THE PROCEDURE HANDBOOK OF ARC WELDING

FOURTEENTH EDITION

This book may be ordered from any dealer or representative of The James F. Lincoln Arc Welding Foundation, or through any recognized book dealer in the world or direct from

The James F. Lincoln Arc Welding Foundation
P.O. Box 17035
Cleveland, OH 44117-0035
(216) 383-2211

FOURTEENTH EDITION May 2000

OVER 1,600,000 Copies
Printed in U.S.A

19 20 21 22 23 QGV 16 15 14 13 12

Published as a service to industry and education by The James F. Lincoln Arc Welding Foundation
P.O. BOX 17035
Cleveland, OH 44117-0035

ACKNOWLEDGEMENTS

The publisher acknowledges with thanks the contributions and cooperation of the following individual and concerns who have aided in the preparation of this and previous editions with information and photographs:

American Hot Dip Galvanizers Association, Inc.
American Institute of Steel Construction
American Iron and Steel Institute
American Petroleum Institute
American Society of Mechanical Engineers
American Society for Metals
  Metals Handbook
  METAL PROGRESS
American Society for Testing and Materials — ASTM Standards
American Welding Society
  Codes, Standards and Specifications
  Welding Handbook
  WELDING JOURNAL
  Welding Metallurgy, George E. Linnert
Arcair Company
Bethlehem Steel Corporation
British Standard Institution
British Welding Association — Welding Processes, P.T. Houldcroft
Bureau of Ships, Navy Department
Hobart Brothers Technical Center
Industrial Publishing Company — Welding Data Book
Jefferson Publications, Inc. — WELDING ENGINEER
Kaiser Aluminum and Chemical Corp., Inc.
The James F. Lincoln Arc Welding Foundation
Linde Division, Union Carbide Corporation
Miller Electric Manufacturing Company
National Aeronautics and Space Administration
National Cylinder Gas Division of Chemetron Corporation
Nelson Stud Welding Company, United-Carr Div., TRW Inc.
Penton Publishing Company — MACHINE DESIGN
Republic Steel Corporation — Republic Alloy Steels
Steel Foundry Research Foundation
Tool and Manufacturing Engineers
WELDING DESIGN & FABRICATION

Special acknowledgement is made to Emmett A. Smith, Robert A. Wilson, Omer W. Blodgett, Jerry Hinkel, Robert E. Greenlee, Jesse Guardado, Ted Bullard, Rosalie Brosilow, David Fink, Charles Hayes, Tom Black, Teresa Melfi, John Gerken, Donald Patton and Mark Hubaker for their technical expertise and professional assistance.

The publisher regrets any omissions from this list which may occur, and would appreciate being advised about them so that the records can be corrected.

THE PROCEDURE HANDBOOK OF ARC WELDING

CONTENTS

Section 1
INTRODUCTION AND FUNDAMENTALS

SECTION 1.1
HISTORICAL DEVELOPMENT OF FUSION JOINING

SECTION 1.2
PROPERTIES OF MATERIALS

SECTION 1.3
ARC-WELDING FUNDAMENTALS

Historical Development of Fusion Joining

EARLY DISCOVERIES

For centuries, the only method man had for metallurgically joining metals was forge welding, a crude and cumbersome blacksmith-type operation in which heated metals were pounded or rammed together until they bonded. Then, within the span of a few years prior to 1900, three new processes came into existence. Arc welding and resistance welding were developed in the late 1880's and put to work in industry a few years later. Oxyacetylene welding was developed during the same period, and was first used industrially in the early 1900's.

No one knows when man first learned to use forge welding. Few implements of iron or steel can survive corrosion over hundreds of years, so there remains little direct evidence of early attempts at the fusion joining of metals.

The working and hardening of steel — advanced arts that doubtless took centuries to evolve — were commonly practiced 30 centuries ago in Greece. But primitive tribes on different continents, and with no apparent means of communication, developed the same basic methods for smelting, shaping, and treating iron. Thus, the principles of welding probably were discovered, lost, and rediscovered repeatedly by ancient peoples.

By the time of the Renaissance, craftsmen were highly skilled in forge welding. Parts to be joined were shaped and then heated in a forge or furnace before being hammered, rolled, or pressed together. Vannoccio Biringuccio's *Pyrotechnia*, published in Venice in 1540, contains several references to such operations. Biringuccio was obviously intrigued by the process, for he wrote, "This seems to me an ingenious thing, little used, but of great usefulness."

For a few centuries thereafter, ordinary fire remained the principal source of heat for welding. The traveling tinker, a familiar figure on the dusty roads of the countryside, carried with him a small charcoal furnace for heating his irons. During this era, tinsmiths and other workers in metal often used the heat of burning gases to braze and solder.

Forge welding of iron developed into a recognized industry. But the joining of large, heavy pieces required great skill and much labor, for they could be brought to the required temperature only if a fire were maintained around them. When the two parts were hot enough, they were forced together by various means, and were often hung from cranes for this operation. The ends were struck repeatedly with a sledge hammer while the heat was maintained. Then the work was withdrawn from the fire and finished on an anvil. Forge welding is still practiced today, but to a very limited extent.

Of the three new processes developed just prior to the Twentieth Century, arc welding has emerged as the most widely used and commercially important method. There is evidence that a Professor G. Lichtenberg may have joined metals by electric fusion as early as 1782 in Germany, but most accounts trace the history of electric welding back to the discovery of the electric arc by Sir Humphrey Davy. In 1801, while experimenting with the infant science of electricity, Davy discovered that an arc could be created with a high-voltage electric circuit by bringing the two terminals near each other. This arc, which cast a bright light and gave off considerable heat, could be struck and maintained at will, and its length and intensity could be varied within limits determined by the circuit voltage and by the type of terminals used. Davy demonstrated the arc at

Fig. 1-1. Carbon-arc welding shop of the late 1800's. Power source is a bank of batteries. Note the dynamo on the right.

the Royal Institute of England in 1808, where his discovery aroused a great deal of interest. For many years, however, it remained a scientific plaything; there appeared to be no practical use for the phenomenon. In fact, Davy did not apply the term "arc" to his discovery until 20 years later.

After the discovery of the arc, the first person known to intentionally join metals by electric welding was an Englishman named Wilde. In the early 1860's he melted together small pieces of iron, and, in 1865, he was granted a patent on his process — the first patent relating to electric welding.

The electric arc, however, remained of scientific interest only until 1881, when the carbon-arc street lamp was introduced. Shortly thereafter, the electric furnace made its appearance in England. One of the earliest was installed in 1886 for the production of aluminum alloys. This particular application of the electric arc was an important step in the early development of the aluminum industry.

NEW WELDING METHODS ARE PUT TO WORK

Probably the first attempt to use the intense heat of the carbon arc for welding was made in 1881 when Auguste de Meritens used a carbon electrode to arc weld lead storage-battery plates. In this experiment, De Meritens connected the work to the positive pole of a current source and attached a carbon rod to the negative pole in such a manner that the distance between the rod and plate could be controlled. Some of the heat developed was lost to the surrounding air, but enough reached the plate to fuse the lead and join the parts. Other early efforts with arc welding employed carbon electrodes arranged similarly to the positions of electrodes in an arc lamp. The heat of the arc was deflected against the work by magnetic fields or by a jet of compressed air.

Two other scientists, Nikolas de Benardos and Stanislav Olszewski, were interested in the De Meritens process and experimented with it. In 1885, they were issued a British patent for a welding process employing carbon electrodes. Benardos, a Russian, also filed for a patent in his homeland. His application described a process in which the work was connected to a negative pole, and the carbon rod was fastened to the positive pole of a DC circuit. The rod was not fixed as in De Meritens' method, but was fitted with an insulated handle so that it could be manipulated by hand. This process was patented in 1887. Thus, Benardos is generally credited as the holder of the first patent on arc welding.

Benardos' carbon-arc process was put to work on a limited scale in England soon after it was developed. In 1887, a shop was using it to make tanks, casks, and iron garden furniture. In the 1890's, another English shop was welding wrought iron pipe up to a foot in diameter. In the United States, the Baldwin Locomotive Works established a shop in 1892, where carbon-arc welding was used extensively for locomotive maintenance. But, in general, acceptance of the carbon-arc process was slow, because the procedures used at that time introduced particles of carbon into the weld metal. These particles made the joint hard and brittle.

In the early days of arc welding, a bank of storage batteries was used as the power source.

12,984. Benardos, N. de, and OLSZEWSKI, S. Oct. 28. *Amended.*

Coating. – Relates to a method of and apparatus for working metals in various ways by electricity, including a method of applying a fused metallic coating for ornamental or other purposes. A voltaic arc is formed by the approach of carbon to the part of the metal operated upon, the carbon usually forming the positive pole and the metal the other pole. The carbon, which may be solid or hollow, is fixed in an apparatus, one form of which is shown in the Figure. The frame A, having a jointed lever B to lower the carbon C, is insulated and supported on the plate or held in the hand. The frame may have wheels running on rails. The work may be supported on an insulated plate electrically connected. Layers of metal are formed by holding an insulated stick of metal in the electric arc. A coloured glass screen is provided to protect the eyes of the workmen.

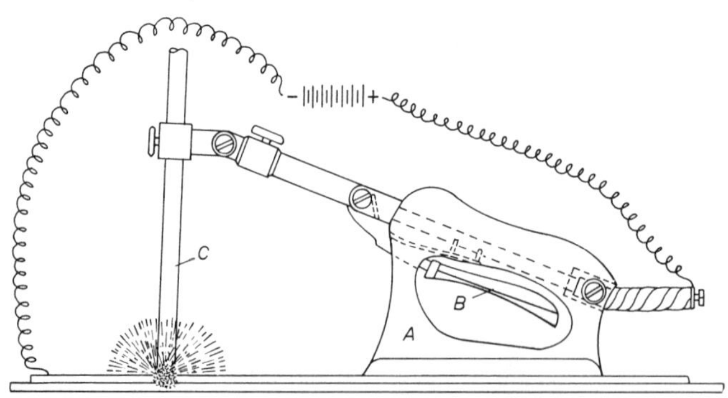

Copy of a British welding patent issued in 1885.

Voltage could be controlled by the number of cells placed in series. Current could be controlled by the number of cells in parallel and by series resistors. The batteries were kept charged by dynamos or generators, powered by the shop steam engine or water wheel.

Two years after Benardos' patent was granted, another Russian, N.G. Slavianoff, announced a process in which the carbon electrode was replaced by a metal rod. After an arc was struck, the rod gradually melted and added fused metal to the weld. In the same year — 1889, unaware of Slavianoff's work, Charles Coffin was granted a U.S. patent on a similar metal-arc welding process.

The metal-arc process simultaneously developed by Coffin and Slavianoff represented a giant step forward, for the metal electrode supplied not only fusing heat, but also added additional filler metal necessary for the joint. In the carbon-arc process, filler metal was supplied by excess metal along the weld line or by a metal rod held in the weldor's hand. Despite this advance in the technology, commercial application of the metal-arc process in the following years was slow because satisfactory metal electrodes were not available.

Fig. 1-3. An early machine for welding the longitudinal seam in a hot water tank with an automatic carbon-arc welding head.

COMMERCIAL ARC WELDING COMES TO AMERICA

Two German weldors who had been working on the metal-arc process in Europe came to the United States in 1907. They formed the Siemund-Wienzell Electric Welding Company and patented a metal-arc welding method. A short time later, another German concern, Enderlein Electric Welding Company, also started operations in the United States. Then a bit of intrigue was attempted. It is reported that Enderlein offered to insure the validity of the Siemund-Wienzell patent by violating it, then putting up a weak defense when Siemund-Wienzell sued. The condition was that the two companies then share the patent rights. Siemund-Wienzell refused the proposal. So when Enderlein began using the process, the firm was promptly and sincerely sued.

Fig. 1-2. A portable arc welder of the early twenties.

In the suit, the patent holders were completely confounded when Enderlein introduced a copy of the *Mechanics Handbook*, published in England in 1888. This handbook contained a woodcut unmistakably showing a shop using the metal-arc process, and its publication date was before any patents had been issued. This revelation cast doubt on the validity of any patents on the process, and, by so doing, opened the field of metal-arc welding in the United States.

By 1917, there were four well-established manufacturers of arc-welding equipment in the United States. One of these was The Lincoln Electric Company, which today is the world's largest producer of arc-welding equipment. Lincoln began experimenting with welding in 1902, and introduced its first machines in 1912.

ELECTRODES THE KEY TO PROGRESS

In the early work with metal-arc welding, it was apparent that the limiting factor was the electrode. The earliest electrodes were bare wire of Norway or Swedish iron, which produced brittle, weak welds. The arcs often overheated the weld metal, and the metal deposited by the electrode was embrittled by reaction with the air. In an attempt to overcome these difficulties, researchers developed a number of electrodes that were lightly coated with various organic or mineral materials. Oscar Kjellborg, of Sweden, who received a patent in 1907, is credited with being one of the pioneer developers of covered electrodes.

The coverings developed during this time, however, did more to stabilize the arc than to shield or purify the weld metal. It was not until 1912, when Strohmenger obtained a U.S. patent for a heavily covered electrode, that industry had an electrode capable of producing weld metal with good mechanical properties. The early covered electrodes, however, were slow in gaining acceptance because of their cost. The covering process required expensive production operations, involving the application of asbestos wrappings, fine aluminum wire, and other materials.

THE IMPETUS ONWARD — WORLD WAR I

The first major increase in the use of welding occurred during World War I. The sudden need for large numbers of transport ships was a contributing factor. At the onset of the war, ships were built by the relatively slow process of riveting. Government officials realized that faster manufacturing methods were needed, and an Emergency Fleet Corporation was set up to find improved shipbuilding methods. Professor Comfort Adams of Harvard was asked to appoint a committee to investigate the problem, and in July 1917 the first committee meeting was held.

Many members of this committee were of the opinion that the key to increased production would be found in resistance welding, a process that had been invented in 1886 by Professor Elihu Thomson, a member of the committee. To gather background information, the committee visited England, where shipbuilders were using welding to some extent. There the committee discovered that it was arc, not resistance, welding that the British were using. England had been forced by gas shortages to curtail gas welding, and was using arc welding with both bare and covered metallic electrodes to produce bombs, mines, and torpedoes. The British had gone

Fig. 1-4. This building was erected in 1928, using arc welding and bare-wire electrodes.

Fig. 1-5. Refineries and oil fields were the earliest beneficiaries of portable gasoline engine-driven welders.

so far as to start construction of a ship with an all-welded hull.

The American committee returned as proponents of the arc-welding method. The various supporters of gas and resistance welding, however, would not accept their findings at face value, and the subsequent debate included the relative merits of carbon and metal electrodes, covered and bare-metal electrodes, and direct and alternating current.

During this discussion, a dramatic incident publicized the capabilities of arc welding. German ships interned in New York Harbor at the outbreak of the war had been scuttled by their crews so that the vessels could not be used in the Allied war effort. Damage was so extensive that revolutionary repair processes were clearly needed if the ships were to be put back into service without long delay. The Navy called in welding experts from two railroad companies, and these men recommended that repairs be made by arc welding. Most of the damaged components were subsequently repaired by this process, and the ships were rapidly returned to service. The potential of the process was clearly established.

In Europe, about the same time, an all-welded cross-channel barge had been put in service. Also, the British launched their all-welded ship, the *Fulagar*, in 1920. Arc welding, thus, became an accepted process for shipbuilding.

The first application of arc welding to aircraft also occurred during World War I. Anthony Fokker, the Dutch airplane manufacturer, used the process to produce fuselages for some German fighter planes.

THE ERA OF SLOW GROWTH

In the years immediately following the war, applications for arc welding did not increase appreciably. In 1919, a patent was granted for a paper-covered electrode that did not leave a slag coating on the joint, yet produced a tough, ductile weld. This welding electrode, was used in 1925 to fabricate heavy pressure vessels for oil refineries. A three-span, 500-ft, all-welded bridge was erected in 1923 in Toronto, Canada. About this time, manufacturers began to use arc welding increasingly for building storage tanks for fuel oil, gasoline, and petroleum distillates. An early application of large proportions was the construction of a million-gallon standpipe that stood 127 feet high.

In 1928, the steel framework for the Upper Carnegie Building in Cleveland, Ohio, was erected, using arc welding in a joint effort by The Austin Company and The Lincoln Electric Company. Construction of this building brought out several important advances in construction techniques. No connection angles or plates were used at intersections, as commonly required with riveted assembly. Since welded lattice joists were used, piping could be concealed between floors. The building was 60 ft by 119 ft and four stories high. The 115 tons of steel required was estimated to be 15% less than required for a riveted design. A factor contributing to this savings was the use of continuous beams, which permitted lighter beams and columns with no sacrifice in strength or rigidity.

In the 1920's, manufacturers were also using arc welding in the production of sheet-steel fabrications, such as blower fans, air conduits, housings for machinery, and bases for machine tools. Foreseeing the potentials, the arc-welding industry began advocating the conversion of cast-iron parts to welded assemblies.

In 1927, the development of an extrusion process for applying a covering to the metal core substantially lowered the cost of covered electrodes. These lower-cost electrodes proved to be one of the most significant developments in the evolution of arc

Fig. 1-6. An all-welded naval vessel that won a major award in a design competition in 1932.

welding. The extrusion process permitted varying the composition of the electrode covering to give desirable operating characteristics and meet specific application requirements. The shielded-arc electrode with its deoxidizers and protective gases and slag became feasible.

YEARS OF RAPID ADVANCE

The applications for arc welding grew rapidly after 1929, and, by the onset of World War II, the process was becoming the dominant welding method. Prior to 1929, the largest undertaking involving welding was the construction of a 5-ft diameter, 90-mile pipeline for carrying water to cities east of San Francisco Bay. It was estimated that this pipeline would have leaked enough water to supply a city of 10,000 if riveted construction had been used. Leakage was minimal with welding.

In the 1930's, welding became increasingly important in shipbuilding. The U.S. Navy, which had contributed much to welding research, turned to the process for practical reasons after the London Naval Treaty of 1930. This treaty imposed limits on the gross tonnages of the major navies of the world, and, thereafter, the Navy often found welding advantageous to minimize weight and thereby maximize the firepower permitted by the tonnage restriction. For the same reason, the Germans used arc welding in their pocket battleships, three of which were launched from 1931 to 1934. Armor plate was difficult to weld and the Germans developed a method for satisfactorily welding it.

Fig. 1-7. Welding steel beer barrel. (1933)

Fig. 1-8. Two all-welded steel presses in an automotive plant. Manufactured by Clearing Machine Corporation, Chicago, Illinois. Capacity 100 tons. (April 1939).

In 1930, the first all-welded merchant ship was built in Charleston, South Carolina. This ship was the forerunner of the thousands of all-welded ships that were to be produced during World War II. Also in the 1930's, the U.S. Army became interested in welding, and a considerable amount of ordinance equipment was redesigned at the Watertown Arsenal for production by welding.

About 1935, improved AC welders were becoming available. These welders offered certain advantages, but AC arcs often proved difficult to maintain. To overcome this difficulty, producers of electrodes developed coverings that ionized more easily and, thus, stabilized the arc. Also during this decade, more stainless steels came into use in metalworking. These materials were relatively difficult to weld because hydrogen in the electrode coverings often caused porosity in the weld. Low-hydrogen electrode coverings were developed to overcome this difficulty. Then, in the early 1940's, it was discovered that these low-hydrogen electrodes also provided good welds in armor plate. Stainless-steel coverings were applied to low-alloy steel electrodes to further improve the quality of welded joints in armor plate.

During the 1930's, numerous attempts were made to bring some degree of mechanization with good shielding to the arc-welding processes. The early attempts at automatic welding were made with

Fig. 1-9. Chicago Bridge and Iron Company Hortonspheroid of 80,000 bbl. capacity in southern Texas. (May 1940)

continuously fed bare wire, with no shielding other than a thin slag flux that was sometimes "painted" on the workpiece. Shielding for automatic carbon-arc welding was provided by passing a flux-impregnated paper string near the arc as it traveled along the seam. Then, in 1932, an innovation was introduced. A heavy layer of flux was placed on the seam ahead of the carbon electrode. The heat of the arc melted the flux into a slag, which provided shielding. The development proved successful, and penstocks for the TVA project and water conduit for the Los Angeles Water Authority were welded by this process.

Use of a granular flux with a continuously fed bare steel electrode led to development in 1935 of the submerged-arc process, which found its first major use in pipe fabrication and shipbuilding. A 521-ft tanker was fabricated by this process in 1936. By 1940, the submerged-arc process was well accepted, but had proved practical primarily on steel plate over 1/4-in. thick. About 1942, the process was improved to accommodate stock down to 3/32-in. thick, and, thus, become feasible for automotive use and for general metal fabrication.

Hand-held, semiautomatic guns were developed for the submerged-arc process in 1946. Voltage and current were controlled automatically, so that weld quality was uniform and results varied less with operator skill. Multiple arcs were introduced in 1948, primarily for manufacturing pipe with 1/4 to 1/2-in. walls in diameters from 18 to 36 inches. Subsequent improvements in submerged-arc welding have been mainly in the areas of improved fluxes and more sophisticated welding equipment and controls.

One problem that continued to defy solution was the joining of the reactive metals aluminum and magnesium. Neither the submerged-arc process nor covered electrodes provided enough shielding to adequately protect these metals from atmospheric contamination. To overcome this difficulty, welding engineers began to use bottled inert gases as shielding agents in the early 1930's. Later in that decade, successful gas-shielded processes powered by DC began to emerge from the aircraft industry in response to a specific need to weld magnesium.

The first gas-shielded process employed a tungsten electrode and helium shielding gas, and became known as the tungsten-arc or tungsten inert-gas (TIG) process. Initially, direct current and a positive electrode were used. It was found, however, that the tungsten electrode tended to overheat and

Fig. 1-10. Four submerged arc passes are made inside the tower of this flame-gouged seam joining the head to the shell. A current of 400 amperes and a speed of 14 inches per minute were used.

Fig. 1-11. Smooth clean lines, without outside stiffeners, carry ramp on sharp radius curve.

transfer particles of tungsten to the weld unless a low current was used. Researchers then discovered that overheating could be avoided by making the electrode negative. This change proved satisfactory for welding stainless steel, but still was not suitable for magnesium or aluminum. The next development was the use of AC with a high-frequency, high-voltage current superimposed over the basic welding current to stabilize the arc. This proved to be the solution to the problem of making good welded joints in aluminum and magnesium. In 1953, the tungsten-arc process was modified by directing the arc through a nozzle, and the resulting method became known as the plasma-arc process.

POSTWAR DEVELOPMENTS CONTINUE

The tungsten-arc process proved unsatisfactory for welding thick sections of highly conductive materials because the workpieces tended to act as heat sinks. To overcome this difficulty, a consumable metal electrode was substituted for the nonconsumable tungsten electrode. The resulting process, announced in 1948, became known as gas metal-arc, metal-inert-gas, or MIG welding. It proved successful for welding aluminum, and was subsequently adapted for other nonferrous materials and for stainless and mild steels. About this time, studies showed that a more stable arc could be obtained by using gas mixtures instead of pure helium or argon.

An important development in manual covered-electrode welding also occurred in this era — namely the use of iron powder in electrode coverings. One benefit of iron powder in the covering was a faster deposition rate and, thus, greater welding speed. Another was that the weldor could simply drag the electrode along the seam without trying to hold it a fixed distance from the work. Thus, less skill was required, and proper welding technique could more easily be taught to beginners. The disadvantage of the iron powder in the covering was the high manufacturing cost. However, by 1953, advances in manufacturing technology and electrode design resulted in cost reductions that made possible the marketing of iron-powder electrodes at acceptable prices. The use of iron-powder electrodes became widespread.

As the GMA and GTA processes gained acceptance in the early 1950's, users found that shielding gases based on argon or helium were often too costly. To lower the material cost of the processes, researchers turned to one of the early developments in arc welding, using carbon dioxide gas as a shielding agent. John C. Lincoln, founder of The Lincoln Electric Company, had applied for a patent on this idea in 1918. Refinements in both the process and equipment for welding steel using carbon dioxide as a shielding gas resulted in a low-cost process. This was immediately adopted by automotive shops and other metalworking plants for applications where the quality of the weld was not exceedingly critical.

Fig. 1-12. Eight-ton transmission housing.

Fig. 1-13. Giant bridge griders for Mississippi River crossing between Dresbach, Minnesota, and Onalaska, Wisconsin, fabricated by Allied Structural Steel.

One of the most significant developments of the period was the Innershield® process, introduced by Lincoln Electric in 1958 for the welding of steel. Prior to its development, self-shielded processes derived protective gases from the decomposition of chemical coverings on the electrode. One could envision possibilities in mechanization with the covered electrode if it could be fed to the arc from a continuous coil. The coverings of such electrodes, however, tended to crack if wound into a coil, and also there was no practical way to feed electric current to a covered continuous electrode. Therefore, self-shielded electrodes as constituted could not be used with automatic or semiautomatic processes.

The Innershield process, also referred to as the self-shielded, flux-cored arc-welding process, solved the problem by incorporating the fluxing and shielding materials inside tubular filler-metal wire. The result was a self-shielded electrode that could be coiled and used with high-speed automatic and semiautomatic equipment. Some of the carbon dioxide-shielded processes also began to employ a fluxed-cored electrode at this time. The concept of a tubular electrode to contain processing ingredients had been employed prior to 1958, but limited to electrodes for surfacing applications.

In 1961, Lincoln Electric introduced an Innershield electrode that provided exceptionally high deposition rates. This electrode referred to as a "fast-fill" electrode is widely used in semiautomatic welding. Because heat input with the fast-fill electrode was considerably less than required by older types, the automatically fed electrode holders, or welding "guns," developed for its use did not require water cooling and, thus, were lightweight and easy to manipulate. The electrode produced welds that had good resistance to cracking and operational characteristics that lessened the amount of care required to fit up workpieces prior to welding.

When first introduced, the fast-fill electrode was limited to single-pass welds in the flat or horizontal positions. By 1962, fast-fill electrodes were available for multiple-pass welds. Thick plates could thus be welded at high deposition rates. In 1967, an all-position electrode was introduced that considerably broadened the application of the process. The American Welding Society has written specifications for flux-cored electrodes. These specifications include both self-shielded electrodes and electrodes requiring gas shielding.

As the arc-welding processes reached a high level of development in the 1960's, research emphasis shifted somewhat. That reliable welds could be

Fig. 1-14. Splicing a column during the erection of a building in Los Angeles, using the semiautomatic self-shielded, flux-cored arc-welding process.

produced was unquestionable, but there was some difficulty in determining whether a given weld made in the plant or field met the metallurgical standards for its particular application. Considerable attention was therefore focused on nondestructive testing — particularly ultrasonic, radiographic, magnetic-particle, and dye-penetrant techniques.

Researchers also exerted considerable effort on the development of exotic joining methods, such as laser welding and electron-beam welding processes that use electricity but do not employ an arc. Although the newer processes do produce welds that were not previously possible, their limitations restrict their use to relatively few specialized applications.

Arc welding continues to serve as the primary means of metal joining. The flash, smoke, and sputter which emanated from the early European laboratories produced one of the most important processes of modern industry.

Fig. 1-15. San Francisco skyline showing building with welded steel frames.

Properties of Materials

The mechanical and physical properties of materials determine their applicability in the design of a product. In the design of weldments, the properties of primary concern are those that indicate the behavior of metallic materials under various condition of loading. These properties are determined in testing laboratories, where standardized procedures and equipment are used to gather data.

MECHANICAL PROPERTIES

Mechanical properties of metals are those that reveal the elastic and inelastic behavior when force is applied. They are:

- ultimate tensile strength
- yield strength
- elongation
- modulus of elasticity
- compressive strength
- shear strength
- fatigue strength
- impact strength
- hardness

All except fatigue and impact strength are determined by steadily applied or static loads. Fatigue and impact properties are determined by pulsating and dynamic loads, respectively.

Tensile Properties

In the standard tensile test, the machined and smoothly finished metal specimen is marked with a centerpunch at two points 2 in. apart, as shown in Fig. 1-16. The specimen is placed in a tensile-testing machine (Fig. 1-17), and an axial load is applied by raising the upper jaw at a slow, constant rate while the lower jaw remains stationary.

As the pulling progresses, the specimen elongates at a uniform rate that is proportional to the rate at which the load or pulling force increases. The load divided by the cross-sectional area of the specimen within the gage marks at the beginning of the test represents the unit stress or the resistance of the material to the pulling or tensile force. The

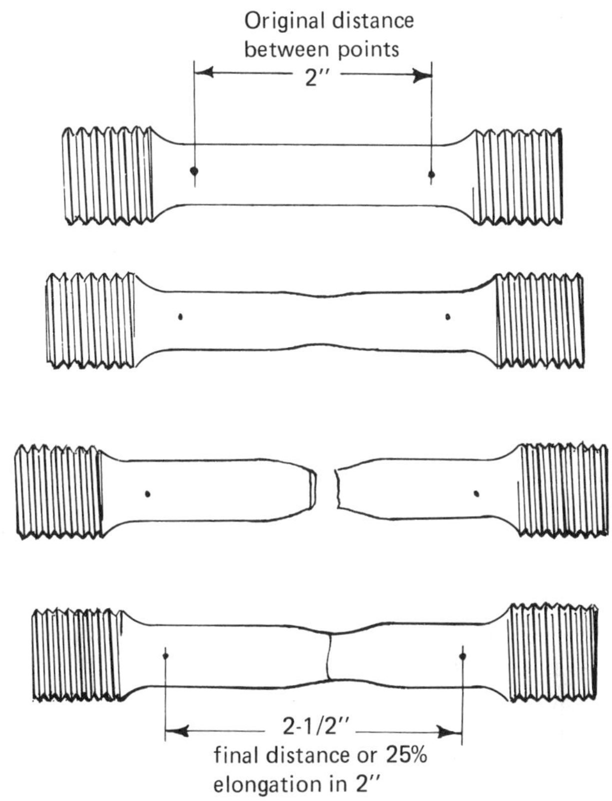

Fig. 1-16. Tensile test specimen before and after testing to failure, showing maximum elongation.

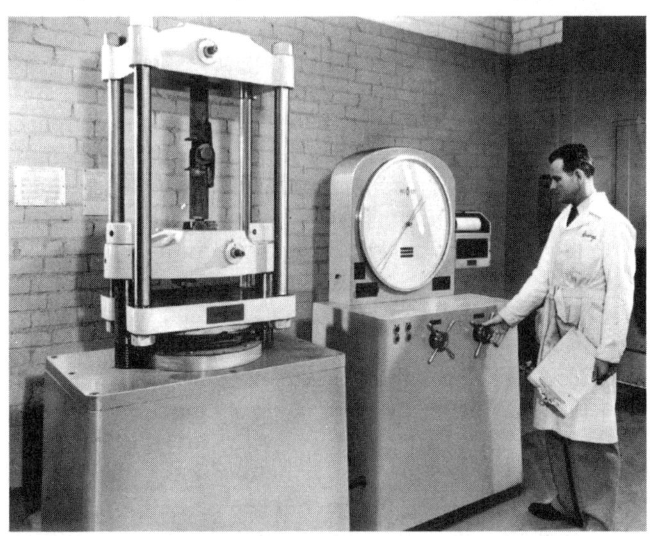

Fig. 1-17. A typical tensile-testing machine. This machine develops the data for the stress-strain diagram.

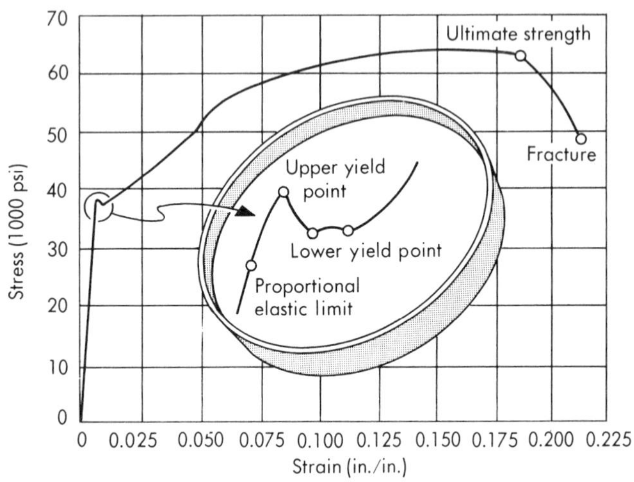

Fig. 1-18. A stress-strain diagram for mild steel. The critical portion of the curve is shown magnified.

stress (σ) is expressed in pounds per square inch, psi. The elongation of the specimen represents the *strain* (ε) induced in the material and is expressed in inches per inch of length, in./in. Stress and strain are plotted in a diagram shown in a simplified form in Fig. 1-18.

The proportional relationship of load to elongation, or of stress to strain, continues until a point is reached where the elongation begins to increase at a faster rate. This point, beyond which the elongation of the specimen no longer is proportional to the loading, is the *proportional elastic limit* of the material. When the load is removed before this point, the specimen returns to its original length and diameter.

Movement of the testing-machine jaw beyond the elastic limit causes a permanent elongation or deformation of the specimen. In the case of low or medium-carbon steels, a point is reached beyond which the metal stretches briefly without an increase in load. This is the yield point. The unit stress at the yield point is considered to be the material's *tensile yield strength* (σ_y).

Beyond the material's elastic limit, continued loading causes the specimen to neck down across its diameter. This action is accompanied by a further acceleration of axial elongation, which is now largely confined to the relatively short necked-down section.

The loading eventually reaches a maximum value, and then falls off rapidly, with little additional elongation of the specimen before fracture occurs. In fracturing, the specimen breaks in two within the necked-down portion. The maximum pulling load in pounds, divided by the original cross section in square inches, is the material's ultimate tensile strength (σ_u).

The standard tensile test specimen is shown in Fig. 1-19. (See *ASTM E 8* for other sizes of round specimens.) The standard test specimen for testing a welded joint transverse to the weld is shown in Fig. 1-20.

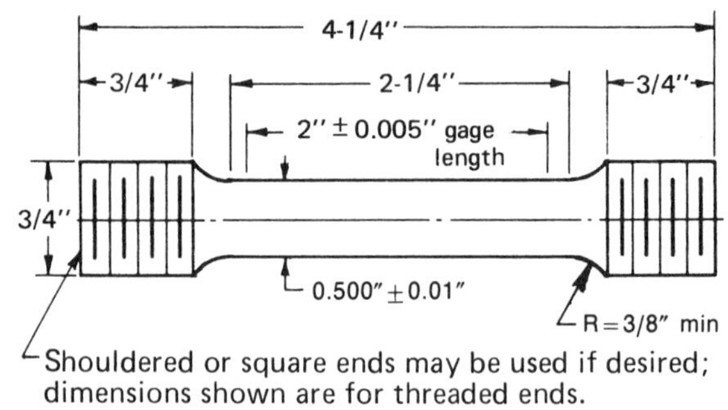

Fig. 1-19. A standard tensile test specimen. The threaded ends may be changed to fit the testing machine.

Ductility and Elasticity

The two halves of the fractured specimen are next fitted together as closely as possible, and the distance between the two punch marks is measured (Fig. 1-16). The increase in length gives the *elongation* of the specimen in 2 in. and is usually expressed as a percentage. The diameter at the point of fracture is measured and the area calculated. The reduction from the original area is calculated. The *reduction in area* is expressed as a percentage. Both the elongation percentage and the reduction in area percentage are measures of *ductility*.

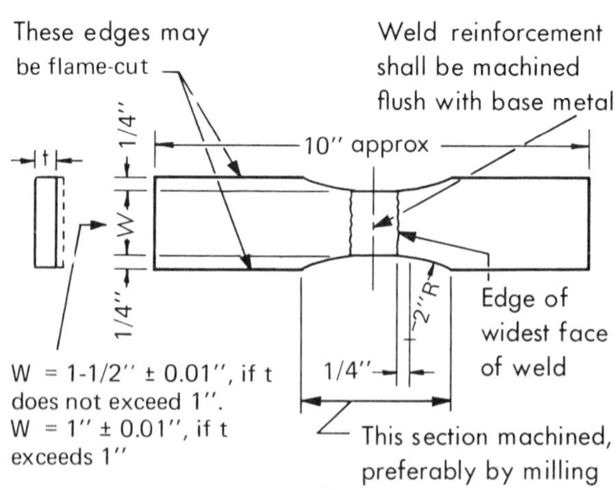

Fig. 1-20. A standard tensile test specimen for transverse test of a welded joint.

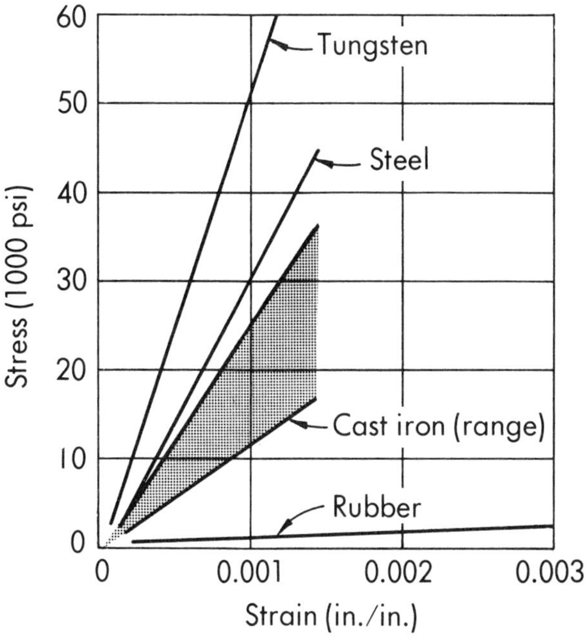

Fig. 1-21. Typical stress-strain curves within the elastic limit of several materials.

In the design of most structural members, it is essential to keep the loading stresses within the elastic range. If the elastic limit (very close to the material's yield strength) is exceeded, permanent deformation takes place due to plastic flow. When this happens, the material is strain-hardened and, thereafter, has a higher effective elastic limit and higher yield strength.

Under the same stress, materials stretch different amounts. The *modulus of elasticity* (E) (also considered as a physical property) of a material simplifies the comparison of its stiffness with that of another material. This property is the ratio of the stress to the strain within the elastic range.

$$\frac{\text{Stress } \sigma}{\text{Strain } \varepsilon} = \text{Modulus of elasticity E}$$

On a stress-strain diagram, the modulus of elasticity is represented visually by the straight portion of the curve where the stress is directly proportional to the strain. The steeper the curve, the higher the modulus of elasticity and the stiffer the material. (See Fig. 1-21.)

Any steel has a modulus of elasticity in tension of approximately 30 million psi. The American Iron and Steel Institute uses a more conservative value of 29 million psi for the modulus of elasticity of steel. The modulus of elasticity will vary for other metals. Steel, however, has the highest value for any commercially available metal used in the structural field.

Compressive Strength

In general design practice, it is assumed that the compressive strength of steel is equal to its tensile strength. This practice is also adhered to in some rigid-design calculations, where the modulus of elasticity of the material in tension is used even though the loading is compressive.

The actual *ultimate compressive strength* of steel is somewhat greater than the ultimate tensile strength. Variations in compressive values are partially dependent on the condition of the steel. The compressive strength of an annealed steel is closer to its tensile strength than would be the case with a cold-worked steel. There is even greater variation between the compressive and tensile strengths of cast iron and nonferrous metals.

The compressive test is conducted in a manner similar to that for tensile properties. A short specimen is subjected to a compressive load, and the ultimate compressive strength is reached when the specimen fails by shattering in the case of a brittle material and deformation in the case of a ductile material.

Shear Strength

There is no recognized standard method for testing the shear strength of a material. Fortunately, pure shear loads are seldom encountered in structural

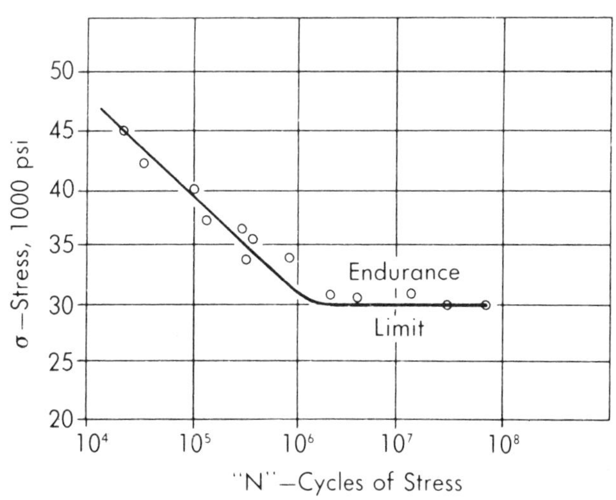

Fig. 1-22. Fatigue test results are plotted to develop a σ-N diagram; stress vs. the number of cycles before failure.

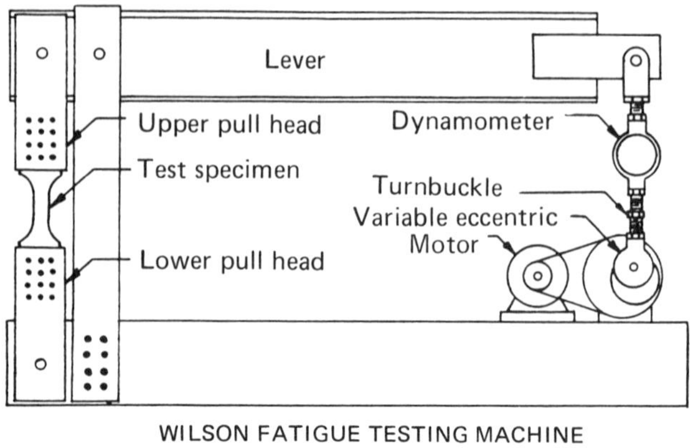

Fig. 1-23. Typical machine for fatigue testing with a pulsating axial stress.

members, but shear stresses frequently develop as a byproduct of principal stresses or the application of transverse forces.

The *ultimate shear strength* (τ) can be obtained by the actual shearing of the metal, usually in a punch-and-die setup, using a ram moving slowly at a constant rate. The maximum load required to punch through is observed and is used to calculate the ultimate shear strength.

Since this is a tedious procedure, the ultimate shear strength, is generally assumed to be 3/4 the ultimate tensile strength for most structural steels.

Fatigue

When the load on the member is constantly varying, is repeated at relatively high frequency, or constitutes a complete reversal of stresses with each operating cycle, the material's fatigue strength must be substituted for ultimate tensile strength where called for by design formulas.

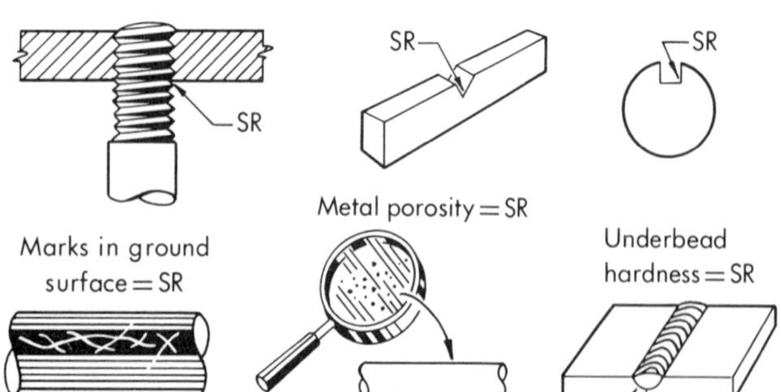

Fig. 1-24. Examples of stress raisers (SR) that lower the fatigue strength.

Under high load, the variable or fatigue mode of loading reduces the material's effective ultimate strength as the number of cycles increases. At a given high stress, the material has a definite service life expressed in "N" cycles of operation.

A series of identical specimens are tested, each under a specific load expressed as unit stress. The unit stress is plotted for each specimen against the number of cycles before failure. The result is a σ-N diagram. (See Fig. 1-22.)

The *endurance limit* is the maximum stress to which the material can be subjected for an indefinite service life. Although the standards vary for various types of members and different industries, it is common practice to accept the assumption that carrying a certain load for several million cycles of stress reversals indicates that the load can be carried for an indefinite time. Theoretically, the load on the test specimen (Fig. 1-23) should be the same type as the load on the proposed member.

Since the geometry of the member, the presence of local areas of high stress concentration, and the condition of the material have considerable influence on the real fatigue strength, prototypes of the member or its section would give the most reliable information as test specimens. However, this is not always practical. The volume *"Design of Welded Structures,"* by Omer W. Blodgett, gives a detailed discussion of fatigue and may be helpful in appraising endurance limits when test data or handbook values are not available.

Local areas of high stress concentration are caused by stress raisers. These are notches, grooves, cracks, tool marks, sharp inside corners, or any other sudden changes in the cross section of the member, as illustrated in Fig. 1-24. Stress raisers can drastically reduce the fatigue life of a member.

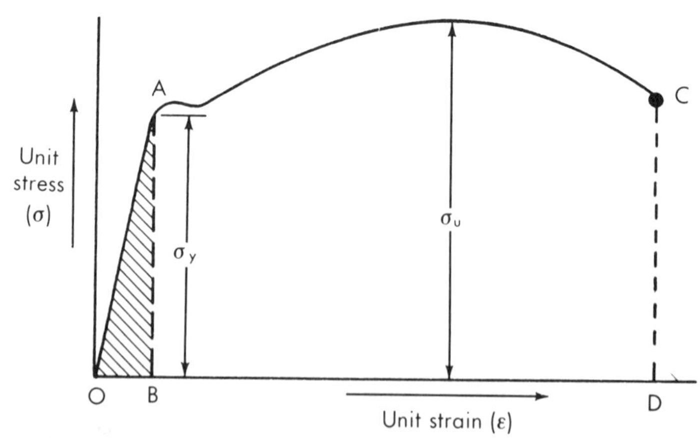

Fig. 1-25. The stress-strain diagram for determining the modulus of resilience and the toughness in terms of ultimate energy resistance.

Fracture Toughness

Fracture toughness generally refers to the ability of a material to resist extension of a crack. Methods of measuring fracture toughness include the Charpy V-notch, the Izod V-notch, the Drop Weight and the Plane Strain Fracture Toughness (K_{IC}) test. The Charpy, Izod and Drop Weight tests measure impact strength or energy absorbed to break a standard specimen. The energy required to cause propagation of a crack in a welded structure may come from the stored energy of residual weld stresses and from static or dynamic loads. The ability of a material to resist crack propagation or its fracture toughness depends on several factors which include yield strength, ductility and elastic modulus.

Impact Strength

Impact strength is the ability of a metal to absorb the energy of a load rapidly applied to the member. A metal may have good tensile strength and good ductility under static loading, yet fracture if subjected to a high-velocity blow.

The two most important properties that indicate the material's resistance to impact loading are obtained from the stress-strain diagram. (See Fig. 1-25.) First is the *modulus of resilience* (u), which is a measure of how well the material absorbs energy when not stressed above the elastic limit or yield point. It indicates the material's resistance to deformation from impact loading. The modulus of resilience (u) is the triangular area OBA under the stress-strain curve, having its apex at the elastic limit. For practicality, let the yield strength (σ_y) be the altitude of the right triangle and the resultant strain (ε_y) be the base. Then

$$u = \frac{\sigma_y \varepsilon_y}{2}$$

$$E = \frac{\sigma_y}{\varepsilon_y} \qquad \varepsilon_y = \frac{\sigma_y}{E}$$

$$u = \frac{\sigma_y^2}{2E}$$

where E is modulus of elasticity
σ_y is yield strength
ε_y is yield strain
u is modulus of resilience

Since the absorption of energy is actually a volumetric property, modulus of resilience is expressed in in.-lb/in.^{3}, or simply lbs/in^2 (psi).

When impact loading exceeds the elastic limit (or yield strength) of the material, it calls for toughness in the material rather than resilience. Toughness, the ability of the metal to resist fracture under impact loading, is indicated by its *ultimate energy resistance* (u_u). This is a measure of how well the material absorbs energy without fracture. The ultimate energy resistance (u_u) is the total area OACD under the stress-strain curve.

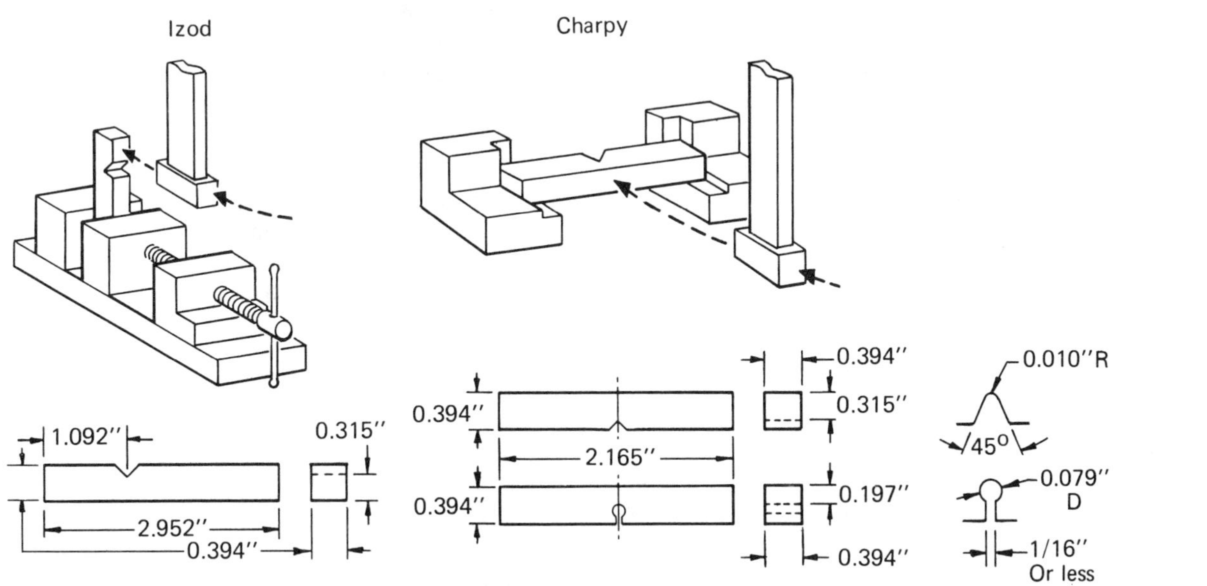

Fig. 1-26. Izod and Charpy impact test specimens showing the method of holding the specimen and applying the load. Two types of Charpy specimens are shown: the upper is V notch and the lower is keyhole notch.

Tests developed for determining the impact strength of metals often give misleading results. Nearly all testing is done with notched specimens. Such tests give results that more accurately describe notch toughness. The two standard tests are Izod and Charpy. In recent years, the Charpy has been replacing the Izod.

In the Charpy test for notch impact strength, there are two types of commonly used specimens, those prepared with the keyhole notch and those with the V notch. Other types of specimens less commonly used are described in *ASTM Standard E 23.* The test specimen (Fig. 1-26) is placed on an anvil, and a heavy pendulum, which swings from a standard height, strikes the specimen on the side opposite the notch. The testing machine indicates the amount of energy in ft-lb required to fracture the specimen. This is a measure of the notch impact strength. Some steels exhibit a considerable loss of notch impact strength at low temperatures, and, for this reason, tests are made at different temperatures to obtain the type of information shown in Fig. 1-27.

Plane Strain in Fracture Toughness

Not all crack propagation occurs on impact loading but will occur when the applied stress at the root of a crack exceeds a certain critical value. This critical value depends on the material and the size or thickness of the component with the crack. For this reason, impact tests with standard size specimens may not give an accurate measure of a material's ability to resist crack propagation in a structural application.

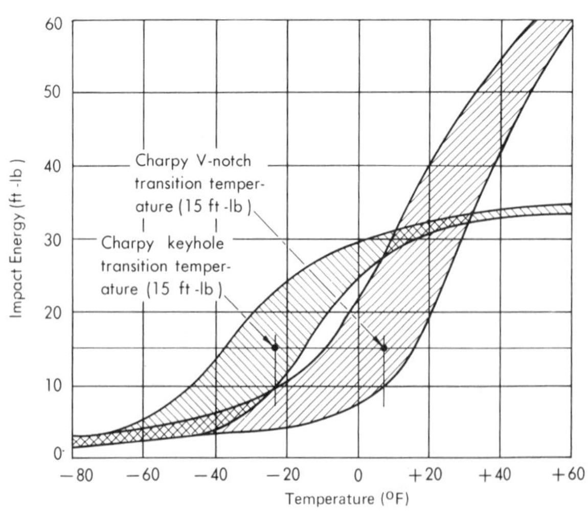

Fig. 1-27. Typical curves for the two types of impact test specimens using the same steel. There is no reliable method to convert V notch data to keyhole values or vice versa.

Plane strain fracture toughness (K_{IC}) is a measure of the resistance to crack extension under conditions of plane strain at the crack tip. Plane strain exists at the tip of a crack in a member of a structure or test specimen when the thickness in both directions normal to the direction of applied load is great enough to restrict deformation or strain in these directions. For this reason, a specimen to measure plane strain fracture toughness (K_{IC}) must be of sufficient size so that the strain in the thickness and width directions is small in comparison to the crack size and specimen dimensions. The size of a specimen for measuring K_{IC} depends on material strength and test temperature where the required size decreases with increasing strength and decreasing test temperature. Thus, there is no one standard specimen for measuring K_{IC} although a number of standard types are available. The validity of a particular test can only be established after the results are calculated. The procedure for conducting K_{IC} tests is expensive and time consuming and normally is not used as a test for welding procedure qualification or material acceptance. This test procedure is described in ASTM E399, "Standard Test Method for Plane Strain Fracture Toughness of Metallic Materials". The test sequence and necessary calculations are summarized in the Welding Handbook, 8th Edition, Volume I page 399. The compact tension specimen, Figure 1-28, is recommended for testing K_{IC} of weldments.

Hardness

Hardness, as related to metals, is the ability of the material to resist indentation or penetration. Two common methods of measuring hardness (Fig. 1-29) are Brinell and Rockwell. Both methods use a penetrator with either a hard sphere or a sharp diamond point. The penetrator is applied to the material under a standardized load, the load removed, and the penetration measured. In the Brinell test, diameter of penetration is measured and converted to hardness by reference to a table. In the Rockwell test, depth of penetration is measured and reads hardness value directly. A numerical value is assigned to the amount of penetration. Another method, Shore Scleroscope, measures the height of rebound of a diamond-tipped hammer when dropped a certain distance. Harder materials cause a higher rebound.

A conversion table appears in Section 16 for the hardness numbers of Brinell and Rockwell. This table is reasonably accurate, but has certain

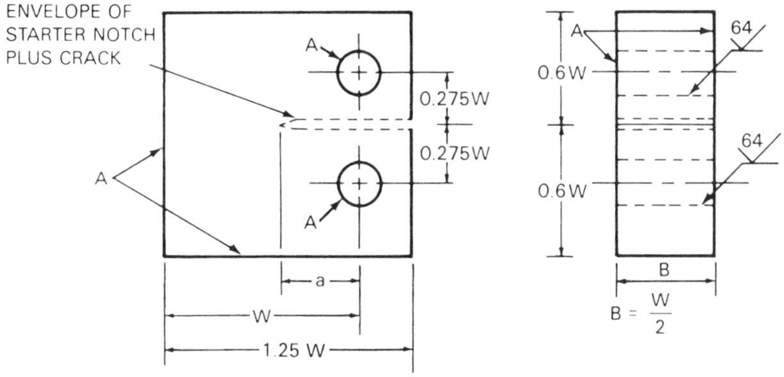

Notes:

1. Dimensions a, B, and W are to be determined in accordance with ASTM E399.
2. Surfaces marked A shall be perpendicular and parallel as applicable to within 0.002W total indicator reading (TIR).
3. The intersection of the crack starter tips with the two specimen faces shall be equally distant from the top and bottom edges of the specimen within 0.005W.
4. Integral or attachable knife edges for clip gage attachment to the crack mouth may be used.

Fig. 1-28. Compact Tension Specimen to Determine Plane Strain Fracture Toughness in Weldments.

limitations. (For limitations, see *ASTM Standard E 140.*)

PHYSICAL PROPERTIES

Physical properties of metals are those other than mechanical and chemical that describe the nature of the metal. They are:

- density
- electrical conductivity
- thermal conductivity
- thermal expansion
- melting temperature

Density

Density of a material is the weight per unit volume. The density of metals is important to the designer, but more important to the weldor is the density of gases. Shielding around the arc is more efficient with a high density gas.

Electrical Conductivity

Electrical conductivity is the efficiency of a material in conducting electrical current. Silver and copper have relatively high electrical conductivities compared to other metals, silver being slightly higher than copper. The conductivity of electrolytic tough pitch copper (ETP) is 101% of the International Annealed Copper Standard (IACS). Other metals compare as follows:

Aluminum (99.99% pure)	65%
Aluminum alloy 5052	35%
Mild steel	15%
Stainless steel type 400	3%
Stainless steel type 300	2.5%

Fig. 1-29. Brinell hardness tester on left measures hardness by the amount of penetration into the metal made by a hard sphere. The Rockwell tester on the right uses either a hard sphere or a sharp diamond point, depending on the hardness of the material.

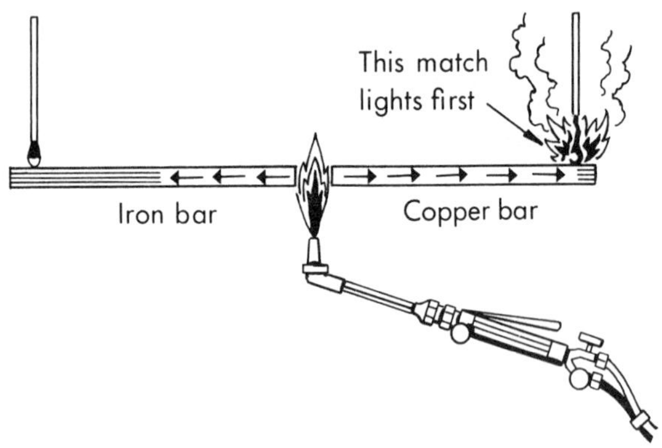

Fig. 1-30. Torch starts heating both the copper bar and the iron bar at the same time. The match in contact with the copper bar ignites first because of the higher thermal conductivity of copper.

Thermal Conductivity

The rate at which heat flows through a material is called *thermal conductivity.* The difference in thermal conductivity between iron and copper can be demonstrated by the arrangement shown in Fig. 1-30. Because the thermal conductivity of copper is about five times that of iron, the match in contact with the copper ignites first.

The thermal conductivity of some of the common metals is shown in Table 1-1. The high thermal conductivity of copper explains why copper is a good material for backup bars. This also explains why copper must be welded with a high heat input or preheat to obtain a satisfactory weld.

TABLE 1-1. THERMAL CONDUCTIVITY OF METALS

Metal	Thermal Conductivity near room temperature cal/cm^2/cm/$^\circ$C/sec
Aluminum EC (99.45%)	0.57
Aluminum 1100	0.53
Aluminum 6061	0.41
Aluminum casting 43	0.34
Copper ETP	0.93
Red brass (15% Zn)	0.38
Cupro-Nickel (30% Ni)	0.070
Nickel (99.95%)	0.22
Monel	0.062
Inconel	0.036
Silver	1.0
Pure iron	0.18
Steel (0.23% C, 0.64% Mn)	0.12
Stainless steel (Type 410)	0.057
Stainless steel (Type 304)	0.036
Manganese steel (14% Mn)	0.032

Data from ASM Handbook Vol. 1

Thermal Expansion

Most metals expand when heated. The change in length is expressed as the *coefficient of linear expansion* and in English units is inches per inch per degree F (in./in./°F). At room temperature, the coefficient for steel is .0000065 in./in./°F, usually expressed 6.5 x 10^{-6} in./in./°F.

Coefficients for thermal expansion are not constant throughout the entire temperature range — for example, from room temperature to the melting point. For this reason, the handbooks give a coefficient within a definite temperature range.

Metals with a high coefficient of expansion present more warping problems, especially if the thermal conductivity is low. The thermal coefficient of linear expansion for several metals is given in Table 1-2.

Melting Point

A pure metal has a definite *melting point* that is the same temperature as its freezing point. Alloys and mixtures of metals start to melt at one temperature (solidus), but the melting is not completed until a higher temperature (liquidus) is reached.

TABLE 1-2. COEFFICIENTS OF LINEAR THERMAL EXPANSION OF SOME METALS AT 68°F

Metal	(in./in./$^\circ$F x 10^{-6})	Metal	(in./in./$^\circ$F x 10^{-6})
Aluminum	13.1	Molybdenum	2.7
Beryllium	6.4	Nickel	7.4
Copper	9.2	Silver	10.9
Iron	6.5	Titanium	4.7
Lead	16.3	Tungsten	2.6
Magnesium	15.1	Zinc	22.0

Data from ASM Handbook Vol. 1

Arc welding a metal with a low melting point or low solidus requires less heat input and more accurate control of the process to prevent melting through, especially if the metal is thin. Melting points of some common metals and other temperatures of interest are shown in Fig. 1-31.

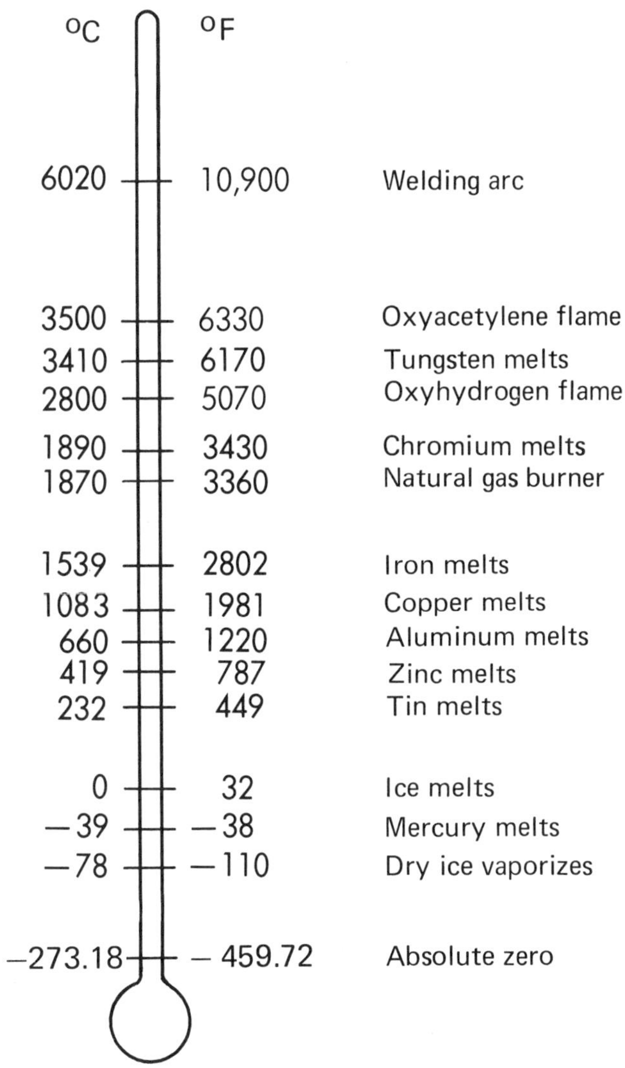

Fig. 1-31. Melting points of some metals and other temperatures of interest.

Arc-Welding Fundamentals

Arc welding is one of several fusion processes for joining metals. By the application of intense heat, metal at the joint between two parts is melted and caused to intermix — directly or, more commonly, with an intermediate molten filler metal. Upon cooling and solidification, a metallurgical bond results. Since the joining is by intermixture of the substance of one part with the substance of the other part, with or without an intermediate of like substance, the final weldment has the potential for exhibiting at the joint the same strength properties as the metal of the parts. This is in sharp contrast to nonfusion processes of joining — such as soldering, brazing, or adhesive bonding — in which the mechanical and physical properties of the base materials cannot be to duplicated at the joint.

In arc welding, the intense heat needed to melt metal is produced by an electric arc. The arc is formed between the work to be welded and an electrode that is manually or mechanically moved along the joint (or the work may be moved under a stationary electrode). The electrode may be a carbon or tungsten rod, of which the sole purpose is to carry the current and sustain the electric arc between its tip and the workpiece. Or, it may be a specially prepared rod or wire that not only conducts the current and sustains the arc but also melts and supplies filler metal to the joint. If the electrode is a carbon or tungsten rod and the joint requires filler metal addition, that metal is supplied by a separately applied filler-metal rod or wire. Most welding in the manufacture of steel products where filler metal is required, however, is accomplished with the second type of electrodes — those that supply filler metal as well as provide the conductor for carrying electric current.

BASIC WELDING CIRCUIT

The basic arc-welding circuit is illustrated in Fig. 1-32. An AC or DC power source, fitted with whatever controls may be needed, is connected by a work cable to the workpiece and by a "hot" cable to an electrode holder of some type, which makes electrical contact with the welding electrode.

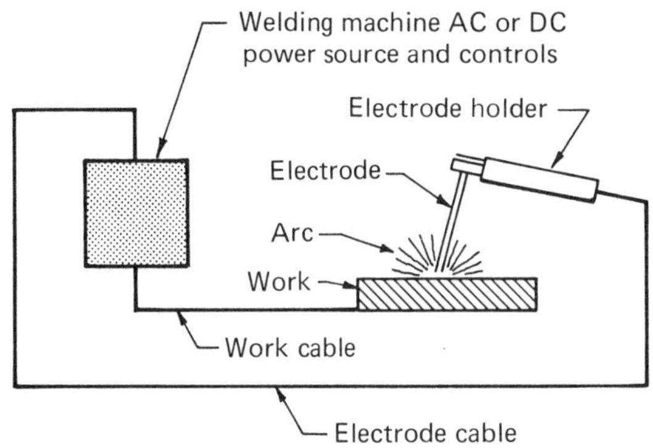

Fig. 1-32. The basic arc-welding circuit.

When the circuit is energized and the electrode tip touched to the workpiece, and then withdrawn and held close to the spot of contact, an arc is created across the gap. The arc produces a temperature of about 6500°F at the tip of the electrode, a temperature more than adequate for melting most metals. The heat produced melts the base metal in the vicinity of the arc and any filler metal supplied by the electrode or by a separately introduced rod or wire. A common pool of molten metal is produced, called a "crater." This crater solidifies behind the electrode as it is moved along the joint being welded. The result is a fusion bond and the metallurgical unification of the workpieces.

ARC SHIELDING

Joining metals with the heat of an electric arc, however, requires more than the moving of the electrode with respect to the weld joint. Metals at high temperatures are reactive chemically with the main constituents of air — oxygen and nitrogen. Should the metal in the molten pool come in contact with air, oxides and nitrides would be formed, which upon solidification of the molten pool would destroy the strength and toughness properties of the weld joint. For this reason, the various arc-welding processes provide some means for covering the arc and the molten pool with a protective shield of gas, vapor, or slag. This is referred to as arc shielding. Arc shielding may be accomplished by various techniques, such as the use of vapor-generating and/or slag forming coverings on filler-metal-type electrodes, the covering of the arc and molten pool with a separately applied inert gas or a granular flux. The use of materials within the core of tubular electrodes also generates shielding vapors and slag.

Whatever the shielding method, the intent is to provide a blanket of gas, vapor, or slag that prevents or minimizes contact of the molten metal with air. The shielding method also affects the stability and other characteristics of the arc. When the shielding is produced by an electrode covering, by electrode core substances, or by separately applied granular flux, a fluxing or metal-improving function is usually also provided. Thus, the core materials in a flux-cored electrode may supply a deoxidizing function as well as a shielding function, and in submerged-arc welding the granular flux applied to the joint ahead of the arc may add alloying elements to the molten pool as well as shielding it and the arc.

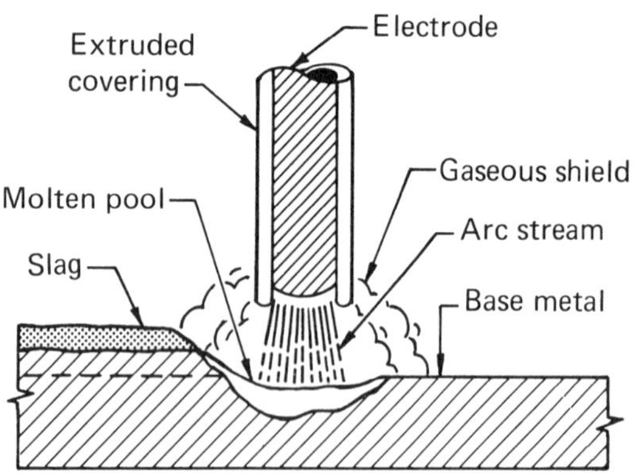

Fig. 1-33. This shows how the coating on a coated (stick) electrode provides a gaseous shield around the arc and a slag covering on the hot weld deposit.

Figure 1-33 illustrates the shielding of the welding arc and molten pool with a covered "stick" electrode, the type of electrode used in most manual arc welding. The extruded covering on the filler metal rod, under the heat of the arc, generates a gaseous shield that prevents air from contacting the molten metal. It also supplies ingredients that react with deleterious substances on the metals, such as oxides and salts, and ties these substances up chemically in a slag that, being lighter than the weld metal, arises to the top of the pool and crusts over the newly solidified metal. This slag, even after solidification, has a protective function; it minimizes contact of the very hot solidified metal with air until the temperature falls to a point where reaction of the metal with air is reduced.

While the main function of the arc is to supply heat, it has other functions that are important to the success of arc-welding processes. It can be adjusted or controlled to transfer molten metal from the electrode to the work, to remove surface films, and to bring about complex gas-slag-metal reactions and various metallurgical changes. The arc, itself, is a very complex phenomenon, which has been intensively studied. In-depth understanding of the physics of the arc is of little value to the weldor, but some knowledge of its general characteristics can be useful.

NATURE OF THE ARC

An arc is an electric current flowing between two electrodes through an ionized column of gas, called a "plasma." The space between the two electrodes — or in arc welding, the space between the electrode and the work — can be divided into three areas of heat generation: the cathode, the anode, and the arc plasma.

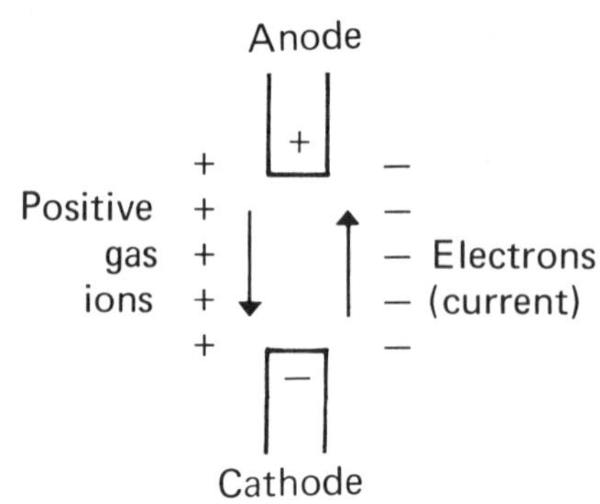

Fig. 1-34. Areas of heat generation in arc.

The welding arc is characterized as a high-current, low-voltage arc that requires a high concentration of electrons to carry the current. Negative electrons are emitted from the cathode and flow — along with the negative ions of the plasma — to the positive anode. Positive ions flow in the reverse direction. A negative ion is an atom that has picked up one or more electrons beyond the number needed to balance the positive charge on its nucleus — thus the negative charge. A positive ion is an atom that has lost one or more electrons — thus the positive charge. However, just as in a solid conductor, the principal flow of current in the arc is by electron travel.

Heat is generated in the cathode area mostly by the positive ions striking the surface of the cathode. Heat at the anode is generated mostly by the electrons. These have been accelerated as they pass through the plasma by the arc voltage, and they give up their energy as heat when striking the anode.

The plasma, or arc column, is a mixture of neutral and excited gas atoms. In the central column of the plasma, electrons, atoms, and ions are in accelerated motion and constantly colliding. The hottest part of the plasma is the central column, where the motion is most intense. The outer portion or the arc flame is somewhat cooler and consists of recombining gas molecules that were disassociated in the central column.

The distribution of heat or voltage drop in the three heat zones can be changed. Changing the arc length has the greatest effect on the arc plasma. Changing the shielding gas can change the heat balance between the anode and cathode. The addition of potassium salts to the plasma reduces the arc voltage because of increased ionization.

The difference in the heat generated between the anode and cathode can determine how certain types of arcs are used. For example, when GTA welding aluminum using argon gas; the electrode as a cathode (negative) can use about 10 times more current without melting than when used as an anode (positive). This indicates that more heat is generated at the anode than at the cathode. The submerged-arc welding process generates more heat at the cathode rather than the anode, as evidenced by the higher melt-off rate when the electrode is negative. The same is also true for EXX10 stick-electrode welding.

In welding, the arc not only provides the heat needed to melt the electrode and the base metal but under certain conditions must also supply the means to transport the molten metal from the tip of the electrode to the work. Several mechanisms for metal transfer exist. In one the molten drop of metal touches the molten metal in the crater and transfer is by surface tension. In another, the drop is ejected from the molten metal at the electrode tip by an electric pinch. It is ejected at high speed and maintains this speed unless slowed by gravitational forces. It may be accelerated by the plasma as in the case of a pinched plasma arc. These forces are the ones that transfer the molten metal in overhead welding. In flat welding, gravity also is a significant force in metal transfer.

If the electrode is consumable, the tip melts under the heat of the arc and molten droplets are detached and transported to the work through the arc column. Any arc-welding system in which the electrode is melted off to become part of the weld is described as "metal-arc." If the electrode is refractory — carbon or tungsten — there are no molten droplets to be forced across the gap and onto the work. Filler metal is melted into the joint from a separate rod or wire.

More of the heat developed by the arc is transferred to the weld pool with consumable electrodes than with nonconsumable electrodes. This produces higher thermal efficiencies and narrower heat-affected zones. Typical thermal efficiencies for metal-arc welding are in the range from 75 to 80 percent; for welding with nonconsumable electrodes, 50 to 60 percent.

Since there must be an ionized path to conduct electricity across a gap, the mere switching on of the welding current with a cold electrode posed over the work will not start the arc. The arc must first be "ignited." This is accomplished either by supplying an initial voltage high enough to cause a discharge or by touching the electrode to the work and then withdrawing it as the contact area becomes heated. High-frequency spark discharges are frequently used for igniting gas-shielded arcs, but the most common method of striking an arc is the touch-and-withdraw method.

Arc welding may be done with direct current (DC) with the electrode either positive or negative or alternating current (AC). The choice of current and polarity depends on the process, the type of electrode, the arc atmosphere, and the metal being welded. Whatever the current, it must be controlled to satisfy the variables — amperage and voltage — which are specified by the welding procedures.

OVERCOMING CURRENT LIMITATIONS

The objective in commercial welding is to get the job done as fast as possible without sacrificing quality so as to lessen the labor time costs of skilled workers. One way to speed the welding process would be to raise the current — use a higher amperage — since the faster electrical energy can be induced in the weld joint, the faster will be the welding rate.

With manual stick-electrode welding, however, there is a practical limit to the current. The covered electrodes are from 9 to 18-in. long, and, if the current is raised too high, electrical resistance heating within the unused length of electrode will become so great that the covering overheats and "breaks down" — the covering ingredients react with each other or oxidize and do not function properly at the arc. Also, the hot core wire increases the melt-off rate and the arc characteristics change. The mechanics of stick-electrode welding is such that electrical contact with the electrode cannot be made immediately above the arc — a technique that would circumvent much of the resistance heating.

Not until semiautomatic guns and automatic welding heads, which are fed by continuous electrode wires, were developed was there a way of solving the resistance-heating problem and, thus, making feasible the use of high currents to speed the welding process. In such guns and heads, electrical contact with the electrode is made close to the arc. The length between the tip of the electrode and the point of electrical contact is, then, inadequate for enough resistance heating to take place to overheat the electrode in advance of the arc, even with currents two or three times those usable with stick-electrode welding.

This solving of the "point-of-contact" problem and circumventing the effects of resistance heating in the electrode was a breakthrough that substantially lowered welding costs and increased the use of arc welding in industrial metals joining. In fact, through the ingenuity of welding equipment manufacturers, the resistance-heating effect has been put to work constructively in a technique known as long-stickout welding. Here, the length of electrode between the point of electrical contact in the welding gun or head and the arc is adjusted so that resistance heating almost — but not quite — overheats the protruding electrode. Thus, when a point on the electrode reaches the arc, the metal at that point is about ready to melt. Thus, less arc heat is required to melt it — and, because of this, still higher welding speeds are possible.

EFFECTS OF ARC ON METAL PROPERTIES

In subsequent sections, also, the effects of the heat of the welding arc on the metallurgy and mechanical properties of weld metal and adjacent base metal will be discussed. Be mindful that what takes place immediately under the welding arc is similar to what happens in an electric furnace for metal production. Electrical furnace steels are premium grades; weld metal from steel electrodes is newly prepared "electric-furnace" steel and also premium grade. Properly executed welds are almost always superior in mechanical properties to the metals they join. In no other metals-joining process is the joint customarily stronger than the metals joined.

Section 2
DESIGNING FOR ARC WELDING

SECTION 2.1
THE SYSTEMS APPROACH TO WELDED DESIGN

Page

SECTION 2.2
THE DESIGN OF WELDED JOINTS

SECTION 2.3
ALLOWABLES FOR WELDS

SECTION 2.4
CODES AND SPECIFICATIONS

LIST OF SYMBOLS

Δ	deflection (in.)
ϵ	unit strain (in./in.)
θ	angular twist (radians; 1 radian = 57.3 degrees)
Σ	sum
σ	tensile strength or stress; compressive strength or stress (psi)
σ_{sr}	allowable range of tensile or compressive stress (psi)
τ	shear strength or stress (psi)
τ_{sr}	allowable range of shear stress (psi)
ϕ	unit angular twist (radians/in.)
ω	leg size of fillet weld (in.)
a	acceleration or deceleration
c	distance from neutral axis to outer fiber (in.)
d	density (lb/in.^{3}); distance between centers of gravity of girder flanges (in.)
f	unit force (lb/in.)
m	mass (lb)
n	number of welds
r	radius (in.); radius of gyration
t	thickness of section; effective throat (in.)
w	uniformly distributed load (lb/in.)
A	area (in.^{2})
C	slenderness factor
℄	center line
E	modulus of elasticity in tension (psi)
E_s	modulus of elasticity in shear (psi)
F	total force (lb)
FS	factor of safety
I	moment of inertia (in.^{4})
J	polar moment of inertia (in.^{4})
K	any specified constant
L	length of member (in.)
M	bending moment (in.-lb)
P	concentrated load (lb)
R	**torsional resistance (in.^{4}); reaction (lb)**
S	section modulus I/c (in.^{3})
T	torque or twisting moment (in.-lb)
U	stored energy (in. lbs)
V	vertical shear load (lb)
W	total width (in.)

GREEK SYMBOLS

Alpha	A	α
Beta	B	β
Gamma	Γ	γ
Delta	Δ	δ
Epsilon	E	ϵ
Zeta	Z	ζ
Eta	H	η
Theta	Θ	θ
Iota	I	ι
Kappa	K	κ
Lambda	Λ	λ
Mu	M	μ
Nu	N	ν
Xi	Ξ	ξ
Omicron	O	o
Pi	Π	π
Rho	P	ρ
Sigma	Σ	σ
Tau	T	τ
Upsilon	Υ	υ
Phi	Φ	ϕ
Chi	X	χ
Psi	Ψ	ψ
Omega	Ω	ω

The Systems Approach To Welded Design

INTRODUCTION

The engineer who has the responsibility for designing a machine part or structural member as a steel weldment frequently operates under a severe handicap. Although he has a mechanical background and academic training in engineering materials and components, it is unlikely that he has adequate information about the specific factors that enter into welded design. He needs to know how to use steel efficiently, how to build stiffness into a beam, how to design for torsional resistance, what weld joints best suit his purpose, etc. He also needs many other bits of practical information — of which few are taught in engineering schools or are found in textbooks.

The art of welded steel design has been evolving gradually, with all the errors that normally accompany an evolutionary process. One of the earlier mistakes in the design of steel weldments — which is still made repeatedly by the novice — is copying the over-all shape and appearance of a casting which is to be replaced by a weldment. Much effort and material can be wasted in the useless attempt to duplicate every flare and offset of the casting. The designer and the management to which he reports should understand that steel weldments are different from castings, and will appear different. There is no reason for shaping a weldment so that it exhibits the protrusions, the separate legs, brackets and housings, and the frills of the casting. A modern steel weldment is an integrated, functional unit that acquires an appearance of its own.

When the change is made from a casting to a weldment, both appearance and function are usually improved, since welded design involves the more conservative and strategic use of materials. But the motivating force for such a change is usually production cost — the desire to fabricate the machine, part, or structure more economically, thus enabling its price to meet the competitive demands of the market. Cost, therefore, must be considered at every step in the design, and the designer must think not only about the obvious production costs, but about all the incidentals from the selection of materials and methods of fabrication down to the final inspection of the finished product and preparation for shipment.

The ability of welded design to make products of superior function and appearance possible at reduced costs arises largely from four advances in fabricating and welding techniques. These are:

1. Machine flame-cutting equipment that produces smoothly cut edges, and machines for shearing thicker plate than could be cut previously.
2. Improved welding electrodes and processes that produce quality welds at high speed.
3. Heavy brakes, which enable the greater use of formed plate, resulting in lower costs, smoother corners, and, because fewer parts are involved, reduced assembly time.
4. Welding positioners which permit more welds to be made in the flat position, resulting in smoother and lower-cost welds.

The use of these advances puts weldments in a favorable light when design changes are considered, whereas, a few decades ago cost restricted the amount of attention weldments received.

GENERAL CONSIDERATIONS

A weldment design program starts with a recognition of a need. The need may be for improving an existing product or for building an entirely new product, using the most advanced design and fabrication techniques. In any event, many factors must be taken into account before a design is finalized. These considerations involve asking numerous questions and doing considerable research into the various areas of marketing, engineering, and production.

Analysis of Present Design

Insofar as possible, when designing an entirely new machine or structural unit, an attempt should be made to gain information about competitive products and their success in the marketplace. If a new machine is to replace an older model, the good points and the deficiencies of the predecessor machine should be understood. The following questions can help in reaching the proper decision:

What do customers and the sales force say about the older machine?

What has been its history of failures?

What features should be retained or added?

What suggestions for improvements have been made?

Was the old model overdesigned?

Determination of Load Conditions

The work the machine is intended to do, or the forces that a structural assembly must sustain, and the conditions of service that might cause overload should be ascertained. From such information, the load on individual members can be determined. As a starting point for calculating loading, the designer may find one or more of the following methods useful:

From the motor horsepower and speed, determine the torque in inch-pounds on a shaft or revolving part.

Calculate the force in pounds on machine members created by the dead weight of parts.

Calculate the load on members of a hoist or lift truck back from the load required to tilt the machine.

Use the maximum strength of critical cables on a shovel or ditch digger that have proved satisfactory in service, to work back to the loads on machine members.

Consider the force required to shear a critical pin as an indication of maximum loading on the machine.

If a satisfactory starting point cannot be found, design for an assumed load and adjust from experience and tests; at least the design will be well proportioned.

Major Design Factors

In developing his design, the designer thinks constantly about how decisions will affect production, manufacturing costs, performance, appearance and customer acceptance. Many factors far removed from engineering considerations per se become major design factors. Some of these are listed below along with other relevant rules:

- The design should satisfy strength and stiffness requirements. Overdesigning is a waste of materials and increases production and shipping costs.
- The safety factor should be adequate but not unrealistically high as indicated by past experience.
- Good appearance has value, but only in areas that are exposed to view. The print could specify the welds that are critical with respect to appearance.
- Deep and symmetrical sections resist bending efficiently.
- Welding the ends of beams rigidly to supports increases strength and stiffness.
- The proper use of stiffeners will provide rigidity at minimum weight of material.
- Closed tubular sections or diagonal bracing should be used for torsion resistance. A closed tubular section may be many times better than an open section.
- Nonpremium grades of steel should be specified wherever possible. Higher carbon and alloy steels require preheating, and frequently postheating, which are added cost items.
- Standard rolled sections should be used wherever possible.
- Standard plate and bar sizes should be used for their economy and availability.
- Accessibility for maintenance should be provided in the design; do not bury a bearing support or other critical wear point in a closed-box weldment.
- The use of standard index tables, way units, heads, and columns should be considered.

Layout

To the designer familiar only with castings, the laying out of a weldment for production may seem complex because of the many possible choices. Variety in the possibilities for layout, however, is one of the advantages of welded design where opportunities for savings exist. Several general rules for effective layout follow:

- Design for easy handling of materials and for inexpensive tooling.
- Check with the shop for ideas that can contribute to cost savings.
- Check the tolerances with the shop. The shop may not be able to hold them, and close tolerances and fits may not be necessary.
- Plan the layout to minimize the number of pieces.

This will reduce assembly time and the amount of welding.

- Lay out parts so as to minimize scrap.
- If possible, modify the shape and size of scrap cutouts, so that such material may be used for pads, stiffeners, gear blanks and other parts.
- If a standard rolled section is not available or suitable, consider forming the desired section from blanks flame-cut from plate. It is also possible to use long flat bar stock welded together, or to place a special order for a rolled-to-shape section.
- In making heavy rings, consider cutting nesting segments from plate to eliminate excessive scrap.

Plate Preparation

Flame cutting, shearing, sawing, punch-press blanking, nibbling, and lathe cutoff are methods for cutting blanks from stock material. The decision relating to method will depend on the equipment available and relative costs. It will be influenced, however, by the quality of the edge for fitup and whether the method also provides a bevel in the case of groove joints. Experience has suggested the following recommendations:

- Dimensioning of the blank may require stock allowance for subsequent edge preparation.
- Not all welds are continuous. This must be borne in mind when proposing to prepare the edge and cut the blank simultaneously.
- Select the type of cutting torch that will allow the cut to be made in one pass. For single-bevel or single-V plate preparation, use a single torch; for double-bevel or double-V, a multiple torch.
- When a plate planer is available, weld metal costs can be reduced with thick plate by making J or U-groove preparations.
- Consider arc gouging, flame gouging, or chipping for back-pass preparations.

Forming and Special Sections

Forming can greatly reduce the cost of a weldment by eliminating welds and, often; machining operations. Thickness of materials, over-all dimensions, production volume, tolerances, and cost influence the choice of forming methods. The suggestions below should be useful in making decisions:

- Create a corner by bending or forming rather than by welding from two pieces.
- Roll a ring instead of cutting from plate in order to effect possible savings.
- Form round or square tubes or rings instead of buying commercial tubing if savings could be effected.
- Put bends in flat plate to increase stiffness.
- Use press indentations in plate to act as ribs, instead of using stiffeners to reduce vibration.
- Use corrugated sheet for extra stiffness.
- The design problem and cost of manufacture may be simplified by incorporating a steel casting or forging into the weldment for a complicated section.
- Use a small amount of hardsurfacing alloy applied by welding, rather than using expensive material throughout the section. (Section 13.7, page 13.7-1).

Welded Joint Design

The type of joint should be selected primarily on the basis of load requirements. Once the type is selected, however, variables in design and layout can substantially affect costs. (See Section 2.2.) Generally, the following rules apply:

- Select the joint requiring the least amount of weld filler metal.
- Where possible, eliminate bevel groove joints by using automatic submerged-arc welding, which has a deep-penetration arc characteristic.
- Use minimum root opening and included angle in order to reduce the amount of filler metal required.
- On thick plate, use double-grooves instead of single-groove joints to reduce the amount of weld metal.
- Use a single weld where possible to join three parts.
- Minimize the convexity of fillet welds; a 45° flat fillet, very slightly convex, is the most economical and reliable shape.
- Avoid joints that create extremely deep grooves.
- Design the joint for easy accessibility for welding.

Size and Amount of Weld

Overwelding is a common error of both design and production. Control begins with design, but must be carried throughout the assembly and welding operations. The following are basic guides:

- Be sure to use the proper amount of welding — not too much and not too little. Excessive weld size is costly.
- Specify that only the needed amount of weld

should be deposited. The allowable limits used by the designer include the safety factor.

- The leg size of fillet welds is especially important, since the amount of weld required increases as the square of the increase in leg size.
- For equivalent strength, longer fillet welds having a small leg size use less filler metal and are less costly than heavy intermittent welds.
- Sometimes, especially under light-load or no-load conditions, cost can be reduced by using intermittent fillet welds in place of a continuous weld of the same leg size.
- To derive maximum advantage from automatic welding, it may be better to convert several short welds into one continuous weld.
- Place the weld in the thinner section and determine the weld size according to the thinner plate.
- Place the weld on the shortest seam. If there is a cutout section, place the welded seam at the cut-out in order to save on the length of welding. On the other hand, in automatic welding it may be better to place the joint away from the cutout area to permit making one continuous weld.
- Stiffeners or diaphragms do not need much welding; reduce the weld leg size or length of weld if possible.
- Keep the amount of welding to a minimum to reduce distortion and internal stress and, therefore, the need and cost for stress relieving and straightening.

Subassemblies

In visualizing assembly procedures, the designer should break the weldment down into subassemblies in several ways to determine which will offer cost savings. The following are points to note:

- Subassemblies spread the work out; more men can work on the job simultaneously.
- Usually, subassemblies provide better access for welding.
- The possibility of distortion or residual stresses in the finished weldment is reduced when the weldment is built from subassemblies.
- Machining to close tolerances before welding into the final assembly is permitted. If necessary, stress relief of certain sections can be performed before welding into the final assembly.
- Leak testing of compartments or chambers and painting before welding into the final assembly are permitted.
- In-process inspection is facilitated allowing errors to be rectified before the job has progressed too far.

Jigs, Fixtures, and Positioners

Jigs, fixtures, and welding positioners should be used to decrease fabrication time. In planning assemblies and subassemblies, the designer should decide if the jig is simply to aid in assembly and tacking or whether the entire welding operation is to be done in the jig. The considerations listed below are significant:

- The jig must provide the rigidity necessary to hold the dimensions of the weldment.
- Tooling must provide easy locating points and be easy to load and unload.
- Camber can be built into the tool for control of distortion.
- The operating factor can be increased by using two jigs, so that a helper can load one while the work in the other is being welded.
- Welding positioners maximize the amount of welding in the flat position, allowing use of larger electrodes and automatic welding.

Assembly

The assembly operations affect the quality of the welds and fabrication costs. Even though the designer may not have control of all the factors entering into good assembly procedures, he should be aware of the following:

- Clean work — parts free from oil, rust, and dirt — reduces trouble.
- Poor fitup can be costly.
- A joint can be preset or prebent to offset expected distortion. (See Section 3-Weldment Distortion).
- Strongbacks are valuable for holding materials in alignment.
- When possible, it is desirable to break the weldment into natural sections, so that the welding of each can be balanced about its own neutral axis.
- Welding the more flexible sections first facilitates any straightening that might be required before final assembly.

Welding Procedures

Although the designer may have little control of welding procedures, he is concerned with what goes on in the shop. Adherence to the following guidelines will help to effect the success of the weldment design:

- Welding helpers and good fixtures and handling equipment improve the operating factor.
- Backup bars increase the speed of welding on the first pass when groove joints are being welded.
- The use of low-hydrogen electrodes eliminates or reduces preheat requirements.
- The welding machine and cable should be large enough for the job.
- The electrode holder should permit the use of high welding currents without overheating.
- Weld in the flat position whenever possible.
- Weld sheet metal 45° downhill.
- If plates are not too thick, consider the possibility of welding from one side only.
- With automatic welding, position fillets so as to obtain maximum penetration into the root of the joint: flat plate, 30° from horizontal; vertical plate, 60° from horizontal.
- Most reinforcements of a weld are unnecessary for a full-strength joint.
- Use a procedure that eliminates arc blow.
- Use optimum welding current and speed for best welding performance. If appearance is not critical and no distortion is being experienced, usual speed frequently can be exceeded.
- Use the recommended current and polarity.
- Consider the use of straight polarity (electrode negative) or long stickout with automatic welding to increase melt-off rate.
- On small fillet welds, a small-diameter electrode may deposit the weld faster by not overwelding.

Distortion Control

Distortion is affected by many factors of design and shop practice. The designer should be aware of some points on shop procedures to control distortion which include:

- High-deposition electrodes, automatic welding, and high welding currents tend to reduce the possibility of distortion.
- The least amount of weld metal, deposited with as few passes as possible, is desirable.
- Welding should progress toward the unrestrained portion of the member, but backstepping may be practical as welding progresses.
- Welds should be balanced about the neutral axis of the member.
- On multipass double-V joints, it may be advisable to weld alternately on both sides of the plate.
- Avoid excessively prestressing members by forcing alignment to obtain better fitup of the parts.
- Joints that may have the greatest contraction on cooling should be welded first.

Cleaning and Inspection

The designer, by his specifications, has some effect on cleaning and inspection costs. He also should recognize the following shop practices that affect these costs:

- Industry now accepts as-welded joints that have uniform appearance as finished; therefore, do not grind the surface of the weld smooth or flush unless required for another reason. This is a very costly operation and usually exceeds the cost of welding.
- Cleaning time may be reduced by use of iron-powder electrodes and automatic welding, which minimize spatter and roughness of surface.
- Spatter resistant films can be applied to the joint to reduce spatter sticking to the plate. Some electrodes and processes produce little or no spatter.
- Sometimes a slightly reduced welding speed or a lower welding current will minimize weld faults. Lower repair costs may result in lower over-all costs.
- Overzealous inspection can increase welding costs; it is possible to be unreasonably strict with inspection.
- Inspection should check for overwelding, which can be costly, and can also contribute to distortion.

WHAT THE DESIGNER NEEDS TO KNOW

The engineer often becomes interested in welded design after having been introduced to it by an isolated instance, such as the use of a steel weldment to solve an unusual vibration or shock-loading problem. The occasion starts him thinking more fully about the exploitation of welded steel. Perhaps, he recognizes that the performance of any member of a structure is dependent on just two basic factors — the properties of its material and the properties of its section. If a design is based on the efficient use of these properties, the weldment is certain to be functionally acceptable and conservative of materials.

When assigned to design a welded steel member, however, the engineer faces many questions. He needs to know how to select the most efficient section; how to determine its dimensions; whether to use stiffeners and, if so, how to size and place them.

Working empirically from past experience was once thought to be the practical approach to welded steel design. This practice turned out to be self-defeating. Guesses and rule-of-thumb methods for selecting configurations and sections almost invariably resulted in excessively heavy designs and excessive costs for materials and fabrication. Not until such "practical" approaches were discarded and fabricators began to use designs based on mathematical calculation did the advantages of weldments come to the fore. Engineers then began to achieve truly efficient use of the properties of steel.

The mathematical formulas for calculating forces and their effects on sections, and for determining the sections needed for resisting such forces, appear quite forbidding to the novice. By proper approach, however, it is possible to simplify design analysis and the use of these formulas. In fact, it is often possible to make correct design decisions merely by examining one or two factors in an equation, without making tedious calculations. On the whole, the mathematics of weld design is no more complex than in other engineering fields; it simply has not reached a comparable degree of formalization and use.

The Design Approach — Part or Whole?

Considerations other than the engineer's wishes may prevail when, say, a machine is to be converted from cast to welded design. Management may favor the redesign of a part or two as a weldment, and conversion over a period of years to an all-welded product. Gradual conversion prevents the abrupt obsolescence of facilities and skills and eases the requirement for new equipment. Capital and personnel considerations often dictate that a company go slowly when changing to welded design. Supplementing these considerations is the need to maintain a smooth production flow and to test the production and market value of the conversion as it is made step by step. When the engineer can redesign a part to improve the existing machine or yield production economies, he is doing his company a service even though he may feel frustrated by the slowness of the conversion.

From the standpoints of performance and ultimate production economies, redesign of the machine as a whole is preferable. The designer is unrestricted by the previous design, and in many cases is able to reduce the number of pieces, the amount of material used, and the labor for assembly. A better, lower-cost product is realized immediately, and the company is in a position to benefit more fully from welded design technology. The benefits almost always include greater market appeal for the product.

Selecting a Basis for Welded Design

The redesign may be based on the previous design or on loading considerations solely.

Following a previous design has the advantage of offering a "safe" starting point; the old design is known to perform satisfactorily. Starting from the old design, however, stifles creative thinking toward developing an entirely new concept to solve the basic problem. Little demand is made on the intelligence or ingenuity of the designer when he models his welded steel design on the previous cast-product. Tables of equivalent sections or nomographs can be used to determine the dimensions for strength and rigidity.

A design based on the loading, however, puts the engineer on his mettle. He starts without preconceived notions. It is up to him to analyze what is wanted and come up with a configuration and selection of materials that best satisfy the need. He must know or determine the value and type of load, and it will be necessary for him to decide on a value for allowable stress in a strength design, or deflection allowable in a rigidity design. Formulas will be required for calculating both strength and rigidity.

DESIGNING FOR STRENGTH AND RIGIDITY

A design may require "strength only" or "strength plus rigidity." All designs must have sufficient strengths so the members will not fail by breaking or yielding when subjected to usual operating loads or reasonable overloads. Strength designs are common in road machinery, farm implements, motor brackets, and various types of structures. If a weldment design is based on calculated loading, design formulas for strength are used to dimension the members.

In certain weldments such as machine tools, rigidity as well as strength is important, since excessive deflection under load would ruin the precision of the product. A design based on loading also requires the use of design formulas for sizing members.

Some parts of a weldment are classed "no load," meaning that they serve their design function without being subjected to loadings much greater than their own weight. Typical no-load members are fenders, dust shields, safety guards, cover plates for access holes, enclosures for esthetic purposes, etc. Only casual attention to strength and rigidity is required in their sizing.

Design Formulas

The design formulas for strength and rigidity always contain terms describing load, member, and stress and strain. If any two of these terms are known, the third can be calculated. All problems of design thus resolve into one of the following:

1. Finding the internal stress or strain caused by an external load on a given member.
2. Finding the external load that may be placed on a given member for any allowable stress or strain.
3. Selecting a member to carry a given load within a given allowable stress or strain.

A load is a force that stresses a member. The result is a strain measured as elongation, contraction, deflection, or angular twist. A useful member must be designed to carry a certain type of load within a certain allowable stress or strain. In designing within the allowable limits, the designer should select the most efficient material and the most efficient section size and shape. The properties of the material and those of the section determine the ability of a member to carry a given load.

The design formulas in use, developed for various conditions and member types, are much too numerous for inclusion here. In the following sections, however, some are used to illustrate specific design problems. Reference material containing formulas applicable to problems encountered may be found in the Suggested Readings at the end of this section. Table 2-1 summarizes the components of design formulas. It should be noted that these components are terms that describe the three basic factors — load, member, and stress and strain. The symbols for values and properties normally used in design formulas are given in the table and in the list of symbols preceding this section.

TABLE 2-1. COMPONENTS OF DESIGN FORMULAS

Load

Application	Type	Value
steady impact fatigue (repeated cyclic variable)	tension compression bending shear torsion	force, lb moment, in.-lb torque, in.lb

Member

Property of material	Property of section
tensile strength, σ compressive strength, σ shear strength, τ fatigue strength modulus of elasticity (tension), E modulus of elasticity (shear), E_s	area, A length, L moment of inertia, I (stiffness factor in bending) section modulus, S (strength factor in bending) torsional resistance, R (stiffness factor in twisting) radius of gyration, r

Stress and strain

Stress	Strain
tensile stress, σ compressive stress, σ shear stress, τ	resulting deformation, elongation or contraction, ϵ vertical deflection, Δ angular twist, θ

The use of design formulas may be illustrated by the problem of obtaining adequate stiffness in a cantilever beam. The problem obviously involves the amount of deflection at the end of the beam under a concentrated load (Fig. 2-1). The following deflection formula may be used:

$$\Delta = \frac{FL^3}{3EI}$$

where F is the given load (force) that would cause deflection, Δ, and L is the given length of the beam. The "member" terms are E, the modulus of elasticity — a property of the material — and I, the moment of inertia — a property of member cross-section.

Since it is desirable to have the least amount of deflection, the equation makes clear that E and I should be as large as possible. The commercial material having the highest modulus of elasticity is steel, with a value of 30×10^6 psi for E. The material thus picked, the only other factor requiring a decision relative to deflection is I, the moment of inertia. This, too, is a "member" factor — a property of section. Desirable, then, is a cross section having a moment of inertia large enough to hold Δ to a permissibly small value. If the designer chooses a section with adequate moment of inertia, he will have satisfied the deflection requirement, whatever the shape of the section.

It should be noted, however, that the designer's task does not end with discovery that steel is the best material and that a large moment of inertia with steel is the key to minimizing deflection. He

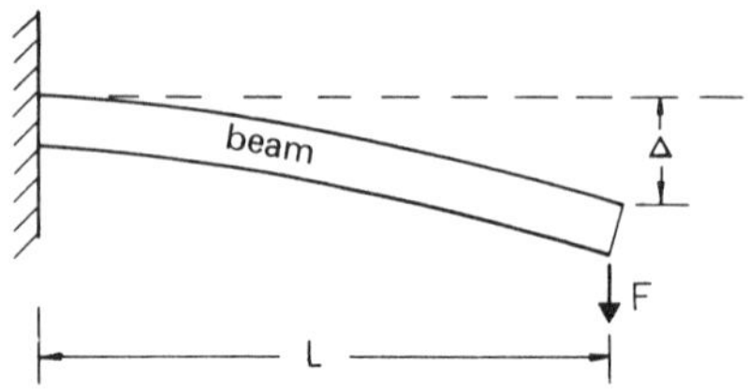

Fig. 2-1. Deflection (Δ) of a cantilever beam under a concentrated load (F).

must decide what shape to use for the best design at lowest cost. Could a standard rolled section be used? Should it be a box section? What are the fabrication costs of sections with the largest moments of inertia? Will the design have sufficient horizontal stability?

Loading

As indicated in Table 2-1, there are five basic types of loads — tension, compression, bending, shear, and torsion. Whatever the type of load, when it is applied to a member, the member becomes stressed. The stresses cause strains, or movements, within the member, the extents of which are governed by the modulus of elasticity of the material. The modulus, E, is defined as the ratio of stress to strain and is a constant value within the elastic limit, which for practical purposes may be regarded as the same as the yield point. Since a load always produces stress and strain, some movement always occurs.

Tension Loading

Tension is the simplest type of loading. It subjects the member to tensile stresses. Figure 2-2 shows a bar under simple axial tension. Here, there is no tendency for the bar to bend. If the force should be applied to a curved or deformed bar as in Fig. 2-3(a), a moment arm would result. The same condition exists when an eccentric load is applied to a straight bar. The axial tensile stress causes axial strains that tend to make the member elongate. The secondary bending stress causes strains that tend to make the member bend, but, in the case of a tensile force, the bending is in a direction that tends to reduce the initial eccentricity, as in Fig. 2-3(c). Thus, the bending moment tends to reduce itself, resulting in a stable condition. There is no danger of a tensile member buckling. Any shape, therefore, can suffice for tensile members. The only requirement is adequacy in the cross-sectional area, A.

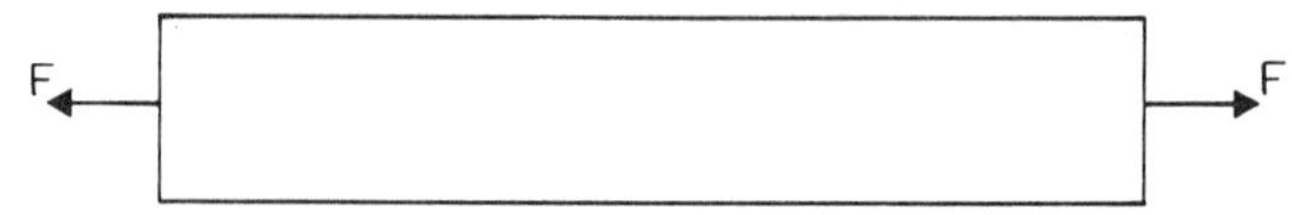

Fig. 2-2. Bar under simple axial tension.

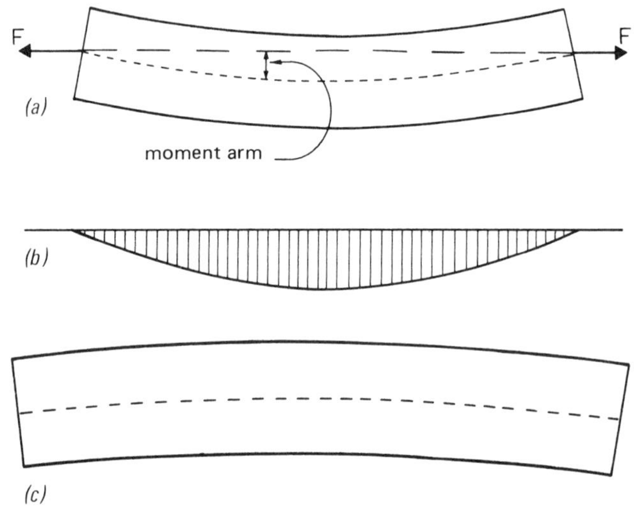

Fig. 2-3. (a) Tensile forces applied on a curved bar result in bending moment diagram as shown in (b); (c) bending moments tend to cause bar to move in the direction shown, thus reducing curvature. (Illustration exaggerates movement).

Compression Loading

A compression force, however, requires designing against buckling. Very few compression members fail by crushing, or exceeding their ultimate compressive strength. If a compression member, such as the column in Fig. 2-4(a), is loaded through its center of gravity, the resulting stresses are simple axial compressive stresses. Because of its slenderness (usually measured by the ratio of its unsupported length to its least radius of gyration and also referred to as its stenderness ratio, L/R), the column will start to move laterally at a stress lower than its yield strength. This movement is shown in Fig.2-4 (b). As a result of this lateral movement, the central portion of the column is then eccentric with the axis of the force, so that a moment arm develops. This causes a bending moment — Fig.2-4 (c) — on the central portion of the column, with resulting bending stresses. The bending stresses cause bending strains, and, as seen in Fig. 2-4(d), these strains cause the column to buckle more. This in turn creates additional eccentricity, a greater moment arm, more moment and still further lateral movement. Finally a point of no return is reached, and the column fails.

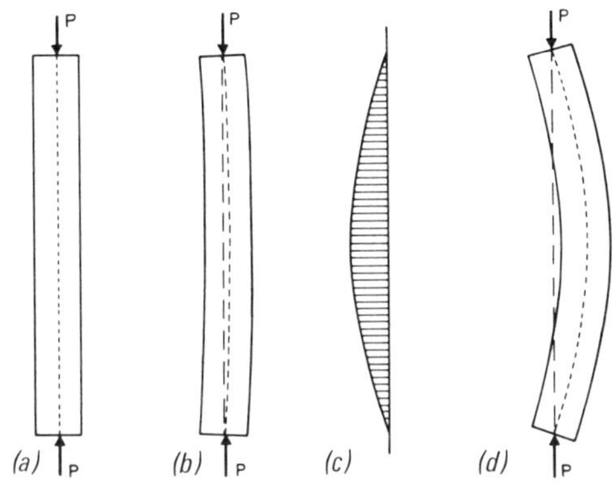

Fig. 2-4. (a) Straight column with concentric load; (b) with increased loading, column tends to move laterally; (c) bending moment diagram as a result of lateral movement; (d) additional buckling as a result of bending moment.

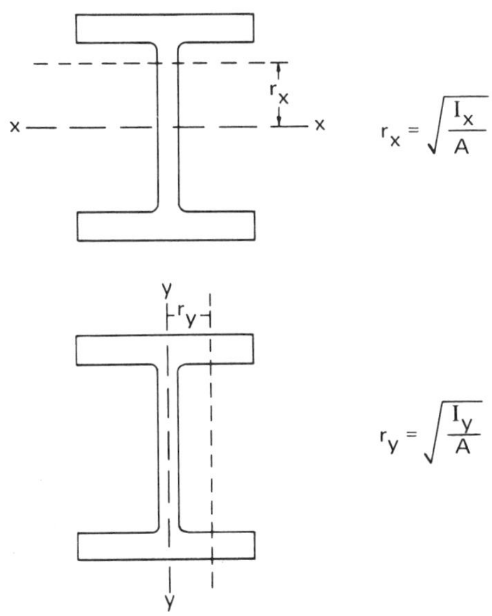

Fig. 2-5. Radius of gyration about the x-x and y-y axes of a column.

Two properties of a column section — area, A, and radius of gyration, r — are important to compressive strength. Area is important because it must be multiplied by the allowable compressive stress to arrive at the compressive load that can be carried. The radius of gyration is important because it indicates to a certain extent the ability of the section to resist buckling. The radius of gyration is the distance from the neutral axis of the section to an imaginary line (Fig. 2-5) in the cross section about which the entire section could be concentrated and still have the same moment of inertia of that section. Since the worst condition is of concern in design work, it is necessary to use the least radius of gyration. Thus, since $r = \sqrt{\frac{I}{A}}$, it is necessary to use the smaller of the two moments of inertia about the x-x and y-y axes to get radius of gyration used in the slenderness ratio.

The design of a compression member or column is by trial and error. A trial section is sketched and its area, A, and the least radius of gyration, r, are determined. A suitable column table will give the allowable compressive stress for the particular slenderness ratio $\left(\frac{L}{r}\right)$. This allowable stress is then multiplied by the area, A, to give the allowable total compressive load that may be placed on the column. If this value is less than that to be applied, the design must be changed to a larger section and tried again. Table 2-2 gives the American Institute of Steel Construction column formulas, and Fig. 2-6 gives allowable compressive stresses with various slenderness ratios, $\frac{L}{r}$.

TABLE 2-2. ALLOWABLE COMPRESSIVE STRESS (AISC)

Range of $\frac{L}{r}$ Values	Average Allowable Compressive Unit Stress (σ)
0 to C_c	$\sigma = \left[1 - \frac{\left(\frac{KL}{r}\right)^2}{2C_c^2}\right] \frac{\sigma_y}{FS}$
C_c to 200	$\sigma = \frac{149 \times 10^6}{\left(\frac{KL}{r}\right)^2}$
where: $C_c = \sqrt{\frac{2\pi^2 E}{\sigma_y}}$ and $FS = \frac{5}{3} + \frac{3\left(\frac{KL}{r}\right)}{8C_c} - \frac{\left(\frac{KL}{r}\right)^3}{8C_c^3}$	

1. For very short columns this provides a factor of safety of 1.67; for longer columns this gradually increases (by 15.0% max) up to 1.92. (Formulas developed earlier were based on a uniform 1.8 factor of safety.)
2. FS = factor of safety; σ_y = yield strength; K = effective length factor.

Bending Loading

Figure 2-7 illustrates bending. When a member is loaded in bending, it is assumed that the bending stresses are zero along the neutral axis and increase linearly, reaching a maximum value at the outer fibers. For a straight beam, the neutral axis is the same as the center of gravity. On one side of the neutral axis, tensile stresses are present; on the other side, there are compressive stresses. These stresses at a given cross section are caused by the bending moment at that particular section. The

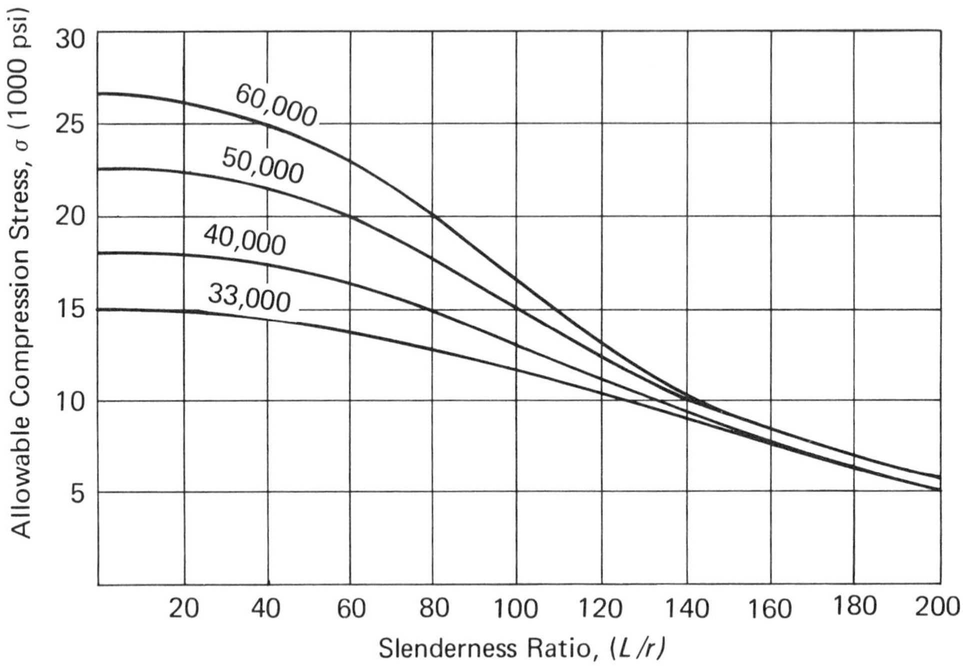

Fig. 2-6. Allowable compressive stress on columns for various slenderness ratios and tensile yield strengths.

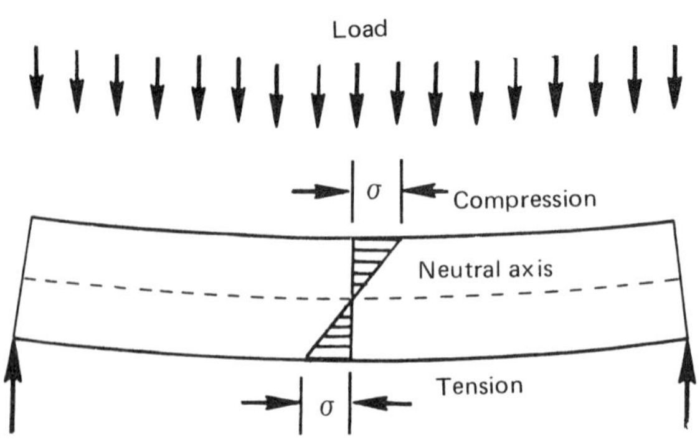

Fig. 2-7. Bending of a beam with uniform loading.

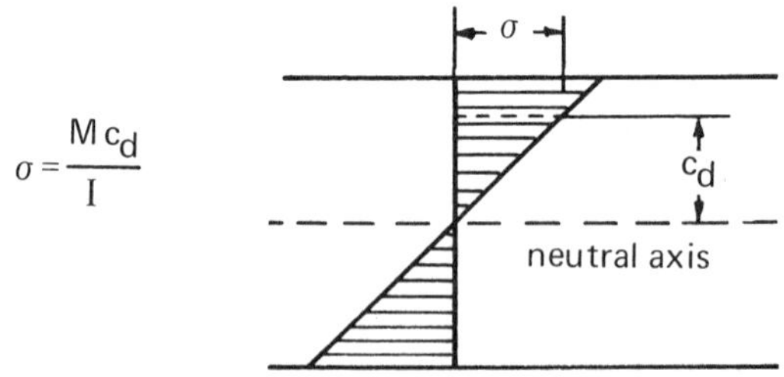

Fig. 2-8. Bending stress at any point (c) in the cross section of a straight beam may be readily calculated.

bending stress at any point in the cross section of a straight beam (Fig. 2-8) may be found by using the formula found above where M is the moment at that point and c_d is the distance from the neutral axis to the point in question.

In most cases, the maximum bending stress (Fig. 2-9) is of greater interest, in which case the formula becomes:

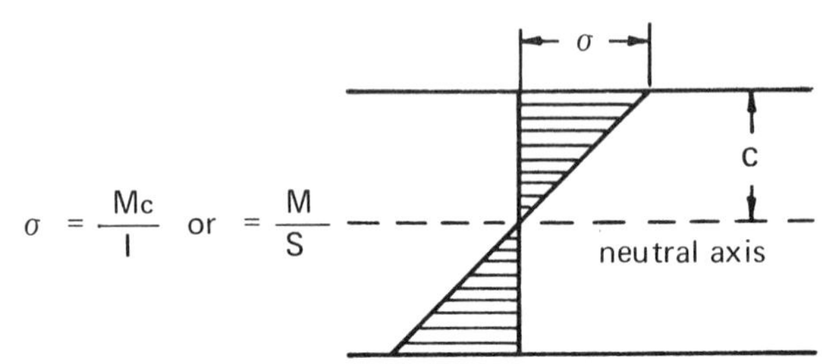

Fig. 2-9. Maximum bending stress in the outside fibers (c) of a beam is of greater interest.

where C is the distance from the neutral axis to the outer fiber and S is section modulus.

As the bending moment decreases along the length of the beam toward its end, the bending stresses in the beam also decrease. This means the bending force in the flange is decreasing as the end of the beam is approached. If a short length of the tension flange within the beam is considered, as shown enlarged in the inset above the beam in Fig. 2-10, a difference in the tensile forces at the two ends is found to exist.

How can a tensile force at one end of a plate

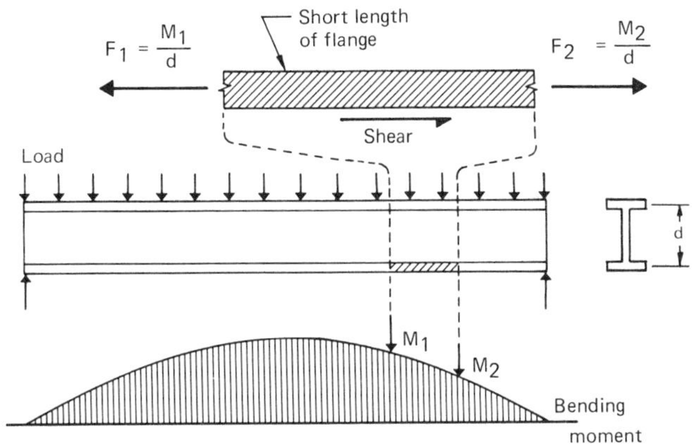

Fig. 2-10. Tensile forces existing in the flange of a beam under load.

be different from that at the other end? The answer is that some of the tensile force has transferred out sideways as shear. This means that, whatever the decrease in the tensile force in the flange, there is a corresponding shear force between the flange and the web through the fillet welds joining the two together. The same thing happens on the upper flange, which is in compression. The decrease in tensile force in the lower flange transfers out as shear up through the web to the other flange and makes up for the decrease in compression in this flange.

Shear Loading

Figure 2-11 illustrates the shear forces in the web of a beam under load. They are both horizontal and vertical and create diagonal tension and diagonal compression. Tension is not a problem, since there is little chance of the tensile stress reaching a value high enough to cause failure. If the diagonal compression reaches a high enough value, however, the web could buckle. For this reason, structural codes limit the shear stress in the web of a beam because there is a possibility of the web buckling, especially if it is very deep or very thin.

The unit shear force on the fillet welds joining the flanges of the beam (Fig. 2-12) to the web can be calculated from the formula for f lbs.:

where

V = external shear force on the member at this cross section, lbs.

a = area held by the connecting welds, in.^{2}

y = distance between the center of gravity of the area held by the welds and the neutral axis of the whole section, in.

I = moment of inertia of the whole section about the neutral axis, in.^{4}

n = number of welds used to hold this area.

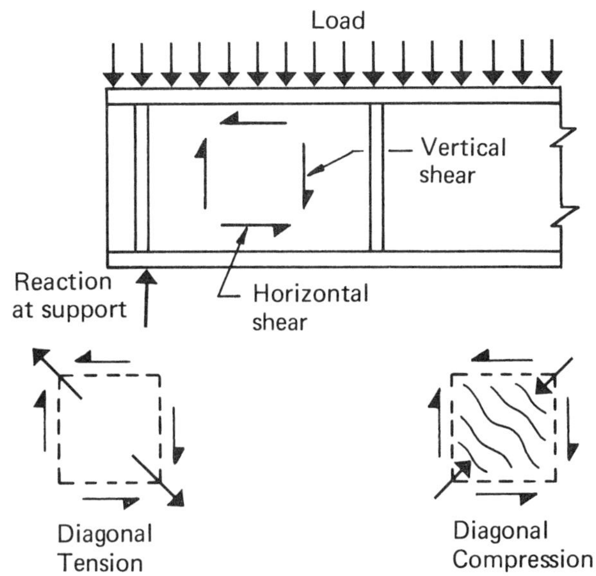

Fig. 2-11. Shear forces in the web of a beam under load.

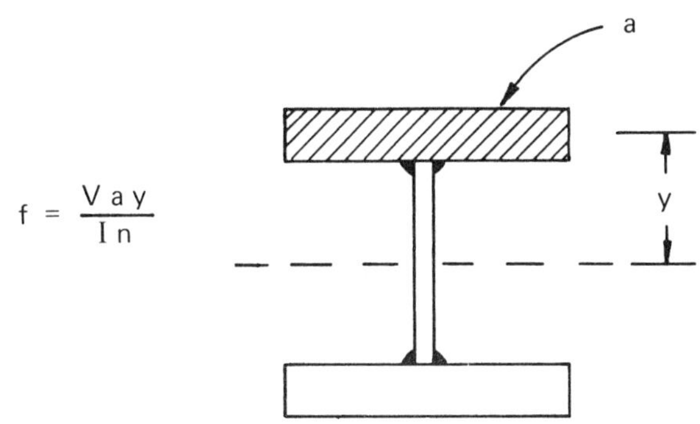

Fig. 2-12. Shear force on the fillet welds connecting the flanges of a beam to the web may be calculated by the formula for (f).

A common bending problem in machinery design involves the deflection of beams. Beam formulas found in many engineering books and nomographs, such as those shown as Figs. 2-13 and 2-14, are useful for quick approximations of the deflection with common types of beams and loads. An example of a typical beam problem, along with applicable formulas, is shown as Fig. 2-15.

To satisfy strength and stiffness requirements, beams should be deep. What was said about a compression member or column can also be said about the compression flange of a beam. It is important that its shape be such that it will not buckle easily, it should not be too wide and too thin. It should have proper lateral or horizontal

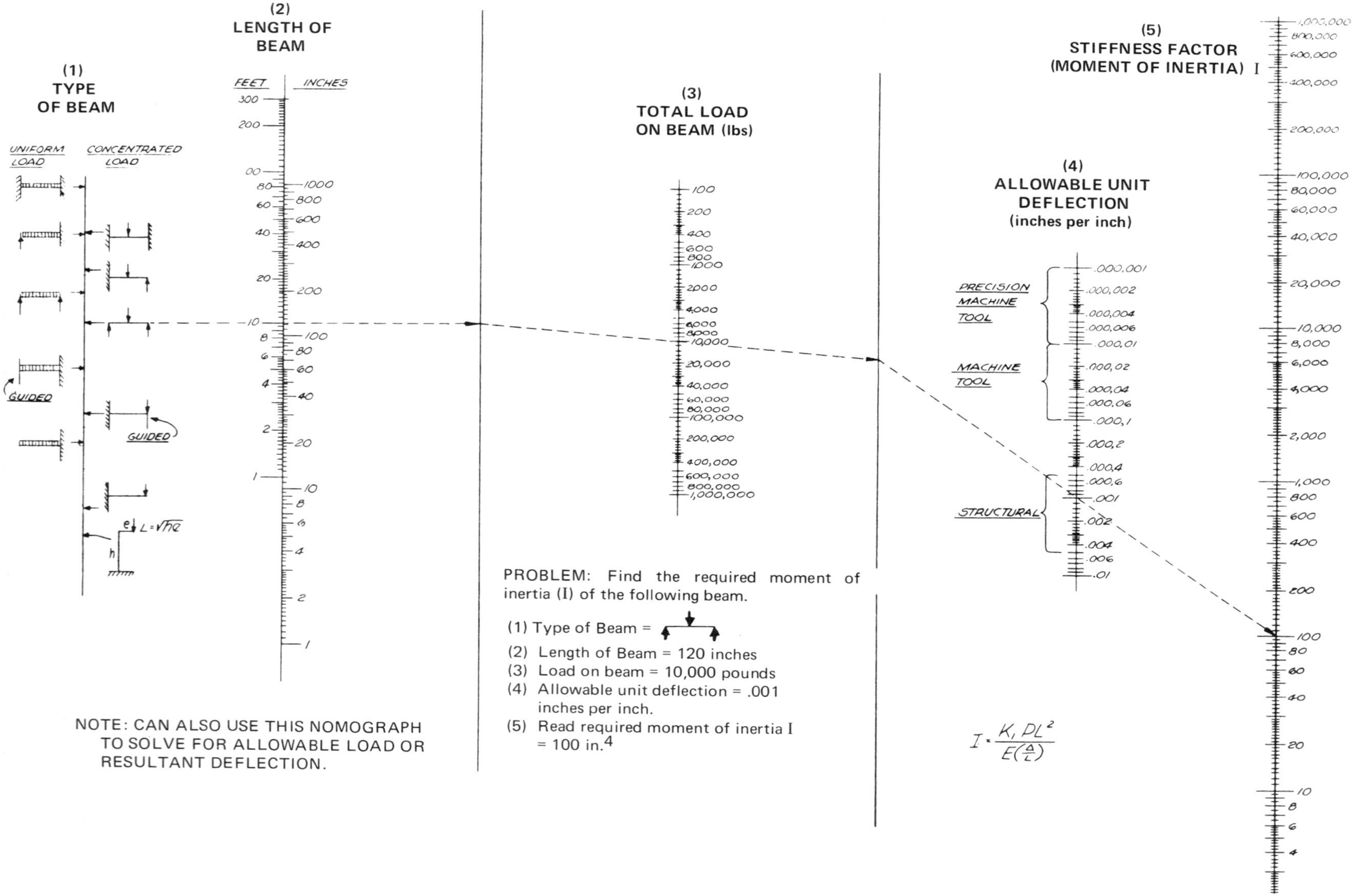

Fig. 2-13. Required moment of inertia of beam under bending load (stiffness nomograph).

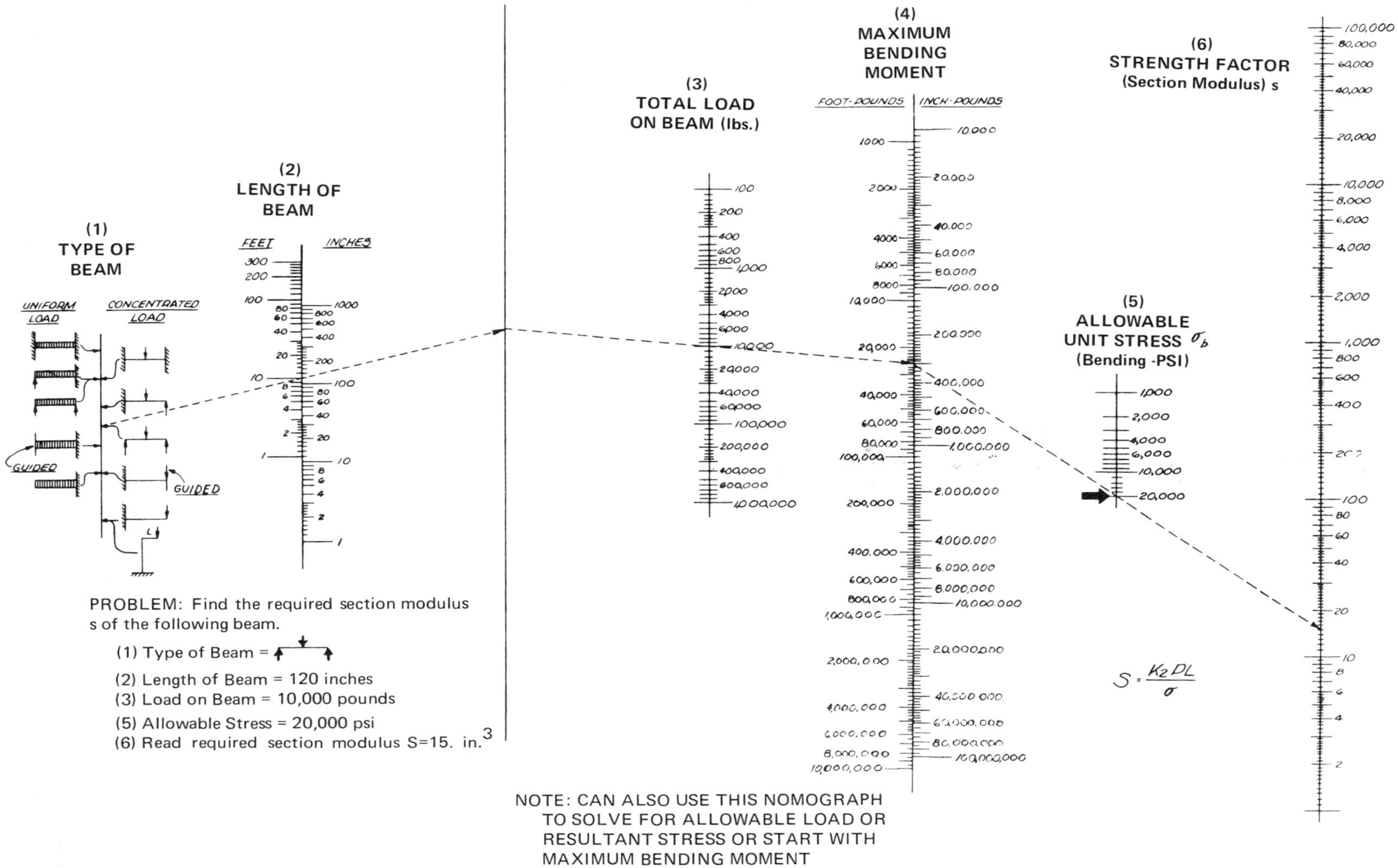

Fig. 2-14. Required section modulus of beam under bending load (strength nomograph).

Bridge girder design with a hinged connection.

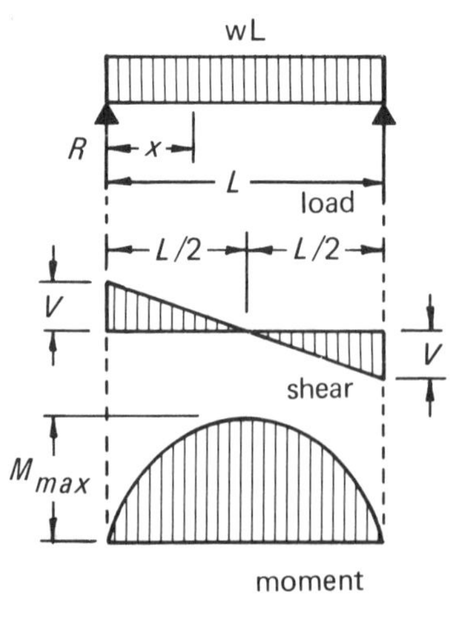

$$R = V = \frac{wL}{2}$$

$$V_x = w\left(\frac{L}{2} - x\right)$$

At center, $$M_{max} = \frac{wL^2}{8}$$

$$M_x = \frac{wx}{2}\ (L - x)$$

At center, $$\Delta\, max = \frac{5wL^4}{384EI}$$

$$\Delta x = \frac{wx}{24EI}\ (L^3 - 2Lx^2 + x^3)$$

At ends, $$\theta = \frac{wL^3}{24EI}$$

Fig. 2-15. Typical beam problem, solvable with nomographs in Figures 2-13 and 2-14.

This welded steel machine base was designed to replace a cast iron base.

support, and the compressive bending stresses must be held within allowable values. It has been found, in testing, that a beam initially fails because the compression flange rotates, thus causing the web to which it is attached to bend outward also and start to buckle, as sketched in Fig. 2-16. Fortunately, most beams have stabilizing members such as crossbeams or flooring, attached to them, and these provide lateral support against buckling. The lower flange, which is not supported, is usually in tension and thus not a source of buckling problems.

Torsional Loading

Torsion — the fifth type of loading — creates problems in the design of bases and frames. A machine with a rotating unit subjects its base to torsional loading. This becomes apparent by the lifting of one corner of the base when the base is not bolted down.

If torsion is a problem, closed tubular sections, as in Fig. 2-17, or diagonal bracing should be used. Closed tubular sections can easily be made from existing open channel or I sections by intermittently welding flat plate to the toes of the rolled sections, thus closing them in. This will increase the torsional resistance several hundred times. An existing frame may be stiffened for torsion by welding in cross bracing at 45° to the axis of the frame. This also increases the torsional resistance by several hundred times.

The torsional resistance of an open section is very poor. The torsional resistance of a built-up section is approximately equal to the sum of the torsional resistances of the individual flat parts that make up the member. The torsional resistance of a flat section, as shown in Fig. 2-18, may be approximated by the formula:

$$R = \frac{Wt^3}{3}$$

where

R = torsional resistance
W = width of the flat section
t = thickness of the flat section

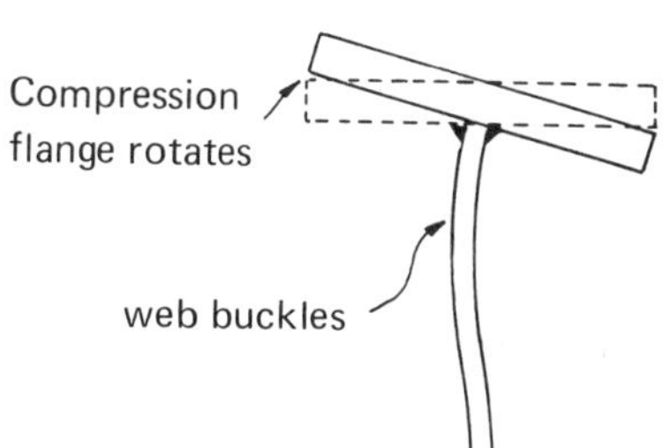

Fig. 2-16. Mode of initial failure of beam in compression flange.

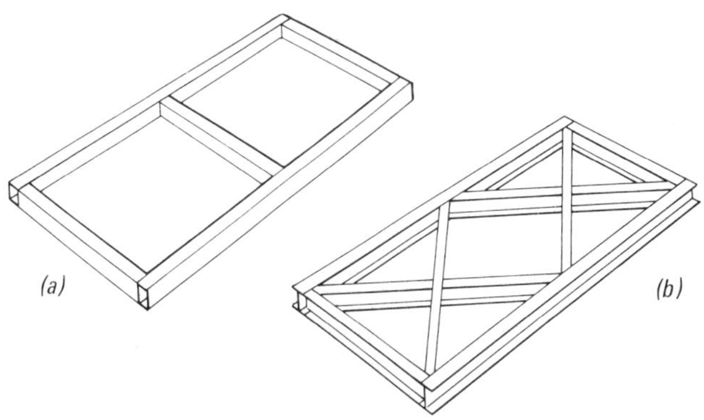

Fig. 2-17. (a) Use of closed box sections to resist torsion; (b) use of diagonal bracing to resist torsion.

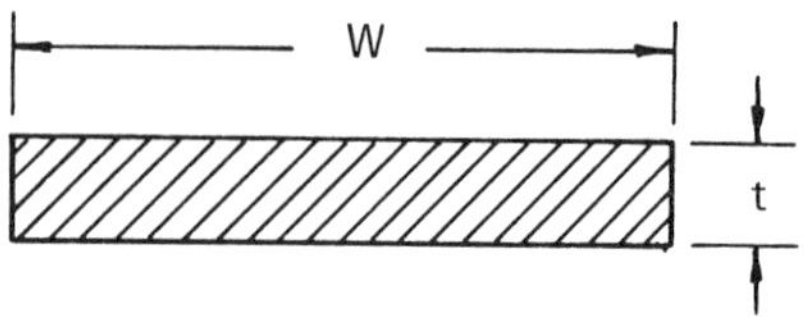

Fig. 2-18. Cross-sectional dimensions of flat section whose torsional resistance, R, is to be determined.

TABLE 2-3. ANGLES OF TWIST

all loadings identical	t = .055 (2)	t = .055 (2 × 2)
Conventional method J polar moment of inertia	.065°	.007°
Method using R Torsional Resistance	21.8°	7.3°
Actual twist	22°	9.5°

Table 2-3 shows the results of twisting a flat section, as well as a small I-beam made of three identical plates. Calculated values of twist by using the conventional polar moment of inertia, J, and torsional resistance, R, are compared with the actual results. This shows the greater accuracy obtained by using the much simpler method — calculating with torsional resistance, R.

No appreciable gain in torsional resistance of individual members occurs until the section is closed. When a flat piece of 16-gage steel was tested for torsional resistance under a given load, it twisted 9 degrees. When formed into a channel, it twisted 9-1/2 degrees; and when a similar piece was rolled into an open-seam tube, the measured twist was 11 degrees. Closed sections made from the same width of material gave, however, torsional resistances from 50 to 100 times as much.

Table 2-4 shows the calculated values of twist by using the conventional polar moment of inertia, J, as well as the torsional resistance, R. The actual values are also shown. Again, this shows the greater accuracy obtainable by using the torsional resistance, R.

TABLE 2-4. ANGLE OF TWIST OF VARIOUS SECTIONS

all loadings identical	t = .060 (3 1/2)	t = .060 (2 1/8, 3/4)	t = .060 (1.15)	t = .060	t = .060
Conventional method J polar moment of inertia	.01°	.006°	.04°	.04°	.045°
Method using R Torsional Resistance	9.5°	9.7°	10°	.04°	.06°
Actual twist	9°	9.5°	11°	too small to measure	too small to measure

The torsional resistance of a frame whose lengthwise members are two channels (Fig. 2-19) would be approximately equal to twice the torsional resistance of each channel section. For the purpose of this example, the distance between these side members is considered to have no effect. Since the closed section is best for resisting twisting, the torsional resistance of a frame can be greatly increased by making the channels into rectangular box sections through the addition of plates.

Once the torsional resistance of an open section has been found, the angular twist may be calculated by the formula:

$$\theta = \frac{TL}{E_S R}$$

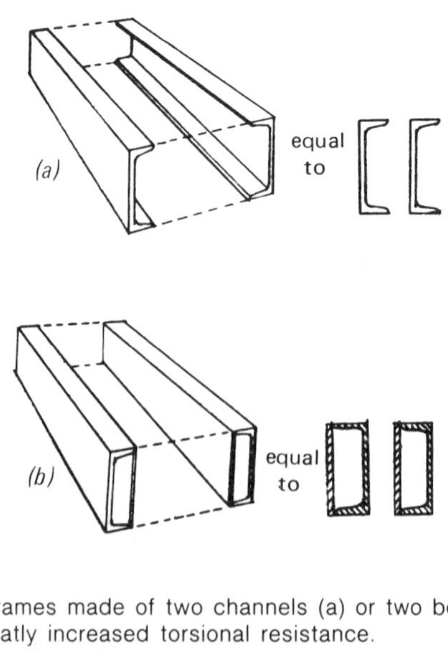

Fig. 2-19. Frames made of two channels (a) or two box sections (b) show greatly increased torsional resistance.

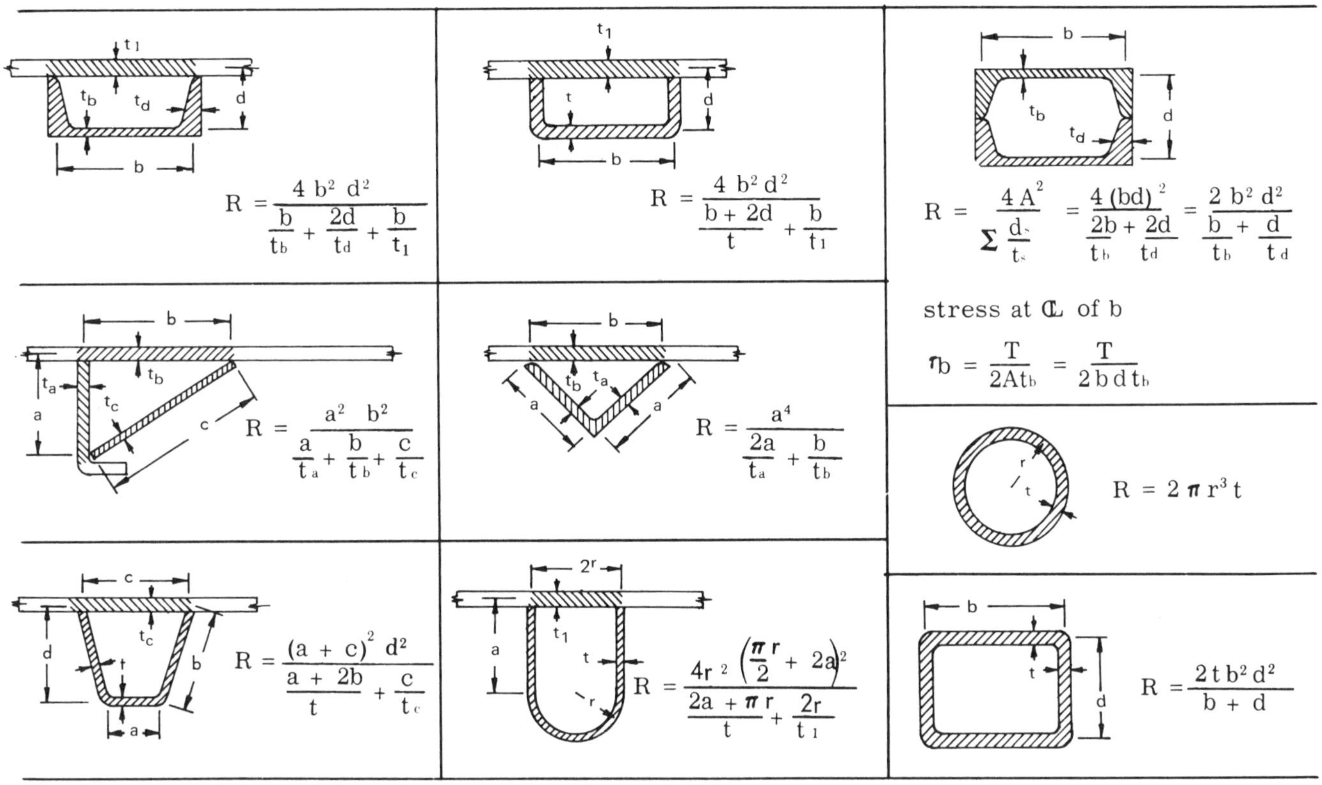

Fig. 2-20. Torsional properties, R, of closed tubular sections.

where E_s equals modulus of elasticity in shear. This formula would also be used for a round shaft.

In the formula for angular twist, in the following formulas, and in Fig. 2-20, the following terms are used:

A	=	area enclosed within mean dimensions (dotted line) (in.^{2})
d_s	=	length of particular segment of section (in.)
t_s	=	corresponding thickness of section (d_s)(in.)
τ_s	=	shear stress at point (s) (psi)
R	=	torsional resistance, (in.^{4})
T	=	torque (in.-lb)
E_s	=	modulus of elasticity in shear (steel = 12,000,000 psi)
θ	=	total angular twist (radians)
φ	=	unit angular twist (radians/in.)
L	=	length of member (in.)
a, b & d	=	mean dimensions of section (in.)
r	=	radius of section (in.)

Each part of an open section will twist the same angle as the whole member. The unit angular twist, φ, is equal to the total angular twist divided by the length of the member.

$$\varphi = \frac{\theta}{L}$$

Knowing the unit angular twist, it is possible with the following formulas to find the resulting shear stress on the surface of the part.

$$\tau = \varphi t E_s = \frac{Tt}{R}$$

The torsional resistance of any closed tubular shape (Fig. 2-21) can be determined by drawing a dotted line through the midthickness all the way around the section. The area enclosed by the dotted line, or mean dimension, is A. Divide this section into convenient lengths. The ratio of these individual lengths divided by their corresponding thicknesses is determined and totaled. Torsional resistance is then obtained from the relation:

$$R = \frac{4A^2}{\Sigma \frac{d_s}{t_s}}$$

Figure 2-20 gives formulas for calculating the torsional resistance of various closed tubular sections. Since most sections resolve themselves into three

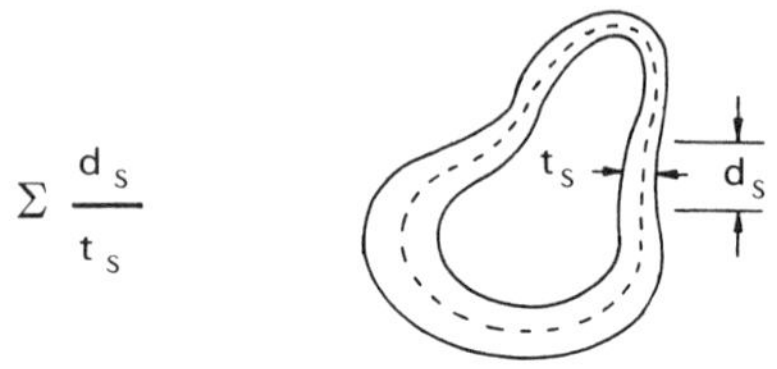

Fig. 2-21. Torsional resistance of any closed tubular section can be calculated by determining mean dimensions and dividing the section into convenient lengths.

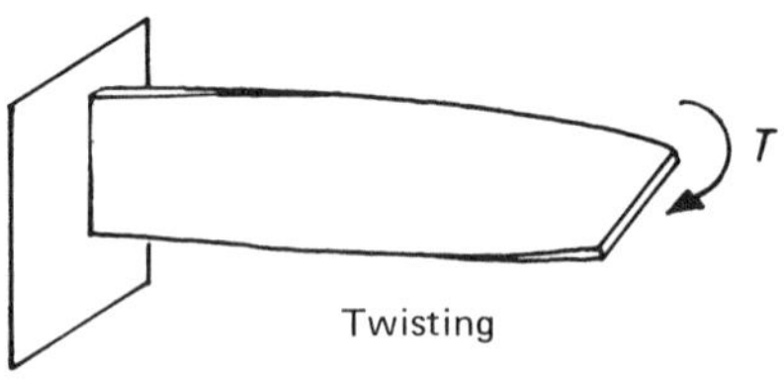

Fig. 2-22. Twisting of flat strip of steel.

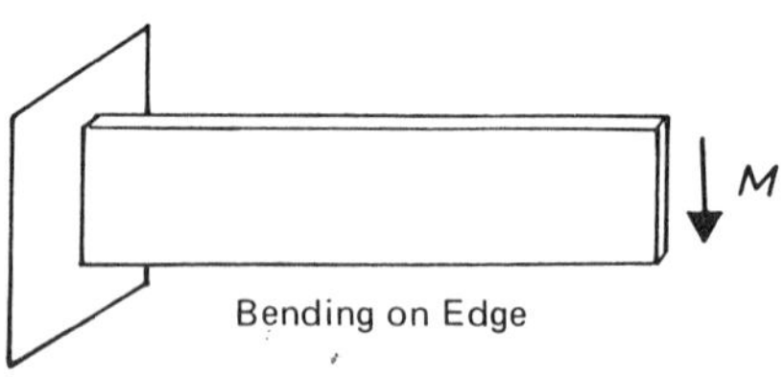

Fig. 2-23. Strip of steel on edge shows exceptional resistance to bending loads.

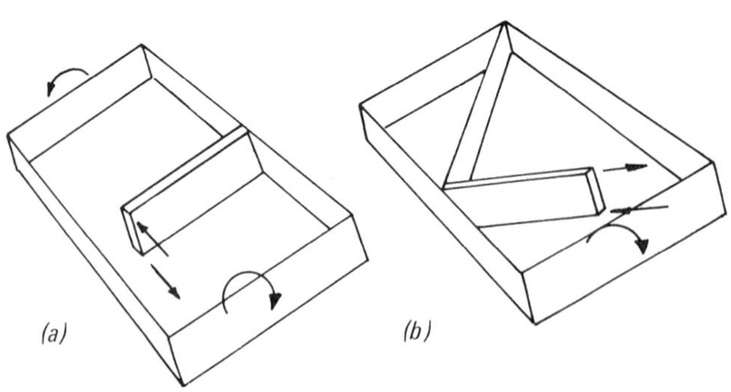

Fig. 2-24. (a) Frame with cross ribs shows little resistance against twisting; (b) diagonal bracing offers exceptional resistance to bending and to twisting of the entire frame.

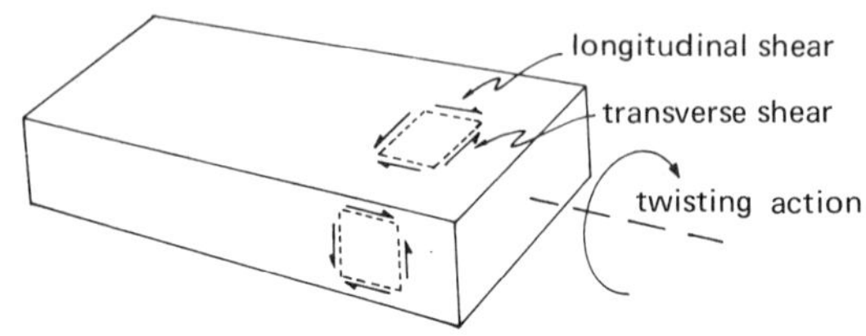

Fig. 2-25. Transverse and longitudinal shear stresses in a frame under torsion.

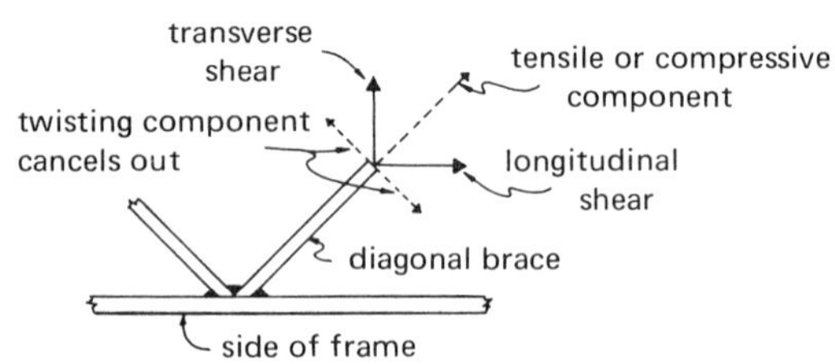

Fig. 2-26. Transverse and longitudinal shear stresses in a frame with diagonal bracing.

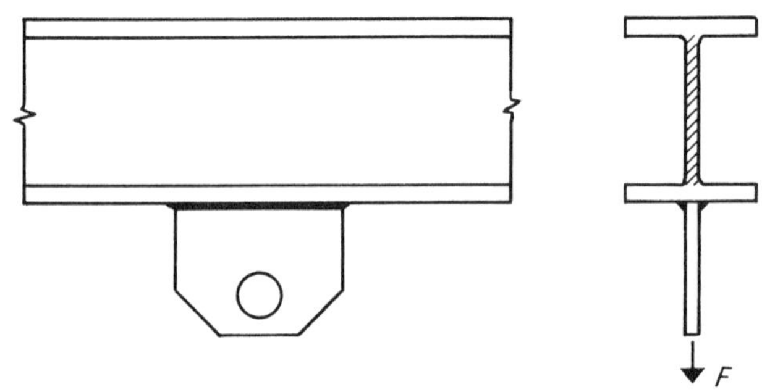

Fig. 2-27. Lug welded parallel to the length of a beam.

or four flat plates, the work required to determine the torsional resistance is greatly simplified.

Diagonal bracing is very effective for preventing twisting. Why is this so? There are several ways to explain the effectiveness of diagonal bracing. A simple explanation arises from an understanding of the direction of the forces involved.

A flat strip of steel (Fig. 2-22) has little resistance to twisting, but when set on edge has exceptional resistance to bending loads (Fig. 2-23).

Consider that a base or frame under torsion has two main stresses: transverse shear stress and longitudinal shear stress. Any steel panel will twist if these stresses are acting at right angles to its axis (transverse), Fig. 2-25.

A frame with cross ribs does not have a great deal of resistance against twisting. This is because, as shown in Fig. 2-24(a) the longitudinal shear stress is applied normal or transverse to the rib, producing torque, and under this torque it will twist just as the flat piece of steel seen in Fig. 2-22.

A consideration of the action of the stresses on a steel panel with 45° diagonal bracing, Fig. 2-26, reveals an entirely different set of conditions. If these shear forces are broken down into their two components, parallel and transverse to the brace, the transverse components cancel out and eliminate the twisting action. The parallel shear components act in the same direction to place one edge of the brace in tension and the other edge in compression, hence the diagonal brace is subjected to a bending action, which it is capable of resisting, and thus the diagonal bracing greatly stiffens the frame.

Transfer of Forces

Loads create forces that must be carried through the structure the engineer is designing to suitable places for counteraction. The designer needs to know how to provide efficient pathways.

One of the basic rules is that a force applied transversely to a member will ultimately enter that portion of the section that lies parallel to the applied force.

Figure 2-27 shows a lug welded parallel to the length of a beam. The portion of the cross section of the beam that lies parallel to the applied force — and thus receives that force — is the web. The force in the lug is easily transferred through the connecting welds into the web, and no additional stiffeners or attaching plates are required. Suppose, however, that the lug were welded to the beam flange at right angles to the length of the beam. The condition shown in Fig. 2-28 then exists: the outer edges of the flange tend to deflect, and the small portion of the weld in line with the web is forced to carry an excessive amount of load.

To prevent this situation, two stiffeners might be welded inside the beam in line with the lug (Fig. 2-29). This will result in an even distribution of force through the welds and through the lug. The stiffeners keep the bottom flange from bending downward, thereby providing a uniform transfer of force through the weld. Since this force enters the web of the beam, there is no reason for welding the stiffener to the top flanges. The three welds (a), (b), and (c) must all be designed to carry the applied force, F.

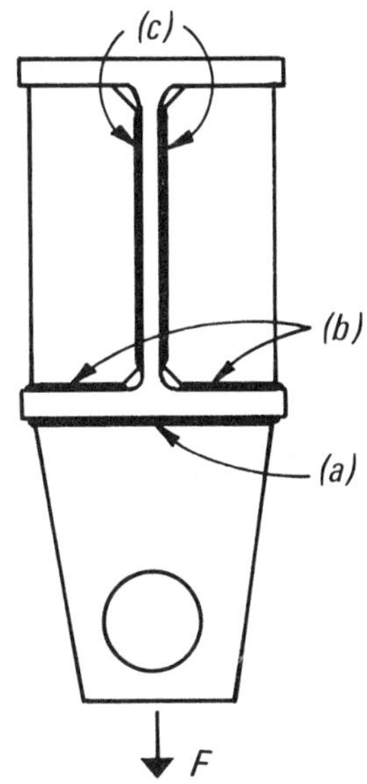

Fig. 2-29. Stiffeners welded inside the beam in line with lug of Fig. 2-28.

If for some reason, the force is to be applied parallel to the flanges, it would not be sufficient to simply attach the lug to the web of the beam. This would cause excessive bending of the web (Fig. 2-30) before it would load up and be able to transfer the force out to the flanges. In such a situation, the lug could be welded into the beam, as in Fig. 2-31, to act as a stiffener, welding to the flanges

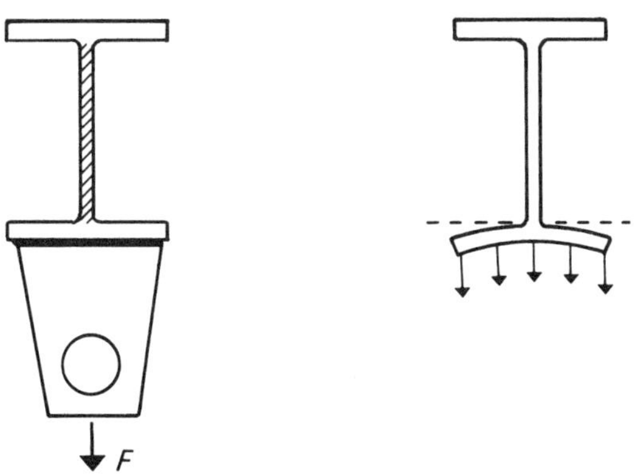

Fig. 2-28. Lug welded at right angles to the length of a beam (left); force causes flange to deflect (right).

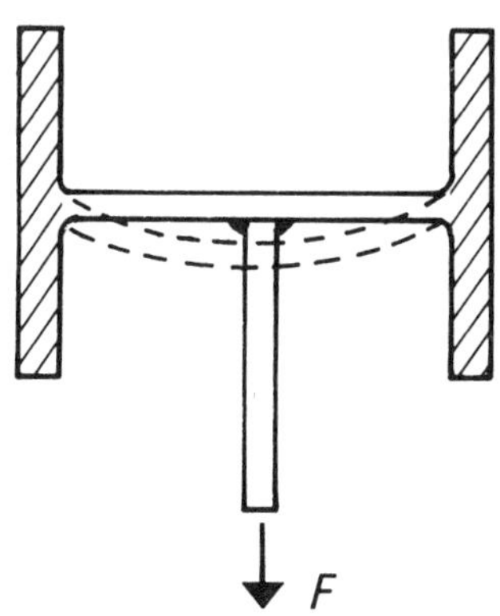

Fig. 2-30. Force applied to web of beam, parallel to flanges, causes excessive bending.

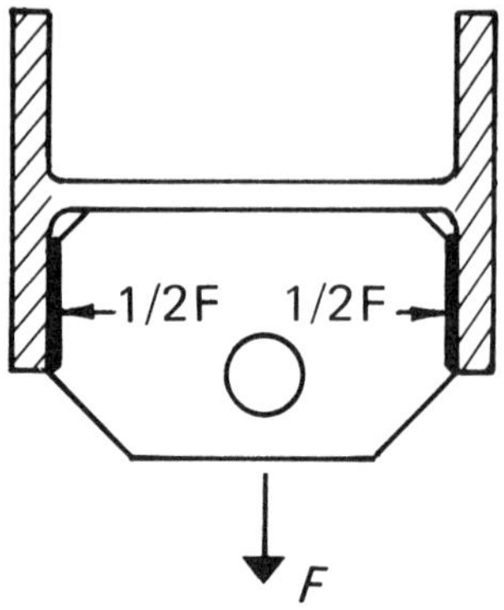

Fig. 2-31. Lug welded to flanges as shown to prevent bending of web.

only would be adequate, since no force is transferred to the web.

For larger loads, it might be necessary to use a stiffener above the web (Fig. 2-32). In this case, both stiffeners would be welded to the web as well as the flanges, since the top stiffener could be loaded only through the welds along the web.

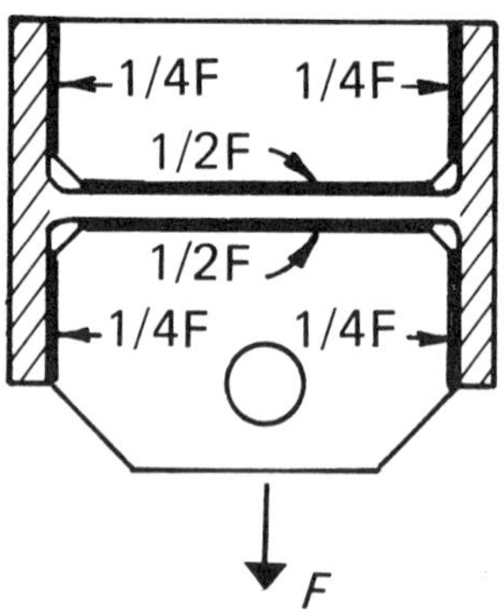

Fig. 2-32. Stiffener used above the web and welded, as shown, to handle larger loads.

Diagonal Bracing

In designing braces, one must be careful to avoid conditions that permit initial movement before the brace loads up. Figure 2-33 shows diagonal bracing made with bent plate. This looks substantial, but the bracing will actually permit some initial twisting of the frame. The reason is brought out by the sketch of the intersection. There is a jog in the dotted line through the center of the diagonal bracing, at the intersection. Under load, tensile forces tend to straighten out one centerline and compressive forces cause additional bending in the other. The result is that some initial movement of the braces occurs, with twisting of the whole frame before the diagonal braces load up. Diagonal braces should always intersect without bends.

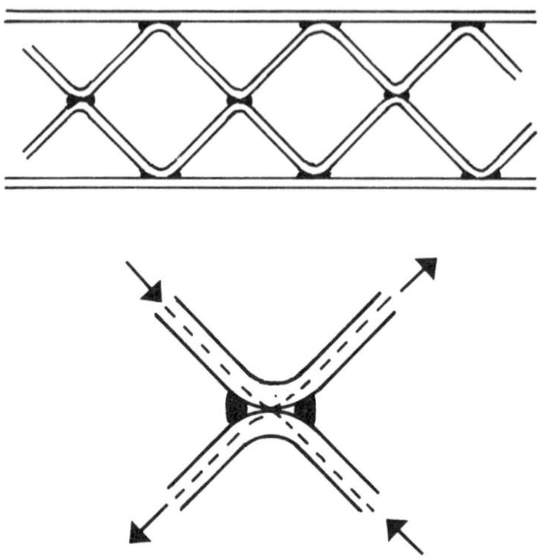

Fig. 2-33. Diagonal bracing made with bent plate (top); enlarged view of intersection of diagonal bracing, showing jog (bottom).

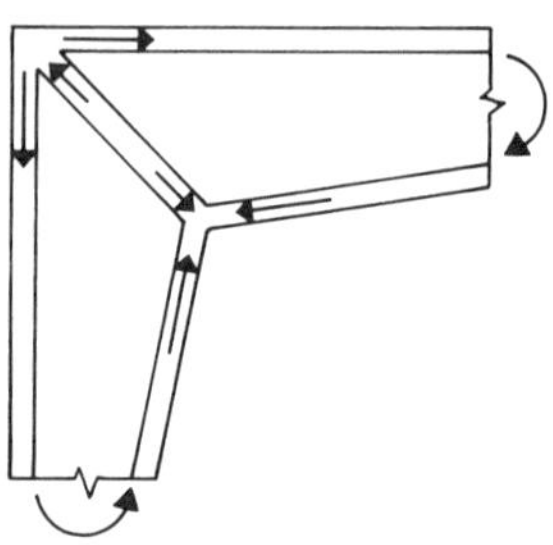

Fig. 2-34. Knee of rigid frame.

Any time a force changes direction, a force component is involved. Figure 2-34 shows a knee of a rigid frame. The compressive force in the lower flange must change direction as it passes into the other flange. To accomplish this, a diagonal stiffener is placed at the intersection of the two flanges. The diagonal force component passes up through the diagonal stiffener and becomes the force component that is also required to change the direction of the tensile forces in the upper or outer flange. Since the force changes direction at a single point, only one diagonal is needed.

THE DESIGN PROCEDURE

When the engineer has acquired a basic understanding of the design approach, the many design and shop factors that should be taken into account, the types of loading he is to deal with, and the use of formulas, he is ready to proceed with his design. Before applying the various formulas, the problem itself should be carefully analyzed and clearly defined.

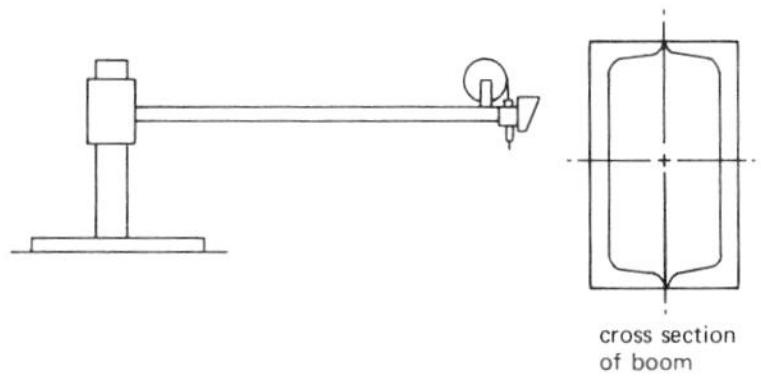

Fig. 2-35. Schematic of automatic welding head (left); cross section of boom (right).

Many uneconomical and unacceptable designs often result from incorrect definitions.

The design procedure may be illustrated by considering the case of a fabricating plant that has set up an automatic welding head on a boom, under which the work moves on a track. At a later date, it becomes necessary to extend the length of the boom so that larger workpieces can be handled (Fig. 2-35).

Defining his problem, the engineer recognizes a simple cantilever beam with a concentrated load at the outer end. The weight of the automatic welding head with its wire reel and flux constitutes the type and amount of loading. This load will produce vertical deflection, which he arbitrarily decides should not exceed 1/8 inch. Even though there is no known horizontal force applied to the beam, he assumes that he should design for a horizontal force of about one quarter the vertical force.

At this stage the problem is fairly well described. The engineer considers the possible use of a box beam of adequate size, a box beam made by welding two channels together toe-to-toe with depth greater than width as shown in Fig. 2-35.

When the welding head it is mounted at the end of the new boom, the boom deflects downward 1/8 in. — just as planned — and there it will remain until it is replaced or moved to another location. Perhaps the real design problem in this instance (the one that was entirely overlooked) concerns movement of the welding head during welding. This can be caused by accidental bumping, or possibly by a crane passing overhead that shakes the building framework.

If this is the case, then, it becomes apparent from investigation that the electrical circuitry of the welding head automatically compensates for limited vertical movement along the y-y axis. Horizontal movement cannot be tolerated, however, since it produces poor bead appearance and can affect the strength of the weld if penetration along the centerline of the joint is important. Obviously the box section configuration in Fig. 2-35 is the wrong design for this situation since the moment of inertia along the vertical y-y axis (I_y) is much less than that about the horizontal x-x axis, (I_x). The solution for this condition would probably call for a beam more nearly square in cross section.

Material Selection

In this case — as in most cases with machine members — there was no question about material selection. Steel was the obvious material. In other instances, after the problem has been defined and load conditions established, it may be necessary to select the best material. Sometimes the designer must establish whether he has a strength problem or a rigidity problem and pick the material that meets his needs most economically.

Here, it is easy to be led astray by wrong premises. One machine tool company experimented with bases of different steels under the assumption that a higher strength steel ought to give a more rigid base. Company engineers were surprised to observe that all the steels tried gave the same deflection under the same load. Had they known that the property of a material that indicates its relative rigidity is its modulus of elasticity — and that all steels have the same modulus and thus the same rigidity — needless experimentation would have been avoided.

Another company was experiencing a deflection difficulty with a lever that operated at very high speeds. The engineers reasoned that the forces were due mainly to inertia, and decided that, if a lighter metal could be used, the mass would be decreased. The inertia formula is:

$$F = ma$$

This in turn would decrease the inertia and reduce the deflection. They had a new lever made from aluminum for testing. At this point, it became relevant to use the following formula for mathematical analysis:

$$\Delta = \frac{KFL^3}{EI} \quad \text{and} \quad F = ma$$

therefore $$\Delta = \frac{KmaL^3}{EI}$$

where

Δ = deflection
K = beam constant
m = mass
a = acceleration or deceleration
E = modulus of elasticity (tension)
I = moment of inertia
L = length

Since the respective densities of aluminum and steel are 0.100 and 0.284 Lb/in.^{3} and the respective moduli of elasticity are 10.3 x 10^6 and 30 x 10^6 psi, inserting the ratios of these values (aluminum to steel) into the above deflection formula, one obtains:

$$\Delta_{aluminum} = \frac{(1)\left(\frac{0.100}{0.284}\right)(1)(1)^3}{\left(\frac{10.3 \times 10^6}{30 \times 10^6}\right)(1)} \Delta_{steel}$$

$$\Delta_{aluminum} = 1.03\,\Delta_{steel}$$

Thus, an aluminum lever designed for equivalent rigidity would actually have a deflection 1.03 times that of a steel lever. The lower modulus of elasticity of aluminum canceled out its lower weight advantage.

Another possible solution for the deflection problem involved seeing if a different section might be used. The deflection formula was put in a slightly different form:

$$\Delta = \frac{KAdaL^4}{EI}$$

where

A = cross-sectional area
d = density

Recalling that property of material and property of section determine the performance of a member, it is noted that $\frac{d}{E}$ in the formula is a material property. This means that for low deflection, $\frac{d}{E}$ should be as small as possible. Neither aluminum nor magnesium would give a value of $\frac{d}{E}$ significantly different from that of steel. This observation confirms the fact that material selection in this example will have little effect on deflection.

Turning away from this marginal approach, it is noted that $\frac{A}{I}$ in the deflection formula is a property of section. For minimum deflection, this ratio should be minimal. Or, the ratio could be inverted to give the opposite of deflection, namely stiffness. A ratio of $\frac{I}{A}$ should therefore be as great as possible. This ratio states that a high value for moment of inertia, I, and a small area. A, are desirable.

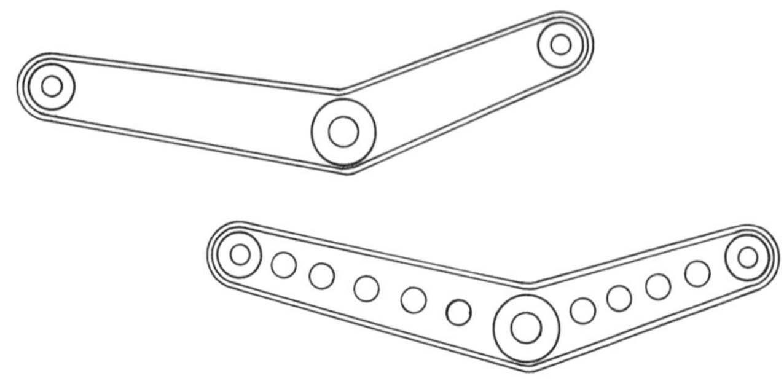

Fig. 2-36. Redesign of lever (left) consisted of cutting holes along neutral axis (right) to reduce sectional area, thus decreasing deflection under inertial forces.

Knowing also that the square root of $\frac{I}{A}$ is the radius of gyration, r, the ratio becomes:

$$\frac{I}{A} = r^2$$

This means that a section having a high radius of gyration should have high stiffness for this inertia load. Cutting holes in the lever along its neutral axis (Fig. 2-36) will reduce area, A, at a faster rate than it reduces moment of inertia, I, thus increasing the radius of gyration and decreasing deflection under the inertial forces. That this function actually occurs becomes increasingly credible when it is noted that it also amounts to reducing the mass in the formula F = ma (the original line of reasoning when aluminum was considered for the lever) without decreasing stiffness, I, to any great extent.

REDESIGN BY EQUIVALENT SECTIONS

When converting from a casting to a weldment, the engineer can avoid complicated computations by using tables of equivalent sections. Tables 2-5 and 2-6 are presented for this purpose. A three-step procedure is used:

1. Determine the type of loading under the basic requirements of strength or rigidity for each member.
2. Determine the critical properties of the original cast members with respect to the loading. The ability of the member to withstand loading is measured by properties of its cross section, such as —

   A = area of cross section
   I = moment of inertia
   S = section modulus, for strength in bending
   R = torsional resistance

TABLE 2-5. EQUIVALENT RIGIDITY FACTORS

STEP 1 – Determine the Type of Loading	Tension	Compression: Short Column	Compression: Long Column	Bending	Torsion
STEP 2 – Determine this property of the cast member.	Area A	Area A	Moment of Inertia I	Moment of Inertia I	Polar Moment of Inertia J
STEP 3	Multiply the above property of the cast member by the following factor to get the equivalent value for steel. *				
EQUIVALENT FACTORS					
Grey Iron ASTM 20	40%	40%	40%	40%	40%
ASTM 30	50	50	50	50	50
ASTM 40	63	63	63	63	63
ASTM 50	67	67	67	67	67
ASTM 60	70	70	70	70	70
Malleable A47–33 35018	83	83	83	83	100
A47–33 32510	83	83	83	83	100
Meehanite Grade GE	40	40	40	40	40
Grade GD	48	48	48	48	48
Grade GC	57	57	57	57	57
Grade GB	60	60	60	60	60
Grade GA	67	67	67	67	67
Cast Steel (.10 – .20%C)	100	100	100	100	100
Magnesium Alloys	22	22	22	22	20
Aluminum Alloys	34	34	34	34	32
	$A_s = \frac{E_c}{E_s} A_c$	$A_s = \frac{E_c}{E_s} A_c$	$I_s = \frac{E_c}{E_s} I_c$	$I_s = \frac{E_c}{E_s} I_c$	$J_s = \frac{E_c}{E_s} J_c$
Subscript "s" is for steel; "c" is for casting					

* The factors above are based on published values of moduli of elasticity.

3. Use equivalent tables to find the corresponding values in steel. It is necessary only to multiply the known properties of the casting by the factor obtained from the appropriate table to get the corresponding value in steel.

The following example illustrates the use of equivalent sections by presenting the problem of redesigning the cast-iron (ASTM 20) base shown in Fig. 2-37 as a steel weldment.

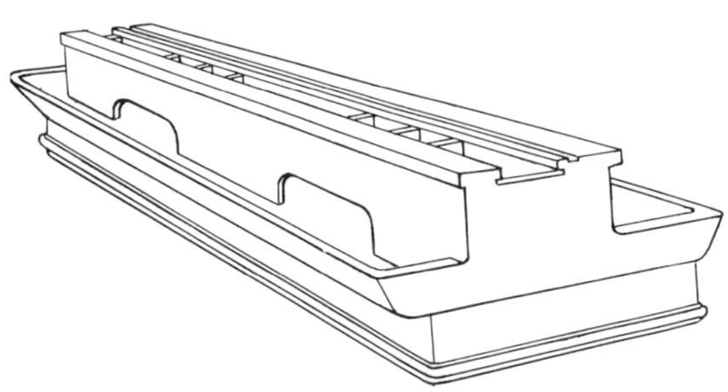

Fig. 2-37. Original cast-iron machine base, ASTM 20 (4900 lbs).

It is desirable that the welded-steel base be as rigid or even more rigid than the cast-iron base. Since it is subject to bending, its resistance to bending must be evaluated.

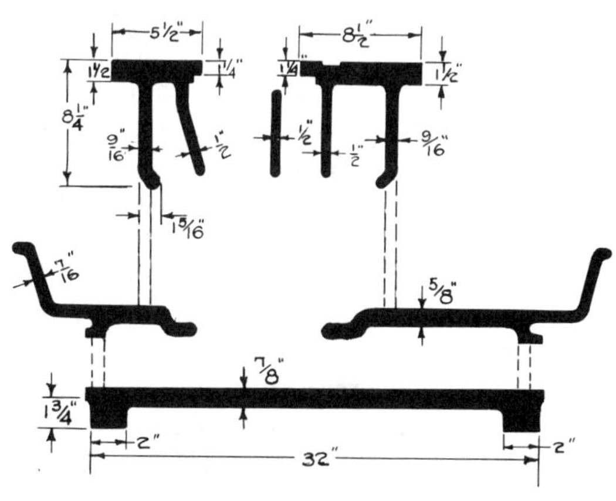

Fig. 2-38. Pattern print — cross section through cast-iron machine base in Fig. 2-37.

TABLE 2-6. EQUIVALENT STRENGTH FACTORS

STEP 1 – Determine the Type of Loading	Tension	Compression Short Column	Bending	Torsion
STEP 2 – Determine this property of the cast member	Area A	Area A	Section Modulus S	Polar Section Modulus $\frac{J}{c}$
STEP 3	Multiply the above property of the cast member by the following factor to get the equivalent value for steel.*			
EQUIVALENT FACTORS				
Grey Iron ASTM 20	21%	94%	21%	28%
ASTM 30	31	123	31	42
ASTM 40	42	136	42	56
ASTM 50	52	156	52	70
ASTM 60	63	167	63	83
Malleable A47 – 33 35018	68		68	76
A47 – 33 32510	54		54	70
Meehanite Grade GE	31	125	31	42
Grade GD	36	136	36	49
Grade GC	44	164	44	58
Grade GB	49	174	49	64
Grade GA	57	199	57	73
Cast Steel (.10 – .20%C)	75	75	75	75
Magnesium H-alloy, AZ63, T6, HTA	50	50	50	33
C-alloy, AZ92, T6, HTA	50	50	50	37
Aluminum Sand Castings 195 T4	40.0	40.0	40.0	43.3
T6	45.0	45.0	45.0	50.0
220 T4	57.5	57.5	57.5	55.0
355 T6	43.7	43.7	43.7	46.6
T7	47.5	47.5	47.5	46.6
356 T6	41.2	41.2	41.2	43.3
T7	42.5	42.5	42.5	40.0
	$A_s = \frac{\sigma_c}{\sigma_s} A_c$	$A_s = \frac{\sigma_c}{\sigma_s} A_c$	$S_s = \frac{\sigma_c}{\sigma_s} S_c$	$\left(\frac{J}{c}\right)_s = \frac{\sigma_c}{\sigma_s}\left(\frac{J}{c}\right)_c$
	Subscript "s" is for steel; "c" is for casting			

* The factors above are based on published values of tensile, compressive, and shear strength, using a safety factor of 3 for mild steel and from 4 to 4.8 for the cast materials, depending upon ductility.

The property of a section that indicates its resistance to bending is the moment of inertia, I. A complete cross-sectional view through the cast base is needed. In the view obtained from the pattern print (Fig. 2-38), the shaded areas indicate the sections that run continuously through the base, acting to resist bending. The moment of inertia, I, about the horizontal and neutral axis must be obtained by calculation; or a ruler specially designed for this purpose may be used as in Fig. 2-39.* The moment of inertia of the casting, I_c is found to be 8640 in.[4]

* For a complete discussion of this method see *"Design of Weldments,"* by Omer W. Blodgett, J.F. Lincoln Arc Welding Foundation (1963).

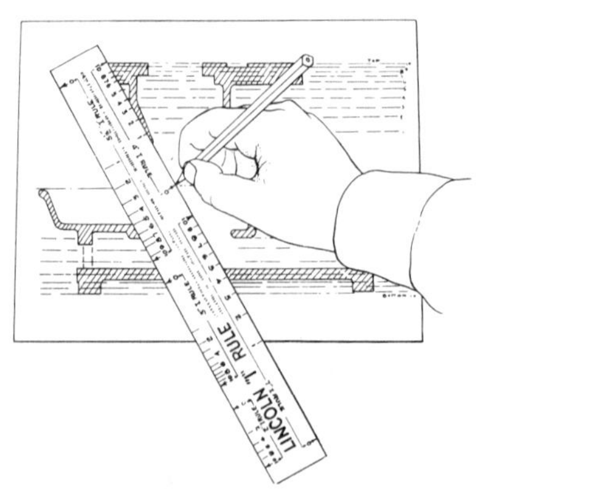

Fig. 2-39. Obtaining the moment of inertia about the horizontal neutral axis of the cast-iron machine base with special I rule.

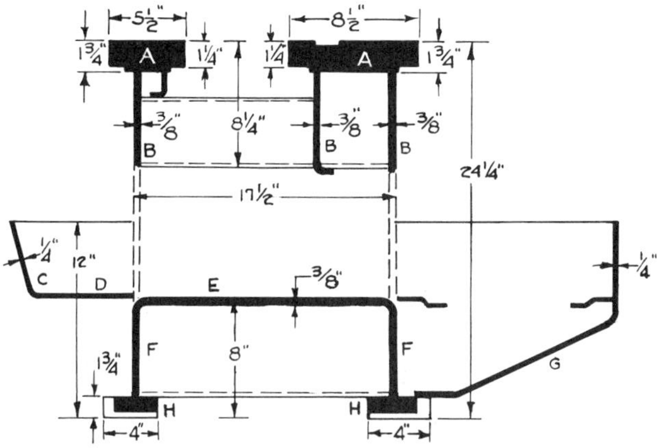

Fig. 2-40. Suggested redesign of cast-iron machine base in Fig. 2-37, for welding.

Table 2-5 shows that the factor for steel replacing an ASTM 20 grey cast-iron in bending is 40% of the moment of inertia, I_c, of the casting. Hence, $I_s = 0.40\ I_c = 3456$ in.^{4}

The problem now is to build up a welded-steel section within the outside dimensions of the cast section having a moment of inertia equaling 3456 in.^{4} The dimensions and location of the two top flange plates must be retained; the design must lend itself to the most economical fabrication methods.

The design in Fig. 2-40 is suggested. Its moment of inertia is quickly found by the method known as "adding areas." Its value is found to be 6280 in.^{4}, or 1.8 times as rigid as the cast-iron base.

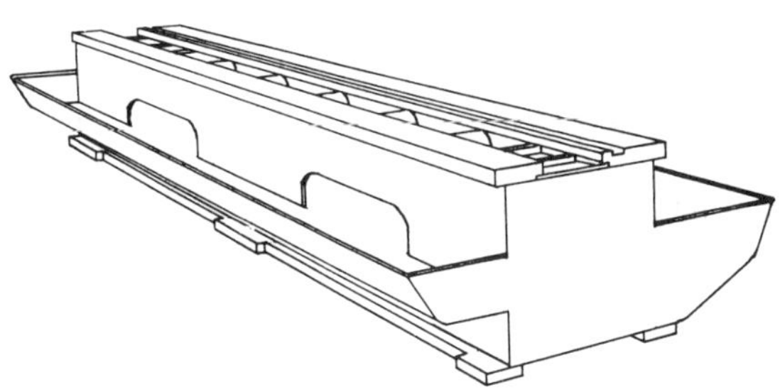

Fig. 2-41. Final design of welded-steel machine base (2500 lbs.)

Once the cross section of the steel base has been designed, other less important components of the cast base are converted to steel. Figure 2-41 shows the final weldment, which has 1.8 times the rigidity but weighs 49% less and costs 38% less.

Nomographs in Conversions

When the same section and over-all dimensions can be used, the conversion to steel is simplified even further by the use of nomographs. Figure 2-42 shows a single-ribbed cast motor base in ASTM

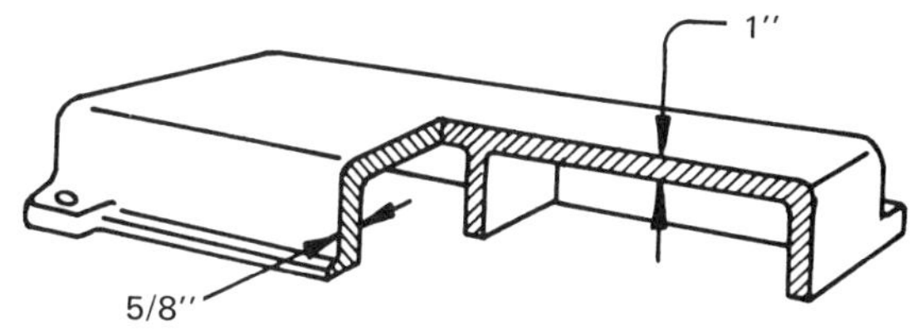

Fig. 2-42. Original cast-iron motor base (681 lbs).

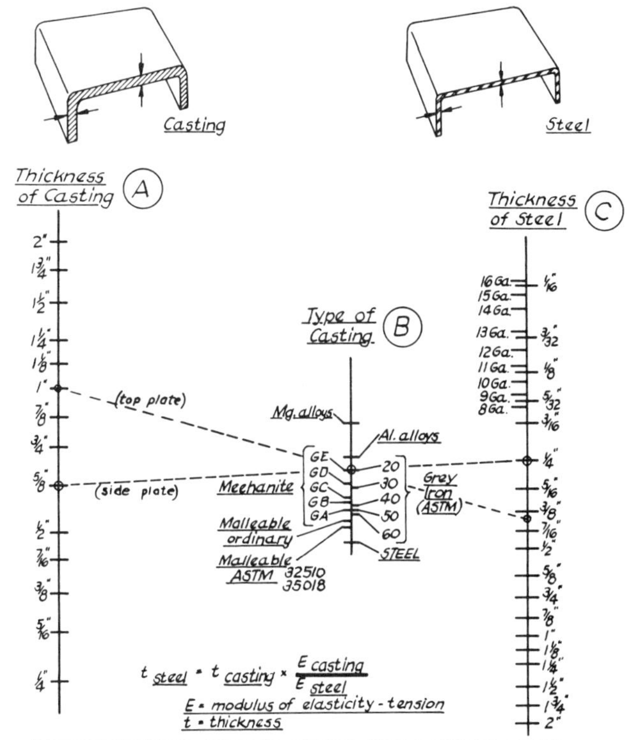

Example: Grey Iron No. 20, 5/8" thick
find thickness of steel section for same rigidty

Construct line from "A" = 5/8" to "B" = No. 20 Grey Iron
read "C" = 1/4" thick steel

Fig. 2-43. Nomograph for determining required thickness of steel section for rigidity equal to cast section.

20 grey iron, 30 in. wide, 60 in. long, and 6 in. deep. The conversion may be accomplished in accordance with the following considerations:

1. The cast base performed satisfactorily in service. The design problem is one of rigidity under a bending load.
2. In a rigidity design, there must be sufficient moment of inertia to resist a bending load. When the shape of the cross section as well as over-all dimensions remain the same, the moment of inertia may be assumed to vary as the specified thickness of the sides and top — the parts that resist bending — and the variance obtained will be accurate within 5 percent.

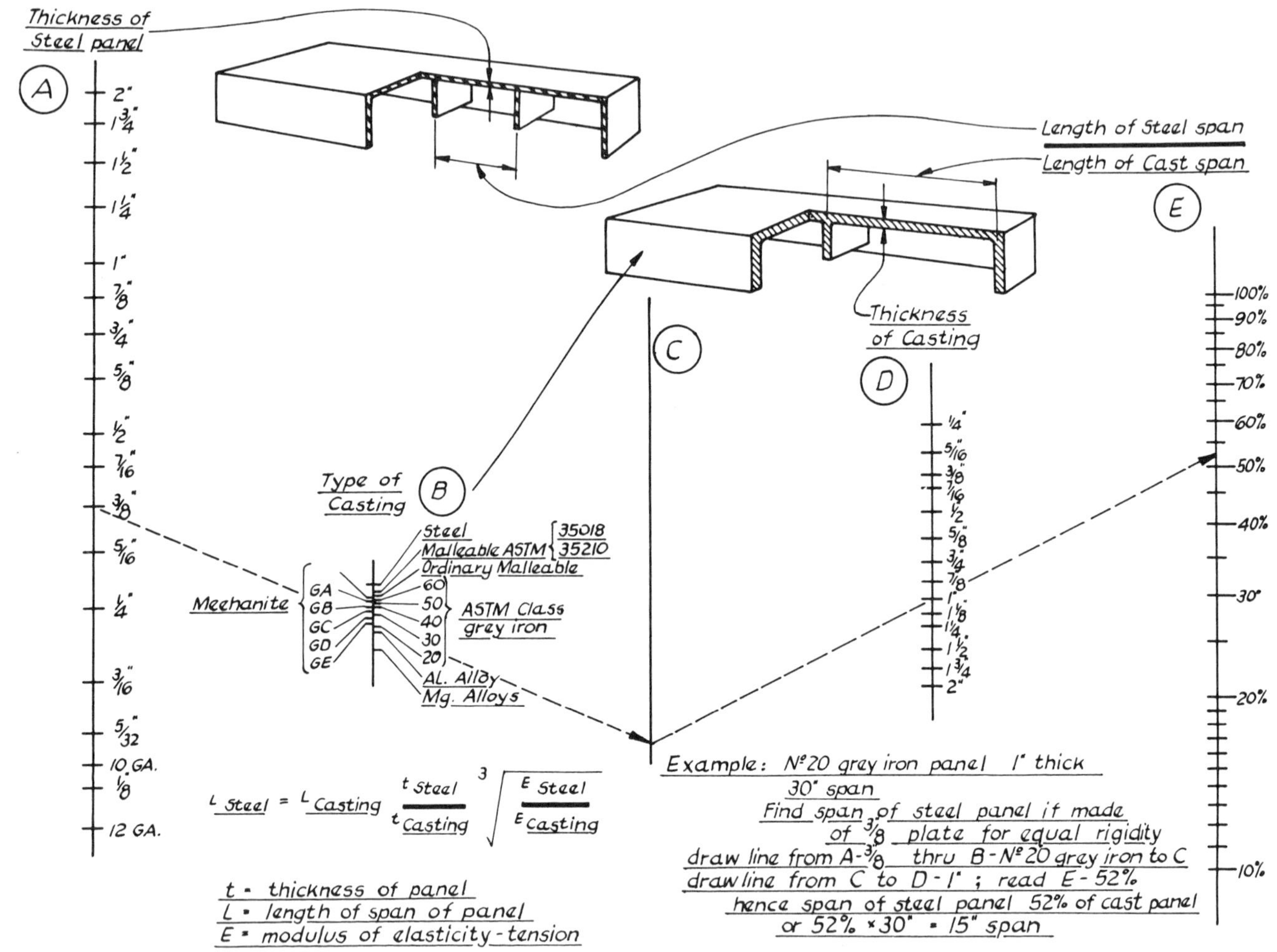

Fig. 2-44. Nomograph for determining required ratio of steel span to cast span for steel section having rigidity equal to cast section.

3. The minimum thicknesses of the top and side panels can be read directly from the nomograph (Fig. 2-43) by using a straight edge and running from line (A) through the point on line (B) indicating an ASTM 20 grey iron casting to line (C). The nomograph shows that a 3/8″ top and 1/4-in. side panels in welded steel should give approximately the same rigidity.

The problem, however, is not solved. The cast base has a rib that serves as a stiffener; thus one or more stiffeners must be provided in the steel base. Reasoning suggests that a thin top panel in steel may require more stiffening than the much thicker cast panel.

The nomograph in Fig. 2-44 is used to determine the maximum span of steel between stiffeners. With a straight edge laid across the point on line (A) indicating the 3/8-in. thickness of the steel top panel and the point on line (B) representing ASTM 20 iron, a reference point on line (C) is obtained. Then with the straight edge repositioned from this point on line (C) through line (D) at the point representing the 1-inch thickness of the cast top panel, a point on line (E) of approximately 52% is obtained. The original casting had a span of 30 in. between its end plate and center rib. A steel top 3/8-in. thick can only have 52% as much span. Three stiffeners about 15 in. apart would thus be required in the steel base to give equivalent rigidity in the top panel.

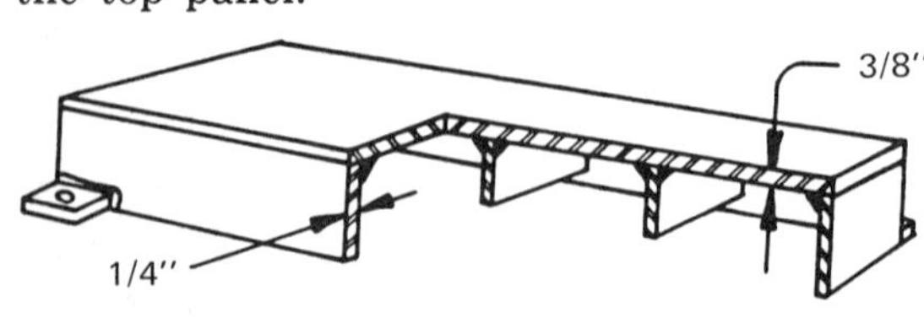

Fig. 2-45. First redesign of casting shown in Fig. 2-42: welded steel base (281 lbs) with three stiffeners on 15-in. centers.

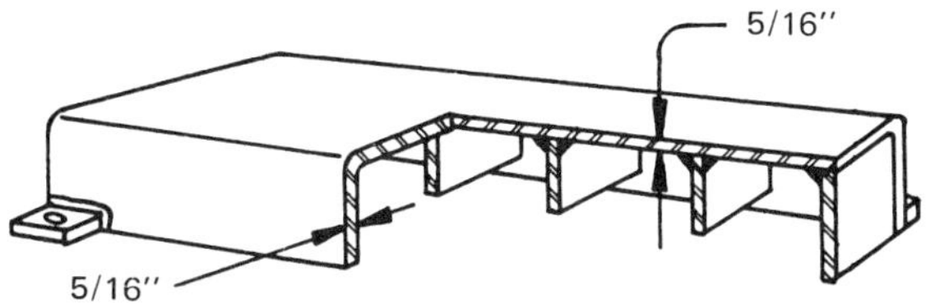

Fig. 2-46. Second redesign of casting in Fig. 2-42: welded steel base (274 lbs) with four stiffeners on 12-in centers.

The redesigned base is shown in Fig. 2-45. Only two operations are required for its fabrication, shearing and welding. The weight has been reduced 41.3% of the cast base and the cost is about 30 to 35% that of the casting.

The redesign program can be taken even further. If the moment of inertia of the cast base is found, it can be multiplied by the equivalent rigidity factor (40 percent). With the redesign based on an equivalent moment of inertia, the base can be made from 5/16-in. steel plate bent into the form of a channel (Fig. 2-46). Bending eliminates preparing the edges for welding, as well as welding three pieces together. A brake-forming operation may increase the cost for one-of-a-kind or small-quantity manufacture, but would result in lower cost if several were to be made at the same time. Note that an additional stiffener is required because of the thinner top panel.

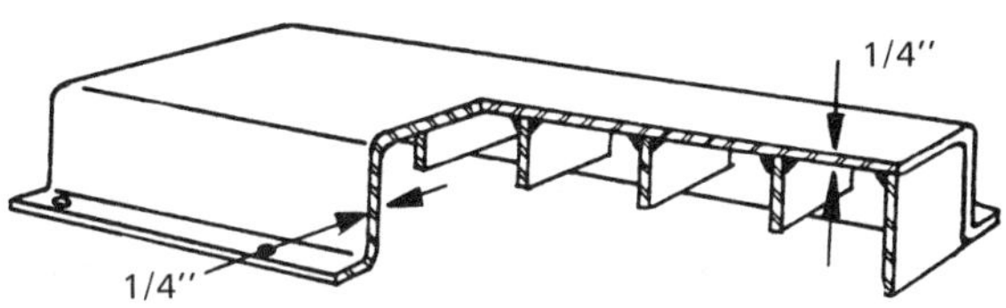

Fig. 2-47. Final design of casting in Fig. 2-42: welded steel base (248 lbs) with five stiffeners on 10-in. centers.

The ultimate design would probably be that shown in Fig. 2-47. Here flanges have been bent into the plate at the bottom to give still greater rigidity. This permits a further reduction in plate thickness, but requires still another stiffener. The weight has now been reduced to 248 pounds.

DESIGNING FROM LOAD CONDITIONS

A motor base, such as the one discussed, may be a member of an entirely new product. In other words, there may be no prior casting to use as the basis for design as a weldment. In this case, the design must be based on analysis of load conditions.

Normally, the calculation of maximum deflection of members subjected to bending loads is very complex. The point of maximum deflection must first be found, and then the deflection at this point determined. Except when there are no more than two loads of equal value at equal distance from the ends of the member, existing beam tables in engineering handbooks are inapplicable to the problem. Most bases have more than two loads, and the maximum deflection usually does not occur in the middle of the member (Fig. 2-48).

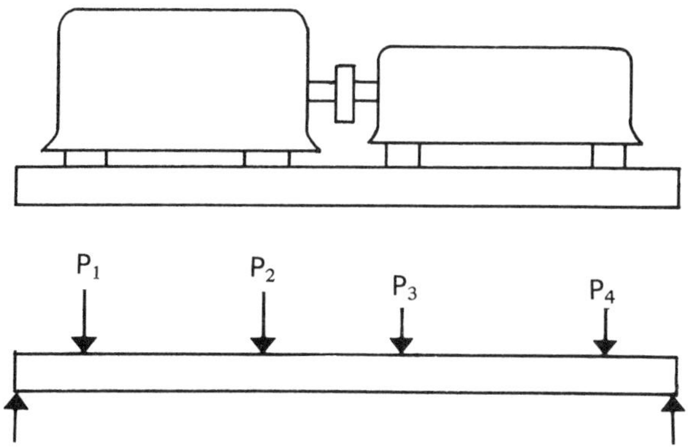

Fig. 2-48. Most machine bases have unsymmetrical loadings; thus maximum deflection is not at center of span.

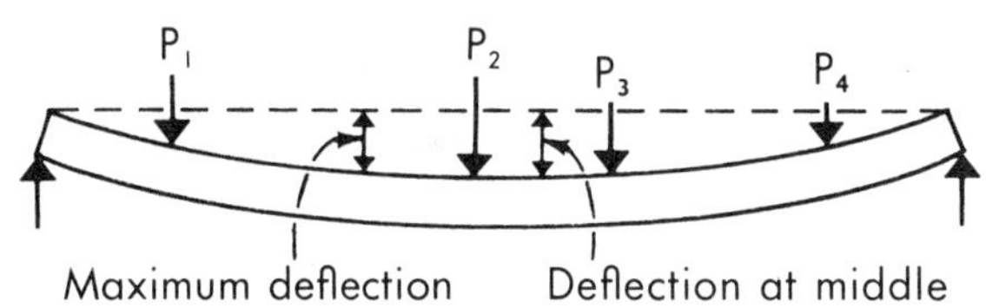

Fig. 2-49. Design based on deflection at middle of beam.

Two things can be done to simplify the problem. First, the deflection at the middle of the member (Fig. 2-49), rather than the maximum deflection at an unknown point, might be used for design purposes. This is justified since the deflection at the midpoint is always within 2% of the maximum deflection. For example, a simply supported beam with a single concentrated load at the one-quarter point has a deflection at the center 98.5% that of the maximum deflection. With a greater number of loads on the member, the error decreases.

Second, a simple method of adding the bending moments of inertia for each individual load can be used. This method may be explained as follows: For a given member, each load will, individually, cause a certain amount of deflection at the center. The total deflection here will equal the sum of the individual deflections (Fig. 2-50).

The principle of adding deflections may be used in a reverse manner to find the required section of

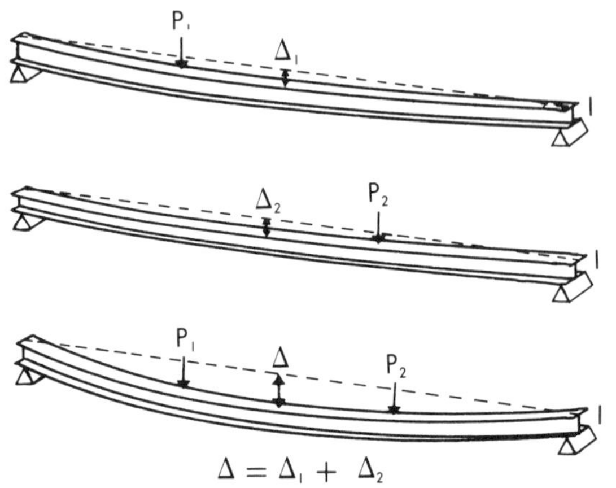

Fig. 2-50. Total deflection, Δ equals sum of individual deflections.

a member, as it is represented by moment of inertia, I. For an allowable deflection, Δ, at the centerline, each load, taken one at a time, will require a section with a moment of inertia (I_1, I_2, I_3, etc.). The moment of inertia, I, of the beam section required to support all the vertical loads within the allowable deflection will equal the sum of the individual moments of inertia (Fig. 2-51).

MAKING USE OF EXPERIENCE

Most types of machines have been in use for many years and continue to perform satisfactorily. Often the actual loads on these machines are unknown, and no effort has been made to determine how forces are transferred through the members. Many machines have come to their present state of development through an evolutionary process. If a casting broke because it was overloaded, the next casting would be made thicker in the weak region.

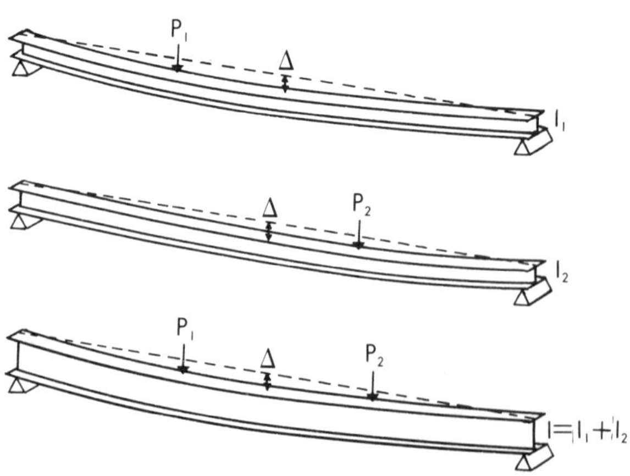

Fig. 2-51. Moment of inertia for beam with several loads may be obtained by adding moments required to handle each load separately.

Eventually a design evolved that gave satisfactory performance. Every machine with a long history of use represents the experience and judgments of a succession of designers.

When given the task of redesigning a machine in welded steel, the weldment designer becomes a theoretician. He knows how to deal with loads, how to transfer forces, size welds, and use materials strategically. He probably has substantial shop experience with the fabrication of weldments, with welding processes, and with cost analysis. But he may not bring to the problem the decades of experience of the machine industry. Through his computations he may produce a welded steel design of apparent excellence, only to find that a critical location for a stiffener is the exact spot where gearing must be located.

Obviously, the only way to be assured of the best design is make certain that the "twain meet" — that theoretical knowledge is joined with practical experience.

Evolution of a Welded Design

Figure 2-52 shows the tedious evolution of a welded design. In (b) in an attempt to convert from the casting (a) and maintain the functional requirements, the designer has taken a narrow approach and considered the problem, piecemeal. He could have saved much effort by looking at the whole. The casting illustrated is a bottom section of a gear housing. It supports the upper part of the housing, holds the oil supply, and provides holes for attaching the upper part with 3/8-in. studs. The performance requirements in the welded steel counterpart are known to the designer, but he approaches his redesign awkwardly.

In weldment (b), the designer reduced the bottom panel thickness to 3/8 in. by going to the equivalent of 1/2-in. cast iron. But to provide adequate width for 3/8-in. studs, he used l-in. plate for the side members of the housing. In order to make the

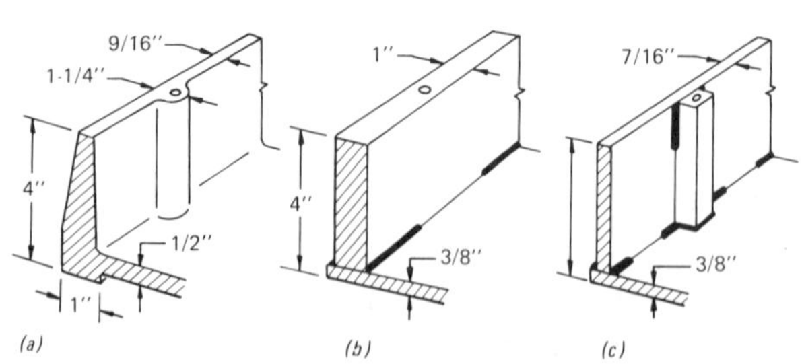

Fig. 2-52. Evolution of welded design of a gear housing. Original casting (a); first redesign — bottom plate thickness reduced (b); thinner member used (c).

housing oiltight, the side members are continuously fillet-welded to the bottom plate on the outside. They are also intermittently fillet-welded on the inside. Welds will be required at all corners, as well. The design satisfied function, but the 1-inch thick side members stand out as a wasteful use of material.

In weldment (c), the side member used is thinner, but still thick enough so that its edge can be machined to provide a bearing surface for the top portion of the gear housing. Bars are welded to the side members to receive the hold-down studs.

The theoretician might note that the design is, after all, nothing but a substitute for a casting. He might suggest starting from scratch to devise a welded steel part to do the job, perhaps listing the functions the cast housing has been serving: (1) providing support for the upper portion of the gear housing; (2) holding oil; and (3) providing a means for access by a machined parting surface.

As he starts to rethink the bottom portion of the housing, he may note that it is basically just an oil pan. Why should an oil pan be the support for the gear housing? Why not use some other way of supporting the gears, their housing, and the loads imposed by the work? For example, in the design illustrated in Fig. 2-53, the gears are held in alignment and proper position by the top portion of the assembly, the same as with the casting. The forces, however, are not transferred to the bottom portion but to an external support through a single plate. Stiffeners are welded at right angles to this plate to help carry the load to the external support. The forces now pass through welded joints rather than bolted joints. The bottom portion is seen to be simply an oil pan, and might be made as a deep-draw stamping from very light-gage material.

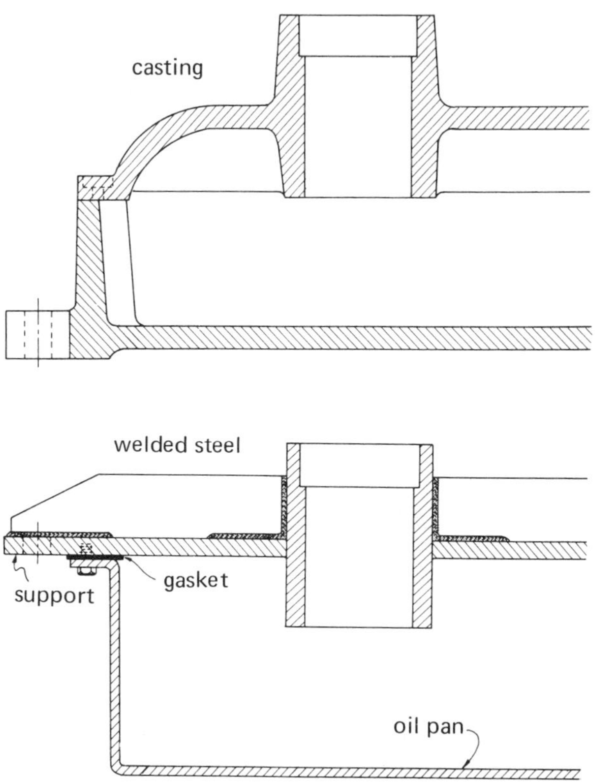

Fig. 2-53. Gear housing. Original grey iron casting (top); suggested weldment replacement (bottom).

The new welded design is enough different than the cast design that in all probability the two designs would not be interchangeable. The next step is to determine the best method to integrate the new design into the over-all machine.

QUALITATIVE VS QUANTITATIVE METHODS

The foregoing has been concerned with the elements of weldment design; how to go about the design procedure; what to do. Assuming that the engineer is ready to attack specific problems, attention will now be turned to examples of practical design problems, to short-cut procedures, design hints, and the "do's" and "don'ts" that are better learned beforehand than acquired by experience.

Chemical analysis can be either qualitative or quantitative. Qualitative analysis answers the question, *What* is the unknown? Quantitative analysis determines *how much* of a substance is in a given sample. It is wasteful to run a long and tedious chemical quantitative determination when only certain elements need to be addressed.

Similarly, in weldment design it is wasteful to make quantitative calculations about stress, deflection, etc., when only qualitative information is wanted. Often the design problem merely asks for

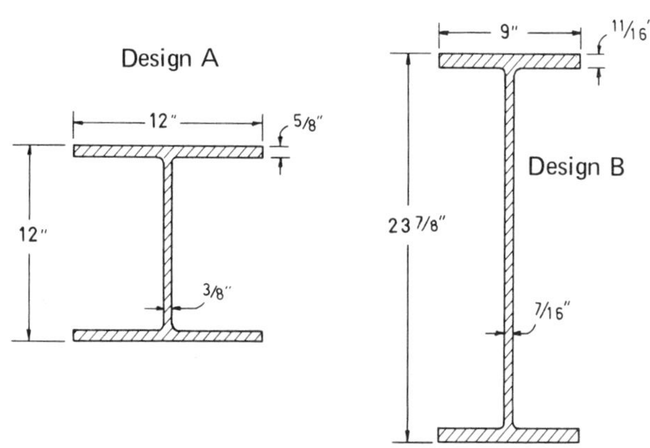

Fig. 2-54. Comparisons of two beams with shock loading.

a "quality" comparison — is one design better or worse than the other? Here, exact quantitative data may not be needed.

The design formulas can provide qualitative answers if they are used to establish ratios which tell whether a design is better or worse without describing the situation quantitatively.

In the example illustrated by Fig. 2-54, the designer has selected a 12-in. WF 65-lb beam to support a load. Reminded that this is a shock load and that a larger beam should be used, he makes what he thinks is a safe selection — a 24-in. WF 76-lb beam. Since the moment of inertia is now four times greater, the designer feels confident that he has made a substantial improvement. But has he?

The situation can be regarded from a qualitative standpoint without becoming involved in numerous calculations. The standard formula for the energy (U) absorbed by a member during shock loading is:

$$U = \frac{\sigma_y 2IL}{6Ec^2}$$

Setting up the ratio of U_B to U_A shows, surprisingly, that the larger beam will not absorb any more energy under the shock load. The ratio is exactly equal to 1.0. The only variables are the moment of inertia, I, and the distance, c, from the neutral axis to the extreme fiber, and their values are such that the ratio becomes 1:1. The designer accomplished nothing by changing his initial selection for the shock load to a 24-in. WF 76-lb beam. A better solution is a 12-in. WF beam with a higher value for I.

Design A	Design B
$I = 533.4$ in.^{4}	$I = 2096.4$ in.^{4}
$c = 6.06$ in.	$c = 11.94$ in.
$E = 30 \times 10^6$ psi	$E = 30 \times 10^6$
$\sigma y = 36{,}000$ psi	$\sigma y = 36{,}000$ psi
$L = 12$ ft.	$L = 12$ ft.

$$\frac{U_B}{U_A} = \frac{\left(\frac{\sigma_B}{\sigma_A}\right)^2 \left(\frac{I_B}{I_A}\right)\left(\frac{L_B}{L_A}\right)}{\left(\frac{E_B}{E_A}\right)\left(\frac{c_B}{c_A}\right)} = \frac{(1)^2\left(\frac{2096.4}{533.4}\right)(1)}{(1)\left(\frac{11.94}{6.06}\right)^2} = 1.0$$

Importance of Correct Analysis

Unless he analyzes correctly, the designer may spend hours or days trying to solve a problem that does not exist. The following example is illustrative.

The welded frame in Fig. 2-55 is made as a box section. How can horizontal diaphragms be inserted inside the box member in line with the top and bottom flanges of the beam section? One solution would be to make the vertical column section by welding the inner flange plates to the two side web plates. Any necessary diaphragms will then be welded into position, and the outside flange of the column welded. The outside flange plate will be cut so that a short length may be inserted and, simultaneously, by means of a single-bevel groove weld, the outside flange will be made continuous and attached to the horizontal diaphragm plates.

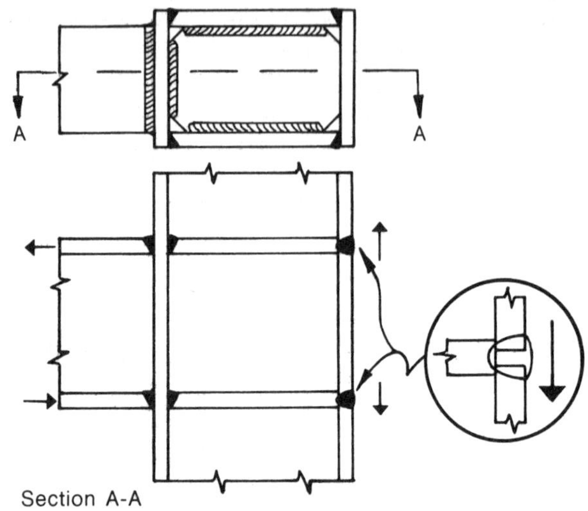

Fig. 2-55. Welded frame box section.

Since the outside flange plate is loaded in tension, and is also subject to fatigue loading, there is some question as to the fatigue strength of this welded joint. Although there are no data on this type of joint — two plates butted together with a

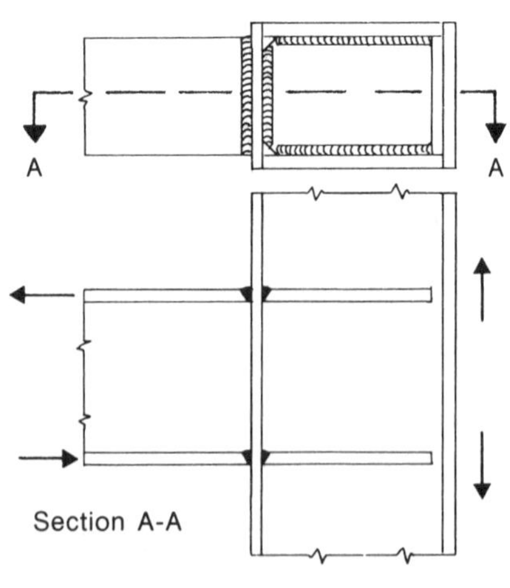

Fig. 2-56. Tension and compression forces enter column through butt welds.

third plate normal to these serving as a backing — it is suggested that it is similar to a butt-welded joint using a backing bar.

An understanding of the purpose of the diaphragms is necessary for full comprehension of the problem. The tensile force in the upper flange of the beam and the compression force in the lower flange enter the column section through the groove weld (Fig. 2-56). These forces then pass directly into the diaphragm plates, and out sideways through the connecting welds into the web plates of the column, where they stop. Since a force applied to a member is transferred ultimately to the portion of the member that lies parallel to the force, it is clear that these forces in the beam flange must enter the two side web plates of the column and go no further. There is no reason for them to enter the outside flange plate. Therefore these diaphragms may be cut short about an inch and need not be welded to the outer flange of the column.

Thus, a situation that appeared to pose a problem, in reality is not a problem. The outside flange plate is now made in one length and does not have to be welded to the diaphragm.

MEETING A DESIGN PROBLEM

Figure 2-57 shows the original design of a steel casting with a very large force, F, exerted on it.

The first weldment (Fig. 2-58) is a copy of the casting. Because of the large force transferred from (c) up through (b) and into (a), all these welds must be complete-penetration groove welds. Much welding is represented, since the plates are thick. The procedures for welding are important because the plates are of quenched and tempered steel. High-strength weld metal must be used.

In view of the fact that a force applied to a member will ultimately be transferred into the part of the member that lies parallel to the force (Fig. 2-59), the force, F, must eventually be found in section (a). Why not, then, use a pair of plates, as in Fig. 2-60? If this is done, the force will pass up through plate (a) without passing through a welded joint. Plates (b), seen in Fig. 2-60, will then be

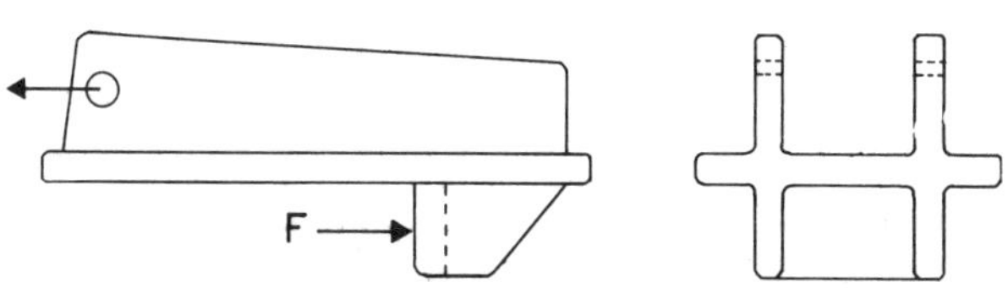

Fig. 2-57. Original steel casting.

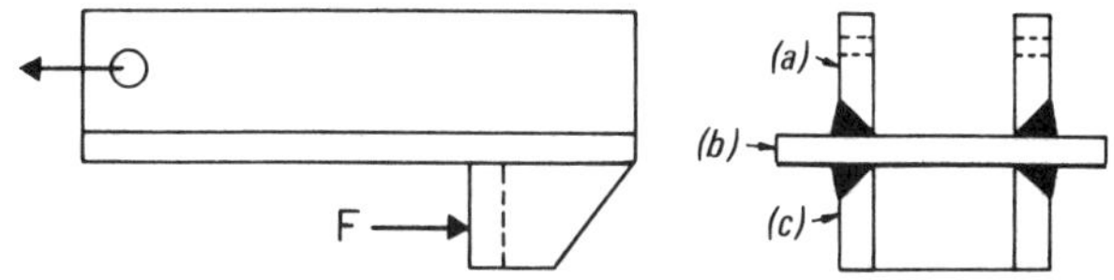

Fig. 2-58. First weldment design of casting in Fig. 57.

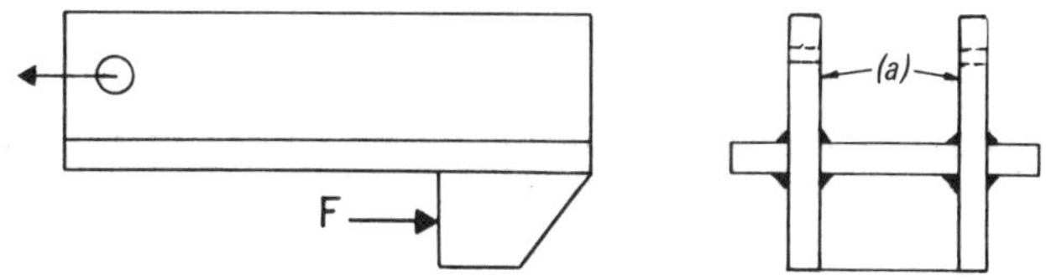

Fig. 2-59. Force, F, is transferred into area (a), which lies parallel to the force.

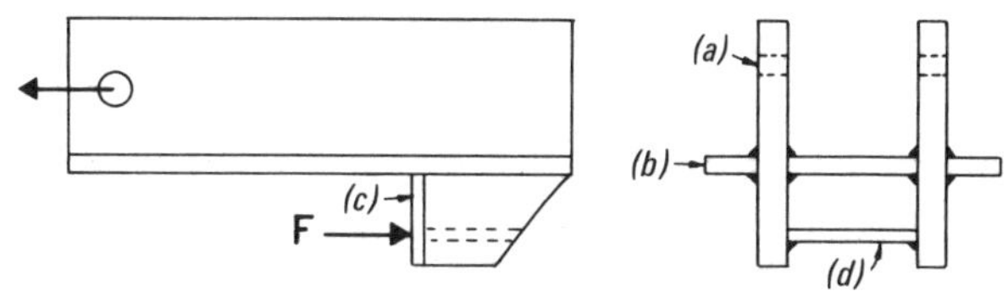

Fig. 2-60. Second weldment design of casting in Fig. 2-57.

added as secondary members, and plate (c) will be used to provide bearing for the force when it is applied to plates (a). Since the force pushes plate (c) against plates (a), the weld is not extremely critical and a smaller weld may be used. Plate (d) is added to provide proper stiffness to plate (c). Again, the load on these welds is not great. Welding costs will be lowered, because welding is used only to connect secondary portions and high-strength weld metal is not required.

Importance of Realistic Specifications

Figure 2-61 shows a section of a welded frame made from 8-in., 11.5-lb channel sections. There are some 1-in.-thick plates welded to the top of the frame to serve as pads between the machinery and the frame. The engineer has indicated a 7/16-in. intermittent fillet-weld, 2 in. in length and on 4-in. centers. The instructions are unrealistic, since the weldor is being asked to put 7/16-in. welds on the toe of a flange only 1/4-in. thick. The engineer, of course, was falling into the easy error of thinking of the thickness of the 1 in. plate.

Ideas from the Shop

Figure 2-62(a) shows a portion of a welded steel gear housing for an earth-moving unit made of formed plates. One boss is located near an inside corner of the housing. When leakage occurred in the joints of the housing behind the box, repairs became difficult and costly.

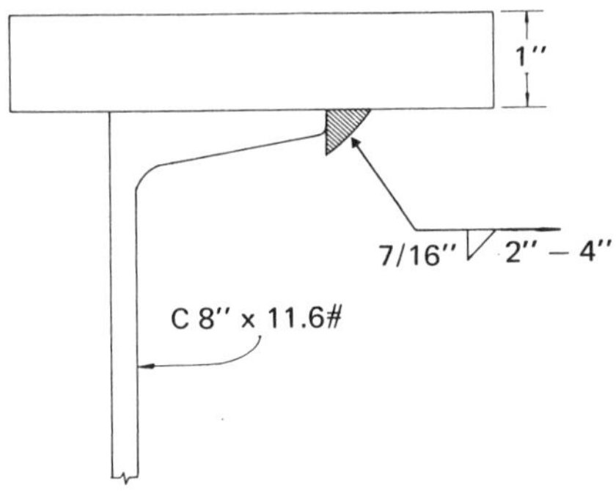

Fig. 2-61. Detail of section of welded frame made of lengths of 8-in., 11.5-lb channel sections.

Finally, a weldor who had frequently made the repair suggested to the engineering department that the boss be changed from the round section to a boss of rectangular section. As shown in Fig. 2-62(b), this could be welded to the housing before the other inside corner welds were made. Thus the weld area was readily accessible. As a result of the change in the shape of the boss, problems of leakage stopped immediately. Should leakage occur, however, repair would be easy.

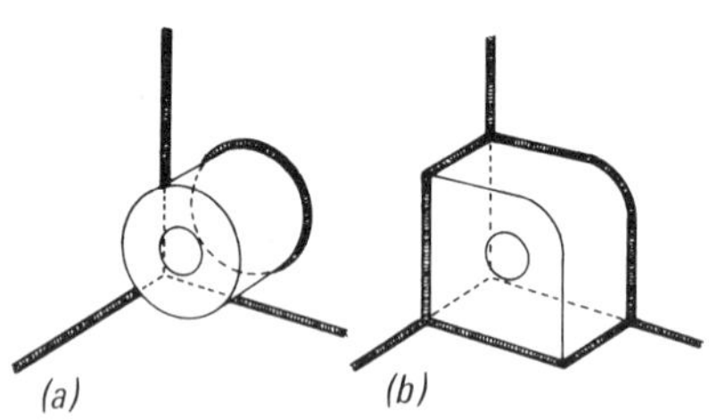

Fig. 2-62. Section of welded gear housing. (a) Original boss; (b) redesigned rectangular boss.

POTENTIAL SOURCES OF TROUBLE

Improper Specifications

When designs are to be altered, the designer should carefully check the changes to be made in the welding. Otherwise, serious errors could occur.

For example, for many years a company had been welding intersecting plates at 90° with a complete-penetration double-bevel groove joint (Fig. 2-63. Full strength was needed, and this represented good practice.

Later, when a job in which plates of the same size intersected at 60° was undertaken, the engineer unthinkingly indicated the same welding symbol. The shop questioned these instructions. Had this not been done, the joint would have required the massive use of weld metal as shown in Fig. 2-64. This would have involved the deposition of 8.67 pounds per foot.

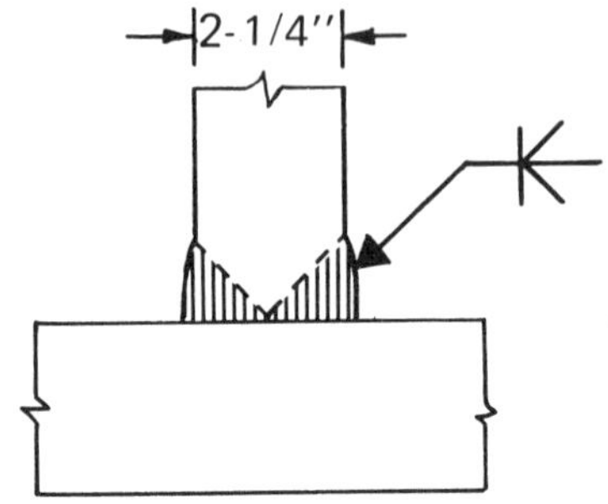

Fig. 2-63. Complete-penetration double-bevel groove joint.

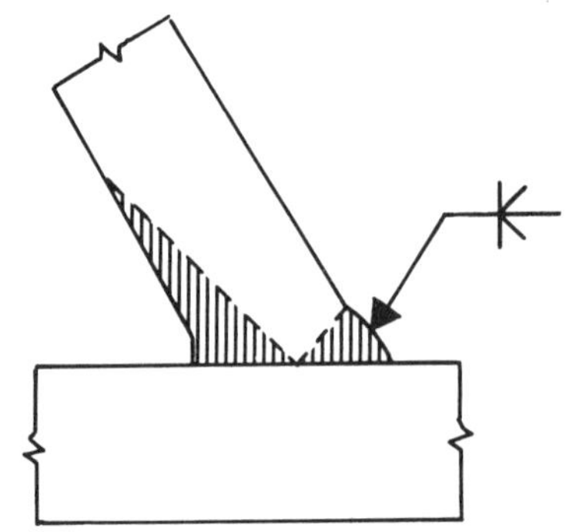

Fig. 2-64. Incorrect specifications would have resulted in the massive use of weld metal.

Figure 2-65 shows the proper way of specifying how plates intersecting at 60° are to be welded with a complete-penetration, double-bevel groove joint. A considerable saving in weld metal is effected simply by shifting the central position of the joint so that, rather than beveling 1-1/8 in. on each side, the bevel is 1/2 in. on the inside face of the joint and 1-3/4 in. on the outside face. This reduces the amount of weld metal (not including reinforcement) from 8.67 pounds per foot to 5.75 pounds per foot.

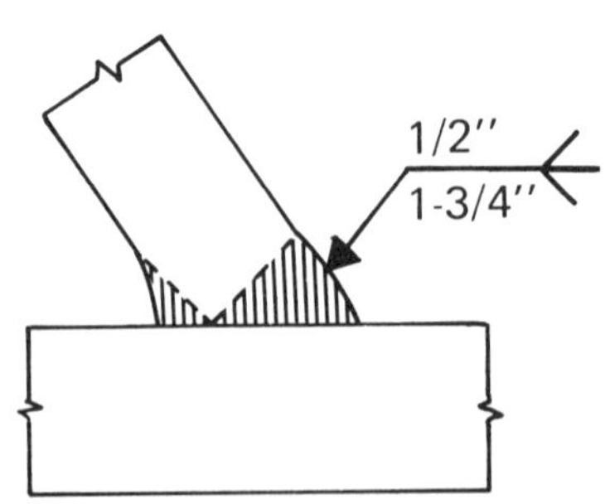

Fig. 2-65. Correct specification for joint on Fig. 2-64.

"Mixing" of Weld Types

The designer may have designed and sized a weld joint correctly — Fig. 2-66(a) — only to have his attention called to a change that must be made. If the change is made without taking into account its effect on the weld joint, a difficult preparation and welding situation may result. Figure 2-66(b) is illustrative. Here, the needed cutout was made on the drawing board, but the alert designer noted that the change would cause trouble by the mixing of weld joints and weld types. The weld in Fig. 2-66(b) changes abruptly from a fillet to a groove weld. The groove weld requires cutting a bevel into the vertical plate. In welding, a smaller electrode would be required to make the stringer bead in the groove-joint portion. A simple method of joining the base to the vertical plate had been destroyed.

The joint was revised to permit continuous fillet welding as seen in Fig. 2-66(c). Not only were fabrication costs reduced, but the second design gave a much better appearance.

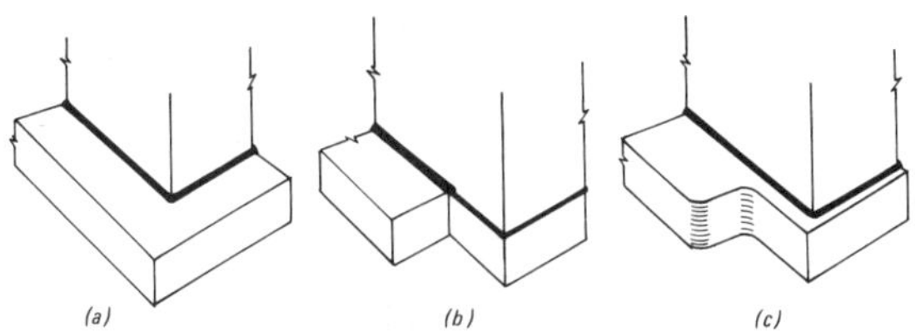

Fig. 2-66. The correctly designed joint shown in (a) could not apply when the part had to be redesigned. If the designer had not been alert, he might have specified the joint shown in (b) when making the change. This would have led to a difficult preparation and welding situation, in which a fillet weld changes to a groove weld. The design in (c) eliminates the difficulty by making possible a continuous fillet weld.

Misuse of Diagnostic Tools

When using diagnostic tools to prove a point, the designer should make certain that what he sets out to prove is correct. Failures in the final product may result if the original premises are erroneous.

An engineer believing that the size of the fillet weld between the rim and the disc of a gear, shown in Fig. 2-67(a), should be increased, set about to prove this point by photoelastic stress analysis of the joint. Taking plastic cross sections of weld (a) and his proposed larger weld (b), he obtained photoelastic stress patterns under polarized light that clearly showed a greater amount of stress at the root of weld (a) when the welds were pulled in the direction indicated by the arrows. To him this was positive proof of the superiority of the proposed larger weld.

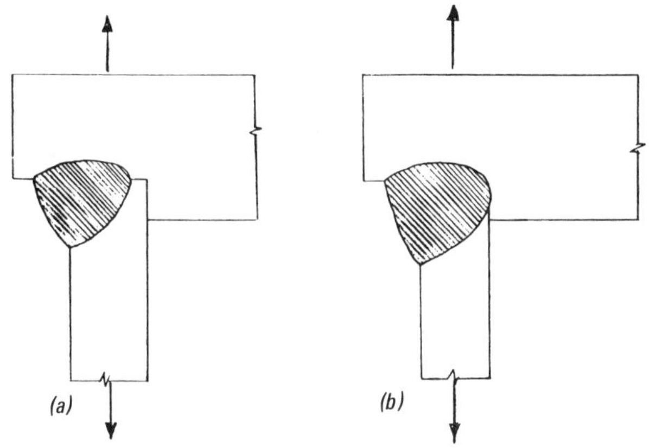

Fig. 2-67. (a) Original fillet weld between rim and disc of a gear; (b) proposed larger weld for same application (which eliminates transverse notch).

The engineer, however, failed to recognize a very important factor. A notch becomes a stress raiser when it is normal or transverse to the flow of stress. In the joint in Fig. 2-67(a), the notch formed by the root of the weld is transverse to the direction of the applied force. This, of course, would constitute a stress raiser and would show a large fringe pattern. In the other joint, the notch at the root of the weld lies parallel to the flow of stress. Thus, its fringe pattern is less. The real difference in the patterns obtained results not from weld size, but from the fact that the notches lie at different angles. Thus, the proof offered by photoelastic stress analysis did not pertain to the really critical factor involved in the problem. The engineer was right, but for a reason he did not perceive.

The Tendency to Overweld

Figure 2-68 shows a weldment used in a machine tool. It is made of 3/4-in.-thick plate, and the top is attached with 1/2-in. fillet welds. There is also welding inside for the attachment of stiffeners. The top serves to tie the four sides of the weldment together and give it stability. During assembly of

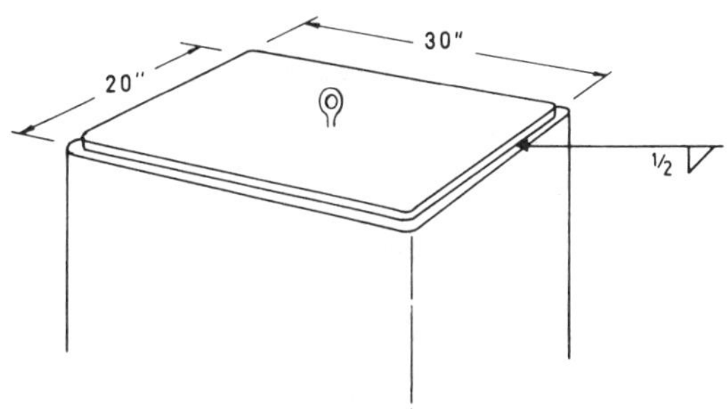

Fig. 2-68. Weldment used in a machine tool.

the machine tool, the weldment will be picked up by means of a lifting lug screwed into the top plate. While the weldment is being lifted, the welds will be subjected to a load of 5000 pounds.

E70 weld metal is used, which has an allowable force of 7,420 pounds-per linear inch of weld for a 1/2-in. fillet weld. There are 100 inches of fillet welding around the top, giving a total allowable lifting capacity of 742,000 pounds, or 371 tons. The weld is thus more than 148 times as strong as it need be. A 3/16-in. fillet, rather the 1/2-in. fillet, would have superfluous load-carrying capacity, and could be produced for about one quarter the cost.

Inside the same weldment, as it appears in Fig. 2-69, are 3/4-in. diaphragms welded to the 3/4-in. side members with 1/2-in. continuous fillets. A full-strength weld would require a fillet leg size equal to three-fourths the plate thickness. Since $3/4 \times 3/4 = 9/16$, the 1/2-in. fillet welds are almost full-strength welds. With this amount of welding on these diaphragms, it should be possible to weld attaching plates on the outside of the weldment in line with the internal diaphragms (Fig. 2-70), place the whole assembly in a huge tensile-testing machine, and pull it apart without breaking the welds.

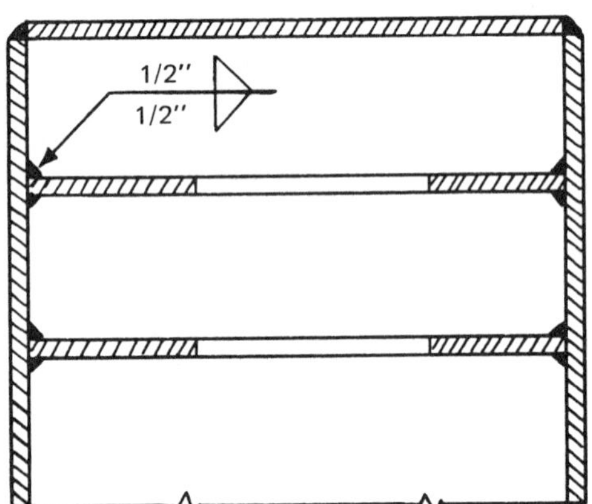

Fig. 2-69. Internal diaphragms welded to side members of weldment shown in Fig. 2-68.

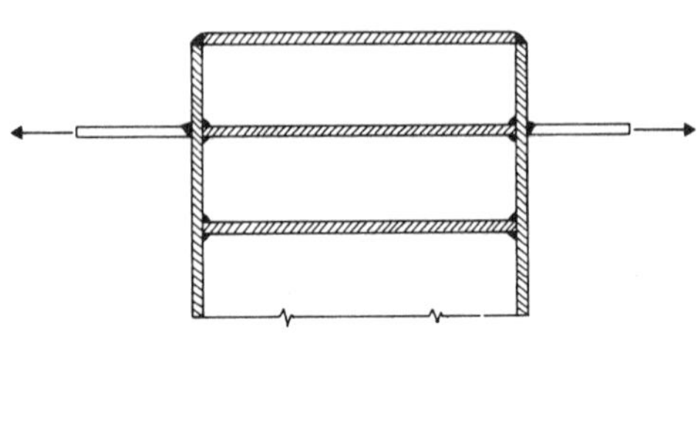

Fig. 2-70. Theoretical test of weldment in Fig. 2-69.

This weldment, of course, will never be subjected to such a load. In fact there is no way to develop a tensile force on this joint because there is no plate adjacent to the diaphragm. Intermittent fillet welds of 1/4-in. size would be sufficient.

Failure To See The Whole Picture

In redesign work, a disadvantage of designing one part at a time is that full advantage may not be taken of the redesign: the over-all picture escapes the engineer.

The weldment illustrated in Fig. 2-71 is a redesign of a cast housing for an earth tamper. Cast designs are usually broken down into units that are bolted together. The original casting had a pair of brackets on the sides of the housing to which handles were bolted. Figure 2-71 reveals that a similar pair of brackets was welded on, imitating unnecessarily a needed provision in the cast design. There is, of course, no point in bolting the handles to the brackets. They can be welded directly to the housing.

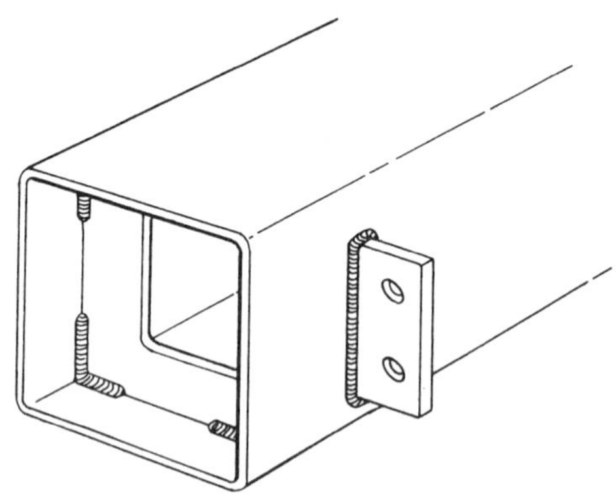

Fig. 2-71. Weldment designed from cast housing for earth tamper.

The Specification of Intermittent Welds

The crown of the press shown in Fig. 2-72 has a variable depth. The designer determines that intermittent fillet welds are adequate for the web-to-flange connections and so specifies; The weldor follows the instructions as to length and distance apart for the intermittent welds. In service, the crown of one of the presses deforms. Investigation reveals that the web has cracked, starting at point (x) in Fig. 2-72(a) where the flange changes direction. This critical point had, by chance, been a skip point in the weldor's sequence, and no intermittent fillets had been placed there.

In service, the upper flange of the crown had pulled away from the web at point (x) and caused the web to undergo substantial plastic yielding. Since

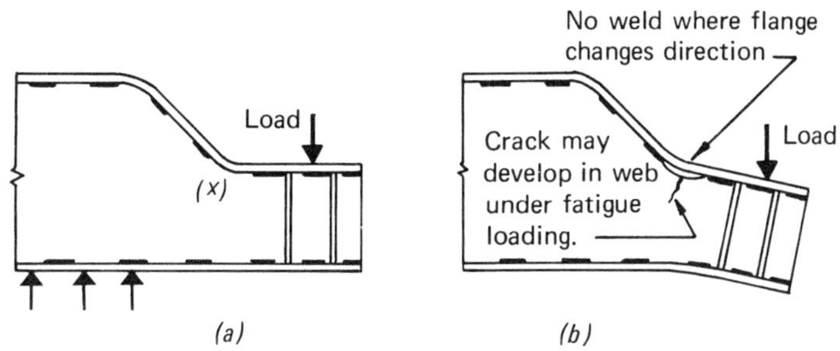

Fig. 2-72. The placement of intermittent fillet welds can be critical.

the flange was free to pull out and could not act to resist bending, the web was forced to carry bending stresses that exceeded its design capacity. Under repeated loading, such high bending stress in the web resulted in a fatigue crack. The presence of a weld of adequate size at-this critical point would have prevented the failure.

Overworked Members

The weldment in Fig. 2-73(a) appeared adequate for the loads. The 6-in.-diameter tube in the center was the backbone of the structure and helped to maintain alignment of parts during fabrication. In service, the weldment performed satisfactorily until it encountered an unexpected torsional load, but one that experience showed could occur occasionally. When this happened, the weldment failed at the fillet welds joining the tube to the ribs.

A designer might have been tempted to increase the size of the fillet — and continue to increase the size until the weldment no longer failed in service. Analysis of the problem, however, indicates that the ribs give little resistance to torsion. Consequently, almost all the torsional load was transferred through the fillet welds into the tube. The tube and the welds were severely overworked.

In the redesign shown in Fig. 2-73(b), the 6-in. tube and the side plates are replaced with two large horizontal channels. These and three intersecting channels form rigid box sections when joined to the flat member.

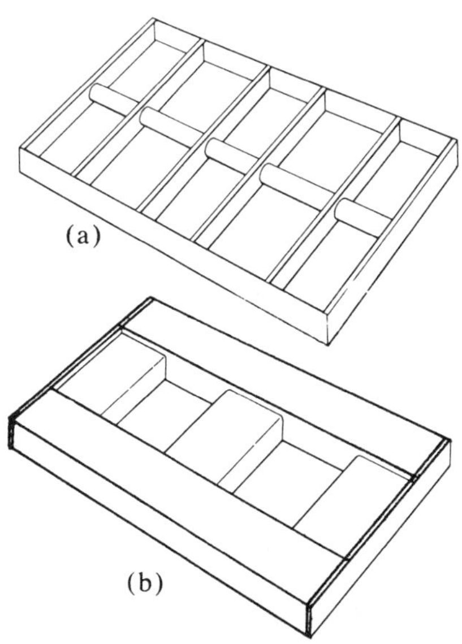

Fig. 2-73. (a) Weldment failed at fillet welds joining tube to ribs under severe torsional load; (b) suggested redesign.

Building in Stress Raisers

There is no such thing as a "secondary" member in a weldment. A supposedly "unimportant" member immediately changes the conditions to which other members are subjected.

Figure 2-74 shows a clamp used to glue laminations together in building wooden arches. A steel rack is welded on to engage the movable jaw. Between the two jaws, the main frame is subjected to bending, with tension on the upper portion. Once the steel rack is welded to the frame, it becomes the upper fiber of the frame. The notches of the teeth then become stress raisers and greatly reduce the strength of the clamp.

If the steel rack is supplied in short lengths, and several of these lengths are used on the clamp, a serious condition arises when a weld crosses a

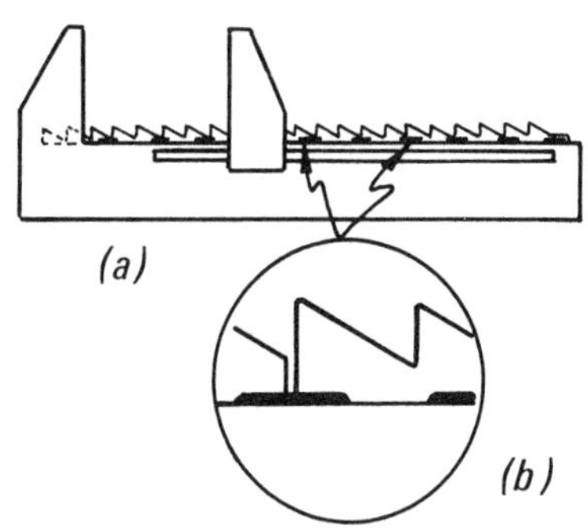

Fig. 2-74. Clamp with welded steel rack.

separation between two segments, as shown in Fig. 2-74(b). A notch is created — a crack virtually built in by the existence of the separation. Since the clamp is loaded repeatedly when in use, a type of fatigue loading may be said to exist. Under this loading, the crack between segments of the rack is propagated through the weld and into the main frame.

To prevent failure, the rack should be supplied in a single length and welded very carefully to the frame on just one end. In this manner, the rack will not become the upper fiber of the frame.

Inefficient Transfer of Forces

As noted previously, to transfer a force efficiently, the force must have a path into the portion of the member that is parallel to the force. The frame depicted in Fig. 2-75(a) is made of channels. Triangular stiffeners are attached to the webs in the corners to help stiffen the frame.

If a horizontal force that would tend to move the frame is applied, it will enter a stiffener and eventually pass into the parts parallel to the horizontal force — the flanges of the channel. But before the transfer is effected, the web will be deflected. The stiffeners, thus, have very little value in keeping the frame rigid. If stiffeners are needed, they should be welded in line with the flanges of the channel.

Directional Change of Forces

Forces in weldment members often change direction. In these cases, a new force component is set up, which must be provided for in the design. Figure 2-76(a) shows an abrupt change in direction of opposing forces. The component F_c is developed, which is concentrated at the point of change. Its axis bisects the angle between the two forces, F. Unless the component force is handled by some type of reinforcement, the member will tend to straighten out under the load.

If the change in direction is more gradual, as in Fig. 2-76(b), either the component force is spread over the length of the change, or a large number of component forces exists. The component can be described in pounds per linear inch. Such a condition is the exact opposite of that existing in a circular pressure vessel, where the internal hydrostatic pressure (radial) causes a tensile force (hoop stress) in the shell (Fig. 2-77).

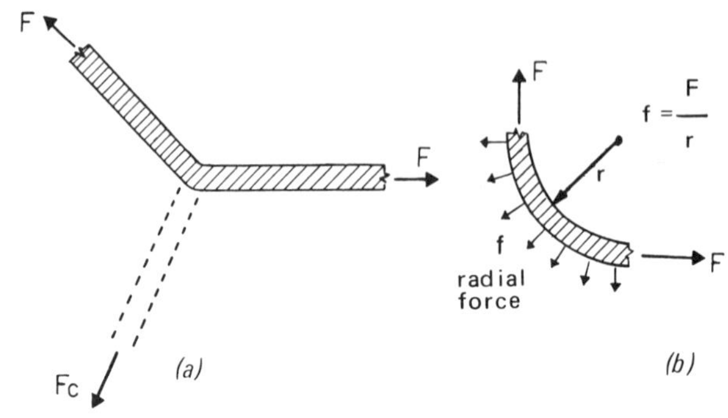

Fig. 2-76. Change of direction of opposing forces. (a) abrupt change; (b) gradual change.

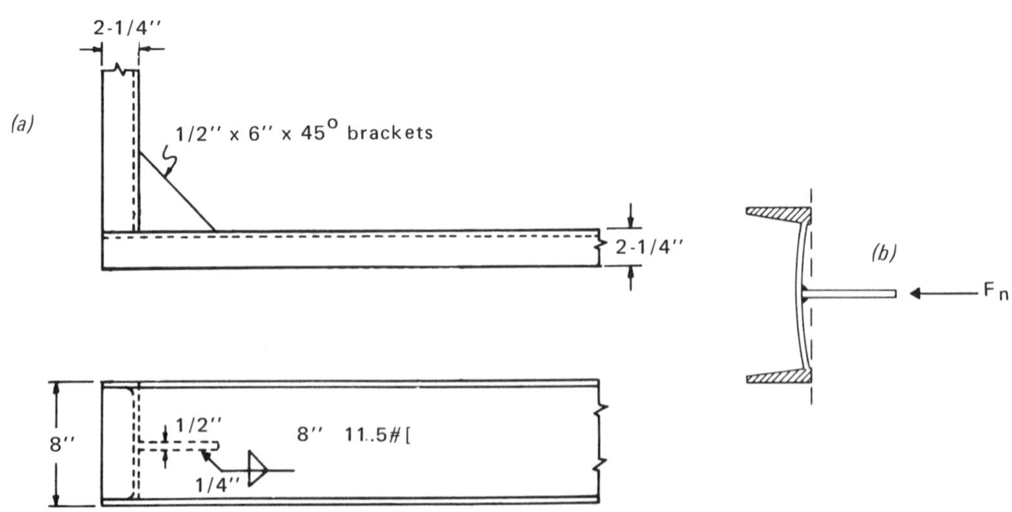

Fig. 2-75. (a) Section of frame made of channels; (b) horizontal force enters stiffener and deflects web of channel.

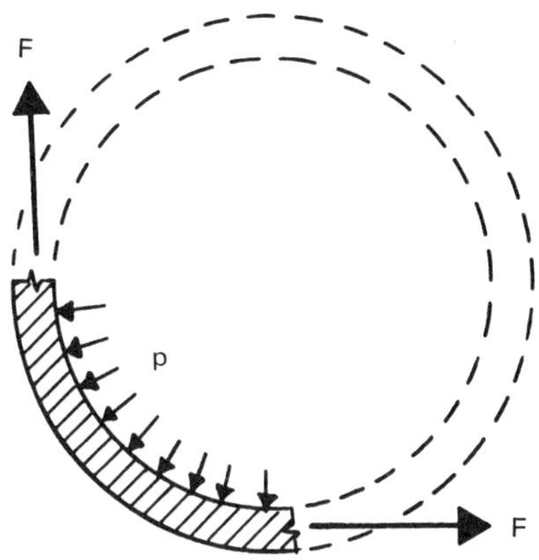

Fig. 2-77. Hoop stress in pressure vessel.

When a curved box beam is used in a press frame, it must be specially designed for the radial force on the welds. The box beam in Fig. 2-78 has tension in the inner flange. The unit radial force, f_r is directed inward away from the supporting web plates. This force tends to pull the connecting welds apart. In addition, the neutral axis shifts inward on a curved beam, increasing both the bending force on the inner flange and the radial force.

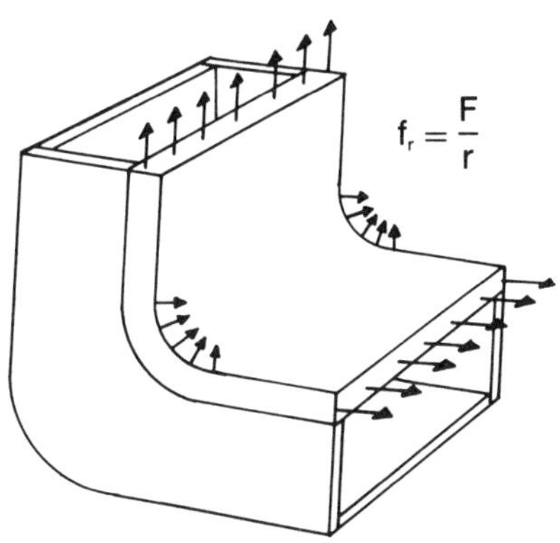

Fig. 2-78. Tension in inner flange of box beam.

Propagating the Cover-Up for an Error

Figure 2-79 shows a gooseneck used in an earth-moving trailer. When the original production unit was field-tested, the gooseneck failed and was repaired with the patch plates. Subsequently, the patch plates were added to the goosenecks on the production floor and became a part of the accepted design. When producers of comparable types of assemblies copied the idea for reinforcing, however, what was originally no more than an expedient method of correcting an error became a method of design.

The patch-plate method of reinforcing the curved beam is not necessarily good, even though it works.

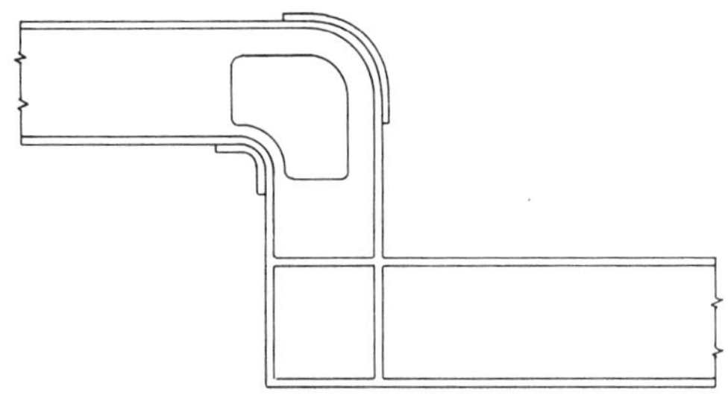

Fig. 2-79. Gooseneck used in earth-moving trailer, reinforced with patch plates.

A better way of attacking the problem is to analyze the piece, discover the cause of weakness, and strengthen the beam in a more professional manner, as in Fig. 2-80. Note that the neutral axis of the beam shifts inward in the curved region, greatly increasing the tensile bending stress in the inner flange. What is needed is a thicker flange in this region. The outer flange has a much lower compressive bending stress and does not need to be increased in thickness.

Incorrect Identification of the Problem

What appears to be the obvious cause of a design failure may have little to do with the problem, and

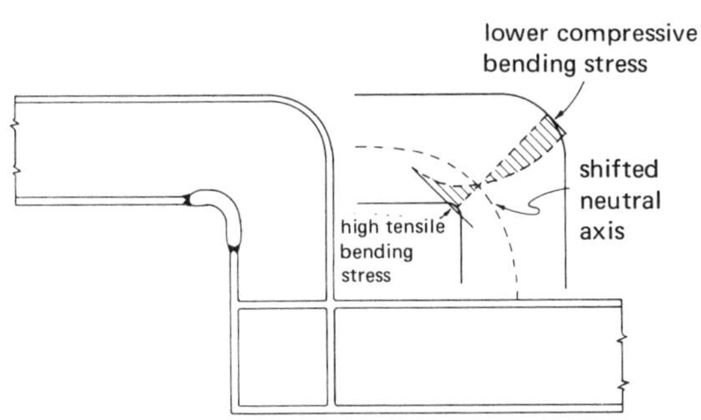

Fig. 2-80. Gooseneck reinforced in accordance with sound engineering practice.

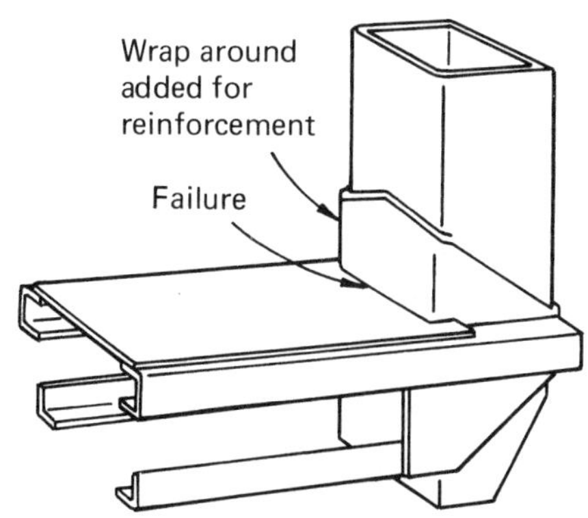

Fig. 2-81. Location of failure in frame supporting rotating truck boom.

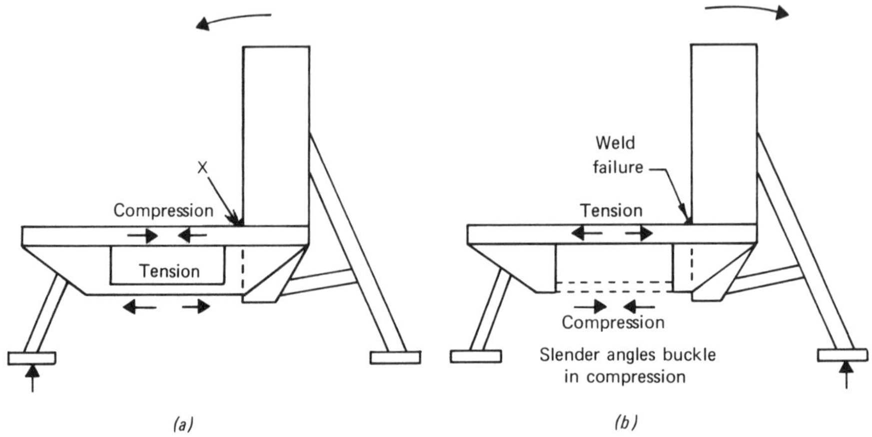

Fig. 2-82. Failure analysis of frame in Fig. 2-81 suggests that lower framing members lack compressive strength. (a) Conditions with boom in far left position; (b) conditions with boom in far right position.

no real progress toward the solution can be made until the real cause of the failure is identified. This is made clear by a consideration of the frame for supporting a rotating truck boom shown in Fig. 2-81.

In tests on the prototype, the frame was required to support a 3000-lb weight at the extreme end of the boom as it rotated through 360 degrees. The frame failed as indicated in Fig. 2-81, and the addition of reinforcing wrap-around plate, plus an increase in weld size, did not correct proneness to failure. An analysis of the varying load conditions was needed.

Figure 2-82 shows the conditions that existed during testing and in service. The two legs of the assembly are extended to the ground to provide a tripod-type support, with the truck acting as one leg. When the boom is in the far left position — Fig. 2-82(a) — the top part of the horizontal member of the frame is in compression and the bottom in tension. The slender angles making up the bottom framework can adequately resist the tensile force. The entire member, top and bottom, makes up a large beam to adequately resist bending. Any giving that might occur would tend to place the weld at (x) under compression. At this position of the boom and all other points along the left semicircle, the frame performs satisfactorily.

When the boom is in the right semicircle of travel, as in Fig. 2-82(b), a different condition exists. The top part of the frame is now in tension and the bottom part in compression. The slender bottom members buckle slightly under the compression load, and only the top portion remains to resist bending. Since this section is small, the bending stresses are transferred to the joint between the vertical and horizontal framing, which tends to open. When these stresses exceed the strength of the weld at this point, the weld cracks.

The place for reinforcement is, thus, at the bottom of the framework. A plate added to the lower slender horizontal members provided the compression strength needed to resist buckling. The point of weakness was not the "obvious" one, but one obscured in casual investigation.

Reinforcements to Prevent Fatigue Cracking

Cover plates have been added to the beam in Fig. 2-83 to provide added strength. As can be seen by the moment diagram, they have been extended to points where the bending stress diminishes to a relatively low value. A fatigue crack that developed in a beam flange at the end of a cover plate would be accessible for repair, and in the next fabrication the cover plate could be extended to a still lower stress region.

The same reinforcement logic would not apply to the press frame design illustrated in Fig. 2-84, since if a fatigue crack develops at a stress raiser (a), bringing the side plates in toward the center of

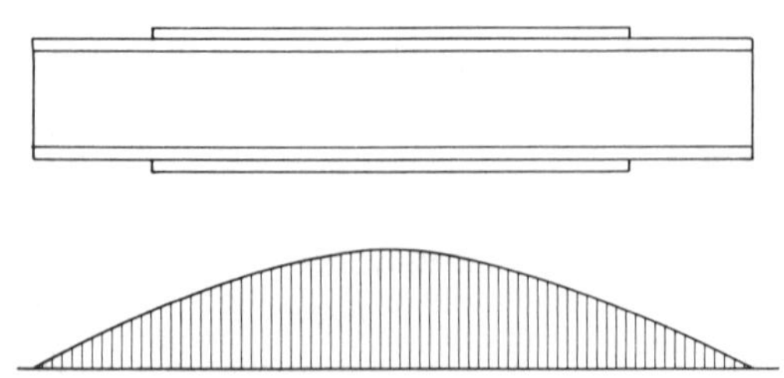

Fig. 2-83. Beam with cover plates added as strengtheners (top); bending moment diagram (bottom).

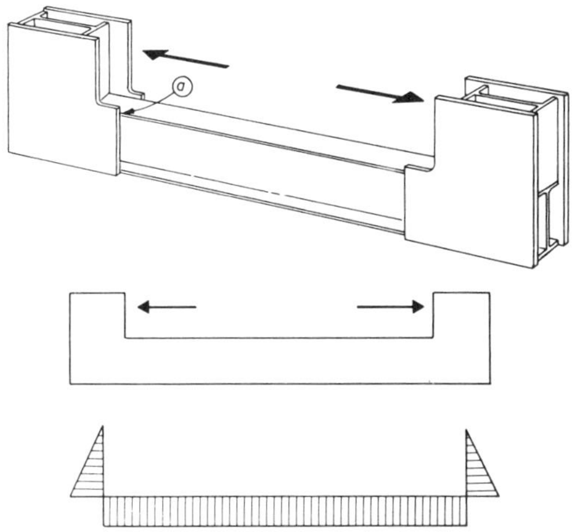

Fig. 2-84. Press frame design (top); simplified diagram showing forces (center); bending moment diagram (bottom).

the beam will not put the stress raisers in regions of lower stress. As the diagram shows, there is a constant bending moment from one end of the beam to the other. Nothing short of carrying the reinforcement the full length of the frame would effectively assure freedom from further fatigue cracking.

Anticipating Trouble

The designer does not always know what problems might develop when the product he is working on is put to use. He may be able to weigh the advantages and disadvantages of various alternatives, one of which may offer the possibility of a design change or a repair, whereas another may not.

For example, tanks frequently leak after welding is completed. Suppose a tank is to be welded with a continuous fillet on one side to make it liquid-tight, and with intermittent fillets on the other side. On which side should the continuous fillet weld be located? If the continuous weld were on the inside of the tank, repair would require rewelding the entire outside joint, should a leak develop. But if the continuous weld were on the outside, the repair could be made at the point of the leak with very little rewelding. The choice is based not on load or service requirements, but on ease of repair, should leakage occur.

Manufacturing Tolerances

The subassembly illustrated in Fig. 2-85(a) typifies composite design and calls for the joining of a formed channel member to a steel casting. The formed member was precut to match the shape of the casting before forming. Naturally, tolerance

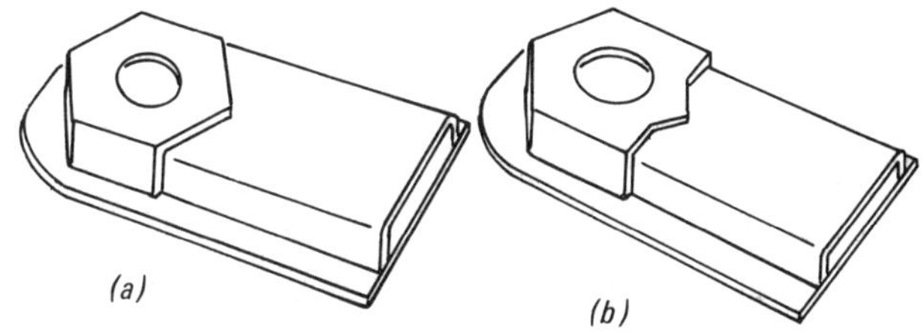

Fig. 2-85. (a) Original design showing channel joined to steel casting; (b) final design to overcome tolerance limitations.

limits apply to the forming operation. If the channel member is formed with a pressbrake and the break occurs a bit too far out, there will be a gap between the vertical leg of the formed section and the casting. If the break is a little too far from the outside edges of the precut plate, the vertical legs will fit tight but an abnormal gap will exist between the horizontal portion and the casting. In either case, a slight "within manufacturing tolerance" variation produces intolerably poor fitup and increases weld costs.

A slight alteration in the shape of the casting, as indicated in Fig. 2-85(b), would solve the problem. Now the position of the break can vary within tolerance limits without affecting the fitup of the joint.

GUIDES TO FABRICATION

When designing a weldment, the designer must keep in mind the equipment, methods, and processes that are available for fabrication. For example, the size and capacity of a bending brake will determine whether a machine base can be made from formed plates or flat plates welded at the corners. If the designer is not familiar with the equipment, a source for this type of information is the personnel of the fabricating department. Keep in mind that changes made at the drawing board before production is started are much less costly than changes made after the design goes into production. On any new design, or any change in design, the shop should be consulted for ideas relative to processing and fabrication.

Ways to Use Material Efficiently

The following are guidelines to more efficient and lower-cost fabrication:

Lay out pieces in a nesting arrangement, as in Fig. 2-86(a).

To shape flat sections, consider the alternatives of cutting from a large plate or cutting and welding bar stock, as in Fig. 2-86(b).

To reduce vibration, bend or press an indentation in the plate to act as a rib, as in Fig. 2-86(c).

A flange on a flat plate increases stiffness, as in Fig. 2-86(d). Stiffeners can be made from plate or welded bar stock, as in Fig. 2-86(e).

Gain a stiffener by bending the edge of a sheet before welding to the next sheet, as in Fig. 2-86(f).

If the section has a cutout, arrange the cutout so the material can be used for pads, stiffeners, gear blanks, etc., as in Fig. 2-86(g) and Fig. 2-88(c).

Build up composite sections by welding to reduce machining and material costs as in Fig. 2-87(a).

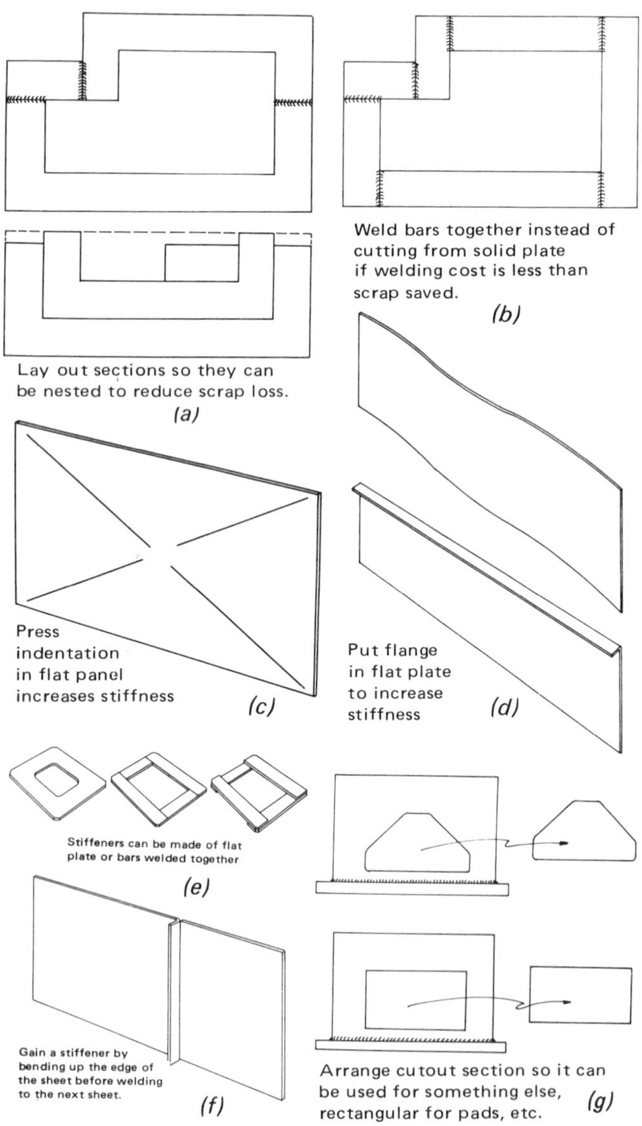

Fig. 2-86. Methods for using material efficiently.

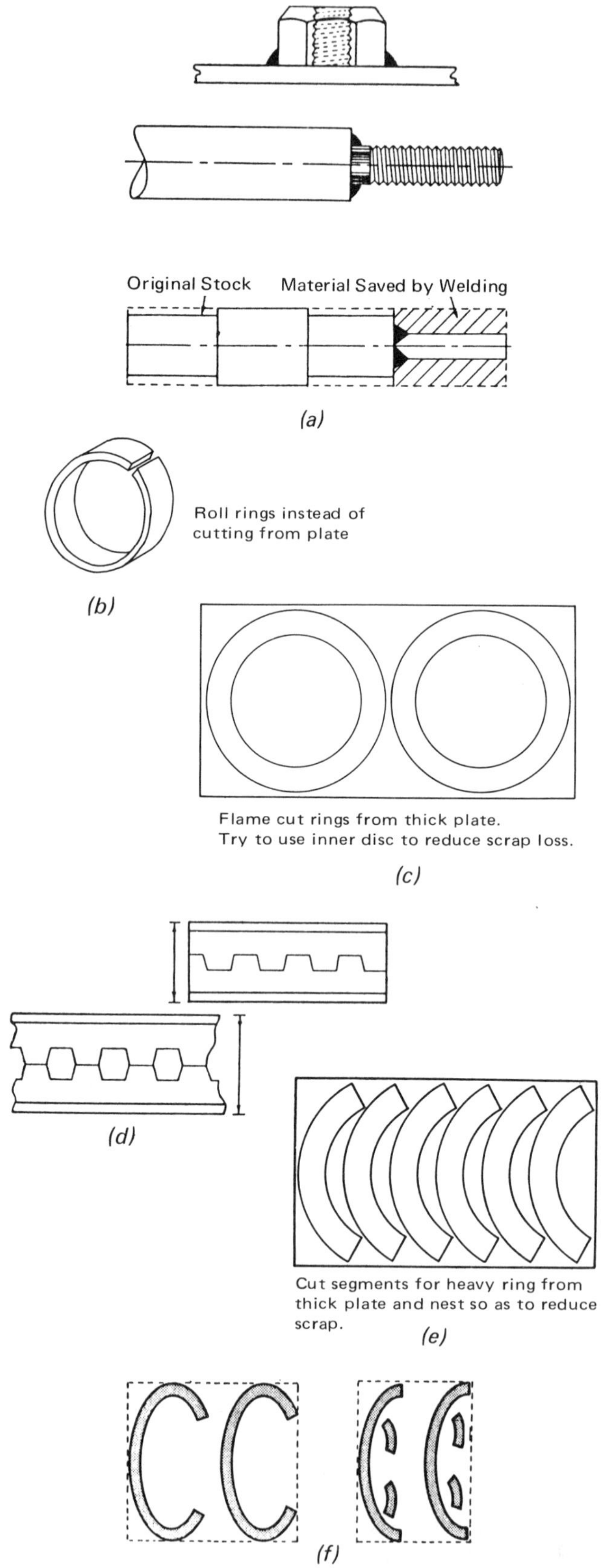

Fig. 2-87. Methods for avoiding wasted material.

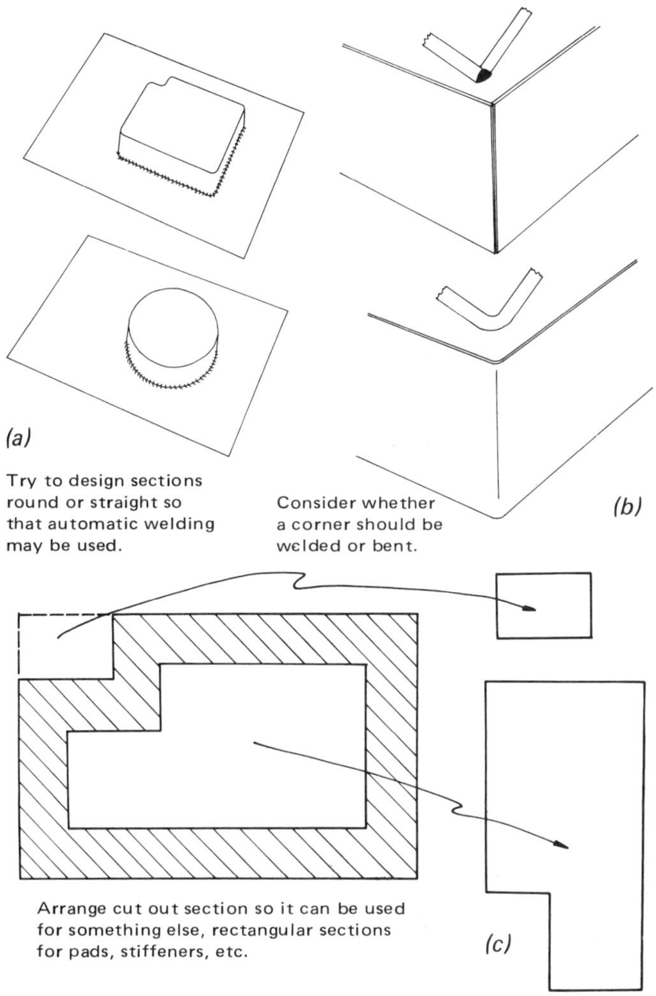

Fig. 2-88. Considerations for design in welding.

Roll rings from bar stock instead of cutting from heavy plate, as in Fig. 2-87(b).

When rings are cut from thick plate, plan to use the inner disc to reduce scrap loss, as in Fig. 2-87(c).

Standard rolled shapes can be cut and welded to produce a more rigid section, as in Fig. 2-87(d).

Cut segments for heavy rings from thick plate and nest to reduce scrap, as in Fig. 2-87(e) and Fig. 2-87(f).

If automatic welding is to be used, design for straight or circular welds, as in Fig. 2-88(a).

Sometimes a bend can be used in place of a weld, as in Fig. 2-88(b), substantially reducing fabrication costs.

Use the minimum amount of weld metal. Shaded areas in Fig. 2-89 indicate the amount of added weld metal.

Design joints for accessibility. Figure 2-90 shows situations where accessibility is a factor in welding efficiency.

Design joints to minimize the problem of melt-through. Figure 2-91 illustrates how to avoid melt-through problems.

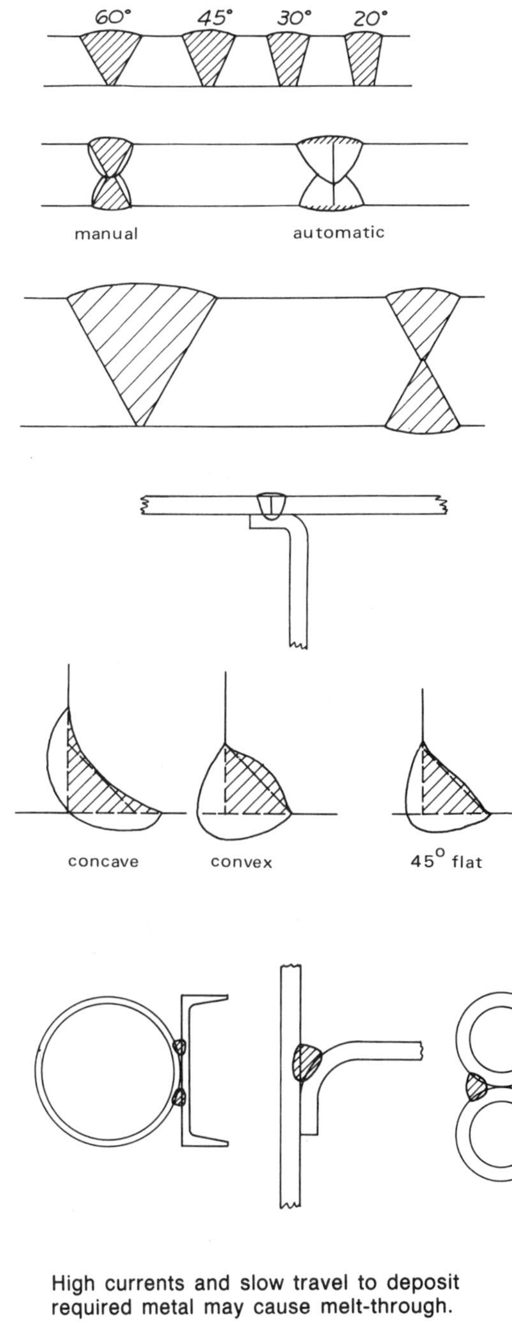

Fig. 2-89. Use minimum amount of weld metal. Shaded areas indicate added metal. Automatic welding eliminates need for beveling.

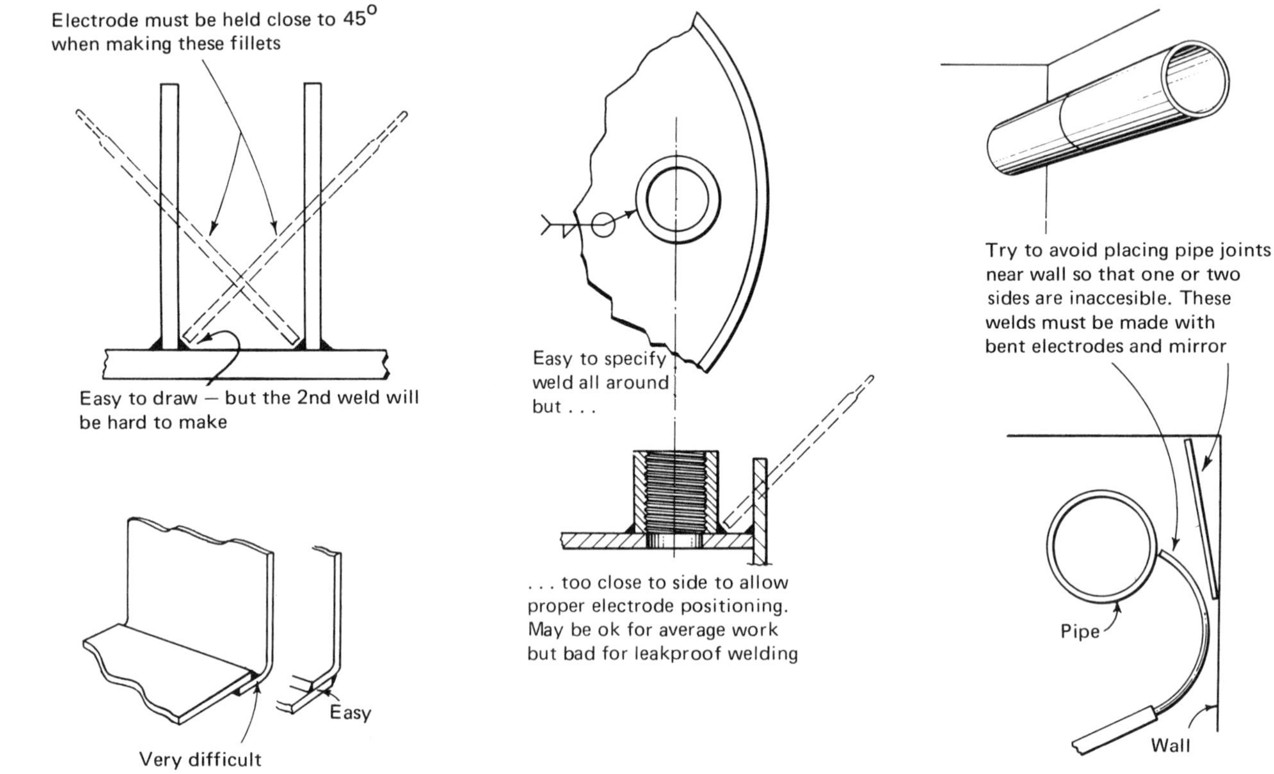

Fig. 2-90. Design for joint accessibility.

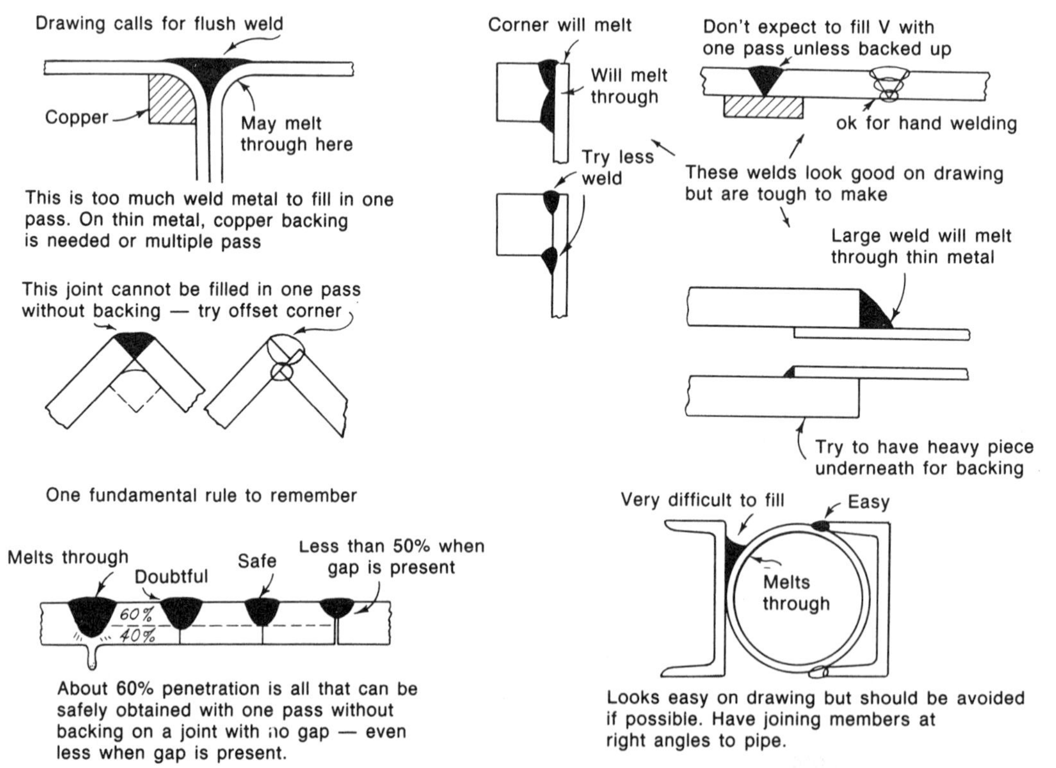

Fig. 2-91. Single-pass welds requiring large amounts of metal tend to melt through, especially with automatic welding.

Jigs and Fixtures

Jigs and fixtures should be used to decrease assembly time. In planning for assemblies and subassemblies, the designer should remember that:

The jig must have adequate rigidity to hold dimensions of weldment.

The assembly must provide easy locating points.

It must be possible to clamp and release quickly.

Jig must be loaded and unloaded easily.

Jig can sometimes be built to precamber weldment to control distortion.

Operating factor can be increased by providing two jigs, so that helper can load one while other is being welded.

It may be better to have separate jigs for tacking and final welding, or it may be better to do the entire job in one jig.

Design for the best possible fitup. Welding joints with gaps larger than necessary is costly.

Provide for clean work. Oil, rust, and dirt make for trouble in welding.

Forming

The proper use of forming can greatly reduce the cost of a weldment by eliminating welds. Several factors determine the best method of forming: thickness, over-all dimensions, number of duplicate parts, and tolerances. Cost is the final factor and is the determining factor if physical or shape requirements do not dictate the method. The cost of forming the part may be offset by a savings in machining.

Consider the following forming methods:

Press brake
Rolling
Tangent bending and contour forming
Flanging and dishing
Die stamping
Nibbling

Finally, consider possible savings by eliminating forming through the use of steel castings or forgings in conjunction with the weldment where very complicated shapes are required.

The Advantages of Subassemblies

Once the product has been designed, the design laid out for production, and the joints selected and designed, the job is ready for assembly. In visualizing assembly procedure, the designer should break the weldment down into subassemblies several different ways to determine which, if any, will offer some of the following cost savings:

A large number of subassemblies spreads the work out so that more people can work on the job.

If stress relief of a portion of the weldment is necessary, it may be easier to do this before welding it into final assembly.

Precision welding possible with modern techniques permits machining to close tolerances before welding assemblies.

Test compartments before welding into final assembly, where required.

Subassemblies facilitate inspection.

Painting before assembly may be more economical.

Repairs, if necessary, are easier on subassemblies.

Subassemblies usually provide better access for welding.

It is easier to control distortion or locked-up stresses in subassemblies than in whole assemblies.

SUGGESTED READING

Welding Handbook, Vol. 1, 8th Ed.: Welding Technology, Chapter 5, "Design For Welding," American Welding Society (1987).

"Design and Fabrication of D6AC Steel Weldments for Aircraft Structures," R.E. Key, J.C. Collins, and H.I. McHenry, *Welding Journal,* 991-1000 (December 1967).

Design of Welded Structures, Omer W. Blodgett, J.F. Lincoln Arc Welding Foundation (1966).

"The Basis of Welding Design," J.E.M. Jubb, *Welding and Metal Fabrication,*Pt. 5, 359-361 (September 1966); Pt. 4, 109-110 (March 1966); Pt. 3, 462-464 (November 1965); Pt. 2 (September 1965); Pt. 1, 342-343 (August 1965).

"The Influence of the Art of Welding on the Creative Concepts of Structural Design," A. Amirikian, *Welding Journal,* 337s-343s (August 1966).

"Root Bead Welding with Y-type Consumable Inserts," C.W. Johnson, *Ibid.,* 227-281 (April 1965).

"Effect of External Geometry on Fatigue Behavior of Welded Joints," W.W. Sanders, Jr.; A.T. Derecho; and W.H. Munse, *Ibid.,* 49s-55s (February 1965).

"Welded Bascule Bridges Return to Florida," J.E. Sawyer and E.R. Estes, Jr., *Ibid.,* 27-33 (January 1964).

Specification for Design, Fabrication and Erection of Structural Steels for Buildings, American Institute of Steel Construction (1969).

Design of Weldments, Omer W. Blodgett, J.F. Lincoln Arc Welding Foundation (1963).

Tubular Steel Structures — Theory and Design, M.S. Troitsky, J.F. Lincoln Arc Welding Foundation (1982).

Prestressed Welded Steel Structures — Theory and Design, M.S. Troitsky, J.F. Lincoln Arc Welding Foundation (1988).

The Aluminum Association. *Specification for aluminum structures.* Washington, D.C.: The Aluminum Association, latest edition.

American Institute of Steel Contruction, Inc. *Manual of steel construction.* Chicago: American Institute of Steel Construction, Inc. (AISC), latest edition.

American Welding Society. ANSI/AWS D1.1, *Structural welding code — steel.* Miami, FL: American Welding Society, latest edition.

American Welding Society. ANSI/AWS D1.2, *Structural welding code — aluminum.* Miami, FL: American Welding Society, latest edition.

Cary, H.B., *Modern welding technology.* Englewood Cliffs, NJ: Prentice-Hall, 1979.

The Lincoln Electric Company. *Solutions to the design of weldments,* D810.17. Cleveland, OH: The Lincoln Electric Co. (Jan. 1975).

Marshall, P.W., *Welding of tubular structures, proceedings of second international conference.* Boston, MA: International Institute of Welding (IIW) (1984).

Rolfe, S.T., Fatigue and fracture control in structures. *AISC Engineering Journal.* 1977: 14 (1).

Rolfe, S.T. and Barsom, J.M. *Fracture and fatigue control in structures: application of fracture mechanics.* Englewood Cliffs, NJ: Prentice-Hall (1977).

Sandus, W.W. and Day, R.H. Fatigue behavior of aluminum alloy weldments. *WRC Bulletin:* No. 286, Aug. (1983).

The Design of Welded Joints

The loads in a welded steel design are transferred from one member to another through welds placed in weld joints. Both the type of joint and the type of weld are specified by the designer.

Figure 2-92 shows the joint and weld types. Specifying a joint does not by itself describe the type of weld to be used. Thus, ten types of welds are shown for making a butt joint. Although all but two welds are illustrated with butt joints here, some may be used with other types of joints. Thus, a single-bevel weld may also be used in a T or corner joint (Fig. 2-93), and a single-V weld may be used in a corner, T, or butt joint.

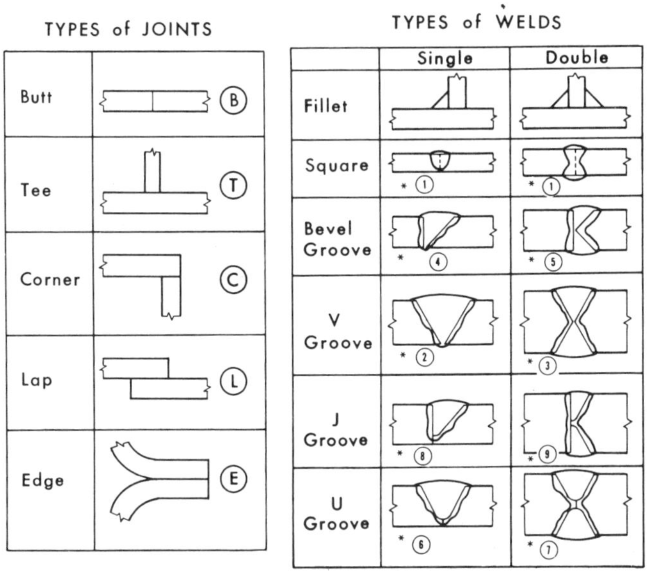

Fig. 2-92. Joint designs (left); weld grooves (right).

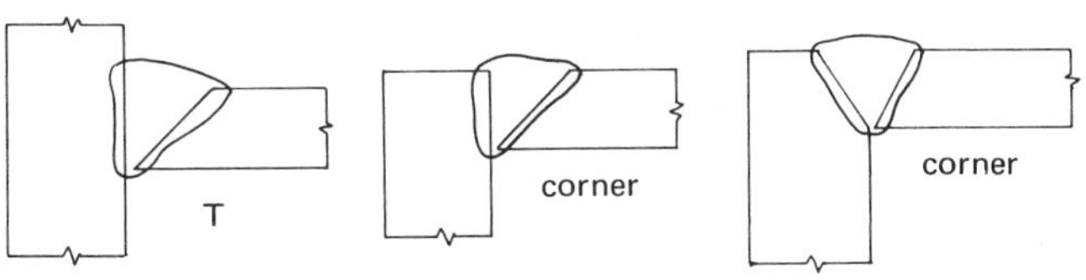

Fig. 2-93. Single-bevel weld used in T joint (left) and corner joint (center); single-V weld in corner joint (right).

FILLET-WELDED JOINTS

The fillet weld, requiring no groove preparation, is one of the most commonly used welds. Corner welds are also widely used in machine design. Various corner arrangements are illustrated in Fig. 2-94. The corner-to-corner joint, as in Fig. 2-94(a), is difficult to assemble because neither plate can be supported by the other. A small electrode with low welding current must be used so that the first welding pass does not melt-through. The joint requires a large amount of metal. The corner joint shown in Fig. 2-94(b) is easy to assemble, does not

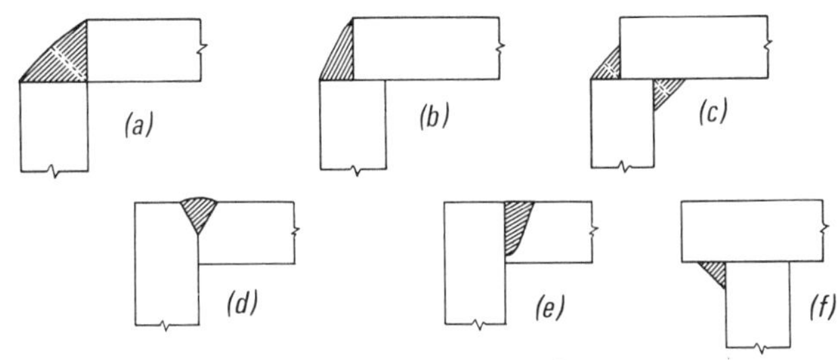

Fig. 2-94. Various corner joints.

easily melt-through, and requires just half the amount of the weld metal as the joint in Fig. 2-94(a). However, by using half the weld size, but placing two welds, one outside and the other inside, as in Fig. 2-94(c), it is possible to obtain the same total throat as with the first weld. Only half the weld metal need be used.

With thick plates, a partial-penetration groove joint, as in Fig. 2-94(d) is often used. This requires beveling. For a deeper joint, a J preparation, as in Fig. 2-94(e), may be used in preference to a bevel. The fillet weld in Fig. 2-94(f) is out of sight and makes a neat and economical corner.

The size of the weld should always be designed with reference to the size of the thinner member. The joint cannot be made any stronger by using the thicker member for the weld size, and much more weld metal will be required, as illustrated in Fig. 2-95.

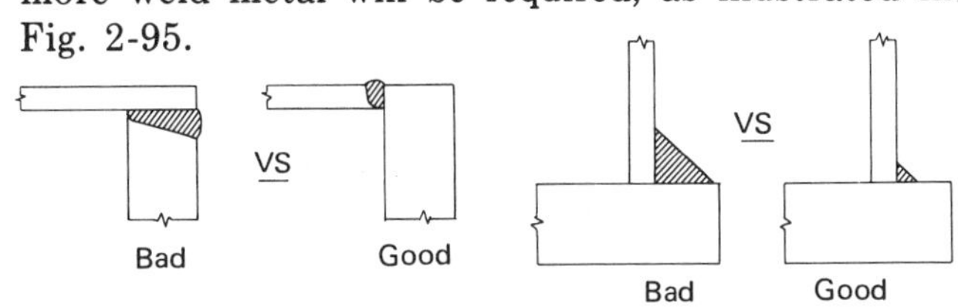

Fig. 2-95. Size of weld should be determined with reference to thinner member.

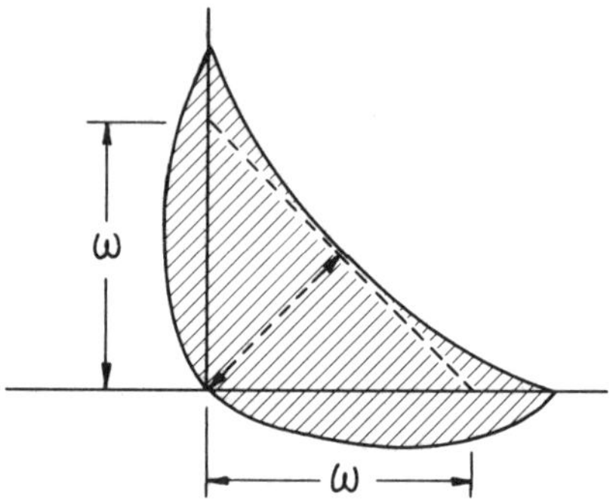

Fig. 2-96. Leg, size, ω, of a fillet weld.

In the United States, a fillet weld is measured by the leg size of the largest right triangle that may be inscribed within the cross-sectional area (Fig. 2-96). The throat; a better index to strength,- is the shortest distance between the root of the joint and the face of the diagrammatical weld. As Figure 2-96 shows, the leg size used may be shorter than the actual leg of the weld. With convex fillets, the actual throat may be longer than the throat of the inscribed triangle.

GROOVE AND FILLET COMBINATIONS

A combination of a partial-penetration groove weld and a fillet weld (Fig. 2-97) is used for many joints. The AWS prequalified, single-bevel groove T joint is reinforced with a fillet weld.

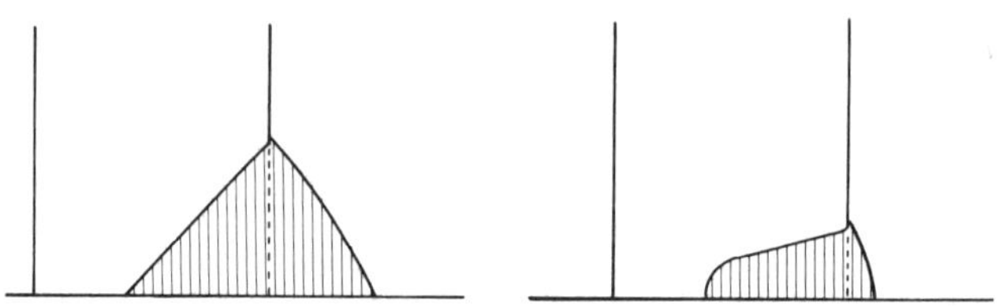

Fig. 2-97. Combined groove and fillet-welded joints.

The designer is frequently faced with the question of whether to use fillet or groove welds (Fig. 2-98). Here cost becomes a major consideration. The fillet welds in Fig. 2-98(a) are easy to apply and require no special plate preparation. They can be made using large-diameter electrodes with high welding currents, and, as a consequence, the deposition rate is high. The cost of the welds increases as the square of the leg size.

In comparison, the double-bevel groove weld in Fig. 2-98(b), has about one-half the weld area of the fillet welds. However, it requires extra preparation and the use of smaller-diameter electrodes with lower welding currents to place the initial pass

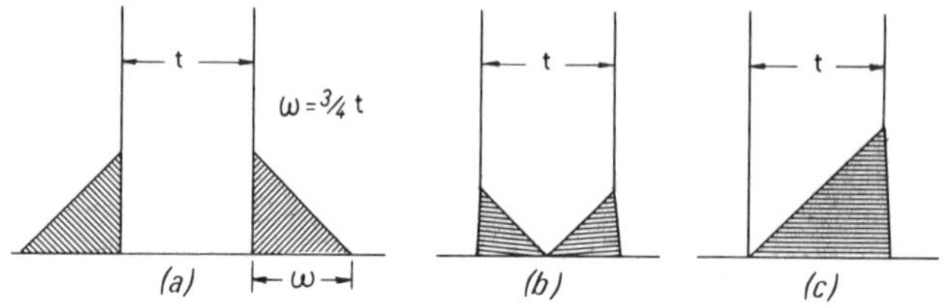

Fig. 2-98. Comparison of fillet welds and groove welds.

without melting through. As plate thickness increases, this initial low-deposition region becomes a less important factor, and the higher cost factor decreases in significance. The construction of a curve based on the best possible determination of the actual cost of welding, cutting, and assembling, such as illustrated in Fig. 2-99, is a possible technique for deciding at what point in the plate thickness the double-bevel groove weld becomes less costly. The point of intersection of the fillet curve with a groove-weld curve is the point of interest. The accuracy of this device is dependent on the accuracy of the cost data used in constructing the curves.

Referring to Fig. 2-98(c), it will be noted that the single-bevel groove weld requires about the same amount of weld metal as the fillet welds deposited in Fig. 2-98(a). Thus, there is no apparent economic advantage. There are some disadvantages, though.

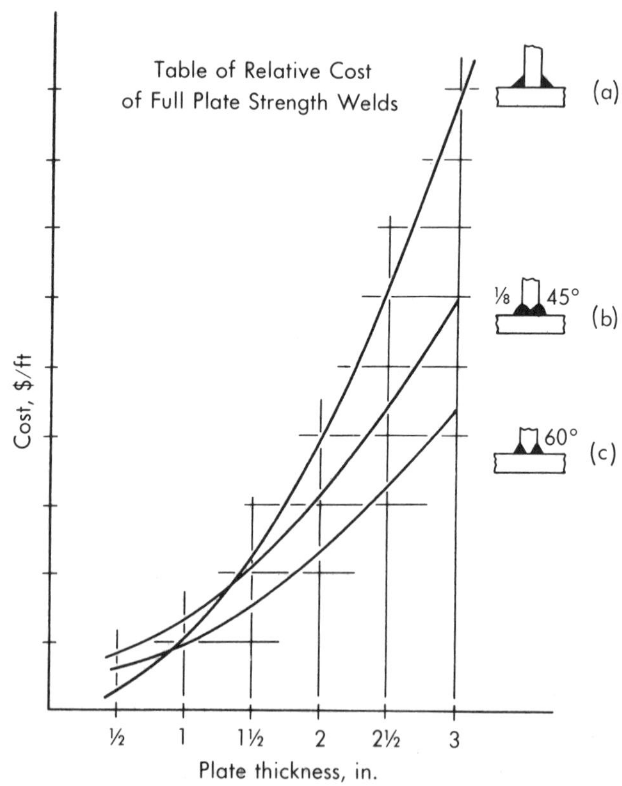

Fig. 2-99. Relative cost of welds having the full strength of the plate.

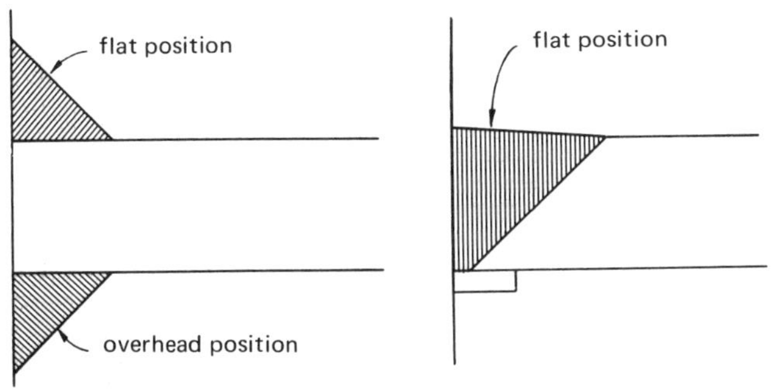

Fig. 2-100. In the flat position, a single-bevel groove joint is less expensive than fillet welds in making a T joint.

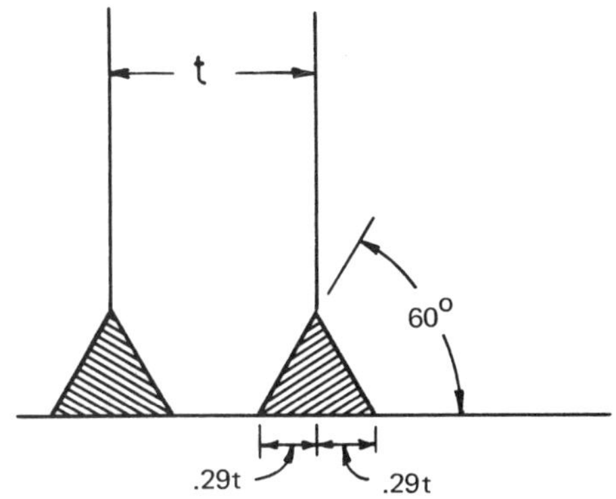

Fig. 2-101. Partial-penetration double-bevel groove joint.

The single-bevel joint requires bevel preparation and initially a lower deposition rate at the root of the joint. From a design standpoint, however, it offers a direct transfer of force through the joint, which means that it is probably better under fatigue loading. Although the illustrated full-strength fillet weld, having leg sizes equal to three-quarters the plate thickness, would be sufficient, some codes have lower allowable limits for fillet welds and may require a leg size equal to the plate thickness. In this case, the cost of the fillet-welded joint may exceed the cost of the single-bevel groove in thicker plates. Also, if the joint is so positioned that the weld can be made in the flat position, a single-bevel groove weld would be less expensive than if fillet welds were specified. As can be seen in Fig. 2-100, one of the fillets would have to be made in the overhead position — a costly operation.

The partial-penetration double-bevel groove joint shown in Fig. 2-101 has been suggested as a full-strength weld. The plate is beveled to 60° on both sides to give a penetration of at least 29% of the thickness of the plate (.29t). After the groove is filled, it is reinforced with a fillet weld of equal cross-sectional area and shape. This partial-penetration double-bevel groove joint has 57.8% the weld metal of the full-strength fillet weld. It requires joint preparation; however, the 60° angle allows the use of large electrodes and high welding current.

Full-strength welds are not always required in the design, and economies can often be achieved by using partial-strength welds where these are applicable and permissible. Referring to Fig. 2-102, it can be seen that on the basis of an unreinforced 1-in. throat, a 45° partial-penetration, single-bevel groove weld requires just one-half the weld area needed for a fillet weld. Such a weld may not be as economical as the same strength fillet weld, however, because of the cost of edge preparation and need to use a smaller electrode and lower current on the initial pass.

If the single-bevel groove joint were reinforced with an equal-leg fillet weld, the cross-sectional area for the same throat size would still be one-half the area of the fillet, and less beveling would be required. The single-bevel 60° groove joint with an equal fillet weld reinforcement for the same throat size would have an area of 57.8% of the simple fillet weld. This joint has the benefit of smaller cross-sectional area — yet the 60° included angle allows the use of higher welding current and larger electrodes. The

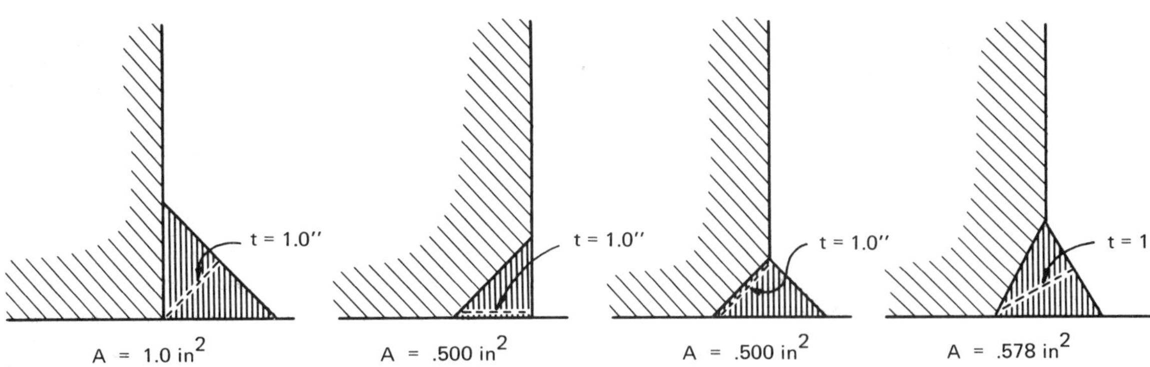

Fig. 2-102. Comparison of weld joints having equal throats.

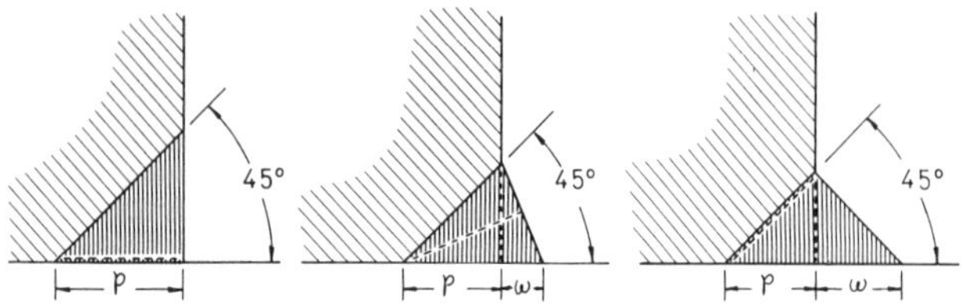

Fig. 2-103. Comparison of weld joints with and without reinforcing fillet welds.

only disadvantage is the extra cost of preparation.

From this discussion, it is apparent that the simple fillet-welded joint is the easiest to make, but may require excessive weld metal for larger sizes. The single-bevel 45°-included-angle joint is a good choice for the larger weld sizes. However, one would miss opportunities by selecting the two extreme conditions of these two joints. The joints between these two should be considered. Referring to Fig. 2-103, one may start with the single-bevel 45 ° joint without the reinforcing fillet weld, gradually add a reinforcement, and finally increase the lower leg of the fillet reinforcement until a full 45° fillet weld is reached. In this figure, p = depth of preparation; ω = leg of reinforcing fillet.

When a partial-penetration groove weld is reinforced with a fillet weld, the minimum throat is used for design purposes, just as a minimum throat of a fillet or partial-penetration groove weld is used. However, as Fig. 2-104 shows, the allowable for this combination weld is not the sum of the allowable limits for each portion of the combination weld. This would result in a total throat much larger than the actual.

Figure 2-105(a) shows the effect of using the incorrect throat in determining the allowable unit force on a combination weld. The allowable for each weld was added separately. In Fig. 2-105(b) weld size is correctly figured upon the minimum throat.

Sizing Fillets

Table 2-7 gives the sizing of fillet welds for rigidity designs at various strengths and plate thicknesses, where the strength of the weld metal matches the plate.

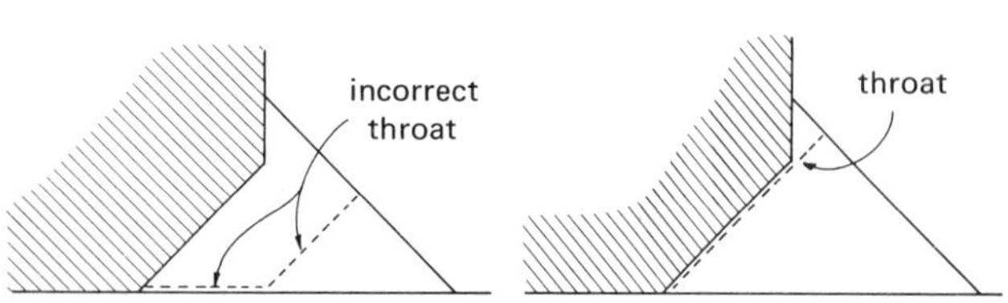

Fig. 2-104. Determining minimum throat: incorrect result (left); correct result (right).

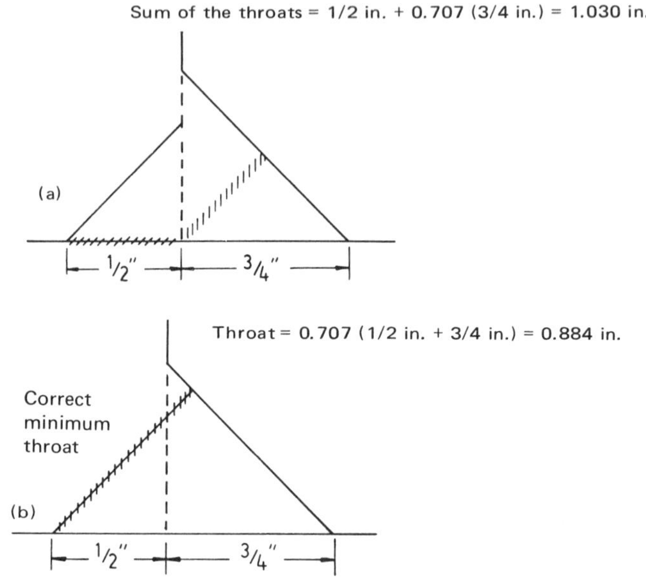

Fig. 2-105. Examples showing effect of correct and incorrect throat dimension in determining allowable load on a combination weld. In (a), the weld allowable is incorrectly figured by adding each weld separately; in (b) weld allowable is correctly figured on the minimum throat.

In machine design work, where the primary design requirement is rigidity, members are often made with extra heavy sections, so that the movement under load will be within very close tolerances. Because of the heavy construction, stresses are very low. Often the allowable stress in tension for mild steel is given as 20,000 psi, yet the welded machine base or frame may have a working stress of only 2000 to 4000 psi. The question arises as how to determine the weld sizes for these types of rigidity designs.

It is not very practical to calculate, first, the stresses resulting in a weldment when the unit is loaded within a predetermined dimensional tolerance, then to use these stresses to determine the forces that must be transferred through the connecting welds. A very practical method, however, is to design the weld for the thinner plate, making it sufficient to carry one-third to one-half the carrying capacity of the plate. This means that if the plate were stressed to one-third to one-half its usual value, the weld would be sufficient. Most rigidity designs are stressed much below these values; however, any reduction in weld size below one-third the full-strength value would give a weld too small an appearance for general acceptance.

GROOVE JOINTS

Figure 2-106 indicates that the root opening (R)

TABLE 2-7. RULE-OF-THUMB FILLET-WELD SIZES WHERE THE STRENGTH OF THE WELD METAL MATCHES THE PLATE

Plate thickness (t)	Strength design	Rigidity design	
	Full-strength weld (ω = 3/4 t)	50% of full-strength weld (ω = 3/8 t)	33% of full-strength weld (ω = 1/4 t)
Less than 1/4	1/8	1/8*	1/8*
1/4	3/16	3/16*	3/16*
5/16	1/4	3/16*	3/16*
3/8	5/16	3/16*	3/16*
7/16	3/8	3/16	3/16*
1/2	3/8	3/16	3/16*
9/16	7/16	1/4	1/4*
5/8	1/2	1/4	1/4*
3/4	9/16	5/16	1/4*
7/8	5/8	3/8	5/16*
1	3/4	3/8	5/16*
1-1/8	7/8	7/16	5/16
1-1/4	1	1/2	5/16
1-3/8	1	1/2	3/8
1-1/2	1-1/8	9/16	3/8
1-5/8	1-1/4	5/8	7/16
1-3/4	1-3/8	3/4	7/16
2	1-1/2	3/4	1/2
2-1/8	1-5/8	7/8	9/16
2-1/4	1-3/4	7/8	9/16
2-3/8	1-3/4	1	5/8
2-1/2	1-7/8	1	5/8
2-5/8	2	1	3/4
2-3/4	2	1	3/4
3	2-1/4	1-1/8	3/4

* These values have been adjusted to comply with AWS-recommended minimums.

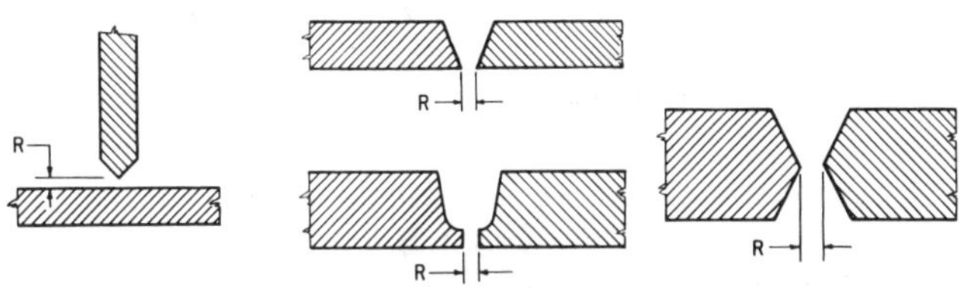

Fig. 2-106.

is the separation between the members to be joined.

A root opening is used for electrode accessibility to the base or root of the joint. The smaller the angle of the bevel, the larger the root opening must be to obtain good fusion at the root.

If the root opening is too small, root fusion is more difficult to obtain, and smaller electrodes must be used, thus slowing down the welding process.

If the root opening is too large, weld quality does not suffer but more weld metal is required; this increases welding cost and will tend to increase distortion.

Figure 2-107 indicates how the root opening must be increased as the included angle of the bevel

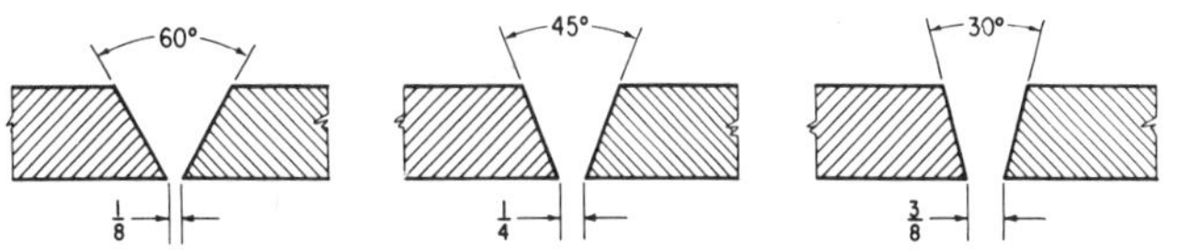

Fig. 2-107.

is decreased. Backup strips are used on larger root openings. All three preparations are acceptable; all are conducive to good welding procedure and good weld quality. Selection, therefore, is usually based on cost.

Root opening and joint preparation will directly affect weld cost (pounds of metal required), and choice should be made with this in mind. Joint preparation involves the work required on plate edges prior to welding and includes beveling and providing a root face.

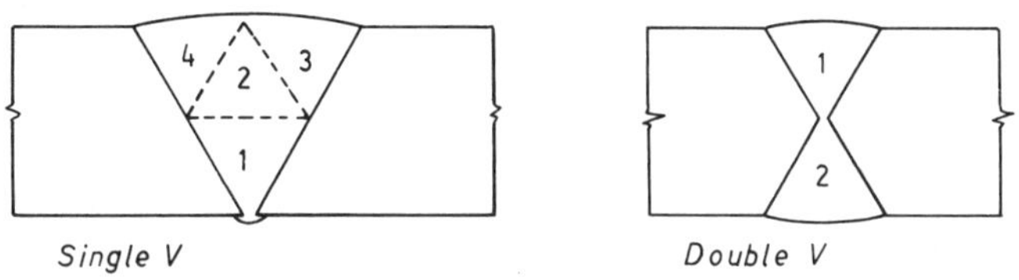

Fig. 2-108. Using double-groove joint in place of single-groove joint reduces amount of welding.

Using a double-groove joint in preference to a single-groove (Fig. 2-108) cuts in half the amount of welding. This reduces distortion and makes possible alternating the weld passes on each side of the joint, again reducing distortion.

In Fig. 2-109(a), if the bevel or gap is too small, the weld will bridge the gap leaving slag at the root. Excessive back-gouging is then required.

Figure 2-109(b) shows how proper joint preparation and procedure will produce good root fusion and will minimize back-gouging.

In Fig. 2-109(c), a large root opening will result in melt-through. Spacer strip may be used, in which case the joint must be back-gouged.

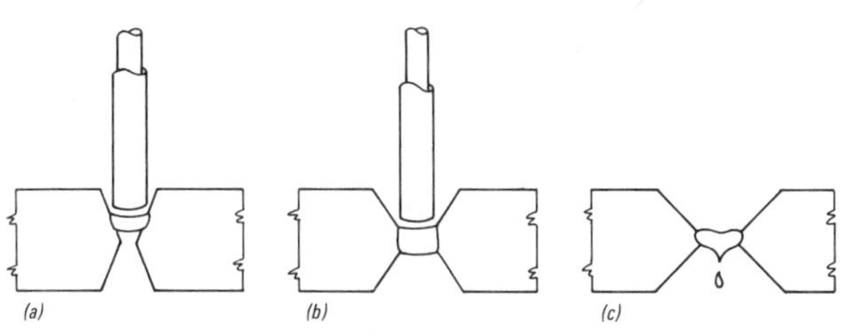

Fig. 2-109. (a) If the gap is too small, the weld will bridge the gap, leaving slag at the root; (b) a proper joint preparation; (c) a root opening too large will result in melt-through.

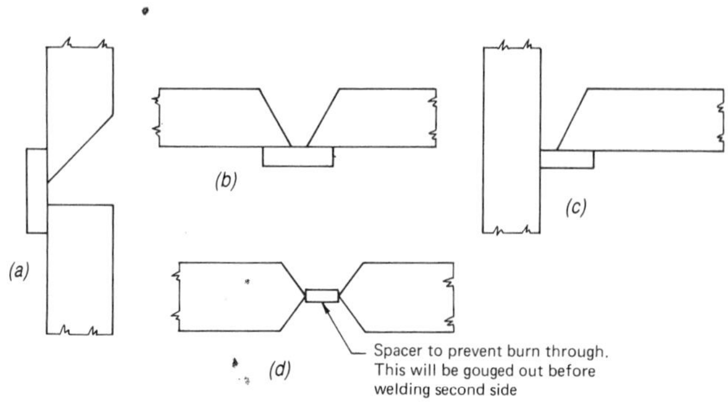

Fig. 2-110. Backup strips — (a), (b), and (c) — are used when all welding is done from one side or when the root opening is excessive, a spacer to prevent melt-through (d) will be gouged out before welding the second side.

Backup strips are commonly used when all welding must be done from one side, or when the root opening is excessive. Backup strips, shown in Fig. 2-110(a), (b), and (c), are generally left in place and become an integral part of the joint.

Spacer strips may be used especially in the case of double-V joints to prevent melt-through. The spacer in Fig. 2-110(d) to prevent melt-through will be gouged out before welding the second side.

Backup Strips

Backup strip material should conform to the base metal. Feather edges of the plate are recommended when using a backup strip.

Short intermittent tack welds should be used to hold the backup strip in place, and these should preferably be staggered to reduce any initial restraint of the joint. They should not be directly opposite one another (Fig. 2-111).

The backup strip should be in intimate contact with both plate edges to avoid trapped slag at the root (Fig. 2-112).

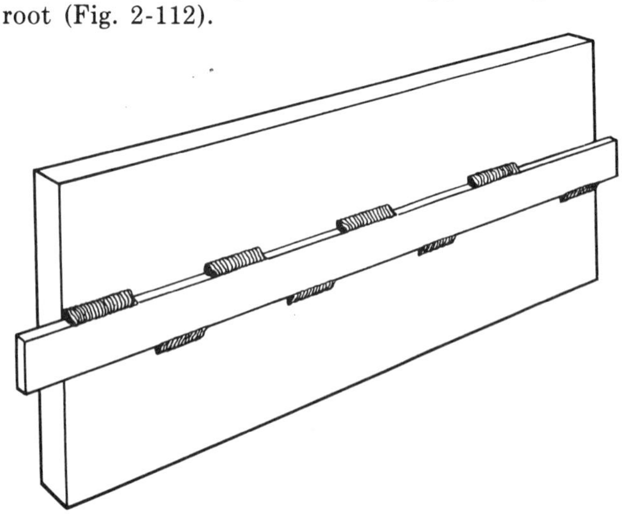

Fig. 2-111. Short intermittent tack welds should be used to hold the backup strip in place.

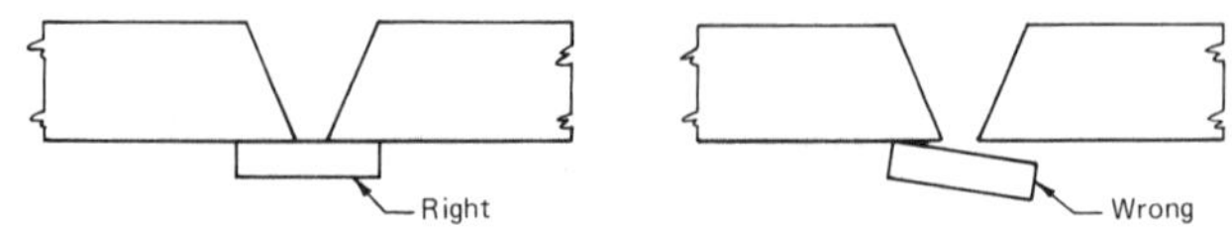

Fig. 2-112. The backup strip should be in intimate contact with both edges of plate.

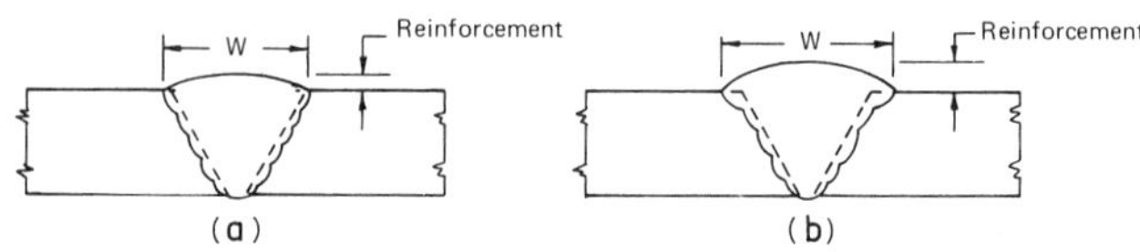

Fig. 2-113. The reinforcement on a butt joint should be minimal, as in (a).

On a butt joint, a nominal weld reinforcement (approximately 1/16″ above flush) is all that is necessary, as in Fig. 2-113(a). Additional buildup, as in Fig. 2-113(b), serves no useful purpose and will increase the weld cost.

Care should be taken to keep both the width and the height of the reinforcement to a minimum.

Edge Preparation

The main purpose of a root face (Fig. 2-114) is to provide an additional thickness of metal, as opposed to a featheredge, in order to minimize any melt-through tendency. A featheredge preparation is more prone to melt-through than a joint with a root face, especially if the root opening becomes a little too large, (Fig. 2-115).

A root face is not as easily obtained as a featheredge. A featheredge is generally a matter of one cut with a torch, while a root face will usually require two cuts or possibly a torch cut plus machining.

A root face usually requires back-gouging if a 100% weld is required. A root face is not recommended when welding, into a backup strip, since a gas pocket would be formed.

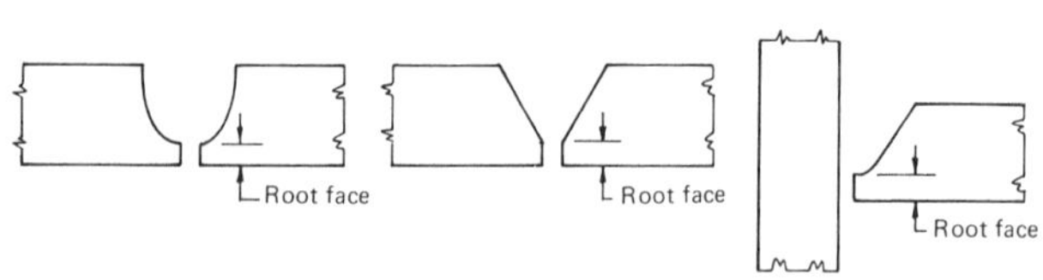

Fig. 2-114. A root face minimizes tendency to melt-through.

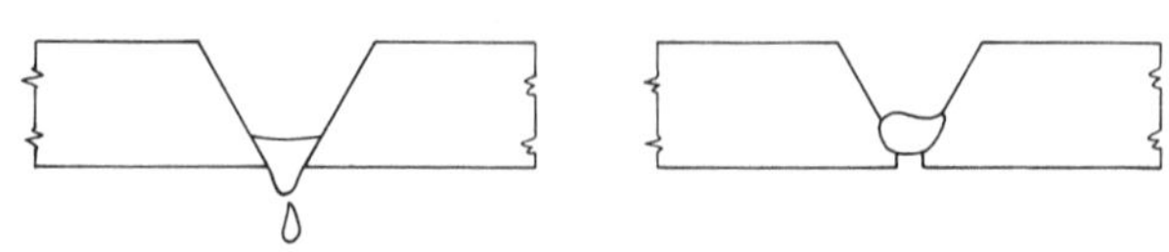

Fig. 2-115. A featheredge is more prone to melt-through than with a root face.

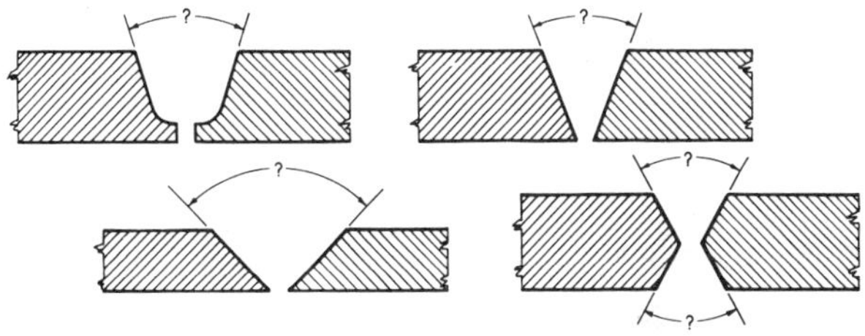

Fig. 2-116. Accessibility is gained by compromising between bevel and root opening.

Plate edges are beveled to permit accessibility to all parts of the joint and insure good fusion throughout the entire weld cross section. Accessibility can be gained by compromising between maximum bevel and minimum root opening (Fig. 2-116).

Degree of bevel may be dictated by the importance of maintaining proper electrode angle in confined quarters (Fig. 2-117). For the joint illustrated, the minimum recommended bevel is 45°.

J and U preparations are excellent to work with, but economically they may have little to offer because preparation requires machining as opposed to simple torch cutting. Also a J or U groove requires a root face (Fig. 2-118) and thus backgouging.

To consistently obtain complete fusion when welding a plate, back-gouging is required.on virtually all joints except bevel joints with feather edge. This may be done by any convenient means, grinding, chipping, or gouging. The latter method is generally the most economical and leaves an ideal contour for subsequent beads.

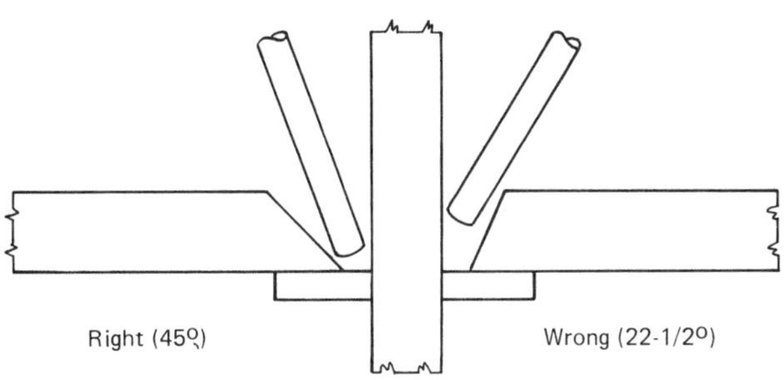

Fig. 2-117. Degree of bevel may be dictated by the need for maintaining proper electrode angle.

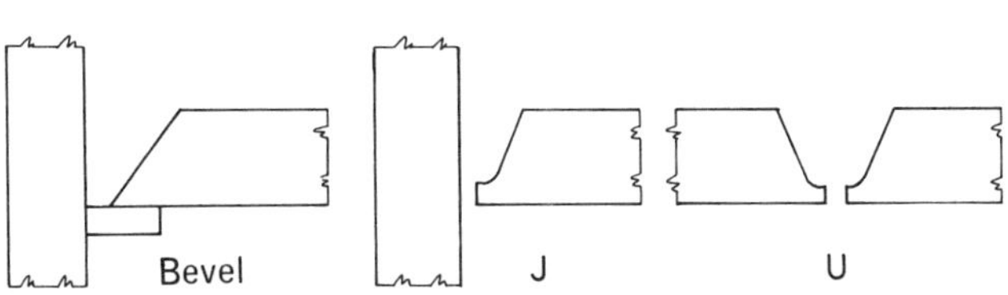

Fig. 2-118. A bevel preparation with a backup strip may be more economical than a J or U groove.

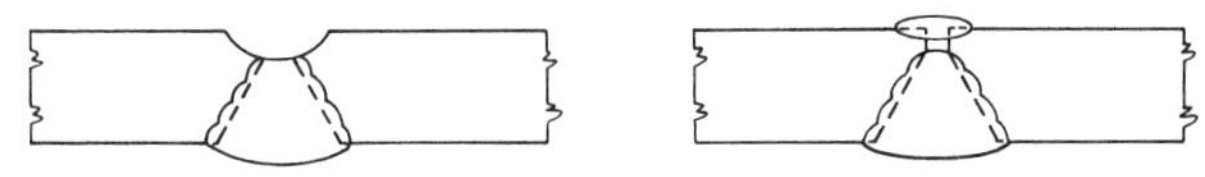

Fig. 2-119. Without back-gouging, penetration is incomplete.

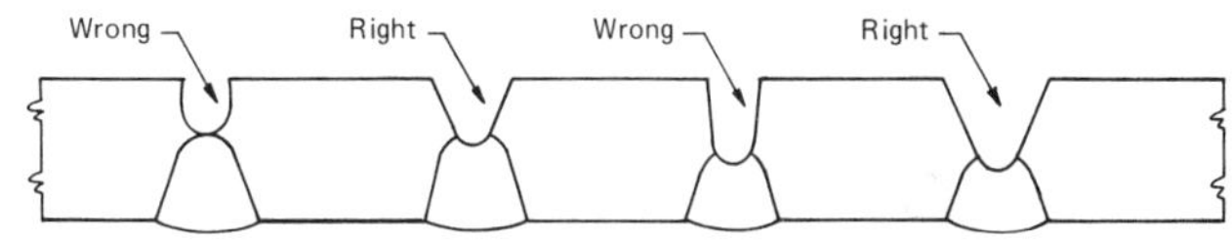

Fig. 2-120. Proper back-gouging should be deep enough to expose sound weld metal.

Without back-gouging, penetration is incomplete (Fig. 2-119). Proper back-chipping should be deep enough to expose sound weld metal, and the contour should permit the electrode complete accessibility (Fig. 2-120).

Joint Preparation After Assembly

The AWS D1.1 Structural Welding Code will be helpful to fabricators who frequently encounter a situation where assembly and tackup before gouge-preparation of U-groove joints for welding would be logical and less costly. This practice has not been forbidden in the past, but some inspectors — not having a clarification of its acceptability — have not permitted it. AWS, by specifically noting that U-groove joints for complete-penetration and partial-penetration welds may be made "prior to or after fitting," removes all doubt regarding acceptability. Figure 2-121 illustrates the joints where gouge-preparation of U grooves after fitup and assembly is applicable and the sequence of fabricating a column in this manner.

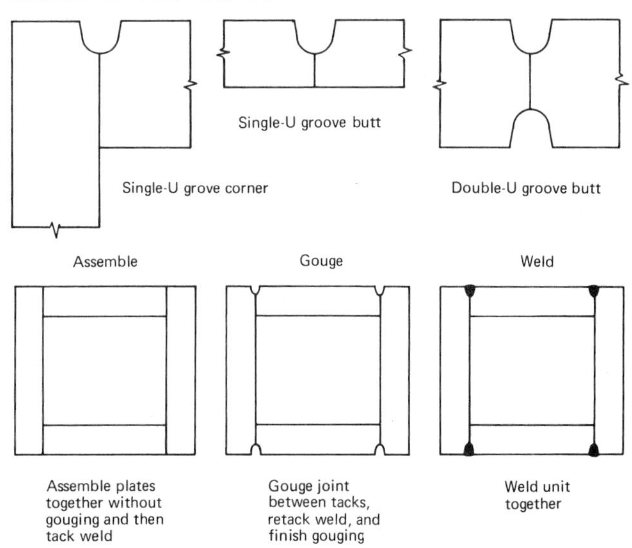

Fig. 2-121. Partial-penetration submerged-arc groove welds. U-groove joint preparation may be done either prior to or after fitting.

Preparing to test beam-to-column connections.

Allowables for Welds

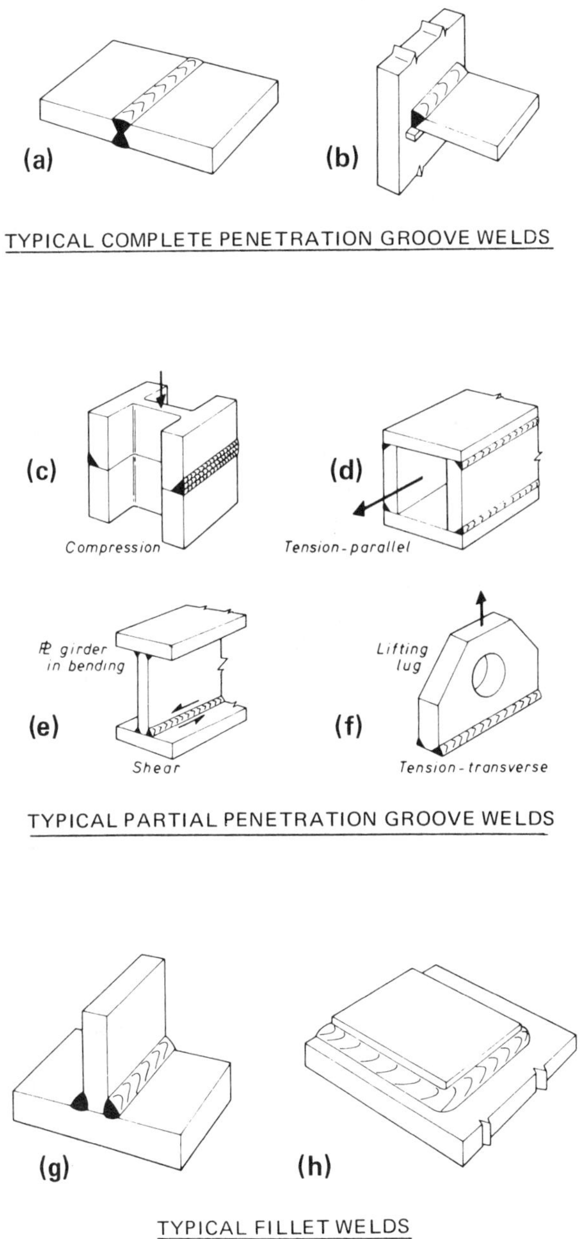

Fig. 2-122. Typical welded joints — a and b, Complete joint penetration groove welds; c to f, Partial joint penetration groove welds; g and h, Fillet welds.

ALLOWABLE STRENGTHS OF WELDS UNDER STEADY LOADS

A structure or weldment is as strong as its weakest point, and "allowable" weld strengths are specified by the American Welding Society (AWS), the American Institute of Steel Construction (AISC), and various other engineering and governmental organizations to make sure that the weld can deliver the mechanical properties of the plate being joined. Allowables are designated for various types of welds under steady and fatigue loads.

In Figs. 2-122a and 2-122b, complete joint penetration groove welds are illustrated. These are considered full-strength welds, since they are capable of transferring the full strength of the members they connect. In calculations, such welds are allowed the same stress as the plate, provided the proper strength level of weld metal is used. In such complete joint penetration welds, the mechanical properties of the weld metal must at least match the strength of the base material. If the plates joined are of different strengths, the weld metal must at least match the strength of the weaker plate.

Figs. 2-122c to 2-122f illustrate partial joint penetration groove welds, which are widely used in the economical welding of very heavy plate. A partial joint penetration groove weld in such heavy material will usually accomplish a savings in weld metal and welding time, while giving the required strength at the joint. The faster cooling and increased restraint, however, justify establishment of a minimum effective throat (t_e). See Table 2-9.

Various factors must be considered in determining the allowable stresses on the throat of partial joint penetration groove welds. Joint configuration is one. If a V, J, or U groove is specified, it is assumed that the weldor can easily reach the bottom of the joint, and the effective weld throat (t_e) equals the depth of the groove. If a bevel groove with an included angle of 45 deg or less is specified and the manual shielded metal arc process is used, 1/8 in. is deducted from the depth of the prepared groove in defining the effective throat. This does not apply to the submerged-arc welding process because of its deeper penetration properties. In the case of gas metal arc welding or flux cored arc welding, the 1/8 in. deduction in throat only applies to bevel groove with an included angle of 45 deg or less in the vertical or overhead positions.

Weld metal subjected to compression in any direction or tension parallel to the axis of the weld, Figs. 2-122c and 2-122d, will have the same allowable as the base material. At least "matching weld metal" (that having strength roughly comparable to the base metal) must be used for compression, but it is not necessary to use matching weld metal for tension parallel loading.

The existence of tension forces transverse to the axis of the weld or shear in any direction, Figs. 2-122e and 2-122f, requires the use of weld metal allowables that are the same as those used for fillet welds. The selected weld metal may have mechanical properties higher or lower than those of the material being joined. If the weld metal has lower strength, however, its allowable must be used for calculating weld size or maximum allowable weld stress. For higher-strength weld metal, the weld allowable may not exceed the allowable of the matching weld metal for the material being joined. The allowable used shall not exceed the shear allowable of the plate.

The AWS has established the allowable shear value for weld metal in a fillet or partial-penetration bevel groove weld at:

$$\tau = 0.30 \text{ (Electrode Min. specified T.S.)} \qquad (1)$$

and has proved its validity with a series of fillet weld tests conducted by a special Task Committee of AISC and AWS.

Table 2-8 presents the allowable shear values for various weld metal strength levels and the more common fillet weld sizes. These values are for equal-leg fillet welds where the effective throat (t_e) equals 0.707 x leg size (ω). With the table, one can calculate the allowable unit force (f) per linear inch for a weld size made with a particular electrode type. For example, calculating the allowable unit force (f) per inch for the 1/2 in. fillet weld made with an E70 electrode:

$$\begin{aligned} f &= 0.707\ \omega\ \tau \\ &= 0.707\ \omega(0.30)\ (EXX) \\ &= 0.707\ (1/2 \text{ in.})\ (0.30)\ (70 \text{ ksi}) \\ &= 7.42 \text{ kips/linear in. (a kip = 1000 lb)} \qquad (2) \end{aligned}$$

The minimum allowable sizes for fillet welds, Figs. 2-122g and 2-122h, are given in Table 2-9. Where materials of different thickness are being joined, the Minimum fillet weld size is governed by the thicker material, but this size does not have to exceed the thickness of the thinner material unless required by the calculated stress.

Table 2-10 summarizes the AWS D1.1 Structural Welding Code and AISC allowables for weld metal. It is intended to provide a ready reference for picking the proper strength levels for the various types of steels. Once this selection has been made, the allowables can be quickly found for the various types of welds that may be required for the specific assembly.

PRIMARY AND SECONDARY WELDS

There are two types of welds, primary and secondary. Primary welds are those that transfer the entire load at the point where they are located. The weld has the same property as the member at this point, and, if the weld fails, the member fails.

Secondary welds are those that simply hold the parts together to form a built up member. In most cases, the forces on these welds are low.

If a full-strength (primary weld) is required, the weld metal is selected to match the mechanical properties of the plate. Generally it is unnecessary to match the chemistry, since the weld will undergo a faster rate of cooling, which increases its strength. An exception to this might be when the weldment is to be heat treated and it is required that the weld metal have the same mechanical properties as the steel after treatment. Although high-strength quenched and tempered plate cools much faster during manufacture than hot-rolled plate because of the quench, the cooling rate is limited somewhat by the large amount of heat contained in the mass. The weld, on the other hand, is quenched by the mass of the plate on which it is deposited, and this rate is very fast — as fast or faster than the cooling rate during the quenching of tempered plate. Of course, any preheating and a high welding heat input will decrease the cooling rate of the weld metal by decreasing the temperature gradient.

In welding high strength alloy steels, do not use full-strength welds unless they are required. High-strength steel may require additional preheat and greater care in welding because there is an increased tendency for cracking, especially if the joint is restrained. Non-full-strength welds or secondary welds can be made with lower-strength weld metal — E70, E80, or E90 electrodes. Since the E70 electrode would be easier to use, it would be the preferable choice. The only design concern is that the weld be sized to give sufficient strength to the joint.

TABLE 2-8. Allowable Load for Various Sizes of Fillet Welds

	Strength Level of Weld Metal (EXX)						
	*60	*70	80	*90	100	*110	120
Allowable Shear Stress on Throat ksi (1000 psi) of Fillet Weld or Partial Penetration Groove Weld							
$\tau =$	18.0	21.0	24.0	27.0	30.0	33.0	36.0
Allowable Unit Force on Fillet Weld kips/linear in.							
f =	12.73ω	14.85ω	16.97ω	19.09ω	21.21ω	23.33ω	25.45ω
Leg Size ω, in.	**Allowable Unit Force for Various Sizes of Fillet Welds kips/linear in.**						
1	12.73	14.85	16.97	19.09	21.21	23.33	25.45
7/8	11.14	12.99	14.85	16.70	18.57	20.41	22.27
3/4	9.55	11.14	12.73	14.32	15.92	17.50	19.09
5/8	7.96	9.28	10.61	11.93	13.27	14.58	15.91
1/2	6.37	7.42	8.48	9.54	10.61	11.67	12.73
7/16	5.57	6.50	7.42	8.35	9.28	10.21	11.14
3/8	4.77	5.57	6.36	7.16	7.95	8.75	9.54
5/16	3.98	4.64	5.30	5.97	6.63	7.29	7.95
1/4	3.18	3.71	4.24	4.77	5.30	5.83	6.36
3/16	2.39	2.78	3.18	3.58	3.98	4.38	4.77
1/8	1.59	1.86	2.12	2.39	2.65	2.92	3.18
1/16	.795	.930	1.06	1.19	1.33	1.46	1.59

*Fillet welds actually tested by the joint AISC-AWS Task Committee.

TABLE 2-9.
Minimum Fillet Weld Size (ω) or Minimum Throat of Partial Penetration Groove Weld (t)

Material Thickness of Thicker Part Jointed (in.)		Minimum Fillet Size, (in.)
** to 1/4 incl.		1/8
over 1/4	to 1/2	3/16
over 1/2	to 3/4	1/4
* over 3/4	to 1-1/2	5/16
over 1-1/2	to 2-1/4	3/8
over 2-1/4	to 6	1/2
over 6		5/8

Not to exceed the thickness of the thinner part.

* For minimum fillet weld size, Table does not go above 5/16″ fillet weld for over 3/4″ material.

** Minimum size for bridge application does not go below 3/16″.

Fig. 2-123 shows a full-strength and non-full-strength weld in A514 (110,000 psi T.S.) steel plate. One weld is transverse to the plate; the other parallel.

In the transverse welded joint, both the plate and the weld will be stressed together, and their behavior can be followed by the stress-strain curve shown in the figure. The plate and weld are fairly well matched. If a test specimen were pulled, it is probable that the plate would neck down and fail first, because of the reinforcement of the weld and slightly higher strength developed in the weld and heat-affected zone due to the rapid cooling following welding.

In the parallel welded joint, both the plate and the weld would be strained together. As the member is loaded, the strain increases from point 1 to point 2, with a corresponding increase in the stress in both the plate and weld from point 1 to point 3. At this point, the E7018 weld metal has reached its 60,000 psi yield. On further loading, the strain is increased to point 4. The weld is now stressed beyond its yield point (point 5) and is plastically yielding. The plate, however, is still below its yield point, point 6. With still further loading, the strain will reach point 7, at which time the elongation of the plate will be exhausted. It will fail before the weld metal, which has a higher elongation. Finally at point 8, the weld will also fail when reaching the limit of its elongation.

From the stress-strain curve for the parallel weld, it can be seen that nothing would be gained

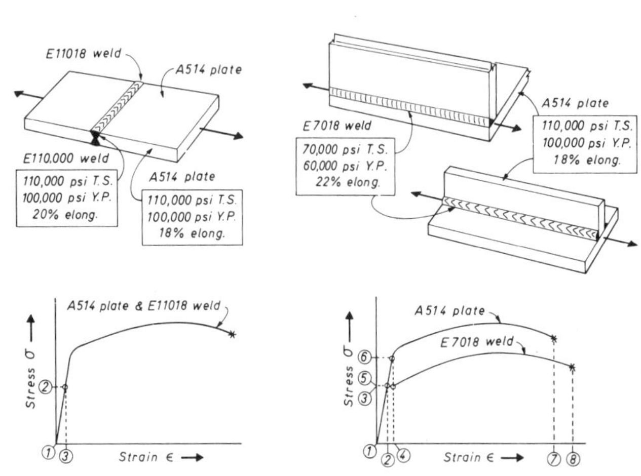

Fig. 2-123. Stress-strain relationships of transverse and parallel welds.

1 (1)

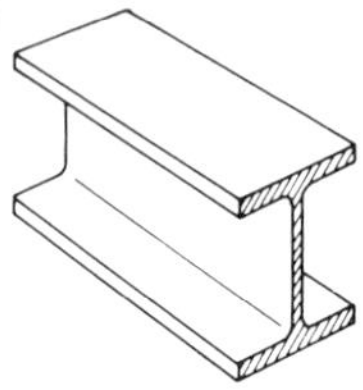

Base metal — no attachments — rolled or clean surfaces

(A)

2 (5)

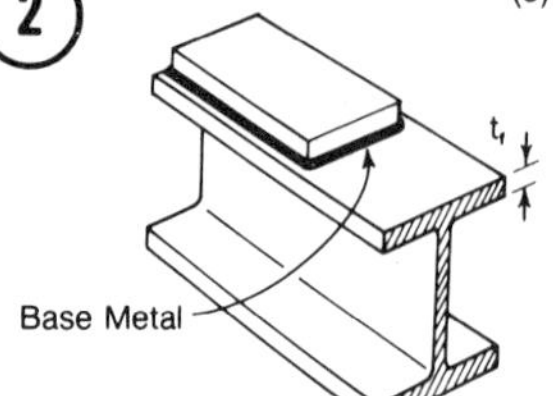

Base metal at ends of partial length welded cover plates narrower than the flange — square or tapered ends — with or without welds across the ends.

$t_f \le 0.8$ in (E)

$t_f > 0.8$ in (E′)

3 (25)

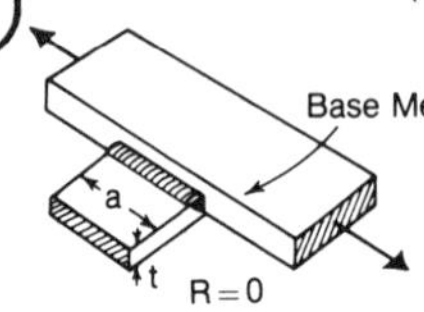

Longitudinal loading
Base metal — full penetration groove weld
Weld termination ground smooth
Weld reinforcement not removed
Not necessarily equal thickness

2 < a < 12 b or 4 in.	(D)
a > 12 b or 4 in. when b ≤ 1 in.	(E)
a > 12 b or 4 in. when b > 1 in.	(E′)

4 (19)

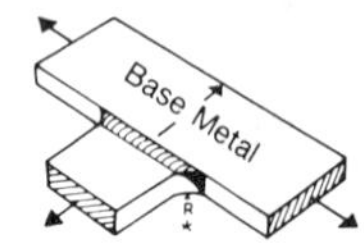

Base metal — full penetration groove weld.
Weld termination ground smooth.
Longitudinal load.
If also transverse load, weld must be inspected by radiography or ultrasound.

R > 24 in. (B)
24 in. > R > 6 in. (C)
6 in. > R > 2 in. (D)
2 in. > R (E)

9 (2)

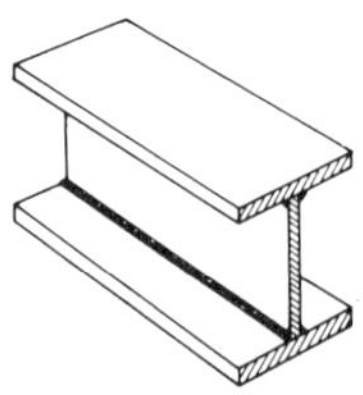

(B)

Base metal — built up plates or shapes — connected by continuous complete penetration groove welds or fillet welds — without attachments. Note: don't use this as a fatigue allowable for the fillet weld to transfer a load — this would be (F).

10 (6)

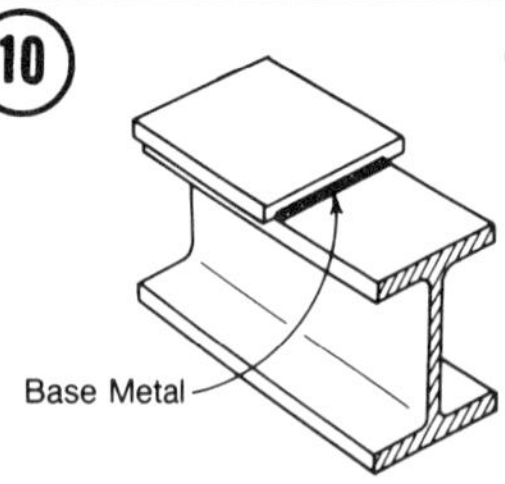

Base metal at ends of partial length welded cover plates wider than the flange.

Weld across end. $t_f \le 0.8$ in (E)

$t_f > 0.8$ in (E′)

No weld across end (E′).

11 (26)

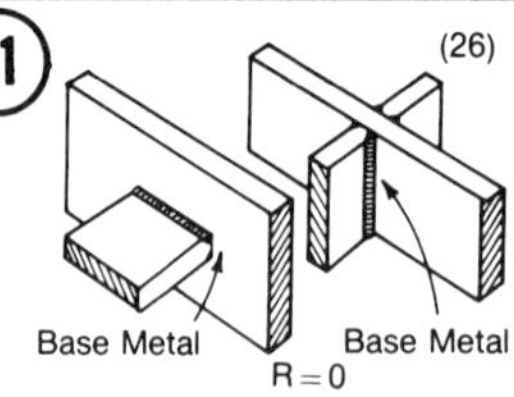

Base metal at detail attached by fillet welds.
Longitudinal loading

a < 2 in.	(C)
2 in. < a < 12 b	(D)
a > 12 b or 4 in. when b ≤ 1 in.	(E)
a > 12 b or 4 in. when b > 1 in.	(E′)

12 (27)

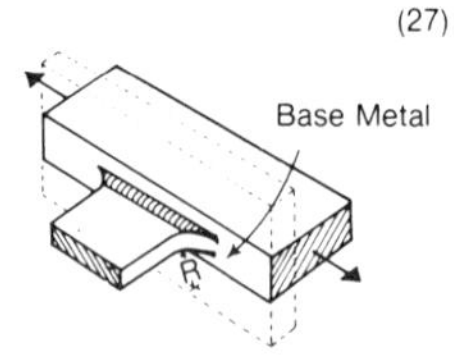

Base metal attached by partial penetration groove welds.
Longitudinal loading.
Weld termination ground smooth
Weld reinforcement left on
Using a transition radius of
R > 2 in. (D)
R ≤ 2 in. (E)

15 (3)

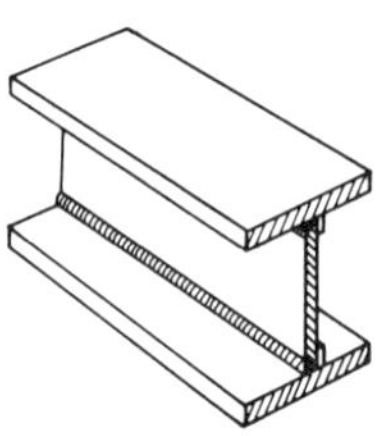

Base metal — built up plates or shapes — connected by continuous complete penetration groove welds with backing bars not removed — or partial penetration groove welds. (B′)

16 (4)

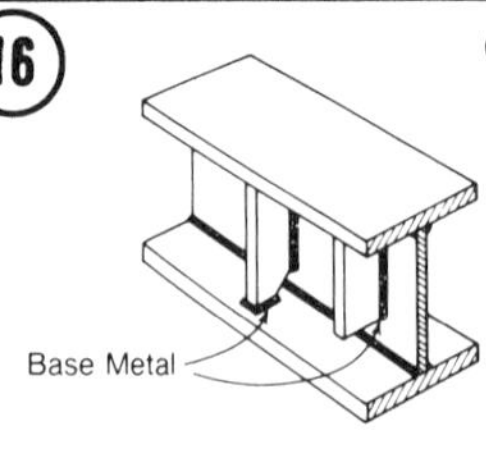

Base metal — at end of welds on transverse stiffeners to web and flanges of girder.

(C)

17 (26)

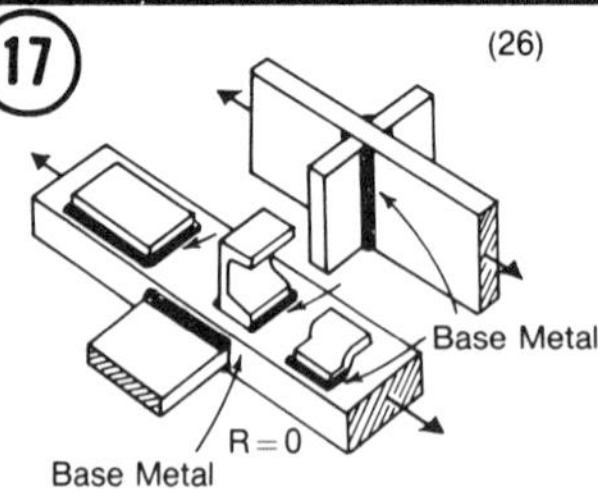

Base metal at detail attached by fillet welds.
Longitudinal loading

a < 2 in.	(C)
2 in. < a < 12 b	(D)
a > 12 b or 4 in. when b ≤ 1 in.	(E)
a > 12 b or 4 in. when b > 1 in.	(E′)

18 (27)

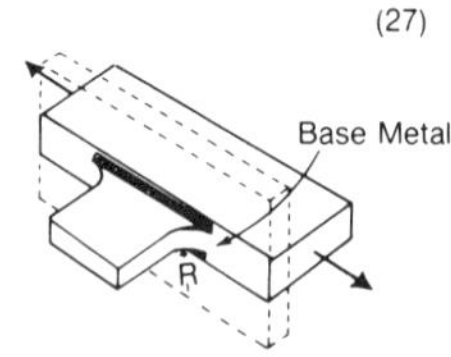

Base metal attached by Fillet welds
Longitudinal Loading
Weld termination ground smooth
Using a transition radius of
R > 2 in. (D)
R ≤ 2 in. (E)

21 (7) (8)

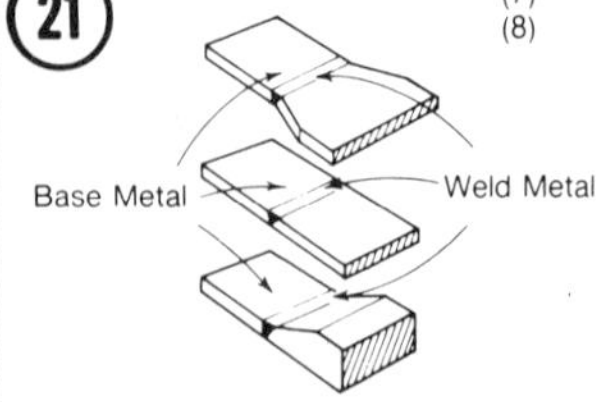

Base metal and weld metal at full penetration groove welds — changes in thickness or width not to exceed a slope of 1 in 2½ (22°). Ground flush and inspected by radiography or ultrasound (B).

For A514 steel (B′)

22 (9)

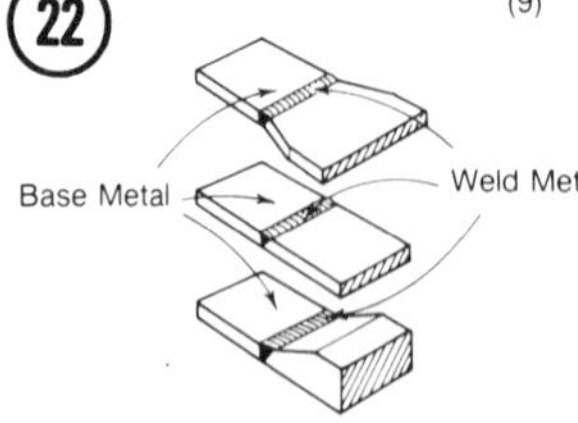

Base metal and weld metal at full penetration groove welds — changes in thickness or width not to exceed a slope of 1 in 2½ (22°). Weld reinforcement not removed. Inspected by radiography or ultrasound.

(C)

23 (12)

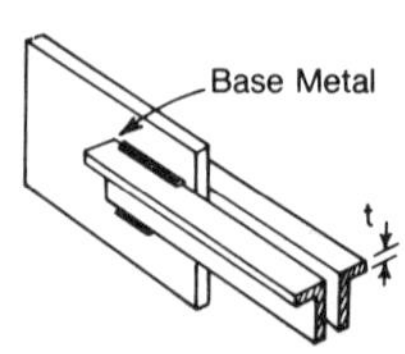

Base metal at junction of axially loaded members with fillet welded end connections. Welds balanced about C.G. of member.

t ≤ 1 in. (E)
t > 1 in. (E′)

24 (13)

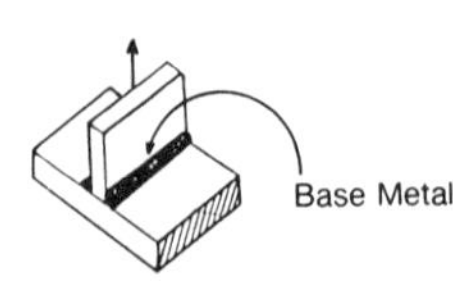

Base metal connected with transverse fillet welds.

t ≤ ½ in. (C)
t > ½ in. (C) See Note C

5 (20) (21)

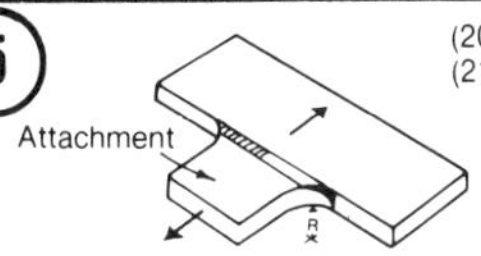

Transverse loading
Attachment — full penetration groove weld.
Weld termination ground smooth.
Equal Thickness

	Weld Reinforcement Removed	Weld Reinforcement Not Removed
R > 24 in.	B	C
24 in. > R > 6 in.	C	C
6 in. > R > 2 in.	D	D
2 in. > R	E	E

6 (22) (23)

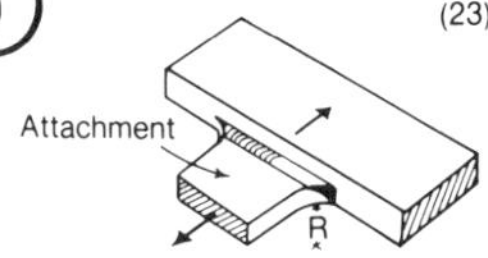

Transverse loading
Attachment — full penetration groove weld.
Weld termination ground smooth
Unequal thickness
Weld reinforcement removed

R > 2 in. (D)
R ≤ 2 in. (E)

Weld reinforcement not removed
any R (E)

7 (24)

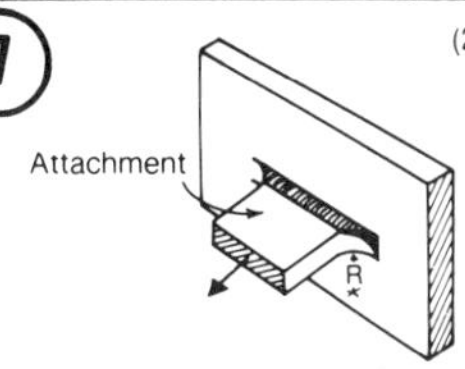

Transverse loading
Attachment — full penetration groove weld
Weld termination ground smooth
Weld reinforcement not removed

R > 6 in. (C)
6 in. > R > 2 in. (D)
R ≤ 2 in. (E)

8 (28)

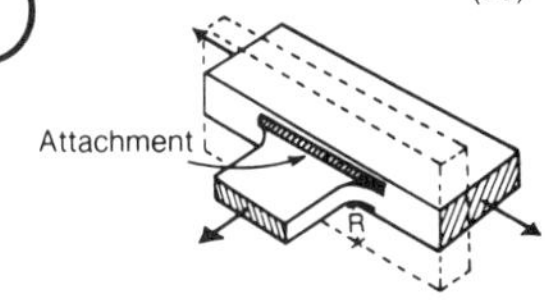

Attachment — Transverse load
Fillet welds
Weld termination ground smooth
Main material longitudinal loaded
Using a transition radius of

R > 2 in. (D)
R ≤ 2 in. (E)

13 (29) (30)

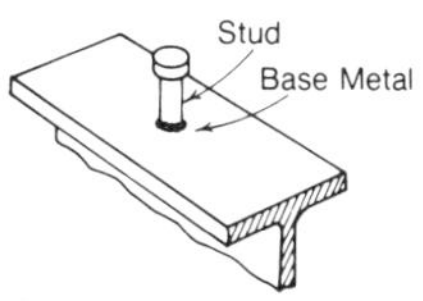

Base metal at stud — type shear connector — fillet weld attachment or automatic end weld. (C)

Shear stress on nominal area of stud-type shear connectors. (F)

14 (15)

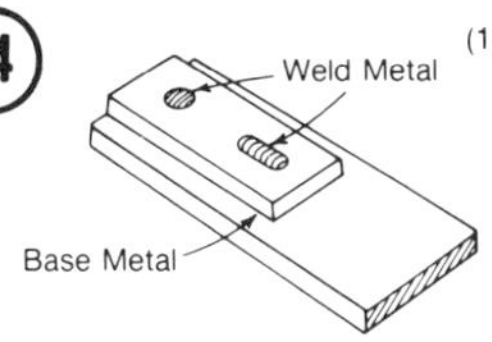

Base metal at plug or slot welds.

(E)

Shear on plug or slot welds.

(F)

19 (11)

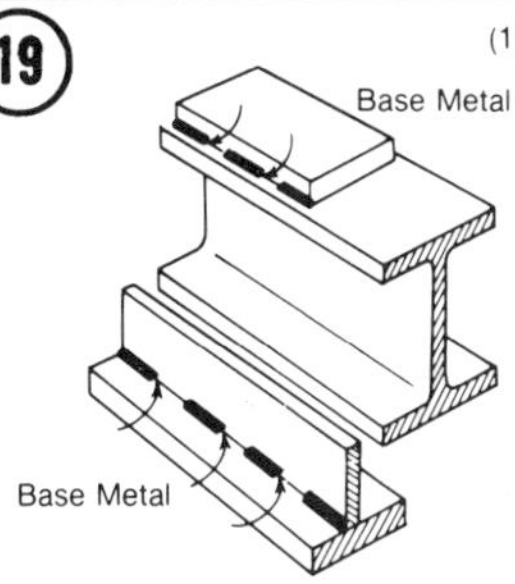

Base metal at intermittent fillet welds.

(E)

20 (16) (17) (18)

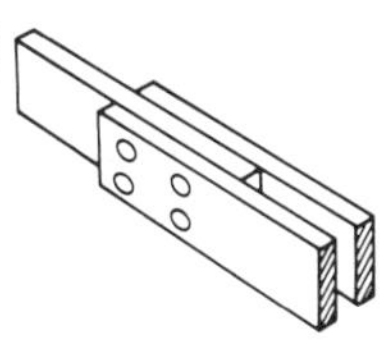

Base metal at *gross section* of high strength bolted *slip-critical* connections — No out of plane bending. (B)

Base metal at *net section* of fully tensioned high strength, bolted *bearing* connections. (B)

Base metal at net section of other mechanical fastened joints. (D)

25 (14)

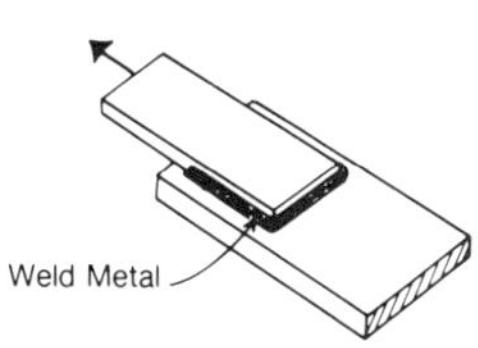

Weld metal of continuous or intermittent longitudinal or transverse fillet welds.

(F)

26 (10)

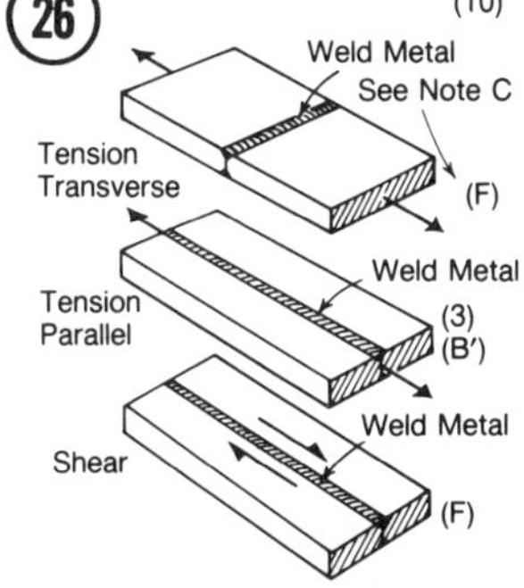

Partial penetration groove welds

Allowable Stress Range, Ksi

Category (from Table A-K4.2)	20,000 to 100,000 ~	100,000 to 500,000 ~	500,000 to 2×10^6 ~	Over 2×10^6 ~
A	63	37	24	24
B	49	29	18	16
B′	39	23	15	12
C	35	21	13	10[a]
D	28	16	10	7
E	22	13	8	5
E′	16	9	6	3
F	15	12	9	8

[a] Flexural stress range of 12 ksi permitted at toe of stiffener welds on flanges.

allowable fatigue stress $\sigma_{max} = \dfrac{\sigma_{sr}}{1 - K}$ for normal stress σ

$\tau_{max} = \dfrac{\tau_{sr}}{1 - K}$ for shear stress τ

but shall not exceed steady allowables

σ_{max} or τ_{max} = maximum allowable fatigue stress

σ_{sr} or τ_{sr} = allowable range of stress from table

$$K = \frac{\sigma_{min}}{\sigma_{max}} = \frac{M_{min}}{M_{max}} = \frac{F_{min}}{F_{max}} = \frac{\tau_{min}}{\tau_{max}} = \frac{V_{min}}{V_{max}}$$

S = shear
T = tension
R = reversal
M = stress in metal
W = stress in weld

NOTE C

$$\sigma_{sr} = \sigma_{sr}^{c}\left(\frac{0.71 - 0.65\dfrac{2a}{t_p} + 0.79\dfrac{w}{t_p}}{1.10\, t_p^{1/6}}\right)$$

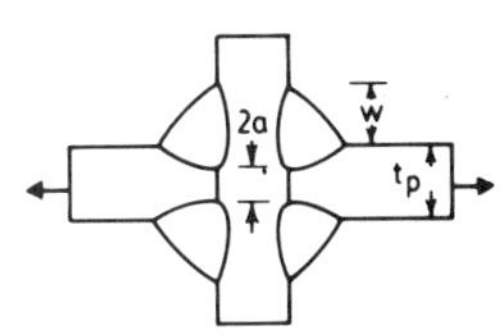

for fillet welds $\dfrac{2a}{t_p} = 1.0$

$$\sigma_{sr} = \sigma_{sr}^{c}\left(\frac{0.06 + 0.79\dfrac{w}{t_p}}{1.10\, t_p^{1/6}}\right)$$

but $\sigma_{xr} \leq \sigma_{xr}^{c}$

σ_{sr} = stress range for this condition
σ_{sr}^{c} = stress range for category (C)
w = leg size of fillet weld
2a = lack of penetration into joint
t_p = plate thickness

Fig. 2-124. AISC allowable range of stress (σ_{sr}, τ_{sr}). (Also used by AWS Structural Code, D1.1-90, Section 9.)

by using an electrode with a strength higher than E70. In this situation, the weld will take any parallel load that the 110,000 psi plate can withstand.

ALLOWABLE FATIGUE STRENGTHS OF WELDS

The performance of a weld under cycles of stress is an important consideration in structures and machinery, and the specifications relating thereto have been developed after extensive research by the AISC.

Although sound weld metal has about the same fatigue strength as unwelded plate, the change in section that a weld induces does lower the fatigue strength. In the case of a complete-penetration groove weld, the reinforcement and any undercut, lack of penetration or crack will act as a notch, which in turn acts as a stress raiser and results in a lower fatigue strength. The very nature of a fillet weld, because it is used in lap and tee joints, provides an abrupt change in section that lowers fatigue strength.

The AISC fatigue specifications cover a wide range of welded joints and also take into consideration the fatigue strength of members attached by welds. Fig. 2-124 shows various types of joints and contains a chart that tabulates the allowable range of stress. Covered are steels having a yield strength of 36,000 psi (A36) up to 110,000 psi (A514) and weld metal from E60XX to E110XX.

In these figures, the fatigue allowables for members are designated as M and the fatigue allowables for welds as W. A tensile load is T, reversal is R, and shear is S. In the chart used for determining values for allowable range of stress (Fig. 2-124), there are four groups representing fatigue life:

1. 20,000 - 100,000 cycles
2. Over 100,000 — 500,000 cycles
3. Over 500,000 — 2,000,000 cycles
4. Over 2,000,000 cycles

and eight different categories (A-F) representing type of joint and member loading. The chart provides the allowable range in stress (σ_{sr} or τ_{sr}) which value may be used in the conventional fatigue formulas:

$$\sigma_{max} = \frac{\sigma_{sr}}{1-K} \quad \text{or} \quad \tau_{max} = \frac{\tau_{sr}}{1-K} \qquad (3)$$

where

$$K = \frac{\text{min. stress}}{\text{max. stress}} \text{ or } \frac{\text{min. force}}{\text{max. force}} \text{ or } \frac{\text{min. moment}}{\text{max. moment}} \text{ or } \frac{\text{min. shear}}{\text{max. shear}} \text{ or } \frac{\text{min. torque}}{\text{max. torque}}$$

In every case, the allowable should not exceed the steady stress allowable for the plate and weld. The fatigue formula:

$$\sigma_{max} = \frac{\sigma_{sr}}{1-K} \qquad (3)$$

is used to reduce the fatigue allowable because of the fatigue condition. As the range of fatigue stress increases, the fatigue allowable decreases. When a high value for K is inserted, it may give a value higher than the steady value, which, of course, must not be exceeded.

Using the information in these figures, it can be seen that the fatigue allowable of a welded member in bending, as shown by 9 in Fig. 2-124, is really determined by the allowable of the plate when connected by fillet welds parallel to the direction of the applied stress. M and W are equal, and the applicable category is B in the chart.

If stiffeners are used in the weldment, as in 16, the fatigue allowable of the web or flange is determined by the allowable in the member at the termination of the weld or adjacent to the weld, category C.

The fatigue allowable of the flange plate at the termination of a cover plate, either square or tapered end, is represented by configuration 2 and 10. The applicable category to use in the chart is E or E′.

If intermittent fillet welds are used on members, the fatigue allowable of the plate adjacent to the termination of the weld is category E, as is the case in 19.

Fatigue allowables for partial-penetration groove welds may be determined by reference to configuration 26.

According to 3, 11 and 17, the fatigue allowable for a member with a transverse attachment increases as the length of the attachment decreases, measured parallel to the axis of the load. Although there may be a similar geometrical notch effect or abrupt change in section in both, it is the stress raiser that is important. The transverse bar in 17 is so short, as far as the axis of the member and the load are concerned, that very little of the force is able to swing up and into the bar and then back down again. Consequently, the stress raiser is not severe. The longer bar attachment, however, is sufficiently long to provide a path for the force through it and the connecting welds. Because of this force transfer into the bar and through the welds, Fig. 2-125, there will be a higher stress raiser and, as a result, a reduction of the fatigue strength of the member.

Since the steady strength of a transverse fillet is about 1/3 stronger than a parallel fillet, one might question why the fatigue allowable for a parallel fillet is the same as the transverse fillet in 25 category F. The actual fatigue strength of the transverse fillet is only slightly higher than the parallel fillet and both have been placed in the same range.

A new change has been the addition of a prime mark which provides a slightly lower fatigue allowable usually for thicker members, see 2, 10, 3, 11, 17 and 23. In general (b) is the thickness of the attachment.

Fatigue values for members loaded parallel, with attachments connected by:

3 and 4 full penetration groove welds
11 and 12 partial penetration groove welds
17 and 18 fillet welds

Even numbers of the above are based upon the radius (R) at the re-entrant corner. Odd numbers are for sharp corners (R=0) and are based upon the longitudinal length (a) of the attachment.

Fatigue values for attachments, loaded transverse to the main member, are shown in:

(5) for full penetration groove welds with equal thickness and are based upon the radius (R) of the re-entrant corner.

(6) for full penetration groove welds with unequal thickness

(7) for full penetration groove welds to the face of the main member.

(8) for fillet welded attachments.

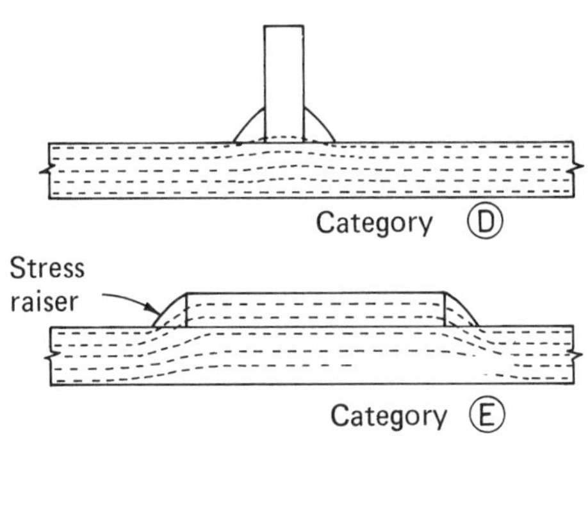

Fig. 2-125. Effect of attachments.

Another new change is the treatment of a fillet welded tee joint, with the additional Note C (Fig. 2-124). The allowable stress range (σ_{sr}) is first determined for category (c). Then a formula using the leg size of the fillet weld and the depth of preparation for the partial penetration groove portion of the joint will result in a factor which is to be multiplied by the initial (σ_{sr}) to provide a modified lower value. This factor should not exceed unity. This recognizes that the usual fillet weld may fail across its throat in fatigue testing. As the depth of level is increased, the form of failure shifts from the throat up the weld to the plate next to the toe of the fillet weld.

Fig. 2-126 is a modified Goodman diagram for a complete-penetration groove butt weld attaching plate with weld reinforcement left on, configuration 22. The category is C, and the life is 500,000 — 2,000,000 cycles. The vertical axis represents the maximum stress (σ_{max}), and the horizontal axis represents the minimum stress (σ_{min}), either positive or negative. A steady load is represented by the 45 deg line to the right, (K = +1.0)and a complete reversal by the 45 deg line to the left. (K = −1.0)

The region to the right of the vertical line (K = 0) represents tensile loading. The fatigue formula (Eq. 3) would apply to welded butt joints in either plates or members. The region to the left of this line represents cycles going into compressive loading.

The fatigue formulas are meant to reduce the allowable stress as fatigue conditions are encountered. Naturally, the resulting allowable fatigue should not exceed the usual steady allowable stress. For this

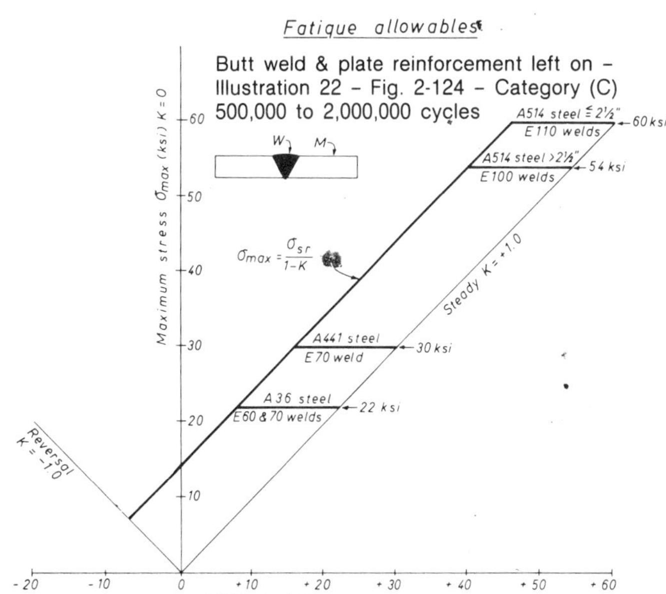

Fig. 2-126. Goodman diagram for butt weld.

reason, these fatigue curves are cut off with horizontal lines representing the steady allowable stress for the particular strength level of steel used. Note that A36 steel and E60 or E70 weld metal is cut off at 22 ksi; A441 steel and E70 weld metal at 30 ksi; and A514 and E110 weld metal at 54 ksi or 60 ksi depending on plate thickness.

The graph in Fig. 1-127 is a modification of that in Fig. 2-126 so that the horizontal axis represents the range, K, of the stress. Here, two additional strength levels of weld metal have been added - E80 and E90 - along with equivalent strength levels of steel. Note that for a small range in stress, K = 0.6 to K = 1.0, higher-strength welds and steels have increased fatigue allowables. However, as the stress range increases - lower values of K - the increase is not as great and below K = +0.05 all of the various strengths of welds and steels have the same fatigue allowable.

Fig. 2-128 represents the same welded joint, but with a lower life of 20,000 — 100,000 cycles. Here, the higher-strength welds and steels have higher fatigue allowables over a wider range of life.

These graphs point up the fact that the wider the range of cycling, the less the usefulness of a high-strength steel. When there is a complete reversal of stress, there is not much advantage in using a high-strength steel.

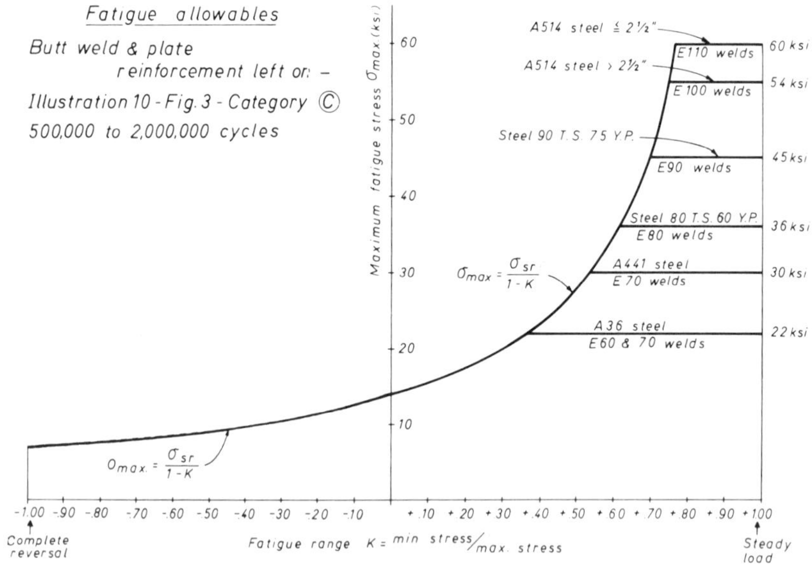

Fig. 2-127. Fatigue allowable for groove weld.

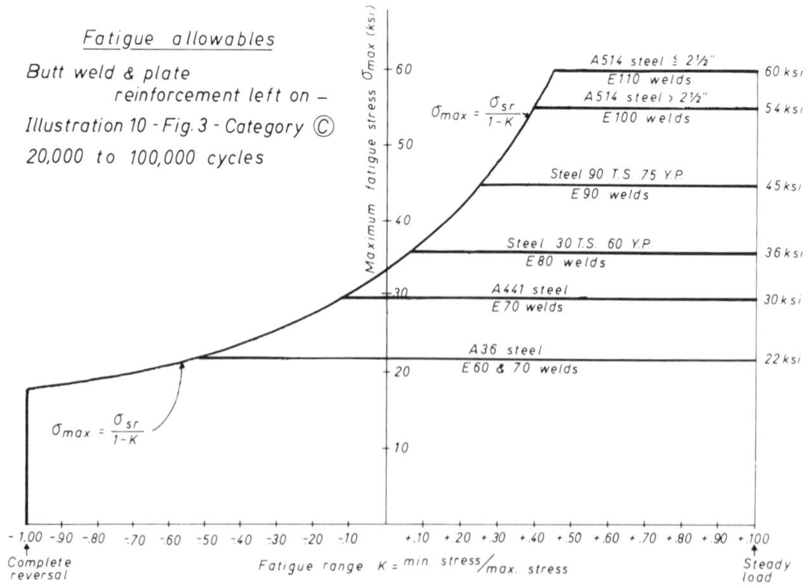

Fig. 2-128. Fatigue allowable for groove weld.

TABLE 2-10. Stress Allowables for Weld Metal

Type of Weld Stress	Permissible Stress[5]	Required Strength Level[1][2]
COMPLETE PENETRATION GROOVE WELDS		
Tension normal to the effective throat.	Same as base metal.	Matching weld metal must be used. See Table below.
Compression normal to the effective throat.	Same as base metal.	Weld metal with a strength level equal to or one classification (10 ksi) less than matching weld metal may be used.
Tension or compression parallel to the axis of the weld.	Same as base metal.	Weld metal with a strength level equal to or less than matching weld metal may be used.
Shear on the effective throat.	.30 × Nominal Tensile strength of weld metal (ksi) except stress on base metal shall not exceed .40 × yield stress of base metal.	
PARTIAL PENETRATION GROOVE WELDS		
Compression normal to effective throat.	Designed not to bear — .50 × Nominal Tensile strength of weld metal (ksi) except stress on base metal shall not exceed .60 × yield stress of base metal. Designed to bear. Same as base metal.	Weld metal with a strength level equal to or less than matching weld metal may be used.
Tension or compression parallel to axis of the weld.[3]	Same as base metal.	
Shear parallel to axis of weld.	.30 × Nominal Tensile strength of weld metal (ksi) except stress on base metal shall not exceed .40 × yield stress of base metal.	
Tension normal to effective throat.[4]	.30 × Nominal Tensile strength of weld metal (ksi) except stress on base metal shall not exceed .60 × yield stress of base metal.	
FILLET WELDS[3]		
Stress on effective throat, regardless of direction of application of load.	.30 × Nominal Tensile strength of weld metal (ksi) except stress on base metal shall not exceed .40 × yield stress of base metal.	Weld metal with a strength level equal to or less than matching weld metal may be used.
Tension or compression parallel to axis of weld.	Same as base metal.	
PLUG AND SLOT WELDS		
Shear parallel to faying surfaces.	.30 × Nominal Tensile strength of weld metal (ksi) except stress on base metal shall not exceed .40 × yield strength of base metal.	Weld metal with a strength level equal to or less than matching weld metal may be used.

(1) For matching weld metal, see AISC Table 1.17.2 or AWS D1.1-90 Table 4.1 or table below.
(2) Weld metal, one strength level (10 ksi) stronger than matching weld metal may be used when using alloy weld metal on A242 or A588 steel to match corrosion resistance or coloring characteristics (Table 4.2 of AWS D1.1-90).
(3) Fillet welds and partial penetration groove welds joining the component elements of built up members (ex. flange to web welds) may be designed without regard to the axial tensile or compressive stress applied to them.
(4) Cannot be used in tension normal to their axis under fatigue loading (AWS (D1.1-90 par. 2.5), AWS D1.1-90 prohibits their use on any butt joint (9.12.1), or any splice in a tension or compression member (9.17), or splice in beams or girders (9.21), however, are allowed on corner joints parallel to axial force of components of built up members (9.12.1.2 (4). Cannot be used in girder splices (AISC 1.10.8).
(5) AWS D1.1-90 Section 9 Dynamically loaded structures — reduce above permissible stress allowables of weld by 10%.

MATCHING WELD METAL AND BASE METAL

Weld Metal	60 or 70	70	80	100	110
Type of Steel	A36; A53, Gr. B; A106, Gr. B; A131, Gr. A, B, C, CS, D, E; A139, Gr. B; A381, Gr. Y35; A500, Gr. A, B; A501; A516, Gr. 55, 60; A524, Gr. I, II; A529; A570, Gr. D, E; A573, Gr. 65; A709, Gr. 36; API 5L, Gr. B; API 5LX, Gr. 42; ABS, Gr. A, B, D, CS, DS, E	A131, Gr. AH32, DH32, EH32, AH36, DH36, EH36; A242; A441; A516, Gr. 65; 70; A537, Class 17; A572, Gr. 42, 45, 50, 55; A588 (4 in. and under); A595, Gr. A, B, C; A606; A607, Gr. 45, 50, 55; A618; A633, Gr. A, B, C, D (2-1/2 in. and under); A709, Gr. 50, 50W; API 2H; ABS Gr. AH32, DH32, EH32, AH36, DH36, EH36.	A572, Gr. 60, 65; A537, Class 2; A63, Gr. E	A514 [over 2-1/2 in. (63 mm)]; A709, Gr. 100, 100W [2-1/2 to 4 in. (63 to 102 mm)]	A514 [2-1/2 in. (63 mm) and under]; A517; A709, Gr. 100, 100W [2-1/2 in. (63 mm) and under]

Column for the San Mateo-Hayward Bridge. Design is attractive yet entirely functional.

Codes and Specifications

Public safety is involved in the design and fabrication of such structures as buildings, pipelines, ships, and pressure vessels, and, to minimize the danger of catastrophic failure or even premature failure, documents are established to regulate the design and construction of these structures. Even if public safety is not involved, some products are built to meet definite requirements that insure a level of quality, uniformity, or interchangeability. These documents are called specifications, codes, standards, and rules. Sometimes the terms are used interchangeably.

Webster's Third International Dictionary (1969) defines the terms as follows:

> **Specification:** "A detailed, precise, explicit presentation (as by numbers, description, or working drawing) of something or a plan or proposal of something."
>
> **Code:** "A set of rules of procedure and standards of materials designed to secure uniformity and to protect the public interest in such matters as building construction and public health, established usually by a public agency."
>
> **Standard:** "Something that is established by authority, custom, or general consent as a model or example to be followed."
>
> **Rule:** "An accepted procedure, custom, or habit having the force of a regulation."

ORGANIZATIONS THAT WRITE CODES

Codes and specifications are generally written by industrial groups, trade or professional organizations, or government bureaus, and each code or specification deals with applications pertaining specifically to the interest of the authoring body. Large manufacturing organizations may prepare their own specifications to meet their specific needs.

Among the major national organizations that write codes that involve arc welding are the American Welding Society (AWS), American Institute of Steel Construction (AISC), American Society for Testing Materials (ASTM), American Society of Mechanical Engineers (ASME), and the American Petroleum Institute (API).

Among government agencies, the Interstate Commerce Commission has rules for the fabrication of over-the-road vehicles and for containers used in interstate commerce. The various branches of the military services also prepare specifications. A list of major agencies involved in code and specification writing is presented in Table 2-11.

Some specifications — for example, those of the Society of Automotive Engineers (SAE) — actually are not standards but are merely guides to recommended practices. Other specifications rigidly call out the design and fabrication procedures to be followed and are legally binding. In any event, neither the design nor fabrication of a welded structure should be undertaken without the full knowledge of all codes and requirements that must be met.

Meeting the requirements of a code does not protect anyone against liability concerning the performance of the welds or structure. Nor, in general, does any code-writing body approve, endorse, guarantee, or in any way attest to the correctness of the procedures, designs, or materials selected for code application.

Some of the major organizations that issue codes pertaining to welding are listed in the following text. These listings do not cover all applications of welding or all code-writing bodies. Many local governments, for example, issue codes. The list is representative of the common applications and of the organizations whose codes are widely used.

American Welding Society: The advancement of the science of welding is a principal aim of the AWS. This organization writes codes for welding buildings and bridges; prepares specifications for welding electrodes, rods, and fluxes; and sets standards for the qualification of welding operators and for the testing and inspection of welds.

American Society of Mechanical Engineers: The Boiler and Pressure Vessel Committee of the ASME establishes standards and rules of safety for the design, construction, and inspection of boilers and other pressure vessels. The Committee also

TABLE 2-11. Major Agencies and Associations Related to the Welding Industry

	Department of the Air Force ORTA-AFWALGLI Wright-Patterson Air Force Base, Ohio 45433 (513) 255-4316
	Department of Navy Naval Supply Depot 5801 Taber Avenue Philadelphia, PA 19120
AA	**Aluminum Association** 900 19th Street, Suite 300 Washington, DC 20006 (202) 862-5100
AASHTO	**American Association of State Highway and Transportation Officials** 444 N. Capitol St., N.W. Suite 225 Washington, DC 20001 (202) 624-5800
ABS	**American Bureau of Shipping** 2 World Trade Center 106 Floor New York, NY 10048 (212) 839-5000
AGA	**American Gas Association** 1515 Wilson Boulevard Arlington, VA 22209 (703) 841-8400
AIME	**American Institute of Mining, Metallurgical, and Petroleum Engineers** 345 E. 47th Street, 14th Floor New York, NY 10017 (212) 705-7695
AIPE	**American Institute of Plant Engineers** 3975 Erie Avenue Cincinnati, OH 45208 (513) 561-6000
AISC	**American Institute of Steel Construction** 1 E Wacker Drive, Suite 3100 Chicago, IL 60601-2001 (312) 670-2400
AISI	**American Iron and Steel Institute** 1101 17th Street, N.W. Washington, DC 20036-4700 (202) 463-6573
ANSI	**American National Standards Institute** 11 W. 42nd Street New York, NY 10036 (212) 642-4900
API	**American Petroleum Institute** 1220 L Street N.W. Washington, DC 20005 (202) 682-8000
ASME	**American Society of Mechanical Engineers** 345 E. 47th Street New York, NY 10017 (212) 705-7722
ASM	**ASM International** Route 87 Metals Park, OH 44073 (216) 338-5151
ASNT	**American Society for Nondestructive Testing** P.O. Box 28518 Columbus, OH 43228-0518 (614) 274-6003
ASSE	**American Society of Safety Engineers** 1800 E. Oakton Street Des Plaines, IL 60018 (708) 692-4121
ASTM	**ASTM** 1916 Race Street Philadelphia, PA 19103 (215) 299-5400
AWWA	**American Water Works Association** 6666 W. Quincy Avenue Denver, CO 80235 (303) 794-7711
AWS	**American Welding Society** P.O. Box 351040 550 N.W. LeJeune Road Miami, FL 33135 (305) 443-9353
AWI	**American Welding Institute** 10628 Dutchtown Road Knoxville, TN 37932 (615) 675-2150
AISE	**Association of Iron and Steel Engineers** Suite 2350 Three Gateway Center Pittsburgh, PA 15222 (412) 281-6323
CISC	**Canadian Institute of Steel Construction** 201 Consumers, Suite 300 Willowdale, Ontario M2J 4G8 (416) 491-4552
CSA	**Cryogenic Society of America** 1033 S. Boulevard Oak Park, IL 60302 (708) 383-6220
CSNT	**Canadian Society for Nondestructive Testing** 17 Wellington Street St. Catherines, Ontario L2R 5P7 (416) 641-0606
CGA	**Compressed Gas Association** Suite 1004 1235 Jefferson Davis Highway Arlington, VA 22202-4100 (703) 979-0900
CDA	**Copper Development Association** Greenwich Office Park #2 P.O. Box 1840 Greenwich, CT 06836-1840 (203) 625-8210
CWB	**Canadian Welding Bureau** 7250 West Credit Avenue Mississauga, Ontario L5N 5N1, Canada
DOT	**U.S. Department of Transportation** 400 7th Street S.W. Washington, DC 20590 (202) 366-4000
DVS	**Deutscher Verband fur Schweissiechnik e.V.** Postfach 27 25, Aachener Strasse 172 4000 Dusseldorf 1 Germany (0211) 15 40 40
EWI	**Edison Welding Institute** 1100 Kinnear Road Columbus, OH 43212-1152 (614) 486-9400
FMA	**Fabricating Manufacturers Association** 5411 East State St. Rockford, IL 61108 (815) 399-8700
GWI	**Grinding Wheel Institute** 30200 Detroit Road Cleveland, OH 44125-1967 (216) 899-0010
ISEA	**Industrial Safety Equipment Association** Suite 808 1901 N. Moore Street Arlington, VA 22209 (703) 525-1695
IIE	**Institute of Industrial Engineers** 25 Technology Park/Atlanta Norcross, GA 30092 (404) 449-0460
IIW	**International Institute of Welding - American Council** P.O. Box 351040 Miami, FL 33135 (305) 443-9353
IMA	**International Magnesium Association** 2010 Corporate Ridge Suite 700 McLean, VA 22102 (703) 442-8888
IPA	**International Tube & Pipe Association** 5411 East State St. Rockford, IL 61108 (815) 399-8700
LRS	**Lloyd's Register of Shipping** 17 Battery Place New York, NY 10004 (212) 425-8050
NACE	**National Association of Corrosion Engineers** P.O. Box 218340 Houston, TX 77218 (713) 492-0535
NBBPVI	**National Board of Boiler and Pressure Vessel Inspectors** 1055 Crupper Avenue Columbus, OH 43229 (614) 888-8320
NEMA	**National Electrical Manufacturers Association** 2101 L Street N.W., Suite 300 Washington, DC 20037 (202) 457-8400
NIST	**National Institute of Science and Technology** Gaithersburg, MD 20899 (301) 975-2000
NSC	**National Safety Council** 444 N. Michigan Avenue Chicago, IL 60611 (312) 527-4800
NWSA	**National Welding Supply Association** 1900 Arch Street Philadelphia, PA 19103 (215) 564-3484
NIDI	**Nickel Development Institute** 15 Toronto Street, Suite 402 Toronto, Ontario M5C 2E3 (416) 362-8850
RWMA	**Resistance Welder Manufacturers' Association** 1900 Arch Street Philadelphia, PA 19103 (215) 564-3484
RI	**Robotics International** One SME Drive P.O. Box 930 Dearborn, MI 48121 (313) 271-1500
RIA	**Robotic Industries Association** 900 Victors Way P.O. Box 3724 Ann Arbor, MI 48106 (313) 994-6088
SAE	**SAE** 400 Commonwealth Drive Warrendale, PA 15096-0001 (412) 776-4841
SAIW	**South African Institute of Welding** Western Boulevard off Main Reef Road City West Johannesburg, South Africa (011) 836-4121/4132
SFSA	**Steel Founders' Society of America** 455 State Street Des Plaines, IL 60016 (708) 299-9160
	Ductile Iron Society 28938 Lorain Road Suite 202 North Olmsted, OH 44070 (216) 734-8040
	National Certified Pipe Welding Bureau 1385 Piccard Drive Rockville, MD 20850-4329 (301) 869-5800
SME	**Society of Manufacturing Engineers** One SME Drive P.O. Box 930 Dearborn, MI 48121 (313) 271-1500
SPFA	**Steel Plate Fabricators Association** 2400 S. Downing Avenue Winchester, IL 60154 (708) 562-8750
SSCI	**Steel Service Center Institute** 1600 Terminal Tower Cleveland, OH 44113 (216) 694-3630
STI	**Steel Tank Institute** 570 Oakwood Road Lake Zurich, IL 60047 (708) 438-8265
TDA	**Titanium Development Association** 11 W. Monument Avenue P.O. Box 2307 Dayton, OH 45401 (513) 223-8432
TWI	**The Welding Institute** Abington Hall Abington, Cambridge UK, CB1 6AL (0223) 891162
WIC	**Welding Institute of Canada** 391 Burnhamthorpe Road E. Oakville, Ontario L6J 6C9 (416) 257-9881
WRC	**Welding Research Council** Room 1301 345 E. 47th Street New York, NY 10017 (212) 705-7956

interprets the rules and considers requests for revisions. Fabricators or manufacturers wishing to produce vessels in accordance with the codes must obtain from the Committee a Certificate of Authorization to use the ASME nameplate. Many states and municipal governments have adopted the ASME Code as a legal requirement for applicable types of construction. Other states have codes similar to, or patterned after, the ASME Code.

American Society for Testing Materials: This national technical society has numerous committees, each of which issues regulations and standards in a prescribed field of materials application. Many of these pertain to construction materials and the methods of testing.

American Petroleum Institute: Preferred practices governing the design and fabrication of welded equipment and structures used in the petroleum industry are issued by the API. Some of the most widely used of these specifications are those for overland-pipelines.

American Institute of Steel Construction: This trade organization issues specifications for design, fabrication, and erection of structural steel for buildings.

Government Agencies: Government specifications consist primarily of two groups: federal and military. Copies of federal specifications are available through the regional offices of the General Services Administration. Copies of military specifications are available through the agencies of the Department of Defense. Distribution of specifications is limited to parties having a contractual relationship with the DOD, or who otherwise need the specifications to fulfill bid requirements. The distribution of specifications having a security classification is, of course, limited.

APPLICATIONS COVERED BY CODES

Pressure Vessels: The construction of welded boilers and pressure vessels is covered by codes and specifications that describe, among other items, the permissible materials, size, configuration, service limitations, fabrication, heat treatment, inspection, and testing requirements. These codes also outline requirements for qualification of welding procedures and operators. Numerous state, city, and other local government agencies also issue codes governing pressure vessels. For equipment to be used outside of the U.S., the applicable foreign government regulations must, of course, be investigated. Commonly applied codes are:

- *ASME Boiler Construction Code*, American Society of Mechanical Engineers
  Section I — Power Boilers
  Section II — Material Specifications
  Section III — Nuclear Vessels & Components
  Section IV — Heating Boilers
  Section V — Nondestructive Examination
  Section VI — Care and Operation of Heating Boilers
  Section VII — Care and Operation of Power Boilers
  Section IX — Welding Qualifications
  Some of the above Sections consist of more than one volume.
- *API-ASME Code — Unfired Pressure Vessels for Petroleum Liquids and Gases*, American petroleum Institute and American Society of Mechanical Engineers.
- *General Specifications for Building Naval Vessels*, United States Navy Department, Bureau of Ships.
- *Marine Engineering Regulations and Material Specifications*, United States Coast Guard.
- *ABS Rules for Building and Classing Steel Vessels*, American Bureau of Shipping.
- *Standards*, Tubular Exchanger Manufacturers' Association, Inc.
- *Lloyd's Rules and Regulations*, Lloyd's Register of Shipping.

Piping: A considerable amount of pipe is welded according to procedures developed by individual contractors without regard to code requirements. Piping for steam and other pressure work, however, is usually welded to code requirements. Many codes for piping are written on the basis of minimum requirements for safety, and some applications may, therefore, require more conservative design and construction practices than stipulated in the codes. A particular application, for example, may be covered by a code and yet require additional allowances for corrosion or erosion, special considerations to prevent distortion or creep, or inspection practices not called for in the code to insure that all joints are leakproof.

The ASME Boiler Construction Code covers piping connected with boilers, and adherence to its requirements has been made mandatory by many state and municipal governments. Sections I and VIII of this code refer to industrial piping.

Reference to piping is also contained in the *API-ASME Code for the Design, Construction, Inspection and Repair of Unfired Pressure Vessels for Petroleum Liquids and Gases.* Although more liberal than Section VIII of the ASME Boiler Code, the API-ASME code includes a mandatory inspection and repair schedule.

The American Welding Society *Standard D10.9-80 Qualification of Welding Procedures and Welders for Piping and Tubing* covers three different levels of weld quality.

The American Standard Code for Pressure Piping (B31.1) serves principally as a guide to state and municipal governments in establishing regulations. It is also used by contractors, manufacturers, architects, and engineers as a reference.

Water Pipelines: Pipe for conveying water is usually purchased to conform to *Specification 7A.3 or 7A.4 of the American Water Works Association.* Pipe purchased under ASTM Specifications A134-85 or A139-88 is satisfactory for water service, as is pipe conforming to API Specification 5L.

For especially large-diameter pipe, having wall thicknesses greater than 1-1/4 in., or for water pipe used above the ground and supported by stiffener rings, it is advisable to use the design practice described- in the section on penstocks in the *AWS Welding Handbook.*

Water pipes usually have a coating inside and outside to protect against corrosion. For mildly corrosive conditions, a good quality coal-tar paint is used. For severely corrosive conditions, either inside or outside, the pipe is usually protected with a coaltar, enamel, tape or epoxy lining or coating applied in accordance with AWWA Specifications C203-86, C209-84 and C210-84.

Field-Welded Storage Tanks: Field-welded storage tanks-are usually constructed in accordance with some code or specification prepared for a particular industry. These codes provide complete specifications for construction of storage tanks for all types of service. Unit stresses for steels will depend upon the service conditions involved and may be found in these industrial specifications:

- *Specifications for Gravity Water Tanks and Steel Towers*; Associated Factory Mutual Fire Insurance Companies
- *Standards for the Construction and Installation of Tanks, Gravity and Pressure, Towers, Etc.*; National Board of Fire Underwriters
- *Specifications for Field Welded Tanks for Storage of Production Liquids*; SPEC 12D-82
- *Specification for Shop welded Tanks for Storage of Production Liquids*; SPEC 12F-88; American Petroleum Institute
- *Standard Specifications for Elevated Steel Water Tanks, Standpipes and Reservoirs, (D100-84; D5.2);* American Water Works Association — American Welding Society
- *Specifications for All-Welded Steel Tanks for Railway Water Service*; American Railway Engineering Association

Aircraft Fabrication: The variety of materials and designs used in aircraft fabrication require the use of most of the welding processes. Because the materials used include aluminum and magnesium alloys, low-alloy steels, austenitic steels, and high-nickel alloys, a wide variation in techniques for most processes is also involved . Typical governing specifications and codes are:

- Weldor Certification
  Test; Aircraft Welding Operators' Certification: AN-T-38
- Filler Metal and Flux
  Electrodes, Welding (covered), Copper-Aluminum-Iron Alloy (Aluminum-Bronze) (for surfacing): JAN-E-278
  Electrodes, Welding (covered), Corrosion-Resisting (Austenitic Type) Steel: BuAer 46E4 (int)
  Electrodes, Welding (covered), Nickel-Copper Alloy: BuAer 17E4
  Nickel - Chromium-Iron Alloy Wire and Welding Rod: AN-N-4 Wire, Iron and Steel, Welding (for aeronautical use): BuAer 46R4, AAF 10286
- Process and Inspection Methods
  Welding, Magnesium: BuAer PW-2
  Welding, Aluminum: BuAer PW-5
  Welding, Steel: BuAer PW-7

Ship Construction: Merchant ships and many merchant-type naval vessels are constructed in accordance with the requirements established by the U.S. Coast Guard and by the American Bureau of Shipping. Naval combat vessels and certain other special types are constructed in accordance with U.S. Navy specifications.

The American Bureau of Shipping requirements can be found in its *Rules for The Building and Classing of Steel Vessels.*

Lloyd's Register of Shipping issues a specification entitled *"Lloyd's Rules and Regulations for the Construction and Classification of Steel Ships.* "This organization, founded in 1760, operates a world-wide service for classifying ships and inspecting their construction.

Coast Guard regulations are a part of the *Code of Federal Regulations* and come under Title 46, Shipping; Chapter I, Coast Guard: Inspection and Navigation. The following sections apply to welding:

TABLE 2-12. Sources of Technical Information

Aluminum Association
900 19th Street
Suite 300
Washington, DC 20006
(202) 862-5100

American Foundrymen's Society
505 State Street
Des Plaines, IL 60016-8399

American National Standards Institute
11 West 42nd Street
New York, NY 10036

American Society for Nondestructive Testing
P.O. Box 28518
Columbus, Ohio 43228-0518

American Welding Society
P.O. Box 351040
550 LeJeune Road
Miami, Florida 33135

ASM International
Route 87
Metals Park, OH 44073

Copper Development Association
Greenwich Office Park #2
P.O. Box 1840
Greenwich, CT 06836-1840
(203) 625-8210

Ductile Iron Society
28938 Lorain Road
Suite 202
North Olmsted, OH 44070

Library of Congress
National Referral Center for Science and Technology
Washington, DC 20540

National Association of Corrosion Engineers
P.O. Box 218340
Houston, TX 77218

National Certified Pipe Welding Bureau
1385 Piccard Drive
Rockville, MD 20850-4329
(301) 869-5800

Steel Founders' Society of America
455 State Street
Des Plaines, IL 60016
(708) 299-9160

Subchapter D — Tank Vessels — Part 31.3-2, 37.2
Subchapter F — Marine-Engineering Regulations, Part 56.20
Subchapter G — Ocean and Coastwise, Parts 59.15 and 59.30
Subchapter H — Great Lakes, Parts 76.15, 76.15a, 76.18, 76.34
Subchapter I — Bays, Sounds, and Lakes, Parts 94.14, 94.14a, 94.17, 94.34
Subchapter J — Rivers, Part 113.23
Subchapter M — Passenger Vessels, Part 144.3
Equipment Lists for Merchant Vessels (includes list of approved electrodes)

U.S. Navy regulations and requirements covering welding are embodied in Chapter 92, Welding and Allied Processes, of the Bureau of Ships Manual and include the following specifications:

General Specifications for Building Vessels of the U.S. Navy, Appendix 5, Specifications for Welding.

General Specifications for Machinery, Section S1-4, Welding and Brazing

General Specifications for Inspection of Material, Appendix VII, Welding.

Although not concerned with the establishment of codes, the *National Certified Pipe Welding Bureau* has a substantial effect on welding practices. This organization of piping contractors has headquarters in Washington, D.C. and local branches throughout the United States. Its purpose is to test and qualify pipe-welding procedures and pipe weldors and eliminate the need for requalifying for each job. NCPWB works within the existing codes and specifications.

General information on metals is published by various technical societies, trade associations, metal producers, and the Federal Government. Some of these organizations are listed in Table 2-12.

The General Services Administration issues the *Index of Federal Specifications and Standards*, which is available from the U.S. Government Printing Office. The actual specifications are available from local GSA Regional Offices.

The Referral Center for Science and Technology Division of the Library of Congress provides a referral service, but does not attempt to answer technical questions or to cite books, journals, or other bibliographic sources. Persons requesting information are provided with names of organizations likely to supply such.

A 300 ft. section of a tunnel being towed to the site after fabrication.

Section 3

VARIABLES IN WELDING FABRICATION

SECTION 3.1

WELDMENT DISTORTION

SECTION 3.2

ARC BLOW

SECTION 3.3

PREHEATING AND STRESS RELIEVING

Weldment Distortion

There are several problems or variables common to welding processes. One of these is distortion.

Distortion in a weldment results from the non-uniform expansion and contraction of the weld metal and adjacent base metal during the heating and cooling cycle of the welding process. During such a cycle, many factors affect shrinkage of the metal and make accurate predictions of distortion difficult. Physical and mechanical properties, upon which calculations must in part be based, change as heat is applied. For example, as the temperature of the weld area increases, yield strength, modulus of elasticity, and thermal conductivity of steel plate decrease, and coefficient of thermal expansion and specific heat increase (Fig. 3-1). These changes, in turn affect heat flow and uniformity of heat distribution. Thus, these variables make a precise calculation of what happens during heating and cooling difficult. Even if the calculations were simple, a practical understanding of the causes of distortion would be of greater value in the design stage and in the shop. This would allow the effects of shrinkage and shrinkage forces in various types of welded assemblies to be used to help control shrinkage.

THE REASONS FOR DISTORTION

To understand how and why distortion occurs during heating and cooling of a metal, consider the bar of steel shown in Fig. 3-2. As the bar is uniformly heated, it expands in all directions, as shown in Fig. 3-2(a). As the metal cools to room temperature it contracts uniformly to its original dimensions.

But if the steel bar is restrained — say, in a vise — while it is heated, as shown in Fig. 3-2(b), lateral expansion cannot take place. Volume expansion must occur, however, so the bar expands a greater amount in the vertical direction (thickness). As the deformed bar returns to room temperature, it will still tend to contract uniformly in all directions, as in Fig. 3-2(c). The bar is now narrower but thicker. It has been permanently deformed, or distorted. For simplification, the sketches show this distortion occurring in thickness only. Actually, of course, length is similarly affected.

In a welded joint, these same expansion and contraction forces act on the weld metal and on the base metal. As the weld metal solidifies and fuses with the base metal, it is in its maximum expanded state — it occupies the greatest possible volume as a solid. On cooling, it attempts to contract to the volume it would normally occupy at the lower temperature, but it is restrained from doing so by the adjacent base metal. Stresses develop within the weld, finally reaching the yield strength of the weld metal. At this point, the weld stretches, or yields, and thins out, thus adjusting to the volume require-

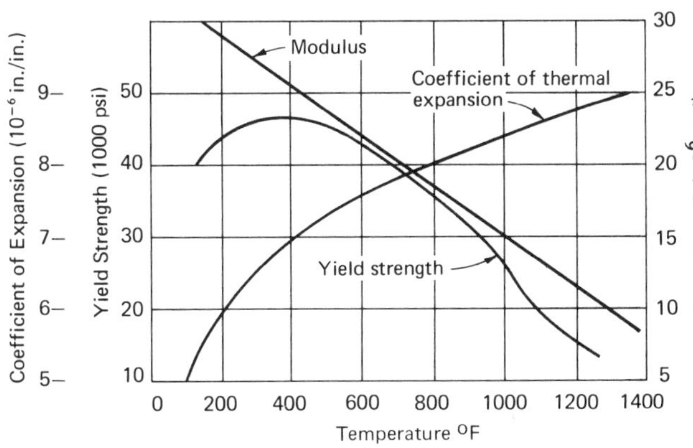

Fig. 3-1. Changes in the properties of steel with increases in temperature complicate analysis of what happens during the welding cycle — and, thus, understanding of the factors contributing to weldment distortion.

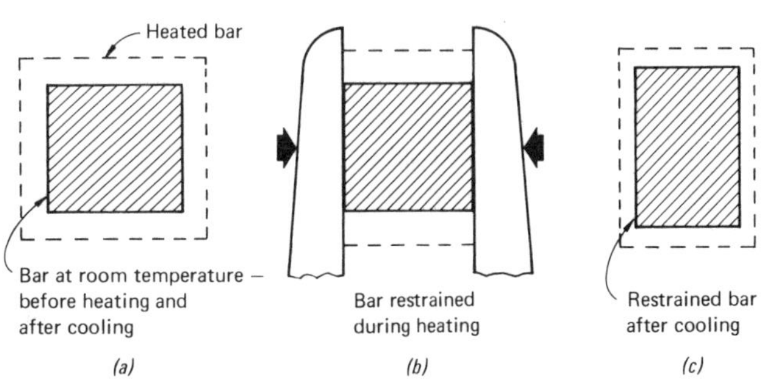

Fig. 3-2. If a steel bar is uniformly heated while unrestrained, as in (a), it will expand in all directions and return to its original dimensions on cooling. If restrained, as in (b), during heating, it can expand only in the vertical direction — become thicker. On cooling, the deformed bar contracts uniformly, as shown in (c), and, thus, is permanently deformed. This is a simplified explanation of a basic cause of distortion in welded assemblies.

ments of the lower temperature. But only those stresses that exceed the yield strength of the weld metal are relieved by this accommodation. By the time the weld reaches room temperature — assuming complete restraint of the base metal so that it cannot move — the weld will contain locked-in tensile stresses approximately equal to the yield strength of the metal. If the restraints (clamps that hold the workpiece, or an opposing shrinkage force) are removed, the locked-in stresses are partially relieved as they cause the base metal to move, thus distorting the weldment.

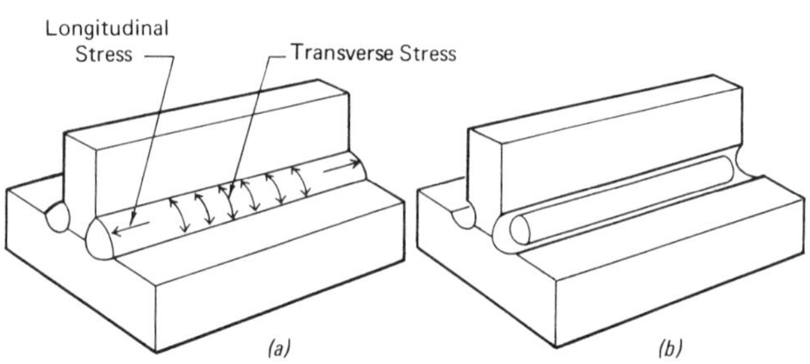

Fig. 3-3. The fillet welds in (a) have internal longitudinal and transverse stresses, and these welds would shrink to the dimensions of those shown in (b) if they could be unattached from the base plate. To re-establish the condition shown in (a), the fillets in (b) would have to be stretched longitudinally and transversely by forces that exceeded their yield strength.

Another approach to understanding internal stresses in a weld is shown in Fig. 3-3. Fillet welds that join two heavy plates contain residual longitudinal and transverse stresses, as indicated in Fig. 3-3(a). To visualize how these stresses got into the welds, imagine the situation depicted in Fig. 3-3(b). Here the fillets have been separated from the base plates. The same amount of weld metal is assumed to exist in both situations. In its unattached condition, the weld metal has shrunk to the volume it would normally occupy at room temperature. It is under no restraint and is stress-free.

To get this unattached weld back to the condition in Fig. 3-3(a), it would be necessary to pull it lengthwise — to impose longitudinal forces — and to stretch it transversely — to impose transverse forces. The weld metal has to give, or yield, in order to stretch, but at the time it reaches the needed dimensions, it is still under stress equivalent to its yield strength. This residual stress attempts to deform the weldment. In the case shown, it is unlikely that the plates would be deformed significantly because they are very rigid, and the weld is relatively small. When the first fillet is laid, however, angular distortion is likely to occur unless the plates are rigidly clamped or tacked.

Shrinkage in the base metal adjacent to the weld adds to the stresses that lead to distortion. During welding, the base metal adjacent to the weld is heated almost to its melting point. The temperature of the base metal a few inches from the weld is substantially lower. This large temperature differential causes nonuniform expansion followed by base metal movement, or metal displacement, if the parts being joined are restrained. As the arc passes on down the joint, the base metal cools and shrinks just like the weld metal. If the surrounding metal restrains the heated base metal from contracting normally, internal stresses develop. These, in combination with the stresses developed in the weld metal, increase the tendency of the weldment to distort.

The volume of adjacent base metal that contributes to distortion can be controlled somewhat by welding procedures. Higher welding speeds, for example, reduce the size of the adjacent base metal zone that shrinks along with the weld. An indication

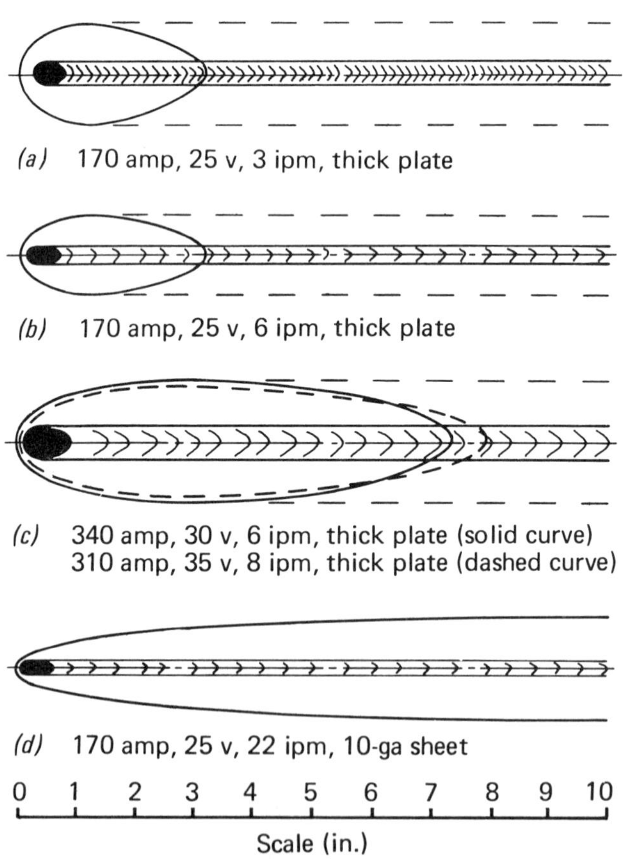

Fig. 3-4. Higher welding speeds reduce the size of the adjacent base metal zone that shrinks along with the weld and help to minimize distortion.

of these effects for some typical welds is shown in Fig. 3-4.

Controlled expansion and contraction is applied usefully in flame-straightening or flame-shrinking of a plate or weldment. For example, to shrink the center portion of a distorted plate, the flame from a torch is directed on a small, centrally located area. The area heats up rapidly and must expand. But the surrounding plate, which is cooler, prevents the spot from expanding along the plane of the plate. The only alternative is for the spot to expand in thickness, Fig. 3-5. In essence, the plate thickens where the heat is applied. Upon cooling, it tends to contract uniformly in all directions. When carefully done, spot heating produces shrinkage that is effective in correcting distortion caused by previous heating and cooling cycles.

Shrinkage of a weld causes various types of distortion and dimensional changes. A butt weld between two pieces of plate, by shrinking transversely, changes the width of the assembly, as in Fig. 3-6(a). It also causes angular distortion, as in Fig. 3-6(b). Here, the greater amount of weld metal and heat at the top of the joint produces greater shrinkage at the upper surface, causing the edges of the plate to lift. Longitudinal shrinkage of the same weld would tend to deform the joined plate, as shown in Fig. 3-6(c).

Angular distortion is also a problem with fillets, as illustrated in Fig. 3-6(d). If fillet welds in a T-shaped assembly are above the neutral axis (center of gravity) of the assembly, the ends of the member tend to bend upward, as in Fig. 3-6(e). If the welds are below the neutral axis, the ends bend down, as in Fig. 3-6(f).

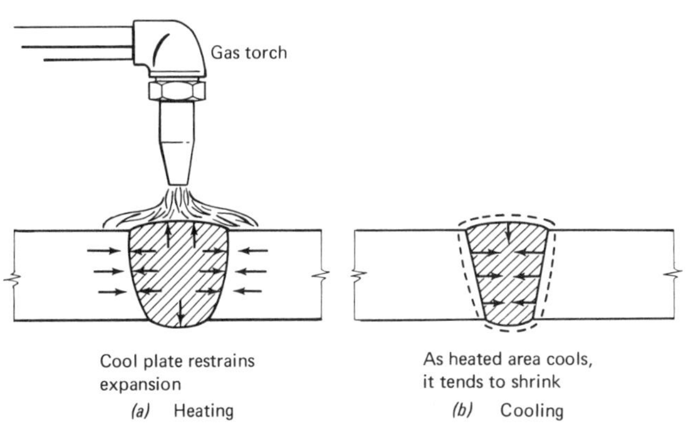

Fig. 3-5. The expansion and shrinkage phenomenon that produces distortion in weldments can be used constructively to remove distortion from steel plate. In (a), the heat from the torch causes a thickening of the spot heated. In (b), the cooled spot has a lesser volume within the thickness of the plate. A buckle that may have existed is now replaced with a slight bulge at the spot that was flame-shrunk.

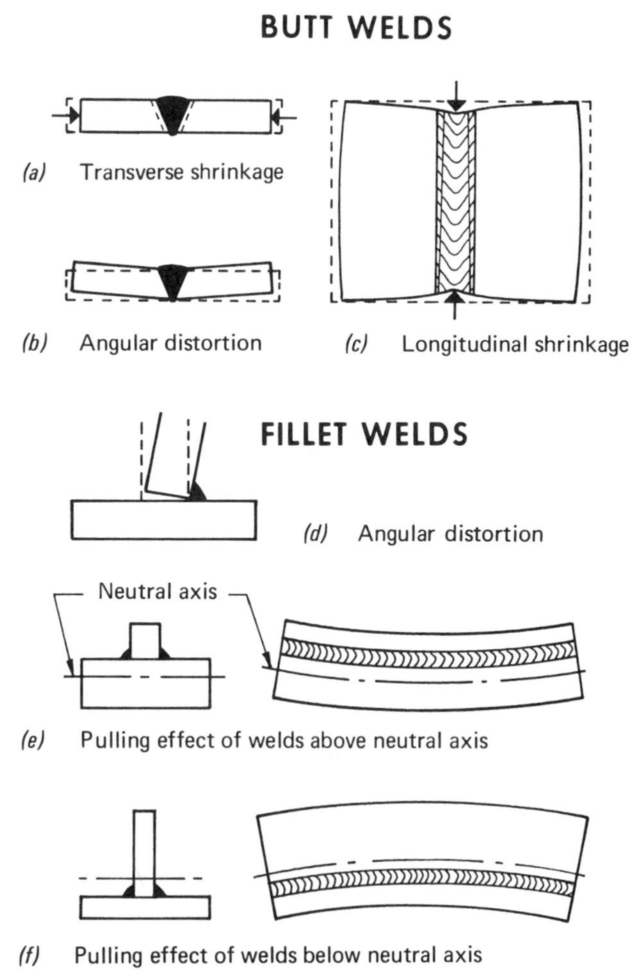

Fig. 3-6. How welds tend to distort and cause dimensional changes in assemblies.

HOW PROPERTIES OF METALS AFFECT DISTORTION

Since distortion is caused by the effects of heating and cooling and involves stiffness and yielding, the related mechanical and physical properties of metals affect the degree of distortion. A knowledge of approximate values of coefficient of thermal expansion, thermal conductivity, modulus of elasticity, and yield strength of the metal in a weldment helps the designer and the weldor to anticipate the relative severity of the distortion problem.

Coefficient of thermal expansion is a measure of the amount of expansion a metal undergoes when it is heated or the amount of contraction that occurs when it is cooled. Metals with high thermal-expansion coefficients expand and contract more than metals with low coefficients for a given temperature change. Because metals with high coefficients have larger shrinkage of both the weld metal and metal adjacent to the weld, the possibility for distortion of the weldment is higher.

Thermal conductivity is a measure of the ease of heat flow through a material. Metals with relatively low thermal conductivity (stainless steels and nickel-base alloys, for example) do not dissipate heat rapidly. Metals with high thermal conductivity (aluminum and copper) dissipate heat rapidly. Welding of low-conductivity metals results in a steep temperature gradient that increases the shrinkage effect in the weld and in the adjacent plate.

Yield strength of the weld metal is another parameter that affects the degree of distortion of a weldment. To accommodate the shrinkage of a weld joint on cooling, stresses must reach the yield strength of the weld metal. After stretching and thinning takes place, the weld and the adjacent base metal are stressed to approximately their yield strength. The higher the yield strength of a material in the weld area, the higher the residual stress that can act to distort the assembly. Conversely, distortion in the lower-strength metals is less likely or less severe.

Yield strength of metals can be changed by thermal or mechanical treatments. Heat treatment of medium-carbon, high-carbon, and alloy steels, for example, can increase yield strength appreciably. Cold working has a similar effect on many stainless-steels and copper and aluminum alloys. To minimize warping, metals should be welded in their annealed (low-strength) condition when possible.

Modulus of elasticity is a measure of stiffness of a material. One with a high modulus is more likely to resist distortion.

Table 3-1 lists these properties that are important in distortion analysis for steel, stainless steel, aluminum, and copper. Following are examples that illustrate how carbon or mild steel compares with other metals of construction with respect to distortion.

Mild Steel vs Stainless Steel: Yield strength and modulus of mild steel and of the commonly used stainless steels are in the same general range, indicating little difference in probable distortion. Thermal conductivity of the stainless grades, however, is only about one-third that of mild steel. This would increase the shrinkage effect. The coefficient of thermal expansion of stainless steel is about 1-1/2 times that of steel; this would also increase shrinkage in the plate adjacent to the weld. Thus, for the same amount of welding and the same size of member, stainless steel would tend to distort more than mild steel.

Mild Steel vs Aluminum: The coefficient of expansion of aluminum is about twice that of steel. If the two metals could be welded at about the same temperature, the shrinkage effect of aluminum would be much higher. But since the fusion temperature of steel is considerably higher than that of aluminum, the expansion factors approximately cancel out. Thermal conductivity of aluminum is about four times that of steel, which means that heat flows out of the aluminum faster, resulting in a lower temperature differential in the plate adjacent to the weld. This should produce less distortion in aluminum. Modulus of aluminum is about one-third that of steel, indicating higher distortion in aluminum for the same residual stress.

Yield strength could vary over a wide range, depending on the aluminum alloy and its heat treatment. The effect, comparatively, on distortion would be minor. Thus, the factors that increase and decrease distortion in aluminum and in steel approximately balance out, indicating that distortion expectancy is nearly equal for the two metals generally. Since the many alloys of both metals vary from the generalities discussed, degree of distortion would depend on the properties of the specific alloys being considered.

Mild Steel vs High-Strength Steel: The only significant difference between properties of these metals affecting distortion is yield strength. This would be higher in the high-strength steel, of course, suggesting increased distortion. Because of the higher strength, a smaller (or thinner) section would probably be used. This would further increase the distortion.

TABLE 3-1. Properties of Typical Metals*

Metal	Modulus of Elasticity (10^6 psi)	Yield Strength (10^3 psi)	Coef of Thermal Expansion (micro-in./in./°F	Thermal Conductivity (cal/cm^2/cm/°C/sec)
Carbon Steels	30	38	7	0.12
Stainless Steels	29	45	10	0.04
Aluminum Alloys	10	20	12	0.50
Copper Alloys	15	10	9	0.90

*Strength and thermal data shown are approximations for comparative use only. Actual data for different grades and alloys of these metals vary widely; for example, small amounts of alloying elements considerably change the thermal conductivity of aluminum and copper.

SHRINKAGE CONTROL

If distortion in a weldment is to be prevented or minimized, methods must be used both in design and in the shop to overcome the effects of the heating and cooling cycle. Shrinkage cannot be prevented, but it can be controlled. Several practical ways can be used to minimize distortion caused by shrinkage:

1. Do not overweld: The more metal placed in a joint, the greater the shrinkage forces. Correctly sizing a weld for the service requirements of the joint not only minimizes distortion, it saves weld metal and time. The amount of weld metal in a fillet can be minimized by use of a flat or slightly convex bead, and in a butt joint by proper edge preparation and fitup. Only the effective throat, dimension T in Fig. 3-7(a), in a conventional fillet can be used in calculating the design strength of the weld. The excess weld metal in a highly convex bead does not increase the allowable strength in code work, but it does increase shrinkage forces.

Proper edge preparation and fitup of butt welds, Fig. 3-7(b), help to force the use of minimum amounts of weld metal. For maximum economy, the plates should be spaced from 1/32 to 1/16 in. apart. A bevel of 30 degrees on each side provides proper fusion at the root of the weld, yet requires minimal weld metal. In relatively thick plates, the angle of bevel can be decreased if the root opening is increased, or a J or U preparation can be used to decrease the amount of weld metal used in the joint. A double-V joint requires about one-half the weld metal of a single-V joint in the same plate thickness.

In general, if distortion is not a problem, select the most economical joint. If distortion is a problem, select either a joint in which the weld stresses balance each other or a joint requiring the least

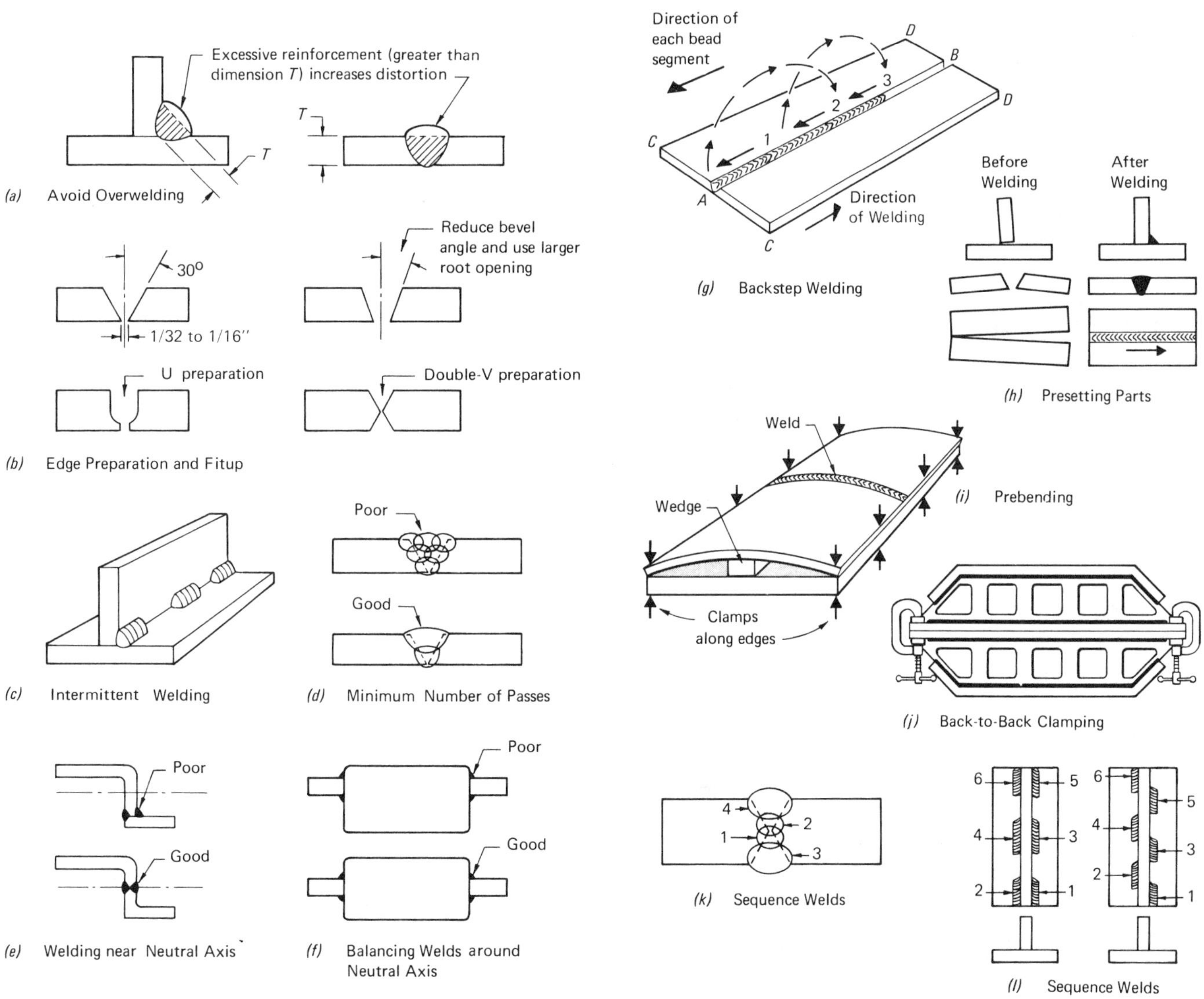

Fig. 3-7. Distortion can be prevented or minimized by techniques that defeat – or use constructively – the effects of the heating and cooling cycle.

amount of weld metal.

2. Use intermittent welding: Another way to minimize weld metal is to use intermittent rather than continuous welds where possible, as in Fig. 3-7(c). For attaching stiffeners to plate, for example, intermittent welds can reduce the weld metal by as much as 75%, yet provide the needed strength.

3. Use as few weld passes as possible: Fewer passes with large electrodes, Fig. 3-7(d), are preferable to a greater number of passes with small electrodes when transverse distortion could be a problem. Shrinkage caused by each pass tends to be cumulative, thereby increasing total shrinkage when many passes are used.

4. Place welds near the neutral axis: Distortion is minimized by providing a smaller leverage for the shrinkage forces to pull the plates out of alignment. Figure 3-7(e) is illustrative. Both design of the weldment and welding sequence can be used effectively to control distortion.

5. Balance welds around the neutral axis: This practice, shown in Fig. 3-7(f), offsets one shrinkage force with another to effectively minimize distortion of the weldment. Here, too, design of the assembly and proper sequence of welding are important factors.

6. Use backstep welding: In the backstep technique, the general progression of welding may be, say, from left to right, but each bead segment is deposited from right to left as in Fig. 3-7(g). As each bead segment is placed, the heated edges expand, which temporarily separates the plates at B. But as the heat moves out across the plate to C, expansion along outer edges CD brings the plates back together. This separation is most pronounced as the first bead is laid. With successive beads, the plates expand less and less because of the restraint of prior welds. Backstepping may not be effective in all applications, and it cannot be used economically in automatic welding.

7. Anticipate the shrinkage forces: Placing parts out of position before welding can make shrinkage perform constructive work. Several assemblies, preset in this manner, are shown in Fig. 3-7(h). The required amount of preset for shrinkage to pull the plates into alignment can be determined from a few trial welds.

Prebending or prespringing the parts to be welded, Fig. 3-7(i), is a simple example of the use of opposing mechanical forces to counteract distortion due to welding. The top of the weld groove — which will contain the bulk of the weld metal — is lengthened when the plates are sprung. Thus the completed weld is slightly longer than it would be if it had been made on the flat plate. When the clamps are released after welding, the plates return to the flat shape, allowing the weld to relieve its longitudinal shrinkage stresses by shortening to a straight line. The two actions coincide, and the welded plates assume the desired flatness.

Another common practice for balancing shrinkage forces is to position identical weldments back to back, Fig. 3-7(j), clamping them tightly together. The welds are completed on both assemblies and allowed to cool before the clamps are released. Prebending can be combined with this method by inserting wedges at suitable positions between the parts before clamping.

In heavy weldments, particularly, the rigidity of the members and their arrangement relative to each other may provide the balancing forces needed. If these natural balancing forces are not present, it is necessary to use other means to counteract the shrinkage forces in the weld metal. This can be accomplished by balancing one shrinkage force against another or by creating an opposing force through the fixturing. The opposing forces may be: other shrinkage forces; restraining forces imposed by clamps, jigs, or fixtures; restraining forces arising from the arrangement of members in the assembly; or the force from the sag in a member due to gravity.

8. Plan the welding sequence: A well-planned welding sequence involves placing weld metal at different points about the assembly so that, as the structure shrinks in one place, it counteracts the shrinkage forces of welds already made. An example of this is welding alternately on both sides of the neutral axis in making a butt weld, as in Fig. 3-7(k). Another example, in a fillet weld, consists of making intermittent welds according to the sequences shown in Fig. 3-7(l). In these examples, the shrinkage in weld No. 1 is balanced by the shrinkage in weld No. 2, and so on.

Clamps, jigs, and fixtures that lock parts into a desired position and hold them until welding is finished are probably the most widely used means for controlling distortion in small assemblies or components. It was mentioned earlier in this section that the restraining force provided by clamps increases internal stresses in the weldment until the yield point of the weld metal is reached. For typical welds on low-carbon plate, this stress level would approximate 45,000 psi. One might expect this stress to cause considerable movement or distortion after the welded part is removed from the jig or clamps. This does not occur, however, since the strain (unit contraction) from this stress is very low compared to the amount of movement that would occur if no restraint were used during welding. For example:

$$\text{Modulus of elasticity (E)} = \frac{\text{Stress } (\sigma)}{\text{Strain } (\epsilon)}$$

$$\epsilon = \frac{\sigma}{\text{E steel}}$$

$$\epsilon = \frac{45{,}000}{30{,}000{,}000}$$

$$\text{Strain} = 0.0015 \text{ in./in.}$$

9. Remove shrinkage forces after welding: Peening is one way to counteract the shrinkage forces of a weld bead as it cools. Essentially, peening the bead stretches it and makes it thinner, thus relieving (by plastic deformation) the stresses induced by contraction as the metal cools. But this method must be used with care. For example, a root bead should never be peened, because of the danger of either concealing a crack or causing one. Generally, peening is not permitted on the final pass, because of the possibility of covering a crack and interfering with inspection, and because of the undesirable work-hardening effect. Thus, the utility of the technique is limited, even though there have been instances where between-pass peening proved to be the only solution for a distortion or cracking problem. Before peening is used on a job, engineering approval should be obtained.

Another method for removing shrinkage forces is by stress relief — controlled heating of the weldment to an elevated temperature, followed by controlled cooling. Sometimes two identical weldments are clamped back to back, welded, and then stress-relieved while being held in this straight condition. The residual stresses that would tend to distort the weldments are thus removed.

10. Minimize welding time: Since complex cycles of heating and cooling take place during welding, and since time is required for heat transmission, the time factor affects distortion. In general, it is desirable to finish the weld quickly, before a large volume of surrounding metal heats up and expands. The welding process used, type and size of electrode, welding current, and speed of travel, thus, affect the degree of shrinkage and distortion of a weldment. The use of iron-powder manual electrodes or mechanized welding equipment reduces welding time and the amount of metal affected by heat and, consequently, distortion. For example, depositing a given-size weld in thick plate with a process operating at 175 amp, 25 v, and 3 ipm requires 87,500 joules of energy per linear inch of weld. The same size weld, produced with a process operating at 310 amp, 35 v, and 8 ipm requires 81,400 joules per linear inch. The difference represents "excessive" heat, which expands the surrounding metal more than necessary.

EQUATIONS FOR CALCULATING SHRINKAGE

Transverse weld shrinkage (shrinkage perpendicular to the axis of a weld) is particularly important when the shrinkage of individual welds is cumulative as, for example, in the beam-to-column connections across the length or width of a large building. Unless allowances are made for transverse weld shrinkage — usually by spreading the joint open by the amount it will contract after welding — the cumulative shrinkage of several beam-to-column connections could be great enough to noticeably shorten the building's dimensions.

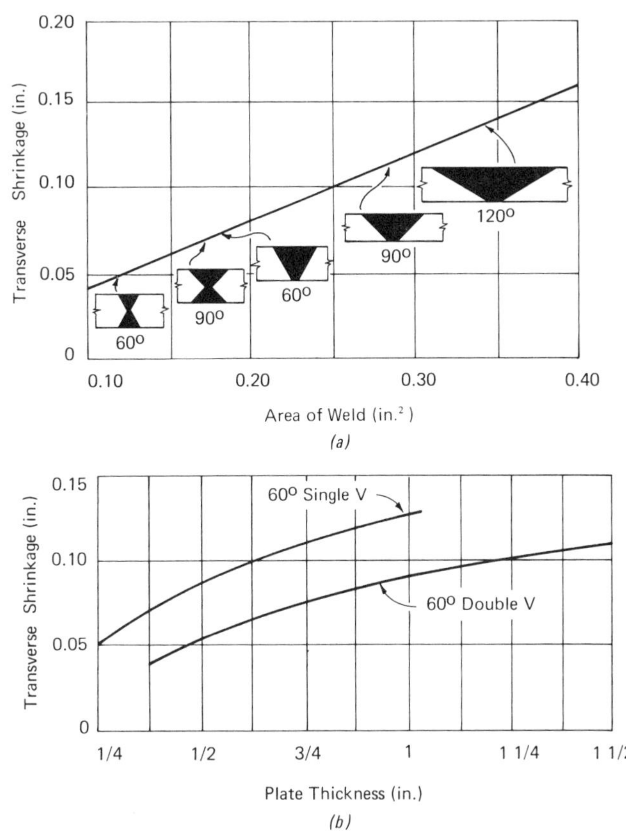

Fig. 3-8. For a given weld thickness, transverse shrinkage increases directly with the cross-sectional area of the weld. The large included angles in (a) are for illustrative purposes only.

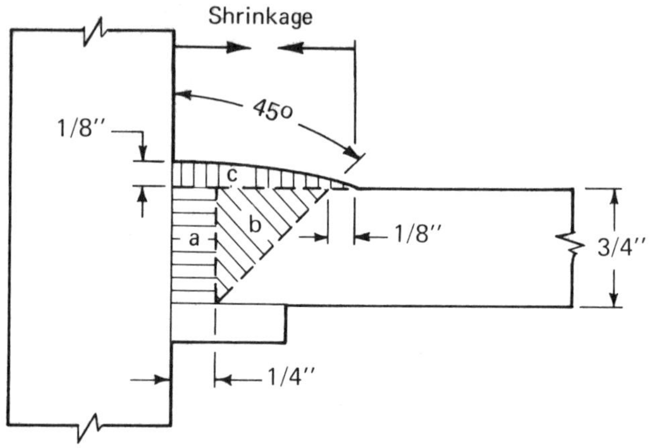

$$A = (1/4)(3/4) + (1/2)(3/4)(3/4) + (2/3)(1/8)(1\text{-}1/8)$$
$$= 0.563 \text{ in.}^2$$

$$\text{Transverse Shrinkage} = 0.10\frac{A}{t} = 0.10\left(\frac{0.563}{0.875}\right) = 0.064 \text{ in.}$$

Fig. 3-9. Transverse shrinkage is calculated by determining the cross-sectional area of the weld and applying it to the transverse-shrinkage formula.

For a given weld thickness, transverse shrinkage of a weld increases directly with cross-sectional area of the weld. Figure 3-8(a) shows this relationship for a 1/2-in. plate. The large included angles shown in this graph are illustrative only; angles above 60° are seldom used in welding. Transverse shrinkage of 60° single-V and double-V joints in several plate thicknesses are shown in Fig. 3-8(b). Shrinkage values shown in both graphs assume that no unusual restraint against transverse shrinkage is imposed.

Approximate transverse shrinkage for other weld angles or sizes can be predicted from:

$$\boxed{\text{Transverse shrinkage} = 0.10\frac{A}{t}} \quad \ldots\ldots(1)$$

where A = cross-sectional area of the weld in in.^{2}, and t = weld thickness in inches.

Another way of stating this relationship is that transverse shrinkage equals one-tenth of the average width of the weld area. A sample calculation using this equation is shown in Fig. 3-9.

Important: When a deep-penetrating welding process (such as submerged-arc) is used, the cross section of the entire fused part of the joint (not only the area of deposited weld metal) must be used in the calculation.

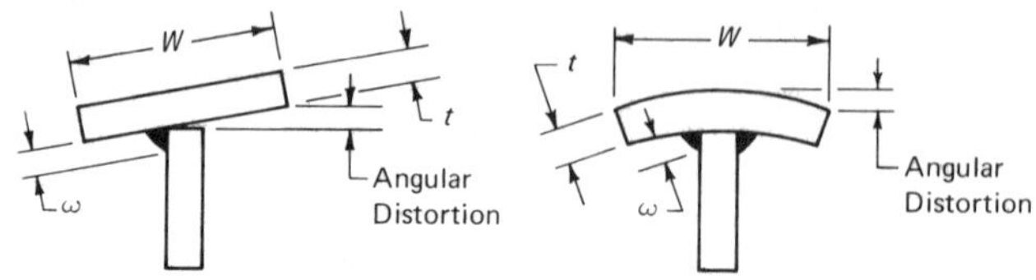

Fig. 3-10. Angular distortion varies directly with the flange width and weld size and inversely with flange thickness.

Angular distortion (Fig. 3-10) varies directly with flange width W and weld size ω and inversely with flange thickness t. The equation is:

$$\boxed{\text{Angular distortion} = \frac{0.02\, W\, \omega^{1.3}}{t^2} \text{ in.}} \quad \ldots\ldots(2)$$

Values of $\omega^{1.3}$ for use in this equation are given in Table 3-2.

TABLE 3-2. VALUES OF $\omega^{1.3}$

Weld Size ω (in.)	Value $\omega^{1.3}$
3/16	0.114
1/4	0.165
5/16	0.220
3/8	0.280
7/16	0.342
1/2	0.406
9/16	0.474
5/8	0.543
3/4	0.688
7/8	0.841
1	1.00

The agreement between measured and calculated values of angular distortion, shown in Fig. 3-11 for eight different flange and web arrangements, attests to the validity of the equation. In only one of the arrangements illustrated does angular distortion approach the AWS allowable limit — 1% of the flange width or 1/4 in., whichever is greater. In this instance, overwelding is obvious.

Longitudinal bending, or cambering, results from a shrinkage force applied at some distance from the neutral axis of a member. Amount of distortion depends on the shrinkage moment and the resistance of the member to bending, as indicated by its moment of inertia. Assuming no unusual initial stresses, the following equation can be used to calcu-

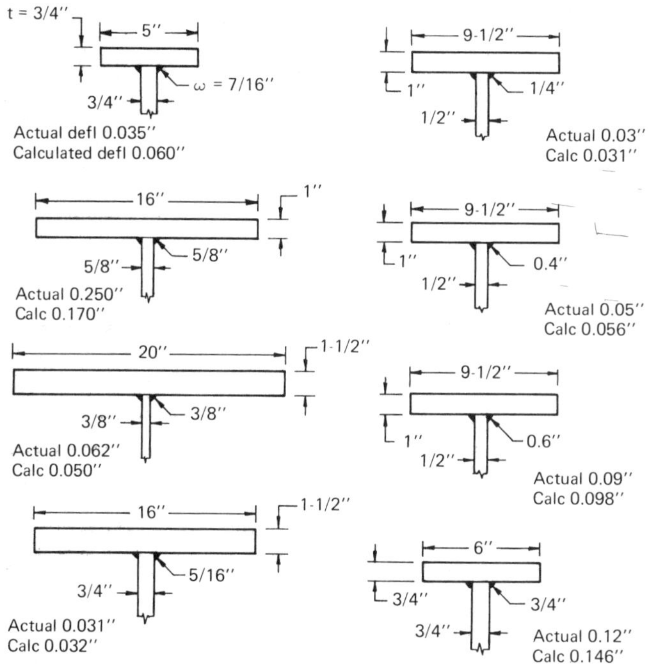

Fig. 3-11. The agreement between measured and calculated values for angular distortion attests to the usefulness of the angular-distortion formula.

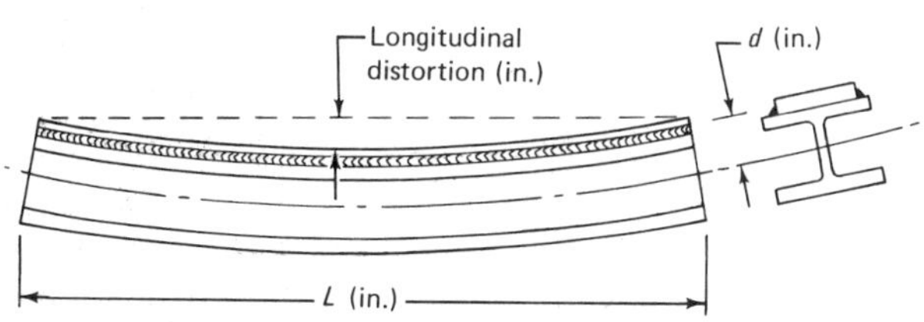

Fig. 3-12. Longitudinal distortion varies directly with the cross-sectional area of weld metal, distance of the center of gravity of the weld from the neutral axis, and length of the member squared, and inversely with the moment of inertia of the member.

late distortion of a member, resulting from longitudinal welds:

$$\text{Longitudinal distortion} = \frac{0.005\ AdL^2}{I} \qquad \ldots\ldots(3)$$

where A = total cross-sectional area of the weld metal and fused base metal in in.^{2}; I = moment of inertia of the member in in.^{4}; and L and d are the length and distance identified in Fig. 3-12. The area A can be estimated from the weld size ω.

Agreement between calculated and measured values for longitudinal distortion is shown for several examples in Fig. 3-13.

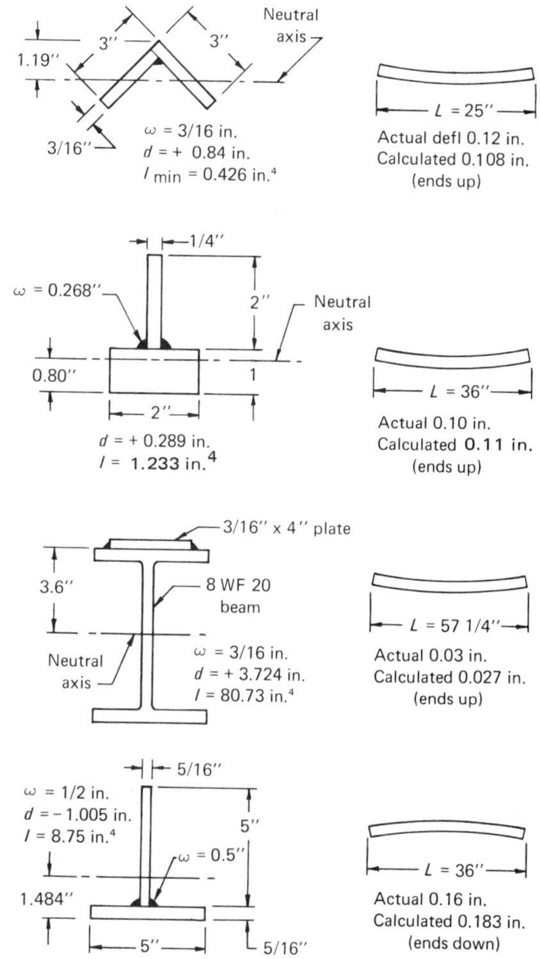

Fig. 3-13. The formula for longitudinal distortion gives values in reasonable agreement with those determined by measurement.

EXAMPLES OF DISTORTION CONTROL

T Section: A manually welded T assembly, Fig. 3-14(a), was distorted laterally after welding, even though the proper size of fillet weld was used. Analysis showed that the center of gravity of the two welds was well above the neutral axis. By changing to deep-penetration, automatic submerged-arc welding, the center of gravity of the welds was lowered, Fig. 3-14(b), substantially reducing the shrinkage moment. Depth, or throat, of the weld is the same but there is now weld metal nearer to the neutral axis. In addition, the higher speed of the automatic welding also reduces distortion.

Three-Member Column: The welds in the lift-truck column shown in Fig. 3-15(a) are balanced and can be made in the flat position by turning the assembly once. But longitudinal distortion proved

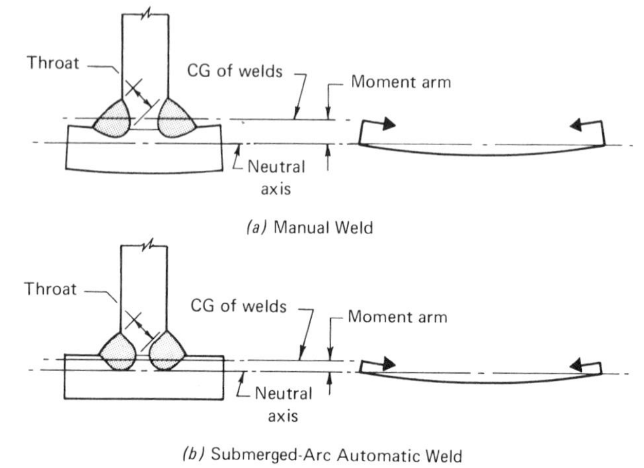

Fig. 3-14. The lateral distortion in (a) resulted from the center of gravity of the two welds being well above the neutral axis. The deep-penetration characteristics of automatic submerged-arc (b) lowered the center of gravity of the welds and helped reduce the distortion.

to be excessive — 0.42 in. in the 100-in. length. Analysis shows that the distance between the center of gravity of the welds and the neutral axis of the section is 0.682 in. If this distance could be reduced by a change in design, less distortion would occur.

One way to put the welds closer to the neutral axis is shown in Fig. 3-15(b). With this design, the distance is reduced to 0.556 in. Calculation for distortion (using Equation 3) shows that the distortion would be reduced to 0.32 in. for the 100-in. length. If this amount of distortion cannot be tolerated, the column members could be prebent about 5/16 in. in the opposite direction so that the assembly would be very near flat after being welded.

Box Section: The lightweight boom section illustrated in Fig. 3-16 exemplifies the importance of method of assembly in minimizing distortion.

One method of assembly would be to tack-weld all pieces together before welding, producing a rigid unit with counterforces to resist those generated by shrinkage. Analysis indicates that the center of gravity of the weld metal would be 1.11 in. from the neutral axis of the section. In a 63-in. length, the deflection calculated from Equation 3 is 0.004 in.

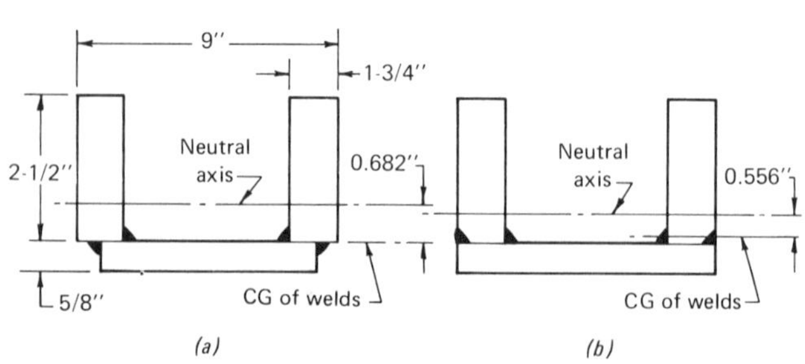

Fig. 3-15. By a design change, the distance of the center of gravity of the welds in this assembly from the neutral axis was reduced from 0.682 in. (a) to 0.556 in. (b). Longitudinal distortion was thus reduced from 0.42 in. to 0.32 in. for the 100-in. length.

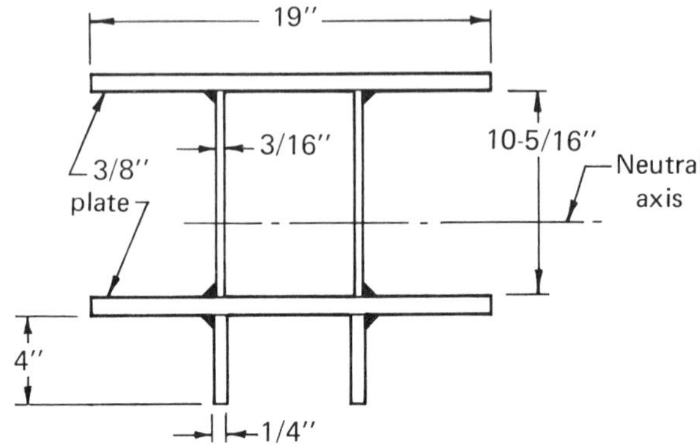

Fig. 3-16. If the stiffeners to the box section were welded on after welding together the box-section members, the longitudinal distortion would be 1-1/2 times that produced by tack-welding the stiffeners at the same time the box section is tacked.

A second method would be to weld the box section first, which should produce no distortion because of the exact coincidence of the center of gravity of weld metal with the neutral axis. The only distortion would then be that developed when the two 1/4-in. stiffener plates are added. The distance between the center of gravity of the stiffener welds and the neutral axis is 3.994 in. Calculating for deflection in a 63-in. length produces a value of 0.006 in. Thus the distortion in the assembly using this sequence would be 1-1/2 times that produced by tack-welding the stiffeners at the same time the box section is tacked.

Unsymmetrical Beam: The welded spandrel beam shown in Fig. 3-17 is to cover a 42-ft span in a structure. How much horizontal deflection, or "sweep," can be expected in this length?

Inspection shows that the welds are balanced about the horizontal (x-x) axis and the section is symmetrical in respect to it. Thus, no vertical distortion, or camber, would be expected as the result of welding. The vertical neutral axis (y-y) is calculated to be 0.242 in. to the right of the centerline of the web plate. The section is thus fairly symmetrical about the vertical axis and, if the welding were balanced about the web plate, there should be little horizontal bending.

However, the welding is not centered about the vertical neutral axis; computation shows the center of gravity of all welds to be 2.63 in. to the right of the centerline of the web. Thus, distance d (in Equation 3) between the neutral axis and the center of gravity of welds is 2.39 in. Horizontal deflection

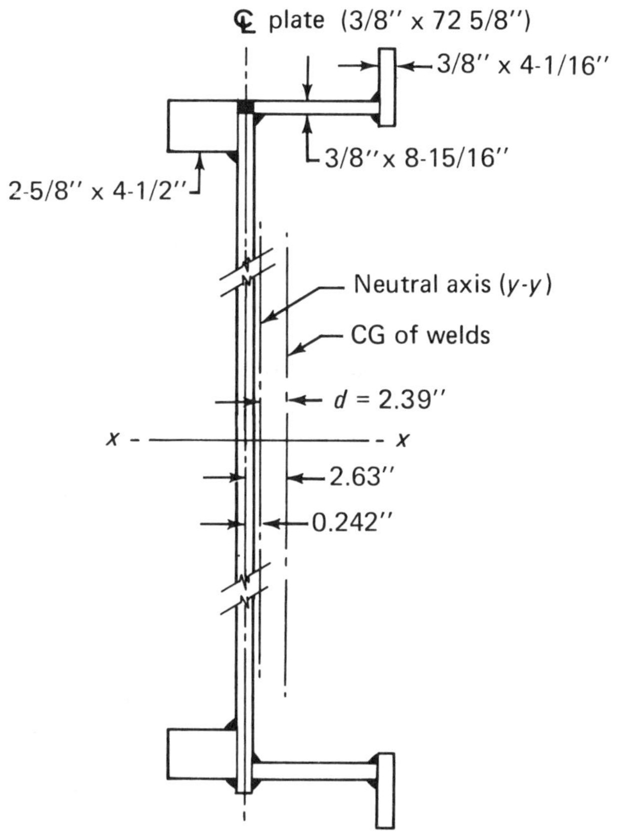

Fig. 3-17. This unsymmetrical beam defies an assembly sequence that lowers horizontal distortion to a point where flame-straightening would be practical. The best solution appears to be one in which the L-shaped subassemblies would be welded in half lengths and then weld-joined into full-length units after flame-straightening.

in the 42-ft length, calculated from Equation 3, is about 2.4 in., if the assembly were all tacked together and welded. This could be too much to straighten by flame-shrinking.

Another possibility might be to weld the two L-shaped parts as subassemblies. As isolated units, these pieces would have a neutral axis approximately 3 in. from the center of gravity of their two 1/4-in. fillets. Calculation shows that the welds would cause the ends of 42-ft lengths to move about 5.1 in. to the right. This would be an excessive amount.

The L-shaped subassemblies could be broken into shorter pieces and then joined by butt welding. Since length L in the deflection equation is a squared term, any reduction in length substantially decreases deflection. If the L-shaped sections were welded in 21-ft lengths, end movement would be less than 1.3 in. — an amount that is easily handled in straightening. After welding the L-shaped sections, they could be assembled into the final girder. Then all of the welding would lie almost along the vertical neutral axis, and no distortion should occur. A fast welding process, avoiding multipass welds, would be recommended.

Welding Sequence: A fabricator was welding frames of different lengths and thicknesses from formed mild-steel channels and plates in the sequence indicated in Fig. 3-18(a). He wanted to know if any other sequence would produce less distortion.

In the analysis of this problem, it should be borne in mind that the shrinkage of the weld metal and adjacent plate will produce a tensile force. If the resulting compressive stress in the member does not exceed the yield strength or is not great enough to produce permanent set, the sequence is immaterial when the welds are not large compared with the rest of the section, and when the section is symmetrical and the welds are balanced about the neutral axis. After the fourth weld, compressive stress would be fairly uniform throughout the section. But, in this example, the question arises as to whether or not the yield point might be exceeded in compression with any of the possible sequences.

After the first weld is completed, Fig. 3-18(b), the weld and some adjacent plate is in tension. Compressive forces exist in the rest of the member, which may be considered as an eccentrically loaded

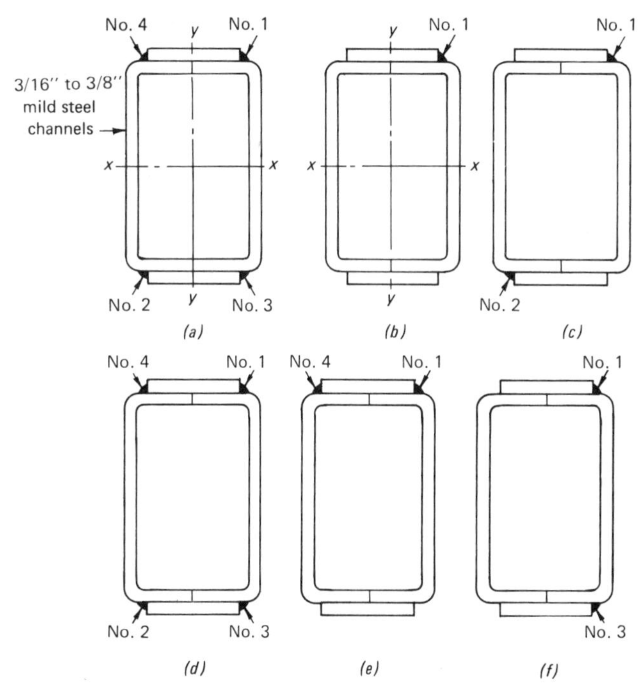

Fig. 3-18. The sequence of placing the welds and the practicality of running two welds simultaneously are important factors in the welding of this assembly.

column. If the weld is large enough, the resulting stress will exceed the yield point of the member and cause permanent deformation. This would produce a larger bend than if the yield point were not exceeded.

When weld No. 2 is made, Fig. 3-18(c), the same but opposite stress distribution should result. However, if the first weld has caused a permanent set, the second weld cannot pull the member back into straight condition.

If both welds are made simultaneously, a uniform stress distribution results, and there is no movement or bending effect. Also, the stress is much lower, and there is very little chance that it would exceed the yield point. Then, when the final two welds are simultaneously made, Fig. 3-18(d), stress distribution would be still more uniform and produce no bending effect.

If two welds are made at the same time in a horizontal position — weld No. 1 and No. 4, in Fig. 3-18(e) — some bending would result because the yield point probably would be exceeded. The remaining two welds would probably not quite pull the member straight.

If instead, the welds across the longer dimension were chosen as pairs, Fig. 3-18(f), bending would be about the y-y axis. Such a member would have little structural utility.

Thus, if it is necessary to position the member to accomodate two adjacent welds; the two welds at the narrower dimension should be made simultaneously. The resulting bending about the x-x axis would be slight, and a camber in this direction is often desirable.

The least amount of distortion would result if welds 1 and 2 were made simultaneously, followed by 3 and 4, also simultaneously. But since this procedure would not be practical in most cases, the next best procedure is to make welds 1, 4, 2, 3 in that order, resulting in some camber.

The girder in Fig. 3-19, made of high strength (A441) steel, needs more flange area on the bottom because the concrete deck on top is attached through shear lugs (composite girder) and will carry some of the bending moment. The strength and allowable stress for this steel decreases with increased thickness. Two 3/4-in. plates are used on the bottom because they have a higher stress-allowable and are stronger than a single 1-3/4-in. plate, thus saving 16% of the steel.

The conventional shop procedure would be to tack all parts of the girder together, then weld the assembly. However, there would be four fillet welds in the bottom portion of the girder and two in the top — an unbalanced condition. In this example, the center of gravity of the six welds is considerably below the neutral axis of the member, and the welding would cause the ends of the girder to move downward, producing camber.

A better procedure might be to weld the cover plate to the bottom flange plate first. If distortion should occur, straightening of the subassembly would not be difficult. Then, when this subassembly and the top flange are welded to the web, only four fillet welds affect distortion, and the center of gravity of the four welds is much closer to the neutral axis. The deflection due to welding will be directly proportional to the amount of weld metal and to the distance d between the center of gravity of the welds and the neutral axis, and inversely proportional to the bending resistance (moment of inertia) of the section.

Changing welding sequence from the first method to the second reduces the number of welds that influence distortion from six to four. In other words, the welds then have only two-thirds the effect on distortion as previously. The distance d is reduced by about half. Bending resistance, of course, remains constant. Theoretically, the second

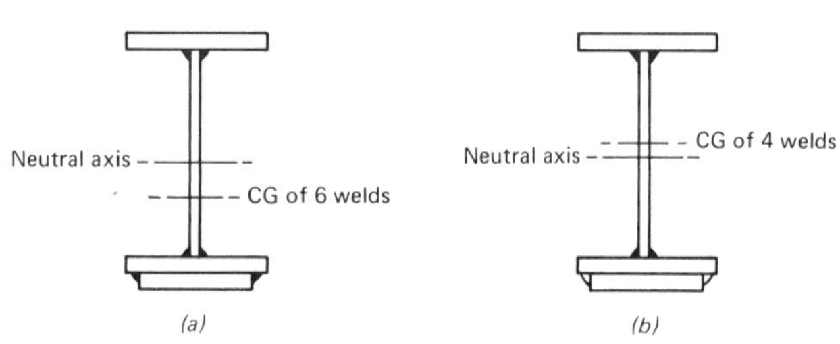

Fig. 3-19. By welding the cover plate to the bottom plate first, as in (b), and straightening it before tacking and welding the beam, the distortion in the final beam would be only about one-third that if all parts were tacked together for welding, as in (a).

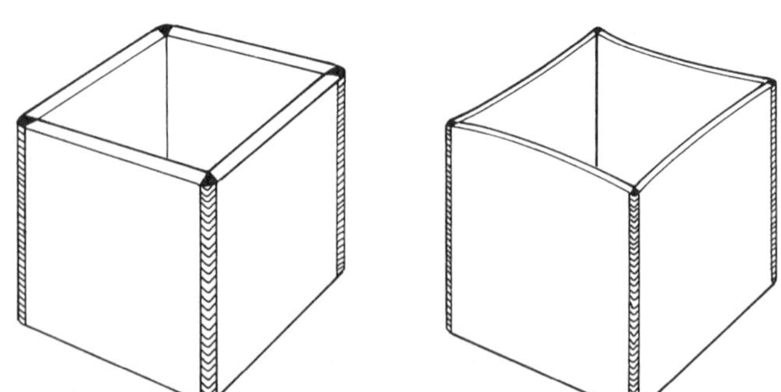

Fig. 3-20. The 1/2-in. plate in the hopper box at the left has about 17 times as much resistance to buckling under the shrinkage forces of the corner welds as the 12-gage sheet material in the box at the right.

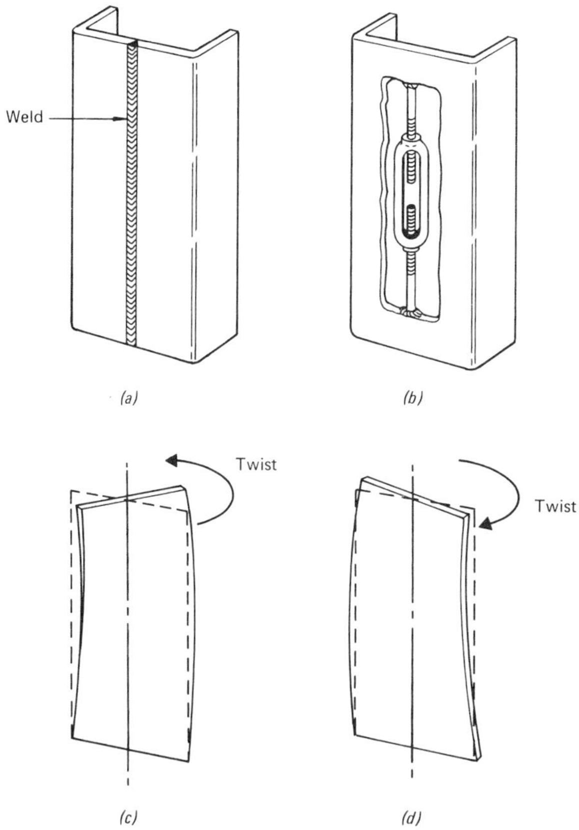

Fig. 3-21. Shrinkage in the centerline weld (a) has the same effect as if a turnbuckle (b) were used to shorten the centerline of the member. A twist, either as shown in (c) or (d), must result when the outer edges of the member are longer than its centerline.

sequence would thus produce only one-third the distortion (2/3 x 1/2 = 1/3), and, since the center of gravity of the welds in this case would be above the neutral axis, the distortion would be in the opposite direction — ends up.

It must be remembered, however, that if the distortion in the first method was not great, it might not be worthwhile to change to the method that would require straightening the flange-cover plate subassembly. Both sequences should be investigated whenever this situation arises.

Buckling and Twisting: Since the shrinkage force of the weld is a function of the square of the thickness of the material and resistance to buckling is a function of the cube of the thickness, it is seen that the buckling due to welding of a panel increases directly as the thickness decreases.

Figure 3-20 illustrates two hopper boxes with welded corners. One is made from 1/2-in. plate and the other, from 12-gage (0.1046-in.) sheet. The welds at the corners shrink and tend to leave excess metal under compression in the central portion of the panels. Distortion by buckling can result. The size of the corner welds in the hopper made from 1/2-in. plate indicates that the shrinkage forces would be much greater than those in the hopper fabricated from 12-gage sheet. However, resistance to buckling is also much greater in the thicker material (because of the t^3 relationship). With the thicknesses used in this example, the plate material has about 17 times more resistance to bending (or buckling) than the sheet material.

Twisting can also be a problem with thin material because of its low torsional resistance. When a weld is made down the center of a member, Fig. 3-21(a), the weld area tends to shrink and become shorter. The effect is the same as if a turnbuckle were attached at this position and tightened, as in Fig. 3-21(b). Under this centrally located tension, a flat rectangle can not exist.

To satisfy the conditions of a member that has outer edges longer than its centerline, the member must twist, as in Fig. 3-21(c). Applying a counter force to untwist the weldment is futile. Once that force is great enough to re-establish the original plane, the material snaps into a twist in the opposite direction, as in Fig. 3-21(d).

Twisting can be prevented or minimized in several ways:

1. Minimize shrinking force by good welding practice — decrease volume of weld metal and weld at the highest practical speed.
2. Keep the length of the welded member as short as practical.
3. Incorporate as much resistance to twisting as feasible. Since the resistance of a plate to twisting is a function of the cube of its thickness, doubling the thickness increases its resistance by a factor of eight. Torsional resistance can also be increased, where

Fig. 3-22. Fillet welds tend to shorten the flanges of a long, thin beam, leading to the twisting shown in (b). Flame-shrinkage of the outer edges of the flanges is a corrective measure.

design permits, by using closed box sections or diagonal bracing.

A twisted weldment can often be corrected by flame-shrinking. The outer edges are heat-shortened to the length of the centerline, and the disparity in dimensions responsible for twisting is eliminated.

Sometimes a very long, thin welded beam twists out of shape after welding. The reason for this is that the member is made up of three parts, as in Fig. 3-22(a), and the fillet welds have shortened the flanges at the centers, while the lengths of the outside edges have remained unchanged. Unless the flanges have adequate torsional resistance, twisting results, as shown in Fig. 3-22(b). Flame-shrinkage of the outer edges of the flanges is the corrective measure for this condition.

Horizontal or Vertical Web: In a shop having automatic equipment with the capability of making two fillet welds simultaneously, beams can be welded with the web either horizontal or vertical (Fig. 3-23). Either way is satisfactory, since distortion would be minimal in each case. When the web is horizontal the beam is more flexible, but since the welds are very close to the neutral axis of the girder, they have practically no bending power. With the web vertical, the welds are farther away from the neutral axis but the girder has considerable rigidity in this direction.

It may be difficult, however, with the web horizontal, to keep the flanges from tilting or rotating to an angle less than 90 degrees. If the welds on the other side of the web do not correct this, the fabricator still has the option of making the first welds with the flanges cocked to an angle greater than 90 degrees. He must ascertain, however, just how much excess angle to use to get 90 degrees after shrinkage

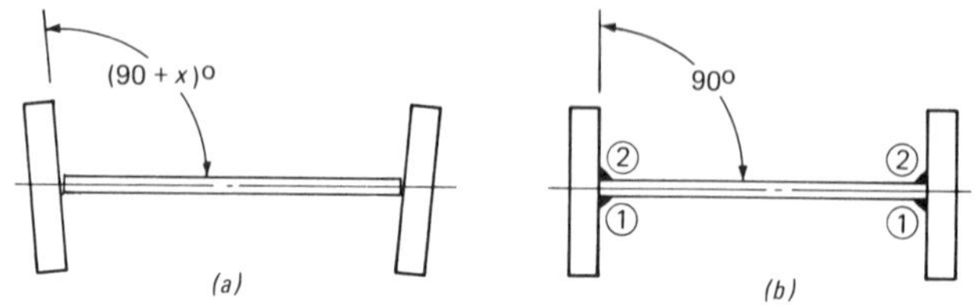

Fig. 3-24. The horizontal web position is often preferred when automatic welding with two heads, since it allows angling the flanges so that shrinkage forces will bring the angle of web-to-flange to 90 degrees in the finished member.

and after completion of the final welds. In addition, he must make certain that his welding procedures remain constant.

In general, when equipment is not available that can deposit two welds simultaneously, the sequence for depositing the four fillet welds on a fabricated plate girder by automatic methods can be varied without a significant effect on distortion. In most cases, the sequence is based on the type of fixture used and the method available for moving the girder from one welding position to another. When a single automatic welder is used, the girder is usually positioned at an angle between 30 and 45 deg, permitting the welds to be deposited in the flat position.

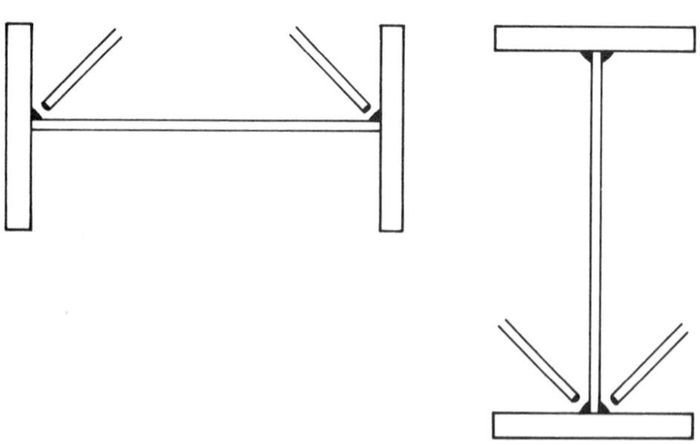

Fig. 3-23. In the automatic welding of flange to web running two fillets simultaneously, positioning the web either horizontal or vertical is acceptable practice.

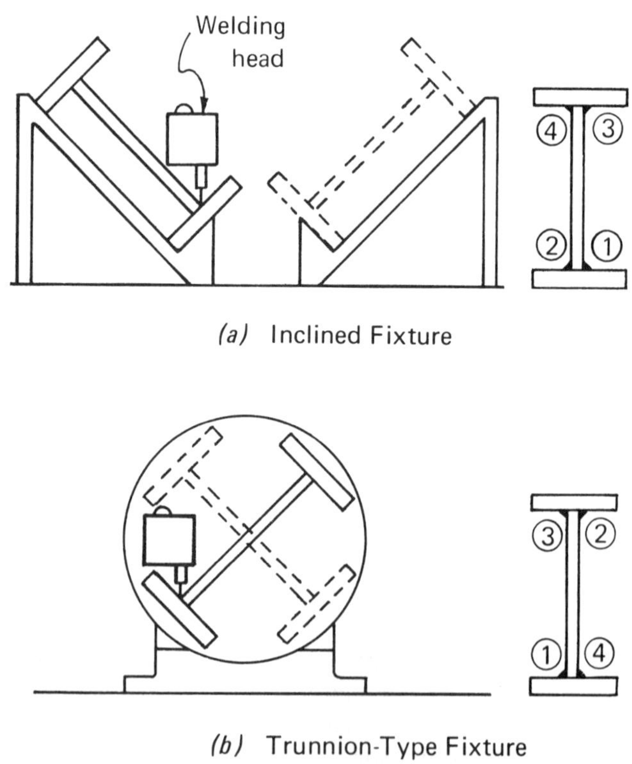

Fig. 3-25. The welding positions and sequences normally used when fabricating a girder, using an inclined or trunnion-type fixture with a single welding head.

This is desirable, since welding is faster. Flat positioning also permits better control of bead shape and allows depositing larger welds when necessary. Figure 3-25 shows the welding positions and sequences for the girder supported by an inclined fixture and by a trunnion-type fixture.

Using jigs to support the girder requires a crane to change position after each weld. Since repositioning takes time and, since it requires the use of equipment that may be serving several welding stations, it should be minimized as much as possible.

A typical setup, using a pair of jigs to support a girder for welding, is shown in Fig. 3-25(a). After weld 1 is made, the crane simply moves the girder to the facing jig where weld 2 is made. After weld 2, the crane rolls the girder over for welds 3 and 4.

A different welding sequence results with the use of a trunnion-type fixture, shown in Fig. 3-25(b). After weld 1 is made, the girder is rolled over 180 deg for weld 2 without changing the position of the welding head. Upon completion of weld 2, the girder is again rolled, and the welding head is repositioned for weld 3. After weld 3, the girder is rolled and weld 4 is made with the head following the same path as for weld 3.

In both cases, welding sequence is not influenced by considerations of distortion control but by the positioning that consumes the least amount of time.

Slender, Light-Gage Columns: Slender columns for lights, signs, or other mounting purposes are commonly fabricated by welding together two light-gage formed channels, Fig. 3-26(a). If the first weld cools before the second weld is made on the opposite side, some bowing results because the

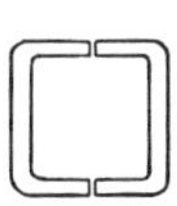

(a) Formed channel column.

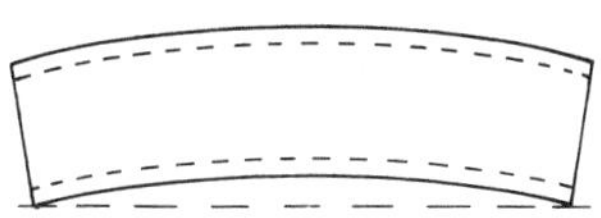

(b) During welding, top expands, center bows up

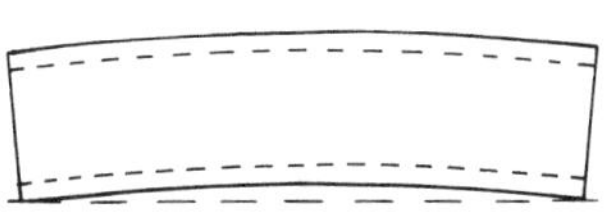

(c) Shortly after welding, still bowed up slightly. Second (bottom) weld is made now.

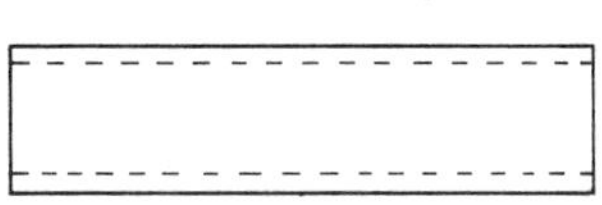

(d) After cooling to room temperature, member is straight.

Fig. 3-26. A technique for welding slender, light-gage columns — so that the end result is a straight column.

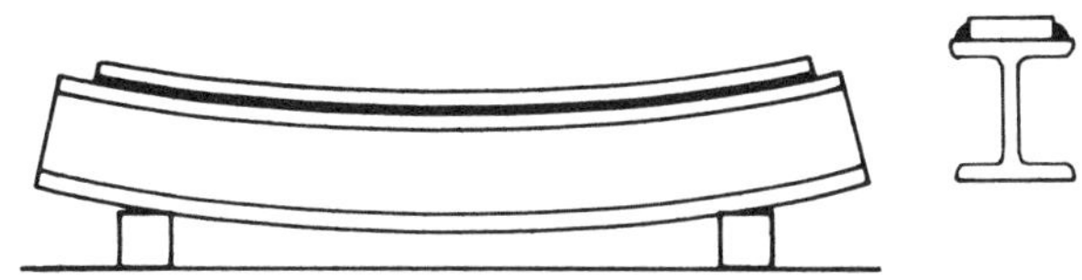

(a) Maximum camber

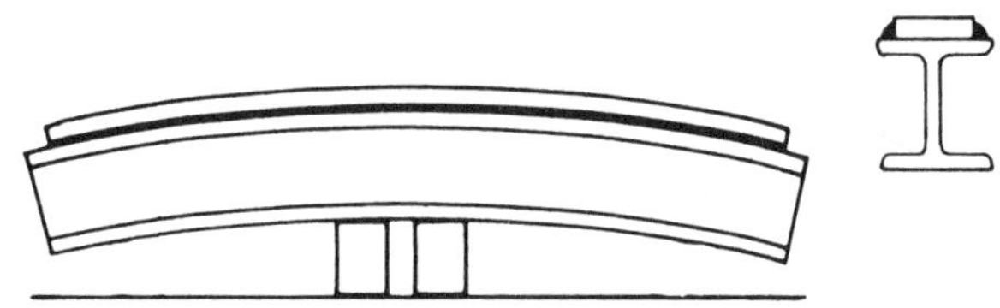

(b) Minimum camber

Fig. 3-27. Weld shrinkage and gravity can be used beneficially to produce and control the camber in a fabricated beam. The supporting arrangement maximizes or minimizes the effects of gravity.

second weld may not be capable of pulling the member straight. The first weld, on the top side of the member, initially causes the center of the column to bow upward, as in Fig. 3-26(b). If the member is turned over quickly and the second weld made immediately, the second weld will initially be slightly shorter than the length of the member. After the welded column cools, it will usually be straight.

Control of Camber: Weld shrinkage can be used beneficially in fabricating a long beam or girder to produce a desired camber. The effects of weld shrinkage and of gravity can be combined to control the amount of camber. The technique involves support of the assembly so that the two effects work in the desired direction — either together, to produce a camber, or in opposition, to produce a straight member or one with minimum camber. Thus, supporting the beam near its ends causes it to sag at the center, and supporting it at the center and letting the ends overhang produces a relatively straight beam.

For example, a beam is to have a cover plate welded to its lower flange and is to have a certain amount of camber (ends down). It is felt that the shrinkage from welding will not provide enough camber. Since the cover plate is narrower than the flange, it must be welded with the beam turned upside down. By supporting the beam near its ends during welding, as in Fig. 3-27(a), the sag supplements weld shrinkage and increases the camber.

If, on the other hand, too much bending will result from the welding, the beam can be supported near the center during welding, as in Fig. 3-27(b).

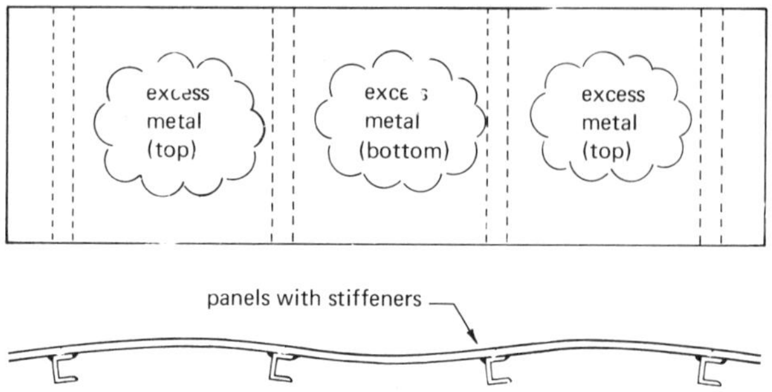

Fig. 3-28. The convex side of the panel is the one with excess metal and the side that requires flame-shrinking.

The effect of gravity will then be subtractive from the bending resulting from welding, thus reducing the camber of the finished beam.

Distortion Correction by Flame-Shrinkage: A high temperature is not required for flame-shrinking, but a large torch is necessary for rapid heat input. The success of the technique depends upon establishing a steep temperature gradient in the member. When a length of metal is to be flame-shrunk, such as the edge of a twisted panel or the flange of a beam, the torch can be moved progressively along the length, or selected spots can be heated and allowed to cool, with intermediate observation of the degree of distortion-removal achieved.

The convex side of a bent beam or buckled panel is obviously the one containing excessive metal and the side that requires shrinking. The buckled panels shown in Fig. 3-28, for example, have too much material in the central areas. If the part of an assembly to be shrunk is restrained — the case in many weldments — overheating could produce residual, or locked-in, stresses. It is good practice, to proceed cautiously in flame-shrinking, periodically allowing the members to cool so that the degree of distortion-removal can be checked.

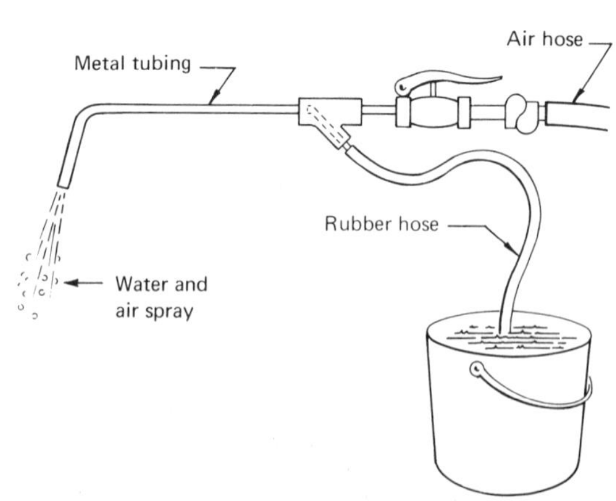

Fig. 3-29. Equipment for providing an atomized spray of water to accelerate cooling of metal when flame-shrinking.

To speed the operation, an atomized spray of water is sometimes used to accelerate cooling. This can be accomplished by simply inserting a Y fitting into a compressed-air line ahead of the valve and running a rubber hose from the fitting to a container of water (Fig. 3-29). When the valve is opened, the rush of air past the orifice in the fitting draws water into the air stream, creating an atomized spray. When the spray strikes the hot plate, it turns into steam, absorbing a substantial amount of heat in so doing. Cooling is rapid and, since all of the sprayed water is vaporized, the work remains dry.

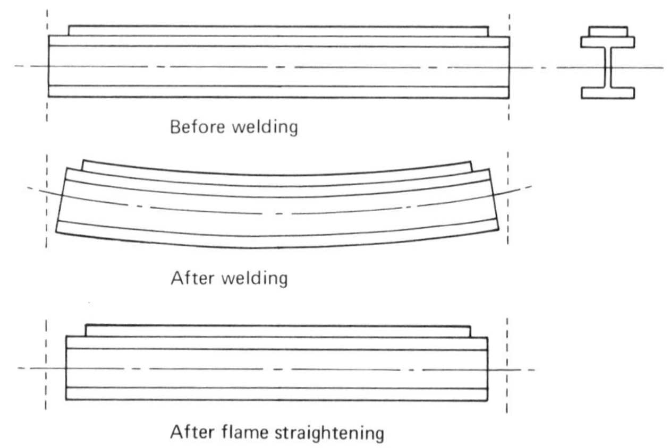

Fig. 3-30. A beam that has been bent when welding on the cover plate can be straightened by flame-shrinking the other flange.

Flame-shrinkage is often applied to long beams and columns that have been bent by welding. A beam with a welded cover plate is likely to bend, as shown in Fig. 3-30, because the welding is not balanced about the neutral axis. The welding on the cover plate produces shrinkage that shortens the length of the flange to which it is welded. Flame-shrinkage shortens the other flange to the same length, thus straightening the beam.

Flame-shrinkage, used properly, can also increase the desired amount of camber in a beam. For example, if the rolled beam with a welded cover plate shown in Fig. 3-30 requires more camber than that produced by the welding of the cover plate, flame is applied to the cover-plate side. However, if the cover plate alone is heated and shrunk, it will pull against the lower beam flange, resulting in a considerable amount of locked-in tensile stress. An accidental overload on the beam in service could

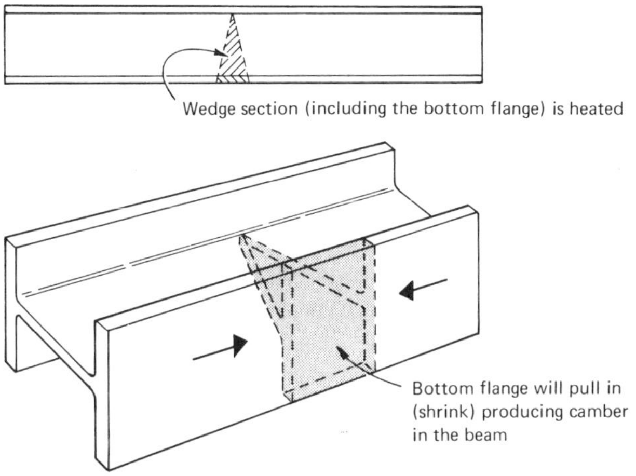

Fig. 3-31. Technique for developing camber in beams with or without cover plates so that little or no locked-in stress results.

exceed the yield stress of the cover plate and cause stretching and the loss of some of the camber. Thus, to minimize the tensile stress developed in the cover plate, so that more of the strength is used for resisting loads and maintaining camber, the beam flange should also be flame-shrunk, along with the cover plate.

Flame-shrinkage can be used to develop camber in beams with or without cover plates so that little or no locked-in stress results. A wedge-shaped area (Fig. 3-31) is first marked off on the web and lower flange with soapstone or other marking material that will withstand high temperature. The flame from one or more heating torches is then applied to the marked region until it reaches a red heat. The flame is removed, the area is allowed to cool, and the camber is measured. Similar areas are then marked off, and the process is repeated until the required camber is obtained. Two beams that were fabricated in normal fashion, then curved by flame-shrinking, are shown in Fig. 3-32.

Fig. 3-32. Two beams that were fabricated in normal fashion and then curved by flame-shrinkage.

SHOP TECHNIQUES FOR DISTORTION CONTROL

Various shop techniques have been developed to control distortion on specific weldments. All make use of the principles discussed relative to restraint, expansion, and contraction.

In sheet-metal welding, for example, a water-cooled jig (Fig. 3-33) is useful to carry heat away from the welded components. Copper tubes are brazed or soldered to copper holding clamps, and water is circulated through the tubes during welding. The restraint of the clamps also helps minimize distortion.

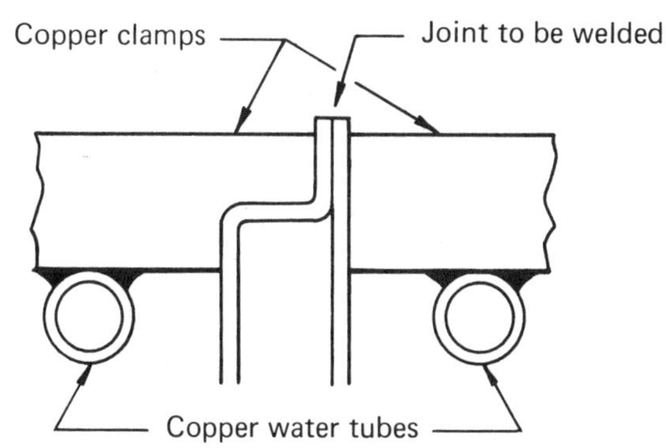

Fig. 3-33. A water-cooled jig for rapid removal of heat when welding sheet metal.

The "strongback" is another useful technique for distortion control during butt welding of plates, as in Fig. 3-34(a). Clips are welded to the edge of one plate and wedges are driven under the clips to force the edges into alignment and to hold them during welding.

In a variation of this method, shown in Fig. 3-34(b), a yoke is welded to form an inverted T section. The yoke is passed through a root-spaced joint from beneath the work. Steel wedges, driven into slots in the yoke, bring the plates into alignment. The thickness of the yoke can be the same as the width of the root opening, thereby serving additionally as a spacer between plate edges.

A yoke can be used in a different way on thicker plates, as in Fig. 3-34(c). Here, the yoke is welded to

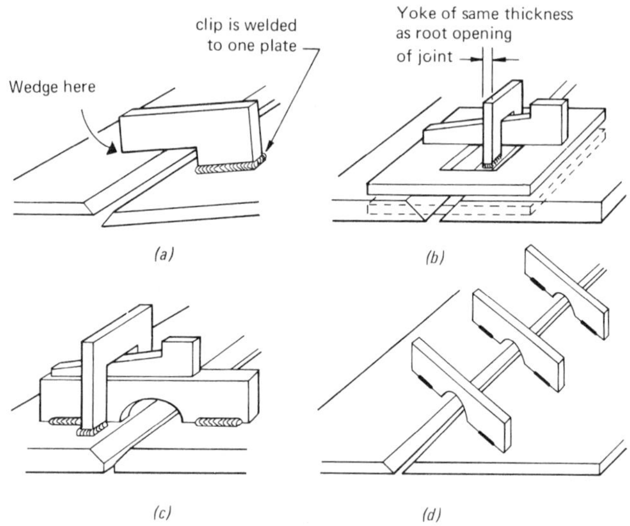

Fig. 3-34. Various strongback arrangements to control distortion during butt-welding.

the top of one plate, a bar is welded to the second plate, and a wedge is driven between the yoke and the bars. After the plates are aligned, the bar is tack-welded to the other plate, and the yoke and wedge are removed and used elsewhere on the joint. When welding is completed, the tack-welded bars are removed.

If there is concern that strongbacks may restrain the joint excessively against transverse movement as the weld shrinks, and thus increase the possibility of weld cracking, the strongbacks can be set at 45° to the joint, as in Fig. 3-34(d). This arrangement allows some transverse movement without loss of restraint in the longitudinal direction.

Welding of very heavy materials may require strongbacks that extend a considerable distance on each side of the joint. The arrangements shown in

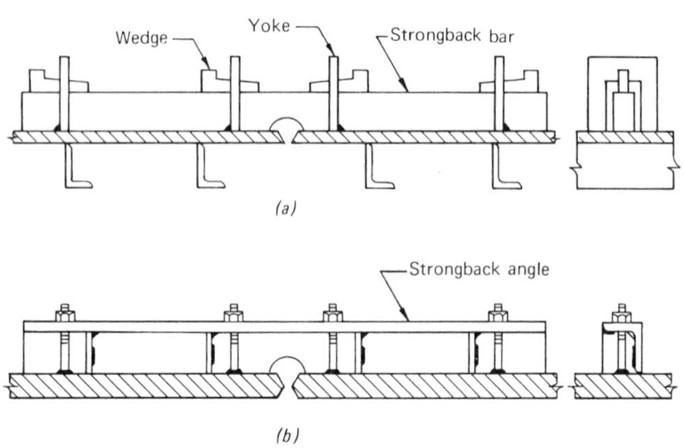

Fig. 3-35. Strongback arrangements that allow pressure to be applied at considerable distance from the joint.

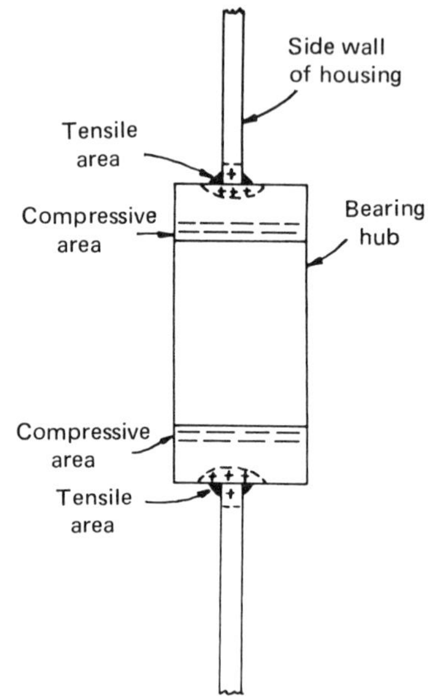

Fig. 3-36. Stress relief before machining would facilitate boring to a specified diameter this hub for a bearing support.

Fig. 3-35 allows pressure to be applied over long spans of material.

Stress Relief: Except in special situations, stress relief by heating is not used for correcting distortion. There are occasions, however, when stress relief is necessary to prevent further distortion from occurring before the weldment is finished.

Residual tensile stresses in the weld area can be of the order of the yield strength of the metal. Compressive stresses exist in other areas to balance the tensile stresses. After sufficient movement (distortion) has taken place to balance the stresses, there should be no further movement in the member. However, if some of the stressed material is subsequently machined out, a new unbalance of stress results. A corresponding movement of the member must then take place to rebalance tensile and compressive stresses. This new movement, or distortion, takes place gradually as machining progresses. To avoid this distortion, weldments are sometimes stress-relieved before machining operations.

An example of a weldment that required stress relief before machining is the hub for a bearing support shown in Fig. 3-36. The hub is welded into the sidewall of a housing member. The two large circumferential welds tend to shrink and assume a smaller circumference and diameter. The inner diameter of the hub resists this movement, and therefore is stressed in compression. Should the hub be bored out without a stress relief, much of the area in compression would be removed. Removal of the balancing compressive forces would allow the welds to shrink to a smaller diameter, and the hub would become smaller as machining progresses. Without preliminary stress relief, it would be necessary to machine out the hub with many light cuts, each successively less, to arrive at the final bore diameter for the bearing. With a stress-relief operation, boring would be straightforward, to the specified diameter.

CHECK LIST FOR MINIMIZING DISTORTION

Do not overweld.

Control the fitup.

Use intermittent welds where possible.

Use the smallest leg size permissible when fillet welding.

Use minimum root opening, included angle, and reinforcement.

Select joints that require minimal weld metal; for example, a double-V joint instead of a single-V joint.

Weld alternately on either side of the joint when possible with multiple-pass welds.

Use minimal weld passes.

Use high-deposition-rate processes.

Use high-speed welding methods — iron-powder-covered electrodes or mechanized welding.

Use methods that give deep penetration and thus reduce the amount of weld metal needed for the same strength and amount of heat input.

Use welding positioners to achieve the maximum amount of flat-position welding. The flat position permits the use of large-diameter electrodes and high-deposition-rate welding procedures.

Balance welds about the neutral axis of the member.

Distribute the welding heat as evenly as possible through a planned welding sequence and weldment positioning.

Weld toward the unrestrained part of the member.

Use clamps, fixtures, and strongbacks to maintain fitup and alignment.

Prebend the members or preset the joints to let shrinkage pull them back into alignment.

Weld those joints that contract most first.

Weld the more flexible sections first. They can be straightened, if necessary, before final assembly.

Sequence subassemblies and final assemblies so that the welds being made continually balance each other around the neutral axis of the section.

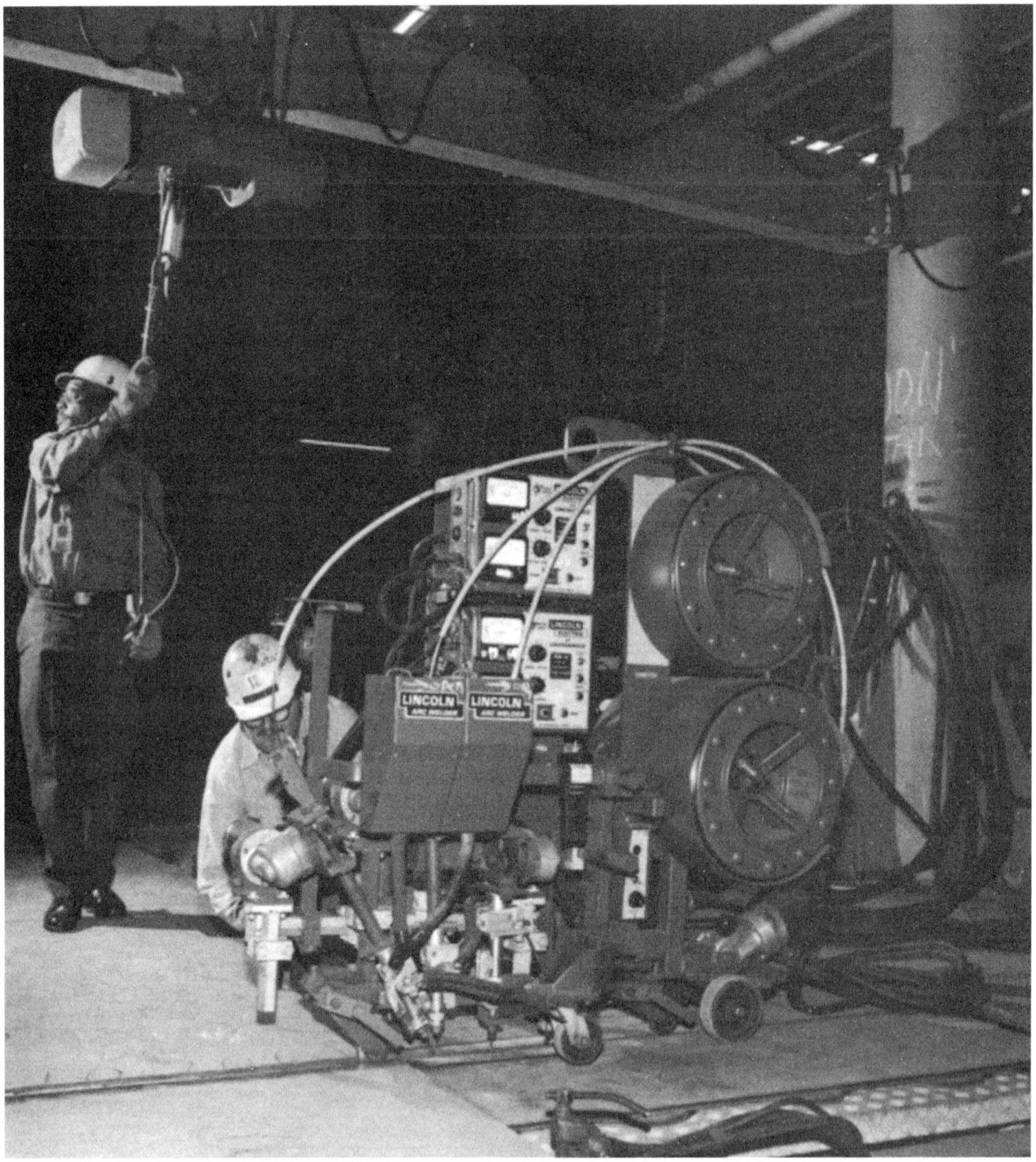

Splicing plate with a one-sided submerged arc welding tractor. The process minimizes distortion and eliminates the need to turn the plate over.

Arc Blow

Arc blow is a phenomenon encountered in DC arc welding when the arc stream does not follow the shortest path between the electrode and the work-piece but is deflected forward or backward from the direction of travel or, less frequently, to one side. Unless controlled, arc blow can be the cause of difficulties in handling the molten pool and slag, excessive spatter, incomplete fusion, reduced welding speed, porosity and lowered weld quality.

Back blow occurs when welding toward the ground connection, end of a joint, or into a corner. Forward blow is encountered when welding away from the ground or at the start of the joint. Forward blow can be especially troublesome with iron-powder or other electrodes that produce large slag coverings, where the effect is to drag the heavy slag or the crater forward and under the arc.

There are two types of arc blow of concern to the weldor. Their designations — magnetic and thermal — are indicative of their origins. Of the two, magnetic arc blow is the type which causes most welding problems.

MAGNETIC ARC BLOW

Magnetic arc blow is caused by an unbalanced condition in the magnetic field surrounding the arc. Unbalanced conditions result from the fact that at most times the arc will be farther from one end of the joint than another and will be at varying distances from the ground connection. Imbalance also always exists because of the change in direction of the current as it flows from the electrode, through the arc, and into and through the workpiece.

To understand arc blow, it is helpful to visualize a magnetic field. Figure 3-37 shows a direct current passing through a conductor, which could be an electrode or the ionized gas stream between an electrode and a weld joint. A magnetic field, or flux, is set up, around the conductor with lines of force that can be represented by concentric circles in planes at right angle to the direction of the current. These circular lines of force diminish in intensity with increasing distance from the electrical conductor.

They remain circular when they can stay in one medium, such as air or steel, which is expansive enough to contain them until they diminish to essentially nothing in intensity. But if the medium changes, for example from steel plate to air, the circular lines of forces are distorted; the forces tend to concentrate in the steel where they encounter less resistance. At a boundary between the edges of a steel plate and air, there is a concentration or squeezing of the magnetic flux lines, with deformation in the circular planes of force. This squeezing can result in a heavy concentration of flux behind or ahead of a welding arc. The arc then tends to move in the direction that would relieve the squeezing and tend to restore flux balance. It deflects away from the side of flux concentration and this deflection is the observed phenomenon of arc blow.

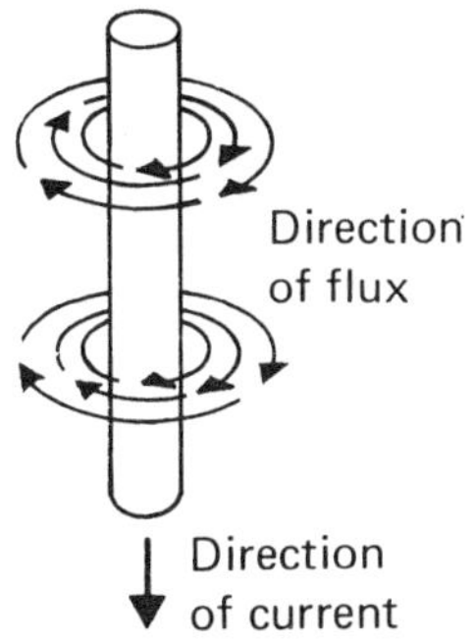

Fig. 3-37. A current through a conductor sets up a magnetic field that may be represented by planes of concentric circles — "flux lines."

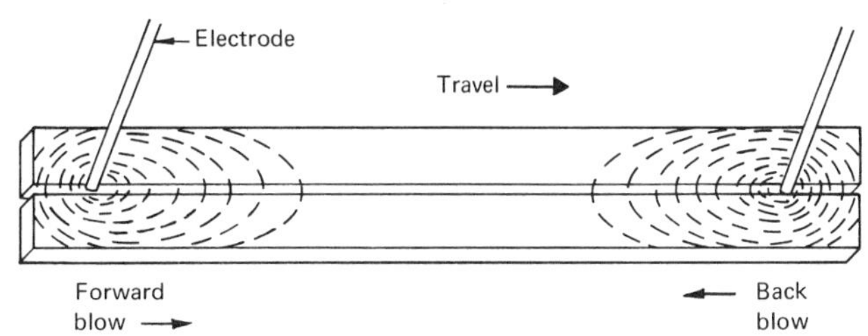

Fig. 3-38. Concentration of magnetic flux behind the arc at start of joint forces the arc forward. Flux concentration ahead of the arc at the end of the joint forces the arc backwards.

Figure 3-38 illustrates the squeezing and distortion of flux fields at the start and finish of a seam weld. At the start, the flux lines are concentrated behind the electrode. (One might say the flux lines balk at leaving the steel plate and moving out into the air.) The arc tries to compensate for this imbalance by moving forward, causing forward arc blow. As the electrode approaches the end of the seam, the squeezing is ahead of the arc, with a resultant movement of the arc backwards, and the development of back blow. At the middle of a seam in two members of the same width, the flux field would be symmetrical, and there would be no back or forward arc blow. If one member should be wide and the other narrow, however, side blow could occur at the midpoint of the weld.

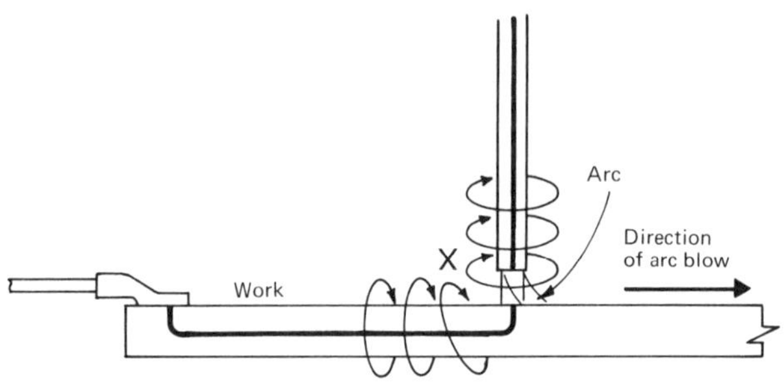

Fig. 3-39. Arc blow caused by ground effect. The magnetic flux set up by the ground current combines with the flux around the electrode causing a high flux concentration at (x) that blows the arc away from the ground connection.

The representations in Fig. 3-38 are only partly descriptive of what really happens. Another "squeezing" phenomenon also has effect on the observed arc blow. This secondary effect results from the ground current within the workpiece. As shown in Fig. 3-39, a magnetic flux is also set up by the electrical current passing through the workpiece to the ground. The heavy line represents the path of the welding current and the light lines the magnetic field set up by the current. As the current changes direction, or turns the corner from the arc to the work, a concentration of flux occurs at x, which causes the arc to blow, as indicated, away from the ground. This is called the "ground effect."

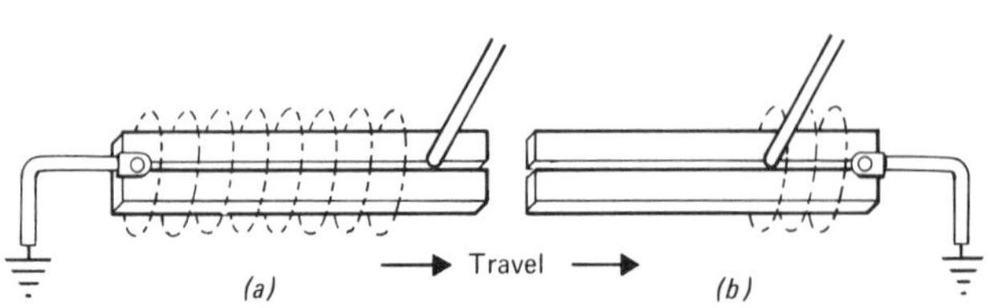

Fig. 3-40. Flux set up by ground current is behind the arc in (a), and ahead of the arc in (b).

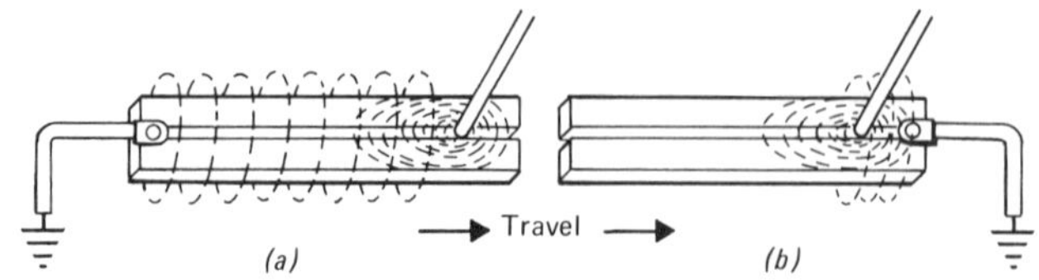

Fig. 3-41. Superimposed magnetic fields. Magnetic blow at the finish end of the joint (a) is reduced because the two flux fields tend to offset one another. At (b) the two fields are additive and cause a strong back blow.

The movement of the arc because of the ground effect will combine with the movement resulting from the concentration previously described to give the observed arc blow. Since the two movements are algebraically additive, ground effect may diminish or increase the arc blow caused by the magnetic flux of the arc. In fact, control of the ground effect is one way to control arc blow, which is especially useful with automatic welding processes.

In Fig. 3-40(a), the ground is connected to the starting end of the seam, and the flux resulting from the ground current in the work is behind the arc. The arc movement resulting from the ground effect would, thus, be forward. At the start of the weld, this would be additive to the arc movement shown in Fig. 3-38. Near the end of the seam, however, the forward movement from the ground effect would diminish the total arc blow by cancelling some of the back blow resulting from concentration of the flux from the arc at the end of the workpiece. Figure 3-41(a) is illustrative.

In Fig. 3-40(b), the ground is connected to the finish end of the seam, and the ground effect results in back blow. Here, it would increase the back blow of the arc flux at the finish of the weld. The combination of "squeezed" magnetic fluxes is illustrated in Fig. 3-41(b). A ground at the finish of the weld, however, may be what the weldor needs to reduce excessive forward blow at the start of the weld.

Because ground effect is less forceful than concentrations of arc-derived magnetic flux at the ends of workpieces, positioning of the ground connection is only moderately effective in controlling arc blow. Other measures must also be used to reduce the difficulties caused by arc blow when welding.

Arc blow is also encountered in corners and in deep V joints. Although analysis of these situations is complicated, the cause is exactly the same as when welding a straight seam — concentrations of lines of magnetic flux and the movement of the arc to relieve such concentrations. Figures 3-42 and

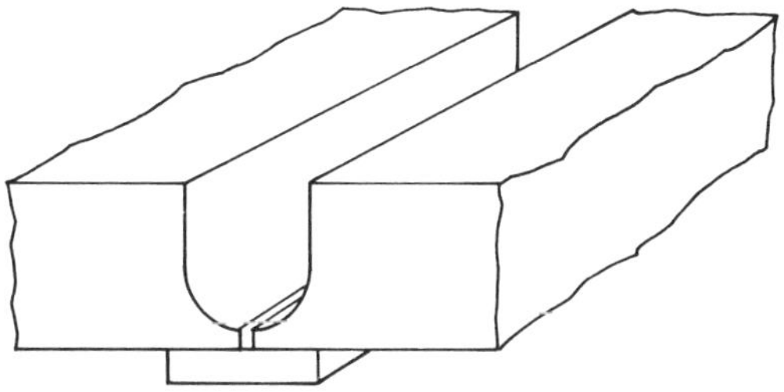

Fig. 3-42. Arc-blow problems are frequently encountered when welding with high DC in deep groove joints such as this. The use of AC may be an expedient solution.

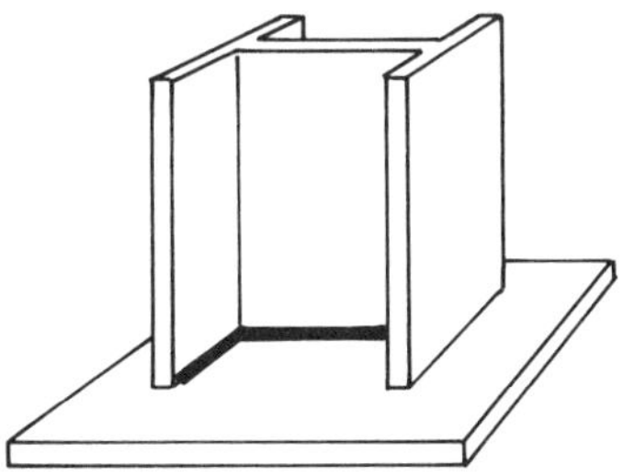

Fig. 3-43. Considerable arc blow can be expected when placing this "inside" fillet, using DC. AC may be the solution.

3-43 illustrate situations in which arc blow with direct current is likely to be a problem.

There is less arc blow with low current than with high current. This is because the intensity of the magnetic field a given distance from the conductor of electric current is proportional to the current. Usually, serious arc-blow problems do not occur when stick-electrode welding with DC up to about 250 amperes, but this is not an exact parameter since joint fitup and geometry could have major influence. With submerged-arc welding, still higher currents can often be used without creating arc-blow problems. The granular flux used with the submerged-arc process tends to dampen the arc-blow caused by magnetic fields.

The use of alternating current markedly reduces arc blow. The rapid reversal of the current induces eddy currents in the base metal, and the fields set up by the eddy currents greatly reduce the strength of the magnetic fields that cause arc blow.

THERMAL ARC BLOW

The physics of the electric arc requires a hot spot on both the electrode and plate to maintain a continuous flow of current in the arc stream. As the electrode is advanced along the work, the arc will tend to lag behind. This natural lag of the arc is caused by the reluctance of the arc to move to the colder plate. The space between the end of the electrode and the hot surface of the molten crater is ionized and, therefore, is a more conductive path than from the electrode to the colder plate. When the welding is done manually, the small amount of "thermal back blow" due to the arc lag is not detrimental, but it may become a problem with the higher speeds of automatic welding or when the thermal back blow is added to magnetic back blow.

ARC BLOW WITH MULTIPLE ARCS

When two arcs are close to each other, their magnetic fields react to cause arc blow on both arcs. Multiple arcs are often used to increase the welding speed of the submerged-arc process, and usually the arcs are less than one inch apart.

When two arcs are close and are of opposite polarities, as in Fig. 3-44(a), the magnetic fields

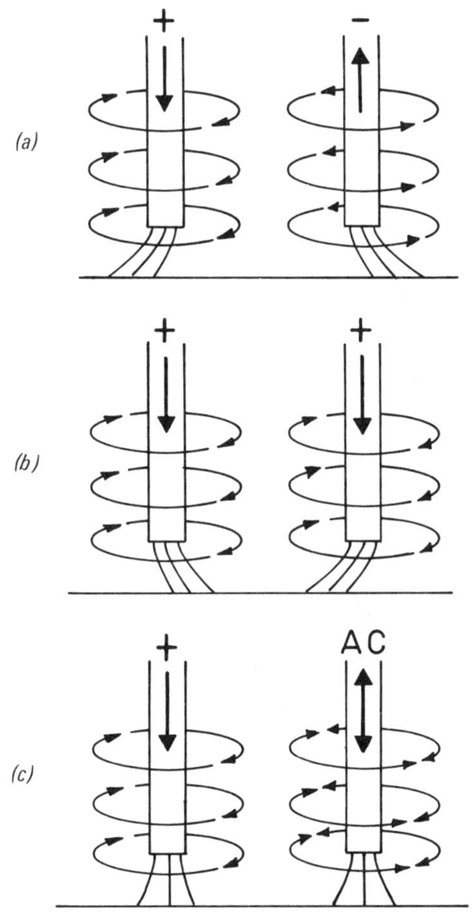

Fig. 3-44. Reactions of the magnetic fields when two arcs are close together: (a) arcs are of different polarities; the magnetic fields are additive and the arcs blow outward; (b) arcs are same polarity; magnetic fields oppose and the arcs blow inward; (c) arcs are DC and AC; little magnetic blow occurs.

between the arcs add to each other. The strong field between the arcs causes the arcs to blow away from each other.

If the arcs are the same polarity, as in Fig. 3-44(b), the magnetic fields between the arcs oppose each other. This results in a weaker field between the arcs, causing the arcs to blow toward each other.

Usually, when two arcs are used, one is DC and the other AC, as shown in Fig. 3-44(c). The flux field of the AC arc completely reverses for each cycle, and the effect on the DC field is small. Very little arc blow results.

Another commonly used arrangement is two AC arcs. Arc-blow interference here is avoided to a large extent by phase-shifting the current of one arc 80 to 90 degrees from the other arc. A so-called "Scott" connection accomplishes this automatically. With the phase shift, the current and magnetic fields of one arc reach a maximum when the current and magnetic fields of the other arc are at or near minimum. As a result, there is very little arc blow.

HOW TO REDUCE ARC BLOW

All arc blow is not detrimental. In fact, a small amount of arc blow can sometimes be used beneficially to help form the bead shape, control molten slag, and control penetration.

When arc blow is causing or contributing to such defects as undercut, inconsistent penetration, crooked beads, beads of irregular width, porosity, wavy beads, and excessive spatter, it must be controlled. Possible corrective measures have already been suggested in the preceding text. In general, here are some methods that might be considered:

- If direct current is being used with the shielded metal-arc process — especially at rates above 250 amperes — a change to alternating current may eliminate problems.
- Hold as short an arc as possible to help the arc force counteract the arc blow.

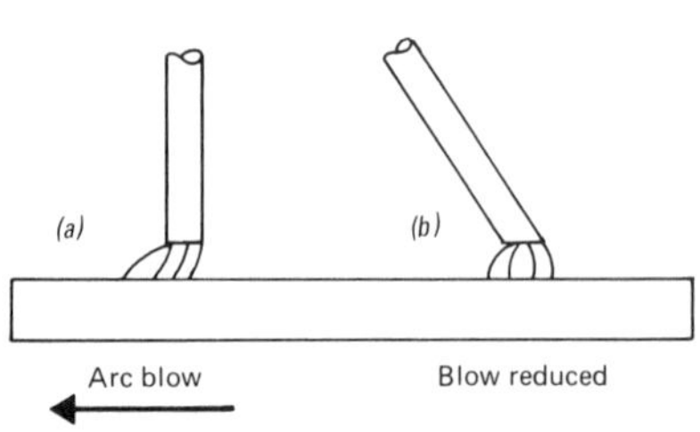

Fig. 3-45. Arc blow, as in (a), can sometimes be corrected by angling the electrode, as in (b).

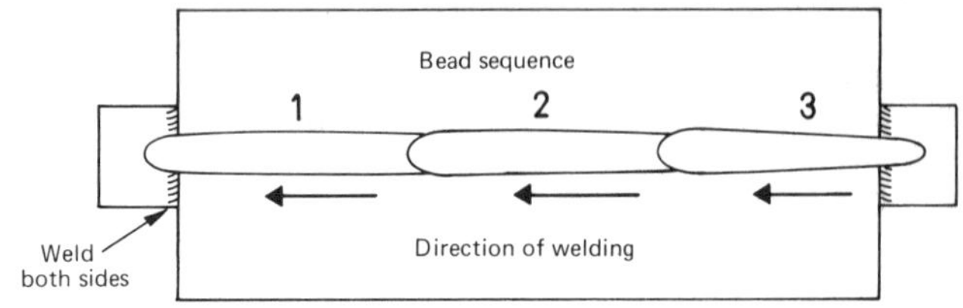

Fig. 3-46. Direction of welding and the sequence of beads for the back-step technique. Note tabs on both ends of the seam. Tabs should be the same thickness as the work.

- Reduce the welding current — which may require a reduction in arc speed.
- Angle the electrode with the work opposite the direction of arc blow, as illustrated in Fig. 3-45.
- Make a heavy tack weld on both ends of the seam; apply frequent tack welds along the seam, especially if the fitup is not tight.
- Weld toward a heavy tack or toward a weld already made.
- Use a back-step welding technique, as shown in Fig. 3-46.
- Weld away from the ground to reduce back blow; weld toward the ground to reduce forward blow.
- With processes where a heavy slag is involved, a small amount of back blow may be desirable; this may be achieved by welding toward the ground.
- Wrap ground cable around the workpiece and pass ground current through it in such a direction that the magnetic field set up will tend to neutralize the magnetic field causing the arc blow.

The direction of the arc blow can be observed with an open-arc process, but with the submerged-arc process must be determined by the type of weld defect.

Back blow is indicated by the following:

Spatter.
Undercut, either continuous or intermittent.
Narrow, high bead, usually with undercut.
An increase in penetration.
Surface porosity at the finish end of welds on sheet metal.

Forward blow is indicated by:

A wide bead, irregular in width.
Wavy bead.
Undercut, usually intermittent.
A decrease in penetration.

THE EFFECTS OF FIXTURING ON ARC BLOW

Steel fixtures for holding the workpieces may have an effect on the magnetic field around the arc and, thus, on arc blow. Usually, the fixturing causes no problem with stick-electrode welding when the current does not exceed 250 amperes. Fixtures for use with higher currents and with mechanized welding should be designed with precautions taken so that an arc-blow-promoting situation is not built into the fixture.

Each fixturing device may require special study to ascertain the best way to prevent the fixture from interfering deleteriously with the magnetic fields. The following are some points to note:

- Fixtures for welding the longitudinal seam of cylinders (Fig. 3-47) should be designed for a minimum of 1-in. clearance between the supporting beam and the work. The clamping fingers or bars that hold the work should be nonmagnetic. Do not attach the ground cable to the copper backup bar; ground directly to the work if possible.
- Fabricate the fixture from low-carbon steel. This is to prevent the buildup of permanent magnetism in the fixture.
- Welding toward the closed end of "horn type" fixtures reduces back blow.
- Design the fixture long enough so that end tabs can be used if necessary.
- The backup bar should be all copper, but relieved under the joint. If the bar is placed in a support, as shown in Fig. 3-48, the support should be a non-magnetic material.
- Provide for continuous or close clamping of parts to be seam-welded. Wide, intermittent clamping may cause seams to gap between clamping points, resulting in arc blow over the gaps.
- Do not build into the fixture large masses of steel on one side of the seam only. Counterbalance with a similar mass on the other side.

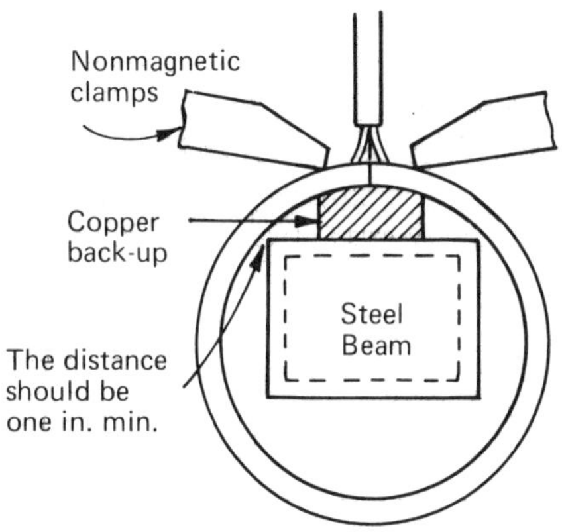

Fig. 3-47. Welding fixtures for clamping cylindrical work should have at least 1 in. of clearance between work and supporting beam. Clamps near the arc should be nonmagnetic.

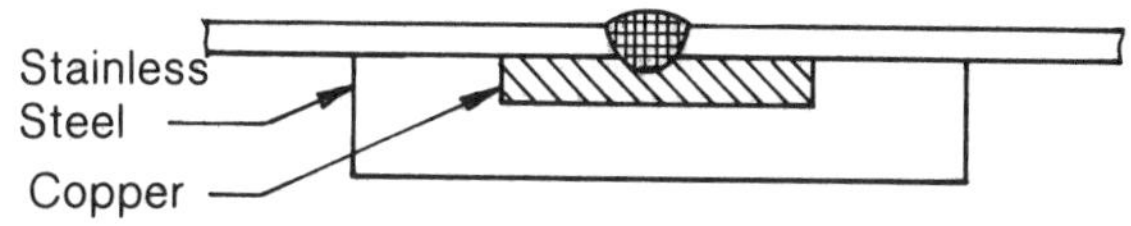

Fig. 3-48. The support for a copper backup bar should be made of a non-magnetic material such as stainless steel or brass where arc blow may be a problem.

An adequate heat source must be used to provide sufficient preheat on large diameter rolls.

Preheating and Stress Relieving

In some welding operations, it is necessary to apply heat to the assembly before starting the welding. In others, a postheat — or application of heat after welding — is needed to relieve the internal stresses that have been developed. With certain weldments, heat may also be applied between welding passes to maintain a required temperature. Each of these applications of heat has a bearing on the quality of weld or the integrity of the finished weldment, and, in code work, control of temperature before, during, and after welding may be rigidly specified.

PREHEATING — WHEN AND WHY

Preheating is used for one of the following reasons:

1. To reduce shrinkage stresses in the weld and adjacent base metal — especially important with highly restrained joints.
2. To provide a slower rate of cooling through the critical temperature range (about 1600°F to 1330°F), preventing excessive hardening and lowering ductility of both the weld and heat-affected area of the base plate.
3. To provide a slower rate of cooling through the 400°F range, allowing more time for any hydrogen that is present to diffuse away from the weld and adjacent plate to avoid underbead cracking.

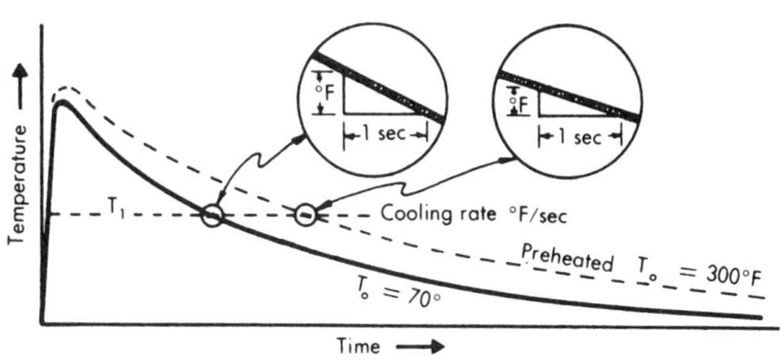

Fig. 3-49. A main purpose of preheat is to slow down the cooling rate. As the insets show, there is a greater temperature drop in one second at a given temperature (T1) when the initial temperature of the plate is 70°F than when the initial temperature is 300°F. In other words, the cooling rate (°F/sec) is slower when preheat is used.

As suggested by the above, a main purpose of preheat is to slow down the cooling rate — to allow more "Time at Temperature," as illustrated in Fig. 3-49. Thus, the amount of heat in the weld area as well as the temperature is important. A thick plate could be preheated to a specified temperature in a localized surface area and yet the heating be ineffective because of rapid heat transfer away from the heated area. Such preheating does not effectively reduce the cooling rate. Having a thin surface area at a preheat temperature is not enough if there is a mass of cold metal beneath it into which the heat can rapidly transfer.

Because of the heat-absorption capacity of a thick plate, the heat-affected zone and the weld metal after cooling may be in a highly quenched condition unless sufficient preheat is provided. What really matters is how long the weld metal and adjacent base metal is maintained in a certain temperature range during the cooling period. This, in turn depends on the amount of heat in the assembly and the heat transfer properties of the material and its configuration. Without adequate preheat, the cooling could be rapid and intolerably high hardness and brittleness could occur in the weld or adjacent area.

Welding at low ambient temperatures or on steel brought in from outside storage on cold winter days greatly increases the need for preheat. It is also true that preheating rids the joint of moisture, but preheating is usually not specified for that purpose.

The Amount of Preheat Required

The amount of preheat required for any application depends on such factors as base metal chemistry, plate thickness, restraint and rigidity of the members, and heat input of the process. Unfortunately, there is no method for metering the **amount** of heat put into an assembly by a preheat torch. The best shop approach for estimating the preheat input is a measure of the temperature at the welding area by temperature-indicating crayon marks or pellets. These give approximate measures

TABLE 3-3. Minimum Preheat and Interpass Temperature. (Degrees F) See also Table 6-21.

Welding Process				
	Shielded Metal-Arc Welding with other than Low-Hydrogen Electrode	Shielded Metal-Arc Welding with Low-Hydrogen Electrodes; Submerged Arc Welding; Gas Metal-Arc Welding; or Flux-Cored Arc Welding		Shielded Metal-Arc Welding with Low-Hydrogen Electrodes; Submerged-Arc Welding with Carbon or Alloy Steel Wire, Neutral Flux; Gas Metal-Arc Welding; or Flux-Cored Arc Welding
Thickness of Thickest Part at Point of Welding — Inches	ASTM A36[4], A53 Gr. B, A106, A131, A139, A375, A381 Gr. Y35, A500, A501, A516 Gr. 55 and 60, A524, A529, A570 Gr. D and E, A573 Gr. 65, API 5L Gr. B; ABS Gr. A, B, C, CS, D, E, R	ASTM A36, A106, A131, A139, A242 Weldable Grade, A375, A381 Gr. Y35, A441, A516 Gr. 65 and 70, A524, A529, A537 Class 1 and 2, A570 Gr. D and E, A572 Gr. 42, 45, 50, A573 Gr. 65, A588, A618, API 5L Gr. B and 5LX Gr. 42; ABS Gr. A, B, C, CS, D, E, R, AH, DH, EH	ASTM A572 Grades 55, 60 and 65	ASTM A514, A517
To 3/4, incl.	None[3]	None[3]	50	50
Over 3/4 to 1-1/2, incl.	150	50	150	125
Over 1-1/2 to 2-1/2, incl.	225	150	225	175
Over 2-1/2	300	225	300	225

[1] Welding shall not be done when the ambient temperature is lower than zero F. When the base metal is below the temperature listed for the welding process being used and the thickness of material being welded, it shall be preheated (except as otherwise provided) in such manner that the surfaces of the parts on which weld metal is being deposited are at or above the specified minimum temperature for a distance equal to the thickness of the part being welded, but not less than 3 in., both laterally and in advance of the welding. Preheat and interpass temperatures must be sufficient to prevent crack formation. Temperature above the minimum shown may be required for highly restrained welds. For quenched and tempered steel the maximum preheat and interpass temperature shall not exceed 400° F for thickness up to 1-1/2 in., inclusive, and 450°F for greater thicknesses. Heat input when welding quenched and tempered steel shall not exceed the steel producer's recommendation.

[2] In joints involving combinations of base metals, preheat shall be as specified for the higher strength steel being welded.

[3] When the base metal temperature is below 32°F, preheat the base metal to at least 70°F and maintain this minimum temperature during welding.

[4] Only low-hydrogen electrodes shall be used for welding A36 steel more than 1 inch thick for bridges.

of temperature at the spots where they are placed. These measurements are taken as indices of the heat input and should be correlated with thickness of metal and chemistry of metal in tables specifying minimum preheat temperatures. Thus, temperature is the gage to preheat inputs, and preheating to specified temperatures is the practical method of obtaining the correct amount of preheat needed to control the cooling rate after welding.

There are various guides which may be used to estimate preheat temperatures, including the recommendation of steel manufacturers. No guide, however, can be completely and universally applicable because of the varying factors of rigidity and restraint in assemblies. Recommendations are, thus, presented as "minimum preheat recommendations," and they should be accepted as such. However, the quenched and tempered steels can be damaged if the preheat is too high and the precautions necessary for these steels are discussed later.

The American Welding Society and the American Institute of Steel Construction have established minimum preheat and interpass temperature requirements for common weldable steels, as shown in Table 3-3 and 6-21. While material thickness, ranges of metal chemistry, and the welding process are taken into account in the minimum requirements, some adjustments may be needed for specific steel chemistry, welding heat input, joint geometry, and other factors.

Generally, the higher the carbon content of a steel, the lower the critical cooling rate and the greater the necessity for preheating and using low-hydrogen electrodes. The Lincoln Preheat and Interpass Temperature Calculator (Fig. 3-50), available from The Lincoln Electric Company, is a convenient tool to use in estimating preheats or adjusting recommended temperatures to specific carbon contents and low alloy additions.

Carbon, however, is not the only element that influences the critical cooling rate. Other elements in the steel are responsible for the hardening and loss of ductility that occur with rapid cooling. Total hardenability is thus a factor to be considered when

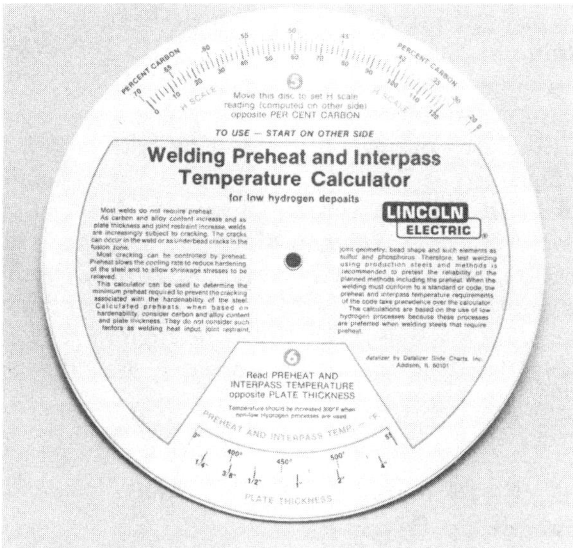

Fig. 3-50. The Lincoln Preheat and Interpass Temperature Calculator is a convenient tool for estimating preheats.

determining preheat requirements. Total hardenability can be expressed in terms of a "carbon equivalent", and this common measure of the effects of carbon and other alloying elements on hardening can be the basis for preheat and interpass temperature estimates.

Carbon equivalents (C_{eq}) are empirical values, determined by various carbon-equivalent formulas that represent the sum of the effects of various elements in steel on its hardenability. The International Institute of Welding (IIW) carbon equivalent equation is:

$$C_{eq} = \%C + \frac{\%Mn}{6} + \frac{\%Ni}{15} + \frac{\%Mo}{5} + \frac{\%Cr}{5} + \frac{\%Cu}{15} + \frac{\%V}{5}$$

This formula is valid only if the alloy contents are less than the following:

0.50% C	3.50% Ni	1.00% Cr
1.60% Mn	0.60% Mo	1.00% Cu

Approximate preheat and interpass temperatures, based on carbon-equivalent values for steels, are:

C_{eq} up to 0.45% .... preheat is optional
C_{eq} = 0.45 to 0.60%...... 200 to 400°F
C_{eq} over 0.60% .... 400 to 700°F

These temperatures are only approximate and are expressed in broad ranges. The carbon-equivalent method of arriving at a preheat range has utility largely when working with steels of unusual chemistries, when the alloy contents fall within the limits specified for the particular formula. Once the carbon-equivalent of such a steel has been determined, it can be correlated to a steel listed in Table 3-3 with a similar carbon-equivalent to judge the effects of plate thickness and the welding process in narrowing the preheat-temperature range.

Whatever the method used to estimate preheat temperature, the value obtained should be confirmed by welding tests on simulated or actual assemblies before it is committed to production welding. Only then can the effects of restraint and welding heat input be taken into account.

Theoretically, it is possible to reduce the preheat temperature requirement below the value listed in preheat tables when using welding currents in the high range of the procedures for semiautomatic and automatic processes. The justification for this is that the welding heat input is likely to be much higher than anticipated by the preheat recommendations. In such cases, heat losses from the assembly might more than be balanced by the welding heat input, bringing the affected metal up to or beyond the minimum preheat and interpass recommendations before it starts to cool.

The heat input during welding for a specific welding procedure is readily calculated by the formula:

$$J = \frac{E\,I\,60}{V\,1000} \qquad \text{.......... } 1$$

where:

J = Heat input in kilojoules/in. or kilowatt-sec/in.

E = Arc voltage in volts

I = Welding current in amperes

V = Arc speed in in./min.

Since all of the welding heat input at the arc does not enter the plate, the following heat inefficiencies are suggested for use with the formula:

75 — 80% for manual welding
90 — 100% for submerged-arc welding

Only after thorough analysis and test of the heat input, transfer, and loss factors should one deviate from recommended practices.

Methods of Preheating

The method of preheating depends on the thickness of the plate, the size of the weldment, and the heating equipment available. In the production welding of small assemblies, preheating in a furnace is the most satisfactory method. Another satisfactory method is torch heating, using natural gas premixed with compressed air. This produces a hot flame and burns clean. Torches can be connected to convenient gas and compressed-air outlets around the shop. Acetylene, propane and oil torches can also be used. On large weldments, banks of heating torches may be used to bring the material up to temperature quickly and uniformly.

Electrical strip heaters are used on longitudinal and girth seams on plate up to 2 in. thick. The heaters are clamped to the plate parallel to the joint and about 6 in. from the seam. After the plate reaches the proper preheat temperature, the heaters may remain in place to add heat if necessary to maintain the proper interpass temperature.

Other means of preheating are induction heating — often used on piping — and radiant heating.

High accuracy is not required in preheating carbon steels. Although it is important that the work be heated to a minimum temperature, no harm is done if this temperature is exceeded by 100°F. This is not true, however, for quenched and tempered steels, since welding on an overheated plate may cause damage in the heat-affected zone. For this reason the temperature should be measured as accurately as possible with such steels.

Temperature-indicating crayons and pellets are available for a wide range of temperatures. A crayon mark for a given temperature on the work will melt suddenly when the work reaches that temperature. Two crayon marks, one for the lower limit and one for the upper limit of temperature, show clearly when the work is heated to the desired temperature range.

Several types of portable pyrometers are available for measuring surface temperature. Properly used, these instruments are sufficiently accurate but must be periodically calibrated to insure reliability.

Thermocouples may be attached to the work and used to measure temperature. Thermocouples, of course, are the temperature-sensing devices in various types of ovens used for preheating small assemblies.

INTERPASS TEMPERATURES

Usually a steel that requires preheating to a specified temperature also must be kept at this temperature between weld passes. With many weldments, the heat input during welding is adequate to maintain the interpass temperature. On a massive weldment, it is not likely that the heat input of the welding process will be sufficient to maintain required interpass temperature. If this is the case, torch heating between passes may be required.

Once an assembly has been preheated and the welding begun, it is desirable to finish the welding as soon as possible so as to avoid the need for interpass heating.

Since the purpose of preheating is to reduce the quench rate, it logically follows that the same slow cooling should be accorded all passes. This can only be accomplished by maintaining an interpass temperature which is at least equal to the preheat temperature. If this is not done, each individual bead will be subjected to the same high quench rate as the first bead of a non-preheated assembly.

PREHEATS FOR QUENCHED AND TEMPERED STEELS

Since the low-alloy quenched and tempered steels are already in a heat-treated condition, any heating beyond a certain temperature will destroy the properties developed in them by the manufacturing process. Some assemblies must be preheated before welding to prevent cracking on rapid cooling, but the preheat must be controlled so as not to destroy throughout the mass of material the high yield strength and toughness that characterize these steels and give them special applications. Yet, during welding the heat-affected zone will be heated far above allowable preheat temperatures. This zone must then cool rapidly enough so as to re-establish the original properties and avoid a brittle structure. As a consequence, preheat temperatures and welding heat inputs must be closely controlled. Narrow limits are thus placed on the procedures.

Through research, welding procedures have been developed that assure high strength and good toughness, ductility, and impact properties in the welded joints. The recommended heat inputs and preheat temperatures are intended to allow sufficiently fast rates to avoid brittle structure. In general, this means a cooling rate of 6°F or more per second through the 900°F temperature range. The chemistry of these steels is such that the carbon equivalent is low enough to minimize the preheat.

In welding quenched and tempered steels, the proper low-hydrogen welding process is selected. Next, the required preheat temperature is determined, based upon the chemistry of the weld metal and plate thickness. Knowing the preheat temperature and the plate thickness, the maximum permissible welding heat input per pass can be found. A welding procedure is then selected that will stay below this maximum value. Welding heat input may be reduced

TABLE 3-4. Comparison Chart of Suggested Preheat Temperatures When Shielded Metal-Arc Welding Representative Quenched and Tempered Alloy Steels

Plate thickness, in.	Minimum preheat or interpass temperature for welding with low-hydrogen electrodes, °F						
	A533B Steel	A517 Steel	A542 Steel	A543 Steel	HY-130 Steel	Mod. A203D Steel	A553 Steel
To 1/2, incl.	50	50	150	100	75	50	50
Over 1/2 to 5/8, incl.	100	50	200	125	75	50	50
Over 5/8 to 3/4, incl.	100	50	200	125	125	50	50
Over 3/4 to 7/8, incl.	100	50	200	150	125	50	50
Over 7/8 to 1, Incl.	100	50	200	150	200	50	50
Over 1 to 1-3/8, incl.	200	150	250	200	200	150	150
Over 1-3/8 to 1-1/2, incl.	200	150	250	200	225	150	150
Over 1-1/2 to 2, incl.	200	150	250	200	225	150	150
Over 2 to 3, incl.	300	200	300	200	225	200	200
Over 3	300	200	300	200	225	200	200

Note: A preheat temperature above the minimum may be required for highly restrained welds. No welding should be done when ambient temperature is below 0°F. Welding of steel at an initial temperature below 100°F may require preheating to remove moisture from the surface of the steel. From ASM Metals Handbook, Volume 6, 8th Edition.

by decreasing the welding current or increasing the arc travel speed. Either change will decrease the amount of weld metal deposited per pass and will result in more passes being used for a given joint. For this reason, stringer beads are used extensively in welding quenched and tempered steels.

Tables 3-4, 3-5, 3-6, 3-7, 3-8, and 3-9 give recommended minimum preheat temperatures and maximum welding heat inputs for various quenched and tempered steels in various thicknesses. Kilojoules per inch of weld in the heat-input table is determined by the Formula (1) in this section. Similar data on other steels of this type are available from the steel producers.

Sometimes, the procedures most desirable from the economic standpoint in welding these steels will lead to a total heat input — preheat plus welding heat — that exceeds the steel manufacturer's recommendations. In such cases, one might question whether the weldment needs maximum notch toughness as well as high yield strength. If it does, the procedures should be modified to reduce the total heat input — not the preheat. Reducing preheat would be too risky, since such action might lead to weld cracking, and maximum toughness in the heat-affected zone would then be of no value. If maximum notch toughness is not required, total heat input limits can be exceeded somewhat without materially reducing the yield strength but there is little information for fatigue and impact properties.

POINTERS ON PREHEAT

- A cardinal rule when welding materials that require preheat is "keep it hot." It is costly to reheat to maintain assembly temperature.
- Preheat requirements can be reduced when running two automatic welding heads a few inches from each other — such as on each side of a web that is being fillet-welded to a flange. The heat input into the flange will be essentially double that resulting from a single head.

TABLE 3-5. Suggested Preheat Temperatures for ASTM A517, Grade B, F and H Steels

Plate thickness, in.	Minimum* preheat or interpass temperature, °F			
			Submerged-arc Process	
	Shielded metal-arc process	Gas metal-arc process	Alloy or carbon-steel wire, neutral flux	Carbon-steel wire, alloy flux
Up to 1/2, incl.	50†	50†	50†	50†
Over 1/2 to 1, incl.	50†	50†	50†	200
Over 1 to 2, incl.	150	150	150	300
Over 2	200	200	200	400

* A preheat temperature above the minimum shown may be required for highly restrained welds.

† Welding at any initial plate temperature below 100°F will require extreme care to minimize moisture on the steel being welded.

TABLE 3-6. Maximum Welding Heat Input in Kilojoules/Inch for Butt Joints in ASTM A533, Grade B Steel

Preheat and interpass temperature, °F	Plate thickness, in.				
	1/4	3/8	1/2	5/8	3/4
70	23.7	35.6	47.4	64.5	88.6
150	20.9	31.4	41.9	57.4	77.4
200	19.2	28.8	38.5	53.0	69.9
300	15.8	23.8	31.9	42.5	55.7
400	12.3	19.1	25.9	33.5	41.9

TABLE 3-7. Maximum Welding Heat Input in Kilojoules/Inch for Butt Joints in ASTM A517, Grade B and H Steels

Preheat and interpass temperature, °F	Plate thickness, in.							
	3/16	1/4	3/8	1/2	5/8	3/4	1	1-1/4 and over
70	17.5	23.7	35.0	47.4	64.5	88.6	Any	Any
150	15.3	20.9	30.7	41.9	57.4	77.4	120.0	Any
200	14.0	19.2	28.0	35.5	53.0	69.9	110.3	154.0
300	11.5	15.8	23.5	31.9	42.5	55.7	86.0	120.0
400	9.0	12.3	18.5	25.9	33.5	41.9	65.6	94.0

TABLE 3-8. Maximum Welding Heat Input in Kilojoules/Inch for Butt Joints in ASTM A517, Grade F Steel

Preheat and interpass temperature, °F	Plate thickness, in.							
	3/16	1/4	1/2	3/4	1	1-1/4	1-1/2	2
70	27.0	36.0	70.0	121.0	Any	Any	Any	Any
200	21.0	29.0	56.0	99.0	173.0	Any	Any	Any
300	17.0	24.0	47.0	82.0	126.0	175.0	Any	Any
400	13.0	19.0	40.0	65.0	93.0	127.0	165.0	Any

Note: Heat-input limits for temperatures and thicknesses included, but not shown, in this table may be obtained by interpolation; 25% higher heat inputs are allowable for fillet welds.

- Don't overlook the value of a preheat to prevent weld cracking in weldments with highly restrained joints — even though the chemistry of the steel does not call for a preheat.
- Heat flow from the joint is faster at a fillet weld that at a butt weld. Heat has three avenues for escape from a conventional fillet (tri-thermal heat flow); two from a conventional butt weld (bi-thermal heat flow). This is made clear by Fig. 3-51.
- Even though adequately preheating a thick section increases the fabrication cost, one experience with field repairs usually teaches that preheating is well worth the cost.
- Consider the use of lower alloy metal — even for highly restrained joints — to minimize the need for preheating.

TABLE 3-9. Suggested Welding Heat Input for Joints in HY-130 Steel

Plate thickness-in.	Heat input, Kilojoules	
	Shielded metal-arc process	Gas metal-arc process
3/8 to 5/8, incl.	40	35
Over 5/8 to 7/8, incl.	45	40
Over 7/8 to 1-3/8, incl.	45	45
Over 1-3/8 to 4, incl.	50	50

STRESS RELIEF

Stress relieving is defined as heating to a suitable temperature (for steel, below the critical); holding long enough to reduce residual stresses; and then cooling slowly enough to minimize the development of new residual stresses. Stress relieving should not be confused with normalizing or annealing, which are done at higher temperatures.

The ASME Code requires certain pressure vessels and power piping to be stress relieved. This is to reduce internal stress. Other weldments such as machine-tool bases are stress relieved to attain dimensional stability after machining.

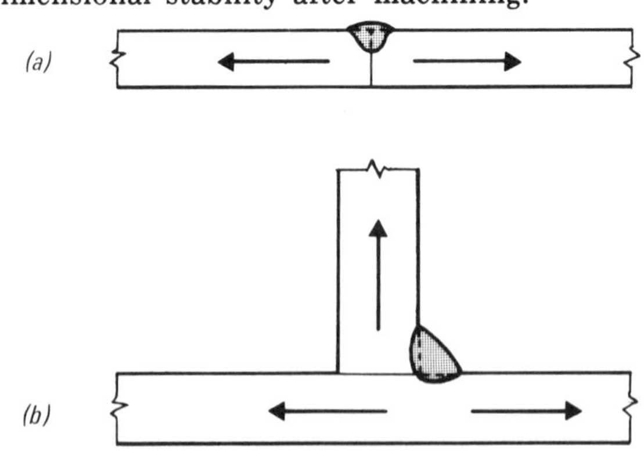

Fig. 3-51. Heat has two avenues of escape from a conventional butt weld (bi-thermal heat flow) and three avenues of escape from a conventional fillet-weld (tri-thermal heat flow).

Heating and cooling must be done slowly and uniformly; uneven cooling could nullify much of the value of the heat treatment or even cause additional stresses in the weldment. In general, the greater the difference between maximum and minimum thickness of the component parts, the slower should be the rate of temperature change. If the ratio of maximum to minimum thickness of the component parts is less than 4 to 1, heating and cooling rates should not exceed 400°F/hour divided by the thickness in inches of the thickest section. However, the heating and cooling rates should not exceed 400°F/hour. If the ratio of the thicknesses of the component parts vary more, rates should be reduced accordingly. For example, with complex structures containing members of widely varying thicknesses, heating and cooling rates should be such that the maximum temperature difference of sections of the same weldment should not exceed 75°F. Temperatures of critical sections can be monitored using thermocouples mounted on the weldment.

The stress-relief range for most carbon steels is 1100 to 1200°F, and the soaking time is usually one hour per inch of thickness.* For the low-alloy chrome-molybdenum steels, with the chromium in the range of 1/2 to 2-1/4% and the molybdenum up to 1%, the stress-relief range is 1250 to 1300°F for one hour. Some of the higher alloy steels require more soaking time. For example, E410, E502, and E505 weld metal is stress relieved at 1550 to 1600°F for two hours, and E430 at 1400 to 1450°F for four hours.

Local stress relieving can be done on girth joints of pipe and pressure vessels when required by codes. The same precautions are necessary as for furnace heating: slow heating, time-at-temperature, and slow cooling.

* AWS proposes to say "per inch of weld thickness when the specified stress-relief is for dimensional stability, the holding time shall be one hour per inch of thickness of the thicker part." However, sometimes ASME Sect. VIII refers also to the "thinner of two adjacent butt-welded plates."

For high production applications electrode can be furnished in 600 lb coils. Large coils eliminate many coil changes and reduce the down time needed for rethreading.

Section 4

CONSUMABLES AND MACHINERY

SECTION 4.1

ARC WELDING CONSUMABLES

Page

SECTION 4.2

POWER SOURCES

SECTION 4.3

WELDING EQUIPMENT

SECTION 4.4

FIXTURES AND MANIPULATORS

SECTION 4.5

ROBOT WELDING

Arc-Welding Consumables

Arc-welding consumables are the materials used up during welding, such as electrodes, filler rods, fluxes, and externally applied shielding gases. With the exception of the gases, all of the commonly used consumables are covered by AWS specifications.

Specifications in the AWS A5.x series prescribe the requirements for welding electrodes, rods, and fluxes. This section briefly reviews some of the important requirements of the A5.x series, with the intent of serving as a guide to the selection of the proper specification. When detailed information is required, the actual AWS specification should be consulted. These specifications are constantly being amended and updated. Care should be taken to obtain the latest publication.

ELECTRODES, RODS, AND FLUXES

The first specification for mild steel covered electrodes, A5.1, was written in 1940. As the welding industry expanded and the number of types of electrodes for welding steel increased, it became necessary to devise a system of electrode classification to avoid confusion. The system used applies to both the mild steel A5.1 and the low-alloy steel A5.5 specifications.

Classifications of mild and low-alloy steel electrodes are based on an "E" prefix and a four or five-digit number. The first two digits (or three, in a five-digit number) indicate the minimum required tensile strength in thousands of pounds per square inch. For example, 60 = 60,000 psi, 70 = 70,000 psi, and 100 = 100,000 psi. The next to the last digit indicates the welding position in which the electrode is capable of making satisfactory welds: 1 = all positions — flat, horizontal, vertical, and overhead; 2 = flat and horizontal fillet welding (see Table 4-1). The last two digits indicate the type of current to be used and the type of covering on the electrode (see Table 4-2).

Originally a color identification system was developed by the National Electrical Manufacturers Association (NEMA) in conjunction with the American Welding Society to identify the electrode's classification. This was a system of color markings applied in a specific relationship on the electrode. The colors

TABLE 4-1. AWS A5.1-91 and AWS A5.5-81 Designations for Manual Electrodes

a. The prefix "E" designates arc-welding electrode.

b. The first two digits of four-digit numbers and the first three digits of five-digit numbers indicate minimum tensile strength:

E60XX	60,000 psi Minimum Tensile Strength
E70XX	70,000 psi Minimum Tensile Strength
E110XX	110,000 psi Minimum Tensile Strength

c. The next-to-last digit indicates position:

EXX1X	All positions
EXX2X	Flat position and horizontal fillets

d. The suffix (Example: EXXXX-A1) indicates the approximate alloy in the weld deposit:

-A1	0.5% Mo
-B1	0.5% Cr, 0.5% Mo
-B2	1.25% Cr, 0.5% Mo
-B3	2.25% Cr, 1% Mo
-B4	2% Cr, 0.5% Mo
-B5	0.5% Cr, 1% Mo
-C1	2.5% Ni
-C2	3.25% Ni
-C3	1% Ni, 0.35% Mo, 0.15% Cr
-D1 and D2	0.25-0.45% Mo, 1.75 Mn
-G	0.5% min. Ni, 0.3% min. Cr, 0.2% min. Mo, 0.1% min. V, 1% min. Mn (only one element required)

TABLE 4-2. AWS A5.1-91 Electrode Designations for Covered Arc-Welding Electrodes

Designation	Current	Covering Type
EXX10	DC(+) only	Organic
EXX11	AC or DC(+)	Organic
EXX12	AC or DC(−)	Rutile
EXX13	AC or DC(±)	Rutile
EXX14	AC or DC(±)	Rutile, iron-powder (approx. 30%)
EXX15	DC(+) only	Low-hydrogen
EXX16	AC or DC(+)	Low-hydrogen
EXX18	AC or DC(+)	Low-hydrogen, iron-powder (approx. 25%)
EXX20	AC or DC(±)	High iron-oxide
EXX24	AC or DC(±)	Rutile, iron-powder (approx. 50%)
EXX27	AC or DC(±)	Mineral, iron-powder (approx. 50%)
EXX28	AC or DC(+)	Low-hydrogen, iron-powder (approx. 50%)

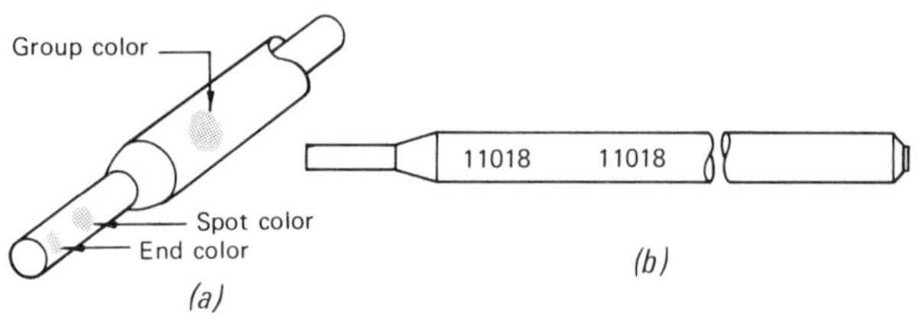

Fig. 4-1. (a) National Electrical Manufacturers Association color-code method to identify an electrode's classification; (b) American Welding Society imprint method.

TABLE 4-3. Color Identification for Covered MILD-STEEL and LOW-ALLOY Steel Electrodes

GROUP COLOR — NO COLOR				
XX10, XX11, XX14, XX24, XX27, XX28, and all 60 XX				
Spot Color \ End Color	**No Color**	**Blue**	**Black**	**Orange**
No Color	E6010	E7010G		EST
White	E6012	E7010-A1		ECI
Brown	E6013		E7014	
Green	E6020			
Blue	E6011	E7011G		
Yellow		E7011-A1	E7024	
Black			E7028	
Silver	E6027			
GROUP COLOR — SILVER				
All XX13 and XX20 except E6013 and E6020				
Brown				
White				
Green		E7020G		
Yellow		E7020-A1		

TABLE 4-4. AWS A5.1-91 Minimum Mechanical Property and Radiographic Requirements for Covered Arc-Welding Electrode Weld Metal

AWS Classification	Tensile Strength, min, psi	Yield Point, min, psi	Elongation in 2 in., min, percent	Radiographic Standard[1]	V-Notch Impact
E60 Series[2]					
E6010	60,000	48,000	22	Grade II	20 ft/lb at −20°F
E6011	60,000	48,000	22	Grade II	20 ft/lb at −20°F
E6012	60,000	48,000	17	Not required	Not required
E6013	60,000	48,000	17	Grade II	Not required
E6020	60,000	48,000	22	Grade I	Not required
E6022[4]	60,000	Not required	Not required	Not required	Not required
E6027	60,000	48,000	22	Grade II	20 ft/lb at −20°F
E70 Series[3]					
E7014			17	Grade II	Not required
E7015			22	Grade I	20 ft/lb at −20°F
E7016			22	Grade I	20 ft/lb at −20°F
E7018	70,000	58,000	22	Grade I	20 ft/lb at −20°F
E7024			17	Grade II	Not required[5]
E7024[5]			17		
E7028			22	Grade II	20 ft/lb at 0°F

[1] See AWS A5.1-91, Fig. 3

[2] For each increase of one percentage point in elongation over the minimum, the yield point or tensile strength, or both, may decrease 1,000 psi to a minimum of 60,000 psi for the tensile strength and 48,000 psi for the yield point for all classifications of the 60-series except E6012 and E6013. For the E6012 and E6013 classifications the yield point and tensile strength may decrease to a minimum of 65,000 psi for the tensile strength and 53,000 psi for the yield point.

[3] For each increase of one percentage point in elongation over the minimum, the yield point or tensile strength, or both, may decrease 1,000 psi to a minimum of 70,000 psi for the tensile strength and 58,000 psi for the yield point.

[4] A transverse tension test and a longitudinal guided bend test are required for classification of $^5/_{32}$, $^3/_{16}$, and $^7/_{32}$ in. (4.0, 4.8, and 5.6 mm) E6022 electrodes. Welding shall be done in the flat position.

[5] Upon agreement between supplier and purchaser, electrodes classified as E7024 may be supplied to a minimum Charpy V-notch impact requirement of 20 ft/lb lb at 0°F and a minimum elongation of 22 percent. Such electrodes shall be identified as E7024-1.

TABLE 4-5. AWS A5.1-91 Standard Covered Arc-Welding Electrode Sizes and Lengths

Core-Wire Diam. (in.)	Standard Lengths (in.)	
	E6010, E6011, E6012, E6013, E7014, E7015, E7016, E7018	E6020 E7024 E6027 E7028
1/16	9	—
5/64	9 or 12	—
3/32	12 or 14	12
1/8	14	14
5/32	14	14
3/16	14	14 or 18
7/32	14 or 18	18
1/4	18	18
5/16	18	18

and their significance are listed in Table 4-3. The NEMA specification also included the choice of imprinting the classification number on the electrode, as in Fig. 4-1.

Starting in 1964, AWS new and revised specifications for covered electrodes required the classification number be imprinted on the covering, as in Fig. 4-1. However, some electrodes can be manufactured faster than the imprinting equipment can mark them and some sizes are too small to be legibly marked with an imprint. Although AWS specifies an imprint, the color code is accepted on electrodes if imprinting is not practical.

Mild Steel Covered Arc-Welding Electrodes, AWS A5.1-91

The scope of this specification prescribes requirements for covered mild steel electrodes for shielded metal-arc welding of carbon and low-alloy steels.

The minimum mechanical property requirements are shown in Table 4-4. Radiographic standard Grade I has less and smaller porosity than Grade II. The actual standards are not contained herein, and, if a comparison is required, the standard in AWS A5.1-91 should be used.

Standard electrode sizes and lengths are given in Table 4-5. Not all classifications, however, are manufactured in all sizes.

Low-Alloy Steel Covered Arc-Welding Electrodes, AWS A5.5-81

This specification prescribes covered electrodes for shielded metal-arc welding of low-alloy steel.

The same classification system is used as for mild steel covered electrodes, with an added suffix that indicates the approximate chemistry of the deposited weld metal (see Table 4-1).

The chemical composition of the deposited weld metal is shown in Table 4-6. The electrodes with the suffix "G" need have only one alloy above the minimum to qualify for the chemical requirements. Electrodes with the suffix "M" will meet or be similar to certain military requirements.

Table 4-7 shows the tensile-strength, yield-strength, and elongation requirements. The preheat, interpass-temperature, and postheat treatments are not the same for all electrodes. For this reason, the complete AWS A5.5-81 specification should be consulted before conducting any tests.

Radiographic requirements are shown in Table 4-8. Grade I has fewer and smaller porosity than Grade II. The radiographic standards can be found in the specification.

Table 4-9 shows the impact requirements. The impact test specimens receive the same heat treatment as the tension test specimens.

Bare Mild Steel and Low Alloy Electrodes and Fluxes for Submerged-Arc Welding, AWS A5.17-89 and A5.23-90

Since the electrode and flux are two separate consumable items, they are classified separately. Electrodes are classified on the basis of chemical composition, as shown in Table 4-10a and b. In the classifying system, the letter "E" indicates an electrode, as in the other classifying systems, but here the similarity stops. The next letter "L," "M," or "H," indicates low, medium, or high-manganese, respectively. The following number or numbers indicate the approximate carbon content in hundredths of one percent. If there is a suffix "K," this indicates a silicon-killed steel.

Table 4-11 gives the standard electrode sizes and tolerances.

Fluxes are classified on the basis of the mechanical properties of the weld deposit made with a particular electrode (see Table 4-12). The classification designation given to a flux wire combination consists of a prefix "F" (indicating a flux) followed by a series of numbers and letters. The significance of the numbering system is explained in Fig. 4.2.

Mild Steel Electrodes for Flux-Cored Arc-Welding, AWS A5.20-79

This specification prescribes requirements for mild steel composite electrodes for flux-cored arc

TABLE 4-6. Composition Requirements of Low-Alloy Weld Metal AWS A5.5-81

AWS Classification[1]	Composition (%)[2]								
	C	Mn	P	S	Si	Ni	Cr	Mo	V
Carbon-Molybdenum Steel Electrodes									
E7010-A1 E7011-A1 E7015-A1 E7016-A1 E7018-A1 E7020-A1 E7027-A1	0.12	0.60 0.60 0.90 0.90 0.90 0.60 1.00	0.03	0.04	0.40 0.40 0.60 0.60 0.80 0.40 0.40	—	—	0.40 to 0.65	—
Chromium-Molybdenum Steel Electrodes									
E8016-B1 E8018-B1	0.05 to 0.12	0.90	0.03	0.04	0.60 0.80	—	0.40 to 0.65	0.40 to 0.65	—
E8015-B2L	0.05	0.90	0.03	0.04	1.00	—	1.00 to 1.50	0.40 to 0.65	—
E8016-B2 E8018-B2	0.05 to 0.12	0.90	0.03	0.04	0.60 0.80	—	1.00 to 1.50	0.40 to 0.65	—
E8018-B2L	0.05	0.90	0.03	0.04	0.80	—	1.00 to 1.50	0.40 to 0.65	—
E9015-B3L	0.05	0.90	0.03	0.04	1.00	—	2.00 to 2.50	0.90 to 1.20	—
E9015-B3 E9016-B3 E9018-B3	0.05 to 0.12	0.90	0.03	0.04	0.60 0.60 0.80	—	2.00 to 2.50	0.90 to 1.20	—
E9018-B3L	0.05	0.90	0.03	0.04	0.80	—	2.00 to 2.50	0.90 to 1.20	—
E8015-B4L E8016-B5	0.05 0.07 to 0.15	0.90 0.40 to 0.70	0.03 0.03	0.04 0.04	1.00 0.30 to 0.60	— —	1.75 to 2.25 0.40 to 0.60	0.40 to 0.65 1.00 to 1.25	— 0.05
Nickel Steel Electrodes									
E8016-C1 E8018-C1	0.12	1.25	0.03	0.04	0.60 0.80	2.00 to 2.75	—	—	—
E7015-C1L E7016-C1L E7018-C1L	0.05	1.25	0.03	0.04	0.50	2.00 to 2.75	—	—	—
E8016-C2 E8018-C2	0.12	1.25	0.03	0.04	0.60 0.80	3.00 to 3.75	—	—	—
E7015-C2L E7016-C2L E7018-C2L	0.05	1.25	0.03	0.04	0.50	3.00 to 3.75	—	—	—
E8016-C3[3] E8018-C3[3]	0.12	0.40 to 1.25	0.03	0.03	0.80	0.80 to 1.10	0.15	0.35	0.05
Nickel-Molybdenum Steel Electrodes									
E8018-NM[4]	0.10	0.80-1.25	0.02	0.03	0.60	0.80-1.10	0.05	0.40-0.65	0.02
Manganese-Molybdenum Steel Electrodes									
E9015-D1 E9018-D1	0.12	1.25 to 1.75	0.03	0.04	0.60 0.80	—	—	0.25 to 0.45	—
E8016-D3 E8018-D3	0.12	1.00 to 1.75	0.03	0.04	0.60 0.80	—	—	0.40 to 0.65	—
E10015-D2 E10016-D2 E10018-D2	0.15	1.65 to 2.00	0.03	0.04	0.60 0.60 0.80	—	—	0.25 to 0.45	—
All Other Low-Alloy Steel Electrodes[5]									
EXX10-G[5] EXX11-G EXX13-G EXX15-G EXX16-G EXX18-G E7020-G	—	1.00 min[6]	—	—	0.80 min[6]	0.50 min[6]	0.30 min[6]	0.20 min[6]	0.10 min[6]
E9018-M[3] E10018-M[3] E11018-M[3] E12018-M[3] E12018-M1[3] E7018-W[7] E8018-W[8]	0.10 0.10 0.10 0.10 0.10 0.12 0.12	0.60 to 1.25 0.75 to 1.70 1.30 to 1.80 1.30 to 2.25 0.80 to 1.60 0.40 to 0.70 0.50 to 1.30	0.030 0.030 0.030 0.030 0.015 0.025 0.03	0.030 0.030 0.030 0.030 0.012 0.025 0.04	0.80 0.60 0.60 0.60 0.65 0.40 to 0.70 0.35 to 0.80	1.40 to 1.80 1.40 to 2.10 1.25 to 2.50 1.75 to 2.50 3.00 to 3.80 0.20 to 0.40 0.40 to 0.80	0.15 0.35 0.40 0.30 to 1.50 0.65 0.15 to 0.30 0.45 to 0.70	0.35 0.25 to 0.50 0.25 to 0.50 0.30 to 0.55 0.20 to 0.30 — —	0.05 0.05 0.05 0.05 0.05 0.08 —

NOTE: Single values shown are maximum percentages, except where otherwise specified.

1 The suffixes A1, B3, C2, etc. designate the chemical composition of the electrode classification.

2 For determining the chemical composition, DCEN (electrode negative) may be used where DC, both polarities, is specified.

3 These classifications are intended to conform to classifications covered by the military specifications for similar compositions.

4 Copper shall be 0.10% max and aluminum shall be 0.05% max for E8018-NM electrodes.

5 The letters "XX" used in the classification designations in this table stand for the various strength levels (70, 80, 90, 100, 110 and 120) of electrodes.

6 In order to meet the alloy requirements of the G group, the weld deposit need have the minimum, as specified in the table, of only one of the elements listed. Additional chemical requirements may be as agreed between supplier and purchaser.

7 Copper shall be 0.30 to 0.60% for E7018-W electrodes.

8 Copper shall be 0.30 to 0.75% for E8018-W electrodes.

TABLE 4-7. AWS A5.5-81 Tensile-Strength, Yield-Strength, and Elongation Requirements for All-Weld-Metal Tension Test[1]

AWS Classification	Tensile Strength, min, psi	Yield Strength at 0.2 percent offset, psi	Elongation in 2 in., min, percent
E7010-X E7011-X E7015-X E7016-X E7018-X E7020-X E7027-X	70,000	57,000	22 22 25 25 25 25 25
E8010-X E8011-X E8013-X E8015-X E8016-X E8018-X	80,000	67,000	19 19 16 19 19 19
E8016-C3 E8018-C3	80,000	68,000 to 80,000	24
E9010-X E9011-X E9013-X E9015-X E9016-X E9018-X	90,000	77,000	17 17 14 17 17 17
E9018-M	90,000	78,000 to 90,000	24
E10010-X E10011-X E10013-X E10015-X E10016-X E10018-X	100,000	87,000	16 16 13 16 16 16
E10018-M	100,000	88,000 to 100,000	20
E11015-X E11016-X E11018-X	110,000	97,000	15
E11018-M	110,000	98,000 to 110,000	20
E12015-X E12016-X E12018-X	120,000	107,000	14
E12018-M E12018-M1	120,000	108,000 to 120,000	18

[1] For the E8016-C3, E8018-C3, E9018-M, E10018-M, E11018-M, and E12018-M electrode classifications the values shown are for specimens tested in the as-welded condition. Specimens tested for all other electrodes are in the stress-relieved condition.

TABLE 4-8. AWS A5.5-81 Radiographic Requirements

AWS Classification	Radiographic Standard
EXX15-X EXX16-X EXX18-X E7020-X	Grade I
EXX10-X EXX11-X EXX13-X E7027-X	Grade II

TABLE 4-9.
AWS A5.5-81 Impact-Property Requirements

AWS Classification	Charpy V-Notch Impact Requirement, min[1]
E8018-NM E8016-C3 E8018-C3	20 ft/lb at −40°F
E8016-D3 E8018-D3 E9015-D1 E9018-D1 E10015-D2 E10016-D2 E10018-D2	20 ft/lb at −60°F[2]
E9018-M E10018-M E11018-M E12018-M	20 ft/lb at −60°F
E12018-M1	50 ft/lb at 0°F
E7018-W E8018-W	20 ft/lb at 0°F
E8016-C1 E8018-C1	20 ft/lb at −75°F[2]
E7015-C1L E7016-C1L E7018-C1L E8016-C2 E8018-C2	20 ft/lb at −100°F[2]
E7015-C2L E7016-C2L E7018-C2L	20 ft/lb at −150°F[2]
All other classifications	Not required

[1] The lowest value obtained, together with the highest value, shall be disregarded for this test.
[2] Stress-relieved impact properties.

welding of mild and low-alloy steels.

Electrodes are classified on the basis of single or multiple-pass operation, chemical composition of the deposited weld metal, mechanical properties, and whether or not carbon dioxide is required as a separate shielding gas. Table 4-13 and 4-14 show the minimum mechanical-property requirements.

Gas-shielded flux-cored electrodes are available for welding the low-alloy high-tensile steels. Self-shielded flux-cored electrodes are available for all-position welding, as in building construction. Fabricators using or anticipating using the flux-cored arc-welding processes should keep in touch with the electrode manufacturers for new or improved electrodes not included in the present specifications.

Mild Steel Electrodes for Gas Metal-Arc Welding, AWS A5.18-79

This specification prescribes requirements for mild steel solid electrodes for gas metal-arc welding of mild and low-alloy steel. The electrodes are classified on the basis of their chemical composition

TABLE 4-10. Chemical Composition Requirements for Solid Electrodes for Submerged-Arc Welding AWS A5.17-89 and A5.23-90

AWS Classi-fication	UNS Number[d]	Wt. Percent[a, b, c]												
		C	Mn	Si	S	P	Cr	Ni	Mo	Cu[e]	V	Al	Ti	Zr
EL8	K01008	0.10	0.25/0.60	0.07	0.030	0.030	—	—	—	0.35				
EL8K	K01009	0.10	0.25/0.60	0.10/0.25	0.030	0.030	—	—	—	0.35				
EL12	K01012	0.05/0.15	0.25/0.60	0.07	0.030	0.030	—	—	—	0.35				
EL12	K01112	0.06/0.15	0.80/1.25	0.10	0.030	0.030	—	—	—	0.35				
EL12K	K01113	0.05/0.15	0.80/1.25	0.10/0.35	0.030	0.030	—	—	—	0.35				
EL13K	K01313	0.07/0.19	0.90/1.40	0.35/0.75	0.030	0.030	—	—	—	0.35				
EL14K	K01314	0.06/0.19	0.90/1.40	0.35/0.75	0.025	0.025	—	—	—	0.35				
EL15K	K01515	0.10/0.20	0.80/1.25	0.10/0.35	0.030	0.030	—	—	—	0.35				
EL11K	K11140	0.07/0.15	1.40/1.85	0.80/1.15	0.030	0.030	—	—	—	0.35				
EL12K	K01213	0.06/0.15	1.50/2.00	0.25/0.65	0.025	0.025	—	—	—	0.35				
EL14	K11585	0.10/0.20	1.70/2.20	0.10	0.030	0.030	—	—	—	0.35		0.03/0.17		
EA1	K11222	0.07-0.17	0.65-1.00	0.20	0.030	0.025	—	—	0.45-0.65	0.35				
EA2	K11223	0.07-0.17	0.95-1.35	0.20	0.030	0.025	—	—	0.45-0.65	0.35				
EA3	K11423	0.07-0.17	1.65-2.20	0.20	0.030	0.025	—	—	0.45-0.65	0.35				
EA3K	K21451	0.07-0.12	1.60-2.10	0.50-0.08	0.025	0.025	—	—	0.40-0.60	0.35				
EA4	K11424	0.07-0.17	1.20-1.70	0.20	0.030	0.025	—	—	0.45-0.65	0.35				
EB1	K11043	0.10	0.40-0.80	0.05-0.30	0.025	0.025	0.40-0.75	—	0.45-0.65	0.35				
EB2	K11172	0.07-0.15	0.45-1.00	0.05-0.30	0.030	0.025	1.00-1.75	—	0.45-0.65	0.35				
EB2H	K23016	0.28-0.33	0.45-0.65	0.55-0.75	0.015	0.015	1.00-1.50	—	0.40-0.65	0.30	0.20-0.30			
EB3	K31115	0.05-0.15	0.40-0.80	0.05-0.30	0.025	0.025	2.25-3.00	—	0.90-1.10	0.35				
EB5	K12187	0.18-0.23	0.40-0.70	0.40-0.60	0.025	0.025	0.45-0.65	—	0.90-1.20	0.30				
EB6	S50280	0.10	0.35-0.70	0.05-0.50	0.025	0.025	4.50-6.50	—	0.45-0.70	0.35				
EB6H	S50180	0.25-0.40	0.75-1.00	0.25-0.50	0.030	0.025	4.80-6.00	—	0.45-0.65	0.35				
EB8	S50480	0.10	0.30-0.65	0.05-0.50	0.030	0.040	8.00-10.00	—	0.80-1.20	0.35				
ENi1	K11040	0.12	0.75-1.25	0.05-0.30	0.020	0.020	0.15	0.75-1.25	0.30	0.35				
ENi2	K21010	0.12	0.75-1.25	0.05-0.30	0.020	0.020	—	2.10-2.90	—	0.35				
ENi3	K31310	0.13	0.60-1.20	0.05-0.30	0.020	0.020	0.15	3.10-3.80	—	0.35				
ENi4	K11485	0.12-0.19	0.60-1.00	0.10-0.30	0.020	0.015	—	1.60-2.10	0.10-0.30	0.35				
ENi1K	K11058	0.12	0.80-1.40	0.40-0.80	0.020	0.020	—	0.75-1.25	—	0.35				
EF1	K11160	0.07-0.15	0.90-1.70	0.15-0.35	0.025	0.025	—	0.95-1.60	0.25-0.55	0.35				
EF2	K21450	0.10-0.18	1.70-2.40	0.20	0.025	0.025	—	0.40-0.80	0.40-0.65	0.35				
EF3	K21485	0.10-0.18	1.70-2.40	0.30	0.025	0.025	—	0.70-1.10	0.40-0.65	0.35				
EF4	K12048	0.16-0.23	0.60-0.90	0.15-0.35	0.035	0.025	0.40-0.60	0.40-0.80	0.15-0.30	0.35				
EF5	K41370	0.10-0.17	1.70-2.20	0.20	0.010	0.010	0.25-0.50	2.30-2.80	0.45-0.65	0.50				
EF6	K21135	0.07-0.15	1.45-1.90	0.10-0.30	0.015	0.015	0.20-0.55	1.75-2.25	0.40-0.65	0.35				
EM2[h]	K10882	0.10	1.25-1.80	0.20-0.60	0.010	0.010	0.30	1.40-2.10	0.25-0.55	0.25	0.05	0.10	0.10	0.10
EM3[h]	K21015	0.10	1.40-1.80	0.20-0.60	0.010	0.010	0.55	1.90-2.60	0.25-0.65	0.25	0.04	0.10	0.10	0.10
EM4[h]	K21030	0.10	1.40-1.80	0.20-0.60	0.010	0.010	0.60	2.00-2.80	0.30-0.65	0.25	0.03	0.10	0.10	0.10
EW	K11245	0.12	0.35-0.65	0.20-0.35	0.040	0.030	0.50-0.80	0.40-0.80	—	0.30-0.80				
EG	Not Specified													

NOTES:

a The filler metal shall be analyzed for the specific elements for which values are shown in this table. If the presence of other elements is indicated in the course of this work, the amount of those elements shall be determined to ensure that their total (excluding iron) does not exceed 0.50 percent.

b Single values are maximum.

c The letter "N" as a suffix to a classification indicates that the electrode is intended for welds in the core belt region of nuclear reactor vessels. This suffic changes the limits on the phosphorus, vanadium and copper, as follows:

P = 0.012% max. V = 0.05% max. Cu = 0.08% max.

"N" electrodes shall not be coated with copper or any material containing copper. The "EF5" and "EW" electrodes shall not be designated as "N" electrodes.

d SAE/ASTM Unified Numbering System for Metals and Alloys.

e The copper limit includes any copper coating that may be applied to the electrode.

f The first 11 electrodes listed are carbon steel wires, the remaining electrodes are low alloy steel wires.

g For the carbon steel electrodes, L = low Mn, M = medium Mn and H = high Mn.

h The composition ranges of classifications with the "EM" prefix for low alloy steels are intended to conform to the ranges for similar electrodes in the military specifications.

TABLE 4-11. AWS A5.17-89 Standard Sizes and Tolerances for Submerged-Arc Electrodes

Standard Electrode Size, dia., in.	Tolerance on Dia., in. ±
1/16 (0.063)	0.0015
5/64 (0.078), 3/32 (0.094)	0.002
1/8 (0.125)	0.003
5/32 (0.156), 3/16 (0.188), 7/32 (0.219), 1/4 (0.250), 5/16 (0.312), 3/8 (0.375)	0.004

and the as-welded mechanical properties of the deposited weld metal (see Tables 4-15 and 4-16). For the chemical-composition requirements of the deposited weld metal, see Table 4-17.

Stainless Steel Electrodes for Shielded Metal Arc Welding, AWS A5.4-92

These electrodes are commonly called the "stainless" or "corrosion-resisting" electrodes and are classified on the basis of the chemical composition of the deposited weld metal and usability characteristics.

Chemical composition requirements are shown in Table 4-18. The specification does not include tests for corrosion resistance. The deposited weld metal can be expected to have the same corrosion resistance as the base metal of the same composition. However, due to the heat of welding or subsequent heat treatment, metallurgical changes can occur that may affect the corrosion resistance of the weld and base metals. For this reason, corrosion tests should be made on critical applications.

Mechanical property requirements are shown in Table 4-19.

The usability of the electrodes is indicated by a suffix to the classification number in Table 4-18. A suffix "-15" indicates the electrode is to be used with DC electrode positive (DC+). If the suffix is "-16" the electrode can be used with AC or DC electrode positive (DC+).

TABLE 4-12A. AWS A5.17-89 Mechanical-Property Requirements for Submerged-Arc Flux Classification

Flux Classification[1]	Tensile Strength		Yield Strength, min.[2]		Elongation %, min[2]
	psi	MPa	psi	MPa	
F6XX-EXXX	60,000-80,000	415-550	48,000	330	22
F7XX-EXXX	70,000-95,000	480-650	58,000	400	22

[1] The letter "X" used in various places in the classification in this table stands for, respectively, the condition of heat treatment, the toughness of the weld metal, and the classification of the electrode (see Fig. 4-2).

[2] Yield strength at 0.2 percent offset and elongation in 2 in. (51 mm) gage length.

Indicates flux.

Indicates the minimum tensile strength (in increments of 10,000 psi [69 MPa]) of weld metal with the flux and some specific classification of electrode deposited according to the welding conditions specified herein (see Table 4-12A).

Designates the condition of heat treatment in which the tests were conducted: "A" for as-welded and "P" for postweld heat treated. The time and temperature of the PWHT are specified herein.

Indicates the lowest temperature at which the impact strength of the weld metal referred to above meets or exceeds 20 ft/lb (27 J). (See Table 4-12B.)

Indicates electrode.

F X X X - E X X X

Classification of the electrode used in depositing the weld metal referred to above.

Example

F7A6-EM12K is a complete designation. It refers to a flux that will produce weld metal which, in the as-welded condition, will have a tensile strength no lower than 70,000 psi and Charpy V-notch impact strength of at least 20 ft/lb at −60°F when deposited with an EM12K electrode under the conditions called for in AWS A5.17-89.

Fig. 4-2. AWS A5.17-89 Classification System for Submerged Arc Electrode and Flux.

Corrosion-Resisting Chromium and Chromium-Nickel Steel Welding Rods and Bare Electrodes, AWS A5.9-81

This specification covers corrosion-resisting chromium and chromium-nickel steel (stainless steel) welding rods for use with atomic hydrogen and gas tungsten-arc processes and bare electrodes for use with submerged-arc and gas metal-arc welding processes.

Rods and electrodes are classified on the basis of the chemical composition. The requirements for solid electrodes and rods are based on the chemical analysis of the filler metal as manufactured. For composite electrodes and rods, the requirements are based on the chemical analysis of a pad of undiluted metal made by melting the filler metal with the GTA process, using argon shielding gas. The analysis of composite electrodes and rods may also be made by any suitable method agreed upon by the purchaser and the supplier. Table 4-20 lists the chemical requirements.

Rods and electrodes are available in a wide variety of diameters and spool sizes. Rods are also available in straight lengths (see Table 4-21).

TABLE 4-12B. AWS A5.17-89 Impact Strength Requirements[1] for Submerged-Arc Flux Classification

Digit[2]	Test Temperature		Energy level, min.
	°F	°C	
Z	—	—	No impact requirement
0	0	−18	20 ft/lb (27 J)
2	−20	−29	
4	−40	−40	
5	−50	−46	
6	−60	−51	
8	−80	−62	

[1] Based on the results of the impact tests of the weld metal, the manufacturer shall insert in the classification (Table 4-12A) the appropriate digit from the table above (Table 4-12B), as indicated in Fig. 4-2. Weld metal from a specific flux-electrode combination that meets impact requirements at a given temperature also meets the requirements at all higher temperatures in this table (i.e., weld metal meeting the requirements for digit 5 also meets the requirements for digits 4, 2, 0, and Z).
[2] 4th Digit, (See Fig. 4-2).

Welding Rods and Covered Electrodes for Welding Cast Iron, AWS A5.15-90

This specification prescribes requirements for welding rods for oxyacetylene and carbon-arc welding and covered electrodes for shielded metal-arc welding of cast irons. These filler metals are suitable for welding gray cast iron, malleable iron, and some alloy cast irons. With the exception of the nickel-base alloys, classification is based on the chemical composition of the bare welding rod and the core wire of the covered electrodes. The chemical composition of the nickel-base alloys ENi-CI, ENiFe-CI, ENiCu-A, and ENiCu-B is based on the composition of the deposited weld metal (see Table 4-22).

Bare Aluminum and Aluminum-Alloy Welding Electrodes and Rods, AWS A5.10-92

This specification prescribes aluminum and aluminum alloy welding rods for use with (GTAW) welding and bare electrodes for use with GMAW welding.

Rods and electrodes are classified on the basis of the chemical composition of the as-manufactured filler metal (see Table 4-23). Electrodes must also meet a usability test. For electrodes 3/32-in. and smaller, a butt joint is welded in the overhead position. For 1/8-in. electrodes, the weld is made in the flat position. The welds are radiographed and must meet an X-ray standard available from AWS. The usability test for rods consists of making a bead on a plate in the flat position with a gas flame or GTA welding. The bead must be uniform in appearance and be free from specified discontinuities.

Copper and Copper Alloy Covered Arc-Welding Electrodes, AWS A5.6-84

This specification covers the requirements for covered copper and copper-alloy arc-welding electrodes for use with the shielded metal arc welding process.

Electrodes are classified on the basis of the chemical composition of deposits of undiluted weld metal for covered electrodes (see Table 4-24). The deposited weld metal must meet the tensile properties shown in Table 4-25.

Covered electrodes are available in sizes from 3/32-in. through 3/16-in.

Copper and Copper-Alloy Welding Rods, AWS A5.7-84

This specification covers copper and copper-alloy bare welding rods and electrodes for the oxyacetylene, gas metal arc, gas tungsten arc, and submerged arc welding processes. Rods are classified on the basis of the chemical composition of the rod and the mechanical properties of the welded joint.

The chemical requirements are shown in Table 4-26. The tensile-strength requirements are shown in Table 4-27. Strength is determined by a transverse test of a welded butt joint. Electrode in coil form is available in diameters from 1/8-in. to 1/4-in. and on spools from 0.020 in. to 0.094 in. Rod in straight lengths is available from 1/16-in. to 3/32-in.

TABLE 4-13. AWS A5.20-79 Mechanical Property Requirements for Flux-Cored Arc-Welding Weld Metal[1]

AWS Classification	Shielding Gas[2]	Current and Polarity	Tensile Strength min.[5], psi	Yield Strength at 0.2% Offset, min.[5], psi	Elongation in 2 inches, min.[5], psi
E60T-7	None	DC, electrode negative	67,000	55,000	22
E60T-8	None	DC electrode positive	62,000	50,000	22
E70T-1	CO_2		72,000	60,000	22
E70T-2			72,000	Not required	
E70T-3	None		72,000	Not required	
E70T-4	None		72,000	60,000	22
E70T-5[6]	CO_2 None		72,000	60,000	22
E70T-6	None		72,000	60,000	22
E70T-G	Not spec.	Not spec.	72,000[3]	Not required	
			72,000[4]	60,000	22[4]

[1] As-welded mechanical properties.
[2] Shielding gases are designated as follows:
CO_2 = carbon dioxide
None = no separate shielding gas
[3] Requirement for single-pass electrodes.
[4] Requirement for multiple-pass electrodes.
[5] For each increase of one percentage point in elongation over the minimum, the minimum required yield strength or the tensile strength, or both, may decrease 1000 psi, for a maximum reduction of 2000 psi in either the required minimum yield strength or the tensile strength, or both.
[6] Where CO_2 and None are indicated as the shielding gases for a given classification, chemical analysis pads and test assemblies shall be prepared using both CO_2 and no separate shielding gas.

TABLE 4-14. AWS A5.20-79 Mechanical Property Requirements in the As-Welded Condition[1]

AWS Classification[2]	Tensile Strength[3] ksi	Yield Strength[3] ksi	Percent Elongation in 2 in. (50 mm)[3]	V-Notch Impact
E6XT-1	62	50	22	20 ft-lb at 0°F
E6XT-4	62	50	22	No requirements
E6XT-5	62	50	22	20 ft-lb at −20°F
E6XT-6	62	50	22	20 ft-lb at −20°F
E6XT-7	62	50	22	No requirements
E6XT-8	62	50	22	20 ft-lb at −20°F
E6XT-11	62	50	22	No requirements
E6XT-G	62	50	22	No requirements
E6XT-GS	62	No requirements	No requirements	No requirements
E7XT-1	72	60	22	20 ft-lb at 0°F
E7XT-2	72	No requirements	No requirements	No requirements
E7XT-3	72	No requirements	No requirements	No requirements
E7XT-4	72	60	22	No requirements
E7XT-5	72	60	22	20 ft-lb at −20°F
E7XT-6	72	60	22	20 ft-lb at −20°F
E7XT-7	72	60	22	No requirements
E7XT-8	72	60	22	20 ft-lb at −20°F
E7XT-10	72	No requirements	No requirements	No requirements
E7XT-11	72	60	22	No requirements
E7XT-G	72	60	22	No requirements
E7XT-GS	72	No requirements	No requirements	No requirements

[1] Single values shown are minimums.
[2] Classifications E6XT-GS, E7XT-3, E7XT-2, E7XT-10, and E7XT-GS are intended for single-pass welding. Only transverse tension and longitudinal guided bend tests are required.
[3] For each increase of one percentage point in elongation over the minimum, the minimum required yield strength or the tensile strength or both may decrease 1000 psi for a maximum reduction of 2000 psi in either the required minimum yield strength or the tensile strength or both.

TABLE 4-15. AWS A5.18-79 Mechanical Property Requirements for Gas Metal-Arc Welding Weld Metal[1]

AWS Classification	Shielding Gas[2]	Tensile Strength[3 4] min., ksi	Yield Strength[3 4] at 0.2% Offset, min., ksi	Elongation in 2 in. min. %
ER70S-2 ER70S-3 ER70S-4 ER70S-5 ER70S-6 ER70S-7	CO_2	72	60	22
ER70S-G	As agreed between supplier and purchaser	72	60	22

[1] As-welded mechanical properties welded with electrode positive.
[2] CO_2 = carbon dioxide shielded gas. The use of CO_2 for classification purposes shall not be construed to restrict the use of Argon-CO_2 or Argon-O_2 shielding gas mixtures. A filler metal classified with CO_2 will also meet the requirements of this specification when used with Argon-CO_2 or Argon-O_2 mixtures.
[3] Mechanical properties as determined from an all-weld-metal tension test specimen.
[4] For each increase of one percentage point in elongation over the minimum, the yield strength, or tensile strength, or both may decrease 1000 psi to a minimum of 70,000 psi for tensile strength and 58,000 psi for the yield strength.

TABLE 4-16. AWS A5.18-79 Impact-Property Requirements for Gas Metal-Arc Welding Weld Metal

AWS Classification	Minimum required Impact Properties[1]
ER70S-2	20 ft-lb at −20°F
ER70S-3	20 ft-lb at 0°F
ER70S-4	Not Required
ER70S-5	Not Required
ER70S-6	20 ft-lb at −20°F
ER70S-7	20 ft-lb at −20°F
ER70S-G	As agreed between supplier and purchaser

[1] The lowest value obtained, together with the highest value obtained, shall be disregarded for this test. Two of the three remaining values shall be greater than the specified 20 ft-lb energy level; one of the three may be lower but shall not be less than 15 ft-lb. The computed average value of the three values shall be equal to or greater than the 20 ft-lb energy level.

TABLE 4-17. AWS A5.18-79 Chemical-Composition Requirements for Gas Metal-Arc Welding Electrode

AWS Classification[1]	Composition (%)												
	C	Mn	Si	P	S	Ni	Cr	Mo	V	Cu[2]	Ti	Zr	Al
ER70S-2	0.07	0.90 to 1.40	0.40 to 0.70								0.05 to 0.15	0.02 to 0.12	0.05 to 0.15
ER70S-3	0.06 to 0.15	0.90 to 1.40	0.45 to 0.70								—	—	—
ER70S-4	0.07 to 0.15	1.00 to 1.50	0.65 to 0.85	0.025	0.035	[3]	[3]	[3]	[3]	0.50	—	—	—
ER70S-5	0.07 to 0.19	0.90 to 1.40	0.30 to 0.60								—	—	0.50 — 0.90
ER70S-6	0.07 to 0.15	1.40 to 1.85	0.80 to 1.15								—	—	—
ER70S-7	0.07 to 0.15	1.50 to 2.00[4]	0.50 to 0.80								—	—	—
ER70S-G	No chemical requirements[5]												

NOTE: Single values shown are maximums.

[1] Electrodes classified E70S-1B in the previous edition of this specification are now classified ER80S-D2 in AWS Specification A5.28, Low Alloy Steel Filler Metals for Gas Shielded Arc Welding.
[2] The maximum weight percent of copper in the rod or electrode due to any coating plus the residual copper content in the steel shall be 0.50.
[3] These elements may be present but are not intentionally added.
[4] In this classification, the maximum Mn may exceed 2.0 percent. If it does, the maximum C must be reduced 0.01 percent for each 0.05 percent increase in Mn or part thereof.
[5] For this classification, there are no chemical requirements for the elements listed, with the exception that there shall be no intentional addition of Ni, Cr, Mo or V.

TABLE 4-18. AWS A5.4-81 Chemical Requirements for Stainless Covered Electrode All-Weld Metal[1] [2] (weight percent)

AWS Classification[3]	Composition (%)									
	C[4]	Cr	Ni	Mo	Cb plus Ta	Mn	Si	P	S	N	Cu
E209[3]	0.06	20.5-24.0	9.5-12.0	1.5-3.0	—	4.0-7.0	0.90	0.03	0.03	0.10-0.30	0.75
E219	0.06	19.0-21.5	5.5-7.0	0.75	—	8.0-10.0	1.00	0.03	0.03	0.10-0.30	0.75
E240	0.06	17.0-19.0	4.0-6.0	0.75	—	10.5-13.5	1.00	0.03	0.03	0.10-0.20	0.75
E307	0.04-0.14	18.0-21.5	9.0-10.7	0.5-1.5	—	3.3-4.75	0.90	0.04	0.03	—	0.75
E308	0.08	18.0-21.0	9.0-11.0	0.75	—	0.5-2.5	0.90	0.04	0.03	—	0.75
E308H	0.04-0.08	18.0-21.0	9.0-11.0	0.75	—	0.5-2.5	0.90	0.04	0.03	—	0.75
E308L	0.04	18.0-21.0	9.0-11.0	0.75	—	0.5-2.5	0.90	0.04	0.03	—	0.75
E308Mo	0.08	18.0-21.0	9.0-12.0	2.0-3.0	—	0.5-2.5	0.90	0.04	0.03	—	0.75
E308-MoL	0.04	18.0-21.0	9.0-12.0	2.0-3.0	—	0.5-2.5	0.90	0.04	0.03	—	0.75
E309	0.15	22.0-25.0	12.0-14.0	0.75	—	0.5-2.5	0.90	0.04	0.03	—	0.75
E309L	0.04	22.0-25.0	12.0-14.0	0.75	—	0.5-2.5	0.90	0.04	0.03	—	0.75
E309Cb	0.12	22.0-25.0	12.0-14.0	0.75	0.70-1.00	0.5-2.5	0.90	0.04	0.03	—	0.75
E309Mo	0.12	22.0-25.0	12.0-14.0	2.0-3.0	—	0.5-2.5	0.90	0.04	0.03	—	0.75
E310	0.08-0.20	25.0-28.0	20.0-22.5	0.75	—	1.0-2.5	0.75	0.03	0.03	—	0.75
E310H	0.35-0.45	25.0-28.0	20.0-22.5	0.75	—	1.0-2.5	0.75	0.03	0.03	—	0.75
E310Cb	0.12	25.0-28.0	20.0-22.0	0.75	0.70-1.00	1.0-2.5	0.75	0.03	0.03	—	0.75
E310Mo	0 12	25.0-28.0	20.0-22.0	2.0-3.0	—	1.0-2.5	0.75	0.03	0.03	—	0.75
E312	0.12	28.0-32.0	8.0-10.5	0.75	—	0.5-2.5	0.90	0.04	0.03	—	0.75
E316	0.08	17.0-20.0	11.0-14.0	2.0-3.0	—	0.5-2.5	0.90	0.04	0.03	—	0.75
E316H	0.04-0.08	17.0-20.0	11.0-14.0	2.0-3.0	—	0.5-2.5	0.90	0.04	0.03	—	0.75
E316L	0.04	17.0-20.0	11.0-14.0	2.0-3.0	—	0.5-2.5	0.90	0.04	0.03	—	0.75
E317	0.08	18.0-21.0	12.0-14.0	3.0-4.0		0.5-2.5	0.90	0.04	0.03	—	0.75
E317L	0.04	18.0-21.0	12.0-14.0	3.0-4.0		0.5-2.5	0.90	0.04	0.03	—	0.75
E318	0.08	17.0-20.0	11.0-14.0	2.0-2.5	6 × C, min to 1.00 max	0.5-2.5	0.90	0.04	0.03	—	0.75
E320	0.07	19.0-21.0	32.0-36.0	2.0-3.0	8 × C, min to 1.00 max	0.5-2.5	0.60	0.04	0.03	—	3.0-4.0
E320LR	0.035	19.0-21.0	32.0-36.0	2.0-3.0	8 × C, min to 0.40 max	1.50-2.50	0.30	0.020	0.015	—	3.0-4.0
E330	0.18-0.25	14.0-17.0	33.0-37.0	0.75	—	1.0-2.5	0.90	0.04	0.03	—	0.75
E330H	0.35-0.45	14.0-17.0	33.0-37.0	0.75	—	1.0-2.5	0.90	0.04	0.03	—	0.75
E347	0.08	18.0-21.0	9.0-11.0	0.75	8 × C, min to 1.00 max	0.5-2.5	0.90	0.04	0.03		0.75
E349[5] [6] [7]	0.13	18.0-21.0	8.0-10.0	0.35-0.65	0.75-1.2	0.5-2.5	0.90	0.04	0.03	—	0.75
E410	0.12	11.0-13.5	0.60	0.75	—	1.0	0.90	0.04	0.03	—	0.75
E410NiMo	0.06	11.0-12.5	4.0-5.0	0.40-0.70	—	1.0	0.90	0.04	0.03	—	0.75
E430	0.10	15.0-18.0	0.60	0.75	—	1.0	0.90	0.04	0.03	—	0.75
E502	0.10	4.0-6.0	0.40	0.45-0.65	—	1.0	0.90	0.04	0.03	—	0.75
E505	0.10	8.0-10.5	0.40	0.85-1.20	—	1.0	0.90	0.04	0.03	—	0.75
E630	0.05	16.0-16.75	4.5-5.0	0.75	0.15-0.30	0.25-0.75	0.75	0.04	0.03	—	3.25-4.00
E16-8-2	0.10	14.5-16.5	7.5-9.5	1.0-2.0	—	0.5-2.5	0.60	0.03	0.03	—	0.75
E7Cr	0.10	6.0-8.0	0.40	0.45-0.65	—	1.0	0.90	0.04	0.03	—	0.75

[1] Analysis shall be made for the elements for which specific values are shown in the table. If, however, the presence of other elements is indicated in the course of routine analysis, further analysis shall be made to determine that the total of these other elements, except iron, is not present in excess of 0.05 percent.
[2] Single values shown are maximum percentages except where otherwise specified.
[3] Suffix-15 electrodes are classified with direct current, electrode positive. Suffix-16 electrodes are classified with alternating current and direct current, electrode positive. Electrodes up to and including 5/32 in. (4.0 mm) in size are usable in all positions. Electrodes 3/16 in. (4.8 mm) and larger are usable only in the flat and horizontal-fillet positions.
[4] Carbon shall be reported to the nearest 0.01 percent except for the classification E320LR for which carbon shall be reported to the nearest 0.005 percent.
[5] Vanadium shall be 0.10 to 0.30 percent.
[6] Titanium shall be 0.15 percent max.
[7] Tungsten shall be from 1.25 to 1.75 percent.

TABLE 4-19. AWS A5.4-81 Mechanical Property Requirements for Stainless Covered Electrode All-Weld Metal

AWS Classification	Tensile Strength, min, ksi	Elongation in 2 in. gage length, min, percent	Heat Treatment
E209	100	15	None
E219	90	15	None
E240	100	15	None
E307	85	30	None
E308	80	35	None
E308H	80	35	None
E308L	75	35	None
E308Mo	80	35	None
E308MoL	75	35	None
E309	80	30	None
E309L	75	30	None
E309Cb	80	30	None
E309Mo	80	30	None
E310	80	30	None
E310H	90	10	None
E310Cb	80	25	None
E310Mo	80	30	None
E312	95	22	None
E316	75	30	None
E316H	75	30	None
E316L	70	30	None
E317	80	30	None
E317L	75	30	None
E318	80	25	None
E320	80	30	None
E320LR	75	30	None
E330	75	25	None
E330H	90	10	None
E347	75	30	None
E349	100	25	None
E410	65	20	1
E410NiMo	110	15	2
E430	65	20	3
E502	60	20	1
E505	60	20	1
E630	135	7	4
E16-8-2	80	35	None
E7Cr	60	20	1

[1] Specimen shall be heated to between 1550 and 1600°F (840 and 870°C), held for 2 hours, furnace-cooled at a rate not exceeding 100°F (55°C) per hour to 1100°F (595°C) and air cooled to ambient.
[2] Specimen shall be heated to between 1100 and 1150°F (595 and 620°C), held for 1 hour, and air cooled to ambient.
[3] Specimen shall be heated to between 1400 and 1450°F (760 and 790°C), held for 2 hours, and furnace-cooled at a rate not exceeding 100°F (55°C) per hour to 100°F (55°C), and air cooled to ambient.
[4] Specimen shall be heated to between 1875 and 1925°F (1025 and 1050°C), held for 1 hour, air cooled to at least 60°F (15°C), and then precipitation hardened at 1135 to 1165°F (610 to 630°C), held for 4 hours, and air cooled to ambient.

Surfacing Welding Rods and Electrodes, AWS A5.13-80

This specification covers the requirements for bare and covered surfacing welding rods for use with oxyacetylene, gas tungsten-arc, carbon-arc, and atomic hydrogen welding processes. The section on electrodes deals with covered electrodes intended for surfacing with the shielded metal-arc welding process. Also, some bare metal-arc welding electrodes are included.

Weld-surfacing applications are extremely diversified, and, as a result, there are a great many different brand-name products available. This specification makes no attempt to classify all filler metals suitable for weld surfacing. Only those filler metals are covered that have gained some degree of industrial standardization and for which technical data are available.

Chemical requirements for bare welding rods and electrodes are given in Table 4-28 and for covered electrodes in Table 4-29. Bare surfacing rods and electrodes are classified on the basis of the chemical composition of the as-manufactured product. All covered electrodes are classified on the basis of the chemical composition of the deposited weld metal.

SHIELDING GASES

Shielding gases are consumables used with the GMAW and GTAW processes. The American Welding Society does not write specifications for gases. There are federal specifications, but the welding industry usually relies on "welding grade" to describe the required purity.

The primary purpose of a shielding gas is to protect the molten weld metal from contamination by the oxygen and nitrogen in air. The factors, in addition to cost, that affect the suitability of a gas include the influence of the gas on the arcing and metal-transfer characteristics during welding, weld penetration, width of fusion and surface shape, welding speed and the tendency to undercut. Among the inert gases — helium, argon, neon, krypton, and xenon — the only ones plentiful enough for practical use in welding are helium and argon. These gases provide satisfactory shielding for the more reactive metals, such as aluminum, magnesium, beryllium, columbium, tantalum, titanium, and zirconium.

Although pure inert gases protect metal at any temperature from reaction with constituents of the air, they are not suitable for all welding applications. Controlled quantities of reactive gases mixed with inert gases improve the arc action and metal-transfer characteristics when welding the steels, but such mixtures are not used for the reactive metals.

Oxygen, nitrogen, and carbon dioxide are reactive gases. With the exception of carbon dioxide, these gases are not generally used alone for arc shielding. Carbon dioxide can be used alone or mixed with an inert gas for welding many carbon and low-alloy

TABLE 4-20. AWS A5.9-81 Chemical Requirements for Bare Stainless Welding Rods and Electrodes

AWS Classification[1]	Composition (%)[1 2]										
	C	Cr	Ni	Mo	Cb plus Ta	Mn	Si	P	S	N	Cu
ER209[5]	0.05	20.5-24.0	9.5-12.0	1.5-3.0	—	4.0-7.0	0.90	0.03	0.03	0.10-0.30	0.75
ER218	0.10	16.0-18.0	8.0-9.0	0.75	—	7.0-9.0	3.5-4.5	0.03	0.03	0.08-0.18	0.75
ER219	0.05	19.0-21.5	5.5-7.0	0.75	—	8.0-10.0	1.00	0.03	0.03	0.10-0.30	0.75
ER240	0.05	17.0-19.0	4.0-6.0	0.75	—	10.5-13.5	1.00	0.03	0.03	0.10-0.20	0.75
ER307	0.04-0.14	19.5-22.0	8.0-10.7	0.5-1.5	—	3.3-4.75	0.30-0.65	0.03	0.03	—	0.75
ER308[4]	0.08	19.5-22.0	9.0-11.0	0.75	—	1.0-2.5	0.30-0.65	0.03	0.03	—	0.75
ER308H	0.04-0.08	19.5-22.0	9.0-11.0	0.75	—	1.0-2.5	0.30-0.65	0.03	0.03	—	0.75
ER308L[4]	0.03	19.5-22.0	9.0-11.0	0.75	—	1.0-2.5	0.30-0.65	0.03	0.03	—	0.75
ER308Mo	0.08	18.0-21.0	9.0-12.0	2.0-3.0	—	1.0-2.5	0.30-0.65	0.03	0.03	—	0.75
ER308MoL	0.04	18.0-21.0	9.0-12.0	2.0-3.0	—	1.0-2.5	0.30-0.65	0.03	0.03	—	0.75
ER309[4]	0.12	23.0-25.0	12.0-14.0	0.75	—	1.0-2.5	0.30-0.65	0.03	0.03	—	0.75
ER309L	0.03	23.0-25.0	12.0-14.0	0.75	—	1.0-2.5	0.30-0.65	0.03	0.03	—	0.75
ER310	0.08-0.15	25.0-28.0	20.0-22.5	0.75	—	1.0-2.5	0.30-0.65	0.03	0.03	—	0.75
ER312	0.15	28.0-32.0	8.0-10.5	0.75	—	1.0-2.5	0.30-0.65	0.03	0.03	—	0.75
ER316[4]	0.08	18.0-20.0	11.0-14.0	2.0-3.0	—	1.0-2.5	0.30-0.65	0.03	0.03	—	0.75
ER316H	0.04-0.08	18.0-20.0	11.0-14.0	2.0-3.0	—	1.0-2.5	0.30-0.65	0.03	0.03	—	0.75
ER316L[4]	0.03	18.0-20.0	11.0-14.0	2.0-3.0	—	1.0-2.5	0.30-0.65	0.03	0.03	—	0.75
ER317	0.08	18.5-20.5	13.0-15.0	3.0-4.0	—	1.0-2.5	0.30-0.65	0.03	0.03	—	0.75
ER317L	0.03	18.5-20.5	13.0-15.0	3.0-4.0	—	1.0-2.5	0.30-0.65	0.03	0.03	—	0.75
ER318	0.08	18.0-20.0	11.0-14.0	2.0-3.0	8 × C min to 1.0 max	1.0-2.5	0.30-0.65	0.03	0.03	—	0.75
ER320	0.07	19.0-21.0	32.0-36.0	2.0-3.0	8 × C min to 1.0 max	2.5	0.60	0.03	0.03	—	3.0-4.0
ER320LR[5]	0.025	19.0-21.0	32.0-36.0	2.0-3.0	8 × C min to 0.40 max	1.5-2.0	0.15	0.015	0.020	—	3.0-4.0
ER321[6]	0.08	18.5-20.5	9.0-10.5	0.75	—	1.0-2.5	0.30-0.65	0.03	0.03	—	0.75
ER330	0.18-0.25	15.0-17.0	34.0-37.0	0.75	—	1.0-2.5	0.30-0.65	0.03	0.03	—	0.75
ER347[4]	0.08	19.0-21.5	9.0-11.0	0.75	10 × C min to 1.0 max	1.0-2.5	0.30-0.65	0.03	0.03	—	0.75
ER349[3 7]	0.07-0.13	19.0-21.5	8.0-9.5	0.35-0.65	1.0-1.4	1.0-2.5	0.30-0.65	0.03	0.03	—	0.75
ER410	0.12	11.5-13.5	0.6	0.75	—	0.6	0.50	0.03	0.03	—	0.75
ER410NiMo	0.06	11.0-12.5	4.0-5.0	0.4-0.7	—	0.6	0.50	0.03	0.03	—	0.75
ER420	0.25-0.40	12.0-14.0	0.6	0.75	—	0.6	0.50	0.03	0.03	—	0.75
ER430	0.10	15.5-17.0	0.6	0.75	—	0.6	0.50	0.03	0.03	—	0.75
ER502	0.10	4.6-6.0	0.6	0.45-0.65	—	0.6	0.50	0.03	0.03	—	0.75
ER505	0.10	8.0-10.5	0.5	0.8-1.2	—	0.6	0.50	0.04	0.03	—	0.75
ER630	0.05	16.0-16.75	4.5-5.0	0.75	0.15-0.30	0.25-0.75	0.75	0.04	0.03	—	3.25-4.00
ER26-1	0.01	25.0-27.5	[8]	0.75-1.50	—	0.40	0.40	0.02	0.02	0.015	0.20[8]
ER16-8-2	0.10	14.5-16.5	7.5-9.5	1.0-2.0	—	1.0-2.5	0.30-0.65	0.03	0.03	—	0.75

[1] Analysis shall be made for the elements for which specific values are shown in this table. If however, the presence of other elements is indicated in the course of routine analysis, further analysis shall be made to determine that the total of these other elements, except iron, is not present in excess of 0.50 percent.
[2] Single values shown are maximum percentages except where otherwise specified.
[3] Vanadium — 0.10-0.30.
[4] These grades are available in high silicon classifications which shall have the same chemical composition requirements as given below with the exception that the silicon content shall be 0.65 to 1.00 percent. These high silicon classifications shall be designated by the addition 'Si' to the standard classification designations indicated below. The fabricator should consider carefully the use of high silicon filler metals in highly restrained fully austenitic welds. A discussion of the problem is presented in paragraphs A5.7, A6.2, and A6.3 in the Appendix of AWS D5.9-81.
[5] Carbon shall be reported to the nearest 0.01 percent except for the classification E320LR for which carbon shall be reported to the nearest 0.005 percent.
[6] Titanium — 9 × C min. to 1.0 max.
[7] Titanium — 0.10 to 0.30. Tungsten is 1.25 to 1.75 percent.
[8] Nickel, max. — 0.5 minus the copper content.

steels. Oxygen is used in small quantities with one of the inert gases — usually argon. Nitrogen is occasionally used alone, but is usually mixed with argon, as a shielding gas to weld copper. The most extensive use of nitrogen is in Europe, where helium is relatively unavailable.

Argon and Helium

As noted, the inert natures of argon and helium are not the only characteristics that make them suitable for gas shielding. Other characteristics are important and are deciding factors in the choice of gas for GTAW or GMAW with specific materials.

For a given arc length and current, arc voltage with helium is higher than with argon. Because more heat is produced with helium than with argon, helium is more effective for welding thick materials, particularly high-conductivity metals such as copper and aluminum alloys. Argon is more suitable for welding thin materials and those with lower heat conductivity, especially in welding positions other than flat.

The heavier a gas, the more effective it is for arc shielding. Helium is very light; argon is about 10 times heavier than helium and about 30% heavier than air. When argon is discharged from the welding

TABLE 4-21. AWS A5.9-81 Standard Sizes

Form	Diameter[1], in.
Welding rods in straight lengths	0.045, 1/16 (0.062), 5/64 (0.078), 3/32 (0.094), 1/8 (0.125), 5/32 (0.156), 3/16 (0.188)
Filler metal in coils with or without support	0.045, 1/16 (0.062), 5/64 (0.078), 3/32 (0.094), 7/64 (0.109), 1/8 (0.125), 5/32 (0.156), 3/16 (0.188), 1/4 (0.250)
Filler metal wound on standard 12 in. O.D. spools	0.030, 0.035, 0.045, 1/16 (0.062), 5/64 (0.078), 3/32 (0.094), 7/64 (0.109)
Electrodes wound on lightweight 1-1/2 and 2-1/2-lb 4-in. O.D. spools	0.020, 0.025, 0.030, 0.035, 0.045

[1] Electrodes and welding rods of diameters up to and including 0.045 in. shall not vary more than ±0.001 from the nominal. Diameters greater than 0.045 in. shall not vary more than ±0.002 from the nominal.

nozzle it forms a protective blanket over the weld area, while helium rises and disperses rapidly. For this reason, higher flow rates are generally required with helium (or with mixtures high in helium) than with argon shielding.

Shape of a weld bead and penetration pattern are determined, to a large extent, by metal-transfer characteristics which, in turn, are affected by the shielding gas used.

Metal is generally deposited either by spray transfer or by globular transfer. Spray transfer (usually the more desirable) produces relatively deep penetration at the center of the bead and shallow penetration at the edges; globular transfer produces a broader and shallower penetration pattern throughout the bead.

Argon generally promotes more spray transfer than helium and at lower current levels. But even with argon shielding, spray transfer cannot always be achieved at usable current levels — one of the problems in welding ferrous metals by the gas metal-arc process.

The physics of metal transfer across an arc is not completely understood. In an argon atmosphere with DCEP, the size of the metal droplet crossing the arc decreases as the current increases. At a critical level of current the mode of transfer changes abruptly. The tip of the electrode becomes pointed, metal transfers from the electrode to the work in a fine spray, the arc becomes very stable, and there is little or no spatter. Figure 4-3 illustrates the appearance of electrode tips in various shielding gases. A degree of spray transfer is possible with 20% argon and 80% helium. Here the argon has a predominate effect because of its higher density.

Inert Gases with Reactive Gas Additions

Improved metal transfer, a more stable arc, and less spatter result from the addition of oxygen or carbon dioxide to an inert shielding gas. These additions, when welding carbon and low-alloy steels,

TABLE 4-22. AWS A5.15-90 Chemical Requirements for Covered Electrodes for Cast Iron

AWS Classification[1]	Composition (%) (Notes 1 and 2)												
	C	Mn	Si	P	S	Fe	Ni[2]	Mo	Cu[3]	Mg	Al	Ce	Total other elements
Cast Iron Filler Metals[4]													
RCI	3.2-3.5	0.60-0.75	2.7-3.0	0.50-0.75	0.10	Remainder	Trace	Trace	—	—	—	—	—
RCI-A	3.2-3.5	0.50-0.70	2.0-2.5	0.20-0.40	0.10	Remainder	1.2-1.6	0.25-0.45	—	—	—	—	—
RCI-B	3.2-4.0	0.10-0.40	3.2-3.8	0.05	0.015	Remainder	0.50	—	—	0.04-0.10	—	0.20	—
Nickel-Base Electrodes[5]													
ENi-CI	2.0	1.0	4.0	—	0.03	8.0	85 min	—	2.5	—	1.0	—	1.0
ENi-CI-A	2.0	1.0	4.0	—	0.03	8.0	85 min	—	2.5	—	1.0-3.0	—	1.0
ENiFe-CI	2.0	1.0	4.0	—	0.03	Remainder	45-60	—	2.5	—	1.0	—	1.0
ENiFe-CI-A	2.0	1.0	4.0	—	0.03	Remainder	45-60	—	2.5	—	1.0-3.0	—	1.0
ENiCu-A	0.35-0.55	2.3	0.75	—	0.025	3.0-6.0	50-60	—	35-45	—	—	—	1.0
ENiCu-B	0.35-0.55	2.3	0.75	—	0.025	3.0-6.0	60-70	—	25-35	—	—	—	1.0
Mild Steel Electrodes[5]													
ESt	0.15	0.60	0.03	0.04	0.04	Remainder	—	—	—	—	—	—	—

NOTE 1 — Analysis shall be made for the elements for which specific values are shown in this table. If the presence of other elements is indicated in the course of routine analysis, further analysis shall be made to determine that the total of these other elements is not present in excess of the limits specified for total other elements in the last column of this table.

NOTE 2 — Single values shown are maximum percentages, except where otherwise specified.

[1] Copper-base filler metals frequently used in the braze-welding of cast iron are no longer included in this specification. For information pertaining to these materials. See AWS A5.6 and A5.27.
[2] Nickel plus incidental cobalt.
[3] Copper plus incidental silver.
[4] Chemical composition requirements for the cast iron filler metals are based on the as-manufactured composition of the bare welding rod.
[5] Chemical composition requirements for the nickel-base and mild steel electrodes are based on weld metal analysis.

TABLE 4-23. AWS A5.10-88 Chemical Requirements for Bare Welding Rods and Electrodes for Aluminum and Aluminum Alloys

AWS Classification	Composition (%) (Notes 1 and 2)											
	Si	Fe	Cu	Mn	Mg	Cr	Ni	Zn	Ti	Percent other[1] elements Each[3]	Percent other[1] elements Total	Al
ER1100	[2]	[2]	0.05-0.20	0.05	—	—	—	0.10	—	0.05	0.15	99.00 min[3]
ER2319[4]	0.20	0.30	5.8-6.8	0.20-0.40	0.02	—	—	0.10	0.10-0.20	0.05	0.15	Remainder
ER4043	4.5-6.0	0.8	0.30	0.05	0.05	—	—	0.10	0.20	0.05	0.15	Remainder
ER4047	11.0-13.0	0.8	0.30	0.15	0.10	—	—	0.20	—	0.05	0.15	Remainder
ER4145	9.3-10.7	0.8	3.3-4.7	0.15	0.15	0.15	—	0.20	—	0.05	0.15	Remainder
ER5183	0.40	0.40	0.10	0.50-1.0	4.3-5.2	0.05-0.25	—	0.25	0.15	0.05	0.15	Remainder
ER5356	0.25	0.40	0.10	0.05-0.20	4.5-5.5	0.05-0.20	—	0.10	0.06-0.20	0.05	0.15	Remainder
ER5554	0.25	0.40	0.10	0.50-1.0	2.4-3.0	0.05-0.20	—	0.25	0.05-0.20	0.05	0.15	Remainder
ER5556	0.25	0.40	0.10	0.50-1.0	4.7-5.5	0.05-0.20	—	0.25	0.05-0.20	0.05	0.15	Remainder
ER5654	[5]	[5]	0.05	0.01	3.1-3.9	0.15-0.35	—	0.20	0.05-0.15	0.05	0.15	Remainder
R242.0[6]	0.7	1.0	3.5-4.5	0.35	1.2-1.8	0.25	1.7-2.3	0.35	0.25	0.05	0.15	Remainder
R295.0[6]	0.7-1.5	1.0	4.0-5.0	0.35	0.03	—	—	0.35	0.25	0.05	0.15	Remainder
R355.0[6]	4.5-5.5	0.6[7]	1.0-1.5	0.50[7]	0.40-0.6	0.25	—	0.35	0.25	0.05	0.15	Remainder
R356.0[6]	6.5-7.5	0.6	0.25	0.35	0.20-0.40	—	—	0.35	0.25	0.05	0.15	Remainder

NOTE 1 — Single values shown are maximum percentages, except where a minimum is specified.

NOTE 2 — Analysis shall be made for the elements of which specific limits are shown. If, however, the presence of other elements is suspected or indicated in the course of routine analysis, further analysis shall be made to determine that these other elements are not in excess of the limits specified for "other elements,".

[1] Beryllium shall not exceed 0.0008 percent.
[2] Silicon plus iron shall not exceed 0.95 percent.
[3] The aluminum content is the difference between 100.00 percent and the sum of all other metallic elements present in amounts of 0.010 percent or more each, expressed to the second decimal before determining the sum.
[4] Vanadium content shall be 0.05-0.15 percent. Zirconium content shall be 0.10-0.25 percent.
[5] Silicon plus iron shall not exceed 0.45 percent.
[6] For repair of castings.
[7] If iron exceeds 0.45 percent, manganese content shall not be less than one-half iron content.

TABLE 4-24. AWS A5.6-84 Chemical Requirements for Undiluted Weld Metal for Copper and Copper Alloy Arc-Welding Electrodes

AWS Classification	UNS Number[3]	Common Name	Composition (%)[1 2]											
			Cu including Ag	Zn	Sn	Mn	Fe	Si	Ni[4]	P	Al	Pb	Tl	Total other elements
ECu	W60189	Copper	Remainder	[6]	[6]	0.10	0.20	0.10	[6]	[6]	0.10	0.02	—	0.50
ECuSi	W60656	Silicon bronze (Copper silicon)	Remainder	[6]	1.5	1.5	0.50	2.4-4.0	[6]	[6]	0.01	0.02	—	0.50
ECuSn-A	W60518	Phosphor bronze (Copper-tin)	Remainder	[6]	4.0-6.0	[6]	0.25	[6]	[6]	0.05-0.35	0.01	0.02	—	0.50
ECuSn-C	W60521		Remainder	[6]	7.0-9.0	[6]	0.25	[6]	[6]	0.05-0.35	0.01	0.02	—	0.50
ECuNi[5]	W60715	Copper nickel (70/30)	Remainder	[6]	[6]	1.00-2.50	0.40-0.75	0.50	29.0-33.0	0.020	—	0.02	0.50	0.50
ECuAl-A2	W60614	Aluminum bronze	Remainder	[6]	[6]	[6]	0.50-5.0	1.5	[6]	—	6.5-9.0	0.02	—	0.50
ECuAl-B	W60619	Aluminum bronze	Remainder	[6]	[6]	[6]	2.5-5.0	1.5	[6]	—	7.5-10.0	0.02	—	0.50
ECuNiAl	W60632	Nickel aluminum bronze	Remainder	[6]	[6]	0.50-3.5	3.0-6.0	1.5	4.0-6.0	—	6.0 to 8.5	0.02	—	0.50
ECuMnNiAl	W60633	Manganese-nickel aluminum bronze	Remainder	[6]	[6]	11.0-13.0	2.0-6.0	1.5	1.0-2.5	—	5.0-7.5	0.02	—	0.50

NOTES:
[1] Analysis shall be made for the elements for which specific values or an "6" are shown in this table. If, however, the presence of other elements is indicated in the course of routine analysis, further analysis, shall be made to determine that the total of these other elements is not present in excess of the limits specified for 'total other elements' in the last column in the table.
[2] Single values shown are maximum.
[3] SAE/ASTM Unified Number System for Metals and Alloys.
[4] Includes cobalt.
[5] Sulfur shall be restricted to 0.015 percent for the ECuNi classification.
[6] Those elements must be included in total of other elements.

also promote wetting and flow of weld metal, thus reducing or eliminating undercut. Effects on penetration and bead shape of oxygen additions are illustrated in Fig. 4-3.

Noticeable change in arc action and metal-transfer characteristics in gas metal-arc welding result from addition of as little as 0.5% oxygen or carbon dioxide to argon. However, 1 to 5% oxygen is generally added. Oxygen or carbon dioxide is seldom added to helium or argon-helium mixtures.

TABLE 4-25. AWS A5.6-84 Tensile-Strength Requirements for Copper and Copper Alloy Weld Metal

AWS Classification	Tensile strength, min		Elongation in. 4 × D gage length, percent, min.
	ksi	MPa	
ECu	25	170	20
ECuSi	50	350	20
ECuSn-A	35	240	20
ECuSn-C	40	280	20
ECuNi	50	350	20
ECuAl-A2	60	410	20
ECuAl-B	65	450	10
ECuNiAl	72	500	10
ECuMnNiAl	75	520	15

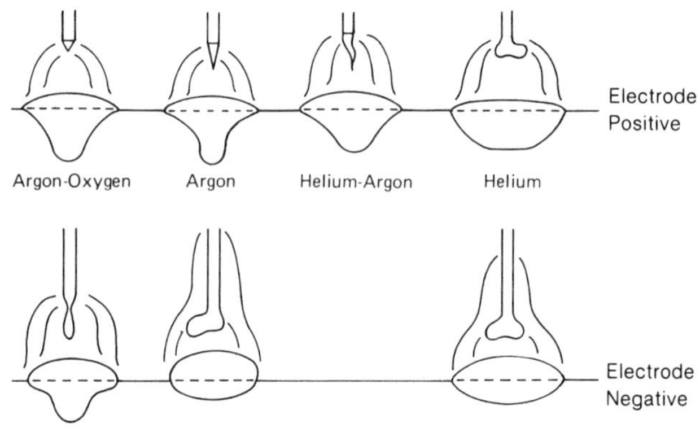

Fig. 4-3. Electrode tip shape, bead contour, and penetration patterns for various shielding gases.

Addition of 5% oxygen or 10 to 25% carbon dioxide to argon produces a significant pinch effect with a DC, electrode negative arc. The filler wire tapers, the metal transfers in the form of a fast-moving stream of droplets, and the penetration pattern approaches that of electrode positive. At the same time, melt-off rate is reduced considerably. With pure argon, melt-off rate with electrode negative is almost double that with electrode positive. However, most GMA welding with an inert gas or carbon dioxide is done with DCEP (see Tables 4-13 and 4-15). Mixtures of 5% oxygen or 25% carbon dioxide with argon are commercially available.

Because of oxidizing effects, addition of oxygen or carbon dioxide to argon may cause porosity in some ferrous metals, as well as loss of such alloying elements as chromium, vanadium, aluminum, titanium, manganese, and silicon. Consequently, filler wires used with oxygen-containing shielding gas require additions of deoxiders to counteract the effects of the oxygen.

TABLE 4-27. AWS A5.7-84 Hardness and Tensile-Strength Requirements for Copper and Copper Alloy Weld Metal

AWS Classification	Brinell Hardness		Minimum tensile strength, psi
ERCu	25 Rockwell F		25,000
ERCuSi-A	80 to 100	(500 kg load)	50,000
ERCuSn-A	70 to 85	(500 kg load)	35,000
ERCuNi	60 to 80	(500 kg load)	50,000
ERCuAl-A1	80 to 110	(500 kg load)	55,000
ERCuAl-A2	130 to 150	(3000 kg load)[1]	60,000
ERCuAl-A3	140 to 180	(3000 kg load)[1]	65,000
ERCuNiAl	160 to 200	(3000 kg load)[1]	72,000
ERCuMnNiAl	160 to 200	(3000 kg load)[1]	75,000

NOTE: Hardness values as listed above are average values for an as-welded deposit made with the filler metal specified. This table is included for information only.

[1] Gas tungsten arc process only.

TABLE 4-26. AWS A5.7-84 Chemical Requirements for Copper and Copper Alloy Arc-Welding Rods

AWS Classification	Composition (%)[1 2 3]													
	UNS Number[4]	Common Name	Cu including Ag	Zn	Sn	Mn	Fe	Si	Ni including Co	P	Al	Pb	Ti	Total other elements
ERCu	C18980	Copper	98.0 min	—	1.0	0.50	—	0.50	—	0.15	0.01	0.02	—	0.50
ERCuSi-A	C65600	Silicon bronze (Copper-Silicon)	Remainder	1.0	1.0	1.5	0.50	2.8-4.0	—	—	0.01	0.02	—	0.50
ERCuSn-A	C51800	Phosphor bronze (Copper-Tin)	Remainder	—	4.0-6.0	—	—	—	—	0.10-0.35	0.01	0.02	—	0.50
ERCuNi[6]	C71580	Copper-nickel	Remainder	—	—	1.00	0.40-0.75	0.25	29.0-32.0	0.02	—	0.02	0.20-0.50	0.50
ERCuAl-A1	C61000	Aluminum bronze	Remainder	0.20	—	0.50	—	0.10	—	—	6.0-8.5	0.02	—	0.50
ERCuAl-A2	C61800	Aluminum bronze	Remainder	0.02	—	—	1.5	0.10	—	—	8.5-11.0	0.02	—	0.50
ERCuAl-A3	C62400	Aluminum bronze	Remainder	0.10	—	—	2.0-4.5	0.10	—	—	10.0-11.5	0.02	—	0.50
ERCuNiAl	C63280	Nickel-aluminum bronze	Remainder	0.10	—	0.60-3.50	3.0-5.0	0.10	4.0-5.50	—	8.50-9.50	0.02	—	0.50
ERCuMnNiAl	C63380	Manganese-nickel aluminum bronze	Remainder	0.15	—	11.0-14.0	2.0-4.0	0.10	1.5-3.0	—	7.0-8.5	0.02	—	0.50

[1] Analysis shall be made for the elements for which specific values are shown in this table. If, however, the presence of other elements is indicated in the course of routine analysis, further analysis shall be made to determine that the total of these other elements is not present in excess of the limits specified for 'total other elements' in the last column in the table.

[2] Single values shown are maximum unless otherwise noted.

[3] Classifications RBCuZn-A, RCuZn-B, RCuZn-C, and RBCuZn-D now are included in A5.27-85. *Specification for Copper and Copper Alloy Gas Welding Rods.*

[4] ASTM-SAE Unified Number System for Metals and Alloys.

[5] Sulfur shall be 0.01 percent for the ERCuNi classification.

TABLE 4-28. AWS A5.13-80 Chemical Requirements for Bare Surfacing Welding Rods and Electrodes

AWS Classification	Composition (%) (Notes 1 and 2)															
	C	Mn	Co	W	Ni	Cr	Mo	Fe	V	Cu	Al	Zn	Si	Pb	Sn	P	Total other elements
RFe5-A	0.7-1.0	0.50	—	5.0-7.0	—	3.0-5.0	4.0-6.0	remainder	1.0-2.5	—	—	—	0.50	—	—	—	1.0
RFe5-B	0.5-0.9	0.50	—	1.0-2.5	—	3.0-5.0	5.0-9.5	remainder	0.8-1.3	—	—	—	0.50	—	—	—	1.0
RFeCr-Al	3.7-5.0	2.0-6.0	—	—	—	27.0-35.0	—	remainder	—	—	—	—	1.10-2.5	—	—	—	1.0
RCoCr-A	0.9-1.4	1.00	remainder	3.0-6.0	3.0	26.0-32.0	1.0	3.0	—	—	—	—	2.0	—	—	—	0.50
RCoCr-B	1.2-1.7	1.00	remainder	7.0-9.5	3.0	26.0-32.0	1.0	3.0	—	—	—	—	2.0	—	—	—	0.50
RCoCr-C	2.0-3.0	1.00	remainder	11.0-14.0	3.0	26.0-33.0	1.0	3.0	—	—	—	—	2.0	—	—	—	0.50
RCuZn-E[2]	—	0.30	—	—	—	—	—	1.50	—	56.0 min	0.01[1]	remainder	0.04-0.25	0.05[1]	2.00-3.00	—	0.50
ERCuSi-A[1]	—	1.5[3]	—	—	[1]	—	—	0.5	—	94.0 min	0.01[1]	1.5[3]	2.8-4.0	0.02[1]	1.5[1]	[1]	0.50
ERCuAl-A2[2]	—	—	—	—	—	—	—	1.5	—	remainder	9.0-11.0	0.02	0.10	0.02[1]	—	—	0.50
ERCuAl-A3	—	—	—	—	—	—	—	3.0-5.0	—	remainder	10.0-11.0	0.10	0.10	0.02[1]	—	—	0.50
RCuAl-C	—	—	—	—	—	—	—	3.0-5.0	—	remainder	12.0-13.0	0.02	0.04	0.02	—	—	0.50
RCuAl-D	—	—	—	—	—	—	—	3.0-5.0	—	remainder	13.0-14.0	0.02	0.04	0.02	—	—	0.50
RCuAl-E	—	—	—	—	—	—	—	3.0-5.0	—	remainder	14.0-15.0	0.02	0.04	0.02	—	—	0.50
ERCuSn-A[2]	—	[1]	—	—	[1]	—	—	[1]	—	93.5 min	0.01[1]	[1]	[1]	0.02[1]	4.0-6.0	0.10-0.35	0.50
RCuSn-D	—	—	—	—	—	—	—	—	—	88.5 min	0.01[1]	—	—	0.05[1]	9.0-11.0	0.10-0.30	0.50
RNiCr-A	0.30-0.60	—	1.50	—	remainder	8.0-14.0	—	1.25-3.25	2.00-3.00 % boron	—	—	—	1.25-3.25	—	—	—	0.50
RNiCr-B	0.40-0.80	—	1.25	—	remainder	10.0-16.0	—	3.00-5.00	2.00-4.00 % boron	—	—	—	3.00-5.00	—	—	—	0.50
RNiCr-C	0.50-1.00	—	1.00	—	remainder	12.0-18.0	—	3.50-5.50	2.50-4.50 % boron	—	—	—	3.50-5.50	—	—	—	0.50

NOTE 1: Analysis shall be made for the elements for which specific values are shown in this table. If, however, the presence of other elements is indicated in the course of routine analysis, further analysis shall be made to determine that the total of these other elements is not present in excess of the limits specified for "total other elements" in the last column in the table.

NOTE 2: Single values shown are maximum percentages, except where otherwise specified.

[1] Total other elements, including the elements marked with footnote a, shall not exceed the value specified.
[2] This AWS classification is intended to be identical with the same classification that appears in the latest edition of AWS A5.7, Specification for Copper and Copper Alloy Bare Welding Rods and Electrodes.
[3] One or more of these elements may be present within the limits specified.

Porosity in aluminum welds can be decreased by adding a small percentage of chlorine to argon or helium. Chlorine combines with dissolved hydrogen to form HCl which comes out of the liquid before freezing. For maximum effectiveness, the chlorine should be introduced separately through the welding torch. Chlorine's disadvantages of being poisonous and corrosive discourage its widespread use. When it is used, extreme caution and all applicable safety rules should be observed.

Carbon Dioxide

Carbon dioxide may be used as a shielding gas for the GMA welding of carbon and low alloy steels, but since it is a reactive gas the electrodes used must contain sufficient deoxidizers to counteract the effects of oxygen. Stainless steel electrodes with high silicon have been developed for use with argon-25% carbon dioxide mixtures.

The low cost of carbon dioxide makes its use as a shielding gas very attractive. With the development of better electrodes, sound weld deposits with good mechanical properties can be made.

Two types of metal transfer occur with carbon dioxide shielding gas — globular and short-circuiting. The spray transfer experienced with argon or argon-oxygen mixture does not occur. Globular transfer produces a harsh arc with excessive spatter. By control of welding conditions, the short-circuiting type of metal transfer is promoted.

To promote the short-circuiting type of transfer when welding carbon and low-alloy steels, argon is often used as the dominant gas in a mixture, with the carbon dioxide content cut to 20 to 30%. Other mixtures with higher percentages of carbon dioxide also give short-circuiting transfer, with its advantages

TABLE 4-29. AWS A5.13-80 Chemical Requirements[1] for Covered Surfacing Electrodes

AWS Classification	Composition (%) (Notes 1 and 2)																
	C	Mn	Co	W	Ni	Cr	Mo	Fe	V	Cu	Al	Zn	Si	Pb	Sn	P	Total other elements
EFe5-A	0.7-1.0	0.60	—	5.0-7.0	—	3.0-5.0	4.0-6.0	remainder	1.0-2.5	—	—	—	0.80	—	—	—	1.0
EFe5-B	0.5-0.9	0.60	—	1.0-2.5	—	3.0-5.0	5.0-9.5	remainder	0.8-1.3	—	—	—	0.80	—	—	—	1.0
EFe5-C	0.3-0.5	0.60	—	1.0-2.5	—	3.0-5.0	5.0-9.0	remainder	0.8-1.2	—	—	—	0.80	—	—	—	1.0
EFeMn-A	0.5-0.9	11.0-16.0	—	—	2.75-6.0	0.50	—	remainder	—	—	—	—	1.3	—	—	0.03	1.0
EFeMn-B	0.5-0.9	11.0-16.0	—	—	—	0.50	0.6-1.4	remainder	—	—	—	—	0.3-1.3	—	—	0.03	1.0
EFeCr-Al	3.0-5.0	4.0-8.0	—	—	—	26.0-32.0	2.0	remainder	—	—	—	—	1.0-2.5	—	—	—	1.0
ECoCr-A	0.7-1.4	2.0	remainder	3.0-6.0	3.0	25.0-32.0	1.0	5.0	—	—	—	—	2.0	—	—	—	0.50
ECoCr-B	1.0-1.7	2.0	remainder	7.0-9.5	3.0	25.0-32.0	1.0	5.0	—	—	—	—	2.0	—	—	—	0.50
ECoCr-C	1.75-3.0	2.0	remainder	11.0-14.0	3.0	25.0-33.0	1.0	5.0	—	—	—	—	2.0	—	—	—	0.50
ECuSi[3]	—	1.5	—	—	[2]	—	—	0.50	—	remainder	0.01	—	2.4-4.0	0.02[2]	1.5	[2]	0.50
ECuAl-A2[3]	—	[2]	—	—	[2]	—	—	0.5-5.0	—	remainder	7.0-9.0	[2]	1.0	0.02[2]	[2]	—	0.60
ECuAl-B[2]	—	[2]	—	—	[2]	—	—	2.5-5.0	—	remainder	8.4-10.0	[2]	1.0	0.02[2]	[2]	—	0.60
ECuAl-C	—	—	—	—	—	—	—	3.0-5.0	—	remainder	12.0-13.0	0.02	0.04	0.02	—	—	0.50
ECuAl-D	—	—	—	—	—	—	—	3.0-5.0	—	remainder	13.0-14.0	0.02	0.04	0.02	—	—	0.50
ECuAl-E	—	—	—	—	—	—	—	3.0-5.0	—	remainder	14.0-15.0	0.02	0.04	0.02	—	—	0.50
ECuSn-A[3]	—	[2]	—	—	[2]	—	—	0.25	—	remainder	0.01	[2]	[2]	0.02[2]	4.0-6.0	0.05-0.35	0.50
ECuSn-C[3]	—	[2]	—	—	[2]	—	—	0.25	—	remainder	0.01	[2]	[2]	0.02[2]	7.0-9.0	0.05-0.35	0.50
ENiCr-A	0.30-0.60	—	1.50	—	remainder	8.0-14.0	—	1.25-3.25	2.00-3.00 % boron	—	—	—	1.25-3.25	—	—	—	0.50
ENiCr-B	0.40-0.80	—	1.25	—	remainder	10.0-16.0	—	3.00-5.00	2.00-4.00 % boron	—	—	—	3.00-5.00	—	—	—	0.50
ENiCr-C	0.50-1.00	—	1.00	—	remainder	12.0-18.0	—	3.50-5.50	2.50-4.50 % boron	—	—	—	3.50-5.50	—	—	—	0.50

NOTE 1: Analysis shall be made for the elements for which specific values are shown in this table. If, however, the presence of other elements is indicated in the course of routine analysis, further analysis shall be made to determine that the total of these other elements is not present in excess of the limits specified for "total other elements" in the last column in the table.

NOTE 2: Single values shown are maximum percentages, except where otherwise specified.

[1] The analysis given is for deposited weld metal.
[2] Total other elements, including the elements marked with footnote 2, shall not exceed the value specified.
[3] This AWS classification is intended to be identical with the same classification that appears in the latest edition of AWS A5.6, Specification for Copper and Copper Alloy Covered Electrodes.

TABLE 4-30. Shielding Gases and Gas Mixtures for Gas Metal-Arc Welding

Shielding Gas	Chemical Behavior	Uses, Remarks
Argon	Inert	For welding most metals, except steel.
Helium	Inert	Al and Cu alloys, for greater heat and to minimize porosity.
A and He (20-80 to 50-50%)	Inert	Al and Cu alloys, for greater heat input and to minimize porosity. Quieter, more stable arc than with He alone.
A and Cl (trace Cl)	Essentially Inert	Al alloys, to minimize porosity.
N_2	Reducing	On Cu, permits very powerful arc; used mostly in Europe.
A + 25-30% N_2	Reducing	On Cu, powerful but smoother operating, more readily controlled arc than N_2 alone; used mostly in Europe.
A + 1-2% O_2	Oxidizing	Stainless and alloy steels, also for some deoxidized copper alloys.
A + 3-5% O_2	Oxidizing	Plain carbon, alloy, and stainless steels; requires deoxidized electrode.
A + 20-30% CO_2	Oxidizing	Various steels, used principally with short-circuiting arc.
A + 5% O_2 + 15% CO_2	Oxidizing	Various steels, requires deoxidized wire; used chiefly in Europe.
CO_2	Oxidizing	Plain-carbon and low-alloy steels; deoxidized electrode is essential.

of low penetration, all-position capability, and the ability of handle poor fitup on light-gage material without melt-through.

In the short-circuiting type of transfer, a globule of molten metal collects on the end of the electrode. The electrode is being fed toward the work and, before the globule detaches from the end of the electrode, it contacts the molten crater and forms a short circuit. The high current due to the short circuit blasts the globule from the electrode into the crater. An arc then forms in the gap between the crater and the tip of the electrode, which starts another globule forming on the tip of the electrode. This cycle of metal transfer is repeated about 20 to 200 times per second.

Shielding Gases for GTAW

Either argon, helium, or a mixture of the two is commonly used in gas tungsten-arc welding. Argon provides the advantages of easier arc starting, smoother arc action, better cleaning action for the AC welding of aluminum and magnesium, and superior resistance to draft. In addition, argon costs less than helium and requires a lower arc voltage for comparable currents and arc lengths.

In the manual welding of thin material, argon is recommended because its lower arc-voltage characteristic reduces the tendency for melt-through. In vertical or overhead welding, this same characteristic reduces the tendency for the metal to sag and run.

Helium's higher arc-voltage characteristic is desirable when welding thick material or metals with high heat conductivity and for the high-speed mechanized welding of stainless-steel tubing. Mixtures of argon and helium are used to balance the arc characteristics.

Mixtures of argon or helium with hydrogen provide higher arc voltage and heat in the welding region than helium alone. This reactive gas, however, can damage many metals and alloys, including aluminum, copper, and magnesium-base materials. Mixtures of inert gas with hydrogen can be used in welding only a few materials, such as certain stainless steels and nickel alloys.

The rate at which some metals are joined by gas tungsten-arc welding and the quality of the resulting welds are significantly affected by gas purity. The reactive metals particularly can be degraded by gas impurities of a few hundredths of one percent. Copper, carbon steel, and stainless steels can tolerate much higher levels of impurities with no adverse effects.

Purity of commercially available argon and helium averages over 99.95%, and in some cases exceeds 99.995%. Impurities in shielding gases usually consist of water vapor, oil, oxygen, or nitrogen — usually from sources other than the original gas supply. Water vapor or atmospheric gases can diffuse through the hose lines, or contaminents can be drawn in at leaks in the lines. Tubing that is not susceptible to gas diffusion should be used to supply shielding gas for welding of materials that are sensitive to impurities.

Shielding Gases for GMAW

The most commonly used gases for gas metal-arc welding are given in Table 4-30.

Initially, only argon, helium, or a mixture of these inert gases were used for gas metal-arc welding. Other gases were not considered, because the primary use of the gas metal-arc process was for welding the more reactive metals, such as aluminum and magnesium, which require an inert gas shield. Today, however, the process is used for welding many metals that do not require inert-gas shielding.

Carbon dioxide shielding is widely used for GMA welding of carbon and low-alloy steels in conjunction with deoxidized electrode. Its advantage over the inert gases is its lower cost.

Filler metals for welding: a. Covered electrodes
b. Solid wire on spools

Power Sources

All arc-welding processes require a continuous supply of electrical current in sufficient amount (amperage) and of proper voltage to maintain an arc. This current may be either alternating (AC) or direct (DC), but it must be supplied to the welding electrode through a device that enables its precise control. Only when the welding current is carefully controlled can the desired welding arc characteristics — and thus maximum welding efficiency — be obtained. The controlling device is called a power source. Current may be supplied to it from utility power lines, or developed within it by generators or alternators driven by close-coupled gasoline or diesel engines.

Various types of power sources provide a range of voltage across the welding arc from 13 to approximately 45 volts. The currents supplied through the power source may range from less than 10 amp to 1500 amp or more, the higher currents for automatic welding.

For efficient welding, the power source must permit control of the arc characteristics needed for a specific job. In one job, a forceful, deeply penetrating arc may be required, while, in another, a soft, less-penetrating arc may be necessary to avoid melt-through. Electrodes are designed for various welding positions and they help compensate for power sources that have no arc characteristic adjustment. The welding process also dictates the type of power source needed. Table 4-31 shows the power source requirements for various processes.

TABLE 4-31. Power Requirements for Arc-Welding Processes

Process	Output Characteristic	Type of Current	Polarity
SMAW Shielded metal-arc, gas tungsten-arc, submerged-arc	Constant Current*	AC or DC	DCEN, DCEP, or AC
FCAW Flux-cored	Constant Voltage	DC	DCEN, DCEP
GMAW Gas metal-arc	Constant Voltage	DC	DCEP

* In some applications, the submerged-arc process can use constant voltage DC.

CLASSIFICATION OF POWER SOURCES

Power sources are classified according to the type of current — AC or DC — and according to their output, which may be either constant current or constant voltage. A further classification designates the method by which energy is supplied to the power source — from a power line directly or through an electric motor, or from a gasoline or diesel engine.

Whatever the type of power source, its main function is to supply the type of current needed for welding. Alternating current direct from the power line goes through a transformer in AC welders that allows control of the current. Thus, a simple AC power source is fed 230-volt single-phase current the same as a kitchen stove, and a selector switch enables the operator to use what AC current he needs for the job — for example, a 225-amp output for 3/16-in. electrodes or 180-amp for 5/32-in. electrodes. A DC power source also gives similar control of the current. Direct current is produced from AC line power by either using the line power to run an electric motor that turns a DC generator (an electric motor-generator set) or running the line power through a transformer and then a rectifier (a rectifier set). Direct current may also be produced by driving a DC generator with any type of fuel-burning engine, such as a gasoline or diesel engine (engine-driven-generator set). A fuel-burning engine may also be used to produce AC for welding by using it to drive an alternator instead of a generator. Combination power sources, producing both AC and DC, are basically transformer-rectifier sets.

Welding power sources of all types are rated according to their current output. This rating is generally set by manufacturers in accordance with standards established by the National Electrical Manufacturers Association (NEMA). These standards are established on a conservative basis, requiring a rating well below the maximum overload capacity of the power source so that it will provide safe operation efficiently over a long period of time.

Ratings are given with a percentage "duty cycle." The duty cycle of a power source is the percentage of a ten-minute period that it can operate at a given output current setting. For example, if a power source is rated 300 amp at a 60% duty cycle, it means that the machine can be operated safely at 300-amp welding current for 6 out of every 10 minutes. If this duty cycle is reduced in actual operation, the maximum permissible current is increased. Thus, at 35% duty cycle, this same 300-amp power source could be operated at 375 amp.

As noted previously, power sources are classified as "constant current" or "constant voltage." A constant current power source is one that delivers a current that changes only slightly with changes in voltage. A constant voltage machine is one that delivers current with the voltage rising or dropping only slightly with changes in current output.

Figure 4-4a shows a typical output curve for a constant current power source. This type of output is used for submerged-arc, gas tungsten-arc, and shielded metal-arc applications. Figure 4-4b shows typical output characteristics of a constant voltage power source. Here, the voltage in the constant voltage curve (a) rises slightly at the low currents and drops at the high currents. Most constant voltage power sources are designed with a small downward slope, as in curve (b), and have adjustments to increase the downward slope, as in curve (c). Some power sources have a rising slope, as in (d), but this type of output is becoming less common.

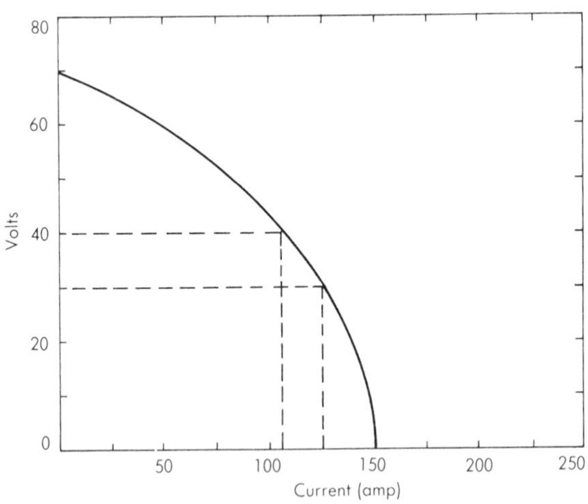

Fig. 4-4a. Typical output curve for a constant-current power source, adjusted for minimum current variation.

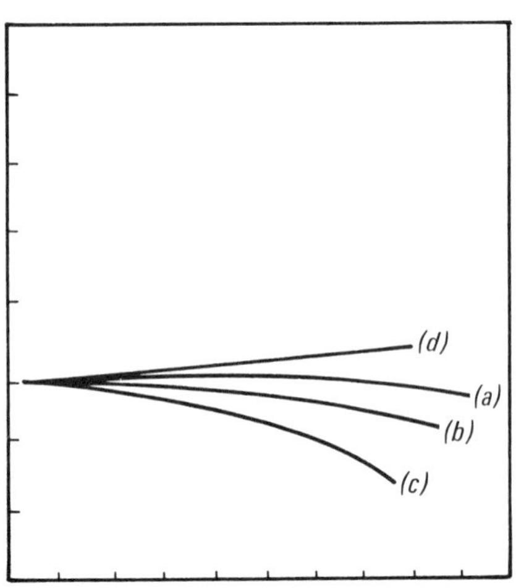

Fig. 4-4b. Typical output curves for a constant voltage power source.

AC POWER SOURCES

Transformer Power Sources: The transformer power source is a voltage step-down transformer that changes high-voltage, low-amperage AC input current to low-voltage, high-amperage AC welding current. AC output transformer power sources usually operate on single-phase power. Most AC power produced in the United States is 60-hertz, and each time the polarity changes the voltage goes through zero, which tends to create an unstable condition in the arc. This problem, however, has been solved by designing better transient characteristics in the power source and better AC electrodes.

Transformer power sources have controls to stabilize and adjust the welding current. A system for controlling the output current is provided either through a series of taps into the secondary windings or by a movable or saturable reactor in the output circuit. The taps provide step control. A reactor provides a continuous stepless control. Various types of starters are used and some are equipped with

Fig. 4-5. A typical small AC transformer power source for light-duty and limited-service welding.

Fig. 4-6. A typical engine-driven power source that supplies AC for welding or power for lights and tools.

low-voltage contactors to reduce open-circuit voltage when the machine is not operating. Some power sources have an "arc booster" that gives an extra surge of current for a few seconds at the start of the arc in order to get deeper penetration at the beginning of the weld. Most welding transformers can be equipped with capacitors to improve power factor and reduce the amount of input current used.

For the inert-gas shielded arc welding processes, transformer power sources are equipped with necessary auxiliary controls. A device is required with GTAW to help establish and maintain the arc.

Small, inexpensive transformer power sources are widely used in light industry, maintenance work, and by farmers. Fig. 4-5 illustrates a typical small (225-amp) AC power source. Rotating the switch at the center of the power source changes taps on the secondary coil, which in turn, changes the welding current.

Small power sources (180-amp or less) are available to meet Rural Electrification Administration input requirements.

Transformer power sources rated at 600 amp or more are used primarily for automatic welding. Available with these power sources are optional accessories required for automatic welding, such as line contactors, remote current control, and DC for control power. With single submerged-arc welding, single-phase power is used. When two AC arcs are used, the power sources are connected to a three phase power system to equalize the load. Three transformers can be used with the primaries connected to the three phase line and the secondaries connected closed delta. Each transformer must have a separate reactor to adjust the welding current and the phase angle between the arc currents. The Scott connection can also be used. Two transformers with a center tap connection on one primary are connected to a three phase power line. The unique connection between the two transformers establishes the proper phase relation between the arcs. With two electrodes,

TABLE 4-32. Typical Ratings and Outputs for AC Constant Current Power Sources

NEMA Rating			Output Current Range (amp)
Rated Current (amp)	Arc Voltage	Duty Cycle (%)	
180	25	20	30-180
225	25	20	40-225
250	30	30	30-300
300	32	60	30-450
400	36	60	40-600
500	40	60	50-750
600	44	60	50-850
1000	44	60	200-1250

NOTE: Input power is single-phase.

TABLE 4-33. Typical Ratings and Outputs for Alternator Power Sources and Auxiliary Power Sources

NEMA Rating			Output		Engine	
Rated Current (amp)	Arc Voltage	Duty Cycle (%)	Current (amp)	Type	Number of Cylinders	HP and Speed (RPM)
130 AC	25	30* 100†	60-130* 3500†	Constant current	1	10.0 3600
225 AC	25	50* 100†	20-225* 5000†	Constant current	2	14.2 2200

* Welding output.
† Watts output when used as auxiliary power source.

Fig. 4-7. A typical industrial AC three-phase input, DC output constant current power source. The heavy duty power sources are available in a wide range of sizes, see Table 4-35.

it is necessary to have approximately 90° out-of-phase operation to prevent interactions between the electrodes that would produce severe arc blow. Reactors are used to adjust the welding current. Details of the connection can be supplied by the equipment manufacturer.

Table 4-32 shows typical AC welder ratings and output currents. A disadvantage of many transformer power sources is that the output current changes with a change in line voltage. In most shops this is not a serious problem, but if the power-line voltage regulation is poor, the welding may not be satisfactory.

The industrial size Idealarc AC 1200 has built-in line voltage compensation that holds the welding current reasonably steady to help maintain weld quality when the input voltage varies within a range of ±10%. This solid-state circuit automatically adjusts the machine so operators do not waste production time resetting the output when input line voltage fluctuates.

Alternators: AC welding current can also be obtained from an engine-driven alternator. A gasoline engine is usually used, and the engine-alternator set serves both as a portable power source and as an auxiliary power supply. Power output — 115 to 230 volts AC — can be used for lights, small tools, or as a standby energy source. A typical machine is illustrated in Fig. 4-6. Table 4-33 shows typical alternator ratings and output currents.

TABLE 4-34. Typical Ratings and Outputs for Transformer-Rectifier Power Sources with Both AC and DC Constant Current Outputs

NEMA Rating			Output Current	
Current (amp)	Voltage	Duty Cycle (%)	AC (amp)	DC (amp)
250	30	30	30-300	30-250
300	32	60	30-450	45-375
400	36	60	40-600	60-500
500	40	60	50-750	75-625
600	44	60	50-850	75-750

TABLE 4-35. Typical Ratings and Outputs for Three-Phase Transformer-Rectifier Power Sources

NEMA Rating			Output Current	
Current (amp)	Voltage	Duty Cycle (%)	DC Current (amp)	Type
300	32	60	45-375	Constant Current
300	32	100	50-375	Constant Voltage
400	36	60	60-500	Constant Current
400	36	80	50-500	Constant Voltage
500	40	60	75-650	Constant Current
600	44	100	70-750	Constant Voltage
600	44	60	75-750	Constant Current
800	44	100	100-1000	Constant Voltage

DC AND AC-DC POWER SOURCES

Transformer-Rectifier Power Sources: Rectifiers for converting AC to DC have been developed to a stage of efficiency and reliability. A result of this development has been the combination of a rectifier with a transformer to form a DC power source. Various semiconducting materials have been used in current rectifiers, however silicon rectifiers have replaced most other types in welding power sources.

In principle, the single-phase rectifier power source is a transformer power source with a rectifier added to obtain a DC output. Adjustment of the welding current is through the AC section, as described for transformer power sources, or through solid state devices controlling output characteristics and providing line voltage compensation. The output characteristic can be either constant current or constant voltage. Some power sources built especially for gas metal-arc welding have adjustments for changing both the slope of the output curve and the reactance in the circuit for better performance when welding with short-circuiting transfer.

Fig. 4-8. An AC-DC welder for gas tungsten arc and covered electrode welding equipped for pulsation and square wave AC.

Transformer-rectifier power sources are often designed with provisions for both AC and DC welding. These power sources are especially convenient for structural work where the vertical welding is done by DC with E7018 electrodes, and flat welding is done by AC with E7028 electrodes. Combination power sources are also convenient for gas tungsten-arc welding; AC is available for welding aluminum, and DC is available for welding stainless and carbon steel. Table 4-34 shows typical ratings and outputs for combination AC-DC transformer-rectifier power sources.

Another type of transformer-rectifier power source is the step-down transformer, in which three-phase AC is fed to rectifier units which, in turn, feed DC to a single output circuit. The output can be either constant current or constant voltage, but only DC is available since the AC is three-phase and cannot be used for welding. Table 4-35 shows typical ratings and outputs for three-phase transformer-rectifier power sources. See Fig. 4-7.

Making optimum use of some welding processes may require that accessory equipment be added to the power source. This is especially true if the process is automated. A good example is the AC-DC Square Wave transformer-rectifier power source built for the gas tungsten-arc process, sometimes called a TIG welder. A typical machine is shown in Fig. 4-8.

This power source can be used for any process using AC or DC constant current, but the accessories are designed primarily for gas tungsten-arc. A high-frequency voltage is superimposed on the Square Wave output voltage so that the arc is established without touching the electrode to the work. The high frequency also stabilizes the arc by igniting the 60-cycle current each time it goes through zero. The intensity of the high-frequency voltage can be adjusted. The welding current is adjusted electronically with provisions to connect a remote current control. The pulse rate feature of this power source can be adjusted from 0.1 pulses per second to 10 pulses per second. Advantages of the adjustable pulse rate include reduction of heat input to minimize distortion and compensation for poor fitup. Standard features include an adjustable crater fill circuit and single range welding current control which can be adjusted from a very low value — 5 amps or less to a maximum of 400 amps. Solenoid valves start and stop the flow of cooling water and gas. The gas valve has an electronic delay so that gas continues to flow after the arc is extinguished — to protect the crater and electrode from oxidation.

The transformer-rectifier power source has the same disadvantages as the transformer power source. A change in voltage on the transformer primary changes the welding current. The transformer-rectifier shown in Fig. 4-8 has line voltage compensation to eliminate the problem.

Inverter Power Source: The state-of-the-art DC transformer power source is an Inverter. The power line voltage can be either single-phase or three-phase and is applied directly to a bridge circuit where it is rectified and filtered to produce a pure DC voltage. Solid-state switches then chop the DC to provide a high frequency AC voltage, typically between 2 and 20 KHz. This high frequency voltage is then applied to the primary of a transformer. The secondary output of the transformer is then rectified and filtered to supply a DC welding voltage and current. Solid-state control circuits are incorporated to provide for output characteristics and line voltage compensation.

Inverter power sources can be designed and manufactured as compact and light weight units. A 300 amp Inverter power source can weigh less than 80 lbs. The corresponding step-down transformer rectifier power source can weigh in excess of 375

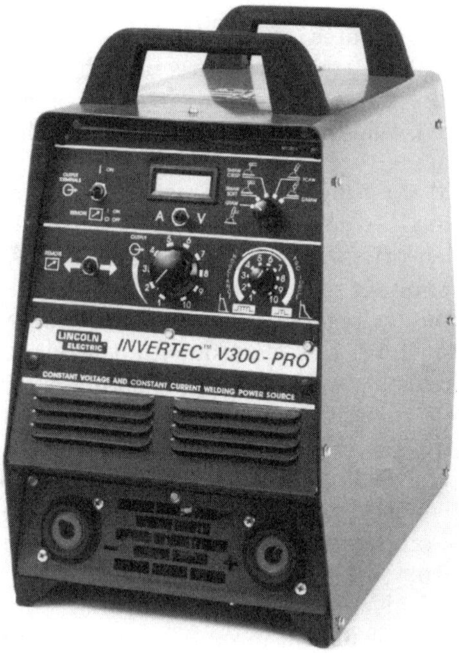

Fig. 4-9. Typical inverter DC transformer welding power supply.

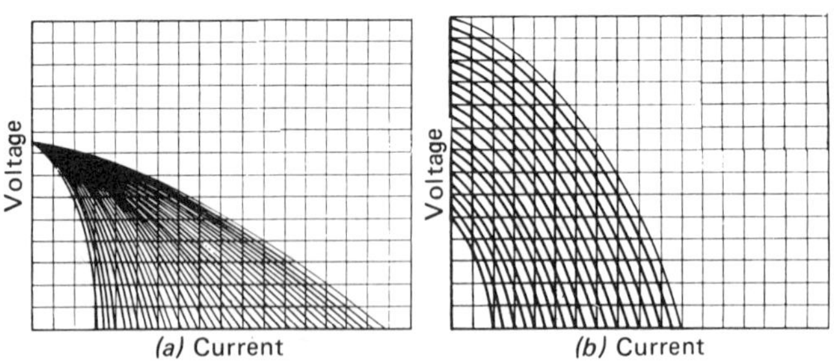

Fig. 4-10. Output for a DC generator power source having adjustments in both the series and shunt fields. Output curves produced by adjusting the series field are shown in (a); curves produced by adjusting the shunt field are shown in (b).

lbs. The light weight design makes an Inverter an excellent choice for jobs where portability is important. See Fig. 4-9.

DC Generators: In the direct-current generator, an armature rotates in a magnetic field. Current is generated in the armature and is taken off for use through a commutator. The armature is rotated either by an electric motor or an internal-combustion engine. The speed of rotation of the armature and the electrical design of the generator change the output characteristics. The arc characteristics of a generator can be precisely controlled. This fact lends DC welding more versatility than AC welding. Polarity of the electrode can be changed with a flip of a switch.

The DC motor-generator power source is driven by AC utility power. It can provide either constant current or constant voltage, or a single unit may provide both types of output. The motor is usually a three-phase induction motor.

The constant current type is a compound generator with a series field that causes the voltage to decrease as the current is increased. Two adjustments can be made to change the welding current:

1. For a given voltage, the output current can be changed by adjusting the series field. This produces an output change as shown in Fig. 4-10 (a), and is sometimes called the "current" control.
2. For a given current control setting, the output can be changed by adjusting the shunt field. This produces an output change as shown in Fig. 4-10 (b).

Combining both adjustments can produce output characteristics similar to those shown in Fig. 4-4a or 4-11b. A typical motor-generator power source is shown in Fig. 4-11a.

DC-generator power sources, in general, have an adjustment that can provide an output of the type shown in Fig. 4-11b. This output is highly suitable for vertical and overhead welding, where the operator uses a whipping motion that alternately raises and lowers the arc voltage. With the flatter characteristic shown in Fig. 4-11b., there is greater change in current for a given change in voltage than with the output in Fig. 4-4a. Since deposition varies with current, the welder can vary deposition and thereby exercise more control of the molten puddle with the flatter output characteristic.

Fig. 4-11a. A typical motor-generator power source.

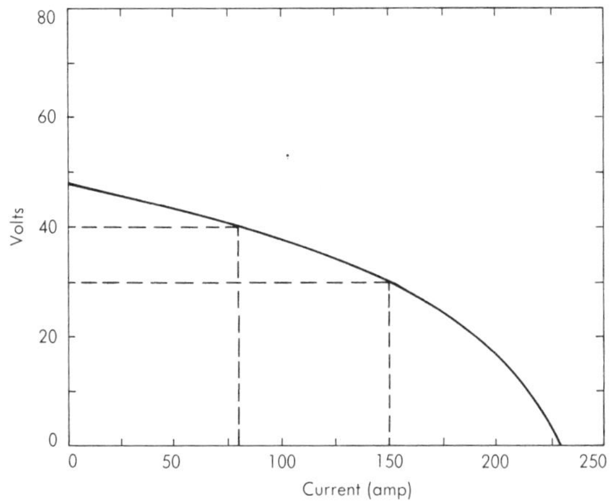

Fig. 4-11b. Typical output curve preferred for vertical and overhead shielded metal-arc welding.

The constant-voltage motor-generator power source is a compound generator with a series field designed to keep the voltage nearly constant within the current capacity of the power source, as in Fig. 4-4b (a). The slope of the output curve can be changed by an adjustment in the series field, as in curves (b) and (c). In some power sources, an output shown by curve (d) can be obtained. These power sources are usually used with automatic or semiautomatic wire-feeding equipment, and the current is changed by changing the speed of the wire feed. The arc voltage is changed by adjusting the shunt field in the generator.

Motor-generator power sources that provide both constant current and constant voltage are gaining wider application, because they can meet a wide range of process requirements. Constant current is used to manually tack weld an assembly, and the welding is then completed with an automatic or semiautomatic process using constant voltage. Table 4-36 shows typical ratings and outputs for these motor-generator power sources.

TABLE 4-36. Typical Ratings and Outputs for Motor-Generator DC Power Sources

Rating			Output	
Current (amp)	Voltage	Duty Cycle (%)	Constant Current (amp)	Constant Voltage (amp)
200	28	60	30-300	—
250	30	30	40-250	—
300	32	60	45-450	—
300	32	60	60-375	60-375
400	36	60	60-600	—
400	36	60	60-600	60-600
600	44	80	—	875 max
800	44	80	100-900	1000 max
1000	44	80	—	1250 max

Every type of DC power source driven by an electric motor can be duplicated with a gasoline or diesel-engine drive. On heavy-duty power sources of 200 amp and larger, the engines are liquid cooled. Gas engines are equipped with governors to maintain constant engine speed and with idling devices to reduce the engine speed when welding is not being done. Power sources with air-cooled engines are available for light-duty work.

Diesel engines cost more than gasoline engines, but the diesel has several advantages. Diesel fuel costs a little less than gasoline, is less hazardous to handle, and is consumed less rapidly. Less maintenance is required with diesels, and engine life is longer.

Multiple-Output Power Source: A multiple-output power source is a single welding power source capable of providing welding current to several operators simultaneously. The use of such power sources is limited to manual welding where several operators are working in a relatively small area. Many factors limit the economic use of these units; when an application appears feasible, the equipment manufacturers should be consulted.

SUGGESTIONS FOR SELECTING A POWER SOURCE

In selecting a power source, two important considerations are its output capacity and its suitability for the particular job.

The size or rated output of a power source required for a given job depends on the thickness of the metal to be welded and the amount of welding to be done. If a conservatively rated power source, made by a reputable manufacturer, is purchased, the selection can be made with confidence in that rating. There is no need to buy more capacity than will be required by the job. Be sure, however, to check the duty cycle. Power Sources with a low duty cycle should be used only for maintenance or intermittent welding. Continued operation of a power source beyond its rated capacity will shorten its service life. Of course, properly made and rated power sources have large overload capacity, which means that higher than rated amperages can be used for shorter periods than the rated duty cycle allows.

In selecting the type of power source, an essential consideration is the energy source available. Motor-generator sets are generally available for only three-phase utility AC power, but can be ordered to different cycles and voltages. They are also available for DC power. AC power sources are generally available for only single-phase power in various cycles, with or without power factor correction in the power source. Fortunately, in most manufacturing, the source of power does not present a limiting factor in the selection of a power source. The decision can be made on the basis of which is the most efficient and economical power source for a given job.

Where utility power service is through a 3 KVA transformer on residential or rural lines, an industrial-type AC power source cannot be used. Here, it is necessary to use a limited-input transformer power source, which is designed so that no more than a specified maximum amount of input current (37.5 amp) can be drawn.

The most important factor to be considered in selecting a power source is performance — what type power source will do the job easiest and enable better welding to be done at lower costs.

There is one best way for every welding job. Sometimes it is AC; sometimes it is DC. For one job, sensitive control may be required for maximum efficiency. For another, certain types of controls may be unnecessary. A power source should be selected, therefore, according to the job to be done.

The following may be used as a guide to select the proper power source based on the type of current.

- DC only
  - Gas metal-arc welding
  - Flux-cored arc-welding
  - Exx10 type electrodes
  - Exx15 type electrodes
- DC preferred
  - The "Out-of-Position" Group
  - The "High Speed" Group
  - Welding stainless steel and other low hydrogen types of electrodes
  - Nonferrous electrodes
  - Surfacing with high alloy electrodes
- AC preferred
  - The "High Deposition" Group
  - Iron powder electrodes except out-of-position welding
  - Where arc blow is a problem
- AC or DC depending on the application
  - Gas tungsten-arc welding
  - Submerged-arc welding

The small transformer power source shown in Fig. 4-5 is widely used on farms, in garages and small machine shops, and by hobbyists. Obviously AC is not always the best type of welding current for such a wide variety of applications. However, the special electrodes and accessories developed for this type of power source make it very versatile even though limited to AC welding current. In this case, the selection of the power source is based on low cost, low power input requirements, and versatility rather than AC or DC.

If a job is entirely flat position in heavy plate, an AC power source will be most efficient. If the job is exclusively sheet metal welding, a DC power source will be most efficient. If the work is a combination of jobs, involving out-of-position welding, as well as straight flat position work, a combination AC-DC power source is the logical choice. These power sources can be adapted to individual job requirements, combining larger AC capacity with smaller DC capacity, or in any way that is required. For most manufacturing situations, both AC and DC are needed for maximum efficiency. The combination power source, therefore, is most efficient for general purpose welding. It gives the weldor the opportunity to select for himself the type of arc and current he can use most efficiently for the job at hand.

Welding Equipment

Since there are several major arc-welding processes and various stages of mechanization in each, the welding equipment, in addition to the power source, involves numerous mechanisms and devices to facilitate laying a weld bead. It is beyond the scope of this handbook to catalog the many hundreds of items that are available. However, by observing the items of equipment required for the simplest form of arc welding — stick-electrode welding — and, then selectively replacing elements of this simple process with the counterparts that give increasing mechanization, it is believed that the reader can be given an understanding of the relation of any item of welding equipment to the basic scheme of fusion-joining by arc heating. In essence, all items of welding equipment other than the protective shields and clothing worn by workers are devices introduced in the basic welding circuit discussed in Section 1.3. Also, the differences between processes at the basic level amount to no more than different ways for accomplishing the same end.

In Fig. 1-31, the basic welding circuit is illustrated. In this circuit is the power source, with cables running from it in one direction to the work and in the other direction to the electrode, from the tip of which the arc is struck. On each side of the power source — extending to the work or the arc-delivering electrode — are other items of equipment needed to accomplish welding. This equipment will vary according to the welding process and its degree of mechanization.

THE WELDING CABLE

Whatever the process, the first item of equipment on both sides of the power source is a cable for carrying electrical current. The cable from the power source to the electrode carries current to the electrode and via the arc to the workpiece. The cable from the power source to the workpiece is the ground cable; it completes the circuit from the workpiece back to the power source.

These conductors are extremely important to the efficiency and success of welding. They must be properly sized to deliver the required current to the welding arc. The size must be adjusted to the length of cable, and the cable assembly must be so designed and built that it performs its current-conducting function efficiently while giving wear resistance where wear is needed and flexibility where flexibility is advantageous.

The conductors in a welding cable are made from strands of fine copper or aluminum wire — preferably copper. The fine wires may be bunched in bundles, with a twisted assembly of bundles enclosed by a paper wrapping. A high-grade insulating material — usually natural or synthetic rubber — over the paper wrapping provides positive insulation. The paper wrapping allows the copper wire to slip within the rubber cover, enhancing the cable's flexibility. A special cable design adds cotton or rayon braiding, followed by a second "outside" covering of rubber. This outside covering is often fluted to increase the cooling area and enhance flexibility.

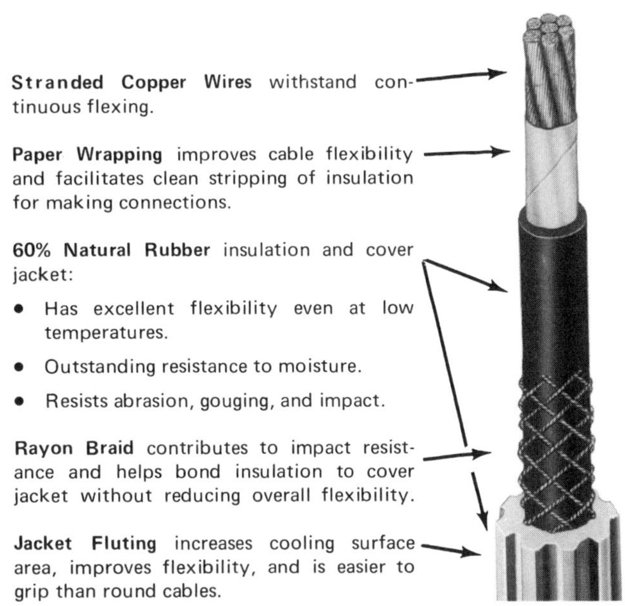

Fig. 4-12. Construction of a typical welding cable designed for good flexibility. A cable of this type would be used as the "whip" section of a cable assembly in manual metal-arc welding.

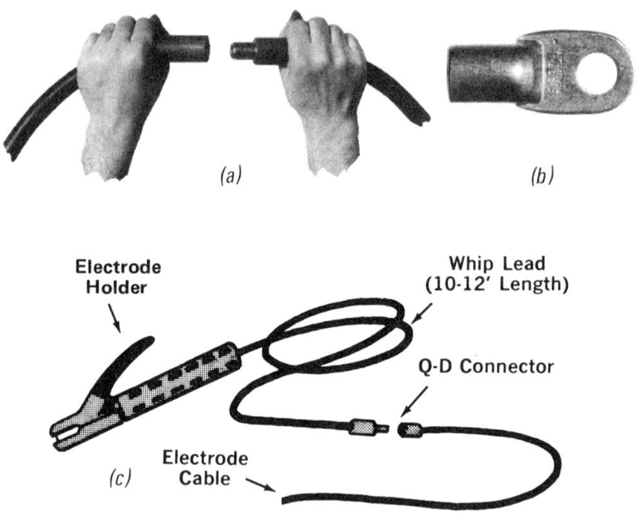

Fig. 4-13. All connections in a cable are important to efficient welding. (a) A quick-detachable connector, used for splicing or to connect (c) the whip section to the main cable. (b) The type of lug used at terminals.

Figure 4-12 shows the construction of a premium-type cable, designed to give good flexibility and long service life. Often a "whip" cable of this flexibility grade would be used within 10 to 12 feet of the electrode holder in manual metal-arc welding, where its flexibility would allow ease in maneuvering the electrode. Such a whip section would use the smallest size permissible for its length and the current to be carried, in order to assure maximum flexibility. A whip section of highly flexible cable is usually connected to a stiffer length of cable that runs to the output terminal of the power source. This section of cable is similar in construction, except that often synthetic rubber is used for insulating coverings to provide toughness and impact and abrasion resistance. The ground cable extending from the power source to the workpiece is also usually made from the stiffer, highly wear-resistant type of cable.

The size of cable used depends on the amperage to be carried and the total length of the electrical circuit. The longer the circuit, the larger the size wire needed to prevent voltage drop and the dissipation of energy by resistance heating within the conductor.

Table 4-37 shows recommended copper cable sizes for manual welding. Using this table, one can determine the size needed for the particular welding station. Thus, if the maximum machine current is 200 amp and the station is 250 ft from the power source, 1/0 cable would be required. This size would allow up to a 60% duty cycle — or welding 6 out of every 10 minutes — without excessive heating of the cable or excessive voltage drop.

Just as the proper construction and size of the cable are important to efficient welding, so are all connections within it or at its terminals. Quick-detachable (Q-D) connectors, Fig. 4-13(a), are provided for splicing cable lengths or for attaching the whip length to the main cable, Fig. 4-13(c). These connectors are designed to lock so they cannot accidentally be pulled apart or work loose. They also are insulated. Cable lugs, Fig. 4-13(b), are needed for connecting welding cables to power sources, panels, switchboards, and some types of ground clamps or electrode holders and are provided in sizes to fit the cables. Lugs and Q-D connectors should be soldered to the cable.

TABLE 4-37. Recommended Cable Sizes for Manual Welding*

Machine Size in Amperes	Duty Cycle (%)	Copper Cable Sizes for Combined Lengths of Electrode plus Ground Cable				
		Up to 50 feet	50-100 feet	100-150 feet	150-200 feet	200-250 feet
100	20	#8	#4	#3	#2	#1
180	20	#5	#4	#3	#2	#1
180	30	#4	#4	#3	#2	#1
200	50	#3	#3	#2	#1	#1/0
200	60	#2	#2	#2	#1	#1/0
225	20	#4	#3	#2	#1	#1/0
250 +	30	#3	#3	#2	#1	#1/0
300	60	#1/0	#1/0	#1/0	#2/0	#3/0
400	60	#2/0	#2/0	#2/0	#3/0	#4/0
500	60	#2/0	#2/0	#3/0	#3/0	#4/0
600	60	#3/0	#3/0	#3/0	#4/0	* * *
650	60	#3/0	#3/0	#4/0	* *	* * *

* For fully automatic welding, use two 4/0 cables for less than 1200 amperes or three 4/0 cables for up to 1500 amperes.

* * Use double strand of #2/0.

* * * Use double strand of #3/0.

\+ For 225-amp 40% duty-cycle machines, use same cable size as 250-amp 30% duty cycle machines.

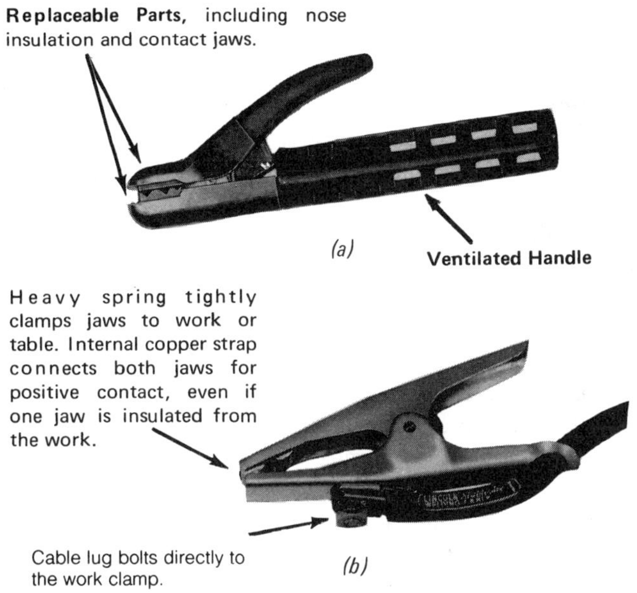

Fig. 4-14. (a) An electrode holder with replaceable jaws, ventilated handle, and nose and arm insulation. (b) A work clamp for attaching the work cable to the workpiece or fixture.

THE ELECTRODE HOLDER

Following the basic welding circuit, there is one more item of welding equipment between the power source and the electrode in manual metal-arc welding — namely, the electrode holder. This is basically no more than a clamping device for grasping and making firm electrical contact with the electrode, with an insulated handle for the operator's hand. It can be designed, however, with a number of refinements, such as replaceable jaws, a ventilated handle, and a hand shield. Figure 4-14(a) shows a type of holder with replaceable jaws, ventilated handle, and nose and arm insulation. The cable conductor may be connected to the holder by a nut-and-bolt arrangement, or the holder may come from the supplier with a whip section of cable permanently attached by some type of fused joint. Solder connections are provided in some designs.

Insulated electrode holders are used almost universally to prevent grounding the holder should it accidentally touch the work or the fixture holding the work.

Electrode holders are rated according to their current capacities, which are usually in the range of 300 to 600 amperes, and are produced in sizes to accommodate the sizes of electrodes required. Since operator comfort is desired, the smallest holder that will do the job is usually chosen. If a holder becomes hot while welding, the operator knows that it is too light for the welding current being used. There should be a margin between the current rating for the holder and the amperage used on the job. Thus, if the welding is done at 275 amp, the holder should be rated for at least 300 amp.

GROUND CONNECTIONS

On the ground side of the power source, there must be a means of connecting the work cable to the workpiece. A work clamp, bolted to the lug at the end of the work cable, is a simple and versatile connector. A work clamp is shown in Fig. 4-14(b). The work clamp may be attached to the workpiece directly or to a fixture that holds the workpiece and makes good electrical contact with it.

The ground connection varies with the process, and, in highly automatic welding installations, may be a permanent stationary connection or a connection made through brushes or rotating or sliding shoes (see Section 4.4). The ground must make a firm, positive connection, and its placement in respect to the layout of the assembly and direction of welding is important for good arc characteristics and the prevention or minimization of arc blow (see Section 3.2). The experienced welder knows that good grounding is essential to good work and that attention to this factor cannot be overemphasized.

In some welding installations, it is the practice of using a steel bar or a reinforcing rod as a common grounding connection between two or more work-holding fixtures. When this is done, care must be taken to assure that the steel grounding bar has adequate cross-sectional area to match the copper welding cable in electrical conductivity. Since the conductivity of copper is almost 7 times that of mild steel, the cross-sectional area of any common steel grounding bar should be at least 7 times the cross section of the welding cable conductor. If the grounding bar is inadequate in cross section, high resistance will develop in the circuit, the voltage will drop, and poor performance will result.

Grounding is discussed repeatedly in subsequent sections, emphasizing its importance to efficient welding. A point to bear in mind is that the ground connection should occupy the total surface of contact and the area of that surface must be at least equal to the cross-sectional area of the conductor. This means that the area of contact must be free from any scale, rust, oil, grease, oxides, or dirt that would act as areas of insulation. Brightening the area of contact with sandpaper or a wire brush before making the connection is good practice.

There is a simple way to test the soundness of the circuit during welding — namely, any time the circuit is off, run the hand over the length of the cable from the power source to the electrode. If a "hot" section or a section warmer than the rest of the cable is felt, one knows that undue resistance heating is occurring in that section. If the hot section is near a terminal, the connection at the terminal is suspect; if any place along the cable, the cause is probably broken strands within the cable. If the entire cable is hot to touch, it is probably undersized for the welding current being used.

With mechanized welding equipment, conductors other than those carrying the welding current may also be required between the power source and the welding gun or head. These conductors are small wires that carry just enough current to operate controlling devices. They are usually in a cable separate from the welding current cable, but, in some instances, may be incorporated in the welding-cable construction.

THE SEMIAUTOMATIC GUN AND WIRE FEEDER

Once any degree of welding mechanization is attempted, the equipment becomes complicated — but still it is merely a modification of the primary elements in the basic welding circuit. Thus, when one moves from stick electrode to semiautomatic welding with continuous electrode, a semiautomatic gun and an electrode wire-feeding mechanism take over the function of the simple electrode holder. A passageway, or conduit, made from plastic tubing or spirally-wound steel wire, is provided for movement of the continuous electrode. This conduit is usually incorporated into the cable carrying the welding current, which may also contain wires for the control circuit that initiates electrode feeding. Additionally, tubing or another arrangement may be required to supply granular flux or shielding gas to the arc, depending on the process.

If the mechanization is carried a step further — to full automation — a welding head with all the auxiliary equipment required is substituted for the simple electrode holder in the basic welding circuit. One of the simplest semiautomatic processes in terms of equipment is the self-shielded flux-cored electrode process, a schematic of which is shown in Fig. 4-15. Here, a wire-feeding mechanism with its controls and a reel of electrode wire has been interposed in that part of the basic welding circuit between the power source and the arc, and the electrode holder used in stick-electrode welding has been replaced with a welding gun.

Figure 4-16(a) depicts a typical wire feeder with the wire drive-roll and control unit built integrally

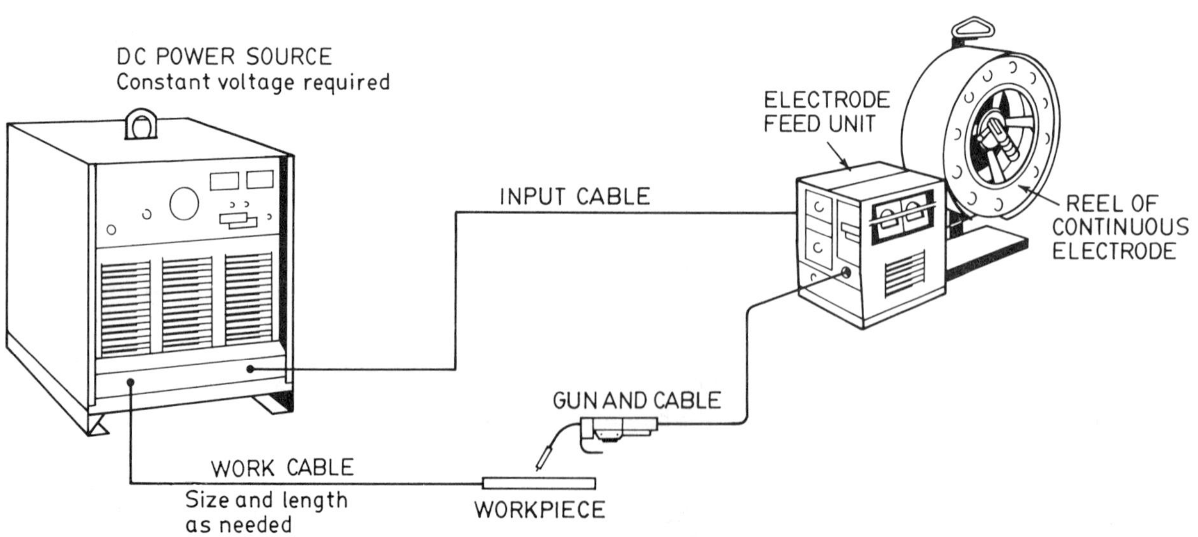

Fig. 4-15. A schematic of the semiautomatic open-arc self-shielded flux-cored electrode process — one of the simplest semiautomatic processes in terms of equipment and the most versatile for steel welding.

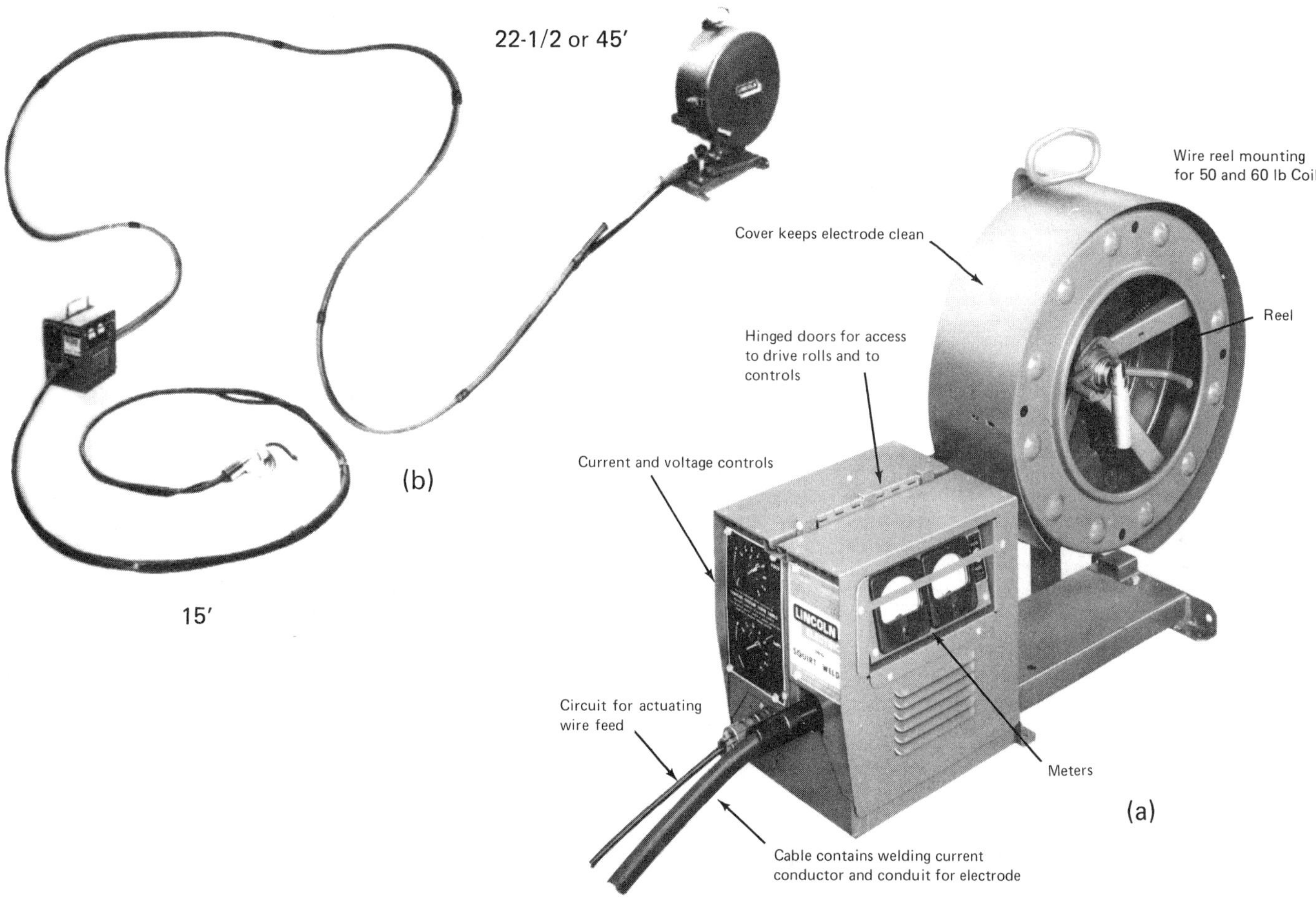

Fig. 4-16. (a) A typical wire-feeding mechanism, as used with semiautomatic self-shielded flux-cored or submerged-arc welding. (b) How the drive-roll and control unit may be separated from the coil of electrode to give even greater flexibility to semiautomatic welding.

with the wire reel. Note that two cables exit from the mechanism. The large one contains both the welding current cable and the conduit passage for the electrode wire. The small cable contains the wires for the circuit that activates wire feed and also activates contactor circuit so electrode is cold when not welding. The electrode feed rate is set by the controls on the control box, and this rate, in turn, regulates the amperage supplied by the constant-voltage power source.

In Fig. 4-16(b) the drive-roll and control unit are seen detached from the reel housing — a modification that permits working farther from the electrode supply and moving the control mechanism into tighter working quarters. Here, the drive mechanism pulls wire from the reel and then pushes it the remaining distance to the welding gun. Wire feeding is initiated and the welding circuit energized when the operator presses a trigger on the welding gun.

Flux-cored electrodes require drive-rolls that will not flatten or otherwise distort the electrode. Various grooved and knurled surfaces are used to advance the wire. Some wire feeders have a single pair of drive-rolls; others, two pairs with at least one roll of each pair powered.

Good, trouble-free wire feeding depends on the size of the drive motor, the diameter of the electrode, the length of cable between feeder and gun, the type of gun — straight-through or curved nozzle — and the surface condition of the electrode. Thus, a system that is satisfactory for driving a 3/32-in. electrode through a 10-ft conduit and a straight-through gun may be unsatisfactory for driving a 7/64-in. electrode 15 ft through a cable and through a 60° curved-nozzle gun.

The semiautomatic welding gun (Fig. 4-17) is designed for comfortable holding and ease in manipulation. The gun provides internal contacts that initiate wire feed and complete the welding

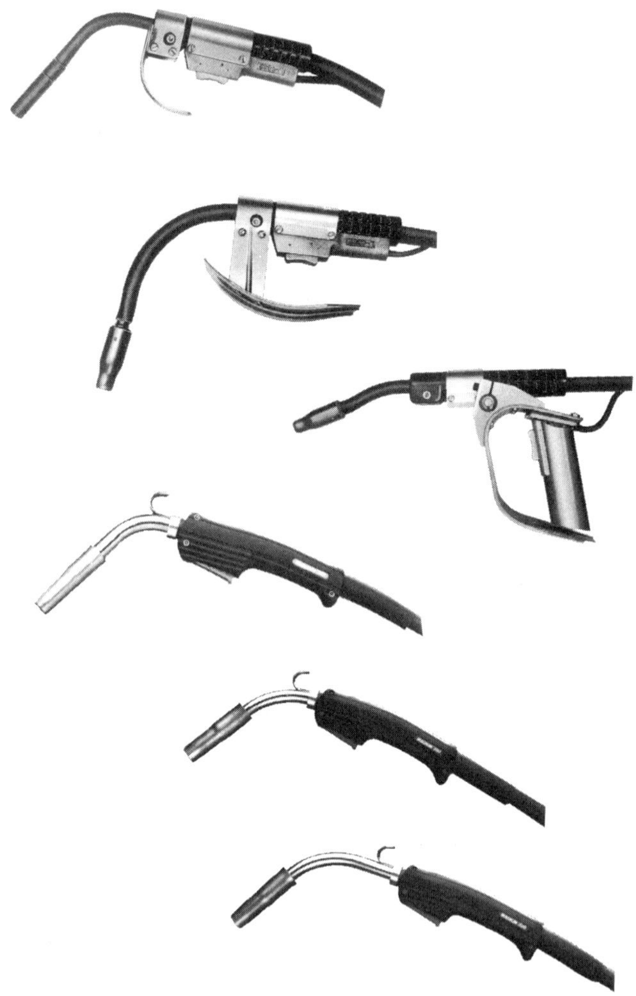

Fig. 4-17. Various designs of guns for semiautomatic welding with continuous wire feed. The ones at top are for self shielded flux-cored electrode welding and are rated at 350, 450 and 600 amps, respectively. The lower three guns, are for gas shielded arc welding and are rated at 200, 300 and 400 amps. respectively.

circuit at the press of the trigger switch. Most semiautomatic guns for the self-shielded, gas-shielded, and submerged-arc processes are designed for air cooling, which permits them to be small, light, and highly maneuverable. Water-cooled guns are also available, but this requires separate tubing for supply and return of water and a bulkier construction to provide water passages; thus, ease of handling and maneuverability are reduced.

Guns may have either straight or curved nozzles. The curved nozzle may vary from 45° to 80°. Straight-through nozzles are favored for fillet welds and when the work flows under the gun. Curved nozzles are preferred for depositing butt welds in heavy sections. The curved nozzle enhances flexibility and ease in manipulation, but requires heavier-duty wire-drive units. Curved nozzles are generally favored with the self-shielded process.

Semiautomatic guns should be of simple construction for minimum maintenance, but must be rugged to withstand physical abuse. Nozzles, contact tips, and switches should be designed for long life and ease of replacement. Nozzles should be rugged, insulated to prevent shorting to the workpiece and resist adherence of spatter. Contact tips should be made of a suitable alloy and should have sufficient cross-sectional area to provide optimum life for the size of electrode and current used.

The equipment for gas-shielded or submerged-arc semiautomatic welding requires additional features — such as equipment for containing, metering, and conducting shielding gas to the welding gun; tanks and hoses for submerged-arc flux; and modifications at the gun for dispensing the shielding material. The sections treating the individual processes discuss some of the specific items of equipment. Several types of semiautomatic guns for submerged-arc welding are shown in Fig. 4-18.

Semiautomatic welding equipment appears quite simple compared to full-automatic welding equipment, but possibly represents even a higher stage in

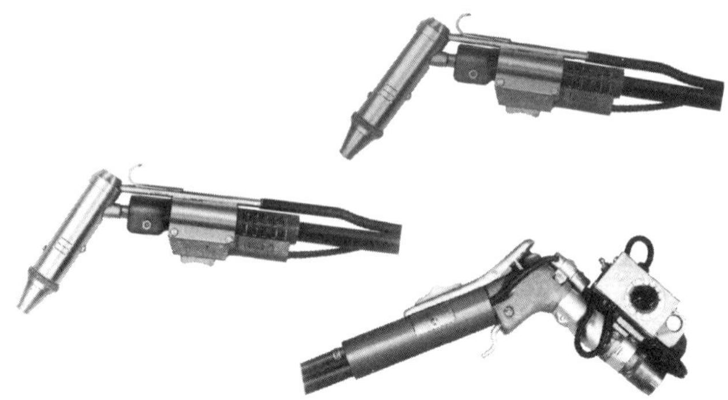

Fig. 4-18. Semiautomatic guns for submerged-arc welding. The concentric nozzle supplies flux as well as electrode wire. The gun at the top is rated at 350 amp; the others at 600 amp. Note the mechanized hand travel unit attached to the bottom gun; this small motorized unit paces the welding travel (see Fig. 4-19).

mechanization ingenuity because of its simplicity Semiautomatic welding attempts to put a degree of mechanization in the great bulk of welding work that is not subject to full mechanization. This work requires extreme flexibility, which is impossible with any fully mechanized equipment. With semiautomatic welding, the operator still holds the equivalence of an electrode holder (the gun) in hand and travels and maneuvers it, but frequent electrode changing

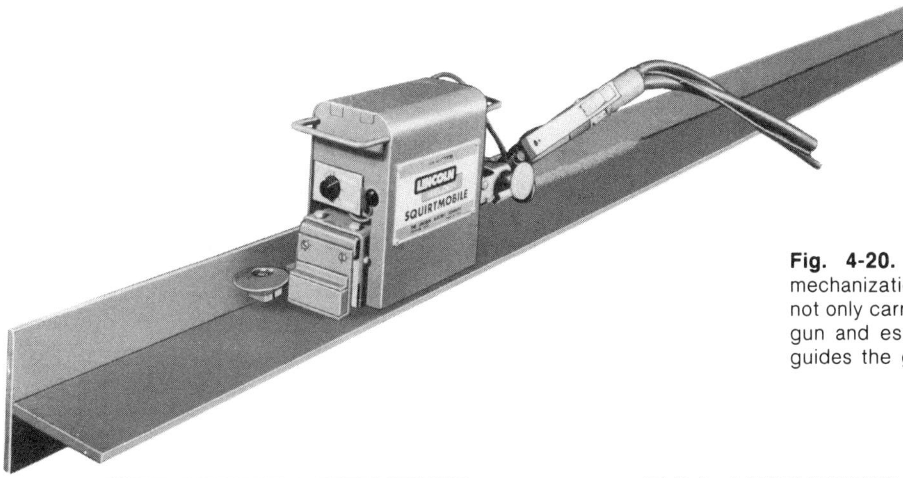

Fig. 4-20. The final step before full mechanization is when a motorized unit not only carries the semiautomatic welding gun and establishes travel rate, but also guides the gun along the weld joint.

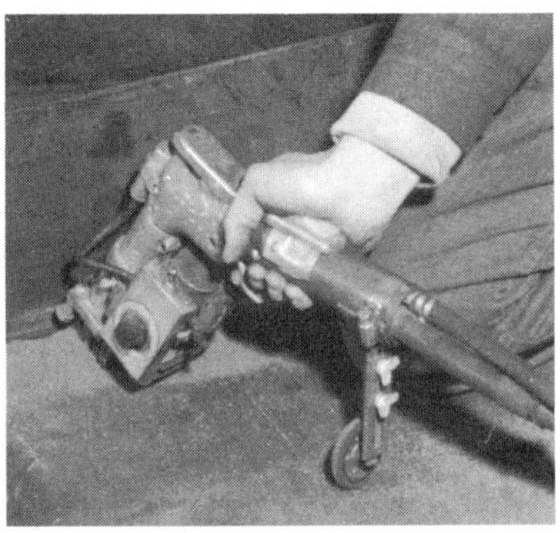

Fig. 4-19. A small motorized unit added to the submerged-arc gun establishes the travel rate and supports the gun as it travels the seam. Here, the weldor merely guides the semiautomatic gun and controls starting and stopping.

is eliminated and the pace of welding is established by the preset electrode feeding rate.

MECHANIZED TRAVEL UNITS

Semiautomatic welding takes a step toward full mechanization when devices and mechanisms are used with the semiautomatic gun to relieve the operator of holding, guiding, and traveling it. Thus, a small mechanized travel attachment mounted on the nozzle of the submerged-arc gun, shown in Fig. 4-19, establishes the rate of travel over a seam and supports the weight of the gun. Or, a small self-propelled trackless carriage, such as shown in Fig. 4-20, relieves the operator of holding and guiding the gun over the joint. His only function left is placing it on the work, starting and stopping its operation, and monitoring its performance. The sketches in Fig. 4-21 show how the spring-loaded guide rolls of the unit are set to maintain the desired path over the assembly.

FULL-AUTOMATIC WELDING HEADS

With fully mechanized welding, a great variety of equipment — much of it custom-designed — is used, but still all of this equipment may be viewed as a substitute for the simple electrode holder used in stick-electrode welding. In a fully mechanized installation, the manual skill of the weldor is replaced entirely. Machines supply the skill — and the operators of such machines provide the instructions and watch over the machines to make sure that instructions are followed.

The welding head replaces the welding gun at the fully mechanized stage of arc-welding development. The head may be stationary with the work moving under it, or it may move across the work.

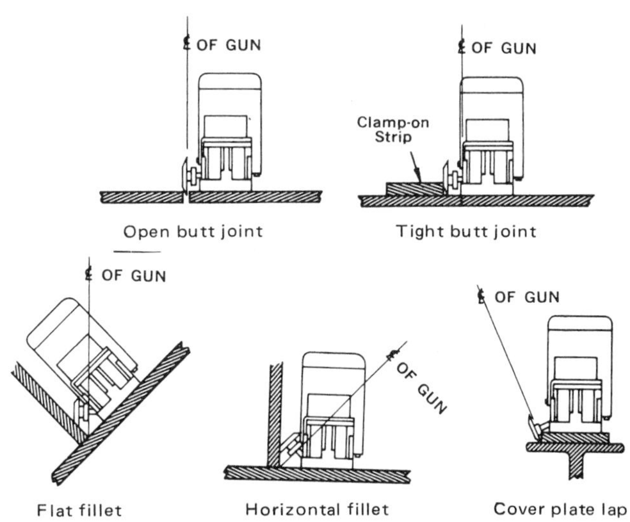

Fig. 4-21. These sketches show how the outboard guide wheels of the motorized unit shown in Fig. 4-20 are set to follow various types of joints.

Fig. 4-22. Schematics of various wire-feeder arrangements for full-automatic submerged-arc welding.

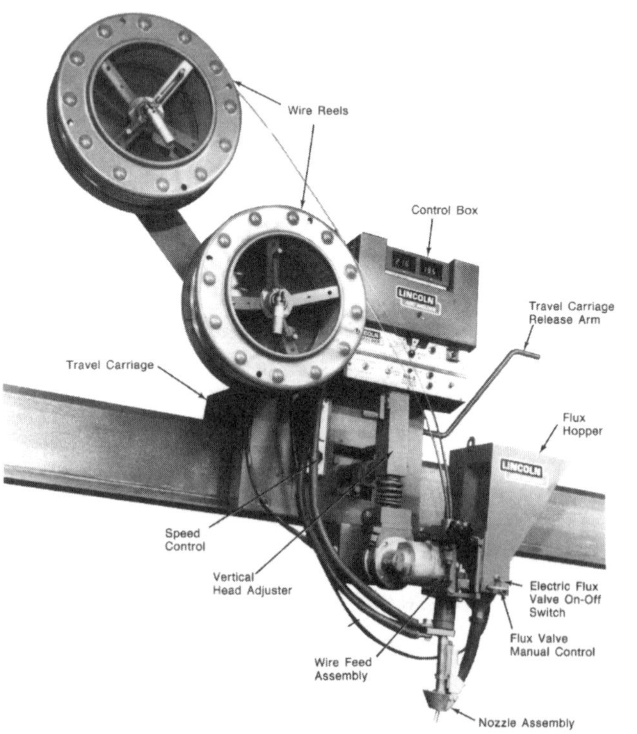

Fig. 4-23. A typical full-automatic welding head for twin submerged-arc welding.

The head may feed one wire (single-arc head) or two wires (Twinarc*), and two or more heads may be used together (tandem-arc and multiarc).

Figure 4-22 shows sketches of various wire-feeder arrangements for full-automatic submerged-arc welding. Here, the heads also meter out the granular flux needed for shielding the arc. With gas-shielded processes, attachments to the head similarly supply the gases needed for the shielding.

A typical full-automatic submerged-arc welding heads is shown in Fig. 4-23. It is depicted as it would ride on a beam over the joint, carrying a reel of electrode wire. The wire feeder is an integral unit, as is the flux hopper. The control circuit automatically varies the wire feed speed to maintain the arc length. The head is adjustable to various positions for different fixtures and applications, as illustrated in Fig. 4-24.

Full-automatic welding is not limited to fixed locations. By use of a self-propelled trackless tractor, the welding equipment may be taken to the work,

* Twinarc is a registered tradename.

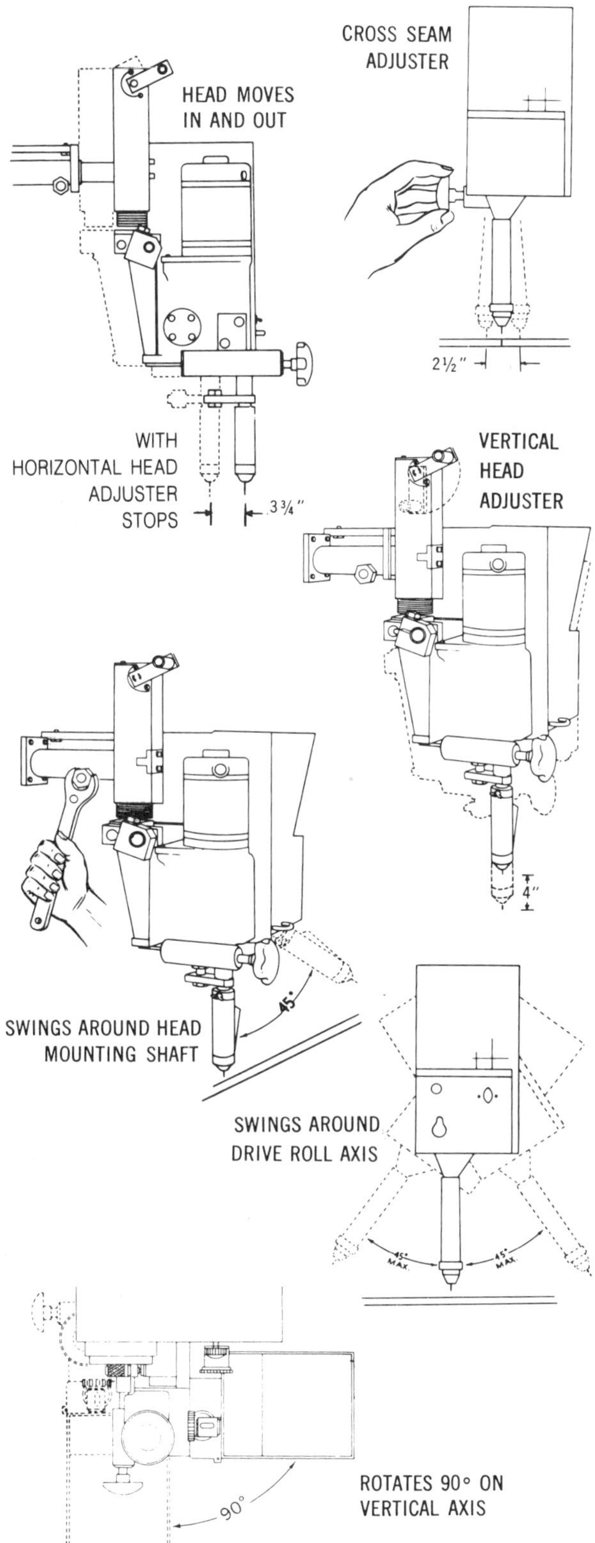

Fig. 4-24. How the head illustrated in Fig. 4-23 can be adjusted to different positions.

Fig. 4-25. A typical self-propelled tractor that takes full-automatic submerged-arc welding to the work, either in the field or in the shop. It can be used with or without a track.

rather than the work brought to the welding equipment. Self-propelled units, such as the DC-AC tandem submerged-arc welder shown in Fig. 4-25, are widely used in ship and barge building; beam, girder, and column fabrication; bridge deck welding;

Fig. 4-26. Twinarc welding on a preformed box beam. Special attachments permit two wires to be fed to the joint through one head. In this application, the use of two wires almost doubled the welding rate over that experienced with one wire.

Fig. 4-27. Welding with two tandem-arc heads simultaneously. Two groove welds in a side member of a foundry flask are deposited in one traverse of the heads. Lead electrodes are operated at 700 amp and 35 volts DC negative and the trail electrodes at 650 amp and 35 volts AC.

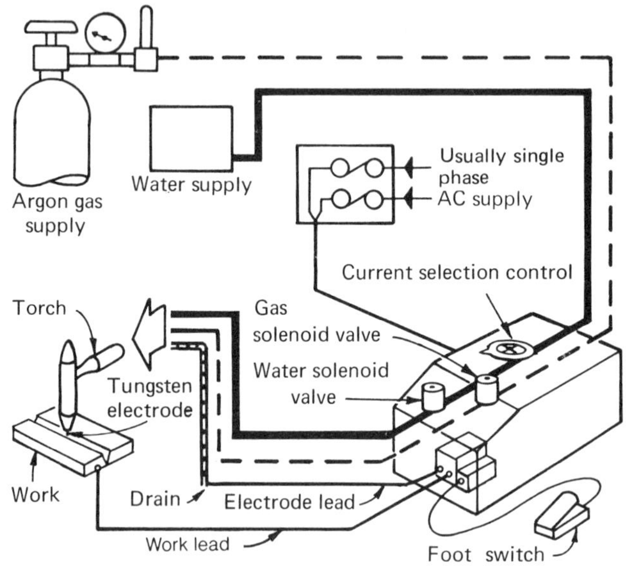

Fig. 4-28. Representative items of equipment used in GTAW.

and for long seams on other weldments, either in the shop or in the field.

The installation for full-automatic welding can become extremely complex — with as many as 20 arcs used simultaneously. Figure 4-26 shows the use of two electrode wires with one head (Twinarc welding) in the welding of a preformed box beam by the submerged-arc process. This is possibly what might be regarded as the first step toward increased complexity in fully mechanized welding. The attachment that permitted the use of two arcs with one head in this particular application almost doubled the welding rate that had been experienced when using a single electrode with the beam-supported head. Normally a 50% increase in speed can be expected.

In Fig. 4-27, an increased degree of complexity is seen. Here, two tandem-arc heads are being used to make two groove welds in one traverse in fabricating foundry flasks. Electrode wire is supplied from four reels. Additional examples of full-automatic welding installations are shown in Section 4.4.

EQUIPMENT FOR ARC HEATING

Heretofore, in this section, welding equipment has been presented as those elements that might be interposed in the basic welding circuit — from the

(b)

Fig. 4-29. A transformer-rectifier welder (a) designed for AC-DC GTA welding or for AC-DC welding with consumable covered electrodes; (b) a high-frequency generator for adapting an AC or DC metal-arc welder for use with the GTAW process.

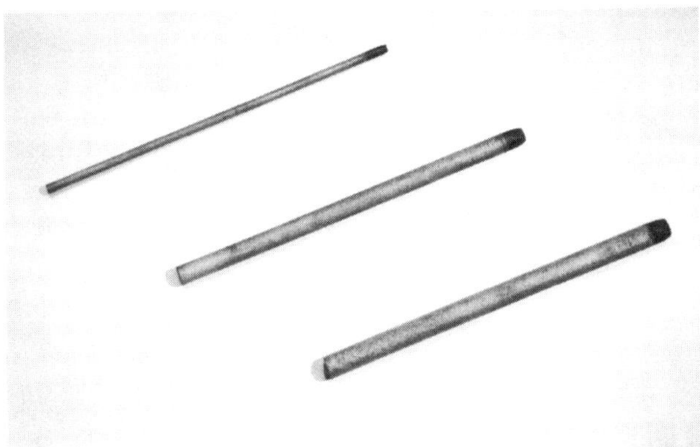

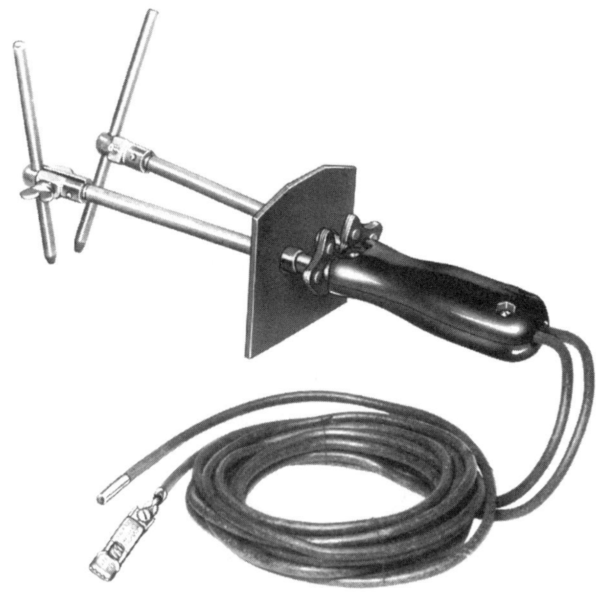

Fig. 4-30. The arc torch is a tool for welding, brazing, soldering, and heating for bending, straightening, and shrinking.

power source to a consumable electrode and from the power source to the workpiece.

As indicated in Section 1.3, the electrode in the basic welding circuit could be nonconsumable, and the arc struck by it could be for the purpose of heating the joint to fuse parts together or to melt separately applied filler metal into the joint. When the electrode is nonconsumable, the arc is much like an acetylene-oxygen torch — a method of heating. Welding may be done with the arc heat — or the heat may be used for brazing, soldering, or such operations as bending and straightening.

A nonconsumable electrode of tungsten allows a weld to be made from a separately applied filler rod or wire and is the basis for the gas tungsten arc (GTAW) process. Figure 4-28 shows representative items of equipment used in GTAW, and Fig. 4-29(a) shows a transformer-rectifier welder that can be used either for AC or DC GTAW process or for AC or DC welding with consumable covered electrodes. The unit illustrated by (b) in Fig. 4-29 is a high-frequency generator for adapting an AC or DC metal-arc welder for use with the GTAW process. Both of the illustrated units have inert-gas and cooling-water valves.

A carbon electrode can also be used as a nonconsumable in the basic welding circuit to develop arc heat and permit welding. Carbon-arc welding, however, has disadvantages and is no longer an accepted industrial process. If the workpiece is eliminated from the circuit, however, and the leads from the power source are run to two carbon electrodes, the tips of which are close to each other, an arc can be established that has practical use in welding, brazing, soldering, and heating for other purposes. Figure 4-30 illustrates the arc torch that results.

Note that the torch has arrangements for adjusting the positions of the rods in the holder and has thumb controls for moving the carbon tips in respect to each other to adjust arc length. The carbon rods are available in several diameters and with a copper coating to increase service life.

The arc torch is used in welding aluminum and copper alloys and in brazing and soldering. Since it can be held any distance above a workpiece, it is also useful for spot or area heating to enable the bending or straightening of metals or for heat-shrinking to remove distortion from a part or assembly.

Tungsten and carbon-arc torches are "open-arc" arc-heating processes. When the arc struck by a nonconsumable electrode is passed through a constricting orifice — such as a water-cooled copper orifice — a plasma jet results. While a conventional gas-tungsten arc is attracted to the nearest ground and is deflected by even low-strength magnetic fields, a plasma jet goes in the direction pointed and is less affected by magnetic fields. High current densities and power concentrations can be produced by arc constriction, and gas temperatures of 18,000°C and higher can be realized. A plasma-arc torch, thus, makes possible the melting of the most refractory materials and the welding of the higher melting-point metals.

Plasma-arc heating and welding is applicable to the more exotic manufacturing endeavors, but is of limited interest to the steel fabrication industries. The AWS Welding Handbook, Eighth Edition, Volume 2, Chapter 10, is a source of information on plasma-arc, heating and welding and the equipment and procedures involved.

PROTECTIVE EQUIPMENT

Arc-welding equipment also includes items worn by welding personnel to protect the face and eyes from the direct rays of the arc and the hands and body from burns. Such items include face shields, head shields, welding lenses, goggles, aprons, gloves and glasses. The requirements for such protective equipment are treated in Section 15.

Welding booths and screens around welding stations may also be looked upon as welding equipment. Such temporary or permanent structures are desirable to protect plant workers from the arc rays and spatter of molten metal and sparks. A ventilation system to remove smoke and fumes is also desirable and is provided by modern factory designs, whether or not welding operations are to be conducted.

Mechanized semiautomatic submerged-arc equipment designed to weld curved seams.

Fixtures and Manipulators

Once the production welding operations reach a stage of mechanization beyond that of the handheld and manually manipulated semiautomatic gun, fixtures, manipulators, and weldment positioners become important items of welding equipment. In fact, even with the hand-held semiautomatic gun some device for rotating the weldment or mechanically positioning it may be essential for realizing the maximum potential of semiautomatic welding.

Thus, the trunnion-type fixture shown in Fig. 4-31 for positioning the weldment so that welding can be in the flat positon with the semiautomatic self-shielded flux-cored arc process is a vital feature of the portrayed welding station. Similarly, in Fig. 4-32, the equipment that rotates cylindrical tanks under the semiautomatic welding gun is important to the total scheme of "almost fully mechanized" welding. In this case, also, a fixture is provided for "resting" the weldor's hand, so — without tension or strain — he can hold the arc an inch or so back of the centerline of the cylinder, so that the molten weld metal solidifies before it begins its downward path.

Fig. 4-31. The trunnion-type fixture used at this station to position the weldment for flat welding under the semiautomatic welding gun gives an approach to full mechanization.

Fig. 4-32. The fixturing equipment being used with the semiautomatic welding gun here in the welding of a head to a cylinder for a storage tank takes the work even a step further than the fixture in Fig. 4-31 toward full mechanization. A roller mechanism is being used to turn the cylinder under the gun, and the weldor's hand rests on a support so he can hold the gun precisely in reference to the centerline of the cylinder.

These two figures illustrate the fact that there is a gradual transition from what is termed "semiautomatic" welding to fully mechanized welding, and that holding and positioning devices are elements in that transition.

When the full-automatic welding head is used, a fixturing device becomes a necessity. It may be nothing more than a support for holding the material to be welded while a self-propelled tractor carries the welding head over the joint, or it may be a

(a) (b) (c) (d)

Fig. 4-33. Some typical work-moving welding fixtures are the (a) rotating positioner, (b) power turning rolls, (c) lathe, and (d) continuous mill.

complex mechanism, electrically or electronically controlled, and capable of such functions as "sensing" and self-correction. In some industries, such as the automotive, the weldment fixtures may cost many thousand of dollars.

TYPES OF FIXTURES

The objective in fixturing is to take maximum advantage of mechanized welding's ability to lower welding costs and increase productivity. The fixtures are usually devices that present the work or welding equipment in a position that assures maximum welding speed. Often these devices also systematize and control the flow of material to and from the welding equipment.

Welding fixtures fall into two broad classifications — those that act on the parts being welded, and those that act on the welding equipment. The type of action may be supporting, clamping, grounding, imparting movement, to name only a few. As an example, a welding fixture for the work may clamp the parts together and support them so that the weld can be made in the flat position. This fixture may hold the work stationary — in which case the welding equipment moves (or is traveled) over it —

(a) (b) (c) (d)

Fig. 4-34. These pictures illustrate four different types of welding equipment fixtures; (a) beam-with-carriage, (b) self-propelled tractor, (c) spud welder, and (d) boom-type manipulator with variable travel of both boom and column.

TABLE 4-38. WELDING EQUIPMENT FIXTURES

	BOOM MANIPULATORS	BOOM WITH CARRIAGE	SPUD WELDING	MILLS	TRACTORS (Full / Semi)
Work Fixtures Used with the Above	Rolls Positioners. (Material is usually pre-assembled)	Rolls. Clamps. Jigs. Positioners.	Centering Clamp with Rotating Head.	Pipe mill. Spiral mill. Forming mill.	Usually no fixtures.
Typical Applications	Tanks (Large). Machine bases. Hardfacing. Structural Beams & Columns.	Tanks. Hardfacing. Box beams. Automotive parts. Hot-water tanks. Refrigerator housings. Light poles.	Spuds in hot-water tanks. Small diameters where piece is too large to rotate.	Pipes. Box sections. Generator frames. Storage racks. Tubing. Structural Beams (Small to Medium).	Structural beams (Medium to Large). Ship building. Inside seams of tank, etc. Any long welds.

or may be mechanized and move the work under the stationary welding equipment. Some typical work moving types of fixtures are rotating positioners, power turning rolls, lathes, and mill-type setups, as illustrated in Fig. 4-33.

Fixtures that act on the welding equipment — some types are called welding manipulators — similarly may provide a stationary support only or both support and move (travel) the welding equipment. The welding equipment may be a welding head or a welding gun, or it may include all of the equipment in the welding circuit, including the power sources and even the flux-recovery unit.

The fixtures for holding the head stationary range from a simple arm and mounting bracket to the very complex installations common in continuous-mill welding. The types of fixtures that both support and travel the welding equipment include the beam-with-carriage, the self-propelled tractor, the spud welder, and a variety of boom-type welding manipulators (Fig. 4-34). Variations in the boom-type manipulators include column-stationary and boom-travel variable and both boom and column movable with variable travel. Table 4-38 charts some of the more frequently used work and welding-equipment fixtures and the weldments on which they are generally used.

Welding fixtures may also perform other functions that are not directly related to the actual welding operation. Thus, the fixturing may be an integral part of a conveyor system or may also position the work for a machining operation, such as drilling or grinding. Sometimes the designer even tries to incorporate part-forming into the welding fixture system. This is not recommended, however, since experience has proved that divorcing the welding fixture from all forming operations leads to more trouble-free operation.

Fig. 4-35. The welding fixture here combines a work-holding device and a beam-with-carriage support for the welding head. The elevator doors being fabricated are held in perfect alignment as a submerged-arc butt weld is made at 150 inches per minute.

Fig. 4-36. Making a circumferential weld with the welding head stationary and the work moving. Here the submerged-arc head is supported by a simple pedestal arrangement.

CONSIDERATIONS IN FIXTURE SELECTION

Elaborate welding equipment or work fixturing usually requires tremendous production volume of a single or reasonably similar parts. If part volume does not exist, the next thing to consider is substantial footage of welding that lends itself to mechanization. In this case, fixture adaptability, simplicity, and cost often are the important fixture-selection factors. This may mean combining a variety of work-supporting or holding devices with a simple beam-with-carriage fixture, as in Fig. 4-35. Even a small self-propelled tractor — if it can be classified as a welding fixture — mounting a semiautomatic welding gun may produce the best answer for a given job, since it features fixturing simplicity and represents a nominal equipment investment.

Joint geometry also plays an important role in the selection of a welding fixture. For example, circumferential welds are made with the welding head stationary and the work moving (Fig. 4-36). Welds in a straight line usually require a fixture that travels the head along the joint with a stationary work fixture providing clamping to maintain part alignment and fitup.

Welding Head Stationary

A stationary welding head calls for work movement and invariably leads to more complex work fixturing. In fact, the work fixturing for a stationary head can be the ultimate in automation (Fig. 4-37). Some fixtures have been designed to have the weight of the workpiece dropping into the fixture activate its positioning and clamping and initiate the welding cycle, thus, even eliminating the simple act of the welding operator pushing a start button.

Sometimes the head mounting of head-stationary fixturing will allow movement up and down or to one side to allow passage of the workpiece into and out of the fixture. It can also permit adjustment of the head position as welding progresses. This is still considered a head-stationary fixture, since welding speed is controlled by the movement of the work under the head.

Although a stationary welding head fixture is often considered the simplest type of fixturing, when it is incorporated as part of a continuous mill welder it becomes part of one of the most complex welding fixtures (Fig. 4-38). In a mill welding fixture, the

Fig. 4-37. The work fixturing for a stationary head can be the ultimate in automation, as this fixture for welding spring pads on axle housings in an automotive plant.

Fig. 4-38. The welding head is stationary in this mill-type fixture for pipe welding. The pipe is formed from plate; then continuously fed into the mill and welded as it passes under the head.

work is continuously fed under the stationary welding head, usually at high speeds. Often, the work is formed just ahead of the welding fixture and machined immediately after welding. Naturally, this complicates the basic work-fixturing operations of supporting, clamping, rounding, and part movement.

The expanding use of multiarc welding — where three, four, and even six welding heads are installed on the same welding fixture — is increasing the number of head-stationary welding installations. The shear weight and bulk of the welding equipment needed for these multihead installations tends to become excessive for a standard beam-with-carriage fixture or boom-type welding manipulator. The stationary head also eliminates the problem of finding ways to travel the control system, power sources, flux supply and recovery system, consumable electrode, and welding cables that must be solved with head-movable welding.

Welding Head Movable

With head-moving welding fixtures, the work remains stationary, and the welding head travels along the joint. Each movement may result from the travel of a carriage on a beam or rail, from a traveling boom mounted on a stationary column, from the travel of the entire manipulator riding on a track in the floor (Fig. 4-39), or from the travel of a tractor riding the surface of the workpiece (Fig. 4-40).

The beam-with-carriage-type welding-equipment fixture usually offers head travel in straight lines that run parallel to and on one side of the support for the beam (Fig. 4-41). The length of the weld that can be made without stopping is limited by the length of the beam.

A boom-type welding manipulator equipped for both boom and column travel can swing the welding

Fig. 4-40. This self-propelled tractor riding on the workpiece simplifies fixturing. In this case both the welding equipment and workpiece are traveling, but in opposite directions to keep the weld puddle in the flat position.

Fig. 4-39. This boom-type welding manipulator moves everything along the joint being welded — equipment, flux, flux recovery system, wire-electrode, power sources, and controls.

Fig. 4-41. The beam-with-carriage fixture usually offers head travel in straight lines parallel to the beam. The length of the weld that can be made without stopping is limited by the length of the beam.

Fig. 4-42. A boom-type welding manipulator can service a number of welding stations. If the loading, unloading, and welding cycles are all properly sequence, the welding operating factor can be substantially increased.

head through a 360° arc. It is capable of welding in a straight line in any direction away from the column to a distance somewhat less than the length of the boom and can service welding stations located anywhere within this area (Fig. 4-42). It duplicates the performance of a beam-with-carriage-type fixture when the boom is held stationary and the entire manipulator is traveled on the floor tracks. In addition, it can weld on either side of the column supporting the boom. Weld length is limited by the length of the tracks on which the manipulator carriage is traveling. Both the beam-with-carriage and the boom-type welding manipulator can serve as welding-head-stationary fixtures.

FIXTURE DESIGN AND PERFORMANCE

The efficient performance of a work station equipped for mechanized welding requires that the work fixture be well integrated with the welding-equipment fixture. In addition, careful attention must be given to finding answers for performance problems related to clamping, weld backup, workpiece grounding, weld-cycle timing, and arc blow.

Clamping and Fitup

Although the work-mounting fixture should not be used to "form" the workpiece, it must achieve complete, uniform joint closure and proper alignment. When possible, the clamping system should also provide follow-up pressure to compensate for weld shrinkage as the bead is run the length of the joint. This means that a pneumatic, hydraulic, or spring-

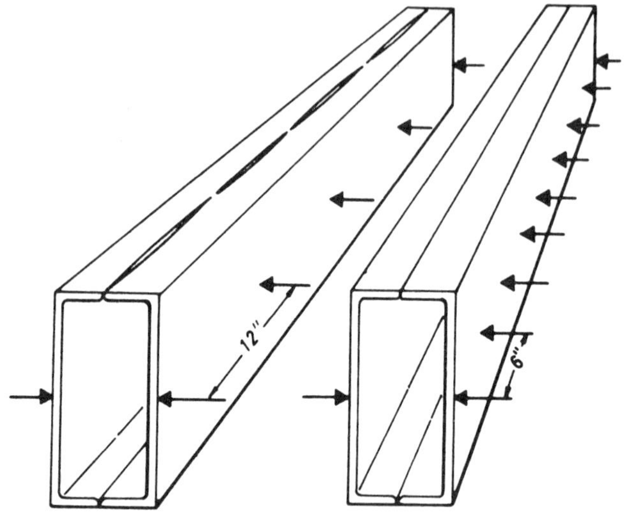

Fig. 4-43. When seam welding two formed sections, numerous pressure points are needed to keep the seam from opening and causing meltthrough as welding progresses.

loaded clamping system is preferred over a solid clamping arrangement.

A good clamping system has many small clamps or pressure-application points rather than a few large clamps or points of pressure. Box beams, for example, are difficult to hold. They should be clamped at closely spaced intervals along their entire length; otherwise gaps will develop during welding, and that promotes meltthrough and causes distortion of the finished part. A 6 or 8-in. channel box section should be clamped every 6 inches. One-foot spacing will let the joint open up ahead of the arc as welding progresses, as illustrated in Fig. 4-43.

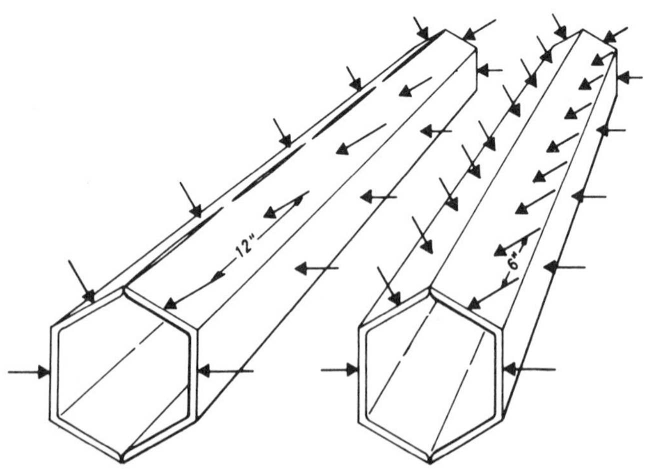

Fig. 4-44. Such a hexagonal box section not only requires clamping in two directions, but also at fairly short intervals to prevent gapping as welding progresses.

A corner weld in a box section is particularly sensitive to gapping. Heat in the edge causes it to expand outward, as illustrated in Fig. 4-44. Usually no backup is feasible for this type of joint, and, unless a constant-pressure clamping system is used, the gap will open and cause meltthrough as well as distortion.

Cylindrical sections likewise tend to open at the seam. The material, if improperly clamped, will lift off the backup bar (Fig. 4-45), causing holes that can be mistaken for porosity. To offset this possibility, the cylinder should be compressed so it is slightly oval-shaped during welding. Sometimes the edges of the tank are preshaped before welding to give the same effect.

If solid clamping is used in lieu of a pneumatic, hydraulic, or spring-loaded mechanical system, consideration should be given to the method for exerting pressure. Much time can often be saved by using wedges where applicable in place of bolts or threaded clamps. A wedge can be quickly hammered into or out of position, whereas tightening or loosening a bolt requires appreciable time. In addition,

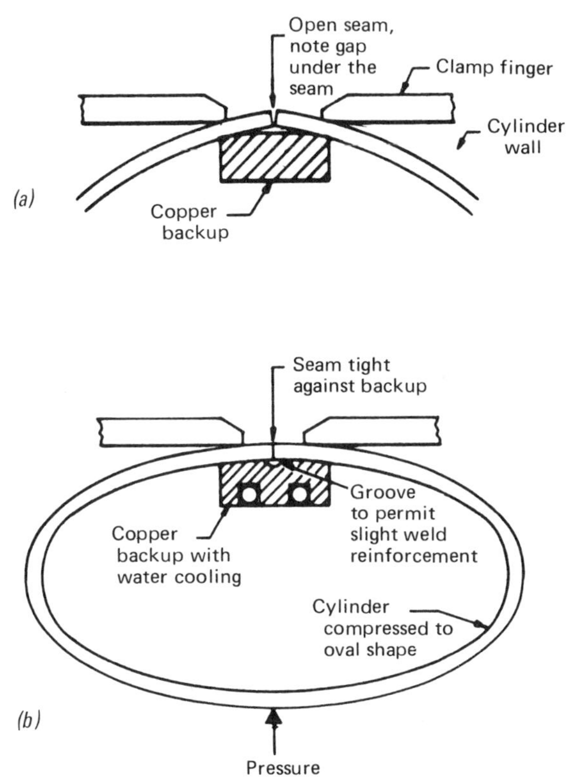

Fig. 4-45. Cylindrical sections tend to open over the copper backup bar (a). Compressing the work so it is slightly oval (b) offsets this tendency. The backup bar (b) has been grooved to establish weld reinforcement inside, and copper tubes are used to permit water cooling.

a modified "follow-up" pressure can be attained while welding is in progress. A tap of a hammer on a wedge can take up slack created by weld shrinkage.

Weld Backup

With some joints, a copper backup strip is needed to achieve high welding speed and good weld quality. The purpose of the backup is to support the molten weld metal during solidification and prevent meltthrough. Copper is used because of its high heat conductivity. This characteristic prevents the weld metal from fusing to the backup strip. Where it is desirable to reinforce the underside of the weld, the backup strip may be grooved to the desired shape of the reinforcement, as illustrated in Fig. 4-45(b).

The backup strip must have enough mass to prevent it from being heated to its melting point beneath the arc and, thus, contaminating the underside of the weld with copper. Sometimes water is passed through the interior of the backup bar to keep it cool. This method of removing heat from the backup bar is usually used for jobs where the production rate is high.

A water-cooled backup bar is easily made by soldering or brazing copper tubing into milled slots in the back of the bar (Fig. 4-45 (b)). A word of caution is necessary concerning the use of a water-cooled backup bar. It is possible to let the backup bar get "too cool." A "too cool" bar will collect condensation on its surface before or even between welds and cause porosity in the weld. The experienced operator will begin welding before turning on the cooling water for the backup bar. He also shuts the water off if the welding operation is interrupted for an extended period of time. It might also be pointed out that the cooling water need not be cold. In fact, warm water will keep a copper backup bar cool, as compared to its melting temperature.

Occasionally, the job may call for a moving backup, such as a sliding shoe or a roll. For example, a mill welder for manufacturing pipe usually uses a spring-loaded copper shoe that is anchored inside the pipe immediately below the arc by an arm extending through the semiclosed crack just ahead of the welding arc. Sliding copper shoes are also used for long beams, and copper rolls on roundabouts.

When welding small diameters, a copper backup roll, positioned inside a pipe or cylindrical tank while welding the circumferential seams, should have a diameter slightly less than the inside diameter being welded. This will make the area of contact as long as possible. It must be small enough, however, to prevent the simultaneous shrinkage of the object

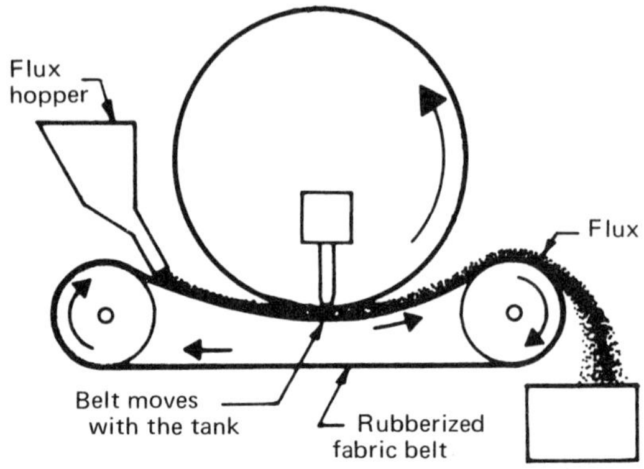

Fig. 4-46. A special adaptation of the principle of flux backup — a flux-fed system sometimes used when circumferentially welding tanks from the inside.

being welded and expansion of the backup roll from freezing it into position. For large cylindrical assemblies, the roll diameter should permit the surface of the roll and inside surface of the weldment to be in contact long enough to allow weld metal solidification.

In general, a copper backup should be pressed tightly against the workpiece to promote maximum heat flow and give a uniform surface appearance to the completed weld.

Flux, under moderate pressure, is sometimes used as a backup material. Its function is the same as the copper backup — to support the molten weld metal until it solidifies. Usually, the loose granular flux rests in a trough on a thin piece of nonconducting material. This nonconducting material, in turn, usually sets on an inflatable rubberized canvas fire hose. The air pressure fed to the hose to develop moderate "flux pressure" on the underside of the weld is usually no more than 5 to 10 psi.

A special adaptation of the principle of flux backup is shown in Fig. 4-46. This "flux-fed belt" is sometimes used when making the inside circumferential weld on large tanks or other types of cylindrical weldments.

Grounding the Workpiece

The method of grounding is an important consideration in fixture design, since it can affect arc action, the quality of the weld, and the speed at which it can be produced. The ground must make good contact with the workpiece through a copper clamp or through copper-graphite brushes or sliding or rotating shoes. Aluminum is too soft, easily melted, and oxidizable to make a good grounding material for welding.

The grounding cable or cables should be neatly organized in reference to the work — not strung about haphazardly. A spring-actuated ground clamp, as used in hand welding, will not assure good contact and doesn't have the capacity needed for mechanized welding. Ground cables should be lugged and C-clamped or bolted to the work.

Fig. 4-47. Fixture used to automatically weld vertical seams with the self-shielded flux-cored electrode process. Note the water cooled movable copper shoe used to contain the molten flux and weld metal.

When grounding is through a sliding shoe, two or more shoes should always be used. This will prevent interruptions of current in case one is lifted out of contact by an unexpected surface protrusion, such as the reinforcement of another weld, a piece of weld spatter, or granular flux.

Preloaded, tapered roller bearings are excellent for rotating grounds and give better performance than sliding brushes. To assure trouble-free performance, the contact area of the bearing should be sufficient to carry the current capacity of the welding cable used in the installation. Since most mechanized welding installations use two 4/0 cables in parallel, the tapered bearings must be fairly large.

A brief examination of the ground cables and connections of an operating welding installation will tell whether the grounding system is adequate and if the ground is making good contact. If the ground cables are uniformly hot their entire length, they are undersized and should be replaced. If the ground and the workpiece are hot near the point of contact, the contact is poor and should be corrected.

Poor ground contact can cause arcing and weld the ground to the workpiece. In addition, a voltage drop will occur across a poor ground, which reduces the voltage available to the arc. In extreme cases, the voltage drop across a poor ground can cause an unstable arc condition and impair weld quality.

Preventing Arc Blow

Properly designed fully mechanized equipment avoids the problem of arc blow (see Section 3.2) by observing the following precautionary measures:

1. All steel other than that being welded should be at least one inch from the arc.
2. The clamping fingers for a "stake-type" fixture should be made from a nonmagnetic material.
3. The ground should be connected directly to the work, not to the backup bar. The point of connection should be as far from the arc as possible. The ground can be split to offset arc blow.
4. Embedding a copper bar in steel to increase the mass of the backup is not good practice, since this will bring a magnetic material too close to the arc. A non-magnetic material should be used.
5. With submerged-arc welding the direction of travel should be away from the ground; toward the ground with open-arc welding. If the weld path crosses over or runs near a large section of steel, it may be necessary to use an auxiliary ground.
6. When possible, the fixture should be built from low-carbon steel and large masses of steel should be normalized to minimize residual magnetism.
7. With horn-type fixtures, the welding should be toward the closed end.
8. The electrode angle should be adjusted to a position that tends to overcome arc blow.

Ingenuity in grounding is often required to eliminate arc blow. For example, long, narrow weldments are more susceptible to arc blow than wider assemblies. An arc-blow problem on these assemblies frequently appears in the form of porosity, which usually occurs at the finish end of the weld. If changing the position or method of grounding doesn't correct the problem, the porosity can usually be eliminated by placing a one-inch tack weld at

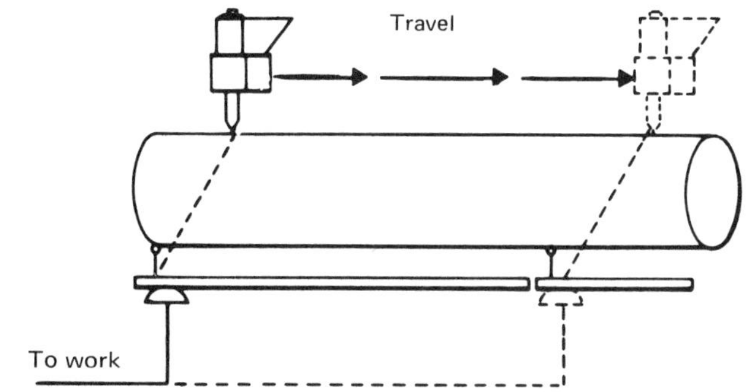

Fig. 4-48. An arrangement for eliminating end arc blow when seam welding hot-water tanks. The shoe slides with the head, switching to a forward position near the end of the weld.

the finish before welding the seam. Figure 4-48 shows a "last resort" remedy for eliminating end arc blow in the seam welding of hot water tanks; Fig. 4-49, a method of coping with arc blow when welding inside a tank.

Arc Starting and Timing

Starting problems are usually caused by a lack of rigidity in either the head or work mounting, or backlash in the drive system. In most cases, rigidizing the system will take care of starting problems when "hot starting." Poor arc starting can also be caused by an electrode diameter too large for the current being used, or an incorrect electrode drag angle causing improper contact between the work and

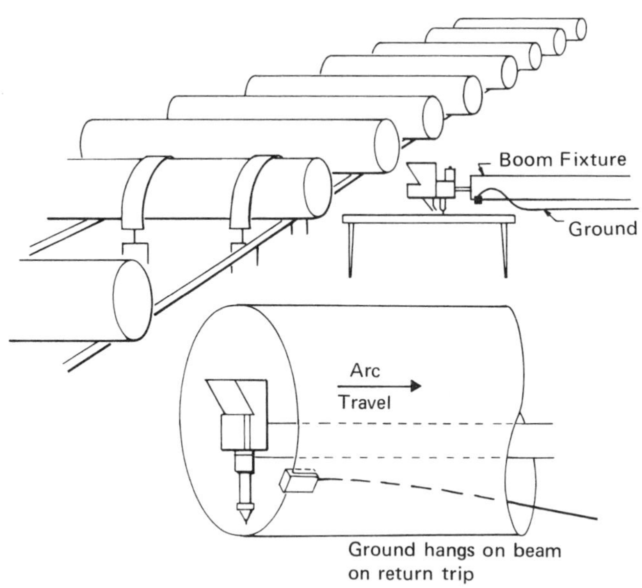

Fig. 4-49. Arc blow is a constant threat when a longitudinal weld is made inside a small tank with a boom fixture. Running the ground cable inside the tank and making the ground connection at the start end is one way to control the problem.

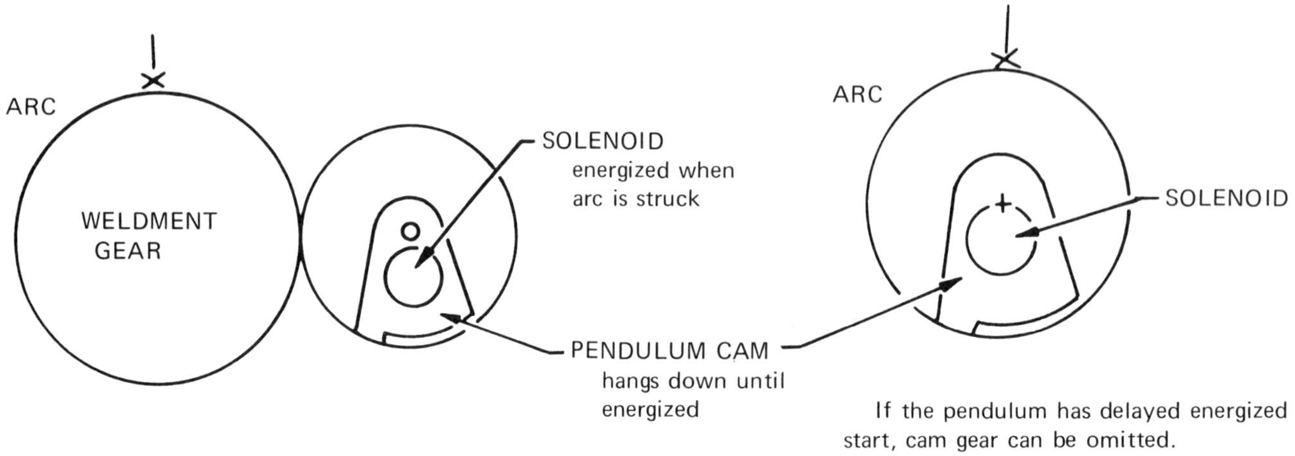

Pendulum-Cam Type
A pendulum cam hangs from the center of a second gear. When the arc is started, a solenoid locks the cam to a face plate and the whole cam assembly rotates around the shaft, which in turn operates the limit switches. The work is rotating when the arc is struck.

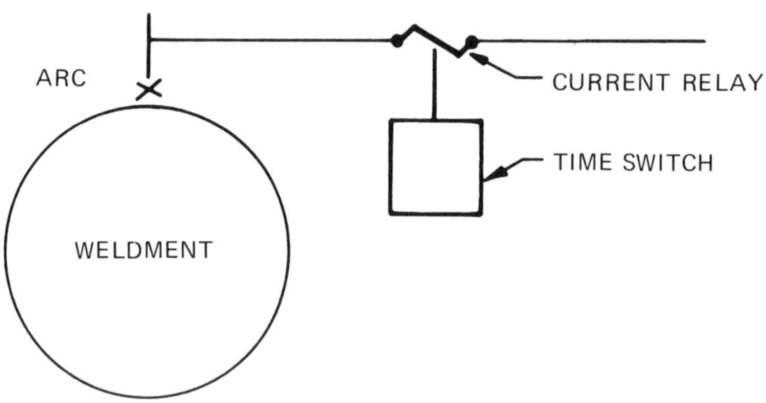

Time-Switch Type
No second gear is required with this type of timer, and the weld can start at any point. The work is rotating prior to striking it with the electrode. This system has advantages where striking with a hot electrode is desired, but requires a close setting of the time switch to match the time of arc revolution plus overlap.

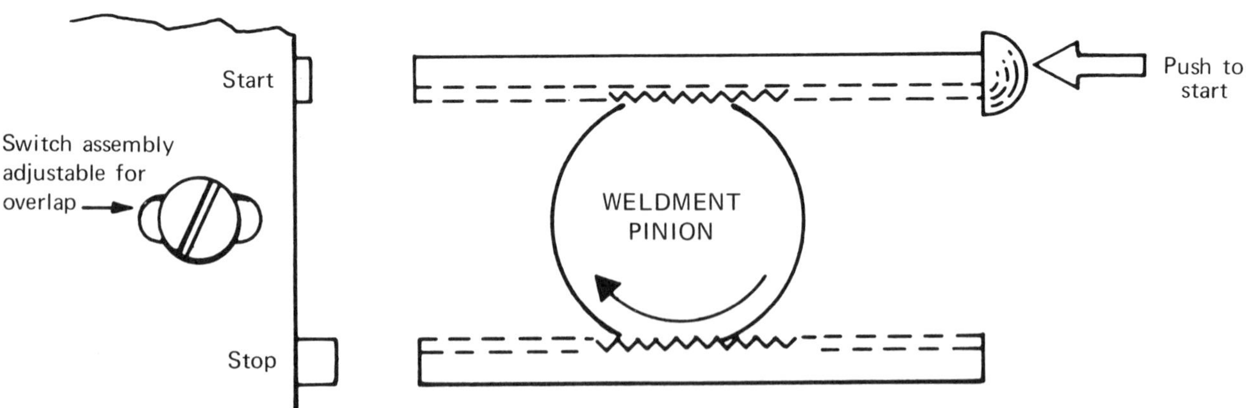

Rack-and-Pinion Type
This is the simplest of the three types. Overlap is simply adjusted, and various diameters can be welded with the least amount of setup time. The starting point can be fixed or initiated by the striking of the arc.

Fig. 4-50. Three types of synchronous timers for circular weldments. Electric timers are used more frequently than mechanized timers.

electrode. Only rarely is there need to resort to high frequency to cure starting problems.

Electric timers are preferred to cams for sequencing action during the welding cycle. A timer can be adjusted easily, whereas a cam must be removed and reshaped. Cams can magnify a backlash problem in gearing that might be tolerated with electric timing. Timers also have the additional advantage of simpler sequencing, permitting any variation of pattern in skip welding.

A timer should have a range of at least 150% of the time required to make the weld. The extra range will allow the use of slower welding speeds if necessary. When multiple heads are used, each should be synchronized with travel through its own timer, otherwise variations in arc striking could result in one weld being incomplete at the end of the cycle. The time-switch, pendulum-cam, and rack-and-pinion types of timers for circular weldments are shown in Fig. 4-50.

Mechanical Considerations

Good fixture design must take into account various mechanical factors that can affect the quality of the weld and operating efficiency.

In any stationary-head fixturing, it is exceedingly important that the head mounting be rigid, since a small movement of the mounting will be magnified at the arc. Any head-positioning adjustments must retain rigidity. A fixture design should always include provisions for head mounting in the original plan, rather than "hanging the head on" after the basic fixturing has been built.

Boom and beam-with-carriage fixtures are built as commercial items, and their manufacturers incorporate the necessary rigidity and stability into them. If the fabricator builds his own equipment, he must make sure that it is designed to minimize deflection. The weight of the welding head at the end of a 12-ft boom requires an exceptionally rigid assembly.

A fixture should be designed so that the welding head is easily accessible to the operator for making procedure adjustments and servicing when necessary. The same applies to the positioning of welding controls. If an operator must climb a ladder to make a change in current settings, welding productivity will suffer. When installed, the equipment should be positioned for easy access on all sides.

Whenever the fixturing must be aligned before or during welding, there should be provision for easy adjustment. Frequently a fixed position is not the optimum, and the ability to make a 1/4-in. adjustment will make the difference between a perfectly functioning machine and one that is performing poorly.

The drive system should be as free as possible from backlash and be designed to minimize acceleration time. V-belt drives should not be used in the final stages of the fixture. There is a time lag as the side of the belt in tension is stretched to the point where it can pick up the load, and this may be accentuated by slippage on the sheaves. Universal joints should be avoided, if possible, since they tend to become sloppy and develop backlash.

If backlash is unavoidable, preloading with a brush or some other method to create a drag in the proper direction may correct the situation. Soft rubber may be used on idler rolls to allow the embedment of seams, but never should be used on drive rolls. Backlash is also minimized by using large-diameter gears in the final stages of the drive system. Table 4-39 lists the types of fixture drive controls commonly used.

In fixturing for submerged-arc welding, due consideration must be given to the methods for handling flux. Most fluxes have an angle of repose of about 45°. This means that tubes or troughs carrying flux by gravity must have a drop in excess of this amount.

TABLE 4-39. TYPES OF FIXTURE DRIVE CONTROLS

Mag-Amp Control
This control permits a wide range of speed. The system can use solid-state components to achieve trouble-free performance. Through the use of feedback windings, small currents are used to control large outputs.

SCR (Silicon Controlled Rectifier)
The SCR control circuit is small, efficient, and reliable. Very useful for AC circuit control, operates on a small control signal and acts as a switch for gate mechanism.

Variable Auto Transformer
This control gives a wide range of speed from one dial, usually from zero to 100%. Instability may exist if the speed is less than 10% of maximum speed. Since it is a transformer, it requires AC input and is subject to line fluctuation.

Shunt Motor (DC, with Field Control)
The control gives steady speed, but the speed range is less than 2 to 1. It usually requires the changing of gears to develop an adequate range of speeds in many welding applications.

Variable-Voltage Generator Control
Very steady speeds and a wide speed range are obtained with this drive. A drawback is the added cost of an extra generator.

Air Cylinder with Hydraulic Control
Provides very uniform travel speed. Should be used for short runs.

Air-Motor Drive
Air-powered equipment has shown merit on some applications. Speed can be controlled through air-pressure gages.

The best bead shape results when the flux system places just enough flux to give good coverage of the arc and weld. There should be little or no relative movement between the flux and the plate. Thus, any enclosure around the arc to retain the flux must not cause the granules to slide over the plate. Poor bead shape and appearance can also be caused by conditions that confine and compact the flux.

Ransome Structural Pipe Mill for Offshore Platforms. Fully mechanized systems for Material Handling and Automatic Welding of Pipe 20″-120″ diameter, 1/2″ to 4″ thickness.

Robot Welding

Industrial robots represent a technological breakthrough that has the potential to improve a company's competitive position if properly applied and supported. Robots have also begun to play a major role in developing the "factory of the future" concept as to key elements in a *flexible factory automation system.* Increased foreign competition, a shrinking labor base, high labor rates, shortage of skilled laborers, and need for greater productivity combine to drive U.S. industries toward automation.

There are thousands of industrial robots in successful use with more than 20,000 estimated to be installed in the U.S. alone. About 50% of the installations are in the automotive industry, with the electronics, off-road vehicle, aerospace, general fabrication, and home appliance industries among the other leading users of robots. Included in the diverse applications of industrial robots are arc and resistance spot welding, material handling, machine tool loading, spray painting, assembly, deburring, grinding, gluing, and polishing. On the world-wide basis, new robots sold each year for arc and resistance spot welding represent about 30% of the total market share for all applications.

Over half of the welding robots are being used today in resistance spot welding applications, mainly for automobile sheet metal parts. However, in the 1990's, this trend is expected to change, as the application need for resistance spot welding robots in the automotive industry becomes saturated. Ongoing development and application of seam finding sensors and adaptive controls will make arc welding robots more popular. This chapter concentrates mainly on the characteristics of arc welding robots.

ROBOT WELDING DEFINITION

Full-automatic welding is defined as "the application of mechanized devices to perform the functions of the basic welding circuit" and "requires not only ingenious devices to move the welding arc or arcs over the work, or to move work under the arc or arcs, but also similarily ingenious fixtures for holding materials and manipulating both the movement of the work and the movement of the welding arcs." Robot welding is a specialized form of automatic welding, which is defined by AWS as "welding with equipment that performs the welding operation without adjustment of controls by a welding operator." However, robot welding differs from "fixed" or "hard" automatic welding using "dedicated" machinery, which is used in high volume production where large numbers of identical parts are produced on a continuous basis. Robot welding is actually a type of *flexible automatic welding* that can handle the welding of many different part designs. This is because the movement of the robot mechanical manipulator is controlled by a computer. Thus, the robot manipulator can be easily reprogrammed for a new part.

The Robot Institute of America officially defines the robot as: "A reprogrammable multifunctional manipulator designed to move material, parts, tools, or specialized devices through variable programmed motion for the performance of a variety of tasks." Three words in the definition are essential to understanding the basic concept of a robot.

- **Reprogrammable** — a robot is controlled by a programmable computer unit with a memory, such as a microprocessor, or a mini-computer. Reprogramming the controller allows a new sequence of movements to be executed by the robot arm.
- **Multifunctional** — a robot can perform a wide variety of tasks, such as the loading/unloading of machine tools, transporting parts, or arc welding, simply by changing the tool attached to the end of the robot manipulator "arm"
- **Manipulator** — a robot has the ability to transport objects or tools through space while, at the same time, reorienting their position. This manipulation is the reason for comparison of the robot to the human worker.

The robots being used in industry today should not be confused with the popular notion of a robot as portrayed in recent movies such as "Star Wars" as having near-human intelligence and sensory capabilities. Most robots in use today are fixed in position and have very little or no sensory capability.

For example, a welding robot is usually taught

to weld a particular part assembly by leading the welding torch or gun attached to the end of the robot manipulator arm along each joint path on a sample part using a hand-held teach pendant with control buttons or a joystick. Points along the joints are programmed and entered into the robot's computer control memory. The robot will repeat this program on the next part assembly, but based on the points in space programmed from some *fixed reference "home" point on the robot.* What happens if the next part assembly to be welded has piece parts that vary in thickness or dimensions, or part holding fixtures with loose tolerances are being used? The welds made by the robot may miss the joints entirely or melt through due to varying joint gaps and part thicknesses! Thus, precision parts and fixtures have been needed for most past robot welding applications to make them a success. This is changing, however, as seam finders and adaptive controls made especially for robotic arc welding are being developed and become more available for production use. These new developments that give welding robots some sensory capability are discussed later in this chapter. However, first the basic characteristics of welding robots will be reviewed.

BASIC COMPONENTS OF A ROBOTIC ARC WELDING SYSTEM

A robot manipulator and its computer control system are only part of a complete robotic arc welding system that can be relatively simple or complex. Listed in Table 4-40 are the basic components of a robotic arc welding system or work cell, which consists of various pieces of equipment integrated under the supervision of the robot computer control. Most robot vendors offer complete *turnkey* robot welding systems to save the buyer the systems engineering time and expense necessary to electronically combine or *interface* the separate components into a workable automated system. A photograph of such a complete turnkey robot welding system is shown in Fig. 4-51.

Robot Manipulator: The manipulator consists of a mechanical *arm* and *wrist* mounted on a support base. An *end effector* or tool (a welding gun or torch in the case of welding) is mounted to the wrist for performing work. Several *mechanical configurations* exist for manipulators. The type of configuration influences the operating space or *work envelope* of the manipulator and the size and shape of the parts to be welded. These configurations are best described in terms of their special coordinate systems.

TABLE 4-40. Basic Components of a Robotic Arc Welding System or Work Cell

Basic Component	Important Considerations
Robot Manipulator	Basic configuration, degrees of freedom, work envelope (range), dexterity, power supply, repeatability, speed, load capacity.
Robot Computer Control	Basic type, programming technique, control feedback system, memory size and backup, program storage, interfacing, special welding software.
Welding Power Source	High duty cycles, interfacing, accuracy.
Arc Welding Equipment	Wire feeder: type control, interfacing to robot control. Guns and torches: accuracy, capacity. Cables: length, mounting.
Work Handling Positioner	Type and control, interfacing, repeatability, load capacity, travel velocity.
Accessory Equipment	Torch or gun mounts and alignment tool, torch or gun nozzle cleaning station, part fixtures, operator station, safety equipment.

The simplest coordinate system is the *rectangular, rectilinear, or cartesian* (Fig. 4-52). All motion is translational along one of the three perpendicular axes. The available work envelope consists of a rectangular box in front of the robot.

The rectangular coordinate configuration is one of the two types that are popular for arc welding robots. An example is shown in Fig. 4-53.

The rectangular manipulators are used on larger workpieces where there is a requirement for long, continuous welds with relatively high arc-on-time. They are also used for hardsurfacing of precision parts, where very accurate, slow speed oscillation during welding is required.

Other coordinate systems incorporate rotational motion, which makes these systems more dexterous, but also more difficult to control. The three main types of rotational systems are *cylindrical, spherical,* and *jointed arm.* The jointed arm (also called *articulated* or *revolute coordinate),* configuration provides three rotational motions (Fig. 4-54) and is the other configuration popular for arc welding. The manipulator resembles a human arm with elbow and shoulder joints connecting the arm components and the base.

An example of an industrial robot with a jointed arm configuration is shown in Fig. 4-55. Compared to the rectangular coordinate robots, the jointed arm configuration allows much higher speed movements between welds and is used for smaller sheet metal parts with short, intermittent welds (low arc-

Fig. 4-51. A complete robotic arc welding system with a robot manipulator, robot computer control, welding power source, welding equipment, and worktable.

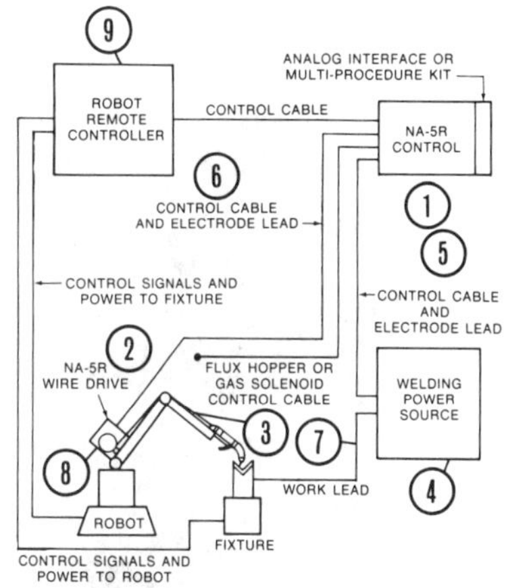

Equipment Requirements and Specifications

① NA-5R Control.

② NA-5R Wire Drive.

③ K397 Gun and Cable assembly (for Innershield), or wire contact assembly when using Automatic Heads.

④ DC Power Source (CV-400, DC-400[1], DC-600[2], DC-1000, DC-1500[3] or PP-500).

⑤ K374 or K215 Input Cables.

⑥ Four ft. control to head electrode cable. (If distance is greater, order K338 or K335.)

⑦ Work Cables.

⑧ K299 Wire Reel Mountings or other electrode supply means.

⑨ User-supplied remote controller.

on-time) that require fast movements between welds to keep up with short production cycle times. Due to its dexterity, the jointed arm provides help in reaching the more inaccessible joints or may even alleviate the need for a positioner to reorient a part to allow the robot to reach all the joints on a workpiece.

The other two types of coordinate systems incorporating rotational motion are the *cylindrical* and *spherical* (also called *polar*) shown, respectively, in Fig. 4-56 and Fig. 4-57. The cylindrical coordinate system includes two translational and one rotational motion. The working envelope for these robots is shaped like a cylinder with a section sliced away. The manipulator consists of a horizontal arm attached to a vertical column which is mounted on a rotating base. The spherical coordinate system consists of one translational and two rotational motions. A boom arm extends and retracts, pivots vertically, and rotates horizontally to trace the partial outline of a sphere.

The cylindrical and spherical coordinate robot manipulators were the first ones developed for

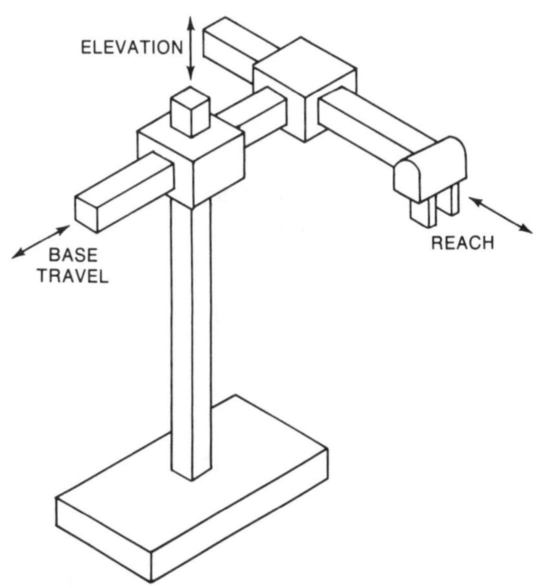

Fig. 4-52. Rectangular coordinate robot configuration.

Fig. 4-53. An industrial welding robot with a rectangular coordinate configuration.

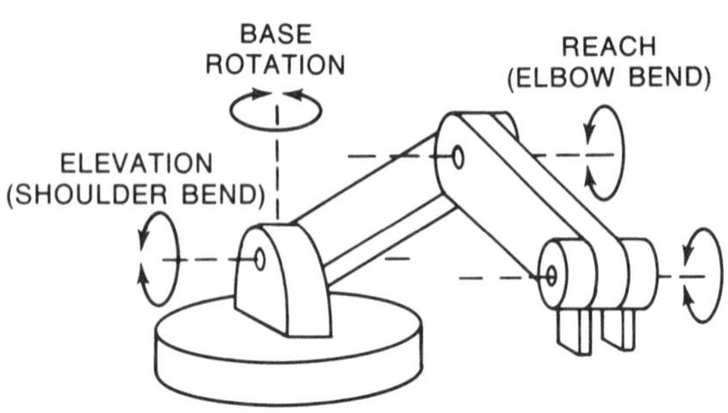

Fig. 4-54. Jointed arm robot configuration.

industrial use in the early 1960's. Their use today for arc welding is very limited, and the great majority of industrial robots being used or sold for welding are the rectangular or jointed arm types.

Each of the four coordinates systems discussed are viewed as having the three basic *degrees of freedom* or *axes of movement.* A *wrist* assembly, added to the lower arm of the manipulator, can provide an additional two or three degrees of freedom to the robot. Typically, industrial robot *welding applications* require a total of *either five or six degrees of freedom.*

Earlier industrial robots used for arc welding were mounted on a rigid base, which was bolted to the floor. More recently, some robot manufacturers

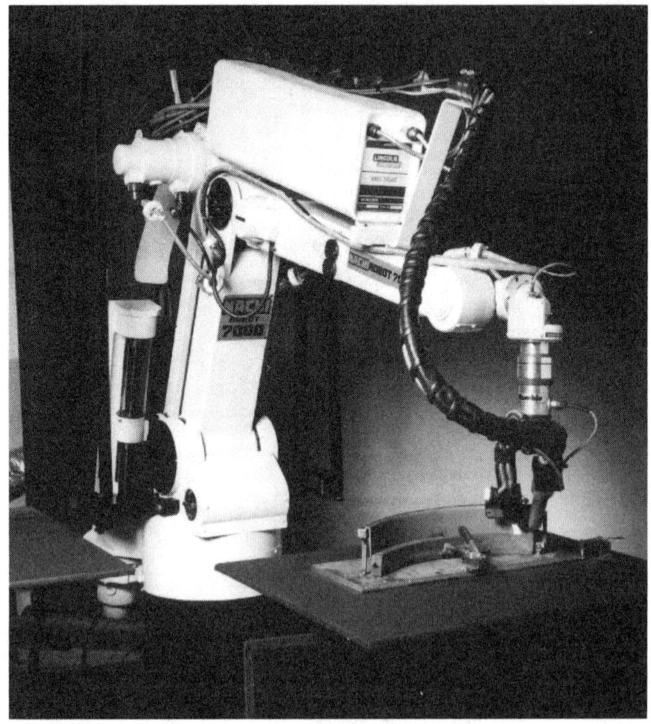

Fig. 4-55. An industrial welding robot with a jointed arm configuration and the following six joints or axes: waist, shoulder, elbow, wrist rotate, wrist flex, hand flex.

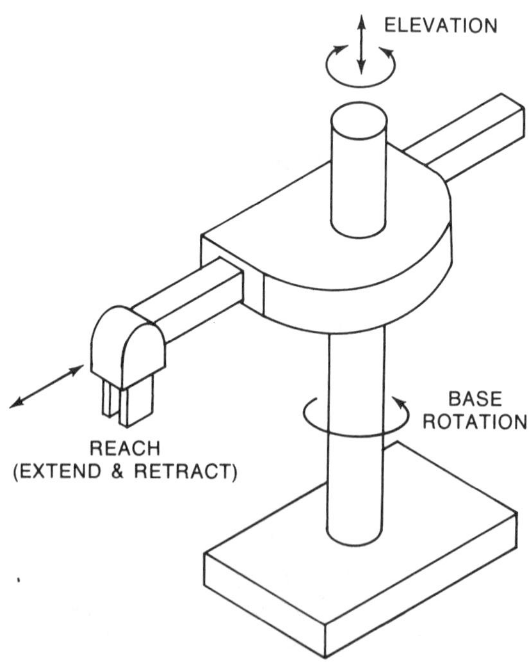

Fig. 4-56. Cylindrical coordinate configuration with two translational and one rotational motion.

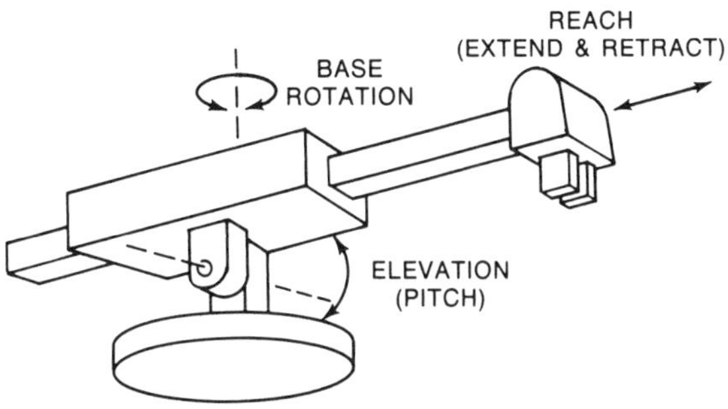

Fig. 4-57. Spherical or polar coordinate system with one translational and two rotational motions.

are offering the option of welding robots that can ride back and forth along a *track* or are mounted upside-down on an *overhead gantry.* These tracked and overhead mountings *provide mobility* and extend the working range of the standard robot configurations already discussed.

The motions of a robot manipulator arm are driven by one of three sources of power: *hydraulic, electric, or pneumatic.* A pneumatic power supply is most often used on the simplest and least expensive type of robot with only point-to-point control used for material handling and machine loading. Pneumatically powered robots are not suitable for arc welding. The early type of welding robots used hydraulic cylinders or hydraulic rotary motors for drives. However, robot manufacturers have substituted electric motors for hydraulic drives in most robots used for arc welding. Attributes of electric over hydraulic drives include reduced maintenance and increased reliability, improved repeatability, and smoother motion control.

Repeatability or repetition accuracy, is a measure of the ability of a robot arm to repeat a previously-programmed position motion at the same location under the same conditions as before. Usually given as a tolerance figure, repeatability of most electrically powered robots is ± 0.004 inch (± 0.1 mm) or better, usually adequate for arc welding with a consumable electrode such as for the GMAW or gas metal arc welding process. The gas tungsten arc (GTAW) process used with a robot to weld precision parts may require tighter repeatability tolerances.

The *load capacity* or *payload* of a robot manipulator arm is the maximum weight of tooling that can be mounted at the end of the arm and handled by the robot within speed and repeatability specifications. Unless specialized equipment such as seam finders or cameras are required to be mounted to the end of the arm, the load capacity is of little importance in arc welding applications, as the weight of a welding gun or torch, collision sensor and mounting bracket is usually less than the rated load capacity. The load capacity of the most common electrically powered robots used for arc welding is 11-22 lb. (5-10 kg.), although some manufacturers offer larger robots for arc welding with up to a 70 lb. (32 kg.) load capacity. (The load capacity does become of prime importance in robots used for resistance spot welding, where a robot has to accurately manipulate welding guns weighing between 75 and 150 lbs.)

Jointed-arm, electrically powered robots used for arc welding can move extremely fast at maximum speeds of about 40 to 60 in./sec (2400 to 3600 in./min, or 1000 to 1500 mm/sec), which is much faster than the actual arc travel speeds used for the various welding processes. Usually, a robot will weld at the same or a slightly faster speed than a human welder. The speeds close to maximum can be used to reduce the time of movements between welds and thus reduce the overall cycle time.

Robot Computer Control: Table 4-41 lists the different types of basic controls used to characterize a robot. Industrial robots used for arc welding must have the *controlled-path* type of control. Controlled-path robots all are *servo-controlled,* which means each axis has a *resolver or encoder* to sense its position and means to feed back the sensed position to the control in such a way that the control can cause a particular path to be followed. Each axis also has a *tachometer* to sense the rotational velocity of the electric motor used to drive the axis, and this information is also fed back to the computer control in order to control the speed of the movements between programmed points.

Robot control requirements are generally much more complex than those of NC (Numerically Controlled) or CNC (Computer Numerically Controlled) machine tools. Robot arm motions consist of rotational as well as linear axis movements which must be coordinated to move the welding gun along the joint path. In addition, the robot control must be capable of interacting with other equipment such as positioners, welding equipment, safety equipment, and sometimes off-line computers. Acceleration/deceleration of the end effector must account for the inertial changes due to the end effector's location relative to the robot base. To control the complex servo, controlled-path robots used for arc welding, it is necessary to use powerful *microprocessors, multiprocessors,* or minicomputers. Multiprocessors or multiple microprocessors are becoming a trend. Each axis is controlled by one microprocessor, and all the microprocessors are coordinated by a central

TABLE 4-41. Basic Type of Robot Controls

Type	Description
Point-to-Point	Able to move from one specified point to another, but cannot stop at arbitrary points not previously designated. Stopping points are often just mechanical stops that must be changed for each new operation (no feedback capability or non-servo).
Continuous Path	Able to stop at any specified number of points along a path, but the path followed between programmed points may not be controlled. Every programmed point must be stored separately in the computer memory of the control.
Controlled Path (computed trajectory)	The path between programmed points is controlled. The path can be defined as linear interpolation (straight line) or circular interpolation. Only the start and finish coordinates and the path definition are required for control.

microprocessor. For example, the main portion of one robot controller is a powerful distributed computer system consisting of seven independent Motorola 6809 microprocessors coordinated by a central 32-bit 68000 microprocessor. Each of the 6809 microprocessors is dedicated to a single task such as controlling a single robot axis motor while the central 68000 microprocessor performs the major computing tasks and coordinates the activity of the 6809 systems.

The process of programming the robot computer control is commonly referred to as *teaching* the robot. There are three basic approaches that can be used to teach an industrial robot to arc weld a particular workpiece. The first two are *on-line* methods and are performed by directly programming the robot. The third can be done on-line or *off-line.*

Walkthrough or *teach-by-guiding* programming is done by the programmer who physically guides the welding gun or torch mounted to the end of the robot arm along a weld joint. Walkthrough programming, which is the method most popular for spray painting robots, is easily and quickly performed by shop people who understand the operation. In the *playback* mode, the robot repeats the path and operations that were stored in memory during the teach operation. There are a few manufacturers that offer robots for arc welding applications with walkthrough programming. However, difficulties with accuracy, incorporating sensory information, synchronization with other operations, and in handling the larger robots have made most robot manufacturers use the *lead-through* programming method using a *teach pendant* control box. Pressing a *control button* or moving a *joystick* on the teach pendant applies a signal to the servo amplifier for a particular axis that causes the arm link to move. After the arm has been moved to the desired point, a *record* button can be pressed to store the coordinates of the point in the computer memory. Examples of two types of teach pendants are shown in Fig. 4-58 and Fig. 4-59.

CRT/keyboard programming is similar to programming used for NC machine tool programming. This method involves development of the program on a computer, sometimes without walkthrough or leadthrough teaching with a teach pendant. The program can be developed *off-line* on a separate computer and downloaded to the robot controller, or it may be programmed directly into the controller. The difficulty in creating CRT/keyboard programs is in accounting for the end effector positioning relative to the workpiece and other objects in the work cell. Off-line programming is an area of ongoing development, the purpose of which is to minimize downtime of robots that are in production usage and need to be reprogrammed for a new workpiece design. Development of more sophisticated programming software will significantly increase the usage of CRT/keyboard or off-line programming, especially for complex robot tasks such as arc welding.

Other important features of a robot computer control are its program memory size, memory backup, program storage, ease of interfacing to other equipment, and availability of special instructions (software) held in memory to assist programming of arc welded workpieces.

- **Memory size or programming capacity.** This usually refers to the number of programmed data points that can be held in the volatile (erasable) memory, called *Random Access Memory (RAM).*

Suppliers of robot welding systems usually list the memory size in the number of positions or data points that can be stored. On standard versions of welding robots, this can vary from about 2,000-3,000 positions, more than sufficient for welding most workpieces. Most robot suppliers also offer expandable memory for welding more complex workpieces.

- **Memory backup.** Most robot suppliers offer as standard or optional equipment a battery backup system which will keep the programmed data and positions from being erased in the event of a sudden power failure.
- **Program storage.** Some medium has to be available to *save* and *store* welding data programs. When a production run of a particular workpiece

Fig. 4-58. Teach pendant with control button control. There is a pair of buttons of each axis. Pressing one button causes the axis to move in one direction; pressing the other button causes the axis to move in the opposite direction.

Fig. 4-59. Teach pendant with joystick control. With the joystick mode switch in the "POSITION" mode, the joystick controls the movement of the base, shoulder, and elbow axes. With the switch in the "ORIENTATION" mode, the joystick controls the movement of the three wrist axes.

assembly is completed, its program can be erased from the robot controller's RAM memory and a new workpiece program *read* into memory from the storage medium. Cassette tapes with recorders and floppy discs with drives are popular storage mediums offered by robot suppliers.

- **Interfacing.** The robot controller must have the ability to *interface* or *electronically communicate* with the other equipment components of a robotic arc welding work cell. Four types of *Input/Output* interfaces are commonly used: *contact closure, Digital-to-Analog (DAC)* interfaces, low voltage AC/DC signals, serial (RS232).

Contact closures are used to turn equipment such as wire feeders, positioners, and shielding gas flow control equipment on and off. For example, an input contact closure signal from a work cell safety gate might be necessary to allow the robot to move at all. A DAC interface allows the wire feeder and power supply to be completely controlled by the robot computer. The computer starts the welding operation, specifies the arc voltage and wire feed speed (which was previously entered into the program by the programmer), changes them anytime required in the program, and stops the welding arc.

- **Software options.** Listed in Table 4-42 are some of the software options available to reduce the time involved in programming and to give the arc welding robots special features. Some work equally well for applications other than welding.

Welding Power Source: Modern, solid-state power sources are usually specified for robot welding systems. Robot welding is usually done at higher duty cycles than manual or semiautomatic welding, and robotic power sources usually are 100% duty cycle rated at an output above the welding parameters to be employed to prevent any danger of overheating the internal components of the power source. Most have remote voltage sensing that measures the voltage across the arc and accurately maintains the programmed arc voltage. An automatic line voltage compensation feature maintains the output voltage constant with fluctuations or spikes in incoming line voltage. Also, the power sources must have the necessary features and controls to interface with the computer control, usually through an intermediate interface control panel. The power source must also be able to consistently start the arc.

TABLE 4-42. Specialized Computer Software Available for Robotic Arc Welding

Software Option	Description
Tool Center Point (TCP)	The tip of the welding gun or torch is defined as the TCP. The robot computer coordinates the movements of all the axes during teaching to move the TCP in a controlled path.
Coordinates Transformation	Allows the programmer to change the type of reference coordinates that the robot arm will move the TCP (torch tip) in, such as cylindrical or rectangular.
Linear Interpolation	The TCP is moved in a straight line between two programmed points.
Circular Interpolation	A circle is taught by three points (for circular welds).
Weave	Gun or torch weaving action is controlled directly through the movement of the robot manipulator.
Three Dimensional Shift	Shifts a welding program that is already taught to any location in space by teaching three representative locations on the fixture or worktable.
Mirror Image	Creates the mirror image of a program. Used on symmetrical workpieces to save about half the programming time.
Offset Displacement	For multipass welds, copies a previously taught weld pass and displaces it in the vertical direction an amount designated by the programmer.
Programmable Welding Parameter Schedules	Welding parameters (arc voltage and wire feed speed) are entered and stored into a table of welding schedules to program into a welding sequence at any location.
Seam Tracking	Allows the robot to search and follow the weld joint as it deviates from the programmed path.
Programmable Positioner	Allows the programming of the axes of a precision robotic work handling welding positioner.

Welding Equipment: Most robotic arc welding is being done using arc welding processes with consumable wire feed, such as the various metal transfer modes (short circuiting, spray, or pulsed arc) of the gas metal arc welding processes and the flux cored arc welding process. The welding current, shielding gas, and metal transfer mode mainly determine the capacity or rating of the *robotic welding gun or torch* used. For gas metal arc welding with short circuiting transfer using carbon dioxide (CO_2) shielding gas at relatively low amperages (<200 amps), *air-cooled* guns or torches have been used. For higher amperages or with mixed argon-base shielding gases, *water-cooled* guns or torches are normally used, thus necessitating a water supply usually in the form of a *water cooler.* All robotic gun or torch parts should be manufactured with close tolerances, so when a part is changed, the position of the gun or torch tip (Tool Center Point) is not changed. Some specialized robotic torches offer features such as a compressed air blast through the center of the torch to aid in weld spatter removal from the gas shielding nozzle.

The *welding cable* length from the gun or torch to the wire feeder should be as short as possible to minimize "drag" and wire feeding difficulties, but be long enough to allow for all the robot arm and wrist orientation movements required. Usually, just the welding torch or gun is actually mounted on the robot manipulator. The wire feeder, wire spool or coil, and welding and torch cables are mounted on some sort of *stand and boom* arrangement over the robot which swings around as the robot rotates through its movements. Thus, the heavy weight of the wire feeder, wire spool or coils, and cables is kept off the robot arm.

Wire feeders for robotic arc welding are usually solid-state controlled and feature a high resolution tachometer feedback system for precise speed control. A separate *interface control panel* and Digital-to-Analog (DAC) hardware and software allows the wire feeder and power supply to be completely controlled by the robot computer control. These controllers usually feature digital readouts of both the "set" and actual (during welding) wire feed speed and arc voltage, and they take the set or programmed values and maintain them through a closed loop feedback system.

Work Handling Positioner: Simpler workpieces with welds on one side may require no work handling positioning equipment; the robot can reach all the joints with the proper gun or torch angle without having to move or reorientate the workpiece. However, some type of work handling positioner may be used for several reasons:

- so the robot can reach all the joints on a particular workpiece that has welds on more than one side.
- to allow welding in the flat position and improve weld quality and speed deposition.
- expand the robot work envelope, enabling it to weld larger workpieces.
- transfer workpieces between the robot welding station and an unload/load station, thus keeping the robot busy welding. While the robot is welding at one station, the work cell operator is unloading a welded workpiece assembly and loading new piece parts at the other station. Then the dual-table positioner rotates or indexes 180° to repeat the operation. This minimizes the cycle time to produce a welded assembly and increases the productivity of the work cell. Also, the operator is safely kept out of the work envelope of the robot arm.

Several types of robotic positioners are listed in Table 4-43. They all have one feature in common which makes them considerably more costly than conventional positioners: *precise repeatability* typically given as ±0.0002 inch per inch of radius. Backlash of conventional positioners is typically more than 1/8 inch, too much for robotic arc welding. Also robotic positioners are specially designed for higher speed moves and for ease of interfacing and programming.

Precision gear-driven positioners with electronic DC servo amplifier/motor systems can be programmed and made extensions of the robot axes. The positioner axes provide positional feedback information to the robot controller, which processes the information and controls the position of each axis on the positioner. If motor speed feedback is also provided along with special software, the positioner axes can be given *coordinated movement* with the axes of the robot. For a weld made on an irregularly-curved surface, the positioner axes could be turning while the robot arm is moving to keep the weld joint in the flat position during welding.

Accessory Equipment: Accessory equipment required in a robotic welding work cell might include the following:

- Welding **torch mount** and **alignment tool.** Collisions between the welding torch and workpiece or fixtures are common occurrences during teaching and editing the program. Special robot welding torch mounts are available that contain small limit switches that break a circuit and put the robot motion in a "hold" or "emergency stop" condition when the torch hits an obstacle during teaching or automatic operation. Also, some mounts contain stiff springs that will bend when an obstacle is hit, or are designed to have the torch and mount shear away completely upon a major collision. (The latter is called a "break-away" torch mount.) These features are intended to prevent damaging the torch in the event of a collision. Also, in case the torch position does shift from a collision or a new torch has to be used, an alignment tool or jig that locates the tip of the welding torch at the originally-defined Tool Center Point from some fixed point is desired. This saves considerable time in not having to reprogram the entire workpiece.

TABLE 4-43. Types of Robotic Work Handling Welding Positioners

Type	Description
A. By Configuration	
1. Turntable	1-axis motion with 360° rotation about fixed vertical axis; usually indexing type with multiple fixtures and indexing stops at 90° or 180° locations.
2. Head stock/tailstock	1-axis motion with 360° rotation about fixed horizontal axis; usually for welding longer workpieces.
3. Tilt/rotate	Single work table with two axes of movement; one 135° tilt from horizontal position, and one with 360° rotation.
4. Dual work table	Two worktables are located at each end of a beam that indexes 180°. The worktables can be fixed, rotate only, or rotate and tilt.
B. By Drive System	
1. Indexing	Simplest, moves to non-adjustable stop locations at a constant speed; usually controlled by robot controller with Input/Output contact closure.
2. Gear-driven	Allows infinite selection of stop positions throughout each axis range and variable speed motion; with resolver and tachometer sensors, each axis is *servo-controlled* and can be programmed and controlled by the robot controller.

- **Torch nozzle cleaning station.** Weld spatter buildup in the torch nozzle will block the flow of shielding gas and could result in weld porosity due to insufficient shielding. Also, spatter may build up sufficiently to short against the contact tip and cause arcing and melting of the tip and nozzle. The operator can clean the torch nozzle at regular intervals, but this stops the operation of the robot. The cleaning action can be accomplished automatically by programming the robot to pause at a cleaning station at regular intervals, possibly while another operation is occurring, such as part unload/load or positioner movement. The torch nozzle is reamed and sprayed with an anti-spatter compound.
- **Operator control console.** Some work cells have an operator's control console which is separate from the robot controller and controls only the automatic cycling of the robot and associated work cell equipment. The control console may only have several buttons such as "cycle start", manual/automatic select, "hold", "emergency stop", and for selecting the job or workpiece program number, if there are multiple work stations.
- **Workpiece fixtures.** Workpiece fixtures are usually custom designed for each application and are usually a major expense item. Fixtures must be designed to locate and hold parts for robot welding accurately and repeatedly in the best position to be welded.
- **Safety equipment.** An industrial robot manipulator arm moves extremely fast and has the potential of causing injury to personnel in the work envelope by collisions or "pinching" them against other equipment in the work cell. Whether a robot malfunctions (it is still just a machine!) or a worker gets careless, careful safety planning will prevent accidents. The most important consideration in preventing accidents is proper indoctrination, familiarization, and training of every person involved with the robot operation: programmers, operators, maintenance technicians, manufacturing engineers, etc. If personnel are familiar with the type and features of the robot, they are less likely to be in a situation to be injured.

The following safety equipment or hardware is used as a means to prevent accidents, especially to keep people out of the work cell when the robot is cycling in automatic mode:

- Warning devices such as floor markings of the robot's work envelope, rotating lights when the robot is energized, ropes or chains, and signs.
- Feedback devices that shut off power to the robot if a person enters the work cell. A typical "interlocked" barrier employed is an 8-foot high fence with gate interlocks, and with two switches needed to reactivate the robot (one switch is located outside of the fence). Other devices intended to prevent intrusion are pressure-sensitive mats, proximity switches, and photocell "eyes" or light curtains.

The operator of the robot work cell should be kept out of the robot's work envelope if possible by the design and arrangement of the work cell components. For example, a dual-table, indexing positioner can be used, so the operator can unload welded assemblies and load new parts completely out of the work envelope. If the operator has to unload and load parts from a positioner in the work envelope, pressure-sensitive mats can be used to deactivate that sector while the operator unloads or loads parts. With the operator on the mat, the positioner cannot move, nor can the robot enter that sector of the work cell. Other precautions might include the placement of emergency stop buttons, that override all other controls, at convenient locations around the work cell, limiting the robot's work envelope to only what is needed by placing limit pins in the base rotation axis, and the use of remote operator control consoles.

The programmer or maintenance personnel that have to enter the robot's work envelope while it is energized are in the most danger. During this time, the robot, may malfunction, or the wrong button could be pushed. As a start, the teach pendant must have an emergency stop button. Also, "pinch points", or places where a programmer or maintenance technician could become trapped between the robot arm and other equipment, should be avoided by designing the work cell with sufficient space for a human to fit between and escape injury.

Another consideration is to protect bystanders and the cell operator from eye damage caused by arc radiation. Weld curtains must be provided; most robotic arc welding work cells employ the transparent, heavy duty vinyl gold curtains that filters 100% of the harmful ultraviolet and blue light radiation from arc welding.

JOINT DESIGN AND TOLERANCES FOR ROBOTIC ARC WELDING

Parts may have to be redesigned for robotic

welding. Welds should be accessible and placed in the flat or horizontal position to produce the highest quality results at the fastest possible speeds. Also, in the design of weldments, the location of the weld is usually not considered critical in relation to the function of the completed part assembly. Dimensional control of overall part length and width, and the location of holes, bosses, and machined surfaces are specified by engineering drawings, not the location of the weld joints. Thus, variance in location of weld joints from part to part by 1/8″ or more and varying joint gaps along the joint are common in welded fabrications. Usually, the weldor can easily compensate for these variations from part to part and along the joint and still make an acceptable quality weld with sufficient penetration, acceptable bead contour, and freedom from undercut and overlap. However, *in robotic or any type of automatic welding, the variation in the weld joint location and gap becomes critical.*

Several items are sources of variability in the joint location and gap:

1. piecepart tolerances.
2. fixture tolerances.
3. repeatability of positioning equipment, if used.
4. distortion during welding.

For a robotic arc welding application to be technically feasible and ultimately successful, two general approaches can be taken:

1. Locate the weld joint seams accurately and repeatedly enough for welds to be made by a standard robot system with no sensory capability; i.e., no seam finding, seam tracking, or adaptive control features.
2. Use available sensors and computer software to allow the robot to find the joint seam, track it, and adapt to varying conditions during welding, if necessary.

Either approach may not be economically justifiable and has to be carefully evaluated. The cost to make precision parts and fixtures, or to purchase additional sensory hardware and software for the robot system may make the "payback" or "return on investment" too low to justify the purchase of the robot system. Before making a decision, make dimensional measurements on the parts and fixtures to see what the present joint tolerances are. Also, the joint location and gap tolerances that would be suitable for robotic arc welding have to be defined by performing weld tests. Published guidelines exist in the literature for determining acceptable joint location and gap tolerances. Some list the acceptable joint location tolerance as plus or minus "one half the welding electrode diameter" or "one welding electrode diameter." However, these guidelines are too simplistic, as besides the electrode diameter, the joint location tolerance also depends on other factors such as the welding position and the welding process and parameters. For example, for a given weld process and electrode diameter, a weld made in the flat position at a high current and slow arc travel speed would have a greater joint location tolerance compared to a small weld made in the overhead position at a low current and fast travel speed. One recent study by the Edison Welding Institute suggests that weld size rather than electrode diameter is a better measure of the joint location tolerance required. The Edison Welding Institute guidelines are given in Table 4-44.

TABLE 4-44. Guidelines for Joint Location Tolerance for Automated Welding (as Published by the Edison Welding Institute).

Weld Pass Size, in.	Joint Location Tolerance, in.
1/8	±0.010 to ±0.020
1/8 to 1/4	±0.020 to ±0.030
over 1/4	±0.035 to ±0.060

SEAM TRACKING AND ADAPTIVE CONTROL

Part assemblies that contain joint seams that are impractical to locate repeatedly and with consistent joint fitup, such as large car body sheet metal assemblies, can be robot welded using modern-day seam tracking and adaptive control sensory systems. Two main types of *non-contacting* seam tracking systems are being sold with arc welding robot systems. Some have adaptive control capability, which means they can adapt to changing joint fitup conditions during "real-time" or while welding.

1. **Through-the-Arc sensing.** These systems incorporate feedback of arc voltage or current as the basic control medium. With the GMAW (MIG) process, the arc current is sensed as the torch is weaved or oscillated across the joint. (Most robot controllers offer a weave software option so the robot arm does the weaving.) This method of seam tracking requires that well-defined joint sidewalls exist as in a T-joint or V-groove butt joint. As the arc reaches a sidewall, the arc length decreases and the current increases. The system samples the current input at each sidewall and adjusts the center of the cross-seam weave pattern.

In a similar manner, the system maintains the torch height and arc length constant, by comparing the current obtained at a certain electrode extension or "stick-out" with a set reference current. More sophisticated versions of the through-the-arc sensors offer an "adaptive fill" capability which adjusts the welding parameters based on the feedback information.

For automatic operation, some other system must be used to find the start location of the joint seam, which can also vary considerably from part to part. Usually, the electrode wire itself is used as a sensor, by accurately cutting the excess wire extension off to some controlled length, then impressing a high voltage low amperage current on the wire. With some type of preprogrammed search routine and varying levels of sophistication, the wire probe can either find the sidewalls and define the joint centerline, find both the joint start and center line, or profile the joint geometry and select welding variables to give an "adaptive fill" capability.

Through-the-Arc sensing has some limitations, in that the torch must be weaved, arc travel speed is relatively slow (about 30 inches per minute maximum), and sidewalls must be present. The other type of sensor that is showing great promise for seam tracking and adaptive control is *computer-aided vision.*

2. **Vision Sensors.** Most vision sensors used for robotic arc welding incorporate a low power laser that projects a stripe or cone of laser light across the joint seam in front of the torch or arc. A solid state camera or optical sensor picks up the light image, and through triangulation calculations, a special sensor system computer compares the actual joint position with the programmed positions and adjusts the robot axes coordinates so the torch finds and follows the joint seam. In the more sophisticated versions with adaptive control, information on joint details such as the joint gap can be sensed, and the welding parameters (wire feed speed, arc voltage, and/or travel speed) can be varied to give a constant weld fill.

Some vision systems are "two-pass", meaning that the robot is first programmed to pass the vision sensor along the joint path with the welding equipment off. The visions system senses the joint position and adjusts the robot's path for the second pass, which is made with the welding equipment on. Other systems are "single-pass," in that the sensing, adjustment of the robot coordinates, and welding are all happening simultaneously in "real-time." Fibre optic cables are sometimes used to carry the laser beam to the generating optics when the laser is remotely housed in the robot controller cabinet and to transmit the images of the weld scene to a remote solid state camera.

More recent versions have tiny electronic light detector chips mounted inside of special welding torches. One prototype system used with the GTAW (TIG) process views the weld seam, arc, and weld pool from inside the torch, coaxially with the tungsten electrode. This later system is closest to the capability of the human eye!

A SYSTEMATIC APPROACH TO ROBOT PURCHASING

As already mentioned, the automation of arc welding processes has made tremendous strides, putting robotics systems within the reach of short run manufacturers and job shop owners. Far from being limited to large plants or high volume production situations, flexible systems, particularly those with vision capability, can be adapted to meet a wide range of needs. These versatile, easily programmable robots with sensory capability can function, in effect, as fixtureless welding systems.

No Miracles. However, all of us have heard horror stories. High tech purchases are sometimes followed by disillusionment verging on despair when an expensive system fails to perform as expected. If your company is considering the purchase of a robotic arc welding system, don't let the excitement of a high tech purchase keep you from following the basic steps for making any good industrial purchasing decision.

In the field of robotics, as in all other sophisticated technologies, there are no miracles. Only a careful plotting of your organization's needs, vs. its current operations, will lead to a purchase which can be justified in terms of reduced costs, improved productivity, and increased quality. You will need a game plan, and of course any game plan is only as effective as its implementation. Let's review the basics to be covered *before* a purchase contract is signed.

Planning. O.K., exactly WHAT will you expect this robotic system to DO? Don't expect a robot to cure the shortcomings of your manufacturing process. If a welding job is difficult for a highly trained, experienced individual to perform, it will be even more difficult to accomplish by automated means. A robot will never understand what you *meant;* at best, it will do *precisely* what it has been instructed to do.

This has important implications. A programmer who does not understand what the human weldor has been doing cannot properly instruct the robot to make the weld. To maximize the programmable capabilities of any robotic welding system, have at least one person who knows how to weld on your project team from the outset. This individual will provide valuable insight, helping to plan and implement the most appropriate initial applications, as well as instructing the programmer on a regular basis. Most programming failures result from the programmer having insufficient knowledge of the application itself, and of how it has been accomplished manually. In fact, the weldor may learn to program a robotic welding system more easily than a programmer could learn to weld. Ignorance of this simple fact has doomed many automated systems to shaky performance from the outset. Don't let it harm the implementation of your system. Listen to the application expert, in this case, the trained weldor.

The remainder of the project team should consist of engineers and technicians who are free to dedicate themselves to the success of the project. The plant foreman, superintendent or plant manager will have ongoing production responsibilities that take precedence over development tasks. While they should receive regular progress reports, it may not be practical to involve them in the day-to-date planning and implementation of the robotic arc welding system.

In summary, then the planning stage will include:

- Making a commitment to an orderly planning process;
- Putting together an appropriate project team, which must include at least one application expert; and
- Analyzing potential applications.

Applications: The Big Picture. For any major investment, you want to be able to calculate the pay-off. Some degree of risk is acceptable and even inevitable, but we all want to minimize the risk as much as possible. In automating a fabrication shop, this comes down to applications. Some recent production applications of automatic arc welding with industrial robots are listed in Table 4-45. Note the wide range of part types robot welded, from bicycle parts to the space shuttle main engine.

The first step in evaluating applications for robot welding is to make a "wish list," detailing all of your shop requirements. Even if an item seems impossible to achieve by automated means, put it on the list. At this stage, keep your expectations high, but don't fail to include the simplest, most obvious requirements. When it is completed, the list should provide a detailed picture of where your operation stands, as well as what you hope to achieve in the future.

Using the Vendor Resource. Next, begin contacting robot vendors. Using your list as a guide, let the vendors do as much detailed spec work as possible; that's part of their job if they want to make a sale. While the vendors are doing some of that leg work, enroll project team members in robotics seminars. Once the vendors have returned with proposals and bids, your team will be in a better position to evaluate the material.

Specific Applications: Initial Decisions. As you approach the crucial stage of selecting the first applications, remind your project team that robots, in order to function effectively, require a more orderly production schedule than do human operators. Then, for each potential application, be sure you can answer the following questions:

- If the robot increases productivity in one area, can the rest of the manufacturing process be scheduled to support the higher level of production?
- Will robot productivity produce a bottleneck at any other stage of your operation, and if so, how do you plan to avoid that problem?

Keep in mind that a robot is a highly visible piece of equipment. Its success or failure will make a lasting impression on both management and shop

TABLE 4-45. Applications of Robotic Arc Welding.

Applications
Automotive car and truck frames (GMAW-pulsed arc)
Automotive car body sheet metal seams (GMAW with vision)
Office and school furniture (GMAW)
Cassette storage system tubular steel racks, computer chassis (GMAW)
Trailer hitches (GMAW)
Bicycle parts (GMAW)
Steam turbine compressor diaphragms (GTAW)
Space shuttle main engine (GTAW)
Aircraft carrier catapult cover assemblies
Snowblower housings, handlebars (GMAW)
Riding lawn mower and garden tractor front axle support, rear deck hanger, cutter deck (GMAW)
Military vehicle aluminum louvered grill assemblies and aluminum frame (GMAW)
U.S. Navy submarine aluminum antenna structures (GMAW)
Nuclear power system heat exchanger tube sheet weld overlays (GMAW)

personnel. Don't attempt to "dazzle" them with a complex operation. The faultless performance of a simple task will be quite dazzling enough!

There will be a learning curve involved in maximizing the potential of an automated system, so start with a simple application, for which every step of the welding process is clearly understood. Remember, a robot is only as good as the instructions it receives; make sure the programmer writing those instructions has acquired a thorough understanding of the welding task to be accomplished.

The Purchase. Now, your team should be ready to select the equipment that best meets your current and future fabrication needs. Let's review the steps that brought you to this point:

1. Committing to an orderly planning process;
2. Building a project team;
3. Compiling a "wish list";
4. Letting the vendors work on preliminary requirements;
5. Educating the team through seminar attendance and ongoing consultation with the expert weldor; and
6. Selecting initial applications that will maximize a robot's chances for success.

Remember: those who buy a robot and *then* try to find appropriate applications for it are often disappointed. By approaching the purchase systematically, you will greatly increase your odds of success.

The Rewards. What kind of payoff can you expect? Once the system is fully implemented, and the learning curve has been successfully negotiated, you may reap the following tangible benefits:

- Increased productivity
- Reduced direct labor costs
- Material cost savings
- Reduced requirements for human-operated equipment
- Increased quality

The intangible advantages of robotic arc welding systems have been shown to include:

- Reduced training or certification costs
- Reduced absenteeism
- A higher degree of safety
- Lower insurance and workman's compensation costs

To the extent that you follow a game plan such as the one outlined here, your shop stands to gain handsomely from its high technology investment.

Bibliography for Robot Welding

1. Critchlow, A.J., *Introduction to Robotics,* MacMillan Publishing Company, 1985.
2. "Engineering, Costs, Quality, and Safety", *Welding Handbook,* Volume 5, 7th Edition, American Welding Society, 1984; Chapter 5, "Automation and Control."
3. "Industry Update," *Robotics Today,* Volume 8, No. 6, pgs. 9-11, December, 1986.
4. Hanright, J., "Selecting Your First Arc Welding Robot — A Guide to Equipment and Features," *Welding Journal,* November, 1984, pgs. 41-45.
5. Gage, J.H., "Identification and Start-Up of a Successful Robotic Arc Welding Application," *Welding Journal,* November, 1983, pgs. 19-26.
6. Schneider, R., "Fixturing for Arc Welding: Guidelines for Success," *Welding Design and Fabrication,* May, 1985, pgs. 60-64.
7. Castner, H., "Joint Tolerances for Automated Welding," *Welding Design and Fabrication,* April, 1987, pgs. 30-32.
8. Pierce, M.F., "A Guide to Robotic Arc Welding Positioners," *Welding Journal,* November, 1985, pgs. 28-30.
9. Kuvin, B., "How to Safeguard Welding Robots," *Welding Design and Fabrication,* May, 1985, pgs. 72-75.
10. Grannan, D., "The New Weld Joint Trackers — Their History and Technology," *Welding Journal,* March, 1986, pgs. 49-51.
11. Hanright, J. "Robotic Arc Welding Under Adaptive Control — A Survey of Current Technology," *Welding Journal,* November, 1986, pgs, 19-24.
12. Hubaker, Mark, "A Systematic Approach to Robot Purchasing," Lincoln Electric literature.

Section 5
WELDING PROCESSES

SECTION 5.0
INTRODUCTION

SECTION 5.1
THE SHIELDED METAL-ARC PROCESS

Page

SECTION 5.2
THE SUBMERGED-ARC PROCESS

SECTION 5.3
THE SELF-SHIELDED FLUX-CORED PROCESS

SECTION 5.4
THE GAS-SHIELDED ARC-WELDING PROCESSES

SECTION 5.5
SPECIALIZED ARC-WELDING PROCESSES

SECTION 5.6
SELECTING A WELDING PROCESS

Introduction

The American Welding Society identifies over 17 metals joining processes which are used commercially. These processes can be grouped into six categories as follows: 1) arc welding, 2) resistance welding, 3) oxyfuel gas welding, 4) solid state welding, 5) high energy density welding, and 6) soldering and brazing. Seven processes are classified under arc welding and these together account for the greatest amount of welding and filler metal deposited commercially.

Resistance welding is used extensively to join sheet metal for items such as automobiles and home appliances. Spot welding, seam welding, projection welding and flash butt welding are variations of this process. Basic equipment is of moderate cost but special tooling and controls can increase costs many fold. Equipment maintenance costs are low. No filler metal is needed with this process.

Oxyfuel welding processes are used for autobody repair and by home and farm mechanics for maintenance because cost of equipment is low. The process is not used to a large extent commercially because weld quality is inferior to that of the arc welding processes. Bare wire is used as filler metal.

Solid state welding includes diffusion welding, friction welding, hot and cold pressure welding, explosive welding and ultrasonic welding. The common feature of these processes is that no melting takes place but that the surfaces to be joined must be brought into intimate contact by the application of pressure and/or heat. Some of these processes are used widely for industrial applications where production rates are high, but the high cost of equipment precludes the use for small production runs.

The high energy density processes include electron beam and laser beam welding. These processes can be used for a wide variety of applications and materials and produce high quality welds. The commercial use of these processes is limited, because of the high cost of equipment, to applications which require the special features of each process.

Soldering and brazing are used for joining a wide variety of alloys where melting of the base material must be avoided to preserve base metal properties. Brazing temperatures can range from about 500°F for lead-tin solders to above 2200°F for nickel base braze alloys used for joining nickel base superalloys for gas turbine applications. Heating for brazing can be done by oxyfuel torch, electric induction, electric resistance, inert or reducing atmosphere furnace or vacuum furnace. The cost of equipment is low for the oxyfuel process and high for the vacuum furnace method. Brazing requires the addition of filler metal which may be different or similar to the base metal composition.

Only the arc welding processes are discussed in depth in this handbook. These include the Shielded Metal Arc Process, Submerged Arc Process, Self-Shielded Flux Cored Process, Gas Metal Arc Process, Gas Tungsten Arc Process, Plasma Arc Process and a few specialized arc processes.

The Shielded Metal-Arc Process

The shielded metal-arc process — commonly called "stick-electrode" welding or "manual" welding — is the most widely used of the various arc-welding processes. Although the semi-automatic welding processes have gained in usage, the "stick electrode" process still accounts for a greater amount of welding than any other single process. In the United States about 40% of the welding is done by the stick electrode process while the factor is more than 50% in other countries. The shielded metal arc process is characterized by application versatility and flexibility and relative simplicity in the equipment. It is the process used by the small welding shop, by the home mechanic, by the farmer for repair of equipment — as well as a process having extensive application in industrial fabrication, structural steel erection and weldment manufacture. Arc welding to persons only casually acquainted with welding usually means shielded metal-arc welding. With this process, an electric arc is struck between the electrically grounded work and a 9 to 18″ length of covered metal rod — the electrode. The electrode is clamped in an electrode holder, which is joined by a flexible conductive cable to the power source. The weldor grips the insulated handle of the electrode holder (see Section 4.3) and manipulates the tip of the electrode with respect to the weld joint. When he touches the tip of the electrode against the work, and then withdraws it to establish the arc, the welding circuit is completed. The heat of the arc melts base metal in the immediate area, the electrode's metal core, and any metal particles that may be in the electrode's covering. It also melts, vaporizes, or breaks down chemically nonmetallic substances incorporated in the covering for arc shielding, metal-protection, or deoxidizing purposes. The mixture of molten base metal, core wire and powdered metal from the covering coalesce to form the weld.

The electrode covering consists of a mixture of silicates binders, and powdered fluorides, carbonates, oxides, metal alloys and cellulose. This mixture is concentrically extended around the core wire and baked to remove the water added to help with binding. As welding progresses, the covered rod becomes shorter and shorter. Finally, the welding must be stopped to remove the stub and replace it with a new electrode. This periodic changing of electrodes is one of the major disadvantages of the process in production welding. It decreases the operating factor, or the percent of the welder's time spent in the actual operation of laying weld beads.

Another disadvantage of shielded metal-arc

welding is the limitation placed on the current that can be used. High amperages, such as those used with semiautomatic guns or automatic welding heads, are impractical because of the long (and varying) length of electrode between the arc and the point of electrical contact in the jaws of the electrode holder. The welding current is limited by the resistance heating of the electrode. The electrode temperature must not exceed the "break down" temperature of the covering. If the temperature is too high the covering chemicals react with each other or with air and therefore do not function properly at the arc. Coverings with organics break down at lower temperatures than mineral or low hydrogen type coverings.

The disadvantages of this process have been overlooked by many in favor of its versatility and simplicity of equipment, even where some degree of mechanized welding was possible. More recently this attitude has changed since semiautomatic self-shielded flux-cored arc welding has been developed to a similar (or even superior) degree of versatility and flexibility in welding (see Section 5.3). Although steel fabricators and erectors have replaced shielded metal arc welding with semiautomatic processes, shielded metal arc welding is expected to remain a primary arc-welding process. It is the one well suited by minimal cost of equipment and broad application possibilities for the home mechanic, the farmer, the repair shop, the garage, the trailer-hitch installer, and many others who are concerned entirely with getting a welding job done.

PRINCIPLES OF OPERATION

Section 1.3 describes the basic welding circuit, and in Section 4.3 various items of equipment are described which when introduced into the basic welding circuit create the shielded metal-arc process. As noted, welding begins when the arc is struck between the work and the tip of the electrode. The heat of the arc melts the end of the electrode and the surface of the work under the arc. Tiny globules of molten metal form on the tip of the electrode and transfer through the arc into the molten weld "pool" on the work surface.

The transfer through the arc stream is brought about by electrical and magnetic forces. As discussed in Section 1.3, movement of the arc along the work (or movement of the work under the arc) accomplishes progressive melting and mixing of molten metal followed by solidification, and, thus, the unification of parts.

When electric arc welding was first developed, bare mild steel electrodes were used. It was not only difficult to weld with bare electrodes because the arc was erratic but the weld deposit contained high levels of porosity, oxides and nitrides from reaction with the air. As may be expected, these welds lacked ductility and toughness. An essential feature of the electrode used in the shielded metal-arc process is a covering or coating, applied to the core metal by extrusion or dipping, that contains ingredients to shield the arc and protect the hot metal from chemical reaction with constituents of the atmosphere.

The shielding ingredients have several functions. One is to shield the arc by providing a dense, impenetrable envelope of vapor or gas around the arc and the molten metal. This is necessary to prevent the pickup of oxygen and nitrogen and the chemical formation of oxides and nitrides in the weld pool. Another is to provide scavengers and deoxidizers to refine the weld metal. A third is to produce a molten slag coating over molten globules of metal during their transfer through the arc stream. The molten slag floats on the surface of the weld pool because it is of lower density than the metal and solidifies after the weld metal. Figure 5-1 illustrates the decomposition of an electrode covering and the manner in which the arc stream and weld metal are shielded from air.

Another function of the shield is to provide the ionization needed for AC welding. With alternating current, the arc goes out 120 times a second. For it to be reignited each time it goes out, an electrically conductive path must be maintained in the arc stream. Potassium compounds in the electrode covering provide ionized gaseous particles that remain

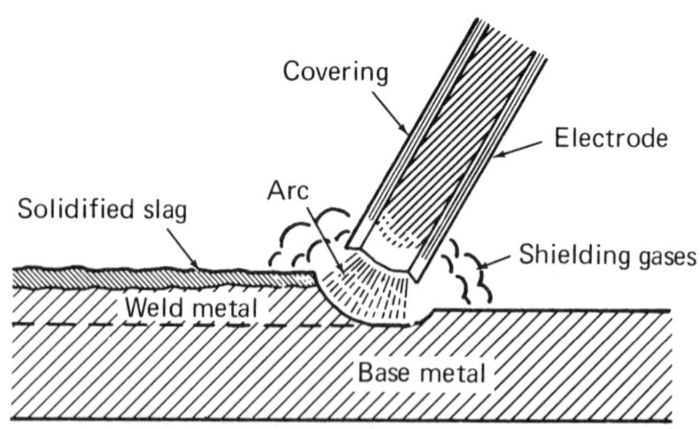

Fig. 5-1. Schematic representation of shielded metal-arc welding. Gases generated by the decomposition and vaporization of materials in the electrode covering — including vaporized slag — provide a dense shield around the arc stream and over the molten puddle. Molten and solidified slag above the newly formed weld metal protects if from the atmosphere while it is hot enough to be chemically reactive with oxygen and nitrogen.

Much of the welding done in building construction is the shielded metal-arc process.

ionized during the fraction of a second that the arc is extinguished with AC cycle reversal. An electrical path for reignition of the arc is thus maintained.

The mechanics of arc shielding vary with the electrode type. Some types of electrodes depend largely on a gaseous shield to protect the arc stream and the weld metal. With these electrodes, only a light covering of slag will be found on the finished weld. Other electrode types depend largely on slag for shielding. The explanation for the protective action is that the tiny globules of metal being transferred in the arc stream are entirely coated with a thin film of molten slag. Presumably, the globules become coated with slag as vaporized slag condenses on them — so the protective action still arises from gasification. In any event, the slag deposit with these types of electrodes is heavy, completely covering the finished weld. Between these extremes are electrodes that depend on various combinations of gas and slag for shielding.

The performance characteristics of the electrode (see Section 4.1) are related to their slag-forming properties. Electrodes with heavy slag formation have high deposition rates and are suitable for making large welds in the flat or horizontal position. Electrodes that develop a gaseous shield that disappears into the atmosphere and give a light slag covering are low-deposition and best suited for making welds in the vertical or overhead positions.

A solid wire core is the main source of filler metal in electrodes for the shielded metal-arc process. However, the so-called iron-powder electrodes also supply filler metal from iron powder contained in the electrode covering or within a tubular core wire. Iron powder in the covering increases the efficiency of use of the arc heat and thus the deposition rate. With thickly covered iron-powder electrodes, it is possible to drag the electrode over the joint without the electrode freezing to the work or shorting out. Even though the heavy covering makes contact with the work, the electrical path through the powder particles is not adequate to short the arc, and any resistance heating that occurs supplements the heat of the arc in melting the electrode. Because heavily-covered iron-powder electrodes can be dragged along the joint, less skill is required in their use.

Some electrodes for the shielded metal-arc process are fabricated with a tubular wire that contains alloying materials in the core. These are used in producing high-alloy deposits. They have an extruded or dipped covering as do some other conventional electrodes.

POWER SOURCE

Shielded metal arc welding can operate over a wide range of currents (10 to 500 amp) and voltages (17 to 45 volts) depending on the type and size

The stick-electrode process does an excellent job field welding pipe, sometimes under adverse conditions.

electrode used. The current may be either AC or DC; thus, the power source may be either AC or DC or a combination AC/DC welder (See Section 4.2). For most work, a variable voltage power source is preferred, since it is difficult for the weldor to hold a constant arc length. With the variable-voltage source and the machine set to give a steep volt-ampere curve, the voltage increases or decreases with variations in the arc length to maintain the current fairly constant. The equipment compensates for the inability of the operator to hold an exact arc length, and he is able to obtain a uniform deposition rate.

In some welding, however, it may be desirable for the weldor to have control over the deposition rate — such as when depositing root passes in joints with varying fitup or in vertical or overhead position work. In these cases variable voltage performance with a flatter volt-ampere curve is desirable, so that the weldor can decrease the deposition rate by increasing his arc length or increase it by shortening the arc length.

Figure 5-2 illustrates typical volt-ampere curves possible with a variable voltage power source. The change from one type of volt-ampere curve to another

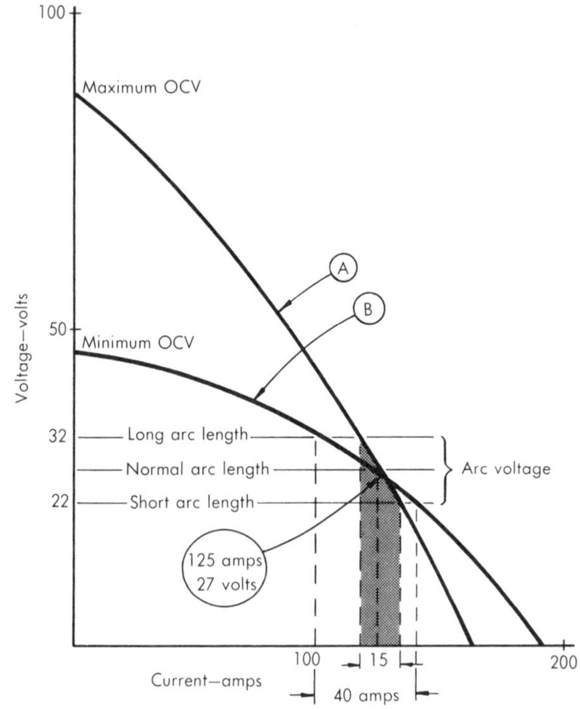

Fig. 5-2. Typical volt-ampere curves possible with a variable voltage power source. The steep curve (A) allows minimum current change. The flatter curve (B) permits the weldor to control current by changing the length of the arc.

is made by changing the open-circuit voltage and current settings of the machine (See Section 4.2 and 6.2).

The fact that the shielded metal-arc process can be used with so many electrode types and sizes — in all positions — on a great variety of materials — and with flexibility in operator control makes it the most versatile of all welding processes. These advantages are enhanced further by the low cost of equipment. The total advantages of the process, however, must be weighed against the cost per foot of weld when a process is to be selected for a particular job. Shielded metal-arc welding is a well recognized way of accomplishing many kinds of welding jobs. However, in selecting a process for any production application several processes should be carefully evaluated with respect to suitability, cost and quality.

Stick-electrode process is used to repair a worn conveyer screw.

The Submerged Arc Process

GENERAL PRINCIPLES OF OPERATION

Heat for the submerged arc welding process is derived from an arc between a bare metal electrode and the work. The arc is shielded by a blanket of granular fusible material called flux which is placed over the joint area ahead of the arc. Filler metal is obtained primarily from the bare electrode wire which is continuously fed through the blanket of flux into the arc and pool of molten flux. Additional filler may be obtained by adding cold wire to the weld pool or from metal powder contained in the granular flux.

The distinguishing feature of submerged arc welding is the granular material which covers the weld area and prevents arc radiation, sparks, spatter and fumes from escaping. The flux is of major importance in achieving the high deposition rates and high quality weld deposit characteristics of this process. In addition to shielding the arc from view, the flux provides a slag which protects the weld metal as it cools, it deoxidizes and refines the weld metal, insulates the weld to reduce the cooling rate and it helps shape the weld contour.

Submerged arc welding can be applied in three general modes: semi-automatic welding, machine welding, and full automatic welding. Each of these modes requires that the joint be in the flat or horizontal position so that flux and molten weld metal will not run out of the joint before solidification takes place.

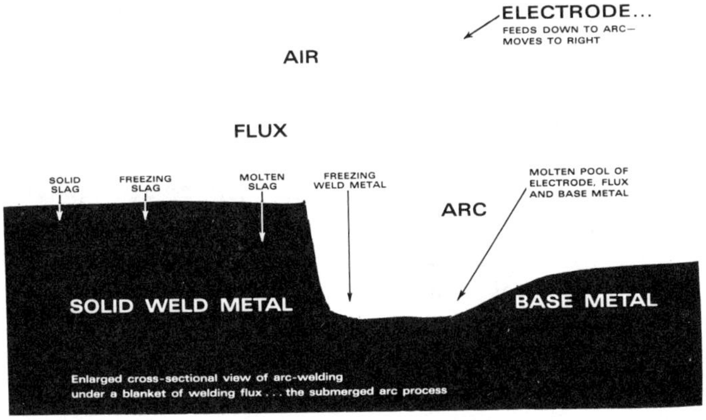

Fig. 5-3. The mechanics of the submerged-arc process. The arc and the molten weld metal are buried in the layer of flux, which protects the weld metal from contamination and concentrates the heat into the joint. The molten flux arises through the pool, deoxidizing and cleansing the molten metal, and forms a protective slag over the newly deposited weld.

Semi-automatic welding uses a hand held welding gun which feeds the wire and flux to the weld area. Wire is pulled from a spool by a wire feeder. Flux may be delivered by gravity from a small hopper attached to the top of the gun or it may be supplied through a hose from a large container using air pressure.

Machine welding uses equipment which performs all the welding operations but requires the operator to start and stop each weld and adjust current, voltage, and joint alignment during the run.

In full automatic welding, flux is deposited ahead of the arc and unfused loose flux is picked up by a vacuum system. Arc current, voltage, travel speed and joint tracking are monitored and adjusted by programmable controllers which maintain these functions at preprogrammed values.

The majority of submerged arc welding is done in the machine mode generally where long straight welds or circular welds are made in steel of heavy thicknesses. These welds require constant monitoring of an operator. Before this process is selected for a production application several factors should be evaluated; length of welds and number of welds to be made, position and accessibility of welds, thickness of joints, required weld mechanical properties, quality and appearance.

During welding, the heat of the arc melts some of the flux along with the tip of the electrode as illustrated in Fig. 5-3. The tip of the electrode and the welding zone are always surrounded and shielded by molten flux covered by a layer of unfused flux. The electrode is held a short distance above the workpiece with the arc between the electrode and the workpiece. As the electrode progresses along the joint, the lighter molten flux rises above the molten metal in the form of a slag. The weld metal which has a higher melting (freezing) point, solidifies while the slag above it is still molten. The slag then freezes over the newly solidified weld metal, continuing to protect the metal from contamination while it is very hot and would react with atmospheric oxygen and nitrogen. After cooling and removing any unfused flux for reuse, the solidified slag may be easily removed from the weld.

Submerged arc automatic welding installation custom fabricates a wide range of nonstandard beams for arc welded steel structures.

ADVANTAGES OF THE PROCESS

High currents can be used in submerged-arc welding and extremely high heat developed. Because the current is applied to the electrode a short distance above its tip, relatively high amperages can be used on small diameter electrodes. This results in extremely high current densities on relatively small cross sections of electrode. Currents as high as 600 amperes can be carried on electrodes as small as 5/64″, giving a density in the order of 100,000 amperes per square inch — six to ten times that carried on stick electrodes.

Because of the high current density, the melt-off rate is much higher for a given electrode diameter than with stick-electrode welding (See Table 6-22). The melt-off rate is affected by the electrode material, the flux, type of current, polarity, and length of wire beyond the point of electrical contact in the gun or head.

Submerged-arc welding may be done with either DC or AC power. Direct current gives better control of bead shape, penetration, and welding speed, and arc starting is easier with it. Bead shape is usually best with DC electrode positive (reverse polarity), which also provides maximum penetration. Highest deposition rates and minimum penetration are obtained with DC electrode negative. Alternating current minimizes arc blow and gives penetration between that of DCEP and DCEN. (See Section 6.3.)

A submerged-arc semiautomatic gun used with a turntable to weld pipe flanges.

The insulating blanket of flux above the arc prevents rapid escape of heat and concentrates it in the welding zone. Not only are the electrode and base metal melted rapidly, but the fusion is deep into the base metal. The deep penetration allows the use of small welding grooves, thus minimizing the amount of filler metal per foot of joint and permitting fast welding speeds. Fast welding, in turn, minimizes the total heat input into the assembly and, thus minimizes problems of heat distortion. Even relatively thick joints can be welded in one pass by submerged-arc.

Welds made under the protective layer of flux have good ductility and impact resistance and uniformity in bead appearance. Mechanical properties at least equal to those of the base metal are consistently obtained. In single-pass welds, the fused base material is large compared to the amount of filler metal used. Thus, in such welds the base metal may greatly influence the chemical and mechanical properties of the weld. For this reason, it is sometimes unnecessary to use electrodes of the same composition as the base metal for welding many of the low-alloy steels.

However, the chemical composition and properties of multipass welds are less affected by the base metal and depend to a greater extent on the

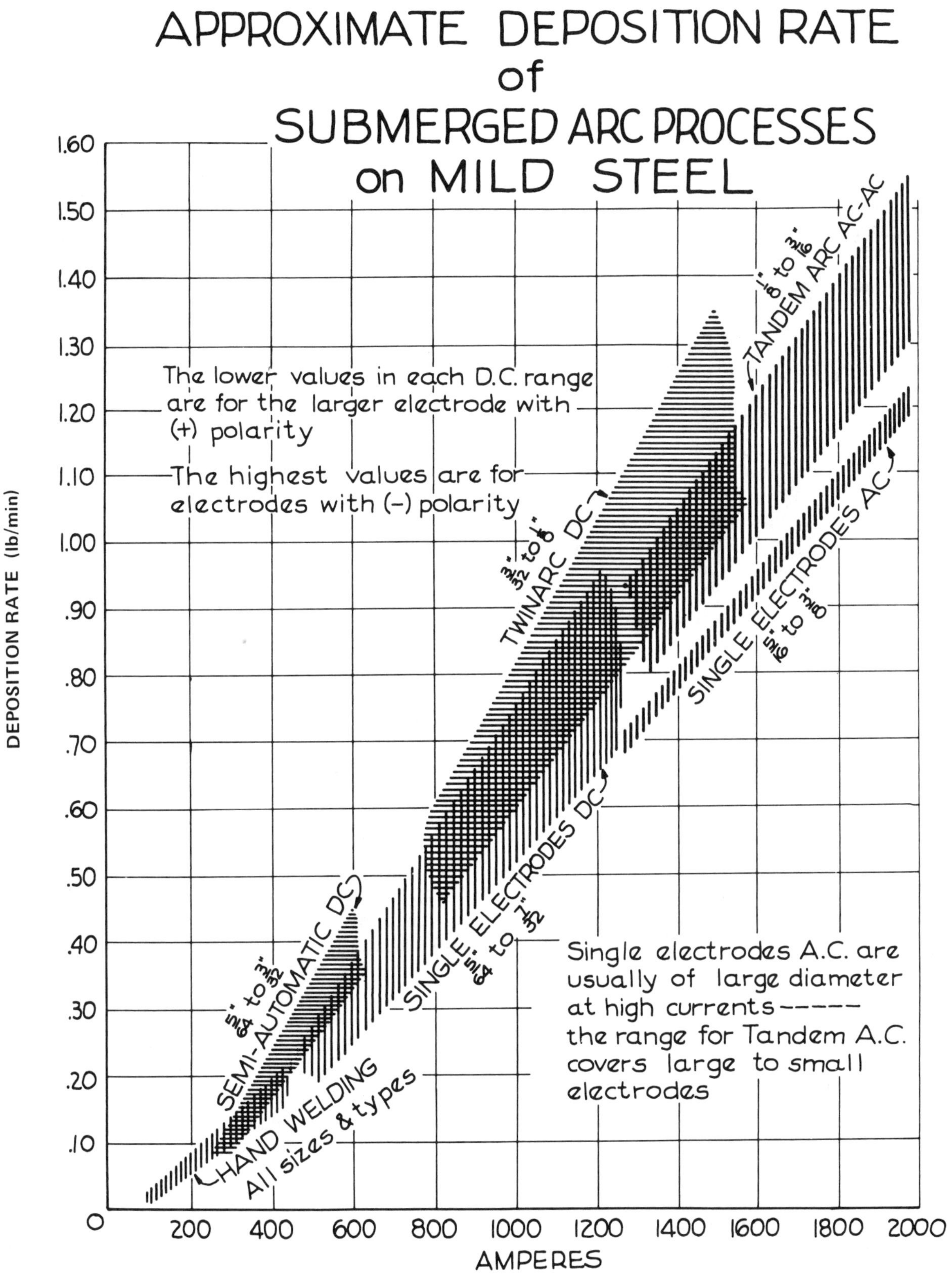

Fig. 5-4. Approximate deposition rates of various submerged-arc arrangements, compared with the deposition rates of stick-electrode welding.

The submerged-arc process joining an elbow to a length of high pressure pipe using fully automatic equipment.

composition of the electrode, the activity of the flux, and the welding conditions.

Through regulation of current, voltage, and travel speed, the operator can exercise close control over penetration to provide any depth ranging from deep and narrow with high-crown reinforcement, to wide, nearly flat beads with shallow penetration. Beads with deep penetration may contain on the order of 70% melted base metal, while shallow beads may contain as little as 10% base metal. In some instances, the deep-penetration properties of submerged-arc can be used to eliminate or reduce the expense of edge preparation.

PROCESS VARIATIONS

A number of multiple electrode systems are available with this process, which permit increased deposition rate and penetration and provide greater flexibility of operation using one or more power sources. Two electrodes may be used with a single power source arranged to follow a joint in tandem so that the arcs produce a single deep penetrating weld pool. The electrode wires normally are fed by a single set of drive rolls with double grooves through a single contact tip with two holes. In this case the wire circuits are in parallel. The same arrangement can be used with the contact tip turned 90° to place the wires side-by-side to obtain a slightly wider weld. If shallow penetration is required, such as for an overlay, the two wires may be connected in series so that the current path is from one electrode to the weld pool then to the other electrode which is connected to the power supply ground. This requires that the drive rolls and contact tips be separated and insulated from each other. If an AC power supply is used with this arrangement the same amount of heat will be generated in each arc and arc-blow will be avoided.

Multiple electrodes may be arranged with each electrode connected to a separate power source; the lead electrode connected to a DC source and the trail electrode to an AC source. Two AC power sources may be connected to a three phase power line using a Scott connection. The Scott connection distributes the AC load over two phases of a three-phase power line. The primary of the lead transformer is connected to lines A and B; the primary of the trail transformer is connected to line C and an intermediate tap on the primary of the lead transformer. By selecting the appropriate tap, the phase angle between the output of the lead and trail transformers can be varied and set for a value of less than 90 degrees which is desirable to minimize arc interaction on tandem arc welding. If the trail transformer were connected to lines C and A or C and B, the phase angle difference would be 120° and magnetic interaction would tend to separate the arcs.

In addition to increasing deposition rate, the use of multiple arcs expands the flexibility of the process. For surfacing applications, two or more arcs may be arranged transverse to the direction of travel to cover a wide area in each pass. Large horizontal fillet welds can be made in one pass using the tandem arc technique. The two arcs are spaced about 4″ apart to produce two separate pools, which will remain in place where a single pool would tend to run out of the corner. Different combination of AC, DC, polarity and phase can be used with multi-electrode welding to control arc deflection to place the deposit in the desired location.

Deposition rates are high — up to ten times those of stick-electrode welding. Figure 5-4 shows approximate deposition rates for various submerged-arc arrangements, with comparable deposition rates for manual welding with covered electrodes.

FLUXES

As mentioned before, the flux serves several functions in submerged arc welding. These include covering the molten weld metal to protect it from the atmosphere and acting as a slag which refines the molten deposit by scavenging oxides and other non-metallic inclusions. Metallic additions to the flux can add to the alloy content of the deposit and can deoxidize it.

There are four types of fluxes based on their method of manufacture; fused, bonded, agglomerated and mechanically mixed.

The ingredients for a fused flux are dry mixed, melted in an electric furnace then poured into chill blocks and allowed to cool. The solidified flux is crushed, screened and packaged. Fused fluxes provide good homogeneity and are not hygroscopic. It is difficult to add alloying elements and deoxidizers to fused fluxes.

The raw materials for a bonded flux, are reduced to powder, are dry mixed then wet bonded with potassium silicate, sodium silicate or a mixture of these. The wet mixture is pelletized and baked at a relatively low temperature. The pellets are broken up, screened and packaged. Alloy elements and deoxidizers can be easily added to bonded fluxes during manufacture. Bonded fluxes have a tendency to absorb moisture and thus must be stored and handled like low hydrogen electrodes.

Agglomerated fluxes are manufactured in much the same way bonded fluxes are made except a ceramic binder is used in place of the silicates. This requires a relatively high temperature to consolidate the mixture which limits the use of deoxidizers and alloy additions.

Mechanically mixed fluxes consist of mixtures of two or more of the fused, bonded, or agglomerated fluxes. This provides a large range of flux compositions for special applications. Since the mixtures are mechanical, these fluxes are subject to segregation during shipment, in handling, and in the feeding system.

Flux particle size and size distribution are important factors to consider both with regard to performance when welding and feeding through the system. Generally particle size should decrease as weld current decreases. However, coarse particle size is preferred when welding rusty steel because the larger spaces between particles allow gases to escape more readily. Fine fluxes tend to pack in feeding and recovery systems.

Where steels require low hydrogen treatment, fluxes must be kept dry. Although fused fluxes do not contain moisture, adsorbed moisture may be attached to the surfaces. Bonded fluxes may contain chemically bonded moisture. Therefore bonded fluxes should be handled and treated the same as low hydrogen electrodes.

When alloy bearing fluxes are used, the amount of alloy obtained in the deposit depends upon the ratio of electrode melted to flux melted. Increasing voltage while holding current constant will increase the amount of flux melted and thus the amount of alloy added to the weld. The amount of wire added will remain constant. It is imperative that all parameters (voltage, wire feed rate, flux and travel speed) be controlled carefully when using an alloy flux to assure that a uniform weld deposit composition is achieved.

Fluxes are also identified as basic, acid, and neutral. Basic fluxes contain oxides of metals which dissociate easily while acidic fluxes contain oxides which dissociate to a small extent. A neutral flux does not add or subtract from the composition of the weld deposit. Fluxes having a ratio of CaO or MnO to SiO_2 which is greater than one are considered basic, those near one are considered neutral, and those less than one are acidic.

PROCESS VARIABLES

Amperage has the greatest effect on melting rate and is controlled by wire diameter and feed rate. Amperage should not be so high as to cause melt-through or excessive buildup or so low as to cause incomplete fusion or lack of penetration.

Varying voltage will cause a change in arc length and arc energy but has little effect on wire melting rate. Increasing voltage with constant current and travel speed will increase the amount of flux melted, produce a flatter weld deposit and reduce porosity caused by rust or scale. If an alloy flux is used, the amount of pickup in the deposit will increase. Decreased voltage will produce a stiffer arc, greater penetration and reduced arc blow. Excessively low voltage will produce a high, narrow bead which tends to prevent easy slag removal.

Welding speed must be adjusted to attain the required weld buildup, penetration and bead shape. As for other processes, penetration and bead size decrease with increased speed and increase with decreased speed. At excessively slow speeds, penetration decreases because the arc force is dissipated by the molten pool. Excessively high speeds cause lack of fusion between deposit and base metal and generally poor bead appearance.

At any fixed current a small size electrode will have a higher deposition rate than a larger electrode because the current density is higher and there is more (I^2R) resistance heating of the wire. However, the larger wire can deposit metal at a higher rate when current is increased proportionally. For high current densities, resistance heating of the electrode may be a significant portion of the total weld heat and this increases with electrode extension. If electrode extension is increased, voltage should be increased to maintain a constant arc length and voltage. As a general rule electrode extension should be approximately eight times the wire diameter.

APPLICATIONS AND ECONOMIES

With proper selection of equipment, submerged-arc is widely applicable to the welding requirements of industry. It can be used with all types of joints, and permits welding a full range of carbon and low-alloy steels, from 16-gage sheet to the thickest plate. It is also applicable to some high-alloy, heat-treated, and stainless steels, and is a favored process for rebuilding and hardsurfacing. Any degree of mechanization can be used — from the hand-held semi-automatic gun to boom or track-carried and fixture-held multiple welding heads.

The high quality of submerged-arc welds, the high deposition rates, the deep penetration, the adaptability of the process to full mechanization, and the comfort characteristics (no glare, sparks, spatter, smoke, or excessive heat radiation) make it a preferred process in steel fabrication. It is used extensively in ship and barge building, railroad car building, pipe manufacture, and in fabricating structural beams, girders, and columns where long welds are required. Automatic submerged-arc installations are also key features of the welding areas of plants turning out mass-produced assemblies joined with repetitive short welds.

The high deposition rates attained with submerged-arc are chiefly responsible for the economies achieved with the process. The cost reductions when changing from the manual shielded metal-arc process to submerged-arc are frequently dramatic. Thus, a hand-held submerged-arc gun with mechanized travel may reduce welding costs more than 50%; with fully automatic multiarc equipment, it is not unusual for the costs to be but 10% of those attained with stick-electrode welding.

Other factors than deposition rates enter into the lowering of welding costs. Continuous electrode feed from coils, ranging in weight from 60 to 1,000 pounds, contributes to a high operating factor. Where the deep-penetration characteristics of the process permit the elimination or reduction of joint preparation, expense is reduced. After the weld has been run, cleaning costs are minimized, because of the elimination of spatter by the protective flux.

Operators work on a tandem fixture to make 3/8 inch fillet welds on low-alloy, high strength "T" sections for a water reservoir structure. They are using a trackless tractor that develops full mechanized welding performance while featuring extreme portability and simplicity.

When submerged-arc equipment is operated properly, the weld beads are smooth and uniform, so that grinding or machining are rarely required. Since the rapid heat input of the process minimizes distortion, the costs for straightening finished assemblies are reduced, especially if a carefully planned welding sequence has been followed. Submerged-arc welding, in fact, often allows the premachining of parts, further adding to fabrication cost savings.

A limitation of submerged-arc welding is that imposed by the force of gravity. In most instances, the joint must be positioned flat or horizontal to hold the granular flux. To deal with this problem, weldment positioners (See Section 4.4) are used to turn assemblies to present joints flat or horizontal — or the assemblies may be turned or rotated by a crane. Substantial capital investments may be necessary to use submerged-arc welding to its fullest extent and gain all of the advantages of the process. As explained in Section 6.3, special fixturing and tooling have been developed for the retention of flux and molten metal in some applications, so that "three-o'clock" and even vertical-up welding is possible.

Problems can arise in the use of submerged-arc resulting from improper joint preparation and improper procedures. Thus, "flash-through" of the arc, melt-through, and weld porosity result from such factors as improper procedures, poor fitup, and joint contaminants. These and other problems and the remedies for them are discussed in Section 6.3.

Vertical-up welding with the self-shielded flux-cored electrode process.

The Self-Shielded Flux-Cored Process

Flux cored arc welding is an arc welding process in which welding heat is obtained from an arc between a continuous flux cored electrode and the work. The flux provides gas shielding for the arc and a slag covering of the weld deposit.

The self-shielded flux-cored arc-welding process is an outgrowth of shielded metal-arc welding. The versatility and maneuverability of stick electrodes in manual welding stimulated efforts to mechanize the process so that a continuous wire electrode could be used. A continuous electrode would eliminate the welding time lost in changing electrodes and would eliminate the material lost in electrode stubs. The main problems with a continuous coated electrode are a means of making electrical contact with the core wire and coiling it without cracking the brittle coating. This need for a continuous arc welding electrode led to the development of the self shielding flux cored electrode where the material needed for shielding is contained inside the core of a hollow electrode.

The "inside-out" construction of the flux cored electrodes solved both of these problems. Continuous electrical contact can be made with these electrodes at any necessary distance from the arc and they can be coiled and packaged on any of the standard spools used for filler wire. The outcome of these efforts was the development of the semiautomatic and full-automatic processes for welding with continuous flux-cored tubular electrode "wires." Such fabricated wires (Figure 5-5) contain in their cores the ingredients for fluxing and deoxidizing molten metal and for generating shielding gases and vapors and slag coverings.

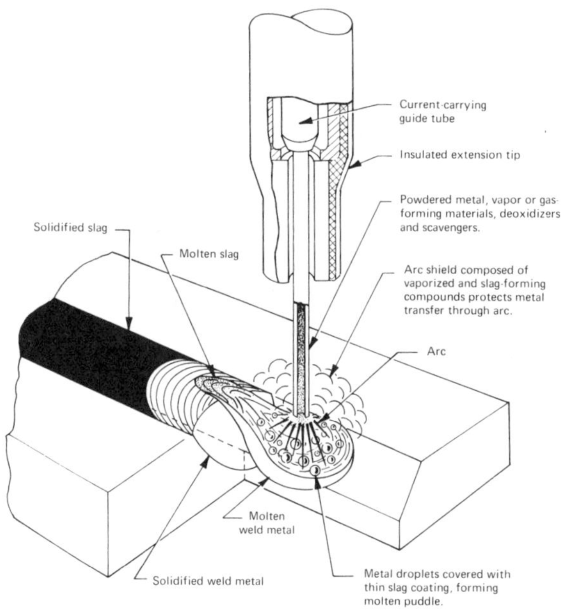

Fig. 5-5. Principles of the self-shielded flux-cored arc-welding process. The electrode may be viewed as an "inside-out" construction of the stick electrode used in shielded metal-arc welding. Putting the shield-generating materials inside the electrode allows the coiling of long, continuous lengths of electrode and gives an outside conductive sheath for carrying the welding current from a point close to the arc.

Fig. 5-6. The operator activates electrode feed when he presses the trigger completing the welding circuit. With the semiautomatic gun he can reach into areas that are inaccessible to the semiautomatic equipment of other processes.

The gas shielding supplied by the flux may be supplemented by a protective gas flow of CO_2 or CO_2– argon mixture. This process is known as the gas shielded flux cored process and is discussed in the next section.

In essence, semiautomatic welding with flux-cored electrodes is equivalent to manual shielded metal-arc welding with an electrode many feet long instead of just a few inches long. By the press of the trigger completing the welding circuit, the operator activates the mechanism that feeds the electrode to the arc (Figure 5-6). He uses a welding gun instead of an electrode holder, but it is similarly light in weight and easy to maneuver. The only other major difference is that the weld metal of the electrode surrounds the shielding and fluxing chemicals, rather than being surrounded by them.

Full-automatic welding with self-shielded flux-cored electrodes is a further step in mechanization — the removal of direct manual manipulation in the utilization of the open-arc process.

In the flux-cored arc welding process, partial shielding is provided by vaporization of ingredients of the flux which form a protective gas covering of the weld pool. When molten weld metal is transferred across the arc, it may not be completely covered by molten flux since the metal surrounds the flux, as both are heated. This means that some oxidation and nitrogen pick-up may occur before the metal enters the weld pool. However, in contrast to the gas-shielded flux cored process, the self-shielded flux-cored process does not depend on shielding alone to provide sound ductile welds. Instead, deoxidizers and denitrifiers are added to the metal and the flux which refine the metal in the molten pool. Since this process does not depend on gas shielding at the arc, it can operate satisfactorily outdoors where strong air currents are encountered.

The self-shielded flux cored process is characterized by long electrode extension beyond the contact tip. Extensions from 3/4″ to 3 3/4″ are used depending upon the application.

As the electrode extension increases, resistance heating of the wire increases in proportion to the extension. For a given weld current this causes a proportional increase in the voltage drop in the wire and a decrease in arc voltage when using a constant potential power supply. For a given electrode feed rate, diameter, and composition, weld current varies inversely with electrode extension. Electrode extension is an additional variable which must be considered and controlled in self-shielded flux cored arc welding. This is necessary to achieve a balance between the melting and activation of the core materials and the metal containment tube. Too long an extension will produce a short unstable arc with excessive spatter, while too short an extension may cause an excessive arc length.

Increasing electrode extension decreases welding current and the heat available for melting base metal. This results in a narrow shallow penetrating weld bead which makes the process suitable for welding thin gage material and bridging gaps caused by poor fitup.

When wire size, composition, voltage, and arc length are held constant, arc current varies directly with wire feed rate and is controlled by the wire feed speed controller. As may be expected, increasing wire feed rate (current) increases electrode deposition rate. Excessive current will produce convex weld beads with irregular appearance, while insufficient current will cause large droplet transfer and excessive spatter. Insufficient current can also result in high nitrogen pickup and weld metal porosity.

Actual arc voltage varies directly with arc length; however, arc voltage is only a part of the total circuit voltage indicated by the meter on the weld power supply. While arc voltage is the largest part of the total voltage, the I^2R drop in the electrode extension is a significant part. Too high an arc voltage will cause excessive spatter, wide irregular weld beads and high nitrogen pickup. This may cause porosity with mild steel electrodes and reduction in ferrite with austenitic stainless steel electrodes. Too low an arc voltage will result in narrow convex beads, excessive spatter and reduced penetration.

The electrode angle has an effect on and can be used to control the weld pool. By tipping the electrode at an angle opposite to the direction of travel (this is known as the drag angle) the arc force can be used to counteract the force of gravity which tends to cause the molten weld pool to run ahead of the arc. For flat and horizontal position welds, the drag angle should be 20° to 45° from vertical. The larger angles are used for thin materials. Smaller angles are used for thicker materials where increased penetration is required. For making fillet welds in the horizontal position the electrode should point at the vertical member at an angle of 40° to 50° from the vertical member in addition to the drag angle.

As noted in Section 5.1, one of the limitations of the stick electrode is the long and varying length of electrode between the point of electrical contact in the electrode holder and the electrode tip. This limits the current that can be used because of electrical resistance heating. High currents — capable of giving high deposition rates — in passing through an electrode length more than a few inches long would develop enough resistance heating to overheat and damage the covering. But when electrical contact can be made close to the arc, as with the inside-out construction of tubular electrodes, relatively high currents can be used even with small-diameter electrode wires.

The inside-out construction of self-shielded electrode, thus, brought to manually operated welding the possibility of using higher amperage than feasible with stick-electrode welding. As a result, much higher deposition rates are possible with the hand-held semiautomatic gun than with the hand-held stick-electrode holder.

Higher deposition rates, plus automatic electrode feed and elimination of lost time for changing electrodes, have resulted in substantial production economies wherever the semiautomatic process has been used to replace stick-electrode welding. Decreases in welding costs as great as 50% have been common, and in some production welding, deposition rates have been increased as much as 400%.

The intent behind the development of self-shielded flux-cored electrode welding was to mechanize and increase the efficiency of manual welding. The semiautomatic use of the process does just that — serves as a direct replacement for stick-electrode welding. The full-automatic use of the process, on the other hand, competes with other fully automatic processes and is used in production where it gives the desired performance characteristics and weld properties, while eliminating problems associated with flux or gas handling. Although the full-automatic process is important to a few industries, the semiautomatic version has the wider application possibilities. In fact, semiautomatic self-shielded flux-cored welding has potential for substantially reducing welding costs when working with steel wherever stick-electrode welding is used to deposit other than minor volumes of weld metal. This has been proved to be true in maintenance and repair welding (Figure 5-7), as well as in production work.

Although the AWS-accepted term for the process is self-shielded flux-cored arc welding, it is often referred to as open-arc "Squirt" welding, vapor-shielded welding, or cored-electrode welding. The prefix "semiautomatic" or "full-automatic" is used to distinguish the degree of mechanization. In shop practice, the proprietary name for the electrode is often the term employed in referring to the process, with the semiautomatic version generally implied.

Fig. 5-7. The semiautomatic self-shielded flux-cored arc-welding process substantially reduces costs in repair, rebuilding, and maintenance work, as well as in manufacturing, fabrication, and structural steel erection. Here, the weldor is using the process to repair and rebuild shovel crawler pads.

Thus, in many shops the trade name "Innershield" is used to mean semiautomatic self-shielded flux-cored welding. Should the full-automatic method be employed in the same shop, it would likely be called "automatic Innershield."

Electrodes for self-shielded flux-cored arc welding are available in standard sizes from 1/16″ (1.58 mm) to 5/32″ (3.97 mm) in mild steel and in some low alloy steel and stainless steel compositions. These electrodes are covered in AWS filler metal specifications, A5.20 for carbon steel, A5.29 for low alloy steels and A5.22 for chromium and chromium-nickel corrosion resisting steels. The latest revision of each of these specifications should be consulted.

ADVANTAGES OF THE PROCESS

The electrode types, performance characteristics, equipment, and operating techniques for the self-shielded flux-cored process are described in Section 6.4. Some of the benefits of the process are brought out or become apparent in the discussion of operating techniques. However, the impact of a mechanized, all-position, manually maneuvered arc-welding process on weldment production and structural steel fabrication and erection is not always obvious and merits mention.

Since the semiautomatic process can be used any place stick-electrode can be used, it makes possible one-process welding in the erection of structural steel in building framing. As proved on major high-rise projects, this factor is possibly as important as the higher deposition rates of semiautomatic welding in reducing erection costs. One process substantially reduces the amount of equipment needed on the job and allows every welding operator to be qualified for every joint. This, in turn, permits the more systematic deployment of men, equipment, and materials and reduces delays and minimizes equipment handling. One-process welding — from tack welding to column-splice and beam-to-column welding — gives erectors the opportunity to take a "sytems approach" to erection logistics, with the result that cost savings above those attributable to the speed of the semiautomatic process are realizable.

Figure 5-8 shows the welding of a beam web-to-column connection in the erection of a major high-rise building. When this weld was made possible by the development of the all-position electrode, one-process mechanized erection welding became feasible.

Fig. 5-8. Before all-position electrodes for self-shielded flux-cored arc welding were developed, beam web-to-column connections in building erection were made with stick electrodes. If the flange-to-column joints were made with semiautomatic equipment, the operator had to change to the stick electrode for the vertical joint, or the connection had to be made by bolting the web to an angle welded to the column. The all-position electrode wires were the development that made possible one-process mechanized welding in building erection, with all of its incidental benefits in the scheduling of erection operations.

The tolerance of the semiautomatic process to poor fitup is also a decisive benefit to steel fabricators. Although every reasonable effort should be made to guard against poor fitup, its avoidance is economically impractical with some weldments, especially those made from heavy material and pieces of complex configuration where appearance or perfection in detail has little or no value in the end product. In manufacturing where fitup has been a major problem, semiautomatic self-shielded flux-cored arc welding reduces rework and repair without affecting final product quality. This factor has been of cost-saving value as important to the company as the cost savings from the increased welding speeds when changing from manual shielded metal-arc welding.

The advantages of the self-shielded flux-cored arc-welding process may be summarized as follows:

1. When compared with stick-electrode welding, gives deposition rates up to four times as great, often decreasing welding costs by as much as 50 to 75%.
2. Eliminates the need for flux-handling and recovery equipment, as in submerged-arc welding, or for gas and gas storage, piping, and metering equipment, as in gas-shielded mechanized welding. The semiautomatic process is applicable where other mechanized processes would be too unwieldy.
3. Has tolerance for elements in steel that normally cause weld cracking when stick-electrode or one of the other mechanized welding processes are used. Produces crack-free welds in medium-carbon steel, using normal welding procedures.
4. Under normal conditions, eliminates the problems of moisture pickup and storage that occur with low-hydrogen electrodes.
5. Eliminates stub losses and the time that would be required for changing electrodes with the stick-electrode process.
6. Eliminates the need for wind shelters required with gas-shielded welding in field erection; permits fans and fast air-flow ventilation systems to be used for worker comfort in the shop.
7. Enables "one-process," and even "one-process, one-electrode," operation in some shop and field applications. This, in turn, simplifies operator training, qualification, and supervision; equipment selection and maintenance; and the logistics of applying men, materials, and equipment to the job efficiently.
8. Enables the application of the long-stickout principle to enhance deposition rates, while allowing easy operator control of penetration.

9. Permits more seams to be welded in one pass, saving welding time and the time that otherwise would be consumed in between-pass cleaning.
10. Is adaptable to a variety of products; permits continuous operation at one welding station, even though a variety of assemblies with widely different joint requirements are run through it.
11. Provides the fast filling of gouged-out voids often required when making repairs to weldments or steel castings.
12. Gives the speed of mechanized welding in close quarters; reaches into spots inaccessible by other semiautomatic processes.
13. Provides mechanized welding where mechanized welding was formerly impossible, such as in the joining of a beam web to a column in building erection (Figure 5-8).
14. Enables the bridging of gaps in fitup by operator control of the penetration without reducing quality of the weld. Minimizes repair, rework, and rejects.

Tubular support assembly for a bulldozer blade is being welding with the self-shielded flux-cored electrode process.

Self-shielded flux-cored arc welding.

The Gas-Shielded Arc-Welding Processes

As noted in the preceding sections, the shielded metal-arc process (stick-electrode) and self-shielded flux-cored electrode process depend in part on gases generated by the heat of the arc to provide arc and puddle shielding. In contrast, the gas-shielded arc-welding processes use either bare or flux-cored filler metal and gas from an external source for shielding. The gas is impinged upon the work from a nozzle that surrounds the electrode. It may be an inert gas — such as argon or helium — or carbon dioxide (CO_2), a much cheaper gas that is suitable for use in the welding of steels. Mixtures of the inert gases, oxygen, and carbon dioxide also are used to produce special arc characteristics.

There are three basic gas-shielded arc-welding processes that have broad application in industry. They are the gas-shielded flux-cored process, the gas tungsten-arc (GTA) process, and the gas metal-arc (GMA) process.

THE GAS-SHIELDED FLUX-CORED PROCESS

The gas-shielded flux-cored process may be looked upon as a hybrid between self-shielded flux-cored arc welding and gas metal-arc welding. Tubular electrode wire is used (Figure 5-9), as in the self-shielded process, but the ingredients in its core are for fluxing, deoxidizing, scavenging, and sometimes alloying additions, rather than for these functions plus the generation of protective vapors. In this respect, the process has similarities to the self-shielded flux-cored electrode process, and the tubular electrodes used are classified by the AWS (See Sections 6.4 and 6.5) along with electrodes used in the self-shielded process. On the other hand, the process is similar to gas metal-arc welding in that a gas is separately applied to act as an arc shield.

The guns and welding heads for semiautomatic and full-automatic welding with the gas-shielded process are of necessity more complex than those used in self-shielded flux-cored welding. Passages must be included for the flow of gases. If the gun is water-cooled, additional passages are required for this purpose. Figure 5-10 shows typical guns for semiautomatic gas-shielded flux-cored arc welding, and Figure 5-11 typical nozzle assemblies for full-automatic operation. The wire feeder is similar to the wire feeder used with the self-shielded flux-cored electrode process. Figure 5-12 shows a schematic

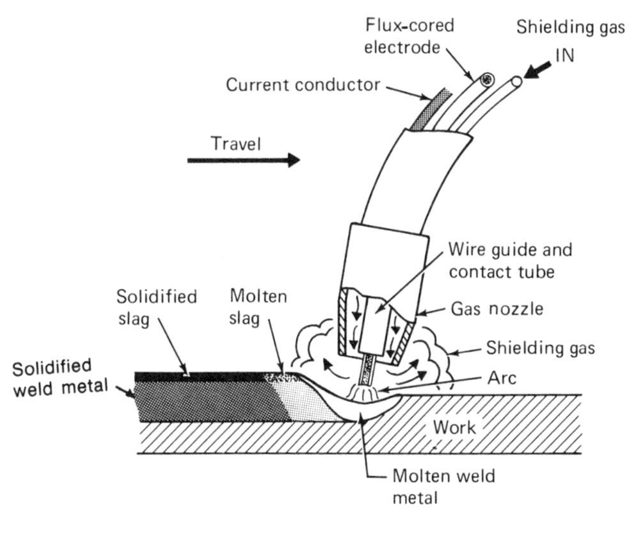

Fig. 5-9. Principles of the gas-shielded flux-cored process. Gas from an external source is used for the shielding; the core ingredients are for fluxing and metal-conditioning purposes.

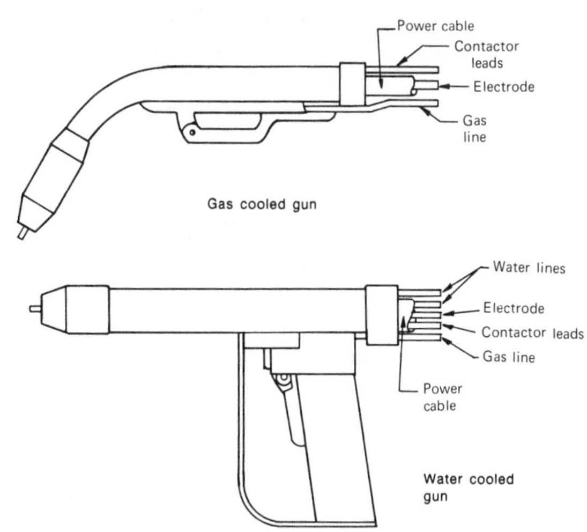

Fig. 5-10. Typical guns for semiautomatic gas-shielded flux-cored arc welding.

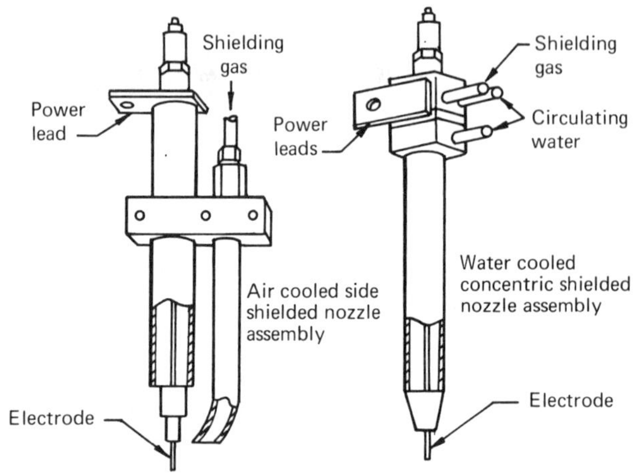

Fig. 5-11. Typical nozzle assemblies for full-automatic welding with gas-shielded flux-cored electrode or with solid-wire electrode (GMA welding).

for a full-automatic welding facility with either self-shielded or gas-shielded flux-cored electrode; the dotted lines indicate the addition required with the gas-shielded version when using CO_2 as the shielding gas.

Although CO_2 is generally used as the shielding gas, mixtures of 25% CO_2 −75% argon and 95% argon −5% oxygen may be used. These gas mixtures can exclude nitrogen from the arc atmosphere and prevent nitrogen pickup by the weld metal but some oxygen can be present in the arc atmosphere from the gas mixture or from dissociation of CO_2. Carbon dioxide is relatively inert at room temperature but when heated to high temperatures, such as in the electric arc, it dissociates to form carbon monoxide and oxygen. Molten iron can react with CO_2 to form iron oxide and CO. Some of the CO can dissociate further at high temperature to form carbon and oxygen. Since the CO_2 shielding gas does not avoid oxygen in the arc atmosphere and does not provide any deoxidizing quality, it is necessary that the core material fulfill this function of deoxidizing and refining the weld pool. Deoxidizers in the wire core react with oxygen in the molten metal to form solid oxide compounds which float to the surface and combine with the slag.

Shielding gas flow rate must be properly adjusted for the type and diameter of the gun nozzle, nozzle to work distance and air movement around the welding operation. An inadequate flow rate will not provide adequate shielding while excessive flow rate will cause turbulence and mixing of air with the shielding gas. Either of these extremes will increase the amount of impurities in the weld metal and deplete the deoxidizers in the core.

The gas-shielded flux-cored process is used for welding mild and low-alloy steels (See Section 6.5). It gives high deposition rates, high deposition efficiencies, and high operating factors. Radiographic-quality welds are easily produced, and the weld metal with mild and low-alloy steels has good ductility and toughness. The process is adaptable to a wide variety of joints and gives the capability for all-position welding.

GAS METAL-ARC WELDING (GMAW)

Gas metal-arc welding, popularly known as MIG welding, uses a continuous electrode for filler metal and an externally supplied gas or gas mixture for shielding. The shielding gas — helium, argon, carbon dioxide, or mixtures thereof — protects the molten metal from reacting with constituents of the atmosphere. Although the gas shield is effective in shielding the molten metal from the air, deoxidizers are usually added as alloys in the electrode. Sometimes light coatings are applied to the electrode for arc stabilizing or other purposes. Lubricating films may also be applied to increase the electrode feeding efficiency in semiautomatic welding equipment. Reactive gases may be included in the gas mixture for arc-conditioning functions. Figure 5-13 illustrates the method of which shielding gas and continuous electrode are supplied to the welding arc.

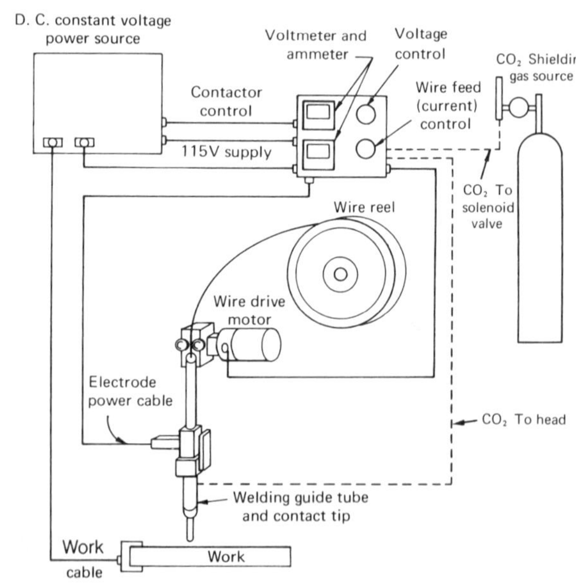

Fig. 5-12. Schematic for a full-automatic welding facility with either self-shielded or gas-shielded flux-cored electrode. The dotted line indicates the additions required with the gas-shielded version when using CO_2 as the shielding gas.

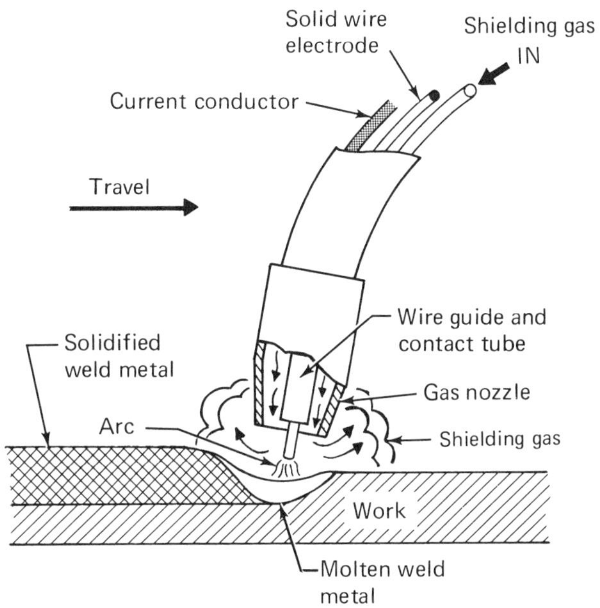

Fig. 5-13. Principle of the gas metal-arc process. Continuous solid-wire electrode is fed to the gas-shielded arc.

Gas metal arc welding was originally developed as a high current density welding process for use with small diameter bare metal electrodes and inert gas shielding. Its primary application was for welding aluminum and aluminum alloys. Since then, the process has been expanded to operate at lower current densities and in a pulsing mode. It can operate with reactive gases such as CO_2 as well as the inert gases.

GMA welding may be used with all of the major commercial metals, including carbon, alloy, and stainless steels and aluminum, magnesium, copper, iron, titanium, and zirconium. It is a preferred process for the welding of aluminum (See Section 9.3), magnesium, copper, and many of the alloys of the reactive metals. Most of the irons and steels can be satisfactorily joined by GMA welding, including the carbon-free irons, the low-carbon and low-alloy steels (See Section 6.6), the high-strength quenched and tempered steels, the chromium irons and steels, the high-nickel steels, and some of the iron and nickel base superalloys. With these various materials, the welding techniques and procedures may vary widely. Thus, carbon dioxide or argon-oxygen mixtures are suitable for arc shielding when welding the low-carbon and low-alloy steels, whereas pure inert gas may be essential when welding highly alloyed steels.

Pure inert gas shielding is essential for welding alloys of aluminum, magnesium, copper, titanium, zirconium, stainless steels and nickel base superalloys.

Welding is either semiautomatic, using a hand-held gun to which electrode is fed automatically, or full-automatic equipment is used. The welding guns or heads are similar to those used with gas-shielded flux-cored welding (Figures 5-10 and 5-11). When the term "manual" gas metal-arc welding is used, the semiautomatic process with its hand-held gun is implied.

Metal transfer with the GMA process is by one of two methods: "spray-arc" or short circuiting. With spray-arc, drops of molten metal detach from the electrode and move through the arc column to the work. With the short-circuiting technique — often referred to as short-arc welding — metal is transferred

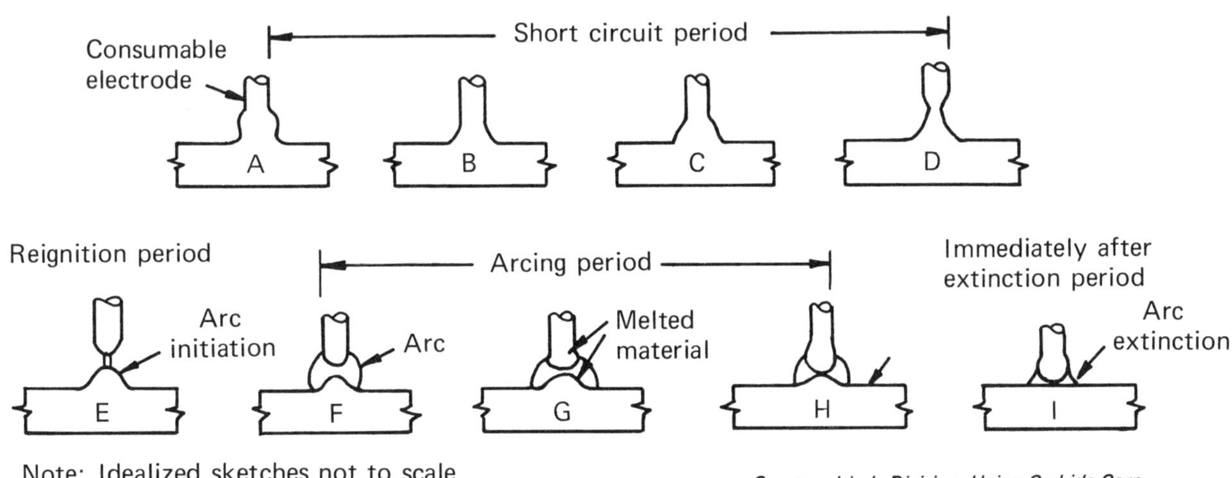

Courtesy Linde Division, Union Carbide Corp.

Fig. 5-14. The short-arc cycle in MIG welding. Short-circuiting starts with (A); the arc is reignited at (E); and is extinguished when the molten metal on the electrode tip touches the molten puddle at (H). Metal is transferred during the short circuit.

to the work when the molten tip of the electrode contacts the molten puddle.

The latter, short-arc welding, uses low currents, low voltages, and small-diameter wires. The molten drop short-circuits the arc an average of 100 times a second and at rates lower and much higher than this average. Metal is transferred with each short circuit, rather than across the arc as in spray-arc welding.

Figure 5-14 illustrates a complete short-arc cycle, starting with (A), the beginning of the short circuit, progressing through arc reignition and the arcing period, and ending with arc extinction immediately prior to the beginning of metal transfer. The technique results in low heat input, which minimizes distortion. It is useful for welding thin-gage materials in all positions and for vertical and overhead welding of heavy sections. Short-arc welding tolerates poor fitup and permits the bridging of wide gaps.

To use short-arc welding efficiently, special power sources with adjustable slope, voltage, and inductance characteristics are required. These power sources produce the predictable and controllable current surges needed for successful use of the short-arc technique.

When the shielding gas is at least 80% argon or helium, metal transfer across the arc changes from globular to spray as current increases above a certain value. This value, which depends upon the wire diameter and composition is known as the globular-to-spray transition current. At this current droplet volume decreases sharply and the rate of droplet transfer increases sharply.

Spray-arc transfer may be subdivided into two different types. When the shielding gas is argon or argon-oxygen mixture the droplets in the spray are very fine and never short circuit the arc. When carbon dioxide or argon-carbon dioxide mixture is used, a molten ball tends to form on the end of the electrode and may grow in size until its diameter is greater than the diameter of the electrode. These droplets, larger in size, may cause short circuits and this mode is known as globular transfer. Under conditions that cause the short-circuiting to occur very rapidly, the mode becomes short-circuiting transfer.

Spray-arc GMAW produces an intensely hot, higher voltage arc, and, thus, gives a higher deposition rate than short-arc welding. A high current density is required for metal transfer through the arc. The spray-arc technique is recommended for 1/8″ and thicker sections, requiring heavy single or multipass welds or for any filler-pass application where high deposition rate is advantageous.

Gas metal arc (GMAW) is normally done with direct current (reverse polarity) electrode positive because it provides a stable arc, smooth metal transfer, relatively low spatter loss and good weld bead characteristics. DC (straight polarity) electrode negative is seldom used because the arc can become unstable and erratic. Melting rate is higher with electrode negative but penetration is less than with electrode positive. AC is not used commercially because arc behavior is erratic as a result of being extinguished every half cycle and not always reigniting.

The gas metal-arc process can be used for spot welding to replace either riveting, electrical resistance, or GTA spot welding. It has proved applicable for spot welding where GTA spot welding is not suitable, such as in the joining of aluminum and rimmed steel. Fitup and cleanliness requirements are not as exacting as with GTA spot welding, and the GMAW process may be applied to thicker materials.

The GMAW process is also adaptable to vertical electrogas welding in a manner similar to that used with the gas-shielded flux-cored electrode process (See Section 6.5).

GAS TUNGSTEN ARC WELDING (GTAW)

The AWS definition of gas tungsten arc welding (GTAW) or tungsten inert gas (TIG), as it is

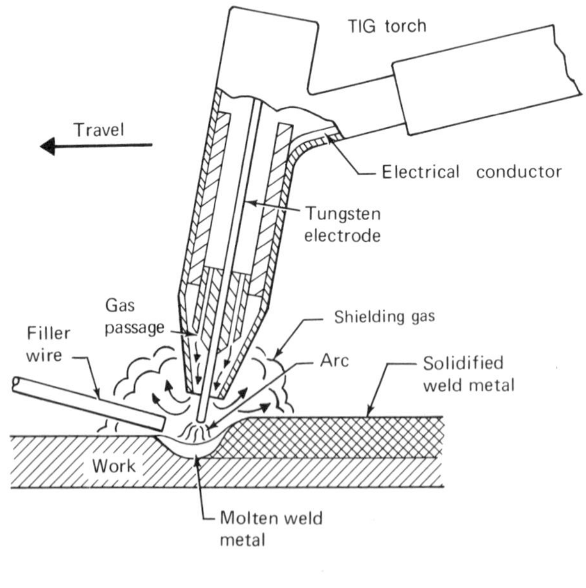

Fig. 5-15. Principles of the gas-tungsten-arc process. If filler metal is required, it is fed into the pool from a separate filler rod.

sometimes known, is "an arc welding process wherein coalescence is produced by heating with an arc between a tungsten electrode and the work." The hot tungsten electrode, arc, and weld pool are shielded by an inert gas or mixture of inert gases. Filler metal may be added, if needed, by feeding rod into the weld pool either manually or automatically.

The tungsten electrode is held in a torch which supplies all the components necessary for GTA welding; arc current, inert gas, and cooling water for water cooled torches. Torches are available in a variety of types and sizes. The smaller ones are usually gas cooled, Figure 5-15, and intended for manual use. The larger torches, which have greater current capability, are water cooled and are made for manual or automatic operation. Torches made for manual operation can be mounted on a support and used for automatic operation; however, torches made for automatic operation cannot be used manually.

ELECTRODES AND POLARITY

Five grades of tungstens are available for GTAW; pure tungsten, tungsten containing 1% and 2% thoria, tungsten containing zirconia, tungsten containing ceria, and tungsten containing a lateral insert of 1-2% thoria. Most of these electrodes are available in sizes from 0.01″ to 0.25″ diameter and lengths up to 24″. The thoria, zirconia, and ceria facilitate arc starting and help maintain a stable arc at low current. The thoriated electrodes with a conical ground tip are recommended for welding with DC electrode negative (straight polarity) because the tip shape produces a concentrated arc, and thus deeper penetration. With DC electrode negative, approximately 70% of the heat is developed at the work and 30% at the electrode so the electrode tip will not melt and will maintain the conical shape when used within the recommended current range. Electrode tip shapes for AC and DC GTA welding are shown in Fig. 5-16.

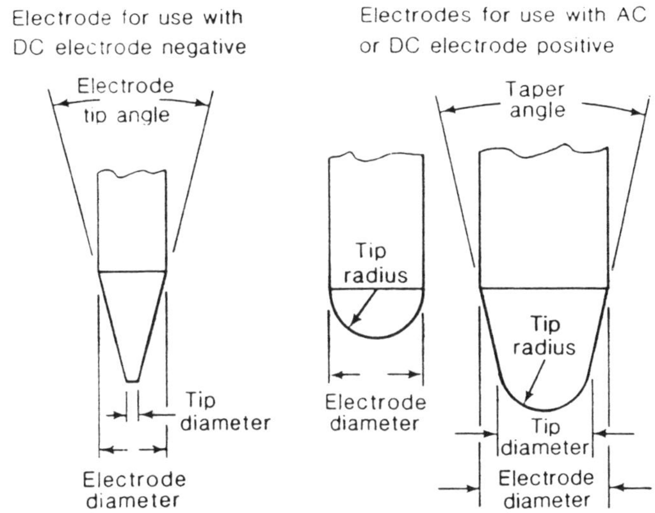

Fig. 5-16. Shapes of Tungsten Electrodes.

TABLE 5-1. Typical Current Ranges for Tungsten Electrodes[1]

Electrode Diameter		Direct Current A		Alternating Current A					
		Negative Polarity	Positive Polarity	Unbalanced Wave			Balanced Wave		
in.	mm	EWP EWTh-1 EWTh-2 EWTh-3	EWP EWTh-1 EWTh-2 EWTh-3	EWP	EWTh-1 EWTh-2 EWZr	EWTh-3	EWP	EWTh-1 EWTh-2 EWZr	EWTh-3
0.010	0.26	up to 15	[2]	up to 15	upto 15	[2]	up to 15	up to 15	[2]
0.020	0.51	5-20	[2]	5-15	5-20	[2]	10-20	5-20	10-20
0.040	1.02	15-80	[2]	10-60	15-80	10-80	20-30	20-60	20-60
1/16	1.59	70-150	10-20	50-100	70-150	50-150	30-80	60-120	30-120
3/32	2.38	150-250	15-30	100-160	140-235	100-235	60-130	100-180	60-180
1/8	3.18	250-400	25-40	150-210	225-325	150-325	100-180	160-250	100-250
5/32	3.97	400-500	40-55	200-275	300-400	200-400	160-240	200-320	160-320
3/16	4.76	500-750	55-80	250-350	400-500	250-500	190-300	290-390	190-390
1/4	6.35	750-1000	80-125	325-450	500-630	325-630	250-400	340-525	250-525

[1] All values are based on the use of argon as the shielding gas. Other current values may be employed depending on the shielding gas, type of equipment, and application.
[2] These combinations are not commonly used.

EWP = Pure Tungsten
EWTh1 = Tungsten + 1% Thoria
EWTh2 = Tungsten + 2% Thoria
EWTh3 = Tungsten + 0.4% Thoria
EWZr = Tungsten + 0.15-0.40% Zirconia

From *AWS Welding Handbook*, 7th Edition, Volume 2.

A hemispherical or balled tip is recommended for welding with DC electrode positive (reverse polarity) or AC because the greater amount of heat is developed at the electrode and causes the end to melt. When operated within the recommended current range the electrode tip will melt and form a stable hemispherical shape. DC electrode positive is rarely used for welding because approximately 70% of the arc heat is developed at the electrode which means that the weld current must be more than doubled to attain the same weld energy as DC electrode negative. It can be used for welding thin sheet where heat input must be minimized. A cleaning action is obtained on the work surface with DC electrode positive which is not obtained with DC electrode negative. This cleaning action is necessary when welding aluminum and magnesium but is not required for most other metals and alloys. AC is preferred to DC electrode positive for welding aluminum because it provides the cleaning action on the positive half cycles and greater heat at the work on the negative half cycles. Tungsten-zirconia electrodes are recommended for AC welding because they provide easier arc starting compared to pure tungsten and the tip will melt and form the hemispherical end more readily than the thoriated tungsten electrodes. Typical current ranges for the various tungsten electrodes are listed in Table 5-1.

SHIELDING GASES

Argon and helium and mixtures of these gases are used for GTAW. For a given set of arc parameters, current, length and electrode, the helium arc will operate at a voltage which is approximately 1.7 times that of the argon arc. This provides greater heat for a given current than the argon arc. However, argon is used for the majority of GTAW because arc starting is easier, the arc is more stable at low currents, cleaning action is better on aluminum, it is about 1/3 the cost of helium and less than half the flow rate is required for effective shielding. Helium is used when the increased arc energy is needed for welding heavy plate or thick materials with high thermal conductivity. Mixtures of argon and helium can be used to obtain arc characteristics between either of the pure gases.

Argon may be supplied as compressed gas in 330 cu. ft. cylinders, as liquid in 5100 cu. ft. portable tanks and in large bulk liquid tanks. The cost of the bulk liquid is about 70% of the cost of argon in cylinders or portable liquid tanks. Helium is usually supplied as compressed gas in cylinders but can be obtained as liquid. Cost of helium is about 3 times the cost of argon. The other inert gases, neon, krypton and xenon, would be satisfactory for GTAW but are not used because of their scarcity and very high cost.

POWER SUPPLIES

Power supplies for use with GTAW should be of the constant current drooping voltage characteristics and may include optional features such as pulsation, up and down slope, and current programming capabilities. For manual welding, the power supply should include a contactor and a means of controlling arc current by the weldor. Foot pedals usually perform these functions of operating the contactor and controlling weld current. A power supply with a single current range is preferred because this allows the weldor to vary the arc current from the maximum to the minimum without changing a range switch. The power supply for GTAW should include solenoid valves to control the flow of inert gas and cooling water for water cooled torches. Timers should be included to turn gas and water on several seconds before the arc can be started and off several seconds after the arc is extinguished. It is important that a power supply for GTAW include a high frequency source for arc starting so that the arc can be started without touching the electrode to the work. It should be possible to use the high frequency for starting a DC arc or continuously for an AC arc.

More advanced power supplies have pulsing capabilities where the pulses approach a square wave shape. This pulsing capability allows root pass welds, welds in thin material, and overhead welds to be made with less chance of melt through or droop. On AC welding, the power supply can vary the positive and negative half cycles to favor either cleaning action or penetration. This feature is useful for welding aluminum where adjustments must be made based on the thickness and cleanliness of the joint.

ADVANTAGES AND DISADVANTAGES

The main desirable feature of GTAW is the high quality welds it is capable of making in almost all metals and alloys. It can be used on these alloys where freedom from atmospheric contamination is critical. Examples are titanium, zirconium, columbium and tantalum where small amounts of oxygen, nitrogen, hydrogen and carbon contamination can cause embrittlement and loss of corrosion resistance.

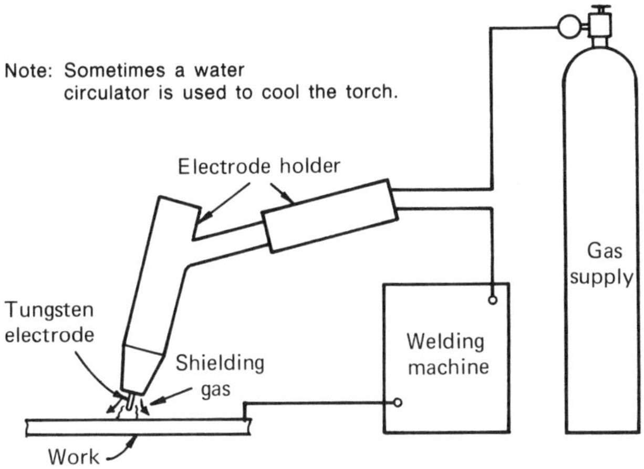

Fig. 5-17. Schematic of manual GTA welding.

GTAW is well suited for welding thin sheet and root passes of all materials. Other features are the ability to add filler metal independently of arc current with very low spatter. The major disadvantages of GTAW are the low rate of filler metal addition and the greater operator skill required. Both of these factors make the process more costly than other arc welding processes. While GTAW can be used for making root passes on carbon steel joints, it would not be economical to use this process for filling the joint. Other processes such as GMAW or SMAW would be better choices.

EQUIPMENT ARRANGEMENT

Figure 5-17 shows the typical arrangement for GTAW for either manual or automatic welding. In manual welding, the torch would be hand-held and arc current would be controlled by a foot pedal or hand control on the torch. The foot pedal or hand control would also operate the secondary contactor.

For automatic welding, the torch would be mounted on a holder which could provide the motion necessary for the intended applications. This would include longitudinal travel with cross slide and vertical adjustments for longitudinal welds. For fully automatic operation, a sensor would actuate the cross slide to track the joint and an arc voltage control would maintain the proper arc length. The same cross slide and arc voltage control could be used for welding circumferential joints in piping and cylinders and the same sensors and controls could be used to track the joint and control arc length.

When water cooled torches are used, flow of tap water is controlled by a solenoid valve in the power supply and this water runs into a drain. When a water cooler circulator is used, the solenoid valve is by-passed and the water is circulated continuously while the pump is running. This conserves water and when distilled water is used in the circulator, prevents mineral deposits from clogging the torch.

MANUAL WELDING TECHNIQUES

The technique for manual GTA welding and adding filler metal to the weld pool is illustrated in Figure 5-18. First the arc is established between the electrode and the work using high frequency arc starting. Touch starting is not recommended. The torch is moved in a small circle to establish a pool. Then the torch is tilted back on about a 75° angle and moved back slightly while sufficient filler metal is dipped into the pool. The filler wire is pulled back and the torch moved in the direction of travel to melt the added filler metal and advance the pool. At this point the sequence of moving the torch backward and adding filler is repeated until the joint is completed. The welder must develop a technique for maintaining a constant arc length and voltage, usually 9-12 volts DCEN with argon, while manipulating the torch in a generally forward direction. The end of the filler wire which is being added to the weld pool should not be moved out of the inert gas shield from the torch. This is to prevent the hot end from oxidizing and subsequently contaminating the pool when it is dipped into it. When helium is used as the shielding gas the arc voltage will be greater, 16-18 volts, producing a hotter arc for a given current. Helium requires greater skill to control the puddle to prevent melt-through and should not be used on thin material.

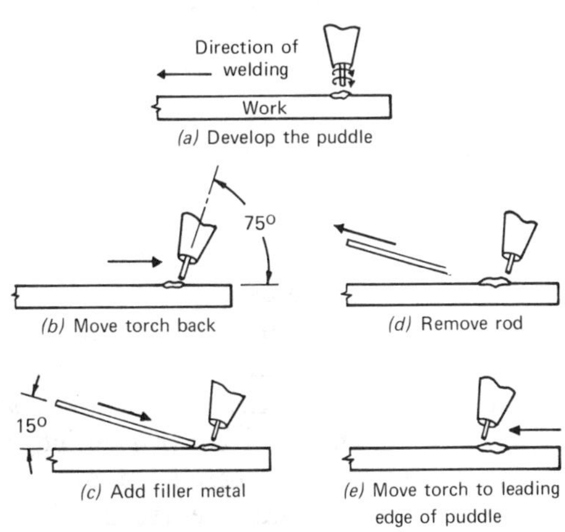

Fig. 5-18. Mode of manually feeding filler metal into the weld puddle when GTA welding.

At the end of the weld the arc current should be decreased gradually to avoid a crater and possible crater cracks at the finish end. It may be necessary to add a small amount of filler metal just before the arc current is decreased.

Many joints in thin material may be welded without addition of filler metal. Examples are edge joints, corner joints, butt joints and flared butt joints. The technique for welding these joints manually is similar to that just described except no filler metal addition is required. The weld is made by fusing together the edges of the sheets. Once the arc is established, the torch is moved at a uniform speed to the finish end of the joint.

Inert gas backup is recommended for both manual and automatic GTAW welding of thin sheet or the root pass of thicker material. The inert gas prevents oxidation of the root of the weld which allows the molten metal to wet and flow together properly and prevents contamination of the weld with oxide inclusions.

AUTOMATIC WELDING TECHNIQUES

In machine or automatic welding, the torch is mounted on a support which provides the necessary adjustments for aligning with the weld joint. Travel along the joint may be accomplished by moving the part or by moving the torch for either longitudinal or circular joints. For machine or semiautomatic welding, the torch must be adjusted continuously by an operator to the correct arc length and center of the joint. If joints are straight and aligned nearly perfectly with travel direction, both vertically and horizontally, these adjustments may not be necessary.

Weld current may be constant for the length of a weld and may include upslope at the start and downslope at the end. Many of the more advanced GTA welding power supplies can be programmed by an external source such as a microprocessor. With this arrangement, weld current can be programmed to compensate for variations in joint thickness or buildup of heat in a component. While current programming would normally be used on a fully automatic set-up, it can be used on a semiautomatic application. In a fully automatic arrangement, all parameters are controlled, arc voltage, arc current, travel speed, and joint tracking. Arc voltage is adjusted by a drive unit and control which senses arc voltage and moves the torch up or down to establish the arc length which corresponds to the desired voltage. Arc current is established by the power supply setting or by an external programmer. Travel speed is controlled by the travel device drive which should be controlled by a feedback type circuit. Joint tracking requires a device which can sense the location of a joint and produce signals which when amplified cause the torch to move in a direction to correct any misalignment with the joint. The sensor may be a mechanical device which contacts the joint, a laser beam, or a visual through-the-torch sensor. All these devices require an electronic circuit to interpret the signals and produce output current which will move the torch in the correct direction.

MATERIALS WHICH CAN BE WELDED

The GTAW process can be used to weld all the common and uncommon metals and alloys except the low melting metals such as lead, tin and zinc base alloys. The arc temperature is too high to successfully weld these alloys.

Although GTAW can be used for welding a wide varieties of steels, it is not economical for use on the plain carbon steels except for root passes. It can be used on low alloy steels, tool steels, stainless steels and nickel base alloys where high quality deposits are necessary. These materials should be welded with DC electrode negative and inert gas backup is recommended for thin sheet and root passes. Alloys of titanium, zirconium, columbium and other reactive and refractory metals should be welded with DC electrode negative. It is imperative that these alloys be completely shielded with inert gas while they are being welded and are cooling. This can be done by inert gas backup from the tooling and a trailer shield attached to the torch to cover the weld until it has cooled to a temperature where oxidation does not occur. Alloys of zirconium, columbium and tantalum are very sensitive to atmospheric contamination and may suffer loss of ductility when welded. In these cases welding should be done in atmosphere chambers or glove boxes which can be filled with inert gas. In standard glove boxes, purging may require 8-10 times the volume of the chamber of inert gas to obtain a sufficiently pure atmosphere. A chamber which can be evacuated to 1×10^{-2} torr or better will contain a satisfactory atmosphere with one filling.

Aluminum and aluminum alloys should be welded with AC using argon or argon-helium shielding gas. Most aluminum alloy GTA welding is done with AC either with balanced wave or square wave AC power supplies. Without these features the current flow will be greater when the electrode is negative than when it is positive and the cleaning action will be inadequate. This unbalance is caused by rectification. On the square wave AC power supplies, the time duration of the negative and positive pulses can be varied so that maximum heating can be obtained on the negative pulses while adequate cleaning is obtained on the positive pulses.

OTHER APPLICATIONS

GTAW is especially adapted to welding thin material and small parts where the requirements for quality and finish are exacting. The ability to operate with a stable arc at a very low amperage makes it one of the few processes that is satisfactory for welding such tiny and thin walled objects as transistor cases, instruments diaphragms and other delicate expansion bellows.

Specialized Arc-Welding Processes

Metals may also be joined by methods other than those utilizing electrical energy. Oxyacetylene, friction, explosive, and thermit welding are examples. There are also various methods of welding that make use of electrical energy that are not "arc" welding. Examples are the variations of electrical-resistance welding, ultrasonic, electronbeam, and electrodeposition welding.

There are also several processes that may appropriately be labeled "arc" welding that do not fall in the categories described in Sections 5.1 to 5.4. In some cases, however, these specialized arc-welding processes may be looked upon as modifications of basic welding processes. Thus, vertical electrogas welding (See Section 6.5) is sometimes listed as a separate welding process, but may also be regarded as a specialized application of the gas-shielded flux-cored process. Similarly, electroslag welding is a variation of the submerged-arc process.

The book *"Welding Metallurgy,"* Volume I, by George E. Linnert, published by the American Welding Society, describes several dozen metals-joining processes, including the 14 processes recognized by the AWS as "arc-welding" processes. This is a recommended reference for more complete information on the specialized arc-welding processes than given in the following text. Volume 2 of the *AWS Welding Handbook,* Eighth Edition, and *"Welding Processes,"* by P.T. Houldcroft, published by the Cambridge University Press, are also recommended references. The descriptions of the specialized arc-welding processes that follow are intended merely to give the reader an indication of what is available and what might be pursued and investigated further should these specialized processes be deemed more suitable than the basic commercial processes for particular applications.

ELECTROSLAG WELDING

Electroslag welding is an adaptation of the submerged-arc welding process for joining thick materials in the vertical position. Figure 5-19 is a diagrammatic sketch of the electroslag process. It will be noted that, whereas some of the principles of submerged-arc welding are applied, in other respects the process resembles a casting operation. Here, a square butt joint in heavy plate is illustrated, but the electroslag process — with modifications in equipment and technique — is also applicable to T joints, corner joints, girth seams in heavy-wall cylinders, and other joints. The process is suited best for materials at least 1″ in thickness, and can be used with multiple electrodes on materials up to 10″ thick without excessive difficulties.

As illustrated by the open, square butt joint in Figure 5-19, the assembly is positioned for the vertical deposition of weld metal. A starting pad at the bottom of the joint prevents the fall-out of the initially deposited weld metal, and, since it is penetrated, assures a full weld at this point. Welding is started at the bottom and progresses upward. Water-cooled dams, which may be looked upon as molds, are placed on each side of the joint. These dams are moved upward as the weld-metal deposition progresses. The joint is filled in one "pass" — a single upward progression — of one or more

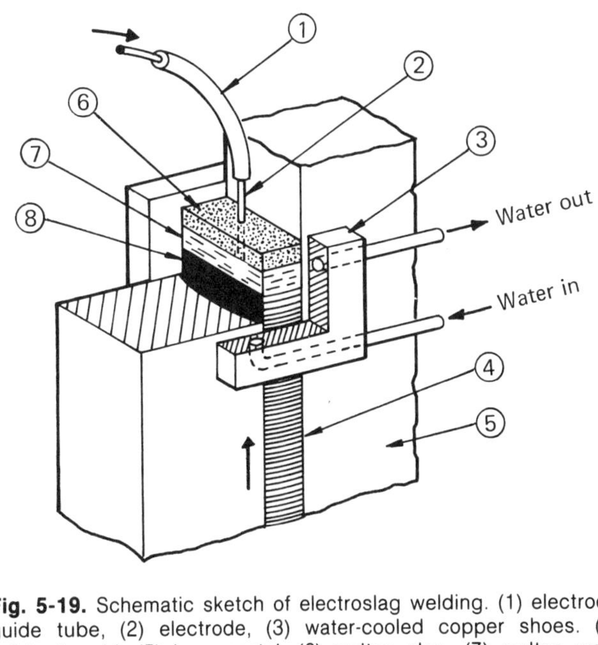

Fig. 5-19. Schematic sketch of electroslag welding. (1) electrode guide tube, (2) electrode, (3) water-cooled copper shoes. (4) finished weld, (5) base metal, (6) molten slag, (7) molten weld metal, (8) solidified weld metal.

consumable electrodes. The electrode or electrodes may be oscillated across the joint if the width of the joint makes this desirable.

At the start of the operation, a layer of flux is placed in the bottom of the joint and an arc is struck between the electrode (or electrodes) and the work. The arc melts the slag, forming a molten layer, which subsequently acts as an electrolytic heating medium. The arc is then quenched or shorted-out by this molten conductive layer. Heat for melting the electrode and the base metal subsequently results from the electrical resistance heating of the electrode section extending from the contact tube and from the resistance heating within the molten slag layer. As the electrode (or electrodes) is consumed, the welding head (or heads) and the cooling dams move upward.

In conventional practice, the weld deposit usually contains about 1/3 melted base metal and 2/3 electrode metal — which means that the base metal substantially contributes to the chemical composition of the weld metal. Flux consumption is low, since the molten flux and the unmelted flux above it "ride" above the progressing weld.

The flux used has a degree of electrical conductivity and low viscosity in the molten condition and a high vaporization temperature. The consumable electrodes may be either solid wire or tubular wire filled with metal powders. Alloying elements may be incorporated into the weld by each of these electrodes.

Weld quality with the electroslag process is generally excellent, due to the protective action of the heavy slag layer. Sometimes, however, the copper dams are provided with orifices just above the slag layer, through which a protective gas — argon or carbon dioxide — is introduced to flush out the air above the weld and, thus give additional assurance against oxidation. Such provisions are sometimes considered worthwhile when welding highly alloyed steels or steels that contain easily oxidized elements.

The electroslag process has various advantages, including no need for special edge preparations, a desirable sequence of cooling that places the outside surfaces of weld metal under compressive rather than tensile stresses, and a relative freedom from porosity problems. Special equipment, however, is required, and few users of welding have enough volume of welding with this process to justify the cost of such equipment.

A variation of this process, Figure 5-20, uses a consumable electrode guide in the form of a thick wall tube which runs the entire length of the joint.

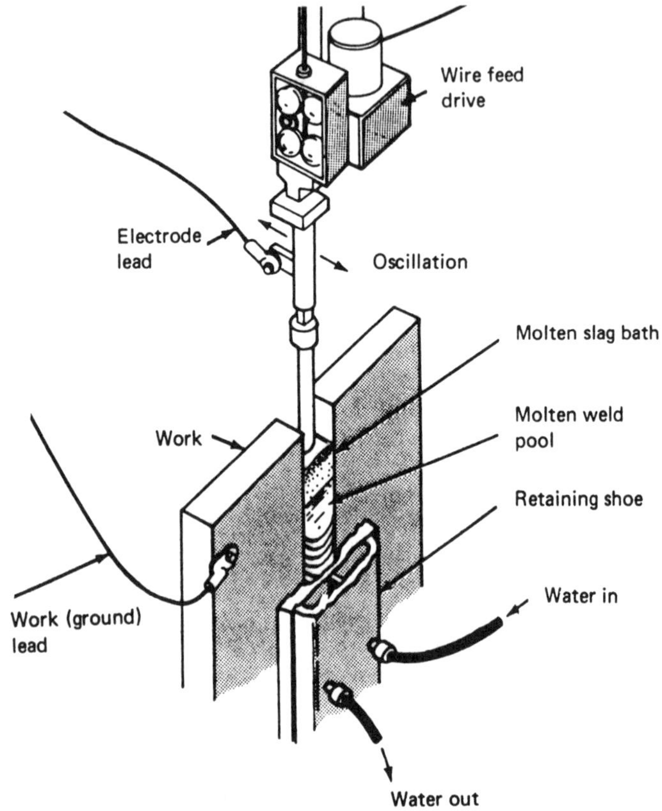

Fig. 5-20. Consumable guide method of electroslag welding.

The electrode is fed through the center hole of the tube and an arc is started at the bottom of the joint as in the standard process. A small amount of granulated flux is added and this melts to form a pool of molten flux which then conducts the weld current.

The consumable tube is heated by the molten slag and this melts and provides 5-10 percent of the filler metal; the other 90-95 percent is supplied by the wire electrode. Since the guide tube is connected to the electrode circuit, it must be insulated from the walls of the joint. This may be accomplished by a slag coating, donut shaped insulaters, fiberglass sleeves, or fiberglass tape. These insulators melt and provide the additional flux needed as the weld progresses upward. One or more electrodes and consumable guide tubes may be used depending on the joint thickness. Each electrode requires a separate power supply and wire feed unit.

Copper water cooled dams must be used on both sides of the joint to contain the molten metal and flux. On short joints these may cover the full length of the joint. On long joints two or more sets of dams may be used. The dams are clamped to the lower section of the joint at the start and as the

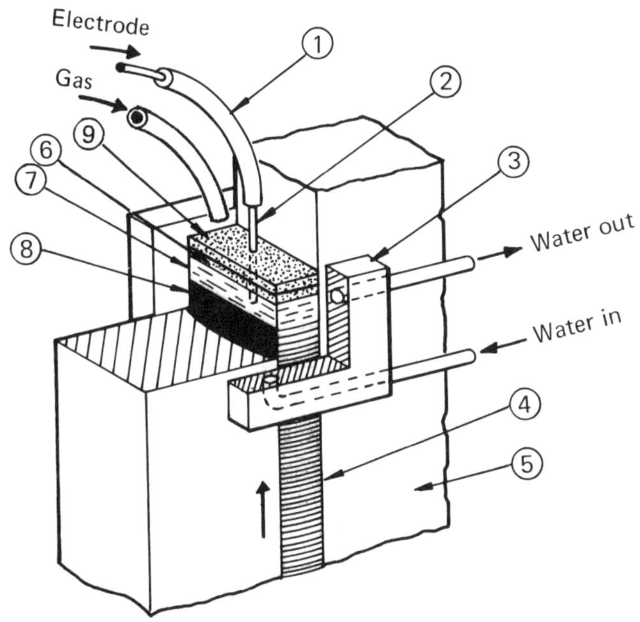

Fig. 5-21. The electro-gas vertical butt-welding process — a special application of gas-shielded flux-cored arc welding. (1) electrode guide tube, (2) electrode, (3) water cooled copper shoes, (4) finished weld, (5) base metal, (6) molten slag, (7) molten weld metal, (8) solidified weld metal, (9) gas shield (CO_2).

welding proceeds upward the lower set of dams are removed and clamped in place above the upper set of dams. This procedure is repeated for the full length of the joint.

ELECTROGAS WELDING

Electrogas welding is very similar to electroslag welding in that the equipment is similar and the joint is in the vertical position. As the name implies, the shielding is by carbon dioxide or an inert gas. A thin layer of slag, supplied by the flux-cored electrode, covers the molten metal, and the heat is supplied by an arc rather than by resistance heating as in the electroslag process.

A disadvantage of the process is that it requires an external source of shielding gas. However, one advantage is that if the welding is stopped the electrogas process can be started again with less difficulty than the electroslag process. Figure 5-21 illustrates the process; additional information on it is given in Section 6.5.

STUD ARC WELDING

Stud arc welding is a variation of the shielded metal-arc process that is widely used for attaching studs, screws, pins, and similar fasteners to a larger workpiece. The stud (or small part), itself — often plus a ceramic ferrule at its tip — is the arc-welding electrode during the brief period of time required for stud welding.

In operation, the stud is held in a portable pistol-shaped tool called a stud gun and positioned by the operator over the spot where it is to be weld-attached. At a press of the trigger, current flows through the stud, which is lifted slightly, creating an arc. After a very short arcing period, the stud is then plunged down into the molten pool created on the base plate. The gun is withdrawn from the stud and the ceramic ferrule — if one has been used — is removed. The timing is controlled automatically and the stud is welded onto the workpiece in less than a second. The fundamentals of the process are illustrated in Figure 5-22.

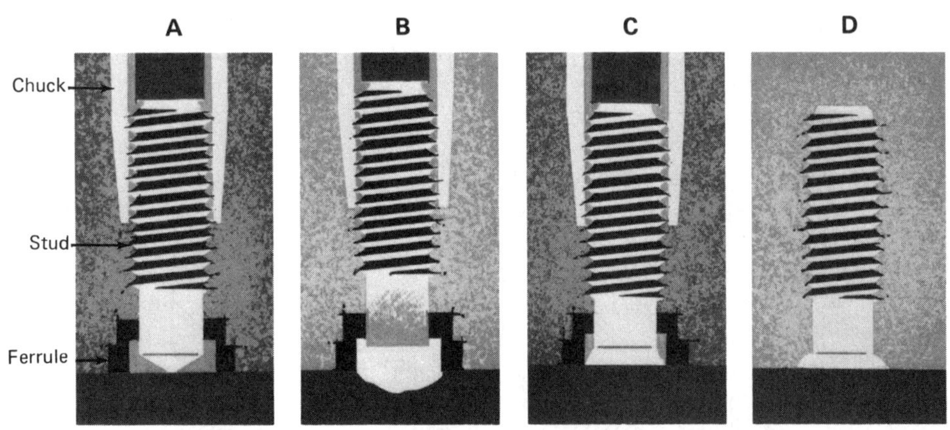

Fig. 5-22. Principles of stud welding, using a ceramic ferrule to shield the pool. (A) The stud with ceramic ferrule is grasped by the chuck of the gun and positioned for welding. (B) The trigger is pressed, the stud is lifted, and the arc is created. (C) With the brief arcing period completed, the stud is plunged into the molten pool on the base plate. (D) The gun is withdrawn from the welded stud and the ferrule removed.

Fig. 5-23. Some of the commonly used fastening devices commonly attached to metal parts and assemblies by stud welding.

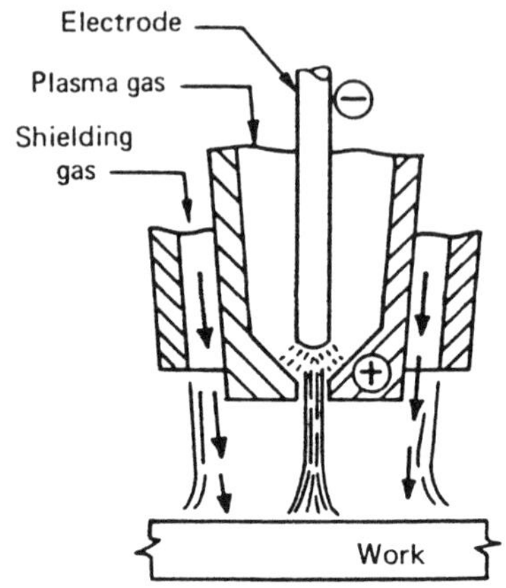

Fig. 5-25. Diagrammatic sketch of the plasma-arc torch.

Studs are of many shapes, as illustrated in Figure 5-23. All may be weld-attached with portable equipment. The stud may be used with a ceramic arc-shielding ferrule, as shown in Figure 5-22, which prevents air infiltration and also acts as a dam to retain the molten metal, or may have a granular flux, flux coating, or solid flux affixed to the welding end, as illustrated in Figure 5-24. The flux may include any of the agents found in a regular electrode covering; most important to stud welding is a deoxidizer to guard against porosity.

Stud welding is a well developed and widely used process. Information needed for its application with the carbon and alloy steels, stainless steels, aluminum, and other nonferrous metals may be found in the *AWS Welding Handbook,* Eighth Edition, Volume 2.

PLASMA-ARC WELDING

Plasma-arc (or plasma-torch) welding is one of the newer welding processes, which is used industrially, frequently as a substitute for the gas tungsten-arc process. In some applications, it offers greater welding speeds, better weld quality, and less sensitivity to process variables than the conventional processes it replaces. With the plasma torch, temperatures as high as 60,000°F are developed and theoretically, temperatures as high as 200,000°F are possible.

The heat in plasma-arc welding originates in an arc, but this arc is not diffused as in an ordinary welding arc. Instead, it is constricted by being forced through a relatively small orifice. The "orifice" or plasma gas (See Figure 5-25) may be supplemented by an auxiliary source of shielding gas.

"Orifice" gas refers to the gas that is directed into the torch to surround the electrode. It becomes ionized in the arc to form the plasma and emerges from the orifice in the torch nozzle as a plasma jet. If a shielding gas is used, it is directed onto the workpiece from an outer shielding ring.

The workpiece may or may not be part of the electrical circuit. In the "transferred-arc" system, the workpiece is a part of the circuit, as in other

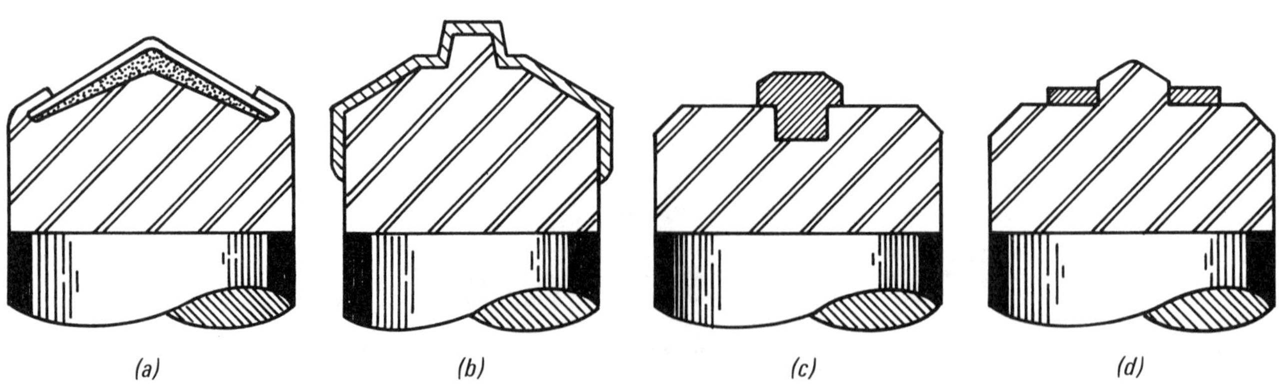

Fig. 5-24. Three methods of containing flux on the end of a welding stud: (a) granular flux; (b) flux coating; (c) and (d) solid flux.

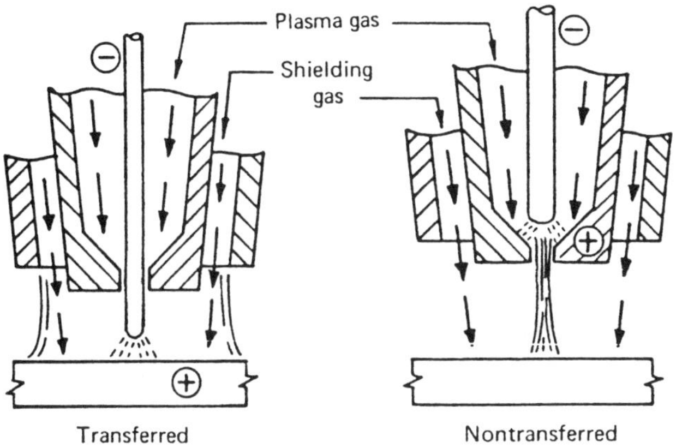

Fig. 5-26. Transferred and non-transferred arcs.

arc-welding processes. The arc "transfers" from the electrode through the orifice to the work. In the "nontransferred" system, the constricting nozzle surrounding the electrode acts as an electrical terminal, and the arc is stuck between it and the electrode tip; the plasma gas then carries the heat to the workpiece. Figure 5-26 illustrates transferred and nontransferred arcs.

The advantages gained by using a constricted-arc process over the gas tungsten-arc process include greater energy concentration, improved arc stability, higher welding speeds, and lower width-to-depth ratio for a given penetration. "Keyhole" welding — or penetrating completely through the workpiece — is possible. Figure 5-27 compares gas tungsten-arc with transferred plasma-arc and schematically illustrates the concentration of energy with the latter and its superior penetration.

Manual and mechanized plasma-arc welding are described in detail in the *AWS Welding Handbook,* Eighth Edition, Volume 2. That volume also lists an extensive bibliography of literature on the subject, with emphasis on applications for the process.

ATOMIC-HYDROGEN WELDING

The atomic-hydrogen process of arc welding may be regarded as a forerunner of inert gas-shielded and plasma arc welding. Although largely displaced by other processes that require less skill and are less costly, it is still preferred in some manual operations where close control of heat input is required.

In the atomic-hydrogen process, an arc is established between two tungsten electrodes in a stream of hydrogen gas, using AC power. As the gas passes through the arc, molecular hydrogen is disassociated into atomic hydrogen under the intense heat. When the stream of hydrogen atoms strikes the workpiece, the environmental temperature is then at a level where recombining into molecules is possible. As a result of the recombining, the heat of disassociation absorbed in the arc is liberated, supplying the heat needed for fusing the base metal and any filler metal that may be introduced.

The atomic-hydrogen process depends on an arc, but is really a heating torch. The arc supplies the heat through the intermediate of the molecular-disassociation, atom-recombination mechanism.

The hydrogen gas, however, does more than provide the mechanism for heat transfer. Before entering the arc, it acts as a shield and a coolant to keep the tungsten electrodes from overheating. At the weld puddle, the gas acts as a shield. Since hydrogen is a powerful reducing agent, any rust in the weld area is reduced to iron, and no oxide can form or exist in the hydrogen atmosphere until it is removed. Weld metal, however, can absorb hydrogen,

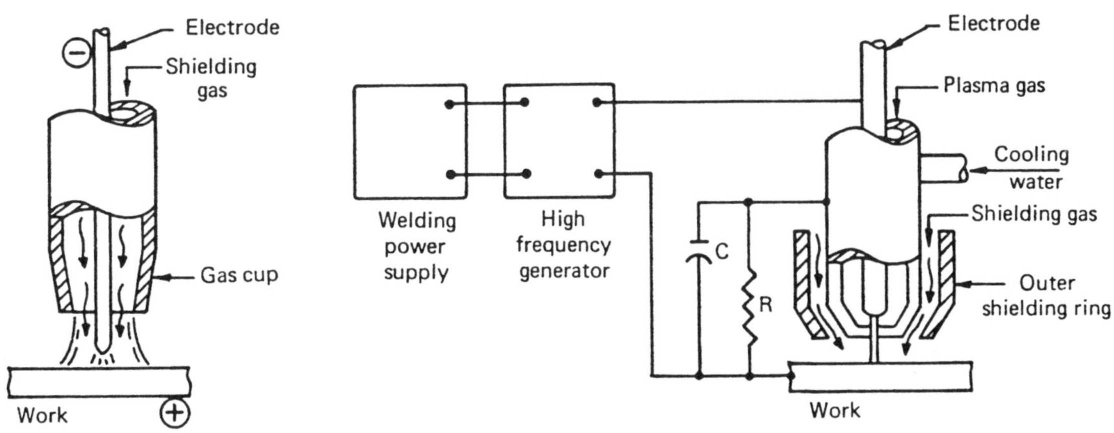

Fig. 5-27. Schematic comparison of gas tungsten-arc and plasma-arc welding torches.

with unfavorable metallurgical effects. For this reason, the process causes difficulties with steels containing sulfur or selenium, since hydrogen reacts with these elements to form hydrogen sulfide or hydrogen selenide gases. These are almost insoluble in molten metal and either bubble out of the weld pool vigorously or become entrapped in the solidifying metal resulting in porosity.

OTHER ADAPTATIONS

Producers of welding equipment and machinery are constantly developing adaptations of conventional arc-welding processes for highly specialized applications. These are usually ingenious techniques to improve the performance of a basic process with a particular type of joint or metal or to solve a welding cost problem, and they frequently are given separate names. Examples are the Linc-Fill* systems for submerged-arc and self-shielded flux-cored welding, a trade term for high-deposition-rate welding using the long electrical stick-out principle.

A similar scheme called "hot-wire welding" increases the deposition rate of submerged-arc or gas tungsten-arc welding. Using a separate power source, the filler wire is preheated to almost the melting point as it enters the arc area (See Figure 7-13). Since very little heat from the arc is required to complete the melting of the wire, it can be fed into the weld crater at a much higher rate than normal (See Section 7-5).

A technique called "narrow-gap" welding is an adaptation of the gas metal-arc process. The equipment is designed to permit welding in a square butt joint with a gap opening of 1/4 to 3/8" wide. Thin weld beads are deposited alternately in the corners, overlapping to fill the gap. The system uses two tandem torches operating together, one filling each side. Special water-cooled contact tubes, only 1/8" thick, are used. Plate up to 6" thick can be welded from one side. In addition to eliminating the need for V or U groove preparations and saving weld metal, the high travel speed used with narrow-gap welding minimizes heat input and, thus, reduces the width of the heat-affected zone and the problems that result from excessive heating of the material. This is especially important when welding high-strength low-alloy steels.

*Linc-Fill is a trade mark of the Lincoln Electric Co.

Selecting a Welding Process

In selecting a process for production welding, a primary consideration is the ability of the process to give the required quality at the lowest cost. Here, the cost factor must include not only the operating cost, but also the amortization of the capital costs of equipment over the job and those jobs that may reasonably be expected to follow. Thus, a fabricator may have a job of running straight fillets where fully mechanized submerged-arc equipment would be the ultimate solution to low-cost, quality welding. But, unless the job involves a great amount of welding, or there is assurance that it will be followed by other jobs of a similar nature, the lowest cost to the fabricator — taking into account the feasibility for amortization — might be with a hand-held semiautomatic gun or even with the stick-electrode process. The job is, thus, the starting point in the selection of a welding process.

The job defines many things. It defines the metals to be joined, the number of pieces to be welded, the total length of welds, the type of joints, the preparations, the quality required, the assembly-positioning requirements — to name but a few. Process selection, thus, considers first the needs of the job at hand. Unless the job will be adequately repetitive or subsequent jobs will be similar in requirements, however, it may be desirable to select a process that is not optimum but which will be most efficient over the long pull.

It is a fact of business life that factors other than technology enter into process selection. Such factors include the need of companies to maintain their working staffs during unproductive periods. Thus, a large company that supplies sand and gravel for the construction industry does the major portion of weld rebuilding and weld resurfacing work on its drag-line buckets and other equipment during very cold weather when concrete pouring is at low ebb. Drag-line, shovel, and bull-dozer operators are accomplished welders who do the rebuilding and resurfacing — working mostly with stick-electrode equipment. The repair and rebuilding work could be done much faster with semiautomatic or full-automatic equipment, but the company cares less about welding cost efficiencies at this stage. A primary objective is to give their operating engineers something productive to do when their products, sand and gravel, are stock-piled beyond all expectancies for customer demand. The alternative would be to lay off skilled equipment operators during such periods — which would lead to greater costs for replacement and retraining than the inefficiencies in repair and rebuilding welding.

The job, thus, may involve intangibles — and these intangibles have a rightful place in process selection. It is because of intangibles that less efficient welding processes are often used. Where intangibles are involved, the company should question their validity. The fact that a weldor who has been using the stick electrode on the job for 20 years vows he will never touch a semiautomatic gun may not be adequate reason for acquiescing to him. Chances are, after a few reluctant days with a semiautomatic gun, he will be acclaiming its virtues with just as much fervor as he acclaims the stick electrode.

Assuming intangible factors do not prevail, what does the job tell that affects process selection?

First, the job tells what metals are to be welded. If the metals are steel, a wide range of processes may be considered. If the metals are aluminum and magnesium, the available processes will be gas tunsten-arc and gas metal-arc. Reference to Sections 5.4 and 9. will be helpful in making a choice. These processes and stick-electrode welding may also be chosen if maximum weld properties are required in high-alloy steels or corrosion-resistant materials. Here, one finds that the gas tungsten-arc process generally provides better properties but is considerably slower than the gas metal-arc process — which is generally preferred in mass production. If thin sections are involved, however, the GTAW process may be preferred.

If the metals to be welded are the carbon or low-alloy steels, selection becomes more difficult. More processes are applicable, and varying degrees of mechanization are possible. The welding is likely to be heavy or involve many feet — or even miles — of weld, with varying types of joints. Furthermore, there may be precise requirements for such joints in terms of fill, follow, freeze, or penetration.

Taking into account the fact that process selection

is almost self-determining when the metals to be welded are aluminum, magnesium, copper, titanium, or the other nonferrous metals — or that the stick electrode is almost certainly to be the choice when welding steel where the immediate cost is of no importance, the following text will be concerned with choosing the optimum process for the *production* welding of steels. Here, intangibles no longer are effective, and only technology and pure economics prevail.

In the *production* welding of steel, mechanization immediately enters the picture. It can't be ignored, since some forms of mechanization are even applicable to tack welding. Here, also, stick-electrode welding is ruled out, since it can't be competitive with the mechanized processes.

All of the mechanized arc-welding processes have been developed as a means of reducing welding costs. Each of these processes has certain capabilities and limitations, and selecting the best process for a particular job can be a difficult, if not confusing, task. Unless the user makes a correct selection, however, he may lose many of the benefits to be gained in moving to mechanization and, thus, not achieve his objective of welds that meet the application requirements at the lowest possible cost.

There are varying degrees of mechanization with all of the mechanized processes — from semiautomatic hand-held equipment to huge mill-type welding installations. Thus, one needs a "common denominator" when processes are to be compared. Semiautomatic equipment possibly provides the best common denominator. Anything added to semi-mechanization is extraneous to the process capability *per se* and merely amounts to "putting the semiautomatic process on wheels."

With semimechanization and the capabilities of the various processes subject to semimechanization in mind, a four-step procedure can be used in deciding which is the preferred process for the particular production welding job.

THE FOUR-STEP PROCEDURE

The four steps involved in process selection are:

1. First, the joint to be welded is analyzed in terms of its requirements.
2. Next, the joint requirements are matched with the capabilities of available processes. One or more of the processes are selected for further examination.
3. A check list of variables is used to determine the capability of the surviving processes to meet the particular shop situation.
4. Finally, the proposed process or processes indicated as most efficient are reviewed with an informed representative of the equipment manufacturer for verification of suitability and for acquisition of subsidiary information having bearing on production economics.

This four-step approach is not mathematically precise, since it requires judgments. However, these *are made in the proper sequence* — which is most important for arriving at correct results. Also, on vital points, such as the joint factors of greatest significance, the judgments are usually clear-cut, with the possibility of error being negligible.

The main problem in applying this approach will be in assessing the capability of various processes to supply the needs of the joint. This will necessitate review of manufacturers' literature, or the directing of pertinent questions to equipment makers — and, to be sure that no process is omitted from consideration, data on each must be available.

STEP I — ANALYSIS OF JOINT REQUIREMENTS

When economy in welding is a prime objective, the needs of any joint can be expressed in four terms:

Fast-fill . . . meaning high deposition rate.

Fast-freeze . . . meaning the joint is out-of-position — overhead or vertical.

Fast-follow . . . being synonymous with high arc speed and very small welds.

Penetration . . . meaning the depth of penetration into the base metal.

Fast-fill is an obvious requirement when a large amount of weld metal is needed to fill the joint. Only by a high deposition rate can a heavy weld bead be laid down in minimum arc time. Fast-fill becomes a minor consideration when the weld is small.

Fast-freeze implies out-of-position — the need for quick solidification of the molten crater. The joint may additionally be a fill-type or penetration-type, but fast freezing is of paramount importance. Not all semiautomatic processes are capable of being used on fast-freeze joints.

Fast-follow implies the ability of the molten metal to follow the arc at rapid travel speed, giving continuous, well-shaped beads, without "skips" or islands. The crater may be said to have low surface

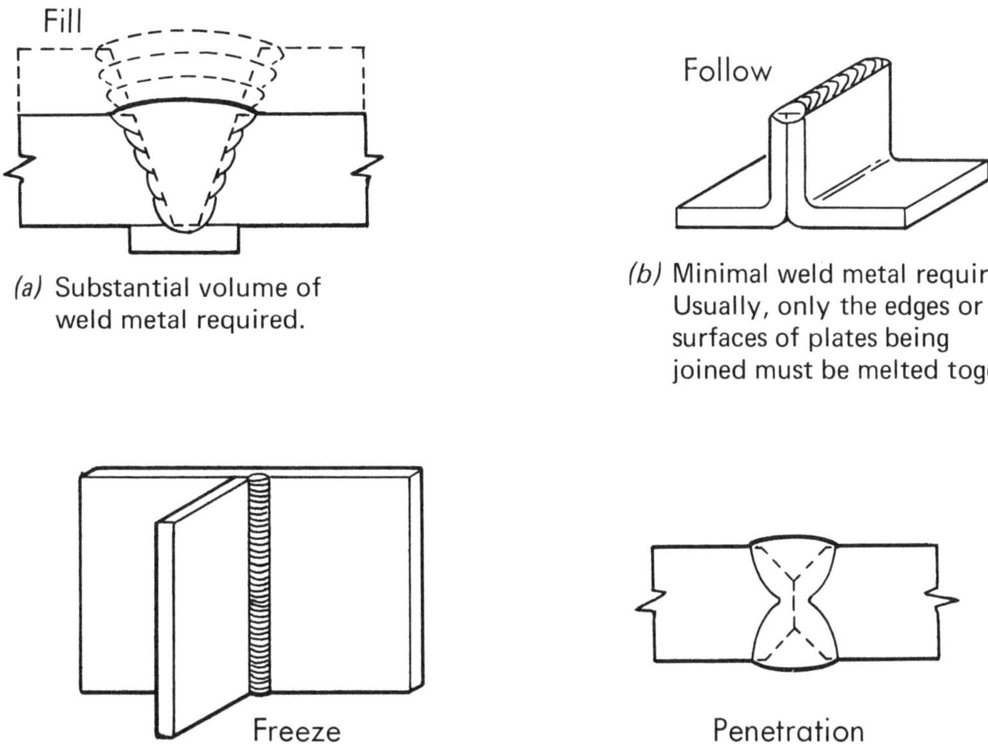

(a) Substantial volume of weld metal required.

(b) Minimal weld metal required. Usually, only the edges or surfaces of plates being joined must be melted together.

(c) Out-of-position joint, where control of molten crater is important.

(d) Joint requires deep penetration into base metal (maximum penetration). Bridging a gap (minimum penetration) may be required if fitup is poor.

Fig. 5-28. Examples of fill, follow, freeze, and penetration joints.

tension; it wets and washes in the joint readily. Fast-follow is especially desirable on relatively small single-pass welds, such as those used in joining sheet metal. It is rarely a requirement of a large or multiple-pass weld; with these, the arc speed automatically will be optimum when the deposition rate is sufficient.

Penetration varies with the joint. With some, it must be deep to provide adequate mixing of the weld and base metal, and with others it must be limited to prevent meltthrough or cracking.

Any joint can be categorized in terms of its needs for these four factors. Examination of the joint detail is all that is required for such categorization.

Figure 5-28 shows examples of fill, follow, freeze, and penetration joints. Note that in the case of the first three types, the adjective "fast" describes what is wanted for maximum efficiency. In the case of penetration, however, either deep or shallow penetration may be desirable, depending upon the joint and the application requirements of the assembly.

In (a) of Figure 5-28, a substantial volume of weld metal is required for fill. The process that will fill most rapidly while meeting other requirements would be given high priority in the four-step selection procedure. In (b), the weld-metal requirements are minimal, and the process that would permit the fastest arc speed would get consideration. In (c), only a process that would permit out-of-position welding would be useful. Here, submerged-arc would be ruled out. But, in (d), deep penetration is indicated as necessary, in which case submerged-arc would definitely be considered.

The examples in Figure 5-28 represent extremes. Although freeze joints are immediately recognizable, many of the joints encountered are not easily categorized in terms of fill, follow, and penetration. Various combinations of these three characteristics may be needed in the joint. Where there is a question, a judgment must be made as to the "weight" of each characteristic in the joint requirements.

Perhaps, follow is of dominating importance, yet

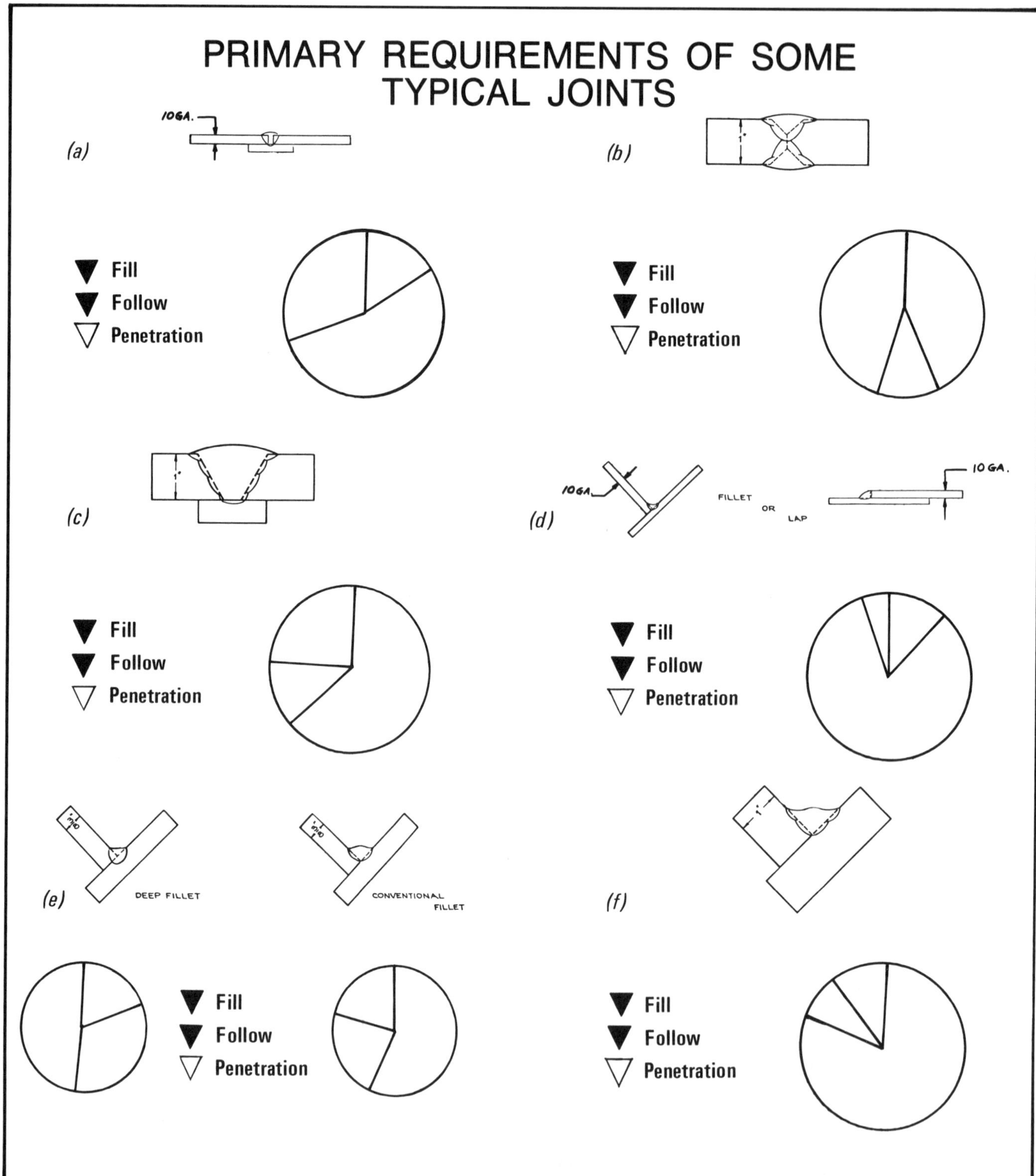

Fig. 5-29. The circular pie chart plots the relative importance of the basic joint requirements.

penetration is also necessary, as in (a) of Figure 5-29. Or, fill may be of equal importance with penetration, as in (b) of the same figure. The pie-chart technique used in Figure 5-29 represents an illustrative effort to "weigh" the relative importance of any joint in terms of fill, follow, and penetration. The persons doing the "weighing" will probably not make pie charts, but they may informally agree that "this joint is mostly fill, with penetration next in importance." With that amount of judgment, the next step toward process selection can be made judiciously.

In analyzing joint requirements, it is virtually impossible for the experienced engineer or welding supervisor to make an error that would substantially affect the selection. The three factors are "weighted" — not quantitatively defined. As long as the order of significance of the three factors is assessed correctly, the information needed for matching up joint requirements with process capabilities has been obtained.

STEP II — MATCHING JOINT REQUIREMENTS WITH PROCESSES

The equipment manufacturers' literature usually will give information on the ability of various processes to deliver the requirements of the joint. Or, a telephone call or letter will bring the needed information. It is virtually impossible to be misled at this point, since the deposition-rate and arc-speed characteristics of each process can be clearly defined. Penetration is related to the current density of the process, but also affected by the thickness of the plate to be welded. In-plant experience with various processes will also be helpful, but care must be exercised that personal prejudices do not lead to destruction of the objectivity of the analysis. Once the capabilities of various semiautomatic processes are known, they can be matched with the joint requirements and tentative selections made.

If the joint in question is the one illustrated by (a) in Figure 5-28, it might be rated 75% fill, 20% penetration, and 5% follow. Fill is obviously all-important; penetration is nominal; and follow is unimportant. No matter what percentages are assigned, as long as they are in the proper order, one has a starting point for matching them with a welding process. The predominant need for fill suggests, possibly, submerged-arc or self-shielded flux-cored welding with a fast-fill electrode. The nominal need for penetration suggests that the latter might be more desirable, since submerged-arc with its deeper penetration could lead to meltthrough.

At this point, however, the welding engineer might ponder the feasibility of using long electrical stickout with submerged-arc, which would reduce excessive penetration while increasing the deposition rate. He now has two processes that look good; other factors, including not only data on deposition rates with the chosen procedures, but also the equipment available, the specifications for the work, the qualification of the weldors, and so on, will affect this final decision. These subsidiary factors come into play in the next step of the four-step selection procedure.

STEP III — THE CHECK LIST

Considerations other than the joint itself have bearing on selection decisions. Many of these will be peculiar to the job or the welding shop. They can be of overriding importance, however, and they offer the means of eliminating alternate possibilities. When organized in a check-list form, these factors can be considered one-by-one, with the assurance that none is overlooked.

Some of the main items to be included on the check list are:

Volume of Production — It must be adequate to justify the cost of the process equipment. Or, if the work volume for one application is not great enough, another application may be found to help defray investment costs.

Weld Specifications — A process under consideration may be ruled out because it does not provide the weld properties specified by the code governing the work. The specified properties may be debatable as far as defining weld quality, but the code still prevails.

Operator Skill — Replacement of manual welding by a semiautomatic process may require a training program. The operators may develop skill with one process more rapidly than with another. Training costs, just as equipment costs, require an adequate volume of work for their amortization.

Auxiliary Equipment — Every process will have its recommended power sources and other items of auxiliary equipment. If a process makes use of existing auxiliary equipment, the initial cost in changing to that process can be substantially reduced.

Accessory Equipment — The availability and the cost of necessary accessory equipment — chipping hammers, deslagging tools, flux lay-down and pickup equipment, exhaust systems, et cetera — should be taken into account.

Base-Metal Condition — Rust, oil, fitup of the joint, weldability of the steel, and other conditions must be considered. Any of these factors could limit the usefulness of a particular process — or give an alternative process a distinct edge.

Arc Visibility — With applications where there is a problem of following irregular seams, open-arc processes are advantageous. On the other hand, when there is no difficulty in correct placement of the weld bead, there are decided "operator comfort" benefits with the submerged-arc process; no headshield is required and the heat radiated from the arc is substantially reduced.

Fixturing Requirements — A change to a semiautomatic process usually requires some fixturing if the ultimate economy is to be realized. The adaptability of a process to fixturing and welding positioners is a consideration to take into account, and this can only be done by realistically appraising the equipment.

Production Bottlenecks — If the process reduces unit fabrication cost, but creates a production bottleneck, its value may be completely lost. For example, highly complicated equipment that requires frequent servicing by skilled technicians may lead to expensive delays and excessive over-all costs.

The completed check list should contain every factor known to affect the economics of the operation. Some may be peculiar to the weld shop or the particular welding job. Other items might include:

Protection Requirements
Application Flexibility
Setup Time Requirements
Cleanliness Requirements

Range of Weld Sizes — Weld Cleaning Costs
Seam Length — Housekeeping Costs
Ability of Follow Seams — Initial Equipment Cost

Each of these items must be evaluated realistically, recognizing the peculiarities of the application as well as those of the process and of the equipment used with it.

Insofar as possible, human prejudices should not enter the selection process; otherwise, objectivity will be lost. At every point, when all other things are equal, the guiding criterion should be welding cost.

The semiautomatic processes, it should be noted, excel over manual welding in respect to the major factors affecting cost — namely, deposition rate, arc speed, and the percentage of the operator's time that is applied to laying the weld bead. With such considerations as setup time, fitup flexibility, application flexibility, range of weld sizes possible, and ability to follow seams, manual welding with the covered electrode will almost always get the highest score.

STEP IV — REVIEW OF THE APPLICATION BY MANUFACTURER'S REPRESENTATIVE

This step in a systematic approach to process selection may seem redundant. However, a basic thesis is that at every step the talents of those who know best should be utilized. Thus, the check list to be used is tailored by the user to his individual situation. The user, and he alone, knows his situation best. It is also true that the manufacturer of the selected equipment knows its capabilities best — will be able to clear up any questions, supply up-to-date information about the process, point out pitfalls in its use, and give practical application tips.

If the analysis of the factors involved has been correct, and the exploration of possible processes thorough, it is almost axiomatic that the equipment manufacturer's representative will confirm the wisdom of the selection. If something important has been overlooked, however, he will be in a position to point it out and make recommendations. This final review before installation of the equipment is to bring to the job all the information bearing on successful application as well as to verify the decision.

SYSTEMIZING THE SYSTEMATIC APPROACH

A system is of no value unless it is used. In a large company where the various welding processes are in use, it is possible to force judicious process selection by requiring analysis of the problem on a standardized process-selection chart. Figure 5-30 shows a process-selection chart that is useful in the analysis of the joint and production requirements — and, thus, in making a process recommendation.

Such a chart recalls to mind the factors that should be considered and prevents oversights. Obviously, it could also be used as a guide for the company that does not have equipment for the major steel-welding processes on its production floors and wishes to make a decision on what to buy for a particular job.

PROCESS SELECTION CHART

DEFINE JOINT REQUIREMENTS AND PERFORMANCE CHARACTERISTICS

STICK Electrode	MIG	Self-Shielded or Gas-Shielded Flux-Cored	SUB-ARC
E ______	Cored Wire ____ Solid Wire ____	______	______

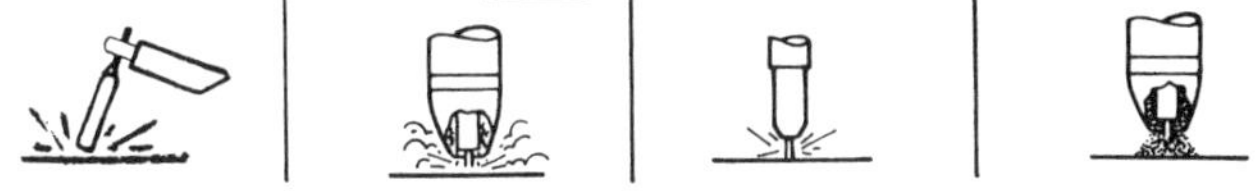

___ FILL
___ FOLLOW
___ FREEZE
___ PENETRATION

(Fast-Fill, Fast-Follow, Fast-Freeze, Penetration)

Sketch Joint Here ↑

PRODUCTION REQUIREMENTS	COMMENTS	PRODUCTION RATING			
PROD. VOLUME REQ'D					
QUALITY, WELD DEPOSIT					
SEAM LENGTH REQ'D					
RANGE, WELD SIZES					
INITIAL EQUIP. COST					
APPLIC. FLEXIBILITY					
ABILITY TO FOLLOW SEAM					
SET UP TIME REQ'D					
FIXTURING COST					
MATERIAL COST					
FITUP FLEXIBILITY					
JOINT PREPAR. FLEXIB.					
CLEANLINESS REQMTS.					
OPERATOR SKILL REQ'D					
WELD CLEANING COST					
HOUSEKEEPING COST					
PROTECTION REQ'D					
CRACK SENSITIVITY (Hard to Weld Steels)					

Fig. 5-30. A process-selection chart that prevents oversights when analyzing the needs of the job.

Bucket-wheel machine for stacking and reclaiming coal. Welding greatly simplified the design.

Section 6

WELDING CARBON AND LOW-ALLOY STEEL

SECTION 6.1

WELDABILITY OF CARBON AND LOW ALLOY STEELS

Page

SECTION 6.2

WELDING CARBON AND LOW-ALLOY STEELS WITH THE SHIELDED METAL-ARC PROCESS

SECTION 6.3

WELDING CARBON AND LOW-ALLOY STEELS WITH THE SUBMERGED-ARC PROCESS

SECTION 6.4

WELDING CARBON AND LOW-ALLOY STEELS WITH THE SELF-SHIELDED FLUX-CORED ELECTRODE PROCESS

SECTION 6.5

WELDING CARBON AND LOW-ALLOY STEELS WITH THE GAS SHIELDED FLUX-CORED ELECTRODE PROCESS

SECTION 6.6

WELDING CARBON AND LOW-ALLOY STEELS WITH THE GAS METAL ARC PROCESS

SECTION 6.7

WELDING CARBON AND LOW-ALLOY STEELS WITH THE GAS TUNGSTEN-ARC PROCESS

Weldability of Carbon and Low Alloy Steels

Carbon and low-alloy steels are the work-horse materials for construction and transportation equipment and for industrial and consumer products of many types. They comprise over 90% of total steel production, and more carbon steel is used in product manufacture than all other metals combined.

Sections 6.1 through 6.7 discuss the weldability of these important materials and the various welding processes that are used for joining them. Selection and operational considerations for each process include details on electrodes, filler wires, welding techniques and procedures, process variables, qualification requirements, welding equipment, fixtures, and other necessary information for designers, welding engineers, and weldors.

Most steels can be welded, but satisfactory joints cannot be produced in all grades with equal ease. A metal is considered to have good weldability if it can be welded without excessive difficulty or the need for special and costly procedures and the weld joints are equal in all necessary respects to a similar piece of base metal. Weldability varies with the grade, chemistry, and mechanical properties of the steel, and when weld joining is to be a major factor in the attachment of steel parts, weldability should be given proper attention in specifying and ordering materials for the job.

In this section, weldability of the various plain carbon, low alloy and alloy steel is discussed. For steels, weldability is related inversely to hardenability and maximum attainable hardness. For example, a steel having high hardenability and hardness will have poor weldability and a tendency to crack when welded. A steel having low hardenability and hardness will have excellent weldability and essentially no tendency to crack when welded.

The maximum hardness attainable in a steel depends upon the carbon equivalent (CE) which is related to the composition and is calculated from the following formula:

$$CE = \%C + \frac{\%Mn + \%Si}{6} + \frac{\%Cr + \%Mo + \%V}{5} + \frac{\%Ni + \%Cu}{15}$$

where the elements are in weight percent. From this, it can be seen that carbon has the greatest effect on hardness and that Mn, Si, Cr, Mo and V have about 1/5 the effect of C. The maximum hardness of a steel will be attained only if it is cooled from an austenitizing temperature (1450-1650°F) at a fast enough rate to obtain an all martensite microstructure. The necessary cooling rate depends on the hardenability of the steel; the higher the hardenability, the slower the cooling rate from the austenite temperature range needed to obtain all martensite. The addition of alloying elements to steel, such as Mn, Cr and Mo, slow down the transformation of austenite on cooling through the transformation temperature range. This prevents transformation to structures such as pearlite which are much softer than martensite and allows the austenite to transform to martensite at a lower temperature (700-300°F). Steels transforming slowest on cooling have the highest hardenability, and those transforming fastest on cooling have the lowest hardenability. Steels having high hardenability can full harden on air cooling except for very thick sections. Steels with low hardenability must be water quenched to full harden except very thin sections.

To prevent cracking when welding a hardenable steel, the cooling rate of the weld and heat affected zone must be slow enough to allow transformation to be complete before it reaches the temperature at which martensite begins to form. Cooling rate can be slowed by preheating, by using a high energy input or by a combination of both.

STEEL SPECIFICATION

Several methods are used to identify and specify steels. These are based on chemistry, on mechanical properties, on an ability to meet a standard specification or industry-accepted practice, or on an ability to be fabricated into a certain type of product.

Specifying by Chemistry

A desired composition can be produced in one of three ways: to a maximum limit, to a minimum limit, or to an acceptable range.

TABLE 6-1. Preferred Analyses for Steels To Be Arc-Welded

Element	Composition (%)	
	Preferred	High*
Carbon	0.06 to 0.25	0.35
Manganese	0.35 to 0.80	1.40
Silicon	0.10 or less	0.30
Sulfur	0.035 or less	0.05
Phosphorus	0.030 or less	0.04

* Additional care is required in welding of steels containing these amounts of the elements listed.

For economical, high-speed welding of carbon-steel plate, the composition of the steel should be within the "preferred-analysis" ranges indicated in Table 6-1. If one or more elements varies from the ranges shown, cost-increasing methods are usually required to produce good welding results. Thus, steels within these ranges should be used whenever extensive welding is to be done unless their properties do not meet service requirements. Published welding procedures generally apply to normal welding conditions and to the more common preferred-analysis mild steels. Low-hydrogen electrodes and processes will generally tolerate a wider range of the elements than shown in Table 6.1.

If the chemical specification of a steel falls outside of the preferred-analysis range, it is usually not necessary to use special welding procedures based on the extremes allowed by the specification. The chemistry of a specific heat, under average mill-production conditions, may be considerably below the top limits indicated in the specification. Thus, for maximum economy, welding procedures for any type of steel should be based on actual rather than allowed chemistry values. A mill test report* can be obtained that gives the analysis of a heat of steel. From this information, a welding procedure can be established that ensures production of quality welds at lowest possible cost.

Standard carbon and alloy steels are identified by AISI (American Iron and Steel Institute), SAE (Society of Automotive Engineers), or ASTM (American Society for Testing Materials) designation systems. In the commonly used four-digit system of

* A mill test report is usually based on a ladle analysis and is an average for an entire heat. Some low-carbon steels are rimmed steels, used because of their excellent forming and deep-drawing properties. The analysis of a rimmed steel varies from the first ingot to the last ingot of a single heat and also from the top to the bottom of a single ingot. Thus, a mill test report is an average and should be interpreted as such.

TABLE 6-2. AISI Designation System for Alloy Steels

Alloy Series	Approximate Alloy Content (%)
13XX	Mn 1.60-1.90
40XX	Mo 0.15-0.30
41XX	Cr 0.40-1.10; Mo 0.08-0.35
43XX	Ni 1.65-2.00; Cr 0.40-0.90; Mo 0.20-0.30
44XX	Mo 0.45-0.60
46XX	Ni 0.70-2.00; Mo 0.15-0.30
47XX	Ni 0.90-1.20; Cr 0.35-0.55; Mo 0.15-0.40
48XX	Ni 3.25-3.75; Mo 0.20-0.30
50XX	Cr 0.30-0.50
51XX	Cr 0.70-1.15
E51100	C 1.00; Cr 0.90-1.15
E52100	C 1.00; Cr 0.90-1.15
61XX	Cr 0.50-1.10; Va 0.10-0.15 (min)
86XX	Ni 0.40-0.70; Cr 0.40-0.60; Mo 0.15-0.25
87XX	Ni 0.40-0.70; Cr 0.40-0.60; Mo 0.20-0.30
88XX	Ni 0.40-0.70; Cr 0.40-0.60; Mo 0.30-0.40
92XX	Si 1.80-2.20

the AISI and SAE (Table 6-2), the last two digits indicate the middle of the carbon range. For example, in grade 1035, the 35 represents a carbon range from 0.32 to 0.38%. The first two digits indicate these carbon-steel grades:

10xx	Nonresulfurized
11xx	Resulfurized
12xx	Resulfurized and rephosphorized

A prefix "B" indicates an acid bessemer steel, an "E" indicates an electric-furnace steel. The E steels are usually alloy or stainless-steel grades. Steels without a prefix designation may be produced by basic open hearth, basic oxygen, or electric-furnace methods.

The letter "L" between the second and third digits indicates a leaded steel. The letter "B" in the same position designates a boron-treated steel. The suffix "H" refers to steels specially produced to narrow chemical and hardenability ranges.

These four-digit AISI or SAE standard steel designations apply primarily to sheet, strip, and bar products. ASTM specifications apply to most plates and structural shapes.

Some of the commonly specified elements and their effects on weldability and other characteristics of steels follow:

Carbon is the principal hardening element in steel. As carbon content increases, hardenability and tensile strength increase, and ductility and weldability decrease. In steels with a carbon content over 0.25%, rapid cooling from the welding temperature may produce a hard, brittle zone adjacent to the weld. Also, if considerable carbon is picked up in the weld puddle through admixture from the metal being welded, the weld deposit itself may be hard. Addition of small amounts of elements other than carbon can produce high tensile strengths without a detrimental effect on weldability. In general, carbon content should be low for best, weldability.

Manganese increases hardenability and strength, but to a lesser extent than carbon. Properties of steels containing manganese depend principally on carbon content. Manganese content of less than 0.30% may promote internal porosity and cracking in the weld bead; cracking can also result if the content is over 0.80%.

For good weldability, the ratio of manganese to sulfur should be at least ten to one. If a steel has a low manganese content in combination with a low carbon content, it may not have been properly deoxidized. In steel, manganese combines with sulfur to form MnS, which is not harmful. However, a steel with a low Mn/S ratio may contain sulfur in the form of FeS, which can cause cracking (a "hot-short" condition) in the weld.

In general, manganese increases the rate of carbon penetration during carburizing and is beneficial to the surface finish of carbon steels.

Sulfur increases the machinability of steels, but reduces transverse ductility, impact toughness, and weldability. Sulfur in any appreciable amount promotes hot shortness in welding, and the tendency increases with increased sulfur. It can be tolerated up to about 0.035% (with sufficient Mn), over 0.050% it can cause serious problems. Sulfur is also detrimental to surface quality in low-carbon and low-manganese steels.

A common cause of poor welding quality that is not apparent from analyses made in the usual way is segregated layers of sulfur in the form of iron sulfide. These layers, which cause cracks or other defects at the fusion line of an arc-welded joint, can be detected by examination of a deep-etched cross section as illustrated in Fig. 6-1.

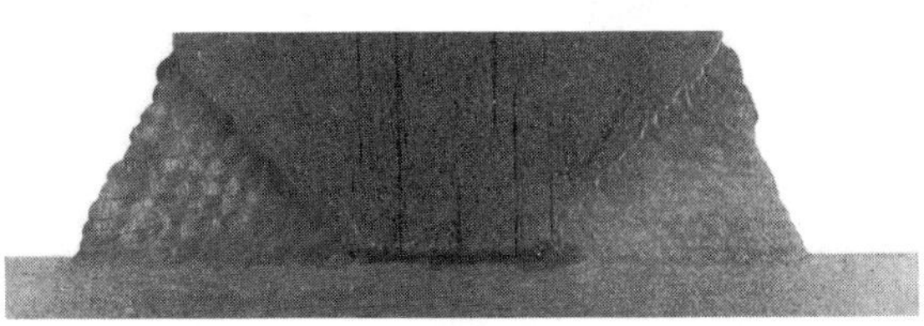

Fig. 6-1. Sulfur segregations. Dark lines in etched section indicate areas of high sulfur concentration.

Silicon is a deoxidizer that is added during the making of steel to improve soundness. Silicon increases strength and hardness, but to a lesser extent than manganese. It is detrimental to surface quality; especially in the low-carbon, resulfurized grades. If carbon content is fairly high, silicon aggravates cracking tendencies. For best welding conditions, silicon content should not exceed 0.10%, but amounts up to 0.30% are not as serious as high sulfur or phosphorus content.

Phosphorus, in large amounts, increases strength and hardness, but reduces ductility and impact strength, particularly in the higher-carbon grades. In low-carbon steels, phosphorus improves machinability and resistance to atmospheric corrosion.

As far as welding is concerned, phosphorus is an impurity, and should be kept as low as possible. Over 0.04% makes welds brittle and increases the tendency to crack. Phosphorus also lowers the surface tension of the molten weld metal, making it difficult to control.

Copper improves atmospheric corrosion resistance when present in excess of 0.15%. (A minimum of 0.20% is usually specified for this purpose.) Most carbon steels contain some copper as a "tramp element," up to about 0.15%. Copper content up to about 1.50% has little or no effect on the acetylene or arc-weldability of a steel, but it affects forge-weldability adversely. Copper content over 0.50% may reduce mechanical properties, if the steel is heat-treated.

Copper content is detrimental to surface quality, particularly in high-sulfur grades.

Specifying by Mechanical Properties

The producer of steels specified by mechanical properties is free to alter the chemistry of the steel (within limits) to obtain the required properties. Mechanical tests are usually specified under one of these conditions: 1. Mechanical test requirements only, with no limits on chemistry. 2. Mechanical test requirements, with limits on one or more elements.

Generally, these tests have been set up according to practices approved by the SAE (Society of Automotive Engineers) or ASTM (American Society for Testing and Materials) or to the requirements of other authorized code-writing organizations, such as the ASME (American Society of Mechanical Engineers) or the API (American Petroleum Institute).

The most common tests are bend tests, hardness tests, and a series of tensile tests that evaluate modulus of elasticity. yield strength, and tensile strength. Section 1.2 discusses some of these tests and the properties they determine. Metallurgical tests are sometimes used to measure grain size, decarburization, or inclusions. Other tests relating to end-use requirements, such as burst tests for pressure tubing, may be included in some specifications.

Most carbon steels are produced to standard specifications established by regulating bodies concerned with public welfare and safety. The largest and most influential body of this type is the ASTM. Other major groups are the SAE, the ASME, the AAR (American Association of Railroads), and the AWWA (American Water Works Association). ASTM specifications are broad, covering requirements of many industries. Most other groups prepare steel specifications for the needs and interests of their particular industries.

Specifying by End-Product

Often more important than exact mechanical properties or chemical analysis is the ability of a steel to be fabricated into a specific end product. Fabricating operations such as welding or deep drawing can change the as-delivered properties of a steel, and more than one chemical analysis or steelmaking method can often produce a suitable material for the product. Consequently, many flat-rolled steel products such as plate, sheet, and strip are specified to have adequate properties for fabrication into an "identified" end product.

A specification for an identified end product tells the steel producer which fabrication processes will be used, finish requirements, and the product's service requirements.

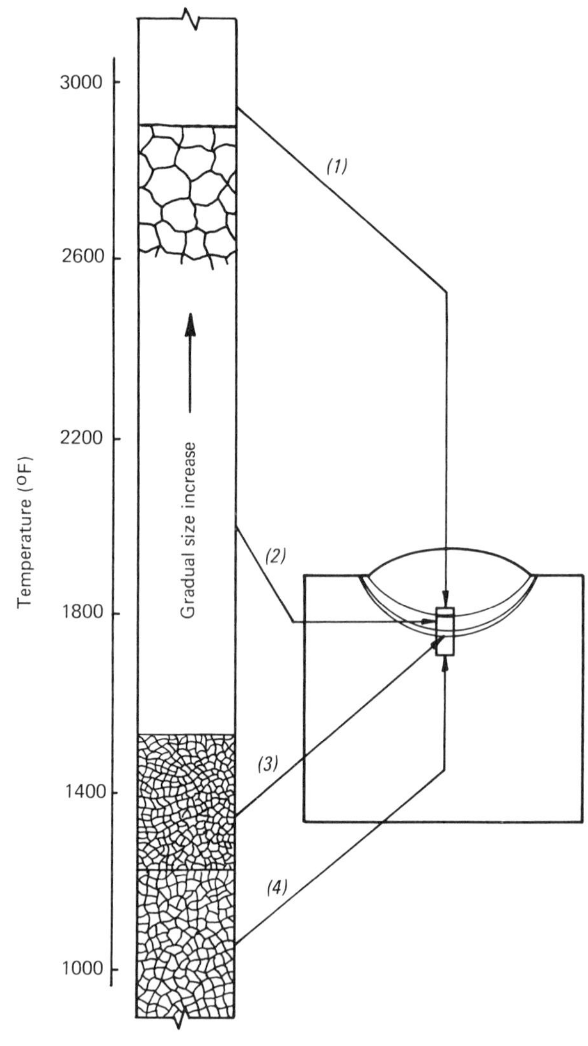

Fig. 6-2. Effect of welding heat on hardness and microstructure of an arc-welded 0.25% carbon steel plate. The schematic diagram represents a strip cut vertically through the weld shown. Significance of the four numbered zones are: 1. Metal that has been melted and resolidified. Grain structure is coarse. 2. Metal that has been heated above the upper critical temperature (1525° for 0.25% carbon steel) but has not been melted. This area of large grain growth is where underbead cracking can occur. 3. Metal that has been heated slightly above the lower critical temperature (1333°F) but not to the upper critical temperature. Grain refinement has taken place. 4. Metal that has been heated and cooled, but not to a high enough temperature for a structural change to occur.

METALLURGY OF A WELD BEAD

The heat of welding brings about certain changes, both in the structure of the steel being welded and in the weld metal. Some of these changes occur during welding; others, after the metal has cooled.

During welding, the temperature of the molten weld metal reaches 3000°F or higher. A short distance from the weld, the temperature of the plate may be only about 600°F. When the steel reaches or exceeds certain critical temperatures between these values, changes occur that affect grain structure, hardness, and mechanical properties. These changes and the temperatures at which they occur are illustrated by Fig. 6-2, a schematic diagram of a section through a weld bead.

The extent of change in structure depends on the maximum temperature to which the metal is subjected, the length of time the temperature is sustained, the composition of the metal, and the rate of cooling. The principal factor that controls these changes is the amount of heat that is put into the plate — both from preheating and from the welding process.

Cooling rate affects properties along with grain size. Rapid cooling rates produce stronger, harder and less ductile steels; slow cooling rates produce the opposite properties. With low-carbon steels, the relatively small differences in cooling rates in normal practice have negligible effects on these properties. However, with steels of higher carbon contents or those with appreciable amounts of alloying elements, the effect can be significant.

Holding the plate material at a high temperature (above the upper critical temperature) for a long time produces a structure with large grain size. During welding, however, the metal adjacent to the weld (Zone 3 in Fig. 6-2) is at the high temperature for a very short time. The result is a slight decrease in grain size and an increase in strength and hardness, compared with the base metal.

In multipass weld joints, each bead produces a grain-refining action on the preceding bead as it is reheated. However, this refining is not likely to be uniform throughout the joint.

CRACKING — CAUSES AND CURES

Except in some weld-surfacing operations, cracks are considered deleterious. Cracking can occur either in the deposited metal or in the heat-affected zone of the base metal adjacent to the weld. The major cause of cracking in the base metal or in the weld metal is a high carbon or alloy content that increases the hardenability. High hardenability, combined with a high cooling rate, produces the brittle condition that leads to cracking. Other causes of weld cracking are: joint restraint that produces high stresses in the weld, improper shape of the weld bead, hydrogen pickup, and contaminants on the plate or electrode.

Factors Causing Underbead Cracking

Subsurface cracks in the base metal, under or near the weld, are known as underbead cracks. Underbead cracking in the heat-affected base metal is caused by: 1. A relatively high carbon or alloy content steel that is allowed to cool too rapidly from the welding temperature. 2. Hydrogen pickup during welding.

Underbead cracking seldom occurs with the preferred-analysis steels (Table 6-1). With carbon steels above 0.35% carbon content and with the low-alloy structural-grade steels, underbead cracking can be minimized by using a low-hydrogen welding process. The problem is most severe with materials such as the heat-treated structural steels having tensile strengths of 100,000 psi and higher. The discussions on specific steels include recommendations for welding these materials.

The second factor that promotes underbead cracking — the pickup and retention of hydrogen

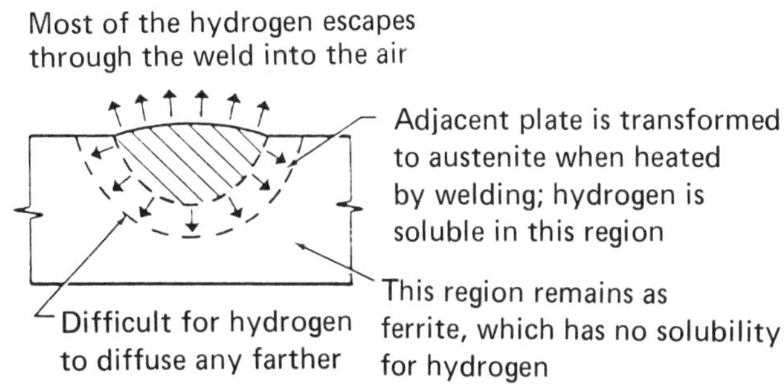

Fig. 6-3. Austenitic heat-affected zone of a weld has high solubility for hydrogen. Upon cooling, the hydrogen builds up pressure that can cause underbead cracking.

— is also influenced by the cooling rate from the welding temperature. During welding, some hydrogen — a decomposition product of moisture from the air, electrode coating, wire, flux, shielding gas, or the surface of the plate — can dissolve in the molten weld metal and from there diffuse into the extremely hot solid base metal. If cooling occurs slowly, the process reverses, and the hydrogen has sufficient time to escape by diffusion through the weld into the air. But if cooling is rapid, some hydrogen may be trapped in the heat-affected zone next to the weld metal, as illustrated by Fig. 6-3. The hydrogen is retained in supersaturated solution and produces a condition of low ductility known as hydrogen embrittlement.

Hydrogen has a high solubility in solid iron at elevated temperatures just under the melting point. This solubility decreases gradually as temperature decreases and decreases abruptly below the transformation temperature range leaving hydrogen atoms in supersaturated solution. According to theory, the hydrogen accumulates in dislocations and micro voids as molecular hydrogen producing very high hydrostatic pressure. This pressure combined with shrinkage stresses and any hardening effect from the chemistry of the steel — causes tiny cracks in the metal immediately under the weld bead (Fig. 6-4). Similar cracks that appear on the plate surface adjacent to the weld are called "toe cracks."

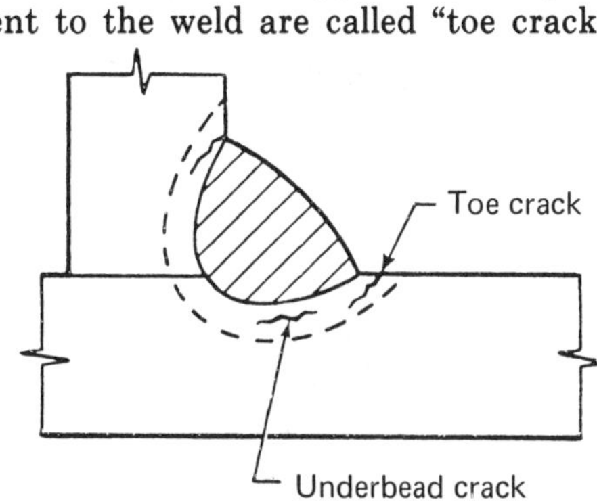

Fig. 6-4. Underbead cracking and toe cracks caused by hydrogen pickup in heat-affected zone of plate.

Slower cooling (by welding slower, or by preheating) allows more of the hydrogen to escape and helps control the problem. In addition, the use of low-hydrogen welding materials eliminates the major source of hydrogen and usually eliminates underbead cracking.

Rapid cooling rates occur when the arc strikes on a cold plate — at the start of a weld with no previous weld bead to preheat the metal. The highest cooling rates occur on thick plate and in short tack welds. The effect of weld length on cooling rate can be illustrated by the time required to cool welds from 1600° to 200°F on a 3/4-in. steel plate:

2-1/2-in. weld . . . . 1.5 min
4-in. weld . . . . 5 min
9-in. weld . . . . 33 min

A 9-in.-long weld made on plate at 70°F has about the same cooling rate as a 3-in. weld on a plate that has been preheated to 300°F.

Welds with large cross sections require greater heat input than smaller ones. High welding current and slow travel rates increase the heat input per inch of weld and reduce the rate of cooling and decrease the likelihood of cracking.

The Effects of Section Thickness

In a steel mill, billets are rolled into plates or shapes while red hot. The rolled members are then placed on finishing tables to cool. Because a thin plate has more surface area in proportion to its mass than a thick plate, it loses heat faster (by radiation) and cools more rapidly.

If a thick plate has the same chemistry as a thin one, its slower cooling rate results in lower tensile and yield strength, lower hardness, and higher elongation. In very thick plates, the cooling rate may be so low that the properties of the steel may not meet minimum specifications. Thus, to meet specified yield-strength levels, the mill increases the carbon or alloy content of the steels that are to be rolled into thick sections.

In welding, cooling rates of thin and thick plates are just the opposite. Because of the larger mass of plate, the weld area in a thick plate cools more rapidly than the weld area in a thin one. The heat input at the weld area is transferred, by conduction, to the large mass of relatively cool steel, thus cooling the weld area relatively rapidly. (Heat is transferred more rapidly by conduction than by radiation.) The thin plate has less mass to absorb the heat, and it cools at a slower rate. The faster cooling rate of the thicker plate produces higher tensile and yield strengths, higher hardness, and lower elongation.

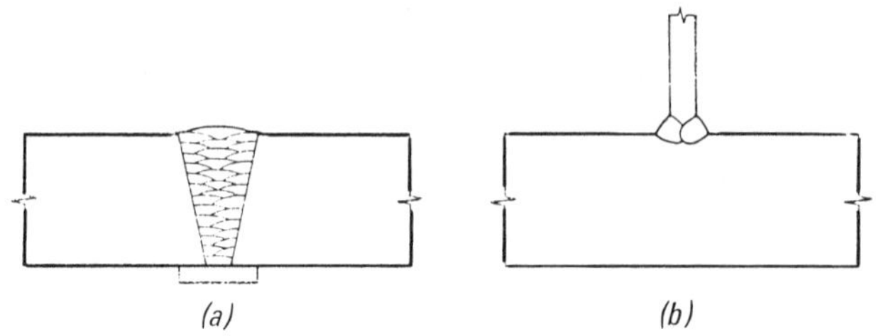

Fig. 6-5. A groove-welded butt joint in thick plate (a) requires a higher preheat, because of joint restraint, than a fillet-welded joint of a thin member and a thick plate (b). See Section 2.3 for the minimum size weld required by AWS.

Welds in structural-steel shapes and plate under 1/2-in. thick have less tendency toward cracking than welds in thicker plate. In addition to the favorable (slower) cooling rate of thinner members, two other factors minimize causes of cracking:

1. Thinner plate weldments usually have a good ratio (high) of weld-throat-to-plate thickness.
2. Because they are less rigid, thinner plates can flex more as the weld cools, thus reducing restraint on the weld metal.

Thicker plates and rolled sections do not have these advantages. Because a weld cools faster on a thick member, and because the thick member probably has a higher carbon or alloy content, welds on a thick section have higher strength and hardness but lower ductility than similar welds on thin plate. If these properties are unacceptable, preheating (especially for the more critical root pass) may be necessary to reduce the cooling rate. (See Section 3.3 for a discussion of preheating.)

Because it increases cost, preheating should be used only when needed. For example, a thin web to be joined to a thick flange plate by fillet welds may not require as much preheat as two highly restrained thick plates joined by a multiple-pass butt weld (Fig. 6-5).

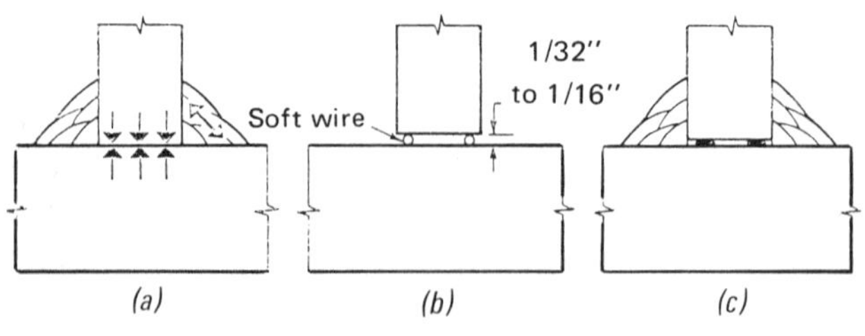

Fig. 6-6. In a restrained joint in thick plates (a), all shrinkage stress must be taken up in the weld. Separating the plates with soft wires (b) allows the plates to move slightly during cooling. The wires flatten (c) and remove most of the stress from the weld metal.

The Effect of Joint Restraint

If metal-to-metal contact exists between thick plates prior to welding, the plates cannot move — the joint is restrained. As the weld cools and contracts, all shrinkage stress must be taken up in the weld, as illustrated in Fig. 6-6(a). This restraint may cause the weld to crack, especially in the first pass on the second side of the plate.

Joint restraint can be minimized by providing a space of 1/32 to 1/16 in. between the two members to allow movement during cooling. Such spaces or gaps can be incorporated by several simple means:

1. Soft steel wire spacers may be placed between the plates, as in Fig. 6-6(b). The wire flattens out as the weld shrinks, as shown in Fig. 6-6(c). (Copper wire should not be used because it may contaminate the weld metal.)
2. Rough flame-cut edges on the plate. The peaks of the cut edge keep the plates apart, yet can deform and flatten out as the weld shrinks.
3. Upsetting the edge of the plate with a heavy center punch. Results are similar to those of the flame-cut edge.

Provision for a space between thick plates to be welded is particularly important for fillet welds.

Fillet Welds: A molten fillet weld starts to solidify, or freeze, along the sides of the joint, as in Fig. 6-7, because the heat is conducted to the adjacent plate, which is at a much lower temperature. Freezing progresses inward until the entire weld is solid. The last material to freeze is that at the center, near the surface of the weld.

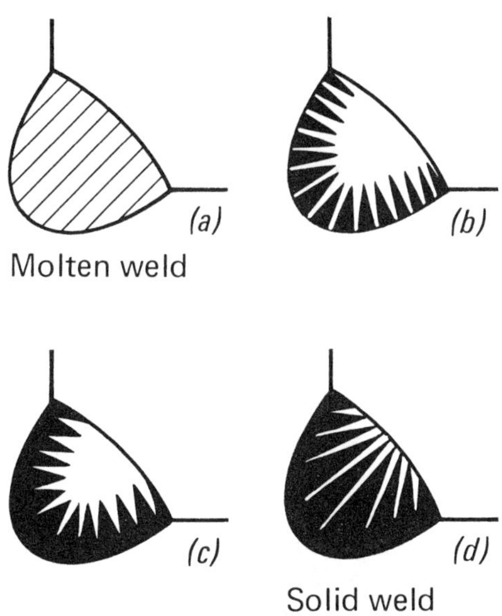

Fig. 6-7. A molten fillet weld (a) starts to solidify along the sides next to the plate (b). Solidification proceeds as shown in (c) and (d).

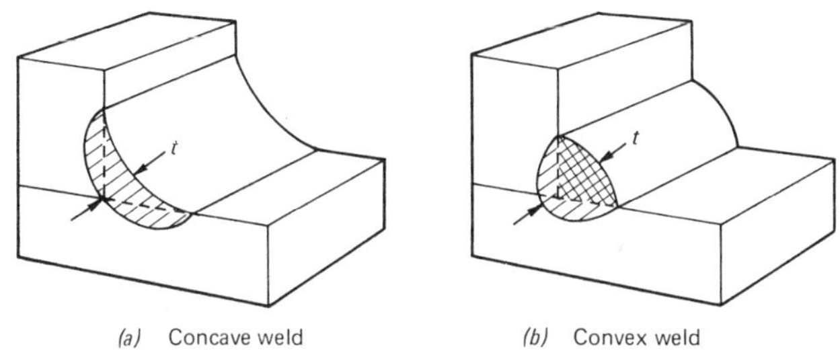

Fig. 6-8. The leg size and the surface of a concave fillet weld (a) may be larger than that of a convex bead (b), but its throat, t, may be considerably smaller.

Although a concave fillet weld may appear to be larger than a convex weld (Fig. 6-8), it may have less penetration into the welded plates and a smaller throat than the convex bead. Thus the convex weld may be the stronger of the two, even though it appears to be smaller.

In the past, the concave weld has been preferred by designers because of the smoother stress flow it offers to resist a load on the joint. Experience has shown however, that single-pass concave fillet welds have a greater tendency to crack during cooling than do convex welds. This disadvantage usually outweighs the effect of improved stress distribution, especially in steels that require special welding procedures.

When a concave bead cools and shrinks, the outer surface is in tension and may crack. A convex bead has considerably reduced shrinkage stresses in the surface area, and the possibility of cracking during cooling is slight. For multiple-pass fillet welds only the first pass need be convex.

When design conditions require concave welds for smooth flow of stresses in thick plate, the first bead (usually three or more passes are required) should be slightly convex. The others are then built up to the required shape.

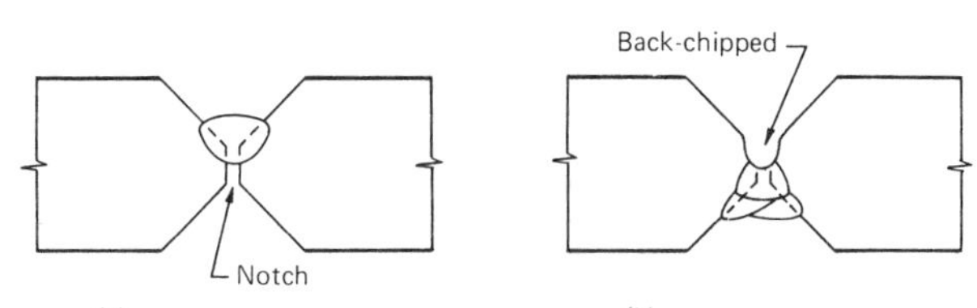

Fig. 6-9. The root pass of a double-V joint is susceptible to cracking because of the notch effect (a). On high-quality work, the notch is minimized by backchipping (b).

Groove Welds: The root pass of a groove weld in heavy plate usually requires special welding procedures. For example, the root pass on the first side of a double-V joint is susceptible to cracking because of the notch, as illustrated in Fig. 6-9(a), which is a crack starter. On high-quality work, this notch is backchipped, as in Fig. 6-9(b), to: 1. Remove slag or oxides from the bottom of the groove. 2. Remove any small cracks that may have occurred in the root bead. 3. Widen the groove at the bottom so that the first bead of the second side is large enough to resist the shrinkage stress that it must withstand due to the rigidity of the joint.

The weld metal tends to shrink in all directions as it cools, and restraint from the heavy plates produces tensile stresses within the weld. The metal yields plastically while hot to accommodate the stresses; if the internal stresses exceed the strength of the weld, it cracks, usually along the centerline.

The problem is greater if the plate material has a higher carbon content than the welding electrode. If this is the case, the weld metal usually picks up additional carbon through admixture with the base metal. Under such conditions, the root bead is usually less ductile than subsequent beads.

A concave root bead in a groove weld, as shown in Fig. 6-10(a), has the same tendency toward cracking as it does in a fillet weld. Increasing the throat dimension of the root pass, as in Fig. 6-10(b), helps to prevent cracking. Electrodes and procedures should be used that produce a convex bead shape. A low-hydrogen process usually reduces cracking tendencies; if not, preheating may be required.

Centerline cracking can also occur in subsequent passes of a multiple-pass weld if the passes are excessively wide or concave. This can be corrected by putting down narrower, slightly convex beads, making the weld two or more beads wide, as in Fig. 6-11.

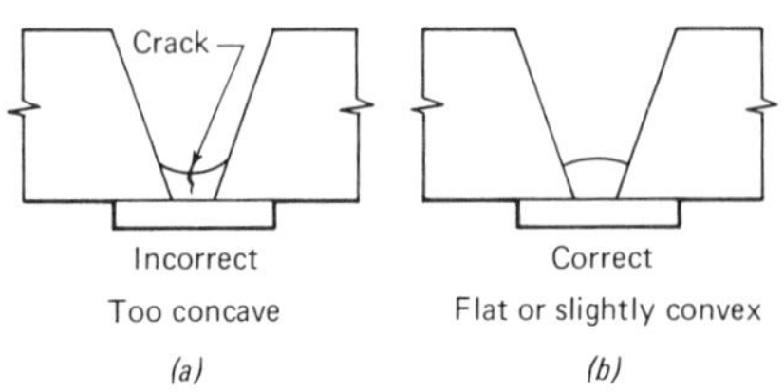

Fig. 6-10. A concave root pass (a) may crack because tensile stresses exceed the strength of the weld metal. A slightly convex root-pass bead (b) helps prevent cracking.

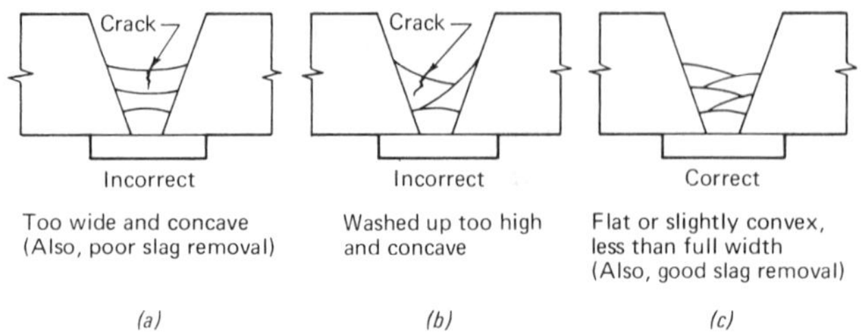

Fig. 6-11. Wide, concave passes (a and b) in a multiple-pass weld may crack. Slightly convex beads (c) are recommended.

Width/Depth Ratio: Cracks caused by joint restraint or material chemistry usually appear at the face of the weld. In some situations, however, internal cracks occur that do not reach the surface. These are usually caused by improper joint design (narrow, deep grooves or fillets) or by misuse of a welding process that can achieve deep penetration.

If the depth of fusion is much greater than the width of the weld face, the surface of the weld may freeze before the center does. When this happens, the shrinkage forces act on the almost-frozen center (the strength of which is lower than that of the frozen surface) and can cause a crack that does not extend to the surface. Figure 6-12(a) is illustrative.

Internal cracks can also be caused by improper joint design or preparation. Results of combining thick plate, a deep-penetrating welding process, and a 45° included angle are shown in Fig. 6-12(b). A similar result on a fillet weld made with deep penetration is shown in Fig. 6-12(c). A too-small bevel, and arc-gouging a groove too narrow for its depth on the second-pass side of a double-V groove weld, can cause the internal crack shown in Fig. 6-12(d).

Internal cracks are serious because they cannot be detected by visual inspection methods. But they can be eliminated if preventive measures are used. Penetration and volume of weld metal deposited in each pass can be controlled by regulating welding speed and current and by using a joint design which establishes reasonable depth-of-fusion requirements. Recommended ratios of width of each individual bead to depth of fusion are between 1.2 to 1 and 2 to 1.

A different type of internal crack occurs in submerged-arc welding when the width-to-depth ratio is too large. Cracks in these so-called "hat-shaped" (Fig. 6-12e) welds are especially dangerous because radiographic inspection may not detect them. The width-to-depth ratio of any individual bead should not exceed 2:1.

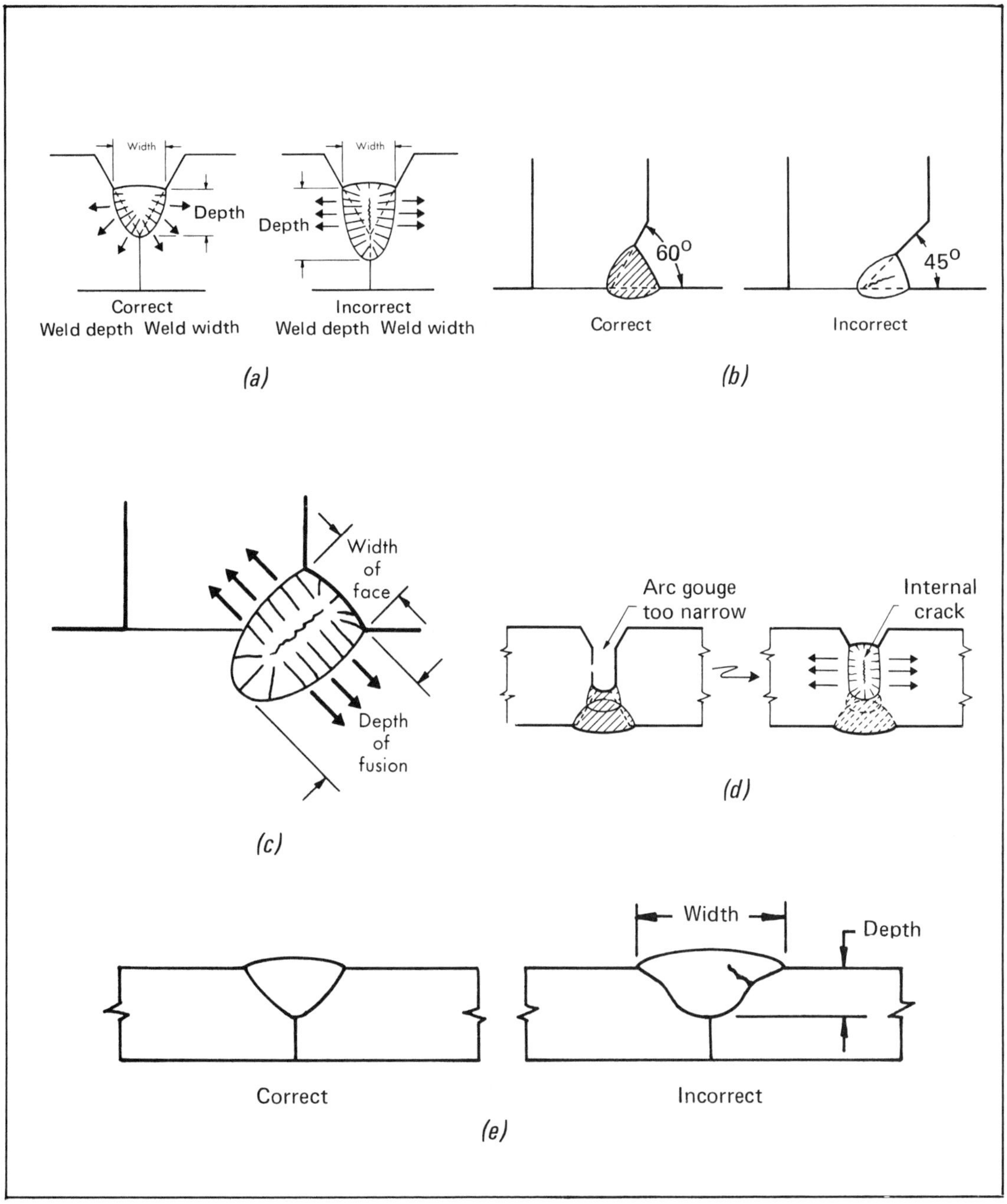

Fig. 6-12. Internal cracking can occur when weld penetration is greater than width. Correct and incorrect proportions are shown in (a), (b), and (c). Arc-gouging a groove too narrow for its depth can cause a similar internal crack (d). Cracks can also occur when depth is too shallow (e). Width of a weld should not exceed twice its depth.

LAMELLAR TEARING

Lamellar tearing is the separation in the parent or base metal caused by thru-thickness stresses induced by weld metal shrinkage. See Fig. 6-13A. Under conditions of high restraint, which is highly typical of thick plate, localized stresses due to weld metal shrinkage may be many times greater than the yield point of the base metal. In the thru-thickness direction, these stresses can produce lamellar tearing.

A lamellar tear occurs only in the parent metal and while it may originate close to the toe or root of a weld, often the tear may originate well beyond the heat affected zone and may not propagate to

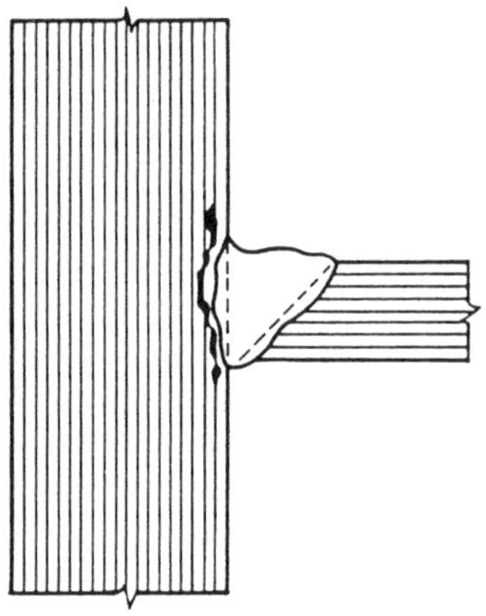

Fig. 6-13A. Because of the directional properties of steel and the shrinkage forces of large volumes of weld metal, this type of connection is highly susceptible to lamellar tearing.

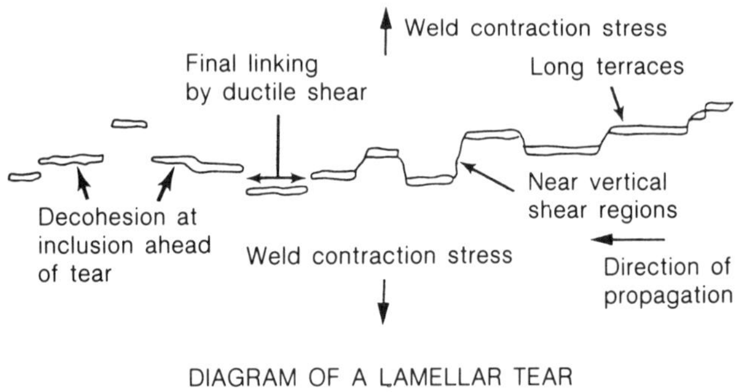

Fig. 6-13B. Diagram of a partially developed lamellar tear.

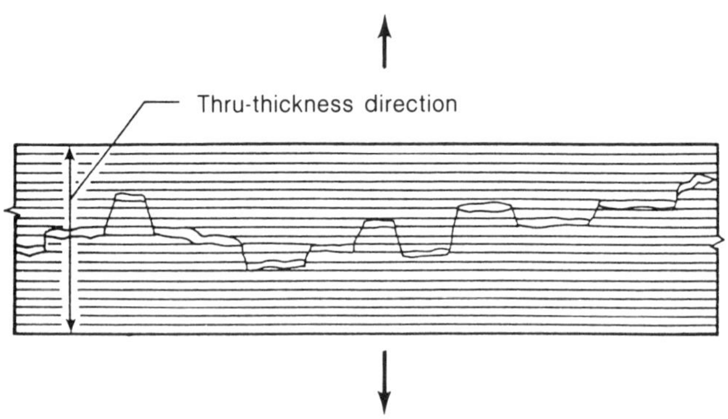

Fig. 6-13C. Cross section of parent material showing complete development of the incomplete tear shown in Fig. 6-13B.

the surface. The cross section of a lamellar tear is steplike, with longitudinal terraces that are markedly longer than the transverse portions. See Figs. 6-13B, 6-13C. The fracture has a fibrous appearance and this characteristic, along with the terraced profile and location within the base metal, is the best way to distinguish lamellar tearing from cracks in the heat affected zone caused by hydrogen.

The tearing occurs principally in tee and corner joints (see Fig. 6-13D) where the degree of restraint is such that the weld shrinkage stresses imposed on the parent metal cannot be accommodated because of the limited ductility of the base metal in the thru-thickness direction. During the progress of welding, after a sufficient number of passes have been deposited, the weld shrinkage stresses increase in magnitude as the weld cools, causing a separation at the interface between microscopic non-metallic inclusions and surrounding matrix. As more weld metal is deposited, additional microscopic tears form. Since the non-metallic inclusions and stresses are dispersed through the steel in an irregular manner, the tear takes the most susceptible path (see Figure 6-13C).

Material Properties

In the hot rolling process used to shape structural steel (see Fig. 6-13E), the greatest strength and ductility occurs in the longitudinal and transverse directions. Elongation and reduction in area values may well be significantly lower in the thru-thickness direction. See Fig. 6-13E. These directional properties are particularly significant on thicker plates since less hot rolling has taken place. As these properties change it places a greater burden on the designer to design accordingly and to avoid those types of welded connections which are highly susceptible to lamellar tearing. In extreme cases the designer may find it necessary to specify the use of a premium grade of steel which will guarantee acceptable thru-thickness properties. Such steels are commercially available.

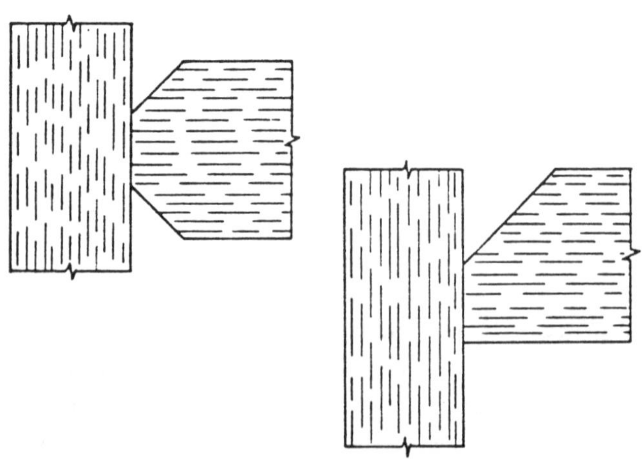

Fig. 6-13D. Tee joints and corner joints, especially on thick plate, are the most susceptible to lamellar tearing.

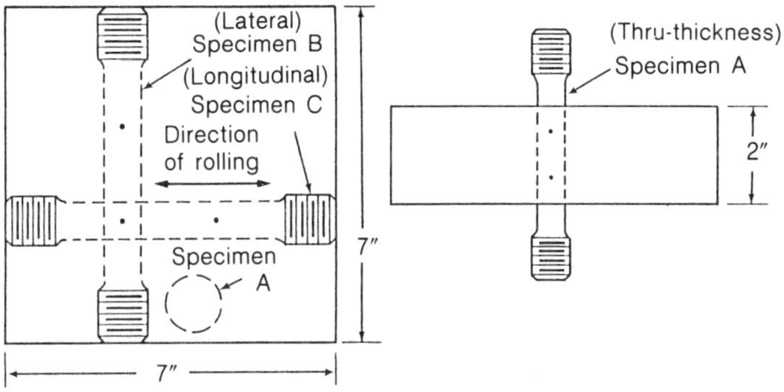

Specimen	Tensile Strength (ksi)	Yield Point	Percent Elongation	Percent Area Reduction
A (thru-thickness)	45.6	None	6.0	12.2
B (lateral)	65.7 68.6	45.3 47.2	21.0 24.5	46.7 46.8
C (longitudinal)	64.3 67.5	46.9 47.5	24.5 20.0	53.1 53.4

Fig. 6-13E. All steels have directional properties, as shown above the lateral and longitudinal properties are considerably better than the thru-thickness properties.

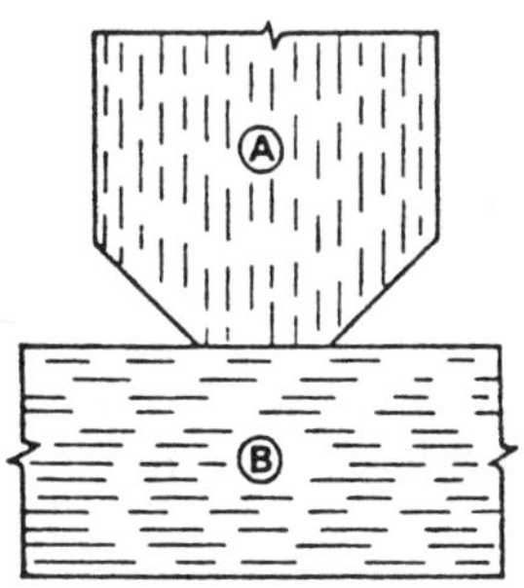

B is much more susceptible to lamellar tearing than A.

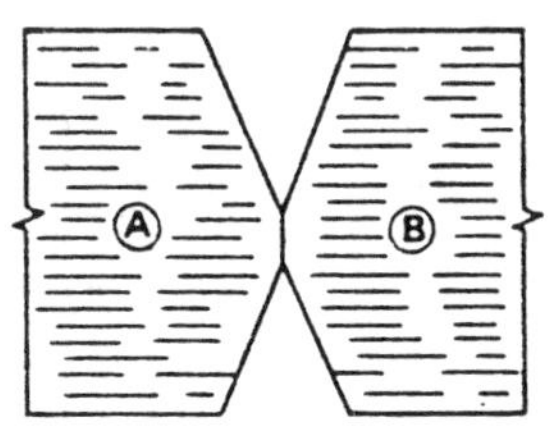

When making butt welds the lamellar orientation is generally parallel with the weld contraction stresses. This condition is not susceptible to lamellar tearing.

Fig. 6-13F.

Recommendations

Successful highly restrained connections require an understanding of the phenomenon and attention to detail on the part of the designer. Good materials and workmanship cannot assure success on poorly conceived and designed or specified connections.

The following are some recommendations presented as representative of possible means to minimize or avoid lamellar tearing:

1. Select electrodes which deposit weld metal with the lowest yield strength adequate to carry design loads. The use of excessively high yield strength weld metal can cause stresses above yield point in the connected material.
2. Where possible, arrange connections so as to avoid welded joints which induce thru-thickness stresses due to weld shrinkage (Fig. 6-13F).
3. Make connections with welds having the minimum throat dimension required to carry the stresses and having a minimum practical volume of weld metal. Lower strength fillet or partial penetration welds may often be used to join higher strength steels when the joint is designed for shear.
4. Design corner joints with proper consideration of edge preparation. See the improved details in Fig. 6-13G.
5. The designer should fully research and utilize available experience and knowledge on specific design details that might be potential sources of lamellar tearing.
6. Do not use larger welds than are necessary to transfer calculated forces.

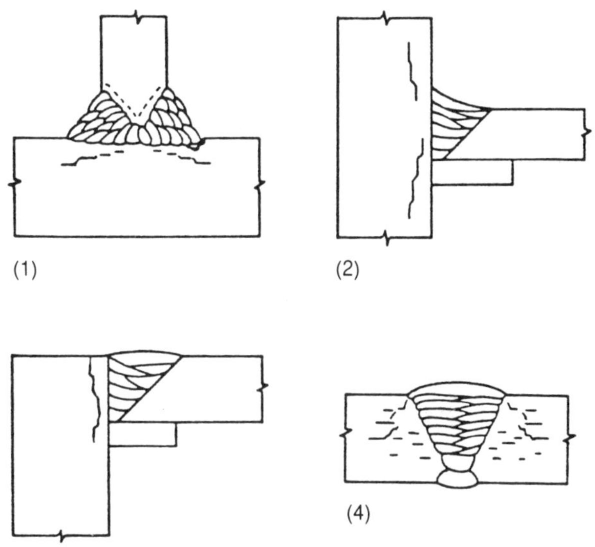

Conditions (1), (2), (3) are susceptible to lamellar tearing. A bevelled butt weld (4) is not.

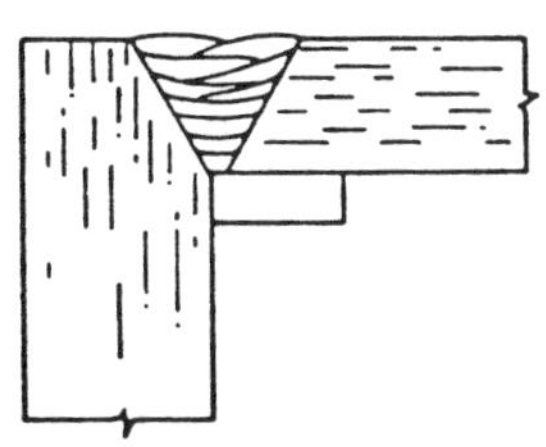

(5)

Bevelling both plates as shown in (5) will greatly minimize the tendency for lamellar tearing.

Fig. 6-13G.

7. Follow the recommended preheat and interpass temperature for the type and thickness of the base metal.
8. The designer should selectively specify ultrasonic inspection *after fabrication and/or erection* of those specific highly restrained welded connections critical to structural integrity that he considers to be subject to lamellar tearing.

Observations on Factors Contributing to Cracking

Two articles[1,2] appearing in the *Welding Journal* in 1964 summarize several of the factors confirmed by research as contributory to weld cracking:

1. The contraction forces of multiple-pass welds tend to cause separations in the base metal and they generally increase with the strength and/or hardenability of the filler metal and base metals. Therefore, softer weld metal would tend to decrease not only weld metal cracks but also heat-affected zone cracks and lamellar tearing.
2. The susceptibility to delayed cracking is proportional to the hydrogen content of the welding atmosphere.
3. Greater crack sensitivity is exhibited by high-chemistry base metal and by heavier plate thicknesses.
4. In general, cracking will initiate in the heat-affected zone of the base metal, except in cases where the weld metal is of higher hardness.
5. With an open-arc or even a shielded-arc manual electrode, it can be assumed that in hot humid weather the arc atmosphere will contain more hydrogen as water vapor than in cool, dry weather. Any tendency to minimize the importance of preheat, of keeping the joint hot, or possibly of postheat in hot summer months, could be at the root of cracking problems on heavy restrained joints. This would be especially true if either the weld metal or the base metal is hardenable because of alloy or carbon content.

[1] *"Welding Cracking Under Hindered Contraction: Comparison of Welding Process,"* Travis, Barry, Moffat, and Adams, MIT. Welding Journal, November, 1964.

[2] *"Delayed Cracking in Steel Weldments,"* Interrante and Stout, Welding Journal, April, 1964.

Low heat input with interruptions in the welding cycle tends to aggravate the problem.

The welding position and its influence on bead size, heat input, number of layers, etc., has a direct influence on the cracking tendency. For example, three-o'clock groove welds are more sensitive to cracking than flat-position groove welds.

STEELS AND WELDING RECOMMENDATIONS

The Carbon Steels

Classification of the carbon steels is based principally on carbon content. The groups are: low-carbon (to 0.30% carbon), medium-carbon (0.30 to 0.45%), and high-carbon (more than 0.45%). The first group is sometimes subdivided into the very-low-carbon steels (to 0.15%) and the mild steels (0.15 to 0.30%). Standard SAE compositions of carbon steels, applicable to structural shapes, plate, strip, sheet, and welded tubing are listed in Table 6-3.

Mechanical properties of hot-finished steels are influenced principally by chemical composition (particularly carbon content), but other factors — finishing temperature, section size, and the presence of residual elements — also affect properties. A 3/4-in. plate, for example, has higher tensile properties and lower elongation than a 1-1/2-in. plate of the same composition. This results from the higher rate of cooling of the 3/4-in. plate from the rolling temperature. Typical tensile properties of hot-rolled and cold-finished low-carbon steels are listed in Table 6-4.

Low-Carbon Steels

In general, steels with carbon contents to 0.30% are readily joined by all common arc-welding processes. These grades account for the greatest tonnage of steels used in welded structures. Typical applications include tanks, structural assemblies, vessels, machine bases. earth-moving and agricultural equipment, and general weldments.

Steels with very low carbon content — to 0.13% — are good welding steels, but they are not the best for high-speed production welding. The low carbon content and the low manganese content (to 0.30%) tend to produce internal porosity. This condition is usually corrected by modifying the welding procedure slightly — usually by using a slower speed. If the presence of some internal porosity has no detrimental effect on service requirements of the assembly standard high-speed welding procedures can be used.

TABLE 6-3. SAE CARBON STEELS
Carbon Steel Compositions

SAE No.	Ladle chemical composition limits, % C	Mn	P, max	S, max	Corresponding AISI No.
1006	0.08 max	0.25-0.40	0.040	0.050	C1006
1008	0.10 max	0.25-0.50	0.040	0.050	C1008
1009	0.15 max	0.60 max	0.040	0.050	C1009
1010	0.08-0.13	0.30-0.60	0.040	0.050	C1010
1012	0.10-0.15	0.30-0.60	0.040	0.050	C1012
1015	0.13-0.18	0.30-0.60	0.040	0.050	C1015
1016	0.13-0.18	0.60-0.90	0.040	0.050	C1016
1017	0.15-0.20	0.30-0.60	0.040	0.050	C1017
1018	0.15-0.20	0.60-0.90	0.040	0.050	C1018
1019	0.15-0.20	0.70-1.00	0.040	0.050	C1019
1020	0.18-0.23	0.30-0.60	0.040	0.050	C1020
1021	0.18-0.23	0.60-0.90	0.040	0.050	C1021
1022	0.18-0.23	0.70-1.00	0.040	0.050	C1022
1023	0.20-0.25	0.30-0.60	0.040	0.050	C1023
1024	0.19-0.25	1.35-1.65	0.040	0.050	C1024
1025	0.22-0.28	0.30-0.60	0.040	0.050	C1025
1026	0.22-0.28	0.60-0.90	0.040	0.050	C1026
1027	0.22-0.29	1.20-1.50	0.040	0.050	C1027
1030	0.28-0.34	0.60-0.90	0.040	0.050	C1030
1033	0.30-0.36	0.70-1.00	0.040	0.050	C1033
1035	0.32-0.38	0.60-0.90	0.040	0.050	C1035
1036	0.30-0.37	1.20-1.50	0.040	0.050	C1036
1037	0.32-0.38	0.70-1.00	0.040	0.050	C1037
1038	0.35-0.42	0.60-0.90	0.040	0.050	C1038
1039	0.37-0.44	0.70-1.00	0.040	0.050	C1039
1040	0.37-0.44	0.60-0.90	0.040	0.050	C1040
1041	0.36-0.44	1.35-1.65	0.040	0.050	C1041
1042	0.40-0.47	0.60-0.90	0.040	0.050	C1042
1043	0.40-0.47	0.70-1.00	0.040	0.050	C1043
1045	0.43-0.50	0.60-0.90	0.040	0.050	C1045
1046	0.43-0.50	0.70-1.00	0.040	0.050	C1046
1048	0.44-0.52	1.10-1.40	0.040	0.050	—
1049	0.46-0.53	0.60-0.90	0.040	0.050	C1049
1050	0.48-0.55	0.60-0.90	0.040	0.050	C1050
1052	0.47-0.55	1.20-1.50	0.040	0.050	C1052
1055	0.50-0.60	0.60-0.90	0.040	0.050	C1055
1060	0.55-0.65	0.60-0.90	0.040	0.050	C1060
1064	0.60-0.70	0.50-0.80	0.040	0.050	C1064
1065	0.60-0.70	0.60-0.90	0.040	0.050	C1065
1070	0.65-0.75	0.60-0.90	0.040	0.050	C1070
1074	0.70-0.80	0.50-0.80	0.040	0.050	C1074
1078	0.72-0.85	0.30-0.60	0.040	0.050	C1078
1080	0.75-0.88	0.60-0.90	0.040	0.050	C1080
1084	0.80-0.93	0.60-0.90	0.040	0.050	C1084
1085	0.80-0.93	0.70-1.00	0.040	0.050	C1085
1086	0.80-0.93	0.30-0.50	0.040	0.050	C1086
1090	0.85-0.98	0.60-0.90	0.040	0.050	C1090
1095	0.90-1.03	0.30-0.50	0.040	0.050	C1095

Steels with very low carbon content are more ductile and easier to form than higher-carbon steels. They are used for applications requiring considerable cold forming, such as stampings or rolled or formed shapes.

TABLE 6-4. Typical Minimum Mechanical Properties of Carbon-Steel Bars

AISI or SAE No.	Condition*	Tensile Strength (1000 psi)	Yield Strength (1000 psi)	Elongation in 2 in. (%)
1010	HR	47	26	28
	CF	53	44	20
1015	HR	50	28	28
	CF	56	47	18
1020	HR	55	30	25
	CF	61	51	15
1025	HR	58	32	25
	CF	64	54	15
1030	HR	68	38	20
	CF	76	64	12
1035	HR	72	40	18
	CF	80	67	12
1040	HR	76	42	18
	CF	85	71	12
1045	HR	82	45	16
	CF	91	77	12
1050	HR	90	50	15
	CF	100	84	10

* HR = hot rolled; CF = cold finished.
Data from ASM Metals Handbook, 8th Ed., Vol. 1.

Steels with 0.15 to 0.20% carbon content have excellent weldability. They seldom require anything beyond standard welding procedures, and they can be welded with all types of mild-steel electrodes. These steels should be used for maximum production speed on assemblies or structures that require extensive welding.

Steels at the upper end of the low-carbon range — the 0.25 to 0.30% carbon grades — have very good weldability, but when one or more of the elements is on the high side of permissible limits, cracking can result, particularly in fillet welds. With slightly reduced speeds and currents, any of the standard electrodes can be used for these steels. In thicknesses to 5/16 in., standard procedures apply.

If some of the elements — particularly carbon, silicon, or sulfur — are on the high side of the limits, surface holes may form. Reducing current and speed minimizes this problem.

Although most welding applications of these steels require no preheating, heavy sections (2-in. or more) and certain joint configurations may require a preheat. Less preheating is required when low-hydrogen processes are used. In general, steels in the 0.25 to 0.30% carbon range should be welded with low-hydrogen electrodes or with a low-hydrogen process if the temperature is below 50°F.

Medium and High-Carbon Steels

Because hardenability of steel increases with carbon content, the medium and high-carbon steels serve where hardness, wear resistance, or higher strength are needed. Important uses for medium-carbon steels (to 0.45%) include wear plates, springs, and components for railroad, agricultural, and earth-moving and materials-handling equipment.

Unfortunately, the same characteristics that make these steels so suitable for use in rugged parts and structures make them more difficult and costly to weld. The medium-carbon steels can be welded successfully, however, provided proper procedures and preheat and interpass temperatures are used. Sometimes, postweld stress relief may be required.

The high-carbon steels are almost always used in a hardened condition. Typical applications are for metalworking and woodworking tools, drills, dies, and knives, and for abrasion-resistant parts such as plowshares and scraper blades. Some farm equipment is built from rerolled rail stock (0.65% C), which is welded in the as-rolled condition, using preheating, interpass heating, and postweld stress relief.

Hardness of these steels can range from dead soft in the annealed condition to Rockwell C 65 (with rapid quench treatment) for the higher-carbon grades. Although an AISI 1020 steel can be made as hard as Rc 50, hardness is very shallow. Increased carbon content increases depth of hardening and maximum attainable hardness to about Rc 65. Alloying elements increase depth of hardening but have little effect on maximum hardness.

It is advisable to make sample weld tests to determine cracking tendencies of steels containing 0.30% or more carbon. If such tendencies are apparent, preheating of the steel may be necessary to retard the cooling rate from the welding temperature. Required preheat temperature varies with analysis, size, and shape of the steel and with the amount of heat input from the welding process. In general, the higher the carbon or alloy content and the thicker the plate, the higher the preheat temperature needed to provide the slow cooling rate required to prevent hardening. For shop calculation, a Preheat Calculator — available from The Lincoln Electric Company at a nominal cost — is a handy tool for determining preheat requirements of various thicknesses of common-analysis steels. (See Section 3.3.)

Use of low-hydrogen processes can minimize the degree of preheating necessary and, may eliminate the need for preheating entirely. As a rule of thumb, preheat temperatures used with low-hydrogen electrodes can be 100 to 200°F lower than those needed for electrodes other than low-hydrogen.

AWS Structural Steels

The American Welding Society does not write specifications for structural steel but does recognize many steels specified by ASTM, API, and ABS as suitable for welded structures with the various arc welding processes. Tables 6-5 and 6-6 show lists of these steels with the mechanical property requirements and the compositions. Since the table does not contain the complete mechanical property or chemical requirements it is suggested the reader consult the original specification for further information.

In general, these steels have maximum limits on carbon, sulfur and phosphorous. Manganese may be specified as a range or in a maximum amount. Small amounts of other alloys may be added in order to meet the mechanical property requirements. All the steels listed in Tables 6-5 and 6-6 have satisfactory weldability characteristics but some may require special procedures or techniques, such as limited heat input or minimum preheat and interpass temperatures. Some structural steels are not intended for arc welding. For example, A440 is intended primarily for riveted or bolted structures.

High-Strength Low-Alloy Structural Steels

Higher mechanical properties and, usually, better corrosion resistance than the structural carbon steels are characteristics of the high-strength low-alloy (HSLA) steels. These improved properties are achieved by additions of small amounts of alloying elements. Some of the HSLA types are carbon-manganese steels; others contain different alloy additions, governed by requirements for weldability, formability, toughness, or economy. Strength of these steels is between those of structural carbon steels and the high-strength quenched-and-tempered steels.

High-strength low-alloy steels are usually used in the as-rolled condition, although some are available that require heat treatment after fabrication. These steels are produced to specific mechanical-property requirements rather than to chemical compositions. Minimum mechanical properties available in the as-rolled condition vary among the grades and, within most grades, with thickness. Ranges of properties available in this group of steels are:

1. Minimum yield point from 42,000 to 70,000 psi.

TABLE 6-5. Minimum Mechanical Properties and Composition for ASTM HSLA Steels Approved for Use by AISC Specifications for the Design, Fabrication and Erection of Structural Steel for Buildings and Bridges, AWS D1.1 – Structural Welding Code – Steel 1992

ASTM GRADE and Descriptive Information	Mechanical Properties			Material Shape	Thickness Group or Grade	Chemical Requirements (Ladle) Percent						
	Tensile Strength† (ksi min)	Yield Point (ksi min)	Elongation in 2 in. % min			C Max	Mn† Max	P Max	S Max	Si†	Cu Min	V Min
A36 Structural Steel	58 to 80	36	23	Shapes		0.26		0.04	0.05	0.40 m	0.20*	
	58 to 80	36	23	Plate	to 3/4″ incl.	0.25		0.04	0.05	0.40 m	0.20*	
	58 to 80	36	23	Plate	over 3/4 to 1-1/2″ incl.	0.25	0.80 – 1.20	0.04	0.05	0.40 m	0.20*	
	58 to 80	36	23	Plate	over 1-1/2 to 2-1/2″ incl.	0.26	0.80 – 1.20	0.04	0.05	0.15 – 0.40	0.20*	
	58 to 80	36	23	Plate	over 2-1/2 to 4″ incl.	0.27	0.85 – 1.20	0.04	0.05	0.15 – 0.40	0.20*	
	58 to 80	36	23	Plate	over 4 to 8″ incl.	0.29	0.85 – 1.20	0.04	0.05	0.15 – 0.40	0.20*	
	58 to 80	36	23	Bars & Bar Shapes	to 3/4″ incl.	0.26		0.04	0.05	0.40 m	0.20*	
	58 to 80	36	23	Bars & Bar Shapes	over 3/4 to 1-1/2″ incl.	0.27	0.60 – 0.90	0.04	0.05	0.40 m	0.20*	
	58 to 80	36	23	Bars & Bar Shapes	over 1-1/2 to 4″ incl.	0.28	0.60 – 0.90	0.04	0.05	0.40 m	0.20*	
	58 to 80	36	23	Bars & Bar Shapes	over 4″	0.29	0.60 – 0.90	0.04	0.05	0.40 m	0.20*	
A53 GRADE B Welded & Seamless Steel Pipe electric resistance or seamless (only chemistry limit is phosphorous)	48	30	**		A	0.25	0.95	0.05	0.06			
	60	35	**		B	0.30	1.20	0.05	0.06			
	45	25	**		F			0.06	0.045			
A242 High Strength Low Alloy Structural Steel	70	50	21	Plates & Bars	to 3/4″ incl.	0.15	1.00	0.15	0.05		0.20	
	67	46	21	Plates & Bars	over 3/4 to 1-1/2″ incl.	0.15	1.00	0.15	0.05		0.20	
(other alloying elements may be added.)	63	42	21	Plates & Bars	over 1-1/2 to 4″ incl.	0.15	1.00	0.15	0.05		0.20	
If Cmax ≤ 0.15, Mn_{max} can be raised to 1.40	70	50	21	Struc. Shapes	1 & 2	0.15	1.00	0.15	0.05		0.20	
	67	46	21	Struc. Shapes	3	0.15	1.00	0.15	0.05		0.20	
	63	42	21	Struc. Shapes	4 & 5	0.15	1.00	0.15	0.05		0.20	
A375 High Strength, Low Alloy Hot Rolled Steel Sheet Strip	70	50	22			0.22	1.25		0.05			
When used for welding, the chemistry shall be checked for weldability based on evidence acceptable to the buyer.												
A441 High Strength Low Alloy Structural Manganese Vanadium Steel	70	50	**	Plates & Bars	to 3/4″ incl.	0.22	.85-1.25	0.04	0.05	0.30	0.20	0.02
	67	46	**	Plates & Bars	over 3/4″ 1-1/2″ incl.	0.22	.85-1.25	0.04	0.05	0.30	0.20	0.02
Replaced by A572/A572M (1989)	63	42	24	Plates & Bars	over 1-1/2″ to 4″ incl.	0.22	.85-1.25	0.04	0.05	0.30	0.20	0.02
	60	40	24	Plates & Bars	over 4″ to 8″ incl.	0.22	.85-1.25	0.04	0.05	0.30	0.20	0.02
	70	50	**	Struc. Shapes	I	0.22	.85-1.25	0.04	0.05	0.30	0.20	0.02
	67	46	**	Struc. Shapes	II	0.22	.85-1.25	0.04	0.05	0.30	0.20	0.02
	63	42	24	Struc. Shapes	III	0.22	.85-1.25	0.04	0.05	0.30	0.20	0.02

* When Specified ** See ASTM Standards for details † Where two figures are given this is a min-max range.

TABLE 6-5. Minimum Mechanical Properties and Composition for ASTM HSLA Steels Approved for Use by AISC Specifications for the Design, Fabrication and Erection of Structural Steel for Buildings and Bridges, AWS D1.1 – Structural Welding Code – Steel 1992 (continued)

ASTM GRADE and Descriptive Information	Mechanical Properties			Material Shape	Thickness Group or Grade	Chemical Requirements (Ladle) Percent						
	Tensile Strength† (ksi min)	Yield Point (ksi min)	Elongation in 2 in. % min			C Max	Mn† Max	P Max	S Max	Si†	Cu Min	V Min
A500 Cold-Formed Welded and Seamless Carbon Steel Structural Tubing in Rounds and Shapes	45	33	25	Round Structural Tubing	A	0.26		0.04	0.05		0.20*	
	58	42	23		B	0.26		0.04	0.05		0.20*	
	62	46	21		C	0.23	1.35m				0.20*	
	58	36	23		D	0.26		0.04	0.05		0.20*	
	45	39	25	Shaped Struc. Tub.	A	0.26		0.04	0.05		0.20*	
	58	46	23		B	0.26		0.04	0.05		0.20*	
	62	50	21		C	0.23	1.35m				0.20*	
	58	36	23		D	0.26		0.04	0.05		0.20*	
A501 Hot-Formed Welded and Seamless Carbon Steel Structural Tubing.	58 min.	36	23			0.26		0.04	0.05		0.20*	
A516 Carbon Steel Plates	55-75	30	27	Plate	55	0.18 0.26	0.60 1.20	0.035	0.035	0.15 0.40		
	60-80	32	25		60	0.21 0.27	0.60 1.20	0.035	0.035	0.15 0.40		
	65-85	35	23		65	0.24 0.29	0.85 1.20	0.035	0.035	0.15 0.40		
	70-90	38	21		70	0.27 0.31	0.85 1.20	0.035	0.035	0.15 0.40		
A529 Structural Steel (42 ksi min yield 1/2 in. max thickness)	60 to 85	42	22			0.27	1.20	0.04	0.05		0.20*	
A570 Hot Rolled Carbon Steel Sheets and Strip, Structure Quality	49	30	19-25		30	0.25	0.90	0.035	0.04		0.20	
	52	33	18-23		33	0.25	0.90	0.035	0.04		0.20	
	53	36	17-22		36	0.25	0.90	0.035	0.04		0.20	
	55	40	16-21		40	0.25	0.90	0.035	0.04		0.20	
	60	45	14-19		45	0.25	1.35	0.035	0.04		0.20	
	65	50	12-17		50	0.25	1.35	0.035	0.04		0.20	
	70	55	10-15		55	0.25	1.35	0.035	0.04		0.20	
A606 High-Strength Low Alloy Sheet and Strip	70	50	22	lengths	as-rolled	0.22	1.25		0.04			
	65	45	22	coils	as-rolled							
	65	45	22	lengths & coils	annealed							
	65	45	22	lengths & coils	cold rolled							
A607 High-Strength Low Alloy Sheet and Strip										Cb	N	
	60 55	45 45	22-25 22-25	Class 1 Class 2	45	0.22 0.15	1.35	0.04	0.04	0.005		0.01
	65 60	50 50	20-22 20-22	Class 1 Class 2	50	0.23 0.15	1.35	0.04	0.05	0.005		0.01
	70 65	55 55	18-20 18-20	Class 1 Class 2	55	0.25 0.15	1.35	0.04	0.05	0.005		0.01
	75 70	60 60	16-18 16-18	Class 1 Class 2	60	0.26 0.15	1.50	0.04	0.05	0.005	 0.020	0.01
	80 75	65 65	14-16 14-16	Class 1 Class 2	65	0.26 0.15	1.50	0.04	0.05	0.005	0.012 0.020	0.01
	85 80	70 70	12-14 12-14	Class 1 Class 2	70	0.26 0.15	1.65	0.04	0.05	0.005	0.012 0.020	0.01

* When Specified

** See ASTM Standards for details

† Where two figures are given this is a min-max range.

TABLE 6-6. Minimum Mechanical Properties and Composition for ASTM HSLA Steels Approved for Use By AISC specification for the Design, Fabrication and Erection of Structural Steel for Buildings and Bridges – AWS-D1.1 Structural Welding Code-Steel 1992

ASTM GRADE and Descriptive Information	Material Shape	Thickness or Group	Mechanical Properties: Tensile Strength† (ksi min)	Mechanical Properties: Yield Point (ksi min)	Mechanical Properties: Elonga. in 2 in. % min	Grade	Chemical Requirements (Ladle) Percent: C† Max	Mn† Max	P Max	S Max	Si Max	Ni† Max	Cr† Max	Mo† Max	Cu† Max	V† Max	Cb Max	Ti† Max	Others Zr
A514 High-Yield Strength Quenched and Tempered Alloy Steel Plate, for Welding	Plate All Grades	to 3/4″ incl.	110 to 130	100	18	A	0.15 0.21	0.80 1.10	0.035	0.035	0.40 0.80		0.50 0.80	0.18 0.28					0.05 0.15
		over 3/4″ to 2-1/2″ incl.	110 to 130	100	18	B	0.12 0.21	0.70 1.00	0.035	0.035	0.40 0.80		0.40 0.65	0.15 0.25		0.03 0.08		0.01 0.03	
		over 2-1/2″ to 4″ incl.	100 to 130	90	16	C	0.10 0.20	1.10 1.50	0.035	0.035	0.15 0.30			0.15 0.30					
						E	0.12 0.20	0.40 0.70	0.035	0.035	0.20 0.40		1.40 2.00	0.40 0.60				0.01 0.10	
						F	0.10 0.20	0.60 1.00	0.035	0.035	0.15 0.35	0.70 1.00	0.40 0.65	0.40 0.60	0.15 0.50	0.03 0.08			
						H	0.12 0.21	0.95 1.30	0.035	0.035	0.20 0.35	0.30 0.70	0.40 0.65	0.20 0.30		0.03 0.08			
						J	0.12 0.21	0.45 0.70	0.035	0.035	0.20 0.35			0.50 0.65					
						M	0.12 0.21	0.45 0.70	0.035	0.035	0.20 0.35	1.20 1.50		0.45 0.60					
These steels contain from 0.0005 to 0.005% B						P	0.12 0.21	0.45 0.70	0.035	0.035	0.20 0.35	1.20 1.50	0.85 1.20	0.45 0.60					
						Q	0.14 0.21	0.95 1.30	0.035	0.035	0.15 0.35	1.20 1.50	1.00 1.50	0.40 0.60		0.03 0.08			
						R	0.15 0.20	0.85 1.15	0.035	0.035	0.20 0.35	0.90 1.10	0.35 0.65	0.15 0.25		0.03 0.08			
						S	0.10 0.20	1.10 1.50	0.035	0.035	0.15 0.35			0.10 0.35			0.06 m.	0.06 m.	
						T	0.08 0.14	1.20 1.50	0.035	0.035	0.40 0.60			0.45 0.60		0.03 0.08			
A517 High Strength Quenched and Tempered Alloy Steel Plates	Plate All Grades	Under 2.5″	115-135	100	16	B	0.15 0.21				0.15 0.35								
						C								0.20 0.30					
		over 2.5″ to 6″	105-135	90	14	E					0.10 0.40								
						H					0.15 0.35								
	Analysis same as A514 except as noted here.					S					0.15 0.40								N
A572 High-Strength Low-Alloy, Columbium-Vanadium Steels of Structural Quality	Shapes and Plates		60 min	42	24	42	0.21	1.35	0.04	0.05	(5)					(2)	(1)(3)		(4)
			65 min	50	21	50	0.23	1.35	0.04	0.05	(5)					(2)	(1)(3)		(4)
			75 min	60	18	60	0.26	1.35	0.04	0.05	(5)					(2)	(1)(3)		(4)
			80 min	65	17	65	0.26	1.35	0.04	0.05	(5)					(2)	(1)(3)		(4)

Alloy content shall be in accordance with one of the following —

- (1) Cb 0.005-0.05
- (2) V 0.01-0.15
- (3) Cb (.05 max)+V 0.02-0.15
- (4) N (with V) . . . . 0.015 max

(4) N (.015 max) when added as a supplement to V shall be reported, and the minimumratio of V to N shall be 4 to 1.

(1)(3) Cb when added either singly or in combination with V unless combined with 0.15 min Si shall be restricted to plate or bar thickness of 1/2″ max, and to shapes of Table A, Group 1 of ASTM Spec. A6.

(5) Si (0.30max) for shapes and plates to 1-1/2″; for grade 42 plates over 1-1/2″ (0.15 – 0.30 max).

TABLE 6-6. Minimum Mechanical Properties and Composition for ASTM HSLA Steels Approved for Use By AISC specification for the Design, Fabrication and Erection of Structural Steel for Buildings and Bridges – AWS-D1.1 Structural Welding Code-Steel 1992 (continued)

ASTM GRADE and Descriptive Information	Material Shape	Thickness or Group	Mechanical Properties: Tensile Strength† (ksi min)	Mechanical Properties: Yield Point (ksi min)	Mechanical Properties: Elonga. in 2 in. % min	Grade	Chemical Requirements (Ladle) Percent: C† Max	Mn† Max	P Max	S Max	Si Max	Ni† Max	Cr† Max	Mo† Max	Cu† Max	V† Max	Cb Max	Ti† Max	Others N
A588 High-Strength Low-Alloy Structural Steel with 50 ksi min. Yield Point to 4 inch Thickness	Plates and Bars	to 4″ incl.	70	50	21	A	0.10- 0.19	0.90- 1.25	0.05	0.05	0.15- 0.30		0.40- 0.63		0.25- 0.40	0.02- 0.10			
	Plates and Bars	over 4″ to 5″ incl.	67	46	21	B	0.10- 0.20	0.75- 1.25	0.04	0.05	0.15- 0.30	0.25- 0.50	0.40- 0.70		0.20- 0.40	0.01- 0.10			
	Plates and Bars	over 5″ to 8″ incl.	63	42	21	C	0.15 max	0.80- 1.35	0.04	0.05	0.15- 0.30	0.25- 0.50	0.30- 0.50		0.20- 0.50	0.01- 0.10			
	Shapes	All Groups	70	50	25	K	0.17 m.	0.50 1.20	0.04	0.05	0.25 0.50	0.40	0.40 0.70	0.10	0.30 0.50		0.005 0.05		
A618 Hot-Formed, Welded and Seamless Low-Alloy Tubing	Tubing Ia Ib II	Walls 3/4″ & Less	70	50	22	Ia	0.15	1.00	0.15	0.05					0.20				
						Ib	0.20	1.35	0.04	0.05					0.20				
		Walls 3/4 to 1-1/2″	67	46	22	II	0.22	0.85 1.25	0.04	0.05	0.30				0.20	0.02			
	Tubing III		65	50	20	III	0.23	1.35	0.04	0.05	0.30					0.02			
A633 Normalized High-Strength Low-Alloy Steel	Plate		63-83	42	23	A	0.18	1.00 1.35	.035	.04	0.15 0.50						0.05		
	Plate	Under 2.5″	70-90	50	23	C	0.20	1.15 1.50	.035	.04	0.15 0.50						0.01 0.05		
	Plate	2.5 to 4″	65-85	46	23	D	0.20	0.70 1.60	.035	.04	0.15 0.50	0.25	0.25	0.08	0.35				N
	Plate	Under 4″	80-100	60	23	E	0.22		.035	.04	0.15 0.50					0.04 0.11			0.01 0.03
	Plate	4-6″	75-95	55	23	E													
A709 Carbon and High Strength Low Alloy Steel Structural Shapes Plates and Bars. Includes quenched and tempered plate.		to 4″	58-80	36	23	36	0.25 0.28	0.60 1.20	0.04	0.05	0.15 0.40				0.20				
		to 4″	65	50	21	50	0.23	1.35	0.04	0.05	0.15 0.40				0.20				
		to 4″	70	50	21	50W	0.15 0.40	0.75 1.35	0.04	0.05	0.15 0.40		0.30 0.70		0.20 0.40				
		to 4″	90-110	70	19	70W	0.19	0.80 1.35	0.04	0.05	0.20 0.65	0.50 m	0.40 0.70		0.20 0.40	0.02 0.10			
		to 2-1/2″	110-130	100	18	100 100W	Compositions of Types A, B, C, E, F, H, J, M, P and Q are equivalent to corresponding types for A514 ASTM-A709 should be consulted for details.												
		2-1/2″ to 4″	100-130	90	16	100 100W													
A710 Low Carbon Age Hardening Ni-Cu-Cr-Mo-Cb, Ni-Cu-Cb, Ni-Cu-Mn-Mo-Cb Steels for Plate	Plate	1/4 to 4″	90	80 85	20	A C/1	0.07	0.40 0.70	.025	.025	0.40	0.70 1.00	0.60 0.90	0.15 0.25	1.00 1.30		0.02 min		
	Plate		60 72	50 65	20	A C/2	0.07	0.40 0.70	.025	.025	0.40	0.70 1.00	0.60 0.90	0.15 0.25	1.00 1.30		0.02 min		
	Plate		70 85	60 75	20	A C/3	0.07	0.40 0.70	.025	.025	0.40	0.70 1.00	0.60 0.90	0.15 0.25	1.00 1.30		0.02 min		
	Plate		88 90	75 85	18	B	0.06	0.40 0.65	.025	.025	0.15 0.40	1.20 1.50			1.00 1.30		0.02 min		
	Plate		100	90	20	C C/1	0.07	1.30 1.65	.025	.025	0.40	0.70 1.00		0.15 0.25	1.00 1.30		0.02 min		
	Plate		90 95	80 85	20	C/3	0.07	1.30 1.65	.025	.025	0.40	0.70 1.00		0.15 0.25	1.00 1.30		0.02 min		
A808 Hot-Rolled High Strength Low Alloy Steel Plate with Improved Notch Toughness	Plate	to 1-1/2″	65	50	22		0.12	1.65	.035	.04	0.05 0.50					0.10 max	0.02 0.10		
	Plate	1-1/2 to 2″	65	46	22		0.12	1.65	.035	.04	0.15 0.50					0.10 max	0.02 0.10		
	Plate	2 to 2-1/2″	60	42	22		0.12	1.65	.035	.04	0.15 0.50					0.10 max	0.02 0.10		

2. Minimum tensile strength from 60,000 to 85,000 psi.
3. Resistance to corrosion, classed as: equal to that of carbon steels, twice that of carbon steels, or four to six times that of carbon steels.

The HSLA steels are available in most commercial wrought forms and are used extensively in products and structures that require higher strength-to-weight ratios than the carbon structural steels offer. Typical applications are supports and panels for truck bodies, railway cars, mobile homes, and other transportation equipment; components for tractors, threshers, fertilizer spreaders, and other agricultural machinery, materials-handling and storage equipment; and buildings, bridge decks, and similar structures.

The high-strength low-alloy steels should not be confused with the high-strength quenched-and-tempered alloy steels. Both groups are sold primarily on a trade-name basis, and they frequently share the same trade-name, with different letters or numbers being used to identify each. The quenched-and-tempered steels are full-alloy steels that are heat-treated at the mill to develop optimum properties. They are generally martensitic in structure, whereas the HSLA steels are mainly ferritic steels; this is the clue to the metallurgical and fabricating differences between the two types. In the as-rolled condition, ferritic steels are composed of relatively soft, ductile constituents; martensitic steels have hard, brittle constituents that require heat treatment to produce their high-strength properties.

Strength in the HSLA steels is achieved instead by relatively small amounts of alloying elements dissolved in a ferritic structure. Carbon content rarely exceeds 0.28% and is usually between 0.15 and 0.22%. Manganese content ranges from 0.85 to 1.60%, depending on grade, and other alloy additions — chromium, nickel, silicon, phosphorus, copper, vanadium, columbium, and nitrogen — are used in amounts less than one percent. Welding, forming, and machining characteristics of most grades do not differ markedly from those of the low-carbon steels.

To be weldable, the high-strength steels must have enough ductility to avoid cracking from the rapid cooling inherent in welding processes. Weldable HSLA steels must be sufficiently low in carbon, manganese, and all "deep-hardening" elements to ensure that appreciable amounts of martensite are not formed upon rapid cooling. Superior strength is provided by solution of the alloying elements in the ferrite of the as-rolled steel. Corrosion resistance is also increased in certain of the HSLA steels by the alloying additions.

Addition of a minimum of 0.20% copper usually produces steels with about twice the atmospheric corrosion resistance of structural carbon steels. Steels with four to six times the atmospheric corrosion resistance of structural carbon steels are obtained in many ways, but, typically, with additions of nickel and/or chromium, often with more than 0.10% phosphorus. These alloys are usually used in addition to the copper.

Standard specifications or recommended practices covering the major types of HSLA steels are available from the American Society for Testing and Materials, the Society of Automotive Engineers, and the Department of Defense.

Other standardizing organizations such as the American Institute of Steel Construction, The American Association of Railroads, and the Department of Transportation have established specifications or practices for the use of HSLA steels in certain industries and applications.

ASTM's specifications are oriented principally to mill form and mechanical properties; SAE's recommended practices include, in addition, information on fabrication characteristics — toughness, weldability, and formability.

ASTM Specifications

Fourteen ASTM specifications cover the plain carbon, high-strength low-alloy, and quenched and tempered structural steels. They are: A36, A53, A242, A440, A441, A514, A517, A572, A588, A606, A607, A618, A633, A709, A710 and A809. Tables 6-5 and 6-6 list the composition and mechanical properties of these steels. Specifications A374 and A375 cover similar steels in sheet and strip form.

ASTM A36 covers carbon steel shapes, plates and bars of structural quality for use in riveted, bolted or welded construction of bridges and buildings and for general structural purposes. Strength requirements are 58 to 80 ksi tensile and 36 ksi min. yield strengths.

ASTM A53 covers seamless and welded black and hot-dipped galvanized steel pipe in NPS 1/8 to 26 and wall thicknesses 0.068 in. to 2.344 in. Strength requirements range from 25 to 35 ksi min. yield and 45 to 60 ksi min. tensile for three types and two grades.

ASTM A242 covers HSLA structural steel shapes, plates, and bars for welded, riveted, or bolted construction. Maximum carbon content of these steels is 0.24%; typical content is from 0.09 to 0.17%. Materials produced to this specification are intended primarily for structural members where light weight and durability are important.

Some producers can supply copper-bearing steels (0.20% minimum copper) with about twice the atmospheric corrosion resistance of carbon steels. Steels meeting the general requirements of ASTM A242 but modified to give four times the atmospheric corrosion resistance of structural steels are also available. These latter grades — sometimes called "weathering steels" — are used for architectural and other structural purposes where it is desirable to avoid painting for either esthetic or economic reasons.

Welding characteristics vary according to the type of steel; producers can recommend the most weldable material and offer welding advice if the conditions under which the welding will be done are known.

ASTM A374 covers cold rolled sheets and strip with 45 ksi min. yield strength. Similar in many respoects to A-242.

ASTM A375 covers hot rolled sheets and strip with 50 ksi min. yield strength. Similar in many respects to A242.

ASTM A440 covers high-strength intermediate-manganese copper-bearing HSLA steels used principally for riveted or bolted structures. These steels are not generally recommended for welding because of their relatively high carbon and manganese contents. ASTM A440 and its companion, A441, have the same minimum mechanical properties as A242.

ASTM A440 steels have about twice the atmospheric corrosion resistance of structural carbon steel and very good abrasion resistance. The high manganese content (typically, about 1.45%) tends to cause weld metal to air harden — a condition that may produce high stresses and cracks in the weld. If these steels must be welded, careful preheating (higher than for A441) is necessary.

ASTM A441 covers the intermediate-manganese HSLA steels that are readily weldable with proper procedures. The specification calls for additions of vanadium and a lower manganese content (1.25% maximum) than ASTM A440. Minimum mechanical properties are the same as A242 and A440 steels, except that plates and bars from 4 to 8-in. thick are covered in A441.

Atmospheric corrosion resistance of this steel is approximately twice that of structural carbon steel. Another property of ASTM A441 steel is its superior toughness at low temperatures. Only shapes, plates, and bars are covered by the specification, but weldable sheets and strip can be supplied by some producers with approximately the same minimum mechanical properties.

ASTM A500 covers cold-formed welded and seamless carbon steel round, square, rectangular or special shape structural tubing for welded, riveted or bolted construction of bridges and buildings and for general structural purposes. This tubing is produced in both seamless and welded sizes with a maximum periphery of 64 in. and a maximum wall thickness of 0.625 in. Minimum strength properties range from 33 ksi yield and 45 ksi tensile for Grade A to 46 ksi yield and 62 ksi tensile for Grade C.

ASTM A501 covers hot-formed welded and seamless carbon steel square, round, rectangular or special shape structural tubing for welded, rivited or bolted construciton of bridges and buildings and for general structural purposes. Square and rectangular tubing may be furnished in sizes from 1 in. to 10 in. across flat sides and wall thicknesses from 0.095 to 1.00 in. Minimum strength requirements are 36 ksi yield and 58 ksi tensile.

ASTM A514 covers quenched and tempered alloy steel plates of structural quality in thicknesses of 6 in. and under intended primarily for use in welded bridges and other structures. Strength requirements range from 100 to 130 ksi tensile and 90-100 ksi yield.

ASTM A516 covers carbon steel plates intended primarily for service in welded pressure vessels where improved notch toughness is important. These plates are furnished in four grades with strength requirements ranging form 55 to 90 ksi tensile and 30 to 38 ksi yield.

ASTM A517 covers high strength quenched and tempered alloy steel plates intended for use in fusion welded boilers and other pressure vessels. Plates are available in 12 grades ranging in thickness from 1.25 in. to 6 in. with strength requirements of 115-135 ksi tensile and 100 ksi minimum yield.

ASTM A529 covers carbon steel plates and bars under 1/2 in. in thickness or diameter and Group 1 shapes shown in Table A of ASTM Specification A6/A6M of structural quality for use in metal building system frames, trusses and related structures for riveted, bolted, or welded construciton. Strength requirements are 60 to 85 ksi tensile and 42 min. ksi yield.

ASTM A570 covers hot-rolled carbon steel sheet and strip of structural quality in cut lengths or coils. Available in thicknesses up to 0.229 in. and seven grades. Strength minimum requirements range from 49 ksi tensile and 30 ksi yield for Grade 30 to 70 ksi tensile and 55 ksi yield for Grade 55.

ASTM A572 includes six grades of high-strength low-alloy structural steels in shapes, plates, and bars. These steels offer a choice of strength levels ranging from 42 to 65 ksi yields (Table 6-6). Proprietary HSLA steels of this type with 70 and 75 ksi yield points are also available. Increasing care is required for welding these steels as strength level increases.

A572 steels are distinguished from other HSLA steels by their columbium, vanadium, and nitrogen content. Copper additions above a minimum of 0.20% may be specified for atmospheric corrosion resistance about double that of structural carbon steels.

A supplementary requirement is included in the specification that permits designating the specific alloying elements required in the steel. Examples are the Type 1 designation, for columbium; Type 2, for vanadium; Type 3, for columbium and vanadium; and Type 4, for vanadium and nitrogen. Specific grade designations must accompany this type of requirement.

ASTM A588 provides for a steel similar in most respects to A242 weathering steel, except that the 50 ksi yield point is available in thicknesses to at least 4 in.

ASTM A606 covers high-strength, low-alloy, hot- and cold-rolled sheet and strip in cut lengths or coils intended for use in structural and miscellaneous purposes where savings in weight and durability are important. They have enhanced atmospheric corrosion resistance and are supplied in two types: Type 2 contains 0.02 minimum copper, Type 4 provides a level of corrosion resistance substantially better than that of carbon steel with or without copper.

ASTM A607 covers high-strength, low-alloy columbium or vanadium hot-rolled sheet and strip, or cold-rolled sheet or combinations thereof, in cut lengths or coils for applications where greater strength and savings in weight are important. Steels are available in two classes similar in strength level. Class 2 offers improved weldability and formability over Class 1. When copper is specified, atmospheric corrosion resistance is twice that of plain carbon steel.

ASTM A618 covers grades of hot-formed, welded and seamless high-strength, low-alloy square, rectangular, round, or special shape structural tubing for welded, riveted or bolted construction of bridges and buildings and for general structural purposes. Grades Ia and Ib have substantially better atmospheric corrosion resistance than Grade II.

ASTM A633 covers normalized high-strength, low-alloy structural steel plates for welded, riveted or bolted construction. It is suited for use at temperatures to −50°F and at higher temperatures where higher than normal notch toughness is expected. Four Grades, A, C, D and E, cover a yield strength range of 42 to 60 ksi.

ASTM A709 covers carbon and high-strength, low-alloy steel structural shapes, plates and bars, and quenched and tempered alloy steel for structural plates intended for use in bridges. Six grades are available in four yield strength levels of 36, 50, 70 and 100 ksi. Grades 50W, 70W and 100W have enhanced atmospheric corrosion resistance.

ASTM A710 covers low-carbon age-hardening nickel-copper-chromium-molybdenum-columbium, nickel-copper-columbium and nickel-copper-manganese-molybdenum-columbium alloy steel plates for general applications. Three different grades and three different conditions provide minimum yield strengths from 50 to 90 ksi.

ASTM A808 covers hot-rolled, high-strength low alloy steel plate with improved notch toughness in the as-rolled condition. Maximum thickness is 2-1/2 in. Strength property requirements range from 60 to 65 ksi min. tensile and 42 to 50 ksi min. yield.

SAE Specifications

High-strength low-alloy steels are also covered in the SAE Recommended Practice J410b. This is not a standard. Rather, it is a recommended practice — a guide or memorandum from SAE to its members to help standardize their engineering practices. SAE J410b was written long before most of the HSLA steels had ASTM specifications. Its content is more general than the ASTM documents, and its intent is to guide material selection in the light of fabrication requirements. Now that ASTM has defined almost all of the HSLA steels in standard specifications, SAE J410b is seldom used as a material specification. But the SAE document is still valuable as a general guide to using the HSLA steels.

The chemical compositions of SAE alloy steels, as contained in SAE standard J404g, are listed in Table 6-7, along with the corresponding AISI numbers.

The SAE document addresses itself primarily to the specific needs of fabricators of automobiles, trucks, trailers, agricultural equipment, and aircraft This is why SAE J410b does not cover the thicker plates and heavier structural shapes. Minimum mechanical properties of commonly used steels covered by SAE J410b are listed in Table 6-8.

TABLE 6-7. SAE Alloy Steel Compositions (SAE J404g)

SAE No.	Composition,* %								Corresponding AISI No.
	C	Mn	P	S	Si	Ni	Cr	Other	
1330	0.28-0.33	1.60-1.90	0.035	0.040	0.20-0.35	—	—	—	1330
1335	0.33-0.38	1.60-1.90	0.035	0.040	0.20-0.35	—	—	—	1335
1340	0.38-0.43	1.60-1.90	0.035	0.040	0.20-0.35	—	—	—	1340
1345	0.43-0.48	1.60-1.90	0.035	0.040	0.20-0.35	—	—	Mo	1345
4012	0.09-0.14	0.75-1.00	0.035	0.040	0.20-0.35	—	—	0.15-0.25	4012
4023	0.20-0.25	0.70-0.90	0.035	0.040	0.20-0.35	—	—	0.20-0.30	4023
4024	0.20-0.25	0.70-0.90	0.035	0.035-0.050	0.20-0.35	—	—	0.20-0.30	4024
4027	0.25-0.30	0.70-0.90	0.035	0.040	0.20-0.35	—	—	0.20-0.30	4027
4028	0.25-0.30	0.70-0.90	0.035	0.035-0.050	0.20-0.35	—	—	0.20-0.30	4028
4032	0.30-0.35	0.70-0.90	0.035	0.040	0.20-0.35	—	—	0.20-0.30	—
4037	0.35-0.40	0.70-0.90	0.035	0.040	0.20-0.35	—	—	0.20-0.30	4037
4042	0.40-0.45	0.70-0.90	0.035	0.040	0.20-0.35	—	—	0.20-0.30	—
4047	0.45-0.50	0.70-0.90	0.035	0.040	0.20-0.35	—	—	0.20-0.30	4047
4118	0.18-0.23	0.70-0.90	0.035	0.040	0.20-0.35	—	0.40-0.60	0.08-0.15	4118
4130	0.28-0.33	0.40-0.60	0.035	0.040	0.20-0.35	—	0.80-1.10	0.15-0.25	4130
4135	0.33-0.38	0.70-0.90	0.035	0.040	0.20-0.35	—	0.80-1.10	0.15-0.25	—
4137	0.35-0.40	0.70-0.90	0.035	0.040	0.20-0.35	—	0.80-1.10	0.15-0.25	4137
4140	0.38-0.43	0.75-1.00	0.035	0.040	0.20-0.35	—	0.80-1.10	0.15-0.25	4140
4142	0.40-0.45	0.75-1.00	0.035	0.040	0.20-0.35	—	0.80-1.10	0.15-0.25	4142
4145	0.43-0.48	0.75-1.00	0.035	0.040	0.20-0.35	—	0.80-1.10	0.15-0.25	4145
4147	0.45-0.50	0.75-1.00	0.035	0.040	0.20-0.35	—	0.80-1.10	0.15-0.25	4147
4150	0.48-0.53	0.75-1.00	0.035	0.040	0.20-0.35	—	0.80-1.10	0.15-0.25	4150
4161	0.56-0.64	0.75-1.00	0.035	0.040	0.20-0.35	—	0.70-0.90	0.25-0.35	4161
4320	0.17-0.22	0.45-0.65	0.035	0.040	0.20-0.35	1.65-2.00	0.40-0.60	0.20-0.30	4320
4340	0.38-0.43	0.60-0.80	0.035	0.040	0.20-0.35	1.65-2.00	0.70-0.90	0.20-0.30	4340
E4340†	0.38-0.43	0.65-0.85	0.025	0.025	0.20-0.35	1.65-2.00	0.70-0.90	0.20-0.30	E4340
4419	0.18-0.23	0.45-0.65	0.035	0.040	0.20-0.35	—	—	0.45-0.60	4419
4422	0.20-0.25	0.70-0.90	0.035	0.040	0.20-0.35	—	—	0.35-0.45	—
4427	0.24-0.29	0.70-0.90	0.035	0.040	0.20-0.35	—	—	0.35-0.45	—
4615	0.13-0.18	0.45-0.65	0.035	0.040	0.20-0.35	1.65-2.00	—	0.20-0.30	4615
4617	0.15-0.20	0.45-0.65	0.035	0.040	0.20-0.35	1.65-2.00	—	0.20-0.30	—
4620	0.17-0.22	0.45-0.65	0.035	0.040	0.20-0.35	1.65-2.00	—	0.20-0.30	4620
4621	0.18-0.23	0.70-0.90	0.035	0.040	0.20-0.35	1.65-2.00	—	0.20-0.30	4621
4626	0.24-0.29	0.45-0.65	0.035	0.04 max.	0.20-0.35	0.70-1.00	—	0.15-0.25	4626
4718	0.16-0.21	0.70-0.90	—	—	—	0.90-1.20	0.35-0.55	0.30-0.40	4718
4720	0.17-0.22	0.50-0.70	0.035	0.040	0.20-0.35	0.90-1.20	0.35-0.55	0.15-0.25	4720
4815	0.13-0.18	0.40-0.60	0.035	0.040	0.20-0.35	3.25-3.75	—	0.20-0.30	4815
4817	0.15-0.20	0.40-0.60	0.035	0.040	0.20-0.35	3.25-3.75	—	0.20-0.30	4817
4820	0.18-0.23	0.50-0.70	0.035	0.040	0.20-0.35	3.25-3.75	—	0.20-0.30	4820
5015	0.12-0.17	0.30-0.50	0.035	0.040	0.20-0.35	—	0.30-0.50	—	5015
50B40‡	0.38-0.43	0.75-1.00	0.035	0.040	0.20-0.35	—	0.40-0.60	—	—
50B44‡	0.43-0.48	0.75-1.00	0.035	0.040	0.20-0.35	—	0.40-0.60	—	50B44
5046	0.43-0.48	0.75-1.00	0.035	0.040	0.20-0.35	—	0.20-0.35	—	—
50B46‡	0.44-0.49	0.75-1.00	0.035	0.040	0.20-0.35	—	0.20-0.35	—	50B46
50B50‡	0.48-0.53	0.75-1.00	0.035	0.040	0.20-0.35	—	0.40-0.60	—	50B50
5060	0.56-0.64	0.75-1.00	0.035	0.040	0.20-0.35	—	0.40-0.60	—	—
50B60‡	0.56-0.64	0.75-1.00	0.035	0.040	0.20-0.35	—	0.40-0.60	—	50B60

(Continued)

* Small quantities of certain elements are present which are not specified or required. Considered as incidental, they are acceptable to the following amounts: 0.35 Cu, 0.25 Ni, 0.20 Cr and 0.06 Mo. † Electric furnace steel. ‡ 0.0005% B min.

TABLE 6-7. SAE Alloy Steel Compositions (SAE J404g) (Continued)

SAE No.	Composition,* %								Corresponding AISI No.
	C	Mn	P	S	Si	Ni	Cr	Other	
5115	0.13-0.18	0.70-0.90	0.035	0.040	0.20-0.35	—	0.70-0.90	—	—
5120	0.17-0.22	0.70-0.90	0.035	0.040	0.20-0.35	—	0.70-0.90	—	5120
5130	0.28-0.33	0.70-0.90	0.035	0.040	0.20-0.35	—	0.80-1.10	—	5130
5132	0.30-0.35	0.60-0.80	0.035	0.040	0.20-0.35	—	0.75-1.00	—	5132
5135	0.33-0.38	0.60-0.80	0.035	0.040	0.20-0.35	—	0.80-1.05	—	5135
5140	0.38-0.43	0.70-0.90	0.035	0.040	0.20-0.35	—	0.70-0.90	—	5140
5145	0.43-0.48	0.70-0.90	0.035	0.040	0.20-0.35	—	0.70-0.90	—	5145
5147	0.46-0.51	0.70-0.95	0.035	0.040	0.20-0.35	—	0.85-1.15	—	5147
5150	0.48-0.53	0.70-0.90	0.035	0.040	0.20-0.35	—	0.70-0.90	—	5150
5155	0.51-0.59	0.70-0.90	0.035	0.040	0.20-0.35	—	0.70-0.90	—	5155
5160	0.56-0.64	0.75-1.00	0.035	0.040	0.20-0.35	—	0.70-0.90	—	5160
51B60‡	0.56-0.64	0.75-1.00	0.035	0.040	0.20-0.35	—	0.70-0.90	—	51B60
50100†	0.98-1.10	0.25-0.45	0.025	0.025	0.20-0.35	—	0.40-0.60	—	—
51100†	0.98-1.10	0.25-0.45	0.025	0.025	0.20-0.35	—	0.90-1.15	—	E51100
52100†	0.98-1.10	0.25-0.45	0.025	0.025	0.20-0.35	—	1.30-1.60	V	E52100
6118	0.16-0.21	0.50-0.70	0.035	0.040	0.20-0.35	—	0.50-0.70	0.10-0.15	6118
6150	0.48-0.53	0.70-0.90	0.035	0.040	0.20-0.35	—	0.80-1.10	0.15 Mo	6150
8115	0.13-0.18	0.70-0.90	0.035	0.040	0.20-0.35	0.20-0.40	0.30-0.50	0.08-0.15	8115
81B45‡	0.43-0.48	0.75-1.00	0.035	0.040	0.20-0.35	0.20-0.40	0.35-0.55	0.08-0.15	81B45
8615	0.13-0.18	0.70-0.90	0.035	0.040	0.20-0.35	0.40-0.70	0.40-0.60	0.15-0.25	8615
8617	0.15-0.20	0.70-0.90	0.035	0.040	0.20-0.35	0.40-0.70	0.40-0.60	0.15-0.25	8617
8620	0.18-0.23	0.70-0.90	0.035	0.040	0.20-0.35	0.40-0.70	0.40-0.60	0.15-0.25	8620
8622	0.20-0.25	0.70-0.90	0.035	0.040	0.20-0.35	0.40-0.70	0.40-0.60	0.15-0.25	8622
8625	0.23-0.28	0.70-0.90	0.035	0.040	0.20-0.35	0.40-0.70	0.40-0.60	0.15-0.25	8625
8627	0.25-0.30	0.70-0.90	0.035	0.040	0.20-0.35	0.40-0.70	0.40-0.60	0.15-0.25	8627
8630	0.28-0.33	0.70-0.90	0.035	0.040	0.20-0.35	0.40-0.70	0.40-0.60	0.15-0.25	8630
8637	0.35-0.40	0.75-1.00	0.035	0.040	0.20-0.35	0.40-0.70	0.40-0.60	0.15-0.25	8637
8640	0.38-0.43	0.75-1.00	0.035	0.040	0.20-0.35	0.40-0.70	0.40-0.60	0.15-0.25	8640
8642	0.40-0.45	0.75-1.00	0.035	0.040	0.20-0.35	0.40-0.70	0.40-0.60	0.15-0.25	8642
8645	0.43-0.48	0.75-1.00	0.035	0.040	0.20-0.35	0.40-0.70	0.40-0.60	0.15-0.25	8645
86B45‡	0.43-0.48	0.75-1.00	0.035	0.040	0.20-0.35	0.40-0.70	0.40-0.60	0.15-0.25	—
8650	0.48-0.53	0.75-1.00	0.035	0.040	0.20-0.35	0.40-0.70	0.40-0.60	0.15-0.25	—
8655	0.51-0.59	0.75-1.00	0.035	0.040	0.20-0.35	0.40-0.70	0.40-0.60	0.15-0.25	8655
8660	0.56-0.64	0.75-1.00	0.035	0.040	0.20-0.35	0.40-0.70	0.40-0.60	0.15-0.25	—
8720	0.18-0.23	0.70-0.90	0.035	0.040	0.20-0.35	0.40-0.70	0.40-0.60	0.20-0.30	8720
8740	0.38-0.43	0.75-1.00	0.035	0.040	0.20-0.35	0.40-0.70	0.40-0.60	0.20-0.30	8740
8822	0.20-0.25	0.75-1.00	0.035	0.040	0.20-0.35	0.40-0.70	0.40-0.60	0.30-0.40	8822
9254	0.51-0.59	0.50-0.80	0.035	0.040	1.20-1.60	—	0.50-0.80	—	—
9255	0.51-0.59	0.70-0.95	0.035	0.040	1.80-2.20	—	—	—	9255
9260	0.56-0.64	0.75-1.00	0.035	0.040	1.80-2.20	—	—	—	9260
9310	0.08-0.13	0.45-0.65	0.025	0.025	0.20-0.35	3.00-3.50	1.00-1.40	0.08-0.15	—
94B15‡	0.13-0.18	0.75-1.00	0.035	0.040	0.20-0.35	0.30-0.60	0.30-0.50	0.08-0.15	—
94B17‡	0.15-0.20	0.75-1.00	0.035	0.040	0.20-0.35	0.30-0.60	0.30-0.50	0.08-0.15	94B17
94B30‡	0.28-0.33	0.75-1.00	0.035	0.040	0.20-0.35	0.30-0.60	0.30-0.50	0.08-0.15	94B30

* Small quantities of certain elements are present which are not specified or required. Considered as incidental, they are acceptable to the following amounts: 0.35 Cu, 0.25 Ni, 0.20 Cr and 0.06 Mo. † Electric furnace steel. ‡ 0.0005% B min.

TABLE 6-8. Minimum Mechanical Properties for SAE J410b HSLA Steels

Grade, Form, and Thickness	Tensile Strength (1000 psi)	Yield Strength 0.2% Offset (1000 psi)	Elongation (%)	
			2 in.	8 in.
945 A, C				
Sheet, strip	60	45	22	. . .
Plate, bar				
To 1/2 in.	65	45	22	18
1/2 to 1-1/2 in.	62	42	24	19
1-1/2 to 3 in.	62	40	24	19
950 A, B, C, D				
Sheet, strip	70	50	22	. . .
Plate, bar				
To 1/2 in.	70	50	22	18
1/2 to 1-1/2 in.	67	45	24	19
1-1/2 in. to 3 in.	63	42	24	19
945X*	60	45	22	18
950X*	65	50	22	18
955X	70	55		
960X	75	60		
965X	80	65		
970X	85	70		

* To 3/8 in. thick.

For mechanical-property data on materials thicker than those listed in the table, suppliers should be consulted. SAE J410b high-strength low-alloy steels may be specified as annealed, normalized, or otherwise specially prepared for forming. When this is done, mechanical properties are agreed upon between supplier and purchaser.

Each grade has chemical composition limits to control welding characteristics in a manner similar to ASTM designations. Table 6-9 lists relative formability, weldability, and toughness of the J410b steels.

Grade 945A has excellent arc and resistance-welding characteristics and the best formability, weldability, and low-temperature notch toughness. It is available in sheets, strip, and light plate.

TABLE 6-9. Fabrication Characteristics for SAE J410b Steels

Formability	Weldability	Toughness
945A	945A	945A
950A	950A	950A
945C, 945X	950D	950B
950B, 950X	945X	950D
950D	950B, 950X	945X, 950X
950C	945C	945C, 950C
	950C	

Alloys are listed in order of decreasing excellence; most formable, most weldable, and toughest alloys at the top.

Source: *Machine Design,* Metals Reference Issue, Dec. 14, 1967.

Grade 945C is a carbon-manganese steel with satisfactory arc-welding properties if proper procedures are used to prevent hardening of the weld metal. Moderate preheat is usually required, especially for thick sections. It is similar to Grade 950C, but has lower carbon and manganese content to improve arc-welding characteristics, formability, and low temperature notch toughness, at some sacrifice in strength.

Grade 945X is a columbium or vanadium-treated carbon-manganese steel similar to 945C except for improved toughness and weldability.

Grade 950A has good weldability, low-temperature notch toughness, and formability. It is normally available only in sheet, strip, and light plate.

Grade 950B has satisfactory arc-welding properties and fairly good low-temperature notch toughness and formability.

Grade 950C is a carbon-manganese steel that can be arc welded if the cooling rate is controlled, but is unsuitable for resistance welding. Formability and toughness are fair.

Grade 950D has good weldability and fairly good formability. Its phosphorus content reduces its low-temperature properties.

Grade 950X is a columbium or vanadium-treated carbon-manganese steel similar to 950C except for somewhat improved welding and forming properties.

Several other grades are also covered by SAE J410b — higher-strength steels that have reduced formability and weldability.

Modifications of standard SAE-grade designations are also available. For example, fully killed steels made to fine-grain practice are indicated by the suffix "K." Thus, 945AK is a fully killed, fine-grain HSLA steel with maximum ladle analysis of 0.15% carbon and a yield strength of about 45,000 psi. All grades made to K practice may not be available from all suppliers. This fine-grain practice is usually specified when low-temperature notch toughness is important.

Steels designated by the suffix "X" contain strengthening elements, such as columbium or vanadium (with or without nitrogen) added singly or in combination. These are usually made semi-killed. However, killed steel may be specified by indicating both suffixes, such as SAE 950XK.

Available HSLA-steel grades often have characteristics in excess of the specification minimums. Literature from producer companies contains information on physical and mechanical property ranges and suggested fabricating and welding practices.

High-Yield Strength Quenched-and-Tempered Alloy Steels

The high-yield-strength quenched-and-tempered construction steels are full-alloy steels that are treated at the steel mill to develop optimum properties. Unlike conventional alloy steels, these grades do not require additional heat treatment by the fabricator except, in some cases, for a stress relief.

These steels are generally low-carbon grades (upper carbon limit of about 0.20%) that have minimum yield strengths from 80,000 to 125,000 psi.

Some high-yield-strength grades are also available in abrasion-resistant modifications (AR steels), produced to a high hardness. Although these steels can have yield strengths to 173,000 psi, hardness (up to 400 Brinell) rather than strength is their key characteristic.

The high-yield-strength quenched-and-tempered alloy steels are used in such widely varying applications as hoist and crane components; end, side, and bottom plates for ore and waste-haulage cars, hopper cars, and gondolas; pressure hulls for submarines; and components for dust-collecting equipment. The AR (abrasion-resistant) modifications are used in applications requiring maximum resistance to abrasive materials — in chutes, hoppers, and dump-truck beds, for example. In such uses, strength properties are secondary and are not usually specified.

Good toughness can be combined with abrasion resistance in these steels, for use in buckets, cutter bars, scraper blades, and impact plates. However, the most abrasion-resistant grades sacrifice impact strength to gain maximum wear resistance.

HY Steels

An important group of high-yield-strength quenched-and-tempered steels is the HY steels. The most common and most available of these is HY80, which has a minimum yield strength of 80,000 psi. Higher-strength grades are HY100 and HY130 of which only HY100 is available commercially.

HY80 is commonly available in plate form. However, it can also be obtained in beams, channels, angles, and tubing. Strength and toughness of HY80 steel and its ability to be welded (under carefully controlled conditions) qualify it for use in critical applications such as pressure hulls for submarines and deep-submergence research and rescue vessels.

Mechanical properties of these steels are influenced by section size. Carbon content is the principal factor that determines maximum attainable strength. Most alloying elements make a small contribution to strength, but their dominant effect is on hardenability — which determines the maximum thickness or depth of steel that can be fully hardened on quenching.

HY80 is normally supplied to the toughness requirements of MIL-S-16216. In plate 1/2 to 1-1/2-in. thick, 50 ft-lb of impact energy absorption is required at minus 120°F with a longitudinal Charpy V-notch specimen.

A typical value for the ductile-to-brittle transition temperature of a 100,000-psi steel in 1/2-in. plate is minus 180°F, as determined with both longitudinal and transverse Charpy V-notch specimens.

Many of the high-yield-strength steels are available in three or four strength or hardness levels. The different levels are achieved by variations in carbon and alloy content, tempering temperature, and tempering time.

In general, the 100,000-psi steels have fatigue strengths in the 50,000 to 70,000-psi range in rotating-beam tests. Higher-strength grades have higher endurance limits — about 60% of their tensile strength.

The compressive yield strength of 100,000-psi steels is usually about the same as tensile yield strength. Shear strength generally ranges from about 85 to 100% of the tensile yield strength.

ASTM Specifications

Two plate specifications, ASTM A514 for welded structures and A517 for boilers and other pressure vessels, allow for the effect of section size on yield strength, tensile strength, and ductility. ASTM A514 requires a minimum yield strength of 100,000 psi for material up to 2-1/2-in. thick, and 90,000 psi for material from 2-1/2 to 4 in. thick. ASTM A517 requires uniform yield strengths of 100,000 psi for all material up to 3/4-in. thick. Representative trade names of the A514 and A517 steels are given in Table 6-10.

TABLE 6-10. Representative ASTM A514/517 Steels

Producer	Trade Name
Bethlehem Steel Corp.	RQ-100A, RQ-100 RQ-100B
National Intergroup Inc.	N-A-XTRA 100 N-A-XTRA 110
USX Corp. and Lukens Steel Corp.*	T-1 T-1 Type A T-1 Type B

* Licensee

Weldability

Most high-yield-strength quenched-and-tempered alloy steels can be welded without preheat or post-heat. If suppliers' recommendations are followed for controlling welding procedures, 100% joint efficiency can be expected in the as-welded condition for the 90,000 and 100,000-psi yield-strength grades.

If the heat-affected zone cools too slowly, the beneficial effects of the original heat treatment (particularly notch toughness) are destroyed. This can be caused by excessive preheat temperature, interpass temperature, or heat input. On the other hand, if the heat-affected zone cools too rapidly, it can become hard and brittle and may crack. This is caused by insufficient preheat or interpass temperature, or insufficient heat input during welding. Producers' recommendations should be followed closely.

The quenched-and-tempered steels can be welded by the shielded metal-arc, submerged-arc, and gas-shielded-arc processes. Weld cooling rates for these processes are relatively rapid, and mechanical properties of the heat-affected zones approach those of the steel in the quenched condition. Reheat-treatment, such as quenching and tempering after welding, is not recommended.

Because of the desirability of relatively rapid cooling after welding, thin sections of these materials can usually be welded without preheating. When preheating is required, both maximum and minimum temperatures are important. If the sections to be welded are warm as a result of preheating and heat input from previous welding passes, it may be necessary to reduce current or increase arc travel speed for subsequent passes, or to wait until the metal cools somewhat. Interpass temperature is just as important as preheat temperature and should be controlled with the same care.

In the ASTM specifications A514 and A517 there are several grades of quenched and tempered constructional steels listed. Welding procedures for all of these steels are similar but no one procedure is right for all grades. Welding procedures are available from the steel manufacturers. When in doubt, consult the steel manufacturer.

The following is a general shielded metal-arc procedure for one of the popular grades of quenched and tempered constructional steels and can be used as a *guide* for all grades or other welding processes.

Use only low hydrogen type electrodes and usually the electrode specified for A514 and A517 steels is E11018M. Under some conditions a lower tensile strength electrode may be used and this will be discussed later. Make sure electrodes are dry. Under normal conditions of humidity electrodes should be returned to the drying ovens after an exposure of four hours maximum. If the humidity is high, reduce the exposure time. Electrodes are shipped in hermetically sealed containers and the contents of any damaged container should be redried before using. See Table 6-17 for drying temperatures.

Clean the joint thoroughly. Remove all rust and scale preferably by grinding. If the base metal has been exposed to moisture, preheat to drive off the moisture. On thin sections, allow the plate to cool, if necessary, before starting to weld.

The amount of preheat and the amount of welding heat put into the weld must be kept within definite boundaries during the actual welding. Usually preheating is not necessary or desirable on thin

TABLE 6-11. Composition of ASTM A-203-90 Nickel-Steel Plate for Pressure Vessels

Element and Plate Thickness	Composition (%)			
	Grade			
	A	B	D	E and F
Carbon, max				
To 2 in.	0.17	0.21	0.17	0.20
2 to 4 in.	0.20	0.24	0.20	0.23
4 to 6 in.	0.23	0.25	—	—
Manganese, max				
To 2 in.	0.70	0.70	0.70	0.70
2 to 4 in.	0.80	0.80	0.80	0.80
4 to 6 in.	0.80	0.80	—	—
Phosphorus, max	0.035	0.035	0.035	0.035
Sulfur, max	0.05	0.05	0.05	0.05
Silicon (heat analysis)	0.15-0.40	0.15-0.40	0.15-0.40	0.15-0.40
Nickel (heat analysis)	2.10-2.50	2.10-2.50	3.25-3.75	3.25-3.75

TABLE 6-12. Suggested Maximum Heat Input in KJ/in.† for USS T-1 steel

Preheat and Interpass Temperature	Plate Thickness							
	3/16″	1/4″	1/2″	3/4″	1″	1-1/4″	1-1/2″	2″
70°F	27	36	70	121	any	any	any	any
200°F	21	29	56	99	173	any	any	any
300°F	17	24	47	82	126	175	any	any
350°F	15	21.5	43.5	73.5	109.5	151	any	any
400°F	13	19	40	65	93	127	165	any

† From the *"Welding Heat Input Calculator"* by the USX Corporation in Kilojoules per inch.

sections but in order to avoid cracks preheating is necessary if:

The joints are highly restrained.

The structure is very rigid.

The weld joint is on thick sections.

Whether or not the base metal is preheated, it is necessary to approximate the heat input before starting to weld. The heat input in watt-seconds (joules) per linear inch of weld is

$$\text{Heat input} = \frac{I \times E \times 60}{V}$$

where I is the arc amperes, E is the arc volts, and V is the welding speed in in./min. Calculation by this formula is only approximate because the heat losses can be large. Also, there are many variables that affect the heat distribution and the maximum temperature of the base metal at the joint but the formula is sufficiently accurate to predict the maximum allowable heat input for a given set of conditions.

In industry, the term "kilo joules per inch" (KJ/in.) is used and is equal to the watt-seconds per linear inch of weld divided by 1000.*

Maximum suggested heat input for USS T-1 steel in kilojoules per linear inch of weld is shown in Table 6-12.

Also see Section 3-3.

Before making a production weld it is recommended to set up a tentative procedure and make a test weld. The tentative procedure includes the preheat, if any, interpass temperature, welding current, voltage, and welding speed. It is important to keep the welding current, speed and interpass temperature under close control.

* A calculator is available from the USX Corporation for quickly determining KJ/in. Also available are tables for maximum KJ/in. when welding T-1, T-1 Type A, and T-1 Type B.

The following are some general rules to follow to promote good weld quality.

Always use stringer beads, never wide weave beads.

Clean thoroughly between passes.

Use the same precautions to prevent cracking as discussed earlier in this section.

Back gouge with arc gouging and remove the scale by grinding. Do not use oxyacetylene to back gouge.

Usually the electrodes used are the E11018M type but lower strength electrodes may be specified where the stress does not require the high yield strength of E11018M. A good example is the lower stress in the web to flange fillet welds. However, if lower strength electrodes are used the same limitations apply as to heat input and interpass temperature.

Low-Alloy Steels

Small amounts of alloying elements such as nickel, chromium, and molybdenum can be added to steels to increase strength, hardness, or toughness, or to improve resistance to heat, corrosion, or other environmental factors. These improvements are sometimes gained with little effect on weldability or other fabricability characteristics. Generally, however, welding of low-alloy steels requires more careful control of procedures and selection of electrodes than welding of the carbon steels.

Nickel Steels

A low nickel addition (2 to 5%) greatly increases strength and hardenability and improves the corrosion resistance of a steel without a proportional reduction in ductility or a significant effect on weldability. The compositions of various thicknesses of nickel-steel plate (ASTM A-203), used principally for pressure vessels, are listed in Table 6-11.

Straight nickel steels are used mainly for low-temperature pressure vessels. The nickel content significantly improves toughness and impact strength at subzero temperatures. Nickel is also very effective in improving the hardenability of steels; heat treatment is easy because nickel lowers the critical cooling rate necessary to produce hardening on quenching.

A nickel steel containing 0.24% carbon and 2.7% nickel can have a tensile strength (normalized and drawn) of over 85,000 psi, an unalloyed steel would require a carbon content of over 0.45% to be that strong. Notch toughness of a 3-1/2% nickel steel, with a tensile strength of 70,000 to 85,000 psi, would be

15 ft-lb at minus 100°F (Charpy V-notch test), whereas a carbon steel of that strength would have a notch toughness of 15 ft-lb at 0°F.

Nickel increases hardenability for a given carbon content. For best weldability and minimum cracking tendency, carbon content should, of course, be low — no more than 0.18% if extensive welding is to be done without preheat.

Chromium Steels

In the low-alloy steels, chromium increases tensile strength, hardenability, and, to some extent, atmospheric corrosion resistance. Chromiun steels with less than 0.18% carbon are readily weldable, using proper precautions against cracking. The combination of chromium and higher carbon increases hardenability and requires preheating and sometimes postheating to prevent brittle weld deposits. Production welding is not recommended for chromium steels containing more than 0.30% carbon.

Nickel-Chromium Steels

The nickel-chromium steels of the AISI series are no longer standard alloys but occasionally there is a need to weld these alloys, especially in maintenance work.

The addition of chromium is intended to increase hardenability and response to heat treatment for a given carbon content over that of the straight nickel low alloy steels. Also a small amount of several alloying elements judiciously chosen may give a greater range of hardenability plus toughness than a larger or more costly amount of a single alloying element.

Chromium is a potent hardening agent and it is necessary to keep the carbon content low for weldability. Thin sections of the lowest carbon content type can usually be welded without preheat but the higher carbon grades require preheat and subsequent stress relief or annealing.

The lower carbon grades of the nickel-chromium steels can be welded with electrodes of the EXX15-16-18 classes and in the as welded condition the weld properties will match the base metal. However, if the weldment must be heat treated after welding, special low-hydrogen type electrodes are required. These electrodes must deposit weld metal that will respond to the same heat treatment as the base metal and match base metal properties.

The higher carbon alloys (above .40%) are not readily welded but, if necessary, a weld can usually be made with stainless E309 (second choice E310) electrodes. The weld will usually be tough and ductile but the fusion zone may be brittle from dilution with the high carbon base metal. The fact that the weld is ductile allows it to give a little without putting too much bending in the brittle zone. Preheat is suggested. See Section 3-3.

Molybdenum Steels

Molybdenum increases the hardenability and high-temperature strength of low-alloy steels. The low-alloy molybdenum steels are of four general types: carbon-molybdenum (AISI 4000 series), chromium-molybdenum (4100 series), nickel-molybdenum (4600 and 4800 series) and nickel-chromium-molybdenum (4300 and 4700 series).

A common use of carbon-moly and chrome-moly steels is in high-pressure piping used at high temperatures. These steels are usually purchased to an ASTM specification. Another typical use of the chrome-moly alloys — usually in the form of tubing — is in highly stressed aircraft parts. Weldability of these thin-section members is good because of the low carbon content. Low-carbon grades of these steels (below 0.18%) can usually be welded without preheat. The higher-carbon nickel and chromium grades of molybdenum steels are air-hardening.

The low carbon grades (below .18%) of carbon-moly steel can be welded much the same as mild steel. E7010-A1, E7018-A1, and E7027-A1 electrodes will give tensile strengths in the same range as plate strength in the as-welded condition. The above electrodes with .5% moly will come close to approximating plate properties and analysis where subsequent heat treatment is required. (See Preheat Table for steels above .18% carbon.)

When carbon content of the carbon-moly alloys is low (approximately .15%), these steels are readily weldable. In pressure vessels, this low carbon content is usually used, but in piping the carbon may be somewhat higher. Where carbon is above .18% preheating is generally required.

In the case of pipe the applicable Code or Specification may limit the choice of welding process and the choice of filler metal. In addition back up rings or chill rings may not be permissible. In this case TIG welding, with a consumable insert, or an appropriate filler rod, would be recommended. The TIG (GTAW) process will produce the very best weld quality and the smoothest and best weld contour on the inside of the pipe. Service conditions may well dictate the use of TIG, at least for the first pass.

If the SMAW process is permitted an experienced weldor, using E-7010A1 can put in an open root first pass with a very small reinforcement on the inside.

However it is important that this weld completely penetrate the joint and that no notch or undercut is left at the root of the joint.

The SMAW process using an E-7010 A1 electrode also works very well if a back-up ring is permitted.

Stress relieving is generally specified when the thickness of the metal is greater than 3/8″. Temperature of 1200° — 1250°F is used with usual procedure as to time of heating (one hour per inch of thickness) and length of pipe heated (6 times thickness on each side of weld).

The cooling rate is from 200° — 250°F per hour down to 150° — 200°F in which case cooling may be done in still air.

For the welding of the steels mentioned herein the use of E7010-A1 electrode is recommended for ease of welding in out-of-position work. The preheat and post heat treatment above is also required when E7010-A1 electrodes are used. Where the work can be positioned for flat position welding or where large welds are required in any position, the low hydrogen electrodes can be used to advantage as they will reduce the preheat temperatures required.

In applications where tensile strength of weld need not be as high as the base metal but where other physical characteristics of the weld should be comparable to the base metal, the regular type of electrode, as used for welding mild steel, can be employed with very satisfactory results. For joining work of this type, E6010 electrodes are recommended.

On light chrome-moly tubing, E6013 electrodes designed especially for aircraft work are often used. These mild steel electrodes usually pick up enough alloy from the base metal to give the required tensile strength in the as-welded condition. When welded on the AISI 4130, their normal 70,000 to 80,000 psi tensile strength is increased by pick-up of alloy and carbon to a satisfactory approximation of the mechanical properties of AISI 4130. The additional thickness of weld due to the usual build-up on light gauge work makes the welded joint stronger than the parent metal.

On the higher carbon and alloy grades where heat treated welds with properties similar to plate properties are necessary, special electrodes can be used that will deposit the proper analysis. A low hydrogen type electrode is used to reduce the tendency for cracking that is quite prevalent on these steels. Preheat and post heat treatment usually will be required.

On the grades over .40% carbon where production welding is not recommended, it is possible to make a weld with E309 type stainless electrode or E310 as a second choice. The weld will be fairly ductile if the proper low penetrating procedure is used; however, the fusion zone may be very brittle depending upon the air hardenability of the alloy. Preheating and slow cooling will tend to reduce this hardness in the fusion zone.

Where molybdenum is added to base metals to increase the resistance to creep at elevated temperatures, the electrode deposit must have a similar amount of molybdenum.

Table 6-13 gives the approximate preheat and interpass temperatures for AISI alloy steel bars when welded with low-hydrogen type electrodes.

TABLE 6-13. Approximate Preheat and Interpass Temperature for AISI Alloy Steel Bars*

AISI Steel	Preheat and Interpass Temperature °F		
	section thickness, in.		
	To 1/2	1/2 – 1	1 – 2
1330	350 – 450	400 – 500	450 – 550
1340	400 – 500	500 – 600	600 – 700
4023	100 min.	200 – 300	250 – 350
4028	200 – 300	250 – 350	400 – 500
4047	400 – 500	450 – 550	500 – 600
4118	200 – 300	350 – 450	400 – 500
4130	300 – 400	400 – 500	450 – 550
4140	400 – 500	600 – 700	600 – 700
4150	600 – 700	600 – 700	600 – 700
4320	200 – 300	350 – 450	400 – 500
4340	600 – 700	600 – 700	600 – 700
4620	100 min.	200 – 300	250 – 350
4640	350 – 450	400 – 500	450 – 550
5120	100 min	200 – 300	250 – 350
5145	400 – 500	450 – 550	500 – 600
8620	100 min	200 – 300	250 – 350
8630	200 – 300	250 – 350	400 – 500
8640	350 – 450	400 – 500	450 – 550

* From ASM Metal Handbook Volume 6, Eighth Edition.

This hopper car has a carbon steel frame and stainless steep hoppers. Weldors are working on the frame.

Welding Carbon and Low-Alloy Steels with the Shielded Metal-Arc Process

A large amount of welding on steel is still done manually with shielded metal-arc (stick) electrodes. As in any manual process, the skill and dexterity of the operator are important for quality work; but equally important is selection of the correct type of electrode.

CONSIDERATIONS IN ELECTRODE SELECTION

Choice of electrode is straightforward when welding high-strength or corrosion-resistant steels. Here, choice is generally limited to one or two electrodes designed specifically to give the correct chemical composition in the weld metal. But most arc welding involves the carbon and low-alloy steels for which many different types of electrodes provide satisfactory chemical compositions in the weld metal. From the many possibilities, the object is to pick an electrode that gives the desired quality of weld at the lowest welding cost. Usually, this means the electrode that allows the highest welding speed with the particular joint. To meet this objective, electrodes are selected according to the design and positioning of the joint.

Electrodes compounded to melt rapidly are called "fast-fill" electrodes, and those compounded to solidify rapidly are called "fast-freeze" electrodes. Some joints and welding positions require a compromise between the fast-fill and fast-freeze characteristics, and electrodes compounded to meet this need are called "fill-freeze" electrodes. There are also electrodes which are classified as "fast follow."

The fill-freeze-follow terminology used to classify types of electrodes is also used to designate types of joints. Overhead or vertical joints that normally require fast-freeze electrodes are thus termed "freeze" joints, while flat joints and some horizontal joints, where rapid deposition is important, are called "fill" joints. Some joints, especially those in sheet metal, require an electrode that permits rapid electrode travel with minimum skips, and are thus called "follow" joints. The fill-freeze electrodes usually are best suited for follow joints, and thus, fill-freeze electrodes are called fast-follow electrodes when the reference is to joints requiring fast electrode travel.

Although the terms fill, freeze, and fill-freeze, are straightforward as applied to electrodes, use of these terms to describe types of joints is not so clear-cut. For example, some overhead "freeze" joints require a fill-freeze, rather than fast-freeze, electrode. By the same token, a "follow" joint in sheet metal may require a fast-freeze, rather than a fill-freeze, electrode. The use of these terms to identify types of joints, and the types of electrodes recommended for these joints, are explained in Fig. 6-14, 6-15,

BUTT WELDS

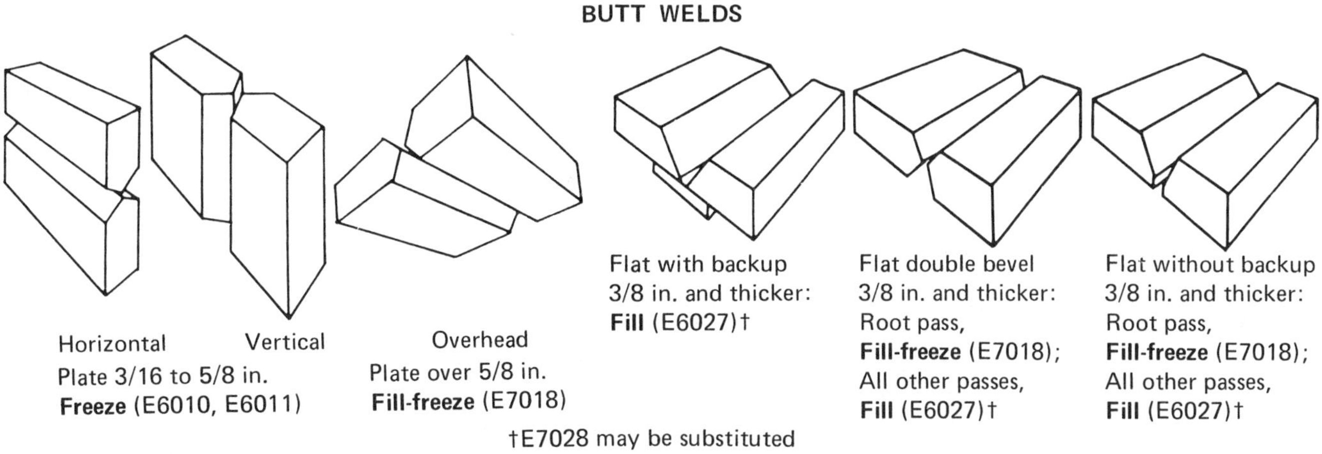

Fig. 6-14. Guide to selection of electrodes for butt welds.

and 6-16, which show butt welds, fillet welds, and sheet-metal welds, respectively.

AWS A5.1-91 is a complete specification for mild-steel electrodes for shielded metal-arc welding (see Section 4.1). Typical mechanical properties of mild-steel deposited weld metal are given in Table 6-14.

A combination of letters and numbers used by the American Welding Society to identify the various classes of electrodes is given in Table 4-1. For a more complete description of this system see Section 4-1. Typical current ranges for all AWS A5.1 electrodes is given in Table 6-15. A guide to the application of electrodes for steels of specific ASTM designations is presented in Table 6-16.

FAST-FREEZE ELECTRODES

Fast-freeze electrodes are compounded to deposit weld metal that solidifies rapidly after being melted by the arc, and are thus intended specifically for welding in the vertical and overhead positions. Although deposition rates are not as high as with other types of electrodes, the fast-freeze type can also be used for flat welding and is, thus, considered an "all-purpose" electrode that can be used for any weld in mild steel. However, welds made with fast-freeze electrodes are slow and require a high degree of operator skill. Therefore, wherever possible, work should be positioned for flat position welding, which permits the use of fast-fill electrodes.

TABLE 6-14. Typical Mechanical Proprerties of Mild-Steel Deposited Weld Metal

Electrode Classification	Conditions							
	As-Welded				Stress-Relieved at 1150°F			
	Tensile Strength (psi)	Yield Strength (psi)	Elong. in 2 in. (%)	Impact* (ft-lb)	Tensile Strength (psi)	Yield Strength (psi)	Elong. in 2 in. (%)	Impact* (ft-lb)
E6010	69,000	60,000	26	55[(1)]	65,000	51,000	32	75
E6011	70,000	63,000	25	50[(1)]	65,000	51,000	30	90
E6012	72,000	64,000	21	43	71,000	62,000	23	47
E6013	74,000	62,000	24	55	74,000	58,000	28	—
E6019	60,000 min	48,000 min	22 min	20 min[(2)]	—	—	—	—
E6020	67,000	57,000	27	50	—	—	—	—
E6022	75,500	—	—	—	—	—	—	—
E6027	66,000	58,000	28	40[(1)]	66,000	57,000	30	80
E7014	73,000	67,000	24	55	73,000	65,000	26	48
E7015	75,000	68,000	27	90	—	—	—	—
E7016	75,000	68,000	27	90	71,000	60,000	32	120
E7018	74,000	65,000	29	80[(1)]	72,000	58,000	31	120
E7018M	70,000	64,000	24 min	50 min[(1)]	—	—	—	—
E7024	86,000	78,000	23	38	80,000	73,000	27	38
E7027	70,000 min	58,000 min	22 min	20 min[(1)]	—	—	—	—
E7028	85,000	78,000	26	26[(2)]	81,000	73,000	26	85
E7048	82,000	70,000	31	85[(1)]	—	—	—	—

* Charpy V-notch at 70°F, except where noted.
(1) Charpy V-notch at −20°F.
(2) Charpy V-notch at 0°F.

TABLE 6-15. Typical Current Ranges for Electrodes

Electrode Diameter (in.)	Current Range (amp)											
	Electrode Type											
	E6010, E6011 DC+	E6012	E6013	E6019	E6020	E6022	E6027 E7027	E7014	E7015, E7016	E7018M E7018	E7024, E7028	E7048
1/16	—	20-40	20-40	—	—	—	—	—	—	—	—	—
5/64	—	25-60	25-60	35-55	—	—	—	—	—	—	—	—
3/32	40-80	35-85	45-90	50-90	—	—	—	80-125	65-110	70-100	100-145*	—
1/8	75-125	80-140	80-130	80-140	100-150	110-160	125-185	110-160	100-150	115-165	140-190	80-140
5/32	110-170	110-190	105-180	130-190	130-190	140-190	160-240	150-210	140-200	150-220	180-250	150-220
3/16	140-215	140-240	150-230	190-250	175-250	170-400	210-300	200-275	180-255	200-275	230-305	210-270
7/32	170-250	200-320	210-300	240-310	225-310	370-520	250-350	260-340	240-320	260-340	275-365	—
1/4	210-320	250-400	250-350	310-360	275-375	—	300-420	330-415	300-390	315-400	335-430	—
5/16	275-425	300-500	320-430	360-410	340-450	—	375-475	390-500	375-475	375-470	400-525*	—

* The E7028 classification is not manufactured in this diameter.

FILLET AND CORNER WELDS

Fillet welds over 10 to 12 in. in length on 3/16-in. or thicker plate

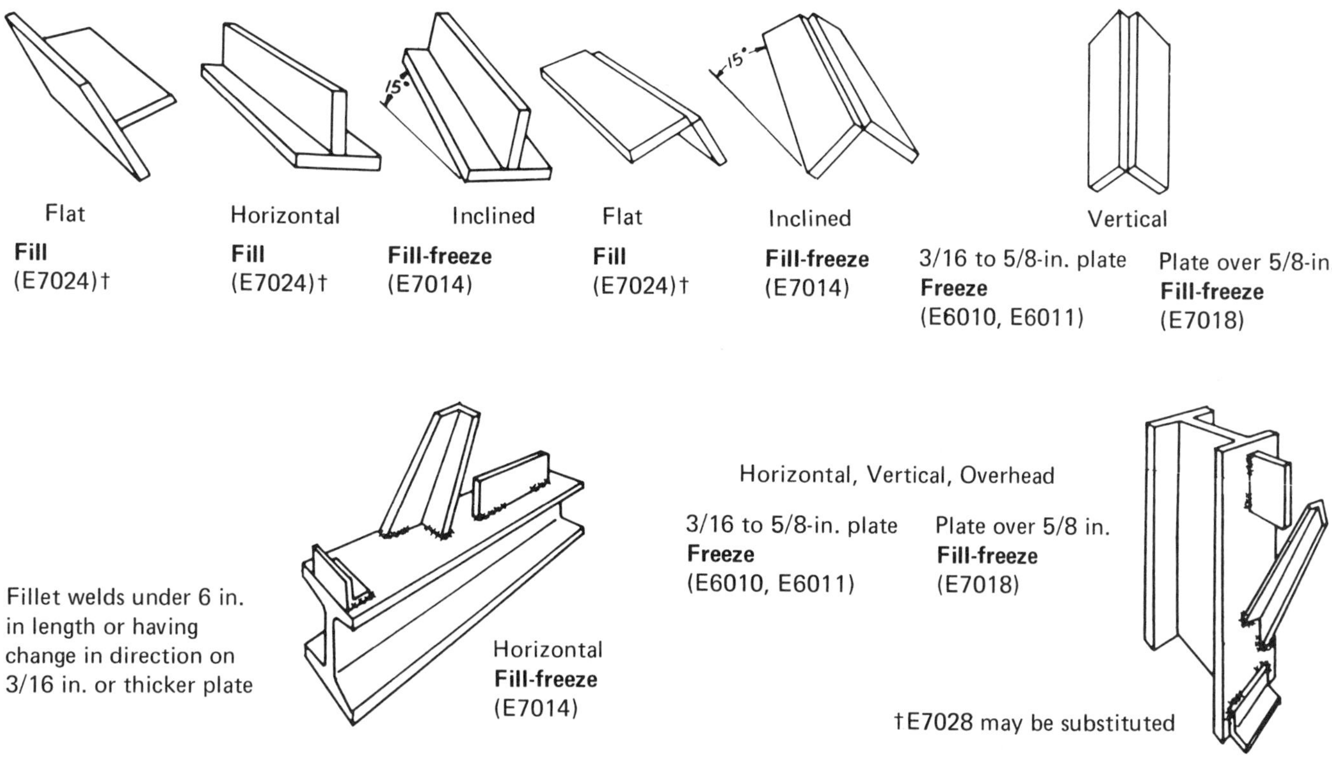

Fig. 6-15. Guide to selection of electrodes for fillet and corner welds.

SHEET METAL JOINTS

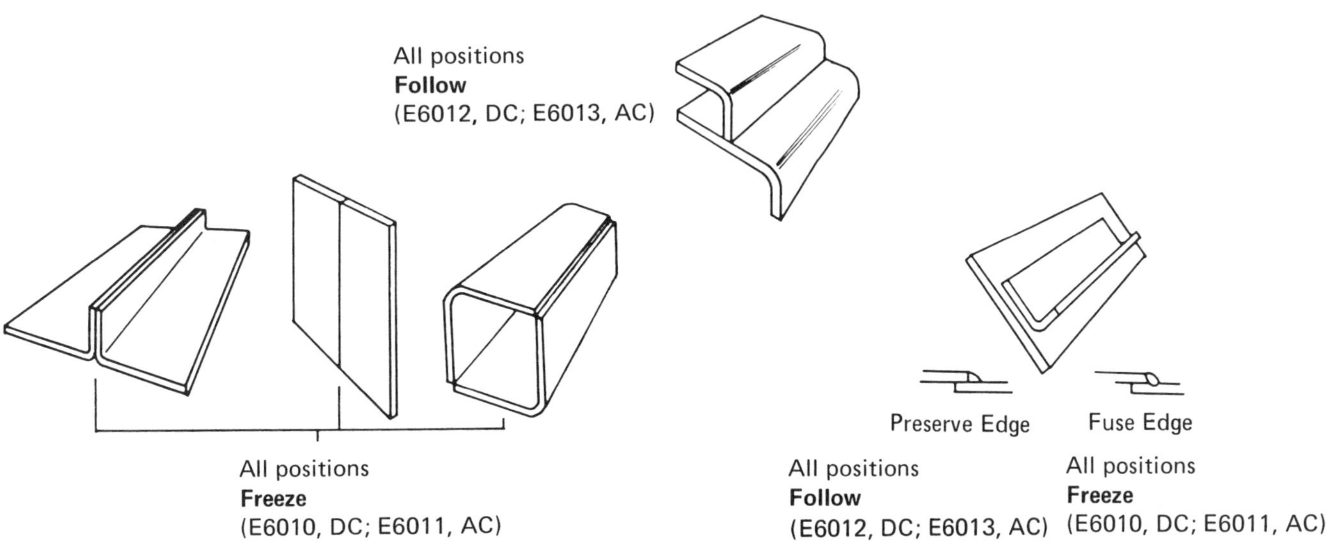

Fig. 6-16. Guide to selection of electrodes for sheet-metal welds.

TABLE 6-16. Recommended Electrodes for Carbon and Low Alloy ASTM Steels (See Note 10) (Plate, Shapes, Forgings and Castings)

ASTM Specifications	Description	Grades	Recommended Electrodes
STEEL PLATES, SHEETS, FORGINGS, SHAPES AND CASTINGS			
A36-91	Structural 36,000 psi Min. YS		Note 1
A113-70a	Railway rolling stock (Discontinued)	All	Note 1
A131-91	Structural for ships	A, B, CS, D, DS & E	Note 1
		AH, DH & EH32	E7015, E7016, E7018, E7028
A148-90	Steel castings for structural use	80-40 & 50	E8018-C3
		90-60	E9018-G
		105-85 & 115-95	E11018-M
A202-90	Boiler & pressure vessel	A & B	E9018-G
A203-90	Pressure vessel	A & B	E8018-C1
		D & E	E8018-C2
A204-90	Boiler & pressure vessel	A & B	E7010-A1 or E7018-A1
		C	E8018-B2
A205-74a	Boiler & pressure vessel (Discontinued)	A & B	E8018-C3
A225-90	Boiler & pressure vessel (Replaced by A233 & A251)	C & D	E8018-C3
A236-74	Forgings, railway (Discontinued)	A & B	E7018 or E7028
		C, D & E	E8018-C3
		F & G	E9018-G
		H	E11018-M
A238-71	Forgings, railway (Discontinued, Replaced by A730)	A	E8018-C3
		B	E9018-G
		C, D & E	E11018-M
A242-91	High strength structural	All	E7018 or E7028, Note 3
A266-90	Drum forgings	1	Note 1
		2 & 4	E7018
		3	E8018-C3
A283-91	Structural plates	All	Note 1
A284-90	Carbon-silicon plates	All	Note 1
A285-90	Flange & firebox plate	All	Note 1
A299-90	Boiler plate	All	E8018-C3
A302-90	Boiler & pressure vessel	All	E8018-C3
A328-90	Steel piling	All	E7018 or E7028
A336-89	Alloy forgings	F1	E7018-A1
		F11, F12, F11A	E8018-B2
		Other grades	Note 9
A352-89	Low-temperature castings	LCA, LCB &LCC	E7018
		LC1	E7018-A1
		LC2	E8018-C1
		LC3	E8018-C2
A356-90	Steam-turbine castings	1, 2, 5, -E	E8018-B1
		6	E8018-B2
		8, 9 & 10	E9018-B3
A361-85(90)	Galvanized sheets		Notes 1 & 8
A366-91	Carbon steel sheets		Note 1
A372-91	Pressure vessel forgings	Type I	E7018 or E7028
		Type II	E8018-C3
		Type III	E9018-G
		Type IV	E11018-M
A374-	Discontinued, Replaced by A606, A607		
A375-	Discontinued, Replaced by A606, A607		
A387-90a	Cr-Mo boiler plate	2, 11, 12	E8018-B2
		D	E9018-B3
A389-89	High-temperature castings	C23	E8018-B2
		C24	E9018-B3
A410-72	Pressure-vessel plate (Discontinued)		E8018-C2
A414-91	Flange & firebox sheet	A, B, C & D	Note 1
		E & F	E7018 or E7028
		G	E8018-C3
A424-91	Sheet for porcelain enameling		E7018
A441-85	High-strength structural	All	E7018 or E7028
	Discontinued, Replaced by A572.		Note 3
A442-90	Fine grain plate	55, 60	E7018 or E7028
A444-89	Galvanized steel sheet	A	Notes 1 & 8
A446-91	Hot dipped galvanized sheet	A, B, C, D, E & F	E7010-A1 Notes 1 & 8
A455-90	C-Mn pressure vessel plate		E8018-C3
A486-84	Highway bridge castings (Discontinued 1989)	70	E7018 or E7028
		90	E9018-G
A487-89a	Castings for pressure service	8A, 9A	E8018-B3
		11N, 16N	Note 1
		BQ & CQ	E8018-C3
A514-91	Quenched & tempered plate	A, B, C, E, F, H, J, M, P, Q, R, S, T	E11018-M Note 4

(continued)

ASTM Specifications	Description	Grades	Recommended Electrodes
STEEL PLATES, SHEETS, FORGINGS, SHAPES AND CASTINGS			
A515-90	High-temperature boiler plate	55, 60 65, 70	E7018, E7028 or E8018-C3
A516-90	Low-temperature pressure vessel plate	55 & 60 65 & 70	E7018 or E7028 E7015, E7016, E7018
A517-90	Quenched & tempered plate	A, B, C, E, F, H, J, M, P, Q, S, T	E11018-M, Note 4
A526-90	Galvanized sheets		Note 8
A528-90	Galvanized sheets		Note 8
A529-89	Structural, 42,000 psi Min. YS		Note 1
A533-90	Quenched & tempered plate	Class 1 Class 2 & 3	E8018-C3 E11018-M
A537-90	Pressure-vessels and structures	Class 1 Class 2	E7018 orE7028 E8018-C3
A541-88a	Pressure-vessel forging	Class 1 & 1A Class 2, 3, 4, 2A, 3A & 11C	E7018 or E7028 E8018-C3
A543-88	Quenched & tempered plate	1, 2 & 3	E11018-M, Note 4
A570-91	Structural sheet & strip	30, 33, 36, 40, 45, 50, 55	Note 1
A572-91	Structural plate	42 50 & 55 60 & 65	E7015-X, E7016-X, E7018-X E7018 or E7028 E8018-C3
E573-90	Structural plate	58, 65 & 70	E7018 or E7028
E588-91a	High-strength structural	A, B, C & K	E7018 or E7028, Note 3
E606-91a	High-strength sheet	All	E7015, E7016, E7018
E607-91	High-strength low-alloy sheet	45, 50 & 55 60, 65 70	E7015, E7016, E7018 E8018-C3 E9018-G
E611-91	Cold rolled sheet	A, B, C, D & E	E8018, Note 1
A615-90	Reinforcement bars	40 60 75	Note 1 E9018-G E11018-M
A616-90	Reinforcement bars	50 60	E8018-C3 E9018-G
E617-90	Reinforcement bars	40 60	Note 1 E9018-G
A633-90a	Normalized high strength sheet	A, C & D E	E7015-X, E7016-X, E7018-X E10015-X, E10016-X, E10018-X
A706-90	Reinforcement bars	60	E9018-G
A709-91	Structural steel plate and shapes	36, 36W 50, 50W, 70W 100, 100W	E60XX, E70XX, E70XX-X E7015-X, E7016-X, E7018-X E10015-X, E10016-X, E10018-X, E11015-X, E11016-X, E11018-X
A710-90	Age hardening alloy steel	A2 A1, A3	E7015-X, E7016-X, E7018-X E10015-X, E10016-X, E10018-X
A808-91	High strength, low alloy plate		E7015-X, E7016-X, E7018-X
ABS		AH32, DH32, EH32, AH36, DH36, EH36	E7015, E7016, E7018
STEEL PIPE, TUBES AND FITTINGS			
A53-90b A106-91 A135-89a A139-90 A179-90 A192-90a A211-75 A214-90a A226-90a A252-90 A523-90 A587-89a A589-89a	Mild-steel pipe	All	Notes 1 and 2
A105-90	High-temperature fittings		E7018
A106-91	High-temperature pipe	A, B, C	E7018
A115-89a	High-temperature pipe	C45, C50, C55 KC & KCF-55, 60 KC & KCF-65 CM65, 70 CM75 CMS75 & CMSH70 1/2, 1 & 1-1/4 Cr 2-1/4 Cr	Note 1 E7018 or E7028 E7018 or E7028 E7010-A1 or E7018-A1 E8018-B2 E8018-C3 E8018-B2 E9018-B3

(continued)

ASTM Specifications	Description	Grades	Recommended Electrodes
STEEL PIPE, TUBES AND FITTINGS			
E161-90a	Still tubes	Low-carbon	Note 1
		T1	E7010-A1 or E7018-A1
E178-90a & E179-90a	Boiler tubes & condenser tubes	A, C, D, All	Note 1
A181-90	General service fittings	60 & 70	E7018 or E7010-A1
A182-91	High-temperature fittings	F1	E7010-A1, E7018-A1
		F2, F11, F12	E8018-B2, Note 9
A199-90a	Heat-exhanger & condenser tubes	T11	E8018-B2, Note 9
A200-90a	Refinery still tubes	T11	E8018-B2, Note 9
A209-90a	Carbon-moly boiler tubes	T1, T1a & T1b	E7010-A1, E7018-A1
A210-90a	Carbon-steel boiler tubes	A1	Note 1 or E7010-A1
		C	E7010-A1
A213-91a	Boiler tubes	T2, T11, T12 & T17	E8018-B2
A214-90a	Condenser tubes		Note 1
A216-89	High-temperature cast fittings	WCA, WCB, WCC	E7018 or E7018-A1
A217-91	High-temperature cast fittings	WC1	E7010-A1
		WC4	E8018-C3
		WC5, WC6, WC11	E8018-B2
A234-91c	Wrought welding fittings	WPA, WPB & WPC	Note 1
		WP1	E7010-A1, Note 2
		WP5, WP7, WP9, WP11, WP22	E8018-B2
A250-91	Carbon-moly boiler tubes	T1, T1a, T1b, T2, T11, T12	E7010-A1, Note 2
A333-91a & A334-91	Low-temperature pipe	1, 6 & 10	E7018 or E8018-C3
		3, 7 & 9	E8018-C2
		7	E8018-C1
A335-91a	High-temperature pipe	P1	E7010-A1, Note 2
		P2, P11 & P12	E8018-B2
		Others	Note 9
A350-91a	Low-temperature fittings	LF1 & LF2	E8018-C1
		LF3	E8018-C2
		LF5	E8018-C3
E369-91a	High-temperature pipe		See A335 & A182
A381-89	High-pressure pipe	Y35, Y42 & Y46	Notes 1 & 2
		Y52 & Y56	Note 5
		Y60 & Y65	Note 5 or E8018-C3
A405-91	High-temperature pipe	P24	E8018-B2
A420-91	Low-temperature pipe	See A203, A333, A334, A350	
A423-90	Low-alloy tube	1 & 2	E8018-C3 or E7018
A426-91	High-temperature cast pipe		See A335
A498-86	Condenser tubes	See A199, A179, A213, A214 & A334	
A500-90a	Structural tubing	A, B, C & D	E7018 & Note 1
A501-89	Structural tubing		E7018 & Note 1
A524-89	Process piping	I & II	E7010-A1 or E7018
A556-90a	Feed water heater tubes	A2, B2 & C2	E7018, Note 1
A557-90a	Feed water heater tubes	A2, B2 & C2	E7018
A618-90a	Structural tubing	I, II & III	E7018, E7015, E7016
API-2H		42 & 50	E7015, E7016, E7018
API-5L		X52	E7015, E7016, E7018

Note 1. Unless restricted by specifications; use any E60XX or E70XX electrode for steel grades with 60,000 psi or less tensile strength, for steel grades with 60,000 to 70,000 psi tensile strength use E70XX electrodes.

Note 2. Use E7010-G, specially designed for field-welding pipe.

Note 3. Use E8018-C1 or E8010-B2 for best color match on unpainted steels with enhanced atmospheric corrosion resistance.Consult the steel supplier.

Note 4. E7018 or E8018-C3 are frequently used for fillet welds.

Note 5. Use special electrode designed for field welding 5LX pipe, Grades X42 thru X65.

Note 6. Do not use E8018-B2 for low-temperature applications.

Note 7. E7018, E7028 for fillets, or E8018-C3 for general-purpose welding, can be used on these steels. If the weldment is to be precipitation-hardened or high weld strength is required, use E8018-B2.

Note 8. Usually E6010 is the most satisfactory electrode for galvanized sheet.

Note 9. Electrode recommendations for other alloy steels may be found in Sections 6.1 and 7.2.

Note 10. These recommendations are based on matching the tensile properties of the weld deposit and the plate, and also the chemical properties of the weld deposit and the plate where chemistry is important. Since it is impossible to foresee all the conditions of every application, other electrodes than those recommended here may also be satisfactory and should be tested before the weldment is started.

Fast-freeze electrodes provide deep penetration and maximum admixture. The weld bead is flat with distinct ripples. Slag formation is light, and the arc is easy to control.

Applications for fast-freeze electrodes are:

- General-purpose fabrication and maintenance welding.
- Vertical-up and overhead plate welds requiring X-ray quality.
- Pipe welding, including cross-country, in-plant, and noncritical small-diameter piping.
- Welds to be made on galvanized, plated, painted, or unclean surfaces.
- Joints requiring deep penetration, such as square-edge butt welds.
- Sheet-metal welds, including edge, corner, and butt welds.

Electrode Characteristics

E6010: This is the basic fast-freeze electrode for general-purpose DC welding. Light slag and good wash-in permit excellent control of the arc. The E6010 electrode is particularly valuable for critical out-of-position applications, such as with pipe welding.

E6011: A general fast-freeze electrode for use with industrial AC welders, E6011 is also the preferred electrode for sheet-metal edge, corner, and butt welds with DCEN. The electrode is also used for vertical-down welding, and for applications requiring exceptionally low silicon deposit. Special grades are available for general-purpose shop use with small, low open-circuit voltage AC welders (not suitable for X-ray quality). E6011 is also available in a special grade producing little slag, that is designed especially for tack welding.

E6022: Electrodes of the E6022 classification are recommended for single pass, high-speed, high current welding of groove welds in the flat welding position, lap joints in the horizontal welding position, and fillet welds on sheet metal. The weld face tends to be more convex and less uniform, especially since the welding speeds are higher.

E7010-A1: This fast-freeze electrode is designed for welding high-strength pipe, such as X52 or X56, and for other out-of-position welding where high strength or control of alloy in the weld are important. It produces a 70,000-psi deposit containing 0.5% molybdenum. Operation is similar to E6010.

E7010-G: This electrode is similar to E7010-Al, but is designed specifically to avoid any surface-hole tendency in fill and cover-pass welds on high-strength pipe. Special grades are available for welding all passes on X60 and X65 high-strength line pipe.

Welding Techniques

Current and Polarity: Unless otherwise specified, use DCEP with Exx10, and use AC with Exx11. Exx11 electrodes can be used on DCEP with a current about 10% below normal AC values. Always adjust current for proper arc action and control of the weld puddle.

Flat Welding: Hold an arc of 1/8 in. or less, or touch the work lightly with the electrode tip. Move fast enough to stay ahead of the molten pool. Use currents in the middle and high portion of the range.

Vertical Welding: Use an electrode of 3/16 in. or smaller. Vertical-down techniques are used by pipeliners and for single-pass welds on thin steel. Vertical-up is used for most plate welding. Make the first vertical-up pass with either a whipping technique for fillet welds (Fig. 6-17), or with a circular motion for V-butt joints. Apply succeeding passes with a weave, pausing slightly at the edges to insure penetration and proper wash-in. Use currents in the low portion of the range.

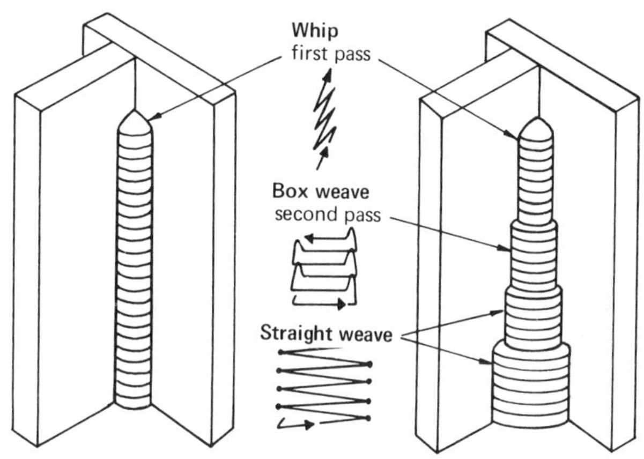

Fig. 6-17. Technique for vertical welding with fast-freeze electrodes.

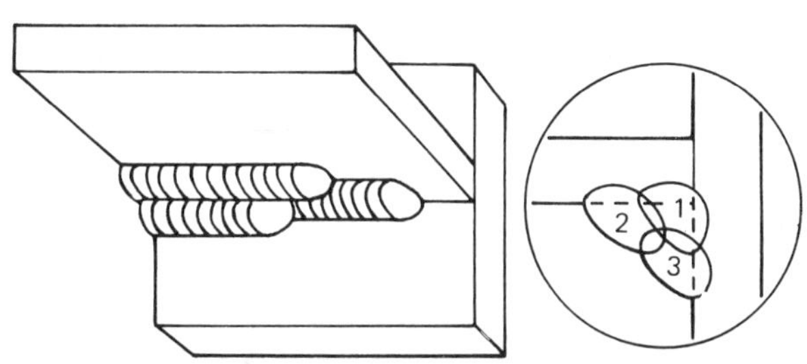

Fig. 6-18. Technique for overhead and horizontal fillet welds with fast-freeze electrodes. These welds are best made with a series of stringer beads.

Overhead and Horizontal Fillet Welds: Use an electrode of 3/16 in. or smaller. These welds (Fig. 6-18) are best made with a series of stringer beads using a technique similar to those described for first-pass vertical-up welds.

Sheet-Metal Edge and Butt Welds: Use DCEN. Hold an arc of 3/16 in. or more. Move as fast as possible while maintaining good fusion. Position the work 45° downhill for fastest welding. Use currents in the middle range.

FAST-FILL ELECTRODES

Fast-fill electrodes are compounded to deposit metal rapidly in the heat of the arc and are, thus, well suited to high-speed welding on horizontal surfaces. The weld metal solidifies somewhat slowly. therefore this type of electrode is not well suited for out-of-position welds. However, a slight downhill positioning is permissible. Joints normally considered fast-fill include butt, fillet, lap, and corner welds in plate 3/16 in. or thicker. These joints are capable of holding a large molten pool of weld metal as it freezes.

Arc penetration is shallow with minimum admixture. The bead is smooth, free of ripples, and flat or slightly convex. Spatter is negligible. Slag formation is heavy, and the slag peels off readily.

Applications for fast-fill electrodes are:

- Production welds on plate having a thickness of 3/16 in. or more.
- Flat and horizontal fillets, laps, and deep-groove butt welds.
- Welds on medium-carbon crack-sensitive steel when low-hydrogen electrodes are not available. (Preheat may be required.)

The coverings of fast-fill electrodes contain approximately 50% iron powder. This powder increases deposition rate by helping to contain the arc heat at the electrode, by melting to add to deposited weld metal, and by permitting currents higher than those permitted by other types of coverings. The thick, iron-bearing covering also facilitates use of the drag technique in welding.

Electrode Characteristics

E7024: This is a general-purpose fast-fill electrode. Special grades provide exceptionally high deposition rates and particularly good operating characteristics. The E7024 electrodes are well suited for making fillet welds in the flat or horizontal position. The weld face is slightly convex to flat, with a very smooth surface and a very fine ripple. These electrodes are characterized by a smooth, quiet arc, very low spatter, and low arc penetration. They can be used with high travel speeds. Electrodes of these classifications can be operated on AC, DCEP, or DCEN.

E7024-1: These electrodes have the same general usability characteristics as E7024 electrodes. They are intended for use in situations requiring greater ductility and a lower transition temperature than normally is available from E7024 electrodes.

E6027: Used principally for flat deep-groove joints and for flat and horizontal fillets, the electrode has excellent wash-in characteristics. A friable slag permits easy slag removal in deep grooves. This electrode is sometimes used as an alternative to E7024 when X-ray quality or high notch toughness are required.

E7020-A1: The electrode is used in place of E6027 when a 70,000-psi strength or 0.5% molybdenum deposit is required.

Welding Techniques

Polarity: Use AC for highest speeds and best operating characteristics. DCEP can be used, but this type of current promotes arc blow and complicates control of the molten puddle.

Flat Welding: Use a drag technique; tip the electrode 10 to 30° in the direction of travel and make stringer beads. Weld with the electrode tip lightly dragging on the work so that molten metal is forced out from under the tip, thereby promoting penetration. The resulting smooth weld is similar in appearance to an automatic weld. Travel rapidly, but not too fast for good slag coverage. Stay about 1/4 to 3/8 in. ahead of the molten slag, as illustrated in Figure 6-19. If travel speed is too slow, a small ball of molten slag may form and roll ahead of the arc, causing spatter, poor penetration, and erratic bead shape. Optimum current usually is 5 to 10 amp above the center of the range for a given electrode. Do not exceed the center of the range if the weld is to be of X-ray quality.

Horizontal Fillets and Laps: Point the electrode into the joint at an angle of 45° from horizontal and use the "flat" technique described above. The tip of the electrode must touch both horizontal and vertical members of the joint. If the 45° angle between plates is not maintained, the fillet legs will be of different sizes. When two passes are needed, deposit the first bead mostly on the bottom plate. To weld the second pass hold the electrode at about 45°, fusing into the vertical plate and the first bead.

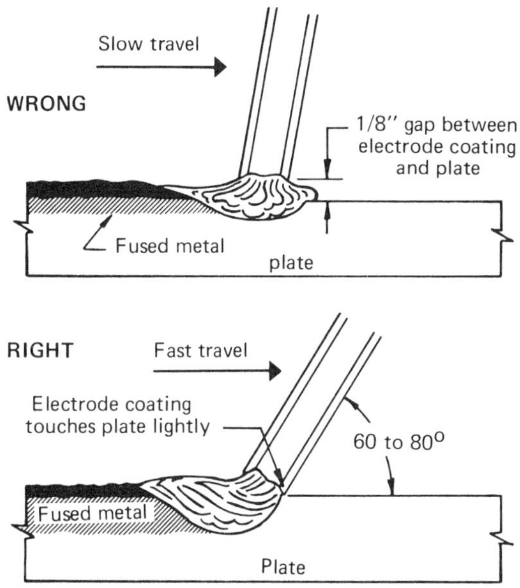

Fig. 6-19. Technique for flat welds with fast-fill electrodes. An incorrect technique is included for comparison.

Make multiple-pass horizontal fillets as shown in Figure 6-20. Put the first bead in the corner with fairly high current, disregarding undercut. Deposit the second bead on the horizontal plate, fusing into the first bead. Hold the electrode angle needed to deposit the filler beads as shown, putting the final bead against the vertical plate.

Deep-Groove Butt Welds: To hold the large pool of molten weld metal produced by fast-fill electrodes, either a backup plate, or a stringer bead made with a deeper-penetrating fast-freeze electrode is required. Deposit fast-fill beads with a stringer technique until a slight weave is required to obtain fusion of both plates. Split-weave welds are better than a wide weave near the top of deep grooves. When welding the second last pass, leave enough room so that the last pass will not exceed a 1/16-in. buildup. A slight undercut on all but the last pass creates no problems, because it is melted out with each succeeding pass.

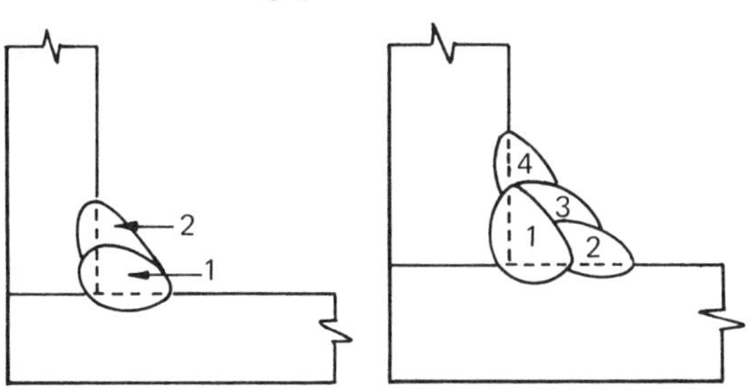

Fig. 6-20. Technique for multi-pass horizontal fillet welds with fast-fill electrodes. Beads should be deposited in the order indicated.

FILL-FREEZE ELECTRODES

Fill-freeze electrodes are compounded to provide a compromise between fast-freeze and fast-fill characteristics, and thus provide medium deposition rates and medium penetration. Since they permit welding at relatively high speed with minimal skip, misses, and undercut, and with minimum slag entrapment, fill-freeze electrodes are also referred to as fast-follow electrodes. The electrode's characteristics are particularly suited to the welding of sheet metal, and fill-freeze electrodes are, thus, often called "sheet-metal" electrodes. Bead appearance with this group of electrodes varies from smooth and ripple-free to wavy with distinct ripples. The fill-freeze electrodes can be used in all welding positions, but are most widely used in the level or downhill positions.

Applications for fill-freeze electrodes include:

- Downhill fillet and lap welds.
- Irregular or short welds that change direction or position.
- Sheet-metal lap and fillet welds.
- Fast-fill joints having poor fitup.
- General-purpose welding in all positions.

Fast-freeze electrodes, particularly E6010 and E6011, are sometimes used for sheet-metal welding when fast-follow electrodes are not available, or when the operator prefers faster solidification. Techniques for sheet-metal welding with these electrodes are discussed in the portion of this section dealing with fast-freeze electrodes.

Electrode Characteristics

E6012: The basic fill-freeze electrode for general-purpose and production welding, this electrode provides a more forceful arc than other electrodes in the group. Special types are available for improved arc stability, minimum spatter, and easier slag removal. Some types contain iron powder in the coatings for greater mileage, better AC operation, and a smoother, quieter arc. These types are excellent for low-current applications, such as sheet-metal welding.

E6013: This electrode is used in place of E6012 for sheet-metal welding where appearance and ease of operation are more important than speed. AC operation is excellent. It is recommended for general-purpose welding with small AC transformer welding machines having low open-circuit voltage.

E6019: These electrodes, although very similar to E6013 and E6020 electrodes in their coverings, have distinct differences. E6019 electrodes, with a rather fluid slag system, provide deeper arc penetration and produce weld metal that meets a 22% minimum elongation requirement, meets the Grade 1 radiographic standards, and has an average impact strength of 20 ft-lb (27J) when tested at 0°F (−18°C). E6019 electrodes are suitable for multipass welding of up to 1 in. (25 mm) thick steel. They are designed for use with AC, DCEN, or DCEP.

E7014: The electrode has highest iron-powder content in the group, and thus provides highest deposition (or maximum fast-fill capability) among the fill-freeze electrodes. It has exceptionally good operating characteristics and is often preferred by weldors. It is frequently used for production welding on short, irregular, or downhill fast-fill types of joints.

Welding Techniques on Steel Plate

Polarity: Use DCEN for best performance on all applications except when arc blow is a problem. To control arc blow, use AC.

Flat and Downhill: Use stringer beads for the first pass except when poor fitup requires a slight weave. Use either stringer or weave beads for succeeding passes. Touch the tip of the electrode to the work or hold an arc length of 1/8 in. or less. Move as fast as possible consistent with desired bead size. Use currents in the middle to higher portion of the range.

Electrode Size: Use electrodes of 3/16-in. or smaller diameter for vertical and overhead welding.

Vertical-Down: Use stringer beads or a slight weave. A drag technique must be used with some E6012 electrodes. Make small beads. Point the electrode upward so that arc force pushes molten metal back up the joint. Move fast enough to stay ahead of the molten pool. Use currents in the higher portion of the range.

Vertical-Up: Use a triangular weave. Weld a shelf at the bottom of the joint and add layer upon layer. Do not whip or take the electrode out of the molten pool. Point the electrode slightly upward so that arc force helps control the puddle. Travel slow enough to maintain the shelf without spilling. Use currents in the lower portion of the range.

Overhead: Make stringer beads using a whipping technique with a slight circular motion in the crater. Do not weave. Travel fast enough to avoid spilling. Use currents in the lower portion of the range.

Welding Techniques with Sheet Metal

The ability to adjust current while welding sheet steel is valuable, particularly when fitup or material thickness varies. Motor-generator welders equipped with foot-operated remote current controls are useful for this purpose.

Generally, use the highest current that does not cause melt-through, does not undercut, or does not melt the edges of lap, corner, or edge welds. For fast welding, the operator must stay precisely on the joint and must travel at a uniform speed. Welding on sheet metal, thus, requires more than average skill, and a good weldor may need a few days of practice when first attempting this type of weld.

For maximum welding speed, minimum distortion, or for welding in the flat position, joints generally should be positioned 45° to 75° downhill. Use copper backup strips where possible to decrease the danger of melt-through. The procedures tables in this handbook assume tight fitup and adequate clamping or tacking. Where poor fitup is encountered:

1. Reduce current.
2. Tilt the electrode into the direction of travel more than normally.
3. With fast-freeze electrodes use a small, quick weave technique to bridge the gap.

Deposit the entire weld in one pass using stringer beads or a slight weave. Drag the electrode on the joint and stay ahead of the molten pool. Tip the electrode well into the direction of travel so the arc force pushes the weld metal back into the joint. Use currents in the high portion of the range.

LOW-HYDROGEN ELECTRODES

Conventional welding electrodes may not be suitable where X-ray quality is required, where the base metal has a tendency to crack, where thick sections are to be welded, or where the base metal has an alloy content higher than that of mild steel. In these applications, a low-hydrogen electrode may be required.

Low-hydrogen electrodes are available with either fast-fill or fill-freeze characteristics. They are compounded to produce sound welds of X-ray quality with excellent notch toughness and high ductility. Low-hydrogen electrodes reduce the danger of underbead and microcracking on thick weldments and on high-carbon and low-alloy steels. Preheat requirements are less than for other electrodes.

Most low-hydrogen electrodes are shipped in hermetically sealed containers, which normally can be stored indefinitely without danger of moisture pickup. But once the container is opened, the electrodes should be used promptly or stored in a heated cabinet. Details on electrode storage and on redrying moisture-contaminated electrodes are presented later in this section.

An "R" designation for an electrode such as E7018R indicates a low-moisture-absorbing coating. To qualify for an "R" designation, electrodes which have been removed from freshly opened packages or which have been reconditioned, are exposed to 80 percent relative humidity and a temperature of 80°F for 9 hours. Moisture content after this exposure shall not be more than 0.4 weight percent.

Applications for low-hydrogen electrodes include:

- X-ray-quality welds or welds requiring high mechanical properties.
- Crack-resistant welds in medium-carbon to high-carbon steels; welds that resist hot-short cracking in phosphorus steels; and welds that minimize porosity in sulfur-bearing steels.
- Welds in thick sections or in restrained joints in mild and alloy steels where shrinkage stresses might promote weld cracking.
- Welds in alloy steel requiring a strength of 70,000 psi or more.
- Multiple-pass, vertical, and overhead welds in mild steel.

Electrode Characteristics

E7018: This electrode has fill-freeze characteristics and is suitable for all-position operation. Iron powder in the electrode coating promotes rapid deposition. Moderately heavy slag is easy to remove. (Weld metal freezes rapidly even though slag remains somewhat fluid.) In addition to their use on carbon steel, the E7018 electrodes are also used for joints involving high-strength, high carbon, or low alloy steels. The fillet welds made in the horizontal and flat welding positions have a slightly convex weld face, with a smooth and finely rippled surface. The electrodes are characterized by a smooth, quiet arc, very low spatter, and medium arc penetration. E7018 electrodes can be used at high travel speeds.

E7018-1: These electrodes have the same usability and weld metal composition as E7018 electrodes, except that the manganese content is set at the high end of the range. They are intended for welds requiring a lower transition temperature than is normally available from E7018 electrodes.

E7028: The electrode has fast-fill characteristics applicable to high-production welds where low-hydrogen quality is required. It performs best on flat fillets and deep-groove joints, but is also suitable for horizontal fillet and lap welds. Excellent restriking qualities permit efficient skip and tack welding.

Welding Techniques

Techniques for E7028 are the same as those described for conventional fast-fill electrodes. However, special care should be taken to clean the slag from every bead on multiple-pass welds to avoid slag inclusions that would appear on X-ray inspection. The ensuing discussion pertains to the techniques recommended for E7018 electrodes.

Polarity: Use DCEP whenever possible if the electrode size is 5/32-in. or less. For larger electrodes, use AC for best operating characteristics (but DCEP can also be used).

Flat Position: Use low current on the first pass, or whenever it is desirable to reduce admixture with a base metal of poor weldability. On succeeding passes, use currents that provide best operating characteristics. Drag the electrode lightly or hold an arc of 1/8-in. or less. Do not use a long arc at any time, since E7018 electrodes rely principally on molten slag for shielding. Stringer beads or small weave passes are preferred to wide weave passes. When starting a new electrode, strike the arc ahead of the crater, move back into the crater, and then proceed in the normal direction. On AC, use currents about 10% higher than those used with DC. Govern travel speed by the desired bead size.

Vertical: Weld vertical-up with electrode sizes of 5/32-in. or less. Use a triangular weave for heavy single-pass welds. For multipass welds, first deposit a stringer bead by using a slight weave. Deposit additional layers with a side-to-side weave, hesitating at the sides long enough to fuse out any small slag pockets and to minimize undercut. Do not use a whip technique or take the electrode out of the molten pool. Travel slowly enough to maintain the shelf without causing metal to spill. Use currents in the lower portion of the range.

Overhead: Use electrodes of 5/32-in. or smaller. Deposit stringer beads by using a slight circular motion in the crater. Maintain a short arc. Motions should be slow and deliberate. Move fast enough to avoid spilling weld metal, but do not be alarmed if some slag spills. Use currents in the lower portion of the range.

Redrying Low-Hydrogen Electrodes

Low-hydrogen electrodes must be dry if they are to perform properly . Electrodes in unopened, hermetically sealed containers remain dry indefinitely in good storage conditions. Opened cans should be stored in a cabinet at 250 to 300°F. Supplying

TABLE 6-17. Procedures for Drying Low-Hydrogen Electrodes

Nature of Moisture Pickup	Drying Temperatures	
	E7015, E7016, E7018, E7028, E7048	E8018-X, E9018-X, E11018-X
Electrodes exposed to air for less than one week; no direct contact with water. Welds not subject to X-ray inspection.	250°F to 300°F	250°F to 300°F
Electrodes exposd to air for less than one week; no direct contact with water. Welds subject to X-ray inspection.	500°F to 800°F	500°F to 800°F
Electrodes have come in direct contact with water, or have been exposed to extremely humid conditions as indicated by core wire rusting at the holder end. Before redrying at 700 - 750°F, predry electrodes in this condition at 180°F for 1 to 2 hours. This minimizes the tendency for coating cracks or oxidation of the alloys in the coating.	500°F to 800°F	500°F to 800°F

NOTE: One to two hours at the listed temperatures is satisfactory. Do not dry electrodes at higher temperature or for more than 8 hours. Several hours at lower temperature are not equivalent to using the specified temperatures. Remove the elecrodes from the can and spread them out in the furnace. Each electrode must reach the drying temperature. (Cardboard can liners char at about 350°F.)

weldors with electrodes twice a shift — at the start of the shift and at lunch, for example — minimizes the danger of moisture pickup. Return electrodes to the heated cabinet for overnight storage.

When containers are punctured or opened so that the electrode is exposed to the air for a few days, or when containers are stored under unusually wet conditions, low-hydrogen electrodes pick up moisture. The moisture, depending upon the amount absorbed, impairs weld quality in the following ways:

1. A small amount of moisture may cause internal porosity. Detection of this porosity requires X-ray inspection or destructive testing. If the base metal has high hardenability, even a small amount of moisture can contribute to underbead cracking.
2. A high amount of moisture causes visible external porosity in addition to internal porosity.
3. Severe moisture pickup can cause weld cracks or underbead cracking in addition to severe porosity.

Redrying completely restores ability to deposit quality welds. The proper redrying temperature depends upon the type of electrode and its condition. Drying procedures are listed in Table 6-17.

TABLE 6-18. Characteristics of Mild-Steel Covered Electrodes*

AWS-ASTM Electrode Classification	Welding Category	General Characteristics
60,000-psi Minimum Tensile Strength		
E6010	Freeze†	Molten weld metal freezes quickly; suitable for welding in all positions with DC electrode positive power; has a low-deposition rate and deeply penetrating arc; can be used to weld all types of joints.
E6011	Freeze†	Similar to E6010, except can be used with AC as well as DC power.
E6012	Follow	Faster travel speed and smaller welds than E6010; AC or DC, electrode negative power; penetration less than E6010. Primary use is for single-pass welding of thin-gage sheet metal in flat, horizontal, and vertical-down positions.
E6013	Follow	Similar to E6012, except can be used with DC (either polarity) or AC power.
E6019	Fill-Freeze	Similar to E6013 provides deeper penetration, meets 22% min. elongation.
E6022	Freeze	For single pass, high-speed groove welds in flat position.
E6027	Fill	Deposition rate high since covering contains about 50% iron powder; primary use is for multipass, deep-groove, and fillet welding in the flat position or horizontal fillets, using DC (either polarity) or AC power.
70,000-psi Minimum Tensile Strength		
E7014	Fill-Freeze	Higher deposition rate than E6010; usable with DC (either polarity) or AC power; primary use is for inclined and short, horizontal fillet welds.
E7018	Fill-Freeze	Suitable for welding low and medium-carbon steels (0.55% C max) in all positions and types of joints. Weld-metal quality and mechanical properties highest of all mild-steel electrodes; usable with DC electrode positive or AC power.
E7018-1	Fill-Freeze	Same as E7018 except Mn content at high end of range to lower transition temperature.
E7024	Fill	Higher deposition rate than E7014; suitable for flat-position welding and horizontal fillets.
E7024-1	Fill	Same usability characteristics as E7024 but has greater ductility and lower transition temperature.
E7028	Fill	Similar to type E7018; used for welding horizontal and flat fillets and grooved butt fillet welds in flat position.

* E6020, E7015 and E7016 are not included because of their limited usage. Only electrodes up to 3/16-in. diameter can be used in all welding positions (flat, horizontal, vertical, and overhead).

† When used for welding sheet metal, these electrodes have follow-freeze characteristics.

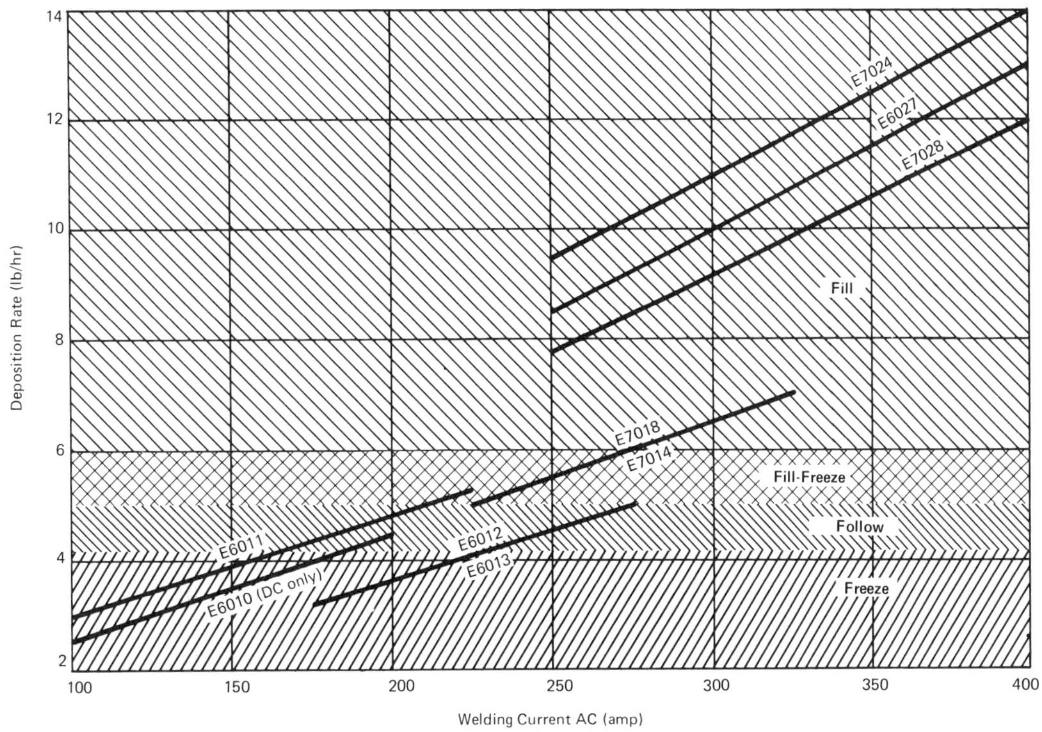

Fig. 6-21. Deposition rates for various mild-steel electrodes.

SUMMARY OF ELECTRODES FOR MILD STEEL

In the AWS specification A5.1-91 there are 17 different classifications of electrodes for welding mild steel. Each classification has different operating characteristics, and a summary of these characteristics is given in Table 6-18. The deposition rates for the electrodes in Table 6-18 are shown in Fig. 6-21.

ALLOY-STEEL ELECTRODES

Alloy content of the weld deposit is not critically important in the welding of common grades of steel. As discussed in the immediately preceding portions of this section, electrode selection for these steels is based largely on whether maximum deposition rates or rapid freeze characteristics are preferred. But for alloy steels — chosen specifically for their high mechanical properties, superior corrosion resistance, or ability to withstand high temperatures — the electrode must be carefully selected so that it provides the specific chemical composition needed to maintain the desired properties of the base metal in the weld deposit.

There are many types of electrodes available for welding low-alloy steels. These types are described completely in AWS A5.5, and a brief summary of typical electrode characteristics and applications is presented in the following paragraphs. The chemical requirements of deposited weld metal are given in Table 4-7. Typical mechanical properties of some of the weld deposits are given in Table 6-19. A guide to the selection of electrodes for welding steels of specific trade names is presented in Table 6-20.

Except for electrodes for welding high-strength line pipe (see Section 13.3), most electrodes for welding low-alloy steel have low-hydrogen, fill-freeze characteristics similar to those of E7018 and are suitable for all-position fabrication and repair welding. Even though these electrodes are suitable for all-position welding, their operating characteristics are quite different from those of fast-freeze electrodes for the common steels. Weld metal from alloy-steel electrodes freezes rapidly even though the slag remains relatively fluid. Deposition rate is high, partially because the coverings contain iron powder.

Beads are flat or slightly convex and have distinct ripples with little spatter. The moderately heavy slag is easy to remove.

Some of the commonly used low-alloy high-strength electrodes include:

E7010-G, E8010-G: These classifications are used for electrodes not meeting the deposit chemistry requirements of existing classifications. The

TABLE 6-19. Typical Mechanical Properties of AWS A5.5-81 Weld Metal

	E7010-A1	E8018-B2	E8018-C3	E8018-C1	E11018-M
As-Welded					
Tensile Strength (psi)	75,000	102,000	86,000	87,000	112,000
Yield Strength (psi)	68,000	90,000	75,000	74,000	102,000
Elongation (% in 2 in.)	24	21	25	22	21
Charpy V Notch (ft-lb)	60 at 70°F	40 at 70°F	48 at −20°F	61 at −75°F	35 at −60°F
Stress Relieved 1150°F					
Tensile Strength (psi)	72,000	93,000 83,000*	81,000	84,000	112,000+
Yield Strength (psi)	60,000	81,000 70,000*	70,000	71,000	96,000+
Elongation (% or 2 in.)	29	20 22*	26	24	22+
Charpy V Notch (ft-lb)	68 at 70°F	65 at 70°F 65 at 70°F*	88 at −20°F	40 at −75°F	35 at −60°F+

* Stress relieved at 1275°F
\+ Stress relieved at 1025°F

E7010-G and E8018-G electrodes are generally designed for welding typical high strength pipe butt welds in the vertical position with downward progression.

E81018-B2: This electrode produces a 1.25%-chromium, 0.5%-molybdenum deposit, commonly required for high-temperature, high-pressure piping. It usually meets requirements of E9018-G for some

TABLE 6-20. Recommended Electrodes for Trade-Name Steels (See Note 10, Table 6-16)

Steel Producer	Steel Trade Name	Recommended Electrodes
Armco Steel Co.	Armco C-42, 45	Note 1
	Armco C-50, 55 Formable 50, 55	E7018 or E7028
	Armco C-60, 65 Formable 70	E8018-C3
	Armco C-70	E8018-B2
Bethlehem Steel Corporation	RQC-60N Mayari R, R-50 Med. Mn	E7018 or E7028
	Mayari R-60	E8018-C3
	RQC-60 Q &T	E8010-C1
	RQC80, 90 RQ100, 100A, 100B	E11018-M
Inland Steel Company	INX-42, 45	Note 1
	INX-50, 55, Cor-Ten A, B Hi-Man, Tri-Steel	Note 7 E7018 or E7028
	INX-60, 65 Cor-Ten C	E8018-C3
	INX-70	E8018-B2
Lukens Steel Company	Lukens 45, 50 LT-75 Lukens A440, A441 Cor-Ten B	E7018 or E7028
	Lukens 55, 60	E8018-C3
	Lukens LT-75QT	E8018-B2
	Lukens T1, T1A, T1B LT-75HS	E11018-M
National Intergroup Inc.	GLX-45W	Note 1
	GLX-50W, 55W NAX-Fine Grain NAX High Tensile	E7018 or E7028
	GLX-60W, 65W	E8018-C3
	GLX-70W	E8018-B2
	N-A-XTRA 80, 90, 100	E11018-M

Steel Producer	Steel Trade Name	Recommended Electrodes
LTV Steel (formally Republic Steel Corp.)	Republic X42W, X45W	Note 1
	Republic X50W, X55W	E7018 or E7028
	Republic X60W, X65W	E8018-C3
	Republic X70W	E8018-B2
	Republic X80W	E11018-M
USX Steel (formerly United States Steel Corp.)	EX-Ten 42, 45	Note 1
	EX-Ten 50, 55 Cor-Ten A, B Par-Ten Tri-Ten	E7018 or E7028
	Cor-Ten C EX-Ten 60, 65	E8018-C3
	Char-Pac	E8018-C1
	EX-Ten 70	E8018-B2
	T-1, T-1A, T-1B	E11018-M
Youngstown Sheet & Tube Company	Yoloy YSW-42, 45	Note 1
	Yoloy YSW-50-55 Yoloy YSW A441 YB-Ten	E7018 or E7028
	Yoloy YSW-60, 65 Yo Man Yoloy HS, HSX	E8018-C3 Note 7
	Yoloy S	E8018-C1
	Yoloy YSW-70	E8018-B2

For notes, refer to Table 6-16.

high-strength (90,000 psi tensile) steels. Since Cr-Mo deposits are air hardenable, both preheat and PWHT (postweld heat treatment) are required for most applications.

E8018-B2L: This electrode produces a weld deposit similar to that of the E8018-B2 electrode, except for the carbon content. This electrode produces a lower carbon content weld metal (.05% maximum) which will improve ductility and lower hardness, but will also reduce the high temperature strength and creep resistance of the weld metal.

E8018-C3: The electrode conforms to MIL 8010-C3 and produces a weld having a tensile strength of 80,000 psi, suitable for general-purpose welding on many high-strength alloys. This type also provides a 1%-nickel deposit for welding alloys that are to be used at low temperatures and which require good notch toughness down to −60°F. The electrode is also used for fillet welds on high-strength (110,000 psi tensile) quenched-and-tempered steels, such as ASTM A514 and A517.

E8018-C1: The type produces a 2.25%-nickel deposit with notch toughness of 50 ft-lb at −75°F and is, thus, commonly required for welding low-temperature alloys. Such alloys are frequently used to fabricate storage, piping, and transportation equipment for liquid ammonia, propane, and other gases. This group of electrodes is also recommended for the best color match on unpainted corrosion-resistant ASTM A242 steels. (Cor-Ten, Mayari-R, and others).

E9018-B3: This electrode produces a 2-1/4% chromium-1% molybdenum deposit, commonly required for certain high temperature service applications, and for matching the properties of basemetals used for these applications. Since Cr-Mo deposits are air hardenable, both preheat and PWHT are required for most applications.

E9018-B3L: This electrode produces a weld deposit similar to that of the E9018-B3 electrode, except for lower carbon content (.05% maximum). This will improve ductility and lower hardness, but will also reduce the high temperature strength and creep resistance of the weld metal.

E11018-M: The electrode conforms to MIL-11018-M and produces a 110,000-psi tensile strength needed for full-strength welds on quenched-and-tempered steels, ASTM A514 and A517 (T-1, SSS-100, HY-80, and others).

GENERAL CONSIDERATIONS IN WELDING

Joint Positions

As noted earlier in this section, joint position is often the primary factor in electrode selection

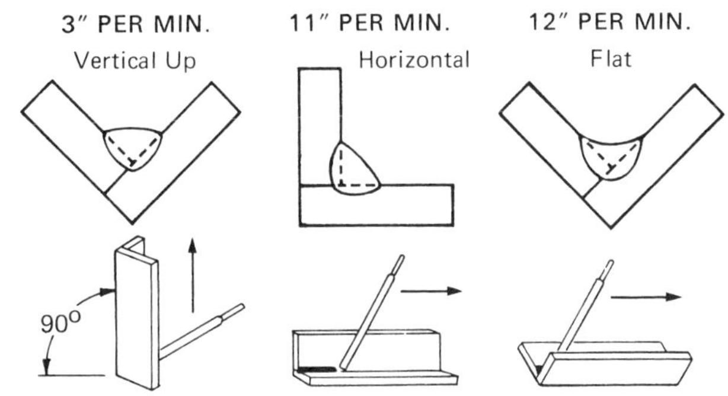

Fig. 6-22. Variations in welding speed with different joint positions.

and is therefore largely responsible for the speed and cost of welding. Where possible, work should be positioned flat for fastest welding speed.

Sheet-Metal Welds: In sheet steel from 10 to 18 gage, welds are usually larger than needed for joint strength. Thus, the primary objective is to avoid melt-through while welding at fast travel speeds with minimum skips and misses. Fastest speeds are obtained with the work positioned 45 to 75° downhill. Refer to the prior portion of this section on fill-freeze electrodes.

Welds on Mild Steel Plate: Plates having a thickness of 3/16-in. or greater are welded most rapidly in the flat position. This position permits easiest manipulation of the electrode and allows use of high-deposition fast-fill electrodes. Variations in welding speed with different joint positions are illustrated in Fig. 6-22. For more information, refer to portions of this section dealing with fast-fill electrodes. If a weld is to be made in the vertical or overhead position, refer to the discussion on fast-freeze electrodes.

Welds on High-Carbon and Low-Alloy Steel: These steels can be welded most readily in the level position. Refer to the discussion on low-hydrogen electrodes.

Joint Geometry and Fitup

Joint dimensions specified in the Procedure Tables are chosen for fast welding speeds consistent with weld quality. Departure from the recommended joint geometry may reduce welding speed or cause welding problems.

Fitup must be consistent for the entire joint. Sheet metal and most fillet and lap joints must be clamped tightly their entire length. Gaps or bevels must be accurately controlled over the entire joint. Any variations in a joint make it necessary for the operator to reduce the welding speed to avoid melt-through and force him to make time-consuming manipulations of the electrodes.

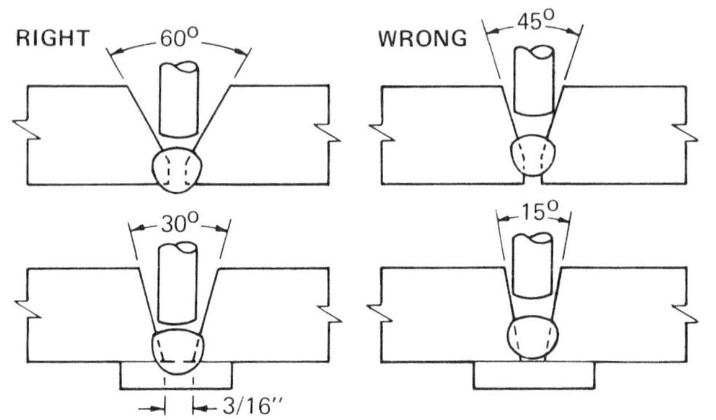

Fig. 6-23. Correct and incorrect bevels for good bead shape and adequate penetration.

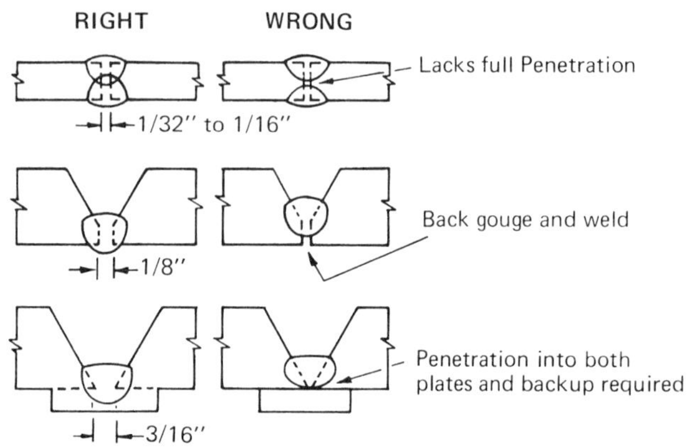

Fig. 6-24. Correct and incorrect gaps for proper penetration.

Sufficient bevel is required for good bead shape and adequate penetration (Fig. 6-23). Insufficient bevel prevents adequate entry of the electrode into the joint. A deep, narrow bead also has a tendency to crack. However, excess bevel wastes material. Sufficient gap is needed for full penetration (Fig. 6-24). Excessive gap wastes metal and slows welding speed. Either a 1/8-in. land or a backup strip is required for fast welding and good quality with thick plate (Fig. 6-25).

Feather-edge preparations require a slow costly seal bead. However, double-V butt joints without a land are practical when the seal bead cost is offset by easier edge preparation and the gap can be limited to about 3/32-in.

Weld seal beads on flat work with 3/16-in. E6010 electrodes at about 150 amp DCEP. Use 1/8-in. electrode at about 90 amp DCEP for vertical, overhead, and horizontal butt welds. Employ a combination whipping technique and circulating motion in the crater.

When low-hydrogen seal beads are required, use the appropriate EXX18 electrode. Weld with the same electrode sizes and about 20 amp higher current than recommended for E6010. Employ stringer bead technique with a slight weave when needed.

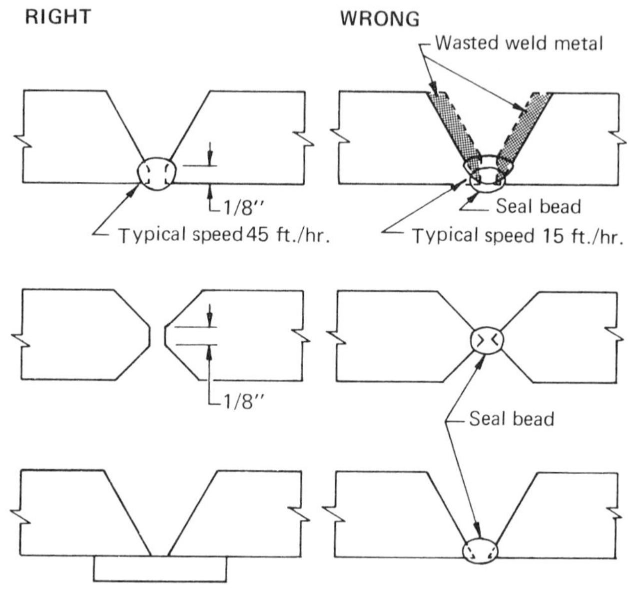

Fig. 6-25. Proper joint geometry for thick-plate welding.

Back-gouging from the second side is needed: 1. For X-ray quality. 2. When irregular gap or poor technique produces a poor bead. 3. When a heavy bead is needed to prevent melt-through of semiautomatic fill beads.

Joint Cleanliness

To avoid porosity and attain the speeds indicated in the Procedure Tables, remove excessive scale, rust, moisture, paint, oil, and grease from the surface of the joints.

If paint, dirt, or rust cannot be removed — as is sometimes the case in maintenance welding — use E6010 or E6011 electrodes to penetrate through the contaminants deeply into the base metal. Slow the travel speed to allow time for gas bubbles to boil out of the molten weld before it freezes.

Electrode Size

Large electrodes permit welding at high currents and high deposition rates. Therefore, use the largest electrode practical consistent with good weld quality. Electrode size is limited by many factors, but the most important considerations usually are:

1. High currents increase penetration. Therefore, electrode size is limited on sheet metal and with root passes where melt-through can occur.

TABLE 6-21. Minimum Preheat and Interpass Temperature[1,2] (AWS D1.1-92, Table 4.3) (see 4.2)

Category	Steel Specification				Welding Process	Thickness of Thickest Part at Point of Welding		Minimum Temperature	
						in.	mm	°F	°C
A	ASTM A36[3]		ASTM A516		Shielded metal arc welding with other than low hydrogen electrodes	Up to 3/4	19 incl.	None[4]	
	ASTM A53	Grade B	ASTM A524	Grades I & II					
	ASTM A106	Grade B	ASTM A529			Over 3/4	19		
	ASTM A131	Grades A, B,	ASTM A570	All grades		thru 1-1/2	38.1 incl.	150	66
		CS, D, DS, E	ASTM A573	Grade 65					
	ASTM A139	Grade B	ASTM A709	Grade 36[3]		Over 1-1/2	38.1		
	ASTM A381	Grade Y35	API 5L	Grade B		thru 2-1/2	63.5	225	107
	ASTM A500	Grade A		Grades X42					
		Grade B	ABS	Grades A, B, D,					
				CS, DS					
	ASTM A501			Grade E		Over 2-1/2	63.5	300	150
B	ASTM A36[3]		ASTM A570	All grades	Shielded metal arc welding with low hydrogen electrodes, submerged arc welding,[5] gas metal arc welding, flux cored arc welding				
	ASTM A53	Grade B	ASTM A572	Grades 42, 50					
	ASTM A106	Grade B	ASTM A573	Grade 65					
	ASTM A131	Grades A, B,	ASTM A588						
		CS, D, DS, E	ASTM A595	Grades A, B, C		Up to 3/4	19 incl.	None[4]	
		AH 32 & 36	ASTM A606						
		DH 32 & 36	ASTM A607	Grades 45, 50, 55					
		EH 32 & 36	ASTM A618						
	ASTM A139	Grade B	ASTM A633	Grades A, B					
	ASTM A242			Grades C, D		Over 3/4	19	50	10
	ASTM A381	Grade Y35	ASTM A709	Grades 36, 50, 50W		thru 1-1/2	38.1 incl.		
			ASTM A808						
	ASTM A441		API 5L	Grade B					
	ASTM A500	Grade A		Grade X42					
		Grade B	API Spec. 2H	Grades 42, 50		Over 1-1/2	38.1	150	66
	ASTM A501		ABS	Grades AH 32 & 36		thru 2-1/2	63.5 incl.		
	ASTM A516	Grades 55 &		DH 32 & 36					
		60, 65 & 70		EH 32 & 36					
	ASTM A524	Grades I & II	ABS	Grades A, B, D,					
	ASTM A529			CS, DS		Over 2-1/2	63.5	225	107
	ASTM A537	Classes 1 & 2		Grade E					
C					Shielded metal arc welding with low hydrogen electrodes, submerged arc welding,[5] gas metal arc welding, flux cored arc welding	Up to 3/4	19 incl.	50	10
	ASTM A572	Grades 60 & 65				Over 3/4 thru 1-1/2	19 38.1 incl.	150	66
	ASTM A633	Grade E				Over 1-1/2 thru 2-1/2	38.1 63.5 incl.	225	107
	API 5L	Grade X52				Over 2-1/2	63.5	300	150
D					Shielded metal arc welding with low hydrogen electrodes, submerged arc welding[5] with carbon or alloy steel wire, neutral flux, gas metal arc welding or flux cored arc welding	Up to 3/4	19 incl.	50	10
	ASTM A514					Over 3/4 thru 1-1/2	19 38.1 incl.	125	50
	ASTM A517								
	ASTM A709	Grades 100 & 100W				Over 1-1/2 thru 2-1/2	38.1 63.5 incl.	175	80
						Over 2-1/2	63.5	225	107
E	ASTM A710	Grade A (All classes)				No preheat is required[6]			

NOTES:

1. Welding shall not be done when the ambient temperature is lower than 0°F (−18°C). Zero °F (−18°C) does not mean the ambient environmental temperature but the temperature in the immediate vicinity of the weld. The ambient environmental temperature may be below 0°F, but a heated structure or shelter around the area being welded could maintain the temperature adjacent to the weldment at 0°F or higher. When the base metal is below the temperature listed for the welding process being used and the thickness of material being welded, it shall be preheated (except as otherwise provided) in such manner that the surfaces of the parts on which weld metal is being deposited are at or above the specified minimum temperature for a distance equal to the thickness of the part being welded, but not less than 3 in. (75 mm) in all directions from the point of welding. Preheat and interpass tempertures must be sufficient to prevent crack formation. Temperature above the minimum shown may be required for highly restrained welds. For A514, A517 and A709 Grades 100 and 100W steel, the maximum preheat and interpass temperture shall not exceed 400°F (205°C) for thickness up to 1-1/2 in. (38.1 mm) inclusive, and 450°F (230°C) for greater thickness. Heat input when welding A514, A517 and A709 Grades 100 and 100W steel shall not exceed the steel producer's recommendations.
2. In joints involving combinations of base metals, preheat shall be as specified for the higher strength steel being welded.
3. Only low hydrogen electrodes shall be used when welding A36 or A709 Grade 36 steel more than 1 in. (25.4 mm) thick for dynamically loaded structures.
4. When the base metal temperature is below 32°F (0°C), the base metal shall be preheated to at least 70°F (21°C) and this minimum temperature maintained during welding.
5. For modificaiton of preheat requirements for submerged arc welding with parallel or multiple electrodes, see 4.10.6 or 4.11.6, AWS D1.1-92.
6. Preheat is not required for the base metal. Preheat for E80XX-X filler metal shall be as for Group C; for higher strength filler metal treat as Group D.

2. The maximum electrode size practical for vertical and overhead welding is 3/16-in. The 5/32-in. electrode is the maximum size for low-hydrogen electrodes.
3. High DC current increases arc blow. When arc blow is a problem, either use AC or limit the current.
4. Joint dimensions sometimes limit the electrode diameter that will fit into the joint.

Preheat and Interpass Temperature

The use of preheat and minimum interpass temperatures may be dictated by the composition of the steel, by the thickness of the material, or by the degree of joint restraint. Preheating may be mandatory if the welding is done according to a code. For example, the preheating requirements in the AISC specification for structural steel buildings are shown in Table 6-21. Other codes, such as the AWS Structural Welding Code D1.1-92 and the AASHTO AWS Bridge Welding Code D1.5-88, have similar requirements. (See Section 3.3)

TROUBLE SHOOTING

Many operating variables can affect the quality and appearance of the weld. The effects produced by the most important of these variables are illustrated in Fig. 6-26. Common undesirable effects are shown in Figs. 6-27 through 6-29. Methods for correcting undesirable characteristics are discussed in the following paragraphs. Not discussed here is arc blow, which is covered in Section 3.2.

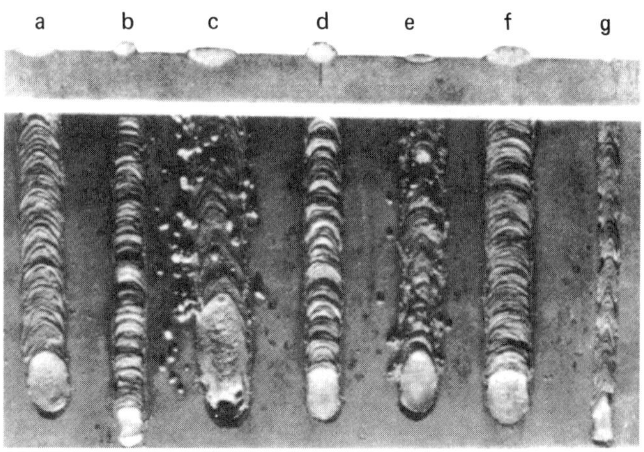

Fig. 6-26. Effect of welding variables on bead characteristics. Proper current, travel speed, and arc length (a). Current too low (b). Current too high (c). Arc length too short (d). Arc length too long (e). Travel speed too slow (f). Travel speed too fast (g).

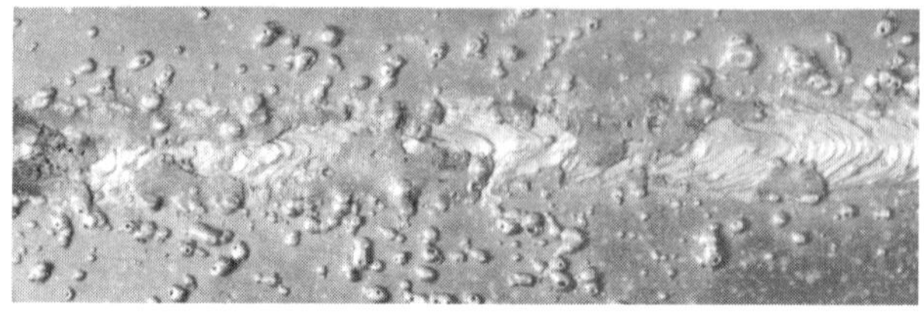

Fig. 6-27. Undesirable bead appearance caused by weld spatter.

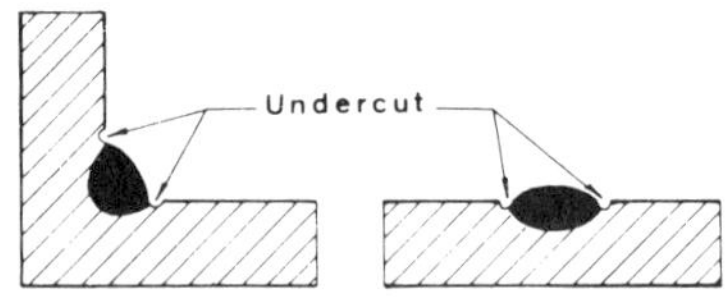

Fig. 6-28. Undercut in a weld. The effect is undesirable from the appearance standpoint and may weaken the joint.

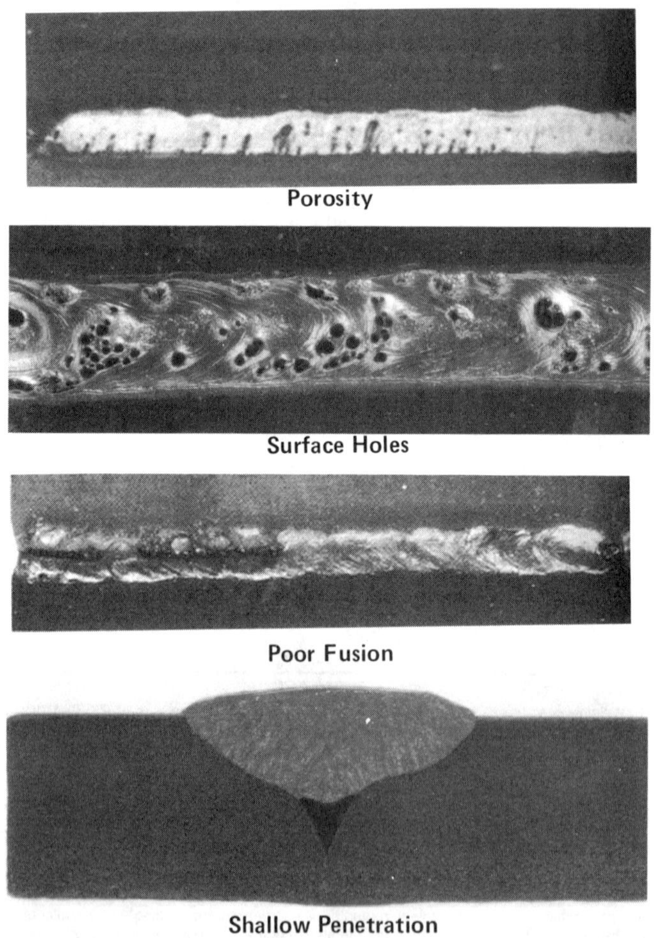

Fig. 6-29. Undesirable effects caused by improper procedures and techniques.

Weld Spatter

Spatter does not affect weld strength but does produce a poor appearance and increases cleaning costs. To control excessive spatter:

1. Try lowering the current. Be sure the current is within the recommended range for the type and size electrode (See Table 6-15.)
2. Be sure the polarity is correct for the electrode type.
3. Try a shorter arc length.
4. If the molten metal is running in front of the arc, change the electrode angle.
5. Watch for arc blow.
6. Be sure the electrode is not too wet.

Undercut

Generally, the only harm from undercutting is impaired appearance. However, undercutting may also impair weld strength, particularly when the weld is loaded in tension or subjected to fatigue. To minimize undercut:

1. Reduce current, travel speed, or electrode size until the puddle is manageable.
2. Change electrode angle so the arc force holds the metal in the corners. Use a uniform travel speed and avoid excessive weaving.

Rough Welding

If polarity and current are within the electrode manufacturer's recommendations but the arc action is rough and erratic, the electrodes may be wet. Try electrodes from a fresh container. If the problem occurs frequently, store open containers of electrodes in a heated cabinet.

Porosity and Surface Holes

Most porosity is not visible. But severe porosity can weaken the weld. The following practices minimize porosity:

1. Remove scale, rust, paint, moisture, or dirt from the joint. Generally use an E6010 or E6011 electrode for dirty steel.
2. Keep the puddle molten for a long time, so that gases may boil out before the metal freezes.
3. Steels very low in carbon or manganese or those high in sulfur or phosphorus should be welded with a low-hydrogen electrode. Minimize admixture of base metal with weld metal by using low currents and fast travel speeds for less penetration.
4. Try using a short arc length; short arcs are required for low-hydrogen electrodes.

Surface holes can be avoided by many of the practices used to minimize porosity.

Poor Fusion

Proper fusion exists when the weld bonds to both walls of the joint and forms a solid bead across the joint. Lack of fusion is often visible and must be avoided for a sound weld. To correct poor fusion:

1. Try a higher current and a stringer-bead technique.
2. Be sure the edges of the joint are clean, or use an E6010 or E6011 electrode.
3. If gap is excessive, provide better fitup or use a weave technique to fill the gap.

Shallow Penetration

Penetration refers to the depth the weld enters into the base metal. For full-strength welds, penetration to the bottom of the joint is required. To overcome shallow penetration:

1. Try higher currents or slower travel.
2. Use small electrodes to reach into deep, narrow grooves.
3. Allow some gap (free space) at the bottom of the joint.

Cracking

Many different types of cracks may occur throughout a weld. Some are visible and some are not. However, all cracks are potentially serious, because they can lead to complete failure of the weld. The following suggestions may help control potential cracking. Practices to minimize cracks are shown in Fig. 6-30.

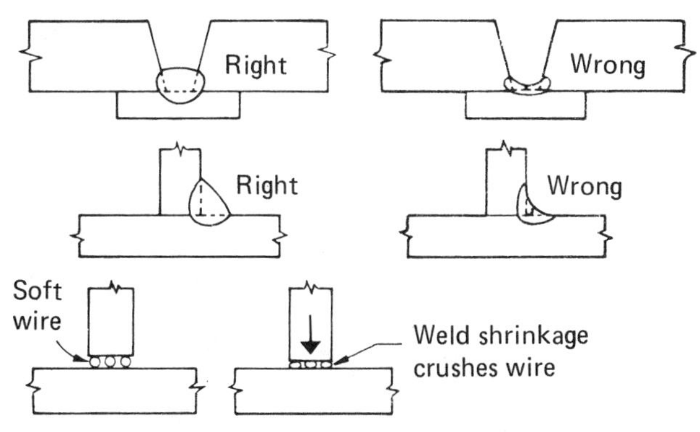

Fig. 6-30. Factors in controlling weld cracking. Illustrated are correct and incorrect joint geometries and bead shapes and a technique to permit stress relaxation in an otherwise rigid joint.

Most cracking is attributed to high-carbon or alloy content or high-sulfur content in the base metal. To control this type of cracking:

1. Use low-hydrogen electrodes.
2. Preheat. Use high preheats for heavier plate and rigid joints.
3. Reduce penetration by using low currents and small electrodes. This reduces the amount of alloy added to the weld from melted base metal.

To control crater cracking, fill each crater before breaking the arc. Use a back-stepping technique so as to end each weld on the crater of the previous weld.

On multiple-pass or fillet welds, be sure the first bead is of sufficient size and of flat or convex shape to resist cracking until the later beads can be added for support. To increase bead size, use slower travel speed, a short arc, or weld 5° uphill. Always continue welding while the plate is hot.

Rigid parts are more prone to cracking. If possible, weld toward the unrestrained end. Leave a 1/32-in. gap between plates for free shrinkage movement as the weld cools. Peen each bead while it is still hot to relieve stresses.

For more on cracking, see Section 6.1.

Arc welding was the only practical means of fabricating this large bull gear for a huge stripper shovel.

INTRODUCTION TO WELDING PROCEDURES

The ideal welding procedure is the one that will produce acceptable quality welds at the lowest overall cost. So many factors influence the optimum welding conditions that it is impossible to write procedures for each set of conditions. In selecting a procedure, the best approach is to study the conditions of the application and then choose the procedure that most nearly accommodates them. The procedures given here are typical, and it may be necessary to make adjustments for a particular application to produce a satisfactory weld.

For some joints, different procedures are offered to suit the weld quality — code quality and commercial quality — that may be required.

Code-Quality Procedures

Code-quality procedures are intended to provide the highest level of quality and appearance. To accomplish this, conservative currents and travel speeds are recommended.

These procedures are aimed at producing welds that will meet the requirements of the commonly used codes: AWS Structural, AISC Buildings and Bridges, ASME Pressure Vessels, AASHTO Bridges, and others. Code-quality welds are intended to be defect-free to the extent that they will measure up to the nondestructive testing requirements normally imposed by these codes. This implies crack-free, pressure-tight welds, with little or no porosity or undercut.

The specific requirements of codes are so numerous and varied that code-quality procedures may not satisfy every detail of a specific code. Procedure qualification tests are recommended to confirm the acceptability of chosen procedures.

All butt welds made to code quality are full-penetration; fillet welds are full-size, as required by most codes. (The theoretical throat, rather than the true throat, is used as the basis of calculating strength.)

Commercial-Quality Procedures

Commercial quality implies a level of quality and appearance that will meet the nominal requirements imposed on most of the welding done commercially. These welds will be pressure-tight and crack-free. They will have good appearance, and they will meet the normal strength requirements of the joint.

Procedures for commercial-quality welds are not as conservative as code-quality procedures; speeds and currents are generally higher. Welds made according to these procedures may have minor imperfections that would be objectionable to the more demanding codes.

It is recommended that appropriate tests be performed to confirm the acceptability of the selected procedure for the application at hand prior to putting it into production.

Weldability of Material

Weldability (see Section 6.1) of a steel has a considerable effect on the welding procedure. For some joints, more than one procedure is offered because of the marginal weldability of the steel.

Good weldability indicates a steel with a composition that is within the preferred range (see Table 6-1) — one whose chemistry does not limit the welding speed.

Fair weldability indicates a steel with one or more elements outside the preferred range or one that contains one or more alloys. These steels require a lower welding speed or a mild preheat, or both, to minimize defects such as porosity, cracking, and undercut.

Poor weldability steels are those with compositions outside the preferred range, alloy additions, segregations, previous heat-treatment, or some other condition that makes them difficult to weld. These steels require still lower welding speeds, preheat, possibly a postheat, and careful electrode selection to obtain a satisfactory weld.

The addition of alloys to steel that enhance the mechanical properties or hardenability usually have an adverse effect on weldability. In general, the weldability of low-alloy steels is never better than "fair."

Procedures Notes

In the following fillet-weld procedures, the fillet

DATA SHEET

Article to be Welded ______________________ Job No. ______________

Plate Specs or Analysis ______________________________________

Welding Process ______________ Submitted by: ______________ Date: ________

JOINT-PREPARATION – BEAD PLACEMENT, SIZE, SEQUENCE

WORK ANGLE – DRAG ANGLE
ELECTRODE ANGLE TO JOINT

Pass No	AC DC (+ –)	Amps	Volts	Travel Speed "/Min	Electrode Type	Elect. Size	Stick-Out	Flux Type					

Special Comments ______________________________________

size is always associated with a particular plate thickness. This relationship is given solely for the purpose of designing a welding procedure and does not imply that a certain size fillet is the only size applicable to that plate thickness. In some of the procedures, the fillet size shown is larger than necessary to meet code requirements for the plate thickness. In such instances, select the procedure for the proper weld size and quality. If the thickness of the plate being welded is appreciably greater than that specified in the procedure, a reduction in welding speed and current will probably be required.

The procedure data given have been developed to provide the most economical procedures for various applications. In some cases, more than one type or size of electrode is recommended for the same joint. In small shops, electrode selection may depend on the available power source; consequently, some joints have procedures for either AC or DC welders.

With some joints, procedures for two different types of electrodes are given — for example, E7014 or E7024, E7018 or E7028. This allows a choice of electrodes so the one with the better usability characteristics can be selected.

Any procedure for a poor or fair welding quality steel may be used on a steel of a better welding quality.

Travel speed is given as a range. The electrode required and the total time are based on the middle of the range.

Unless otherwise indicated, both members of the joint are the same thickness.

Pounds-of-electrode data include all ordinary deposition losses. These values are in terms of pounds of electrode needed to be purchased.

Total time is the arc time only and does not allow for operating factor.

After a satisfactory welding procedure has been established, all the data should be recorded and filed for future reference. This information is invaluable if the same job or a similar job occurs at a later date. A suggested data sheet is shown on the opposite page.

The presented procedures are offered as a starting point and may require changes to meet the requirements of specific applications. Because the many variables in design, fabrication, and erection or assembly affect the results obtained in applying this type of information, the serviceability of the product or structure is the responsibility of the builder.

Welding Parameters, Single Values or a Range

In some of the procedures which follow single values are given for such items as Amps, Volts, and Travel Speed. The purpose of using single values, i.e. 100 amps, is to indicate an optimum or starting place and it is expected that the user will expand these values and show a reasonable range. These values when recorded on a WPS (Welding Procedure Specification) should be shown as a range, not as specific values.

SHIELDED METAL-ARC (MANUAL)

Position: Flat
Weld Quality Level: Commercial
Steel Weldability: Good
Welded From: One side

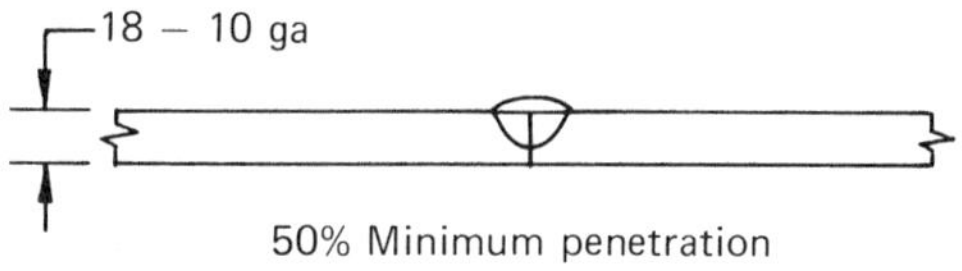

Plate Thickness (in.)	0.048 (18 ga)	0.060 (16 ga)	0.075 (14 ga)	0.105 (12 ga)	0.135 (10 ga)*
Pass	1	1	1	1	1
Electrode Class	E6010	E6010	E6010	E6010	E6010
Size	3/32	1/8	1/8	5/32	3/16
Current (amp) DC(+)	40†	70†	80	120	135
Arc Speed (in./min)	22 – 26	30 – 35	25 – 30	20 – 24	17 – 21
Electrode Req'd (lb/ft)	0.0244	0.0287	0.0262	0.0487	0.0695
Total Time (hr/ft of weld)	0.00833	0.00615	0.00727	0.00909	0.0105

* Use 1/16 in. gap and whip the electrode.
† DC(–)

SHIELDED METAL-ARC (MANUAL)

Position: Flat
Weld Quality Level: Commercial
Steel Weldability: Good
Welded From: One side

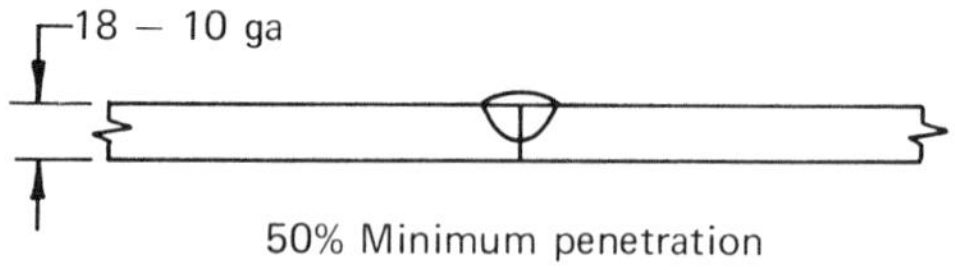

Plate Thickness (in.)	0.048 (18 ga)	0.060 (16 ga)	0.075 (14 ga)	0.105 (12 ga)	0.135 (10 ga)*
Pass	1	1	1	1	1
Electrode Class	E6011	E6011	E6011	E6011	E6011
Size	3/32	1/8	1/8	5/32	3/16
Current (amp) AC	50	100	105	130	145
Arc Speed (in./min)	20 – 24	28 – 33	26 – 31	24 – 29	22 – 27
Electrode Req'd (lb/ft)	0.0251	0.0326	0.0367	0.0527	0.0648
Total Time (hr/ft of weld)	0.00909	0.00656	0.00702	0.00755	0.00817

* Use 1/16 in. gap and whip the electrode.

SHIELDED METAL-ARC (MANUAL)

Position: Flat
Weld Quality Level: Code
Steel Weldability: Good
Welded From: One side

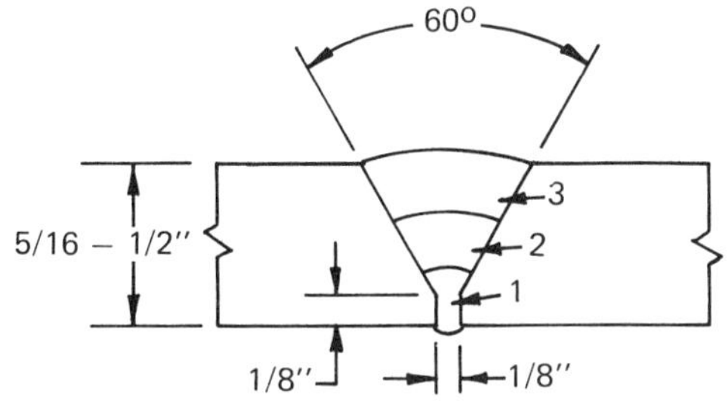

Plate Thickness (in.)	5/16		3/8		1/2		
Pass	1	2	1	2 & 3	1	2	3
Electrode Class	E6011	E6027	E6011	E6027	E6011	E6011	E6027
Size	5/32	5/32	5/32	5/32	5/32	1/4	1/4
Current (amp) AC	135	240	135	240	135	275	400
Arc Speed (in./min)	5.5-6.5	12.0-14.0	5.5-6.5	12.0-14.0	5.5-6.5	8.0-10.0	10.0-12.0
Electrode Req'd (lb/ft)	0.168	0.142	0.168	0.284	0.168	0.228	0.354
Total Time (hr/ft of weld)	0.0487		0.0641		0.0717		

SHIELDED METAL-ARC (MANUAL)

Position: Flat
Weld Quality Level: Code
Steel Weldability: Good
Welded From: One side

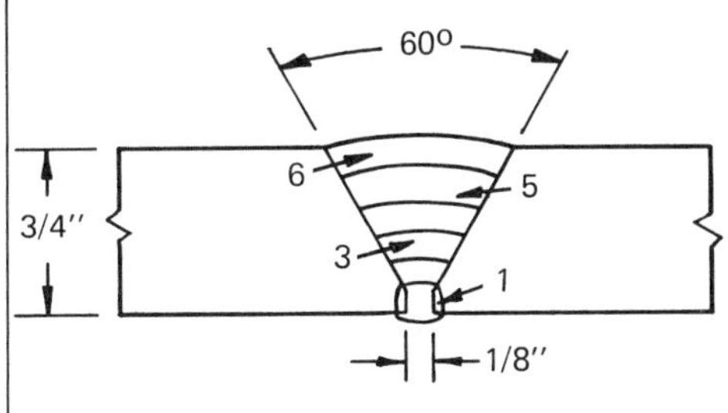

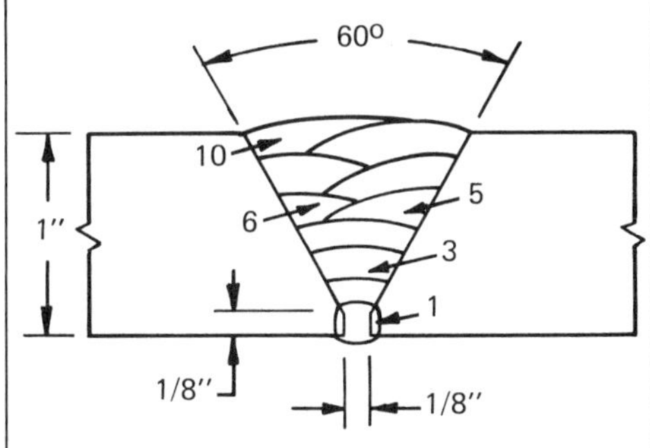

Plate Thickness (in.)	3/4			1		
Pass	1	2	3 – 6	1	2	3 – 10
Electrode Class	E6011	E6011	E6027	E6011	E6011	E6027
Size	5/32	1/4	1/4	5/32	1/4	1/4
Current (amp) AC	135	275	400	135	275	400
Arc Speed (in./min)	5.5 – 6.5	8.0 – 10.0	11.0 – 13.0	5.5 – 6.5	8.0 – 10.0	11.0 – 13.0
Electrode Req'd (lb/ft)	0.168	0.228	1.47	0.168	0.228	2.94
Total Time (hr/ft of weld)	0.122			0.189		

SHIELDED METAL-ARC (MANUAL)

Position: Flat
Weld Quality Level: Commercial
Steel Weldability: Good
Welded From: One side

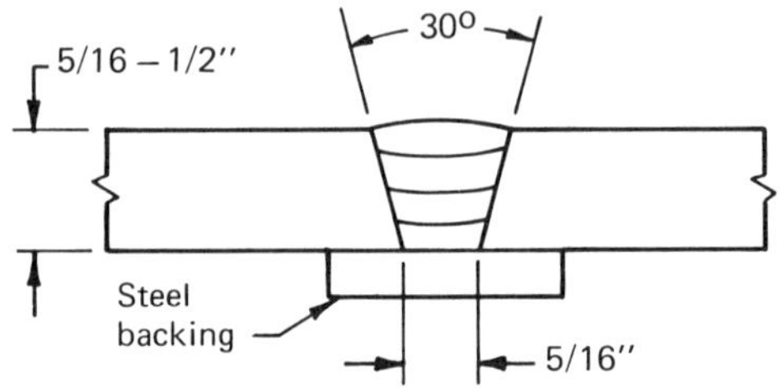

Plate Thickness (in.)	5/16		3/8		1/2	
Pass	1	2 – 3	1	2 – 3	1	2 – 4
Electrode Class	E6027	E6027	E6027	E6027	E6027	E6027
Size	3/16	1/4	3/16	1/4	3/16	1/4
Current (amp) AC	300	400	300	400	300	400
Arc Speed (in./min)	13.0-15.0	15.0-18.0	13.0-15.0	11.5-13.5	13.0-15.0	12.5-14.5
Electrode Req'd (lb/ft)	0.228	0.524	0.228	0.697	0.228	1.00
Total Time (hr/ft of weld)	0.0385		0.0463		0.0605	

SHIELDED METAL-ARC (MANUAL)

Position: Flat
Weld Quality Level: Commercial
Steel Weldability: Good
Welded From: One side

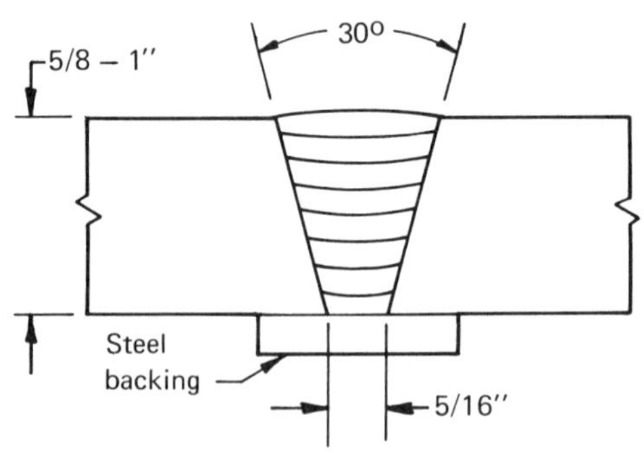

Plate Thickness (in.)	5/8		3/4		1	
Pass	1	2 – 5	1	2 – 6	1	2 – 8
Electrode Class	E6027	E6027	E6027	E6027	E6027	E6027
Size	3/16	1/4	3/16	1/4	3/16	1/4
Current (amp) AC	300	400	300	400	300	400
Arc Speed (in./min)	13.0-15.0	12.5-14.5	13.0-15.0	12.5-14.5	13.0-15.0	12.5-14.5
Electrode Req'd (lb/ft)	0.228	1.35	0.228	1.69	0.228	2.37
Total Time (hr/ft of weld)	0.0759		0.0913		0.122	

SHIELDED METAL-ARC (MANUAL)

Position: Flat
Weld Quality Level: Code
Steel Weldability: Good
Welded From: Two sides

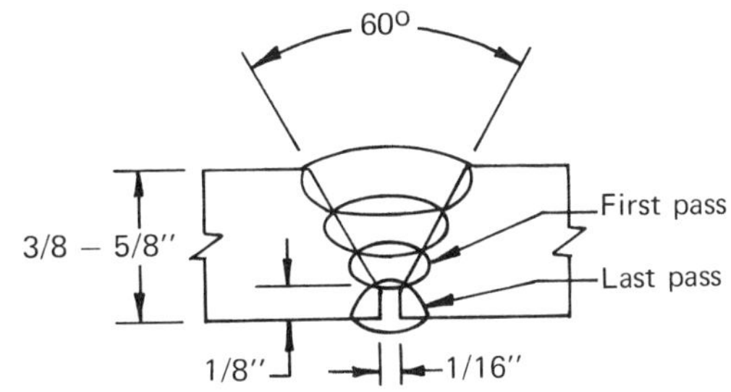

Plate Thickness (in.)	3/8		1/2		5/8		
Pass	1	2 & 3	1	2 & 3	1	2 & 3	4
Electrode Class	E6011	E6027	E6011	E6027	E6011	E6027	E6027
Size	3/16	3/16	1/4	7/32	1/4	1/4	7/32
Current (amp) AC	175	280	275	340	275	375	340
Arc Speed (in./min)	8.0-10.0	14.5-17.5	7.0-9.0	13.5-15.5	7.0-9.0	12.5-14.5	11.5-13.5
Electrode Req'd (lb/ft)	0.148	0.366	0.239	0.480	0.241	0.795	0.235
Total Time (hr/ft of weld)	0.0472		0.0526		0.0706		

Back gouge first pass before welding last pass.

SHIELDED METAL-ARC (MANUAL)

Position: Flat
Weld Quality Level: Code
Steel Weldability: Good
Welded From: Two sides

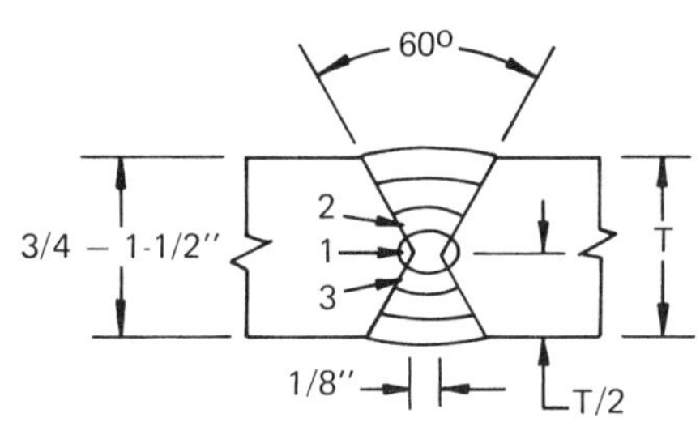

Plate Thickness (in.)	3/4			1			1-1/2		
Pass	1	2 & 3	4 & 5	1	2 & 3	4 – 7	1	2 & 3	4 – 10
Electrode Class	E6011	E6011	E6027	E6011	E6011	E6027	E6011	E6011	E6027
Size	3/16	1/4	1/4	3/16	1/4	1/4	3/16	1/4	1/4
Current (amp) AC	135	275	400	135	275	400	135	275	400
Arc Speed (in./min)	5.5-6.5	8.0-10.0	11.0-13.0	5.5-6.5	8.0-10.0	11.0-13.0	5.5-6.5	8.0-10.0	9.5-11.5
Electrode Req'd (lb/ft)	0.190	0.400	0.728	0.190	0.400	1.45	0.190	0.400	3.04
Total Time (hr/ft of weld)	0.111			0.144			0.211		

Back gouge first pass before welding third pass. Complete third pass side before turning over.

SHIELDED METAL-ARC (MANUAL)

Position: Vertical down
Weld Quality Level: Commercial
Steel Weldability: Good
Welded From: One side

18 – 10 ga

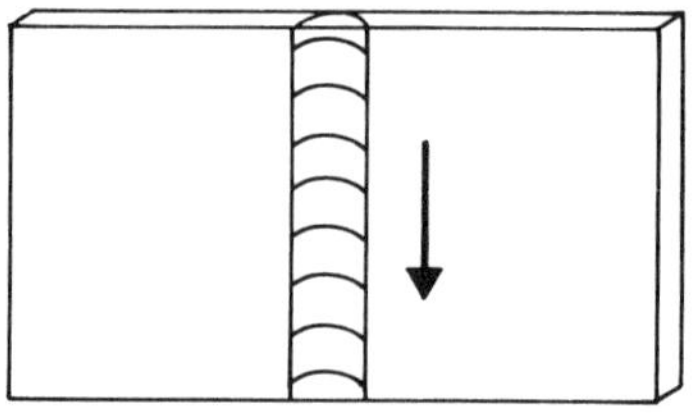

50% Minimum penetration

Plate Thickness (in.)	0.048 (18 ga)	0.060 (16 ga)	0.075 (14 ga)	0.105 (12 ga)	0.135 (10 ga)*
Pass	1	1	1	1	1
Electrode Class	E6010	E6010	E6010	E6010	E6010
Size	3/32	1/8	1/8	5/32	3/16
Current (amp) DC(+)	45†	75†	90	130	150
Arc Speed (in./min)	25 – 30	33 – 38	27 – 32	22 – 27	18 – 22
Electrode Req'd (lb/ft)	0.0234	0.0281	0.0272	0.0478	0.0730
Total Time (hr/ft of weld)	0.00727	0.00555	0.00678	0.00817	0.0100

* Use 1/16 in. gap and whip the electrode.
† Use DC(–)

SHIELDED METAL-ARC (MANUAL)

Position: Vertical down
Weld Quality Level: Commercial
Steel Weldability: Good
Welded From: One side

18 – 10 ga

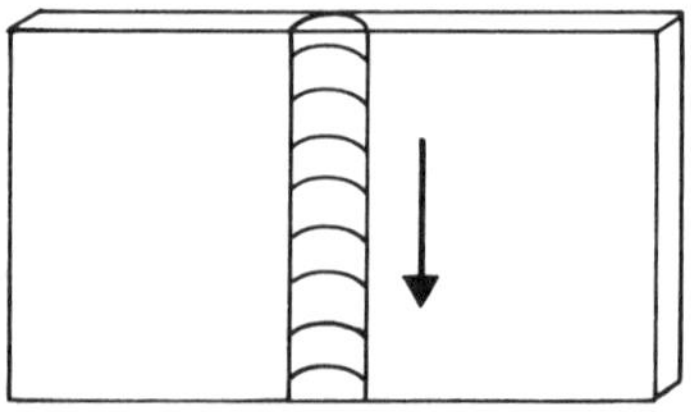

50% Minimum penetration

Plate Thickness (in.)	0.048 (18 ga)	0.060 (16 ga)	0.075 (14 ga)	0.105 (12 ga)	0.135 (10 ga)*
Pass	1	1	1	1	1
Electrode Class	E6011	E6011	E6011	E6011	E6011
Size	3/32	1/8	1/8	5/32	3/16
Current (amp) AC	55	110	115	140	155
Arc Speed (in./min)	23 – 28	29 – 34	27 – 32	26 – 31	24 – 29
Electrode Req'd (lb/ft)	0.0236	0.0345	0.0376	0.0523	0.0640
Total Time (hr/ft of weld)	0.00785	0.00635	0.00678	0.00703	0.00755

* Use 1/16 in. gap and whip the electrode.

SHIELDED METAL-ARC (MANUAL)

Position: Vertical up
Weld Quality Level: Code
Steel Weldability: Good
Welded From: One side

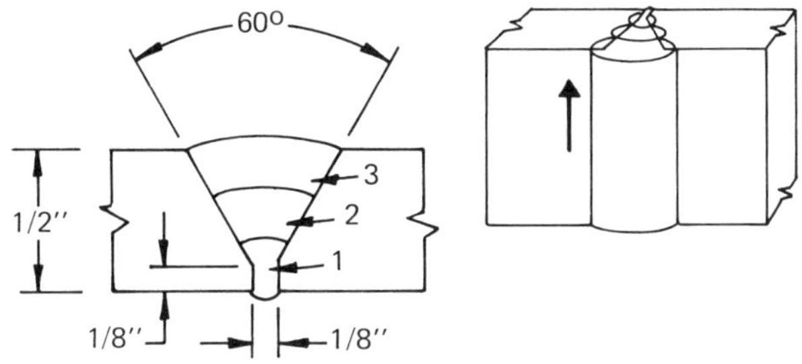

Plate Thickness (in.)	1/4	5/16	3/8	1/2
Pass	1 & 2	1 & 2	1 & 2	1 – 3
Electrode Class	E6010	E6010	E6010	E6010
Size	5/32	5/32	3/16	3/16
Current (amp) DC(+)	110	120	150	170
Arc Speed (in./min)*	5.2-5.8	3.8-4.2	4.8-5.3	3.8-4.2
Electrode Req'd (lb/ft)	0.323	0.440	0.586	0.990
Total Time (hr/ft of weld)	0.0901	0.118	0.130	0.152

* First pass only. Vary speed on succeeding passes to obtain proper weld size.

SHIELDED METAL-ARC (MANUAL)

Position: Vertical up
Weld Quality Level: Code
Steel Weldability: Good
Welded From: One side

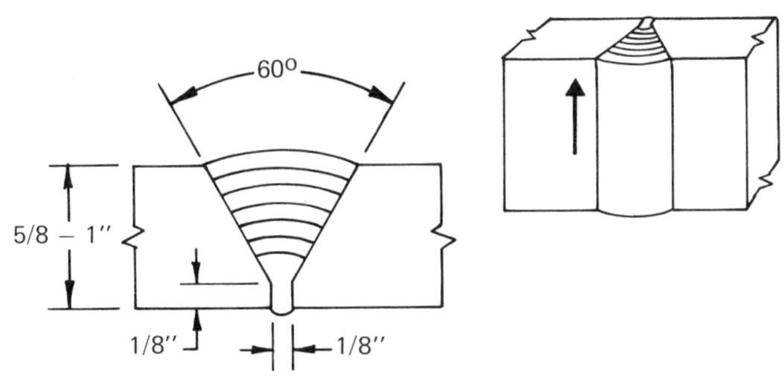

Plate Thickness (in.)	5/8	3/4	1
Pass	1 – 4	1 – 6	1 – 10
Electrode Class	E6010	E6010	E6010
Size	3/16	3/16	3/16
Current (amp) DC(+)	170	170	170
Arc Speed (in./min)*	3.8 – 4.2	3.8 – 4.2	3.8 – 4.2
Electrode Req'd (lb/ft)	1.48	2.08	3.56
Total Time (hr/ft of weld)	0.228	0.318	0.547

* First pass only. Vary speed on succeeding passes to obtain proper weld size.

SHIELDED METAL-ARC (MANUAL)

Position: Vertical up
Weld Quality Level: Code
Steel Weldability: Fair
Welded From: One side

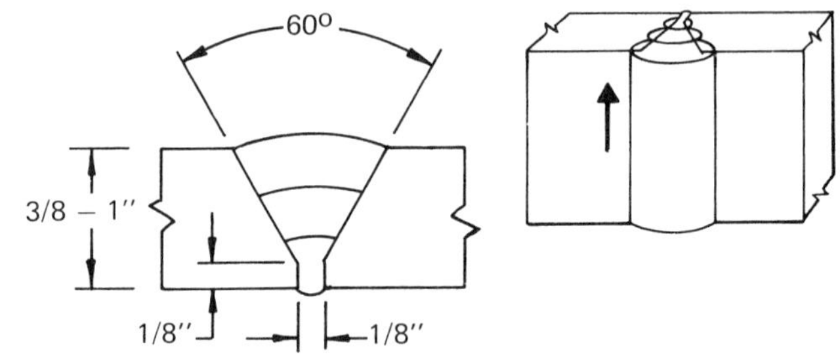

Plate Thickness (in.)	3/8		1/2		3/4		1	
Pass	1	2	1	2 – 3	1	2 – 7	1	2 – 11
Electrode Class	E6010	E7018	E6010	E7018	E6010	E7018	E6010	E7018
Size	5/32	5/32	5/32	5/32	5/32	5/32	5/32	5/32
Current (amp) DC(+)	160	160	160	160	160	160	160	160
Arc Speed (in./min)	4.3-4.7	3.2-3.5*	4.3-4.7	3.2-3.5*	4.3-4.7	3.2-3.5*	4.3-4.7	3.2-3.5*
Electrode Req'd (lb/ft)	0.281	0.341	0.281	0.758	0.281	1.93	0.281	3.52
Total Time (hr/ft of weld)	0.104		0.176		0.381		0.659	

* Second pass only. Vary speed on succeeding passes to obtain proper weld size.

SHIELDED METAL-ARC (MANUAL)

Position: Vertical up
Weld Quality Level: Code
Steel Weldability: Good
Welded From: Two sides

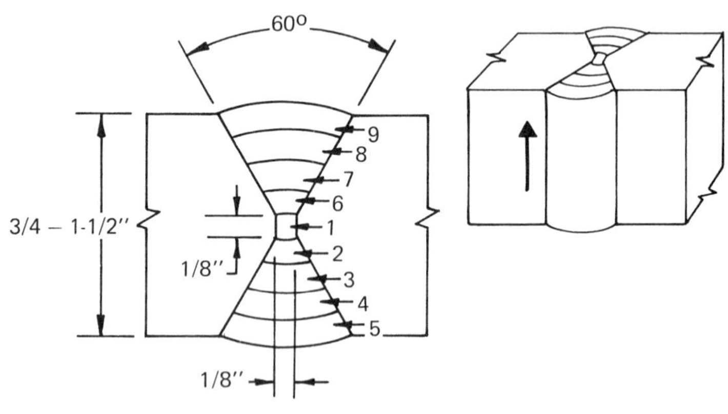

Plate Thickness (in.)	3/4		1		1-1/4		1-1/2	
Pass	1	2 – 5	1	2 – 7	1	2 – 7	1	2 – 9
Electrode Class	E6010	E7018	E6010	E7018	E6010	E7018	E6010	E7018
Size	5/32	5/32	5/32	5/32	5/32	5/32	5/32	5/32
Current (amp) DC(+)	140	160	140	160	140	160	140	160
Arc Speed (in./min)	3.5-4.1	4.1-4.9	3.5-4.1	3.5-4.1	3.5-4.1	2.3-2.9	3.5-4.1	2.4-3.0
Electrode Req'd (lb/ft)	0.240	0.900	0.240	1.66	0.240	2.40	0.240	3.16
Total Time (hr/ft of weld)	0.230		0.367		0.514		0.645	

Gouge out seam for first pass on second side.

SHIELDED METAL-ARC (MANUAL)

Position: Horizontal
Weld Quality Level: Code
Steel Weldability: Fair
Welded From: One side

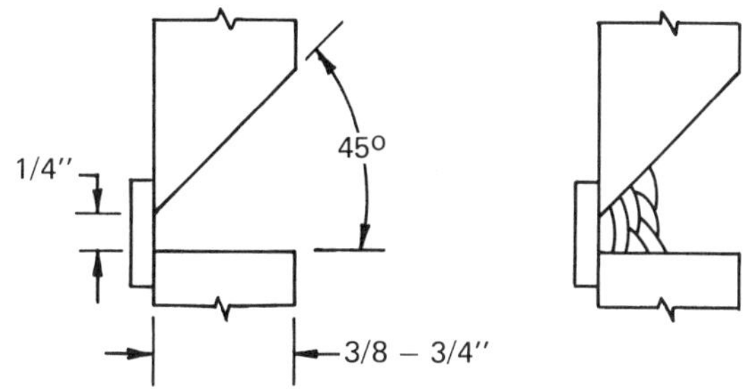

Plate Thickness (in.)	3/8		1/2		5/8		3/4	
Pass	1	2 – 5	1	2 – 7	1	2 – 9	1	2 – 11
Electrode Class	E7018		E7018		E7018		E7018	
Size (in.)	3/16		3/16		3/16		3/16	
Current (amp) DC(+)	240		240		240		240	
Arc Speed (in./min)	4.5-5.5	8.5-9.5	4.5-5.5	7.5-8.5	4.5-5.5	6.7-7.4	5.5-6.5	6.2-6.8
Electrode Req'd (lb/ft)	0.367		1.35		1.75		2.42	
Total Time (hr/ft of weld)	0.118		0.182		0.270		0.345	

SHIELDED METAL-ARC (MANUAL)

Position: Horizontal
Weld Quality Level: Code
Steel Weldability: Fair
Welded From: One side

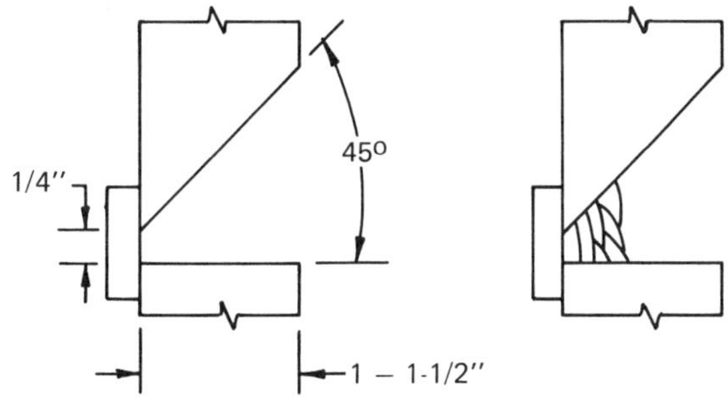

Plate Thickness (in.)		1		1-1/4		1-1/2	
Pass	1*	2 – 13	14 – 19†	2 – 17	18 – 24†	2 – 22	23 – 31†
Electrode Class	E7018	E7018		E7018		E7018	
Size (in.)	3/16	7/32	3/16	7/32	3/16	7/32	3/16
Current (amp) DC(+)	240	280	240	280	240	280	240
Arc Speed (in./min)	5 – 6	6.2-6.8	9.5-10.5	5.7-6.3	9.5-10.5	5.2-5.8	9.5-10.5
Electrode Req'd (lb/ft)		3.39	.994	4.82	1.23	6.40	1.60
Total Time (hr/ft of weld)		0.526		.714		1.00	

* First pass for all thicknesses

† Cover passes.

SHIELDED METAL-ARC (MANUAL)

Position: Horizontal
Weld Quality Level: Commercial
Steel Weldability: Fair
Welded From: Two sides

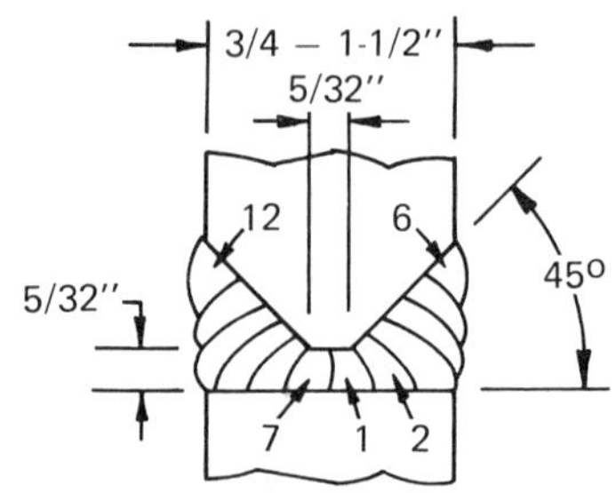

Plate Thickness (in.)	3/4		1		1-1/4		1-1/2	
Pass	1	2 – 6	1	2 – 10	1	2 – 10	1	2 – 12
Electrode Class	E7018	E7018	E7018	E7018	E7018	E7018	E7018	E7018
Size	3/16	3/16	3/16	3/16	3/16	3/16	3/16	3/16
Current (amp) DC(+)	240	240	240	240	240	240	240	240
Arc Speed (in./min)	5.5-6.5	9.0-11.0	4.4-5.2	8.5-10.5	3.8-4.6	5.5-6.5	3.7-4.3	4.6-5.4
Electrode Req'd (lb/ft)	0.956		1.47		2.60		3.84	
Total Time (hr/ft of weld)	0.133		0.230		0.347		0.490	

Fill first pass side. Back gouge as required before welding second side.

SHIELDED METAL-ARC (MANUAL)

Position: Horizontal
Weld Quality Level: Code
Steel Weldability: Fair
Welded From: One side

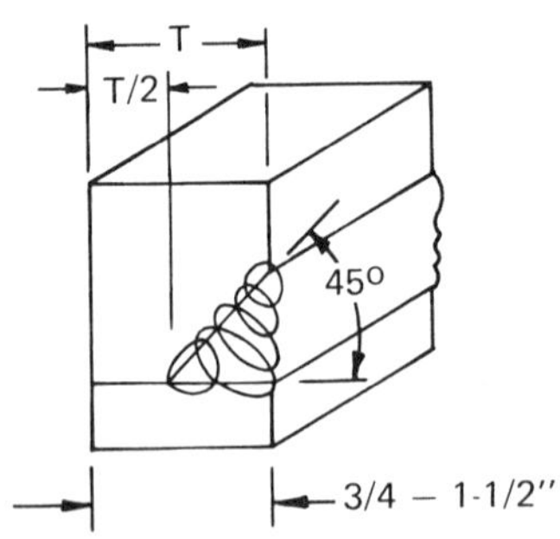

Plate Thickness (in.)		3/4	1	1-1/4	1-1/2
Pass	1*	2 – 3	2 – 5	2 – 5	2 – 6
Electrode Class	E7018	E7018	E7018	E7018	E7018
Size (in.)	3/16	3/16	3/16	3/16	3/16
Current (amp) DC(+)	240	240	240	240	240
Arc Speed (in./min)	4.0-6.0	9.5-10.5	9.0-10.0	5.7-6.3	4.7-5.3
Electrode Req'd (lb/ft)		0.470	0.740	1.80	1.92
Total Time (hr/ft of weld)		0.0800	0.116	0.178	0.250

* First pass for all thicknesses.

SHIELDED METAL-ARC (MANUAL)

Position: Overhead
Weld Quality Level: Code
Steel Weldability: Fair
Welded From: One side

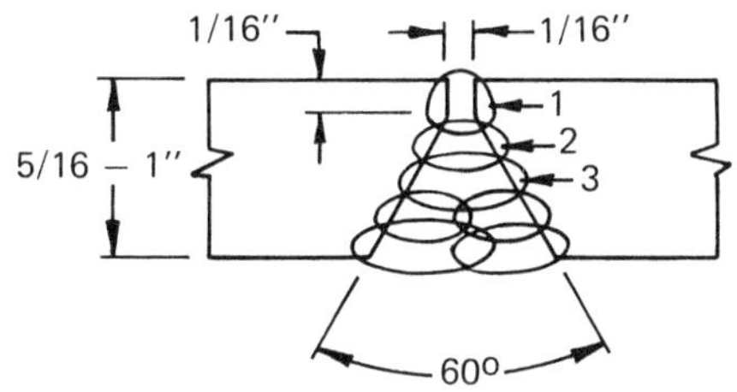

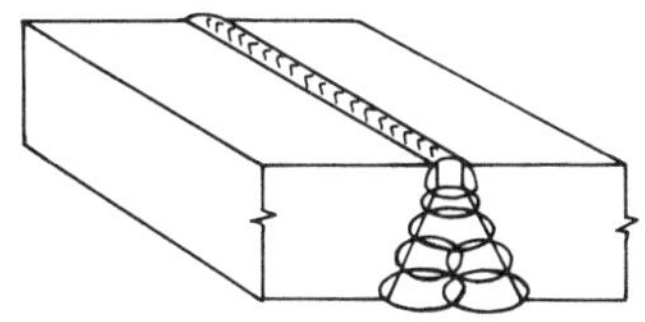

Plate Thickness (in.)	5/16		3/8		1/2		3/4		1	
Pass	1	2	1	2 – 3	1	2 – 5	1	2 – 9	1	2 – 13
Electrode Class	E6010	E7018	E6010	E7018	E6010	E7018	E6010	E7018	E6010	E7018
Size	1/8	5/32	1/8	5/32	1/8	5/32	1/8	5/32	1/8	5/32
Current (amp) DC(+)	110	170	110	170	110	170	110	170	110	170
Arc Speed (in./min)	4.3 – 4.7	3.4 – 3.8	4.3 – 4.7	3.3 – 3.7	4.3 – 4.7	3.6 – 4.0	4.3 – 4.7	4.3 – 4.7	4.3 – 4.7	3.6 – 4.0
Electrode Req'd (lb/ft)	0.155	0.327	0.155	0.671	0.155	0.918	0.155	2.08	0.155	3.70
Total Time (hr/ft of weld)	0.0999		0.158		0.202		0.399		0.575	

Split layers after third pass, as shown in sketch.

SHIELDED METAL-ARC (MANUAL)

Position: Flat and horizontal
Weld Quality Level: Commercial
Steel Weldability: Good

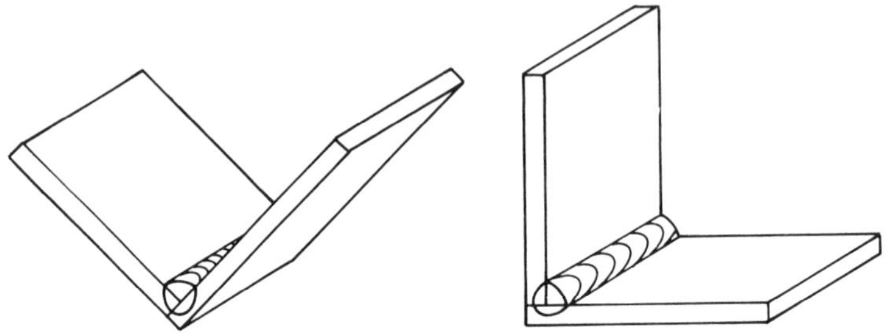

18 – 10 ga

Plate Thickness (in.)	0.048 (18 ga)	0.060 (16 ga)	0.075 (14 ga)	0.105 (12 ga)	0.135 (10 ga)
Pass	1	1	1	1	1
Electrode Class	E6013	E6012	E6012	E6012	E6012
Size	3/32	1/8	5/32	3/16	3/16
Current (amp) DC(–)	70	95	140	190	200
Arc Speed (in./min)	14 – 18	15 – 19	16 – 20	20 – 24	16 – 20
Electrode Req'd (lb/ft)	0.0413	0.0583	0.0848	0.0865	0.112
Total Time (hr/ft of weld)	0.0125	0.0118	0.0111	0.00910	0.0111

SHIELDED METAL-ARC (MANUAL)

Position: Flat and horizontal
Weld Quality Level: Commercial
Steel Weldability: Good

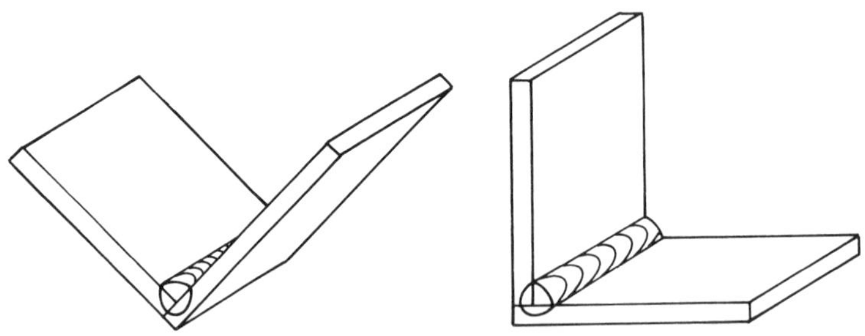

18 – 10 ga

Plate Thickness (in.)	0.048 (18 ga)	0.060 (16 ga)	0.075 (14 ga)	0.105 (12 ga)	0.135 (10 ga)
Pass	1	1	1	1	1
Electrode Class	E6013	E6013	E6013	E6013	E6013
Size	3/32	1/8	5/32	5/32	3/16
Current (amp) AC	70	105	155	160	210
Arc Speed (in./min)	14 – 18	14 – 18	15 – 19	14 – 18	14 – 18
Electrode Req'd (lb/ft)	0.0413	0.0495	0.0670	0.0742	0.0926
Total Time (hr/ft of weld)	0.0125	0.0125	0.0118	0.0125	0.0125

SHIELDED METAL-ARC (MANUAL)

Position: Flat
Weld Quality Level: Commercial
Steel Weldability: Good

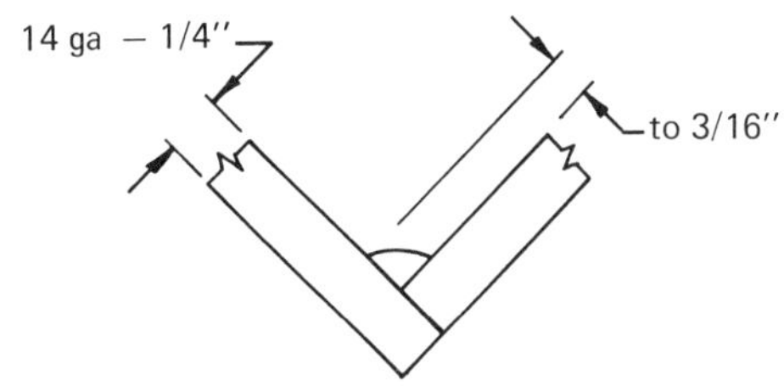

Weld Size, L (in.)				5/32	5/32	3/16	3/16
Plate Thickness (in.)	0.075 (14 ga)	0.105 (12 ga)	0.135 (10 ga)	3/16		1/4	
Pass	1	1	1	1	1	1	1
Electrode Class	E7024	E7024	E7024	E7024	E7024	E7024	E7024
Size	3/32	1/8	1/8	1/8	5/32	5/32	3/16
Current (amp) AC	95	150	160	180	210	230	270
Arc Speed (in./min)	14.5-16.0	16.5-18.5	16.5-18.5	15.0-16.5	16.0-18.0	14.0-15.5	15.5-17.5
Electrode Req'd (lb/ft)	0.0485	0.0760	0.0822	0.102	0.117	0.144	0.162
Total Time (hr/ft of weld)	0.0131	0.0114	0.0114	0.0127	0.0117	0.0136	0.0121

SHIELDED METAL-ARC (MANUAL)

Position: Flat
Weld Quality Level: Commercial
Steel Weldability: Good

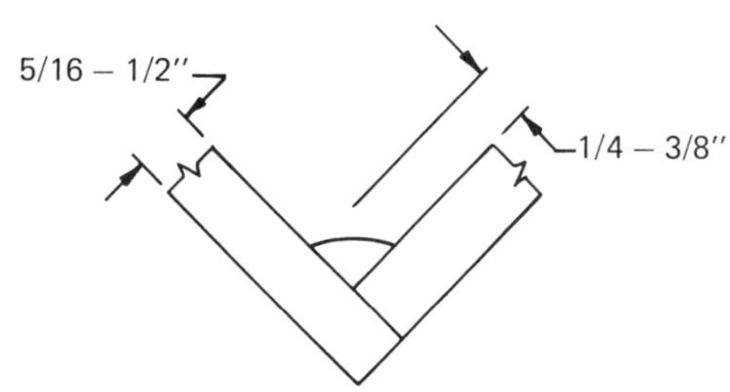

Weld Size, L (in.)	1/4	1/4	9/32	5/16	3/8
Plate Thickness (in.)	5/16		3/8		1/2
Pass	1	1	1	1	1
Electrode Class	E7024	E7024	E7024	E7024	E7024
Size	3/16	7/32	1/4	1/4	5/16
Current (amp) AC	275	325	375	375	475
Arc Speed (in./min)	14.0-16.0	16.0-18.0	17.0-19.0	14.0-15.0	11.0-12.0
Electrode Req'd (lb/ft)	0.19	0.20	0.22	0.29	0.38
Total Time (hr/ft of weld)	0.0133	0.0118	0.0131	0.138	0.174

SHIELDED METAL-ARC (MANUAL)

Position: Flat
Weld Quality Level: Commercial
Weldability: Good

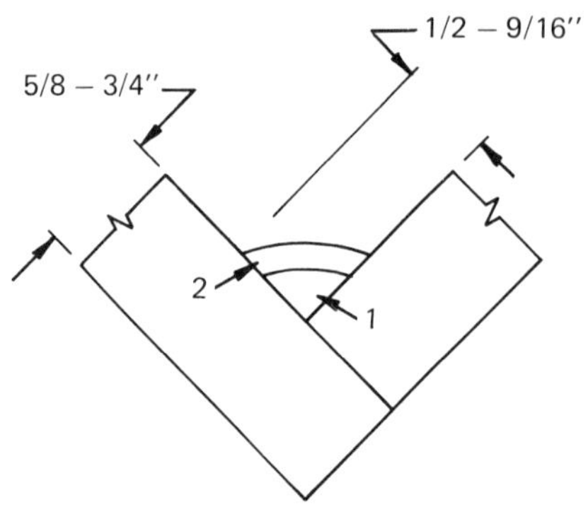

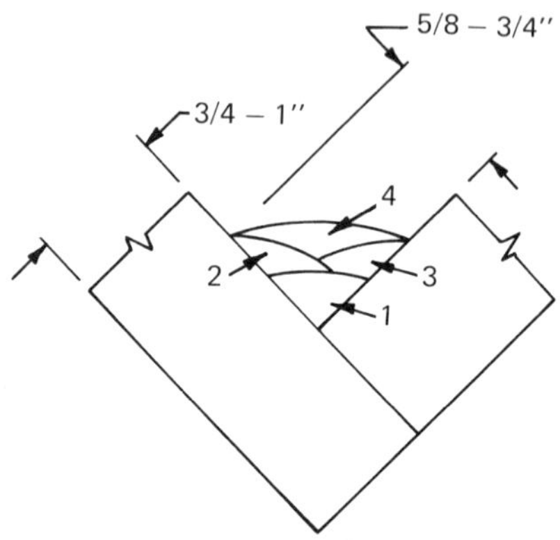

Weld Size, L (in.)	1/2		9/16		5/8		3/4	
Plate Thickness (in.)	5/8		3/4		3/4		1	
Pass	1	2	1	2	1	2 & 3	1	2 – 4
Electrode Class	E7024	E7024	E7024	E7024	E7024	E7024	E7024	E7024
Size	5/16	5/16	5/16	5/16	5/16	5/16	5/16	5/16
Current (amp) AC	475	550	475	550	475	550	475	550
Arc Speed (in./min)	13.0-15.0	14.0-16.0	13.0-15.0	10.0-11.0	13.0-15.0	14.0-15.0	13.0-15.0	13.0-14.0
Electrode Req'd (lb/ft)	0.67		0.85		1.07		1.46	
Total Time (hr/ft of weld)	0.0276		0.0333		0.0429		0.587	

SHIELDED METAL-ARC (MANUAL)

Position: Flat
Weld Quality Level: Commercial
Steel Weldability: Good

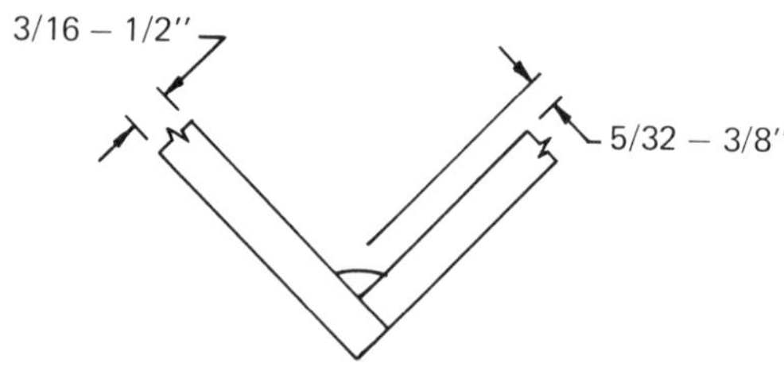

Weld Size, L (in.)	5/32	3/16	1/4	5/16	3/8
Plate Thickness (in.)	3/16	1/4	5/16	3/8	1/2
Pass	1	1	1	1	1
Electrode Class	E7014	E7014	E7014	E7014	E7014
Size	5/32	3/16	7/32	1/4	5/16
Current (amp) AC	200	250	310	370	450
Arc Speed (in./min)	12.5-13.5	12.0-13.0	11.0-12.0	9.0-10.0	7.5-8.5
Electrode Req'd (lb/ft)	0.0980	0.121	0.191	0.270	0.375
Total Time (hr/ft of weld)	0.0154	0.0160	0.0174	0.0211	0.0250

SHIELDED METAL-ARC (MANUAL)

Position: Flat
Weld Quality Level: Code
Steel Weldability: Good

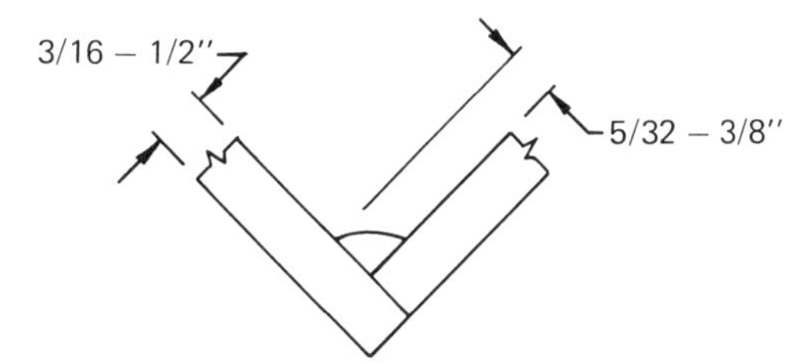

Weld Size, L (in.)	5/32	3/16		1/4		9/32	5/16	3/8
Plate Thickness (in.)	3/16	1/4		5/16		3/8		1/2
Pass	1	1	1	1	1	1	1	1
Electrode Class	E6027	E6027	E6027	E6027	E6027	E6027	E6027	E6027
Size	5/32	5/32	3/16	3/16	7/32	1/4	1/4	1/4
Current (amp) AC	210	220	260	270	335	380	390	400
Arc Speed (in./min)	15.5-17.0	13.5-15.0	15.5-17.0	12.5-14.0	14.5-16.0	14.0-15.5	11.0-12.0	9.5-10.5
Electrode Req'd (lb/ft)	0.119	0.146	0.167	0.215	0.228	0.269	0.343	0.428
Total Time (hr/ft of weld)	0.0123	0.0140	0.0123	0.0151	0.0131	0.0136	0.0174	0.0200

SHIELDED METAL-ARC (MANUAL)

Position: Flat
Weld Quality Level: Code
Steel Weldability: Good

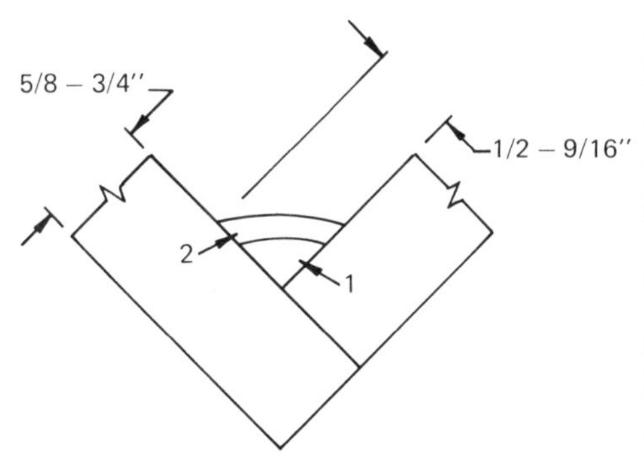

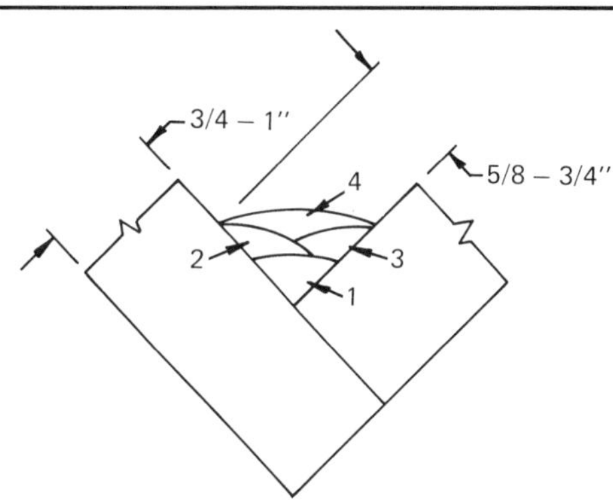

Weld Size, L (in.)	1/2		9/16		5/8		3/4	
Plate Thickness (in.)	5/8		3/4		3/4		1	
Pass	1	2	1	2	1	2 & 3	1	2 – 4
Electrode Class	E6027	E6027	E6027	E6027	E6027	E6027	E6027	E6027
Size	1/4	1/4	1/4	1/4	1/4	1/4	1/4	1/4
Current (amp) AC	400	400	400	400	400	400	400	400
Arc Speed (in./min)	11.5-12.5	11.5-12.5	11.5-12.5	7.5-8.5	11.5-12.5	11.0-12.0	11.5-12.5	10.0-11.0
Electrode Req'd (lb/ft)	0.727		0.936		1.12		1.58	
Total Time (hr/ft of weld)	0.0333		0.0417		0.0512		0.0737	

SHIELDED METAL-ARC (MANUAL)

Position: Flat
Weld Quality Level: Code
Steel Weldability: Poor

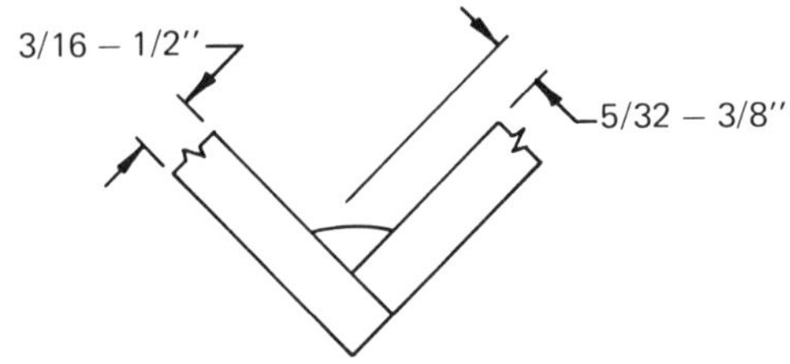

Weld Size, L (in.)	5/32	3/16	1/4	5/16	3/8
Plate Thickness (in.)	3/16	1/4	5/16	3/8	1/2
Pass	1	1	1	1	1
Electrode Class	E7028	E7028	E7028	E7028	E7028
Size	5/32	3/16	3/16	7/32	1/4
Current (amp) AC	215	260	280	330	400
Arc Speed (in./min)	13.5-15.0	13.5-15.0	11.0-12.0	10.0-12.0	8.5-9.5
Electrode Req'd (lb/ft)	0.104	0.147	0.208	0.285	0.437
Total Time (hr/ft of weld)	0.0140	0.0140	0.0175	0.0175	0.222

Preheat may be necessary depending on plate material.

SHIELDED METAL-ARC (MANUAL)

Position: Flat
Weld Quality Level: Code
Steel Weldability: Poor

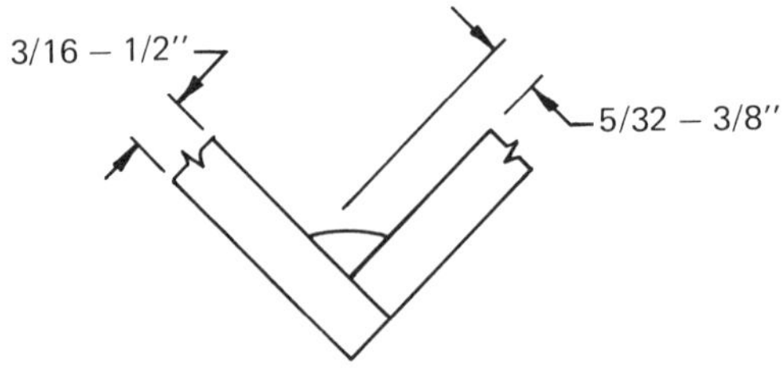

Weld Size, L (in.)	5/32	3/16	1/4	5/16	3/8
Plate Thickness (in.)	3/16	1/4	5/16	3/8	1/2
Pass	1	1	1	1	1
Electrode Class	E7018	E7018	E7018	E7018	E7018
Size	3/16	7/32	7/32	1/4	1/4
Current (amp) AC	240	275	275	350	350
Arc Speed (in./min)	13.5-15.0	13.0-14.0	9.0-10.0	7.0-8.0	6.0-6.8
Electrode Req'd (lb/ft)	0.109	0.132	0.195	0.272	0.409
Total Time (hr/ft of weld)	0.0140	0.0149	0.0202	0.0270	0.0313

Preheat may be necessary depending on plate material.

SHIELDED METAL-ARC (MANUAL)

Position: Flat
Weld Quality Level: Code
Steel Weldability: Poor

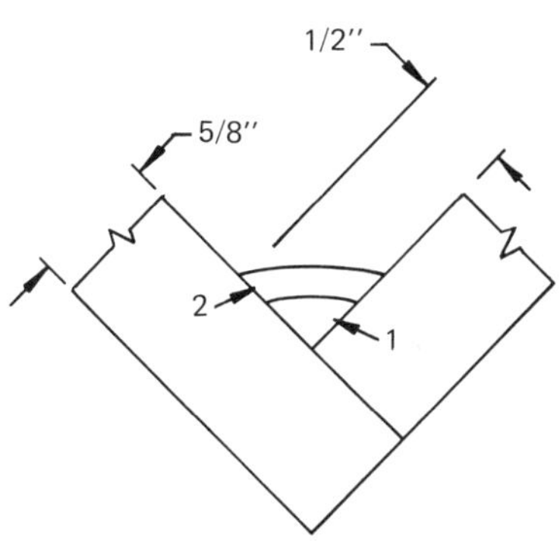

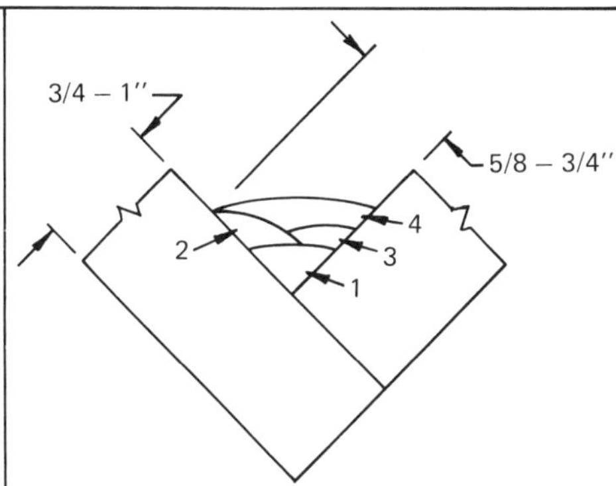

Weld Size, L (in.)	1/2	5/8	3/4
Plate Thickness (in.)	5/8	3/4	1
Pass	1 & 2	1 – 3	1 – 4
Electrode Class	E7028	E7028	E7028
Size	1/4	1/4	1/4
Current (amp) AC	400	400	400
Arc Speed (in./min)	9.5 – 11.5	9.0 – 11.0	9.0 – 11.0
Electrode Req'd (lb/ft)	0.776	1.24	1.79
Total Time (ht/ft of weld)	0.0384	0.0615	0.0887

Preheat may be necessary depending on plate material.

SHIELDED METAL-ARC (MANUAL)

Position: Flat
Weld Quality Level: Code
Steel Weldability: Poor

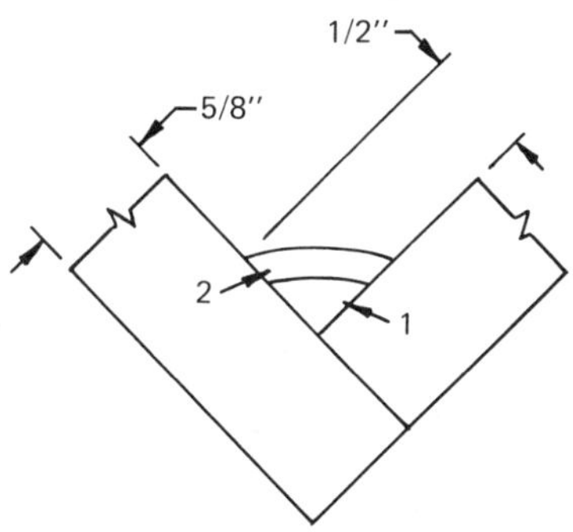

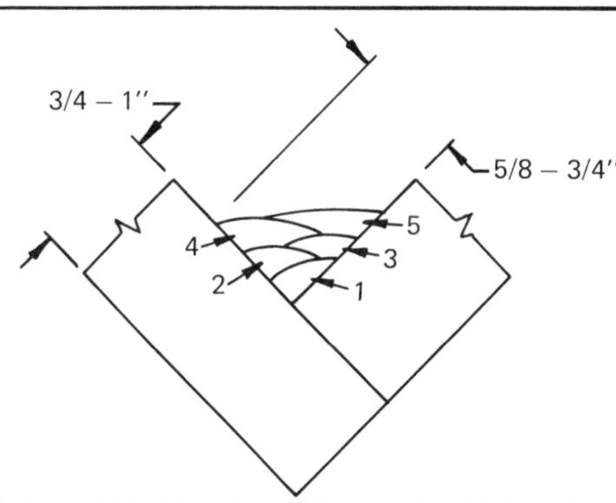

Weld Size, L (in.)	1/2	5/8	3/4
Plate Thickness (in.)	5/8	3/4	1
Pass	1 & 2	1 – 4	1 – 5
Electrode Class	E7018	E7018	E7018
Size	1/4	1/4	1/4
Current (amp) AC	350	350	350
Arc Speed (in./min)	6.9 – 7.6	6.7 – 7.5	6.6 – 7.4
Electrode Req'd (lb/ft)	0.727	1.14	1.50
Total Time (hr/ft of weld)	0.0555	0.114	0.123

Preheat may be necessary depending on plate material.

SHIELDED METAL-ARC (MANUAL)

Position: Vertical down
Weld Quality Level: Commercial
Steel Weldability: Good

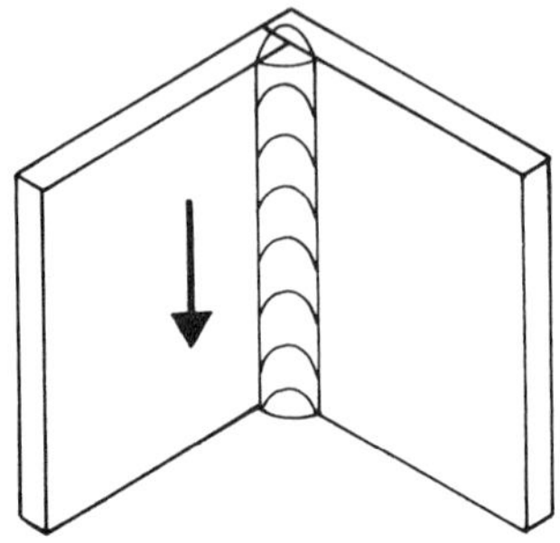

18 – 10 ga

Plate Thickness (in.)	0.048 (18 ga)	0.060 (16 ga)	0.075 (14 ga)	0.105 (12 ga)	0.135 (10 ga)
Pass	1	1	1	1	1
Electrode Class	E6013	E6012	E6012	E6012	E6012
Size	3/32	1/8	5/32	3/16	3/16
Current (amp) DC(–)	70	105	150	200	210
Arc Speed (in./min)	17 – 21	18 – 22	21 – 25	23 – 28	21 – 25
Electrode Req'd (lb/ft)	0.0374	0.0542	0.0713	0.0792	0.0930
Total Time (hr/ft of weld)	0.0105	0.0100	0.00870	0.00785	0.00870

SHIELDED METAL-ARC (MANUAL)

Position: Vertical down
Quality: Commercial
Weldability: Good

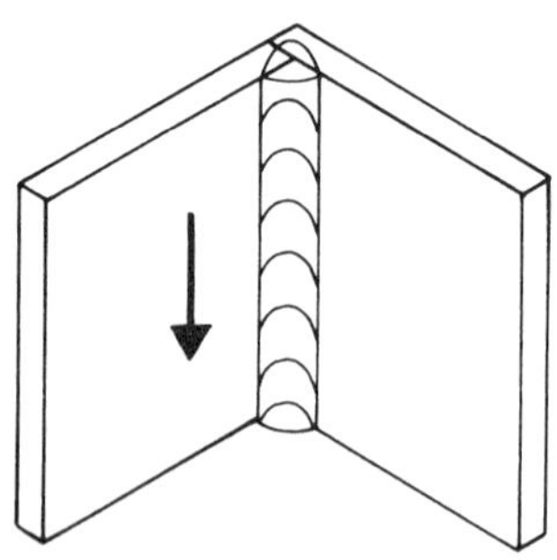

18 – 10 ga

Plate Thickness (in.)	0.048 (18 ga)	0.060 (16 ga)	0.075 (14 ga)	0.105 (12 ga)	0.135 (10 ga)
Pass	1	1	1	1	1
Electrode Class	E6013	E6013	E6013	E6013	E6013
Size	3/32	1/8	5/32	5/32	3/16
Current (amp) AC	75	115	165	170	225
Arc Speed (in./min)	16 – 20	17 – 21	19 – 23	18 – 22	16 – 20
Electrode Req'd (lb/ft)	0.0418	0.0463	0.0583	0.0636	0.0916
Total Time (hr/ft of weld)	0.0111	0.0105	0.00953	0.0100	0.0111

SHIELDED METAL-ARC (MANUAL)

Position: Vertical
Weld Quality Level: Code
Steel Weldability: Good

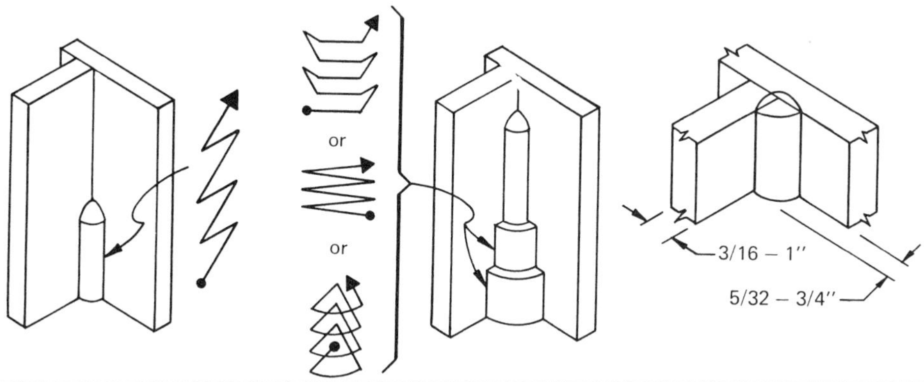

Weld Size, L (in.)	5/32	3/16	1/4	5/16	3/8	1/2	5/8	3/4
Plate Thickness (in.)	3/16	1/4	5/16	3/8	1/2	5/8	3/4	1
Pass	1	1	1	1	1	1 – 2	1 – 3	1 – 4
Electrode Class	E6010	E6010	E6010	E6010	E6010	E6010	E6010	E6010
Size	5/32	3/16	3/16	3/16	3/16	3/16	3/16	3/16
Current (amp) DC(+)	120	150	155	155	155	160	160	160
Arc Speed (in./min)	10.5-11.5	7.4-8.2	5.0-5.5	3.0-3.3	2.0-2.2	4.3-4.7*	4.3-4.7*	4.3-4.7*
Electrode Req'd (lb/ft)	0.0712	0.137	0.211	0.346	0.514	0.850	1.31	1.93
Total Time (hr/ft of weld)	0.0182	0.0256	0.0381	0.0635	0.0952	0.147	0.227	0.333
Direction of welding	Down	Up	Up	Up	Up	Up	Up	Up

* First pass only. Vary speed on succeeding passes to obtain proper weld size.

SHIELDED METAL-ARC (MANUAL)

Position: Vertical
Weld Quality Level: Code
Steel Weldability: Fair

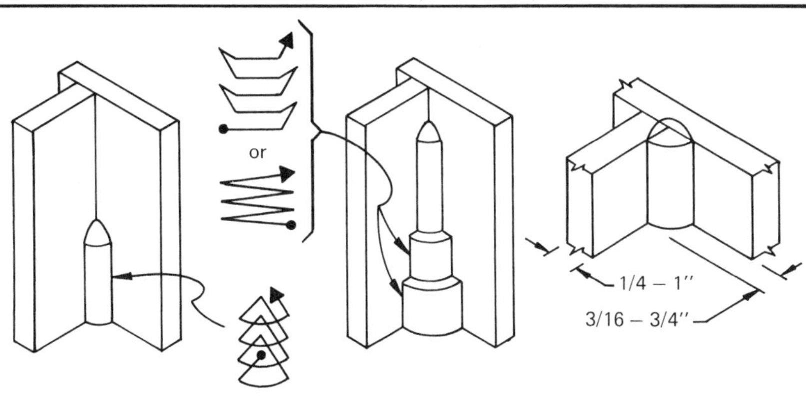

Weld Size, L (in.)	3/16	1/4	5/16	3/8	1/2	5/8	3/4
Plate Thickness (in.)	1/4	5/16	3/8	1/2	5/8	3/4	1
Pass	1	1	1	1	1	1 – 2	1 – 3
Electrode Class	E7018	E7018	E7018	E7018	E7018	E7018	E7018
Size	1/8	1/8	1/8	5/32	5/32	5/32	5/32
Current (amp) DC(+)	135	140	140	150	150	150	150
Arc Speed (in./min)	5.4-5.8	3.8-4.2	2.3-2.5	1.8-2.0	1.1-1.3	1.9-2.1*	1.9-2.1*
Electrode Req'd (lb/ft)	0.155	0.231	0.371	0.556	0.925	1.41	2.11
Total Time (hr/ft of weld)	0.0357	0.0500	0.0833	0.105	0.167	0.261	0.389

* First pass only. Vary speed on succeeding passes to obtain proper size.

SHIELDED METAL-ARC (MANUAL)

Position: Horizontal
Weld Quality Level: Commercial
Steel Weldability: Good

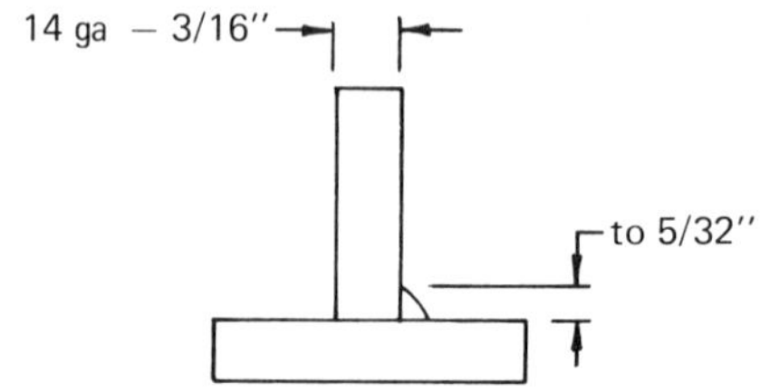

Weld Size, L (in.)				5/32	
Plate Thickness (in.)	0.075 (14 ga)	0.105 (12 ga)	0.135 (10 ga)	3/16	
Pass	1	1	1	1	1
Electrode Class	E7024	E7024	E7024	E7024	E7024
Size	3/32	1/8	1/8	1/8	5/32
Current (amp) AC	95	150	160	180	210
Arc Speed (in./min)	14.0-16.0	16.0-18.5	16.0-18.5	14.5-16.5	15.5-18.0
Electrode Req'd (lb/ft)	0.0495	0.0770	0.0833	0.104	0.119
Total Time (hr/ft of weld)	0.0133	0.0116	0.0116	0.0129	0.0119

SHIELDED METAL-ARC (MANUAL)

Position: Horizontal
Weld Quality Level: Commercial
Steel Weldability: Good

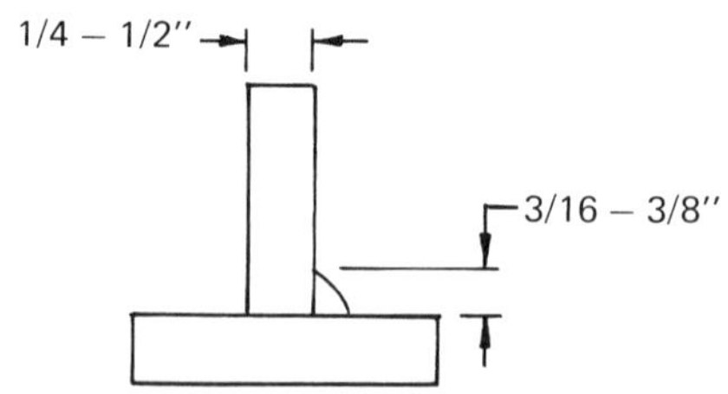

Weld Size, L (in.)	3/16		1/4		9/32	5/16	3/8
Plate Thickness (in.)	1/4		5/16		3/8		1/2
Pass	1	1	1	1	1	1	1
Electrode Class	E7024	E7024	E7024	E7024	E7024	E7024	E7024
Size	5/32	3/16	3/16	7/32	1/4	1/4	1/4
Current (amp) AC	230	270	275	325	375	375	375
Arc Speed (in./min)	13.5-15.0	15.0-17.0	14.0-15.0	16.0-18.0	16.0-18.0	13.0-14.0	10.5-11.5
Electrode Req'd (lb/ft)	0.150	0.166	0.20	0.21	0.23	0.30	0.41
Total Time (hr/ft of weld)	0.0141	0.0125	0.0138	0.0118	0.0118	0.0148	0.0182

SHIELDED METAL-ARC (MANUAL)

Position: Horizontal
Weld Quality Level: Commercial
Steel Weldability: Good

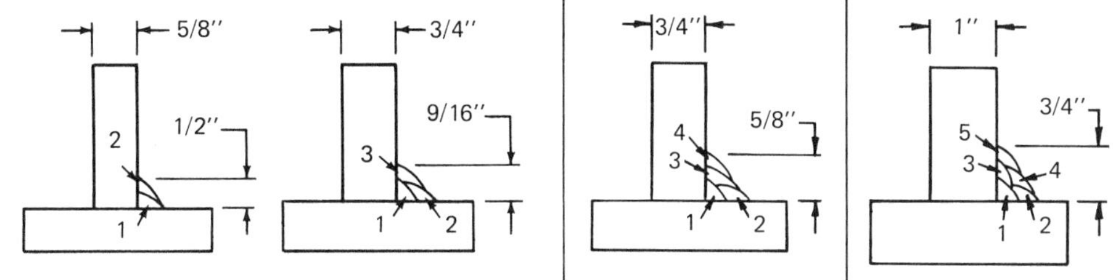

Weld Size, L (in.)	1/2		9/16		5/8		3/4	
Plate Thickness (in.)	5/8		3/4		3/4		1	
Pass	1	2 & 3	1	2 & 3	1	2 – 4	1	2 – 5
Electrode Class	E7024	E7024	E7024	E7024	E7024	E7024	E7024	E7024
Size	1/4	1/4	1/4	1/4	1/4	1/4	1/4	1/4
Current (amp) AC	375	375	375	375	375	375	375	375
Arc Speed (in./min)	10.5-11.5	11.0-12.0	10.5-11.5	14.0-16.0	10.5-11.5	14.0-16.0	10.5-11.5	12.0-13.0
Electrode Req'd (lb/ft)	0.73		0.92		1.15		1.62	
Total Time (hr/ft of weld)	0.0356		0.0449		0.0582		0.0822	

SHIELDED METAL-ARC (MANUAL)

Position: Horizontal
Weld Quality Level: Code
Steel Weldability: Good

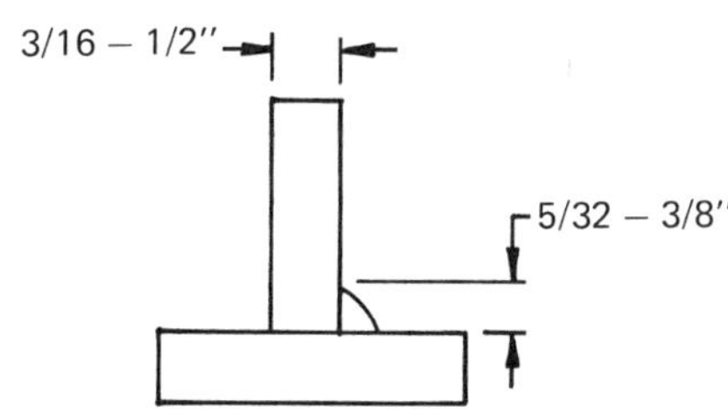

Weld Size, L (in.)	5/32	3/16		1/4		9/32	5/16	3/8
Plate Thickness (in.)	3/16	1/4		5/16		3/8		1/2
Pass	1	1	1	1	1	1	1	1
Electrode Class	E6027	E6027	E6027	E6027	E6027	E6027	E6027	E6027
Size	5/32	5/32	3/16	3/16	7/32	7/32	7/32	1/4
Current (amp) AC	210	220	250	260	320	325	335	360
Arc Speed (in./min)	14.5-16.0	13.0-14.5	14.5-16.0	11.5-12.5	13.0-14.5	11.5-12.5	9.5-10.5	7.5-8.5
Electrode Req'd (lb/ft)	0.128	0.151	0.173	0.224	0.241	0.281	0.356	0.463
Total Time (hr/ft of weld)	0.0131	0.0145	0.0131	0.0167	0.0145	0.0167	0.0200	0.0250

SHIELDED METAL-ARC (MANUAL)

Position: Horizontal
Weld Quality Level: Code
Steel Weldability: Poor

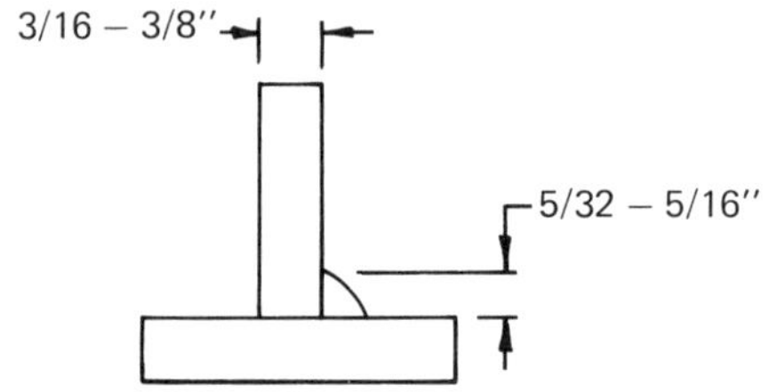

Weld Size, L (in.)	5/32	3/16	1/4		5/16	
Plate Thickness (in.)	3/16	1/4	5/16		3/8	
Pass	1	1	1	1	1	1
Electrode Class	E7028	E7028	E7028	E7028	E7028	E7028
Size	5/32	3/16	3/16	7/32	7/32	1/4
Current (amp) AC	215	260	280	335	335	390
Arc Speed (in./min)	12.5-13.5	11.5-12.5	9.5-10.5	12.0-13.0	9.5-10.5	11.5-12.5
Electrode Req'd (lb/ft)	0.112	0.157	0.235	0.236	0.320	0.330
Total Time (hr/ft of weld)	0.0152	0.0167	0.0200	0.0160	0.0200	0.0167

Preheat may be necessary depending on plate material.

SHIELDED METAL-ARC (MANUAL)

Position: Horizontal
Weld Quality Level: Code
Steel Weldability: Poor

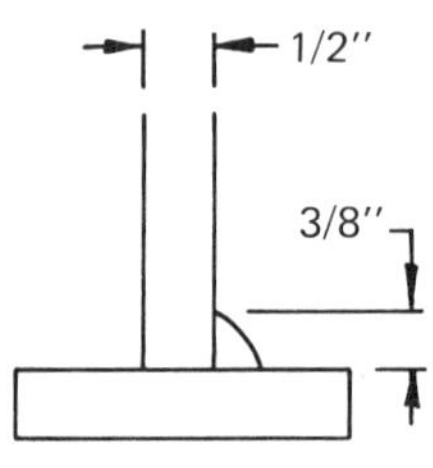

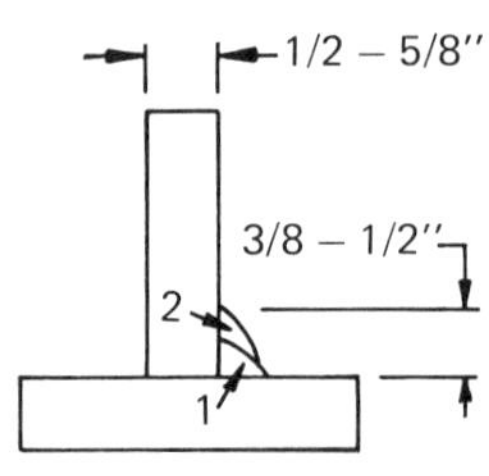

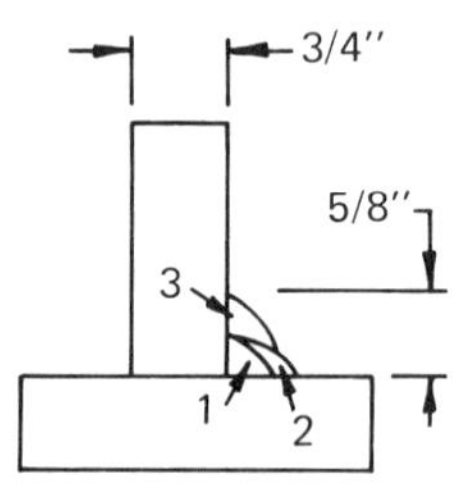

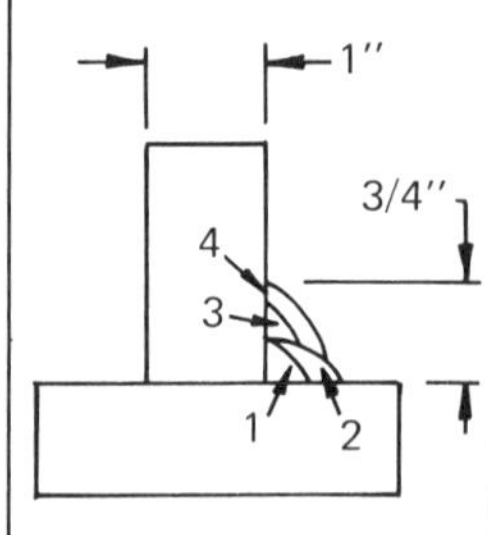

Weld Size, L (in.)	3/8*	3/8	1/2	5/8	3/4
Plate Thickness (in.)	1/2	1/2	5/8	3/4	1
Pass	1	1 – 2	1 – 2	1 – 3	1 – 4
Electrode Class	E7028	E7028	E7028	E7028	E7028
Size	1/4	7/32	1/4	1/4	1/4
Current (amp) AC	390	335	390	390	390
Arc Speed (in./min)	7.5 – 8.5	11.5 – 12.5	9.0 – 10.0	9.0 – 10.0	8.0 – 9.0
Electrode Req'd (lb/ft)	0.483	0.483	0.819	1.28	1.82
Total Time (hr/ft of weld)	0.0250	0.0333	0.0422	0.0633	0.0940

Preheat may be necessary depending on plate material.

* May not be full 3/8 in. on the vertical leg.

SHIELDED METAL-ARC (MANUAL)

Position: Horizontal
Weld Quality Level: Code
Steel Weldability: Poor

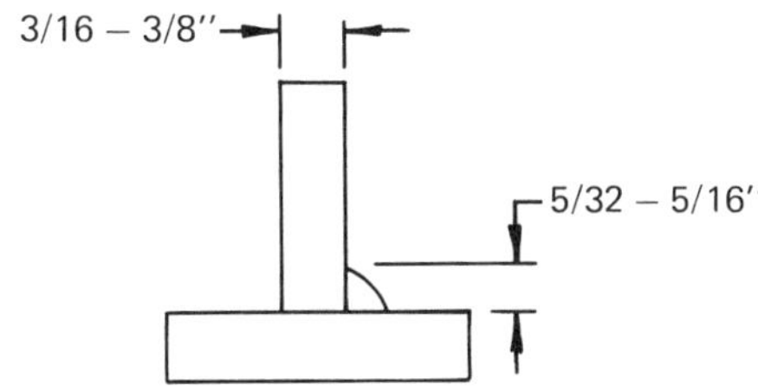

Weld Size, L (in.)	5/32	3/16	1/4	5/16
Plate Thickness (in.)	3/16	1/4	5/16	3/8
Pass	1	1	1	1
Electrode Class	E7018	E7018	E7018	E7018
Size	3/16	7/32	7/32	1/4
Current (amp) AC	240	275	275	350
Arc Speed (in./min)	12.5 – 13.5	11.0 – 12.0	8.5 – 9.5	6.5 – 7.5
Electrode Req'd (lb/ft)	0.111	0.140	0.203	0.335
Total Time (hr/ft of weld)	0.0154	0.0174	0.0222	0.0286

Preheat may be necessary depending on plate material.

SHIELDED METAL-ARC (MANUAL)

Position: Horizontal
Weld Quality Level: Code
Steel Weldability: Poor

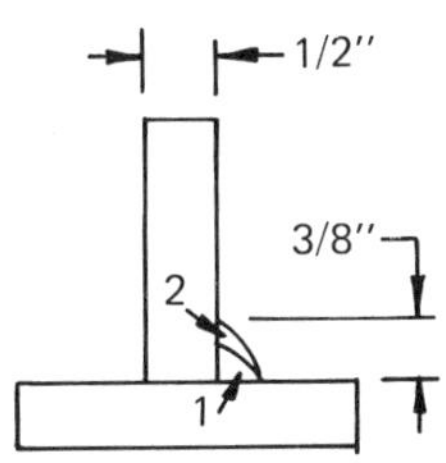

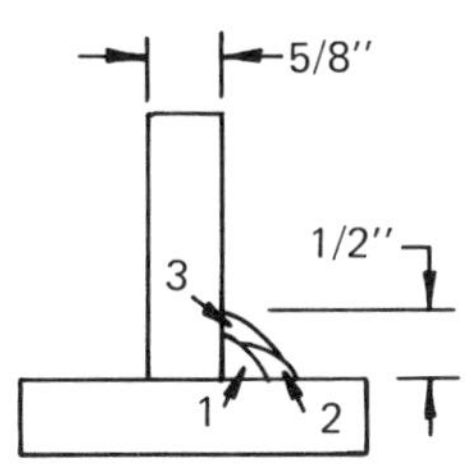

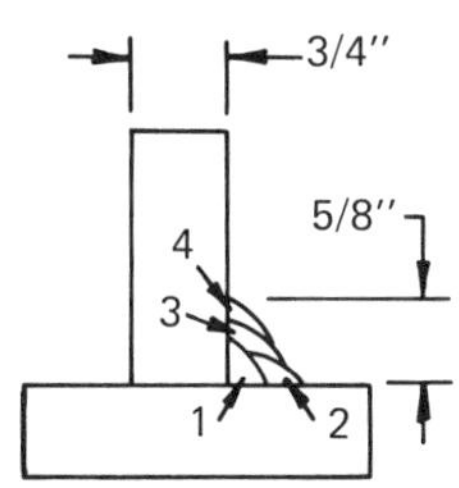

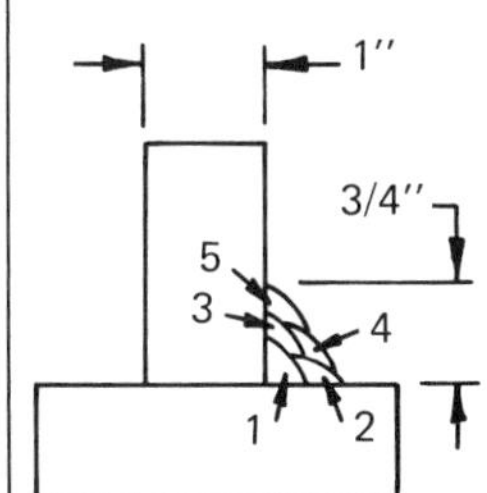

Weld Size, L (in.)	3/8	1/2	5/8	3/4
Plate Thickness (in.)	1/2	5/8	3/4	1
Pass	1 & 2	1 – 3	1 – 4	1 – 5
Electrode Class	E7018	E7018	E7018	E7018
Size	1/4	1/4	1/4	1/4
Current (amp) AC	350	350	350	350
Arc Speed (in./min)	9.5 – 11.5	9.5 – 10.5	8.0 – 9.0	7.0 – 8.0
Electrode Req'd (lb/ft)	0.480	0.785	1.18	1.62
Total Time (hr/ft of weld)	0.0390	0.0600	0.0940	0.133

Preheat may be necessary depending on plate material.

SHIELDED METAL-ARC (MANUAL)

Position: Horizontal
Weld Quality Level: Commercial
Steel Weldability: Good

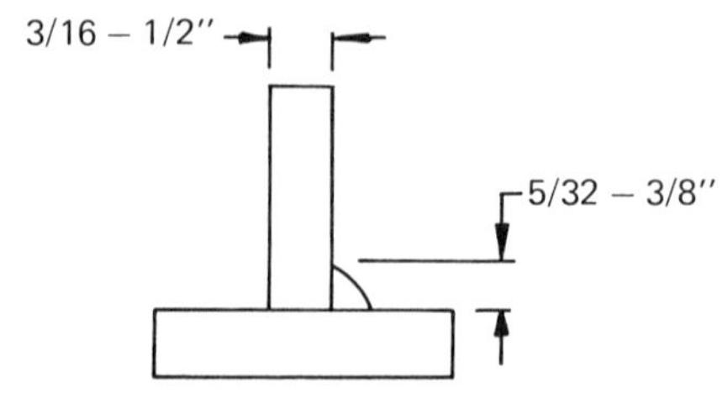

Weld Size, L (in.)	5/32	3/16	1/4	5/16	3/8
Plate Thickness (in.)	3/16	1/4	5/16	3/8	1/2
Pass	1	1	1	1	1
Electrode Class	E7014	E7014	E7014	E7014	E7014
Size	5/32	3/16	7/32	1/4	5/16
Current (amp) AC	200	250	310	370	450
Arc Speed (in./min)	10.5-11.5	11.5-12.5	11.0-12.0	9.0-10.0	7.0-8.0
Electrode Req'd (lb/ft)	0.128	0.127	0.191	0.270	0.388
Total Time (hr/ft of weld)	0.0182	0.0167	0.0174	0.0211	0.0267

SHIELDED METAL-ARC (MANUAL)

Position: Overhead
Weld Quality Level: Code
Steel Weldability: Good

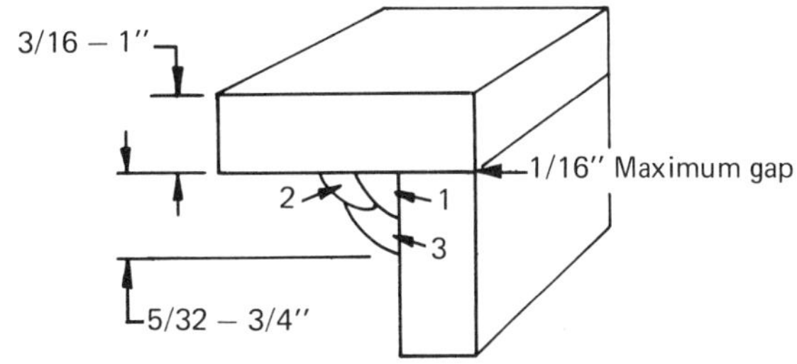

Weld Size, L (in.)	5/32	3/16	1/4	5/16	3/8	1/2	5/8	3/4
Plate Thickness (in.)	3/16	1/4	5/16	3/8	1/2	5/8	3/4	1
Pass	1	1	1	1 – 2	1 – 3	1 – 6	1 – 10	1 – 15
Electrode Class	E6010	E6010	E6010	E6010	E6010	E6010	E6010	E6010
Size	5/32	3/16	3/16	3/16	3/16	3/16	3/16	3/16
Current (amp) DC(+)	130	170	170	170	170	170	170	170
Arc Speed (in./min)*	7.0-7.7	8.5-9.4	4.8-5.3	6.6-7.3	6.6-7.3	6.6-7.3	6.6-7.3	6.6-7.3
Electrode Req'd (lb/ft)	0.100	0.145	0.253	0.369	0.532	0.945	1.48	2.13
Total Time (hr/ft of weld)	0.0272	0.0223	0.0396	0.0567	0.0820	0.145	0.228	0.328

On 1/2 in. plate and thicker, place the first pass of each layer on the top plate.

* First pass only. Vary speed on succeeding passes to obtain proper weld size.

SHIELDED METAL-ARC (MANUAL)

Position: Overhead
Weld Quality Level: Code
Steel Weldability: Fair

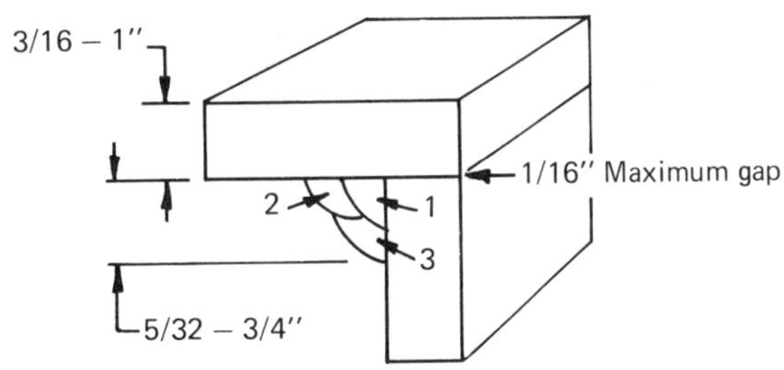

Weld Size, L (in.)	5/32	3/16	1/4	5/16	3/8	1/2	5/8	3/4
Plate Thickness (in.)	3/16	1/4	5/16	3/8	1/2	5/8	3/4	1
Pass	1	1	1 – 2	1 – 3	1 – 4	1 – 6	1 – 10	1 – 15
Electrode Class	E7018	E7018	E7018	E7018	E7018	E7018	E7018	E7018
Size	5/32	5/32	5/32	5/32	5/32	5/32	5/32	5/32
Current (amp) DC(+)	170	170	170	170	170	170	170	170
Arc Speed (in./min)*	10.5-11.5	7.2-8.0	8.2-9.1	8.2-9.1	8.5-9.4	7.0-7.7	7.2-8.0	8.1-8.9
Electrode Req'd (lb/ft)	0.107	0.155	0.277	0.394	0.570	1.01	1.59	2.29
Total Time (hr/ft of weld)	0.0182	0.0264	0.0463	0.0670	0.0967	0.172	0.269	0.388

On 3/8 in. plate and thicker place the first pass of each layer on the top plate.

* First pass only. Vary succeeding passes to obtain proper weld size.

SHIELDED METAL-ARC (MANUAL)

Position: Horizontal
Weld Quality Level: Commercial
Steel Weldability: Good

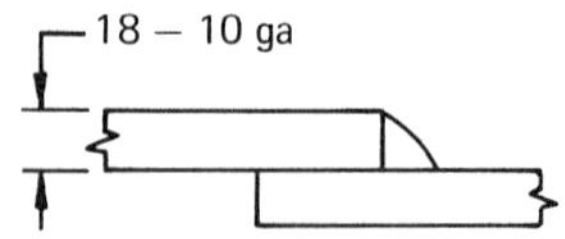

Plate Thickness (in.)	0.048 (18 ga)	0.060 (16 ga)	0.075 (14 ga)	0.105 (12 ga)	0.135 (10 ga)
Pass	1	1	1	1	1
Electrode Class	E6013	E6012	E6012	E6012	E6012
Size	3/32	1/8	5/32	3/16	3/16
Current (amp) DC(–)	70	105	145	200	210
Arc Speed (in./min)	19 – 23	21 – 26	20 – 24	18 – 22	14 – 18
Electrode Req'd (lb/ft)	0.0339	0.0427	0.0717	0.101	0.134
Total Time (hr/ft of weld)	0.00953	0.00851	0.00910	0.0100	0.0125

SHIELDED METAL-ARC (MANUAL)

Position: Horizontal
Weld Quality Level: Commercial
Steel Weldability: Good

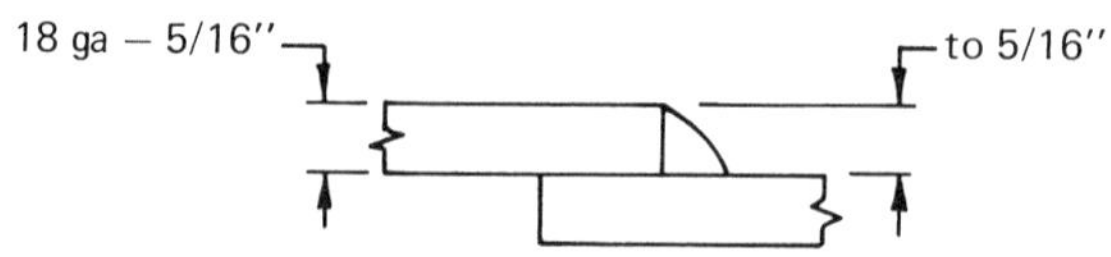

Weld Size, L (in.)						3/16	1/4	5/16
Plate Thickness (in.)	0.048 (18 ga)	0.060 (16 ga)	0.075 (14 ga)	0.105 (12 ga)	0.135 (10 ga)	3/16	1/4	5/16
Pass	1	1	1	1	1	1	1	1
Electrode Class	E6013	E6013	E6013	E6013	E6013	E7024	E7024	E7024
Size	3/32	1/8	5/32	5/32	3/16	3/16	7/32	7/32
Current (amp) AC	75	115	160	165	215	290	360	360
Arc Speed (in./min)	15 – 19	16 – 20	16 – 20	14 – 18	13 – 17	14 – 17	14 – 16	12 – 14
Electrode Req'd (lb/ft)	0.0389	0.0490	0.0667	0.0773	0.103	0.170	0.211	0.253
Total Time (hr/ft of weld)	0.0118	0.0111	0.0111	0.0125	0.0133	0.0129	0.0133	0.0155

SHIELDED METAL-ARC (MANUAL)

Position: Vertical down
Weld Quality Level: Commercial
Steel Weldability: Good

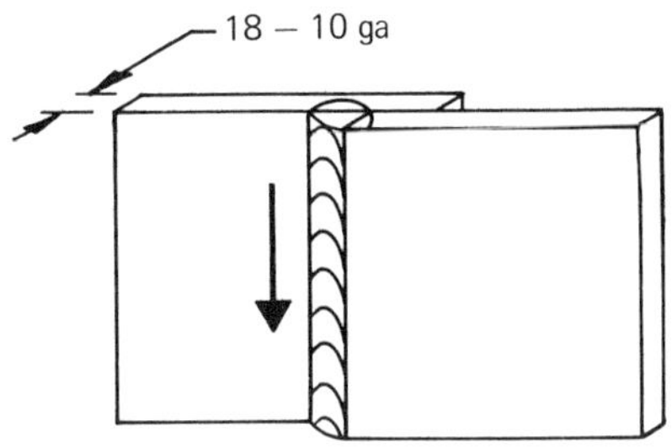

Plate Thickness (in.)	0.048 (18 ga)	0.060 (16 ga)	0.075 (14 ga)	0.105 (12 ga)	0.135 (10 ga)
Pass	1	1	1	1	1
Electrode Class	E6013	E6012	E6012	E6012	E6012
Size	3/32	1/8	5/32	3/16	3/16
Current (amp) DC(–)	75	115	155	210	220
Arc Speed (in./min)	22 – 27	27 – 32	27 – 32	25 – 30	22 – 27
Electrode Req'd (lb/ft)	0.0316	0.0375	0.0576	0.0781	0.0930
Total Time (hr/ft of weld)	0.00817	0.00678	0.00678	0.00728	0.00817

SHIELDED METAL-ARC (MANUAL)

Position: Vertical down
Weld Quality Level: Commercial
Steel Weldability: Good

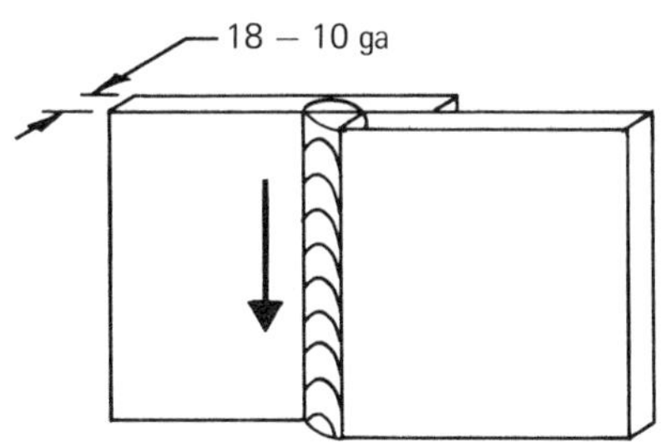

Plate Thickness (in.)	0.048 (18 ga)	0.060 (16 ga)	0.075 (14 ga)	0.105 (12 ga)	0.135 (10 ga)
Pass	1	1	1	1	1
Electrode Class	E6013	E6013	E6013	E6013	E6013
Size	3/32	1/8	5/32	5/32	3/16
Current (amp) AC	85	125	170	175	225
Arc Speed (in./min)	19 – 23	20 – 24	21 – 26	19 – 23	16 – 20
Electrode Req'd (lb/ft)	0.0358	0.0444	0.0546	0.0631	0.0922
Total Time (hr/ft of weld)	0.00953	0.00910	0.00850	0.00953	0.0111

SHIELDED METAL-ARC (MANUAL)

Position: Flat
Weld Quality Level: Commercial
Steel Weldability: Good
Welded From: One side

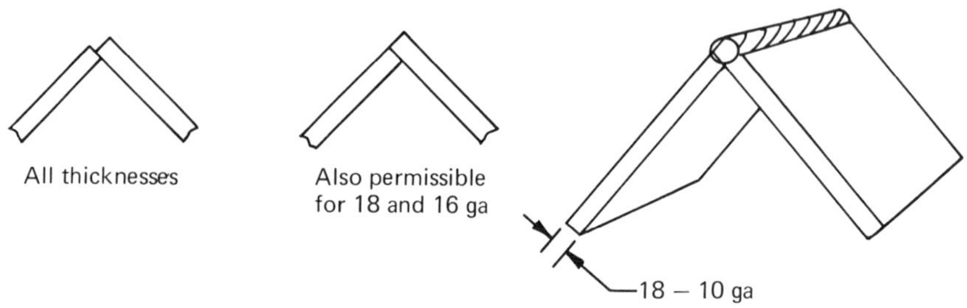

Plate Thickness (in.)	0.048 (18 ga)	0.060 (16 ga)	0.075 (14 ga)	0.105 (12 ga)	0.135 (10 ga)
Pass	1	1	1	1	1
Electrode Class	E6010	E6010	E6010	E6010	E6010
Size	3/32	1/8	1/8	5/32	3/16
Current (amp) DC(–)	45	80	85	110	155*
Arc Speed (in./min)	30 – 35	35 – 40	35 – 40	33 – 38	27 – 32
Electrode Req'd (lb/ft)	0.0197	0.0282	0.0300	0.0432	0.0505
Total Time (hr/ft of weld)	0.00616	0.00533	0.00533	0.00563	0.00678

* Use DC(+)

SHIELDED METAL-ARC (MANUAL)

Position: Flat
Weld Quality Level: Commercial
Steel Weldability: Good
Welded From: One side

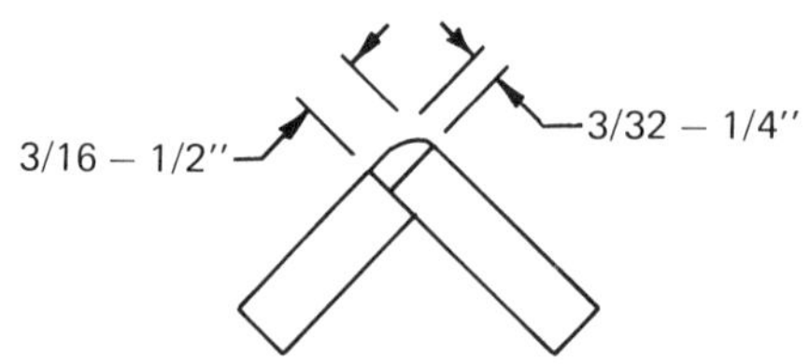

Weld Size, L (in.)	3/32	1/8	5/32	3/16	1/4
Plate Thickness (in.)	3/16	1/4	5/16	3/8	1/2
Pass	1	1	1	1	1
Electrode Class	E7024	E7024	E7024	E7024	E7024
Size	5/32	3/16	7/32	7/32	1/4
Current (amp) AC	215	275	350	360	410
Arc Speed (in./min)	22.0-27.0	19.0-23.0	18.5-22.5	16.5-19.5	14 – 17
Electrode Req'd (lb/ft)	0.0750	0.114	0.152	0.175	0.250
Total Time (hr/ft of weld)	0.00820	0.00952	0.00975	0.0111	0.0130

SHIELDED METAL-ARC (MANUAL)

Position: Flat
Weld Quality Level: Commercial
Steel Weldability: Good
Welded From: One side

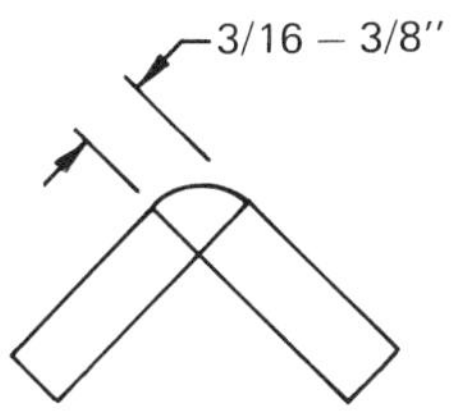

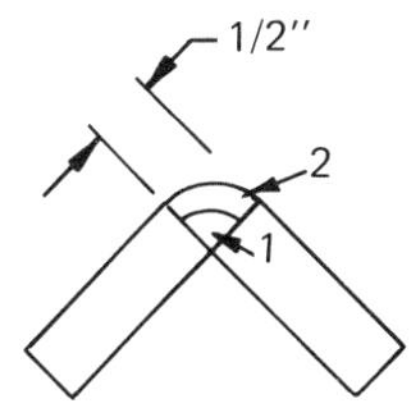

Weld Size, L (in.)	3/16	1/4	5/16	3/8	1/2
Plate Thickness (in.)	3/16	1/4	5/16	3/8	1/2
Pass	1	1	1	1	1 & 2
Electrode Class	E7024	E7024	E7024	E7024	E7024
Size	3/16	7/32	7/32	1/4	1/4
Current (amp) AC	250	320	350	400	410
Arc Speed (in./min)	21.0 – 25.0	18.0 – 22.0	14.5 – 17.5	13.0 – 16.0	11.5 – 14.5
Electrode Req'd (lb/ft)	0.101	0.133	0.198	0.240	0.530
Total Time (hr/ft of weld)	0.00870	0.0100	0.0125	0.0139	0.0308

SHIELDED METAL-ARC (MANUAL)

Position: Vertical down
Weld Quality Level: Commercial
Steel Weldability: Good
Welded From: One side

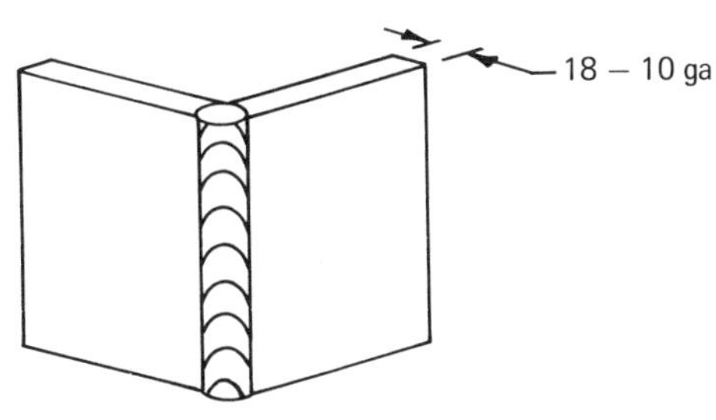

Plate Thickness (in.)	0.048 (18 ga)	0.060 (16 ga)	0.075 (14 ga)	0.105 (12 ga)	0.135 (10 ga)
Pass	1	1	1	1	1
Electrode Class	E6010	E6010	E6010	E6010	E6010
Size	3/32	1/8	1/8	5/32	3/16
Current (amp) DC(–)	50	90	95	120	170*
Arc Speed (in./min)	35 – 40	40 – 45	40 – 45	37 – 42	33 – 38
Electrode Req'd (lb/ft)	0.0184	0.0278	0.0293	0.0436	0.0461
Total Time (hr/ft of weld)	0.00533	0.00471	0.00471	0.00507	0.00563

* DC(+)

SHIELDED METAL-ARC (MANUAL)

Position: Flat
Weld Quality Level: Commercial
Steel Weldability: Good

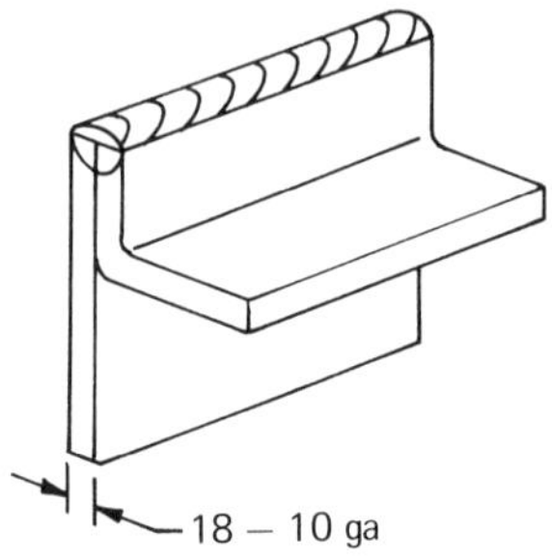

Plate Thickness (in.)	0.048 (18 ga)	0.060 (16 ga)	0.075 (14 ga)	0.105 (12 ga)	0.135 (10 ga)
Pass	1	1	1	1	1
Electrode Class	E6010	E6010	E6010	E6010	E6010
Size	3/32	1/8	1/8	5/32	3/16
Current (amp) DC(–)	50	80	85	115	140
Arc Speed (in/min)	45 – 50	43 – 48	40 – 45	40 – 45	37 – 42
Electrode Req'd (lb/ft)	0.0145	0.0232	0.0263	0.0382	0.0476
Total Time (hr/ft of weld)	0.00421	0.00439	0.00471	0.00471	0.00505

SHIELDED METAL-ARC (MANUAL)

Position: Vertical down
Weld Quality Level: Commercial
Steel Weldability: Good

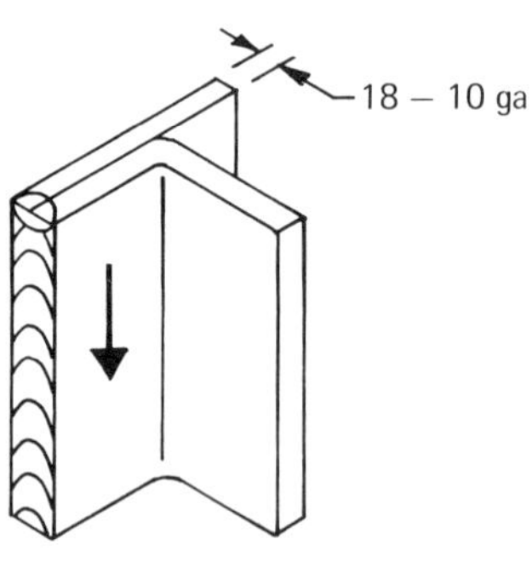

Plate Thickness (in.)	0.048 (18 ga)	0.060 (16 ga)	0.075 (14 ga)	0.105 (12 ga)	0.135 (10 ga)
Pass	1	1	1	1	1
Electrode Class	E6010	E6010	E6010	E6010	E6010
Size	3/32	1/8	1/8	5/32	3/16
Current (amp) DC(–)	55	90	95	125	155
Arc Speed (in./min)	53 – 58	50 – 55	47 – 52	47 – 52	43 – 48
Electrode Req'd (lb/ft)	0.0141	0.0225	0.0251	0.0358	0.0473
Total Time (hr/ft of weld)	0.00361	0.00381	0.00404	0.00404	0.00439

SHIELDED METAL-ARC (MANUAL)

Positon: Flat
Weld Quality Level: Commercial
Steel Weldability: Good

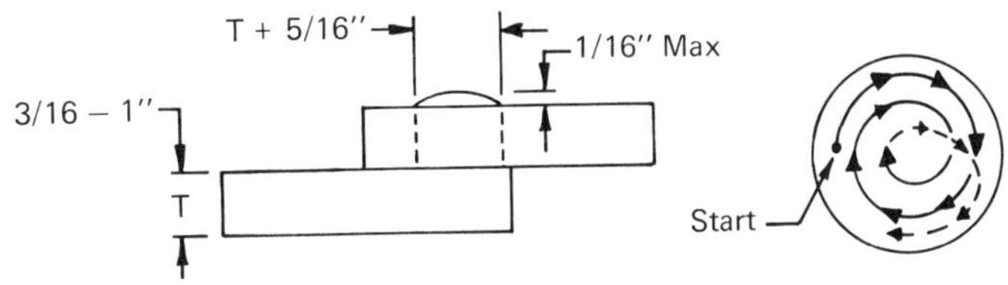

Plate Thickness (in.)	3/16	5/16	3/8	1/2	3/4†	1†
Electrode Class	E7018	E7018	E7018	E7018	E7018	E7018
Size	1/8	3/16	3/16	7/32	1/4	1/4
Current (amp) AC	140	250	250	300	350	350
Arc Speed (in./min)						
Electrode Req'd*	0.0154	0.0440	0.0642	0.113	0.300	0.605
Total Time* (hr)	0.00417	0.00500	0.00731	0.0118	0.0236	0.0475

Weld with spiral motion and continue as long as slag can be kept molten or until the weld is completed.

* Per weld

† Thickness of the weld may be reduced to 5/8 inch per AWS Structural Welding Code D1.1-92 par 2.8.8.

SHIELDED METAL-ARC (MANUAL)
Special Procedures for ASTM A203 and A537 Steels

Position: Flat
Weld Quality Level: Code
Steel Weldability: Poor
Welded From: Two sides

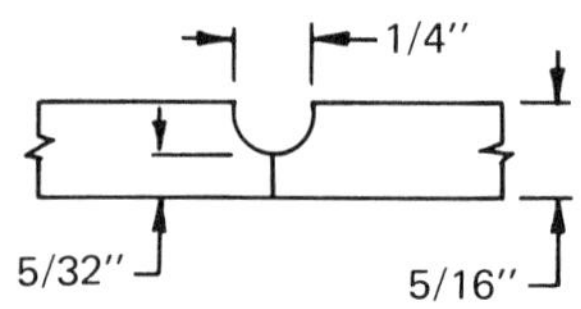

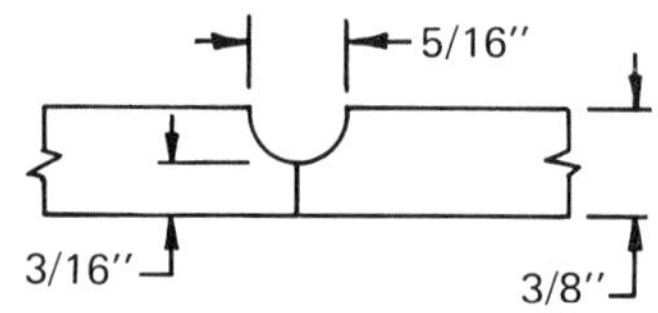

Plate Thickness (in.)	5/16		3/8	
Pass	1 & 2	3 & 4*	1 – 3	4 – 6*
Electrode Class†				
Size	5/32	5/32	5/32	5/32
Current (amp) DC(+)	150	150	150	150
Arc Speed (in./min)	9 – 11	8 – 10	9 – 11	8 – 10
Electrode Req'd (lb/ft)	0.48		0.65	
Total Time (hr/ft of weld)	0.0844		0.127	
Interpass Temperature, Max. (°F)	150		150	

Position: Flat
Weld Quality Level: Code
Steel Weldability: Poor
Welded From: Two sides

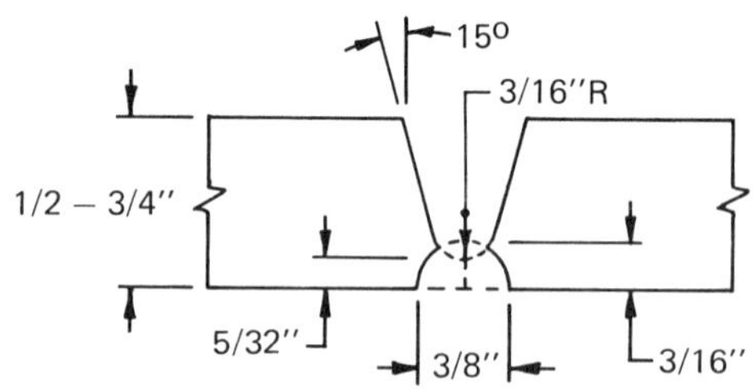

Plate Thickness (in.)	1/2		5/8		3/4	
Pass	1 – 5	6 – 8*	1 – 7	8 – 10*	1 – 10	11 – 13*
Electrode Class†						
Size	5/32	5/32	5/32	5/32	5/32	5/32
Current (amp) DC(+)	150	150	150	150	150	150
Arc Speed (in./min)	7 – 9	8 – 10	7 – 9	8 – 10	7 – 9	8 – 10
Electrode Req'd (lb/ft)	1.40		1.79		2.25	
Total Time (hr/ft of weld)	0.188		0.238		0.313	
Interpass Temperature, Max. (°F)	175		200		225	

* Second side is gouged after first side is completed.

† See Tables 6-16 and 6-20.

Welding Carbon and Low-Alloy Steels with the Submerged-Arc Process

The submerged-arc process is typically some three to ten times faster than stick-electrode welding, yet provides joints of exceedingly high quality and good appearance. The process has various desirable performance features (see Section 5.2) and should be considered whenever the welding is to be done by automatic or semiautomatic means. An indication of the relative speeds of submerged-arc and other processes is given in Table 6-22.

TABLE 6-22. Comparison of Welding Speeds

Welding Process	Arc Speed* (in./min)
Manual shielded metal-arc	4-1/4
Semiautomatic self-shielded flux-cored	10
Semiautomatic submerged-arc	15
Automatic submerged-arc, one wire	18
Twin-arc submerged-arc	21
Tandem submerged-arc, two wires	30
Tandem submerged-arc, three wires	45

* Arc speed for 3/8-in. butt weld.

FLUX AND ELECTRODE REQUIREMENTS

The characteristics of electrodes are discussed in Section 4.1. Electrodes and flux combinations are selected to serve a specific purpose. They may, for example, be chosen to serve general-purpose needs at lowest cost, or they may be selected to meet special metallurgical requirements. Some combinations provide maximum resistance to weld cracking or to porosity from rusty plate, while other combinations provide fast-fill or fast-follow characteristics. Specifications for fluxes and electrodes are listed in Section 4.1. Since hundreds of combinations of flux and electrode are possible, prospective users of the submerged-arc process should consult with suppliers of these materials before choosing a particular combination for a given application. Mechanical properties requirements for submerged-arc weld-metal deposits are given in Table 6-23.

TABLE 6-23. Mechanical Property Requirements for Submerged Arc Weld Metal Deposits

AWS(1) Classification Flux-Wire	Tensile Strength ksi	Yield Strength ksi min.	Elongation in 2 in. % min.	Temperature for 20 ft. lbs.
F6AX-EXXX	60-80	48	22	See Note 2
F7AX-EXXX	70-95	58	22	See Note 2
F8AX-EXXX	80-100	68	20	See Note 2
F9AX-EXXX	90-110	78	17	See Note 2
F10AX-EXXX	100-120	88	16	See Note 2
F11AX-EXXX	110-130	98	15	See Note 2
F12AX-EXXX	120-140	108	14	See Note 2

(1) The A in the first group of characters indicates tests made in the as-welded condition, a P indicates post weld heat treated.

(2) The X in the first group indicates the highest temperature at which an impact strength of at least 20 ft. lbs. is obtained: 0=0°F, 2=−20°F, 4=−40°F, 5=−50°F, 6=−60°F, 8=−80°F, 10=−100°F, 15=−150°F.

(3) The letters and numbers following the E in the second group of characters identify the composition of the electrode wire used to make the welds for testing. Compositions for carbon steel electrodes and for low alloy steel electrodes are listed in Table 4-10 in Section 4.

Electrode Size

Welding equipment for semiautomatic submerged-arc welding can accommodate only a limited range of electrode sizes. Change of electrode may require the purchase of a new gun and the altering of the feed mechanisms. Therefore, the job requirements should be considered carefully before the first equipment is purchased.

Generally, only the 1/16, 5/64, and 3/32-in. electrodes are used for semiautomatic welding. The 1/16-in. wire is used for making high-speed fillet welds on steel ranging from 14-gage to 1/4 in. thick. The 5/64-in. electrode is used for fillet, lap, and butt welds on plate 12-gage or more when the welding gun is hand-held. The 3/32-in. wire is used primarily when the gun is carried mechanically. Electrodes of this diameter can be used for hand-held operation, but the stiffness of the wire tends to make the cable rigid and thereby decreases the maneuverability of the gun.

Fully automatic submerged-arc welding generally employs electrodes of from 5/64 to 7/32 in. in diameter. There are exceptional conditions where wire as small as 0.035 or as large as 3/8-in. has been satisfactory, but most applications require electrodes of 5/64-in. or larger. Current ranges for the various electrode sizes are given in Table 6-24.

When all other conditions are kept constant, an increase in wire size increases the bead width and decreases the penetration and the deposition rate as illustrated in Fig. 6-31. When current is held

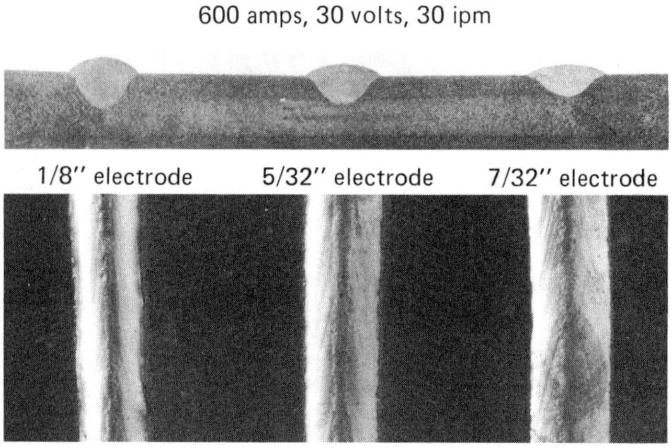

Fig. 6-31. Effect of electrode size on weld characteristics. With the current and voltage held constant as electrode diameter is increased, bead width increases, but penetration and deposition rate decrease.

constant as wire size is increased, wire feed rate must be reduced. A larger wire size, however, increases the maximum current capacity and, thus, permits increased current and a higher deposition rate. A change to a larger wire size, with no increase in current, bridges gaps better if poor fitup is encountered.

WELDING CURRENTS AND VOLTAGES

Welding current determines the rate at which the electrode is melted, the depth of penetration of the weld pool into the base metal, and the amount of base metal fused. An increase in current increases penetration and melt-off rate, but an excessively high current produces a high, narrow bead, an erratic arc, and undercut. Excessively low current produces an unstable arc. Figure 6-32 shows the effect of current variation when other conditions are maintained constant.

TABLE 6-24. Current Ranges for Submerged-Arc Welding Electrodes

Wire Diameter (in.)	Current (amp)
5/64	200 — 600
3/32	230 — 700
1/8	300 — 900
5/32	420 — 1000
3/16	480 — 1100
7/32	600 — 1200
1/4	700 — 1600
5/16	1000 — 2500
3/8	1500 — 4000

Current Type and Polarity

The various attributes of AC and DC welding, and of positive and negative polarity, were discussed in Sections 3.2 and 5.2. Positive-polarity DC is recommended for most submerged-arc welding where fast-follow or deep penetration are important. It also gives welds that exhibit better resistance to porosity, except on high-sulfur and high-phosphorus steel. Negative polarity gives a melt-off rate about one-third greater than that of positive polarity (see

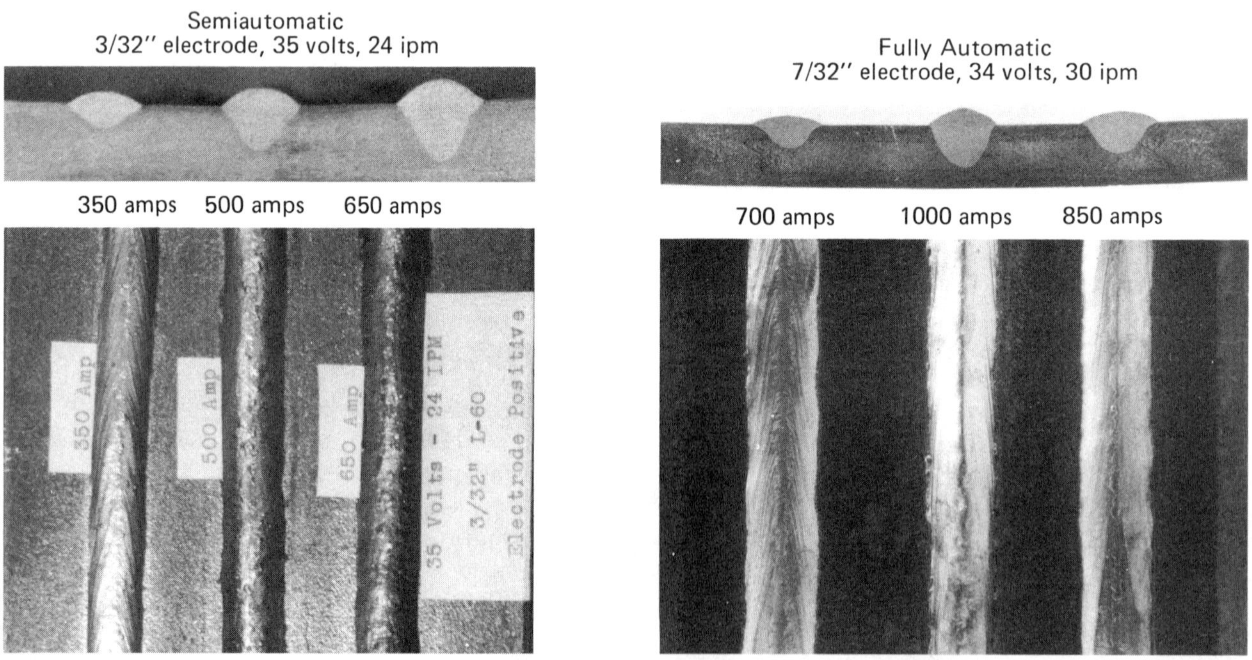

Fig. 6-32. Effect of current level on weld characteristics. Penetration and melt-off rate increase with an increase in current.

TABLE 6-25. Typical Melt-Off Rates (lb/hr) for EL12-Class Submerged-Arc Electrodes

Welding Current (amp)	3/32-in Electrode				5/32-in. Electrode				7/32-in. Electrode			
	1-in. Stickout		3-1/4-in. Stickout		1-in. Stickout		5-in. Stickout		1-in. Stickout		5-in. Stickout	
	DC(+)	DC(−)	DC(+)	DC(−)	DC(+)	DC(−)	DC(+)	DC(−)	DC(+)	DC(−)	DC(+)	DC(−)
300	6.6	10.5	10.0	15.8								
400	9.5	15.0	16.0	23.2	7.9	13.0	11.9	18.0				
500	12.9	19.5	22.5	32.4	11.2	17.5	19.7	24.4				
600	17.3				14.7	21.8	22.5	31.4	13.1	19.0	18.7	27.0
700					18.3	26.1	30.6	38.9	16.2	23.0	23.7	32.5
800					22.0	31.0	39.1	47.1	19.5	26.5	28.8	38.4
900					26.2				22.9	30.0	34.5	44.7
1000									26.4	34.5	40.0	50.7
1100									32.2	38.5	46.5	57.5
1200									34.8			
1300									40.1			

Table 6-25), but negative polarity also produces less penetration. It is used:

1. For conventional fillets where the plate is clean and free from rust.
2. On applications such as hardfacing where greater melt-off is beneficial.
3. Where less penetration is needed to reduce admixture and thereby control cracking or porosity on hard-to-weld steels.
4. Where greater build-up and less penetration helps to prevent cracking in the first passes in deep-groove work.

When changing from positive to negative polarity with no change in current, voltage should be increased about 4 volts if a similar bead shape is to be maintained.

The cost of fillet welds can be reduced through use of electrode positive polarity in place of electrode negative polarity because of the additional penetration provided by positive polarity. The strength of a fillet weld depends on the effective throat size, which in turn, depends on the depth of penetration.

AC is recommended for two specific submerged-arc applications:

1. For the trail electrode when tandem-arc welding.
2. Occasional single-arc applications where arc blow is limiting DC current and slowing travel speed.

For good arc stability a higher current density is needed for AC than for DC. Unstable are conditions can usually be corrected by an increase in current or a reduction in wire size.

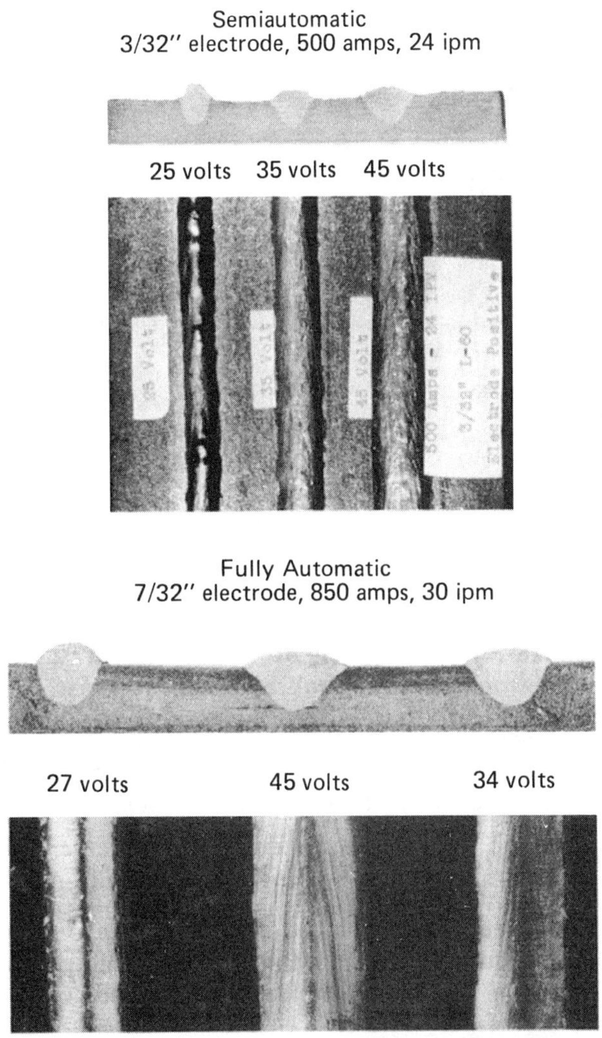

Fig. 6-33. Effect of voltage on weld characteristics. An increase in voltage produces a wider, flatter bead that bridges gaps more readily.

Welding Voltage

Welding voltage influences the shape of the weld cross section and the external appearance of the weld. Figure 6-33 shows the effect of voltage variation when other conditions are maintained constant.

Increasing voltage:

1. Produces a flatter and wider bead.
2. Increases flux consumption.
3. Increases resistance to porosity caused by rust or scale.
4. Helps bridge gaps when fitup is poor.
5. Increases pickup of alloy from the flux. (This can be used to advantage to raise the alloy content of the deposit when welding with alloy or hardfacing fluxes. It can reduce ductility and increase crack sensitivity, particularly with multiple-pass welds.)

Excessively high voltage:

1. Produce a "hat-shaped" bead that is subject to cracking (Fig. 6-34).
2. Make slag removal difficult.
3. In multiple-pass welds, increase the alloy content, thereby producing a crack-sensitive weld.
4. Produce a concave fillet weld that is subject to cracking.

Lowering the voltage produces a "stiffer" arc needed for getting penetration into a deep groove and necessary for resisting arc blow on high-speed work. An excessively low voltage produces a high, narrow bead with poor slag removal.

TRAVEL SPEED

Travel speed is used primarily to control bead size and penetration. It is interdependent with current. In single-pass welds, the current and travel speed should be set to get the desired penetration without melt-through. For multiple-pass welds, the current and travel speed should be set to get the desired bead size.

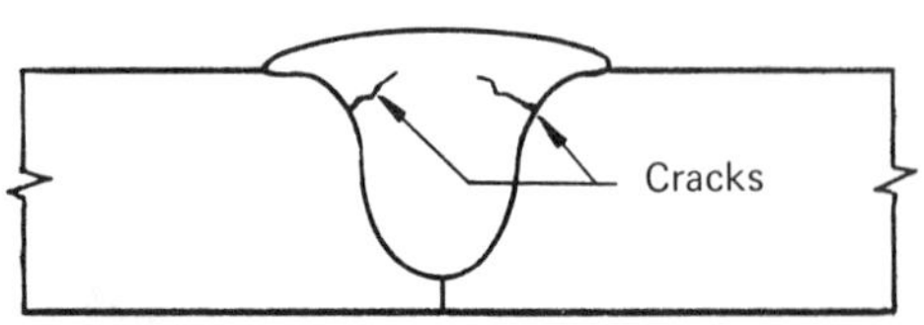

Fig. 6-34. A hat-shaped bead produced by excessive voltage. A similar effect is produced by travel speeds that are too slow. Such beads have a tendency to crack in locations indicated by the arrows.

Figure 6-35 shows the effect of variations in travel speed when other conditions are maintained constant. Changes in travel speed have the following effects:

1. Excessively high travel speed decreases wetting action and increases the tendency for undercut, arc blow, porosity, cracking, and uneven bead shapes. Slow travel speed tends to reduce porosity, because gaseous material has time to boil out of the molten weld.
2. Excessively slow speed produces:
   a. "Hat-shaped" beads that are subject to cracking.

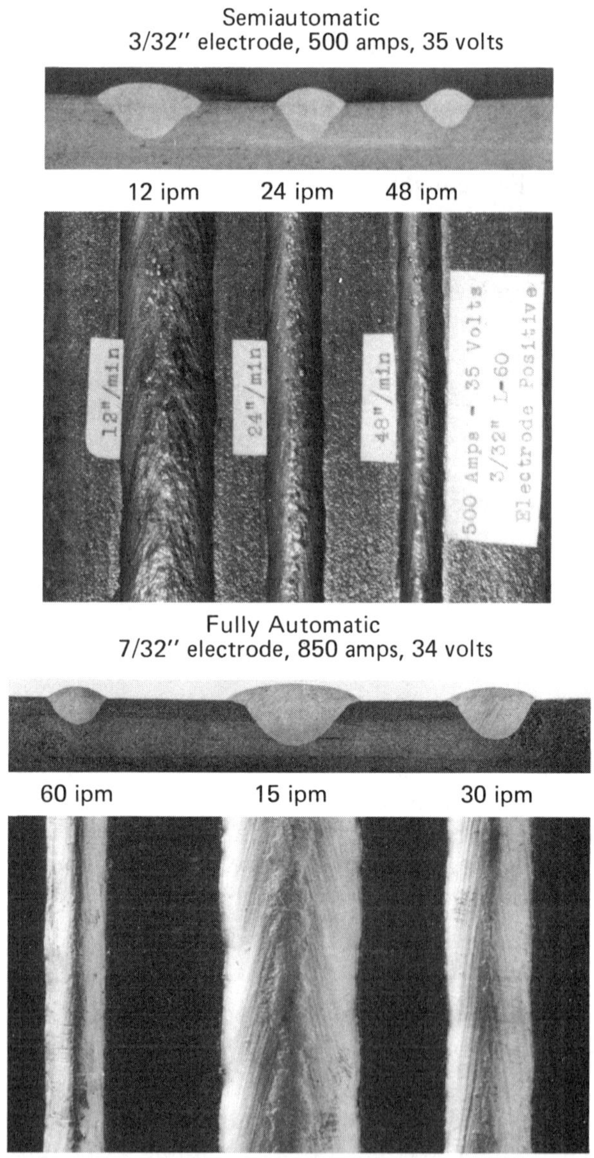

Fig. 6-35. Effect of travel speed on weld characteristics. Penetration and width of the bead decrease with an increase in travel speed.

b. Excessive flash-through, which is uncomfortable for the operator.

c. A large molten pool that flows around the arc resulting in a rough bead, spatter and, slag inclusions.

LONG-STICKOUT WELDING

Electric current is fed into the electrode at the point of electrical contact within the gun or nozzle. The current must then travel to the tip of the electrode to reach the arc. The distance between the point of electrical contact and the electrode tip is referred to as "stickout" or "electrical stickout" (Fig. 6-36).

This entire length of electrode — not just the visible portion protruding from the nozzle — is subject to resistance heating as the current passes through it. The longer the projection of the electrode from the point of electrical contact, the greater the heat build-up within it. This heat can be used to good advantage to increase the melting rate (see Table 6-22) and reduce penetration. Special attachments can be added to standard welding equipment to take advantage of the higher melt-off rates and attendant increase in deposition rate.

Deposition rates with long-stickout welding are typically increased some 25% to 50% with no increase in welding current. With single-electrode, fully automatic submerged-arc welding, the deposition rate may approach that of two-wire welding with a multiple power source.

There are, however, side effects that must be anticipated. A change to long-stickout welding is similar to a change from electrode positive to electrode negative polarity. Increase in deposition rate is accompanied by a decrease in penetration. This decrease generally is not large enough to prevent the use of long stickout except on applications where positive polarity is used for the express purpose of producing maximum penetration. Long-stickout welding also results in less tendency for melt-through, which can be advantageous when the fitup is imperfect, and reduces the heat-affected zone.

A voltage drop accompanies the resistance heating of the extra length of projecting electrode. When the operator sets the voltage control of the wire-feeding equipment, he must compensate for this voltage drop to avoid having the arc operate at less than optimum voltage.

Long-stickout welding provides the greatest economies in fast-fill applications that require a large volume of weld metal, as, for example, in flat-

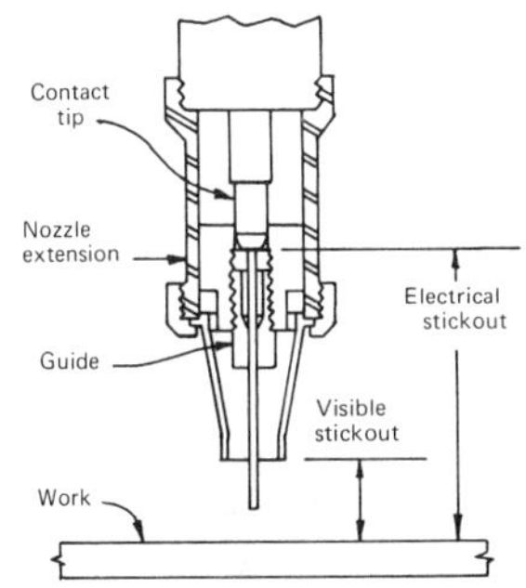

Fig. 6-36. Long electrical stickout — used to increase melt-off rate through resistance heating of the electrode. The term, applied to the distance between the point of electrical contact within the nozzle and the tip of the electrode, is distinct from visible stickout, which denotes the length of electrode extending beyond the guide tip.

fillet and groove butt welds. The reduction in heat-affected zone is particularly beneficial in the welding of quenched and tempered steels. In some instances, the heat-affected zone is reduced 25%.

Long-stickout techniques can also be used to good advantage in build-up and hardfacing applications. Here the benefits include greater speed, improved quality of deposit, and better appearance.

Flux and electrode types are selected for long stickout on the same basis as for standard stickout. Wire sizes from 5/64 through 7/32-in. are used.

The suggested stickout limitations with solid electrode wires for submerged-arc welding are:

5/64, 3/32,1/8-in. electrodes	3 in. maximum
5/32, 3/16, 7/32-in. electrodes	5 in. maximum

These recommendations are based on practical considerations, such as ease of striking the arc, maintaining good wash-in, bead shape control, and avoiding undue electrode "wandering" within the nozzle extension. While these recommendations apply on most applications, longer stickouts are permissible, especially with larger electrodes.

INFLUENCE OF JOINT DESIGN ON SUBMERGED-ARC PROCEDURES

Butt Welds

Butt joints are commonly welded by the submerged-arc process in material ranging from 18-gage sheet metal to plate several inches thick. Procedures and weld design for butt joints depend upon whether the weld is to be made on sheet, on plate, or on extremely thick plate. In the latter case, the weld is called a deep-groove butt.

Butt Welds in Sheet Metal: Principal considerations are control of distortion and prevention of melt-through. For control of distortion, the work must be rigidly supported. A copper or steel backup bar is often used for this purpose as illustrated in Fig. 6-37. Backup bars also help prevent melt-through.

When a steel backup bar is used, a small gap is left between the plates to be joined. The weld penetrates into the steel bar, making it an integral part of the weldment. Where this addition of material is undesirable, a copper bar is used instead of steel. Because of its high thermal conductivity, the copper does not melt and thus does not fuse to the steel. The copper may be flat or may have a small groove, depending on the shape desired on the backside of the bead and the thickness of the sheet. The groove should be wider than the back bead to avoid undercut at the bead edges. When extremely smooth back-beads are desired, the weld is backed by a layer of flux placed in the groove.

Butt Welds on Plate: Complete (100%) penetration without melt-through is usually required in square-edge butt welds for full strength. Plates up to 5/8-in. thick can be butted tightly and then welded with one pass from each side. With normal sheared or flame-cut edges, 60% penetration can usually be achieved from the first side. Penetrations up to 80% are possible if the edges are machined and fitted together tightly.

The built-up bead on top of the joint may be large and have irregular edges if two square edges are butted together tightly, with steel 5/8 to 3/4-in. or thicker. Excessive build-up or irregular beads can be corrected by beveling the edges of the plates, or by leaving a gap between plates as shown in Fig. 6-38.

Gaps of any kind increase penetration. As a rule, either a backup strip or a seal bead is required

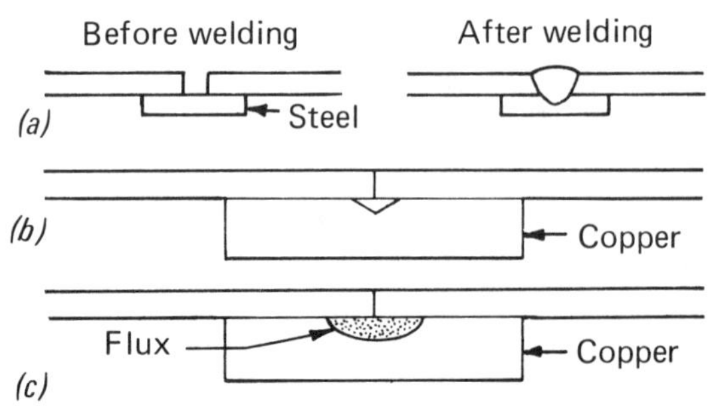

Fig. 6-37. Joints with backup bars. The steel bar in (a) becomes an integral part of the welded assembly. The notched copper bar (b) conducts heat away and does not fuse to the joint. Flux backup with a copper bar (c) provides a smooth back bead.

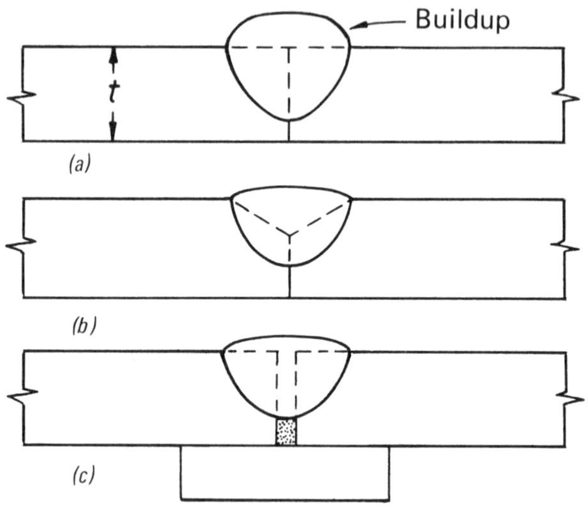

Fig. 6-38. Excessive build-up on thick-plate square-butt joints (a) can be corrected by a bevel (b) or by a gap (c). The maximum penetration possible without a backup strip in a square-butt joint is approximately 60 to 80%.

to support flux if the gap is large enough to permit loose flux to spill through (Fig. 39). Seal beads can be made either manually or with semiautomatic equipment. The beads should be on the second pass side for workpieces more than 1/2-in. thick. On material thinner than 1/2-in., seal beads should be on the first-pass side (Fig. 6-40).

On steel up to 1 /2-in . thick, full-penetration welds can be made from one side if the workpieces are positioned with a gap, and a steel or copper backup bar is used (Fig. 6-41). As with sheet-metal welds, the steel backup bar remains as a permanent part of the weldment. Avoid making deep, narrow beads, or beads with a wide surface flair — referred to as "hat-shaped" beads — because these types tend to develop internal cracks (Figs. 6-34 and 6-42).

Deep-Groove Welds: The first pass in a deep-groove weld requires the same considerations for penetration and melt-through as square-edge butt welds on plate. The only difference is that the

Fig. 6-39. Method for determining whether flux support is needed. A backup bar or seal bead is required if loose flux spills through the gap.

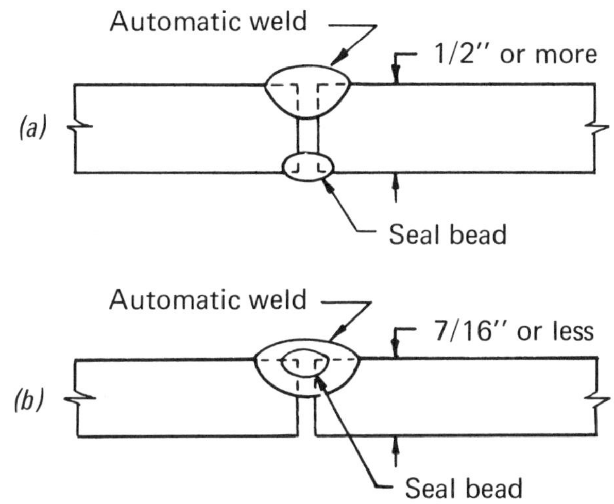

Fig. 6-40. Placement of seal beads depends on material thickness.

thickness of the land, rather than the entire material thickness, is important. In addition, flux depth, plate levelness, and slag removal are all important in deep-groove welding.

Flux has a tendency to pile up to depths greater than 1 in. in deep-groove joints. Since excessive flux causes poor bead shapes, only enough flux to prevent excessive arc flashing should be used.

The multiple-pass welds normally required with deep-groove welds generally create large molten pools that are quite fluid and "run" out of the joint easily if special precautions are not used. Therefore, the work should be level or inclined slightly for an uphill weld. (Welding on inclined plates is discussed later in this section.)

Slag removal is sometimes difficult in deep-groove welding. Small beads with slightly convex

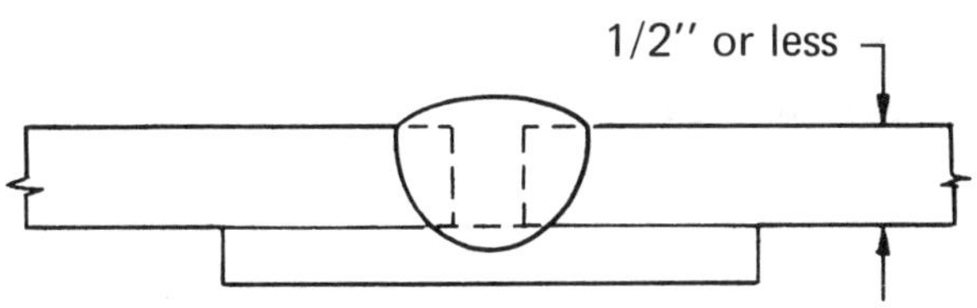

Fig. 6-41. Full-penetration weld made from one side with a backup bar.

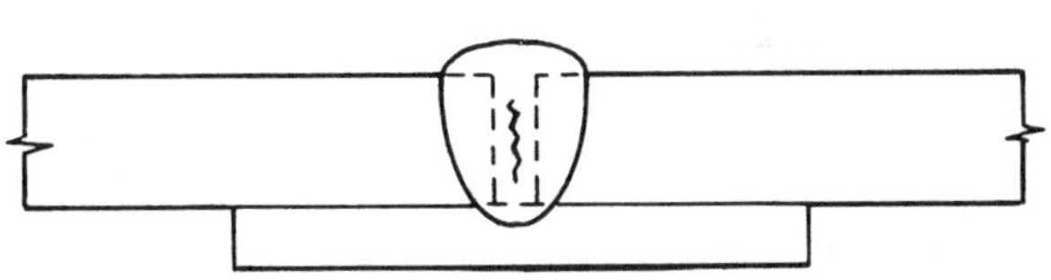

Fig. 6-42. An undesirable bead shape. The deep, narrow configuration promotes internal cracks.

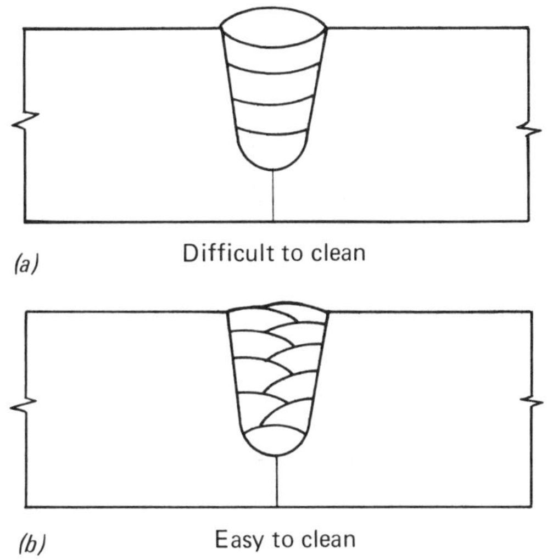

Fig. 6-43. Effect of multiple-pass welding techniques on ease of cleaning deep-groove welds. Slag generally is difficult to remove from large, concave beads (a), but can be removed easily from small, convex beads (b).

curvatures are easier to clean than large beads with concave surfaces, as illustrated in Fig. 6-43. Therefore, there is usually a cost saving in making small beads that are easy to clean, instead of making fewer large beads that are difficult to clean.

The cost of joint preparation is an important factor. V-type plate edges, which can be made at low cost by flame-cutting, are inexpensive (Fig. 6-44). U-type joints, on the other hand, are expensive to prepare.

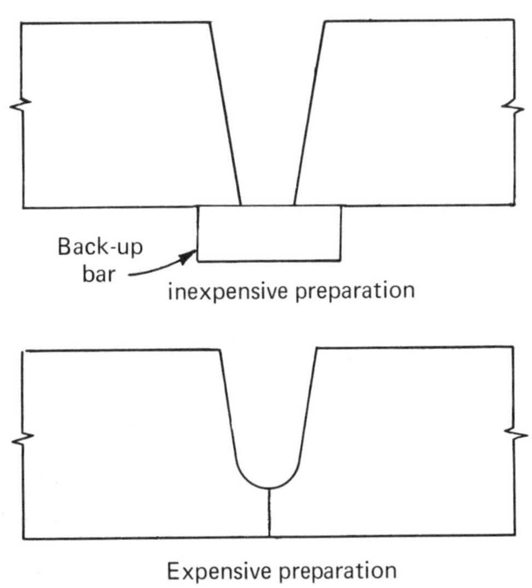

Fig. 6-44. Joint preparation for deep-groove welding. Bevel edges (a) can be prepared at less cost than the machined edges of U joints (b).

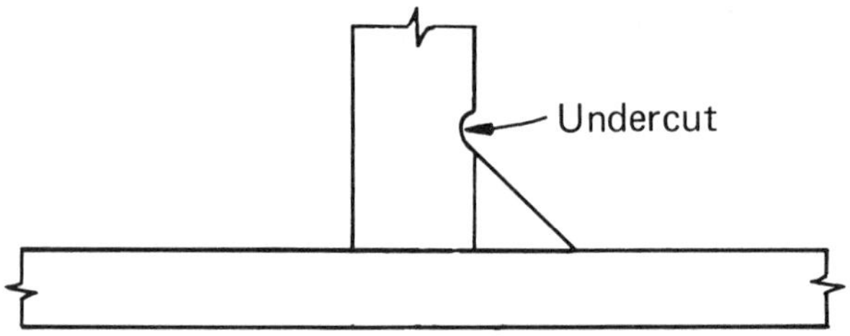

Fig. 6-45. Undercut may occur in single-pass fillet welds having legs larger than 3/8 in.

Fillet Welds

A 3/8-in. leg is the largest single-pass fillet weld that can be made in the horizontal position with a single electrode. Attempts to make larger beads generally cause undercuts, as illustrated in Fig. 6-45. However, horizontal fillets up to 5/8 in. can be made with multiple electrodes.

Bead width should be at least 25% greater than bead depth, since narrow beads are subject to cracking. However, even when fillet welds are properly shaped, they are generally more likely to crack than other types of welds because of the highly restrained nature of fillet joints. This restraint often prevents the small deflections that normally absorb thermal stresses set up as the weld solidifies. The proper measures to observe for joint design are discussed more fully later in this section under **Preventing Weld Cracking.**

The maximum penetration of the arc tends to be aligned with the axis of the electrode. Electrode inclination can thus be used to control the leg length of the bead and the degree of penetration into the joint. This is illustrated in Fig. 6-46. In multiple-pass welds, succeeding passes should be placed so that the two legs are built up equally, as shown in Fig. 6-47. A protruding bead can result if inappropriate welding variables or procedures are used, as shown in Fig. 6-48.

Porosity caused by arc blow is especially prevalent in fillet welds because of the magnetic effects set up by intersecting workpieces. Techniques for minimizing arc blow are discussed later in this section under **Preventing Weld Porosity.** Section 3.2 also discusses arc blow.

A fillet-lap attachment is recommended if horizontal fillet welds are to be made with fully automatic equipment. This attachment keeps the head in a vertical position while turning the electrode to the proper angle for the weld. The standard straight contact assembly should be used for flat or trough-position fillet welds. For any application, the head and flux hopper should not be tilted more

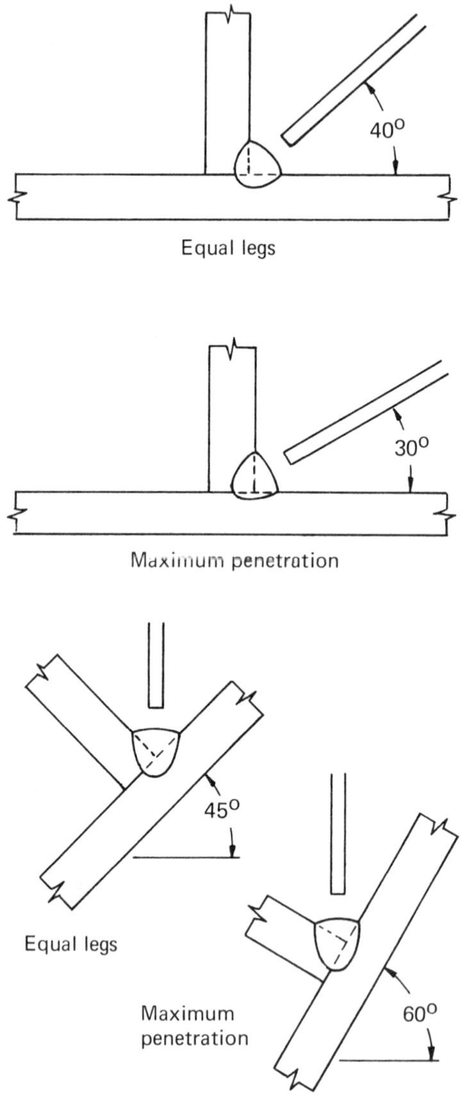

Fig. 6-46. Influence of electrode alignment on penetration and leg size. In horizontal fillet welds, a 40° inclination of the electrode (a) produces equal legs, while a 30° inclination (b) maximizes penetration. In flat fillet welds, positioning the work 45° (c) provides equal legs, while positioning the work at 60° (d) maximizes penetration.

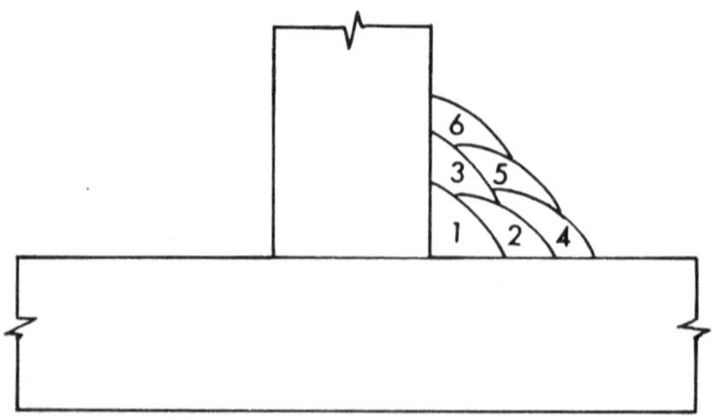

Fig. 6-47. Preferred sequence of bead deposit in multiple-pass welding. The two legs should be built up equally, so that leg lengths do not differ greatly as each layer is completed.

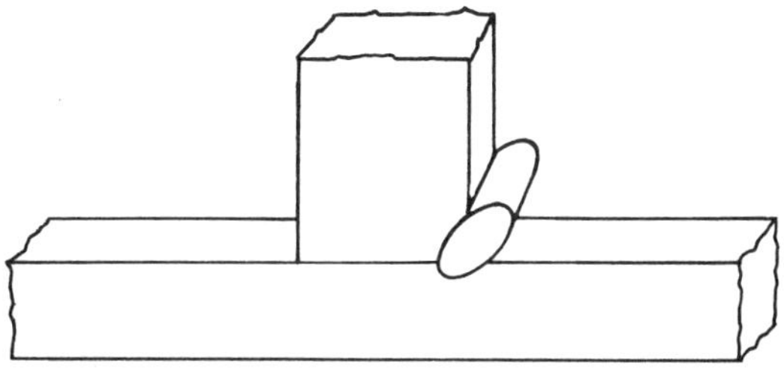

Fig. 6-48. Undesirable bead shape caused by too rapid travel speed, deep flux pile, excessively high current, excessively low voltage, or welding uphill.

than 40° from the vertical. A greater tilt can cause irregular flux flow.

Lap Welds

Tight fitup and proper electrode alignment are principal considerations in making lap welds. The plates must be held together tightly; gaps promote poor bead shape and unsound welds. Since a tight fit inhibits the escape of welding gases, lap welds are particularly susceptible to porosity. Therefore, care must be taken to ensure that the joint is free of moisture, paint, and other contaminants.

Through-laps are preferred to edge-laps on 10-gage or thinner material where good appearance is not essential. When the workpieces are of differing thickness, the recommended procedure is to weld the thin plate to the thick plate (Fig. 6-49).

Position of the electrode is an important factor in determining lap-weld quality. The weld may not properly fuse to the bottom plate if the electrode is inclined too much toward the top plate. If it is tilted too far from the top plate, the weld may not properly fuse to the top plate and may melt through the bottom plate.

Plug Welds

The principal consideration in making plug welds is to be sure that the weld fuses to both plates and completely fills the hole. A weld of 3/4-in. diameter

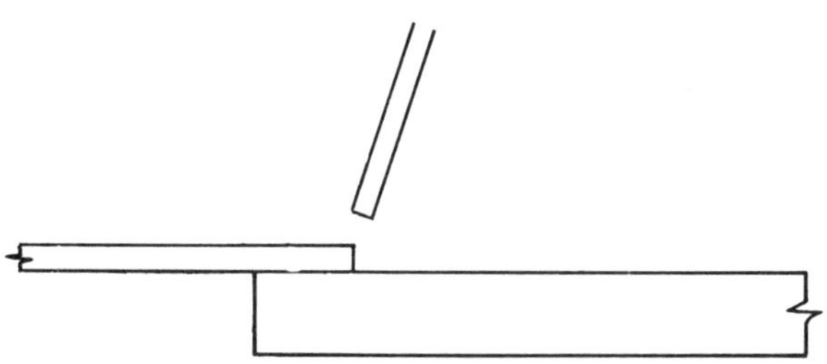

Fig. 6-49. Recommending positioning of workpieces for lap welding. The thin plate should be welded to the thick plate. The electrode should be inclined 0 to 10° from the vertical for material 10-gage and thinner. Inclination should be 45° for plate 3/16 in. and thicker.

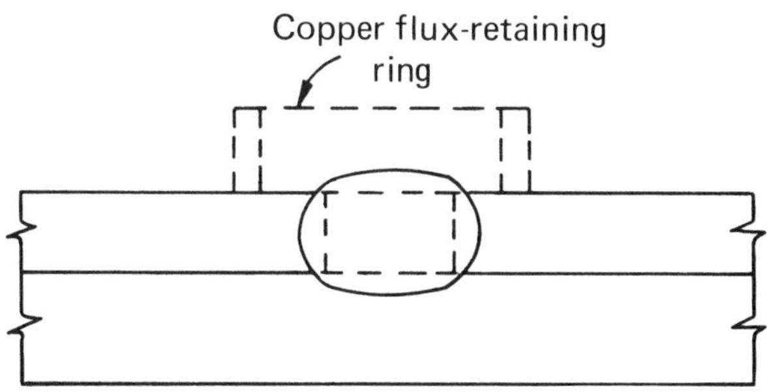

Fig. 6-50. Preferred geometry for plug welds. The weld must fuse into both plates and fill the hole completely.

is normally the maximum size used for plug welding. Larger plugs are sometimes made with semiautomatic equipment by circulating the arc around the plug to completely fill the hole. Plugs up to 2 in. in diameter and 2 in. deep are made in this manner.

It is difficult to tell when the plug is filled, because the weld is always covered completely by flux. Therefore, plug welds are usually timed. The current is reduced for the last few seconds before the arc is broken to reduce the size of the crater at the top of the plug. A typical arrangement for the flux-retaining ring and workpiece is shown in Fig. 6-50.

Edge Welds

Edge welds are made at high travel speeds on sheet metal from 10-gage to about 16-gage. The major considerations involve guiding the electrode and supporting the flux.

The electrode must be guided accurately so that it is always positioned directly over the joint. Support is necessary to keep the flux from spilling over the edge and to hold the molten metal while it freezes. For best results, the weld should be made about 10° downhill. The positioning of workpieces, flux, and electrode for edge welding is shown in Fig. 6-51.

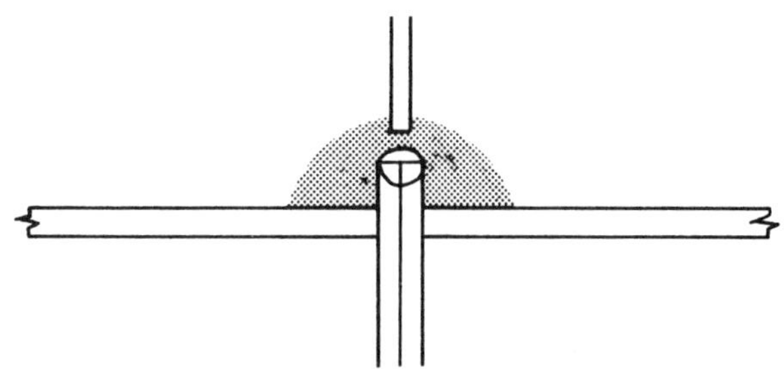

Fig. 6-51. Position of workpiece and flux in an edge weld. The flux supports should be placed as near to the weld as possible without interfering with the weld bead.

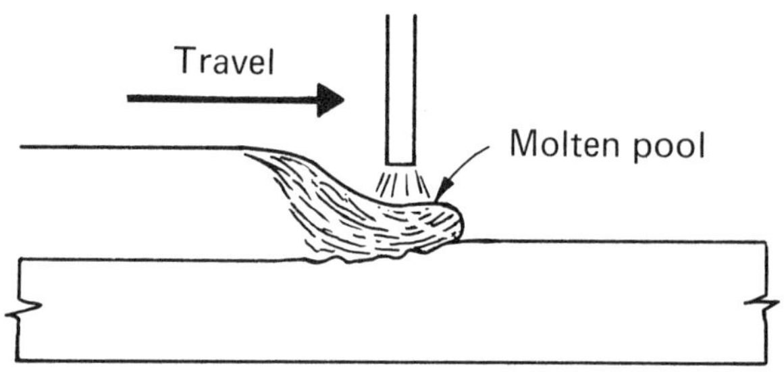

Fig. 6-52. Undesirable flow of molten weld metal in downhill welding. The work should be positioned so that the molten pool does not run ahead of the arc.

Welds on Inclined Plates

The workpieces in submerged-arc welding normally should be positioned level, since welding on inclined plate may result in improper penetration and may produce a distorted bead. Welding uphill increases penetration, and welding downhill reduces penetration. But there are two situations where inclining the work is advantageous — when welding sheet metal and when using multiple arcs on heavy, deep-groove joints.

Since excessive penetration is undesirable when welding sheet metal, a 10 to 20° downhill angle reduces penetration and permits use of high currents and rapid welding speeds. Inclines of more than 20° may produce distorted beads.

On heavy, deep-groove joints, particularly when made with multiple arcs, the molten pool is large and fluid. If the work is slightly downhill, this large pool rolls ahead of the arc and may promote an unstable arc, wavy bead, or shallow penetration, as illustrated in Fig. 6-52. Running the bead 2 to 5° uphill solves the problem.

Circumferential Welds

Circumferential or roundabout welds differ from those made in the flat position in two important ways. First, the pool of molten flux and weld metal tends to spill off the work. Second, slag removal on multipass welds is a factor that controls the progress of the welding.

In flat work, the pool of molten flux and metal is kept beneath the electrode without too much difficulty, but on the curved surface of a circumferential weld tends to sag and run. In severe cases, the metal spills off the work. In less severe cases, the molten pool starts to run, but freezes before it spills, and thereby produces a distorted bead.

To avoid spillage or bead distortion, the weld metal must solidify as it passes over the vertical center of the part. This requires that the electrode be positioned ahead of vertical center — be advanced from the center line opposite the direction of rotation. The bead shape and welding characteristics can be controlled by varying the displacement or "advance" of the electrode along the circumference as shown in Fig. 6-53. Inadequate displacement on an an outer weld or excessive displacement on an interior weld produce deep penetration and a tendency for melt-through. It also results in a narrow, humped bead. Excessive displacement on an outer weld or

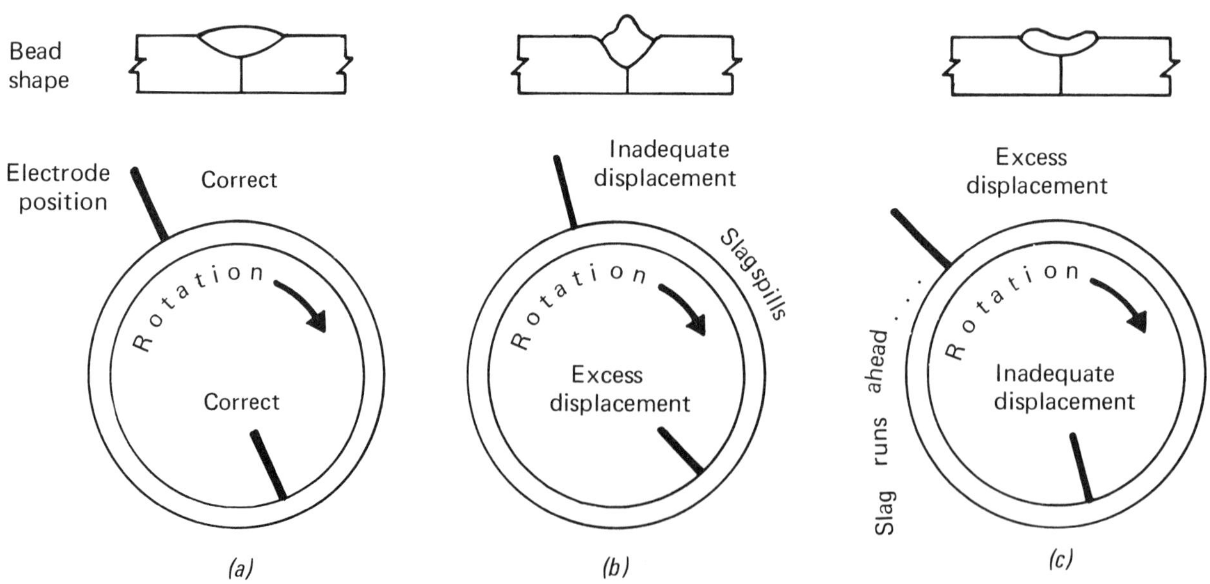

Fig. 6-53. Effect of electrode displacement on bead shape in circumferential welds. A displacement that avoids spillage of the molten pool produces a correctly shaped bead (a). Excess or inadequate displacement produces a distorted bead (b) and (c).

TABLE 6-26. Recommended Electrode Lead For Circumferential Welding

Workpiece Diameter (in.)	Electrode Displacement Ahead of Vertical Center (in.)
1– 3	3/8–3/4
3–18	3/4–1
18–36	1-1/4–1-1/2
36–42	1-1/2–1-3/4
42–48	1-3/4–2
48–72	2–2-1/2
72	3

inadequate displacement on an inner weld produce a shallow, concave bead.

Suggested off-center distances for the electrode are listed in Table 6-26. The electrode should be positioned nearly perpendicular to the surface of the work, particularly on small circumferences. Positioning the electrode at too great an angle of incidence may produce a distorted bead and may cause difficulty in striking the arc.

Regardless of electrode position, the metal will spill simply because it cannot freeze fast enough if the molten pool is too big in relation to the size of the workpiece circumference. The size of the molten pool can be controlled by limiting bead size, which, in turn, depends on the amperage and travel speed. Lowering the current and increasing travel speed reduce bead size.

The size of the bead is limited by the diameter of the work. Approximate maximum sizes for single-pass welds on small-diameter work are shown in Fig. 6-54. Small-diameter work generally should be set up for three-o'clock welding. For large-diameter work, procedures that give good welds with minimum difficulty are presented in Fig. 6-55.

Since the flux is granular, it, too, will spill off small-diameter work if not supported. One method to prevent flux spillage employs a nozzle assembly that pours the flux directly above the arc. Another method employs a wire brush or other flexible heat-resistant material attached to the nozzle assembly ahead of the arc, as shown in Fig. 6-56.

Flux support is also necessary at the edges of the work. Sheet-metal rings tack-welded to the edge, or flexible retainers that ride the edge, are used for this purpose. These devices support the flux so that beads can be made at the very edge of the work (Fig. 6-57). This technique is commonly used in building up or hardfacing rolls.

Excess flux promotes sagging or spilling. The flux should be just deep enough to cover the arc, but should not be so deep that it interferes with the solidification of the weld metal.

Slag Removal: The removal of slag on multiple pass circumferential welds is particularly important, since all slag must be removed before the work makes a complete revolution if the weld is to run continuously. The ease with which the slag is removed is determined largely by the size and shape of the bead. Because small beads cool quickly, slag generally does not stick to them readily. On two-pass welds, the first pass should be flat or slightly concave and blend upward to the top edges of the joint to promote easy slag removal. Convex or undercut beads tend to lock the slag to the weld at the edges of the bead and thereby impede slag removal (Fig. 6-58). However, on deep-groove welds requiring several passes, it is better to stagger the passes as in Fig. 6-59(a). For good slag removal, make the beads less than the width of the groove and slightly convex. Wide, concave beads, as in Fig. 6-59(b), make slag removal difficult.

In heavy work, such as that encountered in thick pressure vessels, weldors have a tendency to put down as large a bead as possible with each pass. This practice, however, is not economical. Circumferential welds can usually be made most rapidly by using many small beads that permit continuous welding because the slag falls away readily. With large beads, slag tends to stick to the bead, and the process must be stopped frequently for slag removal.

Another way to improve slag removal, particularly on small-diameter work that may overheat during welding, is to direct an air jet against the weld. Locating the air nozzle between the two and three-o'clock positions usually provides the best results.

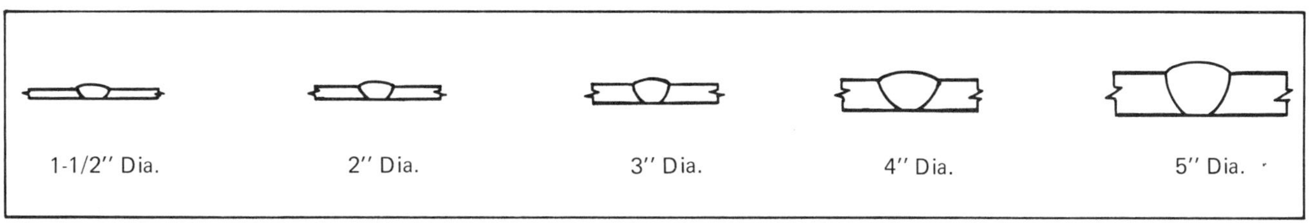

Fig. 6-54. Approximate size of largest beads attainable on small-diameter circumferential welds. Backups are required on thin sections and special flux supports may be necessary.

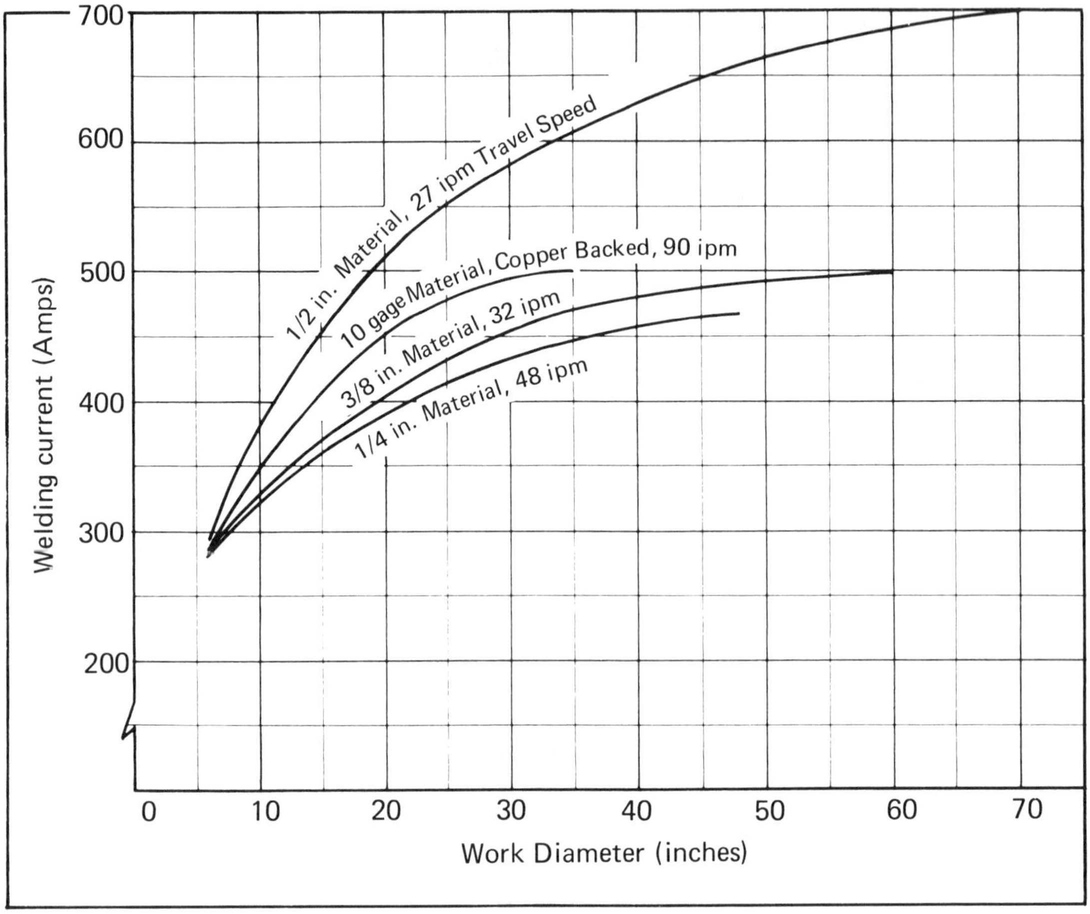

Fig. 6-55. Recommended welding currents for various workpiece diameters and thicknesses. Special provisions to control spillage should not be necessary. For diameters larger than those listed, use standard flat-position procedures.

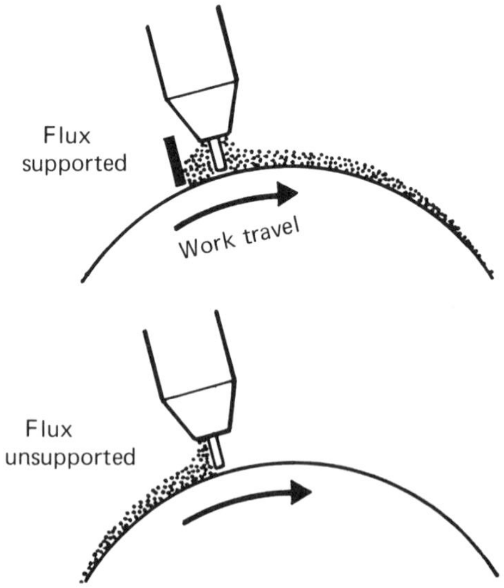

Fig. 6-56. Placement of flux support to avoid flux spillage on circumferential welds.

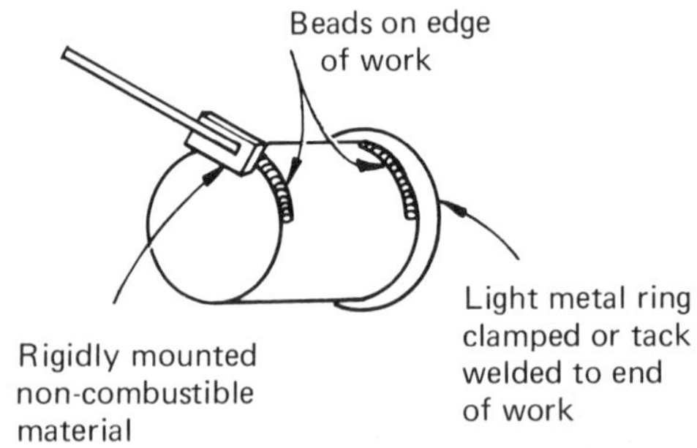

Fig. 6-57. Two types of flux supports for circumferential welds made at the edges of a roll.

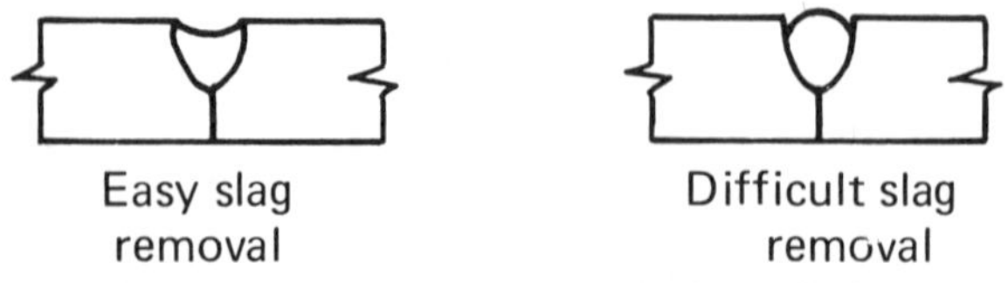

Fig. 6-58. Preferred and undesirable bead shapes for the initial pass of a two-pass weld. The bead should wash up to the edges of the groove to facilitate slag removal.

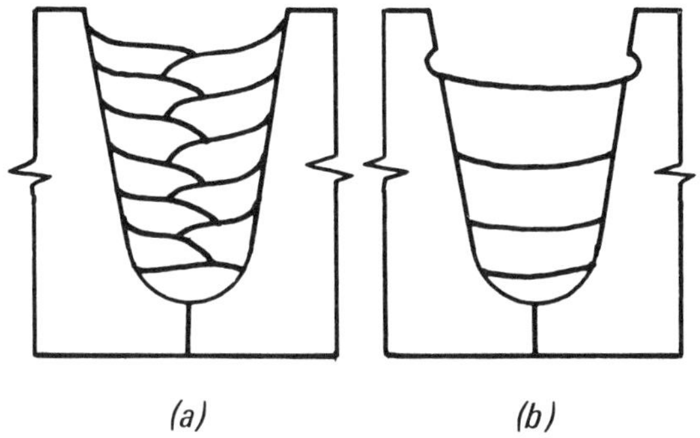

Fig. 6-59. Narrow and slightly convex beads (a) are easier to clean than wide concave beads (b).

Special care should be exercised with flux collection on long, continuous beads. Clean flux not only promotes sound, smooth welds. but also helps avoid clogging in the gun and flux-feed mechanism. The flux should be collected before it reaches the floor. A vacuum recovery unit will remove dust, and a magnetic separator removes particles of steel.

WELDING WITH MULTIPLE ARCS

Although the general rules governing operating variables for single-arc welding also pertain to multiple-arc welding, there are additional considerations that must be kept in mind when welding with multiple arcs. The use of two or three welding heads provides better control of bead shape so that faster speeds are possible. But multiple heads also introduce additional variables that must be controlled properly for satisfactory results.

Since multiple electric arcs interact magnetically, the type of power fed to each must be chosen to minimize arc blow and other detrimental effects from this interaction. In two-arc welding, combining a DC and an AC arc is usually preferred, because the arcs are easier to control. There are, however, many successful applications where both arcs are AC. Where three arcs are used, the lead arc is usually DC, and the two trailing arcs are usually AC.

Welding Current: Penetration and melt-off rate are determined by the current level. The lead arc has the greatest effect on penetration and usually is fed the most current. Current in the trailing arc is adjusted to provide the desired build-up. Excessive current in the trailing arc causes molten metal to splash out of the joint, while too little current may allow the electrode to stick to the workpiece.

Arc Voltage: Bead width and penetration are determined by arc voltage. If the electrode spacing is less than 2 in., the lead arc is usually set at the minimum possible voltage to provide maximum penetration. The voltage in the trailing arc is then set to provide the desired bead shape and width. On the trailing arc, voltage may be set as high as 35 to 42 volts to provide good wash-in on the edges of the weld.

Electrode Size: Electrode diameter also affects penetration. Small diameters maximize penetration, but it may be necessary to select a large electrode for the lead arc to carry the required current. With DC-AC-AC three-arc welding, the third (AC) electrode is usually the same size or smaller than the first AC (second-arc) electrode. Electrode spacing and current for the trailing arc are usually the same or less than that of the lead AC arc, but voltage is usually higher for the trailing arc.

Polarity: On the lead arc, DC electrode positive provides maximum penetration, and DC electrode negative provides maximum deposition rate. When DC is used on the lead arc, AC on the trailing arc provides the best control over magnetic interaction.

Travel Speed: Bead width and penetration are determined by travel speed. Increasing speed decreases penetration and produces a narrow bead with a tendency toward undercutting. Excessively slow speed may allow the trailing arc to wash metal under the lead arc, resulting in an erratic arc and rollover at the bead edges, as illustrated in Fig. 6-60.

Type of Flux: With proper procedures, most mild-steel fluxes can be used in multiple-arc welding. However, some fluxes are designed for high-speed welding, while others work better on slow-speed, deep-groove welding. Consult the flux supplier for advice.

Flux Depth: In addition to protecting the molten weld metal against contamination, flux also serves as a mechanical dam that prevents molten slag and metal from flowing forward where it would interfere with the arcs. Flux should be deposited to a depth where the trailing arc barely breaks through occasionally. Excessive flux depth may promote a

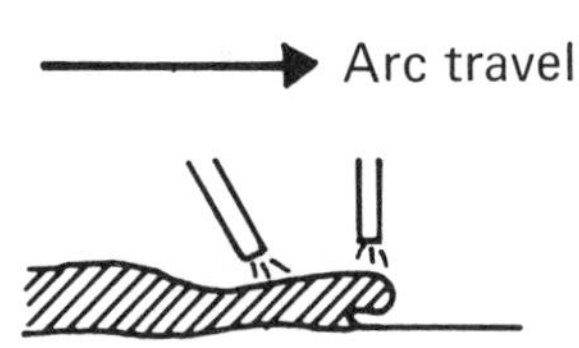

Fig. 6-60. Effect of excessively slow travel speed in multiple-arc welding. The trailing arc may wash metal under the lead arc and cause a wavy bead.

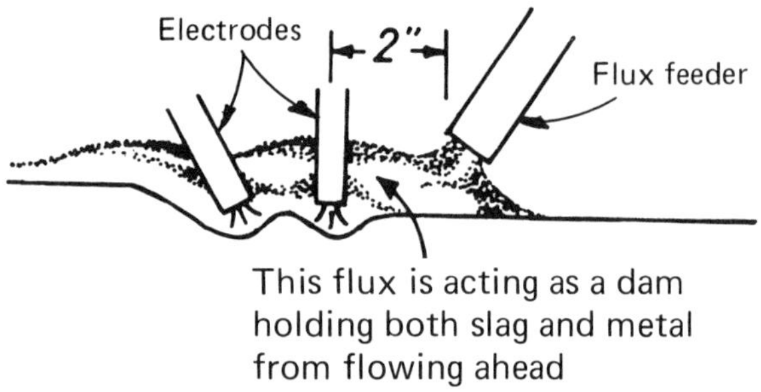

Fig. 6-61. Recommended placement of flux in multiple-arc welding. Flux should be deposited at least 2 in. in front of the lead arc so that the flux serves as a dam for molten metal and slag.

narrow, rough bead. The flux should be applied far enough in front of the arc — about 2 in. minimum — to prevent molten flux or metal from running ahead of the arc (Fig. 6-61).

Electrode Spacing: The distance between electrodes affects penetration and bead width. A spacing of 5/8 to 7/8 in. usually is preferred except on horizontal fillet welds where the spacing may be as much as 5 in.

Electrode Stick-out: The nozzles should be 1-1/4 to 1-1/2 in. above the work, or should just clear the top of the flux. When using the long stickout technique, a wire guide is necessary below the contact tip. (See Fig. 6-36.)

Electrode Alignment: The electrodes should be kept directly in line, one behind the other, to within 1/32 in. The plane of alignment should be perpendicular to the surface of the work. Improper alignment produces undesirable bead shapes.

Grounding: The location of ground connections is important, especially on short welds. Both AC and DC ground connections should be at the start of the weld, unless back blow is desired to keep weld metal from running ahead of the arc. Where back blow is desired, the weld should be made toward the ground connection.

Nozzles: An insulated clamp on the nozzles helps maintain proper electrode alignment. Worn tips should be replaced, because they cause misalignment and poor contact, which can produce erratic arc action.

Deep-Groove Welding with Multiple Arcs

Alignment: Maintaining alignment of the welding heads is exceptionally important when multiple arcs are used for deep-groove welding. The following rules should be observed to maintain alignment:

1. The seam should be aligned exactly with the line of travel of the welding heads so that little or no adjustments are needed after welding begins.
2. The work should be level across the seam, and level or slightly uphill (2 to 5 degrees) along the length of the joint. Leveling across the seam permits proper alignment between the wall of the joint and the electrodes. Welding slightly uphill helps prevent molten flux and metal from running ahead of the arc. Since multiple-pass welds tend to build up faster at the start of the weld, filling the crater at the finish end by stopping travel before breaking the arc helps maintain an even build-up.
3. When making weld beads as in Fig. 6-43(b), electrodes should be spaced away from the wall of the groove a distance approximately equal to the width of the larger electrode. The heads should have provisions for horizontal adjustment so that they can be easily moved across the joint for alignment.
4. Electrode stickout should be the same for all passes. Maintaining this constant stickout usually requires raising the heads slightly after each pass to compensate for the buildup of the previous bead. Constant stickout also ensures proper spacing between electrodes. The heads can be raised easily if the equipment includes a vertical lift adjustment.

Joint preparation: Side-wall angles should not be less than 7° after the plates have pulled together from contraction. Perpendicular sides are easily undercut and thus difficult to clean. But too large an angle wastes material. Run-off tabs are necessary on most jobs, except for the welding of girth seams. Tabs must be used at both ends of the joint and should be shaped exactly like the joint. They should be heavy enough to minimize arc disturbance, and should be welded to both sides of the joint and the steel backup bar if one is used. Close the finish end to provide a magnetic bypass and minimize arc blow.

Preheat and Postheat: The appropriate preheat and interpass temperatures can be determined from tables or special slide rules. The entire workpiece should be preheated if possible. After welding, all pieces should be cooled slowly to room temperature. Postheat can be used to avoid cooling cracks on rigid workpieces.

Root Passes: To ensure adequate penetration, the first few passes in the root are quite critical. For these root passes, a single arc with DCEP (electrode positive) is recommended, except where arc blow occurs, in which case use AC.

Bead Shape: Beads should be made slightly convex to inhibit longitudinal cracking. The beads should also be positioned almost flat, rather than being inclined against the wall as with a fillet.

Control of Molten Weld Metal: Preventing the molten flux and metal from running ahead of the arc is most important. The following practices, summarized from the prior discussion, are helpful in keeping the molten pool properly contained:

Use high speeds.

Weld slightly uphill, not downhill.

Keep flux piled well ahead of the arc.

Do not start on a high spot in the previous bead.

Maintain proper alignment of electrodes and joint.

Use high voltage on the trailing arc to flatten the bead.

PREVENTING WELD POROSITY

Porosity due to the entrapment of gas in the weld metal is not desirable, although the strength of the weld is not lowered appreciably unless the porosity is very severe. Porosity may be evident on the surface, as in Fig. 6-62, or may occur beneath a sound surface. Various factors are causes of weld porosity. These factors and their control are discussed in the following text.

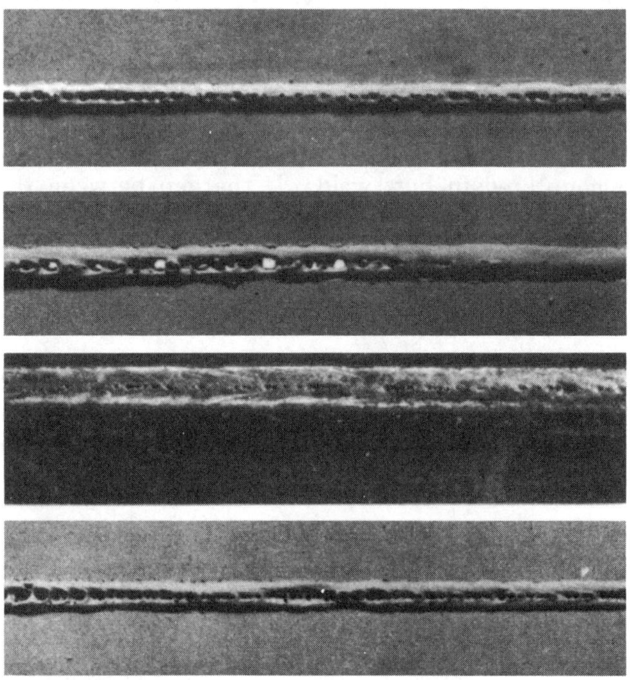

Fig. 6-62. Severe porosity apparent on the surface of the weld bead. Causes of the porosity are (from the top down) oily workpiece, dirty flux, insufficient flux, and backward arc blow.

Contaminants in the Joint: The most common cause of porosity is the presence of organic materials or other gas-producing contaminants in the joint. Weld joints, thus, must be free of foreign matter, such as rust, dirt, oil, and moisture for porosity-free welds. Although cleaning the exterior surfaces (top and bottom of the plate) helps reduce porosity, the abutting edges of the workpieces must be cleaned for best results.

It is not necessary, however, to clean every edge that is to be welded by submerged-arc. Edges that have been prepared by machining or flame-cutting can be welded without further cleaning if they are not rusty or oil-coated. Also, edges with ordinary mill scale can be welded if the scale is not loose or flaky. Even heavy "blue" mill scale can be welded satisfactorily if a silicon-killed electrode is used. Blue mill scale is dark gray or black in color and has no detrimental effect in welding. "Red" mill scale, however, which is reddish brown in color, has the same detrimental effect as rust.

Power wire-brushing and torch-heating are used to clean rust and red mill scale from edges (Fig. 6-63). Either method, used individually, greatly reduces the porosity normally produced by these contaminants. For best results, however, both cleaning techniques should be used in combination.

Joints should be brushed clean before they are fitted together, then a flame torch should be played on the joint about a foot or two in front of the arc during welding to drive off residual moisture. If there is red mill scale or rust on the plate, the surface of the plate edges must be heated to about 400 to 600°F to drive off the moisture. Some moisture will remain to cause porosity if this temperature range is not reached.

Oil, grease, and die lubricants should be removed by degreasing and washing operations. The washing compound must be rinsed away completely, and the work must be dry before welding begins.

Wire Contamination: Welding electrodes sometimes become rusty during storage. This rust may cause porosity, especially in high-speed welds on light-gage sheet metal. Rusty wire also may not feed properly through the feed cables of semiautomatic welders, and the rust causes excessive wear and arcing at the contact nozzle. Do not attempt to use rusty wire.

Wire that has become contaminated with oil, grease, or dirt should be cleaned prior to welding.

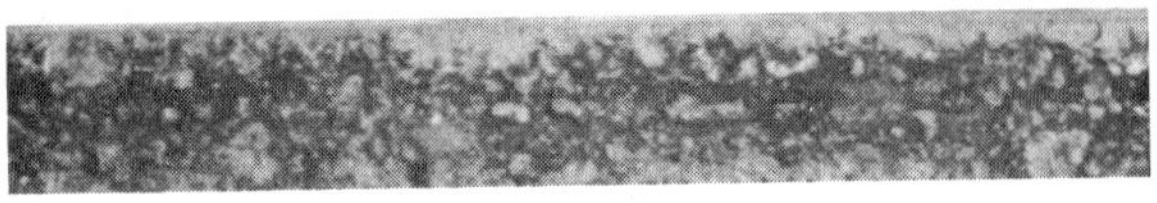

Rusty plate before welding

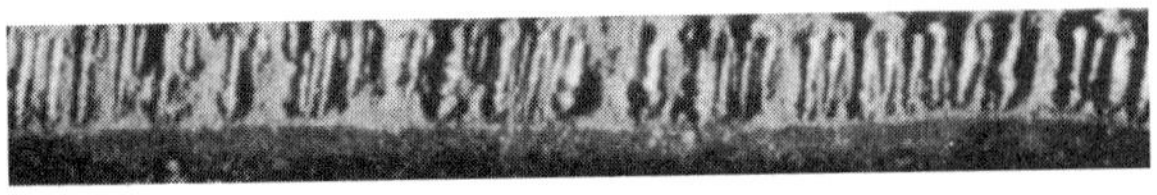

Weld on uncleaned rusty plate

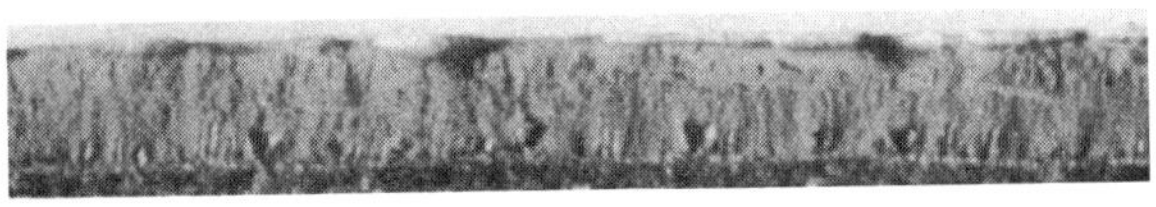

Weld on torch-heated rusty plate

Weld on rusty plate wire-brushed before tacking

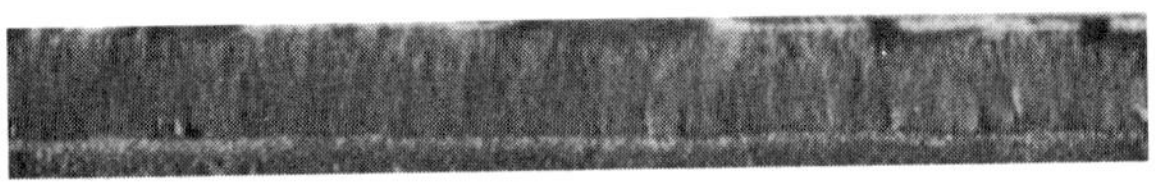

Weld on rusty plate wire-brushed before tacking and torch-heated

Fig. 6-63. Effect of various rust-removal methods on weld porosity. Power wire-brushing combined with torch-heating results in minimum porosity.

Sometimes, a small amount of lubricant is put on the wire for semiautomatic welding to improve feed through the cable. If too much is applied, it will cause porosity.

Insufficient Flux: If an inadequate amount of flux is used, the arc flashes through the flux and causes scattered surface porosity. On the other hand, too much flux causes an undesirable bead shape. The correct amount is indicated when the light of the arc reflects on the wire.

Insufficient flux coverage is more common on circumferential welds than on flat welds. On small circumferential welds, the flux must be contained around the arc by mechanical support. The weld may contain surface porosity if the slag spills off the weld before it has solidified. The danger of slag spillage is especially great with corner welds and with multiple-pass horizontal fillet welds.

Contaminants in the Flux: Contaminants may be picked up by the flux-recovery system and deposited in subsequent welds. The flux should be discarded if these contaminants cannot be removed by the recovery equipment (by magnetic separation or heating, for example).

Welding flux in factory-sealed moisture-resistant bags can be stored for several months in a dry area without picking up moisture. Flux exposed to the atmosphere will pick up moisture by condensation especially in a humid location. Flux can be dried by heating to 500°F for a period long enough to ensure that the entire bulk has reached that temperature. For a long exposure period in a humid atmosphere, 750°F may be necessary.

Dirt must be kept out of the flux. The flux-recovery system should be equipped to remove dust from the flux. Collected dust should be removed from the recovery equipment regularly. If it is not removed frequently, slugs of dust may fall into the recovered flux and cause large holes in the weld.

On some joints, such as those inside tanks, pieces of mill scale may fall onto the joint and contaminate the weld. This scale also may reach the weld through the flux-recovery system. However, mill scale can easily be removed from recovered flux by a magnetic separator.

Entrapped Flux: There is a possibility of porosity wherever there is an opportunity for flux to become trapped between the bottom of the bead and the opposite side of the joint. This porosity may either be subsurface or may come through to the surface as large holes. The solution to this problem is to leave at least 5/32 in. between the root of the weld and the other side, or to penetrate completely into the other side so that no unwelded space remains (Fig. 6-64).

Porosity from entrapped flux is most commonly encountered in butt welds. If the gap between the plate edges is 1/32 in. or more, flux may spill into the gap ahead of the arc. The remedy is to back the joint with a manual or semiautomatic weld or

(a)

(b)

Fig. 6-64. Porosity caused by entrapped flux (a), and sound weld (b) produced by reducing penetration. Penetration in (a) came within 5/32 in. of the backup bead, but missed the backup bead in (b) by 3/16 in.

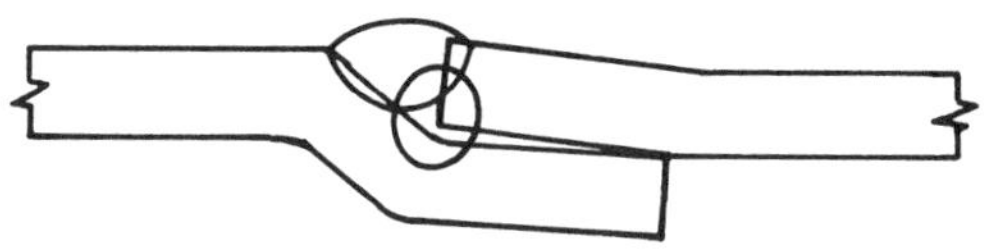

Fig. 6-65. An offset lap weld, which tends to trap flux at the bottom of the joint. The first pass may be porous, but the second pass will be sound.

a backup strip. The bead must penetrate the backup, or clear the backup by at least 5/32 in.

Offset lap welds, as illustrated in Fig. 6-65, are vulnerable to flux entrapment. Here, it is necessary to penetrate beyond the corner of the joint to produce porosity-free welds. Such penetration frequently cannot be obtained with a single pass, so two passes must be used. The second pass remelts most of the first, and the resultant deposit is clean and porosity-free.

Segregation: The composition of the base metal has a bearing on the porosity occurring in the weld. Even when composition is within acceptable limits, possible local segregation of constituents may promote porosity. Sulfur in particular tends to segregate within steel alloys. These sulfur segregations can be detected by a deep etch of the cross section, or by other chemical processes. Severe segregations may appear as laminations and cause large holes in the weld.

When it is necessary to weld steels with sulfur segregation, use procedures that give minimum admixture (low current, negative polarity, and large electrode diameters), and travel as slowly as possible to allow gases to escape through the molten pool. On joints that are normally square-edge butts, scarf the joint; and on joints that are normally beveled, increase the scarf. Also use multiple passes where possible.

Press-Fit Lubricants: Parts to be joined by a press-fit are usually coated with a lubricant, such as white lead, prior to mating. If a weld is then made in the vicinity of the press fit, the lubricant gives off gases that cause porosity, usually large holes at or near the end of the weld. Preferred practice is to avoid press fits in joints to be welded and to allow a gap of up to 1/32 in. But if press fits must be used, one piece should be knurled to allow a path for gases to escape.

Inappropriate Polarity: Though the effect of polarity on porosity is slight and not normally an important factor, positive polarity generally produces less porosity than negative polarity. Negative polarity, however, results in less porosity with sulphur-bearing steels or with other alloys where penetration and admixture are to be minimized.

Excessive Travel Speed: Reducing travel speed reduces porosity, since slow speeds allow gaseous materials to boil out of the molten weld metal. But a reduction in speed generally increases costs, so other solutions should be investigated first. The effect of speed on porosity is particularly strong with light-gage sheet, where rapid travel tends to increase porosity by increasing arc blow. Reducing speed may substantially reduce porosity here, if other means of controlling arc blow are not effective.

Slag Residue from Tack Welds: Slag from some types of electrodes may cause porosity where tack welds are covered with a submerged-arc bead. To avoid this problem, the manual electrodes used for tack welding should be of the E6010, E6011, E7016, or E7018 classes. These electrodes do not leave a residue that causes porosity in subsequent cover welds.

PREVENTING WELD CRACKING

Weld cracking with mild steel is seldom a problem in material less than 3/8 in. thick. With thicker material, welds are subjected to rapid cooling rates, which frequently induce high stresses and lead to cracking. Low ambient temperature increases the cooling rate and produces a similar effect. Constrained shapes not free to flex or deflect under thermal stresses also have a tendency to crack, as do steels of high hardenability or "hot-shortness." The term "weldability," used to indicate the relative ease with which steels can be welded without cracking, is discussed in Section 6.1.

Cracking in Fillet Welds

Cracking is more common in fillet welds than in butt welds because both legs of a fillet joint are rigidly fixed. This rigidity prevents deflections that normally absorb thermal stresses. The following factors influence the crack resistance of fillet welds:

Gap: When any part to be welded is more than 1 in. thick, a gap of 1/32 to 1/16 in. should be used to allow the weld to shrink during cooling. Grooving the plate edges or inserting a compressible material between the workpieces minimize the shrinkage stresses.

Polarity: Positive polarity is normally recommended for fillet welds to obtain greatest penetration and minimum tendency for porosity at high speeds. But if the chemical composition of the workpiece promotes cracking, negative polarity should be used to reduce penetration and minimize admixture with plate metal. Also, the 20 to 30% increase in melt-

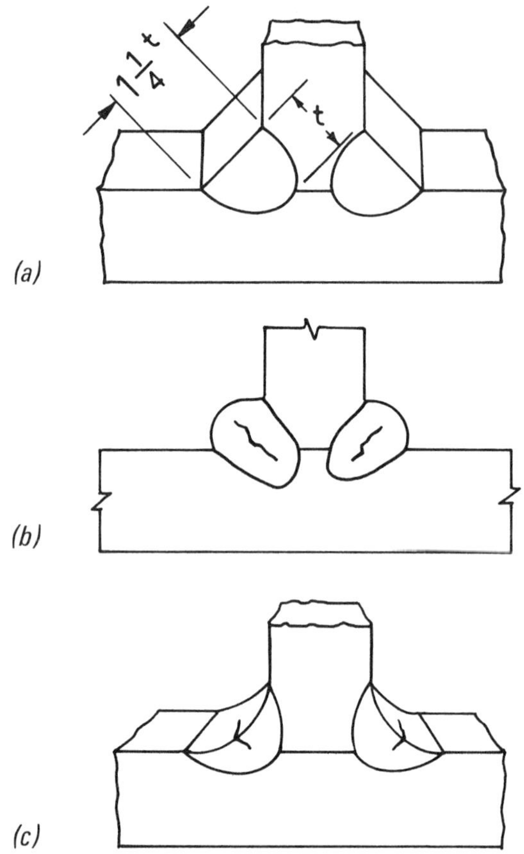

Fig. 6-66. Influence of bead shape on weld cracking in fillet welds. Welds least likely to crack are those (a) with a width that exceeds the depth. The minimum recommended ratio of width to depth is 1.25 to 1. Beads with greater depth than width (b) may crack internally. Regardless of bead shape, concave bead surfaces (c) should be avoided, because they promote surface cracks.

off-rate with negative polarity helps build up an adequate bead with the preferred convex shape.

Electrode Size: Large-diameter electrodes should be used when cracking is a problem to reduce penetration and decrease admixture with the parent plate.

Flux Coverage: Flux thickness should be just adequate to cover the arc.

Number of Arcs: Twin electrodes, frequently used on flat fillets, produce less penetration, less admixture, and more melt-off compared to a single arc. For a given arc speed, twin electrodes thus reduce tendencies toward porosity and cracking. Twin electrodes also produce less arc blow.

Type of First Pass: A manual first pass with an E7018 electrode reduces admixture and thereby minimizes cracking tendencies. This practice is also good assurance against melt-through where poor fitup is encountered.

Bead Shape: Width of the weld cross section should be 1-1/4 times the depth to reduce stresses caused by internal shrinkage, as illustrated in Fig. 6-66(a). This rule-of-thumb is especially important on steels with high cracking tendencies. Beads that are too deep may crack internally, Fig. 6-66(b). Slag removal also is difficult with deep beads. The bead surface must not be concave. Although concave welds have a pleasing appearance, the resulting surface stresses promote surface cracks.

Edge Preparation: Internal cracking and slag inclusions are possible if the prepared angle on single-pass welds is too acute. Bevel joints should not be used unless they have included angles of at least 60°, so that the width is greater than the depth. If 100% penetration T joints are specified on heavy plate, the first pass on each side should be made with E7018 electrode, or the joint should be prepared in a manner to avoid excessive penetration and the resultant possibility of beads having greater depth than width.

Angle of Electrode: When two steels to be welded are of differing chemical analyses, arc movement should be toward the more weldable alloy to minimize admixture with the less weldable alloy.

Electrode Stickout: As electrical stickout (distance from point of electrical contact to electrode tip) is increased, melt-off rate increases while penetration and admixture decrease. Increased stickout thus reduces cracking tendencies, but also increases the difficulty of controlling the bead shape.

Grounding: The workpiece should be grounded at the "start" end of the weld, except on short welds, which should be grounded at both ends of the joint.

Speed and Current: Travel speed and welding current should be decreased as the proportion of carbon and other alloy constituents in the steel increase. This measure reduces cracking tendencies by reducing penetration and minimizing the size of the molten puddle.

Cracking Butt Welds

Cracking in butt welds is less common than in fillet welds because butt joints are less constrained and can generally deflect enough to absorb thermal stresses. But when cracking is encountered in butt welds, the following factors should be reviewed:

Bead Shape: "Hat-shaped" beads, caused by excessively high voltage or slow travel speed, may promote cracks at the change in bead contour where the "crown" joins the "brim" (Fig. 6-34). A reduction in voltage is usually sufficient to avoid such cracks,

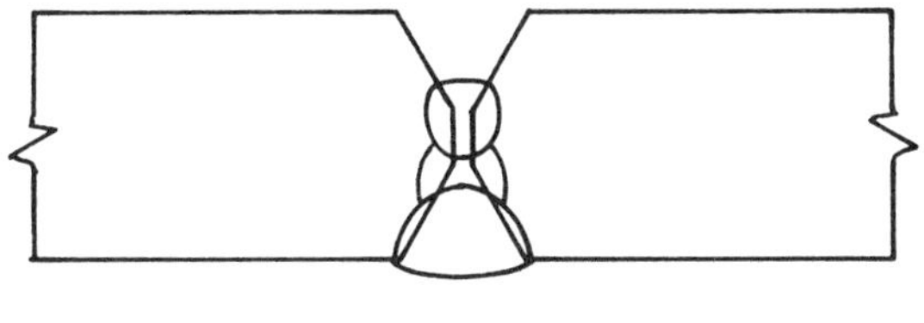

Fig. 6-67. Deep, narrow beads in a back-chipped butt joint are prone to cracking. Beads should be wider than they are deep.

but an increase in travel speed may also be required in some cases. These measures may require that the weld be made with two passes instead of one.

Bead shape may also promote cracking in seams that have been back-chipped or back-gouged. Cracking, therefore, is often a concern in the first pass on the second side of a double-beveled joint. Since the groove produced by back-gouging or back-chipping is deep and narrow, the bead placed in this trough often has a tendency to crack (Fig. 6-67). The appropriate measure in this case is to produce a groove that is wider than it is deep.

Admixture: Butt joints made in steels of poor weldability may have a tendency to crack. The appropriate remedy, as described in the prior paragraphs pertaining to fillet welds, is to reduce admixture with the base metal. This is usually best accomplished by using negative polarity, large electrodes, twin electrodes, a manual first pass, long electrode stickout, slow travel speeds, or low currents.

Alloy Pick-Up from the Flux: Excessively high arc voltages cause substantial pick-up of manganese and silicon from some fluxes. This pick-up usually can be avoided by not exceeding voltages recommended in standard procedures tables.

WIRE FEEDING EQUIPMENT AND CONTROL SYSTEMS

The submerged-arc process is inherently a mechanized process — either semiautomatic or automatic. The electrode-feeding equipment varies with the degree of mechanization.

Semiautomatic Welding: Semiautomatic equipment, commonly called manual "Squirt," maintains a preset wire feed speed (current) and voltage. The operator must strike the arc, guide the welding gun along the seam, and manually pace the travel speed. A remote electrode-feeding mechanism (Fig. 6-68) feeds the electrode and automatically controls the current and voltage. A typical gun, shown in Fig. 6-69, is hand-held by the operator. In Fig. 6-70, a travel-speed device is added to the gun, but the operator must hold the gun and guide it along the

Fig. 6-68. A typical electrode-feeding unit for semiautomatic submerged-arc welding. The flux tank holds approximately 100 lbs of flux, which feeds by dry fluidization through a hose to the gun.

Fig. 6-69. Typical gun for semiautomatic submerged-arc welding. The gun is designed for fluidized flux feeding.

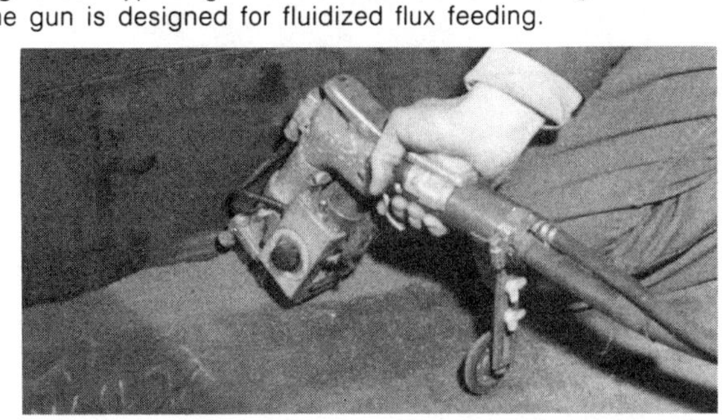

Fig. 6-70. Gun with a travel mechanism attached.

Fig. 6-71. Gun mounted on a small tractor, which relieves the operator of holding and guiding the gun.

seam. In Fig. 6-71, the gun is mounted on a small tractor that rides on the work and follows the joint. This relieves the operator from holding and guiding the gun, and reduces his function to setting the guide rolls for the joint, starting the welding, and monitoring the progress as the tractor rides the seam. Equipment shown in Fig. 6-70 and 6-71 is called mechanized squirt-welding equipment.

Automatic Welding: A typical fully automatic head is shown in Fig. 6-72. The head shown is equipped to feed a single electrode. The same head

Fig. 6-72. Typical submerged-arc welding head for fully automatic welding.

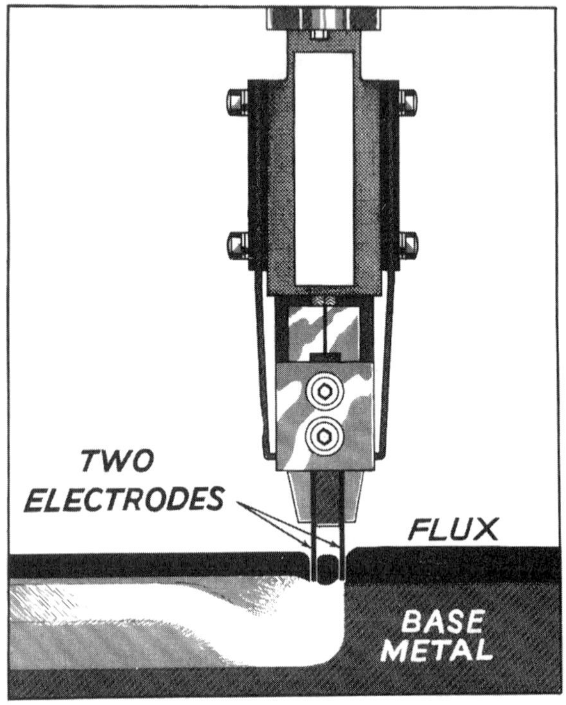

Fig. 6-73. Twin-electrode nozzle for converting a single-arc head for welding with two wires simultaneously.

will feed two electrodes by changing to a contact nozzle as shown in Fig. 6-73. and changing the wire drive rolls. This modification is known as twin-electrode submerged-arc welding. The electrical control system maintains the preset current and voltage, strikes the arc, starts the travel and controls the the travel speed. Either the head is stationary and the works moves, or the work is stationary and the head moves.

Multiple Automatic Heads: For higher welding speeds two or three heads may be combined to weld simultaneously on the same joint. This is known as multiple-electrode or tandem-arc welding. Each welding head is powered by a separate power source, as distinguished from twin-electrode welding, in which both electrodes are powered by the same power source.

Control Systems: The success of any automatic submerged-arc operation depends on the control of the arc length, or the distance from the tip of the electrode to the work. This can be accomplished by varying the wire-feed speed or the rate of melt-off to compensate for irregularities that increase or decrease the distance of the welding head from the work.

Changes in arc length that result from such movement result in arc voltage changes. With a constant current (variable-voltage) control system, a change in arc voltage is detected and the wire-feed speed is automatically increased or decreased to shorten or lengthen the arc and restore it to the desired length. With such a system, by varying the wire feed the arc voltage is kept within a narrow range. The welding current may be AC or DC, but the power source must be of the constant current type (see Section 4.2).

The arc length can also be maintained by current control. This method depends on high short-circuit current characteristics of the power supply to keep the arc length constant. If the arc length becomes shorter than desired, the current increases to melt the electrode more rapidly and lengthen the arc. With this method — the constant-voltage system of arc-length control — the wire-feed speed is usually constant. (Modified constant-voltage control can be used in conjunction with arc-voltage control for greater sensitivity.) The arc voltage is preset at the power source. Only DC is used with the constant-voltage system, and the power source must be the constant-voltage type.

The application determines the type of arc-length control that should be used for optimum results — and thus the type of submerged-arc process.

Constant Current Submerged-Arc: The range of recommended applications include all fast-fill joints, such as multiple-pass welds and heavy single-pass welds where a high deposition rate is a primary requirement. Constant current submerged-arc is also used on high-speed single-pass welds on 1/4-in. and thicker plate. It is not the best, but is, at least, useable on fast-follow applications on 14-gage and thicker steel. When the work consists mostly of 3/16-in. and smaller welds, constant-voltage DC processes are preferred for consistent weld quality at high speeds.

The **single electrode and twin electrode**, constant current submerged-arc:

1. Has flexibility for a wide range of applications, such as multiple-pass welds, single-pass welds on 14-gage and thicker material, and circumferential seams 2 in. and larger in diameter.
2. Is excellent for deep-penetration (DC+) and conventional (DC−) fillets.
3. With long stickout, and operating DC negative, approaches the deposition rates of DC-AC tandem arc.
4. Is a good choice for hardfacing and welding alloys.
5. Is easy to set up and use and easy to convert to long stickout, or to convert the single-electrode head to twin-electrode.

Constant current submerged-arc, however, is not as fast as constant-voltage in fast-follow sheet-metal welding.

Constant current **DC-AC tandem arc** gives deposition rates and speeds 25 to 100% over single-wire DC on many applications, such as multiple-pass welds, single-pass butt welds on 10-gage and thicker material, single-pass horizontal fillets 1/4 to 1/2 in. in leg size, single-pass flat fillets 1/4 to 3/4 in. in leg size, and large-diameter roundabouts (40 in. minimum). It is the best choice for making butt welds up to 3/4 in. from one side into a flux backing. Electrode spacing and controls are easy to set for different applications. Long stickout techniques can be used for increased deposition rates.

Constant current **DC-AC-AC three-wire tandem arc** gives still higher deposition rates and arc speeds, but on a more limited range of applications, such as multiple-pass welds, horizontal fillets 5/16 to 1/2 in. in leg size, flat fillets 1/2 to 3/4 in. in leg size, and single-pass butt welds. Three-wire tandem is widely used on the longitudinal seams in the fabrication of large-diameter pipe. Setting up is more complicated with three arcs than with two arcs, but once the set-up is established, repeat welds can be run without resetting. The long weld crater produced with three arcs minimizes porosity.

Constant-Voltage Submerged-Arc: With constant-voltage DC, small-diameter electrodes are used. The welds are characterized by uniform penetration, which results in minimal melt-throughs, skips, and misses when laying small beads at fast travel speeds, providing the fitup is good. If the fitup is not good or butt joints not tight, backup strips are needed. Recommended applications are fast-follow single-pass welds on sheet metal and plate 5/16-in. and thinner. On thicker plate, constant-voltage submerged-arc tends to produce undercut and ropey beads.

Constant-voltage **single-electrode submerged-arc** permits high travel speeds on 1/8 to 1/4-in. fillets and lap welds, gapped butt welds in 18-gage sheet to 1/4-in. plate, tight butt welds in 18-gage to 3/16-in. material, and on edge and corner welds in sheet metal. Small roundabouts — down to 1/2-in. diameter — may be welded efficiently.

Constant-voltage **twin-electrode submerged-arc** enables the staggering of electrodes to make wide beads and control undercut. It is often used as a replacement for single or tandem-arc welding where melt-through problems are encountered. It permits high speeds on 1/8 to 1/4-in. fillets, gapped butt welds on 14-gage to 1/4-in. material, tight butt welds on 14 to 10-gage sheet, and on gage-size joggled laps and edge and corner welds. The penetration, however, is not adequate for through lap welds.

Electroslag Welding

Electroslag welding is a variation of the submerged arc process and is described in Section 5-5 and illustrated in Figure 5-19 for the standard process and Figure 5-20 for the consumable electrode guide process.

The composition of an electroslag weld depends on the composition of the filler wire electrode, the consumable guide tube, if used, and the base plate. Typically from 30 to 50 percent of the fusion zone comes from base metal which melts and dilutes the added filler metal. Many types of plain carbon steels can be welded by this process without post weld heat treatment. High strength low alloy and heat treatable steels can be welded but normally require a grain refining post weld heat treatment to develop required properties.

Filler wire for electroslag welding may be solid, metal cored or flux cored. Solid wire is more widely used and is available in a variety of compositions. The metal cored wire, through ferro alloy additions to the core, provide for a wider variety of compositions than the solid wire. Flux cored wire is used for a similar process called self-shielded vertical-up butt welding, which is described in Section 6-4.

Fillet welds made by the submerged arc process.

INTRODUCTION TO WELDING PROCEDURES

The ideal welding procedure is the one that will produce acceptable quality welds at the lowest overall cost. So many factors influence the optimum welding conditions that it is impossible to write procedures for each set of conditions. In selecting a procedure, the best approach is to study the conditions of the application and then choose the procedure that most nearly accommodates them. The procedures given here are typical, and it may be necessary to make adjustments for a particular application to produce a satisfactory weld.

For some joints, different procedures are offered to suit the weld quality — code quality, commercial quality, and strength only — that may be required.

Code-Quality Procedures

Code-quality procedures are intended to provide the highest level of quality and appearance. To accomplish this, conservative currents and travel speeds are recommended.

These procedures are aimed at producing welds that will meet the requirements of the commonly used codes: AWS Structural Code, AISC Buildings and Bridges, ASME Pressure Vessels, AASHTO Bridges, and others. Code-quality welds are intended to be defect-free to the extent that they will measure up to the nondestructive testing requirements normally imposed by these codes. This implies crack-free, pressure-tight welds, with little or no porosity or undercut.

The specific requirements of these codes are so numerous and varied that code-quality procedures may not satisfy every detail of a specific code. Procedure qualification tests are recommended to confirm the acceptability of chosen procedures.

All butt welds made to code-quality are full penetration; fillet welds are full size, as required by most codes. (The theoretical throat rather than the true throat is used as the basis of calculating strength.)

Commercial-Quality Procedures

Commercial quality implies a level of quality and appearance that will meet the nominal requirements imposed on most of the welding done commercially. These welds will be pressure-tight and crack-free. They will have good appearance, and they will meet the normal strength requirements of the joint.

Procedures for commercial-quality welds are not as conservative as code quality procedures; speeds and currents are generally higher. Welds made according to these procedures may have minor discontinuities that would be objectionable to the more demanding codes.

It is recommended that appropriate tests be performed to confirm the acceptability of the selected procedure for the application at hand prior to putting it into production.

Strength Only Procedures

The purpose of strength-only procedures is to produce the highest speed welding at the lowest possible cost. To accomplish this, the weld appearance and quality need only be good enough to do the specific job for which it is intended. In this category, imperfections may be acceptable provided that the welds will perform satisfactorily under service conditions.

This quality level may apply to seal beads, partial-penetration welds, some lap welds, high-speed edge welds, and to many low-stress welded connections which have no real need for better quality at higher cost. On fillet welds the joint must be welded from both sides.

Since this is a compromise quality level, realistic tests should be made to confirm the satisfactory performance of welds made according to these procedures. Welds of strength-only quality will be free of cracks but may not be pressure-tight.

Weldability of Material

Weldability, (see Section 6.1) of a steel has a considerable effect on the welding procedure. For some joints, more than one procedure is offered because of the marginal weldability of the steel.

Good weldability indicates a steel with a composition that is within the preferred range (see Table 6-1) — one whose chemistry does not limit the welding speed.

Fair weldability indicates a steel with one or more elements outside the preferred range or one that contains one or more alloys. These steels require a lower welding speed or a mild preheat, or both to minimize discontinuities such as porosity, cracking, and undercut.

The addition of alloys to steel that enhance the mechanical properties or hardenability usually have an adverse affect on weldability. In general, the weldability of low-alloy steels is never better than "fair." Procedures designed for fair weldability can, of course, be used on good weldability steels.

Procedure Notes

In the following fillet weld procedures, the fillet size is always associated with a particular plate thickness. This relationship is given solely for the purpose of designing a welding procedure and does not imply that a certain size fillet is the only size applicable to that plate thickness. In some of the procedures, the fillet size shown is larger than necessary to meet code requirements for the plate thickness. In such instances, select the procedure for the proper weld size and quality. If the thickness of the plate being welded is appreciably greater than that specified in the procedure, a reduction in welding speed and current will probably be required.

Travel speed is given as a range for semiautomatic procedures; the electrode required and the total time are based on the middle of the range.

Unless otherwise indicated, both members of the joint are the same thickness.

Pounds-of-electrode data include all ordinary deposition losses. These values are in terms of pounds of electrode needed to be purchased.

Total time is the arc time only and does not allow for operating factor.

The procedures do not specify the grade of flux or electrode to be used. Manufacturers of flux and electrode will guarantee mechanical properties of flux-electrode combinations but the industry is not standardized on other characteristics such as fill, freeze, and follow. Consult the supplier for a recommendation on the best flux-electrode combination for the application.

Electrode position affects penetration, the placement, and shape of the weld bead. Maximum penetration is in the direction which the electrode points. Thus, for symmetrical weld beads in the flat position, the electrode should be vertical and centered on the joint. For welds in other positions and for unsymmetrical beads, the electrode should be positioned so it controls the point of maximum penetration and the effect of the arc force on the bead shape.

The flux pile should be only deep enough to cover the arc. Excessive flux depth can cause poor bead shape. Insufficient flux causes porosity.

Stickout (SO) is the distance from the tip of the electrode to the contact jaw or contact tip. Increasing the stickout, especially if the current density is high, increases the melt-off rate at the expense of both penetration and accuracy in weld placement. Except where used to purposely increase deposition rate, the stickout should be only enough to clear the flux pile. In general, this is about 3/8-in. for a 5/64-in. electrode, and up to 1-1/2-in. for a 7/32-in. electrode.

Seams should be tight but if gaps do occur, seal them with a fast, stick electrode seal bead. Use a 5/32-in. E6010 or E7018 electrode, clean the slag thoroughly, and place the seal beads as follows:

> On groove joints, place the seal bead on the first-pass side.
>
> On square butt joints, place the seal bead on the first-pass side if the plate is less than 1/2-in. thick. On 1/2-in. and thicker plate, seal on the second-pass side.

Seal beads can also be made with submerged-arc semiautomatic equipment using 200-250 amp, 25-30 v, 45-50 ipm, with electrode negative. Apply the seal beads in the same side as described above for stick electrode.

Where stick electrode welds are used in combination with submerged-arc welds, the procedures data does not specify the electrode or indicate time required for the stick electrode weld. Select an appropriate welding procedure from the manual shielded metal-arc welding section.

After a satisfactory welding procedure has been established, all the data should be recorded and filed for future reference. This information is invaluable if the same job or a similar job occurs at a later date. A suggested data sheet is shown opposite.

The presented procedures are offered as a starting point and may require changes to meet the requirements of specific applications. Because the many variables in design, fabrication, and erection or assembly affect the results obtained in applying this type of information, the serviceability of the product or structure is the responsibility of the builder.

Welding Parameters, Single Values or a Range

In some of the procedures which follow single values are given for such items as Amps, Volts, and Travel Speed. The purpose of using single values, i.e. 100 amps, is to indicate an optimum or starting place and it is expected that the user will expand these values and show a reasonable range. These values when recorded on a WPS (Welding Procedure Specification) should be shown as a range, not as specific values.

DATA SHEET

Article to be Welded ______________________ Job No. ______________

Plate Specs or Analysis ______________________________________

Welding Process ______________ Submitted by: ______________ Date: __________

JOINT-PREPARATION – BEAD PLACEMENT, SIZE, SEQUENCE

WORK ANGLE – DRAG ANGLE
ELECTRODE ANGLE TO JOINT

Pass No	AC DC (+ –)	Amps	Volts	Travel Speed "/Min	Electrode Type	Elect. Size	Stick-Out	Flux Type					

Special Comments ______________________________________

TABLE 6-27. Submerged-Arc Troubleshooting Guide

Joint	Problem	Full Automatic Semiautomatic, Single Electrode, Twin Electrodes Corrective Action – In Order of Importance	
Any	Low Penetration	1. Increase welding current. 2. Use electrode positive. 3. Lower voltage on fillets or V-joints.	4. Use short stickout. 5. Decrease arc speed. 6. Increase included angle on V-joints.
Fillet	Cracking	1. Use EM12K electrode. 2. Use electrode negative. 3. Lower voltage.	4. Decrease welding speed. 5. Preheat joint. 6. Increase electrode diameter and lower voltage.
Root Pass In Groove	Cracking	1. Lower current and voltage. 2. Use electrode negative. 3. Increase root opening or included angle.	4. Preheat joint. 5. Make sure back gouging is not narrow and deep.
Multiple Pass Weld	Transverse Cracking	1. Increase interpass temperature. 2. Decrease welding speed.	3. Decrease voltage. 4. Decrease current and voltage.
Square Butt Weld	Cracking	1. Check fixture for plate movomont. 2. Decrease welding speed.	3. Check for copper pick-up from backup.
Fillet Lap or Sq. Butt	Pock Marking or Slag Sticking	1. Use EM12K electrode. 2. Increase voltage. 3. Decrease current. 4. Decrease speed.	5. Position fillet, if possible. 6. Heavier plate than normal will cause pocking 7. Clean all mill scale, rust and oil off plate.
Spud or Deep Groove	Slag Sticking	1. Decrease voltage. 2. Decrease current and voltage.	
Spud	Not Overlapping	1. Decrease voltage. 2. Decrease current and voltage.	
Any	Undercutting	1. Use electrode negative. 2. Decrease voltage. 3. Decrease current.	4. Increase electrode diameter and lower voltage. 5. Decrease speed.
Any	Rust Porosity	1. Use EM12K or EM13K electrode. 2. Increase voltage. 3. Lower current 4. Use electrode positive.	5. Use torches in front of arc. 6. Clean joint completely (butting edges also). 7. Decrease speed.
Any	Organic Porosity	1. Use EL12 electrode. 2. Use electrode positive.	3. Decrease speed. 4. Degrease joint and dry completely.
Any	Arc Blow Porosity	1. Use EL12 electrode. 2. Use electrode positive. 3. Lower voltage.	4. Lower current and voltage. 5. Increase electrode diameter and lower voltage.
Any	2nd Pass Side Porosity	1. Usually caused by improper tie-in. 2. Increase welding current to tie-in.	3. Decrease welding speed to tie-in. 4. If 100% joint not required, then decrease penetration.
Any	Metal Spots	1. Lower voltage. 2. Use electrode negative.	3. Decrease current and voltage. 4. Increase arc speed.
Out of Position	Metal Spillage	1. On roundabouts, move further off center opposite to direction of travel. 2. Lower voltage. 3. Lower current and voltage.	4. Increase speed on horizontal fillets. 5. On roundabouts, increase speed – lower current and voltage.
Any	Bead Shape	1. Increase voltage to get wider. flatter bead. 2. Decrease current to get flatter bead. 3. Decrease speed to get flatter bead on fillets.	4. Use electrode diameter that is proper for welding current. 5. Use electrode positive on square butt welds and fillets smaller than 1/4".

TABLE 6-28. Submerged-Arc Troubleshooting Guide

Joint	Problem	Full Automatic Multiple Electrodes Corrective Action — In Order of Importance	
Any	Low Penetration	1. Increase lead arc current. 2. Decrease lead arc voltage. 3. Decrease speed.	4. Increase included angle or root opening. 5. Use DC(+) — AC setup.
Any	Undercutting	1. Decrease electrode spacing. 2. Lower trail arc voltage. 3. Raise lead arc voltage.	4. Decrease arc speed. 5. Possibly raise trail arc current. 6. Use proper electrode size for current.
Any	Rust Porosity	Best Combination DC(+) AC 1. Use EM12K or EM13K 2. Increase voltage. 3. Lower current. 4. Use electrode positive.	5. Use torches in front of arc. 6. Clean joint completely (butting edges also). 7. Decrease speed.
Any	Organic Porosity	Best Combination DC(+) AC 1. Use EL12 electrode. 2. Use electrode positive.	3. Decrease speed. 4. Degrease joint and dry completely.
Any	2nd Pass Side Porosity	Best Combination DC(+) AC 1. Usually caused by improper tie-in. 2. Increase welding current to tie-in.	3. Decrease welding speed to tie-in. 4. If 100% joint not required, then decrease penetration.
Any	Metal Spots	Best Combination DC(—) AC 1. Lower voltage. 2. Use electrode negative.	3. Decrease current and voltage. 4. Increase arc speed.
Any	Arc Stability	1. Unstable welding is commonly thought to be poor AC stability. Usually the problem is too much forward blow by the trail arc causing the puddle to get under the lead arc. This is best corrected by increasing the electrode spacing. This is especially true in groove welds. 2. If the problem is really AC stability, then: a. Use a smaller AC electrode. b. Increase the arc voltage. c. Put electrodes closer together.	
Any	Poor Bead Shape	1. Increase lead arc voltage to flatten and widen bead. 2. Decrease trail current to flatten bead.	3. Decrease electrode spacing to flatten and widen bead.
Any	Wavy Edges	Best Combination DC(+) AC 1. Increase diameter of lead arc electrode and decrease diameter of trail arc electrode. 2. Lower trail arc current. 3. Lower lead arc voltage.	4. Increase lead arc current. 5. Increase welding speed.
Fillet	Cracking	Best Combination DC(—) AC 1. Use EM12K 2. Lower lead arc voltage. 3. Lower trail arc current and voltage.	4. Decrease welding speed. 5. Preheat joint. 6. Increase electrode diameter and lower voltage.
Root Pass in Groove	Cracking	1. Lower current and voltage on both arcs. 2. Use electrode negative on lead arc. 3. Increase root opening or included angle.	4. Preheat joint. 5. Make sure back gouging is not narrow and deep.
Multiple Pass Weld	Transverse Cracking	1. Decrease welding speed. 2. Lower voltage on both arcs.	3. Lower trail arc current and voltage. 4. Increase interpass temperature.
Square Butt Weld	Cracking	1. Raise lead arc voltage. 2. Increase trail arc current.	4. Decrease electrode spacing. 5. Check fixture for plate movement.
Fillet, Lap or Square Butt	Pock Marking or Slag Sticking	1. Use EM12K electrode. 2. Lower trail arc current 3. Raise lead arc voltage. 4. Raise trail arc voltage.	5. Decrease welding speed. 6. Clean all mill scale, rust and oil off plate. 7. Position fillets, if possible.
Deep Groove	Slag Sticking	1. Lower lead arc and/or trail arc voltage.	2. Decrease current and voltage.

SUBMERGED-ARC (FULL AUTOMATIC) SINGLE ELECTRODE

Welding Position: Flat
Weld Quality Level: Commercial
Steel Weldability: Good
Welded from: One side

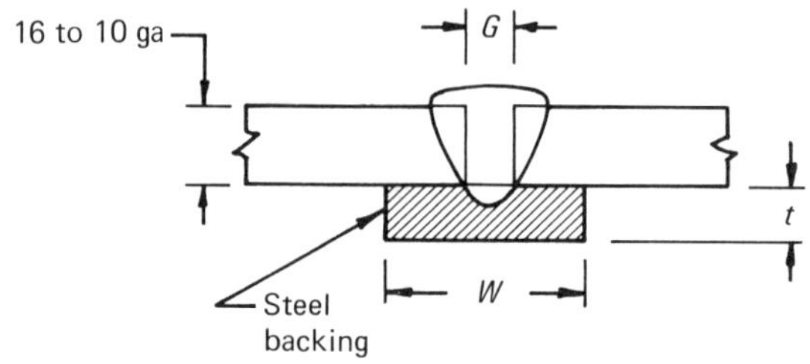

Plate Thickness (in.)	0.060 (16 ga)	0.075 (14 ga)	0.105 (12 ga)	0.135 (10 ga)
Pass	1	1	1	1
Electrode Size (in.)	1/8	1/8	1/8	1/8
Current (amp) DC(+)	450	500	550	650
Volts	25	27	27	28
Arc Speed (in./min)	110	80	65	55
Electrode Req'd (lb/ft)	0.017	0.027	0.038	0.057
Flux Req'd (lb/ft)	0.015 – 0.021	0.023 – 0.031	0.032 – 0.044	0.049 – 0.065
Total Time (hr/ft of weld)	0.00182	0.00250	0.00308	0.00364
G (in.)	1/32	1/32	1/16	1/16
t, min	14 ga	12 ga	12 ga	1/8″
W, min (in.)	3/8	3/8	1/2	5/8

SUBMERGED-ARC (FULL AUTOMATIC) SINGLE ELECTRODE

Welding Position: Flat
Weld Quality Level: Commercial
Steel Weldability: Good
Welded from: One side

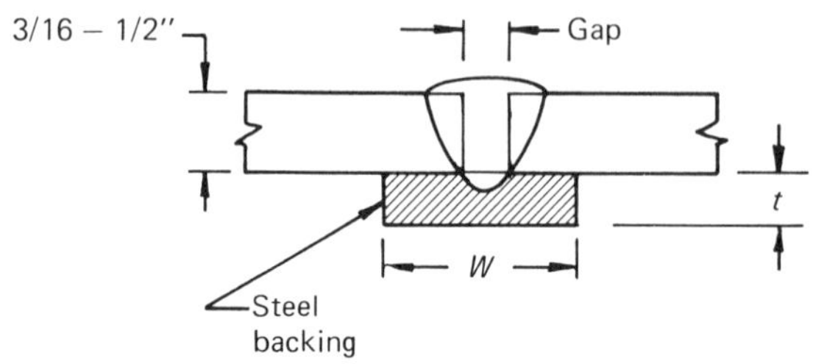

Plate Thickness (in.)	3/16	1/4	3/8	1/2
Pass	1	1	1	1
Electrode Size	3/16	3/16	3/16	3/16
Current (amp) DC(+)	800	850	900	1000
Volts	32	33	34	35
Arc Speed (in./min)	50	33	24	17
Electrode Req'd (lb/ft)	0.087	0.14	0.23	0.35
Flux Req'd (lb/ft)	0.076 – 0.094	0.11 – 0.15	0.18 – 0.24	0.27 – 0.37
Total Time (hr/ft of weld)	0.00400	0.00606	0.00833	0.0118
Backing, minimum size (in.)	3/16 x 3/4	1/4 x 1	5/16 x 1	3/8 x 1
Gap (in.)	3/32	1/8	5/32	3/16
W, min (in.)	3/4	1	1	1
t, min (in.)	3/16	1/4	5/16	3/8

See introductory notes.

SUBMERGED-ARC (FULL AUTOMATIC) SINGLE ELECTRODE

Welding Position: Flat
Weld Quality Level: Code
Steel Weldability: Good or fair
Welded from: One side

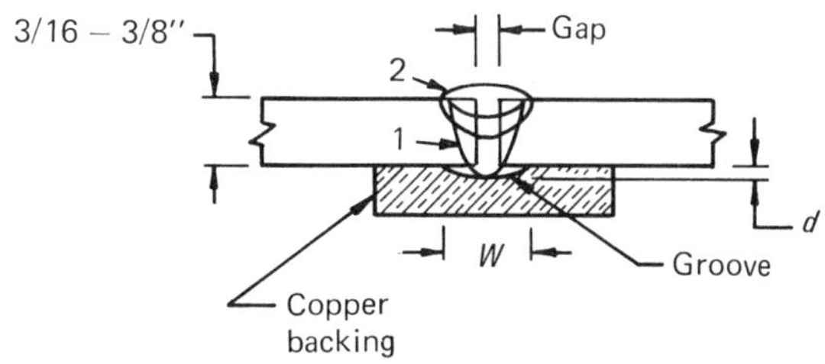

Plate Thickness (in.)	3/16		1/4		5/16*		3/8*	
Pass	1	2	1	2	1	2	1	2
Electrode Size	5/32	5/32	5/32	5/32	5/32	5/32	5/32	5/32
Current (amp) DC(+)	550	500	650	600	650	600	750	700
Volts	25	30	30	35	30	35	30	35
Arc Speed (in./min)	40	40	36	36	30	30	26	26
Electrode Req'd (lb/ft)	0.12		0.17		0.20		0.28	
Flux Req'd (lb/ft)	0.091 – 0.12		0.13 – 0.18		0.14 – 0.20		0.20 – 0.28	
Total Time (hr/ft of weld)	0.0100		0.0111		0.0133		0.0154	
Gap (in.)	1/16 – 3/32		1/16 – 3/32		3/32 – 1/8		1/8 – 5/32	
Groove in backing bar								
Width, W (in.)	11/16		11/16		11/16		11/16	
Depth, d (in.)	3/32		3/32		1/8		5/32	

* Fill groove in backing bar with flux before welding.

See introductory notes.

SUBMERGED-ARC (FULL AUTOMATIC) SINGLE ELECTRODE

Welding Position: Flat
Weld Quality Level: Commercial
Steel Weldability: Good
Welded from: One side

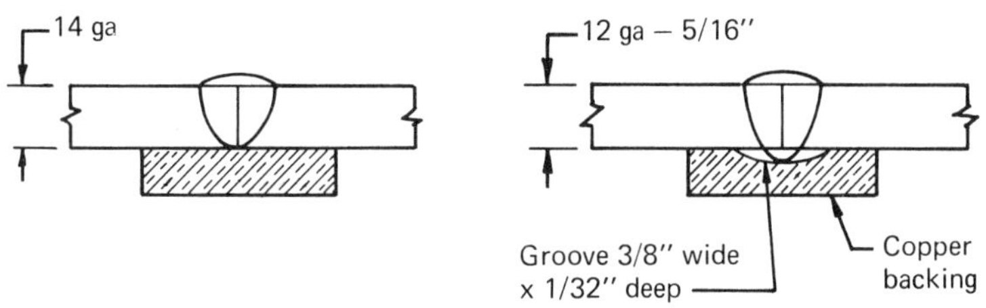

Plate Thickness (in.)	0.075 (14 ga)	0.105 (12 ga)	0.135 (10 ga)	3/16	1/4	5/16
Pass	1	1	1	1	1	1
Electrode Size	1/8	5/32	5/32	5/32	3/16 or 7/32	3/16 or 7/32
Current (amp) DC(+)	400	550	650	725	800	950
Volts	24	30	31	31	34	36
Arc Speed (in./min)	100	110	95	75	45	35
Electrode Req'd (lb/ft)	0.016	0.022	0.033	0.049	0.091	0.15
Flux Req'd (lb/ft)	0.015 – 0.020	0.021 – 0.028	0.029 – 0.040	0.044 – 0.059	0.082 – 0.11	0.13 – 0.16
Total Time (hr/ft of weld)	0.00200	0.00182	0.00211	0.00267	0.00444	0.00571

SUBMERGED-ARC (FULL AUTOMATIC) SINGLE ELECTRODE

Weld Position: Flat
Weld Quality Level: Strength only
Steel Weldability: Good
Welded from: One side

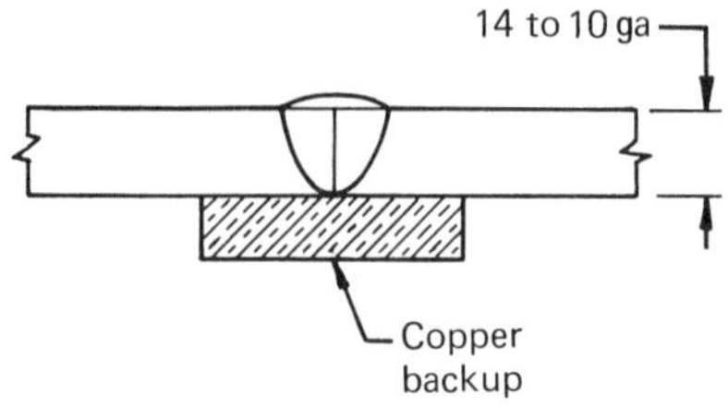

Plate Thickness (in.)	0.075 (14 ga)	0.105 (12 ga)	0.135 (10 ga)
Pass	1	1	1
Electrode Size	1/8	1/8	1/8
Current (amp) DC(+)	475	625	675
Volts	26	28	28
Arc Speed (in./min)	145	135	120
Electrode Req'd (lb/ft)	0.014	0.023	0.029
Flux Req'd (lb/ft)	0.013 – 0.018	0.021 – 0.028	0.026 – 0.035
Total Time (hr/ft of weld)	0.00138	0.00148	0.00167

SUBMERGED-ARC (FULL AUTOMATIC) SINGLE ELECTRODE

Welding Position: Flat
Weld Quality Level: Commercial
Steel Weldability: Good
Welded from: Two sides

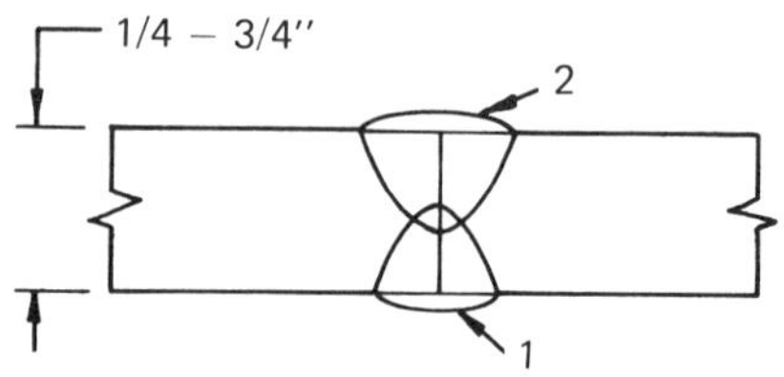

Plate Thickness (in.)	1/4		3/8		1/2		5/8		3/4	
Pass	1*	2	1	2	1	2	1	2	1	2
Electrode Size	3/16		3/16		3/16		3/16		3/16	
Current (amp) DC(+)	600	750	650	800	750	850	750	850	800	900
Volts	31	33	33	35	35	36	35	36	36	37
Arc Speed (in./min)	70	70	48	48	35	35	24	24	22	22
Electrode Req'd (lb/ft)	0.088		0.15		0.23		0.37		0.42	
Flux Req'd (lb/ft)	0.092 – 0.11		0.13 – 0.17		0.20 – 0.26		0.32 – 0.43		0.36 – 0.48	
Total Time (hr/ft of weld)	0.00571		0.00833		0.0114		0.0167		0.0182	

* Must be down flat on a platen.

See introductory notes.

SUBMERGED-ARC (FULL AUTOMATIC) SINGLE ELECTRODE

Weld Position: Flat
Weld Quality Level: Code
Steel Weldability: Good
Welded from: Two sides

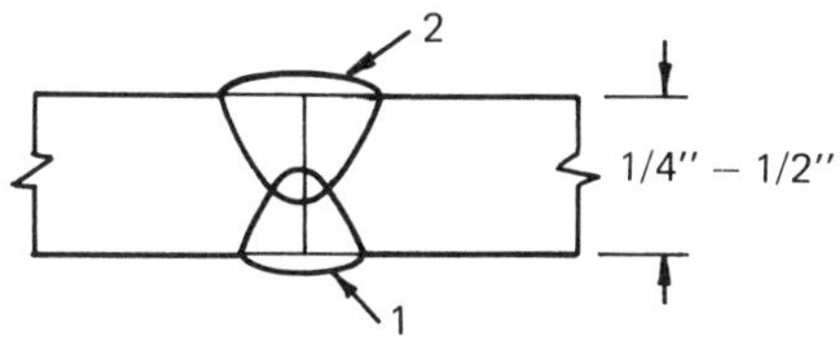

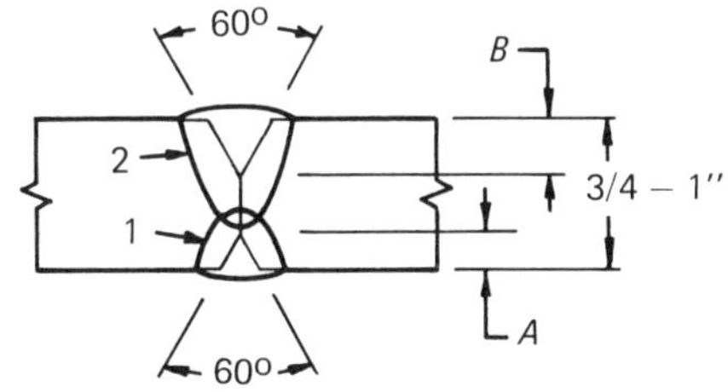

Plate Thickness (in.)	1/4		1/2		3/4		1	
Pass	1*	2	1	2	1	2	1	2
Electrode Size	5/32	5/32	3/16	3/16	3/16	3/16	3/16	3/16
Current (amp) DC(+)	475	575	700	950	700	950	850	1000
Volts	29	32	35	36	35	36	35	36
Arc Speed (in./min)	48	48	27	27	30	16	13.5	17
Electrode Req'd (lb/ft)	0.11		0.34		0.46		0.74	
Flux Req'd (lb/ft)	0.11 – 0.16		0.22 – 0.28		0.36 – 0.48		0.64 – 0.86	
Total Time (hr/ft of weld)	0.00833		0.0148		0.0192		0.0266	
Depth, A (in.)					1/8		3/8	
Depth, B (in.)					3/8		3/8	

* Must be down flat on a platen.

See introductory notes.

SUBMERGED-ARC (FULL AUTOMATIC) SINGLE ELECTRODE

Weld Position: Flat
Weld Quality Level: Code
Steel Weldability: Good
Welded from: Two sides

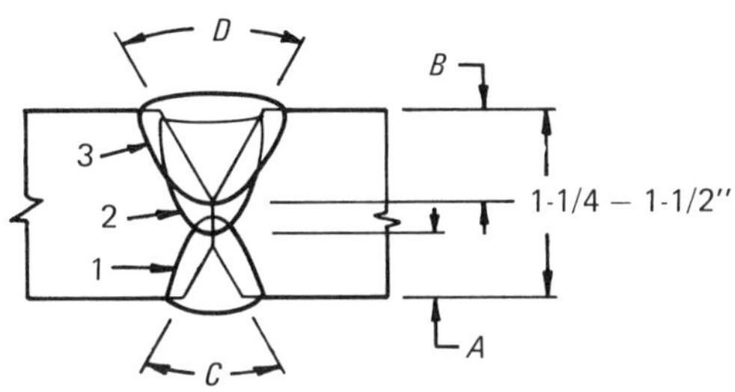

Plate Thickness (in.)	1-1/4			1-1/2		
Pass	1	2	3	1	2	3
Electrode Size	3/16	3/16	3/16	3/16	3/16	3/16
Current (amp) DC(+)	850	1000	850	1000	1000	950
Volts	35	36	35	36	36	34
Arc Speed (in./min)	13.5	12	9	9	10	7
Electrode Req'd (lb/ft)	1.44			2.26		
Flux Req'd (lb/ft)	0.95 – 1.25			1.45 – 2.00		
Total Time (hr/ft of weld)	0.0537			0.0708		
Depth, A (in.)	3/8			1/2		
Depth, B (in.)	5/8			5/8		
Angle, C (deg)	60			70		
Angle, D (deg)	70			90		

See introductory notes.

SUBMERGED-ARC (FULL AUTOMATIC) SINGLE ELECTRODE

Welding Position: Flat
Weld Quality Level: Code
Steel Weldability: Good
Welded from: Two sides*

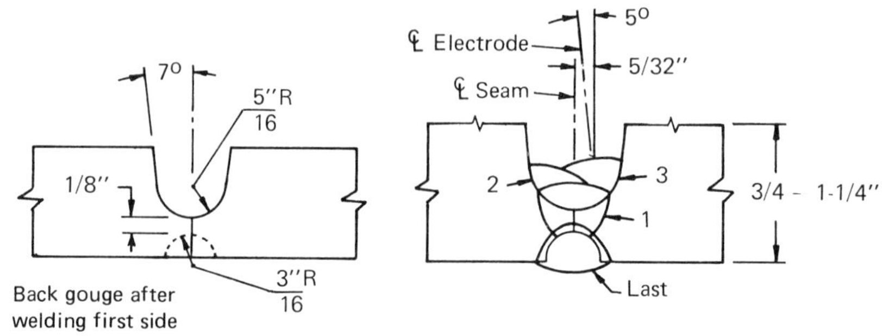

Plate Thickness (in.)	3/4			1			1-1/4		
Pass	1	2 – 5	6	1	2 – 8	9	1	2 – 11	12
Electrode Size	3/16			3/16			3/16		
Current (amp) DC(+)	600	800	900	600	800	900	600	800	900
Volts	27	30	36	27	30	36	28	31	36
Arc Speed (in./min)	20	21	22	20	21.5	22	20	21.5	22
Electrode Req'd (lb/ft)	1.17			1.75			2.40		
Flux Req'd (lb/ft)	0.85 – 1.20			1.25 – 1.80			1.75 – 2.45		
Total Time (hr/ft of weld)	0.0572			0.0842			0.112		

* Back bead can be stick, semiautomatic, or automatic.

See intordoctory notes.

Intended primarily for large girth seams.

SUBMERGED-ARC (FULL AUTOMATIC) SINGLE ELECTRODE

Welding Position: Flat
Weld Quality Level: Code
Steel Weldability: Good
Welded from: Two sides*

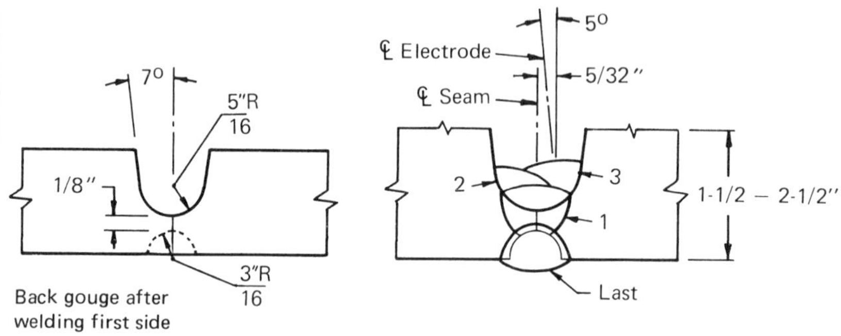

Plate Thickness (in.)	1-1/2			2			2-1/2		
Pass	1	2 – 13	14	1	2 – 21	22	1	2 – 27	28
Electrode Size	3/16	3/16	3/16	3/16	3/16	3/16	3/16	3/16	3/16
Current (amp) DC(+)	600	800	900	600	800	900	600	800	900
Volts	28	31	36	28	31	36	29	32	36
Arc Speed (in./min)	20	20	22	20	23	22	20	23	22
Electrode Req'd (lb/ft)	3.08			4.64			6.32		
Flux Req'd (lb/ft)	2.25 – 3.15			3.40 – 4.75			4.60 – 6.50		
Total Time (hr/ft of weld)	0.139			0.193			0.245		

* Back bead can be stick, semiautomatic, or automatic.

See introductory notes.

Intended primarily for large girth seams

SUBMERGED-ARC (FULL AUTOMATIC) SINGLE ELECTRODE

Welding Position: Flat
Weld Quality Level: Code
Steel Weldability: Good, fair, or poor
Welded from: Two sides

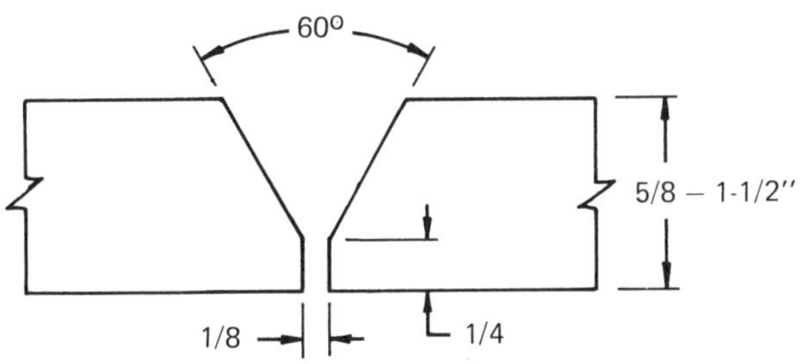

Plate Thickness (in.)	5/8	1	1-1/2
Pass *	3 – 7	3 – 14	3 – 28
Electrode Size (in.)	5/32	5/32	5/32
Current (amp) DC(+)	550	550	550
Volts	28	28	28
Arc Speed (in./min)	16	16	16
Electrode Req'd (lb/ft)†	0.80	1.90	4.13
Flux Req'd (lb/ft)	0.71 – 0.90	1.45 – 2.05	2.70 – 4.15
Total Time (hr/ft of weld)†	0.0625	0.150	0.325

* Passes 1 and 2 are made with stick electrode, preferably with 5/32 in. E7018 electrode (160 amp AC on pass 1, 225 amp AC on pass 2). The back should then be gouged to sound metal and finished as given in the procedure.

† Stick electrode, stick welding time, and gouging time are not included in these figures.

These procedures are conservative and are designed for code quality welding of steels which have tensile strengths of 70,000 psi and limited weldability such as ASTM A515-70.

SUBMERGED-ARC (FULL AUTOMATIC) SINGLE ELECTRODE

Welding Position: Flat
Weld Quality Level: Code
Steel Weldability: Good
Welded from: Two sides

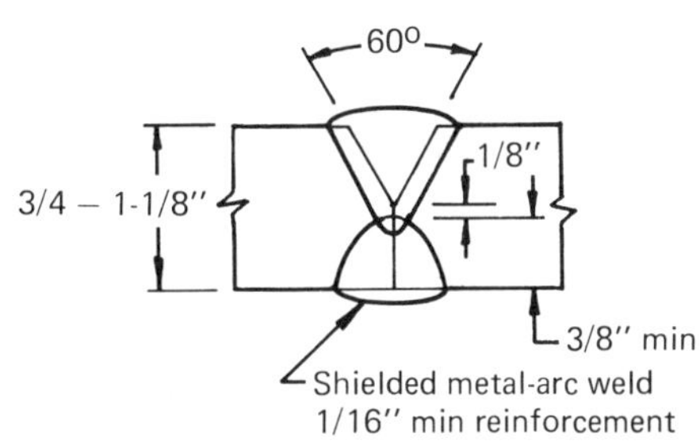

Plate Thickness (in.)	3/4	7/8	1	1-1/8
Pass	1	1	1	1
Electrode Size	7/32	7/32	7/32	7/32
Current (amp) DC(+)	900	1000	1050	1100
Volts	35	35	37	37
Arc Speed (in./min)	18	12	9	7.5
Electrode Req'd (lb/ft)*	0.29	0.51	0.73	0.95
Flux Req'd (lb/ft)	0.22 – 0.31	0.43 – 0.60	0.59 – 0.83	0.82 – 1.2
Total Time (hr/ft of weld)*	0.0111	0.0167	0.0222	0.0267

* Does not include stick electrode weld.

See introductory notes.

Make stick electrode weld first. For plate preparation and electrode selection, see procedures for shielded metal-arc welding.

SUBMERGED-ARC (FULL AUTOMATIC) SINGLE ELECTRODE

Welding Position: Flat
Weld Quality Level: Code
Steel Weldability: Good or fair

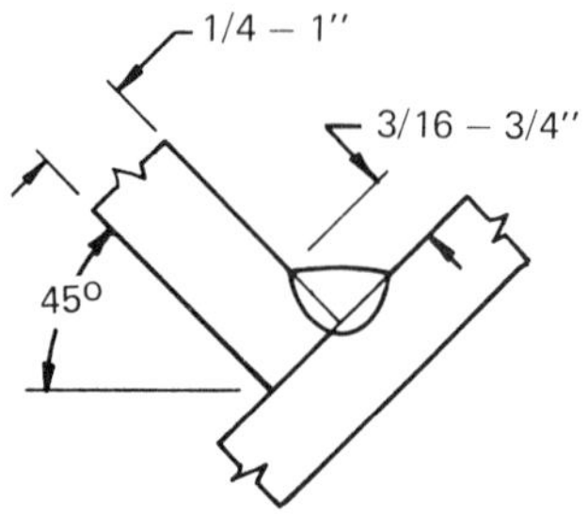

Weld Size (in.)	3/16	1/4	5/16	3/8	1/2	5/8	3/4
Plate Thickness (in.)	1/4	5/16	3/8	1/2	5/8	3/4	1
Pass	1	1	1	1	1	1	1
Electrode Size	1/8	5/32	3/16	3/16	3/16	3/16	3/16
Current (amp) DC(–)	525	625	725	800	875	925	975
Volts	31	32	34	36	38	39	40
Arc Speed (in./min)	55	39	27	22	14	10	7.5
Electrode Req'd (lb/ft)	0.066	0.12	0.18	0.26	0.47	0.73	1.05
Flux Req'd (lb/ft)	0.03 – 0.05	0.05 – 0.08	0.09 – 0.12	0.12 – 0.18	0.22 – 0.30	0.35 – 0.50	0.55 – 0.80
Total Time (hr/ft of weld)	0.00364	0.00513	0.00741	0.00909	0.0143	0.0200	0.0267

For fillets where conventional leg size is required.

See introductory notes.

SUBMERGED-ARC (FULL AUTOMATIC) SINGLE ELECTRODE

Welding Position: Flat
Weld Quality Level: Code
Steel Weldability: Good

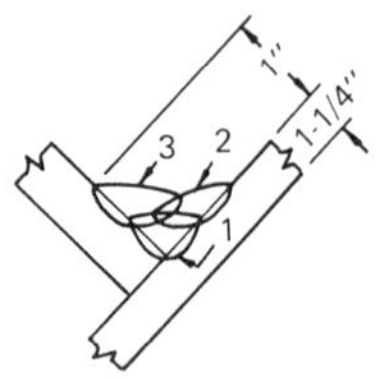

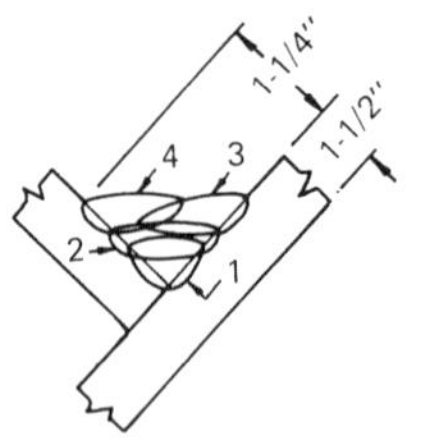

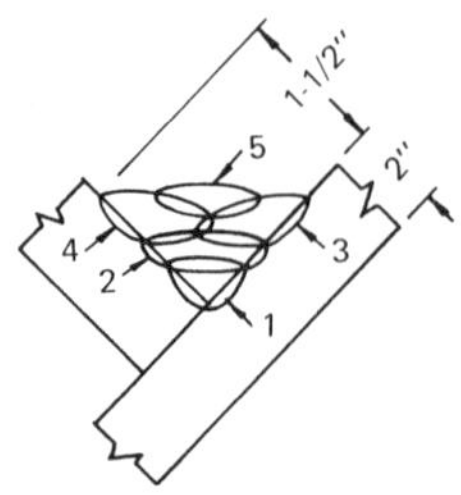

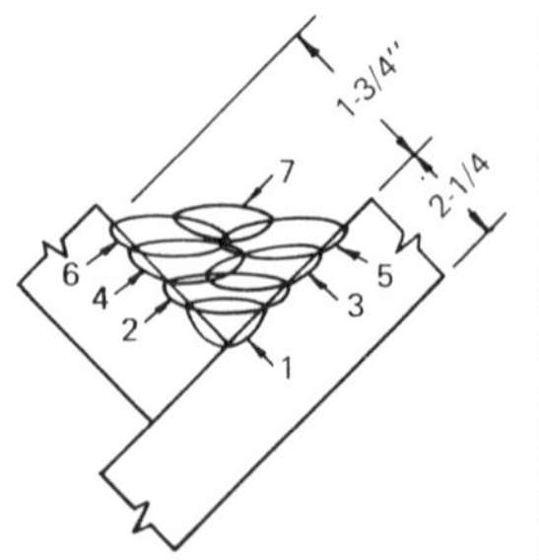

Weld Size (in.)	1		1-1/4			1-1/2		1-3/4		
Plate Thickness (in.)	1-1/4		1-1/2			2		2-1/4		
Pass	1	2 – 3	1	2	3 – 4	1 – 2	3 – 5	1 – 2	3 – 4	5 – 7
Electrode Size	3/16		3/16			3/16		3/16		
Current (amp) DC(–)	1000	900	1000	1000	900	1000	900	1000	1000	900
Volts	38	38	38	38	38	39	38	38	38	38
Arc Speed (in./min)	12	11	10	10	10	9	9	9.5	9.5	9.0
Electrode Req'd (lb/ft)	1.88		2.93			4.20		5.76		
Flux Req'd (lb/ft)	0.91 – 1.25		1.45 – 1.95			2.00 – 2.70		2.80 – 3.75		
Total Time (hr/ft of weld)	0.0531		0.0800			0.111		0.151		

See introductory notes.

SUBMERGED-ARC (FULL AUTOMATIC) SINGLE ELECTRODE

Welding Position: Flat
Weld Quality Level: Commercial
Steel Weldability: Good

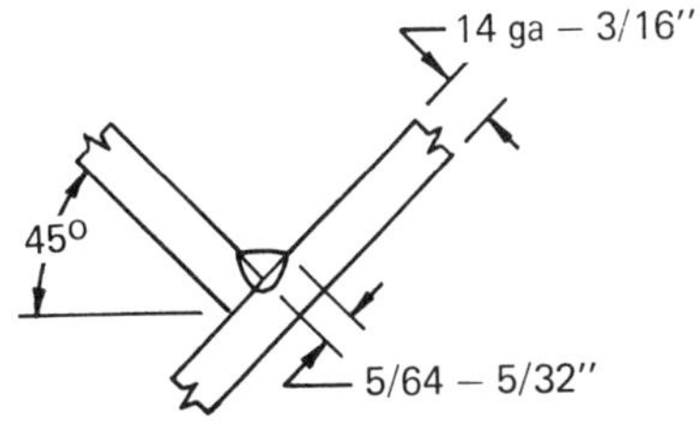

Weld Size (in.)	5/64	3/32(–)	1/8(–)	5/32(–)
Plate Thickness (in.)	0.075 (14ga)	0.105 (12 ga)	0.135 (10 ga)	3/16
Pass	1	1	1	1
Electrode Size	3/32	3/32	1/8	1/8
Current (amp) DC(+)	375	400	450	500
Volts	28	29	27	28
Arc Speed (in./min)	125	110	80	65
Electrode Req'd (lb/ft)	0.013	0.017	0.025	0.040
Flux Req'd (lb/ft)	0.0089 – 0.012	0.012 – 0.016	0.016 – 0.022	0.026 – 0.035
Total Time (hr/ft of weld)	0.00160	0.00182	0.00250	0.00308

A constant-voltage power source is recommended.

Tight copper backup bars are required behind both plates on 14 and 12 ga.

See introductory notes.

SUBMERGED-ARC (FULL AUTOMATIC) SINGLE ELECTRODE

Welding Position: Flat
Weld Quality Level: Commercial
Steel Weldability: Good

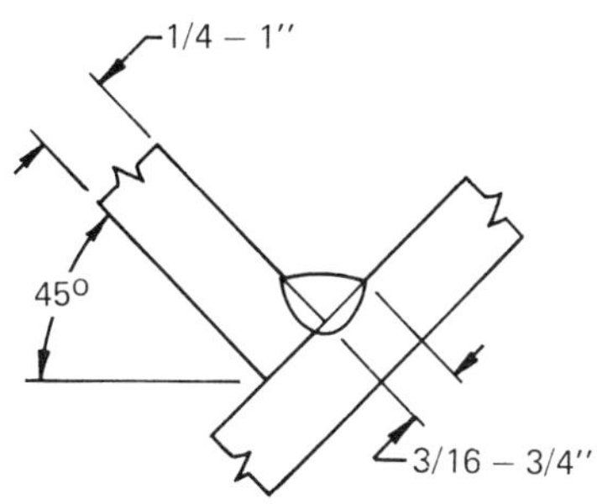

Weld Size, (in.)	3/16	1/4	5/16	3/8	1/2	5/8	3/4
Plate Thickness (in.)	1/4	5/16	3/8	1/2	5/8	3/4	1
Pass	1	1	1	1	1	1	1
Electrode Size	1/8	5/32	3/16	3/16	3/16	3/16	3/16
Current (amp) DC(–)	600	725	850	950	1000	1050	1050
Volts	33	34	36	38	39	40	41
Arc Speed (in./min)	65	48	34	28	17	12	8.5
Electrode Req'd (lb/ft)	0.066	0.12	0.18	0.26	0.47	0.73	1.05
Flux Req'd (lb/ft)	0.035 – 0.047	0.055 – 0.074	0.085 – 0.12	0.13 – 0.17	0.22 – 0.30	0.38 – 0.46	0.48 – 0.66
Total Time (hr/ft of weld)	0.00308	0.00417	0.00588	0.00714	0.0118	0.0167	0.0235

See introductory notes.

SUBMERGED-ARC (FULL AUTOMATIC) SINGLE ELECTRODE

Welding Position: Flat
Weld Quality Level: Strength only
Steel Weldability: Good

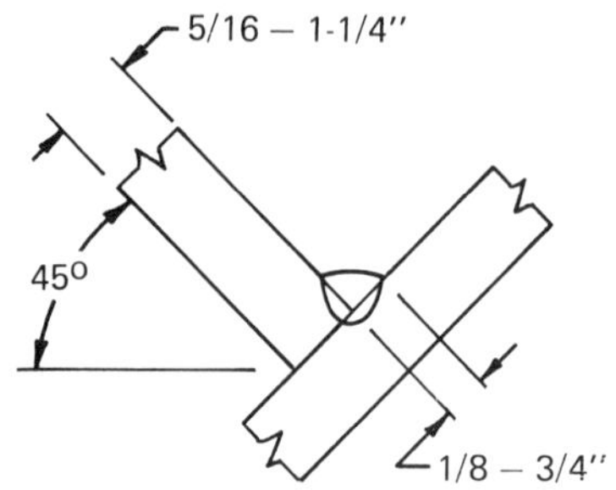

Weld Size (in.)	1/8	3/16	1/4	5/16	3/8	1/2	5/8	3/4
Plate Thickness (in.)	5/16	3/8	1/2	5/8	3/4	7/8	1	1-1/4
Pass	1	1	1	1	1	1	1	1
Electrode Size	1/8	5/32	3/16	3/16	3/16	3/16	3/16	3/16
Current (amp) DC(+)	425	575	675	775	850	950	1000	1000
Volts	26	28	31	34	35	36	37	38
Arc Speed (in./min)	60	42	30	23	18	12	8.5	6.5
Electrode Req'd (lb/ft)	0.028	0.063	0.11	0.17	0.25	0.45	0.71	1.01
Flux Req'd (lb/ft)	0.020 – 0.026	0.041 – 0.055	0.065 – 0.087	0.10 – 0.14	0.15 – 0.20	0.26 – 0.35	0.40 – 0.54	0.57 – 0.78
Total Time (hr/ft of weld)	0.00333	0.00476	0.00667	0.00870	0.0111	0.0167	0.0235	0.0308

See introductory notes.

SUBMERGED-ARC (FULL AUTOMATIC) SINGLE ELECTRODE

Welding Position: Horizontal
Weld Quality Level: Code
Steel Weldability: Good or fair

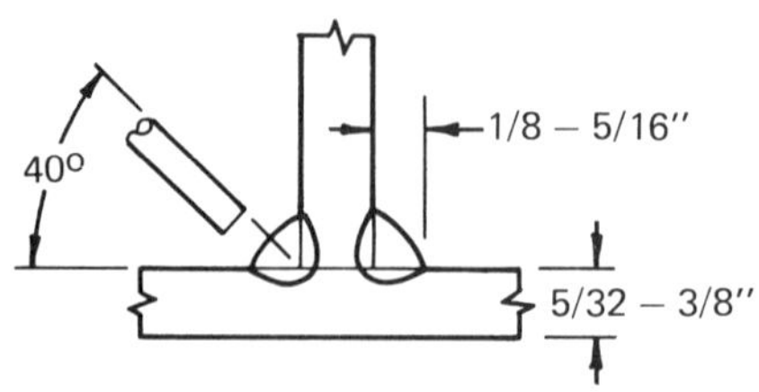

Weld Size (in.)	1/8	5/32	3/16	1/4	5/16
Plate Thickness (in.)	5/32	3/16	1/4	5/16	3/8
Pass	1	1	1	1	1
Electrode Size	1/8	1/8	1/8	1/8	5/32
Current (amp) DC	400(+)	475(+)	550(+)	450 (–)	500(–)
Volts	24	26	28	29	30
Arc Speed (in./min)	54	44	36	25	18
Electrode Req'd (lb/ft)	0.029	0.046	0.066	0.12	0.18
Flux Req'd (lb/ft)	0.024 – 0.032	0.034 – 0.048	0.053 – 0.069	0.070 – 0.095	0.11 – 0.15
Total Time (hr/ft of weld)	0.00370	0.00455	0.00555	0.00800	0.0111

See introductory notes.

SUBMERGED-ARC (FULL AUTOMATIC) SINGLE ELECTRODE

Welding Position: Horizontal
Weld Quality Level: Commercial
Steel Weldability: Good

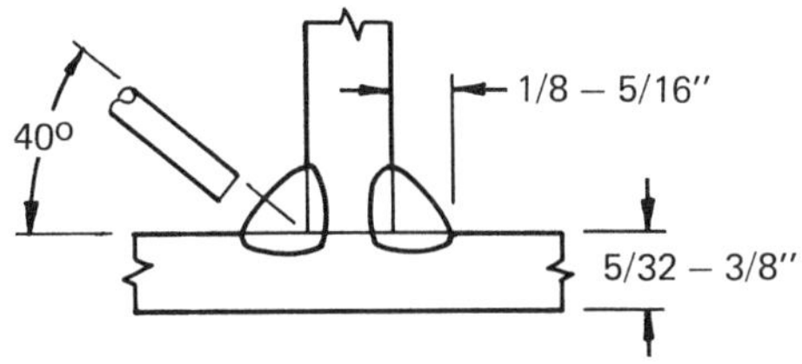

Weld Size (in.)	1/8	5/32	3/16	1/4	5/16
Plate Thickness (in.)	5/32	3/16	1/4	5/16	3/8
Pass	1	1	1	1	1
Electrode Size	1/8	1/8	5/32	5/32	5/32
Current (amp) DC	450(+)	525(+)	600(+)	525(–)	575(–)
Volts	25	27	29	31	34
Arc Speed (in./min)	62	54	42	30	22
Electrode Req'd (lb/ft)	0.029	0.046	0.066	0.12	0.18
Flux Req'd (lb/ft)	0.024 – 0.032	0.034 – 0.048	0.053 – 0.069	0.070 – 0.095	0.11 – 0.15
Total Time (hr/ft of weld)	0.00323	0.00370	0.00476	0.00667	0.00909

See introductory notes.

SUBMERGED-ARC (FULL AUTOMATIC) SINGLE ELECTRODE

Welding Position: Horizontal
Weld Quality Level: Strength only
Steel Weldability: Good

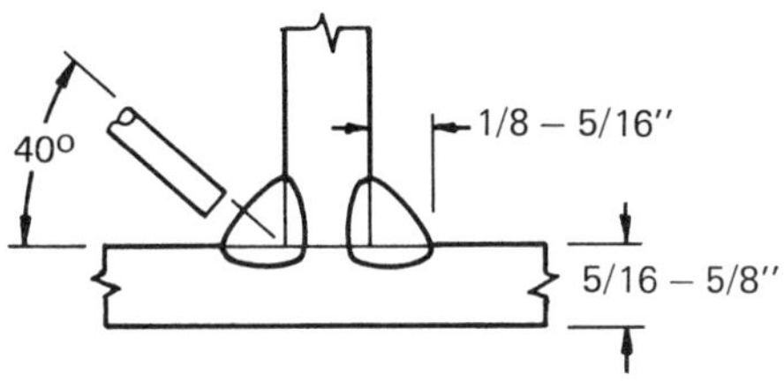

Weld Size (in.)	1/8	5/32	3/16	1/4	5/16
Plate Thickness (in.)	5/16	3/8	7/16	1/2	5/8
Pass	1	1	1	1	1
Electrode Size	1/8	1/8	1/8	5/32	5/32
Current (amp) DC(+)	425	420	450	525	575
Volts	23	25	27	28	30
Arc Speed (in./min)	50	42	35	24	16
Electrode Req'd (lb/ft)	0.028	0.044	0.063	0.11	0.17
Flux Req'd (lb/ft)	0.020 – 0.027	0.029 – 0.040	0.043 – 0.059	0.068 – 0.095	0.10 – 0.15
Total Time (hr/ft of weld)	0.00400	0.00475	0.00571	0.00833	0.0125

See introductory notes.

SUBMERGED-ARC (FULL AUTOMATIC) SINGLE ELECTRODE

Welding Position: Horizontal
Weld Quality Level: Commercial
Steel Weldability: Good

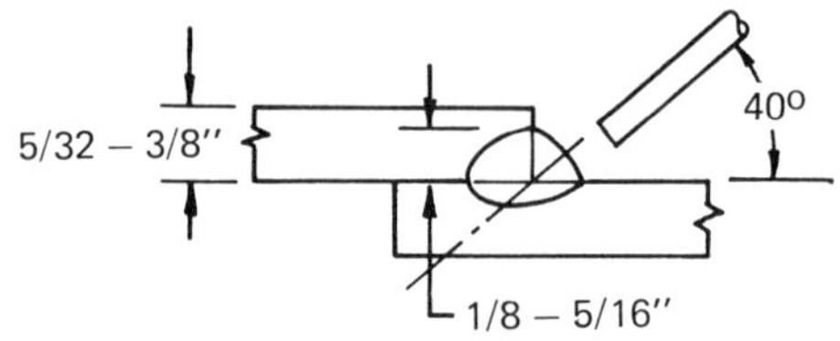

Weld Size (in.)	1/8	3/16	1/4	5/16
Plate Thickness (in.)	5/32	1/4	5/16	3/8
Pass	1	1	1	1
Electrode Size	1/8	1/8	5/32	5/32
Current (amp) DC	450(+)	450(–)	525(–)	575(–)
Volts	25	27	31	31
Arc Speed (in./min)	62	42	30	22
Electrode Req'd (lb/ft)	0.032	0.070	0.12	0.19
Flux Req'd (lb/ft)	0.024 – 0.032	0.042 – 0.056	0.070 – 0.095	0.11 – 0.15
Total Time (hr/ft of weld)	0.00323	0.00476	0.00667	0.00909

See introductory notes.

SUBMERGED-ARC (FULL AUTOMATIC) SINGLE ELECTRODE

Welding Position: Flat
Weld Quality Level: Commercial
Steel Weldability: Good

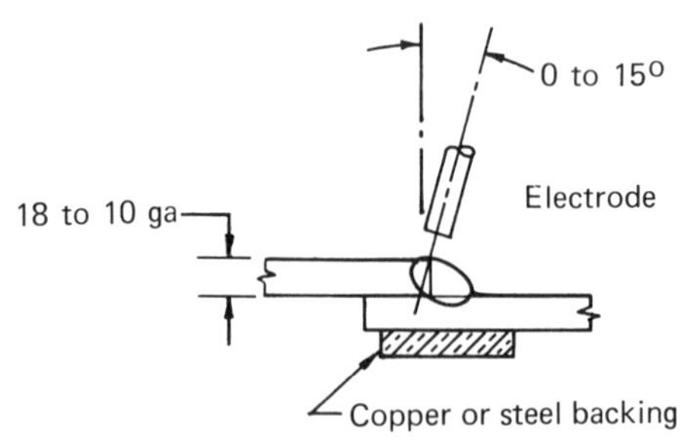

Plate Thickness (in.)	0.048 (18 ga)	0.060 (16 ga)	0.075 (14 ga)	0.105 (12 ga)	0.135 (10 ga)
Pass	1	1	1	1	1
Electrode Size	1/8	1/8	1/8	1/8	1/8
Current (amp) DC(+)	380	425	475	525	575
Volts	23	26	28	29	30
Arc Speed (in./min)	120	120	120	90	70
Electrode Req'd (lb/ft)	0.012	0.014	0.017	0.027	0.040
Flux Req'd (lb/ft)	0.010 – 0.012	0.011 – 0.014	0.013 – 0.017	0.020 – 0.027	0.029 – 0.040
Total Time (hr/ft of weld)	0.00167	0.00167	0.00167	0.00222	0.00286

See introductory notes.

SUBMERGED-ARC (FULL AUTOMATIC) SINGLE ELECTRODE

Welding Position: Flat
Weld Quality Level: Commercial
Steel Weldability: Good

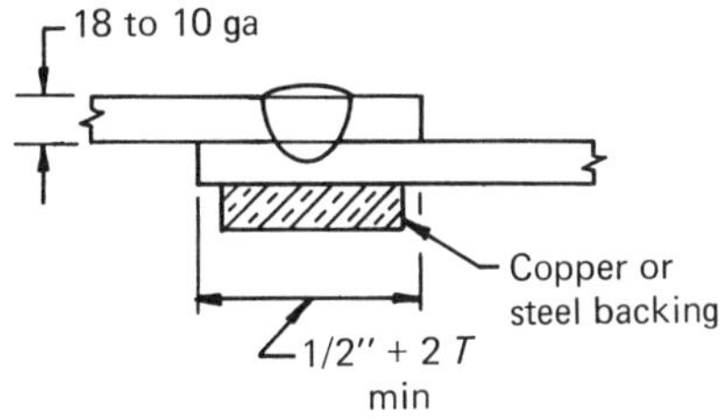

Plate Thickness (in.)	0.048 (18 ga)	0.060 (16 ga)	0.075 (14 ga)	0.105 (12 ga)	0.135 (10 ga)
Pass	1	1	1	1	1
Electrode Size	1/8	1/8	1/8	5/32	5/32
Current (amp) DC(+)	450	500	550	650	750
Volts	23	26	28	30	32
Arc Speed (in./min)	120	120	120	100	80
Electrode Req'd (lb/ft)	0.016	0.019	0.022	0.034	0.051
Flux Req'd (lb/ft)	0.015 – 0.020	0.018 – 0.024	0.021 – 0.028	0.030 – 0.041	0.046 – 0.061
Total Time (hr/ft of weld)	0.00167	0.00167	0.00167	0.00200	0.00250

See introductory notes.

SUBMERGED-ARC (FULL AUTOMATIC) SINGLE ELECTRODE

Welding Position: Flat
Weld Quality Level: Commercial
Steel Weldability: Good

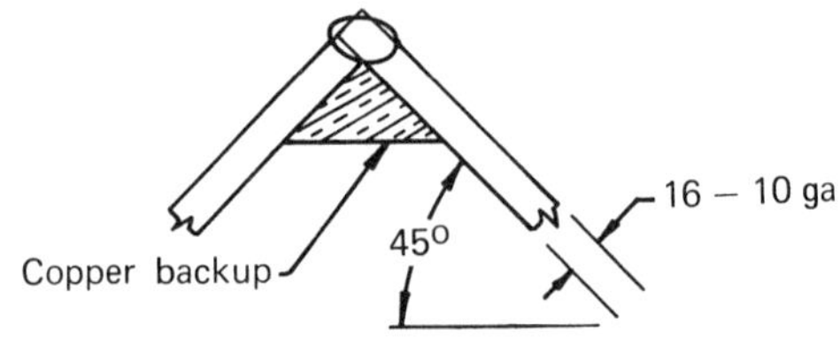

With backing – 80 to 100% penetration

Plate Thickness (in.)	0.060 (16 ga)	0.075 (14 ga)	0.105 (12 ga)	0.135 (10 ga)
Pass	1	1	1	1
Electrode Size	3/32	1/8	1/8	1/8
Current (amp) DC(+)	375	400	500	625
Volts	20	24	30	31
Arc Speed (in./min)	170	130	110	90
Electrode Req'd (lb/ft)	0.0097	0.013	0.020	0.033
Flux Req'd (lb/ft)	0.011 – 0.015	0.013 – 0.018	0.019 – 0.026	0.030 – 0.041
Total Time (hr/ft of weld)	0.00118	0.00154	0.00182	0.00222

A flux retainer is necessary to avoid spillage.

See introductory notes.

SUBMERGED-ARC (FULL AUTOMATIC) SINGLE ELECTRODE

Welding Position: Flat
Weld Quality Level: Commercial
Steel Weldability: Good

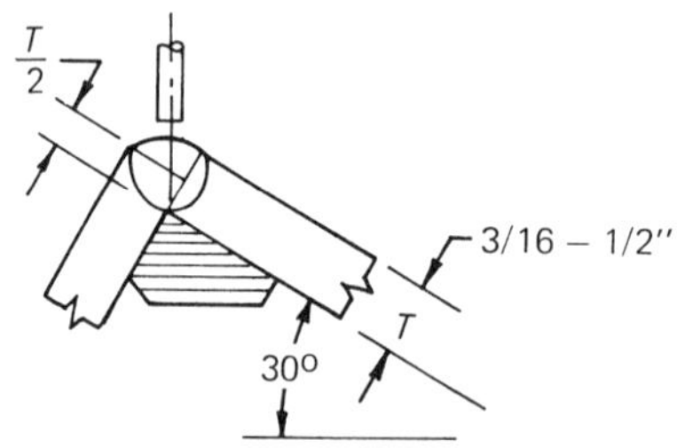

With backing – 80 to 100% penetration

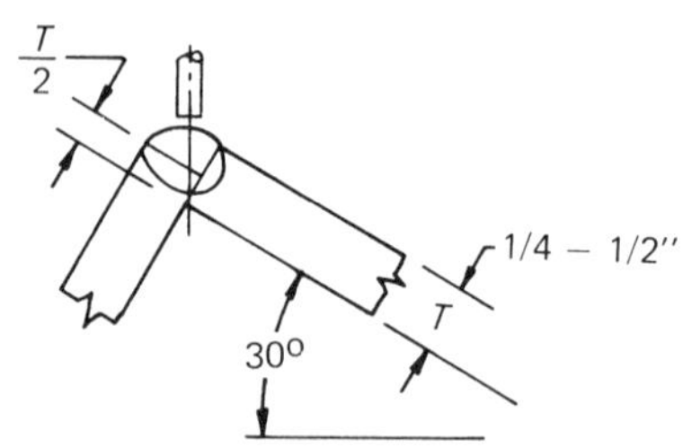

Without backing – 50 to 70% penetration

Plate Thickness (in.)	3/16	1/4	3/8	1/2	1/4	3/8	1/2
Pass	1	1	1	1	1	1	1
Electrode Size	5/32	3/16	3/16	3/16	5/32	3/16	3/16
Current (amp) DC(+)	625	700	750	850	500	650	750
Volts	31	32	35	36	31	33	35
Arc Speed (in./min)	70	56	42	32	60	48	36
Electrode Req'd (lb/ft)	0.046	0.062	0.089	0.13	0.035	0.061	0.10
Flux Req'd (lb/ft)	0.038 – 0.051	0.051 – 0.069	0.074 – 0.10	0.10 – 0.14	0.029 – 0.039	0.047 – 0.065	0.080 – 0.11
Total Time (hr/ft of weld)	0.00286	0.00357	0.00477	0.00625	0.00333	0.00417	0.00555

A flux retainer is necessary to avoid spillage.

See introductory notes.

SUBMERGED-ARC (FULL AUTOMATIC) SINGLE ELECTRODE

Welding Position: Flat
Weld Quality Level: Commercial
Steel Weldability: Good

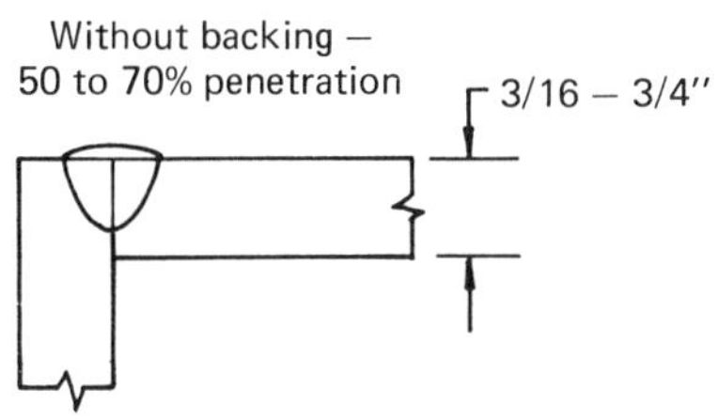

Plate Thickness (in.)	3/16	1/4	3/8	1/2	5/8	3/4
Pass	1	1	1	1	1	1
Electrode Size	5/32	5/32	3/16	3/16	3/16	3/16
Current (amp) DC(+)	450	550	650	750	850	900
Volts	28	30	31	35	35	37
Arc Speed (in./min)	72	60	48	38	32	22
Electrode Req'd (lb/ft)	0.025	0.041	0.061	0.095	0.14	0.22
Flux Req'd (lb/ft)	0.023 – 0.031	0.038 – 0.051	0.053 – 0.072	0.083 – 0.11	0.12 – 0.16	0.18 – 0.25
Total Time (hr/ft of weld)	0.00278	0.00333	0.00417	0.00526	0.00625	0.00908

A flux retainer is necessary to avoid spillage.

See introductory notes.

SUBMERGED-ARC (FULL AUTOMATIC) SINGLE ELECTRODE

Welding Position: Flat
Weld Quality Level: Commercial
Steel Weldability: Good

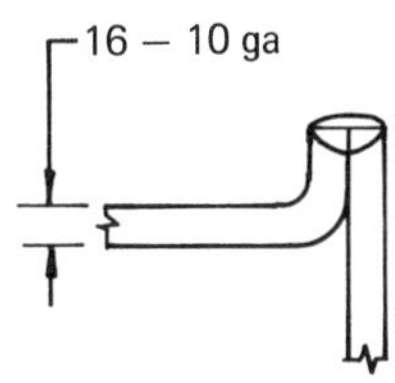

Plate Thickness (in.)	0.060 (16 ga)	0.075 (14 ga)	0.105 (12 ga)	0.135 (10 ga)
Pass	1	1	1	1
Electrode Size	3/32	3/32	1/8	5/32
Current (amp) DC(+)	350	375	450	600
Volts	20	23	24	30
Arc Speed (in./min)	170	130	110	90
Electrode Req'd (lb/ft)	0.0089	0.012	0.018	0.031
Flux Req'd (lb/ft)	0.0090 – 0.012	0.011 – 0.014	0.017 – 0.023	0.028 – 0.038
Total Time (hr/ft of weld)	0.00118	0.00154	0.00182	0.00222

A flux retainer is necessary to avoid spillage.

See introductory notes.

SUBMERGED-ARC (FULL AUTOMATIC) TWIN ELECTRODES

Welding Position: Flat
Weld Quality Level: Commercial
Steel Weldability: Good

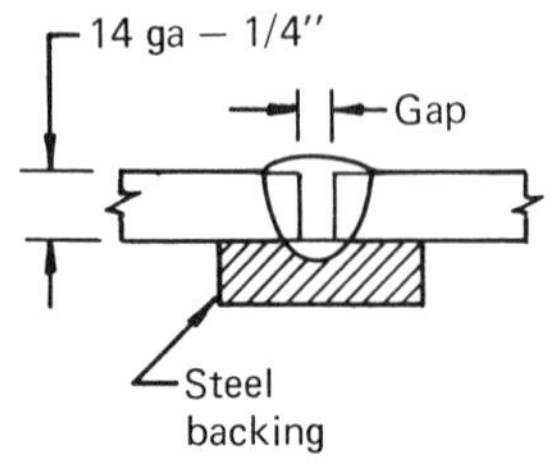

Plate Thickness (in.)	0.075 (14 ga)	0.105 (12 ga)	0.135 (10 ga)	3/16	1/4
Pass	1	1	1	1	1
Electrode Size	1/16 (two)	1/16 (two)	5/64 (two)	5/64 (two)	5/64 (two)
Current (amp) DC	900 (+)	950 (+)	950 (–)	975 (–)	1000 (–)
Volts	27	27	27	30	34
Arc Speed (in./min)	200	170	135	95	60
Electrode Req'd (lb/ft)	0.033	0.043	0.061	0.096	0.152
Flux Req'd (lb/ft)	0.04 – 0.05	0.04 – 0.06	0.05 – 0.07	0.07 – 0.10	0.12 – 0.15
Total Time (hr/ft of weld)	0.00100	0.00118	0.00148	0.00211	0.00333
Backing, minimum size (in.)	12 ga x 3/8	12 ga x 1/2	10 ga x 5/8	3/16 x 3/4	1/4 x 1
Gap (in.)	1/16	1/16	3/32	1/8	5/32
Electrode spacing, 5/16 in.					

Constant voltage power source is recommended.

See introductory notes.

SUBMERGED-ARC (FULL AUTOMATIC) TWIN ELECTRODES

Welding Position: Flat
Weld Quality Level: Commercial
Steel Weldability: Good

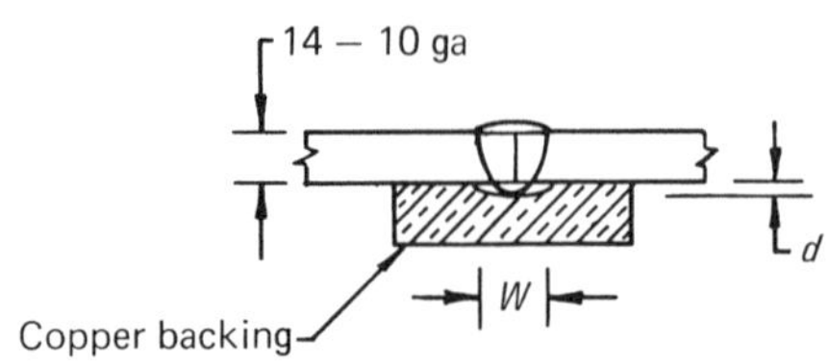

Plate Thickness (in.)	0.075 (14 ga)	0.105 (12 ga)	0.135 (10 ga)
Pass	1	1	1
Electrode Size	0.045 (two)	0.045 (two)	1/16 (two)
Current (amp) DC(+)	550	600	925
Volts	24	27	28
Arc Speed (in./min)	200	170	160
Electrode Req'd (lb/ft)	0.018	0.023	0.043
Flux Req'd (lb/ft)	0.02 – 0.03	0.03 – 0.04	0.04 – 0.05
Total Time (hr/ft of weld)	0.00100	0.00118	0.00125
Electrode Spacing (in.)	0.275	0.275	5/16
Groove in backing bar Width, W (in.)	1/4	3/8	3/8
Depth, d (in.)	1/32	1/32	1/32

Constant potential power source is recommended.

See introductory notes.

SUBMERGED-ARC (FULL AUTOMATIC) TWIN ELECTRODES

Welding Position: Flat
Weld Quality Level: Code
Steel Weldability: Good

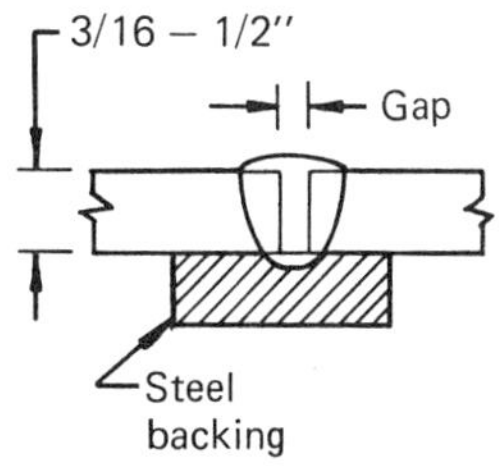

Plate Thickness (in.)	3/16	1/4	3/8	1/2
Pass	1	1	1	1
Electrode Size	3/32 (two)	3/32 (two)	3/32 (two)	3/32 (two)
Current (amp) DC(+)	900	950	1250	1400
Volts	39	39	40	41
Arc Speed (in./min)	48	39	36	24
Electrode Req'd (lb/ft)	0.15	0.19	0.31	0.58
Flux Req'd (lb/ft)	0.10 – 0.14	0.13 – 0.18	0.22 – 0.30	0.40 – 0.56
Total Time (hr/ft of weld)	0.00417	0.00513	0.00555	0.00833
Backing, minimum size (in.)	3/16 x 3/4	1/4 x 1	1/4 x 1	3/8 x 1
Gap (in.)	1/16 – 1/8	1/8 – 3/16	5/32 – 7/32	3/16 – 1/4
Electrode spacing, 3/8 in.				

Only for steels with less than 60,000 psi tensile strength.

See introductory notes.

SUBMERGED-ARC (FULL AUTOMATIC) TWIN ELECTRODES

Welding Position: Flat
Weld Quality Level: Code
Steel Weldability: Fair

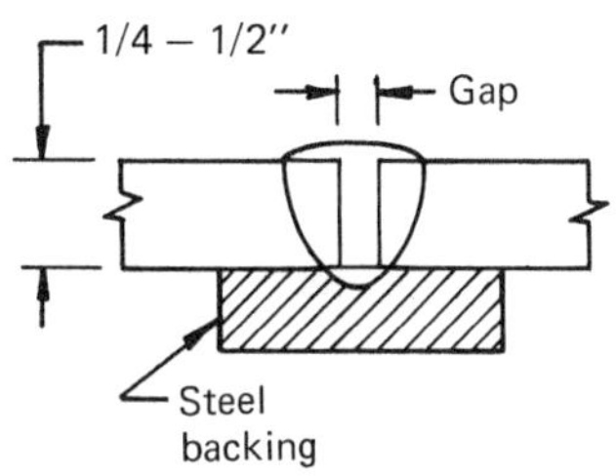

Plate Thickness (in.)	1/4	3/8	1/2
Pass	1	1	1
Electrode Size	3/32 (two)	3/32 (two)	3/32 (two)
Current (amp) DC(+)	670	880	980
Volts	37	39	40
Arc Speed (in./min)	28	27	17
Electrode Req'd (lb/ft)	0.16	0.24	0.46
Flux Req'd (lb/ft)	0.11 – 0.15	0.18 – 0.23	0.34 – 0.44
Total Time (hr/ft of weld)	0.00714	0.00741	0.0118
Backing, minimum size (in.)	1/4 x 1	1/4 x 1	3/8 x 1
Gap (in.)	1/8 – 3/16	5/32 – 7/32	3/16 – 1/4
Electrode spacing, 3/8 in.			

For steels with tensile strength of 60,000 psi or greater.

See introductory notes.

SUBMERGED-ARC (FULL AUTOMATIC) TWIN ELECTRODES

Welding Position: Flat
Weld Quality Level: Code
Steel Weldability: Good
Welded from: Two sides

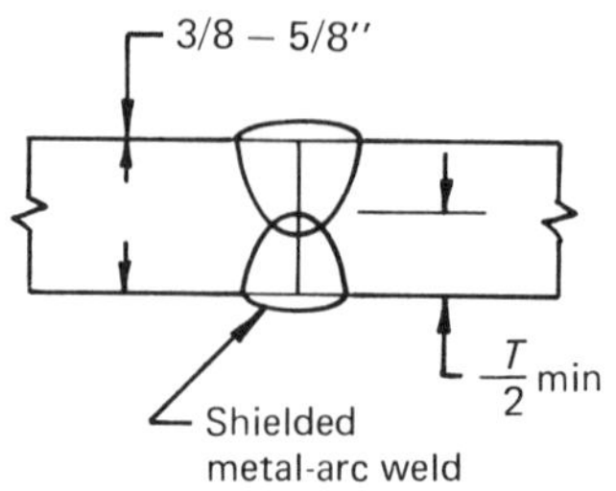

Plate Thickness (in.)	3/8	1/2	5/8
Pass	1	1	1
Electrode Size	3/32 (two)	3/32 (two)	3/32 (two)
Current (amp) DC(+)	1000	1200	1300
Volts	36	38	38
Arc Speed (in./min)	42	40	34
Electrode Req'd (lb/ft)*	0.19	0.26	0.35
Flux Req'd (lb/ft)	0.13 – 0.19	0.16 – 0.25	0.23 – 0.32
Total Time (hr/ft of weld)*	0.00476	0.00500	0.00588
Electrode spacing, 3/8 in.			

Make stick electrode weld first with 1/16 in. minimum buildup.

For plate preparation, see shielded metal-arc procedures.

* Does not include shielded metal-arc weld.

See introductory notes.

SUBMERGED-ARC (FULL AUTOMATIC) TWIN ELECTRODES

Welding Position: Flat
Weld Quality Level: Code
Steel Weldability: Good
Welded from: Two sides

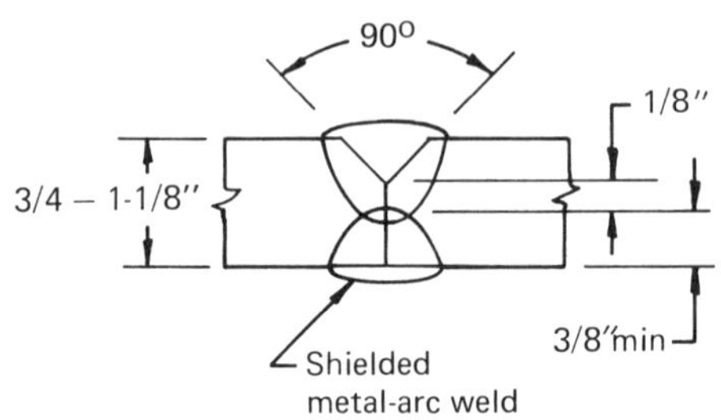

Plate Thickness (in.)	3/4	7/8	1	1-1/8
Pass	1	1	1	1
Electrode Size	3/32 (two)	3/32 (two)	3/32 (two)	3/32 (two)
Current (amp) DC(+)	1200	1300	1500	1500
Volts	37	38	38	39
Arc Speed (in./min)	27	22	16	11
Electrode Req'd (lb/ft)*	0.38	0.55	1.0	1.5
Flux Req'd (lb/ft)	0.24 – 0.34	0.35 – 0.50	0.65 – 0.90	0.95 – 1.35
Total Time (hr/ft of weld)*	0.00741	0.00909	0.0125	0.0182
Electrode spacing (in.)	3/8	3/8	5/8	5/8

Make stick electrode weld first with 1/16 in. minimum buildup.

For plate preparation see shielded metal-arc procedures.

* Does not include manual weld.

See introductory notes.

SUBMERGED-ARC (FULL AUTOMATIC) TWIN ELECTRODES

Welding Position: Flat
Weld Quality Level: Code
Steel Weldability: Good
Welded from: Two sides

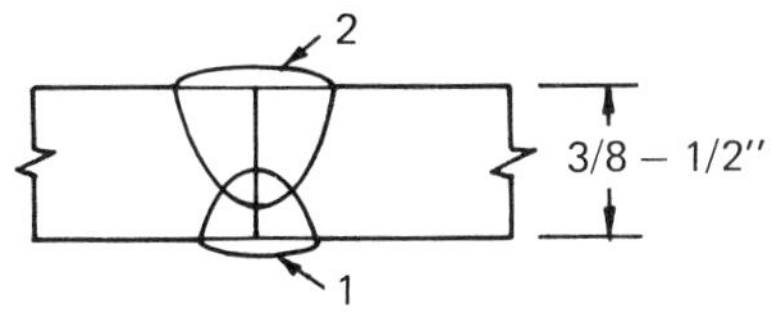

Plate Thickness (in.)	3/8″		1/2″	
Pass	1	2	1	2
Electrode Size	3/32 (two)	3/32 (two)	3/32 (two)	3/32 (two)
Current (amp) DC(+)	750	1000	1000	1200
Volts	34	36	36	38
Arc Speed (in./min)	42	42	40	40
Electrode Req'd (lb/ft)	0.32		0.46	
Flux Req'd (lb/ft)	0.25 – 0.35		0.32 – 0.45	
Total Time (hr/ft of weld)	0.00953		0.0100	
Electrode spacing, 3/8 in.				

See introductory notes.

SUBMERGED-ARC (FULL AUTOMATIC) TWIN ELECTRODES

Welding Position: Flat
Weld Quality Level: Code
Steel Weldability: Good
Welded from: Two sides

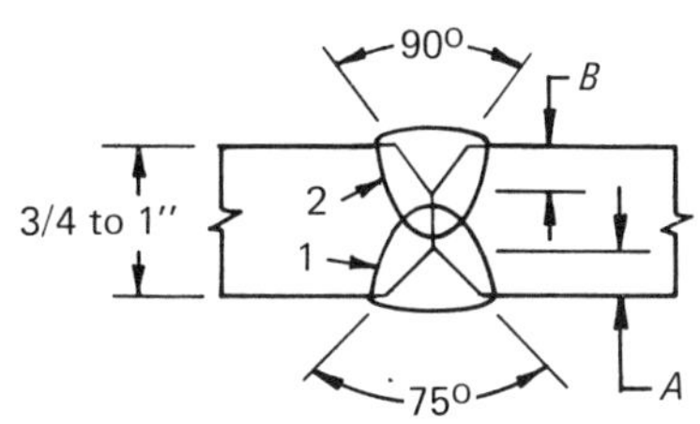

Plate Thickness (in.)	3/4		7/8		1	
Pass	1	2	1	2	1	2
Electrode Size	3/32 (two)	3/32 (two)	3/32 (two)	3/32 (two)	3/32 (two)	3/32 (two)
Current (amp) DC(+)	1000	1300	1000	1400	1200	1400
Volts	39	39	39	38	38	38
Arc Speed (in./min)	32	22	32	18	23	18
Electrode Req'd (lb/ft)	0.80		1.02		1.22	
Flux Req'd (lb/ft)	0.48 – 0.67		0.61 – 0.81		0.85 – 1.1	
Total Time (hr/ft of weld)	0.0154		0.0174		0.0198	
Depth, A (in.)	3/16		1/4		3/8	
Depth, B (in.)	3/8		7/16		7/16	
Electrode spacing, (in.)	3/8		3/8		3/8	

SUBMERGED-ARC (FULL AUTOMATIC) TWIN ELECTRODES

Welding Position: Flat
Weld Quality Level: Code
Steel Weldability: Good
Welded from: Two sides

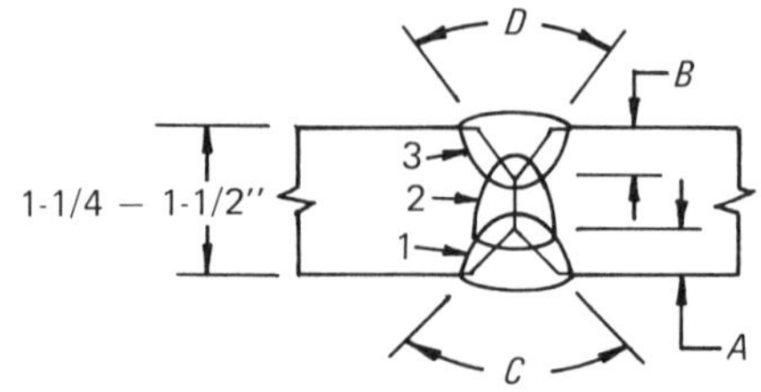

Plate Thickness (in.)	1-1/4			1-3/8			1-1/2		
Pass	1	2	3	1	2	3	1	2	3
Electrode Size	1/8 (two)	1/8 (two)	1/8 (two)	1/8 (two)	1/8 (two)	1/8 (two)	1/8 (two)	1/8 (two)	1/8 (two)
Current (amp) DC(+)	1400	1500	1300	1500	1500	1300	1500	1500	1500
Volts	34	33	35	34	33	35	34	33	35
Arc Speed (in./min)	19	16	17	14	16	16	14	14	13
Electrode Req'd (lb/ft)	1.89			2.24			2.68		
Flux Req'd (lb/ft)	1.0 – 1.6			1.2 – 1.9			1.4 – 2.3		
Total Time (hr/ft of weld)	0.0348			0.0393			0.0440		
Depth, A (in.)	7/16			1/2			1/2		
Depth, B (in.)	9/16			5/8			3/4		
Angle, C (deg)	75			80			80		
Angle, D (deg)	90			80			80		
Electrode spacing, 1/2 in.									

See introductory notes.

SUBMERGED-ARC (FULL AUTOMATIC) TWIN ELECTRODES

Welding Position: Flat
Weld Quality Level: Code
Steel Weldability: Good

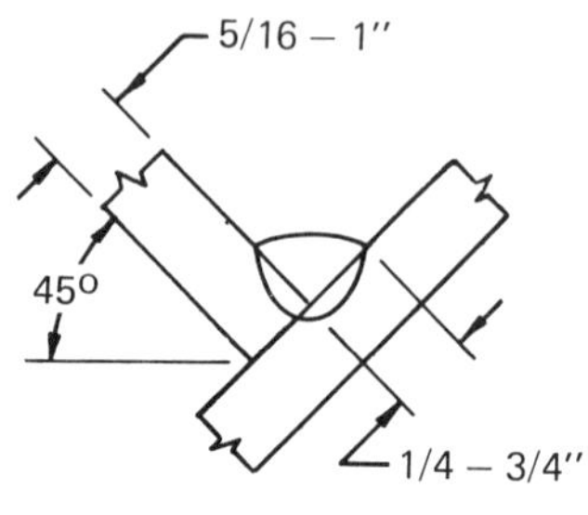

Weld Size, (in.)	1/4	5/16	3/8	1/2	5/8	3/4
Plate Thickness (in.)	5/16	3/8	7/16	5/8	3/4	1
Pass	1	1	1	1	1	1
Electrode Size	3/32 (two)	3/32 (two)	3/32 (two)	3/32 (two)	3/32 (two)	3/32 (two)
Current (amp) DC(+)	900	1050	1200	1350	1400	1500
Volts	35	35	37	39	40	40
Arc Speed (in./min)	55	40	39	25	18	13
Electrode Req'd (lb/ft)	0.11	0.18	0.24	0.47	0.69	1.1
Flux Req'd (lb/ft)	0.07 – 0.10	0.13 – 0.18	0.14 – 0.20	0.27 – 0.38	0.40 – 0.52	0.7 – 1.0
Total Time (hr/ft of weld)	0.00364	0.00500	0.00513	0.00800	0.0111	0.0154
Electrode spacing (in.)	3/8	3/8	3/8	3/8	3/8	5/8

Position electrodes parallel to direction of travel for all except 3/4 in. weld size. For 3/4 in. weld, position electrodes 90° to direction of travel.

See introductory notes.

SUBMERGED-ARC (FULL AUTOMATIC) TWIN ELECTRODES

Welding Position: Flat
Weld Quality Level: Commercial
Steel Weldability: Good

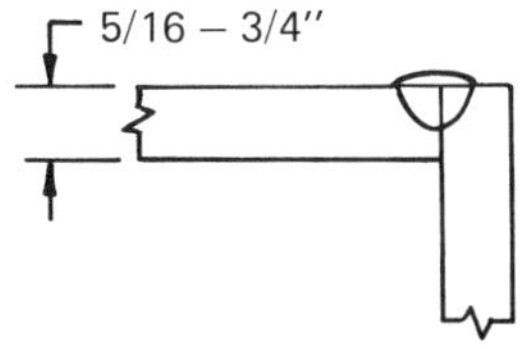

Plate Thickness (in.)	5/16	3/8	1/2	5/8	3/4
Pass	1	1	1	1	1
Electrode Size	3/32 (two)	3/32 (two)	3/32 (two)	3/32 (two)	1/8 (two)
Current (amp) DC(+)	700	900	1100	1200	1300
Volts	34	34	36	38	35
Arc Speed (in./min)	45	42	42	35	33
Electrode Req'd (lb/ft)	0.11	0.17	0.22	0.30	0.35
Flux Req'd (lb/ft)	0.07 – 0.11	0.11 – 0.17	0.15 – 0.22	0.20 – 0.30	0.25 – 0.35
Total Time (hr/ft of weld)	0.00444	0.00476	0.00476	0.00571	0.00606
Electrode spacing, 3/8 in.					

50 to 70% penetration can be achieved without backing.

See introductory notes.

SUBMERGED-ARC (FULL AUTOMATIC) TWIN ELECTRODES

Welding Position: Flat
Weld Quality Level: Commercial
Steel Weldability: Good

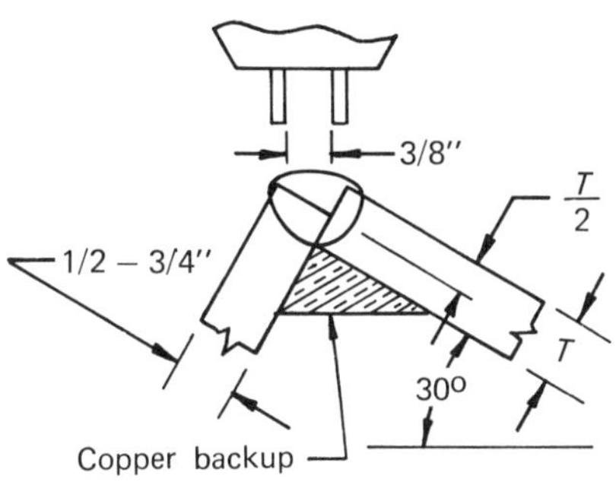

Plate Thickness (in.)	1/2	5/8	3/4
Pass	1	1	1
Electrode Size	3/32 (two)	3/32 (two)	3/32 (two)
Current (amp) DC(–)	1300	1400	1400
Volts	38	39	40
Arc Speed (in./min)	48	40	30
Electrode Req'd (lb/ft)	0.25	0.34	0.46
Flux Req'd (lb/ft)	0.17 – 0.25	0.23 – 0.36	0.31 – 0.46
Total Time (hr/ft of weld)	0.00417	0.00500	0.00667

See introductory notes.

SUBMERGED-ARC (FULL AUTOMATIC) TWIN ELECTRODES

Welding Position: Flat
Weld Quality Level: Commercial
Steel Weldability: Good

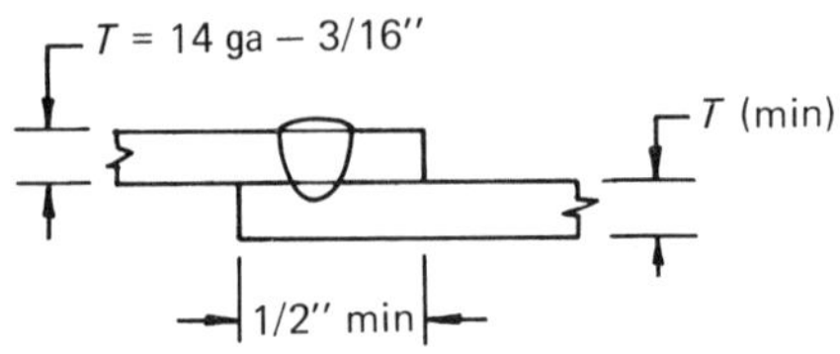

Plate Thickness, T (in.)	0.075 (14 ga)	0.105 (12 ga)	0.135 (10 ga)	3/16
Pass	1	1	1	1
Electrode Size	1/16 (two)	1/16 (two)	1/16 (two)	5/64 (two)
Current (amp) DC(+)	800	900	1000	1100
Volts	27	28	29	30
Arc Speed (in./min)	200	170	130	75
Electrode Req'd (lb/ft)	0.027	0.038	0.058	0.115
Flux Req'd (lb/ft)	0.03 – 0.04	0.03 – 0.05	0.05 – 0.06	0.09 – 0.11
Total Time (hr/ft of weld)	0.00100	0.00118	0.00154	0.00267
Electrode Spacing, 5/16 in.				

A constant-voltage power source is recommended.

See introductory notes.

SUBMERGED-ARC (FULL AUTOMATIC) MULTIPLE ELECTRODES

Welding Position: Flat
Weld Quality Level: Code
Steel Weldability: Good
Welded from: One side

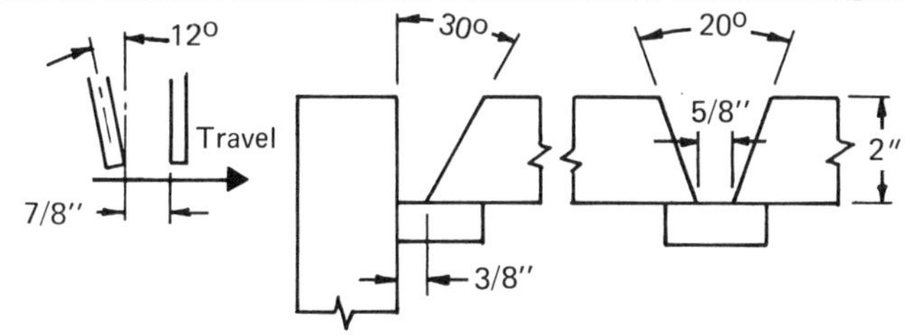

Plate Thickness (in.)			2			
Pass	1 – 2	3 – 18	1 – 2	3 – 16	1 – 2	3 – 14
Electrode Size: Lead (No.1)	3/16	3/16	3/16	3/16	3/16	3/16
Trail (No.2)		3/16		3/16		3/16
Current (amp): Lead (No.1) DC(–)	750	950	750	950	750	950
Trail (No.2) AC		800		800		800
Volts: Lead (No.1)	34	32	38	39	44	45
Trail (No.2)		35		41		47
Arc Speed (in./min)	16	30	25	34	30	36
Electrode Req'd (lb/ft)	7.04		7.04		7.04	
Flux Req'd (lb/ft)	5.5 – 6.3		5.0 – 5.8		4.6 – 5.4	
Total Time (hr/ft of weld)	0.123		0.0963		0.0789	
Stickout (in.)	1-1/2		4		6	
Backing, Minimum thickness (in.)	1/2		3/8		5/16	
Electrode spacing, 7/8 in.						

Slightly higher current on the first pass is preferred but 750 amps is the maximum permitted by the AWS Structural Code.
This procedure can be applied directly or with modifications to several prequalified joints (see Section 11.3) other than those shown here.
See introductory notes.

SUBMERGED-ARC (FULL AUTOMATIC) MULTIPLE ELECTRODES

Welding Position: Flat
Weld Quality Level: Commercial
Steel Weldability: Good

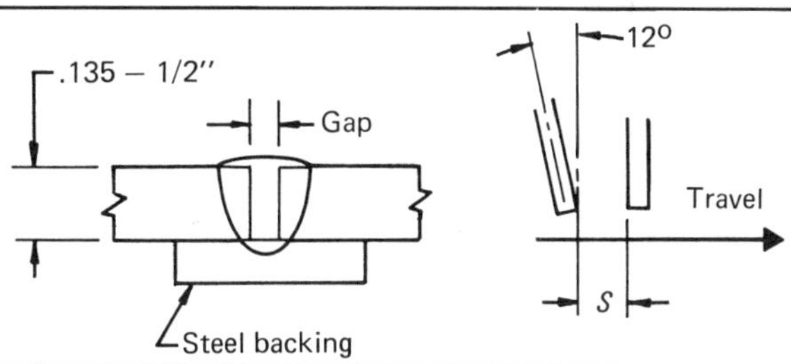

Plate Thickness (in.)	0.135 (10 ga)	3/16	1/4	3/8	1/2
Pass	1	1	1	1	1
Electrode Size: Lead (No.1)	3/16	3/16	3/16	3/16	3/16
Trail (No.2)	5/32	5/32	5/32	5/32	5/32
Current (amp): Lead (No.1) DC(+)	750	975	1000	1000	1000
Trail (No.2) AC	550	725	775	775	800
Volts: Lead (No.1)	31	32	33	33	35
Trail (No.2)	33	36	37	37	39
Arc Speed (in./min)	96	70	50	45	30
Electrode Req'd (lb/ft)	0.07	0.14	0.22	0.24	0.38
Flux Req'd (lb/ft)	0.05 – 0.07	0.10 – 0.15	0.16 – 0.23	0.18 – 0.25	0.26 – 0.40
Total Time (hr/ft of weld)	0.00208	0.00286	0.00400	0.00444	0.00667
Backing, minimum size (in.)	1/8 x 5/8	3/16 x 3/4	1/4 x 1	1/4 x 1	3/8 x 1
Gap (in.)	1/16	1/8	5/32	5/32	3/16
Electrode spacing, S (in.)	5/8	5/8	3/4	3/4	7/8

See introductory notes.

SUBMERGED-ARC (FULL AUTOMATIC) MULTIPLE ELECTRODES

Welding Position: Flat
Weld Quality Level: Code
Steel Weldability: Good
Welded from: Two sides

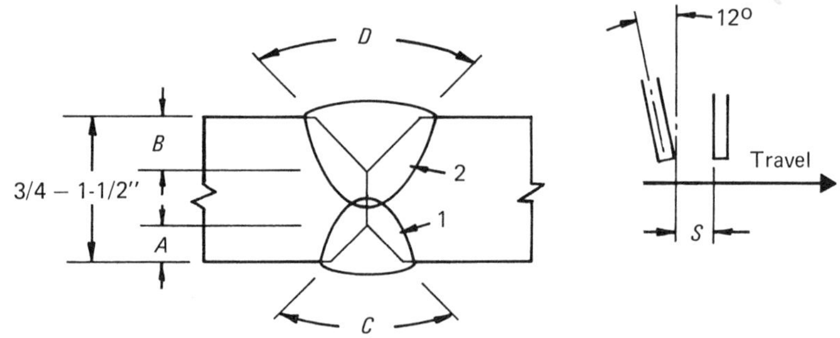

Plate Thickness (in.)	3/4		1		1-1/4		1-1/2	
Pass	1	2	1	2	1	2	1	2
Electrode Size: Lead (No. 1)	3/16	3/16	3/16	3/16	3/16	3/16	3/16	3/16
Trail (No. 2)	3/16	3/16	3/16	3/16	3/16	3/16	3/16	3/16
Current (amp): Lead (No. 1) DC(+)	950	1050	950	1050	950	1075	1075	1150
Trail (No. 2) AC	700	800	700	850	750	850	850	900
Volts: Lead (No. 1)	35	36	36	38	36	38	36	40
Trail (No. 2)	40	41	41	43	42	43	43	44
Arc Speed (in./min)	40	30	26	22	20	16.5	20	15
Electrode Req'd (lb/ft)	0.67		0.97		1.32		1.67	
Flux Req'd (lb/ft)	0.48 – 0.64		0.60 – 0.85		0.92 – 1.30		1.00 – 1.40	
Total Time (hr/ft of weld)	0.0117		0.0168		0.0220		0.0233	
Depth, A (in.)	1/8		1/4		3/8		7/16	
Depth, B (in.)	1/4		3/8		1/2		5/8	
Angle, C (deg)	90		80		70		60	
Angle, D (deg)	90		80		70		70	
Electrode spacing, S (in.)	7/8		1		1-1/8		1-1/4	

See introductory notes.

SUBMERGED-ARC (FULL AUTOMATIC) MULTIPLE ELECTRODES

Welding Position: Flat
Weld Quality Level: Code
Steel Weldability: Good or fair

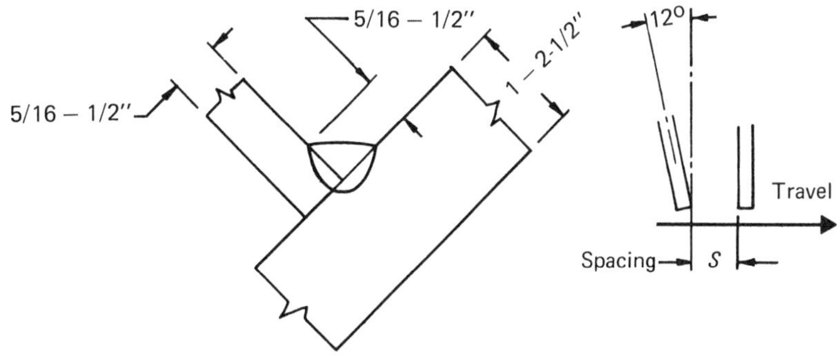

Weld Size, L (in.)	5/16	3/8	1/2
Plate Thickness (in.)	5/16 to 1	3/8 to 1-3/4	1/2 to 2-1/2
Pass	1	1	1
Electrode Size: Lead (No.1)	3/16	3/16	3/16
Trail (No.2)	3/16	3/16	3/16
Current (amp): Lead (No.1) DC(–)	550	600	750
Trail (No.2) AC	420	520	550
Volts: Lead (No.1)	29	29	32
Trail (No.2)	31	34	36
Arc Speed (in./min)	35	30	20
Electrode Req'd (lb/ft)	0.18	0.26	0.47
Flux Req'd (lb/ft)	0.11 – 0.15	0.16 – 0.21	0.27 – 0.36
Total Time (hr/ft of weld)	0.00571	0.00667	0.0100
Electrode spacing, S (in.)	5/8	5/8	3/4

See introductory notes.

SUBMERGED-ARC (FULL AUTOMATIC) MULTIPLE ELECTRODES

Welding Position: Flat
Weld Quality Level: Commercial
Steel Weldability: Good

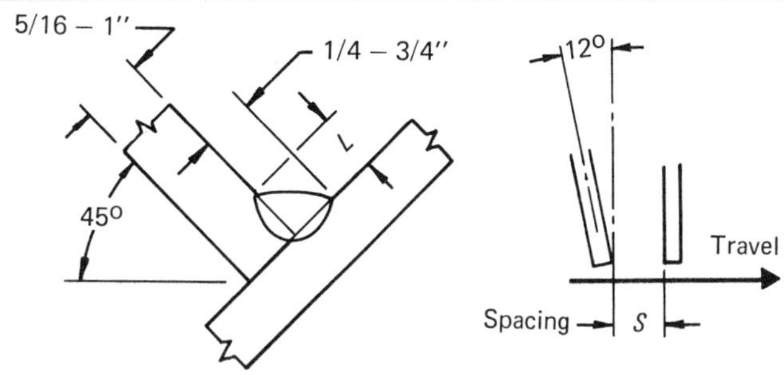

Weld Size, (in.) (L)	1/4	5/16	3/8	1/2	5/8	3/4
Plate Thickness (in.)	5/16	3/8	7/16	5/8	3/4	1
Pass	1	1	1	1	1	1
Electrode Size: Lead (No. 1)	3/16	3/16	3/16	3/16	3/16	3/16
Trail (No. 2)	3/16	3/16	3/16	3/16	3/16	3/16
Current (amp): Lead (No. 1) DC(+)	750	825	900	1075	1100	1100
Trail (No. 2) AC	550	600	700	750	850	850
Volts: Lead (No. 1)	28	30	32	34	37	37
Trail (No. 2)	30	33	34	36	39	39
Arc Speed (in./min)	68	50	40	29	21	14.5
Electrode Req'd (lb/ft)	0.12	0.18	0.26	0.47	0.73	1.05
Flux Req'd (lb/ft)	0.07 – 0.09	0.12 – 0.15	0.15 – 0.20	0.27 – 0.36	0.42 – 0.55	0.61 – 0.80
Total Time (hr/ft of weld)	0.00294	0.00400	0.00500	0.00690	0.00952	0.0138
Electrode Spacing, S (in.)	5/8	5/8	3/4	3/4	3/4	7/8

See introductory notes.

SUBMERGED-ARC (FULL AUTOMATIC) MULTIPLE ELECTRODES

Weld Position: Flat
Weld Quality Level: Commercial
Steel Weldability: Good
Welded from: Two sides

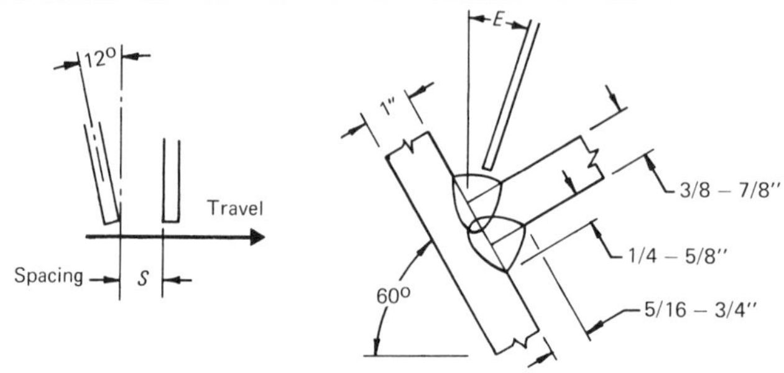

Weld Size, L (in.)*	5/16*	7/16*	1/2*	5/8*	3/4*
Plate Thickness (in.)	3/8 to 1	1/2 to 1	5/8 to 1	3/4 to 1	7/8 to 1
Pass	1,2	1,2	1,2	1,2	1,2
Electrode Size: Lead (No.1)	3/16	3/16	3/16	3/16	3/16
Electrode Size: Trail (No.2)	5/32	3/16	3/16	3/16	3/16
Current (amp): Lead (No.1) DC(+)	850	925	1025	1075	1100
Current (amp): Trail (No.2) AC	575	700	750	800	850
Volts: Lead (No.1)	30	32	34	36	37
Volts: Trail (No.2)	32	34	36	38	39
Arc Speed (in/min)	50	40	30	24	17.5
Electrode Req'd (lb/ft)	0.30	0.51	0.82	1.32	1.76
Flux Req'd (lb/ft)	0.20 – 0.27	0.32 – 0.42	0.52 – 0.67	0.85 – 1.10	1.00 – 1.35
Total Time (hr/ft of weld)	0.00800	0.0100	0.0133	0.0167	0.0228
Electrode angle, E (deg)	0 – 15	0 – 15	0 – 15	0 – 12	0 – 10
Electrode spacing, S (in.)	5/8	3/4	3/4	3/4	3/4
*Minor leg size (in.)	1/4	5/16	7/16	9/16	5/8

See introductory notes.

SUBMERGED-ARC (FULL AUTOMATIC) MULTIPLE ELECTRODES

Welding Position: Flat
Weld Quality Level: Commercial
Steel Weldability: Good
Welded from: Two sides

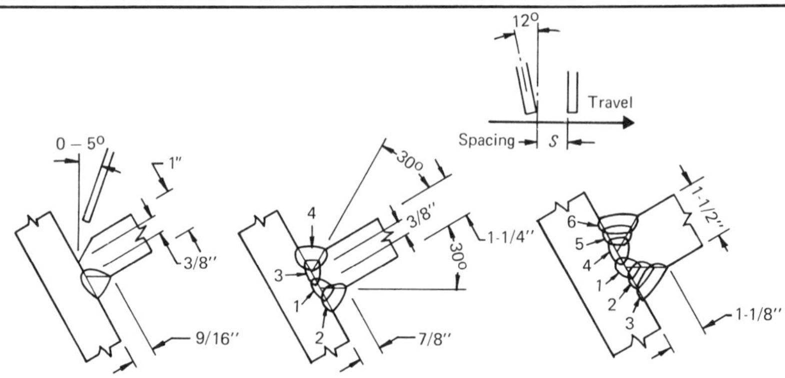

Weld Size, L (in.)	9/16*	7/8*		1-1/8*		
Plate Thickness (in.)	1	1-1/4		1-1/2		
Pass	1, 2	1, 3	2, 4	1, 4	2, 5	3, 6
Electrode Size: Lead (No.1)	3/16	3/16	3/16	3/16	3/16	3/16
Electrode Size: Trail (No.2)	3/16	3/16	3/16	3/16	3/16	3/16
Current (amp): Lead (No.1) DC(–)	900	875	875	900	900	900
Current (amp): Trail (No.2) AC	800	700	700	750	750	750
Volts: Lead (No.1)	34	32	34	33	35	35
Volts: Trail (No.2)	36	35	37	37	39	39
Arc Speed (in./min)	20	18	18	18	18	18
Electrode Req'd (lb/ft)	1.16	2.19		3.51		
Flux Req'd (lb/ft)	0.72 – 0.93	1.40 – 1.80		2.30 – 3.00		
Total Time (hr/ft of weld)	0.0200	0.0444		0.0667		
*L = Minor leg size (in.)	1/4	3/8		1/2		
Electrode spacing, 7/8 in.						

See introductory notes.

SUBMERGED-ARC (FULL AUTOMATIC) MULTIPLE ELECTRODES

Welding Position: Horizontal
Weld Quality Level: Commercial
Steel Weldability: Good

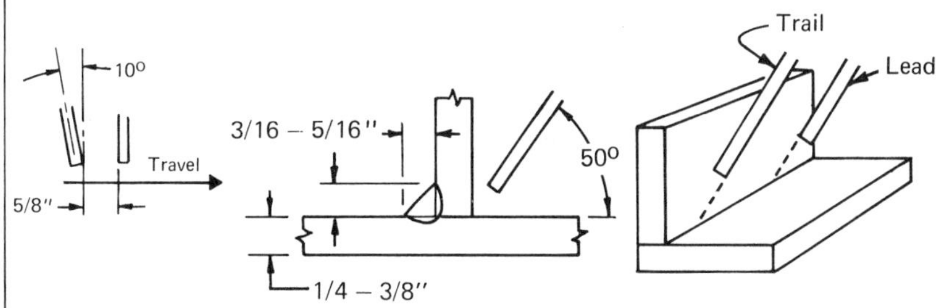

Weld Size, (in.)	3/16	1/4	5/16
Plate Thickness (in.)	1/4	5/16	3/8
Pass	1	1	1
Electrode Size: Lead (No. 1)	5/32	3/16	3/16
Trail (No. 2)	1/8	5/32	5/32
Current (amp): Lead (No. 1) DC(+)	600	800	875
Trail (No. 2) AC	350	550	600
Volts: Lead (No. 1)	26	26	28
Trail (No. 2)	28	28	31
Arc Speed (in./min)	75	60	45
Electrode Req'd (lb/ft)	0.066	0.120	0.181
Flux Req'd (lb/ft)	0.052 – 0.068	0.089 – 0.11	0.13 – 0.17
Total Time (hr/ft of weld)	0.00267	0.00333	0.00444

See introductory notes.

SUBMERGED-ARC (FULL AUTOMATIC) MULTIPLE ELECTRODES

Welding Position: Horizontal
Weld Quality Level: Commercial
Steel Weldability: Good

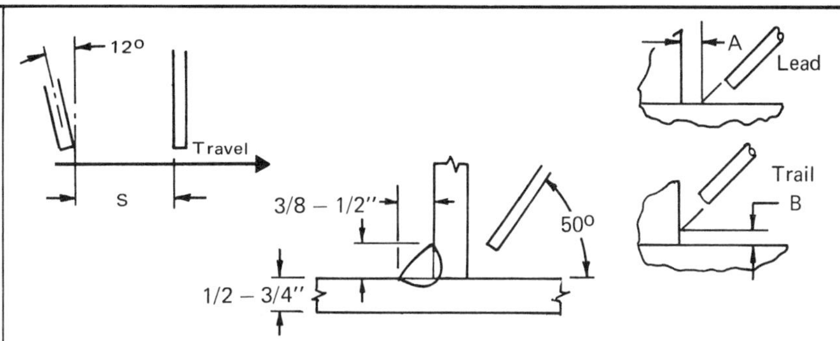

Weld Size, (in.)	3/8	7/16	1/2
Plate Thickness (in.)	1/2	5/8	3/4
Pass	1	1	1
Electrode Size: Lead (No. 1)	3/16	3/16	3/16
Trail (No. 2)	1/8	1/8	1/8
Current (amp): Lead (No. 1) DC(–)	750	750	750
Trail (No. 2) AC	400	375	350
Volts: Lead (No. 1)	30	32	34
Trail (No. 2)	28	27	26
Arc Speed (in./min)	28	21	16
Electrode Req'd (lb/ft)	0.262	0.358	0.480
Flux Req'd (lb/ft)	0.19 – 0.25	0.27 – 0.35	0.35 – 0.46
Total Time (hr/ft of weld)	0.00715	0.00953	0.0125
Electrode Location (in.) A	0	1/16	1/8
B	3/32	1/8	1/8
Electrode Spacing, S (in.)	3	4	5

See introductory notes.

SUBMERGED-ARC (FULL AUTOMATIC) MULTIPLE ELECTRODES

Welding Position: Flat
Weld Quality Level: Code
Steel Weldability: Good

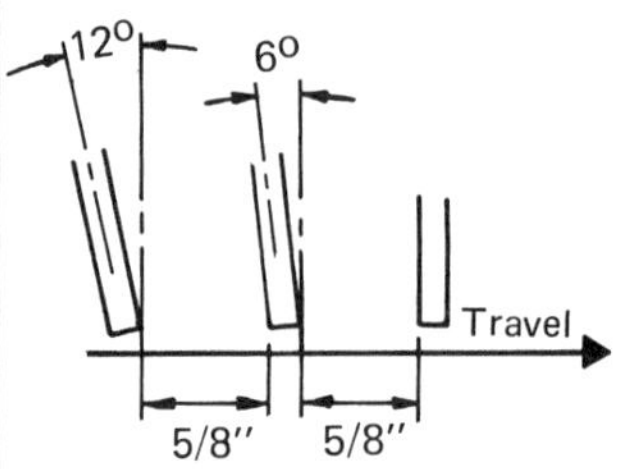

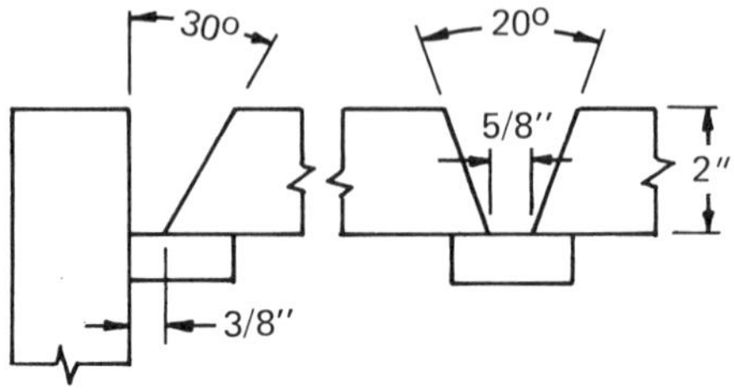

Plate Thickness (in.)	2					
Pass	1 – 2	3 – 18	1 – 2	3 – 16	1 – 2	3 – 14
Electrode Size: Lead (No. 1)	3/16	3/16	3/16	3/16	3/16	3/16
Trail (No. 2&3)	3/16	3/16	3/16	3/16	3/16	3/16
Current (amp): Lead (No. 1) DC(–)	750	850	750	850	750	850
Trail (No. 2) AC		700		700		700
Trail (No. 3) AC		650		650		650
Volts: Lead (No. 1)	34	31	38	37	44	43
Trail (No. 2)		30		36		42
Trail (No. 3)		33		39		46
Arc Speed (in./min)	16	36	25	40	30	43
Electrode Req'd (lb/ft)	7.04		7.04		7.04	
Flux Req'd (lb/ft)	5.1 – 6.8		4.8 – 6.3		4.4 – 5.7	
Total Time (hr/ft of weld)	0.109		0.0859		0.0695	
Stickout (in.)	1-1/2		4		6	
Backing, min thickness (in.)	1/2		3/8		5/16	

Slightly higher current on the first pass is preferred but 750 amps is the maximum permitted by the AWS Structural Codes.

This procedure can be applied directly or with modifications to several prequalified joints (see Section 11.3) other than those shown here.

See introductory notes.

SUBMERGED-ARC (FULL AUTOMATIC) MULTIPLE ELECTRODES

Welding Position: Flat
Weld Quality Level: Code
Steel Weldability: Good
Welded from: Two sides

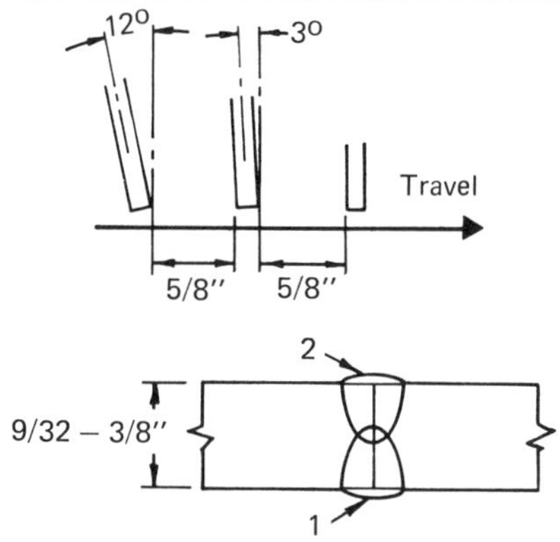

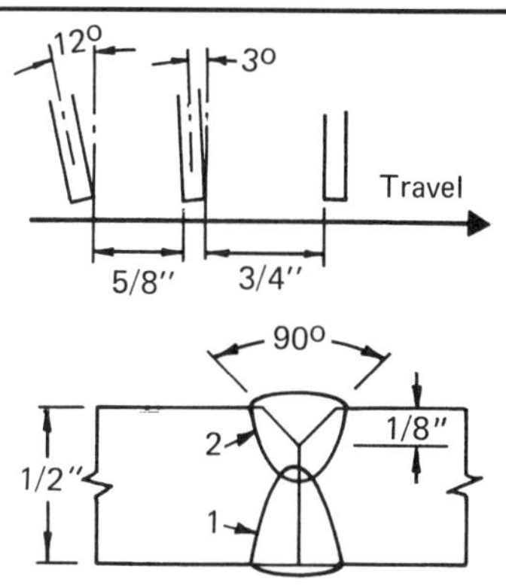

Plate Thickness (in.)	9/32		3/8		1/2	
Pass	1	2	1	2	1	2
Electrode Size: Lead (No. 1)	3/16	3/16	3/16	3/16	3/16	3/16
Trail (No. 2&3)	1/8	1/8	5/32	5/32	5/32	5/32
Current (amp): Lead (No. 1) DC(+)	1000	1100	1050	1100	1100	1100
Trail (No. 2) AC	600	800	700	800	800	800
Trail (No. 3) AC	600	600	700	700	800	800
Volts: Lead (No. 1)	29	31	30	32	31	32
Trail (No. 2)	31	34	32	33	33	33
Trail (No. 3)	37	37	37	37	38	38
Arc Speed (in./min)	104	104	90	90	80	70
Electrode Req'd (lb/ft)	0.27		0.36		0.47	
Flux Req'd (lb/ft)	0.25 – 0.32		0.33 – 0.43		0.42 – 0.53	
Total Time (hr/ft of weld)	0.00384		0.00444		0.00536	

See introductory notes.

SUBMERGED-ARC (FULL AUTOMATIC) MULTIPLE ELECTRODES

Welding Position: Flat
Weld Quality Level: Code
Steel Weldability: Good
Welded from: Two sides

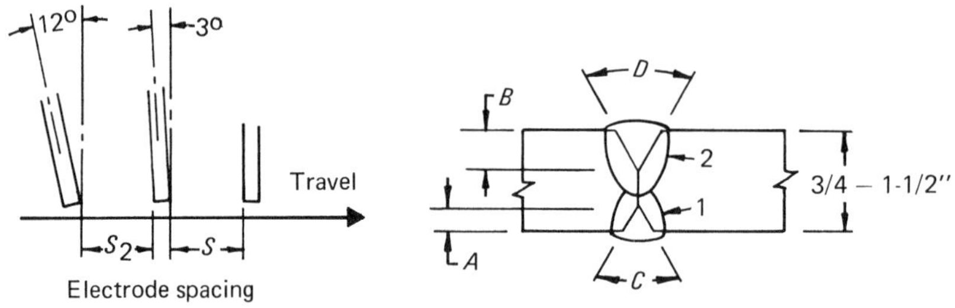

Plate Thickness (in.)	3/4		1		1-1/4		1-1/2	
Pass	1	2	1	2	1	2	1	2
Electrode Size: Lead No. 1)	3/16	3/16	3/16	3/16	3/16	3/16	3/16	3/16
Trail (No. 2&3)	3/16	3/16	3/16	3/16	3/16	3/16	3/16	3/16
Current (amp): Lead (No. 1) DC(+)	1000	1100	1050	1150	1150	1150	1175	1175
Trail (No. 2) AC	700	800	700	825	800	800	800	800
Trail (No. 3) AC	700	800	700	825	800	800	800	800
Volts: Lead (No. 1)	31	32	36	36	36	36	37	37
Trail (No. 2)	33	33	33	33	33	33	33	36
Trail (No. 3)	35	35	36	36	36	36	36	36
Arc Speed (in./min)	55	42	42	34	36	26	34	22
Electrode Req'd (lb/ft)	0.69		0.94		1.26		1.52	
Flux Req'd (lb/ft)	0.56 – 0.73		0.75 – 0.99		1.00 – 1.30		1.20 – 1.60	
Total Time (hr/ft of weld)	0.00840		0.0106		0.0133		0.0150	
Depth, A (in.)	1/8		1/4		3/8		7/16	
Depth, B (in.)	1/4		3/8		1/2		5/8	
Angle, C (deg)	90		80		70		60	
Angle, D (deg)	90		80		70		70	
Spacing, S (in.)	3/4		3/4		7/8		1	
Spacing, S_2 (in.)	3/4		3/4		7/8		1	

See introductory notes.

SUBMERGED-ARC (FULL AUTOMATIC) MULTIPLE ELECTRODES

Welding Position: Horizontal
Weld Quality Level: Commercial
Steel Weldability: Good

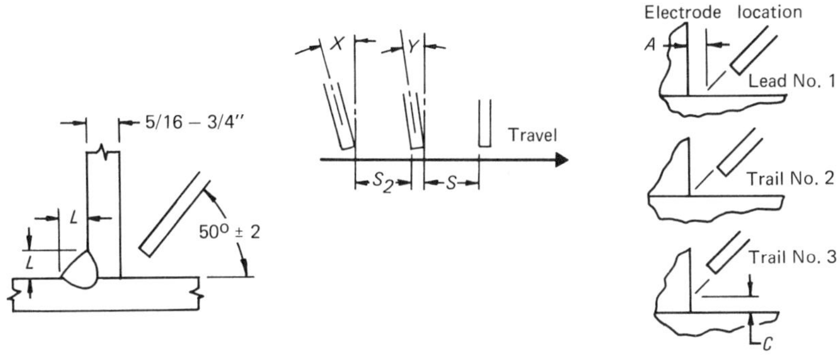

Weld Size, L (in.)		1/4	5/16	3/8	1/2	5/8
Plate Thickness (in.)		5/16	3/8	1/2	5/8	3/4
Pass		1	1	1	1	1
Electrode Size:	Lead (No. 1)	5/32	5/32	3/16	3/16	3/16
	Trail (No. 2)	1/8	1/8	1/8	1/8	1/8
	Trail (No. 3)	3/32	3/32	3/32	3/32	3/32
Current (amp):	Lead (No. 1) DC	800(+)	875(+)	750(–)	750(–)	750(–)
	Trail (No. 2) AC	600	650	400	400	400
	Trail (No. 3) AC	325	350	350	350	300
Volts:	Lead (No. 1)	28	30	31	32	33
	Trail (No. 2)	29	31	29	30	31
	Trail (No. 3)	30	31	31	31	29
Arc Speed (in./min)		80	64	37	21	13.5
Electrode Req'd (lb/ft)		0.12	0.18	0.26	0.47	0.73
Flux Req'd (lb/ft)		0.10 – 0.13	0.13 – 0.18	0.19 – 0.24	0.32 – 0.42	0.51 – 0.66
Total Time (hr/ft of weld)		0.00250	0.00313	0.00541	0.0095	0.0148
Electrode location, A (in.)		0	0	1/8	5/32	3/16
Electrode location, C (in.)		3/32	1/8	1/8	5/32	3/16
Electrode spacing, S (in.)		3/4	3/4	3	4	5
Electrode spacing, S_2 (in.)		7/8	7/8	3	3	3
Electrode angle, X (deg)		8	8	3	3	3
Electrode angle, Y (deg)		15	15	7	7	7

See introductory notes.

SUBMERGED-ARC (SEMIAUTOMATIC) MANUAL

Welding Position: Flat
Weld Quality Level: Commercial
Steel Weldability: Good

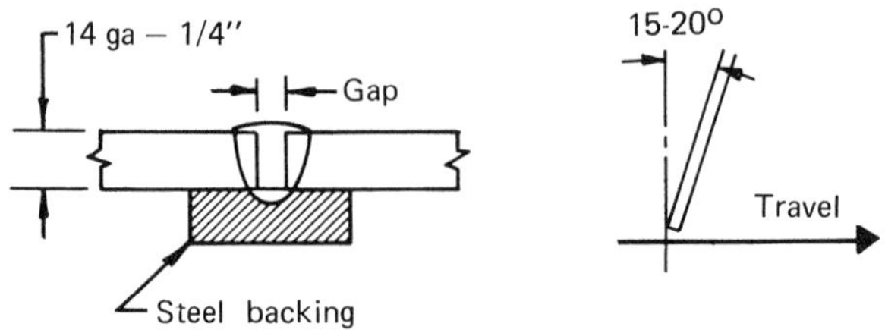

Plate Thickness (in.)	0.075 (14 ga)	0.105 (12 ga)	0.135 (10 ga)	3/16	1/4
Pass	1	1	1	1	1
Electrode Size	1/16	1/16	1/16	1/16	1/16
Current (amp) DC(+)	275	325	375	425	425
Volts	25	27	29	33	36
Arc Speed (in./min)	44 – 50	40 – 46	35 – 40	26 – 30	14 – 18
Electrode Req'd (lb/ft)	0.037	0.047	0.065	0.108	0.190
Flux Req'd (lb/ft)	0.09 – 0.13	0.10 – 0.14	0.12 – 0.16	0.14 – 0.18	0.15 – 0.21
Total Time (hr/ft of weld)	0.00426	0.00465	0.00534	0.00715	0.0125
Backing, minimum size (in.)	12 ga x 3/8	12 ga x 1/2	10 ga x 5/8	3/16 x 3/4	1/4 x 1
Gap (in.)	1/16	1/16	3/32	3/32	1/8

See introductory notes.

SUBMERGED-ARC (SEMIAUTOMATIC) MANUAL

Welding Position: Flat
Weld Quality Level: Commercial
Steel Weldability: Good

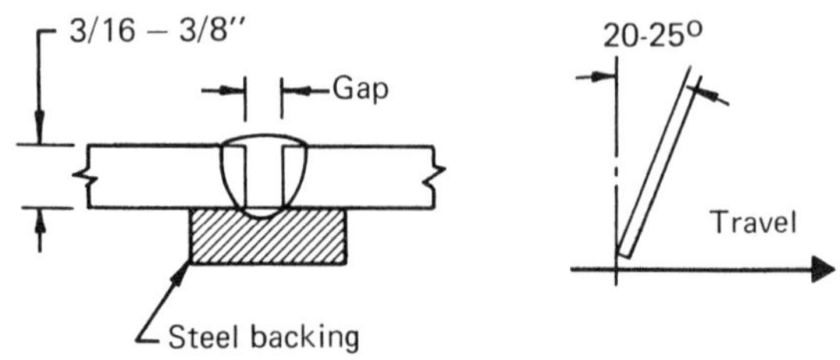

Plate Thickness (in.)	3/16	1/4	5/16	3/8
Pass	1	1	1	1
Electrode Size	5/64	5/64	5/64	5/64
Current (amp) DC(+)	425	450	475	500
Volts	31	32	34	35
Arc Speed (in./min)	20 – 22	15 – 17	13 – 15	10 – 12
Electrode Req'd (lb/ft)	0.12	0.18	0.22	0.30
Flux Req'd (lb/ft)	0.13 – 0.17	0.21 – 0.27	0.25 – 0.32	0.34 – 0.43
Total Time (hr/ft of weld)	0.00952	0.0125	0.0143	0.0182
Backing, minimum size (in.)	3/16 x 3/4	1/4 x 3/4	1/4 x 1	1/4 x 1
Gap (in.)	1/8	5/32	5/32	3/16

See introductory notes.

SUBMERGED-ARC (SEMIAUTOMATIC) MANUAL

Welding Position: Flat
Weld Quality Level: Commercial
Steel Weldability: Good
Welded from: Two sides

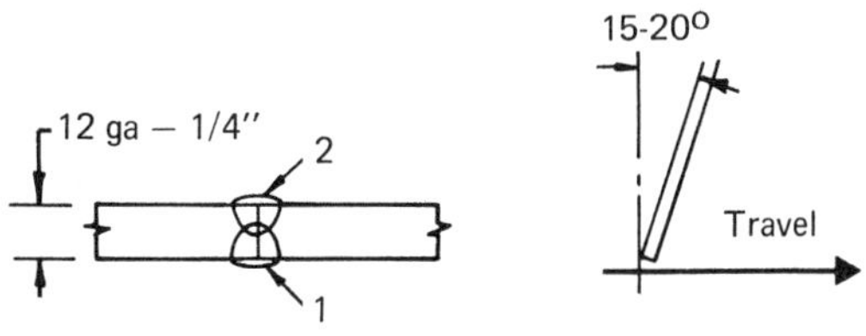

Plate Thickness (in.)	0.105 (12 ga)		0.135 (10 ga)		3/16		1/4	
Pass	1	2	1	2	1	2	1	2
Electrode Size	1/16	1/16	1/16	1/16	1/16	1/16	1/16	1/16
Current (amp) DC(+)	200	275	250	325	300	350	325	375
Volts	23	25	25	27	29	32	31	33
Arc Speed (in./min)	44 – 49		47 – 52		43 – 48		40 – 44	
Electrode Req'd (lb/ft)	0.060		0.070		0.088		0.106	
Flux Req'd (lb/ft)	0.11 – 0.15		0.12 – 0.16		0.14 – 0.18		0.15 – 0.21	
Total Time (hr/ft of weld)	0.00860		0.00808		0.00880		0.00952	

See introductory notes.

SUBMERGED-ARC (SEMIAUTOMATIC) MANUAL

Welding Position: Flat
Weld Quality Level: Commercial
Steel Weldability: Good
Welded from: Two sides

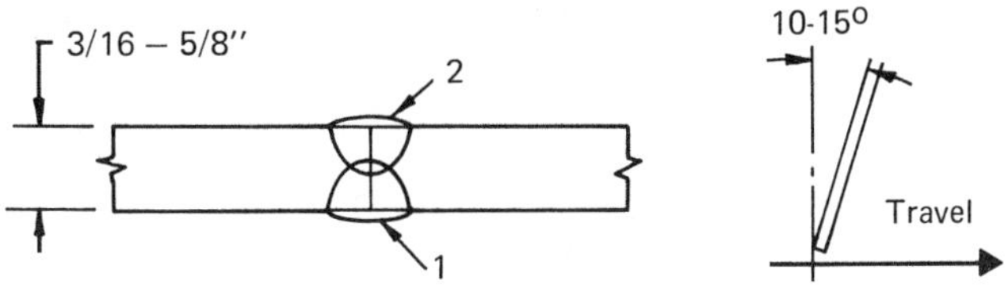

Plate Thickness (in.)	3/16		1/4		3/8		1/2		5/8	
Pass	1	2	1	2	1	2	1	2	1	2
Electrode Size	5/64	5/64	5/64	5/64	5/64	5/64	5/64	5/64	5/64	5/64
Current (amp) DC(+)	300	350	350	400	425	475	475	500	500	500
Volts	30	33	31	34	33	35	34	36	35	37
Arc Speed (in./min)	40 – 44		37 – 41		27 – 30		20 – 23		15 – 17	
Electrode Req'd (lb/ft)	0.072		0.10		0.18		0.28		0.39	
Flux Req'd (lb/ft)	0.14 – 0.18		0.15 – 0.19		0.25 – 0.31		0.42 – 0.50		0.55 – 0.65	
Total Time (hr/ft of weld)	0.00952		0.0103		0.0141		0.0186		0.0250	

See introductory notes.

SUBMERGED-ARC (SEMIAUTOMATIC) MANUAL

Welding Position: Flat
Weld Quality Level: Commercial
Steel Weldability: Good
Welded from: Two sides

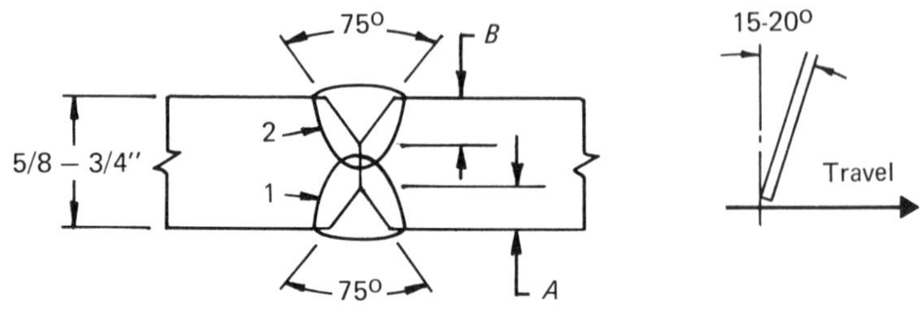

Plate Thickness (in.)	5/8		3/4	
Pass	1	2	1	2
Electrode Size	5/64	5/64	5/64	5/64
Current (amp) DC(+)	475	500	475	500
Volts	35	36	35	36
Arc Speed (in./min)	16 – 18	16 – 18	11 – 13	11 – 13
Electrode Req'd (lb/ft)	0.37		0.53	
Flux Req'd (lb/ft)	0.60 – 0.75		0.75 – 0.96	
Total Time (hr/ft of weld)	0.0235		0.0333	
Depth, A (in.)	7/32		9/32	
Depth, B (in.)	7/32		9/32	

SUBMERGED-ARC (SEMIAUTOMATIC) MANUAL

Welding Position: Flat
Weld Quality Level: Commercial
Steel Weldability: Good
Welded from: Two sides

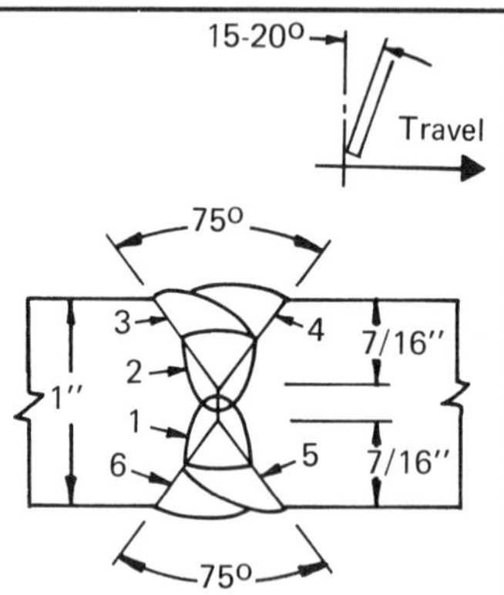

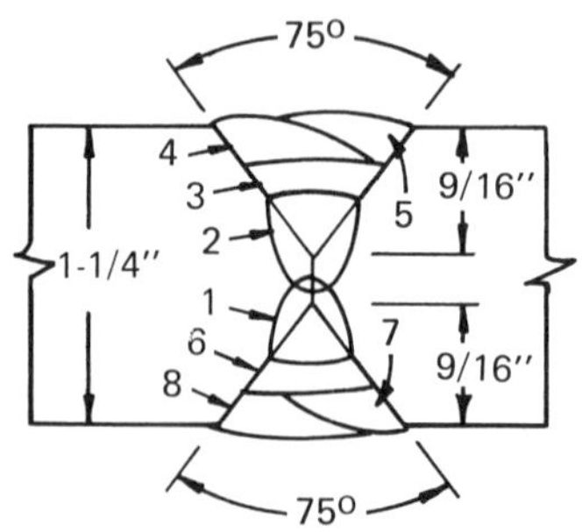

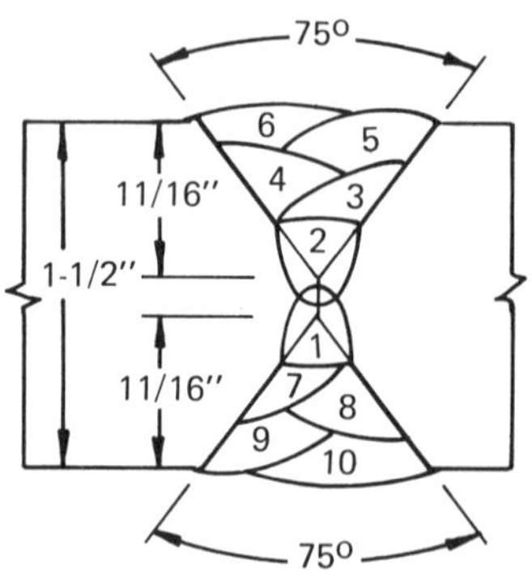

Plate Thickness (in.)	1			1-1/4			1-1/2		
Pass	1	2	3 – 6	1	2	3 – 8	1	2	3 – 10
Electrode Size	5/64	5/64	5/64	5/64	5/64	5/64	5/64	5/64	5/64
Current (amp) DC(–)	450	500	500	450	500	500	450	500	500
Volts	34	34	38	34	34	38	34	34	38
Arc Speed (in./min)	15 – 17	15 – 17	21 – 23	17 – 19	17 – 19	18 – 20	15 – 17	15 – 17	15 – 17
Electrode Req'd (lb/ft)	1.34			1.88			2.75		
Flux Req'd (lb/ft)	2.70 – 3.35			3.80 – 4.45			5.20 – 5.80		
Total Time (hr/ft of weld)	0.0614			0.0852			0.125		

Seam must be tight. Seal all gaps with small bead on first-pass side.

See introductory notes.

SUBMERGED-ARC (SEMIAUTOMATIC) MANUAL

Welding Position: Flat
Weld Quality Level: Commercial
Steel Weldability: Good

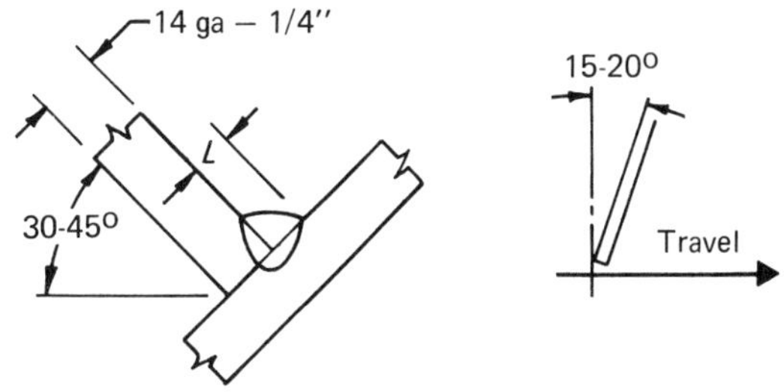

Weld Size, L (in.)	3/32–	3/32	1/8–	5/32–	3/16
Plate Thickness (in.)	0.075 (14 ga)	0.105 (12 ga)	0.135 (10 ga)	3/16	1/4
Pass	1	1	1	1	1
Electrode Size	1/16	1/16	1/16	1/16	1/16
Current (amp) DC(+)	185	250	325	360	400
Volts	23	24.5	28	30	37
Arc Speed (in./min)	35 – 39	47 – 52	57 – 63	47 – 52	31 – 34
Electrode Req'd (lb/ft)	0.028	0.030	0.034	0.047	0.081
Flux Req'd (lb/ft)	0.07 – 0.09	0.08 – 0.10	0.10 – 0.12	0.10 – 0.13	0.11 – 0.14
Total Time (hr/ft of weld)	0.00540	0.00404	0.00333	0.00404	0.00615

See introductory notes.

SUBMERGED-ARC (SEMIAUTOMATIC) MANUAL

Welding Position: Flat
Weld Quality Level: Commercial
Steel Weldability: Good

Procedure for Conventional Fillets with 1-in. Electrical Stickout

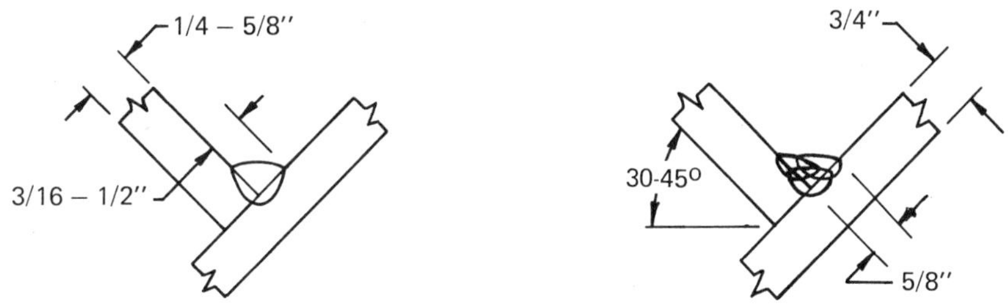

Weld Size, L (in.)	3/16	1/4	5/16	3/8	1/2	5/8
Plate Thickness (in.)	1/4	5/16	3/8	1/2	5/8	3/4
Pass	1	1	1	1	1	1, 2 & 3
Electrode Size	5/64	5/64	5/64	5/64	5/64	5/64
Current (amp) DC(–)	375	400	425	425	425	450
Volts	36	37	38	39	39	40
Arc Speed (in./min)	31 – 32	22 – 24	17 – 19	13.5 – 14.5	7.5 – 8.5	14 – 15
Electrode Req'd (lb/ft)	0.08	0.13	0.19	0.26	0.45	0.74
Flux Req'd (lb/ft)	0.07 – 0.11	0.12 – 0.16	0.17 – 0.24	0.25 – 0.33	0.44 – 0.54	0.80 – 0.95
Total Time (hr/ft of weld)	0.00615	0.00870	0.0111	0.0143	0.0250	0.0404

See introductory notes.

SUBMERGED-ARC (SEMIAUTOMATIC) MANUAL

Welding Position: Flat
Weld Quality Level: Commercial
Steel Weldability: Good

Procedure for Long Stickout (2-1/4 in.)

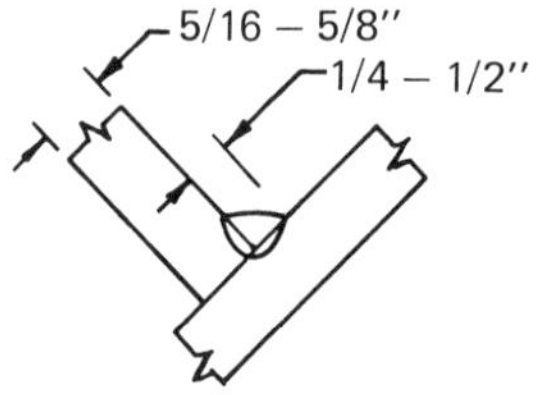

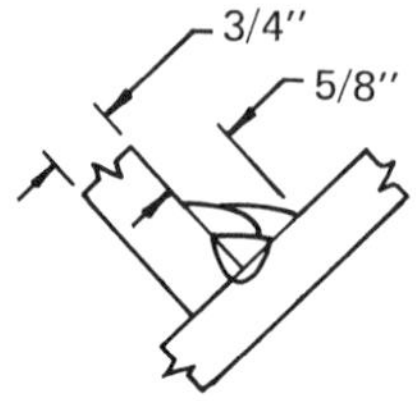

Weld Size, L (in.)	1/4	5/16	3/8	1/2	5/8
Plate Thickness (in.)	5/16	3/8	1/2	5/8	3/4
Pass	1	1	1	1	1, 2 & 3
Electrode Size	5/64	5/64	5/64	5/64	5/64
Current (amp) DC(–)	425	450	450	450	450
Volts	45	47	47	47	47
Arc Speed (in./min)	28 – 31	22 – 24	17 – 19	9 – 11	17 – 19
Electrode Req'd (lb/ft)	0.14	0.22	0.30	0.52	0.81
Flux Req'd (lb/ft)	0.10 – 0.15	0.16 – 0.22	0.26 – 0.35	0.42 – 0.52	0.80 – 0.95
Total Time (hr/ft of weld)	0.00678	0.00870	0.0111	0.0200	0.0333
Electrical Stickout, 2-1/4 in.					

SUBMERGED-ARC (SEMIAUTOMATIC) MANUAL

Welding Position: Flat
Weld Quality Level: Strength only
Steel Weldability: Good

Procedure for Penetration Welds with 1-in. Electrical Stickout

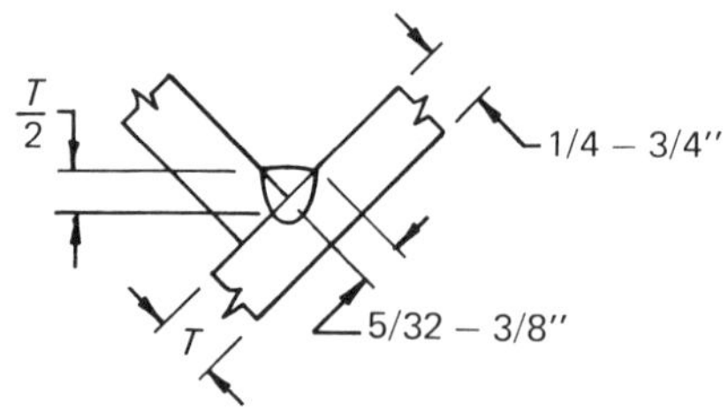

Weld Size, L (in.)	5/32	3/16(–)	3/16	1/4	5/16	3/8
Plate Thickness, T (in.)	1/4	5/16	3/8	1/2	5/8	3/4
Pass	1	1	1	1	1	1
Electrode Size	5/64	5/64	5/64	5/64	5/64	5/64
Current (amp) DC(+)	350	400	410	425	425	425
Volts	30	33	34	35	35	36
Arc Speed (in./min)	36 – 40	33 – 36	24 – 27	16 – 18	11.5 – 12.5	8.5 – 9.5
Electrode Req'd (lb/ft)	0.046	0.061	0.084	0.13	0.19	0.25
Flux Req'd (lb/ft)	0.05 – 0.08	0.09 – 0.13	0.14 – 0.18	0.20 – 0.25	0.26 – 0.32	0.32 – 0.40
Total Time (hr/ft of weld)	0.00526	0.00580	0.00784	0.0118	0.0167	0.0222

See introductory notes.

SUBMERGED-ARC (SEMIAUTOMATIC) MANUAL

Welding Position: Horizontal
Weld Quality Level: Commercial
Steel Weldability: Good

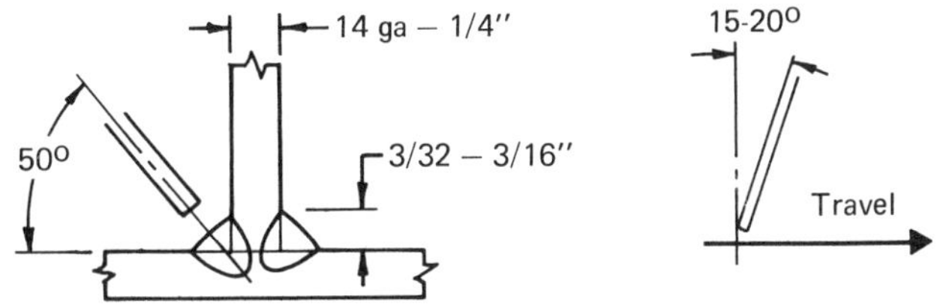

Weld Size, L (in.)	3/32–	3/32	1/8–	5/32–	3/16
Plate Thickness (in.)	0.075 (14 ga)	0.105 (12 ga)	0.135 (10 ga)	3/16	1/4
Pass	1	1	1	1	1
Electrode Size	1/16	1/16	1/16	1/16	1/16
Current (amp) DC(+)	185	250	325	350	375
Volts	23	24.5	28	30	34
Arc Speed (in./min)	35 – 39	47 – 52	52 – 58	43 – 47	27 – 30
Electrode Req'd (lb/ft)	0.029	0.030	0.037	0.050	0.078
Flux Req'd (lb/ft)	0.08 – 0.10	0.08 – 0.12	0.08 – 0.12	0.09 – 0.13	0.09 – 0.13
Total Time (hr/ft of weld)	0.00540	0.00404	0.00364	0.00444	0.00702

See introductory notes.

SUBMERGED-ARC (SEMIAUTOMATIC) MANUAL

Welding Position: Horizontal
Weld Quality Level: Commercial
Steel Weldability: Good

Procedure for Conventional Fillets with 1-in. Electrical Stickout

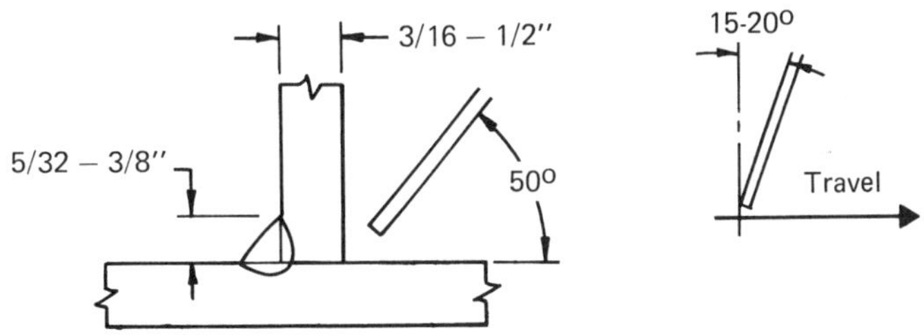

Weld Size, L (in.)	5/32	3/16	1/4	5/16	3/8
Plate Thickness (in.)	3/16	1/4	5/16	3/8	1/2
Pass	1	1	1	1	1
Electrode Size	5/64	5/64	5/64	5/64	5/64
Current (amp) DC(–)	310	340	375	400	425
Volts	31	33	34	35	37
Arc Speed (in./min)	40 – 44	31 – 34	21 – 23	15 – 17	11.5 – 12.5
Electrode Req'd (lb/ft)	0.058	0.078	0.13	0.20	0.30
Flux Req'd (lb/ft)	0.06 – 0.08	0.08 – 0.12	0.14 – 0.18	0.19 – 0.25	0.28 – 0.34
Total Time (hr/ft of weld)	0.00476	0.00615	0.00909	0.0125	0.0167

See introductory notes.

SUBMERGED-ARC (SEMIAUTOMATIC) MANUAL

Welding Position: Horizontal
Weld Quality Level: Commercial
Steel Weldability: Good

Procedure for Long Electrical Stickout (2-1/4")

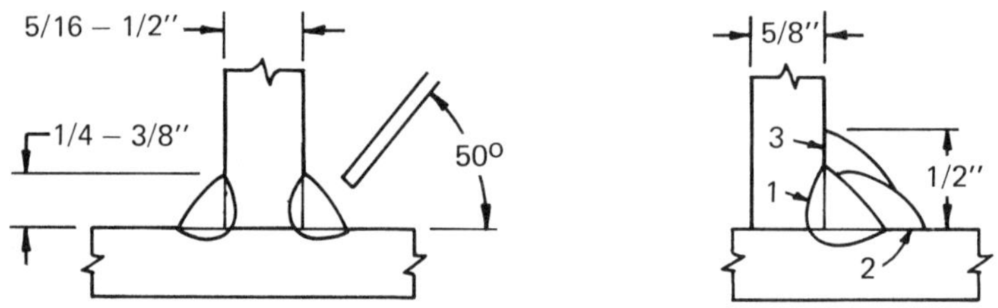

Weld Size, L (in.)	1/4	5/16	3/8	1/2
Plate Thickness (in.)	5/16	3/8	1/2	5/8
Pass	1	1	1	1, 2 & 3
Electrode Size	5/64	5/64	5/64	5/64
Current (amp) DC(–)	450	450	400	400
Volts	48	48	42	42
Arc Speed (in./min)	31 - 34	21 - 23	13 – 14	23 – 25*
Electrode Req'd (lb/ft)	0.16	0.23	0.32	0.56
Flux Req'd (lb/ft)	0.10 – 0.14	0.16 – 0.21	0.37 – 0.43	0.58 – 0.64
Total Time (hr/ft of weld)	0.00615	0.00909	0.0148	0.0250
Electrical Stickout, 2-1/4 in.				

* Each pass

See introductory notes.

SUBMERGED-ARC (SEMIAUTOMATIC) MANUAL

Welding Position: Horizontal
Weld Quality Level: Strength only
Steel Weldability: Good

Procedure for Penetration Fillets with 1-in. Electrical Stickout

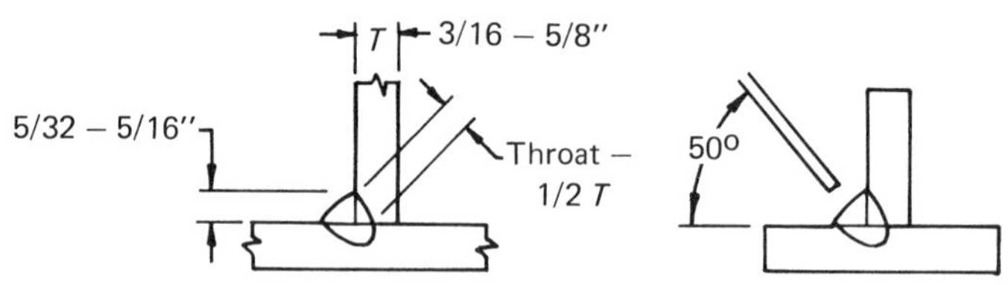

Weld Size, L (in.)	5/32(–)	3/16	3/16(+)	1/4(–)	1/4(+)	5/16(+)
Plate Thickness, T (in.)	3/16	1/4	5/16	3/8	1/2	5/8
Pass	1	1	1	1	1	1
Electrode Size	5/64	5/64	5/64	5/64	5/64	5/64
Current (amp) DC(+)	325	325	350	380	400	400
Volts	26	26	28	31	32	32
Arc Speed (in./min)	37 – 41	32 – 35	25 – 28	21 – 23	13 – 14	8 – 9
Electrode Req'd (lb/ft)	0.038	0.044	0.064	0.081	0.15	0.21
Flux Req'd (lb/ft)	0.05 – 0.07	0.06 – 0.09	0.10 – 0.14	0.15 – 0.20	0.22 – 0.28	0.30 – 0.40
Total Time (hr/ft of weld)	0.00513	0.00597	0.00755	0.00909	0.0148	0.0235

See Introductory notes.

SUBMERGED-ARC (SEMIAUTOMATIC) MANUAL

Welding Position: Horizontal
Weld Quality Level: Commercial
Steel Weldability: Good

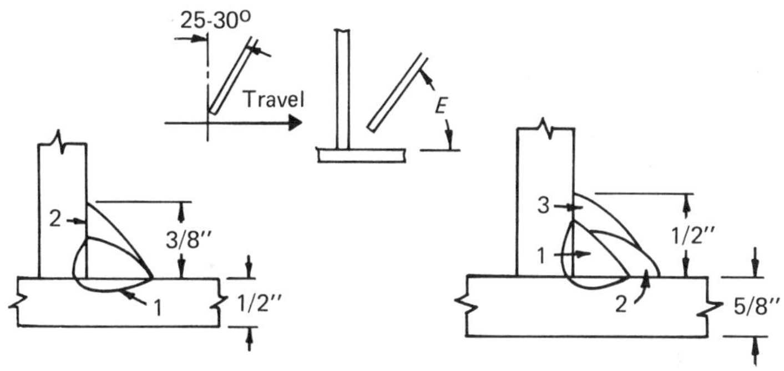

Weld Size, L (in.)	3/8		1/2		
Plate Thickness (in.)	1/2		5/8		
Pass	1	2	1	2	3
Electrode Size	5/64	5/64	5/64	5/64	5/64
Current (amp) DC(−)	425	425	425	425	425
Volts	35	35	35	35	35
Arc Speed (in./min)	23 – 25	23 – 25	20 – 22	20 – 22	20 – 22
Electrode Req'd (lb/ft)	0.31		0.53		
Flux Req'd (lb/ft)	0.32 – 0.50		0.55 – 0.80		
Total Time (hr/ft of weld)	0.0167		0.0286		
Electrode Angle, E (deg)	65	50	50	60	40

See introductory notes.

SUBMERGED-ARC (SEMIAUTOMATIC) MANUAL

Welding Position: Horizontal
Weld Quality Level: Commercial
Steel Weldability: Good

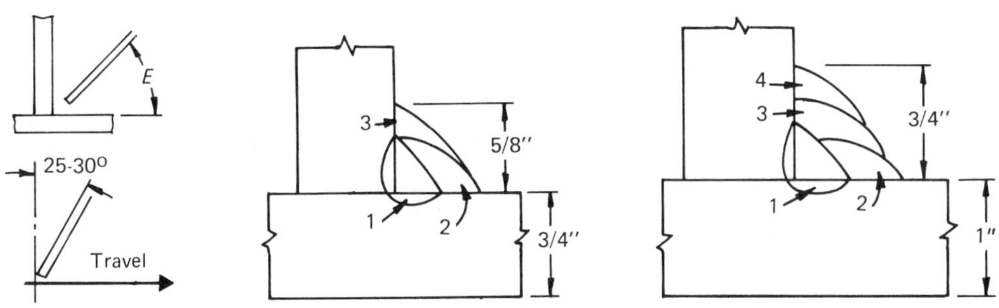

Weld Size, L (in.)	5/8			3/4			
Plate Thickness (in.)	3/4			1			
Pass	1	2	3	1	2	3	4
Electrode Size	5/64	5/64	5/64	5/64	5/64	5/64	5/64
Current (amp) DC(−)	425	425	425	425	425	425	425
Volts	35	35	35	35	35	35	35
Arc Speed (in./min)	13.5 – 14.5	13.5 – 14.5	12.5 – 13.5	12.5 – 13.5	12.5 – 13.5	12.5 – 13.5	12.5 – 13.5
Electrode Req'd (lbs/ft)	0.84			1.14			
Flux Req'd (lb/ft)	1.0 – 1.4			1.3 – 1.8			
Total Time (hr/ft of weld)	0.0440			0.0616			
Electrode angle, E (deg)	50	60	40	50	65	50	40

See introductory notes.

SUBMERGED-ARC (SEMIAUTOMATIC) MANUAL

Welding Position: Horizontal
Weld Quality Level: Commercial
Steel Weldability: Good

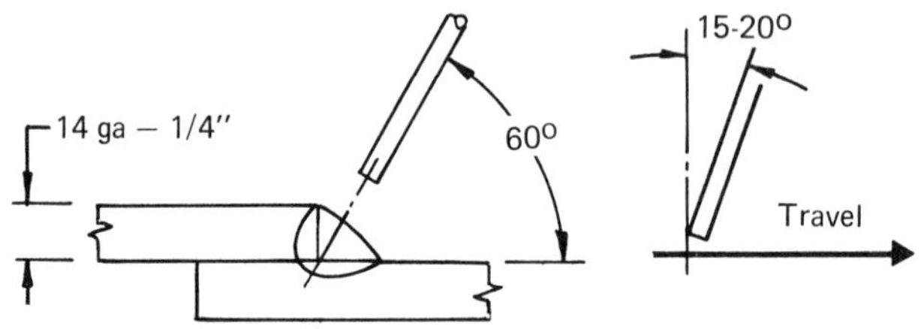

Weld Size, L (in.)	3/32(–)	3/32	1/8(–)	3/16	1/4
Plate Thickness (in.)	0.075 (14 ga)	0.105 (12 ga)	0.135 (10 ga)	3/16	1/4
Pass	1	1	1	1	1
Electrode Size	1/16	1/16	1/16	1/16	1/16
Current (amp) DC(+)	185	250	325	360	400
Volts	23	24.5	28.5	30.5	37
Arc Speed (in./min)	47 – 52	47 – 52	52 – 58	42 – 47	27 – 30
Electrode Req'd (lb/ft)	0.022	0.030	0.037	0.055	0.086
Flux Req'd (lb/ft)	0.09 – 0.13	0.11 – 0.15	0.11 – 0.15	0.12 – 0.16	0.13 – 0.17
Total Time (hr/ft of Weld)	0.00404	0.00404	0.00364	0.00449	0.00702

See introductory notes.

SUBMERGED-ARC (SEMIAUTOMATIC) MANUAL

Welding Position: Horizontal
Weld Quality Level: Commercial
Steel Weldability: Good

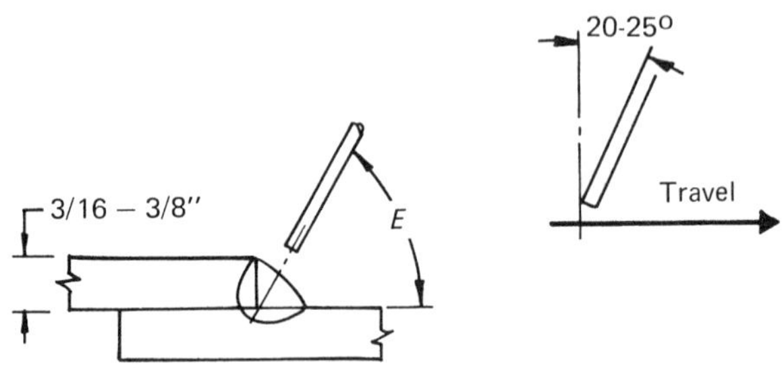

Weld Size, L (in.)	3/16	1/4	5/16(–)	5/16(+)
Plate Thickness (in.)	3/16	1/4	5/16	3/8
Pass	1	1	1	1
Electrode Size	5/64	5/64	5/64	5/64
Current (amp) DC(–)	400	425	450	450
Volts	36	37	38	38
Arc Speed (in./min.)	40 – 44	31 – 34	22 – 24	16 – 18
Electrode Req'd (lb/ft)	0.080	0.12	0.17	0.22
Flux Req'd (lb/ft)	0.10 – 0.13	0.14 – 0.18	0.19 – 0.25	0.22 – 0.29
Total Time (hr/ft of weld)	0.00476	0.00615	0.00870	0.0118
Electrode angle, E (deg)	65	60	55	55

See introductory notes.

SUBMERGED-ARC (SEMIAUTOMATIC) MECHANIZED

Welding Position: Flat
Weld Quality Level: Commercial
Steel Weldability: Good

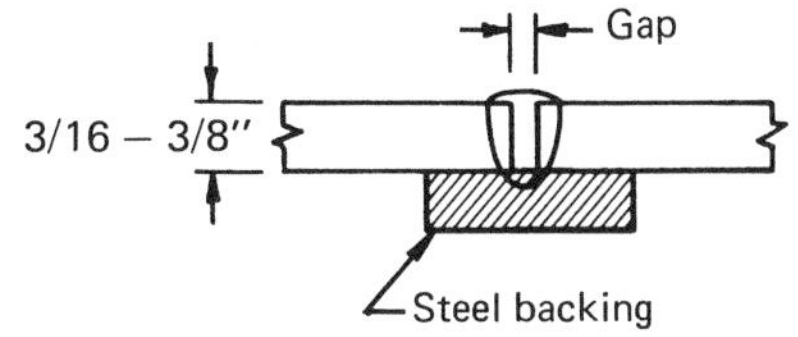

Plate Thickness (in.)	3/16	1/4	5/16	3/8
Pass	1	1	1	1
Electrode Size	3/32	3/32	3/32	3/32
Current (amp) DC(+)	500	550	575	600
Volts	32	34	35	36
Arc Speed (in./min)	26	20	18	14
Electrode Req'd (lb/ft)	0.12	0.18	0.22	0.32
Flux Req'd (lb/ft)	0.14 – 0.18	0.22 – 0.28	0.27 – 0.35	0.35 – 0.48
Total Time (hr/ft of weld)	0.00769	0.0100	0.0111	0.0143
Backing, minimum size (in.)	3/16 x 3/4	1/4 x 3/4	1/4 x 1	1/4 x 1
Gap (in.)	1/8	5/32	5/32	3/16

See introductory notes.

SUBMERGED-ARC (SEMIAUTOMATIC) MECHANIZED

Welding Position: Flat
Weld Quality Level: Commercial
Steel Weldability: Good
Welded from: Two sides

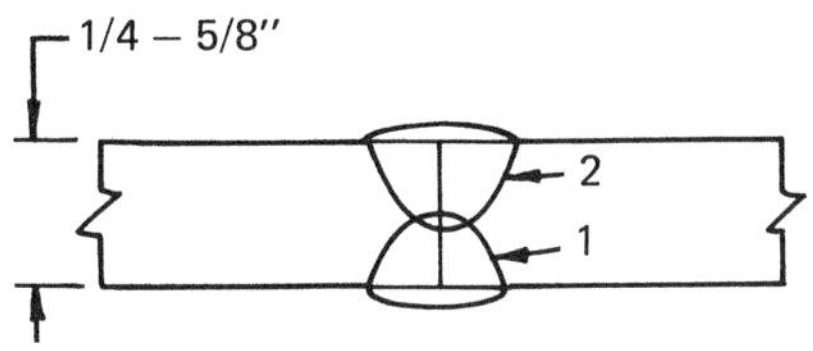

Plate Thickness (in.)	1/4		5/16		3/8		1/2		5/8	
Pass	1	2	1	2	1	2	1	2	1	2
Electrode Size	3/32	3/32	3/32	3/32	3/32	3/32	3/32	3/32	3/32	3/32
Current (amp) DC(+)	350	450	400	475	450	520	520	600	600	650
Volts	33	37	35	38	36	39	38	40	38	40
Arc Speed (in./min)	45	45	38	38	32	32	25	25	20	20
Electrode Req'd (lb/ft)	0.10		0.13		0.18		0.28		0.43	
Flux Req'd (lb/ft)	0.15 – 0.19		0.20 – 0.25		0.26 – 0.32		0.44 – 0.53		0.58 – 0.70	
Total Time (hr/ft of weld)	0.00889		0.0105		0.0125		0.0160		0.0200	

See introductory notes.

SUBMERGED-ARC (SEMIAUTOMATIC) MECHANIZED

Welding Position: Flat
Weld Quality Level: Commercial
Steel Weldability: Good
Welded from: Two sides

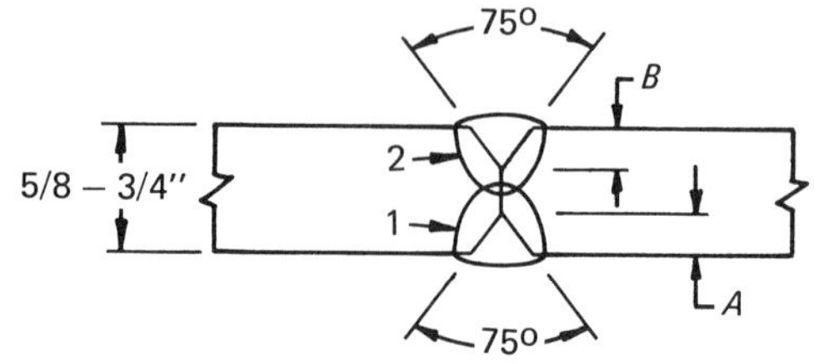

Plate Thickness (in.)	5/8		3/4	
Pass	1	2	1	2
Electrode Size	3/32	3/32	3/32	3/32
Current (amp) DC(+)	500	525	500	525
Volts	34	35	34	35
Arc Speed (in./min)	17.0	18.0	12.0	12.5
Electrode Req'd (lb/ft)	0.37		0.56	
Flux Req'd (lb/ft)	0.64 – 0.71		0.81 – 0.90	
Total Time (hr/ft of weld)	0.0229		0.0327	
Depth, A (in.)	7/32		9/32	
Depth, B (in.)	7/32		9/32	

See introductory notes.

SUBMERGED-ARC (SEMIAUTOMATIC) MECHANIZED

Welding Position: Flat
Weld Quality Level: Commercial
Steel Weldability: Good
Welded from: Two sides

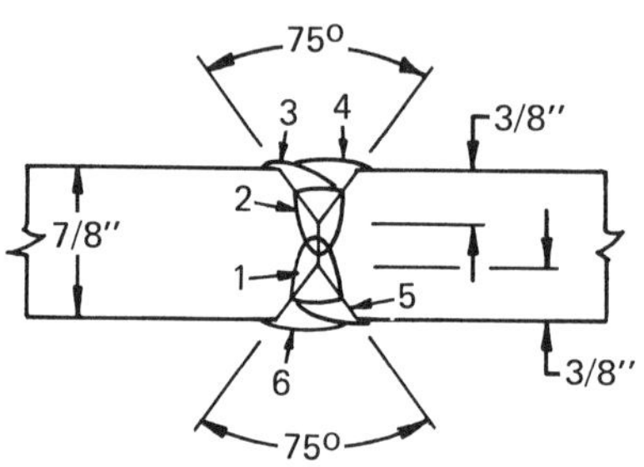

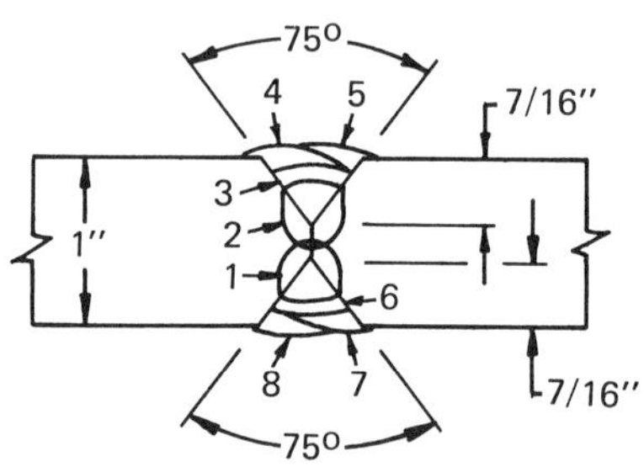

Plate Thickness (in.)	7/8			1		
Pass	1	2	3 – 6	1	2	3 – 8
Electrode Size	3/32	3/32	3/32	3/32	3/32	3/32
Current (amp) DC(–)	500	550	550	500	550	550
Volts	33	33	38	34	34	38
Arc Speed (in./min)	25	25	30	25	25	30
Electrode Req'd (lb/ft)	0.98			1.28		
Flux Req'd (lb/ft)	2.23 – 2.52			2.82 – 3.05		
Total Time (hr/ft of weld)	0.0427			0.0560		

See introductory notes.

SUBMERGED-ARC (SEMIAUTOMATIC) MECHANIZED

Welding Position: Flat
Weld Quality Level: Commercial
Steel Weldability: Good
Welded from: Two sides

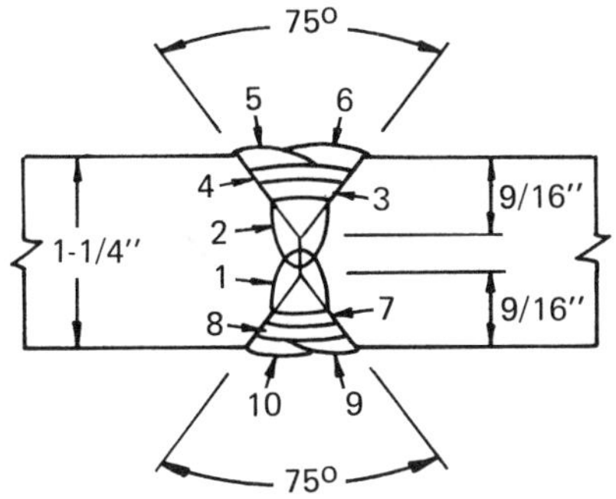

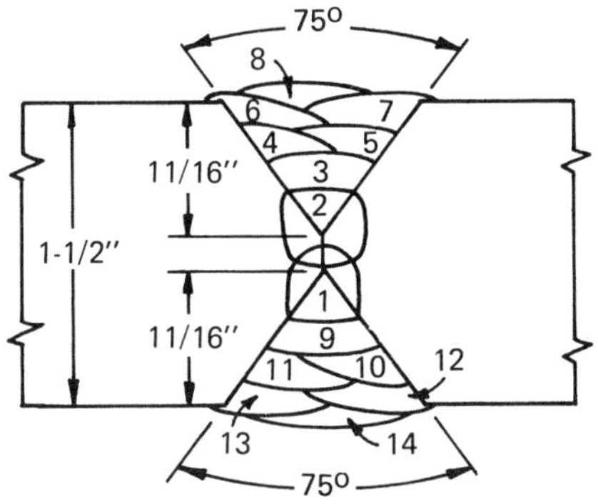

Plate Thickness (in.)	1-1/4			1-1/2		
Pass	1	2	3 – 10	1	2	3 – 14
Electrode Size	3/32	3/32	3/32	3/32	3/32	3/32
Current (amp) DC(–)	500	550	550	500	550	550
Volts	34	34	39	34	34	39
Arc Speed (in.min)	23	23	24	21	21	25
Electrode Req'd (lb/ft)	1.90			2.58		
Flux Req'd (lb/ft)	4.00 – 4.25			5.42 – 5.75		
Total Time (hr/ft of weld)	0.0841			0.115		

See introductory notes.

SUBMERGED-ARC (SEMIAUTOMATIC) MECHANIZED

Welding Position: Flat
Weld Quality Level: Commercial
Steel Weldability: Good

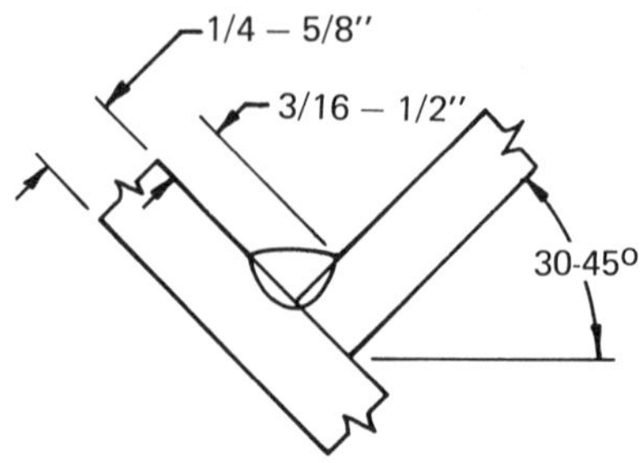

Weld Size, L (in.)	3/16	1/4	5/16	3/8	1/2
Plate Thickness (in.)	1/4	5/16	3/8	1/2	5/8
Pass	1	1	1	1	1
Electrode Size	3/32	3/32	3/32	3/32	3/32
Current (amp) DC(–)	475	550	600	600	600
Volts	37	40	42	42	42
Arc Speed (in./min)	50	36	28	19	11
Electrode Req'd (lb/ft)	0.08	0.13	0.18	0.26	0.46
Flux Req'd (lb/ft)	0.11 – 0.15	0.15 – 0.20	0.20 – 0.27	0.28 – 0.37	0.50 – 0.62
Total Time (hr/ft of weld)	0.00400	0.00556	0.00714	0.0105	0.0182

See introductory notes.

SUBMERGED-ARC (SEMIAUTOMATIC) MECHANIZED

Welding Position: Flat
Weld Quality Level: Commercial
Steel Weldability: Good

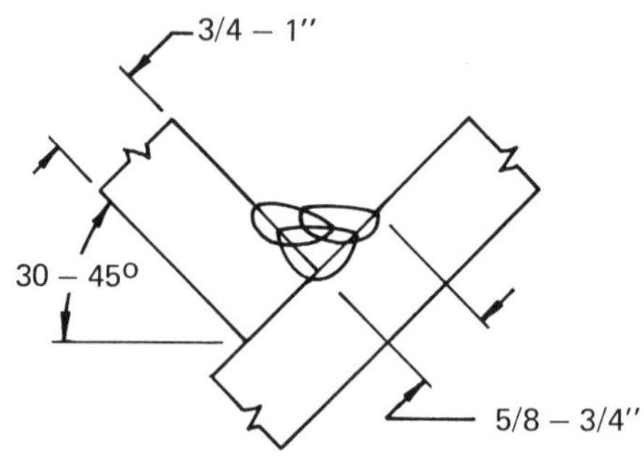

Weld Size, L (in.)	5/8	3/4
Plate Thickness (in.)	3/4	1
Pass	1 – 3	1 – 3
Electrode Size	3/32	3/32
Current (amp) DC(–)	600	600
Volts	42	42
Arc Speed (in./min)	21	15
Electrode Req'd (lb/ft)	0.74	1.05
Flux Req'd (lb/ft)	1.0 – 1.2	1.3 – 1.6
Total Time (hr/ft of weld)	0.0286	0.0400

See introductory notes.

SUBMERGED-ARC (SEMIAUTOMATIC) MECHANIZED

Welding Position: Horizontal
Weld Quality Level: Commercial
Steel Weldability: Good

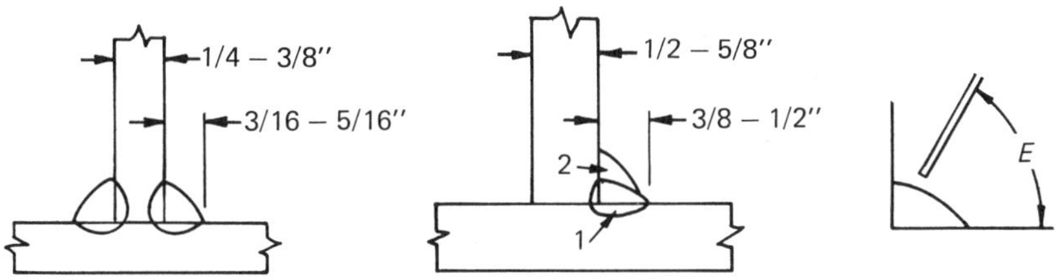

Weld Size, L (in.)	3/16	1/4	5/16	3/8		1/2	
Plate Thickness (in.)	1/4	5/16	3/8	1/2		5/8	
Pass	1	1	1	1	2	1	2
Electrode Size	3/32	3/32	3/32	3/32	3/32	3/32	3/32
Current (amp) DC(–)	450	500	550	500	500	500	500
Volts	34	36	38	36	36	36	36
Arc Speed (in./min)	45	32	24	28	28	16.5	16.5
Electrode Req'd (lb/ft)	0.08	0.13	0.20	0.29		0.49	
Flux Req'd (lb/ft)	0.13 – 0.17	0.18 – 0.23	0.24 – 0.30	0.35 – 0.45		0.60 – 0.75	
Total Time (hr/ft of weld)	0.00445	0.00625	0.00833	0.0143		0.0243	
Electrode Angle, E (deg)	50	50	50	65	40	65	40

See introductory notes.

SUBMERGED-ARC (SEMIAUTOMATIC) MECHANIZED

Welding Position: Horizontal
Weld Quality Level:
Commercial
Steel Weldability: Good

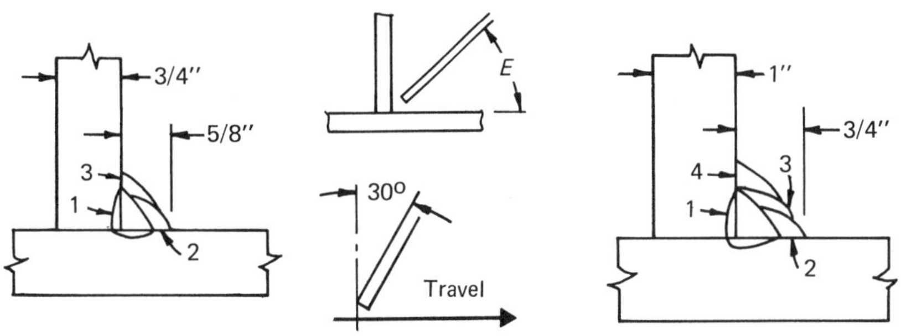

Weld Size, L (in.)	5/8			3/4			
Plate Thickness (in.)	3/4			1			
Pass	1	2	3	1	2	3	4
Electrode Size	3/32	3/32	3/32	3/32	3/32	3/32	3/32
Current (amp) DC(–)	500	500	500	500	500	500	500
Volts	36	36	36	36	36	36	36
Arc Speed (in./min)	17	17	17	16	16	16	16
Electrode Req'd (lb/ft)	0.78			1.10			
Flux Req'd (lb/ft)	1.1 – 1.3			1.4 – 1.7			
Total Time (hr/ft of weld)	0.0353			0.0500			
Electrode angle, E (deg)	50	60	40	50	65	50	40

See introductory notes.

SUBMERGED-ARC (SEMIAUTOMATIC) MECHANIZED

Welding Position: Horizontal
Weld Quality Level: Commercial
Weldability: Good

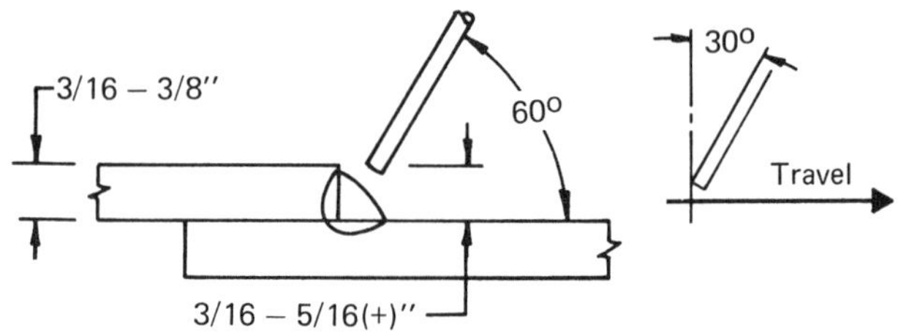

Weld Size, L (in.)	3/16	1/4	5/16(–)	5/16(+)
Plate Thickness (in.)	3/16	1/4	5/16	3/8
Pass	1	1	1	1
Electrode Size	3/32	3/32	3/32	3/32
Current (amp) DC(–)	500	500	500	500
Volts	36	36	36	36
Arc Speed (in./min)	50	32	24	20
Electrode Req'd (lb/ft)	0.08	0.13	0.19	0.23
Flux Req'd (lb/ft)	0.14 – 0.18	0.20 – 0.26	0.26 – 0.34	0.30 – 0.38
Total Time (hr/ft of weld)	0.00400	0.00625	0.00833	0.0100

A field installation to double-end pipe with the submerged-arc process.

Welding Carbon and Low-Alloy Steels with the Self-Shielded Flux-Cored Electrode Process

The self-shielded flux-cored electrode process is described in Section 5.3. As noted, the semiautomatic version of it has been finding increased use as a replacement for the manual shielded metal-arc process in the welding of mild and low-alloy steels. Since the process is of more recent origin, weldors are less experienced with it, and sometimes are apprehensive about their abilities to manually weld with a mechanized welding gun. Where the semiautomatic process has been introduced in welding shops, however, it has been found that an experienced weldor learns to handle the gun with a degree of proficiency after about a day of instruction and practice. In a week, he has mastered the art and is capable of bringing to the work the full cost-reduction benefits of semimechanization.

Operators of full-automatic equipment are usually already acquainted with full-mechanization and merely have to adapt to the specific performance characteristics of the process.

Because the process is relatively new, data on the process are not so rigidly standardized as with other processes. This is especially true in respect to performance characteristics of the electrodes, which vary not only from classification to classification, but also from manufacturer to manufacturer. For this reason, it is important that the manufacturer's performance data be consulted in making a selection of an electrode for a specific type of joint and welding position. Some electrodes can be used in the flat and horizontal positions only; others perform satisfactorily in all positions.

ELECTRODE CLASSIFICATION

The AWS A5.20-79 and A5.29-80 specifications respectively for carbon steels and low alloy steels describe flux-cored arc welding electrodes of both the self-shielded and gas-shielded types which follow the general pattern used in other AWS filler-metal classifications. Some variations are necessary, however, to accommodate the composite nature of the electrodes.

In a typical designation, E60T-7, the prefix "E" indicates an electrode for electric arc welding, as in other classification systems. The number "60" indicates the minimum as-welded tensile strength is 62,000 psi. The letter "T" indicates that the electrode is of tubular construction. The suffix "7" designates a particular grouping based upon chemical composition of deposited weld metal, type of current, polarity of operation, whether it can be used with or without gas, and other specific information for the category. The classification system does not define performance characteristics.

MILD STEEL ELECTRODES (AWS A5.20-79)

The following AWS classifications cover flux-cored electrodes, both self-shielding and those used with gas (see Section 6.5). Electrodes intended to be used in the self-shielding mode should not be used with shielding gas because this will prevent the deoxidzes from being burned out to a low level and reduce weld toughness. Electrodes requiring a gas shield should not be used without a gas shield because the lower level of deoxidizer in the wire may not be enough to prevent porosity. Required chemical compositions for these electrodes for welding mild steels are shown in Table 6-29; required tests for mechanical properties, soundness, and usability are indicated in Table 6-30; and mechanical-property requirements are listed in Table 6-31.

This specification classifies twelve different types of flux cored electrodes. Each suffix (T-1, T-2, T-3, T-4, T-5, T-6, T-7, T-8, T-10, T-11, T-G, and T-GS) indicates a general grouping of electrodes which have similar usability characteristics.

EXXT-1 Electrode Classification. Electrodes of the T-1 group are classified with CO_2 shielding gas by this specification. However, gas mixtures of argon-CO_2 are also used to improve usability, especially for out-of-position applications. Decreasing amounts of CO_2 in the argon-CO_2 mixture will increase manganese and silicon in the deposit and may improve the impact properties. These electrodes are designed for single- and multiple-pass welding. The larger diameters (usually 5/64 in. and larger) are used for welding in the flat position and for horizontal fillets. The smaller diameters (usually

1/16 in. and smaller) are used for welding in all positions. The T-1 electrodes are characterized by a spray transfer, low spatter loss, flat to slightly convex bead configuration, and a moderate volume of slag which completely covers the weld bead. Most electrodes in this group have a rutile base slag.

EXXT-2 Electrode Classification. Electrodes of this classification are essentially T-1 electrodes with higher manganese or silicon or both and are designed primarily for single-pass welding in the flat position and for horizontal fillets. The higher levels of deoxidizers in these electrodes allow single-pass welding over scaled or rimmed steel. This specification does not impose chemical composition requirements for single-pass electrodes since checking the undiluted deposit chemistry will not demonstrate their normal single-pass deposit chemistry. The two-run technique (one pass from each side on the butt welds) is equivalent to the single-pass applications because of the similar weld metal dilution obtained. T-2 electrodes that use manganese as the principal deoxidizing element give good mechanical properties in both single- and multiple-pass applications. However, the manganese content and tensile strength will be high in multiple-pass applications. These electrodes can be used for welding material which has heavier mill scale, rust, or other foreign materials on its surface than can be tolerated by some electrodes of the T-1 classification and still produce welds of radiographic quality. The arc characteristics and deposition rates are similar to those of the T-1 electrodes.

EXXT-3 Electrode Classification. Electrodes of the T-3 classification are self-shielded, are used on dc with positive polarity, and have a spray type transfer. The slag systems is designed to give characteristics which make possible very high welding speeds. They are used to make single-pass welds in the flat, horizontal, and (up to 20°) downhill positions on sheet metal up to 3/16 in. thick. They should not be used for welding material thicker than 3/16 in. or for making multiple-pass welds.

EXXT-4 Electrode Classification. Electrodes of the T-4 classification are self-shielded, operate on dc with positive polarity, and have a globular type transfer. The slag system is designed to give characteristics which make possible very high deposition rates and to desulfurize the weld metal to a very low level, which helps make the weld deposit very resistant to cracking. These electrodes are designed for low penetration beyond the root of the weld, enabling them to be used on weld joints with poor fit-up, and for single- and multiple-pass welding in the flat and horizontal positions.

EXXT-5 Electrode Classification. Electrodes of the T-5 group are designed to be used with CO_2 shielding gas (argon-CO_2 mixtures may be used as in the T-1 types) for single- and multiple-pass welding in the flat position and for horizontal fillets. These electrodes are characterized by a globular transfer, slightly convex bead configuration, and a thin slag which may not completely cover the weld bead. Electrodes in this group have a lime-fluoride base slag. Weld deposits produced by electrodes of

TABLE 6-29. Chemical Composition Requirements[a,b] (AWS A5.20-79)

AWS Classification[c]	Chemical Composition, Percent									
	Carbon	Phosphorus	Sulfur	Vanadium[d]	Silicon	Nickel[d]	Chromium[d]	Molybdenum[d]	Manganese	Aluminum[e]
EXXT-1 EXXT-4 EXXT-5 EXXT-6 EXXT-7 EXXT-8 EXXT-11 EXXT-G	f	0.04	0.03	0.08	0.90	0.50	0.20	0.30	1.75	1.8
EXXT-GS EXXT-2 EXXT-3 EXXT-10	No chemical requirements[g]									

[a] Values are maximums. Composition limits are intended to insure a plain carbon steel deposit.
[b] Chemical composition requirements for electrodes are based on the analysis of deposited weld metal.
[c] EXXT-9 designation is intentionally omitted.
[d] The amounts of these elements shall be reported if added intentionally.
[e] For self-shielded electrodes only.
[f] The amount of this element shall be determined.
[g] Analyses of undiluted weld metal are not meaningful, since these classifications are intended for single-pass welding with resultant high dilution.

TABLE 6-30. Required Tests (soundness, mechanical and usability)[a] of Welds Made by Carbon Steel Flux-Cored Electrodes (AWS A5.20-79)

AWS Classification[b]	Charpy V-Notch Impact Test	Radiographic Soundness Test	All-Weld-Metal Tension Test (Flat Position)	Fillet Weld Test	Transverse Tension Test	Longitudinal Guided Bend Test
EX0T-1	Req'd	Req'd	Req'd	—	—	—
EX0T-4	—					
EX0T-5	Req'd					
EX0T-6	Req'd					
EX0T-7	—					
EX0T-8	Req'd					
EX0T-11	—					
EX0T-G	—					
EX1T-1	Req'd	Req'd	Req'd	Req'd	—	—
EX1T-4	—					
EX1T-5	Req'd					
EX1T-6	Req'd					
EX1T-7	—					
EX1T-8	Req'd					
EX1T-11	—					
EX1T-G	—					
EXXT-2	—	—	—	—	Req'd	Req'd
EXXT-3						
EXXT-10						
EXXT-GS						

[a] The letters "Req'd" indicate that the test is required. Where no indication is shown, the test is not required.
[b] The 0 and 1 before the "T" refer to the usability of the electrode.
EX0T = horizontal and flat position. EX1T = all position. Classifications EXXT-2, EXXT-3, EXXT-10, and EXXT-GS are intended for single-pass welding.

TABLE 6-31. Mechanical Property Requirements for Welds Made with Carbon Steel Flux-Cored Electrodes in the As-Welded Condition[a] (AWS A5.20-79)

AWS Classification[b]	Tensile Strength[c] ksi	Yield Strength[c] ksi	Percent Elongation in 2 in.[c]	V-Notch Impact*
E6XT-1	62	50	22	20 ft-lb at 0°F
E6XT-4	62	50	22	No Requirements
E6XT-5	62	50	22	20 ft-lb at −20°F
E6XT-6	62	50	22	20 ft-lb at −20°F
E6XT-7	62	50	22	No Requirements
E6XT-8	62	50	22	20 ft-lb at −20°F
E6XT-11	62	50	22	No Requirements
E6XT-G	62	50	22	No Requirements
E6XT-GS	62	No Requirements	No Requirements	No Requirements
E7XT-1	72	60	22	20 ft-lb at 0°F
E7XT-2	72	No Requirements	No Requirements	No Requirements
E7XT-3	72	No Requirements	No Requirements	No Requirements
E7XT-4	72	60	22	No Requirements
E7XT-5	72	60	22	20 ft-lb at −20°F
E7XT-6	72	60	22	20 ft-lb at −20°F
E7XT-7	72	60	22	No Requirements
E7XT-8	72	60	22	20 ft-lb at −20°F
E7XT-10	72	No Requirements	No Requirements	No Requirements
E7XT-11	72	60	22	No Requirements
E7XT-G	72	60	22	No Requirements
E7XT-GS	72	No Requirements	No Requirements	No Requirements

[a] Single values shown are minimums.
[b] Classifications E6XT-GS, E7XT-2, E7XT-3, E7XT-10, and E7XT-GS are intended for single-pass welding. Only transverse tension and longitudinal guided bend tests are required.
[c] For each increase of one percentage point in elongation over the minimum, the minimum required yield strength or the tensile strength or both may decrease 1000 psi for a maximum reduction of 2000 psi in either the required minimum yield strength or the tensile strength or both.

this group have improved impact properties and crack resistance in comparison to the rutile types.

EXXT-6 Electrode Classification. Electrodes of the T-6 classification are self-shielded, operate on dc with positive polarity, and have a spray type transfer. The slag system is designed to give very good low-temperature impact properties, deep penetration beyond the root of the weld, and excellent deep groove slag removal. They are used for single- and multiple-pass welding in the flat and horizontal positions.

EXXT-7 Electrode Classification. Electrodes of the T-7 classification are self-shielded and operate on dc with negative polarity. The slag system is designed to give characteristics which allow the larger sizes of these electrodes to be used for high deposition rates and the smaller sizes to be used for all-position welding. The slag system is also designed to desulfurize the weld metal to a very low level, which helps make the weld deposit resistant to cracking. They are used for single- and multiple-pass welding.

EXXT-8 Electrode Classification. Electrodes of the T-8 classification are self-shielded and operate on dc with negative polarity. The slag system is designed to give characteristics which make it possible to use these electrodes for all-position welding. The slag system is also designed to produce very good low-temperature impact properties in the weld metal and to desulfurize the weld metal to a very low level which helps resist weld cracking. They are used for single- and multiple-pass welding.

EXXT-10 Electrode Classification. Electrodes of the T-10 classification are self-shielded and operate on dc with negative polarity. The slag system is designed to give characteristics that enable welds to be made at high travel speeds. They are used for making single-pass welds on material of any thickness in the flat, horizontal, and (up to 20°) downhill positions.

EXXT-11 Electrode Classification. Electrodes of the T-11 classification are self-shielded and operate on DC with negative polarity, and have a smooth spray type arc. The slag system is designed with characteristics that enable these electrodes to be used in all-position welding and to make welds at high travel speeds. They are used as general purpose electrodes for single- and multiple-pass welding in all positions.

EXXT-G Electrode Classification. The EXXT-G classification is for new multiple-pass electrodes which are not covered under any of the presently defined classifications. The multiple-pass properties can be anything covered by these specifications. The slag system, arc characteristics, weld appearance, and polarity are not defined.

EXXT-GS Electrode Classification. The EXXT-GS classification is for new single-pass electrodes which are not covered under any other presently defined classification. The single-pass properties can be anything covered by these specifications. The slag system, arc characteristics, weld appearance, and polarity are not defined.

Low-Alloy Flux-Cored Electrodes (AWS A5.29-80)

Flux Cored arc welding electrodes suitable for welding Low Alloy Steels are available as described in AWS Specification A5.29-80. The specific weld deposit chemistry, mechanical properties and impact properties for these electrodes are given in Tables 6-32 through 6-35. Both self-shielded and gas shielded electrodes are available in this group.

It should be noted that low-alloy flux-cored electrodes are produced in five general classifications, T1 (gas-shielded), T4 (self-shielded), T5 (gas-shielded), T8 (self-shielded, and TG (general, not falling within one of the other types).

Flux-cored electrode are usually packaged in 60-lb or smaller coils for use on semiautomatic or automatic wire feeders, or in 600-lb drums and 300 and 600-lb reels for use with high-production full-automatic welding equipment.

PERFORMANCE CHARACTERISTICS OF SELF-SHIELDED FLUX-CORED ELECTRODES

Performance characteristics include such operational features as welding-position capability, deposition rate, penetration, slag covering, and weld spatter. As noted in the introduction, performance characteristics of flux-cored electrodes within a classification can vary with the manufacturer.

Welding-position capability is influenced by weld-puddle size, slag volume, electrode diameter, and current capacity. All self-shielded flux-cored electrodes are capable of flat-position performance. Those with high deposition rates usually produce flat-position welds most economically. As with covered manual electrodes, the small-diameter flux-cored electrodes suitable for flat position welding can also be used to make vertical and overhead welds, but their performance and weld quality will not match that achieved with electrodes specifically designed for out-of-position work.

Typical deposition rates for E70T-4, E70T-7

TABLE 6-32. Chemical Requirements for Deposited Weld Metal[a] (AWS 5.29-80)

AWS Classification	Chemical composition, percent									
	C	Mn	P	S	Si	Ni	Cr	Mo	V	Al[b]	Cu
Carbon-molybdenum steel electrodes											
E70T5-A1 E80T1-A1 E81T1-A1	0.12	1.25	0.03	0.03	0.80	—	—	0.40/0.65	—	—	—
Chromium-molybdenum steel electrodes											
E81T1-B1	0.12	1.25	0.03	0.03	0.80	—	0.40/0.65	0.40/0.65	—	—	—
E80T5-B2L	0.05	1.25	0.03	0.03	0.80	—	1.00/1.50	0.40/0.65	—	—	—
E80T1-B2 E81T1-B2 E80T5-B2	0.12	1.25	0.03	0.03	0.80	—	1.00/1.50	0.40/0.65	—	—	—
E80T1-B2H	0.10/0.15	1.25	0.03	0.03	0.80	—	1.00/1.50	0.40/0.65	—	—	—
E90T1-B3L	0.05	1.25	0.03	0.03	0.80	—	2.00/2.50	0.90/1.20	—	—	—
E90T1-B3 E91T1-B3 E90T5-B3 E100T1-B3	0.12	1.25	0.03	0.03	0.80	—	2.00/2.50	0.90/1.20	—	—	—
E90T1-B3H	0.10/0.15	1.25	0.03	0.03	0.80	—	2.00/2.50	0.90/1.20	—	—	—
Nickel-steel electrodes											
E71T8-Ni1 E80T1-Ni1 E81T1-Ni1 E80T5-Ni1	0.12	1.50	0.03	0.03	0.80	0.80/1.10	0.15	0.35	0.05	1.8	—
E71T8-Ni2 E80T1-Ni2 E81T1-Ni2 E80T5-Ni2 E90T1-Ni2 E91T1-Ni2	0.12	1.50	0.03	0.03	0.80	1.75/2.75	—	—	—	1.8	—
E80T5-Ni3 E90T5-Ni3	0.12	1.50	0.03	0.03	0.80	2.75/3.75	—	—	—	—	—
Manganese-molybdenum steel electrodes											
E91T1-D1	0.12	1.25/2.00	0.03	0.03	0.80	—	—	0.25/0.55	—	—	—
E90T5-D2 E100T5-D2	0.15	1.65/2.25	0.03	0.03	0.80	—	—	0.25/0.55	—	—	—
E90T1-D3	0.12	1.00/1.75	0.03	0.03	0.80	—	—	0.40/0.65	—	—	—
All other low alloy steel electrodes											
E80T5-K1	0.15	0.80/1.40	0.03	0.03	0.80	0.80/1.10	0.15	0.20/0.65	0.05	—	—
E70T4-K2 E71T8-K2 E80T1-K2 E90T1-K2 E91T1-K2 E80T5-K2 E90T5-K2	0.15	0.50/1.75	0.03	0.03	0.80	1.00/2.00	0.15	0.35	0.05	1.8	—
E100T1-K3 E110T1-K3 E100T5-K3 E110T5-K3	0.15	0.75/2.25	0.03	0.03	0.80	1.25/2.60	0.15	0.25/0.65	0.05	—	—
E110T5-K4 E111T1-K4 E120T5-K4	0.15	1.20/2.25	0.03	0.03	0.80	1.75/2.60	0.20/0.60	0.30/0.65	0.05	—	—
E120T1-K5	0.10/0.25	0.60/1.60	0.03	0.03	0.80	0.75/2.00	0.20/0.70	0.15/0.55	0.05	—	—
E61T8-K6 E71T8-K6	0.15	0.50/1.50	0.03	0.03	0.80	0.40/1.10	0.15	0.15	0.05	1.8	—
E101T1-K7	0.15	1.00/1.75	0.03	0.03	0.80	2.00/2.75	—	—	—	—	—
EXXXTX-G	—	1.00 min*	0.03	0.03	0.80 min*	0.50 min*	0.30 min*	0.20 min*	0.10 min*	1.8	—
E80T1-W	0.12	0.50/1.30	0.03	0.03	0.35/0.80	0.40/0.80	0.45/0.70	—	—	—	0.30/0.75

a. single values are maximums unless otherwise noted.
b. For self-shielded electrodes only.

* In order to meet the alloy requirements of the G group, the weld deposit need have the minimum, as specified in the table, of only one of the elements listed.

TABLE 6-33. Required Tests (Soundness, Mechanical and Usability)[a] (AWS A5.29-80)

AWS Classification[b]	Radiographic Soundness Test	All-Weld-Metal Tension Test (flat position)	Vertical and Overhead Fillet Weld Test	Charpy V-notch Impact Test
EX0TX-X	req	req	—	[c]
EX1TX-X	req	req	req	[c]
EXXTX-G	—	—	—	—

[a] The letters "req" indicate that the test is required. Where no indication is shown, the test is not required.
[b] The 0 and 1 before the "T" refer to the usability of the electrodes: EX0T = horizontal and flat position; EX1TX = all position. The "X's" indicate the tensile strength, slag system, and chemical compositon designations of Table 6-32 and Table 6-34.
[c] The Charpy V-notch impact test is required when the classification in accordance with Table 6-35 indicates toughness performance.

TABLE 6-34. Mechanical Property Requirements[a] (AWS A5.29-80)

AWS Classification	Tensile Strength Range ksi	Yield Strength at 0.2% Offset, min - ksi	Percent Elongation in 2 in., min
E6XTX-X	60 to 80	50	22
E7XTX-X	70 to 90	58	20
E8XTX-X	80 to 100	68	19
E9XTX-X	90 to 110	78	17
E10XT-X	100 to 120	88	16
E11XT-X	110 to 130	98	15
E12XT-X	120 to 140	108	14
EXXXTX-G	Properties as agreed upon between supplier and purchaser		

[a] Properties of electrodes that use external gas shielding (EXXT1-X and EXXT5-X) vary with gas mixutres. Electrodes classified under this specification should not be used with gases other than those listed below without first consulting the manufacturer:
EXXT1-X — CO_2
EXXT4-X — None
EXXT5-X — CO_2
EXXT8-X — None
EXXTX-G — Not specified
All for multipass welds.

and E71T-8 self-shielded flux-cored electrodes are given in Table 6-36. The composition of the core and current-carrying capacity of the mild steel sheath influence deposition rates. The electrode's ability to tolerate extended electrical stickout while maintaining the required mechanical properties and quality of weld metal is also a factor that affects deposition rate.

The arc characteristics of the self-shielded flux-cored electrode are primarily responsible for penetration. Changing welding current and electrical stickout within the tolerance limits of the specific electrode will also affect penetration, but the effect is minimal compared to the difference in penetration of electrodes of different classifications. Electrodes that have a confined arc invariably give deep penetration, as compared to electrodes having the globular, soft-type arc. The need for penetration varies with joint design and precision in fitup.

Slag covering differs primarily in volume and type with the various flux-cored electrodes. Large slag volume is associated with limited-position operation. The type of slag produced by electrodes that give high volume is usually a "friable-porous"

TABLE 6-35. Impact Requirements (AWS A5.29-80)

Classification	Condition[a]	Impact Strength
E80T1-A1	PWHT	Not required
E81T1-A1	PWHT	Not required
E70T5-A1	PWHT	20 ft-lb @ −20°F (27J @ −30°C)
E81T1-B1	PWHT	Not required
E81T1-B2	PWHT	Not required
E80T1-B2	PWHT	Not required
E80T5-B2	PWHT	Not required
E80T1-B2H	PWHT	Not required
E80T5-B2L	PWHT	Not required
E90T1-B3	PWHT	Not required
E91T1-B3	PWHT	Not required
E90T5-B3	PWHT	Not required
E100T1-B3	PWHT	Not required
E90T1-B3L	PWHT	Not required
E90T1-B3H	PWHT	Not required
E71T8-Ni1	AW	20 ft-lb @ −20°F (27J @ −30°C)
E80T1-Ni1	AW	20 ft-lb @ −20°F (27J @ −30°C)
E81T1-Ni1	AW	20 ft-lb @ −20°F (27J @ −30°C)
E80T5-Ni1	PWHT	20 ft-lb @ −60°F (27J @ −51°C)
E71T8-Ni2	AW	20 ft-lb @ −20°F (27J @ −30°C)
E80T1-Ni2	AW	20 ft-lb @ −40°F (27J @ −40°C)
E81T1-Ni2	AW	20 ft-lb @ −40°F (27J @ −40°C)
E80T5-Ni2[b]	PWHT	20 ft-lb @ −75°F (27J @ −60°C)
E90T1-Ni2	AW	20 ft-lb @ −40°F (27J @ −40°C)
E91T1-Ni2	AW	20 ft-lb @ −40°F (27J @ −40°C)
E80T5-Ni3[b]	PWHT	20 ft-lb @ −100°F (27J @ −73°C)
E90T5-Ni3[b]	PWHT	20 ft-lb @ −100°F (27J @ −73°C)
E91T1-D1	AW	20 ft-lb @ −40°F (27J @ −40°C)
E90T5-D2	PWHT	20 ft-lb @ −60°F (27J @ −51°C)
E100T5-D2	PWHT	20 ft-lb @ −40°F (27J @ −40°C)
E90T1-D3	AW	20 ft-lb @ −20°F (27J @ −30°C)
E80T5-K1	AW	20 ft-lb @ −40°F (27J @ −40°C)
E70T4-K2	AW	20 ft-lb @ −0°F (27J @ −18°C)
E71T8-K2	AW	20 ft-lb @ −20°F (27J @ −30°C)
E80T1-K2	AW	20 ft-lb @ −20°F (27J @ −30°C)
E90T1-K2	AW	20 ft-lb @ −0°F (27J @ −18°C)
E91T1-K2	AW	20 ft-lb @ −0°F (27J @ −18°C)
E80T5-K2	AW	20 ft-lb @ −20°F (27J @ −30°C)
E90T5-K2	AW	20 ft-lb @ −60°F (27J @ −51°C)
E100T1-K3	AW	20 ft-lb @ −0°F (27J @ −18°C)
E110T1-K3	AW	20 ft-lb @ −0°F (27J @ −18°C)
E100T5-K3	AW	20 ft-lb @ −60°F (27J @ −51°C)
E110T5-K3	AW	20 ft-lb @ −60°F (27J @ −51°C)
E110T5-K4	AW	20 ft-lb @ −60°F (27J @ −51°C)
E111T1-K4	AW	20 ft-lb @ −60°F (27J @ −51°C)
E120T5-K4	AW	20 ft-lb @ −60°F (27J @ −51°C)
E120T1-K5	AW	Not required
E61T8-K6	AW	20 ft-lb @ −20°F (27J @ −30°C)
E71T8-K6	AW	20 ft-lb @ −20°F (27J @ −30°C)
E101T1-K7	AW	20 ft-lb @ −60°F (27J @ −51°C)
E80T1-W	AW	20 ft-lb @ −20°F (27J @ −30°C)
EXXXTX-G	Properties as agreed upon between supplier and purchaser	

[a] AW = As welded.
PWHT = Postweld heat treated in accordance with 3.6 of A5.29-80.
[b] PWHT temperatures in excess of 1150°F will decrease the impact value.

TABLE 6-36. Typical Deposition Rates for Flux-Cored Electrodes

Current (amp)	Wire Feed Rate (ipm)	Deposition Rate (lb/hr)
E70T-4 Fast-Fill Electrode — DC(+)		
.120-in. dia electrode (0.0279 lb/ft), 2-3/4-in. electrical stickout		
350	130	14.5
400	150	17.5
450	175	20.0
500	200	23.0
550	225	26.5
600	250	30.0
.120-in. dia electrode (0.0279 lb/ft), 3-3/4-in. electrical stickout		
450	205	25.0
500	260	29.5
550	305	34.5
600	355	39.5
3/32-in. dia electrode (0.0183 lb/ft), 2-3/4-in. electrical stickout		
250	120	10.0
300	150	12.5
350	195	15.5
400	235	18.5
E70T-7 Fast-Follow Electrode* — DC(−)		
7/64-in. dia electrode (0.0244 lb/ft), 1-1/2-in. electrical stickout		
325	100	10.0
400	145	14.5
450	175	18.0
550	240	25.5
625	300	33.0
3/32-in. dia electrode (0.0190 lb/ft), 1-1/2-in. electrical stickout		
200	75	5.4
300	135	10.2
325	150	11.4
400	210	16.5
450	270	22.0
E71T-8 Fast-Freeze Electrode* — DC(−)		
3/32-in. dia electrode (0.0198 lb/ft), 7/8-in. electrical stickout		
225	50	4.5
250	60	5.3
270	67	6.0
325	86	7.8
5/64-in. dia electrode (0.0138 lb/ft), 7/8-in. electrical stickout		
170	55	3.2
220	75	4.5
240	86	5.2
280	102	6.2

* A versatile electrode suitable for a wide variety of applications. for additional information, request "Innershield® Production Guide" from the Lincoln Electric Company, 22801 St. Clair Avenue, Cleveland, Ohio 44117.

slag or a "dense-solid" slag. Both types perform the desired action of blanketing the molten weld metal, preventing atmospheric contamination as it solidifies, and protecting the solidified weld metal as it cools. Both types are easily removed from welds.

Self-shielded flux-cored electrodes producing low-volume slags are either those particularly suited to vertical and overhead welding or those designed for the high-speed joining of sheet materials. Electrodes designed for out-of-position performance produce a low volume of slag that sets up quickly and is easily removed. The slag produced by the electrodes designed for high-speed sheet-metal welding promotes good wetting action and uniform fusion between the deposit and the base material along the edges of the weld. Such slag has a very dense, glasslike appearance and structure and is difficult to remove.

Weld spatter volume and size vary with arc characteristics. The electrodes that have a confined arc usually have a smaller droplet transfer in the arc stream. The spatter from this type is finer than that produced by the electrodes that have a globular, soft-arc characteristic. Since the deposition efficiencies of flux-cored electrodes are very similar, the spatter volume, although differing considerably in appearance, can be assumed to be approximately the same.

As a group, the self-shielded flux-cored electrodes are practically immune to moisture pickup. No special storage facilities under normal usage are required — a factor that is especially important in field welding where jobsite control of the moisture content is a difficult task with low-hydrogen stick electrodes. Welds produced in the field with flux-cored electrode wire taken directly from the shipping carton are as low or lower in hydrogen as those produced with carefully stored low-hydrogen electrodes. If the electrode wire is rusty, however, the result is the same as if it had picked up moisture, and excessive spatter and weld porosity can occur. In addition, wire feeding becomes difficult and contact tips wear rapidly.

EQUIPMENT FOR WELDING

The equipment required for self-shielded flux-cored arc welding consists of a power source, a wire-feeder mechanism, and a welding gun or head (see Sections 4.2, 4.3, 4.4, and 5.3).

Power Sources: Constant-voltage DC power sources — either transformer-rectifiers or generators — are used for welding with self-shielded flux-cored electrodes. The transformer-rectifier is preferably of the flat-slope type; the DC generator can be either electric-motor-driven or engine-driven (see Section 4.2).

Wire Feeders: The wire feeder (see Section 4.3) may be located either at a distance from the arc, as semiautomatic welding, or just above the nozzle, as in full-automatic welding. In semiautomatic welding, by the use of a wire-feed extension, the feeder may be as much as 45 ft ahead of the coil of electrode and 15 ft behind the welding gun. In

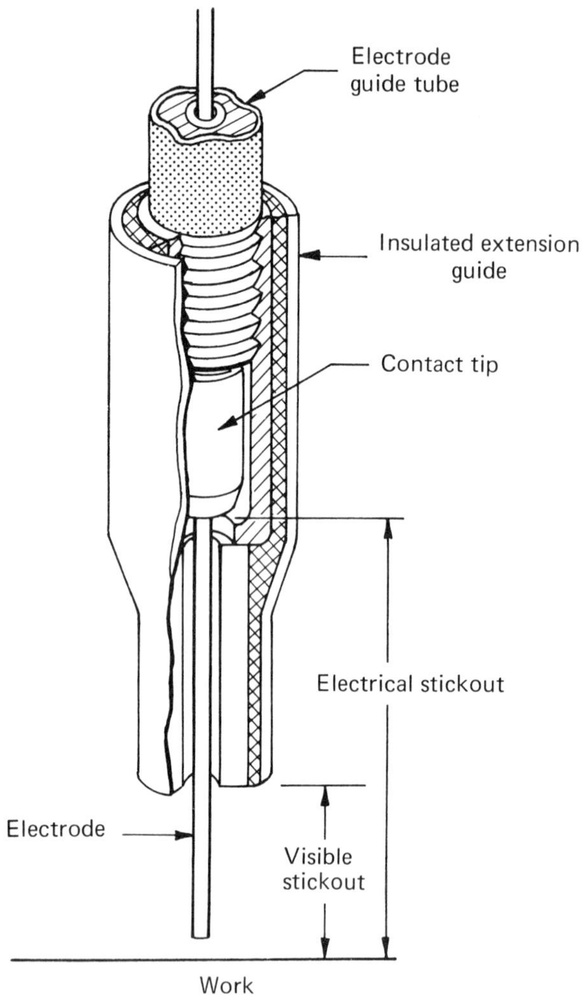

Fig. 6-74. Electrical stickout is the length of electrode wire between the point of electrical contact in a gun or welding head and the arc.

this arrangement, the feeder pulls electrode from the coil and pushes it another 15 ft to the gun. In standard operations, electrode is pushed to the semiautomatic gun and pulled to the full-automatic welding head. The wire-feed motor and rolls in a typical full-automatic head are directly above the nozzle, and the electrode is pulled from a reel that rides on the travel carriage with the welding head.

Some wire-feeding units for semiautomatic flux-cored electrode welding can also be used with semiautomatic submerged-arc welding — or, vice versa. Manufacturers of such equipment provide the basic units with an optional flux-tank, so that they are usable either for the submerged-arc or open-arc process.

Welding Guns: Welding guns for semiautomatic work (see Section 4.3) should be light and maneuverable to facilitate sustained high-speed work. They should be equipped with a small guide tip for reaching into deep grooves. Guns are made in light, medium, and heavy-duty models, with different ampere ratings and to accommodate different electrode diameters. Electrical contact is made within the nozzle of the gun, and the electrode is electrically "cold" until the trigger is pressed. A shield protects the operator's hand from excessive heat and sparks on medium and heavy-duty guns.

Nozzles for full-automatic self-shielded arc welding are also available in light, medium, and heavy-duty types, with optional water-cooling attachments for heavy work.

Electrical Stickout Guide Tips: The procedures for self-shielded flux-cored arc welding specify electrical stickout along with electrode type and diameter and current and voltage ranges. Thus, the procedures for a typical 1/8-in. E70T-4 electrode wire may prescribe a current range of 325 to 600 amp at 28 to 35 volts with a 2-3/4-in. electrical stickout, or a current range of 400 to 600 amp at 34 to 37 volts with a 3-3/4-in. electrical stickout. Amount of electrical stickout must, therefore, be selected in advance of the welding, along with the type and size of electrode.

A long electrical stickout (Fig. 6-74) — the distance from the arc to the point of electrical contact — increases deposition rate by preheating the wire before it is melted at the arc. Effects on welding costs can be significant; deposition rates can increase by as much as 50%. With the various flux-cored electrodes, the electrical stickouts prescribed in procedures range from 3/4 to 3-3/4-in. Use of the 3-3/4-in. stickout is generally limited to 5/16-in. and larger leg-size flat fillets, multiple-pass flat fillets, and flat deep-groove butt welds.

The specified electrical stickouts are obtained by using the proper guide tip and visible stickout on the welding gun. Thus, a medium-length guide tip and a 1-3/8-in. visible stickout provide a 2-3/4-in. electrical stickout, and a long (3-3/8-in.) guide tip and a 1-3/8-in. visible stickout produce a 3-3/4-in. stickout. When changing from a medium guide tip to a long guide tip for a 3-3/4-in. stickout with E70T-4 electrode, the voltage must be increased two to three volts to obtain a good flat bead, along with an increase in the current setting, which increases the wire-feed speed. The need for increases in current and voltage is shown by the higher ranges specified in the procedures.

A guide tip is not usually used with fill-freeze and fast-freeze electrodes, and electrical stickout is limited. The normal stickout with these electrodes is 3/4 to 1-in., although sound welds can be made with the fill-freeze types with stickouts from 5/8 to

1-1/2-in. The longer stickouts are used for making horizontal butt welds and for handling poor fitup.

SEMIAUTOMATIC OPERATING TECHNIQUES

Before welding, control settings should be checked carefully. Control settings should be within the range specified by the procedures, and adjusted according to past experience with the specific joint. Drive rolls and wire guide tubes should be correct for the wire size, and drive-roll pressure should be adjusted according to the manufacturer's instruction. The wire feeder and power source should be set for constant-voltage output. The gun, cable, and nozzle contact tip should be correct for the wire size and for the stickout.

Starting the Arc: To start the arc, the electrode is "inched" out beyond the nozzle to the visible stickout recommended for the electrode size and guide tip. The tip of the electrode is positioned just off, or lightly touching, the work, and the trigger is pressed to start the arc. The electrode should not be pushed into the joint as it melts away, as in stick-electrode welding, since the mechanical feed will take care of advancing the electrode. Welding is stopped by releasing the trigger or quickly pulling the gun from the work. The instruction manual with the specific wire feeder usually gives specific recommendations on setting feed speed and open-circuit voltage to facilitate starting.

When a long electrical stickout is to be used, it is best to start with a visible stickout of about 1/2-in., and increase the visible stickout to the specified amount after the arc has been established.

Accommodating Poor Fitup: As noted in Section 5.3, one of the advantages of flux-cored electrode welding is the ability to handle poor fitup. With a fast-fill electrode poor fitup can be accommodated by increasing the visible stickout to as much as 3-in. Pulling the gun away from the work to increase the visible stickout reduces the current, and thus the penetration, and helps to avoid melt-through. After a poor fitup area has been traversed, normal stickout should be used for the remainder of the joint. This method of controlling penetration should be used only with the short guide tip.

With the fill-freeze electrodes, poor fitup can be handled by reducing the welding current to the minimum value specified in the procedures. Increasing the stickout to 1-1/2-in. also helps to reduce penetration and melt-through.

Removing Slag: Slag removal is easy in most self-shielded flux-cored electrode welding. In heavy fast-fill work, the slag often curls up and peels off behind the welding gun. Otherwise, a light scrape with a chipping hammer or wire brush is usually all that is needed to dislodge the slag.

Slag is occasionally trapped on 90° vertical welds or in a flat position convex bead. Entrapment can be avoided by proper bead location and drag angle and by using a smooth, even travel speed to insure good bead shape.

Electrode Position: The drag angle is the angle between the electrode center line and the seam center line in the direction of travel, as illustrated in Fig. 6-75(a). The desired drag angle is approximately the same as in stick-electrode welding. If slag tends to run ahead of the arc, the drag angle should be decreased.

For best bead shape on most 5/16-in. and larger horizontal fillets, the electrode should point at the bottom plate, as illustrated in Fig. 6-75(b), and the angle between the electrode and bottom plate should be less than 45°. With this arrangement, the molten metal washes up onto the vertical plate. Pointing the electrode directly into the joint and using a 45

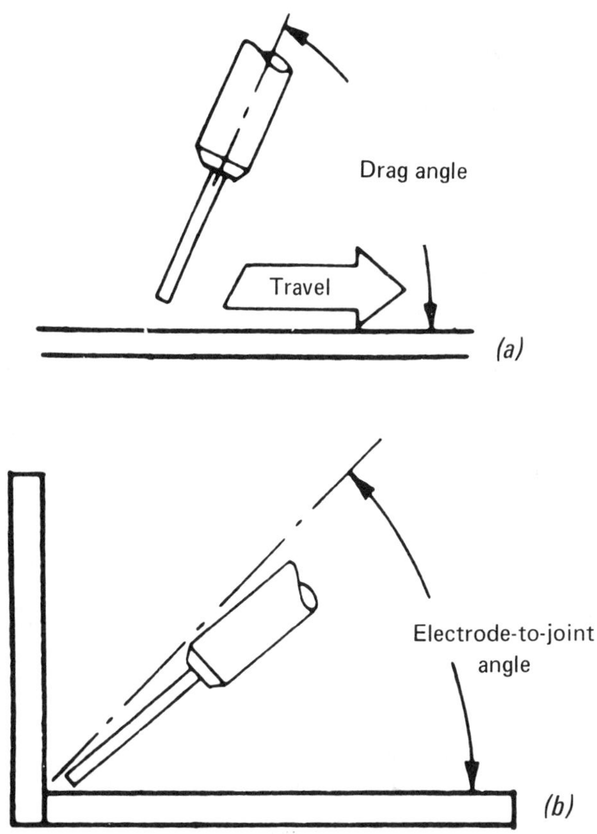

Fig. 6-75. The drag angle (a) and the electrode-to-joint angle (b) are variables that affect performance and weld appearance.

Butt Welds, Including Pipe First Two Passes with Stick Electrode

1. Make a distinct hesitation at the outer edges of bevel.
2. Minimize each upward step. Do not step up at the edges; come straight out from the hesitation point, and move up across the weld.

Fillet and Lap Welds 5/16-in. and Larger

1. FIRST PASS: Use a triangular weave with a hesitation at the outer edges.
2. SECOND PASS: Use a side-to-side weave similar to that used for butt welds. The previous bead should have a face width of 5/16 to 3/8-in. before weave is started.
3. Make welds smaller than 5/16-in. with vertical-down techniques.

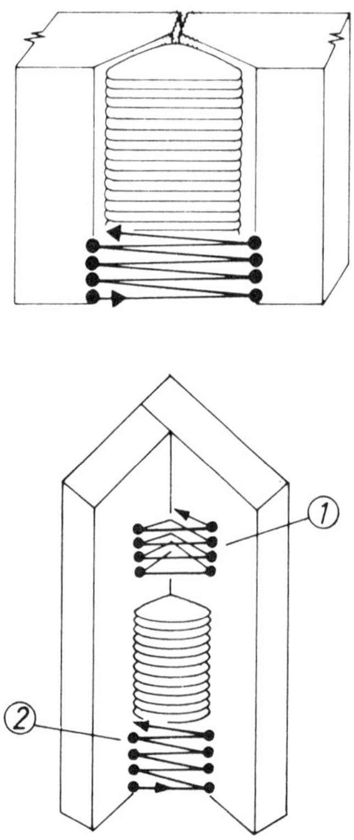

Fig. 6-76. Techniques for vertical-up and overhead welding with E71T-7 electrodes.

to 55° angle will decrease root-porosity problems, if they occur, but may produce spatter and a convex bead.

For 1/4-in. and smaller fillets, the wire should be pointed directly into the joint and the electrode angle held at about 40°.

When a follow-freeze electrode is used on sheet or thin plate, arc striking is best accomplished by starting with a 1/2-in. visible stickout and increasing it slowly to 1-in. Welding should be done with a stringer bead technique in all positions, with a steady travel speed. A steady speed is important; hesitation is likely to cause melt-through, with metal sag developing on the underside of the joint and porosity in the weld.

Fitup with light material should be tight, although a small gap on 12-gage to 3/16-in. steel can be handled by reducing the current and voltage about 10%. A steady whip or weave may help, but these motions should not be overdone.

The light slag on thin-gage material adheres tightly, but does not have to be removed. A wipe with a weak acid solution before painting will neutralize the weld area and remove smoke deposits.

For out-of-position welding with E71T-7 and E71T-11 electrode, best results are obtained by positioning the work downhill or vertical-down. Stringer beads should be used, with the current settings in the middle to high portion of the range. The gun should be tipped in the direction of travel so that the arc force helps hold the molten metal in the joint.

In vertical-up and overhead welding with E71T-7 electrodes, low-hydrogen techniques, as opposed to E6010 techniques, should be used. Whipping, breaking the arc, moving out of the puddle, or moving too fast in any direction should be avoided. Currents should be in the low portion of the range. Figure 6-76 shows general techniques for vertical-up and overhead work.

For vertical and overhead welding with E71T-8 type electrodes, similar techniques are used. Vertical up, multiple pass welding should generally be with a split weave pass sequence, with each bead approximately 3/4″ in width. Some E71T-8 type electrodes are capable of vertical down operation; for those products, a straight progression technique is normally recommended.

In full-automatic welding with self-shielded flux-cored electrodes, the work position angle, electrode-to-joint angle, and electrode drag angle are critical. Figure 6-77 shows how electrode angles and off-center circumferential distances vary for roundabout butt and lap welds in cylindrical assemblies from 3 to 30-in. in diameter.

OPERATING VARIABLES

Four major variables affect welding performance with flux-cored electrodes: arc voltage, wire feed speed rate, travel speed, and electrical stickout. These variables are interdependent, and, if one is changed, one or more of the other three usually require adjustment.

Arc voltage variations, with wire feed speed (current), travel speed, and electrical stickout held constant, produce these effects:

1. A high arc voltage produces a wider and flatter bead.
2. An excessive arc voltage may produce a porous weld.
3. A low voltage tends to cause a convex, ropey bead.
4. Extremely low voltage may produce a tendency for the wire to stub on the plate. The wire may dive through the molten metal and strike the joint bottom, pushing the gun up.

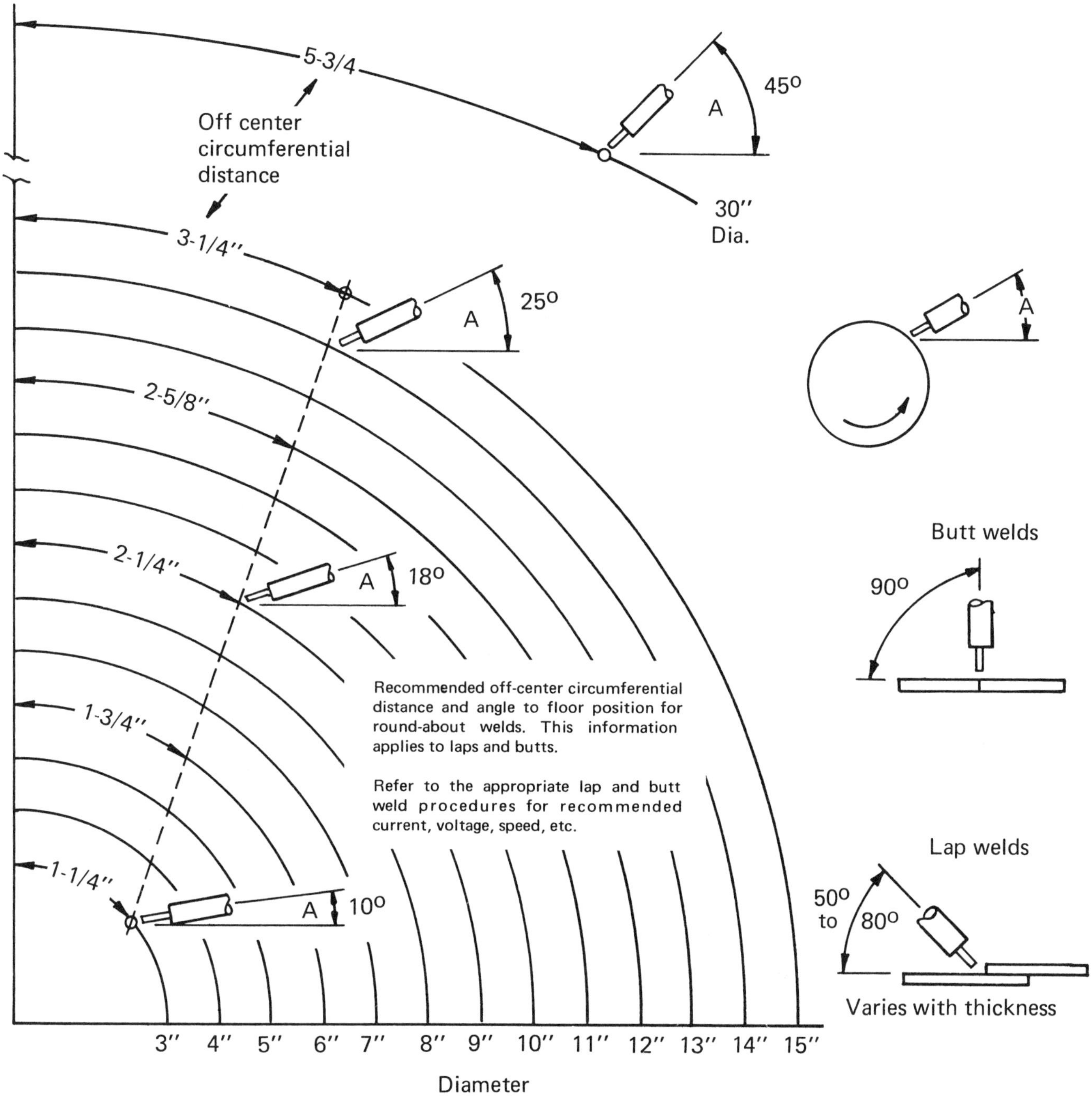

Fig. 6-77. Variations in electrode angle (a) and off-center circumferential distance when making roundabout butt and lap welds in cylindrical assemblies by full-automatic flux-cored welding.

In most applications, a good bead shape is obtained by using the highest voltage possible without causing porosity. With higher wire feed rates, higher voltages can be used without causing porosity.

Wire Feed Speed variations, when arc voltage, travel speed, and electrical stickout are held constant, have the following effects:

1. Increasing wire feed speed increases melt-off and deposition rates.
2. Excessive wire feed speed produces convex beads, resulting in poor appearance and wasted weld metal.
3. Too low wire feed speed gives a droplet transfer, with reduced penetration.

As wire feed speed is increased, arc voltage must also be increased to maintain good bead shape. Increased wire feed speed also increases the maximum voltage that can be used without causing porosity.

Travel Speed, assuming the other variables are held constant, can have the following effects:

1. Excessive travel speed increases the convexity of the bead and causes uneven edges.
2. Too slow a travel speed results in slag interference and inclusions and a rough, uneven bead.

Travel speed is always faster with self-shielded flux-cored electrodes than with stick electrodes — to accommodate the higher deposition rates. The beginning operator with the semiautomatic process will tend to move too slow because of his experience with stick-electrode welding. As in all other welding processes, travel speed should be that necessary to handle the molten metal and slag and produce the desired weld size.

Uniformity in travel speed is important. It is accomplished by maintaining a uniform distance between the wire and the molten slag behind the wire.

Electrical stickout variations, with the other variables held constant, have the following effects:

1. Increasing the stickout decreases the welding current and vice versa.
2. Increasing the stickout lowers the actual voltage across the arc. Lower arc voltage increases the convexity of the bead and reduces the tendency for porosity.
3. Short stickout gives greater penetration than long stickout.

TABLE 6-37. Troubleshooting Adjustments for Welding with Flux-Cored Electrodes

Problem	Solution*				
	Wire Feed Speed	Voltage	Speed	Stickout	Drag Angle
Porosity	5↑	1↓	4↓	2↑	3↑
Spatter	4↓†	1↑	5↓	3↓	2↓
Convexity	4↓	1↑	5↓	2↓	3↑
Back Arc Blow	4↓	3↓	5↓	2↑	1↑
Insufficient Penetration	2↑	3↓	4↑	1↓	5↑
Not Enough Follow	4↑	1↓	5↓	2↑	3↑
Stubbing	4↓	1↑		3↓	2↓

* Arrows indicate the need to increase or decrease the setting to correct the problem. Numbers indicated order of importance.

† Increasing the wire feed speed reduces droplet size and decreases spatter.

Obviously, a proper balance of the four variables is necessary for the best performance. Table 6-37 summarizes the adjustments to be made when trouble is encountered.

FULL-AUTOMATIC OPERATING TECHNIQUES

In full-automatic welding, the weld puddle provides a good index to performance. The puddle should be formed as illustrated in Fig. 6-78 at the start of the weld. Amount of molten metal present and amount and direction of the arc force are two principal factors that influence puddle shape. Sufficient molten metal must be available to form the puddle, and the puddle must follow smoothly behind the arc. A puddle with a sharp leading edge (as shown in the figure) is to be expected in downhill welding as a consequence of the arc force plus the force of gravity.

Following smoothly requires that the distance between the arc and the front edge of the puddle remain constant as welding progresses. With a smooth follow, metal freezes at the back edge at the same rate that new metal is added to the leading edge, and the finished weld is uniform, as illustrated in Fig. 6-79(a).

Lack of follow occurs when a stable leading edge of good shape does not form. When this happens, the puddle follows the arc irregularly, and the weld is thin, irregular, and may have open spots.

In some instances of lack of follow, the leading edge of the weld puddle begins to form after the weld progresses, but the arc force then blows it back into the puddle. The weld is irregular in appearance, as shown in Fig. 6-79(b). The weld builds up from a low, thin bead to a lump at the point where the leading edge was blown back into the puddle. The lump is followed by an open spot in the weld.

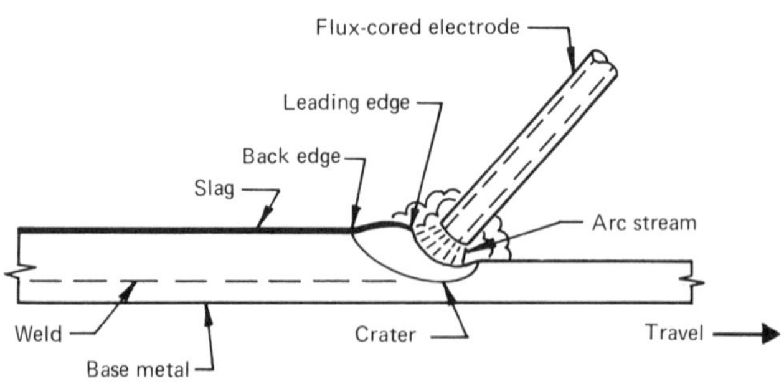

Fig. 6-78. A properly formed weld puddle when full-automatic welding.

(a) (b) (c) (d)

Fig. 6-79. Weld beads formed by full-automatic flux-cored welding. Properly formed bead is shown at (a). The bead in (b) shows lack of follow; the leading edge has been blown back into the puddle. The bead at (c) shows what happens when lack of follow results from excessively high speed or low current. In (d), follow has been too close, as a result of slow speed, high current, or high downhill angle.

The primary causes for lack of follow are insufficient molten metal or too small a downhill angle. The downhill angle is the angle between the horizontal and the joint being welded. With the E70T-3 electrodes in full-automatic welding, welding downhill greatly aids in the proper formation of the weld puddle. Fixtures for full-automatic welding with these electrodes should be made so that the work angle is adjustable. Because of gravity, the weld puddle tends to flow down the workpiece when welding downhill. The arc force holds the metal back in the puddle. Proper interaction of the two forces provides a good weld puddle with a sharp leading edge.

Lack of follow can often be corrected by increasing the downhill angle. Reducing the angle helps keep the puddle from following the arc too closely. Lack of follow can also result from excessively high speed or low current. Figure 6-79(c) illustrates the effects on the weld bead.

Following too closely occurs when the weld puddle flows into the arc and touches the end of the electrode. The resulting short-circuit current blasts the metal away in a shower of spatter, leaving unfilled spots in the weld. The finished weld is wide and irregular, as shown in Fig. 6-79(d). Too much molten metal or insufficient arc force to hold the leading edge in place, as well as excessive downhill angle, can also cause this condition.

The **drag angle** (Fig. 6-75) also affects welding performance and bead shape in full-automatic welding. The proper drag angle prevents spatter at high travel speeds; the higher the travel speed, within limits, the lower the angle. Reducing the drag angle causes a small increase in the arc force and reduced penetration. Changing the drag angle, however, to correct welding defects should be regarded as a fine adjustment, to be used only after travel speed, work angle, and current are adjusted. Figure 6-80 illustrates the effect of drag angle on bead appearance.

Travel speed is important in full-automatic welding with fast-follow self-shielded flux-cored electrodes to obtain proper formation of the weld puddle. Lack of follow is corrected by decreasing travel speed. This permits the build-up of the larger pool of molten metal necessary to form a weld puddle with a sharp leading edge. Overfollow is corrected by increasing travel speed, which reduces the puddle size so that the leading edge does not become so heavy that it overpowers the arc force.

Travel speed also affects penetration. Generally, the slower the travel speed, the deeper the penetration. When travel speed is reduced, the welding wire feed speed may also have to be reduced to prevent melt-through, particularly with fast-follow electrodes.

Welding wire feed speed affects both penetration and melt-off rate. Wire feed speed should be sufficient to produce the desired penetration without melting through. Where penetration is not critical, the wire feed speed can be adjusted to form a good weld puddle. Increasing wire feed speed increases the melt-off rate and, therefore, the amount of metal supplied to the puddle. Higher wire feed speed also increases the arc force. The wire feed speed used must, however, be within the prescribed

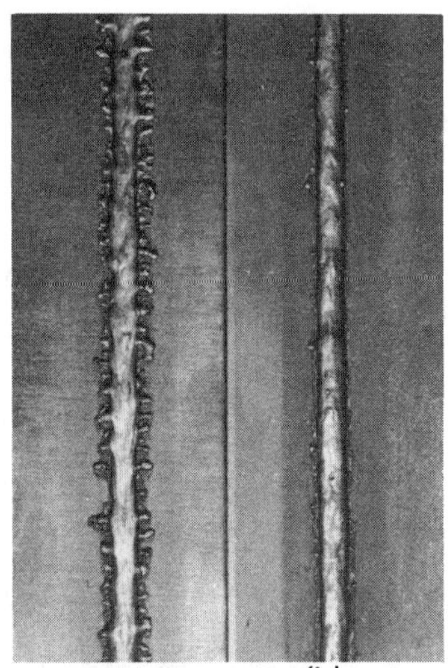

(a) (b)

Fig. 6-80. Effect of drag angle on bead appearance. Bead at (a) was produced by proper drag angle; bead at (b) resulted from too large a drag angle.

range for the electrode type and size. Excessively high wire feed speeds produce undercut and an erratic arc. Low wire feed speeds produce an unstable arc.

If porosity occurs in welds made with full-automatic equipment, lowering the voltage to several volts below the specified minimum will show whether excessive voltage is the cause. If the porosity does not disappear, back blow or contamination of the joint or electrode wire by water, rust, oil, or dirt should be checked as possible causes.

Vertical-Up Automatic Welding Technique

Techniques have been developed to weld steel plates 3/8 to 3-in. thick in the vertical position with self-shielded flux-cored electrode.

The plates are assembled with a simple square edge butt joint and welded in the vertical position. The edges of the plates are spaced about 1/2-in. or more to allow entry of the welding electrode. Water-cooled copper dams are applied to each side of the joint so the dams and the edges of the plates form a cavity into which the weld metal is deposited. As the cavity fills, the electrode nozzle, copper dams, and all the electrode feeding mechanism and controls move upward. For plates up to 1-1/4-in. thick, the electrode position remains fixed but for thicker plate the electrode oscillates back and forth across the seam to fuse the entire cross-section of the joint.

The square edge preparation can be modified to allow a small included angle. This permits easier access for the electrode and nozzle. Another modification is to use a steel or copper backing on one side and only one sliding dam.

A special self-shielded flux-cored electrode is required for the vertical-up technique in order to obtain sound welds with mechanical properties that will meet the code requirements. Consult the electrode supplier for the proper electrode.

The advantages of this technique are:

- High deposition rates with no interruptions to change electrodes.
- Welding speeds of 6-in. per minute on 3/4-in. plate and 4-1/2-in. per minute on 1-1/4-in. plate.
- Electrode is self-shielding, no external gas supply needed.
- Inexpensive plate preparation.
- X-ray quality welds.
- Good mechanical properties.

Welding gondola cars with the self-shielded flux-cored electrode process.

INTRODUCTION TO WELDING PROCEDURES

The ideal welding procedure is the one that will produce acceptable quality welds at the lowest overall cost. So many factors influence the optimum welding conditions that it is impossible to write procedures for each set of conditions. In selecting a procedure, the best approach is to study the conditions of the application and then choose the procedure that most nearly accommodates them. The procedures given here are typical, and it may be necessary to make adjustments for a particular application to produce a satisfactory weld.

For some joints, different procedures are offered to suit the weld quality — code quality, commercial quality, and strength only — that may be required.

Code-Quality Procedures

Code quality procedures are intended to provide the highest level of quality and appearance. To accomplish this, conservative currents and travel speeds are recommended.

These procedures are aimed at producing welds that will meet the requirements of the commonly used codes: AWS Structural, AISC Buildings and Bridges, ASME Pressure Vessels, AASHTO Bridges, and others. Code quality welds are intended to be defect-free to the extent that they will measure up to the nondestructive testing requirements normally imposed by these codes. This implies crack-free, pressure-tight welds, with little or no porosity or undercut.

The specific requirements of these codes are so numerous and varied that code quality procedures may not satisfy every detail of a specific code. Procedure qualification tests are recommended to confirm the acceptability of chosen procedures.

All butt welds made to code quality are full penetration; fillet welds are full size, as required by most codes. (The theoretical throat rather than the true throat is used as the basis of calculating strength.)

Commercial-Quality Procedures

Commercial quality implies a level of quality and appearance that will meet the nominal requirements imposed on most of the welding done commercially. These welds will be pressure-tight and crack-free. They will have good appearance, and they will meet the normal strength requirements of the joint.

Procedures for commercial-quality welds are not as conservative as code quality procedures; speeds and currents are generally higher. Welds made according to these procedures may have minor discontinuities that would be objectionable to the more demanding codes.

It is recommended that appropriate tests be performed to confirm the acceptability of the selected procedure for the application at hand prior to putting it into production.

Strength Only Procedures

The purpose of strength-only procedures is to produce the highest speed welding at the lowest possible cost. To accomplish this, the weld appearance and quality need only be good enough to do the specific job for which it is intended. In this category, imperfections may be acceptable provided that the welds will perform satisfactorily under service conditions.

This quality level may apply to seal beads, partial-penetration welds, some lap welds, high-speed edge welds, and to many low-stress welded connections which have no real need for better quality at higher cost. On fillet welds the joint must be welded from both sides.

Since this is a compromise quality level, realistic tests should be made to confirm the satisfactory performance of welds made according to these procedures. Welds of strength-only quality will be free of cracks but may not be pressure-tight.

Weldability of Material

Weldability (see Section 6.1) of a steel has a considerable effect on the welding procedure.

Good weldability indicates a steel with a composition that is within the preferred range (see Table 6-1) — one whose chemistry does not limit the welding speed.

Fair weldability indicates a steel with one or more elements outside the preferred range or one that contains one or more alloys. These steels require a lower welding speed or a mild preheat, or both, to minimize defects such as porosity, cracking, and undercut.

The addition of alloys to steel that enhance the mechanical properties or hardenability usually have an adverse effect on weldability. In general, the weldability of low-alloy steels is never better than "fair."

Procedure Notes

In the following tables, procedures for some joints are given for two different electrode sizes. The smaller electrode provides greater flexibility and operator comfort at some sacrifice in welding speed.

It is important to control the arc voltage within the range specified in the procedures. These voltages do not allow for poor electrical or ground connections or for undersize welding cables. Such faulty conditions should be corrected rather than compensating by increasing the arc voltage.

The following procedures have been developed with electrodes having these specific characteristics:

E70T-3. A high degree of fast-follow characteristic is necessary in order to obtain the travel speeds specified in the procedures. In general, the E70T-3 electrodes are designed for single pass welds on 10 gage and thinner steel. On 3/16-in. and thicker steel the ductility of the E70T-3 weld metal is reduced and should be thoroughly tested to verify the suitability of the weld metal for service requirements.

E70T-4. This electrode is capable of operating with long electrical stickout and, as a result, has a high deposition rate. Penetration is low and will tolerate a large degree of poor fitup. The deposit is very crack resistant on high sulfur steel and can be used where other electrodes are not satisfactory.

E70T-7. This electrode provides good wetting for fast travel speed and high deposition rates and good penetration on fillet and lap welds. Good penetration and easy slag removal permit use of narrow gap and small bevel angle in butt welds.Can be used in the flat, horizontal and downhill positions and can be used for single or multipass applications. Desulfurizes weld deposit to resist weld cracking in restrained joints.

E71T-7. This electrode can be used in all positions for single or multipass welds. Recommended for vertical up and overhead welding. Desulfurizes weld deposit to resist weld cracking in restrained joints.

E70T-8. This electrode will operate in the flat, horizontal and vertical up positions. The slag system will provide very good low temperature impact properties and desulfurize the weld metal to resist weld cracking.

E71T-8. This electrode can be used in all positions except vertical down. The slag system will provide very good low temperature impact properties and desulfurize the deposit to help avoid weld cracking. It can handle poor fitup and make root passes without backup bars or back gouging.

E70T-10. This electrode is intended for single pass welding of any thickness material in the flat, horizontal and maximum of 20° downhill positions. Slag system allows welds to be made at high travel speeds and easy slag removal. Best for fillet and lap joints and also good for corner, edge and butt joints.

E71T-11. This electrode can be used in all positions. A smooth spray type arc combined with the all-position slag system permits high travel speeds. It can be used for single and multipass welding.

Although the self-shielded flux-cored arc welding process has been in use for many years, the development of new electrodes with improved properties and operating characteristics has continued and some are available commercially. Therefore, when selecting and ordering electrodes for a particular application, the supplier should be consulted regarding the latest availability of electrodes to the AWS specification being used.

In the following fillet weld procedures, the fillet size is always associated with a particular plate thickness. This relationship is given solely for the purpose of designing a welding procedure and does not imply that a certain size fillet is the only size applicable to that plate thickness. In some of the procedures, the fillet size shown is larger than necessary to meet code requirements for the plate thickness. In such instances, select the procedure for the proper weld size and quality. If the thickness of the plate being welded is appreciably greater than that specified in the procedure, a reduction in welding speed and current will probably be required.

Travel speed is given as a range; however, the electrode required and the total time are based on the middle of the range.

Unless otherwise indicated, both members of the joint are the same thickness.

Pounds-of-electrode data include all ordinary deposition losses. These values are in terms of pounds of electrode needed to be purchased.

DATA SHEET

Article to be Welded ______________________ Job No. ______________

Plate Specs or Analysis ______________________________________

Welding Process ______________ Submitted by: ______________ Date: __________

JOINT-PREPARATION – BEAD PLACEMENT, SIZE, SEQUENCE

WORK ANGLE – DRAG ANGLE
ELECTRODE ANGLE TO JOINT

Pass No.	AC DC (+ –)	Amps	Volts	Travel Speed in./min.	Electrode Type	Elect. Size	Stick-out	Flux Type	Wire Speed Feed in./min.				

Special Comments ______________________________________

Total time is the arc time only and does not allow for operating factor.

Each procedure specifies the recommended stickout. This stickout is always the electrical stickout as discussed earlier in this section.

The drag angle is the angle between the center line of the electrode and a line perpendicular to the seam and inclined toward the direction of travel.

After a satisfactory welding procedure has been established, all the data should be recorded and filed for future reference. This information is invaluable if the same job or a similar job occurs at a later date. A suggested data sheet is shown on the previous page.

The presented procedures are offered as a starting point and may require changes to meet the requirements of specific applications. Because the many variables in design, fabrication, and erection or assembly affect the results obtained in applying this type of information, the serviceability of the product or structure is the responsibility of the builder.

Welding Parameters, Single Values or a Range

In some of the procedures which follow single values are given for such items as Amps, Volts, and Travel Speed. The purpose of using single values, i.e. 100 amps, is to indicate an optimum or starting place and it is expected that the user will expand these values and show a reasonable range. These values when recorded on a WPS (Welding Procedure Specification) should be shown as a range, not as specific values.

FLUX-CORED ARC WELDING (FULL AUTOMATIC) SELF-SHIELDED

Welding Position: Flat
Weld Quality Level: Strength only
Steel Weldability: Good

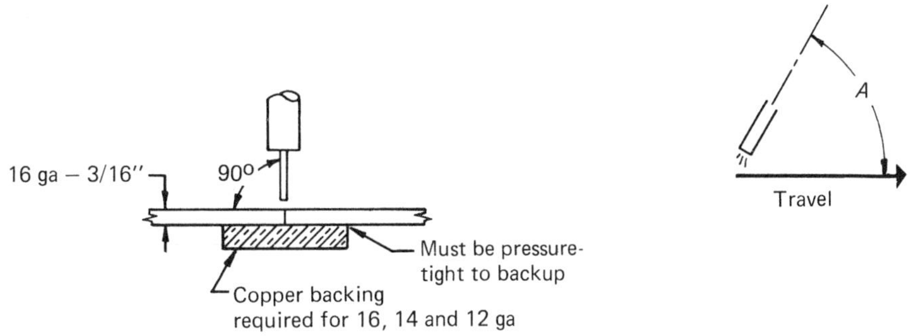

Plate Thickness (in.)	0.060 (16 ga)	0.075 (14 ga)	0.105 (12 ga)	0.135 (10 ga)	3/16
Pass	1	1	1	1	1
Electrode Class	E70T-3	E70T-3	E70T-3	E70T-3	E70T-3
Size	3/32	.120	.120	5/32	5/32
Wire Feed Speed (in./min)	145	155	175	130	135
Current (amp) DC(+) (Approx.)	425	575	625	850	875
Volts	25 - 26	25 - 26	25 - 26	25 - 26	25 - 26
Arc Speed (in./min)	210	150	125	110	95
Electrode Req'd (lb/ft)	0.016	0.025	0.037	0.064	0.078
Total Time (hr/ft of weld)	0.000953	0.00134	0.00160	0.00182	0.00210
Electrical Stickout (in.)	1	1-1/8	1-1/8	1-1/4	1-1/4
Angle, *A* (deg)	55	60	60	60	60

FLUX-CORED ARC WELDING (FULL AUTOMATIC) SELF-SHIELDED

Welding Position: Flat
Weld Quality Level: Commercial
Steel Weldability: Good
Welded from: One side

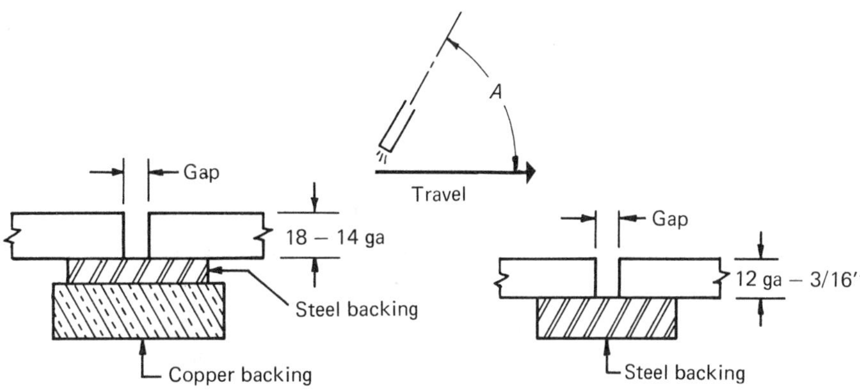

Plate Thickness (in.)	0.048 (18 ga)	0.060 (16 ga)	0.075 (14 ga)	0.105 (12 ga)	0.135 (10 ga)	3/16
Pass	1	1	1	1	1	1
Electrode Class	E70T-3	E70T-3	E70T-3	E70T-3	E70T-3	E70T-3
Size	3/32	3/32	.120	.120	5/32	5/32
Wire Feed Speed (in./min)	145	175	185	205	140	150
Current (amp) DC(+) (Approx.)	425	475	650	700	900	950
Volts	24 - 25	25 - 26	25 - 26	25 - 26	25 - 26	26 - 27
Arc Speed (in./min)	225	200	150	120	100	75
Electrode Req'd (lb/ft)	0.015	0.020	0.032	0.046	0.078	0.117
Total Time (hr/ft of weld)	0.000890	0.00100	0.00134	0.00167	0.00200	0.00267
Electrical Stickout (in.)	1	1	1-1/8	1-1/8	1-1/4	1-1/4
Steel Backing, minimum	16 ga	16 ga	14 ga	12 ga	10 ga	3/16″
Gap (in.)	1/32	3/64	1/16	1/16	3/32	1/8
Angle, *A* (deg)	50	50	55	60	60	60

FLUX-CORED ARC WELDING (FULL AUTOMATIC) SELF-SHIELDED

Welding Position: Horizontal
Weld Quality Level: Commercial
Steel Weldability: Good

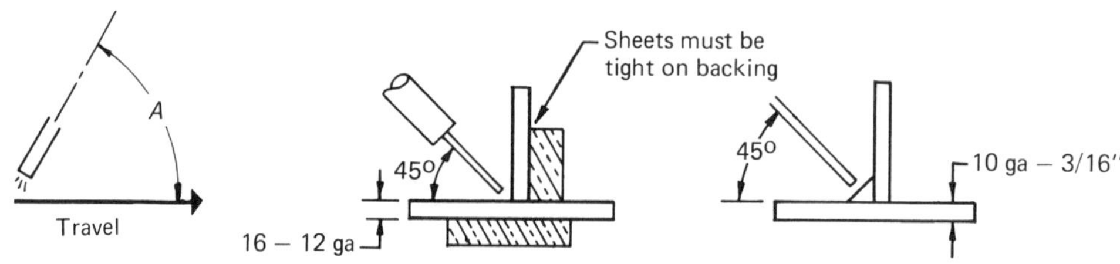

Plate Thickness (in.)	0.060 (16 ga)	0.075 (14 ga)	0.105 (12 ga)	0.135 (10 ga)	3/16
Pass	1	1	1	1	1
Electrode Class	E70T-3	E70T-3	E70T-3	E70T-3	E70T-3
Size	3/32	3/32	.120	5/32	5/32
Wire Feed Speed (in./min)	120	145	165	115	125
Current (amp) DC(+)	375	425	600	775	825
Volts	24 - 25	25 - 26	24 - 25	25 - 26	26 - 27
Arc Speed (in./min)	160	125	90	90	90
Electrode Req'd (lb/ft)	0.018	0.027	0.046	0.068	0.074
Total Time (hr/ft of weld)	0.00125	0.00160	0.00222	0.00222	0.00222
Electrical Stickout (in.)	1	1	1-1/8	1-1/4	1-1/4
Angle, *A* (deg)	45	50	50	60	60

FLUX-CORED ARC WELDING (FULL AUTOMATIC) SELF-SHIELDED

Welding Position: Horizontal
Weld Quality Level: Commercial
Steel Weldability: Good

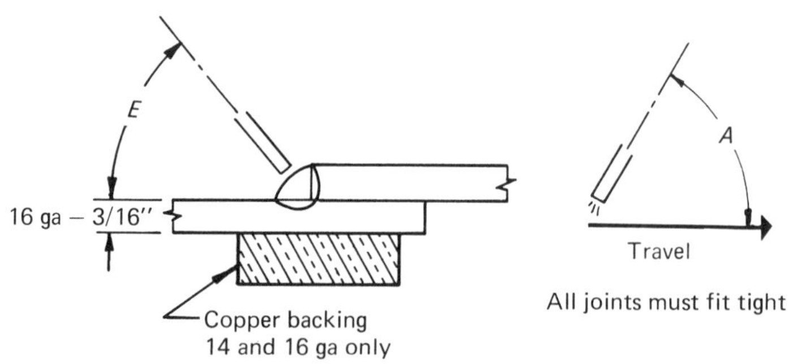

Plate Thickness (in.)	0.060 (16 ga)	0.075 (14 ga)	0.105 (12 ga)	0.135 (10 ga)	3/16
Pass	1	1	1	1	1
Electrode Class	E70T-3	E70T-3	E70T-3	E70T-3	E70T-3
Size	3/32	.120	.120	5/32	5/32
Wire Feed Speed (in./min)	145	165	175	120	125
Current (amp) DC(+) (Approx.)	475	600	625	800	825
Volts	25 - 26	25 - 26	25 - 26	25 - 26	25 - 26
Arc Speed (in./min)	200	150	110	95	80
Electrode Req'd (lb/ft)	0.019	0.028	0.041	0.066	0.083
Total Time (hr/ft of weld)	0.00100	0.00134	0.00182	0.00210	0.00250
Electrical Stickout (in.)	1	1-1/8	1-1/8	1-1/4	1-1/4
Angle, *E* (deg)	70	65	60	50	45
Angle, *A* (deg)	55	55	55	60	60

FLUX-CORED ARC WELDING (FULL AUTOMATIC) SELF-SHIELDED

Welding Position: Vertical up
Weld Quality Level: Code
Steel Weldability: Good or fair

Consult electrode supplier for design of copper shoes.

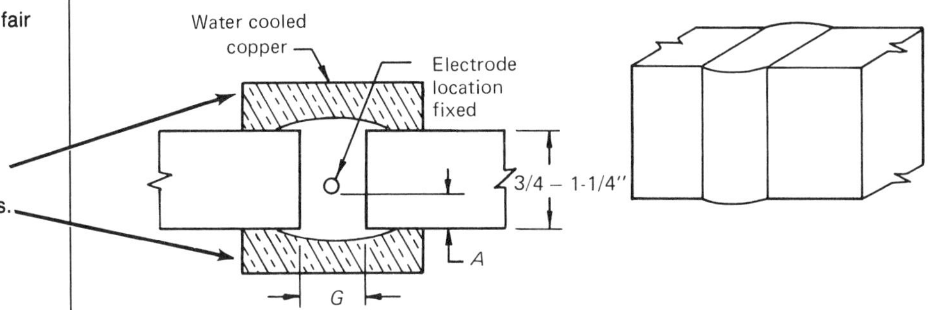

Plate Thickness (in.)	3/4	1	1-1/4
Pass	1	1	1
Electrode Class	EG72T-1	EG72T-1	EG72T-1
Size	.120	.120	.120
Wire Speed Feed (in./min)	350	350	350
Current (amp) DC(+)	625 – 675	625 – 675	625 – 675
Volts	40 – 42	40 – 42	40 – 42
Welding Speed (in./min)	4.1	3.4	2.6
Electrode Req'd (lb/ft)	2.7	3.25	4.25
Total Time (hr/ft of weld)	0.0488	0.0588	0.0769
Electrode Location, *A* (in.)	1/4	5/16	3/8 – 1/2
Gap, G (in.)	7/8	3/4 – 7/8	3/4 – 7/8
Electrical Stickout, in.		3	

FLUX-CORED ARC WELDING (FULL AUTOMATIC) SELF-SHIELDED

Welding Position: Vertical up
Weld Quality Level: Code
Steel Weldability: Good or fair

Consult electrode supplier for design of copper shoes.

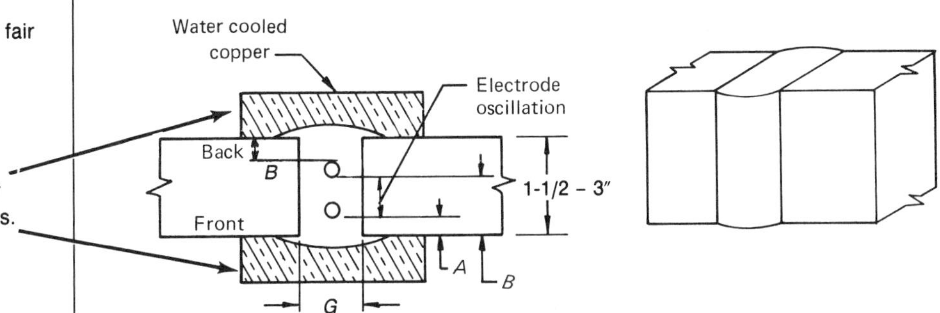

Plate Thickness (in.)	1-1/2	2	3
Pass	1	1	1
Electrode Class	EG72T-1	EG72T-1	EG72T-1
Size	.120	.120	.120
Wire Feed Speed (in./min)	350	350	350
Current (amp) DC(+)	625 – 675	625 – 675	625 – 675
Volts	40 – 42	43 – 45	43 – 45
Welding Speed (in./min)	1.8	1.4	1.0
Electrode Req'd (lb/ft)	6.14	7.91	11.06
Total Time (hr/ft of weld)	0.111	0.143	0.200
Oscillation Time (sec)†	1/2	1	2
Electrode Location (in.) *A*	1/4 – 3/8	1/4 – 3/8	1/4 – 3/8
B	3/8 – 1/2	3/8 – 1/2	3/8 — 1/2
Gap (in.)	7/8 – 1	1 – 1-1/8	1 – 1-1/8
Dwell, 2 sec (front) 2 sec (back)			
Electrical Stickout, in.		3	

† Time for one direction only.

Welding Position: Vertical up
Weld Quality Level: Code
Steel Weldability: Good or fair

FLUX-CORED ARC WELDING (FULL AUTOMATIC) SELF-SHIELDED

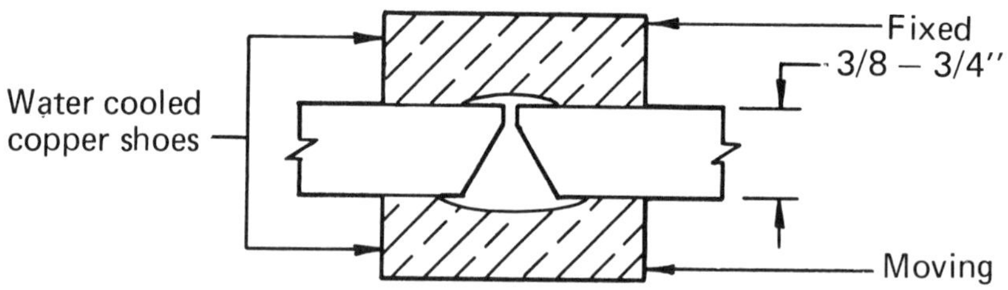

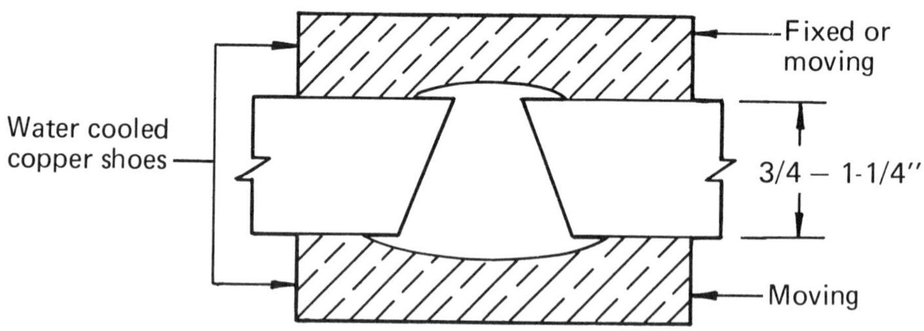

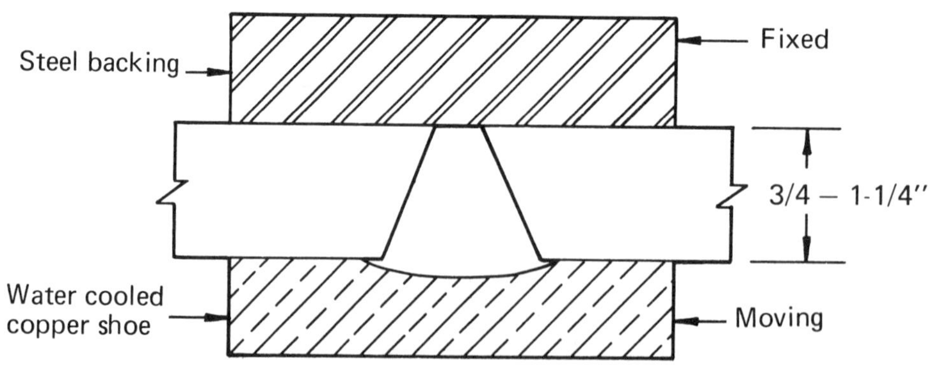

Consult the electrode supplier for welding procedures and design of the water cooled copper shoes.

FLUX-CORED ARC WELDING (SEMIAUTOMATIC) SELF-SHIELDED

Welding Position: Flat
Weld Quality Level: Commercial
Steel Weldability: Good or fair

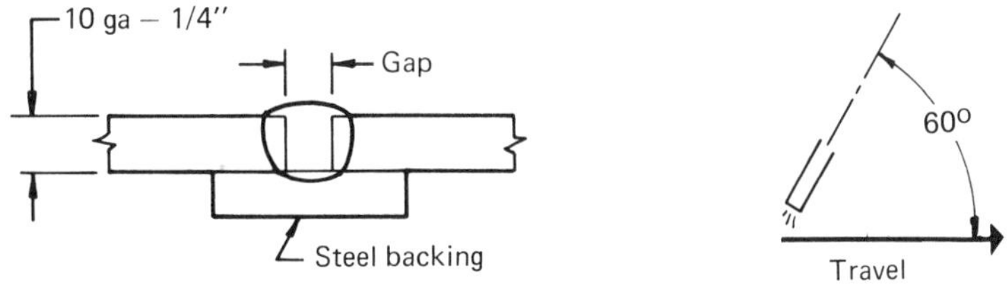

Plate Thickness (in.)	0.135 (10 ga)	3/16	1/4
Pass	1	1	1
Electrode Class	E70T-4	E70T-4	E70T-4
Size	3/32	3/32	3/32
Wire Feed Speed (in./min)	185	225	225
Current (amp) DC(+) (Approx.)	350	400	400
Volts	29 – 30	30 – 31	30 – 31
Arc Speed (in./min)	21 – 23	16 – 18	12.5 – 13.5
Electrode Req'd (lb/ft)	0.16	0.24	0.32
Total Time (hr/ft of weld)	0.00909	0.0118	0.0154
Backing, minimum thickness (in.)	10 ga	3/16	3/16
Gap (in.)	5/32	3/16	7/32
Electrical Stickout, in.	2-3/4		

FLUX-CORED ARC WELDING (SEMIAUTOMATIC) SELF-SHIELDED

Welding Position: Flat
Weld Quality Level: Commercial
Steel Weldability: Good or fair

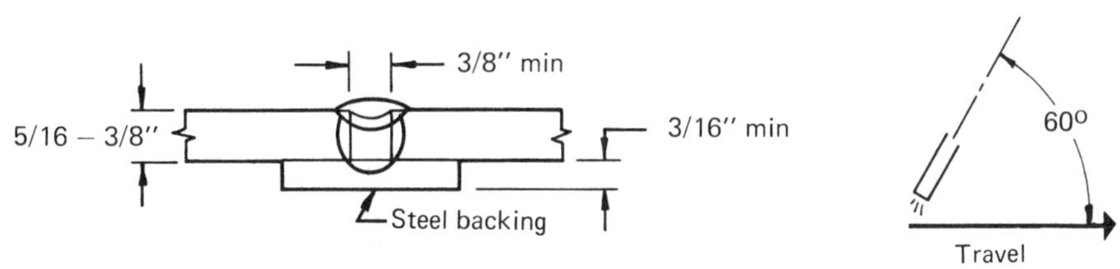

Plate Thickness (in.)	5/16				3/8			
Pass	1	2	1	2	1	2	1	2
Electrode Class	E70T-4		E70T-4		E70T-4		E70T-4	
Size	3/32		.120		3/32		.120	
Wire Feed Speed (in./min)	205		185		205		195	
Current (amp) DC(+) (Approx.)	375		475		375		500	
Volts	33 – 34		32 – 33		33 – 34		33 – 34	
Arc Speed (in./min)	11 – 13	20 – 22	14 – 16	22 – 24	9.5 – 11.5	17 – 19	12 – 14	19 – 21
Electrode Req'd (lb/ft of weld)	0.480		0.593		0.566		0.735	
Total Time (hr/ft of weld)	0.0262		0.0220		0.0301		0.0254	
Electrical Stickout, in.	2-3/4							

FLUX-CORED ARC WELDING (SEMIAUTOMATIC) SELF-SHIELDED

Welding Position: Flat
Weld Quality Level: Commercial
Steel Weldability: Good or fair

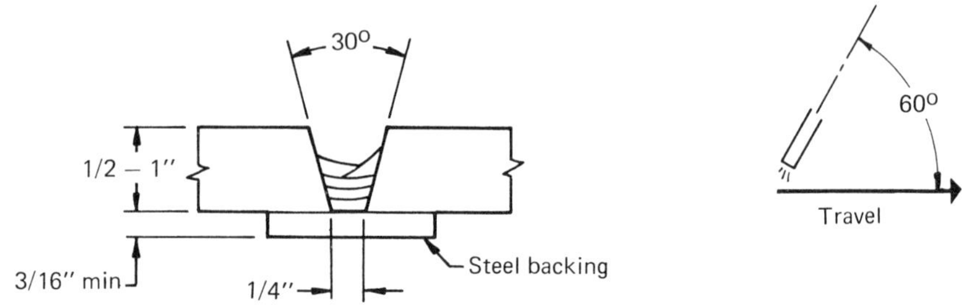

Plate Thickness (in.)	1/2	5/8	3/4	1
Pass	1 - 3	1 - 4	1 - 6	1 - 8
Electrode Class	E70T-4	E70T-4	E70T-4	E70T-4
Size	3/32	3/32	3/32	3/32
Wire Feed Speed (in./min)	185	185	185	185
Current (amp) DC(+)	350	350	350	350
Volts	30 - 31	30 - 31	30 - 31	30 - 31
Arc Speed (in./min)	11.5 - 13.5	11.0 - 13.0	12.5 - 14.5	11.0 - 13.0
Electrode Req'd (lb/ft)	0.830	1.16	1.53	2.29
Total Time (hr/ft of weld)	0.0480	0.0668	0.0888	0.134
Electrical Stickout, in.	2-3/4			

FLUX-CORED ARC WELDING (SEMIAUTOMATIC) SELF-SHIELDED

Welding Position: Flat
Weld Quality Level: Commercial
Steel Weldability: Good or fair

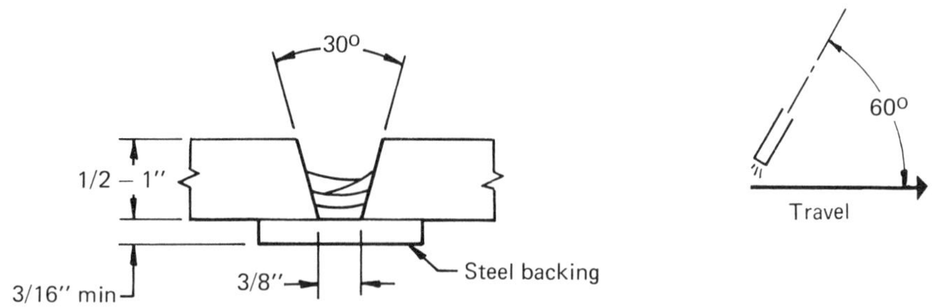

Plate Thickness (in.)	1/2	5/8	3/4	1	
Pass	1 - 3	1 - 4	1 - 5	1	2 - 6*
Electrode Class	E70T-4	E70T-4	E70T-4	E70T-4	E70T-4
Size	.120	.120	.120	.120	.120
Wire Feed Speed (in./min)	190	190	190	190	300
Current (amp) DC(+)	500	500	500	500	550
Volts	31 - 32	31 - 32	31 - 32	31 - 32	35 - 36
Arc Speed (in./min)	13.5 - 16.5	13.5 - 15.5	13.0 - 15.0	14.5 - 17.5	14.5 - 17.5
Electrode Req'd (lb/ft)	1.14	1.56	2.02	3.03	
Total Time (hr/ft of weld)	0.0400	0.0552	0.0715	0.0750	
Electrical Stickout (in.)	2-3/4	2-3/4	2-3/4	2-3/4	3-3/4

* Change to 3-3/4 in. stickout after first pass.

FLUX-CORED ARC WELDING (SEMIAUTOMATIC) SELF-SHIELDED

Welding Position: Flat
Weld Quality Level: Commercial
Steel Weldability: Good or fair
Welded from: One side

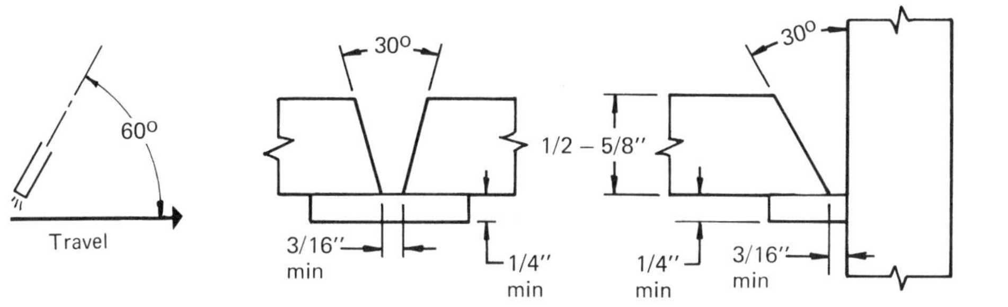

Plate Thickness (in.)	1/2		5/8		
Pass	1	2	1	2	3
Electrode Class*	E70T-7		E70T-7		
Size	7/64		7/64		
Wire Feed Speed (in./min)	160		175		
Current (amp) DC(−) (Approx.)	425		450		
Volts	25 - 26		26 - 27		
Arc Speed (in./min)	12 - 13	8 - 9	13 - 14	12 - 13	11 - 12
Electrode Req'd (lb/ft)	0.75		1.00		
Total Time (hr/ft of weld)	0.0395		0.0482		
Electrical Stickout, in.	1-1/2				

* A fast-follow type E70T-7 with penetration capability. See introductory notes.
This joint is similar to some of the prequalified joints in the AWS Structural Codes.

FLUX-CORED ARC WELDING (SEMIAUTOMATIC) SELF-SHIELDED

Welding Position: Flat
Weld Quality Level: Commercial
Steel Weldability: Good or fair

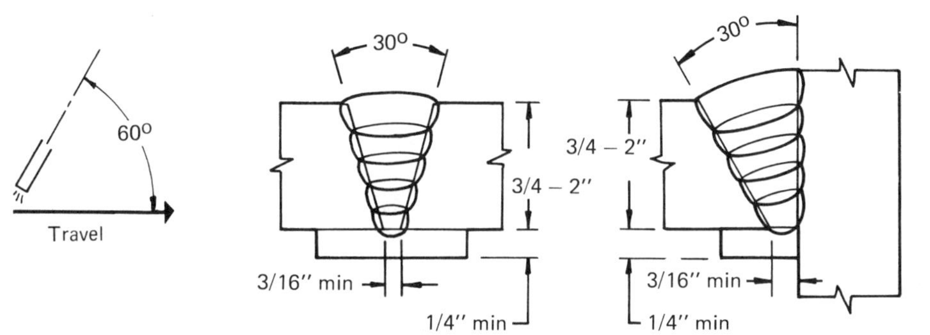

Plate Thickness (in.)	3/4	1	2*	
Pass	1 - 3	1 - 4	1 - 2	3 - 10†
Electrode Class	E70T-7	E70T-7	E70T-7	E70T-4
Size	7/64	7/64	7/64	.120
Wire Feed Speed (in./min)	205	205	205	300
Current (amp) (Approx.)	500 DC(−)	500 DC(−)	500 DC(−)	550 DC(+)
Volts	27 - 28	27 - 28	27 - 28	35 - 36
Arc Speed (in./min)	10 - 11	9 - 10	12.5 - 13.5	11.5 - 12.5
Electrode Req'd (lb/ft)	1.43	2.10	0.72	5.70
Total Time (hr/ft of weld)	0.0570	0.0844	0.164	
Electrical Stickout (in.)	1-1/2	1-1/2	1-1/2	3-3/4

* Typical two electrode procedure. Can be finished with 7/64 E70T-7 but with an increase in total arc time.
† Layers should be split when the previous bead face is 3/4-in. wide. This occurs after pass 4.
This joint is similar to some of the prequalified joints in the AWS Structural Codes.

FLUX-CORED ARC WELDING (SEMIAUTOMATIC) SELF-SHIELDED

Welding Position: Flat
Weld Quality Level: Commercial
Steel Weldability: Good or fair
Welded from: One side

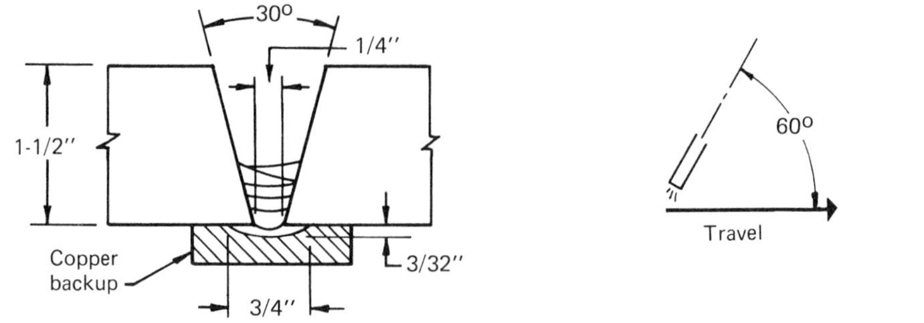

Plate Thickness (in.)	1-1/2			
Pass	1	2 – 14	1	2 – 8*
Electrode Class	E70T-4	E70T-4	E70T-4	E70T-4
Size	3/32	3/32	.120	.120
Wire Feed Speed (in./min)	185	185	190	300
Current (amp) DC(+)	350	350	500	550
Volts	30 – 31	30 – 31	31 – 32	35 – 36
Arc Speed (in./min)	7.0 – 8.5	11.0 – 13.5	11.0 – 13.0	13.0 – 16.0
Electrode Req'd (lb/ft)	4.48		4.59	
Total Time (hr/ft of weld)	0.239		0.114	
Electrical Stickout, in.	2-3/4			

* Change to 3-3/4 in. stickout after first pass.

FLUX-CORED ARC WELDING (SEMIAUTOMATIC) SELF-SHIELDED

Welding Position: Flat
Weld Quality Level: Commercial
Steel Weldability: Good or fair

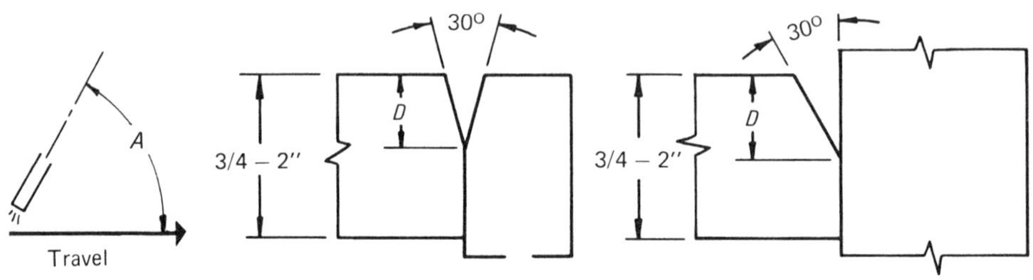

Depth, *D* (in.)	3/4		1-1/4		2		
Pass	1	2 – 3	1 – 2	3 – 5	1 – 2	3 – 4	5 – 9
Electrode Class	E70T-7		E70T-7	E70T-4	E70T-7	E70T-4	E70T-4
Size	3/32		3/32	.120	3/32	.120	.120
Wire Feed Speed (in./min)	160		145	190	145	190	300
Current (amp) (Approx.)	425 DC(−)		400 DC(−)	500 DC(+)	400 DC(−)	500 DC(+)	550 DC(+)
Volts	26 – 27		26 – 27	31 – 32	26 – 27	31 – 32	35 – 36
Arc Speed (in./min)	11 – 12	7.5 – 8.5	10 – 11	13 – 14	10 – 11	12 – 13	13 – 14
Electrode Req'd (lb/ft)	0.71		0.58	1.27	0.58	4.07	
Total Time (hr/ft of weld)	0.0674		0.0822		0.144		
Electrical Stickout (in.)	1-1/4		1-1/2	2-3/4	1-1/2	2-3/4	3-3/4
Drag Angle, *A* (deg)	75		75	60	75	60	

This joint is similar to some of the prequalified joints in the AWS Structural Codes.

FLUX-CORED ARC WELDING (SEMIAUTOMATIC) SELF-SHIELDED

Welding Position: Flat
Weld Quality Level: Commercial
Steel Weldability: Good or fair
Welded from: Two sides

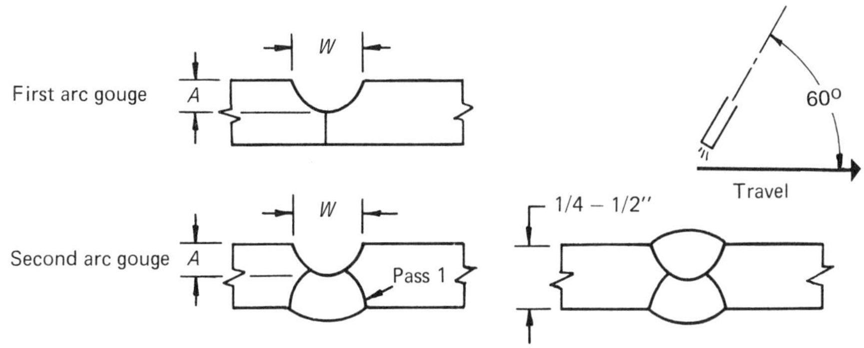

Plate Thickness (in.)	1/4	5/16		3/8		1/2
Pass	1, 2	1, 2	1, 2	1, 2	1, 2	1, 2
Electrode Class	E70T-4	E70T-4	E70T-4	E70T-4	E70T-4	E70T-4
Size	3/32	3/32	.120	3/32	.120	.120
Wire Feed Speed (in./min)	185	185	190	185	190	190
Current (amp) DC(+) Approx.	350	350	500	350	500	500
Volts	30 – 31	30 – 31	31 – 32	30 – 31	31 – 32	31 – 32
Arc Speed (in./min)	19 – 22	14 – 16	23 – 27	11 – 13	19 – 22	13 – 15
Electrode Req'd (lb/ft)	0.338	0.480	0.490	0.595	0.613	0.912
Total Time (hr/ft of weld)*	0.0195	0.0266	0.0160	0.0334	0.0195	0.0286
Depth, *A* minimum (in.)	1/8	5/32	5/32	3/16	3/16	1/4
Width, *W* minimum (in.)	5/16	3/8	3/8	7/16	7/16	1/2
Electrical Stickout, in.	2-3/4					

* Gouging time not included.

FLUX-CORED ARC WELDING (SEMIAUTOMATIC) SELF-SHIELDED

Welding Position: Flat
Weld Quality Level: Code
Steel Weldability: Good or fair
Welded from: Two sides

Where Code Quality is indicated normal practice requires a procedure qualification test.

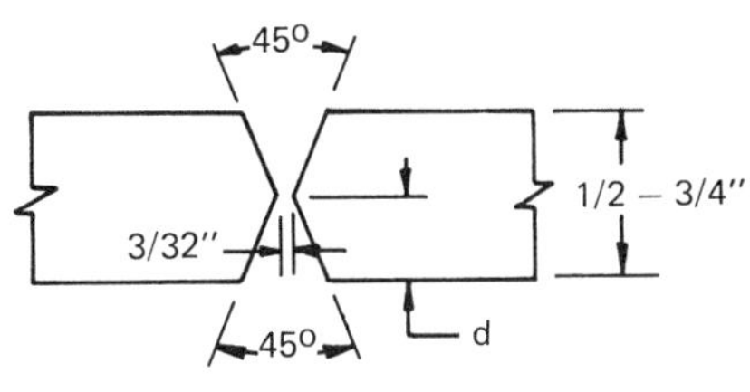

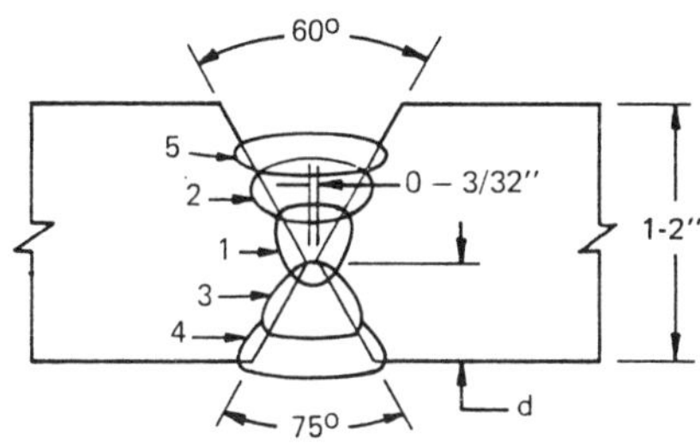

Plate Thickness (in.)	1/2		3/4		1	1-1/2	2
Pass*	1	2 – 4	1	2 – 4	1 – 4	1 – 7	1 – 12
Electrode Class†	E70T-7		E70T-7		E70T-7	E70T-7	E70T-7
Size	3/32		3/32		7/64	7/64	7/64
Wire Feed Speed (in./min)	85	150	85	235	205	205	205
Current (amp) DC(–) (Approx.)	225	325	225	425	500	500	500
Volts	23 – 24	23 – 24	23 – 24	26 – 27	28 – 29	28 – 29	28 – 29
Arc Speed (in./min)	7.6 – 8.4	16 – 18	5.7 – 6.3	12.5 – 13.5	16.5 – 18.5	16.5 – 18.5	16.5 – 18.5
Electrode Req'd (lb/ft)	0.58		1.02		1.51	3.35	6.38
Total Time (hr/ft of weld)	0.0602		0.0793		0.0635	0.141	0.268
Electrical Stickout	1-1/4		1-1/4		1-1/2	1-1/2	1-1/2
Width, *d* (in.)	1/4		3/8		1/2	1/2	1/2

* Layers are split when bead face width exceeds 3/4 in.

FLUX-CORED ARC WELDING (SEMIAUTOMATIC) SELF-SHIELDED

Welding Position: Horizontal
Weld Quality Level: Commercial
Steel Weldability: Good
Welded from: One side

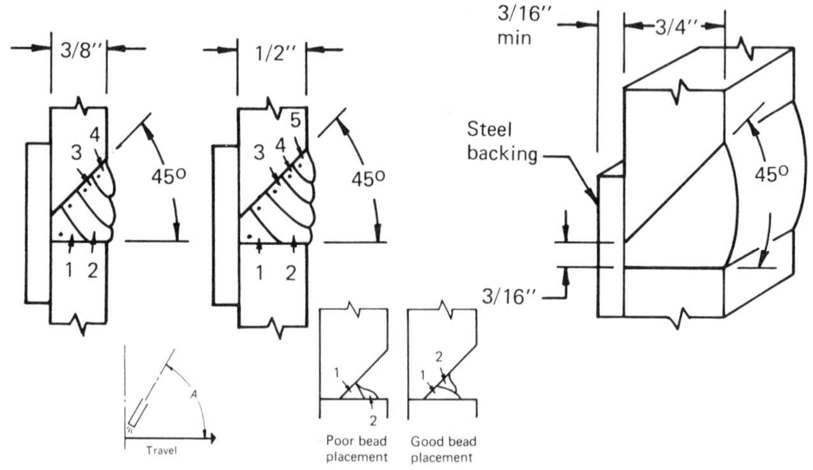

Plate Thickness (in.)	3/8		1/2		3/4			
Pass	1 – 3	4	1 – 4	5	1 – 9	10	1 – 7	8
Electrode Class	E70T-4	E70T-4	E70T-4	E70T-4	E70T-4	E70T-4	E70T-4	E70T-4
Size	3/32	3/32	3/32	3/32	3/32	3/32	.120	3/32
Wire Feed Speed (in./min)	145	110	145	110	145	110	150	110
Current (amp) DC(+) (Approx.)	300	250	300	250	300	250	400	250
Volts	28	25	28	26	28	26	28	26
Arc Speed (in./min)	12.5 – 14.5	12.5 – 14.5	11.0 – 13.0	10.0 – 12.0	12.0 – 14.0	10.0 – 12.0	12.5 – 14.5	10.0 – 12.0
Electrode Req'd (lb/ft)	0.746		1.11		2.11		2.39	
Total Time (hr/ft of weld)	0.0592		0.0840		0.157		0.122	
Drag angle, *A* (deg)	60		60		60		60	
Electrical Stickout, in.	2-3/4							

FLUX-CORED ARC WELDING (SEMIAUTOMATIC) SELF-SHIELDED

Welding Position: Horizontal
Weld Quality Level: Commercial
Steel Weldability: Good
Welded from: One side

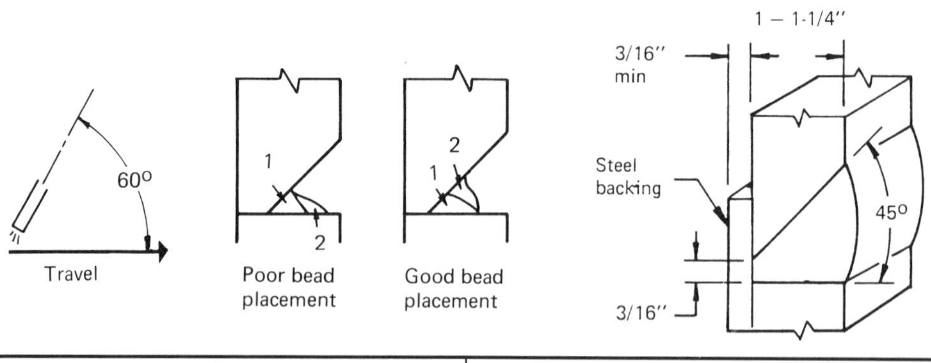

Plate Thickness (in.)	1				1-1/4			
Pass	1 – 14	15	1 – 11	12	1 – 21	22	1 – 17	18
Electrode Class	E70T-4	E70T-4	E70T-4	E70T-4	E70T-4	E70T-4	E70T-4	E70T-4
Size	3/32	3/32	.120	3/32	3/32	3/32	.120	3/32
Wire Feed Speed (in./min)	145	145	150	145	165	145	150	145
Current (amp) DC(+) (Approx.)	300	300	400	300	325	300	400	300
Volts	30	28	29	28	30	28	29	28
Arc Speed (in./min)	11.5 – 13.5	10.0 – 12.0	14 – 16	10.0 – 12.0	11.5 – 13.5	10.0 – 12.0	12.0 – 14.0	10.0 – 12.0
Electrode Req'd (lb/ft)	3.43		3.34		5.01		4.71	
Total Time (hr/ft of weld)	.219		.165		.309		.231	
Electrical Stickout, in.	2-3/4							

FLUX-CORED ARC WELDING (SEMIAUTOMATIC) SELF-SHIELDED

Welding Position: Horizontal
Weld Quality Level: Commercial
Steel Weldability: Good or fair
Welded from: One side

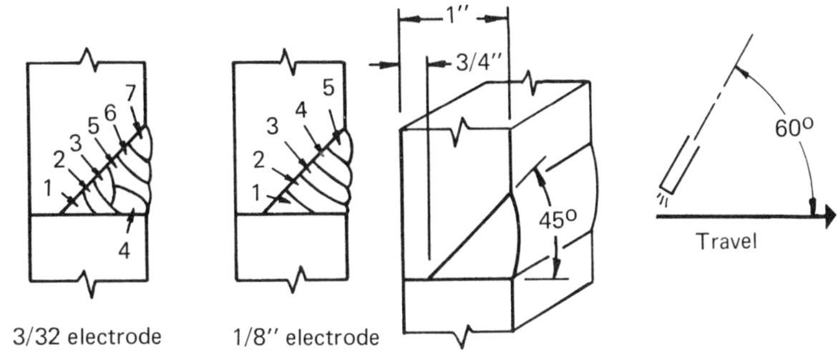

3/32 electrode 1/8" electrode

Plate Thickness (in.)	1				1			
Pass	1	2 – 3	4 – 6	7	1*	2	3 – 4	5**
Electrode Class	E70T-4	E70T-4	E70T-4	E70T-4	E70T-4	E70T-4	E70T-4	E70T-4
Size	3/32	3/32	3/32	3/32	.120	.120	.120	3/32**
Wire Feed Speed (in./min)	165	165	165	125	150	150	150	125
Current (amp) DC(+) (Approx.)	325	325	325	275	400	400	400	275
Volts	28	28	28	26	28*	28	28	26
Arc Speed (in./min)	11.0 – 13.0	12.5 – 14.5	13.0 – 15.0	10.0 – 12.0	11.5 – 13.5	13.5 – 16.5	13.0 – 15.0	10.0 – 12.0
Electrode Req'd (lb/ft)	1.44				1.56			
Total Time (hr/ft of weld)	0.108				0.0762			
Electrical Stickout, in.	2-3/4							

* With tight fitup use 30-32 v or 7/32 in. E7018 stick electrode for the first pass.
** An alternate to changing electrode size is to run the final pass with 3/16 E7018 stick electrode.

FLUX-CORED ARC WELDING (SEMIAUTOMATIC) SELF-SHIELDED

Welding Position: Horizontal
Weld Quality Level: Commercial
Steel Weldability: Good or fair

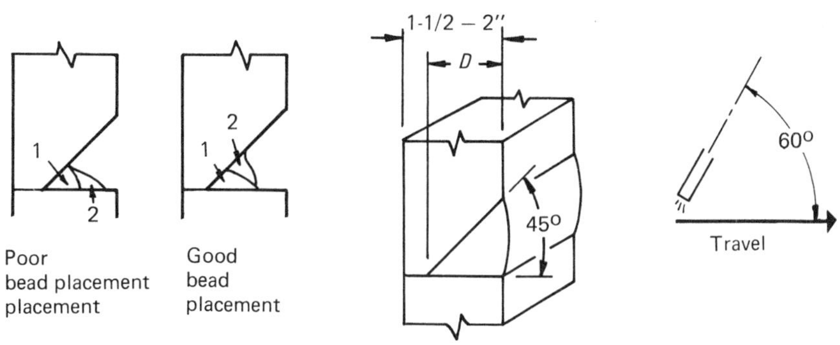

Poor bead placement placement — Good bead placement

Plate Thickness (in.)	1-1/2		1-1/2		2		2	
Pass	1 – 14	15	1* – 9	10	1 – 24	25	1* – 16	17
Electrode Class	E70T-4	E70T-4	E70T-4	E70T-4	E70T-4	E70T-4	E70T-4	E70T-4
Size	3/32	3/32	.120	3/32	3/32	3/32	.120	3/32
Wire Feed Speed (in./min)	145	110	150	110	145	110	150	110
Current (amp) DC(+) (Approx.)	300	250	400	250	300	250	400	250
Volts	28	26	28*	26	28	26	28*	26
Arc Speed (in./min)	12.0 – 14.0	10.0 – 12.0	12.5 – 14.5	10.0 – 12.0	12.0 – 14.0	10.0 – 12.0	12.5 – 14.5	10.0 – 12.0
Electrode Req'd (lb/ft)	3.75		3.25		6.16		5.54	
Total Time (hr/ft of weld)	0.234		0.151		0.388		0.255	
Depth, *D* (in.)	1-1/8		1-1/8		1-1/2		1-1/2	
Electrical Stickout, in.	2-3/4							

* With tight fitup use 30-32 v on the first pass or 7/32 E7018 stick electrode.

FLUX-CORED ARC WELDING (SEMIAUTOMATIC) SELF-SHIELDED

Welding Position: Horizontal
Weld Quality Level: Commercial
Steel Weldability: Good or fair
Welded from: Two sides

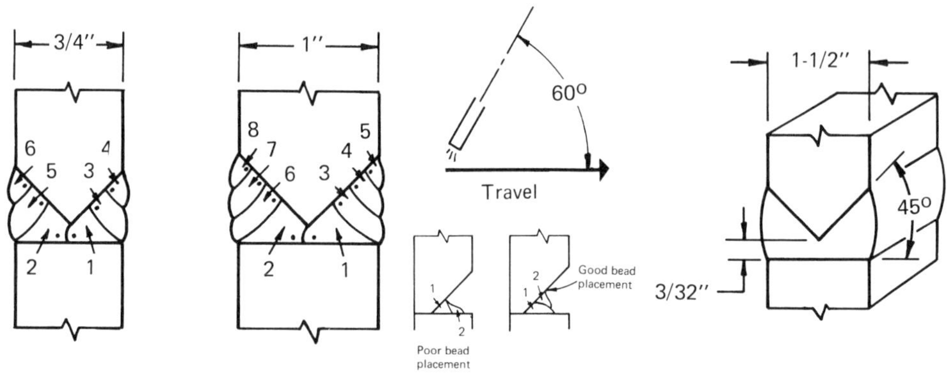

Plate Thickness (in.)	3/4		1		1-1/2			
Pass	1, 2, 3 & 5	4 & 6	1-4 & 6-7	5 & 8	1-7 & 9-15	8 & 16	1-5 & 7-11	6 & 12
Electrode Class	E70T-4	E70T-4	E70T-4	E70T-4	E70T-4	E70T-4	E70T-4	E70T-4
Size	3/32	3/32	3/32	3/32	3/32	3/32	.120	3/32
Wire Feed Speed (in./min)	145	110	145	110	145	110	150	110
Current (amp) DC(+) (Approx.)	300	250	300	250	300	250	400	250
Volts	28	26	28	26	28	26	28	26
Arc Speed (in./min)	11.5 – 13.5	10.0 – 12.0	9.0 – 11.0	10.0 – 12.0	10.0 – 12.0	10.0 – 12.0	11.0 – 13.0	10.0 – 12.0
Electrode Req'd (lb/ft)	1.25		1.98		3.94		4.09	
Total Time (hr/ft of weld)	0.0996		0.154		0.291		0.203	
Electrical Stickout, in.	2-3/4							

First pass should be back-gouged or ground. Using 5/32 E7018 for the first pass reduces the amount of gouging required.

FLUX-CORED ARC WELDING (SEMIAUTOMATIC) SELF-SHIELDED

Welding Position: Horizontal
Weld Quality Level: Code
Steel Weldability: Good or fair
Welded from: Two sides

Where Code Quality is indicated normal practice requires a procedure qualification test.

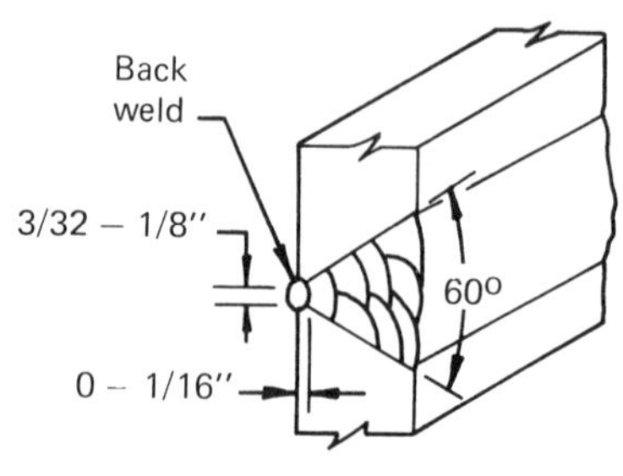

Plate Thickness (in.)	All	3/8	1/2		3/4	
Pass	1	2 – 4	2 – 3	4 – 6	2 – 6	7 – 9
Electrode Class	E71T-7	E71T-7	E71T-7		E71T-7	
Size	5/64	5/64	5/64		5/64	
Wire Feed Speed (in./min)	65	65	110	70	110	70
Current (amp) DC(−) (Approx.)	175	175	250	185	250	185
Volts	18 – 19	18 – 19	19 – 20	18 – 19	19 – 20	18 – 19
Arc Speed (in./min)	5.0 – 5.5	5.1 – 5.6	6.4 – 7.1	5.4 – 6.0	4.9 – 5.4	4.0 – 4.4
Electrode Req'd (lb/ft)*	—	0.53	0.88		1.83	
Total Time (hr/ft of weld)*	—	0.150	0.203		0.375	
Electrical Stickout, in.	7/8					

* Does not include electrode or time to gouge and weld the back weld.
Back weld may be made by stick electrode or semiautomatic welding.

FLUX-CORED ARC WELDING (SEMIAUTOMATIC) SELF-SHIELDED

Welding Position: Flat and Horizontal
Weld Quality Level: Code
Steel Weldability: Good or fair
Welded from: One side

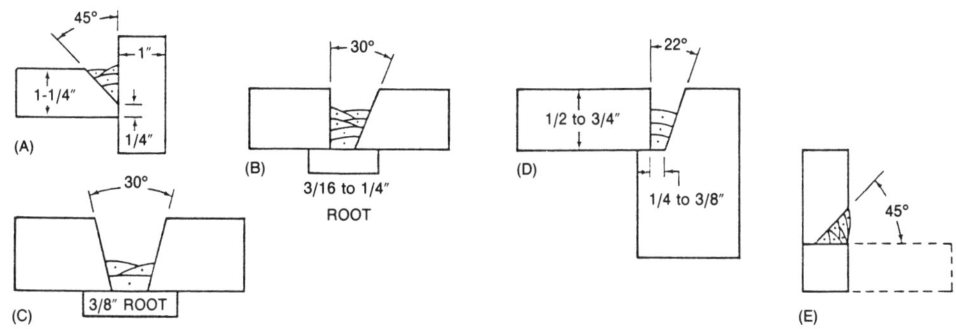

Plate Thickness (in.)	—	A	B	C	D	E
Pass Number	Root	2 & Up	2 & Up	1 & Up	1 & Up	1 & Up
Electrode Class	E70T-6	E70T-6	E70T-6	E70T-6	E70T-6	E70T-6
Size	5/64	5/64	5/64	5/64	5/64	5/64
Power [AC or DC(±)]	DC(+)	DC(+)	DC(+)	DC(+)	DC(+)	DC(+)
Electrical Stickout (in.)	3/4 – 1-1/4				1/2 – 1	3/4 – 1-1/4
Wire Feed Speed (in./min)	380	380	380	380	290	250
Amps (Approx.)	480	480	480	480	410	370
Arc Volts	27 – 29	27 – 29	27 – 29	27 – 29	25 – 27	23 – 25
Travel Speed (in./min)	—	10 – 15	10 – 15	10 – 15	13 – 16	15 – 20
Preheat (°F)	70					
Interpass (°F)	150					
Electrode Angle to Joint (deg)	Bisect Joint					
Electrode Drag Angle (deg)	30 – 45	10 – 45				
Deposition Rate (lbs/hr)	18.5	18.5			14	12

FLUX-CORED ARC WELDING (SEMIAUTOMATIC) SELF-SHIELDED

Welding Position: Flat and Horizontal
Weld Quality Level: Code
Steel Weldability: Good or fair
Welded from: One side

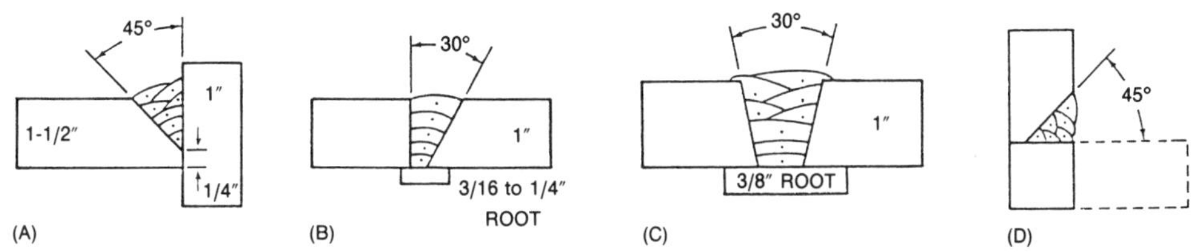

Plate Thickness (in.)	—	A	B	C	D
Pass Number	Root	2 & Up	2 & Up	1 & Up	2 & Up
Electrode Class	E70T-6	E70T-6	E70T-6	E70T-6	E70T-6
Size	3/32	3/32	3/32	3/32	3/32
Power [AC or DC(±)]	DC(+)	DC(+)			
Electrical Stickout (in.)	1-1/2 – 2-1/2				
Wire Feed Speed (in./min)	300	240			200
Amps (Approx.)	475	425			375
Arc Volts	26 – 28	26 – 28			23 – 25
Travel Speed (in./min)	—	10 – 25			15 – 25
Preheat (°F)	70				
Interpass (°F)	150				
Electrode Angle to Joint (deg)	Bisect Joint				
Electrode Drag Angle (deg)	15 – 30				
Deposition Rate (lbs/hr)	21.0	16.7			14.0

FLUX-CORED ARC WELDING (SEMIAUTOMATIC) SELF-SHIELDED

Welding Position: Vertical up
Weld Quality Level: Commercial
Steel Weldability: Good or fair
Welded from: One side

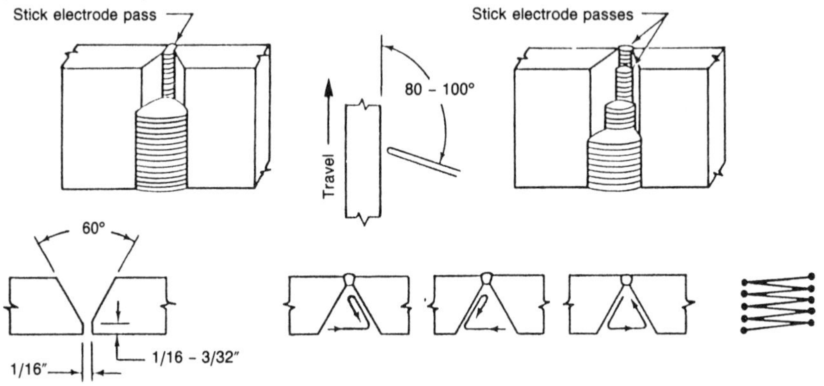

Plate Thickness (in.)	3/8-1/2-5/8	3/8	1/2	5/8	3/4 - 1	3/4		1	
Pass	1	2	2	2	1 & 2	3	4	3	4
Electrode Class	E6010 E6011	E71T-7	E71T-7	E71T-7	E6010 E6011	E71T-7		E71T-7	
Size	1/8	1/16	1/16	1/16	1/8	1/16		1/16	
Wire Feed Speed (in./min)	—	75	75	75	—	90		105	
Current (amp) DC	80 (+)	160 (−)	160 (−)	160 (−)	80 (+)	180 (−)		195 (−)	
Volts	—	19 – 20	19 – 20	19 – 20	—	20 – 21		20.5 – 21.5	
Arc Speed (in./min)	3.3 – 3.7	2.4 – 2.6	1.3 – 1.4	0.8 – 0.9	3.3 – 3.7	1.5 – 1.7	1.3 – 1.4	1.2 – 1.3	0.7 – 0.8
Electrode Req'd (lb/ft)*	0.227	0.342	0.625	1.02	0.454	1.37		2.49	
Total Time (hr/ft of weld)*	0.0571	0.0800	0.147	0.235	0.114	0.273		0.427	
Electrical Stickout, in.	1/2 – 1								

* Add stick electrode to semiautomatic passes.

FLUX-CORED ARC WELDING (SEMIAUTOMATIC) SELF-SHIELDED

Welding Position: Vertical
Weld Quality Level: Code
Steel Weldability: Good or fair
Welded from: One side

> Where Code Quality is indicated normal practice requires a procedure qualification test.

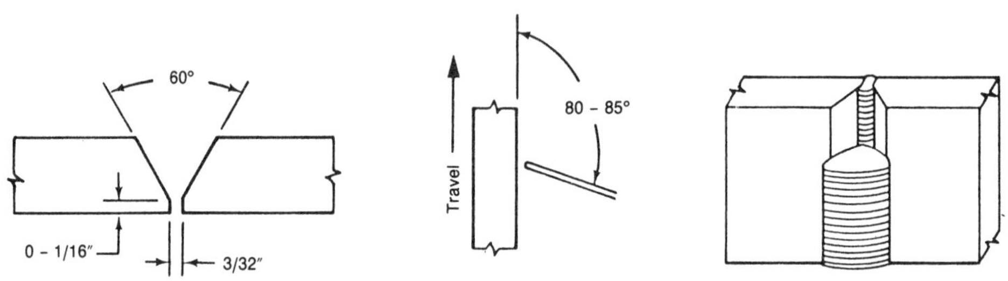

Plate Thickness (in.)	3/8	1/2	3/4	1
Pass	1 – 2	1 – 3	1 – 5	1 – 8
Electrode Class	E71T-7	E71T-7	E71T-7	E71T-7
Size	5/64	5/64	5/64	5/64
Wire Feed Speed (in./min)	65	65	65	65
Current (amp) DC(−) (Approx.)	175	175	175	175
Volts	18 – 19	18 – 19	18 – 19	18 – 19
Arc Speed (in./min)*	2.8 – 3.1	2.8 – 3.1	2.8 – 3.1	2.8 – 3.1
Electrode Req'd (lb/ft)	0.64	1.07	2.21	3.74
Total Time (hr/ft of weld)	0.155	0.259	0.531	0.900
Electrical Stickout, in.	7/8			

* For first pass only. Subsequent full weave passes are slower.

FLUX-CORED ARC WELDING (SEMIAUTOMATIC) SELF-SHIELDED

Welding Position: Vertical up
Weld Quality Level: Code
Steel Weldability: Good or fair
Welded from: One side

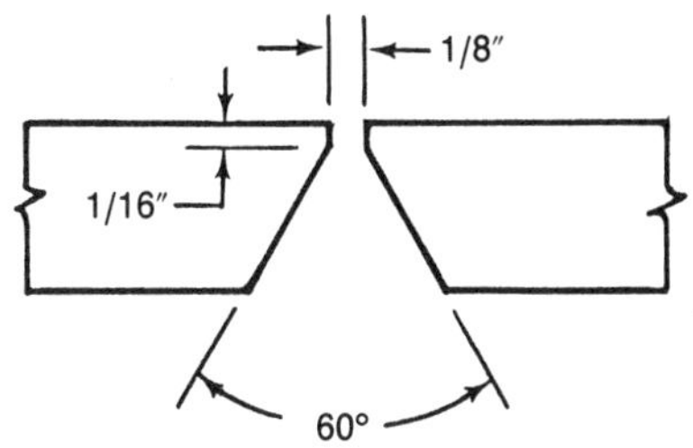

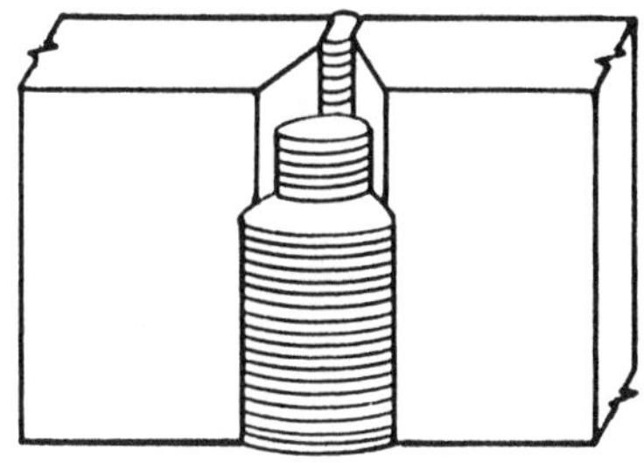

Plate Thickness (in.)	1/2		over 1/2	5/8	3/4	7/8	1	1-1/4 & up	
Pass Number	1	2 & 3	1	2 & 3	2 – 4	2 – 5	2 – 7	2 – 7	8
Electrode Class	E71T-8								
Size	5/64								
Electrical Stickout (in.)	1-1/4	3/4	1-1/4	3/4					
Power — AC or DC(±)	DC(−)								
Wire Feed Speed (in./min)	60	69	69	87					106
Amps	140	175	160	220					250
Volts	16	17	17	18					19
Travel speed (in./min)	2.3 – 2.5	1.1 – 1.2	2.2 – 2.4	.86 – .95	.63 – .69	.48 – .53	.37 – .41	—	
Preheat (°F)	150	—	150	—					
Interpass (°F)	—	250	—	250					
Electrode Location	—								
Electrode Angle to Joint (deg)	90								
Electrode Drag Angle (deg)	0 – 15								
Lb of Elect/Ft of Weld	.75		—	1.14	1.60	2.13	2.73	—	
Hrs/ft of weld	.174		—	.221	.301	.396	.513	—	
Deposition Rate (lbs/hr)	3.1	3.5	3.5	4.3					5.3

FLUX-CORED ARC WELDING (SEMIAUTOMATIC) SELF-SHIELDED

Welding Position: Horizontal, Vertical up and Flat
Weld Quality Level: Code
Steel Weldability: Good or fair
Welded from: One side

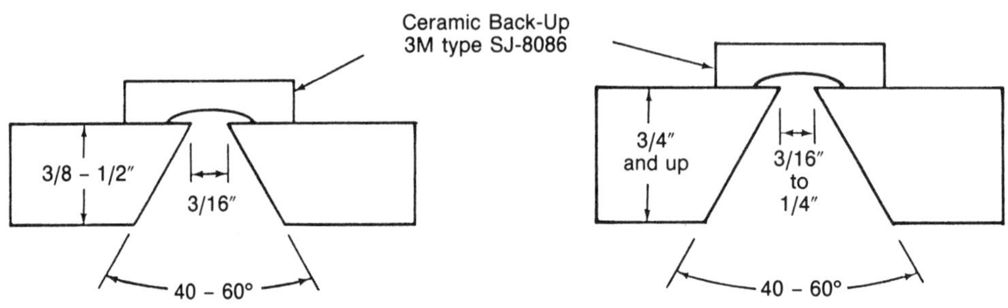

Position	Vertical up	Horizontal	Flat	Vertical up and Horizontal	Flat
Plate Thickness (in.)	3/8 to 1/2			3/4 and Up	
Pass Number	1 (Root)			1 (Root)	
Flux and Electrode Class	E71T-8			E71T-8	
Electrode Size	.072			.072	
Power [AC or DC(±)]	DC(−)			DC(−)	
Electrical Stickout (in.)	3/4 - 1			1/2 - 3/4	
Wire Feed Speed (in./min)	90	90	110	110-120	120-155
Amps (Approx.)	150	150	180	180-190	190-240
Arc Volts	17 - 18	17 - 18	17 - 18	17 - 18	19 - 21
Travel Speed (in./min)	2.3 - 2.7	2.8 - 3.2	3.4 - 3.8	As Required	
Preheat (°F)	70			70	
Interpass (°F)	—			—	
Electrode Angle to Joint (deg)	90	90	90	90	90
Electrode Drag Angle (deg)	0 - 10	20 - 30	20 - 30	0 - 30	5 - 30
Deposition Rate (lbs/hr)	3.6	3.6	4.3	4.3 - 4.7	4.7 - 6.0

FLUX-CORED ARC WELDING (SEMIAUTOMATIC) SELF-SHIELDED

Welding Position: Vertical up
Weld Quality Level: Code
Steel Weldability: Good or fair
Welded from: One side

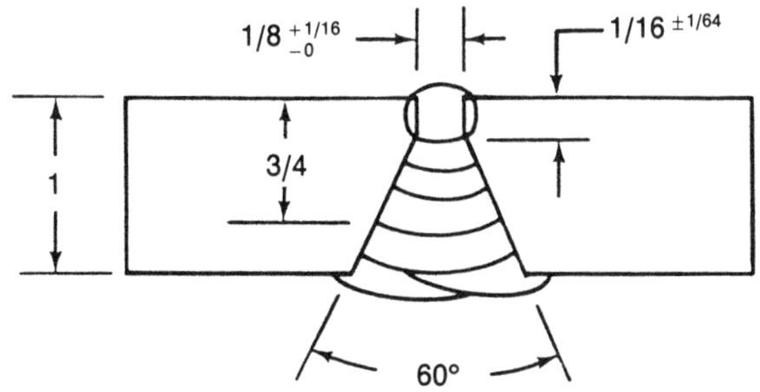

Plate Thickness (in.)	1/2	Over 1/2	1/2	5/8	3/4	7/8	1	1-1/4 and Up	
Pass Number	Root	Root	2 & 3	2 & 3	2 – 4	2 – 5	2 – 7	2 – 7	8 –
Electrode Class	E71T8-Ni1		E71T8-Ni1 or E61T8-K6						
Electrode Size	5/64		5/64						
Power [AC or DC(±)]	DC(−)		DC(−)						
Electrical Stickout (in.)	1-1/4		3/4						
Wire Feed Speed (in./min)	60	69	69	87					106
Amps	140	160	190	225					260
Arc Volts	16	17.5	17.5	19.5					21
Travel Speed (in./min)	2.3 – 2.5	2.2 – 2.5	2.0 – 3.0	2.0 – 3.0	2.0 – 3.5				
Preheat (°F)	150		—						
Interpass (°F)	—		250						
Electrode Location	—								
Electrode Angle to Joint (deg)	90								
Electrode Drag Angle (deg)	0 – 15								
Total Joint Speed (ft/hr)	—		3.6	3.5	2.7	2.2	1.8	—	
Elect. Used (lbs/ft of joint)	—		1.25	1.55	2.07	2.54	3.13	—	
Deposition Rate (lbs/hr)	3.1	3.5	3.5	4.3					5.3

FLUX-CORED ARC WELDING (SEMIAUTOMATIC) SELF-SHIELDED

Welding Position: Horizontal
Weld Quality Level: Code
Steel Weldability: Good or fair
Welded from: One side

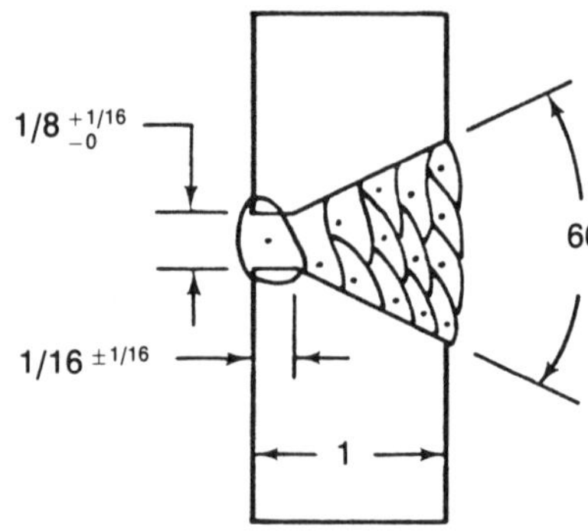

Plate Thickness (in.)	All	3/4	7/8	1	1-1/4 & Up
Pass Number	1	2 – 10	2 – 10	2 – 14	2 & Up
Electrode Class	E71T-8				
Electrode Size	3/32				
Power [AC or DC±)]	DC(−)				
Electrical Stickout (in.)	1-1/4	3/4			
Wire Feed Speed (in./min)	55	95			
Amps	175	290			
Arc Volts	17.5	20.5			
Travel Speed (in./min)	3.4 – 3.8	5.5 – 8.5			
Preheat (°F)	150	—			
Interpass (°F)	—	250			
Electrode Angle to Joint (deg)	90				
Electrode Drag Angle (deg)	0 – 30				
Total Joint Speed (ft/hr)	—	3.4	2.8	2.4	—
Elect. Used (lbs/ft of joint)	—	2.21	2.74	3.29	—
Deposition Rate (lbs/hr)	3.8	6.7			

FLUX-CORED ARC WELDING (SEMIAUTOMATIC) SELF-SHIELDED

Welding Position: Vertical up
Weld Quality Level: Code
Steel Weldability: Good or fair
Welded from: One or two sides

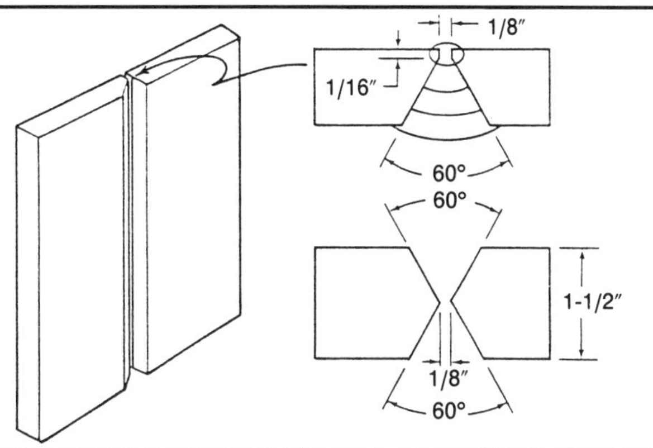

Plate Thickness (in.)	3/4 & Up	3/4	7/8	1	1-1/4 & Up	
Pass Number	1	2 – 4	2 – 5	2 – 7	2 – 7	8 –
Electrode Class	E71T8-Ni1, E61T8-K6, E71T8-K6					
Electrode Size	5/64					
Power [AC or DC±)]	DC(–)					
Electrical Stickout (in.)	1-1/4	3/4				
Wire Feed Speed (in./min)	70	85				105
Amps	160	220				250
Arc Volts	18	19				20
Travel Speed (in./min)	2.2 – 2.5	2.0 – 3.5				
Preheat (°F)	150	—				
Interpass (°F)	—	250				
Electrode Location	—					
Electrode Angle to Joint (deg)	90					
Electrode Drag Angle (deg)	0 – 15					
Deposition Rate (lbs/hr)	3.5	4.3				5.3

FLUX-CORED ARC WELDING (SEMIAUTOMATIC) SELF-SHIELDED

Welding Position: 6G
Weld Quality Level: Code
Steel Weldability: Good
Welded from: One side

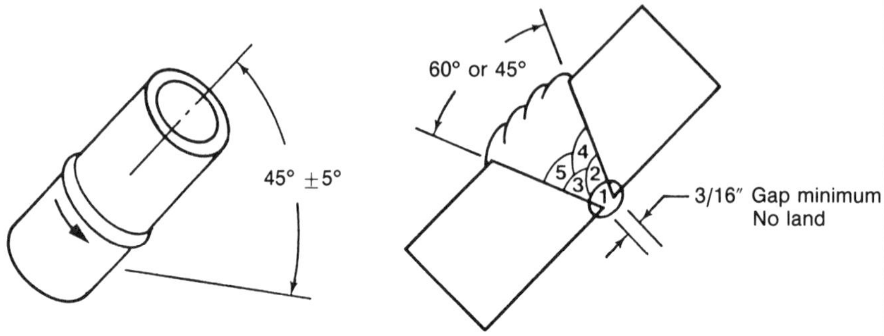

	Root Pass Vertical Up	Fill Passes Vertical Down		
Plate Thickness (in.)	3/4 and Up			
Pass Number	1	2 & 3	Fill	Cap
Electrode Class	E61T8-K6			
Electrode Size	5/64			
Power [AC or DC±)]	DC(−)			
Electrical Stickout (in.)	1 - 1-1/4	3/4 - 1		
Wire Feed Speed (in./min)	50 - 60	70 - 80	70 - 90	70 - 90
Amps	160	195 - 225	195 - 240	195 - 240
Arc Volts	16 - 17	17 - 18	17 - 19	17 - 19
Travel Speed (in./min)	2 - 2-1/2	10 - 12	10 - 12	10 - 12
Preheat (°F)	200			
Interpass (°F)	300			
Electrode Location	—			
Electrode Angle to Joint (deg)	—			
Electrode Drag Angle (deg)	—			
Deposition Rate (lbs/hr)	3.1	3.9	4.4	

FLUX-CORED ARC WELDING (SEMIAUTOMATIC) SELF-SHIELDED

Welding Position: Vertical up and Overhead
Weld Quality Level: Code
Steel Weldability: Good or fair
Welded from: One side

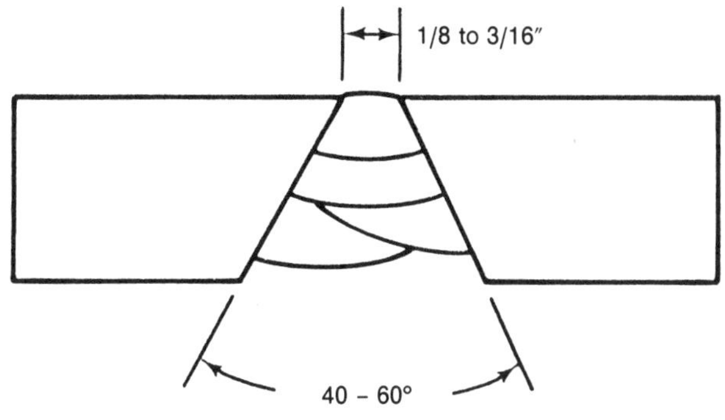

Plate Thickness (in.)	3/8 to 1/2	3/4 and Up		3/4 and Up			
Pass Number	Root			2 and Up		Cap	
Electrode Class	E71T-8						
Electrode Size	.068						
Power [AC or DC(±)]	DC(−)						
Electrical Stickout (in.)	3/4 - 1-1/4			1/2 - 3/4			
Wire Feed Speed (in./min)	90 — 100	110 — 135		150 — 170		150 — 170	
Amps (Approx.)	160	170 - 205		225 - 250		225 - 250	
Arc Volts	17 - 19	18 - 20	19 - 21	19 - 21	20 - 22	19 - 21	20 - 22
Travel Speed (in./min)	As required			As required			
Preheat (°F)	70			—			
Interpass (°F)	—			150			
Electrode Location	—			As required			
Electrode Angle to Joint (deg)	90			90			
Electrode Drag Angle (deg)	5 - 30			0 - 30			
Deposition Rate (lbs/hr)	3.1 - 3.4	3.8 - 4.6		5.2 - 6.0			

FLUX-CORED ARC WELDING (SEMIAUTOMATIC) SELF-SHIELDED

Welding Position: All Position
Weld Quality Level: Code
Steel Weldability: Good or fair
Welded from: One side

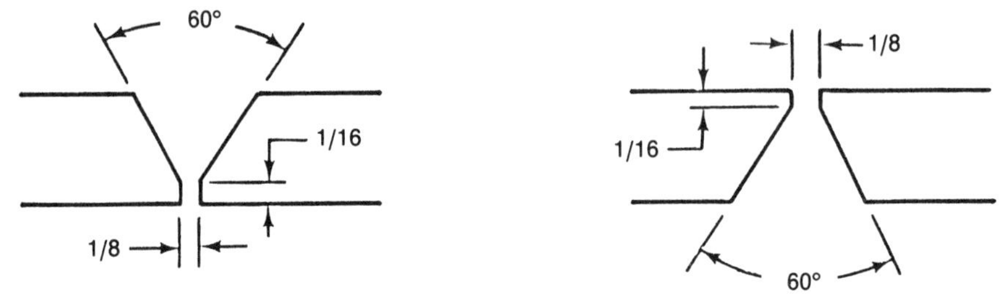

Plate Thickness (in.)	1/2		5/8 & Up		3/4	7/8	1	Over 1
Pass Number	1	2 & 3	1	2 & 3	4	4 – 5	4 – 7	4 –
Electrode Class	E61T8-K6, E71T8-Ni1, E71T8-K6							
Electrode Size	5/64							
Electrical Stickout (in.)	1-1/4	3/4	1-1/4	3/4				
Power [AC or DC(±)]	DC(–)							
Wire Feed Speed (in./min)	60	69	69	69		80		
Amps	140	170	160	170		190		
Volts	16	17.5	17.5	17.5		19		
Travel Speed (in./min)	3.0 – 3.3	1.1 – 1.2	3.0 – 3.3	.72 – .80	.53 – .59	.44 – .49	.33 – .36	—
Preheat (°F)	150	—	150	—				
Interpass (°F)	—	250	—	250				
Electrode Location	—							
Electrode Angle to Joint	—							
Electrode Drag Angle	—							
Lb of Elect/Ft of Weld	.75		—	1.14	1.60	2.13	2.73	—
Hrs/Ft of Weld	.174		—	.263	.357	.430	.580	—
Deposition Rate (lbs/hr)	3.1	3.5	3.5	3.5		3.9		

FLUX-CORED ARC WELDING (SEMIAUTOMATIC) SELF-SHIELDED

Welding Position: Flat,
Horizontal, Vertical down
Weld Quality Level: Commercial
Steel Weldability: Good or fair

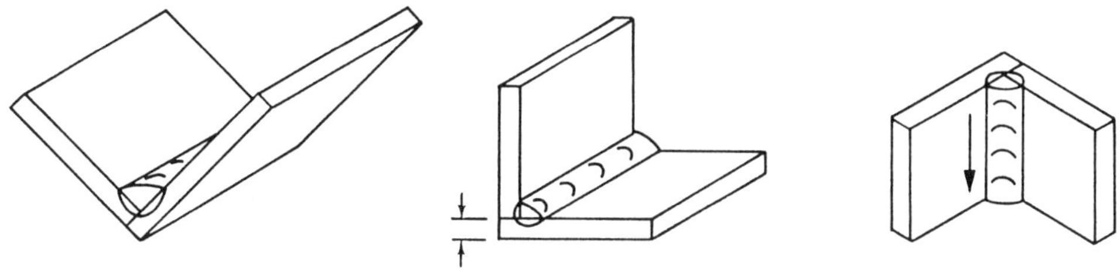

Flat Horizontal Vertical

Plate Thickness (T)	12 ga			10 ga		
Pass Number	1	1	1	1	1	1
Electrode Class	E70T-10					
Electrode Size (in.)	3/32					
Electrical Stickout (in.)	1-1/2					
Power [AC or DC(±)]	DC(−)					
Wire Feed Speed (in./min)	200	250	200	200	250	200
Amps	400	470	400	400	470	400
Volts (Arc)	24	26	24	24	26	24
Travel Speed (in./min)	45 – 50	63 – 70	45 – 50	38 – 42	54 – 60	45 – 50
Preheat (°F)	—					
Interpass (°F)	—					
Electrode Location	—					
Electrode Angle to Joint (deg)	45 – 60					
Electrode Drag Angle (deg)	5					
Lb of Elec/Ft of Weld	.075	.069	.075	.088	.081	.075
Hrs/Ft of Weld	.00421	.00301	.00421	.00500	.00351	.00421
Deposition Rate (lbs/hr)	5.5	7.9	5.5	5.5	7.9	5.5
Degrees from Horizontal	20	30 – 60	90	0	30 – 60	90

FLUX-CORED ARC WELDING (SEMIAUTOMATIC) SELF-SHIELDED

Welding Position: Flat, Horizontal, Vertical down
Weld Quality Level: Commercial
Steel Weldability: Good or fair

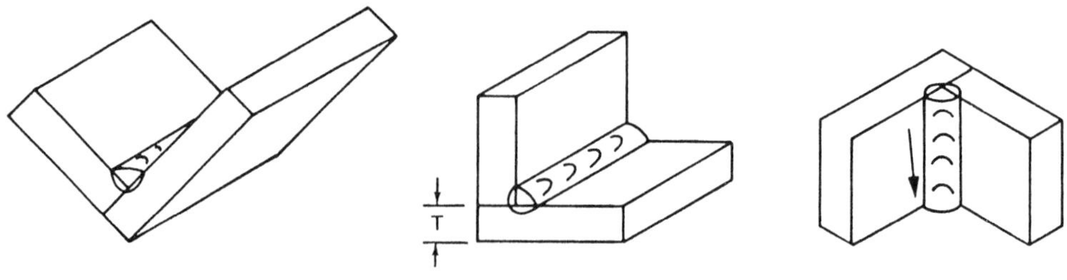

Flat Horizontal Vertical

Fillet Leg Size (in.)	5/32			3/16		
Plate Thickness (T) (in.)	3/16			1/4		
Pass Number	1	1	1	1	1	1
Electrode Class	E70T-10					
Electrode Size (in.)	3/32					
Electrical Stickout (in.)	1-1/2					
Power [AC or DC(±)]	DC(−)					
Wire Feed Speed (in./min)	250	250	200	250	250	200
Amps	470	470	400	470	470	400
Volts (Arc)	26	26	24	26	26	24
Travel Speed (in./min)	32 – 35	45 – 50	45 – 50	32 – 35	45 – 50	36 – 40
Preheat (°F)	—					
Interpass (°F)	—					
Electrode Location	—					
Electrode Angle to Joint (deg)	45 – 60					
Electrode Drag Angle (deg)	5					
Lb of Elec/Ft of Weld	.12	.096	.075	.12	.096	.093
Hrs/Ft of Weld	.00597	.00421	.00421	.00597	.00421	.00526
Deposition Rate (lbs/hr)	5.5	7.9	5.5	5.5	7.9	5.5
Degrees from Horizontal	0	30 – 60	90	0	30 – 60	90

FLUX-CORED ARC WELDING (SEMIAUTOMATIC) SELF-SHIELDED

Welding Position: Flat
Weld Quality Level: Strength only
Steel Weldability: Good

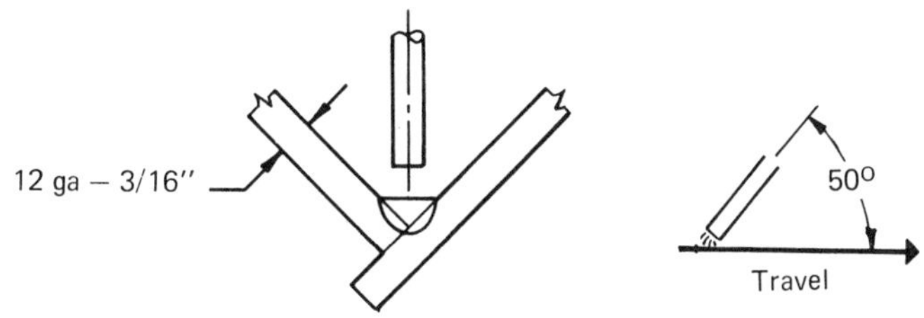

Plate Thickness (in.)	0.105 (12 ga)	0.135 (10 ga)	3/16
Pass	1	1	1
Electrode Class	E70T-3	E70T-3	E70T-3
Size	3/32	3/32	3/32
Wire Feed Speed (in./min)	160	190	190
Current (amp) DC(+)	450	500	500
Volts	27 – 29	28 – 30	28 – 30
Arc Speed (in./min)	90 – 100	65 – 75	50 – 60
Electrode Req'd (lb/ft)	0.0342	0.0535	0.0961
Total Time (hr/ft of weld)	0.00211	0.00286	0.00364
Electrical Stickout, in.	1-1/4		

FLUX-CORED ARC WELDING (SEMIAUTOMATIC) SELF-SHIELDED

Welding Position: Flat and Horizontal
Weld Quality Level: Commercial
Steel Weldability: Good or fair

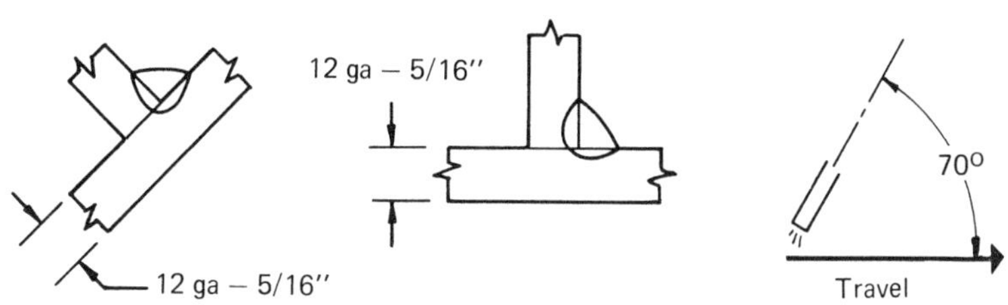

Weld Size, L (in.)	—	—	3/16	3/16+	1/4
Plate Thickness (in.)	0.105 (12 ga)	0.135 (10 ga)	3/16	1/4	5/16
Pass	1	1	1	1	1
Electrode Class	E71T-11	E71T-11	E71T-11	E71T-11	E71T-11
Size	5/64	5/64	3/32	3/32	3/32
Wire Feed Speed (in./min)	80	95	75	100	130
Current (amp) DC(−) (Approx.)	235	260	315	325	350
Volts	19 – 20	20 – 21	23 – 24	24 – 25	25 – 26
Arc Speed (in./min)	25 – 28	23 – 25	20 – 22	17 – 19	15 – 17
Electrode Req'd (lb/ft)	0.053	0.070	0.11	0.13	0.17
Total Time (hr/ft of weld)	0.00755	0.00833	0.00953	0.0111	0.0125
Electrical Stickout (in.)	3/4	3/4	1	1	1

FLUX-CORED ARC WELDING (SEMIAUTOMATIC) SELF-SHIELDED

Welding Position: Flat
Weld Quality Level: Commercial
Steel Weldability: Good or fair

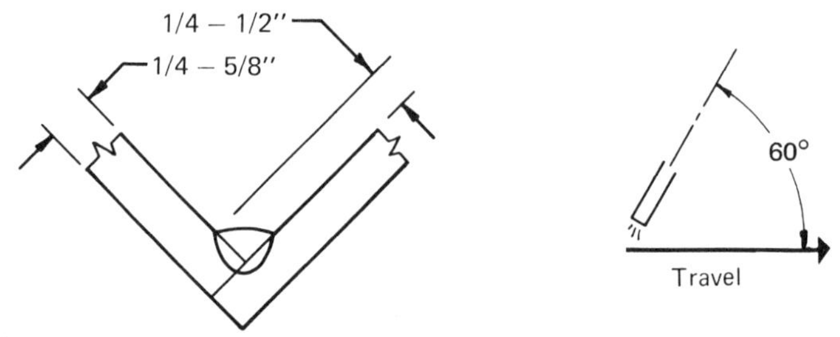

Weld Size, L (in.)	1/4 (–)	1/4	5/16	3/8	1/2
Plate Thickness (in.)	1/4	5/16	3/8	1/2	5/8
Pass	1	1	1	1	1*
Electrode Class	E70T-4	E70T-4	E70T-4	E70T-4	E70T-4
Size	3/32	3/32	3/32	3/32	3/32
Wire Feed Speed (in./min)	190	205	205	205	205
Current (amp) DC(+)	360	375	375	375	375
Volts	29 – 30	30 – 31	30 – 31	30 – 31	30 – 31
Arc Speed (in./min)	21 – 23	18 – 20	13 – 15	10 – 11	5.2 – 5.8
Electrode Req'd (lb/ft)	0.17	0.21	0.28	0.39	0.67
Total Time (hr/ft of weld)	0.00909	0.0105	0.0143	0.0190	0.0364
Electrical Stickout, in.	2-3/4				

* For better penetration use two passes. Overall speed is the same.

FLUX-CORED ARC WELDING (SEMIAUTOMATIC) SELF-SHIELDED

Welding Position: Flat
Weld Quality Level: Commercial
Steel Weldability: Good or fair

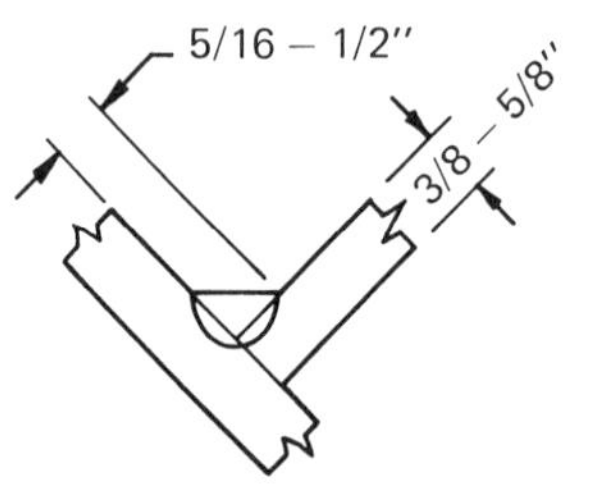

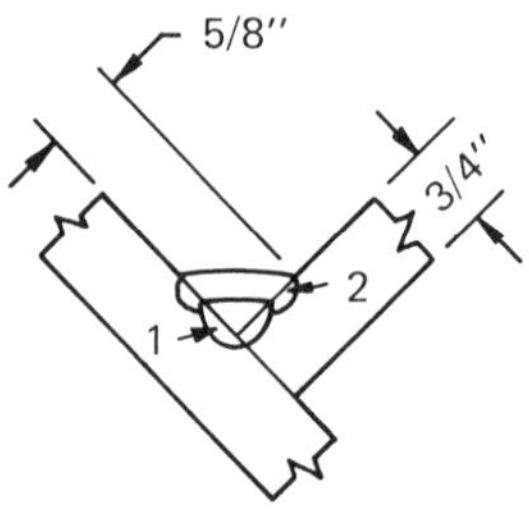

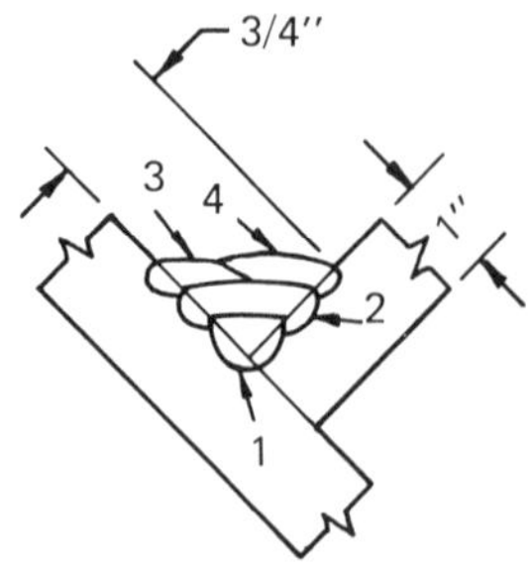

Weld Size, L (in.)	5/16	3/8	1/2	5/8		3/4	
Plate Thickness (in.)	3/8	1/2	5/8	3/4		1	
Pass	1	1	1	1	2	1, 3 & 4	2
Electrode Class	E70T-4	E70T-4	E70T-4	E70T-4	E70T-4	E70T-4	E70T-4
Size	.120	.120	.120	.120	.120	.120	.120
Wire Feed Speed (in./min)	275	310	325	325	325	325	325
Current (amp) DC(+)	525	565	580	580	580	580	580
Volts	35 – 36	36 – 37	37 – 38	37 – 38	37 – 38	37 – 38	37 – 38
Arc Speed (in./min)	23 – 25	20 – 22	13 – 15	21 – 23	16 – 18	26 – 28	17 – 19
Electrode Req'd (lb/ft)	0.313	0.404	0.617	1.02		1.47	
Total Time (hr/ft of weld)	0.00833	0.00952	0.0143	0.0209		0.0333	
Electrical Stickout, in.	3-3/4						

FLUX-CORED ARC WELDING (SEMIAUTOMATIC) SELF-SHIELDED

Welding Position: Flat
Weld Quality Level: Commercial*
Steel Weldability: Good or fair

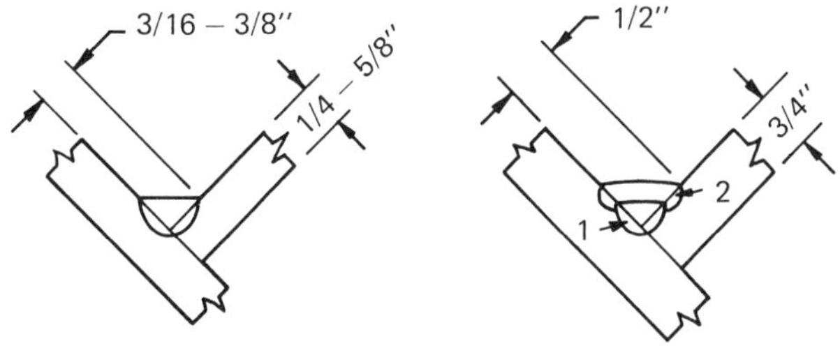

Weld Size, L (in.)	3/16	3/16(+)	1/4	5/16	3/8	1/2
Plate Thickness (in.)	1/4	5/16	3/8	1/2	5/8	3/4
Pass	1	1	1	1	1	1 – 2
Electrode Class*	E70T-7	E70T-7	E70T-7	E70T-7	E70T-7	E70T-7
Size	7/64	7/64	7/64	7/64	7/64	7/64
Wire Feed Speed (in./min)	175	205	205	225	225	225
Current (amp) DC(−) (Approx.)	450	500	500	525	525	525
Volts	27 – 28	28 – 29	28 – 29	29 – 30	29 – 30	29 – 30
Arc Speed (in./min)	32 – 35	26 – 29	22 – 24	18 – 20	13 – 14	15 – 17
Electrode Req'd (lb/ft)	0.12	0.17	0.21	0.27	0.38	0.64
Total Time (hr/ft of weld)	0.00597	0.00727	0.00870	0.0105	0.0148	0.0250
Electrical Stickout, in.	1-1/2					

* Classified "Commercial" because the fillets are smaller than required by most codes. The welds have strength in excess of requirements if an E70T electrode is used which has penetration capabilities.

FLUX-CORED ARC WELDING (SEMIAUTOMATIC) SELF-SHIELDED

Welding Position: Horizontal
Weld Quality Level: Strength only
Steel Weldability: Good or fair

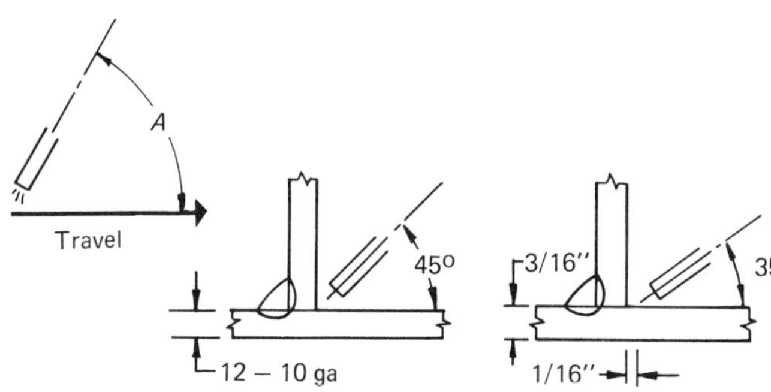

Plate Thickness (in.)	0.105 (12 ga)	0.135 (10 ga)	3/16
Pass	1	1	1
Electrode Class	E70T-3	E70T-3	E70T-3
Size	3/32	3/32	3/32
Wire Feed Speed (in./min)	175	195	195
Current (amp) DC(+) (Approx.)	450	500	500
Volts	27 – 29	28 – 30	28 – 30
Arc Speed (in./min)	85 – 95	55 – 65	30 – 34
Electrode Req'd (lb/ft)	0.0370	0.0627	0.118
Total Time (hr/ft of weld)	0.00222	0.00333	0.00625
Drag Angle, *A* (deg)	45	50	55
Electrical Stickout, in.	1-1/4		

FLUX-CORED ARC WELDING (SEMIAUTOMATIC) SELF-SHIELDED

Welding Position: Horizontal
Weld Quality Level: Commercial
Steel Weldability: Good or fair

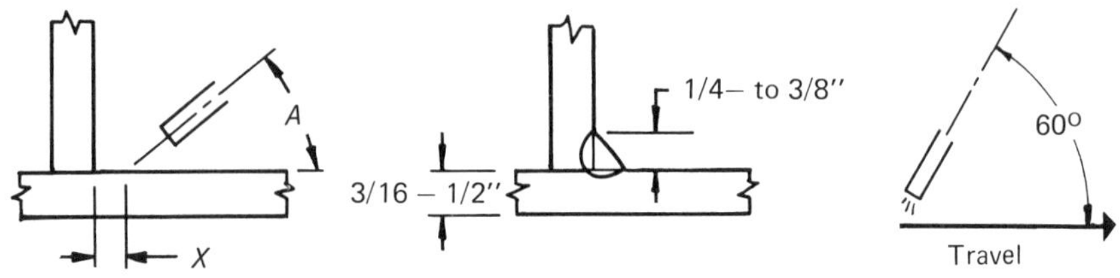

Weld Size, L (in.)	1/4 (−)	1/4 (−)	1/4		5/16		3/8
Plate Thickness (in.)	3/16	1/4	5/16		3.8		1/2
Pass	1	1	1	1	1	1	1
Electrode Class	E70T-4	E70T-4	E70T-4	E70T-4	E70T-4	E70T-4	E70T-4
Size	3/32	3/32	3/32	.120	3/32	.120	.120
Wire Feed Speed (in./min)	165	185	205	160	205	160	175
Current (amp) DC(+)	325	350	375	425	375	425	450
Volts	28 – 29	29 – 30	30 – 31	27 – 30	30 – 31	27 – 30	28 – 31
Arc Speed (in./min)	18 – 20	17 – 19	15 – 17	18 – 20	14 – 16	16 – 18	12 – 13
Electrode Req'd (lb/ft)	0.17	0.20	0.25	0.25	0.27	0.27	0.39
Total Time (hr/ft of weld)	0.0105	0.0111	0.0125	0.0105	0.0133	0.0118	0.0154
Electrode location, *X* (in.)	0	1/16	3/32	1/8	1/8	5/32	5/32
Electrode Angle, *A* (deg)	40	35	35	30	30	25	25
Electrical Stickout, in.	2-3/4						

FLUX-CORED ARC WELDING (SEMIAUTOMATIC) SELF-SHIELDED

Welding Position: Horizontal
Weld Quality Level: Commercial
Steel Weldability: Good or fair

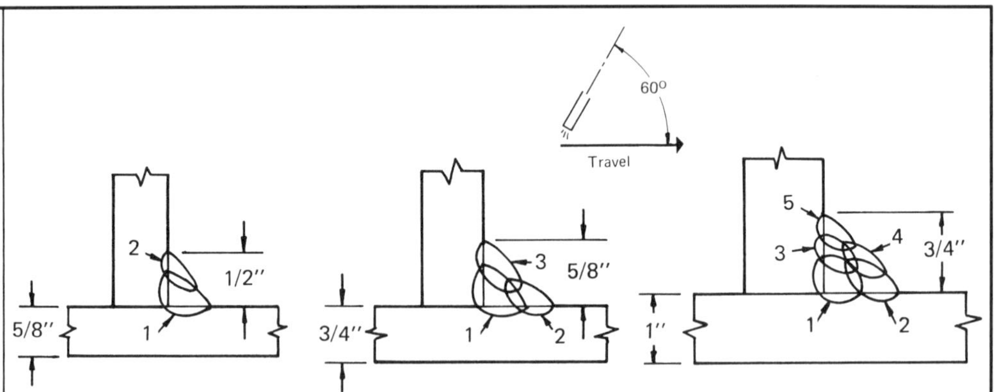

Weld Size, L (in.)	1/2		5/8			3/4		
Plate Thickness (in.)	5/8		3/4			1		
Pass	1	2	1	2	3	1 & 5	2	3 & 4
Electrode Class	E70T-4	E70T-4	E70T-4	E70T-4	E70T-4	E70T-4	E70T-4	E70T-4
Size	.120	.120	.120	.120	.120	.120	.120	.120
Wire Feed Speed (in./min)	175	175	175	175	175	175	175	175
Current (amp) DC(+) (Approx.)	450	450	450	450	450	450	450	450
Volts	28 – 31	28 – 31	28 – 31	28 – 31	28 – 31	28 – 31	28 – 31	28 – 31
Arc Speed (in./min)	13 – 15	16 – 18	15 – 17	16 – 18	13 – 15	15 – 17	14 – 16	18 – 20
Electrode Req'd (lb/ft)	0.670		0.962			1.42		
Total Time (hr/ft of weld)	0.0259		0.0386			0.0593		
Electrical Stickout, in.	2-3/4							

FLUX-CORED ARC WELDING (SEMIAUTOMATIC) SELF-SHIELDED

Welding Position: Horizontal
Weld Quality Level: Commercial*
Steel Weldability: Good or fair

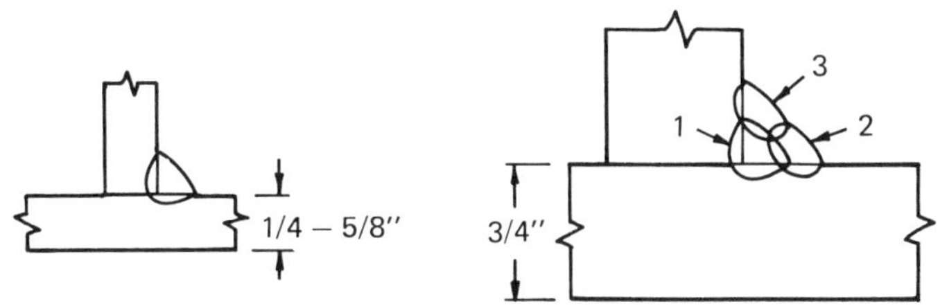

Weld Size, L (in.)	3/16	3/16(+)	1/4	5/16	3/8	1/2
Plate Thickness (in.)	1/4	5/16	3/8	1/2	5/8	3/4
Pass	1	1	1	1	1	3
Electrode Class	E70T-7	E70T-7	E70T-7	E70T-7	E70T-7	E70T-7
Size	7/64	7/64	7/64	7/64	7/64	7/64
Wire Feed Speed (in./min)	160	190	205	205	205	205
Current (amp) DC(−) (Approx.)	425	475	500	500	500	500
Volts	26 - 27	27 - 28	28 - 29	28 - 29	28 - 29	28 - 29
Arc Speed (in./min)	30 - 33	24 - 27	19 - 21	15 - 17	11 - 12	20 - 22
Electrode Req'd (lb/ft)	0.12	0.18	0.24	0.30	0.42	0.69
Total Time (hr/ft of weld)	0.00635	0.00784	0.0100	0.0125	0.0174	0.0286
Electrical Stickout, in.	1-1/2					

* Classified "Commercial" because the fillets are smaller than required by most codes. The welds have strength in excess of requirements if an E70T electrode is used which has penetration capabilities.

FLUX-CORED ARC WELDING (SEMIAUTOMATIC) SELF-SHIELDED

Welding Position: Vertical down
Weld Quality Level: Commercial
Steel Weldability: Good or fair

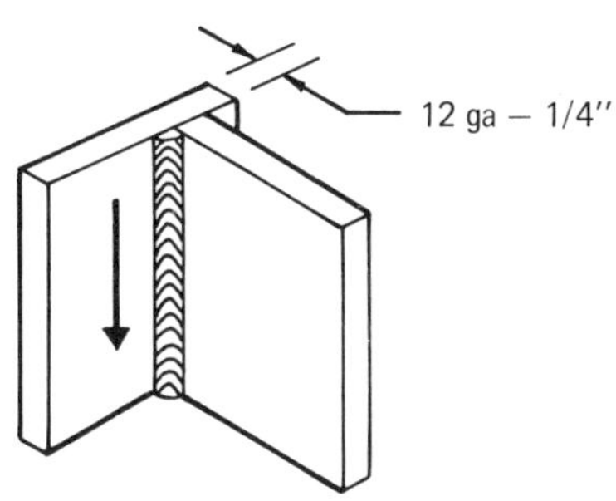

Weld Size, L (in.)	—	—	3/16	3/16+
Plate Thickness (in.)	0.105 (12 ga)	0.135 (10 ga)	3/16	1/4
Pass	1	1	1	1
Electrode Class	E70T-11	E70T-11	E70T-11	E70T-11
Size	5/64	5/64	3/32	3/32
Wire Feed Speed (in./min)	60	75	75	75
Current (amp) DC(−)	205	250	305	305
Volts	19 - 20	20 - 21	21 - 22	21 - 22
Arc Speed (in./min)	30 - 33	26 - 29	25 - 28	22 - 24
Electrode Req'd (lb/ft)	0.045	0.061	0.082	0.10
Total Time (hr/ft of weld)	0.00635	0.00727	0.00755	0.00870
Electrical Stickout, in.	3/4			

See introductory notes.

FLUX-CORED ARC WELDING (SEMIAUTOMATIC) SELF-SHIELDED

Welding Position: Vertical up
Weld Quality Level: Commercial
Steel Weldability: Good

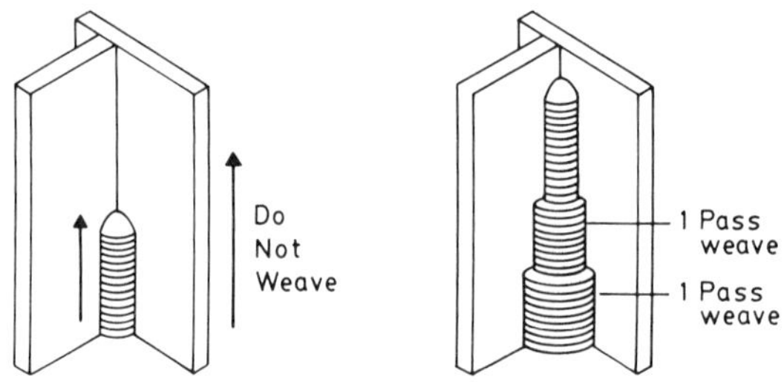

Weld Size, L (in.)	1/4 (+)	5/16	3/8
Plate Thickness (in.)	1/4	3/8	1/2 & Up
Pass	1	1	1
Electrode Class	E70T-8	E70T-8	E70T-8
Size	.068	.068	.068
Wire Feed Speed (in./min)	130	160	160
Current (amp) DC(−)	225	260	260
Volts	19.0	20.0	20.0
Arc Speed (in./min)	6.5 – 8.6	5.5 – 6.5	As Required
Electrode Req'd (lb/ft)	0.180	0.285	0.345
Total Time (hr/ft of weld)	0.0500	0.0607	0.0833
Electrical Stickout, in.	1 ± 1/4		

FLUX-CORED ARC WELDING (SEMIAUTOMATIC) SELF-SHIELDED

Welding Position: Vertical up
Weld Quality Level: Code
Steel Weldability: Good or fair

> Where Code Quality is indicated normal practice requires a procedure qualification test.

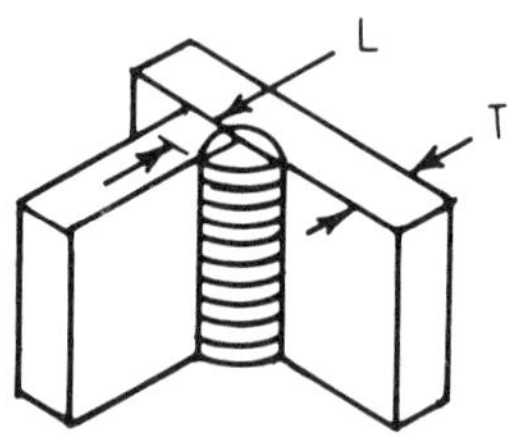

Weld Size, L (in.)	1/4	5/16	3/8	1/2	5/8	3/4
Plate Thickness (in.)	5/16	3/8	1/2	5/8	3/4	1
Pass	1	1	1	1 & 2	1 – 3	1 – 4
Electrode Class	E71T-8	E71T-8	E71T-8	E71T-8	E71T-8	E71T-8
Size	5/64	5/64	5/64	5/64	5/64	5/64
Wire Feed Speed (in./min)	75	87	98	98	98	98
Current (amp) DC(−) (Approx.)	185	220	235	235	235	235
Volts	18 – 19	18 – 19	18 – 19	19 – 20	19 – 20	19 – 20
Arc Speed (in./min)	4.3 – 4.8	4.1 – 4.5	3.4 – 3.8	1.8 – 2.0	1.2 – 1.3	.85 – .95
Electrode Req'd (lb/ft)	.21	.26	.36	.64	.95	1.40
Total Time (hr/ft of weld)	.044	.047	.055	.105	.160	.210
Electrical Stickout, in.	3/4					

FLUX-CORED ARC WELDING (SEMIAUTOMATIC) SELF-SHIELDED

Welding Position: Overhead
Weld Quality Level: Code
Steel Weldability: Good or fair

> Where Code Quality is indicated normal practice requires a procedure qualification test.

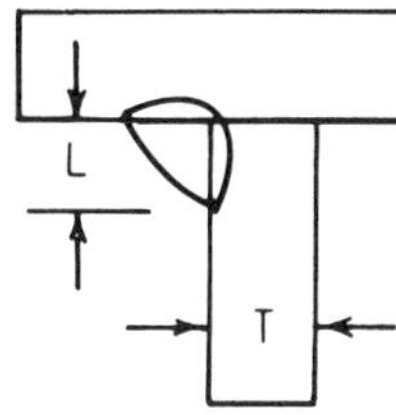

Weld Size, L (in.)	1/4	5/16	3/8
Plate Thickness (in.)	5/16	3/8	1/2
Pass	1	1	1 – 2
Electrode Class	E71T-8	E71T-8	E71T-8
Size	5/64	5/64	5/64
Wire Feed Speed (in./min)	75	87	87
Current (amp)	185	220	220
Volts	18	19	19
Arc Speed (in./min)	4.3 – 4.8	4.1 – 4.5	3.0 – 3.3
Electrode Req'd (lb/ft)	0.20	0.26	0.36
Total Time (hr/ft of weld)	.0440	.0465	.0635
Electrical Stickout, in.		3/4	

FLUX-CORED ARC WELDING (SEMIAUTOMATIC) SELF-SHIELDED

Welding Position: Flat
Weld Quality Level: Strength only
Steel Weldability: Good

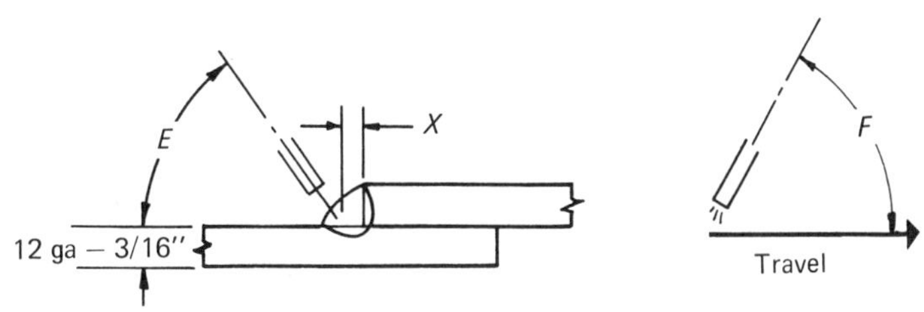

Plate Thickness (in.)	0.105 (12 ga)	0.135 (10 ga)	3/16
Pass	1	1	1
Electrode Class	E70T-3	E70T-3	E70T-3
Size	3/32	3/32	3/32
Wire Feed Speed (in./min)	175	195	195
Current (amp) DC(+) (Approx.)	450	500	500
Volts	27 – 29	28 – 30	28 – 30
Arc Speed (in./min)	85 – 95	65 – 75	38 – 42
Electrode Req'd (lb/ft)	0.0370	0.0502	0.0940
Total Time (hr/ft of weld)	0.00222	0.00286	0.00500
Angle, *E* (deg)	55	50	45
Angle, *F* (deg)	45	50	55
Electrode Location, *X* (in.)	3/64	1/16	5/64
Electrical Stickout, in.		1-1/4	

FLUX-CORED ARC WELDING (SEMIAUTOMATIC) SELF-SHIELDED

Welding Position: Horizontal
Weld Quality Level: Commercial
Steel Weldability: Good or fair

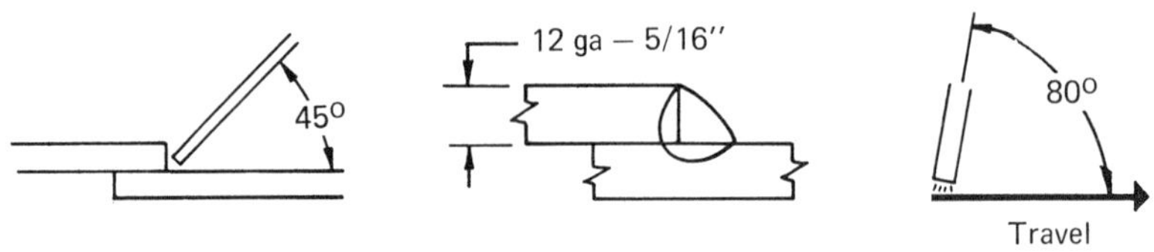

Weld Size, L (in.)	—	—	3/16 (−)	1/4
Plate Thickness (in.)	0.105 (12 ga)	0.135 (10 ga)	3/16	5/16
Pass	1	1	1	1
Electrode Class	E70T-10	E70T-10	E70T-10	E70T-10
Size	3/32	3/32	3/32	3/32
Wire Feed Speed (in./min)	200	200	250	250
Current (amp) DC(−)	400	400	470	470
Volts	24	24	26	26
Arc Speed (in./min)	45 – 50	38 – 42	45 – 50	36 – 40
Electrode Req'd (lb/ft)	.075	.088	.096	.096
Total Time (hr/ft of weld)	.00421	.00500	.00421	.00526
Electrical Stickout, in.	1-1/2			

FLUX-CORED ARC WELDING (SEMIAUTOMATIC) SELF-SHIELDED

Welding Position: Horizontal
Weld Quality Level: Commercial
Steel Weldability: Good or fair

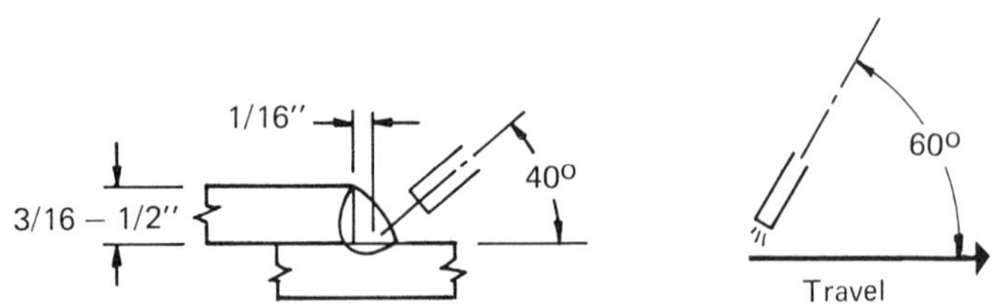

Weld Size, L (in.)	3/16	1/4 (−)	1/4 (+)	5/16	3/8
Plate Thickness (in.)	3/16	1/4	5/16	3/8	1/2
Pass	1	1	1	1	1
Electrode Class	E70T-7	E70T-7	E70T-7	E70T-7	E70T-7
Size	7/64	7/64	7/64	7/64	7/64
Wire Feed Speed (in./min)	115	145	160	175	190
Current (amp) DC(−) (Approx.)	350	400	425	450	475
Volts	25 – 26	26 – 27	27 – 28	28 – 29	29 – 30
Arc Speed (in./min)	21 – 23	20 – 22	19 – 21	15 – 17	11 – 12
Electrode Req'd (lb/ft)	0.13	0.16	0.19	0.26	0.39
Total Time (hr/ft of weld)	0.00909	0.00952	0.0100	0.0125	0.0174
Electrical Stickout, in.	1-1/2				

FLUX-CORED ARC WELDING (SEMIAUTOMATIC) SELF-SHIELDED

Welding Position: Horizontal
Weld Quality Level: Commercial
Steel Weldability: Good

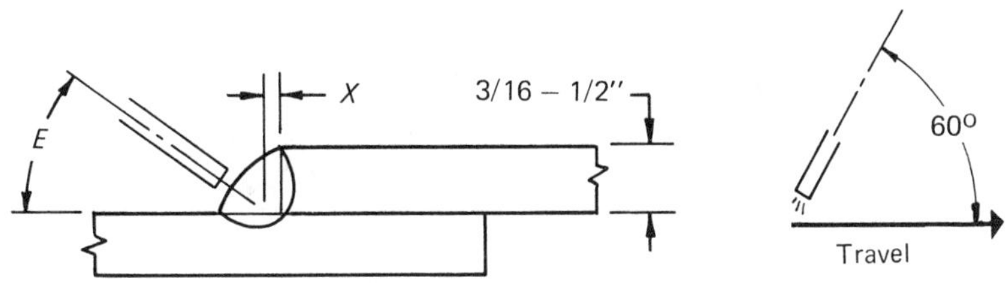

Weld Size, L (in.)	3/16	1/4		5/16		3/8 (–)		1/2 (–)
Plate Thickness (in.)	3/16	1/4		5/16		3/8		1/2
Pass	1	1	1	1	1	1	1	1
Electrode Class	E70T-4	E70T-4	E70T-4	E70T-4	E70T-4	E70T-4	E70T-4	E70T-4
Size	3/32	3/32	.120	3/32	.120	3/32	.120	.120
Wire Feed Speed (in./min)	135	185	175	185	175	185	175	175
Current (amp) DC(+) (Approx.)	275	350	450	350	450	350	450	450
Volts	28 – 29	30 – 31	28 – 31	30 – 31	28 – 31	30 – 31	28 – 31	28 – 31
Arc Speed (in./min)	19 – 21	18 – 20	24 – 26	15 – 17	19 – 21	14 – 16	16 – 18	12 – 14
Electrode Req'd (lb/ft)	0.113	0.171	0.189	0.202	0.231	0.214	0.269	0.346
Total Time (hr/ft of weld)	0.0100	0.0105	0.00800	0.0125	0.0100	0.0133	0.0118	0.0154
Angle, *E* (deg)	35 – 40	35 – 40	30 – 35	30 – 35	30 – 35	30 – 35	25 – 30	25 – 30
Electrode Location, *X* (in.)	0 – 1/16	0 – 1/16	3/32 – 1/8	3/32 – 1/8	3/32 – 1/8	3/32 – 1/8	5/32	5/32
Electrical Stickout, in.	2-3/4							

FLUX-CORED ARC WELDING (SEMIAUTOMATIC) SELF-SHIELDED

Welding Position: Horizontal
Weld Quality Level: Strength only
Steel Weldability: Good

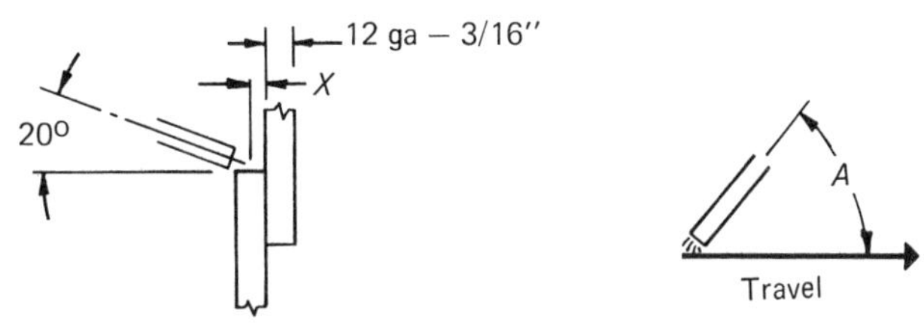

Plate Thickness (in.)	0.105 (12 ga)	0.135 (10 ga)	3/16
Pass	1	1	1
Electrode Class	E70T-3	E70T-3	E70T-3
Size	3/32	3/32	3/32
Wire Feed Speed (in./min)	160	190	190
Current (amp) DC(+) (Approx.)	450	500	500
Volts	27 – 29	28 – 30	28 – 30
Arc Speed (in./min)	85 – 95	70 – 80	40 – 46
Electrode Req'd (lb/ft)	0.0370	0.0502	0.0874
Total Time (hr/ft of weld)	0.00222	0.00267	0.00465
Drag Angle, *A* (deg)	45	45	50
Electrode Location, *X* (in.)	3/64	1/16	5/64
Electrical Stickout, in.	1-1/4		

Self-shielded flux-cored electrode welding a structural assembly for a concrete batch-mixing plant.

Welding Carbon and Low-Alloy Steels with the Gas Shielded Flux-Cored Electrode Process

As noted in Section 5.4, gas-shielded flux-cored arc welding has similarities both to self-shielded flux-cored welding and to gas metal-arc welding. It differs from the former in that a separately applied gas, usually carbon dioxide, acts as the shield, and it differs from GMA welding in that the electrode is tubular instead of a solid wire. The tubular electrode contains fluxing and deoxidizing substances for mild steel, and, in addition, may contain alloying elements for low-alloy steels.

Electrodes are specified in AWS A5.20-79 (reaffirmed 1988) for carbon steel electrodes and in AWS A5.29-80 for low alloy steel electrodes. Table 6-29 (Section 6.4) gives chemical requirements of the deposited metal, Table 6-30 lists test requirements and Table 6-31 gives mechanical property requirements of carbon steel deposited weld metal. Table 6-32 gives chemical requirements of the deposited metal, Table 6-33 lists test requirements, Table 6-34 gives tensile property requirements and Table 6-35 gives impact test requirements of low alloy steel deposited weld metal.

Flux cored electrodes are available for carbon steels and low alloy steels in the following standard diameters: 0.045 in., 0.052 in., 1/16 in., 5/64 in., 3/32 in., 7/64 in., 0.120 in., 1/8 in., 5/32 in. This includes low alloy steels such as 1-1/2 Mn-1/2 Mo, 1 Cr-1/2 Mo, 2-1/4 Cr-1 Mo, 1% Ni, and the Q-and-T steels.

Electrodes 3/32-in. and larger are usually limited to flat position welding; electrodes 5/64-in. and smaller may be used to weld out-of-position by operating at the low end of the welding-current range.

Typical mechanical properties of mild-steel and low-alloy steel weld deposits are shown in Table 6-38. Typical current ranges are listed in Table 6-39.

Figure 6-81 shows deposition rates for typical current ranges for various diameters of E70T1 electrode with CO_2 shielding. Deposition efficiencies of gas-shielded flux-cored electrodes generally range from 85 to 90%.

In the gas shielded flux-cored arc welding process the shielding gas protects the molten metal being transferred across the arc from atmospheric oxygen

TABLE 6-38. Typical Mechanical Properties of Weld Metal Depositied with CO_2-Shielded Flux-Cored Electrodes

Alloy Type or AWS Class	AW (As Welded) or SR (Stress Relieved)	Yield Strength (1000 psi)	Tensile Strength (1000 psi)	Elongation in 2 in. (%)	Impact Strength (ft-lb) Charpy V-Notch				
					Test Temperature (°F)				
					+75	−25	−50	−60	−75
E70T-1	AW	77.5	85	30	98	41	25	21	—
	SR(1150°F)	62	74	34	86	42	25	19	—
E70T-2*	AW	81	86	30	—	—	—	—	—
1-1/2% Mn-1/2% Mo	AW	97	108.5	16.5	50	30	24	—	—
	SR(1150°F)	96.5	107.8	17	—	—	—	—	—
1% Cr-1/2% Mo	SR (1275°F)	78.5	88	25	—	—	—	—	—
1-1/4% Cr-1/2% Mo	AW	94	106	14	—	—	—	—	—
	SR(1275°F)	75	88	22	—	—	—	—	—
	SR(1350°F)	71	85	22	—	—	—	—	—
2-1/4% Cr-1% Mo	AW	109.5	128	12	—	—	—	—	—
	SR(1275°F)	82.5	96	20	—	—	—	—	—
	SR(1350°F)	77.6	92	20	—	—	—	—	—
1% Ni	AW	79.3	88.6	24	64	—	—	32	26
	SR(1150°F)	76.7	85.8	29	74	—	—	29	25
Q and T	AW	107.5	119	20	47	—	—	27	24
	SR(1050°F)	109	121	20	40	—	—	22	19

* Single-pass weld.

TABLE 6-39. Typical Current Ranges for CO_2-Shielded Flux-Cored Electrodes

Electrode Diameter (in.)	Current Range (amp)
1/16, 5/64	150 to 450
3/32	300 to 550
7/64, 1/8	400 to 650
5/32	500 to 750

and nitrogen by forming a protective envelope around the arc and weld pool. There is little need to denitrify the weld metal with the flux because most of the nitrogen is excluded by the gas shield. The shielding gas is usually CO_2 (dew point should be at least — 45°F), and, for most applications, the gas flow is 25 to 45 cfh. Higher flow rates may be necessary when welding in drafts. Mixtures of argon with oxygen and argon with CO_2 reduce spatter and penetration and are sometimes preferred for welding thin material.

EQUIPMENT FOR WELDING

Wire-feeding equipment is essentially the same for the cored electrode as for solid electrode. However, since the diameter of cored wire is larger than a solid electrode for a comparable application, the feeding equipment must be able to feed larger diameter wire. Power sources are the constant-voltage type — the same as used with self-shielded flux-cored welding and gas metal arc welding.

In addition to the wire feeding equipment and welding power supply needed for self-shielded flux cored arc welding, the gas shielded process requires a source of shielding gas and a means of delivering it to the arc zone. The wire feed gun must have a gas nozzle which directs the shielding gas to the weld zone and this must be connected through a tube in the gun cable harness to the gas supply.

Shielding gases may be supplied from cylinders, banks of cylinders manifolded together or from liquified bulk tanks. The latter two would be connected to the weld stations by copper piping which must be clean and leak tight. Regulators and flowmeters are required to control gas pressure and flow rate.

WELDING VARIABLES

Generally the same operating techniques and precautions as recommended for self-shielded flux-cored arc welding should be observed for gas-shielded flux-cored arc welding. The use of a shielding gas allows a wider variation in the metal transfer mode than attainable with the self-shielded process.

The appearance and quality of welds are affected by such variables as arc voltage, travel rate, electrode angle, electrical stickout, nozzle-to-work distance, shielding gas, gas-flow rate and drafts or winds.

Arc Voltage: If the arc voltage is too high, the bead tends to widen in an irregular manner, with excessive spatter. Too low an arc voltage results in a narrow, high bead with excessive spatter and reduced penetration.

Travel Rate: As with other mechanized processes, travel rate affects the build-up of molten metal and penetration into the base material. Slow travel increases penetration, but excessively slow travel can lead to excessive build-up of molten metal, overheating of the weld area, and a rough-appearing bead. Too high travel rates may result in inadequate penetration and a ropy, irregular bead. Travel rates between 12 and 30 inches per minute usually give satisfactory results.

Electrode Angle: Figure 6-82 shows the correct gun positions when welding with gas-shielded flux-cored electrodes. With a butt weld, the electrode should be perpendicular to the sides of the joint and slanted from 2 to 15° in the direction of travel. The leading angle results in a "lagging" gas shield, with much of the gas flowing back over the newly deposited weld metal. With a fillet weld, the electrode is dropped off center of the joint approximately half the diameter of the electrode, as shown in Fig. 6-82, and a leading angle of 2 to 15° used.

Electrical Stickout: Varying the electrical stickout — as with self-shielded flux-cored welding and submerged-arc welding — offers a method of controlling deposition rate and penetration. At a given rate of wire feed, a short stickout results in deeper penetration than a long stickout. With gas-shielded flux-cored electrodes, an electrical stickout of 3/4 to 1-1/4-in. is usually recommended, depending on the type of nozzle. If the stickout is excessive, spatter occurs, and arc shielding is lost. With increased stickout, the wire feed rate should be increased — which gives an increased deposition rate.

Distance to Nozzle: The distance of the nozzle to the work, as well as the electrical stickout, influences the performance (Fig. 6-83). The recommended nozzle-to-work distance is 3/4 to 1 in., which, with concentric-type nozzles, will give an electrical stickout of about 1 to 1-1/4 in. If the nozzle-to-work distance is too short, spatter may rapidly build up on the nozzle and contact tube.

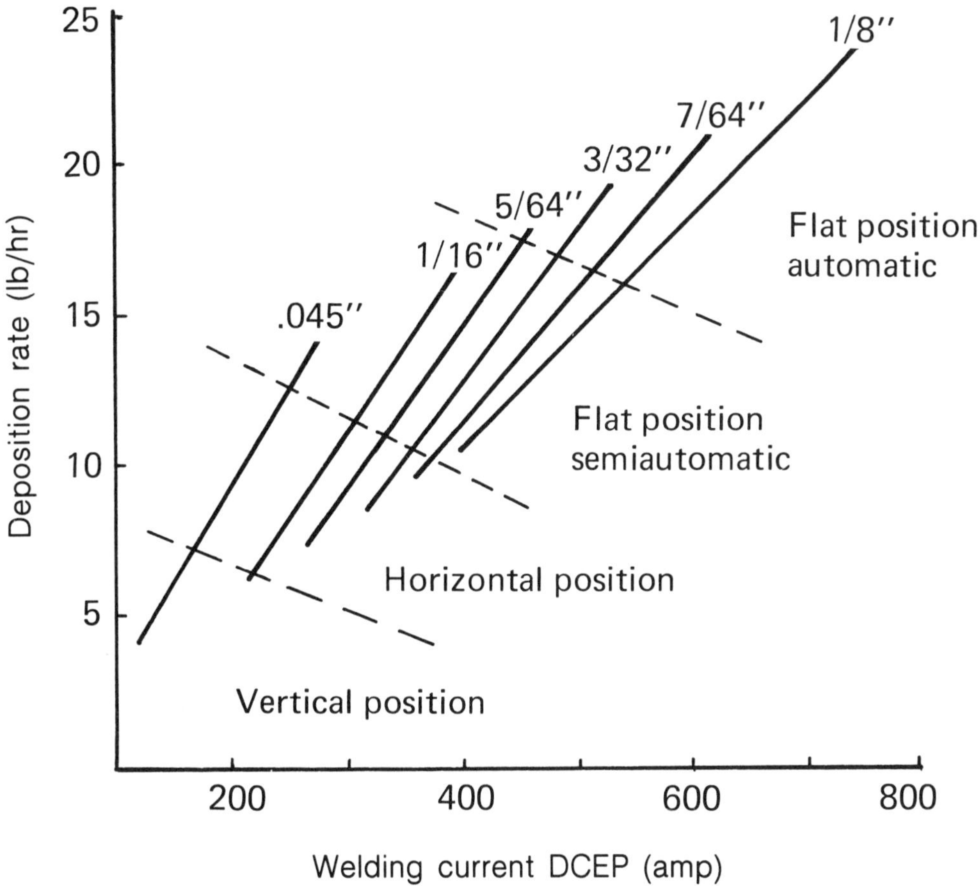

Fig. 6-81. Deposition rate versus current — for various diameters of E70T-1 or E71T1 electrodes with CO_2 shielding.

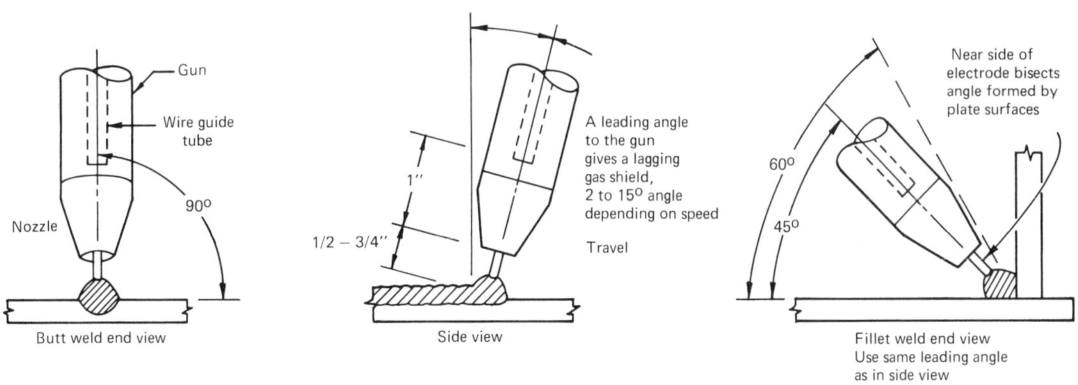

Fig. 6-82. Correct gun positions for gas-shielded flux-cored arc welding.

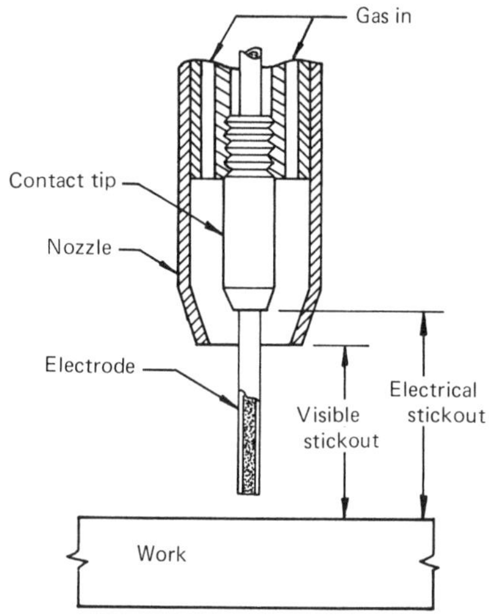

Fig. 6-83. The nozzle-to-work distance influences performance characteristics when welding with flux-cored electrodes.

With side-shielded nozzles, electrical stickout is normally set at 3/4 to 1-1/4 in.

SHIELDING GASES

Carbon dioxide is the most widely used shielding gas for gas-shielded flux-cored arc welding because of its low cost and deep arc penetration characteristics. Metal transfer is of the globular type. Although carbon dioxide is relatively inert at room temperature, it dissociates at arc temperatures to form CO and O_2.

This oxygen in the arc atmosphere will react with the more reactive elements in the weld pool to form oxides which float to the surface and are absorbed in the slag. Elements such as Mn, Al and Si are contained in the core as deoxidizers to compensate for the oxidizing effect of the CO_2.

Molten iron can also react with CO_2 to form iron oxide and CO and some of the CO can dissociate to form carbon and oxygen. Under these circumstances, the molten iron can either lose or pickup carbon. If the carbon level is 0.010 percent or more, it will be lost by oxidation. If it is below 0.05 percent, it will increase by absorbing carbon from the CO. When deoxidizers are present they will react with oxygen to form solid oxides in preference to forming CO so that the carbon level in the deposit may increase over the 0.010 percent. If CO forms in the molten weld metal, it may be trapped on solidification and appear as porosity.

Gas mixtures are used with gas shielded flux cored arc welding to utilize the advantages of each of the individual gases. The higher the percentage of an inert gas in a mixture with CO_2 the greater the transfer percentage of deoxidizers in the deposit.

Argon will protect the molten weld pool from oxidation and contamination at all welding temperatures. When used with CO_2 in adequate percentages it provides improved oxidation resistance over all CO_2. A common shielding gas mixture for this process is 75 percent argon and 25 percent carbon dioxide. Welds made with this gas mixture generally have higher tensile and yield strengths than welds made with 100 percent CO_2. Electrode transfer approaches the spray mode with this gas mixture. The A-CO_2 mixture is preferred by many weldors for out-of-position welding because of its favorable arc characteristics and operator appeal.

As was mentioned previously, the use of high percentages of inert gas in gas mixtures with filler materials designed for use with CO_2 shielding may produce deposited weld metal with high levels of deoxidizing elements. Excessive amounts of Mn, Si or Al will change the mechanical properties, especially reducing the toughness and ductility. Manufacturers of filler wire should be consulted regarding the effect of various gas mixtures of the mechanical properties of their wires. If data are not available, it may be necessary to make welds with a particular gas mixture and run mechanical tests.

Gas Flow: The gas flow will depend on the diameter of the nozzle, distance of nozzle from the work, and air movements in the welding region. Inadequate flow will result in poor shielding of the arc and molten metal, leading to weld porosity. Excessive gas flow can cause turbulence at the arc, resulting in porosity and irregular beads. Most welding with gas-shielded flux-cored electrodes is done with the gas flow in the range of 35 to 45 cubic feet per hour. When welding in moving air or when the stickout is longer than normal, flow rates up to 55 cubic feet per hour may be required. To maintain desired flow rates, it is necessary to keep the nozzle opening free from adhering spatter.

VERTICAL ELECTRO-GAS WELDING

Figure 6-84 illustrates a specialized application of the gas-shielded flux-cored process — automatic vertical electro-gas welding. With it, plates 1/2 to 3 in. in thicknesses may be welded together in the vertical position. The edges of the vertical plate are butted with a gap of about 3/4 in. and water-cooled copper shoes are used on each side of the vertical joint to bridge the gap and prevent weld metal from flowing from the joint as it is deposited. The wire-

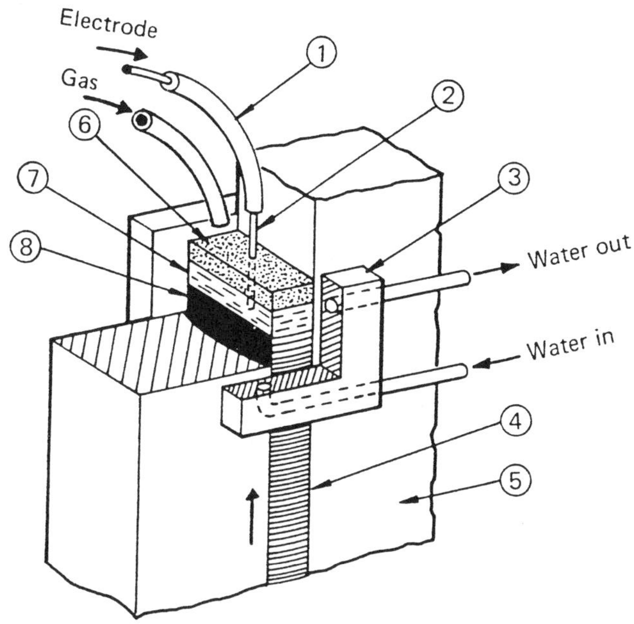

Fig. 6-84. The electro-gas vertical butt-welding process — a special application of gas-shielded flux-cored arc welding. (1) Electrode guide tube, (2) electrode, (3) water cooled copper shoes, (4) finished weld, (5) base metal, (6) slag and gas shield, (7) molten weld metal, (8) solidified weld metal.

feed mechanism, controls, and copper shoes move vertically during operation on a vertical track. The entire operation is automated, and once the setup has been made, a continuous weld up to 40-ft. long can be run without further adjustment.

A special flux-cored electrode, with CO_2 gas, is used. Its flux produces a thin layer of slag between the weld metal and copper shoes, without accumulating excessive slag above the weld puddle. The nonmetallic materials in the core are proportionately less than in conventional gas-shielded and self-shielded flux-cored electrodes. This process can operate with solid filler wire and CO_2 gas shielding. In this case no slag is formed and the molten weld pool is protected only by the CO_2.

The electro-gas process operates on direct current, with the electrode negative. With plate greater than 3/4-in. thick, oscillation of the electrode across the width of the joint is required.

This is accomplished by mounting the head on an oscillating carriage. The welding power source is constant-voltage, rated at 750 amp., 100% duty cycle.

Square, V-groove, and U-groove butt welds may be made by the electro-gas process. The equipment is relatively expensive, and only those having need for repetitively butt-joining plate use this automated variation of gas-shielded flux-cored welding.

INTRODUCTION TO WELDING PROCEDURES

The ideal welding procedure is the one that will produce acceptable quality welds at the lowest overall cost. So many factors influence the optimum welding conditions that it is impossible to write procedures for each set of conditions. In selecting a procedure, the best approach is to study the conditions of the application and then choose the procedure that most nearly accommodates them. The procedures given here are typical, and it may be necessary to make adjustments for a particular application to produce a satisfactory weld.

For some joints, different procedures are offered to suit the weld quality — code quality, commercial quality, and strength only — that may be required.

Code-Quality Procedures

Code quality procedures are intended to provide the highest level of quality and appearance. To accomplish this, conservative currents and travel speeds are recommended.

These procedures are aimed at producing welds that will meet the requirements of the commonly used codes: AWS Structural, AISC Buildings and Bridges, ASME Pressure Vessels, AASHTO Bridges, and others. Code quality welds are intended to be defect-free to the extent that they will measure up to the nondestructive testing requirements normally imposed by these codes. This implies crack-free, pressure-tight welds, with little or no porosity or undercut.

The specific requirements of these codes are so numerous and varied that code quality procedures may not satisfy every detail of a specific code. Procedure qualification tests are recommended to confirm the acceptability of chosen procedures.

All butt welds made to code quality are full penetration; fillet welds are full size, as required by most codes. (The theoretical throat rather than the true throat is used as the basis of calculating strength.)

Commercial-Quality Procedures

Commercial quality implies a level of quality and appearance that will meet the nominal requirements imposed on most of the welding done commercially. These welds will be pressure-tight and crack-free. They will have good appearance, and they will meet the normal strength requirements of the joint.

Procedures for commercial-quality welds are not as conservative as code quality procedures; speeds and currents are generally higher. Welds made according to these procedures may have minor defects that would be objectionable to the more demanding codes.

It is recommended that appropriate tests be performed to confirm the acceptability of the selected procedure for the application at hand prior to putting it into production.

Strength Only Procedures

The purpose of strength-only procedures is to produce the highest speed welding at the lowest possible cost. To accomplish this, the weld appearance and quality need only be good enough to do the specific job for which it is intended. In this category, imperfections may be acceptable provided that the welds will perform satisfactorily under service conditions.

This quality level may apply to seal beads, partial-penetration welds, some lap welds, high-speed edge welds, and to many low-stress welded connections which have no real need for better quality at higher cost. On fillet welds the joint must be welded from both sides.

Since this is a compromise quality level, realistic tests should be made to confirm the satisfactory performance of welds made according to these procedures. Welds of strength-only quality will be free of cracks but may not be pressure-tight.

Weldability of Material

Weldability (see Section 6.1) of a steel has a considerable effect on the welding procedure.

Good weldability indicates a steel with a composition that is within the preferred range (see Table 6-1) — one whose chemistry does not limit the welding speed.

Fair weldability indicates a steel with one or more elements outside the preferred range or one that contains one or more alloys. These steels require a lower welding speed or a mild preheat, or both, to minimize defects such as porosity, cracking, and undercut.

The addition of alloys to steel that enhance the mechanical properties or hardenability usually have an adverse effect on weldability. In general, the weldability of low-alloy steels is never better than "fair."

Procedure Notes

In the following tables, procedures for some joints are given for two different electrode sizes. The smaller electrode provides greater flexibility and operator comfort at some sacrifice in welding speed.

It is important to control the arc voltage within the range specified in the procedures. These voltages do not allow for poor electrical or ground connections or for undersize welding cables. Such faulty conditions should be corrected rather than compensating by increasing the arc voltage.

In the following fillet weld procedures, the fillet size is always associated with a particular plate thickness. This relationship is given solely for the purpose of designing a welding procedure and does not imply that a certain size fillet is the only size applicable to that plate thickness. In some of the procedures, the fillet size shown is larger than necessary to meet code requirements for the plate thickness. In such instances, select the procedure for the proper weld size and quality. If the thickness of the plate being welded is appreciably greater than that specified in the procedure, a reduction in welding speed and current will probably be required.

Travel speed is given as a range; however, the electrode required and the total time are based on the middle of the range.

Unless otherwise indicated, both members of the joint are the same thickness.

Pounds-of-electrode data include all ordinary deposition losses. These values are in terms of pounds of electrode needed to be purchased.

Total time is the arc time only and does not allow for operating factor.

Each procedure specifies the recommended stickout. This stickout is always the electrical stickout as discussed earlier in this section.

The drag angle is the angle between the center line of the electrode and a line perpendicular to the seam and inclined toward the direction of travel.

After a satisfactory welding procedure has been established, all the data should be recorded and filed for future reference. This information is invaluable if the same job or a similar job occurs at a later date. A suggested data sheet is shown on the following page.

The presented procedures are offered as a starting point and may require changes to meet the requirements of specific applications. Because the many variables in design, fabrication, and erection or assembly affect the results obtained in applying this type of information, the serviceability of the product or structure is the responsibility of the builder.

The following procedures have been developed with electrodes having these specific characteristics:

E70T-1. This electrode is designed for use with a CO_2 shielding gas and for welding in the flat position and for horizontal position fillets. It can be used for single and multipass welding. T-1 type electrodes are characterized by a spray deposit mode and low spatter loss. The bead configuration is flat to slightly convex with a moderate covering of slag.

E71T-1. This electrode has similar characteristics to the E70T-1 but is designed for all position welding. It can be used with CO_2 shielding gas, however, argon-CO_2 gas mixtures improve useability especially for out-of-position applications. The use of the argon CO_2 mixture also enhances the spray transfer characteristics of this electrode.

Welding procedures and parameters for the EXXT-1, EXXT-2, EXXT-5 and EXXT-X electrodes will be similar to the procedures and parameters listed here for the same joint designs. When steels of higher strength and hardenability and therefore poorer weldability are used, with higher strength filler metals, the need for preheat should be considered and factored into the welding procedure and verified by mechanical tests.

Welding Parameters, Single Values or a Range

In some of the procedures which follow single values are given for such items as Amps, Volts, and Travel Speed. The purpose of using single values, i.e. 100 amps, is to indicate an optimum or starting place and it is expected that the user will expand these values and show a reasonable range. These values when recorded on a WPS (Welding Procedure Specification) should be shown as a range, not as specific values.

DATA SHEET

Article to be Welded ______________________________ Job No. ______________

Plate Specs or Analysis ____

Welding Process ____________________ Submitted by: ______________ Date: ________

JOINT-PREPARATION – BEAD PLACEMENT, SIZE, SEQUENCE

WORK ANGLE – DRAG ANGLE
ELECTRODE ANGLE TO JOINT

Pass No	AC DC (+ –)	Amps	Volts	Travel Speed "/Min	Electrode Type	Elect. Size	Stick-out	Flux Type					

Special Comments ____

FLUX-CORED ARC WELDING (SEMIAUTOMATIC) GAS-SHIELDED

Welding Position: Flat
Weld Quality Level: Commercial
Steel Weldability: Good or fair

Shield Gas: CO_2
Flow Rate: 35-45 cfh

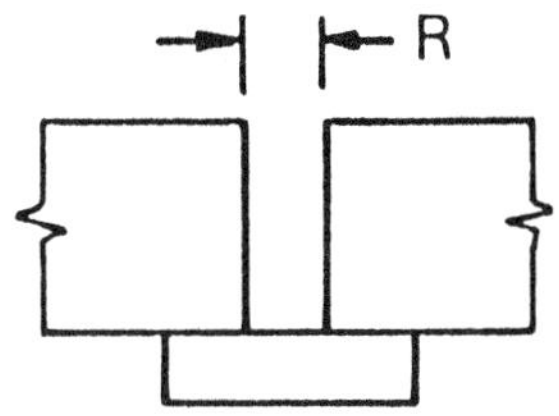

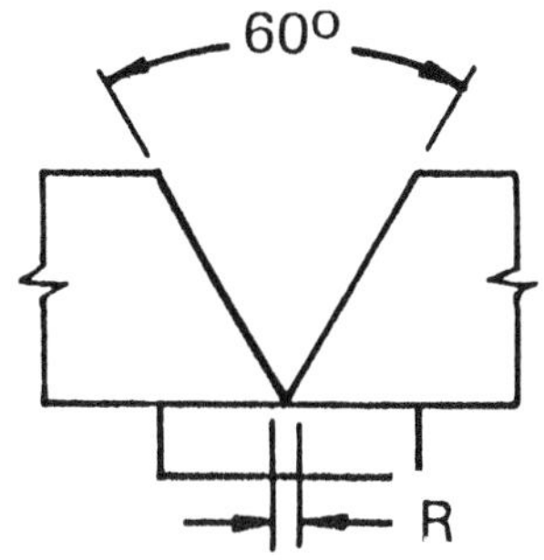

Plate Thickness (in.)	1/8	1/2	1/2	1
Pass	1	1 – 2	1 – 2	1 – 6
Electrode Class	E70T-1	E70T-1	E70T-1	E70T-1
Size	5/64	3/32	3/32	7/64
Wire Feed Speed (in./min)	175	190	200	180
Current (amp) DC(+)	325	450	480	525
Volts	28	32	30	32
Arc Speed (in./min)	40	12	14	11
Electrode Req'd (lb/ft)	0.05	0.50	0.49	2.052
Total Time (hr/ft of weld)	0.005	0.0333	0.0286	0.108
Electrical Stickout (in.)	1	1-1/8	1-1/8	1-1/4
Root Opening (in.)	1/16	1/4	0	0

FLUX-CORED ARC WELDING (SEMIAUTOMATIC) GAS-SHIELDED

Welding Position: Flat
Weld Quality Level: Commercial
Steel Weldability: Good or fair

Shield Gas: CO_2
Flow Rate: 35-45 cfh

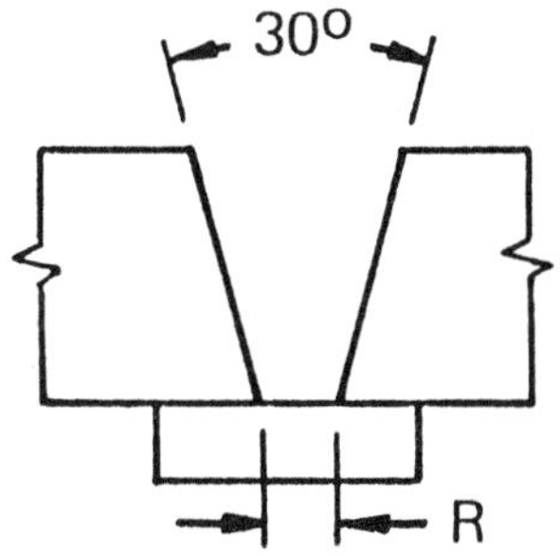

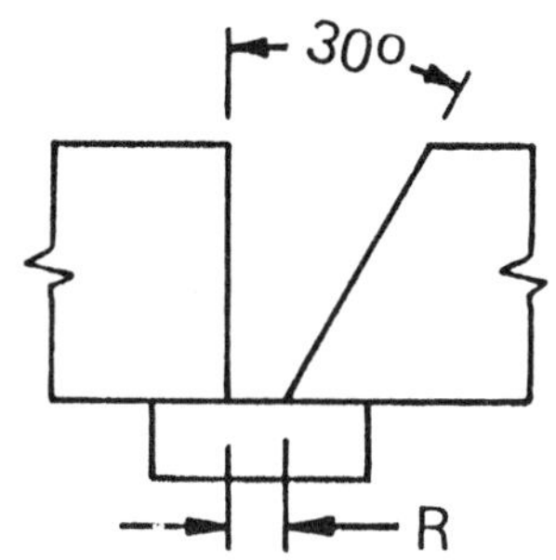

Plate Thickness (in.)	5/8	1	5/8	1
Pass	1 – 3	1 – 6	1 – 3	1 – 6
Electrode Class	E70T-1	E70T-1	E70T-1	E70T-1
Size	7/64	7/64	7/64	7/64
Wire Feed Speed (in./min)	180	180	180	180
Current (amp) DC(+)	525	525	525	525
Volts	32	32	32	32
Arc Speed (in./min)	14	14	16	14
Electrode Req'd (lb/ft)	0.813	1.625	0.711	1.625
Total Time (hr/ft of weld)	0.0428	0.0855	0.0374	0.0855
Electrical Stickout (in.)	1-1/4	1-1/4	1-1/4	1-1/4
Root Opening (in.)	3/16	3/16	1/8	1/8

FLUX-CORED ARC WELDING (SEMIAUTOMATIC) GAS-SHIELDED

Welding Position: Flat
Weld Quality Level: Commercial
Steel Weldability: Good or fair

Shield Gas: CO_2
Flow Rate: 35-45 cfh

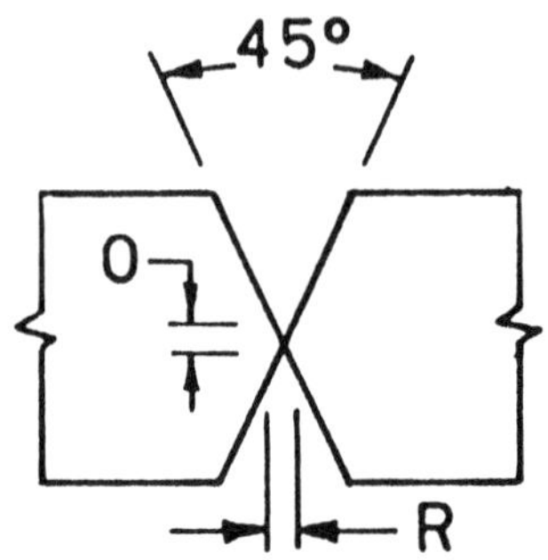

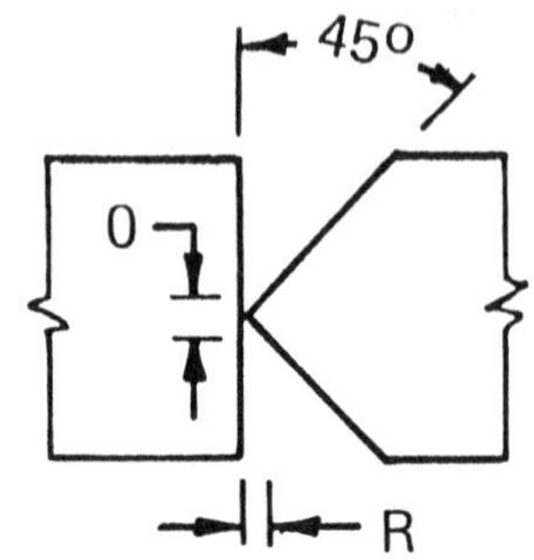

Plate Thickness (in.)	1	2	1	2
Pass	1 – 6	1 – 20	1 – 4	1 – 12
Electrode Class	E70T-1	E70T-1	E70T-1	E70T-1
Size	7/64	7/64	7/64	7/64
Wire Feed Speed (in./min)	180	180	170	170
Current (amp) DC(+)	525	525	500	500
Volts	32	32	32	32
Arc Speed (in./min)	31	26	17	13
Electrode Req'd (lb/ft)	0.674	2.695	0.765	3.006
Total Time (hr/ft of weld)	0.0385	0.154	0.0471	0.185
Electrical Stickout (in.)	1-1/4	1-1/4	1-1/4	1-1/4
Root Opening (in.)	0	0	0	0

FLUX-CORED ARC WELDING (SEMIAUTOMATIC) GAS-SHIELDED

Welding Position: Horizontal
Weld Quality Level: Commercial
Steel Weldability: Good or fair

Shield Gas: CO_2
Flow Rate: 35-45 cfh

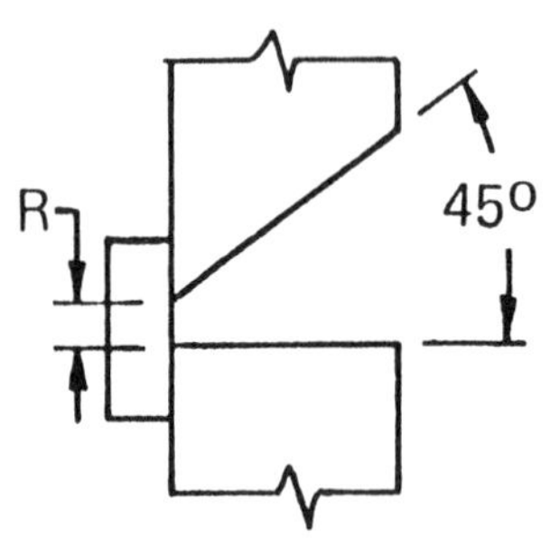

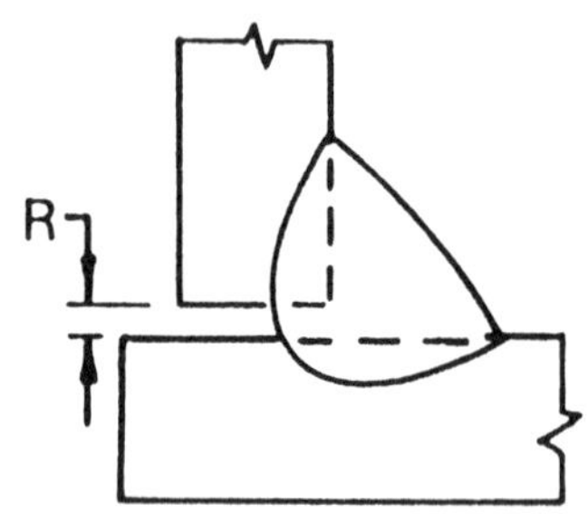

Plate Thickness (in.)	1/2	1	3/16	1/2
Pass	1 – 6	1 – 18	1	1 – 3
Electrode Class	E70T-1	E70T-1	E70T-1	E70T-1
Size	5/64	5/64	5/64	7/64
Wire Feed Speed (in./min)	200	200	200	140
Current (amp) DC(+)	350	350	350	450
Volts	28	28	28	20
Arc Speed (in./min)	20	18	36	18
Electrode Req'd (lb/ft)	0.6888	2.300	0.064	0.466
Total Time (hr/ft of weld)	0.0599	0.200	0.00555	0.0333
Electrical Stickout (in.)	1	1	1	1-1/4
Root Opening (in.)	1/8	1/8	0	0

FLUX-CORED ARC WELDING (SEMIAUTOMATIC) GAS-SHIELDED

Welding Position: Overhead
Weld Quality Level: Commercial
Steel Weldability: Good or fair

Shield Gas: CO_2
Flow Rate: 35-45 cfh

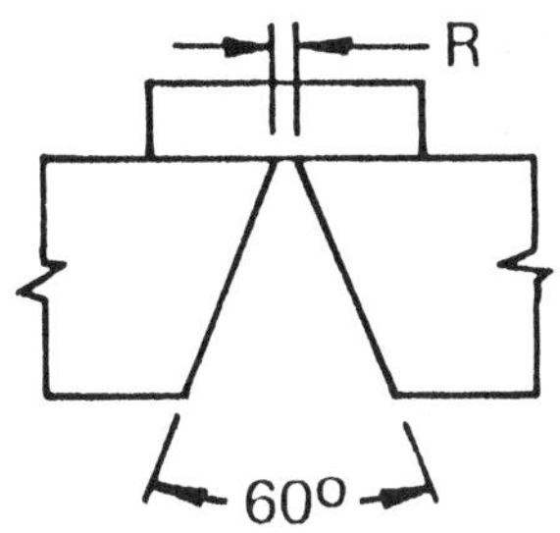

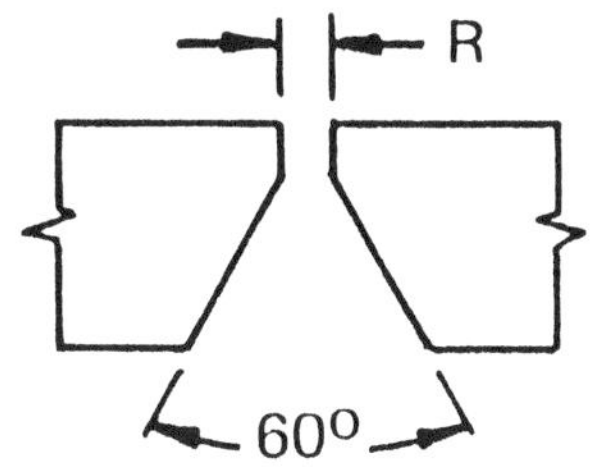

Plate Thickness (in.)	3/8	1/2	1/2	1
Pass	1 – 2	1 – 3	1 – 4	1 – 9
Electrode Class	E70T-1	E70T-1	E70T-1	E70T-1
Size	0.045	0.045	0.045	0.045
Wire Feed Speed (in./min)	275	275	275	275
Current (amp) DC(+)	180	180	180	180
Volts	22	22	22	22
Arc Speed (in./min)	8	8	8	5
Electrode Req'd (lb/ft)	0.260	0.390	0.520	1.851
Total Time (hr/ft of weld)	0.050	0.075	0.100	0.356
Electrical Stickout (in.)	3/4	3/4	3/4	3/4
Root Opening (in.)	0	0	3/32	3/32

FLUX-CORED ARC WELDING (SEMIAUTOMATIC) GAS-SHIELDED

Welding Position:
Weld Quality Level: Commercial
Steel Weldability: Good or fair

Shield Gas: CO_2
Flow Rate: 35-45 cfh

Flat

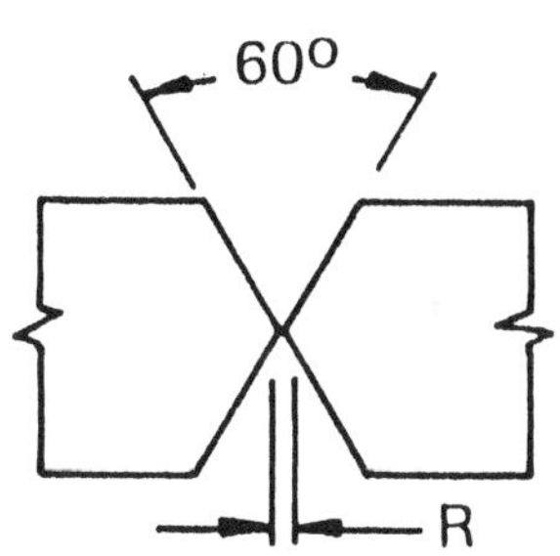

Overhead

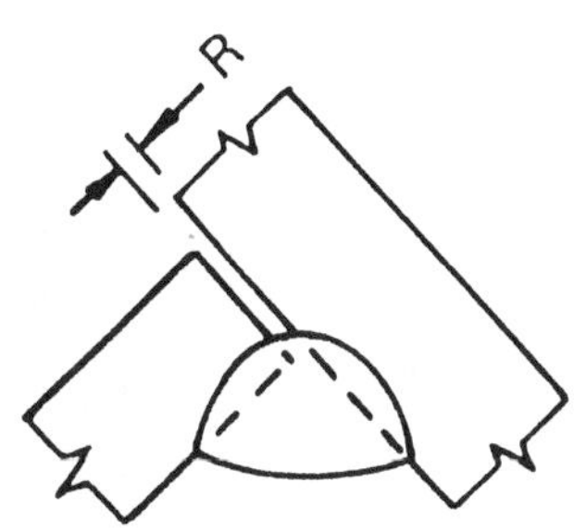

Plate Thickness (in.)	1	2	1/8	1/2
Pass	1 – 6	1 – 16	1	1 – 2
Electrode Class	E70T-1	E70T-1	E70T-1	E70T-1
Size	0.045	0.045	0.045	0.045
Wire Feed Speed (in./min)	275	275	275	275
Current (amp) DC(+)	180	180	180	180
Volts	22	22	21	21
Arc Speed (in./min)	8	5	36	5
Electrode Req'd (lb/ft)	0.775	3.26	0.0289	0.4337
Total Time (hr/ft of weld)	0.149	0.627	0.00555	0.0834
Electrical Stickout (in.)	3/4	3/4	3/4	3/4
Root Opening (in.)	1/16	1/16	0	0

FLUX-CORED ARC WELDING (SEMIAUTOMATIC) GAS-SHIELDED

Welding Position:
Flat, Horizontal
Weld Quality Level: Commercial
Steel Weldability: Good or fair

Shield Gas: CO_2
Flow Rate: 35-45 cfh

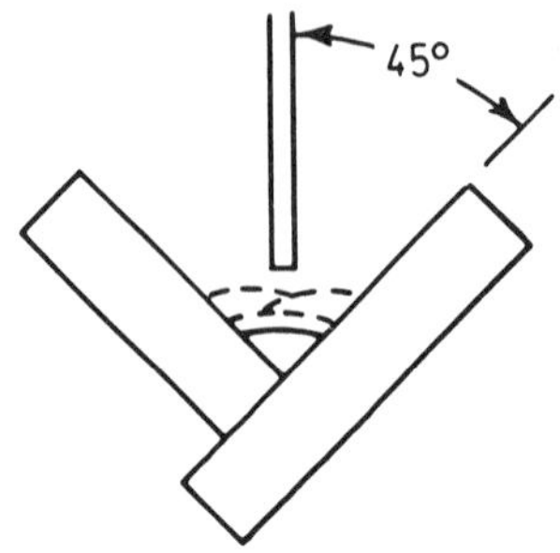

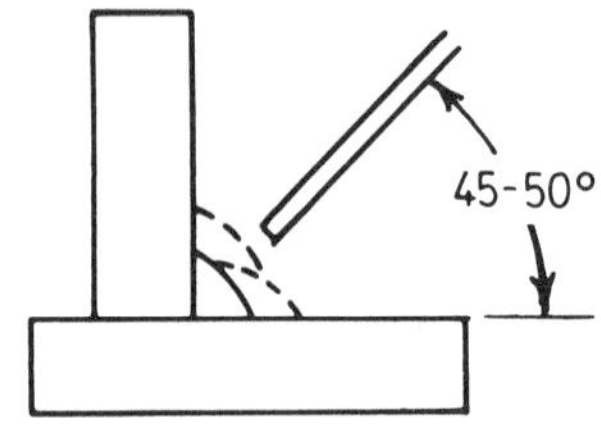

Plate Thickness (in.)	1/4+	1/4+	1/4+	1/4+
Pass	1 – X	1 – X	1	1 – X
Electrode Class	E70T-1	E70T-1	E70T-1	E70T-1
Size	1/16	5/64	3/32	3/32
Wire Feed Speed (in./min)	250 – 350	225 – 300	175 – 190	190 – 250
Current (amp) DC(+)	290 – 365	385 – 455	435 – 460	460 – 550
Volts	25 – 29	29 – 32	27 – 31	28 – 32
Arc Speed (in./min)	12 – 13	17 – 18	15 – 17	15 – 17
Electrode Req'd (lb/ft)/ pass	0.177	0.175	0.196	0.237
Time (hr/ft of weld)/pass	0.016	0.0114	0.0125	0.0125
Electrical Stickout (in.)	3/4	1	1-1/8	1-1/8
Angle, A (deg) Drag	0 – 20	0 – 30	0 – 30	0 – 30

From Lincoln GSP71.

FLUX-CORED ARC WELDING (SEMIAUTOMATIC) GAS-SHIELDED

Welding Position: Flat butt
Weld Quality Level: Commercial
Steel Weldability: Good or fair

Shield Gas: CO_2
Flow Rate: 35 – 45 cfh

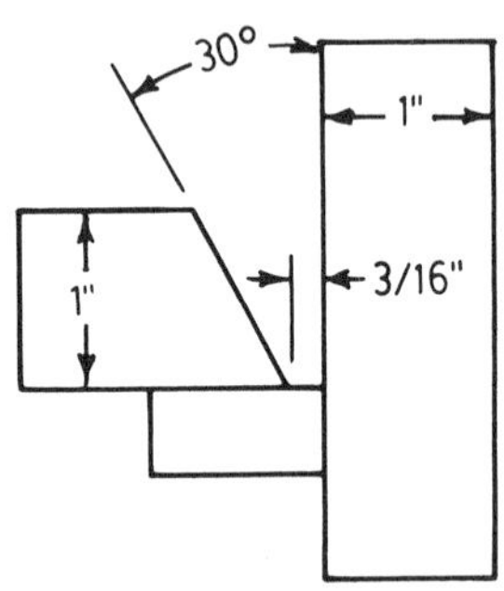

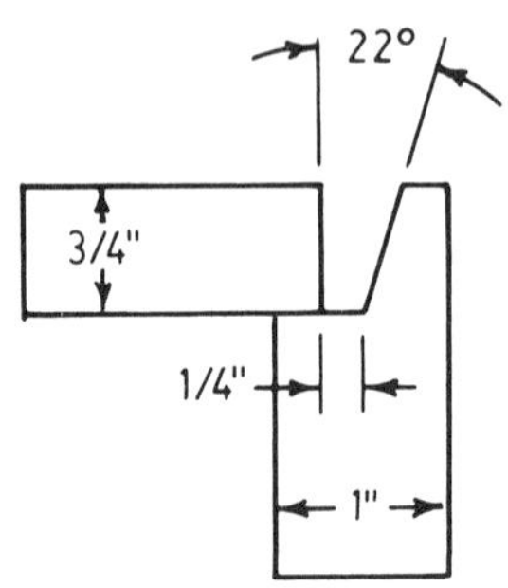

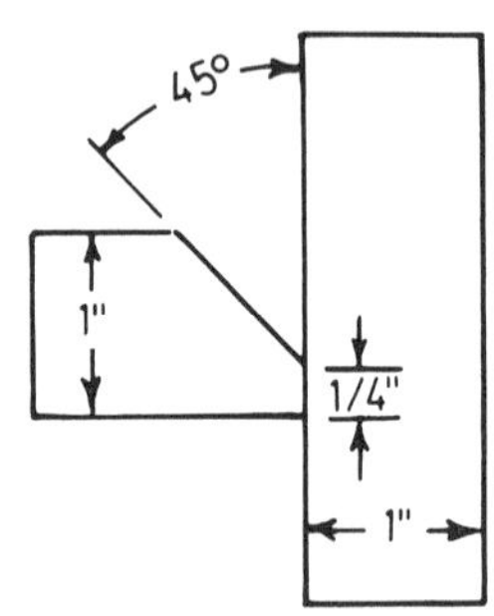

Plate Thickness (in.)	1	1	1
Pass	1+	1+	1+
Electrode Class	E70T-1	E70T-1	E70T-1
Size	1/16	5/64	3/32
Wire Feed Speed (in./min)	350 – 375	225 – 300	250 – 280
Current (amp) DC(+)(Approx.)	365 – 380	385 – 455	550 – 585
Volts	29 – 30	29 – 32	32 – 34
Arc Speed (in./min)	15 – 16	15 – 20	22 – 26
Electrode Req'd (lb/ft)/pass	0.169 – 0.168	0.169 – 0.174	0.193 – 0.168
Time (hr/ft of weld)/pass	0.013 – 0.012	0.013 – 0.010	0.009 – 0.007
Electrical Stickout (in.)	3/4	1	1-1/8
Angle, A (deg) Drag	10 – 20	20 – 30	20 – 30

From Lincoln GSP70.

FLUX-CORED ARC WELDING (SEMIAUTOMATIC) GAS-SHIELDED

Welding Position:
Flat, Horizontal
Weld Quality Level: Commercial
Steel Weldability: Good or fair

Shield Gas: CO_2
Flow Rate: 35 – 45 cfh

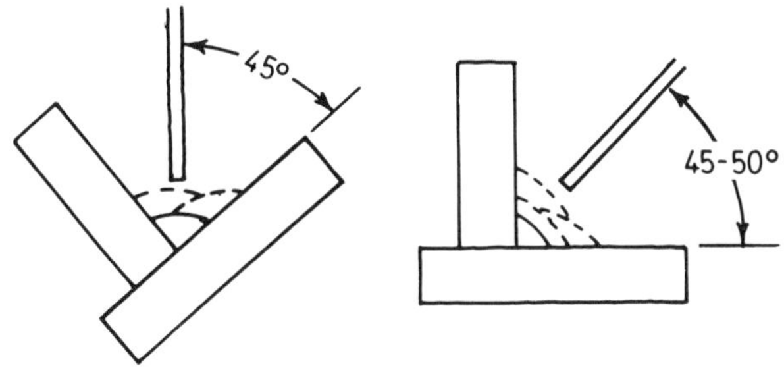

Plate Thickness (in.)	1/4+	1/4+	1/4+
Pass	1+	1+	1+
Electrode Class	E71T-1	E71T-1	E71T-1
Size	0.045	0.052	1/16
Wire Feed Speed (in./min)	340 – 525	250 – 425	275 – 375
Current (amp) DC(+)	220 – 280	225 – 300	310 – 365
Volts	29 – 30	26 – 29	23 – 30
Arc Speed (in./min)	10 – 11	11 – 12	13 – 14
Electrode Req'd (lb/ft)/pass	0.1558	0.1444	0.176
Time (hr/ft of weld)/pass	0.0190	0.0174	0.0148
Electrical Stickout (in.)	3/4	3/4	1
Angle, A (deg) Drag	15	15	15

From Lincoln GSP71.

FLUX-CORED ARC WELDING (SEMIAUTOMATIC) GAS-SHIELDED

Welding Position: Overhead
and Vertical up
Weld Quality Level: Commercial
Steel Weldability: Good or fair

Shield Gas: CO_2
Flow Rate: 35 – 45 cfh

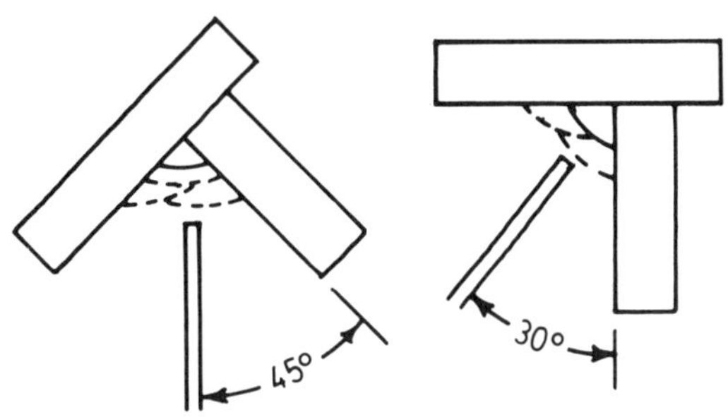

Plate Thickness (in.)	1/4+	1/4+	1/4+
Pass	1+	1+	1+
Electrode Class	E71T-1	E71T-1	E71T-1
Size	0.045	0.052	1/16
Wire Feed Speed (in./min)	275 – 340	240 – 290	150 – 165
Current (amp) DC(+)(Approx.)	180 – 220	210 – 250	210 – 225
Volts	25 – 27	24 – 25	23 – 24
Arc Speed (in./min)	6 – 6.5	6.5 – 7	4.5 – 5.5
Electrode Req'd (lb/ft)/pass	0.187	0.1925	0.230
Time (hr/ft of weld)/pass	0.032	0.0296	0.04
Electrical Stickout (in.)	3/4	3/4	1
Angle, *E* (deg) V up push	5	5	5
Angle, A (deg) Drag (OH)	50	50	50

From Lincoln GSP71.

FLUX-CORED ARC WELDING (SEMIAUTOMATIC) GAS-SHIELDED

Welding Position: Flat, Horizontal
Weld Quality Level: Commercial
Steel Weldability: Good or fair

Shield Gas: CO_2
Flow Rate: 35 – 45 cfh

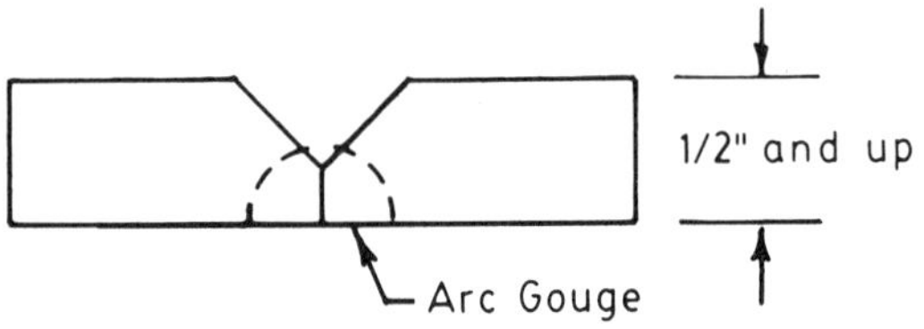

Plate Thickness (in.)	1/2+	1/2+	1/2+
Pass	1+	1+	1+
Electrode Class	E71T-1	E71T-1	E71T-1
Size	0.045	0.052	1/16
Wire Feed Speed (in./min)	340 – 500	290 – 425	275 – 375
Current (amp) DC(+)	220 – 275	250 – 300	310 – 365
Volts	26 – 30	25 – 29	23 – 90
Arc Speed (in./min)	7.4 – 11.1	8.1 – 11.9	11.4 – 15.8
Electrode Req'd (lb/ft)/pass	0.175 – 0.174	0.177 – 0.177	0.170 – 0.179
Time (hr/ft of weld)/pass	0.027 – 0.018	0.025 – 0.017	0.017 – 0.013
Electrical Stickout (in.)	3/4	3/4	1
Angle, A (deg) Drag	15	15	15

From Lincoln GSP71.

FLUX-CORED ARC WELDING (SEMIAUTOMATIC) GAS-SHIELDED

Welding Position: Vertical up and Overhead
Weld Quality Level: Commercial
Steel Weldability: Good or fair

Shield Gas: CO_2
Flow Rate: 35 – 45 cfh

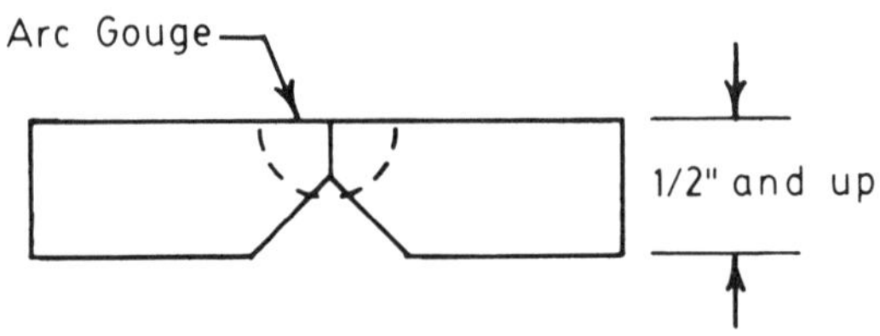

Plate Thickness (in.)	1/2	1/2	1/2
Pass	1+	1+	1+
Electrode Class	E71T-1	E71T-1	E71T-1
Size	0.045	0.052	1/16
Wire Feed Speed (in./min)	300 – 340	240 – 290	150 – 165
Current (amp) DC(+)(Approx.)	190 – 220	210 – 250	210 – 225
Volts	27 – 28	24 – 25	22 – 24
Arc Speed (in./min)	6.5 – 7.4	6.8 – 8.1	6.3 – 6.9
Electrode Req'd (lb/ft)/pass	0.177 – 0.176	0.174 – 0.178	0.176 – 0.174
Total Time (hr/ft of weld)	0.031 – 0.027	0.029 – 0.025	0.032 – 0.029
Electrical Stickout (in.)	3/4	3/4	1
Angle, *E* (deg) Drag	0 – 5	5	5

From Lincon GSP71.

FLUX-CORED ARC WELDING (SEMIAUTOMATIC) GAS-SHIELDED

Welding Position: Flat, Horizontal and Vertical up
Weld Quality Level: Commercial
Steel Weldability: Good or fair

Shield Gas: CO_2
Flow Rate: 35 - 45 cfh

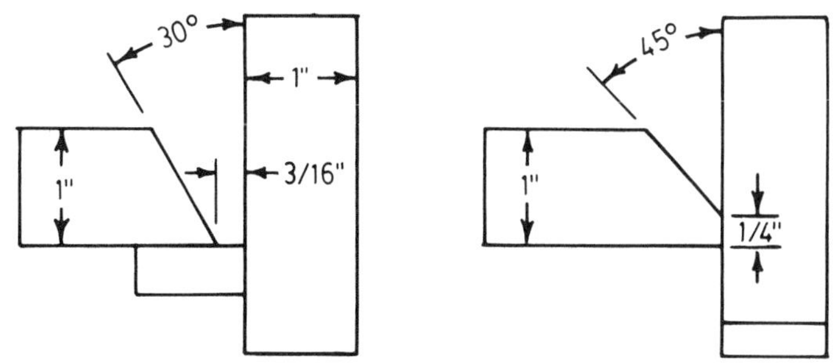

Plate Thickness (in.)	1	1	1	1	1
Pass	1+	1+	1+ Flat	1+ Horizontal	1+ Vertical Up
Electrode Class	E71T-1	E71T-1	E71T-1	E71T-1	E71T-1
Size	0.045	0.052	1/16	1/16	1/16
Wire Feed Speed (in./min)	340 - 525	250 - 425	275 - 375	165 - 275	150 - 275
Current (amp) DC(+)	220 - 290	220 - 300	310 - 370	225 - 310	210 - 310
Volts	26 - 31	24 - 30	23 - 30	23 - 27	22 - 26
Arc Speed (in./min)	7.4 - 11.4	7.1 - 11.9	11.4 - 15.8	6.8 - 11.4	6.3 - 11.4
Electrode Req'd (lb/ft)/pass	0.175 - 0.170	0.174 - 0.177	0.170 - 0.179	0.174 - 0.180	0.175 - 0.180
Total Time (hr/ft of weld)/pass	0.027 - 0.017	0.028 - 0.017	0.017 - 0.013	0.029 - 0.018	0.032 - 0.018
Electrical Stickout (in.)	3/4	3/4	3/4	1	1
Angle, A (deg) Drag	15	15	15	15	5

From Lincoln GSP71.

FLUX-CORED ARC WELDING (SEMIAUTOMATIC) GAS-SHIELDED

Welding Position: Flat, Horizontal
Weld Quality Level: Commercial
Steel Weldability: Good or fair

Shield Gas: CO_2
Flow Rate: 35 - 45 cfh

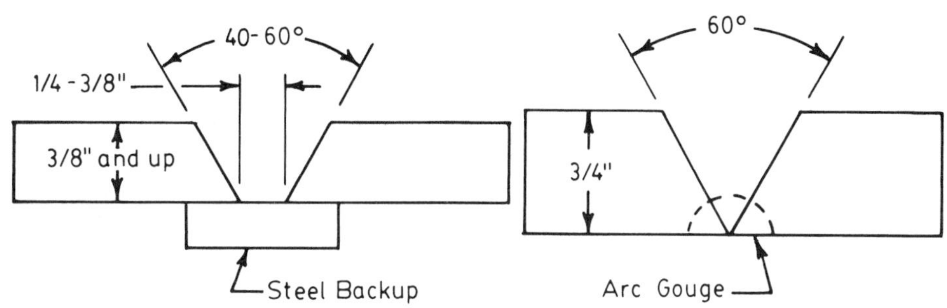

Plate Thickness (in.)	3/8+	3/8+	3/8+
Pass	1+	1+	1+
Electrode Class	E71T-1	E71T-1	E71T-1
Size	0.045	0.052	1/16
Wire Feed Speed (in./min)	340 - 500	290 - 425	275 - 375
Current (amp) DC(+)(Approx.)	220 - 275	250 - 300	310 - 365
Volts	26 - 30	25 - 29	23 - 90
Arc Speed (in./min)	7.4 - 11.1	8.1 - 11.9	11.4 - 15.8
Electrode Req'd (lb/ft)/pass	0.176 - 0.175	0.178 - 0.177	0.180 - 0.179
Time (hr/ft of weld)/pass	0.027 - 0.018	0.025 - 0.017	0.018 - 0.013
Electrical Stickout (in.)	3/4	3/4	1
Angle, A (deg) Drag	15	15	15

From Lincoln GSP71.

FLUX-CORED ARC WELDING (SEMIAUTOMATIC) GAS-SHIELDED

Welding Position: Vertical up, Overhead
Weld Quality Level: Commercial
Steel Weldability: Good or fair

Shield Gas: CO_2
Flow Rate: 35 – 45 cfh

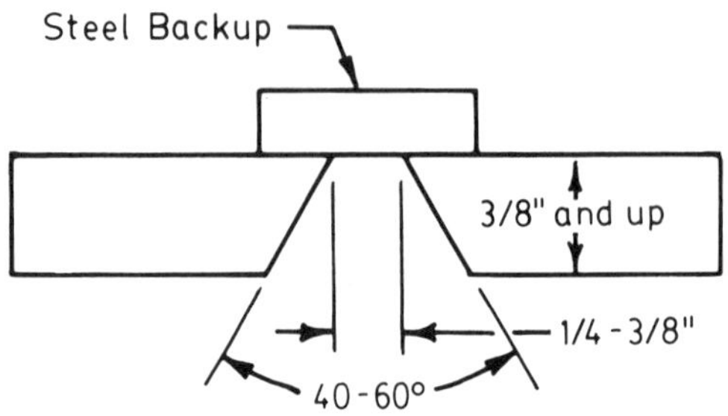

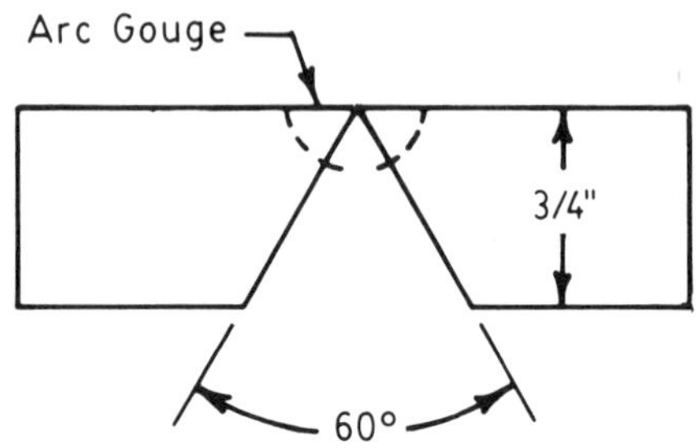

Plate Thickness (in.)	3/8+	3/8+	3/8+
Pass	1+	1+	1+
Electrode Class	E71T-1	E71T-1	E71T-1
Size	0.045	0.052	1/16
Wire Feed Speed (in./min)	300 – 340	240 – 290	150 – 165
Current (amp) DC(+)(Approx.)	190 – 220	210 – 250	210 – 225
Volts	27 – 28	24 – 25	22 – 24
Arc Speed (in./min)	6.5 – 7.4	6.8 – 8.1	6.3 – 6.9
Electrode Req'd (lb/ft)/pass	0.176 – 0.176	0.174 – 0.177	0.176 – 0.174
Time (hr/ft of weld)/pass	0.031 – 0.027	0.029 – 0.025	0.032 – 0.029
Electrical Stickout (in.)	3/4	3/4	3/4
Angle, A (deg) Drag	5	5	5

From Lincoln GSP71.

FLUX-CORED ARC WELDING (SEMIAUTOMATIC) GAS-SHIELDED

Welding Position: Vertical up
Weld Quality Level: Commercial
Steel Weldability: Good or fair

Shield Gas: CO_2
Flow Rate: 35 – 45 cfh

For Butts with 3M SJ-8086 Ceramic Backup Tile

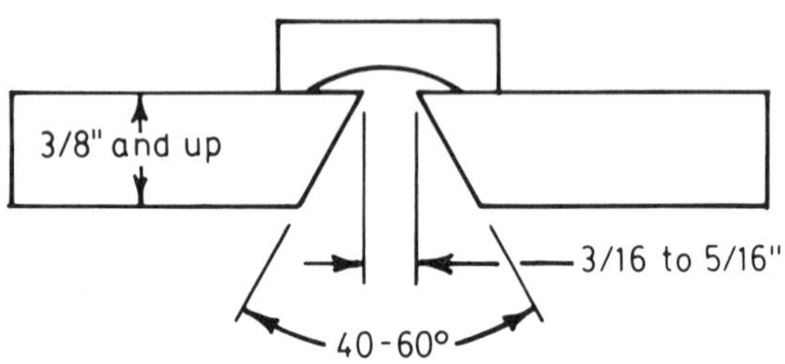

Plate Thickness (in.)	3/8+	3/8+	3/8+
Pass	1+	1+	1+
Electrode Class	E71T-1	E71T-1	E71T-1
Size	0.045	0.052	1/16
Wire Feed Speed (in./min)	300 – 340	240 – 290	150 – 165
Current (amp) DC(+)(Approx.)	190 – 220	210 – 250	210 – 225
Volts	27 – 28	24 – 25	22 – 24
Arc Speed (in./min)	6.5 – 7.4	6.8 – 8.1	6.3 – 6.9
Electrode Req'd (lb/ft)/pass	0.177 – 0.176	0.174 – 0.177	0.176 – 0.174
Time (hr/ft of weld)/pass	0.031 – 0.027	0.029 – 0.025	0.032 – 0.029
Electrical Stickout (in.)	3/4	3/4	3/4
Angle, A (deg) Drag	0	0	0

From Lincoln GSP71.

FLUX-CORED ARC WELDING (SEMIAUTOMATIC) SELF-SHIELDED

Welding Position:
Flat, Horizontal
Weld Quality Level: Commercial
Steel Weldability: Good or fair

Shield Gas: CO_2
Flow Rate: 35 – 45 cfh

For Butts with 3M SJ-8086 Ceramic Backup Tile

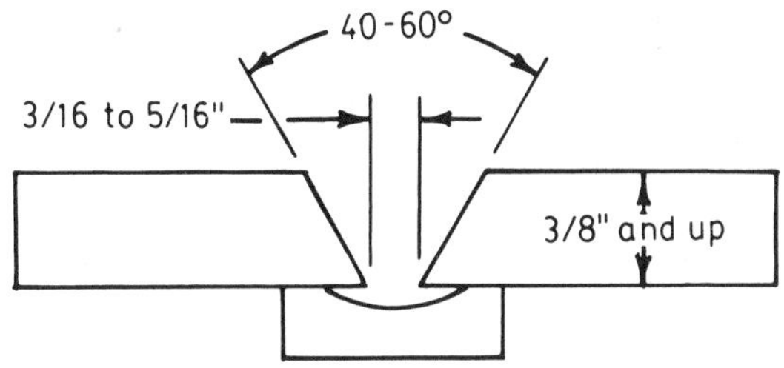

Plate Thickness (in.)	3/8+	3/8+	3/8+
Pass	1+	1+	1+
Electrode Class	E71T-1	E71T-1	E71T-1
Size	0.045	0.052	1/16
Wire Feed Speed (in./min)	340 – 500	290 – 425	275 – 375
Current (amp) DC(+)(Approx.)	220 – 275	250 – 300	310 – 365
Volts	26 – 30	25 – 29	23 – 30
Arc Speed (in./min)	7.4 – 11.1	8.1 – 11.5	11.4 – 15.8
Electrode Req'd (lb/ft)/pass	0.176 – 0.175	0.178 – 0.177	0.170 – 0.179
Time (hr/ft of weld)/pass	0.027 – 0.018	0.025 – 0.017	0.017 – 0.013
Electrical Stickout (in.)	3/4	3/4	1
Angle, *E* (deg) Drag	15	15	15

From Lincoln GSP71.

Field welding a splice on a large column connection. Semiautomatic welding is replacing shielded metal-arc welding on these applications.

Welding Carbon and Low-Alloy Steels with the Gas Metal Arc Process

The gas metal arc process, as described in Section 5.4, is an arc welding process which incorporates mechanized feeding of a continuous consumable wire electrode that is shielded by an externally supplied gas. The process can be operated in the fully automatic or semiautomatic modes. Since the equipment provides for automatic self-regulation of the electrical characteristics of the arc and deposition rate, the only manual controls required by the weldor for semiautomatic operation are gun positioning, guidance, and travel speed. The arc length and the current level are automatically maintained.

The gas metal arc process can be operated in four modes of metal transfer which depend on the current density, the alloy being deposited and the shielding gas. These modes include spray, pulsed spray, globular and short circuiting types of transfer. Distinct droplets are transferred across the arc with the spray, pulsed spray and globular modes of operation. In the short circuiting mode of transfer, the electrode metal is deposited during frequent short circuiting of the wire into the molten weld pool.

Spray Type Transfer. Spray type transfer welding of carbon steel is most commonly done with a shielding gas mixture of argon and 2 to 5 percent oxygen. A characteristic of spray type transfer welding with argon-oxygen shielding gas is the smooth arc plasma through which hundreds of droplets per second are transferred axially from the electrode to the weld pool. When CO_2 is used as the shielding gas, spray type transfer can be achieved only at very high current densities.

Axial spray transfer with argon-oxygen shielding gas is related mainly to the current density and polarity of the arc current and electrical resistance heating of the electrode. The high rate of approximately 250 droplets per second develops abruptly above a critical current level referred to as the transition current. This current depends on the wire size and composition and the shielding gas. Below the transition current, transfer occurs in drops or globules much larger than the wire diameter. For example, the transition current for 1/16 diameter carbon steel electrode is approximately 270 amperes DCEP with argon 2% oxygen shielding.

Pulsed Spray Type Transfer. Metal transfer in pulsed spray arc welding is similar to the spray arc described above but occurs at lower average currents. Lower current welding is made possible by rapid pulsing of the current between a high level where metal will transfer in the spray mode and a low level where no transfer takes place.

At a normal pulsing frequency of about 60 pulses per second, a molten drop forms by the low current arc, then is pinched off and is propelled to the weld pool by the high current pulse. The lower average current of the pulsing mode allows all position welding and welding of thin materials with the spray transfer process.

Globular Type Transfer. The method of transfer which characterizes welding with CO_2 shielding gas is globular and mon-axial in nature. Common practice with plain carbon and low alloy steel globular transfer is to use low arc voltage to cause a "buried arc" which produces deep penetration and minimizes spatter. For this type of transfer, 0.045 and 1/16 in. diameter electrodes are normally used at welding currents in a range of 275-400 amperes DCEP.

Between 20 and 70 globules per second are transferred across this type arc depending on electrode size, current and voltage.

Short Circuiting Type Transfer. This method of gas metal arc welding is generally done with 0.030 to 0.045 in. diameter electrodes, using lower arc voltages and amperages than spray arc welding. A power source designed for short circuiting transfer with the correct amount of inductance should be used for this process.

The electrode short-circuits to the work piece, usually at a rate from 50 to 200 times per second. Metal is transferred with each short circuit and not across the arc. Short circuiting gas metal arc welding of carbon steel is most commonly done with shielding gas mixtures of argon-CO_2 or with 100 percent welding grade CO_2. Penetration is greater with 100

TABLE 6-40. Selection of Gases for GMAW

Metal	Shielding Gas	Advantages
Carbon steel (short circuiting transfer)	75% argon + 25% CO_2	Less than 1/8 in. (3.2 mm) thick; high welding speeds without melt-thru; minimum distortion and spatter.
	75% argon + 25% CO_2	More than 1/8 in. (3.2 mm) thick; minimum spatter, clean weld appearance; good puddle control in vertical and overhead positions.
	CO_2	Deeper penetration; faster welding speeds.
Low alloy steel (short circuiting transfer)	60-70% helium + 25-35% argon + 4-5% CO_2	Minimum reactivity; excellent toughness; excellent arc stability, wetting characteristics, and bead contour; little spatter.
	75% argon + 25% CO_2	Fair toughness; excellent arc stability, wetting characteristics, and bead contour; little spatter.
Carbon steel (spray transfer)	Argon + 1-5% oxygen	Improves arc stability; produces a more fluid and controllable weld puddle; good coalescence and bead contour; minimizes undercutting; permits higher speeds than pure argon.
	Argon + 3-10% CO_2	Good bead shape; minimizes spatter; reduces chance of cold lapping; can not weld out of position.
Low-alloy steel (spray transfer)	Argon + 2% oxygen	Minimizes undercutting; provides good toughness.

percent CO_2 than with argon CO_2 mixtures. However mixtures of 50 to 80 percent argon with CO_2 result in higher short circuiting rates and lower minimum currents which can be useful for welding thin plate and sheet. Table 6-40 lists the gases and gas mixtures used for GMAW and the characteristics of each.

FILLER METALS FOR WELDING CARBON AND LOW ALLOY STEELS

Electrodes for welding carbon steels are covered in AWS A5.18-79,"Specifications for Carbon Steel Filler Metals for Gas Shielded Arc Welding." Chemical compositions of the wire are given in Table 4-18 and mechanical properties of the deposited weld metal are given in Tables 4-16 and 4-17. The alloy additions indicated in this specification are for deoxidation and useability enhancement purposes and not for upgrading the strength of the weld deposit. Electrodes for welding low alloy steels are covered in AWS A5.28-79, "Specification for Low Alloy Steel Filler Metals for Gas Shielded Arc Welding."

Chemical composition requirements of the low alloy solid wires are given in Table 6-41 and in Table 6-42 for deposits from composite or stranded wires. Mechanical property requirements are given in Table 6-43, 6-44 and 6-45. Chemical composition requirements for the bare solid electrodes is based on analysis of the as-manufactured wire. Analysis of the composite metal cored and stranded electrodes are based on analysis of the deposited weld metal.

Descriptions of the electrodes included in AWS A5.18 for carbon steels follow:

ER70S-2 Classification. This classification covers multiple deoxidized steel filler metals which contain a nominal combined total of 0.20 percent zirconium, titanium, and aluminum in addition to the silicon and manganese contents. These filler metals are capable of producing sound welds in semikilled and rimmed steels as well as in killed steels of various carbon levels. Because of the added deoxidants, these filler metals can be used for welding steels which have a rusty or dirty surface, with a possible sacrifice of weld quality depending on the degree of surface contamination. They can be used with a shielding gas of argon-oxygen mixtures, CO_2, or argon-CO_2 mixtures and are preferred for out-of-position welding with the short circuiting type of transfer because of their ease of operation.

ER70S-3 Classification. These filler metals will meet the requirements of this specification with either CO_2 or argon-oxygen as a shielding gas. They are used primarily on single-pass welds, but can be used on multiple-pass welds, especially when welding killed or semikilled steel. Small diameter electrodes can be used for out-of-position welding and for short circuiting type transfer with argon-CO_2 mixtures or CO_2 shielding gases. However, it should be noted that the use of CO_2 shielding gas in conjunction with excessively high heat inputs may result in failure to meet the minimum specified tensile and yield strength.

ER70S-4 Classification. These filler metals contain slightly higher manganese and silicon contents than those of the ER70S-3 classification and produce a weld deposit of higher tensile strength. The primary use of these filler metals is for CO_2 shielded welding applications where a slightly longer arc or other conditions require more deoxidation than provided by the ER70S-3 filler metals. These filler metals are not required to demonstrate impact properties.

ER70S-5 Classification. This classification covers filler metals which contain aluminum in addition to manganese and silicon as deoxidizers. These filler metals can be used when welding rimmed, killed, or semikilled steels with CO_2 shielding gas and high welding currents. The relatively large amount of aluminum assures the deposition of well deoxidized and sound weld metal. Because of the aluminum, they are not used for the short circuiting type transfer, but can be used for welding steels which have a rusty or dirty surface, with a possible sacrifice of weld quality, depending on the degree of surface contamination. These filler metals are not required to demonstrate impact properties.

ER70S-6 Classification. Filler metals of this classification have the highest combination of manganese and silicon, permitting high current welding with CO_2 gas shielding even in rimmed steels. These filler metals may also be used to weld sheet metal where smooth weld beads are desired and steels which have moderate amounts of rust and mill scale. The quality of the weld will depend on the degree of surface impurities. This fuller metal is also usable out-of-position with short circuiting transfer.

ER70S-7 Classification. These filler metals have a substantially greater manganese content (essentially equal to that of ER70S-6) than those of the ER70S-3 classification. This provides slightly better wetting and weld appearance with slightly higher tensile and yield strengths and may permit increased speeds compared with ER70S-3 filler metals. These filler metals are generally recommended for use with argon-oxygen shielding gas mixtures, but they are usable with argon-CO_2 mixtures and CO_2 under the same general conditions as for the ER70S-3 classification. Under equivalent welding conditions, weld hardness will be lower than ER70S-6 weld metal but higher than ER70S-3 deposits.

ER70S-G Classification. This classification includes those solid filler metals which are not included in the preceding classes. The filler metal supplier should be consulted for the characteristics and intended use. This specification does not list specific chemical composition or impact requirements. These are subject to agreement between supplier and purchaser. However, any filler metal classified

TABLE 6-41. Chemical Composition Requiremenets for Bare Solid Electrodes and Welding Rods, percent[a] (AWS A5.28-79)

AWS Classificaiton[b]	Carbon	Manganese	Silicon	Phos-phorus	Sulfur	Nickel	Chromium	Molyb-denum	Vana-dium	Tita-nium	Zirco-nium	Alumi-num	Copper[c]	Total Other Elements[d]
Chromium-molybdenum steel electrodes and rods														
ER80S-B2	0.07-0.12	0.40-0.70	0.40-0.70	0.025	0.025	0.20	1.20-1.50	0.40-0.65	—	—	—	—	0.35	0.50
ER80S-B2L	0.05	0.40-0.70	0.40-0.70	0.025	0.025	0.20	1.20-1.50	0.40-0.65	—	—	—	—	0.35	0.50
ER90S-B3	0.07-0.12	0.40-0.70	0.40-0.70	0.025	0.025	0.20	2.30-2.70	0.90-1.20	—	—	—	—	0.35	0.50
ER90S-B3L	0.05	0.40-0.70	0.40-0.70	0.025	0.025	0.20	2.30-2.70	0.90-1.20	—	—	—	—	0.35	0.50
Nickel steel electrodes and rods														
ER80S-Ni1	0.12	1.25	0.40-0.80	0.025	0.025	0.80-1.10	0.15	0.35	0.05	—	—	—	0.35	0.50
ER80S-Ni2	0.12	1.25	0.40-0.80	0.025	0.025	2.00-2.75	—	—	—	—	—	—	0.35	0.50
ER80S-Ni3	0.12	1.25	0.40-0.80	0.025	0.025	3.00-3.75	—	—	—	—	—	—	0.35	0.50
Manganese-molybdenum steel electrodes and rods														
ER80S-D2[e]	0.07-0.12	1.60-2.10	0.50-0.80	0.025	0.025	0.15	—	0.40-0.60	—	—	—	—	0.50	0.50
Other low alloy steel electrodes and rods														
ER100S-1	0.08	1.25-1.80	0.20-0.50	0.010	0.010	1.40-2.10	0.30	0.25-0.55	0.05	0.10	0.10	0.10	0.25	0.50
ER100S-2	0.12	1.25-1.80	0.20-0.60	0.010	0.010	0.80-1.25	0.30	0.20-0.55	0.05	0.10	0.10	0.10	0.35-0.65	0.50
ER110S-1	0.09	1.40-1.80	0.20-0.55	0.010	0.010	1.90-2.60	0.50	0.25-0.55	0.04	0.10	0.10	0.10	0.25	0.50
ER120S-1	0.10	1.40-1.80	0.25-0.60	0.010	0.010	2.00-2.80	0.60	0.30-0.65	0.03	0.10	0.10	0.10	0.25	0.50
ERXXS-G	Subject to agreement between supplier and purchaser[f]													

NOTES:

1. Single values shown are maximums.
2. Analysis shall be made for the elements for which specific values are shown in this table. If, however, the presence of other elements is indicated in the course of routine analysis, further analysis shall be made to determine that the total of these other elements, except iron, is not present in excess of the limits specified for "Total other elements" in the last column of this table.

a. Chemical requirements for solid electrodes are based on as-manufactured composition.

b. The suffixes B2, Ni1, etc., designate the chemical composition of the electrode and rod classification.

c. The maximum weight percent of copper in the rod or electrode due to any coating plus the residual copper content in the steel shall comply with the stated value.

d. Other elecments, if intentionally added, shall be reported.

e. This composition was formerly classified E70S-1B in AWS Specification A5.18-69.

f. In order to meet the requirements of the G classification, the electrode must have as a minimum one of either 0.50 percent nickel, 0.30 percent chromium, or 0.20 percent molybdenum.

ER70S-G must meet all other requirements of this specification.

Descriptions of the electrodes included in AWS A5.28 for low alloy steels follow:

ER80S-B2 and E80C-B2 Classifications. Filler metals of these classifications are used to weld 1/2Cr-1/2Mo, 1Cr-1/2Mo, and 1-1/4Cr-1/2Mo steels for elevated temperatures and corrosive service. They are also used for joining dissimilar combinations of Cr-Mo and carbon steels. The spray transfer, short circuiting, or pulsed power modes of the GMAW process may be used. Careful control of preheat, interpass temperatures, and postheat is essential to avoid cracking.

ER80S-B2L and E80C-B2L Classifications. These filler metals are identical to the types ER80S-B2 and E80C-B2 except for the low carbon content (0.05 percent max). This alloy exhibits greater resistance to cracking and is more suitable for welds to be left in the as-welded condition or when the accuracy of the PWHT operation is questionable.

ER80S-Ni1 and E80C-Ni1 Classifications. These filler metals deposit weld metal similar to 8018-C3 covered electrodes and are used for welding low alloy high-strength steels requiring good toughness at temperatures as low as −40°F.

ER80S-Ni2 and E80C-Ni2 Classifications. These filler metals deposit weld metals similar to 8018-C1 electrodes. Typically, they are used for welding 3-1/2 percent nickel steels and other materials requiring a tensile strength of 80 ksi and good toughness at temperatures as low as −75°F.

ER80S-Ni3 and E80C-Ni3 Classifications. These filler metals deposit weld metal similar to 8018-C2 electrodes. Typically they are used for welding 3-1/2 percent nickel steels for low temperature service where a tensile strength of 90 ksi is required.

ER80S-D2 Classification (same as E70S-1B of A5.18-79). Filler metals of this classification contain a high level of deoxidizers (Mn and Si), to control porosity when welding with CO_2 as the shielding gas, and molybdenum for increased strength. They will give radiographic quality welds with excellent bead appearance in both ordinary and difficult-to-weld carbon and low alloy steels. They exhibit excellent out-of-position welding characteristics with the short circuiting and pulsed arc processes. The combination of weld soundness and strength makes filler metal of this classification suitable for single- and multiple-pass welding of a variety of carbon and low alloy steels.

ER90S-B3 and E90C-B3 Classifications. Filler metals of these classifications are used to weld the 2-1/4Cr-1Mo steels used for high temperature-high pressure piping and pressure vessels. These may also be used for joining combinations of Cr-Mo and carbon steel. All gas metal arc welding modes may be used. Careful control of preheat, interpass temperatures, and postheat are essential to avoid cracking.

ER90S-B3L and E90C-B3L Classifications. These filler metals are identical to the types ER90S-B3 and E90C-B3 except for the low carbon content (0.05 percent max). These alloys exhibit greater

TABLE 6-42. Chemical Composition Requiremenets for Deposited Weld Metal With Composite Metal Cored and Stranded Electrodes, percent[a] (AWS A5.28-79)

AWS Classificaiton[b]	Carbon	Manganese	Silicon	Phos-phorus	Sulfur	Nickel	Chromium	Molyb-denum	Vana-dium	Tita-nium	Zirco-nium	Alumi-num	Copper[c]	Total Other Elements[d]
Chromium-molybdenum														
E80C-B2L	0.05	0.40-1.00	0.25-0.60	0.025	0.030	0.20	1.00-1.50	0.40-0.65	—	—	—	—	0.35	0.50
E80C-B2	0.07-0.12	0.40-1.00	0.25-0.60	0.025	0.030	0.20	1.00-1.50	0.40-0.65	—	—	—	—	0.35	0.50
E90C-B3L	0.05	0.40-1.00	0.25-0.60	0.025	0.030	0.20	2.00-2.50	0.90-1.20	—	—	—	—	0.35	0.50
E90C-B3	0.07-0.12	0.40-1.00	0.25-0.60	0.025	0.030	0.20	2.00-2.50	0.90-1.20	—	—	—	—	0.35	0.50
Nickel steel														
E80C-Ni1	0.12	1.25	0.60	0.025	0.030	0.80-1.10	—	0.65	0.05	—	—	—	0.35	0.50
E80C-Ni2	0.12	1.25	0.60	0.025	0.030	2.00-2.75	—	—	—	—	—	—	0.35	0.50
E80C-Ni3	0.12	1.25	0.60	0.025	0.030	3.00-3.75	—	—	—	—	—	—	0.35	0.50
Other low alloy														
EXXC-G	Subject to agreement between supplier and purchaser[e]													

NOTES:

1. Single values shown are maximums.
2. Composite electrodes are not recommended for gas tungsten arc welding (GTAW) or plasma arc welding (PAW).

a. Chemical requirements for composite electrodes are based on analysis of their weld metal in the as-welded condition and using the shielding gas specified in Table 6-43.

b. The suffixes B2, Ni1, etc., designate the chemical composition of the electrode classification.

c. The maximum weight percent of copper in the electrode due to any coating plus the residual copper content in the steel shall comply with the stated value.

d. Other elecments, if intentionally added, shall be reported.

e. In order to meet the requirements of the G classification, the electrode must have as a minimum one of either 0.50 percent nickel, 0.30 percent chromium, or 0.20 percent molybdenum.

TABLE 6-43. Mechanical and Soundness Tests Required for Classification (AWS A5.28-79)

AWS Classification	Soundness Test	All-Weld-Metal Tension Test	Charpy V-Notch Impact Test	Condition
Chromium-Molybdenum Steel Electrodes and Rods				
ER80S-B2 ER80S-B2L E80C-B2 E80C-B2L	Required	Required	Not required	PWHT 1150±25°F
ER90S-B3 ER90S-B3L E90C-B3 E90C-B3L	Required	Required	Not required	PWHT 1275±25°F
Nickel-Steel Electrodes and Rods				
ER80S-Ni1 ER80S-Ni2 ER80S-Ni3	Required	Required	Required	As-Welded PWHT 1150±25°F
E80C-Ni1 E80C-Ni2 E80C-Ni3	Required	Required	Required	As-Welded PWHT 1150±25°F
Manganese-Molybdenum Steel Electrodes and Rods				
ER80S-D2	Required	Required	Required	As-Welded
All Other Low-Alloy Steel Electrodes and Rods				
ER100S-1 ER100S-2 ER110S-1 ER120S-1	Required	Required	Required	As-Welded
ERXXS-G EXXC-G	Required	Required	Optional[a]	Optional[a]

[a] The purchaser or supplier may agree to include this test and determine jointly the impact properties required for acceptance.

TABLE 6-44. — Tensile Strength, Yield Strength and Elongation Requirements for All-Weld-Metal Tension Test (AWS A5.28-79)

AWS Classification	Shielding Gas	Current and Polarity	Tensile Strength (min) ksi	Yield Strength at 0.2 Percent Offset (min) ksi	Elongation in 2 in. (min) percent	Condition
ER80S-B2 ER80S-B2L E80C-B2 E80C-B2L	Ar plus 1-5% O_2	DC electrode positive	80	68	19	PWHT 1150±25°F
ER90S-B3 ER90S-B3L E90C-B3 E90C-B3L			90	78	17	PWHT 1275±25°F
ER80S-Ni1			80	68	24	As-welded
ER80S-Ni2 ER80S-Ni3						PWHT 1150±25°F
E80C-Ni1						As-welded
E80C-Ni2 E80C-Ni3						PWHT 1150±25°F
ER80S-D2	CO_2		80	68	17	As-welded
ER100S-1 ER100S-2 ER110S-1 ER120S-1	Ar plus 2% O_2		100 100 110 120	88 to 102 88 to 102 95 to 107 105 to 122	16 16 15 14	As-welded
ERXXS-G EXXC-G	See note b		See note a	See note b	See note b	As-welded

a Tensile strength shall be consistent with the level placed after the "ER" or "E" prefix; e.g., ER90S-G shall have 90,000 psi minimum ultimate tensile strength.
b Subject to agreement between supplier and purchaser.

TABLE 6-45. Impact Properties (AWS A5.28-79)

AWS Classification	Minimum Required Impact Properties
ER80S-B2 ER80S-B2L ER90S-B3 ER90S-B3L	Not required
ER80S-Ni1 ER80S-Ni2 ER80S-Ni3	20 ft-lb @ −50°F[a,b] 20 ft-lb @ −80°F[a,c] 20 ft-lb @ −100°F[a,c]
ER80S-D2	20 ft-lb @ −20°F[a,b]
ER100S-1 ER100S-2 ER110S-1 ER120S-1	50 ft-lb @ −60°F[d]
E80C-B2L E80C-B2 E90C-B3L E90C-B3	Not required
E80C-Ni1 E80C-Ni2 E80C-Ni3	20 ft-lb @ −50°F[a,b] 20 ft-lb @ −80°F[a,c] 20 ft-lb @ −100°F[a,c]
ERXXS-G EXXC-G	As agreed between supplier and purchaser

[a] The lowest and the highest values obtained shall be disregarded for this test. Two of the three remaining values shall be greater than the specified 20 ft-lb energy level; one of the three may be lower but shall not be less than 15 ft-lb. The computed average value of the three values shall be equal to or greater than the 20 ft-lb energy level.

[b] As-welded impact properties.

[c] Required impact properties after postweld heat treatment. (See Table 6-43.)

[d] Impact properties for the ER100S-1, ER100S-2, ER110S-1 and ER120S-1 shall be obtained at test temperature specified in the table, ±3°F. The lowest and the highest impact values thus obtined shall be disregarded for this test. Two of the three remaining values shall be greater than the specified 50 ft-lb energy level; one of the three may be lower but shall not be less than 40 ft-lb; the computed average values of these three shall be greater than the specified 50 ft-lb level.

resistance to cracking and are more suitable for welds to be left in the as-welded condition.

ER100S-1, ER100S-2, ER110S-1 and ER120S-1 Classifications. These filler metals deposit high-strength, very tough weld metal for critical applications. Originally developed for welding HY80 and HY100 steels for military applications, they are also used for a variety of structural applications where tensile strength requirements exceed 100 ksi and excellent toughness is required to temperatures as low as −60°.

ERXXS-G and EXXC-G Classifications. These classifications include those solid electrodes and rods and composite metal cored and stranded electrodes which are not included in the preceding classes. The supplier should be consulted for the characteristics and intended use of these filler metals. This specification does not list specific chemical composition or impact requirements. These are subject to agreement between supplier and purchaser. However, any filler metal classified ERXXS-G or EXXC-G must meet all other requirements of this specification.

SELECTION OF FILLER WIRE

The selection of a filler wire for a specific application will depend upon the base alloy or alloy combinations, the material thickness, the surface cleanliness and deoxidation level of the steel.

In general, electrodes with lower amounts of deoxidizer are intended for welding killed steels with argon-oxygen gas, but they can be used with other steels and CO_2 gas if quality requirements of the weld metal in regard to porosity are not critical. Those electrodes with higher amounts of deoxidizers produce quality welds on semikilled and rimmed steels as well as on killed steels.

Porosity: In order to avoid porosity, the proper electrode must be used. Other ways to minimize porosity in welding carbon steel are:

1. Avoid excessive current densities. Change to a larger wire or lower the current.
2. Check the welding speed with recommended welding procedures. Too high a speed sometimes causes porosity.
3. Maintain a flow of shielding gas of 40 to 60 cfh when argon with 5% oxygen is used with spray transfer, and 10 to 15 cfh with short-circuiting transfer.
4. Keep the electrode centered in the gas flow from the gun.
5. When welding from both sides of the plate, be sure the first pass on the second side penetrates well into the weld on the first side. Back-chip if necessary.

Deposition Rates: Weld-metal deposition rates for three different wire sizes are indicated in Fig. 6-85. These rates are for a triple-deoxidized wire (E70S-2) used with a mixture of argon with 5% oxygen and spray-type transfer. Figure 6-86 shows similar information for E70S-2 wire used with a mixture of argon and 25% CO_2 and a short-circuiting-type transfer. These figures, along with the procedures, can be used for estimating the amount of wire required and welding costs.

Equipment: The wire-feeding equipment for solid electrode is essentially the same as that used for flux-cored electrode. The electrode is fed at a constant speed, and the current (DCEP) is supplied by a constant-voltage power source. The current is adjusted by changing the electrode feed speed, and

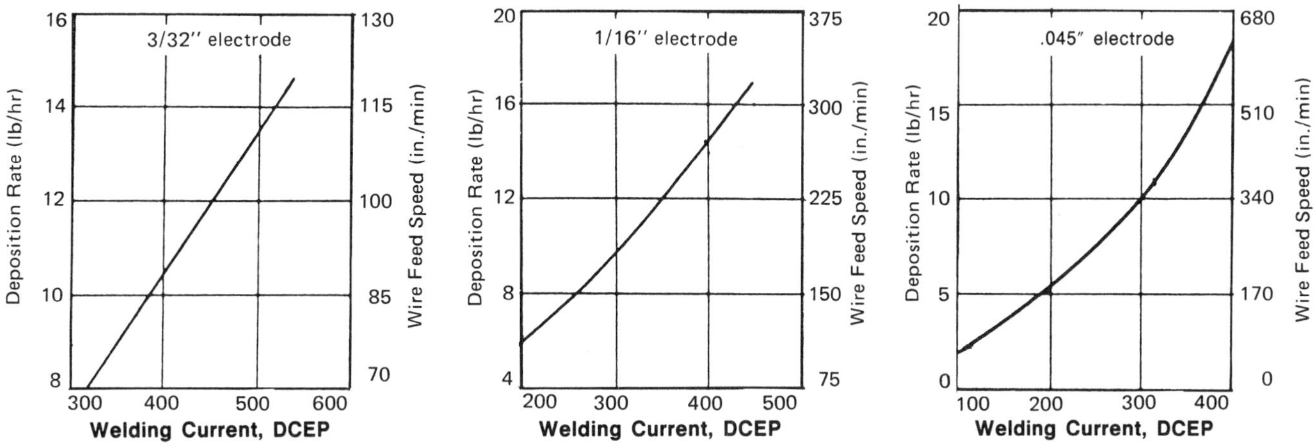

Fig. 6-85. Deposition rates of 3/32, 1/16, and 3/64-in. E70S-2 electrodes with argon-5% oxygen shielding gas.

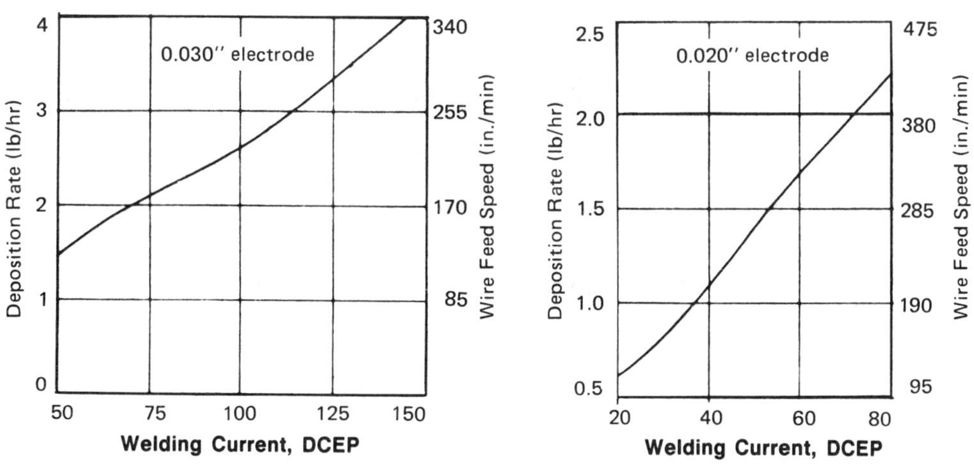

Fig. 6-86. Deposition rate of 0.030 and 0.20-in. E70S-2 electrodes with argon-25% CO_2 shielding gas, short-circuiting-type transfer.

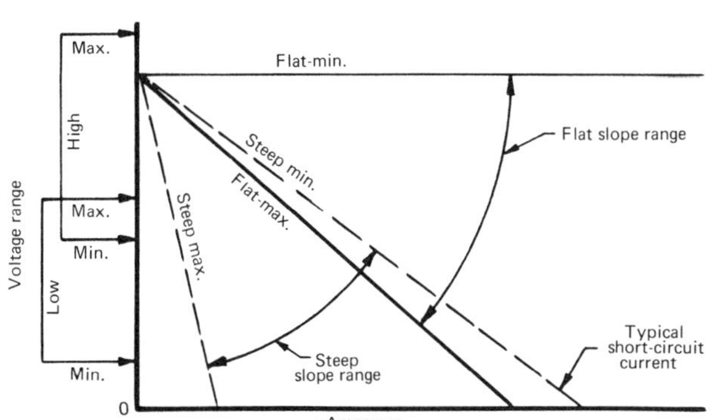

Courtesy Linde Division, Union Carbide Corp.

Fig. 6-87. Slope and voltage ranges typical of power supplies that give control over slope, voltage, and inductance.

the voltage is adjusted at the power source.

Conventional constant-voltage power sources are discussed in Section 4.2. More sophisticated power sources have been developed especially for gas metal-arc welding. These have adjustable slope control and adjustable inductance control. When a short circuit occurs in short-circuiting-type transfer, the slope determines the maximum short-circuit current, and the inductance determines the rate at which the current rises to the maximum. The proper combination of slope and inductance provides the optimum welding conditions for gas metal-arc welding. Figure 6-87 illustrates the slope and voltage ranges that are typical with such power supplies.

JOINT DESIGN

Plate 1/8-in . thick and thinner (down to .035-in.) may be butt-welded with square edges, if an opening of zero to 1/16 in. is used. The short-circuiting technique should be used for wider openings. Short circuiting bridges relatively wide gaps and eliminates excessive penetration.

Plate thicknesses of 3/16-in. may be welded with square edges with a 1/16 to 3/32-in. opening. Two passes are usually necessary — one from each side. A bead overlap greater than the original root spacing is desired to prevent centerline porosity and cold laps. Maximum overlap is attainable if the backhand technique is used on the second side.

For the best quality welds, some bevel should be provided. A 60° single or double-V joint is recommended. A root opening of zero to 1/16-in. should be maintained. Double-V welds can tolerate wider openings than single-V welds. In single-V grooves, a sealing pass from the reverse side will generally be required unless the fitup is uniform.

Plates 1/4-in. thick and thicker generally require single or double-V grooves with 60° included angles. It is generally advantageous to employ a double V. Less metal is needed to fill the joint and less distortion results when welding from alternate sides.

For both types of joints, zero root face is recommended. This assures overlap of the root passes. A root face should not be considered unless an opening larger than normal can be maintained to insure penetration. Uniform penetration is obtainable in joints having no root face if the opening is held less than 3/32-in.

Poor fitup and root overlap should be avoided. Where variations in spacing greater than 1/16 in. are encountered, copper backup strips will prevent excessive penetration.

U grooves should be employed on plate thicker than 1 in. They require considerably less weld metal. Root spacings should be maintained between 1/32 and 3/32 in. Root face should be kept below 3/32 in. to assure adequate penetration.

Welding Parameters, Single Values or a Range

In some of the procedures which follow single values are given for such items as Amps, Volts, and Travel Speed. The purpose of using single values, i.e. 100 amps, is to indicate an optimum or starting place and it is expected that the user will expand these values and show a reasonable range. These values when recorded on a WPS (Welding Procedure Specification) should be shown as a range, not as specific values.

PROCEDURES FOR GAS METAL-ARC WELDING CARBON AND LOW ALLOY STEEL BUTT AND FILLET JOINTS BY SHORT-CIRCUITING TRANSFER

Welding Position:
Flat, Horizontal
Weld Quality Level: Commercial
Steel Weldability: Good or fair

Shield Gas: CO_2
Flow Rate: 25-35 cfh

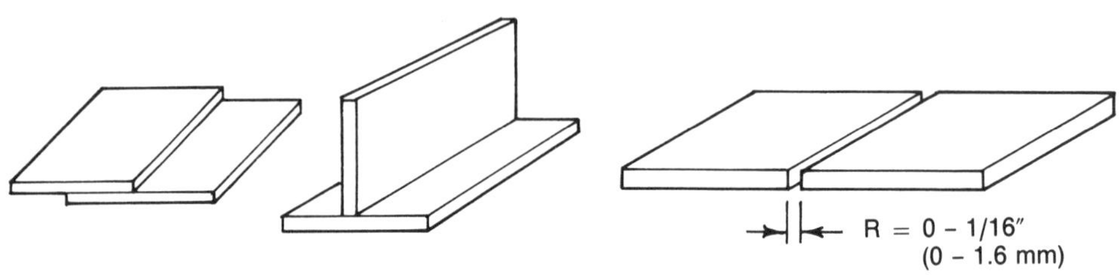

Plate Thickness (in.)	0.024	0.036	0.036	0.060	0.060
Pass	1	1	1	1	1
Electrode Class	EXXS-X	EXXS-X	EXXS-X	EXXS-X	EXXS-X
Size	0.030	0.030	0.035	0.030	0.035
Wire Feed Speed (in./min)	75	125	100	175	150
Current (amp) DC(+)	35	55	80	80	120
Volts	17	18	18	19	19
Arc Speed (in./min)	10	14	13	13	20
Electrode Req'd (lb/ft)	0.016	0.021	0.025	0.030	0.025
Total Time (hr/ft or weld)	0.02	0.014	0.015	0.015	0.010
Electrical Stickout (in.)	1/4 - 1/2	1/4 - 1/2	1/4 - 1/2	1/4 - 1/2	1/4 - 1/2

From Lincoln GS100.

PROCEDURES FOR GAS METAL-ARC WELDING CARBON AND LOW ALLOY STEEL BUTT AND FILLET JOINTS BY SHORT-CIRCUITING TRANSFER

Welding Position:
Flat, Horizontal
Weld Quality Level: Commercial
Steel Weldability: Good or fair

Shield Gas: CO_2
Flow Rate: 25-35 cfh

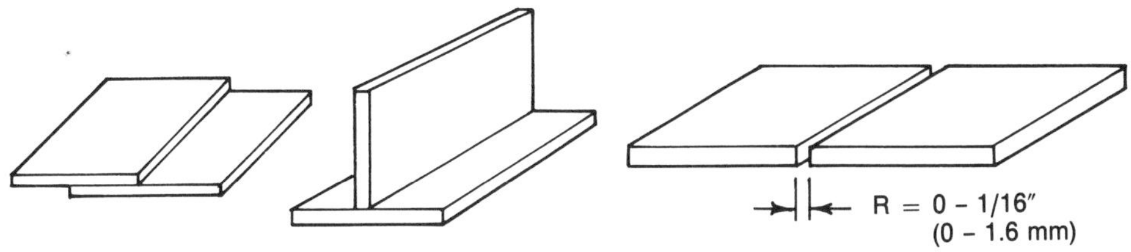

Plate Thickness (in.)	0.075	0.075	0.105	0.105
Pass	1	1	1	1
Electrode Class	EXXS-X	EXXS-X	EXXS-X	EXXS-X
Size	0.030	0.035	0.030	0.035
Wire Feed Speed (in./min)	225	175	275	225
Current (amp) DC(+) (Approx.)	100	130	115	160
Volts	20	20	21	21
Arc Speed (in./min)	18	18	20	20
Electrode Req'd (lb/ft)	0.030	0.032	0.032	0.037
Total Time (hr/ft of weld)	0.011	0.011	0.010	0.010
Electrical Stickout (in.)	1/4 - 1/2	1/4 - 1/2	1/4 - 1/2	1/4 - 1/2

From Lincoln GS100.

PROCEDURES FOR GAS METAL-ARC WELDING CARBON AND LOW ALLOY STEEL BUTT AND FILLET JOINTS BY SHORT-CIRCUITING TRANSFER

Welding Position:
Flat, Horizontal
Weld Quality Level: Commercial
Steel Weldability: Good or fair

Shield Gas: CO_2
Flow Rate: 25-35 cfh

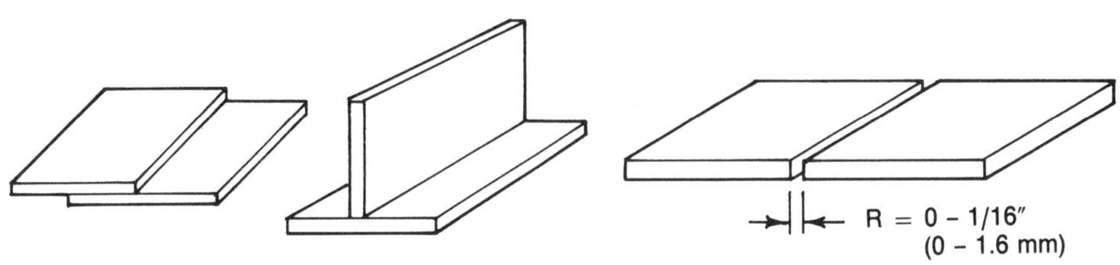

Plate Thickness (in.)	0.135	0.135	0.135	3/16	1/4
Pass	1	1	1	1	1+
Electrode Class	EXXS-X	EXXS-X	EXXS-X	EXXS-X	EXXS-X
Size	0.030	0.035	0.045	0.045	0.045
Wire Feed Speed (in./min)	300	250	125	150	200
Current (amp) DC(+)	130	175	145	165	200
Volts	22	22	18 - 20	19 - 21	20 - 22
Arc Speed (in./min)	17	20	18	15	13
Electrode Req'd (lb/ft)	0.043	0.041	0.039	0.053	0.079
Total Time (hr/ft of weld)	0.012	0.010	0.011	0.013	0.015
Electrical Stickout (in.)	1/4 - 1/2	1/4 - 1/2	1/4 - 1/2	1/4 - 1/2	1/4 - 1/2

From Lincoln GS100.

PROCEDURES FOR GAS METAL-ARC WELDING CARBON AND LOW ALLOY STEEL BUTT JOINTS BY SHORT-CIRCUITING TRANSFER

Welding Position: Flat
Weld Quality Level: Commercial
Steel Weldability: Good or fair

Shield Gas: Argon + 25% CO_2
Flow Rate: 10-15 cfh

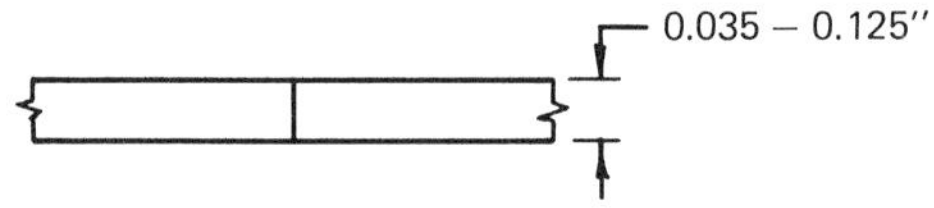

Plate Thickness (in.)	0.035	0.047	0.063	0.078	0.100	0.125
Pass	1	1	1	1	1	1
Electrode Class	EXXS-X	EXXS-X	EXXS-X	EXXS-X	EXXS-X	EXXS-X
Size (in.)	0.030	0.030	0.030	0.030	0.030	0.030
Wire Feed Speed (in./min)	117	140	170	225	235	300
Current (amp) DC(+) (Approx.)	55	65	85	105	110	130
Volts	16	17	17	18	18	19
Arc Speed (in./min)	13	15	15	15	15	16
Electrode Req'd (lb/ft)/pass	0.021	0.021	0.025	0.035	0.038	0.043
Total Time (hr/ft of weld)/pass	0.015	0.013	0.013	0.013	0.013	0.012
Electrical Stickout (in.)	1/4 – 1/2	1/4 – 1/2	1/4 – 1/2	1/4 – 1/2	1/4 – 1/2	1/4 – 1/2

PROCEDURES FOR GAS METAL-ARC WELDING CARBON AND LOW ALLOY STEEL BUTT AND FILLET JOINTS BY SHORT-CIRCUITING TRANSFER

Welding Position:
Vertical down
Weld Quality Level: Commercial
Steel Weldability: Good or fair

Shield Gas: CO_2
Flow Rate: 25-35 cfh

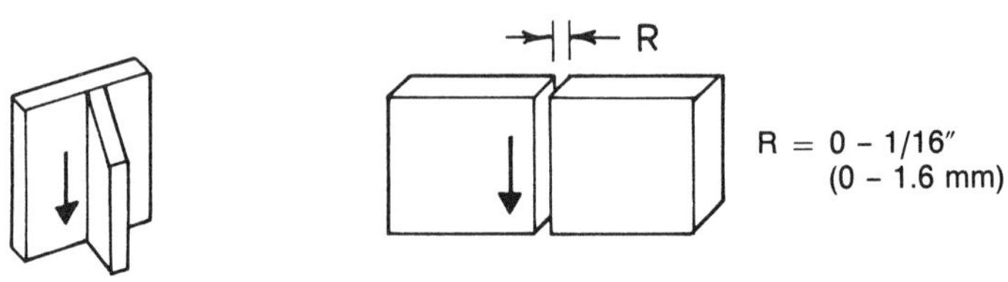

Plate Thickness (in.)	0.024	0.048	0.048	0.075	0.075
Pass	1	1	1	1	1
Electrode Class	EXXS-X	EXXS-X	EXXS-X	EXXS-X	EXXS-X
Size	0.030	0.030	0.035	0.030	0.035
Wire Feed Speed (in./min)	75	150	125	225	175
Current (amp) DC(+)	35	70	100	100	130
Volts	17	18	18	20	20
Arc Speed (in./min)	10	15	19	20	20
Electrode Req'd (lb/ft)/pass	0.016	0.023	0.024	0.027	0.031
Total Time (hr/ft of weld)/pass	0.02	0.013	0.011	0.010	0.010
Electrical Stickout (in.)	1/4 – 1/2	1/4 – 1/2	1/4 – 1/2	1/4 – 1/2	1/4 – 1/2

From Lincoln GS100.

PROCEDURES FOR GAS METAL-ARC WELDING CARBON AND LOW ALLOY STEEL BUTT AND FILLET JOINTS BY SHORT-CIRCUITING TRANSFER

Welding Position: Vertical down
Weld Quality Level: Commercial
Steel Weldability: Good or fair

Shield Gas: CO_2
Flow Rate: 25-35 cfh

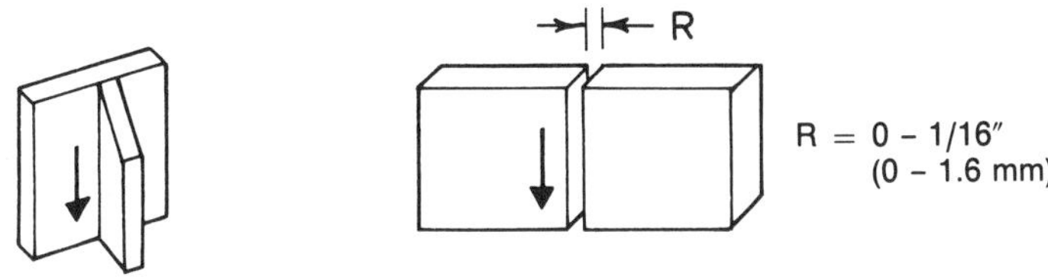

Plate Thickness (in.)	0.135	0.135	0.135	3/16	1/4
Pass	1	1	1	1+	1+
Electrode Class	EXXS-X	EXXS-X	EXXS-X	EXXS-X	EXXS-X
Size	0.030	0.035	0.045	0.045	0.045
Wire Feed Speed (in./min)	300	250	125	150	200
Current (amp) DC(+) (Approx.)	130	175	145	165	200
Volts	22	22	19	20	21
Arc Speed (in./min)	20	20	20	17	17
Electrode Req'd (lb/ft)/pass	0.036	0.041	0.035	0.049	0.063
Total Time (hr/ft or weld)/pass	0.010	0.010	0.010	0.012	0.012
Electrical Stickout (in.)	1/4 – 1/2	1/4 – 1/2	1/4 – 1/2	1/4 – 1/2	1/4 – 1/2

From Lincoln GS100.

PROCEDURES FOR GAS METAL-ARC WELDING CARBON AND LOW ALLOY STEEL FILLETS BY SHORT-CIRCUITING TRANSFER

Welding Position: Vertical up
Weld Quality Level: Commercial
Steel Weldability: Good or fair

Shield Gas: 75%A – 25% CO_2
Flow Rate: 25-35 cfh

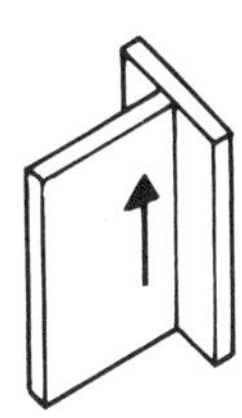

Leg Size	1/4	1/4	5/16	5/16
Plate Thickness (in.)	5/16	5/16	3/8	3/8
Pass	1	1	1	1
Electrode Class	EXXS-X	EXXS-X	EXXS-X	EXXS-X
Size	0.035	0.045	0.035	0.045
Wire Feed Speed (in./min)	225	150	250	150
Current (amp) DC(+)	160	165	175	165
Volts	18	19	20	19
Arc Speed (in./min)	5 – 6	4 – 5	4 – 4.5	4 – 5
Electrode Req'd (lb/ft)	0.156 – 0.129	0.215 – 0.172	0.217 – 0.191	0.215 – 0.172
TotalTime (hr/ft or weld)	0.040 – 0.033	0.050 – 0.044	0.050 – 0.044	0.050 – 0.04
Electrical Stickout (in.)	1/4 – 1/2	1/4 – 1/2	1/4 – 1/2	1/4 – 1/2

From Lincoln GS100.

PROCEDURES FOR GAS METAL-ARC WELDING CARBON AND LOW ALLOY STEEL BUTT JOINTS BY SPRAY TRANSFER

Welding Position: Flat
Weld Quality Level: Commercial
Steel Weldability: Good or fair

Shield Gas: Argon + 2% O_2
Flow Rate: 40-45 cfh

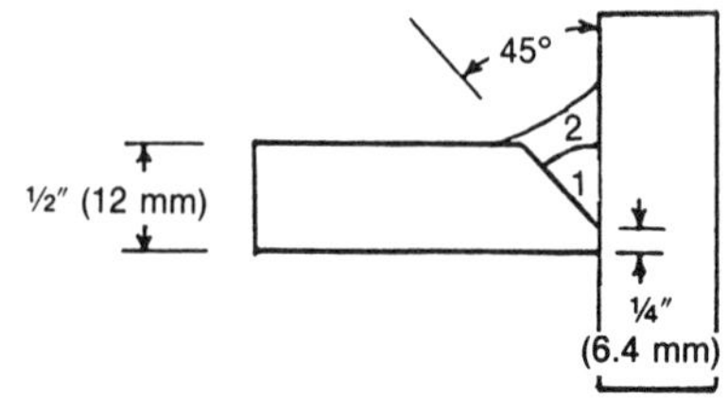

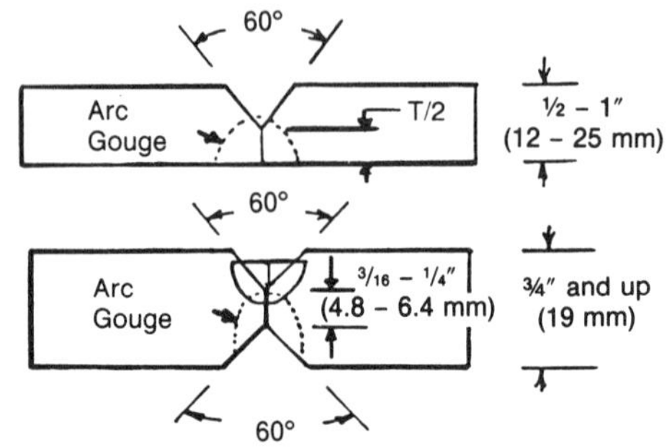

Plate Thickness (in.)	1/2+	1/2+	1/2+	1/2+
Pass	1+	1+	1+	1+
Electrode Class	EXXS-X	EXXS-X	EXXS-X	EXXS-X
Size	0.035	0.045	0.052	1/16
Wire Feed Speed (in./min)	500 – 600	375 – 500	300 – 485	210 – 290
Current (amp) DC(+)	290 – 300	300 – 340	320 – 430	325 – 430
Volts	27 – 28	27 – 28	29 – 31	24 – 27
Arc Speed (in./min)	10 – 15	12 – 18	14 – 24	14 – 23
Electrode Req'd (lb/ft)/pass	0.164 – 0.127	0.172 – 0.148	0.151 – 0.143	0.154 – 0.136
TotalTime (hr/ft of weld)/pass	0.020 – 0.013	0.017 – 0.011	0.014 – 0.008	0.014 – 0.009
Electrical Stickout (in.)	3/4 – 1	3/4 – 1	3/4 – 1	3/4 – 1

From Lincoln GS100.

PROCEDURES FOR GAS METAL-ARC WELDING OF CARBON AND LOW ALLOY STEEL FILLETS BY SPRAY TRANSFER

Welding Position:
Flat, Horizontal
Weld Quality Level: Commercial
Steel Weldability: Good or fair

Shield Gas: Argon + 5% O_2
Flow Rate: 40-45 cfh

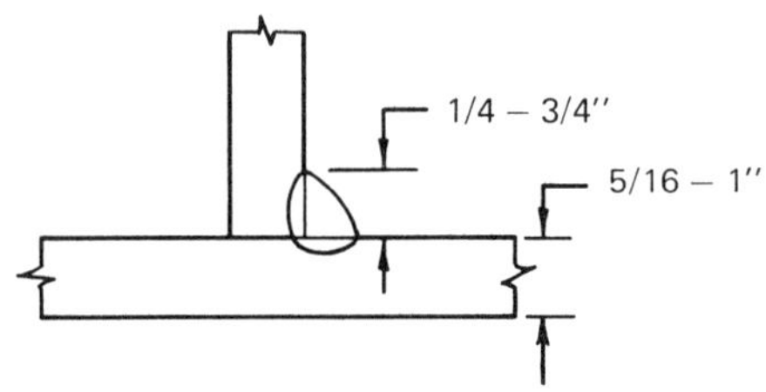

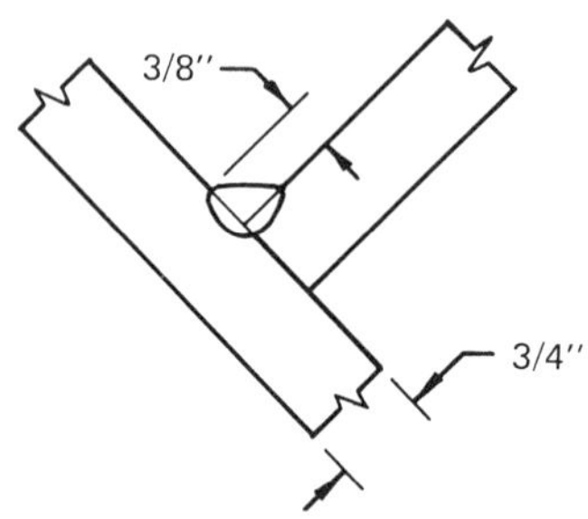

Weld Size (in.)	1/4	5/16	3/4	3/8
Plate Thickness (in.)	5/16	1/2	1	3/4
Pass	1	1	4	1
Electrode Class	EXXS-X	EXXS-X	EXXS-X	EXXS-X
Size	3/32	3/32	3/32	3/32
Wire Feed Speed (in./min)	87	100	100	110
Current (amp) DC(+) (Approx.)	400	450	450	475
Volts	27	28	28	30
Arc Speed (in./min)	16	12	7	9
Electrode Req'd (lb/ft)/pass	0.130	0.200	0.343	0.289
Total Time (hr/ft of weld)/pass	0.0125	0.0167	0.0286	0.022
Electrical Stickout (in.)	3/4 – 1	3/4 – 1	3/4 – 1	3/4 – 1

PROCEDURES FOR GAS METAL-ARC WELDING CARBON AND LOW ALLOY STEEL BUTT JOINTS BY SPRAY TRANSFER

Welding Position: Flat
Weld Quality Level: Commercial
Steel Weldability: Good or fair

Shield Gas: Argon + 5% O_2
Flow Rate: 40-45 cfh

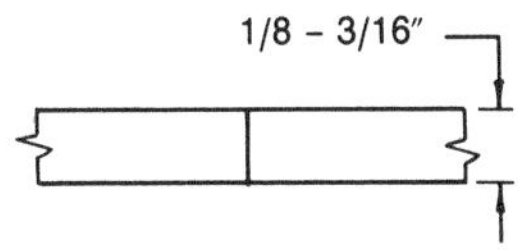

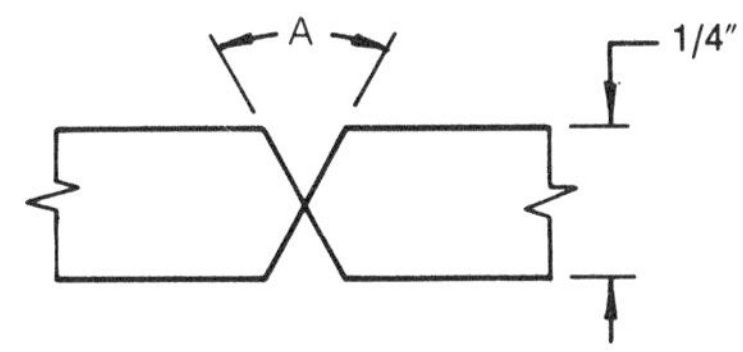

Plate Thickness (in.)	1/8	3/16	1/4	1/4
Pass	1	2	1	2
Electrode Class	EXXS-X	EXXS-X	EXXS-X	EXXS-X
Size	1/16	1/16	3/32	3/32
Wire Feed speed (in./min)	165	260	83	95
Current (amp) DC(+)	280	375	375	430
Volts	24	26	27	27
Arc Speed (in./min)	20	24	24	24
Electrode Req'd (lb/ft)	0.090	0.235	0.255	
Total Time (hr/ft or weld)	0.010	0.017	0.017	
Electrical Stickout (in.)	3/4 – 1	3/4 – 1	3/4 – 1	3/4 – 1
Angle, A (deg)			60	60

PROCEDURES FOR GAS METAL-ARC WELDING CARBON AND LOW ALLOY STEEL BUTT JOINTS BY SPRAY TRANSFER

Welding Position: Flat
Weld Quality Level: Commercial
Steel Weldability: Good or fair

Shield Gas: Argon + 5% O_2
Flow Rate: 40-45 cfh

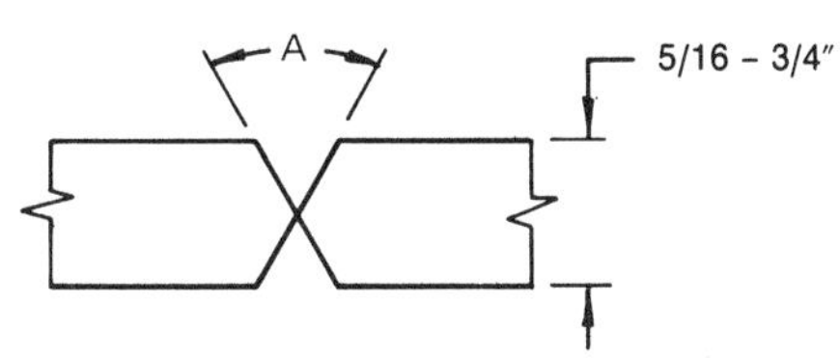

Plate Thickness (in.)	5/16	5/16	1/2	1/2	3/4
Pass	1	2	1	2	1 – 4
Electrode Class	EXXS-X	EXXS-X	EXXS-X	EXXS-X	EXXS-X
Size	3/32	3/32	3/32	3/32	3/32
Wire Feed Speed (in./min)	87	92	87	100	100
Current (amp) DC(+)(Approx.)	400	420	400	450	450
Volts	28	28	28	28	29
Arc Speed (in./min)	20	20	14	14	12
Electrode Req'd (lb/ft)	0.214		0.325		0.804
Total Time (hr/ft of weld)	0.020		0.029		0.067
Electrical Stickout (in.)	3/4 – 1	3/4 – 1	3/4 – 1	3/4 – 1	3/4 – 1
Angle, A (deg)	60	60	60	60	90

PROCEDURES FOR GAS METAL-ARC WELDING CARBON AND LOW ALLOY STEEL FILLETS BY SPRAY TRANSFER

Welding Position:
Flat, Horizontal
Weld Quality Level: Commercial
Steel Weldability: Good or fair

Shield Gas: Argon + 2% O_2
Flow Rate: 35-45 cfh

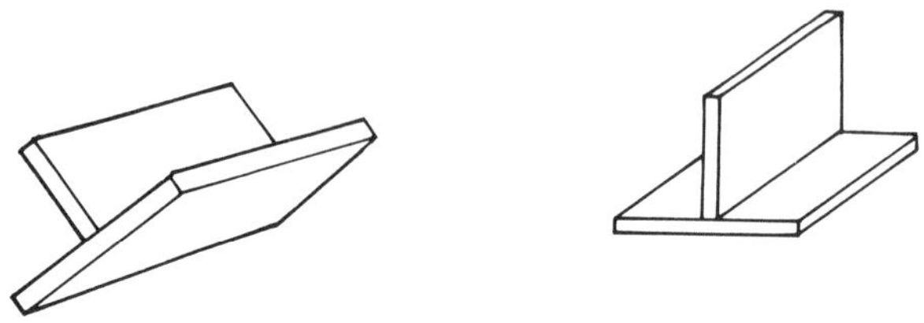

Leg Size	5/32	3/16	3/16	1/4	1/4
Plate Thickness (in.)	3/16	1/4	1/4	5/16	5/16
Pass	1	1	1	1	1
Electrode Class	EXXS-X	EXXS-X	EXXS-X	EXXS-X	EXXS-X
Size	0.035	0.035	0.045	0.035	0.045
Wire Feed Speed (in./min)	375	400	350	500	375
Current (amp) DC(+) (Approx.)	195	200	285	290	300
Volts	21	22	26	27	27
Arc Speed (in./min)	24	19	25	14	18
Electrode Req'd (lb/ft)	0.049	0.072	0.076	0.115	0.111
Total Time (hr/ft of weld)	0.008	0.011	0.008	0.014	0.011
Electrical Stickout (in.)	3/4 – 1	3/4 – 1	3/4 – 1	3/4 – 1	3/4 – 1

From Lincoln GS100.

PROCEDURES FOR GAS METAL-ARC WELDING CARBON AND LOW ALLOY STEEL FILLETS BY SPRAY TRANSFER

Welding Position:
Flat, Horizontal
Weld Quality Level: Commercial
Steel Weldability: Good or fair

Shield Gas: Argon + 2% O_2
Flow Rate: 35-45 cfh

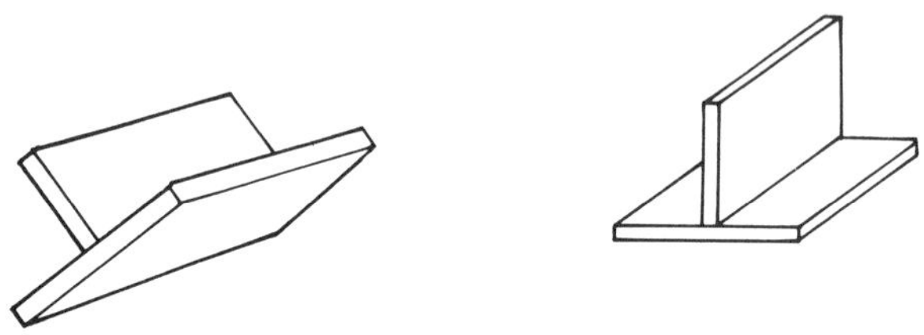

Leg Size	1/4	1/4	5/16	5/16	5/16
Plate Thickness (in.)	5/16	5/16	3/8	3/8	3/8
Pass	1	1	1	1	1
Electrode Class	EXXS-X	EXXS-X	EXXS-X	EXXS-X	EXXS-X
Size	0.052	1/16	0.035	0.045	1/16
Wire Feed Speed (in./min)	320	235	600	475	235
Current (amp) DC(+) (Approx.)	320	350	300	335	350
Volts	26	25	28	28	25
Arc Speed (in./min)	18	19	10	13	12
Electrode Req'd (lb/ft)	0.128	0.135	0.196	0.192	0.209
Total Time (hr/ft of weld)	0.011	0.011	0.020	0.015	0.017
Electrical Stickout (in.)	3/4 – 1	3/4 – 1	3/4 – 1	3/4 – 1	3/4 – 1

From Lincoln GS100.

PROCEDURES FOR GAS METAL-ARC WELDING CARBON AND LOW ALLOY STEEL FILLETS BY SPRAY TRANSFER

Welding Position:
Flat, Horizontal
Weld Quality Level: Commercial
Steel Weldability: Good or fair

Shield Gas: Argon + 2% O_2
Flow Rate: 35-45 cfh

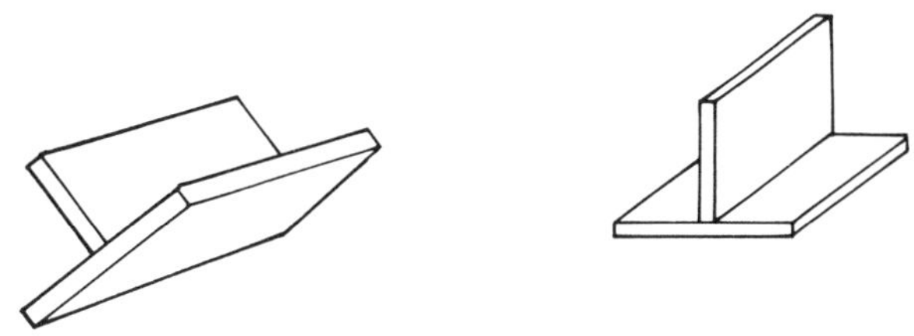

Leg Size	3/8	3/8
Plate Thickness (in.)	1/2	1/2
Pass	1	1
Electrode Class	EXXS-X	EXXS-X
Size	0.052	1/16
Wire Feed Speed (in./min)	485	235
Current (amp) DC(+)	430	350
Volts	30	25
Arc Speed (in./min)	13	9
Electrode Req'd (lb/ft)	0.263	0.271
Total Time (hr/ft or weld)	0.015	0.022
Electrical Stickout (in.)	3/4 – 1	3/4 – 1

From Lincoln GS100.

PROCEDURES FOR GAS METAL-ARC WELDING CARBON AND LOW ALLOY STEEL FILLETS BY PULSED SPRAY TRANSFER

Welding Position: Vertical up
Weld Quality Level: Commercial
Steel Weldability: Good or fair

Shield Gas: Argon + 2% O_2
a. A + 5% CO_2
b. A + 10% CO_2
Flow Rate: 30-40 cfh

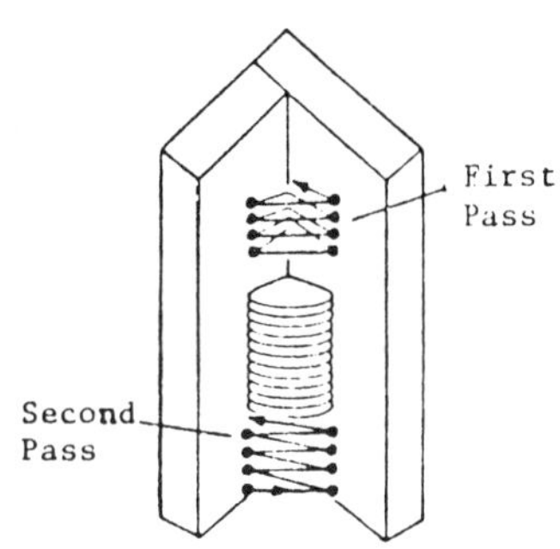

Plate Thickness (in.)	3/8	3/8	1/2	1/2
Pass	1	1		
Electrode Class	EXXS-X	EXXS-X	EXXS-X	EXXS-X
Size (in.)	0.045	0.045	0.045	0.045
Wire Feed Speed (in./min)	125	125	130 – 145	130 – 145
Current (amp) DC(+)	135	135	140 – 155	140 – 155
Volts	17 – 18 a	19 –20 b	18.5 – 19.5 a	20 – 21 b
Arc Speed (in./min)	4 – 5	4 – 5	4 – 5	4 – 5
Electrode Req'd (lb/ft)/pass	0.150 – 0.120	0.150 – 0.120	0.195 – 0.136	0.195 – 0.136
Total Time (hr/ft of weld)/pass	0.050 – 0.040	0.050 – 0.040	0.050 – 0.040	0.050 – 0.040
Electrical Stickout (in.)	1/2 – 3/4	1/2 – 3/4	1/2 – 3/4	1/2 – 3/4

From Lincoln GS100.

Gas metal arc welding of steel.

Welding Carbon and Low-Alloy Steels with the Gas Tungsten-Arc Process

The gas tungsten arc welding process is characterized as a process which produces high quality welds and low filler metal deposition rates. It can be used to weld sheet or plate together without the addition of filler material (autogenous welding) and in this case welding speed may approach the travel speed of other arc welding processes. The GTAW process is described in Section 5.4 along with necessary equipment and welding techniques.

Basically the processes employs an arc between a non-consumable tungsten electrode and the metal to be welded. An inert shielding gas (argon, helium or a mixture) flows from a nozzle on the torch which holds the tungsten electrode. The inert gas becomes ionized to conduct the arc current and protects the electrode and molten weld pool from oxidation and atmospheric contamination. The arc energy input parameters (current, voltage and travel speed) are controlled independently of the rate of filler metal addition, unlike most other arc welding processes where energy input rate is a direct function of rate of filler metal addition.

Because of the inert gas shield, this process produces a very clean weld pool and top surface. In critical applications where the hot weld metal must be kept free of atmospheric contamination, an inert gas backup shield and trailer shield must be used. Only inert gas shields can be used with this process because CO_2 will disassociate in the arc to release oxygen which will erode the tungsten electrode by forming volatile tungsten oxide. Mixtures of inert gas with O_2 or CO_2 should not be used because these will have the same eroding effect as CO_2.

Gas tungsten arc welding is normally performed with direct current electrode negative which develops approximately 70 percent of the arc heat at the work. Tungsten electrodes containing additions of thorium oxide, zirconium oxide, cerium oxide or lanthanum oxide are used for welding carbon and low alloy steels because these provide easy arc starting and a more stable arc than pure tungsten.

GTA welding is used extensively for fusion welding most reactive metals such as aluminum and titanium alloys and other alloys where high quality welds are required. Because deposition rates are low with GTAW, the use of this process for volume welding of plain carbon and low alloy steels is limited to thin sheet, small parts and root passes of multipass butt joints. Subsequent passes are normally deposited by another process such as shielded metal arc or flux cored arc welding.

All of the plain carbon and low alloy steels can be readily welded by the GTAW process; however, rimmed steels and steels which are not fully deoxidized may present a porosity problem when welded without filler metal containing deoxidizers. When deoxidizers are not present in the filler, oxygen in the steel can react with carbon to form CO which is a cause of porosity. Fully deoxidized steels should not present a porosity problem provided the molten weld pool is completely shielded by inert gas until solidification is complete. More thorough deoxidation practice is used in the production of the low alloy steels, thus weld porosity should be minimal with these steels.

WELDING OF STEELS

The parameters listed here should be considered as starting points for establishing actual working parameters which should be developed by welding sample parts and thorough examination.

The weldability of both the plain carbon and low alloy steels depends upon the hardenability of the steel, and this depends upon the carbon content and amount of those elements which increase hardenability. Carbon determines the maximum hardness attainable, and the hardenability elements such as chromium and manganese determine the cooling rate needed to transform austenite to 100 percent martensite for maximum hardness. The weld heat affected zone and fusion zone heat and cool rapidly and usually will cool fast enough to transform to martensite if the carbon and alloy content are high enough. Since the higher carbon martensites are hard and of low ductility, their formation in a weld can lead to cracking. Preheat and use of a high energy input are two methods for slowing the cooling rate to prevent the formation of martensite and thus prevent cracking.

PLAIN CARBON STEEL

Plain carbon steels are classified according to carbon level as: Low Carbon steels up to 0.15% C; Mild Carbon steels, 0.15-0.29% C; Medium Carbon steels, 0.30-0.50% C; and High carbon steels, 0.50-1.00% C.

The Low Carbon steels may be supplied as rimmed, semi-killed or killed and are readily welded without preheat. The rimmed steels are not deoxidized and have a high tendency to form porosity in a weld if a filler metal containing a deoxidizer is not used.

The Mild Carbon steels normally can be GTA welded without preheat but proper welding procedures must be used. Approximate parameters for welding these steels are listed in the procedures at the end of this section. The Medium Carbon steels have higher hardenability and generally will require a preheat of 400-600°F and may require a post heat, tempering or stress relief treatment. High carbon steels can develop heat treated hardness up to 63 R_C and generally are used for items such as tools and dies which do not require welding. They can be welded but require a preheat of 500-600°F, an interpass temperature of 450-500°F and a post weld tempering treatment.

GTAW is normally not used for welding plain carbon steels except for thin sheet and root passes on heavy plate and piping. Filler passes in joints in thick plate or pipe are made by another process because the deposition rate for the GTAW process is too slow and not economical. When making root passes or single fusion passes in thin material it is recommended that the bottom be shielded with argon or CO_2 to prevent oxide formation and provide a smooth root surface free of porosity and wrinkles.

Carbon and low alloy steels are susceptible to hydrogen induced cracking, and this susceptibility increases with increasing carbon content. This effect also increases with increasing material thickness. Therefore, low-hydrogen welding procedures must be used with the susceptible materials by assuring that the base material and filler material are clean and the atmosphere does not contain moisture or other contaminates.

LOW ALLOY STEELS

The low alloy steels represent three groups of steels which normally contain a maximum of 6% alloy content although a few contain more than this.

The high strength, low alloy steels (HSLA) are a group corresponding to ASTM specifications for plate, structural shapes and piping. They are usually furnished in the as-rolled, controlled-rolled or normalized condition. Carbon content is below 0.30%. These steels are usually used for construction of large structures such as buildings, bridges and pressure vessels, and welding processes other than GTAW are used to join them. However, root passes on joints in pressure vessels and piping may be done by the GTAW process. Low hydrogen welding practice and preheat of up to 300°F should be used with these steels. The preheat temperature depends on the particular steel and increases with material thickness. Filler wires should match the base metal composition as close a possible. The approximate welding parameters given in the procedures at the end of this section can be used for these steels.

QUENCHED AND TEMPERED STEELS

The Quenched and Tempered Steels (Q&T) contain less than 0.30% carbon and have compositions which are similar to the HSLA steels but are somewhat higher in Cr and Mo. The main difference is that these steels are furnished in the heat treated condition which consists of austenitizing, quenching and tempering. Yield strengths from 50 to 150 ksi with good notch toughness, ductility and corrosion resistance are attainable. These steels, like the HSLA steels, are used mainly in construction and pressure vessel applications and in these cases are welded by processes other than GTAW. For these steels with yield strengths over 100 ksi, the GMAW and GTAW processes are preferred, especially for smaller components.

The Q&T steels require careful control of the weld cooling rate to avoid loss of strength in the weld and heat affected zone. Too fast a cooling rate would produce a hard brittle martensitic structure, while too slow a cooling rate would produce a mixed structure of ferrite, high carbon martensite, coarse bainite and overtempered martensite. This combination produces a heat affected zone of reduced strength and toughness. Cooling rate can be controlled by choosing the proper preheat temperature and weld energy input, and this will depend upon the steel and thickness. Preheat temperature normally falls between 50 and 300°F, where the 300°F is only used for thicknesses over 2″. Low hydrogen procedures should be used to avoid underbead cracking, thus a minimum preheat of 100°F is recommended to remove moisture from the joint area. Handbook data and recommendations from the steel producer should be used to establish preheat, filler metal and weld parameters for welding each steel and thickness. Filler wire will normally match or closely match base metal.

HEAT TREATABLE LOW ALLOY STEELS

The Heat Treatable Low Alloy Steels (HTLA) range in carbon content from about 0.25-0.45 percent and contain small amounts of Cr, Ni, Mo to provide hardenability. These steels can be heat treated to strength levels of over 200 ksi and hardness of up to 60 R_C. They are normally welded in the annealed or overtempered condition, then heat treated to the required hardness after welding. Because these steels have high hardenability, the proper preheat, energy input and interpass temperature must be used to prevent cold cracking. Since these steels are used mainly for relatively small components, it is possible to heat treat the welded components after welding. The preheat temperature will depend on the particular alloy and thickness and can range from 50 to 550°F. Extra care must be exercised to assure that the joint area is clean and free of rust, oxides, dirt, grease or any organics and that low hydrogen procedures are used. These steels usually are vacuum melted or vacuum arc remelted which produces alloys with very low impurities. To produce welds in these steels with equivalent cleanliness, the filler wires and the weld atmosphere must be clean. The GTAW process is the preferred arc welding process for the HTLA steels because it is capable of producing welds with the lowest hydrogen content. Recommendations for welding procedures, parameters and filler materials can be obtained from handbooks and from producers of each steel. However, the parameters given in the procedures at the end of this section can be used as a starting point in establishing the proper parameters.

HIGH ALLOY STEELS

The category of high alloy steels normally includes steels containing more than 5 percent alloy additions except stainless steels, 9 percent nickel steels and tool steels. Four classes of high alloy steels are nickel-cobalt steels, 5 percent chromium-molybdenum-vanadium steels, maraging steels and austenitic manganese steels. These steels provide substantially higher tensile and yield strengths than the standard low alloy steels. Many of these steels have limited weldability and they require careful adherance to specified procedures and processes. To obtain base metal mechanical properties and toughness in the high alloy steels, weld metal must contain very low levels of impurities. Small amounts of non-metallic compounds of oxygen, nitrogen, carbon, sulfur or phosphorus in grain boundaries can drastically reduce toughness and strength. These elements must be kept low by using filler metals low in these elements and assuring that the surfaces of the filler wire and weld joint are clean. The molten weld pool and surrounding hot metal must be protected from atmospheric contamination. For this reason, the inert gas shielded arc welding processes are recommended for welding the high alloy steels and the gas tungsten arc process is preferred for welding all but the austenitic manganese steels because it produces the highest quality welds and the rates of filler metal addition and power input can be controlled independently of each other. Chemical compositions of representative high alloy steels are listed in Table 6-46. Detailed procedures for welding these and other high alloy steels should be obtained from the manufacturer or suppliers. Information is also available from the AWS Welding Handbook, Volume Four Seventh Edition and Volume Three Eighth Edition.

The nickel-cobalt steels have the best weldability of the high alloy group and some such as the HP-9-4-20 can be welded without preheat or post weld heat treatment and still have base metal properties. The chromium-molybdenum-vanadium steels have high hardenability and thus poor weldability. They must be welded with a carefully controlled preheat, interpass temperature and post weld heat treatment. Too high a preheat can cause oxidation of the joint faces and weld porosity; too low a preheat can lead to martensite formation and weld cracking.

The maraging steels have a low sensitivity to cold cracking and underbead cracking because the low carbon martensite is relatively soft and the low martensite transformation temperature puts the fusion zone in compression. The gas tungsten arc process is recommended for welding these three groups of high alloy steels.

The austenitic manganese steels are sensitive to carbide precipitation on reheating which reduces ductility. Therefore these steels should be welded with processes which result in minimum heat input such as the gas metal arc or shielded metal arc processes. Gas tungsten arc welding is not recommended for these steels.

TABLE 6-46. Chemical Composition of Representative High Alloy Steels

Designation	Composition, Percent											
	C	Mn	P	S	Si	Ni	Cr	Mo	V	Co	Al	Ti
AF1410	0.16	0.05	0.01	0.01	0.1	10	2	1	0.1	14	0.015	0.01
HP9-4-20	0.16 0.23	0.20 0.40	0.01	0.01	0.10	8.50 9.50	0.65 0.85	0.90 1.10	0.06 0.12	4.25 4.75		
HP9-4-30	0.29 0.34	0.10 0.35	0.01	0.01	0.10	7.25 8.00	0.90 1.10	0.90 1.10	0.06 0.12	4.25 4.75		
HY-180	0.09 0.13	0.05 0.25	0.01	0.006	0.15	9.5 10.5	1.8 2.2	0.09 1.10		7.5 8.5	0.025	0.02
H-11	0.33 0.43	0.20 0.50	0.03	0.03	0.80 1.20	0.30	4.75 5.50	1.10 1.60	0.30 0.60			
H-13	0.32 0.45	0.20 0.50	0.03	0.03	0.80 1.20		4.75 5.50	1.10 1.75	0.08 1.20			
Maraging 200 A538, Gr A	0.03	0.10	0.01	0.01	0.10	18		4.0		8.0	0.1	0.2
Maraging 300 A538, Gr C	0.03	0.10	0.01	0.01	0.10	18		5.0		9.0	0.1	0.7
Manganese Casting A128	1.05 1.35	11.5 14.0	0.07		1.00							

FILLER MATERIALS

The filler materials specified for GTA welding of plain carbon and low alloy steels are the same as for GMA welding and are listed in AWS specifications A5.18 and A5.28. A description of each classification is included in Section 6.6, Welding Carbon and Low Alloy Steels with the Gas Metal Arc Process. Only classifications for solid wires should be used for GTA welding. Metal cored and stranded filler metals are not recommended for GTA welding.

Filler metals for the high alloy steels should generally match the composition of the base alloy but should be lower in impurity elements such as P, S and Si to match base metal mechanical properties. If lower strength properties are acceptable in the joints, low alloy steel filler metals may be used to weld high alloy steels. This will provide tougher, more ductile welds with less tendency for cracking.

Procedures for welding plain carbon, low alloy and some high alloy steels with the gas tungsten arc process are given in the charts on the following pages.

Welding Parameters, Single Values or a Range

In some of the procedures which follow single values are given for such items as Amps, Volts, and Travel Speed. The purpose of using single values, i.e. 100 amps, is to indicate an optimum or starting place and it is expected that the user will expand these values and show a reasonable range. These values when recorded on a WPS (Welding Procedure Specification) should be shown as a range, not as specific values.

APPROXIMATE PARAMETERS FOR GTA WELDING PLAIN CARBON AND LOW ALLOY STEELS FOR FLAT POSITION BUTT JOINTS, LAP JOINTS

Electrodes: EWTh2
Electrode Negative
Shielding Gas: 100% argon

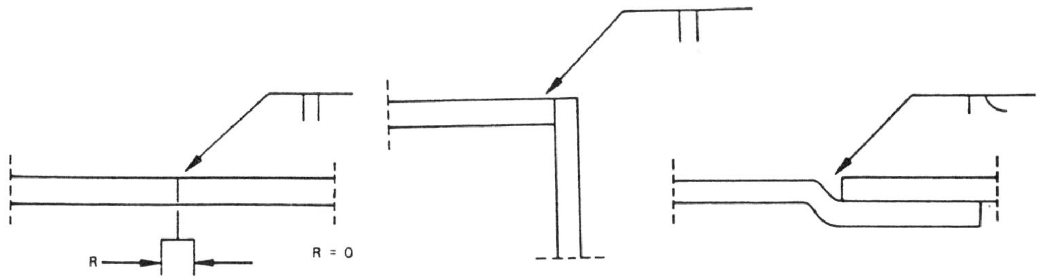

Material Thickness (in.)	.048	.054	.060	.067	.075	.090
Electrode Diameter (in.)	1/16	1/16	1/16	1/16	3/32	3/32
Electrode Angle[1]	30°	30°	30°	30°	45°	45°
Electrode Tip Diameter[1] (in.)	0.03	0.03	0.03	0.03	0.045	0.045
Arc Current, Amps	45 – 65	48 – 69	51 – 73	54 – 77	57 – 81	60 – 85
Arc Volts, Volts	10 – 12	10 – 12	10 – 12	10 – 12	10 – 12	10 – 12
Filler Wire Diameter (in.)	1/16	1/16	1/16	1/16	1/16	3/32
Travel Speed, ipm	10 – 15	10 – 15	10 – 15	10 – 15	10 – 15	10 – 15
Cup Diameter (in.)	3/8	3/8	3/8	3/8	3/8	3/8
Gas Flow Rate, cfh	20	20	20	20	20	20

[1] See Figure 5-15. *(From AWS B2.1008)*
Filler wire should be ER70S-3 or higher strength filler to match strength of base metal. Low carbon steels will not required a preheat over 100°F; as carbon increases and alloy increases preheat will increase up to 600°F.

APPROXIMATE PARAMETERS FOR GTA WELDING PLAIN CARBON AND LOW ALLOY STEELS FOR FLAT POSITION BUTT JOINTS, LAP JOINTS

Electrodes: EWTh2
Electrode Negative
Shielding Gas: 100% argon

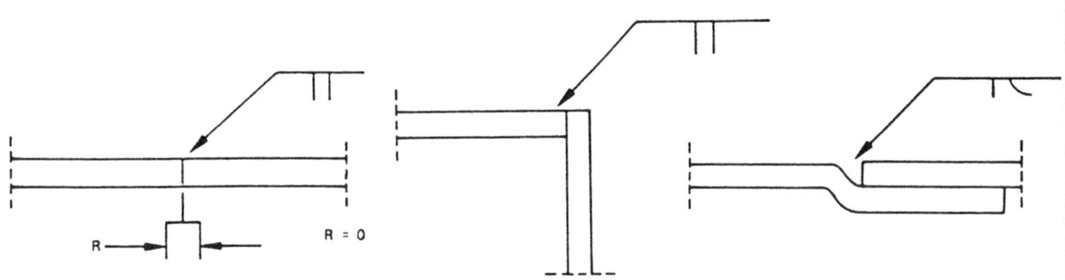

Material Thickness (in.)	.105	.120	.135
Electrode Diameter (in.)	3/32	3/32	3/32
Electrode Angle[1]	45°	45°	45°
Electrode Tip Diameter[1] (in.)	0.045	0.045	0.045
Arc Current, Amps	63 – 90	66 – 95	70 – 100
Arc Volts, Volts	10 – 12	10 – 12	10 – 12
Filler Wire Diameter (in.)	3/32	3/32	3/32
Travel Speed, ipm	10 – 12	10 – 12	8 – 10
Cup Diameter (in.)	3/8	1/2	1/2
Gas Flow Rate, cfh	20	25	25

[1] See Figure 5-15. *(From AWS B2.1008)*
Filler wire should be ER70S-3 or higher strength filler to match strength of base metal. Low carbon steels will not required a preheat over 100°F; as carbon increases and alloy increases preheat will increase up to 600°F.

APPROXIMATE PARAMETERS FOR GTA WELDING PLAIN CARBON AND LOW ALLOY STEELS FOR FLAT POSITION LAP JOINTS AND HORIZONTAL FILLETS

Electrodes: EWTh2
Electrode Negative
Shielding Gas: 100% argon

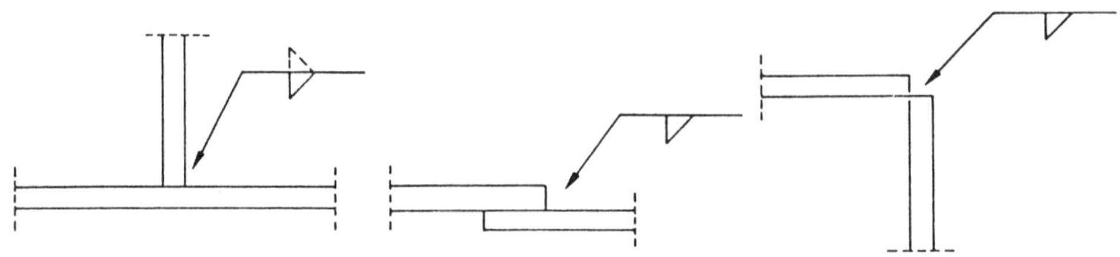

Material Thickness (in.)	.048	.054	.060	.067	.075	.090
Electrode Diameter (in.)	1/16	1/16	1/16	1/16	3/32	3/32
Electrode Angle[(1)]	30°	30°	30°	30°	45°	45°
Electrode Tip Diameter[(1)] (in.)	0.03	0.03	0.03	0.03	0.045	0.045
Arc Current, Amps	50 – 55	59 – 64	68 – 73	77 – 82	86 – 91	95 – 100
Arc Volts, Volts	10 – 12	10 – 12	10 – 12	10 – 12	10 – 12	10 – 12
Filler Wire Diameter (in.)	1/16	1/16	1/16	1/16	3/32	3/32
Travel Speed, ipm	10 – 15	10 – 15	10 – 15	10 – 15	10 – 15	10 – 15
Cup Diameter (in.)	3/8	3/8	3/8	3/8	3/8	3/8
Gas Flow Rate, cfh	20	20	20	20	20	20

(1) See Figure 5-15. *(From AWS B2.1008)*

Filler wire should be ER70S-3 or higher strength filler to match strength of base metal. Low carbon steels will not required a preheat over 100°F; as carbon increases and alloy increases preheat will increase up to 600°F.

APPROXIMATE PARAMETERS FOR GTA WELDING PLAIN CARBON AND LOW ALLOY STEELS FOR FLAT POSITION LAP JOINTS AND HORIZONTAL FILLETS

Electrodes: EWTh2
Electrode Negative
Shielding Gas: 100% argon

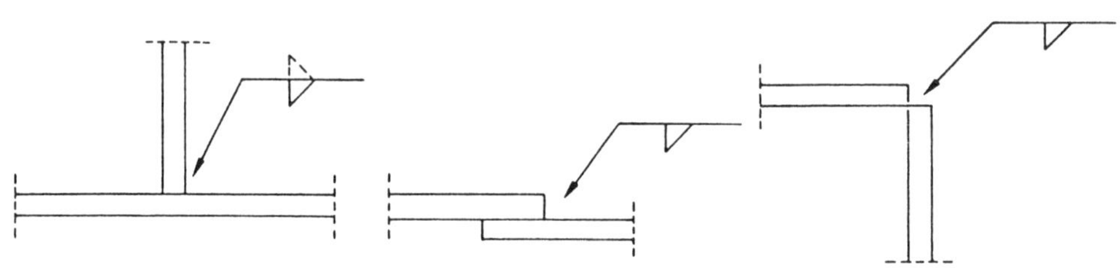

Material Thickness (in.)	.105	.120	.135
Electrode Diameter (in.)	3/32	3/32	3/32
Electrode Angle[(1)]	45°	45°	45°
Electrode Tip Diameter[(1)] (in.)	0.045	0.045	0.045
Arc Current, Amps	105 – 110	115 – 120	125 – 130
Arc Volts, Volts	10 – 12	10 – 12	10 – 12
Filler Wire Diameter (in.)	3/32	3/32	3/32
Travel Speed, ipm	10 – 12	10 – 12	8 – 10
Cup Diameter (in.)	3/8	1/2	1/2
Gas Flow Rate, cfh	20	25	25

(1) See Figure 5-15. *(From AWS B2.1008)*

Filler wire should be ER70S-3 or higher strength filler to match strength of base metal. Low carbon steels will not required a preheat over 100°F; as carbon increases and alloy increases preheat will increase up to 600°F.

APPROXIMATE PARAMETERS FOR GTA WELDING PLAIN CARBON AND LOW ALLOY STEELS FOR FLAT POSITION BUTT JOINTS

Electrodes: EWTh2
Electrode Negative
Shielding Gas: 100% argon

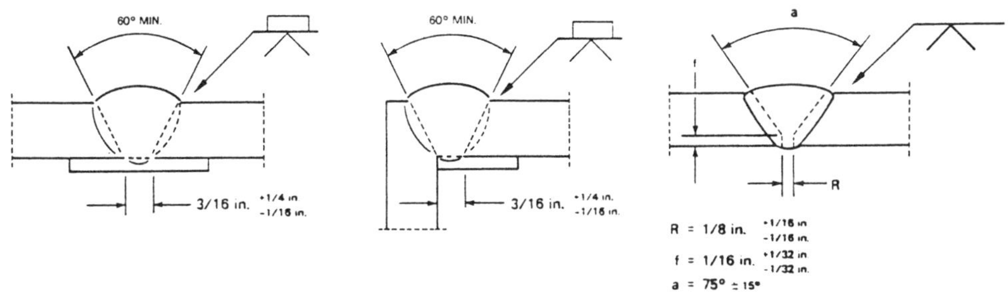

	Pass 1	Pass 2, 3	Pass 1	Pass 2 – 4	Pass 1	Pass 2 – 5
Material Thickness (in.)	3/16	3/16	1/4	1/4	3/8	3/8
Electrode Diameter (in.)	3/32	3/32	3/32	3/32	3/32	3/32
Electrode Angle[1]	45°	45°	45°	45°	45°	45°
Electrode Tip Diameter[1] (in.)	0.045	0.045	0.045	0.045	0.045	0.045
Arc Current, Amps	60 – 130	60 – 130	60 – 130	60 – 130	60 – 130	60 – 130
Arc Volts, Volts	9 – 11	10 – 12	9 – 11	10 – 12	9 – 11	10 – 12
Filler Wire Diameter (in.)	1/16	1/16	1/16	1/16	3/32	3/32
Travel Speed, ipm	8 – 10	8 – 10	8 – 10	8 – 10	7 – 9	7 – 9
Cup Diameter (in.)	1/2	1/2	1/2	1/2	1/2	1/2
Gas Flow Rate, cfh	15 – 25	15 – 25	15 – 25	15 – 25	15 – 25	15 – 25

[1] See Figure 5-15. *(From AWS B2.1002)*
Filler wire should be ER70S-3 or higher strength filler to match strength of base metal. Low carbon steels will not required a preheat over 100°F; as carbon increases and alloy increases preheat will increase up to 600°F.

APPROXIMATE PARAMETERS FOR GTA WELDING PLAIN CARBON AND LOW ALLOY STEELS FOR FLAT POSITION BUTT JOINTS

Electrodes: EWTh2
Electrode Negative
Shielding Gas: 100% argon

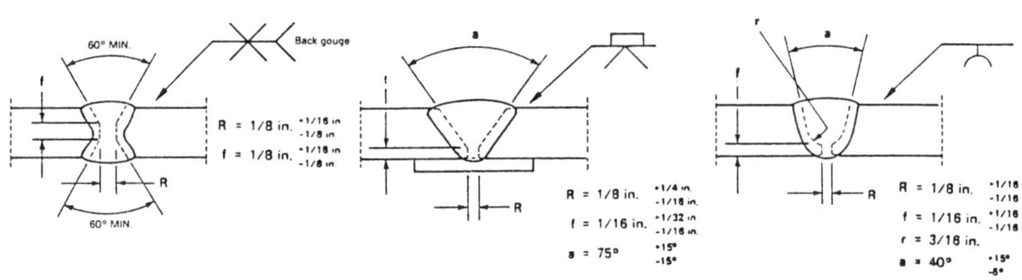

	Pass 1, 2	Pass 3 – 10	Pass 1, 2	Pass 3 – 15	Pass 1, 2	Pass 3+
Material Thickness (in.)	1/2	1/2	5/8	5/8	3/4	3/4
Electrode Diameter (in.)	3/32	3/32	1/8	1/8	1/8	1/8
Electrode Angle[1]	45°	45°	60°	60°	60°	60°
Electrode Tip Diameter[1] (in.)	0.045	0.045	0.045	0.045	0.045	0.045
Arc Current, Amps	60 – 130	60 – 130	90 – 180	90 – 180	90 – 180	90 – 180
Arc Volts, Volts	9 – 11	10 – 12	10 – 12	11 – 13	10 – 12	11 – 13
Filler Wire Diameter (in.)	3/32	3/32	3/32	3/32	3/32	3/32
Travel Speed, ipm	7 – 9	7 – 9	7 – 9	7 – 9	7 – 9	7 – 9
Cup Diameter (in.)	1/2	1/2	1/2	1/2	1/2	1/2
Gas Flow Rate, cfh	15 – 25	15 – 25	15 – 25	15 – 25	15 – 25	15 – 25

[1] See Figure 5-15. *(From AWS B2.1002)*
Filler wire should be ER70S-3 or higher strength filler to match strength of base metal. Low carbon steels will not required a preheat over 100°F; as carbon increases and alloy increases preheat will increase up to 600°F.

APPROXIMATE PARAMETERS FOR GAS TUNGSTEN ARC WELDING MARAGING STEEL FOR FLAT POSITION BUTT JOINTS

Electrodes: EWTh2 Electrode Negative Shielding Gas: 100% argon	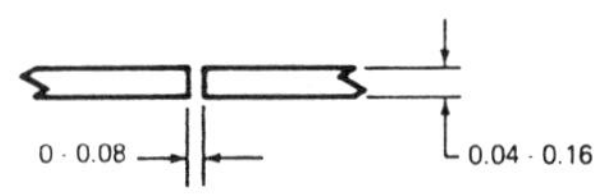	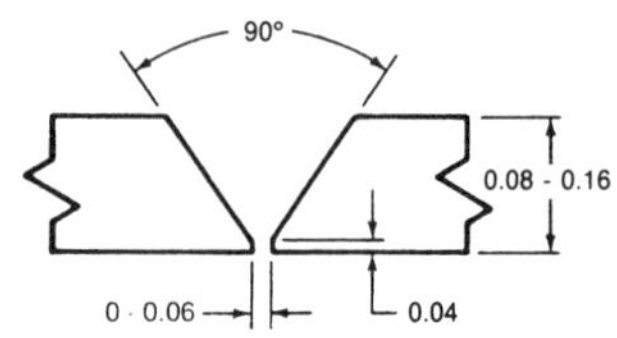		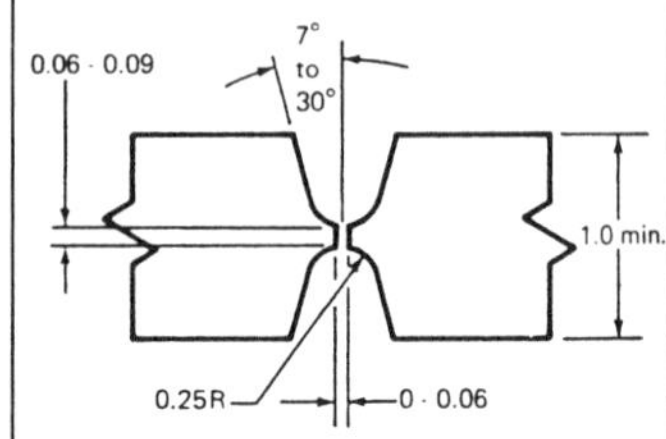	
	Pass 1	**Pass 1**	**Pass 2**	**Pass 1**	**Pass 2 — 10**
Material Thickness (in.)	0.080	0.080	0.080	0.75	0.75
Electrode Diameter (in.)	3/32	3/32	3/32	1/8	1/8
Electrode Angle[(1)]	45°	45°	45°	60°	60°
Electrode Tip Diameter[(1)] (in.)	0.045	0.045	0.045	0.045	0.045
Arc Current, Amps	150	120	150	340	340 - 400
Arc Volts, Volts	8	8	8	10	10
Filler Wire Diameter (in.)	1/16	None	1/16	1/16	1/16
Feed Rate (in./min)	10		10	60	60 - 70
Travel Speed, ipm	7	11	11	12	12 - 15
Cup Diameter (in.)	3/8	3/8	3/8	1/2	1/2
Gas Flow Rate, cfh	20	20	20	30	30

[(1)] See Figure 5-15.
Filler wire should match composition of base metal.

(From AWS WHB Vol. 4, 7th Ed)

APPROXIMATE PARAMETERS FOR GAS TUNGSTEN ARC WELDING MARAGING STEEL FOR FLAT POSITION BUTT JOINTS

Electrodes: EWTh2
Electrode Negative
Shielding Gas: 100% argon

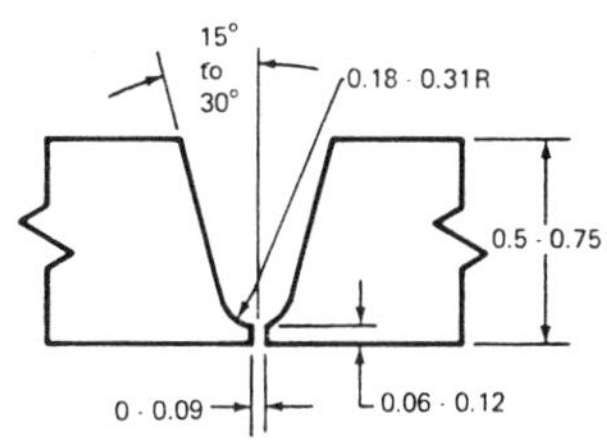

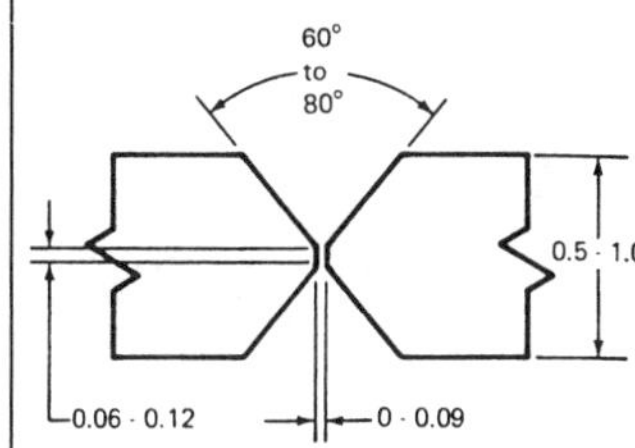

	Pass 1	Pass 2	Pass 3 – 5	Pass 6 – 12	Pass 1	Pass 2 – 30
Material Thickness (in.)	0.5	0.5	0.5	0.5	1.0	1.0
Electrode Diameter (in.)	3/32	3/32	3/32	3/32	3/32	3/32
Electrode Angle[(1)]	45°	45°	45°	45°	45°	45°
Electrode Tip Diameter[(1)] (in.)	0.045	0.045	0.045	0.045	0.045	0.045
Arc Current, Amps	100	150	200	225	230	210 – 230
Arc Volts, Volts	9 – 10	9 – 10	9 – 10	9 – 10	10 – 11	10 – 12
Filler Wire Diameter (in.)	1/16	1/16	1/16	1/16	1/16	1/16
Feed Rate, (in./min)	10	10	20	30	20	20 – 24
Travel Speed, ipm	8	8	8	8	4	4 – 6
Cup Diameter (in.)	1/2	1/2	1/2	1/2	1/2	1/2
Gas Flow Rate, cfh	25[(2)]	25[(2)]	25[(2)]	25[(2)]	25	25

(1) See Figure 5-15.
(2) 80% He – 20% A
Filler wire should match composition of base metal.

(From AWS WHB Vol. 4, 7th Ed.)

The lip on a dragline bucket wears thin and must be repaired. Here the worn areas were cut out and replaced with new plate. The patches are being welded in with a semiautomatic process.

Section 7
WELDING STAINLESS STEEL

SECTION 7.1
WELDABILITY OF STAINLESS STEEL

SECTION 7.2
WELDING STAINLESS STEELS WITH THE SHIELDED METAL-ARC PROCESS

SECTION 7.3
WELDING STAINLESS STEELS WITH THE SUBMERGED-ARC PROCESS

SECTION 7.4
WELDING STAINLESS STEELS WITH THE GAS METAL-ARC PROCESS

SECTION 7.5
WELDING STAINLESS STEELS WITH THE GAS TUNGSTEN-ARC PROCESS

Weldability of Stainless Steel

As a family, the stainless steels offer a combination of mechanical properties, corrosion resistance, and heat resistance unmatched by other commercial metals. Applications range from pots and pans and other kitchenware to complex and sophisticated aircraft landing-gear members. More than 40 wrought compositions of the stainless steels are designated by the American Iron and Steel Institute, and many other stainless alloys — notably, various precipitation-hardening grades — are marketed under proprietary designations.

This section discusses the properties and characteristics of stainless steels, particularly as they affect weldability of these metals. It is a prelude to succeeding sections, in which the selection of welding processes and electrodes is discussed and other information on the welding of stainless steels is presented.

CHEMICAL COMPOSITIONS AND NOMENCLATURE

Chromium is the element that makes stainless steels stainless, or corrosion-resistant. Stainless steels are iron-chromium alloys (most have a low-carbon content) that contain at least 11.5% chromium. The 11.5% level is the content at which effective resistance to atmospheric corrosion begins. Many stainless alloys contain larger amounts of chromium to further improve corrosion resistance and to increase resistance to oxidation at high temperatures.

Addition of elements other than chromium to improve various properties has provided a large range of available mechanical and physical properties. Nickel, for example, in excess of about 6%, increases corrosion resistance slightly and greatly improves mechanical and fabricating properties. Manganese, in conjunction with small amounts of nitrogen, is sometimes used as a replacement for part of the higher price nickel. Small amounts of molydenum increase resistance to pitting type corrosion and general resistance to certain corrosive media. Molydenum also improves high-temperature strength. Carbon increases strength, but in amounts above about 0.04 percent, causes chromium carbide precipitation when exposed to temperatures between 800 and 1600°F. This depletes the matrix in chromium and reduces the corrosion resistance in local areas leading to intergranular corrosion.

Silicon, in larger amounts than used in other alloys, increases oxidation resistance at high temperatures. Sulfur and selenium impart free-machining characteristics. Niobium, (also called columbium) titanium, and tantalum additions stabilize carbides and reduce susceptibility to intergranular corrosion.

AISI STAINLESS STEEL GRADES

Initially, stainless steels were named according to their chromium and nickel contents. Thus, one of the first types developed, which contained 18% chromium and 8% nickel, was called 18-8 stainless. As more stainless steel alloys were developed, this nomenclature became unwieldy. The American Iron and Steel Institute, then, established a numbering system to classify the nonproprietary and generally accepted alloys in the stainless steel family. At the time of publication, AISI recognized as standard alloys 50 grades of stainless steel compositions.

In the AISI system, the chromium stainless steels are assigned numbers in the 400 series, and the chromium-nickel compositions have 300 series designations. The old 18-8 stainless became type 302, and a 17% chromium grade became type 430.

Variations of the standard classification are accommodated in the AISI system by letters following the numbers. For example, low-carbon versions of type 304 and 316 are designated as 304L and 316L. Free machining variations of standard alloys are indicated by the letter F or the symbol for an added element, after the number. Examples are types 430F and 303Se.

Some producers identify their stainless grades by the old alloy designation; some use the AISI numbers. The AISI recognized grades are available from most producers, but grades that are designated simply as 303 or 430 (the same numbers as AISI types) do not necessarily have the same compositions of the corresponding number AISI types. The reason for this is that some producers have developed modifications of certain standard AISI compositions to improve a property or the fabricability of an alloy for a specific application. If such a producer has a good market for this modified grade, it becomes standard with him. But he still may apply the AISI grade number to it, because it is very close to the standard composition. Thus, when the designer specifies a stainless steel grade, he should either specify by the AISI type or he should investigate the suitability of the slightly modified grades available from various producers. Most producers supply the alloys to either specification.

The AISI classifies stainless steels by their metallurgical structures. This system is useful because the structure (austenitic, ferritic, or martensitic) indicates the general range of mechanical and physical properties, formability, weldability, and hardenability. The latest and most up-to-date numbering system for stainless steels incorporates the SAE-ASTM Unified Number System (UNS). Wrought and cast stainless steels are identified by the letters S and J respectively followed by five digits. See Table 7-1.

Austenitic Stainless Steels: The highest tonnage of weldable stainless steels produced are of the austenitic grades. These are the chromium nickel steels of the AISI 200 and 300 series in which chromium usually exceeds 17% and nickel, with a few exceptions, exceeds 7% (Table 7-2). The principal characteristic of this group is the retention of an austenitic structure during cooling from elevated temperatures. The basic grade — type 302 — contains 18% chromium and 8% nickel. More or less of these elements, as well as other alloying additions, have been used to create various alloys in this family for specific end use or fabrication requirements (Fig. 7-1). In the 200 series alloys, manganese and nitrogen are substituted for part of the nickel.

The austenitic types generally have high ductility, low yield strength, and high ultimate strength characteristics that make them suitable for forming and deep drawing operations. These steels can be work hardened to high levels, although not as high as can be obtained by heat treating the hardenable types of the 400 series. In the annealed condition, austenitic grades are nonmagnetic and, depending on composition, they may become slightly magnetic when cold worked. They have excellent properties at cryogenic temperatures and have higher strengths at 1000°F than the 400 series grades. Typical room temperature mechanical properties of austenitic steels in the annealed condition are listed in Table 7-3.

TABLE 7-1. Correlation of Specifications for Stainless Steels

UNS	Wrought Alloys			Cast Alloys		
	AISI	SAE	ASTM	SAE	ACI	ASTM
MARTENSITIC						
S40300	403	51403	Plate, sheet, strip: A176; bar, billet: A276	—	—	—
S41000	410	51410	Plate, sheet, strip: A176, A240; bar, billet: A276; pipe, tube: A268	60410	CA-15	A296
S41400	414	51414	Bar, billet: A276	—	—	—
S41600, S41623	416, 416Se	51416, 51416Se	Bar, billet, rod: A276	—	—	—
S42000	420	51420	Bar, billet: A276	60420	CA-40 —	
S42020	420F	51420F	Bar, wire, rod: A276	—	—	—
S43100	431	51431	Bar, billet, rod: A276	—	—	—
S44002, S44003, S44004	440A, 440B, 440C	—	Bar, billet: A276	—	—	A296

TABLE 7-1. Correlation of Specifications for Stainless Steels (Continued)

UNS	Wrought Alloys			Cast Alloys		
	AISI	SAE	ASTM	SAE	ACI	ASTM
FERRITIC						
S40500	405	51405	Plate, sheet, strip: A240; bar, billet: A276; pipe, tube: A268	—	—	A351
S40900	409	—	Sheet, strip: A176	—	—	—
S42200	422	—	Bar, billet, rod: A276	—	—	—
S42900	429	—	Plate, sheet, strip: A176, A240	—	—	—
S43000, S43020, S43023	430, 430F, 430SeF	51430, 51430F, 51430SeF	Plate, sheet, strip: A176, A240; bar, billet: A276; pipe, tube: A268	—	—	—
S43400	434	—	Sheet, strip, wire: A176	—	—	—
S43600	436	—	Sheet, strip, wire: A176	—	—	—
S44200	442	51442	Plate, sheet, strip: A176; bar, billet: A276; pipe, tube: A268	60442	CB-30	A296
S44600	446	51446	Plate, sheet, strip: A176; bar, billet: A276; pipe, tube: A268	70446	HC	A297
AUSTENITIC						
S20100	201	30201	Sheet, strip: A167; bar, rod: A314	—	—	—
S20200	202	30202	Sheet, strip: A167; bar, rod: A314	—	—	—
S20500	205	30205	Plate, sheet, strip: A167	—	—	—
S30100	301	30301	Plate, sheet, strip, rod: A167, A264 (clad)	—	—	—
S30200	302	30302	Plate, sheet, strip: A167, A240, A264 (clad); rod, bar, billet: A276, A314	—	CF-20	A296
S30215	302B	30302B	Plate, sheet, strip: A167; bar, billet: A276, A314	—	HF	A297
S30300, S30323	303, 303Se	30303, 30303Se	Bar, billet: A276, A314; nuts, bolts: A194, A320	—	CF-16F	A296
S30400, S30403	304, 304L	30304, 30304L	Plate, sheet, strip: A167, A177, A240, A264; bar, billet: A276, A314; pipe, tube: A213, A249, A269, A270, A271, A312, A358, A376; nuts, bolts: A193, A194, A320; forgings, fittings: A182, A336	60304, 60304L	CF-8, CF-3	A296, A351
S30451	304N	30304N	Sheet, strip: A167, A240; bar: A276	—	—	—
S30500	305	30305	Plate, sheet, strip: A240; bar, billet: A314; pipe, tube: A249	—	—	—
S30800	308	30308	Plate, sheet, strip: A167, A264; bar, billet: A276, A314; welding electrodes: A298	—	—	—
S30900, S30908	309, 309S	30309, 30309S	Plate, sheet, strip: A167, A240, A264; bar, billet: A276, A314; pipe, tube: A249, A312, A358; welding electrodes: A298	60309, 70309	CH-20, HH	A296, A297, A351
S31000, S30908	310, 310S	30310, 30310S	Plate, sheet, strip: A167, A240, A264; bar, billet: A276, A314; pipe, tube: A213, A249, A312, A358; forgings, fittings: A182, A336; welding electrodes: A298	60310, 70310	CK-20, HK	A296, A297, A351
S31400	314	30314	Bar, billet: A314	—	—	—
S31600, S31603	316, 316L	30316, 30316L	Plate, sheet, strip: A167, A240, A264; bar, billet: A276, A314; pipe, tube: A213, A249, A269, A312, A358, A376; nuts, bolts: A193: forgings, fittings: A182, A336; welding electrodes: A298	60316, 60316L	CF-3M, CF-8M, CF-12M	A296, A351
S31620	316F	30316F	Sheet: A167, A240; bar: A314	—	—	—
S31651	316N	30316N	Sheet: A167, A240; bar: A276, A314	—	—	—
S31700	317	30317	Plate, sheet, strip: A240; bar, billet: A314; pipe, tube: A249, A269, A312; welding electrodes: A298	60317	CG-8M	A296
S31703	317L	30317L	Sheet, plate: A167, A240; tube: A249, A269	—	—	—
S32100	321	30321	Plate, sheet, strip: A167, A240, A264; bar, billets: A276, A314; pipe, tube: A213, A249, A269, A271, A312, A358, A276; nuts, bolts: A193, A194, A320; forgings, fittings: A182, A336	—	—	—
S32900	329	30329	Strip: A167, A240, A264; bar: A276, A314	—	—	—
S34700	347	30347	Plate, sheet, strip: A167, A240, A264; bar, billet: A276, A314; pipe, tube: A213, A249, A269, A271, A298, A358, A376; nuts, bolts: A193, A194, A320; forgings, fittings: A182, A336; welding electrodes: A298	60347	CF-8C	A296, A351
S34800	348	30348	Plate, sheet, strip: A240; bar, billet: A276, A314; pipe, tube: A213, A249, A269, A358, A376; nuts, bolts: A320	—	—	—
S38400	384	30384	Wire: A313	—	—	—

TABLE 7-2. Typical Compositions of Austenitic Stainless Steels

AISI Type	Composition* (%)			
	Carbon	Chromium	Nickel	Other†
201	0.15	16.0 – 18.0	3.5 – 5.5	0.25 N, 5.5 – 7.5 Mn, 0.060 P
202	0.15	17.0 – 19.0	4.0 – 6.0	0.25 N, 7.5 – 10.0 Mn, 0.060 P
301	0.15	16.0 – 18.0	6.0 – 8.0	—
302	0.15	17.0 – 19.0	8.0 – 10.0	—
302B	0.15	17.0 – 19.0	8.0 – 10.0	2.0 – 3.0 Si
303	0.15	17.0 – 19.0	8.0 – 10.0	0.20 P, 0.15 S (min), 0.60 Mo (opt)
303Se	0.15	17.0 – 19.0	8.0 – 10.0	0.20 P, 0.06 S, 0.15 Se (min)
304	0.08	18.0 – 20.0	8.0 – 12.0	—
304L	0.03	18.0 – 20.0	8.0 – 12.0	—
305	0.12	17.0 – 19.0	10.0 – 13.0	—
308	0.08	19.0 – 21.0	10.0 – 12.0	—
309	0.20	22.0 – 24.0	12.0 – 15.0	—
309S	0.08	22.0 – 24.0	12.0 – 15.0	—
310	0.25	24.0 – 26.0	19.0 – 22.0	1.5 Si
310S	0.08	24.0 – 26.0	19.0 – 22.0	1.5 Si
314	0.25	23.0 – 26.0	19.0 – 22.0	1.5 – 3.0 Si
316	0.08	16.0 – 18.0	10.0 – 14.0	2.0 – 3.0 Mo
316L	0.03	16.0 – 18.0	10.0 – 14.0	2.0 – 3.0 Mo
316N	0.08	16.0 – 18.0	10.0 – 14.0	1.0 Si, 2.0 Mn, 2.0 – 3.0 Mo, 0.10 – 0.16 N
317	0.08	18.0 – 20.0	11.0 – 15.0	3.0 – 4.0 Mo
317L	0.03	18.0 – 20.0	11.0 – 15.0	3.0 – 4.0 Mo
321	0.08	17.0 – 19.0	9.0 – 12.0	Ti (5 × %C min)
329	0.10	25.0 – 30.0	3.0 – 6.0	1.0 – 2.0 Mo
330	0.08	17.0 – 20.0	34.0 – 37.0	0.75 – 1.5 Si, 0.04 P
347	0.08	17.0 – 19.0	9.0 – 13.0	Cb + Ta (10 × %C min)
347M	0.03	19.0	10.0	Cb 13 × C max .65
348	0.08	17.0 – 19.0	9.0 – 13.0	Cb + Ta (10 × %C min but 0.10 Ta max), 0.20 Co

* Single values denote maximum percentage unless otherwise noted.
† Unless otherwise noted, other elements of all alloys listed include maximum contents of 2.0% Mn, 1.0% Si, 0.045% P, and 0.030% S. Balance is Fe.

TABLE 7-3. Typical Properties* of Austenitic Stainless Steels

AISI Type	Room Temperature (annealed)				Max Service Temp (°F) in Air	
	Tensile Strength (1000 psi)	Yield Strength, 0.2% (1000 psi)	Elongation, 2 in. (%)	Hardness Rockwell B	Continuous	Intermittent
201	115	55	55	90	1550	1450
202	105	55	55	90	1550	1450
301	110	40	60	85	1650	1500
302	90	40	50	85	1650	1500
302B	95	40	55	85	1750	1600
304	84	42	55	80	1650	1550
304L	81	39	55	79	1650	1550
305	85	38	50	80	1650	—
308	85	35	50	80	1700	1550
309	90	45	45	85	1950	1850
310	95	45	45	85	2050	1900
314	100	50	40	85	—	—
316	84	42	50	79	1650	1550
316L	81	42	50	79	1650	1550
317	90	40	45	85	1700	1600
317L	85	35	50	80	1700	1600
321	90	35	45	80	1650	1550
347	95	40	45	85	1650	1550

* Mechanical property values are for sheet and strip materials. Properties of bar and plate may vary from these values slightly.

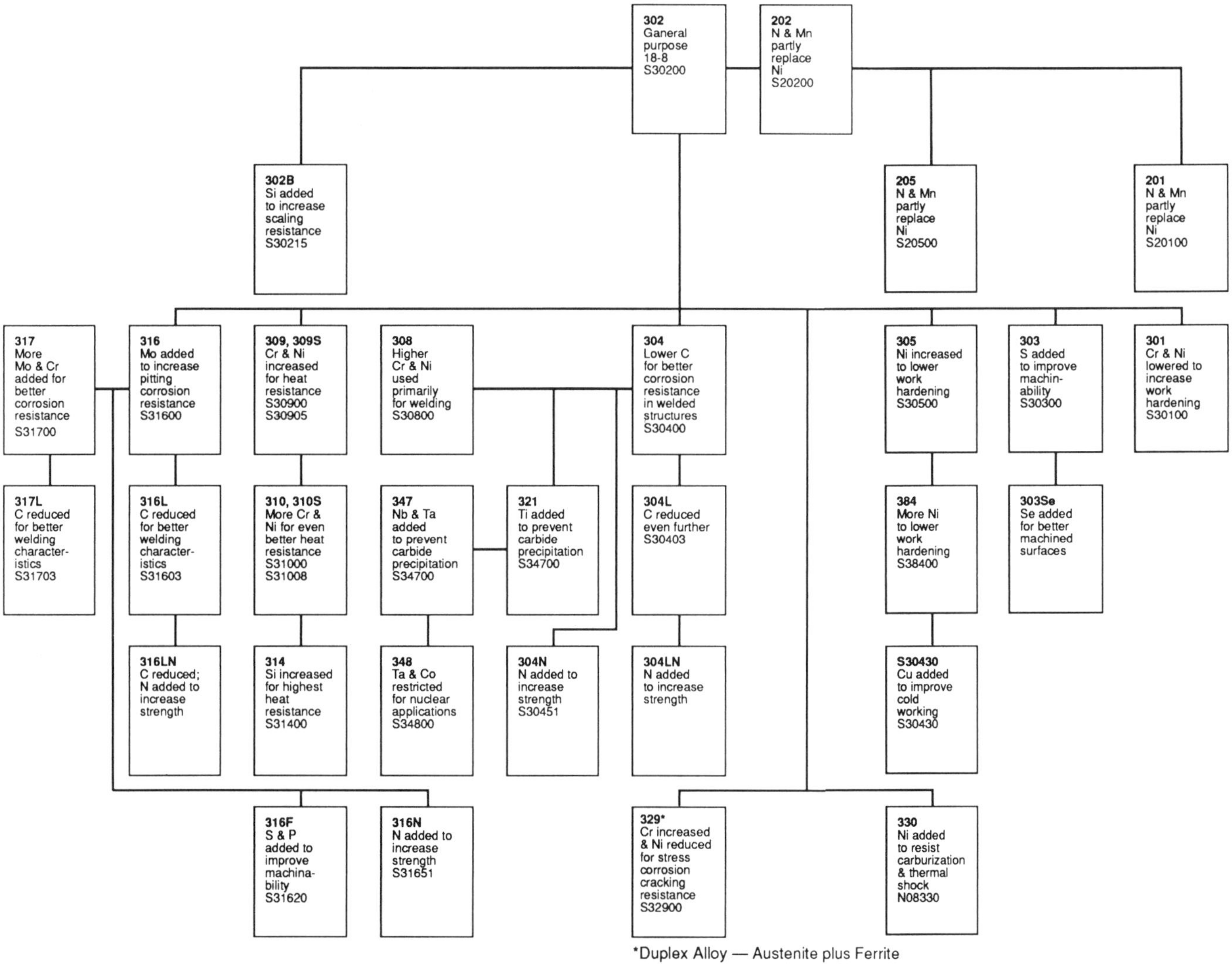

Fig. 7-1. The AISI austenitic stainless steels.

Austentic stainless steels have the highest corrosion resistance of all the stainless steels, particularly when they have been annealed to dissolve the chromium carbides, then quenched to retain the carbon in solution.

Ferritic Stainless Steels: These are the nonhardenable grades of the AISI 400 series that contain from about 14 to 27% chromium and no nickel (Table 7-4). These steels are characterized by high levels of chromium and low carbon, plus additions of titanium and columbium to decrease soluble carbon and additions of powerful ferrite formers such as aluminum that suppresses formation of austenite at high temperatures. Since little or no austenite is present, these grades do not transform to martensite upon cooling, but remain ferritic throughout their normal operating temperature range. For all practical purposes, they are nonhardenable by heat treatment. Typical properties of the ferritic steels are listed in Table 7-5.

Principal applications of the ferritic types are automotive and appliance trim, chemical processing equipment, and products requiring resistance to corrosion and scaling at elevated temperatures, rather than high strength. They are magnetic in all conditions. Figure 7-2 shows how the ferritic stainless steel relate to the basic alloy, type 430, of this group.

Martensitic Stainless Steels: These steels, also in the AISI 400 series, are the iron chromium alloys capable of being heat treated to a wide range of hardness and strength levels. Chromium content is lower and carbon content higher than in the ferritic group. This allows the high temperature ferrite to transform to austenite which subsequently will transform to hard martensite on further cooling. Hardnesses to about Rockwell C 62 and tensile strengths to about 285,000 psi can be developed. These grades are magnetic in all conditions.

Martensitic grades are used to resist abrasion in steam and gas turbine components and for such applications as cutlery and bearings. Also, a modified low carbon 420 stainless is used to repair continuous caster rolls in the steel industry. They are not as corrosion resistant as the austenitic and ferritic types. They have fair cold forming characteristics and can be welded, but usually require preheat and tempering to prevent cracking. The tempered metal suffers a depletion of chromium in solution throughout the structure by precipitation of chromium carbide. This allows the steel to corrode as though it contained a much lower chromium content. Heating to the hardening temperature dissolves the chromium carbides, and rapid quenching prevents their recombining with the carbon and keeps the chromium in solution.

Eleven grades of martensitic stainless steels are classified by the AISI, as shown in Fig. 7-3. Base alloy for the group is type 410. Compositions of these alloys are listed in Table 7-6, and room temperature mechanical properties are given in Table 7-7.

TABLE 7-4. Typical Compositions of Ferritic Stainless Steels

AISI Type	Composition* (%)			
	Carbon	Chromium	Manganese	Other†
405	0.08	11.5 - 14.5	1.0	0.1 - 0.3 Al
409	0.08	10.5 - 11.75	1.0 Max	1.0 Si Max., Ti min. 6×C, 0.045 S, 0.045 P
429	0.12	14.0 - 16.0	1.0	1.0 Si
430	0.12	14.0 - 18.0	1.0	—
430F	0.12	14.0 - 18.0	1.25	0.060 P, 0.15 S (min), 0.60 Mo (opt)
430FSe	0.12	14.0 - 18.0	1.25	0.60 P, 0.060 S, 0.15 Se (min)
434	0.12	16.0 - 18.0	1.0	0.75 - 1.25 Mo, 1.0 Si
436	0.12	16.0 - 18.0	1.0	0.75 - 1.25 Mo, (Cb+Ta) Min 5×%C, 1.0 Si
439	0.07	17.75 - 18.75	1.0	0.6 Si, −0.5 Ni, 0.15 Al, Ti 12×C (1.0 Max)
442	0.20	18.0 - 23.0	1.0	
444	0.025	17.5 - 19.5	1.0	1.0 Ni, 1.75 - 2.5 Mo, 0.035 N max. (Cb+Ta) min, 0.2+4 (% C+ %N)
446	0.20	23.0 - 27.0	1.5	0.25 N
26-1	0.06	25.0 - 27.00	0.75	0.75 Ni, 0.75 - 1.5 Mo, 0.2 - 1.0 Ti 0.04 N, 0.2 Cu, 0.025, 0.75 Si
29-4	0.01	28.0 - 30.0	0.3	0.15 Ni, 3.5 - 4.2 Mo, 0.15 Cu 0.02 N, 0.2 Si, 0.025 P, 0.02 S
29-4-2	0.01	28.0 - 30.0	0.3	2.0 - 2.5 Ni, 3.5 - 4.2 Mo, 0.15 Cu 0.02N, 0.2 Si, 0.025P, 0.02S

* single values denote maximum percentage unless otherwise noted.
† Unless otherwise noted, other elements of all alloys listed include maximum contents of 1.0% Si, 0.040% P, and 0.030% S, Balance is Fe.

TABLE 7-5. Typical Properties* of Ferritic Stainless Steels

AISI Type	Room Temperature (annealed)				Max Service Temp (°F) in Air	
	Tensile Strength (1000 psi)	Yield Strength, 0.2% (1000 psi)	Elongation, 2 in. (%)	Hardness Rockwell B	Continuous	Intermittent
405	70	40	30	81	1400	1450
409	65	35	32	72		
429	60 - 85	35 - 55	20 - 35	70 - 85		
430	75	45	30	83	1550	1650
430F, 430FSe	80	55	25	87	1500	1600
434	75 - 85	50 - 60	22 - 31	78 - 86		
436	65 - 75	40 - 50	25 - 32	70 - 80		
439	78	50	30	200 Brinell		
442	80	45	20	90	1800	1900
446	80	50	25	86	1950	2050

* Mechanical property values are for bar materials. Properties of sheet, strip or plate may vary from these values slightly.

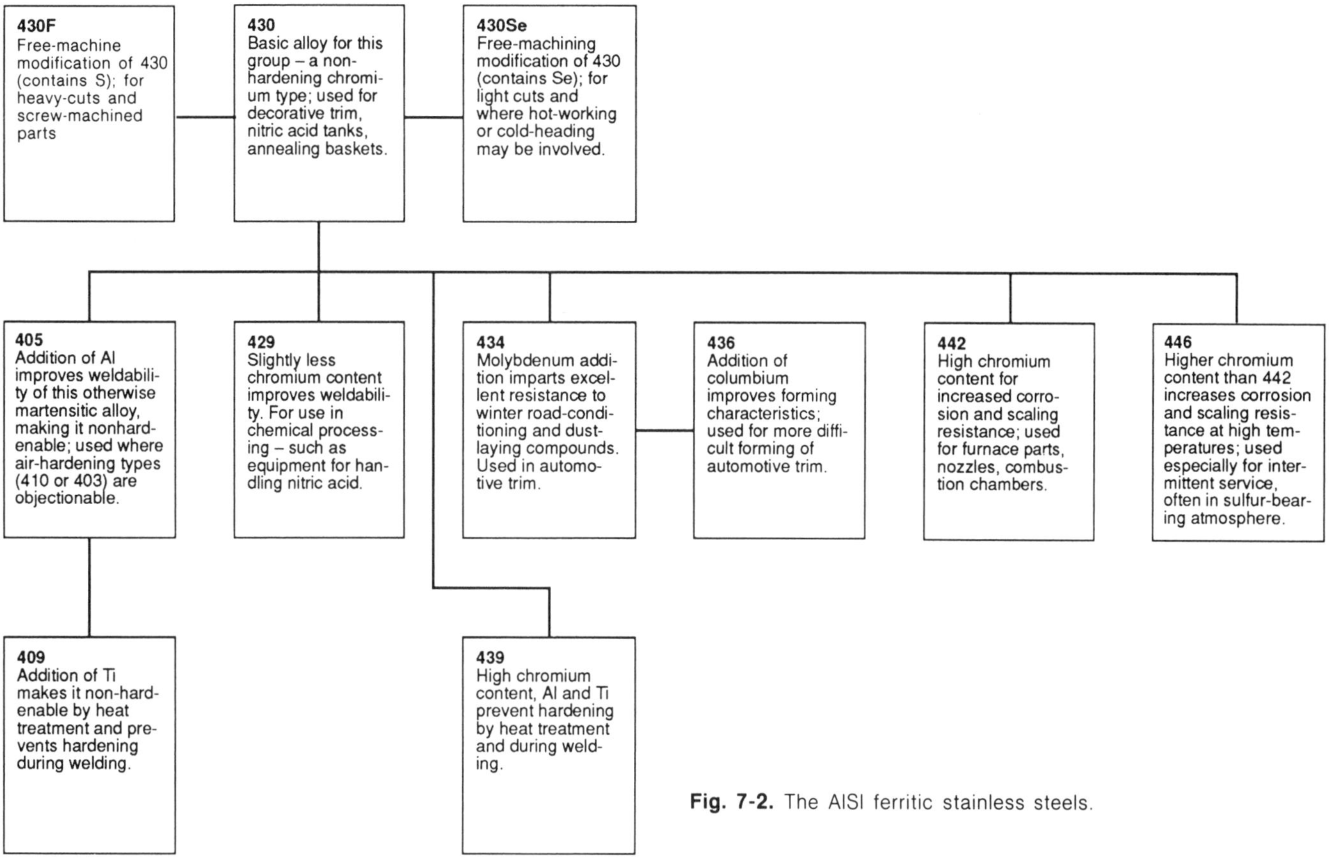

Fig. 7-2. The AISI ferritic stainless steels.

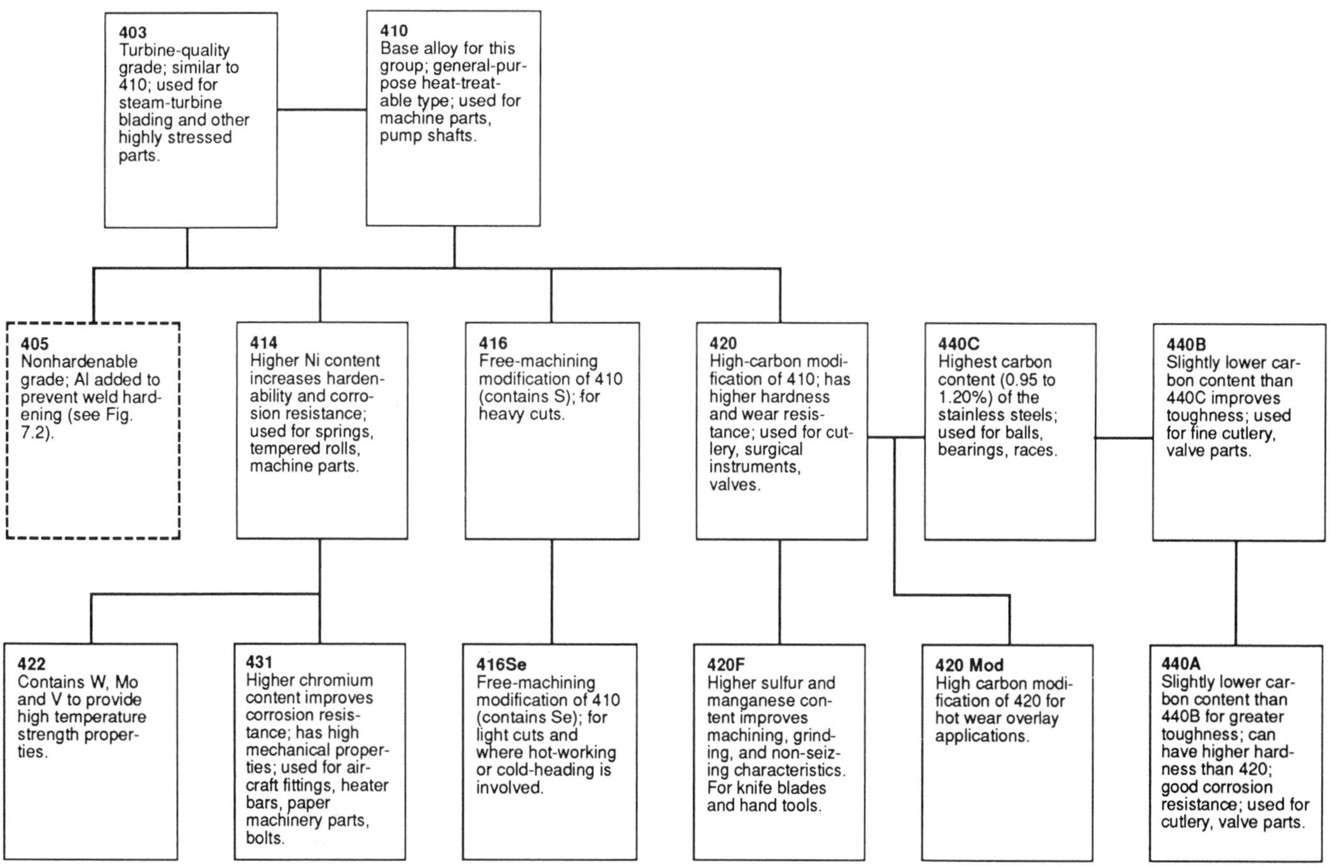

Fig. 7-3. The AISI martensitic stainless steels.

TABLE 7-6. Typical Compositions of Martensitic Stainless Steels

AISI Type	Composition* (%)		
	Carbon	Chromium	Other
403	0.15	11.5 - 13.0	0.5 Si
410	0.15	11.5 - 13.5	—
414	0.15	11.5 - 13.5	1.25 - 2.5 Ni
416	0.15	12.0 - 14.0	1.25 Mn, 0.15 S (min), 0.060 P, 0.60 Mo (opt)
416Se	0.15	12.0 - 14.0	1.25 Mn, 0.060 P, 0.15 Se (min)
420	0.15 min	12.0 - 14.0	—
422	0.2 - 0.25	11.0 - 13.0	1.0 Mn, 0.5 - 1.0 Ni, 0.75 Si, 0.75 - 1.25 Mo, 0.75 - 1.25 W, 0.15 - 0.3 V, 0.025 P, 0.025 S
431	0.20	15.0 - 17.0	1.25 - 2.5 Ni
440A	0.60 - 0.75	16.0 - 18.0	0.75 Mo
440B	0.75 - 0.95	16.0 - 18.0	0.75 Mo
440C	0.95 - 1.20	16.0 - 18.0	0.75 Mo

* Single values denote maximum percentage unless otherwise noted.
† Unless otherwise noted, other elements of all alloys listed include maximum contents of 1.0% Mn, 1.0% Si, 0.040% P, and 0.030% S. Balance is Fe.

TABLE 7-7. Typical Properties* of Martensitic Stainless Steels

AISI Type	Room Temperature (annealed)				Max Service Temp (°F) in Air	
	Tensile Strength (1000 psi)	Yield Strength, 0.2% (1000 psi)	Elongation, 2 in. (%)	Hardness Rockwell B†	Continuous	Intermittent
403	75	40	35	82	1300	1450
410	75	40	35	82	1300	1450
414	115	90	20	97	1300	1450
416, 416Se	75	40	30	82	1250	1400
420	95	50	25	92	1200	1400
431	125	95	20	24 (Rc)	1500	1600
440A	105	60	20	95	1400	1500
440B	107	62	18	96	1400	1500
440C	110	65	14	97	1400	1500

* Mechanical property values are for bar materials. Properties of sheet, strip or plate may vary from these values slightly.
† Rockwell B unless otherwise noted.

Duplex Stainless Steels: The duplex stainless steels typically consist of a microstructure of about 50 percent ferrite and 50 percent austenite but may range from 20 to 80 volume percent ferrite. They were developed to provide a higher strength corrosion-resistant alternative to the 300 series austenitic stainless steels. Compositions are modified to favor the high ferrite levels by increasing the chromium to 22-26%, increasing molybdenum to 2-5%, decreasing the nickel to 4-8% and adding copper up to 2%. These compositions, Table 7-8, provide excellent resistance to pitting, crevice corrosion and stress corrosion cracking. These compositions also provide useful mechanical properties, Table 7-9, over a temperature range from as low as −50°F to 500°F with yield strength nearly twice that of wrought austenitic stainless steel and ductility and toughness approaching those of austenitic stainless steel. The preferred 50 percent ferrite — 50 percent austenite microstructure of the duplex stainless steels is obtained by heat treatment, hot working or a combination of heat treatment and hot working a structure which is initially much higher in ferrite content. The normal heat treatment is in the temperature range of 1900 to 2100°F. This causes part of the original delta ferrite to transform to the more stable austenite form by a diffusion reaction.

A weld made with matching filler metal composition or a fusion weld made without filler metal addition will exist primarily as ferrite unless followed by heat treatment or hot working. The high ferrite welds tend to be very brittle and generally will crack. A post weld annealing treatment is required to produce a balanced ferrite-austenite microstructure and restore ductility. It should be noted that while increasing ferrite content in the 5 to 10 percent level reduces hot cracking tendency for the 300 series steels, ferrite levels over 60 percent increase hot cracking susceptibility.

The unusual combination of mechanical prop-

TABLE 7-8. Composition of Duplex Stainless Steels

Alloy	C	Cr	Ni	Mn	Mo	N	Si	Other
255 (Ferralium)	0.04 m	24.0 - 27.0	4.5 - 6.5	1.5 m	2.0 - 4.0	0.10 - 0.25	1.0 m	1.5 - 2.5 Cu, 0.04 m P, 0.03 m S
2205 (Sandvik)	0.03 m	21.0 - 23.0	4.5 - 6.5	2.0 m	2.5 - 3.5	0.08 - 0.20	1.0 m	0.03 m P, 0.02 m S
2304 (Sandvik)	0.03 m	23.0	4.0	2.5 m	—	0.1	1.0 m	
NU744LN (Uddeholm)	0.067	21.6	4.9	1.7	2.4	0.10	0.44	0.2 Cu, 0.001 S, 0.024 P
21-9 (Experimental)	0.024	20.8	9.0	1.7	0.07	0.007	0.62	0.008 S, 0.025 P
23-7 (Experimental)	0.033	22.8	7.1	1.8	0.06	0.006	0.54	0.009 S, 0.027 P

TABLE 7-9. Typical Properties of Duplex Stainless Steels

Alloy Type	Tensile Strength 1000 psi	Yield Strength 1000 psi	Elongation % in 2 in.	Hardness Vickers
255 (Ferralium)	110 min - 125	80 min - 97	15 min - 27	
2205 (Sandvik)	98 - 130	65 min	25 min	260
2304 (Sandvik)	87 - 120	58 min	25 min	230

erties and corrosion resistance of the duplex stainless steels makes them attractive for many industrial applications including piping for chemical process plants, oil and gas transmission lines and structures for marine environments. Since welding is required as a process for fabricating components of these steels, a knowledge of the factors effecting their weldability and successful techniques for welding them is critical to their industrial application.

A considerable amount of investigation has been done on the weldability of both commercial and experimental duplex stainless steels to determine the effects of filler metal composition and welding procedure on weld properties and hot cracking tendency. Metallographic examination, tensile tests, Charpy impact tests and Varestraint tests were used to evaluate the welds by several investigators. Mechanical test results showed that when ferrite content is below 60 EFN (approximately 60%) in nearly matching weld deposits for alloys 255 and 2205, sufficient ductility and toughness would be attained to pass a side bend test and provide a Charpy "V" notch energy of 20 ft-lbs at −50°F. Weld deposits of nearly matching compositions with ferrite contents of over 30 EFN (approximately 30%) provided tensile strength and yield strength equivalent to that of base metal.

While molybdenum is added to provide resistance to pitting and crevice corrosion, high levels (4.0% and over) cause embrittlement of the deposit, even for low ferrite levels.

Varestraint test results showed that the duplex stainless steels are more susceptible to cracking than a type 304L stainless steel of FN 4.5 but less susceptible than a type 304L stainless steel of FN 0. The two experimental duplex alloys identified as 21-9 and 23-7 in Table 7-8 were less susceptible to weld metal hot cracking than the commercial duplex stainless. Both of these experimental alloys are essentially free of Mo and Cu. This leads to the belief that Mo and Cu form low melting phases with iron which could cause the hot cracking susceptibility.

WELDING THE AISI STAINLESS STEELS

Weldability of stainless steels takes into account not only the usual mechanical properties, but also the chemical characteristics that affect corrosion resistance. Thus, the choice of welding processes is limited because of possible reactions of chromium with carbon and oxygen at welding temperatures. Presently, more welding of stainless steels is done by manual shielded-arc methods than any other process. However, with the availability of semiautomatic and automatic equipment and electrodes, the trend is toward automatic welding. Electrodes for welding stainless steels are available in a wide variety of alloys.

Austenitic Grades

Most weldable of the AISI stainless steels are the austenitic grades, but even these alloys have limitations that require careful attention during processing. Austenitic steels have a high coefficient of thermal expansion — over 50% higher than that of carbon steel or of the 400-series alloys — which demands maximum care to minimize distortion and warping of welded parts. Some of these alloys are susceptible to the formation of sigma phase when exposed to certain high temperature ranges, which can cause cracking and reduce corrosion resistance under certain conditions. Welding can also cause carbide precipitation in some stainless grades which decreases the corrosion resistance in some chemical media.

Ferrite and Sigma Phase

Austenitic stainless steels may be susceptible to hot cracking (micro-fissuring) if the ferrite content in the weld metal is not properly controlled. This problem is corrected by using electrodes that deposit weld metal containing a small amount of ferrite. Thus, recommended electrodes for many standard austenitic grades may deposit weld metal that contains 3 to 15% ferrite even though the same grade base metal contains no ferrite. Since ferrite is magnetic, it is easily detected in an otherwise nonmagnetic weldment.

Ferrite is best determined by measurement with a magnetic instrument calibrated to AWS A4.2. It can also be estimated from the composition of the base material and filler material with the use of any of several constitution diagrams. The oldest of these is the 1948 Schaeffler Diagram. The Cr equivalent (% Cr + % Mo + 1.5 × % Si + 0.5 × % Cb) is plotted on the horizontal axis and the nickel equivalent (% Ni + 30 × % C + 0.5 × % Mn) on the vertical axis. Despite long use, the Schaeffler Diagram is now outdated because it does not consider nitrogen effects and because it has not proven possible to establish agreement among several measures as to the ferrite percent in a given weld metal.

An improvement on the Schaeffler Diagram is the 1973 WRC-DeLong Diagram (Fig. 7-4), which can be used to estimate ferrite level. The main differences are that the DeLong diagram includes nitrogen (N) in the Ni equivalent (% Ni + 30 × % C × 30 × % N + 0.5 × % Mn) and shows Ferrite Numbers in addition to "percent ferrite." Ferrite Numbers at low levels may approximate "percent ferrite." The most recent diagram, the WRC-1988 Diagram (Fig. 7-5), is considered to be the most accurate predicting diagram at present. Its Ni equivalent (% Ni + 35 × % C + 20 × % N) and Cr equivalent (% Cr + % Mo + 0.7 × % Cb) differ from those of Schaeffler and WRC-DeLong.

Ferrite Number may be estimated by drawing a horizontal line across the diagram from the nickel equivalent number and a vertical line from the chromium equivalent number. The Ferrite Number is indicated by the diagonal line which passes through the intersection of the horizontal and vertical lines.

Predictions by the WRC-1988 and WRC-DeLong Diagrams for common grades like 308 are similar, but the WRC-1988 diagram generally is more accurate for less common grades like high manganese austenitic or duplex austenitic-ferritic stainless steels.

Ferrite Number can be measured directly on weld deposits from the magnetic properties of the ferrite. Several instruments are available commercially, including the Magne gage, the Severn gage, the Inspector gage and the Ferritescope which can be calibrated to AWS A4.2 and provide readings in Ferrite Number.

The amount of ferrite normally should not be greater than necessary to prevent hot cracking with some margin of safety. The presence of ferrite can cause formation of sigma phase.

Sigma phase is a crystallographic constituent that forms slowly at high temperatures in high-chromium (20% or more) ferrites in the ferritic portions of unbalanced austenitic alloys and sometimes in austenite itself. Sigma phase increases hardness, but it decreases ductility, notch-toughness and corrosion resistance of stainless steels.

Because of its slow development (in the range of approximately 1000 to 1600°F) sigma phase is primarily a service problem where long exposures at high temperatures are involved. It does not normally develop during welding or even during heat treating. The phase can be redissolved by heating to above 1650°F for a relatively short time.

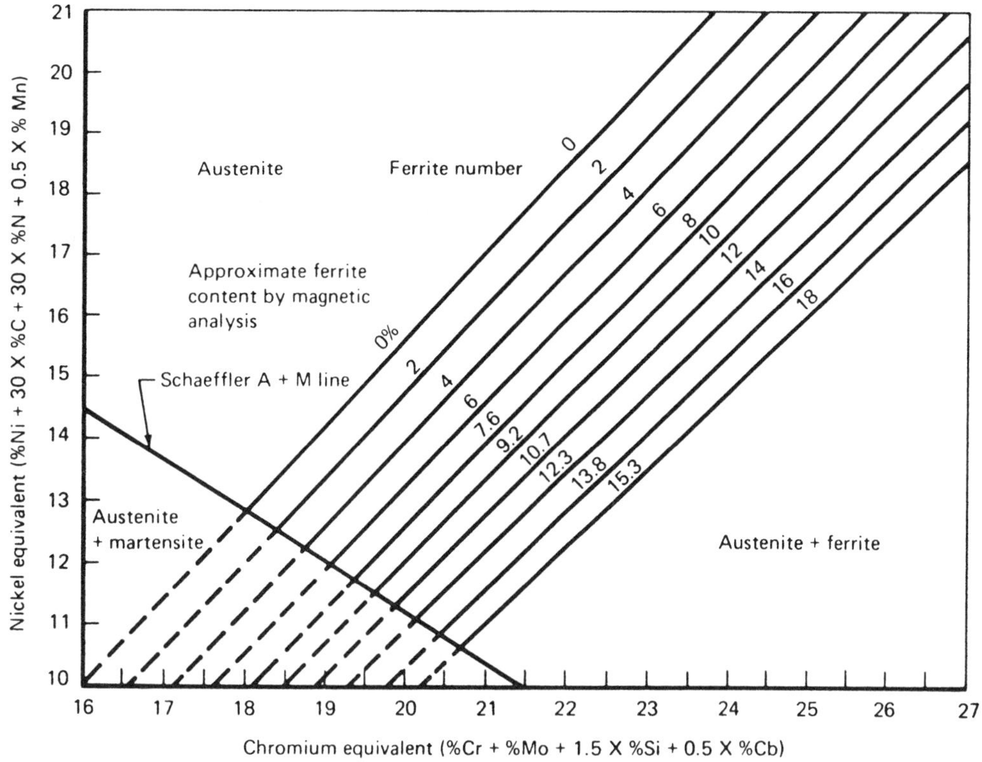

Fig. 7-4. The DeLong constitution diagram for austenitic stainless steel weld metal.

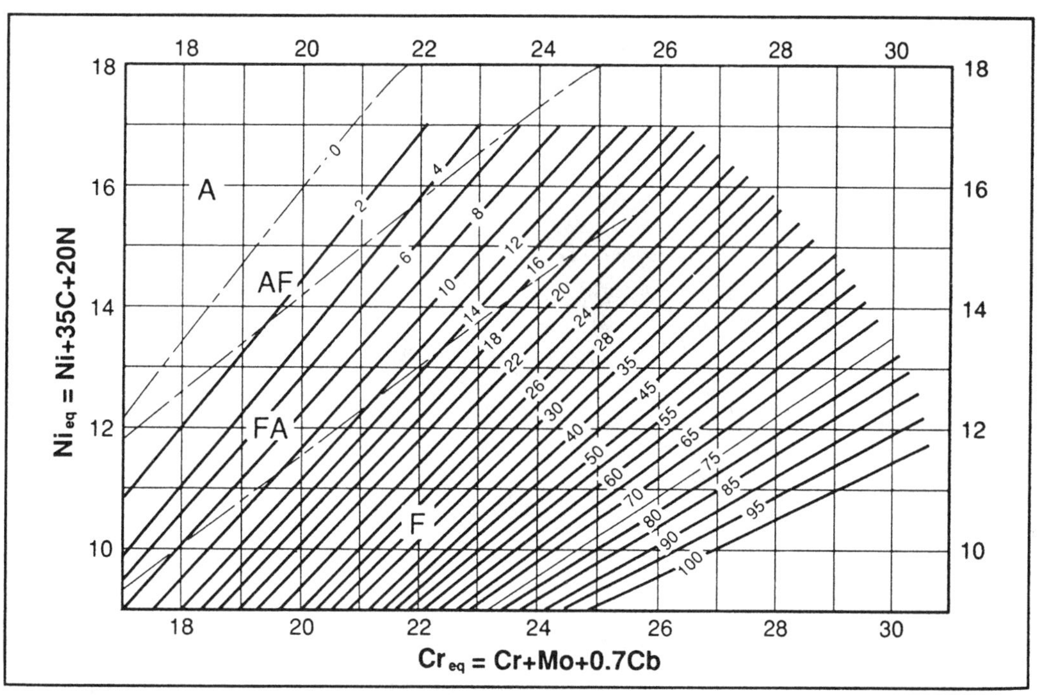

Fig. 7-5. New 1988 WRC Diagram including solidification mode boundaries. (From T.A. Stewert, C.N. McCowan and D.L. Olson, *Welding Journal*, Dec. 1988).

Carbide Precipitation

Although mechanically satisfactory welds can be made on the chromium-nickel (austenitic) stainless steels, heating these materials sometimes promotes intergranular corrosion. When the austenitic steels are heated in the range of 800 to 1600°F (the sensitizing temperature range), or cooled slowly through that range, carbon is precipitated from solid solution (mainly at grain boundaries) and unites with chromium to form chromium-rich carbides. These chromium carbides may contain as much as 90% chromium, taken from the metal immediately adjacent to the grain boundary. The chromium content of the adjacent metal is thus reduced, and corrosion resistance is seriously impaired. This phenomenon is termed "carbide precipitation," and the type of corrosion it promotes is known as "intergranular corrosion" (Fig. 7-6). Stainless steels having chromium contents to about 18% are most susceptible to carbide precipitation.

The thermal conditions that produce carbide precipitation can occur during welding — particularly in multiple-pass welding, welding of heavy beads, or when two beads cross.

The amount of carbide precipitation is reduced by rapid cooling. Fortunately, the copper chill bars, skip welding and other techniques needed to control distortion in sheet metal help reduce carbide precipitation.

Thermal Treatments: Various methods are used to reduce or prevent carbide precipitation in austenitic stainless steels. One is to heat the weldment to the range of 1850 to 2100°F and rapidly cool it (by quenching) through the 1600 to 800°F range. This thermal treatment redissolves the precipitated carbides (puts the carbon back into solution in the austenite and restores the chromium at the grain boundaries), and the rapid quench prevents precipitation from recurring. This method has disadvantages, however. The high-temperature treatment may cause distortion in welded assemblies, and large welded structures cannot be accommodated in heating furnaces.

However, if the weldment is to be used in the sensitization temperature range, carbide precipitation can occur in service after the heat treatment. Therefore, when weldments operate in severe corrosive applications or within the sensitizing temperature range, either ELC or stablilized grades are needed.

Low-Carbon Stainless Steels: Another remedy to the carbide-precipitation problem is to use stainless steels and electrodes having such a low carbon content that no carbides can be precipitated. The 18-8 austenitic steels retain about 0.02% carbon in solid solution under all conditions. With carbon content increased to about 0.08%, the amount of carbon that can be precipitated increases slowly; above that amount, precipitation can occur rapidly (when the material is exposed to the sensitizing temperature).

Extra-low-carbon plate and electrodes cost slightly more than the regular grades, but results are well worth the cost difference, particularly when weldments are to be used in the as-welded condition. Carbide precipitation decreases significantly in changing, from AISI type 304 (0.08% C) to 304L (0.03% C). Low-carbon welding electrodes are also available.

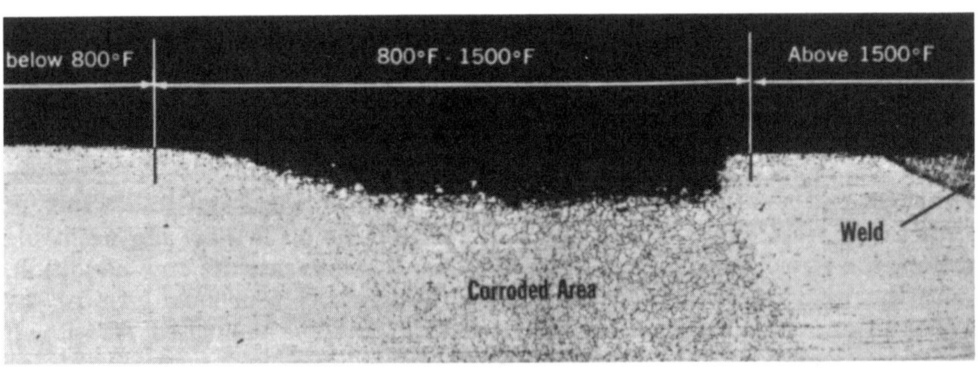

Fig. 7-6. The temperature gradient naturally produced at a weld makes an area adjacent to the bead subject to the 800°F to 1600°F sensitizing temperature. Heating in the sensitizing range causes chromium carbides to precipitate in the grain boundaries. Precipitation of the carbides reduces the chromium content in solid solution in the area adjacent to the grain boundaries making the steel subject to inter-granular corrosion.

The amount of carbide precipitation is reduced by promoting rapid cooling. Fortunately, the copper chill bars, skip welding and other techniques needed to control distortion in sheet metal help reduce carbide precipitation. Annealing the weldment eliminates carbide precipitation, but this is an expensive and often impractical procedure. Therefore, when weldments operate in severe corrosive applications or within the sensitizing temperature range, either ELC or stabilized grades are needed.

Use of low-carbon plate and electrodes may produce a small amount of carbide precipitation, but usually not enough to be harmful. However, when the temperature (either during welding or in service) and corrosion conditions are severe, stabilized electrodes and base metal should be used.

Stabilized Steels

Steels in which chromium carbide forms readily upon heating to the sensitizing temperature range (800 to 1600°F) are called unstabilized steels. Examples of these are AISI 301, 302, 302B, 303, 304, and 308. Since carbide precipitation increases with carbon content, grades 304 and 308 are the best of the 18-8 steels for welding. Maximum carbon content in these two grades is 0.08%; the others listed have a maximum limit of 0.15%.

If types 304 and 308 stainless steels are single-pass welded, the time in the 800 to 1600°F range is so short that very little carbide precipitation can take place. Thus, intergranular corrosion is not likely to occur in these materials (provided, of course, that service temperature of the weldment is not in the sensitizing range).

If the steel is thick enough to require multiple-pass welds, or if the finished product will operate between 800 and 1600°F (1200°F is the most critical point in the range), carbide precipitation is likely to occur, even in grades 304 and 308. Intergranular corrosion will then follow if the product is subjected to a corrosive medium.

The most common way of preventing intergranular corrosion, especially where critical temperatures will be reached in service or where environmental conditions are severe, is to prevent the formation of chromium carbides by using stainless steels that contain columbium or titanium. These elements have a greater affinity for carbon than chromium does; thus they form columbium carbide or titanium carbide, leaving the chromium in solution where it can do its intended job of providing corrosion resistance.

Stainless steels that contain columbium or titanium (AISI types 347 and 321) are called "stabilized" steels since they are not made susceptible to intergranular corrosion by heating. They can be used in the 800 to 1600°F range with no effect on corrosion resistance, and no heat treating is required.

Welding of stabilized stainless steels should be done with stabilized electrodes for best results. Since columbium transfers through an arc much more effectively than titanium, stabilization of electrodes is achieved with columbium additions.

Ferritic Grades

These grades have what is metallurgically known as a ferritic microstructure. They are magnetic and nonhardenable by heat treatment. Typical applications include automobile trim and mufflers, interior building trim, and industrial equipment requiring a low-cost material with good corrosion resistance built with limited welding.

The heat of welding causes embrittlement associated with severe grain growth in the heat-affected zone and fusion zone. When ferritic-type stainless steel weldments are to be annealed, one should weld with the straight chromium-type electrodes listed in Table 7-16.

Martensitic Grades

These grades have what is metallurgically known as a martensitic microstructure. They are magnetic and can be heat treated to a wide range of mechanical properties. Some grades have good strength and scaling resistance to about 1100°F. Typical applications include razor blades, surgical instruments, and industrial equipment requiring specific hardnesses or high strength and good corrosion resistance. Continuous caster rolls are repaired with a modified 420 stainless electrode.

Rapid cooling from welding temperatures produces a hard, brittle zone next to the weld. As hardness of the zone increases, the tendency for weld cracking also increases. To control cracking, particularly on steel over 1/8-in. thick and a carbon content above .10%, preheating is needed.

When martensitic steel weldments are to be heat treated to a specific strength level, they must be welded with straight chromium-type electrodes, as listed in Table 7-16.

Electrode Selection Without Heat Treating

Ferritic or martensitic steel weldments to be used in the "as-welded" condition should be welded with E308, E309, or E310 electrodes. The ductile chromium-nickel welds resist cracking from deformation and impact better than if the weld and heat-affected zone were both brittle. However, differences in thermal expansion rates, weld and base metal color, and other physical properties may require selection of a straight chromium-type electrode.

Duplex Stainless Steels

The duplex stainless steels attain their duplex structure of roughly 50% ferrite and 50% austenite after an annealing heat treatment or hot working. Weld deposits in these alloys with matching filler metals solidify and remain nearly 100% ferrite. This single phase deposit is much more susceptible to hot cracking than a mixed deposit of ferrite and austenite. A post weld annealing treatment would cause the ferrite to partly transform to austenite and improve the ductility. However, hot cracking normally would occur before the heat treatment could be accomplished. In many cases, it would not be acceptable or feasible to heat treat the entire welded assembly.

The recommended approach is to use filler metals which nearly match the base metal but provide an as-welded deposit of a duplex structure such as 50% ferrite and 50% austenite. This is accomplished by increasing the nickel content from the base level of 4.5-6.5% to 8-10%. For alloy 255, flux cored wire of about 10% nickel with remainder of matching composition is recommended to provide appropriate ferrite and excellent mechanical properties in the as-welded condition.

For alloy 2205, flux cored wire of about 8.5% Ni with remainder of matching composition is recommended to provide appropriate ferrite and excellent mechanical properties in the as-welded condition.

Preheating and Postheating

Austenitic stainless-steels are best welded without preheat except to reduce shrinkage stresses on thick sections or restrained joints. No preheat, low interpass temperature, or a stringer-bead technique reduce the time the heat-affected zone is in the sensitizing range (800-1600°F), thereby reducing the amount of carbide precipitation. Usually no postheating is required for the austenitic stainless steels except when an anneal is needed to dissolve the chromium carbides.

The martensitic grades can be welded, although the weldability is not as good as that of the austenitic grades. In the as-welded condition, there is a hard martensitic heat-affected zone, and usually a preheat is required. A rule of thumb is up to 0.10% carbon requires no preheat or postheat. In the 0.10 to 0.20% carbon range, 300°F preheat is required, and the weld should be cooled slowly. Over 0.20% and up to 0.50% carbon requires a preheat of 500°F and a tempering after welding. If a hardening and tempering operation follows the welding, this should be started immediately after the welding is finished. Some reduction in hardness can be had by heating to 900-1200°F and air cooling. For full anneal, heat to 1500-1600°F, furnace cool to 1100°F, and then air cool.

The welding quality of the ferritic-type stainless steels is the poorest of the AISI types. The welding heats a zone in the base metal above a critical temperature (1750°F) and causes rapid grain growth of the ferrite. This coarse grain zone lacks ductility and toughness, and a small amount of martensite may be present, which adds hardness. Postheating reduces the hardness, but the objectional properties due to the coarse grain structure still remain. Any preheating is done only to reduce shrinkage stresses and avoid cracking on cooling. Do not preheat thin sections. preheat thick sections, 1/4-in. or more, 200 to 400°F, depending on the thickness and the amount of restraint. Postheat at 1450°F and furnace cool to 1100°F; then cool rapidly.

Free-machining modifications of the above three stainless categories tend to lead toward porosity and segregation during welding. This can often be avoided by proper electrode selection and taking precautions to make sure the electrode covering is completely dry.

Precipitation-Hardening Stainless Steels

The next category of stainless steels — the precipitation-hardening grades — offers a combination of properties not attainable in other alloys. There are stronger steels; there are steels that have greater corrosion resistance; and there are steels that are easier to fabricate. But few materials combine all of these characteristics as do the precipitation-hardening (also called "age-hardening") stainless steels.

These proprietary steels are used in applications requiring high strength (from 125,000 to nearly 300,000 psi, depending on heat treatment), good fracture toughness, resistance to corrosion (corrosion resistance in most environments is about equal to that of AISI types 302 and 304), and resistance to oxidation at elevated temperatures. They are among the most fabricable of the high-strength materials. They can be worked by most conventional methods while in a very low-strength condition, then strengthened by a low-temperature (900 to 1150°F) heat treatment.

Precipitation hardening is essentially a method of increasing hardness and strength of a metal. Although some variations apply to specific grades, precipitation hardening is generally accomplished by a three-step heat treatment consisting of solution treatment, rapid cooling, and controlled reheating (aging).

The first step (solution heat treating) dissolves certain elements such as copper, titanium, niobium, and aluminum, that are normally insoluable at room temperature. This mechanism might be compared with the ability of hot water to dissolve more salt than can cold water.

The second step (quenching) cools the metal rapidly to retain the solution effect at room temperature. This condition is sometimes called a super-saturated solid solution.

The third step involves reheating of the super-saturated metal to a relatively low aging temperature (about 900°F for some grades) for about an hour. This effects a uniform, submicroscopic precipitation of the special elements (or compounds) throughout the structure. This hardens and strengthens the metal.

The three classes of precipitation-hardenable grades of stainless steels are martensitic, austenitic, and semiaustenitic. Composition of representative materials are listed in Table 7-10, and descriptions and typical applications appear in Table 7-11.

The semiaustenitic grades are basically sheet materials, because their austenitic structure (before heat treatment) provides good formability. The martensitic and austenitic types are generally classed as bar and plate materials — used mainly for machined parts and weldments. These distinctions are not entirely clearcut, however; all three classes have been produced in all basic forms.

Semiaustenitic Grades

Compositions of the semiaustenitic grades are balanced so that the structure is austenitic in the solution-treated condition. This ductile structure permits forming by the conventional processes used for the 18-8 stainless steels. These steels work-harden rapidly, however, and intermediate anneals may be required for complex forming or deep drawing. Following fabrication, the austenite structure is transformed to martensite by a thermal treatment. Maximum strength is then developed by an aging treatment, resulting in precipitation hardening and tempering of the martensite.

Weldability of the semiaustenitic precipitation-hardening steels is good; preheating and postheating are not required. These steels are readily joined by most arc and resistance welding methods. However, inert-gas shielding should be used for fusion-welding methods to overcome hot shortness in single-pass welds and to minimize loss of aluminum in the weld metal.

TABLE 7-10. Composition of Precipitation-Hardening Stainless Steels

Alloy	Composition (%)							
	C	Mn	Si	Cr	Ni	Al	Mo	Other
SEMIAUSTENITIC								
17-7 PH	0.09 max	1.0 max	1.0 max	16.0 - 18.0	6.5 - 7.75	0.75 - 1.50	—	—
PH 15-7 Mo	0.09 max	1.0 max	1.0 max	14.0 - 16.0	6.5 - 7.75	0.75 - 1.50	2.0 - 3.0	—
AM-350	0.12 max	0.90	0.50 max	16.0 - 17.0	4.0 - 5.0	—	2.5 - 3.25	0.10 N
AM-355	0.15 max	0.95	0.50 max	15.0 - 16.0	4.0 - 5.0	—	2.5 - 3.25	0.10 N
PH 14-8 Mo*	0.04	—	—	14.0	8.0	1.2	2.3	—
MARTENSITIC								
Stainless W	0.12 max	1.0 max	1.0 max	16.0 - 18.0	6.0 - 8.0	1.0 max	—	1.0 Ti max, 0.2 N max
17-4 PH	0.07 max	1.0 max	1.0 max	15.5 - 17.5	3.0 - 5.0	—	—	3.0-5.0 Cu, 0.15 - 0.45 Nb+Ta
15-5 PH	0.07 max	1.0 max	1.0 max	14.0 - 15.5	3.5 - 5.5	—	—	2.5-5.5 Cu, 0.15 - 0.45 Nb+Ta
414 Ti	0.08 max	1.0 max	0.75 max	10.5 - 12.5	1.5 - 3.5	—	—	0.75 Ti max
Almar 363	0.05 max	0.3 max	0.15 max	11.0 - 12.0	4.0 - 5.0	—	—	10×C Ti min
PH 13-8 Mo*	0.03	—	—	12.5	8.0	1.1	2.2	—
AM-362*	0.03	—	—	14.5	6.5	—	—	0.8 Ti
Custom 450	0.03	0.25	0.25	15.0	6.0	—	0.8	1.5 Cu, 0.3 Nb
Custom 455*	0.03	—	—	12.0	8.5	—	—	0.40 Cb+Ta, 1.2 Ti, 2.0 Cu
AUSTENITIC								
A-286	0.08 max	1.0 - 2.0	0.40 - 1.00	13.5 - 16.0	24.0 - 28.0	0.35 max	—	1.0 - 1.5 Ti, 0.10 - 0.50 V
17-10 P	0.10 - 0.14	0.50 - 1.00	0.60 max	16.5 - 17.5	9.75 - 10.75	—	—	0.25 - 0.30 P
HNM	0.30	3.50	0.50	18.50	9.50	—	—	0.25 P

* These are typical compositions.

TABLE 7-11. Characteristics and Applications of Precipitation-Hardenable Stainless Steel

Characteristics	Applications
Martensitic	
Corrosion resistance is superior to that of regular martensitic stainless; is magnetic in both solution-treated and precipitation-hardened conditions; does not require preheating or postannealing for welding.	Valves, fasteners, gears, splines, propeller shafts, roller-chain pins.
Austenitic	
Superior corrosion resistance to most chromium-nickel stainless grades; excellent high-temperature (over 1200°F) notch toughness; nonmagnetic in annealed and precipitation-hardened conditions; weldable in annealed condition, but requires extreme care.	Jet-engine frames and hardware, fasteners, turbine wheels and blades.
Semiaustenitic	
Corrosion resistance is superior to regular martensitic stainless grades; is austenitic in annealed condition, martensitic in hardened condition; requires no preheating or postannealing for welding.	Airframes, pressure tanks, conveyor chains, surgical instruments.

The heat of welding austenitizes and solution-treats (anneals) the area adjacent to the weld. Thus, regardless of base-metal condition, the structure of both the heat-affected zone and the weld metal is austenitic in the as-welded condition. For this reason, the semiaustenitic steels can be welded in any conditions without preheating and without requiring control of interpass temperature or cooling rate. Strength of fusion welds, made with welding rods of similar composition and using recommended procedures can be expected to be 90 to 100% of base-metal strength.

Martensitic Grades

In these alloys, a martensitic structure (untempered) forms upon cooling from solution-treating temperature (1500 to 1900°F, depending on alloy). This form is suitable for machining and other fabrication operations. Subsequent aging (precipitation hardening) between 900 and 1150°F increases hardness and strength. Finish-machining operations are seldom required after heat treatment.

Weldability of the martensitic alloys is generally excellent. They are not crack-sensitive (carbon content is low) or susceptible to the hot cracking associated with other copper-bearing stainless steels. Hot cracking due to carbon may occur, however, if the martensitic steels are welded to carbon or low-alloy steels. In such cases, welding methods that minimize weld-metal dilution should be used.

Preheating is not necessary to prevent cracks and excessive hardness in the martensitic steels. Heavy sections or members of highly restrained joints are often heat treated to a tough, overly aged condition, however, before welding.

The superior notch toughness of grades such as 15-5 PH and PH 13-8 Mo make these alloys suitable for restrained-joint weldments in any condition. However, regardless of the toughness level of the steel, notch effects should be eliminated. As with all high-strength steels, residual stresses, unfused notches, and sharp threads should be avoided.

Austenitic Grades

These grades of precipitation-hardening steels retain their austenitic structure at room temperature following solution heat treatment and are hardened by a precipitation treatment.

The austenitic grades are the most difficult of the three types of precipitation-hardening alloys to weld. The most successful welds have been made by the gas tungsten-arc method, keeping heat input low. The metal should be in the solution-treated condition, and multipass welding (light stringer beads) is recommended.

Welding 4-6% Chromium Steels: Steels containing 4 to 6% chromium are not truly stainless though sometimes classified as such. They do have a resistance to corrosion of four to ten times that of mild steel and a resistance to oxidation from three to eight times that of mild steel. They have considerable resistance to oxidation and sulfur corrosion at temperatures up to 1100 — 1200°F. They are therefore used extensively for hot oil transfer lines and other oil refinery equipment operating at elevated temperatures, furnace tubes, heat exchangers, valves and high temperature steam lines.

The most noticeable property of the 4-6% Cr steels is their intense air hardening which is proportional to the chromium and carbon contents. This affects the weldability as both the weld and heat affected zone are apt to be very hard unless preheating and subsequent annealing or stress relieving are used.

The most commonly used additional alloy is molybdenum (AISI 502), usually in an amount of approximately 0.5%. This addition of molybdenum does not appreciably affect the physical qualities of the steels at room temperatures, and after the different heat treatments, but it does increase to a considerable degree the strength of these steels at elevated temperatures and, according to some tests, increases the resistance to certain types of corrosion. Molybdenum also increases the creep strength.

Where comparable properties in weld and plate are needed to meet service requirements, the general rule is to use an AWS E502-15 or E502-16 electrode that will deposit 4-6% chromium of the approximate analysis as the plate and to preheat and then anneal or stress relieve. The use of AWS E309 or E310 is recommended where preheating or postheating cannot be used. In this case the heat affected zone will be hard and brittle but the joint will withstand some impact or bending stress as the weld metal deposited by these electrodes will be relatively soft and ductile. The hard and brittle weld and fusion zone obtained when using the 4-6% chromium electrode without preheat or annealing will not stand bending stresses as well as the combination of the hard fusion zone but softer and more ductile weld metal as obtained with E309 or E310 electrodes.

When using either the 4-6% chromium or stainless electrode, DC electrode positive is used. Hold as short an arc as possible using the standard stainless procedure. E502 electrodes contain molybdenum so that they can be used on all types of 4-6% chromium steels.

References and Bibliography

1. E.S. Kopecki — "Stainless Steels," *Machine Design Metals Reference Issue,* Feb. 17, 1972, P. 26-28.
2. J.J. Vagi, R.M. Evans, and D.C. Martin — *Welding of Precipitation-Hardening Stainless Steels,* NASA-SP-5087, Clearinghouse for Federal Scientific and Technical Information, U.S. Dept. of Commerce, Springfield, VA.
3. Kenneth L. White — "Precipitation-Hardening Stainless Steels," *Machine Design,* Feb. 12, 1970, P. 40-42.

Welded stainless steel processing vessels.

Welding Stainless Steels with the Shielded Metal-Arc Process

TABLE 7-12. Chemical Composition Requirements for Undiluted Weld Metal (AWS A5.4-81)

AWS Classification[c]	UNS Number[d]	Element, Amount Percent by Weight[a,b]										
		C	Cr	Ni	Mo	Cb plus Ta	Mn	Si	P	S	N	Cu
E209[e]	W32210	0.06	20.5-24.0	9.5-12.0	1.5-3.0	—	4.0-7.0	0.90	0.04	0.03	0.10-0.30	0.75
E219	W32310	0.06	19.0-21.5	5.5-7.0	0.75	—	8.0-10.0	1.00	0.04	0.03	0.10-0.30	0.75
E240	W32410	0.06	17.0-19.0	4.0-6.0	0.75	—	10.5-13.5	1.00	0.04	0.03	0.10-0.20	0.75
E307	W30710	0.04-0.14	18.0-21.5	9.0-10.7	0.5-1.5	—	3.3-4.75	0.90	0.04	0.03	—	0.75
E308	W30810	0.08	18.0-21.0	9.0-11.0	0.75	—	0.5-2.5	0.90	0.04	0.03	—	0.75
E308H	W30810	0.04-0.08	18.0-21.0	9.0-11.0	0.75	—	0.5-2.5	0.90	0.04	0.03	—	0.75
E308L	W30810	0.04	18.0-21.0	9.0-11.0	0.75	—	0.5-2.5	0.90	0.04	0.03	—	0.75
E308Mo	W30820	0.08	18.0-21.0	9.0-12.0	2.0-3.0	—	0.5-2.5	0.90	0.04	0.03	—	0.75
E308MoL	W30823	0.04	18.0-21.0	9.0-12.0	2.0-3.0	—	0.5-2.5	0.90	0.04	0.03	—	0.75
E309	W30910	0.15	22.0-25.0	12.0-14.0	0.75	—	0.5-2.5	0.90	0.04	0.03	—	0.75
E309L	W30913	0.04	22.0-25.0	12.0-14.0	0.75	—	0.5-2.5	0.90	0.04	0.03	—	0.75
E309Cb	W30917	0.12	22.0-25.0	12.0-14.0	0.75	0.70-1.00	0.5-2.5	0.90	0.04	0.03	—	0.75
E309Mo	W30920	0.12	22.0-25.0	12.0-14.0	2.0-3.0	—	0.5-2.5	0.90	0.04	0.03	—	0.75
E309MoL	—	0.04	22.0-25.0	12.0-14.0	2.0-3.0	—	0.5-2.5	0.90	0.04	0.03	—	0.75
E310	W31010	0.08-0.20	25.0-28.0	20.0-22.5	0.75	—	1.0-2.5	0.75	0.03	0.03	—	0.75
E310H	W31015	0.35-0.45	25.0-28.0	20.0-22.5	0.75	—	1.0-2.5	0.75	0.03	0.03	—	0.75
E310Cb	W31017	0.12	25.0-28.0	20.0-22.0	0.75	0.70-1.00	1.0-2.5	0.75	0.03	0.03	—	0.75
E310Mo	W31020	0.12	25.0-28.0	20.0-22.0	2.0-3.0	—	1.0-2.5	0.75	0.03	0.03	—	0.75
E312	W31310	0.15	28.0-32.0	8.0-10.5	0.75	—	0.5-2.5	0.90	0.04	0.03	—	0.75
E316	W31610	0.08	17.0-20.0	11.0-14.0	2.0-3.0	—	0.5-2.5	0.90	0.04	0.03	—	0.75
E316H	W31610	0.04-0.08	17.0-20.0	11.0-14.0	2.0-3.0	—	0.5-2.5	0.90	0.04	0.03	—	0.75
E316L	W31613	0.04	17.0-20.0	11.0-14.0	2.0-3.0	—	0.5-2.5	0.90	0.04	0.03	—	0.75
E317	W31710	0.08	18.0-21.0	12.0-14.0	3.0-4.0	—	0.5-2.5	0.90	0.04	0.03	—	0.75
E317L	—	0.04	18.0-21.0	12.0-14.0	3.0-4.0	—	0.5-2.5	0.90	0.04	0.03	—	0.75
E318	W31910	0.04	17.0-20.0	11.0-14.0	2.0-3.0	6 × C, min to 1.00 max	0.5-2.5	0.90	0.04	0.03	—	0.75
E320	W88021	0.07	19.0-21.0	32.0-36.0	2.0-3.0	8 × C, min to 1.00 max	0.5-2.5	0.60	0.04	0.03	—	3.0-4.0
E320LR	W88022	0.03	19.0-21.0	32.0-36.0	2.0-3.0	8 × C, min to 0.40 max	1.50-2.50	0.30	0.020	0.015	—	3.0-4.0
E330	W88331	0.18-0.25	14.0-17.0	33.0-37.0	0.75	—	1.0-2.5	0.90	0.04	0.03	—	0.75
E330H	W88335	0.35-0.45	14.0-17.0	33.0-37.0	0.75	—	1.0-2.5	0.90	0.04	0.03	—	0.75
E347	W34710	0.08	18.0-21.0	9.0-11.0	0.75	8 × C, min to 1.00 max	0.5-2.5	0.90	0.04	0.03	—	0.75
E349[e,f,g]	W34910	0.13	18.0-21.0	8.0-10.0	0.35-0.65	0.75-1.2	0.5-2.5	0.90	0.04	0.03	—	0.75
E383	W88028	0.03	26.5-29.0	30.0-33.0	3.2-4.2	—	0.5-2.5	0.90	0.02	0.02	—	0.6-1.5
E385L	W88904	0.03	19.5-21.5	24.0-26.0	4.2-5.2	—	1.0-2.5	0.75	0.03	0.02	—	1.2-2.0
E410	W41010	0.12	11.0-13.5	0.75	0.75	—	1.0	0.90	0.04	0.03	—	0.75
E410NiMo	W41016	0.06	11.0-12.5	4.0-5.0	0.40-0.70	—	1.0	0.90	0.04	0.03	—	0.75
E430	W43010	0.10	15.0-18.0	0.60	0.75	—	1.0	0.90	0.04	0.03	—	0.75
E630	W37410	0.05	16.0-16.75	4.5-5.0	0.75	0.15-0.30	0.25-0.75	0.75	0.04	0.03	—	3.25-4.00
E16-8-2	W36810	0.10	14.5-16.5	7.5-9.5	1.0-2.0	—	0.5-2.5	0.60	0.03	0.03	—	0.75
E2209	—	0.04	21.5-23.5	8.5-10.5	2.5-3.5	—	0.5-2.0	0.90	0.04	0.03	0.08-0.20	0.75
E2553	—	0.06	24.0-27.0	6.5-8.5	2.9-3.9	—	0.5-1.5	1.0	0.04	0.03	0.10-0.25	1.5-2.5
E502[h]	W50210	0.10	4.0-6.0	0.40	0.45-0.65	—	1.0	0.90	0.04	0.03	—	0.75
E505[h]	W50410	0.10	8.0-10.5	0.40	0.85-1.20	—	1.0	0.90	0.04	0.03	—	0.75
E7Cr[h]	W50310	0.10	6.0-8.0	0.40	0.45-0.65	—	1.0	0.90	0.04	0.03	—	0.75

[a] Analysis shall be made for the elements for which specific values are shown in the table. If, however, the presence of other elements is indicated by the course of routine analysis, further analysis shall be made to determine that the total of these other elements, except iron, is not present in excess of 0.50 percent.
[b] Single values are maximum percentages.
[c] Classification suffix may be -15, -16, -25 or -26.
[d] SAE/ASTM Unified Number System for Metals and Alloys.
[e] Vanadium shall be 0.10 to 0.30 percent.
[f] Titanium shall be 0.15 percent max.
[g] Tungsten shall be from 1.25 to 1.75 percent.
[h] This grade is not a true stainless steel because it contains less than 11.5% Cr.

AWS specification A5.4-81, Corrosion-Resisting Chromium and Chromium-Nickel Covered Welding Electrodes, classes electrodes on the basis of chemistry, mechanical properties of the weld metal, and the type of welding current. Chemical requirements of the deposited weld metal are listed in Table 7-12, and mechanical property requirements are listed in Table 7-13.

The system of classification is similar to that used for other electrode specifications: The letter E indicates an electrode; the last two digits indicate the type of current (-15 is for DC only and -16 is for AC or DC); and digits and letters between identify the alloy.

Most stainless-steel electrodes are imprinted near the holder end with the classification number.

All stainless-steel shielded metal-arc electrode coverings are of the low-hydrogen type and must be protected from moisture pickup. Normally, electrodes packaged in hermetically sealed containers can be stored for several months without deteriorating. However, after the container is opened, the coating begins to absorb moisture and, depending on the ambient air condition, may need to be reconditioned after only four hours of exposure.

Usually, redrying at 500°F for 1 hr restores the electrode to its original condition, and storing in a holding oven at 300°F is satisfactory. Due to differences in materials and processing, the supplier should be consulted if large amounts of electrodes are involved.

SELECTING ELECTRODES

Mechanical properties of stainless-steel weld metal usually are not as important as the chemical composition (Table 7-14) and the heat treatment. There is little difference in mechanical properties or chemical composition between the DC (EXXX-15) and the AC-DC (EXXX-16) weld deposits made with electrodes of the same class. Typical mechanical properties of stainless-steel weld metal are given in Table 7-15.

DC electrodes operate on DC only, have good penetration, produce fillets with a slightly convex profile, and are recommended for:

- Vertical and overhead welding and all-position applications such as pipe. The slag has a fast-freeze characteristic.
- Root passes on heavy plate. The larger throat section of the convex bead helps prevent cracking.

TABLE 7-13.
All-Weld-Metal Mechanical Property Requirements

AWS Classification	Tensile Strength, min. ksi	Elongation min Percent	Heat Treatment
E209	100	15	None
E219	90	15	None
E240	100	15	None
E307	85	30	None
E308	80	35	None
E308H	80	35	None
E308L	75	35	None
E308Mo	80	35	None
E308MoL	75	35	None
E309	80	30	None
E309L	75	30	None
E309Cb	80	30	None
E309Mo	80	30	None
E309MoL	75	30	None
E310	80	30	None
E310H	90	10	None
E310Cb	80	25	None
E310Mo	80	30	None
E312	95	22	None
E316	75	30	None
E316H	75	30	None
E316L	70	30	None
E317	80	30	None
E317L	75	30	None
E318	80	25	None
E320	80	30	None
E320LR	75	30	None
E330	75	25	None
E330H	90	10	None
E347	75	30	None
E349	100	25	None
E383	75	30	None
E385L	75	30	None
E410	75	20	[a]
E410NiMo	110	15	[c]
E430	65	20	[d]
E502	60	20	[b]
E505	60	20	[b]
E630	135	7	[e]
E16-8-2	80	35	None
E7Cr	60	20	6
E2209	100	20	None
E2553	100	15	None

[a] Specimen shall be heated to between 1350 to 1400°F (730 to 760°C), held for 1 hour, furnace-cooled at a rate of 100°F (60°C) per hour to 600°F (315°C) and air-cooled to ambient.

[b] Specimen shall be heated to between 1550 and 1600°F (840 and 870°C), held for 2 hours, furnace-cooled at a rate not exceeding 100°F (55°C) per hour to 1100°F (595°C) and air cooled to ambient.

[c] Specimen shall be heated to between 1100 and 1150°F (595 and 620°C), held for 1 hour, and air cooled to ambient.

[d] Specimen shall be heated to between 1400 and 1450°F (760 and 790°C), held for 2 hours, and furnace-cooled at a rate not exceeding 100°F (55°C) per hour to 100°F (55°C), and air cooled to ambient.

[e] Specimen shall be heated to between 1875 and 1925°F (1025 and 1050°C), held for 1 hour, air cooled to at least 60°F (15°C), and then precipitation hardened at 1135 to 1165°F (610 to 630°C), held for 4 hours, and air cooled to ambient.

AC-DC electrodes are always used on DC when this type of power is available. The fillet profile is flat to concave, the weld surface is smoother, and the penetration is less than with EXXX-15 (DC only) electrodes. The larger amount of slag requires

TABLE 7-14. Typical Compositions of Stainless-Steel Weld Metal

AWS Type	Composition (%)				
	C	Cr	Ni	Mo	Nb (Cb)
308	0.06	20.1	9.6	—	—
308L	0.03	20.5	9.8	—	—
309	0.09	23.8	13.0	—	—
310	0.17	27.0	20.5	—	—
312	0.12	29.0	9.0	—	—
316	0.06	19.0	12.2	2.2	—
316L	0.03	19.2	12.4	2.2	—
318	0.05	19.5	12.5	2.3	0.60
347	0.06	19.9	9.6	—	0.80

TABLE 7-15. Typical Mechanical Properties of Stainless-Steel Weld Metal

AWS Type	Tensile Strength (1000 psi)	Elongation in 2 in. (%)
308	90	41
308L	82	41
309	81	40
310	91	36
312	112	32
316	89	38
316L	82	41
318	85	35
347	96	35

more care to avoid slag inclusions. These electrodes are recommended for horizontal fillets and for all flat-position welding. EXXX-16 electrodes are also used in all positions by skilled weldors.

Selecting the proper electrode must be done with care because of the large number of stainless-steel alloys. All AISI stainless alloys are not available as electrodes. The titanium in type 321 does not transfer efficiently to the weld metal, and free-machining types such as 303Se crack readily. Thus electrodes of these compositions are not produced.

Table 7-16 lists the electrodes commonly used with the various types of stainless alloys. When two different types are to be joined, Table 7-17 can be used as a guide. These tables should be used only as general guides, however — particularly where service conditions are severely corrosive or where high temperatures may cause scaling.

The free-machining grades of stainless steel are not recommended for welding because the sulfur, selenium, or phosphorus cause severe porosity and hot-short cracking. If these steels must be welded,

TABLE 7-16. Typical Filler Metals for Welding Stainless Steels

Base Metal		Service Condition	Covered Electrode	Bare Rod or Filler Wire
Wrought	Cast[(1)]			
201 202 301 302 304 305 308	CF-8 CF-20	As welded or annealed	E308	ER308
302B		As welded	E309, E310	ER309
303 303Se		As welded or annealed	E312, E309	ER314
304L	CF-3	As welded	E308L, E347	ER308L, ER347
308L		As welded	E308L	ER308L
309	CH-20	As welded	E309	ER309
309S			E309, E309Cb	ER309
310	CK-20	As welded	E310	ER310
310S		As welded	E310, E310Cb	ER310
316	CF-8M CF-12M	As welded or annealed	E316, E309Cb†	ER316‡
316L	CF-3M	As welded or stress rel'd	E316L, E309Cb†	ER316L‡
317	CG-8M	As welded or annealed	E317†	ER317
321 321H		As welded	E347	ER321, ER347
347 347H 348 348H		As welded	E347	ER347
403 410		Annealed or hardened As welded	E410 E308, E309, E310	ER410 ER308, ER309, ER310
405		Annealed As welded	E430 E308, E309, E310	ER430 ER308, ER309, ER310
420		Annealed or hardened As welded	E420 E308, E309, E310	ER420 ER308, ER309, ER310
430		Annealed As welded	E430 E308, E309, E310	ER430 ER308, ER309, ER310
430Ti		As welded	E430	430Ti§, ER430
431		Annealed or hardened As welded	431§ E308, E309, E310	431§ ER308, ER309, ER310
442		Annealed As welded	442§ E308, E309, E310	442§ ER308, ER309, ER310
446		Annealed As welded	446§ E308, E309, E310	446§ ER308, ER309, ER310
16-8-2		As welded	E16-8-2	ER16-8-2
A584 Type 630		As welded Heat treated	E630	
2205		As welded	E2209	ER2209
255		As welded		ER2553

(1) Castings higher in carbon but otherwise of generally corresponding composition are available and are designated by the prefix "H". Filler metals for these alloys are high-carbon versions of the listed filler metals.

† Joints deposited by 316, 316L, 317 electrodes and filler wires may have poor corrosion resistance in the as-welded condition particularly in hot acids. The use of 309 or 309Cb filler metals often provides a more suitable weld metal. Corrosion resistance of the molybdenum-bearing stainless steels can be restored by heat treating:
For Types 316 and 317 - full anneal at 1950-2050°F
For Types 316L and 317L - stress relieve at 1600°F

‡ Same comments as (†). However, 309Cb bare wire is not regularly available. ER310 is the best substitute.

§ No standard AWS designation exists, although this filler wire or electrode is available commercially.

From *Tool and Manufacturing Engineer*

TABLE 7-17. Typical Filler Metals for Dissimilar Metal Joints in Austenitic Stainless Steel*

	304L	308	309	309S	310	310S	316 316H	316L	317	321 321H	347,347H 348,348H
304H, 305, 304	308	308	308 309	308 309	308 309 310	308 309 310	308 316	308 316	308 316 317	308	308
304L		308	308 309	308 309	308 309 310	308 309 310	308 316	308L 316L	308 316 317	308L 347	308L 347
308			308 309	308 309	308 309 310	308 309 310	308 316	308 316	308 316 317	308	308 347
309				309	309 310	309 310	309 316	309 316	309 316	309 347	309 347
309S					309 310	309S 310S	309 316	309S 316L	309 316	309 347	309 347
310							316 310 310Mo	316 310Mo 310	317 310Mo 310	308 310	308 310
310S							316 310Mo	316 310Mo	317 310Mo	308 310	308 310
316H, 316								316	317 316	308 316	308 316 347
316L									317	316L	316L 347
317										308 317	308 317 347
321H, 321											308L 347

Electrodes and welding rods listed for the same combination of stainless steels are not in any preferred order.
* Abstracted from NASA *Welding of Stainless Steels*

use E312 or E309 electrodes and techniques that reduce the admixture of the base metal into the weld metal. Grades 416 and 440A, B and C are also not recommended for welding due to severe brittleness caused by the formation of high carbon martensite.

CONSIDERATIONS IN WELDING

Cleaning: For high-quality welds, joints must be clean and dry. The choice of power brushing, degreasing, pickling, grinding, or merely wiping depends upon the kind and amount of dirt. Some specific recommendations are:

1. Remove moisture by heating or by blowing with dry air (beware of moisture in the air line). Moisture can collect on a weldment overnight in high-humidity conditions.
2. Eliminate organic contaminants such as paints, antispatter compounds, grease-pencil marks, cutting compounds, adhesive from protective paper, and soap used for leak-testing.
3. Flame-beveling and machining may leave contaminants or oxide films that must be removed.
4. Avoid zinc contamination from brushes or tools that have been used on galvanized steel. Use only stainless-steel wire brushes that have been used only on stainless steel.

Welding Procedure

Joint Design: Accurate fitup and good joint preparation are necessary for good weld quality and minimum distortion. Joint designs are similar to those described for mild steel. For butt welds on plate to 1/2-in. thick, the bevel should be a 60° included angle for good penetration and easy slag removal. On plate over 1/2-in. thick and up to 1-1/2 in. thick, a double bevel is recommended if the welding can be done from both sides. For butt welds over 1-in. thick that must be done from one side, a U-groove is used. For butt welds in plate over 1-1/2 in. thick that can be welded from both sides, a double U-groove is recommended.

Welding Techniques: Welding with stainless-steel electrodes requires techniques similar to those used for mild-steel low-hydrogen electrodes. Use a short arc, but keep the coating from touching the puddle. Flat beads with good wash-in promote easy slag removal in deep grooves. Fill each crater before breaking the arc, to avoid crater cracks. Clean the slag thoroughly from the finish of the bead before starting another electrode, and clean the complete weld before starting the next pass. On deep-groove butt joints, the root pass should penetrate only enough to fuse to both plates and seal the opening. More penetration may cause cracks.

For vertical and overhead positions never use an electrode larger than 5/32-in. The DC electrodes (EXXX-15) are preferred, but the AC-DC electrodes (EXXX-16) can be used for welding vertical-up (using DC). On thick plate, use the triangular weave technique, welding vertical-up. On thin plate, use small beads, vertical-down.

TABLE 7-18. Recommended Current Ranges for Austenitic Stainless-Steel Electrodes (DCEP)

Electrode Size	Recommended Current (amp)	
	E3XX-15 Electrodes	E3XX-16 Electrodes
3/32	30-70	30-65
1/8	45-95	55-95
5/32	75-130	80-135
3/16	95-165	120-185
1/4	150-225	200-275
	Optimum current for flat position is about 10% below maximum; optimum for vertical-up welding, about 20% below maximum; optimum for vertical-down welding, about maximum.	Optimum current for flat position is about 10% below maximum; AC range is about 10% higher.

Welding techniques can help control distortion. Weld with low current consistent with sufficient penetration to reduce the heat input to the work (Table 7-18). Use stringer beads at a higher speed rather than wide beads at a slower speed. If weave beads must be made, limit the weave to 2-1/2 times the electrode diameter.

Other means to control distortion are:

- Use rigid fixtures to hold parts in alignment.
- Use chill bars near the weld and backing bars under the weld. Rapid cooling of austenitic stainless steels is beneficial rather than harmful. If Cu is used as the chill bar material, care must be exercised to prevent copper grain boundary penetration. This can be done by applying a thin Ni plate to the Cu.

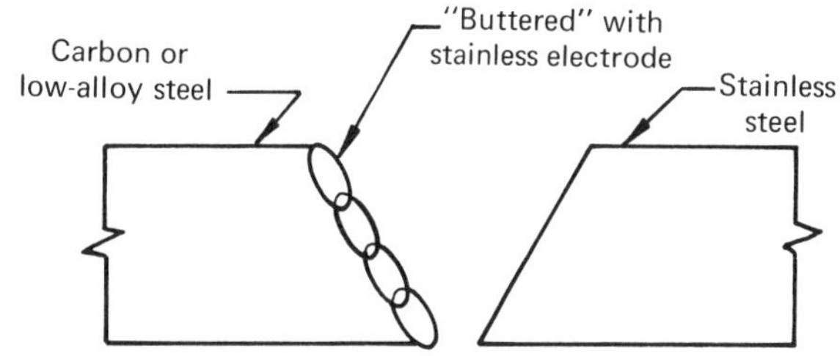

Fig. 7-5. "Buttering" technique for joining mild steel to stainless steel.

- Plan the sequence of welding, using the same techniques as with mild steel, such as skip welding and back-step welding.

For preheating and postheating information see Section 7-1.

Joining Stainless and Other Steels: In some applications, stainless-steel weld-metal is applied to mild steel: for example, lining mild-steel vessels or container with stainless steel. For such applications, stainless electrodes with higher alloy content are used so the admixture of the mild steel into the stainless weld deposit does not form an unsatisfactory alloy.

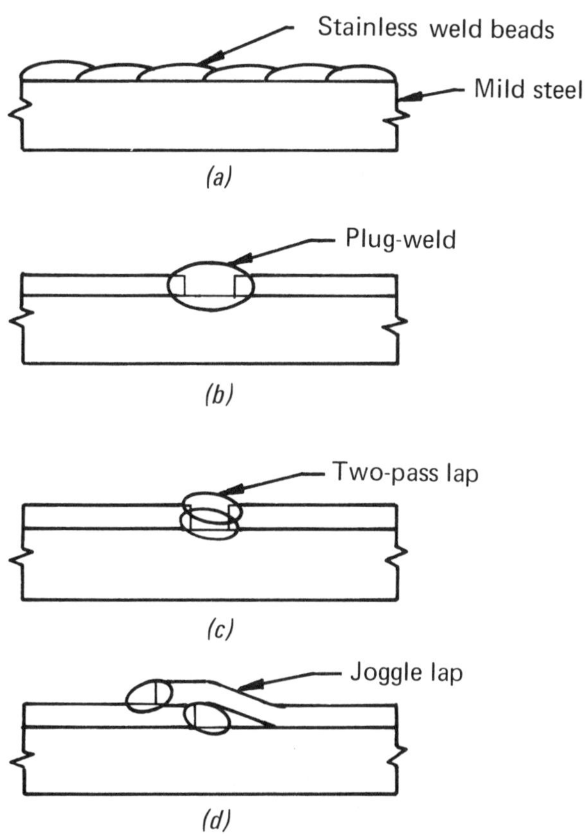

Fig. 7-6. Methods of applying a stainless steel surface to mild steel.

SHIELDED METAL-ARC (MANUAL)
AISI 300 Series Stainless Steels

Welding Position: Vertical and Overhead
Weld Quality Level: Code
Steel Weldability: Good

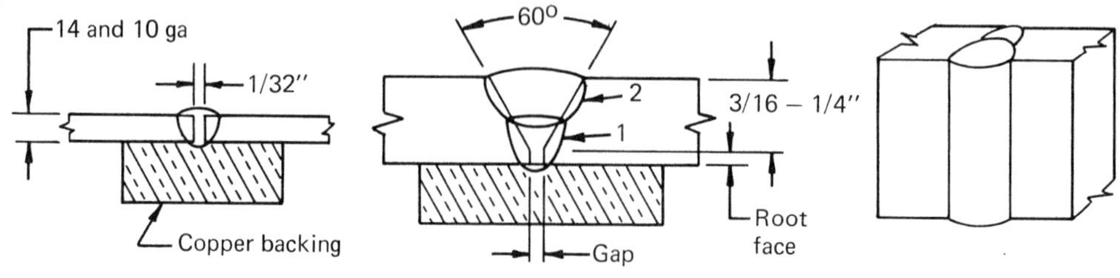

Plate Thickness (in.)	0.078 (14 ga)*	0.140 (10 ga)	3/16	1/4	
Pass	1	1	1	1	2
Electrode Class	E3XX-15	E3XX-15	E3XX-15	E3XX-15	
Size	3/32	1/8	5/32	5/32	
Current (amp) DC(+)	50	75	110	110	
Arc Speed (in./min)	14 – 16	6.7 – 7.3	5.2 – 5.8	5.2 – 5.8	4.3 – 4.7
Electrode Req'd (lb/ft)	0.030	0.091	0.160	0.370	
Total Time (hr/ft of weld)	0.0133	0.0286	0.0364	0.0808	
Gap (in.)			1/16	3/32	
Root face (in.)			1/16	1/16	

* Vertical down, all others vertical up.

See Procedure Notes in Section 6.2.

SHIELDED METAL-ARC (MANUAL)
AISI 300 Series Stainless Steels

Welding Position: Flat or Horizontal*
Weld Quality Level: Code
Steel Weldability: Good

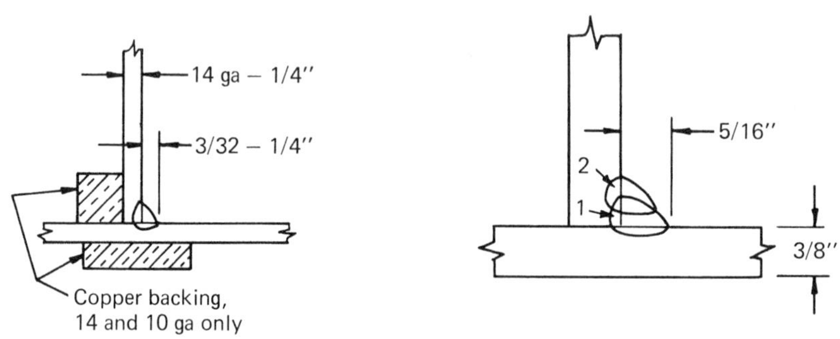

Weld Size (in.)	3/32	1/8	3/16	1/4	5/16	
Plate Thickness (in.)	0.078 (14 ga)	0.140 (10 ga)	3/16	1/4	3/8	
Pass	1	1	1	1	1	2
Electrode Class	E3XX-16	E3XX-16	E3XX-16	E3XX-16	E3XX-16	
Size	3/32	1/8	5/32	3/16	3/16	
Current (amp) DC(+)	60	85	120	160	170	
Arc Speed (in./min)	12.5 – 13.5	12.5 – 13.5	8.6 – 9.4	6.2 – 6.8	6.2 – 6.8	6.7 – 7.3
Electrode Req'd (lb/ft)	0.036	0.056	0.120	0.220	0.430	
Total Time (hr/ft of weld)	0.0154	0.0154	0.0222	0.0308	0.0594	

* For vertical and overhead use same procedures as for vertical and overhead butt welds.

Note: AC can be used with a 10% increase in current. E3XX-15 can be used with a 10% decrease in current.

See Procedure Notes in Section 6.2

SHIELDED METAL-ARC (MANUAL)

AISI 300 Series Stainless Steels

Welding Position: Horizontal
Weld Quality Level: Code
Steel Weldability: Good

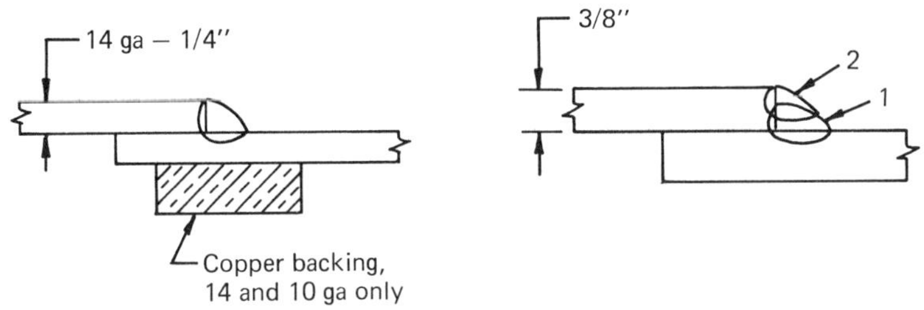

Plate Thickness (in.)	0.078 (14 ga)	0.140 (10 ga)	3/16	1/4	3/8	
Pass	1	1	1	1	1	2
Electrode Class	E3XX-16	E3XX-16	E3XX-16	E3XX-16	E3XX-16	
Size	3/32	1/8	5/32	3/16	3/16	
Current (amp) DC(+)	60	90	125	170	175	
Arc Speed (in./min)	12.5 – 13.5	12.5 – 13.5	8.6 – 9.4	6.2 – 6.8	6.2 – 6.8	6.7 – 7.3
Electrode Req'd (lb/ft)	0.036	0.056	0.130	0.240	0.460	
Total Time (hr/ft of weld)	0.0154	0.0154	0.0222	0.0308	0.0594	

The notes to fillet weld procedure also apply here.
See Procedure Notes in Section 6-2.

SHIELDED METAL-ARC (MANUAL)

AISI Series 300 Stainless Steels

Welding Position: Flat
Weld Quality Level: Code
Steel Weldability: Good

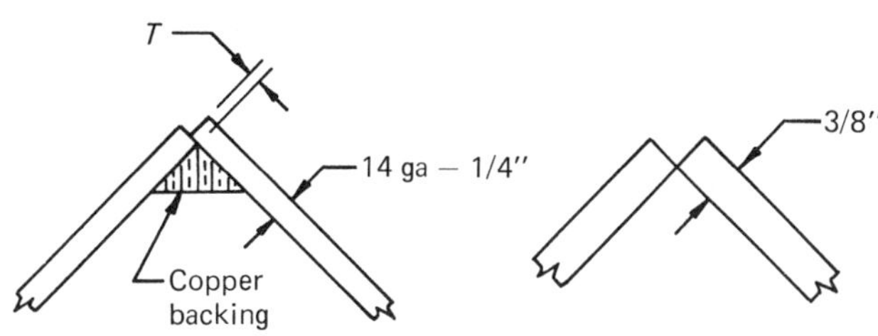

Plate Thickness (in.)	0.078 (14 ga)	0.140 (10 ga)	3/16	1/4	3/8	
Pass	1	1	1	1	1	2
Electrode Class	E3XX-16	E3XX-16	E3XX-16	E3XX-16	E3XX-16	
Size	3/32	1/8	5/32	3/16	3/16	
Current (amp) DC(+)	60	85	125	160	160	175
Arc Speed (in./min)	14 – 16	12.5 – 13.5	10.5 – 11.5	6.2 – 6.8	6.2 – 6.8	5.7 – 6.3
Electrode Req'd (lb/ft)	0.028	0.056	0.094	0.22	0.45	
Total Time (hr/ft of weld)	0.0133	0.0154	0.0182	0.0308	0.0641	
T (in.)	0.04	1/32	3/64	1/16	0	

AC can be used with a 10% increase in current. E3XX-15 electrode can be used with a 10% decrease in current.
See Procedure Notes in Section 6.2.

When stainless-steel is joined to mild steel, the mild steel is "buttered" with stainless steel. This technique consists of depositing a layer of stainless on the surface of the mild steel, then completing the joint with stainless electrode, as illustrated in Fig. 7-5. The electrode commonly used for buttering is E309. This technique is also used for joining hard-to-weld or high-carbon steels that cannot be preheated.

E308 electrode is used for joining austenitic manganese steel to carbon steel or to manganese steel. However, for components that must be replaced periodically, such as dipper teeth, a manganese electrode is recommended because the stainless weld is more difficult to torch cut.

There are several methods for applying a stainless surface to mild steel. For a small area, overlapping weld beads are used, as illustrated in Fig. 7-6(a). For larger areas, stainless sheets are plug-welded to the mild steel [Fig. 7-6(b), or stainless strips are attached as shown in Fig. 7-6(c) and (d)].

Power Sources: The open-circuit voltage of light-duty transformer welders may not be high enough for some brands of EXXX-16 electrodes; otherwise the same power sources used with steel electrodes are satisfactory for stainless electrodes.

Parameters and procedures for welding stainless steel in thicknesses from 18 gauge to 1/2 inch are given on pages 7.2-6 — 7.2-8. These show joint designs and backup bars for butt, T, lap, and 90 degree edge joints.

SHIELDED METAL-ARC (MANUAL)
AISI 300 Series Stainless Steels

Welding Position: Flat
Weld Quality Level: Code
Steel Weldability: Good

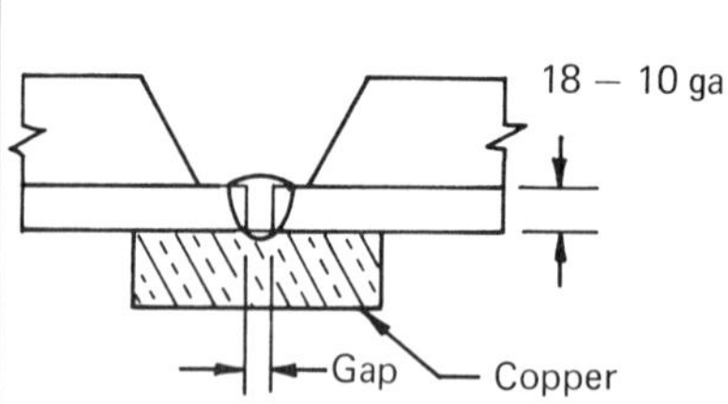

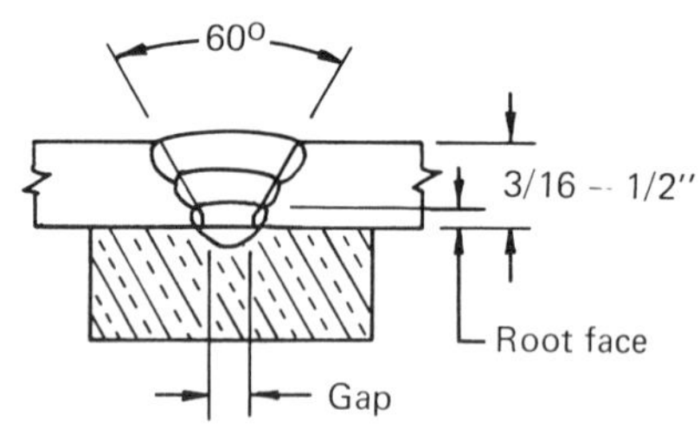

Plate Thickness (in.)	0.050 (18 ga)	0.078 (14 ga)	0.140 (10 ga)	3/16	1/4		3/8		1/2	
Pass	1	1	1	1	1	2	1	2 – 3	1	2 – 5
Electrode Class	E3XX-16	E3XX-16	E3XX-16	E3XX-16	E3XX-16		E3XX-16		E3XX-16	
Size	5/64	3/32	1/8	5/32	5/32	3/16	5/32	3/16	5/32	3/16
Current (amp) DC(+)	40*	60	85	125	125	160	125	160	125	160
Arc Speed (in./min)	14 – 16	11.5 – 12.5	8.5 – 9.5	6.7 – 7.3	5.7 – 6.3	7.6 – 8.4	5.7 – 6.3	5.7 – 6.3	5.7 – 6.3	5.7 – 6.3
Electrode Req'd (lb/ft)	0.020	0.038	0.080	0.150	0.340		0.650		1.06	
Total Time (hr/ft of weld)	0.0133	0.0167	0.0222	0.0286	0.0583		0.100		0.167	
Gap (in.)	0	1/32	1/32	1/16	3/32		3/32		3/32	
Root Face (in.)				1/16	1/16		1/16		1/16	

* Use DC(−)

Note: AC can be used with a 10% increase in current. E3XX-15 electrode can be used with a 10% decrease in current.
See Procedure Notes in Section 6-2.

Welding Stainless Steels with the Submerged-Arc Process

Submerged-arc is applicable to the welding of stainless steels where the higher heat input and slower solidification are tolerable. With submerged-arc welding the silicon content may be much higher than with other processes, a factor that may promote hot short cracking or fissuring.

The submerged-arc process is not recommended where a weld deposit is needed that is fully austenitic or is controlled to a low ferrite content (below 4%). However, high-quality welds may be produced for applications in which more than 4% ferrite in weld deposits is allowable.

JOINT DESIGN

Figure 7-7 shows the type of butt joint designs that can be used for submerged-arc welding.

Good quality single-pass welds up to 5/16-in. thick can be made using the square groove butt joint (Figure 7-7a) without root opening and with suitable backing. Two-pass welds up to 5/8-in. thick are also made without root opening. It is essential on two-pass welds, however, that the edges be closely butted since weld backing is not used. The advantage of this joint design is that it requires a minimum of edge preparation, yet produces welds of good quality having adequate penetration.

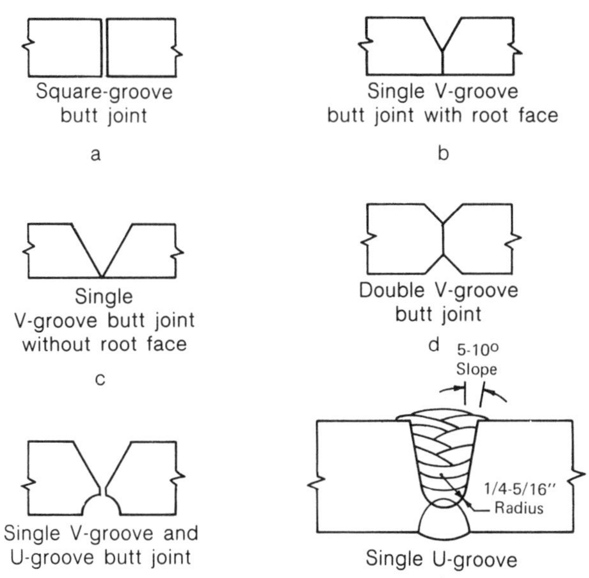

Fig. 7-7. Butt joint designs for submerged-arc welding.

Single-V groove welds with a root face, Figure 7-7b, are used with nonfusible backing for single-pass butt welds of 5/16-in. thickness or greater. For most industrial applications, the maximum thickness is of the order of 1-1/4 to 1-1/2 in. Root face dimensions are 1/8 to 3/16 in. This joint design is also used for two-pass welds without backing where plate thickness exceeds 5/8 in. The first pass is made in the V of the joint, Figure 7-7b. The work is then turned over and the first pass becomes the backing pass. In this position, the finishing pass is made on the flat side of the joint penetrating into the root of the first pass. The root face is approximately 3/8 in. for two-pass welds.

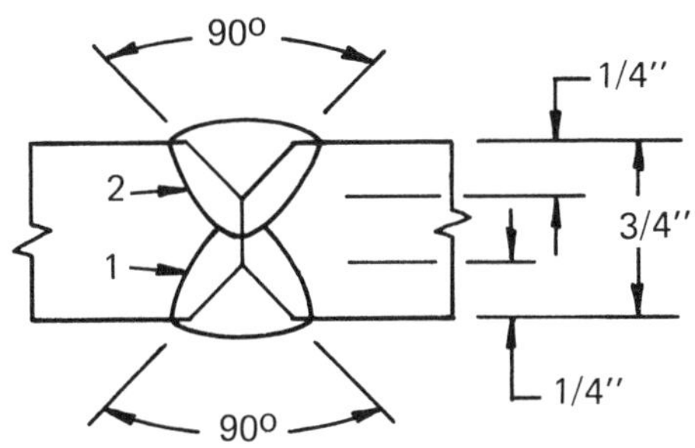

Fig. 7-8. A typical double-V weld in Type 304 plate. Pass 1 was made at 700 amp, 33 volts, 16 ipm; pass 2 at 950 amp, 35 volts, 12 ipm. The power was DCEP; electrode 3/16-in. Type 308; neutral flux.

The double-V groove butt, Figure 7-7d, is the basic joint design for submerged-arc welding. A large root face is generally used with this design. Figure

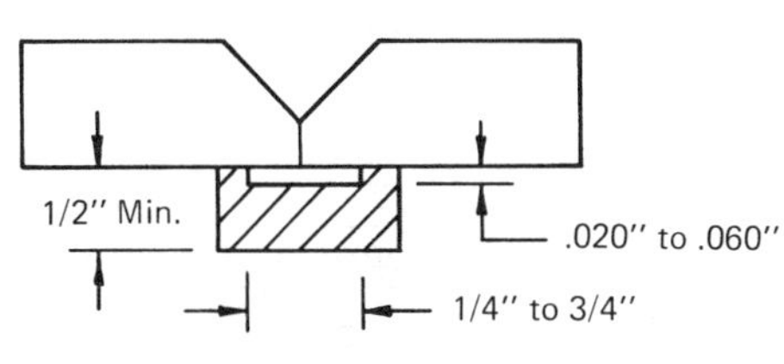

Fig. 7-9. Recommended groove dimensions for copper backing bars in the submerged-arc welding of stainless steels.

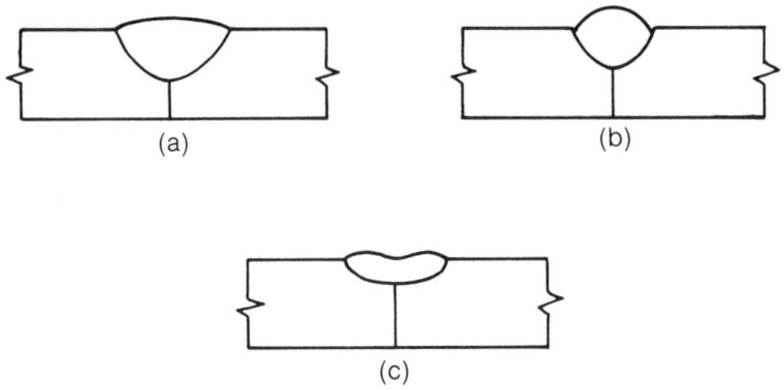

Fig. 7-10. (a) Contour of a weld bead in the flat position with the work horizontal; (b) welding slightly uphill; (c) welding slightly downhill.

7-8 shows a typical double-V groove weld in 3/4 in. 304 plate and dsscribes the welding sequence.

A single-U groove butt joint design, Figure 7-7f, is also commonly used. A small manually produced backing weld is often made from the reverse side of the joint.

WELDING PROCEDURE

For stainless-steel welding, DC power is generally used on thin sections and either AC or DC for heavier sections. However DC is generally the preferred method. Currents are about 80% of those for carbon steels. Single-pass techniques usually result in dilution levels of 40 to 60%. This may be decreased by using multi-pass welds.

Weld Backup

Submerged-arc welding creates a large volume of molten metal that remains fluid for appreciable time. It is essential that this molten metal be supported and contained until it has solidified. The two most common means of weld backing are nonfusible backing and fusible backing.

Copper backing is the most frequently used nonfusible backing in the welding of stainless steel. When copper is used as a chill bar, care must be taken to prevent copper grain boundary penetration. Recommended groove dimensions are shown in Fig. 7-9.

With a fusible metallic backing, the weld penetrates into and fuses with the stainless backing, which either temporarily or permanently becomes an integral part of the assembly.

Inclination of Work

Most submerged-arc welding is done in the flat position. This results in the best bead contour and ease of operation. Occasionally, welding is done on circumferential seams. Figure 7-10 illustrates the effect of various inclinations.

Welding Flux

Submerged-arc fluxes are available as proprietary materials for welding stainless steel. Composition materials fall into two categories — fused type and bonded type. The fused type is glass-like and is produced by melting the ingredients at high temperatures followed by crushing to granulate the flux. The bonded type is produced by mixing the ingredients with a suitable binder and baking the mixture.

Neutral fluxes are generally recommended for the welding of stainless steel. Neutral fluxes are defined as these which will not produce any significant change in the all-weld metal chemistry as a result of a large change in the arc voltage. These fluxes usually contain little or no active alloy. When intended for the welding of stainless steel, however, neutral fluxes may contain small amounts of chromium, nickel, moly, or other alloy ingredients. These alloy additions are intended primarily to compensate for the loss of alloy when the stainless electrode is melted in the welding process.

Welding Electrodes

Twenty-five standard grades of austenitic stainless steel bare welding electrodes are listed in AWS Specification A5.9, covering bare electrodes for submerged-arc and the gas metal-arc welding processes. It also includes chemical composition requirements for the filler metal, but does not specify composition requirements for deposited metal.

The composition ranges listed in this specification are broad. Since composition profoundly affects weld

TABLE 7-19.
Standard Sizes for Stainless Electrodes

Form	Diameter, in.
Electrode in coils, with or without support	0.045, 1/16, 5/64, 3/32, 7/64, 1/8, 5/32, 3/16, 1/4
Electrode wound on standard 12-in. O.D. spools	0.030, 0.035, 0.045, 1/16, 5/64, 3/32, 7/64
Electrodes wound on light-weight 1-1/2 and 2-1/2-lb 4-in. O.D. spools	0.020, 0.025, 0.030, 0.035, 0.045

quality and serviceability, similar variations cannot always be tolerated in the deposit. To maintain control, the welding technique, alloy content of the flux, or other appropriate changes should be made for variations in filler-metal composition. See Table 7-16 for typical electrode type for stainless steels.

Filler-metal specifications are frequently reviewed and periodically revised or expanded. Information about special grades and new or proprietary alloys are also available for submerged-arc welding from producers and suppliers.

Table 7-19 lists various standard electrode sizes and the spools and coils available.

Welding Technique

The method used to start the weld will depend on such factors as the time required for starting relative to the total setup and welding time, the number of pieces to be welded, and the importance of starting the weld on a particular place on the workpiece. The several methods of starting the weld that are commonly in use include:

Scratch Start. In a scratch start, the wire is fed toward the work and the carriage travel also is started. When the wire touches the work, it will not fuse to the workpiece because of the relative motion of the carriage. This type of starting is also called a "flying start."

Retract Starting. The wire is "inched" toward the work and covered with flux. When the weld is started, the wire retracts momentarily and then reverses to feed forward. This method is not recommended for light-gage stainless steel.

Once the arc is initiated, it is important to monitor the various parameters. Welding current is the most influential variable. Next in importance is welding voltage. Welding-speed changes conform to a pattern; if the speed is increased, there is less weld reinforcement; if decreased, there is more weld reinforcement. In addition, weld speed can affect depth of penetration.

An index of typical weld faults is given in Table 7-20.

TABLE 7-20. Weld Faults and Possible Causes

Fault	Possible Cause
Melt-through	Current too high Bevel angle too large Root face too small Excessive plate spacing
"Pear"-shaped fusion zone	Current too high Voltage too low
Penetration too deep or too shallow	Current too high or low Voltage too high or low Incorrect plate spacing
Porosity and gas pockets	Shallow composition burden Improper cleaning Contaminated manual weld backing Insufficient penetration in double welds Improper fitup in welds with manual backing Moisture on the plate or in the flux
Reinforcement narrow and steep-sided	Composition layer too deep Too low voltage Welding up-hill
Reinforcement overlapped	Current too high Speed too slow Voltage too low
Reinforcement ropey	Composition layer too deep
Reinforcement too high or too low	Current too high or low Voltage too high or low Speed too high Improper spacing in backed welds Improper backing Workpiece not level
Undercutting	Speed too high Improper wire alignment in fillet welding Improper backing
Voids and cracks	Improper cooling Improper fitup Concave reinforcement in fillet weld

Welded stainless steel hopper car.

Welding Stainless Steels with the Gas Metal-Arc Process

Stainless steels may be welded by the gas metal-arc process, using either spray-arc, short-circuiting, or pulsed-arc transfer.

Copper backup strips are necessary for welding stainless-steel sections up to 1/16-in. thick. Backup is also needed when welding 1/4-in. and thicker plate from one side only.

No air must be permitted to reach the underside of the weld while the weld puddle is solidifying. Oxygen and nitrogen will weaken molten and cooling stainless steel. If the jig or fixture members permit an appreciable quantity of air to contact the underside of the weld, argon backup gas should be used.

SPRAY-ARC TRANSFER

Filler metals for gas metal arc welding are specified in AWS-A5.9-81 and their chemical compositions are shown in Table 4-20.

Electrode diameters as great as 3/32-in., but usually around 1/16-in., are used with relatively high currents to create the spray-arc transfer. A current of approximately 300-350 amperes is required for a 1/16-in. electrode, depending on the shielding gas and type of stainless wire being used. The degree of spatter is dependent upon the composition and flow rate of the shielding gas, wire-feed speed, and the characteristics of the welding power supply. DCEP is used for most stainless-steel welding. A 1 or 2% argon-oxygen mixture is recommended for most stainless-steel welding.

On square butt welds, a backup strip should be used to prevent weld-metal dropthrough. When fitup is poor or copper backing cannot be used, dropthrough may be minimized by short-circuiting transfer welding the first pass.

When welding with the semiautomatic gun, forehand techniques are beneficial. Although the operator's hand is exposed to more radiated heat, better visibility is obtained.

For welding plate 1/4-in. and thicker, the gun should be moved back and forth in the direction of the joint and at the same time moved slightly from

TABLE 7-21. Gas Metal-Arc Welding (Semiautomatic) General Welding Conditions for Spray-Arc Transfer AISI 200 and 300 Series Stainless Steels

Gas-Argon + 1% Oxygen. Gas flow 35 cfh.	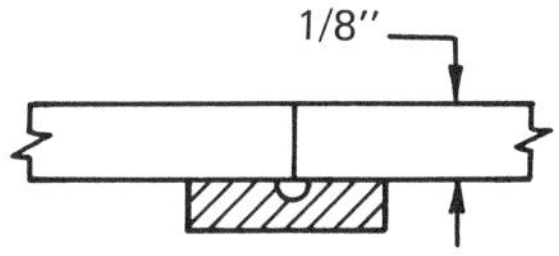	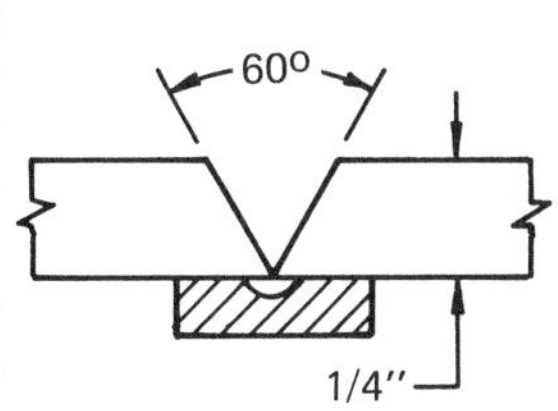	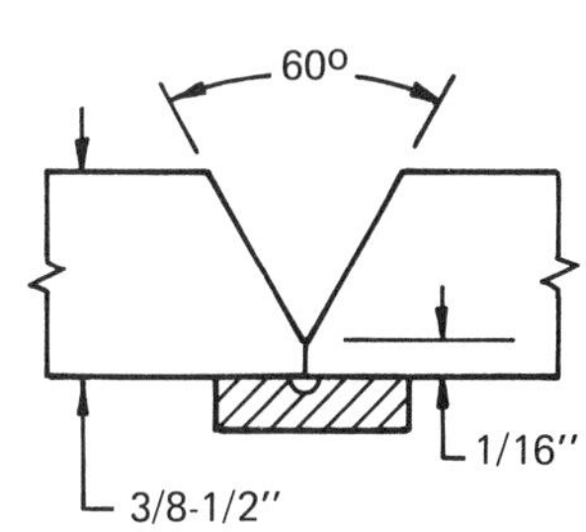	
Plate Thickness (in.) Electrode Size	1/8 1/16	1/4 1/16	3/8 1/16	1/2 3/32
Passes Current DCEP Wire Feed Speed (ipm)	1 225 140	2 275 175	2 300 200	4 325 225
Arc Speed (ipm) Electrode Required (lb/ft) Total Time (hr/ft of weld)	19 – 21 0.075 0.010	19 – 21 0.189 0.020	15 – 17 0.272 0.025	15 – 17 0.495 0.050

side to side. On thinner metal, however, only back and forth motion along the joint is used. Table 7-21 summarizes the welding procedures normally used for the spray-arc welding of stainless steel.

The more economical short-circuiting transfer process for thinner material should be employed in the overhead and horizontal position for, at least, the root and first passes. Although some operators use a short digging spray arc to control the puddle, the weld is apt to be unduly porous.

SHORT-CIRCUITING TRANSFER

Power-supply units with slope, voltage, and inductance controls are recommended for the welding of stainless steel with short-circuiting transfer. Inductance, in particular, plays an important part in obtaining proper puddle fluidity.

The shielding gas recommended for short-circuiting welding of stainless steel contains 90% helium, 7.5% argon, and 2.5% carbon dioxide. The gas gives the most desirable bead contour while keeping the CO_2 level low enough so that it does not influence the corrosion resistance of the metal. High inductance in the output is beneficial when using this gas mixture.

Single-pass welds may also be made using argon-CO_2 gas. The CO_2 in the shielding gas will affect the corrosion resistance of multipass welds made with short-circuiting transfer.

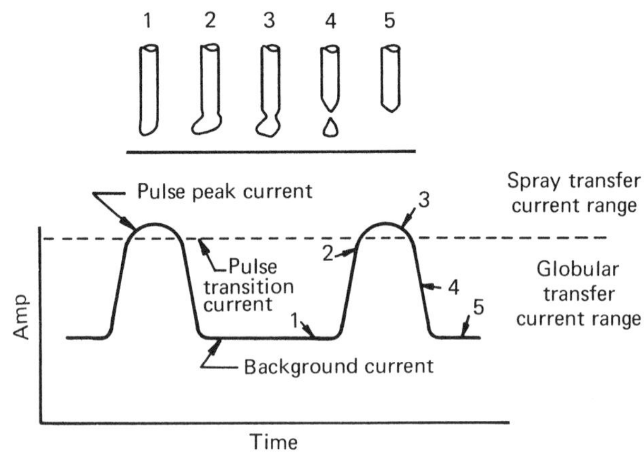

Fig. 7-11. Output current waveform of the pulsed-current power supply. Numbers on curve relate to the metal-transfer cycle shown at top of the figure.

TABLE 7-22. Gas Metal-Arc Welding (Semiautomatic) General Welding Conditions for Short-Circuiting Transfer AISI 200 and 300 Series Stainless Steels

Gas-Argon + 25% CO_2 or Argon + 2% Oxygen. Gas flow 15 to 20 cfh. Electrode 0.030 in. dia. *For Argon + 25% CO_2 or Argon + 2% Oxygen. For Helium, + 7-1/2% Argon, + 2-1/2% CO_2, voltage is 6 to 7 volts higher.	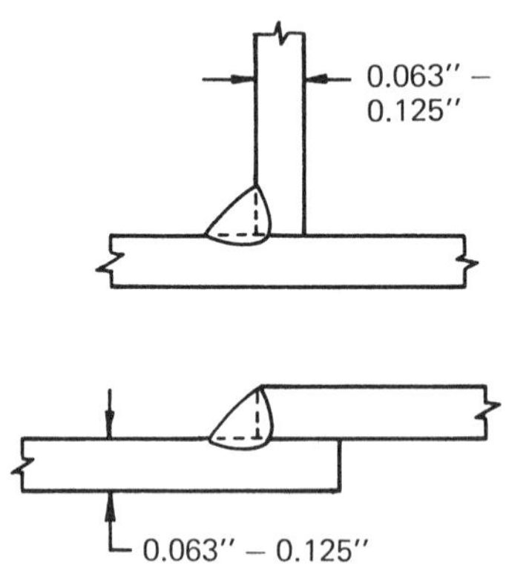 				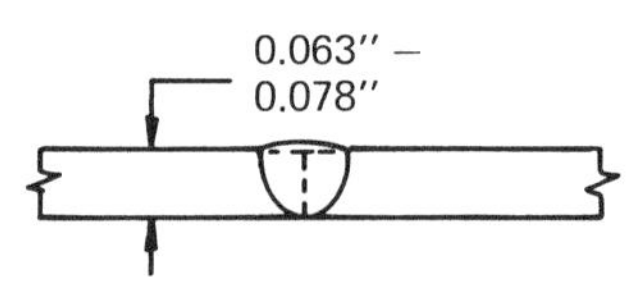 	
Plate Thickness (in.)	0.063	0.078	0.093	0.125	0.063	0.078
Electrode Size	0.030	0.030	0.030	0.030	0.030	0.030
Current DCEP	85	90	105	125	85	90
Voltage*	15	15	17	17	15	15
Wire Feed Speed (ipm)	184	192	232	280	184	192
Welding Speed (ipm)	17 – 19	13 – 15	14 – 16	14 – 16	19 – 21	11.5 – 12.5
Electrode Required (lb/ft)	0.025	0.034	0.039	0.046	0.023	0.039
Total Time (hr/ft of weld)	0.0111	0.0143	0.0133	0.0133	0.0100	0.0167

Wire extension or stickout should be kept as short as possible. Backhand welding is usually easier on fillet welds and will result in a neater weld. Forehand welding should be used for butt welds. Outside corner welds may be made with a straight motion.

A slight backward and forward motion along the axis of the joint should be used. Table 7-22 summarizes the welding procedures normally used for the short-circuiting transfer welding of stainless steel.

Short-circuiting transfer welds on stainless steel made with a shielding gas of 90% He, 7-1/2% A, 2-1/2% CO_2 show good corrosion resistance and coalescence. Butt, lap, and single fillet welds in material ranging from .060-in. to .125-in. in 321, 310, 316, 347, 304, 410, and similar stainless steels can be successfully made.

PULSED-ARC TRANSFER

The pulsed arc process, as normally used, is a spray transfer process wherein one small drop of molten metal is transferred across the arc for each high current pulse of weld current. The high current pulse must be of sufficient magnitude and duration to cause at least one small drop of molten metal to form and be propelled by the pinch effect from the end of the wire to the weld puddle. During the low current portion of the weld cycle the arc is maintained and the wire is heated, but the heat developed is not adequate to transfer any metal. For this reason, the time duration at the low current value must be limited otherwise metal would be transferred in the globular mode.

In earlier pulse type power supplies, the pulsing action was obtained by combining the output of two power supplies mounted in one cabinet.

The low or background current was provided by a DC power supply, and the peak or pulse current was provided by a half wave rectified AC power supply connected in series with the DC power supply. With this arrangement, both background and peak current values can be adjusted independently of each other. However, the pulsing frequency is fixed since it is established by the power line frequency which is normally 60 Hertz in the United States and 50 Hertz in Europe. With the development of solid state circuitry, three phase power supplies are able to attain pulsing frequencies up to 360 Hz which are multiples of 60 Hz.

Wire diameters of 0.045 and 0.060 inches are most commonly used with this process. Gases for pulsed arc welding are argon plus 1% oxygen, the same as used for spray arc welding. These and other wire sizes can be welded in the spray transfer mode at a lower average current with pulsed current than with continuous weld current. The advantage of this is that thin material can be welded in the spray transfer mode which produces a smooth weld with less spatter than the short circuiting transfer mode. Another advantage is that for a given average current, spray transfer can be obtained with a larger diameter wire than could be obtained with continuous currents. Larger diameter wires are less costly than smaller sizes, and the lower ratio of surface to volume reduces the possibility of weld contamination from surface oxides.

The pulsing operation is obtained by the use of a special pulsing power supply. Presently, there are four types available:

1. Two transformers — one for the pulse and one for the background.
2. One transformer producing a pulse and background current. The background current is tied to the pulse by means of an electronic circuit.
3. An inverter transformer is used to produce a rippled DC current. The output current is pulsed to produce a large number (2000 hz) of pulses per second.
4. A single transformer is combined with a chopper circuit to produce as many as 20,000 hz for very high frequency pulses.

Pulsed GMA welding is a modification of short arc welding and spray transfer welding. See Fig. 7-11. The shielding gas is basically argon for non-ferrous materials with a mixture of O_2 or CO_2 for ferrous materials.

Pulsed GMA welding characteristics are excellent transfer with lower currents. There are many advantages with this process including low spatter, penetration without melt-through and excellent operator appeal.

WELDING ELECTRODES

The electrodes diameters for gas metal-arc welding are generally between 0.030 and 3/32 in. For each electrode diameter, there is a certain minimum welding current that must be exceeded to achieve spray transfer. For example, when welding stainless steel in an argon-oxygen atmosphere with 3/64-in. diameter stainless steel electrode, spray transfer will be obtained at a welding current of about 220 amp DCEP. Figure 7-12 illustrates typical deposition rate curves for various stainless steel wire sizes using argon-1% oxygen as the shielding gas. It must be kept in mind that along with the minimum current a minimum arc voltage must also be obtained. This is generally between 24 and 30 v. It increases with an increase in current and is higher for helium shielding than for argon or argon CO_2 mixtures.

Table 7-16 lists austenitic base metals and the recommended electrodes to be used for welding. Electrodes come on spools varying in weight between 2 and 60 lb and sizes from 0.015 to 0.125 in. Also available are electrodes for welding the straight chromium stainless steels and austenitic electrodes that contain more than the usual amount of silicon. The latter have particularly good wetting characteristics when used with the short-circuiting transfer process.

SPECIAL CONSIDERATIONS

Some stainless steels during welding have a tendency toward hot cracking or tearing, type 347, for example. When welding these, more welding passes than indicated in the procedures may be needed. Stringer-bead techniques are also recommended rather than weaving or oscillating from side to side. Hot cracking may be eliminated by stringer-bead techniques since there is a reduction in contraction stresses, hence cooling is more rapid through the hot-short temperature range.

Preheating to about 500°F helps to improve bead contour and reduce hot cracking when using the stringer-bead technique on sections 1-in. or thicker. Weld-metal hot cracking may also be reduced by short-circuiting transfer welding, because of the lower dilution of the base metal. Excessive dilution produces a completely austenitic weld metal having strong cracking characteristics.

When welding magnetic stainless steels (some 400 series) to the relatively nonmagnetic types, it is desirable to:

1. Use a single-bevel joint to obtain minimum joint reinforcement.
2. Use low-heat input short-circuiting transfer to minimize the arc deflection encountered when welding magnetic to nonmagnetic steels.
3. For uniform fusion, be sure the wire is kept centered over the nonbeveled edge of the joint.

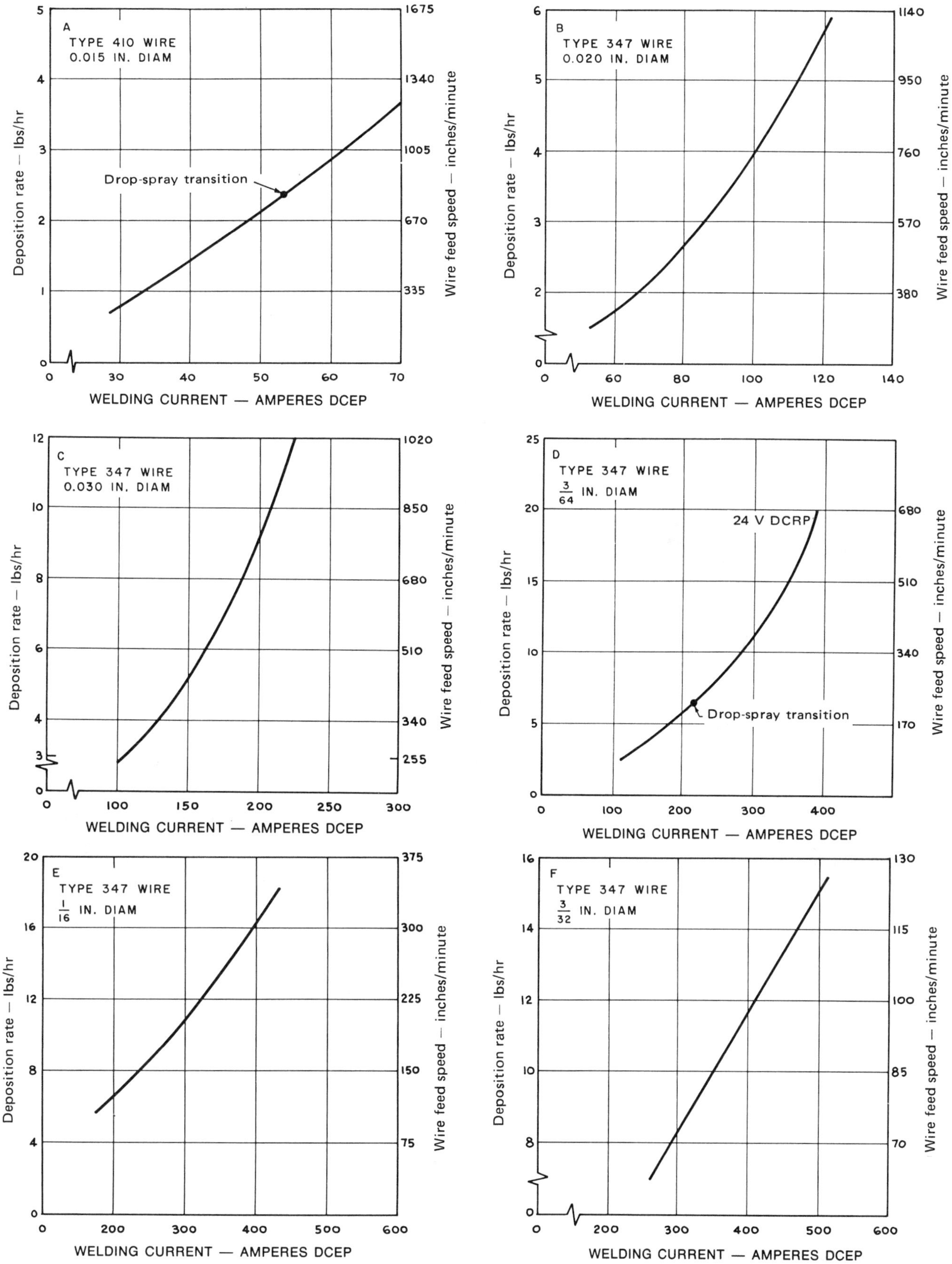

Fig. 7-12. Representative deposition rates of stainless steel electrodes.

Gas metal arc welding stainless steel.

Welding Stainless Steels with the Gas Tungsten-Arc Process

Electrodes and Gases

All stainless-steel alloys that are considered weldable can be welded readily with the gas tungsten-arc process (GTAW).

The preferred electrode is thoriated tungsten as specified in AWS A5.12-80, Class EWTh-2, although other tungsten electrodes can be used. The advantage of the thoriated electrodes is that they have a more stable arc and can be used with higher currents than pure tungsten electrodes.

The shielding gas is usually argon, but helium or mixtures of argon and helium are used on heavy sections. The advantages of argon are that flow rates can be lower, and the arc voltage is somewhat less than with helium. The lower voltage makes it possible to weld thin sheet without meltthrough.

Welding Rods

Welding rods (filler metal) for gas tungsten-arc welding are specified in AWS A5.9-81, and their chemical compositions are shown in Table 4-20. There is no specification for the mechanical properties of the weld metal. According to AWS A5.9-81 the classification is on the basis of chemical composition of the filler metal as manufactured. Mechanical tests have not been included but reference is made in the appendix AWS 5.4-81 which lists mechanical properties for welds made with stainless steel coated electrodes. This is contained as Table 7-13 in this section. It can be stated, however, that joints made with these filler metals under properly controlled welding procedures and using the process prescribed herein will have mechanical properties at least equal

TABLE 7-23. Typical Procedures for Welding Stainless Steels with the Gas Tungsten-Arc Process

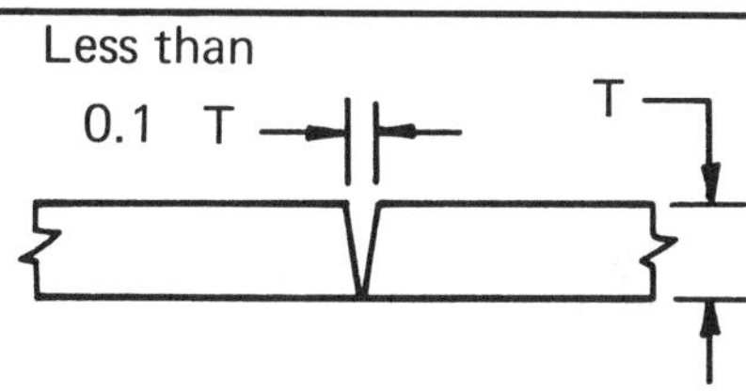

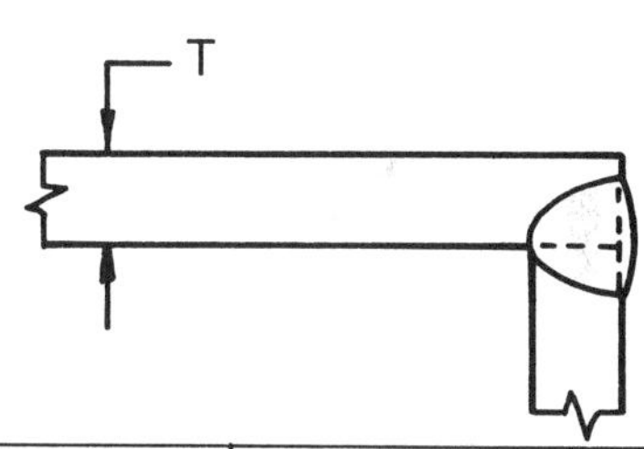

Plate Thickness, T (in.)	1/16	3/32	1/8	3/16
Current, DCEN	80 – 100	100 – 120	120 – 140	200 – 250
Electrode Diameter (in.)	1/16	1/16	1/16	3/32
Gas Flow, Argon (cfh)	10	10	10	15
Filler-Rod Diameter (in.)	1/16	1/16	3/32	1/8
Arc Speed (ipm)	12	12	12	10
Total Time (hr/ft of weld)	0.0167	0.0167	0.0167	0.0200

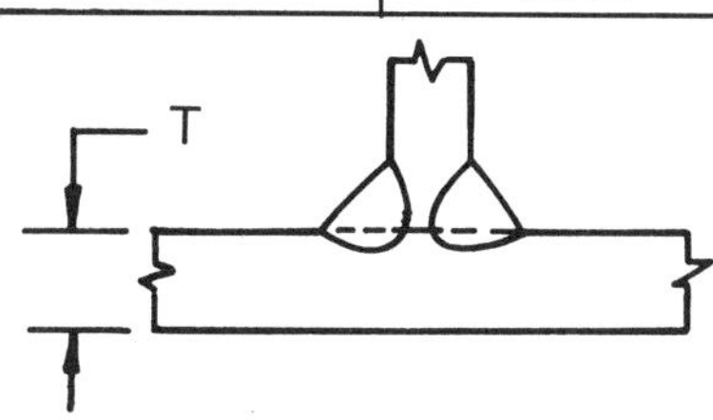

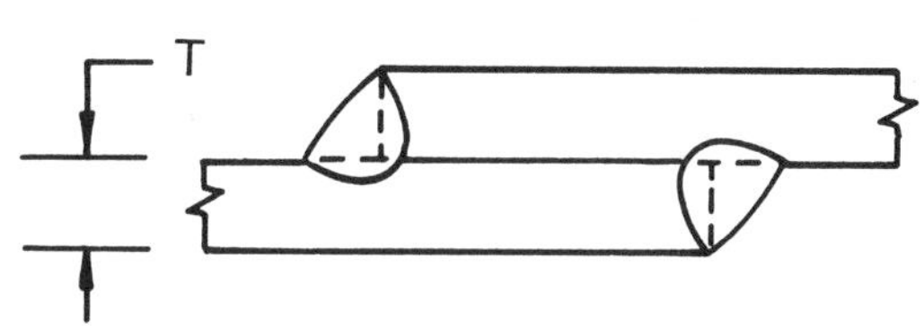

Plate Thickness, T (in.)	1/16	3/32	1/8	3/16
Current, DCEN	90 – 110	110 – 130	130 – 150	225 – 275
Electrode Diameter (in.)	1/16	1/16	1/16	3/32
Gas Flow, Argon (cfh)	10	10	10	15
Filler-Rod Diameter (in.)	1/16	1/16	3/32	1/8
Arc Speed (ipm)	10	10	10	8
Total Time (hr/ft of weld)	0.0200	0.0200	0.0200	0.0250

For vertical-up and overhead, decrease the current 10 to 20%.

to the base metal of corresponding chemical composition.

Welding Procedures

Typical procedures for gas tungsten-arc welding of stainless steels are given in Table 7-23.

Filler-metal selection is very important in GTA welding and is usually specified by the design engineer. If no class of filler metal is specified, Table 7-16 can be used as a general guide to filler-metal selection. When two different types of stainless steels are to be joined, Table 7-17 can be used as a guide. However, the tables should be used with caution. Where corrosive conditions are severe, the filler-metal selection can be very critical, as indicated in the footnote to Table 7-16.

The DC power source for gas tungsten-arc welding must be a variable-voltage type, and it is recommended that a high-frequency voltage be superimposed on the welding circuit. The high frequency need be on only to start the arc. As the electrode is brought close to the work, the high-frequency jumps the gap from the tungsten to the work and ignites the welding arc. Since the tungsten electrode does not actually touch the work, the possibility of contaminating the stainless steel with tungsten is greatly reduced. Electrode negative polarity should be used — which produces deep penetration.

A "scratch" start may be used in lieu of a high-frequency start, although there is some possibility of tungsten pickup. The arc should not be struck on a carbon block because of the likelihood of carbon contamination. Good jigging is essential for good performance. The problems likely to be encountered when GTA welding the austenitic type stainless steels and the possible remedies are listed in Table 7-24.

The pulsed GTA process may be used for welding thin sheet and for root pass welding with inert gas backup. The pulsing rate, which is controlled by a pulsing power supply with dropping voltage characteristics, is much lower than used in constant current power supplies used for GMA welding. The high current pulses are enough to cause fused penetration of the joint but the duration is short enough to prevent excessive drop-through. The low current portion maintains the arc so that it can

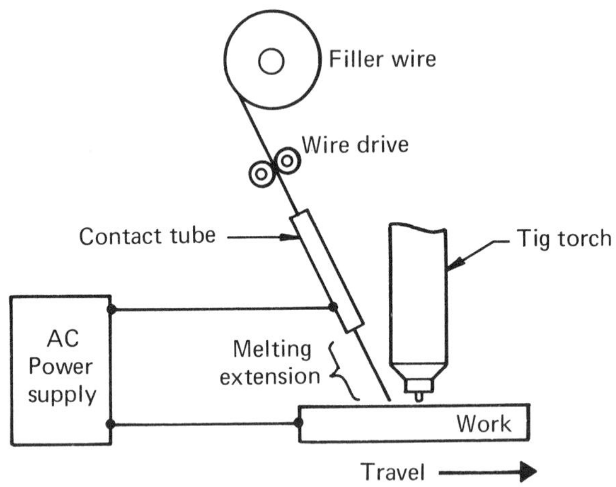

Fig. 7-13. Schematic of the hot-wire system for the automatic GTA welding of stainless steels.

TABLE 7-24. Problems That May Be Encountered in GTA Welding of Austenitic Types of Stainless Steels

Problem	Possible Cause	Possible Remedy
Occasional hot tearing	Improper joint design; sequence of welding; jigging	Change joint design; change sequence; check polarity
Decreased ductility	Excessive heat input	Use lower current; use multipass technique with narrow stringer beads made at high speed with high current.
Decreased corrosion resistance near weld	Carbide precipitation	Use stabilized or ELC grade of base metal, such as 347 or 304L and columbium-stabilized rod; or heat entire assembly to 1900°F and cool rapidly after welding
Arc wanders	Low current density	Use smaller electrode; raise current
	Carbon contamination	Do not strike arc on carbon block
	Air drafts	Shield torch with windbreak; raise gas flow
	Magnetic effects	Alter ground lead; use nonmagnetic hold-down clamps

Condensed from *"Joining of Stainless Steels"* published by ASM increase to the high current of the next pulse without delay. The pulsing GTAW process can produce a more uniform weld in thin material and root passes, than a constant current of the same average current.

AUTOMATIC GTAW WELDING

Stainless steels are readily welded with automatic GTAW. Arc voltage is proportional to arc length — thus, a reliable signal can be generated to operate automatic arc-voltage control equipment. Filler metal may be used, or light-gage material may be joined by simple fusion of the joint edges. When "cold" filler metal is used, it is always added to the front of the puddle.

Hot-Wire Welding

The so-called "hot-wire" method of welding gives greatly increased deposition rates and welding speeds. The wire — which trails the torch, as illustrated in Fig. 7-13 — is resistance-heated by a separate AC power supply. It is fed through a contact tube and extends beyond the tube. The extension is resistance-heated so that it approaches or reaches the melting point before it contacts the weld puddle. Thus, the tungsten electrode furnishes the heat to melt the base metal and the AC power supply furnishes a large portion of the energy needed to resistance-melt the filler wire. The hot-wire method is, in effect, an adaptation of the long-stickout principle used in submerged-arc and self-shielded flux-cored arc welding. The wire used for hot-wire GTAW welding is usually 0.045-in. diameter. Since the wire is melted — or very nearly melted — by its own power sources, the deposition rate can be controlled almost independently of the arc.

Using the GTAW hot-wire method, deposition rates up to 18 lb/hr can be achieved when welding at 400 to 500 amp DCEN (Table 7-25). Still greater deposition rates can be obtained using an automatic oscillated welding technique. Voltage control is essential to achieve control of the large puddle when welding at high deposition rates. For this reason, GTA hot-wire welding requires the use of voltage-control equipment.

Multiple-Electrode Welding

By using closely spaced multiple tungsten electrodes, the welding speed can also be increased substantially when GTAW welding stainless-steel tubing or sheet. Multiple electrode practically eliminate the problem of undercutting at high speeds.

TABLE 7-25. Typical Speeds and Deposition Rates with GTAW Hot-Wire

Wire size: 0.045-in.
Shielding gas: 75% He, 25% A
Electrode: 5/32-3/16-in. diam, 2% Th

Arc current, amp	Voltage	Speed, in./min	Wire, in./min	Deposition rate, lb/hr
300	10-12	4-10	110-370	3-10
400	11-13	6-14	185-445	5-12
500	12-15	8-20	295-665	8-18

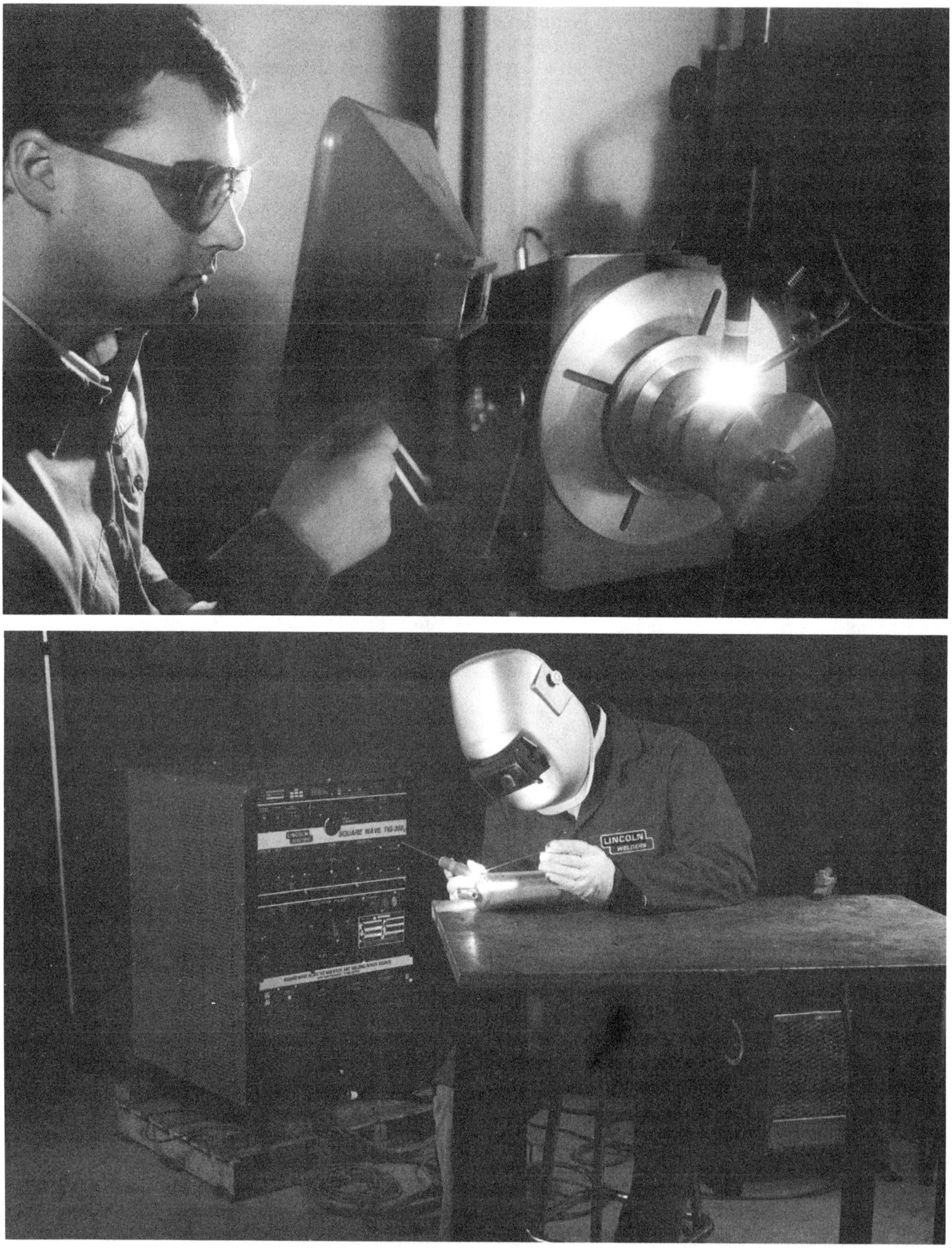

a. Automatic gas tungsten arc welding of stainless steel pipe with internal gas shield.
b. Manual gas tungsten arc welding of stainless steel assembly.

Section 8

WELDING CAST IRON AND CAST STEEL

SECTION 8.1

WELDING CAST IRON

SECTION 8.2

WELDING CAST STEEL

Welding Cast Iron

Types of Cast Iron

There are four major types of cast iron — gray iron, malleable iron, ductile (nodular) iron, and white (or chilled) iron. Most cast-iron parts are made in gray iron. Gray cast iron, malleable iron and ductile (nodular) iron are difficult to weld but they can be welded when the proper procedures are used. Malleable iron and ductile iron require long post heat treatment to restore their normal strength and ductility. White iron, on the other hand, is so difficult to weld that it is generally considered to be unweldable. Approximate compositions of these materials are listed in Table 8-1.

Gray Iron: This material is the most common type of cast iron, and it can be welded if certain precautions are observed concerning preheat and cooling rates after welding. Gray iron contains considerable free graphite, which gives broken surfaces a characteristic gray color. Combined carbon in gray iron is usually less than 0.8%; the remaining carbon is in the form of graphite.

White Iron: This type of iron is an exceptionally hard material, produced by pouring the molten metal against metal or graphite cooling surfaces (called chills or chillers) inserted in the mold. The chills cause the molten metal to solidify and cool very rapidly. When this happens, much of the carbon in the metal remains in solution rather than precipitating out as free graphite, as in gray iron. The casting thus retains a high proportion of iron carbide and is therefore harder and more brittle than ordinary gray iron, even though it has the same chemical analysis. (White iron, however, usually has less silicon than gray iron.) The rapidly chilled portions are extremely hard and are unmachinable. White iron has a fine grain structure and a characteristic white appearance upon fracture.

White iron is seldom used as cast, except for parts such as dies, rollers, and molds. It is usually cast as the first step in the making of malleable iron, a much more useful material.

Malleable Iron: This material is a white cast iron that has been heat treated by a long annealing process (usually lasting 2 or 3 days) that changes the combined carbon (iron carbide or cementite) into iron and a uniformly distributed "temper carbon" (ferrite and free graphite). This heat treatment increases both strength and ductility. Malleable iron is very much tougher than white or gray cast iron and can be bent an appreciable amount. Gray cast iron will break in a brittle fashion with very little bending. White cast iron will break like glass with essentially no bending. When malleable iron is welded, the beneficial effects of the heat treatment are destroyed in the vicinity of the weld by the rapid chilling from the welding temperature. The fusion zone reverts back to chilled cast iron and must again be heat-treated to return to the malleable condition.

Ductile Iron: Although quite similar to gray cast iron in composition, ductile iron possesses much of the strength and ductility of steel. The internal notch effect of flake graphite in gray iron is a prime factor in reducing its tensile strength, toughness, and ductility. In ductile iron, this graphite is converted into spheroids that minimize the notch effect and thereby produce a product combining the strength and ductility of cast steel with the low cost, wear resistance, and damping characteristics of cast iron.

An essential difference in composition between gray cast iron and ductile iron is a small amount of magnesium or cerium in ductile iron. These elements form fine oxide particles which serve as nuclei for graphite to form into spheroids. Also, in gray iron the silicon content is held within lower limits, since silicon has a coarsening effect on flake

TABLE 8-1. Compositions of Cast Irons (% of Constuents)

	Fe	Total C	Si	S	P	Mn
GrayIron	Balance	2.0 - 4.0	1.0 min	0.2	0.6	1.0 max.
Malleable Iron	Balance	2.0 - 3.0	0.9 - 1.8	0.2 max	0.2 max	0.25 - 1.25
Ductile Iron	Balance	3.2 - 4.1	1.8 - 2.8	0.03 max	0.1 max	0.80 max.
White Iron	Balance	2.5 - 4.0	0.4 - 1.6	0.15	0.4	0.3 - 0.8

TABLE 8-2. Recommended Current Ranges (amp) for Welding Cast Ferrous Metals

AWS Class	Electrode Size (in.)			Polarity
	1/8	5/32	3/16	
ECI	150 - 175		175 - 200	DCEP or AC
ESt	80 - 100			DCEP or AC
ENi-CI	60 - 110	100 - 135		DCEP
	65 - 120	110 - 150		AC
ECuSn-C	50 - 125	70 - 170	90 - 220	DCEP

graphite. This effect has not been noted with graphite nodules.

ELECTRODES FOR CAST IRON

Electrodes for welding cast irons are specified in AWS A5.15-90, as shown in Section 4.1, Table 4-22. Recommended current ranges are shown in Table 8-2. These electrodes are classified in the following manner.

RCI Group: Intended for oxyacetylene welding.

ECI Group: A covered electrode with cast-iron core wire.

ESt Group: A covered electrode with steel core wire. The weld deposit is not machinable.

ENi Group: A covered electrode with nickel-base core wire. Welds are machinable.

Copper Base Group: Intended for braze welding and surfacing cast iron.

Details concerning electrode application are presented in the following text. In brief, the ECI group provides a weld most similar to that of the base metal of the casting. This electrode cannot be used, however, unless the casting can be moved to a furnace and subjected to an extremely high (red-hot) preheat. The ESt group is used where lowest cost or maximum strength are desirable, and a nonmachinable deposit is acceptable. The ENi group provides a soft, ductile, machinable deposit and fusion zone with a slight sacrifice in strength and an increase in cost.

WELDING GRAY AND MALLEABLE IRONS

A gray-iron casting is made by pouring molten iron into a mold. The metal is allowed to cool slowly. As it cools most of the carbon in the metal precipitates out as a flaky form of graphite. This graphite gives the iron its gray color and some of its properties.

When the casting is welded, the area adjacent to the weld is raised above the critical temperature — about 1450°F. The mass of the casting around the weld tends to draw the heat from the weld area rapidly. If this cooling of the heat-affected zone and the weld is more rapid than it was when the casting was originally made, a highly brittle, crack-sensitive zone develops.

The best method to slow the cooling rate is to preheat the casting to prevent the iron from rapidly absorbing heat from the weld area. Some castings, however, cannot be preheated. Welding techniques and electrode selection must therefore take into account whether or not the casting can be preheated. White iron is generally considered unweldable and is not discussed in the following.

Electrode Selection and Welding Guidelines

Practically the same welding procedures are used for both gray iron and malleable iron. The prime difference is the need to heat-treat malleable iron to restore its properties after welding.

Small Parts: A considerable amount of welding on cast iron is for the repair of parts, such as brackets, sprockets, and levers. Such parts can be readily preheated and welded on the work bench, and no subsequent machining is required. They are usually welded with ESt electrodes for good strength and low cost. This type of electrode is best for penetrating into corners, V grooves, and other hard-to-get-at places and for use on "dirty" castings with sand inclusions or for castings that are soaked with oil or water. ENi electrodes provide a better bond on clean castings and provide a softer and more ductile weld and fusion zone.

The ESt deposit and fusion zone are not machinable unless several passes are used. ENi deposits are machinable regardless of the number of passes. A preheat (500°F) for these small parts is recommended to retard the cooling. The preheat also provides the added advantage of removing oil and water from the cast surface.

Large Parts with Preheat: Many parts such as flywheels and machine frames, although large, can be moved for preheat and for controlled cooling into a furnace or temporary firebrick oven. Preheat is especially recommended for large parts subjected to extreme shock or severe loading. Recommended procedure is to heat the casting to the red-hot condition and weld with cast-iron filler metal and a carbon arc, or with a cast-iron electrode (ECI).

These procedures produce a weld most nearly approaching the characteristics of the cast metal.

The weld is machinable if the casting is cooled slowly.

Large Parts without Preheat: When no furnace is available, or when the part is too large to move to a furnace, the weld must be made without benefit of preheat. Also, preheat may not be necessary if only light loads are applied to the casting. ENi electrodes are normally used in these applications, since they provide a soft, ductile deposit without preheat.

Liquid-Tight Welds: Occasionally, parts such as water jackets, cylinder heads, valve ports, or gear cases must be repair-welded, and the need to maintain liquid-tightness is a primary concern. In many such cases, preheat cannot be used because it may warp critical mating surfaces. Here, the ENi electrode will produce a weld least likely to crack or check if the casting is clean (not oil or water soaked), and if the preparation can be beveled and the weld made in a flat position. The deposit is machinable, and the fusion zone also may be machinable if a second pass is made to reheat and soften this zone.

An ESt electrode is recommended if the weld is hard to reach or not in the flat position, and if the casting is dirty. Both deposit and fusion zone will then be hard and unmachinable. A typical repair weld in a cylinder head is shown in Fig. 8-1.

Correcting Defects: Welding is often used to repair defects at the foundry or to correct machining errors. The usual practice is to weld with an ENi electrode. Where possible, a local preheat should be used, or more than one pass should be made to retard chilling and produce a fusion zone that is as soft as possible. A typical weld of this type, made with an ENi electrode, is shown in Fig. 8-2.

Welding Steel to Iron: Damaged iron castings are sometimes repaired by the addition of a welded patch of mild steel. Also, steel is sometimes welded to cast iron in production. The welding of steel to cast iron never produces a joint of the same toughness as that between two pieces of mild steel, but in some applications a joint with satisfactory strength can be made with the ESt electrode. The bond to the steel will be strong, but the bond to the cast iron will usually be brittle. The use of the ENi electrode will make a somewhat stronger bond to the cast-iron part at higher cost. The use of lower-cost ESt electrode produces a strongly caustic slag that must be removed thoroughly before painting. This extra cleaning cost should be considered in choosing between the two electrodes if painting is used.

On some production applications where thin sheet metal is to be joined to cast iron, small-diameter, low-hydrogen E7018 electrodes with low currents can be used. The fusion zone in the cast iron will be brittle, so this method of joining should not be used where impact or bending loads are involved. Nor is it recommended for liquid-tight welds. It can, however, be tried experimentally as a lower-cost alternative to ENi electrodes. Welds made with E7018 can be painted after the usual cleaning.

Preparations for Repair Welding

Locating Cracks: Much of the welding on cast iron is to repair cracks that have developed in service. A simple method for locating surface cracks in castings is to swab the damaged area with a kerosene-soaked rag to remove grease and dust.

Fig. 8-1. Repair of a crack in a cast-iron cylinder head with an ESt electrode.

Fig. 8-2. A machinable repair weld made in cast iron with an ENi electrode.

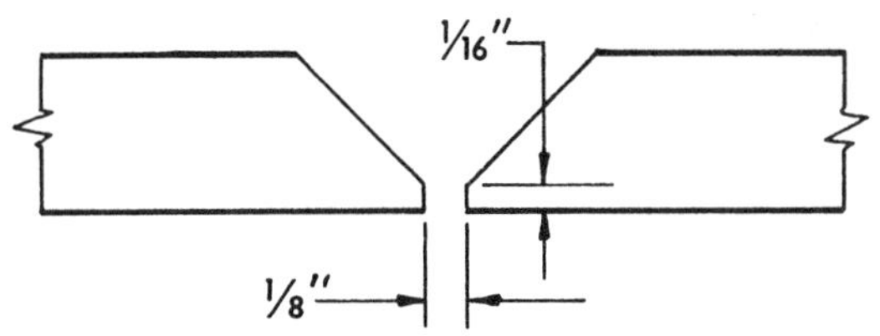

Fig. 8-3. Recommended preparation for repairing cracks extending through thick cast-iron parts.

Allow the kerosene to penetrate any cracks; then wipe the area dry. Immediately chalk over the area with common blackboard chalk. After a few minutes, even cracks not visible to the naked eye become evident by the kerosene in them bleeding back through the chalk. Dye-penetrant procedures may also be used for detecting cracks in castings (see Section 11.2)

Cleaning the Joint Area: Make certain the casting is clean. Remove surface oil, rust, water, and other foreign material. If possible, grind off the "skin" of the casting, which may contain sand or other impurities from the mold. To prevent porosity, especially when using the ENi electrode, remove all water and oil that may have penetrated deeply into the pores of the casting. This can best be done by heating to almost red heat for a short interval, or heating to a lower temperature (400°F) and holding for about an hour. The first pass on a contaminated casting may have porosity resulting from imbedded contaminants, but subsequent passes usually will be satisfactory due to the heating by the first bead.

Preparing the Joint: Prepare the joint with a V groove so that a weld with a cross section at least equal to material cross section can be made without deep penetration and admixture of the base metal into the weld deposit. Sometimes when repairing a crack the heat of the arc will create stresses that cause the crack to propagate. To avoid this, drill a small hole at each end of the crack. Grind, chip, machine, or saw the crack to create the V groove. Gas-cutting or arc-gouging can be used on castings that are preheated for welding. Be sure the groove extends to the bottom of the crack. On sections more than 3/16-in. thick, bevel the edges so the root of the joint is 1/8-in. to 3/16-in. wide. If the crack extends through the section, leave about a 1/8-in. gap and 1/16-in. root face, as illustrated in Fig. 8-3. Remove surface scale by grinding wherever welds are to be placed. Be sure the work is clean and dry.

Preheat and Postheat: A softer and more ductile weld and fusion zone is obtained if the casting is preheated before welding and cooled very slowly. Preheating to 400°F is satisfactory, but higher temperatures (up to 900°F) are usually more effective. For best results, keep the casting at this temperature throughout the welding. If the casting drops below the preheat temperature during welding, reheat before allowing to cool. For slow cooling, the welded casting should immediately be placed in lime, dry sand, or other insulating material to keep the heat from being dissipated rapidly.

Welding without Preheat: Most small castings are welded satisfactorily without preheating, although preheat usually improves the toughness of the weld. Slow cooling, however, is always essential. Where preheating is not used, a two-pass weld is recommended if a soft fusion zone is desirable. The first bead serves as a preheating for the second bead, which tends to slow the cooling and produce an annealing that makes a softer, more machinable fusion zone.

On large parts that must be welded without preheat, the primary goal is to keep the casting as cool as possible. Use small-diameter electrodes and low currents. Deposit short welds spaced apart so that one may cool while the other is being deposited. It may be necessary to stop welding at intervals to allow the casting to cool. The principal reason for keeping the temperature down is to prevent warpage or possible breakage from uneven heating and expansion.

Admixture of Base Metal: Use just enough current to obtain good fusion without voids or slag inclusions. Every attempt should be made to minimize the amount of cast iron melted into the weld. When ESt electrodes are used, the deposited metal becomes high-carbon steel due to admixture of the high-carbon cast iron. If this deposit cools quickly, it becomes extremely hard and brittle, and may crack. Moreover, it is not machinable.

The use of ENi electrodes usually solves this problem. As contrasted with steel rod, nonferrous alloys do not harden appreciably when deposited on cast-iron base metal. The weld deposits are therefore machinable. However, the hardening of cast iron in the casting adjacent to the line of fusion (due to the quenching action of the mass of cold metal under it) remains much the same as in the case of welding with a steel electrode.

Controlling Bead Shrinkage: The shrinkage or contraction of steel as it solidifies from the molten state is greater than that of cast iron. Therefore, when the molten steel from a steel electrode is deposited on cast iron, the steel shrinks more than the cast iron on which it is deposited.

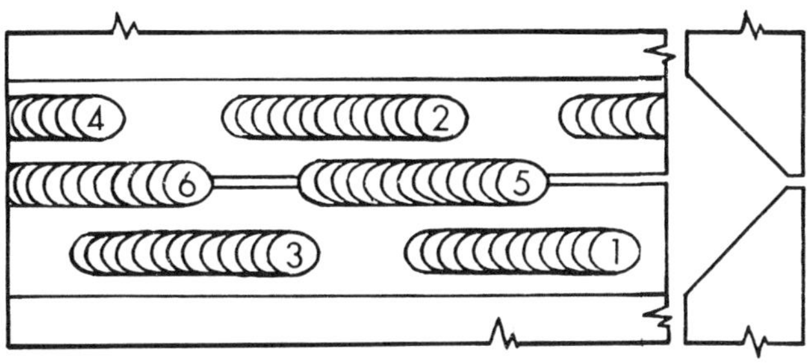

Fig. 8-4. A sequence of short welds helps avoid the cumulative stresses that would occur with long weld beads. Peening these short welds before they cool stretches the metal and reduces stress.

This differential shrinkage causes a residual stress in both the weld metal and the cast iron. A bead of weld metal deposited on cast iron will be in tension if allowed to cool without further treatment. At the same time the cast iron on which the weld metal has been deposited will be stressed due to the shrinkage of the weld metal on the cast iron. Since the cast iron is weaker, failure may occur (particularly when the bead of weld metal is long) as a break in the cast iron just outside of the line of fusion. If the weld is thin in cross section, failure may occur as cross cracks in the bead.

Since the stress is cumulative, it increases with the straight length of weld. To reduce the stress, deposit the weld metal in short lengths and allow each to cool (Fig. 8-4). For example, weld 15 to 30 seconds and then allow the weld to cool for 3 to 5 minutes. By depositing small welds in various parts of the work one weld is allowed to cool and contract while the operator is depositing metal in another location. An alternative is to upset or peen the deposited weld metal lightly before it has a chance to cool and contract. This causes the weld metal to stretch. In many cases, it is best to use a combination of short welds and peening. Use small-ball (1/2-in. — 3/4-in.) peening hammers.

Castings with High-Graphite Content: Castings having a great amount of large, free graphite flakes are difficult to weld, since the weld metal does not fuse to the graphite. Instead, it tends to ball-up and resists "working" into the melted surface of the casting. This behavior is most common on castings that have been heated repeatedly, such as old furnace grates or exhaust manifolds. The ESt electrodes for cast iron have a coating designed to flux out some of this flaky graphite, thereby improving fusion.

"Buttering:" Where large V's are to be welded, it is sometimes beneficial to line the sides with an ESt electrode, and then fill in the remaining void with mild-steel electrode E7018. This process is called "buttering." Shrinkage is reduced since the layers of deposited metal on each side of the groove can cool without restraint, and a smaller volume of deposit subject to restraint is required for filling the final groove.

Studding: As explained previously, the chilling action of the cast iron increases the combined carbon and, in turn, increases hardness and brittleness, thereby weakening the strength of the cast iron just outside of the line of fusion. Welds in cast iron, if of sufficient thickness, may be strengthened mechanically by a process called studding. Studs of steel approximately 1/4 in. to 3/8 in. in diameter should be used. The crack in the casting should be V-cut, drilled, and tapped so that the studs may be screwed into the V or the area of reinforcement. The studs should project about 3/16 in. to 1/3 in. above the surface, and should be long enough to be screwed into the casting to a depth of at least the diameter of the studs. Figure 8-5(a) illustrates the practice of studding.

The cross-sectional area of the studs should be about 25% to 35% of the area of the weld surface. In such cases the strength of the weld may safely and conservatively be taken as the strength of the studs. It is considered good practice to first weld one or two beads around each stud, making sure that fusion is obtained both with the stud and cast-iron base metal. Straight lines of weld metal should be avoided if possible. Welds should be deposited intermittently, and each bead should be peened before cooling.

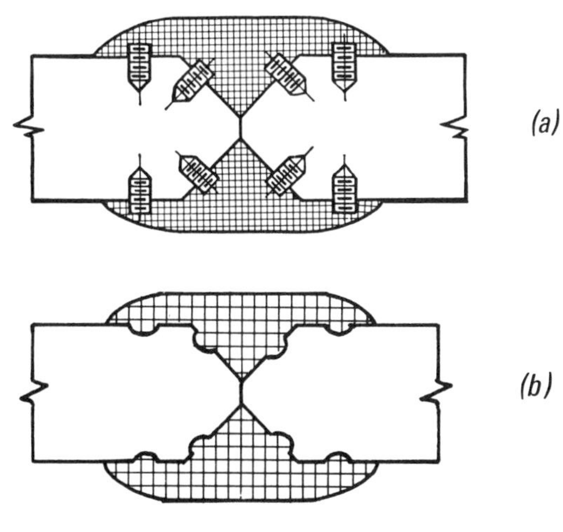

Fig. 8-5. A typical studded joint (a). An alternative to studding is to tool round-bottom grooves or locks into the joint preparation (b).

It is advisable, where the casting is of sufficient thickness, to use V grooves on both sides. In many cases it may be desirable to produce complete penetration at the fracture. In some cases it is better practice not to stud, but to shape grooves or locks into the casting with a round-nose tool. This method of preparation is shown in Fig. 8-5(b).

WELDING DUCTILE IRON

Ductile iron, sometimes referred to as nodular iron, is available in several different metallurgical conditions. One ferritic form provides high ductility; another ferritic form provides high strength. A pearlitic form provides high strength, and an austenitic form provides maximum resistance to corrosion, high temperatures, and thermal shock.

Ductile iron is readily welded when in the ferritic and pearlitic state. When maximum strength and ductility are desired, the part should be in the fully annealed condition. The main concern in welding, then, is to reduce the rates of heating and cooling to prevent cracking. The more complex and massive the assembly, the longer the time and higher the temperature required for heat treatment. It is necessary to preheat the part to 550-750°F and to reduce cooling by external heating. To insure good penetration, it is desirable to remove the cast-iron skin.

Ductile iron can be welded by most commercial welding processes. Regardless of the process used, preferred practice is to use electrodes manufactured specifically for nodular iron. These electrodes have a composition of approximately 60% Ni and 40% iron, with a carboniferous lime-spar flux or coating. With metal-arc welding, the E310-15, E310-16, and E7018 electrodes have been used successfully. Amperage should be low to insure minimum dilution and minimum local heating. Recommended amperages with various sizes of E7018 electrodes are:

Electrode Diameter	Recommended Current (amp)
3/32-in.	60- 70
1/8-in.	90-100
5/32-in.	120-130
3/16-in.	140-150

Parts should be covered with an insulating material to assure slow cooling. For best results, postheat treatments to 850°F for four hours, or 1000°F for two hours should be used for the relief of stress.

Welding Cast Steel

Arc welding is used both to make repairs to steel castings and as a means of joining castings together in the manufacture of cast-steel structures. Castings may also be welded to steel plate or to wrought steel in the fabrication of machine or structural parts. Figure 8-6 is an example of the use of submerged-arc welding to join the two cast halves of a large structural unit. Here, weld-joining of two castings makes practical a final assembly without the heavy flanges that would otherwise be required for bolt-joining.

Despite its usefulness in joining cast parts, or in joining castings to plate or wrought steel, the main use of welding in the casting industry is for the repair of defects. Defects, such as surface blemishes and voids, are consequences of the complexity of the casting operation, and the more complex the casting, the more likely it is to have defects. Welding is also used extensively to repair cracks and breakages that occur with cast-steel parts in service.

The metallurgical problems and chill effects encountered in welding cast steel are not as critical as those encountered when welding cast iron. Steel castings can, in fact, be welded as readily as wrought steels of the same composition and heat treatment. Cast steel, however, is usually somewhat more difficult to weld than wrought steel because of the thermal stresses and resulting tendencies for cracking that normally occur in thick sections typical of castings.

Manual shielded metal-arc welding is used in most repair work on steel castings. Automatic and semiautomatic welding processes give cost advantages with large-volume welds and in repetitive operations. The most widely used of the mechanized processes are gas metal-arc and submerged-arc. Electrode composition can sometimes be matched closely to the composition of the casting. Alloying elements may be supplied by the flux when using the submerged-arc process.

PREPARATION FOR WELDING

The methods for removing defects in steel castings are similar to those used for cast iron (see Section 8.1), and include chipping (with a pneumatic or hand chisel), grinding, arc gouging, and flame gouging. Within reasonable composition limits, chipping, grinding, and arc gouging may be used regardless of the temperature of the casting. Flame gouging usually requires that the casting be preheated to avoid the formation of checks or cracks in the surface of the casting.

All traces of a defect must be removed from a casting prior to repair welding. Defect removal is sometimes difficult with deep cracks, since the operation of removal may "smear" metal across the crack, obscuring it. In such cases, magnetic-particle or dye-penetrant examination, or deep acid etching, are needed to determine whether the defect has been removed completely. Where possible, the sides of the prepared surface should taper toward the bottom of the groove, and all changes in contour should be made gradually.

Fig. 8-6. The two cast halves of this structural unit are being joined by submerged-arc welding. Although weld-joining of cast-steel parts simplifies complex castings by eliminating design provisions for bolt-joining, most of the welding on castings is for repair of defects and blemishes.

Sometimes, two or more discontinuities may be located so close together that one is not detected during inspection. After the known defects are removed, the repair weld may penetrate through the sound metal between the defects to expose the undetected defect. When this occurs, welding should be stopped until the entire defective area can be removed.

When defects extend entirely through a casting, the recommended practice is to turn the casting over and lay one or more sealing passes from the root side of the repair weld. If this cannot be done, a backing should be used wherever possible. Backings can be made from refractory materials such as fire-brick, silica brick, mullite brick, or magnesite. The backings must be thoroughly dry, since water vapor can cause underbead cracking. Metals, such as copper, carbon steels, or stainless steel may also be used as backing material.

EFFECTS OF WELDING

Quench Hardness: The rapid cooling of the high-temperature zone of the base metal next to the weld quench-hardens that zone. Just as in conventional heat treating, the maximum hardness in the heat-affected zone depends primarily on the carbon content of the casting. Other elements may increase maximum hardness somewhat, but the primary effect of these elements is to prevent the transformation of austenite at high temperatures and thus promote deeper and more complete hardening.

It was once thought that underbead cracking was directly related to hardness, because underbead cracking was always found in zones of high hardness. It has since been shown that factors other than hardness can produce underbead cracking. When free from cracks, the principal drawback of a hardened heat-affected zone is that it makes machining difficult. Also, this zone may have low notch toughness.

Fortunately, most castings with welds are, at least, stress-relieved after welding. This eliminates detrimental hardness in the heat-affected zone.

Thermal Stresses: The rapid heating and cooling of the weld may produce high thermal stresses. At times, these stresses are of no concern and are virtually eliminated by heat treatments ordinarily given castings after welding. But these stresses sometimes are high enough to cause hot cracks or permanent distortion, or they may propagate cold cracks.

Hot Cracking: Hot cracking occurs while the heat-affected base metal or the solidified weld metal is still at such high temperature that it cannot withstand even relatively low internal stresses. If stress is developed while the metal is in this weak condition, the casting tears instead of deforming as it would at a slightly lower temperature.

Hot cracks are more likely to occur in the weld metal than in the base metal. One reason for this is that the weld metal is hotter than the base metal and cools last. Also, because of directional freezing, weld metal has columnar grains. Most theories propose that hot cracks are initiated at temperatures above the solidus temperature of the lowest melting phase present in the alloy — that is, when the lowest melting phase is partially molten. However, metallurgists do not completely agree on the mechanism of hot cracking.

Cold Cracking: The term "cold cracking," referring to cracking occurring at temperatures well below the critical temperature, has until recently been applied exclusively to underbead cracking of the base metal in the heat-affected zone. This type of cracking is discussed in Section 6.1. Welds in steel castings are generally not subject to cold cracking if they are properly preheated.

Porosity: Another problem sometimes encountered in welding castings is weld-metal porosity. Porosity can be minimized by slowing down the freezing of the molten weld metal. This allows the gases coming out of solution to escape to the surface of the weld metal. Where severe porosity is found, the weld should be chipped out and the groove rewelded.

THE WELDING OPERATION

Preheat: The best way to minimize possible difficulties from cold cracking, hot cracking, permanent deformation, and porosity is to slow the heating and cooling rates of the weld metal and weld heat-affected zone. This is done by selecting suitable preheating temperatures, interpass temperatures, and postwelding heat treatments.

Two types of preheat are used — a general preheat, in which the entire casting is heated, usually in a furnace; or a local preheat where only a section around the weld is heated, usually by a gas torch or electric-resistance heater. General preheat is preferred, since it minimizes local stresses.

Steels low in carbon and alloy contents do not tend to quench-harden excessively, nor are they prone to cold-crack. They require either no preheat, or require much less preheat than needed for a high-carbon, high-alloy casting. This fact also applies to the deposited weld metal, particularly in multipass welds. In general, carbon and low-alloy castings with carbon contents below 0.30% do not require preheat except on thick sections.

For castings with carbon contents between 0.30 and 0.50%, preheats of 200 to 400°F should be used. The higher side of this range should be used with the higher carbon and alloy contents and for thick sections and complex castings. In some cases, where both the carbon and alloy contents are high, preheats over 400°F may be necessary.

Thick castings have greater heat-absorbing capacities, and, therefore, in effect, greater quenching powers. As a rule, the thicker the casting, the higher the preheat should be. Simple shapes with fairly uniform cross sections cool uniformly, and require minimum preheat. Complex castings consisting of alternating thin and thick sections cool more rapidly in the thin sections, and this differential cooling may lead to severe internal stresses. In such cases, a higher preheat should be used than for a more uniform shape. Also, such castings should be protected from drafts during welding and should be cooled more slowly than less complex castings of the same composition. Simple calculators for determining preheats are available (see Section 3.3).

Another factor influencing the selection of a preheating temperature is the size of the defect in relation to the thickness of the casting. A small weld cools more rapidly than a large one. Thus, tack welding on a crack-sensitive steel can be a dangerous procedure. Many casting failures have resulted from cracks started from a tack weld made without preheat.

Interpass Temperature: The interpass temperature — the temperature between passes in multipass welds — should be considered along with preheat. To maintain the desirable conditions developed by preheat, the interpass temperature should never be below the preheat temperature. The interpass temperature can safely exceed the preheat temperature by 100 to 200°F, depending on the particular casting. The interpass temperature should not be allowed to go so high that the side walls of the weld groove are severely undercut.

With large castings, the welding heat may not be high enough to maintain the preheat temperature between passes. In this case, additional heat is supplied by reheating the entire casting or locally heating the weld area.

Peening: Distortion is sometimes minimized by peening the weld. Heavy peening after each pass reduces distortion by spreading the weld metal, thus counterbalancing its natural shrinkage. However, peening must be carefully controlled, because over-peening may cause weld-metal cracking. The permissible amount of peening depends somewhat on the mass of the casting, heavier peening forces being required for heavier castings. Peening should always be done while the weld metal is hot.

Postwelding Heat Treatment: Most castings can be cooled to room temperature in still air after welding, but those of high alloy content — which are more sensitive to cracking — should be cooled gradually from the welding temperature to give greater assurance against cold cracking. Cooling may be sufficiently delayed by heating with a flame or by burying the casting in some insulating material, However, inasmuch as crack-sensitive steels require preheating as well as slow cooling, one of the best practices is to put the casting back in the preheating furnace as soon as it is welded.

All welded castings, except some of the low-carbon unalloyed steels, should be given a postweld heat treatment. The minimum treatment should be a stress relief, which consists of heating to 1100 to 1250°F. Such a treatment removes practically all residual stresses and helps prevent subsequent cold cracking. But, most important, the heat treatment reduces the hardness of the heat-affected zone to the point where machining or chipping present no problem.

Many castings are given their regular final heat treatment, such as normalizing, or quenching and tempering, after welding. Normalizing eliminates stresses in the heat-affected zone and makes stress-relieving unnecessary. Quenching from the austenitizing temperature may re-introduce stresses. However these stresses will be removed completely or partially by the tempering treatment depending on the tempering temperature and time. Though normalizing or quenching and tempering generally improve the base metal, these operations may reduce the weld-metal properties if proper electrodes are not chosen for the welding operation.

Electrode Selection: Use only a process and an electrode that produces a low-hydrogen deposit. In choosing an electrode, the welding supervisor must know whether there are requirements for specific mechanical properties or for a specific chemical composition in the weld, and he must know whether there is to be a postweld heat treatment.

If mechanical properties only are important and the casting is to receive no postweld heat treatment, for stick welding select a low-hydrogen electrode from Table 4-4. Refer to Table 4-2 for the low-hydrogen types.

If mechanical properties only are important and the casting is to be stress-relieved (postheated to 1250°F or less), for stick welding select a low-hydrogen electrode from Table 4-7. This table is for stress-relieved weld properties. Refer to Table 4-2 for the low-hydrogen types. In selecting an electrode for welding processes other than manual shielded metal-arc, refer to the appropriate table in Section 6 or consult a supplier of electrodes. Also consult the supplier if the casting is to be normalized, fully annealed, or quenched and tempered.

If a specific chemical composition must be provided in the weld, an appropriate electrode may be selected from the AWS specifications. However, not all chemical compositions existing in castings are available as electrodes.

Welded subassemblies for the San Francisco-Oakland Transbay Tube.

Section 9
WELDING ALUMINUM AND ALUMINUM ALLOYS

SECTION 9.1
WELDABILITY OF ALUMINUM AND ALUMINUM ALLOYS

SECTION 9.2
WELDING ALUMINUM AND ALUMINUM ALLOYS WITH THE SHIELDED METAL-ARC PROCESS

SECTION 9.3
WELDING ALUMINUM AND ALUMINUM ALLOYS WITH THE GAS METAL-ARC PROCESS

SECTION 9.4
WELDING ALUMINUM AND ALUMINUM ALLOYS WITH THE GAS TUNGSTEN-ARC PROCESS

Weldability of Aluminum and Aluminum Alloys*

ALLOY TYPES

Wrought aluminum alloys are generally produced in five basic categories: a) flat plate or sheet; b) rod, bar, or wire; c) tubing; d) extruded shapes; and e) forgings. Aluminum alloys are further classified as heat treatable and non-heat treatable. The strength of the non-heat treatable alloys depends upon the solid solution strengthening of elements such as silicon, manganese, iron and magnesium. Additional strengthening is achieved by work hardening from the forming process such as cold rolling. Examples of non-heat treatable alloys are 1100, 3003, 4043, and 5052.

Heat treatable alloys contain additions of copper, magnesium, zinc and silicon either singly or in combination. The solubility of these elements in aluminum increases with temperature providing a capability for precipitation hardening (age hardening). These alloys can also be work hardened then hardened further by heat treatment. Examples of these alloys are 2014, 2219, 6061, 7039 and 7075. Work hardening in all alloys will be diminished by the heat of welding in the heat-affected zone and cannot be recovered by heat treatment. Heat treated hardness can usually be recovered by subsequent re-heat treatment.

The newer aluminum alloys and welding processes has enhanced the use of aluminum usage in welded structures. Among these alloys is the heat-treatable 7000-series, in which zinc is the major alloying element, with magnesium the other principal addition. Outstanding in this group is 7075, one of the strongest aluminum alloys available, used principally in airframe structures and highly stressed parts. Though not easily welded by GMAW or GTAW processes, it responds well to electrical resistance welding.

Alloy 7039 is of somewhat lower strength, but is readily welded by any of the fusion processes. In the T61 temper, this alloy, in thicknesses of 3/16 in. and over, is suitable for welded pressure vessels for use at −200° to −300°F under ASME Boiler and Pressure Vessel Code, Section VIII, Unfired Pressure Vessels. Alloy 7039 is also used for armor plate under MIL Spec. A-46063. Weldments using 7039 filler develop 48,000-psi tensile and 30,000-psi yield strengths after 15 to 30 days of aging at room temperature. Formability and corrosion resistance are excellent. Though more difficult to extrude than 6061 alloy, 7039 is more easily extruded than 2014. Alloy 7005, an extrudable AlMnZn alloy in the Series, also has good weldability.

It is suggested that the designer discuss potential applications with one of the major aluminum suppliers before decision is made on their use.

The aluminum-magnesium-manganese alloys — the 5000 series — are also widely used for welded structures, particularly where corrosion is a problem, as in marine applications. Weld ductility of these alloys is excellent, and they are well suited for applications subjected to impact and shock loading. The 5000-series alloys generally regress slowly to the zero temper when subjected to weld heat.

Heat-treatable alloys 6061 and 6063 also are widely used for welded structures. Their as-welded yield strengths are good, and 6063 is so easily extruded that it is a prime candidate for shapes for general architectural use.

The two welding processes used primarily for aluminum are GMAW (gas metal-arc welding) and GTAW (gas tungsten-arc welding). Both are recommended for most aluminum alloys: generally, GTAW for thicknesses to 1/8 in., and GMAW for heavier material, although this is not a hard and fast rule.

HEAT EFFECTS OF WELDING

Welding heat adversely affects weld-zone mechanical properties of all aluminum alloys. Strength of the weld area is less than that of the unaffected base metal. Although welds in sections that are in the annealed condition do not change the temper, the cast-structure of the weld metal usually has lower strength than the base metal. However, upgrading the weld metal with a suitable filler alloy, increasing the welding speed, and bead reinforcement usually strengthen the welded joint.

Some heat-treatable alloys in the T6 temper

* Section 9 extracted from *Welding Kaiser Aluminum* with permission form Kaiser Aluminum & Chemical Sales, Inc.

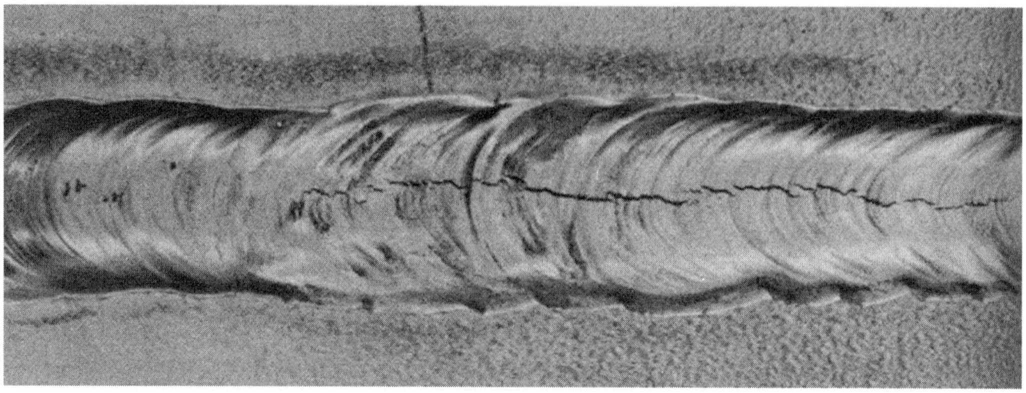

Fig. 9-1. Hot-short crack in aluminum weld metal.

(solution heat-treated, then artificially aged) are used in welded structural applications, where it is customary to proportion members on the basis of yield strength. The minimum tensile and compressive yield strength of 6061-T6 alloy sheet and plate in the heat affected zone adjacent to a butt weld is 20,000 psi with 5356 or 5556 filler alloy. This compares favorably with heat-affected zone yield strengths of other high-strength aluminum alloys.

Hot Cracking of Weld Metal

Hot cracking is caused by stresses induced by metal contraction as the weld pool cools. Cracking of this type usually develops in the weld metal itself, but a similar type of cracking sometimes occurs in the heat-affected zone adjacent to the weld.

Cracking of welds during arc-welding operations can also be caused by a hot-short** characteristic of some resultant weld compositions, as illustrated in Fig. 9-1. Hot-short compositions are weak at elevated temperatures. Hot cracking is a function of contraction stress, and of the hot-short tendency of the alloy. Hot cracking can also be caused by restraint of the joint during welding. Hot cracking does not normally occur in pure aluminum or in deposits of a eutectic composition.

Other factors that affect the amount of hot cracking that occurs in the weld include the solidification temperature range of the alloy, the coefficient of expansion, the thermal gradient, the ratio of the volume of liquid remaining to the volume of solidified material, and the amount of restraint across the weld joint.

Hot-short cracking can be reduced by several methods:

Use Higher Welding Speed. As welding speed is increased, the heat input in the weld zone is decreased. Thus, the stresses created by temperature differentials between the hot and room-temperature metal are less severe. Also, at high welding speeds shorter lengths of weld are within the hot-short range at any one time, and the already solidified metal takes a portion of the load that would otherwise be imposed upon the hot-short weld metal. The same effect is also advantageous in welding aluminum alloys that are not hot-short. In addition, the faster cooling rate that occurs with higher welding speeds results in a finer dendritic structure, which is less susceptible to cracking.

Use Preheat. Preheating reduces the stress on the solidifying weld metal by reducing the temperature gradient across the weld zone and by permitting faster joining rates. Preheating should not be used, however, unless the joint is unrestrained. The mechanical properties of the parent alloy are decreased by excessive preheating, except where the material is already in the annealed condition. This is particularly true of heat-treatable alloys such as 6061.

Changing the Joint Design. The chemical composition of an aluminum alloy affects its hot-short characteristics. The chemical composition of the weld metal is the product of a mixture of the base alloy and the filler metal, and this mixture may be highly susceptible to hot-short cracking. To reduce this susceptibility, the joint design can be changed to either increase or decrease the amount of parent alloy in the final weld metal.

Selection of the proper joint design usually requires that sample weldments be prepared to evaluate crack susceptibility in cases where data on comparable welds are not available.

Select A More Compatible Filler Alloy. A filler alloy that will minimize hot-short cracking is one that, when dilution is considered, will give a resultant weld composition having an individual alloy content above or below known hot-short ranges.

** Hot-shortness is defined as brittleness in metal in the hot forming range.

A proper aluminum filler selection or recommendation, using hot-short cracking tendency as a criterion, requires consideration of the relative cracking tendencies of various weld-metal compositions. For example, in binary alloys of aluminum, maximum cracking occurs in these composition ranges:

Alloying Element	**Amount** (%)
Silicon	0.5 to 1.2
Copper	2.0 to 4.0
Manganese	1.5 to 2.5
Magnesium	2.0 to 5.0
Zinc	4.0 to 5.0
Iron	1.0 to 1.5

FILLER METALS FOR WELDING ALUMINUM ALLOYS

The metal produced in the weld pool is a combination of metals that must have the strength, ductility, freedom from cracking, and the corrosion resistance required by the application. Correct choice of a filler alloy minimizes the presence of intermetallic compounds and brittleness in aluminum fusion welds.

Strength: Some filler metals commonly used for gas tungsten-arc welding, arranged in decreasing order of as-welded strength, reading left to right, are:

5039, 5556, 5183, 5356, 5654, 5554, 4043, and 1100.

This order of strength is generally true for these fillers when used to weld any alloy of the 5000, 3000, or 1100 series. This sequence does not necessarily apply to some combinations of dissimilar alloys or to assemblies subjected to postweld heat treatment.

Ductility: An arrangement of the same filler metals in decreasing order of ductility is almost the reverse of the arrangement by strength:

1100, 4043, 5654, 5356, 5183, 5554, 5556, and 5039.

Up to 50% elongation can be expected with 1100 filler, whereas 15% to 20% is normal for 5183 or 5356 filler. Ductility of the weld is often affected by dilution of the filler metal with the base material. For example, 4043 can vary from 5 to 17% elongation, depending on the type and quantity of alloy with which it is diluted. Table 9-1 lists recommended filler metals for various aluminum alloys, both for maximum strength and for maximum elongation. Table 9-2 contains more complete information on selecting filler metals for the general-purpose welding of aluminum.

TABLE 9-1. RECOMMENDED FILLER METALS FOR VARIOUS ALUMINUM ALLOYS

Base Metal	Recommended Filler Metal[1]	
	For Maximum As-Welded Strength	For Maximum Elongation
EC	1100	EC, 1260
1100	1100, 4043	1100, 4043
2219	2319	(2)
3003	5183, 5356	1100, 4043
3004	5554, 5356	5183, 4043
5005	5183, 4043, 5356	5183, 4043
5050	5356	5183, 4043
5052	5356, 5183	5183, 4043, 5356
5083	5183, 5356	5183, 5356
5086	5183, 5356	5183, 5356
5154	5356, 5183	5183, 5356, 5654
5357	5554, 5356	5356
5454	5356,5554	5554, 5356
5456	5556	5183, 5356
6061	4043, 5183	5356[3]
6063	4043, 5183	5356[3]
7005	5039	5183, 5356
7039	5039	5183, 5356

Notes:
1. Recommendations are for plate of "0" temper.
2. Ductility of weldments of these base metals is not appreciably affected by filler metal. Elongation of these base metals is generally lower than that of other alloys listed.
3. For welded joints in 6061 and 6063 requiring maximum electrical conductivity, use 4043 filler metal. However, if both strength and conductivity are required, use 5356 filler metal and increase the weld reinforcement to compensate for the lower conductivity of 5356.

Availability of Filler Wire: A relatively small number of filler alloys can be used to weld a wide range of aluminum alloys. Certain filler alloys — 5356 or 5183 for example — can be used for practically all aluminum fusion-welding applications.

Filler wire of alloys EC, 1100, 4043, 5039, 5654, 5356, 5183, 5554, and 5556 is available on 1 or 12-1/2-lb spools for use with the GMAW and automatic GTAW processes. Wire of 0.030, 3/64, 1/16, 3/32, and 1/8-in. diameters is readily available. For certain alloys, welding wire of 3/16, 7/32, and 1/4-in. is also available. Wire of alloys other than these is available upon request.

Straight lengths of bare aluminum wire in EC, 1100, 4043, 5039, 5654, 5183, 5356, and 5556 alloys are available in diameters of 1/16, 3/32, 1/8, 5/32, 3/16, and 1/4-in. This bare rod stock is used primarily for GTA welding.

Maximum rate of deposition is obtained with filler wire or rod of the largest practical diameter while welding at the maximum practical welding current. Wire diameter best suited for a specific application depends upon the current that can be used to make the weld. The current, in turn, is

TABLE 9-2. GUIDE TO THE CHOICE OF FILLER METAL FOR GENERAL-PURPOSE WELDING

Base Metal	319, 333 354, 355 C355	13, 43, 344 356, A356 A357, 359	214, A214 B214, F214	7039 A612, C612 D612, 7005 k	6070	6061, 6063 6101, 6151 6201, 6951	5456	5454	5154 5254 a	5086	5083	5052 5652 a	5005 5050	3004 Alc. 3004	2219	2014 2024	1100 3003 Alc 3003	1060 EC
1060, EC	ER4145 c,i	ER4043 i,f	ER4043 e,i	ER4043 i	ER4043 i	ER4043 i	ER5356 c	ER4043 e,i	ER4043 e,i	ER5356 c	ER5356 c	ER4043 i	ER1100 c	ER4043	ER4145	ER4145	ER1100 c	ER1260 c,j
1100, 3003 Alclad 3003	ER4145 c,i	ER4043 i,f	ER4043 e,i	ER4043 i	ER4043 i	ER4043 i	ER5356 c	ER4043 e,i	ER4043 e,i	ER5356 c	ER5356 c	ER4043 e,i	ER4043 e	ER4043 e	ER4145	ER4145	ER1100 c	
2014, 2024	ER4145 g	ER4145	....	....	ER4145	ER4145	....	....	....	....	....	....	....	....	ER4145 g	ER4145 g		
2219	ER4145 g,c,i	ER4145 c,i	ER4043 i	ER4043 i	ER4043 f,i	ER4043 f,i	ER4043	ER4043 i	ER4043 i	ER4043	ER4043	ER4043 i	ER4043	ER4043	ER2319 c,f,i			
3004 Alclad 3004	ER4043 i	ER4043 i	ER5654 b	ER5356 e	ER4043 e	ER4043 b	ER5356 e	ER5654 b	ER5654 b	ER5356 e	ER5356 e	ER4043 e,i	ER4043 e	ER4043 e				
5005, 5050	ER4043 i	ER4043 i	ER5654 b	ER5356 e	ER4043 e	ER4043 b	ER5356 e	ER5654 b	ER5654 b	ER5356 e	ER5356 e	ER4043 e,i	ER4043 d,e					
5052, 5652 a	ER4043 i	ER4043 b,i	ER5654 b	ER5356 e,h	ER5356 b,c	ER5356 b,c	ER5356 b	ER5654 b	ER5654 b	ER5356 e	ER5356 e	ER5654 a,b,e						
5083	....	ER5356 c,e,i	ER5356 e	ER5183 e,h	ER5356 e	ER5356 e	ER5183 e	ER5356 e	ER5356 e	ER5356 e	ER5183 e							
5086	....	ER5356 c,e,i	ER5356 e	ER5356 e,h	ER5356 e	ER5356 e	ER5356 e	ER5356 b	ER5356 b	ER5356 e								
5154, 5254 a	....	ER4043 b,i	ER5654 b	ER5356 b,h	ER5356 b,c	ER5356 b,c	ER5356 b	ER5654 b	ER5654 a,b									
5454	ER4043 i	ER4043 b,i	ER5654 b	ER5356 b,h	ER5356 b,c	ER5356 b,c	ER5356 b	ER5554 c,e										
5456	....	ER5356 c,e,i	ER5356 e	ER5556 e,h	ER5356 e	ER5356 e	ER5556 e											
6061, 6063, 6101 6201, 6151, 6951	ER4145 c,i	ER4043 b,i	ER5356 b,c	ER5356 b,c,h,i	ER4043 b,i	ER4043 b,i												
6070	ER4145 c,i	ER4043 e,i	ER5356 c,e	ER5356 c,e,h,i	ER4043 e,i													
7039 A612, C612 D612, 7005 k	ER4043 i	ER4043 b,h,i	ER5356 b,h	ER5039 e														
214, A214 B214, F214	....	ER4043 b,i	ER5654 b,d															
13, 43, 344 356, A356 A357, 359	ER4145 c,i	ER4043 d,i																
319, 333 354, 355, C355	ER4145 d,c,i																	

Notes:

1. Service conditions such as immersion in fresh or salt water, exposure to specific chemicals, or a sustained high temperature (over 150°F) may limit the choice of filler metals.
2. Recommendations in this table apply to gas shielded-arc welding processes. For gas welding, only R1100, R1260, and R4043 filler metals are ordinarily used.
3. Filler metals designated with ER prefix are listed in AWS specification A5.10.
   - a Base metal alloys 5652 and 5254 are used for hydrogen peroxide service. ER5654 filler metal is used for welding both alloys for low-temperature service (150°F and below).
   - b ER5183, ER5356, ER5554, ER5556, and ER5654 may be used. In some cases they provide: (1) improved color match after anodizing treatment, (2) highest weld ductility, and (3) higher weld strength. ER5554 is suitable for elevated temperature service.
   - c ER4043 may be used for some applications.
   - d Filler metal with the same analysis as the base metal is sometimes used.
   - e ER5183, ER5356, or ER5556 may be used.
   - f ER4145 may be used for some applications.
   - g ER2319 may be used for some applications.
   - h ER5039 may be used for some applications.
   - i ER4047 may be used for some applications.
   - j ER1100 may be used for some applications.
   - k This refers to 7005 extrusions only.
4. Where no filler metal is listed, the base metal combination is not recommended for welding.

governed by the available power supply, joint design, alloy type and thickness, and the welding position. For detailed information on filler sizes, refer to the discussions on GTA and GMA welding.

Quality of Weld Deposit: Good weld quality is obtained only if the filler wire is clean and of high quality. If the wire is not clean, a large amount of contaminant may be introduced into the weld pool, particularly in GMA welding, because of the relatively large surface area of the filler wire with respect to the amount of weld metal being deposited.

Contaminants on the filler wire are most often an oil or a hydrated oxide. The heat of welding releases the hydrogen from these sources, causing porosity in the weld. High-quality aluminum welding wire is manufactured under rigorous control to exacting standards and is packaged to prevent contamination during storage. Since filler wire is alloyed, or diluted, with base metal in the weld pool, the compositions of both the filler wire and the base metal affect the quality of the weld.

Care of Filler Wire and Rod: The best quality welds are obtained by using welding wire as soon as possible after it is removed from the package. Even a high-quality wire can produce welds of inferior quality if the wire surface is contaminated by oil, dust, or other foreign material. Consequently, spools of wire should be protected by a cover during use. Components of a wire-drive system (drive rolls, guides, and liners) can also contaminate the wire if they are not clean. Welding wire packages with desiccant in plastic bags should remain unopened until the wire is to be used.

When maximum weld quality is essential or the cost of repairing welds is prohibitive, questionable spools should be pretested before use in fabrication. Several test methods may be used, all of which involve making a sample weld that is checked by radiography or by examination of the weld fracture. A triple-pass horizontal fillet weld is the simplest for evaluation. A radiograph from a multipass flat-position butt weld is a valid check, provided that at least three passes are used. Such a test may save costly repair cutouts, particularly in welds that require radiographic inspection.

SURFACE PREPARATION FOR WELDING

Aluminum sheet and plate that is to be welded should be carefully stored and handled. Clean, dry storage is preferred. If outdoor storage is necessary, the plate should be covered or stored on edge with interspacing to prevent staining. Aluminum should not be primed or painted before welding.

Need for Cleaning

Pieces to be welded are usually formed, sheared, sawed, or machined prior to the welding operation. Complete removal of all lubricants from these operations is a prerequisite for high-quality welds. Particular care must be taken to remove all oil, other hydrocarbons, and loose particles from sawed or sheared edges prior to welding. Sheared edges should be clean and smooth, not ragged. For ease of cleaning, lubricants used in fabrication should be promptly removed.

To reduce the possibility of porosity and dross in welds, cleanliness of the welding surfaces cannot be overemphasized. Hydrogen can cause porosity, and oxygen can cause dross in welds. Oxides, greases, and oil films contain oxygen and hydrogen that, if left on the edges to be welded, will cause unsound welds with poor mechanical and electrical properties. Cleaning should be done just prior to welding. A summary of general cleaning procedures is given in Table 9-3.

TABLE 9-3. COMMON METHODS FOR CLEANING ALUMINUM SURFACES FOR WELDING

Compounds Removed	Type of Cleaning	
	Welding Surfaces Only	**Complete Piece**
Oil, grease, moisture and dust (Use any method listed.)	Wipe with mild alkaline solution. Wipe with hydrocarbon solvent such as acetone or alcohol for moisture. Wipe with proprietary solvents. Dip edges, using any of above.	Vapor degrease. Spray degrease. Steam degrease. Immerse in alkaline solvent. Immerse in proprietary solvents.
Oxides (Use any method listed.)	Dip edges in strong alkaline solution, then water, then nitric acid. Finish with water rinse. Wipe with proprietary deoxidizers. Remove mechanically, such as by wire-brushing, filing, or grinding. For critical applications, scrape all joint and adjacent surfaces immediately prior to welding, using no lubricant. Remove about one mil of metal.	Immerse in strong alkaline solution, then water, then nitric acid. Finish with water rinse. Immerse in proprietary solutions. Phosphoric acid and water rinse.

Oil or grease films are usually removed chemically by dipping, spraying, or wiping the aluminum sheet or plate with a solvent. Mildly alkaline solutions and commercial degreasers that do not emit toxic fumes during welding may be used.

All welding surfaces should be dried before welding. Cloth or absorbent paper is usually effective. Drying with compressed dry air is often used in difficult-to-reach areas. Some cleaning solutions — especially those containing alcohol or acetone — evaporate much faster than others.

Aluminum and its alloys rapidly develop a self-limiting oxide surface film upon exposure to air. The melting point of this oxide is above 3600°F, or about 2400°F above the melting point of pure aluminum. This oxide film can prevent wetting between the filler metal and base plate, or flakes of oxide or dross may become entrapped within the weld metal, reducing the ductility of the weld.

It is usually necessary to remove surface oxides mechanically just prior to welding. Chemically deoxidized parts may also require mechanical removal of oxide if more than a day has elapsed since deoxidation was performed. Welding should be done as soon as practical after oxide removal or cleaning operations, usually within 24 hours when resistance welding. When fusion welding, two or three days may be allowed to elapse after oxide removal, providing the joint area is wire-brushed immediately before welding. Excessively black or dirty-appearing inert gas welds may be an indication of excessive plate oxide.

Wire brushing, scraping, filing, and grinding are common mechanical cleaning methods used on aluminum plates. In wire brushing, stainless-steel bristles are preferred, but are not mandatory. The main disadvantage of using carbon-steel brushes is that iron deposits left by the bristles may rust if moisture is present. Large amounts of iron oxide may result in weld-metal inclusions. Carbon-steel brushes are satisfactory if they are grease-free, and both they and the work are kept dry. A moisture film that would rule out the use of carbon-steel brushes would also require recleaning the aluminum plate.

Examples of Cleaning Procedures

Example 1: Local cleaning of plate for gas tungsten-arc welding. Wipe all welding surfaces with rags dipped in acetone to remove oil and grease. Wipe surfaces dry with clean rags. Keep solvent-wet rags and solutions in closed containers to reduce fire hazard.

Wire-brush welding surfaces to remove the oxide. Power-driven stainless-steel wire brushes are preferable. Other mechanical methods of oxide removal may be substituted for wire brushing, such as filing, grinding, milling (without lubricant), rubbing with steel wool, and scraping.

Example 2: Local cleaning of sheet for gas tungsten-arc welding. Wipe the welding surfaces of the sheet with rags dipped in acetone or alcohol to remove grease and oil. Wipe dry with clean rags.

Scrape or file (with files designed for aluminum) edges of the sheet to remove entrapped contaminants. If high-quality welding is required, also scrape the top surface of the sheet to remove the oxide film from the area where weld metal will be deposited.

Example 3: Over-all cleaning of sheet or plate for gas tungsten-arc (or spot) welding. The sheet or plate is passed through a spray tunnel. Travel rate is determined by the final cleanliness required.

The tunnel is divided into four zones. In the first zone, grease is removed by spraying the plate with a 160°F water solution containing a compound such as Oakite 161, in the proportion of 1 oz/gal for this particular proprietary preparation. In the second zone, the plate is rinsed with a water spray at room temperature. In the third zone, the oxide film is chemically removed by spraying the plate with a water solution at room temperature containing 0.5 oz/gal of Oakite 34. In the fourth zone, the plate is again rinsed with fresh water at room temperature. An oil-free dry-air blast dries the metal after it comes from the tunnel.

The chemical solutions are periodically checked for strength. Fresh solution is added as required. The solutions are periodically discarded and new batches made up. Intervals for checking solution strength and replacing solutions depend upon the quantity of material being cleaned.

Example 4: Over-all cleaning of sheet and plate for gas tungsten-arc welding. When the sheet or plate is very oily, it should be degreased in a tank of acetone, alcohol, or in a suitable alkaline solution.

If the material is only slightly oily, the following procedure is recommended. Dip the plate into a tank of 5% alkaline solution (0.45 lb of sodium hydroxide per gallon of water) at 160°F for 20 to 60 sec. Next, rinse in a tank of fresh, cold water. Follow by dipping the plate into a tank of concentrated nitric acid at room temperature for 30 sec. Rinse again in the cold water tank and follow with a hot-water rinse. Dry the plate in an oil-free air blast.

This cleaning method is satisfactory for most aluminum welding, but not for electrical-resistance spot welding because the degree of cleanliness is not sufficiently reproducible.

Welding Aluminum and Aluminum Alloys with the Shielded Metal-Arc Process

Aluminum can be welded by the shielded metal-arc (stick-electrode) process, using a conventional DC motor generator, or a rectifier power supply such as that used for the stick-electrode welding of steel. An aluminum electrode, flux-covered and held in a standard holder, is used with a direct-current, electrode positive (DCEP) circuit. The aluminum electrode is the positive pole.

The repetitive quality of shielded metal-arc welds in aluminum is today generally considered below that required for ASME code application. In the United States, stick-electrode welding of aluminum is confined principally to small shops that have only occasional welding jobs and do not wish to invest in inert-gas welding equipment.

Important factors to be considered in the metal-arc welding of aluminum are moisture, preheating, flux, and the cleanliness of electrode and work. Presence of moisture in the electrode covering is a major cause of weld porosity. Dry, clean storage facilities for electrodes are necessary for producing sound welds. Preheating of the work is necessary, especially in the heavier gages. The flux that covers the electrode is very tenacious, and considerable operator skill is required to keep from entrapping it in the weld. Thoroughness in postweld removal of flux is also important.

Although metal-arc welded joints are as strong as oxyacetylene welded joints, it is difficult in metal-arc welding aluminum to obtain liquid-tight or gas-tight welds in material lighter than 1/4-in. thick. Soundness and surface smoothness are not as good as obtained with the gas-shielded arc-welding methods.

One difficulty encountered in metal-arc welding, caused by interruption of the arc, is the formation of a fused flux coating over the end of the electrode. Re-establishing a satisfactory arc is impossible until this formation is removed, usually by striking the rod against the work or other surface.

Square butt joints for metal-arc welding in aluminum can be used on plates up to 3/16 in. thick. No edge preparation other than a relatively smooth, square cut is required. Material of heavier gage, should have a single V included angle of 60° to 90°. Depending upon the gage of the base metal, land thicknesses range between 1/16 and 1/4 in. A root opening of 1/32 to 1/16 in. works well.

ELECTRODE SELECTION

The two types of covered electrodes for metal-arc welding of aluminum, covered by AWS Specification A5.3 and generally available from stock, are the 1100 and 4043 alloys. However, any aluminum alloy available in wire form may be obtained with a flux covering upon special order.

Alloy 1100 is a commercially pure aluminum (99% aluminum), giving a weld deposit with a minimum tensile strength of 12,000 psi. Alloy 4043, which is suitable for most general-purpose work, contains approximately 95% aluminum and 5% silicon and has a weld strength of 14,000 psi. In applications requiring resistance to corrosion, electrodes should be selected with compositions as close as possible to those of the base metal.

Most porous weld structures are caused by moisture in the electrode covering. To minimize this condition, the electrodes are supplied in moisture-proof packages. The electrodes should be stored in a dry location. Deterioration of the flux covering can be rapid when electrodes are exposed to moist air. The entire covering can be affected in only a few hours of exposure to a humid atmosphere.

Because of the necessity for coverings to be completely dry, it is advisable to bake all "doubtful" electrodes, or those taken from previously opened packages, at 350° to 400°F for an hour before welding. After baking, they should be stored in a heated cabinet until used. An old refrigerator or a similar, tightly sealed cabinet having shelves and fitted with a 100-watt light bulb, makes a convenient and economical storage chest.

PREHEATING

Preheating of aluminum members to 400°F is desirable prior to welding and is necessary in welding plate. This helps maintain the weld puddle and results in a more stable arc. Preheating may be done by torch, using oxyacetylene or other suitable gas, or by electrical resistance, using a tungsten electrode, silver-soldered to the top side of the contact jaw of an electrode holder. Current is adjusted to produce the desired temperature when the tungsten contact and ground clamp are connected to the work. This method is not practical for large workpieces.

WELDING TECHNIQUES

The method of striking the arc in (SMAW) stick-electrode welding of aluminum differs from that used for steel. Because both the aluminum electrode and base metal melt and resolidify almost immediately as welding progresses, electrode sticking can be a problem. This can be avoided if the arc is struck by moving the electrode over the surface of the base metal, using a brushing motion.

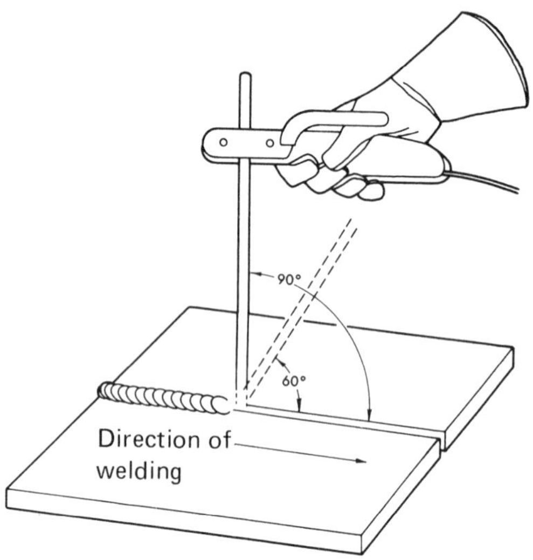

Fig. 9-2. After the arc is started (vertical position), the electrode can be slanted 20° to 30° in the direction of travel. Excessive slant can cause spatter and porosity.

Initially, the electrode should be held in a nearly vertical position, as shown in Fig. 9-2. After the arc has been established, a 20° to 30° forward slant can be used (electrode pointed toward the completed bead). Too much slant causes excessive spatter and porosity. The electrode is moved along the seam at an even speed to produce a uniform bead. Care should be taken in electrode manipulation to "float out" the flux in the molten metal as welding progresses. In-and-out whipping of electrode should be avoided; steady, side-to-side weaving or steady progression is recommended.

TABLE 9-4. RECOMMENDED WELDING CURRENTS FOR DCEP METAL-ARC WELDING OF ALUMINUM

Electrode Diam (in.)	Welding Current (amp)*
1/8	80 - 120
5/32	100 - 150
3/16	125 - 200

*With base metal preheated to 300 - 400°F.

Typical welding currents and corresponding electrode diameters for DCEP metal-arc welding of aluminum are shown in Table 9-4.

A short arc is best — 1/8 to 3/16-in. — because of its greater stability; a long arc may spatter excessively and is more difficult to maintain.

Cleaning: One-pass welding should be used whenever possible. When multiple passes are required, thorough cleaning between passes is essential for optimum results. After the completion of any welding, the bead and work should be thoroughly cleaned of flux. Most of the flux can be removed by mechanical means, such as a rotary wire brush, slag hammer, or peening hammer, and the rest by steaming or by a hot-water rinse. To test for complete flux removal, swab a solution of 5% silver nitrate on weld areas; foaming will occur if flux is present.

Positions: Welding in the flat position is recommended whenever possible. Fillet welds in a horizontal plane are best made by positioning the work to be welded about 20° downhill. Where overhead welding is necessary, a series of stringer beads should be used. Lap and fillet welds can be produced with the angle between the electrode and the horizontal at about 45°.

Jigs, Backup Plates, Tacking: Jigs and backup plates are especially useful in welding the thinner materials. Need for jigs, however, is not as critical in metal-arc welding as in oxyacetylene welding aluminum, largely because of the smaller area heated by the arc, which causes less distortion. Generally, the use of a backup is advisable in butt-welding aluminum by the metal-arc method to obtain complete penetration while avoiding holes or dripping metal from the underside. Graphite, copper, or steel may be used. To allow space for the flux and bead, the backup strip is provided with a half-oval groove about 1/2-in. wide and 1/8-in. deep, which is placed directly beneath the joint.

Welding Aluminum and Aluminum Alloys with the Gas Metal-Arc Process

Process and Equipment

Aluminum alloys are almost always fusion welded with the gas metal arc welding (GMAW) process. Weld properties generally are at least equal to those of the base metal at zero temper. Welding speeds are higher than those obtainable with any other arc or gas process. Heat-affected zones are narrower than those with oxyacetylene or covered electrode arc welding. A DC (electrode positive) electric arc, established in an envelope of inert gas between a consumable electrode and the workpiece, is used for welding aluminum by the GMAW process.

GMAW and another inert-gas process, gas tungsten-arc (GTAW or TIG), are the principal methods for welding aluminum. The two processes are similar in that an inert gas is used to shield the arc and the weld pool, making flux unnecessary. The chief differences are in the electrodes and the characteristics of the power used.

In GMA welding, the electrode is aluminum filler wire fed continuously from a reel into the weld pool. GTA welding uses a nonconsumable tungsten electrode, with aluminum-alloy filler material added separately either from a hand-held rod or from a reel.

The equipment required for GMA welding includes a drive system that pulls the electrode from a reel and pushes it through the welding gun. In the gun, it becomes energized by passing through a contact tube connected to the power supply. While simple in principle, accurate controls are required to initiate and stop the shielding gas, cooling water, and electrode and to operate the contactor in the welding circuit.

The GMAW process is used for either semiautomatic or automatic welding. It is referred to as "semiautomatic" when the gun is operated manually. Once the gun is started and the arc struck, the welding conditions are automatically controlled (on constant-voltage equipment, particularly) by the machine settings. This leaves the operator free to concentrate on placement of the weld bead.

With a direct-current, electrode-positive (DCEP) arc, used for GMA welding aluminum, the electrode is positive, the work negative. Both the consumable filler metal and weld zone of the workpiece are melted.

With argon shielding, the DCEP arc also breaks up the surface oxide on the base metal ahead of the weld pool. This cleaning action is believed to be caused by the electrons leaving the base plate or the inert-gas ions striking the plate, or a combination of the two phenomena.

The DCEP action propels the filler metal across the arc to the workpiece in line with the axis of the electrode, regardless of the orientation of the electrode. Because of this and aluminum's density, surface tension, and cooling rate, welds can be made with relative ease in the horizontal, vertical, and overhead positions. High deposition rates are practical, resulting in less distortion, greater weld strength, and lower welding costs for a given job than other fusion-welding processes.

The efficient use of energy, characteristic of the GMAW process, makes preheating of aluminum work-pieces unnecessary in most cases. Consequently, the process is widely accepted for welding heavy sections of aluminum. GMA welding is also applicable to joining sheet thicknesses of aluminum at high speeds. It is practical to GMA-weld aluminum as thin as 0.062 in. in regular production.

MAINTAINING GMA WELD QUALITY

While a limited amount of porosity and dross can be tolerated in some welded joints, ductility, fatigue strength, and tensile strength are adversely affected by porosity, and weld metal — just as base metal — should be sound in all respects.

The main cause of porosity in aluminum welds is entrapped gas in the weld puddle. Porosity occurs when metal freezes before the gas has a chance to escape. Gases may originate from contaminants in the shielding gas, from air and water, from contaminants that get into the puddle from dirty base or filler metal, from too long an arc, or as a result of violent arc action. The amount of gas that remains in the weld pool is a function of the cooling rate of the weld puddle. Various causes of porosity and some corrective measures are listed in Table 9-5.

Entrapment: Shielding gas, air, or gaseous

contaminants can be entrapped in the weld puddle as a result of the violent arc action. Weld porosity is similar to the air bubbles that are entrapped in a rapidly frozen ice cube.

Turbulence in a weld pool is related to droplet transfer. If too low a welding current is used, so that large globules of metal transfer across the arc, more turbulent puddle reaction occurs than if small, well-formed droplets transfer in a fine-spray pattern.

Excessive welding currents can entrap gas by depositing metal over a gas bubble in the weld pool. Such metal freezes before the entire bubble has escaped. This type of porosity is generally irregular in shape.

Hydrogen: In addition to air entrapment, the generation of hydrogen from contaminants in the weld is another cause of gas porosity in aluminum. Molten aluminum has a high solubility for atomic hydrogen (see Fig. 9-3). On the other hand, solid aluminum can contain very little hydrogen. Therefore, hydrogen gas is evolved as the weld puddle freezes. If the cooling rate of the weld puddle is too fast, the gas cannot rise to the surface, but remains within the metal, causing porosity.

Any compound containing hydrogen and contaminating the filler wire or work surface can be the source of hydrogen gas. For example, if a film of drawing compound (containing hydrocarbons) or

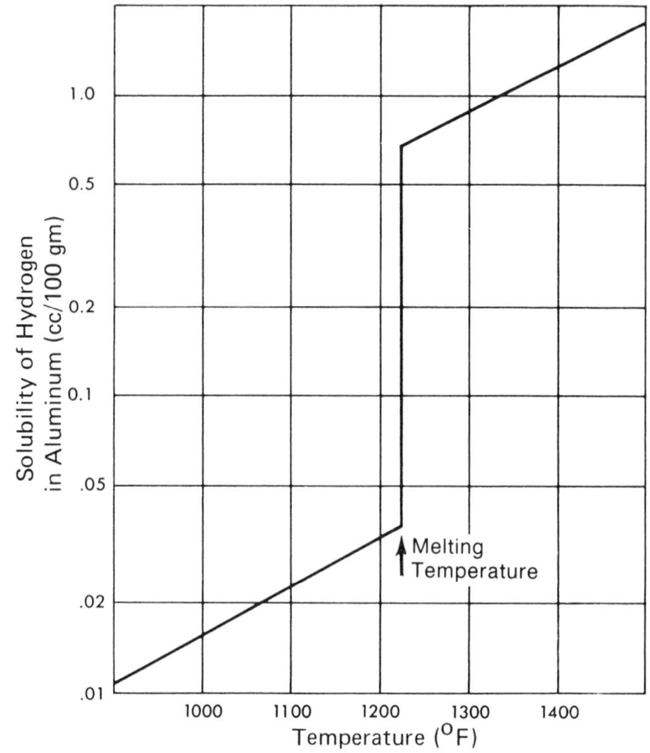

Fig. 9-3. Solubility of hydrogen in aluminum at various temperatures.

TABLE 9-5. CAUSES AND CURES FOR POROSITY IN GMA WELDS

Cause	Contributing Factors	Corrective Measures
Gas entrapment	Turbulence of weld pool Excessive current entraps gas.	Use adequate welding current to stabilize droplet transfer, but avoid excessive current. Reduce travel speed.
Hydrogen	Oil or other contaminants on electrode. Hydrated oxide film on electrode. Oily drive rolls or liner in GMAW gun. Wet shielding gas. Water leaks in GMAW gun. Oily workpiece. Spatter particles ahead of puddle.	Change to clean, high-quality electrode. Keep electrode supply covered. Do not open desicated plastic packaging until ready to use. Do not use electrode that has been kept for any length of time outside of desicated storage. Clean rolls with solvent; change liner. Check dewpoint of gas; reject bottles above minus 40°F dewpoint. Repair guns that have been overheated due to water failure. Clean plate. Adjust welding conditions to minimize spatter.
Rapid cooling rate of weld pool	Low rate of heat input into the weld. Too rapid a rate of heat extraction from the weld. Low temperature of backup bar, if used. Groove configuration of backup bar.	Use higher welding current or a slower speed. Preheat sometimes helps. Hot backup bars reduce root porosity. Shallow, wide grooves are better than deep, narrow grooves in backup bars.
Erratic wire feeding	Drive-roll slip. Excessive bending of guide liner. Bent contact tube puts kink in electrode. Galling in contact tube. Electronic "hunting." Wrong size liner.	Increase pressure on rolls, change to knurled roll; change to V-groove roll rather than U-groove. Change position of GMAW unit to reduce necessity of bending cable to gun. Replace thyratron tube or bias (balance) battery. (old units)
Other	Partial-penetration joint. Multiple-pass welds.	Use full-penetration joint. Increase fillet size to compensate for line of root porosity in critical applications. Use high-current-density single-pass-per-side technique.

hydrated aluminum oxide remains on the filler wire, extreme porosity will occur in the welds made with that wire.

Other possible sources of hydrogen-forming contaminants include the base material and the shielding gas. A bottle of wet gas will cause contaminated welds, as will moisture leaks in the welding torch or hose line. Gas having a dewpoint higher than minus 40°F is unsuitable for GMA welding of aluminum.

Poor welds also result from using electrodes stored under humid conditions, where atmospheric moisture can combine chemically or be absorbed by the aluminum oxide film.

Cooling System: Before welding, the GMAW gun should be checked in regard to the following points:

Are all O-rings seated properly?

Are all fittings drawn up tight?

Are all solder joints tight?
(Guns that have been overheated may have developed leaks in soldered joints.)

Are all ceramic parts in good condition?
(Broken or distorted ceramics should be replaced.)

Is the gas hose clean and in good condition?
(Do not use leaky or deteriorated hose. Do not use hose that previously has carried water, or liquids or gases containing compounds such as hydrocarbons, which will cause weld porosity. A vinyl or other suitable plastic hose is recommended.)

Most GMAW equipment can be adjusted for either "continuous" or "weld-only" cooling-water flow. In humid weather, water flow should not be continuous. Water vapor from the atmosphere may condense on the gas cup and internal parts of the gun if the cold water flows during idle periods. When welding is resumed, the water condensate will vaporize and be carried to the weld area by the shielding gas, causing weld porosity. Therefore, particularly in humid weather, cooling-water flow should be stopped when welding is stopped. Residual heat in the gun will keep the gas cup and other critical parts warm, preventing condensation.

Cooling Rate of Weld Pool: The amount of gas remaining in the weld pool is a function of how fast the weld pool solidifies. If most of the gas has time to escape before the molten metal solidifies, the weld will have minimal porosity and be of generally good radiographic quality. The length of time that the aluminum weld pool is molten is determined by the heat input into the weld, the rate of heat extraction from the weld, the freezing range (solidification temperature) of the alloy, and the temperature and mass of the surrounding material. The temperature and groove configuration of the backup bar (if used) also affect the rate of heat loss.

To retard the cooling rate of the weld puddle, the heat input generally must be increased, since the factors affecting heat loss are usually fixed. This can be done by increasing the current, decreasing the travel speed, or both. Changes in either of these must always be regulated, however, within operating limits conducive to proper bead deposition and contour and to the thickness limitations of the base material.

Use of higher arc voltage also increases the heat input. Helium may be added to argon in amounts greater than 10% to increase arc voltage and reduce porosity.

The rate of heat extracted from the weld is generally high, due to the high thermal conductivity of aluminum. Obviously, heavier or large sections can absorb more heat than thinner sections. Thus, heavy or large sections act as better heat sinks than light sections and extract or conduct heat from the weld faster than thin materials.

When welding speed is reduced to retard the cooling rate, the effect is to allow more heat to be

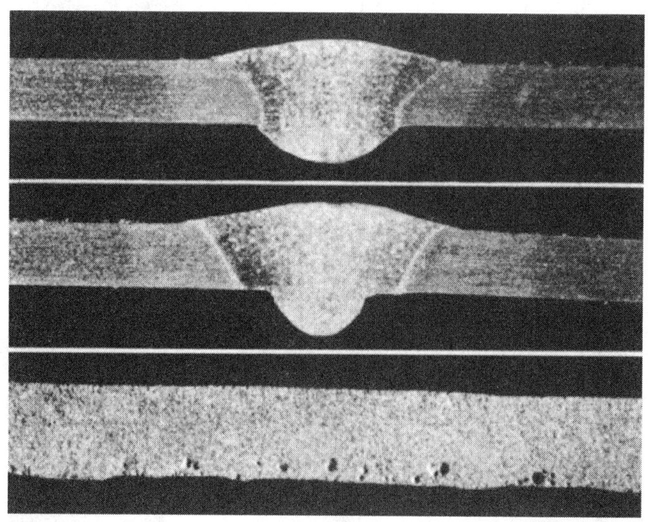

Fig. 9-4. Effects of backup-bar groove configuration on porosity in welds in aluminum sheet. The wide, shallow groove (top) minimizes porosity because it produces a large mass of metal at the root. The deep, narrow groove (center) promotes root porosity. The lower specimen shows root porosity in a GMA weld, resulting from a narrow backup-bar groove. Magnification, 3X.

lost to the heat sink, thus warming the assembly being welded. Consideration must be given to the fact that such heating may decrease the mechanical properties in the metal outside the weld zone, which, itself, is annealed by the arc.

Weld metal cools from the outer edges toward the center of the weld nugget. One frequent condition is that in which a wide weld bead, having a thin edge at its width extremities, has a line of porosity at the edges due to the faster freezing of the thin section of weld metal in those areas. This shows as line porosity in a radiograph and generally is associated with multiple-pass welding.

Root porosity, which is similar to fusion-line porosity, occurs particularly in welds made on a backup bar having an unsuitable groove configuration. A deep, narrow groove in a backup bar is conducive to root porosity, whereas a wide, shallow groove tends to minimize porosity, because of the larger mass of weld metal at the root (see Fig. 9-4). Preheating of the backup bar has been found to reduce further the tendency to form root porosity.

Electrode Feeding: Erratic electrode feeding can cause porosity in weld metal, due to increased turbulence of the weld pool. Erratic feeding can be caused by a number of things. The electrode may slip in the drive rolls because of insufficient pressure, excessive bending in the electrode transport tube, excessive bending in the guide tube of the gun, or excessive bending of the cable between the drive unit and the welding gun. Kinks in the electrode will jam in the gun and cause a burn-back. If the kinks manage to feed through, they usually produce a momentary slowing of the electrode. Improperly wound spools with crossovers or edge-bound strands also cause erratic wire feeding.

Fig. 9-5. Fractured surface of a tensile test bar showing "worm hole" porosity in a GMA weld, caused by spongy cast base metal.

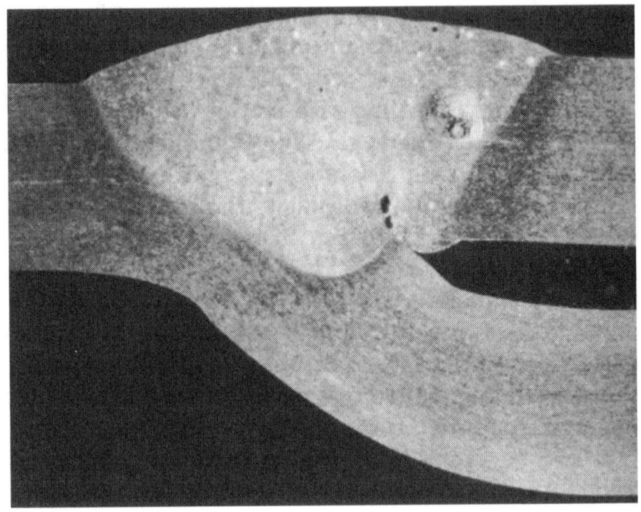

Fig. 9-6. A GMA weld in aluminum sheet showing porosity caused by lamination.

Aluminum electrode can seize and gall inside the contact tube, causing an erratic or stitching arc action. Replacing the contact tube corrects the situation.

Quality of Base Material: Weld porosity is also dependent on the quality of the base material particularly in castings. The presence of large pores in cast base material can cause an elongated type of porosity in the weld, as shown in Fig. 9-5. In the case illustrated, a "boiler effect" occurred where the weld metal heated a volume of gas from the pore in the base material. The gas expanded along a channel in the molten metal to form a long tube, until the diminishing gas pressure was equalized by the weight of the weld metal. A similar effect can occur when the base metal sheet is laminated, as shown in Fig. 9-6.

Welding partial-penetration joints (as in making a welded lap joint) can also cause a boiler effect, as shown by the photomacrograph in Fig. 9-7. In this case, the source of gas was the line at the root, where sufficient pressure was built up from the heat of welding (on the second side to be welded) to cause porosity to form in the molten metal.

Multiple-Pass Welding: Porosity is more prevalent in multiple-pass welds than in single-pass welds. Thus, in welds where porosity exists, top passes in heavy plate have been observed generally to contain more porosity than root passes. Wherever practical, welds with a minimum number of passes should be used. With high current-density techniques, GMA butt welding with a single pass per side can be done in aluminum plate as thick as 1-1/4 in.

Effect of Porosity on Properties: Just what

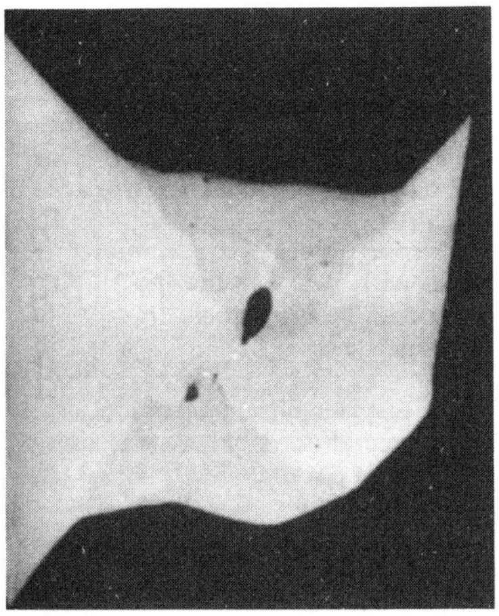

Fig. 9-7. Photomacrograph showing root porosity in a transverse wafer cut from a lap weld.

effect does weld porosity have on the properties of aluminum welds? As shown in Fig. 9-8, some porosity can be tolerated in a weld before mechanical properties are reduced, but tensile properties are reduced appreciably when the amount of porosity is high. Data shown are for GMA weldments in 5083 plate; different porosity levels were obtained by contaminant additions to the welding arc.

Dross in GMA Welds: Most dross consists of broken particles of aluminum oxide or aluminum nitride films, which can be scattered throughout the weld bead. The presence of oxygen in the shielding gas or on the surface of the solidifying weld pool causes dross. Dross has the appearance of pepper on the fractured metal surface.

Dross also may be present as a film in "cold shut" areas, where the arc length was too long, producing temperatures too low to fuse the weld bead into the underlying metal. Dross, unlike gas porosity, has a markedly adverse effect upon the mechanical properties of aluminum welds.

Shielding: Oxygen may enter the weld pool from several sources. The primary cause of oxygen entry from the atmosphere is poor shielding. Drafts in the welding area can cause a momentary loss of shielding, allowing air to contaminate the surface of the molten metal. Also, if the gas flow is too low, inadequate shielding of the weld pool results. Entrapped air entering the contact tube may also cause dross to form. Use of too small a gas cup for wire diameter, current, and size of puddle can result in inefficient shielding, thus allowing excessive air to get to the weld puddle.

Electrode Surface: Just as it can be a source of porosity-causing hydrogen, the electrode can also be a source of dross-causing oxygen. If an electrode has more than a minimal oxide coating, it is a potential source of oxygen and is unsuitable for GMA welding of aluminum. Oxygen may also be present in the form of hydrocarbons on the electrode, particularly if the spools of filler metal have been exposed to the shop atmosphere without protective covers. Unused spools should be kept in their packages, partly used spools on welding machines should be covered between jobs.

Base Metal and Weld Oxide: Welding over base metal surfaces that have not been deoxidized or properly wire-brushed prior to welding can result in oxygen pickup, particularly with sheet material. Preceding weld beads, if not cleaned or brushed, also are a source of oxygen, as well as of hydrogen. All beads must be properly wire-brushed before subsequent beads are deposited.

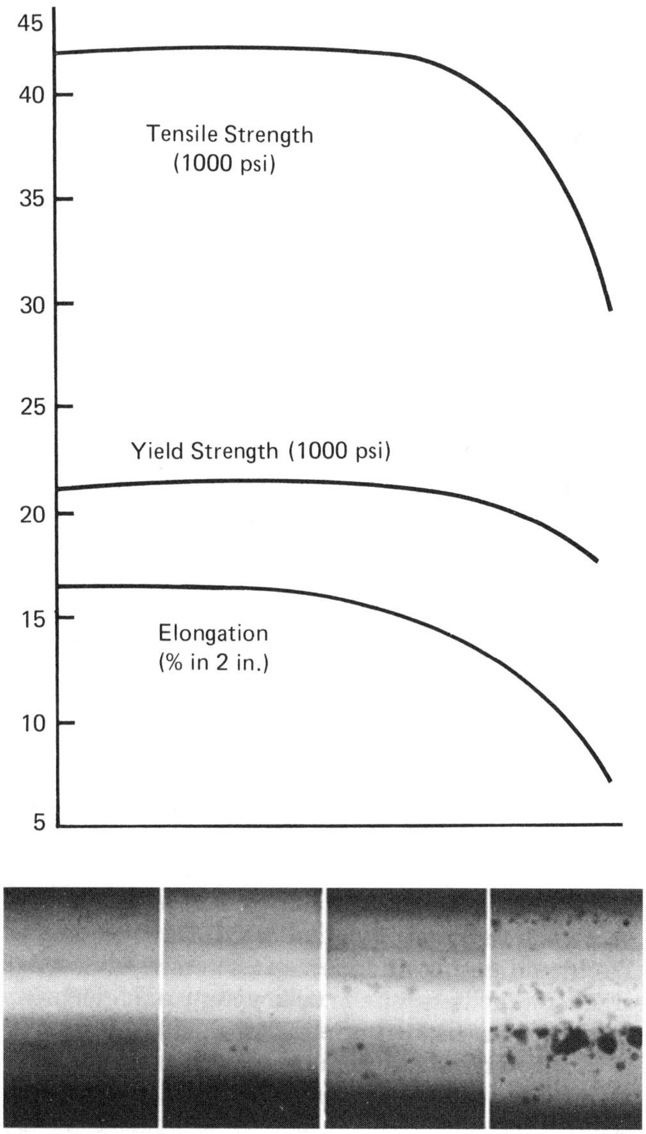

Fig. 9-8. Effect of porosity on GMA weld properties of 5083 plate, welded with 5356 filler.

TABLE 9-6. CAUSES AND CURES FOR DROSS IN GMA WELDS

Cause	Contributing Factors	Corrective Measures
Poor shielding	Drafts, causing loss of shielding. Gas flow too low. Entrapped air in contact tube. Gas cup too small.	Provide protection against drafts in welding area. Increase shielding-gas flow rate. Pressurize contact tube with shielding gas. Use a larger gas cup.
Contaminated electrode surface	Oxygen from excessive oxide or from hydrocarbon contaminant.	Use electrode with clean surface and minimum oxide film. Protect electrode in shop with plastic covers.
Contaminated base-metal surface	Heavy oxide film. Spatter particles.	Wire-brush or chemically remove oxide. Adjust welding conditions so spatter does not occur ahead of puddle.
Contaminated weld	Oxide film on preceding beads. Dross on underside of root passes.	Wire-brush each bead before depositing next bead. Chip or mill underside to sound metal.

The underside of root beads in joints that are to be welded from both sides should be chipped or milled to remove the dross prior to depositing weld metal on the second side. No back-chipping is required on welds made by high current-density techniques in properly prepared square butt or heavy-root-face butt joints. A summary of causes and cures for dross in GMA weld metal is given in Table 9-6.

INERT GASES FOR GMA WELDING

The inert gas has three primary functions. First, it provides a controlled ionized path, which aids in the smooth transfer of current. Second, it surrounds the arc and weld pool with an atmosphere that is nonreactive with molten aluminum. Third, it has a cleaning action, which partially removes aluminum oxide from the plate during DCEP welding. (Only argon shielding produces such cleaning action in DCEP welding, however.) It is believed that this cleaning action is a result of gas ions striking the surface, or electrons leaving the base plate, or a combination of each.

In addition to the inert gases commonly used — argon and helium — neon or krypton could probably be used with the GMAW process for shielding purposes. However, argon and helium are the gases generally used because they are more readily available and much less expensive.

Shielding Efficiency: Any disturbance of the inert gas shield around the arc and the weld pool reduces its shielding efficiency. If insufficient gas is used or if the flow of gas is disturbed, poor quality welding will result. Such disturbances may be caused by improper gas flow, drafts, erratic arcs, gas stoppage, or turbulence due to spatter in the nozzle.

The flow of inert gas necessary for good-quality GMA welding depends upon a number of factors including type of gas used, welding current, diameter of gun nozzle, joint design, welding position, speed of welding, torch lead angle, and the amount of draft in the welding area (see Table 9-7).

Drafts can affect gas usage and weld quality considerably. It is important that the welding area be essentially draft-free. In field welding, suitable shielding with tarpaulins or other types of windbreaks should be provided to prevent air currents from interfering with the gas shield.

TABLE 9-7. FACTORS THAT AFFECT FLOW RATE OF SHIELDING GAS IN GMA WELDING

Factor or Condition	Higher Flow Required	Lower Flow Required
Draft:		
No draft		X
Increasing draft	X	
Gas:		
Argon		X
Helium	X	
Torch lead angle:		
10° Lead		X
0	X	
10° Lag (not recom.)		
Welding current:		
High	X	
Low		X
Diameter of nozzle:		
Large	X	
Small		X
Joint design:		
Flat surface	X	
In a groove or fillet		X
Welding speed:		
Fast	X	
Slow		X
Position		
Flat, overhead		X
Vertical, horizontal	X	

TABLE 9-8. SUGGESTED FLOW RATES FOR GMA WELDING OF ALUMINUM BUTT JOINTS IN VARIOUS POSITIONS

Position	Shielding Gas Flow (cfh)		
	Argon	Helium	75% He, 25% A
Flat	50[1] - 70[2]	70 - 90	60 - 80
Horizontal	50 - 70	70 - 90	60 - 80
Vertical	50 - 70	70 - 90	60 - 80
Overhead	70 - 90	80 - 100	70 - 90

Values are for 1/32 to 1/16-in. electrode size and 5/8 or 3/4-in. cup size.

[1] With 0.030-in. and 3/64-in. electrode, flow can be reduced by 25% under normal conditions.

[2] Highest flow rates indicated apply to thicker base materials and to alloys containing greater amounts of magnesium.

Overhead welding usually requires higher gas flow rates than flat-position welding, regardless of gas used. As a basis for determining correct gas flow for a given welding job, the gas flow rates in Table 9-8 are suggested as starting points. Lower gas flows can frequently be used. The correct flow for any job is determined by gradually reducing the flow rate used in making sample weldments until the weld quality is adversely affected. The gas flow should then be increased to provide a safety margin.

Only high-purity, dry inert gas ("welding grade" gas) should be used in GMA welding. Impure or wet gas will degrade weld quality.

Argon Shielding: Argon is preferred for flat, vertical, and horizontal-position manual GMA welding by most operators, because argon produces a more stable arc than helium. Less argon is required than helium for shielding in GMA welding because its density is greater (argon, 1.4; helium, 0.14; air, 1.0).

Argon is produced by fractionating air and is readily available. Argon is sold in compressed gas form in a variety of cylinder sizes. Where large volumes are consumed, liquid argon installations may be economical.

Helium Shielding: At any given (DCEP) arc length, a helium-shielded arc has a higher voltage than an argon arc. This results in a deeper penetration with helium, which, in turn, can be used to increase welding speed. However, the arc is slightly more erratic in helium than in argon, and, since helium shielding provides no cleaning action, the resultant weld bead appearance is usually inferior.

Helium is usually preferred for welding very heavy material because it provides deeper weld penetration. The lightness of helium makes it preferable for overhead welding. The high welding speeds obtainable with a helium arc make it attractive for automatic welding, where arc length can be controlled precisely. Helium, however, produces more spatter and requires more frequent nozzle cleaning than does argon, a deterrent to the use of helium in certain automatic GMAW operations.

Argon-Helium Mixtures: Mixtures of argon and helium are used to obtain advantages not obtained with either gas alone. Blends of 75% helium and 25% argon are commercially available. Other ratios can be mixed by combining flows from separate tanks of helium and argon. Helium additions of over 10% to argon change the arc characteristics markedly. With increasing helium, arc voltage, spatter, and penetration increase.

PROCEDURES FOR GMA WELDING

Principal factors for consideration in the GMA welding of aluminum are thickness of plate, alloy, and type of equipment available. Typical procedures for conventional GMA welding of various joint designs in aluminum sheet and plate are given in Tables 9-9 through 9-13. These data are approximate and are intended to serve only as starting points. For each application, an optimum set of welding conditions can be established from these procedures.

It is considered good practice to prepare prototype weldments in advance of the actual production so that welding conditions can be determined on the prototype. It is further recommended that weldors practice beforehand under simulated production conditions. This helps avoid mistakes caused by lack of experience.

Where intermittent welding is to be used, one deviation from the regular pattern of gun travel is recommended. GMA welding of aluminum normally leaves a crater at the end of the weld, as illustrated in Fig. 9-9. This crater is prone to cracking which, in turn, could initiate fracture in the intermittent weld.

One method of avoiding this problem is to reverse the direction of welding at the end of each tack or intermittent weld, so that the crater is filled, as shown in Fig. 9-10. Other techniques for eliminating problems of cracking of the crater area are: 1. Use run-on and run-off tabs. 2. Break the arc and restrike to to fill the crater. 3. Use special circuitry and power source control to produce a specific rate of arc decay.

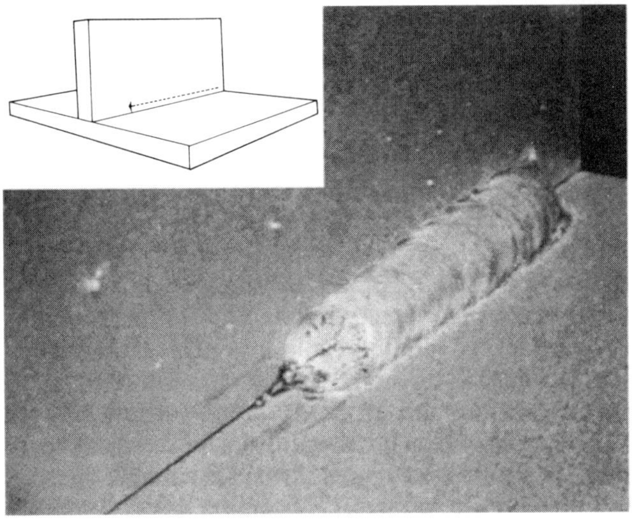

Fig. 9-9. The finish of a GMA weld in aluminum leaves a crater that is very susceptible to cracking.

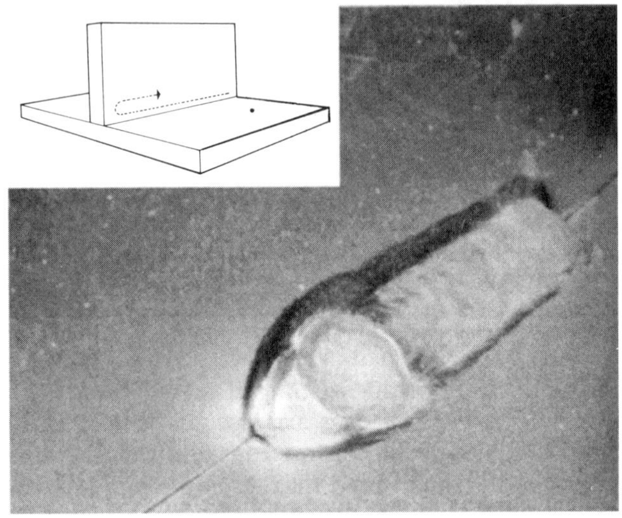

Fig. 9-10. Doubling back at the end of a GMA weld eliminates the crater and the cracking problems that usually accompany it.

TABLE 9-9. PROCEDURES FOR GAS METAL-ARC WELDING OF ALUMINUM (Semiautomatic)

Process: Gas Metal-Arc
Type of Joint: Butt
Plate Thickness: 1/8 - 3/4
Position: Flat
Welded from: Two sides
Polarity: DCEP

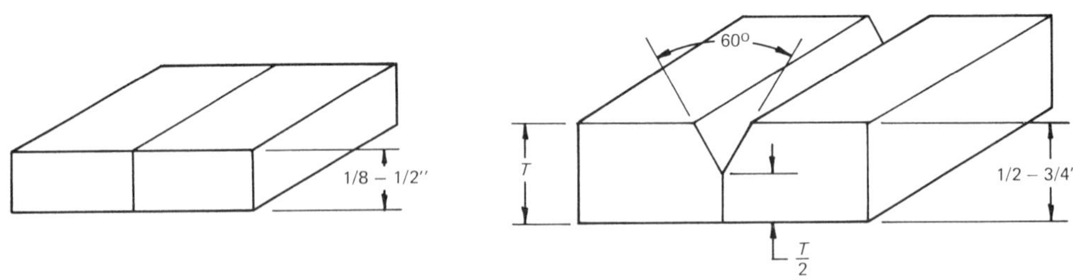

Plate Thickness, T (in.)	1/8	3/16	1/4	3/8	1/2	1/2	5/8	3/4
Pass Face	1	1	1	1	1	1	1	1
Pass Back		1	1	1	1	1	1	1
Electrode Size (in.)	3/64	3/64 or 1/16	1/16	1/16	1/16	1/16	1/16	1/16
Current (amp)	90 – 120	120 – 150	180 – 220	220 – 260	240 – 280	260 – 320	260 – 320	300 – 400
Volts	19 – 21	22 – 24	27 – 29	28 – 30	28 – 30	29 – 31	29 – 31	33 – 37
Arc Speed (ipm)	30 – 34	25 – 29	20 – 24	16 – 20	14 – 18	12 – 16	10 – 14	10 – 14
Argon Gas (cfh)	30 – 40	30 – 40	30 – 40	40 – 60	40 – 60	40 – 60	60 – 80	60 – 80
Gas Cup Opening (in.)	1/2	1/2	1/2	1/2	1/2	1/2	1/2	1/2
Total Time (hr/ft. of weld)	0.00625	0.0148	0.0182	0.0222	0.0250	0.0286	0.0333	0.0333

TABLE 9-10. PROCEDURES FOR GAS METAL-ARC WELDING OF ALUMINUM (Semiautomatic)

Process: Gas Metal-Arc
Type of Joint: Butt
Plate Thickness: 1/2 - 1-1/2
Position: Flat
Welded from: Two sides
Polarity: DCEP

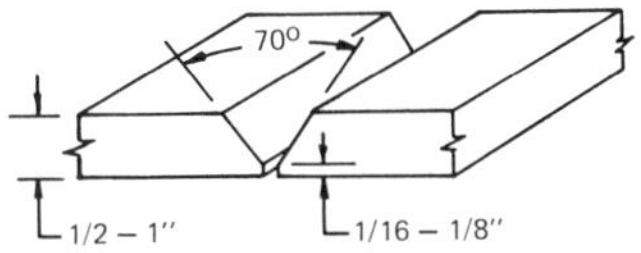

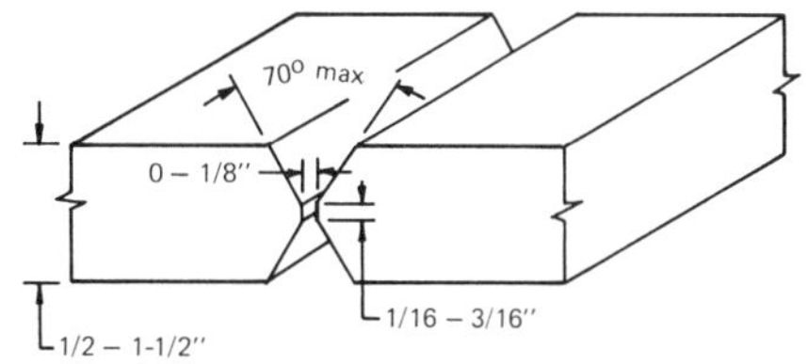

Plate Thickness (in.)	1/2	3/4	1	1/2	3/4	1	1-1/4	1-1/2
Pass Face	4	5	7	1	2	3	3	4
Pass Back*	1	1	1	1	2	3	3	4
Electrode Size (in.)	1/16	3/32	3/32	1/16	1/16	1/16	1/16	1/16
Current (amp)	200 – 230	260 – 280	260 – 300	220 – 250	240 – 280	240 – 280	240 – 280	240 – 280
Volts	27 – 29	27 – 29	27 – 29	28 – 30	28 – 30	28 – 30	28 – 30	28 – 30
Arc Speed (ipm)	16 – 20	12 – 16	10 – 14	16 – 20	16 – 20	14 – 20	14 – 20	14 – 20
Argon Gas (cfh)	40 – 60	40 - 60	40 – 60	40 – 60	40 – 60	60 – 80	60 – 80	60 – 80
Gas Cup Opening (in.)	1/2	5/8	5/8	5/8	5/8	5/8	5/8	5/8
Total Time (hr/ft. of weld)	0.0555	0.0858	0.133	0.0222	0.0444	0.0708	0.0708	0.0944

*Back chip before first pass on back side optional.

TABLE 9-11. PROCEDURES FOR GAS METAL-ARC WELDING OF ALUMINUM (Semiautomatic)

Process: Gas Metal-Arc
Type of Joint: Butt
Plate Thickness: 1/8 - 1
Position: Horizontal
Welded from: Two sides
Polarity: DCEP

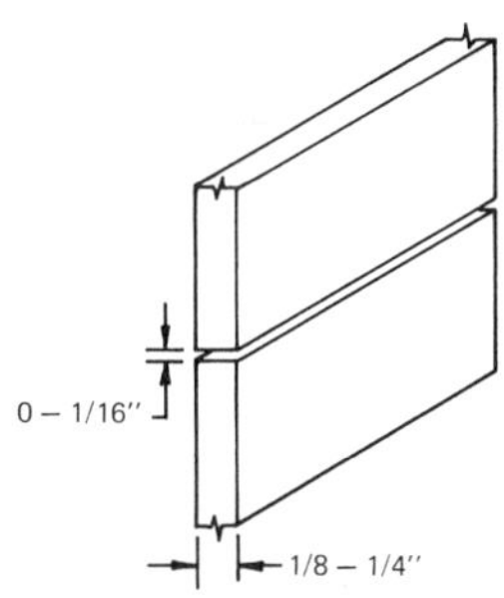

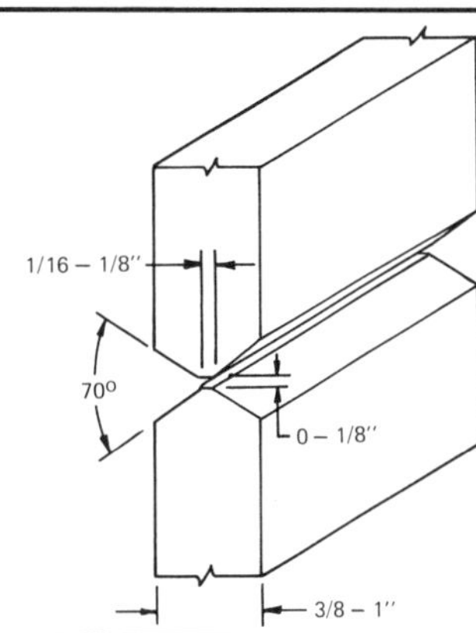

Plate Thickness (in.)	1/8	3/16	1/4	3/8	1/2	3/4	1
Pass Face	1	1	1	2	2	3	5
Pass Back*	1	1	1	2	2	3	5
Electrode Size (in.)	3/64	3/64	3/64	1/16	1/16	1/16	1/16
Current (amp)	100 – 120	120 – 140	160 – 200	180 – 200	200 – 220	220 – 260	240 – 280
Volts	19 – 22	20 – 23	26 – 28	27 – 29	27 – 29	28 – 30	28 – 30
Arc Speed (ipm)	24 – 30	24 – 30	22 – 26	24 – 30	20 – 24	20 – 24	16 – 20
Argon Gas (cfh)	30 – 40	30 – 40	30 – 40	40 – 60	40 – 60	60 – 80	60 – 80
Gas Cup Opening (in.)	1/2	1/2	1/2	1/2	5/8	5/8	5/8
Total Time (hr/ft. of weld)	0.0148	0.0148	0.0167	0.0296	0.0364	0.0545	0.111

*Back chip before first pass on back side optional.

TABLE 9-12. PROCEDURES FOR GAS METAL-ARC WELDING OF ALUMINUM (Semiautomatic)

Process: Gas Metal-Arc
Type of Joint: Butt
Plate Thickness: 1/8 - 1
Position: Vertical
Welded from: Two sides
Polarity: DCEP

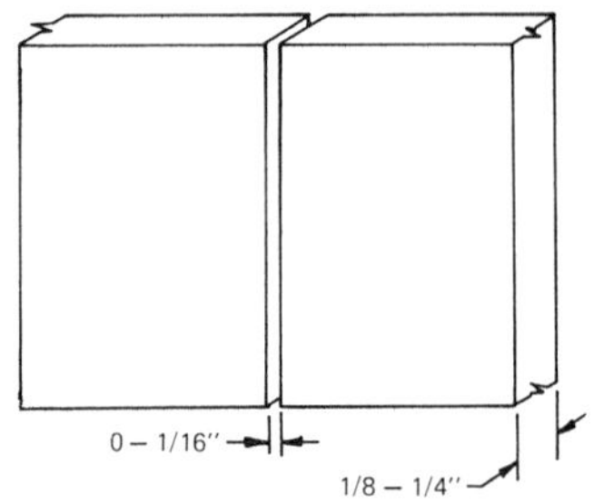

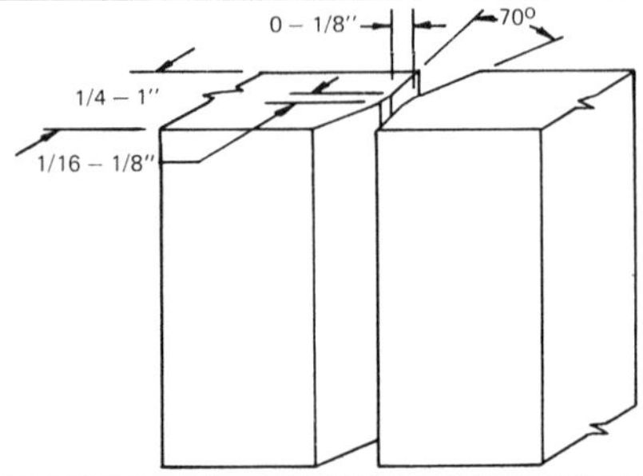

Plate Thickness (in.)	1/8	3/16	1/4	1/4	3/8	1/2	3/4	1
Pass Face	1	1	1	1	1	2	4	6
Pass Back*	1	1	1	1	1	2	4	6
Electrode Size (in.)	3/64	1/16	1/16	1/16	1/16	1/16	1/16	1/16
Current (amp)	90 – 120	100 – 140	140 – 190	160 – 180	180 – 200	200 – 220	220 – 240	220 – 240
Volts	19 – 22	20 – 23	26 – 28	23 – 25	27 – 29	27 – 29	28 – 30	28 – 30
Arc Speed (ipm)	30 – 36	28 – 32	24 – 30	24 – 30	20 – 24	14 – 18	12 – 16	8 – 12
Argon gas (cfh)	30 – 40	30 – 40	30 – 40	30 – 40	40 – 60	40 – 60	60 – 80	60 – 80
Gas Cup Opening (in.)	1/2	1/2	1/2	1/2	1/2	5/8	5/8	5/8
Total Time (hr/ft. of weld)	0.0121	0.0133	0.0148	0.0148	0.0182	0.0500	0.114	0.240

* Backchip before first pass on back side optional.

TABLE 9-13. PROCEDURES FOR GAS METAL-ARC WELDING OF ALUMINUM (Semiautomatic)

Process: Gas Metal-Arc
Weld Size (in.): 1/4 - 1/2
Type of Joint: Fillet
Positions: Horizontal, Vertical and Overhead
Polarity: DCEP

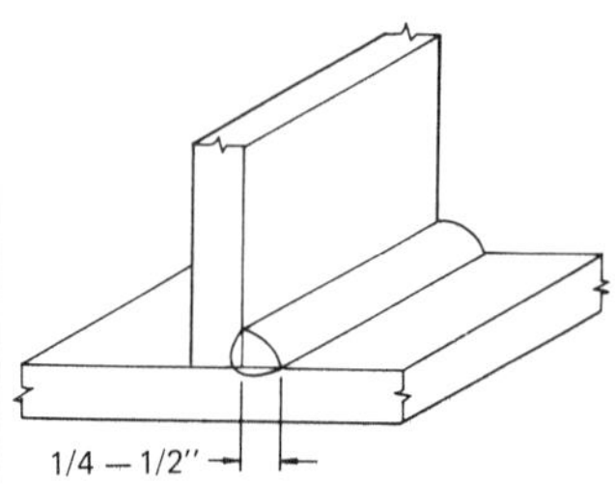

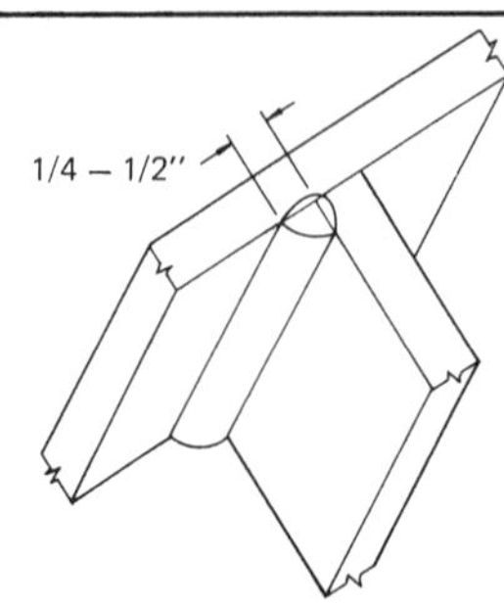

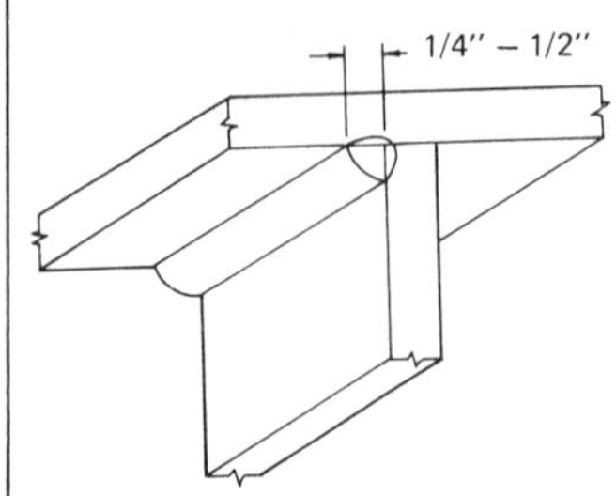

Fillet Size, L (in.)	1/4	3/8	1/2	1/4	3/8	1/2	1/4	3/8	1/2
Pass	1	1	3	1	2	3	1	2	3
Electrode Size (in.)	1/16	1/16	3/32	1/16	1/16	1/16	1/16	1/16	1/16
Current (amp)	160 – 180	200 – 230	240 – 280	160 – 180	180 – 200	180 – 220	160 – 180	180 – 200	180 – 220
Volts	24 – 26	27 – 29	26 – 28	24 – 26	27 – 29	27 – 29	23 – 25	27 – 29	27 – 29
Arc Speed (ipm)	24 – 28	20 – 24	16 – 20	16 – 20	14 – 18	12 – 16	18 – 22	18 – 22	16 – 20
Argon Gas (cfh)	30 – 40	30 – 40	40 – 60	30 – 40	40 – 60	40 – 60	40 – 60	40 – 60	60 – 80
Gas Cup Opening (in.)	1/2	1/2	5/8	1/2	1/2	5/8	1/2	1/2	5/8
Total Time (hr/ft. of weld)	0.00769	0.00909	0.0333	0.0111	0.0250	0.0429	0.0100	0.0200	0.0333

Welding Aluminum and Aluminum Alloys with the Gas Tungsten-Arc Process

GTA Welding Arcs

Aluminum can be welded by the gas tungsten arc (GTA) process using direct-current electrode-negative (DCEN), direct-current electrode-positive (DCEP), or alternating current (AC). Generally, DCEP is used for very thin sheet, DCEN for relatively thick sections with mechanized torch travel and controls; and AC, for thin sheet and light plate. Typical characteristics of each process are listed in Table 9-14. GTA welding is faster than manual shielded metal-arc welding (SMAW), but slower than GMA welding.

GTA welding is accomplished with the heat of an electric arc operating between a tungsten electrode and the work in a shield of inert gas. The inert atmosphere prevents oxidation of the molten aluminum so no flux is required. The gas is usually argon for AC and DCEP GTA, and helium for DCEN GTA welding.

In the inert atmosphere, the tungsten electrode is practically nonconsumable. Actually, it is dissipated slowly by the arc and requires occasional dressing to maintain a hemispherical tip contour. For all practical purposes, however, the tungsten electrode imparts no metal to the weld pool under normal operating conditions.

Each of the three types of power produces different arc characteristics in GTA welding (Table 9-15) and has different penetration characteristics (Fig. 9-11). Appearance of welds made by the three methods is described in Table 9-16. Suggestions for correcting conditions that cause porosity or cracking in GTA welds are given in Table 9-17.

Weldable aluminum in thin gages can be manually GTA-welded without filler metal, although cracking

TABLE 9-14. CHARACTERISTICS OF THE GTA WELDING PROCESS FOR ALUMINUM (AT SAME CURRENT)

Process Characteristic	GTA Welding Process		
	AC	**DC Electrode-Negative (electrode −, work +)**	**DC Electrode-Positive (electrode +, work −)**
Current	Alternates	Flows from electrode to work.	Flows from work to electrode.
Arc Heat Concentration	50% at each cycle (with balanced power source)	70% at work; 30% at electrode.	30% at work; 70% at electrode.
Effects of Heat Concentration	1. Electrode current capacity only slightly less than for DCEN (for AC about 0.001 in.-diameter for each 1.25 amp welding current). 2. Medium penetration, greater than DCEP and somewhat less than DCEN. 3. Midrange welding speeds (5 ipm to 30 fpm is typical range). NOTE: Helium increases penetration and welding speeds as compared to argon; argon gives best cleaning action with AC.	1. Higher currents with smaller electrodes. 2. Narrow and deep penetration of arc heat into base metal. 3. Welding speed: 5 ipm to 120 ipm is typical range.	1. Larger electrode required. 2. Wide and shallow penetration. 3. Much slower welding speeds than DCEN.
Arc Stability	Stable (balanced AC, or AC with continuous high frequency - ACHF).	Stable	Stable
Cleaning Action of Arc and/or Gas	Good, with argon, on positive portion of cycle. (Satisfactory welds also can be made with pure helium or a mixture of argon and helium.)	None; however, satisfactory DCEN welds can be made with helium (maximum current about 600 amp; although with special techniques that help eliminate weld-pool turbulence, this current can be exceeded.) Argon is sometimes used in automatic DCEN GTAW.	Continuous - excellent, with argon; DCEP has best cleaning action of the three, but penetration is poorest.
Size of Tungsten Electrode	Medium (3/32 in. at 125 amp)	Small (1/16 in. at 125 amp)	Large (1/4 in. at 125 amp)
Applications	General-purpose automatic and manual welding.	Automatic.	Thin sheet welding; also root pass in pipe.
Maximum Work Thickness	1/8 in. (full-penetration)	1/4 to 3/4 in. (full-penetration)	Very thin - below 0.050 in.

may occur in some of the heat-treatable alloys such as 6061-T6. With grooved joints and the addition of filler metal, aluminum plate to 1-in. thick can be efficiently welded manually with the GTA process. With DCEN automatic equipment, sections to 1-1/4 in. thick can be welded without the addition of filler metal. Thicker sections that require edge preparation are usually welded by the GMAW process, because larger grooves can be filled more efficiently by GMAW than GTAW.

The four types of electrodes used for GTA welding are pure tungsten, thoriated tungsten, zirconiated tungsten and ceriated tungsten.

Pure tungsten electrodes are usually used for AC GTA welding. Advantages claimed for zirconiated tungsten electrodes for AC GTA welding are that they can carry slightly higher currents and have a longer life than pure tungsten electrodes of equal size. Also, zirconiated tungsten retains the balled tip well, is less susceptible to contamination by molten aluminum, and is a lesser source of tungsten contamination in the weld than is pure tungsten.

Thoriated tungsten does not produce as stable an arc with AC GTA as pure tungsten when welding aluminum. Thoriated tungsten is usually preferred for automatic GTA welding with DCEP.

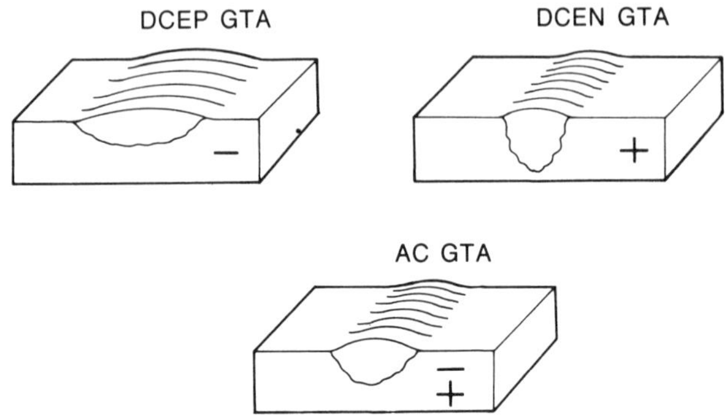

Fig. 9-11. Penetration characteristics of the three types of GTA welding processes.

ALTERNATING-CURRENT GTA WELDING

Alternating current is preferred by many users for both manual and automatic GTA welding of aluminum. This is because AC GTA achieves an efficient balance between penetration and cleaning action, which are the chief advantages, respectively, of electrode-negative and electrode-positive direct-current GTA welding.

TABLE 9-15. CHARACTERISTICS OF GTA WELDING ARCS

Arc Characteristic	GTA Welding Process		
	ACHF or Balanced AC	DCEN	DCEP
Arc Starting and Operation	With high frequency, arc starts without touching work. Strike on starting block; when electrode is thoroughly heated, re-ignite arc in the joint. This reduces tungsten inclusions at start of weld. Hold at starting point until weld pool is established. Add filler at leading edge of pool, but to one side of centerline.	Hot arc forms puddle immediately. Strike arc on work within weld width to avoid marking work. Runout tabs or striking plates are often desirable. Use of high frequency to strike arc is recommended.	Same as AC, but touch electrode to work.
Stability	Smooth arcing; no snapping or cracking sounds (argon).	Stable in helium.	Stable (argon).
Instability: Electrode contamination	Oxides resulting from contact of hot electrode with air or from contact with oxides or metal in weld pool cause arc instability. Dress or cut off tip or replace electrode.	Same as AC.	Same as AC.
Electrode too large	May result in unstable arc and excessive bead width. Use smallest electrode that will carry current.	Same as AC.	Same as AC.
Excessive arc length	Unstable arc results; shorten arc length.	Same as AC.	Same as AC.
Narrow joints	May cause arc to jump from side to side. Widen groove where possible or bring electrode closer to work.	—	—
Length	Maintain short arc (about equal to electrode diameter; approximately 1/8 in.). If arc is too long, root penetration will not be complete, particularly in fillet welding; possible undercutting, excessive bead width, and poor weld contour also may result.	Short, about 1/16 in.	Long, 3/16 to 1/4 in.
Breaking the arc	Gradually reduce arc length or current to minimize weld craters. Continue adding filler at the same time where applicable. Snap torch to horizontal position to break arc.	Same as AC.	Same as AC.

TABLE 9-16. APPEARANCE OF GTA WELD BEADS

	GTA Process and Gas					
	AC		DCEN		DCEP	
Characteristic	**Argon**	**Helium**	**Argon**	**Helium**	**Argon**	**Helium**
Typical Surface Appearance	Medium-width bead, medium rippling, light gray (etched) surface, shiny when brushed.	Medium-width bead, medium rippling, medium to dark gray surface, bright to dull when brushed.	Usually limited to automatic welding. Very narrow bead with finely rippled, bright appearance; smoky at edges.	Narrow bead; smoky film usually dulls surface but is easily brushed off. This is only surface oxide, not porosity, inclusions, or lack of fusion. Surface is shiny and smooth after brushing.	Wide, flat bead; shiny with slightly dull edges when brushed; very slight rippling.	Dirty beads, dull rippled surface.
Bead Faults Stitching	—	—	Very dirty surface. Bead consists of row of alternate deep holes and rough-rippled sections. Reduce current, improve arc-length control.	Turbulent-appearing, uneven bead. Same corrective measures as with argon.	—	—
Cratering	Shrink craters at end of weld bead may contain stress-raiser cracks and are also generally undesirable corrosion crevices. Craters can be eliminated by gradually reducing current, or gradually lengthening the arc, while backtracking (overlapping on circumferential welds) over weld bead.					

Depth of penetration results from the heat produced by the heavy current flow into the work during the electrode-negative portion of the AC cycle. Because cleaning action is inherent in electrode-positive welding, any oxide film on the work is broken up during the electrode-positive portion of the AC cycle. This balance between penetration and cleaning action is particularly important in tack welding and in making the first pass of pipe welds.

AC GTAW can produce excellent quality aluminum welds in all positions. The process achieves complete single-pass weld penetration in thicknesses to 1/2 in. with properly prepared joints and recommended techniques. Particular care should be taken in GTA welding to avoid "bridging" of the molten pool at the root of the joint. Such bridging consists of a blanket of molten metal that barricades the arc at the root opening, preventing full root penetration. Full root penetration can be achieved by maintaining a short arc, assuming arc current and other conditions are adequate for the job.

AC GTAW is used for welding relatively thin aluminum sections. Material from 0.001 to 0.375 in. thick can be welded without filler metal and without preheat; heavier gages require preheat. Of the three types of GTA welding, DCEN is most effective at the heavy end of the thickness range, AC in the middle, and DCEP at the light end.

For efficient AC GTA welding, balanced power that produces a stable arc is required. It is not difficult to recognize an arc where a large amount of undesirable rectification is present. The arc will have an erratic, snapping sound, not the smooth buzzing or humming sound of a stable AC arc. In addition, the cleaning action of the arc and the wetting action of the molten weld metal are reduced if the arc is not stable.

Power Supplies: Alternating-current welding transformers, delivering a balanced wave and incorporating high-frequency stabilization, are designed specifically for AC GTA welding of aluminum (see Section 4). This equipment is recommended for high-quality work. Power supplies intended for AC GTA welding should be smoothly adjustable through the entire range of current required for various thicknesses.

A conventional AC transformer can be adapted to serve as a power supply. One addition should be a high-frequency unit. A wave-balancing component

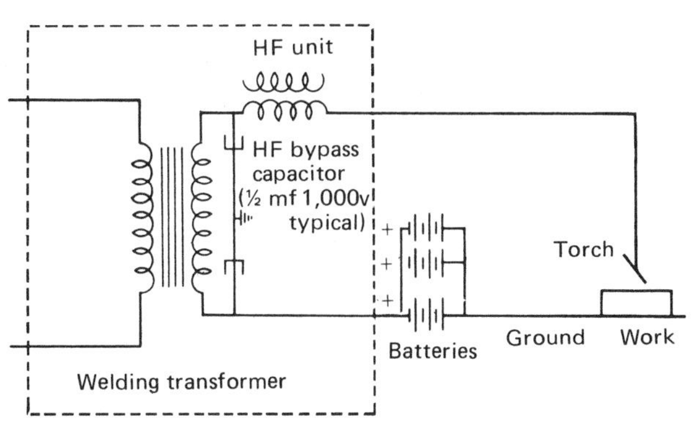

Fig. 9-12. Battery method for obtaining balanced AC power for GTA welding.

is also desirable for certain aluminum welding work, but is not always necessary.

There are several ways to obtain balanced AC power for GTA welding aluminum. A method commonly used is to connect automobile batteries (one 6-v battery per 50 to 75 amp of welding current) in the welding circuit, as shown in Fig. 9-12.

In another system for obtaining balanced current, the output from the secondary windings of the transformer is fed into two half-wave rectifiers of opposite polarity. This output is then combined to give balanced-wave AC power.

When AC transformers are adapted for GTA welding aluminum, care should be taken to install the high-frequency units according to the manufacturer's instructions. Welding torch and ground cables should be kept as short as possible. This helps avoid radio interference and aids in complying with the regulations of the Federal Communications Commission (see current issues of FCC regulations, Part 18, Industrial, Scientific and Medical Service). In addition, power companies may require installation of capacitors for power-factor correction. Most modern AC welding power supplies have these built in as standard equipment.

GTA Welding Torches: Light-duty electrode holders (called "torches") for GTA welding are air-cooled, (Fig. 9-13); others are water-cooled . The GTA torch carries the welding current to the work through the tungsten electrode and directs the inert gas to the weld area. Commercial torches are insulated for maximum welding current ranges to ensure operational safety. They are available in a range of sizes to suit light, medium, and heavy-duty operating conditions. GTA torches should not be used over their rated capacities.

The gas cup or nozzle of the torch can be either ceramic or metal. A ceramic cup is satisfactory for light-duty welding, but for welding at high-current levels ceramic cups are generally unsatisfactory since the nozzle may melt at the tip, partially closing the opening. On the other hand, metal nozzles of too small a diameter will short out the high-frequency current if the work is touched by the nozzle. Torch manufacturers usually recommend a specific type and size of nozzle for different current ranges. Generally, the diameter of the nozzle should be equal to or slightly greater than that of the molten weld pool.

For currents above about 200 amps, water-cooling of the torch is necessary to dissipate excess heat generated by the arc and by the current. In addition, water-cooling the power cable permits use of a more flexible, lighter-weight electrical conductor.

Cooling water should be clean to prevent restricting the flow or clogging the water line or

TABLE 9-17. CAUSES AND CURES FOR POROSITY AND CRACKING IN GTA WELDS

Cause	Contributing Factors	Corrective Measures
Hydrogen	Dirt containing oils or other hydrocarbons; moisture in atmosphere or on metal, or a hydrated oxide film on metal; moisture in gas or gas lines. Base metal may be source of entrapped hydrogen (the thicker the metal, the greater the possibility for hydrogen). Spatter.	Degrease and mechanically or chemically remove oxide from weld area. Avoid humidity; use dry metal or wipe dry. Reduce moisture content of gas. Check gas and water lines for leaks. Increase gas flow to compensate for increased hydrogen in thicker sections. To minimize spatter, adjust welding conditions.
Impurities	Cleaning or other compounds, especially those containing calcium.	Use recommended cleaning compounds; keep work free of contaminants.
Incomplete Root Penetration	Incomplete penetration in heavy sections	Preheat; use higher welding current, or redesign joint geometry.
Temperature	Running too cool tends to increase porosity due to premature solidification of molten metal.	Maintain proper current, arc-length and torch-travel speed relationship.
Welding Speed	Too great a welding speed may increase porosity.	Decrease welding speed and establish and maintain proper arc-length and current relationship.
Solidification Time	Quick freezing of weld pool entraps any gases present, causing porosity.	Establish correct welding current and speed. If work is appreciably below room temperature, use supplemental heating.
Chemical Composition of Weld Metal	Pure aluminum weld metal is more susceptible to porosity than is an aluminum alloy.	If porosity is excessive, try an alloy filler material.
Cracking	Such causes of porosity as temperature, welding time, and solidification may also be contributing causes of cracking. Other causes may be discontinuous welds, welds that intersect, repair welds, cold-working either before or after welding, and weld-metal composition. In general, crack-sensitive alloys include those containing 0.4 to 0.6% Si, or 1.5 to 3.0% Mg, or 1.0% Cu.	Lower current and faster speeds often prevent cracking. However, a change to a filler alloy that brings weld-metal composition out of cracking range is recommended where possible.

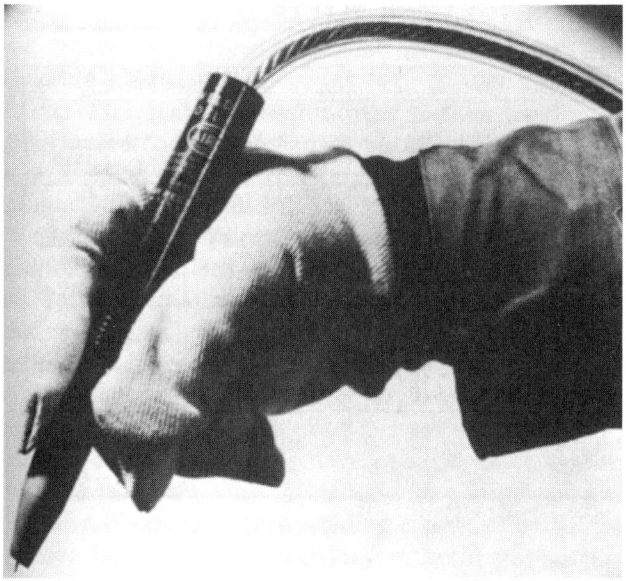

Courtesy Airco Welding Products Division

Fig. 9-13. Light-duty, pencil-type, 100-amp electrode holder for GTA welding.

valves. A filter should be installed in the water system ahead of the welding equipment to remove any foreign material. The water-supply pressure should not be higher than that recommended by the torch manufacturer. If necessary, a reducing valve should be used to lower the pressure sufficiently so that it will not damage the torch.

Overheating a GTA torch can melt the silver-brazed metal joints in the torch handle or melt the plastic tube that sheaths the power cable. A control device is available that does not allow the welding current to start unless the water is flowing.

Water flow should be shut off when welding is stopped to prevent excessive cooling and resultant moisture condensation inside the torch. Moisture can contaminate the tungsten electrode and, being a source of hydrogen, can also cause porosity in the weld metal. The most obvious effect of the presence of water is the loss of the cleaning action of the arc (AC or DCEP). This condition continues until the torch is dry.

Electrodes: The size, tip condition, and position in the torch of the tungsten electrode are important factors that determine the quality of GTA welding. Size of electrode is chosen so that the tip is maintained at a temperature near its melting point when welding is done in any given AC range. The maximum electrode extension beyond the end of the gas cup should be about equal to the diameter of the cup for butt welding, and slightly greater for fillet welding. However, extension of the electrode somewhat less than these maximums is always preferable, where practical, because it helps improve weld shielding.

Electrodes for AC GTA welding aluminum are available in sizes ranging from 0.010 to 1/4 inch. Selecting the right size of electrode for each job is important to prevent poor welds and electrode damage caused by using too high or too low a current for the size electrode used. When electrode size is matched with the correct current, the tip will become a molten hemisphere. Recommended electrode sizes for various ranges of welding current are shown in Table 9-18.

Electrode Spitting: Particles of the electrode being ejected across the arc (electrode "spitting") cannot be tolerated. One result of spitting is that the electrode particles contaminate the weld metal and reduce the mechanical properties of the weld. Another is unnecessary and wasteful consumption of the electrode. Also, the time required to reset the electrode tip and eventually to replace it adds unnecessarily to welding time.

In AC GTA welding, the transfer of electrode particles into the weld pool can be caused by partial rectification of the welding current. The use of batteries in the circuit as a means of reducing partial rectification was discussed earlier.

In all three methods of GTA welding, the transfer of electrode particles can be caused by using too small an electrode for the welding current. This overheats the molten hemisphere at the electrode tip, causing it to drop off.

Spitting of electrode particles across the arc can also occur if too large an electrode is used. In this case, the path of the current or focal point of the arc wanders over the tip of the electrode, and particles of tungsten fly off. It is easy to detect whether the tungsten particles are flying off or whether the tip is becoming overheated and dropping off.

Electrode Care: The electrode should have a clean, silvery appearance. A dirty, rough electrode usually means the inert gas was shut off before the electrode cooled, that there was air leakage in the gas-supply system or in the torch, or that the electrode tip was contaminated by touching metal.

A dirty tungsten electrode can sometimes be cleaned satisfactorily with a fine emery cloth. The contamination can also be removed by using a high current while welding on scrap material until the tip has a sizable molten section; then quickly twisting the torch to flip off a drop of molten metal. If severely contaminated, the electrode should be replaced or the tip broken off and dressed.

TABLE 9-18. TYPICAL CURRENT RANGES FOR TUNGSTEN ELECTRODES[a]

			AC with High Frequency			
	DCEN	DCEP	Balanced Wave		Unbalanced Wave	
Electrode Diameter (in.)	Pure and Thoriated Tungsten	Pure and Thoriated Tungsten	Pure Tungsten	Thoriated Tungsten[b]	Pure Tungsten	Thoriated Tungsten[b]
0.010	up to 15	[c]	up to 15	up to 15	up to 15	up to 15
0.020	5 - 20	[c]	10 - 20	5 - 20	5 - 15	5 - 20
0.040	15 - 80	[c]	20 - 30	20 - 60	10 - 60	15 - 80
1/16	70 - 150	10 - 20	30 - 80	60 - 120	50 - 100	70 - 150
3/32	150 - 250	15 - 30	60 - 130	100 - 180	100 - 160	140 - 235
1/8	250 - 400	25 - 40	100 - 180	160 - 250	150 - 210	225 - 325
5/32	400 - 500	40 - 55	160 - 240	200 - 320	200 - 275	300 - 400
5/16	500 - 750	55 - 80	190 - 300	290 - 390	250 - 350	400 - 500
1/4	750 - 1000	80 - 125	250 - 400	340 - 525	325 - 450	500 - 630

[a] With argon gas.
[b] This range may be lowered when using electrodes with high-thoria cores.
[c] DCEP is not commonly used in these sizes.

In AC GTA welding, the tip should be hemispherical in shape. Needle-point tips are used only for tungsten electrodes when DC GTA welding. In general, tips should be tapered to a blunt end (Fig. 9-14) for AC GTA welding of aluminum.

A grinding wheel is a good tool to make the taper. Some authorities feel that grinding the tip to a sharp point to facilitate arc starting is worthwhile, but others no longer deem this necessary or advisable.

AC GTA Welding Techniques: Before welding is started electrode size, current settings, and gas flow, as detailed in Table 9-19, should be selected to suit the metal thickness and the welding position.

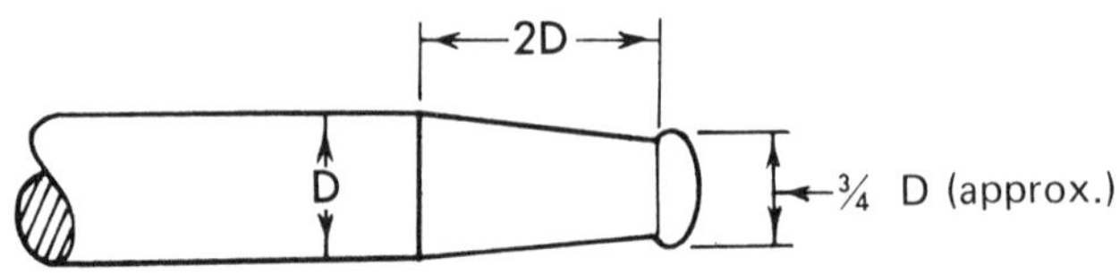

Fig. 9-14. Recommended shape for tip of electrode for AC GTA welding.

Cleaning the Welding Surfaces: For routine GTA jobs cleanliness is about as critical as for GMA welding, but, for high-quality GTA welding, maximum cleanliness is necessary. (For cleaning procedures, see Section 9.1.)

TABLE 9-19. RECOMMENDED PRACTICES FOR MANUAL AC GTA WELDING OF ALUMINUM

Plate Thickness (in.)	Welding Position[a]	Joint Type	Alternating Current (amp)	Electrode[b] Diameter (in.)	Argon Gas Flow[c] (cfh)	Filler Rod Diameter (in.)	Number of Passes
1/16	F	Square butt	70 - 100	1/16	20	3/32	1
	H, V	Square butt	70 - 100	1/16	20	3/32	1
	O	Square butt	60 - 90	1/16	25	3/32	1
1/8	F	Square butt	125 - 160	3/32	20	1/8	1
	H, V	Square butt	115 - 150	3/32	20	1/8	1
	O	Square butt	115 - 150	3/32	25	1/8	1
1/4	F	60° single V groove	225 - 275	5/32	30	3/16	2
	H, V	60° single V groove	200 - 240	5/32	30	3/16	2
	O	100° single V groove	210 - 260	5/32	35	3/16	2
3/8	F	60° single V groove	325 - 400	1/4	35	1/4	2
	H, V	60° single V groove	250 - 320	3/16	35	1/4	3
	O	100° single V groove	275 - 350	3/16	40	1/4	3
1/2	F	60° single V groove	375 - 450	1/4	35	1/4	3
	H, V	60° single V groove	250 - 320	3/16	35	1/4	3
	O	100° single V groove	275 - 340	3/16	40	1/4	4
1	F	60° single V groove	500 - 600	5/16 - 3/8	35 - 45	1/4 - 3/8	8 - 10

[a] F=flat; H=horizontal; V=vertical; O=overhead.
[b] Diameters are for standard pure or zirconium tungsten electrodes. Thoriated tungsten electrodes not generally used for AC GTA.
[c] Helium is not generally used in AC GTA welding aluminum. When helium is used, however, flow rates are about twice those used for argon.

Manual AC GTA welding is slower than GMAW; consequently, the weld pool is molten longer and porosity-causing elements have more time to rise to the surface of the aluminum and escape into the air. However, cleaning before welding is always excellent practice because the possibility of weld porosity is reduced when weld areas are clean. Water or oils usually found in entrapped dirt, hydrated oxide deposits, or hydrogen from any source are potential causes of poor weld quality.

Striking the Arc: In AC GTA welding, the high frequency or the high welding voltage (depending upon type of power supply) establishes an ionized path through the shielding gas for the welding current. When the electrode is brought within 1/8 to 1/16 in. of the plate, the welding arc is initiated. The arc is then adjusted to the desired length.

After the arc has been established, it is held stationary until a molten pool is formed. The area of the pool is determined by electrode size and welding current. Extent of the pool can be increased by rotating the arc with a slight circular motion of the torch. The proper way to hold the GTA torch during welding is shown in Fig. 9-15.

If desired, the arc can be started on a separate plate of aluminum, copper, or steel and then carried to the workpiece. A carbon starting block should not be used. This method was used in the past, but is no longer recommended, because tungsten carbide is formed when carbon impinges on tungsten. The carbide has a lower melting point than pure tungsten and forms a molten ball on the end of the electrode, causing poor current distribution.

Arc Initiation Difficulties: When high-frequency current is used to start the AC welding arc, an arc gap of 1/8 to 1/16 in. in an inert-gas atmosphere is sufficient. The high-frequency arc should not jump a gap of more than 3/16 in. Inability to jump a 1/16-in. gap may be caused by a poor electrical connection in the high-frequency circuit, improper arc gap in the high-frequency oscillator, or excessive oxide film on the base material. Standard procedure should include checking the ground-cable-to-work connection, the high-frequency-unit spark-gap adjustment, and cleanliness of base material. A gap of 0.004 to 0.008 in. in each of the series gaps of the high-frequency oscillator is usually satisfactory, provided that the high-frequency leads to the torch are not excessively long. However, the manufacturer's gap-setting recommendations should be followed.

Before the main weld bead is started, the joint should be tack-welded unless hold-down fixtures are used to secure the work.

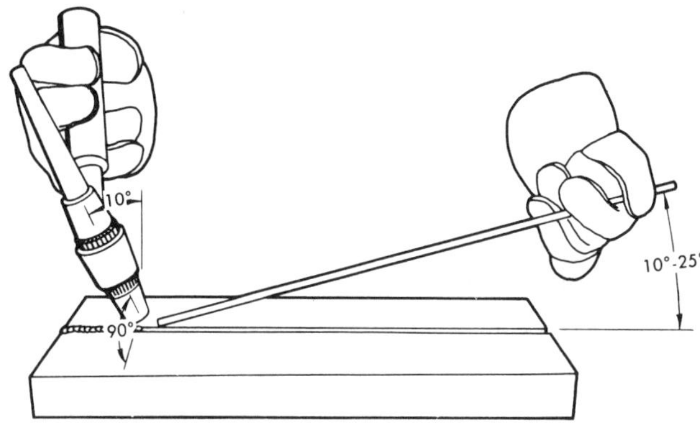

Fig. 9-15. Proper positions of torch and filler rod for manual GTA welding.

Feeding Filler Metal: Filler rod is added manually to the leading edge of the weld pool as needed. Most operators develop a technique of moving the torch and filler rod intermittently. This consists of holding the torch at a point until the pool is formed, then adding filler metal. The filler rod is then withdrawn and the torch moved forward. This sequence of feeding, moving the torch, and holding until a new pool forms, is repeated until the weld is completed. Practice with this technique produces a smooth weld bead.

It is important to feed the filler rod into the arc at as low an angle as possible, as illustrated in Fig. 9-15. This assures good gas coverage of the molten filler metal and also reduces the possibility of touching the tungsten electrode with the filler rod.

In automatic GTA welding, the torch moves steadily forward, and the filler metal, when used, feeds spooled wire continuously to the leading edge of the pool.

Completing the Weld: The manner in which the arc is broken at the end of a weld is important. One method is to withdraw quickly or twist the torch tip up from the work. This rapid breaking of the GTA arc is not recommended, because it leaves a shrink crater. It also promotes porosity, especially when subsequent cover or overlap passes are made over the craters. Craters also may be objectionable in appearance, and reduced weld strength or leakage may result.

One method for eliminating craters is to finish welding with a decreasing current. This technique frequently called "decaying arc" and is usually done by means of a foot or hand-actuated control switch. Backtracking (or overlapping, where applicable) for about 1-1/2 in. on the bead before removing the torch or cutting off the current is also good practice

Still another method is to use runoff tabs made from scrap material.

Preheating: Preheating the work for AC GTA welding of aluminum is necessary for heavier sections, where the heat is conducted away from the joint so fast that the welding arc cannot produce fusion. Preheating can be done with a gas torch or any other suitable means. Unless the ambient temperature is below 40°F, no preheating is used on plate up to 3/8 in. thick and is optional between 3/8 and 1/2 in. thick. Plates 3/4 in. and thicker may require preheating, but the preheat temperature should not exceed 350°F. Maximum temperatures of heat-treatable alloys are particularly critical because of the narrow temperature range that must be maintained to avoid adverse effects of the heat of welding on mechanical properties of the joint.

Rough and Dirty Weld Beads: Various factors can affect the surface appearance of a weld; principal appearance characteristics are described in Table 9-16. As discussed previously, cleanliness is extremely important. Contaminated shielding gas, base material, or filler rod can cause porosity and dross in welds. Insufficient shielding gas can do the same. Poor arc-cleaning action in AC GTA welding can be the result of an insufficient amount of the electrode-positive portion of the current. Reduction of the half cycle of the current that does the cleaning (DCEP) can be due to poor re-ignition of the arc as the current cycle passes zero. It can also be caused by partial rectification, so that the DCEP portion of the AC wave is reduced. The first is caused by the high-frequency unit, and can be corrected by adjusting it. The second, partial rectification, can be alleviated by installing capacitors or a battery "balance" in the welding circuit.

Rough and dirty weld beads can also be caused by poor welding technique. Examples of this are welding with too little current for the travel speed, sloppy torch manipulation, erratic arc due to letting the electrode touch the aluminum, welding in a drafty area, or leaks in the shielding-gas system.

Welding Speed: Within the practical limits of the welding conditions on a given job, the faster the welding speed, the greater the savings in time and the less extensive the heat-affected area in the work. The best speed can be achieved with the highest-practical welding current and use of a flat position. Often, the increase in efficiency obtainable warrants repositioning of the work.

Shielding Gas: Although any inert gas could be used for shielding the arc in the GTA process, argon and helium are used because of their availability, economy, and efficiency. A higher gas flow is required with helium than with argon, but helium permits faster welding speeds and greater penetration. Differentials in rates of gas flow are often about equalized in welding thicker sections. Deeper penetration is obtained with helium, because the AC arc in a helium atmosphere is hotter than in argon.

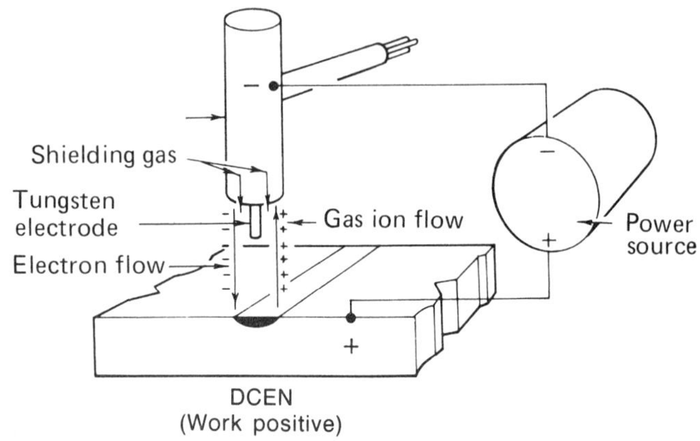

Fig. 9-16. Machine connections for a DCEN GTA welding system; the electrode is negative and the work positive.

Many operators prefer argon for manual AC GTA welding, because the arc has greater stability. Additives to argon shielding gas increase the arc temperature and gain some of the advantages of helium shielding gas. Generally, when 10% or more helium is added to argon, arc penetration increases significantly.

Purity of the inert gas is very important. Only welding-grade gases should be used. Furthermore, contamination of the gas must be prevented during its passage from the cylinder to the torch. Moisture, oil residues or fumes, and dust are the common contaminants to guard against. Hoses previously used for acetylene, compressed air, or water are not suitable for GTA welding.

Inert gas for GTA welding is available in pressurized cylinders. For consistent results, a pressure-reducing regulator-flowmeter of the proper type for the gas being used should be included in the system. Shutoff can be manual, effected by hanging the torch on a hook arm which actuates a shutoff valve, but automatic control is preferred. This can be accomplished by connecting a time-delay solenoid valve (controlled by the welding circuit) in the gas line. Production GTA welding equipment normally has automatic gas and water-control valves built into the welding unit.

DIRECT-CURRENT, ELECTRODE NEGATIVE GTA WELDING

One of the most reliable and useful methods for welding aluminum is the DCEN GTA process. It can be used in the flat, horizontal, and vertical welding positions, and is supplanting GMA welding in many applications.

In DCEN GTA welding, the electrode is negative and the work positive; the electrons go from the electrode to the plate. This electron flow (Fig. 9-16) results in considerable heating of the base plate, while the electrode stays relatively cool.

DCEN GTA produces a narrow, deep weld. Because the best results on aluminum are obtained with very short arc lengths, this process is best suited for automatic welding. Arc lengths are about 1/16 in. for manual and as little as 1/64 in. for machine welding. This critical arc length is somewhat difficult to control in manual DCEN GTA welding.

Since a DCEN system provides no arc cleaning action to remove the oxide film on aluminum, thorough cleaning and oxide removal are necessary.

Equipment: A DC welding generator or rectifier is used for DCEN. (Some welding engineers feel that three-phase power supplies are superior for this process because they provide a smoother current flow.) In addition to the power supply, the same equipment used for AC GTA welding aluminum is employed.

Electrode Selection: Electrode requirements are different for DCEN than for AC GTA welding. For DCEN, most weldors prefer thoriated tungsten electrodes over pure tungsten. The thoriated tungsten electrode has a higher heat resistance than the pure tungsten and, therefore, lasts longer. In addition, the thoriated tungsten appears to repel the "skin" on the weld pool and thus permits welding with the electrode in closer proximity to the molten aluminum. Other weldors prefer pure tungsten electrodes, expressing the opinion that thoriated electrodes tend to shatter or splinter.

Whether the preference is for thoriated or pure tungsten, there is general agreement to use oversized electrodes and to taper the point. Recommended electrode sizes for DCEP, DCEN, and AC GTA welding are shown in Table 9-18.

Shielding Gas: A helium-argon mixture (7 to 35% argon) or helium alone is preferred as the shielding gas in DCEN GTA welding aluminum. Although a helium-shielded arc is hotter than an argon-shielded arc, this does not entirely account for the apparent superiority of helium (or of a

TABLE 9-20. RECOMMENDED PRACTICES FOR DCEN GTA WELDING ALUMINUM IN THE FLAT POSITION

Material Thickness (in.)	Joint Design	Current[a] (amp)	Volts	Diameter of Electrode[b] (in.)	Helium Gas Flow[c] (cfh)	Travel (ipm)	Filler Rod or Wire Diameter (in.)	Number of Passes
0.010	Standing edge	10 - 15		0.020	20 - 50			1
0.020	Square butt	15 - 30		0.020	20 - 50		0.020	1
0.030	Square butt	20 - 50		0.020 or 0.040	20 - 50		0.020 or 0.047	1
0.032	Square butt	65 - 70	10	3/32	20 - 50	52	None	1
0.040	Square butt	25 - 65		3/64	20 - 50		3/64	1
0.050	Square butt	35 - 95		3/64	20 - 50		3/64	1
0.050	Square butt	70 - 80	10	3/32	20 - 50	36	None	1
0.060	Square butt	45 - 120		3/64 - 1/16	20 - 50		3/64 or 1/16	1
0.070	Square butt	55 - 145		1/16	20 - 50		1/16	1
0.080	Square butt	80 - 175		1/16	20 - 50		1/16	1
0.090	Square butt	90 - 185	12.5	1/16	20 - 50	30	1/16	1
1/8	Square butt	120 - 220		1/8	20 - 50		1/8	1
1/8	Square butt	180 - 200		1/8	20 - 50	24	None	1
1/4	Square butt	230 - 340		1/8	25 - 60		1/8 or 3/16	1
1/4	Square butt	220 - 240	12.5	1/8	25 - 60	22	None	1
1/2	60° V bevel, 1/4-in. root face	300 - 450		3/16	25 - 60		1/8 or 1/4	1
1/2	Square butt	260 - 300	13	5/32	25 - 60	20	None	2
3/4	60° V or double-V bevel, 3/16-in. root face	300 - 450		3/16	25 - 60		1/8 or 1/4	3 (single-V) 2 (double-V)
3/4	Square butt	450 - 470	9.5	3/16	40 - 60	6	None	2
1	60° V or double-V bevel, 3/16-in. root face	300 - 450		3/16	25 - 60		1/8 or 1/4	4 (single-V) 2 (double-V)
1	Square butt	550 - 570	9.5	1/4	40 - 60	5	None	2

[a] Automatic welding is required for the higher amperages. Manual welding can be done at the lower amperages.
[b] In lighter gages, it is common to use larger diameter electrodes than recommended and to taper the tip.
[c] Helium-argon mixtures may be substituted. In automatic welding, the arc can be started in argon, and the helium added to the shielded gas when welding begins. The best ratio of He-A is usually determined by experimentation. Gas flow depends in part on welding speed.

helium-argon mixture) over pure argon in DCEN GTA welding.

Argon shielding gas has been used for automatic DCEN GTA welding where the length of arc can be controlled closely. However, welding speed drops considerably. The welding voltage does not vary with the arc length as much in helium as it does in argon.

Joint Design: Use of the DCEN method of GTA welding to its best advantage requires careful joint design. Square butt joints in aluminum are readily welded by this process. When the plate thickness is so great that V joints are required for DCEN welding, the land can be increased and the included angle of the V decreased. Both changes permit faster welding and require less filler metal.

Distortion is less with DCEN than with AC GTAW because of the lower heat input per unit length of weld.

Welding Techniques: High-frequency current is not needed to maintain a stable DCEN arc as in AC GTA welding. It is desirable for initiating the arc, although the arc can be started without it. Thus, the arc can be initiated by touching the work with the electrode tip and then retracting to the proper arc length. The DCEN arc is quieter and smoother than the AC arc, and the weld pool is formed more rapidly. There is no need to play the arc on the base plate to form the pool if sufficient welding current is used.

Because the DCEN arc is so much hotter than the AC arc, manual welding techniques for the two processes differ. In DCEN GTA welding, many operators prefer to hold the filler rod in contact with the joint and to move the arc steadily forward. The rod is not intermittently fed into the weld pool.

Excessive weaving is not desirable in welding aluminum. Instead, multiple stringer beads are recommended for large welds. Interpass brushing is mandatory to remove the oxide from previously completed beads. Standard techniques for eliminating or reducing craters at ends of welds are applicable to this welding method. Recommended joint designs and other details for DCEN GTA welding are given in Table 9-20.

Welds can be made with or without filler addition, depending upon the alloy, thickness, joint design, and whether force is exerted to push the two weld edges together during fusion. Edge undercutting sometimes occurs, due to the highly concentrated heat of the DCEN arc.

In high-speed welding, the DCEN torch is sometimes followed by an AC GTA torch to smooth the weld bead. Also, an AC torch can precede the DC torch, acting as a preheater and surface cleaner.

Weld Appearance: A dull oxide film always covers the DCEN bead at the edge and may cover the entire surface. This film is removed easily by wire brushing. A particularly shiny bead, with dark edges, is obtained in DCEN welding aluminum when helium is used.

Because of the high welding speed possible with DCEN, the heat-affected zones adjacent to the weld are usually narrower than with the AC GTA process. This is more apparent in the heat-treatable alloys, such as 6061 or 6063, than in the nonheat-treatable alloys, such as 5086 or 5083.

DIRECT-CURRENT ELECTRODE POSITIVE GTA WELDING

Equipment required for the DCEP GTAW process is the same as that used for DCEN welding. The machine connections for torch and work cables, however, are reversed, as shown in Fig. 9-17.

In DCEP welding, the tungsten electrode is heated and the plate stays relatively cool. For this reason, electrodes used with DCEP welding must be considerably larger in diameter than those used with AC or DCEN GTA welding. For example, a 1/4-in.-diameter electrode is required for 125 amp of DCEP welding current, whereas a 1/16-in. electrode will handle the same current in DCEN (see Table 9-18). If an electrode is too small, the tip will overheat and drop off, contaminating the weld.

The DCEP arc produces an excellent cleaning action on an aluminum surface, breaking up any oxide present. But the bead produced by this process is shallow (because the plate stays relatively cool),

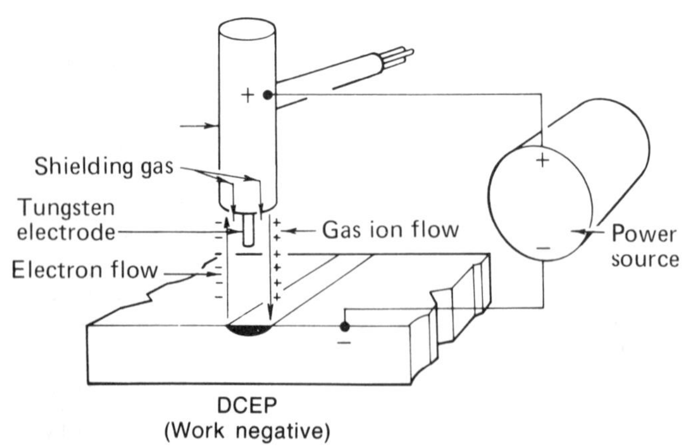

Fig. 9-17. Machine connections for a DCEP GTA welding system; the electrode is positive and the work negative.

which limits DCEP welding of aluminum to thin sheet — to about 0.050 in. maximum.

The shielding gas usually used with DCEP GTA welding is argon. Recommended flow rates, along with suggested procedures for DCEP welding of aluminum, are listed in Table 9-21.

TABLE 9-21. RECOMMENDED PRACTICES FOR DCEP GTA WELDING OF ALUMINUM IN THE FLAT POSITION

Material Thickness t (in.)	Joint Design[a]	Electrode Diameter[b] (in.)	Welding Current[b] (amp)	Gas Flow (cfh)	Filler Diameter (in.)
0.010	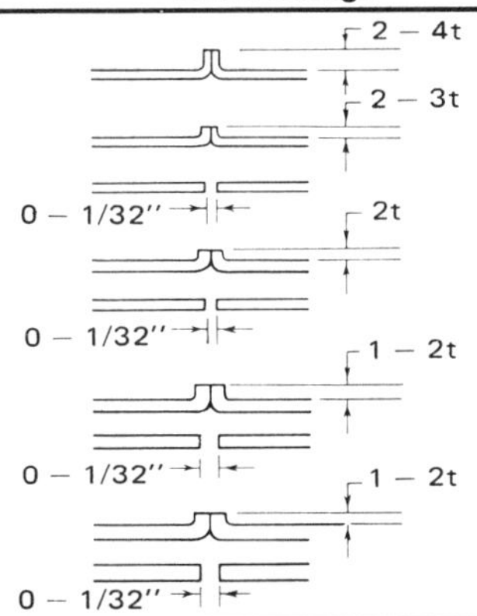	3/32	15 - 25	15 - 25	—
0.020		1/8 - 5/32	40 - 55	15 - 25	—
		1/8 - 5/32	40 - 55	15 - 20	0.020
0.030		3/16	50 - 65	15 - 25	—
		3/16	50 - 65	15 - 20	0.020 or 3/64
0.040		3/16	60 - 80	20 - 30	—
		3/16	60 - 80	25 - 30	3/64
0.050		3/16	70 - 90	20 - 30	—
		3/16	70 - 90	25 - 30	3/64 or 1/16

[a] Use of a backup with a generous groove is recommended in all cases.
[b] High currents, with correspondingly larger electrodes, can be used for automatic welding.

Welded support structure for a 230 KV power line.

Section 10
WELDING COPPER AND COPPER ALLOYS

SECTION 10.1
WELDING COPPER AND COPPER ALLOYS

Page

Welding Copper and Copper Alloys

Copper, one of the best metals for conducting heat and electricity, is also fairly reactive chemically, particularly in respect to oxygen. Yet, it is produced and sold in one of the purest forms of any commercial metal. Ordinary tough-pitch or electrolytic copper is 99.9+% pure, with about 0.05% copper oxide distributed throughout the metal as a copper-oxide eutectic. During welding, the migration of copper oxide to the grain boundaries tends to cause loss of ductility of the otherwise highly ductile material.

The absorption of carbon monoxide and hydrogen at temperatures above 1300°F, with resultant reaction with copper oxide to form carbon dioxide and water vapor, also complicates the welding of copper by causing internal cracking and embrittlement. As indicated in Table 10-1 — Applicability of Four Welding Processes to Copper and Its Alloys, only poor to fair weld quality is obtainable with the manual shielded metal-arc electrode, gas metal-arc, gas tungsten-arc, and carbon-arc processes when welding electrolytic tough-pitch copper.

The addition of a small amount of silicon, phosphorus, or other deoxidizers to copper results in deoxidized copper, which can be welded with the gas metal-arc, gas tungsten-arc, or carbon-arc processes with substantially improved results. The deoxidizing element ties up the oxygen, inhibiting the formation of copper oxide at the grain boundaries.

One form of deoxidized copper is known commercially as "oxygen-free." This is copper that has been remelted and cast in a hydrogen atmosphere to remove oxygen. Oxygen-free copper contains no deoxidizing elements, the absence of which improves some of the mechanical properties, but makes the metal susceptible to internal oxidation on prolonged heating in air at elevated temperatures. Oxygen-free copper, thus, tends to pick up oxygen during welding, forming copper oxide. As a result, the welds are usually inferior in quality to those in deoxidized copper. For this reason, oxygen-free copper can be classed with electrolytic copper in summarizing its weldability — as done in Table 10-1. Only when using the carbon-arc process with its highly reducing atmosphere do welds in oxygen-free copper exhibit significant improvements in quality over welds in electrolytic copper.

The affinity of copper for oxygen is only one of several characteristics that make welding difficult. Even though copper melts at a much lower temperature than steel (1980°F versus 2775°F), it takes longer to melt a section of copper than an equivalent section of steel with the electric arc. This results from the high thermal conductivity of copper — about five times that of steel. When welding copper, the metal adjacent to the weld area draws away heat about as fast as it is introduced. Not until the adjacent metal is adequately heated does the spot being welded reach the melting point. The high thermal conductivity of copper, thus, usually makes preheating necessary when welding.

Copper also has a higher coefficient of expansion than steel. This means that it expands more when heated and contracts more when cooled, accentuating the shrinkage and weld-cracking problem that welding operators are accustomed to dealing with when welding steel. This problem is further heightened by the rapid decrease of the tensile strength of copper above 500°F.

WELDABILITY OF COPPER ALLOYS

There are numerous copper alloys designed for various mechanical, electrical, and architectural applications. The alloying elements used are lead, zinc, tin, nickel, aluminum, iron, and beryllium, in varying admixtures and percentages. All of the alloys are weldable, although not to the same degree of ease or success in respect to the quality of weld. The approximate compositions and properties of the six representative alloys commonly used in products fabricated by welding are listed in Table 10-2 — Properties of Copper and Copper Alloys, along with the properties of electrolytic and deoxidized copper. The weldabilities of these six alloys are summarized by chart fashion in Table 10-1 and discussed in the following text.

TABLE 10-1. APPLICABILITY OF FOUR WELDING PROCESSES TO COPPER AND ITS ALLOYS

Common Name	Oxygen-Free or Electrolytic Tough-Pitch Copper	Deoxidized Copper	Red Brass	Muntz Metal	Silicon Bronze	Phosphor Bronze	Cupro-nickel	Aluminum Bronze
MANUAL SHIELDED METAL ARC — DCEP								
Weld Quality	Fair or poor	Fair. Prefer gas metal-arc or gas tungsten-arc.	Good	Good	Fair. Prefer gas metal-arc or gas tungsten-arc.	Good	Good	Fair. Prefer gas metal-arc or gas tungsten-arc.
Electrode	ECuSn-A or ECuSn-C	ECuSn-A or ECuSn-C	ECuSn-A or ECuSn-C or ECuAl-A2 for higher strength.	ECuSn-A or ECuSnC or ECuAl-A2 for higher strength	ECuSi ECuAl-A2	ECuSn-A ECuSn-C	ECuNi	ECuAl-A2 ECuAl-B for higher strength.
Preheat °F	Up to 900	Up to 900	400 to 500	500 to 700	Not required. Do not exceed 200 interpass.	300 to 400	Not required.	400, up to 1150 for the high-aluminum grades.
GAS METAL-ARC — DCEP								
Weld Quality	Fair to poor	Good	Good	Good	Good	Fair	Good	Good
Electrode	ERCu for best, conductivity. ERCuSi for best strength.	ERCu or ERCuSi	ERCuSn-C best color match. ERCuSi best operation.	ERCuAl-A2 for strength. ERCuSi for best operation.	ERCuSi	ERCuSn-A ERCuSn-C for higher strength.	ERCuNi	ERCuAl-A1 ERCuAl-A2 ERCuAl-B
Preheat °F	Up to 1000, depending on the thickness.	400 for 1/4 in. and thicker.	Up to 400 on heavy sections.	Not required. Can be used to reduce zinc fuming.	Not required. Do not exceed 200 interpass.	200 to 400	Usually not required.	On heavy sections only. Do not exceed 500.
Gas	Argon for 1 in. or thinner. Over 1 in. use 65% helium, 35% argon mixture.	Argon for 1 in. or thinner. Helium-argon mixture over 1 in.	Helium	Helium	Argon or argon-helium mixture.	Helium	Argon or helium. Argon preferred.	Argon or helium or a mixture.
GAS TUNGSTEN-ARC — DCEN EXCEPT ON ALUMINUM BRONZE AC IS PREFERRED								
Weld Quality	Fair	Good	Fair	Fair	Good	Fair	Good	Good
Rod if used	ERCuSi-A or ERCuSn-A for strength ERCu for best conductivity.	ERCuSi-A or ERCuSn-A	ERCuSn-A ERCuSn-C ERCuSi-A	ERCuSi-A ERCuAl-A2	ERCuSi-A	ERCuSn-A	ERCuNi	ERCuAl-A2 ERCuAl-B
Preheat °F	Up to 1000, depending on thickness.	Up to 1000, depending on thickness.	Up to 400, depending on thickness.	Up to 400.	Not required. Do not exceed 200 interpass.	Up to 400 followed by slow cool.	Usually not required.	On heavy sections only. Do not exceed 500.
Gas	Helium or mixture of 75% helium and 25% argon.	Argon 1/8 in. and thinner. Helium over 1/8 in.	Argon	Argon	Argon or argon-helium mixture.	Helium. Argon on thin sheet.	Argon	Argon
CARBON-ARC — DCEN								
Weld Quality	Tough-pitch: Fair to poor Oxygen-free: Fair	Good	Fair	Fair	Fair	Fair	Fair	Fair
Rod if used	ERCu for best conductivity. ERCuSn-A for best strength.	ERCu for best conductivity. ERCuSn-A for best strength.	ERCuSi-A, ERCuSn-C for single pass only.	ERCuSi-A	ERCuSi-A	ERCuSi-A	ERCuNi	Use an ERCuAl-A2 as filler rod.
Preheat °F	Up to 600.	Up to 600.	Up to 400.	Up to 300.	Not required.	Usually not required.	Not required.	On heavy sections only.

WELDING COPPER AND ITS ALLOYS WITH MANUAL SHIELDED METAL-ARC ELECTRODES

Commercial arc welding of copper and its alloys — with several exotic exceptions — is limited to the four processes shown in Table 10-1. The shielded metal-arc process is the simplest to use and has advantages over the gas tungsten-arc and carbon-arc processes. The heating cycle is relatively short, which is desirable with some alloys. With plate over 1/4 in. thick, preheating is usually necessary to obtain proper fusion. As indicated in Table 10-1,

TABLE 10-2. PROPERTIES OF COPPER AND COPPER ALLOYS

Common Name	Tough-Pitch (Electrolytic)		Deoxidized Copper		Red Brass		Muntz Metal		Silicon Bronze		Phosphor Bronze		Cupro-Nickel		Aluminum Bronze	
Approximate composition (%)	Copper 99.95		Copper 99.95 Phos. 0.02		Copper 85 Zinc 15		Copper 60 Zinc 40		Copper 96 Silicon 3		Copper 92 Tin 7 Phos. 0.2		Copper 70 Nickel 30		Copper 91 Aluminum 7 Iron 2	
Temper	Hard	Soft	Hard	Soft	Hard	Soft	Hard	Soft	Hard	Soft	Hard	Soft	Hard	Soft	Hard	Soft
Tensile strength (1000 psi)	50	32	55	32	62	40	74	54	94	60	91	56	84	55	80	72
Yield strength (1000 psi)	45	10	50	10	57	12	60	21	58	25	86	25	79	18	50	38
Elongation in 2 in. (%)	12	55	8	45	5	47	8	45	8	60	17	73	3	36	30	42
Rockwell hardness	50B	40F	60B	40F	73B	59F	80B	80F	93B	60B	94B	43B	85B	40B	83B	74B
Electrical conductivity 68°F (% IACS)	101		85		37		28		7		13		4.6		14	
Thermal conductivity 68°F (Btu/sq ft/ft/hr/°F)	226		196		92		69		21		36		17		44	

the best welds with the manual shielded metal-arc are obtained when welding the copper-zinc, copper-tin, and copper-nickel alloys. Full-strength joints are not possible on electrolytic, oxygen-free, or deoxidized copper.

Electrode selection depends on the chemical composition of the base metal. Section 4.1 gives the AWS classifications and chemical-composition requirements of electrodes for welding copper and its alloys. The minimum specified tensile strengths for welds deposited by these electrodes are also listed in Section 4.1.

Table 10-1 gives the electrode classifications recommended for the metal and six alloys. Thus, from this table it can be seen that an ECuNi electrode would be chosen for welding cupro-nickel alloys and either an ECuSn-A, ECuSn-C, or ECuAl-A2 would give good results with the low-and high-zinc brasses (red brass and Muntz metal).

Direct current, electrode-positive polarity is used with most copper and copper-alloy covered electrodes. The welding operations differ from those of steel as necessitated by the marked differences in physical properties. In general, wider root openings and groove angles are required in the joint preparations, along with a greater amount of tack welding, higher preheat and interpass temperatures, and higher current densities. Figure 10-1 shows the AWS-recommended edge preparations for light-gage and plate material.

The preheat requirements are specified in Table 10-1. With the exception noted for silicon bronze, phosphor bronze, and cupro-nickel, preheat and interpass temperatures of 400°F or more are necessary to obtain adequate flow of molten metal. Joints are usually backed with the same material as the base metal, although carbon or copper backings may be used with most of the alloys. By using a grooved backup of carbon or copper, a small bead may be formed on the underside of the joint.

Although unsatisfactory for quality, full-strength joints in electrolytic copper, and producing only fair-quality welds in deoxidized copper, there are occasions when it is practical to weld these metals

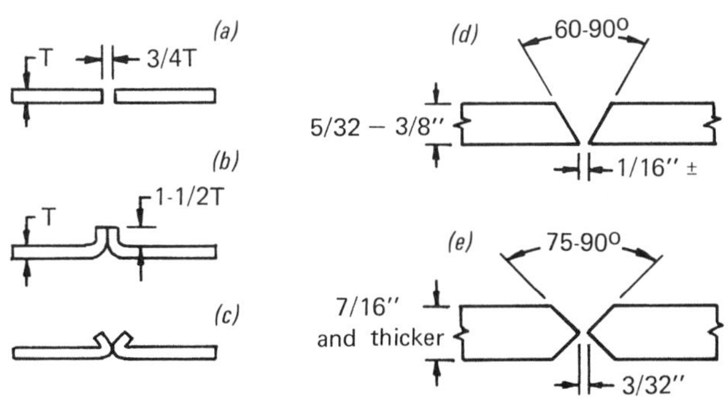

Fig. 10-1. Edge preparations commonly employed when designing copper and copper alloy joints for welding. Preparations (a), (b) and (c) are usually used for light-gage materials, while (d) and (e) may be used for heavier sections.

with the manual shielded metal-arc process, using ECuSn-A or ECuSn-C electrodes. A grooved backing strip of copper with a wide-angle groove should be used, with the welding done in the flat position. The largest diameter electrode possible for the plate thickness should be selected. A high current density is necessary to assure good metal flow and minimize slag entrapment. Weave or stringer beads may be used, depending on the width of the groove. Clamping of the joint should be moderate to minimize the possibility of weld cracking.

Referring to Table 10-1, it can be seen that good results are obtained when welding the brasses (red brass and Muntz metal) by the manual shielded metal-arc process, using ECuSn-A, ECuSn-C, or ECuAl-A2 electrodes. The latter electrode is used when high tensile and fatigue strength and good corrosion resistance are requirements. The preheat and interpass temperatures should be not less than 400°F for red brass nor less than 500°F for Muntz metal. Backing strips of matching plate or copper should be used, and groove preparations should be wide to assure penetration and prevent slag inclusions.

The zinc content of the brasses is a cause of welding difficulties. Welding temperatures tend to volatilize the zinc, leading to zinc oxide fumes that constitute health hazards unless the area is well ventilated. The zinc oxide fumes obscure vision and make fusion difficult by reducing the wetting properties of the molten metal. Vaporization can cause excessive loss of zinc, changing the composition of the metal in the weld area, with resultant changes in physical and mechanical properties from that of the base metal.

To give the best results and minimize zinc volatilization, all welding should be performed in the flat position with the largest size electrode practical. Playing the arc on a small puddle when using the ECuAl-A2 electrode and progressing slowly with a weaving motion minimizes the loss of zinc. The current should be on the high side of the range. With the ECuSn-A and ECuSn-C electrodes, currents on the high side of the range and fast welding speeds should be used with thin stringer beads.

Although gas-shielded arc-welding processes are preferred for the silicon bronzes, welds of fair quality can be obtained with the ECuSi and ECuAl-A2 manual metal-arc electrodes. The silicon bronzes have a lower thermal conductivity than other copper alloys, which makes them easier to weld. Because of this, the groove preparations can be similar to those used with steel. Square butt joints may be used in thicknesses up to 5/32 in., and material of greater thickness may be joined by single or double-V grooves with a 60° included angle.

The silicon bronzes are hot-short and preheats are undesirable. Care should be taken that interpass temperatures do not exceed 200°F. The weld metal should be deposited rapidly in the form of stringer beads, with the current setting in the middle of the recommended range for the electrode size. Maintaining a short arc and a small puddle helps prevent overheating. It is preferable that the work be planned for welding in the flat position.

The manual shielded metal-arc process usually gives better quality welds on the phosphor bronzes than any other arc-welding process. ECuSn-A and ECuSn-C electrodes are specified. Preheat and interpass temperatures up to 400°F and no lower than 300°F are required to give adequate flow to the molten metal. Wide grooves facilitate washing the side walls. A backing strip is helpful, as is flat position welding, although out-of-position work is possible where required by the shape or size of the weldment.

A groove joint may be welded by depositing the first two layers with a slight weaving technique, followed by thin stringer beads. The width of the preliminary weaves should be no more than twice the electrode diameter. This technique of depositing the weld metal facilitates the formation of a fine-grain, dense structure with the best as-welded mechanical properties. The coarse dendritic-grain structure that tends to characterize the phosphor bronzes can also be avoided when using stringer beads exclusively by exercising care in maintaining the interpass temperature in the 300°F to 400°F range. A postheat treatment of 900°F, with fast cooling, is recommended to give the deposit maximum ductility, when necessary.

Shielded metal-arc welding with the ECuNi electrode gives quality deposits with cupro-nickel alloys. The flux of the covering is designed to handle the highly refractory nickel oxides formed under the heat of the arc. Smooth weld deposits, free from slag or porosity, are possible with good workmanship. Welding in the flat position is recommended, although welds can be made out-of-position.

ECuAl-A2 and ECuAl-B electrodes are used to join commercial grades of aluminum bronze sheet and plate by the manual shielded metal-arc process. As noted in Table 10-1, the welds are rated fair in quality, compared to the good ratings by the gas-shielded processes. Welding is made difficult because of the highly refractive aluminum oxides formed, which must be removed by fluxing. A wide groove

— with an included angle of 70° to 90° — should be used with heavier sections, together with a backup strip of the same material. A preheat and interpass temperature of 400°F is recommended for the low-aluminum grades and temperatures up to 1150°F, with postweld fan cooling, for aluminum bronzes of higher aluminum content.

Welding techniques similar to those used with steel may be employed. Beads may be either weave or stringer. The deposits have exceptionally good hot strength and hot ductility, allowing much leeway in bead thickness. Flat position welding is recommended. An anneal at 1150°F, followed by fan cooling, is required as a postheat treatment for the high-aluminum grades.

WELDING COPPER AND COPPER ALLOYS BY THE GAS METAL-ARC PROCESS

Copper and copper alloys may be welded with the semiautomatic and fully automatic gas metal-arc process, using solid or stranded bare electrode wire meeting the chemical compositions shown in Section 4.1 and helium or argon, or mixture thereof, as the shielding gas. The weld quality is especially good with the brasses and bronzes, with the exception of the phosphor bronzes, where the manual shielded metal-arc gives superior mechanical properties in the deposited metal. Gas metal-arc welding also gives good welds when joining deoxidized copper.

The obvious advantage of a semiautomatic or fully automatic process is the higher deposition rates obtainable by the mechanized feeding of the electrode to the arc. Other advantages are excellent weld appearance, good mechanical properties, uniform weld quality, dense deposits, low distortion, and low preheat and interpass temperature requirements.

Gas metal-arc welding may be used to join electrolytic and oxygen-free copper, but the welds are rated only poor to fair in quality. Table 10-1 shows the electrode types, preheat, and shielding gases recommended. The ERCu and ERCuSi electrode wires give good groove welds with various grades of deoxidized copper when the preparations are wide enough to permit proper manipulation. A 90° included angle should be used when plates are 1/2 in. or more in thickness, and a preheat of 400°F is recommended for 1/4-in. or thicker plate. Peening, followed by annealing, is frequently used to refine the weld grain structure.

As Table 10-1 indicates, ERCuSn-C and ERCuSi electrodes give good quality welds with the red brasses. The former produces deposits with the better color match; ERCuSi gives better operating characteristics. The properties of the weld metal are similar. Helium is the preferred gas with the ERCuSn-C electrode, and a preheat of 400°F is recommended when welding plate 3/16 in. or more in thickness. Argon may be used with the ERCuSi electrode. Specifications frequently call for an included angle of 70° in heavier plate grooves welded with the ERCuSn-C electrode and 60° in grooves welded with ERCuSi wire.

The higher-zinc alloys, as typified by Muntz metal, may be welded with ERCuAl-A2 or ERCuSi filler wires, using helium as the shielding gas. Preheating is not required, but is helpful in reducing zinc volatilization by making lower welding currents practical. ERCuAl-A2 wire gives welds with the better strength; ERCuSi wire exhibits better welding characteristics.

The gas metal-arc process is often preferred for the welding of silicon bronzes. It gives good quality welds when used with the ERCuSi electrode and argon gas or a mixture of argon and helium. Preheating is not required, and care should be taken to avoid overheating by welding rapidly. A small weld puddle should be maintained, with the weld progressing at a rate of about 8 in. per minute. The oxide film formed on the weld bead should be removed by wire brushing in multipass work.

Only fair-quality welds can be expected with the gas metal-arc process when welding the phosphor bronzes with the electrodes, preheat ranges, and gas indicated in Table 10-1. These bronzes tend to be hot-short, and they are characterized by thermal conductivities greater than those of the silicon bronzes. Consequently, if the process is used with phosphor bronzes and wider groove preparations, higher current densities are required. The rate of travel must be rapid and the puddle kept small. Heavier sections require a preheat approaching 400°F.

The gas metal-arc process gives good results when welding the copper-nickel alloys — possibly the best welds of any applicable process. Either argon or a mixture of argon and helium may be used with the ERCuNi electrode wire. Square butt joints are used with light-gage sheet, and single or double-V grooves with an included angle of 60° with plate material. Welding may be done out-of-position as well as in the preferred flat positions.

Good-quality welds are also obtained with gas metal-arc welding when joining the aluminum bronzes. Here, also, out-of-position welding is practical, although the best weld quality is attained in joints presented in the flat position. Table 10-1 shows the electrode types and shielding gases recommended. Note that a preheat is not required, except for heavy sections. Stringer beads and small-to-moderate weave beads are preferred.

WELDING COPPER AND COPPER ALLOYS BY THE GAS TUNGSTEN-ARC PROCESS

Copper and its alloys can also be welded by the gas tungsten-arc process, in which the arc is between a nonconsumable tungsten electrode and the work, and the filler metal is obtained from a consumable rod fed into the arc. Welding is done using direct current, electrode-negative, except in the case of welding the aluminum bronzes, where alternating current is preferred.

The chemical composition of welding rods used with the process, and also with carbon-arc welding of copper and copper alloys, are specified by the AWS as shown in Section 4.1. Note that rods carry the initial "ER" letters in the AWS classification to distinguish them from electrodes, which are denoted by "E." The tensile-strength requirements for deposits from the various classifications of welding rods are also shown in Section 4.1.

The gas tungsten-arc process gives dense, high-quality welds on deoxidized copper, silicon bronze, cupro-nickel, and aluminum bronze, and fair-quality welds on electrolytic and deoxidized copper, red brass, Muntz metal, and phosphor bronze. Argon, helium, or a mixture of the inert gases are used for shielding, depending on the material welded. Wide grooves, with 60° to 90° included angles, are required for plate material for good performance, and root gaps ensure complete penetration.

When welding copper, the performance requirements of the weldment may dictate the selection of welding rod. Thus, if it is important to maintain good thermal or electrical conductivity at the joint, the ERCu rod would be selected when welding oxygen-free or electrolytic copper. If strength is a more important consideration, the ERCuSi-A or ERCuSn-A rod would be used. These latter rods are recommended for welding deoxidized copper. Table 10-1 shows the shielding gases that may be used. Note that preheats up to 1000°F may be required to secure proper fusion. The gas tungsten-arc process becomes less attractive as plate thickness increases because of the need for high preheat and a 90° included angle for good welding performance.

GAS METAL-ARC WELDING OF COPPER

Process: Gas Metal-Arc
Plate Material: Oxygen-Free Copper
Plate Thickness: 1/8 - 1/2 in.

Electrode Class: ERCu
Position: Flat
Polarity: DCEP
Gas: Argon 50 cfh

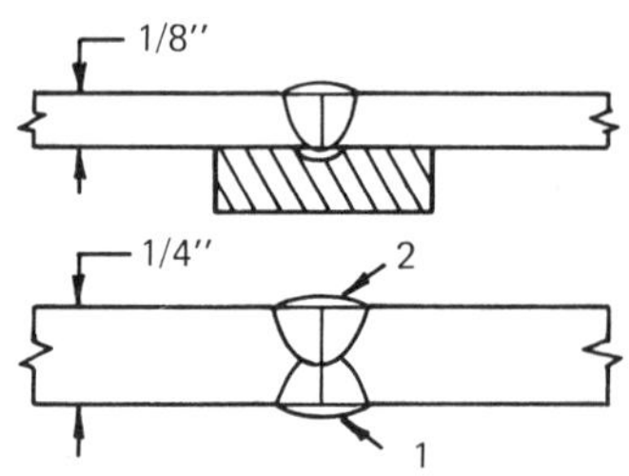

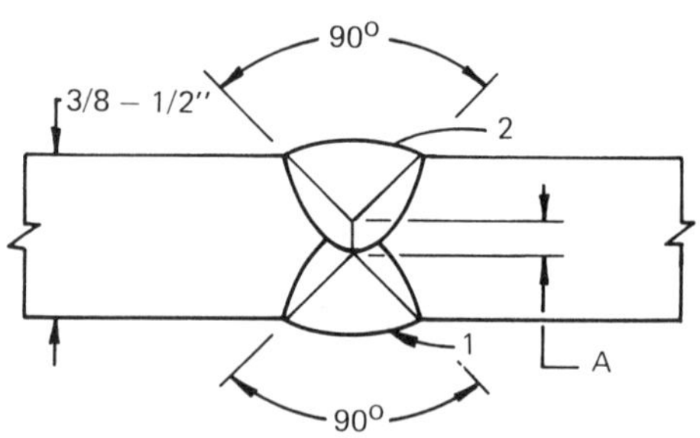

Plate Thickness T (in.)	1/8	1/4		3/8		1/2	
Electrode Size	1/16	3/32		3/32		3/32	
Pass	1	1	2	1	2	1	2
Current (amp)	310	460	500	500	550	540	600
Wire Feed Speed (in./min)	200	135	150	150	170	165	180
Volts	27	26		27		27	
Arc Speed (in./min)	30	20		14		12	10
Total Time (hr/ft of weld)	0.00667	0.0200		0.0286		0.0367	
A (in.)				3/16		1/4	

Even though weld quality is rated only "fair," the gas tungsten-arc process is used extensively in the fabrication and repair of the brasses. The appropriate rod classifications, shielding gases, and preheats are shown in Table 10-1. Sheet brasses are weldable without a preheat.

The silicon bronzes are readily welded with the gas tungsten-arc process. Good-quality welds are obtained with ERCuSi-A electrode. Direct current, electrode-negative is preferred, although alternating current may be used. Note from Table 10-1 that a preheat is not required and that interpass temperatures should not exceed 200°F. Sheet material is frequently joined without an edge preparation or use of a filler rod by the fusing of the base metal. With thicker material, a 60° included angle is required, together with filler rod of appropriate diameter.

The ERCuSn-A welding rod gives a fair-quality weld by the process with the phosphor bronzes. In practice, the process is usually limited to repair work on castings of such materials. Welding should be done as rapidly as possible, using a preheat up to 400°F. Slow cooling is desirable, and hot peening is often used to refine the grain structure and reduce stress and distortion.

Both the cupro-nickels and aluminum bronzes are readily welded with the gas tungsten-arc process. ERCuNi filler rod gives good-quality weld metal with the copper-nickel alloys, using argon as the shielding gas. No preheat is required.

The process is highly recommended for the aluminum bronzes, on which it produces quality welds of excellent appearance. Light-gage sheet can be welded without filler rod; ERCuAl-A2 rod should be chosen for plate joining; and ERCuAl-B is typical of the filler rods recommended for the repair of castings. Preheats are required only with heavy sections. Alternating current usually gives welds of highest quality, although direct current, electrode-negative polarity can be used in many applications. Argon is the preferred shielding gas. No special welding techniques, postweld peening, or heat treatments are normally required.

WELDING COPPER AND COPPER ALLOYS BY THE CARBON-ARC PROCESS

The carbon-arc process was formerly used extensively for the welding of copper and its alloys, but has been largely obsoleted by the gas-shielded and manual shielded metal-arc processes. As noted by Table 10-1, it produces weld metal of good quality only in the case of deoxidized copper. The reducing action of the carbon-arc process tends to cause embrittlement in oxygen-bearing coppers.

Carbon-arc welding procedures call for pointed carbon electrodes in diameters from 5/32″ to 1/2″

GAS TUNGSTEN-ARC WELDING OF SILICON BRONZE

Process: Gas Tungsten-Arc
Plate Material: Silicon Bronze
Plate Thickness: 1/16 - 1/8 in.
Passes: 1
Rod Class: ERCuSi-A
Position: Flat
Polarity: DCEN

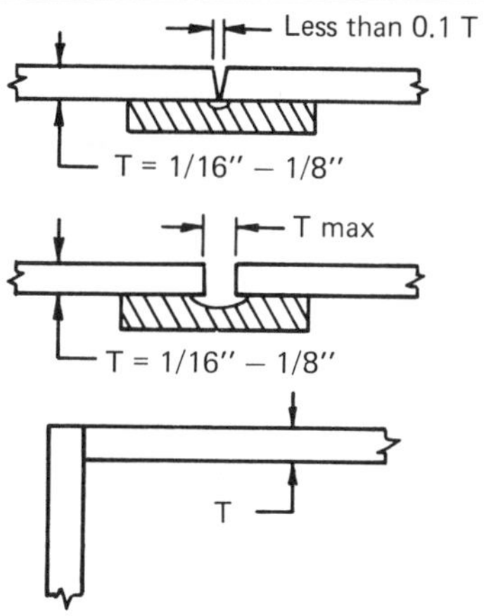

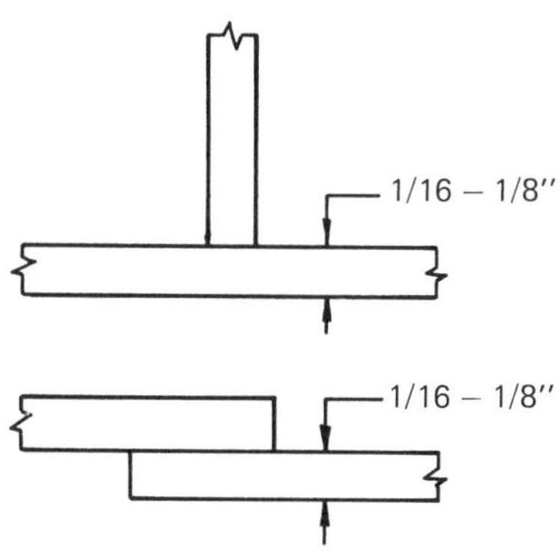

Joint	Butt or Corner		Fillet or Lap	
Plate Thickness, T (in.)	1/16	1/8	1/16	1/8
Rod Size	1/16	3/32	1/16	3/32
Electrode Size	1/16	3/32	1/16	3/32
Current (amp)	100 – 120	130 – 150	110 – 130	140 – 160
Gas (cfh)	Argon 15	Argon 15	Argon 15	Argon 15
Arc Speed (in./min)	12	12	10	10
Total Time (hr/ft of weld)	0.0167	0.0167	0.0200	0.0200

to carry the current and strike the arc. The carbon should be gripped about three inches from the arc to prevent excessive burning away. Electrode-negative polarity is mandatory. The arc is generally played on the filler metal rather than on the base metal, after bringing the base metal up to preheat temperature. Joints in thin material can be fused together without the use of a filler rod.

Table 10-1 shows the recommended filler rods and preheat requirements — should there be reason for selection of this process. In most instances, however, it is recommended that the gas-shielded or covered-electrode processes be used to avoid the difficulties or inferior welds that result from welding copper and copper alloys with the carbon-arc process.

GAS TUNGSTEN-ARC WELDING OF COPPER

Process: Gas Tungsten-Arc
Plate Material: Deoxidized Copper
Plate Thickness: 1/16 - 1/4 in.
Passes: 1
Rod Class: ERCu
Position: Flat

Polarity: DCEN

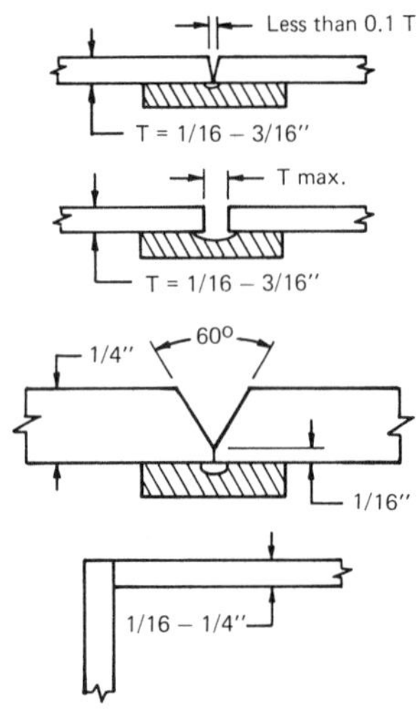

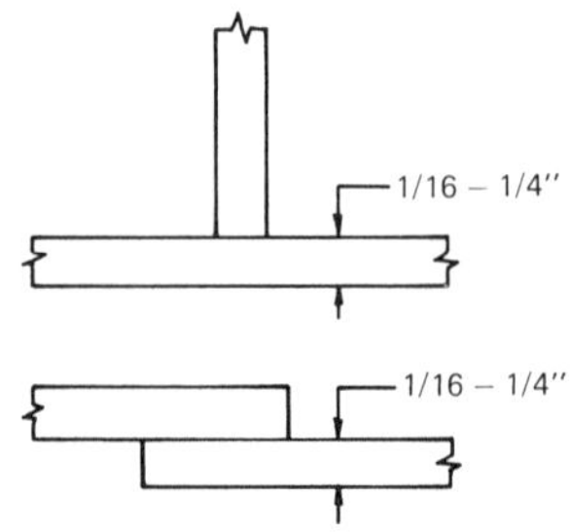

Joint	Butt or Corner				Fillet or Lap			
Plate Thickness T (in.)	1/16	1/8	3/16*	1/4†	1/16	1/8	3/16*	1/4†
Rod Size	1/16	3/32	1/8	1/8	1/16	3/32	1/8	1/8
Electrode Size	1/16	3/32	1/8	1/8	1/16	3/32	1/8	1/8
Current (amp)	110 – 140	175 – 225	190 – 225	225 – 260	130 – 150	200 – 250	200 – 250	250 – 280
Gas (cfh)	Argon 15	Argon 15	Helium 30	Helium 30	Argon 15	Argon 15	Helium 30	Helium 30
Arc Speed (in./min)	12	11	10	9	10	9	8	7
Total Time (hr/ft of weld)	0.0167	0.0182	0.0200	0.0222	0.0200	0.0222	0.0250	0.0286

* Preheat 200°F
† Preheat 300°F

Section 11
QUALITY CONTROL

SECTION 11.1
WELD QUALITY

Page

SECTION 11.2
INSPECTION AND TESTING

SECTION 11.3
WELDING PROCEDURE QUALIFICATION

SECTION 11.4
WELDOR QUALIFICATION TESTS

Weld Quality

Quality in a weld can be defined as conforming to a specification which lists the limits of discontinuities which are acceptable for a particular application. It must be realized that there are no perfect welds. If any apparently perfect weld is examined closely enough, fine discontinuities can be found, but that does not make the weld defective or rejectable. If a weld has fewer discontinuities or rejectable characteristics than allowed by the specification, it is acceptable and considered a quality weld. Quality specifications are not the same for all weld applications. Thus a weld acceptable for static loading may not be acceptable for a dynamic loading application. If the discontinuities are larger or greater in number than allowed, they are considered defects and the weld is defective or rejectable.

While all people involved should strive to attain the highest quality reasonably possible, the cost of obtaining unnecessarily high quality should be considered.

Welds in many of the structural steels may contain small amounts of fine porosity which are within limits allowed by various welding codes such as D1.1. These levels of quality are readily attained by qualified weldors using the standard arc welding processes, good joint fitup and proper edge preparation. A higher level of quality would entail increased cost by more testing to verify the quality. On highly critical applications such as high pressure piping, nuclear power systems and aircraft structures, these high quality welds are necessary. On structural welds, quality should comfortably meet or exceed the code requirements, but extreme measures to obtain near perfect welds would entail an unnecessary extra cost. However, it should be emphasized that all standard procedures should be followed to attain the highest quality possible. No procedures or workmanship which produces welds which are just "good enough" should be tolerated when a little more care will produce welds which exceed the quality requirements.

For each weld application, the required quality and properties should be established based on sound engineering analysis and any applicable code. Then the welding procedure and level of inspection and testing can be determined. From this, a standard of quality can be established which can include appearance and a factor of safety in the design. Once the standard has been set, it is important that established procedure be followed and the required quality level be met consistently.

On the low side, the predetermined standard of quality should never be compromised. On the high side, there is no objection to extra quality, providing it has been obtained at no penalty in cost. If tests repeatedly show that the welds are exhibiting a degree of quality far greater than required by the standard, a cost reduction through modification of weldment design or procedures is possible.

Frequently, the standards are preset by prevailing specifications or engineering and legal codes. Sometimes such standards are ultraconservative, but when they apply they must be honored. The engineer can do his company or the customer a service by pointing out unrealistic specifications and the opportunities for cost savings by communicating with code writing committees and recommending changes. However, the specifications must be adhered to rigidly until revised.

SIX ITEMS THAT ASSURE QUALITY

After the quality standard has been established, the most important step toward its achievement is the selection of the best process and procedures. By giving attention to six items, weld quality will be assured almost automatically. The six items are:

(1) **Process Selection** — the process must be right for the job.

(2) **Joint Preparation** — the joint configuration must be right and compatible with the welding process.

(3) **Procedures** — to assure uniform results the procedures must be spelled out in detail and followed religiously during welding.

(4) **Pretesting** — by full-scale mockups or simulated specimens the process and pro-

cedures are proved to give the desired standard of quality.

(5) **Personnel** — qualified people must be assigned to the job.

(6) **In-Process Monitoring** — during production each step is monitored for correct procedure.

1. Process Selection

In view of the various automatic and semiautomatic processes available, process selection imposes a challenging decision. Of all the processes, manual welding is the most versatile, but economic considerations necessitate its being ruled out in favor of a partially or fully mechanized process wherever such is applicable. Process selection is tied up integrally with the fill, follow, freeze, and penetration needs of the joint. Each process has its advantages and limitations, and each introduces problems affecting joint preparation, welding procedures, and operator training. The process that gives the correct balance between the needs of the joint in terms of fill, follow, freeze, and penetration is likely to be the one that gives optimum weld quality.

2. Joint Preparation

A good joint design will provide access for the weldor or welding operator, adequate root opening to permit full fused penetration and the groove angle or shape to minimize volume of weld metal. Prequalified joints are available within strength and thickness limitations for shielded metal arc, submerged arc, gas metal arc and flux cored arc welding processes. (Refer to AWS Structural Welding Code D1.1 and AWS Bridge Welding Code D1.5.) Deviations from these prequalified joints may be accomplished by running prequalification tests, since practically all codes state that "other welding processes and procedures may be used, provided the contractor qualifies them in accordance with the prescribed requirements."

The type and requirements of a joint are normally made by the designer. These include complete joint penetration groove welds (CJP), partial joint penetration groove welds (PJP) and fillet welds. The decision as to joint preparation, welding process and procedure is usually made by the fabricator. Some joint preparations are standardized and specified by applicable codes. Acceptable groove weld preparations are nothing more than a compromise between the included bevel angle, root face and the root-spacing dimensions. A large included angle will permit a smaller root spacing and, conversely, a small included angle requires a larger root spacing. These and other factors are taken into consideration in prequalified joints. The joint detail influences process selection, electrode size, and welding position. The joint preparation must be correct before welding is started, not only to meet specifications, but to give assurance of weld quality.

3. Procedures

Reliable welding procedures are developed through firsthand experience. They should be completely detailed in advance of production work. A full-scale mockup of the joint, using the same type steel, sizes, and shapes that will be used on the job, should be made to test the procedures if such is possible. If a full-scale mockup is infeasible, a simulated setup should be used to produce specimens that can be destructively and nondestructively tested. By trial and error, all the procedure details for making an acceptable weld can be determined.

4. Pretesting

Pretesting production-size or simulated production specimens also tests the process as well as the procedures. Once it has been ascertained that the procedures and the process give the desired quality, such test specimens can also be used as a final check on the qualifications of the operator.

The closer the conditions of test approach conditions of service, the more meaningful the results. The ideal would be service life tests under slightly exaggerated conditions. Since this is usually impractical, simulated service tests are the next best choice. Specimen assemblies may be subjected to radiographic, ultrasonic, or other nondestructive inspection procedures to evaluate the weld quality Or they may be prooftested or submitted to destructive tests to determine ultimate limits.

By proper use of process-procedure qualification — backed up by convincing test evidence — final inspection takes on the nature of a quality control activity. This is desirable; the intent should be to make the welding so highly precise in giving the desired quality that all subsequent inspection is for the detection of the unexpected and unexplained, rather than the explainable discontinuities.

5. Personnel Qualification

The fifth item can be evaluated in a preliminary way by the AWS Operator Qualification Test and the contractor's judgment. If a semiautomatic process is to be used, some experience with it is desirable, or the weldor may require training. As mentioned previously, the welding of test specimens with the selected process and procedures will affirm the operator's capability.

6. In-Process Monitoring

In-process monitoring at each step of the procedure can help assure the required quality by finding problems or mistakes and correcting them before they become major problems. This monitoring can be done by the weldor, welding operator, foreman or inspector.

QUALITY VERSUS ACCEPTABILITY

It is possible for a weld to be of marginal quality by standards of inspection and judgement, yet perform its job successfully. If that weld is produced at a fraction of the cost of a weld that would be rated high-quality, there is no logical reason for increasing the manufacturing cost to obtain the high-quality weld. Acceptability is, thus, a consideration that must be balanced against the concept of quality in order to control fabrication costs and the cost of the product to the buyer. If a weld of inferior appearance and strength can be made at a fraction of the cost that serves the function just as well, it is acceptable and just as good for the purpose.

The allowing of "acceptability" however, does not mean the toleration of sloppy or careless work. Rather, it means giving the joint what it needs by the most economical method at the best workmanship likely to prevail under the circumstances.

The formed undercarriage channel shown at the top of Fig. 11-1 is illustrative. Here, the strength requirements are nominal and appearance requirements nonexistent. Intermittent welds are adequate, freedom from discontinuities is unimportant, and weld sizing is not specified. The procedures selected give welds that satisfy requirements, yet these welds are imperfect by conventional standards. They are produced at a very low unit cost — and no additional value would be realized if better welds were produced at a higher cost.

Similarly, the welds on the liquid storage tank in Fig. 11-1 have undercut and are not of best appearance. They are leakproof, however — and that, plus minimal cost, is all the chicken farmer who purchases the tank to hold kerosene for supplying the heating equipment to a brood house is concerned about. On the other hand, the storage cabinet in Fig. 11-1 must have welds of good appearance. These are obtained at a slower welding speed and an increase in cost.

Even in welds where strength is important something less than perfection may be acceptable. The double-V weld developed by Procedure A in Fig. 11-2 costs $15.90 per foot and meets code tests. A tensile specimen incorporating it fails in the base

	Common Procedure	Typical Arc Speed	Good Appearance	Free of Surface Holes	Leak-Proof	Free of Undercut	Is Weld Size Considered	Does Weld Meet Service Requirements
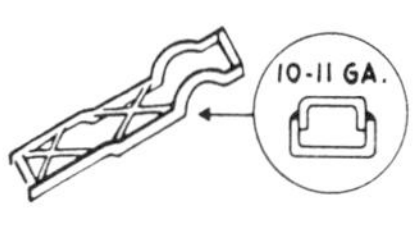 Formed Channels Welded For Undercarriages	3/16 E6012 @ 300-325 Amps	60 in./min	No	No	No	No	No	Yes Note: Members are so designed that intermittent welds of poor quality are sufficient to meet nominal strength requirements'
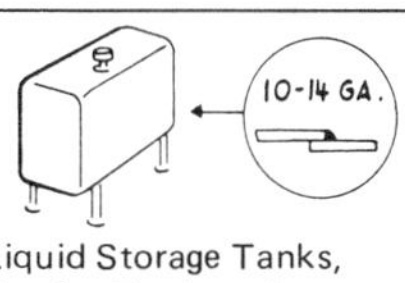 Liquid Storage Tanks, Gravity Feed, no Pressure	5/32 E6010-11 @ 150-175 Amps	30 in./min	No	Yes	Yes	No	No	Yes Note: Welds must be leakproof. All other factors are of secondary importance.
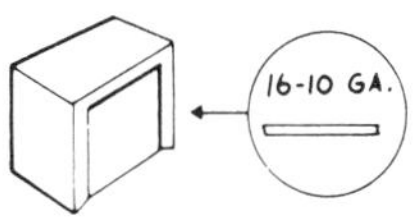 Storage Cabinets — Control Panels, etc.	5/32 E6013 @ 125-135 Amps	20 in./min	Yes	Yes	Yes	Yes	No	Yes Note: Nature of weldment demands good appearance. Any visible defects would not be acceptable.

Fig. 11-1. Each of these welding procedures meets the service requirements of the product and is therefore acceptable. If a low-cost weld does the job, there is no point in demanding a standard of quality that has no value in the product.

		Typical Speed	Cost $/ft	
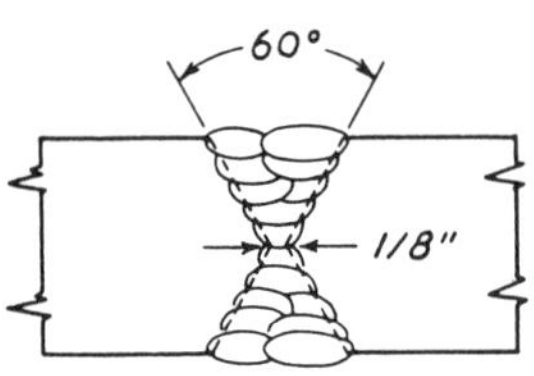	**Procedure "A"** 5/32 & 3/16 E7018 130 – 225 amps Back-gouged before welding second side (2.8 # elect./ft)	1.25 ft/hr	$15.90/ft	1. Clear X-Ray, no defects 2. Face, root & side bends will meet code requirements 3. Reduced-Section Tensile tests fail in base metal.
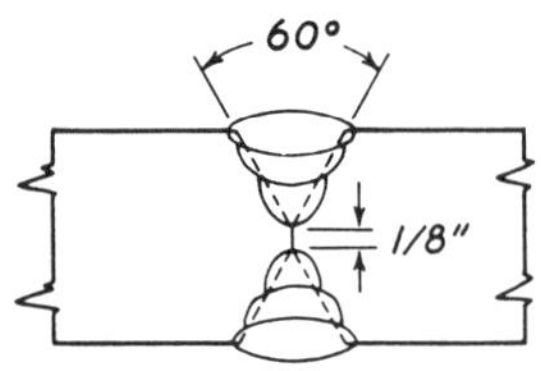	**Procedure "B"** 5/32 & 3/16 E7024 225 – 325 amps No Back-gouging (1.62 # elect./ft)	4.0 ft/hr	$5.34/ft	1. X-Ray shows lack of fusion & scattered porosity. 2. Face & root bends O.K. side bend has 1/8" defect before bending. 3. Reduced-Section Tensile tests fail in base metal.

Fig. 11-2. There is no point in paying $15.90 per foot for a weld when one produced at $5.34 a foot is stronger than the base metal — unless codes or specifications require the higher, quality of the more expensive weld.

metal — proving that the weld is stronger than the metal it joins. But the weld developed by Procedure B at $5.34 per foot also gives a tensile specimen that fails in the base metal. It doesn't meet code tests — yet there is no evidence to indicate that it should give lesser service than the more costly weld.

SOME OBSERVATIONS ON WELD DISCONTINUITIES

Much of what an inspector is required to pass judgment on does not resolve itself into a *go* or a *no-go* situation. Most X-ray inspection is based on standards that give the inspector a means of comparison to serve as a basis for acceptance or rejection and usually are not unrealistic in their demands. The quality demanded is obtainable. On the other hand, these specifications are not loose enough to permit discontinuities that would likely lead to weld failure in service.

Problems arise, however, when a demand is made for much clearer X-rays than the code requires. In some cases, inspectors have insisted that they want perfect X-rays regardless of code requirements. There is doubt that this attitude would prevail if these same people were aware of the facts illustrated in Figs. 11-3 and 11-4. It is an established fact, and one worthy of inspection recognition, that scattered porosity well in excess of that permitted by some codes does not detract from the strength of the joint.

The individual welding operator has a very great responsibility when X-ray quality is required. In addition to being certain he uses the prescribed procedures capable of producing X-ray quality, he must be meticulous in his efforts to make every bead as sound as possible, that ordinary effort is not good enough. Closer supervision and higher costs are natural consequences. It literally requires a change of attitude on the part of the weldor to convert from routine commercial welding to welding for X-ray quality. This has been confirmed through expensive experience, especially by contractors who occasionally contract for X-ray work.

One discontinuity an inspector is required to look for is undercutting. Generally this is not

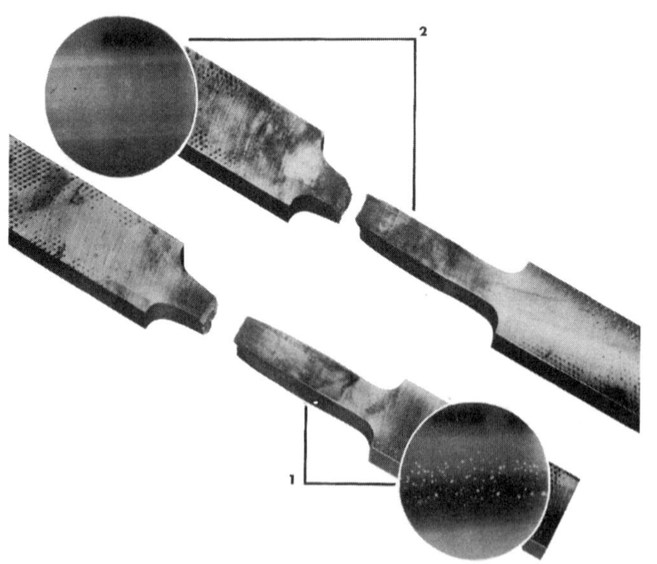

Fig. 11-3. Weld 1 has considerable porosity, as shown by the radiograph (inset). Weld 2 has no porosity. When the specimens were pulled, both broke in the plate at approximately 60,000 psi. In each instance, the weld was stronger than the plate, despite the porosity in the one weld.

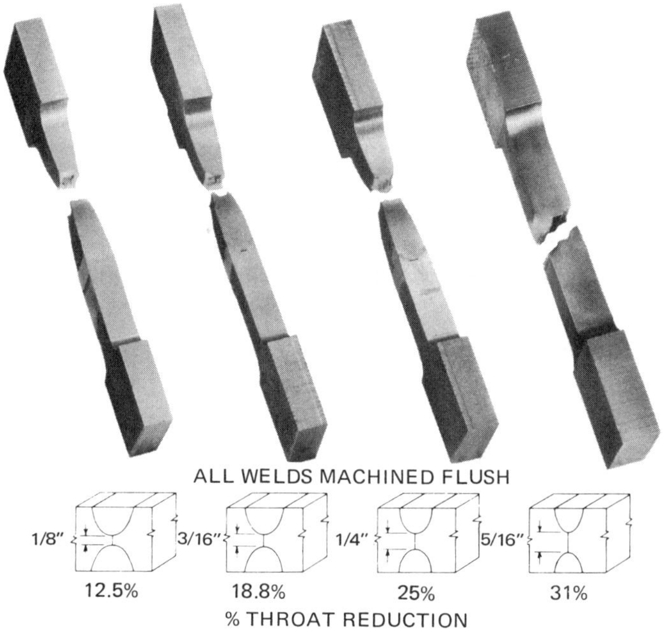

Fig. 11-4. Weld samples made with progressive degrees of lack of fusion. All welds were machined flush before tensile testing. Weld failure finally occurred when the unpenetrated throat dimension reached 31% of the total joint throat. All specimens were made with 1-in. plate, A-36 steel, and E7018 electrode.

troublesome. It is almost always prevalent to some degree on the vertical leg of large horizontal fillets. But the mere fact of its existence someplace along the top edge of a 5/16-in. or larger fillet does not mean that the weld is totally unacceptable.

Undercut serves no useful purpose. Therefore, no reasonable effort should be spared to eliminate it or at least minimize it. The greatest control generally lies within the area of procedure details or in the skill of the individual weldor. But despite its undesirable nature, the presence of undercut, except under fatigue loading, is not nearly as damaging as one might suspect, as illustrated by Fig. 11-5. Here undercut has been deliberately increased to an excessive degree on both sides of a fillet weld. Yet the strength of the joint under static loading has not been impaired. Obviously, it does not make sense to demand repair of a little undercut on the average weldment. However, excessive undercut should not be tolerated.

In many instances, undercut is so oriented that it can only affect appearance and not weld performance. The weld joining the cover plate and the flange of a beam in Fig. 11-6 is an example. Stiffeners welded to the web of a beam provide another example. This weld is almost certain to be made with the beam positioned as illustrated in Fig. 11-7. There is good reason to discourage undercutting the web. The weld position, however, forces any undercut to occur in the stiffener and not the web. Undercut would detract from appearance, but would not interfere with the stiffener's performance.

Another discontinuity controlled by standards and sometimes questionable, is the weld crater. Any arc-welding process or procedure produces a weld crater. The ability to fill this crater at the termination of a weld varies considerably with the welding current, weld size, and other factors. If these limitations are understood, and service requirements of the joint are considered, it should not be difficult to decide whether a crater must be filled.

Most codes make a general statement regarding crater filling. The AWS Structural Welding Code D1.1 states in paragraphs 8.15.1.3 and 9.25.1.3: "All craters shall be filled to the full cross section of

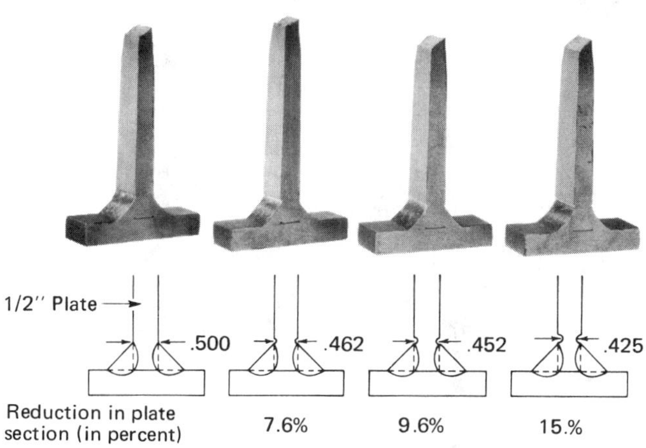

Fig. 11-5. Samples were deliberately prepared to show the effect of undercut. All samples were pulled in tension under a static load. In all cases, failure occurred in the plate and not in the weld.

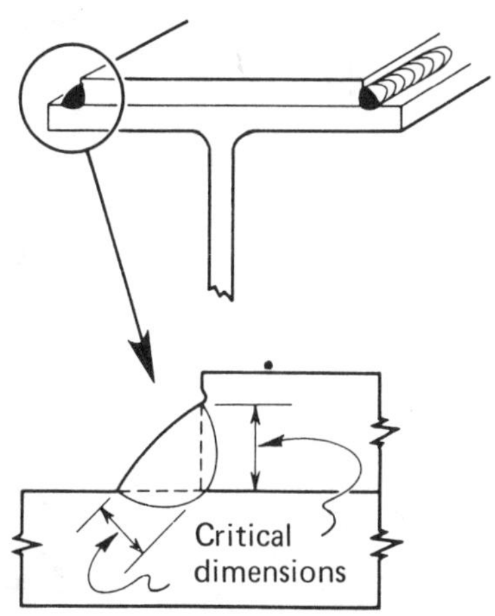

Fig. 11-6. Sketch showing how undercut may occur on a welded cover plate with little or no effect on the strength of the member. The procedure should be controlled to hold undercut to a minimum for reasons of painting and general appearance, but weld performance does not demand complete elimination of undercut.

the weld except for the ends of intermittent fillet welds outside their effective length."

On long continuous welds, the start of each weld fills the crater left by the previous weld — a technique that any weldor can easily master. The only crater requiring special attention would be the final one at the end of the joint.

What happens on intermittent welds is entirely different. An example is the production of 5/16-in. fillet welds, each 3 in. long and spaced on 12-in. centers. A typical production welding procedure would specify the use of 1/4-in. electrode. But any procedure that will efficiently produce a sound weld of the required size will most certainly develop a pronounced crater that is very difficult to eliminate.

Strict adherence by the inspector to the code will require the weldor to change his procedure to one that is more conservative and calls for smaller-diameter electrodes, lower currents, slower travel speeds, or even changing to another electrode type having a slower deposition rate. An alternate approach would require the weldor to use a back-stepping method to fill the crater, which may not produce a sound crater. In either case, the net result has a detrimental affect on weld speed and weld cost.

Where intermittent welds are specified and acceptable, the code now allows that the effective weld length be the total length exclusive of the crater. Thus, if 6-in. welds are required, they should measure 6 in. plus the crater. Then, the inspector's only concern with respect to craters would be to make sure they are free of cracks and undercut.

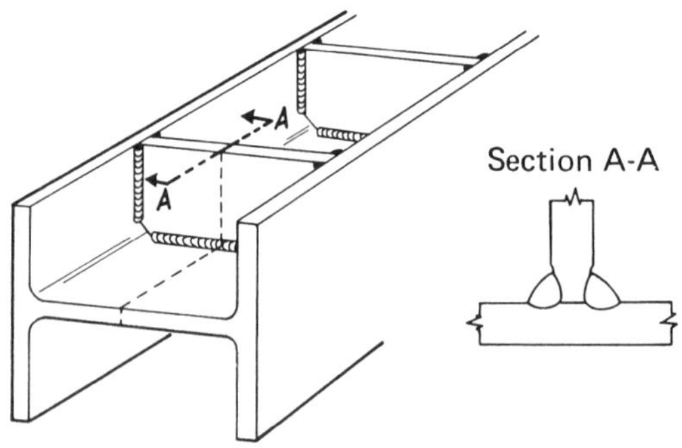

Fig. 11-7. In shop-fabrication, stiffeners are normally positioned horizontal during welding. Any undercut is in the stiffeners — not in the web — where it is not detrimental to weld performance. Yet, some inspection standards demand 100% elimination of undercut — a stipulation difficult to justify technically, but one that adds to fabrication cost.

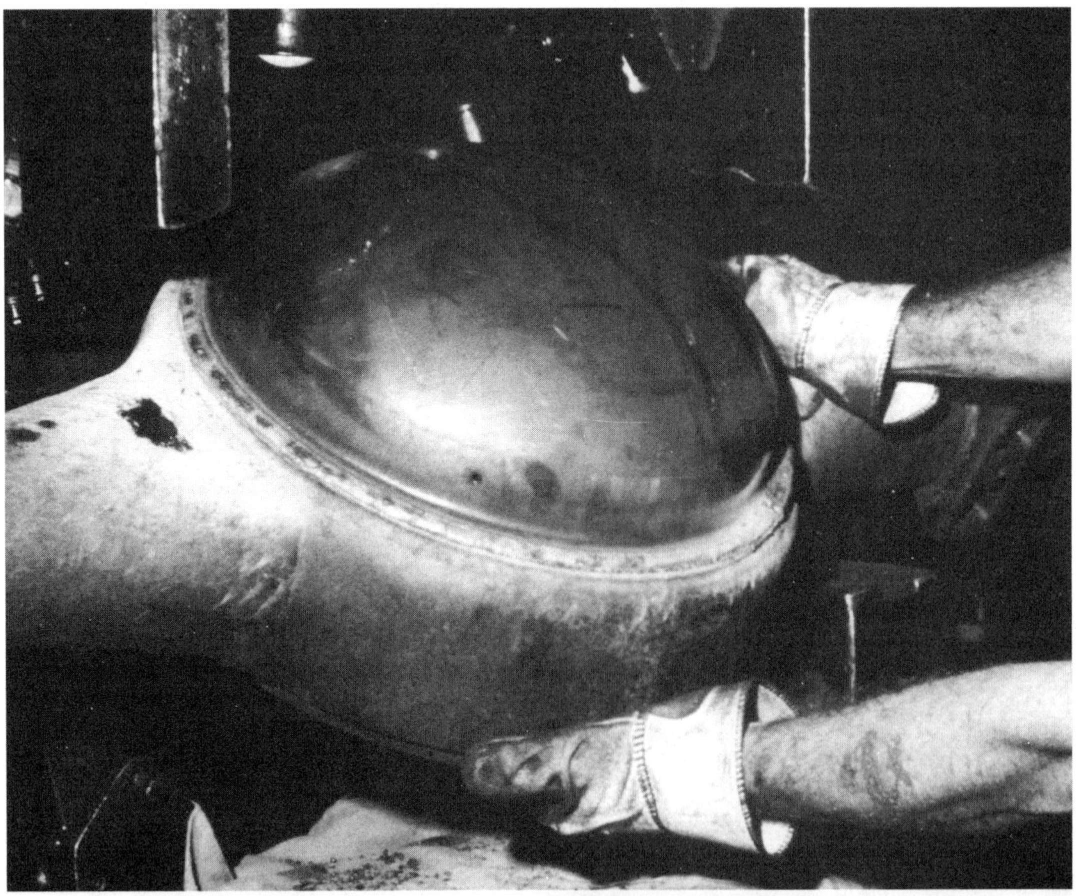

Visual inspection of this self-shielded flux-cored electrode automatic weld indicates good quality.

Inspection and Testing

Whatever the standard of quality, all welds should be inspected, even if the inspection involves no more than the weldor glancing back over his work after running a bead. Testing may also be required, even in relatively rough weldments — such as leak-testing a container for liquids.

Inspection indicates whether the prescribed standard of quality has been met. This function may be the responsibility of the welding supervisor or foreman, a special employee of the company doing the welding, or a representative of the purchasing organization. The formal welding inspector may have a variety of duties. These may begin with interpretation of drawings and specifications and follow each step to the analysis of test results. His operations are both productive and nonproductive — depending on where they are applied.

The American Welding Society has established a program for certification of welding inspectors, and conducts educational courses and tests several times a year in various major cities. The program and requirements to become certified as an inspector are described in AWS-QC-1 "Standard for AWS Certification of Welding Inspectors."

Inspection after the job is finished is a policing action, rather than a productive function. Important as it is to assure quality, it is a burden added to the over-all production cost. No amount of after-the-job inspection will improve the weld; it merely tells what is acceptable and what must be reworked or rejected.

Inspection as the job progresses is a different matter. It detects errors in practice and discontinuities while correction is feasible. It prevents minor discontinuities from piling up into major discontinuities and leading to ultimate rejection. Inspection while weld quality is in the making and can be controlled may justifiably be looked upon as a productive phase of cost, rather than an overburden.

Any program for assuring weld quality should, therefore, emphasize productive inspection and attempt to minimize the nonproductive type. This should be the guiding philosophy, even though its implementation may fall short. In most cases, such a philosophy means that visual inspection will be the main method of ascertaining quality, since it is the one method that can be applied routinely while the job is in progress.

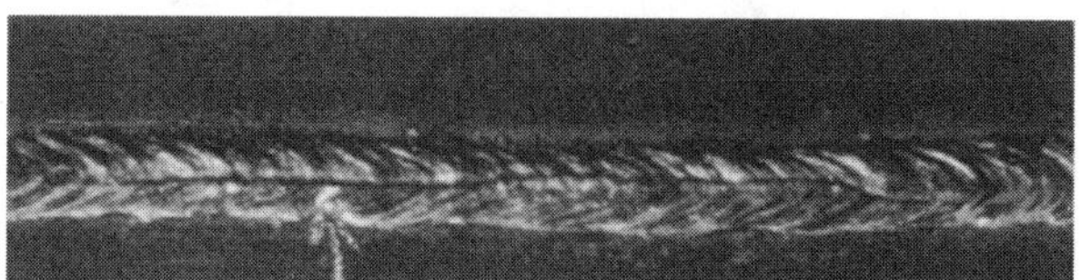

Cracking — can be detected visually unless the cracks are internal or are very fine. It is usually associated with excessively high current, insufficient preheat, insufficient bevel angle and/or highly restrained joints.

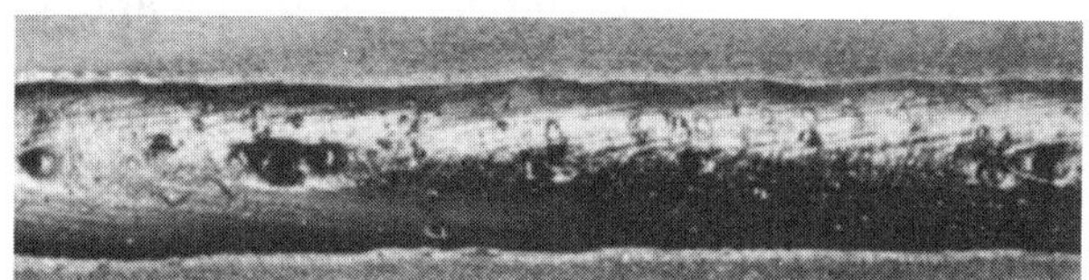

Surface Porosity — is often caused by excessive speed, rusty or dirty plate, wet electrode or flux, insufficient flux coverage, or critical arc-blow conditions.

Surface Slag Inclusions — are usually clues to faulty technique. Improper electrode manipulation, improper electrode size or too steep a downhill angle are typical possibilities.

Undercut — results from poor procedural selections — may be due to too large an electrode or excessive current, voltage, or arc speed.

Fig. 11-8. Several types of discontinuities that can be recognized visually.

VISUAL INSPECTION FROM START TO FINISH

Visual inspection is the best buy in NDT but it must take place constantly; prior to, during and after welding. In a sense, everyone connected with the job, as well as the formal inspector, participates in visual inspection. A conscientious worker does not knowingly pass on work in which he recognizes discontinuities of his making. Nevertheless, it is usually desirable that someone be assigned responsibility for quality checking each operation. In addition to good eyesight and good lighting, the tools for visual inspection are simple — a pocket rule, a weld-size gage, a magnifying glass, and sometimes a straight edge and square for determining straightness, alignment, and perpendicularity.

Visual inspection should begin before the first arc is struck. The materials should be examined to see if they meet specifications for quality, type, size, cleanliness, and freedom from discontinuities. Foreign matter — grease, paint, oil, oxide film, heavy scale — that could be detrimental to the weld should be removed. The pieces to be joined should be checked for straightness, flatness, and dimensions. Warped, bent, improperly cut, or damaged pieces should be ordered for repair or rejected. Alignment and fitup of parts and the fixturing should be scrutinized. Joint preparation should be checked. Often, little more than a passing glance is required in this preliminary inspection, but, despite its almost casual nature, such inspection can be a significant factor in weld quality.

Inspection prior to welding also includes verification that the correct process and procedures are to be employed — that the electrode type and size and the equipment settings for voltage and amperage are as specified — and that provisions are made for the required preheat or postheat.

Assuming the preliminary requirements are in good order, the most productive inspection will take place while the weldment is being fabricated. Examination of a weld bead and the end crater may reveal quality deficiencies such as cracks, inadequate penetration, and gas and slag inclusions to a competent inspector. Figure 11-8 shows several types of discontinuities that can be recognized visually.

On simple welds, inspection of a specimen at the beginning of the operation and periodically as the work progresses may be adequate. When more than one layer of filler metal is deposited, however, it may be desirable to inspect each layer before a subsequent layer is placed.

The root pass in a multipass weld is the most critical one from the standpoint of weld soundness. It is especially susceptible to cracking (Fig. 11-9), and, because it tends to solidify quickly, is prone to trap gas and slag. Subsequent passes are subject to a variety of conditions which may create weld discontinuities. These result from the shape of the weld bead or change in the configuration of the joint (Fig. 11-10). These can be visually detected by the weldor and repair cost minimized if the problem is corrected before welding progresses.

A workmanship standard, constructed for the specific purpose, can be helpful both to the weldor and the inspector in visually appraising the production weld during the stages of its formation. Representative workmanship standards are shown in Fig. 11-11.

Visual inspection at an early stage of the fabrication will also detect underwelding and overwelding Underwelding (Fig. 11-12) is in violation of specifications and cannot be tolerated. Overwelding should be of as much concern to the purchaser's

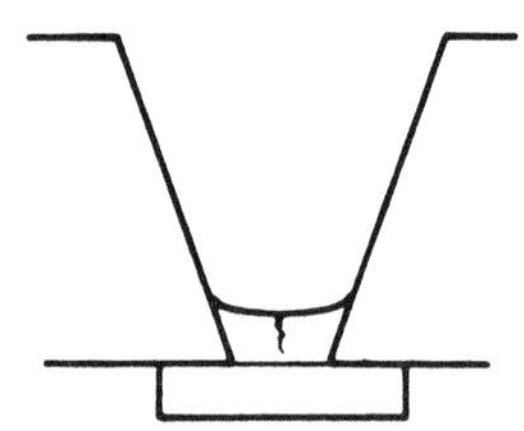

Fig. 11-9. The root pass in a multipass groove weld is especially susceptible to cracking because of the high shrinkage stresses.

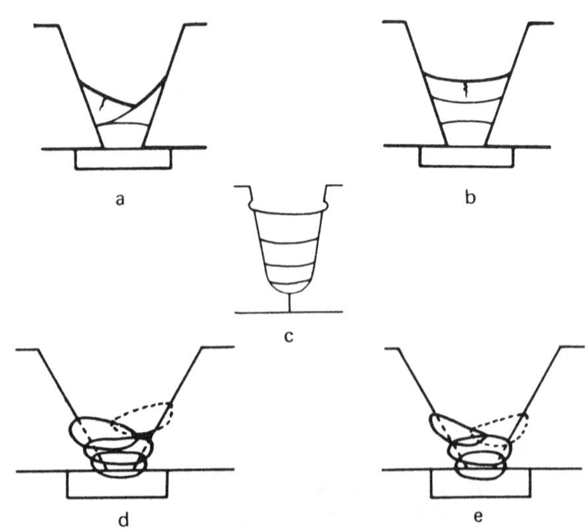

Fig. 11-10. The concave bead shape in (a) and (b) is partly responsible for the cracks. Groove welds are also prone to undercut along the edges, (c), which can trap slag when the next pass is made. Slag trapping can also be caused by a convex bead when there is too little space between the edges of the bead and the side of the joint (d). The correct bead shape is illustrated in (e). These flaws should be corrected before welding progresses.

Fig. 11-11. Mockup welds constructed as workmanships standards. Such mockups serve a variety of useful purposes. They are helpful to the weldor and the inspector in visually appraising the production weld during the stages of its formation. In addition, they give a check on the welding procedures before the procedures are put to work on the production floor, and they provide assemblies on which non-destructive tests can be made, or from which destructive test specimens can be taken, to assure that the procedures give the required weld quality.

inspector as to those members of the shop responsible for monitoring costs, since it is a major cause of distortion. Usually the designer has specified a weld size approaching the limit possible in good practice. The weldor — perhaps wanting to make certain that the joint is strong enough, or having been criticized for making undersize welds — takes it upon himself to add 1/16 in. to a 1/4-in. fillet. Since the weld metal deposited increases as the square of the size, the 1/16-in. increase in leg size increases the amount of weld metal deposited 56%, and has the same effect on shrinkage stress and cost. Figure 11-12 illustrates how overwelding gives this exponential increase.

VISUAL INSPECTION AFTER WELDING

Visual inspection after the weldment has been completed is also useful in evaluating quality, even if ultrasonic, radiographic, or other methods are to be employed. Here, as with visual inspection as welding progresses, surface flaws such as cracks, porosity, and unfilled craters can be detected, and may be of such consequence that repairs are required or the work is rejected without use of subsequent inspection procedures. There is no point in submitting an obviously bad weld to sophisticated inspection methods.

Dimensional variations from tolerances, warpage, and faults in appearance are detected visually at

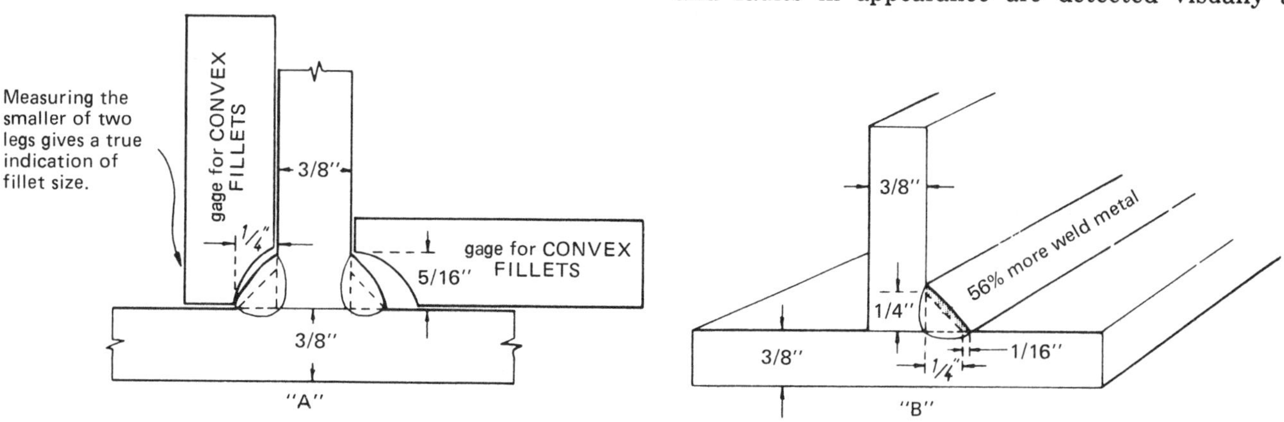

Fig. 11-12. Underwelding is a violation of specifications and cannot be tolerated, whereas overwelding is costly and serves no useful purpose. If 5/16-in. fillet welds were specified in "A", these welds would be undersize. If 1/4-in. fillet welds were specified, these welds would be overwelded. If 1/16-in. were added to both legs, as in "B", the weld volume would increase 56% and increase the cost of welding.

TABLE 11-1. Reference Guide to Major Methods for the Nondestructive Testing of Welds

INSPECTION METHOD	EQUIPMENT REQUIRED	ENABLES DETECTION OF	ADVANTAGES	LIMITATIONS	REMARKS
VISUAL	Magnifying Glass Weld-size gauge Pocket rule Straight edge Workmanship standards	Surface flaws — cracks, porosity, unfilled craters, slag inclusions. Warpage, underwelding, overwelding, poorly formed beads, misalignments, improper fitup.	Low cost Can be applied while work is in process, permitting correction of faults. Gives indication of incorrect procedures.	Applicable to surface defects only. Provides no permanent record.	Should always be the primary method of inspection, no matter what other techniques are required. Is the only "productive" type of inspection. Is the necessary function of everyone who in any way contributes to the making of the weld.
RADIO-GRAPHIC	Commercial X-ray or gamma units, made especially for inspecting welds, castings, and forgings. Film and processing facilities. Fluoroscopic viewing equipment.	Interior macroscopic flaws — cracks, porosity, blow holes, nonmetallic inclusions, incomplete root penetration, undercutting, icicles, and melt-through.	When the indications are recorded on film, gives a permanent record. When viewed on a fluoroscopic screen, a low-cost method of internal inspection.	Requires skill in choosing angles of exposure, operating equipment, and interpreting indications. Requires safety precautions. Not generally suitable for fillet-weld inspection.	X-ray inspection is required by many codes and specifications. Useful in qualification of weldors and welding processes. Because of cost, its use should be limited to those areas where other methods will not provide the assurance required.
MAGNETIC-PARTICLE	Special commercial equipment. Magnetic powders — dry or wet form; may be fluorescent for viewing under ultraviolet light.	Excellent for detecting surface discontinuities — especially surface cracks.	Simpler to use than radiographic inspection. Permits controlled sensitivity. Relatively low-cost method.	Applicable to ferromagnetic materials only. Requires skill in interpretation of indications and recognition of irrelevant patterns. Difficult to use on rough surfaces.	Elongated defects parallel to the magnetic field may not give pattern; for this reason the field should be applied from two directions at or near right angles to each other.
LIQUID-PENETRANT	Commercial kits, containing fluorescent or dye penetrants and developers. Application equipment for the developer. A source of ultraviolet light — if fluorescent method is used.	Surface cracks not readily visible to the unaided eye. Excellent for locating leaks in weldments.	Applicable to magnetic, nonmagnetic materials. Easy to use. Low cost.	Only surace flaws are detectable. Cannot be used effectively on hot assemblies.	In thin-walled vessels, will reveal leaks not ordinarily located by usual air tests. Irrelevant surface conditions (smoke, slag) may give misleading indications.
ULTRASONIC	Special commercial equipment, either of the pulse-echo or transmission type. Standard reference patterns for interpretation of RF or video patterns.	Surface and subsurface flaws, including those too small to be detected by other methods. Especially for detecting subsurface lamination-like defects.	Very sensitive. Permits probing of joints inaccessible to radiography.	Requires high degree of skill in interpreting pulse-echo patterns. Permanent record is not readily obtained.	Pulse-echo equipment is highly developed for weld inspection purposes. The transmission-type equipment simplifies pattern interpretation where it is applicable.

this stage. The extent and continuity of the weld, its size, and the length of segments in intermittent welds can be readily measured or noted.

Welds must be cleaned of slag to make inspection for surface flaws possible. A glass with a magnification of up to 10 diameters is helpful in detecting fine cracks and other faults. Shotblasting should not be used in preparing the weld for examination, since the peening action may seal fine cracks and make them invisible.

The objective of visual inspection at this stage is not only to seek faults not permissible under the quality standard, but also to give clues to what may be wrong in the entire fabrication process. If the inspector has a sound knowledge of welding, he can read much from what he sees. Thus, the presence of excessive porosity and slag inclusions may be an indication that the current is not adequate, regardless of the dial readings on the machine. Subsequent tests will also give clues to faults in equipment or procedures, but the information acquired through visual examination permits corrections to be made before the results from complicated tests are available.

OTHER INSPECTION METHODS

Only the surface discontinuities in welds are visible to the eye, and the specifications or applicable codes may require that the internal portion of the weld and the adjacent metal zones also be examined. Additionally, the application may require assurance that the chemical composition of the weld metal has not changed beyond specified limits. Thus, other inspection methods capable of gathering such information may be necessary.

These methods may be destructive or nondestructive. Destructive methods, obviously, cannot apply to production fabrication, other than for the testing-to-destruction of prototypes, "first units," or sparsely selected samples. In large weldments — building framing, for example — their use would be out of the question, but in an airframe component a periodic test-to-destruction may be regarded as essential to assurance of quality.

Nondestructive methods of testing welds include radiographic, ultrasonic, magnetic-particle, and penetrant techniques. Prooftesting is a mechanical method of determining whether the weld and other parts of the fabrication will withstand certain stresses encountered in service.

Chemical and metallographic methods of inspection may be completely nondestructive or destructive in a very minor way and to a repairable extent. If a sample for chemical or metallographic analysis is taken from a run-out portion of the weld bead, there is no damage; if a core drill is used to remove the sample from the weld proper, a hole results that must be repaired.

Table 11-1 is a condensed reference guide to the use of various nondestructive inspection methods on steel weldments, and Table 11-2 shows a simplified scheme for determining the inspection method recommended for fillet and butt joints.

It should be noted that nondestructive proof of the existence of a flaw does not measure its influence on the serviceability of the product. Only destructive tests have the potential for giving such information. Presumably, the specification or code requiring the nondestructive test is based on a correlation between the flaw and some characteristic affecting service. If such correlations have not been established, **the nondestructive test may well give nothing more than someone's opinion as to what is good or bad.**

The tendency in all matters pertaining to welded structures is to overdo factors affecting strength and safety, which means that flaws unlikely to influence serviceability may be the cause for rejection of the weld. In highly critical engineering structures, this overly cautious approach is justifiable, but in less critical applications some relaxation in the demands for flawless welds may be in order. Both laboratory destructive tests of segments of fabricated structures and tests-to-destruction of whole units where size and cost permit are useful in determining how much significance should be attributed to nondestructively revealed flaws. Also the judgments

TABLE 11-2.
Inspection Methods for Fillet and Butt Joints

RECOMMENDED INSPECTION FOR:		
TYPE OF DEFECT	FILLET JOINTS	BUTT JOINTS
Undersize Weld	Visual (1)	Visual
Surface Porosity	Visual	Visual
Internal Porosity	Destructive	Radiographic
Undercut	Visual	Visual
Cracks	Magnetic-Particle Dye-Penetrant Visual Destructive (2)	Magnetic-Particle Dye-Penetrant Visual Ultrasonic Radiographic (3)
Lack of Penetration	Destructive Ultrasonic	Radiographic Ultrasonic
Slag Inclusions	Destructive Ultrasonic	Radiographic Ultrasonic

(1) Use fillet gages.
(2) Destructive test will reveal subsurface cracks.
(3) Radiographic inspection has its limitations in revealing crack-type defects.

of skilled and experienced interpreters of nondestructive data may be invaluable when a correlation between the flaw and serviceability has not been established.

If, however, the code or specification demands that a weld be rejected or repaired because of a discovered flaw, that code or specification must be followed.

RADIOGRAPHIC INSPECTION

Radiographic inspection is one of the most widely used techniques for showing the presence and nature of macroscopic defects and other discontinuities in the interior of welds. This test method is based on the ability of X-rays and gamma rays to penetrate metal and other opaque materials and produce an image on sensitized film or on a fluorescent screen. The term "X-ray quality," widely used to imply high quality in welds, arises from the inspection method.

X-rays, which have a wavelength 1/10,000 that of visible light, are produced by high-voltage generators. The depth to which the opaque material can be penetrated depends on the power of the generator. Portable units rated up to 300 kilovolts are available and used in weld inspection. The higher-power machines, operating between 1,000 and 2,000 kilovolts, will penetrate from five to nine inches of steel.

Gamma rays are produced by the atomic disintegration of radioisotopes. They are similar to X-rays, except the wavelength is usually shorter than those X-rays produced with lower-voltage equipment. X-rays generated at 1,000 kilovolts or higher, for all practical purposes, appear to be identical to the gamma rays produced by radioisotopes. While the short wave lengths of gamma rays allow penetration to considerable depth, exposure times required to obtain an interpretable picture are usually many times longer than with X-rays, because of the lower intensity of the radiation. The level of radiation is directly proportional to the amount of the radioisotope used.

When X-rays or gamma rays are directed at a section of a weldment, not all of the radiation goes through the metal and to the film placed behind it. Some is absorbed, and the denser or thicker the metal, the greater the absorption. Should there be a cavity in the weld interior, such as a blow-hole or internal crack, the radiation will have less metal to pass through than in a sound weld. Consequently, there will be less absorption in the defective area and a greater amount of radiation striking the film.

Photos Courtesy of Picker X-Ray Corporation

Fig. 11-13. Top photograph, an X-ray image intensifier used to inspect the linear welds in a pipe 40-ft. long. The pipe diameter can be up to approximately 36-in. in diameter. The operator is seated in his radiation-shielded, air-conditioned cabin and views the pipe as it passes at speeds up to 40-ft. per minute.

Center photograph, a portable 160-KV Andrex unit being used in the field to radiograph a circumferential butt weld on a natural gas line. The power for the unit is supplied from a generator mounted on a truck.

Bottom photograph, a mobile Cobalt-60 unit being used to radiograph circumferential welds in sections of special pipes. The unit requires no power and can hold up to 5 curies of Cobalt 60.

After development, the discontinuity will show up in its shape and size as a black or darker area on the film.

The film image is called a radiograph. Radiographs made with X-rays are referred to as exographs; those made with gamma rays, as gammagraphs.

Radiographic equipment typical of that used for inspection purposes in production welding is shown in Fig. 11-13. These commercially available units require careful following of instructions for their proper use. Persons responsible for radiographic inspection should make themselves thoroughly familiar with the equipment, reading carefully the manufacturer's literature and giving due attention to recommended usage, details of operation, applicable films, and safety precautions. An understanding of principles of X-ray generation, radioisotope decay, and film sensitization and development should be acquired by the professional inspector or radiographic technician and will be helpful in his work.

In the shop or field use of radiographic equipment, the parameters affecting the reliability and interpretative value of the image are sharpness and contrast. Radiographic methods should give a "sensitivity" of at least 2%. The term sensitivity is used to denote the least percentage of difference in weld thickness that can be detected visually on a radiograph. The ability of an observer to detect such a thickness difference depends on the sharpness of outline of the image and its contrast with the background. Thus, if a fine crack is to be observable as a crack, its image must be outlined enough to give a recognizable form and must so contrast with the background that no question arises as to the presence of a discontinuity.

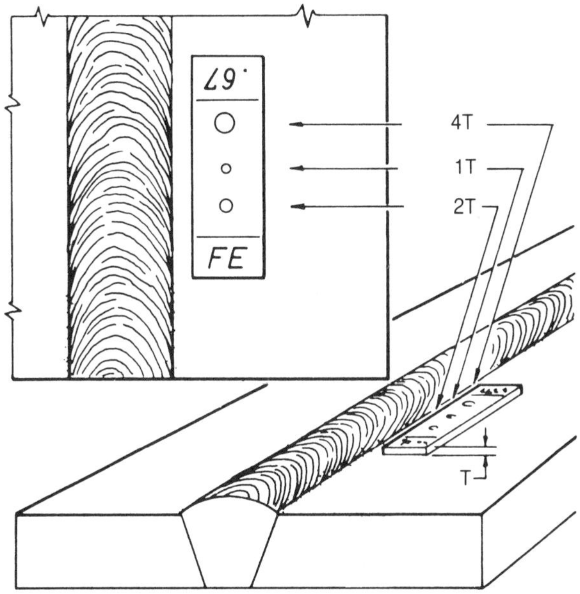

Fig. 11-15. Placement of a penetrameter alongside the weld before subjecting to the X-ray source gives assurance that radiographic image quality is obtained. The shadow of the penetrameter on the developed film is positive proof that 2% sensitivity is achieved, since it adds 2% or less to the thickness of metal penetrated by the radiation. The lead identification number on the penetrameter will stand out plainly on the film, since lead is highly impermeable to X-rays. The holes in the penetrameter have diameters of 1T, 2T and 4T to evaluate film contrast and definition. T = Penetrameter thickness = 2% of metal thickness.

Fig. 11-14. Penetrameters or image quality indicators (IQIs), such as these, are tools that the X-ray technician uses in his trade. The lead letters and numbers are used to permanently identify the radiograph. Penetrameters, available in various thicknesses, are matched to the thickness of the material being examined and establish the sensitivity of the radiograph.

It can be easily seen that a radiograph could be so much off in sharpness and contrast that it would fail to reveal a flaw. To be sure that this doesn't happen, a gage known as an Image QualityIndicator (IQI) or penetrameter is used on the side of the weld away from the film. Figure 11-14 shows typical penetrameters. An ASME Boiler Code IQI is a thin strip of metal with the same absorption characteristics as the weld metal. When a weld is to be X-rayed, a penetrameter with a thickness equal or less than 2% of the weld thickness is selected (see Fig. 11-15). A lead numeral at one end shows the thickness of weld for which the penetrameter is to be used. For reference purposes, three holes are drilled in the face of the gage, the diameters of which are multiples of the penetrameter thickness. The appearance of the image of the penetrameter on the radiographic film tells the observer whether he has the minimum of 2% sensitivity and adequate sharpness and contrast for meaningful interpretation.

Wire type penetrometers may also be used to assure the quality of radiographs, and this is described in ASTM E747, "Standard Method for Controlling Quality of Radiographic Testing Using Wire Pene-

TABLE 11-3. Radiographs of Weld Discontinuities

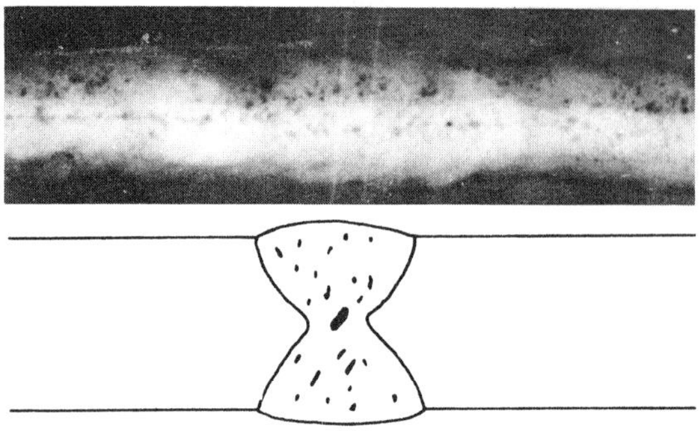

Porosity is shown as rounded shadows of varying size and density, occurring singly, in clusters, or randomly scattered.

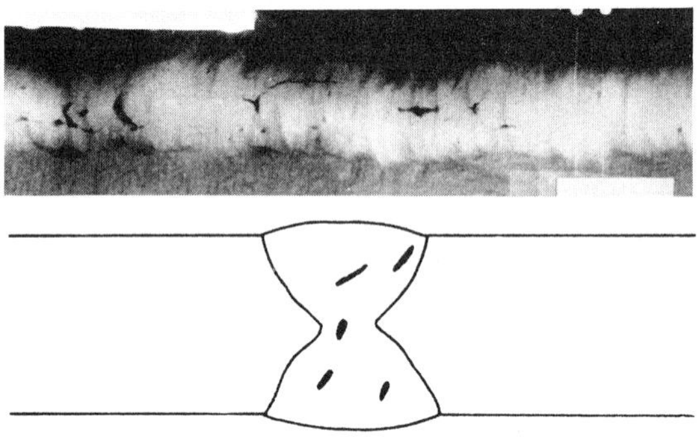

Nonmetallic Inclusions are usually indicated by elongated shadows of irregular shape, occurring singly, in a linear distribution, or scattered randomly.

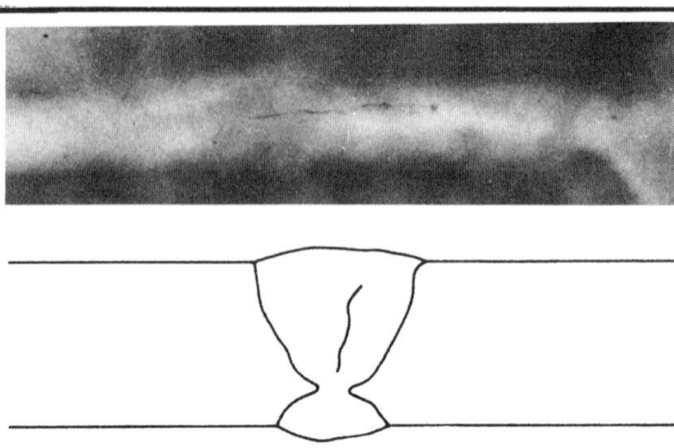

Cracks appear as fine, dark lines, which may be straight or wandering.

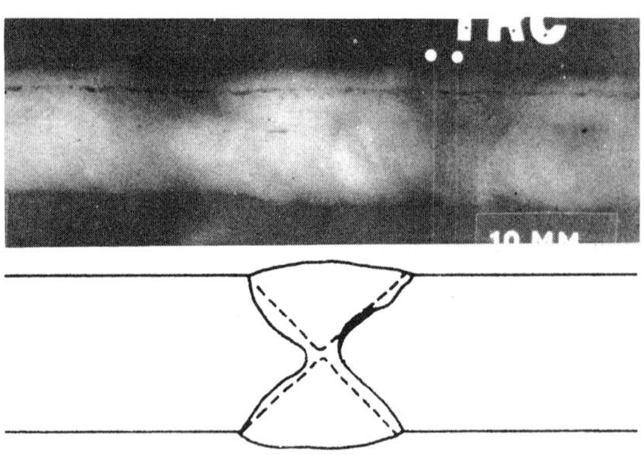

Incomplete Fusion gives dark shadows, usually of elongated shape.

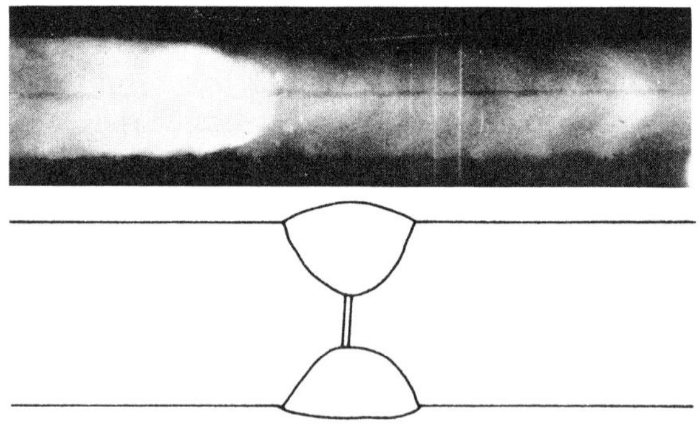

Incomplete Root Penetrtion is usually indicated as a straight, dark continuous or intermittent line, often at the center of the weld.

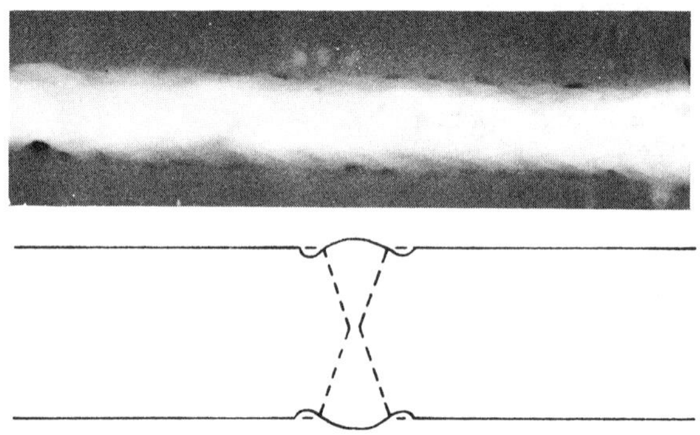

Undercutting shows up as a dark, linear shadow of wavy contour, occurring adjacent to the edge of the weld. This discontinuity is usually detected visually, but its correct identification on the radiograph is needed to prevent misinterpretation as another type of discontinuity.

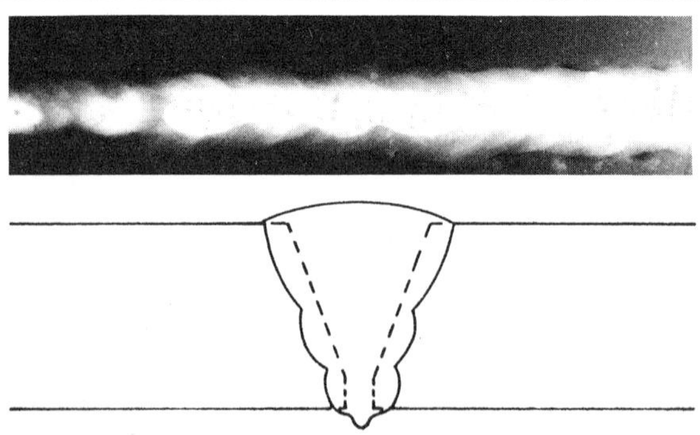

Icicles and Melt-through give individual light circular indications or darkened areas of elongated or rounded contour that may be surrounded by light rings.

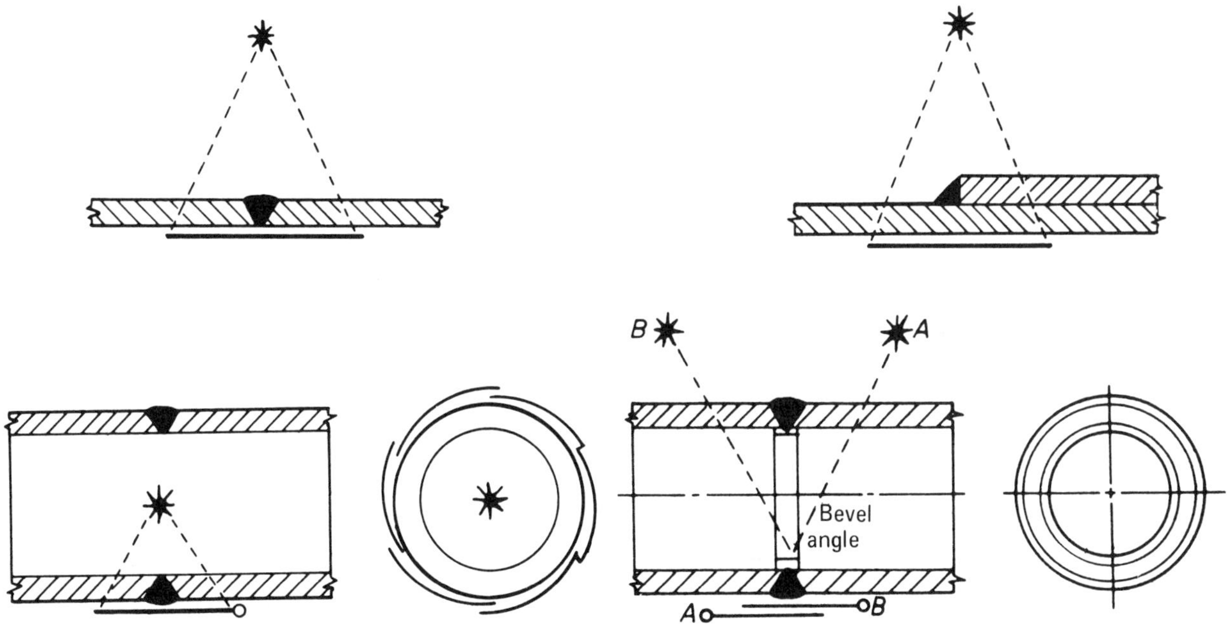

Fig. 11-16. Typical arrangements of X-rays source and film in radiography of welds. The angle of exposure and the geometry of the weld influence interpretation of the radiograph. Note that multiple exposure may be necessary for pipe welds. Of the several welds shown above, fillet welds are the most difficult to x-ray and the most difficult to interpret.

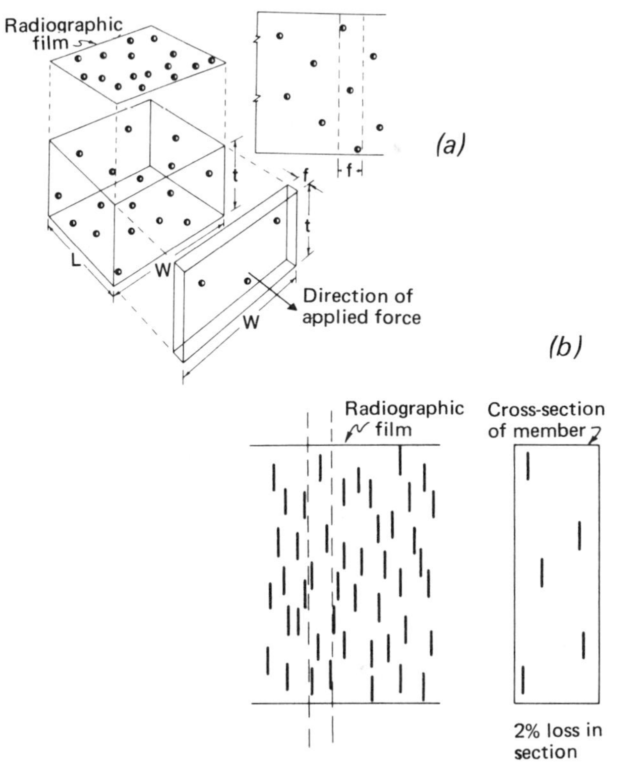

Fig. 11-17. Because a radiographic film accumulates all indications of defects occurring throughout a weld in one plane, its interpretation can be deceptive. The radiographic patterns here look bad, but if a slice, such as Wtf, is examined it is seen that it contains only a few small faults. Final judge of acceptance, in any case, is the standard to which the weldment is made. These standards vary with plate thickness.

trometers." Wires of the same composition as the material being inspected and diameters from 1.5 to 4 percent of the material thickness are used.

In the hands of a skilled inspector, use of the penetrameter also gives other items of information. Sharp image delineation gives assurance that the radiographic procedure is correct. The presence of a penetrameter image is also evidence that can be presented at any time later to prove that the weld was radiographed properly. **A penetrameter is not an indicator or gauge to measure the size of a discontinuity or the minimum detectable flaw size. It is an image quality indicator of the radiographic technique.**

Radiographs show a variety of weld discontinuities Some of the most common ones are illustrated by the reproduced radiographs in Table 11-3

An inspector's ability to recognize discontinuities and identify them is largely a matter of experience. Film-handling marks and streaks, fog, and spots caused by errors in the film-developing procedure complicate identification. Surface discontinuities will also show on the film and must be recognized.

The angle of exposure has an influence on the radiograph. Some typical arrangements for radiography of welds are shown in Fig. 11-16.

A radiographic image of the interior of a weld may be viewed on a fluorescent screen as well as on developed film. This makes possible rapid, low-cost inspection, but definition is poorer than on film.

A radiograph compresses into one plane all the indications of discontinuities that occur throughout the weld. Thus, the radiograph tends to give an exaggerated impression of scattered types of discontinuities, such as porosity or inclusions. Unless allowance is made for this fact, particularly on thick plates, a weld that is entirely adequate for its function could be ruled defective.

In Fig. 11-17(a), the volume LWt contains a number of distributed particles or spots of porosity. On the radiographic film, these appear to exist in a single plane — which could be interpreted as evidence of an excessive number of particles or voids in some section of the volume. When the slice Wtf, however, is cut from the volume, it is seen that it contains only three spots. Any failure due to an applied force would have to occur in some finite section. The three tiny particles or voids in this sample are unlikely to predetermine or accelerate failure. Thus, the radiographic picture of this condition is misleading.

Figure 11-17(b) is another illustration of the same effect. Because the streaks throughout the member are accumulated on the radiographic film, one might gain the impression that they constitute a sizeable proportion of the material. But when a section is removed and viewed from a different direction, it is seen that the streaks take up only a minute percentage of the cross section.

Radiography is useful in the qualification of weldors. The individual's ability to produce welds conforming to specification requirements is readily determined by an examination of welds produced on test plates. The procedure is also useful in evaluating processes proposed for a particular application. When radiography is required by the specifications governing the welding, written procedures for radiography are usually included. In this way, the purchaser exercises control over the inspection and specifies exactly what is wanted to meet his requirements.

A good NDT inspection program must recognize the inherent limitations and peculiarities of each of the NDT processes. For example both radiography and ultrasound have distinct orientation factors which must be understood if they are to be properly applied. In many ways these two processes tend to complement each other, what one is particularly good at the other may be relatively poor. A classic example of this is the inability of radiography to reliably detect lamination-like discontinuities. Conversely ultrasound is much better suited to detecting these types of discontinuities. Ultrasound on the other hand is poorly suited to the detection of scattered porosity while radiography is very good.

REAL-TIME IMAGING FOR WELD INSPECTION

Much work is being done and has been done with X-ray, fluoroscopy and ultrasound to produce immediate and on-the-spot NDT information and interpretations of welds as they are being made. Improvements of this nature are badly needed to eliminate the time that is normally lost between the completion of the weld and the evaluation of the test results. Thus making it possible to effect

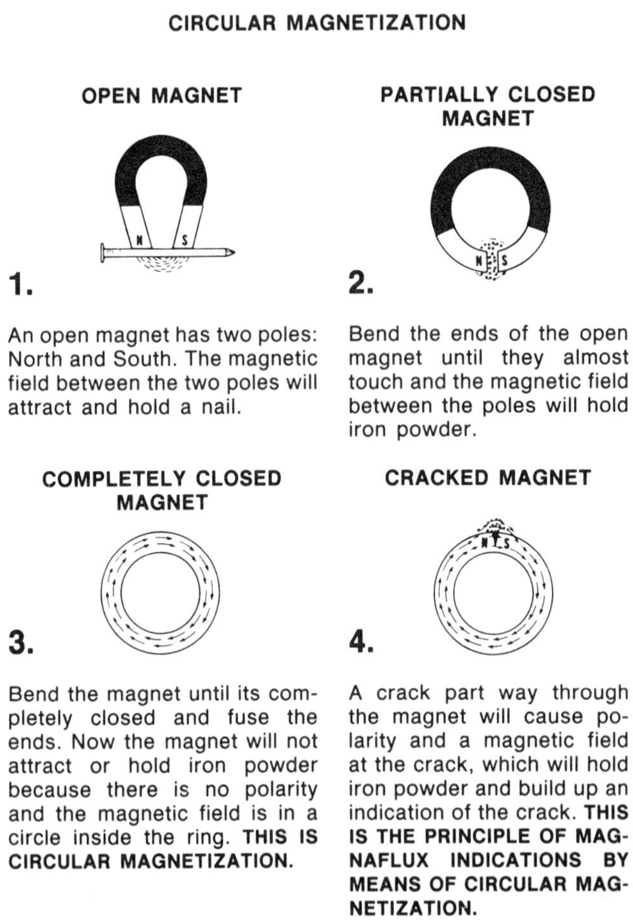

Photo courtesy of Magnaflux Corporation

Fig. 11-18. A discontinuity, such as a crack, in a completely closed magnet causes polarity and the development of a magnetic field, which will hold iron powder and give an indication of the discontinuity.

immediate corrections rather than continuing to make unacceptable welds.

Since radiation from X-ray machines or radioisotope sources can be damaging to body tissue when the exposure is excessive, safety precautions must be taken. The American Standard Safety Code for the Industrial Use of X-rays should be consulted for this purpose.

MAGNETIC-PARTICLE INSPECTION

Magnetic-particle inspection is a method of locating and defining discontinuities in magnetic materials. It is excellent for detecting surface defects in welds, revealing discontinuities that are too fine to be seen with the naked eye. With special equipment, it can also be used to detect discontinuities that are close to the surface. Figure 11-18 gives a simple explanation of the principles of magnetic-particle testing.

Circular magnetization results from longitudinal current transmission. When using this method for weld inspection probes are usually placed on each side of the area to be inspected and a high amperage passed through the workpiece. A magnetic flux is produced at right angles to the flow of current, which may be represented, as in Fig. 11-19, by circular lines of force within the workpiece. When these lines of force encounter a discontinuity, such as a longitudinal crack, they are diverted and leak

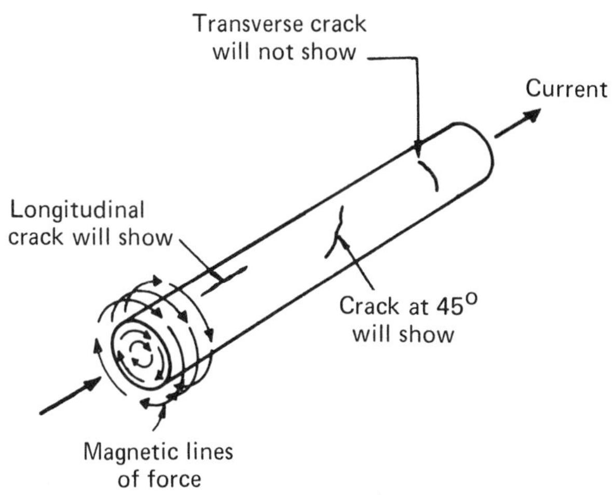

Fig. 11-19. When a current is passed through the workpiece, the magnetic lines of force are at right angles to the current, and discontinuities that are angled against the lines of force will create the diversion needed to produce magnetic poles on the surface. Under the illustrated arrangement, a transverse crack would not give an indication, but by changing the position of the probes 90 degrees it would be at right angle to the lines of force and would show.

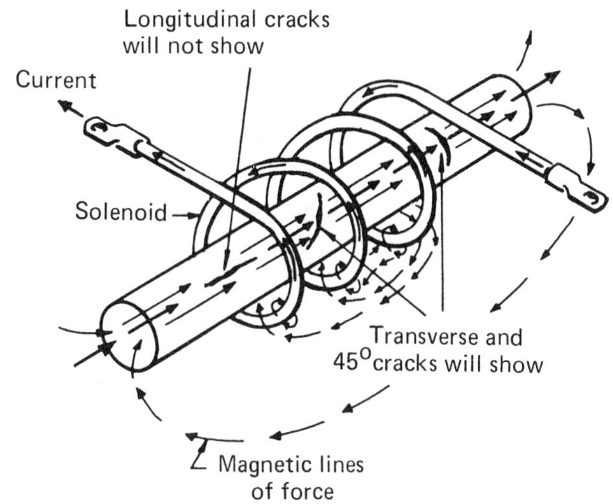

Fig. 11-20. When the magnetic flux is produced with a solenoid coil, the lines of force are parallel and longitudinal. A longitudinal crack will not show, but a transverse crack or one angled against the lines of force will produce an indication.

through the surface cut into the air, creating magnetic poles or points of attraction. A magnetic powder dusted onto the surface will thus cling to the leakage area more tenaciously than elsewhere, forming an indication of the discontinuity.

A workpiece can also be magnetized by putting it inside a solenoid. In this case, the magnetic lines of force are longitudinal and parallel with the workpiece. Transverse cracks show up under this arrangement. Figure 11-20 makes clear why the discontinuity must be angled against the magnetic lines of force for an indication to be developed. If the discontinuity is parallel to the lines of force, the diversion needed to break through the surface and create poles will not occur.

The magnetic-particle inspection method is much simpler to use than radiographic inspection, but has its limitations. It is applicable to ferromagnetic materials only. It cannot be used with austenitic steels. A joint between a base metal and a weld metal of different magnetic characteristics will create magnetic discontinuities, which may produce indications interpretable as unsoundness even though the joint is entirely sound. On the other hand, a true discontinuity can be obscured by the powder clinging over the harmless magnetic discontinuity. The sensitivity of the method decreases with decrease in size of the discontinuity. Sensitivity is less with round forms, such as gas pockets, and best with elongated forms, such as cracks.

To have external leakage, the magnetic field must be distorted sufficiently. Fine, elongated discontinuities, such as hairline cracks, seams, or

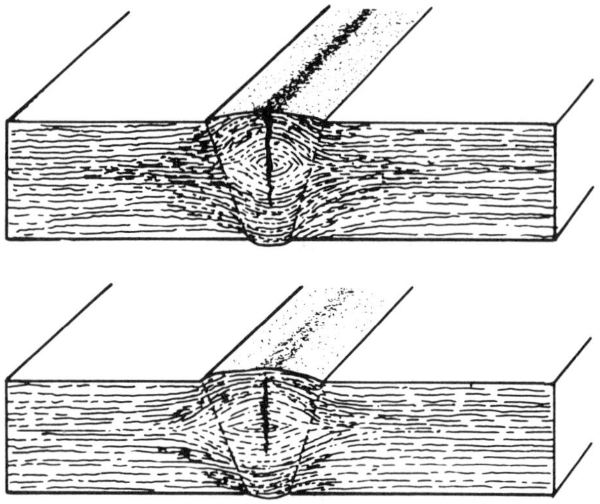

Surface Cracks give powder patterns that are sharply defined, tightly held, and usually built up heavily. The deeper the crack, the heavier the buildup of powder.

Subsurface Cracks produce a less sharply defined, fuzzy pattern. The powder is also less tightly adherent.

Crater Cracks may give a single line, running in almost any direction, or a pattern of lines. The pattern is small and occurs near the end of the weld.

Subsurface Porosity gives weak, poorly defined patterns, as do **Slag Inclusions.**

Fig. 11-21. Typical powder patterns and their interpretation in terms of weld discontinuities.

inclusions, that are "strung" parallel with the magnetic field will not distort it sufficiently to develop leakage. Thus, no indication will result. By changing the direction of the field, however, the indications can be developed. To be certain that discontinuities are detected, it is advisable to apply the field from two directions, preferably at right angles to each other.

Pieces to be inspected must be clean and dry. Wire- brushing or sand-blasting are satisfactory methods for cleaning welds. Surface roughness decreases the sensitivity, tends to distort the magnetic field, and interferes mechanically with formation of the powder pattern.

The shape, sharpness of outline, and width and height to which the particles have built up are features used for identifying discontinuities. When unusual patterns are produced, other test methods may be required to establish identity. Once a pattern has been interpreted by correlating it with the identification established by other methods, the interpretation can be applied to similar indications on other parts. Representative indications are shown in Fig. 11-21.

Since a powder pattern results from various types of discontinuities in the magnetic field, it is easy to mistake an irrelevant indication for a defect. As noted previously, a change in the magnetic characteristics of materials can create an irrelevant indication in the area of the interface. A change in section or a hole drilled in the part will tend to produce indications that have no significance in respect to weld soundness. These patterns are usually readily recognized for what they represent by their shapes and locations in the part. Abrupt changes in magnetic properties may occur at the edge of the heat-affected zone. The pattern will be fuzzy and diffused, running along the base metal close to the edge of the weld. It resembles the pattern caused by undercutting, a difference being that the particles are much less adherent. If a part is heat-treated or stress-relieved before subjecting to inspection, the magnetic characteristics of the heat-affected zone are restored and no indications will appear.

In the shop use of magnetic-particle inspection, judgment as to the significance of patterns can be guided by rules or standards based on experience or laboratory tests. Assuming that the inspector has ruled out the possibility of a false indication and has properly identified the indicated faults, he must then decide whether to pass or reject the part or require its repair. All faults do not affect the integrity of the part in service. Thus, slag inclusions and porosity may have no bearing on the serviceability of the weld. Surface cracks revealed by magnetic-particle inspection, however, should be considered potential stress raisers or focal points for fatigue and corrosion.

The equipment for magnetic-particle inspection is relatively simple (Fig. 11-22). Commercial units, portable and stationary, provide for nearly every situation where the method is applicable. The manufacturer's instructions and recommendations for use of the equipment should be read carefully and thoroughly understood. The units may provide magnetization by direct, alternating, or rectified currents, or combinations thereof. Portable equipment, making use of electromagnets and permanent magnets, is also available.

Direct or rectified current is required for deep penetration. Alternating current magnetizes the

Fig. 11-22. Checking the fillet welds joining connection plates to a large box column with a portable magnetic-particle inspection unit.

surface only, and thus is limited to surface inspection. Full-wave, three-phase rectified current produces results comparable to those produced by battery direct current. Half-wave, single-phase rectified current gives maximum sensitivity. High amperage, low voltage is normally used in all magnetic-particle testing to limit arcing or burning on the test piece.

Magnetic powders may be applied either by the dry or wet methods. Dry powder is dusted uniformly over the work with a spray gun, dusting bag, or atomizer. The finely divided magnetic particles are coated to give them greater mobility and are available in gray, black, and red colors. It is desirable that the particles impinge on the surface at low velocity and with just enough residual force after impact to move them to possible sites of leakage. Excess powder is removed with a light stream of air.

In the wet method, very fine red or black particles are suspended in water or light petroleum distillate. Powders for liquid suspension come from the manufacturer in either paste or dry form, prepared for use in water or oil baths. After the suspension has been made — in accordance with the manufacturer's instructions — it is flowed or sprayed onto the surface to be inspected, or the piece may be dipped into the liquid. The wet method is more sensitive than the dry method since extremely fine particles may be used; this enables detection of exceedingly fine faults. Red particles improve visibility on dark surfaces. When the particle coating is a dye that fluoresces under ultraviolet light, sensitivity is further increased. Fluorescent powders are excellent for locating discontinuities in corners, keyways, splines, deep holes, and similar locations.

The techniques for creating a magnetic field in workpieces of various sizes and shapes, the sequence of operations in magnetizing and applying magnetic particles, adjustment of current to bring out desired results, the practical modes for orienting magnetic fields to produce or better delineate indications — all are important to the successful use of this inspection method. The equipment manufacturers' literature and various ASTM and other specifications (such as ASTM E-109 "Method for Dry Powder Magnetic-Particle Inspection" and ASTM E-138 "Method for Wet Magnetic-Particle Inspection") should be consulted for operational details.

Magnetic-particle inspection is applied to many types of weldments in production practice. The dry-powder method is especially popular for heavy weldments. Many steel weldments in aircraft manufacture are inspected by the wet method, using direct current. Airframe parts are subjected to fatigue conditions, which means that surface cracks cannot be tolerated. Since these weldments are relatively thin, magnetic-particle inspection will usually detect subsurface discontinuities as well as show up the finest surface cracks.

LIQUID-PENETRANT INSPECTION

Liquid-penetrant inspection is a nondestructive method for locating surface cracks and pinholes that are not visible to the naked eye. It is a favored technique for locating leaks in welds, and it can be applied where magnetic-particle inspection is ineffective, such as with austenitic steels or nonferrous metals. Two types of penetrant inspection are used — fluorescent and dye — which define the penetrating substance.

With fluorescent-penetrant inspection, a highly fluorescent liquid with good penetrating qualities is applied to the surface of the part to be examined. Capillary action draws the liquid into the surface openings. The excess liquid is then removed from the part, a so-called "developer" is used to draw the penetrant to the surface, and the resulting indication is viewed by ultraviolet (black) light. The high contrast between the fluorescent material and the background makes possible the detection of minute traces of penetrant.

Since penetration of minute openings is involved, the part to be inspected must be thoroughly clean and dry. Any foreign matter could close the openings leading to false conclusions. The penetrant is applied by dipping, spraying, or brushing. Time must be allowed for absorption of the material into the discontinuities — up to an hour or more in very exacting work.

When penetration is complete — or assumed to be adequate in accordance with testing specifications — the excess material is removed from the surface. If the penetrant is designed for water wash, a low-pressure water spray is used. Some commercial penetrant systems, however, require a solvent wash or what is called a "post emulsifier." With the latter, an emulsifier is applied to the part and allowed to remain on it for one to four minutes before the water spray. The instructions for use of emulsifiers and for the washing operation must be followed closely, since it is necessary that only excess penetrant be removed.

After the wash, the parts are dried if a dry developer is to be used. Hot air may be used to accelerate drying. A dry developer is applied with a powder gun, spray bulb, or by dipping the part into the powder. The developer draws the penetrant from the faults, making it accessible for viewing by ultraviolet light. If only large discontinuities are sought, a developer may not be required to make the indications visible. If a wet developer is to be used, drying of the part after removal of excess penetrant is not required. Wet developer is applied in the form of a colloidal water suspension by dipping or spraying, after which the part is dried by hot air.

Under ultraviolet light, the indications fluoresce brilliantly. The extent and depth of the discontinuity can be gaged by the width and length of the indication and the amount of penetrant bleeding to the surface. The darker the room in which viewing is done, the more brilliant the fluorescence and more easily very small indications are observed.

Dye-penetrant inspection is similar to fluorescent-penetrant inspection except that dyes visible under ordinary light are used. By eliminating the need for ultraviolet light, greater portability in equipment is achieved.

Liquid-penetrant inspection is widely used for leak detection. A common procedure is to apply fluorescent material to one side of a joint by brushing or spraying, allow adequate time for capillary action, and then view the other side of the joint with ultraviolet light. Dry developer may be used on the side being inspected to intensify the indications. In thin-walled vessels, this technique will show up leaks that are not ordinarily located by the usual air test with pressures of 5 to 20 psi. The sensitivity of the leak test decreases, however, when the wall thickness is over 1/4-in.

Fig. 11-23. This hair-line intermittent crack typifies the type of weld discontinuity for which dye-penetrant inspection techniques are particularly suited.

Penetrant inspection is also widely used in the inspection of large and small weldments for cracks and porosity when the materials are nonmagnetic. It must be remembered that only surface discontinuities are revealed by this method. Figure 11-23 shows a typical crack indication using dye penetrant.

ULTRASONIC INSPECTION

Ultrasonic inspection is a supersensitive method of detecting, locating, and measuring both surface and subsurface faults in metals. Flaws that cannot be discovered by other methods, and even cracks small enough to be termed microseparations, may be detected. In the practical inspection of welds, the sensitivity of the process is often curbed by designing or setting the equipment to give a response equivalent to a sensitivity of 2% of the metal thickness, thus giving results comparable with those obtained in radiographic inspection.

Ultrasonic inspection is based on the fact that a discontinuity or density change will act as a reflector for high frequency vibrations propagated through the metal. The searching unit of the pulse-echo type ultrasonic equipment contains a crystal of quartz (or other piezoelectric material). When a voltage is applied, the crystal vibrates rapidly, and imparts mechanical vibrations of the same frequency into materials it contacts.

When an ultrasonic probe is held against metal, the vibrational waves are propagated through the material until a discontinuity or change of density is reached. At these points, some of the vibrational energy is reflected back. If, in the meantime, the current that caused the vibration has been shut off, the quartz crystal can now act as a receiver to pick up the reflected energy. The reflected vibration causes pressure on the quartz crystal, which results

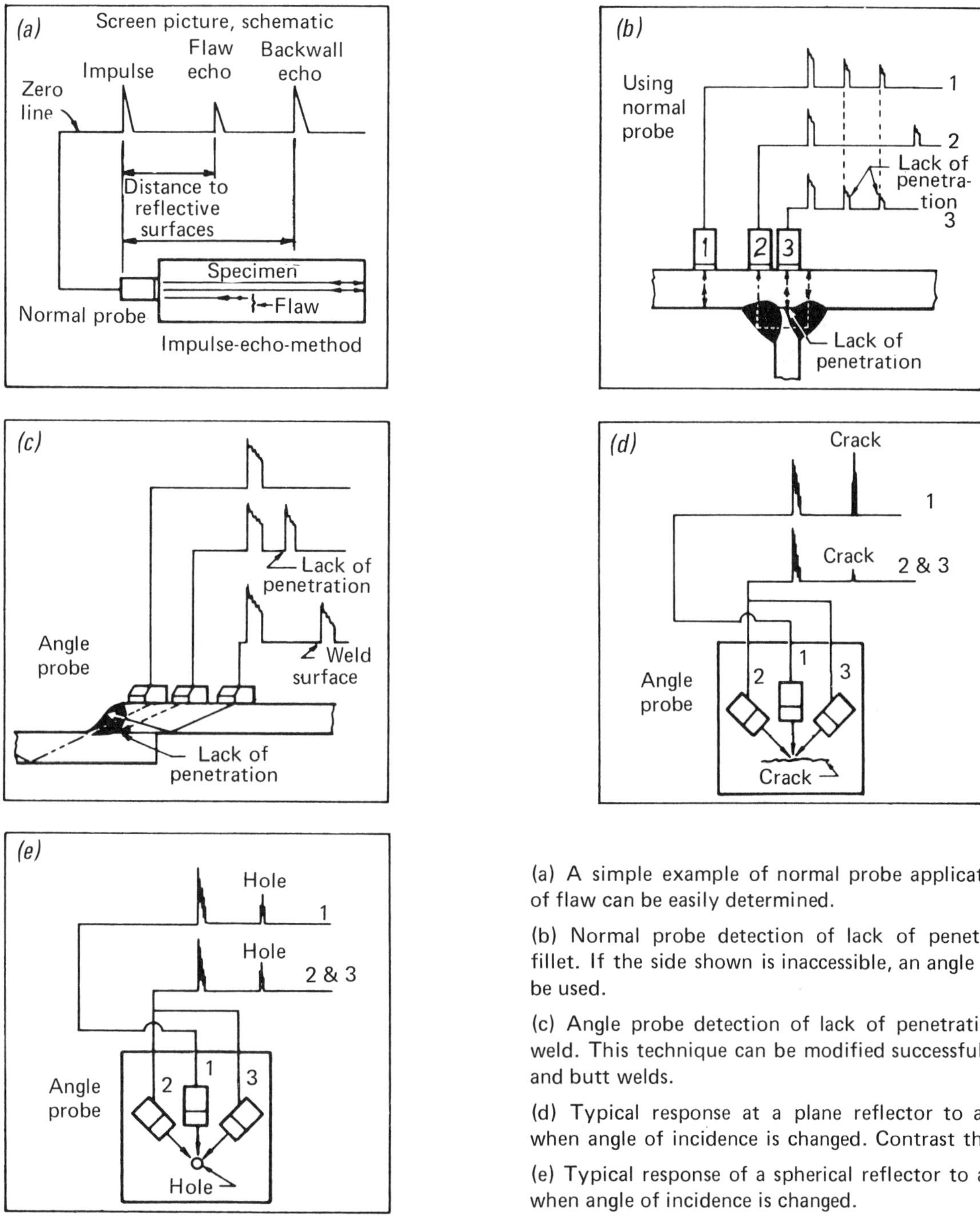

(a) A simple example of normal probe application. Depth of flaw can be easily determined.

(b) Normal probe detection of lack of penetration in a fillet. If the side shown is inaccessible, an angle probe must be used.

(c) Angle probe detection of lack of penetration in a lap weld. This technique can be modified successfully for fillet and butt welds.

(d) Typical response at a plane reflector to angle probe when angle of incidence is changed. Contrast this with (e).

(e) Typical response of a spherical reflector to angle probe when angle of incidence is changed.

Fig. 11-24. Video patterns produced on the cathode-ray tube by different types of discontinuities.

in the generation of an electric current. Fed to a cathode-ray tube, this current produces vertical deflections in the horizontal base line. The pattern on the face of the tube is thus a representation of the reflected signal — and of the discontinuity. The cycle of transmitting and receiving is repeated at a rate of 60 to 1,000 times per second.

Two types of cathode-ray-tube presentation are available — radio frequency (RF) and video. Most commercial units present the video pattern, although many inspectors believe RF provides more useful information on flaw identification. Figure 11-24 shows the video patterns produced on the cathode ray tube by different types of discontinuitiess. It can be readily seen that expertness in interpretation of the deviations from the horizontal oscilloscope

Fig. 11-25. Examination of box-column welds with a portable ultrasonic unit. Inspection of this assembly with radiographic techniques would be very difficult.

line is required, and a similar degree of expertness is needed for operation of the equipment. Interpretations are based on standard patterns made from reference plates, which are prepared in accordance with procedures approved by the IIW, ASTM, ASME, and AWS (D1.1, Section 6, Part C).

Pulse-echo ultrasonic equipment is available in compact form for field use. Figure 11-25 shows a battery-powered portable unit. In addition to the greater sensitivity over radiographic equipment, ultrasonic inspection enables the probing of joints that are inaccessible to radiographic methods and is more dependable in locating all flaws. Operational costs are also less.

Another type of ultrasonic inspection equipment gives what is similar to a "television picture" of the flaw. Although not as popular as pulse-echo equipment, it is believed to have the potential for simplifying pattern interpretation where applicable, since the pattern is essentially a plane image. With this type of equipment, the ultrasonic energy from a generator is transmitted through the specimen much as X-rays are transmitted in radiographic inspection. The transmitted energy impinges on a piezolectric target, producing varying potentials on its face that are characteristics of the flaws in the specimen. The target is scanned with an electron beam, which is modulated by the varying surface potentials. This enables the development of a picture — a representation of the flaw — on a monitoring screen.

Many codes (e.g., The AWS Structural Code D1.1) specify where and when U.T. inspection shall be used to evaluate welds. They also provide detailed equipment setup, calibration, scanning techniques and evaluation requirements. Many codes also state that when ultrasonic testing produces questionable indications of discontinuities, a more detailed evaluation of the discontinuity by other means shall be required (e.g., alternate ultrasonic techniques, radiography, grinding or gouging for visual inspection).

Ultrasonic testing is less suitable for determining porosity in welds than other NDT methods because round gas pores respond to ultrasonic tests as a series of single point reflectors. This results in low amplitude responses. Many codes require very high levels of energy be induced into the part when scanning welds. The resultant low amplitude signals from porosity are often confused with "base line noise" inherent with the testing parameters.

Certain technologically advanced U.T. instruments can reduce base line noise without adversely affecting amplitude linearity through "linear reject control". This advance makes it much easier to detect porosity using the ultrasonic method.

Portable ultrasonic flaw detectors are available that have digital operation and microprocessor controls. These instruments may have built in memory capability, provide hard copy documentation printouts or video monitoring and recording of tests and are computer interfaceable.

CHEMICAL AND METALLOGRAPHIC EXAMINATION

Chemical analysis is not usually required in routine weld inspection. When a certain chemistry is specified for the weld metal, it is known beforehand and becomes an important consideration during procedure qualification. If trouble is encountered in producing acceptable welds, chemical analysis may help reveal the source and suggest changes in procedure to eliminate the difficulty. The AWS electrode identification system (letter, number, and suffix) designates the chemistry of the electrode used. Thus, chemical analysis of the deposited weld metal is a verification procedure rather than an inspection method. A specification, however, may require that a sample be taken from the weld and submitted for analysis.

Metallographic tests are more frequently required by specifications. They are helpful in determining the distribution of nonmetallic inclusions, the grain structure in the weld and fusion zone, and the extent and structure of the heat-affected zone. The tests may involve visual examination, in which case macro-specimens are etched to bring out the structure and are examined by the unaided eye or with a magnifying glass. If microscopic examination is required, the specimens are polished and etched by standard procedures for examination at high magnifications.

Samples for either metallographic or chemical analysis may be taken without destroying the weld by trepanning or core drilling. After the sample has been taken, the hole or cut is repaired by rewelding.

MECHANICAL TESTING

Mechanical testing includes prooftesting and testing-to-destruction. Prooftesting involves applying a load or pressure equal to or greater than that expected in service but not great enough to damage the product. It is a method of assuring that the product is serviceably sound at the time of the test, and it may be required by the specifications for all the production units. Destructive testing, obviously, is applied only to specimens, usually prototypes or samples taken from production in some predetermined statistical sequence.

Prooftesting is common with welded vessels. The test may be run for tightness of the enclosure, in which case the working pressure will likely be used. If the test is run to confirm strength or safety, the test pressure will usually exceed working pressure. Judgment enters into specifying the pressure intended to reveal faults that later may cause failure in service. Usually, some multiple of the design value is selected. The passing of a prooftest in which the weldment is subjected to pressures several times those of normal service does not mean that working pressures of that magnitude can be used. Experience indicates that failures can occur later at much lower pressures as a result of fatigue, corrosion, or stress raisers.

To avoid damage to the weldment, a prooftest must always be run at something less than the yield strength of the weakest component. It is usually desirable to limit the pressure or force to 75% of the calculated yield point. Sometimes the combination of residual stresses with the prooftest stress may cause some local yielding without damage. The test. in fact, may act as a stress-relief operation to a limited extent.

Air, water, oil, and gases may be used to supply the pressure when prooftesting a vessel. Leaks are shown by pressure drops or detected visually. Light-viscosity oil will penetrate leaks that will not pass water, and air will leak where oil will not. On a clean weld, leaking oil or water is easily observed. Air or gas pressure will show a drop if a leak exists. Soap solution, spread over a weld, will reveal the presence of a leak by bubble formation.

Destructive testing gives an absolute measure of the strength of the sample tested. Assuming that materials and the method of fabrication are uniform, it is reasonable to infer that the sample is representative of all the units. Periodic testing of specimens tend to lend validity to this inference, providing the results are similar. Usually the destructive force is applied in a way that simulates service conditions.

Any testing procedure based on sampling may permit defective work to get through inspection. Weight must be given to the penalty resulting therefrom. When safety is an important consideration, destructive techniques with samples do not give assurance of the quality needed. On the other hand, when safety is not a factor and the unit cost of the product is low, destructive testing with samples may be more economical and informative than nondestructive testing.

SPOT CHECKING AND OTHER TESTS

If quality control, rather than the screening out of parts with defects, is the main objective, spot checking may be employed. Any applicable inspection method might be used in spot checking, but in shop practice the term usually alludes to the "sectioning" — taking of samples for metallographic study and tensile, impact, or other mechanical tests. A small circular sample is removed with a hole saw or trepanning tool for metallographic etching, or a boat-shaped specimen taken with a special machine, using a hemispherical saw, for mechanical tests. While it is necessary to repair the weld, the procedure is classed as nondestructive.

Spot checking by sectioning gives an indication of the average quality. Since relatively inexpensive equipment is required and the testing techniques are those standard to all metallurgical laboratories, it is one of the cheaper methods for controlling, even though not assuring, quality.

Hydrostatic testing is a method for determining the tightness of welds in fabricated vessels. It differs from prooftesting in that the pressure built up within the test vessel is usually only moderately greater than the external pressure. Various liquids and gases are used which, when escaping in minute amounts, are easily detected by chemical or physical means. Thus, the vessel may be pressurized with chlorine gas, and a halogen-sensitive leak detector used to spot a leak so small that only 1/100 of an ounce of the gas could pass through the opening in a year. Or helium gas may be used in connection with a commercial mass-spectrometer helium-leak detector capable of detecting one part of helium in 200,000 parts of air.

Hardness tests may also be called for by the specifications governing the welding. Hardness measurements give an indication of the approximate tensile strengths of metals and will show whether or not the base and weld-metal strengths are correctly matched. (See the Reference Data Section for a table relating Rockwell hardness numbers to tensile strength.)

Hardness data also give indications of the metallurgical effects of the welding process on the weld metal and heat-affected zones. Limitations are often placed on hardness, since it reflects the ductility of the joint and to some extent its ability to resist impact loads. Hardness measurements, as other mechanical properties, are usually made by the laboratory rather than by inspection personnel.

SOME POINTS TO REMEMBER ABOUT WELD INSPECTION

- Visual inspection tells most. Everyone can participate in it.
- Radiographic inspection permits looking into the weld for discontinuities that fall within the sensitivity range of the process. It is best for finding discontinuities within the volume of the weld and provides a permanent record of the results.
- Magnetic-particle inspection is outstanding for detecting surface cracks and subsurface cracks which are very close to the surface. It is used to advantage on heavy weldments and assemblies but applies only to magnetic materials.
- Dye-penetrant is easy to use in detecting surface cracks. Its indications are readily interpreted. It may be applied to magnetic and non-magnetic materials.
- Ultrasonic inspection is excellent for detecting subsurface planer type discontinuities, such as cracks and lack of fusion. It requires expert interpretation and is operator sensitive.

Welding Procedure Qualification

INTRODUCTION

Various codes and specifications require that the welding procedures for a job be proved to give satisfactory welds by prescribed tests. In addition, the governing code or specification may demand that the weldors and welding-machine operators be proved by tests as qualified for the job. All such tests are referred to as welding qualification tests.

Although any code-writing body may prescribe its own welding qualification tests, most codes and specifications concerned with welded steel structures in the United States incorporate the welding qualification tests of the American Welding Society, the American Society of Mechanical Engineers, or the American Petroleum Institute in their specifications. Thus, the American Association of State Highway Officials in its specifications for highway bridges incorporates the AWS welding qualification tests for procedures and personnel, as those qualification tests are given in the AASHTO-AWS D1.5 Bridge Welding Code. Many other code-writing organizations, including private companies, industrial groups, or governmental agencies, also refer to AWS, ASME, or API specifications for welding qualification tests.

AWS tests are most widely recognized, but are not universally applicable. In certain types of work, other qualification tests may be required. The contractor doing the welding must make certain that the qualification of procedures and personnel he uses meet the requirements of the governing code.

There are two basic welding qualification tests; procedure qualification tests and personnel qualification tests. Procedure qualification tests evaluate joint design, position, base material, filler material, process and parameters. Personnel qualification tests are divided into weldor qualification tests, tacker qualification tests and welding operator qualification tests. The specific requirements for each of these tests are set forth in codes prepared by recognized organizations such as the American Welding Society (AWS), the American Society of Mechanical Engineers (ASME) and the American Petroleum Institute (API). AWS codes deal with welded structures; AWS D1.1 Structural Welding Code-Steel, AASHTO-AWS D1.5 Bridge Welding Code and AWS D1.2 Structural Welding Code Aluminum. ASME Section IX contains codes which deal with welding high pressure piping and both fired and unfired pressure vessels. The API Code covers welding of piping used in petroleum transmission lines.

In the following text, these tests are described in a condensed form — with the intent to give the reader a concept of what is involved. **Before qualification tests are attempted, a copy of the applicable code or specification should be obtained and studied.**

The AWS Structural Welding Code — Steel-D1.1, Structural Welding Code — Aluminum-D1.2, Structural Welding Code - Sheet Steel-D1.3, Structural Welding Code - Reinforcing Steel-D1.4 and the Bridge Welding Code-D1.5 are sources of information on qualifying procedures, weldors, and welding operators for the welding associated with building and bridge construction. Copies may be obtained from the American Welding Society, Inc., 550 N.W. Le Jeune Road, P.O. Box 351040, Miami, Florida 33135.

The ASME Section IX document may be obtained from the:

American Society of Mechanical Engineers
United Engineering Center
345 East 47th Street
New York, NY 10017

The API Code may be obtained from the:

American Petroleum Institute
1220 L Street, Northwest
Washington, DC 20005

Good weld soundness and strength are dependent on a variety of interacting variables. To ensure that only sound practices are employed in fabrication, procedures must be established and tested to prove their suitability for a specific code and application.

The requirements may include tensile strength, yield strength, elongation, reduction in area, impact strength, fracture toughness and X-ray soundness.

The specific requirements will depend on the code involved. For example, the bridge welding code may include impact strength while the building code may only include strength and ductility requirements.

A welding procedure specification (WPS) must be established for each type of joint, position and process to be used in a particular structure or weldment. A procedure includes joint design, joint preparation, position of welding, base material, filler material, welding process, welding parameters, preheat, interpass temperature and post weld heat treatment. Once established and approved, a procedure (WPS) is used to qualify a weldor, welding operator or tack weldor and that procedure (WPS) must be used for welding that particular joint by any weldor qualified to that process.

A welding procedure qualification test is performed by the manufacturer or contractor in accordance with the provisions of a particular code. The results of the welding procedure qualification test are recorded as a Procedure Qualification Record (PQR) and is used as a basis for preparing a welding procedure specification (WPS). The AASHTO-AWS D1.5 Bridge Welding Code also allows procedure qualification by a pretest-verification route. The pretest is a welding procedure test performed by someone other than the contractor as a basis for preparing a welding procedure specification (WPS). The verification test is a simplified version of a procedure qualification test which is performed by a contractor when verifying a procedure qualification record (PQR) which was obtained from a third party that had performed a pretest. This verification test can be used as a mechanical test basis for preparing a welding procedure specification (WPS).

The American Welding Society recently started publishing Standard Welding Procedures as ANSI/AWS B-2 documents. The publication of these documents will continue on an on-going basis.

TECHNIQUES TO EMPLOY DURING PROCEDURE QUALIFICATION TESTING

To perform a welding procedure qualification test, a number of weld joints must be prepared and welded using samples of base material and filler materials representative of that to be used in the actual structure. The welding process, joint design, joint preparation, welding position, preheat and interpass temperature are selected based on available information and experience such as similar prequalified procedures.

Sample welds are made under these selected conditions over a range of parameters (arc current, arc voltage and travel speed) in base material samples of sufficient size to obtain the necessary test specimens. The rolling direction of the plate is generally normal to the joint axis unless specified otherwise. These welds are cut up and machined into the number and types of test specimens required by the code being used. They are tested according to the code procedure and the results are recorded and compared with the code requirements. A range of welding parameters are selected from this information which provides weld properties well within the range specified by the code. Additional welds may be made near the limits of the parameter range and tested to verify that the required properties are attained. A complete record of all set up, weld parameters and test data is kept for future reference to verify that the procedure qualification test was valid. Where prequalified procedures are available, they require that a verification test of the prequalified procedure be performed by the contractor for bridge application.

SPECIMEN SIZE, SHAPE AND TESTING

Weld qualification tests for the AWS Structural Welding Code D1.1 and the Bridge Welding Code D1.5 require tests in flat plate, either groove welds in butt joints or fillet welds in "T" or lap joints. Butt welds are cut into sections which are machined into tensile or bend test specimens. Fillet welds in "T" joints are cut into sections which are bent to show the completeness of root penetration or are examined by grinding flat and macroetching to show root penetration.

The procedure qualification according to AWS D10.9 Piping and Tubing, the ASME Boiler and Pressure Vessel code and the API Pipe Welding code require weld specimens on pipe and tubing. Sections of pipe of adequate length are butt welded in the position to be qualified and longitudinal sections are cut and machined into tensile and bend specimens. The size and shape of the test specimens are specified by each of the codes. These specimens usually conform to AWS or ASME standard test specimens.

Descriptions of the procedure qualification requirements for the AWS, ASME and API welding codes are covered in the following sections. Details relative to the acceptability standards for mechanical tests are given within the specification or code.

PROCEDURE QUALIFICATION – AWS D1.1 Structural Welding Code

The welding procedures for work done in

accordance with the AWS Structural Welding Code, D1.1, must be "qualified" before fabrication can begin, except when the procedures are prequalified. Prequalified procedures are those that conform to Section 2 — Design, Section 3 — Workmanship, and Section 4 — Technique, of the Code. In many instances, the contractor will find that procedures he intends to use with the specified joints are already prequalified — in which case no procedure qualification tests will be needed.

However, if the contractor wishes to deviate from these requirements, he must qualify the procedure. If the building commissioner or engineers accept evidence of previous qualification, the testing can be waived.

The procedure qualification tests with fillet and groove welds are illustrated and described in Figs. 11-26 and 11-27.

For steels having a specified yield point of 50,000 psi or less, qualification on any one also qualifies for all, or for any combination of the others (5.5.1.1) having a minimum specified yield point equal to or less than that of the base metal used in the test. If the steel has a specified yield point in excess of 50,000 psi, the procedure qualification test on this steel will only qualify this particular steel, grade, or type, and the same specified yield point. A reduction in yield point because of an increase in thickness of the particular steel is permitted. For example, a procedure qualified with 1-in.-thick 100,000-psi yield steel also qualifies for 3-in.-thick 90,000-psi yield steel of the same ASTM specification (5.5.1.3, D1.1).

Qualification on either A242, A441, A572 Gr. 50, A588, API 5L Gr. X52, or ABS Gr. AH, DH or EH will also qualify for any of these, combinations of them or with other approved steels having lower specified yield points. (5.5.1.2, D1.1)

If a steel having a specified yield point in excess of 50,000 psi is qualified to be welded to a steel having a specified yield point of 50,000 psi or less, it is qualified to be welded to any other of the steels having a specified yield point of 50,000 psi or less (5.5.1.4, D1.1).

The procedure qualification for groove welds joining material over 3/4-in.-thick states that the test plate shall be the same thickness as the material but need not exceed 1 in. Should the application in question involve 7/8-in.-thick material, the logical recommendation for procedure qualification would still call for 1-in.-thick plate, since the test is no more difficult than 7/8-in., and passing it would give "unlimited" qualification. Should 1-in. test plates be performed on any thickness greater than 1-in. and still give unlimited qualification.

A test joint for partial-penetration groove welds shall be made for each type of joint and weld position, except that the depth of groove does not have to exceed 1 in. A macroetched cross section of the completed joint shall be made to show that the designated effective throat thickness is obtained and shall conform to the requirements of the Procedure Specification (5.10.2). The test plates for groove welds with gas metal-arc, flux-cored arc, electroslag, and electrogas welding shall be radiographically or ultrasonically tested. (5.10.1.3) (5.12.5, D1.1)

When required by filler-metal specifications applicable to the weld metal being tested, test specimens may be aged at 200° to 220°F for 48 ± 2 hours (5.10.4, D1.1).

PROCEDURE QUALIFICATION — AASHTO-AWS D1.5 BRIDGE WELDING CODE

The AWS D1.5 Bridge Welding Code is based on the 1988 edition of the AWS D1.1 Structural Welding Code where applicable; however extensive revisions were necessary to meet all AASHTO requirements for the construction of highway bridges by welding.

Only five grades of steel are approved for bridge construction: A36, A572-Grade 50, A588, A514/A517, and A852. A852 can be described as an A588 grade steel that has been quenched and tempered. This is similar to the A514, but exhibits the weathering characteristics of A588 material. (See Section 1.2.2 of D1.5.)

D1.5 permits the use of low-hydrogen processes only. Under D1.1, the use of cellulose electrodes is still permitted for certain grades of steel. Their use is prohibited under D1.5. (See 4.5.1.)

D1.5 specified that all welding done in the vertical plane shall proceed upward. For repairing undercut and fillet welds, D1.1 permits vertical down welding. A major concern of the members of the D1.5 committee was that welding vertical down with a high rate of travel may cause a too-rapid cooling rate. This can result in a very hard heat-affected zone surrounding the vertical down beads. Since the vertical up procedures yield higher heat input, the upward direction is the only one permitted for vertical welding. (See 4.68.)

Regarding procedure qualification, a major difference exists between D1.5 and D1.1. Under D1.1, the fabricator has the option of using "prequalified" procedures providing certain conditions are met.

Fillet Welds
when required

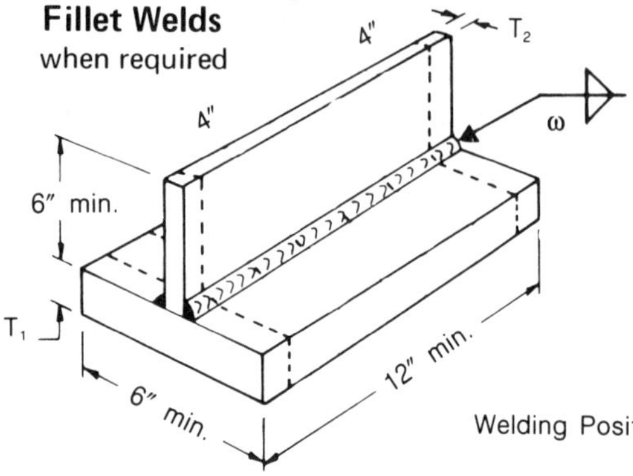

weld size ω	T_1	T_2
3/16	1/2	3/16
1/4	3/4	1/4
5/16	1	5/16
3/8	1	3/8
1/2	1	1/2
5/8	1	5/8
3/4	1	3/4
>3/4	1	1

Two test welds shall be made for each position to be used; one with the maximum size single pass fillet weld and one with the minimum size multiple pass fillet weld to be used. (5.10.3)

Welding Positions

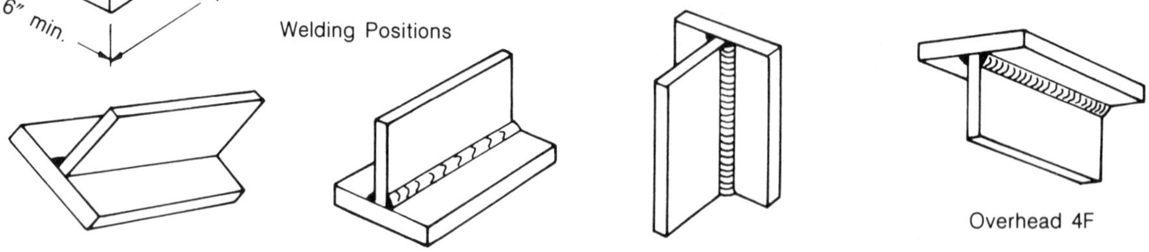

Macroetch Test — The fillet weld shall conform to the quality requirements of 8.15, 9.25, 10.17, whichever is applicable. They shall show fusion to the root, but not necessarily beyond the root, and shall be free from cracks. Convexity and concavity shall not exceed the limits of 3.6.1. (5.12.3)

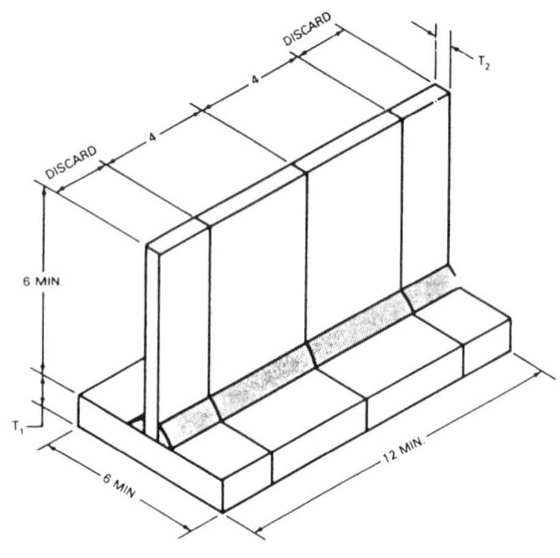

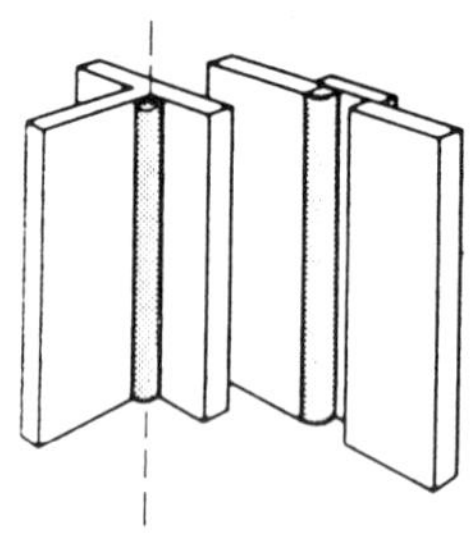

(C) VERTICAL POSITION 3F

AXIS OF WELD VERTICAL

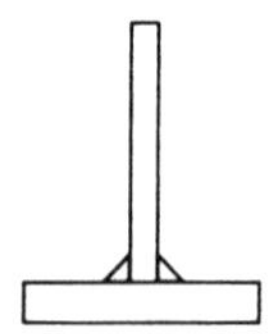

MACROETCH TEST SPECIMEN

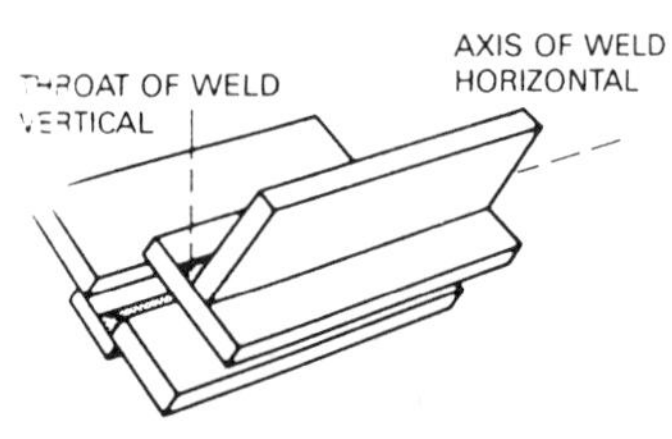

(A) FLAT POSITION 1F

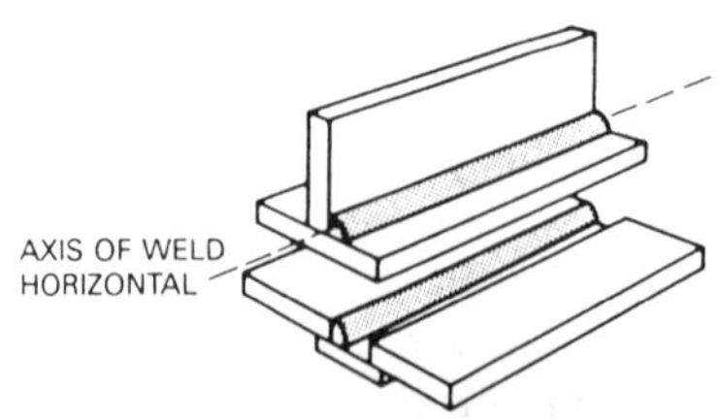

NOTE: ONE PLATE MUST BE HORIZONTAL.

(B) HORIZONTAL POSITION 2F

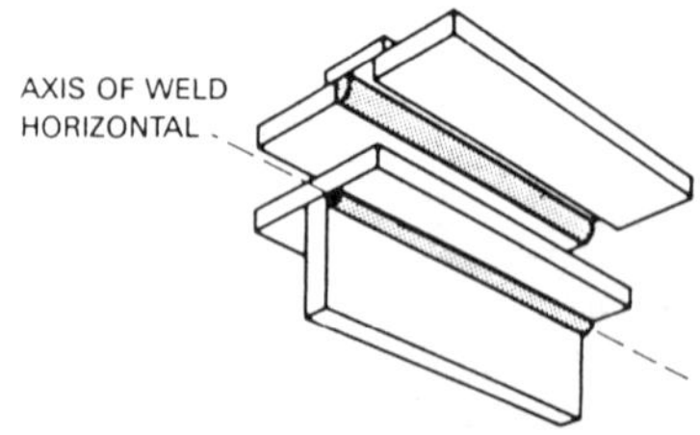

NOTE: ONE PLATE MUST BE HORIZONTAL.

(D) OVERHEAD POSITION 4F

Fig. 11-26. Details of the AWS procedure qualification test (D1.1). *(Continued in Fig. 11-27)*

Groove Welds – Plate

Qualifies for material up to and including 3/4" thick.

Qualifies for material over 3/4" in thickness.

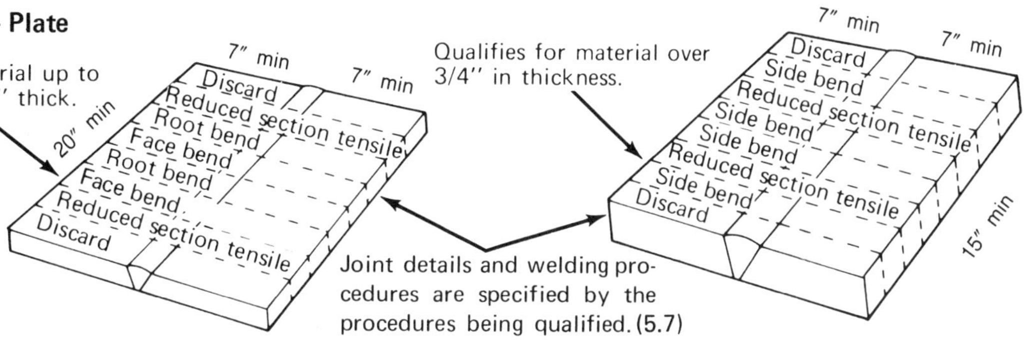

Joint details and welding procedures are specified by the procedures being qualified. (5.7)

Make 3/8" test plate for each position

For material over 3/4" thick a test weld shall be made in material of the maximum thickness, but need not exceed 1".

Welding positions

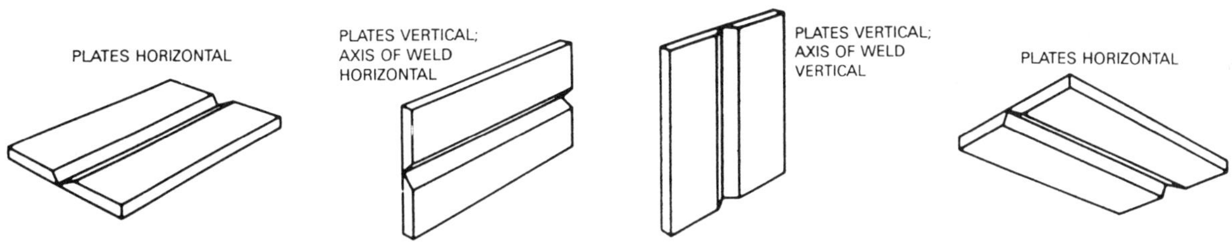

(A) TEST POSITION 1G — Flat

(B) TEST POSITION 2G — Horizontal

(C) TEST POSITION 3G — Vertical

(D) TEST POSITION 4G — Overhead

Reduced Section Tensile

These edges may be flame-cut

Machine weld reinforcement flush with base metal

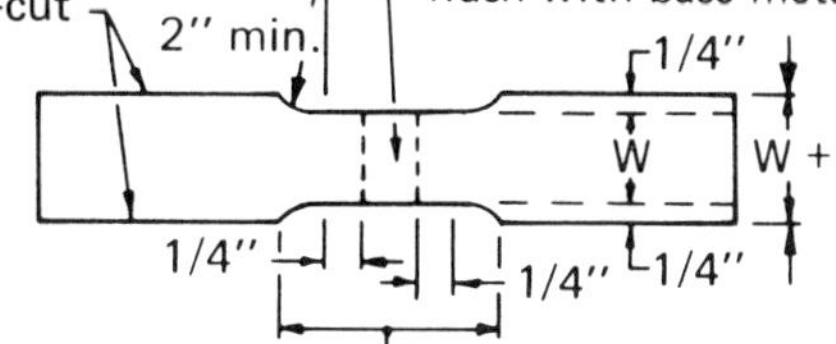

Reduced section machined preferably by milling

Tensile strength shall not be less than the minimum specified tensile strength of the base metal used. (may break in the weld)

$w = 1\text{-}1/2"$ if $t \leqq 1"$

$w = 1"$ if $t > 1"$

Face and Root Bend

These edges may be flame-cut and may or may not be machined

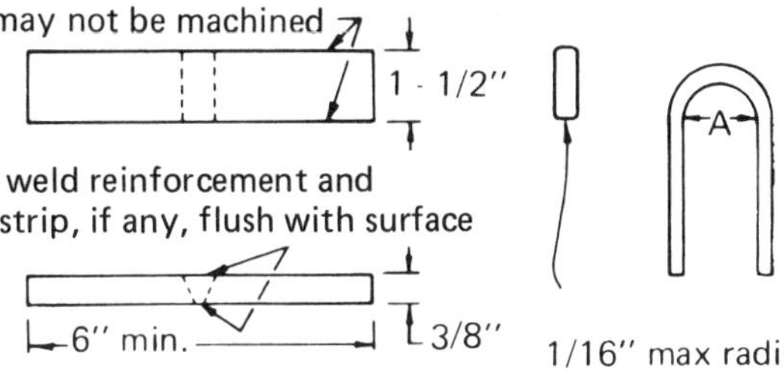

Remove weld reinforcement and backing strip, if any, flush with surface

1/16" max radius on corners

Side Bend

If flame-cut, machine at least 1/8" from edge

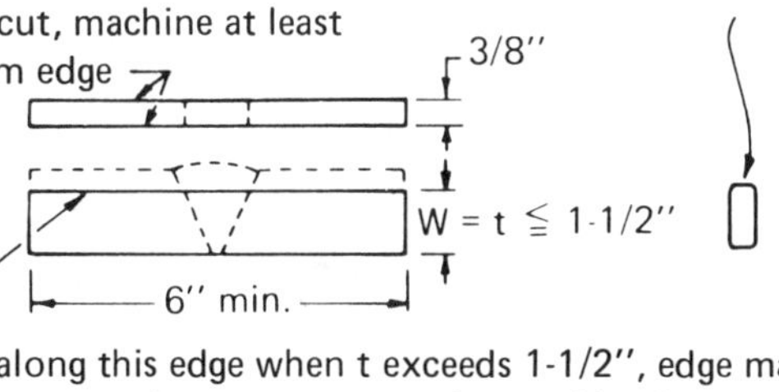

cut along this edge when t exceeds 1-1/2", edge may be flame-cut and may or may not be machined.

YP	A
50 ksi & under	1-1/2"
55 to 90 ksi	2"
over 90 ksi	2-1/2"

NOTE: For plates over 1-1/2" thick, cut specimens into a minimum number of approximately equal strips not exceeding 1-1/2" in width.

Bend specimen passes if crack or other opening does not exceed 1/8" after bending through an angle of 180°. Cracks at corners are not considered. (5.12.2)

Fig. 11-27. Details of the AWS procedure qualification tests. *(Continued from Fig. 11-26.)*

Under D1.5, this is an option only when using SMAW procedures with an approved electrode that deposits weld metal with a minimum specified yield strength of less than 90 ksi. For example, procedures using E7018 can be prequalified under D1.5. Procedures using E11018-M *cannot* be prequalified because the minimum specified yield strength (98 ksi) exceeds the 90 ksi limit. (See 5.1.1, 5.1.2.) Furthermore, procedures for all processes other than SMAW (flux cored, submerged arc, gas metal arc, electroslag and electrogas) must be qualified.

Welding Procedure Approaches — Two approaches may be used for procedure qualification: a Procedure Qualification Test; or a Pretest used in conjunction with a Verification Test. The Procedure Qualification Test described in 5.2.1 is to be performed by the fabricator. The Pretest described in 5.2.2 is identical to the Procedure Qualification Test, but is to be performed by someone other than the fabricator.The somewhat simpler Verification Test is required when the fabricator is using the procedure of the Pretest. Its purpose is to verify that the fabricator can use the Pretest to obtain the desired results.

Test welds for procedure qualification and pretest are made in the position to be qualified in 3/4 in.

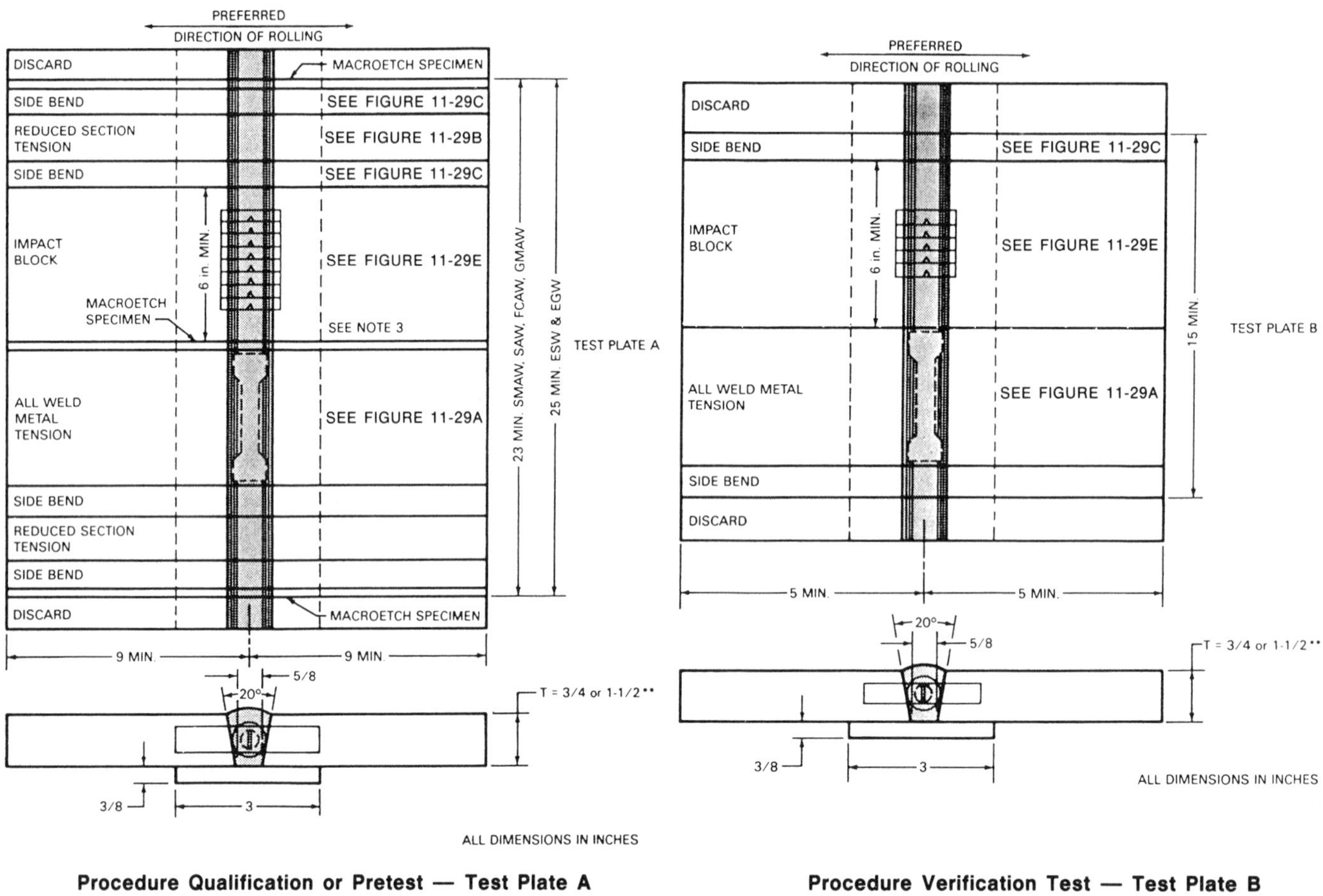

NOTES:

1. Welding and machining shall be witnessed by a state representative or an independent third party acceptable to the state.
2. Test specimens and the Procedure Qualification Record* showing all welding parameters used for the test shall be available to the Engineer.
3. When requested, macroetch specimens 3/8 thick and "T" wide* shall be removed for testing. At least one cut face of each specimen shall be polished and etched for macroscopic examination by the Engineer for test plate A.
4. Testing shall be done by an approved testing laboratory. Such testing need not be witnessed by another agency.

* Test specimens need only be retained for examination by the original approving authority.

** or larger, See 5.3.7 D1.5

Fig. 11-28. Test welds for procedure qualifiction and verification as required for AWS D1.5 Bridge Welding Code.

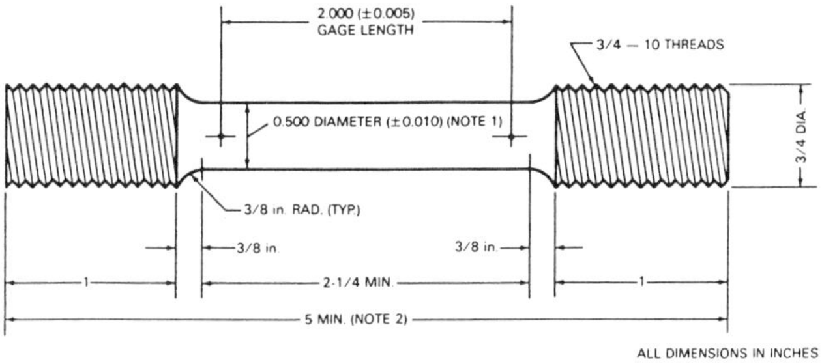

Notes

1. The reduced section may have a gradual taper from the ends toward the center with the ends not more than 0.005 inch larger in diameter than the center
2. Specimen taken from the center of a weld.

A – Standard Round All-Weld-Metal Tension Specimen

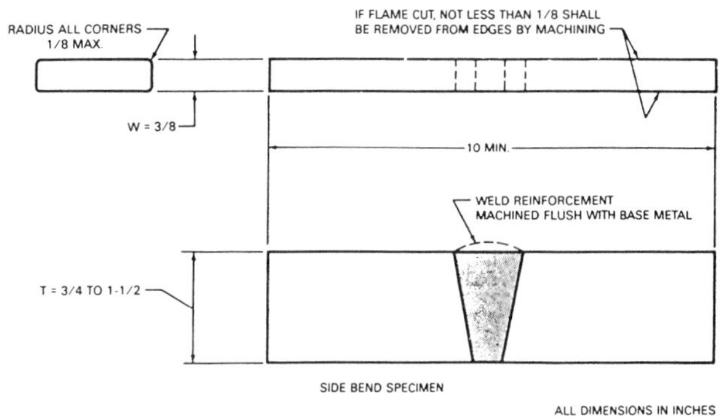

Note: T depends on the thickness of test plate

C – Side Bend Specimen

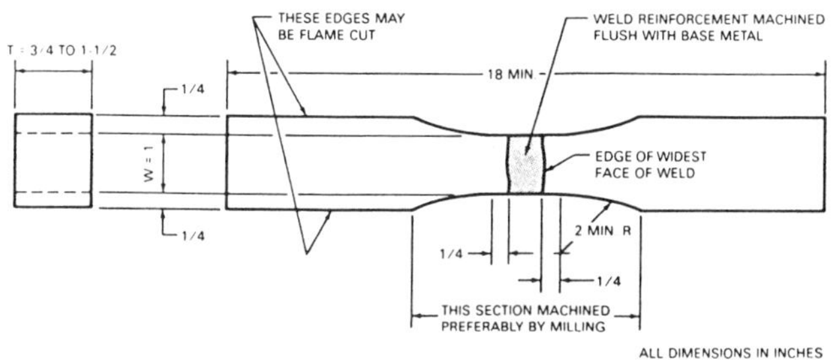

Note: T depends on the thickness of test plate 5.1A or 5.1C

B – Reduced Section Tension Specimen

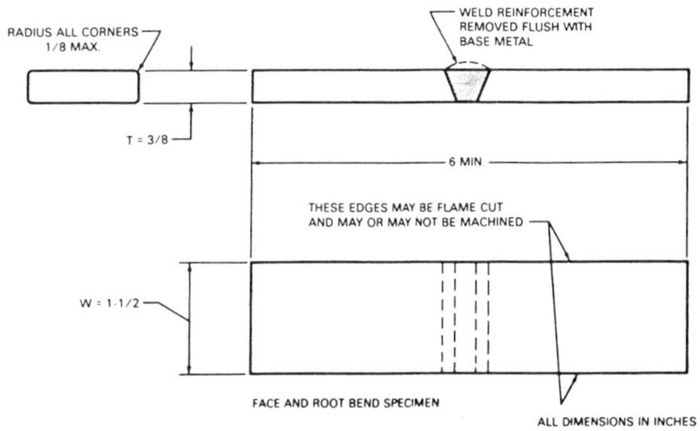

D – Face- and Root-Bend Specimen

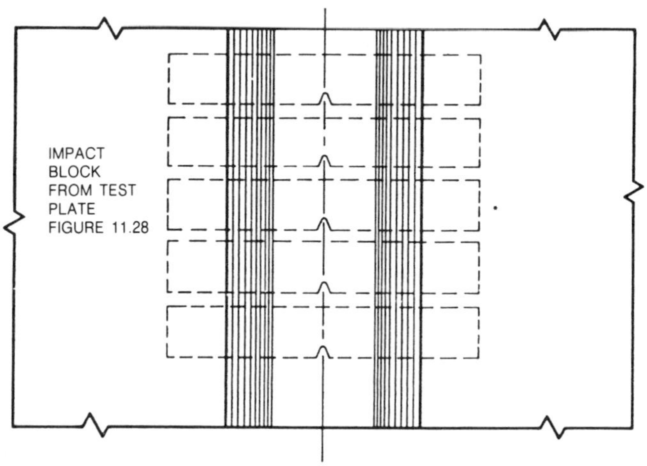

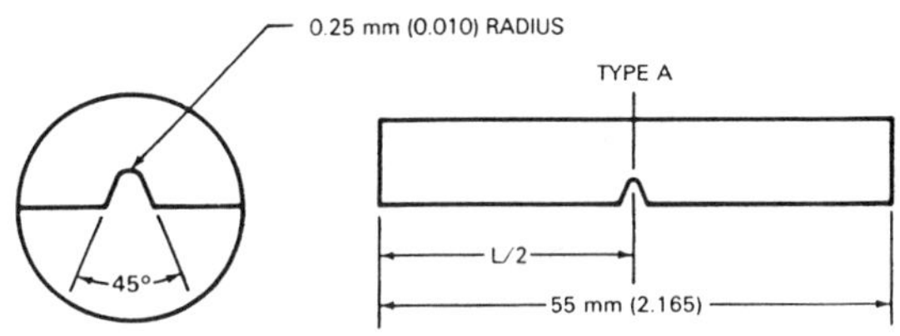

E – Charpy V-Notch Impact Test Specimen

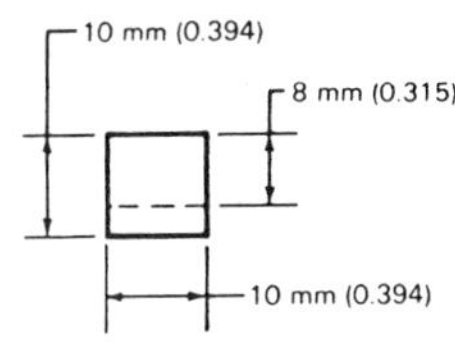

Note — Permissible variations shall be as follows:

Adjacent sides shall be at	90° ± 10 Min.
Cross-section dimensions	±0.025 mm (±0.001 in.)
Length of specimen (L)	±0, −2.5 mm (0.100 in.)
Centering of notch (L/2)	±1 mm (0.039 in.): When an end-centering device is necessary to center the specimen in the anvil, see 11.2.2**, it is necessary that the notch be accurately centered to ensure compliance with 4.3.1**
Angle of notch	±1°
Radius of notch	±0.025 mm (0.001 in.)
Dimensions to bottom of notch: Type A specimen	2 ± 0.025 mm (0.315 ± 0.001 in.)
Finish requirements	2 μm (63 μin.) on notched surface and opposite face, 4 μm (125 μin.) on other two surfaces

*Five test specimens shown, eight required for electroslag and electrogas welds.
**ASTM E23, *Standard Methods for Notched Bar Impact Testing of Metallic Materials.*

Fig. 11-29. Test specimens for procedure qualification tests AWS D1.5.

or 1-1/2 in. plate as shown in Figure 11.28a. The positions of welding are shown in Figure 11.26 and 11.27. The test weld for procedure verification is also made in 3/4 in. or 1-1/2 in. plate as shown in Fig. 11.28b. Details of the specimens to be machined from the test welds are shown in Fig. 11.29.

The following provisions cover the base metal to be used for qualification, pretest, and verification tests.

Test plates used to qualify welding procedures to join steel with a minimum specified yield strength of 50 ksi (345 MPa) or less shall be ASTM A36 (AASHTO M183), A572 Grade 50 (M223), or A588 (M222) steel. A36 (M183) steel qualifies for welding A36 steel. A572 Grade 50 (M223) steel qualifies for welding A36 and A572 Grade 50 steels. A588 (M222) steel qualifies for welding all AASHTO approved steels with a minimum specified yield strength of 50 ksi (345 MPa) or less.

When ASTM A588/AASHTO M222 test plate and backing steel is used to qualify all AASHTO steels having a specified minimum yield strength of 50 ksi (345 MPa), or less, the A588 steel shall have the following chemical composition:

Element	Composition, min, %
Carbon	0.15
Manganese	1.00
Silicon	0.25
Chromium	0.50
Vanadium	0.03

Test plate and backing steel that does not have a chemical composition that conforms to the above limits may be used, provided the steel has equivalent hardenability determined as follows:

(1) The Carbon Equivalent shall be 0.45 minimum as determined by the formula

$$CE = C + \frac{(Mn + Si)}{6} + \frac{(Cr + Mo + V)}{5} + \frac{(Ni + Cu)}{15}$$

or

(2) The hardenability shall be equivalent to steel of specified chemical composition when computed based upon an ideal critical diameter, whether calculated or experimental.

All test plate material should have a minimum Charpy V-notch (CVN) impact toughness of 20 ft-lb (27 J) at 40°F.

When the base metal to be welded has minimum specified yield strength greater than 50 ksi (345 MPa), the base metal used in the qualification test shall be of the same specification and grade as the steel to be used in the work.

Qualification tests for welds in high strength steel [steels with minimum specified yield strength of 70 ksi (480 MPa) or greater] shall measure strength, ductility, toughness, and soundness of the weld metal. When specified in the contract documents, qualification tests for high strength steels shall also measure the Charpy V-notch toughness of the coarse grained area of the heat affected zones. The minimum Charpy V-notch energy, test temperature, orientation of the notch, and other necessary details are to be specified in the contract document when heat-affected zone testing is required.

Maximum Heat Input Tests. Tests to qualify SAW or FCAW welding procedures for filler metals (AWS D1.5 Table 4.1) shall be conducted using Fig. 11-28A, and welded in a manner calculated to produce the highest welding heat input and, therefore, the slowest cooling rate of the weld and base metal heat affected zone. Tests to verify a WPS shall use Fig. 11-28A, or Fig. 11-28B.

Welding conditions shall be as follows:[1]

(1) The number of electrodes shall be as described in the WPS. (Size of the electrode is not an essential variable in this method.)

(2) Current type, polarity, and electrode extension (electrical stickout) shall be as described in the WPS. A change in electrode extension of 3/4 in. (19 mm) or more shall require requalification.

(3) Current shall be WPS maximum.

(4) Voltage shall be WPS maximum.

(5) Travel speed shall be WPS minimum.

(6) Shielding gas flow rate shall be WPS minimum.

(7) Preheat and interpass temperature shall be the maximum expected to be encountered in the work.

Qualifying the maximum heat input permits the fabricator to use procedures which result in heat inputs from 60 to 100 percent of the maximum qualified. The maximum heat input test requires only one test to be performed and opens an acceptable heat input "envelope" of 60 to 100 percent.

To gain greater procedural latitude, the maximum-minimum heat input test may be used with the consumables listed in Table 4.1 D1.5. Under this method, the maximum heat input test portion is conducted as previously described. Next, the minimum heat input test is performed. Again, Plate A is used for procedure qualification tests and pretests, with plate B used for verificiton testing. While this method requires an additional test, it opens the operational "enveolope" to the full range of heat input tested.

Maximum-Minimum Heat Input Tests. Testing to establish the maximum-minimum heat input envelope for approval of SAW, and FCAW, procedures for filler metals (AWS D1.5 Table 4.1) shall be done using Fig. 11-28A, for qualification and Fig. 11-28A or 11-28B for verification. A group of welding procedures that use the same basic welding process, *i.e.,* same electrode type, manufacturer, or classification, same current type and polarity, same electrode flux/shielding-gas combination, same electrode extension, but possibly different electrode diameters and operating variables, may be qualified using the WPS, modified to produce both the maximum and minimum heat input as follows:

Conduct the maximum heat input test as described above.

Conduct the minimum heat input test. Welding variables shall be selected to produce the lowest heat input and therefore, the fastest cooling rates in the weld and heat affected zone as follows:

(1) Number of electrodes shall be as described in the WPS. (Size of electrodes is not an essential variable in this method.)

(2) Current type, polarity, and electrode extension shall be as described in the WPS. A change in electrode extension of 3/4 in. (19 mm) or more shall require requalification.

(3) Current shall be WPS minimum.

(4) Voltage shall be WPS minimum.

(5) The travel speed shall be WPS maximum.

(6) Shielding gas shall be WPS maximum.

(7) Preheat and interpass temperature shall be the minimum permitted by the Code for the thickness of the test plate used. The test plate may be cooled artificially between weld passes. See Footnote 1.

[1] Care should be taken to prevent the build-up of interpass temperature beyond the maximum expected in the work since this may adversely effect the weld metal toughness measured in the test.

The interpass temperature may be controlled by waiting for the test plate to cool between weld passes or by directing a jet of compressed air against the back of the test assembly between passes as necessary.

Production Procedure Test. The Production Procedure Test applies to the consumables listed in Table 4.2 of D1.5 (self-shielded flux cored, gas metal arc, electroslag and electrogas) or as an option to the filler materials listed in Table 4.1 of D1.5 (shielded metal arc, submerged arc and gas shielded flux cored). For Table 4.2 consumables, qualification must be made to Section 5.7 of the D1.5 code and for Table 4.1 consumables, qualification must be made to Section 5.6 of the D1.5 code. The "Limitations of Variables" pertinent to procedures qualified in this manner is described in Section 5.10. While this approach requires only one test, it may yield less production flexibility than the previous two methods.

PROCEDURE QUALIFICATION — AWS D10.9 SPECIFICATION FOR QUALIFICATION OF WELDING PROCEDURES AND WELDERS FOR PIPING AND TUBING

AWS D10.9 "Specification for qualification of Welding Procedures and Welders for Piping and Tubing" covers the qualification of procedures and personnel for the welding of pipe, tubing, and associated components. The standard provides two levels of qualification, termed Acceptance Requirements 1 and 3 — abbreviated AR-1 and AR-3, respectively. AR-1 is the highest level of qualification, followed by AR-3. The differences lie in the type of tests and the acceptance requirements for the sample welds.

Each of the two levels of welding procedure qualification has a corresponding level of weldor qualification. Qualification to the higher level automatically qualifies for the lower level. For example, qualification to AR-1 also qualifies for AR-3.

The relationships between welding procedure qualification and weldor qualification are shown below:

Level Required	Level of Procedure Qualification	Level of Weldor Qualification
AR-1	AR-1	AR-1
AR-3	AR-3 or AR-1	AR-3 or AR-1

Typical applications for the two levels are as follows:

AR-1 — This level applies to systems where the highest degree of weld quality is required. Use of the AR-1 level of quality is intended to provide the confidence required for pipelines that may be found in nuclear-energy, space, high-pressure, high-temperature, chemical, or gas systems.

Some contract specifications may have requirements even more stringent than those required by AR-1. In such cases the qualification must comply with AR-1, plus the additional tests, inspection methods, and acceptance standards required by the contract specifications.

AR-3 — This level applies to systems where a nominal degree of weld quality is required. Use of the AR-3 level of quality is intended to provide the confidence adequate for lines such as low-pressure heating, air-conditioning, and sanitary water.

The standard provides for qualification with groove and fillet welds. Procedure and weldor qualification with groove welds automatically qualifies for fillet welds. However, if fillet welds alone are required for the specific job, the welding procedure and weldor can be qualified for fillet welds only.

AR-1 Welding Procedure Qualification

Table 11-4 shows the welds and tests required for AR-1 procedure qualification. The test positions are shown in Fig. 11-30.

The notes (a) through (k) for Table 11-4 are important to use of the table. They are too voluminous for inclusion in the table, itself, and, in lieu of this customary arrangement, are presented here in the text. The applicable notes to the table state that:

(a) The pipe sizes in the table may not be appropriate for the job-size pipe to be welded. In such instances, actual job-size pipe may be substituted.

(b) For pipe having a wall thickness less than 0.063 in., individual qualification is required for each thickness by visual, penetrant, X-ray, and tensile tests.

(c) The maximum thickness qualified using the oxyacetylene process is the wall thickness of the base metal of the sample weld plus 20 percent. (This note is not applicable to arc welding.)

(d) Procedures qualified prior to issuance of AWS D10.9 may be updated by performing any additional tests required by the standard.

(e) For pipe used at low temperature, impact tests may be required. The purchaser and fabricator should agree in advance on the type and number of test specimens, areas from which the specimens are to be taken, testing temperature, and the acceptance requirements.

(f) Visual inspection before welding shall determine that the proper base metal is used and the fitup is proper.

Visual inspection during welding shall include determining that the proper filler metal is used and inspection for cracks, undercut, porosity, incomplete fusion, and proper cleaning between passes.

TABLE 11-4. AR-1 Level of Welding Procedure Qualification

SAMPLE WELDS					Tests Required						
Wall Thickness 3/4″ and Under											
Pipe Size (a) of Sample Weld	Pipe or Tube Size Qualified	Pipe or (b)(c) Tube Wall Thickness Qualified Min.	Max.	Number of Sample Welds per Position	Visual (f) Inspection	Penetrant (g) Inspection Optional	Radio-graphy (h)	Tensile (i) Test Number of Specimens	Bend Tests (j)(k) Number of Specimens Face	Root	Side
1/2″ Sch. 40	Through 1-1/2″	.063″	.400″	1	Yes	Yes	Yes	1	0	0	0
2″ Sch. 80	1″ Through 4″	.063″	.674″	1	Yes	Yes	Yes	1	1	1	0
6″ Sch. 40	Over 4″	.109″	.750″	1	Yes	Yes	Yes	2	2	2	0
Wall Thickness Over 3/4″											
8″ Sch. 120	6″ and Over	.315″	Any	1	Yes	Yes	Yes	2	0	0	4

NOTES: See Text ("AR-1 Welding Procedure Qualification") for explanation of notes (a) through (k).

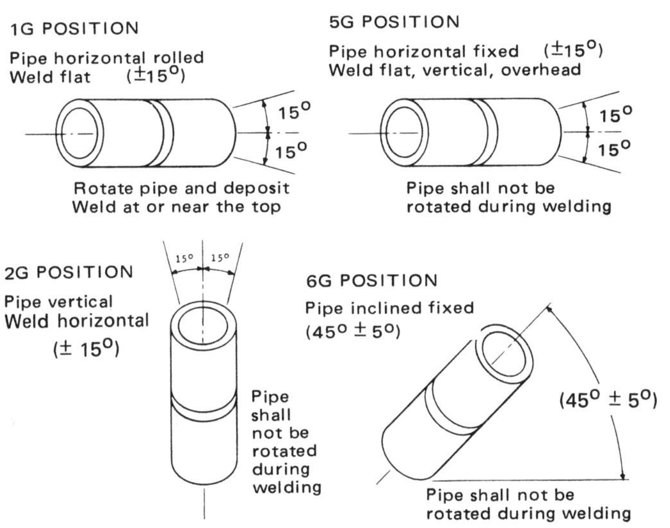

The limits of qualification in a given position are shown below:

1. Qualification in the 1G position qualifies only for this position.
2. Qualification in the 2G position qualifies for welding in the 1G and 2G positions only.
3. Qualification in the 5G position qualifies for welding in the 1G and 5G positions only.
4. Qualification in both the 2G and 5G positions qualifies for welding in all positions.
5. Qualification in the 6G position qualifies for welding in all positions, because it includes elements of those positions.
6. Qualification in a position other than the four standard positions described above is valid only for that position (plus or minus 15 degrees).

Fig. 11-30. Test positions used in qualifying procedures for welding pipe and tubing under AWS D10.9.

Visual inspection after welding and heat treatment, if any, shall determine that the weld is free from cracks, surface porosity, and unfilled craters, and that the weld face is at least flush with the outside surface of the pipe. The weld reinforcement shall not exceed the following:

Pipe Wall Thickness, in.	Reinforcement max., in.
3/8 and less	3/32
Over 3/8 to 3/4 incl.	1/8
Over 3/4	3/16

The root of the weld shall show no evidence of cracks or incomplete fusion. A concave root surface is permitted providing the weld throat is equal to or greater than the base metal and the concavity or convexity of the root surface does not exceed the following:

Sample Weld Pipe Size	Maximum Root-Surface Concavity, in.	Maximum Root-Surface Convexity, in.
1/2″ Sch 40	1/32	1/16
2″ Sch. 80	1/16	1/16
5″ Sch. 80	1/16	3/32
8″ Sch. 120	1/16	1/8

(g) Penetrant inspection is required only if this type of examination is used in production. Each sample weld shall be examined by liquid penetrant in accordance with A-2 or B-3 of ASTM E165. The weld shall be inspected at the completion of welding, including any postheating. Dry-powder magnetic-particle inspection in accordance with ASTM E109 may be used in lieu of liquid-penetrant inspection.

(h) Radiographic inspection is required for all welds of the AR-1 level. For specific techniques and acceptance requirements, see AWS D10.9.

(i) For 1/2, 3/4, 1 and 1-1/2-in. pipe and all thin-wall pipe, the entire joint is tested in tension. Remove the weld reinforcement until it is flush with the base metal. Measure the outside diameter and the wall thickness before testing, and calculate the pipe cross section. Test the pipe in tension using a fixture as shown in Fig. 11-31.

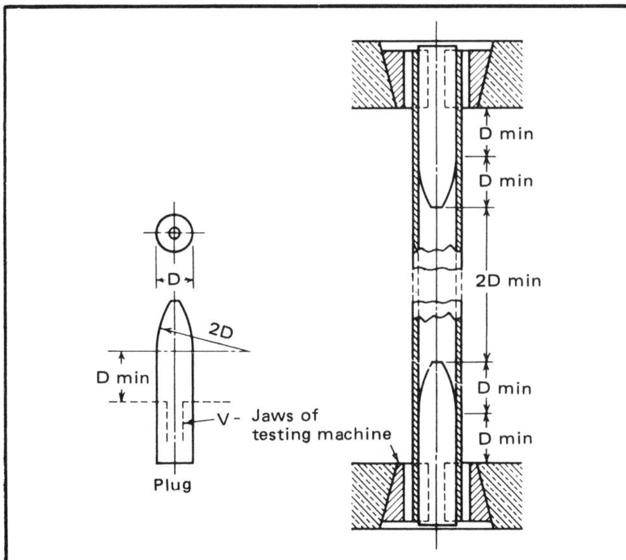

Fig. 11-31. Fixture used for the tensile testing of small-diameter and thin-wall pipe AWS D10.

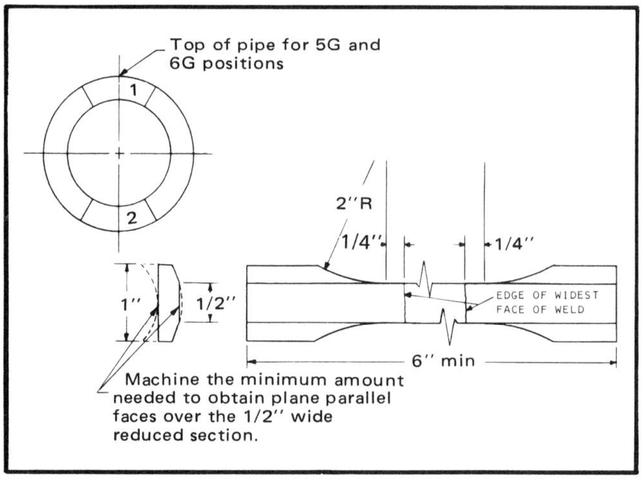

Fig. 11-32. Method of preparing tensile specimens from 2-in. pipe for procedure qualification tests AWS D10.9.

From the 2-in. sample welds, two tensile specimens are cut from one sample weld and prepared as in Fig. 11-32.

From the 5- and 8-in. sample welds, tensile specimens are prepared as in Fig. 11-33.

The tensile test specimens may fail in the base metal, heat-affected zone, or in the weld metal, and are acceptable if the tensile strength is equal to or greater than the minimum specified for the plate.

(j) From 2-in. sample welds, four bend specimens are cut from one sample weld and are prepared as in Fig. 11-34. From 5- and 8-in. sample welds, bend specimens are cut and prepared as in Fig. 11-33. Side bends may be used for wall thicknesses over 3/8 in.

Bend specimens are tested in a guided-bend jig or a wrap-around jig, as shown in Fig. 11-35.

When using the guided-bend jig, the specimen shall be placed on the die with the weld at mid-span. Face-bend specimens shall be placed with the face of the weld directed toward the gap; root-bend specimens shall be placed with the root of the weld directed toward the gap, and side-bend specimens with the surface having the greatest discontinuities, if any, toward the gap.

The plunger shall force the specimen into the die until the specimen becomes U-shaped and a 1/8-in.-diameter wire cannot be inserted between the die and the specimen. The weld and heat-affected zones shall be centered and completely within the bent portion of the specimen after testing.

When using the wrap-around jig, the specimen shall be firmly clamped on one end so that there is no sliding of the specimen during the bending operation. The weld and heat-affected zones shall be completely in the bent portion of the specimen after testing. Test specimens shall be removed from the jig when the outer roll has been moved 180° from the starting point.

In order to pass the test, the bent specimens shall have no cracks or other open discontinuities exceeding 1/8-in. measured in any direction on the convex surface of the specimen, except that cracks occurring at the corners shall not be considered unless there is definite evidence that they result from slag inclusions or other internal discontinuities.

(k) For more ductile materials, a higher sensitivity to bending may be required. Any changes in the bend test should be agreed upon in advance between the purchaser and the fabricator.

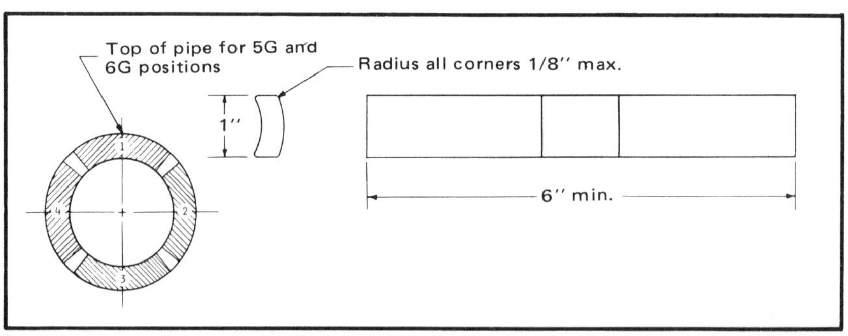

Fig. 11-34. Method of preparing bend specimens from 2-in. pipe for procedure qualification tests AWS D10.9.

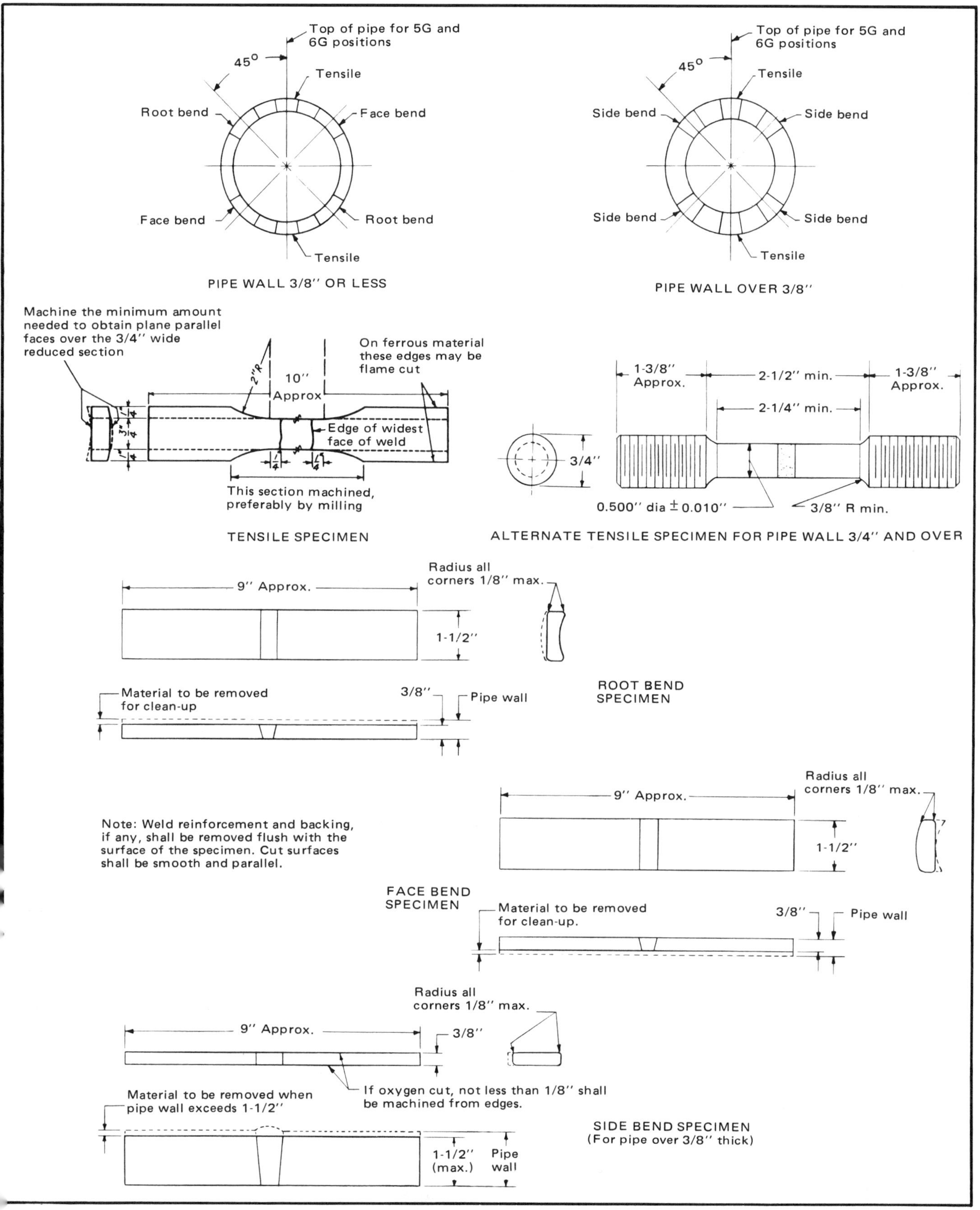

Fig. 11-33. Method of preparing test specimens from 5 and 8-in. pipe for procedure qualification tests AWS D10.9.

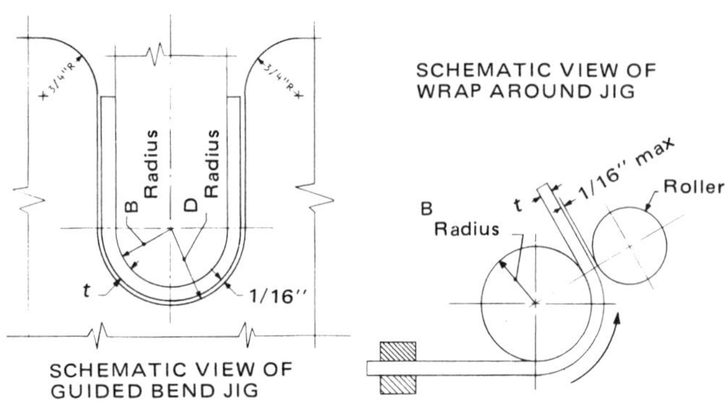

Sample Thickness and Jig Dimensions			
Material	Sample Thickness, in.	Radius of Bend, "B," in.	Radius of Die, "D," in.
Steel	t 0.218 0.375	3-1/3 t 3/4 1-1/4	4-1/3 t + 1/16 1-1/32 1-11/16

Fig. 11-35. Jigs for the bend tests in procedure qualification. For metals other than steel, see AWS D10.9, Table 4.

TABLE 11-5. AR-3 Level of Welding Procedure Qualification
LIMITED TO PIPE UP TO AND INCLUDING 20″ DIAMETER AND 3/8″ WALL THICKNESS

SAMPLE WELDS					TESTS REQUIRED		
Wall Thickness 3/8″ and Under							
Pipe Size of Sample Weld	Pipe or Tube Size Qualified	Pipe or Tube (b) Wall Thickness, (t) Qualified		Number of Sample Welds per Position	Visual (c) Inspection	Pressure (d) Bend	Macrostructure (e) No. of Specimens
		Min.	Max.			No. or Tests	
1/2, 3/4, 1, 1-1/4, 1-1/2″ Sch. 40	1-1/2″ and Under	0.005	0.154	1	Yes	1	0
2, 4, 6, 8, 10 Sch. 40 (a)	2″ through 20″	0.154	0.375	1	Yes	0	4

NOTES: See Text ("AR-3 Welding Procedure Qualification") for explanation of (a) through (e) AWS D10.9.

AR-3 Welding Procedure Qualification

The qualifying positions are as shown in Fig. 11-30. The sample welds and tests required are given in Table 11-5.

The applicable notes are:

(a) The same as note (a) under AR-1 procedure qualification.

(b) The same as Note (c) under AR-1 procedure qualification.

(c) The same as Note (f) under AR-1 procedure qualification, except that an undercut not exceeding 1/32-in. is permissible and the concavity or convexity of the root surface may not exceed the following:

Sample Weld Pipe Size	Maximum Root-Surface Concavity, in.	Maximum Root-Surface Convexity, in.
1/2″ Sch. 40	1/32	1/16
2″ Sch. 40	1/16	1/8
5″ Sch. 40	1/16	1/8

(d) Prepare one sample weld for each position to be qualified for the 1/2-in. Schedule 40 and 2-in. Schedule 40 pipe as shown in Fig. 11-36. Fill with water at city pressure (25 psig minimum) and close the valve. Bend around a 12-in.-diameter mandrel as shown. The weld is acceptable if there are no leaks after a 30° bend.

(e) Section, polish, and etch the 2-in. and over Schedule 40 specimen as shown in Fig. 11-37. The etched sections shall show no cracks, inadequate penetration, or incomplete fusion. Concavity shall not exceed 1/16 in., and the accumulated slag and porosity shall not exceed 1/8 in. from root to face of the weld.

In lieu of this test, the 2-in. and over weld sample may be tested by preparing four bend samples as shown in Fig. 11-34, except the sample width shall be 1-1/2 in. wide. Bend two face and two root

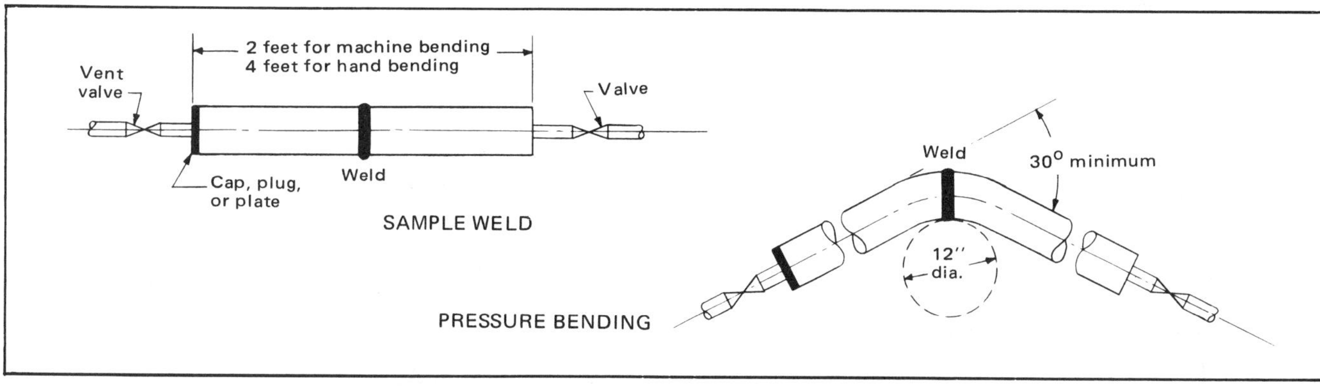

Fig. 11-36. The pressure bend test used in AR-3 procedure qualification AWS D10.9.

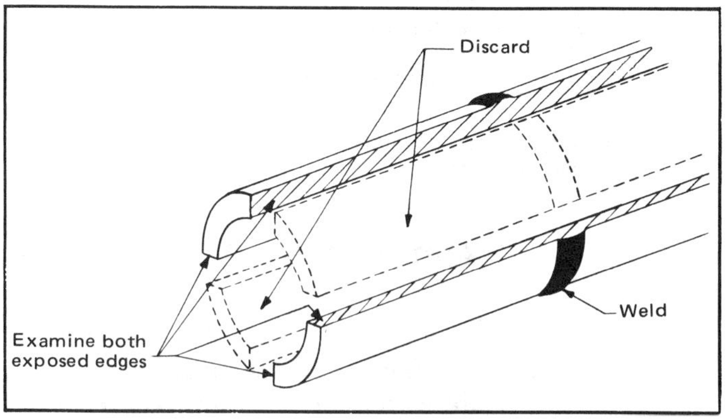

Fig. 11-37. Specimen for macrostructure examination as permitted in AR-3 procedure qualification AWS D10.9.

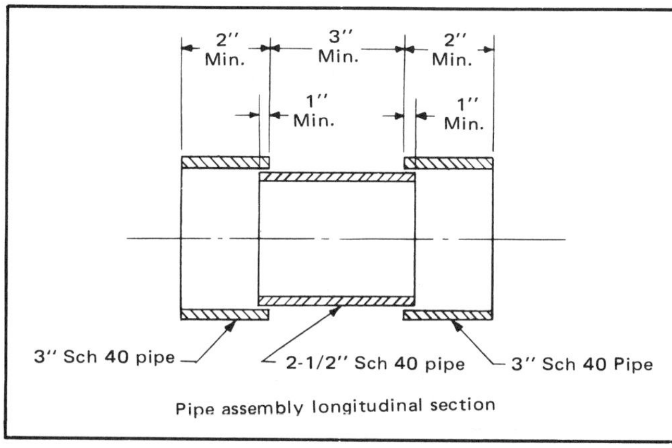

Fig. 11-38. The pipe assembly used for qualifying procedures and weldors for fillet welds AWS D10.9.

specimens with a fixture shown in Fig. 11-35, with "t" equal to 0.375 in. The weld is acceptable if none of the specimens undergo complete failure.

Procedure and Weldor Qualification for Fillet Welds

As mentioned earlier, welding procedure qualification and weldor qualification using groove welds automatically qualifies the procedure or the weldor for fillet welds. however, if only fillet welds are required in production, the welding procedure and the weldor can be qualified for fillet welds only. The sample welds, tests, and acceptance requirements are the same for both procedure and weldor qualification.

A pipe assembly is prepared as shown in Fig. 11-38. Test welds are made as shown in Fig. 11-39 to qualify for all positions. Weld No. 1 is completed with the axis of the assembly in the horizontal position and without rotation. Weld No. 2 is completed with axis of the assembly in the vertical position and without rotation.

Each sample joint must be inspected prior to welding to assure the material is correct, the parts are clean, and the end preparation and fitup are correct. During welding, inspection is made to assure the proper filler metal is used and for cracks, lack of fusion, surface roughness, and poor crater conditions. After welding, and heat treating if any, the weld is inspected for surface discontinuities.

This test may be terminated at any stage of the testing whenever it becomes apparent to the supervisor conducting the test that (1) the weldor is not following the welding procedure or (2) the weldor does not have the necessary skill to produce satisfactory results.

The test assembly is then cut into eight 3/4-in.-wide specimens, as shown in Fig. 11-40. Each specimen must have an acceptable or desirable profile, as illustrated in Fig. 11-41.

Each specimen is fractured in a vise, as shown in Fig. 11-40. The fractured surface must show no evidence of cracks, incomplete fusion, or inadequate penetration at the root of the weld.

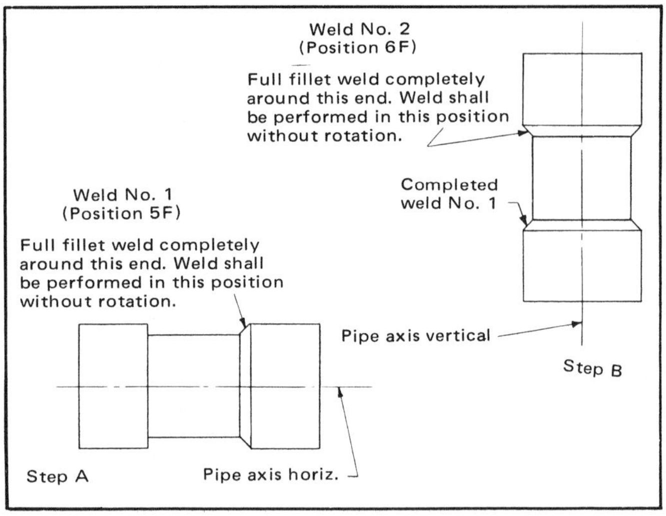

Fig. 11-39. Positions for the fillet-weld qualification tests AWS D10.9.

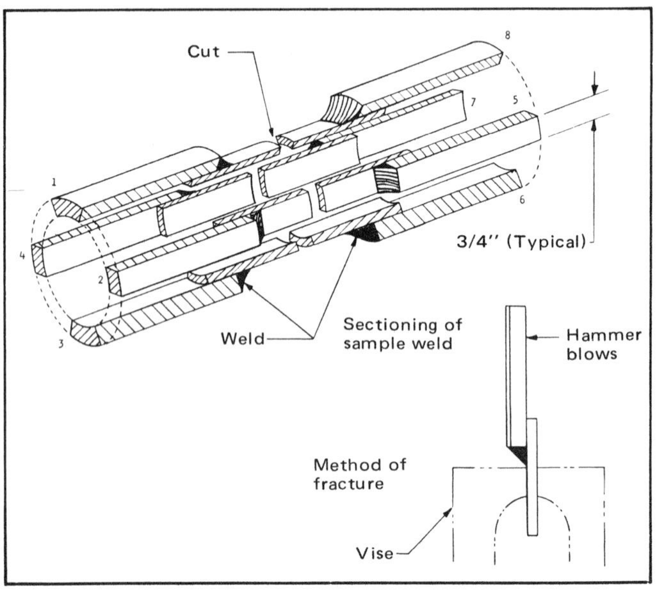

Fig. 11-40. The sampling technique and vise-fracturing method for examining fillet welds AWS D10.9.

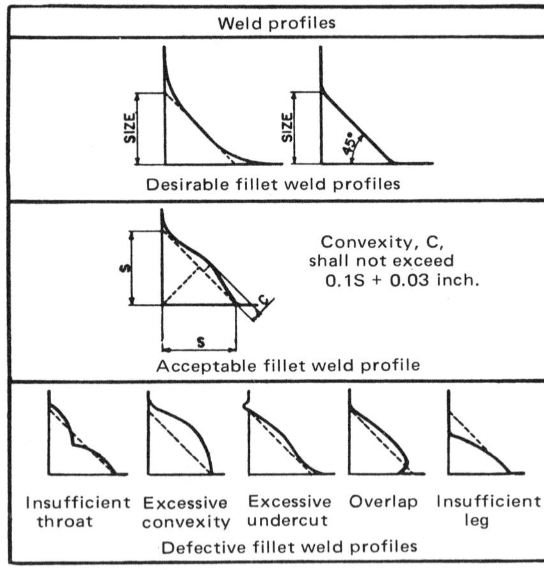

Fig. 11-41. The profile standards that sectioned fillet welds must meet in qualification testing AWS D10.9.

AWS PREQUALIFIED JOINTS

In the discussion of procedure qualification, mention was made of prequalified joints (AWS D1.1 and AWS D1.5) and procedures. These are the joints and procedures recognized and approved by AWS specifications as applicable to the welding of structural members in buildings and bridges. Figures 11-42, 43 and 44 illustrate prequalified joints, and Fig. 11-45 the procedures for welding such joints.

When the work involves these joints and welding procedures, the AWS Structural Welding Code does not require procedure qualification tests.

ASME BOILER AND PRESSURE VESSEL CODE QUALIFICATIONS

Section IX of the Boiler and Pressure Vessel Code of the American Society of Mechanical Engineers provides for qualification tests of procedures and personnel for welding under that code. The following is a condensed and incomplete description of these qualifications. The condensation does not cover qualification tests for steels of 85,000 psi or more and for quenched and tempered steels. Before making tests, a copy of the complete code should be obtained from the American Society of Mechanical Engineers, Order Department, United Engineering Center, 345 East 47th Street, New York, NY 10017.

Welding Procedure Qualification, ASME Code

Each manufacturer or contractor shall list the parameters applicable to welding that he performs in construction of weldments built in accordance with this Code. These parameters shall be listed in a document known as a Welding Procedure Specification (WPS).

Each manufacturer or contractor shall qualify the WPS by the welding of test coupons and the testing of specimens (as required in this Code), and the recording of the welding data and test results in a document known as a Procedure Qualification Record (PQR). The weldors or welding operators used to produce weldments to be tested for qualification of procedures shall be under the full supervision and control of the manufacturer or

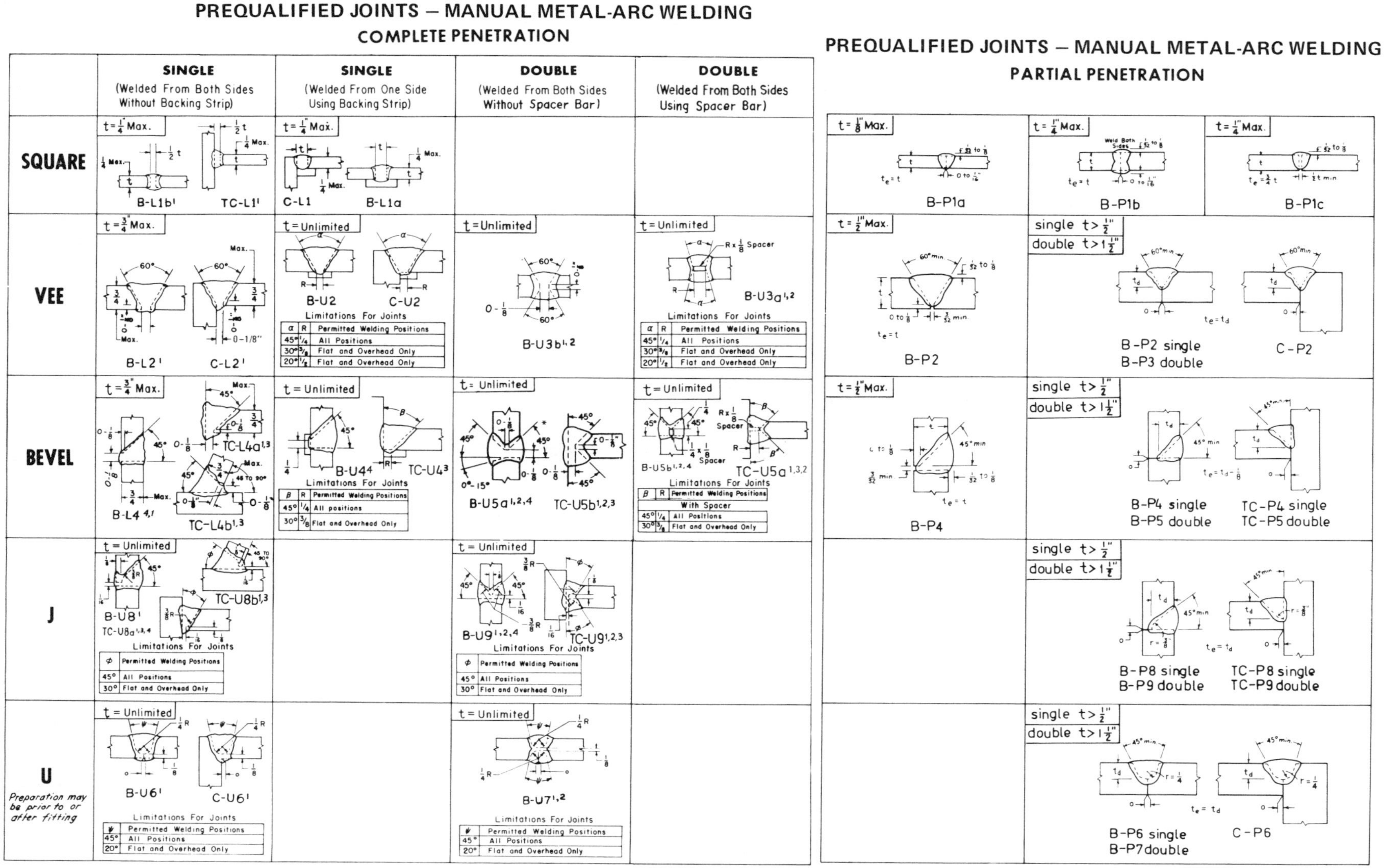

NOTES:
If fillet welds are used to reinforce groove welds in Tee and corner joints, they shall be equal to t/4 but shall be 3/8″ max. (Section 9 – D1.1 requires this reinforcement.)

1. Gouge root before welding second side.
2. Use of this weld preferably limited to base metal thickness 5/8″ or larger.
3. Angle of plate intersection for these joints may range from 90 to 45°. See example TC-L4b.
4. Section 9 – D1.1 in horizontal position only.

* When lower plate is bevelled, first weld root pass this side.

NOTES:
- No back gouging is required for partial penetration welds.
- Minimum effective throat $t_e = \sqrt{t/6}$
- Preparation of second side for double groove joints should be the same as the first side.
- Section 9 – D1.1 allows partial penetration groove welds only in corner joints.

Fig. 11-42. AWS prequalified joints for the manual metal-arc welding process AWS D1.1.

PREQUALIFIED JOINTS – SUBMERGED-ARC WELDING COMPLETE PENETRATION

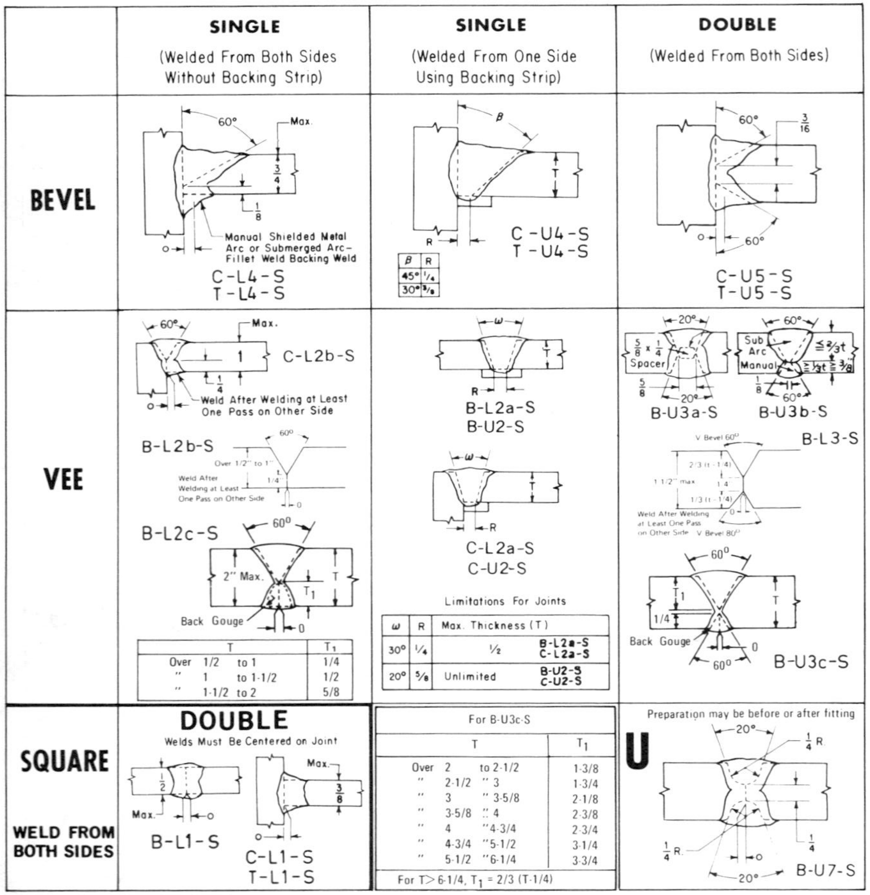

NOTES: Manual weld with low-hydrogen electrode first. Back gouge root than submerged arc weld.
Angle of plate intersection for the joint may be from 90 to 45°
If fillet welds are used to reinforce groove welds in tee and corner joints, they shall be equal to T/4 but not more than 3/8". (Section 9 requires this reinforcement.)

PREQUALIFIED JOINTS – SUBMERGED-ARC WELDING PARTIAL PENETRATION

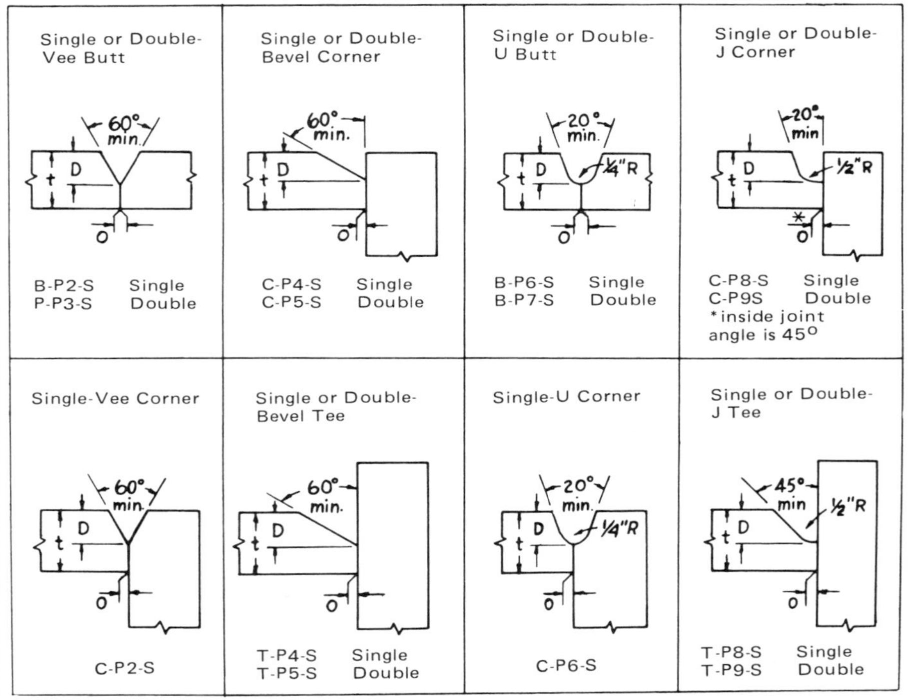

NOTES:
- Welded in the flat position.
- If root face is less than 1/4", there should be at least one manual bead to prevent melt-through.
- Minimum effective throat = $\sqrt{t/6}$ where t is thickness of thinner part.
- Effective throat = D
- Plate thickness single groove joint t ≥ 3/4", double groove joint t ≥ 1-1/4"
- For U groove joints, preparation may be before or after fitting.
- Bevel angle for second side should be same as first side except for C P9 S as indicated.
- Section 9 allows partial penetration groove welds only in corner joints.

Fig. 11-43. AWS prequalified joints for the submerged-arc welding process AWS D1.1.

PREQUALIFIED JOINTS
GAS METAL-ARC AND FLUX-CORED ARC WELDING

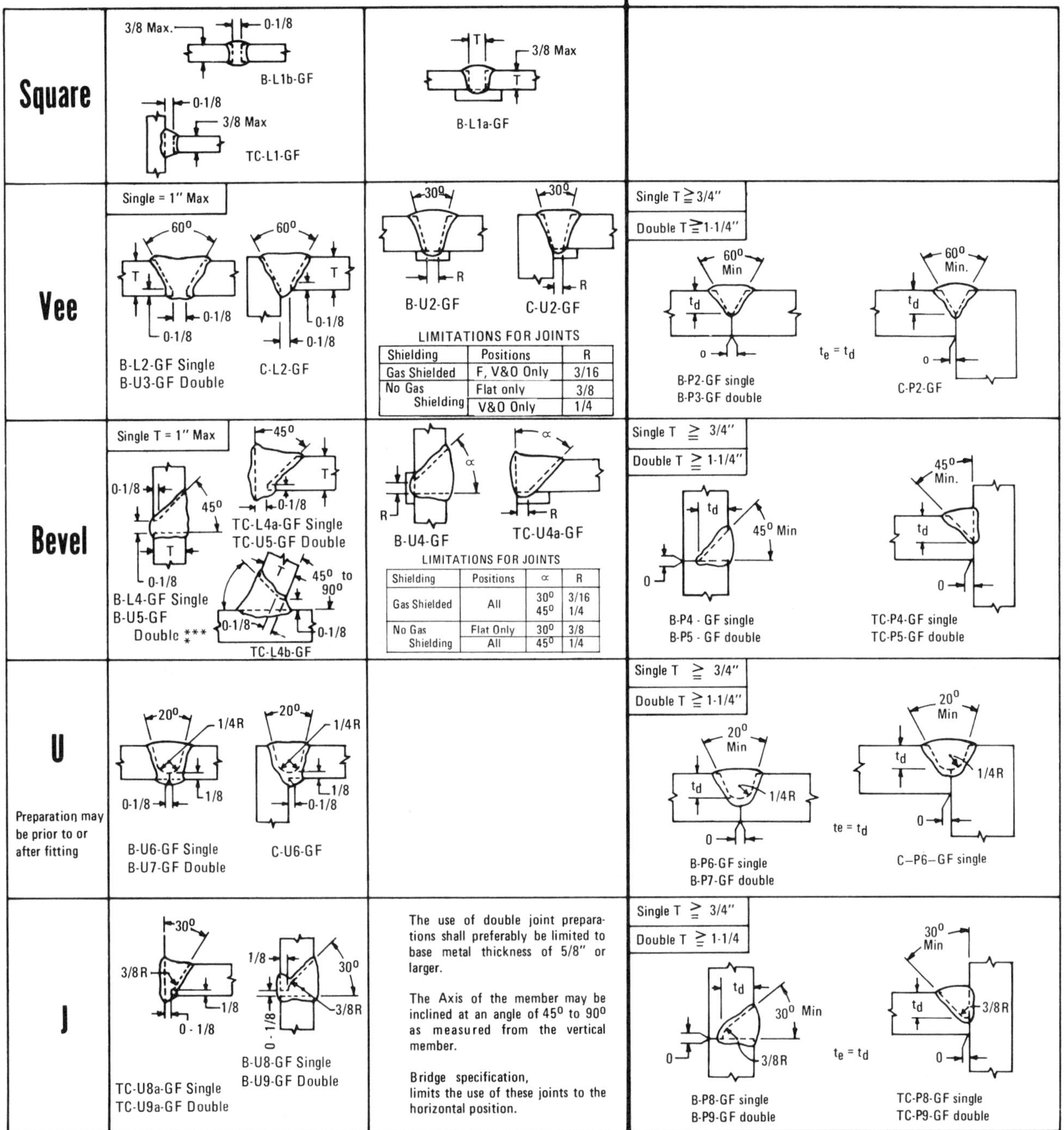

LIMITATIONS FOR JOINTS (B-U2-GF, C-U2-GF)

Shielding	Positions	R
Gas Shielded	F, V&O Only	3/16
No Gas Shielding	Flat only	3/8
	V&O Only	1/4

LIMITATIONS FOR JOINTS (B-U4-GF, TC-U4a-GF)

Shielding	Positions	∝	R
Gas Shielded	All	30° 45°	3/16 1/4
No Gas Shielding	Flat Only	30°	3/8
	All	45°	1/4

NOTES:

1. Gouge roots of joints without backing before welding other side.
2. See 3.5 for allowable variation of dimensions and 3.3.4 for workmanship tolerances.
3. If fillet welds are used to reinforce groove welds in tee and corner joints, they shall be equal to T/4 but not more than 3/8 in.; groove welds in tee and corner joint for bridges shall be reinforced with fillet welds equal to T/4 but not more than 3/8 in.

NOTES:

Effective throat = D for welds made in flat or horizontal position or D–1/8" for welds made in vertical or overhead position.

Minimum effective throat = $\sqrt{T/6}$

AWS spec allows partial penetration groove welds only in corner joints

Fig. 11-44. AWS prequalified joints for the gas metal-arc and flux-cored arc welding processes AWS D1.1.

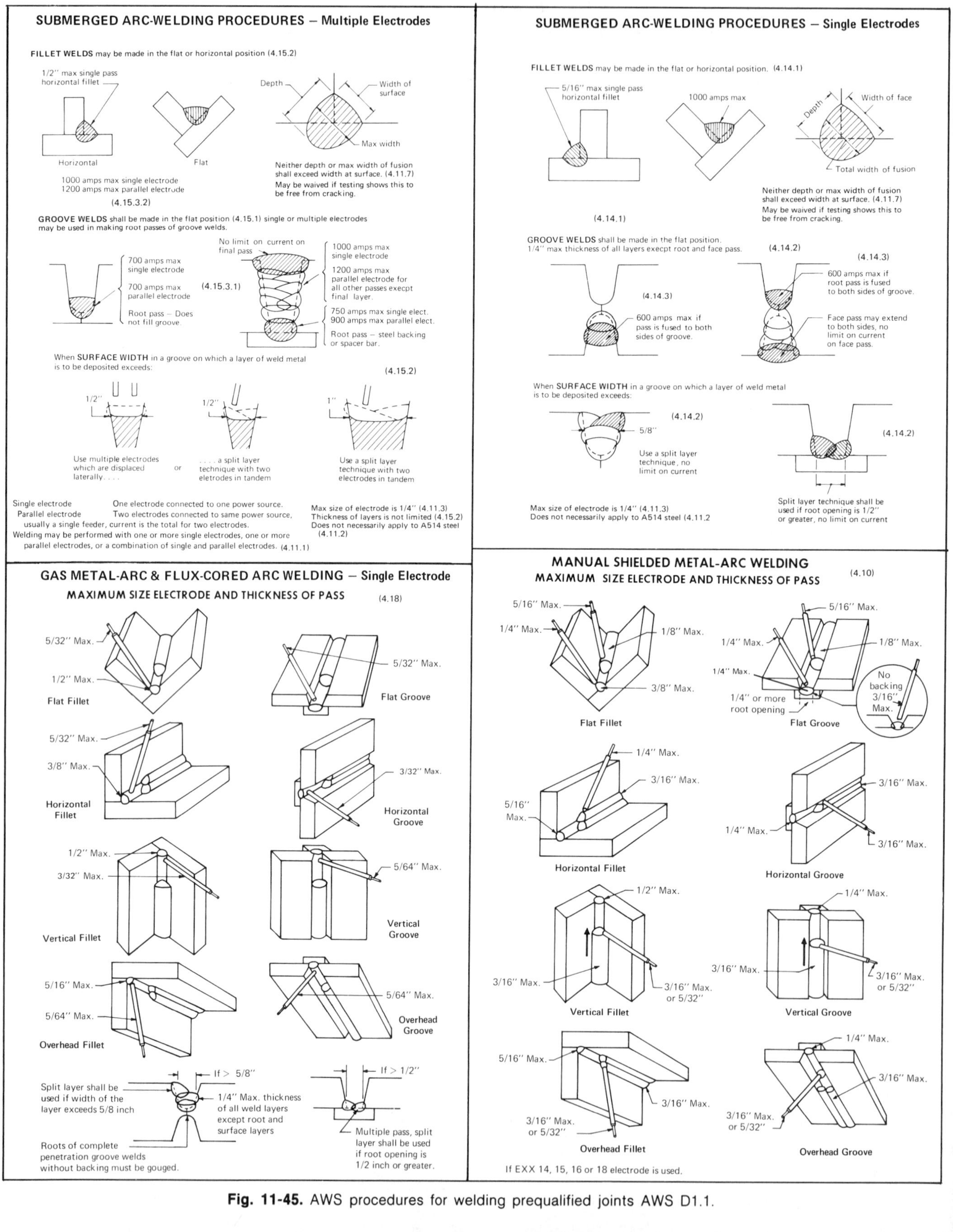

Fig. 11-45. AWS procedures for welding prequalified joints AWS D1.1.

contractor during the production of these test weldments. It is not permissible for the manufacturer or contractor to have the welding of the test weldments performed by another organization. It is permissible, however, to subcontract any or all of the work of preparation of test metal for welding and subsequent work on preparation of test specimens from the completed weldment, performance of nondestructive examination, and mechanical tests, provided the manufacturer or contractor accepts the responsibility for any such work.

Unless specifically required otherwise by the welding variables, a qualification in any position qualifies the procedure for all positions. The welding process must be compatible; and the welding rods, electrodes, and filler metals must be suitable for their use in specific positions. A weldor or welding operator making and passing the WPS qualification test is thereby qualified for the position tested.

To reduce the number of welding procedure qualifications required, ASME established a system where base metals have been assigned P-Numbers, and for ferrous base metals which have specified impact test requirements, Group Numbers within P-Numbers. P-Numbers 1 through 11 are assigned to ferrous base alloys. These assignments are based essentially on comparable base metal characteristics, such as composition, weldability, and mechanical properties, where this can logically be done. These assignments do not imply that base metals may be indiscriminately substituted for a base metal which was used in the qualification test without consideration of compatibility from the standpoint of metallurgical properties, post-weld heat treatment, design, mechanical properties, and service requirements. Where notch toughness is a consideration, it is presupposed that the base metals meet the specific requirements.

For the same reason, F-Number grouping of electrodes and welding rods was established based essentially on their usability characteristics, which fundamentally determine the ability of weldors to make satisfactory welds with a given filler metal. F-Numbers 1 through 6 apply to steels and iron base alloys. The grouping does not imply that base metals or filler metals within a group may be indiscriminately substituted for a metal which was used in the qualification test without consideration of the compatibility of the base and filler metals from the standpoint of metallurgical properties, postweld heat treatment design and service requirements, and mechanical properties.

The base metals and filler metals shall be one or more of those listed in the WPS. The dimensions of the test assembly shall be sufficient to provide the required test specimens.

The base metals may consist of either plate, pipe, or other product forms. Qualification in plate also qualifies for pipe welding and vice versa.

Preparation of the Test Joint. The base material, filler metal, and the welding procedure for the test joint must comply with the job specification. The recommended pipe size is 5 in. in diameter and 3/8 in. in the wall thickness, although larger pipe may be used. Smaller pipe size (job size) may be used, but in such cases the procedure must be qualified for thicknesses between 1/2 and two times the wall thickness of the test pipe, but not over 3/4 in.

The type and number of test specimens for procedure qualification are shown in Table 11-6 and Table 11-7. Also shown is the range of thickness that is qualified for use in construction by a given thickness of test plate or pipe used in making the qualification. Test specimens are to be removed as shown in Fig. 11-46 or Fig. 11-47 for plate and Fig. 11-48 for pipe.

The notes to Tables 11-6 and 11-7 are essential to the use of the tables. The applicable notes are:

(1) The maximum thickness qualified in gas welding is the thickness of the test plate or pipe.

(2) The maximum thickness qualified for pipe smaller than 5 in. is two times the thickness of the pipe wall but not more than 3/4-in.

(3) For submerged-arc welding and gas metal-arc welding, the thickness limitation for production welding, based on plate thickness t, shall be as follows:

  (a) For single-pass welding with no backing strip or against a metal or ceramic (flux) backing, the maximum thickness welded in production shall not exceed the thickness of the test plate or pipe wall.

  (b) If the test plate is welded by a procedure involving one pass from each side, the maximum thickness that may be welded in production shall be 2t, where 2t shall not exceed 2 in. If sections heavier than 2 in. are to be welded in production, a separate test plate shall be prepared with the thickness not less than the thickness to be used in production.

TABLE 11-6. Procedure Qualification. Type, Number of Test Specimens, and Range of Thickness Qualified

Thickness t of Test Plate or Pipe Wall (in.)	Range of Thickness Qualified (in.)		Type and Number of Tests Required			
				Transverse Bend Tests (5)		
	Min. (6)	Max. (1, 3, 6)	Tension	Side Bend	Face Bend	Root Bend
Less than 1/16″	t	2 t	2	—	2	2
1/16 to 3/8″ inclusive	1/16	2 t (2)	2	—	2	2
Over 3/8 to 3/4″	3/16	2 t	2	—	2	2
3/4 to 1-1/2″	3/16	2 t	2	4 (4)	—	—
Over 1-1/2″	3/16	8	2	4	—	—

NOTES: See Text ("Welding Procedure Qualificaiton, ASME Code") for explanation of (1) through (6).

TABLE 11-7. Procedure Qualificaiton. Type, Number of test Specimens, and Range of Thickness Qualified

Thickness t of Test Plate (in.)	Range of Thickness Qualified (in.)		Type and Number of Tests Required		
				Longitudinal Bend Tests (5)	
	Min. (6)	Max. (1, 3, 6)	Tension	Face Bend	Root Bend
Less than 1/16″	t	2 t	2	2	2
1/16 to 3/8″	1/16	2 t	2	2	2
Over 3/8″	3/16	2 t	2	2	2

NOTES: See Text ("Welding Procedure Qualificaiton, ASME Code") for explanation of (1) through (6).

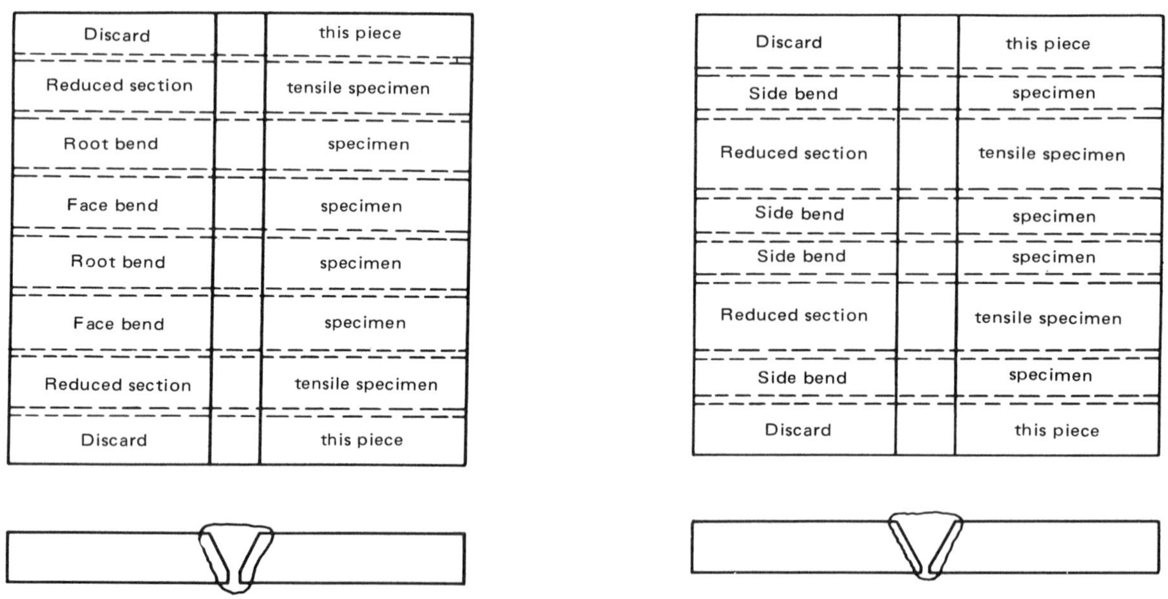

For plate 1/16 to 3/4 in. thick.

For plate over 3/4 in. thick. May be used also for thicknesses from 3/8 to 3/4 in.

Fig. 11-46. Method of taking test specimens from plate for procedure qualification (ASME).

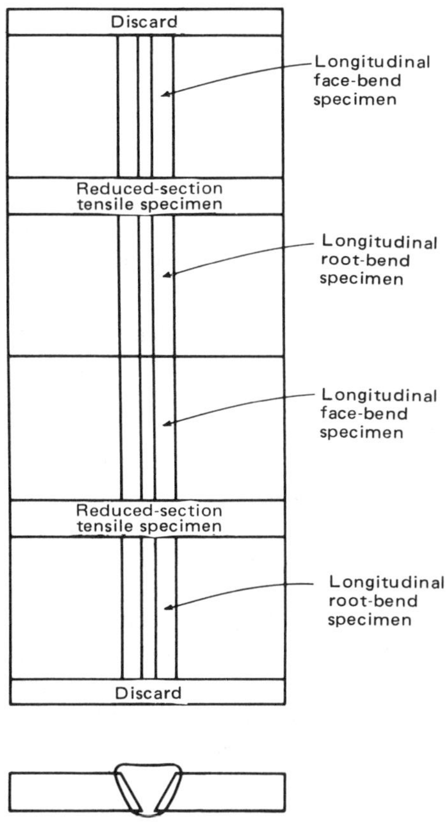

Fig. 11-47. Order of removal of test specimens from welded test plate for longitudinal bend tests (ASME).

(c) For multiple-pass welding, the thickness limitations in Tables 11-6 and 11-7 apply.

(4) Either face and root bends or side bends may be used for thicknesses from 3/8 to 3/4 in.

(5) Longitudinal bend tests may be used in lieu of transverse bend tests in Table 11-6 only for testing material combinations differing markedly in mechanical bending properties between (a) the two base materials or (b) the weld metal and base materials. (See Fig. 11-47 for weld test plate.)

(6) For quenched and tempered steels (tensile strength 95,000 psi or higher) of thicknesses less than 5/8 in., the thickness of the test plate or pipe is the minimum thickness qualified. For test plates or pipe receiving a postweld heat treatment in which the lower critical temperature is exceeded, the maximum thickness qualified is the thickness of the test plate or pipe.

Tension Tests: Reduced-section tension-test specimens shall conform to Fig. 11-49 for plate and Fig. 11-50 for pipe of all wall thicknesses having an outside diameter greater than 3 in. Tension test specimens conforming to Fig. 11-51 may be used for thicknesses to and including 1-1/4 in. Over 1-1/4 in. multiple specimens shall be cut through the full thickness of the weld with the centers parallel to the material surface and not over 1 in. apart.

Tensile specimens conforming to Fig. 11-52 may be used for testing pipe or tube with an outside diameter 3 in. or less.

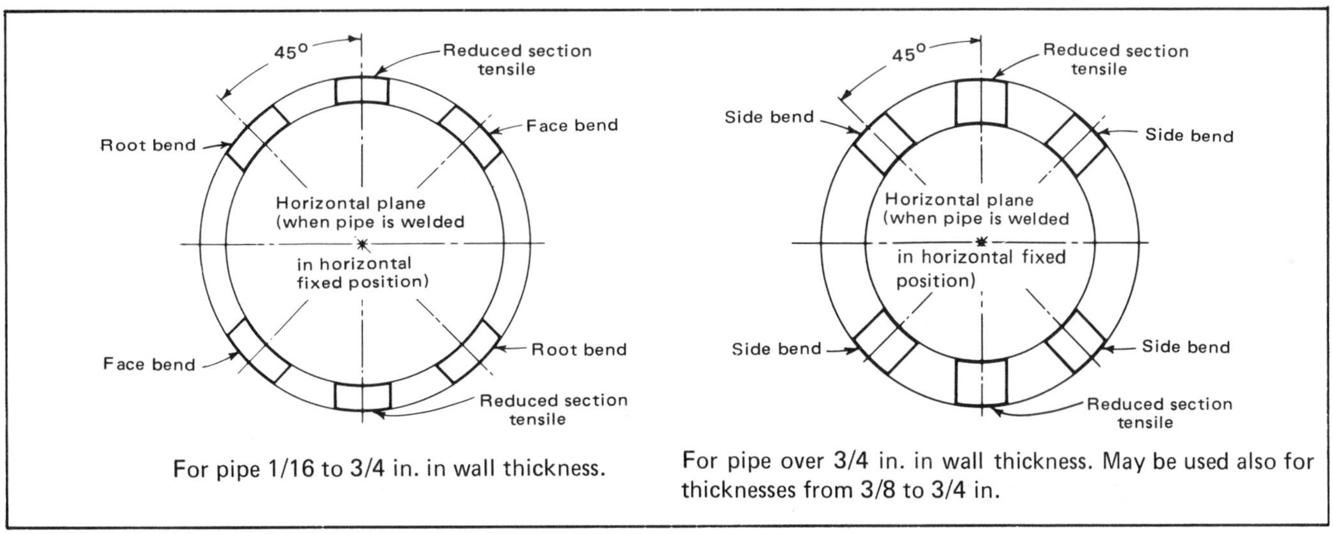

Fig. 11-48. Method of taking test specimens from pipe for procedure qualification (ASME).

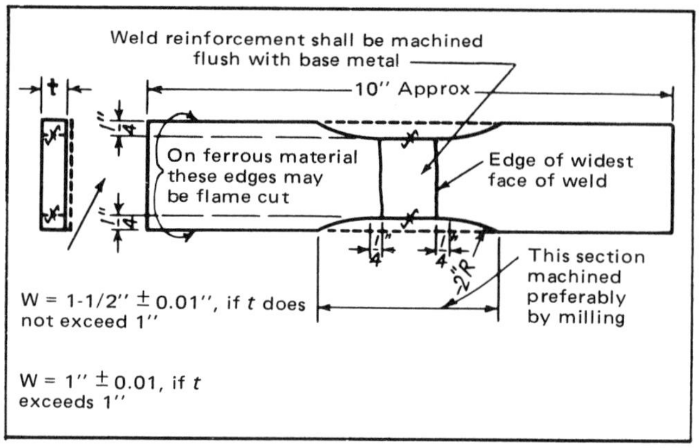

Fig. 11-49. Reduced-section tensile specimen from plate (ASME).

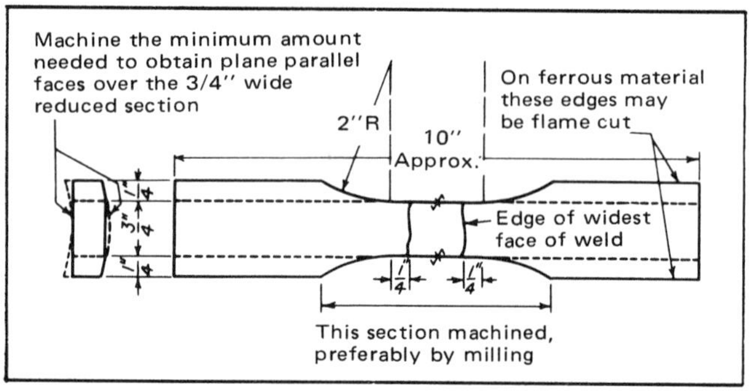

Fig. 11-50. Reduced-section tensile specimen from pipe (ASME).

Tensile Strength Requirements: The tensile strength must not be less than:

(1) The minimum specified for the base material, or

(2) The minimum specified for the weaker of the two, if materials of different specified minimum tensile strengths are used, or

(3) The minimum specified for the weld metal where the weld metal has lower room-temperature strength than the base metal.

If the specimen breaks in the base metal, the test shall be accepted if the tensile strength is not more than 5% below the minimum specified for the base metal.

Guided Bend Tests: Transverse side bend specimens shall conform to the dimensions shown in Fig. 11-53.

Transverse root and face bend specimens shall conform to the dimensions shown in Fig. 11-54.

Longitudinal root and face bend specimens shall conform to the dimensions shown in Fig. 11-55. This type of specimen is used only for testing material combinations differing markedly in mechanical bending properties.

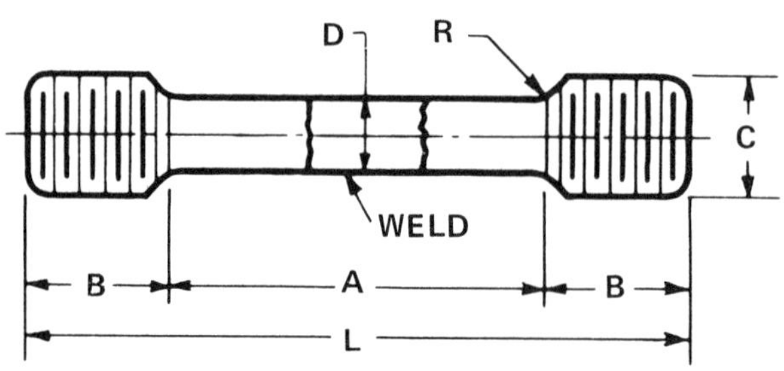

	Standard Dimensions, In.			
	(a) 0.505 Specimen	(b) 0.353 Specimen	(c) 0.252 Specimen	(d) 0.188 Specimen
A-Length reduced section	See Note 4	See Note 4	See Note 4	See Note 4
D-Diameter	0.500 ± 0.010	0.350 ± 0.007	0.250 ± 0.005	0.188 ± 0.003
R-Radius of fillet	3/8, min	1/4, min	3/16, min	1/8, min
B-Length of end section	1-3/8, approx	1-1/8, approx	7/8, approx	1/2, approx
C-Diameter of end section	3/4	1/2	3/8	1/4

NOTES:
1. Use maximum diameter specimen (a), (b), (c), or (d) that can be cut from the section.
2. Weld should be in center of reduced section.
3. Where only a single specimen is required the center of the specimen should be midway between the surfaces.
4. Reduced section "A" should not be less than width of weld plus two "D."
5. The ends may be of any shape to fit the holders of the testing machine in such a way that the load is applied axially.

Fig. 11-51. Alternate reduced-section tensile specimen (ASME).

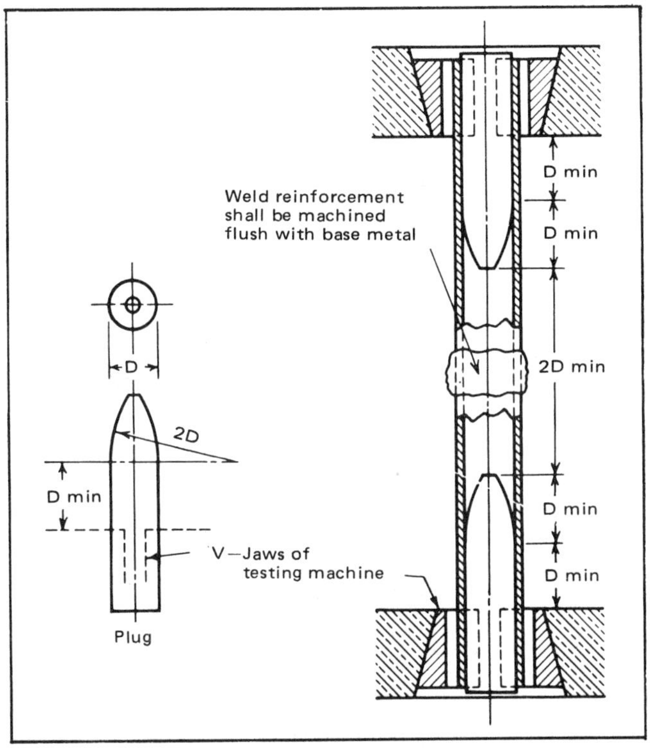

Fig. 11-52. Full-section tensile specimen for small-diameter pipe (ASME).

The guided-bend specimens shall be bent in test jigs that are substantially in accordance with Figs. 11-56, 11-57, and 11-58.

When using jigs in accordance with Figs. 11-56 or 11-57, the side of the specimen turned toward the gap shall be the face for face bends, root for root bends, and the side with greater discontinuities, if any, for side bends. The specimen shall be forced into the die until the curvature of the specimen is

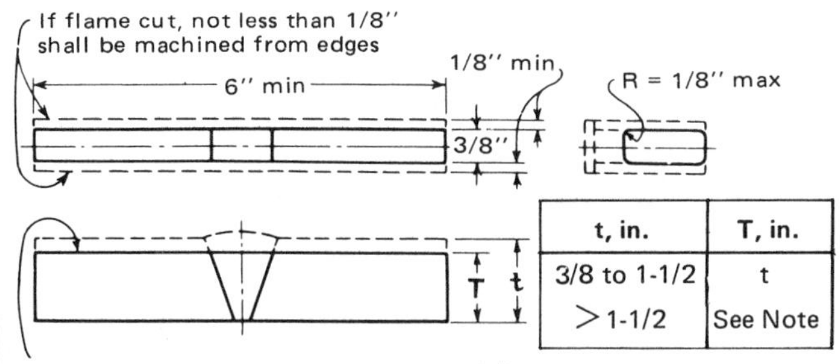

t, in.	T, in.
3/8 to 1-1/2	t
>1-1/2	See Note

Cut along this line when *t* exceeds 1-1/2; Edge may be Flame-Cut and may or may not be Machined.

Note: For Plates over 1-1/2 in. thick, cut Specimen into approximately equal Strips between 3/4 in. and 1-1/2 in. wide and test each Strip.

Fig. 11-53. Transverse side bend specimens, pipe and plate (ASME).

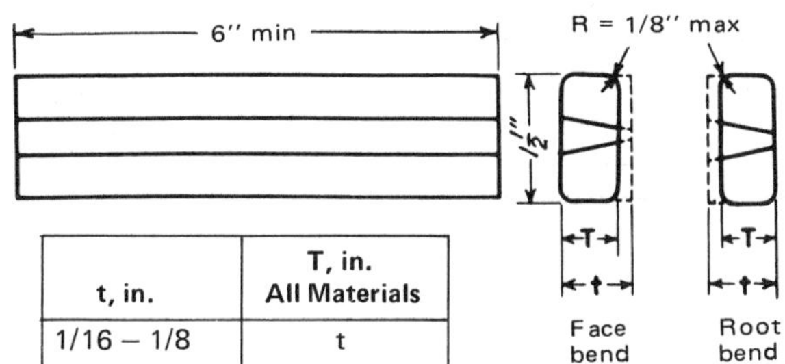

t, in.	T, in. All Materials
1/16 – 1/8	t
1/8 – 3/8	t
>3/8	3/8

Note: Weld reinforcement and backing strip, if any, shall be removed flush with the surface of the specimen. If a recessed strip is used this surface of the specimen may be machined to a depth not exceeding the depth of the recess to remove the strip, except that in such cases the thickness of the finished specimen shall be that specified above.

Fig. 11-55. Longitudinal root and face bend specimens. (ASME)

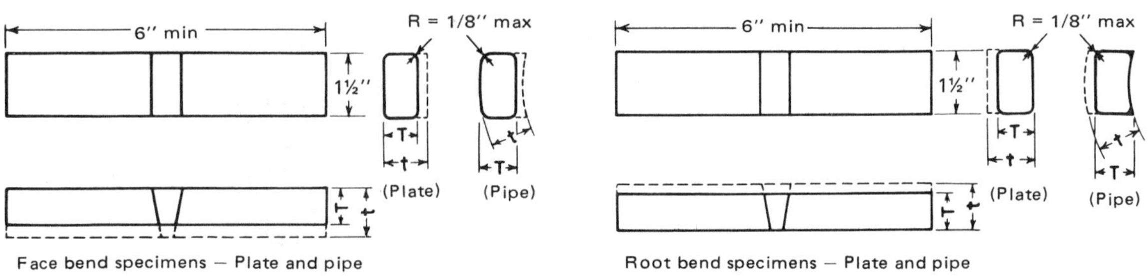

t, in.	T, in. All Ferrous Materials
1/16 – 1/8	t
1/8 – 3/8	t
>3/8	3/8

Note: Weld reinforcement and backing strip or backing ring, if any, shall be removed flush with the surface of the specimen. If a recessed ring is used, this surface of the specimen may be machined to a depth not exceeding the depth of the recess to remove the ring, except that in such cases the thickness of the finished specimen shall be that specified above.

Fig. 11-54. Transverse root and face bend specimens, pipe and plate (ASME).

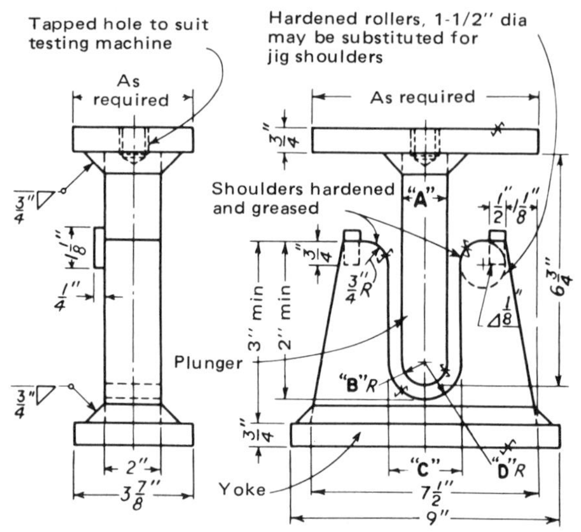

Thickness of Specimens, In.	A In.	B In.	C In.	D In.
3/8	1-1/2	3/4	2-3/8	1-3/16
t	4t	2t	6t + 1/8	3t + 1/16

Fig. 11-56. Guided-bend test jig (ASME).

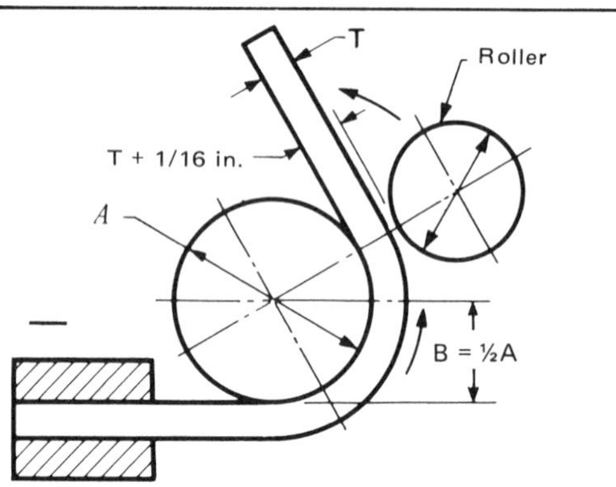

Thickness of Specimens, In.	A In.	B In.
3/8	1-1/2	3/4
t	4t	2t

NOTES:

1. Dimensions not shown are the option of the designer. The essential consideration is to have adequate rigidity so that the jig parts will not spring.
2. The specimen shall be firmly clamped on one end so that there is no sliding of the specimen during the bending operation. The weld and heat-affected zone in the case of transverse weld bend specimens shall be completely within the bent portion of the specimen after testing.
3. Test specimens shall be removed from the jig when the outer roll has been removed 180 degrees from the starting point.

Fig. 11-58. Wrap-around guided-bend jig (ASME).

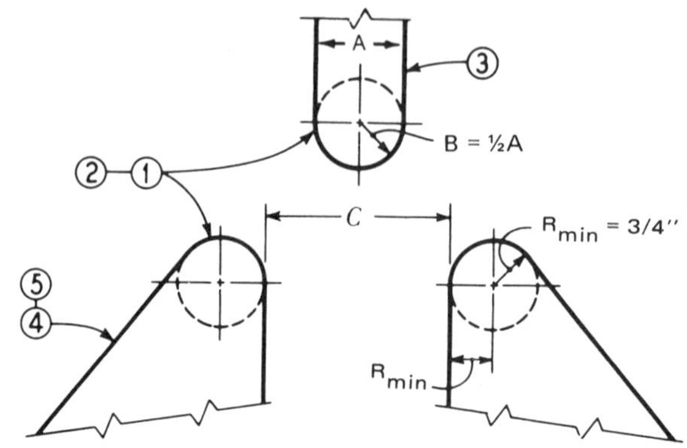

Thickness of Specimens, In.	A In.	B In.	C In.
3/8	1-1/2	3/4	2-3/8
t	4t	2t	6t + 1/8

NOTES:

1. Either hardened and greased shoulders or hardened rollers free to rotate shall be used.
2. The shoulders or rollers shall have a minimum bearing surface of 2 in. for placement of the specimen. The rollers shall be high enough above the bottom of the jig so that the specimens will clear the rollers when the ram is in the low position.
3. The ram shall be fitted with an appropriate base and provision made for attachment to the testing machine, and shall be designed to minimize deflection and misalignment. The ram to be used with the roller jig shall be of identical dimensions to the ram shown in Fig. 11-56.
4. If desired, either the rollers or the roller supports may be made adjustable in the horizontal direction so that specimens of *t* thickness may be tested on the same jig.
5. The roller supports shall be fitted with an appropriate base designed to safeguard against deflection or misalignment and equipped with means for maintaining the rollers centered, midpoint and aligned with respect to the ram.

Fig. 11-57. Roller-equipped guided-bend jig for bottom ejection of test specimen (ASME).

such that a 1/8-in.-diameter wire cannot be inserted between the die and the specimen.

When using the roller-type jig, Fig. 11-57, the specimen shall be bottom-ejected. When using the wrap-around jig, Fig. 11-58, the side of the specimen toward the roller shall be face for face bends, root for root bends, and the side with the greater discontinuities, if any, for the side bends.

Bend Test Requirements: After bending, the specimen shall have no cracks or other open discontinuities exceeding 1/8 in. measured in any direction on the convex surface. Cracks on the corners of the specimen shall not be considered unless there is evidence of slag inclusions or other internal discontinuities.

If the wall thickness of the tube or pipe is less than 3/8 in., or the diameter-to-thickness ration does not permit the preparation of a full-size rectangular guided-bend specimen, the 1-1/2-in.-wide

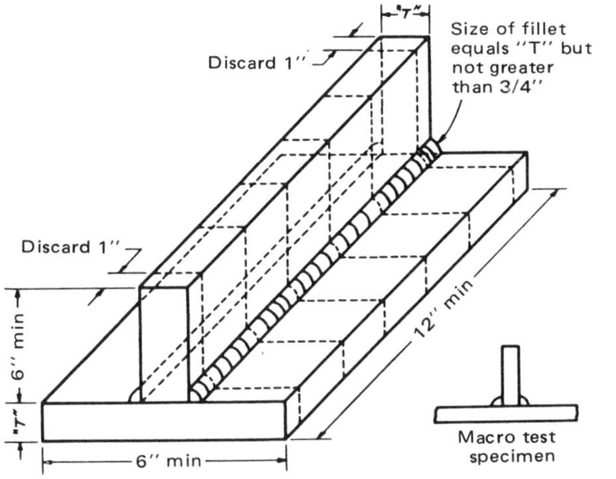

"T" maximum thickness of base material in the vessel at point of welding or 1", whichever is smaller.

Macro test: The fillet shall show fusion at the root of the weld but not necessarily beyond the root. The weld metal and heat-affected zone shall be free of cracks. Both legs of the fillet shall be equal to within 1/8 inch.

Fig. 11-59. Fillet-weld soundness test for procedure qualification (ASME).

standard bend specimen shown in Fig. 11-54 may be replaced by three subsize specimens having a width of 3/8 in. or 4t, whichever is less. The weld reinforcement and backing ring shall be removed flush with the pipe or tube surface, and the corners may be provided with a radius not to exceed 1/3t.

Fillet Weld Tests: Fillet-weld test specimens for procedure qualification shall conform to Fig. 11-59. The weld shall not contain any visible cracks. The specimen shall be cut transversely in five sections approximately 2 in. long, and one face of each section shall be smoothed and etched with a suitable etchant to give a clear definition of the structure. Visual examination shall show:

(1) Complete fusion at the root and no cracks.

(2) Not more than 1/8 in. difference in the length of the legs of the fillet.

API QUALIFICATION TESTS FOR PIPELINE WELDING

The American Petroleum Institute (API) sets specifications for welding procedures and personnel employed on pipeline welding in its "*Standard for Welding Pipelines and Related Facilities*" (API Standard 1104). The standard is available from the American Petroleum Institute, 1220 L Street, N.W., Washington, D.C. 20005. The following is a highly condensed resume of the portions of the standard pertaining to procedure and weldor qualification.

API Qualification of Welding Procedures

Prior to the start of production welding, a procedure specification must be established and qualified to demonstrate that welds having suitable mechanical properties and that soundness will result from the procedure. The quality of the weld is determined by destructive testing.

Table 11-8 shows the type and number of butt-weld test specimens required for procedure qualification, and Figs. 11-60, 11-61, and 11-62 show how specimens are taken and prepared for testing. Fig. 11-63 shows the jig used for guided bend tests.

In the tensile test, the tensile strength of the weld, including the fusion zone, shall be equal to or greater than the minimum specified tensile strength of the pipe material.

In the nick-break test, the exposed surfaces of each specimen shall show complete penetration and fusion and no more than six gas pockets per sq. in. of surface area, with the greatest dimension not to exceed 1/16 in. Slag inclusions shall not be more than 1/32 in. in depth nor 1/8 in. or one-half the nominal wall thickness in length, whichever is shorter, and there shall be at least 1/2 in. of sound weld metal between adjacent inclusions.

The bend tests shall be considered acceptable if no crack or other discontinuity exceeding 1/8 in. or one-half the nominal wall thickness, whichever is smaller, in any direction is present in the weld or between the weld and the fusion zone after bending. Cracks which originate along the edges of the specimen during testing, and which are less than 1/4 in. measured in any direction, shall not be considered unless discontinuities are observed.

The type and number of fillet-weld test specimens for procedure qualification are as follows:

Pipe Size, OD in.	Number of Specimens
Under 2-3/8	4 (Obtain from 2 welds)
2-3/8 to 12-3/4 inclusive	4
Over 12-3/4	6

The methods of taking and preparing fillet-weld specimens are illustrated in Figs. 11-64 and 11-65. The test requirements are similar to those for butt welds.

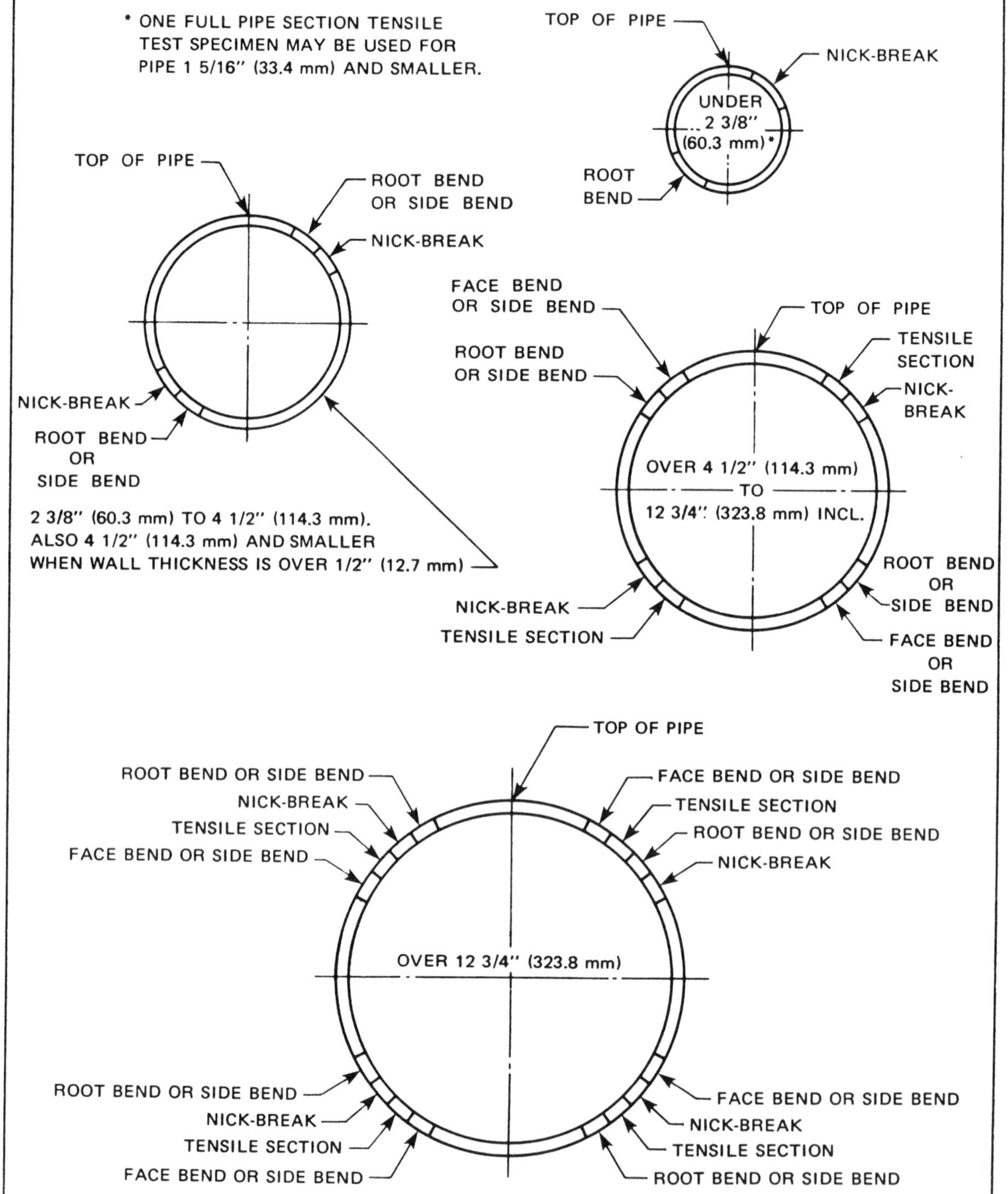

NOTE: AT THE COMPANY'S OPTION, THE LOCATIONS MAY BE ROTATED PROVIDED THEY ARE EQUALLY SPACED AROUND THE PIPE EXCEPT SPECIMENS SHALL NOT INCLUDE THE LONGITUDINAL WELD.

FIGURE 1

LOCATION OF TEST SPECIMENS - BUTT WELD PROCEDURE QUALIFICATION TEST WELD

Fig. 11-60. Location of butt-weld test specimens for API procedure qualification tests.

TABLE 11-8. Type and Number of Butt-Weld Test Specimens for Procedure Qualification Test

Pipe Size, Outside Diameter — In.	Number of Specimens					
	Tensile	**Nick Break**	**Root Bend**	**Face Bend**	**Side Bend**	**Total**
Wall Thickness — 1/2 in. and Under						
Under 2-3/8″	0	2	2	0	0	4*
2-3/8 to 4-1/2″ inclusive	0	2	2	0	0	4
Over 4-1/2 to 12-3/4″ inclusive	2	2	2	2	0	8
Over 12-3/4″	4	4	4	4	0	16
Wall Thickness — Over 1/2 In.						
4-1/2″ and smaller	0	2	0	0	2	4
Over 4-1/2 to 12-3/4″ inclusive	2	2	0	0	4	8
Over 12-3/4″	4	4	0	0	8	16

* One nick-break and one root bend specimen from each of two test welds, or for pipe 1-5/16 in. and smaller, one full pipe section tensile specimen (API).

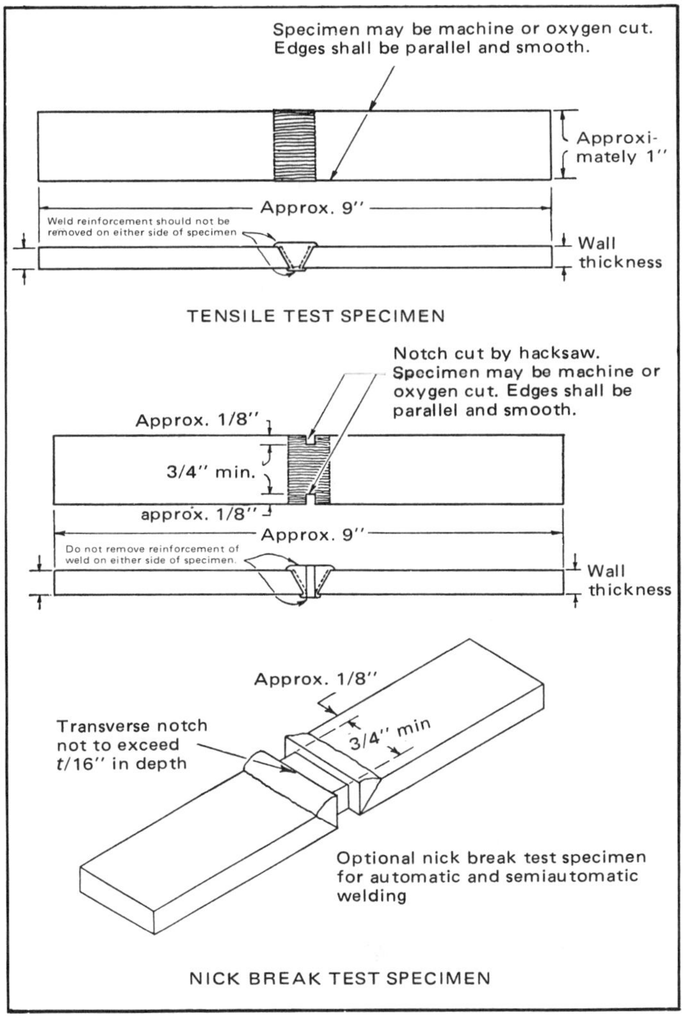

Fig. 11-61. How tensile and nick-break specimens of butt welds are prepared for API procedure qualification tests.

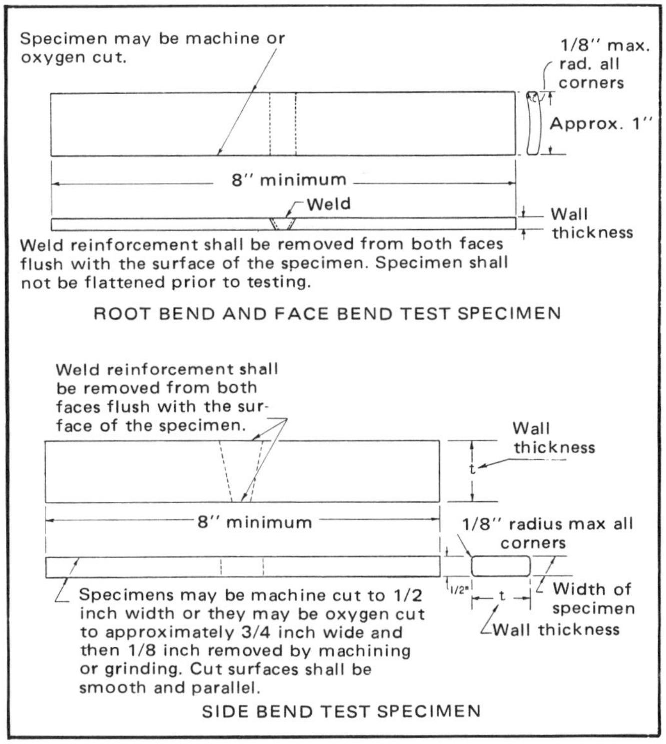

Fig. 11-62. How root-bend, face-bend, and side-bend specimens are prepared (API).

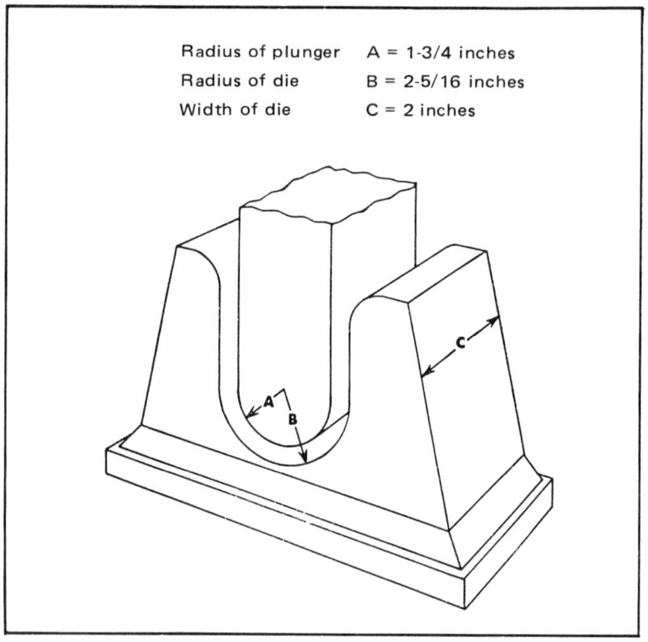

Fig. 11-63. Jig for the API guided bend tests.

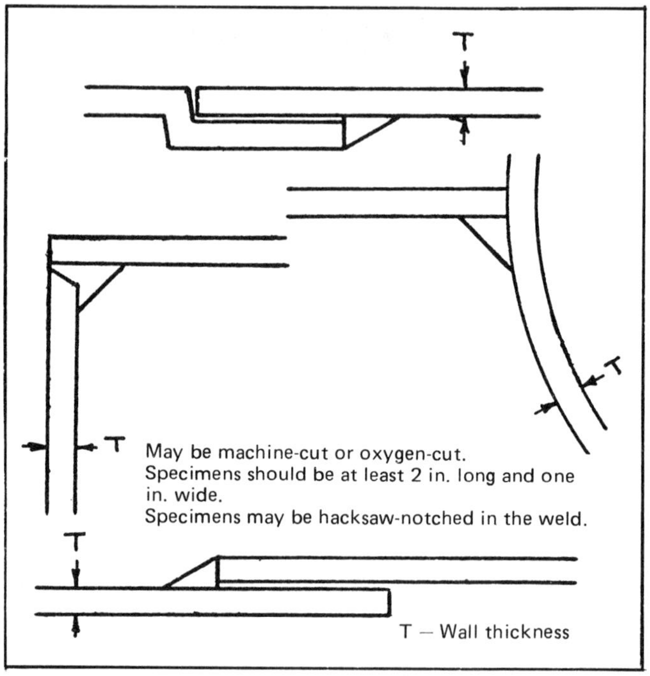

Fig. 11-65. Preparation of fillet-weld specimens (API).

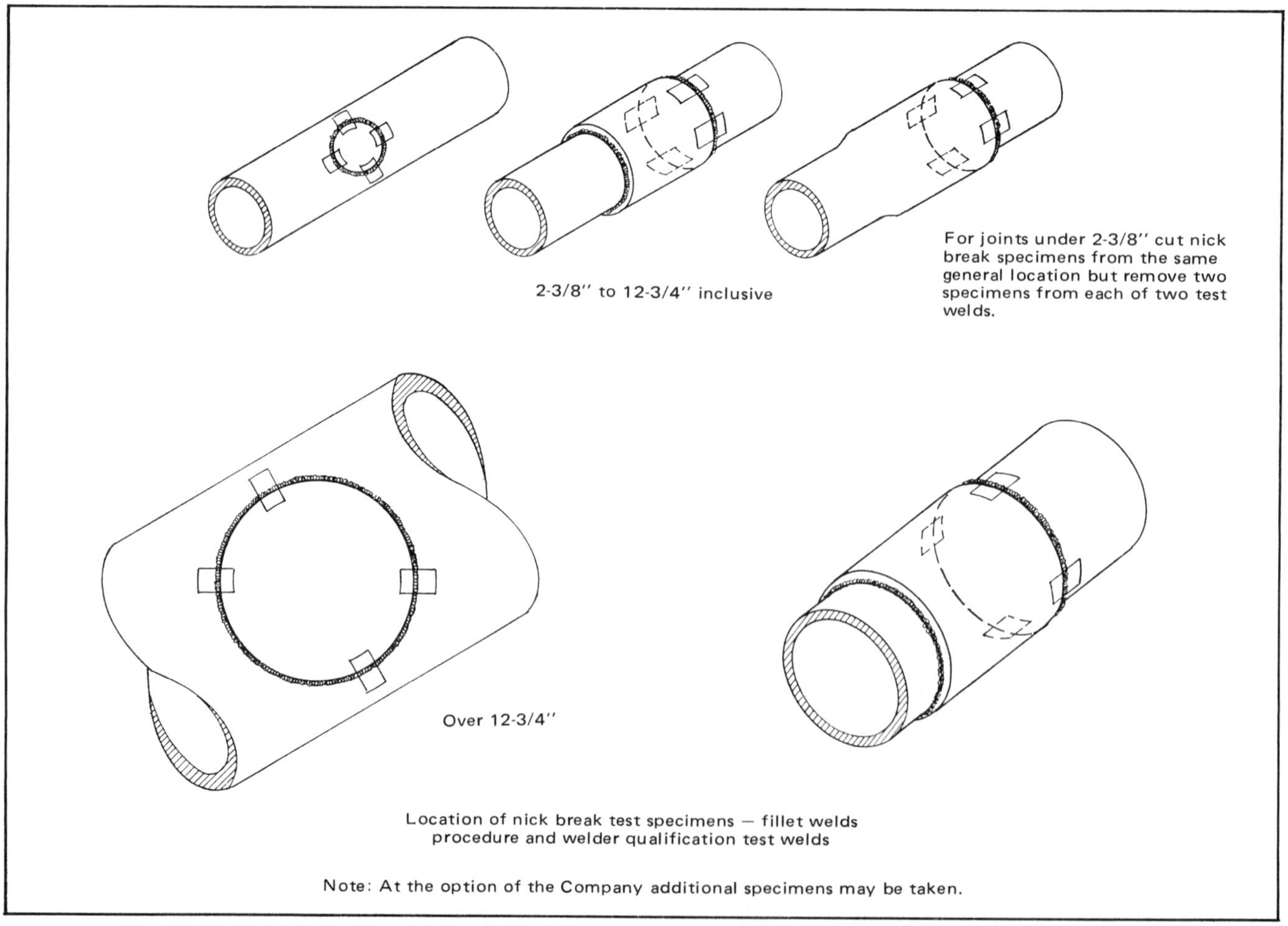

Fig. 11-64. Method of taking nick-break fillet-weld test specimens for API procedure and weldor qualifications.

Weldor Qualification Tests

INTRODUCTION

Weldor qualification tests are performed by an individual to test that person's skill to perform or produce a particular kind of weld according to a specific set of requirements. These requirements may be based on any of several recognized national welding codes and welding procedure specifications (WPS) developed and qualified for the code. The WPS's are developed from welding procedure qualification tests as described in Section 11.3 for several welding codes.

The qualification tests are especially designed to determine the ability of the weldors, welding operators and tack weldors to produce sound welds by following a welding procedure specification. The code does not imply that anyone who satisfactorily completes qualification tests can do the welding for which they are qualified for all conditions that might be encountered during production welding. It is essential that weldors, welding operators and tack weldors have some degree of training for these differences.

The level of skill required to pass a weldor qualification test varies significantly with the process, welding position and base metal. For example, the skill required to pass a welding test by gas tungsten arc welding aluminum in the overhead position will require considerably more skill than shielded metal arc welding mild steel in the flat position. For this reason, a person is qualified only in the process, material, thickness and position in which tested except that the more difficult positions also qualify for the less difficult positions and that thicker plates qualify for thinner plates.

GENERAL TEST REQUIREMENTS

To perform and pass a weldor qualification test, a weldor must make welds according to a qualified procedure. The procedure may be one developed and qualified by the company or a prequalified and verified procedure. The test must be taken on material in a thickness and with filler material which is representative of those to be used in the welded fabrication involved. The welding process and position also must be representative of the application. It follows that the procedure for which the weldor is qualifying is representative of the application. If several types of joints, positions, processes and materials are involved in a structure or weldment, it would be necessary to qualify a weldor in each combination of these items. Some positions, material thicknesses and types of weld may qualify for other positions, thicknesses or types of weld. For example, qualification in the vertical and overhead positions qualifies for all positions and groove welds qualify for the same position fillet welds.

Generally, welds in thick steel plate qualify for any thinner plate, and welds in one inch plate qualify for unlimited thickness. It is usually more economical to take tests under the more difficult conditions such as vertical and overhead and on one inch thick plate because the two tests will qualify the weldor for all positions and unlimited thickness in a specific process. If pipe welding is required, the weldor should qualify on pipe in the vertical fixed and horizontal fixed positions or 45 degree fixed position because these will also qualify for pipe and plate in all positions.

These tests must be witnessed to verify that the tests were performed by the individual to be qualified and according to all the requirements of the procedure. This will include welding process, position, base material, filler material, preheat, joint design and parameters. The cutting, preparation and identifying of specimens prior to submitting for testing should be witnessed. Testing should be performed by a certified testing facility.

Most weldor qualification tests require that the test weld be evaluated by visual inspection for shape of the weld deposit and by bend tests. These are root and face bend tests or side bend tests depending on the thickness of the plate. The test specimens must bend the required amount without breaking or without developing cracks in the weld over a prescribed maximum length and number.

At the contractor's option, radiographic testing of weldor qualification test welds may be performed in lieu of bend tests and tensile tests. However, radiographic testing is not authorized for GMAW-S test welds for bridge code qualification of weldors.

TECHNIQUES TO EMPLOY DURING WELDOR QUALIFICATION TESTING

While the AWS tests are designed to separate capable weldors from amateurs, many professional weldors have failed for reasons not related to their welding ability. The various codes have minimized such possibilities by making the tests more positive demonstrations of weld quality and welding skill. However, it is still possible to fail the test because of poor plate quality, improperly prepared samples, or incorrect interpretation of the results. This is discouraging to the weldor and costly for his employer. It is to everyone's advantage to see that the tests are run properly so the results are an accurate measure of welding ability.

Before taking a test, one should make certain that he has selected the test details that will qualify for the type steel, plate thickness, joint, and welding position to be used. (See Figs. 11-26 through 11-30.) Then, the strength of the plate should be checked. Face and root bend tests are designed so both plate and weld stretch during the test. If the plate has strength substantially higher than the weld metal (and this is possible with steel purchased as mild steel), the plate will not stretch sufficiently, thereby forcing the weld metal to stretch beyond its yield and crack.

As indicated previously, one vertical and one overhead test plate can qualify a manual-electrode or semiautomatic weldor for unlimited welding under the AWS Code. Since it permits using the weldor thoughout the shop without restrictions, most testing is done in these two positions.

For unlimited manual-electrode qualification, low-hydrogen electrode (usually E7018) is required, EXX10 or EXX11 electrodes are also frequently used, but do not qualify the weldor to weld with low-hydrogen electrodes. Low-hydrogen electrodes require different techniques than EXX10 or EXX11, and weldors should be trained with low-hydrogen techniques before testing, even if they are expert EXX10 weldors.

Critical to passing any of the tests is getting good penetration and sound weld metal in the root passes. To obtain good penetration, generally it is best to use the highest current one can handle within the recommended range for the electrode. Electrode data tables should be consulted for current ranges for electrodes.

Many operators worry too much about appearance on the first beads. Appearance means nothing to the testing machine. But poor penetration and unsound weld metal cause most failures.

When manual electrode welding, vertical welds should be made with the vertical-up technique. For gas metal-arc or flux-cored arc welding, vertical welds can be made either in the vertical-up or vertical-down direction. However, the weldor must use the same direction on the job as he used on the qualification test.

No preheat or postheat treatment is necessary or permissible to pass the test. It is, however, helpful to keep the plates hot during welding, and allow them to cool slowly after completion. Interpass temperatures of 300-400°F are ideal and can usually be maintained by operating at a normal pace. Never quench the plates in cold water or accelerate the cooling rate in any other way.

Poor specimen preparation can cause sound weld metal to fail. Specimen size, location,and preparation are specified for each test. Even a slight nick across the sample may open up under the severe bending stress of the test, causing failure. Therefore, always grind or machine lengthwise on the specimen, as indicated in Fig. 11-66. Always grind or machine both faces of the specimen until the entire bend area is even, leaving no indentations or irregular spots.

Remove all reinforcement. This is part of the test requirement and, more important to the weldor; failure to do so can cause failure of a good weld (see Fig. 11-67). Be sure the edges are rounded to a smooth 1/16-in. radius. This can quickly be done

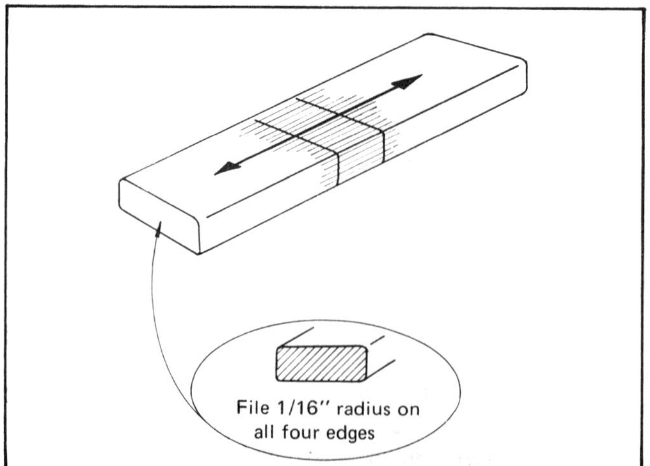

Fig. 11-66. Grinding marks must parallel specimen edges. Otherwise, they may have a notch effect that could cause failure.

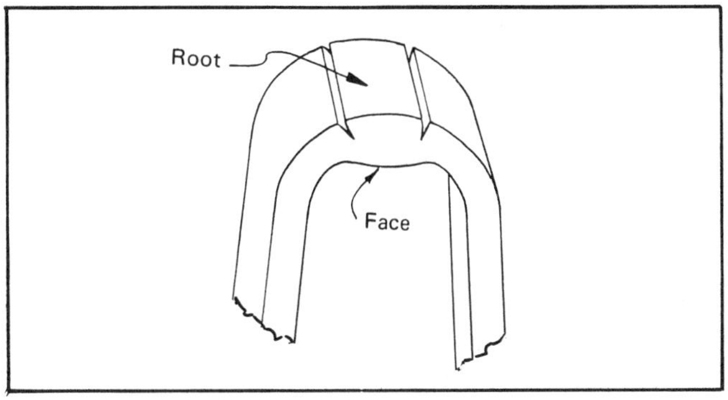

Fig. 11-67. If weld reinforcement is not removed, stretching is concentrated in two places and failure results.

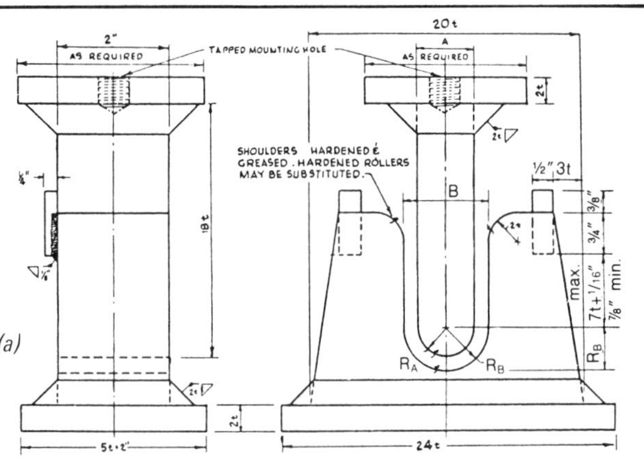

NOTE: "t" refers to specimen thickness
"t" for AWS test is 3/8"
"t" for API Std. 1104 is tabulated wall thickness of pipe

Jig Dimensions	AWS TEST For Mild Steel Min. Yield Strength-psi			API Std. 1104 For All Pipe Grades
	50,000 & under	55-90,000	90,000 & over	
Radius of plunger R_A	3/4	1	1-1/4	1-3/4
Radius of die R_B	1-3/16	1-7/16	1-11/16	2-5/16

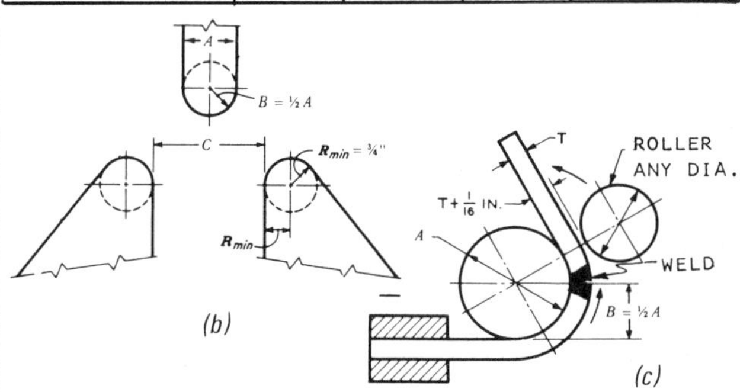

A	B	C	Yield Point-psi*	A	B
1-1/2	3/4	2-3/8	50,000 & Under	1-1/2	3/4
2	1	2-7/8	55,000 to 90,000	2	1
2-1/2	1-1/4	3-3/8	90,000 & Over	2-1/2	1-1/4

Minimum Specified

Fig. 11-68. (a) Jig for guided bend test used to qualify operators for work done under AWS and API specifications. (b) Alternate roller-equipped test jig for bottom ejection. (c) Alternate wrap-around test jig.

with a file and is good insurance against failure caused by cracks starting at the sharp corner.

When grinding specimens, do not water-quench them when they are hot. Quenching may create tiny surface cracks that become larger during the bend test.

The jig for testing weld specimens is shown in Fig. 11-68(a). Note that three plunger diameters are required, with the larger diameter used for the higher strength steels to allow for their lower ductility. Alternate test jigs are shown in Fig. 11-68(b) and (c).

SPECIMEN SIZE, SHAPE AND TESTING

The specimen size and number required for a weldor qualification test depends upon the particular code and type of weld. Generally one or two face bends and one or two root bends or two or four side bend specimens are tested from each qualification test weld. The purpose is to test the soundness and completeness of weld penetration. Radiographic examination may be substituted for bend tests. Tensile, impact and other types of tests normally are not necessary but may be required by some codes and users. These tests are based on the premise that the qualification tests have proved the mechanical properties of welds made according to the procedure and that sound full penetration welds will provide the necessary strength and ductility.

WELDOR QUALIFICATION – AWS D1.1 STRUCTURAL WELDING CODE

Manual-electrode and semiautomatic weldors must be qualified for work covered by the AWS Structural Welding Code. The requirements are also sometimes used by individual companies or specific agencies to prove the skill of welding operators.

A weldor is qualified for any of the approved steels if he qualifies on any one of them. For manual shielded metal-arc welding the proper electrode classification must be used. (5.16.1, D1.1)

With manual shielded metal-arc welding, qualification with any of the following electrodes will qualify for any electrode having the same or lower group designation, where XX represents the strength level of the electrode — 60, 70, 80, 90, 100, 110, and 120. (5.16.3, D1.1)

Group	AWS Electrode Classification
F4	EXX15 EXX16 EXX18 EXX15-X EXX16-X EXX18-X
F3	EXX10 EXX11 EXX10-X EXX11-X
F2	EXX12 EXX13 EXX14 EXX13-X
F1	EXX20 EXX24 EXX27 EXX28 EXX20-X EXX27-X

A weldor qualified with an approved combination of electrode and shielding medium is also qualified for any other approved combination of electrode and shielding medium for the same semiautomatic process as used in the qualification test (5.16.4, D1.1).

The types of joints and welding positions qualified with each test are indicated in the following table:

Test Position	Unlimited(1) & Limited(2) Thickness Test		Fillet Weld Tests(3)
Overhead OH	F &OH groove	F, H & OH fillet	F,H &OH fillet
Vertical V	F, H & V groove	F, H & V fillet	F, H & V fillet
Horizontal H	F & H groove	F & H fillet	F & H fillet
Flat F	F groove	F & H fillet	F fillet

(1) Qualifies for groove and fillet welds on material of unlimited thickness.
(2) Qualifies for groove welds in material not over 3/4-in. thick and fillet welds on material of unlimited thickness.
(3) Qualifies for fillet welds on material of unlimited thickness.

Figures 11-69 and 11-70 detail the qualification tests. The weldor making and passing a procedure qualification test is also qualified for that process and test position for plates of thicknesses equal to or less than the thickness of the test plate welded. In the case of 1-in. or over in thickness, he will be qualified for all thicknesses.

A weldor could, thus, achieve unlimited qualification — would be qualified for making any type weld, in any welding position, on any thickness of plate, on any of the approved weldable steels — by performing satisfactorily with just two weld test plates. Involved would be two groove welds having the prescribed bevel angle and root opening and joining 1-in-thick material of any of the twelve weldable types, using an EXX15, 16, or 18 electrode. One weld would be made vertical and the other overhead. Radiographic inspection would eliminate any machining and bending. If no overhead welding will be required on the job, just one test plate in the vertical position could be made.

Weldor qualification tests remain in effect indefinitely (5.30 D1.1) unless:

1. The weldor does not work with the welding process for which he is qualified for a period exceeding six months. In this case a requalification test is required on 3/8-in.-thick plate.
2. There is specific reason to question his ability.

If a weldor fails his test, he may retest (5.29 D1.1) as follows:

1. An immediate retest shall consist of two test welds of each type he failed, and all test specimens must pass.
2. A retest can be made if the weldor has had further training or practice. In this case, a complete retest shall be made.

Welding Operator Qualification

Qualification of the man who operates automatic welding equipment is covered in Part D of Section 5, D1.1. Here, also, the operator is qualified to weld any of the approved steels if he qualifies on any one of them (5.33.1). An operator qualified for any approved combination of electrode and shielding medium is also approved for any other approved combination of electrode and shielding medium for the same process as used in the qualification test (5.33.2). For other than electroslag or electrogas welding, a welding operator qualified to weld with multiple electrodes shall be qualified to weld with single electrodes but not visa-versa (5.33.3).

Figure 11-71 describes the test. The welding procedure shall be the same as that specified by the Procedure Specification (5.36). The welding operator making and passing a procedure qualification test is also qualified for that process and test position for plates of thicknesses equal to or less than the thickness of the test plate welded. In the case of 1 in. or over in thickness, he will be qualified for all thickness. (5.34.2.1, 5.34.2.2, D1.1)

Radiographic Examination of Welds

Radiographic examination of test plates for weldor and welding operator qualification may be used in place of the guided bend test. Radiographic examination, in addition to being more fair (in that it gives a more positive appraisal of the weld), could easily reduce the cost of qualification tests to but 10% of what they were before revision of the specifications in 1969 by eliminating the machining of test specimens. Furthermore, this option permits a weldor to be tested in the morning and put to work in the afternoon, avoiding the time delays that have frustrated employers and employees alike.

Limited thickness will qualify for groove welds in material not over 3/4" thickness and fillet welds on material of unlimited thickness.

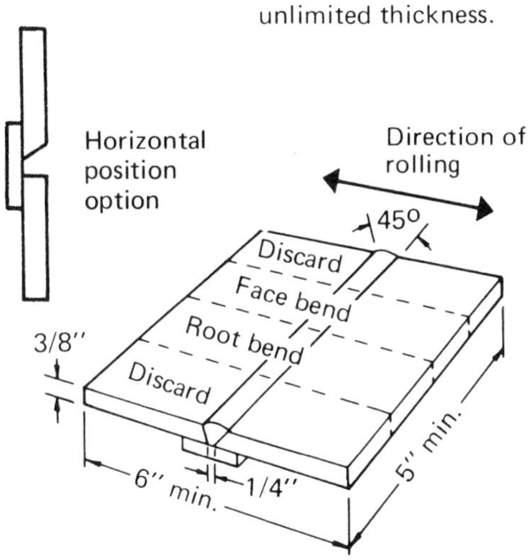

Backing bar is at least 3/8" x 1". If radiography is used then use at least a 3/8" x 3" backing bar.

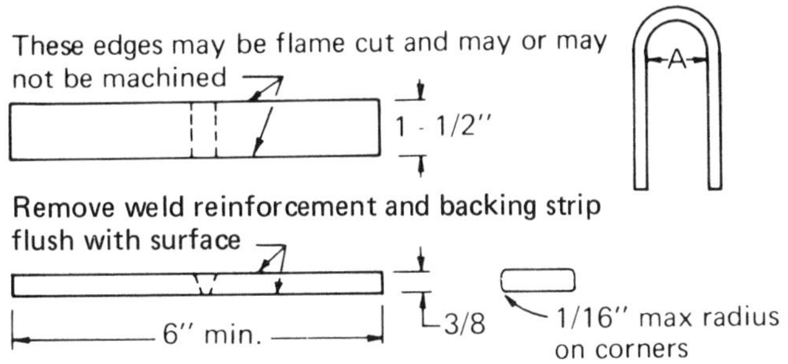

Bend specimen passes if any crack or open defect does not exceed 1/8" after bending through an angle of 180°. Cracks at corners are not considered.

Radiographic inspection of test plate may be used instead of guided bend test. (5.3.2)

Unlimited thickness will qualify for groove or fillet welds in unlimited thickness

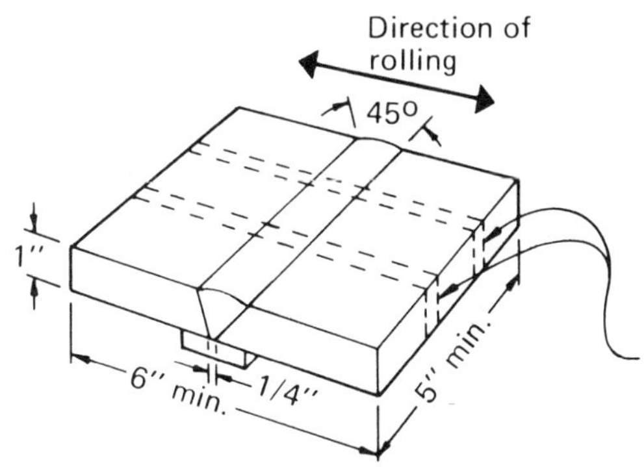

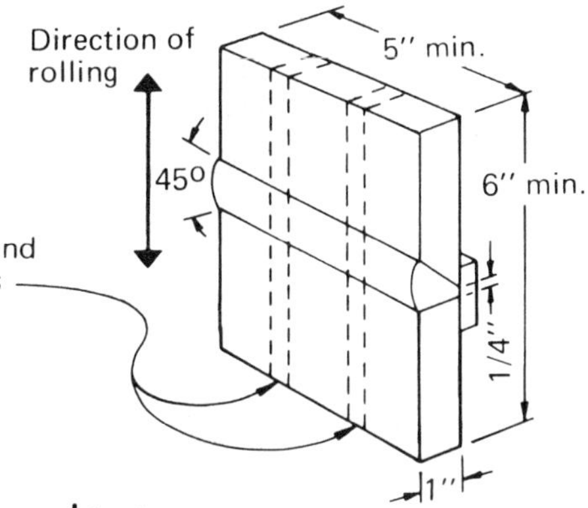

Backing bar is at least 3/8" x 1". If radiography is used then use a 3/8" x 3" backing bar.

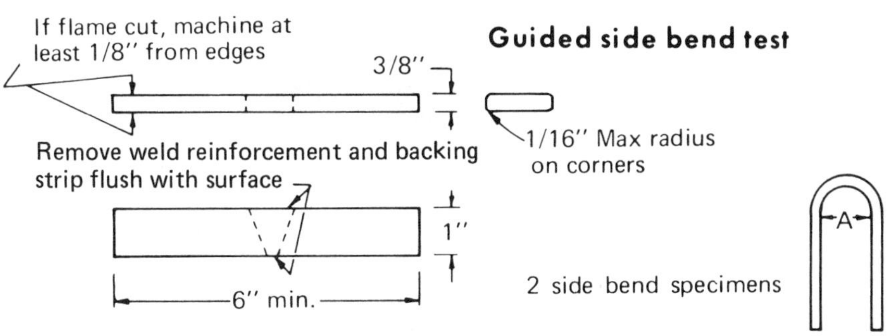

Bend specimen passes if any crack or other open defect does not exceed 1/8" after bending through an angle at 180°. Cracks at corners are not considered.

Radiographic inspection of test plate may be used instead of guided bend test. (5.3.2)

YP	A
50 ksi & under	1-1/2"
55 to 90 ksi	2"
over 90 ksi	2-1/2"

Fig. 11-69. Details of the AWS weldor qualification tests (D1.1). (Groove Welds)

Fillet Welds will qualify for unlimited thickness of material.

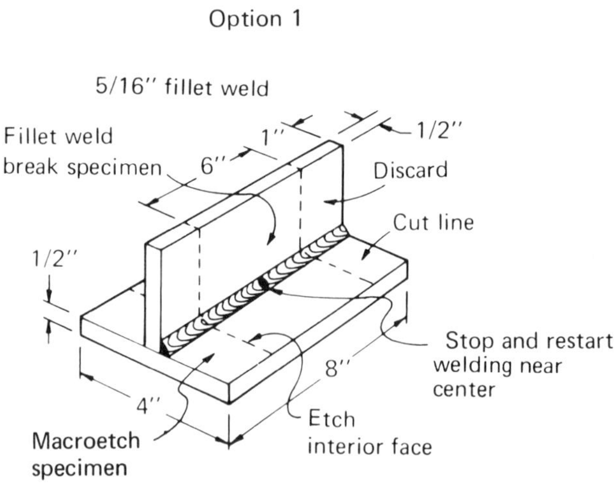

Visual inspection – a reasonably uniform appearance; free of overlap, cracks and excessive undercut; no surface porosity.

Next the 6" section is broken with the root of the weld in tension. The broken section of the weld shall show complete penetration into the root of the joint and fusion into the plate. No inclusion or porosity larger than 3/32". Sum of greatest dimension of all inclusions and porosity shall not exceed 3/8" in the 6" length. If plates are bent upon themselves without weld breaking, will pass.

Macroetch test – shall show fusion to the root but not necessarily beyond root and no cracks. Convexity and concavity shall not exceed 1/16". Legs are equal to within 1/8".

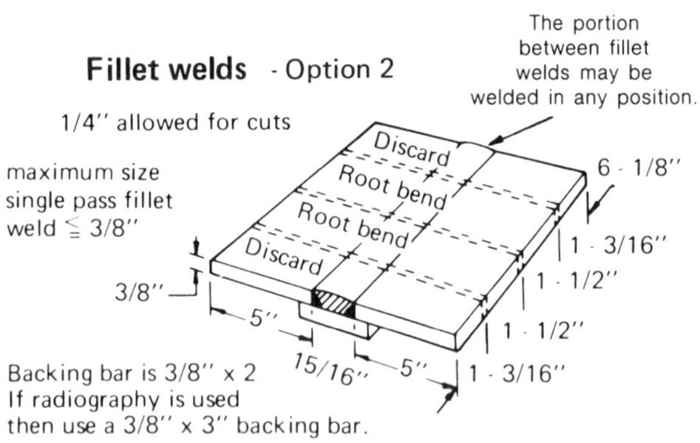

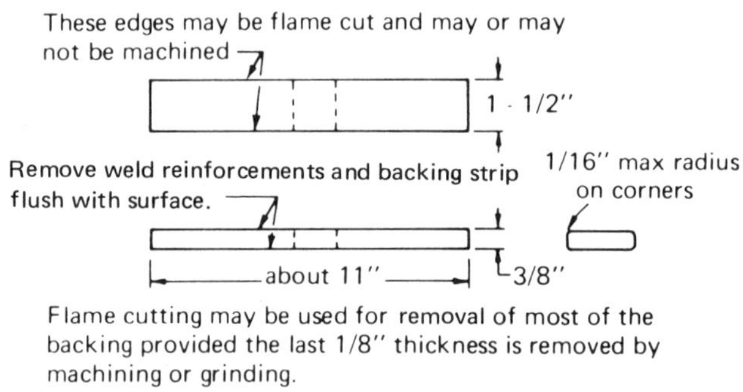

Radiographic inspection of test plate may be used instead of guided bend test. (5.3.2)

Bend specimen passes if any crack or other opening does not exceed 1/8" after bending through an angle of 180°. Cracks at corners are not considered.

Fig. 11-70. Details of the AWS weldor qualification tests (D1.1). (Fillet Welds)

AWS Welding Operator Qualification Tests.

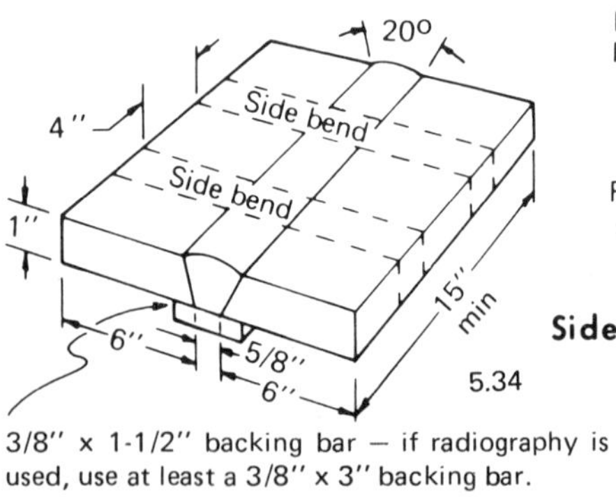

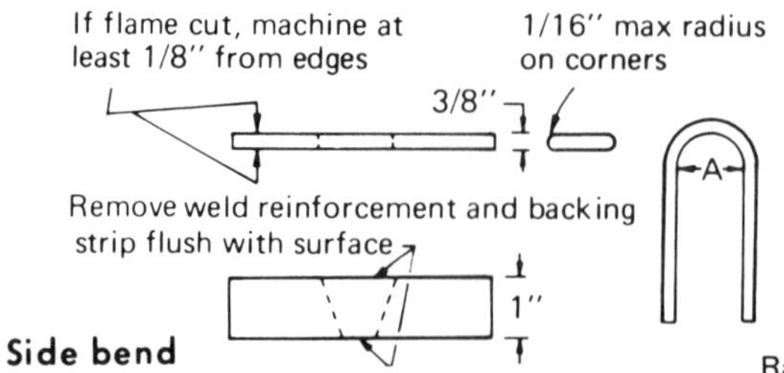

YP	A
50 ksi & under	1-1/2"
55 to 90 ksi	2"
over 90 ksi	2-1/2"

Bend specimen passes if any crack or other opening does not exceed 1/8" after bending through an angle of 180°. Cracks at corners are (5.39.1) not considered.

Radiographic inspection of test plate may be used instead of guided bend test (5.37.1). Will qualify for plate of unlimited thickness with groove and fillet welds.

Fig. 11-71. Details of the AWS welding operator qualification tests (D1.1).

The AWS Structural Welding Code — Steel-D1.1 standards for radiographic weld inspection are applicable to the examination of qualification welds by radiography. Section 8 — Statically Loaded Structures, (8.15) specifies that porosity or fusion-type discontinuities less than 3/32 in. in their greatest dimension are acceptable if the sum of the greatest dimension, when scattered, does not exceed 3/8 in. in any linear inch of weld. Larger porosity of fusion discontinuities are acceptable when:

1. The greatest dimension of the discontinuity does not exceed 2/3t (t = joint or weld throat in inches) or 3/4 in., and:
2. The sum of the greatest dimension of discontinuities in line does not exceed (t) in a length of (6t), or the space between each pair of adjacent discontinuities exceeds three times the greatest dimension of the larger discontinuities. When the length of weld being examined is less than (6t), the permissible sum of the greatest dimension of all such discontinuities shall be proportionately less than (t), and:
3. The discontinuity does not lie closer than three times its greatest dimension to the end of a groove-welded joint carrying primary tensile stress.

Piping porosity is acceptable if the sum of the diameters does not exceed 3/8 in. in any linear inch of weld nor 3/4 in. in any 12-in. length of weld.

Figure 11-72 illustrates graphically these limitations in a groove-welded joint carrying primary tensile stress.

Figures 11-73(a) and (b) illustrate acceptable discontinuities in a 1-in.-thick joint, according to the AWS D1.1, Section 8 — Statically Loaded Structures. In (a) are shown discontinuity sizes and minimum spacings that would be acceptable on a 1-in. plate . Since the minimum spacings are not violated, it is possible for the sum of the discontinuities to exceed 1 in. (t) in any 6-in. (6t) length. In (b), the sum of the discontinuities is less than 1 in. (t) in a 6-in. (6t) length; hence the minimum spacings may be violated.

The discontinuity limitations for dynamically loaded structures is discussed under D1.5 "Bridge Welding Code".

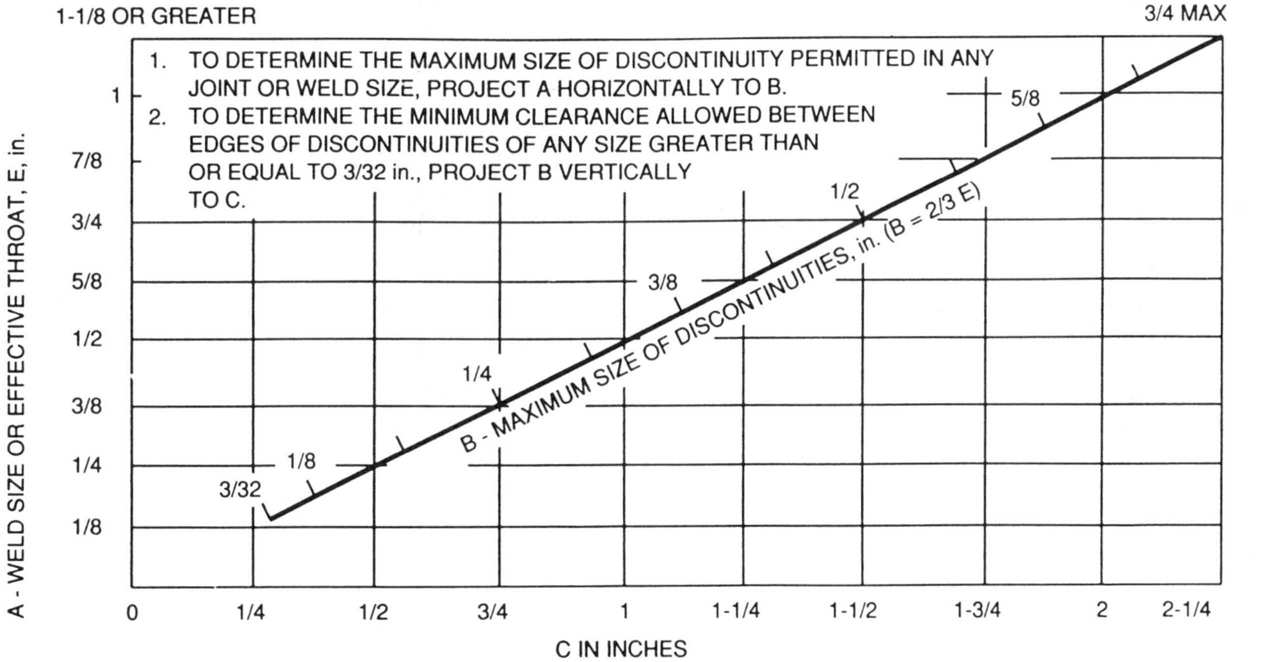

Fig. 11-72. Graphic representation of the discontinuity limitations allowed by the AWS Structural Welding Code in a groove weld carrying primary tensile stress.

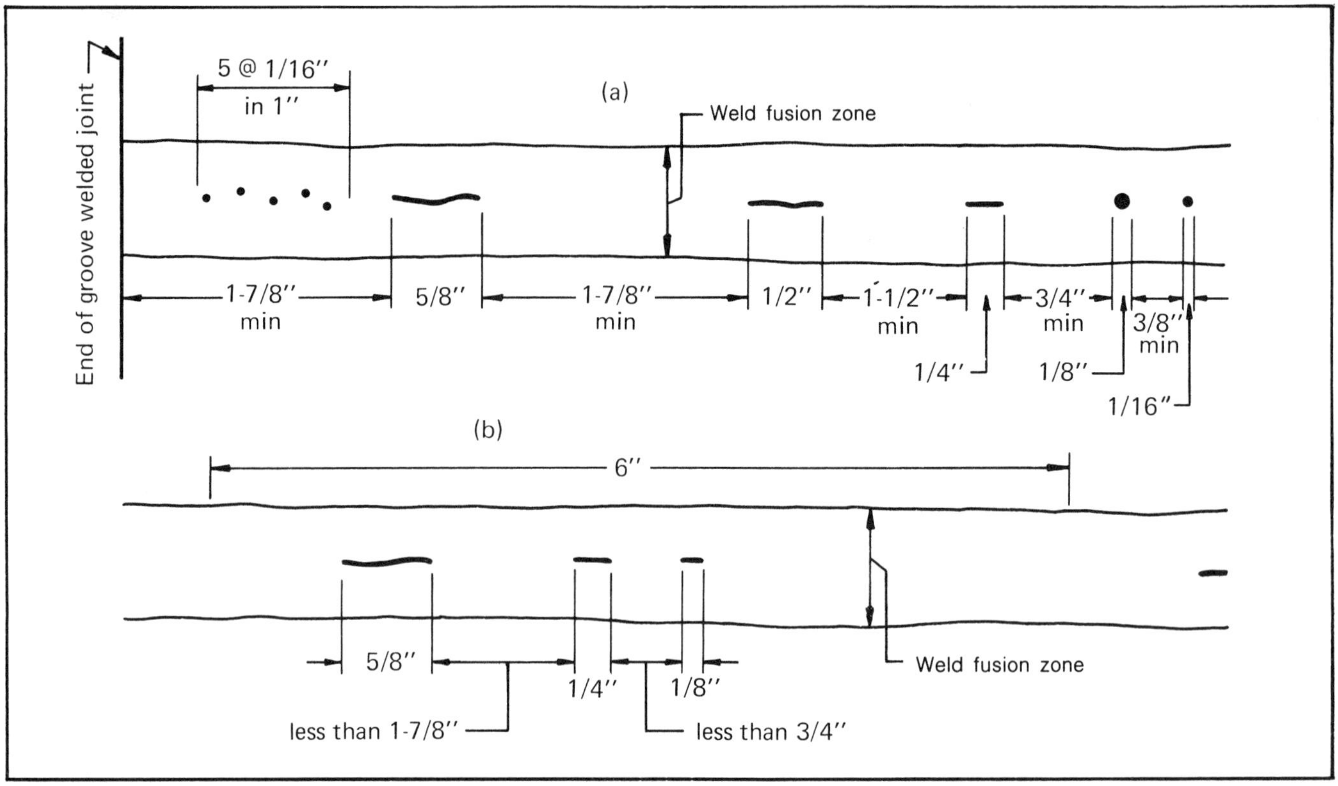

Fig. 11-73. Acceptable discontinuities allowed by the AWS Structural Welding Code in a 1-in.-thick joint.

WELDOR QUALIFICATION - AASHTO-AWS D1.5 BRIDGE WELDING CODE

The weldor qualification tests required for the AASHTO-AWS D1.5 Bridge Welding Code are basically the same as required for the AWS D1.1 Structural Welding Code, Seciton 9, "Dynamically Loaded Structures". To assure that the correct procedure is followed in performing weldor qualification tests it is essential that the AASHTO-AWS D1.5 Bridge Welding Code be obtained and studied.

The preparation of test welds, types of test specimens and methods of testing them are the same as for the D1.1 Structural Welding Code. However, some limitations are related to base material strength, as the D1.5 code states.

Qualification established with any one of the steels permitted by the D1.5 Code shall be considered as qualification to weld or tack weld any of the other steels with the following exception:

> Qualification to weld or tack weld steel with a minimum specified yield strength of 90,000 psi (620 MPa) or greater shall be established with steel having the same specifications as the steel for the project.

A weldor qualified for shielded metal arc welding using EXX18 electrodes shall be qualified to weld with all SMAW electrodes permitted under the D1.5 Code. Weldors required to use an electrode classification of E100XX or higher to joint metals with a minimum specified yield strength of 90,000 psi (620 MPa) or greater shall be tested using E10018 or E11018 electrodes as necessary to match the yield strength of the base metal to be used in the work.

The D1.5 Bridge Welding Code covers welds loaded in tension and would be applicable to the radiographic examination of weldor qualification welds when the job pertains to a highway or railway bridge. It specifies that regardless of the method of inspection welds shall have no cracks, and, shall have no other discontinuities exceeding the following limits in size or frequency of occurrence:

1. The greatest dimension of any porosity or fusion-type discontinuity that is 1/16 in. or larger in greatest dimension shall not exceed the size of discontinuity (B) in Fig. 11-74 for the effective throat thickness or weld size involved. The distance from any porosity or fusion-type discontinuity described above to

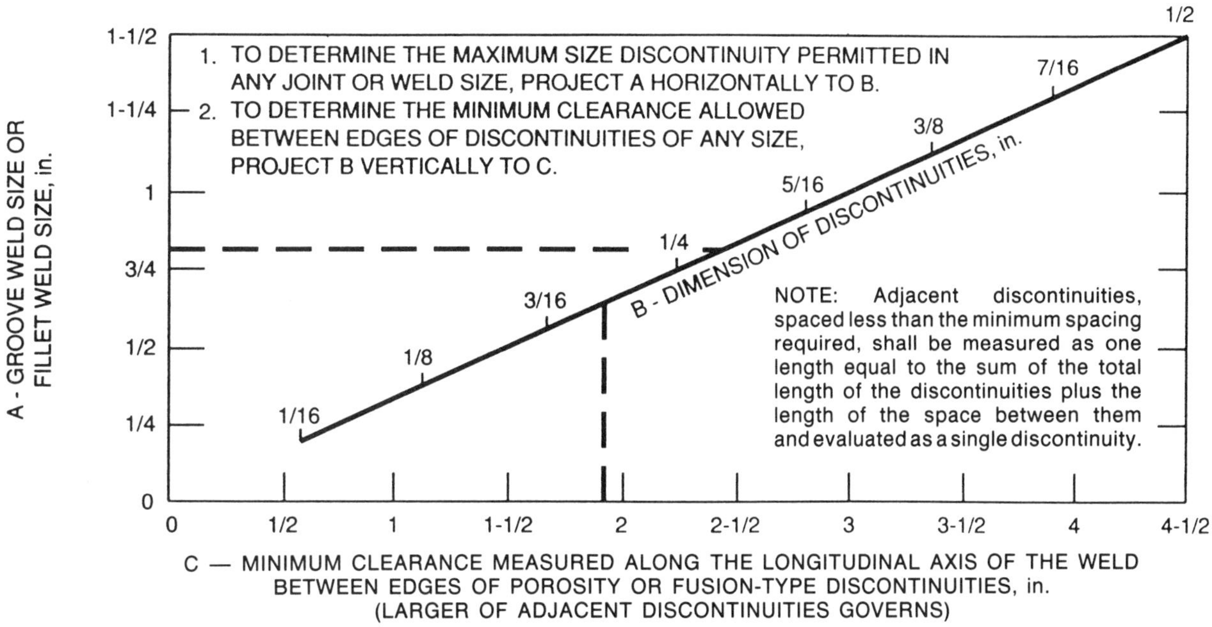

Fig. 11-74. Weld discontinuity limitations of the AWS Bridge Welding Code.

another such discontinuity, to an edge, or to any intersecting weld shall not be less than the minimum clearance allowed (C) in Fig. 11-74 for the size of discontinuity under examination. The limitations given by Fig. 11-74 for a 1-1/2-in. joint or weld-throat thickness shall apply to all joints or weld throats of greater thickness.

2. Independent of the above requirements, the sum of the greatest dimensions of porosity and fusion-type discontinuities less than 1/16 in. in greatest dimension shall not exceed 3/8 in. in any linear inch of weld.
3. The frequency of piping porosity in fillet welds shall not exceed one in each four inches of length, and the maximum diameter shall not exceed 3/32 in. For fillet welds on web stiffeners the sum of diameters of piping porosity shall not exceed 3/8 in. per linear inch of weld, nor 3/4 in. per foot of weld.

The limits for welds loaded statically are covered in the discussion of D1.1 and shown in Fig. 11-72.

WELDOR QUALIFICATION – AWS D10.9 SPECIFICATION FOR QUALIFICATION OF WELDING PROCEDURES AND WELDERS FOR PIPING AND TUBING

AR-1 Weldor Qualification

The welding positions for weldor qualification are the same as for procedure qualification (Fig. 11-30).

The sample welds and the tests required are shown in Table 11-9. The notes to this table are as follows:

TABLE 11-9. AR-1 Level of Weldor Qualification

SAMPLE WELDS					TESTS REQUIRED ON SAMPLE WELDS (c)							
		Pipe or Tube Wall (b) Thickness Qualified, In.					Band Tests (f) Number of Specimens					
							All Positions except 5G & 6G			5G & 6G Positions Only		
Pipe Size (a) of Sample Weld	Pipe or Tube Size Qualified	Min.	Max.	Number of Sample Welds per Position	Visual (d) Inspection	Radiog-raphy (e)	Face	Root	Side	Face	Root	Side
2″ Sch. 80	Through 4″	.063	.674	1	Yes	Yes	1	1	0	2	2	0
5″ Sch. 80	Over 4″	.187	.750	1	Yes	Yes	1	1	0	2	2	0
8″ Sch. 120	5″ and Over	.187	Any	1	Yes	Yes	0	0	2	0	0	4

NOTES: See Text ("AR-1 Weldor Qualification") for explanation of (a) through (f).

(a) The same as Note (a) under AR-1 procedure qualification.

(b) The same as Note (c) under AR-1 procedure qualification.

(c) Weldors qualified prior to the issuance of Standard AWS D10.9 may be updated by performing any additional tests required by the standard.

(d) The same as Note (f) under AR-1 procedure qualification, with the exception that —

Burring or grinding may be used on stops and starts and to remove local discontinuities from bead to bead, but dressing of the completed sample weld is limited to removing local discontinuities appearing on the outer surface of the weld reinforcement. Repair of any root discontinuities is prohibited.

The weldor qualification test may be terminated at any stage of the testing whenever it becomes apparent to the supervisor conducting the test that the weldor (a) is not following the welding procedure or (2) does not have the skill required to produce satisfactory results.

(e) Radiographic inspection is required by all weldor qualification tests for AR-1 level. For specific techniques and acceptance requirements, see AWS D10.9. Radiographic inspection shall not be used to locate sound or defective areas and thereafter make destructive tests to qualify or disqualify a weldor.

(f) For 1G and 2G positions, remove two specimens at random approximately 180° apart, as shown in Fig. 11-75.

For the 5G and 6G positions, remove four specimens as shown in Fig. 11-76.

Machine bend specimens from 2-in. sample welds as shown in Fig. 11-34.

Machine bend specimens from 5 in. and 8 in. sample welds as shown in Fig. 11.33

The bend jigs (Fig. 11-35), methods of testing, and the acceptance requirements are the same as note (j) to Table 11-4, in reference to procedure qualification.

AR-3 Weldor Qualification

The qualifying welding positions are as shown in Fig. 11-30. The sample welds and tests are shown in Table 11-10.

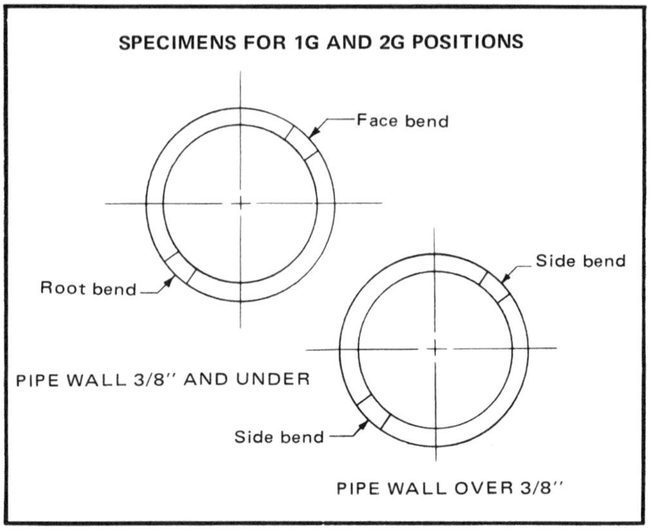

Fig. 11-75. Location of bend specimens for the 1G and 2G positions in weldor qualification tests AWS D10.9.

The notes are:

(a) The same as Note (a) under AR-1 procedure qualification.

(b) The same as Note (c) under AR-1 procedure qualification.

(c) The same as Note (c) under AR-1 weldor qualification.

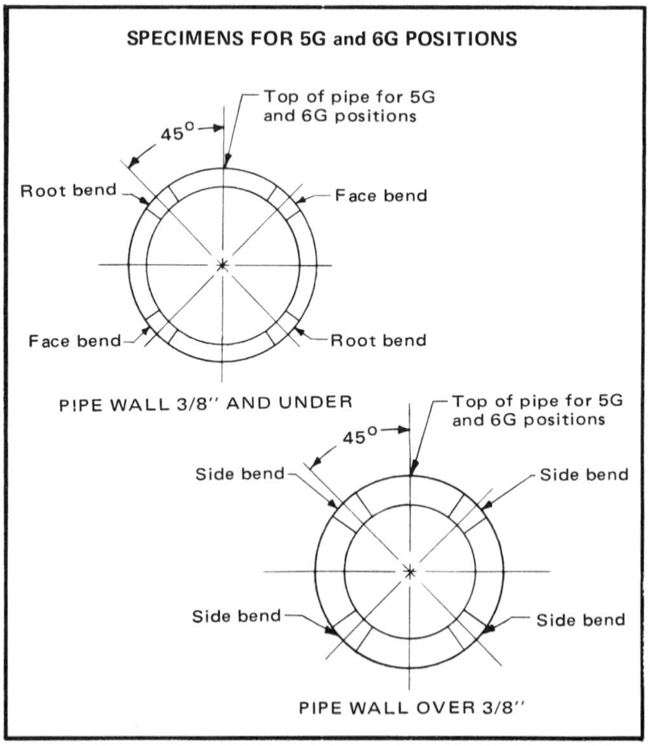

Fig. 11-76. Location of bend specimens for the 5G and 6G positions in weldor qualification tests AWS D10.9.

TABLE 11-10. AR-3 Level of Weldor Qualification

SAMPLE WELDS					TESTS REQUIRED ON SAMPLE WELDS (c)		
Pipe Size (a) of Sample Weld	Pipe or Tube Size Qualified	Pipe or Tube (b) Wall Thickness Qualified, In. Min.	Max.	Number of Sample Welds per Position	Visual (d) Inspection	Pressure Bend, (e) Number of Sample Welds Tested	Macrostructure (f) Examination, No. of Etched Specimens
2″ Sch. 40	Through 4″	.063	.375	1	Yes	1	0
5″ Sch. 40	4-1/2 Through 10″	.187	.375	1	Yes	0	4

NOTES: See Text ("AR-3 Weldor Qualification") for explanation of (a) through (f) AWS D10.9.

(d) The same as Note (d) under AR-1 weldor qualification.

(e) The same as Note (d) under AR-3 procedure qualification, except that the 1/2-in. Schedule 40 sample is omitted.

(f) The same as Note (e) under AR-3 procedure qualification.

Weldor Qualification for Fillet Welds

As mentioned earlier, welding procedure qualification and weldor qualification using groove welds automatically qualifies the procedure or the weldor for fillet welds. However, if only fillet welds are required in production, the welding procedure and the weldor can be qualified for fillet welds only. The sample welds, tests, and acceptance requirements are the same for both procedure and weldor qualification. This is discussed under the section "AWS Procedure Tests for Piping and Tubing" and Figs. 11-38, 11-39, 11-40 and 11-41.

WELDOR QUALIFICATION, ASME CODE

Weldor qualification (called "performance qualification" in the ASME code) is to demonstrate that a weldor can make sound welds when using a previously qualified procedure The code provides that the weldor who prepares the welding procedure qualification test specimens meeting the requirements of the code is thereby qualified without further testing.

Weldor qualification tests are made on groove welds or on fillet welds. Weldors qualified on groove welds are automatically qualified for fillet welds in all thicknesses. Weldors qualified on fillet welds only are qualified to make fillet welds only.

All the requirements for welding positions for procedure qualification also apply to weldor qualification.

The type, number of test specimens, and the ranges of thickness qualified are given in Tables 11-11 and 11-12. The notes to these tables are as follows:

(1) A total of four specimens are required to qualify for position 5G.

(2) The maximum thickness qualified in gas welding is the thickness of the test plate or pipe.

(3) The base material may consist of either plate or pipe. The minimum nominal diameter of 5 in. is recommended for pipe used as base material.

(4) Either face and root bends or side bends may be used for thicknesses from 3/8 to 3/4 in.

(5) Longitudinal bend tests may be used in lieu of the transverse bend tests in Table 11-11 only for testing material combinations differing markedly in mechanical bending properties between (a) the two base materials or (b) the weld metal and base materials. See Fig. 11-47 for the weld test plate.

The base material for the test specimens may be either plate or pipe. The minimum nominal diameter of 5 in. is recommended for pipe used as base material.

The dimensions of the welding groove for the test joint used in making weldor qualification tests, on double-welded butt joints and single-welded butt joints with backing strip, shall be the same as those for procedure qualification, or shall be as shown in Fig. 11-77.

Qualification on a single-welded plate with a backing strip shall also qualify for pipe with a backing strip and vice versa in positions 1G and 2G only (Figs. 11-27 and 11-30).

Qualification on a single-welded plate without a backing strip shall also qualify for single-welded pipe without a backing strip and vice versa in positions 1G and 2G only.

For all other positions qualification on pipe shall qualify for plate but not vice versa.

Test Specimens: Test welds made in plate — the type and number of bend test specimens specified

TABLE 11-11. Welder Qualificaiton. Type, Number of test Specimens, and Ranges of Thickness Qualified

Type of Joint	Thickness t of Test Plate or Pipe As Welded, In. (3)	Range of Thickness of Material Qualified by Test Plate, In.		Type and Number of Tests Required (1)			
				Transverse Bend Tests (5)			
		Min.	Max. (2, 3)	Side Bend	Face Bend	Root Bend	T Joint
Groove	1/16 to 3/8 inclusive	1/16	2 t		1	1	
Groove (4)	Over 3/8 but less than 3/4	3/16	2 t		1	1	
Groove (4)	Over 3/8 but less than 3/4	3/16	2 t	2			
Groove	3/4 and over	3/16	Max to be welded	2 (4)			
Fillet	See Fig. 11-78	All thicknesses					1

NOTES: See Text ("Weldor Qualification, ASME Code") for explanation of (1) through (4).

TABLE 11-12. Weldor Qualification. Type, Number of Test Specimens, and Ranges of Thickness Qualified (5)

Type of Joint	Thickness t of Test Plate or Pipe As Welded, In.	Range of Thickness of Material Qualified by Test Plate, In.		Type and Number of Tests Required		
				Longitudinal Bend Tests		
		Min.	Max. (2)	Face Bend	Root Bend	T Joint
Groove	1/16 to 3/8 inclusive	1/16	2 t	1	1	
Groove	Over 3/8	3/16	2 t	1	1	
Fillet	See Fig. 11-78	All thicknesses				1

NOTES: See Text ("Weldor Qualifications, ASME Code") for explanation of (2) through (5).

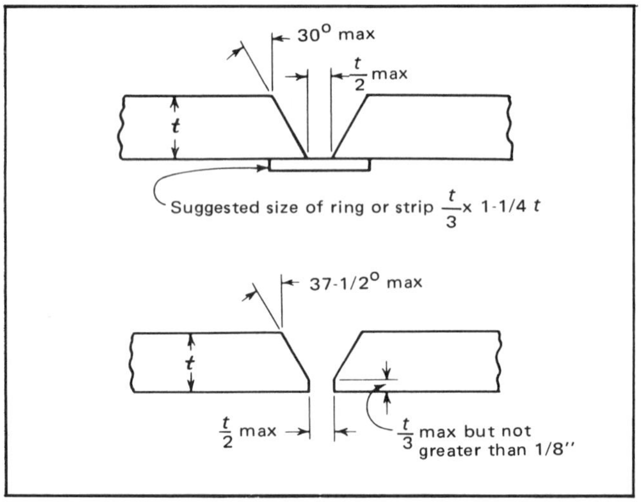

Fig. 11-77. Butt joints for welder qualification tests, with and without backing strip (ASME).

in Tables 11-11 and 11-12 — may be removed from the test plate in any order, but in a manner similar to that shown in Figs. 11-46, 11-47 and 11-48.

For test welds made in pipe in positions 1G or 2G, specimens shall be removed as shown for bend specimens in Fig. 11-48, at approximately 90° apart, omitting the tension specimens. For test welds made on a pipe in position 5G, specimens shall be removed in accordance with Fig. 11-48, and all four specimens shall pass the test.

All bend specimens shall be machined, tested, and meet the same test requirements as the procedure qualification tests.

Fillet-Weld Tests: Dimensions and preparation of the fillet-weld test specimens shall conform to the requirements of Fig. 11-78. The test specimen shall contain no visible cracks. The specimen shall be cut transversely to provide a center section 10 in. long and two end sections each approximately 1 in. long.

The stem of the 10-in. center section shall be located laterally in such a way that the root of the weld is in tension. The load shall be applied until the specimen fractures or bends flat upon itself. In order to pass the test —

(1) The specimen shall not fracture; or

(2) If it fractures, the fractured surface shall show no evidence of cracks or incomplete fusion, and the sum of the lengths of the inclusions and gas pockets shall not exceed 2 in.

One of the end sections shall be smoothed and etched with a suitable etchant to give a clear definition of the structure of the weld metal. In order to pass the test —

(1) The section shall show complete fusion at the root and no cracks;

(2) The weld shall not have a concavity or convexity

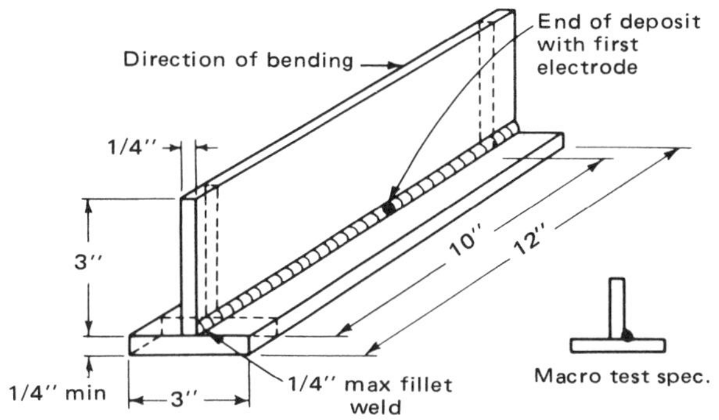

FILLET WELD SOUNDNESS TEST
FOR WELDOR QUALIFICATION

FRACTURE TEST: Maximum permissible flaws such as slag, non-fusion, etc. - 20% or 2 in. Evidence of cracking of fillet shall constitute grounds for rejection.

MACRO TEST: The fillet shall show fusion to the root of the weld but not necessarily beyond the root. Convexity and/or concavity of the fillet shall not exceed 1/16 in. Both legs of the fillet shall be equal to within 1/16 in.

Fig. 11-78. Fillet-weld specimen for weldor qualification test (ASME).

greater than 1/16 in.; and

(3) There shall be not more than 1/16 in. difference in the lengths of the legs of the fillet.

API QUALIFICATION OF WELDORS

The purpose of the weldor qualification tests is to determine the ability of weldors to make sound welds using a previously qualified procedure. Fillet welds are evaluated by destructive tests. Butt welds may be tested by destructive tests or by radiography. API Standard 1104 prescribes the radiographic standards of acceptability.

Table 11-13 gives the type and number of butt-weld specimens for weldor qualification testing, and Fig. 11-79 shows how specimens are to be taken.

Tensile test requirements are the same as for procedure qualification, except that if the specimen should break in the weld, it must meet the nick-break test requirements. The nick-break test requirements are the same as for procedure qualification except that fish eyes are not considered cause for rejection.

The bend test requirements are also the same as for procedure qualification. Welds on high-test pipe (API Std. 5LX) may not bend the full U shape. These shall be considered acceptable if the specimens that crack are broken apart and the exposed surfaces meet the nick-break test requirements.

Fillet-weld specimens are cut from each test weld. If the test weld is a complete circumferential weld, the locations from which the specimens are to be removed are as shown in Fig. 11-64. If the test weld consists of segments of pipe nipples, an approximately equal number of specimens must be removed from each segment. The specimens should be air-cooled to ambient temperature before testing.

The preparation of the fillet-weld test specimens, the testing procedures, and the test requirements are the same as for procedure qualification.

TABLE 11-13. Type and Number of Butt-Weld Test Specimens for Weldor Qualificaiton Test and for Destructive Testing of Production Welds

Pipe Size, Outside Diameter - In.	Number of Specimens					
	Tensile**	Nick Break	Root Bend	Face Bend	Side Bend	Total
Wall Thickness — 1/2 In. and Under						
Under 2-3/8	0	2	2	0	0	4*
2-3/8 to 4-1/2 inclusive	0	2	2	0	0	4
Over 4-1/2 to 12-3/4 inclusive	2	2	2	0	0	6
Over 12-3/4	4	4	2	2	0	12
Wall Thickness — Over 1/2 In.						
4-1/2 and smaller	0	2	0	0	2	4
Over 4-1/2 to 12-3/4 inclusive	2	2	0	0	2	6
Over 12-3/4	4	4	0	0	4	12

* Obtain from two welds, or one full pipe section tensile specimen for pipe 1-5/16 in. and smaller.
** The tensile test may be omitted, in which case the specimens designated for this test shall be subjected to the nick-break test (API).

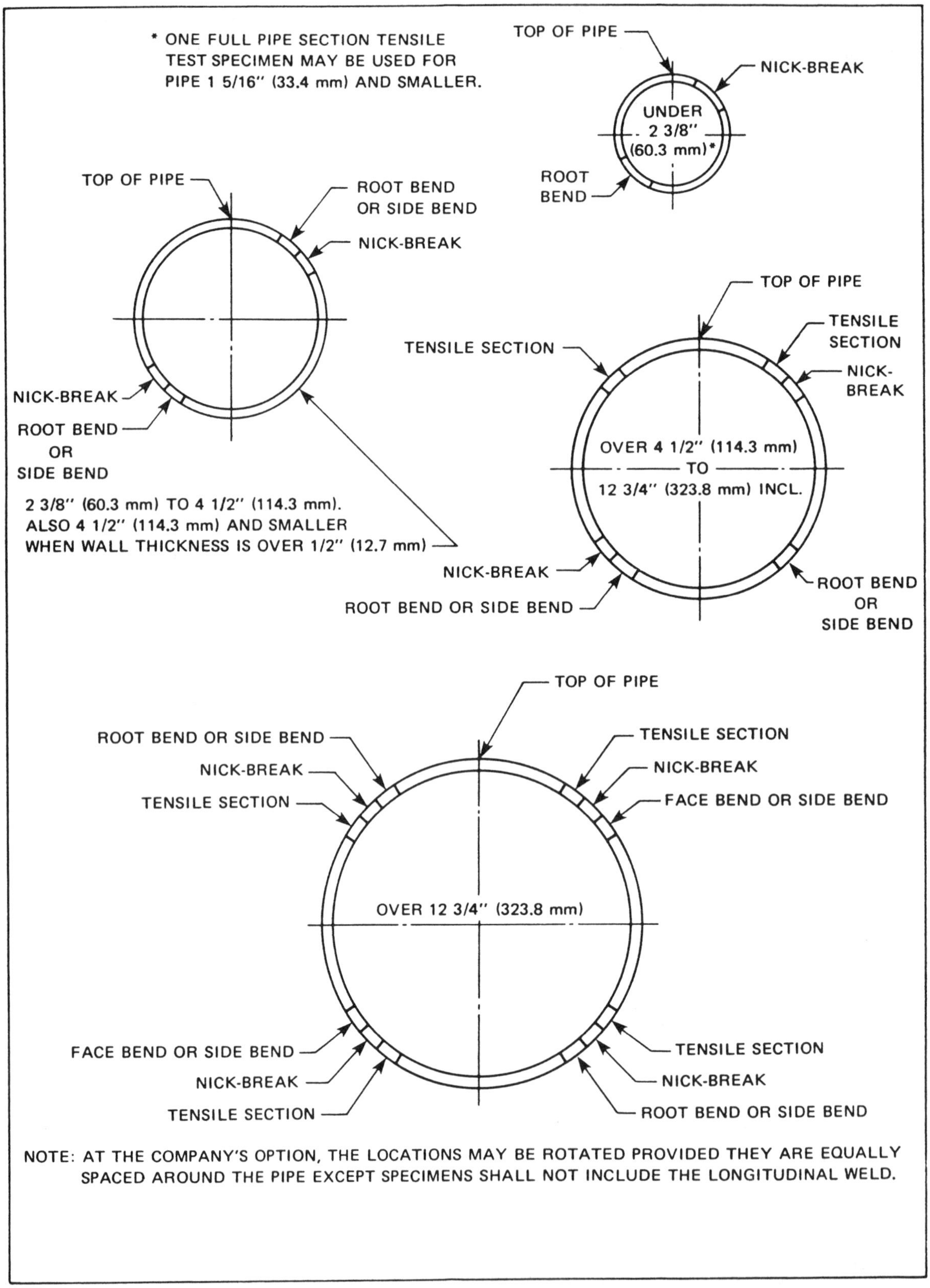

Fig. 11-79. Location of test specimens — butt weld. Welder Qualification Test Weld (API).

Section 12
WELDING COSTS

SECTION 12.1

ESTIMATING WELDING COSTS

Estimating Welding Costs

THE APPROACH TO COST ESTIMATION

All users of welding are, at times, confronted with the problem of how to determine welding costs. A cost analysis may be required in advance of actual welding — in which case the results would constitute a cost estimate — or the cost analysis may be made after welding is in progress for various evaluation or management purposes. The reasons for costing welding are varied, but most often are to:

... provide data needed for bidding on a job

... compare the economics of welding with some other method of fabricating or manufacturing

... establish information required in making a decision between alternate designs

... evaluate proposed changes in procedures

... compare the economic advantages of competing welding processes.

Estimators, designers, shop foremen, and others tend to be apprehensive about the task of cost estimation because of the many factors that must be taken into account and the possibility for errors. Also, the mathematical formulas that have been developed for cost determination look forbidding — even though they only involve simple arithmetic processes. It is the intent here to give practical methods for costing welding that can be applied by anyone and that will lead to usable results. Once one gets beyond rule-of-thumb techniques for approximate estimates, extensive tables of formulas are required, but the use of these formulas is easy.

In fact, by their construction, the formulas eliminate most problems in dealing with dimensions — feet, inches; minutes, hours; cents, dollars — and allow the user to put the data he has into the equation that fits and come up promptly with an answer.

A word of caution, however, is desirable. Although the techniques for determining or estimating welding costs are sound and have been proved in practice, the results of an analysis will only be as good as the data inserted in the cost equations. Some of the data to be used will be dependent on the human factor; other data will be dependent upon corporate accounting practices. Where unrealistic figures are assigned — say, to welding speeds or to the operating factor — the cost formulas will give unrealistic results. It, thus, behooves the person making the cost analysis to verify the validity of his data before applying it to a cost formula. That sound data can be obtained is proved by the experience of large companies that, through years of experience, are able to estimate figuratively to the cent on repetitive welding. Also, the ability of an experienced shop foreman to "guess" the cost of a job with high accuracy suggests that sound judgments about intangible factors can be developed.

FACTORS INVOLVED IN WELDING COSTS

Before attempting to calculate welding costs, it is desirable to have a clear understanding of the factors that are involved in welding costs. Every operation made necessary by the decision to weld, rather than join by another method, is legitimately chargeable to weld fabrication. Thus, weld joint preparation, although it may be a cutting or machining operation, is a cost made necessary by the decision to weld. The cost of it would have to be added to the cost of the welding operation itself to give a true picture of overall cost when making a comparison with the cost of joining by another method. By the same token, any drilling or bending to accommodate bolt assembly would have to be added to the cost of joining by bolts to make a cost comparison with weld joining.

Assuming the parts coming to the welding shop are correctly sized, the costs that might be chargeable to weld fabrication are:

1. Time for weld joint preparation
2. Time for assembly
3. Time for tack-up
4. Time for positioning
5. Time for welding
6. Downtime for changing electrodes, moving location, changing settings
7. Time spent by personnel for personal purposes
8. Time for cleaning
9. Time for removal of the assembly

10. Time for stress relieving (if required)
11. Cost of electrodes
12. Cost of shielding materials
13. Cost of electric power
14. Overhead costs

In production practice, all of these costs may or may not be chargeable to the welding shop for costing purposes. Some plants, for instance, may figure that whatever the method of assembly, parts must be brought together and positioned correctly, with some preparation for that method of assembly. Thus, the costs of joint preparation and positioning may be absorbed in another operation. When, as is often the case, crews are used for handling and preparing material, assembly, and positioning, the operating factor and the economics of welding show up most favorably. In comparing the economics of methods for joining — or even of different welding processes — management should make certain that the practice of cost accounting does not lead to wrong conclusions.

At this point, it might be well to note that the work done in advance by the designer, and the decisions made as to process, procedure, and electrode selection, profoundly affect welding costs. Here, the assumption is made that estimation of welding cost will be based on an economical design, using efficient processes, procedures, and materials. In previous sections of this book, attention has been given to the achievement of economies through the proper design of joints and the sizing of welds. Later in this section, design practices affecting welding costs will be treated and data given that will help the designer minimize production costs. In the immediately following text, the design is considered finalized, the process, procedures, and materials have been selected, and the problem is to determine the costs incurred in the welding shop.

THE OPERATING FACTOR

Review of the factors listed as involved in welding costs shows that there are various "times" listed other than the time of welding. Anytime the weldor's arc is not struck is time when the actual joining process is not in progress. While assembling, positioning, and tacking, the weldor is "preparing" to join. When removing an assembly from the positioner or cleaning a weld, he is executing necessary "follow-up" operations. When changing electrodes, moving from one location to another, or taking a coffee break, he is not joining. The total hours he works are always more than the hours he welds, and the ratio of the hours spent in welding to the total hours worked is known as the operating factor. This factor is one of the most basic factors in cost formulas, and must be determined accurately in order to make sound evaluations of costs.

The operating factor may be defined by the equation:

$$OF = \frac{\text{Arc Time}}{\text{Total Time}}$$

when total time is arc time plus the time the weldor spends in operations other than welding, including time for personal purposes. Since arc time is always divided by a larger number, the ratio is always less than 1.0, and thus a decimal. For convenience in referring to operation factors, the ratio is multiplied by 100 and expressed as a percentage. Thus, one speaks of an operating factor of 30, 40, or 50%. When using an operating factor in a cost formula, however, it must be kept in the decimal form. Thus, a 40% operating factor would be written .40 in a cost formula.

It is desirable, of course, to have the operating factor as high as possible, since a high operating factor means that a high percentage of the weldor's time is being used in joining — and joining is what gets the weldment out of the welding booth and on to the next production station. But when comparisons are made between operating factors, one must be sure that "oranges are being compared with oranges and apples with apples." Thus, in the fabrication of a particular assembly using a specific stick electrode welding procedure, one would be justified in concluding that an operating factor of 50% would be better than an operating factor of 40%. With the 50% factor, the shop might turn out 25 assemblies a day, against 20 assemblies with the 40% operating factor. If, on the other hand, we were to introduce a mechanized welding process on this same job that tripled the welding speed while the non-arc time remained the same, the arc time would be reduced to 3.2 minutes/assembly and the operating factor would drop to 18%. The reduced operating factor in this case, however, would not mean that the number of assemblies welded per day would drop. Actually the output would increase to over 27 assemblies a day.

Figure 12-1, where arc time (T_u) is plotted against down time (T_d), shows that an increase in the operating factor (OF) is not always required to achieve an increase in production. As an example, consider a job where unit completion time was thus:

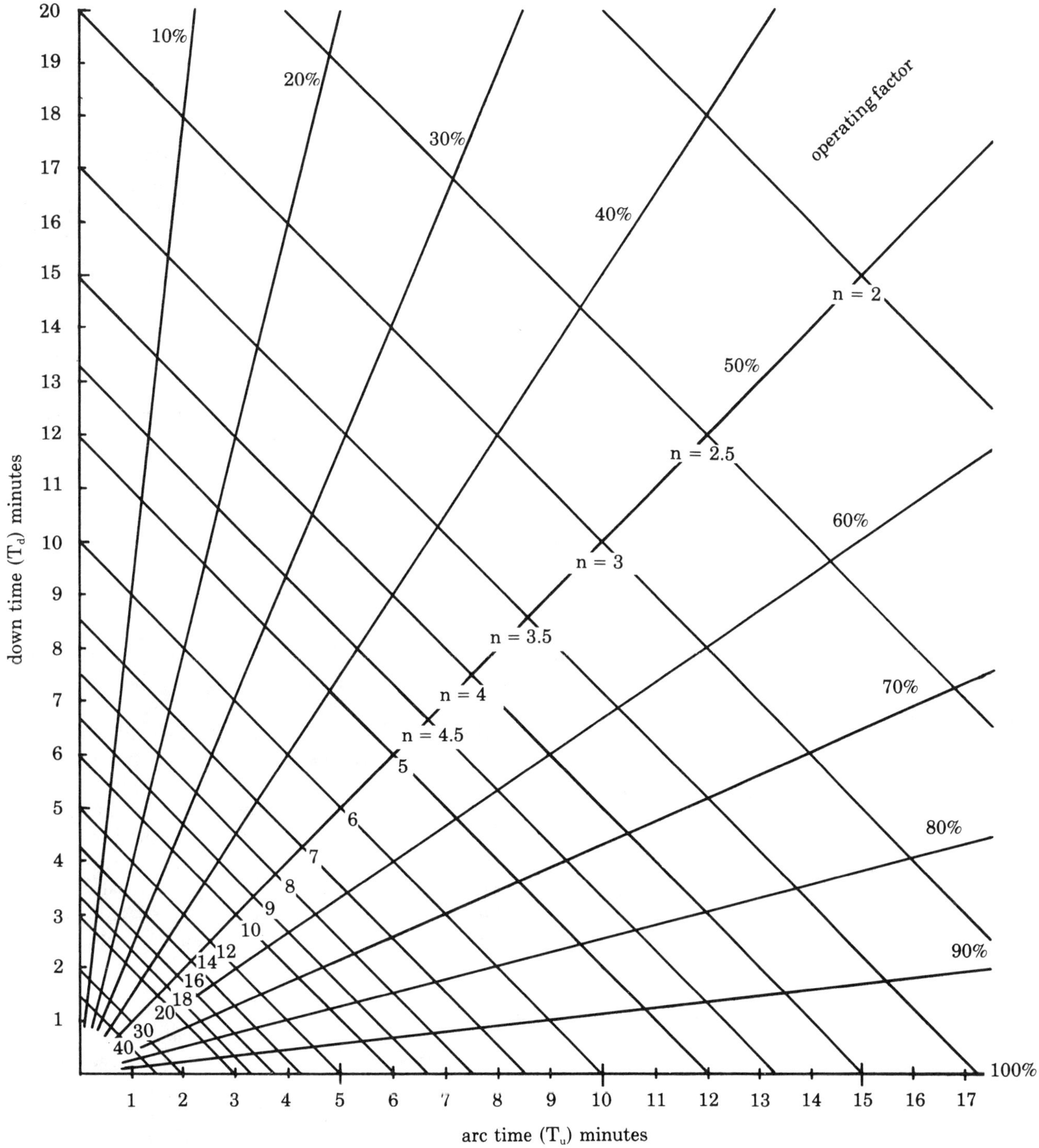

$$\text{OPERATING FACTOR (OF)} = \frac{T_u}{T_u + T_d} \times 100\%$$

$$\text{UNIT PRODUCTION (n)} = \frac{60}{T_u + T_d} \text{ (Units/Hour)}$$

Fig. 12-1. Plot of operating factor as a function of arc time and down time.

Arc Time (T_u) = 16 min

Down Time (T_d) = 16 min

$$\text{Operating Factor (O F)} = \frac{T_u}{T_u + T_d} \times 100 = 50\%$$

$$\frac{\text{Unit Production (n)}}{\text{(8-hour day)}} = \frac{60}{16 + 16} \times 8 =$$

$$= 15 \text{ units/day}$$

Suppose a welding method is introduced that produces a 100% increase in welding speed, with down time remaining the same. Now:

T_u = 8 min
T_d = 16 min

$$\text{O F} = \frac{8}{8 + 16} \times 100 = 33\%$$

$$\frac{\text{Unit Production (n)}}{\text{(8-hour day)}} = \frac{60}{8 + 16} \times 8 =$$

$$= 20 \text{ units/day}$$

Because of a doubling of arc speed, even with a lowered operating factor, production has increased.

Operating factors must always be compared against operating factors derived under basically similar conditions for the comparison to be meaningful. It should be pointed out, however, that improved fixturing and methods that reduce non-arc time should be employed with the introduction on welding mechanization.

Once a job is in progress, the operating factor can be determined by establishing the arc time and dividing it by the worker's total time. The average for several measurements made at a number of welding stations will give a more reliable percentage than one measurement made at one station. For cost estimating purposes in advance of the welding on the particular job, the company's records and the experience of the shop superintendent or foreman may be used to establish the operating factor, or, if the unit is small, specimens may be run through the shop for determining what percentage of the weldor's time is arc time.

From their past experience, most shops have a fairly good idea of what their operating factors are at various stations with the types of weldments fabricated at these stations. One quick, easy method of verifying the accuracy of an adopted figure for operating factor is to compare actual electrode consumption in pounds per hour, day or week, against a calculated value. This calculation would include consideration of the welding procedure, electrode deposition rate, and deposition efficiency, as well as operating factor. All of these terms and the theoretical approach are defined later.

BASIC COSTING PROCEDURES

In many situations, welding costs can be estimated without resorting to use of detailed cost formulas. A few basic equations, with some assistance from prepared nomographs, may be all that is needed.

It is obvious that certain basic factors make up the total cost of welding. These are:

... direct labor costs

... overhead costs (which include amortization of plant and equipment)

... welding consumables costs (electrode, flux, wire, shielding gas, etc.)

... power costs

Total welding costs can, thus, be expressed by the equation:

Total Welding Costs =

(Labor and Overhead Costs) +
(Welding Consumables Costs) + (Power) .... (1)

When the job is in progress, labor costs can be taken directly from the records and multiplied by whatever overhead percentage the shop uses to get "Labor and Overhead Costs." If the cost determination is being made in advance of the job, however, labor and overhead costs will have to be calculated through the use of another equation, using data available on the drawings, or that has been preset by the company's charge practices, or pre-established by the company's experience, or taken from standard welding procedures. "Labor and Overhead Costs" may be calculated by the following equation:

Labor and Overhead Costs per job =

$$\frac{\text{(Total Feet of Weld)} \times \text{(L. \& O. Charge/Hr.)}}{\text{(Welding Speed-Feet of Joint/Hr.)} \times \text{(Operating Factor)}} \quad \ldots \quad (2)$$

"Total Feet of Weld" is no problem — it can be taken directly from the drawings. "Hourly Labor & Overhead Charge" will be the rate that the company has preset to cover direct labor costs and overhead costs. "Operating Factor" will be that established from experience, records, or a time study. Only "Welding Speed" is likely to be the unknown at the time the calculation is attempted. This can be found in three ways: first, and preferably, by drawing upon past production-floor experience with the same type of joint, same procedures, and same material; secondly, by running a test weld and obtaining a figure in feet per hour; or thirdly, by obtaining the value from a handbook, if the joint is one for which there are standard procedures.

The importance of the operating factor should be noted. An error in its evaluation can change the values for labor and overhead costs considerably. The extent of the change will depend on whether the error is in the low or high range of the operating-factor scale. Thus, an error of 0.10 at 0.30 on the operator-factor scale would be an error of 33⅓%, while the same error at 0.80 would be an error of 12½%. Operating factors often lie in the 0.15 to 0.35 range, where the values obtained for labor and overhead by equation (2) can be seriously affected by an error in the operating factor. In highly mechanized automatic welding where the operating factor may run as much as 75 to 80%, an error of 5 to 10% in the operating-factor would be much less significant.

After calculating the value for Labor and Overhead Costs to go into Equation (1), "Welding Consumables" must be determined. The following equation applies:

Welding Consumables Cost =

(Electrode Consumption per Foot of Weld) × (Total Feet of Weld) × (Electrode Price) + (Shielding Materials used per pound of electrode) × (Total pounds of electrode consumed) × (Shielding Materials Price) . . . . (3)

"Electrode Consumption per Foot of Weld" is preferably taken from floor experience, or can be calculated by computing the weight of metal required to make the joint and modifying the value obtained to take into account stub losses, the weight of coating, and spatter. "Total Feet of Weld" has already been taken from the drawings, and "Electrode Price" is a set value.

Computing electrode consumption from known weld metal requires the use of an electrode "Deposition Efficiency" figure. This figure gives the pounds of weld metal that can be expected from each pound of electrode consumed, expressed in percent. Deposition Efficiency accounts for coating and spatter, but makes no consideration of stub losses. The Deposition Efficiency for most hand welding electrodes will fall within the range of 65 to 70%. The flux-cored wire electrodes for mechanized welding approximate 80%, and the solid wires can be considered 100% efficient.

Stub loss ranges from 2 inches up to 4 inches, depending on shop practices. A typical stub loss length considered reasonable for stick electrode welding is 2 inches. Stub loss doesn't enter the picture when mechanized processes are used.

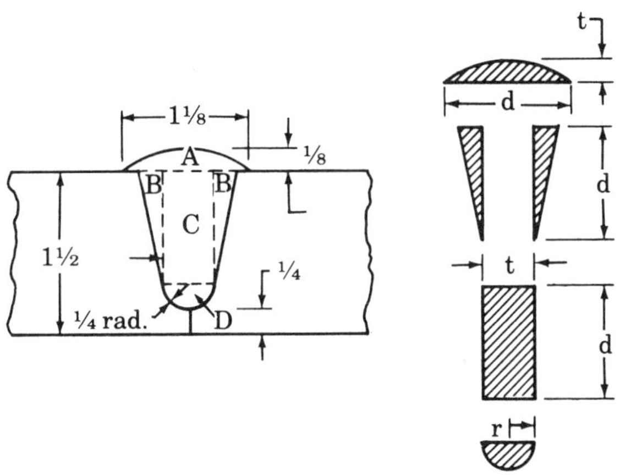

A Since t = ⅛″ and d = 1⅛″ read from Table 1: .318 lbs/ft
B Since included angle is 14° and d = 1″ read from Table 1: .417 lbs/ft
C Since t = ½″ and d = 1″ read from Table 1: 1.7 lbs/ft
D Since r = ¼″ read from Table 1: .334 lbs/ft

Total: 2.77 lbs/ft

Fig. 12-2. Components of a groove weld.

TABLE 12-1. Weight of Weld Metal (Pounds Per Foot of Joint)

(Flat, convex, concave: values below are for leg size 10% oversize, consistent with normal shop practices.)

d or r dimension	t 1/16	t 1/8	t 3/16	t 1/4	t 3/8	t 1/2	t 1/16	t 1/8	t 3/16	t 1/4	r	14°	20°	30°	45°	60°	70°	75°	80°	90°	30°	45°	60°	flat w	convex w	concave w	d or r dimension
1/16											.021																1/16
1/8	.027	.053	.080	.106	.159	.213					.083	.0065	.0094	.014	.022	.031	.037	.041	.045	.053	.015	.027	.046	.032	.041	.036	1/8
3/16	.040	.080	.120	.159	.239	.319	.027				.188	.015	.021	.032	.050	.069	.084	.092	.100	.120	.035	.060	.104	.072	.093	.081	3/16
1/4	.053	.106	.159	.213	.319	.425	.035				.334	.026	.037	.057	.088	.123	.149	.163	.178	.213	.061	.106	.184	.129	.165	.145	1/4
5/16	.066	.133	.199	.266	.398	.531	.044				.522	.041	.059	.089	.138	.192	.232	.255	.279	.332	.096	.166	.288	.201	.258	.226	5/16
3/8	.080	.159	.239	.319	.478	.638	.053	.106			.751	.059	.084	.128	.198	.276	.335	.367	.401	.478	.138	.239	.414	.289	.371	.325	3/8
7/16	.093	.186	.279	.372	.558	.744	.062	.124	.186		1.02	.080	.115	.174	.270	.376	.456	.499	.546	.651	.188	.325	.564	.394	.505	.443	7/16
1/2	.106	.213	.319	.425	.638	.850	.071	.142	.213		1.34	.104	.150	.228	.352	.491	.595	.652	.713	.850	.245	.425	.736	.514	.659	.578	1/2
9/16	.120	.239	.359	.478	.717	.956	.080	.159	.239			.132	.190	.288	.446	.621	.753	.825	.903	1.08	.311	.538	.932	.651	.834	.732	9/16
5/8	.133	.266	.398	.531	.797	1.06	.089	.177	.266	.354		.163	.234	.356	.550	.767	.930	1.02	1.11	1.33	.383	.664	1.15	.804	1.03	.904	5/8
11/16	.146	.292	.438	.584	.877	1.17	.097	.195	.292	.390		.197	.283	.431	.666	.928	1.13	1.23	1.35	1.61	.464	.804	1.39	.972	1.25	1.09	11/16
3/4	.159	.319	.478	.638	.956	1.28	.106	.213	.319	.425		.235	.337	.512	.792	1.10	1.34	1.47	1.60	1.91	.552	.956	1.66	1.16	1.48	1.30	3/4
13/16	.173	.345	.518	.691	1.04	1.38	.115	.230	.345	.460		.276	.396	.601	.930	1.30	1.57	1.72	1.88	2.24	.648	1.12	1.94				13/16
7/8	.186	.372	.558	.744	1.12	1.49	.124	.248	.372	.496		.320	.459	.698	1.08	1.50	1.82	2.00	2.18	2.60	.751	1.30	2.25	1.57	2.02	1.77	7/8
15/16	.199	.398	.598	.797	1.20	1.59	.133	.266	.398	.531		.367	.527	.801	1.24	1.73	2.09	2.29	2.51	2.99	.863	1.49	2.59				15/16
1	.213	.425	.638	.850	1.28	1.70	.142	.283	.425	.567		.417	.600	.911	1.41	1.96	2.38	2.61	2.85	3.40	.981	1.70	2.94	2.06	2.64	2.31	1
1-1/16	.226	.452	.677	.903	1.35	1.81	.151	.301	.452	.602		.471	.677	1.03	1.59	2.22					1.11	1.92	3.32				1-1/16
1-1/8	.239	.478	.717	.956	1.43	1.91	.159	.319	.478	.638		.528	.759	1.15	1.78	2.48					1.24	2.15	3.73	2.60	3.34	2.93	1-1/8
1-3/16	.252	.505	.757	1.01	1.51	2.02	.168	.336	.505	.673		.589	.845	1.28	1.99	2.77					1.38	2.40	4.15				1-3/16
1-1/4	.266	.531	.797	1.06	1.59	2.13	.177	.354	.531	.708		.652	.937	1.42	2.20	3.07					1.53	2.66	4.60	3.21	4.12	3.61	1-1/4
1-5/16	.279	.558	.837	1.12	1.67	2.23	.186	.372	.558	.744		.719	1.03	1.57	2.43	3.38					1.69	2.93	5.07				1-5/16
1-3/8	.292	.584	.877	1.17	1.75	2.34	.195	.390	.584	.779		.789	1.13	1.72	2.66	3.71					1.86	3.21	5.57	3.89	4.99	4.37	1-3/8
1-7/16	.305	.611	.916	1.22	1.83	2.44	.204	.407	.611	.815		.863	1.24	1.88	2.91	4.06					2.03	3.51	6.08				1-7/16
1-1/2	.319	.638	.956	1.28	1.91	2.55	.213	.425	.638	.850		.939	1.35	2.05	3.17	4.42					2.21	3.83	6.63	4.63	5.93	5.20	1-1/2
1-9/16	.332	.664	.996	1.33	1.99	2.66	.221	.443	.664	.885		1.02	1.46	2.22	3.44	4.79					2.40	4.15	7.19				1-9/16
1-5/8	.345	.691	1.04	1.38	2.07	2.76	.230	.460	.691	.921		1.10	1.58	2.41	3.72	5.18					2.59	4.49	7.78	5.43	6.96	6.11	1-5/8
1-11/16	.359	.717	1.08	1.43	2.15	2.87	.239	.478	.717	.956		1.19	1.71	2.59	4.01	5.59					2.79	4.84	8.38				1-11/16
1-3/4	.372	.744	1.12	1.49	2.23	2.98	.248	.496	.744	.992		1.28	1.84	2.79	4.31	6.01					3.01	5.21	9.02	6.30	8.08	7.08	1-3/4
1-13/16	.385	.770	1.16	1.54	2.31	3.08	.257	.514	.770	1.03		1.37	1.97	2.99	4.63	6.45					3.22	5.58	9.67				1-13/16
1-7/8	.398	.797	1.20	1.59	2.39	3.19	.266	.531	.797	1.06		1.47	2.11	3.20	4.95	6.90					3.45	5.98	10.35	7.23	9.27	8.13	1-7/8
1-15/16	.412	.823	1.24	1.65	2.47	3.29	.274	.549	.823	1.10		1.57	2.25	3.42	5.29	7.37					3.68	6.38	11.05				1-15/16
2	.425	.850	1.28	1.70	2.55	3.40	.283	.567	.850	1.13		1.67	2.40	3.64	5.63	7.85					3.93	6.80	11.78	8.23	10.55	9.25	2
2-1/8	.452	.903	1.35	1.81	2.71	3.61	.301	.602	.903	1.20		1.89	2.71	4.11	6.36	8.86					4.43	7.68	13.30	9.29	11.91	10.45	2-1/8
2-1/4	.478	.956	1.43	1.91	2.87	3.83	.319	.638	.956	1.28		2.11	3.04	4.61	7.13	9.94					4.97	8.61	14.91	10.41	13.35	11.71	2-1/4
2-3/8	.505	1.01	1.51	2.02	3.03	4.04	.336	.673	1.01	1.35		2.35	3.38	5.14	7.94	11.07					5.54	9.59	16.61	11.60	14.87	13.05	2-3/8
2-1/2	.531	1.06	1.59	2.13	3.19	4.25	.354	.708	1.06	1.42		2.61	3.75	5.69	8.80	12.27					6.13	10.63	18.40	12.86	16.48	14.46	2-1/2
2-5/8	.558	1.12	1.67	2.23	3.35	4.46	.372	.744	1.12	1.49		2.88	4.13	6.28	9.70	13.53					6.76	11.71	20.29	14.17	18.17	15.94	2-5/8
2-3/4	.584	1.17	1.75	2.34	3.51	4.68	.390	.779	1.17	1.56		3.16	4.53	6.89	10.65	14.85					7.42	12.86	22.27	15.56	19.94	17.49	2-3/4
2-7/8	.611	1.22	1.83	2.44	3.67	4.89	.407	.815	1.22	1.63		3.45	4.96	7.53	11.64	16.23					8.11	14.05	24.34	17.00	21.80	19.12	2-7/8
3	.638	1.28	1.91	2.55	3.83	5.10	.425	.850	1.28	1.70		3.76	5.40	8.20	12.67	17.67					8.83	15.30	26.50	18.51	23.73	20.82	3
3-1/8	.664	1.33	1.99	2.66	3.98	5.31	.443	.885	1.33	1.77		4.08	5.85	8.90	13.75	19.17					9.58	16.60	28.75				3-1/8
3-1/4	.691	1.38	2.07	2.76	4.14	5.53	.460	.921	1.38	1.84		4.41	6.33	9.62	14.88	20.73					10.37	17.96	31.10	21.73	27.85	24.43	3-1/4
3-3/8	.717	1.43	2.15	2.87	4.30	5.74	.478	.956	1.43	1.91		4.76	6.83	10.38	16.04	22.36					11.18	19.36	33.54				3-3/8
3-1/2	.744	1.49	2.23	2.98	4.46	5.95	.496	.992	1.49	1.98		5.11	7.34	11.16	17.25	24.05					12.02	20.88	36.07	25.20	32.30	28.34	3-1/2
3-5/8	.770	1.54	2.31	3.08	4.62	6.16	.514	1.03	1.54	2.05		5.49	7.88	11.97	18.51	25.79					12.90	22.34	38.69				3-5/8
3-3/4	.797	1.59	2.39	3.19	4.78	6.38	.531	1.06	1.59	2.13		5.87	8.43	12.81	19.80	27.60					13.80	23.91	41.41	28.93	37.08	32.53	3-3/4
3-7/8	.823	1.65	2.47	3.29	4.94	6.59	.549	1.10	1.65	2.20		6.27	9.00	13.68	21.15	29.48					14.74	25.53	44.21				3-7/8
4	.850	1.70	2.55	3.40	5.10	6.80	.567	1.13	1.70	2.27		6.68	9.59	14.58	22.53	31.41					15.70	27.20	47.11	32.91	42.19	37.01	4
4-1/8	.877	1.75	2.63	3.51	5.26	7.01	.584	1.17	1.75	2.34		7.10	10.20	15.50	23.96	33.40					16.70	28.93	50.10				4-1/8
4-1/4	.903	1.81	2.71	3.61	5.42	7.23	.602	1.20	1.81	2.41		7.54	10.83	16.46	25.44	35.46					17.73	30.71	53.18				4-1/4
4-3/8	.930	1.86	2.79	3.72	5.58	7.44	.620	1.24	1.86	2.48		7.99	11.48	17.44	26.96	37.57					18.79	32.54	56.36				4-3/8
4-1/2	.956	1.91	2.87	3.83	5.74	7.65	.638	1.28	1.91	2.55		8.45	12.14	18.45	28.52	39.75					19.88	34.43	59.63				4-1/2
4-5/8	.983	1.97	2.95	3.93	5.90	7.86	.655	1.31	1.97	2.62		8.93	12.82	19.49	30.12	41.99					20.99	36.36	62.98				4-5/8
4-3/4	1.01	2.02	3.03	4.04	6.06	8.08	.673	1.35	2.02	2.69		9.42	13.53	20.56	31.78	44.29					22.14	38.36	66.43				4-3/4
4-7/8	1.04	2.07	3.11	4.14	6.22	8.29	.691	1.38	2.07	2.76		9.92	14.25	21.65	33.47	46.65					23.33	40.40	69.98				4-7/8
5	1.06	2.13	3.19	4.25	6.38	8.50	.708	1.42	2.13	2.83		10.44	14.99	22.78	35.21	49.07					24.54	42.50	73.61				5

With stick-electrode welding or with self-shielded flux-cored welding, there will be no value for "Shielding Material Costs," since the shielding materials are part of the cost of the electrode. When using submerged arc welding or a welding process requiring a gas shield, a cost factor for the shielding material will be needed. The amount of the submerged arc flux consumed per pound of electrode will depend upon the recovery system used, as well as the welding procedure and other factors. It is best established by shop experience, or purchasing records, although data from standard procedures may be used. The same applies to the amount of gas consumed per pound of weld metal deposited when using gas-shielded processes. Once the poundage of flux, or cubic footage of gas, per pound of electrode deposited is known, it can be multiplied by the total pounds of electrode required, then by the cost per pound, or per cubic foot, to give the shielding material cost. Usually, power cost runs about 3 cents per pound of weld metal — a safe figure to use in cost estimation.

Once equations (2) and (3) have been solved, and a figure has been set for "Power Costs", the values needed for solving equation (1) are at hand and "Total Welding Costs" can be calculated.

COMPUTING THE WEIGHT OF WELD METAL

In the preceding explanation of how to obtain the value for "Electrode Consumption per Foot" for insertion in the "Electrode Cost" equation, it was stated that the value could be obtained by computing the weight of weld metal required to make the joint and modifying that value appropriately. The weight of weld metal (lbs) is calculated by determining the cross-sectional area of the weld, multiplying by the length of the weld, and then multiplying this product by 0.283. The product of cross-sectional area and weld length is, of course, the volume of weld metal in cubic inches, and the constant .283 is the weight in pounds of one cubic inch of steel. For fillet welds, it is desirable to increase the leg size by 10% over the theoretical size because the operator usually makes the fillets slightly oversized.

The only difficulty one is likely to encounter when computing the weight of weld metal is in determining the cross-sectional area of the weld. The weld joint may be of such a shape that the cross-sectional area can be calculated only after breaking it down into simpler areas. Even these simpler areas may involve complex calculations, and to avoid such, the chart presented in Table 12-1 is handy. The following example illustrates how the chart is used.

The joint is the groove joint shown in Figure 12-2. The shape of the weld cross section is first broken down into geometric areas, indicated by A, B, C, and D. It is now possible to ascribe values for (t), (d), and (r) and read the weight of a linear foot of such shapes from Table 12-1. Thus, for A, when the (t) dimension is 1/8", and the (d) dimension is 1 1/8" the chart indicates that the slice of weld metal will weigh 0.318 pounds per foot. The weight per foot of other shapes are similarly found and the total weight of the weld per foot determined by adding the weight of the shapes, as shown in the figure. Multiplying this value by the total length of weld in feet gives the weight of weld metal required for the job.

Computer programming has enabled the tabulation of weights of weld metal for various types of joints and varying thicknesses of material. In the body of this section are tables giving the weight of weld metal in pounds per linear foot for the various prequalified joints of the AWS Structural Welding Code D.1.1 and also tables giving the weights of pipe welds in pounds per joint for various diameters and thicknesses of pipe wall. Where applicable, reference to these tables will eliminate computation by the use of Table 12-1 and the method described.

An electrode, however, may or may not be all weld metal. Some of its weight may result from its coating, or from the flux and other materials incorporated in flux-cored electrode wire. Also, as a result of stub losses and spatter, all the electrode weight will not be available as weld metal to fill the joint. As explained earlier, the Deposition Efficiency of an electrode (65 to 70 percent for manual welding) determines how much electrode weight can be expected to be converted to weld metal. Thus in the example, 2.769 divided by 0.70 indicates that 4.0 pounds of electrode is required. In addition, if a 3 inch electrode stub is typical for 18-inch electrode, a final upwards adjustment of (18/15) of 4.0 pounds indicates that 4.8 pounds of electrode will be required to produce 1 foot of welded joint.

It should be noted that this method for determining the amount of electrode required per foot of weld is also useful in calculating the amount of electrode to be purchased for a job.

When welding procedures are already established, such as with many of the standard joints welded by submerged arc, it is possible to compute the weight of weld metal required through the meltoff rate and travel speed. Using the submerged-arc nomograph

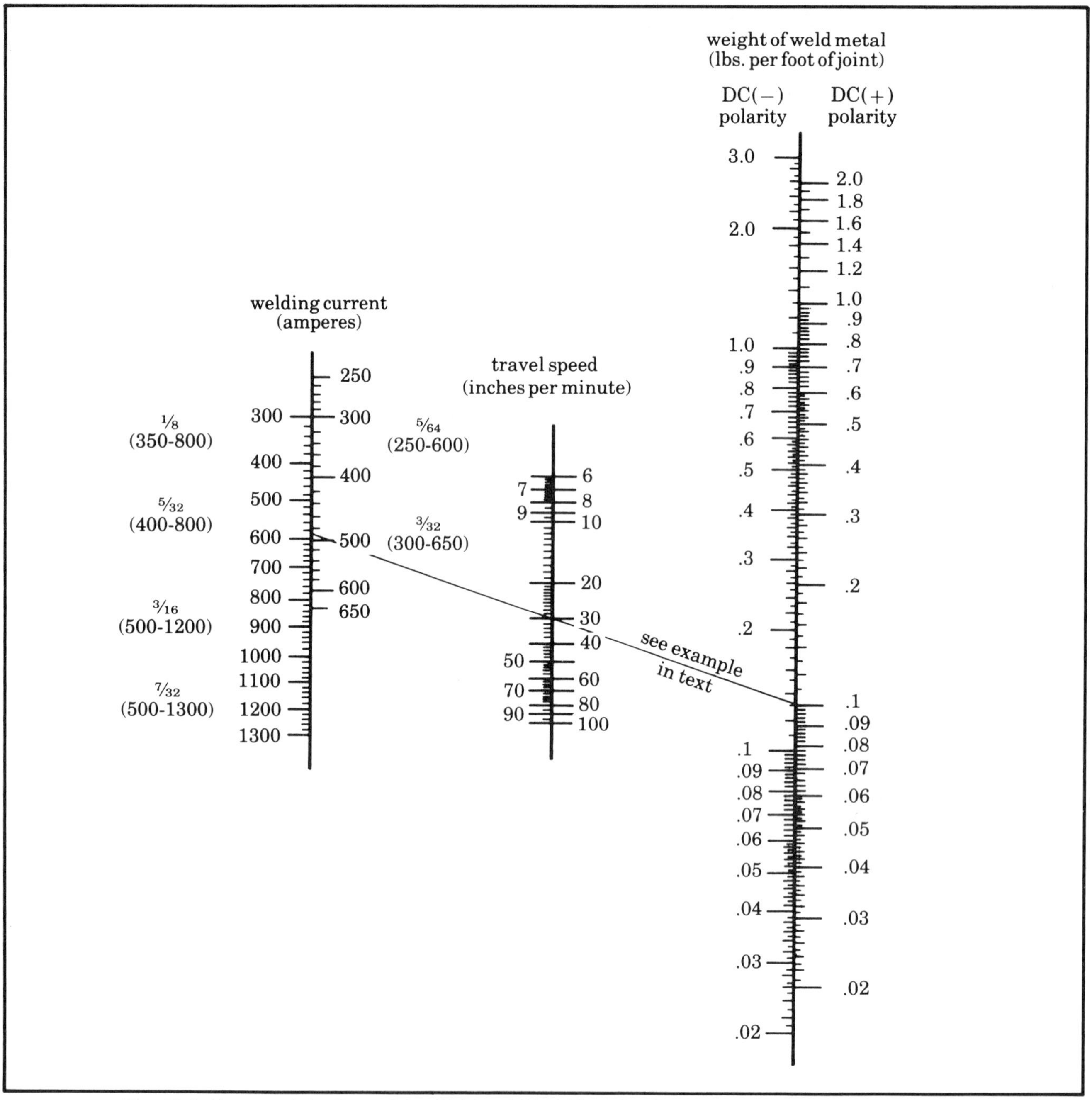

Fig. 12-3. Nomograph for Determining Weight of Weld Metal Per Foot of Joint — Based on Procedures.

in Figure 12-3 eliminates the need for tedious computations. A straight edge is aligned to pass through the value for welding current and travel speed, making sure that the proper side of the welding-current scale is used, depending on electrode size. An extension of the line established to the weight-of-weld-metal scale permits reading weight of weld metal per foot of joint from the proper polarity side of the scale.

In Figure 12-3, the line on the nomograph represents a procedure that uses 590 amperes with a 1/8″ electrode at a travel speed of 30 inches per minute. The projected line shows that 0.10 pounds of weld metal per foot of joint will be deposited at DC positive polarity.

This nomograph, incidentally, is useful in determining the procedures to use with an unconventional joint. As an example, the actual joint might be as shown in Figure 12-4 (a). The joint for which there are established procedures most closely approximating this one would be as shown in Figure 12-4 (b). Inspection suggests that 670 amperes for joint (a) should be about right and that slowing down the welding speed should be a way to get the additional fill required.

The nomograph can be used to determine the welding speed for this unconventional joint, but a second point, the amount of weld metal required per foot, is needed. This figure is determined by the area method described previously, and, in this case, amounts to 0.404 pounds per foot. Locating 670 amperes on the left-hand side of the welding scale (for 5/32″ electrode) and 0.404 pounds per foot on the DC positive polarity side of the weld-metal scale gives a line that intersects the travel speed scale at 9 inches per minute. With this much of the procedure fixed, it is usually a simple matter to adjust voltage to give the desired bead shape.

ROUGH AND READY COST ESTIMATION

Sometimes the person making the cost estimation seeks a "quick" cost figure, not for purpose of a bid but to make comparison with another cost value. Thus, a designer may wish to get a rough idea of how a proposed change in the detail of a joint will affect weld cost. For this purpose, the electrode manufacturer may supply a chart, such as that included here as Table 12-2.

In Table 12-2, a labor and overhead cost of $10.00 per hour is assumed, and the current has been pegged at a value representative of most welding. Operating factors from 20% to 60% are shown. Since the data in this table are dependent on electrode costs and labor and overhead costs that are not likely to apply at a given time, the table must be regarded as illustrative only. To make use of such rough and ready cost estimation, one would need a table based on current costs for electrodes, labor, and overhead.

A table of this type, however, is particularly helpful in suggesting possible joints, welding positions, and processes. As an example, the job might involve field welding where a low operating factor would exist. If a 5/32″ E6010 electrode were to be used at, say, 20% (0.20) operating factor, the approximate overall cost, under the assumptions incorporated in the table, would be $16.14 per pound of weld metal deposited. A study of the job might show that it could be done in a flat or downhand position with 7/32″ E7028 electrode, in which case it would be reasonable to expect an improved factor. Using 30% as the operating factor, the chart shows that a cost of $4.24 per pound of weld metal deposited is the expectancy. The difference is striking — and suggests the move to E7028 electrode should be made if feasible.

In the shop, the same welding would possibly be done with 1/4″ iron powder E7024 electrode, at an operating factor of 40%. The cost would then be reduced to $2.15 per pound. The potential for decreasing overall welding costs by downhand positioning and proper electrode selection is clearly brought out by the chart.

While it is a mistake to equate deposition rate to overall welding costs when considering the various processes and procedures because of the wide differences in the accompanying mechanical operations, deposition rate is very important when consideration is limited to one process. It is, thus, desirable to choose the electrode and current that maximize deposition rate.

The effects of current setting are shown in the following comparison:

3/16″ E7024 at 245 AC		**3/16″ E7024 at 275 AC**
7.0 lbs/hr	Electrode Deposition Rate (100% OF)	8.1 lbs/hr
2.1 lbs/hr	Electrode Deposition Rate (30% OF)	2.43 lbs/hr
47.62 hrs	Deposit Time — 100 lbs Weld Metal	41.15 hrs
$476.20	Labor Cost for 100 lbs Weld Metal at $10.00/hr	$411.50

The higher amperage setting would, thus, bring about a cost savings of $64.70 for each 100 pounds of electrode consumed.

By using a slightly larger electrode at a still higher current setting, the cost savings can be increased additionally:

3/16″ E7024 at 275 AC		**7/32″ E7024 at 340 AC**
8.1 lbs/hr	Electrode Deposition Rate (100% OF)	10.5 lbs/hr
2.43 lbs/hr	Electrode Deposition Rate (30% OF)	3.15 lbs/hr
41.15 hrs	Deposit Time — 100 lbs Weld Metal	31.75 hrs
$411.50	Labor Cost for 100 lbs Weld Metal at $10.00/hr	$317.50

A change of 1/32″ in the size of electrode and use of the higher current reduce the costs by another $94.00 per 100 pounds of electrode.

In general, cost reductions are achieved by using the largest size electrode possible. Choice of size

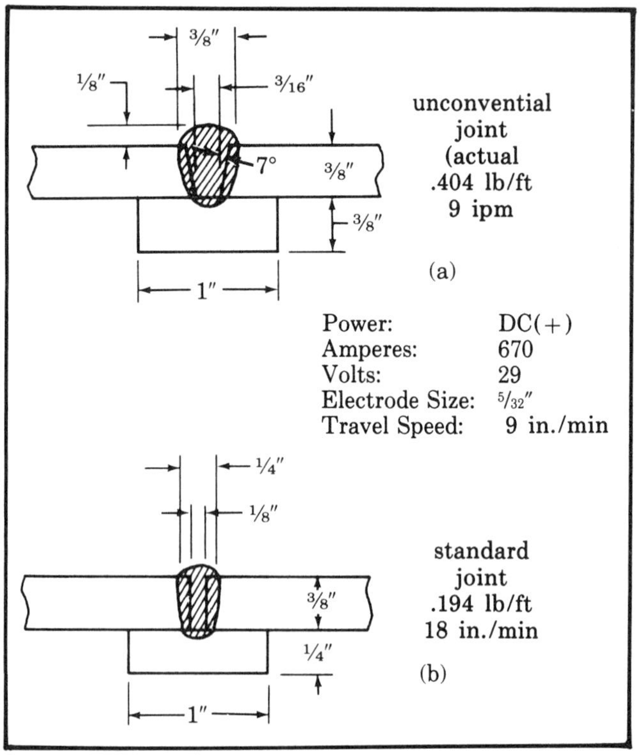

Fig. 12-4. Parameters for making butt joints.

TABLE 12-2. Approximate Cost of Deposited Weld Metal, \$/lb (Includes Electrode Cost and \$10.00/hr Labor and Overhead)

Size	Current	Operating Factor				
		60%	50%	40%	30%	20%
E6010 Fleetweld® 5-P						
1/8″	100 DC(+)	8.47	10.06	12.44	16.41	24.34
5/32″	130 DC(+)	5.73	6.77	8.33	10.93	16.14
3/16″	180 DC(+)	4.79	5.65	6.93	9.06	13.34
Low Hydrogen E7018 Jetweld® LH-70						
1/8″	125 DC(+)	8.09	9.61	11.88	15.67	23.24
5/32″	160 DC(+)	5.88	6.96	8.57	11.26	16.63
3/16″	250 DC(+)	3.65	4.28	5.22	6.79	9.94
1/4″	360 DC(+)	2.55	2.95	3.56	4.58	6.61
Low Hydrogen E7028 Jetweld LH-3800						
5/32″	215 AC	3.75	4.39	5.35	6.95	10.16
3/16″	280 AC	2.85	3.32	4.01	5.17	7.48
7/32″	330 AC	2.39	2.76	3.32	4.24	6.10
1/4″	415 AC	1.73	1.97	2.33	2.92	4.11
E6012 Fleetweld 7						
3/16″	235 AC	4.28	5.04	6.18	8.07	11.86
7/32″	290 AC	3.57	4.19	5.12	6.66	9.75
Iron Powder E7024 Jetweld 3						
7/32″	340 AC	1.83	2.10	2.49	3.16	4.48
1/4″	400 AC	1.61	1.83	2.15	2.70	3.80
5/16″*	525 AC	1.46	1.65	1.94	2.41	3.36
Submerged Arc Automatic						
Full	1000 DC(+)	1.05	1.16	1.32	1.60	2.14
Semi	500 DC(+)	1.54	1.75	2.06	2.57	3.60

* 5/16″ Jetweld 3 does not conform to any AWS/ASME classification.

will depend on the manipulative skill of the weldor, the position of the joint, the type of joint, and the edge preparation. A corollary to increased size of electrode is the potential for increased current, higher energy input, melt-off, and enhanced deposition rate.

The nomograph in Figure 12-3 gives an easy method of "rough-and-ready" estimation of welding costs. The problem included in the figure illustrates how the nomograph is used. Note that the values for (W_m) — the amount of electrode used for each inch of rod melted — found in Table 12-3, make it possible to estimate the cost of electrode as well as the labor and overhead per foot of weld.

THE USE OF WELDING COST FORMULAS

As noted in the introduction, carefully constructed mathematical formulas are available for determining welding costs. These formulas apply to most welding situations and permit the calculation of costs in terms of per-foot-of-weld, per-unit-of-production, and per-pound-of-deposit. Because so many welding situations exist and there are many knowns, unknowns, and variables, a compilation of welding cost formulas extensive enough to be useful becomes sizeable and tends to "frighten off" the user. One should bear in mind, however, that the use of these formulas is quite easy — in fact often much easier than application of the basic methods previously described. The formulas are set up in such a fashion that one merely jots down his known data and then finds the formula into which the data will fit — either directly or after making conversions. All problems of adjusting dimensions are taken care of automatically by the specifications for dimensions and by conversion factors in the equations. In essence, one can calculate welding cost through the use of these formulas without "reasoning out" the approach. After the data for the various values has been jotted down — and expressed in the specified units of measurement — the task resolves itself into searching through the formula tables for the applicable formula, followed by simple computations.

Some of the data needed in the formulas will be preset or can be established by plant records, floor experience, and experiment. Other values are obtainable from standard welding procedures or from the manufacturer's literature.

To cover all the factors that enter into cost formulas, it is necessary to have tables of standardized symbols, with each symbol carefully defined. Table 12-3 gives the symbols that are used in welding cost formulas. Table 12-4 gives formulas that may be required in calculating the values of some of the symbols in Table 12-3 from known values.

If you purchase 50 lbs. of welding electrode and after welding have 10 lbs. left, you have *used* 40 lbs.

If there are 5 lbs. of stubs left, this means that 35 lbs. of electrode were *melted.*

Of this, perhaps 25 lbs. was *deposited* as weld metal.

E_1 = deposition efficiency =

$$= \frac{25\,\text{lbs deposited}}{35\,\text{lbs melted}} \times 100 = 71.4\%$$

E_2 = overall efficiency =

$$= \frac{25\,\text{lbs deposited}}{40\,\text{lbs used}} \times 100 = 62.5\%$$

Note that the following tables — Tables 12-3 through 12-7 — are presented both in English and metric units where such are applicable. Thus, there is a Table 12-3 that defines symbols used in welding cost formulas in terms of pounds, feet, inches, etc., and a Table 12-3-M that defines symbols in terms of millimeters, liters, etc. Likewise, there is a Table 12-4 and a Table 12-4-M.

Table 12-5 is the first cost formula table. With the equations given, labor and overhead, welding materials, handling, and cleaning costs can be determined in terms of cents per foot of weld. Multiplying the cost per foot by the length of welds gives the welding cost, with the exception of electric power. Adding 3 cents for every pound of weld metal to be deposited to cover power, gives the total cost of the welding.

Table 12-6 gives formulas for calculating welding cost in terms of units of production. Note that when the cost per foot of weld is known, the use of these formulas can be avoided merely by multiplying cost per foot of weld by either of the two factors given at the beginning of the table. In using the tables, one should not forget to add on the power costs. Table 12-7 gives formulas for determining costs per pound of deposit.

An Example of the Use of Cost Formulas

The best way to become acquainted with the use of cost formulas is to follow examples step-by-step through the computation process.

In this first example, a ¼″ horizontal fillet weld is to be made with 7/32″ E7024 electrode at 350 amperes. The company uses a charge rate of $10.00 per hour to cover its labor and overhead costs. The electrode is priced at 34 cents per pound. The company wants to know what the cost per foot of weld will be.

The obvious costs will be labor and overhead, electrode, and power. No shielding material costs will be involved, since the electrode coating supplies the shielding. The time for handling materials, weldment positioning, and weld cleaning is included in the 35% operating factor that has been established by floor experience. The company has a standard policy of adding 2% to welding materials and labor and overhead costs to cover power costs — a practice that is commonly used in the fabrication industry. Thus, the only costs to be calculated are labor and overhead and electrode — and these are to be calculated in terms of costs per foot.

Reference to the cost formula table shows that Table 12-5 gives formulas for determining costs in terms of dollars per foot. Inspection of the table shows that the first formula will enable calculation of labor and overhead costs, providing the values for certain symbols are known:

$$\textbf{Labor and Overhead} = \frac{L}{5\,S\,(O\,F)}\ \text{dollars/foot}$$

To find what the symbols mean, the person making the cost analysis is referred to Table 12-3, the table defining symbols used in welding cost formulas. He finds that:

L = charge rate for labor and overhead in dollars per hour
S = inches of weld deposited per minute
O F = operating factor

At this stage, he has the values for (L), $10.00 per hour, and for (O F), 0.35. But because the weld is to be made by a standard procedure, he can get (S) directly from the literature. It turns out to be 15.5 inches per minute. All the values needed for calculating labor and overhead costs are, thus, established. Substituting these values in the cost formula:

$$\textbf{Labor and Overhead} = \frac{(10.00)}{5\,(15.5)\,(0.35)} = \$.369/\text{foot}$$

The next task will be to calculate electrode cost per foot. Again, referring to Table 12-5 the person making the analysis notes seven different formulas will give the cost per foot of manual electrode. He jots down the factors for which he has known values:

O F = 35 percent (.35)
S = 15.5 in. weld/min
L = \$10.00/hr
E = \$0.34/lb

Only (S) and (E) are knowns that would be helpful in solving his problem. But, perhaps, some of the other factors in the seven formulas are ones where values can be found in standard references. Using Table 12-3 to find the meaning of various symbols and Table 12-4 to ascertain whether he could convert known factors into the factors called for by the seven cost formulas, he finally decides that the formula:

$$\textbf{Electrode Cost} = \frac{1200\,M\,E}{N\,l_m\,S} \text{ dollars/foot}$$

gives him the best approach for calculating electrode cost. It uses (S) and (E); (M), or inches of electrode melted per minute, can be found from standard procedures; and (N), or number of electrodes per 100 pounds, is available in the manufacturer's literature. The other factor (l_m), or inches of electrode melted, is just the difference between an 18″ electrode and the stub loss — or about 16″. The values for substitution in the formula are found to be:

M = 12.3 in. electrode/min
E = \$0.34/lb
N = 262 electrodes/100 lbs
l_m = 16.0 in. electrode (2″ stub)
S = 15.5 in. weld/min

The electrode cost in dollars per foot is thus:

$$\textbf{Electrode Cost} = \frac{(1200)\,(12.3)\,(0.34)}{(262)\,(16)\,(15.5)} = \$.077/\text{ft}$$

TABLE 12-3. Symbols Used in Welding Cost Formulas (English System)

(This table defines the symbols used in the welding cost formulas given on these pages. To use the formulas, the known values should first be translated into the appropriate symbols. Then inspection of the tables of formulas will show which one is usable for calculating the desired unknown value.)

- A_w = cross section area of weld in square inches
- C_u = minutes cleaning per unit
- C_f = minutes cleaning per foot $= \dfrac{12\,C_u}{l_x}$
- d = diameter solid electrode (in.)
- D_f = pounds of weld metal deposited per foot of weld = $3.4\,A_w$
- D_h = pounds of weld metal deposited per hour at 100% O.F.
- E = electrode cost in dollars/lb.
- E_1 = deposition efficiency
- E_2 = overall efficiency
- F = flux cost in dollars per pound
- F_f = flux consumption lbs./ft
- F_h = flux consumption lbs./hr.
- G = gas cost in dollars per hour
- g = gas cost in dollars per cubic foot
- H_u = handling minutes per unit.
- I = welding current in amperes
- L = labor and overhead \$/hr.
- l_c = circumference ($\pi \times$ dia.) in.
- l_e = electrode length in inches
- l_m = inches of meltable electrode $= l_e - l_s$
- l_o = overlap in inches
- l_s = stub length in inches
- l_u = inches of electrode used per unit
- l_w = inches of weld per electrode
- l_x = inches of weld per unit. For circumferential welds = $l_c + l_o$
- M = inches of electrode melted per minute (melting rate).
- M_f = pounds of electrode melted per foot of weld
- M_h = electrode melt-off rate in lbs. per hour at 100% OF
- N = number of electrodes per 100 pounds
- n = number of units per hour $= \dfrac{60}{T_u + T_d}$
- OF = operating factor (used as decimal) $= \dfrac{T_u}{T_u + T_d} \times 100$
- R = ratio of flux to electrode weight = F_f / M_f
- r = cubic feet of gas flow per hour during welding
- S = inches of weld deposited per minute (arc speed). For multiple pass welds. $S = \dfrac{1}{\dfrac{1}{S_1} + \dfrac{1}{S_2} + \dfrac{1}{S_3}}$
- $S_{1,2}$ = inches of weld deposited per minute (arc speed) for that pass
- T_d = minutes per unit (downtime)
- T_e = minutes per electrode (arc time)
- T_r = minutes per revolution
- T_u = minutes per unit (arc time). For circumferential welds $= \dfrac{(l_c + l_o)}{l_c}\,T_r$
- U_f = pounds of electrode used per foot of weld
- U_h = pounds of electrode used per hour hour at 100% operating factor $= \dfrac{60 W_e}{T_e}$
- WFS = wire feed speed (in/min)
- W_e = weight of one electrode $= \dfrac{100}{N}$ pounds
- W_m = weight of electrode used per inch of electrode melted in lbs./in. $= \dfrac{W_e}{l_m} = \dfrac{W_e}{l_e - l_s}$
- W_s = weight of one stub in pounds
- W_t = weight of 1000 inches of tubular electrode (lbs)

Adding this cost to labor and overhead cost, the two costs become 0.369 plus 0.077, or .45 per foot. Adding an additional 2% to this to cover power, the total cost per foot for the ¼″ fillet weld using 7/32″ electrode and operating at 350 amperes becomes: $0.46.

Total Cost = $0.46 per foot

TABLE 12-3M. Symbols used in welding cost formulas (Metric System)

This table defines the symbols used in the welding costs formulas given on these pages. To use the formulas, the known values should first be translated into the appropriate symbols. Then, inspection of the tables of formulas will show which one is usable for calculating the desired unknown value.)

A_w = cross sectional area of weld, mm^2

C_u = cleaning time, minutes/unit

C_m = cleaning time, minutes/m

$$= \frac{1000\, C_u}{l_x}$$

D_m = weld metal deposited, Kg/m = $.007833 A_w$

D_h = weld metal deposited, Kg/hr at 100% (OF)

$$= \frac{d^2\, M\, E_1}{2700}$$

d = diameter of solid core wire, mm (cannot be used for iron powder or tubular electrodes)

E = electrode cost, dollars/Kg

E_1 = arc transfer efficiency (spatter)

E_2 = core wire to total electrode efficiency

E_3 = effect of stub efficiency = l_m/l_e

F = flux cost, dollars/Kg

F_m = flux consumption, Kg/m

F_h = flux consumption, Kg/hr

G = gas cost, dollars/hr

g = gas cost, dollars/liter

H_u = handling time, minutes/unit

I = welding current, amperes

L = labor and overhead, dollars/hr

l_c = circumference ($\pi \times$ dia), mm

l_e = electrode length, mm

l_m = length electrode melted, mm = $l_e - l_s$

l_o = weld overlap, mm

l_s = stub length, mm

l_u = length electrode used, mm/unit

l_w = length weld/electrode used, mm

$$= \frac{10^5\, E_1\, E_2\, E_3}{N\, D_m} = \frac{S l_m}{M}$$

l_x = length weld, mm/unit

M = electrode melted, mm/min (melting rate or wire feed speed)

$$= \frac{5 \times 10^4\, U_h}{3\, W_u} = \frac{5 \times 10^4\, M_h}{3\, W_m}$$

$$= \frac{2700\, D_h}{d^2\, E_1} \qquad \frac{M_h\, N\, l_e}{6000}$$

M_m = electrode melted, Kg/m of weld

M_h = electrode melted, Kg/hr

N = number electrodes/100 Kg

n = number units/hr

$$= \frac{60}{T_u + T_d}$$

OF = operating factor (decimal)

$$= \frac{T_u}{T_u + T_d}$$

R = ratio flux to electrode mass

$$= \frac{F_m}{M_m} = \frac{F_h}{M_h}$$

r = gas flow, liters/hr

S = arc speed, mm/min For multiple pass welds:

$$S = \frac{1}{\frac{1}{S_1} + \frac{1}{S_2} + \frac{1}{S_3} + \ldots}$$

$S_{1,2}$ = arc speed of individual passes, mm/min

T_d = downtime, min/unit

T_e = arc time, min/elect = l_m/M

T_r = time 1 revolution, min.

T_u = arc time, min/unit = l_x / S

U_m = electrode used, Kg/m of weld

U_h = electrode used, Kg/hr

$$= \frac{60\, W_e}{T_e}$$

W_e = mass one electrode, grams

$$= \frac{10^5}{N}$$

W_m = grams of electrode melted/m of electrode, g/m

$$= \frac{10^3\, W_e}{l_e}$$

W_u = grams of electrode used/m of electrode, g/m

$$= \frac{10^3\, W_e}{l_m}$$

W_s = mass of stub, grams

TABLE 12-4. Formulas For Calculating Values That May Be Needed When Applying Cost Formulas (English System)

(With certain values known, others required in the cost formulas may be obtained with the following conversion formulas.)

APPROX. MELT-OFF RATE	DEPOSITION EFFICIENCY	OVERALL EFFICIENCY	JOINT SPEED
M_h (approx.) = $\frac{\left(\text{arc volt}\right)\left(\text{welding current}\right)}{1000}$ = lbs/hr.	$E_1 = \frac{D_h}{M_h} = \frac{D_f}{M_f}$	$E_2 = \frac{D_h}{U_h} = \frac{D_f}{U_f}$	$S = \frac{D_h}{5D_f} = \frac{U_h}{5U_f} = \frac{D_h}{17A_w}$

POUNDS OF ELECTRODE USED PER FOOT OF WELD	POUNDS OF WELD METAL DEPOSITED PER FOOT OF WELD	POUNDS OF ELECTRODE USED PER HOUR AT 100% OPERATING FACTOR	POUNDS OF WELD METAL DEPOSITED PER HOUR AT 100% OPERATING FACTOR
$U_f = U_f$ $U_f = \frac{1200\ M}{N l_m S}$ $= \frac{1200}{NST_e}$ $= \frac{1200}{N l_w}$ $= \frac{D_f}{E_2}$ $= \frac{U_h}{5S}$ $= \frac{3.4\ A_w}{E_2}$ $= \frac{12M\ W_m}{S}$	$D_f = E_2\ U_f$ $= \frac{1200\ ME_2}{N l_m S}$ $= \frac{1200\ E_2}{NST_e}$ $= \frac{1200\ E_2}{N l_w}$ $= \frac{U_h\ E_2}{5S}$ $= 3.4\ A_w$ $= \frac{12M\ W_m\ E_2}{S}$	$U_h = 5SU_f$ $= \frac{6000\ M}{N l_m}$ $= \frac{6000}{NT_e}$ $= \frac{6000\ S}{N l_w}$ $= \frac{5SD_f}{E_2}$ $= \frac{17.0\ A_w S}{E_2}$ $= 60M\ W_m$	$D_h = 5SE_2U_f$ $= \frac{6000\ ME_2}{N l_m}$ $= \frac{6000\ E_2}{NT_e}$ $= \frac{6000\ SE_2}{N l_w}$ $= 5SD_f$ $= U_hE_2$ $= 17.0\ A_wS$ $= 60M\ W_mE_2$

Deposition Rate D_h (lbs/hr) — English system

SMAW $\quad D_h = \frac{6000\ M\ DE}{N\ l_e}$

solid electrode

SAW / GMAW $\quad D_h = \frac{d^2\ WFS\ DE}{.075}$

tubular electrode

FCAW $\quad D_h = .06\ W_t\ WFS\ DE$

TABLE 12-4M. Formulas For Calculating Values That May Be Needed When Applying Cost Formulas (Metric System)

(With certain values known, others required in the cost formulas may be obtained with the following conversion formulas.)

$E_1 = \frac{2700\ D_h}{d^2\ M}$ $E_2 = \frac{d^2\ N\ 1_e}{162 \times 10^5}$ $E_3 = 1_m/1_e$	$E_1\ E_2 = \frac{D_h}{M_h} = \frac{D_m}{M_m} = \frac{3\ S\ D_m}{50\ M_h}$ $= \frac{D_h\ N\ 1_e}{M} = \frac{5 \times 10^4\ D_h}{3M\ W_m}$ $= \frac{10^3\ S\ D_m}{W_m}$	$E_1\ E_2\ E_3 = \frac{D_h}{U_h} = \frac{D_m}{U_m}$ $= \frac{5 \times 10^4\ D_h}{3\ M\ W_u}$	$S = \frac{50\ D_h}{3\ D_m} = \frac{50\ U_h}{3\ U_m} = \frac{2120\ D_h}{A_w}$
KILOGRAMS OF ELECTRODE USED PER METER OF WELD	KILOGRAMS OF WELD METAL DEPOSITED PER METER OF WELD	KILOGRAMS OF ELECTRODE USED PER HOUR AT 100% OPERATING FACTOR	KILOGRAMS OF WELD METAL DEPOSITED PER HOUR AT 100% OPERATING FACTOR
$U_m = \frac{10^5\ M}{N\ 1_m\ S}$ $= \frac{10^5}{N\ S\ T_e}$ $= \frac{10^5}{N\ 1_w}$ $= \frac{D_m}{E_1\ E_2\ E_3}$ $= \frac{50\ U_h}{3\ S}$ $= \frac{.007833\ A_w}{E_1\ E_2\ E_3}$ $= \frac{M\ W_u}{10^3\ S}$	$D_m = U_m\ E_1\ E_2\ E_3$ $= \frac{10^5\ M\ E_1\ E_2\ E_3}{N\ 1_m\ S}$ $= \frac{10^5 v\ E_1\ E_2\ E_3}{N\ S\ T_e}$ $= \frac{10^5\ E_1\ E_2\ E_3}{N\ 1_w}$ $= \frac{50\ U_h\ E_1\ E_2\ E_3}{3\ S}$ $= .007833\ A_w$ $= \frac{M\ W_u\ E_1\ E_2\ E_3}{10^3\ S}$	$U_h = \frac{3\ S\ U_m}{50}$ $= \frac{6000\ M}{N\ 1_m}$ $= \frac{6000}{N\ T_e}$ $= \frac{6000\ S}{N\ 1_w}$ $= \frac{3\ S\ D_m}{50\ E_1\ E_2\ E_3}$ $= \frac{A_w\ S}{2120\ E_1\ E_2\ E_3}$ $= \frac{3\ M\ W_u}{5 \times 10^4}$	$D_h = \frac{3\ S\ U_m\ E_1\ E_2\ E_3}{50}$ $= \frac{6000\ M\ E_1\ E_2\ E_3}{N\ 1_m}$ $= \frac{6000\ E_1\ E_2\ E_3}{N\ T_e}$ $= \frac{6000\ S\ E_1\ E_2\ E_3}{N\ 1_w}$ $= \frac{3\ S\ D_m}{50}$ $= U_h\ E_1\ E_2\ E_3$ $= \frac{A_w\ S}{2120}$ $= \frac{3\ M\ W_u\ E_1\ E_2\ E_3}{5 \times 10^4}$

TABLE 12-5. Cost Per Foot Of Weld — Dollars/Foot (English System)

LABOR AND OVERHEAD

$$\frac{L}{5S(OF)} = \frac{3.4\ L\ A_w}{D_h(OF)} = \frac{D_f L}{D_h(OH)}$$

MANUAL ELECTRODE

$$\frac{1200\ M\ E}{N\ 1_m\ S} = \frac{1200\ E}{N\ S\ T_e} = \frac{1200\ E}{N\ 1_w} = \frac{3.4\ A_w\ E}{E_2} = \frac{D_h E}{5\ SE_2} = \frac{D_f E}{E_2} = U_f E = \frac{12\ M\ W_m\ E}{S}$$

AUTOMATIC ELECTRODE AND FLUX

$$M_f(E + RF) = \frac{M_h(E + RF)}{5\ S} = \frac{D_f(E + RF)}{E_1} = \frac{12\ M\ W_m(E + RF)}{S} = \frac{3.4\ A_w(E + RF)}{E_1} = \frac{D_h(E + RF)}{5\ S\ E_1} = ED_f + EF_f$$

GAS

$$\frac{G}{5S} = \frac{gr}{5S}$$

HANDLING

$$\frac{L\ H_u}{5\ 1_x}$$

CLEANING

$$\frac{L\ C_u}{5\ 1_x} = \frac{L\ C_f}{60}$$

TABLE 12-5M. Cost Per Meter Of Weld — Dollars/Meter (Metric System)

LABOR AND OVERHEAD

$$\frac{50L}{3\ S\ (OF)} = \frac{.007833\ A_w}{D_h\ (OF)} = \frac{D_m\ L}{D_h\ (OF)} = \frac{10^3\ L\ A_w}{15\pi d^2\ M\ (OF)\ E_1}$$

MANUAL ELECTRODE

$$\frac{10^5\ M\ E}{N\ 1_m\ S} = \frac{10^5\ E}{N\ S\ T_e} = \frac{10^5\ E}{N\ 1_w} = \frac{.007833 A_w\ E}{E_1\ E_2\ E_3} = \frac{50\ D_h\ E}{3\ S\ E_1\ E_2\ E_3} = \frac{D_m\ E}{E_1\ E_2\ E_3} = U_m\ E = \frac{M\ W_m\ E}{10^3\ S}$$

AUTOMATIC ELECTRODE AND FLUX

$$M_m(E + RF) = \frac{50\ M_h(E + RF)}{3\ S} = \frac{D_m(E + RF)}{E_1\ E_2} = \frac{M\ W_m(E + RF)}{10^3\ S} = \frac{.007833\ A_w(E + RF)}{E_1\ E_2} = \frac{50\ D_h(E + RF)}{3\ S\ E_1\ E_2} =$$

$$= E\ D_m + F\ F_m = \frac{d^2\ M(E + RF)}{162\ S\ E_2}$$

GAS

$$\frac{50\ G}{3\ S} = \frac{50\ gr}{3S}$$

HANDLING

$$\frac{50\ L\ H_u}{3\ 1_x}$$

CLEANING

$$\frac{50\ L\ C_u}{3\ 1_x} = \frac{L\ C_m}{60}$$

TABLE 12-6. Cost Per Unit or Dollars/Unit (English System)

Multiply Cost/Foot of Weld by $\frac{1_x}{12}$ or $\frac{ST_u}{12}$

LABOR AND OVERHEAD

$$\frac{L\,T_u}{60\,(OF)} = \frac{L\,1_x}{60\,S\,(OF)} = \frac{17.0\,L\,1_x\,A_w}{60\,(OF)\,D_h}$$

MANUAL ELECTRODE

$$\frac{100\,M\,1_x\,E}{N\,1_m S} = \frac{10\,1_x E}{N\,S\,T_e} = \frac{100\,1_x E}{N\,1_w} = \frac{.283\,A_w\,1_x E}{E_2} = \frac{D_h\,1_x E}{60\,S\,E_2}$$

$$= \frac{100\,M\,T_u\,E}{N\,1_m} = \frac{100\,T_u\,E}{N\,T_e} = \frac{100\,S\,T_u\,E}{N\,1_w} = \frac{.283\,A_w\,S\,T_u\,E}{E_2} = \frac{D_h\,T_u\,E}{60\,E_2}$$

AUTOMATIC ELECTRODE AND FLUX

$$\frac{M_f\,1_x(E + RF)}{12} = \frac{M_h\,1_x(E + RF)}{60\,S} = \frac{D_f\,1_x(E + RF)}{12\,E_1} = \frac{D_h\,1_x(E + RF)}{60\,S\,E_1} = \frac{M\,W_m\,1_x(E + RF)}{S}$$

$$= \frac{.283\,A_w\,1_x(E + RF)}{E_1} = \frac{M_f\,S\,T_u(E + RF)}{12} = \frac{M_h\,T_u(E + RF)}{60} = \frac{D_f\,S\,T_u(E + RF)}{12\,E_1}$$

$$= \frac{D_h\,T_u\,(E + RF)}{60\,E_1} = M\,W_m\,T_u(E + RF) = \frac{.283\,A_w\,S\,T_u(E + RF)}{E_1}$$

GAS

$$\frac{GT_u}{60} = \frac{gr\,T_u}{60} = \frac{G\,1_x}{60S} = \frac{gr\,1_x}{60S}$$

HANDLING

$$\frac{L\,H_u}{60}$$

CLEANING

$$\frac{L\,C_u}{60}$$

TABLE 12-6M. Cost Per Unit or Dollars/Unit (Metric System)

Multiply Dollars/Meter of Weld by $\frac{l_x}{10^3}$ or $\frac{S\,T_u}{10^3}$

LABOR AND OVERHEAD

$$\frac{L\,T_u}{60\,(OF)} = \frac{L\,l_x}{60\,S\,(OF)} = \frac{.007833\,L\,l_x\,A_w}{D_h\,(OF)} = \frac{L\,l_x\,A_w}{15\pi d^2\,M\,(OF)\,E_1}$$

MANUAL ELECTRODE

$$\frac{10^2\,M\,l_x\,E}{N\,l_m\,S} = \frac{10^2\,l_x\,E}{N\,S\,T_e} = \frac{10^2\,l_x\,E}{N\,l_w} = \frac{783\times10^{-8}\,A_w\,l_x\,E}{E_1\,E_2\,E_3} = \frac{D_h\,l_x\,E}{60\,S\,E_1\,E_2\,E_3}$$

$$= \frac{10^2\,M\,T_u\,E}{N\,l_m} = \frac{10^2\,T_u\,E}{N\,T_e} = \frac{10^2\,S\,T_u\,E}{N\,l_w} = \frac{783\times10^{-8}\,A_w\,S\,T_u\,E}{E_1\,E_2\,E_3} = \frac{D_h\,T_u\,E}{60\,E_1\,E_2\,E_3}$$

AUTOMATIC ELECTRODE AND FLUX

$$\frac{M_m\,l_x(E+RF)}{10^3} = \frac{M_h\,l_x(E+RF)}{60\,S} = \frac{D_m\,l_x(E+RF)}{10^3\,E_1\,E_2} = \frac{D_h\,l_x(E+RF)}{60\,S\,E_1\,E_2} = \frac{M\,W_m\,l_x(E+RF)}{10^6\,S}$$

$$= \frac{783\times10^{-8}\,A_w\,l_x(E+RF)}{E_1\,E_2} = \frac{M_m\,S\,T_u(E+RF)}{10^3} = \frac{M_h\,T_u(E+RF)}{60} = \frac{D_m\,S\,T_u(E+RF)}{10^3\,E_1\,E_2}$$

$$= \frac{D_h\,T_u(E+RF)}{60\,E_1\,E_2} = \frac{M\,W_m\,T_u(E+RF)}{10^6} = \frac{783\times10^{-8}\,A_w\,S\,T_u(E+RF)}{E_1\,E_2} = \frac{d^2\,M(E+RF)}{162000\,S\,E_2}$$

GAS

$$\frac{G\,T_u}{60} = \frac{gr\,T_u}{60} = \frac{G\,l_x}{60\,S} = \frac{gr\,l_x}{60\,S}$$

HANDLING

$$\frac{L\,H_u}{60}$$

CLEANING

$$\frac{L\,C_u}{60}$$

PROBLEM 1

A butt weld is to be used to join 1″ plates. The joint is prepared as a double-V, as shown in the figure below. The process is submerged-arc with single electrode. The problem is to determine the primary costs of weld per foot of joint. The data given or available from standard procedures are:

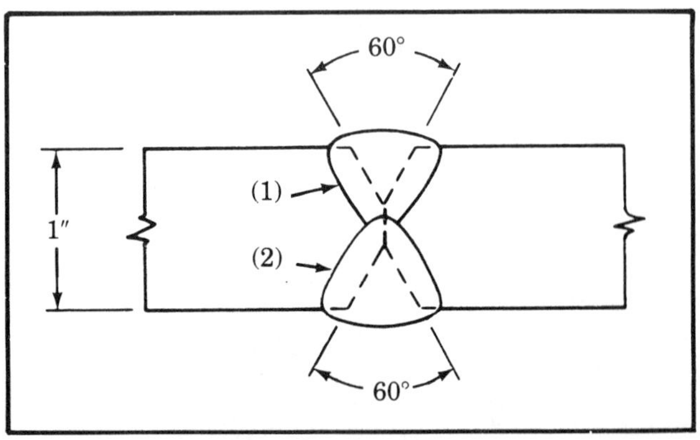

- L = \$10.00 per hour
- O F = 40% (.40)
- E = \$0.31 per pound (electrode cost)
- F = \$0.20 per pound (flux cost)
- M_f = .93 pounds/foot (pounds of electrode melted per foot)
- F_f = .80 pounds/foot (pounds of flux per foot)

$$R = \frac{F_f}{M_f} = \frac{.80}{.93} = .86$$

(ratio of flux-to-electrode weight)

The procedures are:

First Weld — 900 amperes, DC(+), 35 volts, 7/32″ electrode. This procedure will give

TABLE 12-7. Cost Per Pound of Weld Metal — Dollars/Pound (English System)

Multiply Cost/Foot of Weld by $\frac{5S}{D_h}$

LABOR AND OVERHEAD

$$\frac{L}{D_h\,(OF)} = \frac{L}{E_1\,M_h\,(OF)} = \frac{L}{E_2\,U_h\,(OF)}$$

MANUAL ELECTRODE

$$\frac{6000\,M\,E}{N\,1_m\,D_h} = \frac{6000\,E}{N\,T_e\,D_h} = \frac{6000\,S\,E}{N\,1_w\,D_h} = \frac{17.0\,A_2\,S\,E}{E_2\,D_h} = \frac{E}{E_2}$$

AUTOMATIC ELECTRODE AND FLUX

$$\frac{60\,M\,W_m(E\ +\ RF)}{D_h} = \frac{17.0\,A_w\,S(E\ +\ RF)}{E_1\,D_h} = \frac{E\ +\ RF}{E_1}$$

GAS

$$\frac{G}{D_h} = \frac{gr}{D_h}$$

HANDLING

$$\frac{L\,H_u\,S}{1_x\,D_h}$$

CLEANING

$$\frac{L\,C_u\,S}{1_x\,D_h}$$

TABLE 12-7M. Cost Per Kilogram of Weld Metal — Dollars/Kg (Metric System)

Multiply Dollars/Meter of Weld by $\frac{3S}{50\,D_h}$

LABOR AND OVERHEAD

$$\frac{L}{D_h\,(OF)} = \frac{L}{M_h\,(OF)\,E_1\,E_2} = \frac{L}{U_h\,(OF)\,E_1\,E_2\,E_3} = \frac{2700\,L}{d^2\,M\,(OF)\,E_1}$$

MANUAL ELECTRODE

$$\frac{6000\,M\,E}{N\,1_m\,D_h} = \frac{6000\,E}{N\,T_e\,D_h} = \frac{6000\,S\,E}{N\,1_w\,D_h} = \frac{A_w\,S\,E}{2120\,D_h\,E_1\,E_2\,E_3} = \frac{E}{E_1\,E_2\,E_3}$$

AUTOMATIC ELECTRODE AND FLUX

$$\frac{6\,M\,W_m(E\ +\ RF)}{10^5\,D_h} = \frac{E\ +\ RF}{E_1\,E_2}$$

GAS

$$\frac{G}{D_h} = \frac{gr}{D_h}$$

HANDLING

$$\frac{L\,H_u\,S}{1_x\,D_h}$$

CLEANING

$$\frac{L\,C_u\,S}{1_x\,D_h}$$

an arc speed (S_1) of 14 inches per minute.

Second Weld — 1050 amperes, DC(+), 36 volts, 7/32″ electrode. This procedure will give an arc speed (S_2) of 12 inches per minute.

Since the problem seeks an answer in terms of per foot of weld, Table 12-5 is used. The primary components of cost will be labor and overhead and the costs of electrode and flux. The "Labor and Overhead" and "Automatic Electrode and Flux" formulas in Table 12-5 apply.

Using the formula:

$$\textbf{Labor and Overhead} = \frac{L}{5\,S\,(OF)} = \text{dollars/foot}$$

for the Weld (1), the value is found to be:

$$\textbf{Labor and Overhead} = \frac{(10.00)}{5\,(14)\,(.40)} = \$0.357/\text{foot}$$

and for Weld (2):

$$\textbf{Labor and Overhead} = \frac{(10.00)}{5\,(12)\,(.40)} = \$0.417/\text{foot}$$

Material cost is then determined with the formula:

$$\textbf{Auto. Elec. \& Flux} = M_f\,(E+RF)\ \text{dollars/foot}$$

$= (.93)\ (0.31\ +\ .86\ \times\ 0.20)$

$= \$0.448/\text{foot}$

The labor, overhead, and consumables costs in making the butt weld are thus:

$= 0.357\ +\ 0.417\ +\ 0.448$

$= \$1.22/\text{foot}$

Another way to calculate labor and overhead with this double-V weld would use the formula from Table 12-3.

$$\textbf{Joint Speed} = \frac{1}{\frac{1}{S_1} + \frac{1}{S_2}} = \text{inches/min}$$

This gives a value that could be substituted for (S) in the labor and overhead formula. The value obtained (6.46 in./min) enables labor and overhead to be calculated with one application of the formula.

Still another way to solve the problem would be to use the "Approximate Cost of Submerged-Arc

Automatic Weld" formula, since all the factors in it are known.

Approximate Cost

$$= \frac{.0065 I (E + F) + \dfrac{L}{5(OF)}}{S} = \frac{\$}{ft}$$

$$\text{Weld (1)} = \frac{(.0065)(900)(0.31 + 0.20) + \dfrac{10}{5(.40)}}{14} =$$

$$= \$0.570/\text{foot}$$

$$\text{Weld (2)} = \frac{(.0065)(1050)(0.31 + 0.20) + \dfrac{10}{5(.40)}}{12} =$$

$$= \$0.707/\text{foot}$$

The labor, overhead, and consumables costs determined with the approximation formula would be:

$= 0.570 + 0.707$
$= \$1.28/\text{foot}$

PROBLEM 2

The problem is to determine the main costs in welding the automobile axle housing shown in Figure 12-5. Welding is to be done by automatic submerged-arc. Since cost-per-unit information is wanted, the formulas in Table 12-6 apply. The applicable data are:

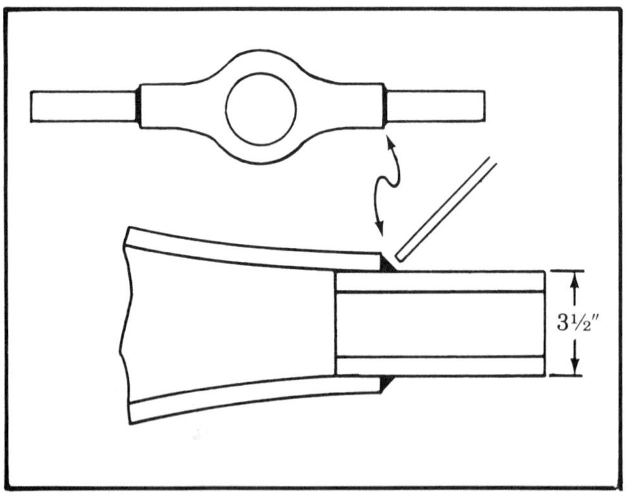

Fig. 12-5. Weld in automobile axil housing.

Procedure: 400 amperes, DC(+), 26 volts, ⅛″ L-61 electrode, 780 flux.

- L = \$10.00 per hour
- $O\ F$ = 50% (.50)
- R = 1.00 (ratio flux weight to electrode weight)
- T_r = 22 seconds or 0.366 minutes per revolution
- l_c = π (diameter) = π (3.5″) = 11.0″
- l_o = ½″
- E = \$0.31/pound, electrode cost
- F = \$0.20/pound, flux cost

From a table of melt-off rates for the electrode or by direct measurement, M_h is found to be 10.2 pounds per hour.

The applicable formulas from Table 12-6 are:

$$\textbf{Labor \& Overhead}_{unit} = \frac{L\,T_u}{60(OF)}\ \$/\text{unit}$$

Automatic Electrode & Flux =

$$= \frac{M_h T_u (E + RF)}{60}\ \text{dollars/unit}$$

T_u must first be calculated from the formula defining it, found in Table 12-3. This increases the weld time per unit to take care of the weld overlap l_o.

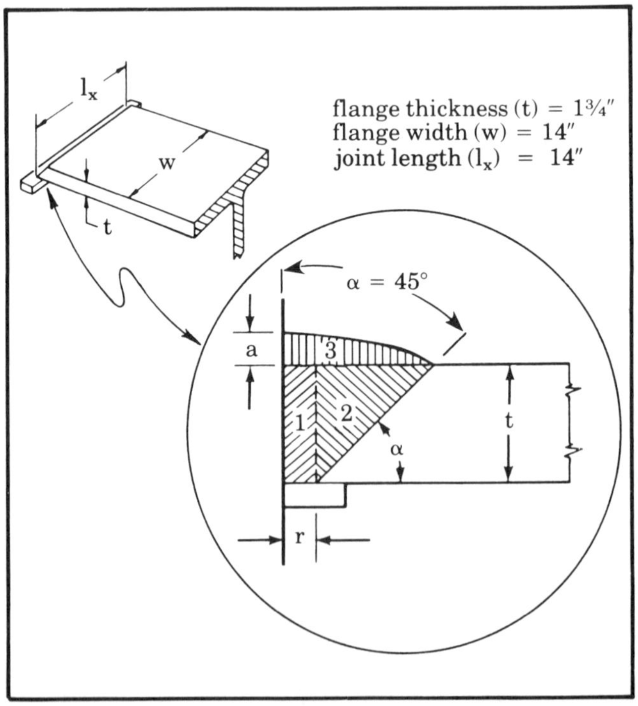

Fig. 12-6. Flange weld in beam to column connection.

$$T_u = \frac{l_c + l_o}{l_c} T_r \text{ min/unit}$$

$$= \frac{(11.0 + .5)(.366)}{11.0}$$

$$= .383 \text{ min./unit}$$

Substituting the various values in the above formulas:

Labor & Overhead/unit =

$$= \frac{L\,T_u}{60\,(O\,F)} = \frac{(10.00)\,(.383)}{60\,(.50)} = \$0.128/\text{unit}$$

Automatic Electrode & Flux =

$$= \frac{M_h\,T_u\,(E + RF)}{60} =$$

$$= \frac{(10.2)\,(.383)\,(.031 + 1.00 \times 0.20)}{60} = \$0.038/\text{weld}$$

The total cost of welding the joint (not taking into account power and maintenance) would, thus, be $0.17 per weld, assuming only one circumferential weld was made at a time. If both welds were made simultaneously, the welding cost would be:

Material for 2 welds = 2 × 0.038 = $0.076/unit

Labor & Overhead + Material/unit = 0.128 + 0.076 = $0.204/unit or $0.102/weld

PROBLEM 3

The problem is to find the cost of the flange weld in a beam-to-column connection. The joint is illustrated in Figure 12-6, where:

thickness (t) = 1¾″

bevel angle (α) = 45°

root spacing (r) = ¼″

In addition, the reinforcement dimension (a) is controlled by the AWS Building Code and Bridge Specifications, which state that 'the reinforcement for a groove weld in tee or corner joints shall equal t/4 but not to exceed ⅜″.' Since:

$$a = \frac{t}{4} = \frac{1\tfrac{3}{4}''}{4} = \frac{7''}{16}$$

therefore, the reinforcement (a) is ⅜″.

The determination of the weld cost can be based on an "area calculations" approach where:

$$A_w = (1) + (2) + (3)$$

$$A_w = r\,t + \frac{t^2}{2} + \frac{2}{3} a\,(r + t)$$

[Since it would be difficult to achieve a flat face to the reinforcement, the formula for a parabola is used to establish the area of (3)].

$$A_w = \left(\frac{1}{4}\right)(1\tfrac{3}{4}) + \frac{1}{2}\left(1\tfrac{3}{4}''\right)^2 + \frac{2}{3}\left(\frac{3}{8}\right)\left(\frac{1}{4} + 1\tfrac{3}{4}\right)$$

$$A_w = 2.47$$

The welding procedure for this joint specified 3/16″ E7027 electrode at 280 amps AC, which will deposit weld metal at:

D_h = 7.8 lb./hr.,

have an overall efficiency of

E_2 = 66% (.66),

and cost

E = $0.35/lb.

From data obtained from production records, the labor and overhead rate and operating factor is:

L = $10.00/hr.
O F = 30% (.30)

Since unit cost is the objective, Tables 12-4 and 12-6 are sources for formulas needed for calculating:

Welding Speed

Electrode Cost per Joint

Labor and Overhead Cost per Joint

Applicable is the formula:

$$\textbf{Joint Speed}\quad S = \frac{D_h}{17 A_w} \text{ inches/minute}$$

So: $S = \dfrac{D_h}{17\,A_w}$

$$S = \frac{7.8}{(17)\,(2.47)}$$

$$S = .186 \text{ in./min.}$$

Now, with known values for (A_w), (D_h), (E_2), (L), (OF), and (l_x length of joint), the formula from Table 12-6 for calculating unit cost can be used.

$$\textbf{Labor \& Overhead}_{unit} = \frac{L\,l_x}{60S\,(O\,F)}\ \$/\text{unit}$$

$$= \frac{(10.00)\,(14)}{(60)\,(.186)\,(.30)}$$

$$= \$41.82/\text{unit}$$

MANUAL ELECTRODE

$$\textbf{Manual Electrode} = \frac{.283\,A_w\,l x\,E}{E_2}\ \text{dollars/unit}$$

$$= \frac{(.283)\,(2.47)\,(14)\,(.35)}{.66}$$

$$= \$5.19/\text{unit}$$

The labor and overhead and consumables costs for the beam-to-column connection would be:

$$= 41.82 + 5.19$$

$$= \$47.01/\text{unit}$$

PRECAUTIONS WHEN USING COST FORMULAS

The formulas in the tables are carefully worked out so the user will not go astray when making computations. In all cases, the dimensions specified should be used, and, if an unknown is to be calculated from existing data, the formulas in Table 12-4 must be followed precisely. The following example shows how the user can go astray by shortcutting the conversion formulas found in Table 12-4.

The arc speeds for the two welds shown in Figure 12-7 are known. Rather than making two calculations requiring different factors (S) for arc speed, the person making the computation decides to take a shortcut by using an average value for (S) and dividing by the number of passes to get the overall joint speed. Thus, if (S_1) were 16 inches per minute and (S_2) 10 inches per minute, he might conclude that the joint speed would be one half of the average (13) of (S_1) and (S_2), or 6.5 inches per minute.

Checking this reasoning with the formula for determining joint speed found in Table 12-4, the true joint speed is found to be:

$$S = \frac{1}{\frac{1}{S_1} + \frac{1}{S_2}}$$

$$= \frac{1}{\frac{1}{16} + \frac{1}{10}}$$

$$= \frac{1}{.0625 + .1000}$$

$$= 6.15 \text{ inches per minute}$$

In this case the error is not great, but it enlarges as the difference between arc speeds increases. Only when arc speeds are equal will the value for joint speed be correct when the average arc speed is divided by the number of passes.

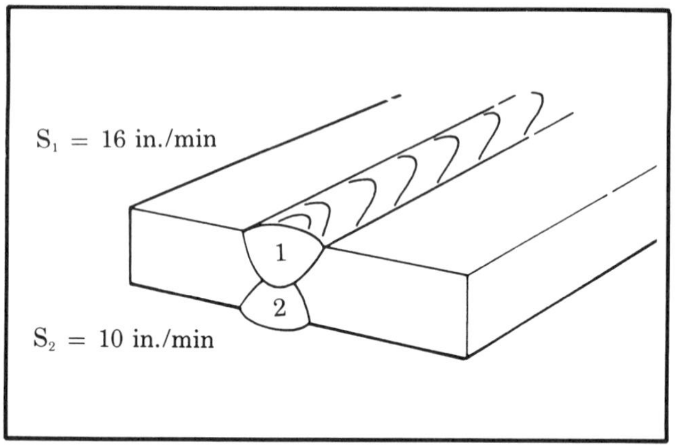

Fig. 12-7. Two-pass butt weld

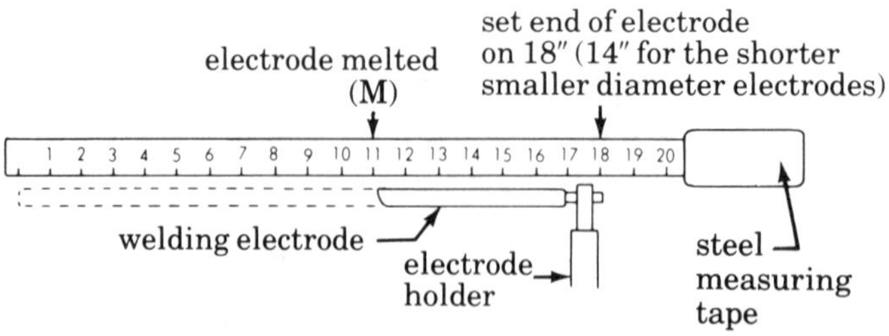

Fig. 12-8. Method of measuring length of electrode melted.

It should be noted that the (S) appearing in the formulas can be either arc speed or as in the case of a multiple pass weld, the actual joint speed. If arc speed were used for determining the cost of a multipass weld, however, a cost calculation would have to be made for each pass and then added together to obtain the actual cost of the joint. If the joint speed is used, only one calculation of cost is necessary.

HOW TO DETERMINE ACTUAL VALUES FOR COST FORMULA FACTORS

In the foregoing examples, the reader was referred to standard procedures to determine values for the electrode melt-off rate (M) and the arc speed (S). However, as repeatedly stressed in the text, the best values to use in cost calculations are those taken directly from the production floor, since they are measurements of the actual performance. In many instances, the procedures being used on the floor will vary somewhat from standard procedures — and there is always the human element to take into account.

A simple way to determine melting rate and arc speed is to measure the length of electrode melted, and the length of weld produced during a 60-second interval, timed by a watch. The melting rate (M) in inches per minute is the inches of electrode melted in that 60-second interval, and the length of the weld in inches per minute is the inches of weld produced, or arc speed (S). To measure the inches of electrode consumed, simply place a steel tape along the electrode stub still held in the holder. With the end of the stub lined up with the number representing the initial length of the electrode, read the number opposite the other end of the electrode. See Figure 12-8.

From the melting rate (M) and arc speed (S), the actual shop values for other factors needed in the cost equation can be calculated from the conversion formulas given in Table 12-4. Thus, if it is necessary to know the value for U_h — pounds of electrode used per hour at 100% operating factor — to solve an equation, it could be found by the formula:

$$\mathbf{U}_h = \frac{6000\,M}{N\,l_m}\ \text{lbs/hour}$$

(M) has already been determined with the watch. (N), the number of electrodes per 100 pounds, can be taken from the manufacturer's literature — or calculated by taking the average weight per electrode of ten electrodes (W_e) and dividing it into 100. The remaining factor (l_m) is the inches of meltable electrode, per electrode, or the length of the original electrode minus the stub length (usually 16 inches for an 18 inch electrode).

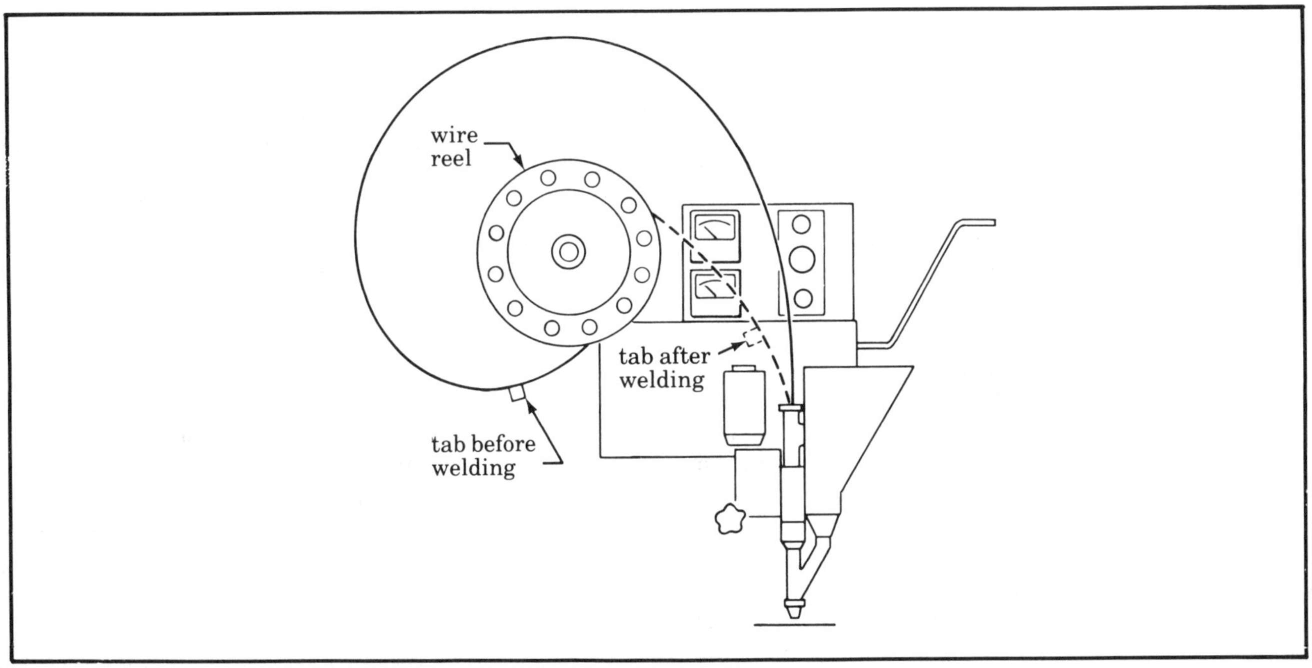

Fig. 12-9. Method of measuring length of wire consumed.

When using automatic or semiautomatic welding processes, the inches of wire melted per minute (M), if not known, can be determined by sticking a tab of adhesive tape to the electrode wire some distance from the top of the wire straightener on the welding head, measuring the distance to the marker tab before welding, and subtracting this latter distance from the former. See Figure 12-9. To place the marker tab and permit measurement of its distance from the head before start of the welding, an adequate length of electrode will have to be unwound from the wire reel. The wire can be measured with a steel or cloth tape.

A more sophisticated way to measure the electrode wire used, however, is to use a wire feed speed device. This employs a spring-loaded guide to keep the pickup shaft on the electrode and give accurate measurements in inches/minute. Many automatic welders have a digital read out for wire feed speed.

These methods for getting actual shop values for use in the cost formulas, rather than depending on the values from standard procedures, will give the highest accuracy in cost estimation.

There is one rule-of-thumb method of approximating melting rate in terms of pounds of electrode melted per hour (M_h) that is convenient to use in the rapid estimation work.

Approximate Melting Rate:*

$$M_{h\,(approx.)} = \frac{(\text{arc voltage})\,(\text{welding current})}{1000}$$

lbs./hour

* Will be high by 2-4 lbs/hr if using the submerged-arc process.

Since (arc voltage) × (welding current) ÷ 1000 is the power across the arc in kilowatts (kw) this means that melting rate is approximately equal to the kw at the arc. This method has been proven fairly accurate for stick and solid-wire electrode. It is not satisfactory for flux-cored electrodes.

Thus, if the arc voltage is 30 and the welding current 300 amperes, the melting rate is approximately 9 pounds of electrode per hour. By this quick calculation, it is possible to get an approximation of the deposition rate in manual welding at any operating factor. The deposition rate is roughly two thirds of the electrode melting rate. At 100% operating factor it would, thus, be 6 pounds per hour for the voltage and current given. If the operating factor is 40%, it would be .40 × 6, or 2.4 pounds per hour.

THE DISCIPLINARY VALUE OF COSTING STUDIES

A smooth-running shop that has all the work it needs and is making profit may look at cost analysis as unnecessary and a useless struggle with mathematical formulas. But this attitude is comparable to the attitude that research and development on new and improved products is not needed since "we have a good product now." Unless the welding fabricator keeps on his toes, he may soon not have all the work he needs or be making a profit.

Cost analysis is a basic prelude to cost-reduction research. It shows up waste in design, shop practices, and the use of personnel. It points up the need for changes in processes and procedures. The analyst working with the formula:

$$\text{Labor and Overhead Cost (\$ per foot)} = \frac{L}{5\,S\,(O\,F)}$$

is likely to ponder if he couldn't increase (S) or (OF), and thus reduce his costs. To increase (S), the arc speed, he begins thinking about larger electrodes, higher operating currents, or semiautomatic and full-automatic processes. This thinking leads to the discovery that the shop is using processes and procedures that are a carryover from earlier days when welding technology was not so advanced. To increase (OF), the operating factor, he envisions better ways of materials handling — and can even prove mathematically that the positioner a foreman has been requesting can be paid off in a few months because of the improved operating factor it will permit.

Cost analysis studies also point up excesses in weld sizes, root openings, included angles, and reinforcements, all of which increase overall welding costs substantially while giving the purchaser of the product no value. In fact, such analyses often lead to the discovery that some welds cost twice as much as necessary because of archaic or ill-advised joint design practices. A classic example is the common use of a V-groove joint with a 60-degree included angle. The overall cost of making the weld in thick plate is almost double the cost of making a V-groove weld with a 20-degree included angle — and the latter joint is readily made by present processes and procedures.

THE DESIGNER'S ROLE IN MINIMIZING WELD COSTS

Once the fabricator has become involved in cost analysis studies, his first attempts at cost reduction will probably be in reducing wastes in his shop, followed by a systematic upgrading of welding practices and modernization of equipment. Eventually, he will turn to his weldments and question: Does this design use weld metal as efficiently as possible? Are there opportunities for conservation?

By conserving weld metal in the design one automatically reduces electrode, flux, labor, and power costs. The designer prescribes how much weld metal is to be used, and, unless the shop is wasteful, has overriding control of weld metal consumption.

A starting point when scrutinizing a design to find opportunities for weld metal conservation is in the joint detail. Often the designer is experienced in specifying joint details and knows what insures good performance. He gives no thought to the possibility of an alternative detail that could give the same performance, perhaps at lower cost. As an example, with a bevel joint, the designer might consider the feasibility of using a smaller included angle with a backup strip with the welding process to be used. Decreasing the included angle in a bevel joint (see Table 12-8) can make a sizeable difference in weld metal requirements for plates thicker than one inch. In the case of 2-inch plate, using a 20° included angle instead of a 60° angle, reduces the weld metal requirement from 9.43 lbs/ft to 6.07 lbs/ft, and the thicker the plate the greater the reduction.

Designers are well aware of how the volume of a fillet weld increases by the square of the increase in leg size. Yet it is difficult for the conscientious designer to specify the very minimum requirement because the weld "just might come from the shop a bit undersized." Although this is unlikely to happen, the designer may feel more comfortable in specifying fillet weld size now that the AWS permits a negative tolerance of 1/16″ in 10% of a long continuous weld.

Savings for weld metal by selecting the optimum joint detail are easily recognized. More subtle are the savings that could occur by consideration of elements in the basic design. Too often weld metal is used where it has no engineering function, and sometimes steel is used where it also serves no purpose.

Figure 12-10 illustrates a case of using both weld metal and steel without useful purpose. A weldment is being substituted for a casting. The designer — knowing that the casting served its function well — decides to duplicate the casting in welded steel. He adds 12 unneeded stiffeners in weldment design — substantially increasing both the amount of weld metal and the amount of steel. The weldment doesn't need the stiffeners that greatly increase its cost, but which could be added to the casting at little extra cost.

Even in some of the more sophisticated designs are found the extravagant uses of weld metal

TABLE 12-8. How the Included Angle Affects Weld Metal Requirements

	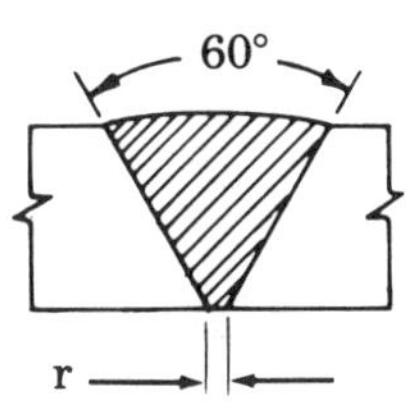	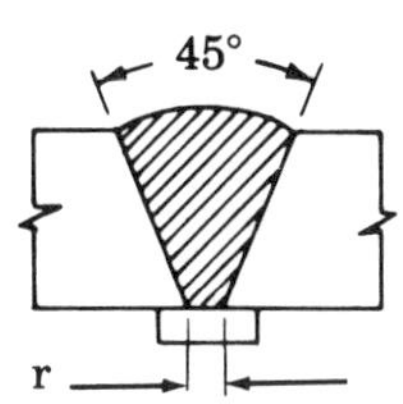	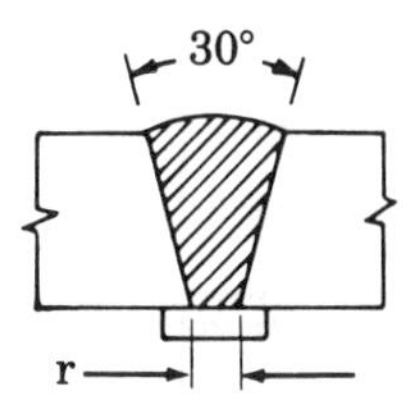	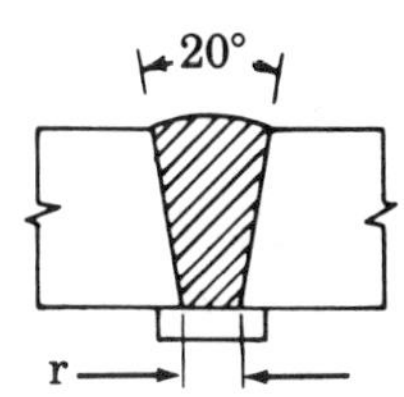
Root Opening r, in.(mm)	**⅛(3.18)**	**¼ (6.35)**	**⅜ (9.52)**	**½ (12.70)**
Plate Thickness, in. (mm)	**Weld Metal Needed, lb/ft (kg/m)**			
½ (12.70)	0.84 (1.25)	0.99 (1.34)	0.99 (1.47)	1.13 (1.68)
⅝ (15.88)	1.21 (1.80)	1.23 (1.83)	1.29 (1.92)	1.44 (2.14)
¾ (19.05)	1.64 (2.44)	1.61 (2.40)	1.62 (2.41)	1.76 (2.62)
1 (25.40)	2.69 (4.00)	2.50 (3.72)	2.38 (3.54)	2.47 (3.68)
1½ (38.10)	5.55 (8.26)	4.81 (7.16)	4.23 (6.29)	4.12 (6.13)
2 (50.80)	9.43 (14.03)	7.85 (11.68)	6.54 (9.73)	6.07 (9.03)
3 (76.20)	20.21 (30.07)	16.08 (23.93)	12.57 (18.70)	10.88 (16.19)
4 (101.6)	35.03 (52.12)	27.19 (40.46)	20.44 (30.41)	16.91 (25.16)
6 (152.4)	76.71 (114.14)	58.01 (86.32)	41.72 (62.08)	32.58 (48.48)
8 (203.2)	134.43 (220.03)	100.29 (149.23)	70.37 (104.71)	53.10 (79.01)
10 (254.0)	208.14 (309.71)	154.01 (229.17)	106.40 (158.32)	78.45 (116.73)

illustrated by the sketches in Figure 12-11. There is no point in placing a weld where force is not transferred — unless enclosure is needed to protect the interior of the member from the weather or for the sake of appearance, in which case, a small bead with no load-carrying capacity would suffice.

Another wasteful design practice is to use continuous welds where intermittent welds would be adequate. Sometimes continuous welds may be justified for appearance, but, as Figure 12-12 shows, when the intermittent welds can be hidden, appearance is even improved.

Weld metal can also be conserved by placing welds in the most effective load-carrying positions. The same amount of welding is used in both of the examples in Figure 12-13, but section modulus calculations reveal the positioning at the right will carry three times the moment of the one at the left. Weld size can be reduced by selecting the more effective position.

In sizing welds, it's also necessary to consider whether loading is parallel or transverse. Extensive test data show a one-third increase in weld strength

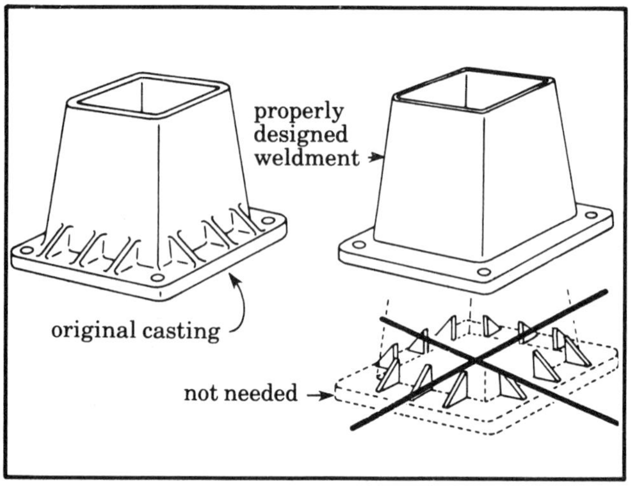

Fig. 12-10. Weldment replacement for casting.

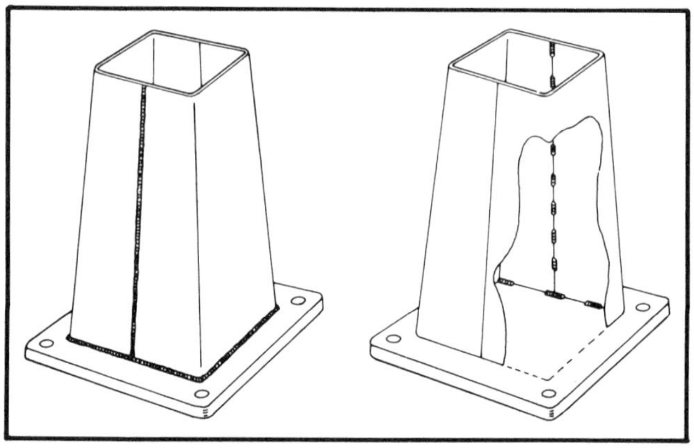

Fig. 12-12. Intermittent use of welds.

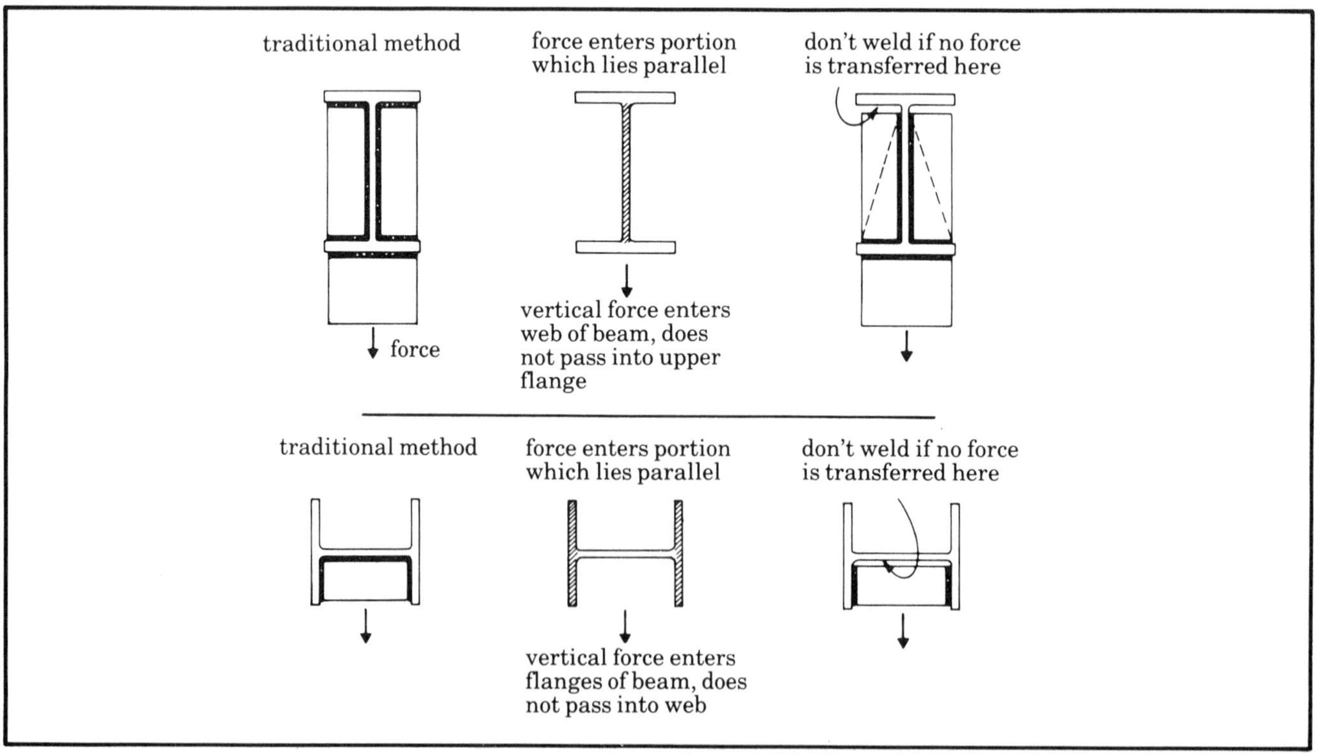

Fig. 12-11. Essential use of weld.

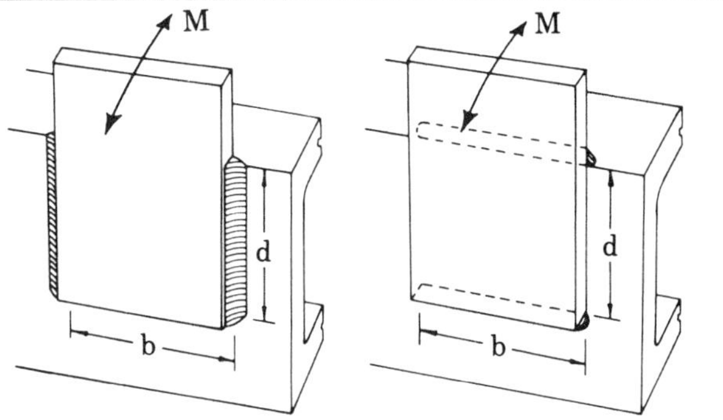

Weld section modulus $S_w = d^2/3$ — Weld section modulus $S_w = bd$

if b and d = 4 in. (101.6 mm);

$S_w = 5\frac{1}{3}$ in.^{2} (3.44×10^3 mm^2) — $S_w = 16$ in.^{2} (10.32×10^3 mm^2)

Fig. 12-13. Effective placement of welds.

under transverse loading. As a result, it may be possible to reduce weld size under transverse loading conditions. Table 12-9 gives the allowable unit forces on fillets of varying sizes under both loading conditions.

Traditionally, a fillet weld has been considered to be a "full strength" weld if its leg size was of the plate thickness. Calculations were based on thinner plate welded on both sides, with the welds extending full length. This ratio (0.75) was true for E60 weld metal and A7 mild steel under the old allowables. However, the value can't be used when working with today's wide ranges of available steel and weld metal strength levels.

TABLE 12-9. Allowable Unit Forces on Fillet Welds, kips/in. (KN/mm)

Leg Size, in. (mm)	E60 or F-60 Weld Metal Parallel Load	E60 or F-60 Weld Metal Transverse Load*	E70 or F-70 Weld Metal Parallel Load	E70 or F-70 Weld Metal Transverse Load*
1/16 (1.59)	0.80 (0.14)	0.93 (0.16)	0.93 (0.16)	1.09 (0.19)
1/8 (3.18)	1.59 (0.28)	1.86 (0.33)	1.86 (0.33)	2.17 (0.38)
3/16 (4.76)	2.39 (0.42)	2.80 (0.49)	2.78 (0.49)	3.26 (0.57)
1/4 (6.35)	3.18 (0.56)	3.73 (0.65)	3.71 (0.65)	4.35 (0.76)
5/16 (7.94)	3.98 (0.70)	4.66 (0.82)	4.64 (0.81)	5.44 (0.95)
3/8 (9.52)	4.77 (0.84)	5.59 (0.98)	5.57 (0.98)	6.52 (1.14)
7/16 (11.11)	5.57 (0.98)	6.52 (1.14)	6.50 (1.14)	7.61 (1.33)
1/2 (12.70)	6.36 (1.11)	7.46 (1.31)	7.42 (1.30)	8.70 (1.52)
5/8 (15.88)	7.95 (1.39)	9.32 (1.63)	9.28 (1.63)	10.87 (1.90)
3/4 (19.05)	9.54 (1.67)	11.16 (1.95)	11.14 (1.95)	13.05 (2.29)
7/8 (22.22)	11.14 (1.95)	13.05 (2.29)	13.00 (2.28)	15.22 (2.67)
1 (25.40)	12.73 (2.23)	14.91 (2.61)	14.85 (2.60)	17.40 (3.05)

*Non-code Work

Table 12-10 shows how the "full strength" factor (w/t) varies with weld metal and plate strength levels for a number of steel grades.

Also shown (at the bottom) are suggested weld-size-to-plate-thickness factors for "rigidity" designs.

When the designer is conscious of costs, he will find numerous methods to achieve cost reduction objectives. Some points to keep in mind:

- Avoid unequal leg sizes on fillet welds.
- Avoid specifying high-strength weld metal on other than primary load-carrying welds when joining high-strength materials.
- Avoid unnecessary use of the all-around symbol.

TABLE 12-10. Weld Size Factors for "Full Strength" and "Rigidity" Fillet Welds

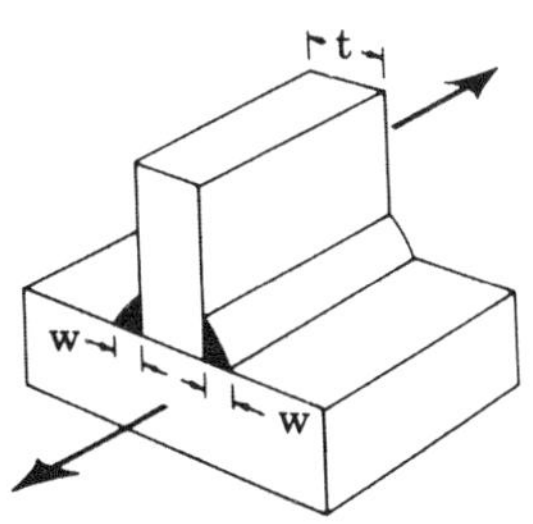

Weld vs Plate

$2(0.707)w\,\tau_w \geq t\,\tau_p$

$2(0.707)w\,\tau_w \geq t(3/4\sigma_p)$*

$\frac{w}{t} \geq 0.5304\,\frac{\sigma_p}{\tau_w}$

*assume $\tau_p = 3/4\sigma_p$

"Full Strength" Weld Size Factor (w/t)

ASTM Steel	Yield Strength θ_y, 10^3 psi (MPa)	E60	E70	E80	E90	E100
A36	36 (248)	0.65	0.56	—	—	—
A441	50 (345)	—	0.76	—	—	—
A572-65	65 (448)	—	1.00	0.86	0.77	—
A514	100 (690)	—	1.52	1.33	1.18	1.06

(E60–E100: Weld Metal Strength Level and Factor)

"Rigidity Design" (Non-Full Strength) Weld Size Factor (w/t) †

ASTM Steel	Yield Strength θ_y, 10^3 psi (MPa)	E60	E70	E80	E90	E100
A36	36 (248)	0.22-0.32	0.19-0.28	—	—	—
A441	50 (345)	—	0.25-0.38	—	—	—
A572-65	65 (448)	—	0.33-0.49	0.29-0.43	0.26-0.38	—
A514	100 (690)	—	0.51-0.76	0.44-0.66	0.39-0.59	0.35-0.53

(E60–E100: Weld Metal Strength Level and Factor)

† First w/t value will develop one-third of plate allowable; second value will develop one-half of plate allowable.

The philosophy of welding "everything that touches" makes no engineering sense and wastes weld metal.

- Consider deep penetration processes, such as submerged-arc welding, to reduce the volume of weld metal needed.

Cost-reduction techniques should, of course, never affect the integrity of the weld or the weldment. The objective is to use weld metal strategically, which usually means increasing the quality of the product rather than degrading it.

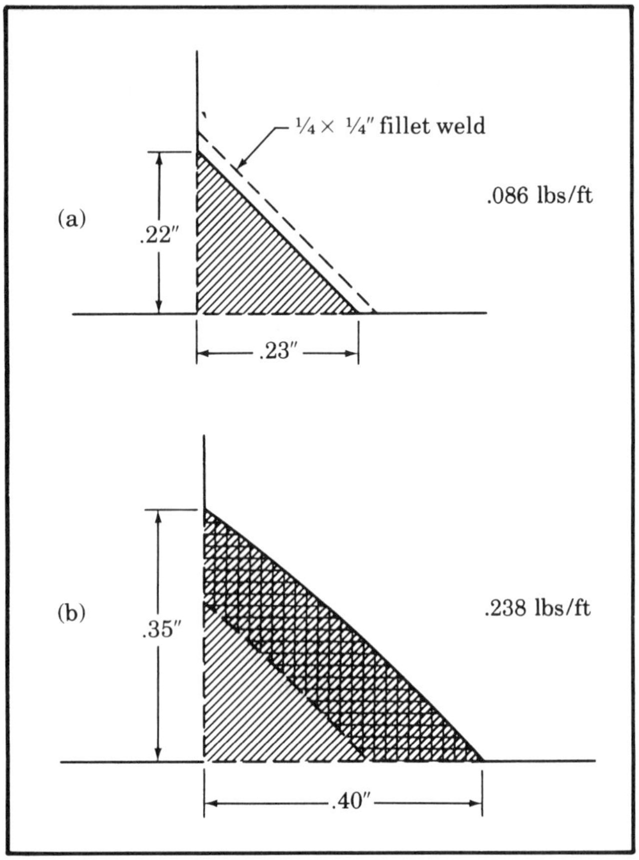

Fig. 12-14 Volume of weld metal vs. leg size.

THE QUESTION OF NEGATIVE TOLERANCE

As previously noted, the American Welding Society does permit a fillet weld to be 1/16″ undersize, providing the undersized portion does not exceed 10% of the length of the weld. On web-to-flange welds on girders, undersize is not permitted at the ends of the weld for a length equal to twice the width of the flange. Since the AWS has granted this concession with fillet welds made to code, the question arises as to whether a negative tolerance might be appropriate in some noncode work. While no one would suggest deliberately undersizing, once a minor undersize weld has been produced in a noncritical area, is the cost of correcting it by additional welding justified?

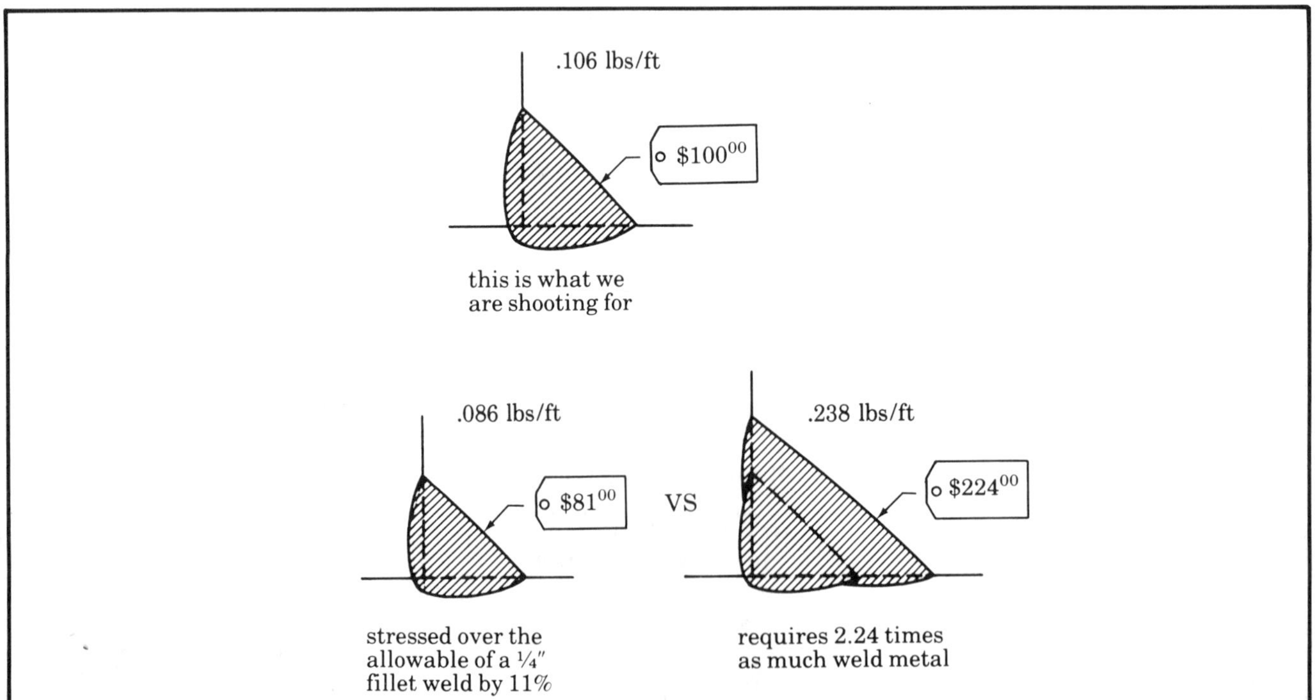

Fig. 12-15 Cost of extra fillet weld pass.

The answer seems to be sometimes definitely not. Consider the following example:

The fillet in Figure 12-14 (a) was specified to be ¼″ leg size, which would require .106 pounds of weld metal per foot. Since it's only by chance that a weld is laid down the exact size specified, the tendency is to use a slight amount of excess metal to make sure that the weld is not undersize. If the fillet leg is 10% oversize, the cross section area will be increased by 21% in the subject weld, or .129 pounds per foot would be used.

But the weld in Figure 12-14 (a) is neither made exactly ¼S, or 10% over. The inspector finds it to be undersize to the extent indicated in the figure, which means .086 pounds of weld metal would be used per foot. He marks it undersize and requires that another pass be added.

After the pass, the section looks as in Figure 12-14 (b). The weld now contains .238 pounds of weld metal per foot, or 2¼ times that called for by a ¼″ fillet. Was it worthwhile to correct for undersize?

Figure 12-15 illustrates the economic considerations to be taken into account.

Let's say the length of a true ¼″ fillet would cost exactly $100. In that case, the undersized weld would have cost $81 and the corrected fillet $224. Calculations show that the undersize fillet would have been stressed 11% over the allowable. But, chances are, the calculations never *really* called for a ¼″ leg size to begin with, but something slightly less — say .225″ — which one would round off to the next larger fraction, namely ¼″, In other words, the probability is high that the undersize weld is not undersize for the stress but oversize only in respect to the designer's detailing. And since weld metal is stronger than credited, there is even a probability that the "undersize" weld is more than adequate for the job it is intended to do.

In this case, it might be desirable to check back to the designer's calculations. The likelihood is great that if the corrective pass is made, $124 will have been expended for no purpose whatsoever.

An example of a different type is illustrated in Figure 12-16. Here, in joining the web to the flanges of the girder, ½″ fillets are prescribed. Calculation shows that fillets of .176″ leg size would be adequate

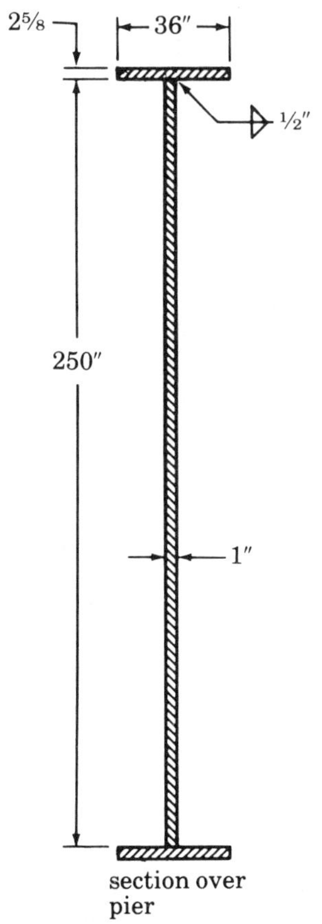

$M_{max} = 60{,}000$ ft kips
$V_{max} = 965$ k
$I = 4{,}322{,}500$ in.^{4}

Horizontal shear force on weld —

$$f = \frac{V\,a\,y}{I\,n} = \frac{(965\,k)\,(2\tfrac{5}{8} \times 36)\,(126\tfrac{3}{8})}{(4{,}322{,}500)\,(2\ \text{welds})} = 1330\ \text{lbs/inch}$$

Fatigue allowable for E70 fillet weld — Category F

Assume over 2×10^6 cycles and $K = +\tfrac{1}{4}$

$\tau_{SR} = 8$ ksi

so $$\tau_{max} = \frac{\tau_{SR}}{1 - K} = \frac{8\,\text{ksi}}{1 - \tfrac{1}{4}} = 10.67\ \text{ksi}$$

But not to exceed 21 ksi

Hence leg size =

$$\omega = \frac{1.33\,\text{k/in.}}{.707\,(10.67\,\text{ksi})} = .176\ \text{in. or}\ {}^{3}/_{16}''$$

but because of 2⅝″ flange,
must use ½″ ◺

Fig. 12-16 Calculated and code requirements for fillet welds.

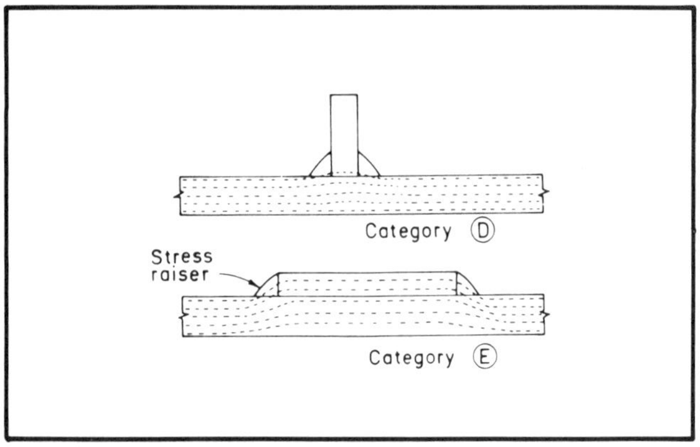

Fig. 12-17 Stress pattern in fillet welds.

for the loading, but because of the thickness of the flanges the code requires that ½″ fillets be used. Assume that after the weld has been run, they were found to be 7/16″ leg size instead of 8/16″. Should they be rewelded when they are 3.4 times as strong as they need be? . . . Meeting the code to the letter, in this case, would accomplish no good — merely waste time and materials.

REINFORCEMENTS AND WELD COSTS

At one time, butt welds had to have a reinforcement of at least 3/16″, and those that did not come up to this, had to be subjected to additional weld passes to build up to 3/16″. Gradually, the realization has come about that reinforcements are unnecessary — in fact do harm — some specifications call for reinforcements not to exceed ⅛″.

In static loading, the reinforcement does neither good nor harm in the engineering sense. It does, however, increase the welding cost unnecessarily.

Under fatigue loading, the reinforcement is damaging. It reduces the strength of the joint by setting up stress raisers at the edges of the reinforcement. See the photoelastic stress patterns in Figure 12-17. Some specifications take this into account and will permit about the same fatigue allowable as the plate if the weld reinforcement is removed.

Regardless of what specifications may say, the best welded butt joint will have no reinforcement. This does not mean that reinforcements should be machined off with most work, but that effort should be made to run the welds so as to finish the joint with little or no reinforcement.

If a butt weld should be made that did not quite come up to full size, as (a) in Figure 12-18, it would be better to leave the joint as it is than to have more welding added to provide a reinforcement, as in (b).

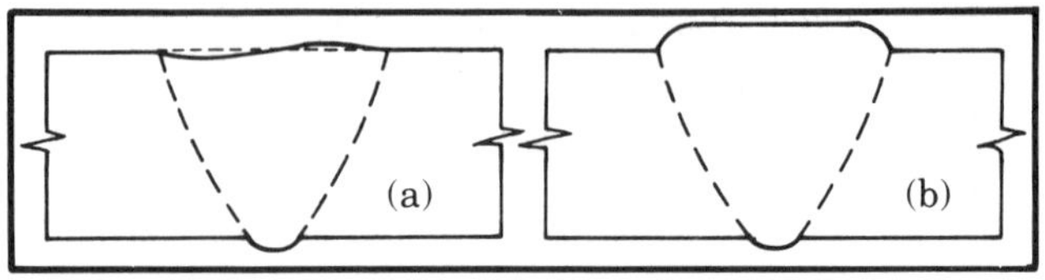

Fig. 12-18 If a butt weld should be made that did not quite come up to full size as shown in (a) it would be better to leave the joint as it is than to have more welding added to provide a reinforcement as in (b).

The slight loss in cross section in (a) would only increase the stress at this point by a few percent, depending upon the percentage reduction in section. Since the weld metal has greater strength than the plate, this would more than make up for the slight loss in section. Also, there is no notch in this joint to produce a stress raiser. The reinforcement in (b) sets up a stress raiser, which will increase the stress at this point by several times.

COMPUTER USAGE IN THE WELDING INDUSTRY

Computers have worked their way into almost every aspect of the workplace — and the welding shop is no exception. Whether it's a decision on which consumable to use to repair a child's wagon, or the preheat necessary to weld nuclear submarines, an answer can probably be garnered from a computer.

Since the advent of the "personal" computer, companies have been developing software to help people design and construct weldments. These same computers, as well as larger "mainframe" computers, are capable of controlling robots in expert welding systems, gathering data from product testing, and implementing quality control and inventory systems.

Computers as an Aid in Base Metal Selection

The same materials selection processes that designers have used for years are still in use today. Specific physical, electrical, mechanical and chemical properties are determined to be necessary, based upon the design. Computers allow these requirements to then be taken to a database; a collection of data on metal properties, chemistries, trade names and specifications that it meets. From this, the computer can extract a list of metals that meet these requirements. For example, a designer may need an alloy of 80,000 psi yield strength, with an 18% elongation that must be suitable for service in marine environments. The computer will very quickly go through its list of metals and determine if there are any that would meet those design criteria. If so, it could print a list, complete with trade names and any information on conformance to code bodies. Many professional organizations and corporations have developed this type of database containing information on frequently used or in-stock materials.[1]

Software developers have further structured materials selection programs so designers can determine what type of metal to specify if they are ordering it in Japan, Belgium, Australia, or under the Unified Number System.[2] Many databases have information on cast metals as well as tool steels, light metals, structural steels and wear-resistant alloys.[3]

Computers as an Aid in Design

Computer programs have been written to determine the distortion that takes place when welding, the cracking tendency of a weld or base plate, and the heat transfer expected on any weldment.[4] There are stress and strain calculating packages, and those that will calculate fatigue as well as analyze fracture mechanics.[5] These sorts of programs are extremely useful for evaluating a design. If a failure is likely under the expected service conditions, the "failure" will take place under computer simulation before the piece is ever manufactured or put into service.

Based on material selection, computer programs can determine the necessary size of beams, gears, plates, columns, catilevered spans, as well as odd shapes.[6] Moments of inertia and resonant frequencies can then be calculated by the computer.[7] Conversely, these programs will calculate material strength requirements, or the type of materials to use, from a specification of size, shape and service conditions. There are special programs written for structural members, piping systems, architectural and special applications.[8] Again, larger companies may develop their own materials databases as a part of their computer aided design (CAD) system.

Computers as an Aid in Welding Consumable Selection

There are several databases available to assist the weldor or welding engineer in selecting the proper consumables for an application.[9] By specifying the base metals, property requirements, weldor qualifications, and code requirements, a selection of the "best" consumables can be made by the computer. Most programs written by a professional agency or university specify the consumable by class, where those written by a manufacturer of consumables will specify these by brand name. Some of these consumables selection programs go on to specify a quantity of electrode or flux required to complete a job.[10]

Computers as an Aid in Welding Procedure Selection

Procedure selection programs may accept a description of a weldment as well as the configuration of the joint to be welded.[11] Geometry and size are taken into consideration, as well as the welding process to be used (eg. SMAW, GMA, Submerged Arc), and any code requirements. From these descriptions, a procedure is developed that, when combined with the proper consumables, will yield the desired properties. Cost calculations for a weldment are often output from these programs along with the recommended procedure. In cases where these procedures are automatically input into a welding controller (robot), this is called an "expert system".

Agencies, universities and private companies have put together the data necessary to calculate preheat and post-weld heat treatment of a weldment.[12] Using information on the base plate composition, process selection, consumables used, and code requirements, the computer makes a determination of the required preheat, interpass temperature, and postweld heat treatment.

Computers for Welding Control

Control of "robots" takes place with microprocessors and programmable controls. Several years ago, these microprocessors would have been powerful enough to run a sophisticated computer system. Today, however, robots are considered "computer controlled" when they not only perform the positioning and manipulations necessary to produce sound welds, but also make decisions about what must be done.

This type of control is generally called an "expert" system, or "knowledge-based" system. These control systems have large databases in the computer, which the control mechanism calls on when making a weld. For example, this type of system is capable of sensing the wire feed speed, and will compensate for a change by adjusting the stickout. A further refinement to the "expert" systems is the vision-based guidance system. This system actually senses the joint position and configuration in front of the welding head, either optically or acoustically, and directs and adjusts the positioner to stay on course. These systems can determine whether the joint is being properly filled, and adjust the travel speed or other operating variables to make up for tack welds and variations in joint configuration.

These systems have been developed or marketed by universities, welding and robotic equipment manufacturers, and fabricators. The sensing, control and responses will vary depending upon the information entered into the system while it was being developed. Because of this, great care must be taken to scrutinize the developers of the software as well as the equipment when selecting this type of system.

Computers Used in Testing and Quality Control

As code requirements become more stringent, and companies become more quality-conscious, more and more testing of weldments is being done. To aid in this end, chemical and mechanical testing equipment are being connected to computers to log, compile and track the data.

There is equipment available to test for almost any element's concentration in a metal sample. These instruments are often linked to a computer for control of the equipment, or simply for logging the results of a test. Mechanical testing equipment can even be hooked to computer controls that will log results and let the engineer see more closely what is happening during testing.[13] For example, when an impact specimen is broken, the computer can log exactly at what points energy is transmitted from the hammer to the test specimen. This data helps to analyze exactly how the failure occurred, which helps to determine what changes must be made to the base plate, design, or weld metal to keep this from occuring in field service.

Non-destructive testing is being more widely used, as refinements to its sensitivity are made. Most NDT systems developed today are computer controlled. These systems will log data, store and analyze pictures, as well as generate testing procedures.[14] Sophisticated controls and decision-making capabilities are necessary for real-time radiography systems. These tools, used as aids in assuring high quality, can be stored in a time-base on the computer, so a manager can see how the quality of weldments is (hopefully) improving as time goes on. Computer systems can also track a period of lower quality, and determine that it conincided with a procedural change that was implemented. The systems can record data and correlate all the variables to determine which ones are critical and should be very closely controlled. Others use these systems as a tracker of employee or supplier performance.

Computers as an Aid in Ordering and Inventory Control

Some of the programs that determine welding procedures (described above) will also ascertain the quantity of a consumable necessary to complete a job. These programs could be interfaced with a company's MRP or JIT inventory systems to place orders for consumables at the proper time, taking into account productivity on that job and lead time to receive the consumables. These types of programs are the cornerstone of companies' efforts who are trying to control their costs by reducing the raw materials inventory they have on hand. This is an especially effective program for welding consumables, since some must be stored under controlled conditions to prevent deterioration of quality.

Computer Usage in the Design of Welding Consumables

Computers are not only used in fabricating and repair shops, but are heavily relied upon in the development of quality welding consumables. As the demands from the field are heard for better quality and lower cost consumables, computers are used to determine what changes must be made. Computer programs estimate the cost of newly designed consumables, and also determine the ultimate cost to the consumer by determining the deposition rate of these new designs. Computer programs are used to determine final weld metal chemistry based on wire and fluxing agents used; from this, mechanical properties can be estimated. Computer programs are also used to give consumable manufacturers an idea of how difficult it will be to make a new product, as well as the sales potential for a particular type of product.

Computers are used to predict weld-pool shape based upon the type of power source, operating parameters, type of consumables, weldment shape, and placement of the electrical ground.[15] From this, the weld bead appearance can be simulated. The resistance to convexity, bead roughness and undercut can be predicted, with some idea as to ease of slag removal. The "follow" tendency and susceptibility to arc blow can be determined once the computer sets up the lines of magnetic flux caused by the electric arc.

Computer Usage in the Design of Welding Equipment

Welding equipment today is a complex blend of transformers, controllers, shunts, gears and P.C. boards. Many of the components that go into a welder, wire feeder, gun or controller are designed with the aid of a computer. The way all of these components fit and work together is also simulated using a computer.

The layout of printed circuit boards is largely done on computers, as is the design of transformers. Computers can help a designer to achieve the proper output characteristics, at the proper temperature and duty cycle, with the lowest possible cost. Finite element analysis — tedious by hand, but simple on a computer, can help to determine optimum conductor size and type, gearing, and even enters into safety considerations. The design and sizing of the equipment cases and design of packaging and crating can be more easily done with the aid of a computer.

Computer aided design (CAD) helps to bring everything together. Most CAD systems will even generate a list of materials that must be available in order to manufacture a machine. Computer integrated manufacturing facilities can often extract this data from a CAD system to order parts on time and then program the production equipment to wind, stamp or mold the parts.

Warning

Every computer has to have programs that utilize its rapid computing and decision making resources to be of much value. The computer cannot make decisions until it is equipped with the proper data, and is programmed with "rules" telling it under what circumstances it should choose one route or another. All programs are written by someone, somewhere. Thus, they are all somewhat subjective. Before relying on a computer or computer system to make the calls for your company, be sure to check it out **thoroughly**. This may mean putting in some normal cases, as well as some very extreme cases, and determining whether the computer makes the same decisions you would have, or if the decisions made by the computer are logical for your situation. To forgo this step is to trust the fate of your welding system to a computer programmer — usually unknown to you, and maybe very unfamiliar with your type of operation. The only way to get around this testing of the programmer's logic would be to program the computer yourself, or within your company; a very time consuming and costly step, indeed.

Sources for Computer Usage in Welding

[1] *MetSel/2,* Engineered by PSI/Systems; Produced and available through ASM International, Metals Park, OH 44073.

[2] *Worldwide Guide to Equivalent Alloys Series;* Available through ASM International, Metals Park, OH 44073.

UNSearch; Available through ASTM and SAE, Department 8406, 1916 Race Street, Philadelphia, PA 19103

[3] *MetSel/2* Data Disks; Available through ASM International, Metals Park, OH 44073.

[4] Blodgett, O.W., The Lincoln Electric Company; "Calculating Cooling Rates by Computer Programming", *Welding Journal,* March, 1984.

Baxter, C.F.G., British Maritime Technology Limited; "Computer Aided Distortion Control", *Computer Technology in Welding,* The Welding Institute, Cambridge, UK, June, 1988.

Michailov, V.G., Bulgarian Academy of Sciences, and Karkhin, V.A., Polytechnical Institute, USSR; "Using the Finite Elements Method in Predicting the Constructional Steels Susceptibility to Cold Cracking During Welding", *Computer Technology in Welding,* The Welding Institute, Cambridge, UK, June, 1988.

[5] *Stress and Strain Software Library,* produced by Engineering Software Company; Available through ASM International, Metals Park, OH 44073.

Fatigue: Life Predictions for Materials Selection and Design, developed by Thomas L. Altshuler; Produced and available through ASM International, Metals Park, OH 44073.

The Fracture Mechanic, developed by Dr. David Broek, FractuREsearch, Inc. Produced and available through ASM International, Metals Park, OH 44073.

[6] *Steel,* Available through Industry Consulting Engineers, 4314 S.W. Washouga Avenue, Portland, OR 97201.

Beams and Frames, developed by Compu-Tec Engineering, Inc.; Produced and available through ASM International, Metals Park, OH 44073.

Geardesign, produced by Engineering Technologies; Available through ASM International, Metals Park, OH 44073.

[7] *Moments of Inertia;* Available through Binary Engineering Associates, P.O. Box 528, Holden, MA 01520.

MSC/CASE, developed by D.V. Schiavello Enterprises, Inc.; Available through The MacNeal-Schwendler Corp., 815 Colorado Blvd., Los Angeles, CA 90041.

Stresses in Circular Components; Available through Binary Engineering Associates, P.O. Box 528, Holden, MA 01520.

[8] *Beam Analysis;* Available through Binary Engineering Associates, P.O. Box 528, Holden, MA 01520.

FLO-SERIES; Available through Engineered Software, 344-L Cleveland Ave., Tumwater, WA 98501.

*FLOW*CALC*SHEET;* Available through Binary Engineering Associates, P.O. Box 528, Holden, MA 01520.

Structural Engineering Library; Available through ENERCALC Engineering Software, 160 Newport Center Dr., Suite 116, Newport Beach, CA 92660.

[9] Hauge M., SINTEF; "The Use of Fracture Mechanics Databank for the Selection of Materials and Welding Procedures", *Computer Technology in Welding,* The Welding Institute, Cambridge, UK, June, 1988.

WELDCONQUEST; Available through The Welding Institute, Cambridge, UK.

[10] *WELDVOL;* Available through The Welding Institute, Cambridge, UK.

WELDCOST; Available through The Welding Institute, Cambridge, UK.

[11] *MAGDATA;* Available through The Welding Institute, Cambridge, UK.

WELDSPEC; Available through The Welding Institute, Cambridge, UK.

[12] *PREHEAT;* Available through The Welding Institute, Cambridge, UK.

[13] *SERIES IX;* Available through Instron Corp., 100 Royall St., Canton, MA 02021.

[14] *Weld Quality-The Role of Computers;* Pergamon Press for the International Institute of Welding, 1988.

[15] Fenn, R., Siores, E., and Egharevba, F., Brunel University, UK; "Computer Imaging of Weld Pool Shapes", *Computer Technology in Welding,* The Welding Institute, Cambridge, UK, June, 1988.

Chandel, R.S., Department of Energy, Mines and Resources, Canada, Goldak, J.A. and Oddy, A.S., Carleton University, Canada; "Computer Prediction of Weld Bead Shapes", *Computer Technology in Welding,* The Welding Institute, Cambridge, UK, June, 1988.

Section 13
SPECIAL APPLICATIONS

SECTION 13.1
UNDERWATER ARC WELDING

SECTION 13.2
WELDING GALVANIZED STEEL AND INGOT IRON

SECTION 13.3
WELDING STEEL PIPE

SECTION 13.4
WELDING CONCRETE/REINFORCING BARS

SECTION 13.5
ARC GOUGING AND CUTTING

SECTION 13.6
WELDING CLAD STEELS

SECTION 13.7
HARDSURFACING BY ARC WELDING

(Continued)

(Section 13.7 Continued)

SECTION 13.8

THAWING FROZEN WATER PIPES

Underwater Arc Welding

Underwater arc welding has been developed to a stage of practical utility largely by the U.S. Navy and commercial salvage and diving companies. Its main use has been in the emergency repair of ships, but, with the expansion of offshore drilling for oil and gas, it is becoming increasingly important in the construction and maintenance of drilling rigs and associated equipment. Although the literature* of the Bureau of Ships, Navy Department, still constitutes the most frequently quoted reference material on methods, techniques, and equipment, this original body of knowledge is being expanded by the experience and innovations of commercials organizations.

It should be emphasized that the electrical hazards associated with underwater welding are greater than welding in air. It should not be attempted by anyone not experienced or trained in underwater welding.

APPLICATIONS

The Navy considers underwater welding suitable for emergency ship repairs of a "semipermanent nature" when no drydock facilities are available. Even then, the welding is to be looked upon as a secondary means of securing a patch to a damaged plate in a ship's hull and of making the repair watertight. The primary means of attachment of an underwater patch plate, other than a very small one, would be by bolting. Small patch plates, however, may be fillet-welded to hulls without other means of fastening as emergency repairs of cracks. Leaky rivets, split seams, and protuberances can also be repaired by underwater welding.

Underwater welding is sometimes the only practical method of making attachments or repairs on the equipment of the offshore oil and gas industry, in which case it falls out of the emergency category prescribed for naval vessels. As an example, anodes are commonly welded to underwater drilling structures to protect them from corrosion, and such welds have proved their durability under hurricane conditions. Commercial transport companies also make repairs to ships and barges underwater and operate repaired vessels for a year or more before checking the repairs during routine dry-docking. While attempting to minimize underwater welding because of its cost, commercial organizations tend to look upon it as a practical way to accomplish certain construction and maintenance ends.

WELD STRENGTH

Underwater welds in mild-steel plate have consistently exhibited tensile strengths of 80% or more of the strength of similar welds made in air in laboratory tests, however, ductility is usually about 50%. The substantial decrease in ductility is explained by the hardening resulting from the drastic quenching of the surrounding water.

Despite the lower strength of underwater welds, the experience of salvage companies and other commercial organizations indicates that these welds can be completely serviceable. One salvage company states that it has made underwater welds that tested stronger than similar topside test welds in the evaluations of an independent testing laboratory. This same company has made underwater welds in pressure vessels in research and development experiments that stood up without failure to pressures as high as 200 psi.

EQUIPMENT

Equipment for underwater welding must be designed for the environmental conditions. The operator usually wears a complete diving outfit, the helmet of which is equipped with a supplementary hinged face plate fitted with No. 6 or 8 welding glass. The Navy recommends that the diver's head be insulated from the helmet by a skull cap and by a piece of rubber tape on the button of the exhaust valve. Although shallow-water work can be done by scuba divers with no protective garb other than

*Bureau of Ships' publication, "*Underwater Cutting and Welding Manual (NAVSHIPS 250-692-9)*," may be obtained by contacting the Superintendent of Documents, U.S. Government Printing Office, Washington, D.C.

rubber gloves and a face plate, this is permitted by the Navy only in extreme emergency.

Since electrode coatings are damaged by water, electrodes must be waterproofed. Airplane dope — cellulose acetate — is suitable for the purpose. Two proprietary products and a solution of celluloid in acetone are also mentioned in Navy literature as materials recommended for dip-waterproofing electrodes. One or two dips are required, depending on the consistency of the solution. After waterproofing, the newly applied coating should be cleaned from the tips of the electrodes to expose enough bare wire for easy starting. The diver should carry only a few electrodes at a time, since the waterproofing offers only temporary protection to the electrode coating. Some brands of electrodes perform satisfactorily without waterproofing if used within a short time after immersion.

Most underwater welding is done with 3/16 and 5/32-in. electrodes. The Navy recommends E6013 electrodes for all-position welding. The 3/16-in. electrode is recommended for all work, except when the metal is too thin for this size. However, recent experiments indicate that the iron-powder electrodes E6027 and E7024 are also satisfactory when used with the same techniques. The iron-powder coverings were treated with paraffin to protect them from water damage.

Electrode holders for underwater welding must be insulated and permit easy changing of electrodes. A plastic holder has been adopted as standard by the Navy. Springless-type topside holders with metal jaws may be used underwater when carefully and completely insulated with rubber tape. Their use should be restricted to emergency situations, however.

The preferred power source is a DC welding generator with at least 300-amp capacity, connected for straight (negative) polarity. A safety switch should be installed in the welding circuit, and this switch must be kept open at all times, except when the diver is actually welding. A single-pole, single-throw switch is commonly used. Care must be taken that the positive action of such a switch is not defeated by a damaged or poorly insulated cable between it and the welding machine, which could allow current to bypass the switch through a deck plate or some other metallic path. AC welders may be used if necessary, but additional special precautions must be taken to protect the diver. Diesel engines are preferred to drive DC generators in offshore work to reduce the fire hazard.

Size 2/0 extra-flexible welding cable (See Section 4.3) is recommended for underwater welding, except for a 10-ft length of 1/0 cable attached to the electrode holder to facilitate maneuverability. Cable is furnished in 50-ft lengths complete with one male and one female connector to minimize voltage drop. A tight wrapping of rubber tape is required to insulate underwater connections.

In addition to the basic equipment, certain accessories will be required by the diver. These include a "C" clamp to which the ground cable is bolted or brazed — for easy attachment of the ground lead to the work. A scraper for removing marine growth, rust, and paint will also be needed, as will a weighted wire brush and a chipping hammer for cleaning the base metal and removing slag.

PREPARING TO WELD

Surface preparation and fitup are extremely important in underwater welding. A sound weld cannot be made over thick paint, rust, or marine growth. The chipping hammer, scraper, and wire brush should be applied vigorously. Pneumatic tools may be of aid in cleaning surfaces badly fouled with marine organisms. Plates to be welded on can be prepared topside with a grinding wheel or sander.

Fitup is often a problem in the weld repair of ship hulls because of the curvature of the surface to which a plate is to be fillet-welded. If possible, the patch plate should be accurately faired to the curvature of the hull. If it is not practical to eliminate a gap entirely, it should be as small as possible. A 1/16-in. gap should be considered the maximum permissible, although gaps as great as 1/8-in. can be handled successfully by the "feeding-in" technique, described later.

A steady platform for the diver to work from is important. If the water is rough and the work can be delayed until calmer weather, that alternative should be taken. Attaching a stage to the object to be welded will provide a more steady work support than attaching to a diving launch, salvage ship, or float.

THE SELF-CONSUMING TECHNIQUE

The so-called "self-consuming" or drag technique is preferred in underwater welding. Once the arc is started, the electrode is pressed against the work with sufficient pressure to allow it to "consume itself." By this technique, a 3/16-in. fillet can be made in a single pass with a 3/16-in. electrode, and the resultant weld has nearly the same strength as a 3-pass weld. The advantages are reduced welding time, no cleaning of weld metal between passes, and the maintenance of a groove along the joint that serves as an easily followed guide for the progressing electrode.

Most underwater welding is fillet welding, and the self-consuming technique is the best way to accomplish it under the poor visibility and unsteady work conditions. Multipass welds are difficult after the first bead is laid because the guiding groove is lost.

Welding Current: Optimum current settings for underwater work are different from those in air, and they vary with length of cable and plate thickness. Sometimes the setting will be as much as 30% higher than that for the same electrode topside, since the surrounding water absorbs heat rapidly. Experiments under actual working conditions are recommended to determine the best settings for the self-consuming technique. Most underwater welding with the 3/16-in. E6013 electrode is done at 220 to 240 amp. Trial welds in the required position may be run with the equipment, using the self-consuming technique, and the settings adjusted to give best performance. The voltage remains nearly constant when using the self-consuming technique, and the rate of melt-off of the electrode is proportional to the current.

Horizontal Fillet Welding: With the safety switch open and the welding generator set to deliver the proper current, a horizontal fillet is made in the following manner:

1. The diver places the electrode against the joint at an angle of from 15 to 45°.
2. He calls for "current on." The arc should start when the topside switch is closed, or starting can be facilitated by tapping or scraping the electrode against the work.
3. Once the arc is started, the electrode is maintained against the work and allowed to "consume itself." An arc is not "held" as when welding in air. The original angle, as preferred by the weldor, is maintained as the bead is run. The bead is run straight — no weaving — until the fillet is finished or the electrode consumed. Constant pressure against the work is maintained. About 8 in. of weld metal is deposited for every 10 in. of electrode consumed.
4. When the electrode is consumed, the weldor calls for "current off." He should keep the stub of the electrode in welding position until verification of "current off" is received from the tender. When assured that the safety switch is open, he is ready to change electrodes.
5. Before starting work with the new electrode, the end of the deposit (or the whole bead, if a multipass weld) should be cleaned. The deposit from the new electrode should slightly overlap the previous deposit.
6. The new electrode must be in position against the work before the call for "current on."

Bubbles generated during the welding tend to interfere with visibility. In horizontal welding, interference is minimized by welding toward, rather than away from, the weldor. Vertical welds should be made from the top down for the same reason.

Vertical and Overhead Fillets: The same technique for horizontal fillets is used for vertical fillets. The weldor may find a slight variation of the electrode-to-work angle advantageous. Some change in current setting may be helpful.

The optimum current range for overhead welding is relatively narrow, and experimentation with samples is usually required. Problems of undercutting and convexity of contour are likely to be encountered when penetration and fusion are good. Maintaining an angle of 35 to 55° between the electrode and line of the weld and a slow, steady rate of progression tend to reduce undercutting and convexity. These techniques require a skill that is obtained only after considerable practice.

Samples of overhead welding, brought to the surface, will indicate errors in procedure. Dripping beads, for example, suggest that the current was too high or that the weldor applied insufficient pressure, or both.

Although poor fitup should be avoided if possible, a satisfactory fillet weld can be made when the gap at the root is as great as 1/8-in. The technique is referred to as "feeding-in" — which simply means that the electrode is fed toward the joint faster than normal with self-consuming welding. Instead of running 8-in. of weld with 10-in. of electrode, about 6-in. is run. The excess weld metal fills up the gap; it solidifies rapidly and does not run out as it would in air.

A HAZARD TO GUARD AGAINST

The need for protection of the diver from contact with electric current is taken into account by the equipment and procedures. Another hazard exists that could be overlooked. This is the possibility of an explosion resulting from the accumulation of hydrogen gas in closed or inadequately ventilated compartments. The bubbles generated in underwater welding have been shown to be about 70% hydrogen. At the time of formation, this hydrogen cannot "burn up" because of the lack of oxygen in the environment. Once the hydrogen rises above water, it continues to ascend until stopped by an overhead surface. Thus, it can accumulate as an overlayer in a closed compartment. Here, it is in contact with oxygen and is subject to explosive ignition by a spark or flame.

Prior to welding operations, closed compartments or corners that could trap and accumulate hydrogen gas should be provided with some means of positive ventilation.

San Mateo-Hayward (California) bridge. Note the old and new structures. See page 13.5-6 for the fixture used to weld the steel columns.

Welding Galvanized Steel and Ingot Iron

Galvanizing — the application of a zinc coating to a ferrous metal — is an effective means for retarding the corrosion of steel. In the galvanizing process, a thin layer of zinc is applied to the steel by electrodeposition, metal spray, sherardizing, or hot-dipping.

The zinc used for galvanizing is nearly 100% pure. The thickness of the coating for galvanized sheet is expressed in ounces per square foot, one oz. per sq. ft. being equivalent to a layer of zinc 0.0017-in. thick. The weight of zinc for galvanized sheet steel is given for both sides; thus a 2-oz.-per-sq.-ft. coating indicates 1 oz. per sq. ft. on each side.

The protective layer on hot-dip galvanized steel is pure zinc on the outer surface, under which are iron-zinc alloys with a progressively increasing iron content. After the pure zinc outer surface has corroded, the underlying iron-zinc alloys continue to protect the steel.

Welding, of course, disrupts the continuity of the protective surface, but investigations indicate that welding affects the corrosion resistance much less than the appearance of the finished joint would suggest. For all practical purposes, the effect is negligible on sheet steel. The arc obviously melts the zinc coating, and some of the coating escapes as zinc oxide fume.

Except for a narrow zone next to the weld bead, however, some zinc-iron alloy remains in the blackened area, and this alloy has essentially the same protective properties as the original zinc coating.

Galvanized sheet with a carbon content of less than 0.2% does not present any special welding problems. However, the lower maintenance costs of galvanized steel structures, such as bridges and storage tanks, have created a demand for galvanized materials in thicker sections and with higher tensile strengths. In general, galvanized plate over 1/4 in. thick has poorer weldability than the equivalent uncoated steel. Welds in galvanized plate have a tendency for porosity and micro-cracks are likely to develop, especially in fillet welds. The porosity is usually within acceptable limits, but the cracking may pose a difficult problem.

Investigations have established that weld-metal cracks in galvanized steel are intercrystalline with a zinc-rich layer lining the surface of the crack. Batch galvanized plates do not have a uniform coating of zinc, and the cracking may, thus, occur to a varying degree associated with the varying thickness of the coating.

Most, if not all, of the cracking can be eliminated by removing the zinc before welding. The following methods, some of which are slow and costly, are recommended:

1. Shear or torch-cut the web members of fillet assemblies after galvanizing.
2. Prepare the edges of butt joints after galvanizing, thus producing an edge surface free of zinc.
3. Burn off the zinc at the faying edges with a carbon arc or an acetylene torch while using an oxidizing flame.
4. Remove the zinc by grinding with a silicon carbide wheel, by gritblasting, or by sandblasting.

Cracking can also be caused by restraint, especially in fillet welds. The tendency for this type of cracking increases as the plate thickness increases. To reduce restraint, fit the joint with a small gap. This practice not only reduces the restraint but also provides a means for the volatilized zinc to escape.

WELDING TECHNIQUES FOR GALVANIZED STEEL

Shielded Metal-Arc Welding: Use E6010 electrodes DCEP, with the same electrode size and current as recommended for uncoated steel (see Section 6.2). Class E6011 electrodes can be used if only AC power sources are available.

Oscillate the arc parallel to the seam with a whipping motion 1/8 to 5/16 in. ahead of the molten crater. This procedure volatilizes the zinc. The weld will contain porosity and possibly cracks if the zinc is not completely volatilized in the area where the weld metal is to be deposited.

On galvanized plate 1/4 in. or more in thickness, insufficient penetration may be a problem with fillet welds. Always use the largest electrode compatible with the plate thickness and welding position. With a whipping motion, concentrate on volatilizing the zinc ahead of the crater.

For complete penetration, butt joints on galvanized plate require slightly more gap compared to those used with uncoated steel plate. Considerable spatter occurs when welding galvanized steel, but an antispatter compound applied in the area of the weld reduces the difficulty of removal.

Welding or cutting galvanized steel always produces zinc fumes. These fumes are toxic, and persons in the vicinity must be protected with proper ventilation or by some other means. (See Section 15.1 and "Safety in Welding and Cutting-ANSI-AWS Standard Z49.1.")

Gas Metal-Arc Welding: Gas metal-arc welding can be used on galvanized sheet and plate. Short-circuiting transfer is used on sheet; typical welding conditions are shown in Table 13-1. Sheet with zinc coatings thicker than those covered in the table require higher currents, lower speeds, or both. A tendency for burn through may require a slight side-to-side oscillation of the torch.

The advantages reported for a shielding-gas mixture of 80% argon and 20% CO_2 for welding uncoated steel is not apparent for galvanized steel. However, there is no disadvantage in using the argon-CO_2 mixture, except for the higher cost.

The same problems of porosity, cracking, and spatter encountered with shielded metal-arc welding are present with gas metal-arc welding, and the same corrective measures are used. With proper adjustment of the power source, the spatter can be reduced but not eliminated.

Carbon-Arc Welding: Galvanized sheet can be welded with a carbon arc and silicon-bronze filler rod. This process is suitable for sheet-metal ductwork, since the weld is satisfactory and there is little distortion. Furthermore, very little of the galvanizing is disturbed.

Carbon electrodes must have a long taper for welding on galvanized metal. A standard carbon is ground with a gradual taper to a sharp point. To maintain this taper, the carbon is held in the electrode holder from 1-1/2 to 2 in. away from the point.

The current range is 50 amp for 16 gage, to 20 amperes for 22 gage. Use a 5/32-in. carbon with 3/32-in. filler rod on 16 and 18 gage. Use 1/4-in. carbon with a 1/16-in. filler rod on 20 and 22 gage.

TABLE 13-1. Typical Short-Circuiting Transfer Welding Conditions for Continuously Galvanized Sheet*

Sheet Thickness (in.)	Joint Type	Position	Current (amps)	Voltage	Travel Speed (ipm)
1/16	Butt	Flat	80	18	12
1/16	Fillet	Flat	110	18	12
1/16	Butt	Vertical down	100	18	14
1/16	Fillet	Vertical down	120	18	14
1/16	Lap	Horizontal	100	18	12
1/16	Lap	Vertical down	100	18	13
1/8	Butt†	Flat	135	20	13
1/8	Butt†	Overhead	135	20	13
1/8	Fillet	Flat	135	20	11
1/8	Fillet	Vertical down	135	20	14
1/8	Fillet	Overhead	135	20	10
1/8	Lap	Horizontal	130	19	10
1/8	Lap	Vertical down	130	19	12
1/8	Lap	Overhead	120	19	8

* For zinc coating on both sides, 1-1/4 to 2-1/2 oz per sq ft; CO_2 gas 40 cfh; 0.040-in.-diameter electrode.
† 1/32 to 1/16-in. gap.

Filler metal is a silicon-bronze AWS A5.7, Class RCuSi-A rod, which contains 3 to 4% silicon.

Lay the filler rod on the seam with its end at an angle of about 10° from the work, as illustrated in Fig. 13-1. Play the arc on the filler metal. A small pool of molten metal forms. The rod is then fed in and melted at the leading edge of the molten pool as the arc is advanced along the seam. The arc is always on the molten pool and never on the galvanized sheet. This process is used on butt, fillet, lap and offset-lap joints, the latter two being the easiest to make.

Carbon-arc welding with silicon-bronze rod can also be used with uncoated sheet steel. The process produces less distortion than manual metal-arc welding and permits bridging a larger gap.

A tin coating on the silicon-bronze filler rod gives a welded joint with improved corrosion resistance. When the tin-coated rod is melted, the

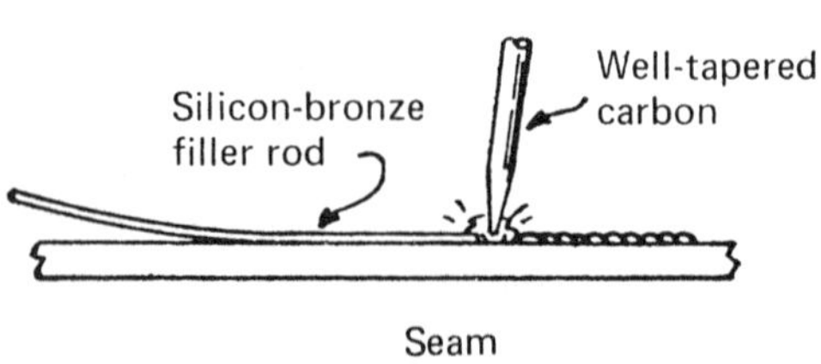

Fig. 13-1. Method of welding galvanized sheet with the carbon arc and a silicon-bronze filler rod.

tin flows over the heat-affected zone, producing a coating of zinc-tin alloy that enhances corrosion resistance.

GALVANIZING OVER WELDS/GALVANIZING KETTLES

To obtain the very best galvanizing results on weldments the following recommendations should be observed.

1. The welded surface and the adjacent area must be completely cleaned to remove all slag, smoke and spatter.
2. Use a welding process and technique which will produce a smooth and uniform surface.
3. The weld deposit should be relatively low in silicon (preferably .30% or less) and low in carbon (preferably .15% and lower) to produce the best and most uniform galvanizing appearance.

   NOTE: Normally E6010-11, 12, 13, type electrodes are acceptable. If in doubt check the code or specification to which the weldment is being built.
4. When welded components are to be galvanized care should be taken that they are not left in the galvanizing tank any longer than necessary.
5. Because of the adverse effects of high silicon, the steel plate and the weld metal used to construct a galvanizing kettle should contain an absolute minimum amount of silicon.
6. When selecting a process to weld a fabricated galvanizing tank or kettle it is important to know that the process is capable of producing the required weld deposit chemistry. (See #3 above).

WELDING INGOT IRON

Ingot iron is 99.9% pure iron. It is made by methods commonly used for making steel, but is more highly refined than steel. The carbon content may be as low as 0.013%, but the iron oxide content is higher than that of steel. A typical analysis is:

C	0.015%
Mn	0.035%
S	0.015%
Si	0.003%

with copper sometimes added for additional corrosion resistance.

Typical mechanical properties are:

Tensile Strength (psi)	45,000
Yield Strength (psi)	18,000
Elongation (% in 2 in.)	47%

Because of its good forming properties and increased resistance to corrosion over mild steel, ingot iron is used for culverts, corrugated building sheets, hot water tanks, vitreous-enameled ware, and for special applications where the magnetic properties of pure iron are wanted.

Ingot iron is readily weldable with the commonly used welding processes. It is nonhardenable because of low carbon content, but its high oxide content causes a tendency for porosity. For manual welding, E6010 or E6011 electrodes are used on light sections, and E6027 electrodes are used on heavy sections except where the work is to be vitreous-enameled. For vitreous-enameled work, low-hydrogen electrodes must be used, or the work must be stress relieved after welding.

Ingot iron can be welded with submerged-arc, self-shielded flux-cored arc, and gas metal-arc processes. However, the current must be slightly lower and travel speed must be slower than that used for steel. Also, electrodes must have more than the usual amount of deoxidizers.

Pipe manufactured in Japan intended for the Prudhoe Bay-Valdez pipe line being stored in Valdez.

Close-up of the above pipe showing the end protector. Pipe is manufactured to API specification X65 (vanadium treated), 48-in. diameter, 0.462-in. wall thickness, 40 ft. 1 in. long, net weight 10,090 lb, tested at 1,190 psi.

Welding Steel Pipe

Pipe welding is a specialized art. It differs so much from plate welding that operators must pass separate tests on pipe techniques to qualify as pipe weldors. Much of the skill required in plate welding, however, is applicable to pipe techniques. This section covers techniques, electrodes, procedures, and qualifications for welding cross-country lines and in-plant systems.

PIPE STEELS AND ELECTRODES

The American Petroleum Institute (API) publishes a standard for welding pipe, which includes procedures, qualification of operators, joint design, testing, and inspection. The standard also includes specifications for the types of pipe steel used for cross-country lines.

The most common steels used for cross-country pipelines are the API 5LX series. The "5L" indicates line pipe, the "X" indicates high-test line pipe, and two additional numbers indicate minimum yield strength in thousands of pounds per square inch. For example, 5LX65 is high-test line pipe with a 65,000-psi minimum yield strength. Table 13-2 lists chemical requirements and Table 13-3 mechanical-property requirements of this series.

For complete specifications on pipe and pipe fittings, see *API Specification for Line Pipe* (API Standard 5L) and *API Specification for High-Test Line Pipe* (API Standard 5LX). Standards (see Section 11.3) for procedure and operator qualification, joint preparation, and weld inspection are included in *Standard for Welding Pipe Lines and Related Facilities* (API Standard 1104). These standards are available from: American Petroleum Institute, 1220 L Street, Northwest Washington, DC 20005.

Weld cracking may occur in some of the pipe steels, particularly when the carbon and manganese contents are near the maximum allowable amounts and when the weather is cold. The X56, X60, and X65 steels may be treated with small amounts of alloying elements to increase their yield strength. With no top limit on tensile strength, this property can, and sometimes does, exceed 100,000 psi. Special techniques are required to prevent cracking in these steels. Some pipe steels have a silicon content up to 0.35%, which tends to produce pinholes unless special pipe electrodes are used.

TABLE 13-3.
Mechanical Property Requirements of Line Pipe

API Grade	Min. Yield Strength (psi)	Min. Tensile Strength (psi)
Grade A	30,000	48,000
Grade B	35,000	60,000
X-42	42,000	60,000
X-46	46,000	63,000
X-52	52,000	66,000(1) 72,000(2)
X-56	56,000	71,000(1) 75,000(2)
X-60	60,000	75,000(1) 78,000(2)
X-65	65,000	80,000

Elongation requirements vary with nominal area of test specimen and minimum specified tensile strength of the steel grade.
(1) For pipe less than 20 in. OD with any wall thickness, and for pipe 20 in. OD and larger with wall thickness greater than 0.375 in.
(2) For pipe 20 in. OD and larger with wall thickness 0.375 in. and less.

TABLE 13-2. Chemical Requirements (ladle analysis) For Welded 5LX Series High-Test Line Pipe

API Grade	Manufacturing Process	Composition, Max (%)			
		C	Mn	P	S
X42	Nonexpanded	0.28	1.25	0.04	0.05
X46, X52	Nonexpanded	0.30	1.35	0.04	0.05
X42, X46, X52	Cold-expanded	0.28	1.25	0.04	0.05
X56*, X60*	Non exp. or cold-exp.	0.26	1.35	0.04	0.05
X65*	Non exp. or cold-exp.	0.26	1.40	0.04	0.05

* These steels may also contain small amounts of columbium, vanadium, or titanium. Other analyses may be furnished by agreement between purchaser and producer. For good weldability, carbon and alloy content should be as low as practical.

TABLE 13-4. Typical Mechanical Properties of Deposited Weld Metal. Pipe-Welding Electrodes of the EXX10 Classes

	E7010-A1	E7010-G for Pipe	High-Yield Pipe Electrode
As-Welded			
Tensile Strength (psi)	75,000	75,000	85,000
Yield Strength (psi)	68,000	61,000	75,000
Elongation (% in 2 in.)	26	25	22
Charpy V Notch (ft-lb @ 70°F)	68	68	55
Stress-Relieved 1150°F			
Tensile Strength (psi)	75,000	72,000	85,000
Yield Strength (psi)	62,000	62,000	80,000
Elongation (% in 2 in.)	33	24	26
Charpy V Notch (ft-lb @ 70°F)	64	68	68

Pipe and tubing for power plants, boiler tubes, and refinery still tubes are usually specified to an ASTM standard. These specifications cover both mild steels and alloy steels for many types of service. Some of the ASTM pipe and tubing steels are listed in Table 6-13 along with recommended electrodes. These electrodes will provide the mechanical properties that meet most industrial code requirements. However, where specific codes apply, check the specification for electrode requirements or recommendations.

Since pipe welding is a specialized application, electrodes have been designed especially for vertical-down welding of pipe joints. These special electrodes are in the EXX10 class, but they may or may not have an AWS classification designation; the electrode supplier must be consulted for the proper product.

Pipe electrodes have several unique operating properties: 1. They produce a steady, concentrated arc, which forms a continuous bead on the inside of the pipe when the first pass is made. 2. They produce a thin covering of slag, which promotes good wetting to the sides of the groove but does not run into the crater and interfere with the arc. 3. They have a minimum tendency to produce pinholes.

The special pipe electrodes of the EXX10 class (Table 13-4) are used on pipe up to 5LX65; low-hydrogen electrodes (Table 6-16) are recommended for pipe with higher specified yield strength and for alloy pipe.

Generally, pipe with a wall thickness of 1/2 in. or less (which includes most cross-country lines) is welded vertical-down with special EXX10 electrodes. Pipe with heavier walls is welded with low-hydrogen electrodes.

VERTICAL-DOWN TECHNIQUES

Joint Preparation: Typical joint design and tolerances for a 30-in. 5LX pipe are shown in Fig. 13-2. Standard practice is to use internal line-up clamps on 16-in. and larger pipe. Often this does not produce a uniform spacing and, since hammering with a sledge is not recommended, the weldor must compensate for the poor fits that usually occur.

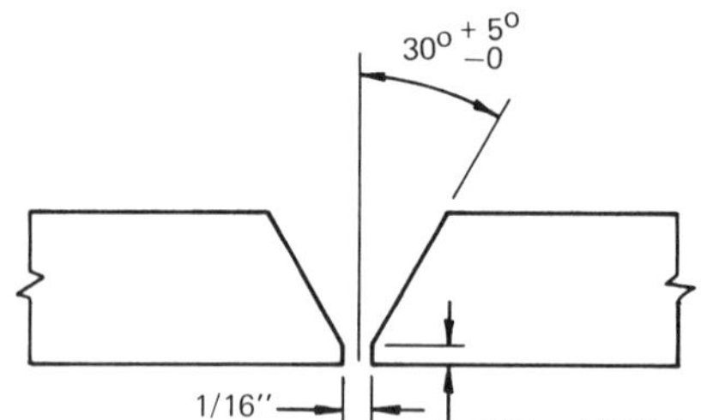

Fig. 13-2. Standard Joint design for a 30-in.-diameter 5LX pipe.

Stringer Bead: The first pass in the bottom of the groove — the stringer bead (Fig. 13-3) — is made by two weldors (more on large-diameter pipe) working on opposite sides of the pipe. The welds are started at the top, and the electrodes are dragged downhill to the bottom of the pipe. Too narrow a gap, too low a welding current, or too large a root face can cause insufficient penetration or lack of fusion. Too high a current, too wide a gap, or too little root face can cause insufficient penetration or lack of fusion. Too high a current, too wide a gap, or too little root face can cause melt-through (called "windows"), globular metal deposits on the inside of the pipe, external undercut (called "wagon tracks"), or internal undercut that can be repaired only from the inside of the pipe. Repairs from the inside of the pipe are, of course, difficult and are impossible on small-diameter pipe.

Hot Pass: The stringer should be cleaned with a power brush. If special pipe-welding electrode is not used, the stringer pass may be very convex and require grinding to remove the slag and to prepare the surface for the hot pass (Fig. 13-3). The hot pass should be applied with sufficient current to melt out

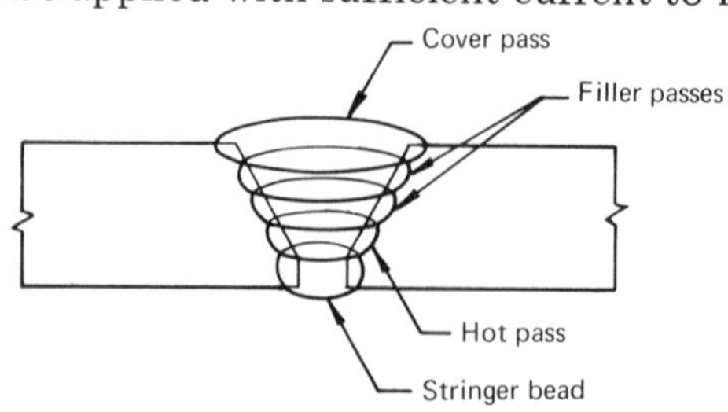

Fig. 13-3. Typical weld cross section on a 3/8-in.-wall pipe.

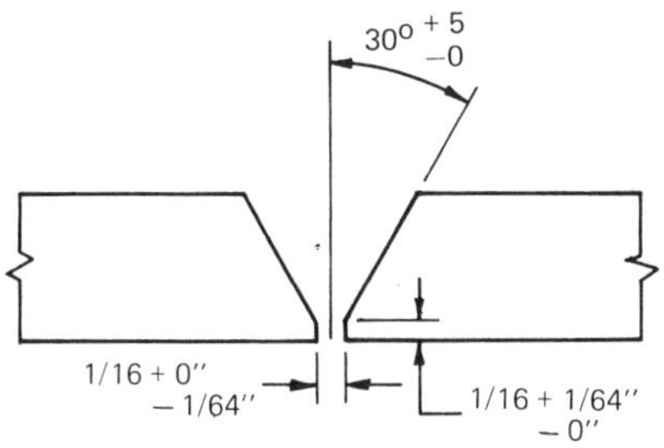

Fig. 13-4. Preferred joint design for X60 and X65 pipe.

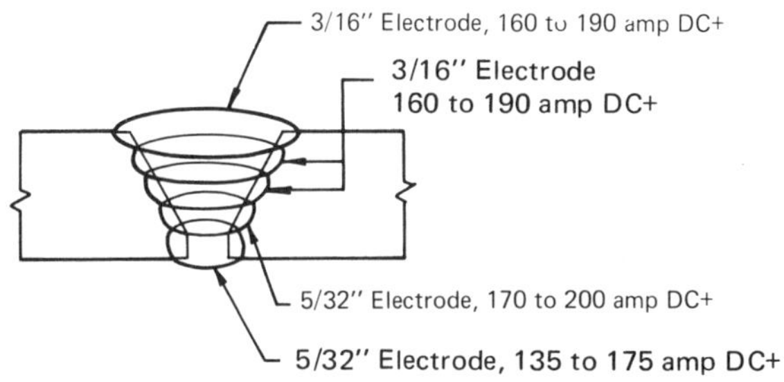

Fig. 13-5. Procedures for vertical-down pipe welding.

all "wagon tracks" and any remaining slag. It should be applied as soon as possible after the stringer is completed — within five minutes.

Filler Passes: A slight weave — only enough to obtain fusion at the sides of the groove — is used to apply the filler passes. Filler passes may fill the groove flush or almost flush at the top and bottom of the pipe, but more weld metal is usually required at the 2 to 5 o'clock positions. These lengths of groove are made flush with short filler passes called "stripper beads."

Cover Pass: The cover pass should be 1/32 to 1/16 in. higher than the pipe surface, and it should overlap the groove about 1/16 in. on each side, as illustrated in Fig. 13-3.

Preventing Cracks: Many factors contribute to weld cracking, but the primary causes are the chemistry, wall thickness, and the temperature of the pipe. Pipeline techniques recommended to prevent cracking are:

- Make the stringer as large as practical.
- Do not remove the line-up clamp until the stringer is completed.
- Make the hot pass as soon as possible after the stringer bead — within five minutes.
- Preheat if the pipe is less than 40°F. Proper preheat temperature depends on the chemistry of the pipe and wall thickness.

For X60 and X65 pipe, additional precautions and more stringent preparation are required:

- Joint preparation and tolerances should be as shown in Fig. 13-4.
- Use only electrodes of the EXX10 class designed for X60 and X65 pipe.
- Preheat to 300°F if the pipe temperature is less than 70°F.
- Always weld the stringer pass and hot pass with two (or more) men working on opposite sides of the pipe.

Vertical-Down Procedures: The procedures illustrated by the accompanying figures and tables are for EXX10-type electrode, designed for welding circumferential pipe joints vertical-down. Joint preparation is shown in Fig. 13-2, and current and electrode size are given in Fig. 13-5.

Table 13-5 shows the number of passes required for various wall thicknesses. This may vary with different operators, especially on thicker walls. One or two stripper beads may be required at the 2 to 5 o'clock location.

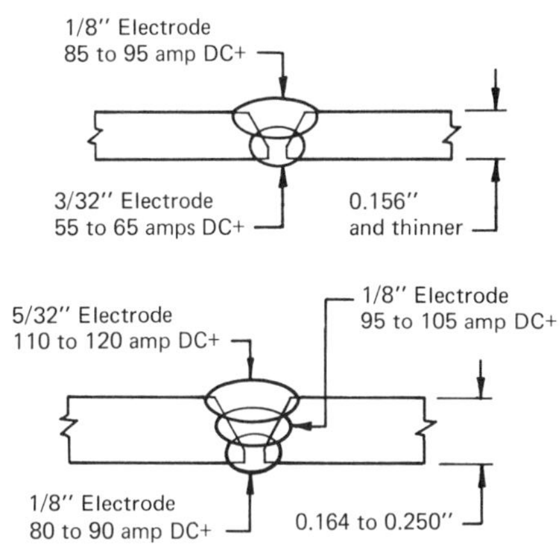

Fig. 13-6. Procedures for vertical-down welding on thin-wall pipe.

Table 13-6 shows the amount of electrode required per joint. Amounts are based on 4-in. stub lengths — about normal for cross-country pipeline welding.

TABLE 13-5. Passes Required for Vertical-Down Welding of Pipe

Wall Thickness of Pipe (in.)	Number of Passes
1/4	3
5/16	4
3/8	5
1/2	7

TABLE 13-6. Type EXX10 Electrode Consumption for Vertical-Down Welding of Pipe

Pipe Diam. (in.)	Wall Thickness of Pipe (in.)											
	1/4			5/16			3/8			1/2		
	Electrode Required* (lb/joint)											
	5/32	3/16	Total	5/32	3/16	Total	5/32	3/16	Total	5/32	5/16	Total
6	0.47	0.24	0.71	0.47	0.47	0.94	0.47	0.75	1.2	....	....	.....
8	0.63	0.32	0.95	0.63	0.63	1.3	0.63	1.0	1.6	0.63	2.0	2.6
10	0.79	0.40	1.2	0.78	0.78	1.6	0.78	1.1	1.9	0.79	2.4	3.2
12	0.95	0.47	1.4	0.95	0.95	1.9	0.95	1.5	2.5	0.95	3.0	4.0
14	1.1	0.55	1.7	1.1	1.1	2.2	1.1	1.8	2.9	1.1	3.5	4.6
16	1.2	0.63	1.8	1.2	1.3	2.5	1.3	2.0	3.3	1.2	4.0	5.2
18	1.4	0.70	2.1	1.4	1.4	2.8	1.4	2.3	3.7	1.4	4.5	5.9
20	1.5	0.76	2.3	1.5	1.6	3.1	1.6	2.4	4.0	1.5	4.8	6.3
22	1.7	0.90	2.6	1.7	1.8	3.5	1.7	2.9	4.6	1.7	5.5	7.2
24	1.9	0.94	2.8	1.9	1.9	3.8	1.9	3.0	4.9	1.9	6.0	7.9
26	2.0	1.1	3.1	2.0	2.1	4.1	2.0	3.3	5.3	2.0	6.5	8.5
28	2.2	1.1	3.3	2.2	2.2	4.4	2.2	3.5	5.7	2.2	7.0	9.2
30	2.3	1.2	3.5	2.3	2.4	4.7	2.3	3.8	6.1	2.4	7.5	9.9
32	2.5	1.3	3.8	2.5	2.5	5.0	2.5	4.1	6.6	2.5	8.0	10.5
34	2.7	1.3	4.0	2.7	2.7	5.4	2.7	4.3	7.0	2.7	8.5	11.2
36	2.8	1.4	4.2	2.8	2.8	5.6	2.8	4.5	7.3	2.8	9.0	11.8

*Values given include 4-in. stubs.

Thin-Wall Pipe Procedures: Electrodes and procedures for thin-wall pipe are much the same as for cross-country pipe. One variation is the occasional use of negative polarity. Where a wide gap causes excessive burnthrough, negative polarity is used on the first pass.

Welding procedures for butt joints in thin-wall pipe are shown in Fig. 13-6. Table 13-7 shows the amount of electrode required per joint. Values are based on 4-in. stub lengths, normal for this type of welding.

Another type of connection used on thin-wall pipe is the bell-and-spigot joint (Fig. 13-7). Procedures for this type of joint are given in Table 13-8, and the amount of electrode required per joint is given in Table 13-9.

VERTICAL-UP TECHNIQUES

Vertical-up welding requires fewer beads than vertical-down and therefore less cleaning. The time saved results in an advantage for vertical-up welding on wall thicknesses of 1/2 in. and thicker.

Vertical-up welding is usually used for in-plant welding of pipe, such as in power stations and refineries; vertical-down welding is used on cross-country pipelines.

Joint Preparation Without Backup Ring: Recommended joint design is shown in Fig. 13-8. The 1/8-in. gap must be maintained accurately all around the joint and tack welded in at least four places on 6-in.-diameter and larger pipe.

Welding: All passes are made starting at the bottom and moving upward. Usually an E6010 or E7010 electrode is used for the first pass, except on alloy pipe, where an alloy electrode must be used. If low-hydrogen electrodes are used, such welding (without backup ring) requires an especially high degree of skill.

After the first pass, the bead should be cleaned thoroughly, and any bumps or imperfections should be ground before additional passes are placed. The finish pass should be about 1/16 in. higher than the pipe and should overlap the original groove by 1/16 to 1/8 in.

Procedures: The procedures shown in Fig. 13-9 are for vertical-up welding of circumferential pipe joints. The first pass is an EXX10-type electrode

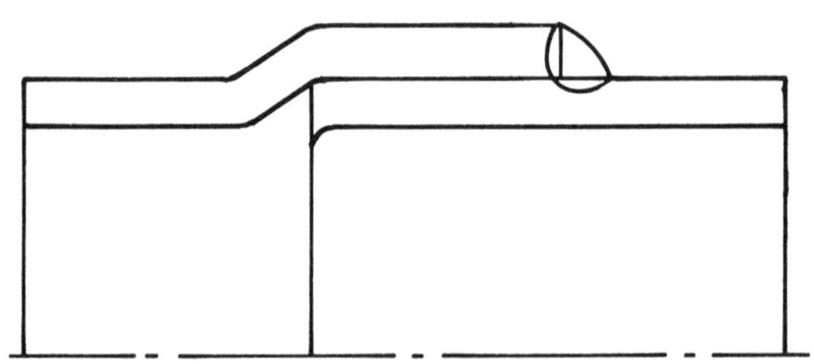

Fig. 13-7. Bell-and-spigot joint for thin-wall pipe.

TABLE 13-7. Type EXX10 Electrode Consumption for Butt Welds in Thin-Wall Pipe (Vertical-Down)

Pipe Diam. (in.)	Wall Thickness of Pipe (in.)								
	1/8			3/16			1/4		
	Electrode Required* (lb/joint)								
	3/32	1/8	Total	1/8	5/32	Total	1/8	5/32	Total
4	0.071	0.092	0.163	0.196	0.146	0.342	0.232	0.177	0.409
4-1/2	0.080	0.102	0.182	0.221	0.164	0.385	0.261	0.199	0.460
6	0.106	0.138	0.244	0.294	0.219	0.513	0.348	0.268	0.616
6-5/8	0.118	0.152	0.270	0.325	0.242	0.567	0.384	0.293	0.677
8	0.142	0.184	0.326	0.392	0.292	0.684	0.464	0.354	0.818
8-5/8	0.153	0.198	0.351	0.422	0.314	0.736	0.500	0.382	0.882
10	0.178	0.230	0.408	0.490	0.365	0.855	0.581	0.442	1.023
10-3/4	0.191	0.247	0.438	0.526	0.392	0.918	0.623	0.475	1.098
12-3/4	0.223	0.289	0.512	0.615	0.458	1.073	0.727	0.552	1.279
14	0.248	0.323	0.571	0.686	0.520	1.206	0.812	0.615	1.427
16	0.284	0.369	0.653	0.784	0.593	1.377	0.929	0.703	1.632
20	0.355	0.461	0.816	0.980	0.741	1.721	1.161	0.879	2.040
24	.....	.....	.....	1.177	0.890	2.067	1.393	1.062	2.455
28	.....	.....	.....	1.372	1.040	2.412	1.624	1.230	2.854
32	.....	.....	.....	1.568	1.186	2.754	1.858	1.406	3.264
36	.....	.....	.....	1.764	1.338	3.102	2.089	1.582	3.671

*Values given include 4-in. stubs.

TABLE 13-8. Type EXX10 Procedures for Bell-and-Spigot Joints in Thin-Wall Pipe

Wall Thickness of Pipe (in.)	Electrode Size (in.)	Current DC + (amp)
0.075	1/8	80 to 100
0.090	1/8	80 to 100
0.105	5/32	120 to 130
0.125	5/32	120 to 130
0.135	5/32	130 to 140

TABLE 13-9. Type EXX10 Electrode Consumption For Bell-and Spigot Joints in Thin-Wall Pipe

Pipe Diam. (in.)	Electrode Required (lb/joint)				
	Wall Thickness of Pipe (in.)				
	0.075	0.090	0.105	0.125	0.135
4	0.041	0.053	0.078	0.095	0.100
4-1/2	0.046	0.059	0.088	0.107	0.113
6	0.062	0.078	0.114	0.142	0.150
6-5/8	0.068	0.088	0.129	0.157	0.166
8	0.082	0.106	0.156	0.190	0.200
8-5/8	0.088	0.114	0.168	0.205	0.215
10	0.098	0.132	0.195	0.238	0.247
10-3/4	0.108	0.142	0.209	0.253	0.264
12-3/4	0.127	0.168	0.245	0.298	0.314
14	0.136	0.185	0.271	0.330	0.345
16	0.155	0.211	0.310	0.377	0.394
20	0.194	0.264	0.387	0.471	0.494

and, for subsequent passes, procedures are given for either EXX10 or EXX18. Table 13-10 gives the approximate number of passes required for different wall thicknesses. This will vary with different operators.

Table 13-11 shows the electrode consumption per joint for vertical-up welding of pipe. Amounts are based on the use of EXX18 electrode and the joint design shown in Fig. 13-8. If the gap is changed, amounts must be revised accordingly.

Joint Preparation With a Backup Ring: Recommended joint design for vertical-up welding with backup ring is shown in Figure 13-10. Gap distance should match the outside diameter of the electrode covering.

TABLE 13-10. Passes Required for Vertical-Up Welding of Pipe

Wall Thickness of Pipe (in.)	Number of Passes
1/4	2
5/16	2
3/8	3
1/2	3
5/8	4
3/4	6
1	7

TABLE 13-11. Type EXX18 Electrode Consumption for Vertical-Up Welding of Pipe

Pipe Diam. (in.)	Wall Thickness of Pipe (in.)														
	3/8			1/2			5/8			3/4			1		
	Electrode Required* (lb/joint)														
	3/32	1/8	Total	3/32	1/8	Total	3/32	1/8	Total	3/32	1/8 & 5/32	Total	3/32	1/8 & 5/32	Total
6	0.51	1.34	1.85	0.51	2.31	2.82	....	....	.....	....	....	.....	....	....	.....
8	0.7	1.8	2.5	0.7	3.1	3.8	0.7	4.7	5.4	....	....	.....	....	....	.....
12	1.0	2.7	3.7	1.0	4.7	5.7	1.0	7.1	8.1	1.0	9.9	10.9	1.0	16.7	17.7
16	1.4	3.6	5.0	1.4	6.1	7.5	1.4	9.8	11.2	1.4	13.1	14.5	1.4	22.1	23.5
20	1.7	4.5	6.2	1.7	7.7	9.4	1.7	11.7	13.4	1.7	16.4	18.1	1.7	27.6	29.3
24	2.0	5.4	7.4	2.0	9.3	11.3	2.0	14.2	16.2	2.0	19.8	21.8	2.0	33.4	35.4
28	2.4	6.2	8.6	2.4	10.8	13.2	2.4	16.5	18.9	2.4	23.0	25.4	2.4	38.8	41.2
32	2.7	7.2	9.9	2.7	12.4	15.1	2.7	19.0	21.7	2.7	26.5	29.2	2.7	44.5	47.2
36	3.1	8.0	11.1	3.1	13.9	17.0	3.1	21.6	24.7	3.1	29.6	32.7	3.1	49.9	53.0
40	3.4	8.9	12.3	3.4	15.4	18.8	3.4	23.5	26.9	3.4	32.8	36.2	3.4	55.3	58.7
48	4.1	14.7	18.8	4.1	18.5	22.6	4.1	28.3	32.4	4.1	39.5	43.6	4.1	66.6	70.7
60	....	....	....	5.1	23.2	28.3	5.1	45.4	50.5	5.1	49.4	54.5	5.1	83.2	88.3

*Values given include 4-in. stubs.

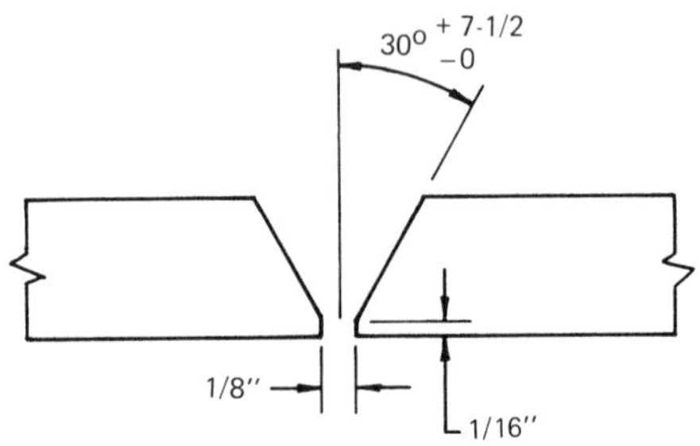

Fig. 13-8. Joint design for vertical-up welding on pipe without a backup ring.

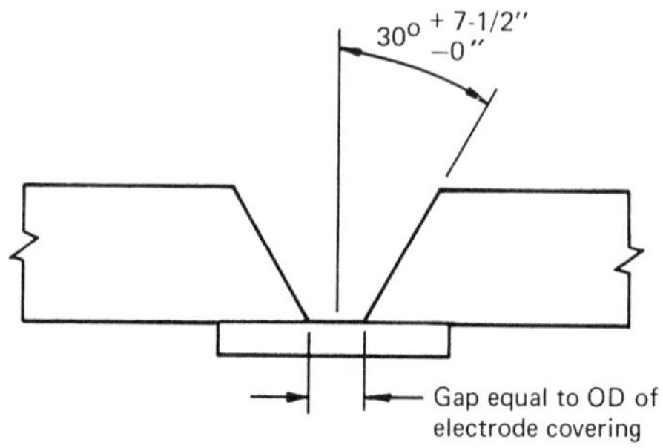

Fig. 13-10. Joint design for vertical-up welding on pipe with backup ring.

Welding: The first pass can be welded with either EXX10 or EXX18 electrode. The root pass must penetrate both lips of the pipe and the backup ring. Any unfused area at the bottom of the root pass is an objectionable defect. Since the gap is wider than that used for welding without a backup ring, the groove is also wider, and split layers are usually necessary after about the fifth layer.

Procedures: Procedures for welding with a backup ring are shown in Figure 13-11. Approximate numbers of passes required are shown in Table 13-12.

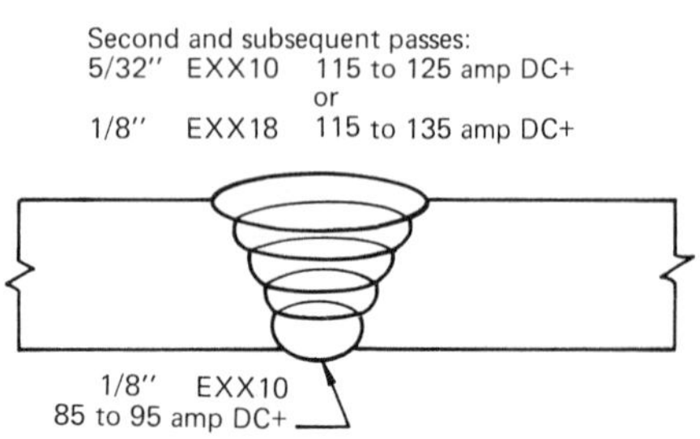

Fig. 13-9. Procedures for vertical-up pipe welding without a backup ring.

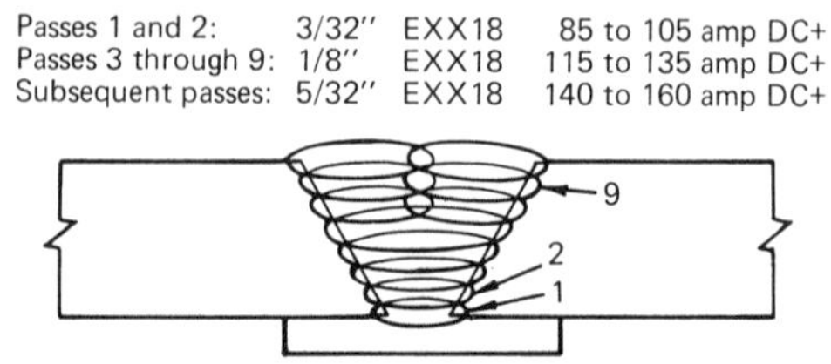

Fig. 13-11. Procedures for vertical-up pipe welding with backup ring.

HORIZONTAL TECHNIQUES

Piping in refineries and plants often requires joints with the axis of the pipe vertical. These horizontal joints require different techniques and procedures than vertical joints, but the designs of horizontal joints are the same as those for vertical-up welding — either with or without the backup ring.

Welding: Without a backup ring, the first pass is usually made with EXX10 electrode, and the weld is completed with either EXX10 or EXX18. With a backup ring, the entire weld can be made with EXX18 electrode.

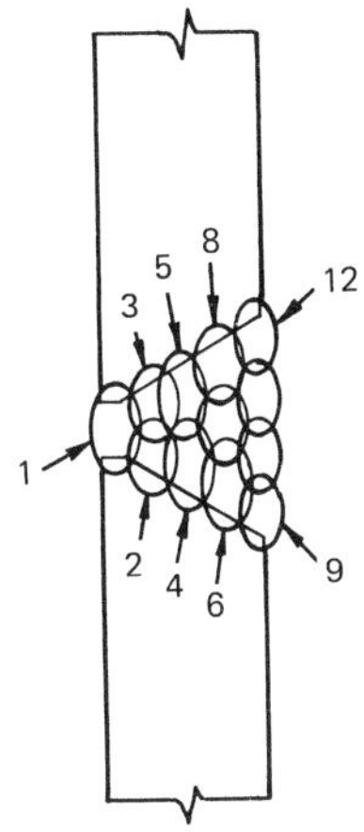

Fig. 13-12. Typical cross section of a horizontal butt weld on pipe, showing bead placement.

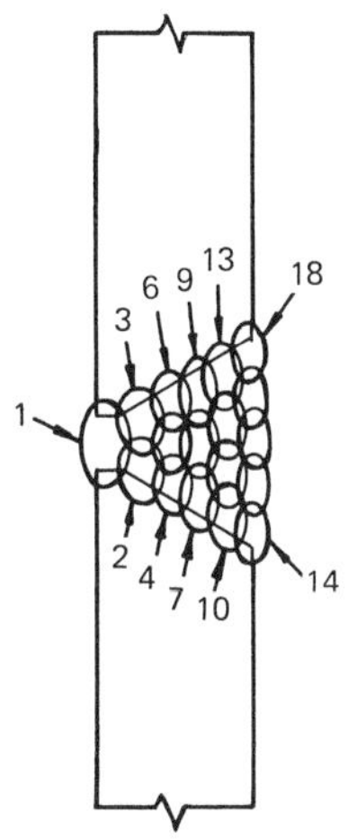

Fig. 13-13. Procedures for a horizontal butt weld on pipe, no-backup ring, EXX10-type electrode.

First pass:	1/8″	EXX10	80 to 110 amp DC+
Subsequent passes:	5/32″	EXX10	120 to 150 amp DC+
			or
	1/8″	EXX18	115 to 135 amp DC+

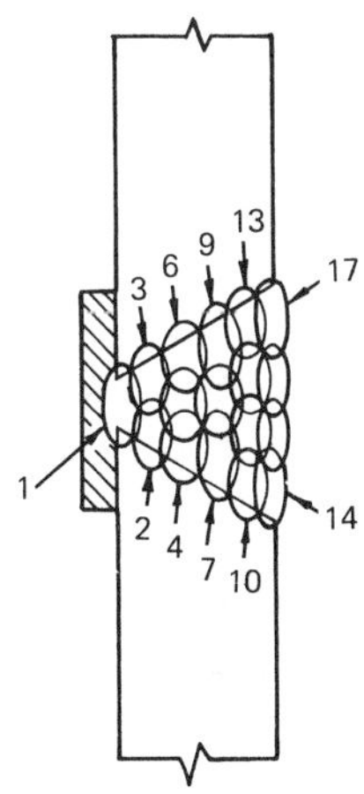

Fig. 13-14. Procedures for a horizontal butt weld on pipe, with backup ring, EXX18 electrode.

Passes 1 through 3:	3/32″	EXX18	85 to 105 amp DC+
Passes 4 through 9:	1/8″	EXX18	115 to 135 amp DC+
Subsequent passes:	5/32″	EXX18	140 to 160 amp DC+

TABLE 13-12. Passes Required for Vertical-Up Welding of Pipe Using Backup Ring

Wall Thickness of Pipe (in.)	Number of Passes
3/8	3
1/2	4
5/8	5
3/4	7
1	10

TABLE 13-13. Procedures for Double-Ending Pipe (manual stringer bead, finish with submerged-arc)

Wall Thickness (in.)	All*	1/4		3/8		1/2	
Pass Number	1	2	3	2	3	2	3 - 5
Electrode Size	5/32	1/4		1/8		1/8	
Current (amp) DC+	160 - 180	350	400	450	600	475	600
Volts	. . . .	27	31	29	33	30	33
Arc Speed (in./min)	13 - 16	26	29	26	24	26	24
Electrode Req'd (lb/ft)	0.090	0.127		0.240		0.513	
Flux Req'd (lb/ft)	. . . .	0.10 - 0.14		0.19 - 0.25		0.40 - 0.54	
Total Time (hr/ft)	0.0138	0.0146		0.0160		0.0326	

*Manual stringer bead is used for all wall thicknesses. For manual electrode, see Fig. 13-5. Add manual time to submerged-arc time to get total time.

After the first pass, the next bead is placed on the lower side of the groove, penetrating both the root pass and the lower pipe member. The following bead is placed immediately above the second bead, so it penetrates both the second bead and the upper pipe member. Similarly, each layer is started at the lower side of the groove and built upward, as shown in Fig. 13-12.

Horizontal Procedures: Figure 13-13 shows procedures, using EXX10 electrodes and a combination of EXX10 and EXX18. Figure 13-14 shows procedures using EXX18 electrodes exclusively.

DOUBLE-ENDING

Double-ending is the welding of two or more lengths of pipe into one longer length. This is done frequently by submerged-arc automatic welding or a combination of submerged-arc and stick-electrode welding. It is practical and economical where the terrain permits hauling the double lengths of pipe to the field site.

One method of double-ending pipe is to assemble two lengths of pipe on a set of power-driven rolls. The joint is lined up with an internal line-up clamp, and the first bead is made manually as described under vertical-down techniques. The second and third beads are made with submerged-arc while the pipe is rotated under an automatic welding

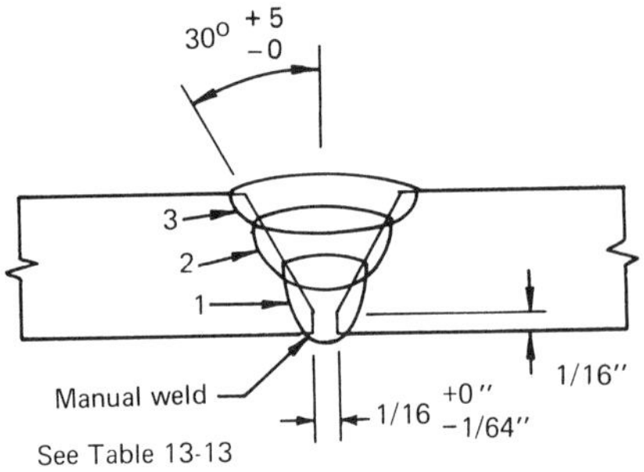

Fig. 13-15. Joint design and sequence for double-ending pipe with submerged-arc (stick-electrode stringer bead).

TABLE 13-14. Submerged-Arc Procedures for Double-Ending Pipe

Wall Thickness (in.)	1/4		3/8		1/2	
Root Face (in.)	5/32		1/4		1/4	
Electrode Size	1/8 or 5/32		1/8 or 5/32		1/8 or 5/32	
Pass Number	1,2	3	1 - 3	4	1 - 4	5
Current (amp) DC+	575	650	700	700	700	700
Volts	32	31	34	33	34	34
Arc Speed (in./min)	60	60	55	55	50	40
Electrode Req'd (lb/ft)	0.167		0.277		0.422	
Flux Req'd (lb/ft)	0.15 - 0.20		0.25 - 0.34		0.38 - 0.50	
Total Time (hr/ft)	0.0100		0.0145		0.0210	

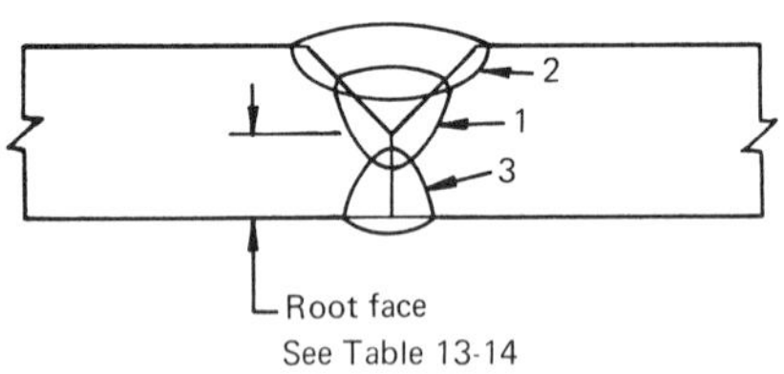

Fig. 13-16. Joint design and welding sequence for double-ending pipe with automatic submerged-arc welding.

head. The joint preparation shown in Fig. 13-15 and the submerged-arc procedures given in Table 13-13 are recommended.

Another method for double-ending pipe is to make all passes automatically. This method requires a joint preparation, as shown in Fig. 13-16. First, the two submerged-arc beads are made on the outside while the pipe is rotated under the automatic head. The third bead is made from the inside of the pipe. The automatic head is mounted at the end of a long boom that positions the head over the seam through the open end of the pipe. Procedures for the automatic welding are given in Table 13-14.

API PROCEDURE AND QUALIFICATION REQUIREMENTS

Procedures and operators for welding cross-country pipe lines are qualified by API Standard 1104. For complete details, the standard should be consulted. The following is a brief resume of the tests and requirements, which are elaborated on with more detail in Section 11.3.

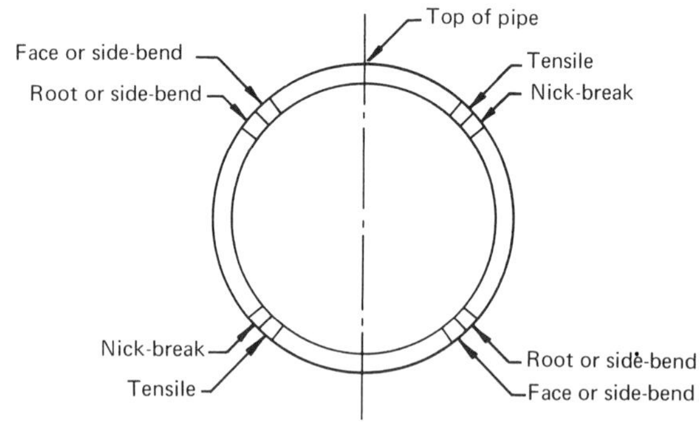

Fig. 13-17. Location of test specimens on pipe from 4-1/2 to 12-3/4 in. in diameter for API qualification tests. For other diameters See Page 11-3-34.

Location of Specimens: Test specimens are cut from the pipe as shown in Fig. 13-17. In the field, the specimens are torch-cut and ground to size. In the shop, specimens are machined rather than ground; machining is the preferred method.

Tensile Requirements: Tensile strength of the weld and the fusion zone must be equal to or greater than the minimum specified tensile strength of the pipe. If the specimen breaks in the plate, the tensile strength must be at least 95% of the minimum specified for the plate.

It is not uncommon for a satisfactory test to break in the weld. This is because there are no maximum strengths specified for the plate, and the tensile strength of the plate may be considerably higher than that of the weld metal.

Nick-Break Test: This test is intended to expose any slag inclusions, gas pockets, or lack of fusion. The specimen is notched with a hack saw, and then broken to expose the interior of the weld. The exposed surfaces must show complete penetration and fusion. There must be no more than six gas pockets per square inch of surface area and no gas pocket larger than 1/16-in. Slag inclusions must not be more than 1/32-in. deep or 1/8-in. (or one-half the wall thickness) in length, and there shall be at least 1/2 in. of sound metal between inclusions.

Bend Tests: Specimens are prepared for bend tests by removing all weld reinforcement from both sides. Bending is done in a fixture as specified in API Standard 1104. The weld is unacceptable if the test produces a crack or defect with the largest dimension greater than 1/8-in. or one-half the wall thickness of the pipe, whichever is smaller. Cracks at the edges of the specimen that are less than 1/4 in. in the greatest dimension can be disregarded.

CODES FOR IN-PLANT PIPING

For consistent high-quality work on in-plant piping, the welding procedure and the weldor should be qualified under some recognized national code. The ASME Boiler and Pressure Vessel Code, Section IX — *Welding Qualifications,* is the code most used to qualify weldors for power piping work. For a detailed review of operator's qualification tests for in-plant welding, see *Pipefitter Welder's Review of Metal-Arc Welding for Qualification Under ASME Code Rules.* This is published by The National Certified Pipe Welding Bureau, 1385 Piccard Drive, Rockville, MD 20850-4329.

Double-ending pipe in the field with the submerged arc process.

Unique design for a pipeline river crossing.

Welding Concrete/Reinforcing Bars

Plain concrete has low tensile strength; concrete members that must withstand appreciable tensile loads are usually reinforced with steel bars or mesh. In this composite construction, the tensile load is carried by the steel, while the concrete carries the compressive load. The amount of steel required for adequate reinforcement varies from about 1% for beams and slabs to about 6% for some columns.

To insure a tight grip between the concrete and the reinforcing bars, the bars are fabricated with small ridges, giving them a raised veined or corrugated surface. These surface deformations vary with different manufacturers, but all must be in accordance with ASTM requirements.

Since the coefficients of thermal expansion are about the same for steel and concrete, temperature changes do not cause appreciable stresses between the two materials.

Most reinforced concrete construction requires that the bars be welded end to end. Bulletin D1.4 of the American Welding Society describes recommended practices for welding of reinforcement steel.

Reinforcing bar is available in sizes from No. 3, which is 3/8 in. in diameter, to No. 18, a special size approximately 2-1/4 in. in diameter. All are "deformed bars." That is, they have lugs or deformations rolled on the surface to provide a grip to the concrete. Number designations for the bars represent eighths of an inch (No. 6, for example, is 3/4-in. in diameter). A number of commonly used sizes of reinforced bar, along with dimensions and weights, are listed in Table 13-15.

TABLE 13-15. Commonly Used Sizes of Deformed Reinforcing Bars

Number	Diam (in.)	Area (sq in.)	Perimeter (in.)	Weight (lb/ft)
3	0.375	0.11	1.178	0.376
4	0.500	0.20	1.571	0.668
5	0.625	0.31	1.963	1.043
6	0.750	0.44	2.356	1.502
7	0.875	0.60	2.749	2.044
8	1.000	0.79	3.142	2.670
9	1.128	1.00	3.544	3.400
10	1.270	1.27	3.990	4.303
11	1.410	1.56	4.430	5.313
14	1.693	2.25	5.320	7.650
18	2.257	4.00	7.090	13.600

STEEL FOR REINFORCING BARS

The steel used to manufacture reinforcing bars comes from several different sources, and the only limitation on chemistry imposed by the ASTM specifications is that the phosphorus shall not exceed 0.05%. When requested by the purchaser an analysis of each lot shall be reported to the purchaser, but there are no restrictions other than phosphorus.

The ASTM specifications clearly state "the weldability of the steel is not a part of this specification but may be subject to agreement between a particular supplier and user." If the bars are to be welded, it is advisable to have some agreement on weldability.

ASTM A615-68 specifies "billet" steel made by the open-hearth, basic-oxygen, or electric-furnace processes. Much of the reinforcing-bar stock used is high-carbon rerolled rail stock; thus, specifications are in terms of railroad materials. A616-68 specifies the bars shall be rolled from standard section T rails. A617-68 specifies the bars shall be rolled from carbon-steel axles for cars and locomotive tenders within certain journal sizes. Table 13-16 lists the grades, sizes, and tensile requirements of some standard bars.

TABLE 13-16. Deformed Bars for Concrete Reinforcement

ASTM Specification	Source of Steel	Range of Sizes	Grade	Yield Strength (min, psi)	Tensile Strength (min, psi)
A615-68	Billets	3-18	40	40,000	70,000
			60	60,000	90,000
			75	75,000	100,000
A616-68	T Rails	3-11	50	50,000	80,000
			60	60,000	90,000
A617-68	Steel Axles	3-11	40	40,000	70,000
			60	60,000	90,000

TABLE 13-17. Preferred Joints for Arc-Welded Splices in Reinforcing Bars

Bar Size	Recommended Type of Joint	
6 and smaller	Square end with plate or angle backing	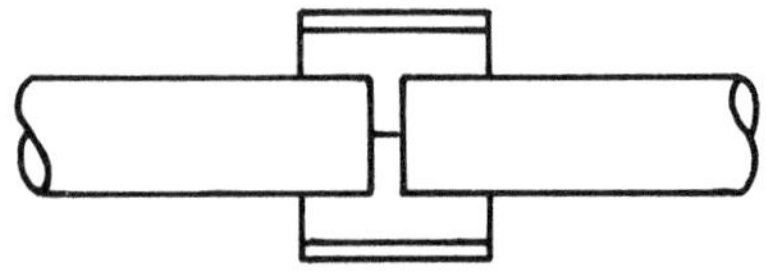
5 to 9	Single bevel with plate or angle backing	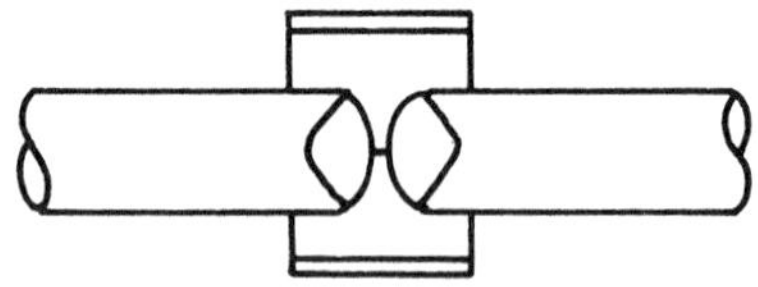
5 to 9	Single bevel without backing	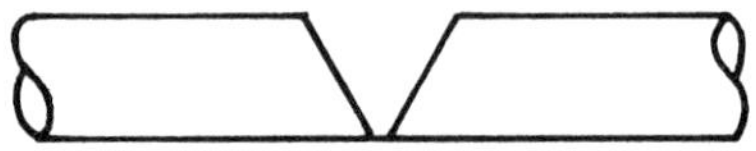
8 to 18	Double bevel	
8 to 18	Double bevel, bars vertical	
11, 14, and 18	Sleeve, bars vertical; compression joints only	

JOINT DESIGN

Joint design is largely determined by the size of the bar and, to a lesser degree, by its position (horizontal or vertical) and by the accessibility of the joint. Straight butt-welded joints are preferred over other arc-welded joints since the stress transfer across the joint is direct and concentric. With the proper electrode, butt-welded joints can develop 100% of the tensile strength specified for the bar.

For small-diameter bars, however, straight butt welds are either very slow or difficult to make. Table 13-17 lists recommended joint types for various bar sizes. Figure 13-18 shows indirect butt splices for bars of size No. 6 and smaller, using a splice plate or angle.

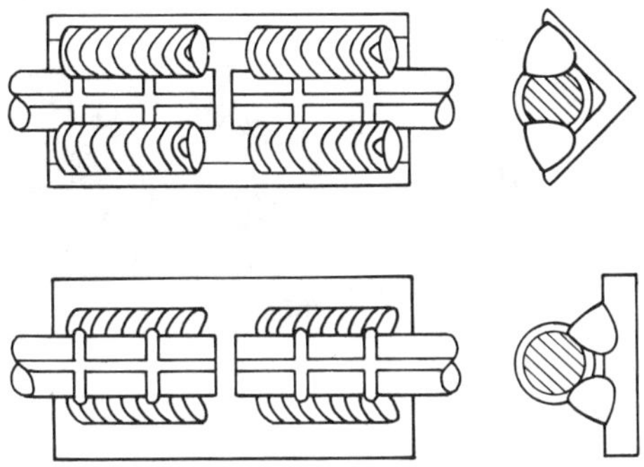

Fig. 13-18. Indirect butt splices for small reinforcing bars (No. 6 and smaller), incorporating a splice plate or angle.

For bar sizes No. 5 through 9, the bar end is beveled and butt-welded along with an angle as shown in Fig. 13-19 or with a plate, Fig. 13-20. Another way to butt-weld reinforcing bars with a backing is shown in Fig. 13-21. A backing strip about 1/8-in. thick is first tacked to the bottom of the joint. After a portion of the weld is made, the backing is red hot and can be easily wrapped partially around the bar with the weldor's slag hammer. Sometimes bar sizes 5 through 9 are butt-welded with a single bevel without a backing, as shown in Fig. 13-22.

Bar sizes No. 8 and larger are prepared with either a single bevel, Fig. 13-22, or a double bevel, Fig. 13-23. When the bars are vertical, the lower bar is cut square and the upper bar is either single or double-beveled, depending on the accessibility of the joint, Figs. 13-24 and 13-25. For heavy column reinforcement where the stress will always be in compression, the sleeve joint, Fig. 13-26 is sometimes used.

The single-lap splice, Fig. 13-27, is subjected to a bending stress when a load is applied, because the

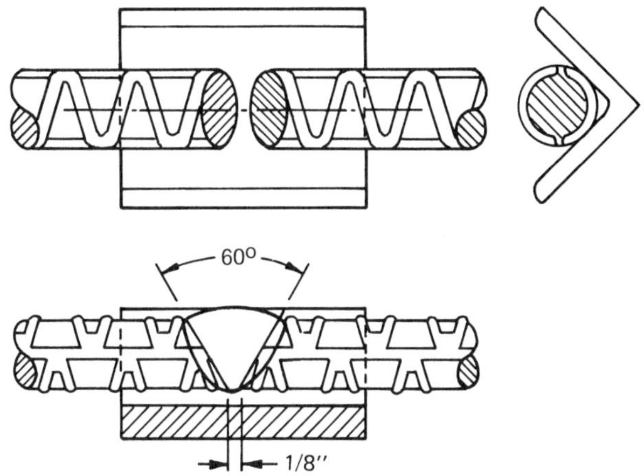

Fig. 13-19. Butt-welded joint with angle backing, for intermediate-size bars.

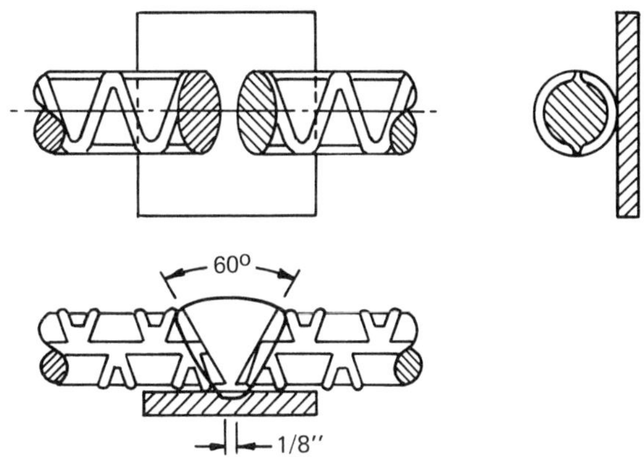

Fig. 13-20. Butt-welded joint with plate backing, for intermediate-size bars.

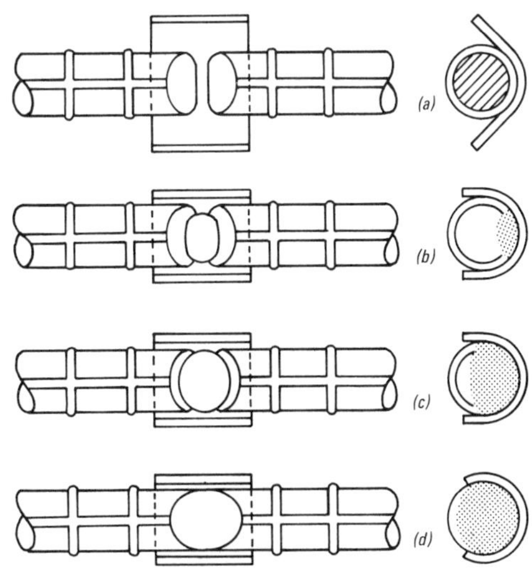

Fig. 13-21. Butt-welded bar with thin backing strip. The strip is tack-welded to the joint (a); then wrapped around the bar (b) as welding progresses. The completed weld is shown at (d).

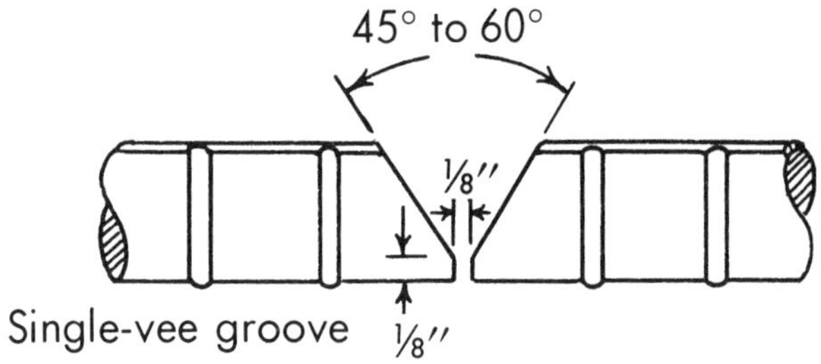

Fig. 13-22. Single-bevel butt joint for size No. 8 and larger.

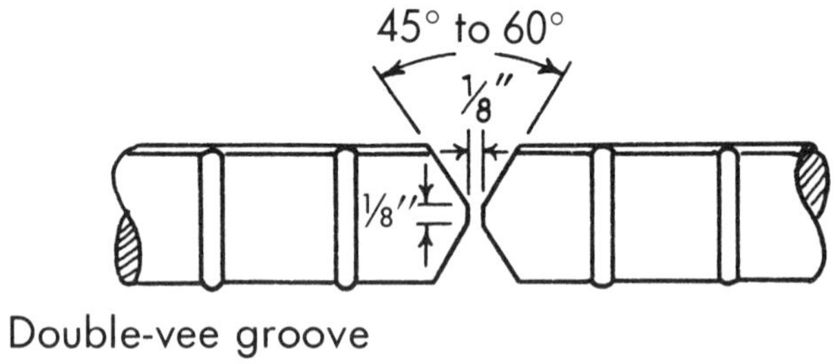

Fig. 13-23. Double-bevel butt joint for size No. 8 and larger.

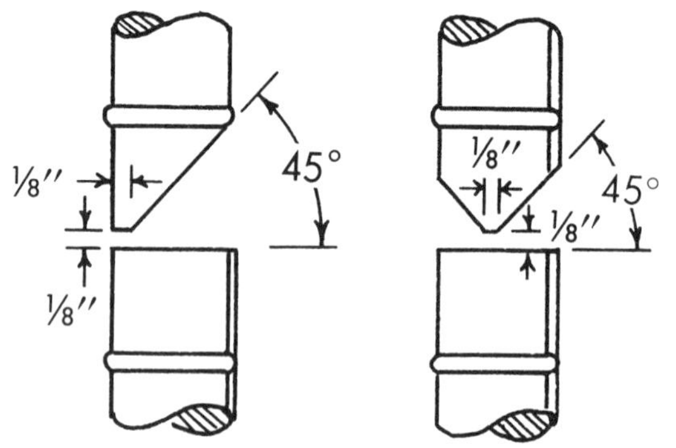

Fig. 13-24. Joint design for butt welds on vertical bars. Design used depends on accessibility of joint when welding.

Fig. 13-25. Welding reinforcing bars in the vertical position. Double bevel is used because joint is accessible from both sides.

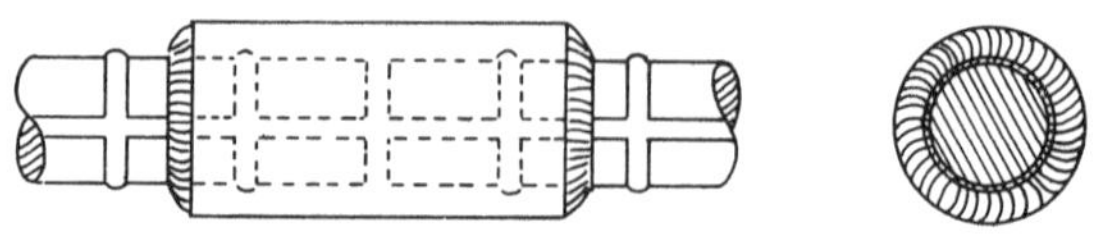

Fig. 13-26. Sleeve joint for bars in compression.

bars are not on the same axis. This type of joint should not be used on large reinforcing bars and on small bars only after the application has been carefully analyzed. A better but more costly splice uses two short splice bars, as illustrated in Fig. 13-28.

WELDING PROCEDURES

When splice plates, angles, or bars are used to join reinforcing bars, there is always a gap at the root of the joint caused by the deformation pattern on the bars. Because of this condition, weldors must acquire specific skills necessary to do quality work on reinforcing bars.

Shielded metal-arc welding electrodes should meet the requirements of AWS specifications and be of the low-hydrogen type (EXX18) and of sufficient strength to match that of the reinforcing bars being welded.

It is important that the composition of the material to be welded be known before the welding starts. This information may be available from the mill or warehouse that supplies the reinforcing bars. Since the composition can vary widely, it is imperative to test sample joints thoroughly to confirm the suitability of the electrode and the skill of the weldor before the job is started.

Specific welding procedures can be extrapolated from the procedures for manual shielded metal-arc welding of a similar joint with low-hydrogen electrodes. Generally, the techniques are the same, but the electrode size is smaller than for steel-plate procedures.

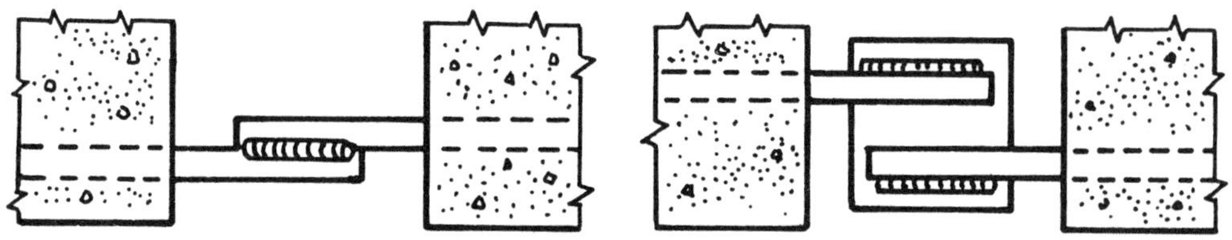

Fig. 13-27. Single lap splices for light loads.

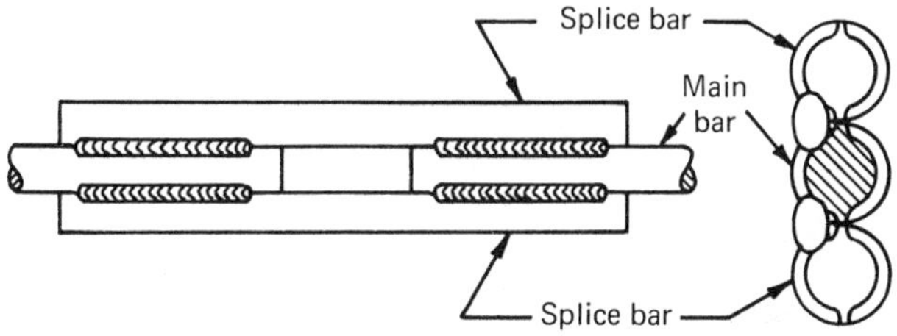

Fig. 13-28. Double lap joint with splice bars.

TABLE 13-18. Preheat and Interpass Temperatures for Welding Reinforcing Bars

Composition (%)		Electrode		Preheat Temperature (°F)
Carbon	Manganese	Low-Hydrogen	Other than Low-Hydrogen	
To 0.30	To 0.60	Not req.	Any E60XX or E70XX	72 if below 32; none if above 32.
0.31-0.35	To 0.90	E70XX	---	100 if below 32; none if above 32.
		– –	E60XX or E70XX	100
0.36-0.50	To 1.30	E70XX	Not recom.	200
0.41-0.50	TTo 1.30	E70XX	Not recom.	400
0.51-0.80	To 1.30	High-tensile types.	Not recom.	Procedure subject to qualification and testing.

Table 13-18 lists recommended preheat and interpass temperatures for bars with varying carbon and manganese content.

TABLE 13-19. Procedures for Gas Metal-Arc Welding Reinforcing Bars

Bar Size	Welding Current (amp)	Arc Voltage
To 8	120	20
9-11	150	22
14, 18	160	22

Procedures are for either horizontal or vertical joints, using CO_2 at 15 cfh, and with 0.035-in.-diam. electrode.

In addition to the shielded metal-arc process, reinforcing bars can also be welded with the gas metal-arc process and the self-shielded cored-wire process. Welding reinforcing bars is almost always done in the field, so when using any gas-shielded process the weldor must be well protected from wind. Typical procedures for gas metal-arc welding are given in Table 13-19. Joints may be either single or double-bevel (use 3/32-in. gap between bars), and the same joint design is used for both horizontal and vertical bars. The same preheat and interpass temperature given in Table 13-18 for low-hydrogen electrode should be used. To insure good bonding to the base metal, large bars (14 and 18) should be preheated regardless of ambient temperature.

Additional information on welding or reinforcing steel can be found in AWS D1.4 *Structural Welding Code, Reinforcing Steel.* AWS 550 N.W. LeJeune Road, Miami, FL 33135.

Control for a rocket launch facility. Welded reinforcing steel bars ready for concrete.

Arc Gouging and Cutting

DESCRIPTION

Air carbon arc cutting (CAC-A) is a physical means of removing base metal or weld metal by using a carbon electrode, an electric arc and compressed air. In the air carbon arc process the intense heat of the arc between the carbon electrode and the workpiece melts a portion of the base metal, or weld. Simultaneously a jet of air is passed through the arc, of sufficient volume and velocity to blow away the molten material. This sequence can be repeated until the required groove or cut has been obtained.

Since the CAC-A process does not depend on oxidation to maintain the cut, it is capable of cutting metals that OFC will not cut; stainless steel, many copper alloys and cast iron.

Material can be removed approximately five times faster by arc gouging than by chipping. A 3/8-in. groove, for example, can be gouged at a speed of more than 2 fpm. Depth of cut can be controlled closely, and welding slag does not deflect or hamper the cutting action as it would with cutting tools. The cost of operating gouging equipment is generally less than for chipping hammers or gas-cutting torches, and the arc-gouging equipment also requires less space. An arc-gouged surface is clean and smooth and can usually be welded without further preparation.

The process has several drawbacks, however. It is not as good as other processes for through-cutting, and large volumes of compressed air are required. Improper operation of the torch may result in carbon pickup (and undesirable metallurgical changes) in some materials. Increased hardness produced on cast iron and air-hardenable materials may be objectionable.

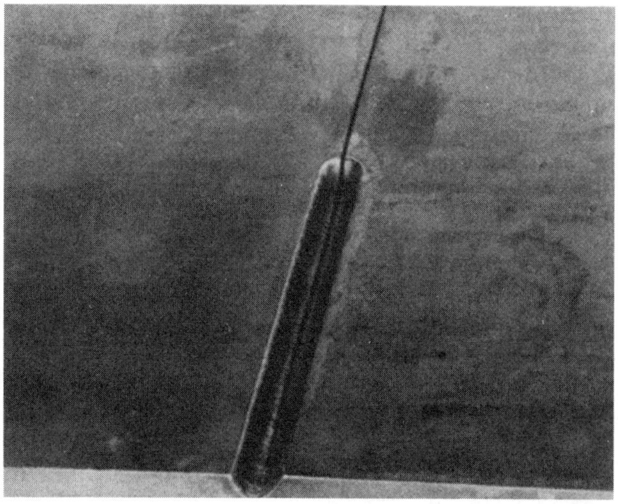

Fig. 13-29. Butt joint that has been partially beveled by arc gouging.

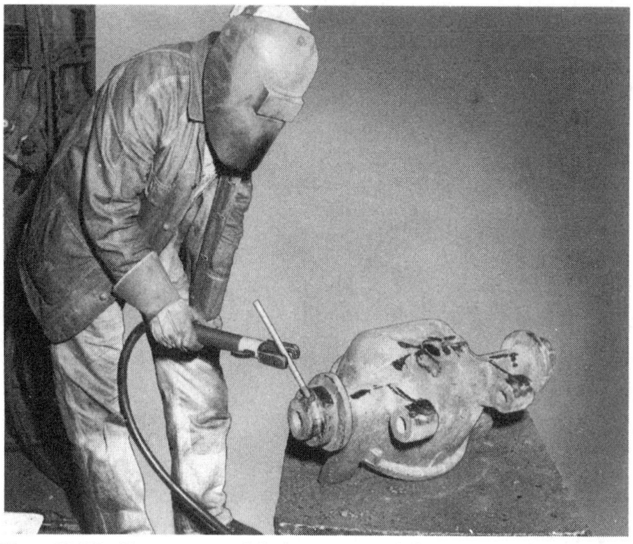

Fig. 13-30. Excess metal being removed from a casting by arc gouging.

APPLICATIONS

The CAC-A process may be used to prepare plates for welding, to provide a suitable bevel or groove, See Fig. 13.29. It may also be used to back gouge a seam prior to welding the second side, Fig. 13.31. It provides an excellent means of removing defective welds or misplaced welds. It has many good applications in industry in metal fabrication, casting finishing, Fig. 13.30, construction, mining, general repair and maintenance.

ELECTRODE HOLDER AND CABLE

While this process may be used manually or mechanized, it is more frequently used as a manual

Fig. 13-31. Back-gouging of a seam welded from inside the tank as preparation for a weld to be made from the outside.

Fig. 13-33. Mechanized torch for arc gouging, providing automatic travel along the seam.

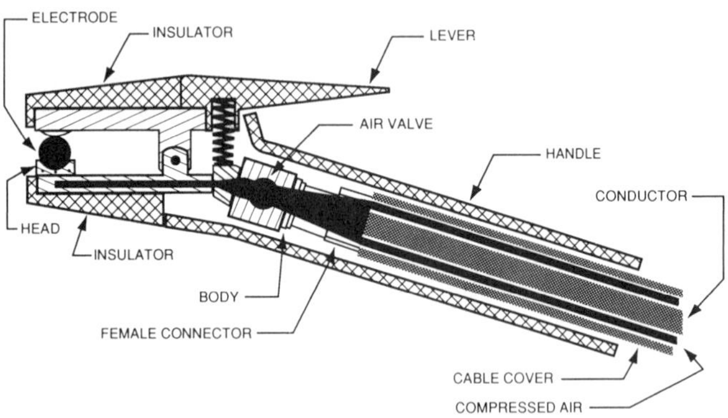

Fig. 13-32. Typical torch for manual arc gouging.

process. This requires a special insulated electrode holder and cable which will hold the copper coated electrode, carry the current, and conduct the compressed air and direct it at the molten metal, See Fig. 13.32. In addition it must have a trigger to release the air when it is needed.

TYPES OF ELECTRODES

Special copper coated carbons must be used; several different types are available. The size of the carbon will influence the size of the groove and the depth of the cut. Flat carbons are also available for washing off weld metal. Carbons are available from 1/8″ to 3/4″ diameter.

The most common type of manual torch is shown in Fig. 13-32, where the electrode is held in a rotatable head. Torches are generally air-cooled, but those intended for application at high currents are water-cooled.

Three types of mechanized torches are available. The simplest type is mounted on a machine carriage, but requires manual feeding of the electrode (Fig. 13-33). A more refined version of the carriage-mounted torch employs a spring-loaded device that maintains a constant distance between the torch and workpiece to provide a uniform groove depth. The third and most sophisticated type has electronic controls that regulate the voltage to provide the correct arc length. This type of torch can make grooves to a depth tolerance of ± 0.025 in.

The specific types of electrodes and current recommended for arc gouging various metals and alloys are summarized in Table 13-20 and discussed in the following paragraphs:

TABLE 13-20. Electrode Type and Polarity Recommended for Arc Gouging

Material	Electrode	Power
Steel	DC	DCEP
	AC	AC
Stainless Steel	DC	DCEP
	AC	AC
Iron (cast iron, ductile) iron, malleable iron)	AC	AC or DCEN
	DC	DCEP (high-amperage)
Copper Alloys	AC	AC or DCEN
	DC	DCEP
Nickel Alloys	AC	AC or DCEN

From AWS Handbook, Sixth Edition, Section 3A.
AC is not the preferred method.

Steels: DC electrodes are used with DC electrode positive for carbon, low-alloy, and stainless steels. If DC is not available, AC (with AC electrodes) can be used. For this application, AC is about 50% as efficient as DC.

Cast Irons: AC electrodes are used with DC electrode negative or with AC at the middle of the amperage range for all cast irons, including malleable and ductile (nodular) iron.

Copper Alloys: DC electrodes are used with DC electrode positive at maximum amperage if the copper content of the alloy is not over 60%. For special marine propeller alloys, use DC electrodes with DC electrode negative. If the copper content is above 60%, or if the workpiece is extremely large, use AC electrodes with AC. Centergrip steel electrodes may be required if copper content is 99.9% or more.

Nickel Alloys: Use AC electrodes and AC power.

Magnesium Alloys: DC electrodes are used with DC electrode positive. Surface of the groove should be wire-brushed prior to welding.

Aluminum Alloys: DC electrodes are used with DC electrode positive. Wire brushing is mandatory prior to welding. Electrode protrusion cannot exceed 4 in. for quality work.

Exotic Alloys: Titanium, zirconium, hafnium, and their alloys can be cut for remelt by arc gouging. A thorough cleaning operation is necessary after arc gouging if these materials are to be welded.

TABLE 13-21. Power Sources for Arc Gouging

Type of Current	Type of Power Source	Remarks
DC	Variable-voltage motor-generator, rectifier, or resistor-grid equipment	Recommended for all electrode sizes.
DC	Constant-voltage motor-generator, or rectifier	Recommended only for electrodes above 1/4-in. diameter.
AC	Transformer	Should be used only with AC electrodes.
AC-DC	Rectifier	DC supplied by three-phase transformer-rectifier is satisfactory. DC from single-phase source not recommended. AC from AC-DC power source is satisfactory if AC electrodes are used.

POWER SOURCE

While it is possible to arc air gouge with AC this is not the preferred method. A DC power source of sufficient capacity and a minimum of 60 open circuit volts, either rectifier or a motor generator will give the best results. With DC it is operated DCEP (electrode positive).

Arc voltages with arc gouging normally range from about 35 to 56 v. Table 13-21 lists recommended power sources, and Table 13-22 lists suggested current ranges.

It is recommended that the power source have overload protection in the output circuit. High current surges of short duration occur with arc gouging, and these surges can overload the power source.

TABLE 13-22. Recommended Range of Current for Arc Gouging

Type of Electrode and Power	Maximum and Minimum Current (amp)					
	Electrode Size (in.)					
	5/32	3/16	1/4	5/16	3/8	1/2
DC Electrodes, DCEP Power	90-150	150-200	200-400	250-450	350-600	600-1000
AC Electrodes, AC Power	- -	150-200	200-300	- - -	300-500	400-600
AC Electrodes, DCEN Power	- -	150-180	200-250	- - -	300-400	400-500

TABLE 13-23. Air Consumption for Arc Gouging

Maximum Electrode Size (in.)	Application	Pressure (psi)	Consumption (cfm)
1/4	Intermittent-duty, manual torch	40	3
1/4	Intermittent-duty, manual torch	80	9
3/8	General-purpose	80	16
3/4	Heavy-duty	80	29
5/8	Semiautomatic mechanized torch	80	25

AIR SUPPLY

The air supply for arc gouging normally should be at a pressure of about 80 to 100 psi. Where compressed air lines are not available, arc-gouging torches suitable for light work can be operated on bottled gas at pressures as low as 40 psi. Pressures above 100 psi are sometimes used, but offer no advantage in metal-removing efficiency.

The air line should have an inside diameter of at least 1/4 in. for both manual and mechanized torches, except for the voltage-control torches. These require a minimum ID of 1/2 in. Table 13-23 list typical requirements for the compressed-air supply.

METALLURGICAL CONSIDERATIONS

The correct use of arc gouging with carbon-based electrodes apparently causes no ill effects insofar as carbon pickup, corrosion resistance, or distortion are concerned. The metallurgical changes produced by the process are similar to those produced by the arc-welding process; that is, a thin, hardened zone may appear in some metals, but subsequent welding remelts this zone and reduces the hardness. Copper contamination from copper-plated electrodes has not been detected.

Heat penetration is shallower with arc gouging than with oxygen cutting, so arc gouging produces less distortion. Machinability of low-carbon and non-hardenable steel is not affected by arc gouging. The surface of cast iron and hardenable steels may be rendered unmachinable by the process. This hard layer, however, is generally only about 0.006-in. thick, and can be easily removed by a cutting tool set to penetrate beyond that depth.

TROUBLESHOOTING

Most problems encountered in arc gouging stem from common oversights. For example, carbon deposits at the beginning of a groove are usually caused by not having the air turned on before the arc is struck. An unsteady arc, which causes the operator to travel slowly, is generally the result of insufficient current. An erratic groove, where the arc wanders from side to side and where the electrode overheats, is normally caused by negative polarity DC.

Intermittent arc action and resulting irregular groove surface is caused by too slow travel speed. Carbon deposits at intervals along the groove are caused by a shorting out of the electrode against the work. Slag adhering to the edges of the groove results form inadequate air pressure. A groove that becomes progressively shallower or deeper is caused by an incorrect arc length.

SAFETY PRECAUTIONS

Any process in which molten metal is spewing about obviously presents hazards. An arc-gouging operator should follow the safety practices for welding in Section 15.1, and particular attention should be paid to possible fire hazards. Inflammable materials should be removed from the work area, protective booths or screens should be set up, and workers should have the proper protective clothing, including ear plugs, especially if they are to work out-of-position.

PLASMA ARC CUTTING AND GOUGING

The plasma arc cutting process is capable of cutting and gouging most metals through the use of a constricted arc which melts the metal in a highly localized area. The molten metal is immediately removed by the high velocity ionized gas or plasma issuing from the constricting orifice. The plasma arcs operate a temperatures in the 18,000-25,000°F range.

This process, which has been in use commercially since the late 1950's is capable of cutting or gouging any metallic material, thus it is particularly attractive for use on stainless steels and non-ferrous alloys which cannot be cut or gouged by the oxyfuel process. Since the plasma arc process operates at a much higher temperatures than the oxyfuel process, it will start cutting instantly without the need for preheat.

The plasma arc cutting process may be used with either a hand held torch or a torch mounted on a travel carriage. The earlier torches were machine mounted and operated at high currents (100-1000 amperes) with argon or nitrogen plasma gas. More recently, hand held torches have been developed which operate at 20-125 amperes and use nitrogen or air as the plasma gas. Power supplies for plasma arc cutting and gouging must have drooping current characteristics and higher voltages than standard welding power supplies. Most power supplies also incorporate circuits for turning the plasma gas on and off, current adjustment and a pilot arc for initiating transfer of the arc to the work. These units also have safety interlock circuits to prevent operator contact with the high voltage if the torch electrode is exposed. Once the machine is set for the desired operating current, the operator holds the torch near the work and pulls the trigger. The pilot arc will come on and within a few seconds establish the main arc and start cutting.

For plate thicknesses under 1/2 inch, the plasma arc process cuts faster than the oxyfuel gas process and produces a narrower kerf with smoother surfaces. When making a cut, the orifice must be held about 1/8 inch away from the surface and the torch moved along at a uniform speed which will produce a through cut. A drag cup is available on some torches which provides the correct standoff distance for cutting when the operator rests the cup on the work and moves the torch along the cut line. The drag cup makes it easier for the operator to maintain the correct standoff distance.

Plasma arc torch consumables include the electrode and orifice. The electrode is copper with a small hafnium insert at the center tip. The arc emanates from the hafnium which gradually erodes with use, and this requires electrode replacement periodically. The torch tip contains the orifice which constricts the plasma arc. Various size orifices are available where the smaller diameters operate at lower amperages and produce a more constricted narrow arc than the larger diameters. The orifices also gradually wear with use and must be replaced when the arc becomes too wide.

The plasma arc cutting torch can be used for arc gouging. The main changes are that the tip orifice is larger than for cutting and the torch is held at an angle of about 30 degrees from horizontal rather than at 90 degrees as in cutting. Plasma gouging can be used on all metals and is particularly suitable for aluminum and stainless steels where oxyfuel cutting is ineffective and carbon arc gouging tends to cause carbon contamination.

When using plasma arcs for cutting or gouging, certain safety precautions must be recognized and observed:

- Operating voltages for plasma arc cutting may range from 150 to 400 volts DC. Most equipment incorporates safety systems which prevent the operator from coming in contact with the high voltages. The operator must see that the equipment is kept in good condition and that the torch cable is not damaged or punctured. Equipment access doors must be kept closed at all times when power is connected.
- Fumes and gases produced by plasma arc cutting can be hazardous to health and should be avoided.
- Sparks from plasma arc cutting can be a fire hazard; therefore, all combustible material and volatile solvents should be kept away from the cutting area.
- The plasma arc emits intense, visible, infrared and ultraviolet radiation which can be harmful to the eyes and skin. The same eye and skin protection normally used for GTA welding at an equivalent current level should be used.

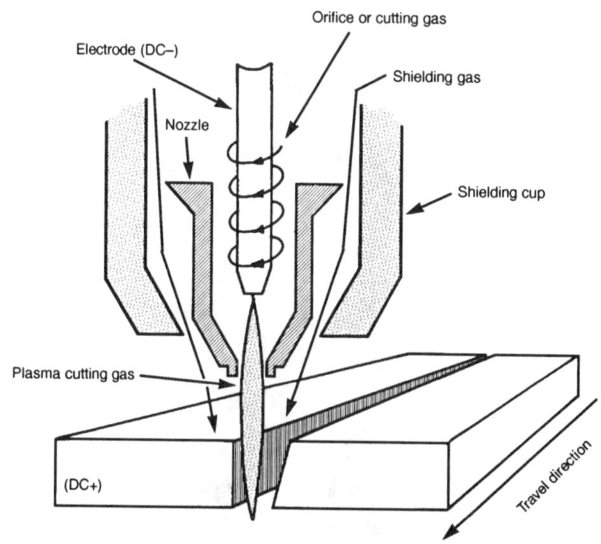

Plasma arc cutting principles.

Typical Plasma Arc Power Source With Torch and Cable

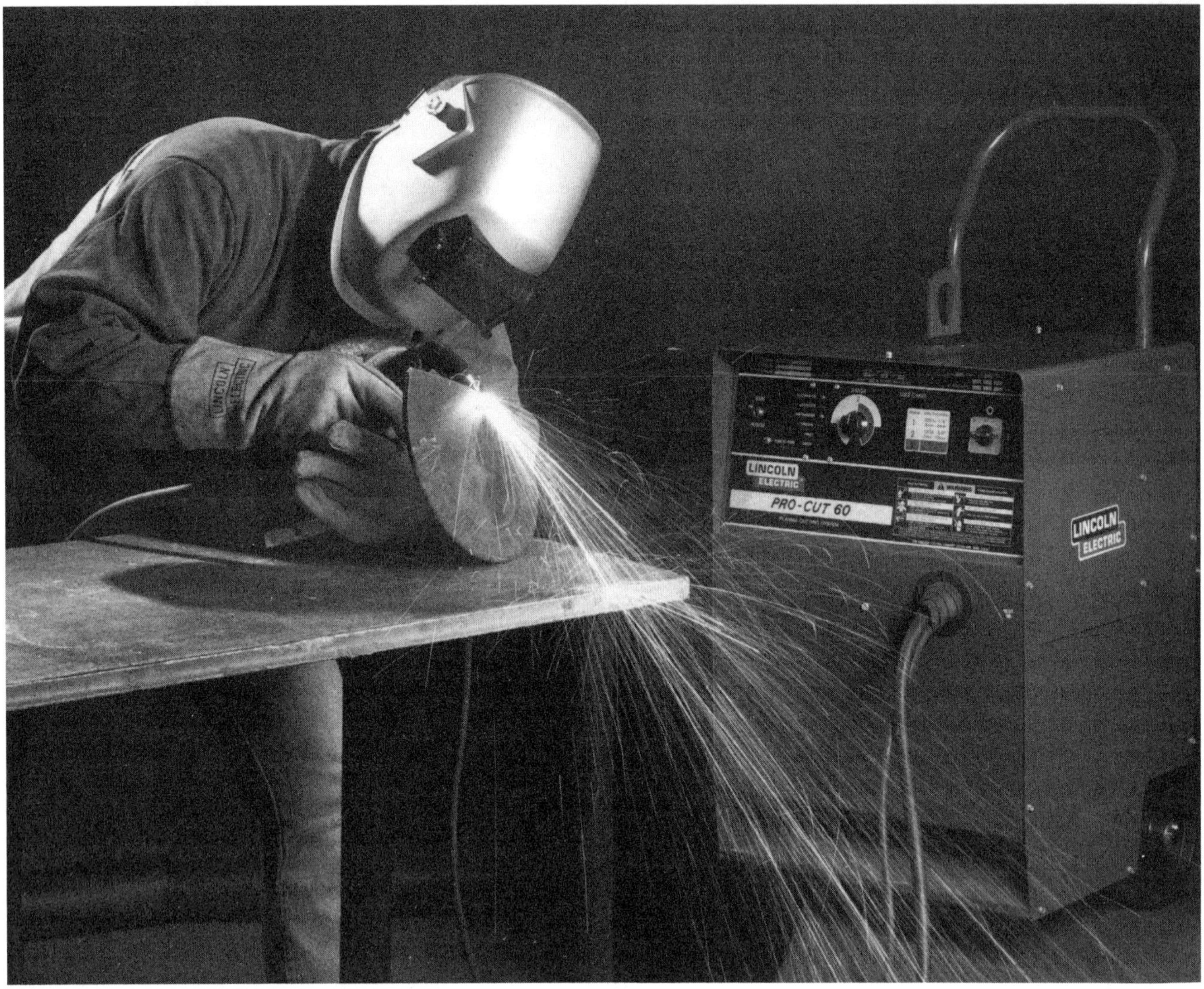

Manual plasma arc cutting of steel plate.

Welding Clad Steels

COMMONLY USED BACKING STEELS

Clad steels are composite materials, made by mill-rolling a thin sheet of a metal that has desired properties over a base ("backing") plate of carbon or alloy steel. The cladding metal may be a stainless-steel alloy, nickel or a nickel alloy, or copper or a copper alloy. Under the heat and pressure of rolling, the dissimilar metals are bonded at the interface. The result is a composite with the strength of the backing steel and with corrosion resistance, abrasion resistance, or other useful properties on the clad face. The cladding may be of any thickness, but is usually from 5 to 20% of the thickness of the composite plate.

Figure 13-34 shows a representative sample of a clad steel. The backing plate is most frequently a plain carbon steel such as A285, but it may be an alloy steel with high-strength or low or elevated-temperature properties. Some of the commonly used backing steels for special applications are:

A285, the most commonly used backing steel.

A515 or A516, where a killed steel is required or for pressure-vessel applications.

A203, for low-temperature uses.

A204, where higher strength than that of A285 at elevated temperature is needed.

A387, where high-temperature characteristics and resistance to graphitization are required.

A302, for high-pressure service at elevated temperatures.

Common cladding materials are the chromium steels, chromium-nickel steels, nickel and nickel-copper and nickel-chromium-iron alloys, cupronickels, and oxygen-free and deoxidized coppers.

CONSIDERATIONS WHEN WELDING

In the weld-joining of the clad steels, it is essential to preserve the properties of the cladding at the weld. Otherwise, the desired corrosion resistance, abrasion resistance, or other property of the clad surface would not be continuous. At the same time, it is also essential that the properties of the backing metal be maintained. These two requirements mean that special joint designs and welding procedures must be used. The backing metal portion of the joint is welded with an appropriate steel electrode and the clad portion with an electrode giving a fill metal with properties compatible with those of the cladding. The joint must be designed to make such "duplex" welding practical.

In general, the procedure is to weld the backing metal portion of the joint first, followed by welding on the clad side. Thus, a high-alloy-clad carbon-steel plate would be welded first on the carbon-steel side with carbon-steel electrode, followed by welding on the clad side with high-alloy electrode. This sequence prevents the formation of hard, brittle areas in the weld, such as would occur if carbon steel weld metal were deposited on high alloy.

Figure 13-35 shows types of edge preparations used. Note the root faces and the stripped-back clad edges — arrangements designed to minimize admixture of the dissimilar weld metals. In joints where

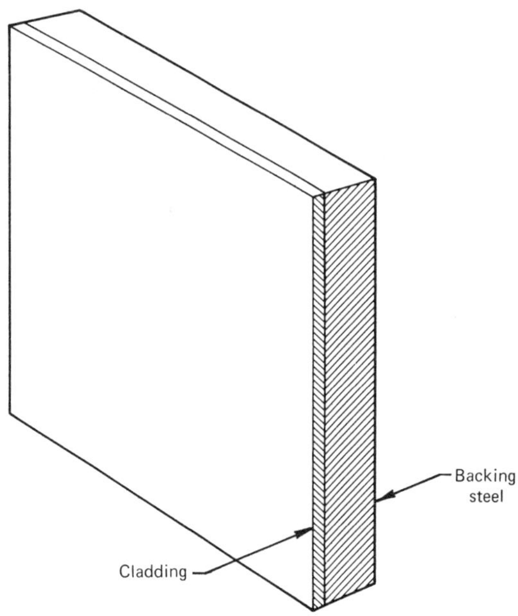

Fig. 13-34. A sample of a clad steel. The cladding is bonded to the backing steel under the heat and pressure of rolling.

full strength is not required, incomplete penetration is sometimes used to avoid admixture. In light-gage plates, an alloy that is compatible with the cladding may be used for the entire joint — a technique referred to as "full-alloy welding."

The clad steels may be welded by the shielded metal-arc, submerged-arc, gas metal-arc, or gas tungsten-arc processes. The shielded metal-arc process (stick-electrode) is most widely used.

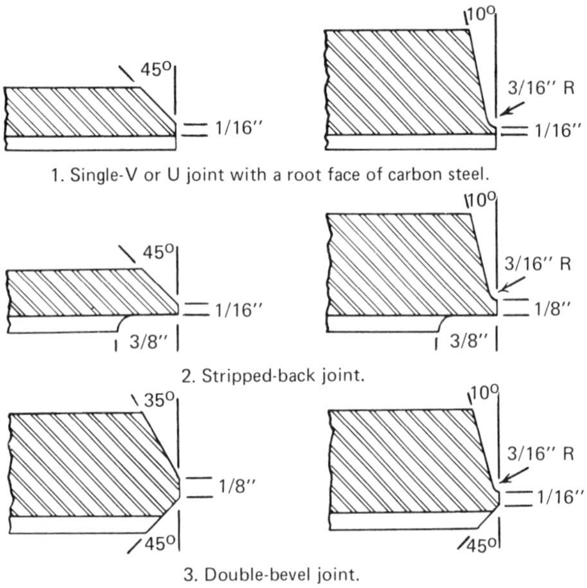

Fig. 13-35. Types of edge preparations used for butt welds in clad steels.

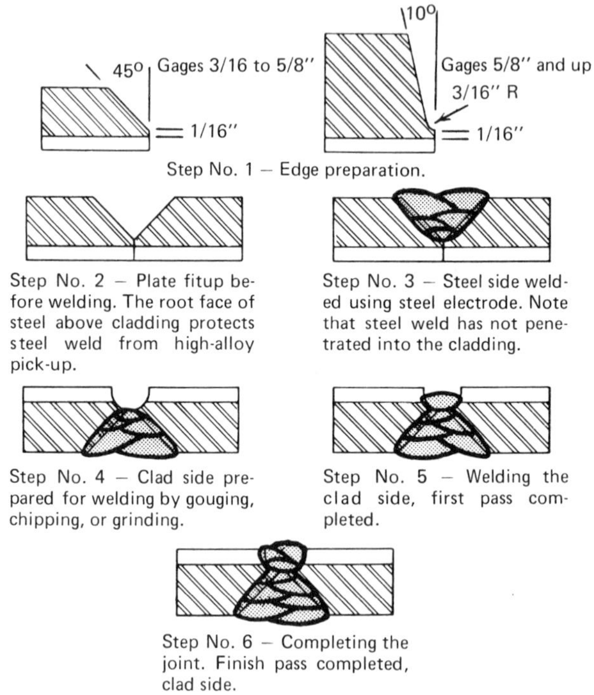

Fig. 13-36. The standard six-step method for making butt welds in clad steels.

SHIELDED METAL-ARC WELDING OF CLAD STEELS

A six-step method for making butt welds in clad steel plate is summarized in Fig. 13-36. The first step — preparing the plate edges — is very important. The 1/16-in. root face of backing steel is a barrier to minimize admixture of the dissimilar weld metals. The fitup must be accurate for the same reason. After assembly and alignment, the parts should be tack welded on the steel side with the same type electrode that will be used for the root bead.

Welding the Steel Side

The first weld is made on the steel side using an electrode that will develop the required characteristics of the backing steel. Care must be used to prevent the root bead from penetrating into the cladding.

If the fitup is good and the plate edges properly prepared, a 1/8-in.-diameter EXX27 type of electrode should give proper penetration. If a wide gap must be bridged or if the root face of the backing metal is thinner than 1/16-in., low-hydrogen electrodes of the EXX16 or EXX18 type should be used. If the cladding is penetrated by the steel root bead in a carbon-steel backing, martensite may be formed in the root pass. If the penetration is too shallow, the amount of metal to be back-gouged from the clad side and the amount of alloy weld metal required for fill will be excessive.

The welding electrodes and procedures for the steel side are the same as would be used with unclad steel (see Section 6.2).

Back-Gouging the Clad Side

The clad side of the joint is back-gouged to sound metal before welding. The operation should remove any unfused area and the slag from the backing-steel weld. If the workmanship has been good, the depth below the cladding to be removed will usually be not more than 1/16-in. The width of the gouged groove should be held to the following approximate dimensions:

Plate Gage	10% Clad	20% Clad
3/16 to 1/2 in.	1/4 in.	5/16 in.
9/16 to 1 in.	3/8 in.	1/2 in.
Over 1 in.	1/2 in.	5/8 in.

If the depth of the gouged groove is excessive, the width of groove should be increased to permit electrode manipulation. The edges of the groove should be rounded to prevent entrapment of slag.

Carbon-arc gouging and chipping are the preferred methods for back-gouging clad materials. When chipping, sharp round-nose chisels must be used.

Welding the Clad Side

The electrode for the clad side should maintain the continuity of the surface properties of that side and contribute to the structural strength of the weld. Table 13-24 gives recommended welding electrodes for various claddings.

TABLE 13-24. Typical Covered Electrodes for Welding the Clad Side of Nonferrous and Stainless Clad-Steel Weld Joints

AISI-Type or Alloy Cladding	AWS-Type Electrode for the First Pass or Layer	AWS-Type Electrode to Complete the Weld[4]
405, 410, 430	E309[1 2],E430[3]	E309[1 2], E430[3]
304	E309, E312	E309, E308
304L	E309, E309Cb	E308L, E309Cb
347, 321	E309, E309Cb	E309, E309Cb
309	E309	E309
316	E309Mo, E309	E309Mo, E316L
316L	E309Mo, E309L	E309Mo, E316L
317	E309Mo, E317	E317
Low-Carbon Nickel	ENi-1	ENi-1
Nickel	ENi-1	ENi-1
Monel	ENiCu-1	ENiCu-1
Inconel	ENiCr-1	ENiCr-1
Cupro-Nickel	ENiCu-1	ENiCu-1

Notes:

[1] Before using this electrode for 400-series stainless cladding, investigate the application to determine if nickel-bearing weld metal can be tolerated.

A preheat of 300°F minimum is suggested for maximum ductility in the heat-affected zone of the cladding, particularly in gages over 1/2-in.

[2] Postheating: Annealing or stress-relief heat treatment after welding is important for maximum ductility and corrosion resistance of heat-affected zones of 430, 410 and 405 cladding and of 430 weld-metal deposits on steel.

Caution: The fact that austenitic weld metal is subject to sensitization when heated in stress-relieving temperature ranges should also be considered.

[3] Preheat of 400°F is required.

[4] Since two layers of deposited metal must be used, it may be necessary to grind off part of the first layer to make room for the second, particularly on light gages or thin claddings.

The straight chromium-clad steels (Types 430, 410, and 405) are commonly welded with austenitic electrodes to avoid the brittle nature of straight chromium weld metal, In some instances, however, the latter must be used. In either case, steps must be taken to overcome detrimental metallurgical effects that occur in the cladding adjacent to the weld metal. When straight chromium electrodes are used, it is necessary to preheat to 300-400°F and to stress relieve after welding. Type 430 stainless steel may become embrittled during welding due to rapid cooling from temperatures above 1700°F. A heat treatment of 1200-1500°F followed by air cooling will correct the conditions.

The electrodes indicated in Table 13-24 for welding claddings of the 300 series of stainless steels have alloy contents that compensate for the iron dilution of the weld metal by the backing steel. The application, however, may require maintaining a specific chemical composition, in which case an electrode with a composition more nearly that of the cladding should be used. If so, small-diameter electrodes should be used at moderate welding current to minimize weld-metal dilution. At least two layers of weld metal should be deposited over the steel, even if it is necessary to chip or grind off part of the first stainless bead to make room for the second pass.

Microfissuring may be encountered with Types 310 and 347 electrodes, and the phenomenon may affect the corrosion resistance of the deposited weld. In such cases, special-order electrodes may be suggested by the electrode supplier.

With nickel claddings, it is virtually impossible to prevent iron pickup. However, the nickel-iron alloys formed are corrosion-resistant — even in caustic applications — when the iron content is as much as 24%.

Small-diameter electrodes should be used with stringer beads for welding nickel claddings. To minimize iron pickup, the top half of the initially deposited beads should be chipped or ground off before succeeding beads are deposited. The greater the number of layers of nickel weld metal, the lower will be the iron content in the top layer — the one in contact with the corrosive environment in the application. As indicated in Table 13-24, ENi-1 electrodes should be used for welding nickel cladding.

Type ENiCu-1 electrodes are used for welding Monel cladding. Developed specially for the purpose, these electrodes prevent the hot-short or brittle conditions that often result from the for-

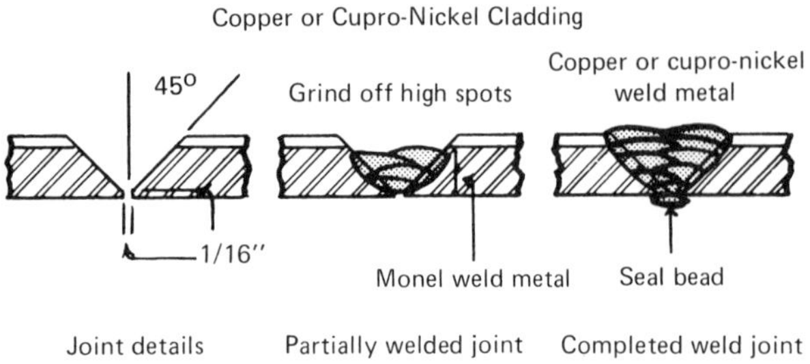

Fig. 13-37. Scheme for welding copper-clad or cupro-nickel-clad plate in gages from 3/16 to 3/8-in.

mation of nickel-copper-iron alloys when regular Monel welding materials are deposited against steel.

When welding Inconel cladding, ENiCr-1 electrodes should be used to compensate for iron dilution. The electrodes should be baked at 500°F for two hours immediately before use to assure X-ray-quality welds.

Copper-clad and cupro-nickel-clad steels present a special problem because of the embrittlement that occurs with iron dilution. This embrittlement can be avoided by depositing a barrier pass of Monel weld metal between the steel and copper or cupro-nickel.

When fabricating copper-clad or cupro-nickel-clad steels in gages from 3/16 to 3/8 in. inclusive, the full-alloy type of weld is recommended with the plate edges beveled as in Fig. 13-37. Welding below the cladding may be done with Monel stick electrodes (ENiCu-1). Generally, 1/8-in. electrodes are used for these plate gages and the edge preparations shown. Care should be taken with the root pass to insure uniform sealing of the gap without excessive melt-through. The seal bead is also made with Monel-coated electrode — after the gap has been wire-brushed with a stainless or other high-alloy brush or lightly grooved by chipping or gouging.

The weld bevel should be filled with Monel metal up to the bond line. The amount of dilution in the deposited weld metal can be minimized with the stringer-bead technique and depositing several layers rather than attempting to deposit a large amount of weld metal in one pass. The surface of the Monel weld metal should be smoothed by grinding so that high points do not affect uniformity of dilution in the welding that follows.

To complete the weld, cupro-nickel or copper is deposited on the Monel weld metal. Copper cladding is preferably put on by the gas-shielded consumable-electrode process, using 1/16-in. bare copper electrode (ECu). Cupro-nickel cladding may be deposited with the stick-electrode process, using 1/8-in. ECuNi covered electrode or by the gas metal-arc process using 1/16-in. ECuNi bare electrodes. To assure a satisfactory chemical composition at the surface of the clad, there should be at least two layers of clad metal above the Monel deposit. It may be necessary to grind off some of the first clad layer to obtain space for the second bead layer.

With copper-clad or cupro-nickel-clad plate 3/8-in.-thick and over, the preparations and welding details shown in Fig. 13-38 are recommended. The heavier gages may be prepared with a U bevel instead of a V, providing the fitup and root face of carbon steel are preserved. The steel side is welded first with a suitable electrode, following which the joint is back-gouged from the clad side to a minimum depth of 1/8-in. below the bond line. Monel metal is then deposited to bring the weld metal surface to the bond line, after which the surface of the Monel metal is ground and the weld finished with copper or cupro-nickel as described in the preceding text.

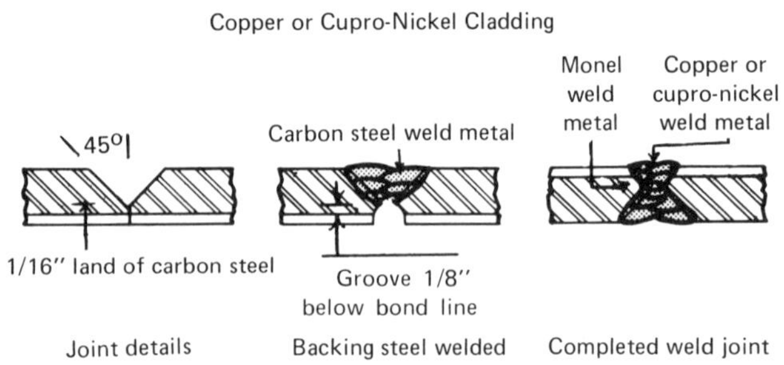

Fig. 13-38. Scheme for welding copper-clad or cupro-nickel-clad plate in gages over 3/8-in.

Full-Alloy Welding

A full-alloy weld is usually more economical than a duplex weld with light-gage clad steels. Full-alloy welding may be practical for gages as thick as 1/2-in. Welding is similar to that used for solid high-alloy plate. Edge preparation is simpler, procedures are less exacting, and labor costs are reduced. The full-alloy technique also minimizes iron pickup when welding materials such as nickel-clad and Monel-clad steels where higher alloy electrodes cannot be used to compensate for iron dilution.

Although the material costs of full-alloy weld metal may be more expensive than combination weld metal, such costs are usually offset by lower labor costs. This is especially true in gages up to 3/8-in. The economical cutoff point depends on the welding process, the joint position, the type of cladding, and specific shop practices.

A bevel to a feather edge is usually prepared for full-alloy welding. The feather edge is then ground to give a 1/16-in. root face. The plates to be joined are spaced approximately 3/32-in. apart to permit full penetration of the high-alloy electrode. The steel side is welded first, using a 1/8-in. electrode for the root pass. The weld is then completed by wire-brushing and depositing a single pass on the cladding side. The wire brush must be made of high-alloy material, such as stainless steel. Figure 13-39 shows the plate edge preparation and the steps used in making a full-alloy weld.

A reverse-type of edge preparation is shown in Fig. 13-40. Here, the beveling is from the cladding side. A 1/16-in. gap is left between the unground feather edges. The root pass is made from the cladding side with 1/8-in. alloy electrode. The opening on the steel side is cleaned before finishing the weld with alloy electrode. This technique makes possible welds with very little iron dilution on the clad steel surface.

Figure 13-41 shows the full-alloy butt welding of light-gage clad steel using a square edge preparation. This is the most economical way of joining light-gage clad materials where a 100% weld is not required. A sealing pass is usually made first from the cladding side. It should not penetrate deeply. The joint is then completed with a deep-penetrating pass from the steel side.

Figure 13-42 shows a method for full-alloy welding of light-gage clad steel in the horizontal position. The top plate edge of the backing is beveled, and the

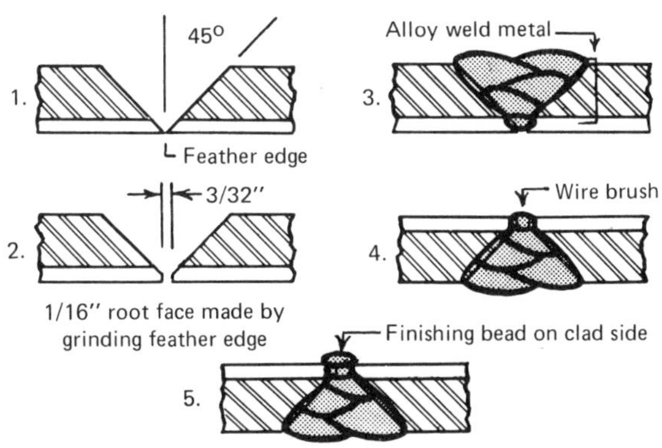

Fig. 13-39. Standard procedures for full-alloy welding of light-gage steel.

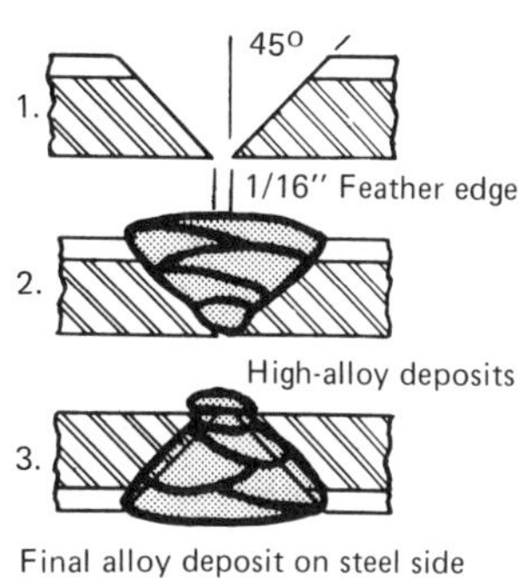

Fig. 13-40. Full-alloy welding of light-gage clad steel using a reverse-bevel edge preparation.

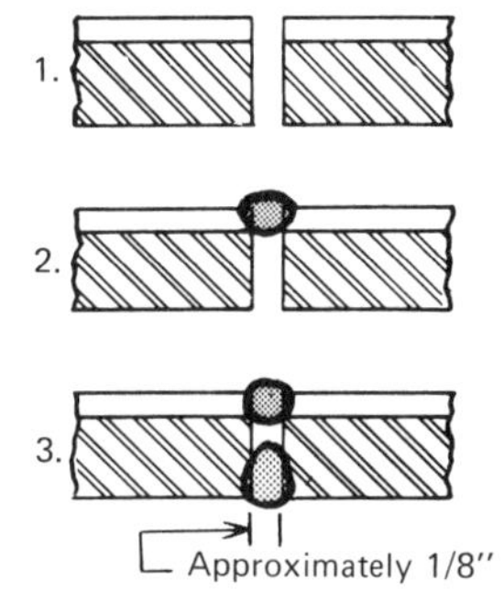

Fig. 13-41. Full-alloy butt welding of light-gage clad steel using a square edge preparation.

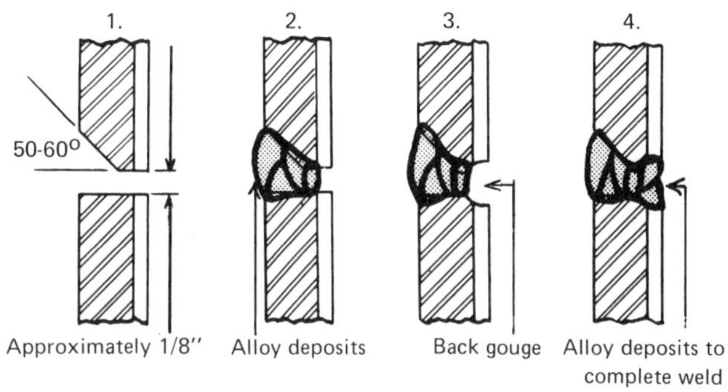

Fig. 13-42. Preparation and scheme of full-alloy butt welding of light-gage steel in the horizontal position.

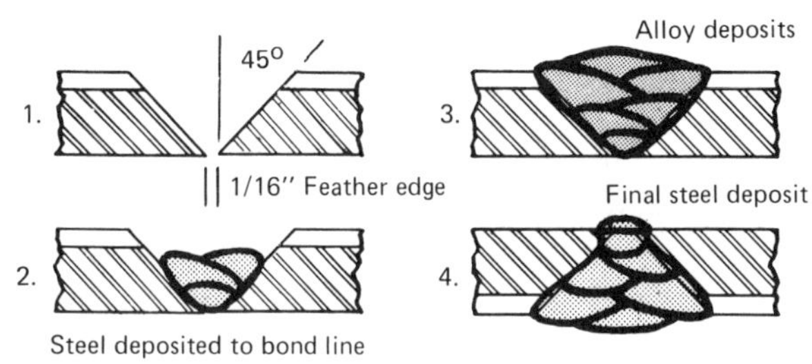

Fig. 13-43. Combination welding of light-gage clad steel using the reverse-bevel technique.

plates are fitted with a root gap of approximately 1/8-in. The steel side is welded first with alloy electrode. The weld is then back-gouged from the cladding side and completed with two or more passes with high-alloy electrode from that side.

In Fig. 13-43, the welding of light-gage clad steel with a combination of steel and high-alloy electrodes is illustrated. A reverse bevel and a 1/16-in. gap between the feather edges are used. The carbon steel electrode is used first, with metal deposited from the cladding side up to the bond line. The weld is then completed on the cladding side with an alloy electrode, followed by cleaning on the steel side and finishing the weld from that side with carbon-steel electrode.

Welding Heavy-Gage Clad-Steel Plate

With clad steel 1-1/2-in. or thicker, a single bevel or a J groove preparation (Fig. 13-44) are frequently used. The steel side of such a joint is usually welded with two manual passes, followed by submerged-arc welding.

Some designs may specify a double-V or double-U edge preparation. In this case it is desirable to minimize the depth of the bevel or U on the clad side, as in Fig. 13-45, to minimize the amount of high-alloy weld metal required.

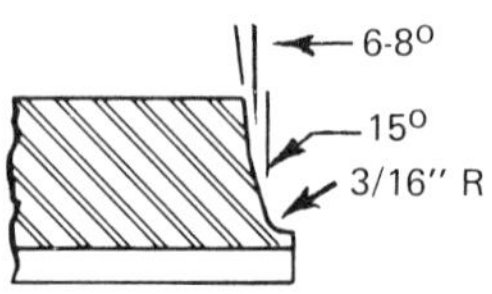

Fig. 13-44. Standard plate edge preparation for welding heavy-gage clad steel. The preparation may be either a bevel or a J groove.

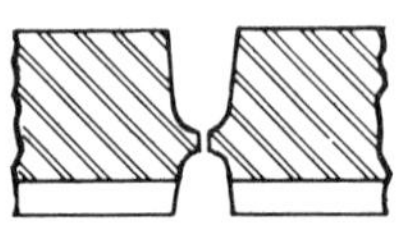

Fig. 13-45. A double-U edge preparation for butt welding heavy-gage clad plate.

GAS-SHIELDED WELDING OF CLAD STEELS

Gas metal-arc welding, either semiautomatic or fully automatic, is an excellent choice for welding clad steels where the volume of work warrants a degree of mechanization. Only relatively low inputs give good penetration, minimizing dilution of weld metal by iron from the clad backing. The amount of dilution is also minimized if the welding arc is directed into the puddle and the molten filler wire is "flowed" into the puddle. Light-gage clad materials are readily welded with a square edge preparation.

The gas tungsten arc process can also be used with all types of clad steels. It, however, is usually considered too slow to be economical. Its use is limited primarily in production to the welding of clad plates with polished surfaces, where the smooth contours achieved by GTA welding save time in polishing the welds to match the surface of the plate.

SUBMERGED-ARC WELDING OF CLAD STEELS

The submerged-arc process gives increased welding speeds, but requires precautions to prevent dilution of the weld deposit with base metal. The plate should be given the conventional preparation on the steel side, with the exception that a steel root face of 3/32-in. minimum thickness should be left above the cladding. One or two beads of steel weld metal should then be run with stick electrode before finishing the weld on the steel side with the submerged-arc process. This provides a backing for the submerged-arc welds and prevents penetration of the cladding.

An alternate method eliminates the manual passes. The plate is prepared with a groove for automatic welding on the steel side, but the cladding is stripped back 1/4-in. on each side of the weld joint to prevent pickup of material from the cladding. A wide root face of steel must be provided for backing the automatic weld.

Table 13-25 gives the welding wires to be used on the clad side with various claddings. Certain precautions must also be used on this side for good results with submerged-arc welding. The groove should be back-gouged to clean, sound metal, and, if it is not possible to compensate for dilution by using a welding wire with higher alloy content than that of the cladding, the alloy weld metal should be deposited in two or more layers. All layers, except the cover pass, should be lightly ground or chipped to assist in minimizing the iron content of weld

TABLE 13-25. Welding Wire for Inert Gas or Submerged-Arc Welding of Clad Side (AWS Classification Numbers)

AISI-Type or Alloy Class	For Covering Exposed Steel with First Pass or Layer	For Complete Balance of Weld
405	E309	E309
410	E309	E309
430	E309	E309
304	E309	E309 or E308
321	E347	E347
347	E347	E347
309	E310	E309
310	E310	E310
316	E316	E316
317	E317	E317
Low Carbon-Nickel	ERNi-3	ERNi-3
Nickel	ERNi-3	ERNi-3
Monel	ERNiCu-7	ERNiCu-7
Inconel	ERNiCrFe-5	ERNiCrFe-5
Cupro-Nickel	ERNiCu-7	ECuNi

surface. Welding current, voltage, and speed of travel should be set to produce a weld with minimum penetration.

HOW TO HANDLE SPECIAL JOINTS

Sometimes it is not possible in a structure to weld clad steels from both sides. When this is the case, backing strips can be used to allow welding through the composite thickness (Fig. 13-46).

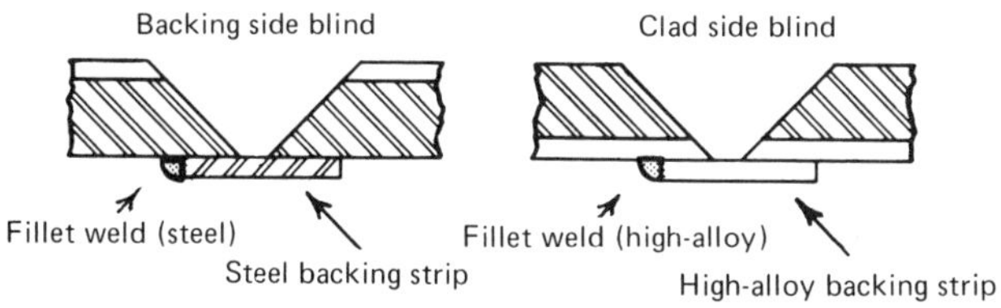

Fig. 13-46. Backing strips enable welding through the composite thickness when one side of the joint is "blind."

When the clad material is on the blind side, a backing of the same analysis as the cladding is used, and when the steel side is blind, a steel backing strip is used. The backing strips are attached to one of the plates by fillet welds from appropriate electrodes.

When the clad side is blind, high-alloy weld metal must be used for the entire joint. When the steel side is blind, steel electrode may be used up to 1/16 to 1/8-in. from the cladding, followed by high-alloy electrode.

Wherever possible, corner joints in clad steels should be eliminated and formed sections as shown in Fig. 13-47(E) used in the design. Separate pieces of clad steels can then be joined by butt welds, which are more desirable from both the fabrication and structural standpoints.

If corner joints must be used, Fig. 13-47 shows the arrangements likely to be encountered. Figure 13-47(A) is the least desirable arrangement, since the continuity of backing is destroyed and an unfused area exists. It should be welded with appropriate high-alloy electrode from both the inside and outside. Such a joint should be permitted only in regions where stresses are negligible.

The joint in Fig. 13-47(B) may be welded from the clad side first with high-alloy electrode, followed by back-chipping the outside and welding the remainder of the joint with high-alloy electrode — or the steel side may be welded first with steel electrode, followed by back-gouging from the inside and finishing the weld with high-alloy electrode. The joints (C) and (D) should be welded throughout with appropriate high-alloy electrode, with the clad side welded first, followed by back-gouging or grinding before welding from the steel side.

WELD DISCONTINUITIES

Since clad steels are most commonly used to give a corrosion-resistant surface to a structural material, the weld will be defective from the application sense if it fails to maintain the corrosion resistance of the cladding or if it fails to maintain the mechanical properties of the backing metal. Defects that would be regarded of little importance in ordinary mild-steel fabrication are of major concern with clad corrosion-resistant materials. Departure from correct procedures can result in welds of both poor mechanical and chemical properties.

Undercutting is a main source of trouble, assuming the proper electrodes are selected and the proper procedures followed. A slight amount of undercut of the cladding can appreciably reduce the thickness of it at that point and drastically reduce the service life of the product. Even a minute area of reduced protection or no protection to a surface that must withstand a corrosive environment is destructive to the whole product. Undercutting, of course, also reduces the fatigue strength. When undercuts are detected, they should be covered with weld metal.

Incomplete penetration and lack of fusion of the weld metal may lead to the initiation of cracks in the unfused area, and porosity on the weld surface may be a focal point for corrosion. Porosity occurring in clusters in the weld may be caused by excessive current, arc instability, inadequate slag removal, moisture on the electrode, improper groove preparation, or improper electrode manipulation.

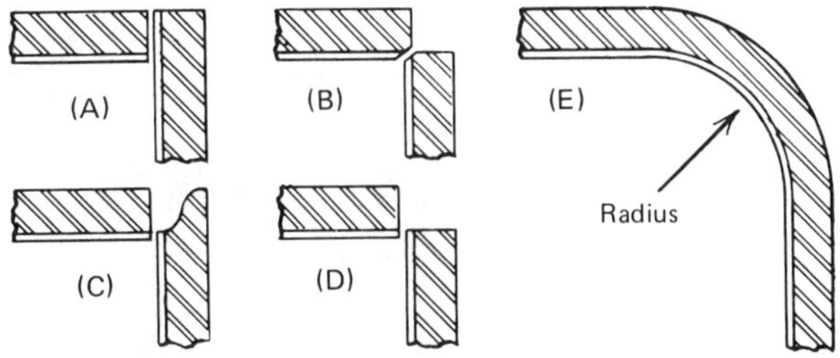

Fig. 13-47. Types of corner joints encountered in clad-steel fabrication.

PREHEATING AND STRESS RELIEF

Preheating is a requirement determined by the nature of the backing steel. It is usually necessary for the successful welding of alloy backing materials, such as A302, A204, and A387 (see Section 3.3). The heating requirements for clad steels are normally the same as for solid plate.

Stress relieving may be done to remove stresses induced by welding, to restore toughness to the backing metal in the heat-affected zone, to impart low-temperature toughness to the deposited weld metal, or to give dimensional stability to the welded structure prior to a subsequent operation, such as machining. Stress relief may be a requirement of the code governing the fabrication; thus, the ASME Unfired Pressure Vessel Code, Section VIII, specifies stress relief based on the gage and specifications of the backing steels.

The 300-series stainless steels are subject to carbide precipitation during a stress-relief heat treatment. The effect can be minimized by the use of the stabilized or extra-low-carbon grades. When the low-carbon grades of Types 304 or 316 are used, the stress-relieving temperature should be no higher than 1050°F.

The 400-series clad stainless steels should be stress relieved in the range of 1100-1350°F. A temperature range of 1150-1200°F may be used to stress relieve clads of Monel, Inconel, nickel, copper, and cupro-nickels. A low-carbon grade of a nickel clad should be used if the weldment is to be stress relieved or used at elevated temperature.

Using a semiautomatic welding process on a dragline bucket.

Hardsurfacing By Arc Welding

Often, the prime requirement of a metal part is that it have exceptionally good resistance to wear, corrosion, or high temperatures. Rather than make the entire part from some expensive wear-resistant, corrosion-resistant or temperature-resistant alloy, it is more economical to make it from an ordinary steel and then cover the critical surfaces with a layer of weld metal capable of withstanding the service conditions. The application of a durable surface layer to a metal is called hardsurfacing, and a simple way to apply hardsurfacing materials is by arc welding.

There are various materials that can be used for hardsurfacing, and selecting the one best suited to a particular application involves balancing cost against performance, as in the selection of any engineering material. Two AWS specifications that pertain to hardsurfacing are *Surfacing Rods and Electrodes, A5.13-80* and *Composite Surfacing Rods and Electrodes, A5.21-80.*

Although some materials are hardsurfaced to improve their resistance to corrosion or high temperatures, by far the most common reason for hardsurfacing is to improve wear resistance. Hardsurfacing for wear resistance will be the main consideration in the following text, with corrosion resistance and elevated-temperature resistance alluded to where they are also factors in the selection of welding materials. Wear is encountered primarily in two forms. An abrasive or galling type of wear occurs when two surfaces rub or slide over one another, and an impact or spalling type of wear occurs when a surface is subjected to repeated blows or impingement.

HARDSURFACING MATERIALS

Hardsurfacing materials tend to be suited to specific types of wear. For example, resistance to abrasive wear generally requires extremely hard materials that also happen to be somewhat brittle and thus not particularly resistant to impact or chipping. On the other hand, high resistance to impact is best provided by ductile or malleable materials that, in turn, are not particularly resistant to severe abrasive wear. There are, however, some hardsurfacing materials that provide a compromise between these two qualities.

The major hardsurfacing materials are discussed in the following text, and, except for copper alloys and the tool and die steels (which have specialized applications), are presented in order of decreasing hardness. This means that, in general, they are also presented in order of decreasing abrasion resistance and increasing impact resistance.

The relative hardness and abrasion resistance of most of these materials are plotted in Fig. 13-48, which provides a graphical indication of their primary application areas. Thus, materials in the upper right portion of the graph are most resistant to abrasive or frictional types of wear, while those in the lower left portion provide maximum resistance to impact or spalling.

Tungsten Carbide

Tungsten carbide is the hardest and generally the most abrasion-resistant of all hardsurfacing materials. It is, thus, best suited to applications requiring exceptional resistance to abrasion and is commonly used on the surfaces of tools for cutting earth or rock. It is particularly well suited to digging in sandy formations, where a coarse form of tungsten carbide is used so that as the hardsurfacing wears it leaves a rough, fast-cutting edge.

This material is also used to protect machine parts that come into contact with severely abrasive material. Screw conveyors, for example, are often hardsurfaced with tungsten carbide. This material is *not* recommended for protecting both surfaces in sliding contact, because the wear pattern is not smooth enough for bearing-type applications. The surface finish is similar to that of fine sandpaper.

For hardsurfacing, tungsten carbide consists of tungsten carbide crystals embedded in an iron-base alloy. This form of tungsten carbide is a cast material and is not to be confused with the sintered (or cemented) carbides used for metal-cutting tools. The sintered form cannot be fusion welded. Hardness of the deposit cannot be measured directly on the Rockwell C scale, because the bonding alloy has

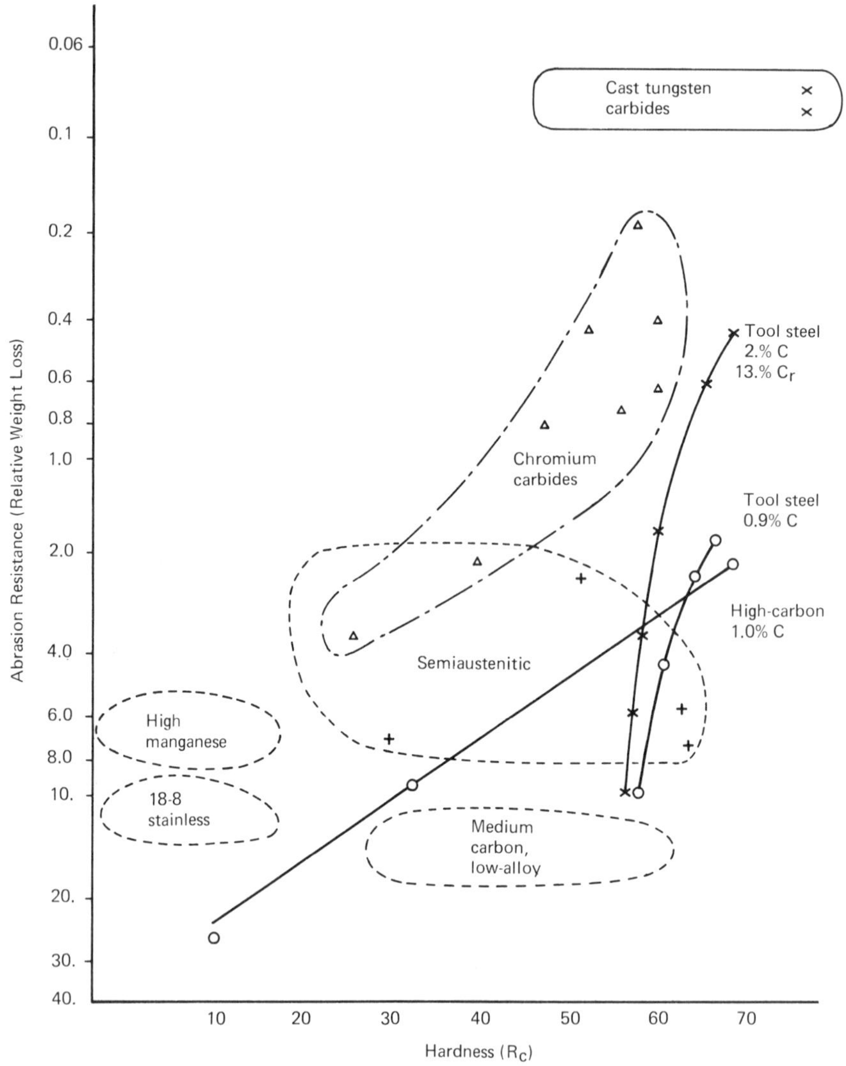

Fig. 13-48. Hardness and abrasion resistance of common hardsurfacing materials.

a relatively low hardness, and the tungsten carbide particles have a high hardness. Metallurgists use a different procedure in evaluating the hardness of such materials, and hardness values when given are on the Rockwell A scale.

Hardsurfacing electrodes are made by encasing the cast tungsten carbide and tungsten-carbon alloy in powder form within mild-steel tubes. During deposition, the steel tube dissolves some carbon and tungsten from the tungsten carbide particles. As deposited, the tungsten carbide is thus embedded in a tungsten-iron or steel.

Chromium Carbides

Chromium-carbide hardsurfacing materials are surpassed in hardness and abrasion resistance only by the tungsten carbides. But the chromium carbides, being slightly softer, are slightly superior to tungsten carbide for impact resistance. Chromium carbides are, nevertheless, still considered to be hard, brittle materials compared to other hardsurfacing materials. They are extremely resistant to wear and abrasion, but do not provide much resistance to chipping, spalling, or damage from impact.

The applications for chromium carbide are generally the same as those for tungsten carbide, except that chromium carbide may also be used where its good corrosion resistance is a desirable property. Typical applications are coatings for earth-cutting tools, for machine parts exposed to abrasive materials, and for chutes and slides that convey ores or other materials not likely to impose significant impact loads.

There are dozens of specific hardsurfacing alloys

TABLE 13-26. Ferrous Chromium Carbide Hardsurfacing Deposits

	Typical Chemistry* (%)					Typical Hardness † (Rc)			
	C	Mn	Si	Cr	Mo	Single Layer Room Temperature	Multiple Layer Room Temperature	Multiple Layer 1000°F	Multiple Layer 1500°F
Manual Electrodes Max. abrasion	4.2	5.4	0.70	22.5	0.25	40 – 52	50 – 58	38	30
Abrasion with mild impact	4.5	1.0	1.0	18.5	6.0	45 – 55	55 – 64	49	38
Submerged-arc, alloy flux	3.8	0.9	0.5	29.0	1.2	53 – 58	58 – 61		

* Two layers in mild steel

† Deposits on mild steel

TABLE 13-27. Nonferrous Chromium Carbide Hardsurfacing Deposits

	Typical Chemistry* (%)					Typical Hot Hardness (Rc)			
	C	Si	Cr	W	Co	Room Temperature	650°F	1050°F	1400°F
Max. abrasion	3.2	0.8	26.0	13.5	53.0	43 – 58	36 – 46	30 – 37	51 – 54*
Abrasion with mild impact	1.1	1.5	29.0	5.5	62.0	23 – 47	26 – 40	19 – 32	15 – 50*

Note: Chemistry and hardness values are for multilayer deposits with manual electrodes

* Rockwell A

TABLE 13-28. Semiaustenitic Steel Hardfacing Deposits

	Typical Chemistry * (%)					Typical Hardness (Rc)	
	C	Mn	Si	Cr	Mo	Single Layer	Double Layer
Impact and abrasion Manual Electrode	2.1	1.1	0.75	6.5	0.4	24 – 53	28 – 53
Submerged-arc, alloy flux	0.95	2.5	1.8	7.0	0.75	36 – 60	32 – 60

* Double Layer

within the chromium carbide group. Some of these are ferrous-based, and some are based on cobalt or nickel. The ferrous chromium carbide alloys (Table 13-26) are too hard to be machined and provide excellent resistance to wear by abrasion. They also maintain high hardness at elevated temperatures. Tong tips for handling hot ingots are often faced with chromium carbide.

The nonferrous chromium carbides (Table 13-27) are often referred to by the term "Stellite," a proprietary trade name. Applications for the nonferrous types are similar to those of the ferrous, but the nonferrous chromium carbides are more resistant to some types of corrosion. However, they also are more expensive than the ferrous types and are not quite as abrasion-resistant. A newer form of nonferrous chromium carbide electrode, having a wrought-cobalt core wire, can be deposited much more easily than the conventional cast electrode.

Chromium carbides generally are nonmagnetic. The deposits are not affected by heat treatment, thus cooling rate has no effect on wear resistance. The materials develop a smooth surface with wear and can, thus, be used to protect surfaces in sliding contact. Bearings for use at high temperatures or in corrosive environments are often faced with chromium carbide. Deposits typically range in hardness from 40 to 63 Rockwell C. For each material, there is a combination of carbon and alloy content that gives both good abrasion resistance and impact strength. An off-balance material, however, may have low resistance to abrasion and also may be brittle. Some electrodes are designed to produce good hardsurfacing deposits when diluted with steel, that is, they provide a one-layer deposit that is resistant to abrasion and to cracking. The best nonferrous chromium carbide electrodes deposit metal that features high hardness (54-63 Rockwell C) and very high abrasion resistance.

Admixture of base metal into the weld metal reduces abrasion resistance and increases impact resistance of the first layer. Thus, for maximum abrasion resistance, two layers should be used, with low current employed on the first layer to minimize admixture. Since the deposits are somewhat brittle, deposit thickness should be limited to no more than two layers to minimize the possibility of spalling. The workpiece surface should be prepared carefully, and underbead cracking should be avoided.

Semiaustenitic Steels

The semiaustenitic alloys can be thought of as intermediate between the hardest and the toughest hardsurfacing materials. They provide a balanced combination of good wear resistance and good impact resistance and probably are the most widely used of all hardsurfacing materials. The deposits are either hard as deposited or become hard with peening impact in service. They are also hardenable by heat treating.

These alloys are iron-based materials that contain up to 20% alloying elements. The varieties most commonly used have from 1 to 2% carbon and from 5 to 12% chromium (Table 13-28). The name for these materials comes from the fact that the alloy, principally austenitic as deposited, remains only semiaustenitic upon cooling. Various metallurgical structures are possible, depending upon the specific composition of the deposit and its cooling rate.

If the deposit cools quickly (small electrode, short beads, and base metal cold before welding), the deposit will be predominantly austenitic and, thus, will be soft and tough. Slow cooling, however, does not impair wear resistance. Also, the surface will work-harden. For example, a material that is 30 Rockwell C as deposited will become more than 50 Rockwell C after moderate peening.

If the deposit cools slowly (large electrode, continuous welding, or preheated base metal), most of the austenite has time to transform to martensite, and the deposit will be less ductile. Both types of deposit work-harden to the same final value.

This group of alloys is machinable with tungsten carbide tools after a long anneal at high temperatures. Some of the alloys can be hot-forged. Although there is a wide range in hardness in this group, abrasion resistance does not vary greatly. Some of the alloys that are hard as deposited are not as resistant to grinding abrasion as some of those having a much lower hardness. The alloys that work-harden are superior for service involving a combination of impact and abrasion.

Two-layer deposits are generally recommended, except with certain submerged-arc processes that produce good wear resistance with one layer. If machining is required, the deposit must be annealed, machined, and then heat treated.

Some alloys of this group may tend to lift or peel away from the substrate upon cooling, particularly when the base metal has less than 0.15% carbon. This is caused by a highly brittle zone that forms between the base metal and weld metal. The transformation and shrinkage stresses then promote cracking and peeling. This effect can be minimized by slowing the cooling rate with a 300°F preheat, and by peening each bead while it is hot. Also, the first layer can be deposited with high current and fast travel as a series of stringer beads separated by small gaps. The high admixture that results avoids a brittle zone. The second layer is then applied conventionally.

Austenitic Manganese Steels

The austenitic manganese-steel hardsurfacing materials (Table 13-29) are tougher and slightly less abrasion-resistant than the semiaustenitic materials,

TABLE 13-29. Austenitic Mn Steel Hardsurfacing Deposits

	Typical Chemistry * (%)					Typical Hardness* (Rc)	
	C	Mn	Si	Ni	Mo	As-Welded	Work Hardened
Severe impact Manual electrode	0.62	14.2	0.15		1.15	17 – 20	43 – 48
Self-shielded flux-cored electrode	0.65	13.5	0.5	2.6	0.5	10 – 15	35 – 45

* Multiple layers on Mn Steel

and are considered to be extremely good alloys for resistance to impact. A typical application might be for coating an ore chute where large rocks may occasionally strike the chute with great force.

The deposit normally contains from 11 to 14% manganese and is work-hardenable. Worn manganese-steel parts often are built up nearly to size with austenitic manganese-steel electrodes, and then are coated with a thin layer of semiaustenitic steel to enhance the abrasion resistance. There is not much variation in the hardness or the abrasion resistance of austenitic manganese deposits. Most deposits contain about 1.0% carbon and some molybdenum or nickel. High-manganese deposits are not recommended for resistance to grinding abrasion but are superior for resistance to a combination of severe impact and abrasion.

The required alloy content of the overlay can be deposited easily without special precautions. Admixture of a manganese-steel base metal into the weld metal has no affect on wear resistance. Admixture of carbon steel into the weld metal can reduce wear resistance slightly. Two-layer deposits are recommended for best performance. Cooling rate has no effect on wear resistance, nor does heat treating have any effect on machinability or wear resistance. Prolonged heating above 500°F, however, may embrittle manganese steel base metal.

Spalling is rarely encountered with these deposits, especially if the surface is prepared properly and if thick deposits are peened after each layer is applied. There is the possibility, however, that prolonged or repeated heating of manganese-steel base metal over 500°F may promote embrittlement and spalling. Thus, heat input should be limited by such techniques as keeping preheat to 70°F (except when a 100 to 200°F preheat may be needed to prevent cracking of a massive part), by using skip welding, or by employing low welding currents. Because weld metal has a lower carbon and silicon content than manganese-steel base metal, the deposit generally does not become embrittled.

Austenitic Stainless Steels

Though not "hard" materials in the generally accepted sense, austenitic stainless steels are grouped with hardsurfacing alloys because this type of stainless steel provides an extremely tough, ductile coating with exceptionally good resistance to chipping from repeated impact forces. These deposits also provide good corrosion resistance. Such coatings typically are used to protect water-turbine blades from corrosion and cavitation erosion. The E300 series stainless steels work-harden, but not to the extent of the manganese steels. Abrasion resistance is higher than that of high-carbon steel of the same hardness.

Stainless steel deposits are sometimes used as base layers for other hardsurfacing materials. For example, an initial layer of stainless avoids a brittle bond that otherwise would develop between a carbon-steel base and a manganese-steel coating. E308 and E312 electrodes are usually used for these base layers. E312, being a higher alloy than E308, is less affected by admixture of the base metal.

Martensitic Stainless Steels

Types 410 and 420 stainless steels (Table (13-30) produce dense, homogeneous weld metal with good resistance to cracking. For best results the deposit should be at least two layers thick, and preferably should be three or more layers thick.

Type 420 must be finished by grinding if a machined finish is required. Type 410 can be machined with carbide tools at slow cutting speeds. The deposits usually are used in the as-welded condition.

These stainless steel deposits are widely used for metal-to-metal wear, such as that encountered with bearings operating at elevated temperatures and with rolls used in steel mills. A typical application is a deposit for a backup roll on a hot-rolling mill.

Heat-Treatable Steels

Several types of heat-treatable steels are used as

TABLE 13-30. Martensitic Stainless Steel Hardsurfacing Deposits

	Typical Chemistry * (%)				Typical Hardness (Rc)		
	C	Mn	Si	Cr	Single Layer	Double Layer	Multiple Layer
Type 420	0.23	1.1	0.9	13.0	45 – 47	49 – 51	52 – 54
Type 410	0.08	1.1	1.0	12.5	38 – 40	37 – 40	30 – 32

* Three-layer deposit by submerged-arc.

hardsurfacing materials, (Table 13-31). These steels produce hard, abrasion-resistant surfaces if the deposit is cooled quickly, or they produce a softer, tougher deposit if allowed to cool slowly. Hard deposits can be softened by annealing. These materials are commonly used for surfacing or for building up parts prior to hardsurfacing by other alloys.

High-Carbon Steels: The cooling rate of high-carbon steel must be rapid to produce maximum abrasion resistance. At best, the high-carbon steels may rival some of the chromium carbides, but if cooled slowly the deposit is less abrasion-resistant.

The abrasion resistance of high-carbon weld deposits increases with hardness. The maximum ductility and impact resistance for a given hardness is obtained by quenching and tempering, however, heat treated properties may be approached by welding alone. To achieve such results, use minimum preheat and deposit the weld metal in small stringer beads or thin layers. Before depositing the next bead or layer, allow sufficient time for the high-carbon weld metal to transform to martensite. Using this procedure, each bead tempers the martensite of the previous bead and the weld as a whole approaches the quenched and tempered condition.

Published hardness values are normally for multilayer deposits applied manually, but the hardness of single-layer welds, which are influenced more by admixture with the base metal, may be higher or lower than that of multilayer deposits. This effect depends on the hardenability of the weld metal. A low-carbon electrode deposited on high-carbon steel produces a medium-carbon weld metal.

It is possible to have a combination of analyses that produces an exceptionally hard deposit. For example, a low-carbon but high-alloy base metal, when surfaced with a high-carbon electrode, produces weld metal that is high in carbon and has absorbed enough alloying elements to produce higher hardenability than either a multilayer high-carbon deposit or the low-carbon high-alloy base metal.

Medium-Carbon Alloy Steels: These materials are most often used to provide protection against metal-to-metal wear, and for building up steels before they are surfaced with the carbide types of hardsurfacing alloys. Though these alloys have low abrasion resistance, some of them have exceptionally high tensile strength and ductility, and thus resist deformation and cracking from impact.

The chemical analyses of this group range from 0.15% carbon with a total of 2.0% chromium, manganese, molybdenum, silicon, and other elements, on up to 0.5% carbon with 5.0% alloying elements. Hardenability increases with alloy content; the as-welded hardness on preheated work is higher than can be obtained with high-carbon deposits. The lower limit on carbon restricts the maximum hardness obtainable, but also increases impact resistance. The metal-arc deposits are air-hardenable and, in the as-welded condition, have high toughness and good abrasion resistance.

A low-alloy content in the deposit reduces abrasion resistance, but increases impact strength. On some deposits, normal procedural variations do not change alloy content significantly, while with other deposits the alloy content can be varied easily by altering procedures.

Admixture has little effect on multiple layer deposits. However, admixture may raise the carbon or alloy content in the first layer on medium to high-carbon steel and on some low-alloy steels. High admixture also may lower the first-layer alloy content with substrates of low-carbon steel.

Slow cooling increases the impact strength of medium-carbon alloy steel deposits and also improves the machinability of some deposits. For best machinability, deposits should be annealed (and

TABLE 13-31. Heat-Treatable Hardsurfacing Deposits

	Typical Chemistry * (%)						Typical Hardness (Rc)		
	C	Mn	Si	Cr	V	Mo	Single Layer	Double Layer	Multiple Layer
Manual †	0.16	1.15	0.60	1.4			15 – 20	18 – 23	23 – 28
Submerged-arc†	0.24	1.20	0.70	1.8		0.30	25 – 32	30 – 43	34 – 45
Submerged-arc ¶	0.10	2.50	0.30	2.8	0.15	0.30	34 – 39	39 – 42	39 – 42
Manual ¶	0.37	2.20	0.15	3.3			40 – 45	48 – 52	50 – 55

* Three-layer deposit on mild steel

† Deposits can be machined with carbide tools without annealing if welded with proper procedures.

¶ Must be annealed before machining

then heat treated after machining).

Tool and Die Steels: Hardsurfacing with these alloys is commonly used to repair worn edges or surfaces of tools and dies and for applying durable wearing surfaces to dies made up of low-cost alloys such as mild steel. Electrodes for these applications generally are of either the chromium-type or molybdenum-type, although some tungsten-types are also available (Table 13-32).

The chromium-type consists of general-purpose alloys that provide a balanced combination of properties. These alloys are generally used when die temperatures are not expected to exceed 600°F. For cutting tools and other high-temperature operations, the molybdenum types are used. These alloys, commonly referred to as high-speed tool steels, retain full hardness at temperatures to about 1000°F. The molybdenum high-speed steel is the type most commonly used as an electrode.

Copper Alloys

Copper alloy overlays are used primarily to provide bearing surfaces for sliding metal-to-metal contact, and for corrosion resistance. The copper-based materials used primarily are aluminum bronze, phosphor bronze, and silicon bronze. Pure copper is seldom used.

Aluminum bronze provides a good bearing surface against steel and has fairly good corrosion resistance. Phosphor bronze, although not as hard as aluminum bronze, also makes a good bearing surface and resists corrosion. Silicon bronze has poor bearing properties and is used only for corrosion resistance.

For best results, if applied to steel the copper-based overlays should be applied in three layers to reduce admixture in the final deposit. Total thickness of the deposit should not be less than 1/4-in. after final machining. Deposits are readily machined, unless there is excessive admixture of iron base metal, which causes hard spots.

SELECTING HARDSURFACING ALLOYS

The choice of hardsurfacing alloy depends in part on the reason for applying the deposit. Nearly all applications can be categorized in one of three classes: 1. A cutting edge is to be maintained. 2. A single surface is to be protected without regard for maintaining a sharp edge. 3. Two surfaces in sliding contact are to be protected.

Selection of an alloy also depends on the nature of the service creating the need for hardsurfacing. Wear-producing conditions usually result from some combination of: 1. Abrasion or erosion from sliding contact. 2. Impact or chipping forces. 3. Loss of hardness or strength from extreme heat. 4. Corrosion or other chemical effects.

Selecting To Maintain a Cutting Edge

Selection of an alloy to maintain a cutting edge depends upon whether the primary concern is to minimize over-all wear while maintaining the cutting edge, or whether the objective is merely to maintain a sharp edge without particular concern for tool wear. A metal-cutting tool, for example, must maintain shape and length as well as sharpness, while the efficiency of earth-cutting tools, on the other hand, is not particularly impaired by nominal dimensional wear so long as the edge stays sharp.

Hardfacings of tool steel and high-speed steel are recommended where dimensional features as well as sharpness are to be maintained. Typical applications are shear blades, punches, and metal-cutting tools where the cutting edge must not only stay sharp, but must maintain its initial size and shape to perform satisfactorily. Failure of these parts usually is by upsetting, chipping, spalling, or galling — caused by high compressive loads and by the flow of metal being cut. Impact and abrasion are seldom problems.

High-speed steel is applied where high temperatures are encountered, while the less expensive tool steels are applied for service below 600°F. Deposits of either of these alloys are homogeneous and fine-grained. They can thus be ground to a sharp thin edge that resists upsetting or chipping.

Entirely different materials are recommended if dimensional wear is not of concern. Typical applications of this type are earth-scraping tools, earth-drilling bits, ensilage knives, and shredder blades.

An important aspect of this class of hardsurfacing is the principle of self sharpening. If a wear-

TABLE 13-32. Typical Tool and Die Steels

	C	Cr	Mo	W	V
Cr-Type	0.35	5.0	1.5	1.5	0.4
W-Type	0.70	4.0	–	18.0	1.0
Mo-Type	0.85	4.0	5.0	6.2	2.0

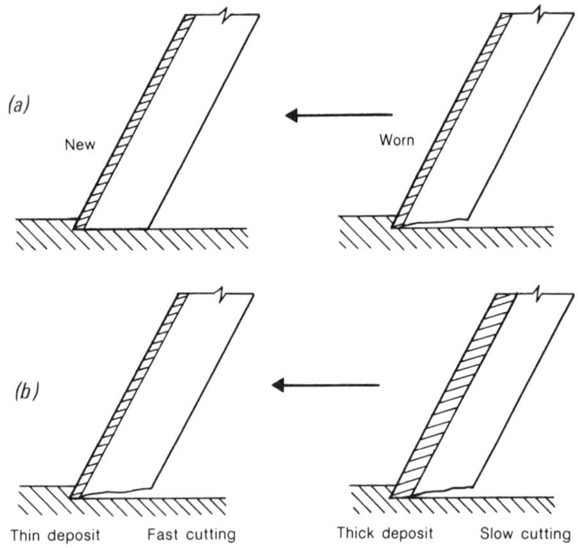

Fig. 13-49. Producing a self-sharpening surface by hardfacing. A thin, wear-resistant advancing surface (a) promotes a wear pattern that tends to maintain a sharp front edge. A thin, hard coating (b) provides fast cutting but short life, while a thick deposit provides longer life but slower cutting.

resistant coating is placed only on the advancing edge of the tool, the back surface wears away at a greater rate and thus tends to promote a sharp cutting edge as shown in Fig. 13-49. Rate of wear and sharpness of the cutting edge are both inversely proportional to the thickness of the hardsurfacing. Thin deposit provides a sharp, fast-cutting tool that wears relatively fast, while a thicker deposit wears more slowly but provides a duller edge. If both front and back sides of the tool are hardsurfaced, the tool tends to dull rapidly and soon begins to slide over rather than cut into the earth or workpiece.

In this class of applications where sharpness rather than dimensional wear is of main concern, carbide materials are often selected. Coarse particles of tungsten carbide provide longest life for earth-cutting tools used in highly abrasive sandy formations. As the bonding material in the hardsurfacing wears away, coarse particles of tungsten carbide are exposed to produce a rough, fast-cutting edge. Fine carbide deposits are preferred where smoother wear patterns are required as with various types of knives.

Chromium carbides are used on cutting edges where abrasive conditions are not particularly severe and where a lower-cost material is preferred. These materials readily self-sharpen and provide smoother, lower-friction edges than those of tungsten carbide.

Neither tungsten carbide nor chromium carbide are suited to applications where high impact loads occur. The semiaustenitic and austenitic manganese steels provide longest life for edges subjected to chipping and impact. Edges protected by these materials are not self-sharpening, but wear to a rounded contour that is tough and ductile. Tungsten carbide and chromium carbide tend to chip away from dipper teeth handling large rock. But austenitic manganese or semiaustenitic deposits work-harden and, thus, become resistant to abrasion, while impact forces are cushioned by the softer metal below the surface.

Selecting to Protect a Single Surface

The need to protect a single surface accounts for the largest number of hardsurfacing applications. The objective is to prevent wear of the surface without regard for a cutting edge or for wear on a contacting surface. It may be desirable to have a low-friction surface that polishes well (a plowshare or moldboard, for example), or a rough high-friction surface (dredge stud clamps) may be preferred.

Chutes for conveying bulk materials are commonly hardsurfaced. In general, hard materials such as chromium carbide are preferred if fine-grained ores or earth are being conveyed without large rocks or other objects that could cause impact. Semi-austenitic or manganese steels are preferred if heavy bulky pieces are being dumped onto a chute or if other service demands are likely to produce high impact forces.

Screw conveyors and earth-drilling tools are normally protected by hard materials such as the carbides. Stainless steels are used to provide corrosion resistance and protection against erosion in water pumps, and for applications requiring good impact resistance.

Selecting to Protect Contacting Surfaces

This type of application normally entails metal-to-metal wear under various combinations of abrasion, impact, friction, and corrosion. The surfacing material must wear smoothly, have a low coefficient of friction, and have little or no tendency to seize or gall. One type of hardsurfacing definitely not suitable for this class of application is tungsten carbide.

Often one surface requires more protection than the other. In such cases, one surface may be coated with bronze while the other surface is not coated. The bronze then wears sacrificially, thereby protecting the other surface.

High-carbon electrodes are used to build up shafts that are to operate with lubrication. The high-carbon deposit when machined or ground provides a smoother surface than is readily obtainable with a mild-steel electrode deposit. Bearings operated at elevated temperatures are surfaced with the chromium carbides, the stainless steels, and the high-chromium and nickel alloys.

Parts operating under corrosive conditions are protected by surfacing with chromium carbides and some of the stainless steels. Chromium carbide facings also give good service on parts operating in sand or mud. Rubber, though obviously not applied by welding, is quite wear-resistant and is a good wearing material for journals and bushings used underwater. Steel shafts give long service when they run in rubber bearings and are lubricated with water to keep out sand and mud.

Galling is the major problem in nonlubricated metal-to-metal contact. Two alloys of identical type are more prone to seize or gall than two different materials, probably because the materials are alike in chemical analysis or in surface condition. Two alloys that provide good service underwater and fair service in sandy mud may soon chew each other up when they are run dry. Soft iron alloys are more likely to gall than harder ones, and rough surfaces gall more readily than smooth ones.

A running-in period with lubricant is recommended for freshly machined or ground surfaces because they often have irregularities that intermesh and produce tears that promote seizing.

Practical Considerations in Selection

A tough, ductile hardsurfacing material may be specified for a new application where no prior experience exists. With a tough material, any failure that may occur is likely to be gradual rather than catastrophic. The semiaustenitic steels are recommended for such initial trials. If these materials hold up without chipping or impact fracture, one of the tougher chromium carbides might be used on the next trial. Progressively harder alloys could then be tried to improve abrasion resistance.

Examination of prior hardsurfacing failures provides valuable clues to selecting a superior material. However, meaningful information can be gained by examining a part that is only partially worn, since a totally worn part often tells very little about the way the wear progressed. For example, it may be impossible to determine whether hardsurfacing was worn away by abrasion or by impact if no hardsurfacing material remains on the part.

Also, it is often less costly to refinish a part while it is only partially worn instead of waiting until it is severely worn. The reason is that rate of wear often increases as wear progresses. A rock that scratches a dipper tooth, for example, will remove only a small amount of surface material if it abrades against hardsurfacing. But if the hardsurfacing is worn away, the same rock will likely create a large gouge in the tooth, thereby requiring expensive rebuilding of underlying metal prior to hardsurfacing. Self-sharpening blades in particular, where hardsurfacing is applied only to the advancing edge, should be hardsurfaced regularly (Fig. 13-50) to avoid excessive wear of the softer blade material.

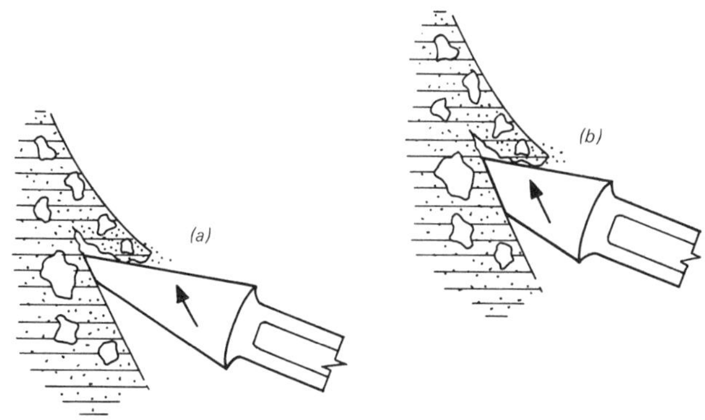

Fig. 13-50. Configuration where rate of wear increases as wear progresses. Minimal bulk material is abraded from slightly worn dipper tooth (a), but increasingly larger amounts of bulk material are removed (b) as advancing edges wears, because more wearing surface is exposed to the abrading action of earth flow.

The hardsurfacing job may have restrictions that narrow the field of possible surfacing materials. Examples are where:

- The deposit must be machinable as welded. This requirement eliminates all materials except the medium-carbon, low-alloy types. The machinability of a medium-carbon deposit depends on the rate of cooling; preheating must be used on large members to provide a deposit that is readily machinable. Deposits of austenitic manganese steel, chromium carbide, and stainless steel can be finished by grinding. The possibility of using an alloy that requires grinding should always be considered, for the increased life may more than offset the higher finishing costs.
- The deposit is to be machined, but may be annealed after welding and then hardened after finishing. This requirement increases the number of deposits that may be used. Possible surfacing materials are high-carbon tool steel and some of the low-carbon, low-alloy types. Materials that should not be used include stainless steel,

chromium carbides, tungsten carbides, and high-manganese steel.

- The part and the deposit are to be hot-forged. This requirement eliminates the tungsten carbides and the chromium carbide groups. The high-carbon and medium-carbon deposits are forgeable at red heat and are also ductile enough to be straightened and forged to some extent after quenching for hardening.These steels form hard martensite when quenched, but the transformation does not take place until the work has cooled to 500 or 400°F. The part may be taken from the quenching bath at 500-600°F and then straightened or forged a small amount while it is still in the ductile austenitic state. If then cooled in air or in the quenching bath, it will be just as hard as if it had been given a complete quench. It may also be tougher, for the martensite is given time to temper.
- A cutting edge is required for cutting or turning steel. This specification necessitates the use of a material with high hardness, high red hardness, and a high edge strength. The field is limited to one of the high-speed tool steels or a tungsten carbide insert.
- The deposit must have maximum resistance to grinding abrasion. Tungsten carbides or chromium carbides are likely to be best for this requirement. Chromium carbides are preferred if low cost is important.

METHODS FOR DEPOSITING HARDSURFACING

Arc welding processes are generally preferred for hardsurfacing for reasons of speed and low cost. However, some hardsurfacing is still done by oxyacetylene processes. The arc processes used are simply counterparts of the methods used to join metals, except that hardsurfacing electrodes are used in place of other welding electrodes or filler wires. In submerged-arc hardsurfacing, alloy elements can be added to the deposit by the flux. Carbon-arc processes are also used.

On steels that are sensitive to overheating, such as austenitic manganese steel, the oxyacetylene and gas tungsten-arc processes should be avoided. Other steels and other surfacing materials may be hardsurfaced by the oxyacetylene or gas tungsten-arc processes as well as the metal-arc processes.

The oxyacetylene and the carbon-arc processes are used for small, delicate parts and where only a thin layer of surfacing is to be applied. Coal cutter bits, for example, often are hardsurfaced by oxyacetylene welding.

Either the oxyacetylene or the carbon-arc process is recommended to deposit nonferrous chromium carbide alloys in cases where the surfacing must be iron-free. By careful manipulation, the alloy may be sweated to the steel base metal and welded without penetration or alloying with the steel. Corrosion resistance and abrasion resistance are lowered when these alloys are diluted with iron.

A possible drawback of metal-arc processes for hardsurfacing is the deep penetration they produce. The more the base metal is melted, the greater the alloying of the deposit with the base metal. Deep penetration is usually a disadvantage since it increases the thickness of the hardsurfaced deposit. The total amount of hardsurfacing is the build-up from the surface plus the penetration below. If the amount of build-up is limited, then deep penetration may be desirable.

In most cases, the first layer of surfacing has lower abrasion resistance but more resistance to impact than the undiluted alloy. This gives a gradual change from the base metal to the hardsurface. But diluted surfacing may produce brittle alloys, as sometimes occurs when carbon steel is deposited on austenitic manganese steel.

Hardsurfacing alloys containing boron should be deposited with an oxyacetylene torch, and care must be taken to prevent the formation of the brittle iron-boride crystals.

Manual hardsurfacing for repair or renewal of worn surfaces is usually done by metal-arc welding and coated or bare electrodes. Semiautomatic and automatic hardsurfacing techniques are often used on large volume production or for protection of large contact areas, such as tread-type tractor idler rolls, or where composite hardsurfacings must be provided on new equipment.

Control of Properties

Once a specific hardsurfacing deposit is selected, the alloy content and cooling rate of the deposit must be carefully tailored to provide the required abrasion resistance and impact properties. Alloy content is determined primarily by the material selected, the procedures employed to lay the deposit, and by the resulting admixture of deposit and base metal. Important procedural considerations are:

Surface Preparations: Remove grease and oil with a solvent, and remove rust and dirt by wire-brushing. If contaminants are not removed, they can cause porosity, cracking, and poor deposit quality. For a good bond between base metal and weld, use arc-gouging or grinding to remove cracks, the remains of old high-alloy hardsurfacing deposits, or badly work-hardened or distorted surfaces. Fill cracks, gouges, and surface depressions by manual welding.

Preheating: Most hardsurfacing does not require preheating, except for bringing the part to a room temperature of 70-100°F. Medium to high-carbon and low-alloy steels may require higher preheat to prevent underbead cracking, weld cracking, spalling, or stress failure. Higher preheat is needed for massive or rigid parts and when cracking actually is encountered. Parts are normally preheated by gas or oil torches, ovens, or electrical heating devices. The particular method used depends upon the size of the part and the equipment available. Preheating is of no benefit if the part cools before welding.

Always be sure the area to be surfaced is at the specified preheat temperature before starting to weld. Sometimes it is advisable to check the temperature of the part during welding to be sure it has not cooled. More heating may be needed if welding is interrupted. Interpass temperature — the temperature of the surface when welding all layers except the first — is just as important as preheat temperature and should usually be as high as the preheat. (Special care should be taken to avoid overheating manganese steel.)

Build-up: Some hardsurfacing deposits are limited to two or three layers, therefore badly worn surfaces are normally rebuilt to within 3/16-3/8 in. or original size before hardsurfacing. The electrode for the preliminary build-up usually should have a strength similar to the base metal.

Pattern of Deposit: Pattern of deposition usually is not important. However, these guidelines can be applied:

- The best pattern is usually the one most economical to apply.
- The pattern affects shrinkage stresses and can thus be used to help control distortion and cracking tendencies.
- A pattern with openings between beads is acceptable if the openings fill with abrasive material in service.
- For roller applications, beads placed on the rolls across the flow of material help pull material through the rolls, for example, rock crusher rolls.
- Beads placed parallel to the flow of abrasive material smooth the flow and thereby reduce wear.

Admixture and Cooling Rate: Small beads made with small electrodes and low currents have fast cooling rates and low admixture. Using two layers reduces admixture in the final layer.

Although cooling rate affects wear resistance of some deposits, it has a much greater effect on the control of spalling, cracking, and distortion. Therefore, a slow cooling rate may be required to reduce these undesirable effects even if wear resistance is reduced as a result.

Methods of controlling cooling rate include:

- Providing adequate preheat.
- Utilizing heat input from the arc or flame during hardsurfacing.
- **Insulating the hot part immediately after hardsurfacing with dry sand, lime, or a fiberglass blanket. This method helps minimize residual cooling stresses, weld cracking, and distortion but does not affect wear resistance of most deposits. Large parts draw heat away from the weld more quickly than small parts and thus cool the weld faster.**

Avoiding Spalling

Spalling is the breaking of weld metal particles away from the base metal or underlying hardsurfacing layers. The broken pieces may vary in size from small chips to large chunks that expose the base metal. To avoid spalling:

Prepare the Surface: As in production welding, hardsurfacing welds must have a sound crack-free bond with the base metal. Such bonds are obtained when the workpiece surface is clean and all cracks and surface damage are repaired.

Avoid Underbead Cracking: Rapid cooling from welding temperature can cause brittle, crack-sensitive heat-affected zones in some base metals.

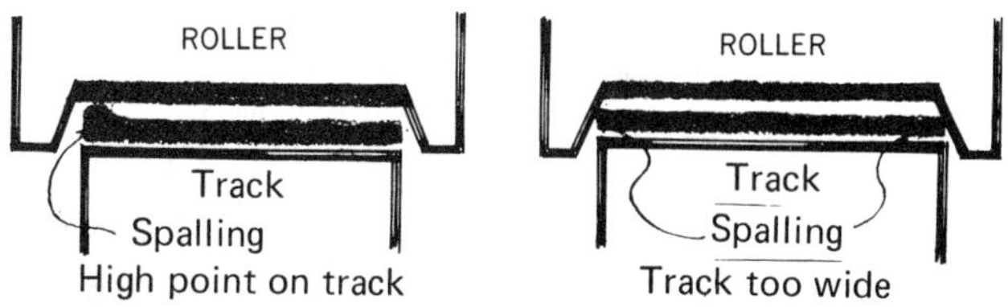

Fig. 13-51. Conditions that promote spalling. High points on a rough deposit or a bead higher than the rest of the deposit (a) absorb a disproportionate amount of load and may thus tend to break away. Mismatched mating surfaces or improper alignment (b) produce load concentrations that promote wear or fracture in the coating.

These zones tend to crack in service and thus lead to spalling. Proper preheating is the surest way to avoid this problem.

Limit Deposit Thickness: Applying hardsurfacing in thick deposits builds up shrinkage stresses that promote spalling. Do not use more hardsurfacing layers than specified for each type of deposit. If thicker welding is required, rebuild before hardsurfacing. Peen each layer of thick build-up to relieve stresses.

Other causes of spalling are presented in Fig. 13-51.

Avoiding Underbead Cracking

Underbead cracks are small cracks that may occur in the heat-affected zone of the base metal under the weld. The cracks do not usually show on the surface but can promote spalling or cracking in the deposit.

The tendency for underbead cracking depends primarily upon the carbon and alloy content of the base metal. It is independent of the type of weld deposit when low-hydrogen processes or electrodes are used. Electrodes that are not the low-hydrogen type may require 100 to 300°F higher preheat than other build-up or hardsurfacing electrodes. However, welding with these electrodes on hot build-up layers usually avoids problems. All build-up and hardsurfacing should be completed without long delays to keep the part hot.

The easiest way to prevent underbead cracking is to slow the cooling rate by preheating. Always be sure the part is up to room temperature (70-100°F) before welding or use higher preheats if specified for the particular base metal.

Recommended Preheats

Low-Carbon Steel: Materials with carbon contents of approximately 0.10 to 0.30% are only slightly hardenable, thus preheating above room temperature is rarely needed. However, heavy, massive, or rigid parts containing more than 0.20% carbon should be preheated to 100 to 300°F.

Medium-Carbon Steel: Materials of 0.30 to 0.45% carbon are moderately hardenable, especially in large parts and heavy sections. Preheats to 300-500°F are recommended for higher carbon contents and for large, rigid, or complex parts.

High-Carbon Steel: Materials of approximately 0.45 to 0.80% carbon are highly hardenable and crack-sensitive in all sizes and shapes. Preheat to 500-800°F, using the higher temperatures for the higher carbon contents and for large, rigid, or complex shapes. When carbon content is near 0.80%, deposit a "buttering" layer with E7018 electrode (or with a mild-steel submerged-arc flux and electrode) prior to depositing build-up or hardsurfacing layers. The buttering layer minimizes the danger of underbead cracking and provides a good bond between base metal and hardsurfacing deposits.

Low-Alloy Steel: These steels vary from medium hardenable to highly hardenable, depending upon carbon and alloy content. Preheat to 100-500°F, using the higher temperatures for higher carbon and alloy contents and for large, rigid, or complex shapes. Preheat temperatures up to 800°F or a buttering layer may be required if the carbon content is over 0.35%.

Manganese Steel: Steels containing 12 to 14% manganese are not particularly hardenable or crack-sensitive. Preheat is not often required. But if parts are massive or highly rigid, preheat to 100-200°F. Prolonged heating over 500°F can cause embrittlement, so avoid high localized heating. On small parts, avoid high localized heating by using a skip welding technique.

Cast Iron: This material is extremely crack sensitive. The heat-affected zone may be full of cracks even with preheat temperatures of 1200-1400°F. Therefore, hardsurfacing cast iron is generally uneconomical. If it must be welded, follow standard cast-iron welding precautions.

Avoiding Stress Failure

Some parts contain high retained internal stresses. When welding stresses are added to these retained stresses, the part can fracture.

Several typical failures are shown in Fig. 13-52. Such failure can occur near the weld or at any weak point in the part, and is most commonly encountered when hardsurfacing:

- Highly rigid parts. (Massive parts and complex shapes are inherently rigid.)
- Shrink-fit parts.
- Some large castings, particularly when they are made of medium to high-carbon steel or medium-carbon low-alloy steels.
- Parts hardened by heat treatment.

The danger of stress failure can be minimized by:

- Preheating slowly to the high side of the temperature range specified for the particular type of base metal. This preheat should be as uniform as possible. Shrink-fit parts should be preheated

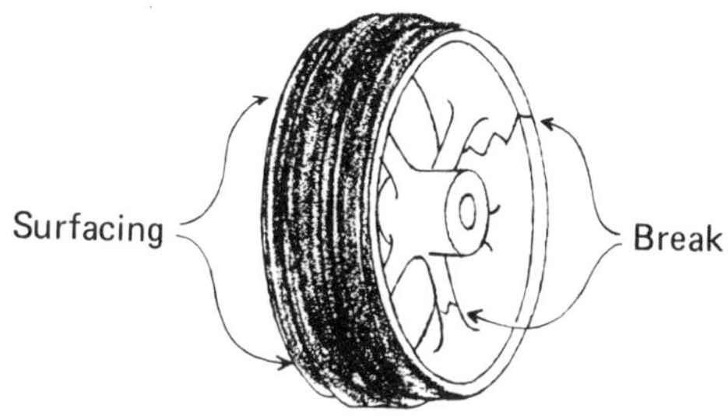

(a)

Spoked tractor idler — complex shape is inherently rigid. Preheat to prevent stress failure at any weak point.

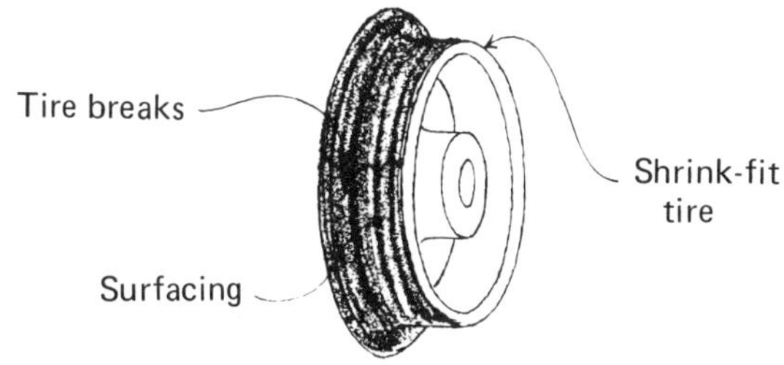

(b)

Mine locomotive wheel — shrink-fit tire is highly prestressed. Expand by preheating to prevent stress failure of the tire.

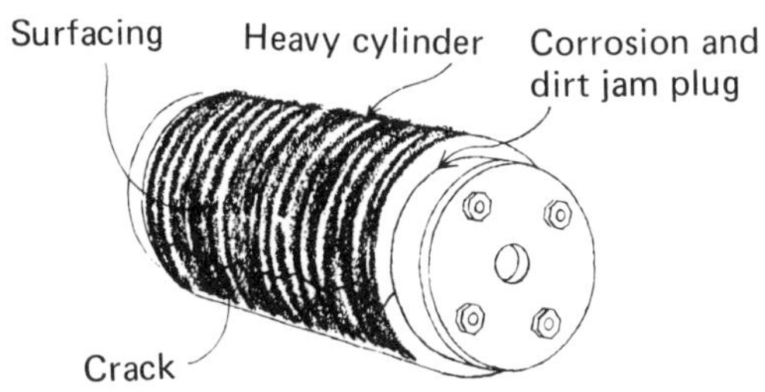

(c)

Crusher roller — massive size and jammed end pieces cause extreme rigidity. Loosen through bolts and expand the roll with preheat to loosen plugs.

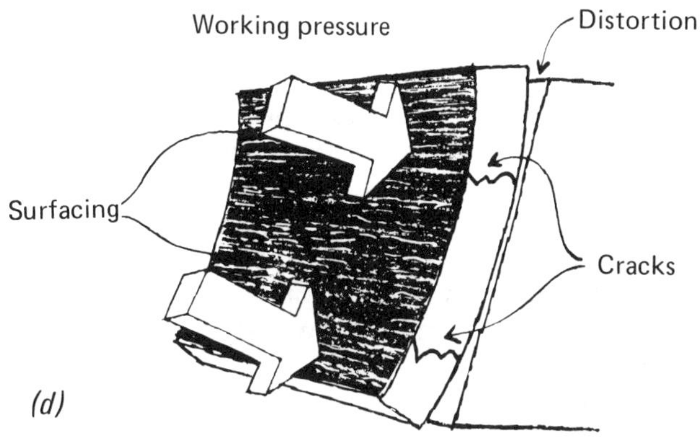

(d)

Jaw crusher — if distortion leaves crusher plate unsupported at the ends, working pressure causes bending stresses that can lead to failure.

Fig. 13-52. Typical part failures encountered with hardsurfacing. Complex shapes (a) frequently have weak sections easily fractured by thermal stresses. Special precautions are advised in shrink-fit parts, (b) Cracks can be avoided in crusher rolls (c) by loosening bolts and expanding the roll by preheat. Distortion may leave working surfaces unsupported (d) so that service stresses produce cracks.

until they are loose. (Manganese steel requires a lower temperature because it has a higher coefficient of expansion.)

- Controlled-atmosphere annealing and slow cooling to the specified preheat temperature parts that have been hardened by heat treatment.
- Arranging the welding schedule so it can be completed without long interruptions.
- Cooling the part slowly and uniformly. This can be done by covering the part with a blanket of glass or some other insulating material, or by cooling in a furnace.

Avoiding Weld Cracking

Transverse or Cross Cracking: Hardsurfacing deposits for resisting severe abrasion commonly cross-crack. This cracking is beneficial because it relieves stresses that could otherwise cause spalling or distortion. The cracking does not impair the wear resistance of the deposit, but it can be minimized by preheating to 1200°F. In other types of deposits, cross cracking can be detrimental. It generally occurs in parts that are massive, rigid, or of complex shape. The preheating used to prevent underbead or stress-failure cracking will also minimize cross cracking.

Longitudinal or Center-Line Cracking: This type of cracking is associated with poor bead shape and is caused by making the bead too flat. Several measures help avoid this problem:

- Use a stringer-bead or minimum weaving technique and low current.
- Be sure fillet welds are slightly convex.
- Use submerged-arc surfacing with the weave-bead technique. (Center-line cracking does not often occur with submerged-arc because the high heat input of the process assures sufficiently slow cooling.)
- With submerged-arc hardsurfacing, decrease the stepover (or increase the bead overlap) enough

to remelt the center of the previous bead.

- When submerged-arc surfacing on roundabouts, set the correct electrode displacement from the center line.

Avoiding Distortion

Distortion in welding is caused principally by the unbalanced stresses that result from expansion of the metal during heating and from contraction during cooling. These stresses and their effects are uneven both in strength and direction. They depend on many factors.

In many cases the amount of workpiece distortion resulting from hardsurfacing is negligible. Consequently, clamping the part in position for welding is the only precaution needed. In some cases the part is hardsurfaced oversize to allow for distortion. It is then machined or ground to size. This method is most often used on parts that must be machined or ground to finish after surfacing. Sometimes the part is allowed to distort and then straightened while still hot enough to bend without cracking.

Where distortion forces must be controlled, the following methods may be used:

- Preforming can be employed with flat pieces and with thin or simply shaped parts. Here the part is bent, formed, or clamped with the proper preset before welding. The distortion forces then pull the part back to its original shape, Fig. 3-7.
- Stresses may be balanced in parts that cannot be straightened after welding (where the increased rigidity does not cause underbead cracking). In this method, two similar parts are welded or clamped back to back. Welding is then alternated back and forth from one part to the other. Or the parts may be welded or clamped to a strongback, fixture, or platen, Fig. 3-7.
- Limit the temperature of the part, particularly when high rigidity could cause cracking. This can be done by a number of methods. One is to distribute the heat evenly by first welding one area, then welding a different area as the first cools. Another is to reduce the heat input by using a procedure with low current. Yet another is to remove some of the heat by blowing air over the part or by circulating water through copper jacket clamped to the surface, Sec. 3.1.
- Relieve stresses, especially with large parts, rigid shapes, shrink-fitted parts, and other highly stressed pieces. This can be done by peening each layer during cooling, a practice most effective with deposits designed for impact or abrasion and impact. Another way to relieve stress is to preheat the entire part slowly and uniformly, Sec. 3.1.
- Use a cross-cracking deposit, namely those designed for resistance to severe abrasion. The cracks minimize distortion by relieving stresses.

HARDSURFACING WITH ALLOY FLUXES

As mentioned previously, alloy fluxes can be used with mild-steel or alloy electrodes when hardsurfacing with the submerged-arc process to give the required alloy composition in the weld deposit. By combining various alloy fluxes with various types of electrodes and changing procedure variables to change the melt-off rates and penetration, a wide range of properties can be developed in the weld metal. The use of an alloy flux gives the most flexible and least expensive method of controlling weld chemistry. In many instances, a mild-steel electrode with an appropriate alloy flux will give the desired hardsurfacing properties at the minimal cost.

Such alloy fluxes are designated as "agglomerated" alloy fluxes. Agglomeration is the process of mixing properly formulated, finely grounded particles together with a binder and drying to give a homogenous mass. The properly sized grains of flux produced from this mass have identical chemical constituency; the alloy particles cannot settle out or separate from the individual grains during shipping or use. Agglomerated alloy fluxes for hardsurfacing are produced in low-alloy, medium-alloy, medium-high alloy, and high-alloy types.

When using these fluxes with mild-steel electrodes, carbon and alloy content of the deposit is increased as:

- The welding current is decreased.
- The arc voltage is increased.
- The carbon or alloying content of the base metal is increased.
- The electrode size is increased.
- Welding speed is decreased.
- The bead overlap is increased.
- The number of layers up to and including three is increased.

Low-Alloy Hardsurfacing Flux

With mild-steel electrode, a typical low-alloy type of hardsurfacing flux produces a machinable deposit of fair to good abrasion resistance and excellent impact resistance. The deposit, depending on the procedure, can be machined either with carbide or high-speed tools. It will produce a dense deposit less sensitive to cracking than a high-carbon deposit of the same abrasion resistance. A typical two-layer deposit made with this flux and 0.12%C mild-steel electrode will have 1.5-2.5 times the abrasion resistance of a machinable hot-rolled medium-carbon steel and up to twice the abrasion resistance of a deposit made with SAE 6150 wire and mild-steel flux.

This flux with mild-steel wire is ideal for both build-up and hardsurfacing. Different deposits can be obtained by varying the arc voltage. Machinable deposits or built-up deposits, as required on tractor rollers and idlers, are easily obtainable at the low-voltage setting (28V). After the work is built up to within two layers of gage, a hardsurfacing deposit of required abrasion resistance and hardness can be made simply by increasing the arc voltage to 32 volts. Thus, it is possible to build-up and hardsurface without flux or wire change. The deposit is magnetic at room temperature and has a consistent hardness. The hardness, depending on the procedure, will range between 33 and 57 Rockwell C. This martensitic-type deposit is a low-carbon alloy steel, which may be annealed or which may be quenched from above the critical temperature to harden.

The weld metal is typically an alloy of carbon, manganese, silicon, chromium, molybdenum, and vanadium.

The flux and mild-steel wire may be used wherever a smooth, dense, crack-free, machinable deposit is required. Typical applications include rolls used in processing steel, kiln rollers, crane wheels, rams, and rollers and idlers on tread-type tractors.

Medium-Alloy Hardsurfacing Flux

The deposit made with mild-steel wire and medium-alloy flux is useful wherever smoothness and minimum porosity are the essential requirements in the particular alloy range. A minimum of two layers is recommended for good abrasion resistance. The deposit is a good selection for applications wherever metal-to-metal wear exists. A typical deposit is magnetic at room temperature, and, with recommended procedures, has a consistent hardness of 44-53 Rockwell C. It is heat-treatable with normal heat-treating practices.

The impact strength may be increased by reheating a multilayer deposit from room temperature to some predetermined temperature up to 1050°F and holding at that temperature 1 hour for each inch of thickness without sacrificing hardness or abrasion resistance at room temperature.

For all practical purposes, the hardness, abrasion resistance, and deposit analysis change only very slightly after the third layer. A typical weld deposit is an alloy of carbon, manganese, silicon, chromium, molybdenum, and vanadium.

This flux and 0.12%C mild-steel electrode are recommended for:

(a) Deposits to be forged, such as plow share blanks.

(b) Metal-to-metal wear, such as tractor rollers, idlers, and extruder rams.

(c) Smooth deposits without porosity and underbead cracking, such as steel-mill roughing rolls that do not require machining.

Medium-High-Alloy Hardsurfacing Flux

The deposit made with 0.12%C mild-steel electrode and medium-high-alloy flux should be used where a single-layer deposit of high abrasion resistance and good impact strength is required. Base metal should not contain over 0.60% carbon. This flux is recommended for fabricated parts, such as scraper blades, and to produce multiple-layer deposits requiring high abrasion resistance, such as required on crusher rolls, whether originally made from carbon steels, manganese steels, or cast iron.

A typical deposit is magnetic, semiaustenitic at room temperature, and has, with recommended procedures, a consistent hardness. A broad range of 36 to 62 Rockwell C can be developed by controlling the weld metal analysis and the cooling rate. It is dense and forgeable, but is nonmachinable and can be finished to size only by grinding. The deposit is heat-treatable by following normal heat-treating practices.

As with medium-alloy flux, the impact strength may be increased by reheating the multilayer deposit from room temperature to some predetermined temperature up to 1050°F and holding at that temperature 1 hour for each inch of thickness without sacrificing hardness or abrasion resistance at room temperature. The hardness, abrasion resistance, and deposit analysis change only very slightly after the third layer.

The deposit is typically an alloy of carbon, manganese, silicon, chromium, and molybdenum.

High-Alloy Hardsurfacing Flux

When used with a mild-steel electrode, a typical

high-alloy flux produces a high-carbon, high-alloy weld deposit. It is recommended for fabrication and maintenance of wearing parts where the service involves severe abrasion under medium impact or high compression loads. It may be used for abrasive service at temperatures up to 1100°F.

The microstructure of the deposit is composed of complex chromium carbides and some secondary carbides in a chromium-iron matrix. Deposit will have a hardness of 53 to 61 Rockwell C, cannot be torch-cut or forged, and retains hardness and resistance to scaling at high temperatures. It will not respond to heat treatment; resists galling and seizing, but will cross-crack on cooling, thereby minimizing locked-up stresses and warpage. It will retain a high polished surface when subjected to metal friction or earth scouring. Hardness and abrasion resistance are unaffected by cooling rates normally encountered in welding.

High-alloy flux and mild-steel wire are used for hardsurfacing crusher rolls and rings of various types used in the coal, rock, cement, and clay industries; mixer parts used in the rubber industry; lining the inside of catalyst towers and piping; lining baffle plates in refinery equipment; lining the inside of ball mills; and in many similar applications in the extractive and processing industries.

Single Submerged-Arc Hardsurfacing Procedures

Most hardsurfacing submerged-arc fluxes are designed to be used with mild steel electrodes. The alloying elements are in the flux, and when the flux is melted the alloys blend with the molten electrode and base metal to produce an alloy deposit.

The recommended power source is DC variable voltage, although DC constant voltage or AC variable voltage can be used.

Almost any automatic or semiautomatic wire feeder can be used, but it should be equipped with good meters.

DC electrode negative (straight polarity) is recommended for the best deposition rate and less penetration into the base metal. AC or DC electrode positive (reverse polarity) can be used, but the deposition rate and buildup are lower and the alloy content of the deposit tends to be slightly higher.

The electrode extension (electrical stickout) should be about 1 in. with a flux depth just enough to prevent flash-through of the arc.

Select a travel speed that produces weld beads that are sound, well shaped, and of the desired width and thickness. With a single arc and stringer bead technique, the travel speed should be within the range listed below:

Flux	Electrode	Travel Speed, ipm
Low to medium-high-alloy	5/64 through 3/16	15-40
High-alloy	3/32 through 5/32	25-35

As discussed earlier, the alloy recovery in the weld deposit can be varied by changing the welding procedure. Once a satisfactory weld deposit has been obtained with a certain procedure, it is necessary to follow strictly to that procedure in order to get consistent results. Accurate meters are necessary to measure voltage and current so the procedures can be reproduced consistently.

Typical procedure guides are given in Fig. 13-53 for high-alloy recovery and low-alloy recovery. Each line of a guide represents a constant alloy recovery for a given electrode size. Further, all lines in the guide provide the same constant alloy recovery. For a small increase in alloy recovery, select a procedure above the line. For a small decrease in alloy recovery, select a procedure below the line. For a large change in alloy content of the deposit, use a different flux or a different procedure guide.

Twin-Electrode and Oscillating Submerged-Arc Hardsurfacing Procedures

Many flat parts are economically hardsurfaced with single submerged-arc stringer beads. However, wider and more economical beads are possible with automatic heads equipped to feed twin electrodes and which have mechanical attachments to oscillate the arc to produce a wide bead.

Automatic heads can be equipped to feed two electrodes (5/64, 3/32, 1/8 in.) through one set of drive rolls into one weld puddle. For the same penetration into the base metal, higher welding current can be used on twin electrodes than on a single large electrode. This can increase the deposition rate as much as 50%.

Twin electrodes used on an oscillating welding head produce smooth beads up to 4 in. wide with a minimum of admixture. Typical procedures are given in Table 13-33 and the welding is illustrated in Fig. 13-54.

Adjust oscillation speed as needed to improve bead appearance. Decrease oscillation speed to increase buildup at the center of the bead. Increase

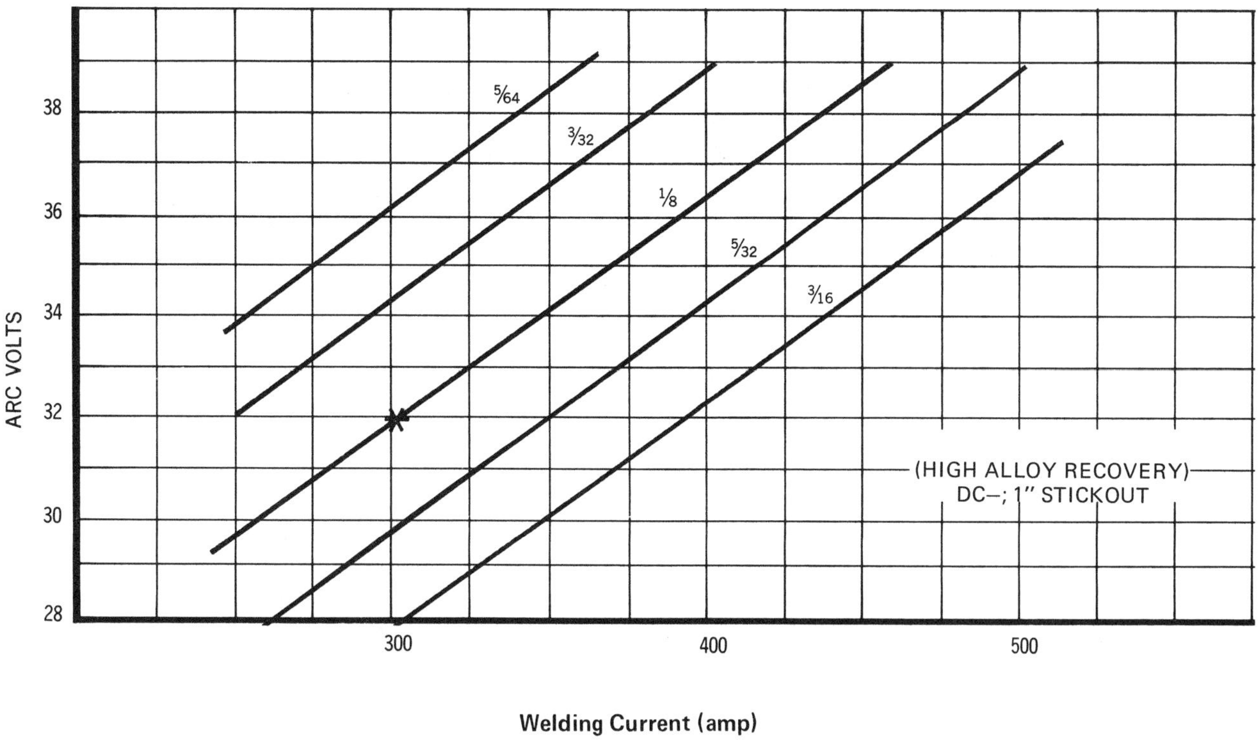

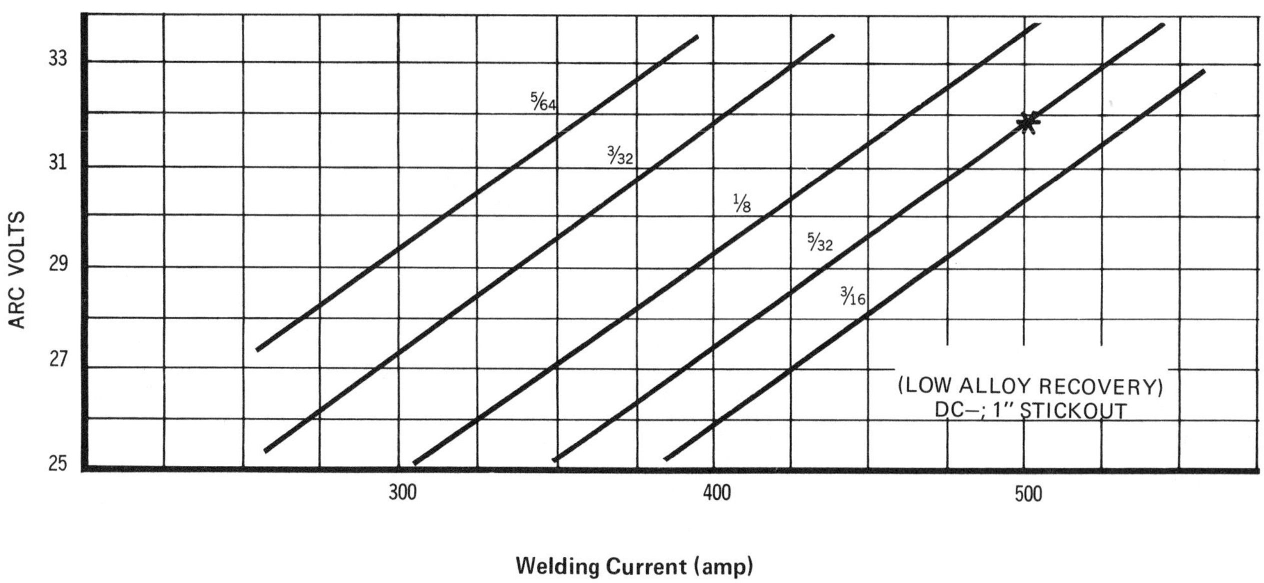

Fig. 13-53. Typical procedure guides for high and low-alloy recovery.

TABLE 13-33. Typical Procedures with Twin-Electrode Welding Head

Flux	Bead Size Width in.	Bead Size Thickness† in.	Wire Spacing in.	*Oscillations per min.	Current DC– amp	Volts	Travel Speed ipm	Min. Plate Thickness in.
Low-Alloy	2	1/8 – 5/32	5/8	30	700	32	8 – 9	5/16
Medium-High-Alloy	4	1/8 – 3/16	5/8	15	800	35	5	3/8
High-Alloy	3	3/16 – 1/4	3/8	20	700 – 750	33	4.5	3/8

* One oscillation is a complete cycle over and back.

† Thickness is for one layer. Use one layer only for high-alloy flux. Two layers can be used for medium-high-alloy flux. More than two layers can be used for low-alloy flux.

DC positive may also be used but the buildup will be less.

Electrode is 1/8-in. mild steel. (See Table 4-11, Classification EL12)

oscillation speed to eliminate rough or scalloped edges. Always be sure the work is level.

Oscillating with a single electrode limits bead width to 3 inches. The bead tends to be rougher and has much greater admixture. With a single electrode, make two complete oscillations before starting forward travel.

Circumferential Submerged-Arc Hardsurfacing Procedures

In circumferential hardsurfacing, 3-in.-diameter and larger cylindrical objects are rotated under the welding head. The welds differ from those made in the flat position in that the flux and molten metal tend to sag or spill off the work.

To prevent spillage or bead distortion, the weld must solidify as it passes the vertical center of the part. (See Fig. 13-55).

Recommended circumferential hardsurfacing voltage, current, and off-center distance ranges are given in the tables below. Speed also must be carefully controlled to make the small beads of proper shape. For a thorough description of circumferential submerged-arc welding, refer to Section 6.3.

Fig. 13-54. Typical hardsurfacing weld bead made with submerged-arc alloy flux, twin electrodes, and oscillating the arc.

Welding Current and Voltage

Diameter (in.)	Current (amp)
3 – 6	250 – 350*
6 – 12	300 – 400
12 – 18	350 – 500
over 18	Standard hardsurfacing procedures (single electrode or twin electrodes).

* "Threading" may be necessary.
Normal Voltage Range is 28 – 32 volts.

Approximate Displacement (d) of the Electrode

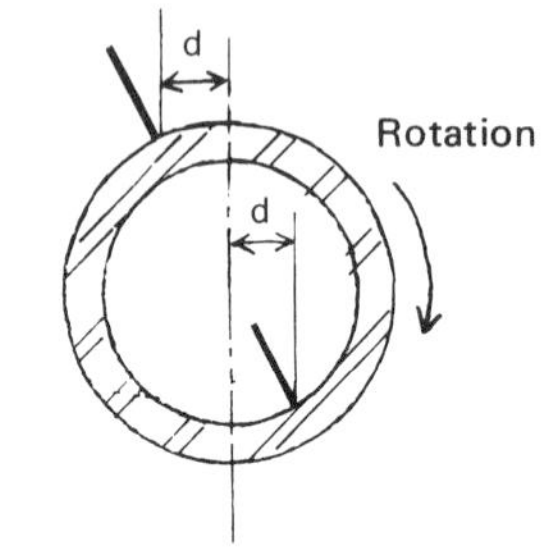

Girth Diameter (in.)	Electrode Position "d" (In. ahead of vertical center)
3 – 18	3/4 – 1
18 – 36	1-1/4 – 1-1/2
36 – 48	1-1/2 – 2
48 – 73	2 – 2-1/2
over	3

Threading: The temperature of the work should be kept below 700°F for easy slag removal and control of spilling. In addition to depositing small beads and using air jets or water cooling (when practical), temperature can be controlled by depositing a stringer bead like a screw thread along the work. Fill the space between beads with successive welds. (See Fig. 13-56).

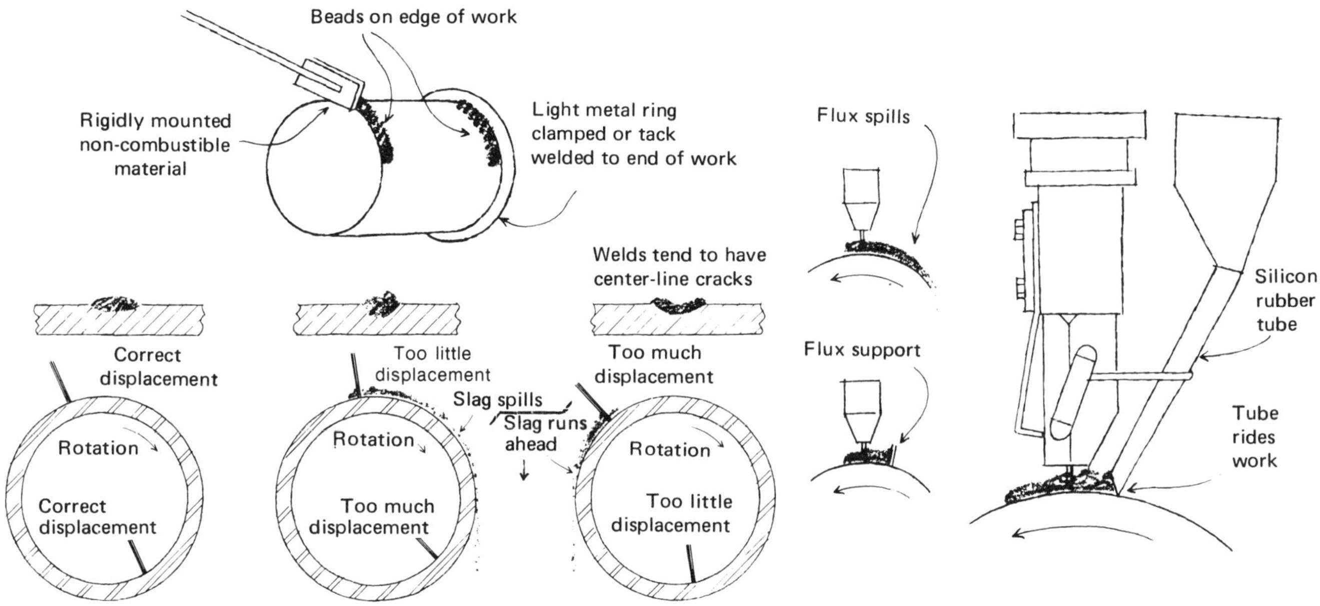

Fig. 13-55. Correct and incorrect methods to prevent spillage and bead distortion.

Overlap: The amount one bead overlaps the adjacent bead affects both admixture of base metal into weld metal — greater overlap reduces admixture — and appearance of the finished weld. Control overlap by adjusting the amount of longitudinal travel with each revolution.

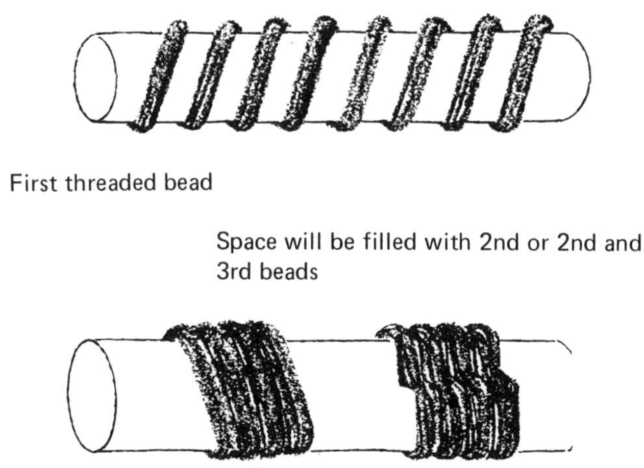

Fig. 13-56. The threading, spiral, and stepover methods of circumferential hardsurfacing.

Longitudinal travel is accomplished either by spiraling the bead or indexing the welding head across the work after each complete revolution (stepover). Unless a lathe with a slow screw feed mechanism is available, the stepover method is recommended because of ease of operation.

For automatic stepover, mount a limit switch that is operated by a cam-type trip on the rotating fixture. Connect the limit switch into the travel carriage motor circuit so the motor runs when the switch is operated. The distance moved is controlled by the size of the cam and speed of the travel motor. A time delay can be used in place of the cam.

Slag must be removed before each bead makes a complete revolution. If slag removal is a problem, the threading technique should be considered.

TESTS FOR ABRASION AND IMPACT RESISTANCE

There are no standardized methods for testing abrasion or impact resistance because of the many different types of wear and service environments. There are, however, various tests that are commonly applied to give relative measures of these properties.

One type of test drags the specimen over a copper slab in a mixture of quartz sand and water; the abrasion factor is the loss of weight of the specimen compared to the loss of weight of a standard annealed SAE 1020 steel specimen. A material that loses the same weight as the SAE 1020 steel will have an abrasion factor of 1.00. The ideal material would not lose any weight and would have a factor of 0.00.

Another test presses the specimen against a notched rubber wheel, which is rotated in a slurry of sand and water; the loss of weight is taken as the abrasion value.

A preferred method is the use of a grinding wheel of aluminum oxide as the abradant. The amount of wheel that is ground away to remove a standard weight of the hardsurfacing material is taken as the abrasion index. This test is excellent for evaluating the cutting ability of hardsurfacing materials used to dig earth. The loss of weight of the specimen when run against the wheel at standard conditions is taken as the abrasion value for the material when it is to be used for wear-resistant purposes. It is important to have both values, because the material that shows the least loss of weight may not be the one that cuts the best on, for example, a rotary well-drilling bit.

The grinding wheel test is preferred to the wet-sand test because the wet sand results do not show as much difference in hardsurfacing materials as is usually found in the field. The grinding wheel method gives a very large difference in weight loss: 0.02 gram loss for a hard grade of sintered tungsten carbide and about 10.0 grams loss for mild steel — a ratio of 1 to 500.

Abrasion values are not directly applicable to field service. Some of the variables that can change the order of abrasion resistance of hardsurfacing are temperature, size, properties of the abradant, and the amount of impact.

Impact tests are commonly made on commercial hardsurfacing materials by pounding weld deposits against each other and against standards. The materials are then arranged according to their resistance to cracking, chipping, or spalling.

Automatic hardsurfacing using submerged-arc alloy flux and twin electrodes.

Thawing Frozen Water Pipes

HOW AN ARC WELDER CAN BE USED

While an arc welder is not specifically designed for this purpose, it can be used for thawing frozen water pipes. It is a good source of low voltage electric current. This low voltage current is a fast and low-cost method of thawing pipes without digging or the need for an open flame.

Under no circumstances should any pipe thawing activity take place without a knowledgeable person in attendance during the entire time. This is true regardless of the number of times a particular joint may have been thawed because conditions change unknowingly. The possibilities of water damage due to melted solder in pipe joints or fire damage due to failure to follow electrical connection instructions can be minimized by proper supervision.

How it Works

When electric current passes through a pipe, wire, or any conductor resistance is encountered. This generates heat. The heat thaws the frozen water. Here are the factors which effect the time required for thawing.

Pipe Material

A material which is a good conductor of electricity has low resistance and vice versa. Steel is a poorer conductor than copper. Therefore, steel has higher resistance and heats more rapidly than copper when a given amount of electric current passes through it.

Size of Pipe

When an electrical cord is overloaded it gets warm. The same principle applies to thawing pipe. A small diameter pipe will heat more rapidly than a large diameter pipe when a given amount of current is passed through it. Thus, it takes higher currents and longer times to thaw large diameter pipes. When using DC current, the time and current required for thawing is about the same for any length of pipe.

Current and Time Required

The amount of thawing time is most dependent upon the amount of electrical current available. The greater the current passed through a pipe the greater the heat generated. With twice as many amperes the heating rate is four times as fast. If the number of amperes is increased three times, the heating rate is nine times as fast.

The time required for thawing different sizes of steel pipe using different currents are given in the following table. Copper pipe requires more current than steel pipe for the same thawing time. To thaw 1/2-inch copper pipe increase the figures for steel pipe by 10%. Increase the current for steel pipe by 25% to thaw 3/4-inch copper pipe. Larger copper pipes require an even larger percentage increase of current.

TABLE 13-34. Amperage* and Time for Thawing Steel Pipe — Pipe Size (in.)

1/2	3/4	1	1-1/4	1-1/2	2	Approx. time (min.)
200	270	400				3 – 4
150	200	300	400			6 – 8
125	170	250	340	400		8 – 12
100	135	200	270	320	440	12 – 16
75	100	150	200	240	330	25 – 30
50	67	100	135	160	220	60

* These figures are actual meter read currents required. The current indicated by the welder current control dial is less than actual thawing current being used.

EQUIPMENT REQUIRED

Large gasoline or diesel engine driven welders (200 amp and larger) are commonly used for thawing because they will produce the high currents required for fast thawing and can be used where there is no electric power available. Most industrial type welders rated 40 volts and 60% duty cycle are suitable for pipe thawing. Also some of the smaller welders rated 25 volts and 30% duty cycle can be used with competent supervision. For specific recommendations for pipe thawing consult the instruction manual for the machine or write to the manufacturer for instructions.

DO NOT USE A CONSTANT VOLTAGE WELDER FOR THAWING PIPE. A constant voltage welder has no control for limiting the current.

Some manufacturers furnish kits as an accessory that is connected to the welder when thawing pipe. These kits, Fig. 13-53, contain an ammeter which indicates the output current, and a fuse to protect the welder. The electrode and ground cables from the welder are connected to one side of the meter-fuse kit and cables connected to the pipe are attached to the other side of the kit. The proper fuse size is important so when ordering a kit always specify the make, model number, and the capacity of the welder.

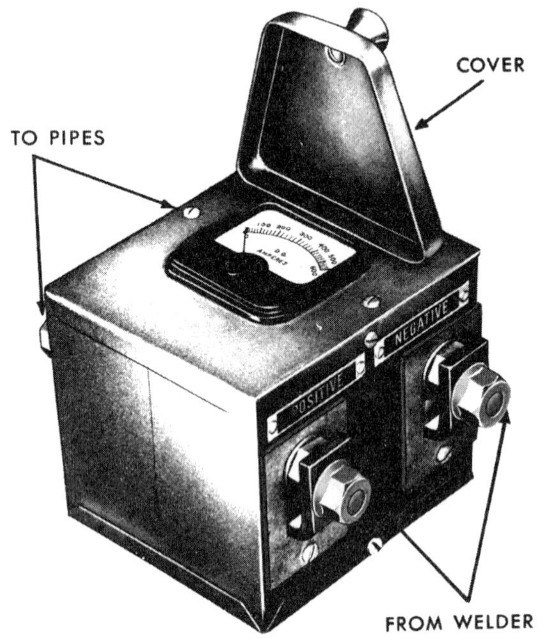

Fig. 13-53. Meter-fuse kit used with a DC welder when thawing frozen water pipe.

PIPE THAWING PROCEDURES

1. Isolate the frozen section of pipe. This can be easily done by turning various faucets on until the specific frozen areas are determined.
2. Be sure the welder or the polarity switch is turned off. Failure to do this may cause arcing when the cables are connected and result in a hole in the pipe.
3. Connect a cable to one end of the pipe. Connect the other cable to the other end. **Be sure this connection makes a complete electrical circuit** from the welder through the frozen section of the pipe, and back to the welder. Sample connections are shown in Fig. 13-54.
4. The connection to the pipe must be good. Clean the pipe, preferably by sanding with coarse emery paper before connecting. Bad connections may cause arcing and result in a hole in the pipe. Poor connections will overheat resulting in loss of thawing current.
5. The cables must be large enough to handle the current and should be kept as short as possible. Recommended cable sizes are shown in Table 13-35.
6. Connect the cables as close to the frozen area as possible. A small empty pipe in the circuit could get hot before a larger frozen pipe thaws. This could melt solder in copper pipes or, in an extreme case, **CAUSE A FIRE.**
7. Always be certain that there are no insulated joints or plastic pipe sections that could prevent

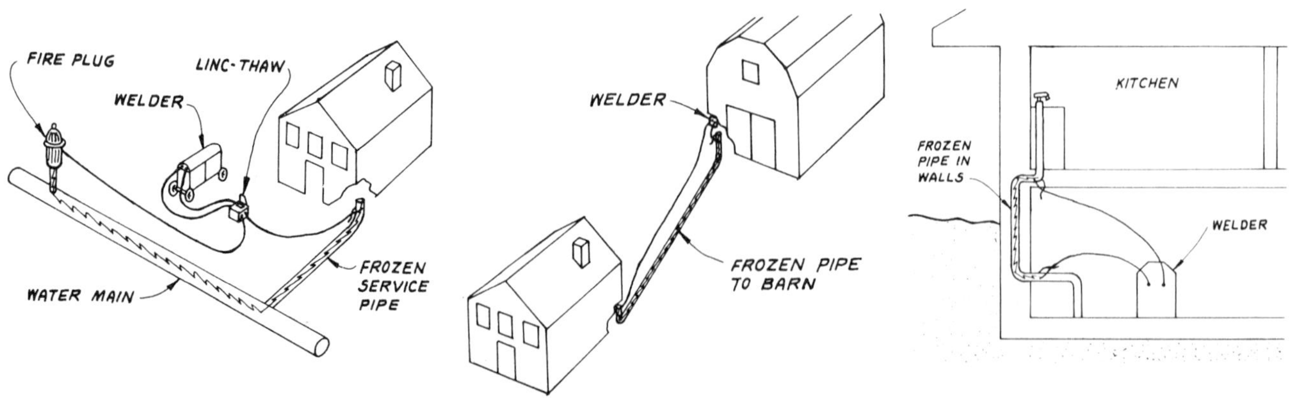

Fig. 13-54. Typical cable connections for thawing frozen water pipe.

the electric current from passing through the frozen pipe between the two cable connection locations. Such interruptions in the electrical circuit could force all of the current to find alternate circuits through electrical grounds connected to the pipe and possibly **CAUSE A FIRE.**

8. Confine all pipe thawing activity within the property boundaries and to fire hydrants or water mains in the immediate vicinity of the property. **DO NOT** attempt to thaw pipes by making connections to neighboring houses.
9. Be sure the valves or faucets at both ends of the line being thawed are open. When the heat melts a thin layer of water next to the pipe wall, the water starts to flow. A small flow of water will help to melt the rest of the ice more rapidly.
10. When the connections are complete, adjust the output current to a value shown in Table 13-34 but do not exceed the capacity of the welder. An ammeter in the output circuit is a great help to prevent overloading.

There are many different types of welders and equally as many types of circuits to adjust output current. Therefore, there are no safe general instructions for all types of welders. Refer to the welder instruction manual for specific directions for adjusting the output current when thawing water pipes. If the instruction manual does not have these directions, request them from the welder manufacturer giving the data on the welder nameplate.

11. Turn the welder on. When the water starts flowing freely, stop the welder.
12. Do not overload the welder. Pipe thawing involves heavier sustained loads than welding. And because pipe thawing establishes close to a short across the welder terminals, actual amperes drawn is always greater than indicated by the current control dial. Therefore, specific pipe thawing instructions for each type machine must be followed. Also, since the welder may overheat if it is overloaded, leaving the machine in a cold place or outside will help to keep it cool.

TABLE 13-35. Recommended Cable Sizes

Amps	Distance in feet from welding Machine to Pipe Connection											
	50	75	100	125	150	175	200	225	250	300	350	400
100	2	2	2	2	1	1/0	1/0	2/0	2/0	3/0	4/0	4/C
150	2	2	1	1/0	2/0	3/0	3/0	4/0	4/0	4/0	2-2/0	2-3/0
200	2	1	1/0	2/0	3/0	4/0	4/0	4/0	2-2/0	2-3/0	2-3/0	2-4/0
250	2	1/0	2/0	3/0	4/0	4/0	2-2/0	2-2/0	2-3/0	2-3/0		
300	1/0	2/0	3/0	4/0	4/0	2-2/0	2-3/0	2-3/0	2-4/0			
350	1/0	2/0	4/0	4/0	2-2/0	2-3/0	2-3/0					
400	2/0	3/0	4/0	2-2/0	2-3/0	2-3/0						

Dragline bucket with weld deposited wear-resistant overlays on surfaces subjected to high wear rate.

Section 14

INSTALLATION AND MAINTENANCE

SECTION 14.1

INSTALLING, MAINTAINING, AND TROUBLESHOOTING EQUIPMENT

Page

Installing, Maintaining, and Troubleshooting Equipment

A welding machine and the accessory equipment used with it are tools required to accomplish a productive goal — and just as the tools of a carpenter or a machinist, they must be used properly and be properly maintained. In the small shop, the user is usually the owner of the equipment and, therefore, is likely to use and maintain it properly. The employed weldor in a large shop, however, seldom has responsibility for equipment maintenance — so it behooves management to set up a system for assuring its care and maintenance. Good welding, the best production rates, and highest efficiencies in operation result when equipment is installed, used, and maintained properly.

INSTALLATION OF WELDING EQUIPMENT

Good welding begins with properly planned and executed installation. Equipment should be installed in uncluttered areas where there is room for handling the materials to be welded without bumping into columns, walls, or adjacent machinery. The area should be reasonably clean, free of litter and regular housekeeping practices should be followed. There should be provisions for a continuous supply of clean air for ventilation. Welding areas should be separated from areas where the production operations generate excessive moisture, dust, or corrosive or flammable fumes. The area temperature should not be high, since the heat of welding would increase the temperature. Equipment installed outdoors should be provided with protection against rain, dew, or snow. The following points should be kept in mind when making a welding installation:

1. Be sure an adequate supply of the right type of electrical power is available. Consult the power company if there is any question.
2. Follow the manufacturer's installation instructions as to input and ground-wire size and fuse size. These will conform to the National Electrical Code. In lieu of such instructions, wire and fuse sizes meeting the National Electrical Code for various types of 60-cycle motor generator sets and transformer welders can be found in Tables 14-1, 14-2. Output and input ratings are taken from the name plate.
3. Provide adequate and even support for the welding equipment.
4. Be sure that there are safeguards from mechanical accidents, such as the striking of equipment by materials being moved by cranes or loaders.
5. Atmospheric conditions should be proper — no corrosive or flammable fumes, no excessive moisture (such as from steam or water spray), no excessive heat, no corrosive or abrasive dust. Adequate fresh air should be available for ventilation and cooling.
6. The frame of the welding machine should be solidly grounded electrically.
7. All electrical connections to and on the welding equipment should be checked for tightness.
8. Be sure the welding cables are of sufficient capacity to handle the current and are well insulated (see Section 4.3).
9. Follow closely the installation instructions that come with the machine.

OPERATING AND MAINTAINING EQUIPMENT

Maximum trouble-free service is obtained from welding equipment with the observance of several precautions and practices. The machine should be kept clean and cool, should not be abused, and should be maintained regularly. **Generally, the manufacturer supplies special instructions on maintenance that are pertinent to the particular machine; these should be followed closely.**

Keeping The Machine Clean and Cool

Because of the large volume of air pulled through welders by the fans to keep the machines cool, the greatest enemies of continuous, efficient performance are air-borne dust and abrasive materials. Where machines are subjected to ordinary dust, they should be blown out at least once a week with dry, clean compressed air at a pressure not over 30 psi. Higher pressures may damage windings.

In foundries or machine shops, where cast-iron or steel dust is present, use vacuum cleaning rather than compressed air. Compressed air at high pressure tends to drive the abrasive dust into the windings. If vacuum-cleaning equipment is not available, compressed air may be used at low pressure. Abrasive material in the atmosphere causes excessive wear on generator commutator and brushes.

Greasy dirt or lint-laden dust quickly clogs air passages between coils on rotating equipment and transformers and causes them to overheat. Since resistance of the coils is raised and the conductivity lowered by heat, it reduces the efficiency and can result in burned-out coils if the machine is not protected against overload. Overheating makes the insulation between coils dry and brittle. This applies to both rotating and transformer-rectifier type power supplies.

Make sure that the air intake or exhaust vents are not blocked to permit the proper flow of air through the machine. Keep the covers on the welder. Removing them destroys the proper path of air flow for ventilation.

General Precautions Against Abuse

Never leave the electrode grounded to the work. This condition creates a "dead" short circuit. The machine is forced to generate much higher currents than it was designed for, which can result in a burned-out machine.

Do not work the machine over its rated capacity. A 200-amp machine will not do the work of a 400-amp machine. Operating above capacity causes overheating; the insulation may be destroyed or the solder in the commutator connections melted.

Use extreme care in operating a machine on a steady load other than arc welding, such as thawing water pipes, supplying current for lighting, running motors, charging batteries, or operating heating equipment. For example, a DC machine, NEMA-rated 300 amp at 32 volts or 9.6 kw should not be used for any continuous load greater than 7.7 kw, and not more than 240 amp. This precaution applies to machines with a duty cycle of at least 60%. Machines with lower duty cycle ratings must be operated at still lower percentages of the rated load.

Do not handle roughly. A welder is a precisely aligned and balanced machine. Mechanical abuse, rough handling, or severe shock may disturb the alignment and balance of the machine, resulting in serious trouble. Misalignment can cause bearing failure, bracket failure, unbalanced air gap, or unbalance in the armature.

TABLE 14-1. Installation Data (Wire and Fuse Size) for Typical Motor-Generator Sets (60 hertz)

WELDER	OUTPUT RATING			INPUT RATING		NATIONAL ELECTRIC CODE SPECIFICATIONS		
MOTOR GENERATORS	Amps	Volts	% Duty Cycle	Voltage	Amperes on Nameplate	Input Wire Size (Type 75°C Copper Conductor in Conduit) AWG	Ground Wire Size (Copper Conductor)	Fuse Size (Super-Lag)
Light production and maintenance. Manual welding.	250	30	30	230 460	32 16	10 14	10 14	60 30
Industrial type for manual welding.	200	40	60	230 460	44 22	8 12	8 10	80 40
Industrial type for manual welding.	300	40	60	230 460	62 31	6 10	8 10	100 50
Combination constant or variable-voltage. Manual and mechanized welding.	300	40	60	230 460	68 34	6 10	8 10	100 50
Industrial type manual or mechanized welding.	400	40	60	230 460	78 39	6 8	6 8	125 70
Combination constant or variable-voltage. Manual and mechanized welding.	400	40	60	230 460	84 42	4 8	6 8	125 70
Combination constant or variable-voltage. Mechanized welding.	675	55	60	230 460	160 80	1 6	3 6	250 125
Constnt-voltage. Mechanized welding.	675	44	60	230 460	124 62	2 6	6 8	200 100
Variable-voltage. Automatic welding only.	900	40	60	230 460	158 79	1 6	3 6	250 125
Constant-voltage. Automatic welding only.	1100	44	60	230 460	198 99	00 4	3 6	300 150

Never pry on the ventilating fan or commutator to try to move the armature. To do so will damage the fan or commutator. If the armature is jammed, inspect the unit for the cause of the trouble. Check for dirt or foreign particles between the armature and frames. Inspect the banding wire on the armature. Look for a frozen bearing.

Do not neglect the engine, if the welder is an engine-driven unit. It deteriorates rapidly if not properly maintained. Follow the engine manufacturer's recommendations. Change oil regularly. Keep air filters and oil strainers clean.

Do not allow grease and oil from the engine to leak back into the generator. Grease quickly accumulates dirt and dust, clogging the air passages between coils.

Maintenance Procedures

Most of the maintenance discussion covered in this section relates to engine drive and motor generator types of welders. This is to be expected because equipment with major rotating components is subject to more wear and requires more maintenance than rectifier type welders with no rotating parts other than the ventilating fan.

TABLE 14-2. Installation Data (Wire and Fuse Size) for Typical Transformer Welders (60 hertz)

WELDER	OUTPUT RATING			INPUT RATING			NATIONAL ELECTRIC CODE SPECIFICATIONS					
					Amperes on Nameplate		Input Wire Size (Type 75°C Copper Conductor in Conduit) AWG		Ground Wire Size (Copper Conductor) AWG		Fuse Size (Super-Lag)	
TRANSFORMERS	Amps	Volts	% Duty Cycle	Voltage	With Cond.	Without Cond.	With Cond.	Without Cond.	With Cond.	Without Cond.	With Cond.	Without Cond.
Single-phase, AC only, VV	180	25	20	230	—	37	—	12	—	12	—	40
Single-phase, AC only, VV	225	25	20	230	—	50	—	10	—	10	—	60
Three-phase, DC only, VV	300	40	60	230 460	— —	56 28	— —	8 10	— —	8 10	— —	80 40
Three-phase, DC only, VV	400	40	60	230 460	— —	75 37.5	— —	6 10	— —	6 10	— —	125 60
Three-phase, DC only, VV	500	40	60	230 460	— —	94 47	— —	4 8	— —	6 8	— —	150 70
Three-phase, DC only, VV	650	40	60	230 460	— —	118 59	— —	3 6	— —	6 8	— —	175 90
Three-phase, DC only, CV	200	28	40	230 460	— —	23.6 11.9	— —	10 12	— —	12 12	— —	40 30
Three-phase, DC only, CV	400	40	100	230 460	— —	68 34	— —	4 8	— —	6 10	— —	110 60
Three-phase, DC only, CV	600	44	100	230 460	— —	100 50	— —	3 6	— —	6 8	— —	150 80
Three-phase, DC only, CV	800	44	100	230 460	— —	142 71	— —	0 4	— —	3 6	— —	225 110
					DC / AC	DC / AC	DC / AC					
Single-phase AC or AC-DC, VV	250	30	30	230 460	50 / 46 25 / 23	70 / 68 35 / 34	8 / 8 12 / 12	6 10	8 12	8 10	70 35	90 45
Single-phase AC or AC-DC, VV	300	40	60	230 460	90-104 / 85-102 45-52 / 42-51	109-116 / 107-116 54-58 / 53-58	4 / 4 8 / 8	4 8	6 8	6 8	150 70	150 80
Single-phase AC or AC-DC, VV	400	40	60	230 460	120 / 112 60 / 56	146 / 142 73 / 71	3 / 3 6 / 8	2 6	6 8	3 6	200 100	225 110
Single-phase AC or AC-DC, VV	500	40	60	230 460	146 / 138 73 / 69	182 / 177 91 / 88.5	2 / 2 6 / 6	0 4	3 6	3 6	250 125	275 125
Single-phase AC or AC-DC, VV	650	40	60	230 460	180 / 172 90 / 86	232 / 226 116 / 113	0 / 0 4 / 4	000 3	3 6	3 6	300 150	350 175
Single-phase AC only	1000	44	1 hr. continuous	230 460	312 156	— —	250 MCM 1	— —	1 3	— —	450 225	— —
Scott-connected AC only	850	44	1 hr. continuous	230 460	312 156	— —	250 MCM 1	— —	1 3	— —	450 225	— —
Single-phase DC CV/VV	200	28	50	230 460	— —	66 33	— —	8 12	— —	8 12	— —	90 45
Three-phase DC CV/VV	400	40	100	230 460	— —	76 38	— —	4 8	— —	6 10	— —	125 60
Three-phase DC CV/VV	600	44	100	230 460	— —	113 56.5	— —	2 6	— —	6 8	— —	175 90
Three-phase DC CV/VV	1000	44	100	230 460	— —	188 94	— —	000 3	— —	4 6	— —	300 150
Three-phase DC CV/VV	1500	60	100	460	—	184	—	000	—	3	—	300
Single-phase AC VV	1200	44	100	460	—	198	—	000	—	3	—	300

Transformer Rectifier Welders

The new generation of welders, mostly transformer rectifiers, depends entirely on solid state PC board control, firing main SCRs for welding output current control. Although the troubleshooting is somewhat simplified by a majority of circuitry being housed upon one or more PC boards, the ability to properly diagnose problems correctly should be left to qualified technicians.

Weldors and service personnel should make use of the excellent instruction manuals and troubleshooting guides supplied with new machines or available from the manufacturer. These instruction manuals provide detailed information on installation, operation, functions of controls, maintenance, troubleshooting and include parts lists. The general maintenance requirements of blowing out dirt, keeping cooling air passages clear, locating away from corrosive or dusty atmosphere and using adequate power leads and welding cables apply to the new power supplies as well.

The troubleshooting guides in the instruction manuals lists troubles with possible causes and remedies which could be handled by weldors or shop maintenance personnel. Remedies for more complicated problems are covered in service manuals, and these can usually be obtained from the manufacturer. Only qualified technicians should attempt to perform the more complicated service procedures described in the service manuals.

Most rectifier type welders intended for gas tungsten arc welding contain high frequency units to provide high frequency for arc starting. The spark gap on these units should be inspected periodically to assure that it is clean and set for the proper gap. Increasing the spark gap increases the high frequency intensity. When spark gap electrodes become pitted, they should be replaced. Dressing of spark gap electrodes is not recommended; however, if they are redressed, surfaces must be as smooth, flat and as parallel as the original surfaces.

Most fan motors have pre-lubricated sealed ball bearings which require no further lubrication. However, some may require periodic oiling which would be indicated in the operating manual.

Inverter Type Welders

Inverter type power supplies are lighter and more compact than standard rectifier type power supplies of the same current rating. They involve high voltage circuits and should be blown out periodically to remove dust and dirt which could cause internal arcing. The service manuals for many of these power supplies contain troubleshooting guides and a detailed test procedure listing check points and corresponding voltage or resistance measurements. These can be performed by qualified technicians to isolate problems. The sheet metal case on these power supplies must be maintained in good condition and all sheet metal screws must be in place and tight to ensure case strength and electrical ground continuity. Breaks or dents must be repaired to assure proper spacing from the high voltage components.

With the main power disconnected and the case removed, all internal components should be inspected for signs of overheating, broken leads or other obvious problems. The interior should be blown out with low pressure air which is filtered to remove oil or moisture. If tests are to be made, follow the procedure set forth in the operating manual or service manual. **Use extreme caution when making these tests because high voltages are present when power is turned on.**

Rotating Type Welders

Bearings. The ball bearings in modern welders have sufficient grease to last the life of the machine under normal conditions. Under severe conditions — heavy usage or in a dirty location — the bearings should be greased about once a year. Add about one ounce (one cubic inch) of grease to each bearing.

Many bearing failures can be traced to over-greasing but dirt is responsible for more bearing failures than any other cause. Before greasing wipe the grease fitting completely clean. Use clean grease. An open container of grease exposed to a factory atmosphere can contaminate the grease and make it unsuitable for ball bearings.

If too little grease is applied, bearings fail. Too light grease will run out. Grease containing solid materials may ruin anti-friction bearings; rancid grease will not lubricate. Dirty grease or dirty fittings or pipes cause bearing failures.

Bearings do not need inspection. They are shielded against dirt, and it is inadvisable to open them unless necessary.

If it is necessary to pull bearings, it should be done with a special puller designed to act against the inner race. Never clean new bearings before installing. Handle them with care. Put them in place by driving against the inner race. Make sure that they fit squarely against the shoulders.

Brackets or end bolts. If it becomes necessary to remove a bracket, to replace a bearing or to disassemble the machine, do so by removing the bolts and tapping lightly and evenly with a babbitt hammer all around the outside diameter of the bracket ring. Do not drive off with a heavy steel hammer.

The bearing housing may become worn oversize, caused by the pounding of the bearing when the armature is out-of-balance. Bracket bearing housings may be checked for size by trying a new bearing for fit. The bearing should slide into the housing with a light drive fit. Replace the bracket if the housing is oversize.

Brushes and brush holders. Set brush holders approximately 0.030 to 0.090 in. above the surface of the commutator. If brush holders have been removed, be certain that they are set squarely in the rocker slot when replaced. Do not force the brush holder into the slot by driving on the insulation. Check to make sure that the brush-holder insulation is squarely set.

Tighten brush holders firmly. When properly set, they are parallel to the mica segments between commutator bars.

Use the grade of brushes recommended by the manufacturer of the welder. Brushes may be too hard or too soft and cause damage to the commutator. Brushes will be damaged by excessive clearance in the brush holder or uneven brush spring pressure. High commutator bars, high mica segments, excessive brush spring pressure, and abrasive dust also will wear out brushes rapidly.

Inspect brushes and holders regularly. A brush may wear down and lose spring tension. It will then start to arc, with damage to the commutator and other brushes.

Keep the brush contact surface of the holder clean and free from pit marks. Brushes must be able to move freely in the holder. Replace them when the pigtails are within 1/4 in. of the commutator or when the limit of spring travel is reached.

New brushes must be sanded in to conform to the shape of the commutator. This may be done by stoning the commutator with a stone or by using fine sandpaper (not emery cloth or paper). Place the sandpaper under the brush, and move it back and forth while holding the brush down in the normal position under slight pressure with the fingers.

See that brush holders and springs seat squarely and firmly against the brushes and that the pigtails are fastened securely.

Commutators. Commutators normally need little care. They will build up a surface film of brown copper oxide, which is highly conductive, hard, and smooth. This surface helps to protect the commutator. Do not try to keep a commutator bright and shiny by constant stoning. The brown copper oxide film prevents the build-up of a black abrasive oxide film that has high resistance and causes excessive brush and commutator wear.

Wipe clean occasionally with a rag or canvas to remove any unnatural films.

If brushes are chattering because of high bars, high mica, or grooves, stone by hand, or remove and turn in a lathe if necessary.

Most commutator troubles start because the wrong grade of brushes is used. Brushes that contain too much abrasive material or have too high a copper content usually scratch the commutator and prevent the desired surface film from building up. A brush that is too soft may smudge the surface with the same result as far as surface film is concerned. In general, brushes that have a low-voltage drop will give poor commutation. Conversely, a brush with high-voltage drop commutates better but may cause overheating of the commutator surface.

If the commutator becomes burned, it may be dressed down by pressing a commutator stone against the surface with the brushes raised. If the surface is badly pitted or out of round, the armature must be removed from the machine and the commutator turned in a lathe. It is good practice for the commutator to run within a radial tolerance of 0.003-in.

The mica separating the bars of the commutator is undercut to a depth of 1/32 to 1/16 in. Mica exposed at the commutator surface causes brush and commutator wear and poor commutation. If the mica is even with the surface, undercut it.

When the commutator is operating properly, there is very little visible sparking. The brush surface is shiny and smooth with no evidence of scratches.

Generator frame. The generator frame and coils need no attention other than inspection to ensure tight connections and cleanliness. Blow out dust and dirt with compressed air. Grease may be cleaned off with naphtha. Keep air gaps between armature and pole pieces clean and even.

Armature. The armature must be kept clean to ensure proper balance. Unbalance in the set will pound out the bearings and wear the bearing housing oversize. Blow out the armature regularly with clean, dry compressed air. Clean out the inside of the armature thoroughly by attaching a long pipe to the compressed air line and reaching into the armature coils.

Motor stator. Keep the stator clean and free from grease. When reconnecting it for use on another voltage, solder all connections. If the set is to be used frequently on different voltages, time may be saved by placing lugs on the ends of all stator leads. This eliminates the necessity for loosening and resoldering to make connections, since the lugs may be safely joined with a screw, nut, and lock washer.

Exciter generator. If the machine has a separate exciter generator, its armature, coils, brushes, and brush holders will need the same general care recommended for the welder set.

Keep the covers over the exciter armature, since the commutator can be damaged easily.

Controls. Inspect every time the welder is used to ensure that the ground and electrode cables are connected tightly to the output terminals. Loose connections cause arcing that destroys the insulation around the terminals.

Do not bump or hit the control handles. Impact damages the controls, resulting in poor electrical contacts. If the handles are tight or jammed, inspect for the cause.

Check the contact fingers of the magnetic starting switch regularly. Keep the fingers free from deep pits or other defects that will interfere with a smooth, sliding contact. Copper fingers may be filed lightly. All fingers should make contact simultaneously.

Keep the switch clean and free from dust. Blow out the entire control box with compressed air.

Connections of the leads from the motor stator to the switch must be tight. Keep the lugs in a vertical position. The line voltage is high enough to jump between the lugs on the stator leads if they are allowed to become loose and cocked to one side or the other.

Keep the cover on the control box at all times.

Capacitors. Capacitors may be placed in an AC welder to raise the power factor if required. When capacitors fail, it is not often apparent from the appearance of the capacitor. Consequently, if it is desired to check to see if they are operating correctly, the following should be done: At rated input voltage and with the welder drawing the rated output load current, the input current reading should correspond to the name-plate amperes. If the reading is 10 to 20 percent more, at least one capacitor has failed.

Never touch the capacitor terminals without first disconnecting the welder from the input power source. Then discharge the capacitor by touching the two terminals with an insulated screwdriver.

Relays. The delay relay contacts may be cleaned by passing a cloth soaked in naphtha between them. Do not force the contact arms or use any abrasives to clean the points. Do not file the silver contacts. The pilot relay is enclosed in a dustproof box and should need no attention. Relays are usually adjusted at the factory and should not be tampered with unless faulty operation is obvious.

Periodic Overhaul and Rebuilding

Rotating type welding machines that are operated day after day in fabrication or field work may require a complete overhaul after a year, or a factory rebuilding job after two years. Whenever downtime for repair becomes frequent, a factory-type overhaul or rebuilding job is indicated as desirable.

It is good practice to schedule periodic overhauls as a part of a regular preventive maintenance program, keeping on hand one or more "extra" welders as substitutes for the one or ones taken out of production for overhaul. In a high-production shop, this is not an extravagance; a bad weld resulting from a machine not operating properly might well lead to as much cost for its repair as the cost of overhauling a welder. Also, such maintenance costs are entirely deductible items for tax purposes, whereas the purchase of a new machine must be depreciated over time. A complete rebuilding job generally costs about 60% of the original cost of the machine — and, for all practical purposes, a factory-rebuilt machine is the same functionally as a new machine.

TROUBLESHOOTING

Welders

The manufacturer of arc-welding equipment usually supplies a troubleshooting guide as well as installation and maintenance instructions in the operating manual that accompanies the machine. Table 14-3 shows a typical troubleshooting guide for a welder. Obviously, there are as many troubleshooting guides as components of welding systems and as brand names for such components, so that no one can be universally applicable.

In general, a troubleshooting guide enables the user to match a recognized trouble with possible causes and possible remedies. The use of the guide shows up errors in settings and oversights as well as faults in the equipment and saves time in making corrections or repairs. **Such data, as well as maintenance data, should be carefully preserved, yet should be readily available for reference by the operator or his foreman.**

The typical troubleshooting guide listed in Table 14-3 is primarily for MG type welders and older mechanically controlled reactor welders. On transformer-rectifier type welders which use PC boards for control and the PC boards are suspected of being defective, the following basic procedures will help to safely and quickly put the machine back in working order.

Procedure for Replacing PC Boards

Before replacing a PC board which is suspected of being defective, visually inspect the PC board in question for any electrical or mechanical damage to any of its components and/or conductors.

A. If there is no visible damage to the PC board, install a new one and see if this remedies the problem. If the problem is remedied, reinstall the old PC board to see if the problem still exists. If it does not exist with the old PC board:

1. Check the PC board harness connector pins for corrosion, contamination or looseness.
2. Check leads in the plug harness for loose or intermittent connection.

B. If PC board is damaged electrically (conductors fused open or components blown apart, burned, etc.), before possibly subjecting the new PC board to the same cause of failure, check the following:

1. Inspect for possible shorts, opens or grounds caused by:
   a. Frayed or pinched lead insulation.
   b. Poor lead termination such as a poor contact or a short to adjacent connection or surface.
   c. Shorted or open control cable leads or other external leads.

TABLE 14-3. Typical Troubleshooting Guide for a Welder

Trouble	Cause	Remedy
Welder will not start.	Line switch not turned "ON". Supply line fuse blown. Power circuit dead. Overload relay tripped. Loose or broken power, electrode or ground lead. Wrong voltage. Polarity switch not centered (AC-DC units only). Open circuit to starter button.	Place line switch in "ON" position. Replace (check reason for blown fuse first). Check input voltage. Let cool, remove cause of overloading. Replace lead or tighten and repair connection. Check input voltage against instructions. Center switch handle on +, −, or AC. Repair.
Welder starts but blows fuse after welding begins.	Short circuit in motor or other connection. Fuse too small.	Check connections and lead insulation. Check instruction manual for correct size.
Welder welds but soon stops welding.	Proper ventilation hindered. Overloading — welding in excess of rating. Fan inopertive. Motor Generator Sets — Wrong direction of rotation. Brushes worn or missing. Wrong driving speed. Excessive dust accumulation in welder.	Make sure all case openings are free for proper circulation of air. Operate welder at rated load and duty cycle. Check leads and connections. Check connection diagram. Check brushes for pressure on commutator. Check nameplate for correct motor speed. Clean thermostat, coils and other components.
Variable or sluggish welding arc.	Current too low. Low line voltage. Welding leads too small. Poor ground, electrode or control-circuit connections. Motor Generator Sets — Brushes improper, weak springs, not properly fitted. Rough or dirty commutator.	Check recommended currents for rod type and size being used. Check with power company. Check instruction manual for recommended cable sizes. Check all connection. Clean, repair or replace as required. Check and repair. Turn down or clean commutator.
Welding arc is loud and spatters excessively.	Current setting is too high. Polarity wrong.	Check setting and output with ammeter. Check polarity; try reversing or an electrode of opposite polarity.
Polarity switch won't turn.	Contacts rough and pitted from improper turning under load.	Replace switch.
Welder won't shut off.	Line switch has failed mechanically.	Replace switch.
Arcing at ground clamp.	Loose connection or weak spring.	Tighten connection or replace clamp.
Electrode holder becomes hot.	Loose connection, loose jaw, inadequate duty cycle.	Tighten connection or replace holder.
Touching welder gives shock.	Frame not grounded.	See instruction manual for proper grounding procedure.

d. Foreign matter of interference behind the PC boards on inside mounting panel.

2. If any rewiring was done or special connections made, make certain it was done precisely per factory wiring diagrams and/or instruction.

C. If a PC board is damaged mechanically, inspect for cause, then remedy before installing a replacement PC board.

Wire and Flux Feeders

Wire and flux-feeding equipment is subject to erratic action as the result of physical difficulties. Some of the common troubles, their causes, and the remedies are given in Table 14-4.

Maintenance Technician/Service

On large fabrication and field jobs, a maintenance technician who is well versed in the operation, maintenance, and repair of welding equipment may be justified as a full-time employee. When not troubleshooting, he can devote his time to the routine maintenance of any items of equipment not in use at that moment. Such a technician can substantially reduce the downtime on a job.

An alternative is to contract for a preventive maintenance service from a firm specializing in welding equipment maintenance. Such services usually provide for the monthly inspection of equipment and the making of any needed repairs or adjustments. Although contracted maintenance services have proved their worth, they do not, however, provide immediate availability for troubleshooting as does the company-employed welding-equipment technician.

TABLE 14-4. Typical Troubleshooting Guide for Wire and Flux-Feeding Equipment

Trouble	Cause	What to Do
WIRE FEEDING		
Wire feeding rough or jerky, or wire feeding stops completely.	Cable liner may be dirty.	Clean cable.
	Electrode may be dirty or rusty.	Replace with clean electrode. If conditions are extremely dirty or electrode is old (solid wire only), put a wiper on the wire before it enters the guide tubes. Use a piece of cloth or felt saturated with "Pyroil B" held around the wire with a light clamp.
	Nozzle contact tip may have been burned or partly melted causing hole to be undersized.	Replace with new tip. Drill contact tip: 1/8 wire — No. 28 drill .120 wire — No. 29 drill 3/32 wire — No. 36 drill 5/64 wire — No. 40 drill
FLUX FEEDING (Submerged-Arc)		
Flux flow stops while welding. If there is flux in the hose right behind the gun (test by squeezing), the stoppage is in the gun.	Flux tube in the gun may be blocked.	Remove the gun from the cable. Check the tube in the gun and the cable handle.
	Magnetic particles may cause bridging at the nozzle tip. Bridging only occurs when welding.	Pass flux over a magnetic separator when filling the flux tank.
Flux stoppage not in the gun. (Be certain there are no kinks or callapsing of the hose.)	There may be a piece of slag in the hose.	Work back along the hose squeezing it until flux can be felt. Shake the hose and feel for slag at this point. Blow out hose with air if necessary.
	Flux tank outlet may be clogged.	If there is no flux in the hose, check the tank outlet for large pieces of slag or paper.
	Pressure in the tank may be set improperly.	Set pressure regulator for 26-30 lbs/sq. in. Set pressure for 55-60 lbs/sq. in. when using a gun-cable extension assembly.
Excessive air blow and uneven flux flow from the gun.	Tank may be almost empty. (May look like the tank is full at the sides but it will be down to the bottom in the center.)	Refill the tank.
	Flux may be falling away from the weld faster than it is being fed.	Alter procedure or make a flux dam.
	Pressure in the tank may be too high.	Set pressure regulator for 26-30 lbs/sq. in. Set pressure for 55-60 lbs/sq. in. when using a gun-cable extension assembly.
Flux in tank gets wet.	There may be water in the air line.	It is possible to get water from air lines when first starting in the morning. Blow out air lines before connecting them to the tank.
	The copper bleeder may be clogged.	Be certain that a slight amount of air is escaping from the crimped end of the copper tube below the flux tank.

Bank of heavy duty welding power supplies installed for production submerged arc welding.

Manual continuous wire feed arc repair welding of large steel casting.

Automatic submerged arc weld build-up of worn roll.

Section 15

SAFETY IN WELDING

Section 15, Safety in Welding, was obtained from Lincoln Bulletin E205. It is strongly recommended that the most recent version of this bulletin be consulted to be aware of current safety practices in welding.

Safety Practices in Welding

INTRODUCTION

Arc welding is a safe occupation when sufficient measures are taken to protect the welder from potential hazards. When these measures are overlooked or ignored, however, welders can encounter such dangers as electric shock, overexposure to fumes and gases, arc radiation, and fire and explosion; which may result in serious, or even fatal injuries.

This bulletin is written with the arc welding operator in mind, containing both mandatory safety practices and those based on shop experience. Be sure to read ANSI Z49.1, and refer to the other publications listed at the end of the bulletin for more detailed information on specific topics of arc welding safety, as well as the manufacturers' instructions and material safety data sheets (MSDS's).

Important Note:
So that you can protect yourself against these hazards, every welder should be familiar with American National Standard ANSI Z49.1, "Safety in Welding and Cutting," and should follow the safety practices in that document. Z49.1 is now available for download at no charge at:

http://www.lincolnelectric.com/community/safety/ or at the AWS website http://www.aws.org.

Download and read it!

PERSONAL PROTECTIVE EQUIPMENT

Protective Clothing

Welders, like firemen, must wear clothing to protect them from being burned. Of all injuries to welders, burns are the most common due to sparks landing on bare skin. Welding arcs are very intense and can cause burns to skin and eyes with just a few minutes of exposure.

The actual gear varies with the job being performed, but generally **protective clothing** must allow freedom of movement while providing adequate coverage against burns from sparks, weld spatter, and arc radiation. Many types of clothing will protect you from ultra-violet radiation exposure, which appears as a skin burn (much like sunburn). Under the worst conditions, however, severe burns and skin cancer may result from excessive radiation.

Because of its durability and resistance to fire, wool clothing is suggested over synthetics (which should never be worn because it melts when exposed to extreme heat) or cotton, unless it is specially treated for fire protection. If possible, keep your clothes clean of grease and oil, as these substances may ignite and burn uncontrollably in the presence of oxygen.

Avoid rolling up your sleeves and pant-cuffs, because sparks or hot metal could deposit in the folds; also, wear your trousers outside your work boots, not tucked in, to keep particles from falling into your boots. While we're on the subject, we suggest leather high-tops with steel toes (especially when doing heavy work).

Other protective wear for heavy work or especially hazardous situations includes: flame-resistant suits, aprons, leggings, leather sleeves/shoulder capes, and caps worn under your helmet.

Heavy, flame-resistant gloves, such as leather, should **always** be worn to protect your hands from burns, cuts, and scratches. In addition, as long as they are dry and in good condition, they will offer some insulation against electric shock.

As to preventing electric shock, the key word is **dry**! We'll have more on the subject later, but for now keep in mind that moisture can increase the potential for and severity of electric shock. When working in wet conditions, or when perspiring heavily, you must be even more careful to insulate your body from electrically "live" parts and work on grounded metal.

WARNING

ARC RAYS can burn.
- **Wear eye, ear and body protection.**

Note To Arc Welding Educators and Trainers:
This Arc Welding Safety brochure may be freely copied for educational purposes if distributed to welders and welding students at no additional charge.

ARC RAYS

It is essential that your **eyes are protected** from radiation exposure. Infrared radiation has been known to cause retinal burning and cataracts. And even a brief exposure to ultraviolet (UV) radiation can cause an eye burn known as "welder's flash." While this condition is not always apparent until several hours after exposure, it causes extreme discomfort, and can result in swelling, fluid excretion, and temporary blindness. Normally, welder's flash is temporary, but repeated or prolonged exposure can lead to permanent injury of the eyes.

Other than simply not looking at an arc, the primary preventive measure you can take is to use the proper shade lens in your helmet. Refer to the lens shade selector chart in Supplement 1 for the recommended shade numbers for various arc welding processes. The general rule is to choose a filter too dark to see the arc, then move to lighter shades without dropping below the minimum rating. The filters are marked as to the manufacturer and shade number, the impact-resistant variety are marked with an "H."

Helmets and hand-held face shields (see Figure A) offer the most complete shading against arc radiation. The shade slips into a window at the front of the shield so that it can be removed and replaced easily. The shields are made from a hard plastic or fiberglass to protect your head, face, ears, and neck from electric shock, heat, sparks, and flames. You should also use safety glasses with side shields or goggles to protect your eyes from flying particles.

Visible light can also be harmful, but it is easy to tell if the light is dangerous: if it hurts to look at, then it's too bright. The same is true for infrared radiation: it can usually be felt as heat. However, there's no real way for you to tell if you're being over exposed to UV radiation, so just don't take chances: always wear eye protection (see Supplement 1 for recommended lens shade numbers).

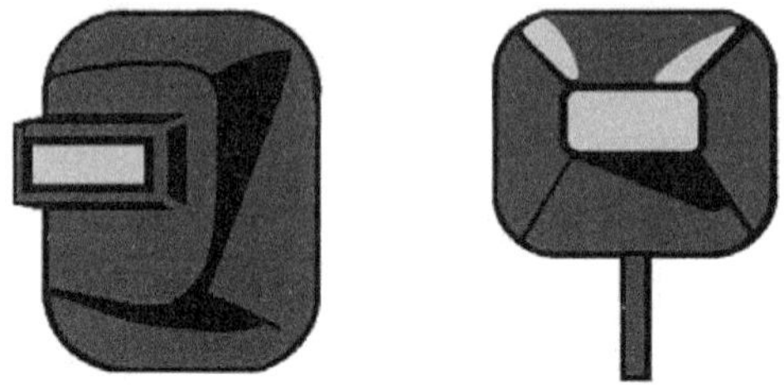

Figure A. A helmet (a) required for protecting the welder's eyes and face and (b) a hand-held face shield that is convenient for the use of foremen, inspectors, and other spectators.

NOISE

There are two good reasons to **wear ear muffs or plugs**:

a) to keep flying sparks or metal out of your ears; and

b) to prevent hearing loss as a result of working around noisy arc welding equipment, power sources, and processes (like air carbon arc cutting or plasma arc cutting).

As with radiation exposure to the eyes, the length and number of times that you are exposed to high levels of noise determine the extent of the damage to your hearing, so be sure to avoid repeated exposure to noise. If it is not possible to reduce the level of noise at the source (by moving either yourself or the equipment, utilizing sound shields, etc.), then you should wear adequate ear protection.

If the noise in your work area becomes uncomfortable, causing a headache or discomfort of the ears, you could be damaging your hearing and should immediately put on ear muffs or plugs.

In fact, the use of ear protection at all times is a good idea, as hearing loss is both gradual and adds up over time. Damage to your hearing may not be noticed until you have a complete hearing test, and then it could be too late.

INSPECTION AND MAINTENANCE OF EQUIPMENT AND WORK

Before starting any arc welding operation, you should make a **complete inspection** of your equipment. All it takes on your part is 5-10 minutes before you turn on your welder; is that too much to spend in preventing injury to yourself or your co-workers?

To begin with:

- Have you read the instruction manual and do you understand the instructions? The instruction manual for your welder is available upon request to your welding distributor or the manufacturer. Manuals for Lincoln Electric welders may be downloaded from lincolnelectric.com at no charge.
- Have you read the warnings and instructions on the equipment nameplates and decals as well as the consumables labels and material safety data sheets? (For older equipment see Supplement 5 to request a FREE Warning Label.)

For the welder:

- Are all the connections tight, including the earth ground?
- OSHA regulations require output terminals to be insulated. Rubber boots are available for that purpose.
- Are the electrode holder and welding cable well insulated and in good condition?
- Are the settings correct for the job you're about to begin?

For an engine-driven welder:

- Is it running OK?
- Are all the hoses on tight?
- Is the fuel cap on tight?
- Is the engine leaking fuel or oil? Some jobsites look for this and may refuse entry if your engine is leaking.
- Is the original enclosure and fan guarding in place? Check with your welding equipment distributor if you are unsure. (See Supplement 6.)

For the work in general: (See also Supplements 4 and 7)

- Are the work area conditions such that normal safety precautions can be observed or must special equipment (i.e., ventilation, exhaust, or respirator, welding equipment, protective equipment, safety equipment) or procedures be used?
- Many jobsites require permits for any welding or cutting. Be sure you have any permits you will need.
- If you will be working in a confined space, many special OSHA regulations and jobsite requirements may apply in addition to the arc welding precautions in this brochure. Understand which of these apply to your jobsite and comply with them.
- Are the cables the right size for your job? Be sure any damaged cable insulation is repaired.
- Are they spread out and run neatly to prevent overheating?
- Is the gas cylinder connected properly?
- Is the cylinder secure?

- Is the work stable and easy to reach from where you're standing?
- Is the Work Lead connected securely?
- Is there enough dry insulation between your body and the work piece?
- Is there adequate ventilation in your work area?

Take some personal responsibility for your own safety. Notify your supervisor if equipment is in need of repair or not working properly or any unsafe condition. You have the most to lose if you get hurt. Don't allow yourself to work in a hazardous situation without taking appropriate safety precautions.

If the hazard is serious and cannot be corrected readily, the machine should be shut down until the needed repairs are made. If the problem is limited to the outside of the welder, such as a loose connection or a damaged cable that needs to be replaced, **disconnect power to the welder** and correct the problem per the manufacturers instructions in the operating/service manual. If the hazard requires repairs to the inside of the welder or to the electrical input supply lines, call a service technician or an electrician. Never attempt to make these repairs if you are untrained.

Important Safety Note:
Consider whether the area in which you will be working creates or increases the level of hazard to you thus requiring special procedures or equipment. Factors such as electrical safety, fume ventilation/exhaust and risk of fire or explosion may be affected. See later sections on those topics and other documents in "Bibliography and Suggested Reading" for further information.

CARE AND CLEANING OF THE WORK AREA

Keeping the area around your work neat is as important as maintaining your equipment. Perhaps even more-so, as the risk of injury is amplified by the larger group of people involved. You may have already inspected your equipment and found it to be OK, but all your caution won't matter when, for example, a co-worker trips over your cable, causing you, and/or the people around you, to be injured by shock, hot metal, or from falling.

Keep all your equipment, cables, hoses, cylinders, etc. out of any traffic routes such as doors, hallways, and ladders. A good practice is to avoid clutter ... and **clean up your work area** when you're done! Not only will it help to protect yourself and others, you'll find it much easier for you to work efficiently.

Also, bear in mind that while you're paying attention to your work, other welders may be preoccupied with their own tasks and not watching where they're going. So be sure that there are protective screens in place, just in case somebody happens to be passing into your work area or walks into a shower of sparks or spatter.

GAS CYLINDERS

Because of the high pressure gas in cylinders, you must pay particularly close attention to their storage and use. Examine the cylinders as you did the rest of your equipment; check the cylinder label to make sure it is the correct shielding gas for the process, and that the regulators, hoses, and fittings are the right ones for that gas and pressure, and are in good condition.

Cylinders must be secured in an upright position, with the valve caps in place, in an area away from combustibles and fuels, and safeguarded from damage, heat, and flames. When in use, keep them out of traffic routes and flying sparks, with all hoses run neatly to the welding area. Never allow the electrode or other "electrically hot" parts of your welder to touch a cylinder. "Crack" the valve open to prevent dirt from entering the regulator; open the cylinder valve only when standing to one side of the cylinder, away from welding or other sources of ignition. Return damaged cylinders to the supplier. Refer to the Compressed Gas Association pamphlet P-1, "Safe Handling of Gas Cylinders," for further information.

⚠ WARNING

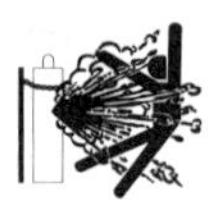

CYLINDER may explode if damaged.
- **Keep cylinder upright and chained to support.**
- **Never allow welding electrode to touch cylinder.**

Electric and Magnetic Fields

Electric current flowing through any conductor causes localized Electric and Magnetic Fields (EMF). Welding current creates EMF fields around welding cables and welding machines. EMF fields may interfere with some pacemakers, and welders having a pacemaker should consult their physician before welding. Exposure to EMF fields in welding may have other health effects which are now not known. All welders should use the following procedures in order to minimize exposure to EMF fields from the welding circuit:

- Route the electrode and work cables together – Secure them with tape when possible.
- Never coil the electrode lead around your body.
- Do not place your body between the electrode and work cables. If the electrode cable is on your right side, the work cable should also be on your right side.
- Connect the work cable to the workpiece as close as possible to the area being welded.
- Do not work next to welding power source.

SPECIFIC CONCERNS

Possible Shock Hazards

The hazard of electric shock is one of the most serious and immediate risks facing you as a welder. Contact with metal parts which are "electrically hot" can cause injury or death because of the effect of the shock upon your body or a fall which may result from your reaction to the shock. The electric shock hazard associated with arc welding may be divided into two categories which are quite different:

- Primary Voltage Shock (i.e., 230, 460 volts); and
- Secondary Voltage Shock (i.e., 20-100 volts).

WARNING

HIGH VOLTAGE can kill.
- **Do not operate with covers removed.**
- **Disconnect input power before servicing.**
- **Do not touch electrically live parts.**

The **primary voltage shock** is very hazardous because it is much greater voltage than the welder secondary voltage. You can receive a shock from the primary (input) voltage if you touch a lead inside the welder with the **power to the welder** "on" while you have your body or hand on the welder case or other grounded metal. Remember that turning the welder power switch "off" does not turn the power off **inside the welder.** To turn the power inside the welder "off", the input power cord must be unplugged or the power disconnect switch turned off. You should never remove fixed panels from your welder; in fact, always have a qualified technician repair your welder if it isn't working properly. Also, your welder should be installed by a qualified electrician so it will be correctly wired for the primary voltage which supplies it power and so the case will be connected to an earth ground. When electrical supply lines are connected to a welder, check the welder capacity nameplate and connection instructions to be sure the input is the correct phase (single phase or three phase) and voltage. Many welders may be set up for single phase or three phase and for multiple input voltages. Be certain the welder is set up for the electrical supply to which it is connected. **Only a qualified electrician should connect input power**. The case must be grounded so that if a problem develops inside the welder a fuse will blow, disconnecting the power and letting you know that repair is required. Never ignore a blown fuse because it is a warning that something is wrong.

WARNING

ELECTRIC SHOCK can kill.
- **Do not touch electrically live parts or electrode with skin or wet clothing.**
- **Insulate yourself from work and ground.**

If welding must be performed under electrically hazardous conditions (in damp locations or while wearing wet clothing; on metal structures such as floors, gratings or scaffolds; when in cramped positions such as sitting, kneeling or lying, if there is a high risk of unavoidable or accidental contact with the work piece or ground) use the following equipment:

- **Semiautomatic DC Constant Voltage Welder**
- **DC Manual (Stick) Welder**
- **AC Welder with Reduced Voltage Control**

A **secondary voltage shock** occurs when you touch a part of the **electrode circuit** — perhaps a bare spot on the electrode cable — at the same time another part of your body is touching the metal upon which you're welding (work). To receive a shock your body must touch both sides of the welding circuit — electrode and work (or welding ground) — at the same time. To prevent secondary voltage shock, you must develop and use safe work habits. Remember the voltage at the electrode is highest when you are **not** welding (open circuit voltage).

- Wear **dry** gloves in good condition when welding.
- Do not touch the electrode or metal parts of the electrode holder with skin or wet clothing.
- Keep **dry** insulation between your body (including arms and legs) and the metal being welded or ground (i.e., metal floor, wet ground).
- Keep your welding cable and electrode holder in good condition. Repair or replace any damaged insulation.

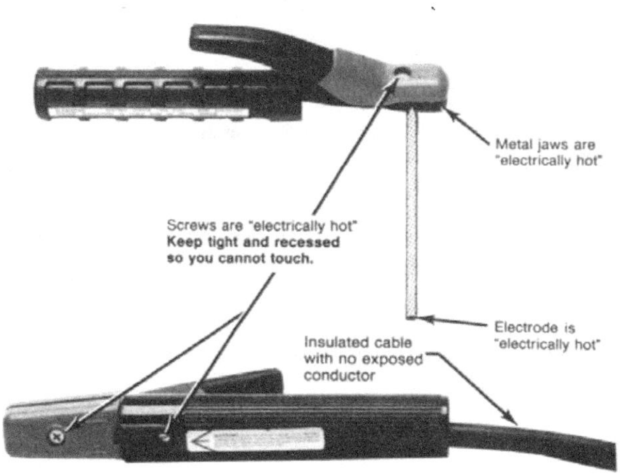

Figure B. Always inspect your electrode holder before turning the welder on.

These rules are basic to welding and you should already know them. Check out the warning on your welder or electrode box next time you weld. You will probably not have a shock while welding if you follow these rules.

Though it may be more difficult to follow the rules under some conditions, the rules still apply. Keep your gloves **dry** even if you have to keep an extra pair. Use plywood, rubber mats, or some other **dry** insulation to stand or lie upon. Insulate your body from the metal you are welding. Don't rest your body, arms, or legs on the workpiece, especially if your clothing is wet or bare skin is exposed (and it should not be if you are dressed properly). In addition to the normal safety precautions, if welding must be performed under electrically hazardous conditions (in damp locations or while wearing wet clothing; on metal structures such as floors, gratings or scaffolds; when in cramped positions such as sitting, kneeling or lying, if there is a high risk of unavoidable or accidental contact with the work piece or ground) use the following equipment:

- Semiautomatic DC Constant Voltage Welder
- DC Manual (Stick) Welder
- AC Welder with Reduced Voltage Control

The condition of your electrode holder and electrode cable is also very important. The plastic or fiber insulation on the electrode holder protects you from touching the metal "electrically hot" parts inside. Always inspect your electrode holder before turning the welder on. Replace the holder if it is damaged — don't try to repair it unless you have replacement parts.

The same is true of the electrode cable except that when not replaced it may be repaired using good electrical tape. If your cable has been repaired, be sure to check and see that the tape is secure before you turn the welder on.

Remember, a stick electrode is always "electrically hot" when the welder is on — treat it with respect. If you do experience a shock, think of it as a warning — check your equipment, work habits and work area to see what is wrong before continuing to weld.

⚠ WARNING

WELDING SPARKS can cause fire or explosion.
• Keep flammable material away.

Fire Hazards

Because of the extreme temperatures associated with any arc welding process, you should always be aware of fire hazards. The heat of the welding arc can reach temperatures of 10,000°F, but this heat in itself is not generally a fire hazard. The danger of fire actually results from the effects of this intense heat upon your work and in the form of **sparks and molten metals**. Because these can spray up to 35 feet from your work, you must recognize and protect combustible materials from the welding arc, sparks and spatter. It is also important to be sure the work is not in contact with any combustible which it may ignite when heated. These materials fall into three categories: liquid (gasoline, oil, paints, and thinners); solid (wood, cardboard, and paper); and gaseous (acetylene and hydrogen).

Watch where the sparks and metals are falling from your work: if there are flammable materials including fuel or hydraulic lines in your work area and you can't move either your work or the combustible substances, put a fire-resistant shield in place. If you're welding above the ground or off a ladder, make sure that there are no combustibles underneath. Also, don't forget about your co-workers, and everybody else who may be in the work area, as they probably wouldn't appreciate being hit with slag or sparks from your work.

Particular care must be taken when welding or cutting in dusty locations. Fine dust particles may readily oxidize (burn) and without warning result in a flash fire or even an explosion when exposed to the welding arc or even sparks.

If you are not sure of the combustible or volatile nature of residue or dust in the work area, no welding or cutting should take place until a responsible person has inspected the area and given approval for the work.

Before you start welding, inspect the surface of your work, looking for flammable coatings or any unknown substances that would ignite when heated. Because of the extreme fire and explosion hazards inherent to welding on or around containers and piping that may have combustible materials, such work should be handled only by experienced welders who review and follow the safety practices recommended in the American Welding Society document F4.1, "Recommended Safe Practices for the Preparation for Welding and Cutting of Containers and Piping Which Had Held Hazardous Substances."

Know where the fire alarms and fire extinguishers are located, and check the pressure gauges so you don't rely upon one that's empty. If there are none in the area, make sure that you have access to fire hoses, sand buckets, fire-resistant blankets, or other fire fighting equipment. If you're welding within 35 feet or so of flammable materials, you should have a fire watcher to see where your sparks are flying, and to grab an extinguisher or alarm if needed. Both you and the fire watcher should wait for a half hour after all welding is finished to find and put out any smoldering fires that may have resulted from your welding.

As with other emergencies that may result from welding accidents, the first rule is: don't panic. Depending on the size of the fire, sound the fire alarm to warn others and call the fire department; shut off your welder; and get to the fire exits as quickly as possible.

⚠ WARNING

FUMES & GASES can be dangerous to your health
- **Keep fumes and gases from your breathing zone and general area.**
- **Keep your head out of the fumes.**
- **Use enough ventilation or exhaust at the arc, or both, to keep fumes and gases from your breathing zone and general area.**

Fumes and Gases

Because of the variables involved in fume and gas generation from arc welding, cutting and allied processes (such as the welding process and electrode, the base metal, coatings on the base metal, and other possible contaminants in the air), we'll have to treat the subject in a rather general way, lumping all but the more hazardous situations together. The precautions we describe will hold true for all arc welding processes.

The **fume plume** contains solid particles from the consumables, base metal, and base metal coating. For common mild steel arc welding, depending on the amount and length of exposure to these fumes, most immediate or short term effects are temporary, and include symptoms of burning eyes and skin, dizziness, nausea, and fever. For example, zinc fumes can cause metal fume fever, a temporary illness that is similar to the flu.

Long-term exposure to welding fumes can lead to siderosis (iron deposits in the lungs) and may affect pulmonary function. Bronchitis and some lung fibrosis have been reported.

Some consumables contain certain compounds in amounts which may require special ventilation and/or exhaust. These Special Ventilation products can be identified by reading the labels on the package. If Special Ventilation products are used indoors, use local exhaust. If Special Ventilation products are used outdoors, a respirator may be required. Various compounds, some of which may be in welding fume, and reported health effects, in summary, are:

Barium: Soluble barium compounds may cause severe stomach pain, slow pulse rate, irregular heart beat, ringing of the ears, convulsions and muscle spasms. In extreme cases can cause death.

Cadmium also requires extra precautions. This toxic metal can be found on some steel and steel fasteners as a plating, or in silver solder. Cadmium fumes can be fatal even under brief overexposures, with symptoms much like those of metal fume fever. These two conditions should **not** be confused. Overexposure to cadmium can be enough to cause fatalities, with symptoms appearing quickly, and, in some circumstances, death a few days later.

Chromium: Chromium is on the IARC (International Agency for Research on Cancer) and NTP (National Toxicology Program) lists chromium as posing a carcinogenic risk to humans. Fumes from the use of stainless steel, hardfacing and other types of consumables contain chromium and/or nickel. Some forms of these metals are known or suspected to cause lung cancer in processes other than welding and asthma has been reported. Therefore, it is recommended that precautions be taken to keep exposures as low as possible. OSHA recently adopted a lower PEL (Permissible Exposure Limit) for chromium (see Supplement 3). The use of local exhaust and/or an approved respirator may be required to avoid overexposure.

Coatings on the metal to be welded, such as paint, may also contain toxic substances, such as lead, chromium and zinc. In general, it is always best to remove coatings from the base metal before welding or cutting.

Cobalt: Exposure to cobalt can cause respiratory disease and pulmonary sensitization. Cobalt in metallic form has been reported to cause lung damage.

Copper: Prolonged exposure to copper fume may cause skin irritation or discoloration of the skin and hair.

Manganese: Manganese overexposure may affect the central nervous system, resulting in poor coordination, difficulty in speaking, and tremor of arms or legs. This condition is considered irreversible.

Nickel: Nickel and its compounds are on the IARC (International Agency for Research on Cancer) and NTP (National Toxicology Program) lists as posing a carcinogenic risk to humans.

Silica: Crystalline silica is present in respirable dust form submerged arc flux. Overexposure can cause severe lung damage (silicosis).

Zinc: Overexposure to zinc (from galvanized metals) may cause metal fume fever with symptoms similar to the common flu.

The **gases** that result from an arc welding process also present potential hazard. Most of the shielding gases (argon, helium, and carbon dioxide) are non-toxic, but, as they are released, they **displace oxygen** in your breathing air, causing dizziness, unconsciousness, and death, the longer your brain is denied the oxygen it needs. Carbon monoxide can also be developed and may pose a hazard if excessive levels are present.

The heat and UV radiation can cause irritation to the eyes and lungs. Some degreasing compounds such as trichlorethylene and perchlorethylene can decompose from the heat and ultraviolet radiation of an arc. Because of the chemical breakdown of vapor-degreasing materials under ultraviolet radiation, arc welding should not be done in the vicinity of a vapor-degreasing operation. Carbon-arc welding, gas tungsten-arc welding and gas metal arc welding should be especially avoided in such areas, because they emit more ultraviolet radiation than other processes. Also, keep in mind that ozone and nitrogen oxides are formed when UV radiation passes through the air. These gases cause headaches, chest pains, irritation of the eyes, and an itchiness in the nose and throat.

There is one easy way to **reduce the risk** of exposure to hazardous fumes and gases: **keep your head out of the fume plume!** As obvious as this sounds, the failure to follow this advice is a common cause of fume and gas overexposure because the concentration of fume and gases is greatest in the plume. Keep fumes and gases from your breathing zone and general area using natural ventilation, mechanical ventilation, fixed or moveable exhaust hoods or local exhaust at the arc. Finally, it may be necessary to wear an approved respirator if adequate ventilation cannot be provided (see Ventilation section).

As a rule of thumb, for many mild steel electrode, if the air is visibly clear and you are comfortable, then the ventilation is generally adequate for your work.The most accurate way to determine if the worker exposure does not exceed the applicable exposure limit for compounds in the fumes and gases is to have an industrial hygienist take and analyze a sample of the air you are breathing. This is particularly important if you are welding with stainless, hardfacing or Special Ventilation products. All Lincoln MSDS have a maximum fume guideline number. If exposure to total fume is kept below that number, exposure to all fume from the electrode (not coatings or plating on the work) will be below the TLV.

There are also steps that you can take to identify hazardous substances in your welding environment. First, read the product label and material safety data sheet for the electrode posted in the work place or in the electrode or flux container to see what fumes can be reasonably expected from use of the product and to determine if special ventilation is needed. Secondly, know what the base metal is, and determine if there is any paint, plating, or coating that could expose you to toxic fumes and/or gases. Remove it from the metal being welded, if possible. If you start to feel uncomfortable, dizzy or nauseous, there is a possibility that you are being overexposed to fumes and gases, or suffering from oxygen deficiency. Stop welding and get some **fresh air** immediately. Notify your supervisor and co-workers so the situation can be corrected and other workers can avoid the hazard. Be sure you are following these safe practices, the consumable labeling and MSDS and improve the ventilation in your area. Do not continue welding until the situation has been corrected.

NOTE: The MSDS for all Lincoln consumables is available on Lincoln's website: www.lincolnelectric.com

Before we turn to the methods available to control welding fume exposure, you should understand a few basic terms:

Natural Ventilation is the movement of air through the workplace caused by natural forces. Outside, this is usually the wind. Inside, this may be the flow of air through open windows and doors.

Mechanical Ventilation is the movement of air through the workplace caused by an electrical device such as a portable fan or permanently mounted fan in the ceiling or wall.

Source Extraction (Local Exhaust) is a mechanical device used to capture welding fume at or near the arc and filter contaminants out of the air.

The ventilation or exhaust needed for your application depends upon many factors such as:

- workspace volume
- workspace configuration
- number of welders
- welding process and current
- consumables used (mild steel, hardfacing, stainless, etc.)
- allowable levels (TLV, PEL, etc.)
- material welded (including paint or plating)
- natural airflow

Your work area has **adequate ventilation** when there is enough ventilation and/or exhaust to control worker exposure to hazardous materials in the welding fumes and gases so the applicable limits for those materials is not exceeded. See Suppement 2 for the legal limits, the OSHA PEL (Permissible Exposure Limit), and the recommended guideline, the ACGIH TLV (Threshold Limit Value), for many compounds found in welding fume.

Ventilation

There are many methods which can be selected by the user to provide adequate ventilation for the specific application. The following section provides general information which may be helpful in evaluating what type of ventilation equipment may be suitable for your application. When ventilation equipment is installed, you should confirm worker exposure is controlled within applicable OSHA PEL and/or ACGIH TLV. According to OSHA regulations, when welding and cutting (mild steels), natural ventilation is usually considered sufficient to meet requirements, provided that:

1. The room or welding area contains at least 10,000 cubic feet (about 22' x 22' x 22') for each welder.
2. The ceiling height is not less than 16 feet.
3. Cross ventilation is not blocked by partitions, equipment, or other structural barriers.
4. Welding is not done in a confined space.

Spaces that do not meet these requirements should be equipped with mechanical ventilating equipment that exhausts at least 2000 cfm of air for each welder, except where local exhaust hoods or booths, or air-line respirators are used.

Important Safety Note:
When welding with electrodes which require special ventilation such as stainless or hardfacing (see instructions on container or MSDS) or on lead or cadmium plated steel and other metals or coatings which produce hazardous fumes, keep exposure as low as possible and below exposure limit values (PEL and TLV) for materials in the fume using local exhaust or mechanical ventilation. In confined spaces or in some circumstances, for example outdoors, a respirator may be required if exposure cannot be controlled to the PEL or TLV. (See MSDS and Supplement 3 of this brochure.) Additional precautions are also required when welding on galvanized steel.

Source Extraction Equipment

Mechanical ventilation is an effective method of fume control for many welding processes. Because it captures fume near the arc or source of the fume, which is more efficient in most cases, local exhaust, also called "source extraction", is a very effective means to control welding fume.

Source extraction of welding fumes can be provided by mobile or stationary, single or multi-station, exhaust and/or filtration equipment designed with adjustable fume extraction arms nozzles or guns, by fixed enclosures, booths or tables with extraction canopies also known as down-draft, or by back-draft or cross-draft tables/booths. Source extraction of weld fume falls into two categories: low vacuum/high volume, or high vacuum/low volume.

Low Vacuum/High Volume

Mobile or stationary, single or multi-station, large centralized exhaust and/or filtration equipment designed with adjustable fume extraction arms are usually low vacuum/high volume systems. When correctly positioned, the capture rate of adjustable fume extraction arms is suitable for all position welding and cutting. For more difficult to reach work areas, flexible hose may be used in place of adjustable fume extraction arms.

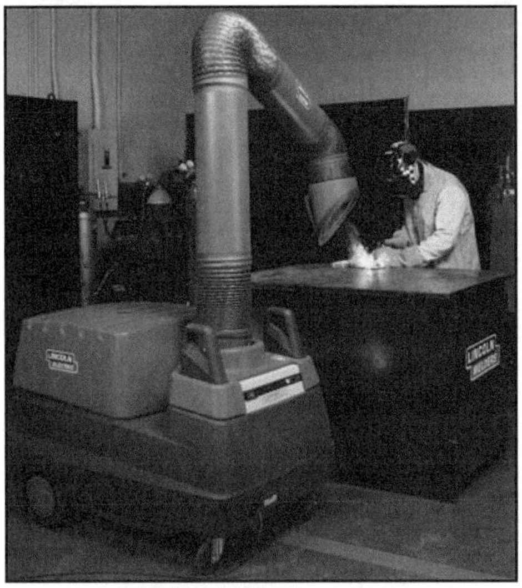

Mobiflex™ 200-M low vacuum mobile fume extraction unit.

These arms generally move between 560 and 860 cubic feet per minute (CFM) (900 – 1400 m^3/hr) of air, but use low vacuum levels (3 to 5 inches water gauge [750 – 1250 Pa]) to minimize power requirements. Water gauge (WG) is a measure of negative pressure: higher numbers mean more negative pressure (more "suction"). With this volume of airflow, the end of the arm can be placed 6 to 15 inches (160 – 375 mm) away from the arc and still effectively capture weld fume.

Fume extraction arms generally use a 6 or 8 inch diameter hose, or hose and tubing combinations. Arm lengths are typically 7, 10, or 13 feet (2, 3, or 4 m), with boom extensions available. The arms may be wall mounted, attached to mobile units, or incorporated into a centralized system.

In general, the farther the extraction hose is from the arc, the more volume of air movement is required to effectively capture welding fume. Overhead hoods (canopies), for example, capture most of the fume, but care must be taken to be sure fume is not pulled through the breathing zone of the operator.

Fixed enclosures, booths or tables with extraction canopies also known as **down-draft, back-draft or cross-draft booths/tables** are a variation of overhead hood technology and can be used as source extraction equipment. A booth is a fixed enclosure that consists of a top and at least two sides that surround the welding operation. These systems use a plenum with openings to the side, back or bottom of the work space rather than above it to capture the weld fume. The weld fume is extracted through the plenum and away from the breathing zone of the operator that is welding or cutting. Down-draft or back-draft booths/tables can be mobile or stationary, single or multi-station, exhaust and/or filtration systems. They are particularly suitable for in-position bench welding or cutting jobs and can be effective when small parts are being welded. The airflow required for effectiveness varies depending upon the installation design, but may be 1,000 CFM or higher.

There are advantages and limitations associated with low vacuum/high volume source extraction systems.

Advantages	*Limitations*
Source extraction with large volume of air being extracted from welder breathing zone.	If not using filtration unit, exhausting air to outside requires make-up air systems and make-up heaters (ie. large volumes of displaced air need to be replaced, resulting in increased utility costs).
Auto-stop delay assists with removal of residual fumes.	Welder must stop to reposition arm over weld area(s).
Low noise level.	Filtration systems larger due to volume of air flow.
Flexible arm for repositioning.	Depending on design, ductwork can be large.
Low installation costs (ductwork).	
Low energy consumption (small fan unit with low rpm).	
Adjustable arms suitable for all-position welding.	

High Vacuum/Low Volume

High vacuum/low volume fume extraction systems are designed for close proximity (2 to 4 inches) positioning. High vacuum/low volume weld fume extraction is achieved with lower airflow rates than those encountered when utilizing low vacuum/high volume systems. There are two methods of high vacuum extraction: welding guns with built-in extraction (fume extraction guns), or separate suction nozzles of various designs.

Fume extraction guns use fume capture nozzles built into the gun tube and handle. The extraction airflow is approximately 35 to 60 CFM (60 – 100 m^3/hr) for integrated fume extraction guns. Therefore, no repositioning is required, since the suction automatically follows the arc. The vacuum level is high (40 to 70 inches WG [9.96 X 103 to 1.74 X 104 Pascal]) permitting the use of hose featuring longer lengths (10 to 25 feet) and smaller diameters (1.25 to 1.75 inches). Fume extraction gun designs have been improved to be more ergonomic and user friendly. Depending upon the type of welding, particularly "in position" welding, extraction guns may be a good solution.

Suction nozzles are positioned near the weld, and commonly use capture distances of less than four inches. Depending upon the design, airflow of suction nozzles is typically between 80 to 100 CFM (135 – 170 m^3/hr). Suction nozzles must be kept near the arc to be used effectively.

The capture rate for fume extraction guns or nozzles is highest when used in flat and horizontal welding positions. High vacuum equipment ranges from small, portable, mobile units to stationary, single or multi-station, large centralized filtration systems.

Miniflex™ high vacuum portable fume extraction unit.

There are advantages and limitations associated with high vacuum/low volume source extraction systems.

Advantages	***Limitations***
When using a fume extraction gun, welder does not need to stop and reposition extraction device.	Required when using a suction nozzle. Welder may need to stop to reposition extraction device.
Low volume of air is displaced-results in energy efficiency and conservation.	High noise level due to increased air velocity and high motor rpm of the fan unit.
Ductwork smaller in diameter (3 to 10 inches) vs. low vacuum systems.	Possible removal of shielding gases affecting weld integrity if nozzle or gun placed too close to source.
Low obstruction of welder vision.	Greater energy consumption (large fan unit with high rpm).
Suitable for heavier particulate (ie. grinding dust).	Residual fumes not extracted.
Suitable option for confined, difficult to reach work spaces.	Less effective in out-of-position welding.
Smaller filter systems due to less volume of airflow.	

Fume extraction is only one component in reducing welding fume. Users should also consider the selection of the welding process, welding procedure, or consumable. Many times a combination of fume extraction, training, process change, and/or consumable change is needed to reduce the amount of fume to acceptable levels. Solutions to a particular application may involve one or all of these factors and the user must determine which solution best fits their application.

OSHA regulations include specific requirements for exhaust systems which should be reviewed when selecting fume extraction systems (see Supplement 2).

Exhaust vs. Filtration

Source extraction exhaust equipment captures and extracts weld fumes from the source and exhausts the fumes to the outside atmosphere. This technique removes welding fume from the breathing zone of the welder but can also displace large volumes of conditioned air which may lead to increased utility and heating costs.

Source extraction filtration equipment captures and extracts weld fumes from the source and filters the fumes by passing them through a cellulose and/or polyester filter cartridge or electrostatic filter. Depending on the weld application, environment, federal or local regulations, and filtration efficiency levels, filtered air may be re-circulated back into the facility or exhausted to the outside atmosphere. By re-circulating filtered air back into the work environment compared to exhausting to the outside, source extraction filtration equipment can be more economical to operate. Particularly in winter months, substantially lower heating costs may be recognized, as less replacement air is required with filtration versus exhaust systems.

Using a cellulose or polyester filter cartridge or electrostatic filter will depend upon the weld application. Electrostatic filters may also be used however, they lose efficiency if they are not frequently washed.

Regardless of the type of mechanical ventilation (exhaust or filtration) source extraction system used, the important factor is that it is a tool designed to control exposure to welding fume and its constituents. All forms of mechanical ventilation or source extraction equipment require routine maintenance. In addition, when using weld fume source extraction equipment, sparks from welding, cutting or grinding processes can cause fire within the equipment. To control this potential fire hazard, operation, service and maintenance instructions for source extraction equipment should be followed.

Note:

It is the equipment owner and operator's responsibility to comply with Occupational Safety, Health Administration (OSHA) Permissable Exposure Limits (PELs) or American Conference of Governments Industrial Hygienists (ACGIH) TLVs for welding fume. It is the responsibility of the equipment owner to research, test and comply with regulations which may apply to filtered air recirculated inside the facility or unfiltered air is exhausted outside of the facility.

Working in Confined Spaces

When arc welding in a confined area, such as a boiler, tank, or the hold of a ship, bear in mind that all the hazards associated with normal arc welding are amplified, so the precautions mentioned here are even more important. This subject is very complicated and only these precautions related to arc welding will be discussed in this brochure. Per OSHA document 29 CFR 1910.146, a particular area is considered a confined space if it:

1) Is large enough and so configured that an employee can bodily enter and perform assigned work; and
2) Has limited or restricted means for entry or exit (for example, tanks, vessels, silos, storage bins, hoppers, vaults, and pits are spaces that may have limited means of entry.); and
3) Is not designed for continuous employee occupancy.

There is a greater danger that enough **flammable gases** may be present in the confined space to cause an explosion. The metal of the enclosure can become part of the welding circuit, so any metal you touch (the walls, floor, ceiling) is **electrically "hot"**. Welding fumes can accumulate more rapidly, with a higher concentration; gases can force out the **breathable air**, suffocating you in the process.

Per OSHA document 29 CFR 1910.146(d)(5)(iii); after an area has been deemed a confined space, the existence of the following atmospheric hazards are to be determined:

1) Test for oxygen
2) Test for combustible gases and vapors
3) Test for toxic gases and vapors

The workplace and OSHA rules regarding confined spaces must be followed. Make sure that your body is insulated from the work-piece using dry insulation. Wear dry gloves and only use a well-insulated electrode holder. Semiautomatic constant voltage welders with cold electrode or stick welders equipped with a device to lower the no-load voltage are recommended, especially when the work area is wet. Make sure that there is adequate ventilation and exhaust (a respirator or an air-supplied respirator may be necessary depending on the application), and that there are no flammable coatings, liquids or gases nearby.

Lastly, you must have someone outside the enclosure trained to handle emergencies, with rescue procedures and a means to disconnect power to your equipment and pull you out if danger arises. We cannot stress this strongly enough: however experienced you are, do not attempt work of this nature without constant communication with the person outside the confined area. When welding within a confined area, problems which arise can immediately become very serious and, in some cases, life-threatening. It is for that reason that OSHA regulations and workplace procedures for confined space work must be followed.

GUIDE FOR SHADE NUMBERS

OPERATION	ELECTRODE SIZE 1/32 in. (mm)	ARC CURRENT (A)	MINIMUM PROTECTIVE SHADE	SUGGESTED(1) SHADE NO. (COMFORT)
Shielded metal arc welding	Less than 3 (2.5) 3-5 (2.5–4) 5-8 (4–6.4) More than 8 (6.4)	Less than 60 60-160 160-250 250-550	7 8 10 11	– 10 12 14
Gas metal arc welding and flux cored arc welding		Less than 60 60-160 160-250 250-500	7 10 10 10	– 11 12 14
Gas tungsten arc welding		Less than 50 50-150 150-500	8 8 10	10 12 14
Air carbon Arc cutting	(Light) (Heavy)	Less than 500 500-1000	10 11	12 14
Plasma arc welding		Less than 20 20-100 100-400 400-800	6 8 10 11	6 to 8 10 12 14
Plasma arc cutting	(Light)[2] (Medium)[2] (Heavy)[2]	Less than 300 300-400 400-800	8 9 10	9 12 14
Torch brazing		–	–	3 or 4
Torch soldering		–	–	2
Carbon arc welding		–	–	14
	PLATE THICKNESS in.	**PLATE THICKNESS mm**		
Gas welding Light Medium Heavy	 Under 1/8 1/8 to 1/2 Over 1/2	 Under 3.2 3.2 to 12.7 Over 12.7		 4 or 5 5 or 6 6 or 8
Oxygen cutting Light Medium Heavy	 Under 1 1 to 6 Over 6	 Under 25 25 to 150 Over 150		 3 or 4 4 or 5 5 or 6

(1) As a rule of thumb, start with a shade that is too dark, then go to a lighter shade which gives sufficient view of the weld zone without going below the minimum. In oxyfuel gas welding or cutting where the torch produces a high yellow light, it is desirable to use a filter lens that absorbs the yellow or sodium line the visible light of the (spectrum) operation

(2) These values apply where the actual arc is clearly seen. Experience has shown that lighter filters may be used when the arc is hidden by the workpiece.

Data from ANSI Z49.1-1999

BIBLIOGRAPHY AND SUGGESTED READING

ANSI Z87.1, *Practice for Occupational and Educational Eye and Face Protection*, American National Standards Institute, 11 West 42nd Street, New York, NY 10036.

Arc Welding and Your Health: A Handbook of Health Information for Welding. Published by The American Industrial Hygiene Association, 2700 Prosperity Avenue, Suite 250, Fairfax, VA 22031-4319.

NFPA Standard 51B, *Cutting and Welding Processes*, National Fire Protection Association, 1 Batterymarch Park, P.O. Box 9146, Quincy, MA 02269-9959.

OSHA General Industry Standard 29 CFR 1910 Subpart Q. OSHA Hazard Communication Standard 29 CFR 1910.1200. Available from the Occupational Safety and Health Administration at http://www.osha.org or contact your local OSHA office.

The following publications are published by The American Welding Society, P.O. Box 351040, Miami, Florida 33135. AWS publications may be purchased from the American Welding society at http://www.aws.org or by contacting the AWS at 800-854-7149.

ANSI, Standard Z49.1, *Safety in Welding, Cutting and Allied Processes*. Z49.1 is now available for download at no charge at http://www.lincolnelectric.com/community/safety/ or at the AWS website http://www.aws.org.

AWS F1.1, *Method for Sampling Airborne Particulates Generated by Welding and Allied Processes.*

AWS F1.2, *Laboratory Method for Measuring Fume Generation Rates and Total Fume Emission of Welding and Allied Processes.*

AWS F1.3, *Evaluating Contaminants in the Welding Environment: A Strategic Sampling Guide.*

AWS F1.5, *Methods for Sampling and Analyzing Gases from Welding and Allied Processes.*

AWS F3.2, Ventilation Guide for Welding Fume Control

AWS F4.1, *Recommended Safe Practices for the Preparation for Welding and Cutting of Containers and Piping That Have Held Hazardous Substances.*

AWS SHF, *Safety and Health Facts Sheets*. Available free of charge from the AWS website at http://www.aws.org.

LISTED BELOW ARE SOME TYPICAL INGREDIENTS IN WELDING ELECTRODES AND THEIR TLV (ACGIH) GUIDELINES AND PEL (OSHA) EXPOSURE LIMITS

INGREDIENTS	CAS No.	TLV mg/m³	PEL mg/m³
Aluminum and/or aluminum alloys (as Al)*****	7429-90-5	10	15
Aluminum oxide and/or Bauxite*****	1344-28-1	10	5**
Barium compounds (as Ba)*****	513-77-9	****	****
Chromium and chromium alloys or compounds (as Cr)*****	7440-47-3	0.5(b)	.005(b)
Fluorides (as F)	7789-75-5	2.5	2.5
Iron	7439-89-6	10*	10*
Limestone and/or calcium carbonate	1317-65-3	10	15
Lithium compounds (as Li)	554-13-2	10*	10*
Magnesite	1309-48-4	10	15
Magnesium and/or magnesium alloys and compounds (as Mg)	7439-95-4	10*	10*
Manganese and/or manganese alloys and compounds (as Mn)*****	7439-96-5	0.2	5.0(c)
Mineral silicates	1332-58-7	5**	5**
Molybdenum alloys (as Mo)	7439-98-7	10	10
Nickel*****	7440-02-0	1.5	1
Silicates and other binders	1344-09-8	10*	10*
Silicon and/or silicon alloys and compounds (as Si)	7440-21-3	10*	10*
Strontium compounds (as Sr)	1633-05-2	10*	10*
Zirconium alloys and compounds (as Zr)	12004-83-0	5	5

Supplemental Information:

(*) Not listed. Nuisance value maximum is 10 milligrams per cubic meter. PEL value for iron oxide is 10 milligrams per cubic meter. TLV value for iron oxide is 5 milligrams per cubic meter.

(**) As respirable dust.

(*****) Subject to the reporting requirements of Sections 311, 312, and 313 of the Emergency Planning and Community Right-to-Know Act of 1986 and of 40CFR 370 and 372.

(b) The PEL for chromium (VI) is .005 milligrams per cubic meter as an 8 hour time weighted average. The TLV for water-soluble chromium (VI) is 0.05 milligrams per cubic meter. The TLV for insoluble chromium (VI) is 0.01 milligrams per cubic meter.

(c) Values are for manganese fume. STEL (Short Term Exposure Limit) is 3.0 milligrams per cubic meter. Values are those proposed by OSHA in 1989. Present PEL is 5.0 milligrams per cubic meter (ceiling value).

(****) There is no listed value for insoluble barium compounds. The TLV for soluble barium compounds is 0.5 mg/m³.

TLV and PEL values are as of April 2006. Always check Material Safety Data Sheet (MSDS) with product or on the Lincoln Electric website at http://www.lincolnelectric.com

⚠ WARNING

⚠ CALIFORNIA PROPOSITION 65 WARNINGS ⚠

Diesel engine exhaust and some of its constituents are known to the State of California to cause cancer, birth defects, and other reproductive harm.

The Above For Diesel Engines

The engine exhaust from this product contains chemicals known to the State of California to cause cancer, birth defects, or other reproductive harm.

The Above For Gasoline Engines

ARC WELDING CAN BE HAZARDOUS. PROTECT YOURSELF AND OTHERS FROM POSSIBLE SERIOUS INJURY OR DEATH. KEEP CHILDREN AWAY. PACEMAKER WEARERS SHOULD CONSULT WITH THEIR DOCTOR BEFORE OPERATING.

Read and understand the following safety highlights. For additional safety information, it is strongly recommended that you purchase a copy of "Safety in Welding & Cutting - ANSI Standard Z49.1" from the American Welding Society, P.O. Box 351040, Miami, Florida 33135 or CSA Standard W117.2-1974. A Free copy of "Arc Welding Safety" booklet E205 is available from the Lincoln Electric Company, 22801 St. Clair Avenue, Cleveland, Ohio 44117-1199.

BE SURE THAT ALL INSTALLATION, OPERATION, MAINTENANCE AND REPAIR PROCEDURES ARE PERFORMED ONLY BY QUALIFIED INDIVIDUALS.

FOR ENGINE powered equipment.

1.a. Turn the engine off before troubleshooting and maintenance work unless the maintenance work requires it to be running.

1.b. Operate engines in open, well-ventilated areas or vent the engine exhaust fumes outdoors.

1.c. Do not add the fuel near an open flame welding arc or when the engine is running. Stop the engine and allow it to cool before refueling to prevent spilled fuel from vaporizing on contact with hot engine parts and igniting. Do not spill fuel when filling tank. If fuel is spilled, wipe it up and do not start engine until fumes have been eliminated.

1.d. Keep all equipment safety guards, covers and devices in position and in good repair. Keep hands, hair, clothing and tools away from V-belts, gears, fans and all other moving parts when starting, operating or repairing equipment.

1.e. In some cases it may be necessary to remove safety guards to perform required maintenance. Remove guards only when necessary and replace them when the maintenance requiring their removal is complete. Always use the greatest care when working near moving parts.

1.f. Do not put your hands near the engine fan. Do not attempt to override the governor or idler by pushing on the throttle control rods while the engine is running.

1.g. To prevent accidentally starting gasoline engines while turning the engine or welding generator during maintenance work, disconnect the spark plug wires, distributor cap or magneto wire as appropriate.

1.h. To avoid scalding, do not remove the radiator pressure cap when the engine is hot.

ELECTRIC AND MAGNETIC FIELDS may be dangerous

2.a. Electric current flowing through any conductor causes localized Electric and Magnetic Fields (EMF). Welding current creates EMF fields around welding cables and welding machines

2.b. EMF fields may interfere with some pacemakers, and welders having a pacemaker should consult their physician before welding.

2.c. Exposure to EMF fields in welding may have other health effects which are now not known.

2.d. All welders should use the following procedures in order to minimize exposure to EMF fields from the welding circuit:

2.d.1. Route the electrode and work cables together - Secure them with tape when possible.

2.d.2. Never coil the electrode lead around your body.

2.d.3. Do not place your body between the electrode and work cables. If the electrode cable is on your right side, the work cable should also be on your right side.

2.d.4. Connect the work cable to the workpiece as close as possible to the area being welded.

2.d.5. Do not work next to welding power source.

ELECTRIC SHOCK can kill.

3.a. The electrode and work (or ground) circuits are electrically "hot" when the welder is on. Do not touch these "hot" parts with your bare skin or wet clothing. Wear dry, hole-free gloves to insulate hands.

3.b. Insulate yourself from work and ground using dry insulation. Make certain the insulation is large enough to cover your full area of physical contact with work and ground.

In addition to the normal safety precautions, if welding must be performed under electrically hazardous conditions (in damp locations or while wearing wet clothing; on metal structures such as floors, gratings or scaffolds; when in cramped positions such as sitting, kneeling or lying, if there is a high risk of unavoidable or accidental contact with the workpiece or ground) use the following equipment:

- **Semiautomatic DC Constant Voltage (Wire) Welder.**
- **DC Manual (Stick) Welder.**
- **AC Welder with Reduced Voltage Control.**

3.c. In semiautomatic or automatic wire welding, the electrode, electrode reel, welding head, nozzle or semiautomatic welding gun are also electrically "hot".

3.d. Always be sure the work cable makes a good electrical connection with the metal being welded. The connection should be as close as possible to the area being welded.

3.e. Ground the work or metal to be welded to a good electrical (earth) ground.

3.f. Maintain the electrode holder, work clamp, welding cable and welding machine in good, safe operating condition. Replace damaged insulation.

3.g. Never dip the electrode in water for cooling.

3.h. Never simultaneously touch electrically "hot" parts of electrode holders connected to two welders because voltage between the two can be the total of the open circuit voltage of both welders.

3.i. When working above floor level, use a safety belt to protect yourself from a fall should you get a shock.

3.j. Also see Items 6.c. and 8.

ARC RAYS can burn.

4.a. Use a shield with the proper filter and cover plates to protect your eyes from sparks and the rays of the arc when welding or observing open arc welding. Headshield and filter lens should conform to ANSI Z87. I standards.

4.b. Use suitable clothing made from durable flame-resistant material to protect your skin and that of your helpers from the arc rays.

4.c. Protect other nearby personnel with suitable, non-flammable screening and/or warn them not to watch the arc nor expose themselves to the arc rays or to hot spatter or metal.

FUMES AND GASES can be dangerous.

5.a. Welding may produce fumes and gases hazardous to health. Avoid breathing these fumes and gases.When welding, keep your head out of the fume. Use enough ventilation and/or exhaust at the arc to keep fumes and gases away from the breathing zone. **When welding with electrodes which require special ventilation such as stainless or hard facing (see instructions on container or MSDS) or on lead or cadmium plated steel and other metals or coatings which produce highly toxic fumes, keep exposure as low as possible and below Threshold Limit Values (TLV) using local exhaust or mechanical ventilation. In confined spaces or in some circumstances, outdoors, a respirator may be required. Additional precautions are also required when welding on galvanized steel.**

5.b. Do not weld in locations near chlorinated hydrocarbon vapors coming from degreasing, cleaning or spraying operations. The heat and rays of the arc can react with solvent vapors to form phosgene, a highly toxic gas, and other irritating products.

5.c. Shielding gases used for arc welding can displace air and cause injury or death. Always use enough ventilation, especially in confined areas, to insure breathing air is safe.

5.d. Read and understand the manufacturer's instructions for this equipment and the consumables to be used, including the material safety data sheet (MSDS) and follow your employer's safety practices. MSDS forms are available from your welding distributor or from the manufacturer.

5.e. Also see item 1.b.

WELDING SPARKS can cause fire or explosion.

6.a. Remove fire hazards from the welding area. If this is not possible, cover them to prevent the welding sparks from starting a fire. Remember that welding sparks and hot materials from welding can easily go through small cracks and openings to adjacent areas. Avoid welding near hydraulic lines. Have a fire extinguisher readily available.

6.b. Where compressed gases are to be used at the job site, special precautions should be used to prevent hazardous situations. Refer to "Safety in Welding and Cutting" (ANSI Standard Z49.1) and the operating information for the equipment being used.

6.c. When not welding, make certain no part of the electrode circuit is touching the work or ground. Accidental contact can cause overheating and create a fire hazard.

6.d. Do not heat, cut or weld tanks, drums or containers until the proper steps have been taken to insure that such procedures will not cause flammable or toxic vapors from substances inside. They can cause an explosion even though they have been "cleaned". For information, purchase "Recommended Safe Practices for the Preparation for Welding and Cutting of Containers and Piping That Have Held Hazardous Substances", AWS F4.1 from the American Welding Society (see address above).

6.e. Vent hollow castings or containers before heating, cutting or welding. They may explode.

6.f. Sparks and spatter are thrown from the welding arc. Wear oil free protective garments such as leather gloves, heavy shirt, cuffless trousers, high shoes and a cap over your hair. Wear ear plugs when welding out of position or in confined places. Always wear safety glasses with side shields when in a welding area.

6.g. Connect the work cable to the work as close to the welding area as practical. Work cables connected to the building framework or other locations away from the welding area increase the possibility of the welding current passing through lifting chains, crane cables or other alternate circuits. This can create fire hazards or overheat lifting chains or cables until they fail.

6.h. Also see item 1.c.

CYLINDER may explode if damaged.

7.a. Use only compressed gas cylinders containing the correct shielding gas for the process used and properly operating regulators designed for the gas and pressure used. All hoses, fittings, etc. should be suitable for the application and maintained in good condition.

7.b. Always keep cylinders in an upright position securely chained to an undercarriage or fixed support.

7.c. Cylinders should be located:

- Away from areas where they may be struck or subjected to physical damage.

- A safe distance from arc welding or cutting operations and any other source of heat, sparks, or flame.

7.d. Never allow the electrode, electrode holder or any other electrically "hot" parts to touch a cylinder.

7.e. Keep your head and face away from the cylinder valve outlet when opening the cylinder valve.

7.f. Valve protection caps should always be in place and hand tight except when the cylinder is in use or connected for use.

7.g. Read and follow the instructions on compressed gas cylinders, associated equipment, and CGA publication P-l, "Precautions for Safe Handling of Compressed Gases in Cylinders," available from the Compressed Gas Association 1235 Jefferson Davis Highway, Arlington, VA 22202.

FOR ELECTRICALLY powered equipment.

8.a. Turn off input power using the disconnect switch at the fuse box before working on the equipment.

8.b. Install equipment in accordance with the U.S. National Electrical Code, all local codes and the manufacturer's recommendations.

8.c. Ground the equipment in accordance with the U.S. National Electrical Code and the manufacturer's recommendations.

PRÉCAUTIONS DE SÛRETÉ

Pour votre propre protection lire et observer toutes les instructions et les précautions de sûreté specifiques qui parraissent dans ce manuel aussi bien que les précautions de sûreté générales suivantes:

Sûreté Pour Soudage A L'Arc

1. Protegez-vous contre la secousse électrique:

   a. Les circuits à l'électrode et à la piéce sont sous tension quand la machine à souder est en marche. Eviter toujours tout contact entre les parties sous tension et la peau nue ou les vétements mouillés. Porter des gants secs et sans trous pour isoler les mains.
   b. Faire trés attention de bien s'isoler de la masse quand on soude dans des endroits humides, ou sur un plancher metallique ou des grilles metalliques, principalement dans les positions assis ou couché pour lesquelles une grande partie du corps peut être en contact avec la masse.
   c. Maintenir le porte-électrode, la pince de masse, le câble de soudage et la machine à souder en bon et sûr état defonctionnement.
   d. Ne jamais plonger le porte-électrode dans l'eau pour le refroidir.
   e. Ne jamais toucher simultanément les parties sous tension des porte-électrodes connectés à deux machines à souder parce que la tension entre les deux pinces peut être le total de la tension à vide des deux machines.
   f. Si on utilise la machine à souder comme une source de courant pour soudage semi-automatique, ces precautions pour le porte-électrode s'applicuent aussi au pistolet de soudage.

2. Dans le cas de travail au dessus du niveau du sol, se protéger contre les chutes dans le cas ou on recoit un choc. Ne jamais enrouler le câble-électrode autour de n'importe quelle partie du corps.

3. Un coup d'arc peut être plus sévère qu'un coup de soliel, donc:

   a. Utiliser un bon masque avec un verre filtrant approprié ainsi qu'un verre blanc afin de se protéger les yeux du rayonnement de l'arc et des projections quand on soude ou quand on regarde l'arc.
   b. Porter des vêtements convenables afin de protéger la peau de soudeur et des aides contre le rayonnement de l'arc.
   c. Protéger l'autre personnel travaillant à proximité au soudage à l'aide d'écrans appropriés et non-inflammables.

4. Des gouttes de laitier en fusion sont émises de l'arc de soudage. Se protéger avec des vêtements de protection libres de l'huile, tels que les gants en cuir, chemise épaisse, pantalons sans revers, et chaussures montantes.

5. Toujours porter des lunettes de sécurité dans la zone de soudage. Utiliser des lunettes avec écrans lateraux dans les zones où l'on pique le laitier.

6. Eloigner les matériaux inflammables ou les recouvrir afin de prévenir tout risque d'incendie dû aux étincelles.

7. Quand on ne soude pas, poser la pince à une endroit isolé de la masse. Un court-circuit accidental peut provoquer un échauffement et un risque d'incendie.

8. S'assurer que la masse est connectée le plus prés possible de la zone de travail qu'il est pratique de le faire. Si on place la masse sur la charpente de la construction ou d'autres endroits éloignés de la zone de travail, on augmente le risque de voir passer le courant de soudage par les chaines de levage, câbles de grue, ou autres circuits. Cela peut provoquer des risques d'incendie ou d'echauffement des chaines et des câbles jusqu'à ce qu'ils se rompent.

9. Assurer une ventilation suffisante dans la zone de soudage. Ceci est particuliérement important pour le soudage de tôles galvanisées plombées, ou cadmiées ou tout autre métal qui produit des fumeés toxiques.

10. Ne pas souder en présence de vapeurs de chlore provenant d'opérations de dégraissage, nettoyage ou pistolage. La chaleur ou les rayons de l'arc peuvent réagir avec les vapeurs du solvant pour produire du phosgéne (gas fortement toxique) ou autres produits irritants.

11. Pour obtenir de plus amples renseignements sur la sûreté, voir le code "Code for safety in welding and cutting" CSA Standard W 117.2-1974.

PRÉCAUTIONS DE SÛRETÉ POUR LES MACHINES À SOUDER À TRANSFORMATEUR ET À REDRESSEUR

1. Relier à la terre le chassis du poste conformement au code de l'électricité et aux recommendations du fabricant. Le dispositif de montage ou la piece à souder doit être branché à une bonne mise à la terre.

2. Autant que possible, l'installation et l'entretien du poste seront effectués par un électricien qualifié.

3. Avant de faires des travaux à l'interieur de poste, la debrancher à l'interrupteur à la boite de fusibles.

4. Garder tous les couvercles et dispositifs de sûreté à leur place.

WARNING LABEL/OPERATING MANUAL REQUEST FORM

NOTE: S18494 WARNING LABELS, SUCH AS THE ONE BELOW FOR LINCOLN ELECTRIC WELDERS, ARE AVAILABLE FREE OF CHARGE to update your welding equipment. Operating manuals are also available upon request. PLEASE write to The Lincoln Electric Company, 22801 St. Clair Ave., Cleveland, Ohio 44117-1199 or visit www.lincolnelectric.com and make the request online.

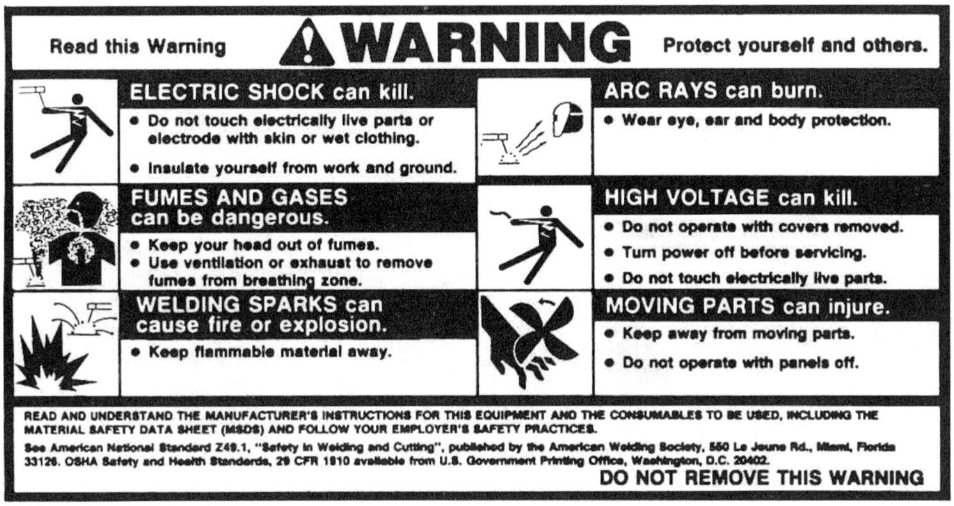

ENGINE WELDER FAN GUARDS

In order to determine whether your engine welder has the proper fan guards, compare your welder with the photo below. If your welder lacks the guards shown, contact your nearest Lincoln Field Service Shop or Distributor for assistance.

NOTE: On some engine welders, including the SA-200, the original fan guard design shown below has been modified to provide added protection and/or to make it more likely to be replaced after maintenance. Check with a Lincoln Field Service Shop or Distributor to determine if updated guarding is available for your welder.

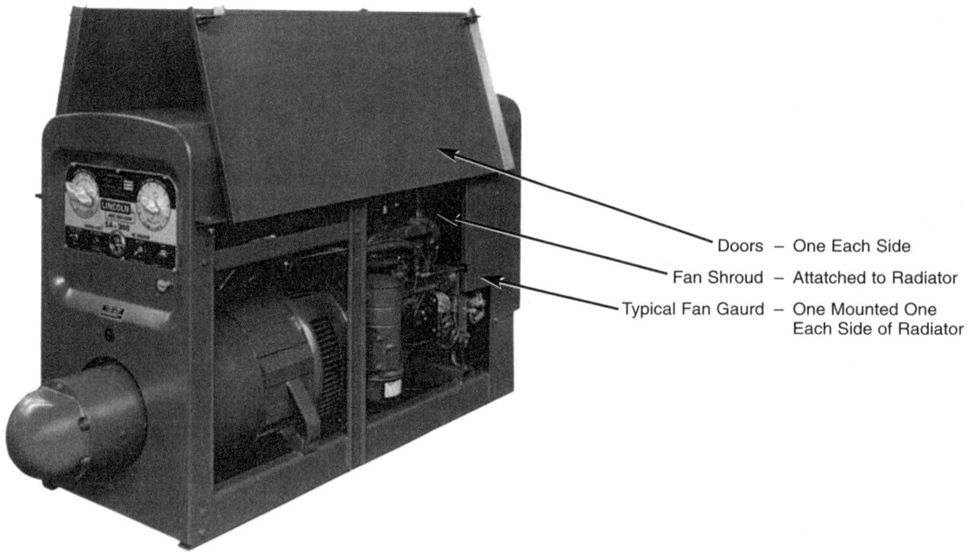

WELDING SAFETY CHECKLIST

HAZARD	FACTORS TO CONSIDER	PRECAUTION SUMMARY
Electric shock can kill 	• **Wetness** • **Welder in or on workpiece** • **Confined space** • **Electrode holder and cable insulation**	• Insulate welder from workpiece and ground using *dry* insulation. Rubber mat or dry wood. • Wear *dry, hole-free* gloves. (Change as necessary to keep dry.) • Do not touch electrically "hot" parts or electrode with bare skin or wet clothing. • If wet area and welder cannot be insulated from workpiece with dry insulation, use a semiautomatic, constant-voltage welder or stick welder with voltage reducing device. • Keep electrode holder and cable insulation in good condition. Do not use if insulation damaged or missing.
Fumes and gases can be dangerous 	• **Confined area** • **Positioning of welder's head** • **Lack of general ventilation** • **Electrode types, i.e., manganese, chromium, etc. See MSDS** • **Base metal coatings, galvanize, paint**	• Use ventilation or exhaust to keep air breathing zone clear, comfortable. • Use helmet and positioning of head to minimize fume in breathing zone. • Read warnings on electrode container and material safety data sheet (MSDS) for electrode, • Provide additional ventilation/exhaust where special ventilation requirements exist. • Use special care when welding in a confined area. • Do not weld unless ventilation is adequate.
Welding sparks can cause fire or explosion 	• **Containers which have held combustibles** • **Flammable materials**	• Do not weld on containers which have held combustible materials (unless strict AWS F4.1 procedures are followed). Check before welding. • Remove flammable materials from welding area or shield from sparks, heat. • Keep a fire watch in area during and after welding. • Keep a fire extinguisher in the welding area. • Wear fire retardant clothing and hat. Use earplugs when welding overhead.
Arc rays can burn eyes and skin 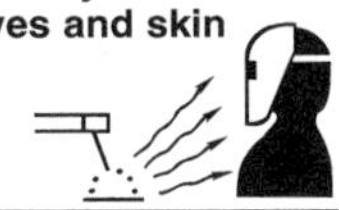	• **Process: gas-shielded arc most severe**	• Select a filter lens which is comfortable for you while welding. • Always use helmet when welding. • Provide non-flammable shielding to protect others. • Wear clothing which protects skin while welding.
Confined space 	• **Metal enclosure** • **Wetness** • **Restricted entry** • **Heavier than air gas** • **Welder inside or on workpiece**	• Carefully evaluate adequacy of ventilation especially where electrode requires special ventilation or where gas may displace breathing air. • If basic electric shock precautions cannot be followed to insulate welder from work and electrode, use semiautomatic, constant-voltage equipment with cold electrode or stick welder with voltage reducing device. • Provide welder helper and method of welder retrieval from outside enclosure.
General work area hazards 	• **Cluttered area**	• Keep cables, materials, tools neatly organized.
	• **Indirect work (welding ground) connection**	• Connect work cable as close as possible to area where welding is being performed. Do *not* allow alternate circuits through scaffold cables, hoist chains, ground leads.
	• **Electrical equipment**	• Use only double insulated or properly grounded equipment. • Always disconnect power to equipment before servicing.
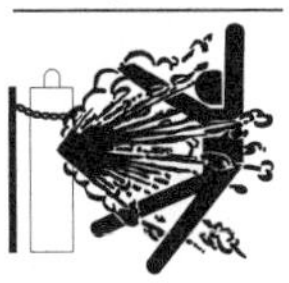	• **Engine-driven equipment**	• Use in only open, well ventilated areas. • Keep enclosure complete and guards in place. • See Lincoln service shop if guards are missing. • Refuel with engine off. • If using auxiliary power, OSHA may require GFI protection or assured grounding program (or isolated windings if less than 5KW).
	• **Gas cylinders**	• Never touch cylinder with the electrode. • Never lift a machine with cylinder attached. • Keep cylinder upright and chained to support.

Section 16
REFERENCE SECTION

SECTION 16.1
REFERENCE SECTION

Terms and Definitions

The following figures and definitions all pertain to arc welding and are taken from the American Welding Society publication "Standard Welding Terms and Definitions," AWS A3.0-89. The complete standard may be ordered from the American Welding Society, 550 N.W. LeJeune Road, P.O. Box 351040, Miami, Florida 33135. The complete standard covers all types of welding and cutting.

A

acceptable weld. A weld that meets all the requirements and the acceptance criteria prescribed by the welding specifications.

actual throat. The shortest distance between the weld root and the face of a fillet weld. See Figs. 18 & 20.

air acetylene welding (AAW). An oxyfuel gas welding process in which coalescence is produced by heating with a gas flame or flames obtained from the combustion of acetylene with air, without the application of pressure, and with or without the use of filler metal.

air carbon arc cutting (AAC). An arc cutting process that melts base metals by the heat of a carbon arc and removes the molten metal by a blast of air.

angle of bevel. See **bevel angle.**

arc blow. The deflection of an electric arc from its normal path because of magnetic forces.

arc brazing (AB). A brazing process that uses an electric arc to provide the heat. See **carbon arc brazing.**

arc cutting (AC). A group of cutting processes that melts the base metal with the heat of an arc between an electrode and the base metal. See **air carbon arc cutting, carbon arc cutting, gas metal arc cutting, gas tungsten arc cutting, metal arc cutting,** and **plasma arc cutting.**

arc cutting gun (gas metal arc cutting). A device used in semiautomatic machine, and automatic arc cutting to transfer current, guide the consumable electrode, and direct the shielding gas.

arc force. The axial force developed by an arc plasma.

arc gouging. An arc cutting process variation used to form a bevel or groove.

arc oxygen cutting. A nonstandard term for oxygen arc cutting.

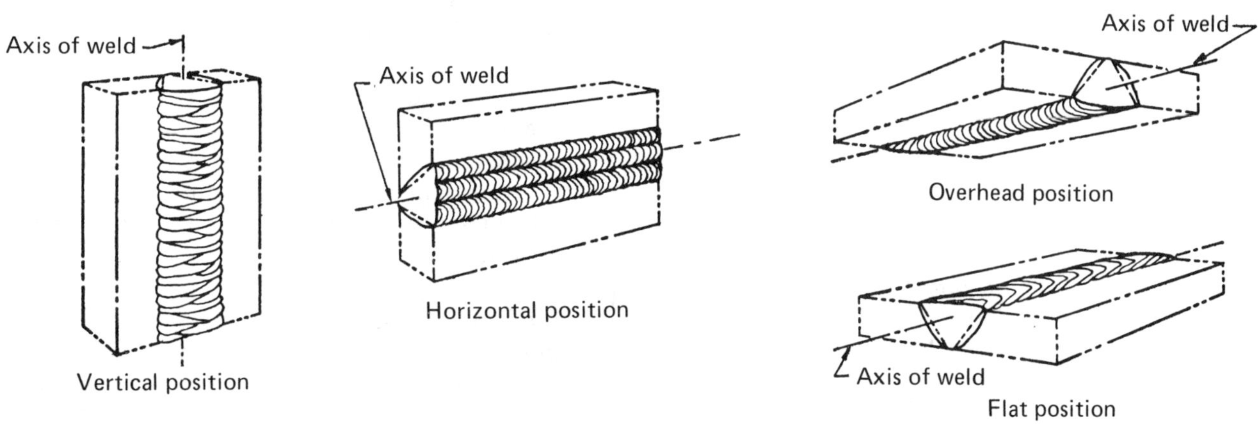

Fig. 1. Positions of Welding for Groove Welds.

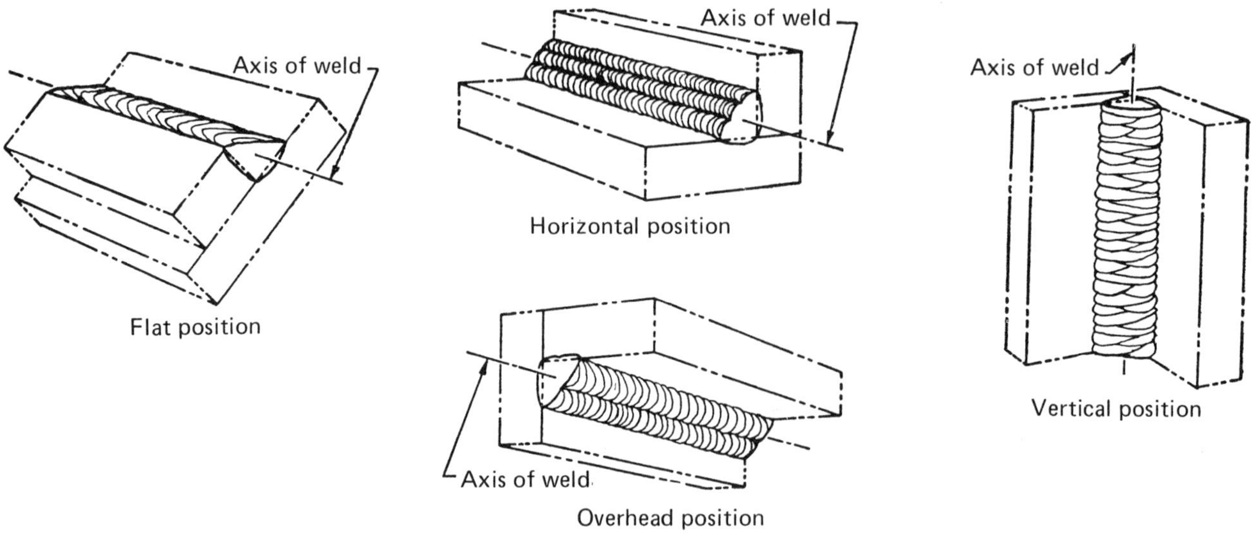

Fig. 2. Positions of Welding for Fillet Welds.

arc plasma. A gas that has been heated by an electric arc to at least a partially ionized conditions, enabling it to conduct an electric current.

arc spot weld. A spot weld made by an arc welding process.

arc spraying (ASP). A thermal spraying process using an arc between two consumable electrodes of surfacing materials as a heat source and a compressed gas to atomize and propel the surfacing material to the substrate.

arc strike. A discontinuity consisting of any localized remelted metal, heat-affected metal, or change in the surface profile of any part of a weld or base metal resulting from an arc.

arc time. The time during which an arc is maintained in making an arc weld.

arc voltage. The voltage across the welding arc.

arc welding (AW). A group of welding processes that produces coalescence of metals by heating them with an arc, with or without the application of pressure, and with or without the use of filler metal.

arc welding electrode. A component of the welding circuit through which current is conducted and which terminates at the arc.

arc welding gun. A device used in semiautomatic, machine, and automatic arc welding to transfer current, guide the consumable electrode, and direct the shielding gas.

as-welded. The condition of weld metal, welded joints, and weldments after welding, but prior to any subsequent thermal, mechanical, or chemical treatments.

atomic hydrogen welding (AHW). An arc welding process that uses an arc between two metal electrodes in a shielding atmosphere of hydrogen and without the application of pressure. This is an obsolete or seldom used process.

autogenous weld. A fusion weld made without the addition of filler metal.

automatic oxygen cutting. Oxygen cutting with equipment that performs the cutting operation without constant observation and adjustment of the controls by an operator. The equipment may or may not load and unload the workpieces. See also **machine oxygen cutting.**

automatic welding. Welding with equipment that performs the welding operation without adjustment of the controls by a welding operator. The equipment may or may not load and unload the workpieces. See also **machine welding.**

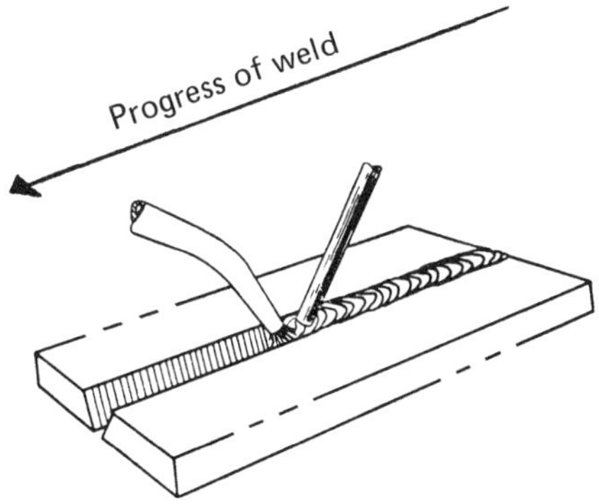

Fig. 3. Backhand Welding.

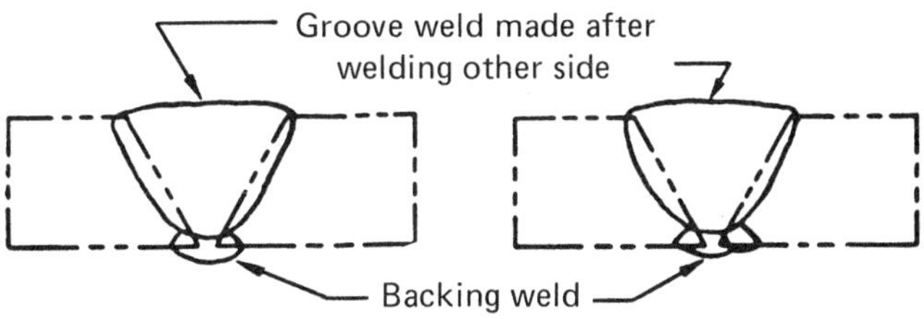

Fig. 4. Backing Welds.

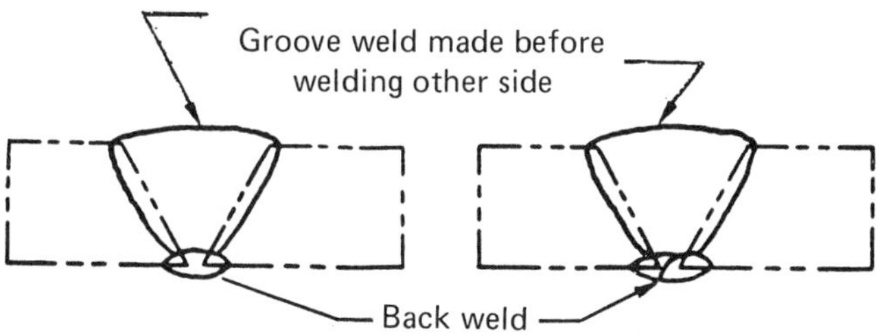

Fig. 5. Back Welds.

B

back bead. A weld bead resulting from a back weld pass.

backfire. The momentary recession of the flame into the welding tip or cutting tip followed by immediate reappearance or complete extinction of the flame.

back gouging. The removal of weld metal and base metal from the other side of a partially welded joint to facilitate complete fusion and complete joint penetration upon subsequent welding from that side.

backhand welding. A welding technique in which the welding torch or gun is directed opposite to the progress of welding. See Fig. 3.

backing. A material or device placed against the back side of the joint, or at both sides of a weld in electroslag and electrogas welding, to support and retain molten weld metal. The material may be partially fused or remain unfused during welding and may be either metal or nonmetal.

backing bead. A weld bead resulting from a backing pass.

backing pass. A weld pass made for a backing weld.

backing ring. Backing in the form of a ring, generally used in the welding of pipe.

backing shoe. A nonconsumable backing device used in electroslag and electrogas welding.

backing weld. Backing in the form of a weld. See Fig. 4.

backstep sequence. A longitudinal sequence in which weld passes are made in the direction opposite to the progress of welding. See Fig. 6.

backweld. A weld made at the back of a single groove weld. See Fig. 5.

balling up. The formation of globules of molten brazing filler metal or flux due to lack of wetting of the base metal.

bare electrode. A filler metal electrode consisting of a single metal or alloy that has been produced into a wire, strip, or bar form and that has had no coating or covering applied to it other than that which was incidental to its manufacture or preservation.

bare metal arc welding (BMAW). An arc welding process that produces coalescence of metals by heating them with an electric arc between a bare or lightly coated metal electrode and the work. Neither shielding nor pressure is used, and filler metal is obtained from the electrode.

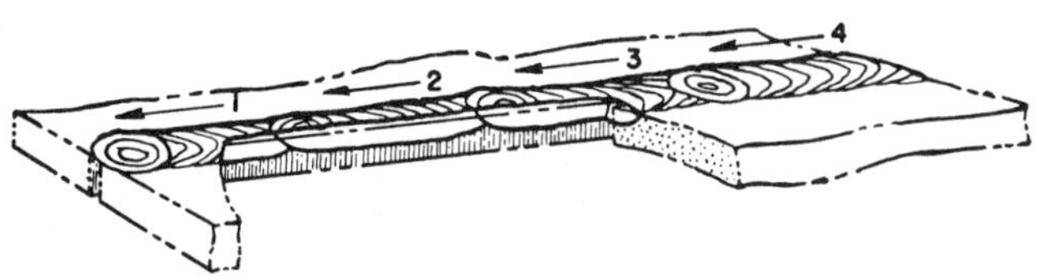

Fig. 6. Backstep Sequence.

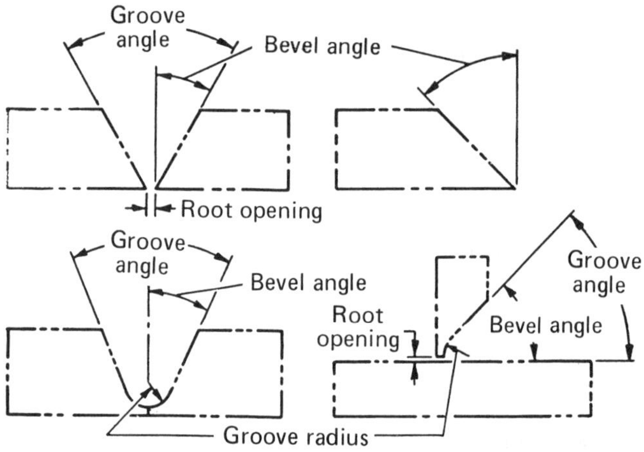

Fig. 7. Bevel Angle, Groove Angle, Groove Radius and Root Openting.

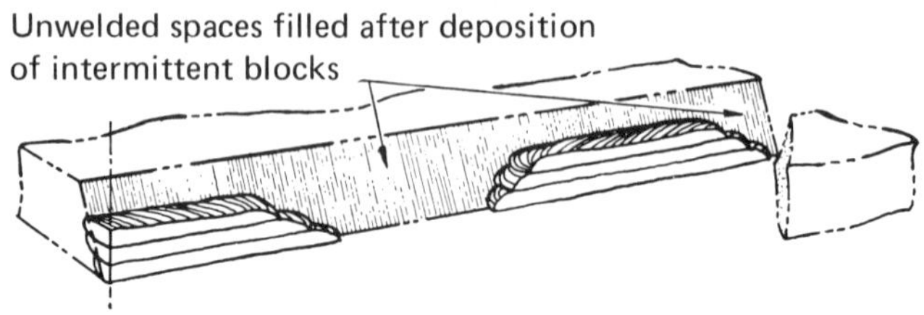

Fig. 8. Block Sequence.

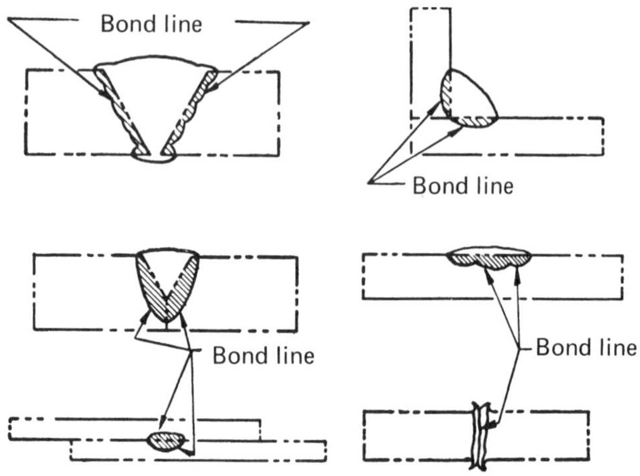

(Fusion zones indicated by cross-hatching)

Fig. 9. Bond Line and Fusion Zone.

base material. The material to be welded, brazed, soldered, or cut. See also **base metal,** and **substrate.**

base metal. The metal to be welded, brazed, soldered, or cut. The use of this term implies that materials other than metals are also referred to, where this is appropriate. See also **base material** and **substrate.**

base metal test specimen. A test specimen composed wholly of base metal.

bead. See **weld bead.**

bead weld. A nonstandard term for **surfacing weld.**

bevel angle. The angle formed between the prepared edge of a member and a plane perpendicular to the surface of the member. See Fig. 7.

bevel groove weld. A type of groove weld.

blacksmith welding. A nonstandard term for **forge welding.**

blasting. A method of cleaning or surface roughening by a forcibly projected stream of sharp angular abrasive.

blind joint. A joint, no portion of which is visible.

block sequence. A combined longitudinal and cross-sectional sequence for a continuous multiple pass weld in which separated increments are completely or partially welded before intervening increments are welded. See Fig. 8.

blowhole. A nonstandard term for **porosity.**

bonding force. The force that holds two atoms together; it results form a decrease in energy as two atoms are brought closer to one another.

bond line. The cross section of the interface between thermal spray deposits and substrate, or between adhesive and adherend in an adhesive bonded joint.

bottle. A nonstandard term for **gas cylinder.**

boxing. The continuation of a fillet weld around a corner of a member as an extension of the principal weld. See Fig. 10.

brazing filler metal. The metal that fills the capillary joint clearance and has a liquidus above 840°F (450°C), but below the solidus of the base metals.

brazing operator. One who operates machine or automatic brazing equipment.

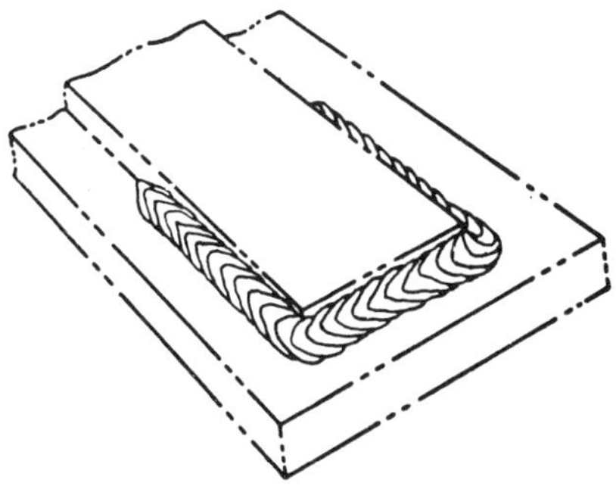

Fig. 10. Boxing.

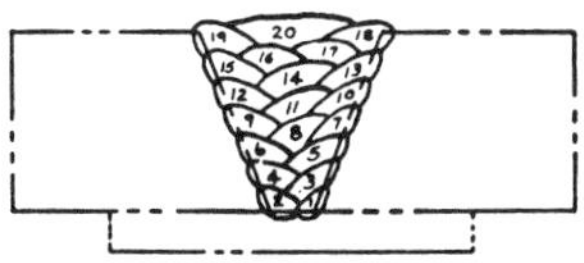

Fig. 11. Buildup Sequence.

buildup. A surfacing variation in which surfacing metal is deposited to achieve the required dimensions. See also **buttering, cladding,** and **hardfacing.**

buildup sequence. A nonstandard term for **cross-sectional sequence.** See Fig. 11.

burnoff rate. A nonstandard term for **melting rate.**

burn through. A nonstandard term for excessive melt through or a hole.

burn through weld. A nonstandard term for a **seam weld** or **spot weld.**

buttering. A surfacing variation that deposits surfacing metal on one or more surfaces to provide metallurgically compatible weld metal for the subsequent completion of the weld. See also **buildup, cladding,** and **hardfacing.**

butt joint. A joint between two members aligned approximately in the same plane. See Fig. 12.

button. That part of a weld, including all or part of the nugget, that tears out in the destructive testing of spot, seam or projection welded specimens.

butt weld. A nonstandard term for a weld in a butt joint.

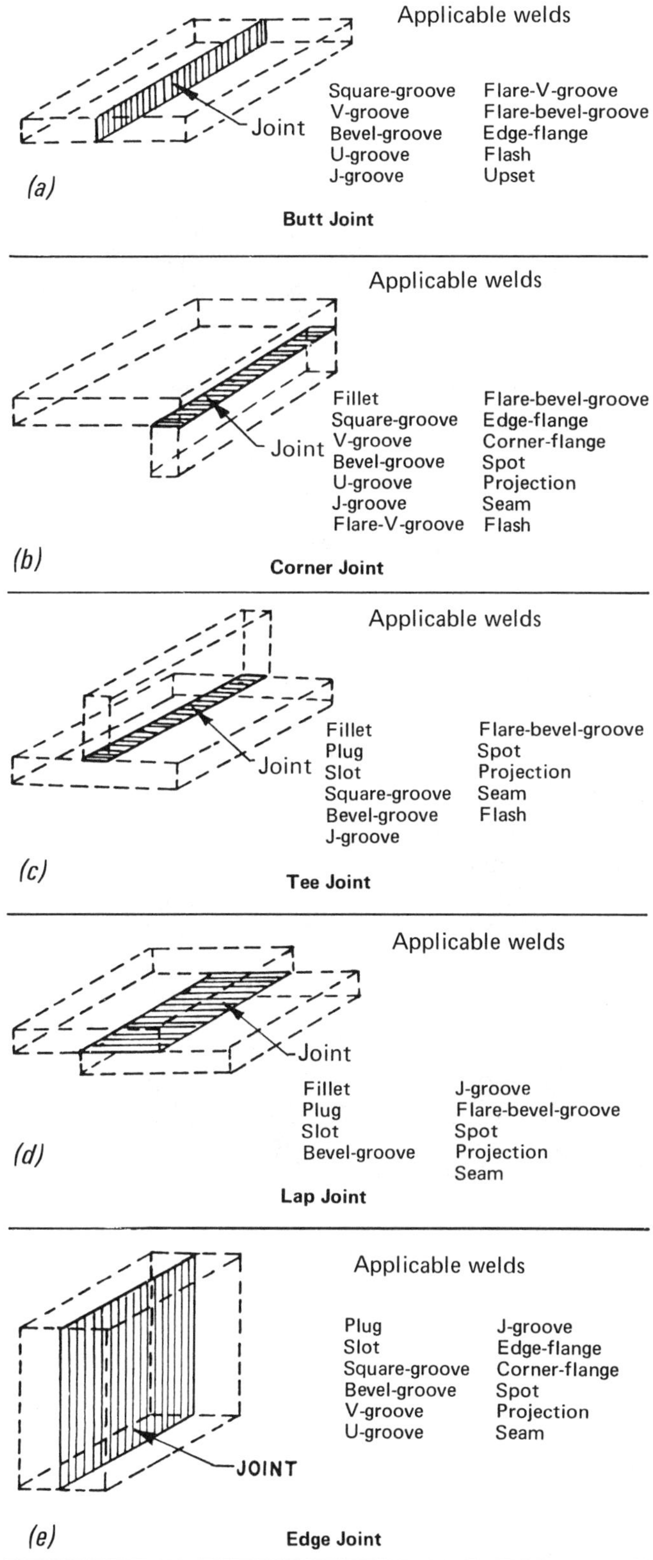

Fig. 12. Types of Joints.

C

capillary action. The force by which liquid, in contact with a solid, is distributed between closely fitted faying surfaces of the joint to be brazed or soldered.

carbon arc brazing (CAB). A brazing process that produces coalescence of metals by heating them with an electric arc between two carbon electrodes. The filler metal is distributed in the joint by capillary attraction.

carbon arc cutting (CAC). An arc cutting process that severs base metals by melting them with the heat of an arc between a carbon electrode and the base metal.

carbon arc welding (CAW). An arc welding process that produces coalescence of metals by heating them with an arc between a carbon electrode and the base metal. No shielding is used. Pressure and filler metal may or may not be used.

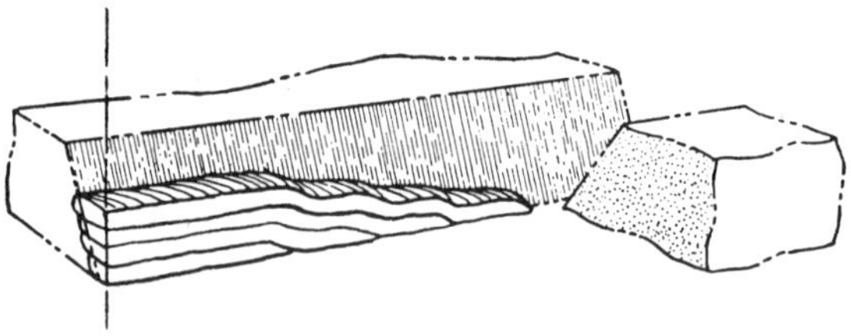

Fig. 13. Cascade Sequence.

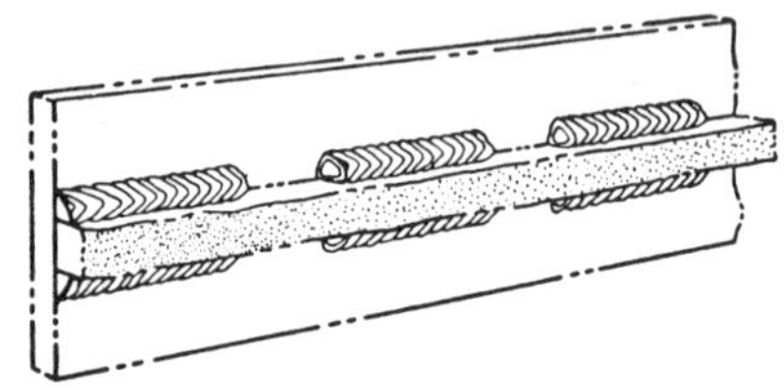

Fig. 14. Chain Intermittent Fillet Welding.

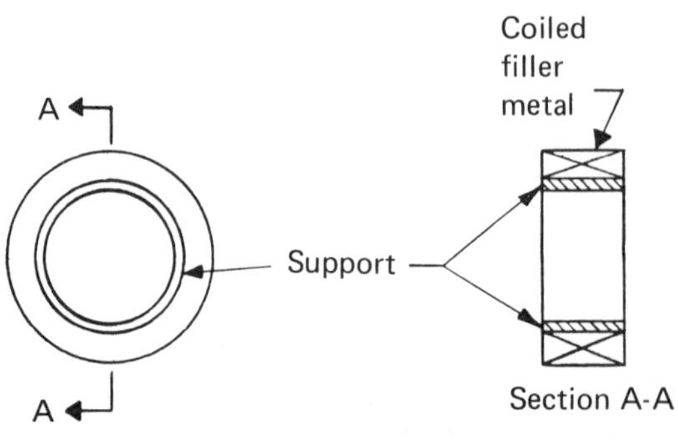

Fig. 15. Coil with Support.

carbon electrode. A nonfiller metal electrode used in arc welding or cutting, consisting of a carbon or graphite rod, which may be coated with copper or other materials.

carbonizing flame. A nonstandard term for **reducing flame.**

carburizing flame. A nonstandard term for **reducing flame.**

cascade sequence. A combined longitudinal and cross-sectional sequence in which weld passes are made in overlapping layers. See Fig. 13.

chain intermittent weld. An intermittent weld on both sides of a joint in which the weld increments on one side are approximately opposite those on the other side. See Fig. 14.

chamfer. A nonstandard term for **bevel.**

chill ring. A nonstandard term for **backing ring.**

cladding. A surfacing variation that deposits or applies surfacing material usually to improve corrosion or heat resistance. See also, **buildup, buttering,** and **hardfacing.**

clad metal. A laminar composite consisting of a metal, with a metal of different chemical composition applied to one or two sides by surfacing, casting, thick chemical deposition, or thick electroplating.

coalescence. The growing together or growth into one body of the materials being welded.

coated electrode. A nonstandard term for **covered electrode** or **lightly covered electrode.**

coil without support. A filler metal package consisting of a continuous length of welding wire in coil form without an internal support. It is appropriately bound to maintain its shape.

coil with support. A filler metal package consisting of a continuous length of welding wire in coil form wound on a simple cylinder without flanges. See Fig. 15.

cold crack. A crack which develops after solidification is complete.

cold soldered joint. A joint with incomplete coalescence caused by insufficient application of heat to the base metal during soldering.

cold welding (CW). A solid-state welding process in which pressure is used at room temperature to produce coalescence of metals with substantial deformation at the weld. See also **hot pressure welding, diffusion welding,** and **forge welding.**

complete fusion. Fusion which has occurred over the entire base metal surfaces intended for welding and between all adjoining weld beads. See Fig. 16.

complete joint penetration. A penetration by weld metal for the full thickness of the base metal in a joint with a groove weld. Secc Fig. 17.

complete penetration. A nonstandard term for **complete joint penetration.**

composite electrode. Any of a number of multi-component filler metal electrodes in various physical forms such as stranded wires, tubes, and covered wire. See also **covered electrode, flux cored electrode, metal cored electrode,** and **stranded electrode.**

concave fillet weld. A fillet weld having a concave face. See Fig. 18.

concave root surface. See Fig. 19.

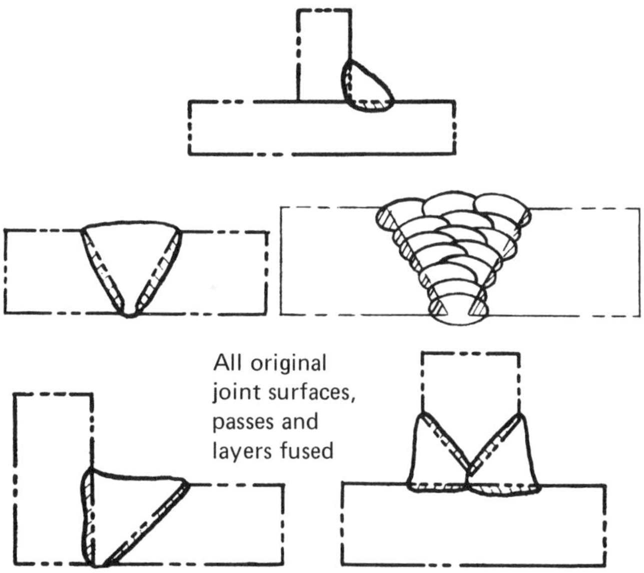

Fig. 16. Complete Fusion.

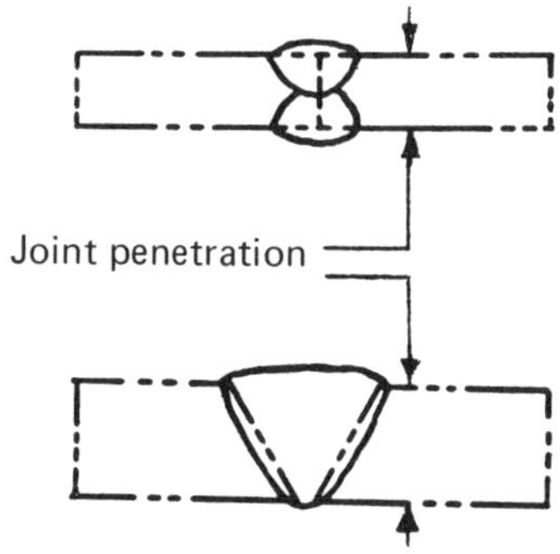

Fig. 17. Complete Joint Penetration.

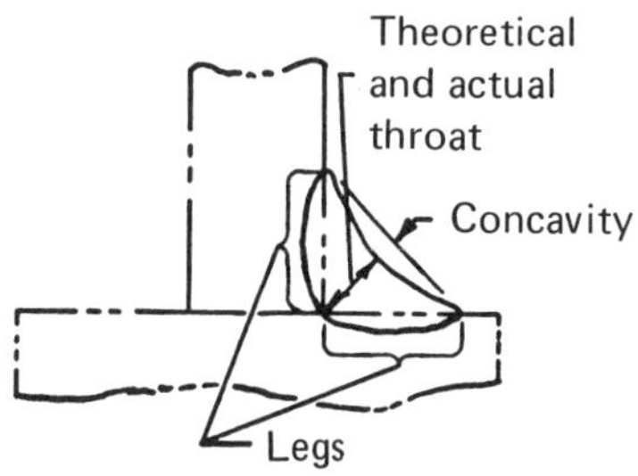

Fig. 18. Concave Fillet Weld.

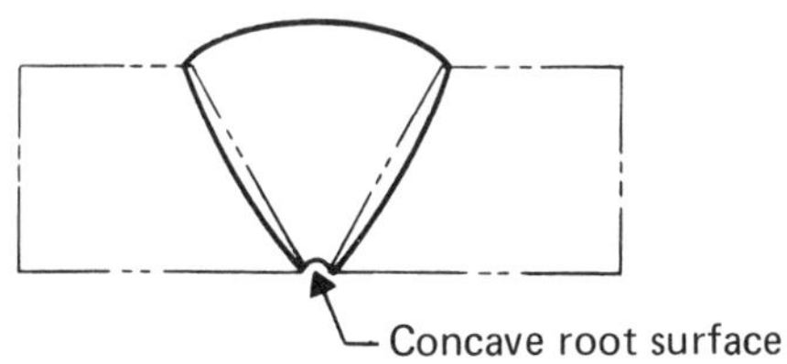

Fig. 19. Concave Root Surface.

concavity. The maximum distance from the face of a concave fillet weld perpendicular to a line joining the weld toes. See Fig. 18.

concurrent heating. The application of supplemental heat to a structure during welding or cutting.

cone. The conical part of an oxyfuel gas flame next to the orifice of the tip.

constricted arc (plasma arc welding and cutting). A plasma arc column that is shaped by a constricting nozzle orifice.

constricting nozzle (plasma arc welding and cutting). A water-cooled copper nozzle surrounding the electrode and containing the constricting orifice.

constricting orifice (plasma arc welding and cutting). The hole in the constricting nozzle through which the arc passes.

consumable guide electroslag welding. An electroslag welding process variation in which filler metal is supplied by an electrode and its guiding member.

consumable insert. Preplaced filler metal that is completely fused into the joint root and becomes part of the weld.

contact resistance (resistance welding). Resistance to the flow of electric current between two workpieces or an electrode and a workpiece.

contact tube. A device which transfers current to a continuous electrode.

continuous sequence. A longitudinal sequence in which each weld pass is made continuously from one end of the weld to the other.

continuous weld. A weld that extends continuously from one end of a joint to the other. Where the joint is essentially circular, it extends completely around the joint.

convex fillet weld. A fillet weld having a convex face. See Fig. 20.

convexity. The maximum distance from the face of a convex fillet weld perpendicular to a line joining the weld toes. See Fig. 20.

cored solder. A solder wire or bar containing flux as a core.

corner-flange weld. A flange weld with only one member flanged at the joint. See Fig. 21.

corner joint. A joint between two members located approximately at right angles to each other. See Fig. 12.

corrective lens (eye protection). A lens ground to the wearer's individual corrective prescription.

corrosive flux. A flux with a residue that chemically attacks the base metal. It may be composed of inorganic salts and acids, organic salts and acids, or activated rosin.

cosmetic pass. A weld pass made primarily to enhance appearance.

CO_2 welding. A nonstandard term for gas metal arc welding.

covered electrode. A composite filler metal electrode consisting of a core of a bare electrode or metal cored electrode to which a covering sufficient to provide a slag layer on the weld metal has been applied. The covering may contain materials providing such functions as shielding from the atmosphere, deoxidation, and arc stabilization, and can serve as a source of metallic additions to the weld.

cover lens (eye protection). A round cover plate.

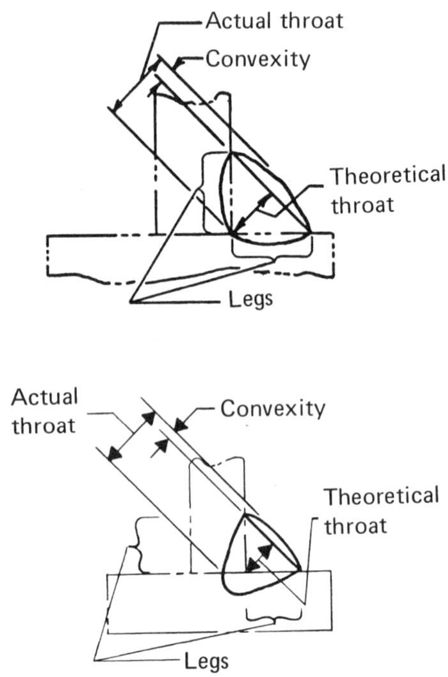

Fig. 20. Convex Fillet Weld.

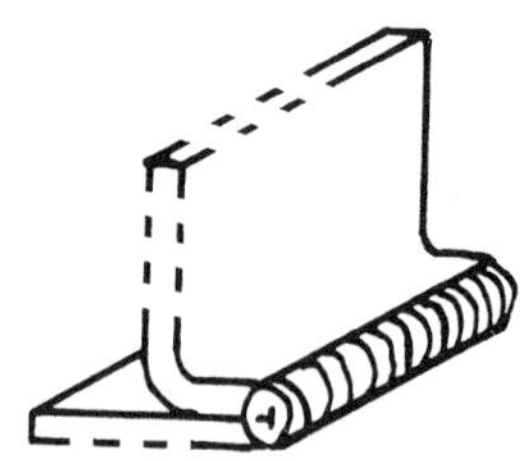

Fig. 21. Corner-Flange Weld.

cover plate (eye protection). A removable pane of colorless glass, plastic-coated glass, or plastic that covers the filter plate and protects it from weld spatter, pitting, or scratching when used in a helmet, hood, or goggles.

crack. A fracture type discontinuity characterized by a sharp tip and high ratio of length and width to opening displacement.

crater. A depression at the termination of a weld bead.

crater crack. A crack in the crater of weld bead.

crater fill current. The current value during crater fill time.

crater fill time. The time interval following weld time but prior to melt-back time during which arc voltage or current reach a preset value greater or less than welding values. Weld travel may or may not stop at this point.

crater fill voltage. The arc voltage value during crater fill time.

cross-sectional sequence. The order in which the weld passes of a multiple pass weld are made with respect to the cross section of the weld. See also **block sequence** and **cascade sequence.**

cross wire welding. A common variation of projection welding wherein the localization of the welding current is achieved by the intersection contact of wires, and is usually accompanied by considerable embedding of one wire into another.

cutting attachment. A device for converting an oxyfuel gas welding torch into an oxygen cutting torch.

cutting head. The part of a cutting machine or automatic cutting equipment in which a cutting torch or tip is incorporated.

cutting nozzle. A nonstandard term for **cutting tip.**

cutting process. A process that results in the severing or removal of metals. See also **arc cutting** and **oxygen cutting.**

cutting tip. The part of an oxygen cutting torch from which the gases issue.

cutting torch (arc). A device used in air carbon arc cutting, gas tungsten arc cutting, and plasma arc cutting to control the position of the electrode, to transfer current, and to control the flow of gases.

cutting torch (oxyfuel gas). A device used for directing the preheating flame produced by the controlled combustion of fuel gases, and to direct and control the cutting oxygen.

cylinder. See **gas cylinder.**

cylinder manifold. A multiple header for interconnection of gas or fluid sources with distribution points.

D

defect. A discontinuity or discontinuities that by nature or accumulated effect (for example, total crack length) render a part or product unable to meet minimum applicable acceptance standards or specifications. This term designates rejectability. See also **discontinuity** and **flaw.**

defective weld. A weld containing one or more defects.

deposited metal. Filler metal that has been added during welding.

deposition efficiency (arc welding). The ratio of the weight of deposited metal to the net weight of filler metal consumed, exclusive of stubs.

deposition efficiency (thermal spraying). The ratio of the weight of spray deposit to the weight of the surfacing material sprayed, usually expressed in percent.

deposition rate. The weight of material deposited in a unit of time.

deposition sequence. A nonstandard term for **weld pass sequence.**

deposit sequence. A nonstandard term for weld pass sequence.

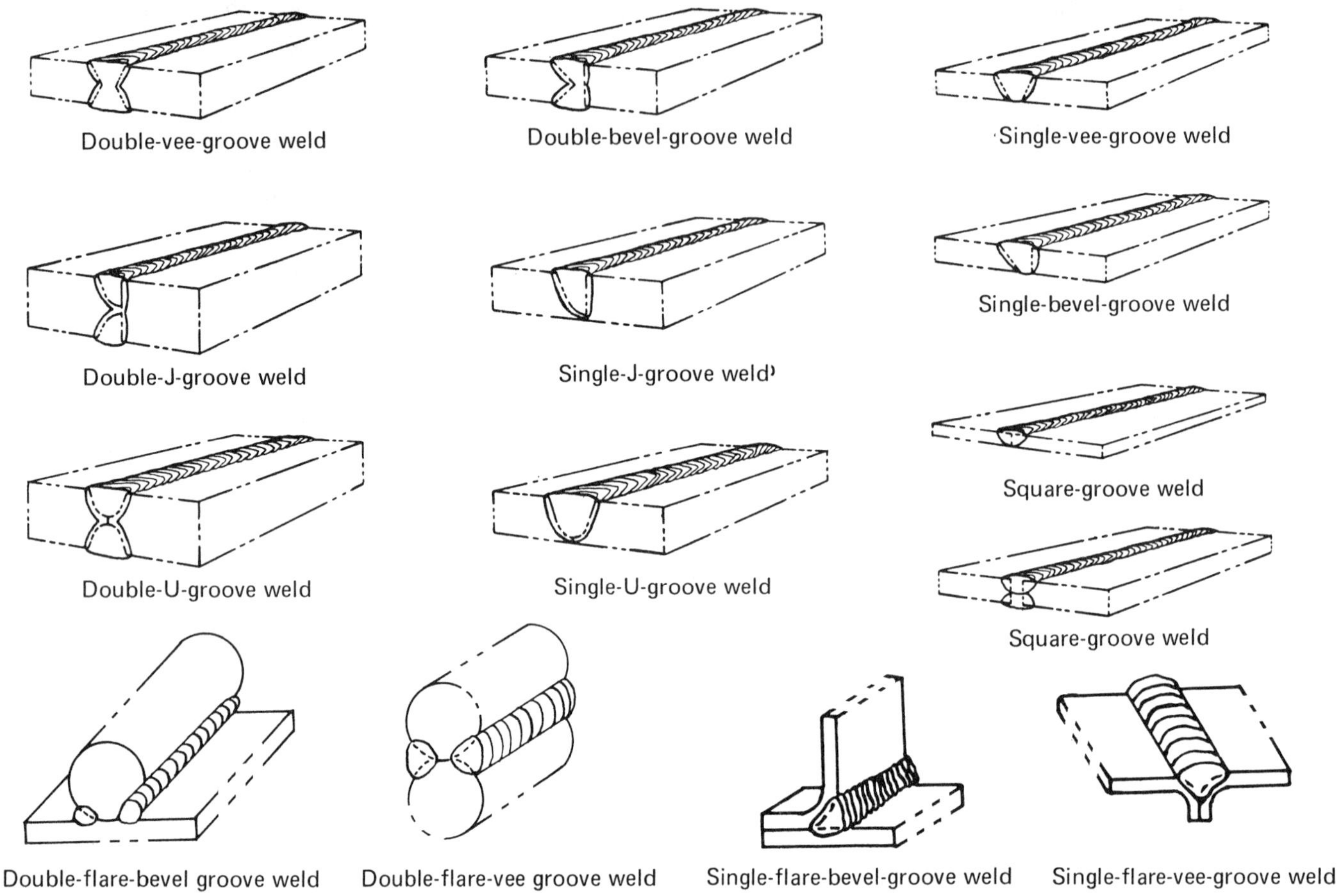

Fig. 22. Types of Groove Welds.

depth of fusion. The distance that fusion extends into the base metal or previous pass from the surface melted during welding. See Fig. 31.

dilution. The change in chemical composition of a welding filler metal caused by the admixture of the base metal or previous weld metal in the weld bead. It is measured by the percentage of base metal or previous weld metal in the weld bead.

dip soldering. A soldering process using the heat furnished by a molten metal bath which provides the solder filler metal.

direct current electrode negative (DCEN). The arrangement of direct current arc welding leads in which the workpiece is the positive pole and the electrode is the negative pole of the welding arc. See Fig. 26.

direct current electrode positive (DCEP). The arrangement of direct current arc welding leads in which the workpiece is the negative pole and the electrode is the positive pole of the welding arc. See Fig. 25.

direct current reverse polarity. A nonstandard term for **direct current electrode positive.**

direct current straight polarity. A nonstandard term for **direct current electrode negative.**

discontinuity. An interruption of the typical structure of a weldment, such as a lack of homogeneity in the mechanical, metallurgical, or physical characteristics of the material or weldment. A discontinuity is not necessarily a defect. See also **defect** and **flaw.**

double-bevel-groove weld. A type of groove weld. See Fig. 22.

double-flare-bevel-groove weld. A weld in grooves formed by a member with a curved surface in contact with a planar member. See Fig. 22.

double-flare-V-groove weld. A weld in grooves formed by two members with curved surface. See Fig. 22.

double-J-groove weld. A type of groove weld. See Fig. 22.

double-square-groove weld. A typeof groove weld. See Fig. 22.

double-U-groove weld. A type of groove weld. See Fig. 22.

double-V-groove weld. A type of groove weld. See Fig. 22.

double-welded joint. A fusion welded joint that is welded from both sides. See Fig. 23.

downhand. A nonstandard term for **flat position.**

downslope time (resistance welding). The time during which the welding current is continuously decreased.

downslope time (automatic arc welding). The time during which the current is changed continuously from final taper current or welding current to final current.

drag angle. The travel angle when the electrode is pointing in a direction opposite to the progression of welding. This angle can also be used to partially define the position of guns, torches, rods and beams. See **backhand welding, push angle, travel angle,** and **work angle.**

drop-through. An undesirable sagging or surface irregularity, usually encountered when brazing or welding near the solidus of the base metal, caused by overheating with rapid diffusion or alloying between the filler metal and the base metal.

drum. A filler metal package consisting of a continuous length of welding wire wound or coiled within an enclosed cylindrical container.

duty cycle. The percentage of time during an arbitrary test period, usually 10 minutes, during which a power supply can be operated at its rated output without overheating.

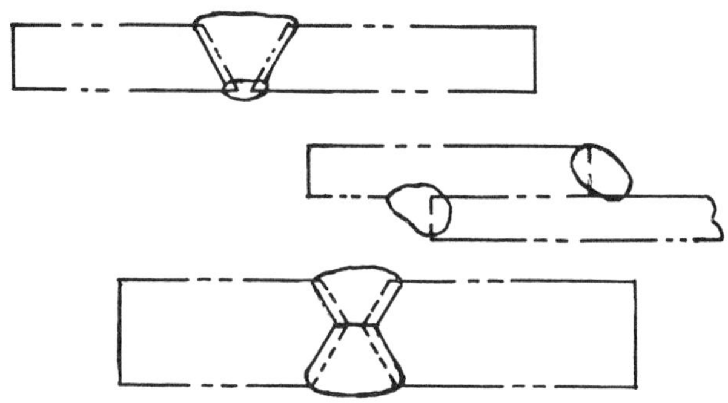

Fig. 23. Double-Welded Joint.

E

edge-flange weld. A flange weld with two members flanged at the location of welding. See Fig. 24.

edge point. A joint between the edges of two or more parallel or nearly parallel members. See Fig. 12.

edge preparation. The surface prepared on the edge of a member for welding.

edge weld. A weld in an edge joint.

edge weld size. The weld metal thickness measured at the weld root.

effective length of weld. The length throughout which the correctly proportioned cross section of the weld exists. In a curved weld, it shall be measured along the weld axis.

effective throat. The minimum distance minus any convexity between the weld root and the face of a fillet weld.

electrode extension. For gas metal arc welding, flux cored arc welding, and submerged arc welding, the length of unmelted electrode extending beyond the end of the contact tube.

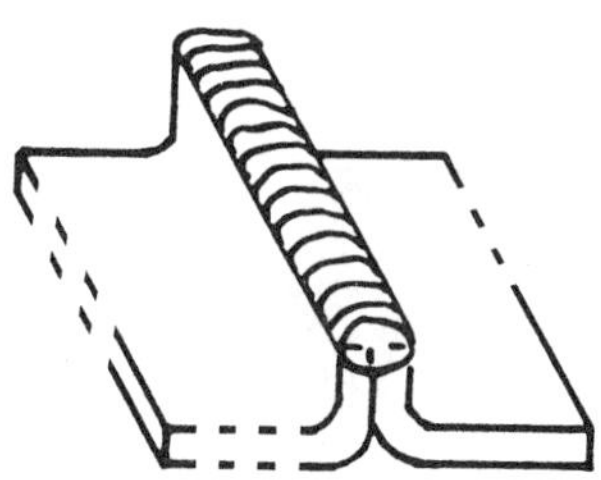

Fig. 24. Edge-Flange Weld.

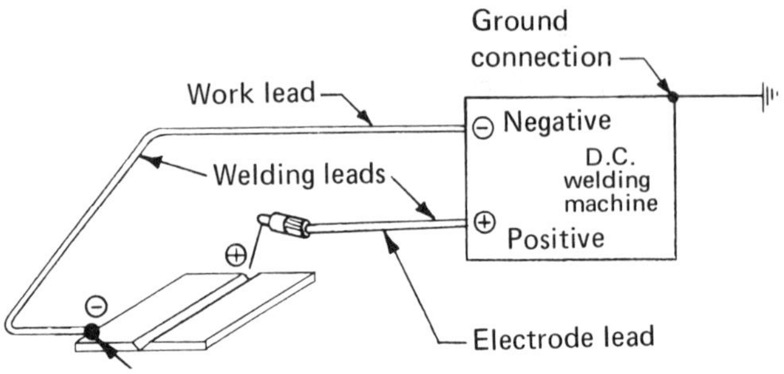

Fig. 25. Electrode Positive Polarity

electrode holder. A device used for mechanically holding the electrode while conducting current to it. See Figs. 25 & 26.

electrode lead. The electrical conductor between the source of arc welding current and the electrode holder. See Figs. 25 & 26.

electrogas welding (EGW). An arc welding process that produces coalescence of metals by heating them with an arc between a continuous filler metal electrode and the work. Molding shoes are used to confine the molten weld metal for vertical position welding. The electrodes may be either flux cored or solid. Shielding may or may not be obtained from an externally supplied gas or mixture.

electron beam cutting (EBC). A cutting process that uses the heat obtained from a concentrated beam composed primarily of high velocity electrons which impinge upon the workpieces; it may or may not use an externally supplied gas.

electron beam gun. A device for producing and accelerating electrons. Typical components include the emitter (also called the filament or cathode) which is heated to produce electrons via thermionic emission, a cup (also called the grid or grid cup), and the anode.

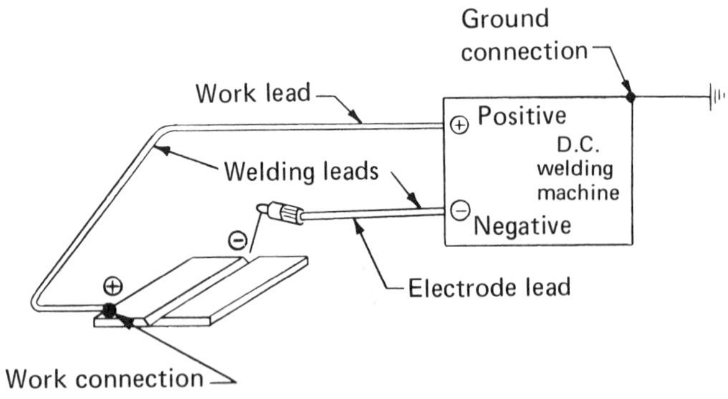

Fig. 26. Electrode Negative Polarity

electron beam gun column. The electron beam gun plus auxiliary mechanical and electrical components which may include beam alignment, focus, and deflection coils.

electron beam welding (EBW). A welding process that produces coalescence of metals with the heat obtained from a concentrated beam composed primarily of high velocity electrons impinging on the joint. See also **electron beam welding — high vacuum, electron beam welding — medium vacuum,** and **electron beam welding — non-vacuum.**

electron beam welding — high vacuum (EBW-HV). An electron beam welding process variation in which welding is accomplished at a pressure of 10^{-4} to 10^{-1} pascals (approximately 10^{-6} to 10^{-3} torr).

electron beam welding — medium vacuum (EBW-MV). An electron beam welding process variation in which welding is accomplished at a pressure of 10^{-1} to 3×10^{3} pascals (approximately 10^{-3} to 25 torr.)

electron beam welding — nonvacuum (EBW-NV). An electron beam welding process variation in which welding is accomplished at atmospheric pressure.

electroslag welding (ESW). A welding process that produces coalescence of metals with molten slag that melts the filler metal and the surfaces of the workpieces. The weld pool is shielded by this slag which moves along the full cross section of the joint as welding progresses. The process is initiated by an arc that heats the slag. The arc is then extinguished by the conductive slag, which is kept molten by its resistance to electric current passing between the electrode and the workpieces, See also **electroslag welding electrode** and **consumable guide electroslag welding.**

electroslag welding electrode. A filler metal component of the welding circuit through which current is conducted from the electrode guiding member to the molten slag.

emissive electrode. A filler metal electrode consisting of a core of a bare electrode or a composite electrode to which a very light coating has been applied to produce a stable arc.

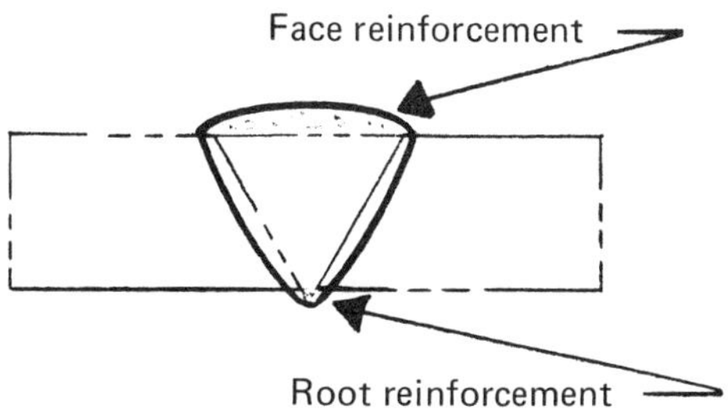

Fig. 27. Reinforcement of Weld.

end return. A nonstandard term for **boxing.**

exhaust booth. A mechanically ventilated, semi-enclosed area in which an air flow across the work area is used to remove fumes, gases, and solid particles.

explosion welding (EXW). A solid-state welding process that effects coalescence by high velocity movement together of the workpieces produced by a controlled detonation.

F

face bend test. A test in which the weld face is on the convex surface of a specified bend radius.

face of weld. See **weld face.** See Fig. 28.

face reinforcement. Weld reinforcement at the side of the joint from which welding was done. See Fig. 27.

face shield (eye protection). A device positioned in front of the eyes and over all or a portion of the face to protect the eyes and face. See also **hand shield** and **helmet.**

faying surface. That mating surface of a member that is in contact with or in close proximity to another member to which it is to be joined.

ferrite number. An arbitrary, standardized value designating the ferrite content of an austenitic stainless steel weld metal. It should be used in place of percent ferrite or volume percent ferrite on a direct replacement basis.

filler metal. The metal to be added in making a welded, brazed, or soldered joint. See also **brazing filler metal, consumable insert, welding electrode, solder, welding rod,** and **welding wire.**

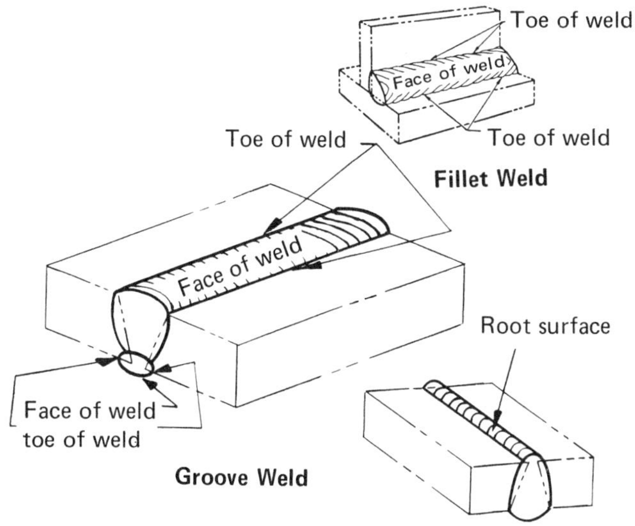

Fig. 28. Face, Root Surface and Toe of Weld.

filler wire. A nonstandard term for **welding wire.**

fillet weld. A weld of approximately triangular cross section joining two surfaces approximately at right angles to each other in a lap joint, T-joint, or corner joint. See Figs. 18, 20 & 29.

fillet weld break test. A test in which the specimen is loaded so that the weld root is in tension.

fillet weld leg. The distance from the joint root to the toe of the fillet weld. See Figs. 18 & 20.

fillet weld size. For equal leg fillet welds, the leg lengths of the largest isosceles right triangle which can be inscribed within the fillet weld cross section. For unequal leg fillet welds, the leg lengths of the largest right triangle that can be inscribed within the fillet weld cross section. See Fig. 29.

fillet weld throat. See **actual throat, effective throat,** and **theoretical throat.**

filter glass. A nonstandard term for **filter plate.**

filter lens. (eye protection). A round filter plate.

filter plate (eye protection). An optical material that protects the eyes against excessive ultraviolet, infrared and visible radiation.

final current. The current after downslope but prior to current shut-off.

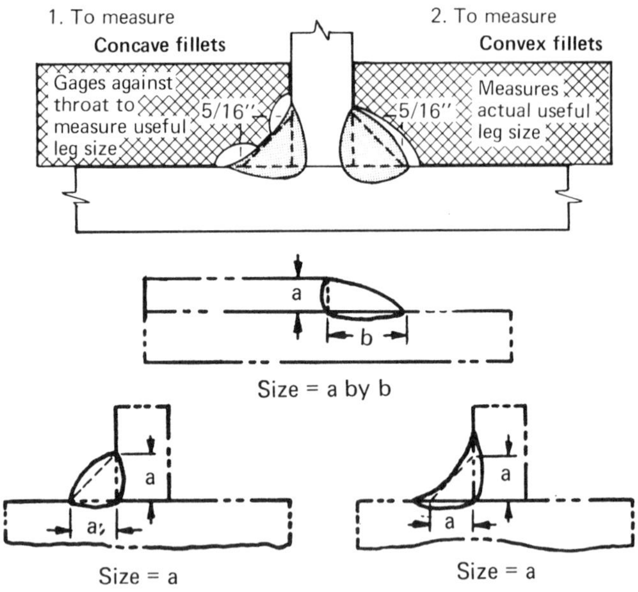

Fig. 29. Size of Fillet Welds.

final taper current. The current at the end of the taper interval prior to downslope.

fines. All material finer than a particular mesh size under consideration.

firecracker welding. A variation of the shielded metal arc welding process in which a length of covered electrode is placed along the joint in contact with the workpieces. During the welding operation, the stationary electrode is consumed as the arc travels the length of the electrode.

fisheye. A discontinuity found on the fracture surface of a weld in steel that consists of a small pore or inclusion surrounded by an approximately round, bright area.

fissure. A nonstandard term for **crack.**

fixture. A device designed to hold parts to be joined in proper relation to each other.

flame cutting. A nonstandard term for **oxygen cutting.**

flange weld. A weld made on the edges of two or more members to be joined, usually light gage metal, at least one of the members being flanged.

flange weld size. The weld metal thickness measured at the weld root. See Fig. 51.

flare-bevel-groove weld. A weld in groove formed by a member with a curved surface in contact with a planar member. See Fig. 22.

flare-V-groove weld. A weld in a groove formed by two members with curved surfaces. See Fig. 22.

flash. Material that is expelled from a flash weld prior to the upset portion of the welding cycle.

flashback. A recession of the flame into or back of the mixing chamber of the torch.

flashback arrester. A device to limit damage form a flashback by preventing propagation of the flame front beyond the location of the arrester.

flat position. The welding position used to weld from the upper side of the joint; the face of the weld is approximately horizontal. See Figs. 1 & 2.

flowability. The ability of molten filler metal to flow or spread over a metal surface.

flux. Material used to prevent, dissolve, or facilitate removal of oxides and other undesirable surface substances.

flux cored arc welding (FCAW). An arc welding process that produces coalescence of metals by heating them with an arc between a continuous filler metal electrode and the work. Shielding is provided by a flux contained within the tubular electrode. Additional shielding may or may not be obtained form an externally supplied gas or gas mixture. See also **flux cored electrode.**

flux cored electrode. A composite filler metal electrode consisting of a metal tube or other hollow configuration containing ingredients to provide such functions as shielding atmosphere, deoxidation, arc stabilization, and slag formation. Minor amounts of alloying materials may be included in the core. External shielding may or may not be used.

flux cover (metal bath dip brazing and dip soldering). A layer of molten flux over the molten filler metal bath.

forehand welding. A welding technique in which the welding torch or gun is directed toward the progress of welding. See Fig. 30.

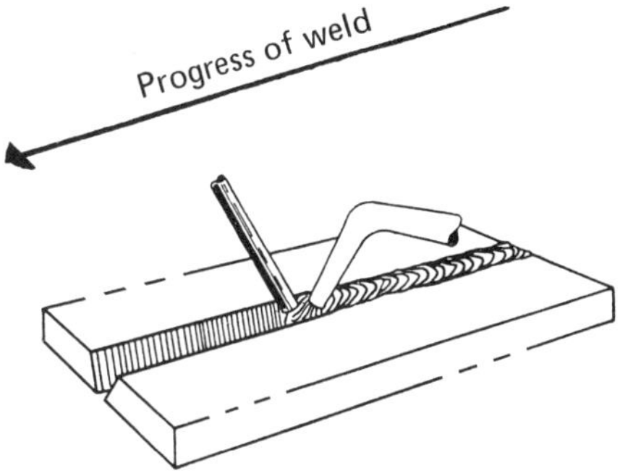

Fig. 30. Forehand Welding.

freezing point. A nonstandard term for **liquidus** and **solidus.**

fuel gas. A gas such as acetylene, natural gas, hydrogen, propane, stabilized methylacetylene propadiene, and other fuels and hydrocarbons usually used with oxygen for heating.

full fillet weld. A fillet weld equal in size to the thickness of the thinner member joined.

furnace brazing (FB). A brazing process in which the workpieces are placed in a furnace and heated to the brazing temperature.

furnace soldering (FS). A soldering process in which the workpieces are placed in a furnace and heated to the soldering temperature.

fused spray deposit (thermal spraying). A self-fluxing spray deposit which is subsequently heated to coaelscence within itself and with the substrate.

fused zone. A nonstandard term for **fusion zone.**

fusing. A nonstandard term for **fusion.**

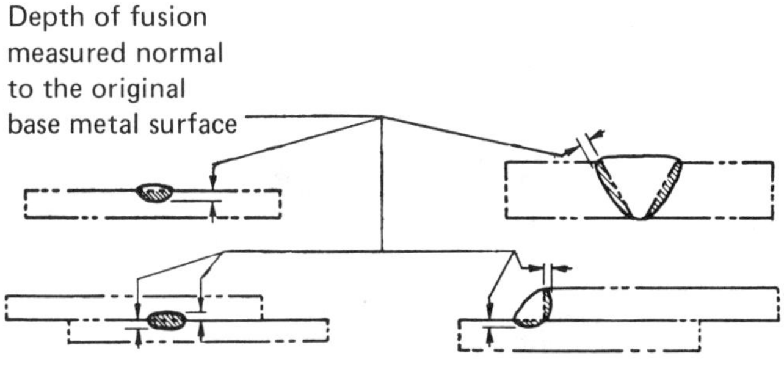

Fig. 31. Depth of Fusion.

fusion. The melting together of filler metal and base metal (substrate), or of base metal only which results in coalescence. See also **depth of fusion.**

fusion face. A surface of the base metal that will be melted during welding.

fusion welding. Any welding process that uses fusion of the base metal to make the weld.

fusion zone. The area of base metal melted as determined on the cross section of a weld. See Fig. 9 & 31.

G

gap. A nonstandard term for **root opening** and **joint clearance.**

gas carbon arc welding (CAW-G). A carbon arc welding process variation that uses a shielding gas. This is an obsolete or seldom used process.

gas cutting. A nonstandard term for **oxygen cutting.**

gas cylinder. A portable container used for transportation and storage of a compressed gas.

gas metal arc cutting (GMAC). An arc cutting process in which metals are severed by melting them with the heat of an arc between a continuous filler metal electrode and the workpiece. Shielding is obtained entirely from an externally supplied gas.

gas metal arc welding (GMAW). An arc welding process that produces coalescence of metals by heating them with an arc between a continuous filler metal electrode and the workpieces. Shielding is obtained entirely from an externally supplied gas.

gas metal arc welding-pulsed arc (GMAW-P). A gas metal arc welding process variation in which the welding current is pulsed. See also **pulsed power welding.**

gas metal arc welding — short circuit arc (GMAW-S). A gas metal arc welding process variation in which the consumable electrode is deposited during repeated short circuits.

gas pocket. A nonstandard term for **porosity.**

gas regulator. A device for controlling the delivery of gas at some substantially constant pressure.

gas shielded arc welding. A general term used to describe **gas metal arc welding, gas tungsten arc welding,** and **flux cored arc welding** when gas shielding is employed.

gas torch. A nonstandard term for **welding torch** and **cutting torch.**

gas tungsten arc cutting (GTAC). An arc cutting process in which metals are severed by melting them with an arc between a single tungsten electrode and the workpiece. Shielding is obtained from a gas.

gas tungsten arc welding (GTAW). An arc welding process that produces coalescence of metals by heating them with an arc between a tungsten electrode (non-consumable) and the workpieces. Shielding is obtained from a gas. Pressure may or may not be used, and filler metal may or may not be used.

gas tungsten arc welding-pulsed arc (GTAW-P). A gas tungsten arc welding process variation in which the current is pulsed. See also **pulsed power welding.**

gas welding. A nonstandard term for **oxyfuel gas welding.**

globular transfer (arc welding). The transfer of molten metal in large drops from a consumable electrode across the arc. See Fig. 32. See also **short circuiting transfer** and **spray transfer.**

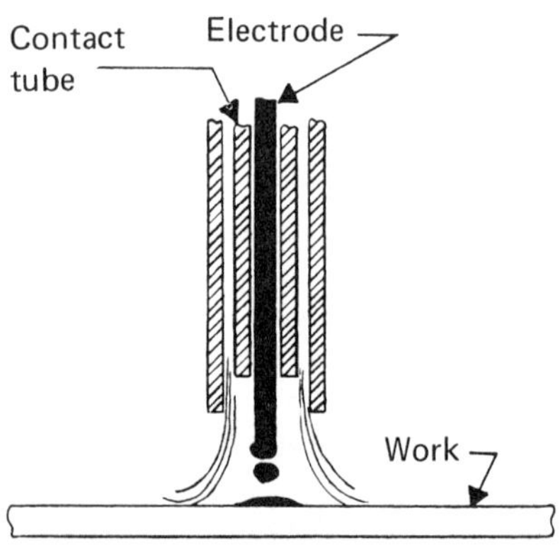

Fig. 32. Globular Transfer.

gouging. The forming of a bevel or groove by material removal. See also **back gouging, arc gouging,** and **oxygen gouging.**

groove. An opening or channel in the surface of a part or between two components which provides space to contain a weld.

groove angle. The total included angle of the groove between workpieces. See Fig. 7.

groove face. The surface of a joint member included in the groove. See Fig. 33.

groove radius. The radius used to form the shape of a J- or U-groove weld. See Fig. 7.

groove type. The geometric configuration of a groove.

groove weld. A weld made in a groove between the workpieces. See Fig. 22.

groove weld size. The joint penetration of a groove weld. See Fig. 45.

groove weld throat. A nonstandard term for **groove weld size.**

ground connection. An electrical connection of the welding machine frame to the earth for safety. See Figs. 25 & 26. See also **workpiece connection** and **workpiece lead.**

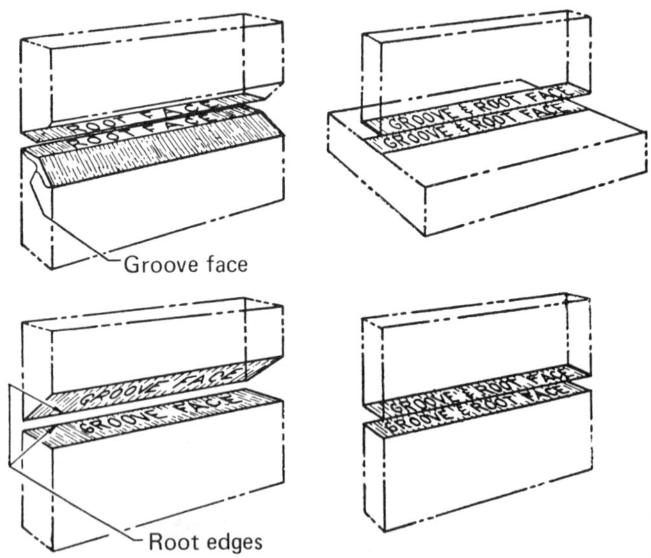

Fig. 33. Groove Face, Root Face and Root Edge.

ground lead. A nonstandard term for **workpiece lead.**

gun. See **arc cutting gun, arc welding gun, electron beam gun, resistance welding gun, soldering gun,** and **thermal spraying gun.**

gun extension. The extension tube attached in front of the thermal spraying gun to permit spraying within confined areas or deep recesses.

H

hand shield. A protective device, used in arc welding, for shielding the eyes, face, and neck. A hand shield is equipped with a filter plate and is designed to be held by hand.

hardfacing. A surfacing variation in which surfacing metal is deposited to reduce wear. See also **buildup, buttering,** and **cladding.**

hard surfacing. A nonstandard term for **hardfacing.**

head. See **welding head** and **cutting head.**

heat-affected zone. That portion of the base metal that has not been melted, but whose mechanical properties or microstructure have been altered by the heat of welding, brazing, soldering, or cutting. See Fig. 34.

heat-affected zone crack. A crack in the heat-affected zone.

helmet. A device designed to be worn on the head to protect eyes, face, and neck from arc radiation, radiated heat, spatter, or other harmful matter expelled during arc welding.

high frequency resistance welding. A group of resistance welding process variations that use high frequency welding current to concentrate the welding heat at the desired location.

high frequency seam welding (RSEW-HF). A resistance seam welding process variation in which high frequency welding current is supplied through electrodes into the workpieces. See Fig. 62. See also **high frequency resistance welding.**

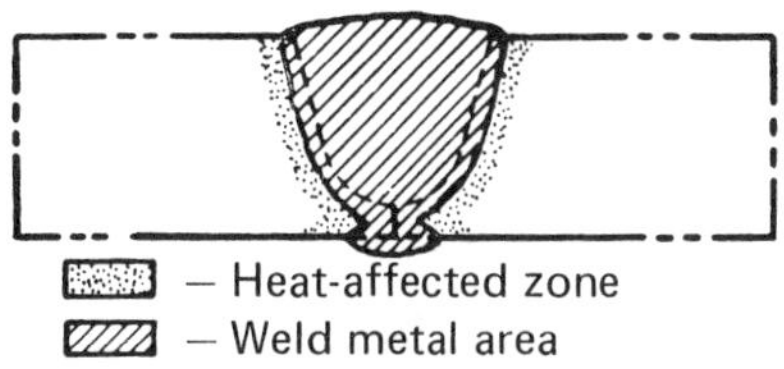

Fig. 34. Heat-Affected Zone and Weld Metal Area.

high pulse current. The current levels during the high pulse time which produces that high heat level.

high pulse time. The duration of the high current pulse time.

horizontal fixed position (pipe welding). The position of a pipe joint in which the axis of the pipe is approximately horizontal, and the pipe is not rotated during welding. See Fig. 35.

horizontal position (fillet weld). The position in which welding is performed on the upper side of an approximately horizontal surface and against an approximately vertical surface. See Fig. 2.

horizontal position (groove weld). The position of welding in which the weld axis lies in an approximately horizontal plane and the weld face lies in an approximately vertical plane. See Fig. 1.

horizontal rolled position (pipe welding). The position of a pipe joint in which the axis of the pipe is approximately horizontal, and welding is performed in the flat position by rotating the pipe. See Fig. 35.

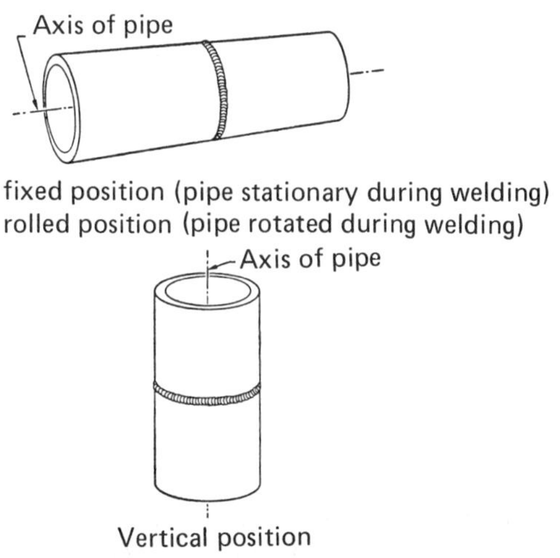

Fig. 35. Positions of Pipe During Welding.

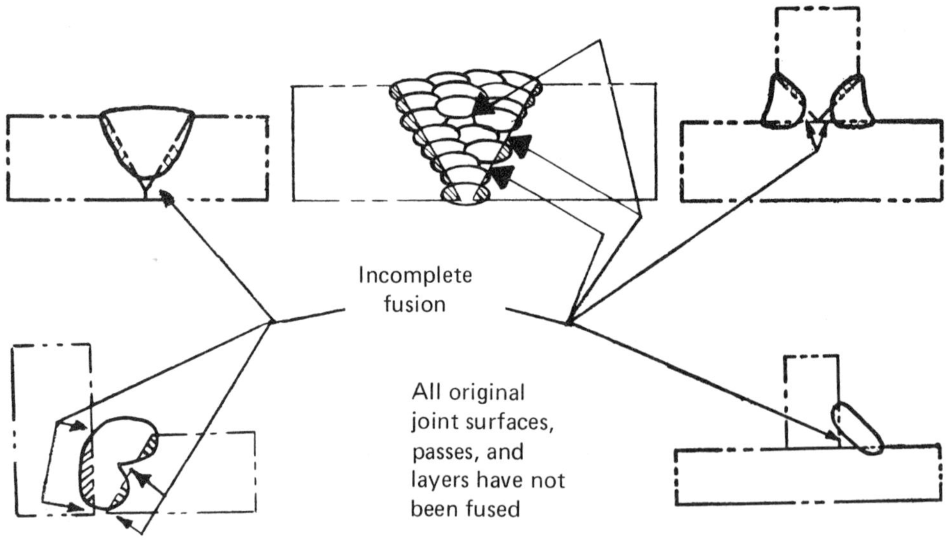

Fig. 36. Incomplete Fusion.

hot crack. A crack that develops during solidification.

hot start current. A very brief current pulse at arc initiation to stabilize the arc quickly.

hot wire welding. A variation of the gas tungsten arc, plasma arc, and submerged arc welding processes in which a filler metal wire is resistance heated by current flowing through the wire as it is fed into the weld pool.

I

inclined position. The position of a pipe joint in which the axis of the pipe is at an angle of approximately 45 degrees to the horizontal, and the pipe is not rotated during welding.

included angle. A nonstandard term for **groove angle.**

incomplete fusion. A weld discontinuity in which fusion did not occur between weld metal and fusion faces or adjoining weld beads. See Fig. 36.

induction brazing (IB). A brazing process in which the heat required is obtained from the resistance of the workpieces to induced electric current.

induction seam welding (RSEW-I). A resistance seam welding process variation in which high frequency welding current is induced in the workpieces. See also **high frequency resistance welding.**

induction soldering (IS). A soldering process in which the heat required is obtained from the resistance of the workpieces to induced electric current.

induction upset welding (UW-I). An upset welding process variation in which high frequency welding current is induced in the workpieces.

induction welding (IW). A welding process that produces coalescence of metals by the heat obtained from the resistance of the workpieces to the flow of induced high frequency welding current with or without the application of pressure. The effect of the high frequency welding current is to concentrate the welding heat at the desired location.

induction work coil. The inductor used when welding, brazing, or soldering with induction heating equipment.

inert gas. A gas that normally does not combine chemically with the base metal or filler metal. See also **protective atmosphere.**

inert gas metal arc welding. A nonstandard term for **gas metal arc welding.**

inert gas tungsten arc welding. A nonstandard term for **gas tungsten arc welding.**

inertia friction welding. A variation of friction welding in which the energy required to make the

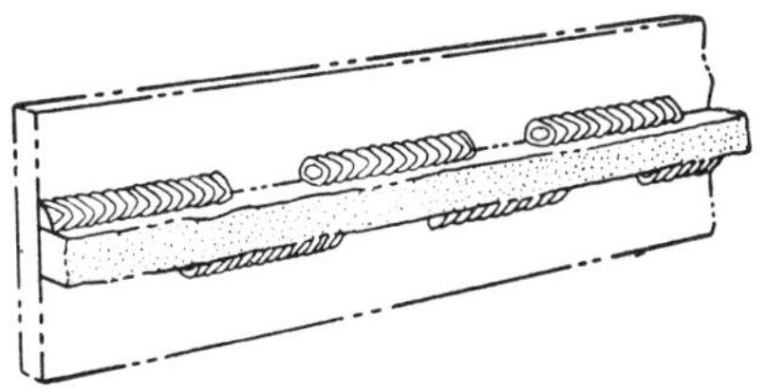

Fig. 37. Staggered Intermittent Fillet Welding.

weld is supplied primarily by the stored rotational kinetic energy of the welding machine.

initial current. The current after starting, but before establishment of welding current.

intermittent weld. A weld in which the continuity is broken by recurring unwelded spaces. See Figs. 14 and 37.

interpass temperature. In a multipass weld, the temperature of the weld metal before the next pass is started.

iron soldering (INS). A soldering process in which the heat required is obtained from a soldering iron.

J

J-groove weld. A type of groove weld. See Fig. 22.

joint. The junction of members or the edges of members which are to be joined or have been joined.

joint buildup sequence. A nonstandard term for **cross-sectional sequence.**

joint design. The joint geometry together with the required dimensions of the welded joint.

joint efficiency. The ratio of the strength of a joint to the strength of the base metal, expressed in percent.

joint geometry. The shape and dimensions of a joint in cross section prior to welding.

joint penetration. The depth a weld extends from its face into a joint, exclusive of reinforcement. See Fig. 38.

joint root. That portion of a joint to be welded where the members approach closest to each other. In cross section, the joint root may be either a point, a line, or an area. See Fig. 47.

joint type. A weld joint classification based on the five basic arrangements of the component parts such as butt joint, corner joint, edge joint, lap joint, and T-joint. See Fig. 12.

joint welding sequence. See **welding sequence.**

K

kerf. The width of the cut produced during a cutting process.

keyhole welding. A technique in which a concentrated heat source penetrates completely through a workpiece, forming a hole at the leading edge of the weld pool. As the heat source progresses, the molten metal fills in behind the hole to form the weld bead.

L

lack of fusion. A nonstandard term for **incomplete fusion.**

lamellar tear. A terrace-like fracture in the base metal with a basic orientation parallel to the wrought surface. It is caused by the high stress in the thickness direction that results from welding.

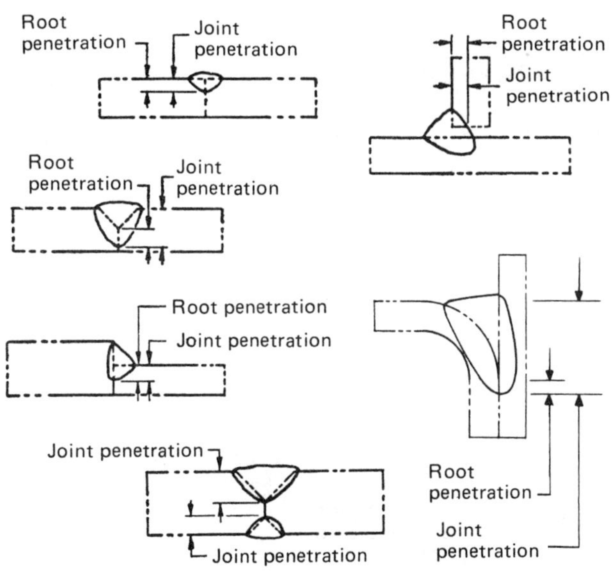

Fig. 38. Root Penetration and Joint Penetration of Groove Welds.

lap joint. A joint between two overlapping members in parallel planes. See Fig. 12.

land. A nonstandard term for **root face.**

laser. A device that produces a concentrated coherent light beam by stimulating electronic or molecular transitions of lower energy levels. Laser is an acronym for light amplification by stimulated emission of radiation.

laser beam air cutting (LBC-A). A laser beam cutting process variation that severs material by melting with the heat obtained from a laser beam combined with an air jet to remove molten and vaporized metal. See also **laser beam cutting.**

laser beam cutting (LBC). A thermal cutting process that severs materials by melting or vaporizing them with the heat obtained from a laser beam, with or without the application of gas jets to augment the removal of material.

laser beam welding (LBW). A welding process that produces coalescence of materials with the heat obtained from the application of a concentrated coherent light beam impinging upon the joint.

layer. A stratum of weld metal, consisting of one or more weld beads. See Fig. 39.

layer level wound. A nonstandard term for **level wound.**

layer wound. A nonstandard term for **level wound.**

lead angle. A nonstandard term for **travel angle.**

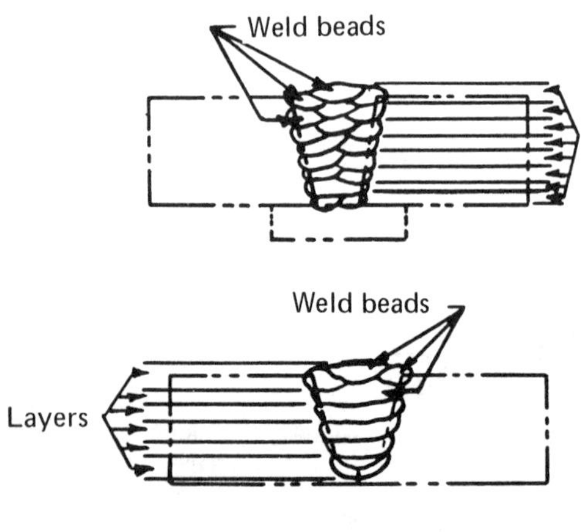

Fig. 39. Weld Beads and Layers.

leg of a fillet weld. See **fillet weld leg.**

lens. See **filter lens.**

level wound. Spooled or coiled filler metal that has been wound in distinct layers such that adjacent turns touch. See also **random wound.**

lightly coated electrode. A filler metal electrode consisting of a metal wire with a light coating applied subsequent to the drawing operation, primarily for stabilizing the arc.

liquidus. The lowest temperature at which a metal or an alloy is completely liquid.

local preheating. Perheating a specific portion of a structure.

local stress relief heat treatment. Stress relief heat treatment of a specific portion of a structure.

locked-up stress. A nonstandard term for **residual stress.**

longitudinal crack. A crack with its major axis orientation approximately parallel to the weld axis.

logitudinal sequence. The order in which the weld passes of a continuous weld are made with respect to its length. See also **backstep sequence, block sequence, cascade sequence, continuous sequence,** and **random sequence.**

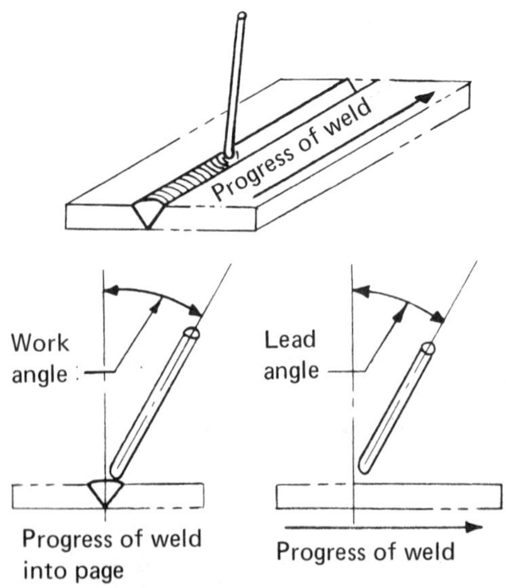

Fig. 40. Lead Angle and Work Angle.

low pulse current. The current levels during the low pulse time which produce the low heat levels.

low pulse time. The duration of the low current pulse.

M

machine oxygen cutting. Oxygen cutting with equipment that performs the cutting operation under the constant observation and control of an oxygen cutting operator. The equipment may or may not load and unload the workpieces. See also **automatic oxygen cutting.**

machine welding. Welding with equipment that performs the welding operation under the constant observation and control of a welding operator. The equipment may or may not load and unload the workpieces. See also **automatic welding.**

macroetch test. A test in which the specimen is prepared with a fine finish and etched to give a clear definition of the weld.

manifold. See **cylinder manifold.**

manual oxygen cutting. A cutting operation peformed and controlled completely by hand. See also **automatic oxygen cutting** and **machine oxygen cutting.**

manual welding. A welding operation performed and controlled completely by hand. See also **automatic welding, machine welding,** and **semiautomatic arc welding.**

mechanical bond (thermal spraying). The adherence of a spray deposit to a roughened surface by the mechanism of particle interlocking.

meltback time. The time interval at the end of crater fill time to arc outage during which electrode feed is stopped.

melting range. The temperature range between solidus and liquidus.

melting rate. The weight or length of electrode melted in a unit of time.

melt-through. Visible root reinforcement produced in a joint welded from one side. See Fig. 41.

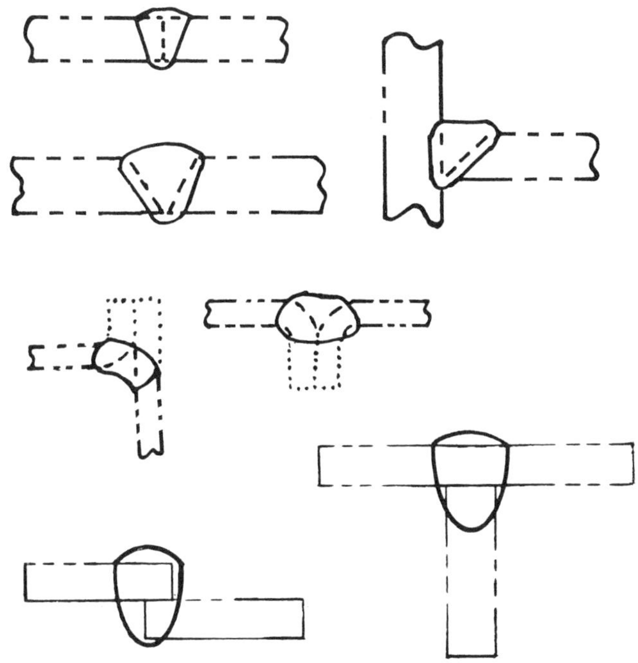

Fig. 41. Melt-Thru.

metal arc cutting (MAC). Any of a group of arc cutting processes that severs metals by melting them with the heat of an arc between a metal electrode and the base metal. See also **gas metal arc cutting** and **shielded metal arc cutting.**

metal arc welding. See **flux cored arc welding, gas metal arc welding, gas tungsten arc welding, plasma arc welding, shielded metal arc welding, stud arc welding,** and **submerged arc welding.**

metal cored electrode. A composite filler metal electrode consisting of a metal tube or other hollow configuration containing alloying materials. Minor amounts of ingredients providing such functions as arc stabilization and fluxing of oxides may be included. External shielding gas may or may not be used.

metal electrode. A filler or nonfiller metal electrode used in arc welding or cutting that consists of a metal wire or rod that has been manufactured by any method and that is either bare or covered.

metallic bond. The principal bond that holds metals together and is formed between base metals and filler metals in all welding processes. This is a primary bond arising form the increased spatial extension of the valence electron wave functions when an aggregate of metal atoms is brought close together.

metallizing. A nonstandard term for **thermal spraying.**

metallurgical bond. A nonstandard term for **metallic bond.**

metal powder cutting (POC). An oxygen cutting process that severs metals through the use of powder, such as iron, to facilitate cutting.

MIG welding. A nonstandard term for **gas metal arc welding** and **flux cored arc welding.**

mixing chamber. That part of a welding or cutting torch in which a fuel gas and oxygen are mixed.

molten weld pool. A nonstandard term for **weld pool.**

N

neutral flame. An oxyfuel gas flame in which the portion used is neither oxidizing nor reducing. See also **oxidizing flame** and **reducing flame.**

nozzle. A device that directs shielding media.

nugget. The weld metal joining the workpieces in spot, roll spot, seam, or projection welds.

O

open-circuit voltage. The voltage between the output terminals of the welding machine when no current is flowing in the welding circuit.

orifice. See **constricting orifice.**

orifice gas. (plasma arc welding and cutting). The gas that is directed into the torch to surround the electrode. It becomes ionized in the arc to form the plasma, and issues from the orifice in the torch nozzle as the plasma jet.

overhead position. The position in which welding is performed from the underside of the joint. See Figs. 1 & 2.

overlap. The protrusion of weld metal beyond the weld toe or weld root. See Fig. 42.

overlaying. A nonstandard term for **surfacing.**

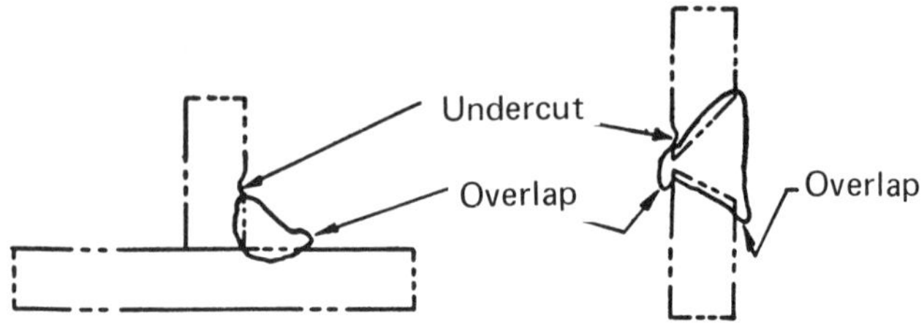

Fig. 42. Undercut and Overlap.

oxidizing flame. An oxyfuel gas flame having an oxidizing effect due to excess oxygen. See also **neutral flame** and **reducing flame.**

oxyacetylene cutting (OFC-A). An oxyfuel gas cutting process used to sever metals by means of the chemical reaction of oxygen with the base metal at elevated temperatures. The necessary temperature is maintained by gas flames resulting from the combustion of acetylene with oxygen.

oxyacetylene welding (OAW). An oxyfuel gas welding process that produces coalescence of metals by heating them with a gas flame or flames obtained from the combustion of acetylene with oxygen. The process may be used with or without the application of pressure and with or without the use of filler metal.

oxyfuel gas cutting (OFC). A group of cutting processes used to sever metals by means of the chemical reaction of oxygen with the base metal at elevated temperatures. The necessary temperature is maintained by means of gas flames obtained from the combustion of a specified fuel gas and oxygen. See also **oxygen cutting, oxyacetylene cutting, oxyhydrogen cutting, oxynatural gas cutting** and **oxypropane cutting.**

oxyfuel gas welding (OFW). A group of welding processes that produces coalescence by heating materials with an oxyfuel gas flame or flames, with or without the application of pressure and with or without the use of filler metal.

oxygen arc cutting (AOC). An oxygen cutting process used to sever metals by means of the chemical reaction of oxygen with the base metal at elevated temperatures. The necessary temperature is maintained by an arc between a consumable tubular electrode and the base metal.

oxygen cutter. One who performs a manual oxygen cutting operation.

oxygen cutting (OC). A group of cutting processes used to sever or remove metals by means of the chemical reaction between oxygen and the base metal at elevated temperatures. In the case of oxidation-resistant metals, the reaction is facilitated by the use of chemical flux or metal powder. See **oxygen arc cutting, oxyfuel gas cutting, oxygen lance cutting, chemical flux cutting,** and **metal powder cutting.**

oxygen cutting operator. One who operates machine or automatic oxygen cutting equipment.

oxygen gouging. An application of oxygen cutting in which a bevel or groove is formed.

oxygen grooving. A nonstandard term for **oxygen gouging.**

oxygen lance. A length of pipe used to convey oxygen to the point of cutting in oxygen lance cutting.

oxygen lance cutting (LOC). An oxygen cutting process used to sever metals with oxygen supplied through a consumable lance. The preheat to start the cutting is obtained by other means.

oxygen lancing. A nonstandard term for **oxygen lance cutting.**

oxyhydrogen cutting (OFC-H). An oxyfuel gas cutting process used to sever metals by means of the chemical reaction of oxygen with the base metal at elevated temperatures. The necessary temperature is maintained by gas flames resulting from the combustion of hydrogen with oxygen.

oxyhydrogen welding (OHW). An oxyfuel gas welding process that produces coalescence of materials by heating them with a gas flame(s) obtained from the combustion of hydrogen with oxygen, without the application of pressure and with or without the use of filler metal.

oxynatural gas cutting (OFC-N). An oxyfuel gas cutting process used to sever metals by means of the chemical reaction of oxygen with the base metal at elevated temperatures. The necessary temperature is maintained by gas flames resulting from the combustion of natural gas with oxygen.

oxypropane cutting (OFC-P). An oxyfuel gas cutting process variation that uses propane as the fuel gas.

P

parallel welding. A resistance welding secondary circuit variation in which the secondary current is divided and conducted through the workpieces and electrodes in parallel electrical paths to simultaneously form multiple resistance spot, seam, or projection welds

parallel wire welding.[1] A system of machine welding which feeds two electrodes, usually of the same diameter, through a single wire feeder at the same speed. Both electrodes share the current supplied by a common power source. This system is sometimes called twin arc parallel wire welding.

parent metal. A nonstandard term for **base metal.**

partial joint penetration. Joint penetration that is intentionally less than complete. See Fig. 43.

pass. See **weld pass.**

pass sequence. See **weld pass sequence.**

peel test. A destructive method of testing that mechanically separates a lap joint by peeling.

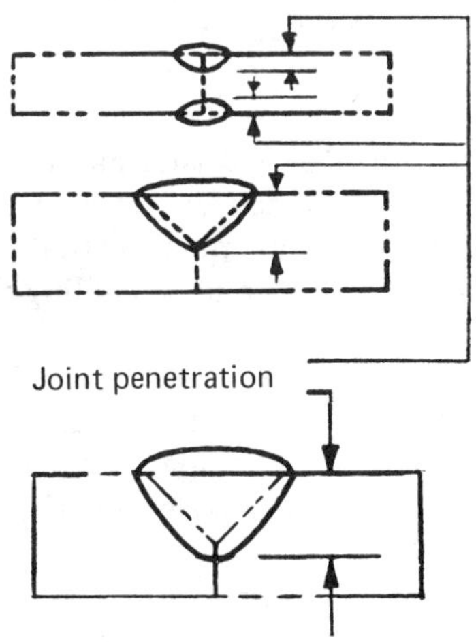

Fig. 43. Partial Joint Penetration.

[1] Not presently defined by AWS A3.0.

peening. The mechanical working of metals using impact blows.

penetration. A nonstandard term for **joint penetration** and **root penetration.**

percent ferrite. A nonstandard term for **ferrite number.**

percussion welding (PEW). A welding process that produces coalescence at the faying surface using the heat from an arc produced by a rapid discharge of electrical energy. Pressure is applied percussively during or immediately following the electrical discharge.

plano lens (eye protection). A lens that does not incorporate correction.

plasma. See **arc plasma.**

plasma arc cutting (PAC). An arc cutting process that severs metal by melting a localized area with a constricted arc and removing the molten material with a high velocity jet of hot, ionized gas issuing from the constricting orifice.

plasma arc welding (PAW). An arc welding process that produces coalescence of metals by heating them with a constricted arc between an electrode and the workpiece (transferred arc) or the electrode and the constricting nozzle (nontransferred arc). Shielding is obtained from the hot, ionized gas issuing from the torch which may be supplemented by an auxiliary source of shielding gas. Shielding gas may be an inert gas or a mixture of gases. Pressure may or may not be used, and filler metal may or may not be supplied.

plasma spraying (PSP). A thermal spraying process in which a nontransferred arc of a plasma torch is utilized to create a gas plasma that acts as the source of heat for melting and propelling the surfacing material to the substrate.

plenum chamber (plasma arc welding and cutting, and plasma spraying). The space between the inside wall of the constricting nozzle and the electrode.

plug weld. A weld made in a circular hole in one member of a joint fusing that member to another member. A fillet-welded hole is not to be construed as conforming to this definition. See Fig. 44.

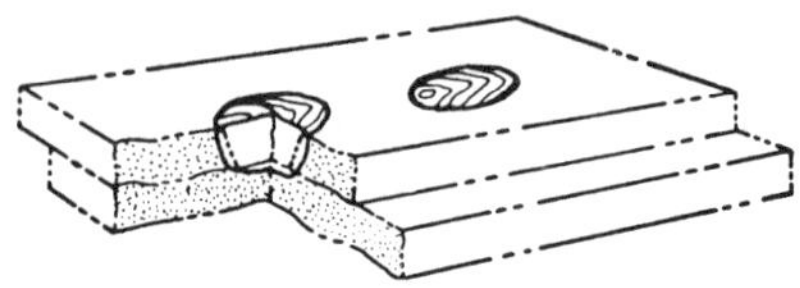

Fig. 44. Plug Welds.

polarity. See **direct current electrode negative** and **direct current electrode positive.**

porosity. Cavity type discontinuities formed by gas entrapment during solidification.

position. See **welding position.**

positioned weld. A weld made in a joint that has been placed to facilitate making the weld.

position of welding. See **welding position.**

postflow time. The time interval from current shut off to either shielding gas or cooling water shut off.

postheating. The application of heat to an assembly after welding, brazing, soldering, thermal spraying, or thermal cutting. See also **postweld heat treatment. weld heat treatment.**

post weld heat treatment. Any heat treatment after welding.

preheat. A nonstandard term for **preheat temperature.**

preheating. The application of heat to the base metal immediately before welding, brazing, soldering, thermal spraying, or cutting.

preheat temperature. A specified temperature that the base metal must attain in the welding, brazing, soldering, thermal spraying, or cutting area immediately before these operations are performed.

pressure gas welding (PGW). An oxyfuel gas welding process that produces coalescence simultaneously over the entire area of faying surfaces by heating them with gas flames obtained from the combustion of a fuel gas and oxygen and by the application of pressure, without the use of filler metal.

pressure welding. A nonstandard term for **solid-state welding, hot pressure welding, forge welding, diffusion welding, pressure gas welding,** and **cold welding.**

procedure. The detailed elements of a process or method used to produce a specific result.

procedure qualification. The demonstration that welds made by a specific procedure can meet prescribed standards.

procedure qualification record (PQR). A document providing the actual welding variables used to produce an acceptable test weld and the results of tests conducted on the weld to qualify a welding procedure specification.

process. A grouping of basic operational elements used in welding, cutting, adhesive bonding, or thermal spraying.

progressive block sequence. A block sequence in which successive blocks are completed progressively along the weld, either from one end to the other or from an intermediate location of the weld toward either end

projection welding (PW). A resistance welding process that produces coalescence by the heat obtained from the resistance to the flow of the welding current. The resulting welds are localized at predetermined points by projections, embossments, or intersections.

protective atmosphere. A gas or vacuum envelope surrounding the workpieces used to prevent or facilitate removal of oxides and other detrimental surface substances.

puddle. A nonstandard term for **weld pool.**

pull gun technique. A nonstandard term for **backhand welding.**

pulsation welding. A nonstandard term for **multiple-impulse welding.**

pulsed power welding. An arc welding process variation in which the power is cyclically programmed to pulse so that effective but short duration values of power can be utilized. Such short duration values are significantly different from the average value of power. Equivalent terms are pulsed voltage or pulsed current welding. See also **pulsed spray welding.**

pulsed spray welding. An arc welding process variation in which the current is pulsed to utilize the advantages of the spray mode of metal transfer at average currents equal to or less than the globular to spray transition current.

pulse start delay time. The time interval from current initiation to the beginning of current pulsation.

push angle. The travel angle when the electrode is pointing in the direction of weld progression. This agle can also be used to partially define the position of guns, torches, rods, and beams. See **drag angle, forehand welding, travel angle**, and **work angle.**

Q

qualification. See **welder performance qualification** and **procedure qualification.**

R

random intermittent welds. Intermittent welds on one or both sides of a joint in which the weld increments are made without regard to spacing.

random sequence. A longitudinal sequence in which the weld bead increments are made at random.

random wound. Spooled or coiled filler metal that has not been wound in distinct layers. See also **level wound.**

rate of deposition. See **deposition rate.**

reaction stress. A stress that cannot exist in a member if the member is isolated as free body without connection to other parts of the structure.

reactor. A device used in arc welding circuits to minimize irregularities in the flow of the welding current.

reducing section tension test. A test in which a transverse section of the weld is located in the center of the reduced section of the specimen.

reducing atmosphere. A chemically active protective atmosphere which at elevated temperature will reduce metal oxides to their metallic state.

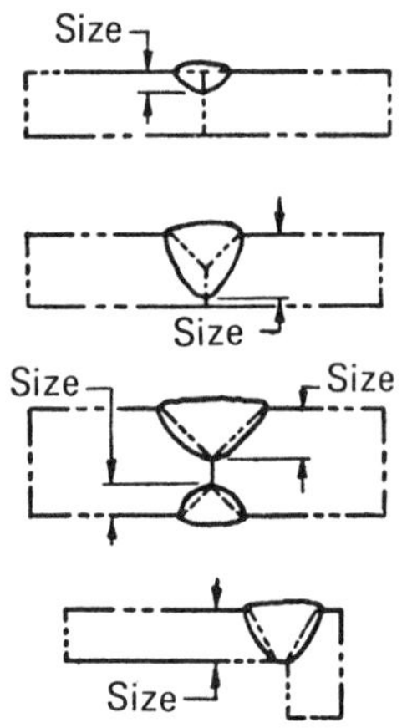

Fig. 45. Size of Groove Welds.

reducing flame. A gas flame having a reducing effect due to excess fuel. See **neutral flame** and **oxidizing flame.**

reinforcement of weld. See **weld reinforcement.**

residual stress. Stress present in a member that is free of external forces or thermal gradients.

resistance brazing (RB). A brazing process in which the heat required is obtained from the resistance to electric current flow in a circuit of which the workpiece is a part.

resistance butt welding. A nonstandard term for **upset welding** and **flash welding.**

resistance seam welding (RSEW). A resistance welding process that produces coalescence at the faying surfaces of overlapped parts progressively along a length of a joint. The weld may be made with overlapping weld nuggets, a continuous weld nugget, or by forging the joint as it is heated to the welding temperature by resistance to the flow of the welding current.

resistance soldering (RS). A soldering process in which the heat required is obtained from the resistance to electric current flow in a circuit of which the workpiece is a part.

resistance spot welding (RSW). A resistance welding process that produces coalescence at the faying surfaces of a joint by the heat obtained from resistance to the flow of welding current through the workpieces from electrodes that serve to concentrate the welding current and pressure at the weld area.

resistance welding (RW). A group of welding processes that produces coalescence of the faying surfaces with the heat obtained from resistance of the work to the flow of the welding current in a circuit of which the work is a part, and by the application of pressure.

resistance welding control. The device, usually electronic, that determines the welding sequence and timing with regard to the welding current pattern, electrode or platen forces or movement, and other operational conditions of a resistance welding machine.

resistance welding electrode. The part or parts of a resistance welding machine through which the welding current and, in most cases, force are applied directly to the work. The electrode may be in the form of a rotating wheel, rotating roll, bar, cylinder, plate, clamp, chuck, or modification thereof.

resistance welding gun. A manipulatable device to transfer current and provide electrode force to the weld area (usually in reference to a portable gun).

reverse polarity. A nonstandard term for **direct current electrode positive.**

rim. A filler metal package type consisting of a continuous length of electrode wound in a circular channel. (This package type is a proposed standard of the International Organization for Standardization

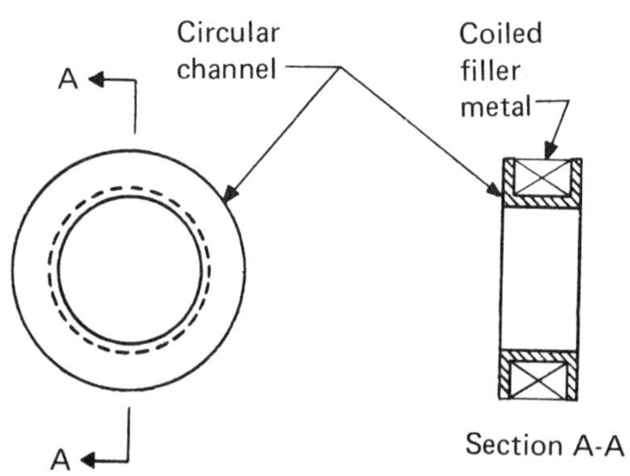

Fig. 46. Rim.

and is not currently used by U.S. industry. See Fig. 46.) This term was previously used to denote coil with support.

roll spot welding. A resistance welding process variation that makes separated spot welds using one or more rotating circular electrodes. The rotation of the electrodes may or may not be stopped during the making or a weld.

roll welding (ROW). A solid-state welding process that produces coalescence of metals by heating and by applying sufficient pressure with rolls to cause deformation at the faying surfaces. See also **forge welding.**

root. A nonstandard term for **joint root** and **weld root.**

root bead. A weld that extends into or includes part or all of the joint root.

root bend test. A test in which the weld root is on the convex surface of a specified bend radius.

root crack. A crack in the weld or heat affected zone occurring at the root of the weld.

root edge. A root face of zero width. See Fig. 33. See also **root face.**

root face. That portion of the groove face adjacent to the joint root. See Fig. 33.

root gap. A nonstandard term for **root opening.**

root of joint. See **joint root.** See also Fig. 47.

root of weld. See **weld root.** See also Fig. 48.

root opening. A separation at the joint root between the workpieces. See Fig. 7.

root penetration. The depth that a weld extends into the joint root. See Fig. 38.

root radius. A nonstandard term for **groove radius.**

root reinforcement. Weld reinforcement opposite the side from which welding was done. See Fig. 27.

root surface. The exposed surface of a weld opposite the side from which welding was done. See Fig. 28.

S

scarf. A nonstandard term for **edge preparation.**

seal weld. Any weld designed primarily to provide a specific degree of tightness against leakage.

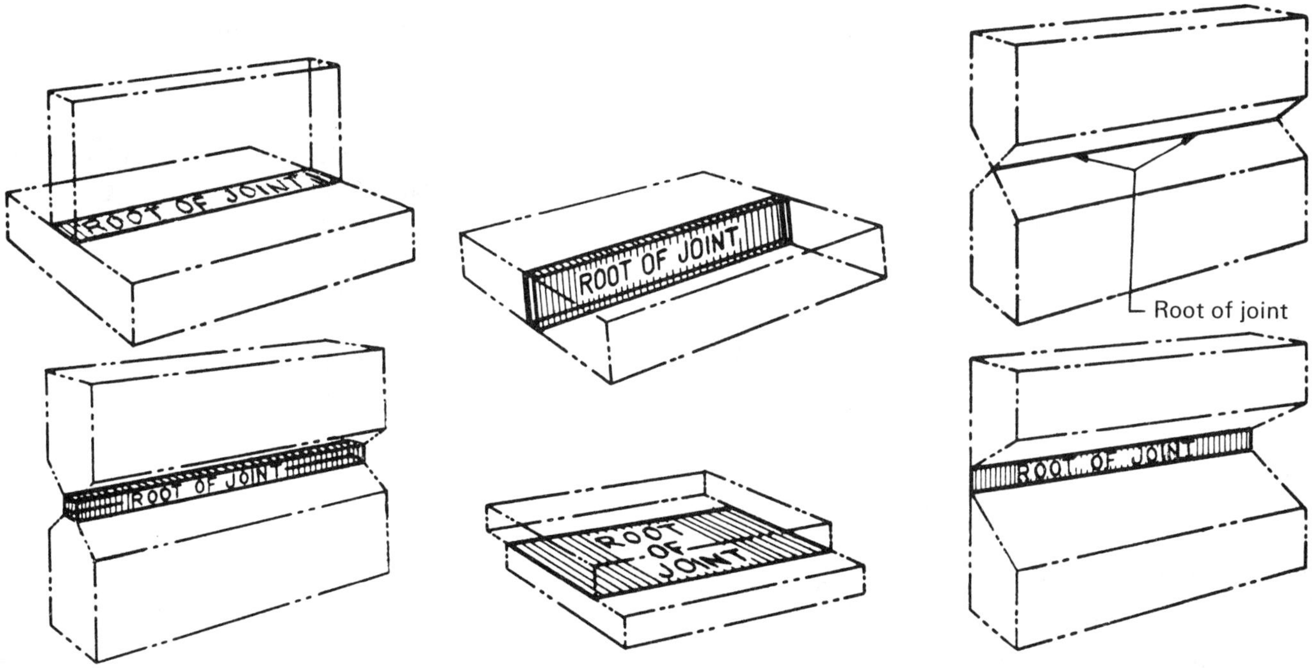

Fig. 47. Root of Joint.

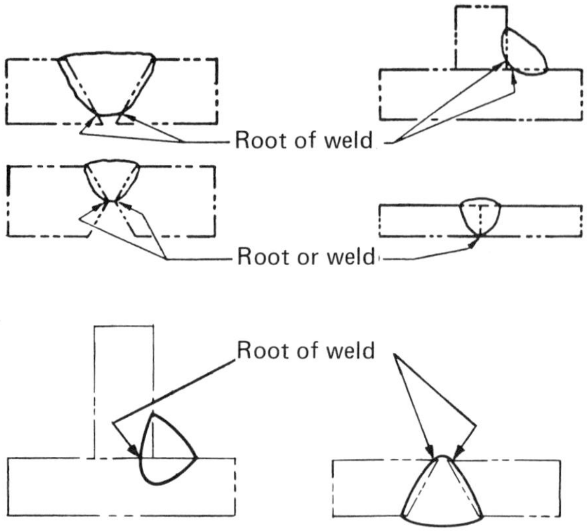

Fig. 48. Root of Weld.

seam weld. A continuous weld made between or upon overlapping members, in which coalescence may start and occur on the faying surfaces, or may have proceeded from the outer surface of one member. The continuous weld may consist of a single weld bead or a series of overlapping spot welds. See **arc seam weld** and **resistance seam welding.**

secondary circuit. That portion of a welding machine that conducts the secondary current between the secondary terminals of the welding transformer and the electrodes, or electrode and workpiece.

selective block sequence. A block sequence in which successive blocks are completed in an order selected to control residual stress and distortion.

semiautomatic arc welding. Arc welding with equipment that controls only the filler metal feed. The advance of the welding is manually controlled.

semiautomatic brazing. Brazing with equipment that controls only the filler metal feed. The advance of the brazing is manually controlled.

semiblind joint. A joint in which one extremity of the joint is not visible.

sequence time. A nonstandard term for **welding cycle.**

series submerged arc welding (SAW-S). A submerged arc welding process variation in which electric current is established between two consumable electrodes that meet just above the surface of the workpieces which are not part of the electrical circuit.

series welding. A resistance welding secondary circuit variation in which the secondary current is conducted through the workpieces and electrodes or wheels in a series electrical path to simultaneously form multiple resistance spot, seam, or projection welds.

shielded carbon arc welding (CAW-S). A carbon arc welding process variation that produces coalescence of metals by heating them with an electric arc between a carbon electrode and the workpieces. Shielding is obtained from the combustion of a solid material fed into the arc or from a blanket of flux on the workpieces, or both. Pressure may or may not be used, and filler metal may or may not be used.

shielding metal arc cutting (SMAC). A metal arc cutting process in which metals are severed by melting them with the heat of an arc between a covered metal electrode and the base metal.

shielded metal arc welding (SMAW). An arc welding process that produces coalescence of metals by heating them with an arc between a covered metal electrode and the workpieces. Shielding is obtained from decomposition of the electrode covering. Pressure is not used, and filler metal is obtained from the electrode.

shielding gas. Protective gas used to prevent atmospheric contamination.

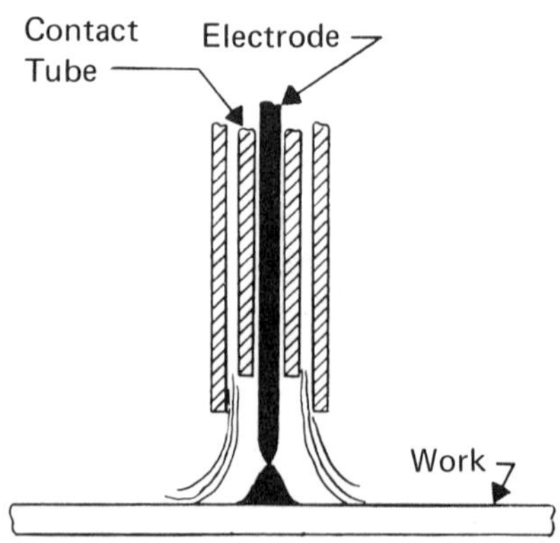

Fig. 49. Short Circuiting Transfer.

short circuiting arc welding. See **gas metal arc welding-short circuit arc.**

short circuiting transfer (arc welding). Metal transfer in which molten metal from a consumable electrode is deposited during repeated short circuits. See Fig. 49. See also **globular transfer** and **spray transfer.**

shoulder. A nonstandard term for **root face.**

shrinkage stress. A nonstandard term for **residual stress.**

shrinkage void. A cavity type discontinuity normally formed by shrinkage during solidification.

sieve analysis. A method of determining particle size distribution, usually expressed as the weight percentage retained upon each of a series of standard screens of decreasing mesh size.

side bend test. A test in which the side of a transverse section of the weld is on the convex surface of a specified bend radius.

silver soldering. A nonstandard term for brazing with a silver-base filler metal.

single-bevel-groove weld. A type of groove weld. See Fig. 22.

single-flare-bevel-groove weld. A weld in a groove formed by a member with a curved surface in contact with a planar member. See Fig. 22.

single-flare-V-groove weld. A weld in a groove formed by two members with curved surfaces. See Fig. 22.

single-J-groove weld. A type of groove weld. See Fig. 22.

single-square-groove weld. A type of groove weld. See Fig. 22.

single-U-groove weld. A type of groove weld. See Fig. 22.

single-V-groove weld. A type of groove weld. See Fig. 22.

single-welded joint. A fusion welded joint that is welded from one side only. See Fig. 50.

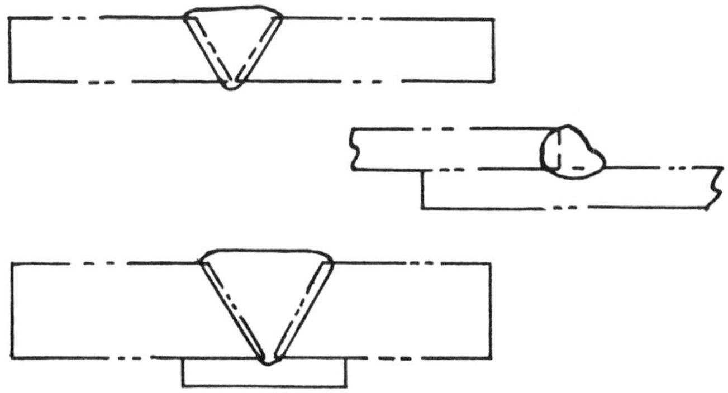

Fig. 50. Single-Welded Joint.

size of weld. See **weld size.**

skip weld. A nonstandard term for **intermittent weld.**

slag inclusion. Nonmetallic solid material entrapped in weld metal or between weld metal and base metal.

slot weld. A weld made in an elongated hole in one member of a joint fusing that member to another member. The hole may be open at one end. A fillet welded slot is not to be construed as conforming to this definition. See Fig. 52

slugging. The act of adding a separate piece or pieces of material in a joint before or during welding that results in a welded joint not complying with design, drawing, or specification requirements.

soft solder. A nonstandard term for **solder.**

solder. A filler metal used in soldering that has a liquidus not exceeding 840°F (450°C).

solderability. The capacity of a material to be soldered under the imposed fabrication conditions into a specific, suitably designed structure and to perform satisfactorily in the intended service.

soldering (S). A group of welding processes that produces coalescence of materials by heating them

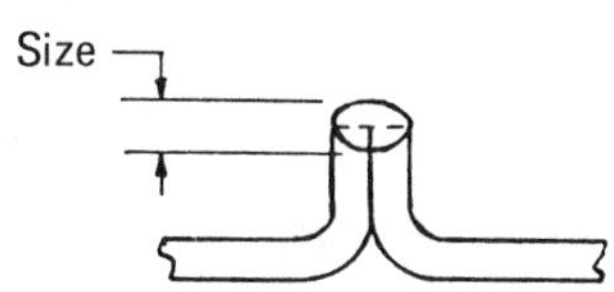

Fig. 51. Size of Flange Weld.

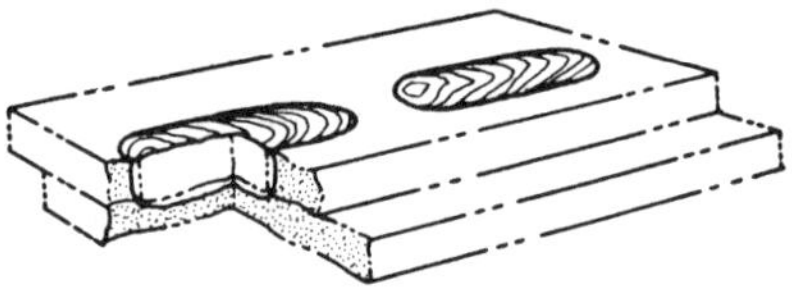

Fig. 52. Slot Welds.

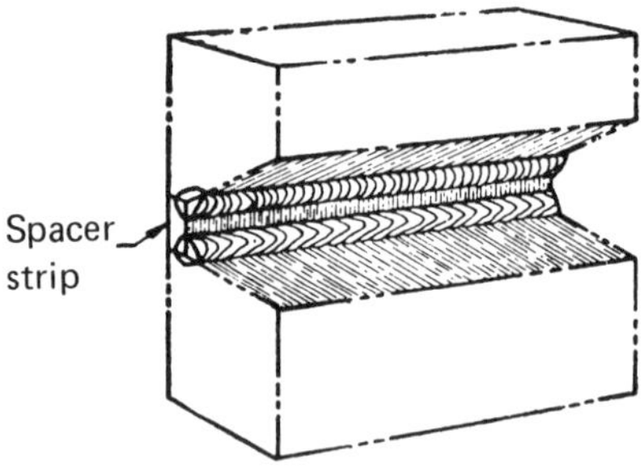

Fig. 53. Spacer Strip.

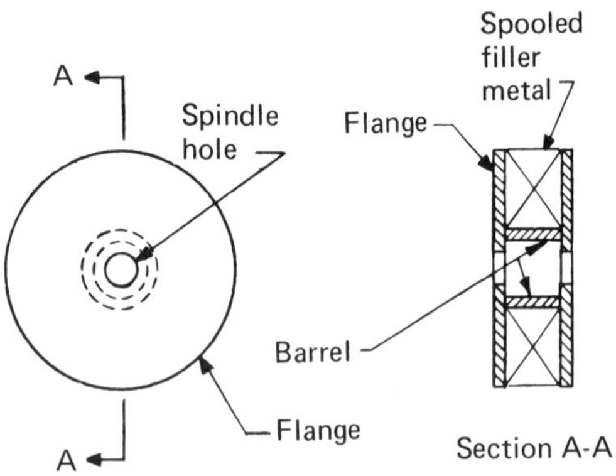

Fig. 54. Spool.

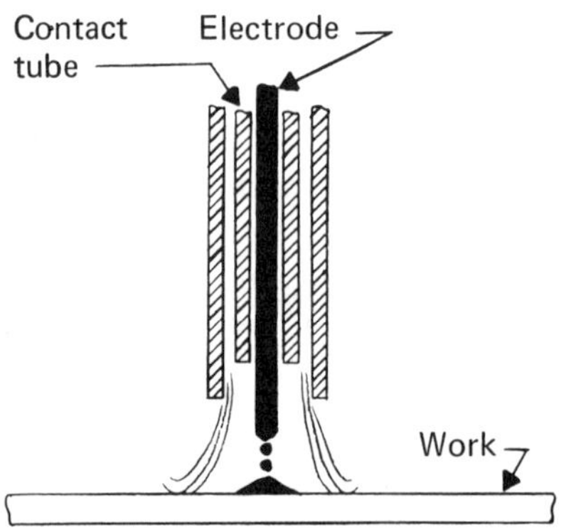

Fig. 55. Spray Transfer.

to the soldering temperature and by using a filler metal having a liquidus not exceeding 840°F (450°C) and below the solidus of the base metals. The filler metal is distributed between the closely fitted faying surfaces of the joint by capillary action.

soldering blowpipe. A device used to obtain a small, accurately directed flame for fine work. A portion of any flame is blown to the desired location by the blowpipe, which is usually mouth operated.

soldering gun. An electrical soldering iron with a pistol grip and a quick heating, relatively small bit.

soldering iron. A soldering tool having an internally or externally heated metal bit usually made of copper.

solder interface. The interface between filler metal and base metal in a soldered joint.

solidus. The highest temperature at which a metal or alloy is completely solid.

spacer strip. A metal strip or bar prepared for a groove weld and inserted in the joint root to serve as a backing and to maintain the root opening during welding. It can also bridge an exceptionally wide root opening due to poor fit. See Fig. 53

spatter. The metal particles expelled during fusion welding that do not form a part of the weld.

spatter loss. Metal lost due to spatter.

spool. A filler metal package consisting of a continuous length of welding wire in coil form wound on a cylinder (called a barrel) which is flanged at both ends. The flange contains a spindle hole of smaller diameter than the inside diameter of the barrel. See Fig. 54

spot weld. A weld made between or upon overlapping members in which coalescence may start and occur on the faying surfaces or may proceed from the outer surface of one member. The weld cross section (plan view) is approximately circular.

spray deposit. The coating or layer of surfacing material applied by a thermal spraying process.

spray rate (thermal spraying). The rate at which surfacing material passes through the gun.

spray transfer (arc welding). Metal transfer in which molten metal from a consumable electrode is propelled axially across the arc in small droplets. See Fig. 55.

stack cutting. Thermal cutting of stacked metal plates arranged so that all the plates are severed by a single cut.

staggered intermittent weld. An intermittent weld on both sides of a joint in which the weld increments on one side are alternated with respect to those on the other side. See Fig. 37

standoff distance. The distance between a nozzle and the workpiece.

start current. The current value during start time interval.

start time. The time interval prior to weld time during which arc voltage and current reach a preset value greater or less than welding values.

start voltage. The arc voltage during the start time.

stepback sequence. A nonstandard term for **backstep sequence.**

stick electrode. A nonstandard term for **covered electrode.**

stick electrode welding. A nonstandard term for **shielded metal arc welding.**

stickout. A nonstandard term for **electrode extension.**

stitch weld. A nonstandard term for **intermittent weld.**

straight polarity. A nonstandard term for **direct current electrode negative.**

stranded electrode. A composite filler metal electrode consisting of stranded wires that may mechanically enclose materials to improve properties, stabilize the arc, or provide shielding.

stress corrosion cracking. Failure of metals by cracking under combined action of corrosion and stress, residual or applied. In brazing, the term applies to the cracking of stressed base metal due to the presence of a liquid filler metal.

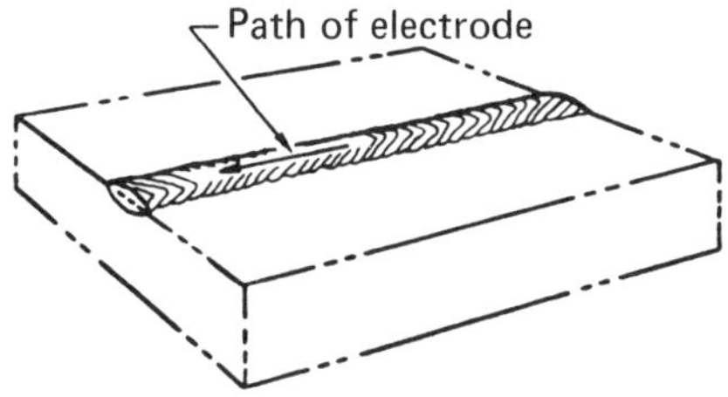

Fig. 56. Stringer Bead.

stress relief cracking. Intergranular cracking in the heat-affected zone or weld metal that occurs during the exposure of weldments to elevated temperatures during postweld heat treatment or high temperature service.

stress relief heat treatment. Uniform heating of a structure or a portion thereof to a sufficient temperature to relieve the major portion of the residual stresses, followed by uniform cooling.

stringer bead. A type of weld bead made without appreciable weaving motion. See Fig. 56

stub. The short length of welding rod or consumable electrode that remains after its use for welding.

stud arc welding (SW). An arc welding process that produces coalescence of metals by heating them with an arc between a metal stud, or similar part, and the other workpiece. When the surfaces to be joined are properly heated, they are brought together under pressure. Partial shielding may be obtained by the use of a ceramic ferrule surrounding the stud. Shielding gas or flux may or may not be used.

stud welding. A general term for joining a metal stud or similar part to a workpiece. Welding may be accomplished by arc, resistance, friction, or other process with or without external gas shielding.

submerged arc welding (SAW). An arc welding process that produces coalescence of metals by heating them with an arc or arcs between a bare metal electrode or electrodes and the workpieces. The arc and molten metal are shielded by a blanket of granular, fusible material on the workpieces. Pressure is not used, and filler metal is obtained from the electrode and sometimes from a supplemental source (welding rod, flux, or metal granules). See also **series submerged arc welding.**

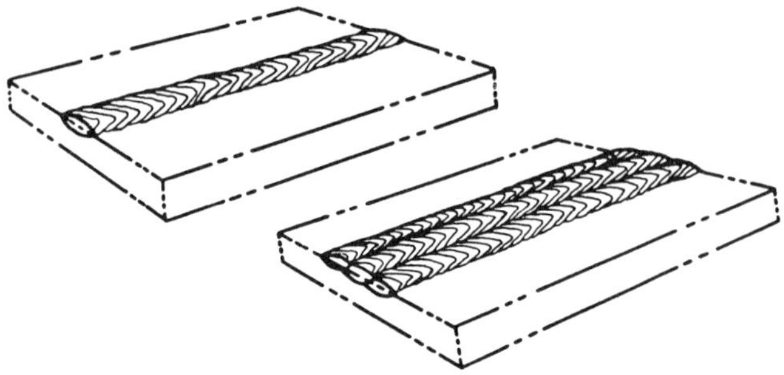

Fig. 57. Surfacing Welds.

suck-back. A nonstandard term for **concave root surface.**

surface preparation. The operations necessary to produce a desired or specified surface condition.

surface roughening. A group of methods for producing irregularities on a surface. See also **dovetailing, groove and rotary roughening, rotary roughening, rough threading,** and **threading and knurling.**

surfacing. The application by welding, brazing, or thermal spraying of a layer(s) of material to a surface to obtain desired properties or dimensions, as opposed to making a joint. See also **buildup, buttering, cladding,** and **hardfacing.**

surfacing material. The material that is applied to a base metal or substrate during surfacing. The use of this term implies that materials other than metals are also included, where this is appropriate.

surfacing metal. The metal or alloy that is applied to a base metal or substrate during surfacing.

surfacing weld. A weld applied to a surface, as opposed to making a joint, to obtain desired properties or dimensions. See Fig. 57

T

tacker. A nonstandard term for a tack welder.

tack weld. A weld made to hold parts of a weldment in proper alignment until the final welds are made.

tandem arc welding.[2] A system of machine welding which uses two or more wire feeders each feeding a single welding electrode. Each wire feeder and welding electrode has its own power source to provide the required amperage, voltage and polarity. Two electrode Tandem Arc systems are considerably more common than three or more electrodes in Tandem.

taper delay time. The time interval after upslope during which the current is constant.

taper time. The time interval when current increases or decreases continuously from the welding current to final taper current.

taps. Connections to a transformer winding that are used to vary the transformer turn ratio, thereby controlling welding voltage and current.

temporary weld. A weld made to attach a piece or pieces to a weldment for temporary use in handling, shipping, or working on the weldment.

theoretical throat. The distance from the beginning of the joint root perpendicular to the hypotenuse of the largest right triangle that can be inscribed within the cross section of a fillet weld. This dimension is based on the assumption that the root opening is equal to zero. See Figs. 18 & 20

thermal cutting. A group of cutting processes that melts the base metal. See also **arc cutting, oxygen cutting, electron beam cutting,** and **laser beam cutting.**

thermal spraying (THSP). A group of processes in which finely divided metallic or nonmetallic surfacing materials are deposited in a molten or semimolten condition on a substrate to form a spray deposit. The surfacing material may be in the form of powder, rod, or wire. See also **arc spraying, flame spraying,** and **plasma spraying.**

thermal spraying gun. A device for heating, feeding, and directing the flow of a surfacing material.

thermal stress. Stress resulting from nonuniform temperature distribution.

thermit mold. A mold formed around the workpieces to receive the molten metal.

[2] Tandem Arc Welding is not presently defined by AWS A3.0.

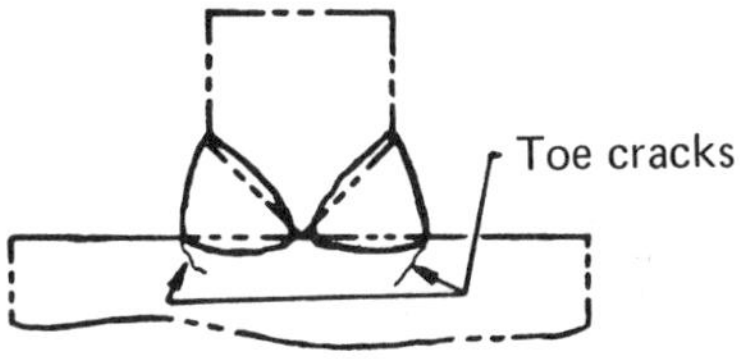

Fig. 58. Toe Cracks.

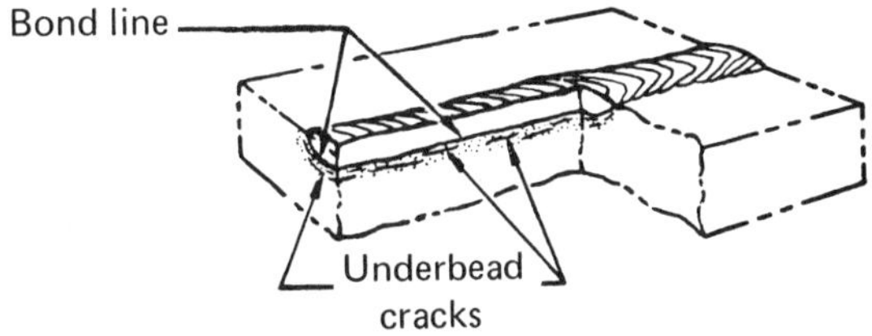

Fig. 59. Underbead Cracks.

thermit reaction. The chemical reaction between metal oxide and aluminum that produces superheated molten metal and a slag containing aluminum oxide.

thermit welding (TW). A welding process that produces coalescence of metals by heating them with superheated liquid metal from a chemical rection between a metal oxide and aluminum, with or without the application of pressure. Filler metal is obtained from the liquid metal.

throat of a fillet weld. See **actual throat, effective throat,** and **theoretical throat.**

throat of a groove weld. A nonstandard term for **groove weld size.**

TIG welding. A nonstandard term for **gas tungsten arc welding.**

T-joint. A joint between two members located approximately at right angles to each other in the form of a T. See Fig. 12

toe cracks. See Fig. 58.

toe of weld. See **weld toe.**

torch. See **cutting torch** and **welding torch.**

torch tip. See **welding tip** and **cutting tip.**

transverse crack. A crack with its major axis oriented approximately perpendicular to the weld axis.

travel angle. The angle less than 90 degrees between the electrode axis and a line perpendicular to the weld axis, in a plane determined by the electrode axis and the weld axis. This angle can also be used to partially define the position of guns, torches, rods, and beams. See **drag angle, push angle,** and **work angle.** See also Fig. 40.

travel angle (pipe). The angle less than 90 degrees between the electrode axis and a line perpendicular to the weld axis at its point of intersection with the extension of the electrode axis, in a plane determined by the electrode axis and a line tangent to the pipe surface at the same point. This angle can also be used to partially define the position of guns, torches, rods, and beams. See **drag angle, push angle,** and **work angle.**

tungsten electrode. A nonfiller metal electrode used in arc welding or cutting, made principally of tungsten.

type of joint. See **joint type.**

U

U-groove weld. A type of groove weld. See Fig. 22

ultrasonic welding (USW). A solid-state welding process that produces coalescence of materials by the local application of high frequency vibratory energy as the workpieces are held together under pressure.

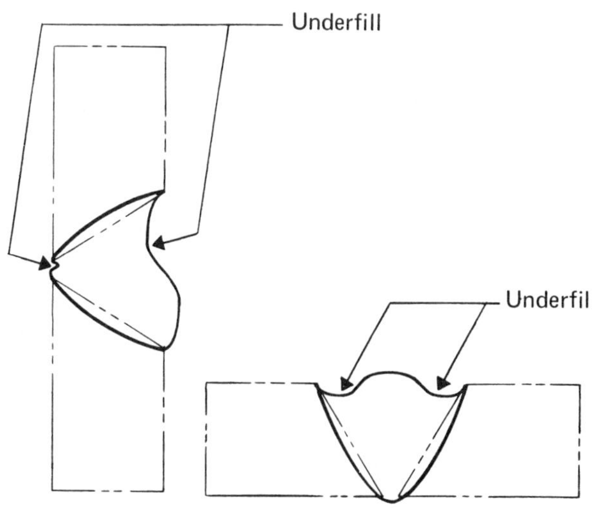

Fig. 60. Underfill.

underbead crack. A crack in the heat-affected zone generally not extending to the surface of the base metal. See Fig. 59

undercut. A groove melted into the base metal adjacent to the weld toe or weld root and left unfilled by weld metal. See Fig. 42

underfill. A depression on the weld face or root surface extending below the adjacent surface of the base metal. See Fig. 60

upslope time (automatic arc welding). The time during which the current changes continuously from initial current value to the welding value.

V

vacuum brazing. A nonstandard term for various brazing processes that take place in a chamber or retort below atmospheric pressure.

vertical position. The position of welding in which the weld axis is approximately vertical. See Figs. 1 & 2

vertical position (pipe welding). The position of a pipe joint in which welding is performed in the horizontal position and the pipe may or may not be rotated. See Fig. 35

V-groove weld. A type of groove weld. See Fig. 22

voltage regulator. An automatic electrical control device for maintaining a constant voltage supply to the primary of a welding transformer.

W

weave bead. A type of weld bead made with transverse oscillation. see Fig. 61

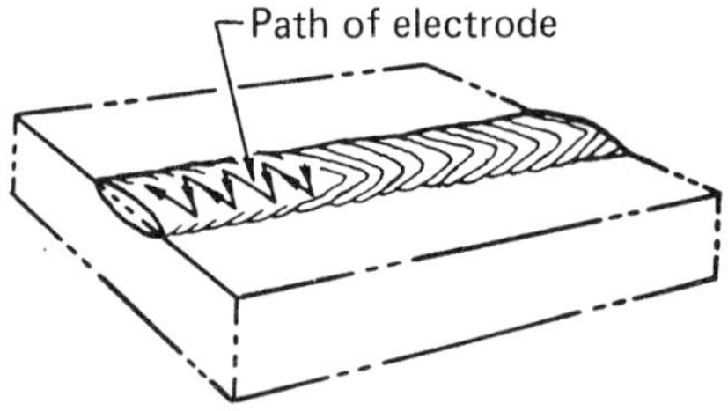

Fig. 61. Weave Bead.

weld. A localized coalescence of metals or nonmetals produced either by heating the materials to the welding temperature, with or without the application of pressure, or by the application of pressure alone and with or without the use of filler metal.

weldability. The capacity of material to be welded under the imposed fabrication conditions into a specific, suitable designed structure and to perform satisfactorily in the intended service.

weld axis. A line through the length of a weld, perpendicular to and at the goemetric center of its cross section. See Figs. 1 & 2

weld bead. A weld resulting from a pass. See Fig. 39. See also **stringer bead** and **weave bead.**

weld crack. A crack located in the weld metal or heat affected zone.

welder[3]. One who is capable of performing a manual or semiautomatic welding operation. Used in this handbook to denote a welding machine.

welder certification. Certification in writing that a welder has produced welds meeting prescribed standards.

welder performance qualification. The demonstration of a welder's ability to produce welds meeting prescribed standards.

welder registration. The act of registering a welder certification or a photostatic copy of the welder certification.

weld face. The exposed surface of a weld on the side from which welding was done. See Fig. 28

weld gauge. A device designed for measuring the shape and size of welds. See Fig. 29

welding current. The current in the welding circuit during the making of a weld.

welding current (automatic arc welding). The current in the welding circuit during the making of a weld, but excluding upslope, downslope, start, and crater fill current.

[3] This handbook deviates from AWS practice by using "welder" to denote a machine and "weldor" to indicate one who is caple of performing a welding operation.

welding current (resistance welding). The current in the welding circuit during the making of a weld, but excluding preweld or postweld current.

welding cycle. The complete series of events involved in the making of a weld.

welding electrode. A component of the welding circuit through which current is conducted and that terminates at the arc, molten conductive slag, or base metal. See also **arc welding electrode, bare electrode, carbon electrode, emissive electrode, flux cored electrode, lightly coated electrode, metal cored electrode, metal electrode, resistance welding electrode, stranded electrode,** and **tungsten electrode.**

welding generator. A generator used for supplying current for welding.

welding ground. A nonstandard term for **workpiece connection.**

welding head. The part of a welding machine or automatic welding equipment in which a welding gun or torch is incorporated.

welding leads. The workpiece lead and electrode lead of an arc welding circuit. See Figs. 25 & 26

welding machine. Equipment used to perform the welding operation. For example. spot welding machine, arc welding machine, and seam welding machine.

welding operator. One who operates machine or automatic welding equipment.

welding position. See **flat position, horizontal position, horizontal fixed position, horizontal rolled position, inclined position, overhead position,** and **vertical position.**

welding procedure. The detailed methods and practices involved in the production of a weldment. See also **welding procedure specification.**

welding procedure specification (WPS). A document providing in detail the required variables for specific application to assure repeatability by properly trained welders and welding operators.

welding process. A joining process that produces coalescence of materials by heating them to the welding temperature, with or without the application of pressure or by the application of pressure alone, and with or without the use of filler metal.

welding rectifier. A device in a welding machine for converting alternating current to direct current.

welding rod. A form of welding filler metal, normally packaged in straight lengths, that does not conduct electrical current.

welding sequence. The order of making the welds in a weldment.

welding technique. The details of a welding procedure that are controlled by the welder or welding operator.

welding tip. The part of an oxyfuel gas welding torch from which the gases issue.

welding torch (arc). A device used in the gas tungsten and plasma arc welding processes to control the position of the electrode, to transfer current to the arc, and to direct the flow of shielding and plasma gas.

welding torch (Oxyfuel gas). A device used in oxyfuel gas welding, torch brazing, and torch soldering for directing the heating flame produced by the controlled combustion of fuel gases.

welding transformer. A transformer used for supplying current for welding.

welding voltage. See **arc voltage.**

welding wire. A form of welding filler metal, normally packaged as coils or spools, that may or may not conduct electrical current depending upon the welding process with which it is used. See also **welding electrode** and **welding rod.**

weld interface. The interface between weld metal and base metal in a fusion weld, between base metals in a solid-state weld without filler metal, or between filler metal and base metal in a solid-state weld with filler metal.

weld length. See **effective length of weld.**

weld line. A nonstandard term for **weld interface.**

weldment. An assembly whose component parts are joined by welding.

weld metal. That portion of a weld that has been melted during welding.

weld metal area. The area of the weld metal as measures on the cross section of a weld.

weld metal crack. See weld crack.

weld pass. A single progression of welding along a joint. The result of a pass is a weld bead, or layer.

weld pass sequence. The order in which the weld passes are made. See also **longitudinal sequence** and **cross-sectional sequence.**

weld penetration. A nonstandard term for **joint penetration** and **root penetration.**

weld pool. The localized volume of molten metal in a weld prior to its solidification as weld metal.

weld puddle. A nonstandard term for **weld pool.**

weld reinforcement. Weld metal in excess of the quantity required to fill a joint. See also **face reinforcement** and **root reinforcement.**

weld root. The points, as shown in cross section, at which the root surface intersects the base metal surfaces. See Fig. 48.

weld size. See **edge weld size, groove weld size, fillet weld size,** and **flange weld size.**

weld tab. Additional material on which the weld may be initiated or terminated.

weld throat. See **actual throat, effective throat,** and **theoretical throat.**

weld time (automatic arc welding). The time interval from the end of start time or end of upslope to beginning of crater fill time or beginning of downslope.

weld toe. The junction of the weld face and the base metal.

weld voltage. See **arc voltage.**

wetting. The phenomenon whereby a liquid filler metal or flux spreads and adheres in a thin continuous layer on a solid base metal.

wiped joint. A joint made with solder having a wide melting range and with the heat supplied by the molten solder poured onto the joint. The solder is manipulated with a hand-held cloth or paddle so as to obtain the required size and contour.

wire feed speed. The rate of speed at which a filler metal is consumed in arc welding or thermal spraying.

wire straightener. A device used for controlling the cast of coiled welding wire to enable it to be easily fed into the gun.

work angle. The angle less than 90 degrees between a line perpendicular to the major workpiece surface and a plane determined by the electrode axis and the weld axis. In a T-joint or a corner joint, the line is perpendicular to the nonbutting member. This angle can also be used to partially define the position of guns, torches, rods, and beams. See Fig. 40. See also **drag angle, push angle**, and **travel angle.**

work angle (pipe). The angle less than 90 degrees between a line, which is perpendicular to the cylindrical pipe surface at the point of intersection of the weld axis and the extension of the electrode axis, and a plane determined by the electrode axis and a line tangent to the pipe at the same point. In a T-joint, the line is perpendicular to the nonbutting member. This angle can also be used to partially define the position of guns, torches, rods, and beams. See **drag angle, push angle**, and **travel angle.**

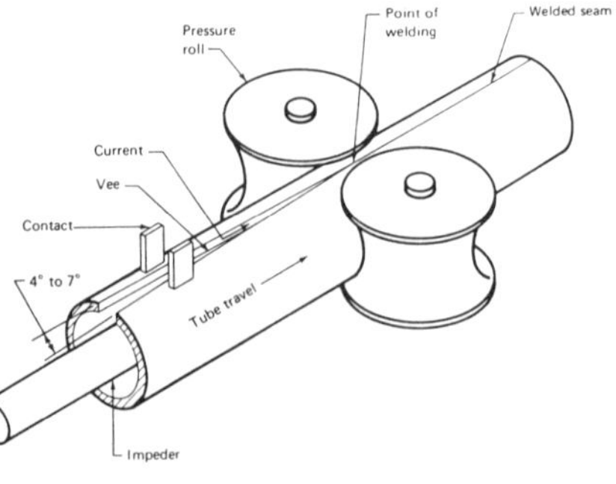

Fig. 62. High Frequency Resistance Seam Welding.

work connection. A nonstandard term for **workpiece connection.**

work lead. A nonstandard term for **workpiece lead.**

workpiece. The part being welded, brazed, soldered or cut.

workpiece connection. The connection of the workpiece lead to the workpiece.

workpiece lead. The electrical conductor between the arc welding current source and the workpiece connection.

Welding Symbols

STANDARD WELDING SYMBOLS*

The weld symbols presented here have been standardized and adopted by the American Welding Society. Like any systematic plan of symbols, they quickly describe to the designer, draftsman, production supervisor, and weldor the weld that will develop the required joint or connection strength and meet the prevailing in-service performance requirements.

A welding symbol consists of basic elements. These elements can be joined in various combinations to denote any type of weld needed for any type of joint. The condensed summary presented here identifies these elements and will serve as a ready reference for those associated with a welding program.

Since fillet welds and simple butt welds comprise 95 percent of most work, it is wise to initially limit the use of the symbols to just these types of welds. Later on, after the draftsmen and shop personnel learn to communicate with the symbols, additional, rarely used symbols can be introduced to aid in detailing special welds.

Adoption of this uniform system of symbols will assure that the correct welding instructions are being transmitted to all concerned with the work. This can help to cut drafting time and reduce shop costs arising from misinterpretation of instructions. In addition, since this system is in general use today, it simplifies the transmission of this type of information to outside suppliers.

In a joint, the adjoining members may contact each other in several ways, as illustrated by the butt, T, corner, lap, and edge joints. These general descriptions of the joint geometry, however, do not define the weld joint configuration, since it can be made in various ways. Thus, a welded butt joint can be made square, double-square, single-bevel, double-bevel, single-V, double-V, or by four other joint configurations. A T connection can be made with a double fillet, as shown: or it may be made with a single or double bevel or single or double J. V and U weld joints are feasible only for butt and corner welds because of the need for the preparation of both surfaces.

TYPES of JOINTS

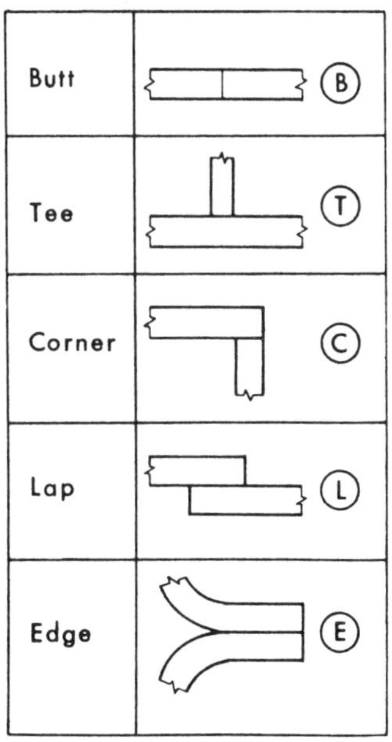

TYPES of WELDS

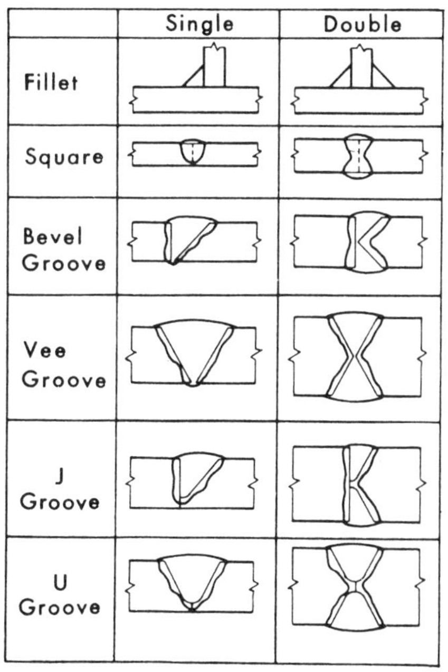

* From the American Welding Society "Standard Symbols for Welding, Brazing and Non-Destructive Examination", A2.4-86. For a complete copy, order from American Welding Society, 550N.W. LeJeune Rd., P.O. Box 351040, Miami, Florida 33135.

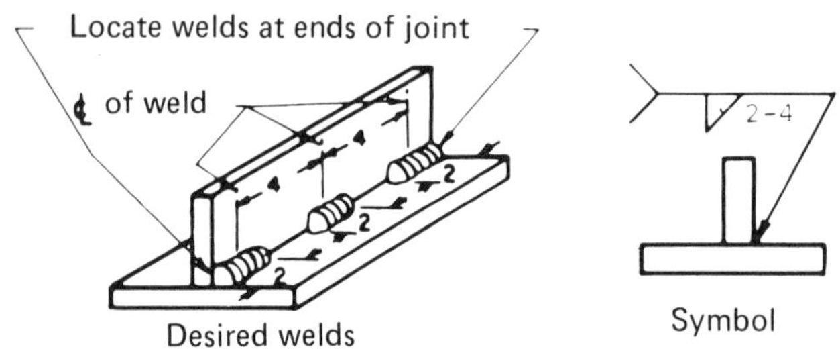

Length and Pitch of Increments of Intermittent Welding

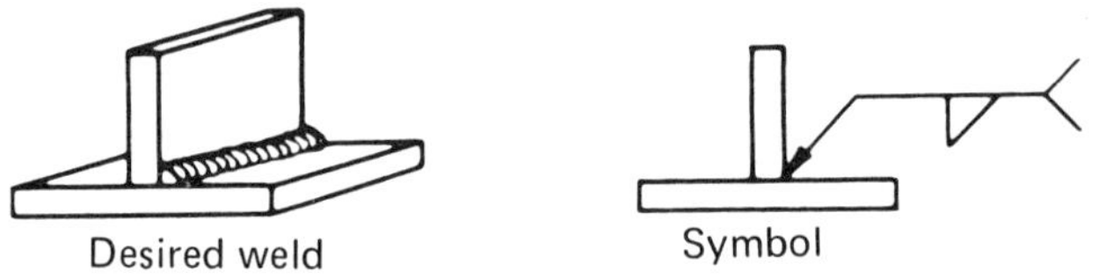

Continuous Fillet Weld

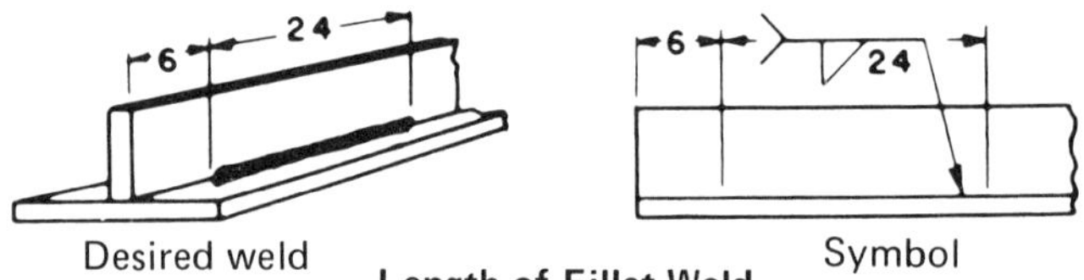

Length of Fillet Weld

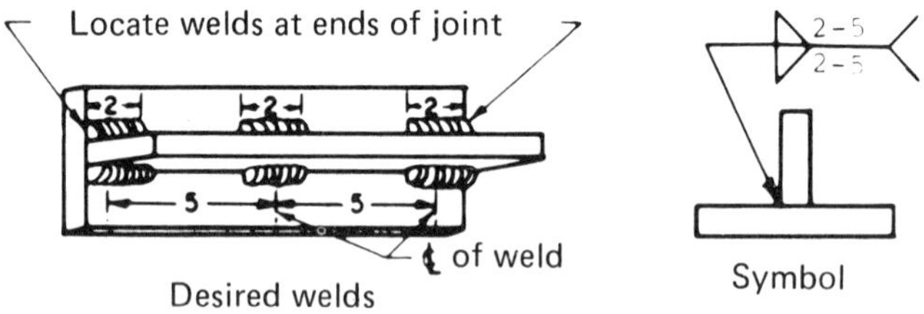

Length and Pitch of Increments of Chain Intermittent Welding

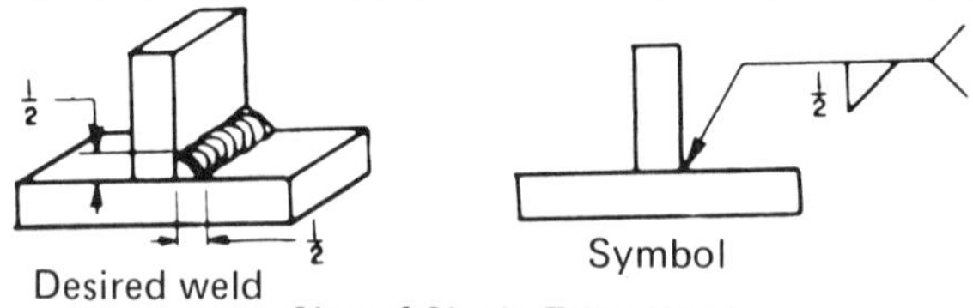

Size of Single-Fillet Weld

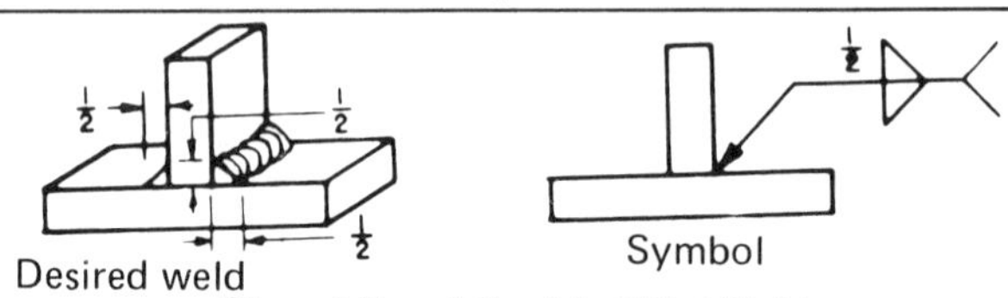

Size of Equal Double-Fillet Welds

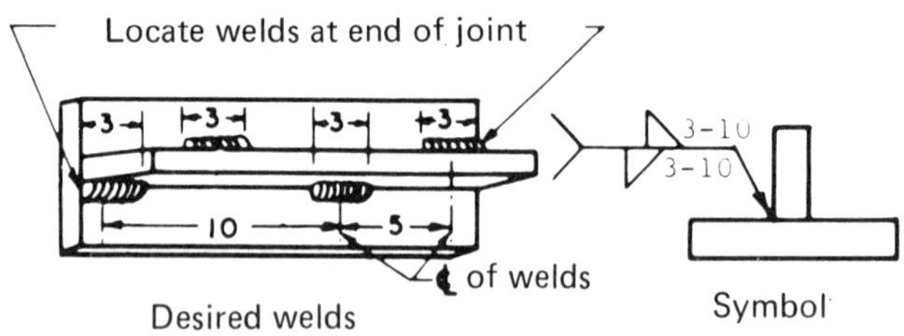

Length and Pitch of Increments of Staggered Intermittent Welding

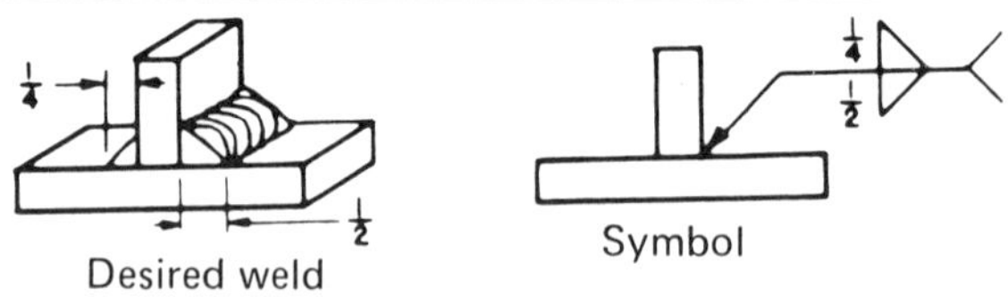

Size of Unequal Double-Fillet Welds

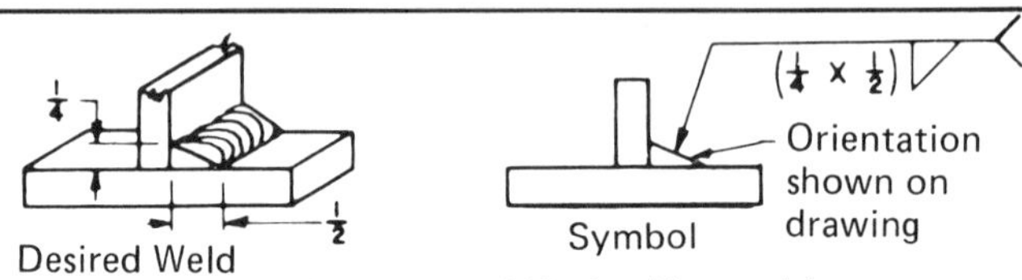

Size of Fillet Weld Having Unequal Legs

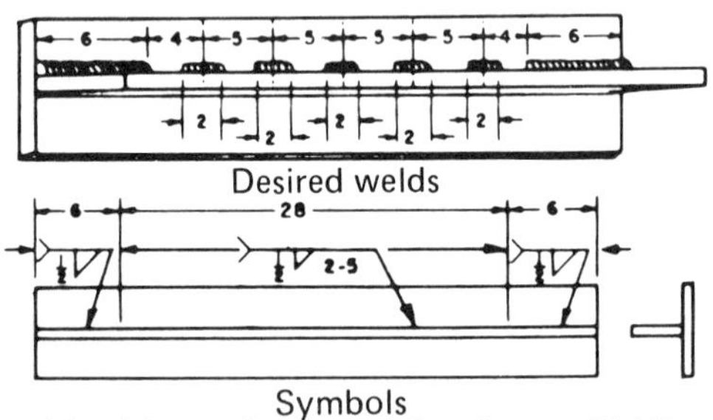

Combined Intermittent and Continuous Welding
(One side of joint)

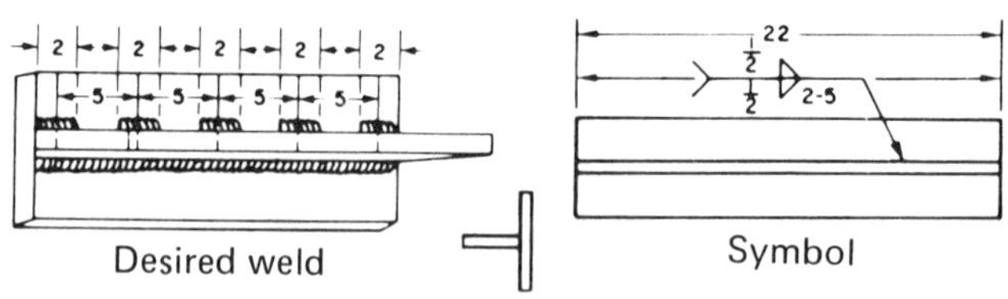

Combined Intermittent and Continuous Welding
(Opposite sides of joint)

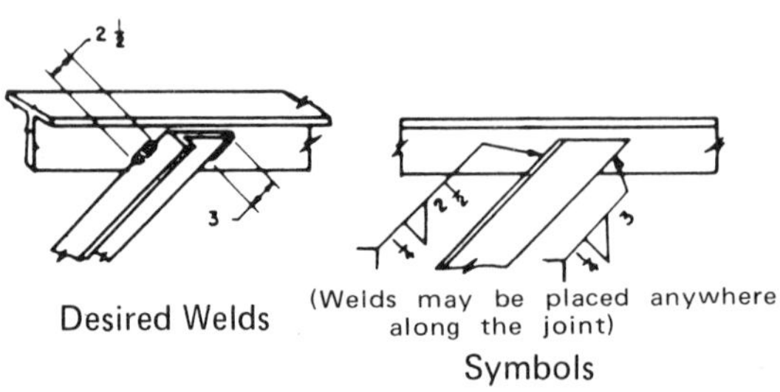

Welds Approximately Located

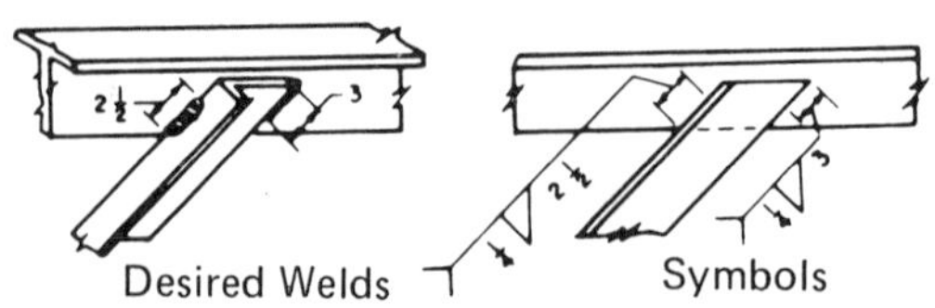

Welds Definitely Located

Basic Joints — Identification of Arrow Side and Other Side of Joint

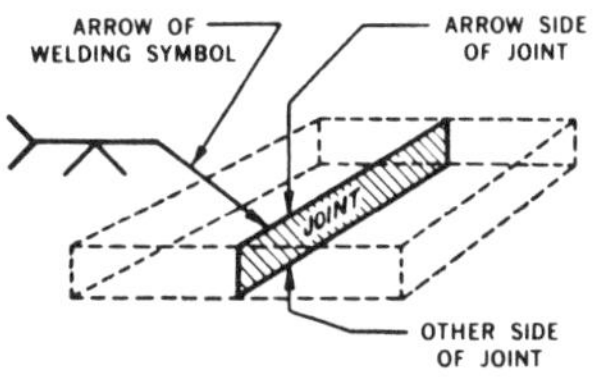

BUTT JOINT

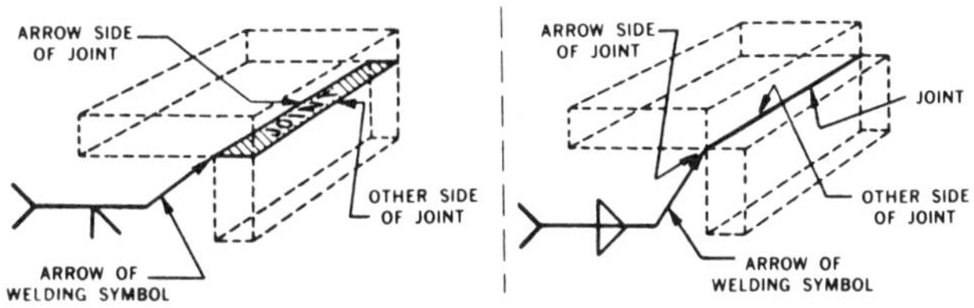

CORNER JOINT

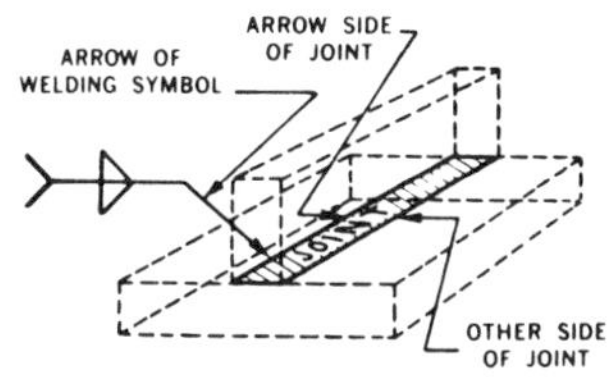

TEE JOINT

Arrow-Side and Other-Side Member of Joint

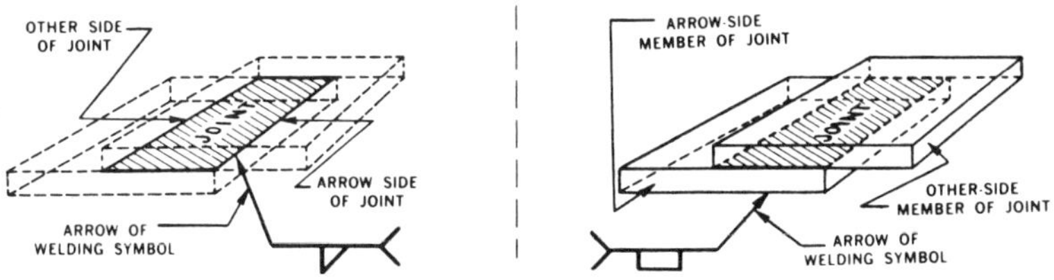

LAP JOINT

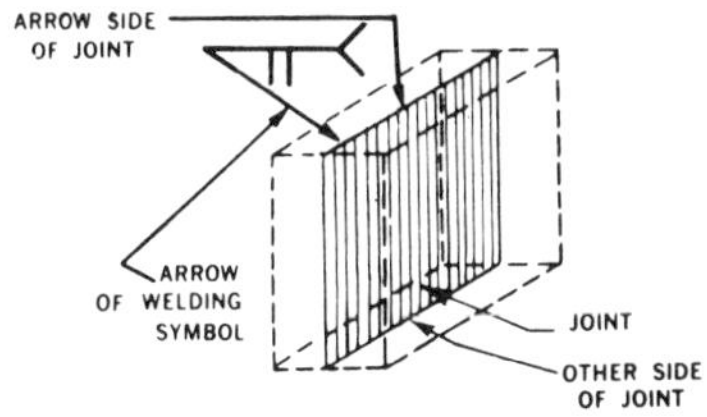

EDGE JOINT

Fillet	Plug or Slot	Spot or Projection	Seam	Groove: Square (Flash or Upset)	Groove: V	Groove: Bevel	Groove: U	Groove: J	Groove: Flare V	Groove: Flare-bevel	Back or Backing	Surfacing	Flange: Edge	Flange: Corner

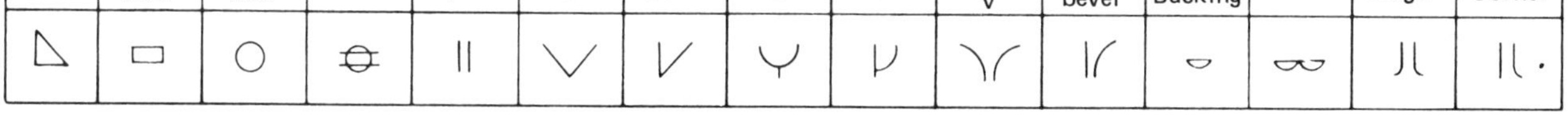

ARC-SEAM OR ARC-SPOT	RESISTANCE SPOT	PROJECTION	RESISTANCE SEAM	FLASH OR UPSET

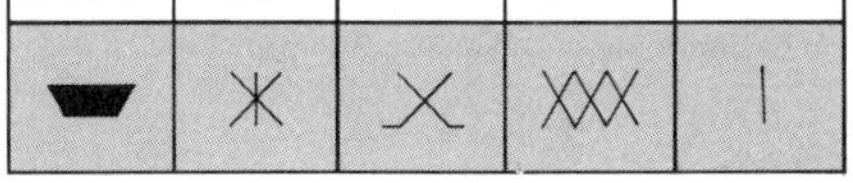

NONPREFERRED SYMBOLS: USE PREFERRED SYMBOL WITH PROCESS REFERENCE IN THE TAIL

Designation of Welding Processes by Letters

- CAW . . . . Carbon-Arc Welding
- CW . . . . Cold Welding
- DB . . . . Dip Brazing
- DFW . . . . Diffusion Welding
- EBW . . . . Electron Beam Welding
- EW . . . . Electroslag Welding
- EXW . . . . Explosion Welding
- FB . . . . Furnace Brazing
- FCAW . . . . . Flux-Cored Arc Welding
- FOW . . . . Forge Welding
- FRW . . . . Friction Welding
- FW . . . . Flash Welding
- GMAW . . . . . Gas Metal-Arc Welding
- GTAW . . . . Gas Tungsten-Arc Welding
- IB . . . . Induction Brazing
- IRB . . . . Infrared Brazing
- IW . . . . Induction Welding
- LBW . . . . Laser Beam Welding
- OAW . . . . Oxyacetylene Welding
- OHW . . . . Oxyhydrogen Welding
- PAW . . . . Plasma-Arc Welding
- PEW . . . . Percussion Welding
- PGW . . . . Pressure Gas Welding
- RB . . . . Resistance Brazing
- RPW . . . . Projection Welding
- RSEW . . . . . Resistance-Seam Welding
- RSW . . . . . . Resistance-Spot Welding
- SAW . . . . Submerged-Arc Welding
- SMAW . . . Shielded Metal-Arc Welding
- SW . . . . Stud Welding
- TB . . . . Torch Brazing
- TW . . . . Thermit Welding
- USW . . . . Ultrasonic Welding
- UW . . . . Upset Welding

Designation of Cutting Processes by Letters

- AAC . . . . Air Carbon-Arc Cutting
- AC . . . . Arc Cutting
- AOC . . . . Oxygen-Arc Cutting
- CAC . . . . Carbon-Arc Cutting
- FOC . . . . Chemical Flux Cutting
- MAC . . . . Metal-Arc Cutting
- OC . . . . Oxygen Cutting
- PAC . . . . Plasma-Arc Cutting
- POC . . . . Metal Powder Cutting

AMERICAN WELDING SOCIETY

Basic Welding Symbols and Their Location Significance

Location Significance	Fillet	Plug or Slot	Spot or Projection	Seam	Back or Backing	Surfacing	Scarf for Brazed Joint	Flange Edge
Arrow Side					Groove weld symbol			
Other Side					Groove weld symbol	Not used		
Both Sides		Not used	Not used	Not used	Not used	Not used		Not used
No Arrow Side or Other Side Significance	Not used	Not used			Not used	Not used	Not used	Not used

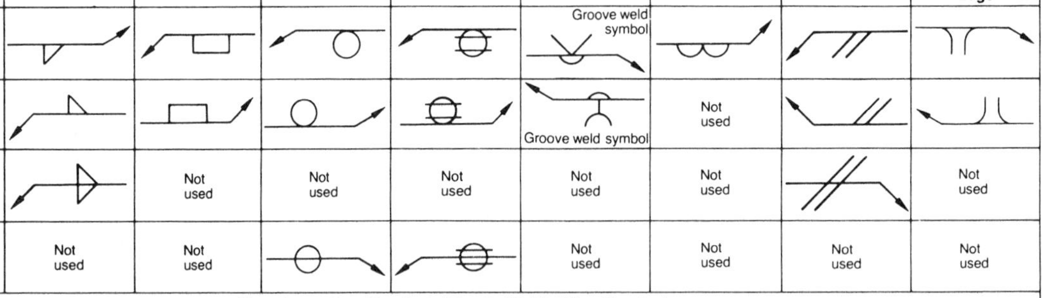

Supplementary Symbols Used with Welding Symbols

Convex Contour Symbol

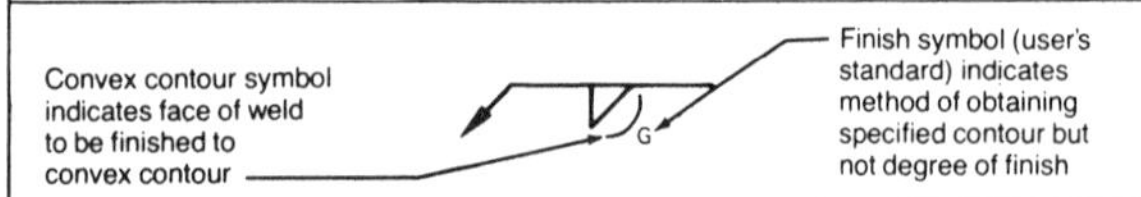

Weld-All-Around Symbol

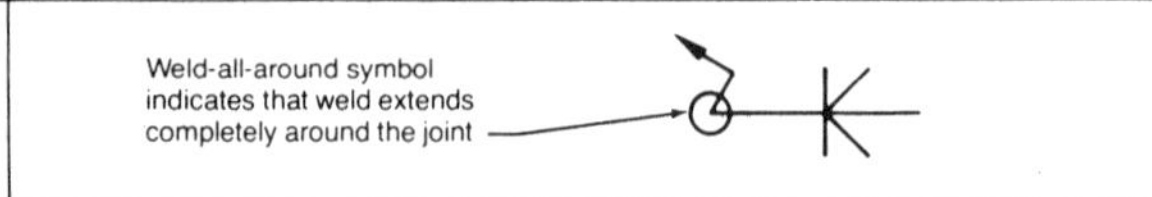

Joint with Backing

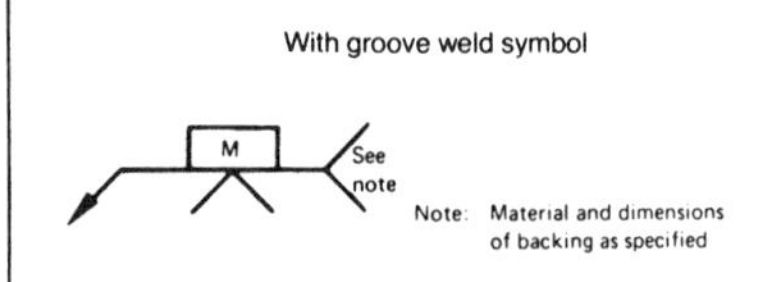

Joint with Spacer

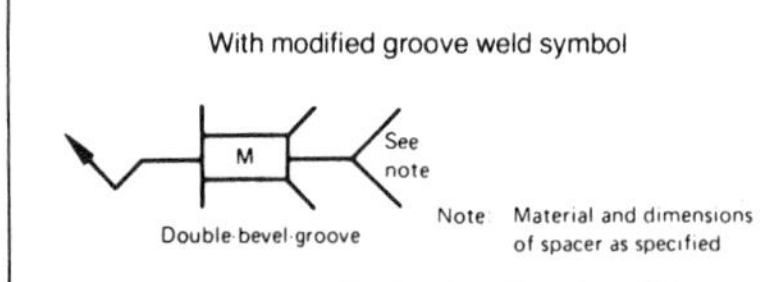

Melt-Thru Symbol

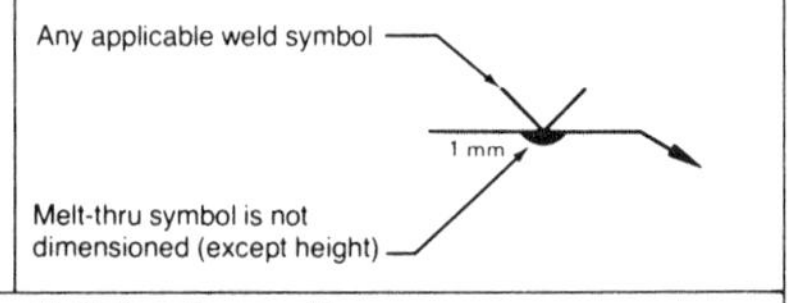

Flush Contour Symbol

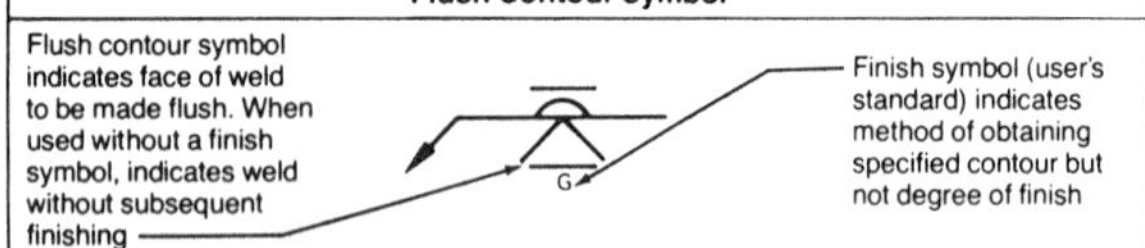

Multiple Reference Lines

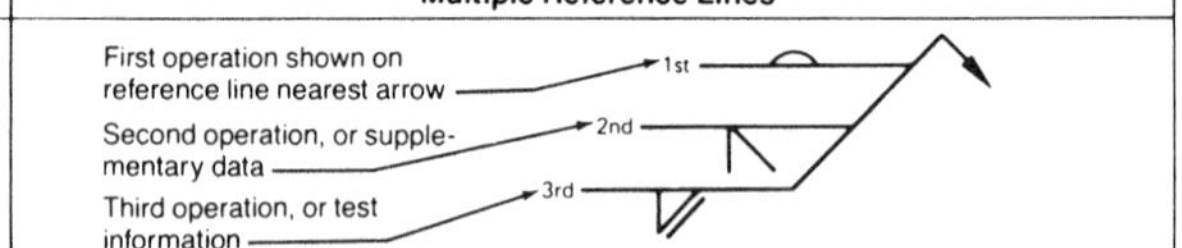

Field Weld Symbol

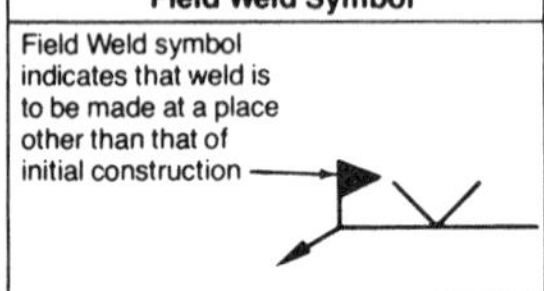

Complete Penetration

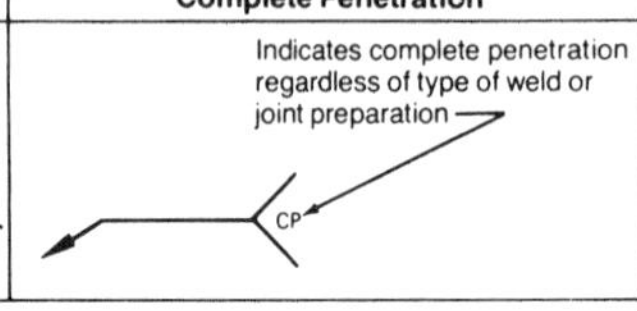

Location of Elements of a Welding Symbol

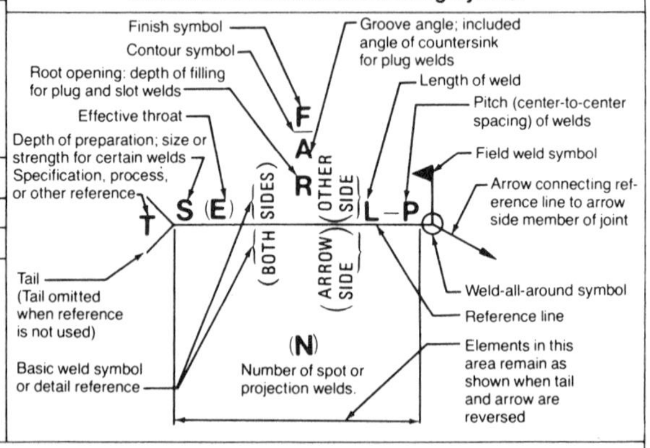

Supplementary Symbols

Weld-All-Around	Field Weld	Melt-Thru	Backing, Spacer	Contour: Flush	Contour: Convex	Contour: Concave

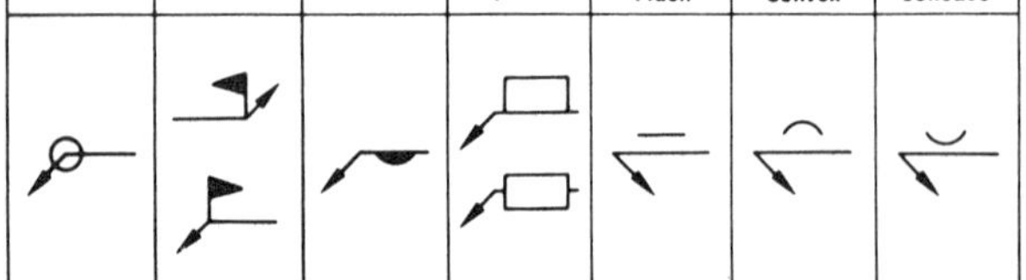

Basic Joints—Identification of Arrow Side and Other Side of Joint

Butt Joint

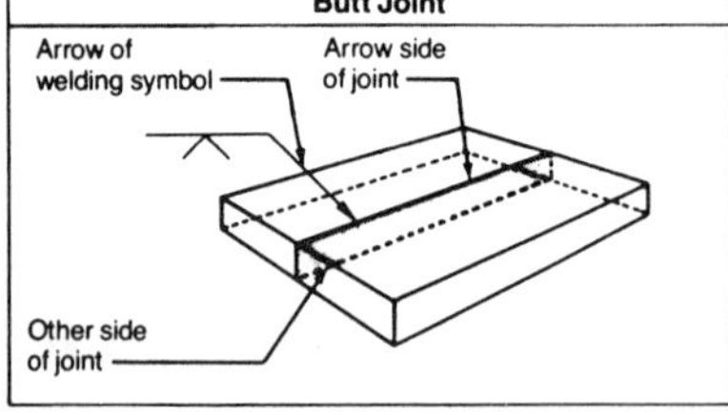

Corner Joint

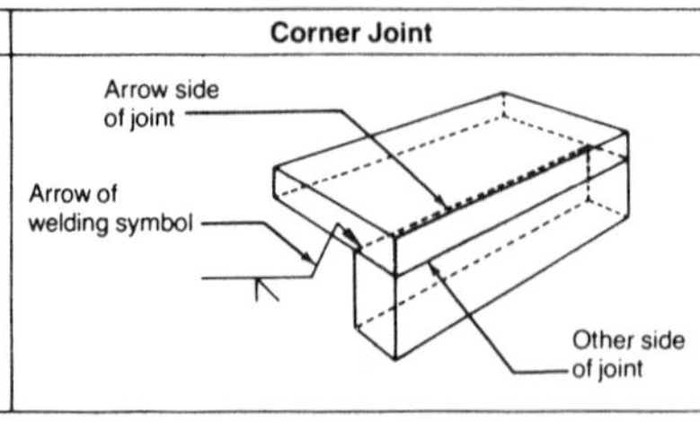

T-Joint

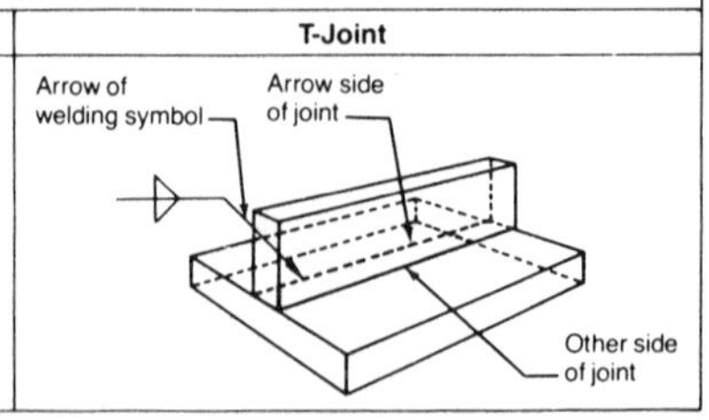

STANDARD WELDING SYMBOLS

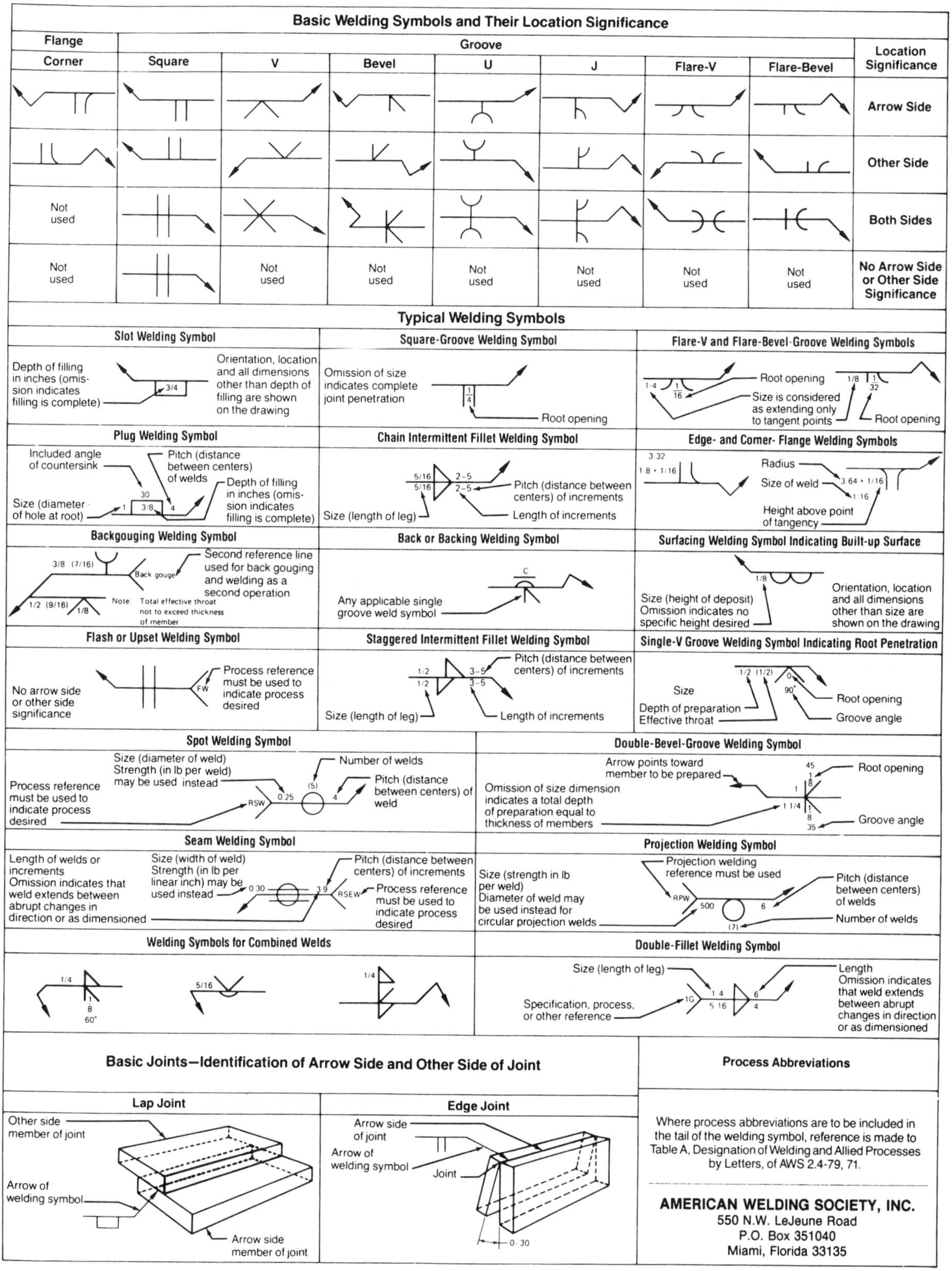

Process Abbreviations

Where process abbreviations are to be included in the tail of the welding symbol, reference is made to Table A, Designation of Welding and Allied Processes by Letters, of AWS 2.4-79, 71.

AMERICAN WELDING SOCIETY, INC.
550 N.W. LeJeune Road
P.O. Box 351040
Miami, Florida 33135

MASTER CHART OF WELDING AND ALLIED PROCESSES

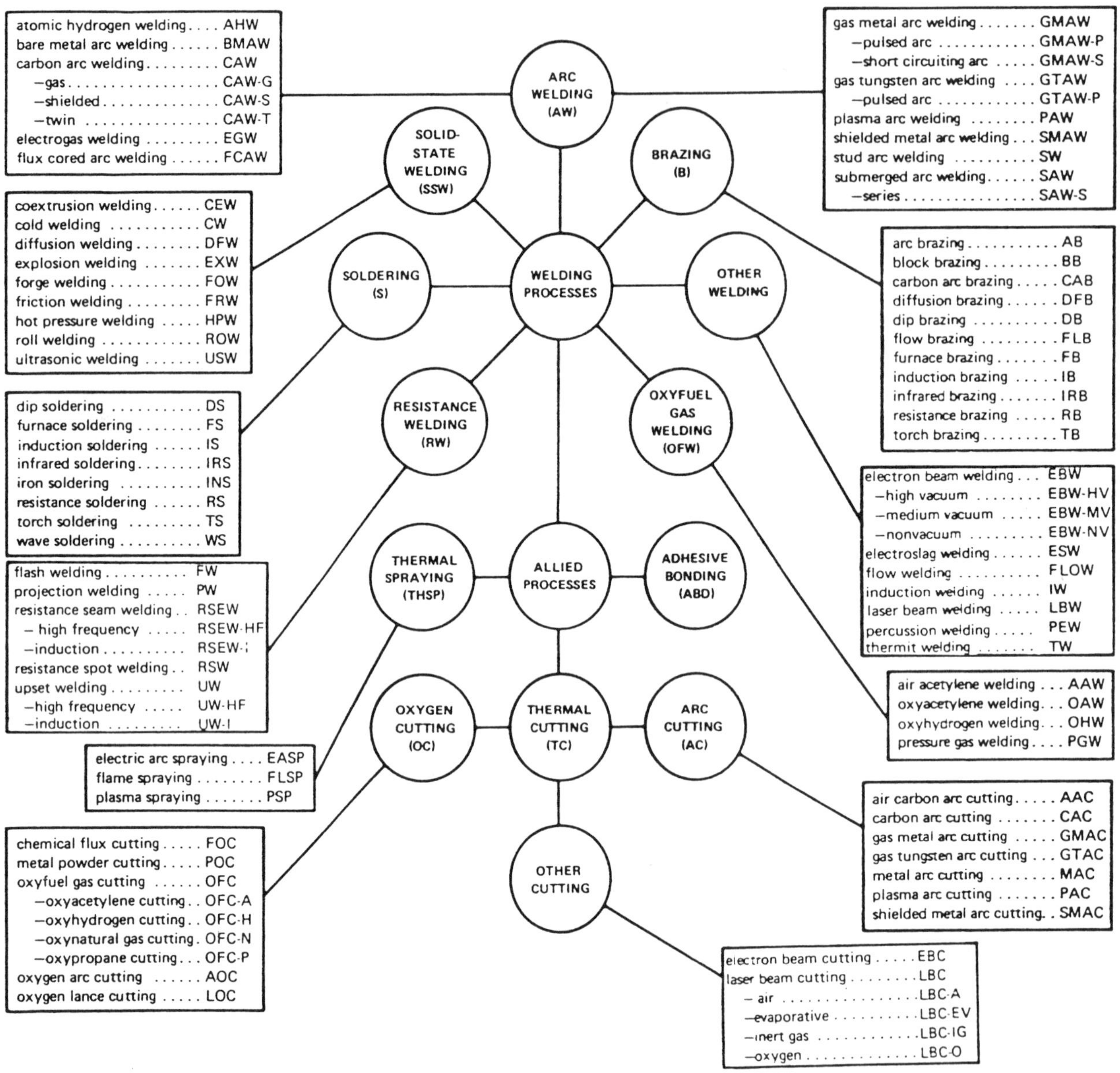

From AWS Welding Handbook, 8th Edition, Volume 1, 1987.

Conversion Factors

COMMON UNITS FOR CONVERTING FROM THE ENGLISH TO THE METRIC (SI) SYSTEM†

The International System of Units (SI for short) is a modernized version of the metric system. It is built upon six base units and two supplementary units. Symbols for units with specific names are given in parentheses. The information in this Data Sheet, adapted from the revised "Metric Practice Guide," Standard E380-68, 1969 Book of ASTM Standards, Part 30, includes a selected list of factors for converting U.S. customary units to SI units.

METRIC CONVERSION FACTORS FOR COMMON ENGINEERING TERMS

Property	To Convert From	To	Multiply By
acceleration (angular)	revolution per minute squared	rad/s^2	$1.745\ 329 \times 10^{-3}$
acceleration (linear)	in./min^2	m/s^2	$7.055\ 556 \times 10^{-6}$
	ft/min^2	m/s^2	$8.466\ 667 \times 10^{-5}$
	in./min^2	mm/s^2	$7.055\ 556 \times 10^{-3}$
	ft/min^2	mm/s^2	$8.466\ 667 \times 10^{-2}$
	ft/s^2	m/s^2	$3.048\ 000 \times 10^{-1}$
angle, plane	degree	rad	$1.745\ 329 \times 10^{-2}$
	minute	rad	$2.908\ 882 \times 10^{-4}$
	second	rad	$4.848\ 137 \times 10^{-6}$
area	in^2	m^2	$6.451\ 600 \times 10^{-4}$
	ft^2	m^2	$9.299\ 304 \times 10^{-2}$
	yd^2	m^2	$8.361\ 274 \times 10^{-1}$
	in^2	mm^2	$6.451\ 600 \times 10^{2}$
	ft^2	mm^2	$9.290\ 304 \times 10^{4}$
	acre (U.S. Survey)	m^2	$4.046\ 873 \times 10^{3}$
density	pound mass per cubic inch	kg/m^3	$2.767\ 990 \times 10^{4}$
	pound mass per cubic foot	kg/m^3	$1.601\ 846 \times 10$
energy, work heat, and impact energy	foot pound force	J	1.355 818
	found poundal	J	$4.214\ 011 \times 10^{-2}$
	Btu*	J	$1.054\ 350 \times 10^{3}$
	calorie*	J	4.184 000
	watt hour		$3.600\ 000 \times 10^{3}$
force	kilogram-force	N	9.806 650
	pound-force	N	4.448 222
impact strength	(see energy)		
length	in.	m	$2.540\ 000 \times 10^{-2}$
	ft	m	$3.048\ 000 \times {}^{-1}$
	yd	m	$9.144\ 000 \times {}^{-1}$
	rod (U.S. Survey)	m	5.029 210
	mile (U.S. Survey)	km	1.609 347
mass	pound mass (avdp)		
	metric ton	kg	$4.535\ 924 \times 10^{-1}$
	ton (short,	kg	$1.000\ 000 \times 10^{3}$
	2000 lbm)	kg	$9.071\ 847 \times 10^{2}$
	slug	kg	$1.459\ 390 \times 10$

Property	To Convert From	To	Multiply By
power	horsepower (550 ft lbf/s)	W	$7.456\ 999 \times 10^{2}$
	horsepower (electric)	W	$7.460\ 000 \times 10^{2}$
	Btu/min*	W	$1.757\ 250 \times 10$
	calorie per minute*	W	$6.973\ 333 \times 10^{-2}$
	foot pound-force per minute	W	$2.259\ 697 \times 10^{-2}$
pressure	pound force per square inch	kPa	6.894 757
	bar	kPa	$1.000\ 000 \times 10^{2}$
	atmosphere	kPa	$1.013\ 250 \times 10^{2}$
	kip/in.^{2}	kPa	$6.894\ 757 \times 10^{3}$
temperature	degree Celsius, $t°_C$	K	$t_K = t°_C + 273.15$
	degree Fahrenheit, $t°_F$	K	$t_K = \frac{(t°_F + 459.67)}{1.8}$
	degree Rankine, $t°_R$	K	$t_K = t°_R/1.8$
	degree Fahrenheit, t_F	°C	$t°_C = (t_F - 32)/1.8$
	kelvin, t_K	°C	$t°_C = t_K - 273.15$
tensile strength (stress)	ksi	MPa	6.894 757
torque	inch pound force	N•m	$1.129\ 848 \times 10^{-1}$
	foot pound force	N•m	1.355 818
velocity (angular)	revolution per minute	rad/s	$1.047\ 198 \times 10^{-1}$
	degree per minute	rad/s	$2.908\ 882 \times 10^{-4}$
	revolution per minute	deg/min	$3.600\ 000 \times 10^{2}$
velocity (linear)	in./min	m/s	$4.233\ 333 \times 10^{-4}$
	ft/min	m/s	$5.080\ 000 \times 10^{-3}$
	in./min	mm/s	$4.233\ 333 \times 10^{-1}$
	ft/min	mm/s	5.080 000
	mile/hour	km/h	1.609 344
volume	in.^{3}	m^3	$1.638\ 706 \times 10^{-5}$
	ft^3	m^3	$2.831\ 685 \times 10^{-2}$
	yd^3	m^3	$7.645\ 549 \times 10^{-1}$
	in.^{3}	mm^3	$1.638\ 706 \times 10^{4}$
	ft^3	mm^3	$2.831\ 685 \times 10^{7}$
	in.^{3}	L	$1.638\ 706 \times 10^{-2}$
	ft^3	L	$2.831\ 685 \times 10$
	gallon	L	3.785 412

From AWS Welding Handbook, 8th Edition, Volume 1, 1987

METRIC CONVERSIONS FOR COMMON WELDING TERMS*

Property	To Convert From	To	Multiply By	Property	To Convert From	To	Multiply By
area dimensions	in.²	mm²	$6.451\ 600 \times 10^2$	power density	W/in.²	W/m²	$1.550\ 003 \times 10^3$
(mm²)	mm²	in.²	$1.550\ 003 \times 10^{-3}$	(W/m²)	W/m²	W/in.²	$6.451\ 600 \times 10^{-4}$
current density	A/in.²	A/mm²	$1.550\ 003 \times 10^{-3}$		psi	Pa	$6.894\ 757 \times 10^3$
(A/mm²)	A/mm²	A/in.²	$6.451\ 600 \times 10^2$	pressure (gas	lb/ft²	Pa	$4.788\ 026 \times 10$
deposition rate**	lb/h	kg/h	0.45**	and liquid)	N/mm²	Pa	$1.000\ 000 \times 10^6$
(kb/h)	kg/h	lb/h	2.2**	(kPa)	kPa	psi	$1.450\ 377 \times 10^{-1}$
electrical resistivity	Ω • cm	Ω • m	$1.000\ 000 \times 10^{-2}$		kPa	lb/ft²	$2.088\ 543 \times 10$
(Ω • m)	Ω • m	Ω • cm	$1.000\ 000 \times 10^2$		kPa	N/mm²	$1.000\ 000 \times 10^{-3}$
electrode force	pound-force	N	4.448 222		torr	kPa	$1.333\ 22 \times 10^{-1}$
(N)	kilogram-force	N	9,806 650		(mm Hg at 0°C)		
	N	lbf	$2.248\ 089 \times 10^{-1}$		micron	kPa	$1.333\ 22 \times 10^{-4}$
flow rate	ft³/h	L/min	$4.719\ 475 \times 10^{-1}$		(μm Hg at 0°C)		
(L/min)	gallon per hour	L/min	$6.309\ 020 \times 10^{-2}$		kPa	torr	7.500 64
	gallon per minute	L/min	3.785 412		kPa	micron	$7.500\ 64 \times 10^3$
	cm³/min	L/min	$1.000\ 000 \times 10^{-3}$	tensile strength	psi	kPa	6.894 757
	L/min	ft³/h	2.118 880	(MPa)	lb/ft²	kPa	$4.788\ 026 \times 10^{-2}$
	cm³/min	ft³/h	$2.118\ 880 \times 10^{-3}$		N/mm²	MPa	1.000 000
fracture toughness	ksi • in.$^{1/2}$	MN • m$^{-3/2}$	1.098 855		MPa	psi	$1.450\ 377 \times 10^2$
(MN • m$^{-3/2}$)	MN • m$^{-3/2}$	ksi • in.$^{1/2}$	0.910 038		MPa	lb/ft²	$2.088\ 543 \times 10^4$
heat input	J/in.	J/m	$3.937\ 008 \times 10$		MPa	N/mm²	1.000 000
(J/m)	J/m	J/in.	$2.540\ 000 \times 10^{-2}$	thermal	cal/(cm • s • °C)	W/(m • K)	$4.184\ 000 \times 10^2$
impact energy	foot pound force	J	1.355 818	conductivity			
linear	in.	mm	$2.540\ 000 \times 10$	(W/m • K)			
measurment	ft	mm	$3.048\ 000 \times 10^2$	travel speed, wire	in./min	mm/s	$4.233\ 333 \times 10^{-1}$
(mm)	mm	in.	$3.937\ 008 \times 10^{-2}$	feed speed	mm/s	in./min	2.362 205
	mm	ft	$3.280\ 840 \times 10^{-3}$	(mm/s)			

* Preferred units are given in parentheses.

** Approximate conversion.

From AWS Welding Handbook, 8th Edition, Volume 1, 1987.

TABLE OF CONVERSION FACTORS*

MULTIPLY	BY	TO OBTAIN	MULTIPLY	BY	TO OBTAIN
Abamperes	10	amperes	bushels	32	quarts (dry)
Abamperes	$3x10^{10}$	statamperes	Centares	1	square meters
abamperes per sq cm	64.52	amperes per sq inch	centigrams	0.01	grams
abampere-turns	10	ampere-turns	centiliters	0.01	liters
abampere-turns	12.57	gilberts	centimeters	0.3937	inches
abampere-turns per cm	25.40	ampere-turns per inch	centimeters	0.01	meters
abcoulombs	10	coulombs	centimeters	393.7	mils
abcoulombs	$3x10^{10}$	statcoulombs	centimeters	10	millimeters
abcoulombs per sq cm	64.52	coulombs per sq inch	centimeter-dynes	$1.020x10^{-3}$	centimeter-grams
abfarads	10^{9}	farads	centimeter-dynes	$1.020x10^{-8}$	meter-kilograms
abfarads	10^{15}	microfarads	centimeter-dynes	$7.367x10^{-9}$	pound-feet
abfarads	$9x10^{20}$	statfarads	centimeter-grams	980.7	centimeter-dynes
abhenries	10^{-9}	henries	centimeter-grams	10^{-5}	meter-kilograms
abhenries	10^{-6}	millihenries	centimeter-grams	$7.233x10^{-5}$	pound-feet
abhenries	$1/9x10^{-20}$	stathenries	centimeters of mercury	0.01316	atmospheres
abmhos per cm cube	$1.662x10^{2}$	mhos per mil foot	centimeters of mercury	0.4461	feet of water
abmhos per cm cube	10^{3}	megmhos per cm cube	centimeters of mercury	136.0	kgs per square meter
abohms	10^{-15}	megohms	centimeters of mercury	27.85	pounds per sq foot
abohms	10^{-3}	microhms	centimeters of mercury	0.1934	pounds per sq inch
abohms	10^{-9}	ohms	centimeters per second	1.969	feet per minute
abohms	$1/9x10^{-20}$	statohms	centimeters per second	0.03281	feet per second
abohms per cm cube	10^{-3}	microhms per cm cube	centimeters per second	0.036	kilometers per hour
abohms per cm cube	$6.015x10^{-3}$	ohms per mil foot	centimeters per second	0.6	meters per minute
abvolts	$1/3x10^{-10}$	statvolts	centimeters per second	0.02237	miles per hour
abvolts	10^{-8}	volts	centimeters per second	$3.728x10^{-4}$	miles per minute
acres	43,560	square feet	cms per sec per sec	0.03281	feet per sec per sec
acres	4047	square meters	cms per sec per sec	0.036	kms per hour per sec
acres	$1.562x10^{-8}$	square miles	cms per sec per sec	0.02237	miles per hour per sec
acres	5645.38	square varas	circular mils	$5.067x10^{-6}$	square centimeters
acres	4840	square yards	circular mils	$7.854x10^{-7}$	square inches
acre-feet	43,560	cubic feet	circular mils	0.7854	square mils
acre-feet	$3.259x10^{5}$	gallons	cord-feet	4 ft x 4 ft x 1 ft	cubic feet
amperes	1/10	abamperes	cords	3 ft x 4 ft x 4 ft	cubic feet
amperes	$3x10^{9}$	statamperes	coulombs	1/10	abcoulombs
amperes per sq cm	6.452	amperes per sq inch	coulombs	$3x9^{9}$	statcoulombs
amperes per sq inch	0.01550	abamperes per sq cm	coulombs per sq inch	0.01550	abcoulombs per sq cm
amperes per sq inch	0.1550	amperes per sq cm	coulombs per sq inch	0.1550	coulombs per sq cm
amperes per sq inch	$4.650x10^{8}$	statamperes per sq cm	coulombs per sq inch	$4.650x10^{8}$	statcouls per sq cm
ampere-turns	1/10	abampere-turns	cubic centimeters	$3.531x10^{-5}$	cubic feet
ampere-turns	1.257	gilberts	cubic centimeters	$6.102x10^{-2}$	cubic inches
ampere-turns per cm	2.540	ampere-turns per in	cubic centimeters	10^{-6}	cubic meters
ampere-turns per inch	0.03937	abampere-turnspercm	cubic centimeters	$1.308x10^{-6}$	cubic yards
ampere-turns per inch	0.3937	ampere-turns per cm	cubic centimeters	$2.642x10^{-4}$	gallons
ampere-turns per inch	0.4950	gilberts per cm	cubic centimeters	10^{-3}	liters
areas	0.02471	acres	cubic centimeters	$2.113x10^{-8}$	pints (liq)
areas	100	square meters	cubic centimeters	$1.057x10^{-8}$	quarts (liq)
atmospheres	76.0	cms of mercury	cubic feet	$2.832x10^{4}$	cubic cms
atmospheres	29.92	inches of mercury	cubic feet	1728	cubic inches
atmospheres	33.90	feet of water	cubic feet	0.02832	cubic meters
atmospheres	10.333	kgs per sq meter	cubic feet	0.03704	cubic yards
atmospheres	14.70	pounds per sq inch	cubic feet	7.481	gallons
atmospheres	1.058	tons per sq foot	cubic feet	28.32	liters
Bars	$9.870x10^{-1}$	atmospheres	cubic feet	59.84	pints (liq)
Bars	1	dynes per sq cm	cubic feet	29.92	quarts (liq)
Bars	0.01020	kgs per square meter	cubic feet per minute	472.0	cubic cms per sec
Bars	$2.089x10^{-8}$	pounds per sq foot	cubic feet per minute	0.1247	gallons per sec
Bars	$1.450x10^{-5}$	pounds per sq inch	cubic feet per minute	0.4720	liters per second
board-feet	0.2530	cubic inches	cubic feet per minute	62.4	lbs of water per min
British thermal units	144 sq. in x 1 in.	kilogram-calories	cubic inches	16.39	cubic centimeters
British thermal units	777.5	foot-pounds	cubic inches	$5.787x10^{-4}$	cubic feet
British thermal units	$3.927x10^{-4}$	horsepower-hours	cubic inches	$1.639x10^{-5}$	cubic meters
British thermal units	1054	joules	cubic inches	$2.143x10^{-5}$	cubic yards
British thermal units	107.5	kilogram-meters	cubic inches	$4.329x10^{-3}$	gallons
British thermal units	$2.928x10^{-4}$	kilowatt-hours	cubic inches	$1.639x10^{-2}$	liters
Btu per min	12.96	foot-pounds per sec	cubic inches	0.03463	pints (liq)
Btu per min	0.2356	horsepower	cubic inches	0.01732	quarts (liq)
Btu per min	0.01757	kilowatts	cubic meters	10^{6}	cubic centimeters
Btu per min	17.57	watts	cubic meters	35.31	cubic feet
Btu per sq ft per min	0.1220	watts per square inch	cubic meters	61,023	cubic inches
bushels	1.244	cubic feet	cubic meters	1.308	cubic yards
bushels	2150	cubic inches	cubic meters	264.2	gallons
bushels	0.03524	cubic meters	cubic meters	10^{9}	liters
bushels	4	pecks	cubic meters	2113	pints (liq)
bushels	64	pints (dry)	cubic meters	1057	quarts (liq)

From the Eugene Dietzen Company

MULTIPLY	BY	TO OBTAIN
cubic yards	7.646x10^{5}	cubic centimeters
cubic yards	27	cubic feet
cubic yards	46,656	cubic inches
cubic yards	0.7646	cubic meters
cubic yards	202.0	gallons
cubic yards	764.6	liters
cubic yards	1616	pints (liq)
cubic yards	807.9	quarts (liq)
cubic yards per minute	0.45	cubic feet per second
cubic yards per minute	3.367	gallons per second
cubic yards per minute	12.74	liters per second
Days	24	hours
Days	1440	minutes
Days	86,400	seconds
decigrams	0.1	grams
deciliters	0.1	liters
decimeters	0.1	meters
degrees (angle)	60	minutes
degrees (angle)	0.01745	radians
degrees (angle)	3600	seconds
degrees per second	0.01745	radians per second
degrees per second	0.1667	revolutions per min
degrees per second	0.002778	revolutions per sec
dekagrams	10	grams
dekaliters	10	liters
dekameters	10	meters
drams	1.772	grams
drams	0.0625	ounces
dynes	1.020x10^{-3}	grams
dynes	7.233x10^{-5}	poundals
dynes	2.248x10^{-6}	pounds
dynes per square cm.	1	bars
Ergs	9.486x10^{-11}	British thermal units
Ergs	1	dyne-centimeters
Ergs	7.376x10^{-8}	foot-pounds
Ergs	1.020x10^{-3}	gram-centimeters
Ergs	10^{-7}	joules
Ergs	2.390x10^{-11}	kilogram-calories
Ergs	1.020x10^{-8}	kilogram-meters
ergs per second	5.692x10^{-9}	Btu per minute
ergs per second	4.426x10^{-6}	foot-pounds per min
ergs per second	7.376x10^{-8}	foot-pounds per sec
ergs per second	1.341x10^{-10}	horsepower
ergs per second	1.434x10^{-9}	kg.-calories per min
ergs per second	10^{-10}	kilowatts
Farads	10^{-9}	abfarads
Farads	10^{6}	microfarads
Farads	9x10^{-11}	statfarads
fathoms	6	feet
feet	30.48	centimeters
feet	12	inches
feet	0.3048	meters
feet	.36	varas
feet	1/3	yards
feet of water	0.02950	atmospheres
feet of water	0.8826	inches of mercury
feet of water	304.8	kgs per square meter
feet of water	62.43	pounds per sq ft
feet of water	0.4335	pounds per sq inch
feet per minute	0.5080	centimeters per sec
feet per minute	0.01667	feet per sec
feet per minute	0.01829	kilometers per hour
feet per minute	0.3048	meters per minute
feet per minute	0.01136	miles per hour
feet per second	30.48	centimeters per sec
feet per second	1.097	kilometers per hour
feet per second	0.5921	knots per hour
feet per second	18.29	meters per minute
feet per second	0.6818	miles per hour

MULTIPLY	BY	TO OBTAIN
feet per second	0.01136	miles per minute
feet per 100 feet	1	per cent grade
feet per sec per sec	30.48	cms per sec per sec
feet per sec per sec	1.097	kms per hr per sc
feet per sec per sec	0.3048	meters per sec per sec
feet per sec per sec	0.6818	miles per hr per sec
foot-pounds	1.286x10^{-3}	British thermal units
foot-pounds	1.356x10^{7}	ergs
foot-pounds	5.050x10^{-7}	horsepower-hours
foot-pounds	1.356	joules
foot-pounds	3.241x10^{-4}	kilogram-calories
foot-pounds	0.1383	kilogram-meters
foot-pounds	3.766x10^{-7}	kilowatt-hours
foot-pounds per minute	1.286x10^{-3}	Btu per minute
foot-pounds per minute	0.01667	foot-pounds per sec
foot-pounds per minute	3.03x10^{-5}	horsepower
foot-pounds per minute	3.241x10^{-4}	kg.-calories per minute
foot-pounds per minute	1.260x10^{-5}	kilowatts
foot-pounds per second	7.717x10^{-2}	Btu per minute
foot-pounds per second	1.818x10^{-3}	horsepower
foot-pounds per second	1.945x10^{-2}	kg.-calories per min
foot-pounds per second	1.365x10^{-3}	kilowatts
furlongs	40	rods
Gallons	3785	cubic centimeters
Gallons	0.1337	cubic feet
Gallons	231	cubic inches
Gallons	3.785x10^{-3}	cubic meters
Gallons	4.951x10^{-3}	cubic yards
Gallons	3.785	liters
Gallons	8	pints (liq)
Gallons	4	quarts (liq)
gallons per minute	2.228x10^{-3}	cubic feet per second
gallons per minute	0.06308	liters per second
gausses	6.452	lines per square inch
gilberts	0.07958	abampere-turns
gilberts	0.7958	ampere-turns
gilberts per centimeter	2.021	ampere-turns per inch
gills	0.1183	liters
gills	0.25	pints (liq)
grains (troy)	1	grains (av)
grains (troy)	0.06480	grams
grains (troy)	0.04167	pennyweights (troy)
grams	980.7	dynes
grams	15.43	grains (troy)
grams	10^{-3}	kilograms
grams	10^{3}	milligrams
grams	0.03527	ounces
grams	0.03215	ounces (troy)
grams	0.07093	poundals
grams	2.205x10^{-3}	pounds
gram-calories	3.968x10^{-3}	British thermal units
gram-centimeters	9.302x10^{-8}	British thermal units
gram-centimeters	980.7	ergs
gram-centimeters	7.233x10^{-5}	foot-pounds
gram-centimeters	2.344x10^{-8}	joules
gram-centimeters	10^{-5}	kilogram-calories
gram-centimeters	9.807x10^{-5}	kilogram-meters
grams per cc	5.600x10^{-3}	pounds per inch
grams per cc	62.43	pounds per cubic foot
grams per cc	0.03613	pounds per cubic inch
grams per cc	3.405x10^{-7}	pounds per mil-foot
Hectares	2.471	acres
Hectares	1.076x10^{5}	square feet
hectograms	100	grams
hectoliters	100	liters
hectometers	100	meters
hectowatts	100	watts
hemispheres (solid angle)	0.5	sphere
hemispheres (solid angle)	4	spherical right angles
hemispheres (solid angle)	6.283	steradians

MULTIPLY	BY	TO OBTAIN
henries	10^{9}	abhenries
henries	10^{3}	millihenries
henries	$1/9 \times 10^{-11}$	stathenries
horse-power	42.44	Btu per min
horse-power	33,000	foot-pounds per min
horse-power	550	foot-pounds per sec
horse-power	1.014	horsepower (metric)
horse-power	10.70	kg.-calories per min
horse-power	0.7457	kilowatts
horse-power	745.7	watts
horse-power (boiler)	33,520	Btu per hour
horse-power (boiler)	9.804	kilowatts
horse-power-hours	2547	British thermal units
horse-power-hours	1.98×10^{6}	foot-pounds
horse-power-hours	2.684×10^{6}	joules
horse-power-hours	641.7	kilogram-calories
horse-power-hours	2.737×10^{5}	kilogram-meters
horse-power-hours	0.7457	kilowatt-hours
hours	60	minutes
hours	3600	seconds
Inches	2.540	centimeters
Inches	10^{3}	mils
Inches	.03	varas
inches of mercury	0.03342	atmospheres
inches of mercury	1.133	feet of water
inches of mercury	345.3	kgs per square meter
inches of mercury	70.73	pounds per square ft
inches of mercury	0.4912	pounds per square in
inches of water	0.002458	atmospheres
inches of water	0.07355	inches of mercury
inches of water	25.40	kgs per square meter
inches of water	0.5781	ounces per square in
inches of water	5.204	pounds per square ft
inches of water	0.03613	pounds per square in
Joules	9.486×10^{-4}	British thermal units
Joules	10^{7}	ergs
Joules	0.7376	foot-pounds
Joules	2.390×10^{-4}	kilogram-calories
Joules	0.1020	kilogram-meters
Joules	2.778×10^{-4}	watt-hours
Kilograms	980,665	dynes
Kilograms	10^{2}	grams
Kilograms	70.93	poundals
Kilograms	2.2046	pounds
Kilograms	1.102×10^{-3}	tons (short)
kilogram-calories	3.968	British thermal units
kilogram-calories	3086	foot-pounds
kilogram-calories	1.558×10^{-2}	horsepower-hours
kilogram-calories	4183	joules
kilogram-calories	426.6	kilogram meters
kilogram-calories	1.162×10^{-3}	kilowatt-hours
kg-calories per min	51.43	foot-pounds per sec
kg-calories per min	0.09351	horsepower
kg-calories per min	0.06972	kilowatts
kgs-cms squared	2.373×10^{-3}	pounds-feet squared
kgs-cms squared	0.3417	pounds-inches squared
kilogram-meters	9.302×10^{-3}	British thermal units
kilogram-meters	9.807×10^{7}	ergs
kilogram-meters	7.233	foot-pounds
kilogram-meters	9.807	joules
kilogram-meters	2.344×10^{-3}	kilogram-calories
kilogram-meters	2.724×10^{-6}	kilowatt-hours
kgs per cubic meter	10^{-2}	grams per cubic cm
kgs per cubic meter	0.06243	pounds per cubic foot
kgs per cubic meter	3.613×10^{-5}	pounds per cubic inch
kgs per cubic meter	3.405×10^{-10}	pounds per mil foot
kgs per meter	0.6720	pounds per foot
kgs per square meter	9.678×10^{-5}	atmospheres
kgs per square meter	98.07	bars
kgs per square meter	3.281×10^{-2}	feet of water
kgs per square meter	2.896×10^{-3}	inches of mercury
kgs per square meter	0.2048	pounds per square ft
kgs per square meter	1.422×10^{-3}	pounds per square in
kgs per sq millimeter	10^{6}	kgs per square meter
kilolines	10^{9}	maxwells
kiloliters	10^{3}	liters
kilometers	10^{5}	centimeters
kilometers	3281	feet
kilometers	10^{3}	meters
kilometers	0.6214	miles
kilometers	1093.6	yards
kilometers per hour	27.78	centimeters per sec
kilometers per hour	54.68	feet per minute
kilometers per hour	0.9113	feet per second
kilometers per hour	0.5396	knots per hour
kilometers per hour	16.67	meters per minute
kilometers per hour	0.6214	miles per hour
kms per hour per sec	27.78	cms per sec per sec
kms per hour per sec	0.9113	ft per sec per sec
kms per hour per sec	0.2778	meters per sec per sec
kms per hour per sec	0.6214	miles per hr per sec
kilometers per min	60	kilometers per hour
kilowatts	56.92	Btu per min
kilowatts	4.425×10^{4}	foot-pounds per min
kilowatts	737.6	foot-pounds per sec
kilowatts	1.341	horsepower
kilowatts	14.34	kg.-calories per min
kilowatts	10^{3}	watts
kilowatt-hours	3415	British thermal units
kilowatt-hours	2.655×10^{6}	foot-pounds
kilowatt-hours	1.341	horsepower-hours
kilowatt-hours	3.6×10^{6}	joules
kilowatt-hours	860.5	kilogram-calories
kilowatt-hours	3.671×10^{5}	kilogram-meters
knots	6080	feet
knots	1.853	kilometers
knots	1.152	miles
knots	2027	yards
knots per hour	51.48	centimeters per sec
knots per hour	1.689	feet per sec
knots per hour	1.853	kilometers per hour
knots per hour	1.152	miles per hour
Lines per square cm	1	gausses
lines per square inch	0.1550	gausses
links (engineer's)	12	inches
links (surveyor's)	7.92	inches
liters	10^{3}	cubic centimeters
liters	0.03531	cubic feet
liters	61.02	cubic inches
liters	10^{-3}	cubic meters
liters	1.308×10^{-3}	cubic yards
liters	0.2642	gallons
liters	2.113	pints (liq)
liters	1.057	quarts (liq)
liters per minute	5.855×10^{-4}	cubic feet per second
liters per minute	4.403×10^{-3}	gallons per second
$\log^{10}$ N	2.303	$\log_e$ N or ln N
$\log_e$ N or ln N	0.4343	$\log^{10}$ N
lumens per sq ft	1	foot-candles
maxwells	10^{-3}	kilolines
megalines	10^{6}	maxwells
megmhos per cm cube	10^{-3}	abmhos per cm cube
megmhos per cm cube	2.540	megmhos per in cube
megmhos per cm cube	0.1662	mhos per mil foot
megmhos per inch cube	0.3937	megmhos per cm cube
megohms	10^{6}	ohms
meters	100	centimeters
meters	3.2808	feet
meters	39.37	inches
meters	10^{-3}	kilometers
meters	10^{3}	millimeters
meters	1.0936	yards
meter-kilograms	9.807×10^{7}	centimeter-dynes

MULTIPLY	BY	TO OBTAIN
meter-kilograms	10^5	centimeter-grams
meter-kilograms	7.233	pound-feet
meters per minute	1.667	centimeters per sec
meters per minute	3.281	feet per minute
meters per minute	0.05468	feet per second
meters per minute	0.06	kilometers per hour
meters per second	0.03728	miles per hour
meters per second	1968	feet per minute
meters per second	3.284	feet per second
meters per second	3.0	kilometers per hour
meters per second	0.06	kilometers per min
meters per second	2.237	miles per hour
meters per sec per sec	0.03728	miles per minute
meters per sec per sec	3.281	feet per sec per sec
meters per sec per sec	3.6	kms per hour per sec
meters per sec per sec	2.237	miles per hour per sec
mhos per mil foot	6.015×10^{-3}	abmhos per cm cube
mhos per mil foot	6.015	megmhos per cm cube
mhos per mil foot	15.28	megmhos per in cube
microfarads	10^{-15}	abfarads
microfarads	10^{-6}	farads
microfarads	9×10^5	staffarads
micrograms	10^{-6}	grams
microliters	10^{-6}	liters
microhms	10^3	abohms
microhms	10^{-12}	megohms
microhms	10^{-6}	ohms
microhms	$1/9\times10^{-17}$	statohms
microhms per cm cube	10^3	abohms per cm cube
microhms per cm cube	0.3937	microhms p in cube
microhms per cm cube	6.015	ohms per mil foot
microhms per inch cube	2.540	microhms p cm cube
microns	10^{-6}	meters
miles	1.609×10^5	centimeters
miles	5280	feet
miles	1.6093	kilometers
miles	1760	yards
miles	1900.8	varas
miles per hour	44.70	centimeters per sec
miles per hour	88	feet per minute
miles per hour	1.467	feet per second
miles per hour	1.6093	kilometers per hour
miles per hour	0.8684	knots per hour
miles per hour	26.82	meters per minute
miles per hour per sec	44.70	cms per sec per sec
miles per hour per sec	1.467	feet per sec per sec
miles per hour per sec	1.6093	kms per hour per sec
miles per hour per sec	0.4470	M per sec per sec
miles per minute	2682	centimeters per sec
miles per minute	88	feet per second
miles per minute	1.6093	kilometers per hour
miles per minute	0.8684	knots per minute
miles per minute	660	miles per hour
milligrams	10^{-2}	grams
millihenries	10^6	aghenries
millihenries	10^{-3}	henries
millihenries	$1/9\times10^{-14}$	stathenries
milliliters	10^{-2}	liters
millimeters	0.1	centimeters
millimeters	0.03937	inches
millimeters	39.37	mils
mils	0.002540	centimeters
mils	10^{-3}	inches
miner's inches	1.5	cubic feet per min
minutes (angle)	2.909×10^{-4}	radians
minutes (angle)	60	seconds (angle)
months	30.42	days
months	730	hours
months	43,800	minutes
months	2.628×10^6	seconds
myriagrams	10	kilograms
myriameters	10	kilowatts
myriawatts	10	kilometers

MULTIPLY	BY	TO OBTAIN
Ohms	10^9	abohms
Ohms	10^{-6}	megohms
Ohms	10^6	microhms
Ohms	$1/9\times10^{-11}$	statohms
ohms per mil foot	166.2	abohms per cm cube
ohms per mil foot	0.1662	microhms per cm cube
ohms per mil foot	0.06524	microhms per in cube
ounces	8	drams
ounces	437.5	grains
ounces	28.35	grams
ounces	0.0625	pounds
ounces (fluid)	1.805	cubic inches
ounces (fluid)	0.02957	liters
ounces (troy)	480	grains (troy)
ounces (troy)	31.10	grams
ounces (troy)	20	pennyweights (troy)
ounces (troy)	0.08333	pounds (troy)
ounces per sq inch	0.0625	pounds per sq inch
Pennyweights (troy)	24	grains (troy)
Pennyweights (troy)	1.555	grams
Pennyweights (troy)	0.05	ounces (troy)
perches (masonry)	24.75	cubic feet
pints (dry)	33.60	cubic inches
pints (liquid)	28.87	cubic inches
poundals	13,826	dynes
poundals	14.10	grams
poundals	0.03108	pounds
pounds	444,823	dynes
pounds	7000	grains
pounds	453.6	grams
pounds	16	ounces
pounds	32.17	poundals
pounds (troy)	0.8229	pounds (av)
pound-feet	1.356×10^7	centimeter-dynes
pound-feet	13,825	centimeter-grams
pound-feet	0.1383	meter-kilograms
pounds-feet squared	421.3	kgs-cms squared
pounds-feet squared	144	pounds-in squared
pounds-in (squared)	2.926	kgs-cms squared
pounds-in (squared)	6.945×10^{-3}	pounds-feet squared
pounds of water	0.01602	cubic feet
pounds of water	27.68	cubic inches
pounds of water	0.1198	gallons
pounds of water per min	2669×10^{-4}	cubic feet per sec
pounds per cubic foot	0.01602	grams per cubic cm
pounds per cubic foot	16.02	kgs per cubic meter
pounds per cubic foot	5.787×10^{-4}	pounds per cubic inch
pounds per cubic foot	5.456×10^{-9}	pounds per mil foot
pounds per cubic inch	27.68	grams per cubic cm
pounds per cubic inch	2.768×10^4	kgs per cubic meter
pounds per cubic inch	1728	pounds per cubic foot
pounds per cubic inch	9.425×10^{-6}	pounds per mil foot
pounds per foot	1.488	kgs per meter
pounds per inch	178.6	grams per cm
pounds per mil foot	2.306×10^6	grams per cubic cm
pounds per square foot	0.01602	feet of water
pounds per square foot	4.882	kgs per square meter
pounds per square foot	6.944×10^{-3}	pounds per sq inch
pounds per sq inch	0.06804	atmospheres
pounds per sq inch	2.307	feet of water
pounds per sq inch	2.036	inches of mercury
pounds per sq inch	703.1	kgs per square meter
pounds per sq inch	144	pounds per sq foot
Quadrants angle)	90	degrees
Quadrants angle)	5400	minutes
Quadrants angle)	1.571	radians
quarts (dry)	67.20	cubic inches
quarts (liq)	57.75	cubic inches
quintals	100	pounds
quires	25	sheets
Radians	57.30	degrees
Radians	3438	minutes
Radians	0.637	quadrants

MULTIPLY	BY	TO OBTAIN
radians per second	57.30	degrees per second
radians per second	0.1592	revolutions per second
radians per second	9.549	revolutions per min
radians per sec per sec	573.0	revs per min per min
radians per sec per sec	9.549	revs per min per sec
radians per sec per sec	0.1592	revs per sec per sec
reams	500	sheets
revolutions	360	degrees
revolutions	4	quadrants
revolutions	6.283	radians
revolutions per minute	6	degrees per second
revolutions per minute	0.1047	radians per second
revolutions per minute	0.01667	revolutions per sec
revs per min per min	1.745x10^{-3}	rads per sec per sec
revs per min per min	0.01667	revs per min per sec
revs per min per min	2.778x10^{-4}	revs per sec per sec
revolutions per second	360	degrees per second
revolutions per second	6.283	radians per second
revolutions per second	60	revs per min
revs per sec per sec	6.283	rads per sec per sec
revs per sec per sec	3600	revs per min per min
revs per sec per sec	60	revs per min per sec
rods	16.5	feet
Seconds (angle)	4.848x10^{-6}	radians
spheres (solid angle)	12.57	steradians
spherical right angles	0.25	hemispheres
spherical right angles	0.125	spheres
spherical right angles	1.571	steradians
square centimeters	1.973x10^{5}	circular mils
square centimeters	1.076x10^{-3}	square feet
square centimeters	0.1550	square inches
square centimeters	10^{-6}	square meters
square centimeters	100	square millimeters
sq cms-cms sqd	0.02402	sq inches-inches sqd
square feet	2.296x10^{-5}	acres
square feet	929.0	square centimeters
square feet	144	square inches
square feet	0.09290	square meters
square feet	3.587x10^{-3}	square miles
square feet	.1296	square varas
square feet	1/9	square yards
sq feet-feet sqd	2.074x10^{4}	sq inches-inches sqd
square inches	1.273x10^{6}	circular mils
square inches	6.452	square centimeters
square inches	6.944x10^{-3}	square feet
square inches	10^{6}	square mils
square inches	645.2	square millimeters
sq inches-inches sqd	41.62	sq cms-cms sqd
sq inches-inches sqd	4.823x10^{-5}	sq ft-feet sqd
square kilometers	247.1	acres
square kilometers	10.76x10^{6}	square feet
square kilometers	10^{6}	square meters
square kilometers	0.3861	square miles
square kilometers	1.196x10^{6}	square yards
square meters	2.471x10^{-4}	acres
square meters	10.764	square feet
square meters	3.861x10^{-7}	square miles
square meters	1.196	square yards
square miles	640	acres
square miles	27.88x10^{6}	square feet
square miles	2.590	square kilometers
square miles	3,613,040.45	square varas
square miles	3.098x16^{6}	square yards
square millimeters	1.973x10^{3}	circular mils
square millimeters	0.01	square centimeters
square millimeters	1.550x10^{-3}	square inches
square mils	1.273	circular mils
square mils	6.452x10^{-6}	square centimeters
square mils	10^{-6}	square inches
square varas	.0001771	acres
square varas	7.716049	square feet
square varas	.0000002765	square miles
square varas	.857339	square yards

MULTIPLY	BY	TO OBTAIN
square yards	2.066x10^{-4}	acres
square yards	9	square feet
square yards	0.8361	square meters
square yards	3.228x10^{-7}	square miles
square yards	1.1664	square varas
statamperes	1/3x10^{-10}	abamperes
statamperes	1/3x10^{-9}	amperes
statcoulombs	1/3x10^{-10}	abcoulombs
statcoulombs	1/3x10^{-9}	coulombs
statfarads	1/9x10^{-20}	abfarads
statfarads	1/9x10^{-11}	farads
statfarads	1/9x10^{-5}	microfarads
stathenries	9x20^{20}	abhenries
stathenries	9x10^{11}	henries
stathenries	9x10^{14}	millihenries
statohms	9x20^{20}	abohms
statohms	9x10^{5}	megohms
statohms	9x10^{17}	microhms
statohms	9x10^{11}	ohms
statvolts	3x10^{10}	abvolts
statvolts	300	volts
steradians	0.1592	hemispheres
steradians	0.07958	spheres
steradians	0.6366	spherical right angles
steres	10^{3}	liters
Temp (degs C) + 17.8	1	abs temp (degs C)
temp (degs) F) + 460	1.8	temp (degs Fahr)
tem (degs F) — 32	1	abs temp (degs F)
Temp (degs C) + 273	5/9	temp (degs Cent)
tons (long)	1016	kilograms
tons (long)	2240	pounds
tons (metric)	10^{3}	kilograms
tons (metric)	2205	pounds
tons (short)	907.2	kilograms
tons (short)	2000	pounds
tons (short) per sq ft	9765	kgs per square meter
tons (short) per sq ft	13.89	pounds per sq inch
tons (short) per sq in	1.406x10^{6}	kgs per square meter
tons (short) per sq in	2000	pounds per sq inch
Varas	2.7777	feet
Varas	33.3333	inches
Varas	.000526	miles
Varas	.9259	yards
volts	10^{8}	abvolts
volts	1/300	statvolts
volts per inch	3.937x10^{7}	abvolts per cm
volts per inch	1.312x10^{-3}	statvolts per cm
Watts	0.05692	Btu per min
Watts	10^{7}	ergs per second
Watts	44.26	foot-pounds per min
Watts	0.7376	foot-pounds per sec
Watts	1.341x10^{-3}	horsepower
Watts	0.01434	kg-calories per min
Watts	10^{-3}	kilowatts
watt-hours	3.415	British thermal units
watt-hours	2655	foot-pounds
watt-hours	1.341x10^{-3}	horsepower-hours
watt-hours	0.8605	kilogram-calories
watt-hours	367.1	kilogram-meters
watt-hours	10^{-3}	kilowatt-hours
webers	10^{8}	maxwells
weeks	168	hours
weeks	10,080	minutes
weeks	6044,800	seconds
Yards	91.44	centimeters
Yards	3	feet
Yards	36	inches
Yards	0.9144	meters
Yards	1.08	varas
years (common)	365	days
years (common)	8760	hours
years (leap)	366	days
years (leap)	8784	hours

IMPACT DATA CONVERSION TABLE

One ft-lb = 0.13826 kg-m

Ft-lb	Kg-m	Ft-lb	Kg-m	Ft-lb	Kg-m
1	.14	34	4.70	68	9.40
2	.28	35	4.84	69	9.54
3	.42	36	4.98	70	9.68
4	.55	37	5.12	71	9.82
5	.69	38	5.25	72	9.95
6	.83	39	5.39	73	10.09
7	.97	40	5.53	74	10.23
8	1.11	41	5.67	75	10.37
9	1.24	42	5.81	76	10.51
10	1.38	43	5.95	77	10.65
11	1.52	44	6.08	78	10.78
12	1.66	45	6.22	79	10.92
13	1.80	46	6.36	80	11.06
14	1.94	47	6.50	81	11.20
15	2.07	48	6.64	82	11.34
16	2.21	49	6.78	83	11.48
17	2.35	50	6.91	84	11.61
18	2.49	51	7.05	85	11.75
19	2.63	52	7.19	86	11.89
20	2.77	53	7.33	87	12.03
21	2.90	54	7.47	88	12.17
22	3.04	55	7.60	89	12.31
23	3.18	56	7.74	90	12.44
24	3.32	57	7.88	91	12.58
25	3.46	58	8.02	92	12.72
26	3.60	59	8.16	93	12.86
27	3.73	60	8.30	94	13.00
28	3.87	61	8.43	95	13.13
29	4.01	62	8.57	96	13.27
30	4.15	63	8.71	97	13.41
31	4.29	64	8.85	98	13.55
32	4.42	65	8.99	99	13.69
33	4.56	66	9.13	100	13.83
		67	9.26		

LOAD CONVERSION TABLE

LONG TONS PER SQUARE INCH TO POUNDS PER SQUARE INCH						KILOGRAMS PER SQUARE MILLIMETER TO POUNDS PER SQUARE INCH					
Long Tons per Sq. in.	Pounds per Sq. in.	Long Tons per Sq. in.	Pounds per Sq. in.	Long Tons per Sq. in.	Pounds per Sq. in.	Kg. per Sq.mm.	Pounds per Sq. in.	Kg. per Sq.mm.	Pounds per Sq. in.	Kg. per Sq.mm.	Pounds per Sq. in.
10.0	22,400	35.0	78,400	70	156,800	10	14,223	60	85,340	110	156,457
10.5	23,520	35.5	79,520	71	159,040	11	15,646	61	86,763	111	157,880
11.0	24,640	36.0	80,640	72	161,280	12	17,068	62	88,185	112	159,302
11.5	25,760	36.5	81,760	73	163,520	13	18,490	63	89,607	113	160,724
12.0	26,880	37.0	82,880	74	165,760	14	19,913	64	91,030	114	162,147
12.5	28,000	37.5	84,000	75	168,000	15	21,335	65	92,452	115	163,569
13.0	29,120	38.0	85,120	76	170,240	16	22,757	66	93,874	116	164,991
13.5	30,240	38.5	86,240	77	172,480	17	24,180	67	95,297	117	166,414
14.0	31,360	39.0	87,360	78	174,720	18	25,602	68	96,719	118	167,836
14.5	32,480	39.5	88,480	79	176,960	19	27,024	69	98,141	119	169,258
15.0	33,600	40.0	89,600	80	179,200	20	28,447	70	99,564	120	170,681
15.5	34,720	40.5	90,720	81	181,440	21	29,869	71	100,986	121	172,103
16.0	35,840	41.0	91,840	82	183,680	22	31,291	72	102,408	122	173,525
16.5	36,960	41.5	92,960	83	185,920	23	32,714	73	103,831	123	174,948
17.0	38,080	42.0	94,080	84	188,160	24	34,136	74	105,253	124	176,370
17.5	39,200	42.5	95,200	85	190,400	25	35,558	75	106,675	125	177,792
18.0	40,320	43.0	96,320	86	192,640	26	36,981	76	108,098	126	179,215
18.5	41,440	43.5	97,440	87	194,880	27	38,403	77	109,520	127	180,637
19.0	42,560	44.0	98,560	88	197,120	28	39,826	78	110,943	128	182,059
19.5	43,680	44.5	99,680	89	199,360	29	41,248	79	112,365	129	183,482
20.0	44,800	45.0	100,800	90	201,600	30	42,670	80	113,787	130	184,904
20.5	45,920	45.5	101,920	91	203,840	31	44,093	81	115,210	131	186,327
21.0	47,040	46.0	103,040	92	206,080	32	45,515	82	116,632	132	187,749
21.5	48,160	46.5	104,160	93	208,320	33	46,937	83	118,054	133	189,171
22.0	49,280	47.0	105,280	94	210,560	34	48,360	84	119,477	134	190,594
22.5	50,400	47.5	106,400	95	212,800	35	49,782	85	120,899	135	192,016
23.0	51,520	48.0	107,520	96	215,040	36	51,204	86	122,321	136	193,438
23.5	52,640	48.5	108,640	97	217,280	37	52,627	87	123,744	137	194,861
24.0	53,760	49.0	109,760	98	219,520	38	54,049	88	125,166	138	196,283
24.5	54,880	49.5	110,880	99	221,760	39	55,471	89	126,588	139	197,705
25.0	56,000	50	112,000	100	224,000	40	56,894	90	128,011	140	199,128
25.5	57,120	51	114,240	101	226,240	41	58,316	91	129,433	141	200,550
26.0	58,240	52	116,480	102	228,480	42	59,738	92	130,855	142	201,972
26.5	59,360	53	118,720	103	230,720	43	61,161	93	132,278	143	203,395
27.0	60,480	54	120,960	104	232,960	44	62,583	94	133,700	144	204,817
27.5	61,600	55	123,200	105	235,200	45	64,005	95	135,122	145	206,239
28.0	62,720	56	125,440	106	237,440	46	65,428	96	136,545	146	207,662
28.5	63,840	57	127,680	107	239,680	47	66,850	97	137,967	147	209,084
29.0	64,960	58	129,920	108	241,920	48	68,272	98	139,389	148	210,506
29.5	66,080	59	132,160	109	244,160	49	69,695	99	140,812	149	211,929
30.0	67,200	60	134,400	110	246,400	50	71,117	100	142,234	150	213,351
30.5	68,320	61	136,640	111	248,640	51	72,539	101	143,656	151	214,773
31.0	69,440	62	138,880	112	250,880	52	73,962	102	145,079	152	216,196
31.5	70,560	63	141,120	113	253,120	53	75,384	103	146,501	153	217,618
32.0	71,680	64	143,360	114	255,360	54	76,806	104	147,923	154	219,040
32.5	72,800	65	145,600	115	257,600	55	78,229	105	149,346	155	220,463
33.0	73,920	66	147,840	116	259,840	56	79,651	106	150,768	156	221,885
33.5	75,040	67	150,080	117	262,080	57	81,073	107	152,190	157	223,307
34.0	76,160	68	152,320	118	264,320	58	82,496	108	153,613	158	224,730
34.5	77,280	69	154,560	119	266,560	59	83,918	109	155,035	159	226,152

WEIGHT COMPARISON TABLE

Avoirdupois Pounds	Kilograms	Short Tons	Long Tons	Metric Tons
1	.45359	.0005	.00044643	.00045359
2	.90718	.0010	.00089286	.00090718
3	1.36078	.0015	.00133929	.00136078
4	1.81437	.0020	.00178571	.00181437
5	2.26796	.0025	.00223214	.00226796
6	2.72155	.0030	.00267857	.00272155
7	3.17515	.0035	.00312500	.00317515
8	3.62874	.0040	.00357143	.00362874
9	4.08233	.0045	.00401786	.00408233
2.20462	1	.00110231	.00098421	.001
4.40924	2	.00220462	.00196841	.002
6.61387	3	.00330693	.00295262	.003
8.81849	4	.00440924	.00393683	.004
11.02311	5	.00551156	.00492103	.005
13.22773	6	.00661387	.00590524	.006
15.43236	7	.00771618	.00688944	.007
17.63698	8	.00881849	.00787365	.008
19.84160	9	.00992080	.00885786	.009
2000	907.18	1	.89287	.90718
4000	1814.37	2	1.78571	1.81437
6000	2721.55	3	2.67857	2.72155
8000	3628.74	4	3.57143	3.62874
10000	4535.92	5	4.46429	4.53592
12000	5443.11	6	5.35714	5.44311
14000	6350.29	7	6.25000	6.35029
16000	7257.48	8	7.14286	7.25748
18000	8164.66	9	8.03571	8.16466
2240	1016.05	1.12	1	1.01605
4480	2032.09	2.24	2	2.03209
6720	3048.14	3.36	3	3.04814
8960	4064.19	4.48	4	4.06419
11200	5080.24	5.60	5	5.08024
13440	6096.28	6.72	6	6.09628
15680	7112.32	7.84	7	7.11232
17920	8128.38	8.96	8	8.12838
20160	9144.42	10.08	9	9.14442
2204.62	1000	1.10231	.98421	1
4409.24	2000	2.20462	1.96841	2
6613.87	3000	3.30693	2.95262	3
8818.49	4000	4.40924	3.93683	4
11023.11	5000	5.51156	4.92103	5
13227.73	6000	6.61887	5.90524	6
15432.36	7000	7.71618	6.88944	7
17636.98	8000	8.81849	7.87365	8
19841.60	9000	9.92080	8.85786	9

From Republic Alloy Steels, Republic Steel Corp.

Areas and Dimensions of Plane Figures

In the following tables are given formulas for the areas of plane figures, together with other formulas relating to their dimensions and properties; the surfaces of solids; and the volumes of solids. The notation used in the formulas is, as far as possible, given in the illustration accompanying them; where this has not been possible, it is given at the beginning of each set of formulas.

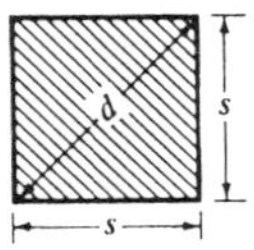 **Square**	A = area. $A = s^2$; $A = \frac{1}{2}d^2$; $s = 0.7071d = \sqrt{A}$; $d = 1.414s = 1.414\sqrt{A}$
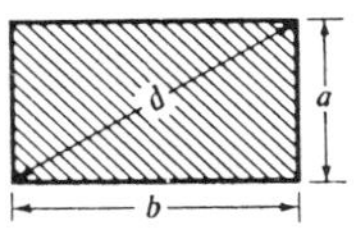 **Rectangle**	A = area. $A = ab$; $A = a\sqrt{d^2 - a^2} = b\sqrt{d^2 - b^2}$; $d = \sqrt{a^2 + b^2}$; $a = \sqrt{d^2 - b^2} = A \div b$; $b = \sqrt{d^2 - a^2} = A \div a$
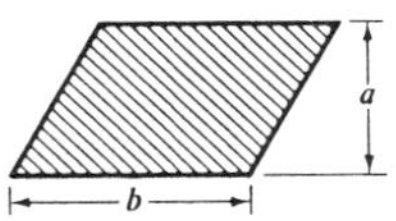 **Parallelogram**	A = area. $A = ab$; $a = A \div b$; $b = A \div a$. Note that dimension a is measured at right angles to line b.
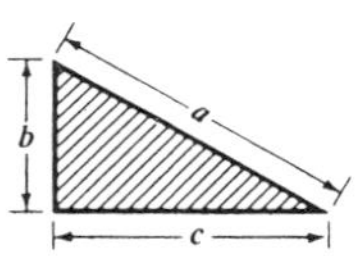 **Right-angled Triangle**	A = area. $A = \frac{bc}{2}$; $a = \sqrt{b^2 + c^2}$; $b = \sqrt{a^2 - c^2}$; $c = \sqrt{a^2 - b^2}$
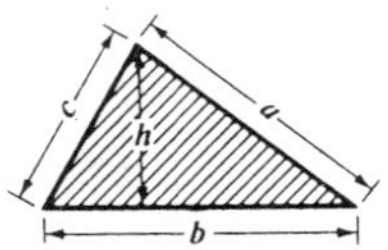 **Acute-angled Triangle**	A = area. $A = \frac{bh}{2} = \frac{b}{2}\sqrt{a^2 - \left(\frac{a^2 + b^2 - c^2}{2b}\right)^2}$. If $S = \frac{1}{2}(a + b + c)$, then $A = \sqrt{S(S-a)(S-b)(S-c)}$

Areas and Dimensions of Plane Figures

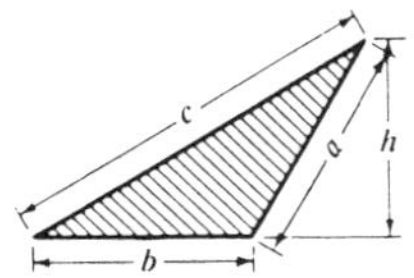 **Obtuse-angled Triangle**	A = area. $A = \frac{bh}{2} = \frac{b}{2}\sqrt{a^2 - \left(\frac{c^2 - a^2 - b^2}{2b}\right)^2}$. If $S = \frac{1}{2}(a + b + c)$, then $A = \sqrt{S(S-a)(S-b)(S-c)}$
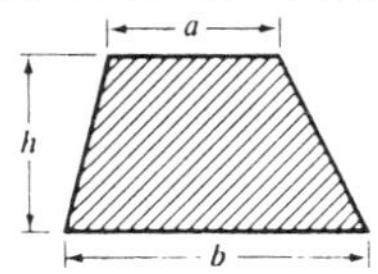 **Trapezoid**	A = area. $A = \frac{(a+b)h}{2}$. *Note:* In England, this figure is called a *trapezium* and the one below it is known as a *trapezoid*, the terms being reversed.
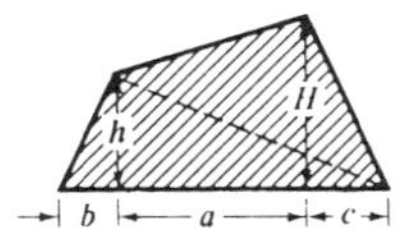 **Trapezium**	A = area. $A = \frac{(H+h)a + bh + cH}{2}$. A trapezium can also be divided into two triangles as indicated by the dotted line. The area of each of these triangles is computed, and the results added to find the area of the trapezium.
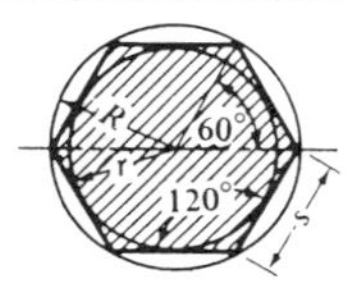 **Regular Hexagon**	A = area; R = radius of circumscribed circle; r = radius of inscribed circle. $A = 2.598s^2 = 2.598R^2 = 3.464r^2$; $R = s = 1.155r$; $r = 0.866s = 0.866R$; $s = R = 1.155r$
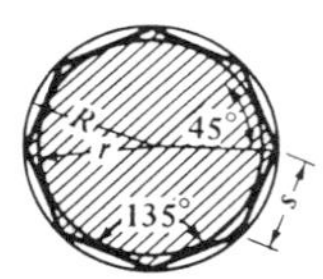 **Regular Octagon**	A = area; R = radius of circumscribed circle; r = radius of inscribed circle. $A = 4.828s^2 = 2.828R^2 = 3.314r^2$; $R = 1.307s = 1.082r$; $r = 1.207s = 0.924R$; $s = 0.765R = 0.828r$
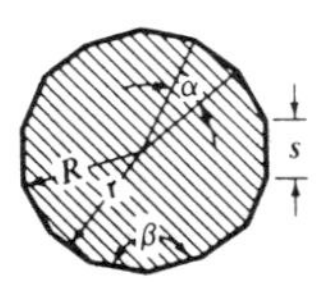 **Regular Polygon**	A = area; n = number of sides. $\alpha = 360° \div n$; $\beta = 180° - \alpha$; $A = \frac{nsr}{2} = \frac{ns}{2}\sqrt{R^2 - \frac{s^2}{4}}$; $R = \sqrt{r^2 + \frac{s^2}{4}}$; $r = \sqrt{R^2 - \frac{s^2}{4}}$; $s = 2\sqrt{R^2 - r^2}$

Formulas for Areas and Volumes

Areas and Dimensions of Plane Figures

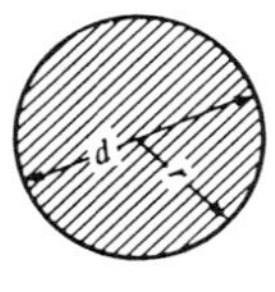 Circle	A = area; C = circumference. $A = \pi r^2 = 3.1416r^2 = 0.7854d^2$ $C = 2\pi r = 6.2832r = 3.1416d$ $r = C \div 6.2832 = \sqrt{A \div 3.1416} = 0.564\sqrt{A}$ $d = C \div 3.1416 = \sqrt{A \div 0.7854} = 1.128\sqrt{A}$ Length of arc for center-angle of 1° = $0.008727d$ Length of arc for center-angle of $n°$ = $0.008727nd$
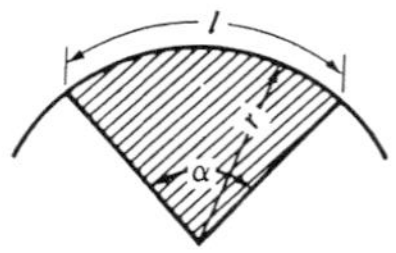 Circular Sector	A = area; l = length of arc; α = angle, in degrees. $l = \dfrac{r \times \alpha \times 3.1416}{180} = 0.01745r\alpha = \dfrac{2A}{r}$ $A = \frac{1}{2}rl = 0.008727\alpha r^2$ $\alpha = \dfrac{57.296\,l}{r}$ $r = \dfrac{2A}{l} = \dfrac{57.296\,l}{\alpha}$
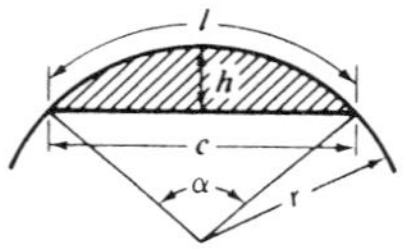 Circular Segment	A = area; l = length of arc; α = angle, in degrees. $c = 2\sqrt{h(2r-h)}$ $A = \frac{1}{2}[rl - c(r-h)]$* $r = \dfrac{c^2 + 4h^2}{8h}$ $l = 0.01745r\alpha$ $h = r - \frac{1}{2}\sqrt{4r^2 - c^2}$ $\alpha = \dfrac{57.296\,l}{r}$ $h = r[1 - \cos(\alpha/2)]$ * See also p. 70.
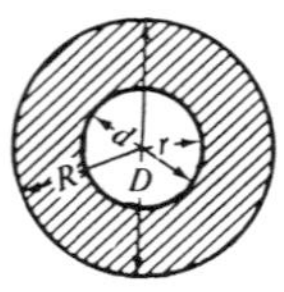 Circular Ring	A = area. $A = \pi(R^2 - r^2) = 3.1416(R^2 - r^2)$ $= 3.1416(R + r)(R - r)$ $= 0.7854(D^2 - d^2) = 0.7854(D + d)(D - d)$
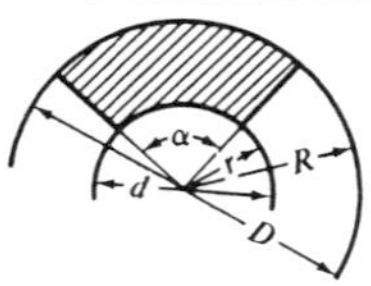 Circular Ring Sector	A = area; α = angle, in degrees. $A = \dfrac{\alpha\pi}{360}(R^2 - r^2) = 0.00873\alpha(R^2 - r^2)$ $= \dfrac{\alpha\pi}{4 \times 360}(D^2 - d^2) = 0.00218\alpha(D^2 - d^2)$
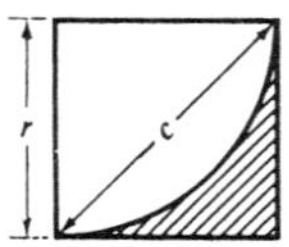 Spandrel or Fillet	A = area. $A = r^2 - \dfrac{\pi r^2}{4} = 0.215r^2$ $= 0.1075c^2$

Areas and Dimensions of Plane Figures

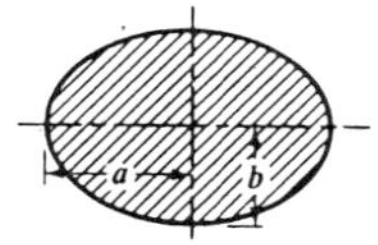 Ellipse	A = area; P = perimeter or circumference. $A = \pi ab = 3.1416ab$. An approximate formula for the perimeter is: $P = 3.1416\sqrt{2(a^2 + b^2)}$ A closer approximation is: $P = 3.1416\sqrt{2(a^2 + b^2) - \dfrac{(a-b)^2}{2.2}}$
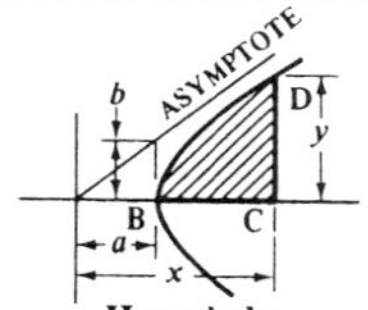 Hyperbola	A = area BCD. $A = \dfrac{xy}{2} - \dfrac{ab}{2}\ln\left(\dfrac{x}{a} + \dfrac{y}{b}\right)$
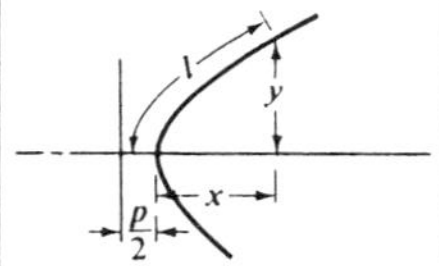 Parabola	l = length of arc. $l = \dfrac{p}{2}\left[\sqrt{\dfrac{2x}{p}\left(1 + \dfrac{2x}{p}\right)} + \ln\left(\sqrt{\dfrac{2x}{p}} + \sqrt{1 + \dfrac{2x}{p}}\right)\right]$ When x is small in proportion to y, the following is a close approximation: $l = y\left[1 + \dfrac{2}{3}\left(\dfrac{x}{y}\right)^2 - \dfrac{2}{5}\left(\dfrac{x}{y}\right)^4\right]$, or $l = \sqrt{y^2 + \dfrac{4}{3}x^2}$
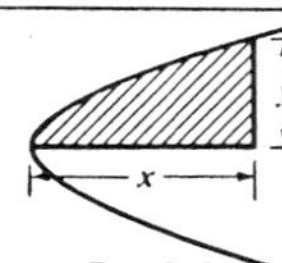 Parabola	A = area. $A = \frac{2}{3}xy$ (The area is equal to two-thirds of a rectangle which has x for its base and y for its height.)
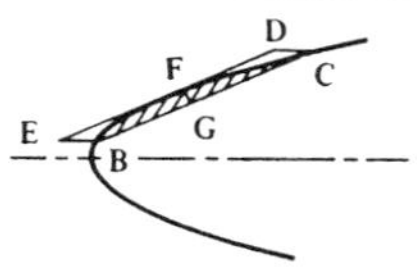 Segment of Parabola	A = area. Area $BFC = A = \frac{2}{3}$ area of parallelogram $BCDE$. If FG is the height of the segment, measured at right angles to BC, then: Area of segment $BFC = \frac{2}{3}BC \times FG$
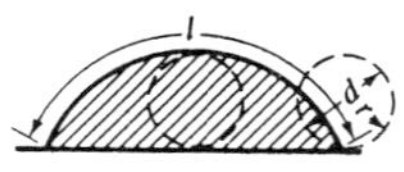 Cycloid	A = area; l = length of cycloid. $A = 3\pi r^2 = 9.4248r^2 = 2.3562d^2$ $= 3 \times$ area of generating circle $l = 8r = 4d$

Volumes of Solids

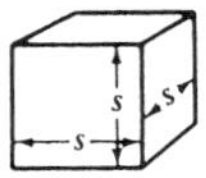 **Cube**	V = volume. $V = s^3$ $s = \sqrt[3]{V}$
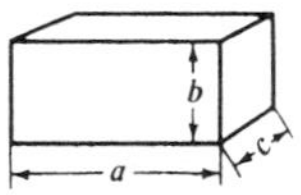 **Square Prism**	V = volume. $V = abc$ $a = \frac{V}{bc}$ $b = \frac{V}{ac}$ $c = \frac{V}{ab}$
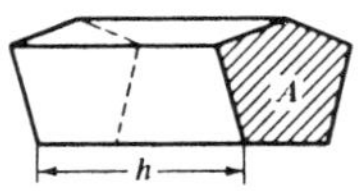 **Prism**	V = volume; A = area of end surface. $V = h \times A$ The area A of the end surface is found by the formulas for areas of plane figures on the preceding pages. Height h must be measured perpendicular to end surface.
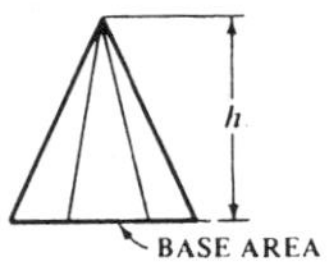 **Pyramid**	V = volume. $V = \frac{1}{3}h \times$ area of base. If the base is a regular polygon with n sides, and s = length of side, r = radius of inscribed circle, and R = radius of circumscribed circle, then: $V = \frac{nsrh}{6} = \frac{nsh}{6}\sqrt{R^2 - \frac{s^2}{4}}$
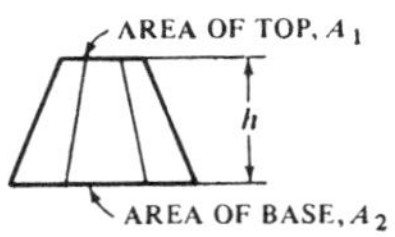 **Frustum of Pyramid**	V = volume. $V = \frac{h}{3}(A_1 + A_2 + \sqrt{A_1 \times A_2})$
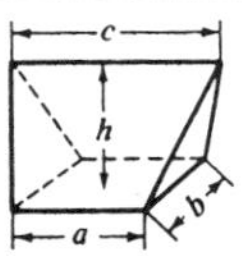 **Wedge**	V = volume. $V = \frac{(2a + c)bh}{6}$

Volumes of Solids

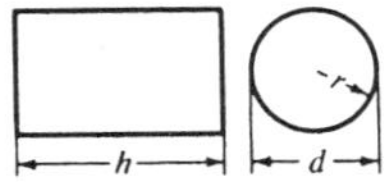 **Cylinder**	V = volume; S = area of cylindrical surface. $V = 3.1416r^2h = 0.7854d^2h$ $S = 6.2832rh = 3.1416dh$ Total area A of cylindrical surface and end surfaces: $A = 6.2832r(r + h) = 3.1416d(\frac{1}{2}d + h)$
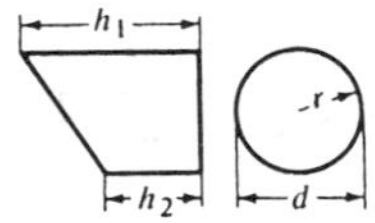 **Portion of Cylinder**	V = volume; S = area of cylindrical surface. $V = 1.5708r^2(h_1 + h_2) = 0.3927d^2(h_1 + h_2)$ $S = 3.1416r(h_1 + h_2) = 1.5708d(h_1 + h_2)$
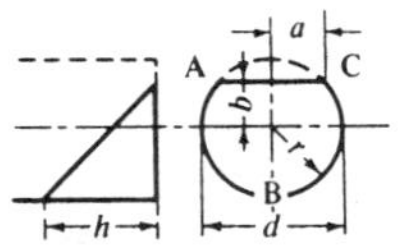 **Portion of Cylinder**	V = volume; S = area of cylindrical surface. $V = (\frac{2}{3}a^3 \pm b \times \text{area } ABC)\frac{h}{r \pm b}$ $S = (ad \pm b \times \text{length of arc } ABC)\frac{h}{r \pm b}$ Use + when base area is larger, and − when base area is less than one-half the base circle.
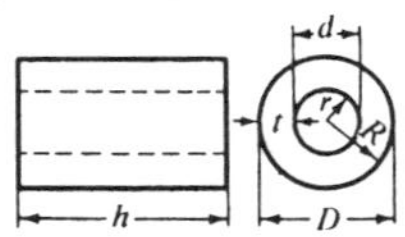 **Hollow Cylinder**	V = volume. $V = 3.1416h(R^2 - r^2) = 0.7854h(D^2 - d^2)$ $= 3.1416ht(2R - t) = 3.1416ht(D - t)$ $= 3.1416ht(2r + t) = 3.1416ht(d + t)$ $= 3.1416ht(R + r) = 1.5708ht(D + d)$
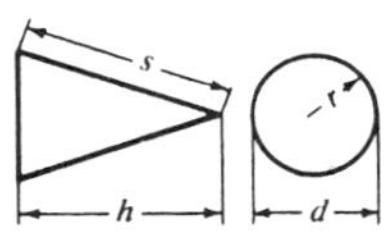 **Cone**	V = volume; A = area of conical surface. $V = \frac{3.1416r^2h}{3} = 1.0472r^2h = 0.2618d^2h$ $A = 3.1416r\sqrt{r^2 + h^2} = 3.1416rs = 1.5708ds$ $s = \sqrt{r^2 + h^2} = \sqrt{\frac{d^2}{4} + h^2}$
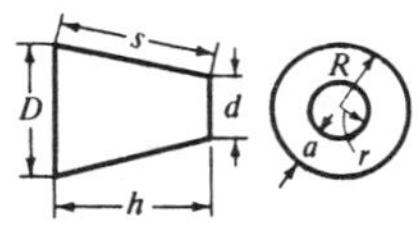 **Frustum of Cone**	V = volume; A = area of conical surface. $V = 1.0472h(R^2 + Rr + r^2) = 0.2618h(D^2 + Dd + d^2)$ $A = 3.1416s(R + r) = 1.5708s(D + d)$ $a = R - r$ $s = \sqrt{a^2 + h^2} = \sqrt{(R - r)^2 + h^2}$

Volumes of Solids

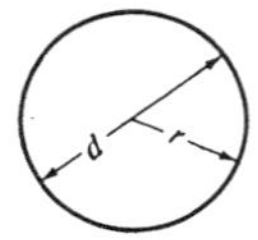 Sphere	V = volume; A = area of surface. $V = \frac{4\pi r^3}{3} = \frac{\pi d^3}{6} = 4.1888r^3 = 0.5236d^3$ $A = 4\pi r^2 = \pi d^2 = 12.5664r^2 = 3.1416d^2$ $r = \sqrt[3]{\frac{3V}{4\pi}} = 0.6204\sqrt[3]{V}$
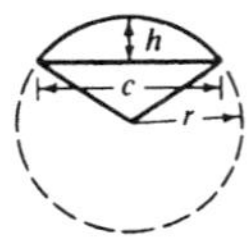 Spherical Sector	V = volume; A = total area of conical and spherical surface. $V = \frac{2\pi r^2 h}{3} = 2.0944r^2h$ $A = 3.1416r(2h + \frac{1}{2}c)$ $c = 2\sqrt{h(2r-h)}$
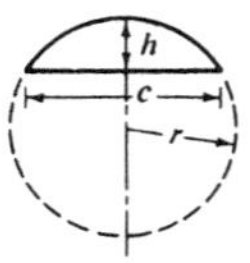 Spherical Segment	V = volume; A = area of spherical surface. $V = 3.1416h^2\left(r - \frac{h}{3}\right) = 3.1416h\left(\frac{c^2}{8} + \frac{h^2}{6}\right)$ $A = 2\pi rh = 6.2832rh = 3.1416\left(\frac{c^2}{4} + h^2\right)$ $c = 2\sqrt{h(2r-h)}$; $r = \frac{c^2 + 4h^2}{8h}$
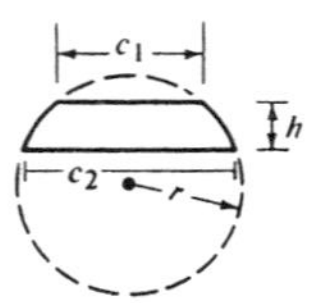 Spherical Zone	V = volume; A = area of spherical surface. $V = 0.5236h\left(\frac{3c_1^2}{4} + \frac{3c_2^2}{4} + h^2\right)$ $A = 2\pi rh = 6.2832rh$ $r = \sqrt{\frac{c_2^2}{4} + \left(\frac{c_2^2 - c_1^2 - 4h^2}{8h}\right)^2}$
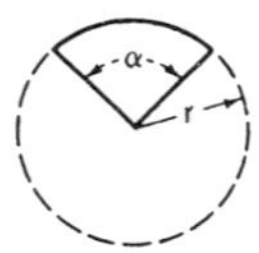 Spherical Wedge	V = volume; A = area of spherical surface; α = center angle in degrees. $V = \frac{\alpha}{360} \times \frac{4\pi r^3}{3} = 0.0116\alpha r^3$ $A = \frac{\alpha}{360} \times 4\pi r^2 = 0.0349\alpha r^2$
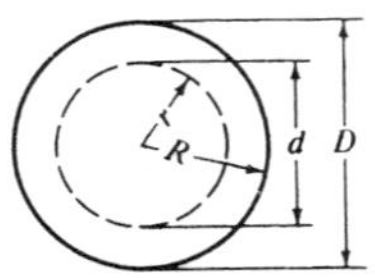 Hollow Sphere	V = volume. $V = \frac{4\pi}{3}(R^3 - r^3) = 4.1888(R^3 - r^3)$ $= \frac{\pi}{6}(D^3 - d^3) = 0.5236(D^3 - d^3)$

Volumes of Solids

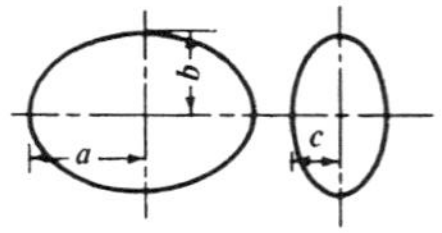 Ellipsoid	V = volume. $V = \frac{4\pi}{3}abc = 4.1888abc$ In an ellipsoid of revolution, or spheroid, where $c = b$: $V = 4.1888ab^2$
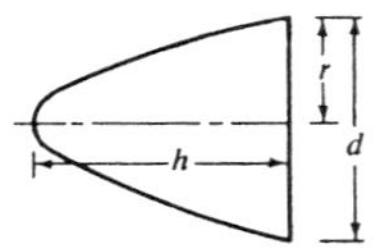 Paraboloid	V = volume; $V = \frac{1}{2}\pi r^2 h = 0.3927d^2h$ A = area; $A = \frac{2\pi}{3p}\left[\sqrt{\left(\frac{d^2}{4} + p^2\right)^3} - p^3\right]$ in which $p = \frac{d^2}{8h}$
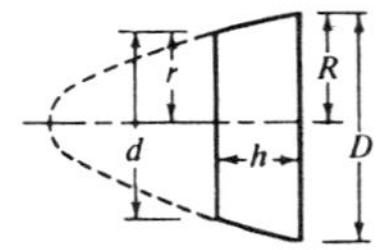 Paraboloidal Segment	V = volume. $V = \frac{\pi}{2}h(R^2 + r^2) = 1.5708h(R^2 + r^2)$ $= \frac{\pi}{8}h(D^2 + d^2) = 0.3927h(D^2 + d^2)$
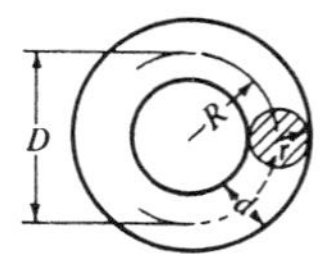 Torus	V = volume; A = area of surface. $V = 2\pi^2Rr^2 = 19.739Rr^2$ $= \frac{\pi^2}{4}Dd^2 = 2.4674Dd^2$ $A = 4\pi^2Rr = 39.478Rr$ $= \pi^2Dd = 9.8696Dd$
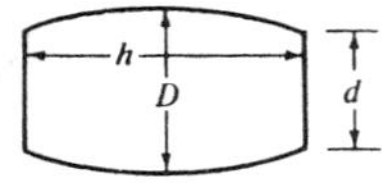 Barrel	V = approximate volume. If the sides are bent to the arc of a circle: $V = \frac{1}{12}\pi h(2D^2 + d^2) = 0.262h(2D^2 + d^2)$ If the sides are bent to the arc of a parabola: $V = 0.209h(2D^2 + Dd + \frac{3}{4}d^2)$
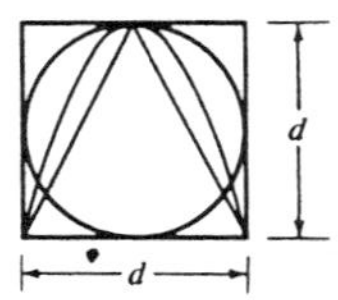	If d = base diameter and height of a cone, a paraboloid and a cylinder, and the diameter of a sphere, then the volumes of these bodies are to each other as below: Cone : paraboloid : sphere : cylinder $= \frac{1}{3} : \frac{1}{2} : \frac{2}{3} : 1$

Wire Data

INCHES PER POUND OF WIRES

Diameter (in.)

	0.020	0.025	0.030	0.035	0.045	0.062	0.078	0.093	0.125
					3/64	1/16	5/64	3/32	1/8
Magnesium	50500	34700	22400	16500	9990	5270	3300	2350	1280
Aluminum	32400	22300	14420	10600	6410	3382	2120	1510	825
Aluminum Bronze	11600	7960	5150	3780	2290	1220	756	538	295
400 series Stainless Steel	11350	7820	5050	3720	2240	1180	742	528	289
Mild Steel	11100	7680	4960	3650	2210	1160	730	519	284
300 series Stainless Steel	10950	7550	4880	3590	2170	1140	718	510	279
Silicon Bronze	10300	7100	4600	3380	2040	1070	675	480	263
Copper Nickel	9950	6850	4430	3260	1970	1040	650	462	253
Nickel	9900	6820	4400	3240	1960	1030	647	460	252
Deoxidized Copper	9800	6750	4360	3200	1940	1200	640	455	249

From Gas Metal-Arc Welding, Miller Electric Manufacturing Co.

TWIST DRILL GAUGE SIZES

Inches

No.	Size	No.	Size	No.	Size	No.	Size	No.	Size
1	0.2280	17	0.1730	33	0.1130	49	0.0730	65	0.0350
2	0.2210	18	0.1695	34	0.1110	50	0.0700	66	0.0330
3	0.2130	19	0.1660	35	0.1100	51	0.0670	67	0.0320
4	0.2090	20	0.1610	36	0.1065	52	0.0635	68	0.0310
5	0.2055	21	0.1590	37	0.1040	53	0.0595	69	0.0292
6	0.2040	22	0.1570	38	0.1015	54	0.0550	70	0.0280
7	0.2010	23	0.1540	39	0.0995	55	0.0520	71	0.0260
8	0.1990	24	0.1520	40	0.0980	56	0.0465	72	0.0250
9	0.1960	25	0.1495	41	0.0960	57	0.0430	73	0.0240
10	0.1935	26	0.1470	42	0.0935	58	0.0420	74	0.0225
11	0.1910	27	0.1440	43	0.0890	59	0.0410	75	0.0210
12	0.1890	28	0.1405	44	0.0860	60	0.0400	76	0.0200
13	0.1850	29	0.1360	45	0.0820	61	0.0390	77	0.0180
14	0.1820	30	0.1285	46	0.0810	62	0.0380	78	0.0160
15	0.1800	31	0.1200	47	0.0785	63	0.0370	79	0.0145
16	0.1770	32	0.1160	48	0.0760	64	0.0360	80	0.0135

Centimeters

No.	Size	No.	Size	No.	Size	No.	Size	No.	Size
1	0.5791	17	0.4394	33	0.2870	49	0.1854	65	0.0889
2	0.5613	18	0.4305	34	0.2819	50	0.1778	66	0.0838
3	0.5410	19	0.4216	35	0.2794	51	0.1702	67	0.0813
4	0.5309	20	0.4089	36	0.2705	52	0.1613	68	0.0787
5	0.5220	21	0.4039	37	0.2642	53	0.1511	69	0.0743
6	0.5182	22	0.3988	38	0.2578	54	0.1397	70	0.0711
7	0.5105	23	0.3912	39	0.2527	55	0.1321	71	0.0660
8	0.5055	24	0.3861	40	0.2489	56	0.1181	72	0.0635
9	0.4978	25	0.3797	41	0.2438	57	0.1092	73	0.0610
10	0.4915	26	0.3734	42	0.2375	58	0.1067	74	0.0572
11	0.4851	27	0.3658	43	0.2261	59	0.1041	75	0.0533
12	0.4801	28	0.3569	44	0.2184	60	0.1016	76	0.0508
13	0.4699	29	0.3454	45	0.2083	61	0.0991	77	0.0457
14	0.4623	30	0.3264	46	0.2057	62	0.0965	78	0.0406
15	0.4572	31	0.3048	47	0.1994	63	0.0940	79	0.0368
16	0.4496	32	0.2946	48	0.1930	64	0.0914	80	0.0343

COMPARATIVE TABLE OF WIRE GAGES

Dimensions of Sizes (in.)

No. of Wire Gage	American or B. & S.	Washburn & Moen, Worcester, Mass.	W. & M. Steel Music Wire	New Am. S. & W. Co's Music Wire Gage	Imperial Wire Gage	Stubs' Steel Wire	Birmingham or Stubs' Iron Wire
00000000	......	......	0.0083	......	......	......	......
0000000	......	0.490	0.0087	......	......	......	......
000000	......	0.4615	0.0095	0.004	0.464	......	......
00000	......	0.4305	0.010	0.005	0.432	......	......
0000	0.460	0.3938	0.011	0.006	0.400	......	0.454
000	0.40964	0.3625	0.012	0.007	0.372	......	0.425
00	0.3648	0.3310	0.0133	0.008	0.348	......	0.380
0	0.32486	0.3065	0.0144	0.009	0.324	......	0.340
1	0.2893	0.2830	0.0156	0.010	0.300	0.227	0.300
2	0.25763	0.2625	0.0166	0.011	0.276	0.219	0.284
3	0.22942	0.2437	0.0178	0.012	0.252	0.212	0.259
4	0.20431	0.2253	0.0188	0.013	0.232	0.207	0.238
5	0.18194	0.2070	0.0202	0.014	0.212	0.204	0.220
6	0.16202	0.1920	0.0215	0.016	0.192	0.201	0.203
7	0.14428	0.1770	0.023	0.018	0.176	0.199	0.180
8	0.12849	0.1620	0.0243	0.020	0.160	0.197	0.165
9	0.11443	0.1483	0.0256	0.022	0.144	0.194	0.148
10	0.10189	0.1350	0.027	0.024	0.128	0.191	0.134
11	0.090742	0.1205	0.0284	0.026	0.116	0.188	0.120
12	0.080808	0.1055	0.0296	0.029	0.104	0.185	0.109
13	0.071961	0.0915	0.0314	0.031	0.092	0.182	0.095
14	0.064804	0.0800	0.0326	0.033	0.080	0.180	0.083
15	0.057068	0.0720	0.0345	0.035	0.072	0.178	0.072
16	0.05082	0.0625	0.036	0.037	0.064	0.175	0.065
17	0.045257	0.0540	0.0377	0.039	0.056	0.172	0.058
18	0.040303	0.0475	0.0395	0.041	0.048	0.168	0.049
19	0.03589	0.0410	0.0414	0.043	0.040	0.164	0.042
20	0.031961	0.0348	0.0434	0.045	0.036	0.161	0.035
21	0.028462	0.03175	0.046	0.047	0.032	0.157	0.032
22	0.025347	0.0286	0.0483	0.049	0.028	0.155	0.028
23	0.022571	0.0258	0.051	0.051	0.024	0.153	0.025
24	0.0201	0.0230	0.055	0.055	0.022	0.151	0.022
25	0.0179	0.0204	0.0586	0.059	0.020	0.148	0.020
26	0.01594	0.0181	0.0626	0.063	0.018	0.146	0.018
27	0.014195	0.0173	0.0658	0.067	0.0164	0.143	0.016
28	0.012641	0.0162	0.072	0.071	0.0149	0.139	0.014
29	0.011257	0.0150	0.076	0.075	0.0136	0.134	0.013
30	0.010025	0.0140	0.080	0.080	0.0124	0.127	0.012
31	0.008928	0.0132	......	0.085	0.0116	0.120	0.010
32	0.00795	0.0128	......	0.090	0.0108	0.115	0.009
33	0.00708	0.0118	......	0.095	0.0100	0.112	0.008
34	0.006304	0.0104	......	......	0.0092	0.110	0.007
35	0.005614	0.0095	......	......	0.0084	0.108	0.005
36	0.005	0.0090	......	......	0.0076	0.106	0.004
37	0.004453	......	......	......	0.0068	0.103	......
38	0.003965	......	......	......	0.0060	0.101	......
39	0.003531	......	......	......	0.0052	0.099	......
40	0.003144	......	......	......	0.0048	0.097	......

From Republic Alloy Steels, Republic Steel Corp. (now LTV Steel)

WIRE TABLES*
Solid Copper Wire

A. W. G. or B. & S. Gage; English Units
100 per cent conductivity; density 8.89 at 20 deg cent

Gage No.	Diameter in Mils	Cross-section		Resistance at 20° C or 68° F		Weight in Pounds		Feet per Pound
		Circular Mils	Square Inches	Ohms per 1000 ft	Ohms per Mile	per 1000 ft	per Mile	
0000	460.0	211,600	0.1662	0.04901	0.259	640.5	3380	1.561
000	409.6	167,800	0.1318	0.06180	0.326	507.9	2680	1.968
00	364.8	133,100	0.1045	0.07793	0.411	402.8	2130	2.482
0	324.9	105,500	0.08289	0.90827	0.519	319.5	1680	3.130
1	289.3	83,690	0.06573	0.1239	0.654	253.3	1340	3.947
2	257.6	66,370	0.05213	0.1563	0.825	200.9	1060	4.977
3	229.4	52,640	0.04134	0.1970	1.04	159.3	841	6.276
4	204.3	41,740	0.03278	0.2485	1.31	126.4	667	7.914
5	181.9	33,100	0.02600	0.3133	1.65	100.2	529	9.980
6	62.0	26,250	0.02062	0.3951	2.09	79.46	420	12.58
7	144.3	20,820	0.01635	0.4982	2.63	63.02	333	15.87
8	1 8.5	16,510	0.01297	0.6282	3.32	49.98	264	20.01
10	181.9	10,380	0.008155	0.9989	5.28	31.43	166	31.82
12	80.81	6,530	0.005129	1.588	8.38	19.77	104	50.59
14	64.08	4,107	0.003225	2.525	13.3	12.43	63.3	80.44
15	57.07	3,257	0.002558	3.184	16.8	9.858	52.0	101.4
16	50.82	2,583	0.002028	4.015	21.2	7.818	41.3	127.9
17	45.26	2,048	0.001609	5.064	26.7	6.200	32.7	161.3
18	40.30	1,624	0.001276	6.385	33.7	4.917	26.0	203.4
19	35.89	1,288	0.001012	8.051	42.5	3.899	20.6	256.5
20	31.96	1,022	0.0008023	10.15	53.6	3.092	16.3	323.4
21	28.46	810.1	0.0006363	12.80	67.6	2.452	12.9	407.8
22	25.35	642.4	0.0005046	16.14	85.2	1.945	10.3	514.2
23	22.57	509.5	0.0004002	20.36	108	1.542	8.14	648.4
24	20.10	404.0	0.0003173	25.67	135	1.223	6.46	817.7
25	17.90	320.4	0.0002517	32.37	171	0.9699	5.12	1,031
26	15.94	254.1	0.0001996	40.82	216	0.7692	4.06	1,300
27	14.20	201.5	0.0001583	51.46	272	0.6100	3.22	1,639
28	12.64	59.8	0.0001255	64.90	343	0.4837	2.55	2,067
29	11.26	126.7	0.00009953	81.84	432	0.3836	2.03	2,607
30	10.03	100.5	0.00007894	103.2	545	0.3042	1.61	3,287
31	8.928	79.70	0.00006260	130.1	687	0.2413	1.27	4,145
32	7.950	63.21	0.00004964	164.1	866	0.1913	1.01	5,227
33	7.080	50.13	0.00003937	206.9	1,090	0.1517	0.814	6,591
34	6.305	39.75	0.00003122	260.9	1,380	0.1203	0.635	8,310
35	5.615	31.52	0.00002476	329.0	1,740	0.09542	0.504	10,480
36	5.000	25.00	0.00001964	414.8	2,190	0.07568	0.400	13,210
37	4.453	19.83	0.00001557	523.1	2,762	0.06001	0.317	16,660
38	3.965	15.72	0.00001235	659.6	3,480	0.04759	0.251	21,010
39	3.531	12.47	0.000009793	831.8	4,392	0.03774	0.199	26,500
40	3.145	9.888	0.000007766	1049	5,540	0.02993	0.158	33,410
41	2.800	7.842	0.000006159	1323	6,983	0.02374	0.125	42,130
42	2.494	6.219	0.000004884	1668	8,806	0.01882	0.0994	53,120
43	2.221	4.932	0.000003873	2103	11,100	0.01493	0.0788	66,990
44	1.978	3.911	0.000003072	2652	14,000	0.01184	0.0625	84,470

*From Electrical Engineer's Handbook, Pender and McIlwain

WIRE TABLES*
Solid Copper Wire

A. W. G. or B. & S. Gage in Metric Units
100 per cent conductivity; density 8.89 at 20 deg cent

Gage No.	Diameter, mm	Cross-section, sq mm	Ohms per Kilometer 20° C	Kilograms per Kilometer
0000	11.68	107.2	0.1608	953.2
000	10.40	85.03	0.2028	755.9
00	9.266	67.43	0.2557	599.5
0	8.252	53.48	0.3224	475.4
1	7.348	42.41	0.4066	377.0
2	6.544	33.63	0.5126	299.0
3	5.827	26.67	0.6464	237.1
4	5.189	21.15	0.8152	188.0
5	4.621	16.77	1.028	149.1
6	4.115	13.30	1.296	118.2
7	3.665	10.55	1.634	93.78
8	3.264	8.366	2.061	74.37
10	2.588	5.261	3.277	46.77
12	2.053	3.309	5.211	29.42
14	1.628	2.081	8.285	18.50
15	1.450	1.650	10.45	14.67
16	1.291	1.309	13.18	11.63
17	1.150	1.038	16.61	9.226
18	1.024	0.8231	20.95	7.317
19	0.9116	0.6527	26.42	5.803
20	0.8118	0.5176	33.31	4.602
21	0.7230	0.4105	42.00	3.649
22	0.6438	0.3255	52.96	2.894
23	0.5733	0.2582	66.79	2.295
24	0.5106	0.2047	84.22	1.820
25	0.4547	0.1624	106.2	1.443
26	0.4049	0.1288	133.9	1.145
27	0.3606	0.1021	168.8	0.9078
28	0.3211	0.08098	212.9	0.7199
29	0.2859	0.06422	268.5	0.5709
30	0.2546	0.05093	338.6	0.4527
31	0.2268	0.04039	426.9	0.3590
32	0.2019	0.03203	538.3	0.2847
33	0.1798	0.02540	678.8	0.2258
34	0.1601	0.02014	856.0	0.1791
35	0.1426	0.01597	1079	0.1420
36	0.1270	0.01267	1361	0.1126
37	0.1131	0.01005	1716	0.08931
38	0.1007	0.007967	2164	0.07083
39	0.08969	0.006318	2729	0.05617
40	0.07987	0.005010	3441	0.04454
41	0.07113	0.003973	4339	0.03532
42	0.06334	0.003151	5472	0.02801
43	0.05641	0.002499	6900	0.02222
44	0.05023	0.001982	8700	0.01762

AISI-SAE Standard Carbon Steels

Free-Machining Grades

AISI No.	Composition*, % C	Mn	P	S	SAE No.
	Resulfurized				
1108	0.08 to 0.13	0.50 to 0.80	0.040 max	0.08 to 0.13	1108
1109	0.08 to 0.13	0.60 to 0.90	0.040 max	0.08 to 0.13	1109
1110	0.08 to 0.13	0.30 to 0.60	0.040 max	0.08 to 0.13	1110
1116	0.14 to 0.20	1.10 to 1.40	0.040 max	0.16 to 0.23	1116
1117	0.14 to 0.20	1.00 to 1.30	0.040 max	0.08 to 0.13	1117
1118	0.14 to 0.20	1.30 to 1.60	0.040 max	0.08 to 0.13	1118
1119	0.14 to 0.20	1.00 to 1.30	0.040 max	0.24 to 0.33	1119
1132	0.27 to 0.34	1.35 to 1.65	0.040 max	0.08 to 0.13	1132
1137	0.32 to 0.39	1.35 to 1.65	0.040 max	0.08 to 0.13	1137
1139	0.35 to 0.43	1.35 to 1.65	0.040 max	0.13 to 0.20	1139
1140	0.37 to 0.44	0.70 to 1.00	0.040 max	0.08 to 0.13	1140
1141	0.37 to 0.45	1.35 to 1.65	0.040 max	0.08 to 0.13	1141
1144	0.40 to 0.48	1.35 to 1.65	0.040 max	0.24 to 0.33	1144
1145	0.42 to 0.49	0.70 to 1.00	0.040 max	0.04 to 0.07	1145
1146	0.42 to 0.49	0.70 to 1.00	0.040 max	0.08 to 0.13	1146
1151	0.48 to 0.55	0.70 to 1.00	0.040 max	0.08 to 0.13	1151
	Resulfurized and Rephosphorized				
1211	0.13 max	0.60 to 0.90	0.07 to 0.12	0.10 to 0.15	1211
1212	0.13 max	0.70 to 1.00	0.07 to 0.12	0.16 to 0.23	1212
1213	0.13 max	0.70 to 1.00	0.07 to 0.12	0.24 to 0.33	1213
1215	0.09 max	0.75 to 1.05	0.04 to 0.09	0.26 to 0.35	1215
12L14†	0.15 max	0.85 to 1.15	0.04 to 0.09	0.26 to 0.35	12L14

*When silicon is required, the following ranges and limits are commonly used: for nonresulfurized steels: up to 1015, 0.10% max; 1015 to 1025, 0.10% max, or 0.10 to 0.20%, or 0.15 to 0.30%, or 0.20 to 0.40%, or 0.30 to 0.60%; over 1025, 0.10 to 0.20%, or 0.15 to 0.30%, or 0.20 to 0.40%, or 0.30 to 0.60%; 1513 to 1524, 0.10% max, or 0.10 to 0.20%, or 0.15 to 0.30%, or 0.20 to 0.40%, or 0.30 to 0.60%; 1525 and over, 0.10 to 0.20%, or 0.15 to 0.30%, or 0.20 to 0.40%, or 0.30 to 0.60%. For resulfurized steels: to 1110, 0.10% max; 1116 and over, 0.10% max, or 0.10 to 0.20%, or 0.15 to 0.30%, or 0.20 to 0.40%, or 0.30 to 0.60%. It is not common practice to produce resulfurized and rephosphorized steels to specified limits for silicon because of its adverse effect on machinability.

Copper can be added to a standard steel.

Standard killed carbon steels, which are generally fine grain, may be produced with a boron addition to improve hardenability. Such steels can be expected to contain 0.0005% B min. These steels are identified by inserting the letter "B" between the second and third numerals of the AISI number — for example, 10B46.

†0.15 to 0.35% Pb. When lead is required as an added element to a standard steel, a range of 0.15 to 0.35%, inclusive, is generally used. Such a steel is identified by inserting the letter "L" between the second and third numeral of the AISI number.

Sources: American Iron & Steel Institute, New York; SAE Standard J403f.

Nonresulfurized Grades

AISI No.	Composition*, % C	Mn	P Max	S Max	SAE No.
1005	0.06 max	0.35 max	0.040	0.050	1005
1006	0.08 max	0.25 to 0.40	0.040	0.050	1006
1008	0.10 max	0.30 to 0.50	0.040	0.050	1008
1010	0.08 to 0.13	0.30 to 0.60	0.040	0.050	1010
1011	0.08 to 0.13	0.60 to 0.90	0.040	0.050	1011
1012	0.10 to 0.15	0.30 to 0.60	0.040	0.050	1012
1013	0.11 to 0.16	0.50 to 0.80	0.040	0.050	1013
1513	0.10 to 0.16	1.10 to 1.40	0.040	0.050	1513
1015	0.13 to 0.18	0.30 to 0.60	0.040	0.050	1015
1016	0.13 to 0.18	0.60 to 0.90	0.040	0.050	1016
1017	0.15 to 0.20	0.30 to 0.60	0.040	0.050	1017
1018	0.15 to 0.20	0.60 to 0.90	0.040	0.050	1018
1518	0.15 to 0.21	1.10 to 1.40	0.040	0.050	1518
1019	0.15 to 0.20	0.70 to 1.00	0.040	0.050	1019
1020	0.18 to 0.23	0.30 to 0.60	0.040	0.050	1020
1021	0.18 to 0.23	0.60 to 0.90	0.040	0.050	1021
1022	0.18 to 0.23	0.70 to 1.00	0.040	0.050	1022
1522	0.18 to 0.24	1.10 to 1.40	0.040	0.050	1522
1023	0.20 to 0.25	0.30 to 0.60	0.040	0.050	1023
1524	0.19 to 0.25	1.35 to 1.65	0.040	0.050	1524
1025	0.22 to 0.28	0.30 to 0.60	0.040	0.050	1025
1525	0.23 to 0.29	0.80 to 1.10	0.040	0.050	1525
1026	0.22 to 0.28	0.60 to 0.90	0.040	0.050	1026
1526	0.22 to 0.29	1.10 to 1.40	0.040	0.050	1526
1527	0.22 to 0.29	1.20 to 1.50	0.040	0.050	1527
1029	0.25 to 0.31	0.60 to 0.90	0.040	0.050	1029
1030	0.28 to 0.34	0.60 to 0.90	0.040	0.050	1030
1033	0.29 to 0.36	0.70 to 1.00	0.040	0.050	1033
1034	0.32 to 0.38	0.50 to 0.80	0.040	0.050	—
1035	0.32 to 0.38	0.60 to 0.90	0.040	0.050	1035
1536	0.30 to 0.37	1.20 to 1.50	0.040	0.050	1536
1037	0.32 to 0.38	0.70 to 1.00	0.040	0.050	1037
1038	0.35 to 0.42	0.60 to 0.90	0.040	0.050	1038
1039	0.37 to 0.44	0.70 to 1.00	0.040	0.050	1039
1040	0.37 to 0.44	0.60 to 0.90	0.040	0.050	1040
1541	0.36 to 0.44	1.35 to 1.65	0.040	0.050	1541
1042	0.40 to 0.47	0.60 to 0.90	0.040	0.050	1042
1043	0.40 to 0.47	0.70 to 1.00	0.040	0.050	1043
1044	0.43 to 0.50	0.30 to 0.60	0.040	0.050	1044
1045	0.43 to 0.50	0.60 to 0.90	0.040	0.050	1045
1046	0.43 to 0.50	0.70 to 1.00	0.040	0.050	1046
1547	0.45 to 0.51	1.35 to 1.65	0.040	0.050	1547
1548	0.44 to 0.52	1.10 to 1.40	0.040	0.050	1548
1049	0.46 to 0.53	0.60 to 0.90	0.040	0.050	1049
1050	0.48 to 0.55	0.60 to 0.90	0.040	0.050	1050
1551	0.45 to 0.56	0.85 to 1.15	0.040	0.050	1551
1552	0.47 to 0.55	1.20 to 1.50	0.040	0.050	1552
1053	0.48 to 0.55	0.70 to 1.00	0.040	0.050	1053
1055	0.50 to 0.60	0.60 to 0.90	0.040	0.050	1055
1059	0.55 to 0.65	0.50 to 0.80	0.040	0.050	—
1060	0.55 to 0.65	0.60 to 0.90	0.040	0.050	1060
1561	0.55 to 0.65	0.75 to 1.05	0.040	0.050	1561
1065	0.60 to 0.70	0.60 to 0.90	0.040	0.050	1065
1566	0.60 to 0.71	0.85 to 1.15	0.040	0.050	1566
1069	0.65 to 0.75	0.40 to 0.70	0.040	0.050	1069
1070	0.65 to 0.75	0.60 to 0.90	0.040	0.050	1070
1572	0.65 to 0.76	1.00 to 1.30	0.040	0.050	1572
1074	0.70 to 0.80	0.50 to 0.80	0.040	0.050	1074
1075	0.70 to 0.80	0.40 to 0.70	0.040	0.050	1075
1078	0.72 to 0.85	0.30 to 0.60	0.040	0.050	1078
1080	0.75 to 0.88	0.60 to 0.90	0.040	0.050	1080
1084	0.80 to 0.93	0.60 to 0.90	0.040	0.050	1084
—	0.80 to 0.93	0.70 to 1.00	0.040	0.050	1085
1086	0.80 to 0.93	0.30 to 0.50	0.040	0.050	1086
1090	0.85 to 0.98	0.60 to 0.90	0.040	0.050	1090
1095	0.90 to 1.03	0.30 to 0.50	0.040	0.050	1095

SAE Alloy Steel Compositions

SAE No.	Composition,* %								Corresponding AISI No.
	C	Mn	P	S	Si	Ni	Cr	Other	
1330	0.28 - 0.33	1.60 - 1.90	0.035	0.040	0.20 - 0.35	—	—	—	1330
1335	0.33 - 0.38	1.60 - 1.90	0.035	0.040	0.20 - 0.35	—	—	—	1335
1340	0.38 - 0.43	1.50 - 1.90	0.035	0.040	0.20 - 0.35	—	—	—	1340
1345	0.43 - 0.48	1.60 - 1.90	0.035	0.040	0.20 - 0.35	—	—	Mo	1345
4012	0.09 - 0.14	0.75 - 1.00	0.035	0.040	0.20 - 0.35	—	—	0.15 - 0.25	4012
4023	0.20 - 0.25	0.70 - 0.90	0.035	0.040	0.20 - 0.35	—	—	0.20 - 0.30	4023
4024	0.20 - 0.25	0.70 - 0.90	0.035	0.035-0.050	0.20 - 0.35	—	—	0.20 - 0.30	4024
4027	0.25 - 0.30	0.70 - 0.90	0.035	0.040	0.20 - 0.35	—	—	0.20 - 0.30	4027
4028	0.25 - 0.30	0.70 - 0.90	0.035	0.035 - 0.050	0.20 - 0.35	—	—	0.20 - 0.30	4028
4032	0.30 - 0.35	0.70 - 0.90	0.035	0.040	0.20 - 0.35	—	—	0.20 - 0.30	—
4037	0.35 - 0.40	0.70 - 0.90	0.035	0.040	0.20 - 0.35	—	—	0.20 - 0.30	4037
4042	0.40 - 0.45	0.70 - 0.90	0.035	0.040	0.20 - 0.35	—	—	0.20 - 0.30	—
4047	0.45 - 0.50	0.70 - 0.90	0.035	0.040	0.20 - 0.35	—	—	0.20 - 0.30	4047
4118	0.18 - 0.23	0.70 - 0.90	0.035	0.040	0.20 - 0.35	—	0.40 - 0.60	0.08 - 0.15	4118
4130	0.28 - 0.33	0.40 - 0.60	0.035	0.040	0.20 - 0.35	—	0.80 - 1.10	0.15 - 0.25	4130
4135	0.33 - 0.38	0.70 - 0.90	0.035	0.040	0.20 - 0.35	—	0.80 - 1.10	0.15 - 0.25	—
4137	0.35 - 0.40	0.70 - 0.90	0.035	0.040	0.20 - 0.35	—	0.80 - 1.10	0.15 - 0.25	4137
4140	0.38 - 0.43	0.75 - 1.00	0.035	0.040	0.20 - 0.35	—	0.80 - 1.10	0.15 - 0.25	4140
4142	0.40 - 0.45	0.75 - 1.00	0.035	0.040	0.20 - 0.35	—	0.80 - 1.10	0.15 - 0.25	4142
4145	0.43 - 0.48	0.75 - 1.00	0.035	0.040	0.20 - 0.35	—	0.80 - 1.10	0.15 - 0.25	4145
4147	0.45 - 0.50	0.75 - 1.00	0.035	0.040	0.20 - 0.35	—	0.80 - 1.10	0.15 - 0.25	4147
4150	0.48 - 0.53	0.75 - 1.00	0.035	0.040	0.20 - 0.35	—	0.80 - 1.10	0.15 - 0.25	4150
4161	0.56 - 0.64	0.75 - 1.00	0.035	0.040	0.20 - 0.35	—	0.70 - 0.90	0.25 - 0.35	4161
4320	0.17 - 0.22	0.45 - 0.65	0.035	0.040	0.20 - 0.35	1.65 - 2.00	0.40 - 0.60	0.20 - 0.30	4320
4340	0.38 - 0.43	0.60 - 0.80	0.035	0.040	0.20 - 0.35	1.65 - 2.00	0.70 - 0.90	0.20 - 0.30	4340
E4340†	0.38 - 0.43	0.65 - 0.85	0.025	0.025	0.20 - 0.35	1.65 - 2.00	0.70 - 0.90	0.20 - 0.30	E4340
4419	0.18 - 0.23	0.45 - 0.65	0.035	0.040	0.20 - 0.35	—	—	0.45 - 0.60	4419
4422	0.20 - 0.25	0.70 - 0.90	0.035	0.040	0.20 - 0.35	—	—	0.35 - 0.45	—
4427	0.24 - 0.29	0.70 - 0.90	0.035	0.040	0.20 - 0.35	—	—	0.35 - 0.45	—
4615	0.13 -0.18	0.45 - 0.65	0.035	0.040	0.20 - 0.35	1.65 - 2.00	—	0.20 - 0.30	4615
4617	0.15 - 0.20	0.45 - 0.65	0.035	0.040	0.20 - 0.35	1.65 - 2.00	—	0.20 - 0.30	—
4620	0.17 - 0.22	0.45 - 0.65	0.035	0.040	0.20 - 0.35	1.65 - 2.00	—	0.20 - 0.30	4620
4621	0.18 - 0.23	0.70 - 0.90	0.035	0.040	0.20 - 0.35	1.65 - 2.00	—	0.20 - 0.30	4621
4626	0.24 - 0.29	0.45 - 0.65	0.035	0.04 max	0.20 - 0.35	0.70 - 1.00	—	0.15 - 0.25	4626
4718	0.16 - 0.21	0.70 - 0.90	—	—	—	0.90 - 1.20	0.35 - 0.55	0.30 - 0.40	4718
4720	0.17 - 0.22	0.50 - 0.70	0.035	0.040	0.20 - 0.35	0.90 - 1.20	0.35 - 0.55	0.15 - 0.25	4720
4815	0.13 - 0.18	0.40 - 0.60	0.035	0.040	0.20 - 0.35	3.25 - 3.75	—	0.20 - 0.30	4815
4817	0.15 - 0.20	0.40 - 0.60	0.035	0.040	0.20 - 0.35	3.25 - 3.75	—	0.20 - 0.30	4817
4820	0.18 - 0.23	0.50 - 0.70	0.035	0.040	0.20 - 0.35	3.25 - 3.75	—	0.20 - 0.30	4820
5015	0.12 - 0.17	0.30 - 0.50	0.035	0.040	0.20 - 0.35	—	0.30 - 0.50	—	5015
50B40‡	0.38 - 0.43	0.75 - 1.00	0.035	0.040	0.20 - 0.35	—	0.40 - 0.60	—	—
50B44‡	0.43 - 0.48	0.75 - 1.00	0.035	0.040	0.20 - 0.35	—	0.40 - 0.60	—	50B44
5046	0.43 - 0.48	0.75 - 1.00	0.035	0.040	0.20 - 0.35	—	0.20 - 0.35	—	—
50B46‡	0.44 - 0.49	0.75 - 1.00	0.035	0.040	0.20 - 0.35	—	0.20 - 0.35	—	50B46
50B50‡	0.48 - 0.53	0.75 - 1.00	0.035	0.040	0.20 - 0.35	—	0.04 - 0.60	—	50B50
5060	0.56 - 0.64	0.75 - 1.00	0.035	0.040	0.20 - 0.35	—	0.40 - 0.60	—	—
50B60‡	0.56 - 0.64	0.75 - 1.00	0.035	0.040	0.20 - 0.35	—	0.40 - 0.60	—	50B60

* Small quantities of certain elements are present which are not specified or required. Considered as incidental, they are acceptable to the following amounts: 0.35 Cu, 0.25 Ni, 0.20 Cr, and 0.06 Mo. † Electric furnace steel. ‡0.0005% B min.

SAE Alloy Steel Compositions

(Continued)

SAE No.	Composition,* %								Corresponding AISI No.
	C	Mn	P	S	Si	Ni	Cr	Other	
5115	0.13 - 0.18	0.70 - 0.90	0.035	0.040	0.20 - 0.35	—	0.70 - 0.90	—	—
5120	0.17 - 0.22	0.70 - 0.90	0.035	0.040	0.20 - 0.35	—	0.70 - 0.90	—	5120
5130	0.28 - 0.33	0.70 - 0.90	0.035	0.040	0.20 - 0.35	—	0.80 - 1.10	—	5130
5132	0.30 - 0.35	0.60 - 0.80	0.035	0.040	0.20 - 0.35	—	0.75 - 1.00	—	5130
5135	0.33 - 0.38	0.60 - 0.80	0.035	0.040	0.20 - 0.35	—	0.80 - 1.05	—	5135
5140	0.38 - 0.43	0.70 - 0.90	0.035	0.040	0.20 - 0.35	—	0.70 - 0.90	—	5140
5145	0.43 - 0.40	0.70 - 0.90	0.035	0.040	0.20 - 0.35	—	0.70 - 0.90	—	5145
5147	0.46 - 0.51	0.70 - 0.95	0.035	0.040	0.20 - 0.35	—	0.85 - 1.15	—	5147
5150	0.48 - 0.53	0.70 - 0.90	0.035	0.040	0.20 - 0.35	—	0.70 - 0.90	—	5150
5155	0.51 - 0.59	0.70 - 0.90	0.035	0.040	0.20 - 0.35	—	0.70 - 0.90	—	5155
5160	0.56 - 0.64	0.75 - 1.00	0.035	0.040	0.20 - 0.35	—	0.70 - 0.90	—	5160
51B60‡	0.56 - 0.64	0.75 - 1.00	0.035	0.040	0.20 - 0.35	—	0.70 - 0.90	—	51B60
50100†	0.98 - 1.10	0.25 - 0.45	0.025	0.025	0.20 - 0.35	—	0.40 - 0.60	—	—
51100†	0.98 - 1.10	0.25 - 0.45	0.025	0.025	0.20 - 0.35	—	0.90 - 1.15	—	E51100
52100†	0.98 - 1.10	0.25 - 0.45	0.025	0.025	0.20 - 0.35	—	1.30 - 1.60	V	E52100
6118	0.16 - 0.21	0.50 - 0.70	0.035	0.040	0.20 - 0.35	—	0.50 - 0.70	0.10 - 0.15	6118
6150	0.48 - 0.53	0.70 - 0.90	0.035	0.040	0.20 - 0.35	—	0.80 - 1.10	0.15	6150
								Mo	
8115	0.13 - 0.18	0.70 - 0.90	0.035	0.040	0.20 - 0.35	0.20 - 0.40	0.30 - 0.50	0.08 - 0.15	8115
81B45‡	0.43 - 0.48	0.75 - 1.00	0.035	0.040	0.20 - 0.35	0.20 - 0.40	0.35 - 0.55	0.08 - 0.15	81B45
8615	0.13 - 0.18	0.70 - 0.90	0.035	0.040	0.20 - 0.35	0.40 - 0.70	0.40 - 0.60	0.15 - 0.25	8615
8617	0.15 - 0.20	0.70 - 0.90	0.035	0.040	0.20 - 0.35	0.40 - 0.70	0.40 - 0.60	0.15 - 0.25	8617
8620	0.18 - 0.23	0.70 - 0.90	0.035	0.040	0.20 - 0.35	0.40 - 0.70	0.40 - 0.60	0.15 - 0.25	8620
8622	0.20 - 0.25	0.70 - 0.90	0.035	0.040	0.20 - 0.35	0.40 - 0.70	0.40 - 0.60	0.15 - 0.25	8622
8625	0.23 - 0.28	0.70 - 0.90	0.035	0.040	0.20 - 0.35	0.40 - 0.70	0.40 - 0.60	0.15 - 0.25	8625
8627	0.25 - 0.30	0.70 - 0.90	0.035	0.040	0.20 - 0.35	0.40 - 0.70	0.40 - 0.60	0.15 - 0.25	8627
8630	0.28 - 0.33	0.70 - 0.90	0.035	0.040	0.20 - 0.35	0.40 - 0.70	0.40 - 0.60	0.15 - 0.25	8630
8637	0.35 - 0.40	0.75 - 1.00	0.035	0.040	0.20 - 0.35	0.40 - 0.70	0.40 - 0.60	0.15 - 0.25	8637
8640	0.38 - 0.43	0.75 - 1.00	0.035	0.040	0.20 - 0.35	0.40 - 0.70	0.40 - 0.60	0.15 - 0.25	8640
8642	0.40 - 0.45	0.75 - 1.00	0.035	0.040	0.20 - 0.35	0.40 - 0.70	0.40 - 0.60	0.15 - 0.25	8642
8645	0.43 - 0.48	0.75 - 1.00	0.035	0.040	0.20 - 0.35	0.40 - 0.70	0.40 - 0.60	0.15 - 0.25	8645
86B45‡	0.43 - 0.48	0.75 - 1.00	0.035	0.040	0.20 - 0.35	0.40 - 0.70	0.40 - 0.60	0.15 - 0.25	—
8650	0.48 - 0.53	0.75 - 1.00	0.035	0.040	0.20- 0.35	0.40 - 0.70	0.40 - 0.60	0.15 - 0.25	—
8655	0.51 - 0.59	0.75 - 1.00	0.035	0.040	0.20 - 0.35	0.40 - 0.70	0.40 - 0.60	0.15 - 0.25	8655
8660	0.56 - 0.64	0.75 - 1.00	0.035	0.040	0.20 - 0.35	0.40 - 0.70	0.40 - 0.60	0.15 - 0.25	—
8720	0.18 - 0.23	0.70 - 0.90	0.035	0.040	0.20 - 0.35	0.40 - 0.70	0.40 - 0.60	0.20 - 0.30	8720
8740	0.38 - 0.43	0.75 - 1.00	0.035	0.040	0.20 - 0.35	0.40 - 0.70	0.40 - 0.60	0.20 - 0.30	8740
8822	0.20 - 0.25	0.75 - 1.00	0.035	0.040	0.20 - 0.35	0.40 - 0.70	0.40 - 0.60	0.30 - 0.40	8822
9254	0.51 - 0.59	0.50 - 0.80	0.035	0.040	1.20 - 1.60	—	0.50 - 0.80	—	—
9255	0.51 - 0.59	0.70 - 0.95	0.035	0.040	1.80 - 2.20	—	—	—	9255
9260	0.56 - 0.64	0.75 - 1.00	0.035	0.040	1.80 - 2.20	—	—	—	9260
9310	0.08 - 0.13	0.45 - 0.65	0.025	0.025	0.20 - 0.35	3.00 - 3.50	1.00 - 1.40	0.08 - 0.15	—
94B15‡	0.13 - 0.18	0.75 - 1.00	0.035	0.040	0.20 - 0.35	0.30 - 0.60	0.30 - 0.50	0.08 - 0.15	—
94B17‡	0.15 - 0.20	0.75 - 1.00	0.035	0.040	0.20 - 0.35	0.30 - 0.60	0.30 - 0.50	0.08 - 0.15	94B17
94B30‡	0.28 - 0.33	0.75 - 1.00	0.035	0.040	0.20 - 0.35	0.30 - 0.60	0.30 - 0.50	0.08 - 0.15	94B30

*Small quantities of certain elements are present which are not specified or required. Considered as incidental, they are acceptable to the following amounts: 0.35 Cu, 0.25 Ni, 0.20 Cr, and 0.06 Mo. † electric furnace steel. ‡0.0005% B min.

Standard Line Pipe Data

Seamless, Electric Weld

Size, Inches		Wall Thickness, Inches	Class	Schedule Number
Nominal Size	OD			
2	2.375	0.154	Std.	40
		0.218	XS	80
		0.344		160
		0.436	XXS	
2½	2.875	0.203	Std.	40
		0.276	XS	80
		0.375		160
		0.552	XXS	
3	3.500	0.216	Std.	40
		0.250		
		0.281		
		0.300	XS	80
		0.438		160
		0.600	XXS	
3½	4.000	0.226	Std.	40
		0.250		
		0.281		
		0.318	XS	80
		0.636		
4	4.500	0.156		
		0.172		
		0.188		
		0.203		
		0.219		
		0.237	Std.	40
		0.250		
		0.281		
		0.312		
		0.337	XS	80
		0.438		120
		0.531		160
		0.674	XXS	
5	5.563	0.156		
		0.188		
		0.219		
		0.258	Std.	40
		0.281		
		0.312		
		0.344		
		0.375	XS	80
		0.500		120
		0.625		160
		0.750	XXS	
		0.875		
		0.938		

Size, Inches		Wall Thickness, Inches	Class	Schedule Number
Nominal Size	OD			
6	6.625	0.156		
		0.172		
		0.188		
		0.203		
		0.219		
		0.250		
		0.280	Std.	40
		0.312		
		0.344		
		0.375		
		0.432	XS	80
		0.500		
		0.562		120
		0.625		
		0.719		160
		0.864	XXS	
		1.000		
		1.125		
		1.188		
8	8.625	0.188		
		0.203		
		0.219		
		0.250		20
		0.277		30
		0.312		
		0.322	Std	40
		0.344		
		0.375		
		0.406		60
		0.438		
		0.500	XS	80
		0.562		
		0.594		100
		0.625		
		0.719		120
		0.812		140
		0.875	XXS	
		0.906		160
		1.000		
		1.125		
		1.250		

Seamless, Electric Weld, Plain End

Size, Inches		Wall Thickness, Inches	Class	Schedule Number
Nominal Size	OD			
10	10.750	0.188		
		0.203		
		0.219		
		0.250		20
		0.279		
		0.307		30
		0.344		
		0.365	Std.	40
		0.438		
		0.500	XS	60
		0.562		
		0.594		
		0.625		
		0.719		100
		0.812		
		0.844		120
		1.000		140
		1.125		160
		1.438		
12	12.750	0.188*		
		0.203*		
		0.219*		
		0.250*		20
		0.281		
		0.312		
		0.330		30
		0.344		
		0.375	Std.	
		0.406		40
		0.438		
		0.500	XS	
		0.562		60
		0.625		
		0.688		80
		0.750		
		0.844		100
		1.000		120
		1.125		140
		1.312		160
		1.500		
		1.594		
		1.625		
		1.750		
		2.000		
		2.375		
		2.500		
14	14.000	0.188*		
		0.203*		
		0.210*		
		0.219*		
		0.250*		
		0.281		
		0.312		20
		0.344		
		0.375	Std.	30
		0.406		
		0.438		40
		0.459		
		0.500	XS	
		0.562		
		0.594		60
		0.625		
		0.688		
		0.750		80
		0.812		
		0.938		100
		1.094		120
		1.250		140
		1.406		160
		1.500		
		2.000		
16	16.000	0.188*		
		0.203*		
		0.219*		
		0.250*		
		0.281		
		0.312		20
		0.344		
		0.375	Std.	30
		0.406		
		0.438		
		0.469		
		0.500	XS	40
		0.562		
		0.625		
		0.656		60
		0.688		
		0.750		
		0.812		
		0.844		80
		1.031		100
		1.219		120
		1.438		140
		1.594		160
		1.618		
		2.000		

NOTE: Some OD's, walls and grades are listed for information only and are not necessarily regular production items. In addition, certain walls lighter than shown above, are available as AWWA Water Pipe.

Some OD's, walls and grades are non-API.

* Not available in seamless.

Seamless, Double Submerged Arc Weld, Plain End

Size, Inches		Wall Thickness, Inches	Class	Schedule Number
Nominal Size	OD			
18	18.000	0.219*		
		0.250*		
		0.281*		
		0.312*		20
		0.344		
		0.375	Std	
		0.406		
		0.438		30
		0.469		
		0.500	XS	
		0.562		40
		0.625		
		0.688		
		0.750		60
		0.812		
		0.938		80
		1.156		100
		1.375		120
		1.500		
		1.562		140
		1.652		
20	20.000	0.219*		
		0.250*		
		0.281		
		0.312		
		0.344		
		0.375	Std.	20
		0.406		
		0.438		
		0.469		
		0.500	XS	30
		0.562		
		0.594		40
		0.625		
		0.688		
		0.750		
		0.812		60
		1.031		80
		1.281		100
		1.375		
22	22.000	0.375	Std.	20
		0.406		
		0.438		
		0.469		
		0.500	XS	30
		0.562		
		0.625		
		0.688		
		0.750		
		0.812		
		0.875		60
		1.125		80
		1.219		

Size, Inches		Wall Thickness, Inches	Class	Schedule Number
Nominal Size	OD			
24	24.000	0.250*		
		0.281		
		0.312		
		0.344		
		0.375	Std.	20
		0.406		
		0.438		
		0.469		
		0.500	XS	
		0.562		30
		0.625		
		0.688		40
		0.750		
		0.812		
		0.875		
		0.938		
		0.969		60
		1.219		90
		1.312		
26	26.000	0.250*		
		0.281*		
		0.312		
		0.344		
		0.375	Std.	
		0.406		
		0.438		
		0.469		
		0.500	XS	20
		0.562		
		0.625		
		0.656		
		0.688		
		0.750		
		0.875		
		1.188		
30	30.000	0.250		
		0.281		
		0.312		
		0.344		
		0.375		
		0.406		
		0.438		
		0.469		
		0.500		
		0.562		
		0.625		
		0.656		
		0.688		
		0.750		
		0.812		
		0.875		

NOTE: Some OD's, walls and grades are listed for information only and are not necessarily regular production items.

Some OD's, walls and grades are non-API.

*Not available in seamless.

Double Submerged Arc Weld, Plain End

Nominal Size (Size, Inches)	OD (Size, Inches)	Wall Thickness, Inches
36	36.000	0.250 0.281 0.312 0.344 0.375 0.406 0.438 0.469 0.500 0.562 0.625 0.656 0.688 0.750 0.812 0.875 0.938 1.000
40	40.000	0.312 0.344 0.375 0.406 0.438 0.469 0.500 0.562 0.625 0.688 0.750 0.812 0.875 0.938 1.000
42	42.000	0.312 0.344 0.375 0.406 0.438 0.469 0.500 0.562 0.625 0.688 0.750 0.812 0.875 0.938 1.000
44	44.000	0.344 0.375 0.406 0.438 0.469 0.500 0.562 0.625 0.688 0.750 0.812 0.875 0.938 1.000
46	46.000	0.344 0.375 0.406 0.438 0.469 0.500 0.562 0.625 0.688 0.750 0.812 0.875 0.938 1.000
48	48.000	0.375 0.406 0.438 0.469 0.500 0.562 0.625 0.688 0.750 0.812 0.875 0.938 1.000

NOTE: Some OD's, walls and grades are listed for information only and are not necessarily regular production items.

Some OD's, walls and grades are non-API.

Pipe and Pipe Fittings

Standard Weights and Dimensions of Welded and Seamless Steel Pipe
(ASA B36.10-1950)

Size (Nom. Inside Diam.) In.	Outside Diam. In.	"Standard Weight" (Standard Wall) Pipe			"Extra Strong" Pipe		"Double Extra Strong" Pipe	
		Threads per Inch	Thick-ness,[2] In.	Weight per Foot, with Couplings, Lb	Thick-ness,[2] In.	Weight per Foot, Plain Ends, Lb	Thick-ness, In.	Weight per Foot, Plain Ends, Lb
1/8	0.405	27	**0.068**	0.24	**0.095**	0.31	—	—
1/4	0.540	18	**0.088**	0.42	**0.119**	0.54	—	—
3/8	0.675	18	**0.091**	0.57	**0.126**	0.74	—	—
1/2	0.840	14	**0.109**	0.85	**0.147**	1.09	0.294	1.71
3/4	1.050	14	**0.113**	1.13	**0.154**	1.47	0.308	2.44
1	1.315	11 1/2	**0.133**	1.68	**0.179**	2.17	0.358	3.66
1 1/4	1.660	11 1/2	**0.140**	2.28	**0.191**	3.00	0.382	5.21
1 1/2	1.900	11 1/2	**0.145**	2.73	**0.200**	3.63	0.400	6.41
2	2.375	11 1/2	**0.154**	3.68	**0.218**	5.02	0.436	9.03
2 1/2	2.875	8	**0.203**	5.82	**0.276**	7.66	0.552	13.70
3	3.500	8	**0.216**	7.62	**0.300**	10.25	0.600	18.58
3 1/2	4.000	8	**0.226**	9.20	**0.318**	12.51	—	—
4	4.500	8	**0.237**	10.89	**0.337**	14.98	0.674	27.54
5	5.563	8	**0.258**	14.81	**0.375**	20.78	0.750	38.55
6	6.625	8	**0.280**	19.18	**0.432**	28.57	0.864	53.16
8	8.625	8	**0.322**	29.35	**0.500**	43.39	0.875	72.42
10	10.750	8	**0.365**	41.85	**0.500**	54.74	—	—
12	12.750	8	0.375	51.15	0.500	65.42	—	—
14	14.000	8	0.375	54.57[1]	0.500	72.09	—	—
16	16.000	8	0.375	62.58[1]	0.500	82.77	—	—
18	18.000	8	0.375	70.59[1]	0.500	93.45	—	—
20	20.000	8	0.375	78.60[1]	0.500	104.13	—	—
24	24.000	8	0.375	94.62[1]	0.500	125.49	—	—

The decimal thicknesses listed for the respective pipe sizes represent their nominal or average wall dimensions.

[1] Denotes pipe with plain ends, i.e., without thread or coupling.

[2] Thicknesses shown in **bold face** type for *Standard Wall* are identical with corresponding thicknesses shown in **bold face** type for Schedule 40 pipe. Those shown in bold face for *Extra Strong Wall* are identical with corresponding thicknesses shown in **bold face** type for Schedules 60 and 80 pipe. *Double Extra Strong Wall* has no corresponding Schedule number.

Welded Wrought Iron Pipe (ASA B36.10-1950) — The nominal sizes are the same as shown for welded and seamless steel pipe in the above table and the weights per foot also correspond to those shown therein. However, because of the difference in weight per cubic foot of steel and wrought iron, the wall thicknesses for wrought iron pipe are slightly greater than those for steel pipes of corresponding sizes, as given in the above table.

American Standard Welded and Seamless Steel Pipe (ASA B36.10-1950)

Thicknesses shown in **bold face** type in Schedule 40 are identical with thicknesses for "standard weight" pipe; those shown in **bold face** in Schedules 60 and 80 are identical with thicknesses for "extra strong" pipe.

Nom. Pipe Size, In.	Out-side Diam., In.	Nominal Wall Thickness, Inches									
		Schedule Number									
		10	20	30	40	60	80	100	120	140	160
⅛	0.405	—	—	—	**0.068**	—	**0.095**	—	—	—	—
¼	0.540	—	—	—	**0.088**	—	**0.119**	—	—	—	—
⅜	0.675	—	—	—	**0.091**	—	**0.126**	—	—	—	—
½	0.840	—	—	—	**0.109**	—	**0.147**	—	—	—	0.187
¾	1.050	—	—	—	**0.113**	—	**0.154**	—	—	—	0.218
1	1.315	—	—	—	**0.133**	—	**0.179**	—	—	—	0.250
1¼	1.660	—	—	—	**0.140**	—	**0.191**	—	—	—	0.250
1½	1.900	—	—	—	**0.145**	—	**0.200**	—	—	—	0.281
2	2.375	—	—	—	**0.154**	—	**0.218**	—	—	—	0.343
2½	2.875	—	—	—	**0.203**	—	**0.276**	—	—	—	0.375
3	3.500	—	—	—	**0.216**	—	**0.300**	—	—	—	0.438
3½	4.000	—	—	—	**0.226**	—	**0.318**	—	—	—	—
4	4.500	—	—	—	**0.237**	—	**0.337**	—	0.438	—	0.531
5	5.563	—	—	—	**0.258**	—	**0.375**	—	0.500	—	0.625
6	6.625	—	—	—	**0.280**	—	**0.432**	—	0.562	—	0.718
8	8.625	—	0.250	0.277	**0.322**	0.406	0.500	0.593	0.718	0.812	0.906
10	10.75	—	0.250	0.307	**0.365**	**0.500**	0.593	0.718	0.843	1.000	1.125
12	12.75	—	0.250	0.330	0.406	0.562	0.687	0.843	1.000	1.125	1.312
*14	14.0	0.250	0.312	0.375	0.438	0.593	0.750	0.937	1.093	1.250	1.406
*16	16.0	0.250	0.312	0.375	0.500	0.656	0.843	1.031	1.218	1.438	1.593
*18	18.0	0.250	0.312	0.438	0.562	0.750	0.937	1.156	1.375	1.562	1.781
*20	20.0	0.250	0.375	0.500	0.593	0.812	1.031	1.281	1.500	1.750	1.968
*24	24.0	0.250	0.375	0.562	0.687	0.968	1.218	1.531	1.812	2.062	2.343
*30	30.0	0.312	0.500	0.625	—	—	—	—	—	—	—

The nominal sizes of pipes larger than 12 inches are the same as the outside diameters. These large sizes commonly are designated as O.D. pipe. The inside diameter depends upon the wall thickness. The standard includes six sizes of O.D. pipe. Each of these sizes has a range of ten wall thicknesses, excepting the 30-inch size which has a range of three thicknesses. These O.D. pipes (marked * in table) all have 8 threads per inch. They will be furnished with plain ends unless otherwise specified.

The decimal thicknesses listed for the respective pipe sizes represent their nominal or average wall dimensions and include an allowance for mill tolerance of 12.5 per cent under the nominal thicknesses.

These specifications cover welded and seamless steel pipe. The "Schedule Numbers" indicate approximate values of the expression 1000 × P.S, in which P = internal pressure in pounds per square inch and S = allowable fiber stress in pounds per square inch.

Weights of Steel

THEORETICAL WEIGHTS OF ROUND SQUARE, AND HEXAGON CARBON STEEL BARS

SQUARE AND ROUND STEEL BARS

Weight of Squares

(Size in in.)^2 X 3.400 Weight, lb/ft
(Size in in.)^2 X 0.2833 Weight, lb/in.
(Size in in.)^2 X 0.00784 Weight, kilo/cm

Weight of Rounds

(Diam in in.)^2 X 2.670 Weight, lb/ft
(Diam in in.)^2 X 0.2225 Weight, lb/in.
(Diam in in.)^2 X 0.00616 Weight, kilo/cm

Thick-ness or Diameter, Inches	ROUND		SQUARE		HEXAGON	
	Weight Pounds Per Inch	Weight Pounds Per Foot	Weight Pounds Per Inch	Weight Pounds Per Foot	Weight Pounds Per Inch	Weight Pounds Per Foot
1/32	0.0002	0.0026	0.0003	0.0033	0.0002	0.0028
1/16	0.0009	0.0104	0.0011	0.0133	0.0010	0.0115
3/32	0.0020	0.0235	0.0025	0.0299	0.0022	0.0259
1/8	0.0035	0.0417	0.0044	0.0531	0.0038	0.0460
5/32	0.0054	0.0652	0.0069	0.0830	0.0060	0.0719
3/16	0.0078	0.0939	0.0100	0.1195	0.0086	0.1035
7/32	0.0106	0.1278	0.0136	0.1627	0.0117	0.1409
1/4	0.0139	0.1669	0.0177	0.2125	0.0153	0.1840
9/32	0.0176	0.2112	0.0224	0.2689	0.0194	0.2329
5/16	0.0217	0.2608	0.0277	0.3320	0.0240	0.2875
11/32	0.0263	0.3155	0.0335	0.4018	0.0290	0.3479
3/8	0.0313	0.3755	0.0398	0.4781	0.0345	0.4141
13/32	0.0367	0.4407	0.0468	0.5611	0.0405	0.4860
7/16	0.0426	0.5111	0.0542	0.6508	0.0470	0.5636
15/32	0.0489	0.5867	0.0623	0.7471	0.0538	0.6470
1/2	0.0556	0.6676	0.0708	0.8500	0.0613	0.7361
17/32	0.0628	0.7536	0.0800	0.9596	0.0693	0.8310
9/16	0.0704	0.8449	0.0896	1.076	0.0776	0.9317
19/32	0.0785	0.9414	0.0999	1.199	0.0865	1.038
5/8	0.0869	1.043	0.1107	1.328	0.0958	1.150
21/32	0.0958	1.150	0.1220	1.464	0.1057	1.268
11/16	0.1052	1.262	0.1339	1.607	0.1160	1.392
23/32	0.1150	1.380	0.1464	1.756	0.1268	1.521
3/4	0.1252	1.502	0.1594	1.913	0.1380	1.656
25/32	0.1358	1.630	0.1729	2.075	0.1498	1.797
13/16	0.1469	1.763	0.1870	2.245	0.1620	1.944
27/32	0.1584	1.901	0.2017	2.421	0.1747	2.096
7/8	0.1704	2.044	0.2169	2.603	0.1879	2.254
29/32	0.1828	2.193	0.2327	2.792	0.2015	2.418
15/16	0.1956	2.347	0.2490	2.988	0.2157	2.588
31/32	0.2088	2.506	0.2659	3.191	0.2303	2.763
1	0.2225	2.670	0.2833	3.400	0.2454	2.944
1-1/16	0.2512	3.015	0.3199	3.838	0.2770	3.324
1-1/8	0.2816	3.380	0.3586	4.303	0.3106	3.727
1-3/16	0.3138	3.766	0.3995	4.795	0.3460	4.152
1-1/4	0.3477	4.172	0.4427	5.313	0.3834	4.601
1-5/16	0.3833	4.600	0.4881	5.857	0.4227	5.072
1-3/8	0.4207	5.049	0.5357	6.428	0.4639	5.567
1-7/16	0.4598	5.518	0.5855	7.026	0.5070	6.085
1-1/2	0.5007	6.008	0.6375	7.650	0.5521	6.625
1-9/16	0.5433	6.519	0.6917	8.301	0.5991	7.189
1-5/8	0.5876	7.051	0.7482	8.978	0.6479	7.775
1-11/16	0.6337	7.604	0.8068	9.682	0.6988	8.385
1-3/4	0.6815	8.178	0.8677	10.41	0.7515	9.018
1-13/16	0.7310	8.773	0.9308	11.17	0.8060	9.67
1-7/8	0.7823	9.388	0.9961	11.95	0.8626	10.35
1-15/16	0.8354	10.02	1.064	12.76	0.9211	11.05
2	0.8901	10.68	1.133	13.60	0.9815	11.78
2-1/16	0.9466	11.36	1.205	14.46	1.044	12.53
2-1/8	1.005	12.06	1.279	15.35	1.108	13.30
2-3/16	1.065	12.78	1.356	16.27	1.174	14.09
2-1/4	1.127	13.52	1.434	17.21	1.242	14.91
2-5/16	1.190	14.28	1.515	18.18	1.312	15.75
2-3/8	1.255	15.06	1.598	19.18	1.384	16.61
2-7/16	1.322	15.87	1.683	20.20	1.458	17.49
2-1/2	1.391	16.69	1.771	21.25	1.534	18.40
2-5/8	1.533	18.40	1.952	23.43	1.691	20.29
2-3/4	1.683	20.19	2.143	25.71	1.856	22.27
2-7/8	1.839	22.07	2.342	28.10	2.028	24.34
3	2.003	24.03	2.550	30.60	2.208	26.50
3-1/8	2.173	26.08	2.767	33.20	2.396	28.75
3-1/4	2.350	28.21	2.993	35.91	2.592	31.10
3-3/8	2.535	30.42	3.227	38.73	2.795	33.54
3-1/2	2.726	32.71	3.471	41.65	3.006	36.07
3-5/8	2.924	35.09	3.723	44.68	3.224	38.69
3-3/4	3.129	37.55	3.984	47.81	3.451	41.41
3-7/8	3.341	40.10	4.254	51.05	3.684	44.21
4	3.560	42.73	4.533	54.40	3.926	47.11
4-1/3	3.786	45.44	4.821	57.85	4.175	50.10
4-1/4	4.019	48.23	5.118	61.41	4.432	53.18
4-3/8	4.259	51.11	5.423	65.08	4.700	56.36
4-1/2	4.506	54.07	5.738	68.85	4.970	59.63
4-5/8	4.760	57.12	6.061	72.73	5.248	62.98
4-3/4	5.021	60.25	6.393	76.71	5.536	66.44
4-7/8	5.289	63.46	6.734	80.80	5.831	69.98
5	5.563	66.76	7.083	85.00	6.134	73.61
5-1/8	5.845	70.14	7.442	89.30	6.445	77.34
5-1/4	6.133	73.60	7.809	93.71	6.763	81.16
5-3/8	6.429	77.15	8.186	98.23	7.089	85.07
5-1/2	6.732	80.78	8.571	102.85	7.422	89.07
5-5/8	7.041	84.49	8.965	107.58	7.763	93.16
5-3/4	7.357	88.29	9.368	112.41	8.112	97.35
5-7/8	7.681	92.17	9.779	117.35	8.470	101.63
6	8.011	96.13	10.200	122.40	8.833	106.00

From the Steel Products Manual, American Iron and Steel Institute

Weights of Flat Steel Bars
Pounds per Linear Foot

Thickness in.	Width, In. 1	2	3	4	5	12
1/32	0.1063	0.213	0.319	0.425	0.531	1.275
1/16	0.2125	0.425	0.638	0.850	1.063	2.550
3/32	0.3188	0.638	0.956	1.275	1.594	3.825
1/8	0.4250	0.850	1.275	1.700	2.125	5.100
5/32	0.5313	1.063	1.594	2.125	2.656	6.375
3/16	0.6375	1.275	1.913	2.550	3.188	7.650
7/32	0.7438	1.488	2.231	2.975	3.719	8.925
1/4	0.8500	1.700	2.550	3.400	4.250	10.200
5/16	1.0625	2.125	3.188	4.250	5.313	12.750
3/8	1.2750	2.550	3.825	5.100	6.375	15.300
7/16	1.4875	2.975	4.463	5.950	7.438	17.850
1/2	1.7000	3.400	5.100	6.800	8.500	20.400
9/16	1.9125	3.825	5.738	7.650	9.563	22.950
5/8	2.1250	4.250	6.375	8.500	10.625	25.500
11/16	2.3375	4.675	7.013	9.350	11.688	28.050
3/4	2.5500	5.100	7.650	10.200	12.750	30.600
13/16	2.7625	5.525	8.288	11.050	13.813	33.150
7/8	2.9750	5.950	8.925	11.900	14.875	35.700
15/16	3.1875	6.375	9.563	12.750	15.938	38.250
1	3.4000	6.800	10.200	13.600	17.000	40.800
1-1/8	3.8250	7.650	11.475	15.300	19.125	45.900
1-1/4	4.2500	8.500	12.750	17.000	21.250	51.000
1-3/8	4.6750	9.350	14.025	18.700	23.375	56.000
1-1/2	5.1000	10.200	15.300	20.400	25.500	61.200
1-5/8	5.5250	11.050	16.575	22.100	27.625	66.300
1-3/4	5.9500	11.900	17.850	23.800	29.750	71.400
1-7/8	6.3750	12.750	19.125	25.500	31.875	76.500
2	6.8000	13.600	20.400	27.200	34.000	81.600

Manufacturers' Standard Gage For Steel Sheets

Manufacturers' Standard Gage No.	Ounces per Square Foot	Pounds per Square Inch	Pound per Square Foot	Inch Equivalent for Steel Sheet Thickness
3	160	0.069444	10.0000	0.2391
4	150	0.065104	9.3750	0.2242
5	140	0.060764	8.7500	0.2092
6	130	0.056424	8.1250	0.1943
7	120	0.052083	7.5000	0.1793
8	110	0.047743	6.8750	0.1644
9	100	0.043403	6.2500	0.1495
10	90	0.039062	5.6250	0.1345
11	80	0.034722	5.0000	0.1196
12	70	0.030382	4.3750	0.1046
13	60	0.026042	3.7500	0.0897
14	50	0.021701	3.1250	0.0747
15	45	0.019531	2.8125	0.0673
16	40	0.017361	2.5000	0.0598
17	36	0.015625	2.2500	0.0538
18	32	0.013889	2.0000	0.0478
19	28	0.012153	1.7500	0.0418
20	24	0.010417	1.5000	0.0359
21	22	0.0095486	1.3750	0.0329
22	20	0.0086806	1.2500	0.0299
23	18	0.0078125	1.1250	0.0269
24	16	0.0069444	1.0000	0.0239
25	14	0.0060764	0.87500	0.0209
26	12	0.0052083	0.75000	0.0179
27	11	0.0047743	0.68750	0.0164
28	10	0.0043403	0.62500	0.0149
29	9	0.0039062	0.56250	0.0135
30	8	0.0034722	0.50000	0.0120
31	7	0.0030382	0.43750	0.0105
32	6.5	0.0028212	0.40625	0.0097
33	6	0.0026042	0.37500	0.0090
34	5.5	0.0023872	0.34375	0.0082
35	5	0.0021701	0.31250	0.0075
36	4.5	0.0019531	0.28125	0.0067
37	4.25	0.0018446	0.26562	0.0064
38	4	0.0017361	0.25000	0.0060

From ASTM, Part 3, A568

Table shows the upper and lower limits for unit weight and thickness for each number of the manufacturers standard gage table. *These ranges are not tolerances.*

Sheets specified to a manufacturer's standard gage number are produced to the inch equivalent for that gage number as shown in the preceding table.

Weight and Thickness Ranges
For Manufacturers' Standard Gage Numbers
For Steel Sheets

Manufacturers' Standard Gage No.	Weight Range, lb/ft^2	Thickness Range, in.
3	10.312 to 9.688	0.2465 to 0.2317
4	9.687 to 9.063	0.2316 to 0.2168
5	9.062 to 8.438	0.2167 to 0.2018
6	8.437 to 7.813	0.2017 to 0.1869
7	7.812 to 7.188	0.1868 to 0.1719
8	7.187 to 6.563	0.1718 to 0.1570
9	6.562 to 5.938	0.1569 to 0.1420
10	5.937 to 5.313	0.1419 to 0.1271
11	5.312 to 4.688	0.1270 to 0.1121
12	4.687 to 4.063	0.1120 to 0.0972
13	4.062 to 3.438	0.0971 to 0.0822
14	3.437 to 2.969	0.0821 to 0.0710
15	2.968 to 2.657	0.0709 to 0.0636
16	2.656 to 2.375	0.0635 to 0.0568
17	2.374 to 2.125	0.0567 to 0.0509
18	2.124 to 1.875	0.0508 to 0.0449
19	1.874 to 1.625	0.0448 to 0.0389
20	1.624 to 1.438	0.0388 to 0.0344
21	1.437 to 1.313	0.0343 to 0.0314
22	1.312 to 1.188	0.0313 to 0.0284
23	1.187 to 1.063	0.0283 to 0.0255
24	1.062 to 0.938	0.0254 to 0.0225
25	0.937 to 0.813	0.0224 to 0.0195
26	0.812 to 0.719	0.0194 to 0.0172
27	0.718 to 0.657	0.0171 to 0.0157
28	0.656 to 0.594	0.0156 to 0.0142
29	0.593 to 0.532	0.0141 to 0.0128
30	0.531 to 0.469	0.0127 to 0.0113
31	0.468 to 0.422	0.0112 to 0.0101
32	0.421 to 0.391	0.0100 to 0.0094
33	0.390 to 0.360	0.0093 to 0.0086
34	0.359 to 0.329	0.0085 to 0.0079
35	0.328 to 0.297	0.0078 to 0.0071
36	0.296 to 0.274	0.0070 to 0.0066
37	0.273 to 0.258	0.0065 to 0.0062
38	0.257 to 0.243	0.0061 to 0.0058

APPROXIMATE THEORETICAL WEIGHTS

Alloy Steel Sheets

Wt. per Sq. Ft. in Lbs.	Approx. Dec. of Inch	USS Gage
7.0813	.17187	8
6.4375	.15625	9
5.7937	.14062	10
5.1500	.125	11
4.5063	.10937	12
3.8625	.09375	13
3.2187	.07812	14
2.8968	.07031	15
2.5750	.0625	16
2.3175	.05625	17
2.0600	.050	18
1.8025	.04375	19
1.5450	.0375	20
1.4160	.03437	21
1.2875	.03125	22
1.1587	.02812	23
1.0300	.025	24
.9013	.02187	25
.7725	.01875	26
.7081	.01718	27
.6438	.01562	28
.5794	.01406	29
.5150	.0125	30
.4506	.01093	31

Alloy Steel Strip

No. BWG	Dec. Equiv. in Inches	Wt. per Sq. Ft. Chr. Iron Alloys
7	.180	7.434
8	.165	6.814
9	.148	6.112
10	.134	5.534
11	.120	4.956
12	.109	4.501
13	.095	3.923
14	.083	3.427
15	.072	2.973
16	.065	2.684
17	.058	2.395
18	.049	2.023
19	.042	1.734
20	.035	1.445
21	.032	1.321
22	.028	1.156
23	.025	1.032
24	.022	0.908
25	.020	0.826
26	.018	0.743
27	.016	0.660
28	.014	0.578
29	.013	0.536
30	.012	0.495
31	.010	0.413
32	.009	0.371
33	.008	0.330
34	.007	0.289
35	.005	0.206
36	.004	0.165

Alloy Steel Plates

No. of Gage	Approx. Dec. Parts of an Inch	Approx. Fractions of an Inch	Approx. Wt. per Sq. Ft. in lbs.
	1.0000	1	40.478
	.9375	15/16	37.949
	.8750	7/8	35.419
	.8125	13/16	32.889
	.7500	3/4	30.359
	.6875	11/16	27.829
	.6250	5/8	25.299
	.5625	9/16	22.769
	.5000	1/2	20.239
	.46875	15/32	18.974
	.4375	7/16	17.709
	.40625	13/32	16.444
	.3750	3/8	15.179
	.34375	11/32	13.914
0	.3125	5/16	12.650
1	.28125	9/32	11.385
2	.26562	17/64	10.752
3	.2500	1/4	10.120
4	.234375	15/64	9.487
5	.21875	7/32	8.855
6	.203125	13/64	8.222
7	.1875	3/16	7.590

From Republic Alloy Steels, Republic Steel Corp.

Hardness and Temperature Conversion Tables

Hardness Conversion Numbers for Steel (ASTM E140)

Diamond Pyramid Hardness Number	Brinell Hardness Number*			Rockwell Hardness Number			Rockwell Superficial Hardness Number			Diamond Pyramid Hardness Number
	10-mm Standard Ball, 3000-kg Load	10-mm Hultgren Ball, 3000-kg Load	10-mm Carbide Ball, 3000-kg Load	C Scale, 150-kg Load, Diamond Cone Penetrator	A Scale, 60-kg Load, Diamond Cone Penetrator	D Scale, 100-kg Load, Diamond Cone Penetrator	15-N Scale, 15-kg Load Superficial Diamond Cone Penetrator	30-N Scale, 30-kg Load, Superficial Diamond Cone Penetrator	45-N Scale, 45-kg Load, Superficial Diamond Cone Penetrator	
940	..	..	..	68.0	85.6	76.9	93.2	84.4	75.4	940
920	..	..	..	67.5	85.3	76.5	93.0	84.0	74.8	920
900	..	..	..	67.0	85.0	76.1	92.9	83.6	74.2	900
880	..	..	..	66.4	84.7	75.7	92.7	83.1	73.6	880
860	..	..	..	65.9	84.4	75.3	92.5	82.7	73.1	860
840	..	..	**745**	65.3	84.1	74.8	92.3	82.2	72.2	840
820	..	..	**733**	64.7	83.8	74.3	92.1	81.7	71.8	820
800	..	..	**722**	64.0	83.4	73.8	91.8	81.1	71.0	800
780	..	..	**710**	63.3	83.0	73.3	91.5	80.4	70.2	780
760	..	..	**698**	62.5	82.6	72.6	91.2	79.7	69.4	760
740	..	..	**684**	61.8	82.2	72.1	91.0	79.1	68.6	740
720	..	..	**670**	61.0	81.8	71.5	90.7	78.4	67.7	720
700	..	**615**	**656**	60.1	81.3	70.8	90.3	77.6	66.7	700
690	..	**610**	**647**	59.7	81.1	70.5	90.1	77.2	66.2	690
680	..	**603**	**638**	59.2	80.8	70.1	89.8	76.8	65.7	680
670	..	**597**	630	58.8	80.6	69.8	89.7	76.4	65.3	670
660	..	**590**	620	58.3	80.3	69.4	89.5	75.9	64.7	660
650	..	**585**	611	57.8	80.0	69.0	89.2	75.5	64.1	650
640	..	**578**	601	57.3	79.8	68.7	89.0	75.1	63.5	640
630	..	**571**	591	56.8	79.5	68.3	88.8	74.6	63.0	630
620	..	**564**	582	56.3	79.2	67.9	88.5	74.2	62.4	620
610	..	**557**	573	55.7	78.9	67.5	88.2	73.6	61.7	610
600	..	**550**	564	55.2	78.6	67.0	88.0	73.2	61.2	600
590	..	**542**	554	54.7	78.4	66.7	87.8	72.7	60.5	590
580	..	**535**	545	54.1	78.0	66.2	87.5	72.1	59.9	580
570	..	**527**	535	53.6	77.8	65.8	87.2	71.7	59.3	570
560	..	**519**	525	53.0	77.4	65.4	86.9	71.2	58.6	560
550	**505**	512	517	52.3	77.0	64.8	86.6	70.5	57.8	550
540	**496**	503	507	51.7	76.7	64.4	86.3	70.0	57.0	540
530	**488**	495	497	51.1	76.4	63.9	86.0	69.5	56.2	530
520	**480**	487	488	50.5	76.1	63.5	85.7	69.0	55.6	520
510	**473**	479	479	49.8	75.7	62.9	85.4	68.3	54.7	510
500	**465**	471	471	49.1	75.3	62.2	85.0	67.7	53.9	500
490	**456**	460	460	48.4	74.9	61.6	84.7	67.1	53.1	490
480	448	452	452	47.7	74.5	61.3	84.3	66.4	52.2	480
470	441	442	442	46.9	74.1	60.7	83.9	65.7	51.3	470
460	433	433	433	46.1	73.6	60.1	83.6	64.9	50.4	460
450	425	425	425	45.3	73.3	59.4	83.2	64.3	49.4	450
440	415	415	415	44.5	72.8	58.8	82.8	63.5	48.4	440
430	405	405	405	43.6	72.3	58.2	82.3	62.7	47.4	430
420	397	397	397	42.7	71.8	57.5	81.8	61.9	46.4	420
410	388	388	388	41.8	71.4	56.8	81.4	61.1	45.3	410
400	379	379	379	40.8	70.8	56.0	80.8	60.2	44.1	400
390	369	369	369	39.8	70.3	55.2	80.3	59.3	42.9	390
380	360	360	360	38.8	69.8	54.4	79.8	58.4	41.7	380
370	350	350	350	37.7	69.2	53.6	79.2	57.4	40.4	370
360	341	341	341	36.6	68.7	52.8	78.6	56.4	39.1	360
350	331	331	331	35.5	68.1	51.9	78.0	55.4	37.8	350
340	322	322	322	34.4	67.6	51.1	77.4	54.4	36.5	340
330	313	313	313	33.3	67.0	50.2	76.8	53.6	35.2	330
320	303	303	303	32.2	66.4	49.4	76.2	52.3	33.9	320
310	294	294	294	31.0	65.8	48.4	75.6	51.3	32.5	310
300	284	284	284	29.8	65.2	47.5	74.9	50.2	31.1	300
295	280	280	280	29.2	64.8	47.1	74.6	49.7	30.4	295
290	275	275	275	28.5	64.5	46.5	74.2	49.0	29.5	290
285	270	270	270	27.8	64.2	46.0	73.8	48.4	28.7	285
280	265	265	265	27.1	63.8	45.3	73.4	47.8	27.9	280
275	261	261	261	26.4	63.5	44.9	73.0	47.2	27.1	275
270	256	256	256	25.6	63.1	44.3	72.6	46.4	26.2	270
265	252	252	252	24.8	62.7	43.7	72.1	45.7	25.2	265
260	247	247	247	24.0	62.4	43.1	71.6	45.0	24.3	260
255	243	243	243	23.1	62.0	42.2	71.1	44.2	23.2	255
250	238	238	238	22.2	61.6	41.7	70.6	43.4	22.2	250
245	233	233	233	21.3	61.2	41.1	70.1	42.5	21.1	245
240	228	228	228	20.3	60.7	40.3	69.6	41.7	19.9	240

*The Brinell hardness numbers in boldface type are outside the range recommended for Brinell hardness testing in Section 5(c) of the Method of Test for Brinell Hardness of Metallic Materials (ASTM E10).

Hardness Conversion Numbers for Steel (ASTM E140)

Rockwell C Hardness Number	Diamond Pyramid Hardness Number	Brinell Hardness Number*			Rockwell Hardness Number		Rockwell Superficial Hardness Number			Rockwell C Hardness Number
		10-mm Standard Ball, 3000-kg Load	10-mm Hultgren Ball, 3000-kg Load	10-mm Carbide Ball, 3000-kg Load	A Scale, 60-kg Load, Diamond Cone Pene-trator	D Scale, 100-kg Load, Diamond Cone Pene-trator	15-N Scale, 15-kg Load, Superfi-cial Diamond Cone Pene-trator	30-N Scale, 30-kg Load Superfi-cial Diamond Cone Pene-trator	45-N Scale, 45-kg Load Superfi-cial Diamond Cone Pene-trator	
68	940	..	..	..	85.6	76.9	93.2	84.4	75.4	68
67	900	..	..	..	85.0	76.1	92.9	83.6	74.2	67
66	865	..	..	..	84.5	75.4	92.5	82.8	73.3	66
65	832	..	..	**739**	83.9	74.5	92.2	81.9	72.0	65
64	800	..	..	**722**	83.4	73.8	91.8	81.1	71.0	64
63	772	..	..	**705**	82.8	73.0	91.4	80.1	69.9	63
62	746	..	..	**688**	82.3	72.2	91.1	79.3	68.8	62
61	720	..	..	**670**	81.8	71.5	90.7	78.4	67.7	61
60	697	..	**613**	**654**	81.2	70.7	90.2	77.5	66.6	60
59	674	..	**599**	**634**	80.7	69.9	89.8	76.6	65.5	59
58	653	..	**587**	615	80.1	69.2	89.3	75.7	64.3	58
57	633	..	**575**	595	79.6	68.5	88.9	74.8	63.2	57
56	613	..	**561**	577	79.0	67.7	88.3	73.9	62.0	56
55	595	..	**546**	560	78.5	66.9	87.9	73.0	60.9	55
54	577	..	**534**	543	78.0	66.1	87.4	72.0	59.8	54
53	560	..	**519**	525	77.4	65.4	86.9	71.2	58.6	53
52	544	**500**	**508**	512	76.8	64.6	86.4	70.2	57.4	52
51	528	**487**	494	496	76.3	63.8	85.9	69.4	56.1	51
50	513	**475**	481	481	75.9	63.1	85.5	68.5	55.0	50
49	498	**464**	469	469	75.2	62.1	85.0	67.6	53.8	49
48	484	451	455	455	74.7	61.4	84.5	66.7	52.5	48
47	471	442	443	443	74.1	60.8	83.9	65.8	51.4	47
46	458	432	432	432	73.6	60.0	83.5	64.8	50.3	46
45	446	421	421	421	73.1	59.2	83.0	64.0	49.0	45
44	434	409	409	409	72.5	58.5	82.5	63.1	47.8	44
43	423	400	400	400	72.0	57.7	82.0	62.2	46.7	43
42	412	390	390	390	71.5	56.9	81.5	61.3	45.5	42
41	402	381	381	381	70.9	56.2	80.9	60.4	44.3	41
40	392	371	371	371	70.4	55.4	80.4	59.5	43.1	40
39	382	362	362	362	69.9	54.6	79.9	58.6	41.9	39
38	372	353	353	353	69.4	53.8	79.4	57.7	40.8	38
37	363	344	344	344	68.9	53.1	78.8	56.8	39.6	37
36	354	336	336	336	68.4	52.3	78.3	55.9	38.4	36
35	345	327	327	327	67.9	51.5	77.7	55.0	37.2	35
34	336	319	319	319	67.4	50.8	77.2	54.2	36.1	34
33	327	311	311	311	66.8	50.0	76.6	53.3	34.9	33
32	318	301	301	301	66.3	49.2	76.1	52.1	33.7	32
31	310	294	294	294	65.8	48.4	75.6	51.3	32.5	31
30	302	286	286	286	65.3	47.7	75.0	50.4	31.3	30
29	294	279	279	279	64.7	47.0	74.5	49.5	30.1	29
28	286	271	271	271	64.3	46.1	73.9	48.6	28.9	28
27	279	264	264	264	63.8	45.2	73.3	47.7	27.8	27
26	272	258	258	258	63.3	44.6	72.8	46.8	26.7	26
25	266	253	253	253	62.8	43.8	72.2	45.9	25.5	25
24	260	247	247	247	62.4	43.1	71.6	45.0	24.3	24
23	254	243	243	243	62.0	42.1	71.0	44.0	23.1	23
22	248	237	237	237	61.5	41.6	70.5	43.2	22.0	22
21	243	231	231	231	61.0	40.9	69.9	42.3	20.7	21
20	238	226	226	226	60.5	40.1	69.4	41.5	19.6	20

*The Brinell hardness numbers in boldface type are outside the range recommended for Brinell hardness testing in Section 5(c) of the Method of Test for Brinell Hardness of Metallic Materials (ASTM Designation: E10).

Hardness Conversion Numbers for Steel (ASTM E140)

Brinell Indentation, Diameter, mm	Brinell Hardness Number*: 10-mm Standard Ball, 3000-kg Load	Brinell Hardness Number*: 10-mm Hultgren Ball, 3000-kg Load[a]	Brinell Hardness Number*: 10-mm Carbide Ball, 3000-kg Load	Diamond Pyramid Hardness Number	Rockwell Hardness Number: C Scale, 150-kg Load, Diamond Cone Penetrator	Rockwell Hardness Number: A Scale, 60-kg Load, Diamond Cone Penetrator	Rockwell Hardness Number: D Scale, 100-kg Load, Diamond Cone Penetrator	Rockwell Superfical Hardness Number: 15-N Scale, 15-kg Load, Superficial Diamond Cone Penetrator	Rockwell Superfical Hardness Number: 30-N Scale, 30-kg Load, Superficial Diamond Cone Penetrator	Rockwell Superfical Hardness Number: 45-N Scale, 45-kg Load Superficial Diamond Cone Penetrator	Brinell Indentation, Diameter, mm
2.35	..	..	**682**	737	61.7	82.2	72.0	91.0	79.0	68.5	2.35
2.40	..	..	**653**	697	60.0	81.2	70.7	90.2	77.5	66.5	2.40
2.45	..	..	627	667	58.7	80.5	69.7	89.6	76.3	65.1	2.45
	..	..	..		..	..	..	..	..	..	
2.50	..	**601**	..	677	59.1	80.7	70.0	89.8	76.8	65.7	2.50
	..	..	601	640	57.3	79.8	68.7	89.0	75.1	63.5	
	..	..	..		..	..	..	..	..	..	
2.55	..	**578**	..	640	57.3	79.8	68.7	89.0	75.1	63.5	2.55
	..	..	578	615	56.0	79.1	67.7	88.4	73.9	62.1	
	..	..	..		..	..	..	..	..	..	
2.60	..	**555**	..	607	55.6	78.8	67.4	88.1	73.5	61.6	2.60
	..	..	555	591	54.7	78.4	66.7	87.8	72.7	60.6	
	..	..	..		..	..	..	..	..	..	
2.65	..	**534**	..	579	54.0	78.0	66.1	87.5	72.0	59.8	2.65
	..	..	534	569	53.5	77.8	65.8	87.2	71.6	59.2	
	..	..	..		..	..	..	..	..	..	
2.70	..	**514**	..	553	52.5	77.1	65.0	86.7	70.7	58.0	2.70
	..	..	514	547	52.1	76.9	64.7	86.5	70.3	57.6	
	495	..	..	539	51.6	76.7	64.3	86.3	69.9	56.9	
2.75	..	495	..	530	51.1	76.4	63.9	86.0	69.5	56.2	2.75
	..	..	495	528	51.0	76.3	63.8	85.9	69.4	56.1	
	477	..	..	516	50.3	75.9	63.2	85.6	68.7	55.2	
2.80	..	477	..	508	49.6	75.6	62.7	85.3	68.2	54.5	2.80
	..	..	477	508	49.6	75.6	62.7	85.3	68.2	54.5	
	461	..	..	495	48.8	75.1	61.9	84.9	67.4	53.5	
2.85	..	461	..	491	48.5	74.9	61.7	84.7	67.2	53.2	2.85
	..	..	461	491	48.5	74.9	61.7	84.7	67.2	53.2	
	444	..	..	474	47.2	74.3	61.0	84.1	66.0	51.7	
2.90	..	444	..	472	47.1	74.2	60.8	84.0	65.8	51.5	2.90
	..	..	**444**	472	47.1	74.2	60.8	84.0	65.8	51.5	
2.95	429	429	429	455	45.7	73.4	59.7	83.4	64.6	49.9	2.95
3.00	415	415	415	440	44.5	72.8	58.8	82.8	63.5	48.4	3.00
3.05	401	401	401	425	43.1	72.0	57.8	82.0	62.3	46.9	3.05
3.10	388	388	388	410	41.8	71.4	56.8	81.4	61.1	45.3	3.10
3.15	375	375	375	396	40.4	70.6	55.7	80.6	59.9	43.6	3.15
3.20	363	363	363	383	39.1	70.0	54.6	80.0	58.7	42.0	3.20
3.25	352	352	352	372	37.9	69.3	53.8	79.3	57.6	40.5	3.25
3.30	341	341	341	360	36.6	68.7	52.8	78.6	56.4	39.1	3.30
3.35	331	331	331	350	35.5	68.1	51.9	78.0	55.4	37.8	3.35
3.40	321	321	321	339	34.3	67.5	51.0	77.3	54.3	36.4	3.40
3.45	311	311	311	328	33.1	66.9	50.0	76.7	53.3	34.4	3.45
3.50	302	302	302	319	32.1	66.3	49.3	76.1	52.2	33.8	3.50
3.55	293	293	293	309	30.9	65.7	48.3	75.5	51.2	32.4	3.55
3.60	285	285	285	301	29.9	65.3	47.6	75.0	50.3	31.2	3.60
3.65	277	277	277	292	28.8	64.6	46.7	74.4	49.3	29.9	3.65
3.70	269	269	269	284	27.6	64.1	45.9	73.7	48.3	28.5	3.70
3.75	262	262	262	276	26.6	63.6	45.0	73.1	47.3	27.3	3.75
3.80	255	255	255	269	25.4	63.0	44.2	72.5	46.2	26.0	3.80
3.85	248	248	248	261	24.2	62.5	43.2	71.7	45.1	24.5	3.85
3.90	241	241	241	253	22.8	61.8	42.0	70.9	43.9	22.8	3.90
3.95	235	235	235	247	21.7	61.4	41.4	70.3	42.9	21.5	3.95
4.00	229	229	229	241	20.5	60.8	40.5	69.7	41.9	20.1	4.00

*The Brinell hardness numbers in boldface type are outside the range recommended for Brinell hardness testing in Section 5(c) of the Method of Test for Brinell Hardness of Metallic Materials (ASTM Designation: E10).

[a] Brinell hardness values are given for the Hultgren ball to 601 Brinell hardness; however, flattening of the ball at the higher hardnesses gives lower Brinell hardness values than obtained with the carbide ball. For example, material having a DPH value of 667 has a carbide ball Brinell hardness of 627, while material having a DPH value of 667 has a Hultgren ball Brinell hardness of only 601.

Hardness Conversion Numbers for Nickel and High Nickel Alloys (ASTM E140)

Diamond Pyramid Hardness Number, DPH	Brinell Hardness Number, BHN	Rockwell Hardness Number								Rockwell Superficial Hardness Number					
Diamond Pyramid Penetrator 1, 5, 10, 30-kg Load	10-mm Standard Ball, 3000-kg Load[a]	A Scale 60-kg Load, Diamond Cone Penetrator	B Scale 100-kg Load, 1/16-in. Ball	C Scale 150-kg Load, Diamond Cone Penetrator	D Scale 100-kg Load, Diamond Cone Penetrator	E Scale 100-kg Load, 1/8-in. Ball	F Scale 60-kg Load, 1/16-in. Ball	G Scale 150-kg Load, 1/16-in. Ball	K Scale 150-kg Load, 1/8-in. Ball	15-N Scale 15-kg Load, Superficial Diamond Cone Penetrator	30-N Scale 30-kg Load, Superficial Diamond Cone Penetrator	45-N Scale 45-kg Load, Superficial Diamond Cone Penetrator	15-T Scale 15-kg Load, 1/16-in. Ball	30-T Scale 30-kg Load, 1/16-in. Ball	45-T Scale 45-kg Load, 1/16-in. Ball
513	**479**	75.5	..	**50.0**	**63.0**	..	..	..	..	**85.5**	**68.0**	54.5	..	..	..
481	**450**	74.5	..	**48.0**	61.5	..	..	..	..	**84.5**	**66.5**	52.5	..	..	..
452	**425**	73.5	..	**46.0**	**60.0**	..	..	..	..	**83.5**	**64.5**	50.0	..	..	..
427	403	72.5	..	**44.0**	58.5	..	..	..	..	**82.5**	**63.0**	47.5	..	..	..
404	382	71.5	..	**42.0**	57.0	..	..	..	..	**81.5**	**61.0**	45.5	..	..	..
382	363	70.5	..	40.0	55.5	..	..	..	..	80.5	59.5	43.0	..	..	..
362	346	69.5	..	38.0	54.0	..	..	..	..	79.5	58.0	41.0	..	..	..
344	329	68.5	..	36.0	52.5	..	..	..	..	78.5	56.0	38.5	..	..	..
326	313	67.5	..	34.0	50.5	..	..	..	..	77.5	54.5	36.0	..	..	..
309	298	66.5	**106**	32.0	49.5	..	**116.5**	94.0	..	76.5	52.5	34.0	94.5	85.5	77.0
285	275	64.5	**104**	28.5	46.5	..	**115.5**	91.0	..	75.0	49.5	30.0	94.0	84.5	75.0
266	258	63.0	**102**	25.5	44.5	..	**114.5**	87.5	..	73.5	47.0	26.5	93.0	83.0	73.0
248	241	61.5	100	22.5	42.0	..	**113.0**	84.5	..	72.0	44.5	23.0	92.5	81.5	71.0
234	228	60.5	98	20.0	40.0	..	**112.0**	81.5	..	70.5	42.0	20.0	92.0	80.5	69.0
220	215	59.0	96	**17.0**	38.0	..	**111.0**	78.5	100.0	69.0	39.5	17.0	91.0	79.0	67.0
209	204	57.5	94	**14.5**	36.0	..	**110.0**	75.5	98.0	68.0	37.5	14.0	90.5	77.5	65.0
198	194	56.5	92	**12.0**	34.0	..	**108.5**	72.0	96.5	66.5	35.5	11.0	89.5	76.0	63.0
188	184	55.0	90	**9.0**	32.0	**108.5**	**107.5**	69.0	94.5	65.0	32.5	7.5	89.0	75.0	61.0
179	176	53.5	88	**6.5**	30.0	**107.0**	**106.5**	65.5	93.0	64.0	30.5	5.0	88.0	73.5	59.5
171	168	52.5	86	**4.0**	28.0	**106.0**	**105.0**	62.5	91.0	62.5	28.5	2.0	87.5	72.0	57.5
164	161	51.5	84	**2.0**	26.5	**104.5**	**104.0**	59.5	89.0	61.5	26.5	**−0.5**	87.0	70.5	55.5
157	155	50.0	82	..	24.5	**103.0**	**103.0**	56.5	87.5	..	..	..	86.0	69.5	53.5
151	149	49.0	80	..	22.5	**102.0**	**101.5**	53.0	85.5	..	..	..	85.5	68.0	51.5
145	144	47.5	78	..	21.0	**100.5**	**100.5**	50.0	83.5	..	..	..	84.5	66.5	49.5
140	139	46.5	76	..	**19.0**	99.5	99.5	47.0	82.0	..	..	..	84.0	65.5	47.5
135	134	45.5	74	..	**17.5**	98.0	98.5	43.5	80.0	..	..	..	83.0	64.0	45.5
130	129	44.0	72	..	**16.0**	97.0	97.0	40.5	78.0	..	..	..	82.5	62.5	43.5
126	125	43.0	70	..	**14.5**	95.5	96.0	37.5	76.5	..	..	..	82.0	61.0	41.5
122	121	42.0	68	..	**13.0**	94.5	95.0	34.5	74.5	..	..	..	81.0	60.0	39.5
119	118	41.0	66	..	**11.5**	93.0	93.5	31.0	72.5	..	..	..	80.5	58.5	37.5
115	114	40.0	64	..	**10.0**	91.5	92.5	..	71.0	..	..	..	79.5	57.0	35.5
112	111	39.0	62	..	**8.0**	90.5	91.5	..	69.0	..	..	..	79.0	56.0	33.5
108	108	..	60	..	..	89.0	90.0	..	67.5	..	..	..	78.5	54.5	31.5
106	106	..	58	..	..	88.0	89.0	..	65.5	..	..	..	77.5	53.0	29.5
103	103	..	56	..	..	86.5	88.0	..	63.5	..	..	..	77.0	51.5	27.5
100	100	..	54	..	..	85.5	87.0	..	62.0	..	..	..	76.0	50.5	25.5
98	98	..	52	..	..	84.0	85.5	..	60.0	..	..	..	75.5	49.0	23.5
95	95	..	50	..	..	83.0	84.5	..	58.0	..	..	..	74.5	47.5	21.5
93	93	..	48	..	..	81.5	83.5	..	56.5	..	..	..	74.0	46.5	19.5
91	91	..	46	..	..	80.5	82.0	..	54.5	..	..	..	73.5	45.0	17.0
89	89	..	44	..	..	79.0	81.0	..	52.5	..	..	..	72.5	43.5	14.5
87	87	..	42	..	..	78.0	80.0	..	51.0	..	..	..	72.0	42.0	12.5
85	85	..	40	..	..	76.5	79.0	..	49.0	..	..	..	71.0	41.0	10.0
83	83	..	38	..	..	75.0	77.5	..	47.0	..	..	..	70.5	39.5	7.5
81	81	..	36	..	..	74.0	76.5	..	45.5	..	..	..	70.0	38.0	5.5
79	79	..	34	..	..	72.5	75.5	..	43.5	..	..	..	69.0	36.5	3.0
78	78	..	32	..	..	71.5	74.0	..	42.0	..	..	..	68.5	35.5	1.0
77	77	..	30	..	..	70.0	73.0	..	40.0	..	..	..	67.5	34.0	**−1.5**

Note — The use of hardness scales for hardness values shown above in boldface type are not recommended since they are beyond the ranges recommended for accuracy. Such values are shown for comparative purposes only, where comparisons may be desired and the recommended machine and scale are not available.

[a]Note that in Section 5(a) of the Method of Test for Brinell Hardness of Metallic Materials (ASTM Designation: E10), the use of a 3000-kg load is recommended (but not mandatory) for material in the hardness range 96 to 600 BHN, and a 1500-kg load is recommended (but not mandatory) for material in the hardness range 48 to 300 BHN. These recommendations are designed to limit impression diameters to the range 2.50 to 6.0 mm. The Brinell hardness numbers in this conversion table are based on tests using a 3000-kg load. When the 1500-kg load is used for the softer nickel and high-nickel alloys these conversion relationships do not apply.

Hardness Conversion Numbers for Cartridge Brass (70 Percent Copper, 30 Percent Zinc Alloy). ASTM E140

Diamond Pyramid Hardness Number	Rockwell Hardness Number: B Scale, 100-kg Load, 1/16-in. Ball	Rockwell Hardness Number: F Scale 60-kg Load, 1/16-in. Ball	Rockwell Superficial Hardness Number: 15-T Scale, 15-kg Load, 1/16-in. Ball	Rockwell Superficial Hardness Number: 30-T Scale, 30-kg Load, 1/16-in. Ball	Rockwell Superficial Hardness Number: 45-T Scale, 45-kg Load, 1/16-in. Ball	Brinell Hardness Number: 500-kg Load, 10-mm Ball	Diamond Pyramid Hardness Number	Rockwell Hardness Number: B Scale, 100-kg Load, 1/16-in. Ball	Rockwell Hardness Number: F Scale, 60-kg Load, 1/16-in. Ball	Rockwell Superficial Hardness Number: 15-T Scale, 15 kg Load, 1/16-in. Ball	Rockwell Superficial Hardness Number: 30-T Scale, 30-kg Load, 1/16-in. Ball	Rockwell Superficial Hardness Number: 45-T Scale, 45-kg Load, 1/16-in. Ball	Brinell Hardness Number: 500-kg Load, 10-mm Ball
45	..	40.0	..	..	..	42	120	67.0	95.5	..	61.0	41.0	106
46	..	43.0	..	..	..	43	122	68.0	96.0	83.0	62.0	42.0	108
47	..	45.0	..	..	..	44	124	69.0	96.5	..	62.5	43.0	110
48	..	47.0	53.5	..	..	45	126	70.0	97.0	83.5	63.0	44.0	112
49	..	49.0	54.5	..	..	46	128	71.0	97.5	..	63.5	45.0	113
50	..	50.5	55.5	..	..	47	130	72.0	98.0	84.0	64.5	45.5	114
52	..	53.5	57.0	..	..	48	132	73.0	98.5	84.5	65.0	46.5	116
54	..	56.5	58.5	12.0	..	50	134	73.5	99.0	..	65.5	47.5	118
56	..	58.8	60.0	15.0	..	52	136	74.5	99.5	85.0	66.0	48.0	120
58	..	61.0	61.0	18.0	..	53	138	75.0	100.0	..	66.5	49.0	121
60	10.0	63.0	62.5	20.5	..	55	140	76.0	100.5	85.5	67.0	50.0	122
62	12.5	65.0	63.5	23.0	..	57	142	77.0	101.0	..	67.5	51.0	124
64	15.5	66.8	65.0	25.5	..	59	144	77.5	101.5	86.0	68.0	51.5	126
66	18.5	68.5	66.0	28.0	..	61	146	78.0	102.0	..	68.5	52.5	128
68	21.5	70.0	67.0	30.0	..	62	148	79.0	102.5	..	69.0	53.0	129
70	24.5	71.8	68.0	32.0	..	63	150	80.0	..	86.5	69.5	53.5	131
72	27.5	73.2	69.0	34.0	..	64	152	80.5	103.0	..	..	54.0	133
74	30.0	74.8	70.0	36.0	1.0	66	154	81.5	103.5	..	70.0	54.5	135
76	32.5	76.0	70.5	38.0	4.5	68	156	82.0	104.0	87.0	70.5	55.5	136
78	35.0	77.4	71.5	39.5	7.5	70	158	83.0	104.5	..	71.0	56.0	138
80	37.5	78.6	72.0	41.0	10.0	72	160	83.5	..	..	71.5	56.5	139
82	40.0	80.0	73.0	43.0	12.5	74	162	84.0	105.0	87.5	..	57.5	141
84	42.0	81.2	73.5	44.0	14.5	76	164	85.0	105.5	..	72.0	58.0	142
86	44.0	82.3	74.5	45.5	17.0	77	166	85.5	..	..	72.5	58.5	144
88	46.0	83.5	75.0	47.0	19.0	79	168	86.0	106.0	88.0	73.0	59.0	146
90	47.5	84.4	75.5	48.0	21.0	80	170	87.0	..	..	..	59.5	147
92	49.5	85.4	76.5	49.0	23.0	82	172	87.5	106.5	..	73.5	60.0	149
94	51.0	86.3	77.0	50.5	24.5	83	174	88.0	..	88.5	74.0	60.5	150
96	53.0	87.2	77.5	51.5	26.5	85	176	88.5	107.0	..	..	61.0	152
98	54.0	88.0	78.0	52.5	28.0	86	178	89.0	..	..	74.5	61.5	154
100	56.0	89.0	78.5	53.5	29.5	88	180	90.0	107.5	..	75.0	62.0	156
102	57.0	89.8	79.0	54.5	30.5	90	182	90.5	108.0	89.0	..	62.5	157
104	58.0	90.5	79.5	55.0	32.0	92	184	91.0	..	..	75.5	63.0	159
106	59.5	91.2	80.0	56.0	33.0	94	186	91.5	108.5	..	76.0	63.5	161
108	61.0	92.0	..	57.0	34.5	95	188	92.0	..	89.5	..	64.0	162
110	62.0	92.6	80.5	58.0	35.5	97	190	92.5	109.0	..	76.5	64.5	164
112	63.0	93.0	81.0	58.5	37.0	99	192	93.0	..	..	77.0	65.0	166
114	64.0	94.0	81.5	59.5	38.0	101	194	..	109.5	..	..	65.5	167
116	65.0	94.5	82.0	60.0	39.0	103	196	93.5	110.0	90.0	77.5	66.0	169
118	66.0	95.0	82.5	60.5	40.0	105							

Approximate Tensile Strength Estimated from Hardness Numbers

Brinell Hardness 3000 KG 10 mm St'd.	Rockwell B Scale	Rockwell C Scale	Wrought Steel Tensile Strength 1000 psi	Steel Weld Metal Tensile Strength 1000 psi
270	–	27.8	134	124
265	(103.5)	27.1	131	–
261	–	26.4	129	121
256	(102.0)	25.6	126	–
252	–	24.8	124	115
247	(101.0)	24.0	121	113
243	–	23.1	119	–
238	99.5	22.2	116	108
233	–	21.3	114	–
228	98.1	20.3	111	104
219	96.7	(18.0)	106	99
209	95.0	(15.7)	101	–
200	93.4	(13.4)	97	89
190	91.5	(11.0)	92	–
181	89.5	(8.5)	88	80
171	87.1	(6.0)	84	76
162	85.0	(3.0)	79	–
152	81.7	0	75	71
143	78.7	–	71	–
133	75.0	–	66	68
124	71.2	–	62	–
114	66.7	–	57	–

Temperature Conversion Tables

—459.4 TO 0

C		F
—273	—459.4	
—268	—450	
—262	—440	
—257	—430	
—251	—420	
—246	—410	
—240	—400	
—234	—390	
—229	—380	
—223	—370	
—218	—360	
—212	—350	
—207	—340	
—201	—330	
—196	—320	
—190	—310	
—184	—300	
—179	—290	
—173	—280	
—169	—273	—459.4
—168	—270	—454
—162	—260	—436
—157	—250	—418
—151	—240	—400
—146	—230	—382
—140	—220	—364
—134	—210	—346
—129	—200	—328
—123	—190	—310
—118	—180	—292
—112	—170	—274
—107	—160	—256
—101	—150	—238
— 96	—140	—220
— 90	—130	—202
— 84	—120	—184
— 79	—110	—166
— 73	—100	—148
— 68	— 90	—130
— 62	— 80	—112
— 57	— 70	— 94
— 51	— 60	— 76
— 46	— 50	— 58
— 40	— 40	— 40
— 34	— 30	— 22
— 29	— 20	— 4
— 23	— 10	14
— 17.7	0	32

0 TO 100

C		F	C		F
—17.7	0	32	9.9	50	122.0
—17.2	1	33.8	10.4	51	123.8
—16.6	2	35.6	11.1	52	125.6
—16.1	3	37.4	11.5	53	127.4
—15.5	4	39.2	12.1	54	129.2
—15.0	5	41.0	12.6	55	131.0
—14.4	6	42.8	13.2	56	132.8
—13.9	7	44.6	13.7	57	134.6
—13.3	8	46.4	14.3	48	136.4
—12.7	9	48.2	14.8	59	138.2
—12.2	10	50.0	15.6	60	140.0
—11.6	11	51.8	16.1	61	141.8
—11.1	12	53.6	16.6	62	143.6
—10.5	13	55.4	17.1	63	145.4
—10.0	14	57.2	17.7	64	147.2
— 9.4	15	59.0	18.2	65	149.0
— 8.8	16	60.8	18.8	66	150.8
— 8.3	17	62.6	19.3	67	152.6
— 7.7	18	64.4	19.9	68	154.4
— 7.2	19	66.2	20.4	69	156.2
— 6.6	20	68.0	21.0	70	158.0
— 6.1	21	69.8	21.5	71	159.8
— 5.5	22	71.6	22.2	72	161.6
— 5.0	23	73.4	22.7	73	163.4
— 4.4	24	75.2	23.3	74	165.2
— 3.9	25	77.0	23.8	75	167.0
— 3.3	26	78.8	24.4	76	168.8
— 2.8	27	80.6	25.0	77	170.6
— 2.2	28	82.4	25.5	78	172.4
— 1.6	29	84.2	26.2	79	174.2
— 1.1	30	86.0	26.8	80	176.0
— .6	31	87.8	27.3	81	177.8
0	32	89.6	27.7	82	179.6
.5	33	91.4	28.2	83	181.4
1.1	34	93.2	28.8	84	183.2
1.6	35	95.0	29.3	85	185.0
2.2	36	96.8	29.9	86	186.8
2.7	37	98.6	30.4	87	188.6
3.3	38	100.4	31.0	88	190.4
3.8	39	102.2	31.5	89	192.2
4.4	40	104.0	32.1	90	194.0
4.9	41	105.8	32.6	91	195.8
5.5	42	107.6	33.3	92	197.6
6.0	43	109.4	33.8	93	199.4
6.6	44	111.2	34.4	94	201.2
7.1	45	113.0	34.9	95	203.0
7.7	46	114.8	35.5	96	204.8
8.2	47	116.6	36.1	97	206.6
8.8	48	118.4	36.6	98	208.4
9.3	49	120.2	37.1	99	210.2
			37.7	100	212.0

100 TO 1000

C		F	C		F
38	100	212	260	500	932
43	110	230	265	510	950
49	120	248	271	520	968
54	130	266	276	530	986
60	140	284	282	540	1004
65	150	302	288	550	1022
71	160	320	293	560	1040
76	170	338	299	570	1058
83	180	356	304	580	1076
88	190	374	310	590	1094
93	200	392	315	600	1112
99	210	410	321	610	1130
100	212	413	326	620	1148
104	220	428	332	630	1166
110	230	446	338	640	1184
115	240	464	343	650	1202
121	250	482	349	660	1220
127	260	500	354	670	1238
132	270	518	360	680	1256
138	280	536	365	690	1274
143	290	554	371	700	1292
149	300	572	376	710	1310
154	310	590	382	720	1328
160	320	608	387	730	1346
165	330	626	393	740	1364
171	340	644	399	750	1382
177	350	662	404	760	1400
182	360	680	410	770	1418
188	370	698	415	780	1436
193	380	716	421	790	1454
199	390	734	426	800	1472
204	400	752	432	810	1490
210	410	770	438	820	1508
215	420	788	443	830	1526
221	430	806	449	840	1544
226	440	824	454	850	1562
232	450	842	460	860	1580
238	460	860	465	870	1598
243	470	878	471	880	1616
249	480	896	476	890	1634
254	490	914	482	900	1652
			487	910	1670
			493	920	1688
			498	930	1706
			504	940	1724
			510	950	1742
			515	960	1760
			520	970	1778
			526	980	1796
			532	990	1814
			538	1000	1832

1000 TO 2000

C		F	C		F
538	1000	1832	815	1500	2732
543	1010	1850	820	1510	2750
549	1020	1868	827	1520	2768
554	1030	1886	831	1530	2786
560	1040	1904	838	1540	2804
565	1050	1922	842	1550	2822
571	1060	1940	849	1560	2840
576	1070	1958	853	1570	2858
582	1080	1976	860	1580	2876
587	1090	1994	864	1590	2894
593	1100	2012	871	1600	2912
598	1110	2030	876	1610	2930
604	1120	2048	882	1620	2948
609	1130	2066	887	1630	2966
615	1140	2084	893	1640	2984
620	1150	2102	898	1650	3002
626	1160	2120	904	1660	3020
631	1170	2138	909	1670	3038
637	1180	2156	915	1680	3056
642	1190	2174	920	1690	3074
648	1200	2192	926	1700	3092
653	1210	2210	931	1710	3110
659	1220	2228	937	1720	3128
664	1230	2246	942	1730	3146
670	1240	2264	948	1740	3164
675	1250	2282	953	1750	3182
681	1260	2300	959	1760	3200
686	1270	2318	964	1770	3218
692	1280	2336	970	1780	3236
697	1290	2354	975	1790	3254
704	1300	2372	981	1800	3272
708	1310	2390	986	1810	3290
715	1320	2408	992	1820	3308
719	1330	2426	997	1830	3326
726	1340	2444	1003	1840	3344
734	1350	2462	1008	1850	3362
737	1360	2480	1014	1860	3380
741	1370	2498	1019	1870	3398
748	1380	2516	1025	1880	3416
752	1390	2534	1030	1890	3434
760	1400	2552	1036	1900	3452
765	1410	2570	1041	1910	3470
771	1420	2588	1047	1920	3488
776	1430	2606	1052	1930	3506
782	1440	2624	1058	1940	3524
787	1450	2642	1063	1950	3542
793	1460	2660	1069	1960	3560
798	1470	2678	1074	1970	3578
804	1480	2696	1080	1980	3596
809	1490	2714	1085	1990	3614
			1093	2000	3632

2000 TO 3000

C		F	C		F
1093	2000	3632	1371	2500	4532
1098	2010	3650	1376	2510	4550
1104	2020	3668	1382	2520	4568
1109	2030	3686	1387	2530	4586
1115	2040	3704	1393	2540	4604
1120	2050	3722	1398	2550	4622
1126	2060	3740	1404	2560	4640
1131	2070	3758	1409	2570	4658
1137	2080	3776	1415	2580	4676
1142	2090	3794	1420	2590	4694
1149	2100	3812	1427	2600	4712
1154	2110	3830	1432	2610	4730
1160	2120	3848	1438	2620	4748
1165	2130	3866	1443	2630	4766
1171	2140	3884	1449	2640	4784
1176	2150	3902	1454	2650	4802
1182	2160	3920	1460	2660	4820
1187	2170	3938	1465	2670	4838
1193	2180	3956	1471	2680	4856
1198	2190	3974	1476	2690	4874
1204	2200	3992	1483	2700	4892
1209	2210	4010	1488	2710	4910
1215	2220	4028	1494	2720	4928
1220	2230	4046	1499	2730	4946
1226	2240	4064	1505	2740	4964
1231	2250	4082	1510	2750	4982
1237	2260	4100	1516	2760	5000
1242	2270	4118	1521	2770	5018
1248	2280	4136	1527	2780	5036
1253	2290	4154	1532	2790	5054
1259	2300	4172	1538	2800	5072
1264	2310	4190	1543	2810	5090
1270	2320	4208	1549	2820	5108
1275	2330	4226	1554	2830	5126
1281	2340	4244	1560	2840	5144
1286	2350	4262	1565	2850	5162
1292	2360	4280	1571	2860	5180
1297	2370	4298	1576	2870	5198
1303	2380	4316	1582	2880	5216
1308	2390	4334	1587	2890	5234
1315	2400	4352	1593	2900	5252
1320	2410	4370	1598	2910	5270
1326	2420	4388	1604	2920	5288
1331	2430	4406	1609	2930	5306
1337	2440	4424	1615	2940	5324
1342	2450	4442	1620	2950	5342
1348	2460	4460	1626	2960	5360
1353	2470	4478	1631	2970	5378
1359	2480	4496	1637	2980	5396
1364	2490	4514	1642	2990	5414
			1649	3000	5432

Note: The numbers in the unmarked column refer to the temperature either in degrees Centigrade or Fahrenheit which it is desired to convert into the other scale. If converting from Fahrenheit degrees to Centigrade degrees the equivalent temperature will be found in the left column, while if converting from degrees Centigrade to degrees Fahrenheit, the answer will be found in the column on the right.

From Republic Alloy Steels, Republic Steel Corp.

Miscellaneous

CLEANING AND PAINTING OVER WELDS

Discoloration, flaking, and blistering of paint over and immediately adjacent to welds are chronic problems. There are actually two problems here, caused by different conditions: 1. Paint flakes off. 2. Paint discolors and may eventually blister and peel. These problems may be caused by surface condition or by chemical action.

The surface can be impaired by dust, smoke film, iron oxide, grease, and similar materials which form a barrier between the paint and the surface, thus preventing the paint from properly bonding to the surface. A surface that has been burnished very smooth with a power wire brush may also prevent the paint from bonding properly.

The smoke and fumes of welding form a film on the surface of the plate which combines with moisture in the air. The resulting alkaline solution reacts with paint to produce discoloration and blistering. The effect increases with higher humidity.

Effect of Processes and Electrodes

Submerged-arc welds produce few paint problems because the slag is nearly always removed, and the process leaves no smoke or iron-oxide film on the adjacent material.

Covered electrodes for welding cast iron and aluminum pick up moisture quickly, thus causing more chemical-reaction problems than do other electrodes. For practical purposes, most shielded metal-arc electrodes produce about the same chemical reaction. Any apparent differences usually result from differences in humidity conditions.

Remedies

Cleaning is the obvious first step. Removal of slag, spatter, smoke film and iron-oxide film eliminates most paint-adhesion problems. Cleaning provides a surface to which the paint can bond, and removes most of the chemical-containing deposits. However, avoid burnishing the surface with a power wire brush.

If discoloration or blistering persists after normal cleaning, substitute a paint that is more alkaline-resistant, such as a vinyl, epoxy, or chlorinated-rubber base type. The alkaline surface can be neutralized by washing with a mild acid solution such as boric acid or phosphoric acid. Commercially available products such as Metal-Prep, Rusticide, Naval Jelly and others will not only neutralize the alkalinity and remove rust but will lightly etch the surface to provide an ideal condition for a coat of primer paint.

Simply wiping the surface with a shop rag removes much of the film and improves paint bonding. Application of the paint with a brush instead of by spraying provides a better wetting action and forms a better bond.

Regardless of application method, the affected area should be treated and painted as soon as possible after welding to reduce the time for moisture pick-up by the deposited film. Less moisture means less alkaline solution formed to attack the paint. Two coats, including an alkaline-resistant primer, are better than a single coat.

British and United States Shielded Metal-Arc Electrode Size Comparison Table

Metric		Gage		Approx. Fraction
mm	in.	No.	in.	in.
2	0.0787	14	0.080	5/64
2.5	0.0984	12	0.104	3/32
3.15	0.1240	10	0.128	1/8
3.25*	0.1280			
4	0.1576	8	0.160	5/32
5	0.1969	6	0.192	3/16
6*	0.2362	4	0.232	7/32
6.3	0.2481			1/4
8	0.3150	0	0.324	5/16
10	0.3937			3/8

*The British Standard Institution in the Specifications for Covered Electrodes (BS 639:1969) proposes that, with the exception of the 3.15 and 6 mm sizes, the metric sizes in the first column become standard.

The 3.25 mm diameter is at present widely used and will be retained until such a time as it can be dropped through its lesser usage.

The 6 mm diameter is not a preferred size but it is included because the existing capacities of some welding equipment may make the use of 6.3 mm impractical.

ETCHING SOLUTIONS FOR FERROUS ALLOYS

Used to Show	Composition of Etching Solution	Remarks
Recommeded Ferric Chloride Solution Etching Agent for Macrostructure of Welds in Steel	$FeCl^3$ - $6H_2O$ — 200 grams HNO_3 conc — 300 ml H_2O — 100 ml	Dissolve the ferric chloride in water, then add the HNO_2 and mix thoroughly.Wet the surface of the specimen with the solution using a medicine dropper contained in a bottle. Etching takes one to five minutes depending on the quality of the surface and composition of metal. It should be done under a fume hood. At end of etching period, wash specimen carefully under running water, let drain and swab with methyl alcohol in order to speed drying and inhibit rusting of th surface. The Macrostructure of the weld is then clearly discernible. This works best when sample is at room temperature. Sample can be merely ground or sanded smooth, it does not hav to b highly polished.
For Macroscopic Examination		
Segregations	1 part conc. hydrochloric acid, HCl 1 part water	Use at 100 to 212°F Etch for 1/2 hour or until defects are brought out.
Segregations	5 g picric acid, $HOC_6H_2 (NO_2)_3$ 100 cc ethyl alcohol, C_2H_5OH	Etch for 4 to 5 hours.
Phosphorus segregations	10 g cuprous ammonium chloride, $CuClNH_4Cl\text{-}6H_2O$ 100 cc water	This solution is more easily prepared than are most copper chloride solutions.
Phosphorus segregations	10 g cupric chloride, $CuCl_2$ 40 g magnesium chloride, $MgCl_2$ 20 cc hydrochloric acid, HCl 180 cc water 1000 cc absolute alcohol, C_2H_5OH	Absolute methyl alcohol, CH_3OH, may be used in place of ethyl alcohol.
Variations in crystal structure, and segregations	20 g ammonium persulfate, $(NH_4)_2S_2O_6$ 100 cc distilled water	Use at room temperature. A very good reagent for the purpose indicated.
To reveal the structure of welds and to distinguish between steel and wrought iron	1 part conc. nitric acid, HNO_3 3 parts of water	Use at about 100°F.
For Microscopic Examination		
The generl structure of normal iron and steel	4 g picric acid and 95 cc of absolute alcohol, or 5 g picric acid and 100 cc of 95% alcohol, C_2H_5OH	Keep bottle well stoppered when absolute alcohol is used.
The general structure of normal iron and steel	5 cc nitic acid, HNO_3 100 cc of ethyl alcohol, C_2H_5OH	For heat treated steels use a more dilute solution.
To establish the difference between hardened and normal steels	4 parts nitric acid, HNO_3 96 parts iso-amyl alcohol, $C_5H_{11}OH$	
Fine Pearlite and tempered Martensite	1 part of a 40% solution of nitric acid in acetic anhydride 1 part methyl alcohol, CH_3OH 1 part ethyl alcohol, C_2H_5OH 1 part iso-amyl alcohol, $C_5H_{11}OH$	Colors fine pearlite and tempered martensite.
Cementite and other carbides, and tungstides	2 g picric acid, $HOC_6H_2 (NO_2)_3$ 24.5 g sodium hydroxide, NaOH 73.5 cc water	Use boiling for 5 to 10 minutes. Tungsten carbide is unaffected, but other carbides and tungstides are colored.
Structure of stainless steels	10 g ferric chloride, $FeCl_2$ 30 cc hydrochloric acid Water to make total volume 120 cc	May be diluted, and applied gently with a piece of cotton. Time required is less than 30 seconds.

From Lang's Handbook of Chemistry

(Continued on next page)

ETCHING SOLUTIONS FOR FERROUS ALLOYS

Used to Show	Composition of etching solution	Remarks
Structure of stainless steels	30 cc hydrochloric acid, HCl 10 cc nitric acid, HNO_3 80 cc glycerol (glycerine)	Use full strength for rapid work, but handle carefully.
Carbides and tungstides in high-speed and tungsten steels	10 g potassium ferricyanide, $K_3Fe(CN)_6$ 10 g potassium hydroxide, KOH 100 cc water	Use hot.
Structure of high-speed steel and tungsten carbide	10 cc commercial hydrogen peroxide, H_2O_2 20 cc of a 10% aqueous sodium hydroxide (NaOH) solution	Use fresh from 10 to 12 minutes. Tungsten carbide is darkened.

ETCHING SOLUTIONS FOR NON-FERROUS ALLOYS

Metal or Alloy	Composition of etching solution	Remarks
Copper Base Alloys		
Copper or Cu-Al alloys	5 g copper-ammonium chloride 120 cc water Add conc. ammonium hydroxide until the precipitate which forms is redissolved, giving a clear blue color.	Etch from 30 to 60 seconds; this is a very good copper etch.
Copper, brasses, Cu-Al alloys, German silver	A 10% solution of ammonium persulfate, $(NH_4)_2 S_2O_8$	Etch from 1 to 2 minutes. Keep the surface of the specimen free from grease.
Copper or brasses	10 cc hydrogen peroxide, H_2O_2 40 cc conc. ammonium hydroxide, NH_4OH	Apply with a swab to the surface of polished specimen. Rub briskly until mirror-like surface is obtained.
Brasses (Used for copper products such as sheets, rods, etc.), Cu-Al alloys, German silver	15 cc of a saturated ferric chloride ($FeCl_3$) solution 30 cc concentrated hydrochloric acid, HCl 200 cc water	Use this solution after the ammoniacal peroxide solution above. Dip the specimen into this ferric chloride solution for a few seconds after following the directions above.
Cu-Al alloys	Nitric acid (HNO_3), sp. gr. 1.42	Time for etching is very short. It is difficult to control the intensity of the attack.
Aluminum and Light Alloys		
Al and Al alloys	0.5 to 20% hydrofluoric acid (HF) acid solutions. The aqueous or alcoholic solutions are used.	Etch for about 15 seconds. Remove the black deposit so formed by immersing for a short time in conc. nitric (HNO_3) acid. $CuAl_2$ and $FeAl_3$ are darkened by HF.
Aluminum	45 cc conc. hydrochloric acid, HCl 15 cc conc. nitric acid, HNO_3 15 cc hydrofluoric acid, HF 25 cc hot water	This etch generally requires only a few seconds.
Al alloys	10 cc conc. nitric acid, HNO_3 3 to 4 cc hydrofluoric acid, HF 86 to 87 cc hot water	
Al and Al alloys	10 to 20% aqueous sulfuric acid (H_2SO_4) solution	Heat to 160°F., immerse specimen for a few seconds, quench in cold water. $FeAl_3$ is stained black.
Al and Al alloys	10 cc conc. hydrofluoric acid, HF 15 cc conc. hydrochloric acid, HCl 90 cc water	Etch from 30 to 90 seconds. Reveals grain size very clearly.
Miscellaneous Metals and Alloys		
Nickel and its alloys	†50 cc of 70% nitric acid, HNO_3 50 cc of 50% acetic acid, $HC_2H_3O_2$	This is *Merica's solution.* It is very good for cast, cold drawn, and annealed nickel.
Nickel and its alloys	†20 to 40 cc of nitric acid, HNO_3 30 to 40 cc of 75% acetic acid, $HC_2H_3O_2$ 30 to 40 cc acetone, $(CH_3)_2 CO$	This solution is very similar in action to the one above.
Magnesium and its alloys	2% aqueous nitric acid solution	Etch for about 10 seconds.

PARALLELING ARC WELDERS

What is Paralleling?

Paralleling consists of connecting the output of two, three, or more arc welding machines of the same model in parallel so the total available rated output equals the sum of the rated output of each machine at rated duty cycle.

Why Parallel?

Machines are paralleled for two primary reasons:

1. For high current application when power sources of sufficient size are not readily obtainable.
2. To utilize existing power sources for operations which require higher output than the rating of any one available machine.

Applications utilizing paralleled welders include automatic and semiautomatic welding, stud welding, arc gouging, resistance heating, high current manual welding and others.

Basic Requirements for Paralleling

1. Since improper paralleling or improper operation of paralleled machines can cause serious damage, the specific instructions for the equipment involved MUST be accurately and completely followed.
2. Welders to be paralleled must be the same size and model and must operate on the same input volts and frequency.
3. The control settings for each machine must be set for the same output so the current draw from each machine is the same.

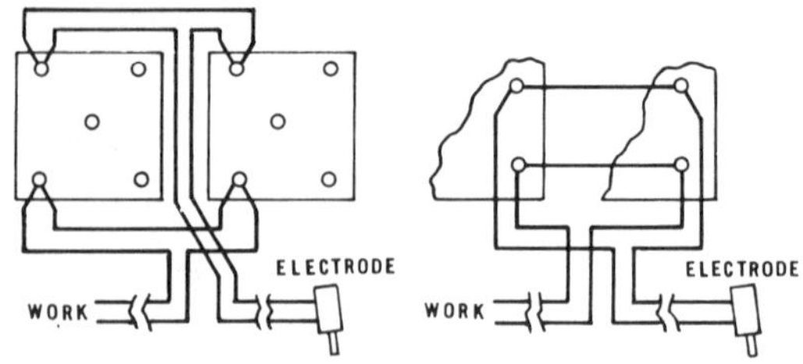

4. Connect "like" output terminals (example: connect "Work" to "Work" and "Min to 525" to "Min to 525") together using equal lengths of welding cable. Use the size cable recommended for the rated output current and duty cycle of the welder. Interconnect motor-generator and engine welders per instructions on the specific connection diagram for the models being used.
5. When they are to be connected to a wire feeder, the paralleled power sources must be interconnected with the wire feeder per the appropriate connection diagram available from the manufacturer. (See sample request form below.)

Paralleling Motor-Generator Welders

CAUTION: The proper paralleling instructions must be followed for the specific equipment involved. Improper paralleling can result in one generator driving the other as an unloaded DC motor. If this occurs, the driven machine increases speed until it literally destroys itself and endangers personnel and nearby equipment. Therefore, special precautions must be taken to insure safe operation. In all instances, contact your local welding equipment representative for assistance before paralleling.

REQUEST FOR PARALLELING ASSISTANCE (Suggested Form)

Company ________________

Required Output Current, Voltage and Operation Factor ________________

Purpose: ☐ Automatic welding — Process ☐ Submerged Arc; ☐ Innershield; or ☐ Other ________________

Wire Feeders: ☐ LAF-3; ☐ LAF-4: ☐ LAF-5; ☐ LT-7; ☐ LT-56: ☐ NA-2; ☐ NA-3; ☐ NA-4;

☐ NA-5; ☐ Other ________________

☐ Semiautomatic Welding — Process ________________

Wire Feeders ☐ LN-8; ☐ LN-9; ☐ Other ________________

☐ Stick Electrode Welding: ☐ Stud Welding; ☐ Arc Gouging; ☐ Resistance Heating

☐ Other ________________

Number of power sources to be paralleled ________________

Description of Lincoln power sources to be paralleled:

	Machine No. 1	Machine No. 2	Machine No. 3
Input volts/ph/hertz	________	________	________
Model and Size	________	________	________
Code Number	________	________	________
Serial Number	________	________	________

Index

— A —

— B —

— C —

— D —

— E —

— F —

— G —

— H —

— I —

— J —

— L —

— M —

— N —

— O —

— P —

— Q —

— R —

— S —

— T —

— U —

— V —

— W —

— X —

— Y —